Biochemistry

INTERNATIONAL ADAPTATION

Biochemistry

An Integrative Approach with Expanded Topics

INTERNATIONAL ADAPTATION

John T. Tansey

University of Massachusetts at Amherst

WILEY

Biochemistry

An Integrative Approach with Expanded Topics

INTERNATIONAL ADAPTATION

Contributing Subject Matter Experts: Dr. Kavishankar G B, Assistant Professor, Department of Life Science, Central University of Karnataka, Kadaganchi, Kalaburagi, Karnataka State, India and Dr. Sorokhaibam Jibankumar Singh, Assistant Professor, Department of Biochemistry, Manipur University, Canchipur, Imphal, Manipur, India

Founded in 1807, John Wiley & Sons, Inc. has been a valued source of knowledge and understanding for more than 200 years, helping people around the world meet their needs and fulfill their aspirations. Our company is built on a foundation of principles that include responsibility to the communities we serve and where we live and work. In 2008, we launched a Corporate Citizenship Initiative, a global effort to address the environmental, social, economic, and ethical challenges we face in our business. Among the issues we are addressing are carbon impact, paper specifications and procurement, ethical conduct within our business and among our vendors, and community and charitable support. For more information, please visit our website: www.wiley.com/go/citizenship.

ISBN: 978-1-119-82080-2

ISBN: 978-1-119-82081-9 (ePub)

ISBN: 978-1-119-82082-6 (ePdf)

Printed and bound by CPI Group (UK) Ltd, Croydon, CR0 4YY

C117806_080322

To Emily, Aidan, and Eli for their love, support, and encouragement. We did it!

To my professors and teachers for all they have taught me. I owe you so much. I can never thank you enough. I hope I've made you proud.

To my students for their curiosity. You continue to inspire me. I hope this work helps future generations of scientists and clinicians.

BRIEF CONTENTS

CONTENTS

4   Proteins II: Enzymes, Allosterism and Receptor–Ligand Interactions   133

5   Membranes and an Introduction to Signal Transduction   193

MEDICAL APPLICATIONS

ABOUT THE AUTHOR

JOHN T. TANSEY received his B.S. in Biochemistry and Molecular Biology from the University of Massachusetts, Amherst, and his PhD in Biochemistry from Wake Forest University. Following his graduate degree, John undertook a postdoctoral fellowship at the National Institutes of Health in the lab of Dr. Constantine Londos, where he led the team that generated and characterized the perilipin 1 knockout mouse. John's scientific interests are a natural extension of his postdoctoral studies and focus on the structure and function of the perilipins, a class of lipid droplet proteins that regulate neutral lipid storage and breakdown. John has been teaching for the past seventeen years in the Chemistry department and Biochemistry and Molecular Biology program at Otterbein University, a primarily undergraduate institution outside Columbus, Ohio. He has also taught at the community college and graduate levels. John was awarded both the Best New Teacher and Master Teacher awards from his university. He is an education fellow of the American Society for Biochemistry and Molecular Biology (ASBMB) and is a member of the editorial board of Biochemistry and Molecular Biology Education (BAMBEd). In addition to his basic science interests, he has studied the use of science fiction or other readings to promote engagement and discussion in science classes, integration of codes of conduct and basic ethics in science courses, and development of curricula and tools for programmatic assessment. When he isn't teaching or in the lab, John enjoys cooking, the outdoors, games, and a good laugh.

Courtesy of Edward Syguda, Otterbein University

Over time, few fields have changed as much as biochemistry. During the past 50 years, advances in computing, technology, and molecular biology have radically changed how we approach biochemical questions. The laboratory tools available to biochemists continue to improve and change. Obviously, what was current in 1959, 1979, or even 1999 will not suffice now. Arguably, the current rate of change in biochemistry is greater than in many other fields of science and mathematics, and the materials faculty choose to use in their courses should reflect those changes. Biochemistry is a living, breathing science.

Perhaps in light of how biochemistry is changing as a science, biochemistry is also changing as an undergraduate course of study. A very different pool of students now takes undergraduate biochemistry courses compared to even a few years ago. These students include biology majors pursuing advanced life science degrees, an expanded pool of pre-professional and pre-med students, and ACS-certified chemistry majors. These changes have arisen partially due to student perceptions of the value of biochemistry as well as changes from both professional societies and revisions to the MCAT.

We would be remiss if we did not mention the students themselves. As students with more diverse scientific backgrounds are enrolling in biochemistry, biochemistry courses must adapt to help these students optimize their strengths and maximize their learning. Students have significant distractions and demands on their time. This often increases demand on both students and faculty. Clearly, there must be new solutions and pedagogies to help meet these emerging challenges.

Biochemistry: An Integrative Approach with Expanded Topics, International Adaptation by John Tansey caters to a variety of majors and backgrounds with relevant, engaging examples while still preserving balance between the classical and modern and between chemistry and biology.

Approach

Biochemistry: An Integrative Approach with Expanded Topics, International Adaptation uses a thematic approach to present concepts with a chemical perspective in a biological context and highlights four important themes:

- **Evolution.**   Biomolecules, like all life, have changed over time through the mutation of genes, selection of properties and traits, and transmission of the genes that yield the most favorable traits.
- **Structure and Function.**   Basic units of structure define the function of biomolecules and all living things.

- **Information Flow, Exchange, and Storage.**   All levels of biomolecular organization depend on molecular interactions and information transfer.
- **Pathways and Transformations of Matter and Energy.** Biomolecules, and all biological systems, change through organic reaction pathways governed by the laws of thermodynamics and kinetics.

Biochemistry: An Integrative Approach with Expanded Topics, International Adaptation emphasizes the core competencies advocated by several professional and governmental agencies. Higher-level thinking skills such as data interpretation and experimental design are then developed from these core competencies.

The text presents material recommended by the American Society for Biochemistry and Molecular Biology in a way that is suitable for a variety of course structures. Material is organized so that the first half is suitable for the one-semester biochemistry course, including descriptive biochemistry, metabolism, elementary enzyme kinetics and mechanisms, allosteric regulation, DNA/RNA structure/function, signal transduction, and supramolecular assemblies. Second semester coverage focuses on specific topics at greater levels of detail. By often placing structure and function of biomolecules in the same chapter as their metabolism, biochemistry is presented as a more integrated subject. The modular nature of the chapters means that instructors can easily customize content and change the order in which topics are covered to suit their classes' needs without a disruption or disconnect in the flow of material.

Although organic chemistry remains as an assumed prerequisite for the course, the text will also draw from physical chemical principles (such as thermodynamics and kinetics) approached from a general chemistry perspective. Presenting the material this way facilitates division of material for those students only taking a single biochemistry course and allows for differentiated learning at higher levels; for example, signal transduction is introduced in the first semester but revisited at a greater level of detail in the second.

Biochemistry: An Integrative Approach with Expanded Topics, International Adaptation provides more continuity with other science courses and texts by emphasizing worked problems throughout and including expanded end-of-chapter problems with a wide range of subject matter and difficulty. The worked problems are both qualitative (conceptual) and quantitative to model for students the biochemical reasoning they need to succeed in the course. Students will often be asked to analyze data and make critical assessments of experiments.

To help students understand biochemistry at a deeper level than rote memorization, topics are integrated and cross-referenced to enhance pattern recognition. By pointing out similarities and differences in major types of biochemical reactions, mechanisms, enzyme controls, metabolic pathways, and signal transduction, students can more easily build conceptual frameworks.

Given the role of biochemistry in modern medicine and the advances that have been made in science over the past few decades, it makes sense to emphasize these topics and use complex examples from animals in teaching biochemistry. Although this may make the material more complex in some instances than simply studying analogous systems in *E.coli*, it provides an important frame of reference for students who are interested in biomedical applications.

Visual literacy is the ability to recognize and understand ideas conveyed through visible actions or images, such as illustrations. In our digital age, the ability of scientists and medical professionals to interpret data, complex illustrations, and images is critical. *Biochemistry: An Integrative Approach with Expanded Topics, International Adaptation* includes clean, crisp images and micrographs that use common themes, structures, and color palettes throughout. This consistent approach creates familiarity for the students such that they immediately understand the type of figure being displayed, its relative importance, and how to put the figure into a broader context.

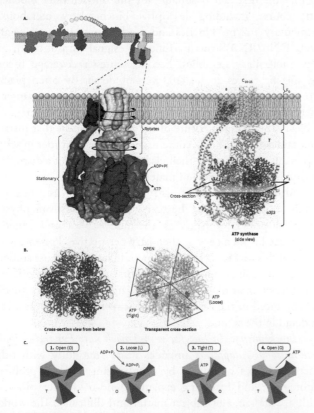

FIGURE 7.25 Reactions of ATP synthase. A. The ATP synthase consists of two larger fragments, the membrane-bound F₀ complex and the catalytic F₁ complex. **B.** As the F₁ complex rotates past the b subunits, the conformation of a and b change, resulting in binding of ADP and Pᵢ and formation of ATP. **C.** The complex consists of three active sites that progress through three different states: open, where nothing is bound; loose, where ADP and Pᵢ bind; and tight, where the change in conformation generates ATP.
(Source: (A) Data from PDB ID SARA Zhou, A., Rohou, A., Schep, D.G., Bason, J.V., Montgomery, M.G., Walker, J.E., Grigorieff, N., Rubinstein, J.L. (2015) Structure and conformational states of the bovine mitochondrial ATP synthase by cryo-EM. *Elife* 4: e10180-e10180 (B) Data from PDB ID 1BMF Abrahams, J.P., Leslie, A.G., Lutter, R., Walker, J.E. (1994) Structure at 2.8 A resolution of F1-ATPase from bovine heart mitochondria. *Nature* 370: 621-628)

Finally, *Biochemistry: An Integrative Approach with Expanded Topics, International Adaptation* emphasizes that biochemistry is an interdisciplinary and collaborative science conducted in a societal context. Each chapter has features that explore current topics and ethical discussions that ask students to engage with

these concepts in a meaningful and fact-based way. There may not be clear answers to some of these questions, but they are clearly worthy of discussion.

Organization

In many texts, students' initial steps into biochemistry are divided into a discussion of the molecules found in biochemical systems and the metabolism of these molecules. This creates two problems. First, students taking a two-semester biochemistry sequence must retain and apply the descriptive biochemistry they learned several months earlier when discussing the metabolism of these molecules. It also often creates situations where material can be encapsulated and segregated in some aspect of a chapter. For example, if enzyme kinetics and mechanisms are only emphasized in a single chapter, students may be under the impression that they exist as one-off topics not pertinent to other discussions instead of part of an integrated whole.

One-Semester Course

Because of its organization and integration, *Biochemistry: An Integrative Approach with Expanded Topics, International Adaptation* can be used for either a one- or a two-semester sequence. The first 12 chapters, suitable for a one-semester course, begin with a review of chemical principles germane to biochemistry. These include free energy, thermodynamics, weak forces, organic functional groups, solubility, and the properties of water. This is followed by a review of the structure and function of nucleic acids and central dogma of biology. This material is presented so that all students have foundational concepts and tools in both chemistry and biology prior to discussion of biochemical systems. The remainder of the introductory chapters covers the basics of amino acid chemistry and protein structure, function, and purification in chapter 3, enzyme kinetics and mechanisms, allosterism, and receptor–ligand interactions in chapter 4, and the structure of the cell and signal transduction in chapter 5. These chapters build on one another but are modular so that material is easily customized.

The first five chapters provide a solid basis for the discussions of metabolism that occur in chapters 6 to 12. Chapter 6 begins with a discussion of carbohydrates followed by glycolysis, gluconeogenesis, and the fates of pyruvate. Chapter 7 explores the citric acid cycle and electron transport and builds from chapter 6. Chapter 8 returns to carbohydrate metabolism to discuss the structure, function, and metabolism of glycogen, but it also discusses the pentose phosphate pathway and several other roles of complex carbohydrates, including glycoproteins and the extracellular matrix. Chapter 9 introduces and describes the numerous classes of lipids, and chapter 10 covers lipoprotein and membrane lipid biosynthesis, cholesterol metabolism, and neutral lipid metabolism. These two chapters parallel the format of chapters 6 and 8 and describe lipids and their metabolism. Chapter 11 discusses amino acid metabolism, the urea cycle, and scavenging of amino acid skeletons. Kidney function and metabolism of xenobiotics are also discussed in chapter 11. Neither of these topics are commonly covered in biochemistry texts but help complete the story of metabolism and the excretion of waste products. Chapter 11 also reiterates aspects of ion transport, protein structure, and regulation of metabolism that have been encountered and built on throughout the text. Finally, chapter 12 integrates all of metabolism and ties fundamental pathways into metabolic disease, including the hormonal integration of pathways and organ systems that helps regulate metabolism. Discussion of a variety of metabolic diseases (for example, diabetes or starvation) helps clarify and define metabolic states for the student. This approach differs from other works in that it builds from aspects of biochemistry discussed in the previous 11 chapters and integrates it with modern diseases and human health.

Two-Semester Course

For two-semester courses, *Biochemistry: An Integrative Approach with Expanded Topics, International Adaptation* includes the first twelve chapters but then develops these themes further. In chapter 13, nucleotide metabolism can be covered as part of either central metabolism or nucleic acid metabolism, allowing instructors to easily place it in either arrangement. Chapters 14, 15, 16, and 17 cover DNA replication, transcription, translation, and the regulation of translation, respectively. These chapters again approach the central dogma from a modern biochemical approach. The most recently solved structures and regulatory mechanisms such as RNAi and riboswitches are discussed alongside discussions of more classical structures such as operons.

The final portion of the text revisits key concepts covered in the first 12 chapters in greater detail. Chapter 18 examines protein structure/function and discusses how structures of proteins are determined using EM, NMR, and X-ray diffraction as well as databases of these structures such as the Protein Data Bank. Chapter 19 discusses protein engineering in greater detail, including techniques ranging from the more biological (epitope tagging and site-directed mutagenesis) to the more biophysical (solid-phase peptide synthesis and amber codon suppression). Chapter 20 discusses -omics, mass spectrometry and the data-driven approaches that are more and more commonly found in the laboratory and clinic. Signal transduction is reexamined in Chapter 21 with an emphasis on some of the signaling pathways that are not commonly encountered in introductory biochemistry but are clearly relevant and applicable. These include the toll-like receptors, MAP kinase, and Akt. This chapter builds from what was first learned in Chapter 5, reemphasized throughout intermediary metabolism and added to in the transitional chapters on gene regulation. Chapter 22 covers macromolecular processes including secretory pathways, protein trafficking, molecular motors, and biochemical aspects of neuroscience and development. This chapter again builds from what has been seen in previous chapters on protein structure/function but applies the concepts to these more complex aspects of cellular or organismal biology. The final chapter of the text (Chapter 23) unites topics and concepts throughout to discuss plant and agricultural biochemistry. Chapter 23 is as much a review of all the basic concepts found in the book as it is a discussion of photosynthesis and how plants function on the biochemical level.

Features

- *Context:* Chapters begin with a short discussion that puts the content of the chapter into a broader context of biochemistry.
- *Common Themes:* Examples of the four guiding themes encountered in the chapter provide students with pertinent examples that ground the concepts in the chapter with the broader themes used throughout the book.

Common Themes

Evolution's outcomes are conserved.	• Many of the pathways we will see in this chapter are conserved throughout evolution. • In contrast to genes discussed in other chapters, such as the genes involved in eicosanoids and steroid metabolism, the genes coding for enzymes involved in glycolysis are found throughout biology and are therefore more ancient in origin. • Because glycolysis is so basic to the existence of an organism, mutations that disrupt this pathway are often lethal or put organisms at such an evolutionary disadvantage that they generally fail to pass on these genes.
Structure determines function.	• Monosaccharides are a family of molecules that are largely related through stereochemistry. • Monosaccharides are frequently depicted in Fischer projections to highlight their stereochemical differences. However, in solution, these molecules adopt a cyclic hemiacetal or hemiketal structure; this structure is best seen in a Haworth projection. • In the broadest sense, polysaccharides can be functionally characterized as being energy stores or as providing structure to the organism. • Energy-storing polysaccharides can either be linear like amylose or highly branched like amylopectin or glycogen. The branched forms enable the mobilization of monosaccharides at a higher rate, because release can happen simultaneously from several sites. • Structural polysaccharides are generally linear molecules, with multiple places where the strands can interact with one another through weak forces (typically hydrogen bonding, dipole–dipole, and electrostatic interactions). • The structure of many enzymes is conserved throughout biology, and many of the same chemical reactions are employed in the breakdown and biosynthesis of carbohydrates in divergent species.
Biochemical information is transferred, exchanged, and stored.	• Carbohydrate metabolism is regulated by several allosteric regulators and kinases that inhibit or activate key enzymes in the glycolytic pathway.
Biomolecules are altered through pathways involving transformations of energy and matter.	• In this as in other chapters, we will see that energetically unfavorable reactions become favorable when coupled to reactions with very negative ΔG° values. Although we typically think of this in terms of ATP hydrolysis, reactions of other groups such as phosphoesters or thioesters are common in biochemistry. Likewise, reactions that have ΔG° values close to zero can be tipped to move in either direction by changing the concentrations of substrates or products. • The synthesis of glucose and the degradation of monosaccharides use similar reactions but different pathways. Key steps are different, and compartmentalization is important. A similar theme is seen in fatty acid metabolism.

- *Worked Problems:* Each section contains qualitative and quantitative worked problems. These problems help students engage with the material and think like a biochemist.

Worked Problem 7.2   Anaplerotic reactions

Acetyl-CoA contributes directly to the formation of citrate, but acetate is not considered an anaplerotic intermediate. Explain why this is the case.

Strategy Review the reactions of the citric acid cycle. What does "anaplerosis" mean?

Solution Anaplerotic reactions "build up" levels of citric acid cycle intermediates, replenishing them. The intermediates most often discussed include α-ketoglutarate, succinyl-CoA, and oxaloacetate. The reactions of the citric acid cycle themselves are not thought of as anaplerotic because they do not lead to an increase in the concentration of that intermediate; the product of the reaction is immediately used by the next step. Because the formation of citrate from acetyl-CoA and oxaloacetate is part of the citric acid cycle, this step does not really elevate the concentrations of intermediates and it is not anaplerotic.

Follow-up question In several neurodegenerative processes, fluctuations of glutamate and glutamine levels lead to low energy levels in the cell by slowing down the citric acid cycle. Suggest how depletion of cellular glutamate could reduce flux through the citric acid cycle.

- *End-of-Chapter Problems:* Each chapter includes nearly 40 problems written by the author. These problems use research and biomedical examples to reinforce key concepts in the chapter. In addition to problems that address content, each chapter includes three additional problem sets—Data Interpretation, Experimental Design, and Ethics and Social Responsibility—requiring students to think at a higher level and defend and support their views. Detailed worked solutions written by the author preserve a consistent problem-solving pedagogy throughout.

- *Relevance, Explorations, and Applications:* Each chapter contains three types of boxed features focusing on current issues germane to the chapter topics:

  - *Medical Biochemistry* illustrates how biochemistry has advanced and informed medicine, offers hope for cures, and indicates where more research is necessary.

  - *Biochemistry* ties key concepts covered in general or organic chemistry to biochemistry and uses biochemical examples to illustrate these points.

  - *Societal and Ethical Biochemistry* examines societal and ethical questions that are currently faced by investigators and members of society. These range from genetically modified organisms to the use of cultured cells in the laboratory.

- *Bioinformatics Exercises:* Written by Rakesh Mogul, California State Polytechnic University–Pomona, these activities are a combination of tutorials and questions that ask students to use digital tools such as PDB, GenBank, BLAST, the KEGG database, and other resources to become familiar with these modern research tools.

Instructor Supplements

- Instructor's Solutions by John Tansey. Detailed explanations to each Worked Problem and End-of-Chapter Problem by the author provide consistent problem-solving pedagogy throughout the text. Students are provided with the answers to odd-numbered problems only.

- Test Bank by Anne Grippo, *Arkansas State University*, contains multiple-choice, text entry, molecular drawing, and short answer questions.

- PowerPoint Lecture Slides by Tanea Reed, *Eastern Kentucky University*, highlight topics to help reinforce students' grasp of essential concepts.

- PowerPoint Art Slides and jpegs. All of the art and tables are available as both slides and jpeg files.

- Personal Response System ("Clicker") Questions are written by Michael Chen, *California State University–Los Angeles*.

ACKNOWLEDGMENTS

This book may have only one author, but it has been far from a solo project. I have been so fortunate to have a great team of people that I have worked with over the years. This began with John Stout, my Wiley textbook representative, who asked me back in 2009 what I liked or disliked about current texts and set me on this journey. From there I was introduced to Joan Kalkut. Joan has served in many roles on this project including project manager, editor, and product designer. Her sustained energy and optimism has been nothing short of amazing. More than anyone, her drive has helped this project along, offering both deadlines and words of encouragement, and many, many edits. Thank you, Joan, for believing in this project; it wouldn't have happened without you. For several years Clifford Mills played the role of project manager and fellow New England sports fan. Cliff was patient and encouraging and kept the project rolling through the first draft of the manuscript. Sadly, Cliff passed on before this project came to fruition. He is still sorely missed. As we completed the first draft of the manuscript, art editor Kathy Naylor joined the team. Kathy's experience in developing art to describe complex scientific principles has been invaluable. Her attention to detail and organizational skills have been incredible and have also helped make this such a better project than it would have been otherwise. As we began to edit the second half of the text, Rebecca Heider joined the team as Development Editor. Rebecca has an eye for the written word that I will never have and has made this a far better work through her edits and suggestions. Finally, at Wiley, I need to thank Sladjana Bruno, Executive Editor; Patricia Gutierrez, Production Editor and Media Specialist; Geraldine Osnato, Director of Content Enablement and Operations; Andrew Moore, Course Content Developer; Maureen Shelburn, Market Development Manager; and Michael Olsen, Marketing Manager.

Other people I need to single out and thank include my accuracy checkers Joseph Provost, Tanea Reed, Leanna Patton, and Hannah Bailey and Brodie Ranzau. Despite their best efforts and the work of hundreds of people I'm sure there are mistakes that have crept through. Any errors are mine, but please gently remind me of them to help make future editions of this a better work. I also need to thank Laura Lynn Gonzalez and her team at Dynamoid as well as Laura Zapanta, Spencer Jordan, Rebekah Dalton, and Erin Hughes for their work on the Dynamic Figures. I also need to thank Rakesh Mogul for his excellent bioinformatics exercises found in each chapter. I would be remiss if I didn't thank the members of my advisory panel (Laura Zapanta, Paul Craig, Emily Westover, Kerry Smith, Cheryl Ingram-Smith, and Ron Gary). These reviewers gave multiple rounds of detailed feedback. Their expertise in teaching and in different aspects of biochemistry helped shape this book in its early stages and throughout its development.

Finally, I would also like to thank all of the faculty members who helped review this work as it was being developed. This project would not be possible without their hard work, input, and suggestions.

Reviewers

Alabama
Jong Kim, Alabama A&M University

Arkansas
Anne Grippo, Arkansas State University
Denise Greathouse, University of Arkansas, Fayetteville

Arizona
Scott Lefler, Arizona State University

California
Jinah Choi, UC–Merced
Douglas McAbee, California State University, Long Beach
Elizabeth Komives, UC San Diego
Rakesh Mogul, California State Polytechnic University, Pomona
Joseph Provost, University of San Diego
Brian Sato, University of California, Irvine
Jay Brewster, Pepperdine University
Ellis Bell, University of San Diego
Pavan Kadandale, University of California, Irvine
Elaine Carter, Los Angeles City College

Florida
Frank Mari, Florida Atlantic University
David Brown, Florida Gulf Coast University
Alberto Haces, Florida Atlantic University
Sulekha Coticone, Florida Gulf Coast University

Georgia
Mary Peek, Georgia Institute of Technology

Illinois
Paul Strieleman, University of Chicago
Keith Gagnon, Southern Illinois University

Indiana
Andrew Kusmierczyk, Indiana University–Purdue University Indianapolis
Ann Kirchmaier, Purdue University
Ann Taylor, Wabash College

Iowa
Charles Brenner, University of Iowa

Kentucky
Tanea Reed, Eastern Kentucky University
Kevin Williams, Western Kentucky University

Louisiana
Patricia Moroney, Louisiana State University
Lucy Robinson, LSU Health Sciences Center–Shreveport

Maryland
Pamela S. Mertz, St. Mary's College of Maryland
Allison Tracy, University of Maryland, Baltimore County

Massachusetts
Emily Westover, Brandeis University
Adele Wolfson, Wellesley College

Michigan
Robert Stach, University of Michigan–Flint
Michael Pikaart, Hope College
Deborah Heyl-Clegg, Eastern Michigan University
Stephen Juris, Central Michigan University

Minnesota
David Mitchell, College of Saint Benedicts, Saint John's University
Henry Jakubowski, College of Saint Benedicts, Saint John's University
Debra Martin, Saint Mary's University of Minnesota

Missouri
Peter Tipton, University of Missouri–Columbia
Ruth Birch, Fontbonne University
Brent Znosko, Saint Louis University
Blythe Janowiak, Saint Louis University

Nebraska
Robert Mackin, Creighton University
Madhavan Soundararajan, University of Nebraska–Lincoln
Rich Lomneth, University of Nebraska–Omaha
Jodi Kreiling, University of Nebraska–Omaha

Nevada
Ronald Gary, University of Nevada, Las Vegas

New Mexico
David Bear, University of New Mexico School of Medicine
Martina Rosenberg, University of New Mexico

New York
Paul Craig, Rochester Institute of Technology
Scott Bello, Rensselaer Polytechnic Institute
Christopher Stoj, Niagara University
Carla Theimer, University at Albany, SUNY

North Carolina
Brooke Christian, Appalachian State University
Megen Culpepper, Appalachian State University
Brian Coggins, Duke University

North Dakota
Daniel Barr, University of Mary

Ohio
Michael Lieberman, University of Cincinnati College of Medicine
Tim Mueser, University of Toledo
Amy Stockert, Ohio Northern University, Raabe College of Pharmacy
John Cogan, The Ohio State University
Stephen Mills, Xavier University

Oklahoma
Donald Ruhl, Oklahoma State University

Pennsylvania
Laura Zapanta, University of Pittsburgh
Daniel Dries, Juniata College
Felicia Barbieri, Gwynedd Mercy University

South Carolina
Kerry Smith, Clemson University
Cheryl Ingram-Smith, Clemson University
Srikripa Chandrasekaran, Clemson University

Texas
Kayunta Johnson-Winters, University of Texas at Arlington
Mian Jiang, University of Houston Downtown

Utah
Bethany Buck-Koehntop, University of Utah

Virginia
Deborah Polayes, George Mason University
Pinky McCoy, Old Dominion University

Washington
Jeff Watson, Gonzaga University
Jeffery Corkill, Eastern Washington University

West Virginia
Michael Gunther, West Virginia University School of Medicine

Wisconsin
Kevin Siebenlist, Marquette University, College of Health Sciences
Jim Lawrence, University of Wisconsin, Stevens Point

Canada
Isabelle Barretee-Ng, University of Calgary
Adrienne Wright, University of Alberta
Odette Laneuville, University of Ottawa
Daman Bawa, University of Toronto Scarborough

United Kingdom
Kevin Gaston, University of Bristol

CHAPTER 1

The Chemical Foundations of Biochemistry

Chemistry in Context

An archeologist who has discovered a lost civilization would be confronted with numerous questions. How did that civilization function? How did it store food and confront challenges to survive? How did it archive and pass information from one generation to the next? If there were some way to translate the writings found into something understandable, we would stand a better chance of discovering what went on in that civilization.

This story is an analogy for where we begin with biochemistry, the study of life at the molecular level. In this analogy, the civilization is any system we choose to study, from the bacterium *Escherichia coli* to humans. From the perspective of the biochemist, the language that unites these systems and explains how and why things occur is chemistry.

This chapter focuses on several topics that will be helpful in defining and explaining biochemical reactions throughout the rest of the text. These topics include thermodynamics, equilibrium, organic reaction mechanisms, and some of the basics of structure and function. The chapter then goes on to discuss water, the properties of acids and bases, and buffers (systems resistant to changes in pH). It is intended to be a review of specific topics pertinent to biochemistry.

The dialect of chemistry that we speak in biochemistry is specialized. Biochemistry deals in aqueous systems; in small, multifunctional organics; and in large polymers. The systems have been optimized by eons of natural selection and will continue to be altered throughout time as conditions and selective pressures change. However, the foundational laws that all systems must abide by are chemical in nature.

Common Themes

Evolution's outcomes are conserved.	• Evolution takes advantage of the basic laws of chemistry, and it must abide by those laws as systems change.
Structure determines function.	• The rules that dictate the behavior of small chemical structures also apply to macromolecules.
	• The properties and reactivity of organic functional groups also apply to biochemical systems.
Biochemical information is transferred, exchanged, and stored.	• Information is stored in polymers of organic compounds.
	• Flow, exchange, and storage of information all follow the basic laws of chemistry.
Biomolecules are altered through pathways involving transformations of energy and matter.	• The basic laws of chemistry and thermodynamics apply to biochemical systems as they do to classical chemical systems.

1.1 General Chemical Principles

Chemistry is the study of matter. We observe molecules and how they behave, we create new ways to synthesize molecules, we characterize molecules using light or other wavelengths of the electromagnetic spectrum, and we determine how molecules form and break down. General chemistry, the introductory subject taught in high school and the first year of college, is a combination of several different subdisciplines of chemistry, including:

- *organic* chemistry—the study of carbon-containing molecules
- *inorganic* chemistry—the study of compounds comprised of elements other than carbon
- *analytical* chemistry—the identification, quantitation, and precise analysis of compounds
- *physical* chemistry—the study of the interactions of matter with energy

General chemistry covers several topics that are highly pertinent to biochemistry and is a good place to begin a review of chemical principles. Most of the topics covered in general chemistry are important to biochemistry and can help to explain many important biological principles, but they are not often discussed in a biochemical context in a general chemistry course. Few, if any, biochemical examples are used, and students are sometimes left with the impression that the material lacks relevance. Nothing could be further from the truth. Biochemical examples can be used to illustrate any chemical principle. Freed from this bias, you can begin to think of biological questions as having answers written in the language of chemistry.

1.1.1 Principles of thermodynamics describe all systems

Thermodynamics is the study of energy, and it is an important concept in biochemistry. Studying the thermodynamics of a reaction or system gives us valuable information about how and why a reaction can occur. In some cases, it can also provide information about the conditions required for a reaction to occur: for example, whether energy is required, if the reaction sits on the knife edge of an equilibrium constant, or if the loss of a group such as carbon dioxide (CO_2) helps to drive the reaction forward. Most important, it can tell us whether a reaction will proceed in the forward direction. Thus, understanding the basics of thermodynamics helps to paint a rich chemical picture of what is happening in a biological process.

Fundamental laws describe energy in thermodynamics
This section briefly reviews the three fundamental laws of thermodynamics in the context of biochemical systems.

The first law of thermodynamics states that all energy is conserved. That is, energy cannot be created or destroyed but only converted from one form to another. This law of thermodynamics is one of the easiest for people to understand, yet it can be one of the most complex when dissected. When we discuss the conservation of energy in a strictly physical sense, we often think of potential and kinetic energy; this can be illustrated by the example of a roller coaster. At the top of a hill, the roller coaster has potential energy, but as it moves down the hill, some of that potential energy is converted to kinetic energy (the speed of the roller coaster). For biological systems, we often think of potential energy in terms of the energy trapped in the chemical bonds of food or other energy-storage molecules such as adenosine triphosphate (ATP) (**Figure 1.1**). The kinetic energy of a biological system could be the locomotion of that organism or the movement of fluids or gases in chambers of the organism, but there are many other ways in which energy can be used. For example, energy in biological systems can be used to generate electrochemical potentials by pumping ions across an otherwise impermeable barrier, heat energy can be released in biological systems to increase the temperature of the organism or the immediate surroundings, and chemical energy can be converted into a photon of light by bioluminescent organisms. There is no shortage of ways that energy can be used, but all of the ways obey the first law of thermodynamics.

When we discuss the energetics of any reaction, we can describe the energy as being exchanged in terms of heat or in terms of work. Heat is the amount of thermal energy that is dissipated or absorbed by a system, whereas work is generally defined as a force applied over a distance or as a change in pressure or volume. Because most biochemical systems operate under constant pressure, the heat of a reaction is a close approximation of the energy of that system. We define this energy as the **enthalpy**, **H**.

*The second law of thermodynamics states that the **entropy, S**, of the universe is always increasing and that the entropy of any system tends toward increasing randomness and disorder.* This definition should be familiar to us on a macroscopic scale. If we consider gas molecules distributed throughout a flask, these molecules are randomly distributed. Similarly, the air we breathe has a statistically random sample of gas molecules. It would be unthinkable to walk down the street and encounter a pocket of carbon dioxide molecules that had randomly and spontaneously assembled out of the mixture. A better definition is the number of potential ways we could arrange those molecules or, to put it more scientifically, "the number of microstates a system can achieve." This definition is somewhat more detailed than disorder, and it can be used to explain a greater variety of phenomena. Imagine a system containing two objects, one of which is cool, the other warm. The atoms that make up the warmer object have more kinetic energy, more molecular motion, and more microstates, while the cooler atoms have less motion and therefore fewer microstates. According to the second law, it is impossible to trap and segregate the thermal energy from the cool object and use it to further heat the warm object. Therefore, it is not possible to use an ice cream cone to cook a hot dog, and a metal bar will not spontaneously segregate energy from one end of the bar to the other, cooling one end while warming the other.

The third law of thermodynamics states that the entropy of a perfect crystalline system at 0 Kelvin is 0. Consider an absolutely pure and perfect crystal of a monomolecular solid. As the crystal cooled to lower and lower temperatures, it would have less and less vibration. As the crystal approached 0 K, the molecular motion would decrease and the number of possible microstates would also approach zero. Of the three laws of thermodynamics, this one seemingly has the least impact on biochemistry because living systems never operate at such low temperatures. However, this

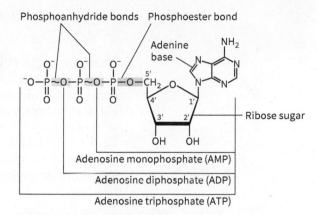

FIGURE 1.1 Structure of ATP. Adenosine triphosphate contains three phosphate groups linked via a phosphoanhydride or phosphoester bond to adenosine.

(Source: Wessner, *Microbiology*, 1e, copyright 2013, John Wiley & Sons. This material is reproduced with permission of John Wiley & Sons, Inc.)

definition of the third law allows us to define entropy (0 K = no entropy) and therefore merits mention here.

Free energy relates both enthalpy and entropy in a single useful term

Building from the three laws of thermodynamics, there are two key parameters to discuss: enthalpy and entropy. Although enthalpy is useful in predicting the direction of a reaction, not all of the energy released can be used as enthalpy because some of it is consumed in the generation of entropy. Likewise, while entropy is important, the amounts of entropy are typically much smaller than enthalpy for a given reaction. While neither term alone can accurately describe the energetics of a reaction, the **Gibbs free energy** or simply **free energy** relates these two parameters in a simple expression:

$$G = H - TS$$

Often, free energies are compared to one another. Due to experimental constraints, these are expressed not as ultimate values (G) but rather as the change in the value (ΔG). The Gibbs free energy expression thus becomes:

$$\Delta G = \Delta H - T\Delta S$$

where the change in the free energy is related to the change in the enthalpy of the reaction, minus the change in the entropy multiplied by the temperature, measured in Kelvin (**Table 1.1**).

The equilibrium constant of a reaction can also be expressed in terms of ΔG. Although a negative value of ΔG is often thought of as meaning that a reaction is favorable, it really means that it is favorable in the forward direction. Hence, a ΔG value of zero indicates that a system is at equilibrium, a negative value indicates that the reaction proceeds in the forward direction, and a positive value indicates that the reaction is favorable in the reverse direction.

Although ΔG can predict the net direction of a reaction, it does not affect $\Delta G^{\ddagger}$, **the free energy of the transition state** or the free energy of activation. In other words, ΔG gives the energy difference between the reactants and products, but not the energetic "hill" that has to be climbed to go from one to the other. Thus, ΔG does not predict reaction rate. Reactions with ΔG values of -30, -0.3, $+0.3$, or $+30$ kJ/mol may all proceed at the same rate.

The Gibbs free energy describes the amount of energy available to do work at constant temperature and pressure, or the amount of energy that is unavailable because it is lost to increasing

TABLE 1.1 Common Reactions with a Strongly Negative $\Delta G^{\circ\prime}$ Found in Biochemistry

Compound and Hydrolysis Reaction	$\Delta G^{\circ\prime}$ (kJ/mol)
Phosphoenolpyruvate → Pyruvate + P_i	−62.2
1,3-bisphosphoglycerate → 3-phosphoglycerate + P_i	−49.6
Creatine phosphate → Creatine + P_i	−43.3
Acetyl phosphate → Acetate + P_i	−43.3
Adenosine-5′-triphosphate → ADP + P_i	−35.7
Adenosine-5′-triphosphate → ADP + P_i (with excess Mg^{2+})	−30.5
Adenosine-5′-diphosphate → AMP + P_i	−35.7
Pyrophosphate → P_i + P_i (in 5 mM Mg^{2+})	−33.6
Adenosine-5′-triphosphate → AMP + PP_i (excess Mg^{2+})	−32.3
Uridine diphosphoglucose → UDP + glucose	−31.9
Acetyl-coenzyme A → Acetate + CoA	−31.5
S-adenosylmethionine → Methionine + adenosine	−25.6
Glucose-1-phosphate → Glucose + P_i	−21.0
Glycerol-3-phosphate → Glycerol + P_i	−9.2
Adenosine-5′-monophosphate → Adenosine + P_i	−9.2

entropy. Because biochemical systems are defined as operating under constant pressure rather than constant volume, the Gibbs free energy is used in most biochemical applications.

The standard state in biochemical systems In all chemical systems, including biochemistry, scientists define reference states or standard states. The **standard state** is defined by a temperature at 25°C (298 K), a pressure of 1 atmosphere, and solutes with an activity of 1 (roughly corresponding to their molar concentration). These criteria are used to define the **free energy of the standard state, $\Delta G°$**. The biochemical standard state is $\Delta G°'$. This includes the activity of water (the value used in free energy calculations) as 1 and the pH as 7.0.

The free energy of the standard state is important to know because it can provide important information regarding the favorability of a reaction, that is, whether or not it will proceed in the forward direction under standard conditions. However, by itself the free energy is not enough to predict the favorability of a reaction. To do so requires that the standard free energy change be related to the conditions in the reaction. Specifically, for any reaction where:

$$a\,A + b\,B \rightarrow c\,C + d\,D$$

$$\Delta G = \Delta G° + RT \ln([C]^c[D]^d/[A]^a[B]^b)$$

For any reaction, the direction of the reaction (the free energy) is a function of the free energy of the standard state, the temperature, T, the ideal gas constant, R, and the natural log of the ratio of the products to reactants in the reaction. While this final term may look similar to an equilibrium constant, recall that these concentrations are the ones found in the cell and are not at equilibrium. The final term is the reaction quotient, sometimes denoted as Q.

Some important biological molecules have a large negative free energy of hydrolysis Some molecules have groups that will hydrolyze, cleaving the bond through the addition of water, with large, negative free energy values. This happens because these molecules have certain properties such as charge–charge repulsion, bond enthalpy, and differences in hydration. Adenosine triphosphate (ATP) is the best known of these molecules (**Figure 1.2**), but there are others. Any of the nucleoside triphosphates (the other building blocks of ribonucleic acid, RNA) can and do serve as molecules that donate energy to reactions. Cleavage of the phosphoanhydride bond also increases the number of available resonance states and results in an overall increase in entropy. When these molecules are hydrolyzed, typically one or two phosphate groups are cleaved away from the nucleoside phosphate. The energy of this reaction is used to drive reactions that would otherwise be energetically unfavorable in the forward direction.

Many other phosphorylated groups, for example, phosphoenolates, also serve as a temporary repository of energy (**Figure 1.3**). The energy liberated in these reactions is also used to drive reactions. Recall that enols are tautomeric forms of a ketone. Phosphoenolates can be generated by

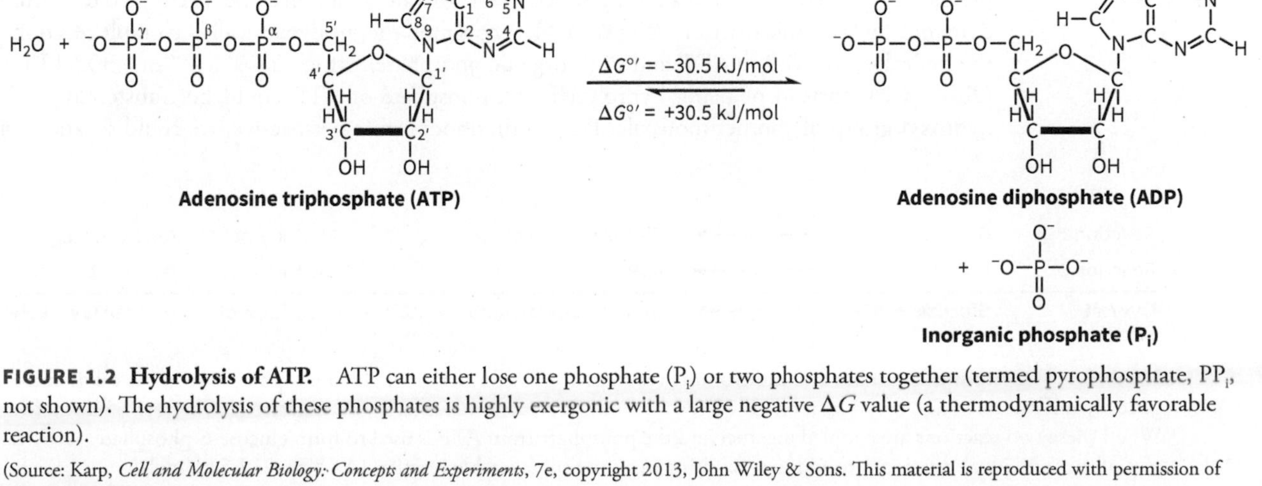

FIGURE 1.2 Hydrolysis of ATP. ATP can either lose one phosphate (P_i) or two phosphates together (termed pyrophosphate, PP_i, not shown). The hydrolysis of these phosphates is highly exergonic with a large negative ΔG value (a thermodynamically favorable reaction).

(Source: Karp, *Cell and Molecular Biology: Concepts and Experiments*, 7e, copyright 2013, John Wiley & Sons. This material is reproduced with permission of John Wiley & Sons, Inc.)

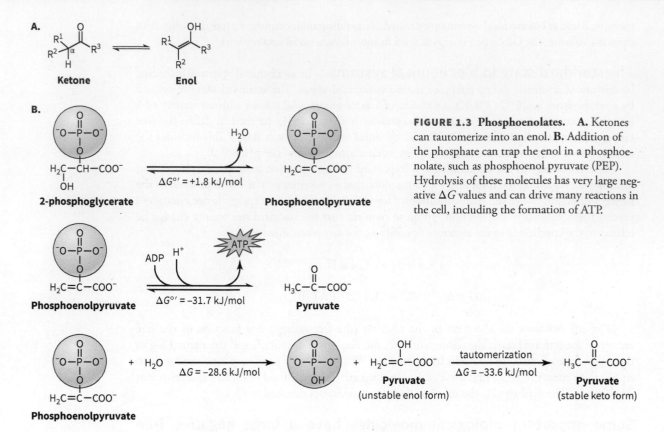

FIGURE 1.3 Phosphoenolates. A. Ketones can tautomerize into an enol. **B.** Addition of the phosphate can trap the enol in a phosphoenolate, such as phosphoenol pyruvate (PEP). Hydrolysis of these molecules has very large negative ΔG values and can drive many reactions in the cell, including the formation of ATP.

phosphorylating a hydroxyl group, followed by generation of a double bond via dehydration of a neighboring hydroxyl group. This phosphate has a much more negative free energy of hydrolysis and therefore a much higher transfer potential. These groups are used by the cell as one means of synthesis of ATP. The free energy of transfer from the enol to ADP is negative. Once the phosphate is lost, the enol tautomerizes into its keto form.

Knowing that these molecules hydrolyze spontaneously and have large negative ΔG values provides no information about rate. In the absence of a **catalyst** to accelerate the rate of breakdown, ATP can be isolated and stored indefinitely. In the instance of ATP, there is a relatively large activation energy barrier ($\Delta G^{\ddagger}$) that must be overcome for these molecules to react.

Unfavorable reactions can be made energetically possible by coupling them to favorable ones
On their own, many biochemical reactions are thermodynamically unfavorable in the forward direction ($\Delta G > 0$). For example, forming an ester between a phosphate group and a hydroxyl group generally has a positive free energy of reaction. In biochemistry, these reactions can still proceed in the forward direction by being coupled to thermodynamically favorable reactions (**Figure 1.4**). A common thermodynamically favorable reaction is the hydrolysis of ATP to ADP and Pi (inorganic phosphate), which has a $\Delta G^{\circ\prime}$ of -30.5 kJ/mol (**Table 1.2**). Instead of being hydrolyzed, the phosphate of ATP could be transferred to the hydroxyl group of another molecule. From a thermodynamic perspective, we could separate this

Reaction 1	Glucose + P$_i$ ⟶	Glucose-6-phosphate	+18.0 kJ/mol	(unfavorable)
Reaction 2	ATP + H$_2$O ⟶	ADP + P$_i$	−30.5 kJ/mol	(favorable)
Overall reaction:	**Glucose + ATP ⟶**	**Glucose-6-phosphate + ADP**	**−12.5 kJ/mol**	**(overall favorable)**

FIGURE 1.4 Coupling favorable and unfavorable reactions. The formation of glucose-6-phosphate in the cell is energetically unfavorable ($\Delta G^{\circ\prime} = 18$ kJ/mol). The hydrolysis of ATP, reaction 2, is large and negative ($\Delta G^{\circ\prime} = -30.5$ kJ/mol). When these two reactions are coupled together and the phosphate from ATP is used to form glucose-6-phosphate, the energies of the individual reactions are additive, resulting in an overall favorable reaction ($\Delta G^{\circ\prime} = -12.5$ kJ/mol).

TABLE 1.2 Common Organic Functional Groups Found in Biochemistry

Functional Group	Structure	Notes
Oxygen-containing groups		
Hydroxyl	R—O—H	Hydroxyl groups are common to all classes of biological molecules. They participate in numerous reactions including esterifications, dehydrations, and oxidations.
Carbonyl (aldehydes and ketones)	R—C—H, ‖O	The polar carbonyl group participates in numerous biochemical reactions. Frequently carbonyl carbons are subjected to nucleophilic attack by hydroxy or amino groups.
Carboxyl	R—C—O⁻, ‖O	Carboxyl groups are weak acids. They are common functional groups and participate in the formation of several other groups, including esters and amides.
Ester	R¹—C—O—R², ‖O	Esters are the result of a hydroxyl and carboxylic acid reacting with one another. Esters are commonly found in many lipids.
Sulfur analogs		
Thiol	R—S—H	Thiol groups can be thought of as hydroxyl analogs. They are found in the amino acid cysteine.
Disulfide	R—S—S—R′	Disulfides result from the reaction of two thiol groups together. Disulfide reducing agents cleave the disulfide and restore the two thiols.
Thioester	R¹—C—S—R², ‖O	Thioesters have a large negative free energy of hydroylsis.
Nitrogen analogs		
Amino	R—N⁺H, with H above and H below	Amino groups are common to all classes of biological molecules. The lone pair of electrons on the amino group can acquire a proton resulting in a positively charged amino group.
Amide	R—C(=O)—N(—H)(—R)	Amides are formed through the reaction of an amine and a carboxylic acid. Peptides are the amide linkages found in proteins, but other amide bonds are found throughout biochemistry.
Phosphorous-containing groups		
Phosphates	O⁻—P(=O)(O⁻)—OH	Phosphate, also known as inorganic phosphate, Pi, is commonly found in all biological systems.
Phosphoesters	R—O—P(=O)(OH)—OH	Phosphoesters are the result of the addition of a phosphate group to a hydroxyl group.
Phosphanhydrides	R¹—O—P(=O)(O⁻)—O—P(=O)(O⁻)—O⁻	Phosphanhydrides are commonly found in nucleoside triphosphates like ATP and have a large negative free energy of hydrolysis.
Other groups		
Enol	R—C(OH)=C(H)(H)	Enols are tautomers of carbonyl groups. In biochemistry they are often trapped as phosphoenolates.

reaction into two separate reactions: the hydrolysis of ATP to ADP and inorganic phosphate (P_i), and the esterification of phosphate to the hydroxyl of the new molecule. The energies of the two reactions are added together to find the net ΔG for the entire reaction.

ATP is used to drive many reactions, but in theory, any molecule with a favorable free energy of hydrolysis ($\Delta G < 0$) can be used.

1.1.2 Equilibrium concepts of specific thermodynamic state

Equilibrium refers to the chemical state when the rates of the forward and reverse state of a reaction are equal. The rates can be fast or slow, provided the net rate of product formed or reactant used is zero. Likewise, the concentrations of reactant and product do not need to be the same (and often are not). While it may seem confusing at first that a system can still be at equilibrium when the ratio of products to reactants is 1×10^9 or 1×10^{-9}, values commonly span this range.

We define the **equilibrium constant, K_{eq}**, as the ratio of products to reactants at equilibrium. For the given reaction at equilibrium:

$$a\,A + b\,B \rightleftharpoons c\,C + d\,D$$

$$K_{eq} = [C]^c\,[D]^d / [A]^a\,[B]^b$$

Combining this equation with our previous definition of the free energy of the reaction,

$$\Delta G = \Delta G° + RT \ln([C]^c[D]^d/[A]^a[B]^b)$$

we obtain an expression that defines the free energy of any reaction as the sum of the standard free energy and the temperature and gas law constant, multiplied by the natural log of the ratio of concentrations in the reaction.

Because equilibrium is defined here as the point at which there is no net reaction, ΔG at equilibrium is zero. As a result, $\Delta G°$ becomes:

$$\Delta G° = -RT \ln K_{eq}$$

Rearranging this equation describes the equilibrium constant in terms of the other parameters:

$$K_{eq} = e^{-\Delta G°/RT}$$

Effectively, this means that the free energy of a reaction can be used in the standard state to describe the equilibrium constant, or vice versa. The free energy is therefore simply another way to describe the equilibrium constant.

All systems tend toward equilibrium, despite the ratio of reactants to products. If reactants are mixed in a beaker in the absence of product, they will begin to react to form product. Likewise, if the products of this same reaction are mixed together in a beaker and allowed to react, reactants will eventually appear. If a system at equilibrium is in some way perturbed, it will react to return to equilibrium. This observation is known as **Le Châtelier's principle** and is a feature of all systems, including biochemical systems. If the concentrations of reactants are increased above the equilibrium concentration, the reaction will proceed in the forward direction to compensate by making more products. Likewise, if the concentration of product is increased, the reaction will proceed in the reverse direction to generate more reactant. A biochemical example of this occurs in erythrocytes (red blood cells). In tissues, erythrocytes acquire carbon dioxide from muscle and quickly convert it into carbonic acid (H_2CO_3) via the enzyme carbonic anhydrase. In the lung, carbon dioxide concentrations are low and the process moves in the reverse direction, generating carbon dioxide from carbonic acid.

Enzymes, or any catalyst for that matter, will not perturb or in any way alter equilibrium concentrations; they will only help a reaction to reach equilibrium more quickly.

Although often used interchangeably in the popular literature, equilibrium should not be confused with **homeostasis**, the ability of an organism to live within its energetic and chemical means. An organism living in homeostasis is not consuming more energy than it is using or using its energy stores; however, it is not at chemical equilibrium with its environment. An organism that has reached equilibrium with its environment is, in all biochemical ways of thinking, dead.

1.1.3 Kinetics approaches of the rate of chemical reactions

Kinetics is the field of study that analyzes the rates of chemical reactions. Some reactions involve a single reactant being transformed to a single product, while others involve multiple reactants leading to multiple products. The rate of a reaction depends on general factors including the concentration of products and reactants, temperature, and factors specific to that individual reaction. The rate of any chemical reaction can be described using a mathematical expression termed a rate law. These laws link the rate, the change in concentration over time, to the concentration of the starting materials multiplied by a constant, k, the rate constant. The rate constant takes into account the temperature and factors such as how often a collision between reactants leads to products.

$$A + B \rightarrow C$$

$$\text{Rate} = d[C]/dt = k[A][B]$$

In the simple example shown above, two molecules, A and B, react to form the molecule C. Based on the rate law given above, the rate of this reaction is linearly dependent on the concentration of both A and B; thus, doubling the concentration of A doubles the rate, and the same is true for B. The rate needs to be experimentally determined: the rate law for that reaction is then derived from experimental data; however, once a rate law has been determined, it is possible to calculate reaction rates from given concentrations, or concentrations from observed rates.

1.1.4 Thermodynamics, equilibrium, and kinetics are used together to describe biochemical reactions and systems

The direction of any reaction depends on the initial and final energies of the starting materials and products (ΔG), and the concentrations of the reactants. The rate of a reaction is related to the activation energy ($\Delta G^{\ddagger}$), not to the initial or final states. When a catalyst acts on a reaction, it lowers the activation energy, affecting the rate constant k. The catalyst does not affect ΔG, nor does it affect the equilibrium concentrations of reactants and products, which are also ΔG dependent. The lowering of the energetic barrier between products and reactants is the same if the reaction is going in the forward or reverse direction; therefore, the rates of *both* reactions are increased. Hence, a catalyst can help a system to reach equilibrium more quickly, but it cannot change equilibrium concentrations.

When considering biochemical systems, the same rules apply, but nearly all biochemical reactions are catalyzed by specific protein catalysts termed **enzymes**. Enzymes lower activation energy by forming more stable states and facilitating collisions between reactants (**Figure 1.5**).

Worked Problem 1.1	Using thermodynamics to examine reactions

Examine the following two reactions.

$$ATP + H_2O \rightarrow ADP + P_i \qquad \Delta G^{\circ\prime} = -30.5 \text{ kJ/mol}$$

$$Glucose + P_i \rightarrow Glucose\text{-}6\text{-phosphate} + H_2O \quad \Delta G^{\circ\prime} = +18.0 \text{ kJ/mol}$$

What is the standard free energy change for the formation of glucose-6-phosphate from glucose and ATP? That is, what is the standard free energy change for the reaction:

$$ATP + glucose \rightarrow Glucose\text{-}6\text{-phosphate} + ADP$$

Strategy The energies of thermodynamic properties, such as free energy, are additive. Combine the energies to find the overall energy of the reaction.

Solution Adding the free energies of the first two reactions, we see that the overall free energy for the two reactions combined (-30.5 kJ/mol $+ 18.0$ kJ/mol) is negative and exergonic (-12.5 kJ/mol). Therefore, the reaction will proceed in the forward direction.

The formation of glucose-6-phosphate from glucose and phosphate is energetically unfavorable in the forward direction. Rather, it is endergonic with a positive $\Delta G^{\circ\prime}$ value. Nevertheless, glucose-6-phosphate is found in the cell because of the presence of ATP. The large negative $\Delta G^{\circ\prime}$ of hydrolysis of ATP allows it to act as a phosphate donor. More important, it provides the energy needed for this reaction to happen.

Follow-up question The ΔG values used in the previous problem were in the standard state ($\Delta G^{\circ\prime}$). What factors will modify ΔG in the cell? How could ΔG be made more negative, and would this affect how quickly the reaction proceeds?

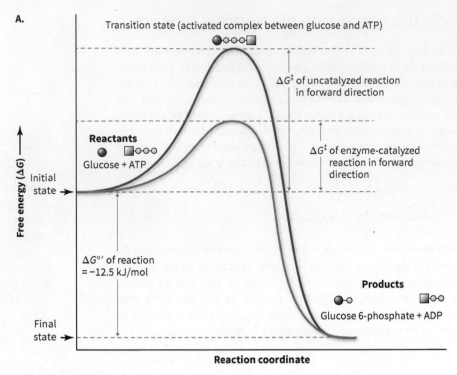

FIGURE 1.5 Reaction coordinates and catalysis. **A.** Enzymes do not alter the initial or final states but rather lower the activation energy by stabilizing the transition state ($\Delta G^{\ddagger}$), to increase rate. **B.** and **C.** This is done in part by constraining the reactants in the active site of the enzyme.

(Source: (A) Karp, *Cell and Molecular Biology: Concepts and Experiments,* 7e, copyright 2013, John Wiley & Sons. This material is reproduced with permission of John Wiley & Sons, Inc.)

It may be hard to conceptualize how "lower activation energy" translates to actual chemistry. Imagine two molecules such as ATP and glucose that may react together to form a glucose phosphate and ADP. Although these molecules could react with each other in the absence of a catalyst, simple random collisions between the molecules will rarely have the proper direction of collision and enough energy for the reaction to occur. Now imagine these two molecules held in an enzyme in the correct orientation to react with one another, or brought together so that a reaction is more favorable; this gives a picture of what the lowering of the transition-state energy looks like on a physical level.

1.1.5 ATP-the energy carrier

Biological systems use many forms of organic molecules as a source of metabolic energy. Among them, ATP is the most used molecule as the "energy carrier". Although other ribonucleotides like GTP contain the same amount of energy, ATP has been preferably used as the "energy currency" by living systems to power metabolism. Living cells synthesize ATP in three ways: substrate-level phosphorylation, oxidative phosphorylation, and photophosphorylation. The synthesis of ATP by transferring a phosphate from a substrate to ADP during a metabolic reaction is known as substrate-level phosphorylation, e.g., ATP production in the reaction catalyzed by pyruvate kinase (Section 6.3). In oxidative phosphorylation, ATP synthesis is coupled to the flow of electrons from reduced coenzymes produced during the oxidation of metabolic fuels via a chain of electron transport complexes (Figure 7.12). Photophosphorylation is the synthesis of ATP using solar energy by photosynthetic cells only. This mode of ATP production is very much like that of oxidative phosphorylation. The significant difference is that photophosphorylation uses sunlight as the ultimate energy source to eject electrons from water.

Summary

- There are three fundamental laws of thermodynamics:
  - Energy is conserved. In general, the energy of a biochemical system can be defined as its enthalpy.
  - The entropy of a spontaneous reaction is always increasing.
  - $S = 0$ at 0 Kelvin.
- Both enthalpy and entropy contribute to biochemical processes. We can aggregate these terms into one expression: the Gibbs free energy, ΔG.
- The free energy of a reaction describes the equilibrium state ($\Delta G = 0$) and whether a reaction will move in the forward ($\Delta G < 0$) or reverse direction ($\Delta G > 0$).
- The equilibrium state is one in which the rates of the forward and reverse reactions are equal; hence, there is no net production or loss of product or starting material. It can be expressed in terms of the equilibrium constant, K_{eq} (a ratio of the equilibrium concentrations) or in terms of free energy.
- Le Châtelier's principle states that all reactions seek to move toward equilibrium from other states.
- Kinetics is the study of chemical rates. Most biochemical reactions employ protein catalysts called enzymes to lower activation energies and increase reaction rates.
- ATP is the common energy molecule for driving metabolic reactions. Cells synthesize ATP through substrate-level phosphorylation, oxidative phosphorylation, or photophosphorylation.

Concept Check

1. Describe the basic concepts of thermodynamics, equilibrium, and kinetics in your own words.
2. Use these concepts to describe a basic biochemical problem.
3. Why is ATP called the energy currency of cells?

1.2 Fundamental Concepts of Organic Chemistry

Biochemistry draws from all fields of chemistry to help answer questions, but organic chemistry has been the classical gateway into this field. This happens largely because most biological molecules contain carbon. This section focuses on four aspects of organic chemistry to prepare for the study of biochemistry: functional groups, solubility and polarity, mechanism, and polymer chemistry.

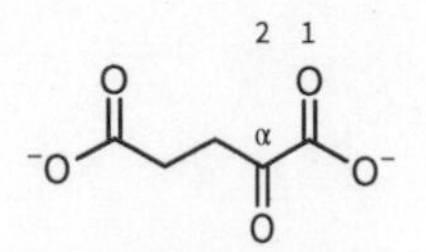

α-ketoglutarate

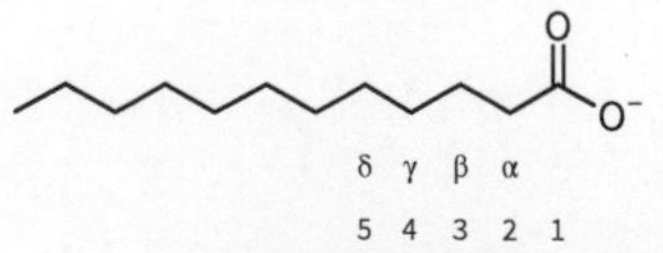

δ γ β α
5 4 3 2 1

Fatty acid

FIGURE 1.6 Organic nomenclature using the Greek lettering system. In α-ketoglutarate the carbonyl (keto) group is α to the higher priority carboxyl group. Fatty acids are broken down through a process termed β-oxidation. The oxidation occurs on the β-carbon.

1.2.1 Chemical properties and reactions can be sorted by functional groups

Functional groups are fragments of a molecule, such as a hydroxyl or carbonyl group. Understanding how each group reacts, as well as its properties, makes it possible to apply that information to other molecules. Likewise, if a functional group does not behave as expected, the hypothesis needs to be re-examined, and theories or understanding of the system may need to change.

Many of the common functional groups found in biochemistry will be familiar (Table 1.2). They include hydroxyl groups, carbonyl groups, amines, and carboxylic acids. Others, such as thiols and phosphates may be less familiar. These groups combine to generate more complex molecules such as amides and disulfides.

Many of the molecules studied in biochemistry were identified before systematic methods of nomenclature were adopted, and therefore they have common names. Other molecules use the older Greek lettering system instead of numbering to identify the location of functional groups in a molecule (**Figure 1.6**). Although this is not the International Union of Pure and Applied Chemistry (IUPAC) nomenclature system that is used in organic chemistry, dissecting the name of a molecule can still help provide clues to its structure and function. For example, as the name implies, α-ketoglutarate contains a carbonyl group (keto) α to a carboxylate (ending in the suffix "-ate"). Without knowing other information, such as the structure of glutarate, it is not possible to draw out the full structure, but the name still provides a guide. Likewise, fatty acids are broken down through a process termed β-oxidation. In this example, the oxidation occurs at the β-carbon.

1.2.2 The solubility and polarity of a molecule can be determined from its structure

All biochemistry occurs in aqueous systems. Water is the common denominator, and it is required in some way and at some level for all life to occur. When the National Aeronautics and Space Administration (NASA) searches for life on other planets, the first molecule they often look for is water. Water is the most common solvent used in biochemistry; hence, most biological molecules need to be soluble in water. Molecules can be made more water soluble by incorporating more polar groups, like hydroxyl groups and thiols, or charged groups, such as amine and carboxylate salts; phosphates; and, in some instances, sulfates.

Although all living systems require water, not all biochemical reactions occur in water. There are microenvironments in the cell or within individual macromolecules that exclude water. These systems are hydrophobic and nonpolar. Reactions that might not occur in an aqueous system can occur within a hydrophobic microenvironment in an enzyme. Likewise, although a cell is largely a watery place, hydrophobic groups on an otherwise water-soluble molecule can tether or anchor that molecule in a hydrophobic location; for example, a water soluble protein can be anchored within a hydrophobic layer of a cell membrane.

In addition to the structure and polarity of a molecule, stereochemistry also plays a key role in biochemical systems. This is discussed in **Societal and Ethical Biochemistry: The relevance of stereochemistry.**

Societal and Ethical Biochemistry

The relevance of stereochemistry

Stereochemistry plays a critical but subtle role in biochemistry. Due to the nature of biological molecules—mostly organic, often with multiple functional groups—many of them are chiral. As we build larger molecules from smaller chiral building blocks, these macromolecules are also chiral. Being chiral themselves, enzymes often catalyze chiral or stereospecific reactions. Overall, most of biochemistry is chiral or contains molecules that are chiral in at least one carbon center. We may refer to these pairs of isomers using the R and S system, in which stereoisomers are assigned by priority of functional groups, or the D and L system, in which chiral molecules rotate plane polarized light.

Fortunately, through evolution, nature uses a conserved set of molecules; with few exceptions, these have the same stereochemistry. Therefore, humans can obtain the amino acids they need to build new proteins from their diet (from animal, plant, or even microbial sources) and not worry about the chirality of those amino acids. The same is not true, however, for molecules made in the laboratory.

It can be difficult and expensive to produce stereochemically pure substances. Instead, when creating molecules where chirality may be important, the effort is spent on removing chiral centers or preventing their formation in the first place. A good illustration of this is in pharmaceutical development. Many pharmaceuticals bind to, and thus block or activate, either enzymes or receptors, leading to downstream biological effects. Although many pharmaceuticals have been designed to avoid chiral centers as far as possible, a lot of drugs still have them.

Having different stereoisomers of a drug can have a variety of effects. For example, the over-the-counter pain reliever ibuprofen has a single chiral center; this drug is supplied as a racemic mixture or mixture of the two stereoisomers, but only the S isomer is biologically active. The R isomer has no known effect. Conversely, for the drug naproxyn, the S enantiomer is the over-the-counter pain reliever, whereas the R enantiomer causes liver toxicity or hepatotoxicity. Therefore, naproxyn is synthesized and sold as the stereochemically pure S enantiomer. Similarly, dextromethorphan is a common over-the-counter cough suppressant, or antitussive, whereas levomethorphan, its enantiomer, is a powerful narcotic opioid analgesic.

Pharmaceutical companies are now aware that racemic drugs are likely to have an active and an inactive component, or one isomer that is causing the desired effect and one that is causing other effects or no effect. Advances in synthetic chemistry have made it possible to produce and test enantiomerically pure compounds, that is, only the R or S isomers. Often, when the pure drug is tested, it is found that a lower dosage of the drug can be used, and adverse effects, due to the other isomer or to metabolites of the drug, can be avoided. For example, citalopram is a selective serotonin reuptake inhibitor (SSRI), an antidepressant supplied as a racemic mixture. Although the adverse effects of this drug are relatively mild, there are common effects of SSRIs, such as weight gain and decreased libido. The stereochemically pure L isomer (escitalopram) minimizes these adverse effects. This raises some interesting scientific questions, such as whether the D isomer is responsible for the adverse effects, and if so, why and how? A smaller amount of the pure drug (the L isomer) is needed, which provides more options for delivery systems, and this puts less of a metabolic burden on the liver to detoxify other stereoisomers.

There are other advantages for pharmaceutical companies in pursuing stereochemically pure compounds. Because the effectiveness and safety of the parent compound has already been demonstrated, it is often easier and quicker to bring these drugs to market. Also, drugs that are stereochemically pure are considered new in terms of patent protection. Thus a drug company can theoretically extend patent protection on a drug, up to another 17 years (the length of a patent on a pharmaceutical compound).

R-Naproxen

S-Naproxen

Dextromethorphan

Levomethorphan

1.2.3 Reaction mechanisms attempt to explain how a reaction occurs

An organic reaction mechanism is an evidence-based explanation of how a reaction occurs. In a mechanism, we follow the flow of electrons from electron-rich nucleophiles to electron-poor nuclei. In reaction mechanisms, the direction of the arrow indicates the direction of electron flow. The headedness of the arrow indicates the number of electrons. A double-headed arrow ($\rightarrow$) means two electrons are flowing, whereas a single-headed arrow ($\rightharpoonup$) indicates one electron, or radical transfer.

The reaction mechanisms covered in organic chemistry are generally relatively simple in comparison to the enzymatically catalyzed reaction mechanisms in biochemistry. Biochemical reaction mechanisms can be quite complex, employing multiple steps, such as the mechanism of triose

Worked Problem 1.2   Properties of a biological molecule

Shown below is the structure of sphingomyelin, a molecule that is involved in cell-to-cell communication and has a structural role in the cell membrane. Which functional groups can you identify in sphingomyelin? Would you anticipate that this molecule would be soluble in water or in a nonpolar solvent? Which functional groups in this molecule would interact with water molecules and how?

$$\text{OH}$$
$$\text{CH}_2\text{—O—P—O—CH}_2\text{—CH}_2\text{—N}^+\text{—(CH}_3)_3$$
$$\text{NH}$$
$$\text{O}$$

Strategy   Compare the figure of sphingomyelin to Table 1.2. Which functional groups can you identify? Are those groups polar or nonpolar?

Solution   There are multiple functional groups in sphingomyelin: a quaternary amine, a phosphate in a phosphodiester, an amide, a hydroxyl, and an alkene. The charged moieties (the phosphate and amine) and the hydroxyl will help make that end of the molecule water soluble. The amide has lone pairs of electrons on both the oxygen and nitrogen; these can participate in hydrogen bonding as hydrogen-bond acceptors but not as hydrogen-bond donors. The hydrocarbon tails, on the other hand, are not going to be water soluble, making the entire molecule **amphipathic**—literally, a molecule with affinities for both water and oil.

$$\text{OH}$$
$$\text{CH}_2\text{—O—P—O—CH}_2\text{—CH}_2\text{—N}^+\text{—(CH}_3)_3$$
$$\text{NH}$$
$$\text{O}$$

Follow-up question   What would be the products if the amide bond of the molecule were hydrolyzed?

phosphate isomerase (**Figure 1.7**). In either instance, the same basic rules apply. A reaction mechanism explains the making and breaking of bonds at the molecular level. Arrows indicate the number of electrons and direction of electron flow, and electrons flow from regions of high electron density to regions of low electron density. Frequently in biochemistry these mechanisms employ nucleophiles and electrophiles similar or identical to ones that you have seen in organic chemistry.

A. Reaction

$$\text{CH}_3\text{Cl} + \text{OH}^- \longrightarrow \text{CH}_3\text{OH} + \text{Cl}^-$$

B. Organic reaction mechanism

$$\text{OH}^- \quad \text{H}\overset{\text{H}}{\underset{\text{H}}{—}}\text{Cl} \longrightarrow \left[\text{HO}\overset{\text{H}}{\underset{\text{H H}}{\cdots}}\text{Cl} \right]^{\ddagger} \longrightarrow \text{HO}\overset{\text{H}}{\underset{\text{H}}{—}}\text{H} \quad \text{Cl}^-$$

Nucleophile attacks   **Transition state**   **Substitution product**
alkyl halide

C. Enzyme reaction mechanism

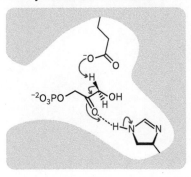

1.2.4 Major constituent biomolecules in living organisms

Many of the molecules discussed so far are small, with molecular weights below 500 g/mol. However, many if not all of these small molecules have functional groups such as alcohols, amines, and carboxylic acids that facilitate the formation of **polymers**. Polymers are macromolecular assemblies of smaller building blocks termed **monomers**,

FIGURE 1.7 Reaction mechanisms.   A. A typical substitution reaction studied in organic chemistry. **B.** The mechanism of the example shown in A (an S_N2 substitution reaction). **C.** A biochemical reaction mechanism, while more complicated, illustrates the underlying principles, such as electron flow, intermediates, and nucleophilic attack, are the same.

and they may be connected in a linear fashion, like the cars of a train, or in a branched fashion, like the branches of a tree.

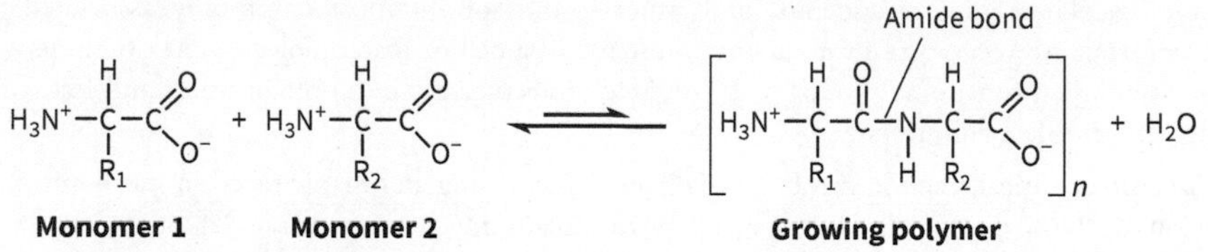

Monomer 1   **Monomer 2**   **Growing polymer**

Worked Problem 1.3   Creating a polymer

The ethylene glycol ester of *N*-formylglycine reacts with tris(hydroxymethyl)aminomethane (referred to as TRIS) to form an amide bond. The reaction is shown below:

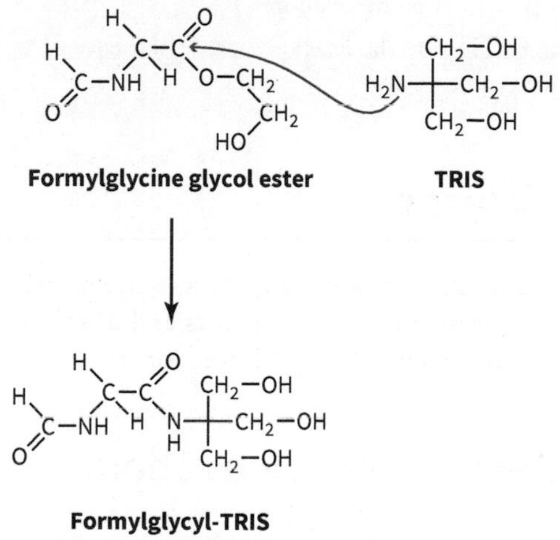

Formylglycine glycol ester   **TRIS**

Formylglycyl-TRIS

Propose a possible mechanism for this reaction. Can these compounds polymerize?

Strategy   Examine the given reaction. What are the reactants and products? How might these starting materials react together to yield the final product?

Solution   The mechanism of amide bond formation in the ribosome, the protein assembly complex in the cell, involves an amino acyl ester as the donor group for adding amino acids to the growing protein. Ester aminolysis is also a common organic synthetic technique.

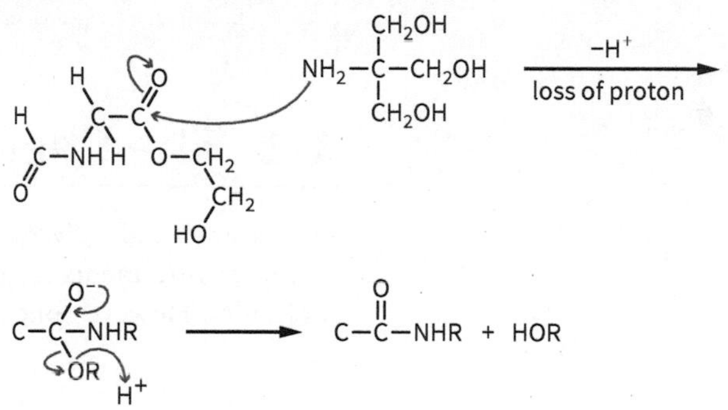

These compounds cannot polymerize. To simplify and facilitate study of this reaction, a formyl protecting group (CHO) is coupled to the amino group of the glycine glycol ester. The formyl group ensures that only one reaction can occur—the reaction between the formylglycine glycol ester and the amino group of TRIS. If that formyl group was missing, the glycine groups could probably polymerize.

Follow-up question   Would you anticipate that this reaction would proceed more favorably at an acidic pH or a basic pH?

While biochemistry is often thought of as the study of small organic molecules, it is equally a study in polymer chemistry and macromolecules. DNA, RNA, proteins, and enzymes are polymers, as are molecules such as glycogen, starch, and cellulose. Other molecules, for instance, some lipids, are not polymers, but they are assembled in a modular fashion, similar to polymers. Common themes are found throughout the biosynthesis and degradation of these molecules.

Polymers in biochemistry can be broadly categorized as those that have been coded for by some sort of template, for example DNA, RNA, and proteins, and those that have their structures determined by other means, such as carbohydrates. In both instances, the bonds that hold the polymer together generating amides, phosphodiesters, or acetals have a higher energy than the starting materials. That is, formation of a polymer requires an input of chemical energy. Polymers are formed using activated intermediates, that is, compounds that can "donate" a monomer to the growing polypeptide chain. Whereas the synthesis of a polymer requires the input of energy, the net reactions of the breakdown of a polymer releases energy. This release of energy comes not from breaking chemical bonds but in the formation of new, more stable products. Therefore, polymers can be used to store energy and drive otherwise unfavorable reactions.

Summary

- In organic chemistry molecules are categorized by functional groups. Many of the same functional groups are used in biochemistry.
- All biological systems are aqueous, and water is the solvent most commonly associated with biochemistry; however, certain microenvironments in a cell or macromolecule may exclude water. Polar or charged functional groups make organic molecules more soluble in water and less soluble in a hydrophobic environment.
- An organic chemical reaction can be partially explained using an organic reaction mechanism. The reaction mechanism shows how electrons flow, to explain the chemistry observed. Such diagrams are useful in dissecting how complex molecules such as enzymes function.
- Many biological molecules are polymers of simpler building blocks. Proteins are made from amino acids, nucleic acids from nucleotides, and complex carbohydrates from simple sugars. Other molecules including some lipids are not polymers but are made in a similar modular fashion.

Concept Check

1. Identify polar and nonpolar functional groups.
2. Identify which functional groups make a molecule more soluble in water, and which make it more soluble in a nonpolar environment, such as hexane or olive oil.
3. Describe the basic guidelines for any organic reaction mechanism.
4. What does a reaction mechanism illustrate? How do the reaction mechanisms covered in organic chemistry compare with those in biochemical reactions?

1.3 The Chemistry of Water

Life is amazingly diverse and robust. It can exist without oxygen, at extreme temperatures, and in environments ranging from dilute to nearly saturated with salts and dissolved inorganic molecules. However, one common denominator to all of life is the need for water.

1.3.1 The structure of water provides clues about its properties

Water is so common on Earth that it is easy to forget that it is unique among solvents (**Figure 1.8**). Water has two lone pairs of electrons on the oxygen atom and two polar H–O bonds. The polarity in the oxygen–hydrogen bond comes from the unequal sharing of electrons between the weakly electronegative hydrogen and the highly electronegative oxygen atom. The lone

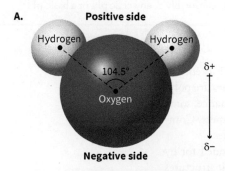

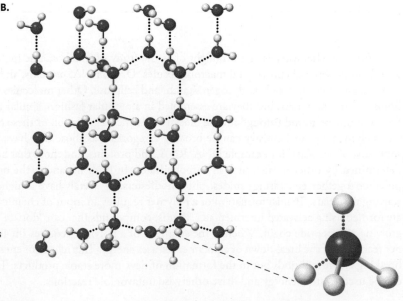

FIGURE 1.8 Structure of water. A. A single molecule of water has a 104.5° bond angle and a dipole moment running through its center. It has a bent geometry due to lone pairs of electrons on oxygen. **B.** In ice, water forms a tetrahedral lattice with each water molecule forming four hydrogen bonds with other molecules.

(Source: (A) Wiley Trefil: *The Sciences: An Integrated Approach*, 7e. copyright 2013, John Wiley & Sons. This material is reproduced with permission of John Wiley & Sons, Inc.

A.

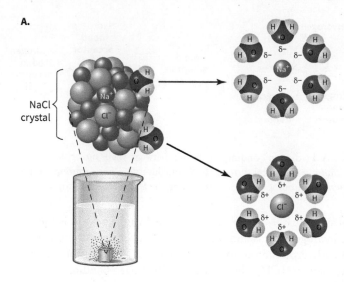

B. Dielectric Constants of Common Laboratory Liquids

Solvent	Dielectric Constant (D)
Water	78.5
Ethyl alcohol	24.3
Acetone	20.7
Acetic acid	6.2
Benzene	2.3
Hexane	1.9

FIGURE 1.9 Coulombic interactions. Coulombic interactions describe the forces that hold ions together. **A.** Sodium and chloride ions in a crystal are disrupted and dissolved by water molecules. The water molecules form hydration shells around each of the ions. **B.** The dielectric constant of some common solvents. Water has one of the highest dielectric constants known.

(Source: (A) Karp, *Cell and Molecular Biology: Concepts and Experiments*, 7e, copyright 2013, John Wiley & Sons. This material is reproduced with permission of John Wiley & Sons, Inc.)

pairs of electrons give water a bent shape, with a nearly tetrahedral bond angle (104.5° in water compared to 109.5° in a truly tetrahedral center). Water's bent shape allows its polar H–O bonds to contribute to an overall dipole moment which runs through the entirety of the water molecule.

The polarity of water influences the strength of ionic interactions

Chemists categorize the polarity of a solvent using a constant known as the **dielectric constant**, the ability of an insulating liquid to carry current, measured in Debye units, D. Vacuum and air both have dielectric constants close to 1 D. Typical nonpolar solvents used in organic chemistry (hexane, benzene, or oils) have dielectric constants of 2 to 4 D (**Figure 1.9**). Ethanol and acetone have dielectric constants in the low 20s. Water, on the other hand, has a dielectric constant of 80 D, one of the highest values known.

The dielectric constant comes into play in coulombic interactions, defined by an equation that describes the interactions between charged species. Coulomb's law is:

$$F = kQ_1Q_2/r^2D$$

where F is the force between the two charges, k is the Coulomb's law constant, Q_1 and Q_2 are the charges, r^2 is the square of the distance between the forces, and D is the dielectric constant of the medium containing the charges.

Coulomb's law states that the strength of an ionic interaction is the product of the charge of the two charges and a constant, divided by the square of the distance between those ions and the dielectric constant. This means that the force holding the two ions together decreases as the distance between the ions increases, or as the strength of the dielectric constant increases.

The high dielectric constant of water means that it can readily solubilize many ionic solids and significantly decrease ionic interactions. As a simple example, table salt can dissolve in water but not in vegetable oil. The interactions of water molecules with the sodium and chloride ions contribute more than the interactions of those ions with one another. Vegetable oil, with a far lower dielectric constant, is unable to overcome the interactions of the ions in the crystal and, as a result, the salt will not dissolve.

Although water can decrease the strength of ionic interactions, this observation should be used cautiously. Biochemistry is rich with microenvironments that may exclude water and strengthen ionic interactions.

1.3.2 Hydrogen bonding among water molecules is one of the most important weak forces in biochemistry

Hydrogen bonds are a weak force resulting from a strongly electronegative atom (typically oxygen, nitrogen, or fluorine) bound to a hydrogen atom (**Figure 1.10**). The polarity of the bond between

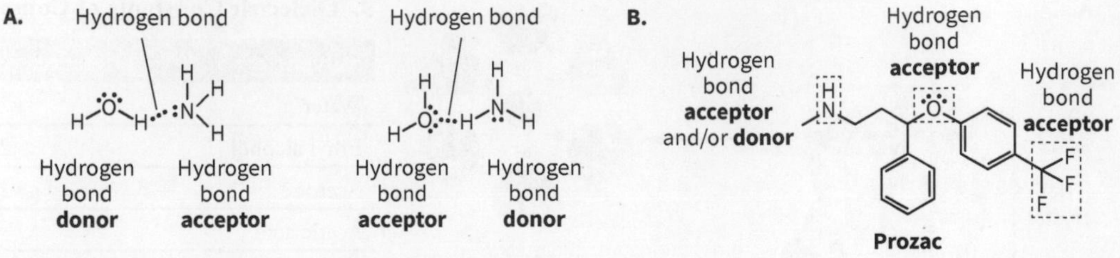

FIGURE 1.10 Hydrogen bonding. A. Hydrogen bonding is common between water molecules or between water and other molecules found in biological systems. **B.** While fluorine is not commonly found in biochemistry, it is an important element in numerous drugs, including the antidepressant Prozac.

these atoms results in a partial positive charge on the hydrogen. The hydrogen bond is the weak interaction between one of these hydrogen atoms and the lone pair of electrons on another electronegative atom. The presence of a hydrogen atom in a molecule is not sufficient for formation of a hydrogen bond; for example, methane (CH_4) does not exhibit hydrogen bonding.

Water has two hydrogen atoms bound to oxygen and two lone pairs of electrons; thus, it can form up to four hydrogen bonds at any one time. Water is unique in this regard, having equal numbers of hydrogen bond donors (hydrogen atoms bound to oxygen) and hydrogen bond acceptors (lone pairs). In liquid water, it has been calculated that any individual water molecule is involved in 3.7 hydrogen bonds at any one time.

The properties of hydrogen bonds   Hydrogen bonds are considered a weak force (10 to 40 kJ/mol in biological systems) when compared to a covalent bond (200 to 600 kJ/mol). The strength of a hydrogen bond is based partially on the angle between the electronegative hydrogen and the lone pair. A linear (180°) alignment produces the strongest bond. As with other bonds, hydrogen bonds have an optimal bond distance. For a hydrogen bond between a hydroxyl proton and the lone pair of electrons on another oxygen atom, the optimal length is 2.0 Å, but this can vary from 1.5 to 2.2 Å. The distance between the two non-hydrogen atoms is 2.5 to 3.2 Å.

Hydrogen bonds are not as strong as covalent bonds, but there are many hydrogen bonds within biological macromolecules. Water molecules and many biological molecules contain functional groups such as hydroxyl, amine, carbonyl, and carboxyl groups that can participate in hydrogen bonding, with tens to billions of hydrogen bonds occurring at all times in biological molecules. For example, hydrogen bonds lend strength to the cellulose in wood and plant fibers and proteins in wool and silk. Hydrogen bonds are also necessary to form the double helix of DNA.

Hydrogen bonds influence the properties of bulk water   Hydrogen bonds influence the properties of water in other ways. Water exhibits a property known as surface tension. The surface tension of water, which is illustrated by its ability to bead on a hydrophobic surface and resist an opposing force, is in part a product of the high degree of hydrogen bonding at the air–water interface. Biological systems take advantage of surface tension, but they also produce molecules to eliminate the tension. For example, water striders use surface tension to skim the surface of water, whereas in the lung, detergent-like molecules disrupt the surface tension making it easier for our lungs to expand, allowing us to breathe.

Water also has a high heat capacity, which is defined as the amount of energy needed to raise 1 gram of material by 1°C. At 4.18 J/g · K, water has a much higher heat capacity than most compounds. This is in part due to the energy needed to disrupt the hydrogen-bonding network of water to generate molecular motion—in other words, to raise the water's temperature. The heat capacity of water means that, in nature, water is relatively slow to heat up or cool down, and water is therefore liquid over a relatively broad temperature range. In effect, the heat capacity of water serves as a natural thermal buffer, helping to regulate the temperature of biochemical systems.

1.3.3 Water can ionize to acids and bases in biochemical systems

Acids, bases, and buffer systems are also important in biochemistry. There are several ways to describe acids and bases. Using the Arrhenius definition, acids are H^+ donors and bases generate an increased concentration of OH⁻. Although limited, this definition works well in many cases, especially when examining strong acids or bases. For compounds containing ammonia or an amine, the Arrhenius definition fails to explain the molecule's behavior. The **Brønsted-Lowry** definition of acids and bases categorizes acids as proton donors and bases as proton acceptors. Using this latter definition, we see that ammonia and most other amines accept a proton from water, making the solution more basic.

When examining biochemical systems, it is important to consider the protonation state of an acid or base in biochemical systems. A protonated amine is not a base; rather, it is an acid because it can act as a proton donor but not an acceptor.

The acid ionization constant and pK_a describe the strength of an acid

The acidity of a proton can be measured using its acid ionization constant, K_a:

$$K_a = [H^+][A^-]/[HA]$$

where H^+ is the proton concentration, A^- is the conjugate base of the acid, and $[HA]$ is the concentration of the acid at equilibrium, expressed in units of molarity.

Strong acids are almost 100% ionized in solution and hence have high K_a values. Conversely, weak acids are poorly ionized, often 1 molecule in 100 to 1 in 10 million, leading to small K_a values. These values are difficult to conceptualize, but taking the negative logarithm gives a whole number, the **pK_a** value, which is about 2.0 to 7.0 for weak acids (Table 1.3). The higher the pK_a is, the less acidic is the proton. Although in general chemistry we often think of the acidic proton in a weak acid as coming from a carboxylic acid, it is possible to determine a pK_a value for any proton. Hydroxyl protons have pK_a values of 14.0 to 16.0, whereas alkane protons may have values in the mid- to upper 30s, which is far beyond the values found in biochemistry. The pK_a value is a logarithmic rather than a linear value, so a change of three pK_a units is actually a 1,000-fold change in value. In addition, pK_a values can be modified significantly by neighboring atoms within a molecule. Introduction of electron withdrawing groups, or resonance states within the same molecule, or alterations to the local environment around a molecule of interest can change

TABLE 1.3 K_a and pK_a Values for Weak Acids Used in Biochemistry

Acid	K_a	pK_a
H_3PO_4	7.08×10^{-3}	2.15 (pK_1)
Succinic acid	6.17×10^{-5}	4.21 (pK_1)
Oxalate⁻	5.37×10^{-5}	4.27 (pK_2)
Acetic acid	1.74×10^{-5}	4.76
Succinate⁻	2.29×10^{-6}	5.64 (pK_2)
H_2CO_3	4.47×10^{-7}	6.35 (pK_1)
$H_2PO_4^-$	1.51×10^{-7}	6.82 (pK_2)
3-(N-Morpholino)propanesulfonic acid (MOPS)	7.08×10^{-8}	7.15
N-2-hydroxyethylpiperazine-N'-2-ethanesulfonic acid (HEPES)	3.39×10^{-8}	7.47
Tris(hydroxymethyl)aminomethane (TRIS)	8.32×10^{-9}	8.08
Boric acid	5.75×10^{-10}	9.24
NH_4^+	5.62×10^{-10}	9.25
Glycine (amino group)	1.66×10^{-10}	9.78
HCO_3^-	4.68×10^{-11}	10.33 (pK_2)
HPO_4^{2-}	4.17×10^{-13}	12.38 (pK_3)

Source: Dawson, R.M.C., Elliott, D.C., Elliott, W.H., and Jones, K.M., *Data for Biochemical Research* (3rd ed.), pp. 424–425, Oxford Science Publications (1986); *and* Good, N.E., Winget, G.D., Winter, W., Connolly, T.N., Izawa, S., and Singh, R.M.M., *Biochemistry* **5**, 467 (1966).

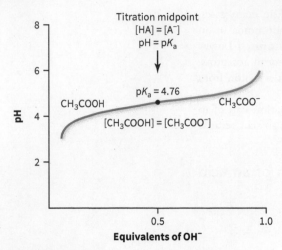

FIGURE 1.11 Buffers. Weak acids and bases will either ionize to release a proton or absorb one. Solutions buffer best over a range of ±1 pH unit around the pK_a value (the pH at which there are equal concentrations of conjugate acid and conjugate base).

a pK_a value radically. For example, molecules that would normally not deprotonate in aqueous solution often do so readily when bound to an enzyme.

The pH scale is used to measure hydrogen ion concentration
The hydrogen ion concentration of a system can range from 1 to 1×10^{-14} M. The pH scale transforms these concentrations into values that are easier to comprehend. Again, because these values are small, taking the logarithm of the value helps to generate numbers that are easier to work with. The **pH** is the negative logarithm of the hydrogen ion concentration, providing a scale that runs from 0 to 14.

In biochemistry, the pH of a reaction is critical; many important macromolecular interactions in biochemistry are dictated by weak forces that may be sensitive to changes in pH. Likewise, many catalytic mechanisms employ simple acid–base chemistry that fail to react outside a narrow pH range.

Biochemical systems are buffered from changes in pH
Buffers are chemical systems that are resistant to changes in pH. They are mixtures of weak acids, and their conjugate base (**Figure 1.11**). The mixture buffers the pH by acting like a proton sponge, releasing or absorbing protons to resist changes in pH. Buffers made with weak acids work best at pH values nearest the pK_a value of the weak acid in the system. The expression that is used to calculate the pH of a buffer system is the **Henderson–Hasselbalch** equation:

$$pH = pK_a + \log[A^-]/[HA]$$

where pH is the pH in question, the pK_a is the pK_a of the weak acid, $[A^-]$ is the molar concentration of the conjugate base of the weak acid, and $[HA]$ is the concentration of the un-ionized weak acid, also expressed in units of molarity, M. The Henderson–Hasselbalch equation illustrates why buffers are only effective at pH values near the pK_a of the acid or base. If the pH is more than 1 pH unit from these values, the logarithmic term begins to dominate the equation, and the ratio of acid to conjugate base in the buffer fails to regulate pH. However, at pH values near the pK_a value, up to a 10-fold change in acid or base concentration will be blunted by the effect of the buffer.

It is often useful to see how an equation is derived. The derivation of the Henderson–Hasselbalch equation is shown in **Biochemistry: The derivation of the Henderson–Hasselbalch equation**.

Biochemistry

The derivation of the Henderson–Hasselbalch equation

The Henderson–Hasselbalch equation is a valuable expression in biochemistry. It relates the pH of a solution to the pK_a value of the weak acid in question, and the ratio of the concentration of weak acid to that of the conjugate base of the weak acid. The equation is useful for performing calculations with buffer solutions and pH; it is also relatively easy to derive. Although it is probably easier to simply know or look up the Henderson–Hasselbalch equation, deriving it illustrates several other important pieces of acid–base chemistry and how the equations are put together, a useful tool when examining or deriving other equations.

A weak acid (HA) will partially ionize in aqueous solution to give a proton (H^+) and the conjugate base of the weak acid (A^-). Expressing these as molar quantities at equilibrium allows us to write an acid ionization constant, K_a:

$$K_a = \frac{[H^+][A^-]}{[HA]}$$

Taking the base-10 logarithm of both sides gives:

$$\log_{10} K_a = \log_{10}\left(\frac{[H^+][A^-]}{[HA]}\right)$$

The right-hand side of the equation can be rearranged to isolate the log of the hydrogen ion concentration. Adding logarithms is the equivalent of multiplying integer values:

$$\log_{10} K_a = \log_{10}[H^+] + \log_{10}\left(\frac{[A^-]}{[HA]}\right)$$

Next, we can simplify the equation. The abbreviation used when taking the negative log of a term, is p (as in pH and pK_a). The above cases use the log (not the negative log), so the equation can be simplified to:

$$-pK_a = -pH + \log_{10}\left(\frac{[A^-]}{[HA]}\right)$$

Rearranging the pK_a and pH terms gives the familiar form of the Henderson–Hasselbalch equation:

$$pH = pK_a + \log_{10}\left(\frac{[A^-]}{[HA]}\right)$$

The pH at which a weak acid or conjugate base system will buffer is a function of its pK_a, but the total amount of acid or base that can be consumed is a function of the concentration. Thus, if two potassium phosphate buffers have been prepared in the laboratory, both at pH 7.3, and one buffer is 10 mM and the other 100 mM, the latter will have the ability to consume more acid or base. This is termed the **buffer capacity** of a system.

Biochemical systems are buffered by several different combinations of weak acids and their conjugate bases or weak bases and their conjugate acids. The most common buffering system in the blood is the bicarbonate/carbonate buffer. Cells use a combination of bicarbonate and phosphate to buffer against pH changes. Common biological buffers used in the laboratory include bicarbonate, phosphate, and acetate. Both carbonate and phosphate are polyprotic acids, acids that can donate more than one proton. Generally, only one of these ionizations plays a role in buffering of biochemical systems. In the laboratory, small organic molecules are often chosen to buffer near a specific pH; these include TRIS, MOPS, and HEPES.

TRIS
tris(hydroxymethyl)aminomethane

MOPS
3-(N-Morpholino)propanesulfonic acid

HEPES
4-(2-hydroxyethyl)-1-piperazineethanesulfonic acid

In an experiment, if there is no means of buffering, it is easy for the pH to change rapidly. In the early days of experimental biochemistry, the lack of control of factors such as buffering and pH made it difficult to obtain consistent results from repeated experiments, and this generated spurious and false results (referred to as artifacts). Buffers help regulate and control biochemical experiments, and standardize results. Biochemical systems are buffered for similar reasons; the inability to regulate pH can result in wide fluctuations, many of which could be harmful to the organism. Altering the pH of a system can have implications on the molecular level such as the ability of a protein to bind to a small molecule. This can have significant effects on the organism, for example, on the ability to carry oxygen to tissues that need it.

The buffering capacity of the blood is discussed in **Medical Biochemistry: Buffers in blood.**

Medical Biochemistry

Buffers in blood

Buffers play a central role in biochemistry. In the laboratory, students are advised to always use a buffered solution for a biochemical experiment. Naturally occurring bodies of water, including streams, lakes, and even oceans, have low millimolar levels of dissolved compounds such as carbon dioxide and phosphate, which provide minor amounts of buffering. The dissolved compounds are also partially responsible for the observed pH. However, the necessity and effect of buffering

are most evident inside an organism. In the blood, the bicarbonate ion (HCO_3^-, pK_a 6.3) is the primary molecule responsible for buffering and maintaining a tight pH range around 7.35 (±0.1 pH units). Other ions, such as phosphate, ammonia, or organic ions, play minor roles in blood buffering but are found at a much lower concentration than bicarbonate (17 to 24 mM in healthy humans), or have pK_a values too far from 7.35 to be effective. At physiological pH we are outside the pH range at which bicarbonate would buffer a solution: ±1 pH unit around the pK_a value or, for bicarbonate,

5.3 to 7.3. At pH 7.35 the ratio of bicarbonate to carbonic acid is over 10:1.

Several factors account for the observed pH. First, the organism can regulate pH through the release or reabsorption of CO_2 either in the lung or kidney. Second, while less important than bicarbonate, other molecules, such as phosphate and proteins, can also participate in buffering. It may seem paradoxical that the organism operates outside the optimal buffering range; however, the bicarbonate buffering system has several advantages. Tissues produce acidic waste products, and this system provides a means by which the organism can rapidly respond to increased acid levels. Carbon dioxide is generated when carbon-containing molecules are oxidized for energy, and this CO_2 must be eliminated from the body. Carbon dioxide typically diffuses out of cells and into the bloodstream. There, dissolved carbon dioxide enters red blood cells or erythrocytes, where it is acted on by the enzyme carbonic anhydrase, catalyzing the formation of carbonic acid (H_2CO_3) from carbon dioxide and water. The carbonic acid generated is mildly acidic ($pK_{a1} = 6.35$), and it rapidly ionizes to form a bicarbonate anion and a hydronium ion. The second pK_a (pK_{a2}) is 10.33 and is not typically encountered in biological systems. The bicarbonate that is generated in this reaction accumulates in the erythrocyte. These reactions would rapidly reach equilibrium were it not for the chloride–bicarbonate exchange protein, which is found in the membrane of the erythrocyte and exchanges a bicarbonate ion for a chloride ion. This process ensures that there is no significant accumulation of bicarbonate in the erythrocyte, and keeps the reaction moving in the forward direction in carbon dioxide generating tissues.

In the lung, there is a low concentration or partial pressure of carbon dioxide. Thus, as blood passes through the lung, the described reactions occur in reverse: chloride rushes out of the erythrocyte and bicarbonate rushes in. The latter protonates and is rapidly degraded to carbon dioxide, which is free to diffuse out of the erythrocyte and blood, and is exhaled and eliminated by the lung. The erythrocyte demonstrates several chemical principles including equilibrium, Le Châtelier's principle, solubility, permeability, and catalysis. In addition, the bicarbonate generated serves as a buffer for the blood, bathing tissues in a constant pH close to 7.35.

In times of excessive acid production, such as during anaerobic muscle exertion or in several disease states, more acid is produced than the buffer system can compensate for, and blood pH levels drop below pH 7.2. This condition, known as acidosis, can have serious consequences. At pH 6.8, the level of acidity is such that it becomes difficult for the blood to deliver oxygen to tissues that need it. Damage can quickly result to tissues, cells, and the molecules that make up those cells. In the clinic, sodium bicarbonate is used to raise the pH of the blood until the underlying cause of the acidosis can be determined.

In other conditions, the blood may become dangerously alkaline (above pH 7.45). This may happen, for example, when a person hyperventilates, in which case too much carbon dioxide is exhaled, depleting the body of its buffering capacity. In this instance, carbon dioxide can be recovered by having a patient breathe an environment enriched in carbon dioxide, such as a paper bag, to increase the amount of carbon dioxide and bicarbonate dissolved in blood.

Worked Problem 1.4   Working with buffers

You have been asked to prepare 1 liter of a 50 mM buffer (pH 5.2) for an experiment. In the laboratory are the following buffers: TRIS, MOPS, phosphate, and acetate.

Which buffer will you choose to make, and how will you make it?

Strategy   Examine the pK_a values for each buffer in Table 1.3. How does that information help in choosing which buffer to use?

Solution   Buffers only work over a range of ± 1 pH unit around the pK_a of the acid or base in the buffer. Of the four buffers available,

only one, acetate, has a pK_a near 5.2 (4.7). To make the buffer, you would measure out enough acetate to make 1 liter of final solution but dilute it into less than the final volume. You would then adjust the pH of this solution with a solution of base like NaOH to the desired pH (5.2). Finally, you would then dilute the buffer to the final volume (1 liter).

Follow-up question   If you have a solution of 50 mM acetic acid and 50 mM sodium acetate, how much of each would you need to make 1 liter of pH 5.5 buffer?

Summary

- Water is:
  - necessary for life
  - the central solvent in biochemistry
  - highly polar
  - particularly effective at making hydrogen bonds—bonds between a hydrogen attached to an electronegative atom and another electronegative atom.
- All biological systems require water, and it is the solvent most commonly associated with biochemistry; however, microenvironments in a cell or macromolecule may exclude water. Polar or charged functional groups make organic molecules more soluble in water and less soluble in a hydrophobic environment.

- Water has the ability to ionize, forming H^+ and OH^-, affecting pH.
- Buffers are pairs of weak acids and their conjugate base (or weak bases and their conjugate acid).
- Buffers act to resist changes in pH. The compositions of a buffer and the pH can be calculated using the Henderson–Hasselbalch equation.

Concept Check

1. Discuss how biological processes can be understood in the context of water and its interaction at molecular level.

2. Describe hydrogen bonding and how it imparts to water many of the properties it exhibits.

3. Perform calculations with both strong and weak acids and bases, including pH calculations. What is the approximate pK_a of a weak acid HA if a solution of 0.1 M HA and 0.3 M A^- has a pH of 6.5?

4. Perform calculations using buffers and the Henderson–Hasselbalch equation. What would be the resulting pH if one drop (0.05 mL) of 1.0 M HCl was added to one liter of pure water?

Bioinformatics Exercises

Exercise 1 The Periodic Table of the Elements and Domains of Life

Exercise 2 Organic Functional Groups and the Three-Dimensional Structure of Vitamin C

Exercise 3 Structure and Solubility

Exercise 4 Amino acids, Ionization, and pK values

Problems

1.1 General Chemical Principles

1. State the three laws of thermodynamics and explain their significance in context of biochemical systems.

2. What is the difference between ΔG, $\Delta G°$, $\Delta G°'$ and $\Delta G^{\ddagger}$? Can any of these be used interchangeably, and if so, when?

3. All biochemical macromolecules are folded into complex shapes held together in part by weak forces. Discuss the thermodynamics of these processes. Are they favorable or unfavorable? Which thermodynamic terms (ΔG, ΔH, ΔS, and others) are important in folding into a stable conformation? What other phenomena (bond breaking or forming, changes in entropy, and so on) contribute to the larger thermodynamic parameters?

4. Phosphoglucomutase catalyzes the reaction in which a phosphate group is transferred from the C-1 of glucose to the C-6 of glucose (G1P → G6P). A student incubates a 0.2 M solution of glucose-1-phosphate overnight with a small amount of the enzyme. At equilibrium the concentration of glucose-1-phosphate is 9.0×10^{-3} M and the concentration of glucose-6-phosphate is 19.1×10^{-2} M. Calculate the equilibrium constant (K_{eq}) and the standard state free energy ($\Delta G°'$) for this reaction at 25°C.

5. If we consider the polymerization of a water soluble monomer into a polymer, is this process entropically positive or negative? What types of interactions need to be considered?

6. In organisms, β-hydroxybutyrate, found in some metabolic states, will spontaneously decarboxylate into acetone and carbon dioxide. Write a rate law for this reaction. What factors will contribute to this rate?

1.2 Fundamental Concepts of Organic Chemistry

7. Examine these amino acids. Which functional groups can you identify? What types of bonds can these groups form?

8. Amino acids are the building blocks of proteins. Can you put these amino acids into order of increasing solubility in water?

9. Which of these molecules would you anticipate has the highest solubility in water? Which would be most soluble in hexane?

10. Examine the structures given in problem 9. If each were dissolved in water, how would the water molecules surround these solutes? What sort of interactions would occur and where?

11. Propose a mechanism for the base catalyzed cleavage of an ester.

12. If a polymer is 1,000 monomers long and includes four different types of monomers, how many different combinations of the monomers are possible?

13. Examine the structure of the polymer shown. What are the monomeric units of that polymer?

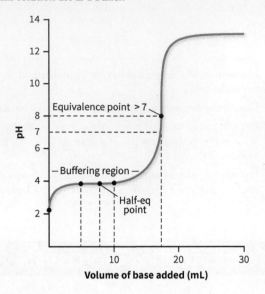

1.3 The Chemistry of Water

14. Examine the functional groups in Table 1.2. Which of these can hydrogen bond with water? How and where will these bonds form?

15. Which of these bonds would be the shortest H bond, and why? OH–O, OH–N, OH–F

16. A molecule of water is said to participate in 3.7 hydrogen bonds at any one time. How could you calculate this number? How could you measure it?

17. What is the pH value of each of these solutions?

a. 0.1 M HCl	**b.** 3 M H_2SO_4
c. 12 mM HCl	**d.** 2 M KOH
e. 100 mM $Ca(OH)_2$	**f.** 0.3 M acetic acid
g. 43 mM H_2CO_3	**h.** 25 mM ammonia
i. 80 mM TRIS base	**j.** 0.5 M KH_2PO_4

18. What percentage of a solution of acetic acid is ionized at pH 5.6? What percentage of a solution of HCl is ionized at the same pH?

19. What is the pH of a solution that contains three parts acetic acid and one part sodium acetate? The pK_a for acetic acid is 4.76.

20. What is the final pH if you have 500 mL of a sodium acetate buffer (50 mM, pH 5.0) and you add 5 mL of 1 M NaOH?

21. How much 6 M HCl do you need to add if you have 800 mL of 100 mM TRIS, pH 10.2 and you wish to lower the pH to 7.0?

22. The pK_a of acetic acid is 4.76. If 20% of the acetic acid in a 0.1 M solution is in the protonated state, what is the pH of the solution?

23. Buffering capacity is the property of a buffer to resist changes in pH. Which has more buffering capacity: a liter of 50 mM MOPS, pH 6.9 or 100 mL of 1 M MOPS at the same pH?

24. Normal blood pH is approximately 7.35 and the concentration of carbonate and bicarbonate ions (total) is approximately 20 mM. What is the ratio of carbonate to bicarbonate ions in the blood under these conditions?

Data Interpretation

25. In the titration curve of a weak acid with a strong base, where is the equivalence point? What is the pK_a of this acid? Over what pH range would this solution act as a buffer?

Experimental Design

26. Based on your knowledge of organic chemistry, how could you increase the solubility of an amine or a carboxylic acid in an aqueous solution?

27. Tyrosine is an amino acid, one of the building blocks of proteins. It has a hydrophobic side chain. Design an experiment to test the solubility of tyrosine in can be changed to 250 mM NaCl.

28. Based on your experiences in organic chemistry, how would you identify an unknown molecule? What organic techniques would you use? What are the limitations of those techniques? Based on your current knowledge of biochemistry, what are the strengths and drawbacks of employing organic techniques to identify biological molecules?

29. If you were asked to make 1 liter of 250 mM KCl, how would you do it? Give a detailed description, including the steps in the process and the pieces of glassware you would use.

30. How would you go about making 500 milliliters of 100 mM TRIS buffer, pH 7.2? Describe the steps involved, including which apparatus and glassware you would use and when.

Ethics and Social Responsibility

31. As you begin your studies in biochemistry, what do you think are the important ethical and social questions in this field?

32. All professional organizations have codes of conduct. These include the codes of conduct for the American Society for Biochemistry and Molecular Biology, the International Union for Biochemistry and Molecular Biology, the American Chemical Society, the American

Medical Association, and others. Using the Internet, examine one of these organization's codes of conduct. Is there anything that you find surprising?

33. Various governmental agencies regulate the sale of certain biomolecules, ranging from drugs to foods to fuels. Is this regulation justified, and if so, why?

34. If you were to examine the effect that table sugar has on the body, you could find many ways in which it elicits physiological and metabolic changes. Why is sugar not considered a drug? Find a reliable resource that defines drugs and pharmaceuticals. What differentiates drugs from other molecules?

35. Drug companies have patent protection for drugs that they have developed and are on the market. If an enantiomerically pure drug can be developed from a racemate, the company is given a new patent on the stereochemically pure compound. Do you feel this situation is justified?

Suggested Readings

1.1 General Chemical Principles

Alberty, R. A. *Thermodynamics of Biochemical Reactions.* New York, NY: John Wiley and Sons, 2003.

Jesperson, N. D., J. E. Brady, and A. Hyslop. *Chemistry: The Molecular Nature of Matter.* 6th ed. New York, NY: John Wiley and Sons, 2012.

Pauling, L., *The Nature of the Chemical Bond and the Structure of Molecules and Crystals.* Ithaca, NY: Cornell University Press, 1960.

Tinoco, I., K. Sauer, J. C. Wang, J. D. Puglisi, G. Harbison, and D. Rovnyak. *Physical Chemistry: Principles and Applications in Biological Sciences.* 5th ed. Englewood Cliffs, NJ: Prentice Hall, 2013.

Van Holde, K. E. *Physical Biochemistry.* 2nd ed. Englewood Cliffs, NJ: Prentice Hall, 1985.

Van Holde, K. E., W. C. Johnson, and P. S. Ho. *Principles of Physical Biochemistry.* 2nd ed. Englewood Cliffs, NJ: Prentice Hall, 2006.

1.2 Fundamental Concepts of Organic Chemistry

Cahn, R. S, C. K. Ingold, and V. Prelog. "Specification of Molecular Chirality." *Angewandte Chemie International Edition 5*, no. 4 (1966): 385–415.

Carey, F. A., and R. J. Sundberg. *Advanced Organic Chemistry, Part A, Structure and Mechanisms.* 5th ed. New York, NY: Springer 2007

Grossman, R. B. *The Art of Writing Reasonable Organic Reaction Mechanisms.* 2nd ed. New York, NY: Springer, 2002.

Klein, D. *Organic Chemistry.* New York, NY: Wiley and Sons, 2012.

Smith, M. B., and J. March. *March's Advanced Organic Chemistry.* 5th ed. New York, NY: Wiley and Sons, 2001.

1.3 The Chemistry of Water

Beynon, R. J., and J. S. Easterby. *Buffer Solutions: The Basics,* New York, NY: IRL Press, 1996.

Halperin, M. L., and M. B. Goldstein. *Fluid, Electrolyte, and Acid-Base Physiology: A Problem-Based Approach.* 3rd ed. Philadelphia, PA: W.B. Saunders, 1999.

Jeffrey, G. A., and W. Saenger. *Hydrogen Bonding in Biological Structures.* New York, NY: Springer, 1994.

Pauling, L., *The Nature of the Chemical Bond and the Structure of Molecules and Crystals.* Ithaca, NY: Cornell University Press, 1960.

Segel, I. H., *Biochemical Calculations.* 2nd ed. New York, NY: Wiley and Sons, 1976.

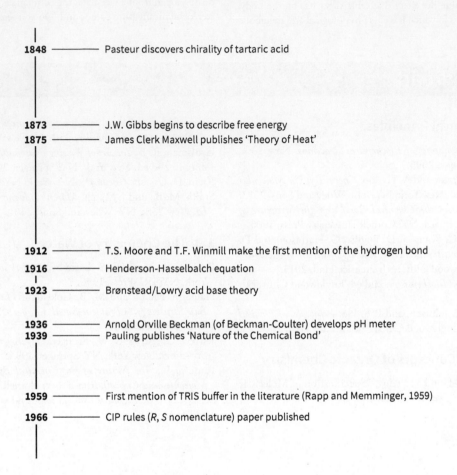

1848 ———— Pasteur discovers chirality of tartaric acid

1873 ———— J.W. Gibbs begins to describe free energy
1875 ———— James Clerk Maxwell publishes 'Theory of Heat'

1912 ———— T.S. Moore and T.F. Winmill make the first mention of the hydrogen bond
1916 ———— Henderson-Hasselbalch equation
1923 ———— Brønstead/Lowry acid base theory

1936 ———— Arnold Orville Beckman (of Beckman-Coulter) develops pH meter
1939 ———— Pauling publishes 'Nature of the Chemical Bond'

1959 ———— First mention of TRIS buffer in the literature (Rapp and Memminger, 1959)

1966 ———— CIP rules (*R*, *S* nomenclature) paper published

CHAPTER 2

Nucleic Acids

Nucleic Acids in Context

It is easy to imagine how even a small amount of information, such as a credit card or social security number, could be misused if it were to get into the wrong hands. In many cases data is often encrypted to keep it safe. We could also imagine how, if there were an error in the code on a credit card number, an e-mail address, or mailing address, it could result in a problem with an online order. It would be easy to charge items to the wrong account or the items may be sent to the wrong place. Depending on where these errors occurred, it may be more difficult to rectify the situation.

Life, too, uses a code; unlike online data, life's code is written in nucleic acids. This code can be duplicated, transcribed into RNA, and finally translated into protein; these processes are referred to as replication, transcription and translation, respectively. Each of these processes occurs in all living creatures, all of which use the same genetic code. Hence, a piece of DNA coding for a human protein can be inserted into other organisms to produce those proteins for research or therapeutic use.

This chapter describes the physical and chemical properties of nucleic acids; reviews the basics of the central dogma, that is, how information in DNA is transcribed into RNA and then translated into protein; and describes how nucleic acids can be manipulated to alter mRNA and protein levels in the organism.

Chapter Outline

2.1 Nucleic acids are the Genetic Materials

2.2 Nucleic Acids Have Distinct Structures

2.3 Nucleic Acids Have Many Cellular Functions

2.4 The Manipulation of Nucleic Acids Has Transformed Biochemistry

Common Themes

Evolution's outcomes are conserved.	• The classic observations and deductions made by Charles Darwin when he formulated the theory of evolution are equally relevant when we examine the molecular aspects of gene transmission.
	• Genetic traits, coded for by deoxyribonucleic acid (DNA) or ribonucleic acid (RNA), are conserved throughout evolution. Small, gradual changes in these coding sequences can be tracked through history to generate an evolutionary family tree.
Structure determines function.	• DNA molecules are made up of base pairs and have information coded within those base pairs.
	• The structure of DNA helps to explain how complex phenomena such as DNA replication occur, and how proteins interact with DNA (either specifically or nonspecifically) to organize DNA and regulate function.
	• RNA molecules can have complex structures that can encode information, regulate gene expression, or act catalytically.
Biochemical information is transferred, exchanged, and stored.	• Both DNA and RNA can serve as repositories of information that can be stored, translated into another type of molecule (RNA or protein), and transmitted to future generations.
Biomolecules are altered through pathways involving transformations of energy and matter.	• The nucleotide building blocks of RNA are used by the organism as a source of energy in numerous other chemical reactions.

2.1 Nucleic Acids are the Genetic Materials

Nucleic acids exist as long linear or circular macromolecules in the form of DNA and RNA made of linked nucleotides. The most striking feature of these molecules is that they carry the essential genetic information that directs the functions of a cell.

Cell nucleus was recognized to play a major role in inheritance during the late 1870s, based on the observations that male nuclei and female reproductive cells fertilize by undergoing fusion. Subsequently, thread-like entities called chromosomes were found inside the cell when stained with certain dyes. In the late 1900s, the chromosomes were found to be the promising material for carrying genes and their features suggested an interrelationship with DNA. The presence of DNA in chromosomes was observed using special stains. In addition, several types of chromosomal proteins with varying amounts were found across various cell types in contrast to the amount of DNA that remained constant. In higher organisms, all of the DNA present in cells was found in the chromosomes. These findings were insufficient to establish DNA as the genetic material due to insufficiency in the chemical diversity that was required in genetic material. Contrastingly, as proteins existed in diverse forms, these were widely accepted as genetic material with an assumption of DNA being destined to function as structural support to the chromosomes. These assumptions were finally rectified by experiments carried out by Griffith and other scientists to prove that DNA is the genetic material.

2.1.1 DNA as genetic material

The experimental proof of DNA as the genetic material came in 1928 when Frederick Griffith demonstrated the transformation of genetic information from one bacterial strain type of pneumococcus (*Streptococcus pneumonia*) into another. Based on the polysaccharide and the capsule type of each strain (the inherited characteristic features), mice were injected with various preparations of rough (R) and smooth (S) surface strains of *Streptococcus pneumonia*. The R strain could not cause pneumonia as it lacked the capsule that facilitates the bacterium to cause pneumonia by curbing the defense mechanisms of the host. However, the S strain that synthesized the gelatinous capsule composed of polysaccharide was able to cause pneumonia. Griffith's critical finding came with the observation that mice injected with avirulent strain combined with heat-killed encapsulated strain die of pneumonia, as did the mice that received live virulent cells. The bacterial strain from the dead mice was noted with encapsulation and genetic characteristics of the heat-killed strain. The injected material from the dead S cells harbored an entity that was capable of getting transferred to the living R cells and in some way restored the virulence. In other words, the R strain did undergo transformation or got changed into S strain. This gave a clue about the inherited transformation and not as a result of avirulent cells being coated with capsule material. The clarification came with the transmission of capsule-type to daughter cells in cultured cells. Until 1944, the chemical entity responsible for the change in the strain was not identified, when Oswald Avery, Colin MacLeod, and Maclyn McCarty in a milestone experiment showed that the entity causing the transformation of R strains to S strain was DNA. They isolated and purified the DNA from S strain and added it into the growing cultures of R strain. In the growing medium, a few cells of new S strains were observed although the preparations of DNA contained minute quantities of RNA and protein. The transformation activity did not have any effect or showed altered changes upon treatment with protease or RNase that destroyed either protein or RNA. However, the transformation activity was eliminated when treatment with DNase destroyed DNA. It was inferred from these experiments that the DNA of the cell was responsible for genetic transformation.

2.1.2 RNA as genetic material

Viruses are known to contain RNA as their genetic material. In the 1930s, experiments on tobacco mosaic virus (TMV) showed the presence of protein and RNA. DNA was not found in the viral particle. Separation of protein from RNA was easily achieved by alkali treatment. In 1956, Gierer and Schramm demonstrated directly and showed that inoculation of RNA alone into the tobacco plants causes infection indicating that it is the RNA that acts as the genetic material in such viral particles. The infective viral particles reconstituted by blending the protein and RNA were determined by Frankel-Conrat and his associates. Exploiting this technique, in 1957 Gierer and Singer reconstituted viral particles by mixing RNA from one strain together with protein produced by a mutant strain or *vice versa*. The production of new viral particles in the infected host by either combination had protein type that was to be produced by the parent RNA molecule. These experiments suggested and proved that RNA is a source of genetic inheritance.

Summary

- Several experiments were performed by Griffith and subsequently by Avery, MacLeod, and McCarty to conclude DNA as genetic material.
- RNA was found to be the genetic material in viral particles that lack DNA.

Concept Check

1. Describe the experiment performed by Frederick Griffith to prove the transformation process.
2. Who proved RNA as a genetic material and how?

2.2 Nucleic Acids Have Distinct Structures

The **nucleic acids** deoxyribonucleic acid (DNA) and ribonucleic acid (RNA) are among the most highly recognized molecules. In many ways, these molecules are biochemistry in a microcosm. While they may be familiar to you, your earlier studies of these may not have included a chemical focus. This section first presents the basic chemical composition of DNA and RNA, and the building blocks used to construct these molecules. It then discusses how weak forces give rise to the complex structures that these molecules adopt.

2.2.1 Nucleic acids can be understood from their chemical constituents

DNA and RNA are anionic polymers that were originally identified in the nucleus, hence the name "nucleic acid," and are constructed from nucleotide monomers.

A **nucleotide** consists of three parts—a nitrogenous base, a carbohydrate, and a phosphate. Each figures prominently in the chemistry of the molecule. This section considers each of these parts.

Nitrogenous bases give rise to specific nucleotides

Nitrogenous bases are heterocyclic structures that contain either one or two rings (**Figure 2.1**). The structures containing a single ring are the **pyrimidines** (named for their structural similarity to pyrimidine), and those containing two rings are **purines** (named for their structural similarity to purine). A simple way to recall and relate these two types of molecule is that the shorter name goes with the larger molecule, for example, purines are bicyclic. Each of these basic structures is modified to contain additional keto, amine, and methyl groups, which can participate in hydrogen bonding. Some of the nitrogens in the ring systems bear hydrogens and are also capable of hydrogen bonding. Although these molecules are referred to as bases and contain several amines within the ring systems, the pK_a values of the nitrogens are outside the physiological range; that is, they are never found in an ionized state.

The common purines are **adenine** (A) and **guanine** (G). These bicyclic bases are found in both RNA and DNA, and they appear in nucleotides that serve other functions in the cell, such as adenosine triphosphate, ATP. Cells also contain purines that are not found in nucleic acids; these include caffeine, theobromine, uric acid, and xanthine. Some of these molecules are involved in the metabolism of nitrogenous bases or nitrogenous waste, whereas others are potent signaling molecules.

The monocyclic pyrimidines that are commonly found are **cytosine** (C), **thymine** (T), and **uracil** (U). Here we see a difference between DNA and RNA. DNA contains the bases C and T, while RNA contains C and U. As is true for the purines, these bases and nucleotides also serve other roles in biochemistry in addition to being found in DNA and RNA.

Both purines and pyrimidines are aromatic structures. The aromatic ring imparts extra stability through resonance stabilization. It also means that nucleotides, and the nucleic acids made from them, will absorb in the ultraviolet (UV) region of the electromagnetic spectrum. This property is commonly used to measure concentrations of nucleic acids.

The aromatic properties and resonance states of purines and pyrimidines also facilitate several instances of tautomerization (**Figure 2.2**). In these forms, migration of protons can result in molecules with the properties of a carbonyl group (the keto form) or a hydroxyl group (the enol form). This could also result in an amino group changing to an imino group. In most instances in nature, the keto or amino form of any molecule will predominate, and nitrogenous bases are no exception. However, there are several instances, usually pH dependent, where enols can also be observed.

The tautomeric switch from keto to enol form is important on several fronts. First, the presence of a ketone versus a hydroxyl will change the hydrogen bonding properties of the molecule. As outlined in the

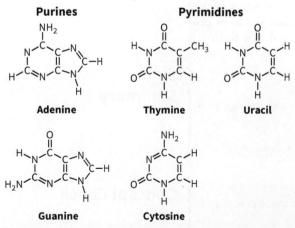

FIGURE 2.1 Nitrogenous bases. The nitrogenous bases that make up DNA are the purines adenine and guanine and the pyrimidines cytosine and thymine. In RNA uracil takes the place of thymine.

next section, this ability to precisely hydrogen bond is of central importance in nucleic acid chemistry. Second, hydroxyl groups are important as catalytic centers in some nucleic acids, and this property may also be affected by tautomerism.

Addition of a carbohydrate to a nitrogenous base produces a nucleoside

Nitrogenous bases are coupled to a carbohydrate structure through a glycosidic bond, that is, a bond to a carbohydrate, to form a **nucleoside** (Figure 2.3). The different carbohydrates used in DNA and RNA give rise to the names of these two molecules. RNA uses the carbohydrate **ribose**. In a ribonucleoside, the base is attached to the 1′ carbon of the five-membered ribose ring; the atoms of the carbohydrate ring are given the designation "′" or "prime" to distinguish them from the atoms of the base. The carbohydrate rings in nucleosides are numbered from the 1′ position around the carbohydrate ring to the 5′ position. Ribose has two characteristic hydroxyl groups on the 2′ and 3′ carbons found on the bottom face of the ring. In contrast, the 5′ carbon is found on the upper face of the ring.

DNA and other deoxyribonucleosides contain **2′-deoxyribose**, often referred to simply as deoxyribose, a ribose molecule that lacks the hydroxyl group on the 2′ position.

Addition of phosphate groups produces a nucleotide

A nucleotide is formed by the addition of a phosphate to a nucleoside. This occurs via a phosphoester linkage to the hydroxyl group on the 5′ carbon of the ribose or 2′-deoxyribose ring. When added to ribose-containing nucleosides, this produces adenosine monophosphate (AMP), guanosine monophosphate (GMP), cytosine monophosphate (CMP), or uracil monophosphate (UMP). Correspondingly, the addition of phosphates to 2′-deoxyribonucleosides results in 2′-deoxyadenosine monophosphate (dAMP) and the other 2′-deoxy analogs: dGMP, dCTP, and dTTP.

Additional phosphate groups can be added through phosphoanhydride linkages. This results in the diphosphate and triphosphate forms of the aforementioned nucleotide monophosphates, that is, ADP and ATP or their deoxy analogs, dADP and dATP, and so on, as noted in Figure 2.3.

The phosphate groups added to these nucleosides and deoxynucleosides produce negatively charged molecules at neutral pH. ATP, for example, has three phosphate groups linked together through phosphoanhydride bonds, giving it a charge of −4 at pH 7.2. Therefore, these molecules are polyprotic acids; that is, they can donate more than one proton. Because of the negative charge on these phosphates, nucleotide triphosphates and deoxynucleotide triphosphates often form complexes with divalent metal ions (metal ions with a 2+ charge), most notably magnesium ions (Mg^{2+}). In these instances, the phosphates wrap around and cradle the metal ion. Often in biochemistry divalent metal ions are required for reactions involving nucleotide triphosphates and deoxynucleotide triphosphates.

The phosphates found on nucleotides and deoxynucleotides will hydrolyze with large, negative ΔG values. This energy can be used to drive reactions in the cell by coupling an otherwise energetically unfavorable reaction, such as the addition of a phosphate group to a hydroxyl, to a favorable one, such as the hydrolysis of ATP. Although ATP is the best known molecule that can be used to drive reactions, all of the nucleotide triphosphates are involved in providing energy to help drive specific reactions in the cell.

Finally, many nucleotide metabolites also play roles in cellular signaling. Compounds such as cyclic AMP (cAMP) that have a 5′ to 3′ phosphodiester linkage are found in signaling pathways ranging from the bacterium *E. coli* to humans.

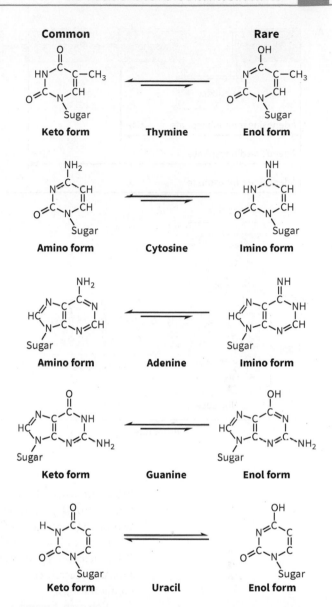

FIGURE 2.2 Tautomeric forms of bases. The bases of DNA can undergo keto-enol tautomerism. While not usually apparent, these forms can lead to mismatches or changes to the DNA sequence if improperly paired.

(Source: Snustad; Simmons, *Principles of Genetics*, 6e, copyright 2012, John Wiley & Sons. This material is reproduced with permission of John Wiley & Sons, Inc.)

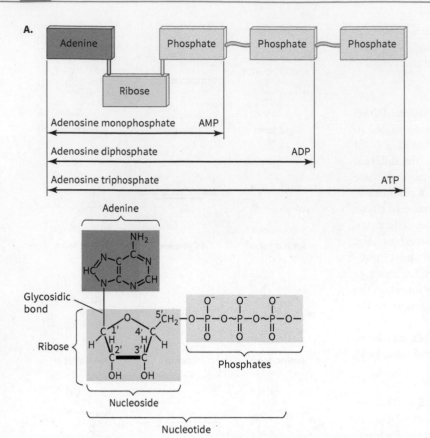

FIGURE 2.3 Nucleosides, ribose, and deoxyribose.
A. Nucleosides are comprised of a nitrogenous base and a ribose sugar. Nucleotides are comprised of the base, sugar, and a phosphate esterified to the 5′ carbon. Nucleotides can have one (AMP), two (ADP), or three (ATP) phosphates **B.** Deoxyribonucleotides use deoxyribose, which is missing an oxygen on the 2′ position of the ribose ring.

(Source: (A) Black, *Microbiology: Principles and Explorations*, 8e, copyright 2012, John Wiley & Sons. This material is reproduced with permission of John Wiley & Sons, Inc. (B) Alters, *Biology: Understanding Life*, 1e, copyright 2006, John Wiley & Sons. This material is reproduced with permission of John Wiley & Sons, Inc.)

Nucleic acids are polymers of nucleotides   Because nucleotides contain both free hydroxyl groups and phosphates, they are capable of forming diester linkages between two or more molecules (**Figure 2.4**). This occurs most commonly between the hydroxyl group on the 3′ carbon of one nucleotide and the phosphate connected to the 5′ carbon of another. Ribonucleotides polymerize to form RNA, and 2-deoxyribonucleotides polymerize to form DNA. DNA is typically comprised of two strands running in opposite directions (termed an **antiparallel** configuration).

Polymers formed of a small number of nucleotides can be referred to by number, for example, a *di*nucleotide or *tri*nucleotide. However, larger polymers, including five to ten or more nucleotides, are often referred to as oligomers (or oligos for short). Nucleic acids can be lengthy, stretching for hundreds of millions of monomers.

Perhaps one of the most fascinating aspects of biochemistry is that these polymers are copied from a template, resulting in an almost perfect copy each time. The information stored in this sequence of bases is the code of life, and is usually represented as a sequence of single letters resulting from the bases A, C, G, and T (or U in the case of RNA). The information coded within these bases provides the cell with the directions for *how* to make protein, *where* or in *which* cells to synthesize these proteins, and to some extent *when* and under which conditions to do so.

2.2.2 The complex shapes of nucleic acids are the result of numerous weak forces

Nucleic acids differ with regard to their structure. RNA consists of a single strand, often several thousand bases or more in length, folded back on itself in a complex configuration that is unique for each molecule. DNA differs from this in that it is typically comprised of two strands which then twist around each other, resulting in the characteristic double helix that is so familiar. This section discusses the various aspects and forces involved in forming nucleic acid structures. Although discussed here in relation to nucleic acids, these same forces shape all other biological molecules as well.

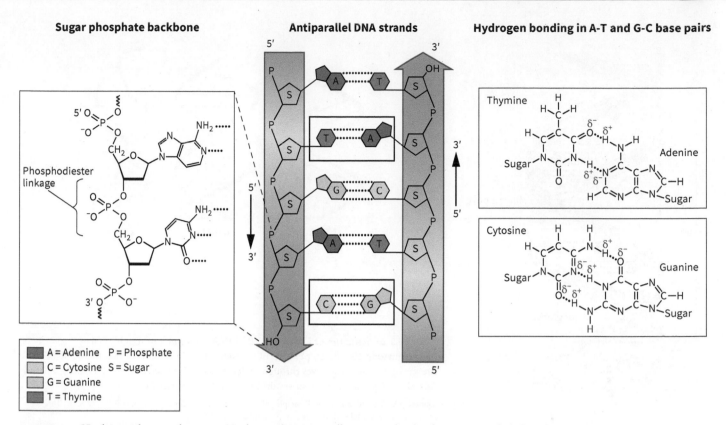

FIGURE 2.4 Nucleic acids are polymers.   Nucleic acids are generally macromolecules, long strings of nucleotides or deoxynucleotides. Nucleotides in each strand are joined via phosphate esters between the 5′ carbon, the phosphate, and the 3′ carbon of the next nucleotide. RNA is single stranded while DNA is double stranded with the strands running in opposite directions. The bases of DNA hydrogen bond with one another forming base pairs.

(Source: Snustad; Simmons, *Principles of Genetics*, 6e, copyright 2012, John Wiley & Sons. This material is reproduced with permission of John Wiley & Sons, Inc.)

The structure of DNA is a double helix   DNA consists of two antiparallel strands,
each of which is a polymer of deoxyribonucleotides. These strands align like a ladder: the bases of DNA pair up (A binding to T, and G binding to C) to form the rungs of the ladder, and the phosphates and deoxyribose (the sugar-phosphate backbone) form the sides of the ladder. Finally, the entire structure twists upon itself to generate the double helix (**Figure 2.5**).

As the DNA helix forms, several details become clear. The bases of DNA stack up on one another. This generates a hydrophobic core to DNA, and helps to stabilize the structure. The helix of DNA has about 10.3 bases per turn; with about 0.34 Å per base pair (bp), the entire length of one turn of DNA is about 3.4 Å. Although short structures of DNA are often depicted as a roughly cylindrical shape, DNA is fairly flexible throughout its length, and can be looped around organizing proteins or regulatory proteins.

DNA is a double helix. Each of the two strands is helical and the strands are slightly offset from one another. Thus, a section of DNA contains two grooves that run the length of the molecule: the larger major groove and the smaller minor groove. The diameter of these grooves, in particular the major groove, makes it easier for proteins to interact with and thus regulate DNA.

There are three known forms of DNA: A DNA, B DNA, and Z DNA. Of these, B DNA was the first type to be characterized and is the type of DNA most commonly found in cells. In contrast, A DNA is thought to be more common under laboratory conditions, for instance, in the absence of water. Both A and B forms of DNA are right-handed helices, but the B DNA helix forms a slightly tighter structure than the A DNA helix. Therefore, B DNA has a wider major groove and a narrower minor groove than A DNA. In contrast, Z DNA is a left-handed helix with a much wider diameter and is thought to form in some DNA-protein complexes. The roles of these different DNA structures in biology is not clear, but it may be that they play roles in the interaction of DNA with proteins.

DNA molecules can be relatively short in the laboratory, but those that are used to transmit genetic information in nature are much larger. For example, the complete collection of DNA,

A.

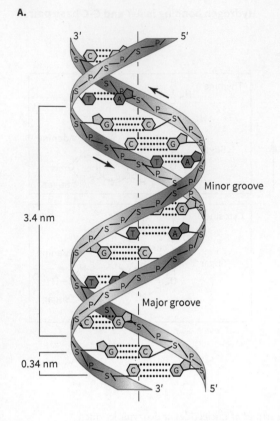

3'   5'

Minor groove

3.4 nm

Major groove

0.34 nm

3'   5'

B.

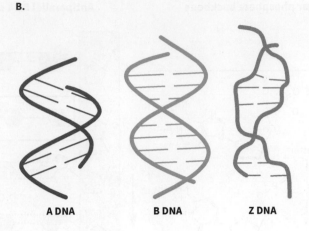

A DNA   **B DNA**   **Z DNA**

FIGURE 2.5 Structure of DNA.   A. "B" DNA has a pitch of roughly 3.4Å (approximately 10.4 bases) per turn of the helix. Twisting results in the double helix and the major and minor grooves through which proteins interact with the DNA. **B.** "A" and "Z" DNA have different twists resulting in a different structure.

(Source: (A) Snustad; Simmons, Principles of Genetics, 6e, copyright 2012, John Wiley & Sons. This material is reproduced with permission of John Wiley & Sons, Inc.)

or genome in an average bacterium, contains about 10 million base pairs, whereas the human genome contains 3.3 billion base pairs. The largest single piece of DNA in the human genome is 249 million base pairs.

Elucidation of the structure of DNA was a watershed moment in biochemistry
Often, when some facet of nature reveals itself to science through experimentation, we struggle for a time to understand it well enough to put the observations into context and learn from what we have found. However, the discovery of the structure of DNA immediately tied together numerous observations, such as the ratios between A and T and between G and C, and laid the groundwork for others, including the elucidation of the replication of DNA. As with most great discoveries in science, answering this key question opened the door to many new questions.

Numerous weak forces are at work in the formation of the double helix
This section examines the forces involved in the formation and stabilization of the double helix. Looking first at a single strand of DNA, the strand has a hydrophilic sugar-phosphate backbone and a less hydrophilic nucleotide base. At this point, an easy assumption would be that hydrogen bonding holds DNA together. Hydrogen bonds between A and T or G and C form the basis of the genetic code, and are how recognition of sequences happens. This scenario is true in that base pairing in DNA involves hydrogen bonding; however, those bonds are slightly skewed, resulting in a weaker bond. Likewise, if the issue were merely hydrogen bonding, why would the bases not form hydrogen bonds with water? After all, there is far more water available than other nucleotides. The secret of helix formation lies in a phenomenon known as the **hydrophobic effect**, whereby hydrophobic molecules, including the bases of DNA, aggregate and group together in an oily pocket within a molecule.

The hydrophobic effect is not due to the strength of the hydrophobic interaction of molecules with one another. The **van der Waals** and **London-dispersion** forces that are found between two hydrophobic molecules are weak, being 10 to 1,000-fold weaker than any single hydrogen bond. The hydrophobic effect occurs because it takes fewer water molecules to surround and cage a larger group than single molecules. Hydrophobic molecules do not actively form groups; instead, they are pushed by water molecules interacting with each other. Freeing water from the need

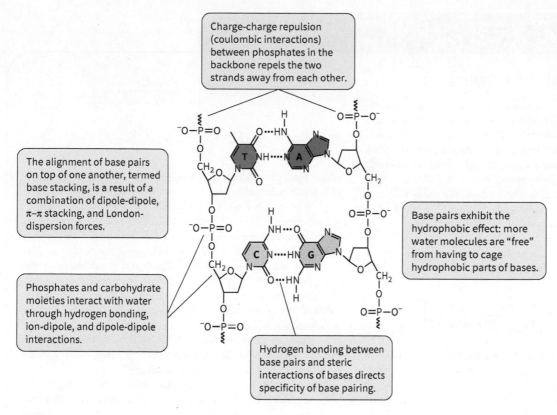

Charge-charge repulsion (coulombic interactions) between phosphates in the backbone repels the two strands away from each other.

The alignment of base pairs on top of one another, termed base stacking, is a result of a combination of dipole-dipole, $\pi-\pi$ stacking, and London-dispersion forces.

Phosphates and carbohydrate moieties interact with water through hydrogen bonding, ion-dipole, and dipole-dipole interactions.

Base pairs exhibit the hydrophobic effect: more water molecules are "free" from having to cage hydrophobic parts of bases.

Hydrogen bonding between base pairs and steric interactions of bases directs specificity of base pairing.

FIGURE 2.6 Analysis of weak forces involved in DNA stabilization. Numerous weak forces are involved in the stabilization of the DNA helix, including hydrogen bonding, coulombic interactions, hydrophobic interactions of base pairs, and the hydrophobic effect.

to cage hydrophobic molecules provides the increased energy to drive hydrophobic molecules together, creating the hydrophobic effect. Thus, helix formation can be attributed to hydrogen bonding, and hydrogen bonding between base pairs clearly drives the specificity of the assembly, but what drives the formation of the DNA duplex is not hydrogen bonds between bases, but the bulk hydrogen bonding of water.

Hydrogen bonding and the hydrophobic effect are not the only forces involved in the stabilization of DNA. In addition to these two, the charged phosphate groups on the backbone repel one another and therefore act to pull the helix apart (**Figure 2.6**). In strictly aqueous conditions, these forces are diminished by the strength and polarity of water; however, in DNA, the hydrophobic base pairs fail to decrease these forces to the same extent. Instead, the charges on the phosphate are partially balanced out by cations, most notably, divalent cations such as Mg^{2+}, and positively charged proteins.

The other forces involved in helix formation are often grouped into a single phenomenon known as **base stacking**. Base stacking involves numerous interactions: the hydrophobic effect, London-dispersion, van der Waals, dipole–dipole, and $\pi-\pi$ stacking. Collectively, these interactions stabilize the helix.

DNA can be melted or denatured, either chemically or by heating. As we might predict from the number of hydrogen bonds between the bases, DNA with a high GC content has a higher melting point than DNA with a high AT content. Buffer components such as salt or divalent metal ions also affect the melting point of DNA. Finally, chaotropes, molecules such as urea or guanidinium hydrochloride that disrupt the hydrogen bonding network of water, can denature DNA. Although these compounds have different structures, they all denature DNA by disrupting the forces that stabilize the helix. Simple means of calculating the melting point of short stretches of DNA can be found in **Biochemistry: Thermodynamics and DNA annealing**.

Urea   **Guanidinium hydrochloride**

Biochemistry

Thermodynamics and DNA annealing

When double-stranded DNA is heated, it loses its secondary structure and denatures or melts, ultimately resulting in two single-stranded pieces of DNA. In the laboratory, the DNA sample is generally heated to above 97°C. When the temperature cools, the DNA will refold and, if this is done slowly enough, the base pairs will realign with one another, reforming double-stranded DNA.

In some of the techniques discussed in this chapter, shorter strands of DNA need to anneal to one another. The same forces apply, but for these experiments to work, the melting point of the short stretch of DNA needs to be calculated.

The simplest approximations for melting temperatures are based on the number of G:C pairs versus the number of A:T pairs in the sequence.

$$T_m = 64.9 + 41.0 \times \left(\frac{y\text{G} + z\text{C} - 16.4}{w\text{A} + x\text{T} + y\text{G} + z\text{C}} \right)$$

In this equation, w, x, y, and z are the number of bases of the respective nucleotides in the sequence, and the values calculated are in degrees Celsius. Because of the higher number of hydrogen bonds in G:C, it would be expected to have a higher melting point; however, other factors such as divalent metal ions or ionic strength also contribute to destabilization of the helix. Taking those parameters into consideration yields the following equation:

$$T_m = 100.5 + 41.0 \times \left(\frac{y\text{G} + z\text{C} - 16.4}{w\text{A} + x\text{T} + y\text{G} + z\text{C}} \right)$$
$$- \left(\frac{820.0}{w\text{A} + x\text{T} + y\text{G} + z\text{C}} \right) + 16.6 \log([\text{Na}^+])$$

where again the values w, x, y, and z are the corresponding number of bases of each nucleotide.

Another factor to consider is the order of the nucleotides in the sequence. If there is a stretch rich in G:C base pairs, this region would have a higher melting temperature than would be the case if the G:C pairs were distributed throughout the sequence. This is termed the "nearest neighbor model" and is preferable to the other methods, although it is a slightly more involved calculation that takes into account the thermodynamic parameters involved in DNA–DNA interactions. Using a computer for these calculations greatly streamlines the process.

There are other models of DNA melting that use statistical mechanics to take into account the degrees of freedom of the molecules and other forces such as base stacking.

Having the correct melting temperature of small oligonucleotides can mean the difference between success and failure in many experiments. Thankfully, the algorithms discussed in this feature have been adapted for online use. Web sites quickly calculate melting temperatures using all three methods to provide a better estimate of the actual value.

RNA is single stranded but can form complex structures   RNA molecules can form many different complex structures, each of which is unique for that particular molecule. Although RNA can form base pairs, in general RNA molecules are single stranded rather than double stranded. Often in RNA, the strand doubles back upon itself, forming double-stranded sections of about 3 to 12 nucleotides in length (**Figure 2.7**). Like DNA molecules, these structures are double helical. Other RNA structures form a stem loop structure comprising a short stretch of nucleotides, a turn, and then a second stretch of nucleotides complementary to the first. The stem part of the stem loop structure can adopt a helical conformation. RNA can also form triplexed structures, loops, and clustered formations known as pseudoknots. Each of these elements then contributes to the larger overall structure by folding into a complex three-dimensional shape. The same weak forces discussed earlier for DNA structure are also at work in RNA, helping to fold the molecule and stabilize its conformation.

In terms of function, RNA is more diverse than DNA. RNA can store information and transmit it in much the same way that DNA stores information. RNA provides information for protein synthesis in all organisms; it also forms the genome of some viruses. RNA can be catalytic (e.g., ribozymes) and the protein synthesis machinery of the cell uses RNA as its catalytic machinery. RNA can bind to small molecules either through a covalent bond by binding to an amino acid in transfer RNA (tRNA) molecules or through noncovalent interactions such as in a riboswitch.

RNA is not as stable as DNA. It breaks down into nucleotides at a greater rate. RNA is also susceptible to hydrolysis under basic conditions, and even weakly alkaline solutions of pH 8 or 9 can rapidly accelerate the degradation of RNA. The reduced stability is due to the ribose ring. The 2′ hydroxyl is capable of attack on the phosphodiester bond of the backbone, resulting in cleavage (**Figure 2.8**). Hence, RNA must be treated and stored somewhat differently

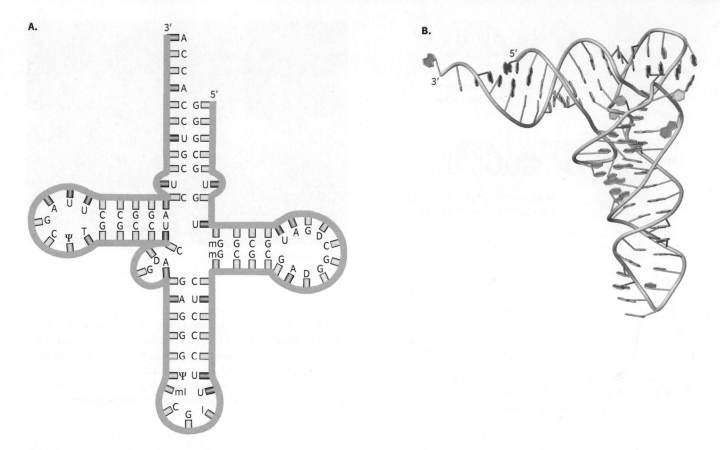

FIGURE 2.7 Complex RNA structures. RNA molecules form complex structures. **A.** In these tRNA structures we can still observe base pairing. **B.** Short helical segments and complex tertiary structures also form. Modified bases are denoted using Greek letters (such as ψ).

(Source: (A) Allison, *Fundamental Molecular Biology*, 2e, copyright 2012, John Wiley & Sons. This material is reproduced with permission of John Wiley & Sons, Inc. (B) Allison, *Fundamental Molecular Biology*, 2e, copyright 2012, John Wiley & Sons. This material is reproduced with permission of John Wiley & Sons, Inc.)

than DNA. While not recommended, DNA can be dissolved in water and left on the laboratory bench with few ill effects, while RNA is typically stored at −80°C in various buffers chosen to stabilize the molecule.

Frequently, fluorescent dyes are used to visualize, measure, and monitor DNA and RNA (**Figure 2.9**). Various factors affect fluorescence, but in the case of DNA and RNA the most important is whether the environment is aqueous or hydrophobic. In a more hydrophobic environment, the intensity of the fluorescence emitted by these molecules increases and the wavelength (color) of the emitted light is shifted further to the red side of the spectrum. Thus, a fluorescent dye can be used as a sensitive probe for hydrophobic environments. Fluorescent dyes, such as ethidium bromide, bind to nucleic acids and intercalate between the stacked base pairs of DNA. In this hydrophobic environment they fluoresce more brightly and at a longer wavelength than dye molecules in solution. Because DNA and RNA differ in the amount and type of base stacking that occurs, they bind to dyes with different affinity; hence, different dyes can be used to differentiate these molecules.

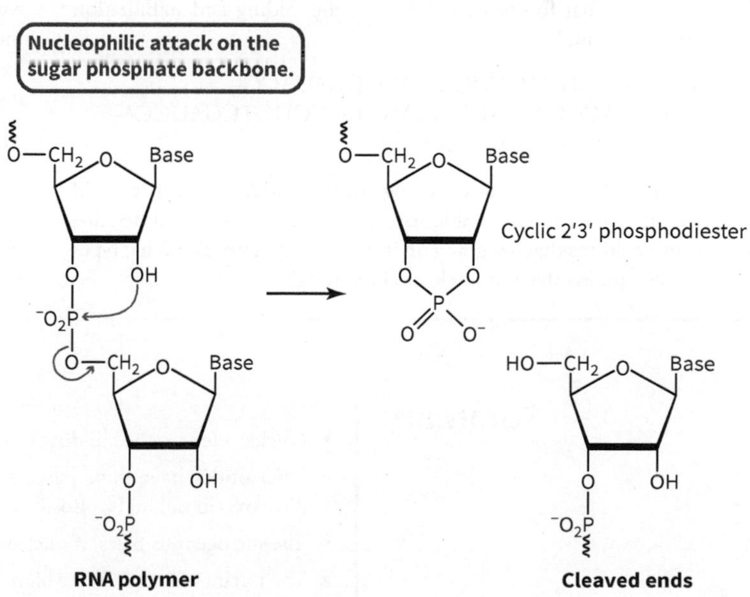

FIGURE 2.8 RNA cleavage. The 2′ hydroxyl group on ribose is in position to undergo nucleophilic attack on the phosphate attached to the 3′ hydroxyl. Basic conditions accelerate this reaction by deprotonating the hydroxyl.

A.

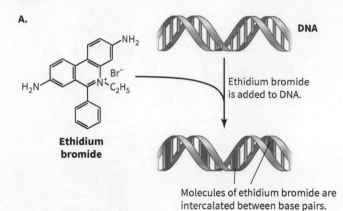

Ethidium bromide

Ethidium bromide is added to DNA.

Molecules of ethidium bromide are intercalated between base pairs.

B.

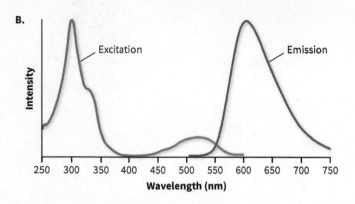

C.

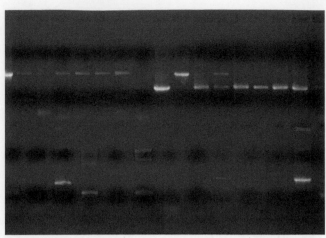

Klaus Guldbrandsen/Science Source

FIGURE 2.9 Fluorescent dyes and DNA. A. Fluorescent dyes, such as ethidium bromide, bind DNA by intercalating in between the base pairs of DNA. **B.** When bound in this hydrophobic environment the spectral properties of the dye change, increasing the amount of light emitted for every photon absorbed and undergoing a shift toward the red end of the spectrum. **C.** Dyes such as these can be used to visualize DNA in a gel.

Worked Problem 2.1   Structure of a nucleic acid

The tRNAs are a family of short nucleic acids that play critical roles in protein synthesis. The structure of the nucleotide sequence of a tRNA is shown in figure 2.7, along with a schematic of the structures it forms and a three-dimensional representation of the final structure. What forces are involved in the folding and stabilization of this molecule?

5'–GCGGAUUUAGCUCAGDDGGGAGAGC-
     CCAGACUGAAYACUGGAGUCCUGUGTCGAUCCA-
     CAGAAUUCGCACCA–3'

Strategy   Based on our knowledge of nucleotides and nucleic acid structures, we should be able to analyze the sequence and describe the interactions that are going on in the folded molecule. What types of forces are involved in nucleic acid folding?

Solution   The forces at work in the folding and stabilization of this molecule are the same as those in DNA or in other RNA molecules. There is electrostatic repulsion of the phosphates in the backbone, hydrogen bonding of the ribose rings and bases with one another or with water, dipole–dipole interactions of various polar parts of the molecule with one another, and London-dispersion forces causing weak interactions throughout the entire molecule. However, the driving force that guides folding of this molecule is the hydrophobic effect.

Although we can explain the forces at work in the folded tRNA molecule or in short stretches of RNA, we cannot yet *predict* the folded structure of macromolecules from the sequence of its building blocks.

Follow-up question   What techniques could you use to probe the structure of tRNA? What information would they provide?

Summary

- Nucleotides are the building blocks of nucleic acids.
- Nucleotides have three parts: a nitrogenous base (A, C, G, T, or U), a carbohydrate (ribose or 2'-deoxyribose) and a phosphate group.
- The nitrogenous bases of nucleotides can be categorized as purines or pyrimidines.
- The purines (A and G) are bicyclic structures.
- The pyrimidines (C, T, and U) are monocyclic.
- A, C, G, and T are the bases found in DNA, whereas A, C, G, and U are found in RNA.
- In addition to forming nucleic acids, nucleotides also play important roles in metabolism and signal transduction due to their ability to transfer phosphate groups to other molecules.

- Numerous weak forces contribute to the double-helical structure of DNA and the more complex forms of RNA, including the hydrophobic effect, hydrogen bonding, coulombic repulsion and base stacking effects.
- Nucleic acids can lose their secondary structure if heated or treated with chemical compounds that will disrupt these weak forces.
- Due to the hydroxyl group on the 2′ carbon of ribose, RNA molecules can be readily degraded under basic conditions.
- Because of their structure, nucleic acids can absorb UV light and bind fluorescent dyes.

Concept Check

1. Draw the structures of the nucleotides.
2. Explain how the chemical structures of nucleotides give rise to the properties of these molecules, for example, UV absorbance and hydrogen bonding.
3. Explain which forces are important in the formation of a double helix or the folding of a complex RNA molecule, and how alterations to the nucleotide sequence might influence this folding.

2.3 Nucleic Acids Have Many Cellular Functions

The list of known functions of nucleic acids in biochemistry grows as research continues, but perhaps the most important function of these molecules is to code and store the information needed to produce many of the molecules necessary for life. Nucleic acids also provide a means of copying and sharing that information when the organism replicates, and of translating that code into other types of biological molecules. Nucleic acids are even involved in the regulating when these processes happen. Because it is useful to have an understanding of these processes before discussing other aspects of biochemistry, this section gives a brief overview of the **replication** of DNA, **transcription** of DNA into RNA, **translation** of RNA into protein, and regulation of these steps. It also briefly discusses how viruses and retroviruses exploit a cell's own processes to infect the cell and reproduce.

2.3.1 Replication is the process by which DNA makes its own replica

The genetic material of the cell is DNA, which is organized in chromosomes. Coded in the sequence of bases of DNA is the information that tells the cell how to make proteins, which proteins to make, and even when to make them. Every time a cell divides, this information must be passed on, as accurately as possible. If an organism has 1 billion base pairs of DNA in its genome and copies it with an error rate of 1 error in a million, this will result in 1,000 errors every time a cell divides. In addition, even slight errors in the DNA sequence, termed **mutations**, can disrupt the message encoded in that piece of DNA. If a mutation has a negative effect, the organism may be unable to survive or propagate and thus will be unable to pass on its DNA. If, however, a mutation gives an organism some form of advantage in survival or propagation, it is more likely that this mutation will be passed on to the offspring of the organism. This is the central theme of evolution, and it is as true for bacteria living in thermal vents as it is for humans and redwood trees. All living organisms need to be able to replicate their DNA with a low error rate.

DNA replication is semiconservative   As described above, the DNA double helix is comprised of two strands of nucleotides, each of which is held together by phosphodiester bonds between 2-deoxyribose sugars (the sugar-phosphate backbone). These strands run in opposite directions. In both strands, the linkage runs between the 5′ carbon of one deoxyribose, through a phosphate, and then to the 3′ carbon of the next deoxyribose; however, the two strands are aligned so that this bonding runs in opposite directions, as shown in Figure 2.5.

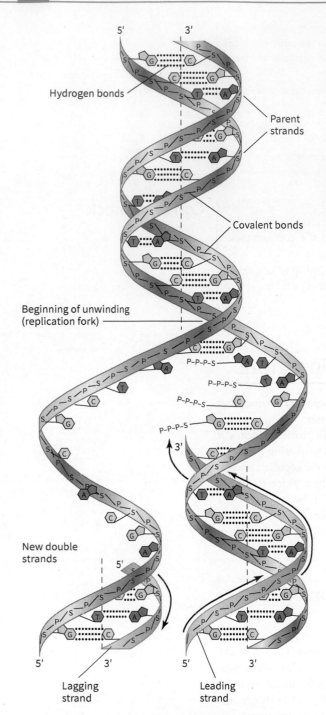

Hydrogen bonds

Parent strands

Covalent bonds

Beginning of unwinding (replication fork)

P-P-P-S

P-P-P-S

P-P-P-S

P-P-P-S

New double strands

Lagging strand

Leading strand

FIGURE 2.10 Semiconservative replication. Shown is a replication fork. In this view, DNA is being replicated moving from top to bottom. The right strand is the leading strand. As the fork unwinds (moving upward) the hydrogen bonds holding the two strands together are broken. The new strand is being synthesized moving into the divide. The left strand (the lagging strand) must be copied in the other direction in short segments measuring several hundred nucleotides in length.

(Source: Snustad; Simmons, *Principles of Genetics*, 6e, copyright 2012, John Wiley & Sons. This material is reproduced with permission of John Wiley & Sons, Inc.)

DNA replication must have a low error rate, but how is it possible to copy billions of bases billions of times yet still minimize errors? The answer partially comes from the structure of DNA itself. Because of its double-stranded structure, when DNA is denatured or unraveled, each strand can serve as a template. This makes the job of copying the sequence much easier; it also provides a way to proofread the sequence. Hence, when a cell divides, each daughter cell receives a copy of DNA that contains one strand from the parent and a second strand that has recently been synthesized. This mechanism of replication whereby each of the progeny cells receives a full set of DNA comprising one strand from the parent and one copy is referred to as **semiconservative replication** (**Figure 2.10**). The process has interesting ramifications because, theoretically, each person has two cells within the body that contain copies of the original DNA inherited when that person was conceived.

DNA replication is bidirectional and occurs in many places at once
The antiparallel structure of DNA presents an interesting dilemma when it comes to replication. DNA polymerases (the enzymatic complexes that replicate DNA) only replicate DNA along a strand in a single direction, from 5′ to 3′. However, a double helix has two strands, each running in opposite directions. How is it possible to replicate both sides in a concerted fashion? The answer is that replication does occur on both strands and the copying is concerted, but not in unison.

As the cell begins to copy its DNA, a bubble is formed where two strands of DNA are unwound, exposing single-stranded DNA to be copied (**Figure 2.11**). This replication bubble results in two replication forks—regions where the bubble and single strands meet. Replication occurs and spreads outward from these forks.

A replication fork contains a large complex of enzymes and proteins, which bind and coordinate the copying of DNA. Along the **leading strand**, new nucleotides are put into position and esterified by the polymerase in the 5′ to 3′ direction. On the **lagging strand**, a region of several hundred nucleotides is looped out from the polymerase complex. Once a large enough loop has been exposed, this loop is fed through a second unit of polymerase in the opposite direction, again synthesizing the DNA 5′ to 3′. Therefore, on the leading strand, the DNA is continually synthesized as the polymerase complex dives into the replication fork, whereas on the lagging strand, a short (200 bp) stretch of nucleotides is exposed and then filled in. Other enzymes on the lagging strand fill in gaps and join, or ligate, the strand fragments together.

DNA replication is highly regulated and coordinated
In bacteria and other prokaryotes, DNA replication is signaled by the ratios of ATP:ADP or the amount of energy available to the cell and the presence of DnaA, an ATP binding protein. DnaA is one of the proteins found in the replication fork that is involved in the unwinding of DNA. When ATP levels are high, DnaA binds to ATP and a specific sequence in the bacterial genome, known as the DNA box. Binding of the DnaA-ATP complex to the DNA box promotes replication of DNA.

In eukaryotes, DNA replication is coordinated with the cell cycle. It is more complicated than in prokaryotes, and it involves a complex signaling cascade. The key members of this cascade are the cyclins and the cyclin dependent kinases (CDKs)—enzymes that phosphorylate

A. Prokaryotic cell

Origin

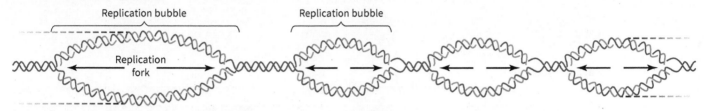

Replication fork

B. Eukaryotic cell

Replication bubble   Replication bubble

Replication
fork

FIGURE 2.11 Replication bubbles.   A. In prokaryotic cells there are relatively few origins of replication, distinct sequences where replication bubbles initiate. **B.** In eukaryotic cells numerous replication bubbles form in each chromosome in response to cell cycle dependent kinases.

(Source: (A) Snustad; Simmons, *Principles of Genetics*, 6e, copyright 2012, John Wiley & Sons. This material is reproduced with permission of John Wiley & Sons, Inc. (B) Alters, *Biology: Understanding Life*, 1e, copyright 2006, John Wiley & Sons. This material is reproduced with permission of John Wiley & Sons, Inc.)

proteins involved in chromosomal organization and replication. Bacteria and yeast contain specific sequences where replication begins, but no such sequences are thought to exist in higher eukaryotes. Instead, the action of several kinases results in the formation of numerous pre-replication complexes, in a process known as **replication licensing**. This process can only occur in the phase of the cell cycle prior to replication; thus, DNA replication is blocked in the other phases.

DNA replication can be rapid but is nevertheless the rate-determining step in cell division. In prokaryotes, DNA polymerases can copy upwards of 1,000 nucleotides per second, moving at a speed that has been said to approach that of a jet engine. This enables a bacterium, such as *Pseudomonas aeruginosa,* to divide every 20 minutes. Eukaryotes, however, have about 1,000-fold more DNA to replicate, and eukaryotic polymerases move 10-fold slower. Replication under these conditions is too much for a single enzyme or complex to accomplish. Instead, replication forks form in thousands of places in a eukaryotic genome. Although replication is slower in eukaryotes than in prokaryotes, eukaryotic cells have astounding propensities for division; for example, some cancer cells or cultured cells in the laboratory can divide every 16 to 20 hours.

2.3.2 Transcription is the copying of DNA into an RNA message

Coded within DNA is the information necessary to synthesize proteins, the diverse molecules that carry out most of the functions of the cell. However, proteins are not made directly from the DNA code. Instead, DNA is first transcribed into RNA, in a process known as transcription. This occurs differently in prokaryotes than it does in eukaryotes (**Figure 2.12**).

Not all DNA contains information that codes for protein, but those regions that do code are known as **genes**. In prokaryotes, genes may be organized into **operons**, regions of a chromosome that code for multiple proteins with related purposes.

In prokaryotes, genes comprise a single message that has no gaps or breaks in it. At the front, or the 5′ end, of the gene there are stretches of DNA known as promoter regions.

A. Prokaryotic cell

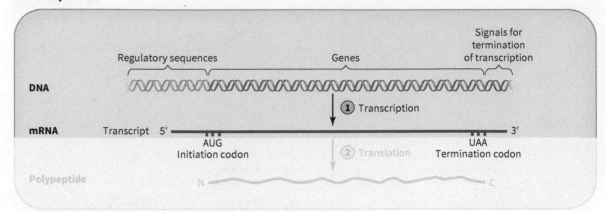

B. Eukaryotic cell

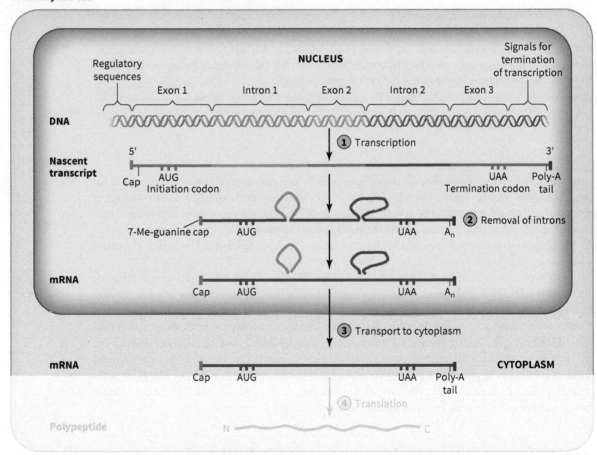

FIGURE 2.12 Transcription. **A.** Prokaryotic genes are transcribed into mRNA. **B.** In eukaryotes, nascent RNA must be spliced and processed to generate a mature mRNA. The translation steps shown shaded are illustrated in Figure 2.13.

(Source: (A) Snustad; Simmons, *Principles of Genetics*, 6e, copyright 2012, John Wiley & Sons. This material is reproduced with permission of John Wiley & Sons, Inc. (B) Snustad; Simmons, *Principles of Genetics*, 6e, copyright 2012, John Wiley & Sons. This material is reproduced with permission of John Wiley & Sons, Inc.)

These regulatory sequences can bind to specific proteins known as transcription factors that will regulate the copying of the gene into a messenger RNA (mRNA) by RNA polymerase.

In eukaryotes, the process is somewhat more complicated. Eukaryotic genes are interrupted and divided into regions that code, known as **exons**, and regions that do not code, known as **introns**. As with prokaryotes, transcription factors bind to DNA in the promoter region 5′ of the coding sequence. This elicits binding of an RNA polymerase and copying of a nascent transcript—essentially an mRNA copy of the gene. The nascent message must then be processed, edited to remove the introns, and modified to lend some stability to this molecule. Introns are spliced out of the gene, resulting in a single, uninterrupted message. The 5′ end of the message is modified with a 7-methyl-guanine cap, which is thought to prevent premature degradation of the message. Other modifications are made to the opposite end of the message, with a poly-adenosine (poly-A) tail being enzymatically added to the 3′ end of most messages. This tail is not coded for in the DNA sequence, but its addition is signaled by a specific poly-A sequence at the end of the message. Like the 7-methyl guanine cap, the poly-A tail is thought to stabilize the message.

In the strictest and most classical sense, each gene codes for a single protein. However, in reality, genes in eukaryotes can be differentially spliced, or proteins can be modified after synthesis, resulting in several highly related proteins that can be made from each gene. While we tend to focus on the big picture—DNA to RNA to protein—gene expression is regulated at multiple levels: DNA itself can be modified, and there are many RNAs that regulate the function of genes and other RNAs. There are also organisms, such as retroviruses, that do not follow the central dogma.

The nearly universal nature of the central dogma means that a piece of human DNA or a human gene can be inserted into another organism for study. Species may differ in how they transcribe genes or splice or modify RNAs, but the ability to study a gene or protein by expressing it in a different organism is one of the more powerful tools of biochemistry.

2.3.3 Translation is the synthesis of proteins by ribosomes using an mRNA code

In translation, proteins are synthesized using the code found in an mRNA (**Figure 2.13**). Two other types of RNA figure prominently in this process: tRNA and ribosomal RNA (rRNA). The tRNA are small RNA molecules (approximately 180 nucleotides) that act as adapters. Each amino acid has its own type of tRNA. At the other end of the tRNA is a loop that can "read" the code of the mRNA. These tRNAs transfer a specific amino acid to a growing protein, depending on which amino acid is called for by the code in the mRNA. The macromolecular complex that brings together the mRNA and the tRNA is the **ribosome**, the protein synthesis factory of the cell. Ribosomes themselves are predominantly made of several RNA molecules termed rRNA or ribosomal RNA. These RNA molecules catalyze the addition of new amino acids to the growing amino acid chain of the new protein.

All of biology, from *E.coli* to elephants uses the same code to make proteins. Hence, it is possible to use bacteria to make human proteins from human genes, although in a bacterium, the final product may not be processed in exactly the same way as it would be in a human. In addition, it is possible to put proteins into particular cells to study their function or to better understand disease states and other problems. When studying a protein or gene using these techniques, it is important to keep in mind that the system being studied is not identical to the original, but it may be sufficiently similar to provide meaningful data.

2.3.4 Regulation of replication, transcription, and translation is critical to an organism's survival and propagation

As we go through the course of a day, our needs will change. For example, we may need to consume food for energy, move from one place to another, or respond to environmental challenges. Likewise, as we grow and mature throughout development, we express different genes or alter the levels of gene expression. To be evolutionarily successful, an organism must be able to respond to its environment, grow and develop, and ultimately reproduce and transmit its genes to the next generation.

The sections above discussed how DNA is replicated and noted that different proteins known as transcription factors bind to promoter sequences in DNA, either upregulating or downregulating gene expression or the production of mRNA. However, there are several other ways in which genes may be regulated (**Figure 2.14**).

Some genes use a riboswitch. There are several different types of riboswitch, but each involves a segment of the mRNA folding into a complex shape that will bind to a small molecule,

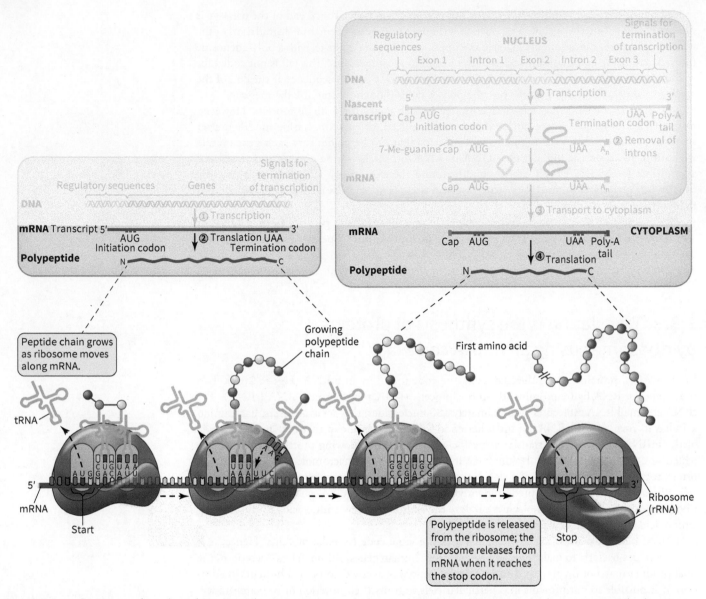

FIGURE 2.13 Protein synthesis (translation). Ribosomes, mRNA, and amino acyl tRNAs act in concert to synthesize proteins. The directions of how to formulate the protein are provided by the mRNA. The tRNA molecules act as adapters and help match the amino acid with the proper code on the mRNA. Ribosomes are assemblies of RNA and protein that catalyze the reaction. The transcription steps shown shaded are illustrated in Figure 2.12.

(Source: (Top) Snustad; Simmons, *Principles of Genetics*, 6e, copyright 2012, John Wiley & Sons. This material is reproduced with permission of John Wiley & Sons, Inc. (Bottom) Wessner, DuPont, Charles, Neufeld, *Microbiology*, 2e, copyright 2017, John Wiley & Sons. This material is reproduced with permission of John Wiley & Sons, Inc.)

affecting the ability of the mRNA to be translated. Through this mechanism, some genes can be regulated after an mRNA is synthesized.

Other genes may be regulated by short RNA molecules that complement the mRNA, forming a double-stranded mRNA structure. These RNAs interfere with transcription or lead to message degradation, in a process known as RNA interference (RNAi). This process lends itself to several laboratory applications, and because of its ease of use and relatively low cost, it is rapidly changing how biochemical experiments are being conducted.

It used to be thought that once a gene was activated, a protein product would result. Now, it is clear that genes can be regulated in several different ways. It is also clear that proteins vary in their stability, with some having lifetimes on the order of the life of the organism and others lasting only a few seconds before being degraded. Additionally, some proteins are synthesized in a futile cycle; that is, they are synthesized and then rapidly degraded unless stabilized by the presence of some other molecule.

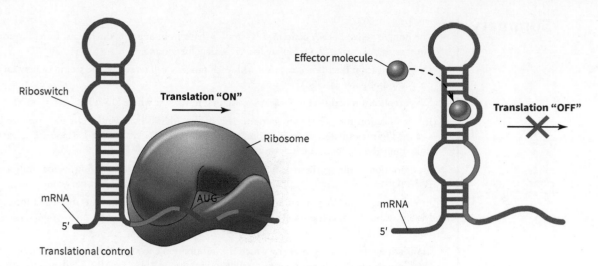

FIGURE 2.14 Riboswitches are one mechanism to regulate gene expression. A riboswitch consists of a complex structure formed by the 5′ end of the mRNA. This structure can bind to a small molecule, resulting in a structural change that alters transcription.

(Source: Wessner, DuPont, Charles, Neufeld, *Microbiology*, 2e, copyright 2017, John Wiley & Sons. This material is reproduced with permission of John Wiley & Sons, Inc.)

2.3.5 Viruses and retroviruses use the cell's own machinery to reproduce

Viruses are small pathogenic assemblies of nucleic acids, proteins, and occasionally lipids that require a cellular host for replication. Those that contain DNA or RNA as their genetic material are typically categorized as **viruses**, whereas those that employ RNA as their genetic material but go through a DNA intermediate at some point in their life cycle are categorized as **retroviruses**. Viruses invade a cell through one of several different mechanisms; once inside a cell, the virus integrates its genetic material with that of the host. The virus then uses the host's own replication, transcription, and translation machinery to produce copies of viral nucleic acids and viral proteins. These assemble into mature virions, which are released from the cell and can then infect other cells in the organism, or other organisms.

Worked Problem 2.2   How do mutations affect the central dogma?

DNA can be chemically damaged by exposure to UV light. As illustrated below, adjacent molecules of thymine in one strand can become cross-linked to one another, forming a cyclobutane ring.

How might this type of mutation affect replication, transcription, and translation?

Strategy Examine the structure shown. Think about what happens during replication, transcription, and translation. How would these processes be affected?

Solution Thymine dimers are a common DNA lesion. The formation of the cyclobutane ring induces strain in the DNA backbone, resulting in a bulge. In theory, the covalent linkage between the thymines could affect replication, in that a polymerase may not be able to read the thymine dimer as well as it could single thymines. Depending on where the mutation occurs in the DNA, it could affect transcription. Again, an RNA polymerase may have difficulty reading through the thymine dimer. Because translation takes place after gene transcription, it is unlikely that the mutation will affect translation.

Follow-up question Some other mutations generate oxidatively damaged or alkylated bases. How might modification to the amine or carbonyl groups of the base give rise to problems in replication, transcription, and translation?

Summary

- The double helix of DNA contains two strands that run in opposite directions. In replication, each of these strands serves as a template for a new daughter strand.
- DNA replication is semiconservative, with each progeny cell receiving one strand of new DNA and one strand of parental DNA.
- DNA replication occurs at replication forks, that is, regions where DNA has been unraveled.
- In a replication fork, the leading strand is copied as the fork unwinds. On the opposite lagging strand, DNA synthesis occurs in short fragments as the DNA is exposed. These are subsequently linked together to form one long strand.
- Transcription is the synthesis of RNA molecules from DNA. Transcription factors indicate to an RNA polymerase where to bind and start synthesizing an mRNA copy of the gene.
- In eukaryotes, mRNA molecules are modified and edited before they are translated into protein. Modifications include the splicing out of introns and the addition of a 7-methyl guanosine cap or poly-A tail.
- Translation refers to the process by which the information encoded in the mRNA molecule is translated into protein. Ribosomes, large, catalytic complexes of RNA, and tRNA, adapter molecules that insert the proper amino acid, are also important in this process.
- Viruses and retroviruses use the cell's own machinery to replicate and reproduce. This often means integration of the viral genome into the host.
- Regulation of replication, transcription, and translation is critical to an organism's survival and propagation.

Concept Check

1. Discuss the enzymes involved in DNA and RNA replication.
2. Describe in basic terms how replication, transcription, and translation are regulated by the organisms.

2.4 The Manipulation of Nucleic Acids Has Transformed Biochemistry

The commonality of the central dogma and the universal nature of the genetic code have had powerful implications for biochemistry and medicine. It has been said that in the first half of the twentieth century, science belonged to physics, but in the second half, the crown was abruptly passed to other fields such as biochemistry. Part of this succession is due to a biochemist's ability to manipulate the environment of the cell, and even multicellular organisms, with molecular precision. This section describes several techniques that can be used to manipulate or alter DNA, RNA, or proteins in the organism. These techniques are the basis of modern biochemistry.

2.4.1 DNA can be easily manipulated and analyzed *in vitro*

DNA is vital to life because it codes for the proteins that comprise the cell. DNA passes from generation to generation and is synonymous with many things we may consider deeply personal. Nevertheless, DNA is in fact just another chemical and, as such, can be manipulated in the laboratory. Advances in organic synthesis, molecular biology, protein expression, and enzymology have taken DNA manipulations from the advanced research laboratory to DNA instruction at the elementary school level in less than a generation. This section discusses how we can manipulate and analyze DNA.

Short DNA fragments can be synthesized chemically   Short pieces of DNA (up to approximately 70 nucleotides long) can be synthesized chemically in a technique known as solid phase oligonucleotide synthesis (**Figure 2.15**). In this technique a nucleotide is bound to a resin or bead that is trapped in a reactor. A second nucleotide is added and reacts with the first nucleotide. Further reactions with the second nucleotide are stopped by a protecting group. Unreacted materials are washed away, and the dinucleotide bound to the column is left behind in the reactor. The protecting group is removed from the dinucleotide, and the reaction is repeated with a new nucleotide. In this way, the oligonucleotide is built up one nucleotide at a time. Initially, this process was undertaken in the laboratory by hand; today, however, robotic devices can perform all of the necessary steps, making it possible to generate these molecules literally at the push of a few buttons.

The oligonucleotides generated through solid phase synthesis have multiple applications in the laboratory.

DNA can be amplified using the polymerase chain reaction

The **polymerase chain reaction** (PCR) is one of the most powerful tools developed for the manipulation of DNA fragments (**Figure 2.16**). PCR makes it possible to simply amplify or create millions of copies of DNA in a short period of time.

Included in the reaction mix for PCR are:

- *template DNA*—the DNA that is going to be copied
- *primers*—short oligonucleotide sequences that flank the region to be copied
- *a polymerase*—the enzyme that replicates the DNA
- *deoxynucleotides*—the building blocks of DNA

The polymerase used in PCR has been isolated from thermophilic organisms that live in hot springs or near volcanic vents at the bottom of the ocean. In contrast to most enzymes, these polymerases will not lose activity when heated, an essential property for this reaction.

The final piece of the PCR reaction is the instrument needed to carry out the experiment. A thermocycler heats and cools the sample to facilitate DNA amplification.

First, DNA is heated to denature it, melting the double helix into two single strands. Next, the sample is cooled to allow the primers to anneal (bind) to the template strand. Then, the sample is warmed again slightly to activate the polymerase, which extends from the primers to generate a copy of the parent DNA. This cycle is repeated 25 to 30 times, resulting in over a million-fold amplification of the original molecule.

A variant of PCR is emulsion PCR (**Figure 2.17**). In this technique, the primers for the reaction are coupled to a microscopic agarose bead. The beads, PCR reaction mix, and DNA are emulsified in oil so that each individual bead will generate a specific PCR product. These reactions are often used to amplify genomic DNA in modern genomic DNA sequencing.

PCR is such a versatile technique that it is commonly coupled with or used as the basis of many other techniques, including DNA sequencing, genotyping, disease identification, and forensic science.

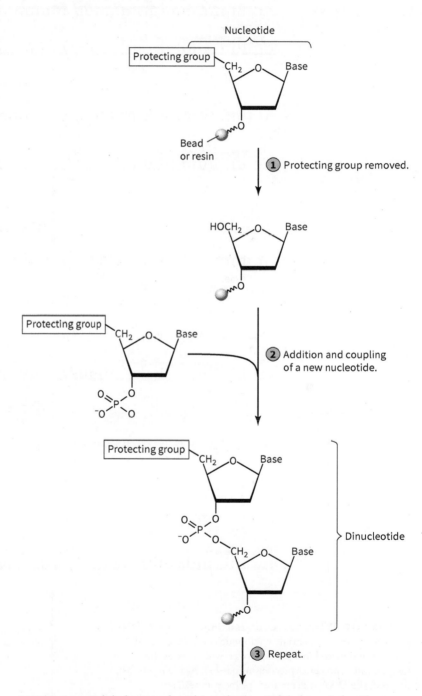

FIGURE 2.15 Solid phase nucleotide synthesis. Solid phase nucleotide synthesis proceeds through the addition of protected bases to a nucleotide affixed to a solid support (a bead). The newly added group is deprotected with mild acid, and the process is repeated with a new nucleotide.

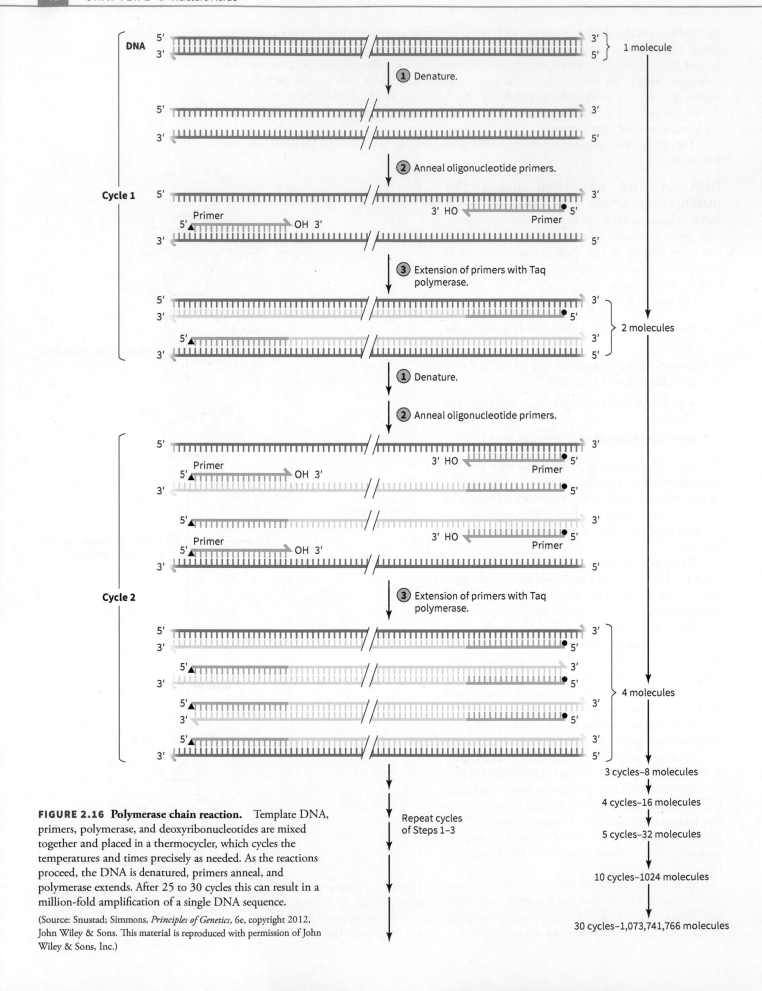

FIGURE 2.16 Polymerase chain reaction. Template DNA, primers, polymerase, and deoxyribonucleotides are mixed together and placed in a thermocycler, which cycles the temperatures and times precisely as needed. As the reactions proceed, the DNA is denatured, primers anneal, and polymerase extends. After 25 to 30 cycles this can result in a million-fold amplification of a single DNA sequence.

(Source: Snustad; Simmons, *Principles of Genetics*, 6e, copyright 2012, John Wiley & Sons. This material is reproduced with permission of John Wiley & Sons, Inc.)

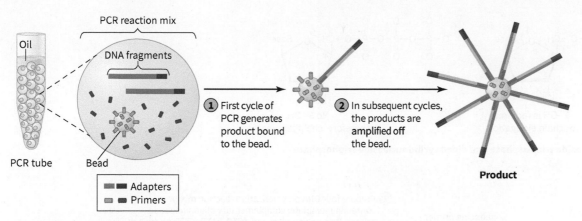

FIGURE 2.17 Emulsion PCR.   Fragments of DNA are ligated to adapters. The fragments are emulsified with beads coated with one of the primers and PCR reagents in such a way to ensure there is only a single copy of DNA per emulsion droplet. The fragments melt and then anneal to the bead through a primer that recognizes the adapter. The PCR reaction copies the strand, and the reaction is repeated, each time generating a copy of the fragment bound to the bead.

First-generation DNA sequencing technologies employed chain termination

Sequencing of a polymer such as a protein or nucleic acid is the determination of the order of the monomers that make up that polymer. One of the earliest successful systems for DNA sequencing was the Sanger method or dideoxy system (**Figure 2.18**). It is comprised of a sequencing mixture containing a DNA polymerase, a template to be sequenced, an oligonucleotide or sequencing primer from which the polymerase can extend, and radiolabeled deoxynucleotide triphosphates. Sequencing primers bind specifically to one region of DNA, ensuring that the region downstream will be the one that is sequenced. This mixture is divided into four reactions, and a small amount of a different 2'3'-dideoxynucleotide triphosphate is added to each. Because these nucleotides lack a hydroxyl group on the 3' carbon of the deoxyribose ring, the polymerase is unable to extend from one of these nucleotides, which effectively becomes a chain terminator, blocking further elongation from that chain. Because we know which dideoxynucleotide is in the reaction, we know which base is found at the end of that chain. The products of the four reactions are separated, based on size, on a polyacrylamide electrophoresis gel. Gel electrophoresis is a technique in which molecules are separated in a gel using an electric current. A film exposed to this gel contains bands that show the order of the bases, one at a time. In the mid-1990s, an accomplished investigator could obtain about 200 to 300 bases per set of reactions.

Fluorescent DNA sequencing permitted automation and facilitated modern genomic studies

The next evolution of DNA sequencing technology employed fluorescently tagged dye terminator nucleotides (**Figure 2.19**). Because four different fluorophores could be used, only one reaction mix was needed. In addition, the strands were separated with a capillary column rather than a gel. The entire system could now be automated, initially running as many as 384 samples at once. Thanks to technical advances, by the early 2000s this method generated 500 to 600 bases per reaction, and, by the end of the decade, the number of bases increased to almost 1,000. This technology greatly advanced sequencing, and it enabled the completion of the human genome project in 2003, two years ahead of schedule.

Modern parallel sequencing technologies sequence millions of samples at once

Modern sequencing technologies enable the sequencing of an entire genome within days or weeks. These technologies employ from hundreds of thousands to billions of parallel reactions, which are then read optically from a microscope slide. The sequence data generated are found in short stretches, which are then assembled into a complete genome using advanced computer algorithms. This section discusses two of these technologies: Illumina sequencing, and SOLiD or two-base encoding.

A.

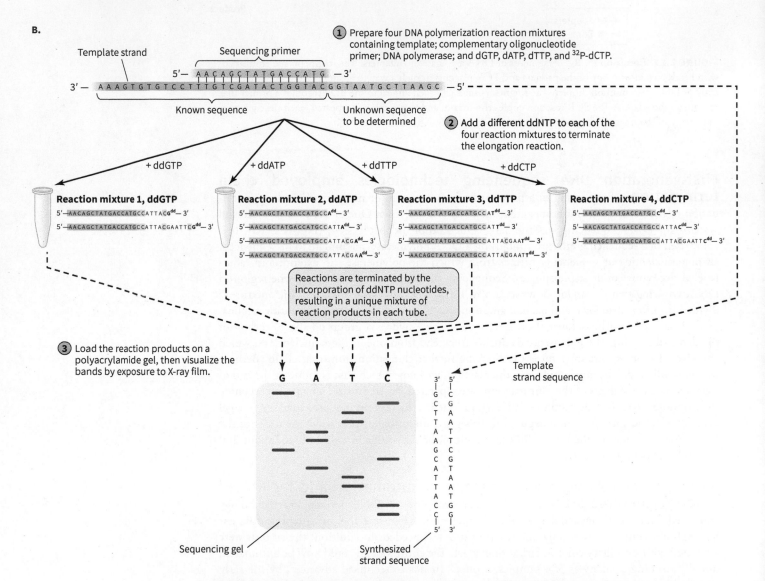

3'-OH is required
for chain elongation.

No 3'-OH;
therefore, chain terminates.

Deoxyribonucleoside triphosphate and dideoxyribonucleoside triphosphate

B.

FIGURE 2.18 First-generation DNA sequencing using the Sanger dideoxy method.   A. Sanger sequencing employs dideoxy nucleotides that lack hydroxyl groups on both the 2′ and 3′ carbons, blocking elongation once one of them has been incorporated. **B.** Sequencing using this method employs four different reactions, each of which contains a strand to be synthesized, a sequencing primer from which the DNA strand can be synthesized, a polymerase to make the new strand, and the four different deoxynucleotides. Each reaction also contains small amounts of a single dideoxynucleotide. The four different reactions are separated on a gel. Each band terminates with a known nucleotide (one of the dideoxynucleotides from the reaction), enabling the reading of the DNA sequence.

(Source: (A) Wessner, DuPont, Charles, Neufeld, *Microbiology*, 2e, copyright 2017, John Wiley & Sons. This material is reproduced with permission of John Wiley & Sons, Inc. (B) Wessner, DuPont, Charles, Neufeld, *Microbiology*, 2e, copyright 2017, John Wiley & Sons. This material is reproduced with permission of John Wiley & Sons, Inc.)

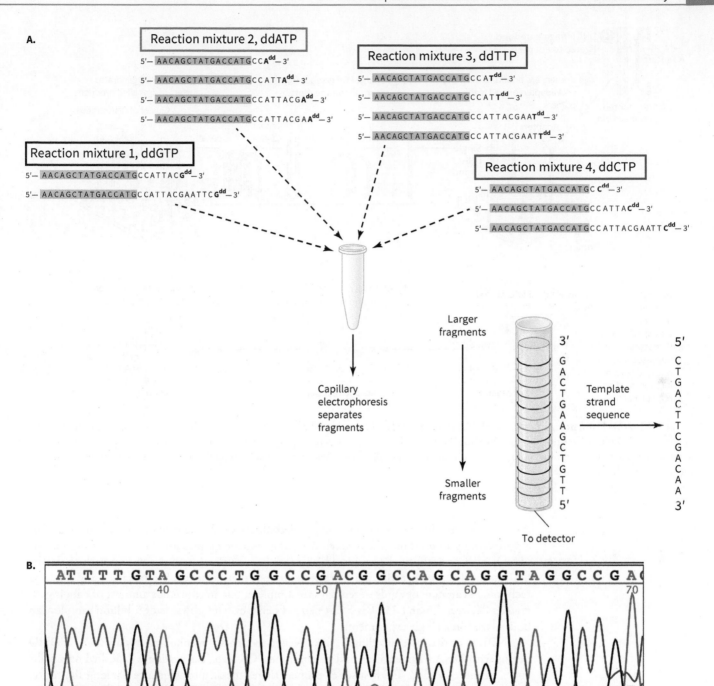

A.

Reaction mixture 2, ddATP

5'— AACAGCTATGACCATGCCA**A**dd — 3'
5'— AACAGCTATGACCATGCCATTA**A**dd — 3'
5'— AACAGCTATGACCATGCCATTACGA**A**dd — 3'
5'— AACAGCTATGACCATGCCATTACGAA**A**dd — 3'

Reaction mixture 3, ddTTP

5'— AACAGCTATGACCATGCCAT**T**dd — 3'
5'— AACAGCTATGACCATGCCATT**T**dd — 3'
5'— AACAGCTATGACCATGCCATTACGAAT**T**dd — 3'
5'— AACAGCTATGACCATGCCATTACGAATT**T**dd — 3'

Reaction mixture 1, ddGTP

5'— AACAGCTATGACCATGCCATTAC**G**dd — 3'
5'— AACAGCTATGACCATGCCATTACGAATTC**G**dd — 3'

Reaction mixture 4, ddCTP

5'— AACAGCTATGACCATGC**C**dd — 3'
5'— AACAGCTATGACCATGCCATTAC**C**dd — 3'
5'— AACAGCTATGACCATGCCATTACGAATT**C**dd — 3'

Capillary electrophoresis separates fragments

Larger fragments

Smaller fragments

3'
G
A
C
T
G
A
A
G
C
T
G
T
T
5'

Template strand sequence

5'
C
T
G
A
C
T
T
C
G
A
C
A
A
3'

To detector

B.

AT TTT GTA GCCCTG GCCG ACG GCCAG CAGG TAGGCCGA

40   50   60   70

FIGURE 2.19 Fluorescent PCR-based sequencing. **A.** Fluorescent sequencing is similar to dideoxy sequencing but employs nucleotides that terminate in a nucleotide coupled to a fluorescent dye. There are four dyes used, one for each different nucleotide. **B.** The products of the reaction are separated based on size and the results automatically detected and presented as a chromatogram corresponding to the nucleotide sequence.

(Source: (A) Wessner, DuPont, Charles, Neufeld, *Microbiology*, 2e, copyright 2017, John Wiley & Sons. This material is reproduced with permission of John Wiley & Sons, Inc. (B) Wessner, DuPont, Charles, Neufeld, *Microbiology*, 2e, copyright 2017, John Wiley & Sons. This material is reproduced with permission of John Wiley & Sons, Inc.)

In Illumina sequencing, genomic DNA is fragmented and ligated to adapters (**Figure 2.20**). These double-stranded fragments are denatured and annealed to the surface of a plate that is densely coated with both adapters. Bridge PCR, or amplification that bridges the two adapters on the surface, is carried out to amplify single copies into small spots termed "clusters." The DNA content of the cluster is identified by sequencing all of the amplified clusters at once, but one nucleotide at a time. Modified versions of fluorescent deoxynucleotides are added;

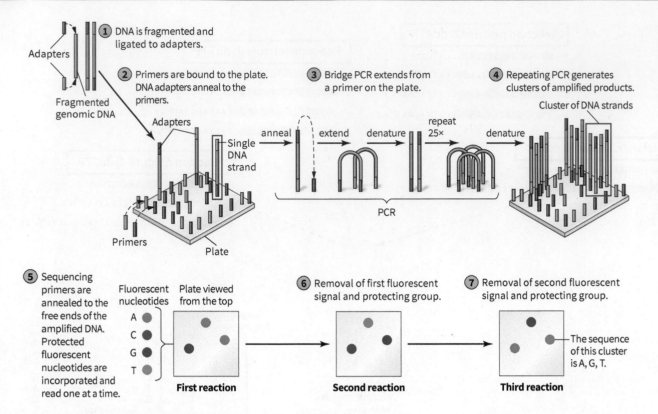

FIGURE 2.20 Illumina sequencing.   Genomic DNA is fragmented and ligated to a linker. The other end of the linker has a different primer and both are bound to the surface of a microscope slide. Next, "bridge" PCR is used to amplify single spots on the slide between the two adapters. Sequencing is done using fluorescent dyes, and the results are read optically. Individual spots will light up in different colors on the slide, corresponding to the four different nucleotides.

each of the four different types of deoxynucleotides has a different fluorophore attached, plus a protecting group on the 3′ hydroxyl. After the sequencing reaction has been performed, the plate is read optically to determine which color (i.e., nucleotide) is present on each spot. The fluorescent dye together with the protecting group is then removed. Sequencing using Illumina technology can take upwards of ten days to complete, but an enormous amount of data is generated. Between 25 and 250 bases can be read per spot, and upwards of 3 billion samples can be analyzed in a single experiment.

SOLiD or two-base encoding is a ligase-based sequencing technology (**Figure 2.21**). SOLiD begins with DNA fragmentation and an emulsion PCR step. The beads are deposited on a slide, and a primer is added. The first set of primers shares the adapter sequence flanking the region to be sequenced; the second is eight bases in length and has a fluorescent group attached to it. If the two nucleotides at the 3′ end are complementary to the DNA template being sequenced, they will anneal. Ligase is added, and if the second primer anneals, it will be ligated with the first. The remaining primers are washed away. Next, fluorescence is detected wherever annealing and ligation has occurred. The fluorescent tags are removed using a nuclease and the cycle is repeated, building up short stretches of sequence, two bases at a time. Use of different primer combinations enables completion of the sequence. Typically, each spot can have somewhere around 50 bases sequenced using this technique, and upwards of 1.5 million spots can be analyzed. Run times are one to two weeks for an entire genome.

Each of these technologies has its strengths and weaknesses. Because of two-base recognition, SOLiD is better than other techniques at detecting single nucleotide polymorphisms and is less expensive per base, but it is slower than comparable techniques. All of these technologies take advantage of computer analysis of data for quality control and analysis, and all use computers to assemble the final genome from the sequences acquired.

High throughput sequencing is already being used in the clinic, as is discussed in **Medical Biochemistry: Diagnosing neonatal diseases with DNA.**

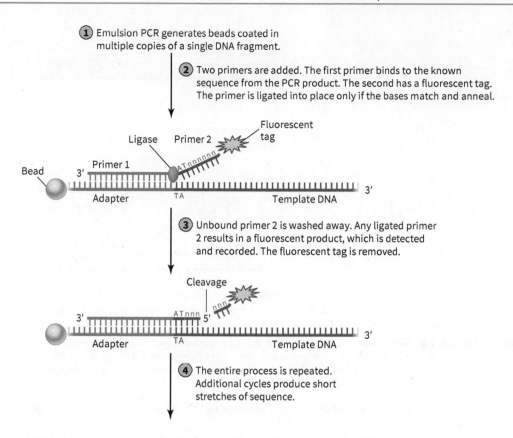

① Emulsion PCR generates beads coated in multiple copies of a single DNA fragment.

② Two primers are added. The first primer binds to the known sequence from the PCR product. The second has a fluorescent tag. The primer is ligated into place only if the bases match and anneal.

③ Unbound primer 2 is washed away. Any ligated primer 2 results in a fluorescent product, which is detected and recorded. The fluorescent tag is removed.

④ The entire process is repeated. Additional cycles produce short stretches of sequence.

FIGURE 2.21 SOLiD sequencing. SOLiD sequencing first employs emulsion PCR to generate beads from which to sequence. Two primers are added. The first is complementary to the template, while the second consists of eight nucleotides and a fluorescent tag. Ligase is added. Should the second primer bind and be ligated, the sequence will now fluoresce with the color of that tag. The samples are deprotected, and the next primer is added. Through successive cycles it becomes possible to sequence short sequences and ultimately sequence the entire genome.

(Source: From Oelkerding, K.V., Dames, S.A., Durtschi, J.Y., Next-Generation Sequencing: From Basic Research to Diagnostics, *Clinical Chemistry* Apr 2009, 55 (4) 641–658; DOI: 10.1373/clinchem.2008.112789. Reproduced with permission from the American Association for Clinical Chemistry.)

Medical Biochemistry

Diagnosing neonatal diseases with DNA

One in 20 children born in the United States has some form of inherited disease. Many diseases in children or adults have a progression to them, but in newborns, an inherited disease often presents suddenly and early, sometimes within hours of birth. Newborns are unable to verbally communicate what is wrong, and their tiny size and fragile nature make invasive procedures undesirable. Classical diagnosis of these diseases, when possible, can often take weeks to months, and requires the patient be seen by multiple specialists who deal in identifying specific classes of illnesses.

Several studies have now used next-generation DNA sequencing to diagnose these patients. In this technique, a small sample of the patient's blood is used to obtain a DNA sample. The DNA is sequenced and a set of specific genes, either mitochondrial genes or those related to inherited disease, are analyzed. This approach has several benefits. First, it simplifies the DNA analysis. Instead of having to reconstruct an entire genome, researchers can focus on the parts more likely to be affected. Second, it avoids the ethical pitfalls of obtaining the entire DNA sequence, because genes linked with other diseases such as cancer will not be analyzed. Finally, and most important, it streamlines the diagnosis. Instead of focusing on every gene, researchers narrow in on those likely to be causal. Using this technology, it is currently possible to sequence and analyze a group of 3,500 genes in a little more than a day.

More than 7,500 inherited diseases are known and, although treatments are still unavailable for most of them, knowing the prognosis offers comfort to some families. DNA sequencing studies are useful in genetic counseling, because they can help in identifying recessive disorders the parent may be carrying or diseases that were previously unknown. Perhaps the greatest benefit of such studies is that they indicate a future direction of medicine, in which personal and individual diagnoses become possible.

DNA fragments can be cut and ligated using a wide array of enzymes

Knowledge of DNA sequences is most valuable when it can be used in the laboratory. DNA can be manipulated using a series of enzymes (**Figure 2.22**) that were originally identified in nature. These enzymes assist in DNA replication or serve as part of a primitive immune system for bacteria, but they are now commercially available. In many ways, this is similar to scrapbooking. In a craft or hobby store there are aisles which have numerous different patterns of scissors or paper cutters and different types of glue or adhesive, all dedicated to scrapbooking. This is analogous

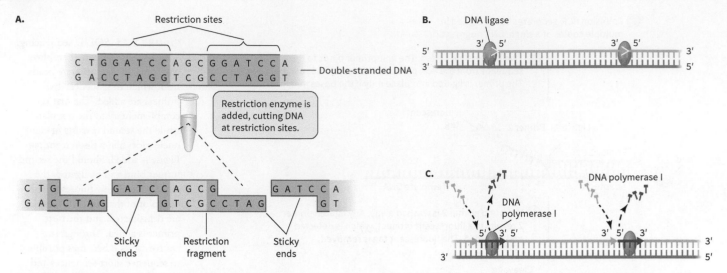

FIGURE 2.22 Enzymes that manipulate DNA.   Numerous enzymes are commercially available to manipulate and synthesize DNA. **A.** Restriction enzymes recognize a four to eight base DNA sequence and cleave both strands of the phosphodiester backbone. **B.** DNA ligase joins the cut ends of DNA. **C.** DNA polymerase synthesizes a new strand of DNA.

(Source: (A) Alters, *Biology: Understanding Life*, 1e, copyright 2006, John Wiley & Sons. This material is reproduced with permission of John Wiley & Sons, Inc.
(B) Wessner, DuPont, Charles, Neufeld, *Microbiology*, 2e, copyright 2017, John Wiley & Sons. This material is reproduced with permission of John Wiley & Sons, Inc.
(C) Wessner, DuPont, Charles, Neufeld, *Microbiology*, 2e, copyright 2017, John Wiley & Sons. This material is reproduced with permission of John Wiley & Sons, Inc.)

to what can be achieved in the laboratory as fragments of DNA are cut up and joined together (**Table 2.1**).

The enzymes that cut DNA are called **restriction enzymes**. They were originally identified as components of bacteria that enabled growth in the presence of some bacteriophages, restricting the impact of the virus. Restriction enzymes are endonucleases; that is, they cut inside a strand of nucleotides. Palindromes, such as RACECAR, are words or phrases that read the same forward as they do backward. Restriction enzymes typically recognize a palindromic sequence 4 to 8 nucleotides long and cleave the sugar-phosphate backbone on both strands, leaving either a blunt end or an overhang, referred to as the sticky end.

DNA ligases are enzymes that seal breaks in the sugar-phosphate backbone. They do not fill in missing bases but can join two blunt or complementary sticky ends together.

Copying DNA in a living organism is termed **cloning**. Subcloning involves copying only a piece of the whole, for example, a fragment of a gene. The term "cloning" is also used to mean creation of a copy of a single cell, such as a bacterial or cultured mammalian cell, or creation of a copy of an organism, such as a sheep from the DNA of a single parent cell. Copying DNA in an organism differs from amplifying a piece of DNA in a PCR reaction. Inside the organism, it is possible to copy DNA with much higher fidelity, that is, with fewer errors, and to copy much larger pieces than is currently possible in a test tube. Bacteria are often used as part of the process of manipulating DNA.

Cutting and piecing DNA together would be easier if we had some means of holding onto the pieces like a sort of cover to the scrapbook. Nature has provided this as well. Plasmids are

TABLE 2.1 Enzymes Commonly Used in Molecular Biology to Manipulate Nucleic Acids

Enzyme	Purpose/Function
Restriction enzymes	Cuts specific DNA sequences
Ligase	Joins two DNA fragments together
Polymerase (Klenow fragment)	Generates a copy of one strand of DNA from a primer and a template
Thermophilic polymerases	Polymerizes enzyme in polymerase chain reaction (PCR)
Topoisomerase	Unwinds DNA in the cell; used to partially join DNA fragments; used in place of ligase in some cloning applications
Reverse transcriptase	Converts RNA into DNA; used in producing cDNA libraries and reverse transcriptase PCR

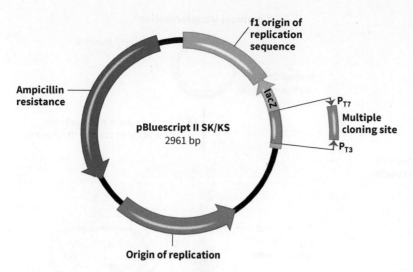

FIGURE 2.23 Plasmid DNA.   Plasmids are short (less than 10 kb) circular pieces of DNA. Shown is pBluescript II, a common plasmid used in cloning. Plasmids all contain an *ori* sequence (an origin of replication) necessary for propagation in bacteria, and at least one selectable marker. Most other plasmids also contain a multiple cloning site (MCS), an engineered stretch of DNA consisting of multiple restriction sites adjacent to one another.

short (2 to 5 kb) loops of bacterial DNA that bacteria use to exchange genes with one another (**Figure 2.23**). In the laboratory, plasmids such as pBluescript II have been intentionally rearranged and mutated to provide a handy tool for DNA manipulation. These plasmids contain:

- an origin of replication sequence, called an *ori* sequence, which enables the plasmid to be replicated in bacteria
- a selectable marker, that is, a gene that provides a trait to the bacteria so that only the bacteria containing the plasmid will grow and reproduce
- a cloning site, that is, an engineered series of restriction sites that facilitate the insertion of the DNA fragment of interest

Plasmids may also have other features, depending on the research in which they are employed. For example, some plasmids have been optimized for cloning DNA, whereas others contain sequences that drive high levels of expression of mRNA, and subsequently of protein.

Plasmids are one method that can be used to clone DNA. Other methods that offer options for cloning larger fragments of DNA referred to as vectors include bacterial artificial chromosomes (BACs), yeast artificial chromosomes (YACs), and cosmids (a fusion of a plasmid with the *cos* sequences of a bacteriophage).

2.4.2 DNA can be used to drive protein expression

This section explains how the information coded within DNA fragments can be used to alter the biochemistry of a cell or organism. For example, a cell or organism can be manipulated to produce proteins it usually lacks, or the information can be altered and tests undertaken to see how the alterations affect the proteins coded for by the gene.

Genes can be expressed or overexpressed in cultured cells   All organisms use basically the same genetic code. Because of this, it is possible to use the pieces of DNA produced in the laboratory to make RNA or protein for subsequent study. The first step is to get the pieces of DNA into the cells, a process known in bacteria as **transformation** and in eukaryotic cells as **transfection**. In several instances, the same technique can be used on either bacterial or eukaryotic cells.

DNA can be introduced into cells through electrical, chemical, or physical methods, or by using an engineered virus (**Figure 2.24**).

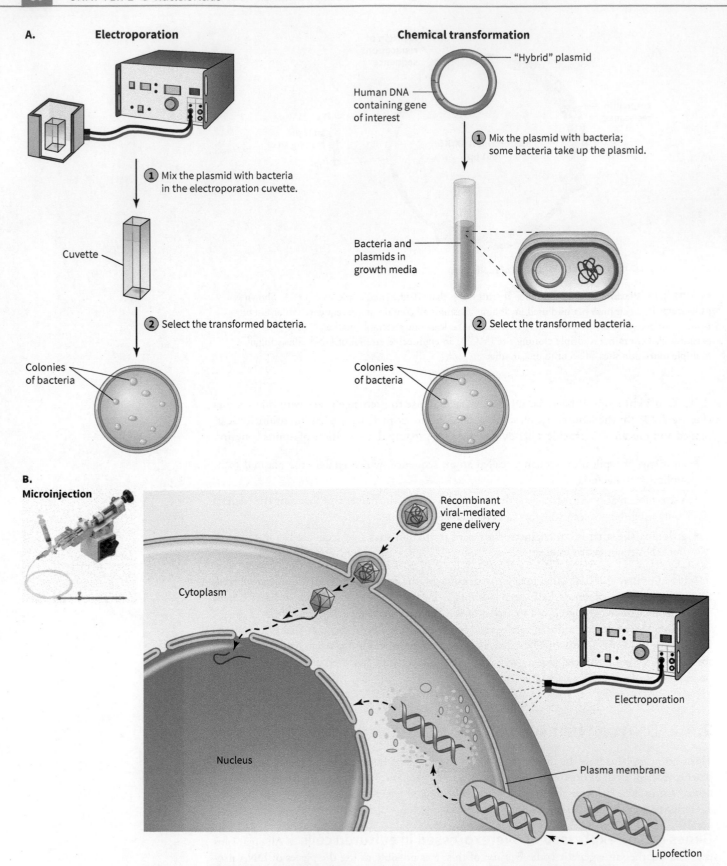

A. Electroporation

Chemical transformation

"Hybrid" plasmid

Human DNA containing gene of interest

1 Mix the plasmid with bacteria; some bacteria take up the plasmid.

1 Mix the plasmid with bacteria in the electroporation cuvette.

Cuvette

Bacteria and plasmids in growth media

2 Select the transformed bacteria.

2 Select the transformed bacteria.

Colonies of bacteria

Colonies of bacteria

B. Microinjection

Recombinant viral-mediated gene delivery

Cytoplasm

Electroporation

Nucleus

Plasma membrane

Lipofection

FIGURE 2.24 Transformation, transfection, and infection.   A. Bacterial cells can be transformed either by electroporation or using chemical means. **B.** Eukaryotic cells can be infected with a virus or transfected using electroporation, cationic lipids, or microinjection.

(Source: (A) Alters, *Biology: Understanding Life*, 1e, copyright 2006, John Wiley & Sons. This material is reproduced with permission of John Wiley & Sons, Inc. (B) Wessner, DuPont, Charles, Neufeld, *Microbiology*, 2e, copyright 2017, John Wiley & Sons. This material is reproduced with permission of John Wiley & Sons, Inc.)

In **electroporation**, cells are suspended in a weak buffer in a cuvette with the DNA to be electroporated. A special power supply known as the electroporator sends short, strong pulses of electricity through the cuvette. These pulses are often several hundred volts and are on the order of milliseconds long. In theory, these pulses temporarily blow holes into the plasma membrane of cells and allow the cells to acquire DNA or RNA. Other molecules, such as drugs and dyes, have also been electroporated. Electroporation can be used with bacteria as well as eukaryotic cells.

In addition to electroporation, bacteria can be treated to make them chemically competent to absorb DNA. This involves treating the bacteria with different salt solutions at low temperature. Transformation efficiencies using this method are typically 10-fold lower than electroporation, but the technique is simple and effective, and does not require additional equipment.

There are several means to transform eukaryotic cells chemically. The earliest used diethylaminoethyl (DEAE) dextran or calcium phosphate. In the DEAE technique, DNA was complexed to microscopic particles of dextran, a high molecular weight carbohydrate coated with DEAE groups. The positively charged groups in DEAE attracted the negatively charged DNA and the molecules formed a complex, which was then applied to cells. These complexes were taken up by cells through endocytosis.

A second early technique that is still commonly used is calcium phosphate precipitation. DNA is mixed with a calcium-containing buffer to form DNA–calcium complexes. Following incubation, a phosphate-containing buffer is added. The calcium and phosphate ions react to form a precipitate of microscopic crystals containing the DNA. As with DEAE-dextran transfection, these crystals are thought to be absorbed by the cell through endocytosis.

The main obstacle to transfection is the insertion of DNA into the cell, that is, getting past the plasma membrane. A more recent innovation uses DNA complexed to cationic lipids to smuggle the DNA into the cell. The plasma membrane of the cell is largely composed of phospholipids. These phospholipids form a hydrophobic layer through which anything entering the cell must pass. Cationic lipids have a positively charged end that binds to the sugar-phosphate backbone of DNA, generating a liposome, or a vesicle of lipid, surrounding the DNA. The liposomes can fuse to the plasma membrane and result in high transfection efficiencies (the percentage of cells that take up the DNA molecule). This technique is sometimes referred to as lipofection.

A third means of introducing DNA is to coat it onto microscopic particles of gold or tungsten and physically blast these particles into the cell using a pulse of helium gas. This technique is referred to as biolistics (from "biological" and "ballistics"). This is often effective on cells that can be otherwise hard to transfect. A handheld version of this device called the Gene Gun is commercially available.

One final method that is commonly used to introduce DNA is to alter the genetic sequence of a virus and use that altered virus to deliver genes; this technique is known as recombinant viral-mediated gene delivery. Several viruses, including lentivirus and adenovirus, are commonly used for this procedure. The technique involves modifying the viral genome so that the DNA lacks the genes coding for proteins the virus would use to replicate, and instead inserting a gene coding for the protein of interest. This new piece of viral DNA is transfected into a packaging cell line (a cell line that has been modified to produce the viral proteins but lacks the viral DNA). When the two come together, the packaging cell can now produce a virus, but the virus lacks the proteins needed for replication and instead produces the protein of interest.

Modified viruses are currently the gold standard in gene delivery. They often have high transfection efficiencies and can transfect cells that are otherwise difficult to work with. However, some of these viruses are human pathogens and could, without proper precautions, contaminate and infect people exposed to the virus.

Once a plasmid of DNA with the proper sequence has been put into a cell, that cell's native polymerases and ribosomes will use the instructions coded within it to produce protein. However, with this modification alone, the effect will be short lived. Unless there is selective pressure on the cell, most cells will stop producing protein from a plasmid relatively quickly. Therefore, the plasmid is constructed to contain a selectable marker, a gene which codes for a protein that enables cells to survive in an environment where others cannot. This gene is often an antibiotic resistance marker but can also be an enzyme that plays an

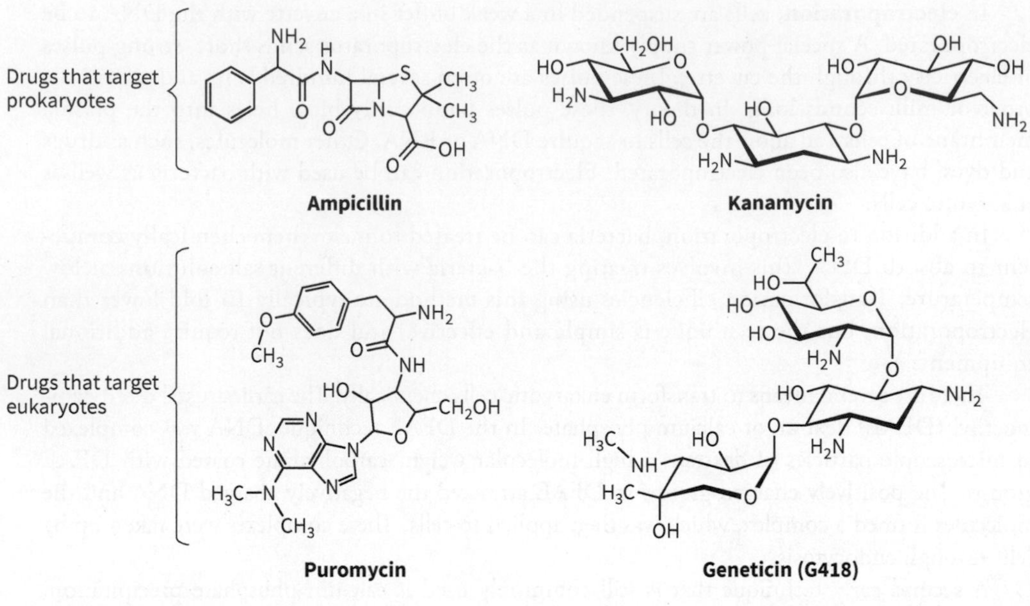

FIGURE 2.25 Antibiotics used in selecting cells. Ampicillin, kanamycin, puromycin, and geneticin are all antibiotics that are used to select transformed cells. Ampicillin and kanamycin are used on prokaryotes, while puromycin and geneticin are used to treat eukaryotic cells.

important role in metabolism, such as in the production of an amino acid (**Figure 2.25**). Cells that lose this piece of DNA or stop producing the proteins coded for by the plasmid will not survive.

In addition to selection, expression in some cell types (typically eukaryotic cells) can be divided broadly into two categories: transient and stable. All transfections begin as transient transfections. In the first 24 to 48 hours following transfection, the transfection rate is highest and the proteins will be strongly expressed. However, as time goes on, the rate of transfection drops, as does the expression rate. To keep expression at a high level, cells are selected using an antibiotic appropriate for the resistance gene. In eukaryotic cells, this is often puromycin or geneticin. After treatment with the antibiotic for several rounds of cell division, the plasmid of interest is presumed to have integrated into the host chromosome and to have passed from one generation to the next. This is referred to as stable transfection.

Entire organisms can be manipulated to express genes of interest

So far, this chapter has focused on bacterial and eukaryotic cells, but technology exists to express genes in more complex organisms. In many animals, zygotes are large enough that DNA can be injected into them (**Figure 2.26**). As with transfections of other cell types, if the piece of DNA in question is able to integrate into the host genome, it will be passed from cell to cell as the organism grows and develops, resulting in an organism that contains this DNA in every cell of its body. Again, depending on the promoter sequence used to drive expression, this can mean expression in all cells or in only a select organ or tissue. This sort of experiment results in a **transgenic organism**.

Transgenic plants can be made using a somewhat different technique, also shown in Figure 2.26. For example, the bacterial plant pathogen *Agrobacter tumefaciens* has sequences that can recombine with the plant's genome. It is possible to generate a plasmid containing these sequences, modify the plasmid, and transform *Agrobacter* using the techniques previously discussed. Cultures of *Agrobacter* are then incubated with cultured plant cells. The bacteria infect some of the plant cells, and the new piece of DNA is incorporated into the plant genome. Using this technology, genes have been incorporated that can slow the ripening process, increase yield, increase resistance to plant diseases, and increase the nutrition found in plants. There is continuing debate and controversy at the ethical, scientific, legal, and political levels as to whether this is a good idea. A discussion of some of the ramifications of genetically modified organisms can be found in **Societal and Ethical Biochemistry: Genetically modified organisms.**

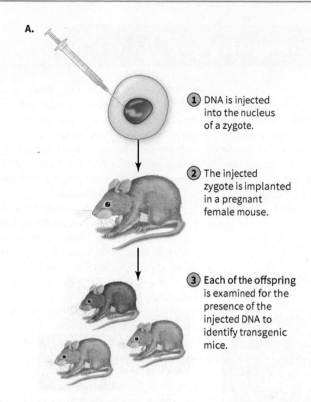

A.

1. DNA is injected into the nucleus of a zygote.

2. The injected zygote is implanted in a pregnant female mouse.

3. Each of the offspring is examined for the presence of the injected DNA to identify transgenic mice.

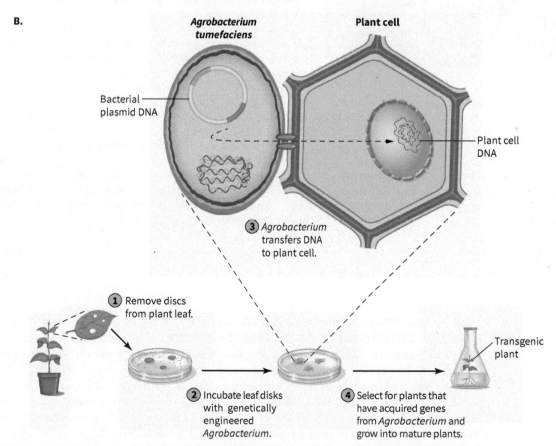

B.

Agrobacterium tumefaciens

Plant cell

Bacterial plasmid DNA

Plant cell DNA

3. *Agrobacterium* transfers DNA to plant cell.

1. Remove discs from plant leaf.

Transgenic plant

2. Incubate leaf disks with genetically engineered *Agrobacterium*.

4. Select for plants that have acquired genes from *Agrobacterium* and grow into mature plants.

FIGURE 2.26 Transgenic organisms. Transgenic organisms can be made by different methods depending on the organism. **A.** Transgenic mice can be made by injecting DNA into a fertilized mouse egg prior to cell division. **B.** Transgenic plants are often made using engineered *Agrobacter* lines. The modification is made to plasmid DNA, the DNA is introduced to *Agrobacter*, and then the bacteria infect and incorporate the DNA into the plant genome.

(Source: (A) Snustad; Simmons, *Principles of Genetics*, 6e, copyright 2012, John Wiley & Sons. This material is reproduced with permission of John Wiley & Sons, Inc. (B) Allison, *Fundamental Molecular Biology*, 2e, copyright 2012, John Wiley & Sons. This material is reproduced with permission of John Wiley & Sons, Inc.)

Societal and Ethical Biochemistry

Genetically modified organisms

Often, science becomes popularized. This probably occurs for a variety of reasons including nationalism, power, or some great success like landing on the Moon. Other times there are causes that we embrace because we feel they personally impact us or impact others. Examples of these causes include things that affect our health or impact the environment. One of the most controversial science issues of the past decade has been genetically modified organisms (GMOs).

dscz/Getty Images

GMOs are organisms that have had their genomes manipulated in the laboratory, usually through the introduction of genes from other organisms or the manipulation of genes to alter expression levels. Since the beginning of agriculture, humanity has been breeding and crossbreeding food crops and livestock to bring out favorable characteristics. Traditionally, this has been done through selective breeding. Although few people look at golden retrievers or tulips as genetic manipulations, they are the result of tens of generations of selective breeding and inbreeding.

Scientists have now identified genes that impart desirable traits and, using the tools of molecular biology, are able to introduce the desired genes into host organisms. This type of genetic engineering has been used to create crops that are resistant to particular diseases or herbicides (Roundup Ready corn or soybeans) or produce their own pesticides (Bt Cotton). Genetic engineering also been used to create animals that have a high growth rate (AquAdvantage salmon) or produce less waste (EnviroPig).

In many ways this sounds ideal. We can grow foods using less energy and chemicals, and feed more people on an increasingly crowded planet. However, the situation is more complex. In theory at least, genes inserted into GMO could be passed on to other related species, for example from corn to grasses, resulting in other species carrying the modified trait, such as resistance to herbicides. In the case of GMO salmon, the concern is that the modified organism could escape from aquaculture and damage the native population.

Another concern is that the organism produced is the property of a company. Farmers who grow a GMO crop must buy new seeds every year from the manufacturer. Currently, it is illegal to take seeds from last year's crops and plant them. This law was enacted to protect companies producing GMO seeds and some view it as controversial.

Generation of resistant pests is another criticism of the GMO industry. This argument is most commonly made with Bt crops, crops expressing an insecticidal protein from the bacterium *Bacillus thuringiensis*, such as Bt cotton. Many are concerned that like any pesticide an overuse will lead to resistant strains of insects. Cultures of *B. thuringiensis* are a standard treatment used by many organic farmers and loss of this would make organic insect control even more difficult.

Finally, there are concerns that the driving force behind production of GMO organisms may be commercial gain rather than public good and that the benefits do not make it worth taking the risk.

Although many people may not have considered how they feel about GMOs, they are likely to have been consuming genetically modified products.

Gene sequences can be altered before expression to facilitate the study of protein function or disease states The combination of solid-phase nucleotide synthesis and PCR make it simple to alter DNA sequences through site-directed mutagenesis (a specific mutation generated at a specific site in the sequence) (**Figure 2.27**). The first step is to design and synthesize a primer that contains the mutation of interest. This primer will have a mismatch with the template DNA at the mutation, but it needs to have enough homology to be able to anneal and allow the polymerase to extend. When this primer is used in PCR, the mutation will be incorporated into the PCR product. Typically, the product created in these reactions is not a full-length copy of the coding region of the gene, known as a cDNA. Instead, a fragment containing the mutated sequence can be cut on the ends with restriction enzymes and cloned in place in the cDNA. This technique is called cassette mutagenesis. Using this approach, numerous mutations can be generated in a single sequence and swapped in or out as desired.

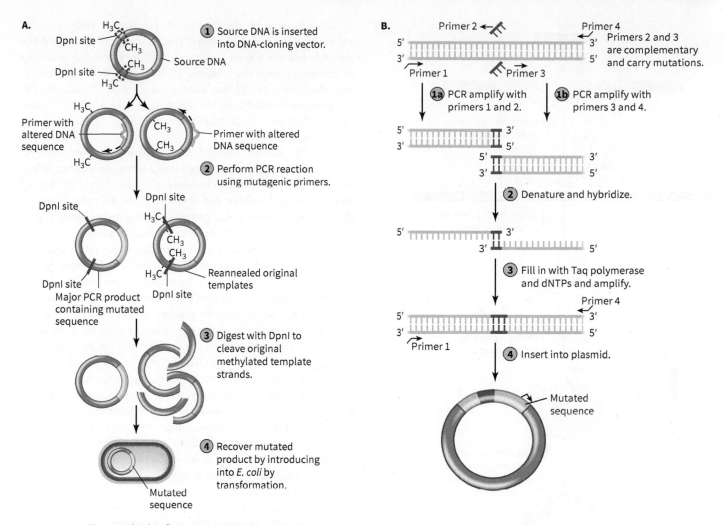

FIGURE 2.27 Two methods of generating site-directed mutants. **A.** In single-step mutagenesis, two primers each containing the mutation are used. These face away from each other and are used to amplify the entire plasmid. The parent plasmid is of bacterial origin and can be degraded by the restriction enzyme DpnI while the mutant PCR products are not cleaved by Dpn I. **B.** In classical PCR-based site-directed mutagenesis PCR is used to generate two separate products, each containing the mutation and a short region that overlaps the mutation site. A second round of PCR is used with flanking primers to generate a fragment that can be cloned into a plasmid.

(Source: (A) Wessner, DuPont, Charles, Neufeld, *Microbiology*, 2e, copyright 2017, John Wiley & Sons. This material is reproduced with permission of John Wiley & Sons, Inc. (B) Allison, *Fundamental Molecular Biology*, 2e, copyright 2012, John Wiley & Sons. This material is reproduced with permission of John Wiley & Sons, Inc.)

2.4.3 Techniques can be used to silence genes in organisms

Although molecular biology techniques to produce a protein in a cell or organism are useful, there are times when it would be helpful to know what would happen if a protein were absent. Two techniques are commonly used for this type of experiment: a **knockout** organism or **gene silencing** using siRNA. A relatively new method is **CRISPR**. All three techniques are explained below.

Knockout technology effectively deletes a gene from the organism

A knockout organism takes advantage of homologous recombination to delete a gene (**Figure 2.28**). The first step is to map the gene of interest to identify intron and exon boundaries. A region of the gene to be deleted is then identified. This region often includes the start codon or results in a frameshift mutation, to ensure that any protein that is accidently produced is nonfunctional. Segments of DNA flanking the region to be deleted are selected and cloned into a knockout vector, which contains sequences to facilitate its manipulation in bacteria, and selectable markers for screening its insertion into the host organism's genome.

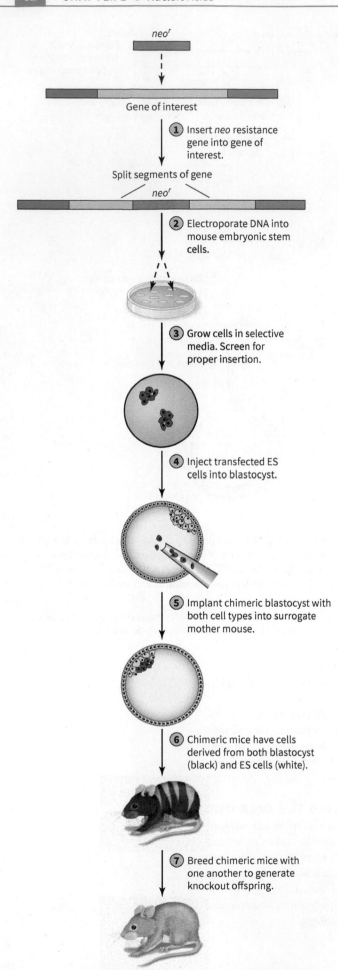

neo^r

Gene of interest

1 Insert *neo* resistance gene into gene of interest.

Split segments of gene

neo^r

2 Electroporate DNA into mouse embryonic stem cells.

3 Grow cells in selective media. Screen for proper insertion.

4 Inject transfected ES cells into blastocyst.

5 Implant chimeric blastocyst with both cell types into surrogate mother mouse.

6 Chimeric mice have cells derived from both blastocyst (black) and ES cells (white).

7 Breed chimeric mice with one another to generate knockout offspring.

Once generated, the knockout construct is electroporated into embryonic stem (ES) cells. The cells are cultured, and the selectable marker is used to identify cells in which the construct has integrated into the genome. In 3%–10% of these integrations, the construct will have homologously recombined with the host genome, resulting in the deletion of one of the cell's two copies of that gene. Cells must be genotyped using PCR or other techniques.

The ES cells are injected into a growing mouse embryo, termed a blastocyst, and implanted into a surrogate mother. When born, the offspring, termed **chimeras**, will have some cells derived from the original embryo and some from the ES cells. If the ES cells have contributed to the gametes of the chimera, these organisms can be interbred to produce a knockout organism.

At one level the ramifications of knocking out a gene are clear, but the broader impacts may not be. If the process is successful, the organism will lack functional copies of the deleted gene and be completely unable to produce the protein coded for by that gene. The inability to produce a particular protein can manifest in three ways. First, the deletion may have no obvious effect, either because other genes help to compensate for the missing gene or because the changes only become evident when specific assays are performed. Second, the gene may be essential for life, so its deletion causes an embryonic lethal mutation, meaning the organism dies early in development or *in utero*. Although this situation provides important data, it makes it difficult to study the system. The third option, which is perhaps the most useful, is the situation in which the knockout organism has an interesting phenotype but is still capable of reproducing and is therefore easy to study.

One final drawback of this type of experiment is the time and cost involved. The process is complex, requiring a team of experts with a range of specialized skills, and takes as long as 18 months.

RNAi offers the ability to quickly and affordably knockdown gene expression

In the late 1990s, it was discovered that eukaryotes have the ability to regulate gene expression after transcription, that is, after mRNA synthesis (**Figure 2.29**). The cell can do this by synthesizing short stretches of mRNA that are the reverse complement of the message.

FIGURE 2.28 Production of knockout mice. The production of a knockout mouse begins with the generation of a knockout construct, a piece of DNA in which a central region of the gene of interest has been replaced by the *neo* selectable marker. The construct is electroporated into mouse embryonic stem cells and selected with geneticin. Surviving cells are screened to ensure that the DNA has integrated properly into the chromosome. These mutant ES cells are injected into blastocysts that are implanted into pregnant female mice. The cells develop into chimeric mice containing some cells derived from the blastocyst and some derived from the ES cells. These heterozygous mice are crossed with one another to generate knockout animals.

(Source: (Steps 1–2) Snustad; Simmons, *Principles of Genetics*, 6e, copyright 2012, John Wiley & Sons. This material is reproduced with permission of John Wiley & Sons, Inc. (Steps 3–7) Karp, *Cell and Molecular Biology: Concepts and Experiments*, 7e, copyright 2013, John Wiley & Sons. This material is reproduced with permission of John Wiley & Sons, Inc.)

① An RNAi construct is produced. This piece of DNA has a fragment of the gene that is being silenced and a reverse complement of that sequence separated by several nucleotides.

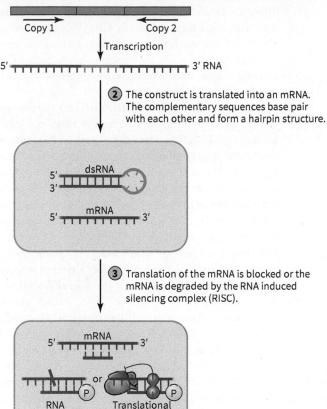

② The construct is translated into an mRNA. The complementary sequences base pair with each other and form a hairpin structure.

③ Translation of the mRNA is blocked or the mRNA is degraded by the RNA induced silencing complex (RISC).

FIGURE 2.29 RNAi regulating gene expression. RNAi is a technique for regulating gene expression post transcriptionally. Short RNAs that are complimentary to the gene being regulated are transcribed and processed in the nucleus. These form part of the gene silencing complex.

(Source: Snustad; Simmons, *Principles of Genetics*, 6e, copyright 2012, John Wiley & Sons. This material is reproduced with permission of John Wiley & Sons, Inc.)

These strands can bind to mRNA and either trigger its destruction or block translation. Soon after the initial discovery, it was found that results could be obtained by introducing artificial complementary mRNA or DNA constructs into the cell. This technique has several variations but is generically known as **RNA interference** (RNAi).

A short RNA molecule can be used to interfere, but use of these molecules is typically limited because they cannot be copied and passed from one cell to the next as the cells divide. Instead, a common approach is to construct a fragment of DNA coding for a short (approximately 20 nucleotide) stretch of the coding sequence, a short loop and a reverse complement of the first sequence. When translated, this forms a short hairpin structure that uses the cells' natural RNA processing mechanisms to produce fragments that will knock down or silence the gene of interest.

There are several key differences between a knockout organism and cells treated with RNAi. Construction of a knockout mouse can take years and cost hundreds of thousands of dollars, whereas RNAi can be done in an afternoon for less than one hundred dollars. Because of the lower cost and simplicity of the technique, hundreds of genes can be knocked down at once. Unfortunately, RNAi is not completely predictable or absolute because some RNAi constructs will silence a gene but others may not. The reasons for this difference are unclear. Hence, RNAi may produce variable phenotypes in which the expression of a gene has been blunted but not shut off. Finally, because knockout organisms do not have a functional copy of the gene, they never produce functional protein, whereas RNAi works from the point at which the gene becomes expressed. This can be valuable if an RNAi can be triggered at a specific time in development or an experiment, but this also assumes that the protein will appear or disappear at a rate that can be effective in the experiment. If a gene is silenced using RNAi, we assume that the protein product of that gene is not produced, and this leads to a phenotype or effect. However, if the half-life of any pre-existing protein is long and lingers after the gene has been silenced, silencing the gene may have little effect.

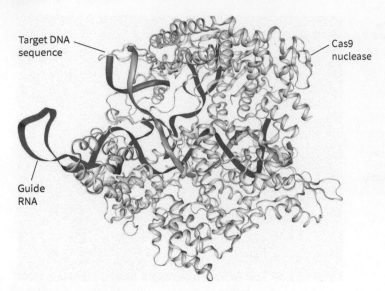

Target DNA sequence

Cas9 nuclease

Guide RNA

FIGURE 2.30 CRISPR gene editing. The CRISPR system consists of a nuclease, in this case the Cas9 nuclease, and a guide RNA. The guide RNA has one end that binds to the nuclease and one that binds to the DNA sequence, guiding the nuclease to the site of cleavage.

(Source: Data from PDB ID 5B2R, Hirano, S., Nishimasu, H., Ishitani, R., Nureki, O. (2016) Structural Basis for the Altered PAM Specificities of Engineered CRISPR-Cas9 *Mol. Cell* 61: 886–894.)

CRISPR/Cas9 is a flexible system used for genomic engineering.

CRISPR stands for *C*lustered *R*egularly *I*nterspaced *S*hort *P*alindromic *R*epeats. These are DNA sequences used by prokaryotes in association with a nuclease, often the CRISPR associated nuclease 9 or Cas9, to form part of the prokaryotic immune response to viruses. In this system, RNA guides Cas9 to a specific location where the enzyme binds and cuts the foreign DNA (**Figure 2.30**).

In the laboratory, the CRISPR/Cas9 system consists of a plasmid of DNA encoding the Cas9 nuclease and a single guide RNA. The single guide RNA consists of two parts: a gene specific sequence and a Cas9 specific sequence. The gene specific RNA guides Cas9 to the specific sequence in the genome to be edited while the other sequence binds to Cas9. Once bound at the specific site, Cas9 cuts both strands of DNA leaving exposed ends. The cell's endogenous DNA repair mechanisms rejoin these ends but often add or delete bases, resulting in the loss of gene function if the addition or deletion occurs in a coding sequence. Variations of the CRISPR system allow deletions of large sequences of DNA, permit generation of point mutations, and facilitate insertion of specific new sequences. CRISPR therefore is not simply a tool to delete or silence genes, but rather a system through which entire genomes can be edited. Furthermore, because the flexibility of the CRISPR system lies within the guide RNA sequence, changes to the system are relatively simple to make; investigators only need to change the DNA that encodes the RNA. Thousands of CRISPR experiments have been performed in just the few years since this technique was developed.

While CRISPR has made it easy to edit organisms in the laboratory, it is evident that this technique can be used to edit any genome, including the human genome. Among many ethicists this has raised concern that this technique will be used not to prevent disease but to make "improvements" to people, such as increasing intelligence or athletic performance. While these alterations still remain in the realm of science fiction, they may not be far off.

Worked Problem 2.3   Genetic engineering

Your research advisor has heard about using a fluorescent protein from coral called mCherry to help identify the lab's protein of interest. She has a sample of the cDNA for mCherry, and you already have a cDNA for your protein. Your professor would like you to express the two joined together as one protein. How would you fuse your protein and the fluorescent protein? What steps would be involved?

Strategy   Break the process down into smaller parts. What are the individual steps that need to be taken into consideration to achieve the ultimate goal, that is, expression of the protein in cells?

Solution   To make a fusion of your protein with mCherry, you should first consider which type of cells will be used to produce the protein. If a large quantity is needed, then bacterial expression is often the best technique to use; however, if the protein has some sort of modification or processing that is specific to eukaryotes, another system, for instance, cultured cells, may be more appropriate.

The next step is to develop a plan for creating a DNA construct that contains cDNAs coding for both the desired protein and mCherry. These must be joined together in frame—that is, ligated in such a way as to generate a single mRNA when transcribed. This may be as simple as cutting both pieces with a restriction enzyme that leaves ends that can join to each other, or it may require the engineering of restriction sites onto each cDNA using PCR. The newly joined piece of cDNA coding for the fusion protein will need to be cloned into an expression vector, a piece of DNA that can be replicated inside the cell and has promoter sequences that will drive high levels of protein expression. Finally, a method is needed to put this piece of DNA into the bacteria. This is commonly accomplished with chemical transformation or electroporation.

Follow-up question   How would this process be different if the aim were to express the protein in mammalian cells?

Summary

- Oligonucleotides up to 70 nucleotides long can be chemically synthesized in the laboratory.
- PCR can be used to amplify fragments of DNA.
- DNA can be cut and ligated, using a variety of commercially available enzymes.
- Vectors are pieces of DNA that can be inserted into different organisms such as bacteria, yeast, or other eukaryotic cells to aid in the study of their function.
- Pieces of DNA can be used to drive protein production in individual cells, or even in whole organisms, to test for function.
- Genes can be removed using homologous recombination, generating a knockout organism.
- CRISPR and RNAi are means of silencing gene expression.

Concept Check

1. Describe the various techniques used in the manipulation of nucleic acids.
2. Explain which of these techniques can be used to construct pieces of DNA to learn more about a biochemical system.
3. How would you generate insertions or deletions in a DNA sequence using PCR?

Bioinformatics Exercises

Exercise 1 Taxonomic Trees

Exercise 2 Drawing and Visualizing Nucleotides

Exercise 3 The DNA Double Helix

Exercise 4 Analysis of Genomic DNA

Exercise 5 Restriction Enzyme Mapping

Problems

2.1 Nucleic Acids are the Genetic Materials

1. How was it proved that DNA is a genetic material?
2. What are the key features of DNA as a genetic material?
3. Why is RNA better for transmission of genetic information?

2.2 Nucleic Acids Have Distinct Structures

4. Draw the structure of five bases that are found in DNA and RNA molecules.
5. Based on the structures of the bases, would you anticipate that they would be water soluble?
6. Each of the five bases contains nitrogen. Why are these amines not drawn in the ionized state?
7. Several bases can tautomerize. Draw the enol form of each base. How would this affect hydrogen bonding and reactivity?
8. Examine the structure of the carbohydrate moiety in a ribonucleotide or deoxyribonucleotide. What shape is it? Where are the hydroxyl groups found relative to the rest of the molecule?
9. Describe the structure of a RNA molecule by listing its key characteristics.
10. Review the forces involved in the folding of DNA and RNA. Would these molecules fold differently in a nonpolar solvent such as hexane?

11. Examine the structure of a nucleotide. What weak forces are involved in interactions with other nucleotides or other molecules we have discussed? What are the relative strengths of these interactions?

12. Would the DNA helix be more or less stable in the absence of Mg^{2+}? Give reason your response

13. DNA can lose a purine base in strong acid. Propose a mechanism for the acid-catalyzed depurination of DNA.

14. A riboswitch is a posttranscriptional regulatory mechanism. In a riboswitch, one region of the mRNA can fold into a complex shape that will block translation. Binding of a small ligand stabilizes this structure. Draw a model for how this occurs. Write an equilibrium expression for the riboswitch in the presence and absence of a ligand.

15. Examine the structure of DNA. What weak forces are involved in interactions with other molecules? What are the relative strengths of these interactions?

2.3 Nucleic Acids Have Many Cellular Functions

16. What does "semiconservative replication" mean?

17. Describe the sequence of events at a replication fork as DNA is being copied.

18. Mutations are changes to DNA sequences or chemical modifications of those sequences. If DNA were methylated on a select cytosine residue, how might this change the DNA sequence.

19. DNA can be damaged by UV light. One sort of damage that can be caused is the formation of thiamine dimers, as shown below. How might this be harmful to the cell? How does the repair mechanism rectify this?

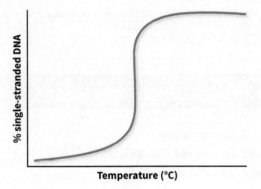

20. Ribosome is the the protein synthesis factory of the cell. List the key functions of the ribosome.

21. Some toxins block the synthesis of mRNA or proteins. Why would this effect be toxic, and how could it be used in the laboratory?

22. Consider the thermodynamics of DNA replication. What are the different thermodynamic parameters (enthalpy, entropy) that need to be considered? Why is this process favorable overall?

23. Retroviruses have an RNA genome instead of a DNA genome, and the RNA must be copied into the host's DNA. How does this process differ from other reactions in biochemistry? What additional steps must happen for the virus to infect a cell?

24. List the places where the expression of a gene and, therefore, the production of a protein can be regulated. Which are exclusively eukaryotic, which are exclusively prokaryotic, and which are both?

2.4 The Manipulation of Nucleic Acids Has Transformed Biochemistry

25. Why do restriction enzymes recognize a four to eight base pair sequence? Why are many of these sequences palindromic?

26. On an average, how many fragments would a restriction enzyme, that recognizes a specific four base sequence in DNA, be expected to cleave a double-stranded bacteriophage with a genome size of 5,000 bp into?

27. How frequently would a six-base cutter cut a random sequence of DNA that is 10,000 base pairs long?

28. In a cloning experiment a fragment of DNA is ligated with a vector such as a plasmid. Following the initial ligation of the sample into a vector, the new DNA construct is used to transform bacteria. Why is this step needed?

29. UV light is frequently used to monitor DNA in molecular biology experiments. Some investigators attempt to minimize exposure of their experiments to UV light. What would be the reason for this?

30. What are the reagents required to perform PCR? List in correct order the sequence of steps in a PCR reaction beginning with double-stranded DNA.

31. What was the mythical beast known as a chimera, and why is the first organism generated in the production of a knockout mouse termed a chimera?

32. How could you engineer a restriction enzyme site into a piece of DNA?

33. What are the strengths and weaknesses of using high throughput sequencing compared to more classical methods, such as fluorescent dye terminator sequencing?

34. DNA sequencing by the chain-termination method uses DNA polymerase I to make a complementary copy of the target or template DNA molecule. A reaction with a 20 bp template and dideoxyadenosine nucleotides as terminators results in the production of a 5 bp fragment. What can be concluded about the template from this result?

35. If PCR were performed on a highly conserved gene from two species, one prokaryotic and one eukaryotic, would you anticipate different results or the same result? Why?

Data Interpretation

Questions 36-39 pertain to the following figure.

Y-axis: % single-stranded DNA; X-axis: Temperature (°C)

36. What does the shape of this curve tell us about the transition from double-stranded to single-stranded DNA?

37. What is the melting temperature of this piece of DNA?

38. What effect would the addition of acid have on the melting point, and what would be the reason for the effect? Draw a curve illustrating melting in the presence of acid.

39. What effect does EDTA (ethylenediaminetetraacetic acid) have, and why would it influence the melting temperature?

40. The sequences at the cloning site of three vectors are given below. The Bam HI (GGATCC) and Hind III (AAGCTT) sites are underlined. Only the sequences around the restriction sites are shown. The symbol "_____" indicates rest of the sequence.

Vector 1: _____ Promoter..ATGGGTCGC<u>GGATCC</u>GGCTGC.. <u>AAGCTT</u> _____

Vector 2: _____Promoter..ATGGGTCG<u>GGATCC</u>GGCTGCT.. <u>AAGCTT</u> _____

Vector 3: _____ Promoter..ATGGGTCG<u>GGATCC</u>GGCTGCTA.. <u>AAGCTT</u> _____

Which one of the above three vectors is appropriate to clone the following ORF for expression?

_____ ATGCCCAACACCCGGATCCCG..TAA <u>AAGCTT</u> _____

41. Draw the restriction map of the plasmid given the following data (the gel pattern shown below is not to scale). The size of each DNA fragment (in kb) is indicated next to it.

EcoRI	SalI	HindIII	EcoRI & HindIII	SalI & Hind II	EcoRI & SalI
5.4–	5.4–				
					3.6–
		2.1–	2.1–		
		1.9–		1.9–	
					1.8–
		1.4–	1.4–	1.4–	
				1.3–	1.2–
				0.6–	0.9–

Experimental Design

42. Using the PCR technique, describe how you could distinguish the knockout mouse from a wild-type mouse.

43. Describe three different ways that could be used to make a cultured eukaryotic cell line produce a recombinant protein.

44. Virtual PCR uses a pair of primers to search through a genomic DNA sequence for amplified fragments. Explain how this technique may be useful in the laboratory.

45. You are investigating a protein that you believe binds to DNA. You hypothesize that binding of this protein increases the melting point of DNA. Design an experiment to test your hypothesis.

46. You are working on designing a primer set for amplifying a target sequence. What parameters should be considered while designing the primer and how would you identify the formation of primer-dimer and avoid misinterpretation about the product.

47. You are studying a protein that is phosphorylated on a specific amino acid. How could you mutate the gene that codes for this protein to another amino acid?

Ethics and Social Responsibility

48. Are you in favor of or against using recombinant DNA technologies to create GMOs? Are there some instances where use of genetic engineering is acceptable and others when it is not?

49. Why do most people find cloning of bacteria or cells acceptable, cloning of larger organisms such as mice or sheep questionable, and cloning of humans unethical?

50. Each of us harbors traits that are beneficial and those we might wish were different. If you could make changes to your genome to give you traits that you find favorable, would you do so? If you could increase your chances that your children would have these favorable traits, would you do so?

51. Your personal DNA sequence may reveal that you are harboring mutations or alleles of genes that could put you at risk for future diseases. Would you like to know or have access to your genomic DNA sequence?

Suggested Readings

2.1 Nucleic Acids are the Genetic Materials

McCarty M. The *Transforming Principle*. WW Norton, 1986.

Rosenberg E. *It's in Your DNA, From Discovery to Structure Function, and Role in Evolution, Cancer, and Aging.* Elsevier Science, 2017.

2.2 Nucleic Acids Have Distinct Structures

Branden, C., and J. Tooze. *Introduction to Protein Structure.* New York, NY: Garland Science, 1999.

Egil, M., and W. Saenger. *Principles of Nucleic Acid Structure.* Berlin, Germany: Springer Verlag, 1988.

Neidle, S. *Principles of Nucleic Acid Structure.* Cambridge, MA: Academic Press, 2007.

Sinden, R. R. *DNA Structure and Function.* Cambridge, MA: Academic Press, 1994.

2.3 Nucleic Acids Have Many Cellular Functions

Bloomfield, V. A., D. M. Crothers, and I. Tinoco, Jr. *Nucleic Acids: Structures, Properties, and Functions.* Herndon, VA: University Science Books, 2000.

Watson, J. D., T. A. Baker, S. P. Bell, A. Gann, M. Levine, and R. Losick. *Molecular Biology of the Gene.* 7th ed. San Francisco, CA: Benjamin Cummings, 2013.

2.4 The Manipulation of Nucleic Acids Has Transformed Biochemistry

Brown, T. A. *Gene Cloning and DNA Analysis: An Introduction.* Hoboken, NJ: Wiley-Blackwell, 2016.

Buckingham, L. *Molecular Diagnostics: Fundamentals, Methods, and Clinical Applications.* 2nd ed. Philadelphia, PA: F. A. Davis Co., 2011.

Dale, J. W., and M. von Schantz. *From Genes to Genomes: Concepts and Applications of DNA Technology.* Hoboken, NJ: Wiley, 2011.

Green, M. R., and J. Sambrook. *Molecular Cloning, A Laboratory Manual.* 4th ed. Oxford, England: Cold Spring Harbor Laboratory Press, 2012.

Lesk, A. M. *Introduction to Genomics.* 2nd ed. Oxford, England: Oxford Press, 2012.

McPherson, M. J., and S. G. Moller. *PCR, The Basics.* 2nd ed. New York, NY: Garland Science, 2006.

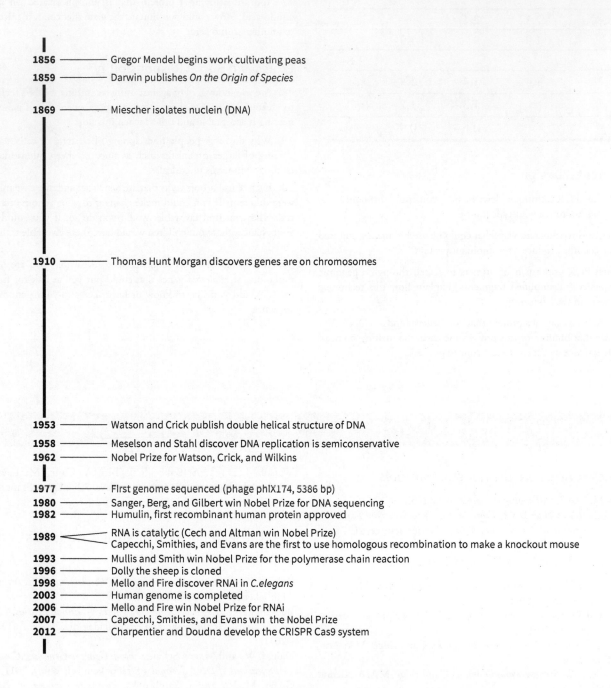

1856 ——————— Gregor Mendel begins work cultivating peas

1859 ——————— Darwin publishes *On the Origin of Species*

1869 ——————— Miescher isolates nuclein (DNA)

1910 ——————— Thomas Hunt Morgan discovers genes are on chromosomes

1953 ——————— Watson and Crick publish double helical structure of DNA

1958 ——————— Meselson and Stahl discover DNA replication is semiconservative

1962 ——————— Nobel Prize for Watson, Crick, and Wilkins

1977 ——————— First genome sequenced (phage phiX174, 5386 bp)

1980 ——————— Sanger, Berg, and Gilbert win Nobel Prize for DNA sequencing

1982 ——————— Humulin, first recombinant human protein approved

1989 ——————— RNA is catalytic (Cech and Altman win Nobel Prize)
Capecchi, Smithies, and Evans are the first to use homologous recombination to make a knockout mouse

1993 ——————— Mullis and Smith win Nobel Prize for the polymerase chain reaction

1996 ——————— Dolly the sheep is cloned

1998 ——————— Mello and Fire discover RNAi in *C.elegans*

2003 ——————— Human genome is completed

2006 ——————— Mello and Fire win Nobel Prize for RNAi

2007 ——————— Capecchi, Smithies, and Evans win the Nobel Prize

2012 ——————— Charpentier and Doudna develop the CRISPR Cas9 system

CHAPTER 3

Proteins I

An Introduction to Protein Structure, Function and Purification

Proteins and their Purification in Context

Throughout history advances in materials have changed tools and how we make them. Ancient civilizations may have used clay pots or made weapons of bronze or iron. In modern times refined alloys of aluminum have been used in applications ranging from cans to aircraft. Perhaps people in the future will mark our current era by the use of plastics or silicon microchips.

In biochemistry, we consider proteins to be a wonder material used in thousands of applications. Proteins are polymers of amino acids which have evolved to have numerous distinct functions: they can be structural or catalytic, transmit information, help protect an organism by binding to foreign molecules, provide motility, store amino acids, or transport molecules within the cell or the organism.

This chapter begins by discussing amino acids, the building blocks of proteins, then moves on to the basics of protein structure and a brief description of how these macromolecules fold into their specific conformations. Several different examples of different proteins are examined next. The chapter then discusses the basic scheme for conducting a purification, isolating a single protein of interest from a crude mixture. The next section discusses size-exclusion chromatography, a common means of achieving that separation, followed by discussion on use of affinity and ion exchange chromatography techniques to achieve separation. These same techniques can also be used to determine many of the properties of proteins and sometimes their functions.

Chapter Outline

3.1 Amino Acid Chemistry

3.2 Proteins Are Polymers of Amino Acids

3.3 Proteins Are Molecules of Defined Shape and Structure

3.4 Examples of Protein Structures and Functions

3.5 Protein Purification Basics

3.6 Size-Exclusion Chromatography

3.7 Affinity Chromatography

Common Themes

Evolution's outcomes are conserved.	• Mutations to DNA sequences may result in alterations to a protein's amino acid sequence. The new protein may function normally, have a new and slightly altered function, or be completely nonfunctional.
	• The amino acid sequence of proteins can help establish the evolutionary relationship between organisms.
	• Protein structures and the amino acids they are made from have undergone billions of years of natural selection and random change (genetic drift).
	• Proteins with conserved amino acid sequences or folds may have similar topology or ligand binding; these attributes can guide the development of purification schemes.
Structure determines function.	• The structures of the 20 amino acids commonly found in proteins partially describe the function of these molecules. The backbone and side chains of these amino acids act in concert to give the structure of the protein.
	• An alteration to the structure of a protein changes the function of that protein.
	• Knowledge of protein structure or function can be useful in the development of purification schemes.
	• If a protein binds to a specific ligand, that ligand can be used in a purification scheme.
Biochemical information is transferred, exchanged, and stored.	• The information encoded in deoxyribonucleic acid (DNA) and transcribed into messenger ribonucleic acid (mRNA) is ultimately translated into proteins.
	• Many proteins are involved in cell-to-cell communication, either between organisms or within the same organism. In such cases the proteins may be the signal, the receptor of the signal, or elements of the signaling cascade that translate the signal into an effect.
	• DNA sequences coding for proteins can be edited or altered, causing cells to produce increased amounts of protein for study, enzymes that can be used to modify molecules in the purified state, or enzymes that can be purified more easily.
Biomolecules are altered through pathways involving transformations of energy and matter.	• Proteins are made from a common set of amino acid building blocks. Specific modifications are sometimes made to particular amino acids within a protein.
	• The synthesis of the covalent chemical bonds in a protein costs energy. In addition, the assembly of hundreds of amino acids into a single protein has a highly negative entropy. Therefore, the free energy (ΔG) of protein formation from isolated amino acids in a test tube is quite positive. In the cell, the overall reaction becomes favorable ($\Delta G < 0$) by coupling these reactions to the hydrolysis of ATP and GTP.
	• Unlike the synthesis of the peptide bonds in a protein, the folding of a protein into its final conformation is largely governed by numerous weak forces. Here, increases in entropy provided by liberation of water molecules drive $\Delta G < 0$ and lead to formation of the final structure.
	• Purified biomolecules can be chemically changed in the same way that any molecule can be changed; however, due to the complexity and general fragility of these molecules, enzymatic modifications are usually employed.

3.1 Amino Acid Chemistry

Amino acids are the building blocks of proteins, and they play critical roles in many metabolic pathways. Because of these diverse roles, amino acids are a good place to start a discussion of protein chemistry.

3.1.1 The structure of amino acids dictates their chemical properties

As their name suggests, amino acids have a single unifying property—they have both an amine and a carboxylic acid moiety (**Figure 3.1**). Amino acids found in proteins are all α-amino acids; that is, they contain an amino group, a central α-carbon, and a carboxyl group. In 19 of the 20 common amino acids found in proteins, the amine is a primary amine bound only to the α-carbon. The remaining amino acid, proline, is technically an imino acid, containing a secondary amine with the side chain joining the backbone amine, forming a ring.

All but one of the 20 common amino acids have four different substituents: an amino group, a carboxyl group, a hydrogen, and a side chain bound to the central carbon. The side chain is referred to as "R" because it can be one of 20 different groups. These amino acids lack a plane of symmetry and are **chiral** (**Figure 3.2**). The remaining amino acid, glycine, has a hydrogen atom as a side chain; because of the plane of symmetry created by the hydrogen atom, glycine is **achiral**. As with other molecules, the three-dimensional shape of an amino acid dictates its reactivity and ability to bind to other molecules. Additionally, nature almost exclusively uses the L-enantiomer of amino acids. Unless otherwise noted, all amino acids discussed here can be assumed to be the L-enantiomer.

Both the amino and the carboxyl group of amino acids are charged in aqueous solution near neutral pH. Molecules possessing both a positive and negative charge are **zwitterions**. In some depictions of amino acids, the amino group is shown as deprotonated (NH_2) and the carboxyl as protonated (COOH). A quick check of the pK_a of these groups indicates that this cannot be the case. The pK_a of the amino group in glycine is 9.6, whereas that of the carboxyl is 2.3. Hence, at any pH between these pK_a values, both groups will be found in the ionized state (**Figure 3.3**). The pK_a values of the carboxyl and amino group may be slightly different from the values expected for a carboxylic acid or primary amine, because the proximity of the two groups to one another can influence the pK_a (causing shifts of almost 2.5 units), as can the side chain of the amino acid or the immediate environment. The change of the pK_a value of a functional group due to its immediate environment is a recurring theme in biochemistry.

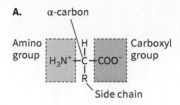

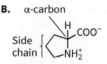

FIGURE 3.1 Structure of an amino acid. A. The general structure of an α-amino acid is shown. The primary amino group, carboxyl group, and side chain are all bound to the α-carbon. **B.** Proline is the only amino acid with a secondary amine.

(Source: (A) Snustad; Simmons, *Principles of Genetics*, 6e, copyright 2012, John Wiley & Sons. This material is reproduced with permission of John Wiley & Sons, Inc.)

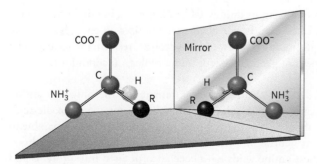

FIGURE 3.2 Chirality of amino acids. All amino acids, with the exception of glycine, have four different substituents on the α carbon and are therefore chiral. Glycine has an H for a side chain and therefore is achiral. Naturally occurring amino acids found in proteins are in the L configuration.

(Source: Karp, *Cell and Molecular Biology: Concepts and Experiments*, 7e, copyright 2013, John Wiley & Sons. This material is reproduced with permission of John Wiley & Sons, Inc.)

Ionized at low pH   **Zwitterionic at neutral pH**   **Ionized at high pH**

FIGURE 3.3 Zwitterionic nature of amino acids. Despite how structures may be depicted on paper, amino acids exist as zwitterions in neutral solutions. Changing the pH may alter the protonation of the carboxyl group or amino group.

It is possible to encounter pH values that would permit the protonation of only one state of zwitterion, such as stomach acid at pH 1, but these situations are the exception in biochemistry.

Amino acids are versatile but relatively simple molecules. The common skeleton provides a central core from which to build an amazingly diverse array of molecules.

3.1.2 The side chains of amino acids impart unique properties

Amino acids are differentiated by their **side chain or residue**, a short extension from the common backbone. The side chains can be as small as a hydrogen atom or as large as the bicyclic ring of the amino acid tryptophan. Collectively, however, amino acids are considered small molecules.

Amino acids can be organized based on the properties of their side chains (**Figure 3.4**). There are various ways to do this. The organization used here is based on the chemical properties of the side chains. This section discusses the properties of the side chain rather than the entire molecule. Keep in mind, in solution all amino acids are zwitterions and are soluble in water.

Nonpolar side chains   Amino acids with a nonpolar (hydrophobic) side chain include glycine, alanine, leucine, isoleucine, valine, proline, methionine, phenylalanine, and tryptophan. Glycine is the simplest amino acid, with a single hydrogen atom as a side chain. Alanine, the next simplest, has a methyl group as a side chain. Valine, leucine, and isoleucine have isopropyl, *sec*-butyl, and *iso*-butyl groups, respectively, as side chains and are collectively known as the branched chain amino acids (BCAAs). Proline has a side chain that connects back to the backbone nitrogen, resulting in a five-membered ring that has constrained geometry. Methionine has an ethyl methyl thioester as a side chain. Phenylalanine is simply alanine with a phenyl group added. Finally, tryptophan, the largest of the amino acids, has an aromatic, heterocyclic indole ring system comprised of both a five-membered and a six-membered ring.

Polar side chains   Amino acids with polar side groups are serine, threonine, tyrosine, cysteine, asparagine, and glutamine. Each of these side chains contains groups that are polar and can form hydrogen bonds, but are not sufficiently polar to ionize at typical physiological pH.

Three of these amino acids are alcohols. Serine is a primary alcohol, while threonine is a secondary alcohol. Tyrosine contains a phenol group with a pK_a of approximately 10. While this side chain is not typically ionized under physiological pH, the acidity of this group means that it is more reactive than other hydroxyl groups.

Cysteine can be thought of as a thiol analog of serine. The thiol in these residues is redox active—two cysteine residues can cross-link to form a disulfide bond, either within a protein or between two proteins. Cysteine can also ionize with a pK_a of 8.

Asparagine and glutamine both contain carboxamides, $CONH_2$ groups, but differ by a methylene group. The amine moiety in the carboxamide side chain is *not* ionized. If enough acid were added to protonate the amine of this group, the group would hydrolyze, resulting in a carboxylic acid and ammonia, something that does not generally occur under biological conditions.

Negatively charged side chains   Aspartate and glutamate are the only two amino acids typically categorized as having negatively charged side chains. Both contain carboxylic acids and are therefore weak acids. The pK_a of these side chains ranges from 2 to 5 and is comparable to that of other carboxylic acids. Hence, at neutral pH, most of these side chains are ionized; however, this may not be the case when these amino acids have been incorporated into proteins and their local environment has changed. Aspartate and glutamate are the ionized forms of aspartic and glutamic acid, respectively. Like asparagine and glutamine, aspartate and glutamate differ by a methylene unit.

Positively charged side chains   Lysine, arginine, and histidine are designated as having positively charged side chains. As with their negatively charged counterparts, they are ionized or partially ionized at physiological pH. Lysine's side chain is a primary amine at the end of a

Nonpolar side chains

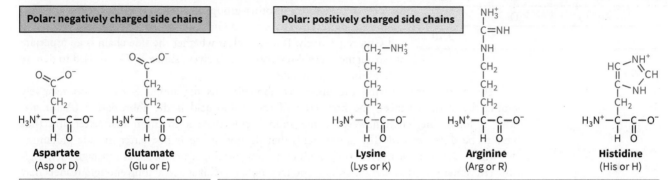

Alanine
(Ala or A)

Valine
(Val or V)

Leucine
(Leu or L)

Isoleucine
(Ile or I)

Methionine
(Met or M)

Phenylalanine
(Phe or F)

Tryptophan
(Trp or W)

Properties of side chains
Hydrophobic side chains are nonpolar and incapable of hydrogen bonding. These side chains may be aliphatic or aromatic.

Polar: negatively charged side chains   ### Polar: positively charged side chains

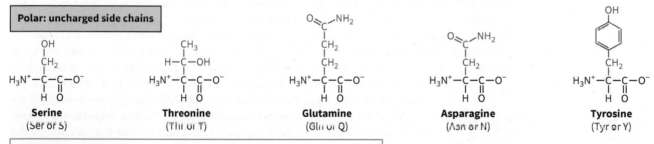

Aspartate
(Asp or D)

Glutamate
(Glu or E)

Lysine
(Lys or K)

Arginine
(Arg or R)

Histidine
(His or H)

Properties of side chains
Negatively charged amino acids contain a carboxyl group and are commonly deprotonated at physiological pH.

Properties of side chains
Positively charged amino acids include lysine, arginine, and histidine. These residues contain amino groups. Lysine and arginine have pK_a values of 10.5 and 12.5 and are protonated and positively charged at physiological pH. The side chain of histidine has a pK_a of 6.0 and is partially protonated at physiological pH.

Polar: uncharged side chains

Serine
(Ser or S)

Threonine
(Thr or T)

Glutamine
(Gln or Q)

Asparagine
(Asn or N)

Tyrosine
(Tyr or Y)

Properties of side chains
Polar side chains are hydrophilic but are not usually ionized at physiological pH.

Side chains with unique properties

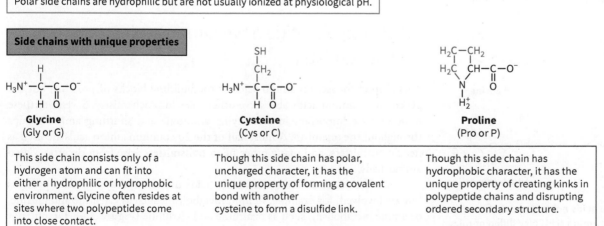

Glycine
(Gly or G)

Cysteine
(Cys or C)

Proline
(Pro or P)

This side chain consists only of a hydrogen atom and can fit into either a hydrophilic or hydrophobic environment. Glycine often resides at sites where two polypeptides come into close contact.

Though this side chain has polar, uncharged character, it has the unique property of forming a covalent bond with another cysteine to form a disulfide link.

Though this side chain has hydrophobic character, it has the unique property of creating kinks in polypeptide chains and disrupting ordered secondary structure.

FIGURE 3.4 Amino acids.   The 20 amino acids are grouped by hydrophobic, polar, positively charged, and negatively charged side chains. Their names are shown in one- and three-letter abbreviations.

TABLE 3.1 The pKₐ of Amino Acid Side Chains

Amino acid	Three-letter abbreviation	One-letter abbreviation	pK_a
Amino			9.30
Arginine	Arg	R	12.48
Lysine	Lys	K	10.53
Tyrosine	Tyr	Y	10.07
Cysteine	Cys	C	8.18
Histidine	His	H	6.00
Glutamate	Glu	E	4.25
Aspartate	Asp	D	3.65
Carboxyl			2.10

4-carbon long chain. Arginine has a guanidinium group ($NHC(NH_2)_2^+$) and uses resonance to distribute the positive charge across this system. Histidine has a heterocyclic imidazole group as a side chain. This group can become protonated and deprotonated with a pK_a near 6.0, making it the only amino acid with a pK_a in the physiological range.

Amino acids are known by either one-letter or three-letter abbreviations

All amino acids have both a one-letter and a three-letter abbreviation. These are shown in Figure 3.4, with the structure of the amino acid side chains. Although most of these abbreviations are self-explanatory, it is worth highlighting some of them. The three-letter abbreviation for isoleucine is Ile rather than Iso (because "iso" is used to describe the chemistry of other groups like *iso*-butyl). Asparagine and glutamine are abbreviated to Asn and Gln, and aspartate and glutamate to Asp and Glu, respectively. If it is unclear whether the side chain is an aspartate or asparagine, the abbreviation Asx is given; similarly, Gsx is used to denote a glutamine or glutamate.

The system for deriving the one-letter abbreviations for amino acids is also relatively simple. In most instances, the first letter of the amino acid is the abbreviation (A for alanine, S for serine, etc.); however, some amino acids share a common first letter and thus have to be differentiated. The amino acids that do not use their first letter include glutamine (Q, which can be remembered as Q-tamine), asparagine (N, remembered as asparagiNne), arginine (R, remembered as Arrrgh!-anine), and tryptophan (W, if stared at long enough, readers may start to see the W in the structure of tryptophan). The other amino acids that have single letter abbreviations that do not match their first letter are aspartate (D) and glutamate (E), which at least fall in alphabetical order. Finally, lysine is K, and there is no easy way to remember that one.

The ionization of amino acid side chains

The side chains of amino acids are partially responsible for the way in which a protein folds and mainly responsible for the chemistry the protein can conduct. The functional groups of the 20 amino acids commonly found in proteins have pK_a values throughout the pH range encountered by living systems (**Table 3.1**). Most of these values are as expected, with carboxylic acids typically in the range of 2 to 5, and amines at the other end of the spectrum in the range of 10 to 12. Others that are worthy of mention include histidine with a pK_a near 6, cysteine with a pK_a near 8 (due to its thiol side chain) and tyrosine with a pK_a near 10 (due to its phenolic group). However, these are the pK_a values of the protons on free amino acids in solution; folded into a protein or in the active site of an enzyme, these values may be changed radically by the local environment in the protein.

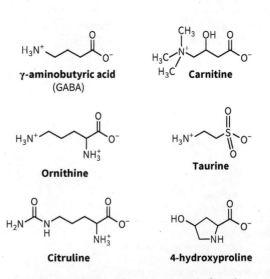

FIGURE 3.5 Amino acids not commonly found in proteins. Other amino acids play different roles in biochemistry. Some of these molecules are found in proteins (4-hydroxyproline, for example) while others are neurotransmitters (GABA) or play roles in metabolism (ornithine and citrulline are components of the urea cycle).

3.1.3 Amino acids have various roles in biochemistry

This chapter focuses on amino acids as the building blocks of proteins, but the 20 common amino acids also serve other roles in biochemistry. Several of these molecules are important in detoxifying ammonia and shuttling amino groups throughout the organism. Almost half of the 20 common amino acids are used as neurotransmitters, and many other neurotransmitters arise from decarboxylated amino acids.

In addition, there are other amino acids not commonly found in proteins but are involved, for example, in lipid metabolism, such as carnitine and taurine, or amine metabolism, such as ornithine and citrulline (**Figure 3.5**). Despite not being incorporated into proteins, these amino acids are clearly important in biochemistry and metabolism.

Often, molecules we consume in our diet have effects that are like those of drugs. This topic is examined in **Societal and Ethical Biochemistry: Are nutritional supplements foods or drugs?**

Societal and Ethical Biochemistry

Are nutritional supplements foods or drugs?

Nutritional products have always captured the attention of those looking to improve their health or physique. These products have progressed from herbal extracts to vitamins, minerals, amino acids, and small organic cofactors, and they are now readily available from a variety of outlets, ranging from specialty supplement shops to corner gas stations. Are these molecules drugs, intended to relieve the symptoms of an illness, or are they foods? Do these molecules do what they claim to do?

We often think of drugs as eliciting some physiological or psychological effect in the body, for example, altering blood pressure, making someone more alert or drowsy, or causing the release of a hormone. Based on this definition, almost any molecule can be defined as a drug.

The U.S. Food and Drug Administration (FDA) defines drugs in the 2006 Food Drug and Cosmetics Act (FD&C Act) as follows:

> ... in part, by their intended use, as "articles intended for use in the diagnosis, cure, mitigation, treatment, or prevention of disease" and "articles (other than food) intended to affect the structure or any function of the body of man or other animals." [FD&C Act, sec. 201(g)(1)]

This legal definition contains loopholes. For example, if a bodybuilder takes amino acid supplements, these would probably "affect the structure and function of the body," but possibly the same could be said if that bodybuilder was eating egg whites. Furthermore, not all molecules found in foods have been isolated and identified. For example, cranberries are eaten as a food, but what if an extract of cranberries was found to be beneficial in treating some cancers. Would that extract not be considered a drug? Likewise, the caffeine in coffee or tea is clearly a drug because it stimulates our nervous system and has no value as a source of energy or building block for other biological molecules.

Therefore, the question of whether a substance is a drug, a pharmaceutical, a supplement, or a food often comes down to intent and dose. These factors need to be determined before a molecule can be marketed as a drug or as a supplement.

Regulations govern the use of dietary supplements. For example, if a new molecule that was not marketed before 1994 is introduced into a supplement, companies are required to notify the FDA of the new ingredient. Also, companies cannot make the type of claims about efficacy with supplements that they can for drugs developed and marketed to treat a specific disorder. The rules for introducing supplements are more lenient and do not need to demonstrate the same level of effectiveness or safety. With supplements, when it comes to safety, the burden of proof is on the FDA rather than on the company marketing the compound. If safety concerns do arise, the FDA can, in some instances, issue warning letters and seize compounds.

Because many nutritional supplements are viewed as "natural" and are available without a prescription, people have an unfounded confidence and trust that these molecules will be at best helpful and at least not harmful. Pharmaceuticals and drugs, on the other hand, are perceived as being synthetic or unnatural in some way and therefore less beneficial. These impressions do not hold up under the test of scientific scrutiny. For instance, many people take an extract of St. John's wort as a nutritional supplement to relieve mild depression. However, this supplement has been shown to alter certain metabolic processes in the liver that result in significantly more rapid clearing of some drugs, including some compounds found in birth control pills. This can reduce efficacy and potentially lead to unplanned pregnancies.

Worked Problem 3.1   The acid–base properties of amino acids

You have a sample of the amino acid histidine at pH 2 and you begin to titrate it with NaOH. Draw a predicted titration curve for histidine.

Strategy Examine the structure of histidine. There are three groups that can ionize: the backbone amino and carboxyl groups, and the side chain amino group. What are the pK_a values of those groups? What happens to each of these groups as the pH increases?

Solution

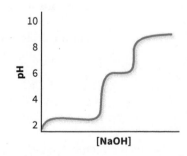

Follow-up question Would alanine act as a buffer at pH 4.1?

Summary

- Amino acids are small organic molecules containing both an amino group and a carboxyl group. Most are chiral and, in nature, the L enantiomer predominates.
- Twenty amino acids are commonly found as components of proteins.
- The amino acids found in proteins can be categorized by their side chains, which have diverse structures and chemistry, and contribute to the overall structure and properties of proteins.
- Amino acids serve other roles in biochemistry. For example, they act as metabolic intermediates and signaling molecules.

Concept Check

1. Categorize the amino acids by property or functional group.
2. Describe how each amino acid is unique with regard to its structure and chemistry.
3. Identify which parts of an amino acid are ionizable and give the pK_a of those groups.
4. Discuss the zwitterionic state of amino acids.
5. Explain on a chemical level why and how the immediate environment contributes to the pK_a of a group.

3.2 Proteins Are Polymers of Amino Acids

Amino acids play several important roles in biochemistry but are chiefly used as building blocks for peptides and proteins. This section begins with a discussion of the peptide bond and its properties, and then discusses cofactors and modifications to amino acids found in proteins.

3.2.1 Both peptides and proteins are polymers of amino acids

Two amino acids can join via an amide linkage between the carboxyl group of one amino acid and the amino group of the next (**Figure 3.6**). This linkage is referred to as a **peptide bond**, and the resulting molecule is termed a dipeptide. Three linked amino acids are referred to as a tripeptide, four as a tetrapeptide, and so on. **Oligopeptides** are generally 4 to 20 amino acids long, and peptides of more than 20 amino acids are usually referred to as **polypeptides**, although these terms are not strictly defined. Short stretches of amino acids are referred to as **peptides**, and longer chains, generally greater than 100 amino acids long, are referred to as **proteins**.

Both peptides and proteins are linear polymers of amino acids, and the amino acids they are made from have both amino and carboxyl groups. Therefore, when assembled into a peptide or protein, the sequence of amino acids can be described, proceeding from the **amino** (or N) **terminus** to the **carboxy** (or C) **terminus**.

Proteins are large, complex molecules. The average protein is a polymer of about 450 amino acids and a molecular weight of nearly 50,000 g/mole (amino acids in a peptide or

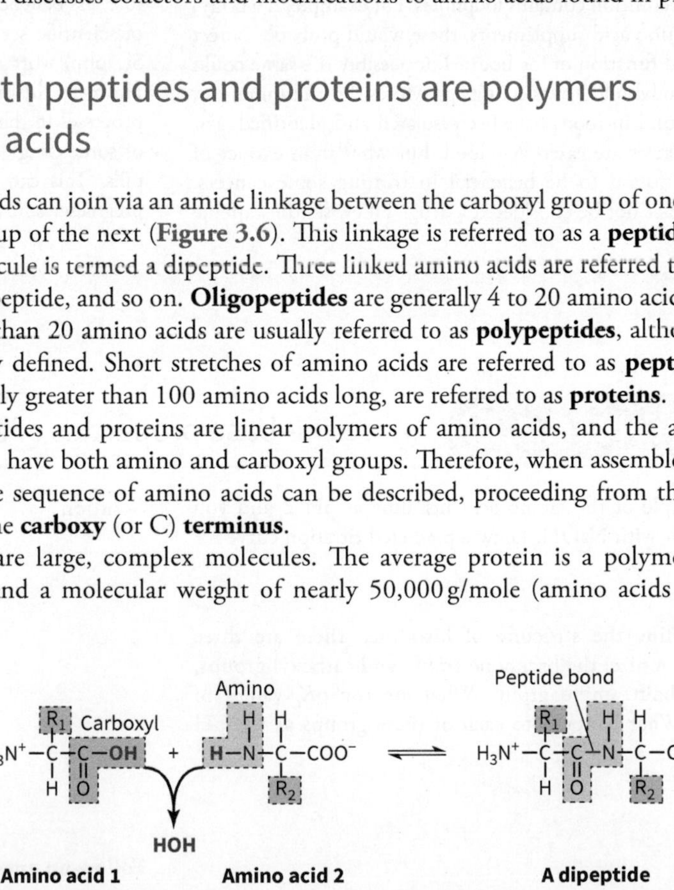

FIGURE 3.6 Peptide bond. The peptide bond is the amide bond that connects the carboxyl group of one amino acid to the amino group of the next. In this process, water is formed.

(Source: Black, *Microbiology: Principles and Explorations*, 8e, copyright 2012, John Wiley & Sons. This material is reproduced with permission of John Wiley & Sons, Inc.)

protein have an average molecular weight of 110 g/mole). Instead of using these unwieldy numbers, biochemists refer to the molecular weight of a protein in terms of the mass of the individual molecule. Hence, a molecular weight of 50,000 g/mol can be expressed in Daltons, or in the case of proteins, kiloDaltons (kDa), as 50 kDa. About 80% of proteins fall between 30 and 80 kDa in mass, although peptides can be as small as a few amino acids. For example, the hormone oxytocin is an octapeptide with a molecular weight of 1 kDa. Proteins can be as large as thousands of amino acids, such as the protein titin, which has 34,350 amino acids and a molecular weight of 3,816 kDa.

We might expect proteins with greater mass to be physically larger. While this is often the case, the shape of the protein also plays an important role. **Figure 3.7** shows some representative proteins and their Stokes radius, a property of their size and shape. As we might expect, if we were to assume that these proteins are spherical, the ratio of their diameters is proportionally less than the ratio of their masses. This is not necessarily the case if the protein adopts some other conformation.

In the simplest scenario, a protein is a **monomer**, comprised of a single chain of amino acids folded into an active protein; however, in reality, proteins are often **multimers** consisting of several subunits. When assembled into its active complex, a protein may have the same subunits repeated several times, for example, a homotetramer with four identical subunits, or several different subunits, like a heterotrimer that has three different subunits. Often, subunits of a protein are labeled with regard to their function, such as regulatory and catalytic, or with Greek letters to assist in identifying them (**Figure 3.8**). The stoichiometry of the subunits is treated as in a chemical formula; thus, a tetrameric protein comprising two copies of each subunit would have the designation $\alpha_2\beta_2$.

Many of the techniques we employ in studying proteins are shared with the ones that are used to study synthetic polymers. These are discussed in **Biochemistry: Polymer chemistry**.

Protein	MW (kDa)	Stokes radius (nm)
IgM pentamer	1048	6.70
Apoferritin band 1	720	5.91
Apoferritin band 2	480	5.17
β-phycoerythrin	242	4.11
Lactate dehydrogenase	146	3.46
Bovine serum albumin	66	2.66
Soybean trypsin inhibitor	20	1.78

FIGURE 3.7 Sizes of proteins.   A comparison of the overall molar mass and relative size (Stokes radius) of proteins of differing sizes. The ratio of the Stokes radii of the IgM pentamer to soybean trypsin inhibitor is 3.7:1, while the ratio of their masses is 52.4:1. Note that this assumes that these proteins are roughly spherical in shape, which is not the case for all proteins.

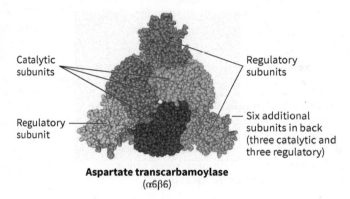

Aspartate transcarbamoylase
($\alpha6\beta6$)

FIGURE 3.8 Multimeric proteins.   Many proteins have multiple subunits or are found in solution as multimers. Aspartate transcarbamoylase has six regulatory and six catalytic subunits in the holoenzyme. The second three catalytic subunits are found behind the three shown in this image.

(Source: Data from PDB ID 1Q95, Huang, J., Lipscomb, W.N. (2004) Aspartate Transcarbamylase (ATCase) of Escherichia coli: A New Crystalline R-State Bound to PALA, or to Product Analogues Citrate and Phosphate, *Biochemistry* 43: 6415–6421)

Biochemistry

Polymer chemistry

Many important molecules in biochemistry are polymers, including proteins and the nucleic acids DNA and RNA. Many carbohydrates can also form polymers that can play structural roles, store energy, or contribute to protein trafficking and cell–cell recognition. Lipids do not typically form polymers, but they are built in a modular fashion from smaller building blocks and aggregate using many of the same weak forces that govern the behavior of larger molecules. Relevant lessons can be learned by comparing polymer chemistry and biological molecules.

Polymers are either linear or branched chains built of smaller molecules termed monomers. In most synthetic polymers, the monomeric units are small, simple organic molecules that join together through addition reactions between alkenes, or through condensation reactions to form polyamides or polyesters. These last two reactions are found in biochemistry, where proteins are an example of a polyamide and nucleic acids are an example of a polyphosphodiester.

The properties of polymers are partially dependent on the monomers that make up the polymer and partially dependent on the size of the polymer. In synthetic polymers, the desired qualities are often the ability to be drawn into filaments, such as nylon or Kevlar; the clarity of the material, for example, polycarbonate or Plexiglas; and the moldability or strength of the material, such as most plastics including polystyrene, polyethylene, and polypropylene. Proteins and carbohydrates display some of the same properties as synthetic polymers. However, because of the variety of side chains in the amino acids of proteins and the different types of weak forces involved, there is a greater variety of conformations that a protein can adopt and different types of roles it can play.

Polymer chemistry tools can be applied to answer biochemical questions. Low resolution techniques to determine the general overall size and shape of a protein include size exclusion chromatography, dynamic light scatter, fluorescence anisotropy, analytical ultracentrif-

ugation, electron microscopy, or soft ionization mass spectrometry. Although such techniques do not provide the same information as X-ray diffraction or nuclear magnetic resonance (NMR) studies, two effective techniques for determining the structure of a molecule, they do provide important information about how the molecule behaves in solution and whether it forms multimeric assemblies.

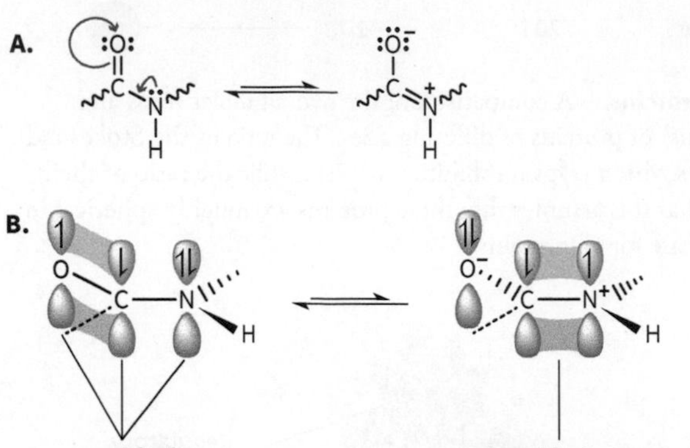

FIGURE 3.9 Nature of the peptide bond. **A.** An analysis of a Lewis structure of the peptide bond reveals resonance states that contribute to the overall structure. **B.** Because of *p* orbital overlap, the C–N bond is shorter and more stable, and the peptide unit is planar, all supportive of the predicted structure.

3.2.2 The peptide bond has special characteristics

The amide bond formed between two amino acids is referred to as a peptide bond. The nature of the peptide bond has several interesting and important characteristics. Although this structure is often depicted as containing a double bond between the oxygen and carbon, it is actually more complex. Analysis of a peptide bond's Lewis structure indicates that the double bond could also be drawn between the carbonyl carbon and nitrogen. This contributes to the peptide bond structure to a small but significant extent (**Figure 3.9**).

Sketching the molecular orbitals forming the peptide unit shows that there is a pi (π) cloud that extends through the carbonyl oxygen and carbon, as well as the nitrogen. Therefore, according to molecular orbital theory, there is double bond-like character to the C–N peptide bond. The peptide bond is shorter than an ordinary C–N bond, and it has a higher bond enthalpy, requiring more energy to break. Finally, because the C–N bond has overlap of these π orbitals, we would anticipate that this would limit rotation around the C–N bond axis. Peptide units are planar with the peptide bond fixed in either a *cis* or *trans* conformation. In most instances, the conformation observed in proteins is *trans*, due in part to steric clashes between the carbonyl oxygen and the hydrogen attached to the peptide nitrogen.

Since the peptide is planar, it can be thought of as a fixed unit (**Figure 3.10**). Based on this information, it is possible to start to determine the crude structure of a protein and tease out how these planar units are interconnected. However, the torsional angles between the peptide units are also required. Imagine laying a series of cards down flat on a table and connecting them with

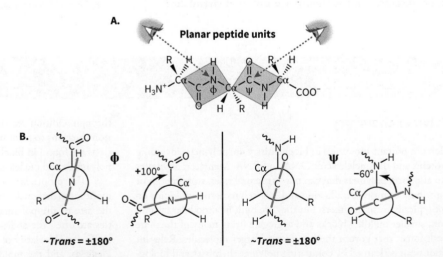

FIGURE 3.10 Phi and psi bond angles. **A.** The phi (Φ) and psi (Ψ) angles are the torsional bond angles located on either side of the peptide bond. **B.** Newman projections show the view down the nitrogen–carbon or carbon–carbon bond. Knowing these angles enables investigators to solve the structure of the protein structure at low resolution.

(Source: Adapted from drawings created by Jason D. Kahn, Univ. Maryland.)

flexible connectors such as pieces of string. If the torsional bond angles (the twist found between two cards) on either side of the card (the peptide unit) angles are known, it is possible to start to see how these flat units can form a three-dimensional structure. These angles are termed the phi (Φ) and psi (Ψ) angles.

Peptide bonds are rather stable. In the absence of enzymes, peptide bonds are generally resistant to base-catalyzed cleavage, although they can be broken with acid. Enzymatic degradation of proteins and peptides is catalyzed by proteases and peptidases, that is, enzymes that employ nucleophilic attack on the peptide carbon to cleave these bonds.

3.2.3 Many proteins require inorganic ions or small organic molecules to function

Many proteins require other groups for binding or activity; these "non-protein" groups are termed **cofactors** or **prosthetic groups**. A protein missing its cofactor is referred to as an **apoprotein**, whereas a **holoprotein** contains both the protein and its cofactor. Apoenzymes or holoenzymes are enzymes that are analogous to apoproteins or holoproteins. Cofactors can be broadly divided into metal ions and small organic groups.

Divalent metal ions are a common cofactor   Many proteins use divalent metal ions as cofactors in enzymatic reactions, either to initiate conformational changes in the protein, to bind and transport electrons, or to facilitate binding of the protein to other proteins or nucleic acids. Such ions include Ca^{2+}, Mg^{2+}, Zn^{2+}, Fe^{2+}, and Cu^{2+}. Other oxidation states of these metals, such as Fe^{3+} or $Cu+$ may also be observed as cofactors; less commonly found metal ions include Sr^{2+}, Cr^{3+}, Ni^{2+}, or Mn^{2+}. Nutritionally, all of these ions are referred to as minerals.

Small organic molecules are often used as cofactors   Another type of cofactor used by proteins is small organic molecules (**Figure 3.11**). As with divalent metal ions, this group includes molecules involved in oxidation–reduction reactions, catalysis, binding, and transport. Examples of small organic cofactors are the redox active nicotine adenine dinucleotide (NAD^+) and flavin adenine dinucleotide (FAD); pyridoxal phosphate, which is used in amino acid metabolism; and heme, a functional group that combines both an organic molecule (the protoporphyrin ring), and an inorganic group (the iron ion found in its center). In the diet, these cofactors or their precursors are often referred to as vitamins.

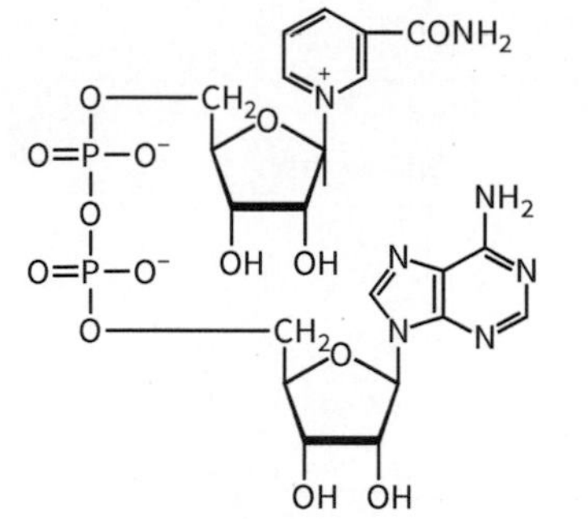

3.2.4 Amino acids can be modified within proteins

Most of the amino acids founds in proteins are the 20 common amino acids discussed above; however, amino acids can be modified within a protein (**Figure 3.12**). Several of the common modifications can have major effects; they can covalently cross-link proteins, alter activity, or change the location of proteins within the cell. Such modifications include covalent attachment of a carbohydrate through a glycosylation, or a lipid by an acylation or acetylation, or by phosphorylation or hydroxylation of key amino acids.

Lipid modifications often change the subcellular localization or function of a protein   Lipids are largely nonpolar hydrophobic molecules. Binding or attachment of a lipid often results in a protein being localized to a lipid bilayer. This is the case when the fatty acids myristate or palmitate are attached covalently to serine residues of a protein through the serine hydroxyl group, generating an ester linkage between the serine hydroxyl and the carboxyl of the lipid, or when the steroid precursor farnesyl couples through a cysteine residue, forming a thioether.

Other lipid modifications can alter the function of a protein. Histones are small, positively charged proteins that organize DNA. Acetylation of lysine residues in histones neutralizes the positive charge on the amino group and decreases the interaction between the histone protein and the DNA backbone, increasing accessibility of enzymes to DNA and subsequently altering gene expression.

NAD⁺

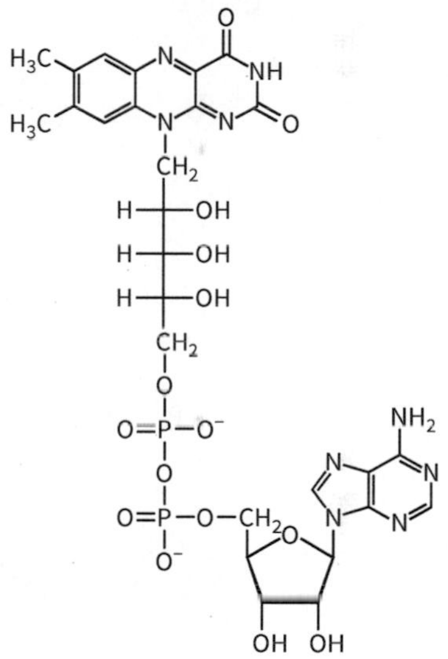

FAD

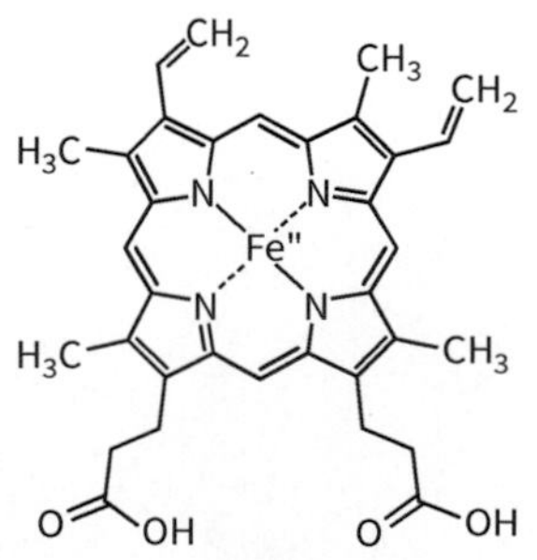

Heme

FIGURE 3.11 Examples of cofactors. These molecules are examples of common cofactors found in proteins. NAD⁺ and FAD are both involved in redox reactions and are important electron carriers. Heme is used in oxygen transport, oxidation reactions, and as an electron carrier.

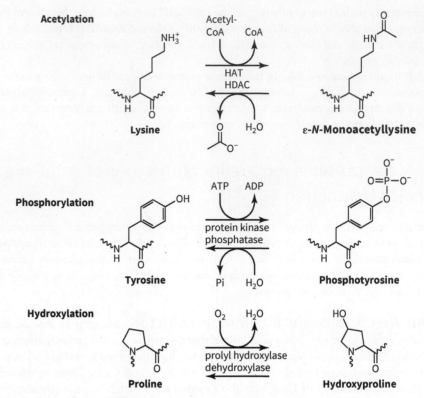

FIGURE 3.12 Amino acid modifications. Many amino acids are regulated through post-translational modifications. Shown are examples of acetylation, phosphorylation, and hydroxylation.

Carbohydrates are often found on secreted proteins

Many secreted proteins have carbohydrate modifications. The modifications may be short stretches of carbohydrate or massive carbohydrate complexes. Carbohydrates link to proteins either through the nitrogen of the asparagine side chain (termed an *N*-linked carbohydrate) or through the oxygen of the hydroxyl groups of serine or threonine (termed an *O*-linked carbohydrate).

Phosphorylation can alter protein function

One of the most common protein modifications is the addition of a phosphate. The phosphorylation is through enzymes known as **kinases**, and ATP is typically the source of the phosphate. Phosphates can be added to the hydroxyl groups of serine, threonine, or tyrosine residues via a phosphoester linkage. The incorporation of a phosphate can dramatically change the activity of an enzyme or cause the translocation or movement of a protein within a cell. Both phosphorylation and dephosphorylation are often thought of as switches in the cell, whereby a pathway or reaction is turned on or off. One example of this is lipolysis, the breakdown of stored fat in the cell. Lipolysis is mediated via protein kinase A, which phosphorylates several key enzymes and regulators of triacylglycerol storage, activating lipid breakdown.

Proteins of the extracellular matrix may be hydroxylated

Hydroxylated amino acids are often contained in the secreted structural proteins of the extracellular matrix, such as collagen. In these proteins, proline or lysine residues may be hydroxylated by prolyl or lysyl hydroxylases. In proline, hydroxylation occurs on either the γ carbon, C-4 of the side chain, or β carbon, C-3 of the side chain; the resulting amino acid is in 4-hydroxy or 3-hydroxyproline. Hydroxylations of lysine occur on the δ carbon. Elemental oxygen, O_2, is the oxygen source in both of these reactions. Ascorbate, also known as vitamin C, is used by the enzymes catalyzing these reactions to complete their catalytic cycle and become trapped in an oxidized state; in the absence of vitamin C, these enzymes are unable to operate and cannot hydroxylate proteins. The dietary-deficiency disease scurvy is marked by weakened connective tissue, bleeding, and tooth loss and results from the body's inability to modify these proteins.

Worked Problem 3.2 — Myoglobin structures

Using the Protein Data Bank (PDB), access a protein structure for myoglobin, the oxygen-binding protein found in muscle. Does myoglobin have a cofactor?

Strategy Access the PDB online (http://www.rcsb.org/pdb/home/home.do). Search for "myoglobin" (it will appear numerous times). Read the page and examine these structures. Do any of the structures have groups that are not comprised of protein?

Solution Myoglobin contains a heme group: a large planar heterocyclic structure with a single iron atom in the center. Heme groups are common prosthetic groups found in many proteins. They are important in the binding of oxygen, electron transport, and redox chemistry.

Follow-up question Is myoglobin a small, average, or large protein?

Summary

- Proteins are polymers of amino acids, averaging 450 amino acids in length (approximately 50 kDa).
- Proteins can have one or more subunits, and these subunits can be repeats of the same polypeptide or multiple different polypeptides.
- The peptide bond joining amino acids in the protein is planar because it has partial double bond character.
- Proteins may contain cofactors in the form of metal ions or small organic prosthetic groups.
- Proteins may have modifications such as acetylations or farnesylations (lengthy hydrocarbon chains), phosphorylations or hydroxylations.

Concept Check

1. Describe what a protein is like chemically.
2. Describe the peptide bond and explain why the peptide unit is planar.
3. Name several different prosthetic groups.
4. Describe the different types of post-translational modifications to proteins.

3.3 Proteins Are Molecules of Defined Shape and Structure

Proteins are varied in terms of both function and structure. As proteins are synthesized in the cell, they spontaneously and, in most cases, independently fold into the proper conformation. This is an amazing phenomenon, given that a molecule with over 5,000 atoms and an equal number of covalent bonds snaps into a single structure with near perfection. This section discusses the complex structures that proteins form. Proteins are organized into four levels of organization: primary, secondary, tertiary, and quaternary.

3.3.1 The primary structure of a protein is its amino acid sequence

At one level, a polymer of amino acids is like a charm bracelet, with each amino acid being one of the charms or beads. There can be multiples of a particular type of charm or bead in the bracelet, but they are placed in a specific order. The order or sequence of the amino acids in a protein is the **primary structure** of that protein (**Figure 3.13**). Just as alterations to the number, type, or position of charms in a bracelet might change the appearance, changes to the number of amino acids, the types of amino acid, or even the specific order of amino acids can alter the properties of a protein.

If a single hydrophobic amino acid valine (Val or V) is substituted for a similar hydrophobic amino acid such as isoleucine (Ile or I), there may be no observed difference in the new protein. This type of substitution is termed a **conservative substitution**. A more radical substitution, such as the bulky hydrophobic tryptophan (Trp or W) for the small negatively charged aspartate (Asp, or D) may result in a more substantial change to protein conformation or function.

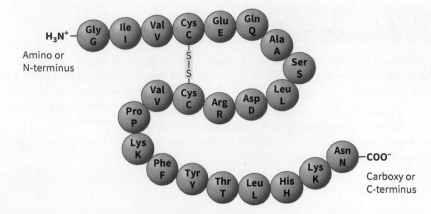

FIGURE 3.13 Primary sequence of a protein. The primary sequence of a protein is the sequence of amino acids from the amino to carboxy terminus (the N-to-C terminus).

Here are two examples of how proteins may vary with regard to amino acid sequence and the ramifications of these changes. The peptide hormone insulin regulates carbohydrate and lipid metabolism and is used to treat type 1 diabetes, also known as diabetes mellitus. It is made as a single polypeptide that is then cleaved into two fragments termed the A and B chains. From its discovery in 1923 until the early 1980s, the insulin used to treat type 1 diabetes was derived from dogs or pigs rather than humans. The insulin from dogs or pigs is similar to the human hormone but differs at several sites (**Figure 3.14**). These differences were not sufficient to render the hormone ineffective but did lead to immunological reactions in some people.

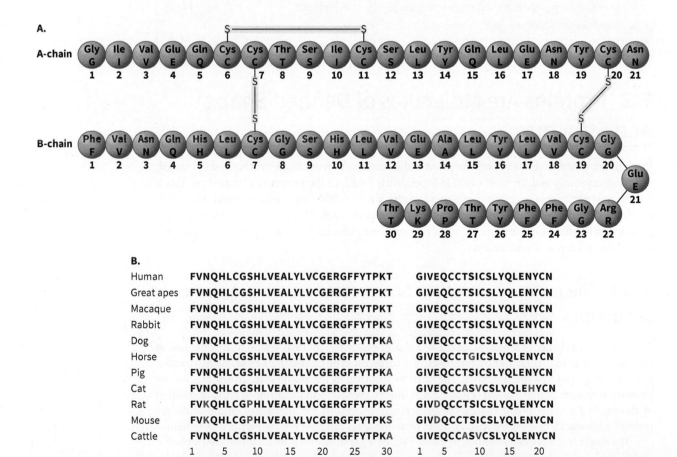

FIGURE 3.14 Differences in insulin sequence. **A.** The sequence of the native human insulin is shown. **B.** An alignment of the A and B chains are compared for different species.

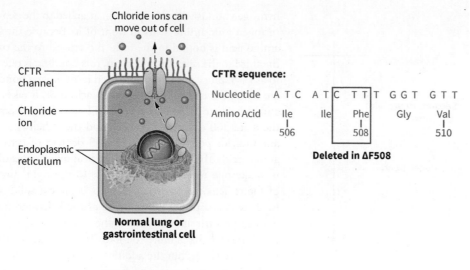

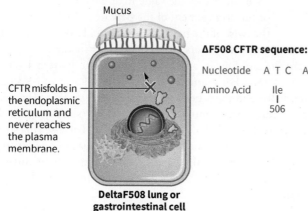

FIGURE 3.15 **Cystic fibrosis.** There are numerous mutations that can lead to cystic fibrosis, all of which affect the CTFR protein, a chloride channel found in lung and gastric mucosa. Mutations to this protein block the secretion of chloride, which results in the formation of thick mucus in the lung and gastrointestinal tract.

An example of a more harmful amino acid change is found in the disease cystic fibrosis. Although many different mutations can lead to cystic fibrosis, one of the most common is ΔF508 in the *CFTR* gene. This gene codes for a protein that allows the passage of water and chloride ions through the plasma membrane of cells (**Figure 3.15**). This mutation results in a protein where the amino acid phenylalanine (F) at position 508 in the protein has been deleted (Δ). As a result of this single change in the amino acid sequence, the chloride channel fails to move to the plasma membrane and is retained in the endoplasmic reticulum following synthesis. This causes a lack of chloride ion transport, leading to the formation of thick, viscous mucus in the lungs of people with this disorder.

The primary structure of a protein is one of the most important aspects of protein structure; however, it is currently not possible to determine the more complex structure of a protein based solely on the amino acid sequence.

3.3.2 The major secondary structures found in proteins are α helix and β sheets

The next level of organization of protein structure is **secondary structure**, which consists of two major structural elements: the α helix and β sheet. In addition, there are **turns**, tight connectors between two other structural elements, and **coils** or **loops**, larger regions without defined secondary structure.

The alpha helix is the most common element of secondary structure

The α **helix** is formed by the peptide backbone of a protein wrapping into a right-handed spiral (**Figure 3.16**). This spiral is held together by lone pairs of electrons on carbonyl oxygens forming

A.

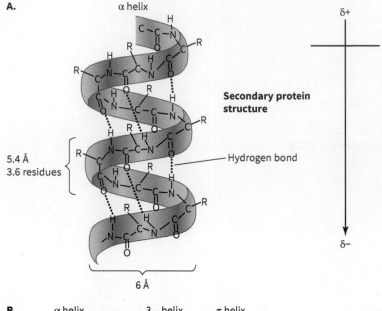

α helix

Secondary protein structure

δ+

δ−

5.4 Å
3.6 residues

Hydrogen bond

6 Å

B.

α helix   3₁₀ helix   π helix

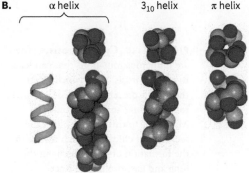

FIGURE 3.16 Alpha helix.   A. The α helices are the most common structure observed in proteins. They are right-handed helices with an i + 4 hydrogen bonding pattern. **B.** The 3:10 helix (i + 3) and the π helix (i + 5) are tighter or looser, respectively.

(Source: (A) Allison: *Fundamental Molecular Biology*, 2e, copyright 2011, John Wiley & Sons. This material is reproduced with permission of John Wiley & Sons, Inc. (B) Data from PDB ID 2QD1 Medlock, A.E., Dailey, T.A., Ross, T.A., Dailey, H.A., Lanzilotta, W.N. (2007) A pi-Helix Switch Selective for Porphyrin Deprotonation and Product Release in Human Ferrochelatase. *J. Mol. Biol.* 373: 1006–1016)

hydrogen bonds with the hydrogen attached to the peptide nitrogen four amino acids in front of it. Because the first amino acid is bound to the fifth, the second to the sixth, the third to the seventh, and so on, this bonding is referred to as i + 4. Since the groups involved in hydrogen bonding are not at the end of a whole amino acid unit, it takes 3.6 amino acids to complete one turn of an α helix. Because there are 360 degrees in a circle, and the α helix has 3.6 amino acids per turn, there are 100 degrees of turn per amino acid. The helix's pitch, or vertical distance required to make one rotation is 5.4Å. Due to the tight turning of the α helix, the core of the helix is almost solid; that is, there is no space in the middle of the helix and nothing can pass through its core. Although often depicted as pipes, the α helix is more rod-like. The diameter of the α helix's core is 6Å, but the addition of the amino acid side chains increases the value to approximately 12Å. Because of the arrangement of the peptide units in an α helix and the twist of the helix, the amino acid side chains are found only on the outside of the central structure.

Although hydrogen bonding in an α helix follows the pattern i + 4, other patterns are found in nature. For example, the **3₁₀ helix** follows an i + 3 pattern, resulting in a tighter helix, and the **π helix** has a looser i + 5 pattern. The π helices are often short, containing seven amino acids on average (slightly more than a single turn) and are embedded within a stretch of α helix. Because this structure introduces a bulge to the helix, it is sometimes referred to as an α aneurism. The length of an average α helix is ten amino acids, encompassing about three turns. The influenza protein hemagglutinin is the largest helix found to date, at 104 amino acids in length.

Most amino acids can be found in α helices, but there are two notable exceptions. Glycine's small side chain, comprised of a single hydrogen atom, is floppy and discourages helix formation. In proline, the side chain links back onto the amino group, preventing the Φ and Ψ bond angles from forming the geometry necessary for an α helix. This leads either to a kink in the helix or, more often, termination of the helix at the proline. Proline also lacks the proton necessary for hydrogen bonding, which would destabilize the helix.

The α helix aligns the carbonyl groups in a single direction, resulting in a dipole moment for the entire structure, which yields a partial positive charge on the amino-terminal end and a partial negative charge on the carboxy-terminal end of the helix. In addition, the handedness of the helix (α helices are right-handed) means that the overall structure is chiral. This chirality can be experimentally measured with a technique called circular dichroism, which provides a percent of residues in an α helix in the protein.

The beta pleated sheet is a major element of secondary structure

The other major element of secondary structure is the β **pleated sheet** or β **sheet**. Unlike the more compact form of the α helix, in a β sheet the individual strands are extended. The sheet is named "β" because it was the second structure to be determined (**Figure 3.17**). As in the α helix, the bonding unit in a β sheet is formed by hydrogen bonding between carbonyl oxygens and the peptide hydrogen and nitrogens, although here it occurs in the plane of the sheet between different strands. This means that the amino acid side chains attached to these backbones alternate on either side of the sheet, perpendicular to the plane created by the backbones. Numerous individual strands can contribute to a single β sheet. Although β sheets are often depicted as being short and flat, in reality, they are often twisted or flowing, and they may even circle back upon themselves to form a barrel-like structure.

There are two distinct types of β sheet: **parallel** β sheets in which the two strands forming the sheet run in the same direction, and **antiparallel** β sheets in which the two strands run in opposite directions. This does not alter the positioning of the side chains of the amino acids, but it does

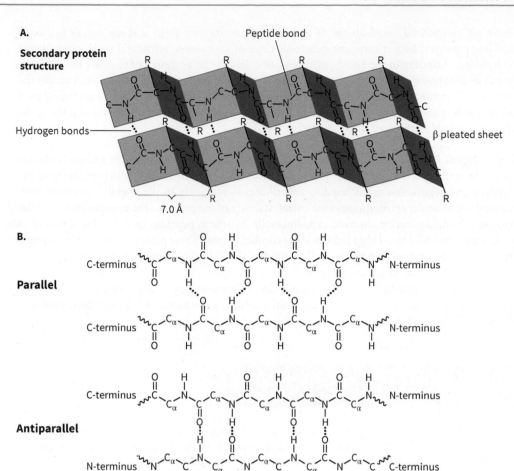

A.

Secondary protein structure

Peptide bond

Hydrogen bonds

β pleated sheet

7.0 Å

B.

Parallel

C-terminus

N-terminus

Antiparallel

C-terminus

N-terminus

FIGURE 3.17 Beta sheet. A. In β sheets there are hydrogen bonds between the backbone carbonyl groups and amino protons of adjacent sheets. Side chains (R) emerge from the top and bottom sides of the sheet. **B.** A β pleated sheet can be broadly categorized as parallel, where the strands run in the same direction, or antiparallel, where the strands run in opposite directions.

(Source: (A) Allison: *Fundamental Molecular Biology*, 2e, copyright 2011, John Wiley & Sons. This material is reproduced with permission of John Wiley & Sons, Inc.)

change the pattern of hydrogen bonding between the backbones. Instead of a simple rectangular pattern of hydrogen bonding seen in parallel β sheets, antiparallel sheets have a trapezoidal shape. There is no apparent difference in hydrogen bond strength or structural stability between these two types of β sheet, and all amino acids can participate in these structures.

Hairpin turns, and loops or coils are minor but important elements of secondary structure

Alpha helices and β sheets are connected to each other using either hairpin turns (or loops) and coils (**Figure 3.18**). Nine different types of hairpin turns exist.

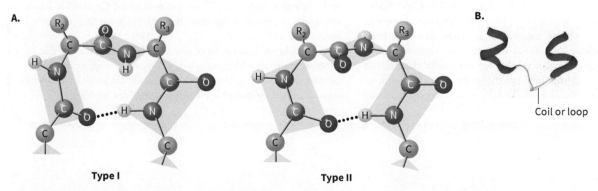

A.

Type I

Type II

B.

Coil or loop

FIGURE 3.18 Turns and coils. Turns and coils link other pieces of secondary structure. **A.** Hairpin turns require a minimum of four residues to complete the turn and are stabilized by a single backbone hydrogen bond. **B.** A coil or loop, which is a longer connection between two elements of secondary structure.

(Source: (B) Data from PDB ID 3WFT Mizohata, E., Morita, Y., Oohora, K., Hirata, K., Ohbayashi, J., Inoue, T., Hisaeda, Y., Hayashi, T. (2014) Co(II)/Co(I) reduction-induced axial histidine-flipping in myoglobin reconstituted with a cobalt tetradehydrocorrin as a methionine synthase model. *Chem. Commun. (Camb.)* 50: 12560–12563)

These are categorized based on the Φ and Ψ bond angles they form and the overall hydrogen bonding pattern. Hairpin turns comprise only three or four amino acids, and again are stabilized by hydrogen bonding in the backbone. Loops or coils may be as short as five or six amino acids, or can be composed of tens of amino acids, but they lack elements of α helix or β sheet and in the past were therefore referred to as "random coils." Today, the term *random coil* is no longer used because it is recognized that despite the lack of regular elements such as α helix and β sheet, the coils are probably organized into a discrete structure.

Hydrogen bonds stabilize secondary structures   Both helix and sheet are stabilized by hydrogen bonds in the peptide backbone. This is a critical feature in protein structure: repeated and regular hydrogen bonding contributes significantly to the overall stability of both elements of secondary structure and the overall stability of the protein. The contribution made by hydrogen bonding can be observed experimentally. Synthetic peptides comprised of a variety of amino acids can fold into either helix or sheet conformations. These proteins denature in response to similar experimental conditions indicating that the side chains are not contributing to the overall stability of the secondary structure.

Changes to secondary structure of a protein can have serious implications on protein folding and protein function. Some of these implications are discussed in **Medical Biochemistry: Protein folding and disease**.

Medical Biochemistry

Protein folding and disease

Despite their remarkable complexity, proteins fold into a single conformation and generally do this consistently, without assistance, and almost instantaneously. However, some proteins have more than one stable conformation.

In the mid-twentieth century there was an epidemic of a neurological disease among the Fore people of eastern Papua New Guinea. This disease is called kuru, also known as the shivering disease. People affected, who were typically the elder women of the group, suffered muscle weakness followed by shaking and tremors or truncal ataxia. As the disease developed over the course of a year, patients gradually lost control of their musculature, eventually becoming unable to take care of themselves or eat. They also lost control of their emotions, undergoing bouts of depression and of uncontrollable laughter. The disease always resulted in death.

Kuru has now been identified as a type of spongiform encephalopathy—a degradation of the brain related to Creutzfeldt–Jakob syndrome; bovine spongiform encephalopathy (also known as BSE or mad cow disease); and scrapie, a degenerative disease of sheep and goats. These diseases are all caused by prions—infectious protein particles that have, for reasons that are not totally clear, folded into an alternative conformation that causes disease. In all of these instances, the protein found in the healthy state is largely α helical,

but in disease states it mostly adopts a β sheet conformation. It is now believed that when it is in the β sheet conformation, the protein polymerizes and forms insoluble plaques that lead to neurodegeneration. Exposure of a properly folded protein to the protein in the prion conformation will stabilize or catalyze the refolding of the healthy protein to the diseased one.

Exactly how the prion enters the body is unclear, but it is clear that consumption of diseased tissue increases the risk. In the case of the Fore, kuru was partially spread through funerary rituals. As a sign of respect for the dead, the Fore practice transumption; that is, cannibalism of the dead by the remaining village members. In this practice, women were more likely to eat neurological tissue and thus had a greater chance of exposure.

The study of this disease may provide important clues to many other disorders. In the United Kingdom in the late 1980s and early 1990s, over 4.4 million cattle were destroyed in an attempt to contain an epidemic of BSE. Over 180,000 cattle were found to be infected, and over 175 people contracted Creutzfeldt-Jakob syndrome as a result of exposure, through consumption of infected meat.

More broadly, the issue of protein folding as a cause of disease is becoming more widely accepted. It may be that other neurological diseases, including Alzheimer's disease, are the result of neuronal damage caused by insoluble plaques of misfolded protein.

3.3.3 The tertiary structure of a protein is the organization of secondary structures into conserved motifs

The **tertiary structure** of a protein often gives the overall shape of the protein (**Figure 3.19**). As the structures of more proteins have been solved, it has become clear that, as with the α helix and β sheet, some elements of tertiary structure are repeated in different proteins. These combinations of secondary structure are termed **motifs**. Motifs should not be confused with **domains**, which are pieces of a protein that retain their structure in the absence of the rest of the protein; that is, a domain is a discretely folding structure whereas a motif may be part of a domain. This section examines several of the most common structural motifs and reviews the forces involved in stabilizing these structures.

Alpha motifs contain predominantly helical domains

Several common motifs are mostly helical (**Figure 3.20**). They include the helix-turn-helix motif, the three-helix bundle, and the four-helix bundle. The globin fold, which comprises at least eight helices, is an example of a more complex motif that has been retained in evolution.

Beta motifs are arrangements of either parallel or antiparallel beta sheets

Other common motifs contain largely or solely β structures (**Figure 3.21**). These include the Greek-key motif, an arrangement of four strands of β sheet; the β barrel, a β sheet that has been closed to form a barrel-like structure; and the β propeller, a β sheet that is twisted from a central point until it has the appearance of a propeller. The immunoglobulin fold is an example of a more complex motif consisting of flattened β barrel structures.

Protein folds are conserved

Protein folds are conserved throughout evolution, although the folds may be made with different combinations of amino acids. Therefore, in addition to other databases you may be familiar with, there are structural databases and tools for proteins, such as Structural Classification of Proteins (SCOP), and the Class, Architecture, Topology, Homologous superfamily (CATH), which can be used to categorize, organize and search for conserved structural motifs. Both of these databases are freely accessible on the internet.

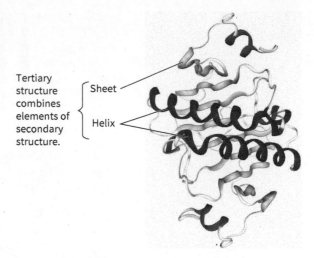

Tertiary structure combines elements of secondary structure.

Sheet

Helix

FIGURE 3.19 Tertiary structures. The tertiary structure of a protein gives the overall shape of the protein, in other words, how elements of the secondary structure (helix and sheet) are joined together to give the structure of a single polypeptide chain in a protein.

(Source: Data from PDB ID 1AV5 Lima, C.D., Klein, M.G., Hendrickson, W.A. (1997) Structure-based analysis of catalysis and substrate definition in the HIT protein family. *Science* 278: 286–290)

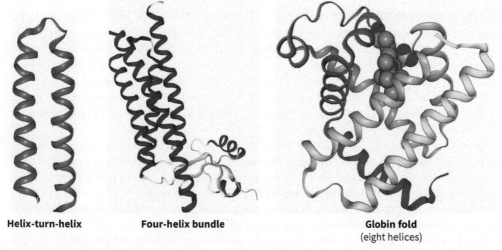

Helix-turn-helix

Four-helix bundle

Globin fold
(eight helices)

FIGURE 3.20 Alpha structures. Examples of all α structures include the helix-turn-helix, four-helix bundle, and globin fold.

(Source: (Left) Data from PDB ID 4DO2 Amprazi, M., Kapetaniou, E.G., Kokkinidis, M. (2013) Structural plasticity of 4-alpha-helical bundles exemplified by the puzzle-like molecular assembly of the Rop protein. *Proc.Natl.Acad.Sci.* USA 111: 11049–11054. (Middle) Data from PDB ID 1SZI Hickenbottom, S.J., Kimmel, A.R., Londos, C., Hurley, J.H. (2004) Structure of a Lipid Droplet Protein: The PAT Family Member TIP47 *Structure* 12: 1199–1207. (Right) Data from PDB ID 3WFT Mizohata, E., Morita, Y., Oohora, K., Hirata, K., Ohbayashi, J., Inoue, T., Hisaeda, Y., Hayashi, T. (2014) Co(II)/Co(I) reduction-induced axial histidine-flipping in myoglobin reconstituted with a cobalt tetradehydrocorrin as a methionine synthase model. *Chem.Commun.(Camb.)* 50: 12560–12563)

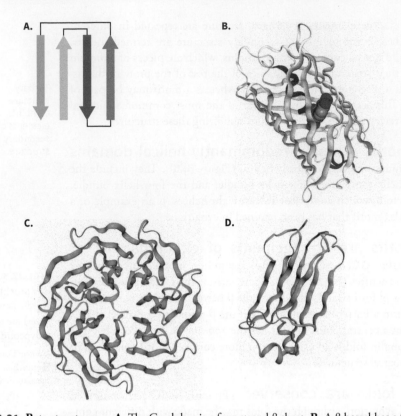

FIGURE 3.21 Beta structures.   A. The Greek key is a four-strand β sheet. **B.** A β barrel has a β sheet wrapped up on itself. **C.** The β propeller structure has from four to eight blades, each of which is twisted like a fan blade. **D.** An immunoglobulin fold has strands of β sheet sandwiched on top of one another.

(Source: (B) Data from PDB ID 5EB6 Close, D.W., Langan, P.S., Bradbury, A.R.M. (2016) Evolution and characterization of a new reversibly photoswitching chromogenic protein, Dathail. *J.Mol.Biol.* 428: 1776–1789. (C) Data from PDB ID 3UVL Zhang, P., Lee, H., Brunzelle, J.S., Couture, J.F. (2012) The plasticity of WDR5 peptide-binding cleft enables the binding of the SET1 family of histone methyltransferases. *Nucleic Acids Res.* 40: 4237–4246. (D) Data from PDB ID 1IGT Harris, L.J., Larson, S.B., Hasel, K.W., McPherson, A. (1997) Refined structure of an intact IgG2a monoclonal antibody. *Biochemistry* 36: 1581–1597)

Numerous weak forces (and one covalent one) dictate tertiary structure

The covalent peptide bond links amino acids together in a protein. However, weak forces dominate the assembly of secondary and tertiary structures (**Figure 3.22**). These are the weak interactions that are studied in both general and organic chemistry; however, when studied as part of biochemistry, there are specific aspects of these forces that need to be considered.

Hydrogen bonding is, in many ways, the most important interaction in protein folding and in many other phenomena, including formation of phospholipid bilayers and the formation of the DNA double helix.

To briefly review, a hydrogen bond forms when a hydrogen attached to a highly electronegative atom, usually an oxygen, nitrogen, or fluorine, is polarized to the point at which it will interact with the lone pair of electrons of a nearby molecule, again often a molecule of oxygen, nitrogen, or fluorine. In biochemistry, hydrogen bonds do not typically occur with hydrogen atoms bound to carbon, sulfur, or any other atoms, because these atoms lack the electronegativity required to sufficiently polarize the hydrogen. Hydrogen bonds are strongest, approximately 5 kcal/mol in biological systems, when the groups are arranged linearly and are within 2.2 Å of one another. Electronegativity also plays a role in hydrogen bond strength. The fluorine-based hydrogen bonds are stronger than those made with oxygen, and the oxygen-based structures are stronger than the structures using nitrogen.

Hydrogen bonds can be considered in one of two ways: first between amino acids within the protein, and second between the solvent molecules.

Hydrogen bonds are common and are found in every protein. Most of the polar amino acids contain groups such as hydroxyl groups, amines, and carboxamides that can participate in hydrogen bonding. In addition, the peptide backbone has lone pairs of electrons on the

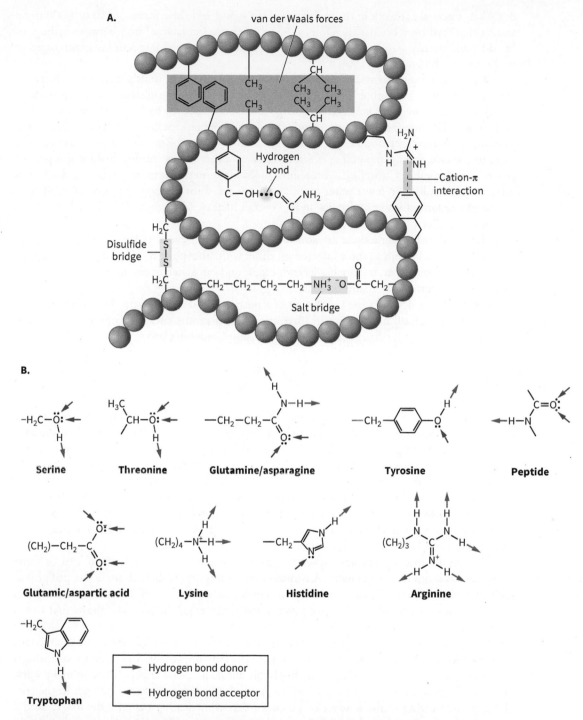

FIGURE 3.22 Forces involved in stabilizing proteins. **A.** The main side chain interactions found in proteins are shown. **B.** The different ways that side chains contribute to hydrogen bonding are shown.

carbonyl carbon and peptide nitrogen that can participate in hydrogen bonding. The tertiary structure of proteins depends mainly on interactions of side chains with each other or with the backbone, whereas secondary structure depends mainly on interactions that occur within the peptide backbone itself.

The second way that hydrogen bonds affect protein structure is by hydrogen bonding within the solvent water molecules. This type of bonding is part of the hydrophobic effect, the phenomena in which hydrophobic groups cluster together. The hydrophobic effect does not depend on hydrophobic residues being attracted to one another. For example, hydrophobic molecules, such as oil or hexane, spread out to coat surfaces; thus, on a rainy day a drop of oil will spread out and form a slick on a puddle. In a hydrophobic liquid, such as hexane, London dispersion and van der Waals forces are at work in the molecules of the liquid, but these forces are 10 to 1,000 times weaker than hydrogen bonds. Therefore, although there are interactive forces between hydrophobic side chains, they are not strong enough to explain how hydrophobic residues group together in the core of the protein.

Water has strong intermolecular interactions, and it can form multiple hydrogen bonds at any time. When a hydrophobic molecule such as hexane or a hydrophobic side chain is introduced into water, the water molecules rearrange to accommodate the hydrophobic group. There is no hydrogen bonding possible between water and a hydrophobic molecule. Instead, water molecules organize into cages around hydrophobic molecules. The organization and formation of these cages to preserve hydrogen bonding in water results in a decrease in entropy, making this process generally unfavorable. Clustering hydrophobic molecules means that more can be caged with less surface area, resulting in fewer water molecules needed to form the cage. The overall result is an increased freedom of water molecules and an overall increase in entropy. This is the basis of the hydrophobic effect.

Dipole–dipole interactions are also found in proteins. Several polar side chains have the potential to participate in a dipole interaction, either with other polar residues or with the peptide backbone. In addition, an α helix has its own dipole and can generate enough of an electric field to interact with other side chains or helices.

Salt bridging describes the interaction of a positively and a negatively charged amino acid. A salt bridge is basically an electrostatic interaction; therefore, the energetics of the interaction of the positive and negative charge can be described using Coulomb's law:

$$F = \frac{kq_1q_2}{r^2D}$$

In this expression, the force of the interaction is dependent on Coulomb's constant (k) and the magnitude of the charges q_1 and q_2, and is inversely dependent on the square of the distance between the two charges (r^2) and the dielectric constant of the media, D. The charge on ionized amino acid side chains is always + or −1; therefore, q_1 and q_2 are constant in proteins, and the strength of the interaction depends on the square of the distance between the charges and the strength of the dielectric constant. The dielectric constant of water is high, almost 80 Debye units, whereas the dielectric constant of a nonpolar solvent such as hexane is only about 1 to 5 Debye units, so the strength of an electrostatic interaction is decreased by a factor of 80 in water, but by a much smaller value in a nonpolar solvent.

Coulombic interactions have several interesting ramifications and tend to lead to some misunderstandings in biochemistry. A common misconception is that adding a charged group, such as a phosphate, will cause a protein to translocate across the cell based on electrostatics alone. A quick analysis of the distance covered and the strength of the force shows that this is not the case.

Biochemical systems are aqueous, leading some to assume that *all* electrostatic interactions are weakened by the dielectric constant of water. The core of proteins where many salt bridges are found is also the location of many hydrophobic amino acid side chains. Thus, in many ways, the core of a protein is an oily droplet, so the strength of a salt bridge may be high in the core of the protein owing to the absence of water. Further complicating matters, the low dielectric constant and the close proximity of charged groups to one another in the core of the protein create the possibility that all charged species in the core of the protein interact with one another. Finally, although salt bridges can be viewed as an attractive force between positive and negative side chains, the rules that apply to salt bridges also apply to side chains with the same charge, which repel one another.

Cation-π interactions occur between the positively charged amino acids arginine, lysine, and histidine and a π bond, as shown in Figure 3.22. Pi bonds are seen in any double-bond system. Although the amino acids typically found in these interactions are tyrosine, tryptophan, phenylalanine, or histidine, weak cation-π interactions have been observed between a positively charged side chain and the peptide backbone carbonyl. Cation-π interactions are akin to a hydrogen bond, in that the bond strength decreases with the square of the distance between the two molecules involved ($1/r^2$). Also, the bond is stronger when the cation and π cloud approach each other at a 90° angle, that is, orthogonal to each other. Cation-π interactions are as strong as hydrogen bonds and as common as salt bridges, with anywhere from 5 to 12 bonds in an average protein.

Disulfide bonds are the only covalent force involved in the formation of tertiary structure. In a disulfide bond, two cysteine residues undergo redox chemistry to form a disulfide linkage. Disulfides are more oxidized than a pair of cysteine residues, and will form spontaneously in oxidizing environments. To keep cysteine residues from reacting with each other, a reducing agent is used. In the laboratory, compounds such as β-mercaptoethanol (βME) or dithiothreitol (DTT) are often used to accomplish this task. In the cell, the cytosol is a reducing environment owing to the presence of the peptide glutathione. These compounds will reduce any disulfides that form. Proteins containing disulfide bonds are synthesized and folded in the endoplasmic reticulum (ER).

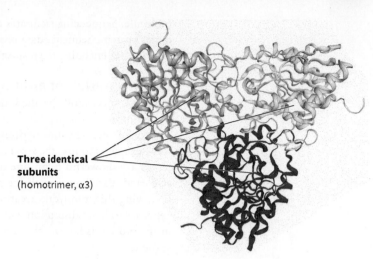

Three identical subunits
(homotrimer, α3)

3.3.4 The quaternary structure of a protein describes how individual subunits interact

Most proteins exist in a complex, either with other subunits of a protein or with different proteins. This is termed the **quaternary** structure of the protein (**Figure 3.23**). The forces that govern the interactions of these individual subunits are the same as those involved in the formation of tertiary structure.

FIGURE 3.23 Quaternary structure. The structure of purine nucleotide phosphorylases is an example of quaternary structure. This protein is a homotrimer with α3 stoichiometry. This enzyme has three identical subunits, each shown here in a different color. Note how the subunits are in close contact with each other in this example.

(Source: Data from PDB ID 4EAR Haapalainen, A.M., Ho, M.C., Suarez, J.J., Almo, S.C., Schramm, V.L. (2013) Catalytic Site Conformations in Human PNP by (19) F-NMR and Crystallography. *Chem.Biol.* 20: 212–222)

Worked Problem 3.3   Protein structure and disease

Sickle cell anemia is a common genetic disorder. The causal mutation in sickle cell anemia is E6V, that is, the substitution of a glutamic acid for valine in the sixth position of the β chain of hemoglobin, a tetrameric protein with the stoichiometry α2β2. In the deoxygenated state, the mutant hemoglobin can polymerize, causing erythrocytes (red blood cells) to form a sickled shape and lyse, leading to the symptoms of this disease. How might this mutation affect all four levels of protein structure?

Strategy   Examine the information we have been given in the question and think about the four levels of protein structure. How could this alteration in amino acid sequence lead to the observed phenotype and disease?

Solution   Hemoglobin is a tetrameric protein. The mutation has altered the primary sequence of the β chains of hemoglobin by replacing a negatively charged glutamic acid with a hydrophobic valine residue. It is not apparent from the information provided in the question how this might affect the secondary and tertiary structure of the protein. However, we do know that the protein is functional in the oxygenated state; that is, the mutation is not lethal, and therefore we may assume that the secondary and tertiary structures have not been significantly disrupted. Functional studies have borne out this hypothesis. However, the quaternary structure of the protein has been somewhat altered. Although the hemoglobin is still able to act as a tetramer, in the deoxygenated state, the hydrophobic valine enables the protein to assemble into polymers, leading to symptoms of the disease.

Follow-up question   A protein has an α helix with an isoleucine residue facing the core of the protein. Compare isoleucine to each of the other amino acids. Hypothesize which mutations would be more deleterious to the protein and which would be less.

3.3.5 Amino acid sequencing in proteins

The amino acid sequence of a protein gives significant insights into the protein's history, structure and function. It will allow for comparison with other known sequences for any possible similarities. Similarity comparison will yield a scope to track the evolutionary pathways. Internal repeats in the protein molecule could be ascertained by amino acid sequencing and some amino acids that serve as the signal sequences help to understand their destinations or regulatory control mechanisms. It will also allow tracking the posttranslational modifications that take place in the protein molecule. Sequencing facilitates the designing and preparing the antibodies that are specific to a target protein. Sequence data provides valuable information to probe DNA that could be used to target genes encoding the respective proteins.

Determination of amino acid composition of proteins

Once the protein of interest is successfully purified, the next step would be to determine the amino acid composition in that protein. The peptide or the protein is placed in a solution of 6M HCl at a temperature of 110°C for one day. This hydrolyzes the peptide bonds resulting in its constituent amino acids. The next step requires the separation of different amino acids, and ion-exchange chromatography serves as a powerful technique in achieving so. A buffer of increasing pH will elute the amino acids and, based on the volume of buffer required, the identity of the amino acid is determined. Following this, ninhydrin treatment of the solution containing a single type of amino acid with exposure to light absorption will determine the concentration. Once the amino acid composition in the protein is known, the next step would be to determine the first amino acid in the specific sequence.

Identification of N-terminus

Identification of a protein's N-terminal residue can be achieved by several methods. Dansyl chloride yields dansylated polypeptide when it reacts with the primary amines and further acid hydrolysis releases the N-terminal residue as a dansylamino acid. This product can be identified chromatographically even at picomolar concentration. Apart from the chemical approach, enzymatic identification of the N-terminus is possible by aminopeptidases which cleave the polypeptide chain from the N-terminus.

Sanger degradation   This method helps in determining the terminal amino acid (the first amino acid in the sequence). The first step involves the breakdown of any disulfide bonds by thiol to obtain individual polypeptide subunits. Next, the polypeptide is treated with 2,4 dinitrofluorobenzene that labels the amino terminus of each polypeptide with the 2,4-dinitro-fluorobenzene group. Hydrolysis of the polypeptides produces the amino acid but only the amino terminus is labeled (**Figure 3.24**). The Sanger degradation is not a very useful method as it offers limited scope to determine the first amino acid in a sequence. It also involves exposing the peptide to hydrolysis, which cleaves all the peptide bonds and scrambles the amino acids. A much more useful method that does not involve hydrolysis is called Edman degradation.

Edman degradation   The approach to protein sequence determination can be achieved directly by sequencing the protein or by sequencing DNA that encodes the protein from which the amino acid sequence could be derived. Edman degradation is a technique in which the N-terminus group of the protein, adsorbed to a solid phase, reacts with phenylisothiocyanate under mild alkaline conditions to form PTC (phenlythiocarbamayl) adduct. This product is treated with trifluoroacetic acid, a strong acid (anhydrous) that cleaves the N-terminal amino acid resulting in thiazolinone derivative without hydrolyzing the other peptide bonds. An intramolecular cyclization occurs and results in the cleavage of N-terminal amino acids. The thiazolinone-amino acid is treated with aqueous acid and converted to a more stable PTH (phenylthiohydantoin) derivative. The resulting PTH-amino acid is identified by comparing the retention time on HPLC with a known PTH-amino acid. The most important feature about Edman degradation is that it keeps the polypeptide chain intact and releases only the N-terminal amino acid residue (**Figure 3.25**).

Edman degradation cannot be used to determine the sequence of polypeptides longer than 50 amino acids in length.

If proteins are larger, it could still be possible to sequence them using specific enzymes called endoproteases such as trypsin (cleaves after Lys and Arg) and chymotrypsin (specifically cleaves after aromatic amino acids like Tyr, Phe, and Trp), which cleave proteins after specific side chains. Specific cleavage can also be achieved by chemical treatment with molecules such as cyanogen bromide (cleaves after methionine residue). The resulting peptides with known sequences can be aligned and compared.

Identification of C-terminus

The C-terminal amino acid can be determined by treating the polypeptide with either chemical agents or with the addition of enzymes that cleave amino

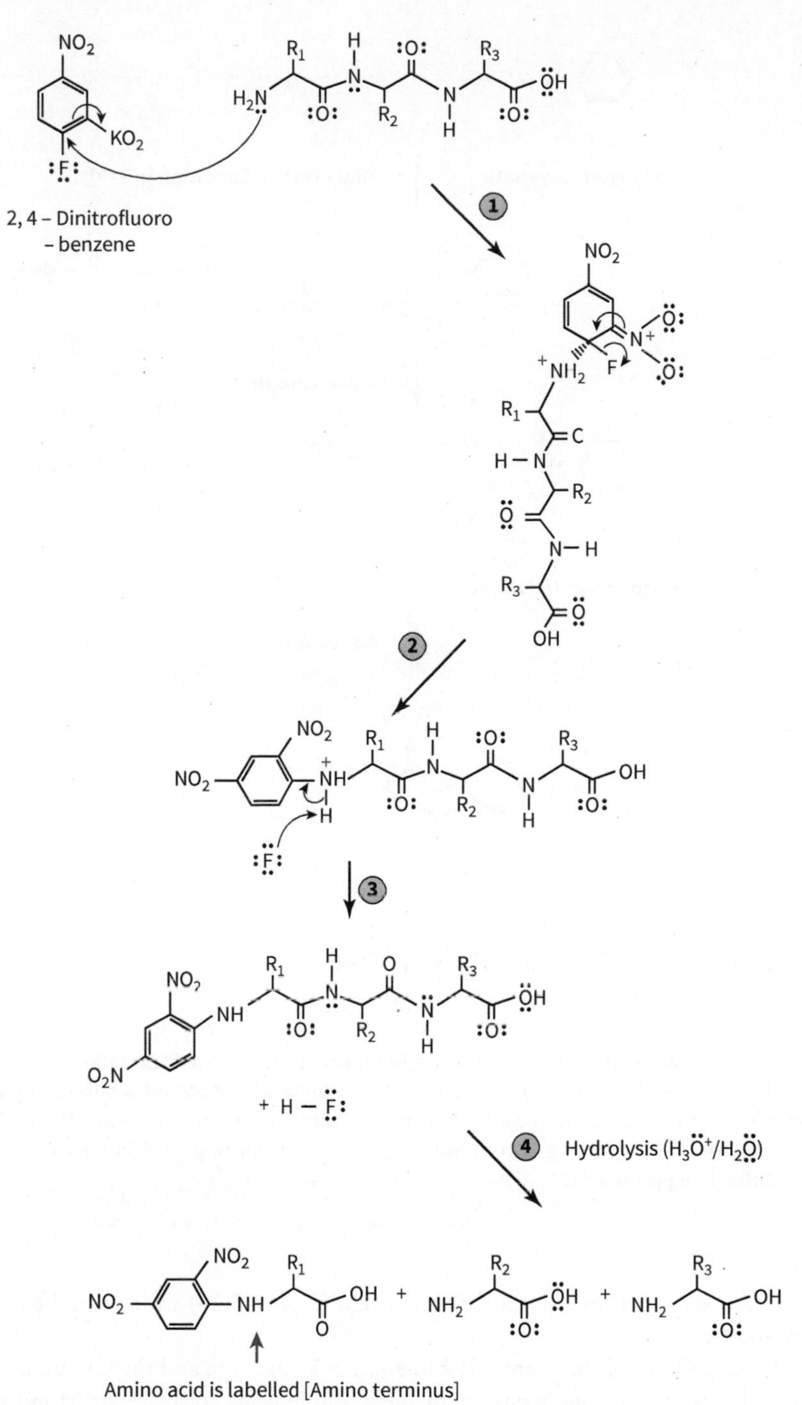

FIGURE 3.24 Determination of amino acid sequence by Sanger degradation. The polypeptide is to be treated with 2,4 dinitrofluorobenzene followed by hydrolysis to yield N-terminal amino acid.

acids from the C-terminal end. The side chain specificity of peptidases (endo- and exopeptidase) commonly used to fragment and identify the sequence are listed in **Table 3.2**. Classes of carboxypeptidase enzymes from different sources have varying specificity. However, carboxypeptidase A and B do not cleave the C-terminal residues with a preceding proline amino acid. In this scenario, chemical methods can be adapted to identify the C-terminal residue. Hydrazinolysis, a reliable chemical method can be used wherein, the polypeptide is treated with hydrazine (anhydrous) at a high temperature (90°C) for 20 to 100 hours in the presence of a catalyst such as mild acidic ion exchange resin. This results in the cleavage of all peptide bonds to yield aminoacyl hydrazides

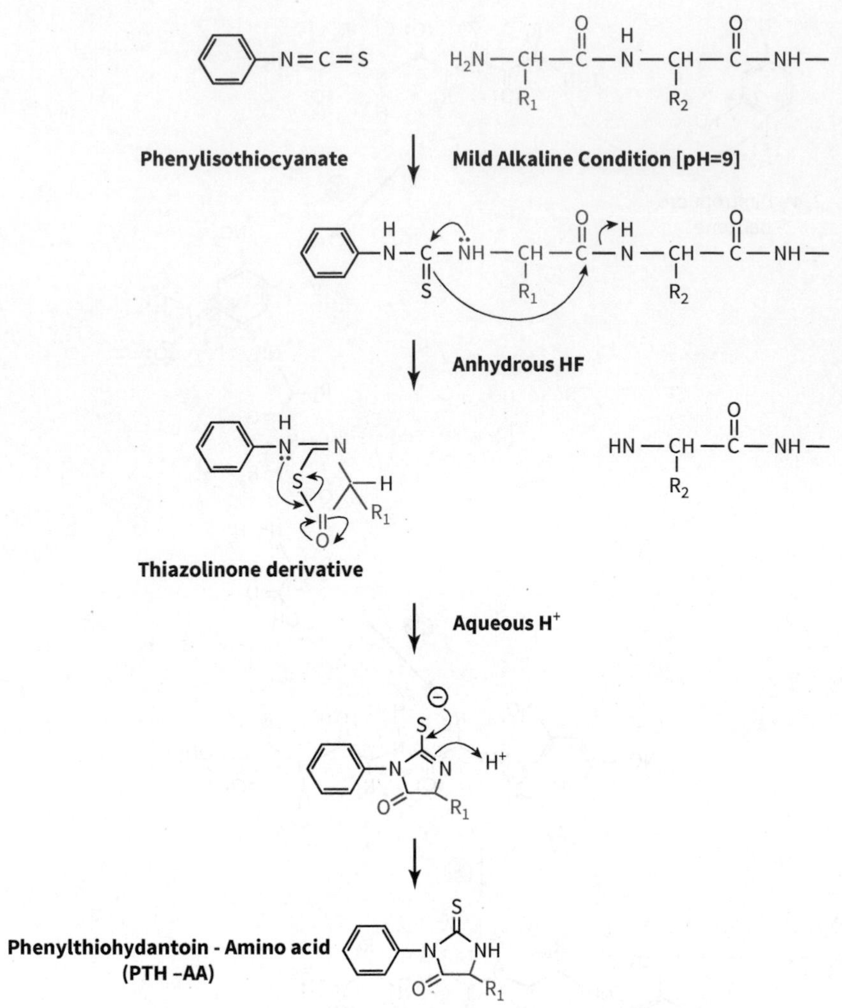

FIGURE 3.25 Determination of protein or amino acid sequence by Edman degradation. The peptide or protein is to be treated with phenylisothiocyanate under the mild alkaline condition to form phenyl-thiocarbamayl adduct. Subsequent treatment with strong anhydrous acid, trifluoroacetic acid results in thiazolinone derivative and eventually gets converted to a more stable form, phenylthiohydantoin-amino acid, on the addition of aqueous acid.

of all amino acids except that of C-terminal amino acid (**Figure 3.26**) which could be identified chromatographically.

Once the composition of the protein is known, that is, the type and the relative amounts of each amino acid, determining the sequence of those amino acids (primary structure) comes as

TABLE 3.2 Cleaving Agents with Their Site of Action

Cleaving Agent	Site of Cleavage	
Cyanogen bromide		Met
Trypsin (Endopeptidase)		Lys and Arg
Chymotrypsin (Endopeptidase)		Tyr, Met, Phe, Trp, Leu
Pepsin (Endopeptidase)	Carboxyl end of	Phe, Tyr, Leu, Trp
Thrombin		Arg
Carboxypeptidase A, B (Exopeptidase)		Lys, Arg, Pro
Carboxypeptidase C, Y (Exopeptidase)	Free C-terminal residues	

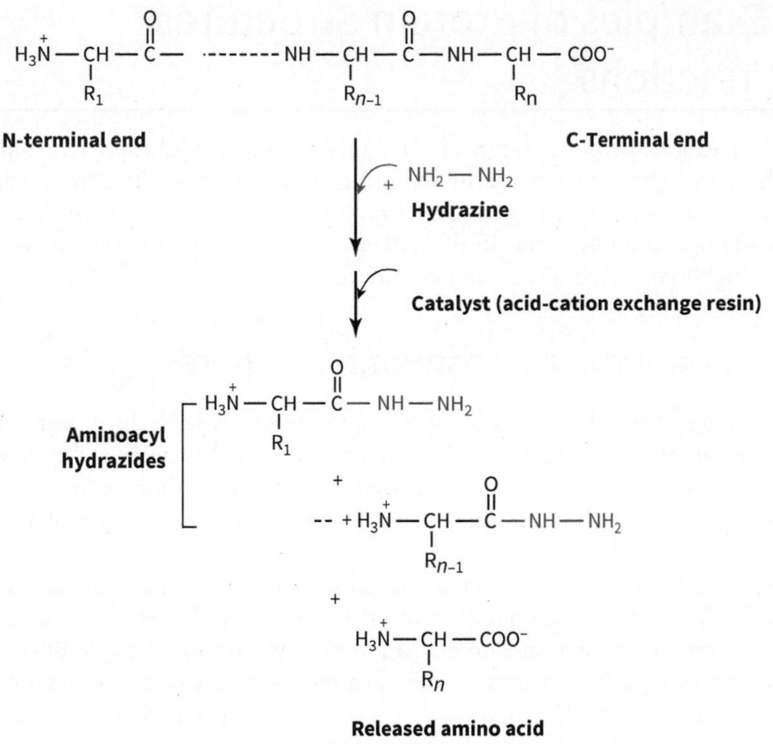

FIGURE 3.26 Determination of C-terminus by using a chemical agent. Hydrazine is used along with a catalyst to form aminoacyl hydrazides and free C-terminal amino acids.

the next step. Upon determining the specific sequence of amino acids in each segment, the order of those segments concerning one another remains a question. To determine the correct order of the sequence, this usually requires another proteolytic agent to cleave the original polypeptide at different locations and eventually overlapping the cleaved regions.

Summary

- Protein structure can be described on four levels—primary through quaternary.
- The basic unit of protein structure is the peptide bond.
- Secondary structure describes higher order structures: α helices, β sheets, turns, and coils.
- Secondary structures are stabilized by hydrogen bonds in the peptide backbone, and can be formed by many different amino acids.
- Tertiary structures describe the interaction of different elements of helix and sheet to form a complex structure.
- Tertiary structure represents the complete folding of a single polypeptide into a functional protein.
- Examples of tertiary motifs include the four-helix bundle and the Greek key motif.
- Tertiary structures are stabilized by a combination of many weak forces; these include hydrogen bonding, dipole–dipole interactions, salt bridges, disulfide bonds, cation–π interactions, and a phenomenon known as the hydrophobic effect.
- Many proteins exhibit quaternary structure interactions between and among different polypeptide chains.
- Amino acid sequencing depends on chemical or enzymatic digestion methods to separate peptides and detect the amount and composition of amino acid residues.

Concept Check

1. Describe the four levels of structure of a protein, and explain the forces involved in the stabilization of each level.
2. Provide examples of secondary and tertiary structure, and draw them in molecular detail.
3. How do secondary structures contribute to the organization of the tertiary structure of proteins?

3.4 Examples of Protein Structures and Functions

As might be expected from a group of molecules composed of 20 different building blocks, proteins have a vast diversity in terms of both structure and function. This section describes seven different proteins, each of which exemplifies some interesting aspect of protein structure-function and plays an important role in the life of a cell or organism. All of the molecules discussed in this section are highly conserved throughout evolution.

3.4.1 Aquaporin, a transmembrane pore

Aquaporins are proteins that act as pores in the membrane, selectively allowing the passage of water into or out of the cell (**Figure 3.27**). Lipid bilayers have low permeability to water, and the aquaporins are necessary to rapidly transport water and for cells to function normally. Aquaporins are so specific that they will permit the passage of water but not the passage of acid (hydronium ion, H_3O^+).

Aquaporins consist of six transmembrane α helices pitched at a slight angle to each other in a single bundle. The center of this bundle forms the pore through which water passes. The outside of the helices are hydrophobic and are exposed to the acyl chains of the lipid bilayer.

Aquaporins are pores or channels rather than transporters. Channels allow for the transport of molecules down either a chemical or electrical gradient. They have much higher transit rates than transporters and do not require the input of energy for transport.

3.4.2 Chymotrypsin, an enzyme

Chymotrypsin is an enzyme; it is a small (26 kDa) globular protein that is produced by the pancreas and secreted into the intestinal lumen, where it cleaves dietary proteins into peptides. It is comprised predominantly of two compressed β barrel structures and one short α helix (**Figure 3.28**).

Chymotrypsin has two structural features that are common to most enzymes. First, its active site, where catalysis occurs, is relatively small compared to the rest of the structure. Only three amino acid side chains are directly involved in the catalytic steps of the reaction, although several

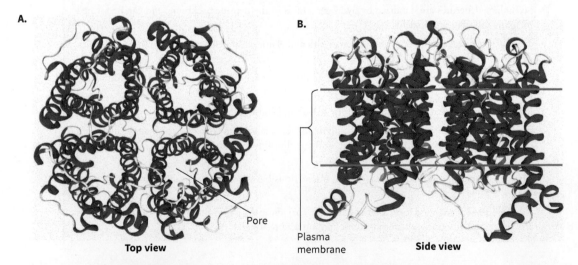

A.

Pore

Top view

B.

Plasma membrane

Side view

FIGURE 3.27 Aquaporin. The aquaporins permit the passage of water through the plasma membrane. **A.** View from above looking down on the four subunits. **B.** In this side view, the plane of the plasma membrane is shown as yellow lines and the subunits of the aquaporin are shown in magenta. Water takes a path through the channel in each subunit formed by the six helices.

(Source: (A, B) Data from PDB ID 3D9S Horsefield, R., Norden, K., Fellert, M., Backmark, A., Tornroth-Horsefield, S., Terwisscha Van Scheltinga, A.C., Kvassman, J., Kjellbom, P., Johanson, U., Neutze, R. (2008) High-resolution x-ray structure of human aquaporin 5. *Proc.Natl.Acad.Sci.* USA 105: 13327–13332)

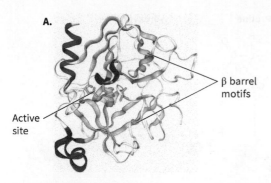
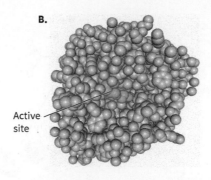

FIGURE 3.28 Chymotrypsin. A. Chymotrypsin is mainly comprised of β sheets. At the interface of the two regions of sheet is the active site of this enzyme. **B.** In this space-filling model the catalytic residues are located in the middle of the structure.

(Source: (A, B) Data from PDB ID 1T7C Czapinska, H., Helland, R., Smalas, A.O., Otlewski, J. (2004) Crystal structures of five bovine chymotrypsin complexes with P1 BPTI variants. *J.Mol.Biol.* 344: 1005–1020)

other side chains are involved in the binding and recognition of the protein that is cleaved. This is in stark contrast to the other 242 amino acids found in the protein which are important for forming the scaffold on which the essential residues reside. These other parts of the protein may also play roles in regulation or localization of an enzyme in the cell. Second, the active site of chymotrypsin is found relatively close to the surface. Active sites are often near the surface of a protein but found in a canyon, trench, or other crevice. Frequently, amino acids found in turns or coils are found in an active site. In the example given here, binding of substrate and catalysis occurs at the interface of the two β structures.

3.4.3 Collagen, a structural protein

Collagen is the protein in greatest abundance in the body (**Figure 3.29**). **Isoforms** are related proteins encoded by the same genes. Several isoforms of collagen are found in the body, and their expression varies in different organs and tissues.

Collagen is secreted by fibroblasts and assembles as part of the extracellular matrix. It is most concentrated in muscle and connective tissue and was one of the first protein structures to be elucidated. However, in several ways, collagen is unlike other proteins discussed in this chapter. Collagen is a fibrous protein forming extended strands and is helical, but not α helical. Instead, it has a structure known as a collagen helix in which three helical filaments twist around each other. In addition, collagen has an unusual amino acid composition, with a high proportion of residues of glycine (up to 33%), hydroxyproline, and proline (16% each). Hydroxyproline is a modified amino acid; select proline residues are modified by the enzyme prolyl hydroxylase. This enzyme uses molecular oxygen to complete the modification but requires ascorbate, vitamin C, to reduce the enzyme back to the active form. In the absence of vitamin C, the enzyme is unable to operate properly, and hydroxylation does not occur. Consequently, collagen lacks hydroxyl groups and lacks structural integrity, leading to the disease scurvy, marked by weakened connective tissue, bleeding, and tooth loss. Glycine-rich patterns of amino acids and proteins containing modified amino acids such as hydroxylproline are also found in other filamentous proteins of the extracellular matrix.

3.4.4 Hemoglobin, a transport protein

Hemoglobin is a tetrameric protein found in erythrocytes or red blood cells (**Figure 3.30**). Each subunit of hemoglobin contains one molecule of heme: a large heterocyclic ring system surrounding a central iron atom. Oxygen binds to this central iron and that binding is at the core of oxygen transport, which is hemoglobin's primary function. A single molecule of oxygen can bind to each molecule of heme; thus, a single molecule of hemoglobin can transport up to four molecules of oxygen. However, the interaction of oxygen with the heme group is not simple; the four subunits of the protein cooperate, altering the affinity of oxygen for the heme group.

FIGURE 3.29 Collagen.
Individual collagen molecules enriched in glycine, proline, and hydroxyproline form a collagen helix, a less compact, more gently spiraling left-handed helix. Three of these molecules twist around each other to form a collagen triple helix. Hydrogen bonding and steric interactions of the side chains stabilize this structure. Five triple helices form a collagen microfibril.

(Source: (Middle) Karp, *Cell and Molecular Biology: Concepts and Experiments*, 7e, copyright 2013, John Wiley & Sons. This material is reproduced with permission of John Wiley & Sons, Inc.)

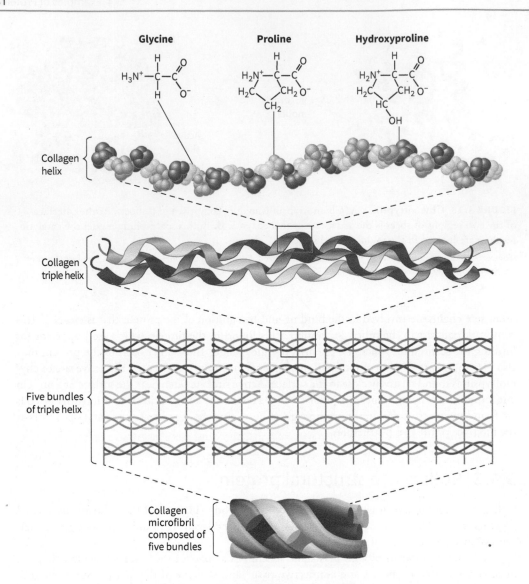

An examination of the structure of hemoglobin reveals that it is a globular protein that is predominantly made up of α helices joined by turns. As might be expected, the four subunits interact tightly, facilitating communication between them. Likewise, there are several places in the hemoglobin holoenzyme where other molecules can bind to alter the binding affinity of oxygen. These other molecules are termed **allosteric regulators**, because they act at sites removed from where the chemistry is occurring.

3.4.5 Immunoglobulins, antigen-binding proteins

Immunoglobulins are molecules of the immune system that recognize and bind to antigens, which are small pieces of a molecule that the body recognizes as foreign (**Figure 3.31**). Usually, an immune reaction is provoked by an infection such as a bacterial or viral infection, but it could be from other sources such as a food (for example, peanuts) or any sort of foreign molecule in the body. In effect, immunoglobulins act as adapter molecules between the foreign molecule and the immune system cells required to neutralize it. To accomplish this, immunoglobulins require an amazing amount of diversity to recognize an array of potential pathogens, but they also require a common portion to communicate with the cells of the immune system; these different parts of the molecule, performing different functions, are located at either end of the immunoglobulin.

An examination of the structure of immunoglobulin gamma (IgG) shows how it accomplishes its tasks. The molecule has four subunits: two identical heavy chains (50 kDa each) and two identical light chains (23 kDa each), all of which are held together by disulfide bonds. IgG can be split into three main fragments: two identical variable fragments (Fab) that bind

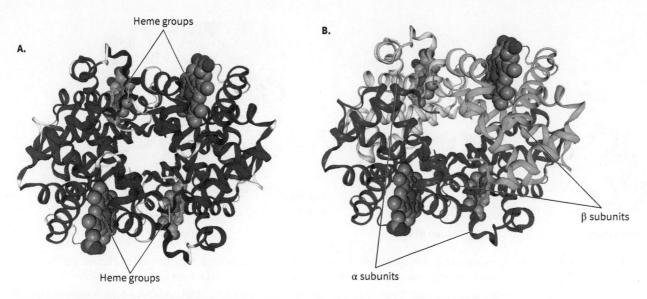

FIGURE 3.30 Hemoglobin. Hemoglobin is depicted in two different ribbon depictions. **A.** Hemoglobin is comprised of eight α helices connected by turns. **B.** Hemoglobin is a tetrameric molecule consisting of two α chains, shown in red and yellow, and two β chains, shown in green and blue, each of which has a heme functional group. Oxygen can bind to each of the heme groups. The four subunits are in close contact with one another, which facilitates communication between the four heme groups.

(Source: (A, B) Data from PDB ID 4YU4 Abubakkar, M.M., Maheshwaran, V., Ponnuswamy, M.N. (2015) Crystal structure of Mongoose (Helogale parvula) hemoglobin at pH 7.0.)

to foreign molecules and a constant fragment (Fc) that binds to cells of the immune system and elicits an immune response. Having multiple variable regions within the same IgG molecule increases the interaction with the antigen, allows for binding to more than one copy of the foreign molecule at once, and increases the overall binding efficiency. The Fab and Fc fragments are connected by a hinge region that allows the molecule to be flexible and thus bind more easily.

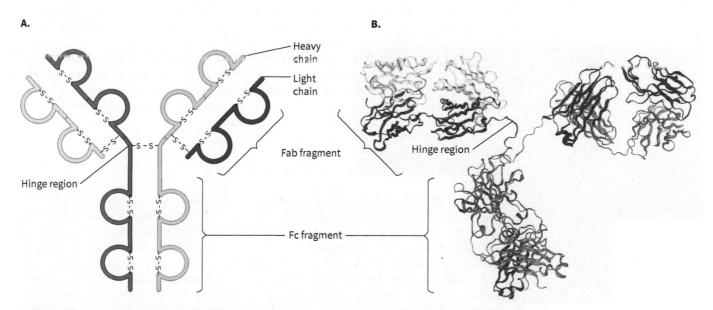

FIGURE 3.31 Immunoglobulins. **A.** The immunoglobulins are a family of proteins that have an immunoglobulin fold, a series of β sheet structures stabilized by four interchain disulfide bonds. **B.** The basic structure seen in immunoglobulin gamma (IgG) is a dimer of dimers or a tetramer. Each dimer is comprised of a heavy chain and a light chain. The Fc and Fab domains are connected by a flexible hinge region.

(Source: (A) Karp, *Cell and Molecular Biology: Concepts and Experiments*, 7e, copyright 2013, John Wiley & Sons. This material is reproduced with permission of John Wiley & Sons, Inc. (B) Data from PDB ID 1IGT Harris, L.J., Larson, S.B., Hasel, K.W., McPherson, A. (1997) Refined structure of an intact IgG2a monoclonal antibody. Biochemistry 36: 1581–1597)

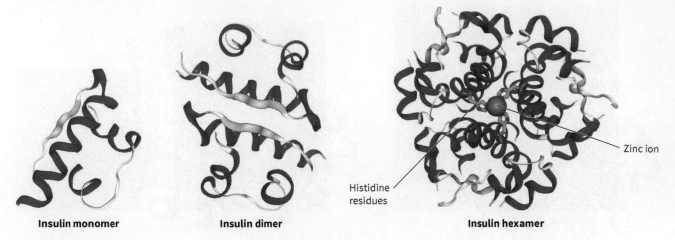

Insulin monomer   **Insulin dimer**   **Insulin hexamer**

Zinc ion

Histidine
residues

FIGURE 3.32 Insulin.   Insulin is a small (51 amino acid) protein hormone. Only the monomeric form of insulin is active, but the hormone will assemble into dimers and hexamers. The central atom in the hexamer is a zinc ion, chelated by histidine residues.

(Source: Data from PDB ID 1ZEH Whittingham, J.L., Edwards, E.J., Antson, A.A., Clarkson, J.M., Dodson, G.G. (1998) Interactions of phenol and m-cresol in the insulin hexamer, and their effect on the association properties of B28 pro → Asp insulin analogues. Biochemistry 37: 11516–11523)

In terms of structure, immunoglobulins are comprised mainly of β sheet domains joined by turns and coils. The generic term **antibody** is often used to describe IgG.

3.4.6 Insulin, a signaling protein

Insulin is a small protein hormone made by the β-cells of the pancreas (**Figure 3.32**). Hormones bind to protein receptors in target cells and tissues to elicit a response. Insulin is a growth factor, signaling cells to store energy and divert energy to growth, but it is best known for its role in diabetes. Insulin signals cells to absorb glucose from the circulation. Interrupted or impaired insulin signaling results in diabetes. An autoimmune attack on the β-cells of the pancreas or on insulin itself results in a lack of the insulin signal, leading to type 1 diabetes. Type 2 diabetes develops in situations where there is an insulin signal, but the response of tissues is blunted or absent due to a problem with the insulin receptor or a post receptor defect.

Insulin is first synthesized as a single polypeptide of 104 amino acids, which is cleaved multiple times in the ER to yield a single protein of 51 amino acids comprised of two polypeptide chains, termed A and B, held together by disulfide bonds. The protein itself adopts an α helical configuration. It is unclear if insulin forms higher order structures (trimers or hexamers) joined through a common metal ion (Zn^{2+}) while in the circulation.

3.4.7 Myosin, a molecular motor

Myosins are a family of proteins that act as molecular motors, using the energy of ATP hydrolysis to elicit muscle contraction and transport vesicles in the cell (**Figure 3.33**). Myosins have heavy and light types of chains. The myosin heavy chain consists of three domains, a bulky globular head, a flexible neck, and a lengthy coiled α helical tail. The head domain interacts with the cytoskeletal protein actin, and it moves the myosin down the strand to the positive end of the actin fiber. The neck of myosin acts as a hinge, providing the head with flexibility. The tail of myosin can interact with adapter or cargo proteins; it can also form large bundles with other myosin tails. The light chains of myosin wrap around the neck region of myosin, potentially to provide stability to this flexible region. The light chains of myosin are structurally conserved with the calcium binding protein calmodulin; they may also play a role in the binding of Ca^{2+}.

The mechanism by which myosin mediates movement starts with the binding of ATP to the active site of myosin; this elicits a conformational change that releases actin from myosin. As the ATP is hydrolyzed, the head of myosin is moved back, resulting in myosin binding elsewhere on the actin molecule. Release of phosphate from myosin causes a conformational change in myosin

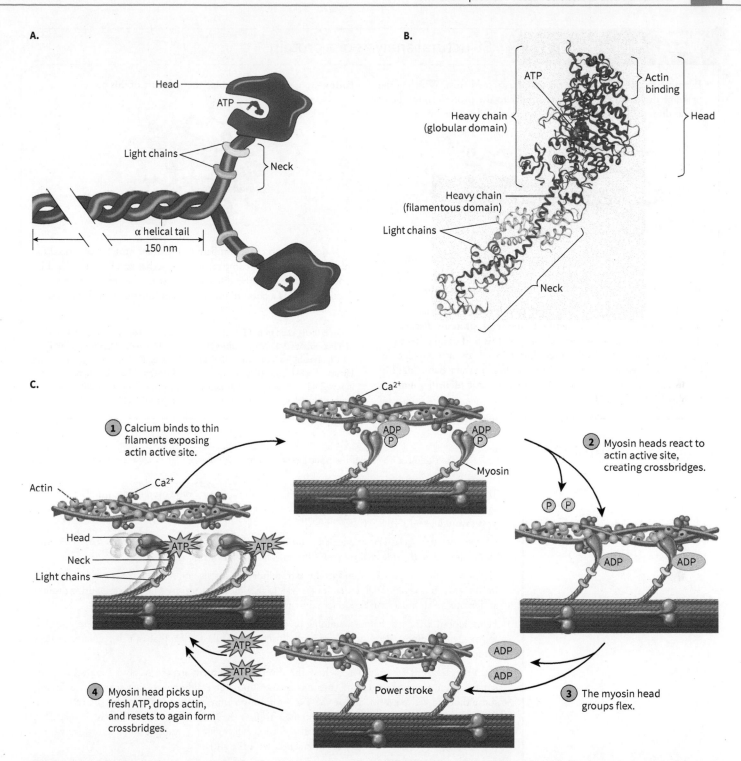

FIGURE 3.33 Myosin.   Myosin is an example of a molecular motor. **A.** and **B.** The heavy chain consists of a globular domain containing the actin binding region and ATP binding site and an extended filamentous domain. Myosin heavy chains bundle together joined by two other subunits (the light chains). **C.** Myosin functions in movement.

(Source: (A) Karp, *Cell and Molecular Biology: Concepts and Experiments*, 7e, copyright 2013, John Wiley & Sons. This material is reproduced with permission of John Wiley & Sons, Inc. (B) Data from PDB ID 1S5G Risal, D., Gourinath, S., Himmel, D.M., Szent-Gyorgyi, A.G., Cohen, C. (2004) Myosin subfragment 1 structures reveal a partially bound nucleotide and a complex salt bridge that helps couple nucleotide and actin binding. *Proc. Natl. Acad. Sci. USA* 101: 8930–8935. (C) Ireland, *Visualizing Human Biology*, 4e, copyright 2012, John Wiley & Sons. This material is reproduced with permission of John Wiley & Sons, Inc.)

so that the head rocks, moving closer to the tail in a movement termed "the power stroke." Finally, ADP dissociates from the active site.

When an animal dies, cells exhaust the supply of ATP. In the absence of ATP, myosin is unable to release, resulting in muscle rigidity (rigor mortis).

Worked Problem 3.4   Structural analysis of a protein

Examine the structure of the protein illustrated below. Which of the proteins discussed above is most similar to the protein shown here? What does this suggest about the function of the protein?

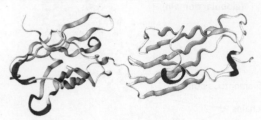

Follow-up question   Examine the structure of this protein.

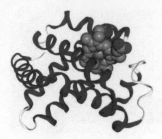

Strategy   How does the overall structure of the protein shown compare to the other proteins depicted in this chapter? Do you recognize any of the elements of secondary or tertiary structure, or any domains?

Solution   This image shows a protein with two domains enriched in β sheet. Comparing this structure to the seven proteins discussed in this section, or using software to compare this fold to all other structures determined to date, we see that it forms an immunoglobulin fold. This structure is deposited in the Protein Data Bank, and like all proteins there it has a four digit alphanumeric identifier number, such as PDB ID 1IGT.

How does this structure compare to the other proteins discussed in this section? Use your comparison to hypothesize a function for this protein. How would the differences in structure from the proteins discussed in this section result in different functions for this protein?

(Source: (Left) Data from PDB ID 1IGT Harris, L.J., Larson, S.B., Hasel, K.W., McPherson, A. (1997) *Biochemistry* 36: 1581–1597. (Right) Data from PDB ID 3WFT Hayashi, T., Morita, Y., Mizohata, E., Oohora, K., Ohbayashi, J., Inoue, T., Hisaeda, Y. (2014) Co(II)/Co(I) reduction-induced axial histidine-flipping in myoglobin reconstituted with a cobalt tetradehydrocorrin as a methionine synthase model. *Chem.Commun.*(Camb.) 50: 12560–12563)

Summary

- Proteins are the most diverse of the basic classes of biological molecules. They vary in size, structure, and function.

- Transmembrane proteins pass through the phospholipid bilayer. Some proteins are simply anchored in the membrane, whereas others, such as the aquaporins, form channels for specific molecules to pass through.

- Chymotrypsin is a classic example of an enzyme. Structural studies have made it possible to identify and characterize the side chains involved in intermolecular interactions and catalysis.

- Collagen, a fibrous protein found in the extracellular matrix, is one of the most prevalent proteins in the body. It is exceptional in that it possesses highly modified amino acids such as hydroxyproline and forms an unusual structure, referred to as the collagen helix.

- Hemoglobin is the red, heme-containing protein found in erythrocytes. Hemoglobin is responsible for oxygen binding and transport and is an illustration of allosterism; that is, it is regulated at sites distant from where the chemistry is occurring.

- Immunoglobulins are examples of how the diversity of protein structure solves interesting challenges. These proteins need to be able to bind to innumerable different foreign antigens, including some that are yet to evolve, and yet have a common portion that can interact with the cells of the immune system to elicit an immune response. To do this, IgG has regions that are highly variable linked to a domain that is common to all IgG molecules. Having multiple variable regions within the same IgG molecule increases the interaction with the antigen.

- Insulin is an example of a protein that acts as a signaling molecule. The relatively small insulin molecule is produced in the pancreas and travels through the bloodstream, signaling cells to increase uptake of glucose; it also acts as a growth factor and plays other roles in cell signaling.

- Myosin is a muscle protein that is partially responsible for muscle contraction. It is a type of molecular motor, using the energy of ATP hydrolysis to generate motion.

Concept Check

1. Described the proteins discussed in this section in terms of both their structure and function on a macromolecular level.

2. Describe the elements of secondary structure seen in these proteins.

3. How does the myosin use the chemical energy stored in ATP to generate motion?

3.5 Protein Purification Basics

Why might it be necessary to purify a protein? First, a pure protein may be needed for research purposes. The choice of whether to study a pure protein or one *in situ* is a fundamental question in biochemistry. The latter approach can provide critical data about cellular localization and interaction partners, but many scientific questions are difficult or impossible to answer using a protein in a mixture. For example, the protein *in situ* may have binding partners or regulators that we are not aware of. Second, purified proteins may be used as drugs or as additives in detergents, foods, and industrial processes. In such instances, it may be necessary to have a pure protein to optimize the effect of the drug or enzyme or for safety reasons.

As a purification proceeds, the number of components and contaminants decreases. Likewise, the steps proceed from crude (a physical separation of tissues) to intermediate (precipitation) to more precise (affinity chromatography). This section describes the basics of a purification scheme and the generalities of purification. The specifics of various chromatographic techniques are given in sections 3.6 and 3.7.

3.5.1 The first step of purifying a protein is to identify a source

The first consideration in protein purification is the source of the protein of interest. As with any purification process, some of the product is lost along the way, and sometimes the losses can be substantial, giving a low overall yield. Hence, if a large amount of protein is needed or the process has a low yield, a large pool of enriched material will be required as the starting point.

A variety of protein sources are used in protein purification, including microorganisms, plants, and animal tissue (**Figure 3.34**). In the past, material from larger organisms was used or large amounts of tissue from animals that were being used for other purposes; for example, for many years insulin was purified from pig pancreas obtained from slaughterhouses. Other proteins and enzymes were isolated from microorganisms that could be grown in large quantities such as *Escherichia coli* or yeast.

Advances in molecular biology have provided a wealth of options for sourcing protein. For example, will discuss in section 19.1, proteins can be overexpressed in bacteria, yeast, or other cultured cells, and the protein can then be purified from the cells themselves or from the growth media. This approach has numerous advantages. Growing microorganisms is often easier than obtaining large amounts of tissue from a large animal. Also, a protein typically found at low levels in a cell such as a transcription factor can be produced in abundance by overexpressing it in culture. To aid purification, the protein can be modified to introduce sequences that facilitate separation by affinity chromatography, discussed later in the chapter. Finally, human proteins that would otherwise be difficult or unethical to obtain can be expressed in culture.

A.

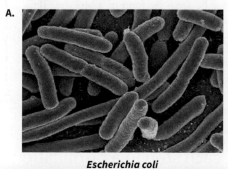

Cultura Creative Ltd / Alamy Stock Photo

Escherichia coli

B.

RBG

Dioscoreaceae mexicalis

C.

Kidneys   Adrenal glands

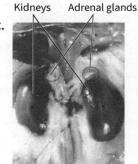

La Perle, K. M. D., & Dintzis, S. M. (2018). Endocrine System. Comparative Anatomy and Histology, 251–273. doi:10.1016/ b978-0-12-802900-8.00015-4

Mus musculus

FIGURE 3.34 Natural sources of biological molecules. **A.** The bacteria *E. coli* is a common source of proteins because it is easy to grow in large quantities. **B.** *Dioscoreaceae mexicalis,* a Mexican yam, is used as a source of steroid hormone precursors. **C.** The kidney and adrenal glands of a mouse are tissues sometimes used as protein sources, but their small size limits what can be obtained.

The use of such proteins in drugs is one of the exciting frontiers in medicine, but working with proteins presents some unique challenges for drug companies (see **Medical Biochemistry: Purifying proteins for use as drugs**).

Medical Biochemistry

Purifying proteins for use as drugs

Chapter 16 discusses the growing use of biologics, or protein molecules used as pharmaceuticals. The earliest and most classic example of a biologic is insulin, a protein hormone that was originally isolated from the pancreatic tissue of dogs, horses, or pigs for the treatment of diabetes. More recently, there has been an explosion in the number of biologics. Some of these molecules are recombinant versions of naturally occurring proteins such as recombinant erythropoietin (EPO) alpha, which stimulates production of red blood cells. Others are proteins that have been engineered in the laboratory such as the monoclonal antibodies Enbrel (etanercept), which treats rheumatoid arthritis, and Remicade (infliximab), used to treat Crohn's disease.

The formulation and purification of biologics has much in common with processes used for organics. When a small organic molecule is synthesized as a drug, common organic purification techniques such as distillation, recrystallization, and chromatography are used to ensure that the molecule is pure. These techniques are coupled with efficient, high-yield reactions to minimize the number of steps involved and maximize the yield. Each of these steps is optimized to produce the compound in the most efficient manner at the lowest cost. No matter how well designed and executed the process, every batch of a compound must undergo extensive quality assurance testing to verify that it has not been modified and is sufficiently pure. Once the active ingredient is produced, the drug is formulated. In this step, a dose of drug is combined with binder, dyes, flavoring agents, and carriers to produce the final product obtained at the pharmacy.

Because biologics are proteins, however, they present a unique set of challenges. First of all, biologics are made in a biological system (cells) using recombinant DNA technology. The DNA constructs that produce biologics are made and introduced into cells using techniques such as cloning, production of site-directed mutants and fusion proteins, or protein overexpression and transfection. Because cells grow, divide, replicate the DNA, and produce the biologics, techniques used to produce organic molecules are not needed.

The challenge for biologics is the scale at which cells must be grown to produce the protein of interest. To meet these demands, cells are grown in large bioreactors ranging from hundreds to thousands of liters in size. In these reactors, cell growth conditions such as concentrations of dissolved oxygen and glucose or pH are closely monitored to maintain optimal cell growth. The biotechnology industry has learned from the technology and practices of the brewing and dairy industries, both of which use microbes to produce a final product.

Gorodenkoff/Shutterstock.com

Because of how they are produced, biologics are not at all pure following production. Tens of thousands of contaminants are present, and the biologic may not even be the most prevalent protein in the sample. Isolating a protein molecule from a 10,000-L reactor does not require a different technique than organics, but it does operate at a different scale. A simple size-exclusion chromatography step might work well in the research laboratory, but it would not be feasible to scale up from either a size or cost perspective. Nevertheless, purification schemes are necessary to isolate the biologic away from the myriad other components of the cell. Even so, it is difficult to isolate a biologic product as pure as a small organic.

Furthermore, because biologics are proteins, they can be modified, for example, glycosylated and phosphorylated. The degree to which these modifications occur may vary based on the cell type and growth conditions. Hence, no two batches of a biologic can be said to be truly identical. It can be difficult, if not impossible, to measure some of these modifications, and there may be other modifications that are as yet unknown. The growth conditions of the cells producing the biologic must be monitored to minimize lot-to-lot variability and thus ensure consistency in drug production, patenting, and marketing.

Biologics are a unique product because of how they are produced and because they are somewhat uncharacterized in nature. As a result, there cannot ever be a generic form of a biologic. Thus, although competing companies may develop their own monoclonal antibodies to bind to tumor necrosis factor α, for example, each company's antibody will differ from others on the market. The FDA has not yet determined how to best regulate biologics from different sources with interchangeable functions.

3.5.2 Purification schemes detail each step in a purification

Purifying a protein requires a detailed plan of action called a purification scheme that summarizes and describes the steps involved (**Figure 3.35**). The investigator plans each step in the purification scheme based on which particular tools and techniques make sense for the protein under investigation, considering, for example, its size, location, charge, and solubility.

Measuring protein purity   As purification proceeds, the number of components and contaminants decreases. Depending on the intended use of the protein, a higher or lower degree of purity may be required. Because each step in the process involves loss of product, it is desirable to use the minimum number of steps to achieve the desired purity.

After each step in a purification, a sample is taken to test its purity. The purity of a protein is generally described in terms of percentages (for example, 99% pure), but protein purification uses a different measure. The **fold purification** is a measure of the amount of the protein of interest in the crude fraction divided by the amount in the final sample. This describes, for example, how a protein's purity may have increased by 100-fold or 1,000-fold during a certain step (not related to the term "protein folding"). Purity can be ascertained using a technique such as sodium dodecyl sulfate polyacrylamide gel electrophoresis (SDS-PAGE). In the case of an enzyme, an easier approach is to divide the activity (units catalyzed per unit time) of the enzyme in the crude fraction by its activity in the final sample. The **specific activity** (units catalyzed per unit time per milligram of enzyme) is another measure of enzyme purity.

Often, a table is constructed that indicates the amount of total protein and the relative purity of the protein of interest at each step of the purification process (Table 3.3). Purification tables are useful for both developing and describing a purification process. If a specific step does not enrich the protein of interest enough or results in a significant loss of activity or low yield, it should be reconsidered as part of the overall purification scheme.

A typical protein purification might enrich a protein more than 100,000-fold above what is found in a cell or biological mixture. In all cases, the fold purification or the yield is expressed relative to the original mixture rather than to the specific step of the purification. If the protein being purified has a large epitope tag that is cleaved in one of the later steps of the purification, resulting in an apparent change in yield or purity, this should also be noted in the table.

3.5.3 Purification begins with crude steps of separation and becomes more refined

Protein purifications typically begin with a series of crude steps, including extracting the protein from the sample and isolating it. This is usually followed by intermediate steps, including precipitation and

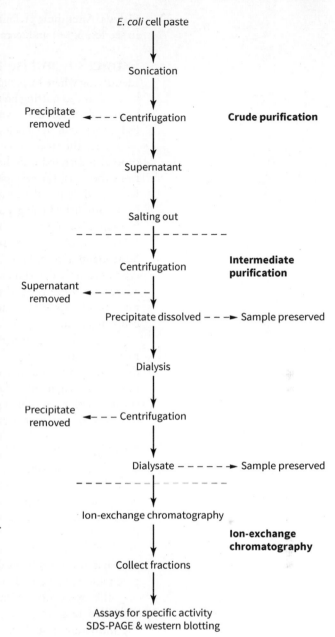

FIGURE 3.35 Protein purification scheme. This example of a protein purification scheme shows a protein expressed in *E. coli*. Among the many steps, the purification includes an intermediate purification step (salting out) and a later, more refined purification step (ion-exchange chromatography).

TABLE 3.3 An Example of a Protein Purification Table

Step	Volume (mL)	Protein (mg)	Activity (units)	Specific Activity (units/mg)	Purification (fold)	Yield (%)
Lysate	6,000	3,360	10,150	3.0	1	100
Ultrafiltration	500	2,374	8,900	3.7	1.2	87.6
Salting out	200	1,685	6,900	4.2	1.4	68.3
Ion exchange	200	312	3,200	10.3	3.4	31.6
Size exclusion	12.5	11	488	45.9	15.2	4.8

dialysis. After these preliminary steps come more refined chromatographic techniques (introduced in section 3.5.4) and, occasionally, a final "polishing" step to remove specific contaminants.

Extraction and isolation

The first step of protein purification is to extract the protein. Identifying where in an organism a protein is found means that the particular tissue or sample can be used as the starting point for the purification. For example, if a plant protein being studied is produced only in the leaves of the plant, the purification process would start with a sample of the leaf rather than the whole plant.

Once the tissue or sample has been obtained, the protein of interest must be extracted. If a protein is secreted into the cell culture medium, the first step would be to separate the medium from the cells, for example, using low-speed centrifugation. However, if the protein of interest is retained within the cells, it is first necessary to lyse the cells to release their contents. This can be accomplished using a process like homogenization or sonication. The various techniques for **homogenization** all involve disrupting the tissue to bring the sample into uniform distribution (**Figure 3.36**). For example, imagine blending various fruits with yogurt to make a smoothie. **Sonication** accomplishes a similar result by bombarding the sample with ultrasonic waves. The type of sample determines the best method of extracting the proteins, and each method has its advantages and drawbacks. Once the cells have been lysed, the cell components and organelles must be physically separated from the soluble proteins by techniques like filtration or low-speed centrifugation.

Centrifugation at higher speeds can be used to separate organelles of different densities that may contain a protein of interest. Common misconceptions are that proteins of different sizes differ in density and that proteins of higher molecular weight can be isolated from those of lower molecular weight through ultracentrifugation. In reality, although centrifugation can be used to isolate fractions, it cannot generally be used to separate proteins of different molecular weights, so this must be undertaken at a later step.

Precipitation and solubilization

In nature, proteins are generally found in a dilute salt solution. Making a substantial change to the salt concentration of a solution affects protein solubility, causing some proteins to precipitate in a process referred to as **salting out** (**Figure 3.37**). A sample of mixed proteins can be treated with a salt such as ammonium sulfate. At different concentrations of salt, different proteins become insoluble and precipitate. The precipitate and the liquid above it (the supernatant) can then be separated using centrifugation. The protein of interest may be in either the precipitate or the supernatant. If the protein of interest is in the supernatant, it can be used directly for the next step of purification. If the protein of interest has precipitated, however, it must first be redissolved.

The exact mechanism of salting out is unclear, but several probable scenarios have been described (**Figure 3.38**). When the salt concentration of a solution of proteins is increased, the individual ions found in the salt solution form hydration spheres, or shells of organized water molecules around each ion. As the salt concentration is further increased, more water is drawn from the overall solution, and less is available to form a shell of hydration around the proteins in solution. These hydrated proteins aggregate together and precipitate. Because proteins have different polar and charged surfaces, different proteins will precipitate at different

FIGURE 3.36 Initial steps in protein purification. Cells from starting materials often need to be broken open to release molecules of interest. This may involve homogenization, which is usually followed by centrifugation steps to remove unbroken cells and large organelles.

(Source: *Karp, Cell and Molecular Biology: Concepts and Experiments*, 8e, copyright 2016, John Wiley & Sons. This material is reproduced with permission of John Wiley & Sons, Inc.)

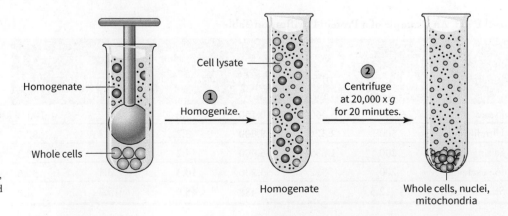

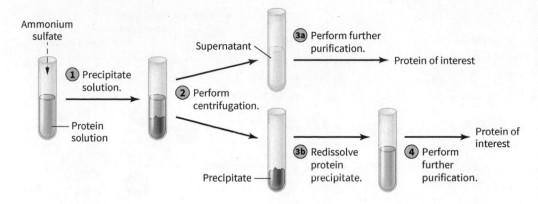

FIGURE 3.37 Salting out. A sample of mixed proteins in solution can be treated with a salt such as ammonium sulfate. As the salt concentration rises, different proteins precipitate according to their solubility. The precipitate can then be separated from the supernatant through centrifugation. If the protein of interest is in the precipitate, it can be dissolved into solution before further purification.

salt concentrations. A similar effect is seen when nucleic acids are precipitated out using ethanol or isopropanol.

Ammonium sulfate is often the salt of choice for precipitating proteins. The **Hofmeister series** orders anions and cations based on how they decrease or increase protein solubility (**Figure 3.39**). Ammonium (NH_4^+) and sulfate (SO_4^{2-}) ions are found at the end of this scale that decreases solubility (salting out), whereas guanidinium ($CH_6N_3^+$) and isothiocyanate (SCN^-) ions are found at the end that increases solubility (salting in).

Salting in (solubilizing) is the reverse process of salting out (precipitating). The addition of certain salts such as guanidinium chloride increases the solubility of some proteins and can assist in dissolving proteins that have already been precipitated. As with salting out, this phenomenon is thought to occur through the interaction of the ions of the salt with water molecules and the protein itself. Ions used for salting in also decrease protein stability and increase denaturation, so care must be taken to ensure that the protein of interest is not denatured through the direct interactions of the salt with the protein.

Dialysis Once a protein of interest (or the contaminating proteins) has been salted out, the salt must be removed from the sample. **Dialysis** is the classic means by which a protein solution can be either desalted or changed from one buffer to another without precipitating the protein of interest. The dialysis bag, typically made of cross-linked cellulose, is porous and retains the protein or macromolecule of interest while allowing smaller components such

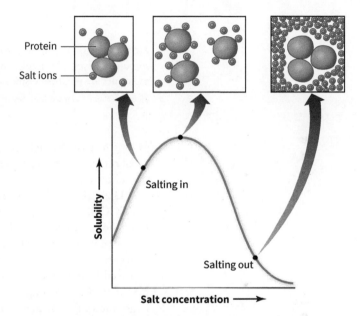

FIGURE 3.38 The theory behind salting out. At low or high concentrations of salt, proteins are less soluble than they are at physiological concentrations. At low concentrations, low ionic strength promotes hydrophobic interactions between the protein molecules, decreasing solubility. At high concentrations, the salt competes with the proteins for water molecules, decreasing the amount of solvent that is available to solubilize the proteins, also resulting in decreased solubility.

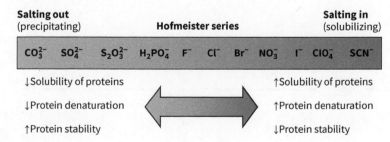

FIGURE 3.39 The Hofmeister series. Franz Hofmeister determined the ability of different cations and anions to salt in or salt out proteins. A partial list of ions is shown here.

(Source: From Bradley A. Rogers, Tye S Thompson, Yanjie Zhang, The journal of physical chemistry. "Hofmeister Anion Effects on Thermodynamics of Caffeine Partitioning between Aqueous and Cyclohexane Phases." Copyright © 2016 American Chemical Society.)

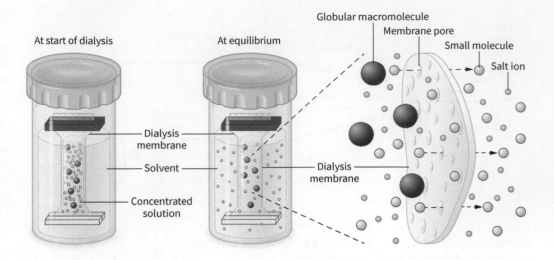

FIGURE 3.40 **Dialysis membranes.**   Dialysis membranes are typically cross-linked cellulose with specific pore sizes. Molecules smaller than the pore size can pass through the membrane, whereas larger molecules are retained. Osmosis drives the passage of water through a dialysis membrane.

(Source: (Right) Adapted from Separation Characteristics of Dialysis Membranes, Thermo Fisher Scientific.)

as ions to pass through the pores (**Figure 3.40**). When the bag is soaked in an excess of buffer, small molecules that can pass through the pores equilibrate across the membrane, diluting the salt concentration.

Dialysis membranes have different pore sizes, referred to as molecular weight cut-offs (MWCOs). For example, a membrane with an MWCO of 10 kDa has a pore size that will retain most molecules greater than 10 kDa but allow the passage of smaller molecules.

A more recent application of dialysis is **ultrafiltration** (**Figure 3.41**). In an ultrafiltration concentrator, a dialysis membrane is stretched across the bottom of a centrifuge tube. The sample is placed in the tube, which is then inserted into a larger catch tube, and the whole assembly is placed into a centrifuge. Water and small molecules pass through the dialysis membrane, but larger molecules are retained, concentrating the protein sample.

3.5.4 Protein chromatography separates proteins based on physical or chemical properties

Once proteins have passed through the crude and intermediate steps of the purification scheme, chromatography is often used to further purify them. The various chromatographic methods

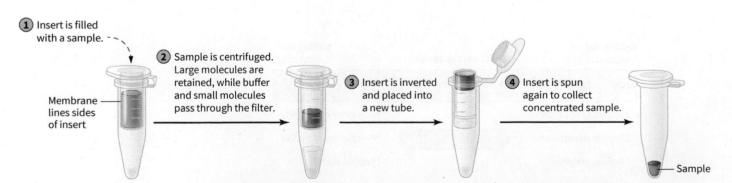

FIGURE 3.41 **Ultrafiltration.**   Ultrafiltration uses the gravitational field produced by a centrifuge to drive solvents and small molecules through a dialysis membrane. This effectively acts as a molecular colander, retaining larger molecules and allowing the passage of smaller ones.

(Source: Amicon Ultra-0.5 mL Centrifugal Filters for DNA and Protein Purification and Concentration, Merck KgaA.)

developed for the isolation of protein molecules are similar to other chromatographic techniques in that they separate components of a solution as they move through a medium at different rates. This section discusses protein chromatography in general; subsequent sections discuss two specific types of chromatography: size-exclusion chromatography and affinity chromatography.

Chromatographic media and columns

At the core of protein chemistry are solid chromatographic media, generally referred to as resins or beads (**Figure 3.42**). Chromatographic media are basically microscopic beads that form a support for separation chemistry to occur. The size of the beads (the mesh size) refers to the diameter of any particle, and the numbers are reciprocally related such that a larger number indicates a smaller particle size. For example, a resin with a mesh size of 100 has a particle size of at least 150 microns, whereas one with a mesh size of 400 has a particle size of at least 37 microns.

Generally, resins are placed in a chromatography column, typically a glass cylinder. At the bottom of the column is a plug of glass wool, a Teflon mesh, or a glass frit (a piece of porous glass) that will retain the chromatographic medium but allow the passage of the buffer and proteins. Columns have fittings at either end to allow the addition of buffer and sample to the top and collection of fluid from the bottom. The dimensions of the column vary depending on the requirements of the experiment.

There are alternatives to columns. For example, if a protein of interest binds to an affinity resin, it can be bound to the resin in a beaker or flask instead of physically being kept in a column, and the unbound proteins can be washed away using filtration. These steps can save time in purification; they can also lead to an increased yield by helping to preserve protein integrity and enzyme activity.

The mesh size of the resin affects the flow characteristics of fluid through the medium, regardless of the column type. Media composed of smaller beads have a greater surface area and therefore potentially a greater area for proteins to interact, but they also have a slower flow rate. A column appropriate for a particular application can be created by balancing mesh size against flow rate. A flow adapter at the top of the column prevents the column from running dry when a low flow rate is used.

The chromatography system

In addition to the resin and columns, a chromatography system also requires a means of feeding buffer and sample through the column, detecting what has been separated, and collecting samples (**Figure 3.43**).

Although it is possible to feed the liquid through a reservoir using gravity, a pump provides greater control over flow rate. These simple systems run at relatively low pressure (<30 psi). Fast protein liquid chromatography (FPLC) runs at somewhat higher pressures (up to 75 psi). The use of higher pressures and flow rates expedites separations, and these systems allow for some level of automation in the purification process. High-pressure liquid chromatography (HPLC) is often used in biochemistry to analyze lipids, metabolites, and small organics. It is less suitable for

A.

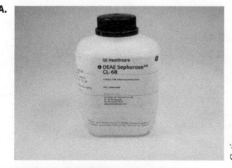

Cytiva

Chromatographic media

B.

Electron micrograph of porous beads

Porous beads from Özlem Bahadir Acikara (April 10th 2013). Ion-Exchange Chromatography and Its Applications, Column Chromatography, Dean F. Martin and Barbara B. Martin, IntechOpen, DOI: 10.5772/55744. Available from: https://www.intechopen.com/books/column-chromatography/ion-exchange-chromatography-and-its-applications

FIGURE 3.42 Chromatographic media. Chromatographic media (also called resin or beads) are microscopic particles of polystyrene, agarose, or acrylamide. **A.** Media are often supplied as a slurry and, if purchased that way, should not be allowed to dry out. **B.** This electron micrograph shows the porous chromatographic media.

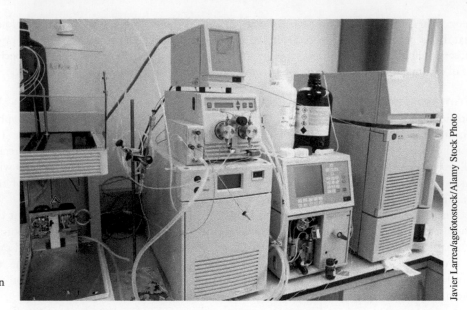

FIGURE 3.43 A chromatography system. A chromatography system uses a series of pumps that feed buffers and samples into the column(s) and a fraction collector to collect the final products. Often the entire system is computer controlled.

protein chemistry because the high pressure (up to 5,000 psi) necessitates the use of materials like metal tubing that are not compatible with protein purifications.

Whatever the method, as sample elutes (washes out) from a column, different fractions are collected in test tubes using a fraction collector, which changes the collection tube based on a set time (e.g., every 2 minutes) or volume (e.g., every 0.5 mL).

The next part of the process is to determine where proteins eluting from the column (particularly the protein of interest) can be found. Rather than running a protein assay on each column fraction, which would add time to the purification, the usual approach is to measure the UV absorbance of the eluent as it comes off the column. Tryptophan and phenylalanine residues, which absorb at 280 nm, can be used as a general marker for proteins, provided there are no other compounds in the buffer (such as some detergents) that absorb in the UV range. A UV spectrophotometer equipped with a flow cell (a cuvette through which the eluent passes as it flows from the column) can facilitate this process.

There are various ways to identify the point at which the protein of interest has eluted. Sometimes, the protein of interest is bound to a chromophore, in which case a visible-range spectrophotometer can be used to identify it. For example, proteins bound to heme or flavin groups and some metal-binding proteins have characteristic absorbances in the visible range. In other cases, a western blot or enzymatic assay of the fractions can be used.

This chapter discusses purification of large amounts of proteins (milligram quantities) for subsequent analysis, but many of these techniques can also be used at the analytical level. For example, an affinity resin used in a purification experiment could be used on a microscale to perform a pulldown assay, where proteins of interest are selectively precipitated with analysis of the sample through immunoblotting. Likewise, small size-exclusion columns can be used for desalting proteins or nucleic acids.

Worked Problem 3.5   Purification schemes

Lysozyme is a 14-kDa enzyme found in the whites of hen eggs. Without any other data, propose a general purification scheme for lysozyme.

Strategy   Where is the protein found in the egg? Based on the information given, what might be your first step in purifying the protein?

Solution   First, crack open the egg and separate the white and any membranes away from the shell and yolk. Next, perform a

salting-out step with 20% ammonium sulfate. Test the precipitate to see whether the lysozyme is there or in the supernatant. Assuming, for the sake of this example, that the lysozyme is still soluble, the next step would be to use chromatography to remove other contaminating proteins.

Follow-up question   How could the salting-out step in this purification scheme be optimized?

Summary

- Protein purification is the isolation of a protein from a complex mixture. Sources of material from which to purify a protein include plant and animal tissues as well as cultured cells.

- A purification scheme is the plan of steps that will be taken to purify the protein; it typically comprises multiple steps beginning from crude and progressing to more refined. A table of purification data is used to denote the purity of the protein at each step in the purification.

- Protein purification begins with the isolation of the organ or tissue containing the protein of interest. Tissues or cells are ground or disrupted to release proteins. Unbroken cells and organelles are removed through centrifugation. Salting out is the treatment of a protein solution with salt to precipitate some proteins selectively, which are then removed by centrifugation. Dialysis and ultrafiltration can be used to remove salts and other small molecules.

- Chromatography is often employed in a purification to separate proteins in solution based on how long it takes them to pass through media. Chromatography systems consist of pumps, columns, UV detectors, and fraction collectors.

Concept Check

1. Explain various methods for obtaining proteins for purification.
2. Describe how data are recorded in a protein purification table and which data are needed.
3. Describe a basic protein purification process, including centrifugation, salting out, dialysis, and ultrafiltration.
4. Describe the steps in chromatography for protein purification, including the components of a chromatography system and how chromatographic media are used.

3.6 Size-Exclusion Chromatography

Size-exclusion chromatography, also called gel filtration, is a common technique used to separate proteins based on size. The process can be complicated if the proteins involved have elongated conformations or awkward geometries that alter the interaction of the gel with the protein. This section describes the theory behind size-exclusion chromatography and introduces the concept of hydrodynamic properties, or the principles that govern how proteins behave in solution. It also discusses some of the practical aspects of this technique.

3.6.1 Size-exclusion chromatography uses the size and shape of a protein to achieve separation

The basis of size-exclusion chromatography is that molecules of different sizes are forced to elute from a medium at different times. The medium (or gel) in size-exclusion chromatography is comprised of microscopic beads containing millions of even smaller pores. The pore size of the gel is about the same size as the proteins being separated.

Molecules too large to enter the pores of the gel pass around the beads and elute in the **void volume**, or the volume of buffer that surrounds the beads. The beads and buffer together are called the **bed volume**. Small molecules are able to diffuse freely into the pores of the gel and therefore have a larger volume through which to move; hence, they are retained and exit the column last. Molecules of intermediate size may be able to enter the pores of the gel partially and are retained but to a lesser extent than the smallest molecules (**Figure 3.44**). The point at which a specific molecule elutes is termed its **elution volume**. This point can be visualized by a peak on a chromatogram.

The chromatographic medium   The medium used in size-exclusion chromatography is typically agarose or acrylamide beads, which are synthesized with a set percentage of polymer or crosslinker to produce differing degrees of porosity in the bead. The higher the percentage of monomer or the amount of crosslinker, the smaller the size of the pores in the bead. Resins with

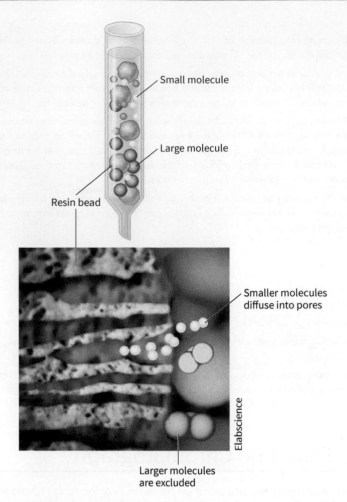

FIGURE 3.44 Size-exclusion chromatography.   Size-exclusion chromatography employs a glass column packed with size-exclusion (resin) beads. These chromatographic beads contain millions of microscopic pores. Smaller molecules diffuse into the pores and take longer to elute, whereas larger molecules are excluded and therefore elute earlier in the process.

small pore sizes will be able to resolve small molecules but not large ones. Therefore, the size of the pore needs to be matched to the size of the protein that is being purified.

The size of the pore is denoted using a numbering system. Small numbers (around 25 or 50) denote small pore sizes appropriate for removing salt, small molecules, or ions from proteins and nucleic acids (desalting them) or for separating single nucleotides away from the products of a polymerase reaction. Large numbers (up to 1,200) denote large pore sizes appropriate for separating macromolecular complexes such as lipoproteins.

Columns and flow rate   The length and shape of the column is another factor that affects the separation. Wider columns hold a greater volume of chromatographic media and can thus resolve more protein. Longer columns provide greater resolution. Typical columns are between 1 and 4 cm wide and between 10 and 75 cm long, although columns as short as 0.5 cm are used for desalting and as long as 2 m for some separations.

Flow rate also influences the separation, with a longer column length leading to a lower flow rate. A flow rate that is too high will limit the diffusion of the proteins of interest into the pores and thus decrease resolution. It can also lead to high pressures that crush the chromatographic medium, although this problem can be overcome by using agarose beads encased with a rigid polymer coating. In contrast, if the flow rate is too low, the sample will diffuse, resulting in broad peaks of eluted protein on the chromatogram.

In some instances, the manufacturers of a specific type of chromatographic medium will provide information about column geometry and flow rate, but usually this needs to be determined empirically.

3.6.2 Size-exclusion chromatography is based on hydrodynamic properties

The properties of a hydrated particle in aqueous solution, or its **hydrodynamic properties**, include viscosity, sedimentation coefficient, diffusion coefficients, and hydrodynamic radius. Size-exclusion chromatography is just one of many techniques that can be used to determine hydrodynamic properties (see **Biochemistry: Quantifying hydrodynamic properties of proteins**).

Biochemistry

Quantifying hydrodynamic properties of proteins

Various techniques are used to determine the hydrodynamic properties of a macromolecule and calculate hydrodynamic radius. In addition to size-exclusion chromatography, common techniques are sedimentation equilibrium (analytical ultracentrifugation), fluorescence anisotropy, and quasi-elastic light scattering.

Sedimentation equilibrium is an analytical ultracentrifugation technique that can be used to determine the molecular mass and dimensions of a macromolecule. In short, a pure sample of a macromolecule such as an enzyme is introduced into an analytical ultracentrifuge. There, an optical system measures how quickly it sediments in a gravitational field. Based on this value, the other properties of the molecule can be calculated, for example, the diffusion coefficient and molecular weight.

Fluorescence anisotropy employs fluorescence and the rotation of molecules to measure rotational diffusion coefficients. In fluorescence, a photon of light excites a fluorophore from the ground state to an excited state. Electrons stay in the excited state for a short period from microseconds to milliseconds, the lifetime of fluorescence. While the electron occupies the excited state, it can release some energy as heat or through other nonradiative processes, moving to a slightly lower excited state. From this lower state it collapses back to the ground state, emitting a new photon with less energy and hence a longer wavelength. If polarized light is used to excite the fluorophore, the emitted light should likewise be polarized. However, if the fluorophore is free to move, some of the polarization will be lost. This is the basis of fluorescence anisotropy. Furthermore, we quickly see that a small fluorophore will tumble and rotate more quickly than a large molecule, leading to differences in the loss of the polarized signal. These differences in polarization can be used to extrapolate back to determine the size of the molecule in solution, which provides a measure of hydrodynamic radius.

Quasi-elastic light scattering (QELS), also known as dynamic light scattering, relies on the ability of particles such as macromolecules to scatter visible light. When light makes contact with a particle, it is scattered randomly in all directions. Due to Brownian motion, particles are constantly moving throughout a sample, giving rise to fluctuations in the signal. As with other techniques mentioned here, smaller particles can diffuse more quickly than larger particles and produce differences in the amount of light scattered and how it is scattered. Data can be extracted from these signals to determine the size of the particle and the ratio of the long and short axes.

One way to envision hydrodynamic properties is to think of a series of balls commonly used in sporting events, such as a golf ball, a tennis ball, a soccer ball, a basketball, a rugby ball, and an American football (**Figure 3.45**). Clearly, these balls have different sizes and shapes and would thus roll or rotate differently. If we were to roll the golf ball and a basketball across a surface, the smaller ball would have to rotate more quickly to cover the same distance in the same time. If we were to roll the soccer ball and the American football across the same surface, they would roll differently due to their shape. The balls would also interact with their surroundings differently. If a baseball and a tennis ball were rolled across a hard surface, the tennis ball would stop before the smooth baseball as its soft felt interacted with the hard surface. This analogy illustrates how the size, shape, and structure of objects in part dictate their interactions with the environment.

FIGURE 3.45 An analogy for hydrodynamic properties. Balls of different sizes and shapes behave differently when rolled across a surface. This is analogous to how particles with different hydrodynamic radii behave differently in solution.

Richard Thomas/123RF

Hydrodynamic radius There are different ways to describe how a sphere might behave. For a sphere in solution, we could describe how it changes the viscosity of the solution through interactions with solvent molecules. For a sphere in an ultracentrifuge, we could describe how rapidly it diffuses through media (translational diffusion) or how quickly it rotates (rotational diffusion). These properties are determined in part by the size and shape of the molecule, which is reflected in a property termed the **hydrodynamic radius**.

A macromolecule can be represented with a spherical model. This is the simplest shape, and some molecules such as globular proteins are roughly spherical. In a sphere with an x, y, and z coordinate system running through the center, those coordinates are equal. However, crushing the sphere along one or two axes produces an oblate spheroid (a disc in which one axis is shorter than the other two) or a prolate spheroid (a cigar shape in which one axis is longer than the other two). By changing the ratios of the three axes, it is possible to manipulate the model of a protein structure into one of the three basic shapes that describe many protein molecules: spherical, oblate, or prolate (**Figure 3.46**). More advanced models can take other shapes into consideration, for example, proteins that are highly filamentous or shaped like a dumbbell or lollipop.

Size-exclusion chromatography does not depend on the size of a protein alone. To return to the ball analogy, if there were pores in the bead that could accommodate the pointed end of a football but not a soccer ball of equal overall size, the football might be retained on the column longer and elute in a greater volume than the soccer ball. Protein conformation as well as shape contribute to the hydrodynamic properties of a protein.

Uses for hydrodynamic data
The hydrodynamic properties of a protein are useful to an investigator. At their most basic level, these properties complement the higher-resolution structure-determination techniques of X-ray diffraction, nuclear magnetic resonance (NMR), and electron microscopy. Because hydrodynamic techniques are performed in solution and often at low concentrations, the data obtained are closer to biological conditions than is the case with a protein crystal. They can also be performed with a far smaller sample.

Hydrodynamic techniques may provide important data that might otherwise be missed. For example, a protein may crystallize as a tetramer but be found as a dimer in solution. In addition, if a protein forms a complex with other interaction partners or regulatory proteins, this is readily apparent in a hydrodynamic experiment, whereas it may not be apparent in other techniques. In another example, if a 60-kDa enzyme in the cell were found bound to a 40-kDa regulatory protein, SDS-PAGE would not be able to indicate whether the regulatory protein interacted with the enzyme or was simply a contaminant in a purification. However, size-exclusion chromatography would indicate that the protein was eluting from the column far earlier than would be anticipated from its molecular weight. Further analysis would be necessary to confirm this finding.

One example of how hydrodynamic techniques inform investigations is the protein bovine serum albumin (BSA), which is widely used in the laboratory. A primary function of BSA is binding and transporting free (unesterified) fatty acids in blood, but it can also transport many other hydrophobic molecules such as drugs, vitamins, and hormones. In the laboratory, BSA is used as a carrier protein to bind fatty acids or hydrophobic molecules or as a protein concentration standard. As an additive to restriction digests, BSA helps prevent degradation of restriction endonucleases.

Based on its amino acid composition, BSA has a molecular weight of 66,463 g/mol (66.4 kDa) and is strongly anionic, with an isoelectric point of 4.5. Multiple high-resolution

FIGURE 3.46 Simple protein shapes. Most proteins can be modeled using three basic shapes. A sphere has equal x, y, and z axes. In an oblate spheroid, one axis is longer than the other two, resulting in a disclike shape. In a prolate spheroid, one axis is shorter than the other two, resulting in a cigarlike shape.

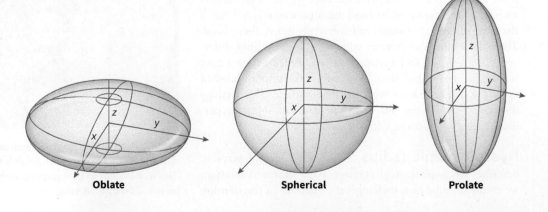

Oblate   **Spherical**   **Prolate**

A.

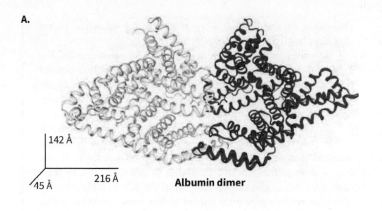

142 Å

45 Å   216 Å

Albumin dimer

B.

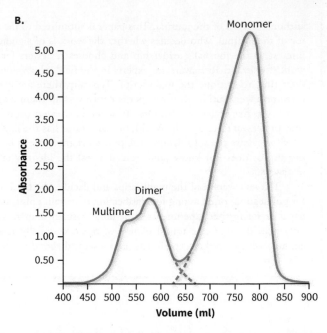

FIGURE 3.47 Size-exclusion chromatography of BSA. **A.** Albumin crystalizes as a dimer in which the two albumin molecules have a staggered overlap. **B.** Chromatography of BSA from a size-exclusion column shows three different peaks corresponding to monomers, dimers, and multimers of BSA.

(Source: Data from PDB ID 3V03 Majorek, K.A., Porebski, P.J., Dayal, A., Zimmerman, M.D., Jablonska, K., Stewart, A.J., Chruszcz, M., Minor, W. (2012) Structural and immunologic characterization of bovine, horse, and rabbit serum albumins. *Mol. Immunol.* **52:** 174–182)

(X-ray and NMR) structures of BSA have been solved, and the X-ray studies indicate BSA is a unit cell with dimensions of $216 \times 45 \times 142$ Å (**Figure 3.47**). Size-exclusion chromatography and sedimentation-velocity experiments give an approximate molecular weight of $66,700 \pm 400$ g/mol, close to the accepted value. However, in size-exclusion chromatography, aggregates of the protein (dimers and trimers) elute later. The concentrations of the monomer, dimer, and trimer can be manipulated in the laboratory by changing the ionic strength of the solution.

Hydrodynamic studies model BSA as a prolate spheroid with a major axis of 140 Å and minor axes of 40 Å ($140 \times 40 \times 40$ Å). When the dimer is examined using these same hydrodynamic techniques, a model can be derived in which two molecules of BSA overlap but are staggered by 50% with respect to one another. This model is not in perfect agreement with the crystal structure, but it is clearly close in one of the three dimensions. Perhaps most fascinating is that these hydrodynamic techniques were used in 1968 to determine many of the various features of the complex (overall size, molecular weight, shape, and predicted structure of the dimer), whereas the X-ray structure was not determined until 44 years later. Examples like this one demonstrate how research builds on past results to gradually create a more complete scientific picture. This picture will only be reliable, however, if all the results are accurate. One check against the publication of inaccurate data is peer review (see **Societal and Ethical Biochemistry: Peer review**).

Societal and Ethical Biochemistry

Peer review

Training in the scientific method often focuses on choosing a hypothesis or experimental design. These are worthy goals and key parts of the scientific method, but one of the final steps of the scientific process is often overlooked. This is the dissemination or sharing of results. It does little good to find a cure for a dread disease if you do not share your findings with the greater scientific community.

Furthermore, you may be wrong. There may be some aspect of your experiment that you have overlooked, or you may have misinterpreted some piece of data. Although not perfect, peer review is the single best means to validate findings.

Peer review is the process through which scientists submit their work to check that it is worthy of publication. Scientists wishing to submit their work to a peer-reviewed journal prepare and format a manuscript (a paper) according to the "instructions to authors" section of the journal. This paper is submitted to the editor of the journal, who decides whether the work is of significant interest to the journal's readership and chooses reviewers for the work if relevant. Reviewers are experts in the field who volunteer their time to critique the manuscript. Typically, there are two or three reviewers, although if a paper is controversial, there may be as many as five reviewers. Authors can suggest reviewers (or people they prefer *not* to review the work), but the editor has the final say in who reviews the submission. The process is anonymous in that the author does not know (and never knows) the identity of the reviewers.

The reviewers read the manuscript and decide whether it is fit for publication, often asking for clarification of certain points in the paper or for further experiments to be done. Usually, a manuscript only goes through one round of review, but occasionally reviewers may ask for further experiments and a second round of review.

Depending on the journal, acceptance rates for submitted papers vary from less than 10% to over 90%.

Modified forms of peer review are used by some journals. The Public Library of Science (PLoS) suite of journals employs a model in which submitted manuscripts are peer reviewed, but the significance of the work is decided in the public domain through comments and citations of the work rather than by the journal editor.

Recently, there have been several examples of the breakdown of the peer review process. For example, manuscripts have been accepted that were simply randomly generated words or, in one case, a short phrase repeated thousands of times. Clearly, in these cases there was negligence on the part of the editors and reviewers. Although reviewing journal articles makes some demands on an investigator's time, it is a professional responsibility and at some level an honor. Peer review only works if reviewers are professional and put forth their best effort in reviewing.

Worked Problem 3.6   Data discrepancy

An enzyme analyzed with SDS-PAGE produces the following data:

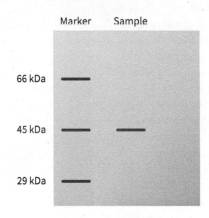

Before being subjected to SDS-PAGE, the sample was isolated with size-exclusion chromatography and produced the following elution data:

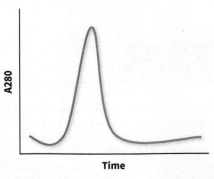

The peak eluted close to a protein with a molecular weight of 275 kDa. Based on these collective data, what can you deduce about the protein?

Strategy   What do each of these techniques actually measure? How is the analysis different in each technique?

Solution   SDS-PAGE shows a single band migrating at about 45 kDa. In contrast, the size-exclusion data indicate that the protein has a molecular weight of about 275 kDa. This is unlikely to be due simply to conformation; instead, it is likely that this protein is a multimer. Five 45-kDa subunits would have a molecular weight of 225 kDa, but six 45-kDa subunits would have an expected molecular weight of 270 kDa, closer to the 275 kDa observed. Based on these data, we would hypothesize that the protein in solution is a hexamer of 45-kDa subunits. Other techniques (light scattering, non-denaturing gradient gel electrophoresis, or X-ray structure determination) would be necessary to substantiate this hypothesis.

Follow-up question   Could size-exclusion chromatography be used to distinguish between folded and unfolded forms of a protein? Explain your answer.

Summary

- Size-exclusion chromatography separates proteins based on size and shape using a matrix of agarose or acrylamide beads with pores of a specific size. Smaller proteins are able to diffuse into the pore and therefore move through the column slowly, whereas larger proteins move through the column more quickly. Variables include pore size, mesh size, column geometry, and flow rate.

- The properties that describe how proteins behave in solution, known as hydrodynamic properties, include viscosity, sedimentation coefficient, rotational diffusion coefficient, and hydrodynamic radius. Hydrodynamic properties provide important information about the overall conformation of a protein in solution. Hydrodynamic techniques can reveal data that otherwise might be missed, such as whether a protein is found in a higher-order structure (a dimer, trimer, or multimer) in solution.

Concept Check

1. Explain how size-exclusion chromatography works, describing the different variables available to modify the experiment.

2. Explain what the hydrodynamic properties of a protein or complex can indicate about that molecule that may be missed using other techniques.

3.7 Affinity Chromatography

Affinity chromatography takes advantage of weak interactions between the molecule of interest and the binding resin. Molecules with little or no interaction with the resin do not bind and instead wash off (elute). The molecule of interest binds to the resin and can then be eluted through one of several different means specific to the interaction.

3.7.1 Affinity chromatography takes advantage of numerous weak forces to separate molecules

Weak forces are critical in affinity chromatography (**Figure 3.48**). These are the same forces that are at work when a protein folds, when an enzyme binds to a substrate, or when a protein interacts with another protein or molecule in a biological system. The different weak forces all contribute to the types of protein–ligand interactions discussed here and throughout this book. If necessary, review the discussion of weak forces from Chapter 3 before continuing.

Hydrogen bonds Perhaps the predominant driving force in biological systems is hydrogen bonding, which occurs between the lone pair of electrons on an electronegative atom (an oxygen or nitrogen) and a polarized hydrogen that is attached to an electronegative atom (an OH or NH group). Recall that hydrogen bonding is strongest when groups are arranged in a linear fashion and have a bond length of 2.2 Å. Water forms hydrogen bonds well, as do most of the polar and charged amino acids and the peptide backbone. Hydrogen bonding of water molecules is in large part what drives the hydrophobic effect, discussed in a later section.

London dispersion forces All atoms employ London dispersion forces, that is, forces that arise

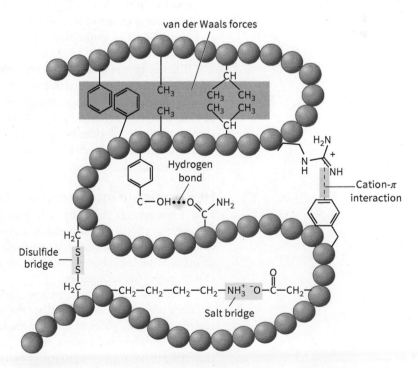

FIGURE 3.48 Forces and bonds that stabilize proteins. Shown are examples of some bonds and forces that stabilize the tertiary and quaternary structure of proteins.

through transient interactions of electron clouds of all molecules with each other. These forces are 100 to 1,000 times weaker than hydrogen bonds, and they are the predominant weak intermolecular force in hydrocarbons, including the tails of membrane lipids.

Salt bridges   Salt bridges, also known as ionic or coulombic interactions, occur between two isolated ions. The strength of these interactions obeys Coulomb's law, meaning that it varies with the inverse square of the distance (r) between the two charges (q_1 and q_2) and the dielectric constant of the media in which the ions are found (D):

$$F = k\left\{\frac{q_1 q_2}{r^2 D}\right\}$$

In this expression, k is a constant that varies with the units used. Because water has one of the highest dielectric constants ($D = 80$), ionic interactions are relatively weak in water compared to nonpolar solvents. However, the interactions can be much stronger if two charges are found on either side of a protein, separated by the hydrophobic core of the protein. Interactions are also stronger when charges are on two different proteins brought together in such a way that water is excluded from the protein–protein interface.

Dipole–dipole interactions   Dipole–dipole interactions occur between two standing dipoles, such as two carbonyl groups. In this case, the electronegative oxygen can be thought of as having a partial negative charge and the carbonyl carbon as having a partial positive charge. These two "charges" attract each other through space like a coulombic interaction, although the charges are standing dipoles rather than integer charge units. Like other coulombic interactions, dipole–dipole interactions are optimized at shorter distances and by less polar solvents. The peptide backbone has several groups that have standing dipoles, as do the side chains for asparagine and glutamine, among others.

Cation–π interactions   A cation–π interaction forms between a positively charged ion and any π-bond system containing double or triple bonds. Triple bonds are rare in biochemistry, but double bonds and cations are common. Tryptophan, tyrosine, and phenylalanine all have delocalized π clouds of electrons and participate. Histidine can participate as a π cloud donor when found in the deprotonated state or as a cation when protonated. On average, a protein has nearly as many cation–π interactions as it does salt bridges (8 to 12 in an average size protein), and over 25% of all tryptophan residues noted in the Protein Data Bank participate in some sort of cation–π interaction.

The strength of a cation–π interaction is similar to that of a hydrogen bond. These bonds are strongest when groups are within 6 Å of one another, which leads to an interesting phenomenon. Because this distance is large enough to accommodate another molecule, cation–π interactions, like salt bridges, can function at a distance. That is to say, an interaction can still be observed when there are other molecules between the cation and the π cloud. This is in contrast to hydrogen bonds, which are only effective over shorter distances, and covalent interactions, which require orbital overlap.

Hydrophobic interactions   Hydrophobic interactions can be viewed in different ways. First, from general chemistry, it is clear that all molecules can exhibit London dispersion forces. Second, the effects of π–π stacking, although poorly understood, are found when molecules with clouds of delocalized π orbitals (aromatic rings) stack together. This is a relatively rare interaction in most of biology, but it may help to stabilize nucleic acids and could be employed in some chromatographic techniques.

Although hydrophobic groups exhibit these interactions, the hydrophobic effect itself is a different phenomenon. It describes the clustering of hydrophobic molecules or amino acid side chains due to the gains in entropy made through freeing water molecules from having to cage multiple hydrophobic groups. Although the term "hydrophobic effect" may bring to mind the clustering of hydrophobic groups, the effect itself is one of water and entropy rather than enthalpy.

Metal chelation effects   Transition metal ions with charges of +1 to +3 are common cofactors found in proteins, but chelation effects can also be used in protein purification.

Histidine, cysteine, and anionic amino acids are all involved in the binding of metal ions. Metal ions can form as many as 12 bonds, although four and six are more common in biological systems, with the geometries being tetrahedral (for a center with four bonds) or octahedral (for a center with six bonds). Heme is an example of a metal ion (Fe^{2+}) that forms six bonds. The central iron ion is chelated in four positions by the porphyrin ring and in the fifth position by a histidine residue, whereas oxygen binds to the sixth position. In each instance, the ligand acts as an electron donor (Lewis base) and the metal ion as the electron acceptor (Lewis acid) in the formation of these complexes. In addition to the binding of metal ions in a protein or cofactor, in some instances amino acid side chains can form complexes with metal ions, a property that is used in metal affinity chromatography.

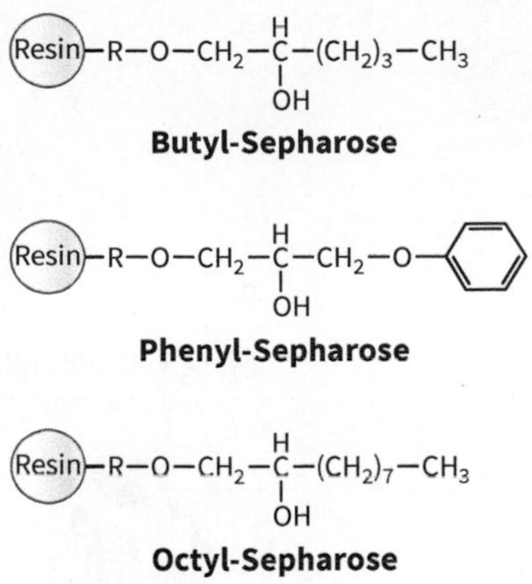

3.7.2 Different resins are used to separate proteins

Proteins have functional groups that can exhibit all of the weak forces noted in the previous section. Resins have been synthesized to have different functional groups capable of forming weak interactions with the amino acid side chains of proteins. No two proteins have the same amino acid sequence or tertiary structure, and chromatographic separations exploit the differences in sequence and structure to separate different proteins.

Hydrophobic interaction resins
In **hydrophobic interaction chromatography (HIC)**, the resin is coated with hydrophobic groups. Typically, these are butyl groups (four carbons long, denoted C-4) or octyl groups (eight carbons long, denoted C-8). Increasing the length of the carbon tail attached to the resin increases the hydrophobicity of the column. Proteins bind differentially to the column, depending on the strength of the hydrophobic interactions between the proteins and the resin. It is not possible to determine precisely which residues or regions of the protein will interact with the column, but it is possible to predict that a hydrophobic peptide or protein will interact more strongly with the resins than a hydrophilic protein. A highly specialized hydrophobic resin used in HIC is a phenyl column (**Figure 3.49**).

Proteins are typically bound to an HIC resin under high-salt conditions (up to 3-M NaCl), and unbound proteins are eluted using the same salt solution. At high ionic strength, water molecules are associated with salt ions, and the hydrophobic groups of the protein have a stronger interaction with the resin. The protein of interest is then eluted using a buffer of decreasing ionic strength. The change in buffer composition can either be an abrupt, stepwise change that is accomplished by simply switching from one buffer to another, or it can be gradual via a gradient. Gradients can be generated using a gradient mixer (a pair of connected chambers containing buffers of different ionic strengths) or a gradient pump system (a pair of computer-controlled pumps that work differentially to manufacture the gradient).

Reverse-phase chromatography is a technique similar to HIC that is used in analytical chemistry and organic chemistry. HIC differs from reverse-phase chromatography in that the density of hydrophobic groups on the resin in HIC is lower, so the interactions are weaker, and proteins can be eluted using less harsh conditions, for example, salt gradients rather than hydrophobic solvents.

Ion-exchange resins
Ion-exchange chromatography (IEX) uses coulombic effects to separate proteins. All proteins have an **isoelectric point**, or a pH at which the protein is electronically neutral, with the positively charged residues equaling the negatively charged ones. This pH value can be close to neutral (7), or acidic (down to pH 4 or 5), or basic (up to pH 9 or 10). A protein with an acidic isoelectric point contains more acidic (anionic) residues than basic (cationic) ones, and at neutral pH it has a net negative charge due to the deprotonated carboxyl groups. Conversely, a protein with a basic isoelectric point will have a net positive charge at neutral pH due to protonated amine groups. Proteins with a net negative charge will bind to a positively charged resin, whereas proteins with a net positive charge will bind to a negatively charged resin.

The most commonly used resins in IEX are those with either a diethylaminoethyl (DEAE) or a carboxymethyl (CM) modification (**Figure 3.50**). The DEAE resin contains a positively charged amino group and is called an **anion-exchange resin** because the anions bind to the cationic resin. The CM resin bears a negative charge due to the carboxyl group and is termed a **cation-exchange resin** because the cations bind to the anionic resin. Both CM and DEAE resins

FIGURE 3.49 Hydrophobic groups. Butyl-, phenyl-, and octyl-Sepharose can be used as column media in hydrophobic interaction chromatography.

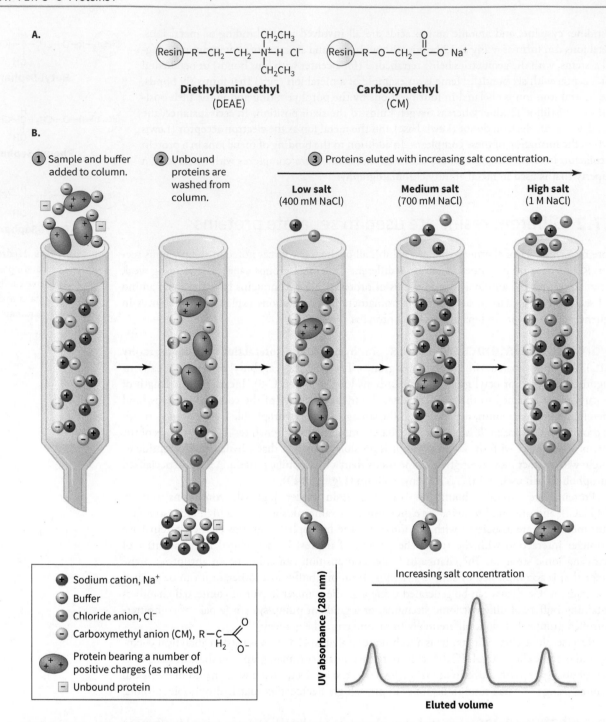

FIGURE 3.50 Ion-exchange chromatography. **A.** Diethylaminoethyl (DEAE) and carboxymethyl (CM) are used as column media in IEX. **B.** In IEX, proteins can either be eluted through stepwise changes in the salt concentration of the buffer or through the use of a continuous gradient of salt concentrations.

(Source: (B) From Protein Purification by Ion-Exchange Chromatography, REACH Devices, LLC. Reproduced with permissions of REACH Devices, LLC.)

are considered weakly binding resins because of their low charge density. Newer, stronger binding resins contain quaternary amines or sulfonates.

Proteins are typically bound to an ion-exchange resin in an aqueous solution with low ionic strength and at a pH value that optimizes binding of the protein of interest to the resin. Elution is accomplished using one of two methods. One approach is to change the pH of the buffer in order to protonate or deprotonate amino acid side chains, resulting in release and elution of the protein of interest. The other approach is to change the ionic strength of the buffer; this results in competition between ions in the buffer and the charged groups on the protein, employing chemistry similar to the common-ion effect. As with other types of affinity chromatography, the buffer can be changed via abrupt, stepwise changes, or it can be altered using a gradient.

Metal affinity resins

The techniques of **metal affinity chromatography (MAC)** or immobilized metal affinity chromatography (IMAC) take advantage of the interactions between specific side chains (most often histidine residues) and metal ions bound to a column (**Figure 3.51**). The resin used is often coated in Ni^{2+}, Co^{2+}, Zn^{2+}, Cu^{2+}, or some combination of metals. The metal ions are bound to the resin through iminodiacetic acid (IDA), carboxymethyl aspartate, or nitrilotriacetic acid (NTA). Native histidine residues found on the surface of a protein may be used in some instances; however, the technique is more commonly used with engineered proteins. Such proteins have been engineered to have a histidine epitope tag (for example, the 6-His tag), and tags as long as 14 histidines have been used successfully.

In these interactions, the basic imidazole ring of the histidine interacts with the metal ion. As noted earlier, this is an example of Lewis acid–base chemistry and a covalent coordinate bond.

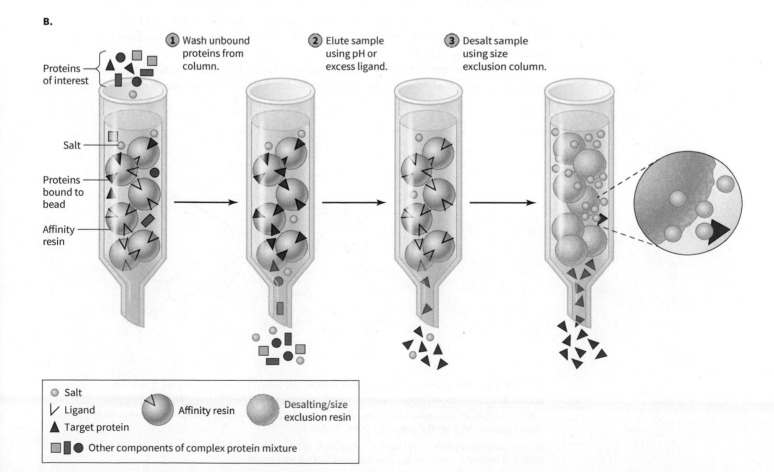

A.

FIGURE 3.51 Metal affinity chromatography. **A.** In metal affinity chromatography, divalent metal ions, such as Ni^{2+} and Cu^{2+}, are chelated to a column through nitrilotriacetic acid (NTA) linking groups. The imidazole groups of histidine side chains can bind to these metals. This interaction is often increased by adding a chain of four to eight histidine residues onto a recombinant protein. **B.** In affinity chromatography, a mixture of molecules is passed over an affinity resin. The molecule of interest binds to the resin, while others are washed off. The protein of interest is then eluted using either a change in pH, salt concentration, or additional ligand. In the case of metal affinity chromatography, the eluent is frequently imidazole. The molecule used to elute the protein can be removed through a desalting column (a gel filtration column).

(Source: (A) From What You Need to Know About NTA and IDA Ligands, Aug 18, 2014, Geno Technology Inc.; (B) From Introduction to Affinity Chromatography, Bio-Rad Laboratories, Inc.)

B.

① Wash unbound proteins from column.

② Elute sample using pH or excess ligand.

③ Desalt sample using size exclusion column.

Proteins of interest

Salt

Proteins bound to bead

Affinity resin

● Salt

⋁ Ligand　Affinity resin

▲ Target protein

■ ▮ ● Other components of complex protein mixture

Desalting/size exclusion resin

The strength of these bonds is partially due to the direct interaction of the two groups (the lone pair on the imidazole nitrogen and the metal ion), and the interaction is modulated by the effects of solvent or groups bound to the interacting pair. These bonds may seem somewhat weak or less stable than other covalent bonds, but they can be relatively strong compared to some of the weak forces used in other forms of affinity chromatography.

Samples are typically eluted from a metal-affinity column using an excess of imidazole (50 mM). The free imidazole outcompetes the protein through mass action, and the protein is released. The column can be regenerated by washing off the imidazole under acidic conditions. Another method for eluting the protein is to lower the pH of the buffer below the pK_a of the histidine side chain. This results in protonation of the imidazole pyrrolic nitrogen and disruption of the imidazole–metal ion complex. Finally, chelating agents that bind metal ions such as EDTA can be used to release the protein, although this strips the metal ions from the column and leaves the chelating agent and bound metal ion as a contaminant in the eluate.

3.7.3 Proteins can be used to bind to a molecule of interest

Interactions between proteins and some small ligands can also be used to bind to molecules of interest. Resins are available that are either already bound to avidin, antibodies, or lectins, or that have chemistry available to bind a protein of the investigator's design.

Biotin and avidin binding pairs   Biotin is a small organic cofactor whose involvement in several different reactions. Biotin binds to carbon dioxide and activates it in several carboxylases, including acetyl-CoA carboxylase (ACC) in the production of malonyl-CoA during fatty acid biosynthesis and pyruvate carboxylase in the production of oxaloacetate in gluconeogenesis. Biotin is bound tightly by two proteins: the avian protein avidin and the bacterial protein streptavidin. It is thought that these proteins bind avidin to act as an inhibitor of bacterial growth.

The binding of avidin to biotin has been well studied and has one of the highest affinities found in nature, with a dissociation constant on the order of 10^{-15} (**Figure 3.52**). The interaction of avidin and biotin is not as simple and straightforward as that found in the affinity resins discussed in the last section. Instead, the pair interacts as an enzyme would bind to its substrate or a ligand to its cognate receptor. Numerous weak forces combined with steric effects contribute to avidin–biotin binding.

There are several ways the avidin–biotin system can be used in a purification scheme. For example, a fusion protein can be created in which a protein of interest is expressed joined to

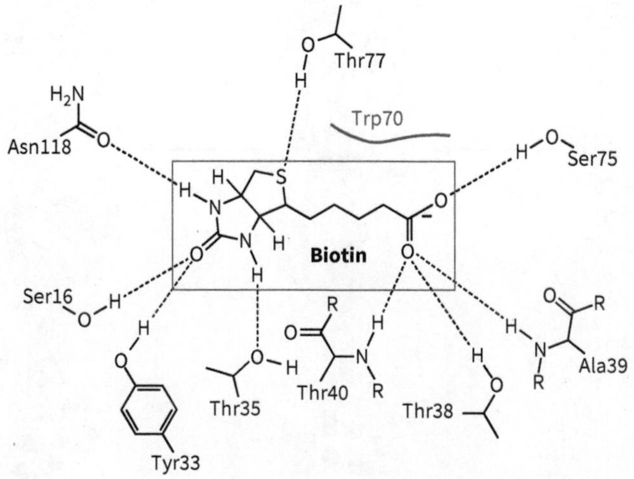

FIGURE 3.52 Interactions of biotin with avidin.   Biotin is an organic cofactor that binds to the protein avidin. The figure shows biotin interacting with amino acid residues in the avidin-binding site.
The binding of biotin and avidin is one of the strongest protein–ligand interactions known, with a dissociation constant (K_d) on the order of 1×10^{-15} M.

(Source: Data from PDB ID 2AVI Livnah, O., Bayer, E.A., Wilchek, M., Sussman, J.L. (1993) Three-dimensional structures of avidin and the avidin-biotin complex. *Proc. Natl. Acad. Sci. USA* **90:** 5076–5080)

avidin; this fusion protein can then bind to a biotin-coated resin. Likewise, it is a relatively easy matter to couple biotin chemically to a protein (termed biotinylation) and then purify the protein using an avidin-coated resin.

The avidin–biotin pair is used in several other instances. For example, the small size of the biotin modification means that proteins with a biotin tag will probably not be too significantly modified and will still interact with other partners in the cell as they would in the native, unmodified state. Avidin beads can then be used in a pulldown assay to separate out selectively the protein of interest and any interacting partners. These proteins can be subsequently identified using immunoblotting or mass spectrometry.

Immunoaffinity chromatography Immunoglobulins, or antibodies, are a class of protein molecules made by the cells of the immune system that recognize foreign structures termed antigens. These highly specific interactions can be exploited in immunoaffinity chromatography (**Figure 3.53**). Immunoglobulins can be assembled into several different higher-order structures, the simplest and most common of which is immunoglobulin γ (IgG).

IgG is a Y-shaped molecule that is composed of two heavy and two light chains joined by disulfide bridges. As mentioned previously, antibodies are very different from other proteins. The base of the Y is common to all antibodies and is termed the Fc region (the constant region). At the upper ends of the Y is the hypervariable region, which varies from one antibody to another. This region (called the Fab region for "antigen binding") is formed through a unique recombination process termed VDJ recombination (discussed in Chapter 14), which results in each cell of the immune system coding for a specific antibody. IgG is largely formed of beta-sheet structures, but the hypervariable region consists of six loops formed from both the heavy and light chains. These loops are short (3 to 12 amino acids in length), and they provide the basis for recognition and binding of the antibody to antigen. Although some level of canonical structure has been

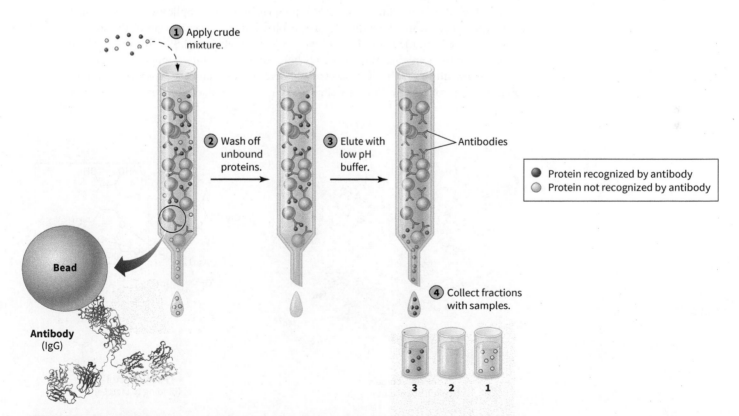

FIGURE 3.53 Immunoaffinity chromatography. Antibodies covalently coupled to agarose beads can be used in an affinity chromatography column. As with other affinity techniques, a mixture of proteins is applied to the column. Proteins that are recognized by the antibody bind to the column, while others are washed away. The protein of interest is typically eluted through a change to a lower pH.

(Source: Data from PDB ID 1IGT Harris, L.J., Larson, S.B., Hasel, K.W., McPherson, A. (1997) Refined structure of an intact IgG2a monoclonal antibody. *Biochemistry* **36:** 1581–1597)

proposed for these loops, their composition is largely random. Therefore, virtually all the weak forces involved in protein structure and protein–protein recognition also contribute to antibody–antigen interactions.

There are numerous permutations of immunoaffinity chromatography. Antibodies can be coupled to agarose or acrylamide beads and used either in column chromatographic separations or (in small amounts) in precipitation assays.

Because of the diverse nature of the antibody–antigen interaction, there are several means of dissociating the pair. The most common method is to decrease the pH. A change from pH 7 to 3 will usually result in release of the protein of interest. This method has the advantage of using relatively gentle conditions and not introducing any contaminants into the purification process. Proteins can also be eluted by competing off the protein of interest with a peptide that the antibody recognizes. Often, this can be the same peptide that was used to generate the antibody in the first place. This elution method has the advantage of using mild conditions but the disadvantage of introducing excess peptide into the elution mix. A final means of elution is to dissociate the antibody using a sulfhydryl reducing agent such as β-mercaptoethanol. Because the structure of the antibody depends on the disulfide bonds holding the heavy and light chains together, dissociation of those bonds will result in loss of antibody structural integrity. This is the most stringent of the dissociation conditions and results in the destruction of the antibody.

Other types of affinity chromatography

Several other types of affinity chromatography are used to varying degrees. Generally, these resins bind through the action of multiple weak forces, similar to an immunoaffinity resin or avidin binding to biotin.

Lectins are a group of plant proteins that bind tightly to specific carbohydrate moieties (**Figure 3.54**). Resins coated with these proteins were commonly used to purify proteins before the increased availability of immunoaffinity columns. Lectin-coated resins are still available and used for specific applications. The ability of these resins to bind to a protein of interest must be determined empirically.

Affi-Gel blue is an affinity resin that is popular in specific applications. It is formed from an agarose bead coated in a blue dye (Cibacron Blue F3GA). Examination of the structure of this dye gives some indication as to how it binds to proteins (**Figure 3.55**). The dye has hydrophobic aromatic groups, is anionic, and potentially binds to proteins using a range of these forces. As with other generic affinity resins, it is necessary to determine empirically whether the protein undergoing purification will bind to the resin; however, Affi-Gel blue is popular for another reason. As

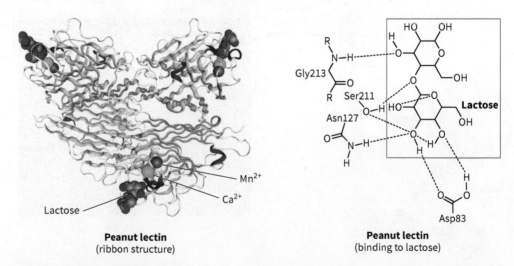

Peanut lectin
(ribbon structure)

Peanut lectin
(binding to lactose)

FIGURE 3.54 Lectin chromatography. Although other techniques have largely replaced lectin chromatography, it is still commonly used in some applications. Peanut lectin is shown binding to lactose as well as calcium and manganese ions.

(Source: Data from PDB ID 2PEL Banerjee, R., Das, K., Ravishankar, R., Suguna, K., Surolia, A., Vijayan, M. (1996) Conformation, protein-carbohydrate interactions and a novel subunit association in the refined structure of peanut lectin-lactose complex. *J.Mol.Biol.* **259:** 281–296)

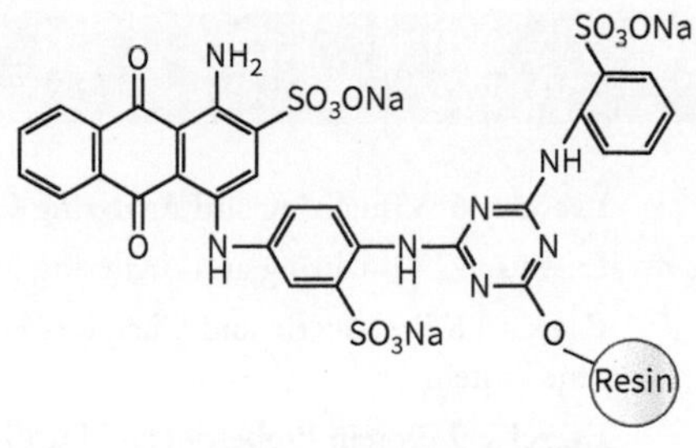

Cibacron Blue F3GA

FIGURE 3.55 Structure of Cibacron Blue F3GA. Cibacron Blue F3GA serves as the dye in Affi-Gel blue. Note its aromatic rings, highly conjugated π-bond system, and anionic nature. All of these factors play a role in how it binds to proteins.

mentioned earlier, albumin is used in the laboratory as a carrier molecule for fatty acids and some drugs, and it is found in most tissue culture media; hence, it is a common contaminant in protein purifications. Albumin binds tightly to Affi-Gel blue; therefore, this resin can be used in some of the final steps of a purification to remove albumin from the protein of interest.

Worked Problem 3.7   Affinity chromatography

Given the following peptide sequence, propose two different affinity chromatography resins that might be used to purify it:

$$NH_3^+\text{-MVKRGRRHKMCGWFAM-COO}^-$$

Strategy Examine the sequence of the peptide. Do the side chains of the amino acids have any distinguishing properties that might be useful in designing a purification scheme?

Solution There are two predominant classes of amino acids in this peptide: hydrophobic residues (Met, Ala, Phe, Val, Gly, and Trp)

and positively charged residues (Lys, Arg, and His). The peptide does have a single histidine residue, which in theory would interact with a MAC resin, but because the peptide is hydrophobic and positively charged, a better choice would either be HIC or cation-exchange chromatography using a CM resin.

Follow-up question If you wanted to separate the peptide shown below from the one given in the first part of the problem, which affinity chromatography techniques could you use?

$$NH_3^+\text{-MVDDGREHDMCGWFAM-COO}^-$$

Summary

- In affinity chromatography, a resin is coated with a molecule that has some capacity to bind a protein of interest. Affinity chromatography takes advantage of numerous weak interactions between the protein and the resin.

- Common resins include hydrophobic-interaction resins, ion-exchange resins, and metal-affinity resins. Molecules that are bound to an affinity resin can be eluted through a variety of techniques, the most common of which are washing with the ligand or altering the pH.

- Interactions between proteins and some small ligands can also be used to bind to molecules of interest. Resins are available that are either already bound to avidin, antibodies, or lectins or that have chemistry available to bind a protein of the investigator's design.

Concept Check

1. Explain the weak interactions used in affinity chromatography.
2. Discuss the different methods by which molecules are eluted from each type of resin.
3. Describe how avidin, antibodies, and lectins can be used in affinity chromatography.
4. What is the principle of ion exchange chromatography?
5. What are the applications of ion exchange chromatography?

Bioinformatics Exercises

Problems

3.1 Amino Acid Chemistry

1. Name the amino acids that have more than one chiral carbon.

2. Why it is that proline is considered to disrupt a helix but histidine does not?

3. Examine the side chains of the amino acids. Based on their structures, which ones will absorb light? In which part of the electromagnetic spectrum might they absorb?

4. Examine the structures of the amino acids. Based on their structure, what might they appear like in the purified state? What would most proteins look like when purified?

5. Like the amino acids from which they are made, proteins are chiral. Circular dichroism is a technique that is used to determine how much a sample of a protein will rotate plane-polarized light. The amount of rotation correlates with the degree of α helicity of a protein. Why does an α helix rotate plane-polarized light but a β sheet does not?

6. The amino acid glycine is a common component in many buffers. Draw a titration curve of 100 mM glycine titrated with 1 M NaOH.

7. Calculate the percentage ionization in these amino acid side chains at the pH values specified:

   i. aspartate at pH 5.4

   ii. glutamate at pH 11.8

   iii. histidine at pH 8.0

   iv. tyrosine at pH 8.0

8. Tyrosine is generally listed as a polar amino acid. How else could this amino acid be categorized?

9. Which amino acids have aromatic side chains? Recall from organic chemistry that organic molecules follow Huckle's rule (the $4n+2$ rule). Show that each of the proposed aromatic amino acids follows this rule.

10. Which amino acids absorb in the UV? Why do other amino acids not absorb in the UV?

3.2 Proteins Are Polymers of Amino Acids

11. Explain why most of the amino acids (with a few exceptions) can form either an α helix structure or a β sheet structure.

12. How many different ways can the 20 common amino acids combine to form a protein 400 amino acids long?

13. The average mass of an amino acid in a protein is 110 g/mol. Dividing the mass of titin (3,816 kDa) by the number of amino acids (34,350) does not give the exact figure of 110. Why?

14. Which level of protein organization is the most important in determining the final conformation?

15. ATP is usually the source of the phosphate in biochemical reactions, such as phosphorylation of proteins. Why is this the case? Could other molecules serve as the source? Which of the phosphates on ATP is the source of the phosphate? Why, in terms of energy and of equilibrium, does ATP have this role?

3.3 Proteins Are Molecules of Defined Shape and Structure

16. The four-helix bundle is one of the most commonly found tertiary motifs, and it is found in a wide variety of proteins. Describe how this motif could arise in two different proteins. Explain your observation first using divergent evolution and then using convergent evolution. Given that the four-helix bundle is found in thousands of different proteins, which instance, convergent or divergent evolution, is more likely?

17. Why are enzymes so large in comparison to the molecules in the reactions they catalyze?

18. Given the importance of structure in biochemistry, it would seem that obtaining the structure of a protein might be a good first step in investigating its properties. Think of other chemistry courses you have taken. If you know the structure of a molecule, what other information does that provide about the molecule?

19. What is the core of a protein generally like compared to the surface? What types of amino acid side chains would you expect to find in these two environments?

20. If a protein were chemically synthesized in hexane instead of water, would it retain the same structure? Of the levels of structure discussed, which is most likely to be affected?

3.4 Examples of Protein Structures and Functions

21. Immunoglobulins all have a similar structure but bind to different antigens. What is it about their structure that makes this possible?

22. Collagen is a fibrous protein found in the extracellular matrix. What challenges does this present when attempting to study collagen? What advantages are present?

23. Sickle cell anemia is a genetic disease resulting from a single amino acid substitution (glutamate to valine in the sixth position, or E6V) in the hemoglobin β chain. This mutation causes hemoglobin to polymerize into strands in the deoxygenated state, leading to the erythrocytes forming a sickled shape. Speculate as to why the mutation

leads to the polymerization of hemoglobin and why this only happens in the deoxygenated state.

24. Before the 1980s, the insulin used to treat diabetes was obtained from pigs (porcine insulin) or cattle (bovine insulin). These molecules differ by only three amino acids in the 104 amino acid sequence, as shown in the table. Based on this comparison, which would you anticipate to be the better treatment for diabetes in humans, porcine or bovine insulin?

Insulin	A chain, position 8	A chain, position 10	B chain, position 30
Human	Threonine	Isoleucine	Threonine
Porcine	Threonine	Isoleucine	Alanine
Bovine	Alanine	Valine	Alanine

25. In chymotrypsin, only three amino acids are involved in catalysis, but there are almost 200 amino acids in the entire protein. What is the function of the other amino acids?

26. What special features of aquaporin permit it to cross the plasma membrane? How does aquaporin allow water to flow through the membrane?

3.5 Protein Purification Basics

27. Compare and contrast the strengths and weaknesses of obtaining a protein from:

  a. a human source versus a nonhuman source

  b. cultured cells versus tissue

  c. a liver versus an adrenal gland

  d. donkeys versus mice

28. If you were interested in isolating large quantities of a rare transcription factor from neurons for crystallographic studies, what options would you have as a source of this protein?

29. A 3-mL protein sample is 1 M in NaCl. What volume would it need to be dialyzed into in order to obtain a final concentration of 145 mM?

30. Polyethylene glycol is a hydrophilic, high-molecular-weight polymer. How could this polymer be used in conjunction with dialysis to purify a protein? Would the sample being purified experience an increase in the salt or buffer concentration?

31. What parameters affect how quickly a dialysis will obtain equilibrium?

32. Explain how dialysis could be used to take a protein out of a 145-mM NaCl solution and into a 20-mM KCl solution.

3.6 Size-Exclusion Chromatography

33. Which would elute from a size-exclusion column first, a low-molecular-weight protein such as glucagon (MW 3.5 kDa) or a high-molecular-weight one such as ferritin (MW 450 kDa)?

34. If a column had a resolution range of 15,000 to 50,000 Da, would a protein such as aldolase (MW 160 kDa) be resolved from ferritin (MW 450 kDa) in this experiment?

35. A column has a resolution range of 15,000 to 50,000 Da. Explain if

  a. increasing the length of the column would allow it to resolve proteins outside this range.

  b. altering the flow rate would affect the resolution.

36. Some small size-exclusion chromatography columns (known as spin columns) can be run in a centrifuge to increase flow rate and decrease time. The protocol for a spin column is to spin 30 sec at $1,000 \times g$. How might the results change if the column were run under the following conditions:

  a. longer time

  b. shorter time

  c. higher speed

  d. lower temperature

37. Describe these objects as oblate or prolate spheroids. Propose an axial ratio for each.

  a. a piece of chalk

  b. a chalkboard eraser

  c. a cell phone

  d. a calculator

38. How might the structure of a protein change as it changes conformation? Could that be detected using the techniques described in this section? Which structural changes could be detected and which could not?

3.7 Affinity Chromatography

39. HIC is often used directly following a salting-out step. Why are purifications done in this order?

40. Why are proteins eluted with a gradient of decreasing ionic strength from an HIC column?

41. Some ion-exchange columns contain quaternary amines or sulfonates. Would these be anion or cation exchangers?

42. Some columns can be run under reducing conditions (in the presence of β-mercaptoethanol) to maintain proteins in a reduced state. Could this be done with immunoaffinity columns or an avidin–biotin binding column? Explain.

43. Why are affinity tags usually placed at the amino or carboxy terminus of a protein instead of somewhere in the middle?

44. Can metal affinity chromatography be used with native proteins? Why are genetically engineered proteins often used?

45. Describe how a polyhistidine tag such as a 6-His tag could be incorporated into a protein of interest.

46. What conditions might you use to elute a molecule from each of these columns?

  a. an immunoaffinity column

  b. a cation-exchange resin

  c. a phenyl-Sepharose column

Data Interpretation

47. A protein is loaded into a gel containing a pH gradient at pH 7. An electric field is applied and the protein migrates as shown. What does this data indicate about the protein?

48. Reaction of an intact peptide with dansyl chloride with subsequent acid hydrolysis results in a methionine derivative. Cleavage at specific sites produces the following sequences:

> Sequence A: His-Gly-Glu-Lys
>
> Sequence B: Met-Ser-Phe-Val
>
> Sequence C: Met-Ala-Leu-Arg

Based on this result, what could be the sequence of the peptide?

49. A protein has a predicted molecular weight of 47 kDa based on the amino acid sequence, an apparent molecular weight of 50 kDa by SDS-PAGE, and an approximate molecular weight of 95 kDa by size exclusion chromatography. Each of these techniques is described in the appendix. Why do the results vary between the different techniques? In particular, why is there such a difference between size exclusion chromatography and the other techniques? Which technique is likely to be most accurate and why?

50. Kwashiorkor is a type of malnutrition characterized by dietary protein deficiency. Examine the map shown below. The darker the color, the higher the prevalence of kwashiorkor. Where is the deficiency most prevalent? What is its likely cause, and how could this theory be tested? Why is the rate higher in France than in other countries in Europe? Why is the rate higher in the United States than in Australia or Canada? What other data would be useful in interpreting this figure and answering these questions? What other questions arise after thinking about this data?

Questions 51–53 relate to the data in the following table:

	Volume (mL)	Total protein (mg)	Activity (μmol/min)	Specific activity (μmol/mg/min)	Fold purification	Yield (%)
Cell lysate	5	4.8	20,400	4,250	1	100
Centrifugation	4.8	4	18,000	4,500	1.1	88.2
Salting out (ammonium persulfate precipitation)	0.5	2.5	13,000	5,200	1.2	63.7
Metal affinity chromatography	0.25	0.1	5,900	59,000	13.9	28.9
Ultrafiltration	0.05	0.1	5,400	54,000	12.7	26.4

51. Which step of the purification was most effective? In other words, which gave the highest fold increase in purification?

52. Which step is likely the most cost-effective step of the purification? In other words, which step cost the least and gave the greatest fold purification?

53. Are there any steps in this purification that were less effective or could be eliminated?

Experimental Design

54. The amino acid proline disrupts the regular pattern of the α helix. How could you demonstrate this experimentally?

55. Most proteins fold into a single conformation and do so almost instantaneously and without assistance. How could you test this statement experimentally?

56. You are studying a protein that consists of two domains: one catalytic and one regulatory. How could you test to see whether these domains can fold independently or whether they are dependent on one another for folding?

57. Develop a purification scheme to isolate a galactose-binding protein expressed in germinating seeds.

58. Design an experiment:

a. to test the hypothesis that a protein is in a multimeric complex in solution.

b. to separate a highly anionic matrix protein away from a complex mixture.

c. to ascertain if a protein has unfolded.

Ethics and Social Responsibility

59. Amino acids are freely available in purified form in vitamin and health food stores. Many amino acids and other naturally occur in molecules have distinct physiological effects in the body. Research the following question: When is a molecule considered a food item and when is it considered a drug? Do you agree with the definition you found in your research?

60. Several prion-based diseases seem to be acquired through the consumption of neurological tissue. Is it ethical to prohibit the consumption of these tissues based on the potential for causing or transmitting disease?

61. In preclinical data, intake of sulfur amino acids in higher quantities is speculated to be associated with aging-related diseases. However, there is a lack of significant data to ascertain the risk factors in humans. Recent findings suggest that diets lower in sulfur amino acids which are close to the estimated average requirement are linked with reduced risk for cardiometabolic disorders. How would you proceed and interpret such a scenario? Discuss.

62. What does it mean to be recused from the review of a work? Ask a professor or post-doctoral student in the sciences if they have ever been recused from a review and if so, for what reason.

63. Do the same rules apply for who can review a manuscript being submitted for publication as for who can review a grant proposal? Examine the instructions for reviewers of a journal or granting agency, and discuss the differences.

64. Where would you go to find the criteria for publishing in a journal and their review policies?

Suggested Readings

3.1, 3.2 Amino Acids and the Basics of Protein Structure

Branden, C., and J. Tooze. *Introduction to Protein Structure*. New York, NY: Garland Science, 1999.

Buxbaum, E. *Fundamentals of Protein Structure and Function*. New York, NY: Springer, 2007.

Kessel, A., and N. Ben-Tal. *Introduction to Proteins: Structure, Function, and Motion*. Boca Raton, FL: CRC press, 2011.

Williamson, M. *How Proteins Work*. New York, NY: Garland Science, 2011.

3.3, 3.4 Protein Architecture and Examples of Protein Structure

Ferscht, A. M. *Structure and Mechanism in Protein Science*. New York, NY: W.H. Freeman, 1998.

Lesk, A. M. *Introduction to Protein Science Architecture, Function, and Genomics*. Oxford, UK: Oxford University Press, 2010.

Lesk, A. M. *Introduction to Proteins: The Structural Biology of Proteins Architecture*. Oxford, UK: Oxford University Press, 2001.

Liska, K. *Drugs and the Human Body*. Upper Saddle River, NJ: Pearson/Prentice Hall, 2004.

Petsko, G. A., and D. Ringe. *Protein Structure and Function*. Sunderland, MA: Sinauer Associates, Inc. 2003.

3.5 Protein Purification Basics

Bonner, P. L. R. *Protein Purification*. New York: Garland Science, 2007.

Burgess, R. R., and M. P. Deutscher, eds. *Guide to Protein Purification*, 2nd ed. (Methods in Enzymology, vol. 436). San Diego, CA: Academic Press, 2009.

Cutler, P. *Protein Purification Protocols* (Methods in Molecular Biology, vol. 244). Totowa, NJ: Humana Press, 2003.

Deutscher, M. P., ed. *Guide to Protein Purification* (Methods in Enzymology, vol. 182). San Diego, CA: Academic Press, 1990.

Doyle, S. A. *High Throughput Protein Expression and Purification: Methods and Protocols* (Methods in Molecular Biology). Totowa, NJ: Humana Press, 2010.

Janson, J.-C. *Protein Purification: Principles, High Resolution Methods, and Applications*. New York: Wiley, 2011.

Scopes, R. K. *Protein Purification: Principles and Practice* (Springer Advanced Texts in Chemistry). New York: Springer, 1993.

3.6 Size-Exclusion Chromatography

Axelsson, I. "Characterization of Proteins and Other Macromolecules by Agarose Gel Chromatography." *Journal of Chromatography A* 152, no. 1 (1978): 21–32.

Bloomfield, V. A. "Survey of Biomolecular Hydrodynamics." Chapter 1 in *On-Line Biophysics Textbook: Separations and Hydrodynamics*, edited by T. M. Schuster. https://www.biophysics.org/Portals/1/PDFs/Education/vbloomfield.pdf.

Carrasco, B., and J. García de la Torre. "Hydrodynamic Properties of Rigid Particles: Comparison of Different Modeling and Computational Procedures." *Biophysical Journal* 76, no. 6 (June 1999): 3044–57.

Fernandes, M. X., A. Ortega, M. C. López Martínez, and J. García de la Torre. "Calculation of Hydrodynamic Properties of Small Nucleic Acids from Their Atomic Structure." *Nucleic Acids Research* 30, no. 8 (2002): 1782–8.

García de la Torre, J., and V. A. Bloomfield. "Hydrodynamic Properties of Complex, Rigid, Biological Macromolecules: Theory and Applications." *Quarterly Reviews of Biophysics* 14, no. 1 (February 1981): 81–139.

García de la Torre, J., and V. A. Bloomfield. "Hydrodynamic Properties of Macromolecular Complexes. I. Translation." *Biopolymers* 16 (1977): 1747–63.

García de la Torre, J., M. L. Huertas, and B. Carrasco "Calculation of Hydrodynamic Properties of Globular Proteins from Their Atomic-Level Structure." *Biophysical Journal* 78, no. 2 (February 2000): 719–30.

Hardin, S. E. "Protein Hydrodynamics." Chapter 7 in *Protein: A Comprehensive Treatise* (vol. 2, pp. 271–305), edited by G. Allen. Stamford, CT: JAI Press, 1999.

López-Corcuera, B., R. Alcántara, J. Vázquez, and C. Aragón. "Hydrodynamic Properties and Immunological Identification of the Sodium- and Chloride-Coupled Glycine Transporter." *Journal of Biological Chemistry* 268, no. 3 (January 1993): 2239–43.

Majorek, K. A., P. J. Porebski, A. Dayal, M. D. Zimmerman, K. Jablonska, A. J. Stewart, M. Chruszcz, and W. Minor. "Structural and Immunologic Characterization of Bovine, Horse, and Rabbit Serum Albumins." *Molecular Immunology* 52 (2012): 174–82.

Monkos, K. "A Comparison of Solution Conformation and Hydrodynamic Properties of Equine, Porcine and Rabbit Serum Albumin Using Viscometric Measurements." *Biochimica et Biophysica Acta* 1748, no. 1 (April 2005): 100–9.

Ortega, A., and J. Garcia de la Torre. "Hydrodynamic Properties of Rodlike and Disklike Particles in Dilute Solution." *Journal of Chemical Physics* 119, no. 18 (2003): 9914–9.

Putnam, F. W. *The Plasma Proteins: Structure, Function and Genetic Control*, vol. I, 2nd ed. San Diego, CA: Academic Press, 1975.

Scherag, H. A., and L. Mandelkern. "Consideration of the Hydrodynamic Properties of Proteins." *Journal of the American Chemical Society* 75, no. 1 (1953): 179–84.

Squire, P. G., P. Moser, and C. T. O'Konski. "Hydrodynamic Properties of Bovine Serum Albumin Monomer and Dimer." *Biochemistry* 7, no. 12 (1968): 4261–72.

Szymczak, P., and M. Cieplak. "Hydrodynamic Effects in Proteins." *Journal of Physics Condensed Matter* 23, no. 3 (2011): 033102.

Wilkins, D. K., S. B. Grimshaw, V. Receveur, C. M. Dobson, J. A. Jones, and L. J. Smith, "Hydrodynamic Radii of Native and Denatured Proteins Measured by Pulse Field Gradient NMR Techniques." *Biochemistry* 38, no. 50 (1999): 16424–31.

Wright, A. K., and M. R. Thompson. "Hydrodynamic Structure of Bovine Serum Albumin Determined by Transient Electric Birefringence." *Biophysical Journal* 15, no. 2, Pt 1 (1975): 137–41.

3.7 Affinity Chromatography

Bornhorst, J. A., and J. J. Falke "Purification of Proteins Using Polyhistidine Affinity Tags." *Methods in Enzymology* 326 (2000): 245–54.

Chakrabarti, P. "Geometry of Interaction of Metal Ions with Histidine Residues in Protein Structures." *Protein Engineering* 4, no. 1 (October 1990): 57–63.

Chothia, C., and A. M. Lesk. "Canonical Structures for the Hypervariable Regions of Immunoglobulins." *Journal of Molecular Biology* 196, no. 4 (1987): 901–17.

Chothia, C., et al. "Conformations of Immunoglobulin Hypervariable Regions." *Nature* 342 (1989): 887–83.

Crowe, J., H. Döbeli, R. Gentz, E. Hochuli, D. Stüber, and K. Henco. "6xHis-Ni-NTA Chromatography as a Superior Technique in Recombinant Protein Expression/Purification." *Methods in Molecular Biology* 31 (1994): 371–87.

Gallivan, J. P., and D. A. Dougherty. "Cation-π Interactions in Structural Biology." *Proceedings of the National Academy of Sciences USA* 96, no. 17 (August 1999): 9459–64.

Leszczynski, J., and M. K. Shukla. *Practical Aspects of Computational Chemistry I: An Overview of the Last Two Decades and Current Trends.* New York: Springer, 2012.

Xiu, X., N. L. Puskar, J. A. Shanata, H. A. Lester, and D. A. Dougherty. "Nicotine Binding to Brain Receptors Requires a Strong Cation–π Interaction." *Nature* 458, no. 7237 (March 2009): 534–7.

Zacharias, N., and D. A. Dougherty. "Cation–π Interactions in Ligand Recognition and Catalysis." *Trends in Pharmacological Sciences* 23, no. 6 (June 2002): 281–7.

1800s	Acetone powders used to precipitate proteins
1838	Berzelius and Mulder describe and name proteins
1851	Funke crystallizes hemoglobin
1888	Hofmeister uses salt series to precipitate proteins
1902	Fischer and Hofmeister propose that proteins are made of amino acid chains
1920–21	Banting identifies insulin as a protein and begins using it to treat diabetics
1923	Banting and McLeod share Noble Prize for insulin
1926	Using urease, Summer determines that enzymes are proteins
1929–1946	Northrup and Stanley isolate proteins and viruses
1930s	Pauling predicts the secondary structures of proteins and Dialysis membranes become commercially available
1945	Kendrew begins work on elucidation of the structure of myoglobin
1946	Northrup and Stanley awarded Nobel Prize for isolation of pure proteins and viruses
1949	Insulin is sequenced by Sanger (Nobel Prize 1958)
1950	Edman first describes protein sequencing through labeling and stepwise amino terminal degradation and Size-exclusion chromatography developed
1957	Structure of myoglobin solved
1958	First protein structure (myoglobin) solved by X-ray diffraction
1962	Nobel Prize awarded to Perutz, Kendrew, and Crowfoot-Hogkins for use of X-ray diffraction in biological structure determination
1963	Merrifield develops solid phase peptide synthesis
1967	First mention of SDS-PAGE in biochemical literature (Shapiro et al.)
1970	Lamelli uses SDS-PAGE to characterize viral proteins
	First biological macromolecular characterization by NMR, late 1960s, early 1970s
	First protein NMR
1977	Antibodies/immunopurification techniques first identified
	First recombinant protein (insulin)
1980	Sanger awarded the Nobel Prize in Chemistry for sequencing DNA
1982	Humulin (recombinant human insulin) goes to market
1985	Soft ionization techniques adapted for protein mass spectrometry
	First recombinant protein (insulin) 1977, Humulin (recombinant human insulin) goes to market
1988	Metal-affinity techniques introduced
1990s	Epitope-tagging techniques developed
2002	Wuthrich, Tanaka, and Fenn share Nobel Prize for NMR structure determination of proteins in solution and mass spectrometry of proteins, respectively
2015	Number of protein structures in the Protein Data Bank surpasses 100,000

CHAPTER 4

Proteins II

Enzymes, Allosterism and Receptor–Ligand Interactions

Enzymes and Allosterism in Context

In a toolbox, there are a wide variety of tools to choose from—screwdrivers, wrenches, hammers, and saws—each with a specific structure and function. If we imagine the tool section of a large hardware or home improvement center, we may encounter unusual tools, such as a giant wrench. Even though this particular tool may be unfamiliar, it is easy to imagine a function for it, based on its appearance (the structure) and how other similar tools act (the mechanism). Looking at the bigger picture, we can see how tools can be grouped according to their function, that is, whether they cut, fasten, bend, hold, and so on.

Just as tools can be grouped, enzymes can be broadly characterized based on the reactions they catalyze and the chemistry they conduct. Alternatively, they can be categorized based on protein conformation or structure. Armed with a knowledge of structure, we can amass other data that can help in determining a possible reaction mechanism for an enzymatically catalyzed reaction. We can then build on this knowledge to probe the types of reactions that an enzyme will catalyze and those it will not, and explain those differences. With these differences in mind, it is then possible to identify different types of inhibitors, alter biochemical processes, develop new drugs, and diagnose and treat disease.

Many enzymes are not regulated, whereas others are regulated by post-translational modifications or by the concentrations of substrates or products .This chapter also discusses another means of control: allosteric regulation. Allosterism describes how structurally different molecules can bind and regulate an enzyme. Hemoglobin,though not an enzyme, is discussed next because its binding of oxygen is allosterically regulated. Finally, because many receptors are also allosterically regulated, this chapter also discusses receptor–ligand binding.

Common Themes

Evolution's outcomes are conserved.	• Mutations to DNA sequences may alter an enzyme's amino acid sequence. The new enzyme may act normally (a silent mutation), have a slightly altered function, such as a change in rate or binding of a different substrate, or be completely nonfunctional. If a mutation benefits the organism, there is an increased likelihood that it will be transmitted to the organism's progeny. • As organisms have become more complex and organs and systems have evolved, enzymes have evolved functions or regulatory mechanisms specific to that organ or tissue. • The mechanisms enzymes employ to catalyze reactions are themselves repeated and conserved throughout nature; this is an example of evolution at the chemical level. • Analysis of mechanisms and structures of enzymes illustrate both convergent and divergent evolution. • The regulatory mechanisms employed by enzymes have coevolved with catalytic mechanisms. • Allosteric sites have evolved over time to bind substrates and catalyze reactions in parallel with the evolution of enzymes. Allosteric regulation is thus an adaptation that gives organisms an advantage.
Structure determines function.	• The structure of an enzyme and the position of amino acid side chains in the active site provide important biochemical information that can be used to help elucidate an enzymatic mechanism. • Regulatory mechanisms (e.g., proteolytic cleavage, phosphorylation, or binding of a small regulatory molecule) lead to changes in the tertiary or quaternary structure of the enzyme, modifying substrate binding or the geometry of the active site, and ultimately altering activity. • Allosteric enzymes have distinct sites where allosteric modulators bind to the enzyme and regulate enzyme activity. • Many allosteric enzymes and receptors are multimeric.
Biochemical information is transferred, exchanged, and stored.	• Many enzymes play important roles in transmitting and carrying out biochemical signals in the cell. • The regulation of enzymes (through either covalent modification or allosteric mechanisms) regulates both the rates of biochemical reactions and flux through biochemical pathways. • Allosteric regulation is one of the most rapidly acting means of enzyme regulation. • Allosteric regulation allows flux through a pathway or signaling through a receptor to be finely tuned by molecules that are neither substrates or products of that reaction or pathway nor ligands for that specific receptor.
Biomolecules are altered through pathways involving transformations of energy and matter.	• Most biochemical reactions are catalyzed by enzymes. • The use of enzymes affords a means to regulate biochemical reactions and a way to couple energetically unfavorable reactions with favorable ones. • Data about the rate at which a catalyzed reaction proceeds can be used to describe the reaction mathematically. • Spectroscopic, structural, and kinetic data can be used to describe the mechanism of an enzymatically catalyzed reaction. • Allosteric regulators have been found to act in almost every known biochemical pathway.

4.1 Regarding Enzymes

Most of the thousands of reactions that occur in the cell require a catalyst to run at biologically significant rates.

At room temperature, a bag of sugar (sucrose) is reacting with oxygen in the air, gradually forming CO_2 and H_2O. This reaction is happening far too slowly for us to see; even so, it is possible to measure this reaction chemically and understand from general chemistry that it is happening. The energetics of the reaction indicates that the reaction is exothermic; that is, heat is given off in the process. Examining other thermodynamic parameters, we find that there is a negative free energy (ΔG) for this reaction, indicating spontaneity in the forward direction. Finally, as this reaction runs, there is an increase in the entropy—the number of possible microstates or the disorder of the system. All of these thermodynamic parameters are true, yet they reveal nothing of the speed or rate of the reaction. If you were to leave a bag of sugar exposed to oxygen in your kitchen and return even several years later, you would see no appreciable change.

If that same bag of sugar were dissolved in water and incubated with yeast, the same net reactions would occur. Sucrose would be consumed, and CO_2 and H_2O would be generated, as would some heat from the making of chemical bonds. The entropy of the system would increase, and the reaction would proceed favorably. However, this reaction would be visible to the naked eye because the microbes would generate CO_2 so quickly that it would bubble out of solution. The same overall reaction is happening in the two cases, and the thermodynamics are unchanged. However, what *has* changed is that a biological system has contributed a catalyst (or, in this case, multiple catalysts) to increase the rates of reactions to the point where the yeast can use these reactions at room temperature. The catalysts are protein molecules termed **enzymes**.

4.1.1 Enzyme nomenclature

Tens of thousands of different reactions can occur in an organism. However, these reactions can be divided into seven different classes, each having different subclasses, based on the type of substrate and the mechanism of action of the enzyme involved (**Table 4.1**). The classes are:

1. Oxidoreductases—catalyze reactions involving the gain or loss of electrons.
2. Transferases—transfer one group to another.
3. Hydrolases—cleave a bond with water.
4. Lyases—break double bonds using some other means than oxidation or hydrolysis.
5. Isomerases—catalyze a rearrangement of the molecule.
6. Ligases—join two molecules.
7. Translocase—catalyze movement of ions across membrane

These groups were decided upon by the Enzyme Commission (EC) of the International Union of Biochemistry and Molecular Biology (IUBMB), and are called **EC numbers (Figure 4.1)**. Every enzyme has a unique EC number, based on its category and further subclassifications.

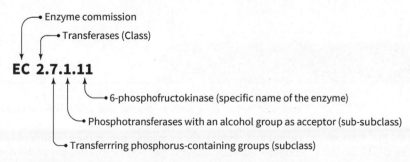

FIGURE 4.1 Enzyme commission number.   Illustration of the enzyme commission number of the enzyme 6-phosphofructokinase

TABLE 4.1 Classes of Enzymes Production

Class	General Reaction	Reaction Catalyzed	Important Subclasses	Examples
1. Oxidoreductases	$A_{Red} + B_{Ox} \leftrightarrows A_{Ox} + B_{Red}$	Transfer of electron(s)	Dehydrogenases Oxidases, peroxidases Reductases Monooxygenases Dioxygenases	Cytochrome oxidase Lactate dehydrogenase
2. Transferases	$A\text{-}B + C \rightarrow A + C\text{-}B$	Transfer of a group from a donor substrate to an acceptor substrate	C-1 Transferases Glycosyltransferases Aminotransferases Phosphotransferases	Acetate kinase Alanine deaminase
3. Hydrolases	$A\text{-}B + H_2O \rightarrow A\text{-}H + B\text{-}OH$	Hydrolysis of a bond	Esterases Glycosidases Peptidases Amidases	Lipase Sucrase
4. Lyases ("synthases")	$A\text{-}B \leftrightarrows A + B$ (reverse reaction: synthases)	Nonhydrolytic cleavage of a bond removing a group and leaving behind a double bond	C-C-Lyases C-O-Lyases C-N-Lyases C-S-Lyases	Oxalate decarboxylase Isocitrate lyase
5. Isomerases	$A\text{-}B\text{-}C \leftrightarrows A\text{-}C\text{-}B$	Interconversion of isomers	Epimerases *cis trans* Isomerases Intramolecular transferases	Glucose-phosphate isomerase Alanine racemase
6. Ligases ("synthetases")	$A + B + ATP \rightarrow A\text{-}B + ADP + P_i$	Condensation of two or more substrates coupled with the hydrolysis of ATP or similar molecules	C-C-Ligases C-O-Ligases C-N-Ligases C-S-Ligases	Acetyl-CoA synthetase DNA ligase
7. Translocases	$AX + B\| = A + X + \|B$ (side 1)　(side 2)	Movement of ions or molecules across membranes or their separation within membranes		Adenine nucleotide translocase Ornithine translocase

The EC number consists of four numbers separated by full stops. The first number represents the class of reaction catalyzed. The second number (the subclass) gives information about the type of compound or group involved. The third number, the sub-subclass, is for the type of reaction involved, and the fourth is a serial number of the individual enzyme within a sub-subclass. To illustrate, the enzyme commission number for 6-phosphofructokinase (EC 2.7.1.11), which catalyzes the following reaction, is expanded in Figure 4.1.

D-fructose 6-phosphate + ATP → D-fructose 1,6-bisphosphate + ADP

One way our understanding of enzyme chemistry has assisted in the clinic through the measurement of enzyme activity in the blood is discussed in **Medical Biochemistry: Enzymes in diagnosis**.

Medical Biochemistry

Enzymes in diagnosis

Physicians often request blood work on a patient to aid in a diagnosis. A comprehensive metabolic panel (referred to as a CHEM-20) is a commonly requested battery of blood tests. Many parameters will be already familiar: pH and the concentrations of different ions and metabolites, such as glucose and urea. These values provide the clinician with a general view of the function of various organs and systems. Other values, such as the activities of several enzymes typically found in the liver or muscle, can probe deeper. Because these enzymes are not typically found in blood, their appearance indicates organ damage. Such enzymes include lactate dehydrogenase, alanine transaminase, aspartate aminotransferase, and creatine phosphokinase.

Lactate dehydrogenase (LDH) is usually localized to specific tissues, including the heart, the liver, the muscle, and/or the kidney. If LDH is released from those tissues, it will accumulate in the blood rather than being excreted by the kidney. LDH is therefore used as a marker for general tissue damage.

Alanine transaminase (ALT) and aspartate aminotransferase (AST) are localized to the liver; hence, as with LDH, elevated levels of these enzymes in blood indicate liver damage. Similarly, the liver enzyme gamma-glutamyl transpeptidase (gamma-GT) is a marker for liver damage. Among its multiple functions, gamma-GT catalyzes the transfer of the gamma-linked glutamate from glutathione to an acceptor molecule.

One final molecule that is often included on chemistry panels is creatine phosphokinase (CPK). This enzyme is specifically found in muscle tissue and the liver, where it functions in creatine metabolism. Elevated levels of CPK indicate muscle damage or, more frequently, heart attack.

4.1.2 Enzymes are biological catalysts

Enzymes are the cell's solution to most of its catalytic problems. Enzymes are usually globular proteins, although some have irregular shapes, or form filaments. Some enzymes are monomeric, meaning they contain a single subunit, while others are multimeric; that is, they have two or more subunits (**Figure 4.2**). Multimeric enzymes may have both **catalytic subunits** and **regulatory subunits**, or may have multiple catalytic subunits that interact and help to regulate one another, an example of allosteric regulation.

The reactants in an enzymatically catalyzed reaction are referred to as **substrates**, and the molecules produced are referred to as **products**. Although this chapter focuses on reactions that have a single substrate, over 60% of reactions use more than one substrate and have more than one product. The substrate binds to the enzyme at the location where catalysis occurs, termed the **active site**. This site is on the surface of the enzyme but is often in a cleft, pocket, or trench. The residues that make up the surface of the active site are responsible for substrate binding and for catalysis, although some residues have more specialized functions. The active site is generally formed by residues on turns or coils, although sometimes sections of helixes or sheets may also participate in the reaction.

Several models are used to describe the interaction of the substrate with the enzyme. Some substrates interact with enzymes like a key in a lock, known as the **lock and key** model (**Figure 4.3**). In this analogy, only the correct key will fit the lock; that is, the correct steric interactions are needed for the substrate to fit into the enzyme's active site. This model explains some aspects of enzyme-substrate interactions particularly well. For example, enzymes are generally quite specific in terms of what substrates they will bind to, being able to distinguish, for example, between glutamate and aspartate or between the carbohydrates glucose and galactose (which differ only in the chirality of a single chiral center). In this regard, enzymes may indeed be like a key in a lock. However, other aspects of enzyme molecules do not fit this model.

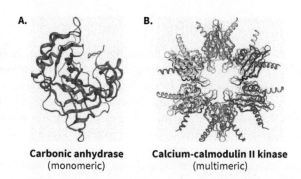

A. Carbonic anhydrase (monomeric)

B. Calcium-calmodulin II kinase (multimeric)

FIGURE 4.2 General topology of enzymes.   Enzymes can be **A.** monomeric with a single subunit or **B.** multimeric with several subunits.

(Source: (A) Data from PDB ID 1CA2 Eriksson, A.E., Jones, T.A., Liljas, A. (1988) Refined structure of human carbonic anhydrase II at 2.0 A resolution. *Proteins* 4: 274–282. (B) Data from PDB ID 2UX0 Rellos, P., Pike, A.C.W., Niesen, F.H., Salah, E., Lee, W.H., von Delft, F., Knapp, S. (2010) Structure of the CaMKIIdelta/Calmodulin Complex Reveals the Molecular Mechanism of CaMKII Kinase Activation. *Plos Biol.* 8: 426)

A.

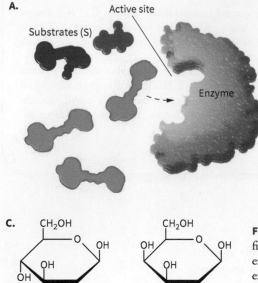

Substrates (S)

Active site

Enzyme

B.

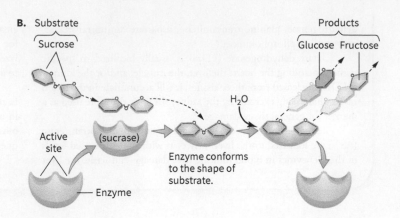

Substrate

Sucrose

Active site

(sucrose)

Enzyme

Enzyme conforms to the shape of substrate.

H_2O

Products

Glucose Fructose

C.

Glucose

Galactose

FIGURE 4.3 Models of substrate binding.   A. In the lock and key model, the substrate fits in the active site like a key fitting into a lock. **B.** In induced fit, the substrate fits into the enzyme like a hand inserting into a glove. **C.** Enzymes are specific enough to discern the difference between very similar structures, such as glucose and galactose.

(Source: (A) Karp, *Cell and Molecular Biology: Concepts and Experiments*, 7e, copyright 2013, John Wiley & Sons. This material is reproduced with permission of John Wiley & Sons, Inc. (B) Alters, *Biology: Understanding Life*, 1e, copyright 2006, John Wiley & Sons. This material is reproduced with permission of John Wiley & Sons, Inc.)

In **induced fit**, the interaction of the substrate with the enzyme helps to form the active site. An analogy for this is a latex or nitrile laboratory glove. The glove itself is roughly hand shaped, but in the absence of a hand, the glove is flat and floppy. The hand filling the glove is like a substrate in an active site. The induced fit interaction helps to describe many enzyme-substrate interactions, as well as some conformational changes the enzyme may undergo as a result of substrate binding.

The active site is a tight fit. Binding of substrate usually fills the site entirely. Even water is excluded. If water molecules are present, they are usually involved in substrate binding or the reaction mechanism rather than being passive bystanders. This observation explains several reactions that would otherwise make no sense. Reactions such as transesterifications occur only because there is no water available to hydrolyze a bond, whereas other reactions add a hydroxyl group stereospecifically, through the specific positioning of a water molecule by the enzyme. The enzyme hexokinase catalyzes the transfer of a phosphate group from ATP to a hydroxyl group in glucose. To accomplish this, the enzyme binds both substrates and closes down around them, excluding water or any other molecule from the active site before catalysis occurs (**Figure 4.4**).

4.1.3 How do enzymes work?

Enzymes are catalysts. All catalysts work by decreasing the **activation energy** (E_a or $\Delta G^{\ddagger}$) of a reaction. As mentioned above, enzymes do not change the thermodynamic parameters of a reaction. Those are equations of state, and they do not depend on *how* reactants are transformed into products, merely that they *are* transformed. Likewise, catalysts do not affect equilibrium. If an equilibrium for a reaction lies heavily to the left or to the right, it will still be that way in the presence of an enzyme; the only difference is that the reaction will achieve that equilibrium more quickly in the presence of a catalyst than without.

Catalysts increase reaction rate by lowering the energy of the transition state (**Figure 4.5**). General chemistry indicates that the number of molecules that can cross the energy barrier from reactants, or substrate, to products is limited by the height of the barrier, known as the activation energy, through the Arrhenius equation:

$$k = Ae^{-\frac{E_a}{RT}}$$

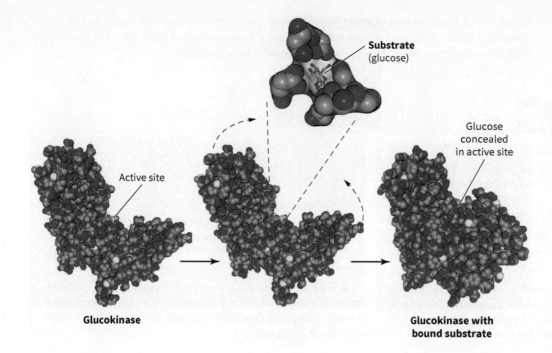

Substrate
(glucose)

Active site

Glucose
concealed
in active site

Glucokinase

**Glucokinase with
bound substrate**

**FIGURE 4.4 Topology of the
active site.** Shown is the enzyme
glucokinase and a close-up view of
its active site. Once the substrate is
bound, the enzyme closes around it
like the two halves of a clam shell,
excluding all molecules except the
two substrates. This is an example
of induced fit.

(Source: Data from PDB ID 1V4T, 1V4S
Kamata, K., Mitsuya, M., Nishimura, T.,
Eiki, J., Nagata, Y. (2004) Structural basis
for allosteric regulation of the monomeric
allosteric enzyme human glucokinas.
Structure 12: 429-438)

In this instance, the higher the barrier (E_a, the activation energy), the slower the reaction.
Catalysts work by lowering this barrier and thus increasing the reaction rate. A quick calculation
shows how significant this can be. A 36 kJ/mol change in the activation energy results in an
astounding 10^6-fold change in the **rate constant** (k) at the same temperature.

As an analogy, think about living in one town in the mountains and wishing to go to a second
town, located in a nearby valley. We would be limited by the height of the mountain we need to
climb to go to the second town; a tunnel through the mountain would facilitate our travel.

Enzymes lower activation energy in both a surprisingly simple and terribly complex way.
Enzymes bind substrates using a combination of weak forces and steric interactions that align
functional groups, polarize bonds, promote electron flow, constrain molecules, and otherwise
place molecules in a favorable geometry for reactions to occur. That is, enzymes literally form

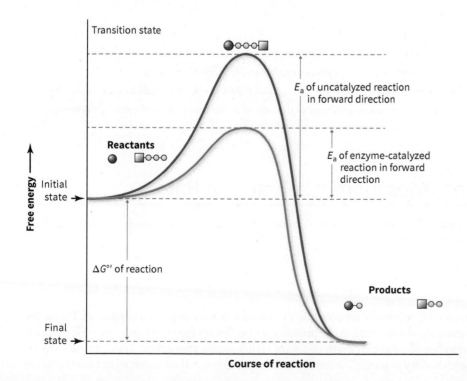

Transition state

E_a of uncatalyzed reaction
in forward direction

Reactants

E_a of enzyme-catalyzed
reaction in forward
direction

Free energy

Initial
state

$\Delta G^{\circ\prime}$ of reaction

Products

Final
state

Course of reaction

**FIGURE 4.5 Enzymes lower activation
energy.** Shown is a reaction coordinate with
energy on the *y*-axis and the progress of the reaction on the *x*-axis. All reactions have an energetic
barrier they need to cross (the activation energy,
E_a). Enzymes work by lowering this barrier. This
makes both the forward and reverse directions
more favorable.

(Source: Karp, *Cell and Molecular Biology: Concepts and
Experiments*, 7e, copyright 2013, John Wiley & Sons.
This material is reproduced with permission of John Wiley
& Sons, Inc.)

a pocket that brings groups together or pulls them apart to form a structure similar to the transition state. Once at the peak of this energetic barrier some fraction of molecules will decompose back into substrates, while the remainder will convert into products.

Worked Problem 4.1   **Describing an enzymatically catalyzed reaction**

Lysozyme is an enzyme that breaks bonds in the carbohydrate chains found in bacterial cell walls. Describe this reaction using the vocabulary found in this section.

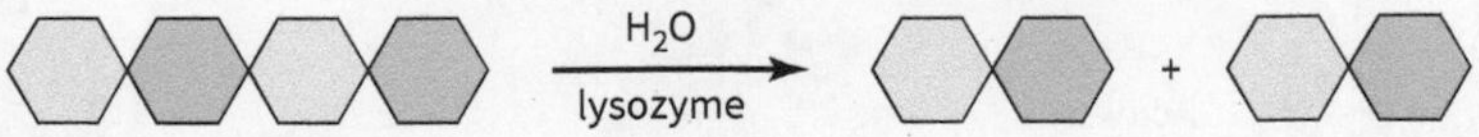

Strategy   Examine the image and the reaction shown. Describe in your own words what is going on. Next, determine whether there are specific terms or names for what you have described, and put what you have written into scientific terms.

Solution   Lysozyme is a monomeric enzyme that catalyzes the hydrolytic cleavage of carbohydrate chains in the bacterial cell wall; thus it belongs to EC classification 3, hydrolases. The carbohydrate chains are the substrate, and they bind to the active site of the

enzyme using induced fit. This binding stabilizes the transition state, lowering the activation energy—the barrier to catalysis. The broken fragments of the carbohydrate chain are the products of the reaction.

Follow-up question   Choose any three enzymatically catalyzed reactions you can think of (either from this text or from other sources). Categorize these reactions based on the six types of reactions found in section 4.1.1.

Summary

- Enzymes are protein catalysts that increase rates of reaction by lowering activation energy.
- Enzymes are often globular proteins that may be monomeric or multimeric.
- Substrates (the reactants) are converted to products in the active site of the enzyme, a surface cleft that specifically binds the substrate.
- The active site binds substrate like a key in a lock or a hand in a glove. The binding is specific and tight, with solvent often being excluded.
- Enzymes can be categorized based on the type of reaction they catalyze.
- Enzymes do not alter equilibrium, but they do allow systems to achieve equilibrium more quickly than would be the case in the absence of a catalyst.

Concept Check

1. Describe what generally happens in an enzymatically catalyzed reaction.
2. Describe the interactions of substrate and enzyme in biochemical terms.
3. Describe the two models (lock and key and induced fit) that are used to describe these interactions.
4. Use a simple thermodynamic argument to describe why enzymes increase the rate of a reaction.

4.2 Enzymes Increase Reaction Rate

Rates of enzymatically catalyzed reactions can be derived from rates of chemical reactions and are powerful tools for describing reactions.

4.2.1 Boltzmann distribution

The rate constant (k) relates the concentration to the rate. It has two main components. First is the frequency factor, A, which describes the percentage of collisions that result in a favorable reaction, that is, the number of times that molecules must collide before they collide in the proper orientation to cause chemistry to occur. The second term is an exponential term that relates the activation energy, E_a; the temperature (measured in Kelvin), T; and the gas law constant, R. Here, e is being raised to the negative activation energy (E_a) power. Hence, the higher the E_a, the lower the value of the exponential, the lower the value of k, and the lower the rate. However, as the temperature increases, it can modify the exponential term to increase k and thus increase the reaction rate. This is a variation of a Boltzman–Maxwell distribution and describes the fraction of molecules with enough energy to overcome the activation energy barrier (**Figure 4.6**). Combining the frequency factor A and the exponential term gives the Arrhenius equation you may recall from general chemistry:

$$k = Ae^{-\frac{E_a}{RT}} \tag{4.1}$$

When a chemical reaction is catalyzed, the catalyst takes part in the reaction but is not consumed. Rather, it provides an alternative path for a reaction to occur, and this alternative mechanism has a lower activation energy.

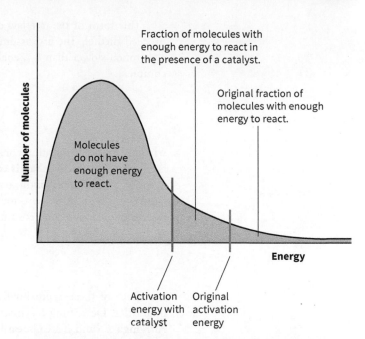

FIGURE 4.6 Boltzmann distribution. A Boltzmann distribution shows the distribution of particles with different energies.

4.2.2 The Michaelis-Menten equation relates enzymatic rates to measurable parameters

Before examining an enzymatically catalyzed reaction, it is helpful to define the terms used. In such a reaction, the starting material is termed the substrate (S), and it is converted by the enzyme (E) to the product (P). The rate of an enzymatically catalyzed reaction is referred to as a velocity (v).

In some of the first experiments to determine enzymatic rate, Adrian Brown, a professor of malting and brewing, determined that invertase (an enzyme that breaks down sucrose) is independent of sucrose concentration provided that the concentration of the substrate is much higher than that of the enzyme.

$$\text{Sucrose} \xrightarrow{\text{invertase}} \text{glucose} + \text{fructose}$$

From this finding, Brown surmised that catalysis must take at least two steps, the formation of an enzyme–substrate complex (ES), and the formation of product, and that the product-forming step does not involve sucrose.

$$E + S \underset{k_{-1}}{\overset{k_1}{\rightleftharpoons}} [ES] \xrightarrow{k_2} E + P \tag{4.2}$$

If there were enough substrate available to bind the entire free enzyme (saturation), then k_2 would become the rate-determining step of this reaction. This is consistent with the observation that increasing substrate concentration past a certain point fails to elicit an increase in reaction rate. When this is the case, and the enzyme is saturated with substrate, the rate law becomes:

$$v = \frac{d[P]}{dt} = k_2[ES] \tag{4.3}$$

Here, the rate of formation of $[ES]$ is the total of the rates that build up the concentration of $[ES]$ minus those that break it down:

$$\frac{d[ES]}{dt} = k_1[E][S] - k_{-1}[ES] - k_2[ES] \tag{4.4}$$

This form of the rate law cannot be integrated simply; instead, assumptions are needed to proceed further. The first assumption, made by Leonor Michaelis and Maude Menten, was that k_2 is much slower than k_{-1}, enabling an equilibrium to be established between E, S, and the ES complex:

$$K_s = \frac{k_{-1}}{k_1} = \frac{[E][S]}{[ES]} \qquad (4.5)$$

where K_s is the dissociation constant of the ES complex. In honor of their work, the ES complex is referred to as the **Michaelis complex**.

The second assumption, termed the **steady state assumption**, assumes that the ES complex forms rapidly from free enzyme and substrate, and it exists at a relatively unchanging concentration as the reaction proceeds until substrate is depleted:

$$\frac{d[ES]}{dt} = 0 \qquad (4.6)$$

Both of these approximations are useful in expressing enzyme rate in terms of measurable quantities. Describing enzymatic catalysis requires the use of terms that can be readily measured. Although ES and E are not easily measured, total enzyme is:

$$[E] + [ES] = [E]_{tot} \qquad (4.7)$$

To calculate rate, we need to be able to rearrange these expressions to determine $[ES]$. Combining equations 4.4 and 4.7 gives:

$$k_1([E]_{tot} - [ES])[S] = (k_{-1} + k_2)[ES] \qquad (4.8)$$

Distributing k_1 and S across the first term yields:

$$k_1[E]_{tot}[S] - k_1[ES][S] = (k_{-1} + k_2)[ES] \qquad (4.9)$$

This can be rearranged to:

$$k_1[E]_{tot}[S] = (k_{-1} + k_2)[ES] + k_1[ES][S] \qquad (4.10)$$

Grouping terms gives:

$$k_1[E]_{tot}[S] = [ES](k_{-1} + k_2 + k_1[S]) \qquad (4.11)$$

Solving for $[ES]$ gives:

$$[ES] = \frac{[E]_{tot}[S]}{\left\{\dfrac{(k_{-1} + k_2)}{k_1}\right\} + [S]} \qquad (4.12)$$

Defining the ratio of the rate constants $(k_{-1} + k_2)/k_1$ as K_M, the Michaelis constant, gives:

$$[ES] = \frac{[E]_{tot}[S]}{K_M + [S]} \qquad (4.13)$$

To express this equation as a velocity or rate instead of simply in terms of $[ES]$, it is necessary to use the equation $v = d[P]dt = k_2[ES]$ (equation 4.3). Substituting gives:

$$v_0 = \frac{d[P]}{dt} = k_2[ES] = \frac{k_2[E]_{tot}[S]}{K_M + [S]} \qquad (4.14)$$

The maximal rate achievable (V_{max}) occurs when the entire enzyme is saturated with substrate, $[ES] = [E]_{tot}$. This can be substituted into equation 4.14 as shown:

$$v_0 = \frac{V_{max}[S]}{K_M + [S]} \tag{4.15}$$

This is the **Michaelis-Menten equation**, the central equation of **enzyme kinetics** that relates substrate concentration $[S]$ to two fundamental properties of the enzyme: the theoretical maximal velocity for a given concentration of enzyme (V_{max}) and the Michaelis constant (K_M). A plot of this curve, substrate concentration versus **initial rate** or **initial velocity**, gives a characteristic saturation kinetic: a rectangular hyperbola rotated 45° (**Figure 4.7**).

4.2.3 The Michaelis constant has several meanings

The Michaelis constant (K_M) has several meanings in biochemistry, all of which are equally valid. First, it can be seen as a ratio of rate constants. This ratio describes the rate constant of formation of the *ES* complex, divided by the rate constants of the breakdown of the complex. This provides a second way of thinking about the Michaelis constant, that is, as an equilibrium constant that describes the interaction of *E*, *S*, and *ES*. Therefore, the Michaelis complex also describes the affinity of substrate for enzyme (how tightly substrate and enzyme interact), which is an important and distinct parameter of enzyme kinetics. The lower the K_M value, the higher the affinity of the pair for each other. There are enzymes that have low K_M values and therefore bind substrate well but also have low k_2 values and do not rapidly catalyze reactions. These enzymes act as molecular timers, binding to substrate but then taking seconds to minutes to catalyze the reaction. The signaling proteins known as G proteins employ this mechanism; while bound to GTP, they activate other proteins but the effect stops once GTP has been hydrolyzed.

Finally, K_M can be also thought of as the concentration at which the enzyme operates at one-half of maximal velocity at a given concentration of enzyme ($V_{max}/2$). While this is a commonly used explanation for introductory students, please be aware that the K_M value for any enzyme is not defined by the V_{max} value and K_M is independent of enzyme concentration. This definition can be seen either on a plot of substrate concentration versus velocity or by analyzing the Michaelis-Menten equation as shown in Figure 4.7.

4.2.4 Kinetic data can be graphically analyzed

It is difficult to obtain meaningful data by looking at a graph of substrate concentration versus initial velocity. As computational tools have advanced, it has become easy to use software to perform a nonlinear curve fit to the dataset, to obtain the asymptote that represents V_{max}, and from there to calculate K_M from the $V_{max}/2$ value.

The classical way to interpret Michaelis-Menten kinetic data is to construct a **Lineweaver-Burke plot**.

$$\frac{1}{v_0} = \left(\frac{K_M}{V_{max}}\right)\frac{1}{[S]} + \frac{1}{V_{max}} \tag{4.16}$$

This is a double reciprocal plot of the data, with the point slope form of the line ($y = mx + b$) (**Figure 4.8**). The inverse of the substrate concentration is plotted against the inverse of the velocity. The resulting data give a straight line with a y-intercept of $1/V_{max}$ and an x-intercept of $-1/K_M$. The slope of this line is K_M/V_{max}.

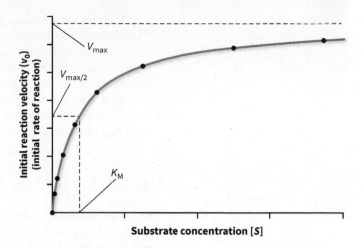

FIGURE 4.7 Saturation kinetic curve.   Shown is a graph of initial reaction velocity (reaction rate) versus substrate concentration. This curve illustrates saturation kinetics, the observation that, above a certain concentration of substrate, the rate cannot increase any further. Many enzymes will obey these kinetics.

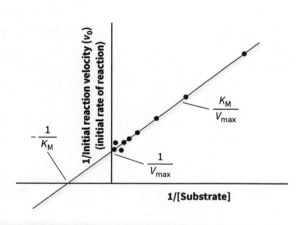

FIGURE 4.8 Lineweaver-Burke plot.
A Lineweaver-Burke plot is a different means of graphing the Michaelis-Menten equation. In this graph, $1/v_0$ on the y-axis is graphed versus $1/[S]$ on the x-axis. The y-intercept is $1/V_{max}$ and the x-intercept is $-1/K_M$.

The Lineweaver-Burke plot is the most common means of representing kinetic data. While it is relatively easy to read compared to a saturation kinetic curve, it has its own shortcomings. Rate data collected over the concentrations typically associated with enzyme kinetics experiments are often compressed near the x-axis. Second, points taken at lower concentrations (and therefore found farther down the x-axis in the reciprocal plot) can distract the viewer and skew the result, although this is not as significant a problem if the curve is being fit by a graphing program. Finally, the reciprocal nature of the axis means that any errors in the experiment become amplified in the graph.

4.2.5 k_{cat}/K_M is a measure of catalytic efficiency

So far we have been working with the approximation that k_2 is the catalytic step of the enzymatic reaction. In the simple reaction model shown in equation 4.2, k_2 is sometimes called k_{cat}. The k_{cat} value is calculated by dividing V_{max} by $[E]_{tot}$. The value k_{cat} is also known as the **turnover number** of the enzyme, that is, the number of reactions that the enzyme can catalyze per unit time (k_{cat} is a first-order rate constant with the units sec^{-1}).

At less than saturating conditions, most of the enzyme is found as free E rather than ES. In this instance, dividing k_{cat}/K_M gives the catalytic efficiency: a second-order rate constant that describes how often a reaction occurs for every encounter of enzyme and substrate.

4.2.6 Enzymatic reactions may be inhibited through one of several different mechanisms

Numerous small molecules can act as **enzyme inhibitors**. This section discusses the three major mechanisms through which such molecules can inhibit an enzyme, and the effect they have on kinetics. There are other means through which enzymes can be inhibited. For example, some enzymes can be directly poisoned through an **irreversible inhibitor**, also termed a **suicide substrate**. These inhibitors covalently modify the active site of the enzyme, irreversibly blocking its further function. In terms of a kinetic study, as more inhibitor is added, the enzyme seems to disappear: in effect, the catalyst has been inactivated. In terms of biology these inhibitors can have very serious effects. For example, many pesticides and nerve agents are potent irreversible inhibitors of acetylcholinesterase (**Figure 4.9**). Acetylcholinesterase is found in the synapse between neurons and muscle cells and functions to inactivate the neurotransmitter acetylcholine following a signal. Blocking this enzyme results in rigid muscle contraction, paralysis, and eventual death.

Competitive inhibition   Many enzymes recognize a single small molecule as their substrate, yet other small molecules with similar shape can also bind in the active site, competing directly with the substrate. These competitors, known as **competitive inhibitors**, may or may not

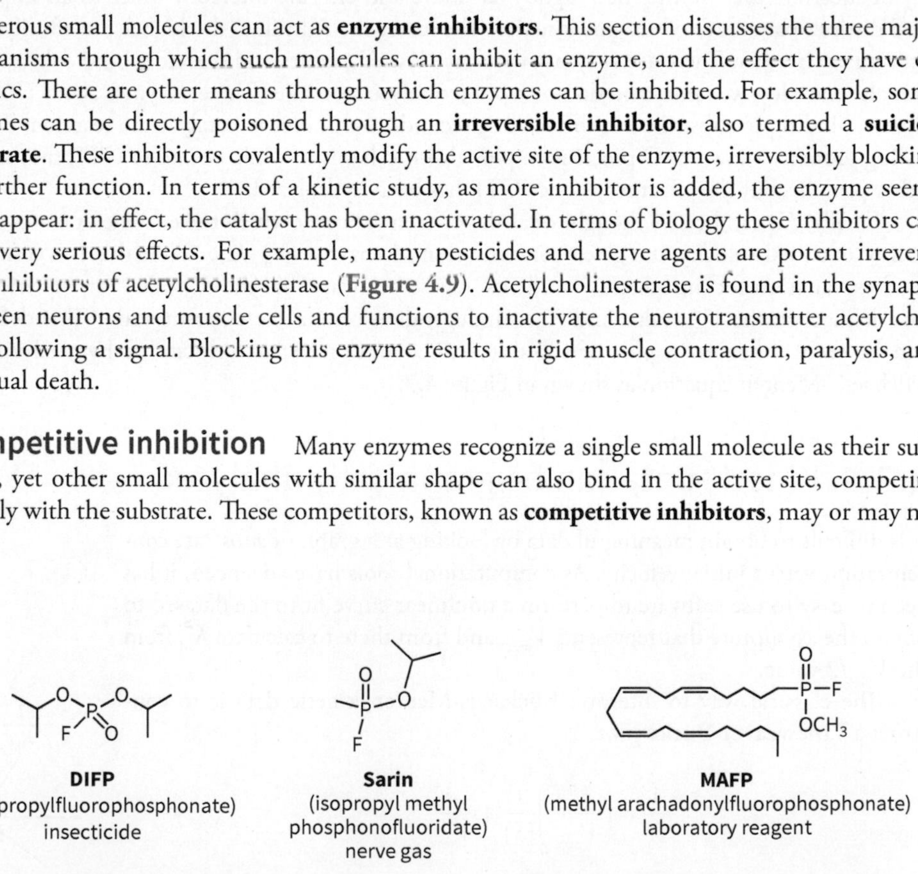

DIFP
(diisopropylfluorophosphonate)
insecticide

Sarin
(isopropyl methyl
phosphonofluoridate)
nerve gas

MAFP
(methyl arachadonylfluorophosphonate)
laboratory reagent

FIGURE 4.9 Irreversible enzyme inhibitors.   Examples of suicide enzyme inhibitors include the nerve gas sarin, the protease inhibitor DIFP (diisopropylfluorophosphate), and MAFP (methyl arachadonyl-fluorophosphonate). Each of these inhibitors includes a fluorophosphonate. In each of the enzymes, there is nucleophilic attack on the phosphorus. The fluoride (a good leaving group) departs, leaving the enzyme forever covalently bound to the active site nucleophile.

be converted to a product by the enzyme, but they tie up the enzyme and prevent the conversion of the substrate into product.

$$E + S \rightleftarrows ES \rightarrow E + P$$
$$+$$
$$I$$
$$\uparrow\downarrow$$
$$EI$$

where I is the inhibitor, and EI is the enzyme inhibitor complex. The competitive inhibitor ties up free enzyme but can be outcompeted with added substrate. In other words, if the substrate concentration is high enough, all of the free E will be saturated with substrate and become ES. Because the inhibitor cannot bind ES, it loses efficacy. Hence, competitive inhibitors do not affect V_{max}, but they do raise K_M.

The Michaelis-Menten equation for a system with a competitive inhibitor is:

$$v_0 = \frac{V_{max}[S]}{\alpha K_M + [S]} \tag{4.17}$$

where α is $1 + [I]/K_I$, $[I]$ is the concentration of I, and K_I is the dissociation constant of the EI complex:

$$K_I = \frac{[E][I]}{[EI]} \tag{4.18}$$

Plots including reactions with these inhibitors illustrate the change they have on V_{max} (**Figure 4.10**). The double reciprocal form of the Michaelis-Menten equation for a competitive inhibitor is:

$$\frac{1}{v_0} = \left(\frac{\alpha K_M}{V_{max}}\right)\frac{1}{[S]} + \frac{1}{V_{max}} \tag{4.19}$$

Uncompetitive inhibition

In contrast to competitive inhibitors, **uncompetitive inhibitors** bind only to the ES complex.

$$E + S \rightleftarrows ES \rightarrow E + P$$
$$+$$
$$I$$
$$\downarrow\uparrow$$
$$ESI$$

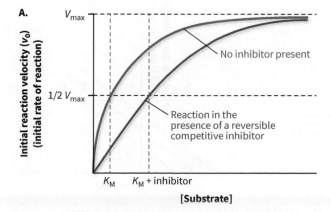

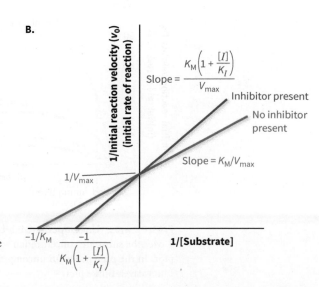

FIGURE 4.10 Competitive inhibition. A. In the presence of a competitive inhibitor, V_{max} is unchanged, and K_M is increased. **B.** In a Lineweaver-Burke plot, the y-intercept is unchanged, but the slope is changed.

By binding only to the *ES* complex, these inhibitors decrease V_{max} because they are lowering *ES*; however, they also decrease K_M by depleting *ES* complex.

The Michaelis-Menten equation for an uncompetitive inhibitor is:

$$v_0 = \frac{V_{max}[S]}{K_M} + \alpha'[S] \tag{4.20}$$

where

$$\alpha' = 1 + \frac{[I]}{K_I'} \tag{4.21}$$

with

$$K_I' = \frac{[ES][I]}{[ESI]}. \tag{4.22}$$

Please note that in uncompetitive inhibition, the inhibitor constant is α' instead of α.

The double reciprocal form of the Michaelis-Menten equation in this instance is:

$$\frac{1}{v_0} = \left(\frac{K_M}{V_{max}}\right)\frac{1}{[S]} + \frac{\alpha'}{V_{max}} \tag{4.23}$$

Here, both $1/V_{max}$ and $-1/K_M$ are altered by α', resulting in a *y*-intercept of α'/V_{max} and an *x*-intercept of $-\alpha'/K_M$ (**Figure 4.11**). When the substrate concentration in an uncompetitive situation is low, there is little *ES* complex and the uncompetitive inhibitor loses effectiveness. This is the opposite of the situation seen in competitive inhibition.

Mixed inhibitors

Mixed inhibitors bind to the enzyme in the presence or absence of substrate:

$$E + S \rightleftarrows ES \rightarrow E + P$$

$$
\begin{array}{ccc}
+ & & + \\
I & & I \\
\downarrow\uparrow & & \downarrow\uparrow \\
EI & & ESI
\end{array}
$$

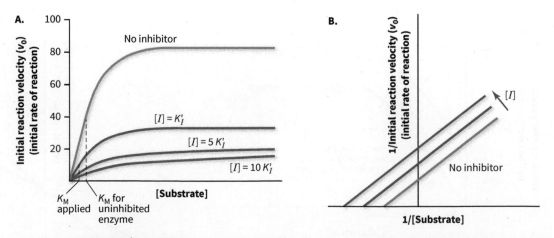

FIGURE 4.11 Uncompetitive inhibition.   A. Shown is a graph of a saturation kinetic plot. **B.** Graphing 1 over the substrate concentration versus 1 over the velocity gives a double reciprocal or Lineweaver-Burke plot. In the presence of an uncompetitive inhibitor, V_{max} and K_M are both altered, leading to parallel lines on a Lineweaver-Burke plot.

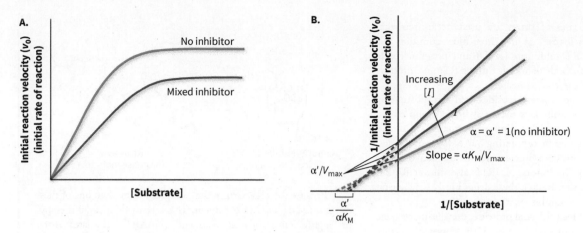

FIGURE 4.12 Mixed inhibition.   A. In the case of a mixed inhibitor, both V_{max} and K_M are altered.
B. Lineweaver-Burke plots of these data show the lines intersecting somewhere in the plane to the left of the y-axis.

The Michaelis-Menten equation for mixed inhibition is:

$$v_0 = \frac{V_{max}[S]}{\alpha \ K_M} + \alpha'[S] \tag{4.24}$$

Mixed inhibitors act as a combination of competitive and uncompetitive inhibitors, and they are effective regardless of substrate concentration.

The Lineweaver-Burke equation for a mixed inhibitor is:

$$\frac{1}{v_0} = \left(\frac{\alpha \ K_M}{V_{max}}\right)\frac{1}{[S]} + \frac{\alpha'}{V_{max}} \tag{4.25}$$

The Lineweaver-Burke plot of this equation gives a series of lines that intersect at a point left of the y-axis but above the x-axis (**Figure 4.12**). In situations where $\alpha = \alpha'$, the line will intersect on the x-axis, in which case the inhibition is referred to as noncompetitive or pure mixed inhibition.

A summary of how these inhibitors act is found in **Table 4.2**.

An example of enzyme inhibition and how it applies to modern medicine can be found in **Societal and Ethical Biochemistry: Enzyme inhibitors as lifesavers-antiretroviral drugs.**

TABLE 4.2 Effects of Enzyme Inhibitors on Kinetic Parameters

Inhibitor	Apparent K_m	Apparent V_{max}
None	K_m	V_{max}
Competitive	αK_m	V_{max}
Uncompetitive	$K_{m/\alpha'}$	$V_{max/\alpha'}$
Mixed	$\alpha K_{m/\alpha'}$	$V_{max/\alpha'}$

Societal and Ethical Biochemistry

Enzyme inhibitors as lifesavers-antiretroviral drugs

This chapter discusses enzyme inhibitors and the different mechanisms of inhibition. Enzyme inhibitors are a useful laboratory tool for probing enzyme function and mechanism, but their benefits are much wider than that; for example, many everyday drugs, including the antiretroviral drugs used to treat human immunodeficiency virus (HIV) are enzyme inhibitors.

HIV (the virus that causes acquired immunodeficiency syndrome or AIDS) is a retrovirus, meaning that it uses RNA rather than DNA as its genetic material. As part of its life cycle, HIV uses reverse transcriptase (RT) to copy the viral genome into a DNA copy inserted into the host's genome. The viral genome contains the code for all viral proteins. Several of these proteins are transcribed as a single protein that is then cleaved into mature proteins by a specific viral protease. Healthy mammalian cells lack both RT and the viral protease; therefore, these two enzymes are suitable targets for pharmaceutical treatments aimed at blocking the spread of the virus.

Several different classes of drugs are used in antiretroviral therapy, but this section focuses on the three most common and well-established classes: RT inhibitors (RTIs) and non-nucleoside RT inhibitors (NNRTIs), both of which inhibit RT; and protease inhibitors, which inhibit the viral protease.

RTIs are analogs of the deoxynucleosides used as a substrate by RT. Nucleoside analogs typically lack the 3′ hydroxyl group on the ribose or deoxyribose ring, making it impossible for RT to extend the sequence of the DNA after incorporating one of these molecules. These analogs compete with deoxynucleosides for the active site; as such, they are competitive inhibitors. NNRTIs also inhibit RT but do not bind in the active site. Rather, these drugs bind elsewhere on RT and are noncompetitive inhibitors.

Protease inhibitors block the viral protease, preventing cleavage of the viral proteins and synthesis of mature virus. A notable feature of these inhibitors are the peptide bonds apparent in their structure. Protease inhibitors bind in the active site of the viral protease and are potent competitive inhibitors of the protease. Newer variants of these drugs have modifications—tetrahedral transition-state analogs or structures that use linkages other than peptides—that make them more resistant to breakdown and metabolism in the body, increasing the half-life and efficacy of the drug.

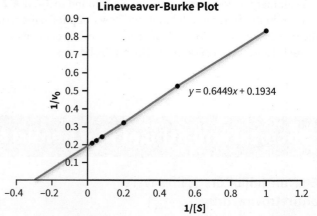

Azidothymidine (AZT)
(nucleoside analog)

Ritonavir
(protease inhibitor)

Because the HIV virus mutates rapidly, more than one of these drugs is usually used in treatment. This is termed "cocktail therapy" or "highly active antiviral treatment" (HAART). The latest drugs used in the treatment of HIV target other aspects of the viral life cycle, such as the entry of the virus into the host cell or integration of the viral genome into the host's genome.

As we move into the fourth decade of the AIDS epidemic, HIV infection has become a disease that can be managed, but it is still a significant health threat. In 2016 the World Health Organization (WHO) reported that over 36.7 million people are infected with HIV, including 2.1 million children. It still causes over a million deaths annually.

4.2.7 Many reactions have more than one substrate

This chapter has focused on **Michaelis-Menten kinetics** with a single substrate and a single product. However, most reactions have more than one substrate and more than one product. Similar equations describe the behavior of enzymes that have two substrates and two products, but the basic parameters (K_M and V_{max}) are still observed. These reactions can be defined by the order in which substrates bind and products leave.

Worked Problem 4.2   Calculating kinetic parameters

An enzyme is reacted with its substrate, A, for 1 minute at 37°C and yields the following rates.

[A], (mM)	Initial velocity, v_0 (µMol/min)
1	1.2
2	1.9
5	3.1
12.5	4.1
20	4.5
40	4.8

What are K_M and V_{max} for this enzyme?

Strategy   Given the kinetic data, how could you transform it to get K_M and V_{max}?

Solution

Lineweaver-Burke Plot

$y = 0.6449x + 0.1934$

1/v_0 vs 1/[S]

A Lineweaver-Burke plot of the data will give the values of V_{max} (the y-intercept is $1/V_{max}$, 5.2) and K_M (the x-intercept is $-1/K_M$, 3.33 mM).

Follow-up question   What other information would you need to calculate k_{cat}?

The equations and concepts outlined in the study of Michaelis-Menten kinetics have been used to describe many situations in which it might be anticipated that these kinetics would not work, and yet they often do. Such situations include the metabolism of ethanol by the liver, protein–protein interactions, and growth of bacteria. Despite the usefulness and flexibility of the Michaelis-Menten equation, caution must be employed to ensure that the assumptions made in the analysis of any system do not confound the actual biochemistry occurring.

Summary

- The rates of enzymatically catalyzed reactions can be examined mathematically.
- Analysis of the rate of a chemical reaction provides important data that are useful in elucidating the mechanism of the reaction.
- Kinetic parameters such as V_{max} and K_M can be obtained from plots of kinetic data.
- The efficiency of an enzyme can be calculated as the ratio of k_{cat}/K_M.
- Some enzymes have achieved catalytic perfection; they carry out a reaction almost every time they encounter a substrate molecule.
- Enzymes can be inhibited by several different mechanisms.
- Competitive inhibitors compete with substrate for the active site; they increase K_M but do not affect V_{max}.
- Uncompetitive inhibitors bind solely to the ES complex; they decrease V_{max} and modify K_M.
- Mixed inhibitors bind to either the free enzyme or the ES complex; they decrease V_{max} and modify K_M.
- Noncompetitive inhibition occurs where both α and α' are equal in a mixed inhibitor.
- Many reactions have more than one substrate and product. These reactions can be classified and their kinetics defined by the number of substrates and products and by the order in which they bind and leave.

Concept Check

1. Describe the Michaelis-Menten equation and the meaning of the various components such as V_{max} and K_M.
2. Explain the behavior of enzymes in terms of parameters V_{max} and K_M.
3. Construct and interpret a double reciprocal plot to extract K_M and V_{max} from raw kinetic data.
4. Using graphs and equations, explain the difference between the three types of inhibition in terms of how these small molecules act on the enzyme and how they affect kinetic parameters.

4.3 The Mechanism of an Enzyme Can Be Deduced from Structural, Kinetic, and Spectral Data

From a chemical perspective, the mechanism of any reaction is a valuable piece of information. The mechanism gives us information about how the reaction proceeds (or does not) at one of the most fundamental levels. It explains why some products and not others are seen. The information found in a chemical mechanism provides basic information that helps in predicting what will happen in future experiments. The same is true for biochemistry. An enzymatic mechanism ties together numerous pieces of chemical and structural data to give a concise explanation of the chemistry that is believed to occur. To clarify, mechanisms are models; although data can be consistent or inconsistent with a mechanism, they cannot be proven. Nevertheless, understanding the fundamental chemistry occurring in an enzyme is valuable when it comes to understanding why some substrates will react and others will not, or in designing enzyme inhibitors or drugs.

One other point of clarification: the term mechanism has more than one use in biochemistry. When we are discussing how an enzyme catalyzes a reaction, we refer to the mechanism by which that enzyme catalyzes the reaction. This is a typical organic or inorganic reaction mechanism shown with arrows pushing electrons. There is, however, a broader use of the term mechanism in biochemistry. When discussing how a drug elicits an effect, a mechanism of action of that drug (for example, signaling through a G protein coupled receptor) may be mentioned. This use of the term mechanism uses a pathway, a series of reactions, or an accepted chain through which proteins interact. In this regard, the term mechanism does not include the atomic level of detail in a reaction mechanism, but it is nevertheless a valid use of the term.

4.3.1 Enzymatically catalyzed reactions have common properties

In thinking about organic reaction mechanisms, we may consider the general categories of mechanisms (S_N1 or E2), or something more involved (4+2 cycloaddition). Other questions arise, such as the effect of leaving groups, solvent selection, bond polarity, electron withdrawing or donating groups, or the effects of stereochemistry. Collectively, all these contribute to the overall reaction and the type of mechanism that occurs.

In examining an enzymatically catalyzed reaction, many of the same conditions arise, but there are several important distinctions. Because enzymes lower activation energy, several general principles and guidelines are followed, and relatively few types of catalysis can occur.

The general principles are as follows:

- Enzymes bind substrate and orient substrates properly for chemistry to occur. Hence, they favor reactions by organizing and positioning groups to prime them for reactions. This lowers entropic contributions (ΔS) to the free energy of the transition state.

- Enzymes bind substrates using a large number of weak forces. This is sometimes referred to as electrostatic catalysis and is the sum total of the weak forces acting on the substrate to effect chemical change.

- Using induced fit, enzymes bind substrates in a way that favors the transition state. If we think of the hand-in-glove analogy, the glove is more of a baseball glove that positions the hand to properly catch the ball.

In relation to general chemical mechanisms, there are three distinct categories for these types of catalysis. However, there are many instances where more than one type of mechanism is employed; for example, a metal ion may activate an acid-base catalyst, or a combination of covalent catalysis and acid-base catalysis may be found in a single mechanism. The general chemical mechanisms are as follows:

- **General acid-base catalysis**, in which an amino acid side chain either donates or accepts a proton, causing reactions to occur. The polar and charged amino acids play important roles in this form of catalysis.

- **Metal ion catalysis**, in which an active site metal ion can act as a redox active center, or as a Lewis acid or base (donating or accepting electrons as needed), or can assist in the polarization of water molecules to generate hydroxide. Metal ions are found in a third of all enzymes; the metals include zinc, magnesium, iron, calcium, copper, nickel, cobalt, and other polyvalent metal ions.

- **Covalent catalysis**, in which nucleophilic or electrophilic attack on an atom results in a covalent intermediate formed between the enzyme and substrate. The mechanistic steps in forming and breaking the intermediate are at lower transition state energies than the uncatalyzed reaction. The completion of the catalytic cycle releases the products and regenerates the active site residues. Serine, histidine, aspartate, cysteine, lysine, tyrosine, and the cofactors thiamine pyrophosphate and pyridoxal phosphate often participate in covalent catalysis.

A review of chemical reaction mechanisms and how they are used to describe enzymatic catalysis can be found in **Biochemistry: Chemical reaction mechanisms**.

Biochemistry

Chemical reaction mechanisms

Chemical reaction mechanisms indicate how scientists believe electrons flow and the details of how a reaction occurs. Mechanisms are valuable because they provide a way to explain how a reaction proceeds, and they connect structural, chemical, and kinetic data to give an elegant picture of what is happening at the molecular level.

Chemical mechanisms will be familiar from organic chemistry. In these mechanisms, the flow of a pair of electrons is depicted with a two-headed arrow:

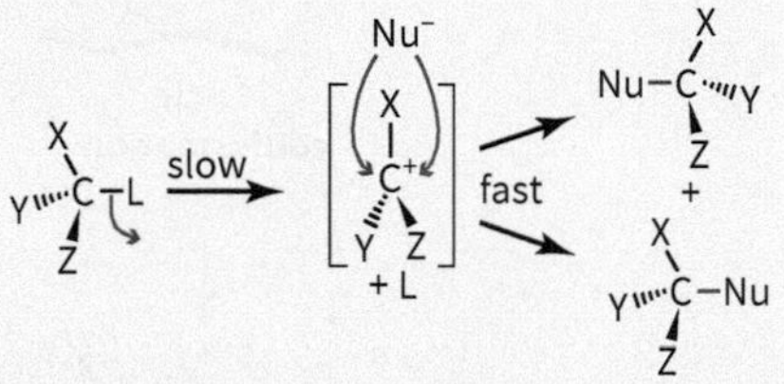

Electrons flow from regions of high-electron density (a double bond, lone pair of electrons, or other electron-rich center) to regions of low-electron density (a polarized hydrogen in a hydroxyl group or a carbonyl carbon):

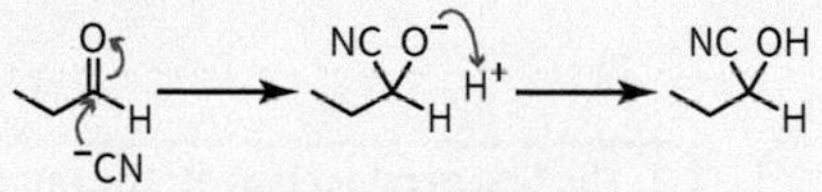

Movement of a single electron can be depicted using a single-headed arrow. The same rules apply for reactions using inorganic centers (metal ions), observed in some oxidation reactions and in the electron transport chain in mitochondria.

Single-headed arrows represent movement of a single electron.

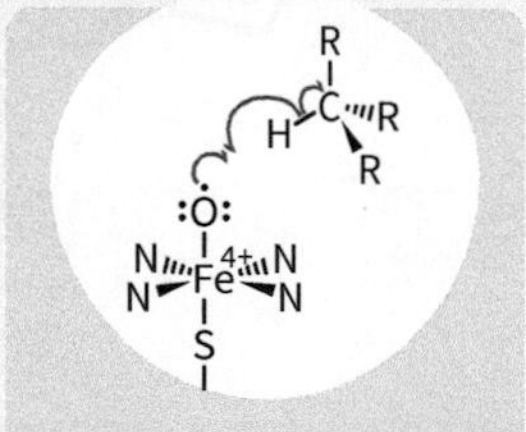

In biochemistry, the same rules apply, but there are several other things to keep in mind. Most reactions in biochemistry are catalyzed; thus, the geometry and topology of the active site plays a key role in the catalytic mechanism. Likewise, because of the positioning of nucleophiles and electrophiles in the active site, pK_a values can be significantly altered within the enzyme. An example of this is water. Activated water molecules that are coordinated and polarized in a mechanism become potent nucleophiles, a phenomenon that is not observed when water is in bulk solution.

A coordinated and polarized water molecule becomes a nucleophile and attacks the phosphorous.

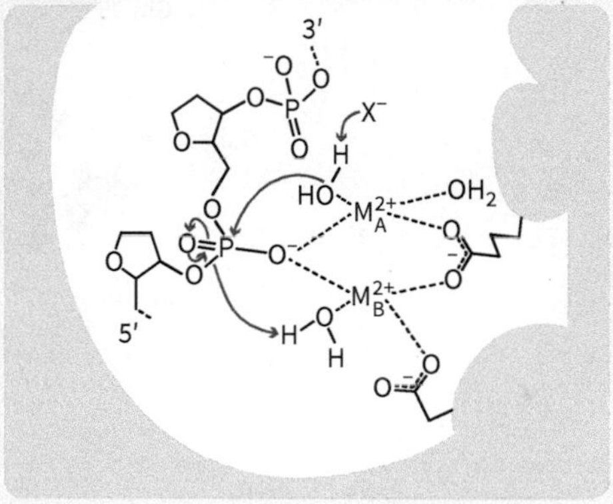

A second consideration for enzymatically catalyzed reactions is their complexity. A biochemically catalyzed mechanism may well have upward of 10 to 15 arrows and multiple steps. Do not be intimidated by this. If you can explain the hydration of an alkene, you can use the same logic and rules (electron flow, electronegativity, bond polarity, and geometry) to explain an enzymatically catalyzed reaction.

4.3.2 Examining examples of enzymatic reaction mechanisms illustrates underpinning principles

Mechanisms are known for many enzymes, and three representative examples are discussed here to illustrate the underpinning principles.

Acid–base catalysis: lysozyme
Lysozyme is a natural antibiotic found in mucus and tears, and at high levels in egg white. Lysozyme is a hydrolase and cleaves a link between

carbohydrate chains found in the cell wall of bacteria. When this cell wall loses structural integrity, the bacteria lyse and die.

Lysozyme is a compact enzyme, only 128 amino acids in length (14.3 kDa). The structure is slightly oblong and consists of two domains, one with several short helical stretches and one with a short portion of β sheet, connected by an α-helical linker. The active site is found in a trench that runs the length of the enzyme between the two domains and accommodates the carbohydrate substrate.

The active site of lysozyme has two negatively charged residues that are necessary for catalysis: an aspartate in position 52 (Asp 52) and a glutamic acid in position 35 (Glu 35) (**Figure 4.13**). In the mechanism shown, the glutamic acid is protonated and the aspartate is not. The mechanism can only proceed if this is the case.

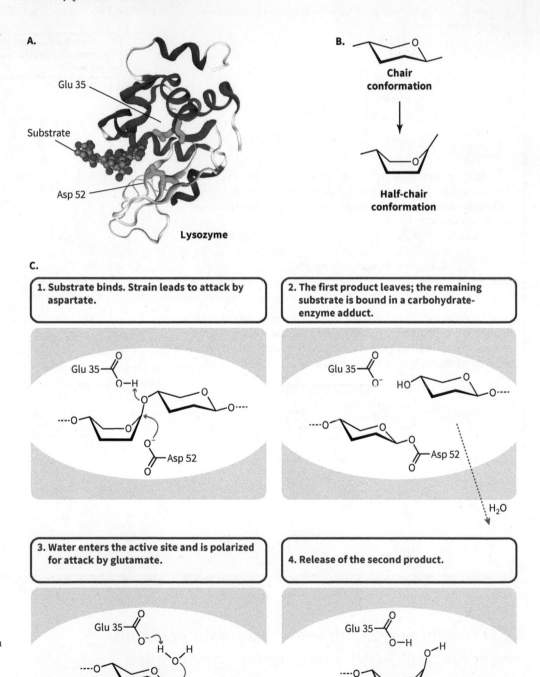

FIGURE 4.13 Mechanism of lysozyme. A. Shown is the structure of lysozyme with the active site residues and substrate highlighted in a ball and stick depiction. **B.** Binding of carbohydrate to lysozyme results in a conformational change in the carbohydrate ring (from chair to half-chair). **C.** Lysozyme induces strain in the molecule, then uses an activated water for catalysis.

(Source: Data from PDB ID 9LYZ Kelly, J.A., Sielecki, A.R., Sykes, B.D., James, M.N., Phillips, D.C. (1979) X-ray crystallography of the binding of the bacterial cell wall trisaccharide NAM-NAG-NAM to lysozyme. *Nature* 282: 875–878)

It took nearly 60 years from the initial discoveries of lysozyme activity to the elucidation of its structure. A mechanism was first proposed in 1966 and was debated for over 35 years until an alternative mechanism was proposed in 2001. Both of these mechanisms have merit and have data in support of them. Indeed, years later they are still being debated.

The currently accepted mechanism for lysozyme proceeds as follows. The carbohydrate substrate of lysozyme is a polymer of six-membered rings. These rings have a similar conformation to cyclohexane and normally exist in a chair conformation. Binding of the carbohydrate to lysozyme distorts the first ring of the carbohydrate, changing it from a chair conformation to a half-chair conformation. The oxygen forming the linkage between the first and second rings attacks the proton of the active site glutamate. This results in cleavage of the bond between the rings and release of the second ring side of the carbohydrate chain. The first ring can exist in several resonance states, including that of an oxonium ion. This transition state is attacked by the active site aspartate, which forms a covalent adduct with the first ring. Water then enters. The active site glutamate orients the water and deprotonates it, generating a hydroxyl ion, which then attacks the intermediate at the linkage between the first ring and the aspartyl group of the enzyme. This causes release of the first-ring side of the carbohydrate chain, and regeneration of the active site.

Metal based catalysis: carbonic anhydrase

Carbonic anhydrase is a relatively small enzyme of 29 kDa (**Figure 4.14**). It is found in numerous cells, including erythrocytes (red blood cells), the kidney, and the stomach where it catalyzes the conversion of CO_2 into H_2CO_3, (carbonic acid), which rapidly ionizes into H^+ and HCO_3^- (bicarbonate). Bicarbonate serves as the most important buffering system in the body, but it is also involved in kidney function and in maintaining the hydrostatic pressure in the eye. The direction of this reaction depends on the concentrations of substrates available.

Carbonic anhydrase is largely spherical in shape. It has one large β sheet that dominates the structure. Part of this sheet is the floor of the active site, which lies in a deep (15 Å) crevice formed by several short stretches of helix, and several turns and coils.

The active site of carbonic anhydrase contains a Zn^{2+} ion, held in place by the imidazole rings of three histidine residues (**Figure 4.15**). The fourth coordination site of the Zn^{2+} binds a water molecule. Two other residues (a glutamate and a threonine, not shown) help to position and polarize the water molecule through hydrogen bonding. Another residue important to the mechanism is histidine 64. While the other residues position either the zinc ion or water, His 64 is important in shuttling protons out of the active site.

The mechanism begins by the zinc ion binding to water. Binding polarizes the water to the point of deprotonation, generating a hydroxyl ion, which is stabilized by the zinc ion. This nucleophile attacks the carbon of carbon dioxide, generating HCO_3^-, which is released. The proton

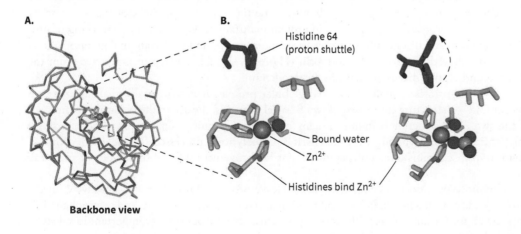

A.

Backbone view

B.

Histidine 64 (proton shuttle)

Bound water

Zn^{2+}

Histidines bind Zn^{2+}

C.

$$CO_2 + H_2O \underset{k_{-1}}{\overset{k_1}{\rightleftharpoons}} \underset{\text{Carbonic acid}}{HOCOOH} \rightleftharpoons \underset{\text{Bicarbonate ion}}{HOCO^-O} + H^+$$

FIGURE 4.14 Carbonic anhydrase structure and reaction. **A.** The active site of carbonic anhydrase is found at the bottom of a trench lined by a β sheet. **B.** His 64 acts as a proton shuttle moving protons out of the active site. **C.** Carbonic anhydrase catalyzes the reaction of water and carbon dioxide to form carbonic acid, which ionizes to form bicarbonate ion.

(Source: (A, B, Left) Data from PDB ID 1CA2 Eriksson, A.E., Jones, T.A., Liljas, A. (1988) Refined structure of human carbonic anhydrase II at 2.0 A resolution. *Proteins* 4: 274–282. (Right) Data from PDB ID 1CAM Xue, Y., Liljas, A., Jonsson, B.H., Lindskog, S. (1993) Structural analysis of the zinc hydroxide-Thr-199-Glu-106 hydrogen-bond network in human carbonic anhydrase II. *Proteins* 17: 93–106)

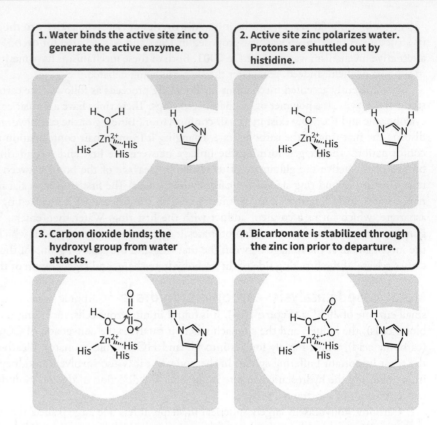

1. Water binds the active site zinc to generate the active enzyme.

2. Active site zinc polarizes water. Protons are shuttled out by histidine.

3. Carbon dioxide binds; the hydroxyl group from water attacks.

4. Bicarbonate is stabilized through the zinc ion prior to departure.

FIGURE 4.15 Mechanism of carbonic anhydrase. Carbonic anhydrase uses an active site zinc complexed to three histidines to polarize a water molecule. Protons from the active site are shuttled through two fixed water molecules to transfer the proton to a histidine residue that shuttles the proton out of the active site.

released from the water molecule in the first step is shuttled out of the active site through binding to a network of two intervening hydrogen-bonded water molecules, ultimately transferring to His 64. The ring of His 64 can pivot on the methylene group to transfer these protons out of the active site cleft to buffers in the surroundings of the enzyme.

Carbonic anhydrase is found in almost all species, from corals to plants to mammals, but in each case it serves different functions ranging from carbonate deposition to carbon fixation in photosynthesis, to acid–base balance. Although these enzymes differ in structure, each employs a Zn^{2+} ion chelated in place by three highly conserved histidine residues. This is an example of convergent evolution—taking different evolutionary paths to arrive at a structure that performs the same chemistry.

Covalent catalysis: chymotrypsin, a serine protease

Proteases are enzymes that degrade proteins, cleaving peptide bonds. Some proteases are specific, cleaving only a select sequence; such proteases are important in protein maturation, blood clotting, trafficking of proteins in the cell, and disease. Other proteases are more generic and important in digestion; among these is the digestive enzyme chymotrypsin (**Figure 4.16**). Chymotrypsin is secreted from the pancreas and cleaves dietary proteins in the duodenum.

Chymotrypsin is a small (25.6 kDa) globular protein. It is structurally characterized as a trypsin fold—a structure comprised of two β barrel domains, the active site being at the interface of the two. This structure is found in many proteases. Chymotrypsin is synthesized as a single polypeptide (chymotrypsinogen), which itself is proteolytically processed. The final active form of chymotrypsin contains three polypeptide chains held together by five disulfide bonds and weak forces.

Chymotrypsin does not indiscriminately cleave proteins; instead, it cuts selectively on the carboxyl side of bulky hydrophobic residues such as tyrosine, phenylalanine, and tryptophan. This provides an important clue as to the topology of the active site but is more important in substrate recognition than in catalysis. This substrate recognition pocket (termed the S1 pocket) is large enough to accommodate the bulky residue, and it is lined with hydrophobic residues that also facilitate binding of the substrate.

Adjacent to the substrate recognition pocket are two regions important in catalysis. One of these is the charge relay system (also known as the catalytic triad), which consists of three key residues: aspartate, histidine, and serine, with the latter being the nucleophile in the catalytic

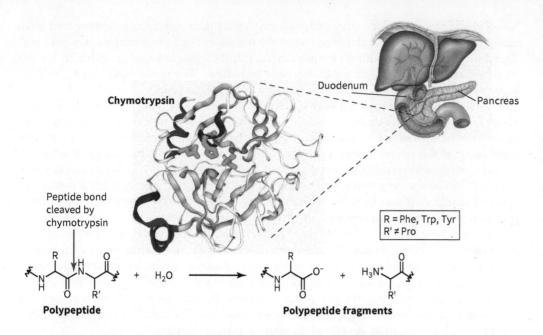

FIGURE 4.16 Chymotrypsin structure and reaction. Chymotrypsin is a protease, an enzyme that cleaves a peptide bond. The enzyme consists of two distinct β sheet domains. The active site is found at the interface of these two domains.

(Source: Tortora; Derrickson, *Principles of Anatomy & Physiology*, 14e, copyright 2013, John Wiley & Sons. This material is reproduced with permission of John Wiley & Sons, Inc. Data from PDB ID 1T7C Czapinska, H., Helland, R., Smalas, A.O., Otlewski, J. (2004) Crystal structures of five bovine chymotrypsin complexes with P1 BPTI variants. *J. Mol. Biol.* 344: 1005–1020)

mechanism. Also nearby is a glycine residue that participates in the oxyanion hole, the other important region which is formed by a gap created by the peptide backbone and glycine which stabilizes intermediates as the reaction proceeds.

The mechanism of chymotrypsin employs a nucleophilic attack and a covalently bound intermediate
The reaction mechanism of chymotrypsin begins with the binding of a protein in the active site (**Figure 4.17**). The hydroxyl group of the serine residue would probably be protonated if this amino acid were on the surface of the protein or free in solution; however, this is not the case in the active site, where the serine is adjacent to the

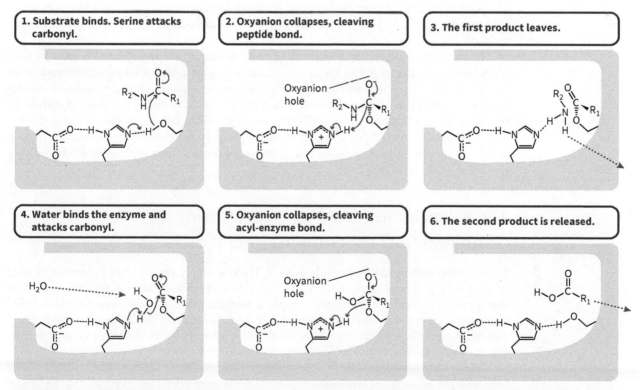

FIGURE 4.17 Mechanism of chymotrypsin.   The active site of chymotrypsin has three critical residues for catalysis (the catalytic triad or charge transfer system). A deprotonated serine attacks the peptide carbon forming an acyl enzyme intermediate. Then water enters the active site and cleaves the acyl enzyme intermediate.

histidine and aspartate of the charge relay system. Rather than existing as a deprotonated imidazole and protonated hydroxyl, the pair share the proton through a low-barrier hydrogen bond. Therefore, the interaction with the histidine means that the serine hydroxyl group has far more of an alkoxide-like character than would otherwise be possible. This deprotonated serine hydroxyl becomes the nucleophile that attacks the carbonyl carbon of the peptide.

As electrons flow from the hydroxyl into the carbonyl carbon and oxygen, the formerly planar sp^2 carbonyl center changes to a tetrahedral sp^3 center. The peptide is now covalently linked to the enzyme through the serine residue. The reaction is stabilized by the oxyanion hole. The stabilization of this transition state is one of the most important steps in the catalytic mechanism.

The tetrahedral oxyanion intermediate is short-lived, however. As it collapses and electrons flow back out of the oxygen through the bond network, the carbon-nitrogen bond breaks, and nitrogen acquires a proton from the histidine of the catalytic triad. The fragment of the substrate protein with the new amino terminus leaves the active site.

Water now enters the active site. The active site histidine deprotonates this water, and the resulting hydroxyl group attacks the acyl-enzyme intermediate (the ester linking the carboxyterminus of the substrate protein to the active site serine). Again, electrons flow up through the carbonyl group, forming a tetrahedral intermediate stabilized by the oxyanion hole. As electrons flow back through this center and this intermediate collapses, the carbon–oxygen bond is broken, and the serine hydroxyl returns to share its proton with histidine. The protein containing the newly generated carboxyl terminus leaves, and the catalytic cycle is complete.

Armed with this information, it is possible to probe the structure and mechanism of chymotrypsin further. If specific site-directed mutants of the enzyme were generated, the effects of these mutations on the kinetics or mechanism of an enzyme could be studied. Creating a mutation in the S1 pocket might affect K_m (affinity of substrate for enzyme) but not catalysis, whereas inserting a bulky residue in place of the glycine in the oxyanion hole might do the reverse, affecting catalysis (V_{max}, k_2, or k_{cat}) but not substrate binding (K_M). In reality, mutations designed to alter the specificity of chymotrypsin by placing charged residues in the S1 pocket resulted in an enzyme that had no catalytic activity. In this instance, the mutation may have affected protein folding, or it may be that the binding of substrate is more complex than is currently thought.

Other proteases share some properties with chymotrypsin. Trypsin and elastase both have substrate-recognition pockets, but these enzymes have different substrate specificities. Trypsin's S1 pocket is specific for basic residues, and its substrate-recognition pocket contains an aspartate residue, whereas the S1 pocket of elastase is specific for short-chain hydrophobic residues and is hydrophobic.

Chymotrypsin is a member of a class of proteins termed **serine proteases** because of their active site serine (**Figure 4.18**). Other enzymes (typically proteases or esterases) also employ an active site serine that is nucleophilically primed by a catalytic triad. Cysteine proteases are among the enzymes that use an active site cysteine as the nucleophile to attack the carbonyl carbon. In this instance, a neighboring amino acid increases the nucleophilicity of the attacking thiolate through deprotonation. Other proteases include aspartyl proteases and metalloproteases, both of which use an activated water molecule (a hydroxyl group) as the nucleophile. In this regard they are similar to carbonic anhydrase.

4.3.3 Mechanisms are elucidated using a combination of experimental techniques

Enzyme mechanisms cannot be truly proven. There is simply the presence or absence of data to support a proposed mechanism. A combination of structural, kinetic, spectral, and other biochemical data combine to validate a proposed mechanism. That mechanism may still be called into question if new data arise, but a combined body of data reduces the likelihood of this happening. As an example, let us briefly review the data in support of the mechanism of chymotrypsin. The main data are based on affinity labeling, structural studies, and kinetics, as outlined below.

Diisopropylfluoridate (DIPF) is an affinity label, that is, a molecule that will specifically label only one of the serine residues of chymotrypsin (**Figure 4.19**). This suggests that the labeled serine is different than the other serine residues in the protein; that is, it is implicated in the catalytic mechanism—a prediction that is corroborated by other data. Another affinity label, tosyl-L-Phenylalanine chloromethyl ketone (TPCK), irreversibly derivatizes only one histidine residue in

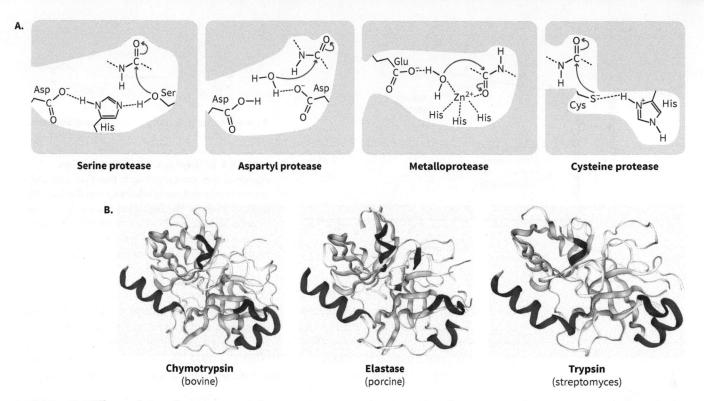

FIGURE 4.18 Different classes of proteases. **A.** Serine proteases, aspartyl proteases, metalloproteases, and cysteine proteases all employ similar chemistry: nucleophilic attack on the carbonyl carbon. **B.** Note that these enzymes not only employ similar nucleophiles, they have a conserved topology and overall structure, although not necessarily the same amino acid sequence.

(Source: (Left) Data from PDB ID 1T7Z Delker, S.L., West Jr., A.P., McDermott, L., Kennedy, M.W., Bjorkman, P.J. (2004) Crystallographic studies of ligand binding by Zn-alpha2-glycoprotein. *J.Struct.Biol.* 148: 205–213. (Middle) Data from PDB ID 1QNJ Wurtele, M., Hahn, M., Hilpert, K., Hohne, W. (2000) Atomic Resolution Structure of Native Porcine Pancreatic Elastase at 1.1 A. *Acta Crystallogr.*, Sect. D 56: 520. (Right) Data from PDB ID 1SGT Read, R.J., James, M.N. (1988) Refined crystal structure of Streptomyces griseus trypsin at 1.7 A resolution. *J.Mol.Biol.* 200: 523–551)

the entire protein, the histidine residue of the active site. Labeling with both of these molecules fails to occur if the protein is denatured, and labeling can be blocked by competitive inhibitors. Now that we have seen the mechanism of chymotrypsin, these data should begin to make sense. Of all of the serine and histidine residues available in chymotrypsin, the two found in the active site are specifically labeled because they participate in the hydrolysis reaction. This does not explicitly lead us to a mechanism, but it does implicate these two residues in the process.

In terms of structural studies, the structure of chymotrypsin has been solved by X-ray diffraction, which provides the coordinates of the atoms used in proposing these mechanisms. Other proteins that interact with chymotrypsin have had their structures solved by either X-ray diffraction or nuclear magnetic resonance (NMR) and can be shown to interact with chymotrypsin as predicted.

Chymotrypsin has an acyl-enzyme intermediate in which the carboxy terminus of the cleaved peptide bond is transiently bound to the active site serine. This can be demonstrated with the artificial substrate paranitrophenylacetate, which is colorless until the acetate moiety is cleaved, giving paranitrophenol, which has a yellow color. When paranitrophenylacetate is incubated with chymotrypsin, there is a rapid release of paranitrophenol as the

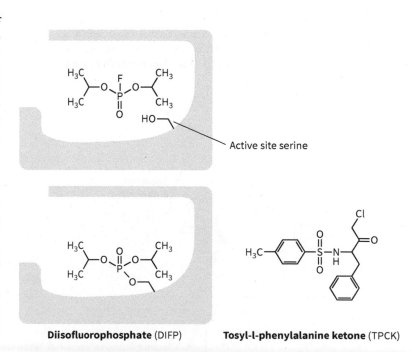

FIGURE 4.19 Affinity labels diisofluorophosphate (DIFP) and tosyl-L-Phenylalanine chloromethyl ketone (TPCK). DIFP and TPCK are both used as affinity labels, molecules that specifically derivatize key amino acid residues in the active site, such as this active site serine.

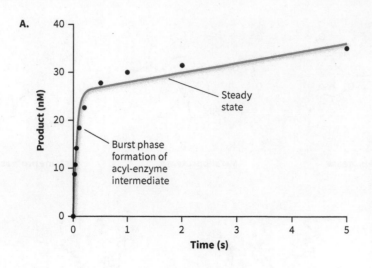

FIGURE 4.20 Biphasic kinetics. A. When examined over short timeframes (less than a second), chymotrypsin and many other enzymes demonstrate biphasic kinetics. The amount of product is measured in units of concentration (nanomolar). **B.** Reactions like this can be studied through the use of a colorimetric substrate, such as paranitrophenyl acetate.

enzyme undergoes the initial step of catalysis. However, to degrade more molecules, the enzyme needs to finish the catalytic cycle and resolve the existing acetyl-enzyme intermediates. This is seen in kinetic studies as a biphasic curve. In the first segment of the curve, the rapid hydrolysis and release of paranitrophenol is shown (**Figure 4.20**). This is followed by a second phase in which the enzyme is operating at the steady state. The slopes of these lines indicate the relative rates of each of the steps of the mechanism with the burst phase having a steeper curve followed by the slower steady state.

There are other ways to probe the mechanism of enzymes. These include examination of spectra as enzymatic reactions proceed, to look for evidence of different states. An examination of the rates of reaction using substrates isotopically labeled in specific positions is termed an isotope effect.

Worked Problem 4.3　A proposed chemical mechanism

You identified a novel metalloprotease and obtained its structure. Propose a mechanism for this reaction.

Strategy Examine the active site. What molecules are present, and how might they participate in catalysis? How is this active site similar to ones we have seen previously? How does a metalloprotease function?

Solution The proposed mechanism is shown. Experimental data would be needed to validate this proposed mechanism.

Follow-up question If the active site glutamate in this mechanism were mutated to an aspartate, how might that effect the mechanism?

Summary

- Enzymes catalyze chemical reactions through transition state stabilization, which is achieved through decreased entropy, increased weak interactions, and induced fit.
- Enzymes use some combination of general acid–base, metal ion, or covalent catalysis to complete reactions.

- Lysozyme, carbonic anhydrase, and chymotrypsin are provided as examples of enzyme mechanisms. Each of these illustrates several of the principles of enzymatic catalysis.
- Proposed enzyme mechanisms are built up from several different sources of data; they present a rich picture of the steps of a chemical reaction in an enzyme. They can also provide important information to guide drug design, explain disease, and further scientific understanding.

Concept Check

1. Describe how enzymes generally catalyze reactions, including the forces involved.
2. Describe how lysozyme, carbonic anhydrase, and chymotrypsin catalyze reactions.
3. Identify the nucleophiles and electrophiles found in each mechanism.
4. Explain the data that support the proposed mechanism for chymotrypsin.

4.4 Examples of Enzyme Regulation

Tools can be useful in accomplishing a job, but only if they can be controlled. Most tools have controls that turn them on or off, or perhaps change their speed. The same applies to enzymes. Evolution has produced complex means of regulating the rate at which enzymes catalyze reactions.

Enzyme activity can be regulated by altering gene expression (altering levels of the enzyme itself), sequestering the enzyme in one compartment of the cell or one organ of the organism, or limiting the access of the enzyme to substrate (potentially also through compartmentalization). This section describes two specific means of regulating enzyme activity: **covalent modification** and **allosteric regulation**.

4.4.1 Covalent modifications are a common means of enzyme regulation

Covalent modification of proteins includes any covalent addition or removal of groups from the proteins. Two common means of activation, discussed here, are **proteolytic cleavage** and **phosphorylation**.

Proteolytic cleavage An example of an enzyme activated through proteolytic cleavage is chymotrypsinogen, the zymogen form of chymotrypsin (**Figure 4.21**). **Zymogens** are inactive enzyme precursors that require proteolytic activation. In the case of chymotrypsinogen, the active form of the protein (chymotrypsin) is a protease that could degrade proteins in the acinar cells of the pancreas in which it is made, harming the cell and the organism. To avoid this, chymotrypsin is synthesized as a zymogen. The full-length peptide (chymotrypsinogen) has 245 amino acids. Once secreted into the digestive tract, it is attacked by trypsin, which cleaves the zymogen between an arginine and an isoleucine residue in the amino terminus, facilitating two important structural changes. First, the peptide nitrogen of the arginine residue is now free to interact with the aspartate of the active site, stabilizing the glycine residue lining the pocket that stabilizes the tetrahedral oxyanion intermediate. This change in conformation allows the oxyanion hole to form and thus allows catalysis to proceed. Second, the cleavage between the arginine and isoleucine alters the conformation of the protein, exposing a sequence that is recognized by other molecules of chymotrypsin and leading to optimally activated enzyme. The three polypeptides found in chymotrypsin are linked together by disulfide bonds.

Phosphorylation **Protein kinases** are enzymes that add phosphates to the hydroxyl groups of serine, threonine, or tyrosine residues of proteins. Often, the phosphate comes from a donor molecule such as ATP or GTP. Phosphates are removed by a second family of enzymes termed **phosphatases**.

A phosphorylation event can be seen as throwing a switch with regard to protein function: turning an enzyme either on or off. This can happen and cause an effect in less than a millisecond,

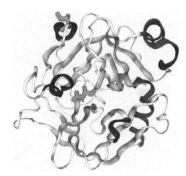

Chymotrypsinogen

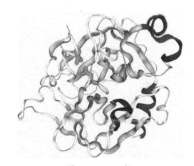

Chymotrypsin

FIGURE 4.21 Chymotrypsinogen and chymotrypsin. Chymotrypsinogen is the zymogen form of chymotrypsin. Proteolytic cleavage leads to activation of the enzyme. Note the movement of isoleucine 16 relative to the active site aspartate 194, both shown highlighted with green spheres.

(Source: (Top) Data from PDB ID 1EX3 Pjura, P.E., Lenhoff, A.M., Leonard, S.A., Gittis, A.G. (2000) Protein crystallization by design: chymotrypsinogen without precipitants. *J.Mol.Biol.* 300: 235–239. (Bottom) Data from PDB ID 1T7C Czapinska, H., Helland, R., Smalas, A.O., Otlewski, J. (2004) Crystal structures of five bovine chymotrypsin complexes with P1 BPTI variants. *J.Mol.Biol.* 344: 1005–1020)

or it can take several hours depending on the process. In considering the chemical changes going on at the surface of the protein that is being phosphorylated, it is possible to hypothesize how and why alterations in enzyme activity occur. Adding a phosphate group to a serine, threonine, or tyrosine residue adds a bulky group. This may cause steric hindrance between domains of a protein, between subunits, or between different polypeptides that need to interact for catalysis to occur. The phosphate also incorporates a negative two charge, which can participate in coulombic interactions, to a previously uncharged part of the protein. However, these interactions are tempered by the solvent between the two charges. The charge added to a protein through phosphorylation would probably interact with and influence many of the other charged groups in that protein, but would be unlikely to attract another charged molecule from across the cell. Nevertheless, if two proteins need to bind together for catalysis to occur, adding a negative charge to one protein could increase the likelihood of an interaction with the second protein. This would be dependent on the latter having a region of positive charge near the surface that could interact with the phosphate. Finally, the oxygen atoms bound to the phosphate have the potential to participate in more hydrogen bonds than the unphosphorylated side chain.

An example of regulation by phosphorylation is the signaling protein src (pronounced sark), a regulatory kinase originally identified and named for the Rous sarcoma virus (**Figure 4.22**).

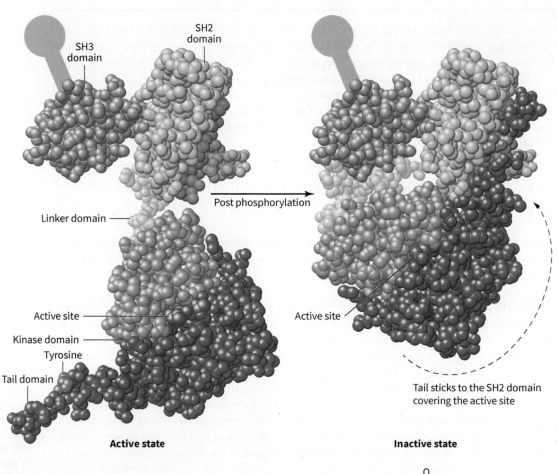

FIGURE 4.22 Src is regulated by phosphorylation.   Phosphorylation of a tyrosine residue in the tail of src causes increased electrostatic interactions with residues in the SH2 domain. This results in binding of the phosphotyrosine in the tail by SH2 domain, generating the inactive state of the protein.

(Source: Image from the RCSB PDB July 2003 Molecule of the Month feature by David Goodsell [doi:10.2210/rcsb_pdb/mom_2003_7])

The regulation of kinases and phosphatases occurs through a series of signaling steps termed a **signaling cascade**. Src is a kinase and an important player in several signaling cascades. The kinase has several domains: two SH domains (src homology domains), SH2 and SH3, that are important in binding to other proteins and substrate recognition; a flexible linker that acts as a hinge; a catalytic kinase domain; and a tail that contains a specific tyrosine residue. In the active state, the tail-domain tyrosine is not phosphorylated, the protein hangs open, and the active site is accessible to substrate. However, when the kinase is phosphorylated, the tail domain wraps around the kinase domain and the phosphotyrosine residue docks to one of the SH domains, closing off the active site to substrate.

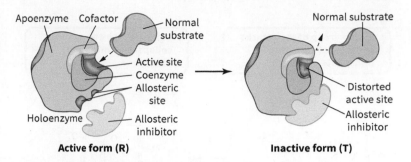

FIGURE 4.23 Allosteric regulation.   Allosteric regulators bind to the enzyme (or complex) at a site called the allosteric site. Binding of inhibitors stabilizes the T state while binding of activators stabilizes the R state. Shown here is a single subunit, although there can be multiple subunits influencing each other

(Source: Black, *Microbiology: Principles and Explorations*, 8e, copyright 2012, John Wiley & Sons. This material is reproduced with permission of John Wiley & Sons, Inc.).

4.4.2 Allosteric regulators bind at sites other than the active site

Allosteric regulators act to increase or decrease the activity of an enzyme by binding at a site other than the active site (**Figure 4.23**). The term "allosteric" comes from Greek and literally means "other place." Allosteric regulators are most often small molecules, in some instances as small as a proton. Proteins that are allosterically regulated are classically thought of as being multimeric and cooperative, although this view is currently being reconsidered. It is clear that allosteric regulation is far more common than was once thought.

In classical allosteric models, there are two states: a tense state (T) in which the enzyme is inactive or inhibited and affinity for substrate is low, and a relaxed state (R) in which the enzyme is active and affinity for the substrate is high. An easy way to remember this is that you do not get any work done when you are tense. The names of these states are slightly misleading, because the name correlates with the *activity* rather than with the *conformation* of the protein. The binding of an allosteric regulator shifts the conformation from one state to the other, altering the activity. Only these two states, T and R, exist—there is no middle ground. Binding of allosteric activators favors the R state, while binding of allosteric inhibitors favors the T state.

Enzymes that are allosterically regulated exhibit sigmoidal (S-shaped) activity curves (**Figure 4.24**). These curves are the result of two different types of binding, T and R. The high-affinity, high-activity R state resembles a saturation curve that is shifted to the left, while the lower affinity, lower activity T curve is shifted to the right. When these curves are summed together, the resulting graph is sigmoidal and represents the mixed population of enzyme molecules found in the cell or test tube. The addition of an activator shifts this curve to the left, meaning that the enzyme is more active at a lower concentration of substrate whereas an inhibitor conversely shifts it to the right. Allosteric regulators work acutely and instantaneously, in contrast to some other forms of regulation that require a cascade of events or induction of gene expression. Allosteric regulation is the most rapid form of regulation and the most direct.

Aspartate transcarbamoylase is an example of an allosterically regulated enzyme   One of the best studied examples of an allosteric enzyme is aspartate transcarbamoylase or ATCase (**Figure 4.25**). This enzyme catalyzes the formation of *N*-Carbamoyl aspartate from aspartate and carbamoyl phosphate—the rate-determining step in pyrimidine biosynthesis in *E. coli*. Rate-determining steps are frequently highly regulated; ATCase is no different, having at least three allosteric regulators.

The ATCase holoenzyme is a dodecamer with six catalytic and six regulatory subunits. If envisioned as a globe, the catalytic subunits are arranged as a trimer on either pole of the enzyme, largely separated by a belt of regulatory

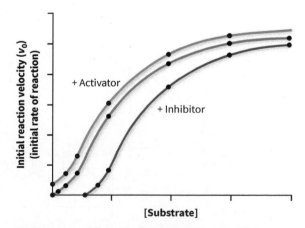

FIGURE 4.24 Allosteric enzymes have sigmoidal kinetics.   Cooperativity between active sites in an allosteric enzyme leads to sigmoidal (S-shaped) kinetics. Addition of an allosteric inhibitor shifts the curve to the right, requiring a higher concentration of substrate for the same level of activity, while addition of an activator shifts the curve to the left.

A.

1. Amino group of asp attacks carbonyl of carbamoyl phosphate.

2. Tetrahedral intermediate collapses, eliminating phosphate.

3. Forming *N*-Carbamoyl aspartate and free phosphate.

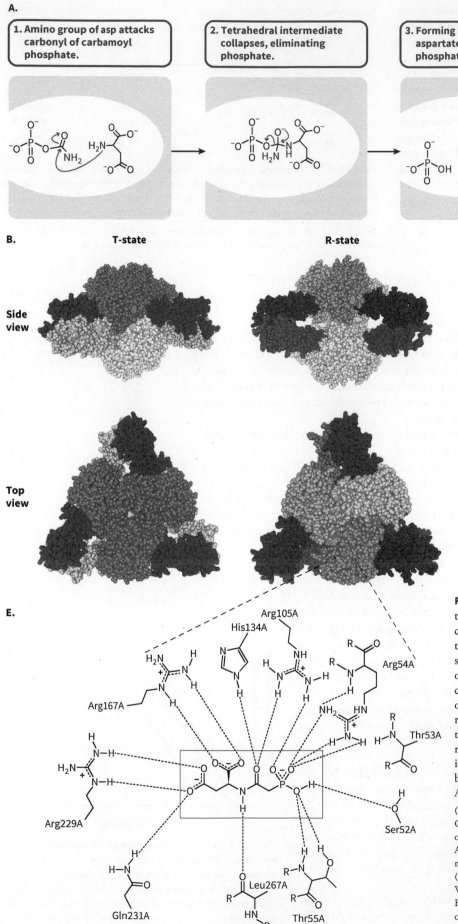

B.   T-state   R-state

Side view

Top view

C.

Catalytic subunits

Regulatory subunits

D.

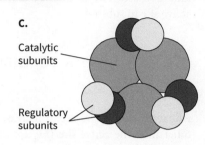

Phosphono linkage

E.

Arg105A

His134A

H_2N Arg54A

Arg167A

Thr53A

H_2N

Arg229A

Ser52A

Gln231A   Leu267A   Thr55A

FIGURE 4.25 ATCase regulation.   A. Aspartate transcarbamoylase catalyzes the addition of a carbamoyl group to the amine moiety of aspartate, the rate-determining step of pyrimidine biosynthesis. **B.** The active enzyme is a dodecamer consisting of six catalytic and six regulatory subunits. **C.** The catalytic subunits form two triads at either end of the molecule largely separated by a band of six regulatory subunits (in three pairs). **D.** Binding of the transition state analog PALA shifts the equilibrium between the T-state and R-state and results in inhibition of the enzyme. **E.** PALA, shown in the box, binds using numerous hydrogen bonds in the ATCase active site.

(Source: (B, Left) Data from PDB ID 1RAH Kosman, R.P., Gouaux, J.E., Lipscomb, W.N. (1994) Crystal structure of CTP-ligated T state aspartate transcarbamoylase at 2.5 A resolution: implications for ATCase mutants and the mechanism of negative cooperativity. *Proteins* 15: 147–176. (B, Right) Data from PDB ID 1Q95 Huang, J., Lipscomb, W.N. (2004) Aspartate Transcarbamylase (ATCase) of *Escherichia coli*: A New Crystalline R-State Bound to PALA, or to Product Analogues Citrate and Phosphate. *Biochemistry* 43: 6415–6421)

subunits at the equator. The active site of ATCase is unusual in that it is found at the interface of two of the catalytic subunits and involves both subunits. Analysis of the mechanism and regulation of ATCase has been facilitated by crystallographic studies, which have been advanced in part through the discovery and use of the transition-state analog N-phosphonoacetyl-L-aspartate (PALA). ATCase is unable to cleave the phosphono linkage in this molecule. Instead, it binds tightly in the active site, stabilizing the enzyme in the R state. This has enabled investigators to obtain structures of ATCase in both the T (unoccupied) and R states, and compare the two.

ATCase is the rate-determining step in pyrimidine biosynthesis, and it is regulated by molecules downstream in the pathway; this is an example of feedback control. The three best studied allosteric regulators of ATCase are ATP, CTP, and UTP. As you may recall, the adenine moiety of ATP is not a pyrimidine but a bicyclic purine. Both cytosine (from CTP) and uracil (from UTP) are downstream products of this pathway. All three nucleoside triphosphates (NTPs) bind to the regulatory subunits, but there are only two **binding sites**: a high-affinity binding site that can only bind to ATP and CTP, and a low-affinity binding site that preferentially binds to UTP. Complicating matters further, when UTP binds to the low-affinity site, it results in preferential binding of CTP at the high-affinity site. Therefore, only ATP and CTP can bind to the high-affinity site, and these molecules compete with one another for the site. However, if concentrations of all three NTPs are high enough, UTP can bind to the low-affinity site, resulting in preferential CTP binding at the high-affinity site.

High levels of CTP indicate that the cell has enough CTP and that further synthesis of pyrimidines should be slowed. When CTP binds to the regulatory subunit, the sigmoidal kinetic curve for the synthesis of *N*-carbamoyl aspartate is shifted to the right and the enzyme's activity falls by 50%–70% (**Figure 4.26**). Binding of CTP favors the T state. If UTP is also added, this inhibition increases to 95%.

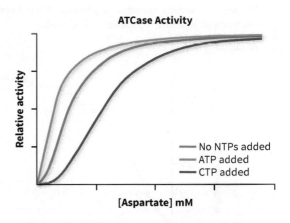

ATCase Activity

— No NTPs added
— ATP added
— CTP added

[Aspartate] mM

FIGURE 4.26 Effect of allosteric regulators on ATCase. ATCase activity in the presence of the allosteric activator ATP shifts the rate curve to the left, while binding of the allosteric inhibitor CTP shifts the curve to the right.

In contrast to the pyrimidine NTPs, binding of ATP activates the enzyme, shifting the curve to the left and favoring the R state. High levels of ATP could indicate that the cell has an excess of energy and can thus promote growth and biosynthesis of DNA and mRNA; alternatively, it could indicate that there is an excess of purine nucleotides compared to pyrimidines. In either case, the regulatory mechanism should make biological and chemical sense to us.

Binding of these allosteric regulators results in a shift to either the T or R state, as noted. In ATCase, this leads to a large conformational change in the enzyme that involves complex interactions between numerous residues. The catalytic subunits of ATCase have a high degree of interaction with each other. Part of this interaction involves a loop of amino acids termed the 240's loop, named for its location within the protein (amino acids 230–245). The switch from T to R causes a large movement of the 240's loop, which decreases the interaction between the adjacent subunits. Furthermore, the 240's loop and two other residues are adjacent to the active site of ATCase. When they move, they change the topology of the active site, moving key residues away from one another, and hence decreasing activity.

The movement and interactions of the 240's loop are only part of the story of how ATCase functions. However, they illustrate how structural information, combined with activity and other data, help to paint a detailed picture of how enzymatic reactions occur and are regulated.

The regulation of enzymes is also under the selective pressures of evolution
Evolution can be illustrated at many different levels in biological systems, as discussed previously throughout this work. Biochemists often focus on the conservation of a particular nucleotide or amino acid sequence, whereas organismal biologists may focus on an aspect of morphology (beak shape or skull size); however, evolution covers all aspects between these extremes. The examples of protein regulation given here have undergone hundreds of millions of years of natural selection. An enzyme does not simply evolve as a catalyst. In some instances, the substrate for the enzyme may also be changing over time, and the regulatory mechanisms that dictate enzymatic rate are also under selective pressure. The selective pressures on a multimeric enzyme complex regulated by several different phosphorylation sites or allosteric regulators illustrates the complexity of this scenario.

Worked Problem 4.4 | Regulation of an enzyme

Lactate dehydrogenase catalyzes the conversion of pyruvate to lactate. It is a tetrameric enzyme that is allosterically activated in some species of bacteria by fructose-1,6-bisphosphate (FBP). What would a graph of pyruvate concentration versus rate look like in the presence and absence of FBP? Would FBP stabilize the T or the R state?

Strategy Describe allosteric regulation. What have our previous examples of allosteric regulation looked like?

Solution

Acting as an activator means that FBP stabilizes the R state and shifts the curve to the left.

Follow-up question The allosteric regulators discussed so far have been heterotropic; that is, they are not substrates for the enzyme in question. In contrast, homotropic allosteric regulators are substrates for the enzyme. How can a substrate for an enzyme also be a regulator?

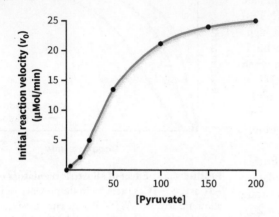

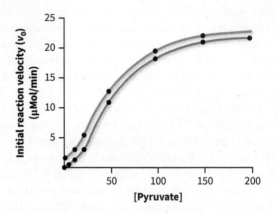

Summary

- In many cases, the rate at which an enzyme operates can be manipulated by the cell, the system, or the organism.
- The many examples of regulation can be grouped into covalent modification and allosteric regulation.
- Two examples of covalent modification are proteolytic cleavage and phosphorylation.
- Proteolytic cleavage works by removing domains of the pro-enzyme that inhibit activity. This type of regulation is irreversible.
- Phosphorylation is the addition of a phosphate group, usually from a high-energy phosphate donor such as ATP, and usually to the hydroxyl group of a serine, threonine, or tyrosine residue.
- Phosphorylations are mediated by a group of enzymes called kinases. These mutations are reversible because the phosphate group can be removed by the action of a phosphatase.
- Allosteric regulation occurs when small molecules bind to sites on the enzyme distal to the active site. The action of these regulators is often cooperative, with multiple regulatory sites acting on multiple active sites.

Concept Check

1. Provide examples of each type of regulation mechanism.
2. Describe how each type of mechanism works.
3. Explain which of the regulatory mechanisms are reversible, which are not, and why this is the case.
4. Describe the advantages and disadvantages of each of these control systems.

4.5 Allosterism and Cooperativity

This section describes allosterism and cooperativity at the chemical level, introduces two different models of cooperativity, and uses several examples to illustrate cooperative allosteric control.

4.5.1 Allosterism is a means of regulating protein function

The term **allosterism** literally means "other shape" or "other site." It is used to describe the regulation of a molecule such as an enzyme, receptor, or transport protein through a conformational change induced by the binding of a regulator. Numerous proteins are regulated by allosteric mechanisms (**Table 4.3**).

In allosteric regulation, molecules (typically small ones) bind to the protein at another site on the enzyme, the allosteric binding site, which is different from the active site (**Figure 4.27**). The opposite of allosteric regulation is **orthosteric regulation**, which means "same site." An example

TABLE 4.3 Common Proteins Regulated by Allosteric Mechanisms

Protein	Function or Pathway
Glucokinase	Glycolysis
Phosphofructokinase	
Isocitrate dehydrogenase	Citric acid cycle
α-ketoglutarate dehydrogenase	
Acetyl-CoA carboxylase I (ACC I)	Fatty acid synthesis
Aspartate transcarbamoylase (ATCase)	Pyrimidine biosynthesis
Chorismate mutase	Aromatic amino acid synthesis
Glycogen phosphorylase	Glycogen breakdown
Glutamine synthetase	Amino acid metabolism
Ribonucleotide reductase	Deoxyribonucleotide synthesis
Immunoglobulins	Immune system function
Hemoglobin	Oxygen transport
Nicotinic acetylcholine receptor	Neurotransmitter binding

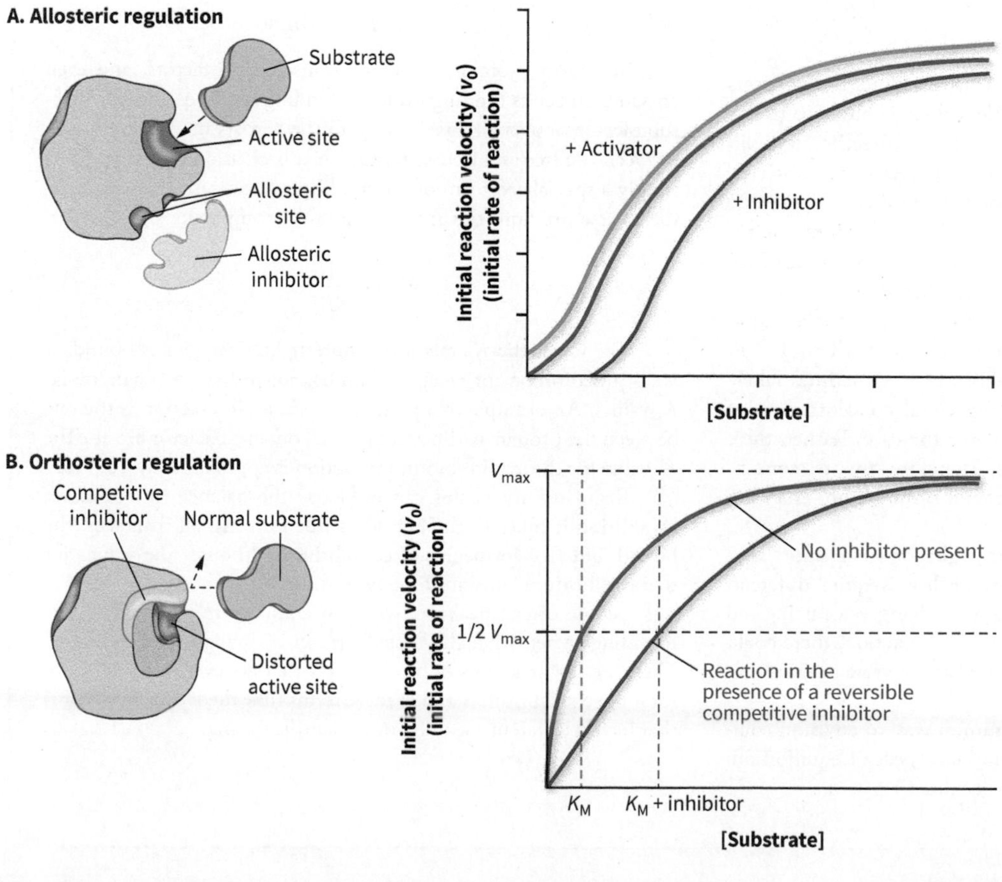

A. Allosteric regulation

Substrate
Active site
Allosteric site
Allosteric inhibitor

+ Activator
+ Inhibitor

Initial reaction velocity (v_0) (initial rate of reaction)
[Substrate]

B. Orthosteric regulation

Competitive inhibitor
Normal substrate
Distorted active site

V_{max}
$1/2\,V_{max}$

No inhibitor present
Reaction in the presence of a reversible competitive inhibitor

Initial reaction velocity (v_0) (initial rate of reaction)
K_M K_M + inhibitor
[Substrate]

FIGURE 4.27 Allosteric and orthosteric regulation.
A. Allosteric regulators bind at a site (the allosteric site) different than the active site of the protein or enzyme. The activity curve of an allosteric enzyme is sigmoidal (S shaped). If inhibited, the curve is shifted to the right; if activated, it is shifted to the left. **B.** In orthosteric regulation, the binding of a competitive, noncompetitive, or mixed inhibitor in the active site leads to changes in the overall kinetic parameters, but the response is still a rectangular parabola (a saturation curve).

(Source: (A, B Left) Black, *Microbiology: Principles and Explorations*, 8e, copyright 2012, John Wiley & Sons. This material is reproduced with permission of John Wiley & Sons, Inc.)

of orthosteric control is competitive inhibition, where a structural analog binds to the active site of the enzyme. Orthosteric regulation requires sufficient homology between the product of the reaction and the binding site. This is not necessarily true in allosteric regulation, where a reaction can be regulated by a molecule with a quite different structure and function, for example, citrate regulating acetyl-CoA carboxylase. Thus, in some ways, an allosteric site is analogous to the active site of the enzyme in terms of how it binds to a small molecule; however, whereas a substrate binds to the enzyme's active site, an allosteric regulator binds to the allosteric site.

The kinetics of enzymes that are allosterically regulated differ from those of enzymes that follow Michaelis–Menten kinetics. With the former, the plot of substrate concentration versus activity is sigmoidal, or S shaped; with the latter, there is a hyperbolic saturation kinetic as seen in Figure 4.27.

The sigmoidal curve observed in allosteric proteins has a particular significance. The simple binding of a ligand to a protein can be expressed using the Langmuir equation:

$$\theta = \frac{[L]}{K_d + [L]}$$

where θ is the percentage of sites bound, L is the concentration of ligand, and K_d is the dissociation constant for the protein–ligand complex. A graph of this equation gives a hyperbolic curve similar to one generated according to Michaelis–Menten kinetics. As the concentration of ligand increases, eventually all the sites will become bound, and the line will approach saturation. However, the sigmoidal curve seen with allosteric enzymes indicates that the binding is cooperative rather than simple. In other words, as one molecule of ligand binds, it increases the affinity of binding for other ligands. Hence, although allosterism means "other site," it is not possible to discuss allosteric regulation without discussing cooperativity among sites. **Biochemistry: Chemical equilibrium** gives a more nuanced discussion of equilibrium and the dissociation constant, K_d.

Biochemistry

Chemical equilibrium

Any chemical reaction can be described as reactants or substrates being converted to products. For example, A being converted to B is a chemical reaction. When a reaction runs in this direction, it is said to run in the forward direction, but this is subject to the observer. The reaction could just as easily be defined as running in the reverse direction (B going to A). For any reaction, the equilibrium point is where the concentrations of A and B no longer change; that is, the rates of the forward and reverse direction are the same.

We often use a seesaw or teeter-totter analogy to describe equilibrium, evoking an image of a system in balance, with equivalent weights on either side. Chemical equilibrium is sometimes falsely conflated with the notion of there being equal concentrations of products and reactants; this is almost never the case. Rather, there is a balance between the reaction rates. Therefore, the net concentrations of substrates and products at equilibrium are not changing ($d[A]/dt = d[B]/dt = 0$).

Equilibrium is therefore often thought of as a static state in which nothing is changing, but the reality is quite different. Although the net change is 0, there are always changes occurring and reactions going on. Depending on the rates of reaction, there could be many or few, but the rates of the forward and reverse reactions are equal to one another.

Thinking about a reaction is a common way to envision equilibrium, but this chapter deals with another aspect of equilibrium; that is, the reversible binding of two molecules:

$$[A] + [B] \leftrightarrows [AB]$$

Transport proteins, receptors, transcription factors, and even enzymes all act as binding proteins, binding to either another macromolecule or a small ligand. Each of these exists in an equilibrium between the free and bound forms. The dissociation constant, K_d, is simply a specialized form of the equilibrium constant that examines the reverse reaction (the dissociation of the complex):

$$K_d = \frac{[A][B]}{[AB]}$$

The lower the K_d value, the more tightly the pair is bound. It is not uncommon for receptors and ligands to have sub-nanomolar K_d values. An example of a particularly strong interaction is the one between the protein avidin and the small organic cofactor biotin. The K_d value for the avidin–biotin interaction is approximately 10^{-15} M.

Regulator molecules can influence the balance in a reaction by shifting it more to the bound or free form. In all instances, the bound and free forms are in equilibrium (although the values for that equilibrium constant have been shifted).

People often speak of "living in equilibrium" with their surroundings when what they usually mean is "living in homeostasis." There are few reactions in biochemistry that are ever in equilibrium for long; equilibrium is one of those terms (like theory or significant) that have a different meaning in a scientific setting.

4.5.2 Multiple models describe cooperative allosteric interactions

Models of cooperativity have evolved as scientific knowledge has advanced. The two most common models are the concerted model (or MWC, proposed by Monod, Wyman, and Changeaux) and the sequential model (or KNF, proposed by Koshland, Nemethy, and Filmer) (**Figure 4.28**). In both models, the subunits exist in one of two states: the **tense (T) state**, which has low affinity, and the **relaxed (R) state**, which has high affinity. Where the two models differ is in the transition between these states. In the **concerted model**, all of the subunits of the complex change state at once, whereas in the **sequential model**, the subunits change one at a time as regulators are bound.

Both models can yield sigmoidal curves, and both accurately reflect what is observed in the laboratory. It is likely that some systems in nature can be better represented by the sequential model and others by the concerted model. Neither model is entirely correct or incorrect, and both faithfully represent aspects of allosterism.

The concerted model   In the concerted (MWC) model, the entire complex exists only in the T or R state. All subunits must convert from one state to the other in a concerted fashion. The concerted model has the advantage of using only three equilibria to describe the binding: the binding constant for the ligand to the T state, the binding constant for the ligand to the R state, and the equilibrium constant for the transition of the unbound protein between T and R states, shown here as K_1, K_2, and K_3. Gamma, γ, is the fraction of subunits that are bound, and X is the concentration of ligand.

$$\gamma = \frac{K_2 \dfrac{X}{K_1}\left(1-\dfrac{X}{K_1}\right)^3 + \dfrac{X}{K_3}\left(1-\dfrac{X}{K_3}\right)^3}{K_2\left(1+\dfrac{X}{K_1}\right)^4 + \left(1+\dfrac{X}{K_3}\right)^4}$$

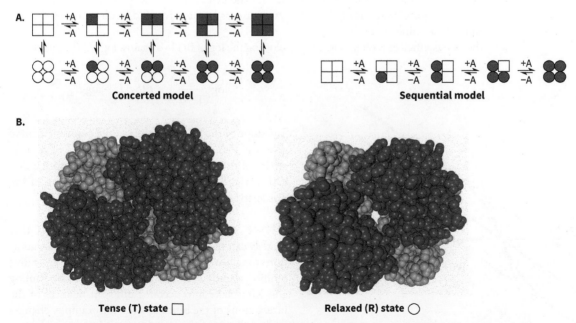

FIGURE 4.28 Concerted and sequential models of allosterism.   A. In both models, subunits exist in either the tense (T) state, represented here by squares, or the relaxed (R) state, represented by circles. Each A represents the addition of a ligand shown by a filled circle or square. This illustration shows four binding sites, but the models can be adapted for other numbers of sites. In the concerted model, all subunits transition between the T and R states at the same time as regulators bind. In the sequential model, subunits transition between T and R states one at a time. **B.** The T and R states of human and horse hemoglobin are shown.

(Source: (B, Left) Data from PDB ID 1THB Waller, D.A., Liddington, R.C. (1990) Refinement of a partially oxygenated T state human haemoglobin at 1.5 A resolution. *Acta Crystallogr.,Sect.B* **46**: 409–418; (B, Right) Data from PDB ID 1IBE Wilson, J., Phillips, K., Luisi, B. (1996) The crystal structure of horse deoxyhaemoglobin trapped in the high-affinity (R) state. *J.Mol.Biol.* **264**: 743–756)

The sequential model   In the sequential (KNF) model, as regulatory molecules bind to the allosteric sites, the activity of the subunit that includes the active site is shifted from the T to the R state, and the subunit is activated. This description fits mathematical models first developed by Adair:

$$\gamma = \frac{\dfrac{1}{4a_1x} + \dfrac{1}{2a_2x^2} + \dfrac{3}{4a_3x^3} + a_4x^4}{1 + a_1x + a_2x^2 + a_3x^3 + a_4x^4}$$

In this equation, γ is the fraction of subunits that are bound, x is the concentration of ligand, and a_1 to a_4 are the equilibrium constants for the binding of ligand to protein. In this specific instance, we are assuming four binding sites, but this number would vary proportionally if there were more or fewer of them. The cooperative nature of the curve necessitates different values for the four binding constants and easier binding of the last ligand than of the first. This is consistent with the sequential model and represents a stepwise switch from T to R states.

The Hill plot   The degree to which the sites in an allosteric protein or complex cooperate can be determined using a plot of the **Hill equation**:

$$\theta = \frac{[L]^n}{K_d + [L]^n}$$

in which θ is the fraction of sites bound, $[L]$ is the concentration of the ligand, K_d is the dissociation constant of the protein–ligand complex, and n is the Hill coefficient. This relationship is similar to the Langmuir equation described earlier, but the exponential terms are more similar to the models that were subsequently developed by other groups. The Hill coefficient describes the cooperativity of the system. If this coefficient is greater than 1, then the interaction of the allosteric sites is positive, and the binding of subsequent ligands is enhanced. If it is less than 1, then the system is negatively cooperative, and the binding of one ligand makes it more difficult for subsequent ligands to bind. Finally, if the Hill coefficient is equal to 1 (in which case it is the same as the Langmuir equation), then the system is not cooperative, and binding to one site will not influence binding at another.

The Hill coefficient is often obtained by plotting the log of $\theta/1 - \theta$ against log $[L]$ (**Figure 4.29**):

$$\log\left(\frac{\theta}{1-\theta}\right) = n\log[L] - \log K_d$$

The slope of the line from the middle part of the curve gives the Hill coefficient (n), or the degree of cooperativity of the different sites. Extrapolating from either end of the line to the x-axis will give the log of the dissociation constant (K_d) for T or R states.

Other models of cooperativity   Other models can be used to describe cooperativity in addition to allosterism and the models of cooperative binding already discussed. Models derived by Pauling and Klotz take into account different nuances of the interactions of the binding sites, for example, changes in the affinity of the sites for a modulator or changes in equilibrium constants as moderators bind. Cooperative binding has been discussed here generically in the context of ligand binding to a protein. These same concepts can be applied specifically to oxygen transport in hemoglobin, ligand binding by receptors, and phenomena such as the phase transitions of membrane lipids and the melting of DNA.

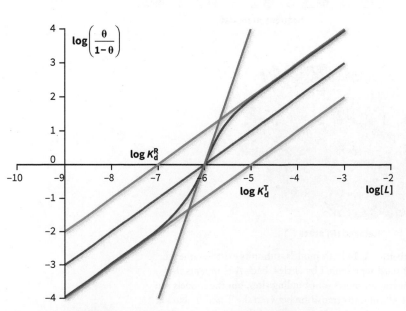

FIGURE 4.29 Hill plot.   A Hill plot is a graphic description of ligand binding. The line is a plot of the log of the fraction of sites bound plotted as a function of the log of the ligand, L. The slope of the central portion of the curve gives the Hill coefficient, which is a measure of cooperativity of the subunits in the complex.

4.5.3 Allosteric molecules can operate in different ways

Molecules that bind the allosteric site can act in one of two ways: they can be allosteric activators, stimulating the enzyme, or they can be allosteric inhibitors, inhibiting the enzyme. Allosteric activators shift a plot of activity versus substrate concentration [S] to the left, whereas allosteric inhibitors shift the curve to the right, which means that more substrate is needed to achieve the same activity (**Figure 4.30**).

Homotropic versus heterotropic regulators

Regulators that bind to an allosteric protein may be homotropic or heterotropic. **Homotropic regulation** (from the Greek for "same" and "change") occurs when a protein is allosterically regulated by its substrate. We may ask, then, how is this different from orthosteric regulation (competitive or noncompetitive inhibition)? The answer lies in the multimeric nature of the protein complex in homotropic regulation. An example of homotropic regulation is oxygen binding to hemoglobin. Because hemoglobin is a tetramer and undergoes allosteric regulation, binding of oxygen to one subunit influences the conformation of the complex (the T or R state) and influences binding (the binding constant) of other oxygen molecules at a *different* oxygen binding site.

Heterotropic regulation (from the Greek for "other" and "change") is when a protein is regulated by a molecule that is not its substrate. Heterotropic regulators do not bind in the active site of the protein of interest, but they do allosterically influence binding at those sites. An example of a heterotropic regulator is 2,3-bisphosphoglycerate (2,3-bPG), which does not bind to hemoglobin at the oxygen-binding sites, but it does allosterically influence oxygen binding.

Allosteric regulation of monomeric enzymes

Most proteins that are regulated through allosteric control are multimeric, but some are monomeric. Glucokinase is a monomeric enzyme with a single active site (**Figure 4.31**). It is regulated by glucokinase regulatory protein (GKRP), which in turn is regulated allosterically by other small molecules, including glucose, fructose-6-phosphate, and fructose-1-phosphate.

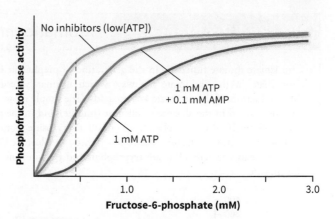

FIGURE 4.30 Allosteric regulators. The graph shows phosphofructokinase activity as a function of substrate concentration (fructose-6-phosphate). The activity curve is sigmoidal, indicating allosteric regulation. In the presence of increasing concentration of the allosteric inhibitor ATP, the curve is shifted to the right.

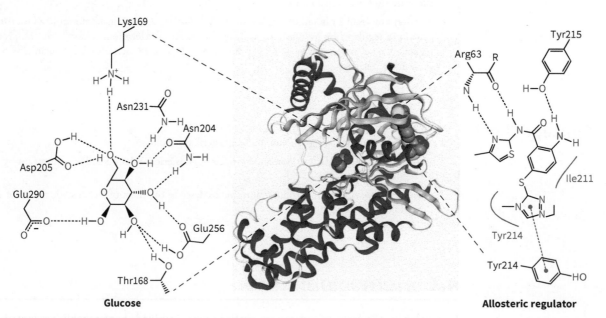

FIGURE 4.31 A monomeric allosteric enzyme. Glucokinase is one of the enzymes that catalyzes the first step of glycolysis (phosphorylation of glucose to glucose-6-phosphate). The structure shown here depicts glucokinase bound to glucose and to an allosteric activator. Key residues involved in the binding of each small molecule are shown.

(Source: Data from PDB ID 3FR0 Nishimura, T., Iino, T., Mitsuya, M., Bamba, M., Watanabe, H., Tsukahara, D., Kamata, K., Sasaki, K., Ohyama, S., Hosaka, H., Futamura, M., Nagata, Y., Eiki, J. (2009) Identification of novel and potent 2-amino benzamide derivatives as allosteric glucokinase activators. *Bioorg. Med. Chem. Lett.* **19**: 1357–1360)

Worked Problem 4.5   Aromatic amino acid synthesis

Chorismate mutase functions in the generation of prephenate from chorismate in the shikimate pathway, which ultimately leads to the production of the aromatic amino acids. The graph shows this enzyme assayed in the absence of effector (middle) and in the presence of either 10 μM tryptophan (top) or 100 μM tyrosine (bottom). Is chorismate mutase obeying Michaelis–Menten or allosteric kinetics? How can you tell? How are tryptophan and tyrosine affecting the enzyme?

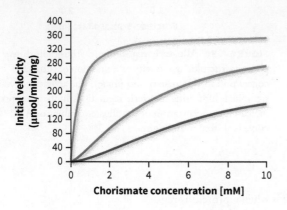

Strategy   Examine the graph. What is the shape of the curve? What does that indicate about the kinetics of the reaction?

Solution   The activity of chorismate mutase is an example of allosterism. This can be seen from the sigmoidal, or S-shaped, curve, in contrast to the rectangular hyperbola observed with Michaelis–Menten kinetics. In the presence of tryptophan, the activity of the enzyme (the y coordinate) is higher at a lower substrate concentration (the x coordinate). Therefore, tryptophan is acting as an allosteric activator. Tyrosine, on the other hand, is acting as an allosteric inhibitor, requiring higher concentrations of substrate to achieve the same level of activity. A clear comparison of this can be seen if we examine the concentration of substrate required to achieve 150 μmol/min/mg of activity for all three curves (<1 mM of substrate for the stimulated case, 3 mM for the case without regulator, and 7 mM for the case with inhibitor).

Follow-up question   How might you calculate the cooperativity of the subunits found in the previous question? What other data would you need?

Summary

- Allosterism is a way to regulate a protein function such as enzyme activity or ligand binding. Allosteric enzymes do not display hyperbolic Michaelis–Menten kinetics; instead, the graph of substrate concentration versus activity is sigmoidal.

- Several models describe allosterism, including the concerted (MWC) model and the sequential (KNF) model. In both models, the subunits exist in either the tense (T) state (low activity) or the relaxed (R) state (high affinity). Hill plots can be used to describe the cooperativity of different allosteric sites with one another.

- Allosteric molecules can either activate or inhibit enzyme activity. In homotropic allosterism, the protein is regulated by its substrate or ligand, whereas in heterotropic allosterism, the protein is regulated by a molecule that is not its substrate or ligand. Allosteric molecules can also bind to monomeric enzymes.

Concept Check

1. Define allosterism, and give examples of allosteric regulation.

2. Describe the sequential and concerted models of allosteric regulation and what is meant by T and R states.

3. Use the terms homotropic and heterotropic to describe the effect of an allosteric regulator on the reaction rate of an enzyme.

4.6 Hemoglobin

Hemoglobin, the brilliant red oxygen carrier found in blood, has probably fascinated humans since antiquity. Because of its abundance and its role in biology, its study has been an important part of the field of biochemistry. Many biochemical phenomena were initially studied with this molecule, including protein–ligand binding, oxygen transport, biological redox chemistry, the characterization of prosthetic groups, allosterism, and macromolecular X-ray crystallography.

This section first reviews the structure of hemoglobin in detail and compares it to its monomeric cousin, myoglobin. It then discusses oxygen transport by hemoglobin in the context of hemoglobin structure and the allosteric models discussed in section 4.5. Finally, we turn to examples of hemoglobin in human diseases, in particular, sickle cell anemia and the thalassemias.

4.6.1 Hemoglobin and myoglobin have related structures

The structures of hemoglobin and the related molecule myoglobin were the first to be solved by X-ray crystallography, and determining the structure of myoglobin helped to solve the structure of hemoglobin (**Figure 4.32**). Found predominantly in erythrocytes (red blood cells), hemoglobin is a tetrameric protein with an $\alpha_2\beta_2$ topology. In contrast, myoglobin is found in muscle and is monomeric, but it has a similar structure to the subunits of hemoglobin. Both molecules contain

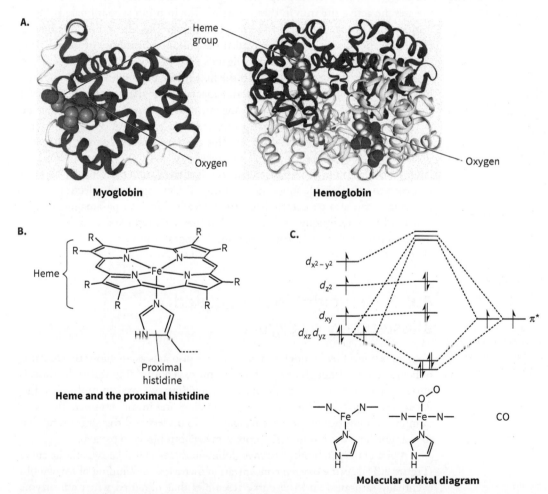

FIGURE 4.32 Structures of myoglobin and hemoglobin.   A. Myoglobin and hemoglobin have similar shapes. Both are nearly all alpha helix with a conserved tertiary structure termed the globin fold. Both are also bound to a cofactor termed heme. In both of these structures, heme is bound to oxygen. Note that myoglobin is monomeric, whereas hemoglobin is a tetramer with $\alpha_2\beta_2$ stoichiometry. **B.** Heme consists of an Fe^{2+} ion chelated by four of the nitrogens of a porphyrin ring. The fifth position of the iron ion is bound by a histidine residue termed the proximal histidine. Without oxygen, the iron is slightly too large to fit into the hole in the middle of the porphyrin ring. When oxygen binds, the electrons rearrange and the ion shrinks, allowing the iron to move into the plane of the ring. **C.** This molecular orbital diagram for heme depicts the participating valence electrons of the iron ion on the left, the participating valence electrons of the oxygen on the right, and the electronic configuration of the complex in the middle. In either atomic orbital (the left or right) the electrons are unpaired. In the complex, the electrons are spin-paired as shown. This is observable using EPR, which will only detect unpaired electrons.

(Source: (A) Data from PDB ID 6BB5 Terrell, J.R., Gumpper, R.H., Luo, M. (2018) Hemoglobin crystals immersed in liquid oxygen reveal diffusion channels. *Biochem. Biophys. Res. Commun.* **495**: 1858–1863)

a single heme prosthetic group per subunit (one for myoglobin and four for hemoglobin), and it is this group that binds to oxygen.

In hemoglobin, the a subunits have 141 amino acids, and the b subunits have 146; both types of subunits have a molecular weight of approximately 16 kDa, giving the tetramer a molecular weight of approximately 64 kDa. The four subunits are arrayed in a tetrahedron, and the subunits are held together by weak forces (salt bridges, hydrogen bonds, and hydrophobic interactions). There are no disulfide bonds in hemoglobin.

The secondary structure of each subunit of hemoglobin is comprised of eight alpha helices connected by turns; there is no beta sheet in this structure. The helices fold up into a tertiary structure known as a **globin fold**. Each subunit of hemoglobin has this retained fold. Of the eight helices in hemoglobin (labeled A to H), four are terminated by a proline residue. These helices are on average 14 amino acids in length (approximately four turns of the helix), giving the entire complex a diameter of approximately 55 Å.

The heme functional group of hemoglobin has two parts: a planar organic ring system (the porphyrin ring) and an iron ion (Fe^{2+}). The porphyrin ring is synthesized from glycine and succinyl-CoA, and it chelates the active site iron in these proteins. The iron ion is coordinated at four positions by nitrogen atoms of the porphyrin ring. The fifth position is bound to a nitrogen atom of a histidine (called the proximal histidine) of helix F, and the sixth position is free to bind to ligands such as oxygen. In the absence of ligand, the iron is slightly larger than the hole in the porphyrin ring. As ligands bind to the iron, one of two things can happen. Either the iron can be oxidized from Fe^{2+} to Fe^{3+}, or the orbital arrangement of electrons in the iron can be altered. In either case, the diameter of the iron ion shrinks, and it is pulled farther into the plane of the ring system. This also changes the spectral properties of hemoglobin in several ways. In the oxygenated state, the protein has a brilliant red color; in the deoxygenated state, it is slightly duller and bluer. This can be observed using visible-range spectroscopy. The change in the electronic state of the iron can also be detected spectroscopically. Electron paramagnetic resonance (EPR), a technique similar to nuclear magnetic resonance, can be used to detect unpaired electrons. The EPR signal changes as the iron binds to oxygen (**Figure 4.33**).

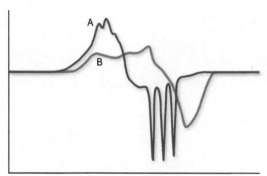

Magnetic field

FIGURE 4.33 EPR of hemoglobin.   EPR spectra of hemoglobin bound to NO in the presence of oxygen (A) and in the absence of oxygen at different pH values (B). Note the shape of the spectra changes in the presence of oxygen.

(Source. Reprinted with permission from Takashi Yonetani, Antonio Tsuneshige, Yuxiang Zhou§, and Xuesi Chen, (1998) Electron Paramagnetic Resonance and Oxygen Binding Studies of a-Nitrosyl Hemoglobin, *The Journal of Biological Chemistry*; 273(32).

4.6.2 Hemoglobin is regulated allosterically by oxygen

Oxygen is bound by hemoglobin, and oxygen also acts as an allosteric regulator of hemoglobin. Although oxygen binds to the oxygen-binding site in this case, it also serves as an allosteric regulator because binding one oxygen molecule influences how oxygen binds to *other* oxygen-binding sites in the tetramer. This is an example of homotropic allosteric regulation. To understand this, it is useful first to examine how the monomeric protein myoglobin binds to oxygen.

A plot of oxygen binding to myoglobin produces a familiar hyperbolic curve (**Figure 4.34**). As the oxygen concentration increases, the binding of myoglobin eventually saturates, and the curve resembles that obtained when an enzyme obeys Michaelis–Menten kinetics.

A plot of oxygen binding to hemoglobin is quite different. Instead of a hyperbola, the curve is sigmoidal, indicative of cooperation among the subunits. Indeed, when oxygen binds to hemoglobin, the second molecule of oxygen binds with greater affinity than the first, the third with greater affinity than the second, and the fourth with greater affinity than the third. This has important implications for biology. In the lungs, where the partial pressure (effectively the concentration) of oxygen is at its highest, hemoglobin binds tightly, and the molecule loads up with oxygen. In tissues where oxygen is being used up by the electron transport chain, there is a low partial pressure of oxygen. Here, oxygen binds to hemoglobin loosely, meaning that hemoglobin can easily release up to 90% of its oxygen in these tissues. Without allosteric regulation, hemoglobin would behave like myoglobin and would at best be able to deliver 38% of the available oxygen.

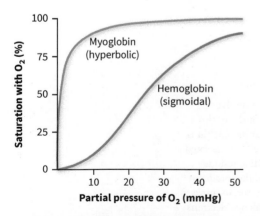

FIGURE 4.34 Oxygen-binding curves of hemoglobin and myoglobin.   Myoglobin binds oxygen with high affinity and saturates with a hyperbolic curve. In contrast, hemoglobin has a sigmoidal curve, indicating cooperativity among the oxygen-binding sites.

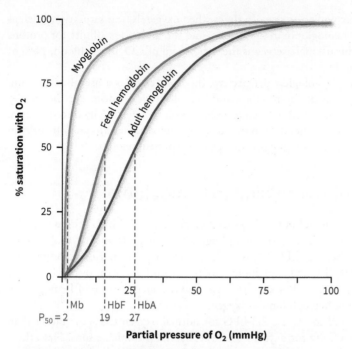

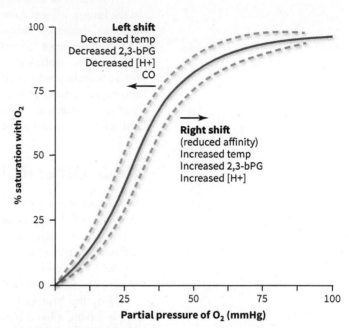

FIGURE 4.35 Fetal hemoglobin. Fetal hemoglobin (HbF) is a tetramer with $\alpha_2\gamma_2$ stoichiometry. It has a higher affinity for oxygen than adult hemoglobin (HbA) and, as a result, can absorb oxygen from the maternal circulation.

FIGURE 4.36 The Bohr effect. Decreases in pH protonate key histidine residues, thereby altering hydrogen bonding in hemoglobin and stabilizing the T state. This, in turn, leads to a decreased affinity of hemoglobin for oxygen.

Oxygen acts as a homotropic allosteric regulator of hemoglobin in that the binding of oxygen to one heme group influences the affinity at the other sites. Using the concerted and sequential models of allosterism, the oxygen-bound form of hemoglobin can be considered to be in either the high-affinity R state or the low-activity T state.

Hemoglobin expression varies over the course of human development. *In utero*, humans produce fetal hemoglobin, a variant that is comprised of two α chains and two γ chains. This difference in structure has several important functional consequences, the greatest of which is that fetal hemoglobin has a greater affinity for oxygen than does maternal hemoglobin. The difference in affinity creates an oxygen gradient, which causes oxygen to be transported from the mother to the growing fetus (**Figure 4.35**).

The Bohr effect

An early hemoglobin chemist was the Danish physiologist Christian Bohr, father of the physicist Niels Bohr. Christian Bohr discovered that as CO_2 increases and pH decreases, oxygen's affinity for hemoglobin also decreases, shifting oxygen curves to the right. This phenomenon is now known as the Bohr effect (**Figure 4.36**).

Structural studies have revealed the chemistry underlying the Bohr effect. In hemoglobin, a key histidine (His146) at the carboxy terminus of the β subunit and the side chain carboxyl group of an aspartate (Asp94) in the β subunits are the important players. When pH is normal or high, these groups do not interact strongly, and the complex is found in R state. When the pH drops, the histidine becomes protonated, and a salt bridge forms between the histidine and the carboxyl group of the aspartate. This interaction stabilizes the low-activity T state.

CO_2 is also transported by hemoglobin (**Figure 4.37**), reacting with the amino terminus of each of hemoglobin's α subunits to form a carbamate. Formation of the carbamate shifts the equilibrium, stabilizing the deoxygenated T state. Likewise, the protons released by the formation of the carbamate favor stabilization of the deoxygenated state, as noted previously.

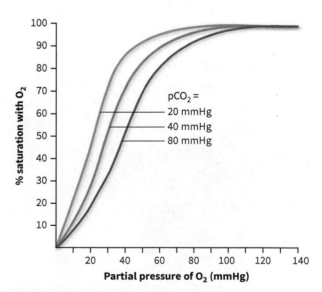

FIGURE 4.37 CO_2 transport by hemoglobin. Hemoglobin can react with the free amino groups of the β chains of hemoglobin to form a carbamate. No enzyme is needed for this reaction. Binding of CO_2 stabilizes the T state and decreases hemoglobin's affinity for oxygen.

The converse of these principles is also true. In times when oxygen is near saturation (such as in the lungs), oxygen stabilizes hemoglobin in the R state, and the molecule's affinity for protons and CO_2 decreases. Although bicarbonate serves as the major pool of CO_2 in the blood, 10% of CO_2 is transported by hemoglobin.

From an evolutionary and physiological perspective, the Bohr effect is a useful adaptation. Active muscle produces CO_2 and acidic by-products and thus allosterically decreases the affinity of oxygen for hemoglobin. This causes hemoglobin to release oxygen preferentially into the active muscle that needs it. Examining the Bohr effect in the context of active tissues that require oxygen and release CO_2 and acidic by-products highlights this property of hemoglobin.

4.6.3 Other molecules also bind to hemoglobin

Another allosteric regulator of hemoglobin is 2,3-bisphosphoglycerate (2,3-bPG) (**Figure 4.38**), an isomer of the glycolytic intermediate 1,3-bisphosphoglycerate, made from that molecule by bisphosphoglycerate mutase. When 2,3-bPG binds in the pocket at the center of the tetrameric complex, it makes contact with all four hemoglobin subunits, forming salt bridges with lysine and histidine residues. Binding of 2,3-bPG stabilizes the deoxygenated T state of hemoglobin, effectively lowering the affinity of hemoglobin for oxygen.

Why has hemoglobin evolved to require 2,3-bPG for normal oxygen transport rather than simply having a lower binding affinity for oxygen? There are several possible reasons. First, there are times when it would be useful to provide more oxygen than is typically available. For example, someone suffering from congestive heart failure or chronic obstructive pulmonary disease (COPD) has a greater need for hemoglobin to deliver all the oxygen it can to tissues because of shortcomings elsewhere in oxygen transport. Similarly, people who live at high altitude breathe air that has a reduced partial pressure of oxygen; hence, they also need extra oxygen. As an individual adapts to these conditions, synthesis of 2,3-bPG is increased, resulting in hemoglobin releasing oxygen in tissues that need it.

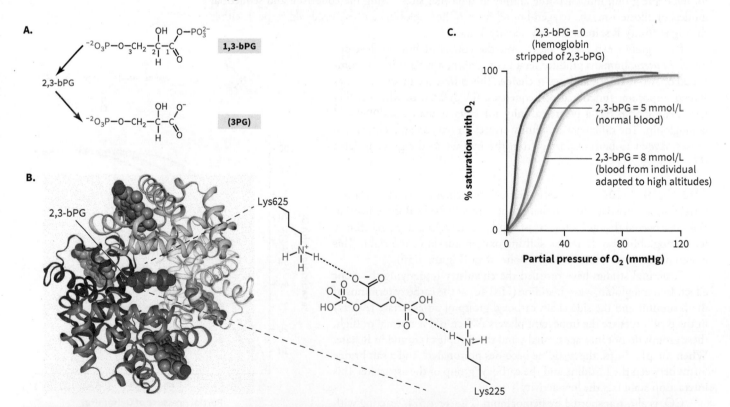

FIGURE 4.38 Effect of 2,3-bisphosphoglycerate on oxygen transport by hemoglobin.   **A.** 2,3-bisphosphoglycerate (2,3-bPG) is made by bisphosphoglycerate mutase from the glycolytic intermediate 1,3 bPG. It can be converted into 3-phosphoglycerate by the action of 2,3-bPG phosphatase. **B.** 2,3bPG binds in the cleft at the center of the four subunits of hemoglobin but interacts specifically with basic residues (His2, Lys82, and His143) of the β chains. **C.** Binding of 2,3-bPG stabilizes the T state and lowers hemoglobin's affinity for oxygen.

(Source: (B) Data from PDB ID 1B86 Richard, V., Dodson, G.G., Mauguen, Y. (1993) Human deoxyhaemoglobin-2,3-diphosphoglycerate complex low-salt structure at 2.5 A resolution. *J.Mol.Biol.* **233**: 270–274)

Another situation that explains the need for 2,3-bPG is pregnancy. As explained earlier, the growing fetus requires oxygen from the maternal blood supply. Fetal hemoglobin, which has γ chains in place of the β chains of hemoglobin (with the structure $\alpha_2\gamma_2$), has a higher affinity for oxygen than maternal hemoglobin. The γ subunits have serine residues in place of the histidine side chains found in the β subunits. This alteration means that 2,3-bPG binds fetal hemoglobin with a far lower affinity than it does maternal hemoglobin. Hence, there is an increased affinity of fetal hemoglobin for oxygen, which generates a gradient whereby oxygen is transported from the mother to the fetus.

Other molecules that bind to hemoglobin

Molecules other than oxygen and allosteric regulators can also bind to hemoglobin, sometimes with dire effects. Nitric oxide (NO) is a small, gaseous, signaling molecule that has various roles, including promoting vasodilation (the widening of blood vessels). NO is transported by hemoglobin, to which it binds by forming an S-nitrosothiol with key thiol groups of cysteine residues in the protein. As oxygen is released, hemoglobin assumes the T state, and NO is also released. These molecules may work together to help deliver oxygen to tissues that need it.

$$\langle^{SH} \quad + NO \longrightarrow \quad \langle^{S-N=O}$$

Cysteine   Nitric oxide   S-Nitrosothiol

Carbon monoxide results from the incomplete combustion of carbon-containing compounds. The oxygen atom of this compound will bind to the iron in the heme group of hemoglobin, but carbon monoxide binds with a 250-fold higher affinity for hemoglobin than molecular oxygen (O_2). This means that carbon monoxide effectively binds hemoglobin irreversibly, leaving the protein unable to transport oxygen. The inability to transport oxygen can lead to suffocation through carbon monoxide poisoning.

Other molecules that can bind to the heme group are hydrogen sulfide (HS), sulfur monoxide (SO), and cyanide (CN^-). In humans, binding of these molecules can have lethal effects owing to their ability to block oxygen transport. However, in certain species, such as bacteria that live near undersea volcanic vents, the ability of heme to bind to sulfur-containing compounds allows them to thrive in environments devoid of O_2.

Carbon dioxide transport in plasma

Most CO_2 is transported not bound to hemoglobin but as bicarbonate (HCO_3^-) (**Figure 4.39**. Bicarbonate is the ionized form of carbonic acid (H_2CO_3), which results from dissolving CO_2 in water. Although CO_2 will dissolve in water on its own, the enzyme carbonic anhydrase increases the rate of the reaction. Carbonic anhydrase is found in erythrocytes and is important in catalyzing both the forward and reverse reactions. In metabolically active tissues where CO_2 is being produced and the partial pressure of CO_2 is high, carbonic anhydrase catalyzes the reaction in the forward direction, producing carbonic acid from water and carbon dioxide. In the lungs, the partial pressure of CO_2 drops, which causes the reaction to run in the reverse direction, and carbonic anhydrase catalyzes the breakdown of bicarbonate into CO_2 and H_2O.

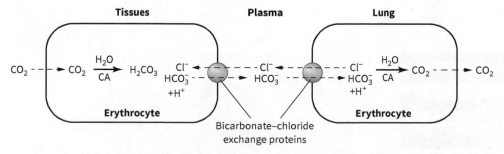

FIGURE 4.39 Carbon dioxide transport in plasma. Most carbon dioxide is transported as bicarbonate. In metabolically active tissues (left), carbonic anhydrase in the erythrocyte generates carbonic acid from CO_2 and water. The bicarbonate–chloride exchange protein swaps chloride in the plasma for bicarbonate, enabling the reaction to proceed in the forward direction. In the lung (right), the reverse reactions occur. Bicarbonate is exchanged into the erythrocyte, broken down into CO_2 by carbonic anhydrase, and lost through respiration.

If these reactions were all occurring in the erythrocyte, the cells would quickly build up a high concentration of bicarbonate, which would limit their ability to carry CO_2 away from active tissues. Instead, the erythrocyte membrane contains the bicarbonate–chloride exchange protein, an antiporter that exchanges these molecules for one another. This exchange protein maintains charge balance across the erythrocyte membrane and enables the erythrocyte to produce bicarbonate without bearing the responsibility of carrying it all to the lung.

In addition to serving as the major means of transport of CO_2, bicarbonate is also the largest contributor to the buffering capacity of the blood.

4.6.4 Several diseases arise from mutations in hemoglobin genes

Several notable mutations are found in the genes coding for hemoglobin. This section discusses types of anemia that can result from mutations to hemoglobin: sickle cell anemia and the thalassemias.

Sickle cell anemia **Sickle cell anemia** is caused by a single point mutation (A to T) in the hemoglobin β gene, which results in the change of a glutamate to a valine residue at the sixth position in the β chains of the hemoglobin tetramer (**Figure 4.40**). This mutation generates

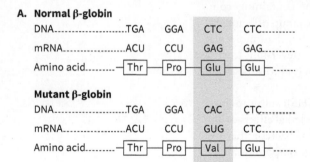

A. Normal β-globin

DNA	......................TGA	GGA	CTC	CTC............
mRNA	......................ACU	CCU	GAG	GAG..........
Amino acid	Thr	Pro	Glu	Glu

Mutant β-globin

DNA	......................TGA	GGA	CAC	CTC..........
mRNA	......................ACU	CCU	GUG	CTC...........
Amino acid	Thr	Pro	Val	Glu

FIGURE 4.40 Sickle cell disease. **A.** Sickle cell anemia results from a single mutation that causes the substitution of a valine for a glutamate residue in the β chains of hemoglobin. **B.** The mutation causes hemoglobin to polymerize into lengthy strands **C.** Polymerized proteins are visible with electron microscopy. **D.** The polymerized hemoglobin causes the erythrocytes to adopt a sickled shape that is less able to pass through capillaries and is more fragile.

(Source: (B) From Sam F. Greenbury, Iain G. Johnston, Ard A. Louis and Sebastian E. Ahnert, A tractable genotype – phenotype map modelling the self-assembly of protein quaternary structure, *Journal of The Royal Society Interface*, 2014. By Creative Commons 3.0. Public Domain.)

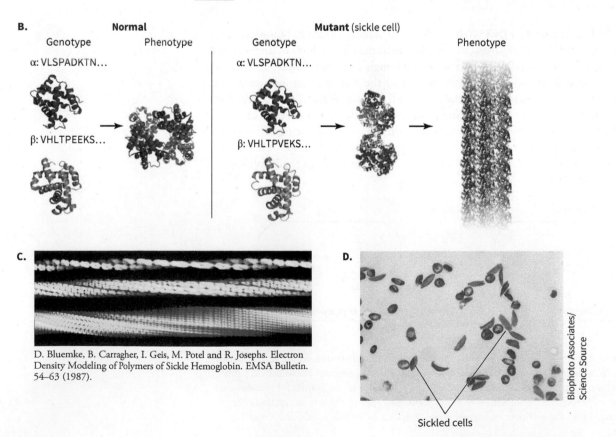

B.

Normal

Genotype   Phenotype

α: VLSPADKTN...

β: VHLTPEEKS...

Mutant (sickle cell)

Genotype   Phenotype

α: VLSPADKTN...

β: VHLTPVEKS...

C.

D. Bluemke, B. Carragher, I. Geis, M. Potel and R. Josephs. Electron Density Modeling of Polymers of Sickle Hemoglobin. EMSA Bulletin. 54–63 (1987).

D.

Biophoto Associates/Science Source

Sickled cells

a hydrophobic patch on the hemoglobin molecule. In the absence of oxygen, the patches can cause the hemoglobin to polymerize into long strands, resulting in deformation of the erythrocyte into a sickle shape. This change in morphology has serious health consequences because the sickle-shaped cells have difficulty passing through capillary beds and can rupture, releasing their contents. This puts a large burden on the spleen and kidneys and causes anemia (low red blood cell count). Sickle cell anemia is extremely painful and can be lethal.

Several treatments are available for sickle cell anemia. Blood transfusions provide healthy erythrocytes that can help relieve some of the symptoms. However, the life span of a healthy erythrocyte is only 120 days, which means the effect of this treatment is temporary. Furthermore, blood transfusions pose a risk of infection by blood-borne pathogens.

A bone marrow transplant is one way to overcome the need for chronic blood transfusions. It involves destroying the native bone marrow of the recipient with chemotherapy or radiation and replacing it with the bone marrow of a donor. The donor marrow has hemopoietic stem cells that give rise to normal erythrocytes. This major procedure carries risks; for example, if the new bone marrow is rejected (called graft versus host disease), the effects can be fatal. It may be that new technologies such as CRISPR or adenovirus can be employed to engineer repairs to the globin genes of a patient's own stem cells, which offers hope for novel treatments or cures. As with any research, scientists working on these breakthroughs in sickle cell treatment must take care to avoid conflicts of interest to protect the integrity of the project as well as their reputations (see **Societal and Ethical Biochemistry: Conflicts of interest**).

Societal and Ethical Biochemistry

Conflicts of interest

Imagine that someone researching treatments for sickle cell anemia had a child with the disease. It makes sense that the researcher would choose to work on a disease that touched her personally—she might be even more motivated to find a cure. However, this personal connection could affect her judgment about how quickly to push a treatment through the approval process so that her child could benefit sooner. This is an example of a conflict of interest (COI), which may be harmless but can be damaging. Failing to disclose a COI may even be a criminal offense. Despite the COI, the researcher in this example can ethically work on sickle cell research—she just needs to acknowledge her connection openly so her team can factor that into their decision making.

Several different COIs can arise in science. The first is a lack of impartiality. Often, scientists are asked to review manuscripts or grant applications submitted by their peers. When this happens, the reviewer needs to consider whether they can be impartial toward the people being reviewed. If the group whose work is under review contains friends or collaborators of the reviewer, the reviewer might give them an unfair advantage, even without intending to. On the other hand, if the group being reviewed is a competitor, the reviewer might judge them in an unduly harsh manner, again generating an unfair situation. In both of these examples, the reviewer has a COI and needs to notify the editor or head of the study section to inform them of the situation. In most cases, the manuscript or grant application will then be reassigned to another reviewer. However, if the reviewer feels she can impartially judge the work, and the journal editor agrees, she may be allowed to continue. Most journals and organizations now have tightly defined rules for when a COI may exist and how to avoid it. In all instances, it is best to notify the editor or head of study section of any possible conflict.

The second type of COI is one where there is a financial interest. This type of COI often does not arise until later in a scientist's or physician's career. If an investigator is taking part in a study and stands to gain financially from the success of that study, for example, by the direct marketing of a product or through owning stock in that company, or if a company specifically paid for a study to be done, the investigator is ethically bound to disclose that relationship. This does not mean that the study has been biased or is not credible, but it lets reviewers and the general public know who helped pay for the work or who stands to gain from its findings. Again, failing to disclose this relationship puts the investigator's reputation in jeopardy and in some instances can leave the person open to criminal investigation.

Another option for treating sickle cell anemia is with the drug hydroxyurea. This drug causes cells to express the γ subunit of hemoglobin, effectively producing fetal rather than adult hemoglobin. Because the $\alpha_2\gamma_2$ assembly of fetal hemoglobin lacks the mutant β subunit, it does not sickle. The mechanism of action of this relatively simple drug is not completely clear, but two pertinent observations have been made. First, hydroxyurea has been shown to induce production of NO, which in turn stimulates production of cyclic GMP (cGMP) by stimulating guanylyl cyclase. This leads to increased expression of the γ subunit of hemoglobin. Second, hydroxyurea breaks down to give hydroxycarbamide, which has been shown to inhibit deoxyribonucleotide biosynthesis. Deoxynucleotides are produced from ribonucleotides by the action of ribonucleotide reductase, an enzyme that employs a tyrosyl radical mechanism (refer to Figure 13.14). By scavenging these radicals, hydroxycarbamide inhibits the formation of deoxynucleotides. It is not clear how this action leads to relief of sickle cell anemia, but it may explain some of the other effects (and side effects) of this drug; for example, hydroxyurea is also used as a chemotherapeutic agent that dampens the immunological response.

Hydroxyurea

Thalassemias   Another common disorder stemming from mutations to the hemoglobin gene is **thalassemia** (**Figure 4.41**). This family of disorders arises from several different mutations to genes coding for the α, β, or γ subunits of hemoglobin. Recall that adult hemoglobin is typically $\alpha_2\beta_2$ but that fetal hemoglobin is $\alpha_2\gamma_2$. In many cases, the mutations alter the splicing of mRNA for the affected subunit and thus lead to formation of a mutant subunit. Sometimes this simply reduces the available number of subunits, but in more severe mutations there is a complete lack of the affected subunit, resulting in disease.

In α-thalassemia, a mutation results in a dysfunctional α subunit, which is recognized by the cell during processing of the protein and degraded by the proteasome. The β subunits therefore increase in concentration and join together in unstable all-β tetramers. Because two genes code for the α subunits, the severity of the disease varies with the number of alleles affected (between one and four).

Mutations to the gene coding for the β subunit result in β-thalassemia. Like α-thalassemias, mutations vary in severity from those that cause only a small change in the total number of β subunits to those that lead to a complete absence of β subunits. The cell responds to a low number of β subunits by decreasing production of α subunits, that is, by downregulating the genes that code for the α subunits. This leads to microcytic anemia, a type of anemia in which the erythrocytes are smaller and lighter in color (hypochromic) than normal.

Treatments for thalassemias are similar to those for sickle cell anemia, namely, blood transfusions and bone marrow transplants.

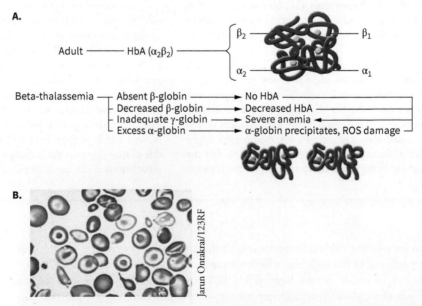

A.

B.

Jarun Ontakrai/123RF

FIGURE 4.41 Thalassemias.   A. Normal hemoglobin consists of two α and β chains. **B.** Thalassemic erythrocytes appear pale and pink due to low levels of hemoglobin in these cells.

(Source: (A) Reprinted with permission from Bank A, Understanding globin regulation in beta-thalassemia: it's as simple as alpha, beta, gamma, delta. Journal of Clinical Investigation,(115)(6), 2005.)

Hidden benefits of hemoglobin mutations

Both sickle cell anemia and thalassemia occur at much higher rates in the general population than would normally be expected for such diseases. Prevalence of these diseases is highest in the Mediterranean, sub-Saharan Africa, the Middle East, and India. Both diseases are autosomal recessive disorders, meaning that they require two mutant copies of the gene for the disease phenotype to present itself. Based on natural selection, a mutant allele should be lost over time if it puts future generations at risk for developing disease without any evolutionary benefit. However, sickle cell anemia and thalassemia are examples of heterozygous advantage, in which a mutant allele does yield some advantage to the organism—in this case, resistance against malaria.

Plasmodium falciparum, a causal parasite in malaria, lives in the erythrocytes of its host (**Figure 4.42**). Having a single mutant copy of a gene for sickle cell anemia (a condition called sickle cell trait) increases the chances of surviving a bout of malaria by increasing the tolerance of the cell to *P. falciparum* by two means. First, the sickle cell hemoglobin induces production of heme oxygenase-1. This enzyme releases carbon monoxide and prevents accumulation of free heme. Second, sickle cell hemoglobin inhibits the activation and proliferation of CD8+/killer T cells, the cells of the immune system that are responsible for destroying infected cells. Inhibition of this aspect of the immune system spares infected cells, which leads to decreased morbidity and mortality. It may seem counterintuitive that preserving infected cells would be beneficial, but many of the complications of malaria are caused by erythrocyte lysis, and so preventing the destruction of these cells leads to some level of resistance to the disease.

Infected erythrocyte

A.
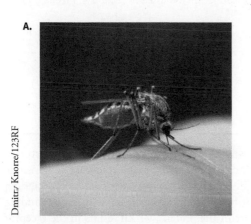
Dmitry Knorre/123RF

B.
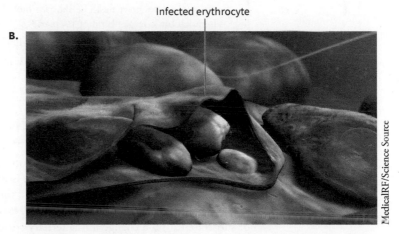
MedicalRF/Science Source

FIGURE 4.42 Plasmodium falciparum and malaria. *Plasmodium falciparum* is a species of protozoan that causes malaria. **A.** It is spread through insect bites, primarily those of mosquitoes (*Anopheles* spp.). **B.** In the course of the protozoan's life cycle, erythrocytes are infected, resulting in erythrocyte lysis and release of the parasite.

Worked Problem 4.6   A mutation to hemoglobin

If hemoglobin had a lysine to glutamate mutation to the residues found in the 2,3-bPG-binding pocket, how might this affect oxygen transport?

Strategy What role does 2,3-bPG play in oxygen transport? How does it bind to hemoglobin?

Solution Hemoglobin's affinity for oxygen is decreased by 2,3-bPG. In the absence of 2,3-bPG, hemoglobin would bind oxygen more tightly and thus have more difficulty delivering oxygen to tissues. Lysine plays an important role in this pocket by providing a positive charge, which can interact with the phosphate groups of 2,3-bPG. Changing this to an anionic glutamate residue would cause charge–charge repulsion and, again, decrease the strength of binding of 2,3-bPG.

Follow-up question How might a mutation to bisphosphoglycerate mutase (the enzyme that makes 2,3-bPG) influence oxygen transport?

Summary

- Hemoglobin is a tetrameric, heme-containing, oxygen-binding protein in erythrocytes, whereas myoglobin is a monomeric, heme-containing, oxygen-binding protein in muscle. Binding of oxygen to heme causes the restructuring of the orbitals in the iron ion, shrinking that ion and its movement into the plane of the porphyrin ring.

- Allosteric regulation of hemoglobin by oxygen enables hemoglobin to deliver more oxygen than would be possible with a monomeric protein. Fetal hemoglobin differs from adult hemoglobin in that γ chains are expressed instead of β chains. This enables the fetus to take oxygen from the maternal blood supply. The Bohr effect describes the decreased affinity of hemoglobin for oxygen under conditions of decreased pH or elevated concentrations of CO_2. These phenomena can be attributed to alterations in salt bridges that stabilize the T state.

- The small molecule 2,3-bisphosphoglycerate stabilizes the T state of hemoglobin, decreasing hemoglobin's affinity for oxygen. Hemoglobin can bind to and transport other molecules, including NO, HS, SO, and CO_2. CO_2 is also transported as bicarbonate, produced by the action of carbonic anhydrase.

- Mutations to the hemoglobin gene are responsible for several diseases. Sickle cell anemia is caused by a single point mutation that results in the substitution of a valine for a glutamate. This causes the protein to polymerize and erythrocytes to adopt a sickled shape. Thalassemias are diseases caused by the inability to produce one of the chains of hemoglobin.

Concept Check

1. Describe the structures of hemoglobin and myoglobin, relating them to the functions and properties of these molecules. Use the two models of allosterism discussed in section 4.5 to describe hemoglobin's oxygen-binding properties.

2. Describe the chemistry of how different allosteric regulators affect hemoglobin.

3. Describe how mutations to hemoglobin can cause disease.

4.7 Receptor–Ligand Interactions

This section provides a detailed description of this class of molecules and the chemistry of receptor–ligand interactions. The discussion is included here because many receptors demonstrate allosterism when binding to ligands. Ligand binding of all receptors is analogous to the binding of substrate in the active site of an enzyme, except that there is no catalysis. Rather, the binding of substrate causes a change in the tertiary structure of the protein, which results in the receptor initiating a signaling cascade.

Broadly speaking, receptors fall into two classes based on where they function: nuclear hormone receptors, which bind ligands within the cell, and membrane-bound receptors. Receptors can also be classified according to their structure and mechanism.

4.7.1 Nuclear hormone receptors bind ligands within the cell

Nuclear hormone receptors are cytosolic or nuclear proteins that bind to largely hydrophobic signal molecules. Examples of the ligands to which these receptors bind include steroid and thyroid hormones and retinoic acid. In contrast to the membrane-bound receptors, nuclear hormone receptors never leave the cell or, in some cases, the nucleus. Instead, the ligand enters the cell or nucleus and binds to the receptor (**Figure 4.43**). If the receptor is initially found in the cytosol, the receptor–ligand complex translocates to the nucleus after binding to the ligand. There the receptor binds to specific stretches of DNA, acting as a transcription factor and altering gene expression. It is currently thought that nuclear receptors that reside in the nucleus are permanently bound to DNA but only activated upon binding of the ligand.

As with many molecules, the structure of receptor proteins provides information about their function. Nuclear hormone receptors have two domains: a ligand-binding domain and a

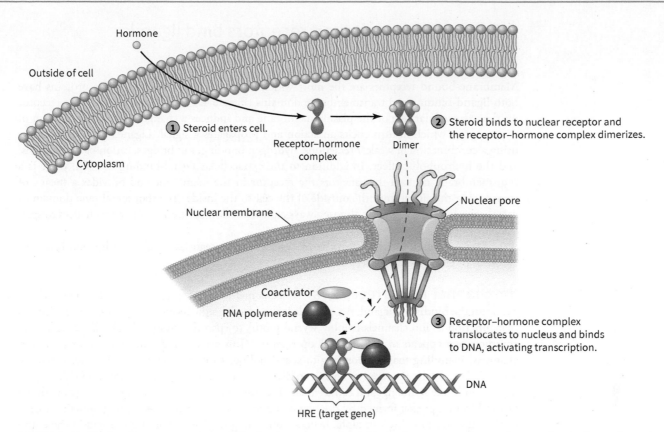

FIGURE 4.43 Nuclear hormone receptors. Nuclear hormone receptors bind to hormones, either in the cytosol or in the nucleus. They translocate to the nucleus, where they bind DNA as a dimer, interacting with coactivators and corepressors. In the nucleus, these receptors influence gene transcription.

DNA-binding domain (**Figure 4.44**). The ligand-binding domain is mostly alpha helical and has a hydrophobic cleft (pocket) in which the ligand binds. Corepressors and coactivators bind to this domain of the receptor to repress or activate transcription, respectively.

Nuclear hormone receptors bind to DNA via a zinc finger motif in the DNA-binding domain. This domain is comprised of a short stretch of alpha helix and some random coil or one strand of alpha helix and one of beta sheet that are held together by a zinc ion chelated to a combination of cysteine and histidine residues in the protein. There are many types of zinc fingers, and a single protein may have multiple zinc finger domains (the domain itself is generally only about 20 to 30 amino acids). Typically, nuclear hormone receptors bind DNA as either a homo- or heterodimer with other nuclear hormone receptors.

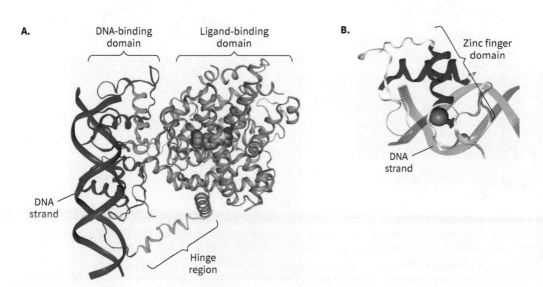

FIGURE 4.44 Structure of a nuclear hormone receptor.
A. A nuclear hormone receptor has two domains: a ligand-binding domain and a DNA-binding domain. These are connected to one another by a flexible hinge.
B. Many nuclear hormone receptors interact with DNA through a zinc finger domain.

(Source: (A, B) Data from PDB ID 3DZY Chandra, V., Huang, P., Hamuro, Y., Raghuram, S., Wang, Y., Burris, T.P., Rastinejad, F. (2008) Structure of the intact PPAR-gamma-RXR- nuclear receptor complex on DNA. *Nature* **456**: 350–356)

4.7.2 Membrane-bound receptors bind ligands outside the cell

Membrane-bound receptors are the most widely known type of receptor. These proteins have both ligand-binding and transmembrane domains (**Figure 4.45**). A ligand binds to the receptor on the extracellular side of the plasma membrane and induces a change in the conformation of the receptor, which in turn elicits an action and propagates a signal. Ligands bind to receptors using a combination of weak interactions: hydrogen bonding, salt bridges, cation–π interactions, and the hydrophobic effect. In addition to the extracellular ligand-binding domain, there is a transmembrane domain that anchors the receptor in the membrane and provides a means of transducing the signal from the outside of the cell to the inside. The transmembrane domain of membrane-bound receptors is almost always alpha helical in nature and uses alpha helices to span the phospholipid bilayer.

Membrane-bound receptors can be divided into three categories: ion channel-linked, G protein-coupled, or enzyme-linked receptors.

Ion channel-linked receptors

Ion channel-linked receptors open or close an ion channel upon the binding of ligand. An example is the acetylcholine receptor (**Figure 4.46**). Receptors that function as ion channels employ several motifs to span the phospholipid bilayer. The most common one appears to be the Cys-loop receptor. This class includes many receptor-gated ion channels, including those responding to acetylcholine, serotonin, glycine, and γ-amino butyric acid (GABA). Here, the receptor consists of five subunits, each of which has three domains: cytosolic, membrane spanning, and extracellular. The stoichiometry of the complex is $\alpha_2\beta\gamma\delta$, with the ligand-binding pocket formed by the two α subunits. The membrane-spanning domain of each of the subunits contains four alpha helices, combining to give a total of 20 helices in the entire complex. When the receptor binds its ligand—for example, the nicotinic acetylcholine receptor binding to acetylcholine—the subunits twist with respect to one another. This twisting opens an ion-selective channel at the core of the complex that allows the passage of ions (in this instance, Na^+) and generates an action potential or muscle contraction.

G protein-coupled receptors

G protein-coupled receptors (GPCRs) act through G proteins. This vast family of receptors uses a heterotrimeric G protein to transmit an extracellular signal to an intracellular second messenger generator (for example, adenylate cyclase) or a phospholipase to generate an effect. Most G protein-coupled receptors signal through the 7-transmembrane helix (7 TM) family of receptors. In this family, the seven transmembrane helices are not directly perpendicular to the membrane but are instead oriented at a slight pitch to

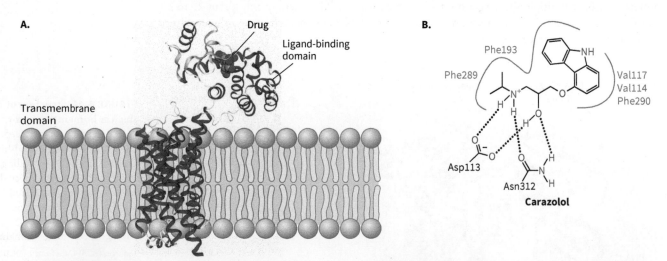

FIGURE 4.45 Structure of a membrane-bound receptor.   **A.** The β-adrenergic receptor is a 7-transmembrane G protein-coupled receptor. The protein has an extracellular domain that is responsible for ligand binding and a transmembrane domain characterized by seven alpha helices. **B.** Molecules such as this drug, carazolol, bind to the receptor through a number of weak interactions.

(Source: Data from PDB ID 2RH1 Cherezov, V., Rosenbaum, D.M., Hanson, M.A., Rasmussen, S.G., Thian, F.S., Kobilka, T.S., Choi, H.J., Kuhn, P., Weis, W.I., Kobilka, B.K., Stevens, R.C. (2007) High-resolution crystal structure of an engineered human beta2-adrenergic G protein-coupled receptor. *Science* **318**: 1258–1265)

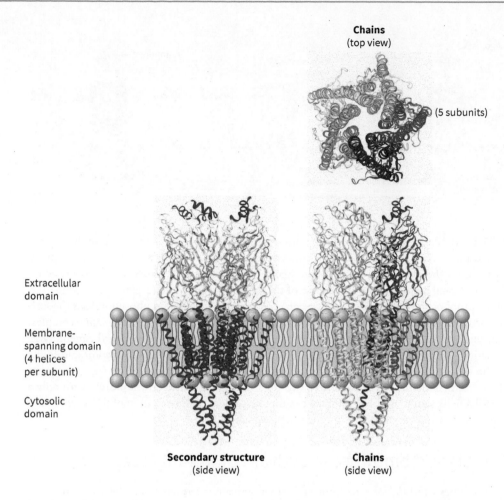

Chains
(top view)

(5 subunits)

Extracellular
domain

Membrane-
spanning domain
(4 helices
per subunit)

Cytosolic
domain

Secondary structure
(side view)

Chains
(side view)

FIGURE 4.46 The acetylcholine receptor, an example of an ion channel. The acetylcholine receptor is comprised of five subunits, each of which has four helical transmembrane segments, a small helical cytosolic domain, and a large extracellular ligand-binding domain, which is largely beta sheet. The holoenzyme has 20 transmembrane helices.

(Source: Data from PDB ID 2BG9 Unwin, N. (2005) Refined Structure of the Nicotinic Acetylcholine Receptor at 4A Resolution *J. Mol. Biol.* **346**: 967)

the plane of the membrane. One might expect that these helices would be hydrophobic in order to span the phospholipid bilayer. In reality, the bundling of the helices means that they are usually amphipathic, having a hydrophobic face exposed to the hydrophobic phospholipid acyl chains and other hydrophilic faces that interact with the other helices, preserving the tertiary structure of the receptor. G protein-coupled receptors offer exciting possibilities in terms of drug development, as described in **Medical Biochemistry: Allosteric regulators as drugs**.

Medical Biochemistry

Allosteric regulators as drugs

Drugs are often thought of as molecules that inhibit enzymes or that bind to receptors and activate or inactivate them. These are accurate pictures of how some drugs function, but drugs that function as allosteric regulators of enzymes or receptors offer exciting new avenues for drug development. In particular, allosteric regulators of G protein-coupled receptors (GPCRs) hold promise as a novel class of pharmaceuticals.

Allosteric regulators have several advantages over traditional orthosteric regulators, which can either activate or inactivate the receptor to some extent. For example, the catecholamines (dopamine, epinephrine, and norepinephrine) bind to multiple receptors, which means they may also activate those receptors and trigger side effects. Another disadvantage is that levels of signaling molecules vary throughout the day, and these drugs are unable to work effectively with such variability. Finally, the interactions

that occur between peptides or proteins are complex to mimic, especially when taking into account how the drug will act in the body.

Allosteric regulators can be tailored specifically for the protein of interest or class of proteins, which means they lack many of the side effects seen with orthosteric regulators. They are often small molecules compared to some of the hormones and ligands involved in signaling. Perhaps most interesting and important, the receptor is still subject to the natural ebb and flow of ligand that occurs throughout the day because allosteric regulators only regulate the response in the presence of the natural ligand.

There are currently no allosteric regulators of GPCRs on the market, but several molecules in development hold promise. These include regulators of glutaminergic and muscarinic acetylcholine receptors. These molecules are being investigated as treatments for difficult-to-treat diseases including Parkinson's, schizophrenia, and Alzheimer's.

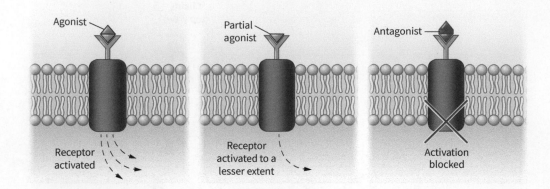

FIGURE 4.47 Agonists and antagonists. Agonists are compounds that bind to receptors and activate them. Partial agonists bind to receptors and activate them but not as well as agonists. Antagonists are molecules that bind to receptors and block their activation.

Enzyme-linked receptors The third subclass of membrane-bound receptors is enzyme-linked receptors. An example is the receptor tyrosine kinase (RTK) class of hormone receptors, which includes the insulin receptor. When ligand binds to these receptors, they become active enzymes and catalyze a reaction; in the case of the insulin receptor, the receptor causes dimerization and autophosphorylation of the other copy of the receptor and then phosphorylates the scaffolding proteins (for example, the insulin receptor substrate), which assemble a signaling complex.

Enzyme-linked receptors often have a single alpha helix that passes through the membrane. Examination of the structure of epidermal growth factor (an enzyme-linked receptor and RTK) reveals how this helix functions in tethering the extracellular ligand-binding domains to the intracellular effector domains. The structure of this protein also illustrates how a protein with only a single alpha helix spanning the membrane can transduce signals; dimerization of the receptor and autophosphorylation are necessary to propagate a signal.

4.7.3 Ligand binding can activate or inhibit receptors

Many molecules can bind to a receptor and either activate the receptor or block its activation. Molecules that bind to and activate a receptor (acting like the natural ligand) are **agonists** (**Figure 4.47**). Those that bind to the receptor but activate it to a lesser extent than the native ligand are **partial agonists**. Molecules that bind to a receptor and block its activation are **antagonists**; this term is used for all molecules that bind and block a receptor, and antagonists are analogous to enzyme inhibitors.

Competitive agonists compete with the native ligand for the ligand-binding site. In contrast, **noncompetitive agonists** do not compete with the native ligand, but they lower the ability of the native ligand to activate the receptor maximally. **Inverse agonists** appear similar to antagonists but often cause a different effect in the cell. The binding of an inverse agonist inhibits the basal level of activity in a receptor, that is, the level of activity the receptor has in the absence of ligand.

4.7.4 Receptor–ligand interactions can involve simple or allosteric binding

A study of how receptors respond to ligands will show that some responses are allosteric in nature. The interactions of a ligand with a receptor follow the models of allosterism introduced in section 4.5. In short, if a receptor has a single ligand-binding site, the Langmuir equation can be used to represent the percentage of sites bound. Plotting the percentage of bound sites on the y-axis and the concentration of ligand [A] on the x-axis gives a hyperbolic curve similar to one produced according to Michaelis–Menten kinetics. The data are often linearized and plotted as a Scatchard plot, in which the percentage bound/[A] is plotted on the y-axis against the percentage bound on the x-axis (**Figure 4.48**).

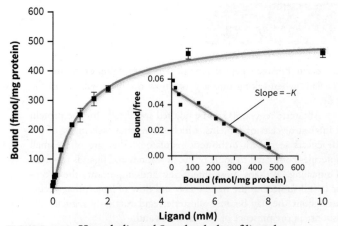

FIGURE 4.48 Hyperbolic and Scatchard plot of ligand binding. The larger plot shows a rectangular hyperbola obtained when the amount of ligand is graphed versus the fraction bound to the protein. The inset shows a Scatchard plot, which is a transformation performed to linearize a rectangular hyperbola. The concentration of bound ligand is plotted against the fraction bound to generate a linear plot. The slope of the line is the negative of the binding constant K.

If there is complex binding—that is, sites that are not equivalent or independent—the binding follows what is observed for an allosteric situation and can frequently fit either the concerted or sequential model for ligand binding.

Worked Problem 4.7   Sweetness!

The sensing of sweet tastes by humans is accomplished by binding of any one of a number of ligands to the human taste receptor hT1R2/hT1R3, a multimeric receptor that is G protein-coupled. The taste receptor was recently expressed in cultured cells in a study by Masuda et al. The cells were exposed to different concentrations of sucralose (a common artificial sweetener), and the binding of the compound was detected through a fluorescence assay. Data were expressed as DRFU, or the change in relative fluorescent units, RFU.

[Sucralose], mM	ΔRFU
0.005	22
0.01	29
0.05	47
0.10	92
0.05	135
1	148
5	151
10	153

Is the sweet taste receptor responding using simple binding or allosteric (cooperative) binding?

Strategy   It can be difficult to see the shape of a curve from a table of numbers. Therefore, graph the data to see the shape of the curve. A linear-log plot may give a clearer indication of the shape of the curve. Based on how this curve looks, we could plan future analysis, like a Langmuir or Scatchard plot.

Solution   These data are from Masuda et al. "Characterization of the Modes of Binding between Human Sweet Taste Receptor and Low-Molecular-Weight Sweet Compounds." *PLoS One* 7, no. 4 (2012): e35380.

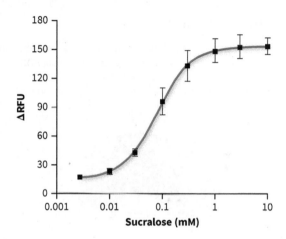

The response of the human taste receptor to the binding of sucralose is allosteric and cooperative in nature (this is more obvious when the data are graphed as a linear-log plot).

Follow-up question   Is this regulation homotropic or heterotropic? Explain your response.

Summary

- Interactions between ligand-binding domains and ligands are analogous to those between substrates and enzymes.
- Receptors can be classified based on whether they bind ligands inside the cell (nuclear hormone receptors) or outside the cell (membrane-bound receptors). Nuclear hormone receptors have two domains: a ligand-binding domain and a DNA-binding domain.
- Membrane-bound receptors have a ligand-binding domain and a transmembrane domain. Membrane-spanning domains are alpha helical. Membrane-bound receptors can further be separated into ion-channel linked, G protein-coupled, or enzyme-linked receptors.
- Agonists bind to and activate receptors, whereas antagonists bind to block the activation of receptors.
- Ligands bind to receptors using both simple Michaelis–Menten kinetics and also allosteric models of interaction and regulation.

Concept Check

1. Compare and contrast receptor–ligand interactions with those between substrates and enzymes.
2. Compare the two major classes of receptors.
3. Describe how membrane-bound receptors can be subdivided based on structure and function.
4. Explain the difference between (a) competitive and noncompetitive agonists, (b) agonists and antagonist.

Bioinformatics Exercises

Problems

4.1 Regarding Enzymes

1. Examine this structure. Based only on this information, where might you find the active site on this enzyme? (Source: Data from PDB ID 1V4T Kamata, K., Mitsuya, M., Nishimura, T., Eiki, J., Nagata, Y. (2004) Structural basis for allosteric regulation of the monomeric allosteric enzyme human glucokinase. *Structure* 12: 429–438)

2. The burning of sugar in air is chemically similar to the enzymatically catalyzed oxidation of sugar. Draw a reaction coordinate diagram for the reaction of sucrose ($C_{12}H_{22}O_{11}$). What are the different thermodynamic parameters of this reaction (such as free energy)? How do these change when the reaction is catalyzed?

3. Like proteins, RNA molecules can be catalytic. Examine the structure of a short sequence of RNA. What types of functional groups are there in RNA molecules that could participate in catalysis in the way that amino acid side chains do?

4. Based on your knowledge of DNA, RNA, and proteins, propose a reason why catalytic RNA and protein molecules have been observed but catalytic DNA molecules have not.

5. Contaminating proteases are a common problem in protein or enzyme purification. These enzymes degrade the very proteins the scientist is attempting to isolate, but they can be inhibited with a mixture of protease inhibitors. What would such a mixture be likely to contain? What would be the properties of that group of inhibitors?

6. Efavirenz is one of the NNRTI class of drugs used to treat HIV infection. Drug-resistant HIV has undergone mutations which lead to drug resistance. The most common mutation observed after efavirenz treatment is K103N, which is also observed with other NNRTIs. Examine the genetic code. Why might this particular mutation predominate rather than others?

7. Azidothymidine (AZT) is an example of an antiretroviral drug used to treat HIV infection. Its structure is shown below. What class of drug is this (NNRTI, RTI, or protease inhibitor)?

Azidothymidine

8. Would the antiretroviral drugs developed and used to treat HIV be useful for treating other viral infections? Why or why not?

9. Norvir is an HIV protease inhibitor. Based on the structure of Norvir, what amino acids might you predict to bind in the active site of the protease?

Norvir

4.2 Enzymes Increase Reaction Rate

10. The Eadie-Hofstee plot is another means of obtaining a linear graph of Michaelis-Menten data, and has the equation $v = K_M(v/[S]) + V_{max}$. Velocity is plotted on the y-axis, and velocity over substrate concentration is plotted on the x-axis. What are the values of the slope and y-intercept of this graph? What is the advantage of this graph over a Lineweaver-Burke or a saturation kinetic (Michaelis-Menten) plot? Can you derive the Eadie-Hofstee form of the equation from the Michaelis-Menten equation?

11. If a Lineweaver-Burk plot gave a line with an equation of $y = 0.25\ x + 0.34$, what are the values of K_M and V_{max} if the substrate concentration is in mM and the velocity in mM/s?

12. Acetazolamide is a diuretic used to treat conditions ranging from glaucoma to altitude sickness. It is an inhibitor of carbonic anhydrase.

Examine the structure of drug given below. What kind of inhibitor might this be and why? What part of acetazolamide might bind carbonic anhydrase?

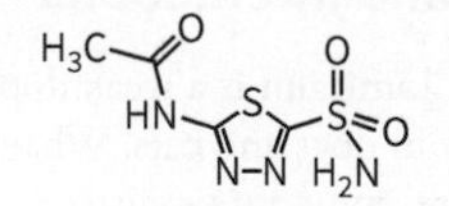

Acetazolamide

13. Bis(sulfosuccinimidyl)suberate (BS3) is a cross-linking reagent. It will react with primary amines to cross-link them, forming new amide linkages. Draw a diagram of how this reaction may occur. How could this reagent be used to help determine whether an enzyme has multiple subunits? Which amino acids would be affected?

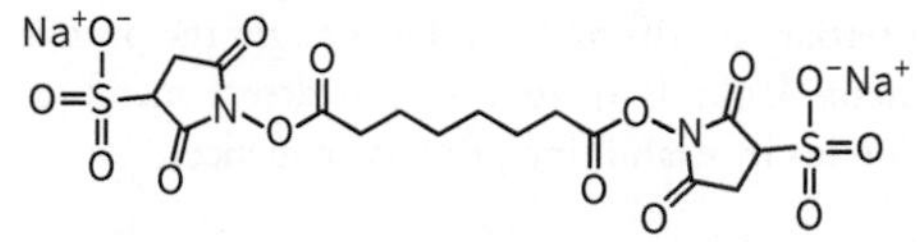

BS3

14. Aspartate transcarbamylase will also bind to succinate, shown below. What would this lead you to predict about the aspartate binding site?

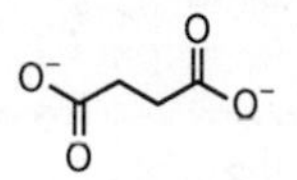

Succinate

15. Determine k_{cat} for a reaction in which V_{max} is 4×10^{-4} mol·min^{-1} and the reaction mixture contains one microgram of enzyme (the molecular weight of the enzyme is 200,000 D).

16. Carbonic anhydrase is said to have achieved catalytic perfection. Could this enzyme be artificially improved? If so, how?

4.3 The Mechanism of an Enzyme Can Be Deduced from Structural, Kinetic, and Spectral Data

17. Examine the mechanism of carbonic anhydrase. If the Zn^{2+} were removed from the active site, and replaced with either Cu^+, Cd^{2+}, or Fe^{3+} would you expect the enzyme to still be functional? Why or why not?

18. Examine the structure and mechanism of lysozyme. If the glutamate and aspartate residues were reversed, would you expect the enzyme to still be active? Why or why not?

19. Examine the substrate binding site of lysozyme. If there were a mutation to the region distant from the bond that is cleaved, would you expect that this mutation would be more or less harmful to the enzyme than a mutation that was closer to the bonds that are cleaved?

20. If the Asp in the chymotrypsin active site was mutated to another amino acid, what would be the invisible mutation in that it is least likely to affect the function of the enzyme?

21. The active site residues in some proteins seem to be "ultimately conserved," meaning that any change to that residue results in a complete lack of activity. Is it possible for the gene coding for that enzyme to evolve? Explain your answer.

22. Some residues of an enzyme are considered more highly conserved than others (for example, the active site serine). Explain what "highly conserved" means and why some residues are more highly conserved than others.

23. If the glycine residue in the oxyanion hole of chymotrypsin were mutated to an alanine residue, how might this affect the kinetics of the reaction? Which parameters would be affected more significantly: K_M or k_{cat}?

24. If the glycine residue in the oxyanion hole of chymotrypsin were mutated to a tryptophan residue, how might this affect the kinetics of the reaction? Which parameters would be affected more significantly: K_M or k_{cat}?

25. If the substrate binding pocket of chymotrypsin were mutated to include a lysine residue, how might this affect the kinetics of the reaction? Which parameters would be affected more significantly: K_M or k_{cat}?

4.4 Examples of Enzyme Regulation

26. In thinking of enzymes and evolution, the focus is often on the binding of substrate or catalysis, with regulation seen as a secondary event. What sorts of selective pressures are at work in regulating enzyme function in enzymes such as the ones discussed in this chapter?

27. Compare the three types of mechanisms for regulation of proteins discussed in this chapter, with regard to level of control. Discuss how quickly does each type of control take effect and how fine-tuned is each.

28. Biochemical pathways are molecular assembly lines where numerous enzymes each catalyze a single step. Where might you anticipate that such a process would be controlled?

29. The strength of an electrostatic interaction can be estimated using a version of Coulomb's law: $\Delta E = A\ k\ q_1 q_2 / D\ r$ where A is Avogadro's number (6.02×10^{23}), k is 9.0×10^9 joules/coulomb^2, q_1 and q_2 are the charges on the groups measured in coulombs $(1.6 \times 10^{-19}$ joules per electron), D is the dielectric constant of the media, and r is the distance in meters. How strong is the coulombic attraction between a phosphoserine residue (charge 2−) and a histidine residue (charge 1+) on the other side of a protein 20 angstroms away? Assume the core of the protein is similar to hexane, with a dielectric constant of 4. What if the histidine were on an adjacent protein 20 angstroms away, with water $(D = 80)$ separating them?

30. PALA competes for binding in the active site of ATCase, yet it acts as an allosteric inhibitor and stabilizes the R state. Explain this phenomenon.

31. Some enzymes have been shown to polymerize or form multimers upon the addition of an allosteric regulator. How could you test whether an enzyme had formed some sort of higher-order structure upon the addition of an allosteric regulator?

32. Small molecules that affect allosteric sites are a growing area of drug development. Speculate as to why development of an allosteric regulator may have advantages over other types of inhibitors.

4.5 Allosterism and Cooperativity

33. The Hill coefficient of a plot for oxygen binding to hemoglobin is approximately 3. How would oxygen transport be influenced if it were a different value such as 1, 1.5, or 2?

34. Can the Hill coefficient ever be 0 or a decimal less than one? Can it be a negative number? Explain.

35. What are the assumptions of the Hill equation?

36. Rearrange the Hill equation to yield the form found in the Hill plot in Figure 4.19.

37. What is the difference between allosteric and orthosteric regulations. List two characteristics of proteins/enzymes that can be regulated allosterically.

4.6 Hemoglobin

38. For each of the mutations provided, predict how it might alter the activity of hemoglobin.

  a. Glu to Val at the 6 position of the β chain

  b. Lys to Trp at the 82 position of the β chain

  c. Lys to Arg at the 82 position of the β chain

  d. Mutation of the proximal His to Ser in the β chain

  e. His to Asp at the 2 position of the β chain (in the 2,3-bPG-binding pocket)

39. The protein sequence of hemoglobin differs slightly among different animals. For each animal, consider what environment it lives in, and describe what you might observe in their hemoglobin based on those environmental conditions.

  a. blue whales

  b. mountain goats

  c. cheetahs

  d. migratory birds

40. What is the key role of the structure of the globin protein in the affinity of hemoglobin for oxygen?

41. Draw a model for allosteric inhibition of glucokinase like the ones found in Figure 4.28.

42. Using equilibrium constants, describe the reactions involved in the regulation of glucokinase.

43. Draw approximate reaction rate curves for glucokinase activity plotted as a function of glucose concentration in the presence and absence of glucokinase regulatory protein.

44. If the life span of an erythrocyte is 120 days, how many times a year would a sickle cell patient need to be given a blood transfusion to ensure a unit of healthy erythrocytes at any given time?

45. Why would hydroxyurea be used as a chemotherapeutic agent? Why would its use lead to a decreased immune response?

46. Is γ-thalassemia a disease? Would it be a problem if you had γ-thalassemia or not?

47. Would someone with sickle cell anemia still be considered a carrier of the disease after receiving a bone marrow transplant?

48. There are two separate genes (loci) for the a subunit of hemoglobin: *HBA1* and *HBA2*. Is this an evolutionary advantage? Explain why or why not.

49. Many drugs that are used to treat thalassemias chelate iron out of the body.

  a. How might these molecules bind iron?

  b. Why is it important to bind iron in the body?

  c. Would taking additional iron be beneficial for these anemic patients?

  d. Why is this a less significant issue in the treatment of sickle cell anemia?

50. Using the literature, compare and contrast hemolytic anemia, microcytic anemia, nutritional anemia, and glucose-6-phosphate dehydrogenase deficiency. In each case, what is defective or missing? How are these diseases similar, and how are they different?

51. Based only on the facts that the γ subunit of hemoglobin is expressed *in utero* and relieves the complications of sickle cell anemia, what can you say about the sequence or structure of γ hemoglobin?

52. Explain how receptor–ligand interactions can be modeled using the mathematical expressions that describe allosteric interactions.

4.7 Receptor–Ligand Interactions

53. Clomipramine or Clomicalm is a weak dopamine receptor antagonist used to treat anxiety in dogs and cats. What does it mean for a drug to be a weak dopamine receptor antagonist?

54. If the ligand-binding pocket of a receptor is largely hydrophobic, what does this tell us about the energetics of binding compared to one that is lined with charged residues? What forces are directing ligand binding?

55. Nicotine is an agonist of the nicotinic class of acetylcholine receptors. Describe what this means in terms of signaling through that receptor.

56. Both a 7-transmembrane helix G protein-coupled receptor (7TM-GPCR) and the insulin receptor span the plasma membrane but use different tertiary motifs to do so. How might the primary sequences of these proteins look? Is there a way to detect membrane-spanning regions of proteins by examining primary sequence?

Data Interpretation

57. An experiment yields the kinetic data shown below:

[Substrate], mM	Rate (µM/min/mg enzyme), substrate, no inhibitor	Rate (µM/min/mg enzyme) [substrate] + 7 nM inhibitor
0.02	0.8	1.5
0.04	1.5	2.9
0.10	3.7	6.9
0.25	8.6	15
1.00	24	36
2.50	41	50

- Graph this data and provide values of V_{max} and K_M for this enzyme.

- How is the inhibitor acting on this enzyme?

- What information would you need to be able to calculate K_I for this inhibitor?

58. You identified a new protein and obtained the amino acid sequence, but you lack structural data. Mutation of a serine of interest to alanine results in a partial loss of activity, but mutation of the same serine to aspartate returns activity to wild-type levels. What explanation could link these observations?

59. Phosphofructokinase is an enzyme in the glycolytic pathway (which the cell uses to break down glucose). Phosphofructokinase uses ATP to add a phosphate group to C-1 of fructose-6-phosphate, generating fructose 1,6-bisphosphate. Explain how ATP and AMP are acting in this graph.

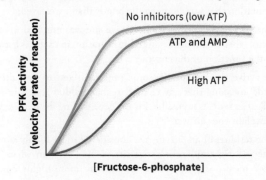

60. A pure sample of chymotrypsin gives three bands that are smaller than the expected 25.6 kDa when examined on SDS-PAGE, but elutes from a size exclusion column somewhere between 25 and 27 kDa. Explain this finding.

61. An enzyme has a molecular weight of 66 kDa when analyzed by SDS-PAGE when run under reducing conditions (with β-mercaptoethanol). This same enzyme has a molecular weight of 200 kDa when run on a native PAGE gel. Propose a stoichiometry for the holoenzyme.

62. A given enzyme has a molecular weight of 40 kDa. It is known to interact with a small regulatory protein with a molecular weight of 20 kDa. The molecular weight of the complex when determined using size-exclusion chromatography is 100 kDa. What are all the possible combinations of enzyme and regulatory protein?

Experimental Design

63. You are studying an enzyme you believe to be regulated by phosphorylation. Design an experiment to test whether this enzyme is more active in the phosphorylated state. Include proper controls in your design.

64. You have identified a previously undiscovered family of enzymes. Examination of related sequences in Genbank shows a group of three histidine residues that are conserved in all related species. Design an experiment to see whether this enzyme employs a metal ion in its active site.

65. You discovered a small molecule you believe may be an inhibitor of the HIV protease. How could you determine the type of inhibitor?

66. ATCase is a dodecameric enzyme comprising six regulatory and six catalytic subunits. Design an experiment to show that ATCase is a functional dodecamer.

67. Electrophoresis is a common technique for the study of proteins. How might you use electrophoresis on native hemoglobin to discern between native and sickle cell hemoglobin?

68. Polymerization of sickle cell hemoglobin is favored only in certain conditions. What techniques might you be able to use to determine if hemoglobin is polymerized?

69. SDS-PAGE is another common technique used to study proteins. How might SDS-PAGE be used to study thalassemia?

70. Allosteric enzymes are nearly always multimers. How can the stoichiometry of an enzyme in its native state be determined?

Ethics and Social Responsibility

71. Most antibiotics are inhibitors of bacterial enzymes. Antibiotics are a hugely successful category of drugs, yet generally, they are not as profitable for pharmaceutical companies as other types of medications. The lack of profitability, coupled with the difficulty of identifying new antibiotics, has limited discoveries of new classes of antibiotics. Should the government create incentives to increase the number of new antibiotics under development?

72. Many enzymes are used commercially in cleaning products. Based on your knowledge of chemistry and biology, is this a "green" approach to cleaning, or would it be better to use synthetic detergents?

73. Some of the older types of contact-lens solutions use hydrogen peroxide to clean and decontaminate lenses and use the enzyme catalase or a metal catalyst to break down the hydrogen peroxide. What is the difference between these two catalysts, and why would you choose one over the other?

74. The study of enzymes and enzyme mechanisms by itself is not of interest to most people, but disease-resistant crops, antibiotics, and antivirals are of interest. Look up the terms "basic research," "applied research," and "translational research." How do these fields differ, and how do they complement each other? Can any of these fields exist without the others?

75. WHO keeps a list of medicines that it considers essential for all healthcare systems. Based on your current knowledge, what types of drugs (for example, pain relievers or antibiotics) would you anticipate might be included in this list? Look up a current copy of the list (it is reviewed biannually) on the WHO Web site. Are there inclusions or omissions you find surprising?

76. Siblings are often the best donors for bone marrow transplants. There have been instances of children being conceived using *in vitro* fertilization using zygotes that have been pre-screened for genetic defects (such as thalassemia) to provide bone marrow for affected siblings. Do you find this ethical or not? Explain.

77. Bone marrow transplants are used to treat multiple diseases, including sickle cell anemia and thalassemia. Would you consider becoming a donor for a bone marrow transplant? Why or why not?

78. You are investigating a novel mutant of hemoglobin found in a large family living in a city in United States. What ethical obligations do you have as a researcher to this family? Would you be bound by the same ethical responsibilities if you were studying a mutation found in a family of natives Peruvian Indians living in village in Peru.

79. If you did not wish to participate in a genetic study of your family, what rights and responsibilities would you have?

Suggested Readings

4.1 Regarding Enzymes

Branden, C., and J. Tooze. *Introduction to Protein Structure*. New York, NY: Garland Science, 1999.

Buxbaum, E. *Fundamentals of Protein Structure and Function*. New York, NY: Springer, 2007.

Kessel, A., and N. Ben-Tal. *Introduction to Proteins: Structure, Function, and Motion*. Boca Raton, FL: CRC press, 2011.

Petsko, G. A., and D. Ringe. *Protein Structure and Function*. Sunderland, MA: Sinauer Associates, Inc., 2003.

Williamson, M. *How Proteins Work*. New York, NY: Garland Science, 2011.

4.2 Enzymes Increase Reaction Rate

Cook, P. F., and W. W. Cleland. *Enzyme Kinetics and Mechanism*. New York, NY: Garland Science, 2007.

Copeland, R. A. *Enzymes: A Practical Introduction to Structure, Mechanism, and Data Analysis.* 2nd ed. Weinheim, Germany: Wiley-VCH, 2000.

Cornish-Bowden, A. *Fundamentals of Enzyme Kinetics.* Hoboken, NJ: Wiley Blackwell, 2012.

4.3 The Mechanism of an Enzyme Can Be Deduced from Structural, Kinetic, and Spectral Data

Bar-Even, A., E. Noor, Y. Savir, W. Liebermeister, D. Davidi, D. S. Tawfik, and R. Milo. "The Moderately Efficient Enzyme: Evolutionary and Physicochemical Trends Shaping Enzyme Parameters." *Biochemistry* 50, no. 21 (2011): 4402–4410.

Eisenthal, R., M. J. Danson, and D. W. Hough. "Catalytic Efficiency and kcat/KM: A Useful Comparator?" *Trends in Biotechnology* 25, no. 6 (June 2007) 247–249.

Fersht, A. M. *Structure and Mechanism in Protein Science.* New York, NY: W.H. Freeman, 1998.

Xie, F., J. M. Briggs, and C. M. Dupureur. "Nucleophile Activation in PD...(D/E)xK Metallonucleases: An Experimental and Computational pKa Study." *Journal of Inorganic Biochemistry* 104, no. 6 (1998): 665–672.

4.4 Examples of Enzyme Regulation

Hammes, G. G. *Enzyme Catalysis and Regulation.* Cambridge, MA: Academic Press, 1982.

Kantrowitz, E. R. "Allostery and Cooperativity in Escherichia Coli Aspartate Transcarbamoylase." *Archives of Biochemistry Biophysics* 519, no. 2 (2012 March): 81–90.

4.5 Allosterism and Cooperativity

Changeux, J. P., and S. J. Edelstein. "Allosteric Mechanisms of Signal Transduction." *Science* 308, no. 5727 (2005): 1424–28.

Ferrer, J. C., C. Favre, R. R. Gomis, et al. "Control of Glycogen Deposition." *FEBS Letters* 546, no. 1 (2003): 127–32.

Jordan, A., and P. Reichard. "Ribonucleotide Reductases." *Annual Review of Biochemistry* 67 (1998): 71–98.

Kenakin, T., ed. *Oligomerization and Allosteric Modulation in G Protein Coupled Receptors* (vol. 115 in *Progress in Molecular Biology and Translational Science*). San Diego, CA: Academic Press, 2012.

Perutz, M. F. *Mechanisms of Cooperativity and Allosteric Regulation in Proteins.* Cambridge, UK: Cambridge University Press, 1990.

Ruderman, N. B., A. K. Saha, D. Vavvas, and L. A. Witters. "Malonyl-CoA, Fuel Sensing, and Insulin Resistance." *American Journal of Physiology-Endocrinology and Metabolism* 276, no. 1 (1999): E1–E18.

Traut, T. W. *Allosteric Regulatory Enzymes.* New York: Springer, 2010.

4.6 Hemoglobin

Cokic V. P., R. D. Smith, B. B, Beleslin-Cokic, et al. "Hydroxyurea Induces Fetal Hemoglobin by the Nitric Oxide-Dependent Activation of Soluble Guanylyl Cyclase." *Journal of Clinical Investigation* 111, no. 2 (2003): 231–9. doi:10.1172/JCI16672. PMC 151872. PMID 12531879.

Dickerson, R. E., and I. Geiss. *Hemoglobin: Structure, Function, Evolution, and Pathology.* Menlo Park, CA: Benjamin Cummings, 1983.

Ferreira, A., I. Marguti, I. Bechmann, V. Jeney, Â. Chora, N. R. Palha, S. Rebelo, A. Henri, Y. Beuzard, and M. P. Soaresemail. "Sickle Hemoglobin Confers Tolerance to Plasmodium Infection." *Cell* 145, no. 3 (April 29, 2011): 398–409. doi: http://dx.doi.org/10.1016/j.cell.2011.03.049.

Perutz, M. *I Wish I'd Made You Angry Earlier.* Oxford, UK: Oxford University Press, 1998.

Taliaferro, W. H., and J. G. Huck. "Inheritance of Sickle Cell Anemia in Man." *Genetics* 8 no. 6 (1923): 594–8.

Vandegriff, K. D., L. Benazzi, M. Ripamonti, M. Perrella, Y. C. Le Tellier, A. Zegna, and R. M. Winslow. "Carbon Dioxide Binding to Human Hemoglobin Cross-Linked between the Alpha Chains." *Journal of Biological Chemistry* 266, no. 5 (February 15, 1991): 2697–700.

4.7 Receptor–Ligand Interactions

Christopoulos A. "Allosteric Binding Sites on Cell Surface Receptors: Novel Targets for Drug Discovery." *Nature Reviews Drug Discovery* 1 (2002): 198–210.

Conn, P. J., A. Christopoulos, and C. W. Lindsley. "Allosteric Modulators of GPCRs, a Novel Approach for the Treatment of CNS Disorders." *Nature Reviews Drug Discovery* 8, no. 1 (January 2009): 41–54.

Conn, P. J., and J. P. Pin. "Pharmacology and Functions of Metabotropic Glutamate Receptors." *Annual Review of Pharmacology and Toxicology* 37 (1997): 205–37. doi: 10.1146/annurev.pharmtox.37.1.205.

Epping-Jordan, M., E. Le Poul, and J.-P. Rocher. "Allosteric Modulation: A Novel Approach to Drug Discovery." *Innovations in Pharmaceutical Technology* 24, no. 24 (December 2007): 22–26.

Garcia, K. C. *Cell Surface Receptors* (vol. 68 in *Advances in Protein Chemistry and Structural Biology*). San Diego, CA: Academic Press, 2004.

Lauffenburger, D. A., and J. J. Linderman. *Receptors: Models for Binding, Trafficking, and Signaling.* Oxford, UK: Oxford University Press, 1993.

May, L. T., K. Leach, P. M. Sexton, and A. Christopoulos. "Allosteric Modulation of G Protein-Coupled Receptors." *Annual Review of Pharmacology and Toxicology* 47 (2007): 1–51.

Nickols, H. H., and P. J. Conn "Development of Allosteric Modulators of GPCRs for Treatment of CNS Disorders." *Neurobiology of Disease* 61 (January 2014): 55–71.

Schlessinger, J., and M. A. Lemmon, eds. *Signaling by Receptor Tyrosine Kinases* (Cold Spring Harbor Perspectives in Biology). Woodbury, NY: Cold Spring Harbor Laboratory Press, 2013.

Tsai, M. J., and B. W. O'Malley. "Molecular Mechanisms of Action of Steroid/Thyroid Receptor Superfamily Members." *Annual Review of Biochemistry* 63 (1994): 451–86. doi: 10.1146/annurev.biochem.63.1.451.

1833 ———— Payen discovers first enzyme diastase

1877 ———— Khune coins term "enzyme"

1902 ———— Brown describes reactions of invertase
1904 ———— Christian Bohr identifies the Bohr effect involving hemoglobin's ability to carry oxygen
1907 ———— Buchner receives Nobel Prize for cell free fermentation (zymase)
1909 ———— Laschtschenko describes lysozyme
1910 ———— James Herring describes sickle cell anemia
           Archibald Hill describes the effect of oxygen binding on hemoglobin
1913 ———— Michaelis and Menten propose equation describing the behavior of enzyme rates

1922 ———— Verne Mason originates the term sickle cell anemia
1923 ———— Huck publishes the term sickle cell anemia and notes its basis of inheritance
1925 ———— Brigs and Haldane describe steady state model of enzymatic catalysis
           Adair generates the first allosteric model

1946 ———— Sumner, Stanley, and Northrup share Nobel Prize for enzyme purification and crystallization

1949 ———— Linus Pauling identifies the molecular cause of sickle cell anemia
           Myoglobin structure solved
1958 ———— Jensen identifies first steroid receptor (estrogen)
1959 ———— First enzyme (a bacterial protease) added to laundry detergent

1963 ———— Polymeric structure of acetyl-CoA carboxylase discovered
1965 ———— First enzyme structure solved by Phillips using X-ray crystallography (lysozyme)
           Monod, Wyman, and Changeaux propose the concerted model of allosterism
1966 ———— Blake proposes first mechanism for lysozyme
           Koshland, Nemethy, and Filmer propose the sequential model of allosterism

1975 ———— New England Biolabs begins selling restriction enzymes (enzymes used to cleave specific DNA sequences)
           First studies of aspartate transcarbamoylase (ATCase) and PALA
1978 ———— Arber and Messleson receive Nobel Prize for their discovery and characterization of restriction enzymes
1981 ———— Captopril, the first angiotensinogen converting enzyme (ACE) inhibitor available to the public. These are
           popular diuretic drugs used to treat high blood pressure.
1987 ———— Reverse transcriptase inhibitor azidothymidine (AZT), the first class of antiviral drugs used to treat HIV
           available to the public

2001 ———— Withers proposes alternate mechanism for lysozyme

2009 ———— Blackburn, Greider and Szostak share Nobel Prize for discovery and characterization of telomerase,
           an enzyme involved in the regulation of chromosome telomere length and cell longevity
2010 ———— Morpheeins first mentioned in literature
2011 ———— Ferreria et al. determine cause of malaria resistance in sickle cell trait
2013 ———— Mechanism discovered for allosteric regulation of glucokinase

CHAPTER 5

Membranes and an Introduction to Signal Transduction

Biochemistry in Context

We live in an age of communication. We can imagine how difficult it would be if we could not communicate with each other. Communicating means that we need to cross some sort of barrier, be it distance, or time, or a physical separation between the two parties.

Cells, too, need to communicate with other cells and respond to their surroundings. All cells have a membrane that delineates and separates the inside from the outside. These membranes define the boundaries of the cell and its organelles, and are important in how these units function, but the cell also needs some way to receive messages across this barrier.

This chapter discusses two aspects of biochemistry that we will use to help understand the regulation of metabolism. It begins with a description of the structure of the plasma membrane and a short introduction of the components of that membrane, including the structures and functions of some associated proteins. The chapter concludes with an introduction to signal transduction, that is, the way that cells respond to chemical signals with changes in enzyme function, flux through a metabolic pathway, or gene expression.

Chapter Outline

5.1 Membrane Structure and Function

5.2 Signal Transduction

Common Themes

Evolution's outcomes are conserved.	• The structure of membranes and the molecules involved in their formation are conserved throughout evolution. As a result, many of the pathways involved in the biosynthesis of these molecules are also conserved.
	• The themes discussed in signal transduction are conserved in other pathways, cells, organs, tissues, and organisms.
	• Many of the signaling pathways in this chapter are conserved throughout evolution.
Structure determines function.	• The hydrophobic nature of the plasma membrane excludes small hydrophilic organic molecules and ions.
	• Molecules are transported across the plasma membrane by protein transporters, channels, or pumps.
	• The highly dynamic nature of the plasma membrane is important in its function.
	• Proteins associated with the plasma membrane are either integral (requiring detergent for dissociation) or peripheral (able to be dislodged without disrupting the lipid bilayer).
	• The hydrophobic effect and other weak forces explain the spontaneous assembly of the lipid bilayer and the association of proteins with the bilayer.
Biochemical information is transferred, exchanged, and stored.	• Signal transduction employs a series of signaling proteins and kinases to transmit information from a chemical signal outside the cell that causes an effect inside the cell.
	• Chemical signals can lead to alterations in metabolic pathways, enzyme activation, or gene regulation.
Biomolecules are altered through pathways involving transformations of energy and matter.	• As with all processes, the laws of thermodynamics apply to both membrane formation and the association of proteins with membranes.

5.1 Membrane Structure and Function

Before studying biochemistry, people often see membranes as something static, such as the skin of an onion or the thin layer of connective tissue on a piece of meat. Those are not membranes in the biochemical sense. Instead, membranes are the submicroscopic barriers that surround and define cells and many organelles. These structures are dynamic, fluid, and only a few molecules thick.

5.1.1 The chemical properties of the membrane components dictate the characteristics of the membrane

Membranes in the cell are comprised of two main components: lipids and proteins. Lipids are hydrophobic molecules. The lipids found in cell membranes are characterized as **polar lipids**, containing polar, water-soluble functional groups in addition to their hydrophobic component.

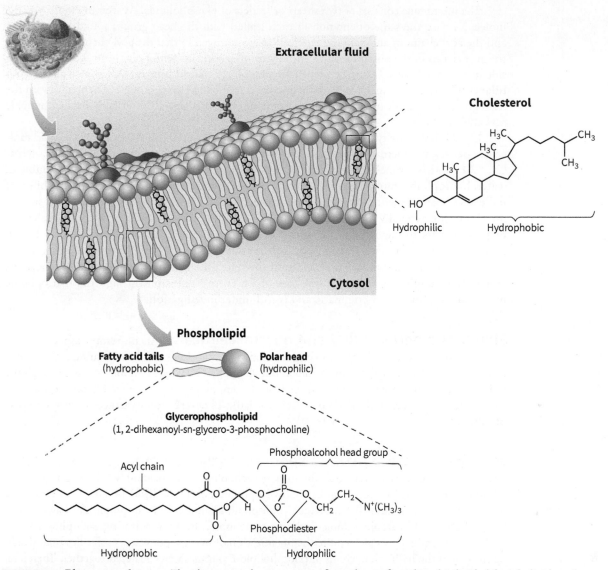

FIGURE 5.1 Plasma membrane. The plasma membrane consists of two sheets of amphipathic lipids (phospholipids and cholesterol) and numerous proteins. Both phospholipids and cholesterol have polar (hydrophilic) and nonpolar (hydrophobic) groups in the same molecule.

(Source: Ireland, *Visualizing Human Biology*, 4e, copyright 2012, John Wiley & Sons. This material is reproduced with permission of John Wiley & Sons, Inc.)

This combination makes these molecules **amphipathic**, having both water soluble and water insoluble properties, an important property in forming a membrane (**Figure 5.1**).

The predominant lipids in membranes are glycerophospholipids. These lipids have two acyl chains, long-chain fatty acids, esterified to a glycerol backbone. The third position on the glycerol backbone has a phosphate group. Finally, an alcohol is linked to the phosphate, forming a phosphodiester linkage. The phosphate group and alcohol are polar hydrophilic groups, whereas the acyl chains are hydrophobic, making the overall molecule amphipathic. Phospholipids vary both in terms of the length and degree of unsaturation of the acyl chains, and the type of alcohol found in the head group.

The membrane lipid **cholesterol** is the most common steroid found in the body. Cholesterol is largely hydrophobic, but it has a single hydroxyl group; within a membrane, this hydroxyl group is oriented away from the core of the membrane. Structurally, cholesterol is planar, a property that allows cholesterol to disrupt interactions of the acyl chains of phospholipids, resulting in a higher degree of membrane fluidity at lower temperatures. However, once the percentage of cholesterol in the membrane increases past a certain point, fluidity decreases. Therefore, the levels of cholesterol found in the membrane must be tightly regulated.

The membrane consists of two sheets or leaflets of phospholipids. As seen in Figure 5.1, the molecules have the same orientation in each leaflet, with the head groups found in one plane and the acyl chains in another. The two leaflets are arranged so that the acyl chains of one adjoin the acyl chains of the other, forming a sandwich of acyl chains with polar head groups on either side of the membrane. This structure is termed a **phospholipid bilayer**, also known as **the lipid bilayer**. Cholesterol is inserted into the bilayer so that its hydroxyl group is oriented toward the head groups of one side or another and the remainder of the molecule is buried in the hydrophobic acyl chains.

Lipid bilayers are typically 45 Å (4.5 nm) thick and contain on the order of 10^6 lipid molecules per square micron (10^{-6} m) of membrane. Lipid bilayers are permeable to gases such as O_2, CO_2, and N_2, which can diffuse through the membrane. Although not readily soluble in water, hydrophobic molecules can pass fairly easily through the lipid bilayer. The solubility of molecules decreases with increasing size such that larger molecules are less soluble than smaller ones. Ions are generally unable to cross the membrane without some type of protein channel or pore to pass through. Likewise, small polar organic molecules including most biological molecules require proteins for transport into or out of the cell. Water and urea can both diffuse through the membrane, into and out of the cell, although it is thought that this transport is mainly through aquaporins. In the case of water, this diffusion through the hydrophobic membrane is particularly enigmatic and is still under investigation.

Membrane permeability and pharmaceuticals

An important aspect of drug design is solubility. A challenge for drug developers is that drugs need to be sufficiently soluble to dissolve and be transported in the blood but also sufficiently hydrophobic to pass through the lipid bilayer. It has been estimated that 80% of compounds that are effective drug candidates (molecules that will elicit the desired effect in the lab) will never become marketable drugs because of solubility issues.

Membranes are dynamic, spontaneously assembling structures

A property that scientists often look for in nanotechnology is self-assembly (the ability to form without external assistance). Due to the hydrophobic effect, the **lipid bilayer** is one of these self-assembling structures. An individual phospholipid molecule is amphipathic, having both hydrophobic and hydrophilic moieties in the same molecule. Thus, solvating each phospholipid molecule individually would require the organization of many more water molecules than would be the case if the hydrophobic tails of phospholipid molecules were clustered together. Therefore, membrane assembly is driven not by the attractive forces between acyl chains, nor by the attractive forces between head groups (in fact, the charged groups often repel one another), but by the phospholipids being "pushed" together as water molecules hydrogen-bond with each other.

Diffusion is the movement of particles down a concentration gradient. An example is a solution of dye or food coloring dropped into a container of water, or a strong smelling molecule in air, but it also could be the lateral movement of proteins in a lipid bilayer. Over time, without any forces to prevent it, the molecule will become randomly distributed throughout the solvent.

The diffusion of any molecule depends on several factors, including the molecule's size and shape, solvent viscosity, and the temperature. These factors combine to give a diffusion constant, a value that varies for each type of molecule. Because the diffusion constant is partially dependent on viscosity, it also varies with regard to solvent. When a molecule such as a protein is constrained in the lipid bilayer, the solvent may be the lipid molecules themselves. Smaller molecules like lipids move more quickly than larger ones, such as proteins, and molecules can move more rapidly through a medium with lower viscosity than one with higher viscosity.

Molecules will diffuse through a medium until they are randomly distributed and entropy has been maximized. This diffusion of molecules down a concentration gradient is an illustration of the second law of thermodynamics—that is, the entropy of a system tends to increase. The reverse process, the segregation or concentration of molecules in one area or compartment, is a violation of the second law, unless energy (ΔH) is expended somewhere else to drive transport.

Lipid bilayers are not static structures; they are constantly in motion. The term "fluid mosaic" has been used to describe the plasma membrane. In this analogy, the lipid molecules in the membrane are analogous to balls floating on the surface of a swimming pool. They are free to rotate and to move laterally but are generally unable to rise out of the pool or sink to the bottom.

The same can be said for the phospholipids of the lipid bilayer. They have high rates of rotational and lateral diffusion (because only weak London-dispersion forces hold these molecules next to one another). A single phospholipid molecule can cross the length of the cell in less than a second. However, without enzymatic assistance, lipids cannot readily leave the lipid bilayer or flip from one side to the other.

Membranes in the cell also contain proteins

While half of the cell membrane's mass is comprised of lipid, the other half is protein. Membrane-associated proteins are diverse in structure and in function, and they give the membrane many of its properties. These proteins can be broadly categorized as **integral** or **peripheral membrane proteins** (**Figure 5.2**).

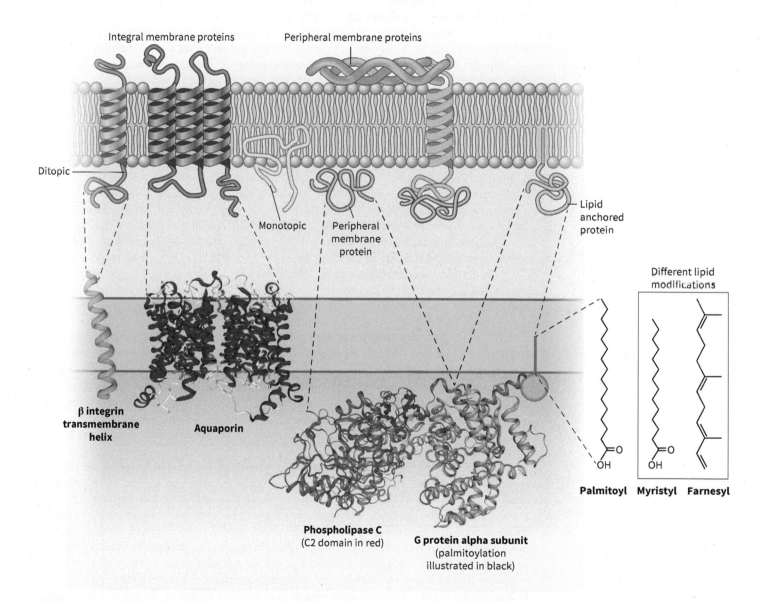

FIGURE 5.2 Membrane proteins and the plasma membrane.   In addition to the mechanisms listed for tethering to the cell membrane, proteins can be conjugated to hydrophobic groups (myristoyl, palmitoyl, or farnesyl, shown in enlarged view). Aquaporins and integrins are examples of integral membrane proteins. Phospholipase C and the G protein alpha subunit are examples of peripheral membrane proteins.

(Source: Karp, *Cell and Molecular Biology: Concepts and Experiments*, 7e, copyright 2013, John Wiley & Sons. This material is reproduced with permission of John Wiley & Sons, Inc. (Left) Data from PDB ID 2M3E Surya, W., Li, Y., Millet, O., Diercks, T., Torres, J. The Integrin Alpha L Transmembrane Domain in Bicelles: Structure and Interaction with Integrin Beta 2. (Middle) Data from PDB ID 3D9S Horsefield, R., Norden, K., Fellert, M., Backmark, A., Tornroth-Horsefield, S., Terwisscha van Scheltinga, A.C., Kvassman, J., Kjellbom, P., Johanson, U., Neutze, R. (2008) High-resolution x-ray structure of human aquaporin 5. *Proc. Natl. Acad. Sci. Usa* 105: 13327–13332. (Right) Data from PDB ID 3OHM Waldo, G.L., Ricks, T.K., Hicks, S.N., Cheever, M.L., Kawano, T., Tsuboi, K., Wang, X., Montell, C., Kozasa, T., Sondek, J., Harden, T.K. (2010) Kinetic Scaffolding Mediated by a Phospholipase C-(beta) and Gq Signaling Complex. *Science* 330: 974–980).

Integral membrane proteins are proteins that are somehow imbedded in the plasma membrane. These proteins can be further subclassified as **monotopic**, inserted into only one leaflet of the lipid bilayer, or **ditopic**, spanning both leaflets. Ditopic integral membrane proteins are also known as **transmembrane proteins**; they may have a single hydrophobic α helix spanning the membrane, or a series of helices or β sheets forming some type of anchor or pore. Other structures that tether monotopic integral membrane proteins to a membrane are hydrophobic "fingers," or loops of hydrophobic amino acids or hydrophobic domains that enter the membrane but do not pass completely through it. By definition, integral membrane proteins require detergent to remove them from the membrane. Detergents are amphipathic molecules with both polar and nonpolar ends to them that can disrupt and solubilize membrane proteins and lipids.

An example of a membrane-spanning protein is the seven-transmembrane (7TM) family of receptors, each of which has seven transmembrane α helices that span the membrane. Extracellular signals bind to the outside of the membrane and transduce signals to the interior of the cell through conformational changes to the receptor.

Integrins are structural proteins that anchor a cell in the extracellular matrix. They have both an α and β subunit, each of which has a single α helix passing through the plasma membrane of the cell.

Aquaporins are an example of pore-forming proteins. They have a series of α helices that span the membrane and twist about each other to form a selective pore through which particular molecules (such as water) can pass.

In contrast to integral membrane proteins, peripheral membrane proteins are associated with the surface of the membrane, either with the polar head groups of specific phospholipids or with other membrane proteins. Peripheral membrane proteins can also be associated with membranes through a hydrophobic anchor such as a myristoyl, palmitoyl, or farnesyl group. The G protein α subunit is associated with the plasma membrane through a hydrophobic anchor. Peripheral membrane proteins are not imbedded in the plasma membrane, and can be dissociated from the membrane by changes to pH, or by treatment with salt or concentrated solutions of molecules like urea or guanidinium hydrochloride that disrupt hydrogen bonding and the hydrophobic effect.

Multiple protein domains are involved in membrane interaction. For example, C1 and C2 domains are found in some kinases (enzymes that join a phosphate to a hydroxyl group) and lipases (enzymes that hydrolyze lipids). These domains recognize specific phospholipid head groups and are important in recognition and binding of the phospholipid bilayer, but not necessarily in binding to individual substrate molecules and catalyzing reactions.

Proteins regulate the passage of most molecules through the membrane
Because lipid bilayers are virtually impenetrable to ions and most biological molecules, some way is needed for these molecules to pass in and out of the cell or organelle. This passage is accomplished through several different types of proteins and mechanisms. If a single ion is pumped or diffuses in a single direction, the protein is termed a **uniporter**. If two or more ions are transported in the same direction across the membrane, the protein is termed a **symporter**. If two or more ions are transported in different directions across a membrane, the protein is termed an **antiporter** (Figure 5.3).

Passage through a membrane can be broadly classified as either active or passive. In **active transport**, energy must be expended to move a molecule from one side of the membrane to another. Usually this is done to transport a molecule up a concentration gradient. In primary active transport, the energy of ATP hydrolysis is used directly to drive ions across a membrane. An example is the gastric proton pump that acidifies the stomach. This protein (the hydrogen/potassium ATPase) was originally named an ATPase because it catalyzes the breakdown of ATP. In **secondary active transport**, a gradient of other molecules (often ions) is used to pump a second molecule across the membrane. An example is the sodium–chloride symporter in the kidney, which helps to reabsorb these ions from glomerular filtrate in the nephron.

In **passive transport**, **transporter proteins** (the glucose transporters or aquaporins), permit the passage of molecules down a concentration gradient or electrochemical gradient. The transporter proteins are largely α helical and have a central pore that selectively permits the passage of a single specific molecule or a closely related group of molecules. Passive transport can also occur via **channels**, which are similar to transporters but differ in that channels are gated and are regulated. As shown in Figure 5.3, channels open and shut either through binding of a ligand (ligand-gated ion channels) or a change in the voltage across a membrane (voltage-gated ion channels). Both transporters and channels are important in the nervous system and in neurotransmission of signals.

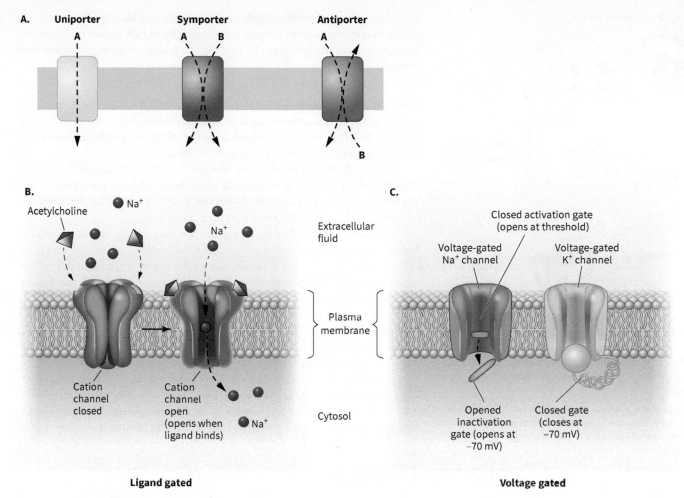

FIGURE 5.3 Transporters, pumps, and channels. **A.** Uniporters enable the transport of a single molecule or ion; symporters use one ion or molecule to drive the transport of a second in the same direction, while antiporters use a gradient of one molecule or ion to drive the transport in the opposite direction. There are two classes of ion channels: ligand gated or voltage gated. **B.** Ligand-gated ion channels open when a ligand binds. **C.** Voltage-gated ion channels open when there is enough of an electrochemical potential across the membrane.

(Source: (B, C) Ireland, *Visualizing Human Biology*, 4e, copyright 2012, John Wiley & Sons. This material is reproduced with permission of John Wiley & Sons, Inc.)

A lipid bilayer devoid of channels is impermeable to ions; therefore, membranes can have an uneven distribution of chemicals on either side. This is a potential energy difference that the cell employs to generate ATP in mitochondria, drive various transport processes, and generate action potentials in neurons. It has been discussed in detail later in Section 5.1.4.

Osmosis is the movement of solvent (water in biochemical systems) through a semipermeable barrier from a region of low solute concentration (the hypotonic side) to a region of high concentration (the hypertonic side). We can visualize osmosis as a system that is moving toward an equilibrium in which solvent and solute concentrations are the same on both sides, although a barrier prevents this from ever truly happening. Like diffusion, the movement of molecules in osmosis is random, does not require the input of energy, and tends toward maximal randomness, driven by entropy.

5.1.2 Other aspects of membrane structure

Membrane proteins also may be glycosylated, that is, modified by conjugation to carbohydrates (**Figure 5.4**). Inside eukaryotic cells, small carbohydrate modifications are important in trafficking of proteins to the correct location in the cell. Secreted proteins or proteins in the extracellular leaflet of the plasma membrane of the cell often have larger carbohydrate modifications that contribute to the function of the protein.

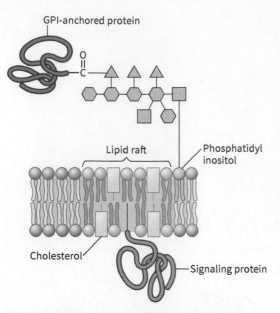

FIGURE 5.4 Other aspects of membrane biology.
Some membrane proteins are clustered in transient segments of membrane, termed lipid rafts. Rafts are enriched in specific phospholipids and cholesterol. Rafts are thought to play roles in cell signaling and membrane transport.

(Source: Karp, *Cell and Molecular Biology: Concepts and Experiments*, 7e, copyright 2013, John Wiley & Sons. This material is reproduced with permission of John Wiley & Sons, Inc.)

Cholesterol and sphingolipids can transiently and loosely aggregate in the plasma membrane to form a structure termed a **lipid raft**, shown in Figure 5.4. These structures are involved in cell signaling and tethering of membrane proteins.

Membranes vary throughout an organism or cell The composition of a membrane is not uniform throughout the cell, or even from one leaflet of the lipid bilayer to another. Rather, membranes are asymmetric and have an uneven distribution of lipids.

Figure 5.5 shows how the lipid composition of membranes differs among different cell types and organelles. This is consistent with the principles of both biochemistry and evolution. Just as specialized cells or organelles contain different proteins, they also have a different lipid composition.

The leaflets of membranes in the cell are also diverse and asymmetrical. The asymmetry helps to maintain the structural integrity of the cell, and it has a role in cell signaling. The cytosolic face of the plasma membrane is enriched in phospholipids (phosphatidylinositol, phosphatidate, and phosphatidylserine) that act in cell signaling pathways.

Asymmetry is generated and maintained in the membrane by the action of phospholipid translocases (also known as flippases). These enzymes alter the composition of lipid bilayers by catalyzing the movement of phospholipids from one leaflet to another.

The structure of the lipid bilayer is so common we may forget that there was a time before this structure was known. How the structure of the plasma membrane was elucidated is discussed in **Biochemistry: Numerous techniques converge to give a picture of the plasma membrane.**

5.1.3 Membrane fusion and membrane budding

Membranes can increase in size through the addition or breakdown of single molecules of phospholipid, but this is not the accepted means by which large aggregates of lipid molecules move between pools within the cell. Membranous structures can either fuse together when a hormone or neurotransmitter is secreted from the cell or they can bud or bleb off from an existing structure (**Figure 5.6**).

Membrane fusion begins with two membranes coming into close proximity; as they approach one another, water is excluded. Some destabilization of one or both membranes causes the formation of a stalklike structure (the fusion pore), which then grows and spreads as the two membranes fuse.

Studies of the plasma membrane and membrane fusion and countless other scientific investigations have been facilitated by the use of cultured cells. The development and use of these cells is discussed in **Societal and Ethical Biochemistry: Cultured cell lines.**

FIGURE 5.5 Composition of lipid bilayers. The lipid composition of a membrane varies from cell type to cell type and from organelle to organelle. Shown is the composition for three different membranes. The lipids shown are cholesterol, phosphatidylethanolamine (PE), phosphatidylcholine (PC), sphingomyelin (SM), phosphatidylserine (PS), cardiolipin (CL), and all other lipids.

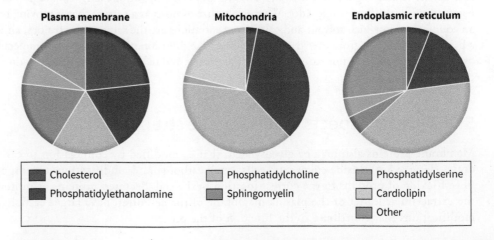

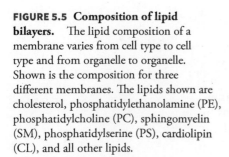

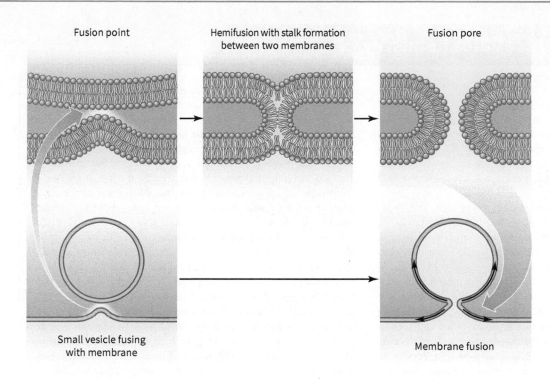

Fusion point

Hemifusion with stalk formation between two membranes

Fusion pore

Small vesicle fusing with membrane

Membrane fusion

FIGURE 5.6 Fusion and budding. When membranes fuse or bud, bends in the membrane facilitate the initial steps. As the structures approach each other, water is excluded between them. Membranes fuse at a single point forming an unstable stalklike structure (hemifusion) and then the two membranes zipper together or apart at a fusion pore. Alterations to the phospholipid composition facilitate these changes in geometry and lipid packing.

Biochemistry

Numerous techniques converge to give a picture of the plasma membrane

The fluid mosaic model of the cell membrane is familiar to those studying the biological sciences. Students are taught that the surface of the cell is a membrane barrier, the surface of which, viewed from above, looks like a mosaic (an art form comprising numerous small pieces of tile). They are also taught that, unlike a mosaic, the membrane is fluid, with its "tiles" (the phospholipids and proteins in the membrane) able to freely rotate and move about within the membrane. Models of the plasma membrane are sometimes used to illustrate this structure, for example, using small balls floating on the surface of water. The term "fluid mosaic" is a wonderful name, in that it is memorable and generates a picture in the mind, but how do we know that this model describes the plasma membrane?

The fluid mosaic model was proposed by Singer and Nicholson in 1972. It had the basic familiar structure of two sheets of phospholipids oriented to form a hydrophobic barrier, with their polar head groups facing away from the hydrophobic core. Singer and Nicholson also proposed that integral membrane proteins were amphipathic and had structures that helped them to anchor in the membrane. Key to this model was the dynamic interactions of the components rather than the components themselves.

Since the late 1800s it has been known that proteins and phospholipids are components of the cell. However, advances in technology, such as chemical analysis, electron microscopy, light microscopy, and ultracentrifugation, were needed before scientists could combine numerous results to produce the currently accepted model. The amphipathic structures of phospholipids—part of the foundation of the fluid mosaic model—were elucidated in the late

1950s and early 1960s. An attractive aspect of the Singer and Nicholson model is that it is consistent with the restrictions imposed by thermodynamics; in other words, it makes chemical sense. The arrangement and orientation of phospholipid molecules to form the membrane, and the way that proteins float in the membrane, are easily explained by chemistry and do not violate any of the basic rules of thermodynamics. Furthermore, knowing the basic structure of the lipid bilayer, it makes sense that the phospholipid and protein molecules in the membrane can rotate and move laterally (there are few weak forces preventing them from doing so) but cannot readily flip from one leaflet to the other without some sort of enzymatic assistance.

Using these basic paradigms, Singer and Nicholson built a case for their model using experiments ranging from the more biophysical (differential calorimetry of membrane lipids) to the more biological (microscopy illustrating the clustering of membrane proteins when tethered to a divalent antibody). Other evidence included studies involving fusion of human and mouse cells and the diffusion of membrane proteins across the resulting hybrid cell. Further, using both their data and the broader literature, Singer and Nicholson showed that these phenomena were not specific to single circumstances, but rather they were found in various species and membranes, with various proteins. They were careful not to be proscriptive, and they pointed out weaknesses or exceptions to their model (including myelin sheaths of neurons), but collectively their data made a compelling case.

Since the original proposal of the fluid mosaic, there have been other techniques that also validate the model, and it is difficult now to imagine a time when the nature of the plasma membrane was considered controversial.

Societal and Ethical Biochemistry

Cultured cell lines

By its very nature, biochemistry involves studying living systems or molecules derived from living systems. Although study of isolated molecules can provide useful and meaningful data, studying a molecule in situ in the cell, tissue, or organism can provide a much richer picture of what the molecule is doing and how it affects overall biology.

Historically, studies were conducted in model organisms: bacteria, yeast, plants, and animals, or in human volunteers. However, such models are not applicable in some situations. For example, imagine attempting to study a disease that affects avocado trees; these trees take years to mature, so they are difficult to study directly. Model organisms such as bacteria or mice might not give meaningful results in this case, but cultured cells could provide a useful alternative.

Cells have been cultured since antiquity in the processes of making fermented foods such as cheese, bread, wine, and beer. In the laboratory, the culture of eukaryotic cells dates back to the early 1900s. These initial experiments helped scientists to develop an understanding of the conditions cells require for growth but were limited in their scope. For example, most of the experiments used tissue obtained from animal sources, a type of culture now referred to as primary culture. Such cells have a preset number of divisions they can undergo before they become senescent and stop dividing, and they often die after several days or weeks in culture. In addition, antibiotics were generally not available at that time, and cultures frequently became contaminated with microbes.

From the late 1940s onward, advances in tissue culture helped to set the stage for the widespread use of cultured cells; these advances included antibiotics, the use of serum as a source of growth factors, and standardized culture conditions. The final advances that made cell culture possible were the cells themselves. Advances in virology and cancer biology led to the discovery that certain cells were transformed or immortalized (able to divide without limits), providing an almost endless source of cells that could be used to develop vaccines, express proteins, and study cellular processes.

The use of cultured cells might seem like a simple bypass around the potential ethical quagmire of using either animals or humans in biomedical research, but actually the use of such cells itself raises ethical concerns. At the time when some of the most popular cell lines were derived, there was little use of the current practice of informed consent, the concept that an experiment can only go ahead if the patients are informed of what is happening to them and give their permission. Patients had little idea that they were donating samples of tumors to be used in biomedical research, and at that time no one—patient, physician, or researcher—had any idea about the potential of these tools. There were no laws governing such decisions and the laws that have been written since are ambiguous at best or vary from jurisdiction to jurisdiction. In addition, science continues to advance and use these tools in research. At present, there are concerns that the genomes of these cell lines contain personal information about the family of the original donor, including relatives and descendants who are still living and may not want that information shared.

The history of the discovery of one cell line (HeLa cells) and the lasting impact this has had on the family of the donor (Henrietta Lacks) is described in *The Immortal Life of Henrietta Lacks* by Rebecca Skloot.

Cell culture has been an important part of biochemistry for many years, and it will continue to be so for the foreseeable future, but researchers should know and take into account the source of cells and the conditions under which they were generated. The extent to which descendants of donors should have a voice in the use of these cell lines is a question that remains to be answered.

Worked Problem 5.1   Interpreting a FRAP experiment

FRAP is a technique used to determine the mobility of a molecule in a membrane. This is accomplished by measuring the time it takes for a fluorescent molecule to diffuse back into a photobleached region of the membrane. Photobleaching is the fading or bleaching of a fluorescent or chromogenic molecule with light.

A membrane-bound receptor is labeled with the fluorescent protein tag mCherry, and functional studies indicate that the labeled receptor still operates normally. The diffusion of the receptor is measured using fluorescence recovery after photobleaching (FRAP), in the presence and absence of ligand, giving the result shown below.

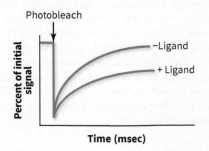

What can be said qualitatively about these results? What quantitative information can be obtained from this type of information?

Strategy   Compare the experimental condition (+ ligand) to the control (– ligand). What is the difference between the two curves? What is happening in the experiment?

Solution   A comparison of the two curves indicates that in the presence of ligand the fluorescent signal returns to the bleached region more slowly than in the unbound state. This correlates to a slower diffusional coefficient for the complex with the bound ligand. The results correlate percentage of original signal as a function of time. The half-life of recovery and the recovery rate can be modeled using several equations that are beyond the scope of this chapter, but they are not difficult to comprehend. Based on these data, a diffusional coefficient (D) can be determined. The diffusional coefficient can be used to calculate parameters such as the hydrodynamic radius or the diameter of the protein or complex if it were modeled as a sphere.

Follow-up question   Why do you think the diffusional constant of the receptor in the previous question might change in the presence of ligand?

5.1.4 Electrical properties of the membranes

There exists a potential difference across the plasma membrane of all cell types. It is known as the **membrane potential** of the cell and expressed in millivolts (mV). This voltage difference is due to the unequal distribution of small ions on the two sides of the membrane. When the cells are at rest, that is not stimulated, the inside of the cell has slightly more negative ions than the outside (**Figure 5.7**). The potential when the cell is at rest is known as the **resting membrane potential**. Different cell types may have values of resting membrane potential ranging from −40 to −90 mV.

If a membrane potential gets less negative than the resting potential, it is said to be **depolarized**. In contrast, a change towards more negative potential than the resting potential is said to be **hyperpolarization**. A return to the resting potential from the depolarized state is called **repolarization**. The underlying molecular basis of the changes in the membrane potential is the movement of ions across the membrane through specific ion channels down their concentration gradients. The ion channels are mostly gated, that is they open or close in response to stimuli. In the case of neurons, the opening and closing of voltage-gated Na^+ and K^+ channels found in the axonal membrane are responsible for the rapid changes in membrane potential during the generation and transmission of a nerve impulse or "action potential" along the axon. Any drug or chemical which can block the movement of ions is a potential neurotoxin. An example of one of the most potent neurotoxins is tetrodotoxin. This toxin found in the ovaries of certain types of pufferfish blocks the movement of Na^+ through the voltage-gated Na^+ channel located in the axonal membrane. A person consuming tetrodotoxin may die within few hours due to paralysis followed by respiratory failure.

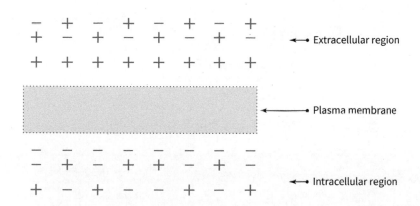

← Extracellular region

← Plasma membrane

← Intracellular region

FIGURE 5.7 Unequal distribution of ions across the membrane. The inside of the cell has more negative ions than the outside when the cell is at rest.

Summary

- Membranes are dynamic semipermeable barriers that compartmentalize the cell and separate it from the external environment.

- The membrane is a lipid bilayer, that is, two sheets of phospholipids arrayed with their hydrophobic acyl chains oriented to the middle of the membrane and their polar head groups oriented away from the membrane core.

- Membranes are about 50% protein by mass. These proteins provide many of the functions of the membrane, including permeability for ions and organic molecules, ligand binding, protein trafficking, endocytosis, and exocytosis.

- Proteins associated with the membrane can be classified as either peripheral or integral. Peripheral membrane proteins can be dissociated from the membrane with alterations to pH or salt concentration, whereas integral membrane proteins require detergent to remove them from the phospholipid bilayer.

- Membranes are dynamic structures and can fuse with one another or bud new membranous structures; this process is highly regulated and involves numerous trafficking proteins.

- A potential difference exists across the plasma membrane, called the membrane potential. It arises due to the unequal distribution of small ions on either side of the membrane.

Concept Check

1. Describe the general topology and behavior of a membrane.
2. Discuss the ways in which proteins bind to membranes and the forces, functional groups, and energies involved.
3. D-Glucose and D-Mannitol are similarly soluble, but D-Glucose is transported through the erythrocyte membrane four times as rapidly as D-Mannitol. Explain.
4. Describe in qualitative terms how membranes fuse or bud.

5.2 Signal Transduction

Imagine that you are planning a party and need to notify your friends. There are numerous ways you could try to contact them: calling, texting, using social media, and so on. You could also ask some friends to tell others. If planning a large event, you could make flyers and hang them around campus. All of these means of communication involve conveying some type of information from a sender to a receiver. This is analogous to signal transduction—the means by which cells and tissues in the body communicate with one another.

5.2.1 General principles underlie signal transduction

In a signal transduction pathway, also known as a cascade, a chemical signal (a hormone or neurotransmitter) binds to a protein receptor found in the plasma membrane of the cell (**Figure 5.8**).

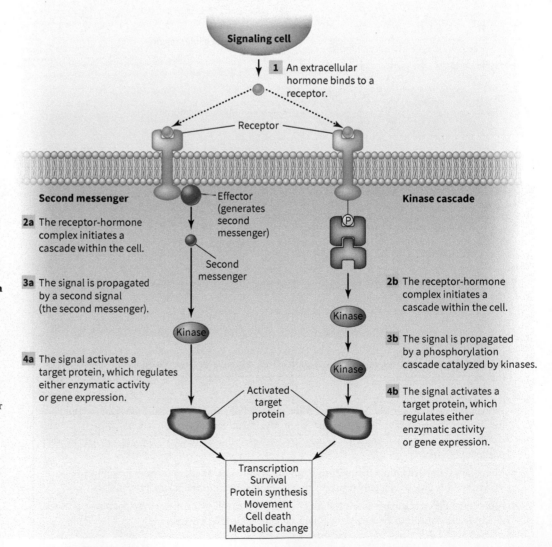

FIGURE 5.8 Signal transduction fundamentals. At its most basic, signal transduction occurs when a signal (a hormone, neurotransmitter, or extracellular stimulus) binds to a receptor. This causes changes inside the cell to produce other signaling molecules (second messengers) or a cascade of kinases. Ultimately, this results in a change to some protein, altering its function in the cell.

(Source: Karp, *Cell and Molecular Biology: Concepts and Experiments*, 7e, copyright 2013, John Wiley & Sons. This material is reproduced with permission of John Wiley & Sons, Inc.)

This causes a conformational change in the receptor that initiates a chain of events on the cytosolic face of the membrane. Often, a second chemical signal, or **second messenger**, is produced inside the cell. This is analogous to a flyer posted on campus being seen inside a dorm; although it was not the original signal, it still propagates the message. A signal transduction cascade activates or inhibits protein kinases, the enzymes that add phosphates to proteins. The addition of phosphate can result in conformational change, activating or inhibiting enzymes. Often, multiple rather than single kinases are acted upon, with each being activated or inhibited either by a specific signaling event or by a specific kinase.

There are many advantages to this system. For example, signal transduction can be at once both general and highly specific. One signal may have multiple different outcomes in different tissues, based on the types of receptor and downstream signaling pathways activated. Conversely, multiple chemical signals may bind to a single receptor, eliciting the same response. The multiple steps in these pathways provide multiple points at which the system can be controlled and fine-tuned, either by other proteins or through allosterism.

The use of a second messenger provides an additional opportunity to amplify the signal. To continue the analogy given above, if you were to text one friend about your party, she would receive your message; however, if you gave her a flyer and she posted it in a dorm, hundreds of people might see it, so the signal has been amplified. Because second messengers are typically small organic compounds that are enzymatically synthesized and degraded, there is an opportunity for significant amplification at this step. One molecule of signal binds to a single receptor, which activates a single enzyme, but this enzyme may rapidly make tens of thousands of copies of a second messenger.

Finally, signal transduction allows for a range of outputs over a range of time frames. The cascades can be as simple as the activation of an enzyme or as complex as the activation of a family of genes. They function in processes ranging from neurotransmission to development of an organism and can therefore operate in time frames ranging from milliseconds to hours or days.

Three common signaling pathways discussed in this section are chosen as representative pathways to illustrate the general principles of signal transduction and for their relevance to intermediary metabolism.

5.2.2 The protein kinase A (PKA) signaling pathway is activated by cyclic AMP

One of the most common signaling pathways in the cell is the **protein kinase A** (also known as **PKA**, cAMP dependent protein kinase, or A-kinase) signaling pathway (**Figure 5.9**). PKA transduces the signal received into actions in numerous cellular pathways, including lipolysis (the breakdown of stored fats), glycogen (stored carbohydrate) metabolism (decreasing glycogen synthesis and increasing glycogen breakdown), and some types of neurotransmission. In addition, PKA can alter gene expression and can target other kinases, activating or inhibiting them.

Multiple ligands and receptors signal via PKA. As an example, the catecholamine hormone epinephrine signals through a β-adrenergic receptor and binds to the receptor on the extracellular side of the plasma membrane. Like the interactions of an enzyme with its substrate, binding of hormone induces a conformational change in the receptor.

On the cytosolic face of the membrane, the receptor is complexed to a heterotrimeric G protein that has three different subunits: α, β, and γ. Both the α and γ subunits are tethered to the plasma membrane by a lipid modification, most typically a palmitoylation on the α subunit and a farnesylation on the γ subunit, although this can vary depending on the subclass of the subunit. These lipid modifications are shown in Figure 5.3. The α subunit is bound to a guanine nucleoside phosphate, either GDP or GTP. In the inactive state, the G protein is associated with GDP and bound to the receptor; however, when the receptor binds to a ligand, the conformational change to the receptor causes the α subunit of the G protein to exchange the GDP for GTP, and the G protein complex dissociates into an active G_α subunit and a dimer consisting of G_β and γ. The G_α subunit bound to GTP is now active, and it can associate with and activate the enzyme adenylate cyclase.

Adenylate cyclase catalyzes the formation of **cyclic AMP (cAMP)** from ATP. This is an illustration of how signals can be amplified. A single G protein activates a single molecule of adenylate

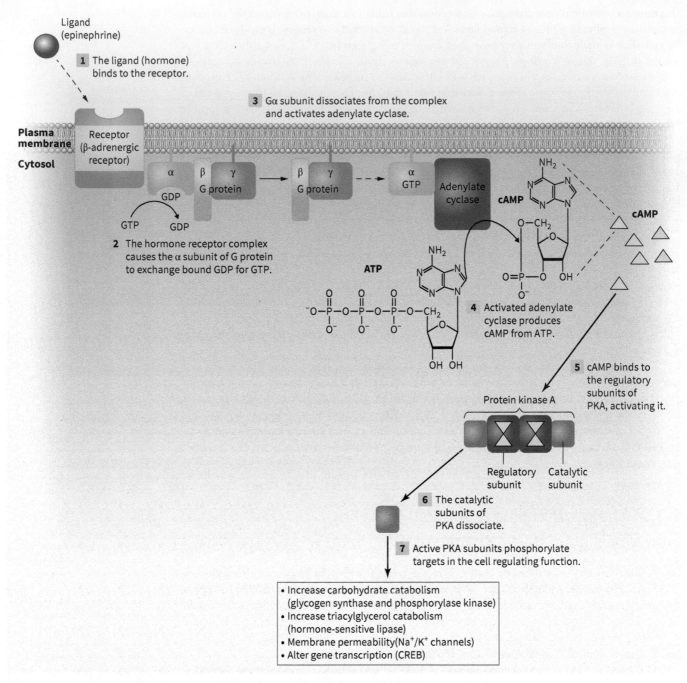

FIGURE 5.9 PKA signaling pathway. The PKA pathway was one of the first elucidated. In this example of the pathway, a hormone signals through a β-adrenergic receptor to stimulate a heterotrimeric G protein. This results in activation of adenylate cyclase and production of cAMP. The cAMP (a second messenger) binds to PKA, activating the enzyme.

cyclase, but this enzyme can rapidly produce tens of thousands of molecules of cAMP. In this instance, cAMP is serving as a second messenger that activates PKA.

PKA is a heterotetrameric enzyme complex comprised of a dimer of regulatory subunits found between two catalytic subunits. Each regulatory subunit has two cAMP binding sites (four sites in the holoenzyme complex), and the binding of cAMP is cooperative and allosteric. Binding of cAMP releases the catalytic subunits, which are then free to phosphorylate their cellular targets. These targets include enzymes which regulate carbohydrate and lipid metabolism, ion channels in the plasma membrane, and the transcription factor CREB (the cAMP response element binding protein).

There are numerous ways in which this pathway is regulated. First, when GTP is hydrolyzed, the G_α subunit needs to re-associate with the $G_{\beta\gamma}$ dimer and then to a ligand-bound receptor in order to reactivate. Therefore, hydrolysis of GTP is one of the fundamental ways this system is regulated. Second, the ligand can be removed from the receptor or degraded, or the entire ligand-receptor complex can be removed from the plasma membrane by receptor-mediated endocytosis. Removal of the signal or the signaling complex brings signaling to a halt. The final means by which signaling can be regulated is by modulation of the concentration of cAMP, which can be enzymatically degraded by phosphodiesterase. This enzyme cleaves cAMP to form AMP. Molecules that inhibit phosphodiesterase lead to elevated levels of cAMP and therefore to increased PKA signaling.

Once the role of G proteins and PKA in β-adrenergic signaling was established, this finding was used to elucidate several aspects of disease. One such example is cholera (**Figure 5.10**), a

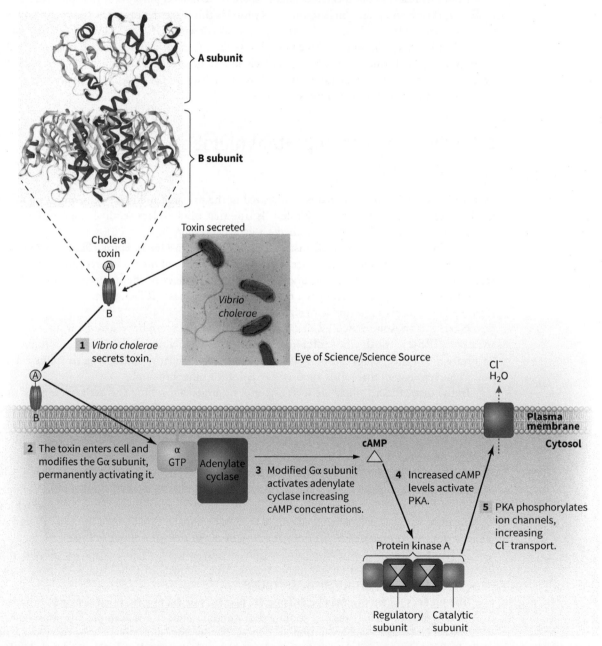

FIGURE 5.10 Cholera toxin. Cholera toxin consists of seven subunits, a two-peptide A subunit and a pentamer of B subunits. The B subunits bind to the surface of the enterocyte and aid in the entry of the toxin to the cell. The A subunit of the toxin is an ADP-ribosylation factor that activates the G_α subunit permanently, resulting in prolonged activation of adenylate cyclase and much higher than normal levels of cAMP. This results in activation of PKA and CFTR, the ion channel that pumps chloride out of the cell, causing loss of water and sodium.

(Source: Data from PDB ID 1LTT Sixma, T.K., Pronk, S.E., Kalk, K.H., van Zanten, B.A., Berghuis, A.M., Hol, W.G. (1992) Lactose binding to heat-labile enterotoxin revealed by X-ray crystallography *Nature* 355: 561–564)

bacterial disease spread by drinking contaminated water. Cholera kills by dehydrating its victims. The toxin produced by these bacteria cause people with cholera to produce as much as a liter per hour of watery diarrhea. The catalytic subunit of the toxin is an enzyme (an ADP-ribosylation factor) that catalyzes the transfer of an ADP-ribose group to the G_α subunit of the G protein. This modification to the G_α subunit locks it in the active conformation, making it chronically active and unable to hydrolyze GTP and switch off. This leads to elevated cAMP levels and overactive PKA, resulting in activation of a chloride channel in the cells lining the intestine. The opening of this channel causes the accumulation of chloride in the lumen of the intestine. As these ions leave the cell, they draw water and sodium with them. The cell is unable to reabsorb these molecules, so dehydration and loss of electrolytes follows.

G proteins are an interesting and unexpected example of enzymatic catalysis and of how evolution can result in unexpected changes. There is a common perception that processes evolve to become faster and more efficient, but in G proteins this is not the case. The G protein complex was first identified as a GTPase, an enzyme that breaks down GTP. The G_α subunit binds strongly to GTP (it has a low K_M value). Once bound to GTP, it is in the active state and does not readily hydrolyze the GTP (the enzyme has a relatively low value of k_{cat}). Therefore, the G_α subunit has evolved to act as a molecular timer. It can bind to substrate (GTP), but the hydrolysis of this substrate can take seconds to minutes to occur.

5.2.3 Insulin is an important metabolic regulator and growth factor

Insulin is a protein hormone that is synthesized in the pancreas and is best known for its role in glucose metabolism and diabetes. While it is true that insulin plays seminal roles in these processes, it is also a growth factor and can affect gene expression.

Insulin signaling begins with insulin binding to the insulin receptor (**Figure 5.11**). This binding causes the receptor to dimerize, binding another copy of the insulin receptor. The insulin receptor is itself a kinase that phosphorylates tyrosine residues in the other copy found in the homodimer. This type of receptor is a **receptor tyrosine kinase (RTK)**, a type of receptor architecture often found in growth factor signaling.

Next, the phosphorylated insulin receptor binds to and phosphorylates the insulin receptor substrate (IRS-1), which is a scaffolding protein. On a building, scaffolding acts a framework around which other things can be positioned and hung. Similarly, **scaffolding proteins** act as a molecular framework that forms an assembly point for other proteins.

In the case of IRS-1, there are several groups of proteins that bind and activate. One of these groups is the series that leads to activation of the Erk proteins of the MAP kinase cascade. This cascade ultimately affects gene expression, regulating cell growth and differentiation.

The other major pathway through which insulin signals is the phosphoinositide or PI cascade. Phosphatidylinositol is a phospholipid found on the cytosolic face of the plasma membrane. One of the proteins that phosphorylated IRS recruits is phosphatidyl inositol-4,5-bis phosphate 3 kinase (PI3K). Phosphatidylinositol is found in the lipid bilayer and has the polar alcohol inositol as its head group. Inositol has six different hydroxyl groups, each of which can be phosphorylated, and alterations to these are important in cellular signaling. The PI cascade is the series of enzymatic modifications that can occur to this phospholipid head group.

PI3K is a kinase that phosphorylates an already phosphorylated form of phosphatidylinositol (phosphatidylinositol-4,5-bisphosphate, also known as PIP2) on C-3 to produce phosphatidylinositol-3,4,5-trisphosphate (PIP3).

The prefixes *bis* and *tris* may be unfamiliar. Bis refers to a compound that has two of the same functional groups added, but not in conjunction with one another. For example, the phosphates in PIP2 are on C-4 and C-5 of the inositol ring. In ADP (adenosine diphosphate), the two phosphates are linked together from the single C-5 of ribose. Tris and tri are analogous. If a molecule has four phosphates on different positions, the prefix used is either "tetra-" or "kis-."

PIP3 activates phosphoinositide dependent kinase 1 (PDK1), an enzyme that has several roles in the cell, and it acts in part to regulate cross talk between several signaling pathways. In this instance, however, PDK1 phosphorylates and activates the kinase Akt.

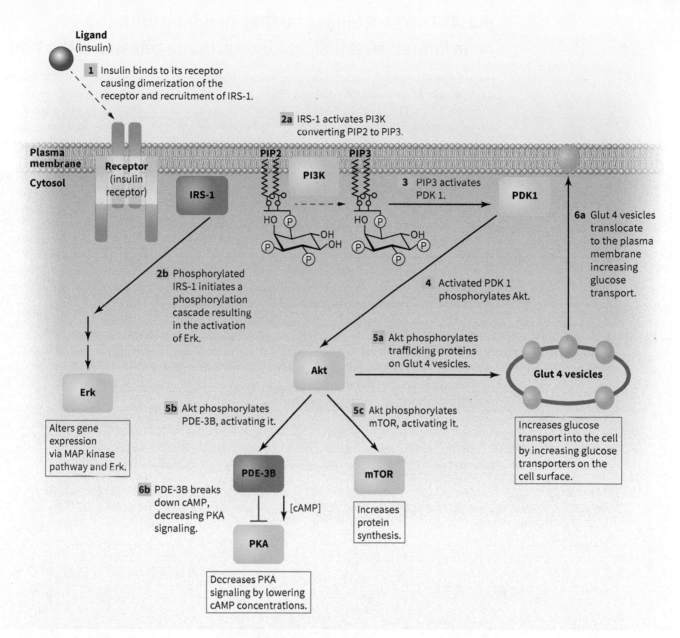

FIGURE 5.11 Insulin signaling pathway.   Insulin binds to the insulin receptor, which undergoes dimerization and autophosphorylation. The receptor, now an active kinase itself, phosphorylates the insulin receptor substrate (IRS). Phosphorylated IRS interacts with the PI3 kinase complex which phosphorylates a specific phospholipid (phosphatidylinositol) into more highly phosphorylated forms (PIP2 and PIP3). The different phosphoinositides produced activate PDK1 (phosphoinositide dependent kinase) which in turn phosphorylates and activates Akt. Akt can affect numerous targets in the cell leading to changes in gene expression, cell growth, and differentiation, and changes to glucose metabolism.

Akt phosphorylates several proteins in a complex that initiates the translocation and fusion of Glut4-coated vesicles to the plasma membrane. Glut4 is a glucose transporter. Therefore, stimulation of Akt via the insulin signaling cascade results in an increased number of glucose transporters being deposited in the plasma membrane, leading to increased glucose transport. Akt can also increase protein expression and production via a signaling cascade involving the protein complex mTOR.

Insulin is the hormone that signals the fed state. As such, it promotes the storage of the carbohydrate glycogen and of fats. Insulin signaling also plays a role in blocking fat breakdown (lipolysis) by stimulating phosphodiesterase 3B (PDE3B). This lowers cAMP concentrations, which in turn lowers PKA activity, blocking breakdown of glycogen and increasing biosynthesis of glycogen and fatty acids. All of these processes are coordinated and regulated through Akt.

5.2.4 The AMP kinase (AMPK) signaling pathway coordinates metabolic pathways in the cell and in the body

A third signaling pathway that plays critical roles in the regulation of metabolism is the **AMP dependent protein kinase (AMPK)** pathway (**Figure 5.12**). AMPK is *not* the PKA pathway, and it is *not* stimulated by cAMP. Rather, AMPK is a cytosolic kinase, which is activated not by a receptor or second messenger but by binding of AMP or phosphorylation by other kinases. In the cell, levels of total adenosine nucleoside phosphates are relatively constant. When the cell has abundant energy, the levels of AMP and ADP will be low, and levels of ATP will be high. Conversely, when the cell has a low energetic state, there is relatively less ATP and more ADP and AMP. AMPK acts as an energy sensor. If the insulin signaling pathway signifies the fed state and promotes growth and energy storage, signaling through the AMPK pathway indicates a low energy level and stimulates the oxidation of molecules for ATP production. Elevated levels of AMP will activate AMPK, activating pathways that increase fatty acid oxidation, and glucose uptake, glycolysis (the breakdown of glucose for energy), and block glucose storage (glycogen synthesis). AMPK also blocks gluconeogenesis (the synthesis of new glucose), steroid biosynthesis, and fatty acid biosynthesis. Overall, AMPK promotes oxidation of fatty acids and carbohydrates for ATP production. The AMPK pathway is one of the principal regulators of homeostasis in the organism.

AMPK is also regulated by several other kinases and signaling pathways including PKA and insulin signaling. PKA phosphorylates a second kinase called liver kinase B-1 (LKB1), which in turn can phosphorylate and activate AMPK. Insulin can act through the insulin receptor to activate the PI cascade, ultimately resulting in the activation of the kinase Akt. Akt can also phosphorylate and activate AMPK. Again, this is an example of how cellular communication pathways intersect and influence one another.

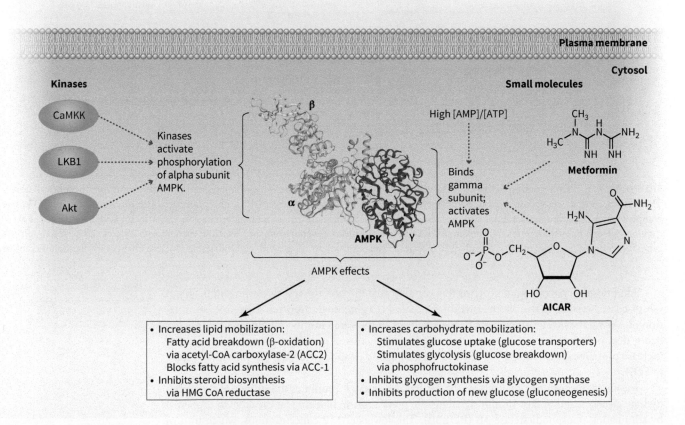

FIGURE 5.12 AMP kinase signaling. AMP kinase is a heterotrimeric protein that is implicated in several metabolic pathways. It is directly stimulated by a high AMP:ATP ratio and the kinases CaMKK, Akt, and LKB1.

(Source: Data from PDB ID 4CFF Xiao, B., Sanders, M.J., Carmena, D., Bright, N.J., Haire, L.F., Underwood, E., Patel, B.R., Heath, R.B., Walker, S., Giordanetto, F., Martin, S.R., Carling, D., Gamblin, S.J. (2013) Structural Basis of Ampk Regulation by Small Molecule Activators. *Nat. Commun.* 4: 3017)

A third means of regulating AMPK is through the calcium calmodulin kinase kinase (CaMKK). This kinase normally regulates the calcium calmodulin kinase (hence the name "kinase kinase"). In the presence of elevated levels of cytosolic calcium ions, CaMKK can phosphorylate AMPK, activating it. Several small molecules are known to interact with AMPK, including the drugs metformin and 5-aminoimidazole-4-carboxamide ribonucleotide (AICAR, pronounced "a-car").

Metformin is a biguanide, an oral antidiabetic agent that acts as an insulin sensitizer, making cells respond to lower levels of insulin. This drug also decreases the hyperglycemia (high blood sugar) in people with diabetes by decreasing hepatic gluconeogenesis. Metformin acts by stimulating AMPK, which in turn increases the response of the cell to insulin and blocks gluconeogenesis. These effects occur in addition to the metabolic benefits of activating AMPK, such as increased fat burning and decreased cholesterol biosynthesis.

ACIAR is a small molecule that acts as an AMPK agonist, and it was developed as a drug to prevent cardiac damage during heart surgery and ischemia (low blood flow). Like metformin, ACIAR acts to stimulate AMPK, but it has found an additional use. In 2011, the World Anti-Doping Agency listed AICAR as a performance-enhancing drug of abuse and a banned substance in international competitions. People who take AICAR as a performance enhancer hope to gain the benefits of exercise in pill form.

Perhaps the most intriguing way of activating AMPK is through exercise itself. It is not clear whether this activation is due to a combination of elevated Ca^{2+} concentrations, increased AMP levels, or activation of kinases, nor is it clear that AMPK is the key regulatory enzyme in this process.

As the examples of metformin and AICAR illustrate, numerous small molecules affect signal transduction and therefore how the cell and organism function. This is discussed in **Medical Biochemistry: Signal transduction and pharmaceuticals**.

Medical Biochemistry

Signal transduction and pharmaceuticals

Signal transduction pathways are not an esoteric aspect of biochemistry. Understanding signal transduction is important and has many practical uses. In any research question involving a complex system such as a cell or organism, a researcher has a duty to understand how signaling pathways work, because simple changes to a system may elicit a signaling response, complicating results. It is often easiest to biochemically shut down some pathways with specific inhibitors to simplify results.

Beyond biomedical research, molecules that affect signal transduction are some of the world's most popular and widely used pharmaceuticals, and at one time it was said that approximately 80% of all drugs affected cell signaling in some way. Although recent estimates suggest that number is closer to 30%, a quick analysis of the most widely used drugs shows just how important these pathways can be.

Of the over-the-counter medications sold in 2016, six of the ten highest selling medications (cough and cold remedies, analgesics, heartburn medications, sleep aids, anti-itch medications, and smoking cessation aids) all affect signaling in some way, as do five of the ten most widely used prescription drugs. These drugs can be as common as aspirin or nicotine, or as cutting edge as recombinant protein hormones.

This section introduces signal transduction and starts to lay the groundwork for how signaling affects metabolism. Each of these pathways plays a significant role in biochemistry and metabolism. The interactions and phosphorylation events presented here are meant to form a foundation from which to build an understanding of the basics of signal transduction and provide several examples relevant to metabolism. The interactions of these proteins are far more complex than presented here, and new interactions and cross talk between existing pathways are being discovered daily.

Worked Problem 5.2   The effect of caffeine on signal transduction

Caffeine (3,5,7-trimethyl xanthine) is the world's most commonly used stimulant; it acts as a phosphodiesterase inhibitor. How would caffeine affect the three signaling pathways discussed in this section?

Strategy   If a compound inhibits phosphodiesterases, how would that affect the products of the phosphodiesterase reaction? Examine the pathways for insulin, PKA, and AMPK signaling. In which of these pathways does phosphodiesterase play a role, and what is that role?

Solution   By inhibiting phosphodiesterase (the enzyme that breaks down cAMP), caffeine allows cAMP levels to increase. This means

that PKA is active with a lower level of stimulation, or that the background activity of PKA is elevated, and PKA signaling is increased.

Phosphodiesterases are found in multiple signaling pathways involving cAMP, and other cyclic nucleotides; however, they are not known to play a central role in either insulin signaling or AMPK signaling (despite structural similarities, AMP and cAMP are involved in very different pathways).

Follow-up question   What are the physiological effects of caffeine on the body? What does this say about the signal process involved in the regulation of those processes?

Summary

- Signal transduction is the study of how cells communicate with each other.
- Most signal transduction pathways are complex, involving chemical signals (hormones or neurotransmitters), receptors, enzymes, kinases, and regulatory proteins.
- The complex nature of signal transduction pathways provides considerable variability, both in terms of how a signal arrives at the cell and the cellular response to that signal. There are also multiple levels of control, multiple places for pathways to interact, and multiple opportunities for amplification of signals.
- PKA is a kinase that is activated by the second messenger cAMP. It figures prominently in several cellular processes including some aspects of neurotransmission, several metabolic pathways, and gene transcription.
- The insulin signaling pathway responds to the protein hormone insulin. This pathway employs a receptor tyrosine kinase (RTK) to transduce signals into the cell. Although insulin signaling is best known for increasing the transport of glucose into the cell, it also regulates several important metabolic pathways and alters gene expression.
- AMPK is another important metabolic regulator. Unlike the first two pathways, no extracellular ligand leads to the activation of AMPK. Rather, the activation is induced by an elevated AMP:ATP ratio, exercise, and the activation of other kinases (calcium-calmodulin kinase kinase and LKB1).

Concept Check

1. Describe the general principles involved in signal transduction.
2. Draw the steps of the three signal transduction pathways discussed in this section.
3. Compare these different pathways in terms of how they regulate metabolism.
4. Predict what would happen if these pathways were altered by disease, such as cholera, or pharmacological intervention, such as caffeine.

Bioinformatics Exercises

Exercise 1 Properties of Membrane Proteins

Exercise 2 Porins: Maltoporin and OmpF

Problems

5.1 Membrane Structure and Function

1. If the ratio of protein mass in the membrane to lipid mass is 1:1 (50/50), what is the approximate molar ratio of the two to each other?

2. Describe the mechanism by which peripheral proteins dissociate from the membrane by changing the pH, increasing the ionic concentration and by adding the chaotrophic agents?

3. Examine the structure of the head group of a phospholipid. What sorts of weak forces could form between those phospholipid head groups and a peripheral membrane protein? What amino acids would need to be involved in those noncovalent bonds?

4. Examine the thermodynamic parameters (ΔG, ΔH, ΔS) behind the association of an integral membrane protein with the plasma membrane. How must these parameters change for a protein to be removed from the plasma membrane? Draw two illustrations, one with a protein in the plasma membrane and one with a protein separate from the plasma membrane. How does your thermodynamic model explain what is going on in the illustration?

5. Analyze the generic structure of a phospholipid, and comment on why it aligns in the plasma membrane as it does.

6. The perilipins (a family of lipid droplet proteins) were originally identified as proteins that stayed associated with the lipid droplet when washed with sodium carbonate (100 mM, pH 9.0). Based on this observation, describe a model for the association of the perilipins with the lipid storage droplet.

7. In a helical protein that spans a cell membrane, what amino acids are likely to be in the center of the membrane and associated with the polar head groups and aqueous environment?

8. If cells are disrupted by physical or chemical means, then sealed spherical compartments surrounded by a membrane forms instead of sheets of membrane because the exposed edges of a membrane are energetically unfavorable. What does "energetically unfavorable" mean here? Use thermodynamic arguments to describe why vesicles will form instead of sheets.

9. Proteins associated with the membrane can be classified as either peripheral or integral. How do these differ in their mode of association and dissociation from the membrane?

10. Generally speaking, how does tethering molecules together in the plasma membrane affect the speed of their diffusion in the membrane? How can this alter their biochemical properties with regard to function, such as binding to a ligand?

11. The presence of two membranes in some organelles has been interpreted as meaning that eukaryotic cells arose through a symbiotic relationship between two prokaryotic cells. Describe in basic terms how this could happen.

12. Describe what happens to membranes when two membranous compartments fuse together or when one membranous compartment pinches off to form a new compartment.

5.2 Signal Transduction

13. What are the steps that result from insulin binding to its receptor?

14. Metformin is an antidiabetic agent but it does not work through effects on insulin signaling. Explain this finding.

15. Why is AICAR a banned substance for athletic competitions but caffeine is not?

16. What are the advantages of protein phosphorylation related to biochemical signaling?

17. This chapter discusses caffeine as a phosphodiesterase inhibitor, but caffeine is also an adenosine receptor antagonist. What does the term "antagonist" mean in this instance? Look up the structure of caffeine. How does the structure of caffeine help to explain these two functions in the body?

18. Give examples of proteins that are targets for protein kinase A (PKA) and phosphorylated by this kinase. List the steps associated with cAMP binding to cAMP-dependent PKA.

19. Many drug companies are interested in formulating compounds that act on the kinase Akt, but not many are interested in compounds that act on the insulin receptor. What are the reasons for different levels of interest in Akt and insulin?

20. If a drug increased the activity of Akt, what would be the outcome in the cell with regard to glucose metabolism?

21. Phosphodiesterase inhibitors increase signaling through PKA. How does this work? Why does the presence of an inhibitor increase signaling, and what does this say about background levels of signaling in this pathway?

Data Interpretation

22. You have identified a rare condition in which, under certain unknown circumstances, a protein (termed "imstudyin") accumulates, leading to increased cognitive development. It is not clear whether this is an inherited condition, although environmental factors (exposure to the library and deprivation from videogames) seem to increase the risk of this condition.

Parts a–e pertain to this scenario.

a. To begin your investigations on imstudyin accumulation, you need a model system. Three have been proposed: a mouse model in which mice have repeatedly solved mazes, leading to increased accumulation of the protein; a sample of cells from an undergraduate's liver (obtained with written permission and informed consent) that has been transformed and is being cultured in the lab; and brain samples from a graduate student cadaver (which has been used in previous studies). Cite the strengths and weaknesses of each of these models for a biochemical study on imstudyin.

b. You would like to know where in the cell imstudyin accumulates. A sample of cells is disrupted by homogenization, and the resulting cellular lysate is subjected to density gradient ultracentrifugation. The results are shown below.

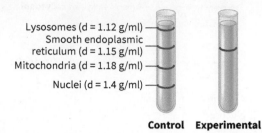

Control   Experimental

Based on the results of this experiment, where would you hypothesize that imstudyin is accumulating? How else could you test this?

c. The fraction containing imstudyin from the previous experiment was treated with either 250 mM NaCl, 50 mM glycine (pH 3.0), or 0.1% Tween 20 (a detergent). The sample density was adjusted to 1.05 g/mL. The samples were centrifuged again and the results are shown below.

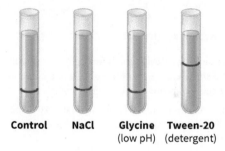

Control   NaCl   Glycine   Tween-20
(low pH)   (detergent)

Is imstudyin found in the lipid bilayer of this organelle? How can you tell? Use the data to support your hypothesis.

d. Samples were prepared as they were for part b. Following isolation of the region corresponding to imstudyin, some of the sample was treated with a protease. Samples were then subjected to immunoblotting (western blotting) and probed with an antibody to imstudyin. The results are shown below.

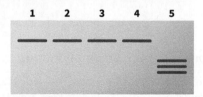

Lane 1, control, no protease
Lane 2, control, with protease
Lane 3, NaCl, with protease
Lane 4, glycine, with protease
Lane 5, Tween-20, with protease

Describe these results as you would if you were writing a scientific manuscript. What conclusions can you draw from the results of this question? How do these data align with your observations from questions b and c?

e. Other studies have noted that treatment with caffeine and glucose helps to increase the effects of imstudyin in some students. Cells were treated with caffeine, glucose, insulin, or some combination of these agents, and the amount of imstudyin found in the organellar fraction was quantitated using immunoblotting. Results are shown below.

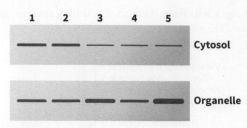

Lane 1, control, no treatment

Lane 2, glucose

Lane 3, caffeine

Lane 4, insulin

Lane 5, insulin, glucose, and caffeine

Top panel-cytosol

Bottom panel-organelle

How is each compound affecting the distribution of the protein? What can this data tell you? Is this protein phosphorylated by either pathway? In other words, what does the data actually show you, and where are we extrapolating or drawing conclusions based on the outcomes shown?

Experimental Design

23. You are studying a novel protein thought to be involved in carbohydrate metabolism. Describe two experiments you could use to determine where in the cell this protein can be found.

24. Membranes are asymmetric. Design an experiment to test this hypothesis. What techniques would you employ, and what would you use as controls?

25. Several factors contribute to the fluidity of a membrane, one of which is the degree of saturation of the lipids comprising the membrane. Membranes made with unsaturated lipids generally have a lower melting point and are more fluid. Design an experiment to test this hypothesis.

26. FRAP is a technique used to determine the mobility of a molecule in a membrane. Design a FRAP experiment to test the mobility of an insulin receptor. Would you expect the receptor to be more mobile before or after insulin binding?

27. The insulin receptor is an example of an RTK and a representative member of the growth factor receptor family of receptors. Design an experiment to test the hypothesis that other growth factors will bind to the insulin receptor. Design a second experiment to determine whether EGF will activate the insulin receptor.

Ethics and Social Responsibility

28. Tens of thousands of people are on waiting lists for donated organs. Are you a registered organ donor? Explain why you are or are not a registered donor.

29. The drug caffeine is discussed in this chapter. Caffeine is generally not considered harmful, and its sale is not restricted in any way to minors, whereas other drugs, such as ethanol and nicotine, are restricted. Briefly research this topic. Why are some substances restricted and others not?

30. Many people believe that cultured cell models can take the place of animals in biomedical research. Based on your background knowledge and what you have learned in this chapter, is that a true statement? Do animals still have a place in biomedical research?

31. Stem cells, especially embryonic stem cells (those derived from embryonic tissue) have been a topic of ethical debate since their discovery. It is now known that many tissues in the body harbor stem cells, and that other cells can be reprogrammed into stem cells. Many people still have a visceral reaction to the term "stem cell," without considering the source of those cells. Is their response valid? Is the use of stem cells (or all cultured cells for that matter) forever tainted by knowledge originally obtained in ways we may currently deem unethical?

Suggested Readings

5.1 Membrane Structure and Function

Ehnholm, C., ed. *Cellular Lipid Metabolism.* New York, NY: Springer, 2009.

Gurr, I. M. *Lipid Biochemistry: An Introduction.* 3rd ed. New York, NY: Springer, 1980.

The AOS lipid library retrieved from http://lipidlibrary.aocs.org/index.html

Vance, D. E., and J. E. Vance. *Biochemistry of Lipids, Lipoproteins, and Membranes.* 5th ed. New York, NY: Elsevier, 2008.

5.2 Signal Transduction

Dennis, E. A., S. G. Rhee, M. M. Billah, and Y. A. Hannun. "Role of Phospholipase in Generating Lipid Second Messengers in Signal Transduction." *FASEB Journal* 5, no. 7 (1991): 2068–77.

Gomperts, B. D., I. M. Kramer, and P. E. R. Tatham. *Signal Transduction.* San Diego, CA: Academic Press, 2003.

Milligan, G., ed. *Signal Transduction: A Practical Approach.* Oxford, NY: Oxford University Press, 1999.

Schönbrunner, N., J. Cooper, and G. Krauss. Translated by G. Weinheim. *Biochemistry of Signal Transduction and Regulation.* New York, NY: Wiley-VCH, 2000.

Sitaramayya, A. ed. *Introduction to Cellular Signal Transduction.* Boston, MA: Birkhauser, 1999.

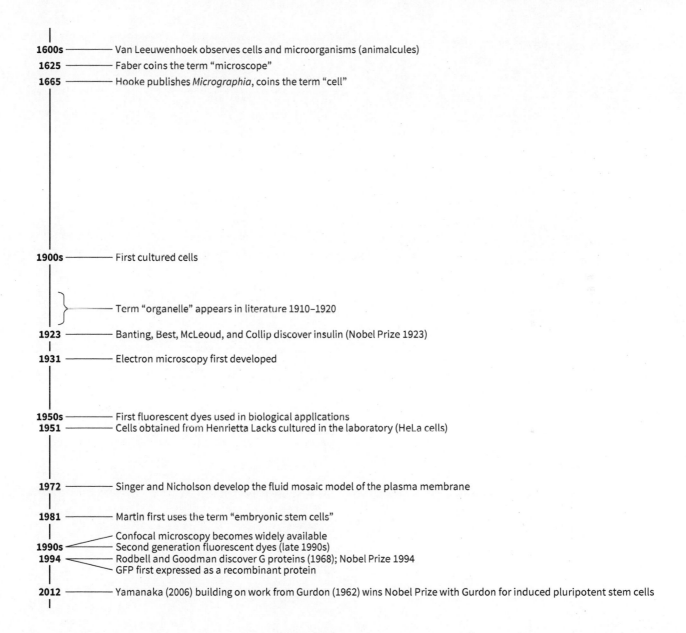

1600s ——————— Van Leeuwenhoek observes cells and microorganisms (animalcules)

1625 ——————— Faber coins the term "microscope"

1665 ——————— Hooke publishes *Micrographia*, coins the term "cell"

1900s ——————— First cultured cells

——————— Term "organelle" appears in literature 1910–1920

1923 ——————— Banting, Best, McLeoud, and Collip discover insulin (Nobel Prize 1923)

1931 ——————— Electron microscopy first developed

1950s ——————— First fluorescent dyes used in biological applications

1951 ——————— Cells obtained from Henrietta Lacks cultured in the laboratory (HeLa cells)

1972 ——————— Singer and Nicholson develop the fluid mosaic model of the plasma membrane

1981 ——————— Martin first uses the term "embryonic stem cells"

——————— Confocal microscopy becomes widely available

1990s ——————— Second generation fluorescent dyes (late 1990s)

1994 ——————— Rodbell and Goodman discover G proteins (1968); Nobel Prize 1994

——————— GFP first expressed as a recombinant protein

2012 ——————— Yamanaka (2006) building on work from Gurdon (1962) wins Nobel Prize with Gurdon for induced pluripotent stem cells

CHAPTER 6

Carbohydrates I

Mono- and Disaccharides, Glycolysis, Gluconeogenesis, and the Fates of Pyruvate

Carbohydrates in Context

Many people without biochemistry backgrounds can name foods rich in carbohydrates. Sugar, rice, bread, and pasta are good examples of carbohydrates, but many other molecules are also carbohydrates. Carbohydrates play structural roles, such as cellulose in cotton and wood, or chitin in the shells of crabs and insects. Carbohydrates make up part of the sugar phosphate backbone of 2-deoxyribose in DNA and ribose in RNA. Polymers of carbohydrate are found in the synovial fluid of joints, in mucus, and in some secreted extracellular matrix molecules. Even the ABO blood group antigen system is caused by carbohydrate modifications to proteins on the surface of cells.

This chapter introduces the study of metabolism—the reactions through which organisms transform one molecule into another and provide the energy and intermediates for other reactions. While many questions in metabolism were solved decades ago, there are current questions, such as the effect of high fructose corn syrup in the diet, that remain unanswered.

This chapter discusses the properties and functions of simple carbohydrates and how they can be joined to one another to form more complex molecules. Also discussed is how glucose can be catabolized into pyruvate when needed for energy through the process known as glycolysis, and how new glucose is synthesized in animals in times of need. The chapter concludes with a discussion of the fates of pyruvate—the three-carbon α-keto acid produced at the end of glycolysis.

Chapter Outline

Common Themes

Evolution's outcomes are conserved.	• Many of the pathways we will see in this chapter are conserved throughout evolution.
	• In contrast to genes discussed in other chapters, such as the genes involved in eicosanoids and steroid metabolism, the genes coding for enzymes involved in glycolysis are found throughout biology and are therefore more ancient in origin.
	• Because glycolysis is so basic to the existence of an organism, mutations that disrupt this pathway are often lethal or put organisms at such an evolutionary disadvantage that they generally fail to pass on these genes.
Structure determines function.	• Monosaccharides are a family of molecules that are largely related through stereochemistry.
	• Monosaccharides are frequently depicted in Fischer projections to highlight their stereochemical differences. However, in solution, these molecules adopt a cyclic hemiacetal or hemiketal structure; this structure is best seen in a Haworth projection.
	• In the broadest sense, polysaccharides can be functionally characterized as being energy stores or as providing structure to the organism.
	• Energy-storing polysaccharides can either be linear like amylose or highly branched like amylopectin or glycogen. The branched forms enable the mobilization of monosaccharides at a higher rate, because release can happen simultaneously from several sites.
	• Structural polysaccharides are generally linear molecules, with multiple places where the strands can interact with one another through weak forces (typically hydrogen bonding, dipole–dipole, and electrostatic interactions).
	• The structure of many enzymes is conserved throughout biology, and many of the same chemical reactions are employed in the breakdown and biosynthesis of carbohydrates in divergent species.
Biochemical information is transferred, exchanged, and stored.	• Carbohydrate metabolism is regulated by several allosteric regulators and kinases that inhibit or activate key enzymes in the glycolytic pathway.
Biomolecules are altered through pathways involving transformations of energy and matter.	• In this as in other chapters, we will see that energetically unfavorable reactions become favorable when coupled to reactions with very negative $\Delta G°$ values. Although we typically think of this in terms of ATP hydrolysis, reactions of other groups such as phosphoesters or thioesters are common in biochemistry. Likewise, reactions that have $\Delta G°$ values close to zero can be tipped to move in either direction by changing the concentrations of substrates or products.
	• The synthesis of glucose and the degradation of monosaccharides use similar reactions but different pathways. Key steps are different, and compartmentalization is important. A similar theme is seen in fatty acid metabolism.

6.1 Properties, Nomenclature, and Biological Functions of Monosaccharides

Carbohydrates are a group of biological molecules with the basic formula $C_x(H_2O)_x$, hence the name "carbo-hydrate." At their most basic level, carbohydrates have the equivalent of one molecule of water for every carbon atom. This means that carbohydrates are polyhydroxy aldehydes or polyhydroxy ketones, that is, molecules in which one carbon bears a carbonyl group (C=O) and the others all carry hydroxyls (—OH). They also contain at least three carbons. We will first discuss the basic structures of carbohydrates and different ways we can represent these molecules on paper, and then we will move into modifications of these molecules.

$C_x(H_2O)_x$	$C_6H_{12}O_6$	Bond-angle depiction	Fischer projection of
Generic formula	**Formula of a carbohydrate**	**of a carbohydrate**	**a carbohydrate**

6.1.1 Monosaccharides are the simplest carbohydrates

This section focuses on the simplest carbohydrates, the **monosaccharides**. Monosaccharides can be linked together to form more complex carbohydrates. Monosaccharides all end in the suffix -ose to designate them as carbohydrates. They can be classified based by the number of carbons in the molecule, the presence of an aldehyde or a ketone, and the stereochemistry at the **penultimate carbon** (the chiral center farthest from the carbonyl). In terms of size, monosaccharides contain three to nine carbons but most contain five, six, or seven carbons (**Table 6.1**). The reasons for this are rooted in organic chemistry. A numbering system can be combined with the "-ose" suffix to provide a generic name for a group of carbohydrates; for example, five-carbon monosaccharides are **pentoses** and six-carbon monosaccharides are **hexoses**. Thanks to the carbonyl moiety, a monosaccharide contains either an aldehyde or a ketone and can therefore be broadly categorized as **aldoses** or **ketoses**. These terms can be combined with the numbering system mentioned previously. Thus, for example, fructose is a ketohexose and ribose is an aldopentose.

Fructose	**Ribose**
(ketohexose)	(aldopentose)

Carbohydrates can also be classified by the stereochemistry of the chiral carbons in the molecule. To do this, we examine the chiral center that is farthest from the carbonyl moiety. This chiral center is termed the penultimate carbon. The carbon farthest from the carbonyl—the last or ultimate carbon—has two hydrogens. Therefore, it has a plane of symmetry and is *achiral*—that is, not chiral. In categorizing monosaccharides, the molecule is first drawn in a **Fischer projection**. Recall that in a Fischer projection bonds are represented as horizontal and vertical lines where the horizontal lines project out toward the viewer and the vertical lines project back through the plane of the illustration. Next, a comparison is made between the penultimate carbon

TABLE 6.1 Names of Monosaccharides

Number of Carbons	Name
3	Triose
4	Tetrose
5	Pentose
6	Hexose
7	Heptose

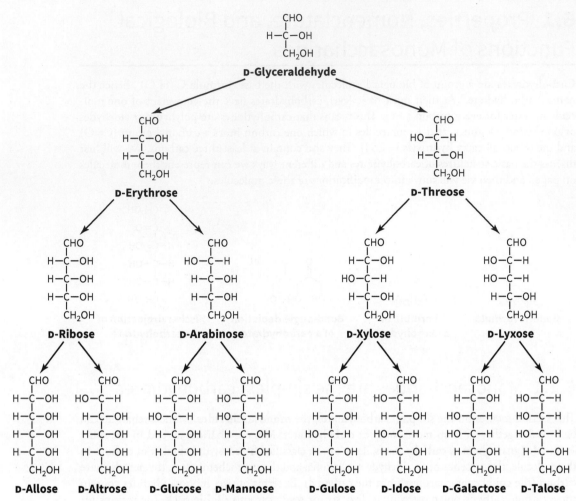

FIGURE 6.1 First fifteen D-aldoses. The most common of these molecules found in biochemistry are glucose, mannose, galactose, ribose, and glyceraldehyde.

in the monosaccharide in question and glyceraldehyde, the simplest monosaccharide. If the hydroxyl group is found on the right, the molecule is given the stereochemical designation D; if it is on the left, it is given the designation L. Nearly all carbohydrates found in nature have D stereochemistry. It may be tempting to relate the D, L nomenclature of monosaccharides to other stereochemical systems, such as the *d, l* dextrorotary/levorotary system or the *R, S* system. However, the D, L system is merely a means of grouping molecules by comparing them to a standard, in this case, glyceraldehyde. The *R, S* names of carbohydrates and their ability to rotate plane polarized light are not related to the D, L system. The smallest ketose (dihydroxyacetone) is symmetric and lacks a chiral center.

Figure 6.1 and Figure 6.2 are Fischer projections showing the degree of relatedness of aldoses and ketoses, respectively. In a Fischer projection, placement of a group on one side of the carbon or the other indicates that these molecules are different stereoisomers. For example, the aldotetroses erythrose and threose are stereoisomers of one another, differing only in the stereochemistry around C-2. Carbohydrates that differ at one stereocenter, for example, the C-4 of the aldohexoses glucose and galactose, are termed **epimers** of one another. Different stereoisomers give the molecule a different final shape or conformation. Because they differ in conformation, these molecules will bind differently to enzymes and produce different products.

Although carbohydrates can be named according to the number of carbons and the presence of an aldehyde or ketone, most have common names. Since these molecules often differ only in their stereochemistry, it becomes problematic to refer to them using systematic names. For example, it is easier to refer to a molecule as "mannose" than as "(3S,4S,5S,6R)-6-(hydroxymethyl) oxane-2,3,4,5-tetrol." Glucose or blood sugar is the most prevalent monosaccharide and the building block of many complex carbohydrates. Because it is the most common D carbohydrate, glucose is also called **dextrose**, particularly in clinical situations.

6.1.2 Monosaccharides form hemiacetals and hemiketals

Fischer projections are used to depict monosaccharides because they make it easier to view the stereochemical differences between the different molecules. However, the projections are not an accurate reflection of these molecules in the cell nor do they provide an easily identifiable shape. One of the hydroxyl moieties in the monosaccharide can participate in nucleophilic attack on the carbonyl carbon to form a new structure—a hemiacetal or hemiketal. Hemiacetals and hemiketals are often unstable. However, tethering both the hydroxyl and carbonyl groups within the same molecule increases the likelihood of bond formation, and leads to a much greater proportion of the molecules existing as hemiacetals or hemiketals than would be the case if the carbonyl and hydroxyl were in separate molecules. In other words, intramolecular hemiacetal and hemiketal bonds form much more readily than intermolecular bonds.

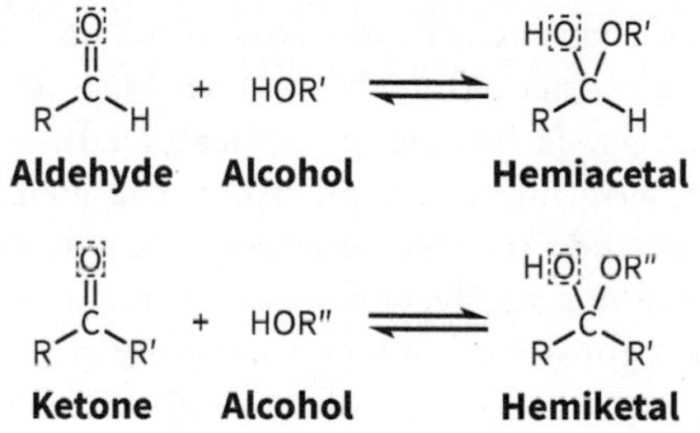

Aldehyde   Alcohol   Hemiacetal

Ketone   Alcohol   Hemiketal

When a hemiacetal or hemiketal forms, the carbonyl group is converted from a trigonal planar sp^2 hybridized carbon to a tetrahedral sp^3 center. This carbon is now a new chiral center, and the carbonyl oxygen is converted to a hydroxyl group (**Figure 6.3**).

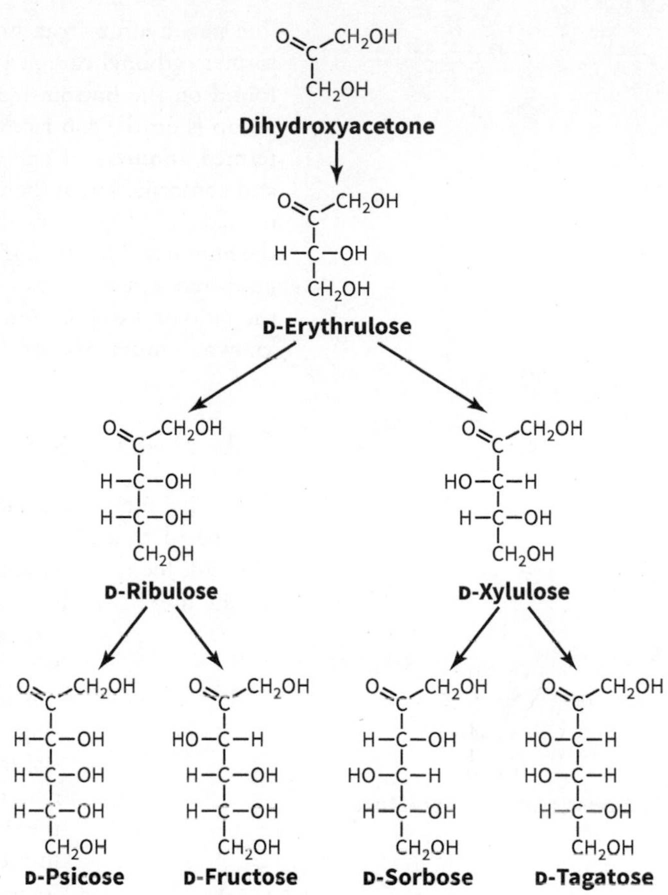

FIGURE 6.2 Seven d-ketoses. The most common of these molecules found in biochemistry are fructose, xylulose, and dihydroxyacetone.

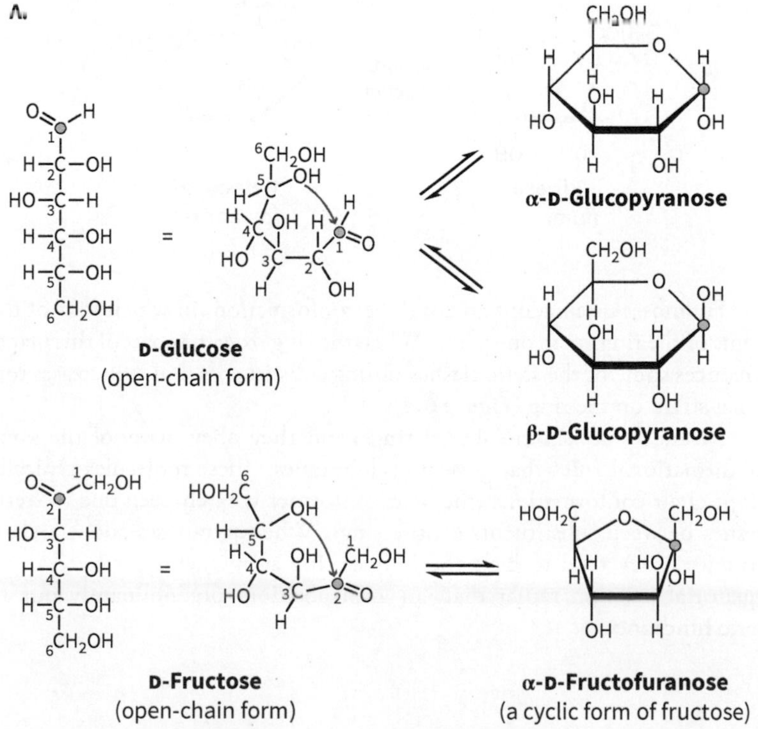

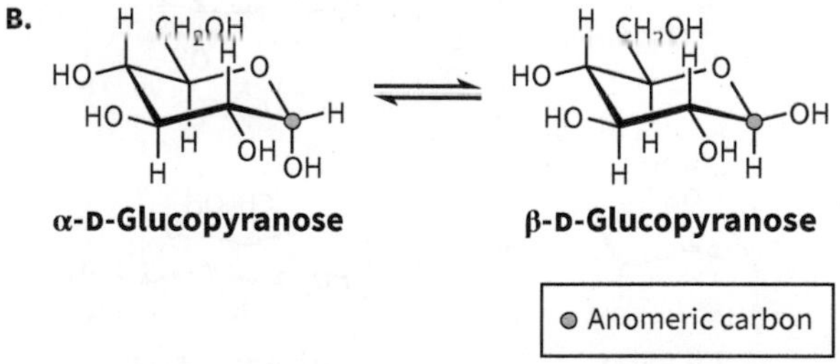

FIGURE 6.3 Most monosaccharides will form hemiacetal or hemiketal structures. **A.** The formation of a new chiral center (from the old carbonyl carbon) results in an uneven mix of α and β anomers. **B.** Cyclic structures can open and reclose, rearranging from one anomer to another, a process known as *mutarotation*.

The new hydroxyl can be found on either the bottom face or the top face of the ring. The former carbonyl carbon is now referred to as the **anomeric carbon**. If the hydroxyl group is found on the bottom face of the ring, the monosaccharide is designated α; if the hydroxyl group is on the top face, it is designated β. The α and β structures of a monosaccharide are termed **anomers** of one another. Although α and β anomers have different stereochemistry and conformation at the C-1, this is not usually a problem in biochemistry. The hemiacetal or hemiketal linkage is highly labile; it opens and recloses, and as it recloses it can form either of the anomers. However, the odds of forming one anomer over another are not 50:50. For any given monosaccharide, a particular ratio of anomers is observed at equilibrium. For glucose, the ratio of α to β is 36:64. The process of converting from one anomeric form to the other is known as **mutarotation**.

6.1.3 Monosaccharides form heterocyclic structures

To depict monosaccharides in their cyclic form, it is best to use **Haworth projections**. Compared to a Fischer projection, a Haworth projection more accurately depicts what the molecule looks like in solution, but it lacks some details of the molecule's conformation. Based on the number of hydroxyl groups present, it would seem that numerous rings of different sizes are possible. In reality, owing to ring strain and steric hindrances, five- and six-membered rings are the most common. Some carbohydrates can adopt either of these structures, each of which exists as a heterocyclic ring containing an oxygen (a hemiacetal or hemiketal). Based on their similarity to cyclic ethers (furan and pyran), a five-membered ring is designated as a **furanose** and a six-membered ring as a **pyranose**. However, the size of the ring is not necessarily a reflection of the number of carbons in the molecule: some pentoses can adopt a pyranose conformation, and some hexoses can be furanoses (because of the oxygen in the ring).

A.

Furanose envelope structure

B.

Chair form   **Boat form**

C.

All axial   **All equatorial**
(not observed)

FIGURE 6.4 Conformation of furanoses and pyranoses. A. Furanoses adopt an envelopelike structure with one member of the ring out of the plane. **B.** Pyranoses can adopt a number of conformations. Shown are the chair and boat forms. In B, "a" indicates axial substituents and "e" indicates equatorial substituents. Note that the boat conformer is rarely seen due to steric clashes between axial substituents on the ring. **C.** When pyranoses adopt a chair structure, equatorial positions are favored to axial ones, again predominantly due to steric hindrance.

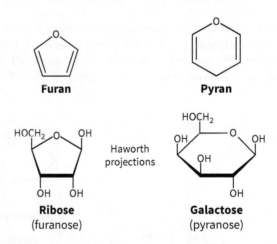

Furan   **Pyran**

Haworth projections

Ribose   **Galactose**
(furanose)   (pyranose)

Furanoses often adopt an envelope conformation, in which four of the atoms of the ring are in one plane. When the ring oxygen is part of this plane, it reduces some of the steric clashes of ring substituents, but it increases torsional strain on the ring (**Figure 6.4**).

Pyranoses are six-membered rings, and they obey many of the same conformational rules that govern cyclohexanes. These molecules typically adopt chair conformations (the boat conformer is rarely seen due to steric clashes between substituents on the ring). When pyranoses adopt a chair structure, they tend to have the bulkier ring substituents arranged in an equatorial position rather than an axial position, predominantly due to steric hindrance.

6.1.4 Monosaccharides can be chemically modified

The monosaccharides discussed so far are polyhydroxy aldehydes and ketones; however, carbohydrates can also undergo chemical modifications through oxidation, reduction, or substitution reactions.

Glucosamine   ***N*-Acetylglucosamine**   **Xylitol**

Glucuronic acid   **2-deoxyribose**

In **amino sugars**, a hydroxyl group has been replaced by an amine. Frequently, these amines are further modified, for example, being acetylated to form an amide. The amino derivative of glucose (2-aminoglucose) is commonly known as **glucosamine**. Both glucosamine and the acetyl derivative (*N*-Acetylglucosamine) are found in important biological polymers, including chitin, extracellular matrix and insect exoskeletons.

In **sugar alcohols**, the carbonyl group is reduced to an alcohol. Without the carbonyl group it is impossible for sugar alcohols to form a hemiacetal or hemiketal linkage and form a cyclic structure.

Sugar alcohols can bind to taste receptors on the tongue and therefore still taste sweet; however, they cannot form cyclic structures and therefore cannot be absorbed by glucose transporters in the intestine. Because of this they have no caloric value; instead, sugar alcohols pass through the gut, where they are either fermented by bacteria or pass undigested. In nature, sugar alcohols are found in seaweed and some other plants, but they are also easy to make in the laboratory through the chemical reduction of the carbonyl group of monosaccharides. Xylitol, mannitol, and glucitol are sugar alcohols that are found in dietetic candy and gum.

Several monosaccharides can be oxidized to form a **sugar acid**, a carboxylic acid modification. Glucuronic acid (glucuronate) is a molecule of glucose in which C-6 has been enzymatically oxidized to a carboxylic acid. The liver employs glucuronic acid to detoxify foreign compounds known as xenobiotics. These compounds range from pharmaceuticals to the pigments and flavoring agents found naturally in all foods. These molecules often require modification to facilitate their removal from the body. An acetal is formed to a hydroxyl or an amino group on the foreign substance, and the complex is eliminated in the urine.

Deoxy sugars are formed by the removal of a hydroxyl moiety on one or more carbons of a monosaccharide. While the namesake carbohydrate in DNA is deoxyribose and is the most common and best known of these, hundreds of different deoxy sugars have been identified in bacteria, several of which are found in important antibiotics. In humans, deoxyribose is not made directly but is rather synthesized from ribonucleotide diphosphates by ribonucleotide reductase.

Monosaccharides can be attached to each other or to proteins, lipids, and peptides forming acetals and ketals. These linkages can occur through amines or hydroxyls and are referred to as **glycosidic bonds**. Compared to hemiacetals and hydroxyls the acetals and ketals found in glycosidic bonds are fairly stable and robust structures that require specific enzymatic degradation or strong acids to disassemble them.

Glycosidic bonds

Methyl α-D-Glucopyranoside   **Methyl β-D-Glucopyranoside**

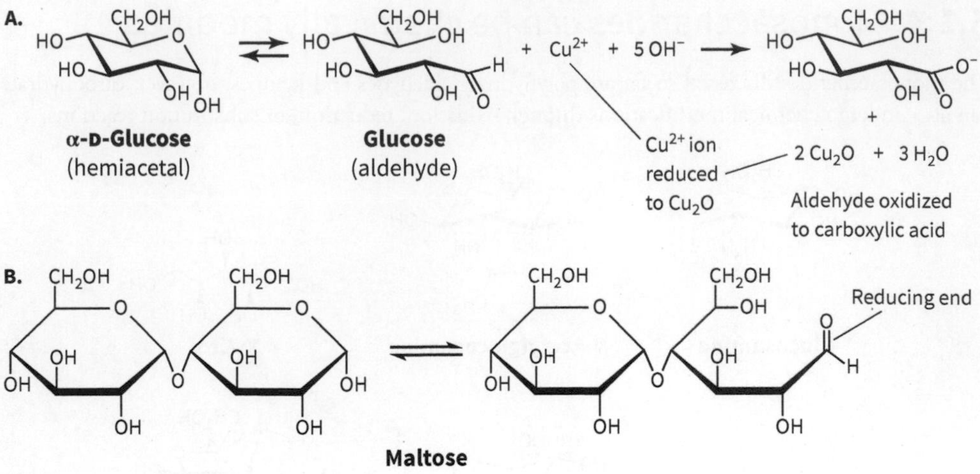

FIGURE 6.5 Reducing sugars.
A. Sugars like glucose that have free aldehyde groups or ketoses that can isomerize to aldoses have the ability to reduce metal ions, such as Cu^{2+}, becoming oxidized in the process. **B.** Disaccharides, such as maltose, are able to open to an aldehyde form and, therefore, also are considered reducing sugars.

A.

α-D-Glucose
(hemiacetal)

Glucose
(aldehyde)

$+ Cu^{2+} + 5 OH^-$ →

Cu^{2+} ion reduced to Cu_2O

$2 Cu_2O + 3 H_2O$

Aldehyde oxidized to carboxylic acid

B.

Reducing end

Maltose

6.1.5 Carbohydrates can be classified as reducing sugars or nonreducing sugars

All aldehydes are prone to oxidation to carboxylic acids. When one molecule is oxidized, another is reduced. Classically, carbohydrates were often categorized by their ability to reduce a particular reagent such as Cu^{2+} or Ag^+. The ability to reduce metals was a useful assay for aldehydes, and this was often used to measure aldose levels. Aldoses are reducing sugars, but some ketoses such as fructose will isomerize under the assay conditions and are also reducing sugars. Complicating matters further, some disaccharides have a free hemiacetal carbon. These hemiacetals are capable of opening, creating an aldehyde moiety that can undergo redox chemistry. Complex carbohydrates generally lack enough free aldehyde groups to be considered reducing sugars (**Figure 6.5**).

The ability of glucose to reduce metals figured prominently in the development of some of the first clinical tests for glucose (see **Medical Biochemistry: The interesting history of glucose testing**).

Medical Biochemistry

The interesting history of glucose testing

Nearly 2,000 years ago, the Greek physician Areteus of Cappadocia described diabetes as "…a melting down of the flesh and limbs into urine…." Indeed, copious urination is one of the hallmarks of diabetes mellitus. In diabetes, cells are unable to absorb glucose from the plasma, owing to either a lack of insulin (type 1 diabetes) or a post-receptor signaling defect (type 2). As glucose levels rise, transporters in the kidney that reabsorb glucose from the filtrate are overwhelmed, and glucose literally "spills over" into the urine. The extra glucose carries water with it, hence the increased urinary output. Since antiquity, it has been known that the urine of people with diabetes contains glucose. By the middle of the nineteenth century it was appreciated that a test for urinary glucose could be beneficial in the clinic, both as a diagnostic test for diabetes and as a way to determine how well the patient is responding to different treatments.

Such tests were available, but they were primitive. Some bold clinicians were apparently willing and able to taste the sugar in the urine of their patients, an unpleasant approach that clearly cannot produce a quantitative result. A more common technique was to collect a liter of urine, boil it dry, and then measure the mass of the residue that was left behind. This approach also had its drawbacks. Several different salts and solids can be found in urine, and these can vary depending on the composition of the diet and the disease state of the patient. The test was also unpleasant; it took some time to boil the sample dry, and it was difficult not to scorch or lose any of it.

The most significant advance in glucose testing at this time was made by Stanley Rossiter Benedict, a clinical biochemist credited with developing Benedict's solution: an alkali solution of carbonate, citrate, and, most importantly, Cu^{2+} ions. Benedict knew both from his work and from that of his competitor Otto Folin that the aldehyde moiety found in glucose could reduce Cu^{2+} to Cu^+. The reagent would change from a brilliant blue solution of Cu^{2+} to a brick red precipitate as the Cu^+ formed Cu_2O. Benedict published his findings in the *Journal of Biological Chemistry* in 1908.

Benedict's test was quickly adopted by clinicians as a rapid, simple, and semiquantitative means of determining urinary glucose concentrations. Perhaps the greatest endorsement came from Dr. Elliot Joslin, director of the Joslin Diabetes Center in Boston, Massachusetts. Joslin directed his patients to use the reagent daily to track their urinary glucose and to record their results in a diary.

This simple test resulted in several major paradigm shifts in medicine. For one of the first times, it gave clinicians the ability to measure the rate of progression of disease and make alterations to a treatment—medicine was evolving from an art to a science. Also, people could now be active participants in their own healthcare. People with diabetes were not subjects treated by the physician, but were partners who could directly affect their urinary glucose levels through diet and who could record the findings.

Benedict's test remained the standard test for more than 50 years. In more recent times, it has been replaced by enzymatic assays based on glucose oxidase. A significant difference between these tests is the sample tested. Older tests assay glucose in urine. This roughly correlates with plasma glucose levels in people with diabetes, but the results can be influenced by diet, volume of urine, and several other factors. The urine tests do not give an accurate picture of blood glucose at a particular moment in time. Enzymatic assays give a snapshot of blood glucose level at the time the blood is drawn. In the clinic, these assays are colorimetric (producing a colored product in the presence of glucose).

In the home setting, testing is now accomplished by a modification of the enzymatic assay, using a glucometer, which uses a disposable test strip impregnated with a stabilized sample of glucose oxidase. When blood is drawn into the strip, the glucose oxidase converts the glucose to glucuronic acid, and the electrons that are captured are transferred to a device that measures changes in current. The more glucose available for oxidation, the more current flows. The current is compared to standard values, and the device gives a concentration of glucose in the blood. One drawback of using glucometers is that blood must be drawn (using a fingerstick); however, the advantage of knowing the blood glucose concentrations in real time outweighs the inconvenience and pain of a fingerstick. As glucometer design improves, the devices require less blood; the newest ones use as little as 0.3 µl of blood.

6.1.6 Identification of monosaccharides using different reagents

Qualitative analysis of carbohydrates (**Figure 6.6**), especially monosaccharides, is specific to the functional groups present in the sugar molecule. The reducing property of the molecule is mainly due to the presence of a free hydroxyl group at the anomeric carbon atom. However, few carbohydrates such as glycogen and starch do not show reducing properties. For identifying

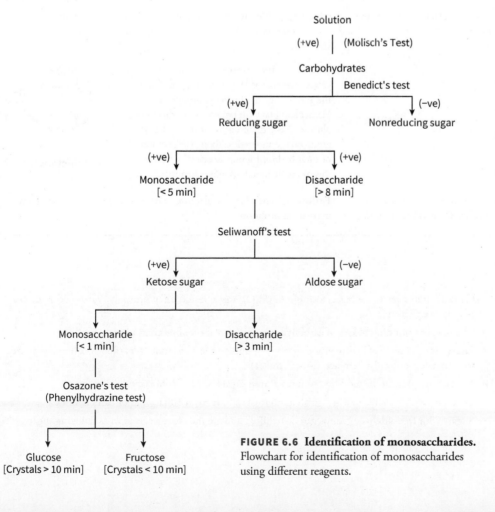

FIGURE 6.6 Identification of monosaccharides.
Flowchart for identification of monosaccharides using different reagents.

carbohydrates, many tests are widely used in clinical biochemistry to diagnose diseases like diabetes and glycosuria. It is important to know that qualitative analysis will facilitate knowing the type of carbohydrate. In contrast, quantitative analysis will determine the number of carbohydrates present.

Molisch's test is a general test to identify presence of carbohydrates. Sugars undergo dehydration in the presence of sulphuric acid to form furfural or hydroxymethylfurfural, which condenses with α-naphthol (1-hydroxy naphthalene), resulting in the formation of a purple-colored ring. *Benedict's* test is useful for identifying the presence or absence of reducing sugars. Cupric (Cu^{2+}) ions present in the alkaline copper sulfate solution are reduced to cuprous (Cu^+) hydroxide by the reducing sugar which endures spontaneous dehydration to yield cupric oxide, appearing as a colored precipitate. The intensity of the color varies from green to red (red, brick-red, orange, green) depending on the concentration of the reducing sugar. The same principle is applicable for *Barfoed's* test, wherein, the reduction of cupric ions occurs under mildly acidic conditions. This test is more rapid for monosaccharides than disaccharides. *Seliwanoff's* test is useful for the identification of keto sugar and to distinguish between aldoses and ketoses. Keto sugars (e.g., fructose) when exposed to acid medium form hydroxymethylfurfural more rapidly than aldohexoses. This furfural condenses with resorcinol (*m*-dihydroxy benzene) to produce a deep pink color molecule. *Bial's* test is used to determine the presence of pentose sugars. The pentose is dehydrated to form furfural in the presence of resorcinol, HCl, and ferric chloride producing a bluish precipitate. *Osazone* test is the confirmatory test in which reducing sugars upon reacting with phenylhydrazine produce osazones, which are the characteristic derivatives of carbohydrates. These osazone derivatives have a definite crystalline shapes that aid in confirming the type of carbohydrate molecule present.

Worked Problem 6.1   The structures of monosaccharides

Shown below is the Fischer projection of D-galactose. Draw the Haworth projection of β-D-galactopyranose.

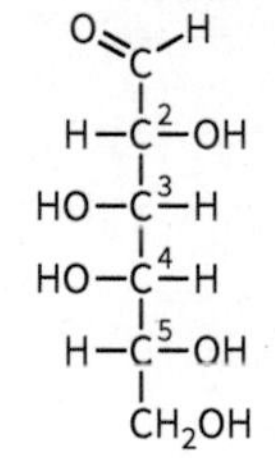

D-Galactose

Strategy   Review the cyclization of glucose discussed in this section. Which hydroxyl groups would be likely to contribute to a ring structure? Which hydroxyl group would have to contribute to form a pyranose? What do the designations α and d tell you about the structure?

Solution   The structure of β-D-galactopyranose is shown below. Compare the position of the hydroxyl groups in the Fischer and Haworth projections to those of glucose. Note the position of the hydroxyl group on the anomeric carbon. The position of each hydroxyl group is essential in how this molecule functions in biochemistry.

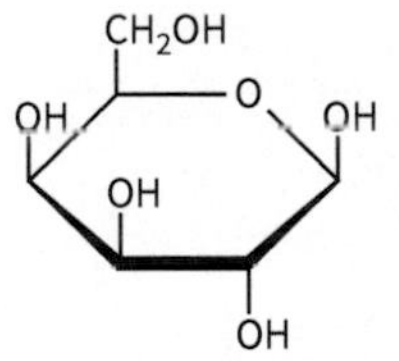

β-D-Galactopyranose

Follow-up question   Galactose can also form a furanose. Draw α-D-galactofuranose.

Summary

- Carbohydrates have the basic formula $C_x(H_2O)_x$; as a result, they are polyhydroxyaldehydes or polyhydroxyketones.
- The simplest building blocks of carbohydrates are termed monosaccharides.
- Due to the presence of hydroxyl and carbonyl groups in the same molecule, monosaccharides cyclize to form intramolecular hemiacetals or hemiketals.
- Monsaccharides can be modified to form amino sugars, sugar alcohols, and sugar acids.
- Qualitative tests enable to distinguish a monosaccharide from other types of carbohydrates.
- Sugars with free aldehyde groups can be oxidized and are therefore referred to as reducing sugars.

6.2 Properties, Nomenclature, and Biological Functions of Complex Carbohydrates

Examining the structures of monosaccharides highlights limitations to the size and diversity of these molecules. Their small size and relative lack of different functional groups present fewer opportunities for interesting chemistry and function than, for example, the amino acids. Fortunately, monosaccharides have several structural features that work in their favor. Most contain a hemiacetal or hemiketal linkage, and all contain multiple hydroxyl groups. These functional groups can combine with other hydroxyl groups to form acetals or ketals. These linkages are more stable than hemiacetal or hemiketal bonds, and they facilitate the synthesis of more complex and diverse structures. Likewise, the multiple hydroxyl groups found on monosaccharides allow multiple acetals and ketals to branch off the same molecule, resulting in linear or branched polymers of the monosaccharide. In biochemistry, we refer to these types of bonds as glycosidic bonds, and molecules with carbohydrates attached are often referred to as glycosides.

Polysaccharides can be classified by the number of monomeric units they contain. **Disaccharides** contain two monosaccharides, **trisaccharides** contain three. If the molecule contains more than three monosaccharides, it may be referred to as an **oligosaccharide**, although larger assemblies are commonly referred to as **polysaccharides**.

The number of monosaccharide units in the polysaccharide is important, but equally important are the bonds joining those monosaccharides. Specific enzymes catalyze the formation and breakdown of these bonds. The presence or absence of these enzymes dictates the ability of an organism to use one of these molecules as a fuel source. Bonds between monosaccharides describe the position of the carbon involved in the linkage as well as the α or β face from which the acetal or ketal involved in the linkage originates. Thus, a β-1,4 linkage connects an acetal or ketal from the β face of the C-1 of one monosaccharide to C-4 of a second monosaccharide. The rest of this section discusses the structure and function of several important disaccharides, trisaccharides, and polysaccharides.

6.2.1 Common disaccharides include lactose, sucrose, and maltose

A disaccharide is composed of two monosaccharides joined together. The types of monosaccharides and the linkage between them dictate some of the properties of the disaccharide. This section discusses three simple examples: lactose, sucrose, and maltose (**Figure 6.7**).

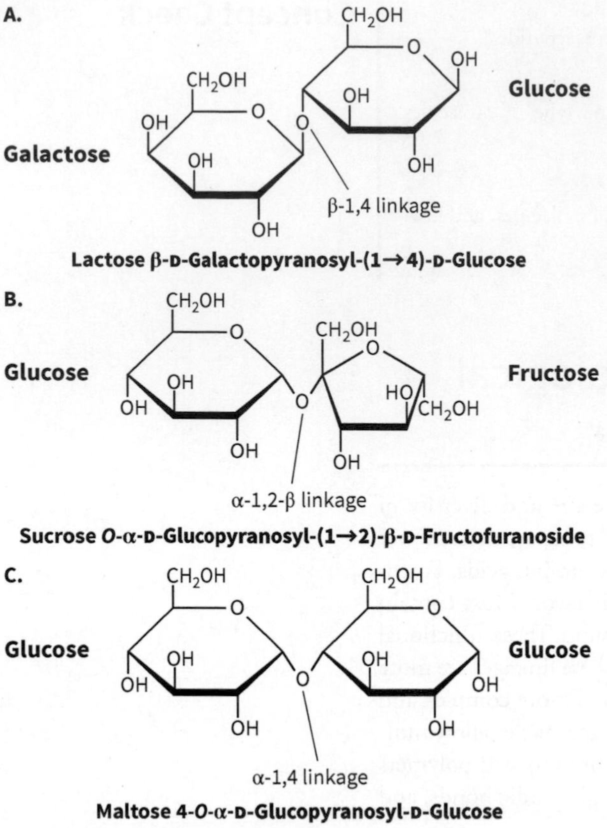

A.

Lactose β-D-Galactopyranosyl-(1→4)-D-Glucose

B.

Sucrose O-α-D-Glucopyranosyl-(1→2)-β-D-Fructofuranoside

C.

Maltose 4-O-α-D-Glucopyranosyl-D-Glucose

FIGURE 6.7 Examples of common disaccharides. The structures of **A.** lactose, **B.** sucrose, and **C.** maltose are shown.

Lactose Lactose (milk sugar) is a disaccharide comprised of one molecule of glucose and one molecule of galactose, connected via a β-1,4 linkage (β-D-Galactopyranosyl-(1→4)-D-glucose). As its name suggests, lactose is commonly found in milk; however, the concentrations of lactose vary depending on the species. For example, the lactose concentration is higher in the milk of humans (9%) than in that of cows (4.7%), goats (4.7%), or sheep (4.6%).

In the stomach, lactose is degraded by lactase, releasing molecules of glucose and galactose. All humans possess this enzyme activity at birth, and some ethnic groups with lengthy histories of dairy farming, for example Europeans and some African and Central American groups, have retained the ability to express this enzyme throughout their lives. Therefore, they can continue to consume milk as adults. Worldwide, however, most people lose this enzyme by the end of childhood, and they are unable to digest the lactose found in milk. Instead, the lactose is fermented by microbes in the digestive tract, resulting in nausea, cramps, bloating, and diarrhea. People who are lactose intolerant have, in the past, needed to avoid most milk products. Recently, however, scientists have developed alternatives for those who are lactose intolerant but wish to consume milk. These include milk that has been treated with lactase and capsules of recombinant lactase, which can be consumed before eating a meal that contains milk. The lactase in these products takes the place of the endogenous enzymes that infants and children produce because it degrades the lactose into harmless monosaccharides for subsequent metabolism.

People who are lactose intolerant do not need to avoid all dairy products. In many products, for instance, those that are fermented such as yogurt and some cheeses, the lactose that was originally present has been catabolized by the microbes that produced the food, or has been lost in the curding process. These foods can be eaten without undesirable side effects.

Sucrose Sucrose (table sugar) is one of the most commonly consumed carbohydrates worldwide. Sucrose is a 1,2-linked disaccharide of glucose and fructose (α-D-Glucopyranosyl-(1→2)-β-D-fructofuranoside). Plants synthesize sucrose as a way to temporarily store energy and to attract organisms to help pollinate a plant or disperse its seeds. The taste of sucrose is characteristically sweet, and it is used as the basis of comparison for all other sweeteners.

Table sugar is crystalized sucrose often obtained from sugar cane or sugar beets. Sucrose can also be found in honey and maple syrup.

The chemistry and controversy of high fructose corn syrup is discussed in **Societal and Ethical Biochemistry: The ongoing debate over high fructose corn syrup.**

Societal and Ethical Biochemistry

The ongoing debate over high fructose corn syrup

Over the past 30 years, increased attention has been focused on the composition of the American diet. One of the molecules under the greatest scrutiny is high fructose corn syrup (HFCS).

HFCS is a sweetener prepared from corn. The corn is dried and milled, and to liberate glucose the resulting corn starch is treated with either dilute hydrochloric acid or amylase from bacteria or mold. The glucose is passed over a solid support to which glucose isomerase has been bound. The enzyme converts approximately 42% of the glucose to fructose. This product is HFCS-42, which is found in numerous processed foods. Some of the fructose is used to make an enriched product, HFCS-55 (55% fructose), which is used in soft drinks.

HFCS is desirable from a food-production perspective for several reasons. First, it is more economical than sucrose on a per-kilogram basis, and is even cheaper if we take into account that fructose is 1.7 times sweeter than sucrose. Second, HFCS is a liquid and is therefore easier than solid sucrose to handle in an industrial setting. Finally, the United States grows a lot of corn. The USDA reports that in 2017 over 14 billion bushels were grown. The conditions that are needed to grow corn and the diseases that affect it are well studied and understood. HFCS is used in a wide range of products, from baked goods to beverages to ketchup. In 2016, the average American consumed over 18 kg of HFCS.

Commercially, HFCS is a success. Unfortunately, as HFCS consumption has increased in the American diet, so have obesity and type 2 diabetes. Is HFCS the cause? The science is unclear. Consumption of HFCS has increased over the past 30 years, but so have other factors that may also be implicated in obesity and diabetes, such as increased consumption of fats (including *trans* fats), decrease in exercise, and an overall increase in the total number of calories consumed due to larger portion sizes. If we include in the analysis changes in how we eat, for example, in the car, on the run, or late at night, we see a complex problem with no simple answer.

Focusing on the biochemistry, we know that fructose is metabolized partially through glycolyis, discussed in section 6.3,

but also through the fructolysis pathway. In this pathway fructose is phosphorylated to fructose-1-phosphate and then cleaved by an aldolase to yield DHAP and glyceraldehyde, both of which ultimately enter glycolysis as GAP. However, other details may add to this story. Consumption of glucose elicits a release of insulin from the pancreas, whereas consumption of fructose does not. Likewise, glucose triggers responses in the brain that blunt the appetite, whereas fructose does not. Finally, consumption of HFCS has been associated with increased levels of uric acid—a by-product of purine metabolism and a causal molecule in the disease gout. It may be that fructose metabolism provides a link between glycolysis and uric acid production.

None of these observations offers a clear answer. As is often the case, complex problems have equally complex solutions.

Maltose   **Maltose** (malt sugar) is a dimer of glucose molecules connected in an α-1,4 linkage (α-D-Glucopyranosyl-(1→4)-β-D-Glucopyranoside). Maltose is the major breakdown product formed when plants catabolize starch. When hearing the term "malt," most people think of malted milk balls or malted milk; however, the molecule plays a central role in the brewing industry.

Grains such as barley store energy as starch (polymers of glucose). When barley grains germinate, they mobilize the energy stored in starch as maltose. An early step in beer production is malting of grain, in which the grains are kept damp and warm and are allowed to germinate. During the malting process, enzymes cleave maltose, a dimer of glucose, from starch polymers. Following malting, the grains are roasted, and the soluble maltose is then removed from the grain by steeping in warm water. This is the actual brewing step. These soluble carbohydrates provide the fuel source for yeast in the subsequent steps in beer production.

Not all grains lend themselves to malting. Barley is the main grain used in beer production and can be malted, but corn cannot because it lacks the necessary enzymes. There are two options for extracting soluble carbohydrates from corn. In North America, the corn is often combined with malted barley or other grains. The enzymes found in the malted barley can hydrolyze the starches found in corn, in a process known as mashing. If the microbes from a previous mash are added to assist in the starch breakdown, this is known as a sour mash. The product of the fermentation of malt and corn is far from palatable, so it cannot be consumed as beer; instead, it is filtered and distilled to produce whiskey. Ancient cultures used different sources of enzymes to obtain similar results. In Central America, fermentable monosaccharides and disaccharides are released from corn by treating it with amylase from human saliva. Corn was communally chewed and spat into a pot; it was then fermented into a beer-like beverage called chicha, which was common in the Inca culture. Modern chicha drinks use other malting techniques to liberate fermentable carbohydrates from grains.

Worked Problem 6.2   Drawing a disaccharide

Draw the disaccharide cellobiose, a dimer of two glucose molecules in a β-1,4 linkage.

Strategy   Think about the structures of the monosaccharides that make up this molecule. Draw both in Haworth projections. Which carbons are involved in the linkage, and which face of the ring (α or β) would be involved in the acetal bond?

Solution   The structure of cellobiose is shown here. Examine the structures of the monosaccharides comprising the disaccharide, and look at how they are connected.

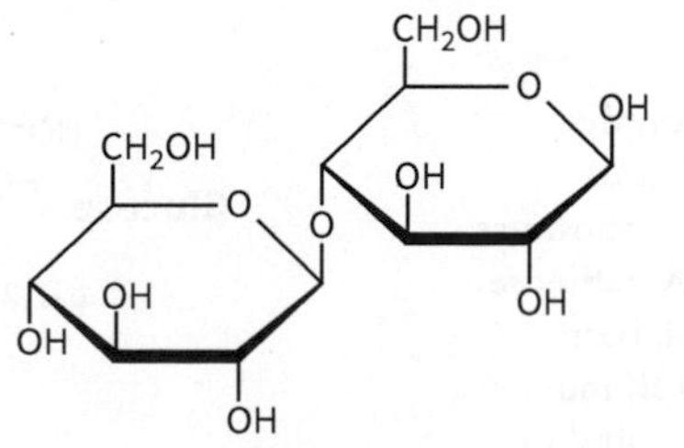

Cellobiose

Follow-up question   How would the structure and properties of this molecule change if it had an α linkage?

6.2.2 Trisaccharides and oligosaccharides contain three or more monosaccharide units bound by glycosidic linkages

Trisaccharides and oligosaccharides are connected in a similar fashion to disaccharides through glycosidic bonds. As with disaccharides, both the composition of the building blocks and the connections between them impart properties to the final molecule (**Figure 6.8**). Oligosaccharides are typically between 7 and 20 monomers in length, although the delineation between larger oligosaccharides and shorter polysaccharides is a point of debate.

Raffinose—a trisaccharide

Raffinose is a trisaccharide composed of linked molecules of galactose, glucose, and fructose. Raffinose is effectively a molecule of galactose linked via an α-1,6 linkage to the glucose moiety of sucrose. It is commonly found in the *brassica* family—broccoli, cabbage, Brussels sprouts and also some legumes. Humans cannot digest raffinose because we lack α-galactosidase activity. Gut microbes can metabolize raffinose, resulting in the production of gaseous by-products (CO_2, H_2, and CH_4) in some individuals who ingest these vegetables. As with people who are lactose intolerant, an enzyme (in this case an α-galactosidase) can be taken in tablet or capsule form together with the food, to help prevent these side effects.

Inulin—an oligosaccharide

Inulin (not to be confused with the peptide hormone *insulin*) is a plant oligosaccharide consisting predominantly of polymers of fructose capped by glucose units, and contains 20–1,000 monomeric units. Plants use inulin as a storage form of energy. As with several other carbohydrates, humans lack the enzymes necessary to mobilize inulin and are therefore unable to digest it. As a result, inulin in the diet passes largely undigested into the intestine, where it is partially digested by gut microbes. Inulin represents one of several molecules broadly characterized as a soluble fiber, and is thought to be beneficial to the health

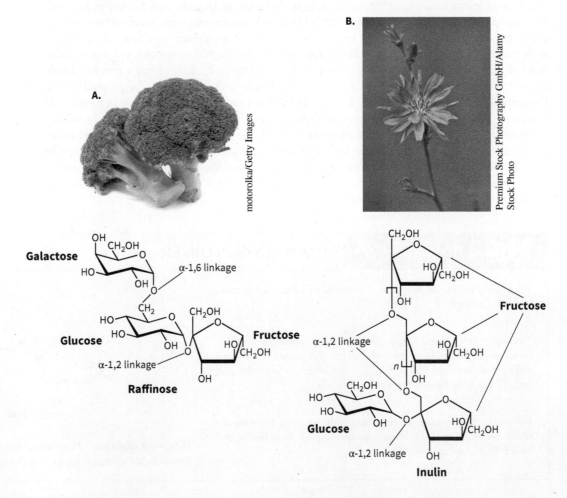

FIGURE 6.8 Examples of a trisaccharide and oligosaccharide. Shown are the structures of **A.** raffinose (found in broccoli, beans, and cabbage) and **B.** inulin (a polysaccharide found in chicory and added to many foods). Neither molecule is digestible by humans due to the linkages between the monosaccharide subunits.

of the gastrointestinal tract. It is found in high concentrations in plants such as chicory, and it is sometimes added to processed foods to elevate their fiber content.

6.2.3 Common polysaccharides function to store energy or provide structure

Polysaccharides are typically made up of simple building blocks with a repeated linkage and are larger than oligosaccharides. Often these molecules act as stores of energy or as a structural matrix for the organism.

Stored sources of energy   Plants use sugar for energy and store this sugar as starch. This starch can be found in roots and rhizomes underground, as in potatoes, or in grains like wheat or rice. Starch takes several different forms.

Amylose is a linear polymer of several hundred to several thousand glucose monomers linked by α-1,4 glycosidic bonds. It adopts a helical shape, but it can also form several other higher order structures. Amylose is relatively insoluble in water.

Amylopectin is a branched polymer of several thousand to several hundred thousand glucose monomers with α-1,4 linkages, with a branch point (linked α-1,6) about every 24–30 residues along the main chain. Because amylopectin is branched, it has numerous ends from which enzymes can degrade it. Hence, glucose monomers can be released from amylopectin more quickly than from amylose.

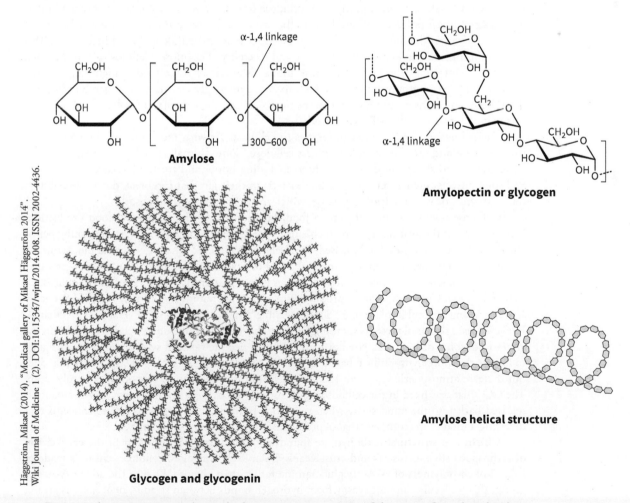

Häggström, Mikael (2014). "Medical gallery of Mikael Häggström 2014". Wiki Journal of Medicine 1 (2). DOI:10.15347/wjm/2014.008. ISSN 2002-4436.

Amylose

Amylopectin or glycogen

Glycogen and glycogenin

Amylose helical structure

Due to its linear nature, amylose, which composes up to 30% of the mass of starch, is harder to metabolize than amylopectin, the other component. Amylases, the enzymes responsible for breaking down amylose, can only attack the molecule from one end, cleaving either single glucose molecules or glucose dimers (maltose) from the molecule.

Differences in the ratio of amylose to amylopectin result in different forms of starch with different properties that are noticeable in food. This explains why some varieties of potato are good for baking and others are good for salads, and why some varieties of rice are good for sushi while others are good for risotto.

Glycogen is the storage form of carbohydrate in animals. Like amylopectin it is a highly branched polymer of glucose with α-1,4 and α-1,6 linkages, but the molecules differ in several important aspects. First, glycogen is generally larger than amylopectin: a single molecule of glycogen can be composed of several tens of thousands of glucose monomers. Second, glycogen is branched at every 6–12 glucose residues instead of every 24–30. This creates even more branch points and results in a molecule that can be rapidly degraded, mobilizing glucose molecules for energy. Third, a glycogen granule has a protein molecule (glycogenin) at its core.

Glycogen is found in granules in the cytosol of cells. Although low levels of glycogen are found in the kidneys and the intestine, the main glycogen stores in animals are in the liver and skeletal muscle. The liver has the highest concentration of glycogen (at least 10% of tissue weight), whereas muscle has only 1%−2%. However, because overall there is much more muscle by weight, muscles contain more than double the mass of glycogen in liver.

Glycogen serves as a reservoir to moderate levels of plasma glucose. When plasma glucose levels drop, liver glycogen is mobilized to provide energy for tissues and cells in need. Chief among these are erythrocytes and the cells of the brain.

Structural polysaccharides

There are numerous examples of secreted carbohydrates that different organisms use to provide structure or to act as an extracellular matrix (**Figure 6.9**).

Cellulose is a structural polysaccharide found in plants. It is the most abundant biological molecule on Earth. Cotton is nearly 90% cellulose fiber by mass. It is similar to amylose in that it consists of linear polymers of hundreds to thousands of glucose monomers connected via a 1,4 linkage. However, the linkage in cellulose is β, rather than the α linkage found in amylose. This has significant ramifications for the structure and reactivity of cellulose. Cellulose forms extended chains (in contrast to the helical structure found in amylose). Due to the orientations of the numerous hydroxyl groups on the cellulose, there are many places where hydrogen bonding can occur between chains, forming strong interactions between aligned adjacent polymers, thus giving this material its tough fiber-like characteristics.

The β linkage imparts other properties to cellulose. As with most animals, humans lack a β-glucosidase and are thus unable to digest cellulose. Some animals, notably ruminants such as cows, goats, and sheep; hind-gut fermenters including horses and turtles; lagomorphs (rabbits); and some insects, such as termites, harbor bacteria in their gut that can break down the cellulose found in grass or wood, releasing glucose, which the host organism can absorb.

Cellulose is not a source of human food, but it is a by-product of most food production. Concerns about the availability of petrochemicals and the environmental impact of burning petrochemicals has resulted in efforts focused on developing cellulosic ethanol—ethanol biochemically derived from cellulose. Current research in the United States is concentrating on these by-products (husks of corn, wheat, and rice) or on other crops that grow quickly and with minimal input (such as switch grass). Currently, ethanol production plays a small but important part in the U.S. energy strategy. Using ethanol in place of gasoline is feasible. Brazil fuels most of its vehicles with ethanol. However, using ethanol presents some engineering challenges. Unlike diesel vehicles, which can use either traditional or biodiesel, ethanol-fueled vehicles need to be built with fuel systems designed to handle the more hydrophilic ethanol. Ethanol has been seen as a "green" energy source that does not increase atmospheric CO_2, but the current form of production of ethanol is not truly clean. The CO_2 that was fixed in the cellulose of a plant grown last summer and converted to ethanol is released back into the atmosphere when the fuel is burned. Additionally, there are concerns that diverting a food crop (corn) to ethanol production results in higher food prices.

Chitin is a structural carbohydrate found in the cell wall of fungi and in the exoskeleton of arthropods such as insects and crustaceans. Chemically similar to cellulose, chitin is made of β-1,4 linked polymers of *N*-Acetylglucosamine rather than glucose. Due to the added *N*-Acetyl group, this molecule can more easily form hydrogen bonds between strands; hence, it has greater strength than other polysaccharides. Chitin is slow to break down in nature, and accumulation of chitin waste is a concern at seafood processing plants. Several commercial products are now made from chitin or its derivatives, including sutures, wound dressings for burn victims, and dietary supplements.

A. Cellulose

β-1,4 linkage

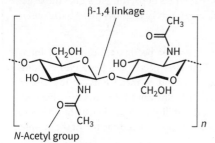

Glucose polymer

Martin Ruegner/Getty Images

B. Chitin

β-1,4 linkage

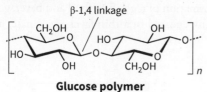

N-Acetyl group

N-Acetylglucosamine polymer

Cyndi Monaghan/ Getty Images

C. Alginate

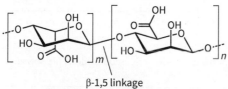

β-1,5 linkage

Guluronate and mannuronate copolymer

Daniel Poloha Underwater/ AlamyStock Photo

D. Cross-linking in alginate

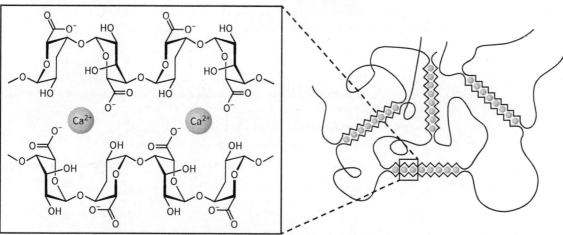

FIGURE 6.9 Structures of cellulose, chitin, and alginate.   Shown are the repeating units of the structural carbohydrates **A.** cellulose, **B.** chitin, and **C.** alginate. Note how these molecules can cross-link with one another through hydrogen bonding (A, B, or C) or **D.** electrostatic interactions with a bridging metal ion.

Alginate is a structural polysaccharide found in the cell walls of brown algae and kelp. Unlike the other polysaccharides discussed here, alginate is a **copolymer**: a combination of two monomeric units. In the case of alginate, these monomers are guluronate and mannuronate (the deprotonated forms of guluronic and mannuronic acid). These molecules are themselves the enzymatically oxidized forms of gulose and mannose, with carboxyl moieties found on the number 6 carbon of each. The resulting alginate polymer is highly anionic and readily absorbs water, one of its most notable properties. Alginate is commercially used as a diet aid, food thickener, and absorbent owing to its ability to absorb water.

Having numerous negative charges, alginate will react with cations, particularly Ca^{2+} ions. The strands of alginate become linked, resulting in the formation of a gel. The food and beverage industry is now creating pearls or spheres of food or drink encased in a thin layer of alginate: a novel application of polysaccharide chemistry.

Worked Problem 6.3   The properties of a polysaccharide

Pectins are a family of structural polysaccharides found in fruit. While there is considerable variability among pectins, most consist of a D-1,4 linked galacturonic acid (with methanol esterified to the carboxyl group in some cases). Pectins can be classified as high ester or low ester, depending on the degree of esterification. They will form gels (highly hydrated networks of carbohydrate and water) and are used as gelling agents in the production of jams, jellies, and candy. Both high- and low-ester pectins can be used in this process, with different pectins used in different applications. Based on your knowledge of polysaccharides and the descriptions of pectins provided, describe the forces that are involved in gelling high- and low-ester pectins.

Strategy   Think about the structures of pectin. How does the structure and what you know of other polysaccharides fit with the observations in the question?

Solution   Both high- and low-ester pectins are capable of interstrand hydrogen bonding. Low-ester pectin also has a greater number of carboxyl groups that can bind to metal ions. In the case of divalent metal ions, such as Ca^{2+}, this can lead to bridging between strands of pectin (increasing the ability to gel). Hence, the ability to gel depends on both the metal ion content and the pH of the sample.

Follow-up question   Acetylated pectins are also found in nature. These varieties fail to act as gelling agents. Propose a chemical reason why this is so.

Summary

- Monosaccharides can be joined to form disaccharides, trisaccharides, oligosaccharides, and polysaccharides.
- The acetal or ketal linkage formed between monosaccharide units can be described based on the positions of the carbons involved in the bond.
- Polysaccharides often serve as stores of energy or provide structure for an organism.
- The properties of complex carbohydrates are based on the monomeric units (glucose or galactose) and the connection between those units (β-1,4).

Concept Check

1. Name and describe at the chemical level several different examples of disaccharides, trisaccharides, oligosaccharides, and polysaccharides.
2. Describe the biochemical function of these molecules.
3. Explain how the structure of these molecules contributes to their function.
4. Explain how these molecules are similar to one another or to other biological molecules, and how they are different.

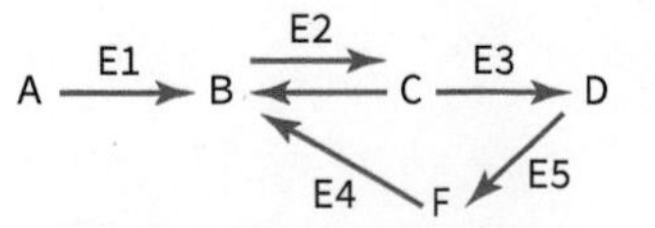

FIGURE 6.10 Overview of metabolic pathways. Any metabolic pathway consists of intermediates (A, B, C, D, and F) converted to one another by enzymes (E1, E2, E3). Some steps may be reversible, others may not. Pathways may branch or be linear.

6.3 Glycolysis and an Introduction to Metabolic Pathways

This section discusses the breakdown of glucose into pyruvate, a process called **glycolysis**. In this pathway, as in any metabolic pathway, a small molecule (a metabolite) is transformed into a different molecule through a series of enzymatic reactions (**Figure 6.10**). This section describes what is meant by a metabolic pathway and some general themes to look for in any pathway, as well as how to apply this information to a discussion of glycolysis.

Because glycolysis is a catabolic pathway, one that breaks molecules down, the process is oxidative, and thus the final products are more oxidized than the starting materials. This means that the overall process gives off energy. However, glycolysis requires an initial energetic investment to initiate the process.

6.3.1 Metabolic pathways describe how molecules are built up or broken down

The sum total of all the chemical reactions in an organism constitutes that organism's metabolism. These reactions can be classified as **anabolic** (buildup molecules) or **catabolic** (breakdown molecules). The molecules in question are small organic molecules generically known as **metabolites**. They include amino acids, carbohydrates, lipids, and nucleotides, and other small molecules that fit somewhere in between these four groups. As a general rule, the overall breakdown of a metabolite yields energy and is oxidative, whereas reactions that build up metabolites require an input of energy and are reductive.

Some of the reactions that enzymes employ may not look familiar at first, but once examined, they reveal themselves to be reactions first encountered in organic chemistry. An example of this is the retro aldol condensation. A review of the retro aldol condensation from organic chemistry and its applications in biochemistry are found in **Biochemistry: The retro aldol reaction**.

Biochemistry

The retro aldol reaction

The controlled and direct formation or cleavage of carbon–carbon bonds presents an interesting conundrum in biochemistry. Given the sheer number of carbon–carbon bonds in nature, it might seem as though these bonds could form in many different ways, but that is not the case. In reality, there are relatively few ways to make or break carbon–carbon bonds. This box discusses the aldol reaction, a key reaction of glycolysis and gluconeogenesis. (This reaction is related to the Claisen condensation, found in organic chemistry.)

Aldols, as the name suggests, are organic molecules containing a carbonyl group such as an aldehyde with a hydroxyl moiety β to the carbonyl. Aldols are formed by the addition of one aldehyde to another:

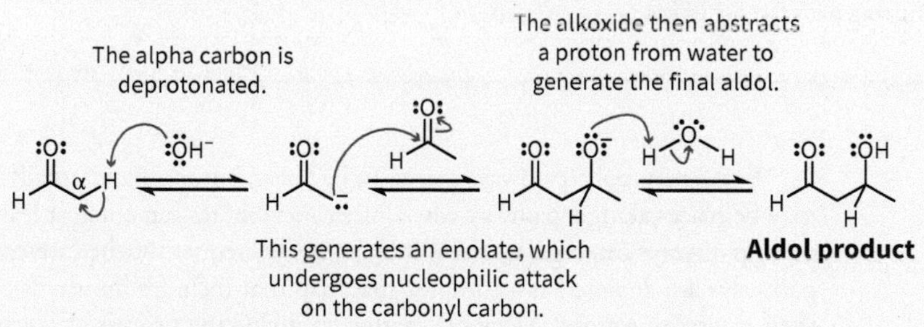

The aldol (a β-hydroxy aldehyde) can then undergo elimination to form an enone. That reaction is known as an aldol condensation, and water is eliminated in the reaction.

In the organic laboratory, an aldol can be made via either an acid- or base-catalyzed mechanism:

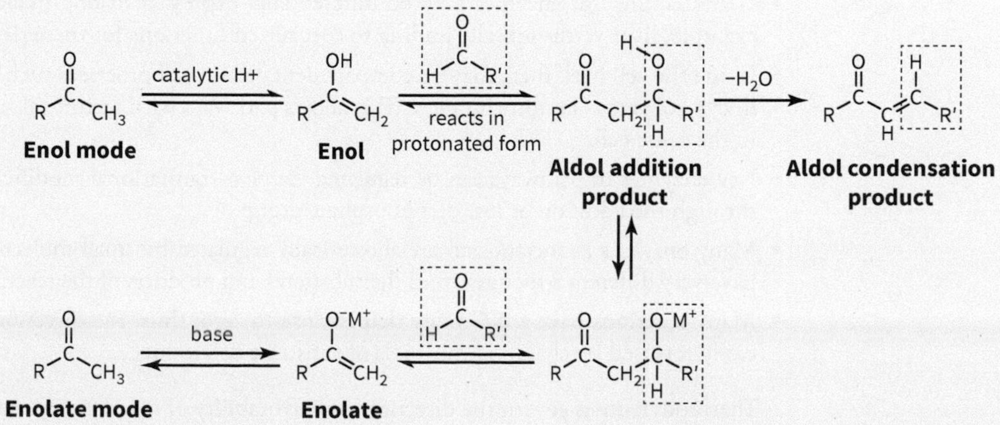

The reverse reaction is a retro-aldol reaction, which results in bond cleavage:

The β-hydroxyl is deprotonated.

Resulting in collapse of the alkoxide, and reformation of the carbonyl.

This leads to loss of the enolate.

In biochemistry, retro-aldol reactions are central to the cleavage of F-1,6-bP in the fourth step of glycolysis, and to the reverse reaction in gluconeogenesis.

If we examine the Fischer projection of F-1,6-bP, we see that it fits the definition of an aldol.

The enzymes that carry out these reactions are known as aldolases, and they employ a Schiff base in carrying out this chemistry. A Schiff base is simply a carbon–nitrogen double bond in which the nitrogen is also bound to one other carbon.

The Schiff base forms between the carbonyl carbon of F-1, 6-bP, and a lysine side chain in the active site of the aldolase. In the retro-aldol cleavage, the Schiff base acts as an electron sink, assisting in electron withdrawal and cleavage of the α–β carbon–carbon bond; the positive charge on this Schiff base makes it drive the reaction to an even greater extent. In the laboratory, the enolate that forms on the carbon α to the carbonyl is stabilized by tautomerism and resonance. Although the reaction may look different, similar chemistry, such as the stabilization of the enolate, is occurring with the Schiff base.

1. F-1,6-bP substrate binds. Lysine forms Schiff base with carbonyl.

2. Aspartate deprotonates hydroxyl. Electrons flow into Schiff base.

Glyceraldehyde-3-phosphate
(product 1)

Some metabolic pathways appear to be linear, but most are actually branched. For example, there may be places along a pathway where other metabolites can enter or leave, or where metabolites from other pathways can regulate the initial pathway. Frequently, the interconnections of these metabolic pathways are depicted as a complex roadmap that includes hundreds or thousands of reactions and their interconnections. Metabolic pathways divide the process of conversion from one molecule to another into discrete steps. Multiple steps provide added places where a pathway can be controlled.

Metabolic pathways are regulated in multiple ways, including:

- The expression of genes encoding for the enzymes and other proteins involved in the pathway can be upregulated or downregulated, leading to higher or lower levels of enzymes.

- Genes coding for enzymes may be differentially expressed in one tissue over another, for example, liver versus muscle, leading to specialized functions for these tissues.

- Inside the cell itself there may be compartmentalization of processes such as those that occur in cytosol versus in mitochondria. This allows pathways with reciprocal functions to operate in the same cell.

- Key enzymes in pathways can be regulated via post-translational modification, for example, through the addition or loss of a phosphate group.

- Many enzymes in metabolism are allosterically regulated by small molecules, some of which have very different structures from the substrates and products of the reaction being catalyzed.

- Many reactions have a ΔG value that is close to zero; thus, the direction of the reaction is easily reversed by changing the levels of substrate or product.

Thermodynamics govern the direction and favorability of reactions in metabolism, and reactions only proceed downhill energetically. The best term for discussing the favorability of these reactions is Gibbs free energy, ΔG. Often, in metabolism, reactions are coupled, with an energetically favorable reaction making an unfavorable reaction favorable. An example of this is the addition of a phosphate group to a molecule. If the phosphate reactant were a free phosphate anion (HPO_4^{2-}) in solution, it would be energetically unfavorable to add that phosphate to an alcohol group, such as a serine or threonine side chain. Hence, this type of reaction does not occur. However, if that phosphate is donated in a reaction from a donor with a greater transfer potential, such as ATP, the overall reaction can proceed favorably with a negative ΔG value.

6.3.2 Glycolysis is the process by which glucose is broken into pyruvate

Glycolysis is the process used by almost all organisms to break down glucose into a three-carbon metabolite called **pyruvate** (**Figure 6.11**). The process provides energy to the organism in terms of both ATP and NADH/H$^+$, but also provides entry points for other molecules, such as fructose and glycerol, to be metabolized. The end-product, pyruvate, is often oxidized into acetyl-CoA and moves on to the citric acid cycle to provide even more energy. This same pyruvate can, however, be used to synthesize new amino acids or new molecules of glucose in other tissues, or it can be converted into acetyl-CoA and reduced into fatty acids. Clearly, glycolysis has a number of different functions.

The fact that glycolysis is found in almost all organisms from *E. coli* to humans indicates that this pathway has been in use for billions of years. Because of its central importance to metabolism, there are relatively few inherited mutations associated with glycolysis. Any major mutations would be extremely harmful to the organism.

Reactions of glycolysis   The glycolytic pathway is composed of ten reactions, which are generally put into two groupings. In the first five reactions of the pathway, energy in the form of two molecules of ATP needs to be "invested" in glucose. This investment energetically primes the energy-yielding reactions that occur in the last five reactions. This situation is analogous to a roller coaster in that the initial investment of energy (towing the train of cars to the top of a hill) provides the energy for the remaining trip.

The steps of glycolysis, shown in **Figure 6.12**, are as follows:

1. **Glucose is phosphorylated to glucose-6-phosphate, trapping it in the cell and maintaining a concentration gradient.** In the first step, glucose is phosphorylated on C-6 (the carbon outside the ring) to form glucose-6-phosphate. One molecule of ATP is consumed in this reaction. Because glucose-6-phosphate has a charge associated with it and a different structure than glucose, this molecule is unable to pass back through the glucose transporter and is thus trapped within the cell. This helps maintain a gradient of glucose, facilitating the diffusion of glucose into the cell.

    The enzyme responsible for catalyzing this reaction is one of several isoforms of the enzyme hexokinase. We might ask why multiple enzymes can be used to catalyze this reaction. The answer has to do with the properties of these enzymes and regulation of glucose metabolism in different tissues (such as the liver versus muscle) and at different metabolic states (for example, immediately after eating).

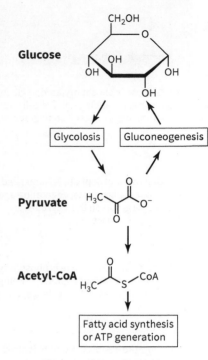

FIGURE 6.11 Pyruvate in metabolism.   Glucose and other monosaccharides are catabolized through glycolysis into pyruvate. Pyruvate can be decarboxylated and enter the citric acid cycle, producing other metabolic intermediates or contributing to ATP production.

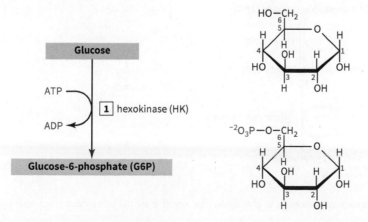

Energy investment stage

1. Glucose is phosphorylated to glucose-6-phosphate, trapping it in the cell and maintaining a concentration gradient.

2. Glucose-6-phosphate is isomerized to fructose-6-phosphate, generating a free hydroxyl group on the 1 carbon for subsequent reactions.

3. Fructose-6-phosphate is phosphorylated to fructose-1,6-bisphosphate, priming the molecule for subsequent cleavage.

4. Fructose-1,6-bisphosphate is cleaved into glyceraldehyde-3-phosphate and dihydroxyacetonephosphate.

5. Dihydroxyacetonephosphate is isomerized to glyceraldehyde-3-phosphate, ensuring that all carbons of glucose are available for oxidation.

Energy yielding stage

6. The 1 carbon of glyceraldehyde-3 phosphate is oxidized to a carboxyl group and phosphorylated with free phosphate to generate a molecule of NADH/H+ and 1,3-bisphosphoglycerate.

7. 1,3-bisphosphoglycerate transfers a phosphate from the 1 carbon to ADP making ATP and generating 3-phosphoglycerate.

8. 3-phosphoglycerate is isomerized to 2-phosphoglycerate, facilitating the subsequent dehydration of the 3 carbon.

9. 2-phosphoglycerate is dehydrated to generate phosphoenolpyruvate.

10. Phosphoenolpyruvate transfers its phosphate to ADP making ATP. The dephosphorylated enol tautomerizes to a keto group.
Pyruvate has multiple fates in the cell.

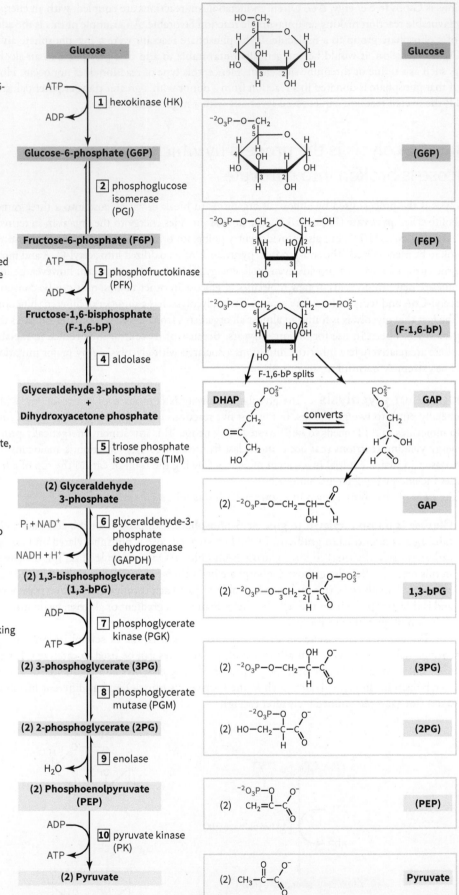

FIGURE 6.12 Reactions of glycolysis. There are ten reactions in glycolysis. The first five are the investment stage, and the final five are where energy is yielded.

2. **Glucose-6-phosphate is isomerized to fructose-6-phosphate, generating a free hydroxyl group on the number 1 carbon for subsequent reactions.** Fructose is an isomer of glucose. In the second reaction, G-6-P is isomerized to F-6-P by phosphoglucose isomerase. The formation of F-6-P facilitates subsequent phosphorylation and cleavage reactions.

3. **Fructose-6-phosphate is phosphorylated to fructose-1,6-bisphosphate, priming the molecule for subsequent cleavage.** Fructose-6-phosphate undergoes another round of phosphorylation, this time on the newly exposed C-1. This process, catalyzed by phosphofructokinase (PFK), is also ATP dependent and generates the product fructose-1,6-bisphosphate.

4. **Fructose-1,6-bisphosphate is cleaved into glyceraldehyde-3-phosphate and dihydroxyacetone phosphate.** In this step, the hexose fructose-1,6-bisphosphate is cleaved by aldolase in a retro aldol reaction into two 3-carbon intermediates: an aldehyde (glyceraldehyde-3-phosphate) and a ketone (dihydroxyacetone phosphate). Both of these triose phosphate intermediates are used as glycolysis proceeds (**Figure 6.13**).

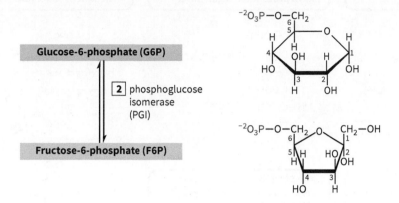

The mechanism of aldolase employs a Schiff base. This structure, which is used in several enzyme mechanisms, forms when a primary amine attacks a carbonyl carbon. The result is a carbon with a double bond to a nitrogen. In aldolase, a protonated Schiff base forms. This acts as an electron sink, withdrawing electrons from adjacent bonds and facilitating bond

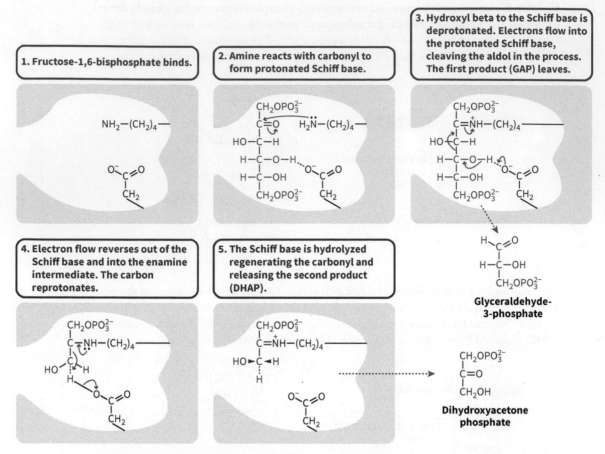

FIGURE 6.13 The mechanism of aldolase. The aldol cleavage reaction begins with the formation of a Schiff base intermediate on C-2 of F-1,6-bP. Following deprotonation of the hydroxyl on the β carbon, the Schiff base acts as an electron sink, withdrawing electrons and facilitating bond cleavage between the α and β carbons.

cleavage, in this case between the α and β carbons. Following bond cleavage, electron flow reverses out of the sink, resulting in protonation of the remaining product. Hydrolysis of the Schiff base forms the new carbonyl and regenerates the free amine form of the enzyme.

5. **Dihydroxyacetone phosphate is isomerized to glyceraldehyde-3-phosphate, ensuring that all carbons of glucose are available for oxidation.** The remainder of the pathway proceeds with only glyceraldehyde-3-phosphate. Keto groups are unable to be oxidized further, and therefore dihydroxyacetone phosphate cannot be oxidized via the same pathway as GAP. Instead, DHAP is isomerized into GAP in a reaction that involves an enediol intermediate and is catalyzed by triose phosphate isomerase (TIM) (**Figure 6.14**). In this mechanism, DHAP is bound in the active site between two key residues: a deprotonated glutamate and a histidine. Abstraction of a proton from the carbon bearing the free hydroxyl results in electron flow into the carbon–carbon bond, forming a new double bond and causing protonation of the old carbonyl oxygen. This molecule containing both a double bond and two adjacent hydroxyl groups is called an *enediol*. In the final step of the mechanism, reversal of electron flow causes deprotonation of the terminal hydroxyl generating the new aldehyde found in the final product.

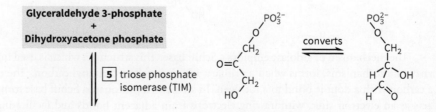

1. **Substrate binds. Glutamate deprotonates carbon, resulting in enediol formation.**

2. **Enzyme changes conformation. Histidine deprotonates primary alcohol.**

3. **Electrons flow back the through a double bond into aspartate to generate aldehyde product.**

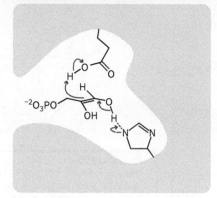
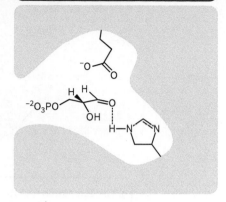

FIGURE 6.14 The mechanism of triose phosphate isomerase. In this mechanism, the isomerization of DHAP to GAP proceeds through an enediol intermediate stabilized through the pyrolic nitrogen of the histidine ring. Both molecules of GAP, the one coming from the aldol cleavage and the one resulting from this reaction (the isomerization of DHAP), continue through glycolysis.

Both molecules of GAP, the one coming from the aldol cleavage and the one resulting from the isomerization of DHAP, continue through glycolysis.

Therefore, from one 6-carbon glucose molecule we generate two 3-carbon intermediates (DHAP and GAP) and the DHAP is converted into a second molecule of GAP, which proceeds through the second phase of glycolysis.

This is the end of the investment phase of glycolysis. To summarize, in this phase glucose is absorbed by the cell, phosphorylated, isomerized, phosphorylated again, and cleaved; the resulting fragments are isomerized to yield two molecules of GAP. The cost is two molecules of ATP. During the second phase of glycolysis (the energy-yielding stage) the cell regains energy expended in the first phase.

6. **The number 1 carbon of glyceraldehyde-3 phosphate is oxidized to a carboxyl group and phosphorylated with free phosphate to generate a molecule of NADH/H⁺ and 1, 3-bisphosphoglycerate.** In this step, C-2 of GAP is oxidized from a hydroxyl to a keto group (**Figure 6.15**). This process yields one molecule of NADH/H⁺. In addition, a phosphoester is formed on C-1 of the molecule. Inorganic phosphate (HPO_4^{2-}) is used in this reaction.

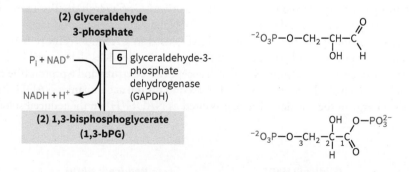

Some aspects of the mechanism of glyceraldehyde-3-phosphate dehydrogenase may look familiar to you. As substrate binds, a base assists in the deprotonation and nucleophilic attack on the carbonyl carbon of the substrate. We have seen similar chemistry in the mechanism of chymotrypsin, but here the nucleophile is the thiolate anion generated by deprotonating cysteine. This results in the formation of a thiohemiacetal intermediate, again analogous to the tetrahedral oxyanion intermediate observed in chymotrypsin. At this point the mechanisms differ. Electrons from the thiohemiacetal intermediate are transferred to NAD⁺, resulting in the formation of a covalently bound acyl thioester. The reduced NADH is exchanged for a new molecule of NAD⁺. The mechanism resolves and the enzyme is

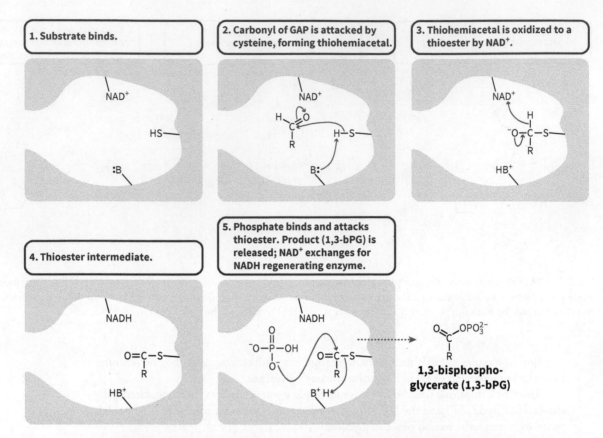

FIGURE 6.15 The mechanism of glyceraldehyde-3-phosphate dehydrogenase. In the mechanism of glyceraldehyde-3-phosphate dehydrogenase, the binding of substrate (glyceraldehyde-3-phosphate) causes nucleophilic attack of the active site cysteine on the carbonyl carbon, resulting in the formation of a thiohemiacetal intermediate. Electrons from this unstable intermediate are transferred to NAD^+, resulting in the formation of an acyl thioester intermediate. The reduced NADH is exchanged for a new molecule of NAD^+ and phosphate attacks the thioester, regenerating the enzyme and releasing 1,3-bisphosphoglycerate.

regenerated through the attack of a phosphate ion on the thioester carbon. In chymotrypsin, water was used to attack the acyl-enzyme intermediate and regenerate the free hydroxyl found in the active site serine. Thioesters are considerably less stable than their oxygen-containing counterparts and have a strongly negative ΔG of hydrolysis. This thermodynamically drives the reaction and generates the 1,3-bisphosphoglycerate product.

This reaction is the first example of NAD^+ used in a reaction (**Figure 6.16**). NAD^+ is a small organic molecule (nicotinamide adenine dinucleotide) that is used by biological systems as an electron acceptor. NAD^+ will accept two electrons and a proton (the chemical equivalent of a hydride) to form NADH. A second proton accompanies NADH, balancing out the charge of the reaction. The pair is cited as $NADH/H^+$ in the reduced state.

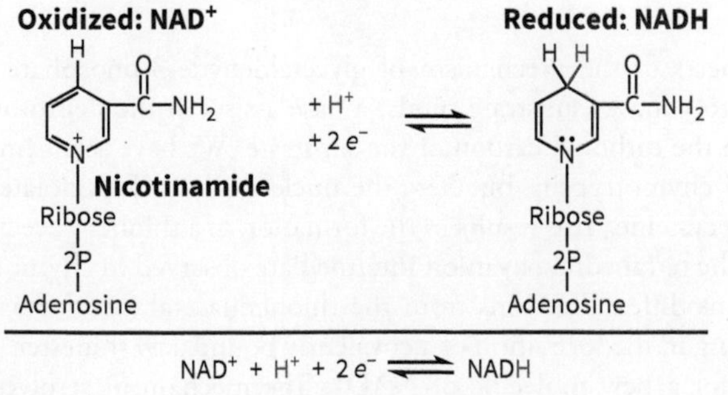

FIGURE 6.16 Numerous molecules participate in redox chemistry in biochemistry. NAD^+ is a small redox active organic molecule that is a critical cofactor in many enzymatic reactions. Electrons transferred to NAD^+ are used in subsequent reactions, such as the electron transport chain.

7. **1,3-bisphosphoglycerate (1,3-bPG) transfers a phosphate from the number one carbon to ADP, making ATP and generating 3-phosphoglycerate.** In this step, phosphoglycerate kinase removes a phosphate from 1,3-bPG and transfers it to ADP to create a molecule of ATP. This is a substrate-level phosphorylation, the direct formation of a nucleotide triphosphate by a coupled reaction, rather than an oxidative phosphorylation as seen in mitochondria. It may seem odd that a kinase removes a phosphate, but this enzyme also can remain true to its name: in the presence of phosphoglycerate and ATP it will generate 1,3-bPG and ADP. This is an example of how an enzyme lowers the activation energy and enhances the rate of both the forward and the reverse reaction. As we will see in gluconeogenesis, this reaction also highlights how altering the concentrations of substrate and product can alter the reaction quotient Q and tip the free energy of a reaction, ΔG, to the negative.

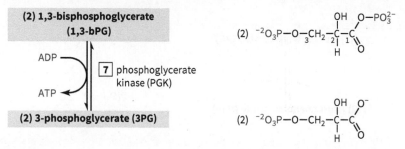

8. **3-phosphoglycerate is isomerized to 2-phosphoglycerate, facilitating the subsequent dehydration of C-3.** This reaction is catalyzed by phosphoglycerate mutase (**Figure 6.17**) and proceeds through a phosphoenzyme intermediate (2,3-bPG), coupled via a histidine residue. As with most isomerizations seen in metabolism, this neither costs nor provides energy.

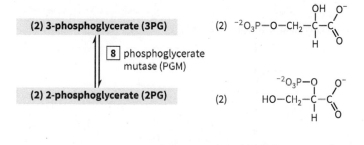

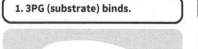

FIGURE 6.17 The mechanism of mutase. Mutase catalyzes the isomerization of 2-phosphoglycerate to 3-phosphoglycerate. Exchange of a phosphate between C-2 or C-3 and the active site histidine results in the formation of 2,3-bisphosphoglycerate (an intermediate in this reaction and an important regulator of hemoglobin).

Mutase catalyzes the isomerization of 2-phosphoglycerate to 3-phosphoglycerate. Exchange of a phosphate between C-2 or C-3 and the active site histidine results in the formation of 2, 3-bisphosphoglycerate, an intermediate in this reaction and an important regulator of hemoglobin.

9. **2-phosphoglycerate (2-PG) is dehydrated to generate phosphoenolpyruvate (PEP).** In this reaction, the hydroxyl group from C-3 is removed by enolase to give the dehydrated form. We often think of ATP as being readily able to transfer a phosphate and participate in chemical reactions because the hydrolysis of the phosphate has a large, negative ΔG value; however, the phosphate bond in PEP has an even more negative ΔG of hydrolysis and can easily transfer this phosphate group to ADP to form ATP in the next step (**Table 6.2**).

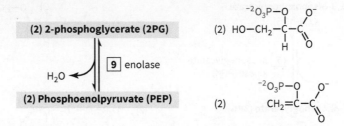

10. **Phosphoenolpyruvate (PEP) transfers its phosphate to ADP, making ATP. The dephosphorylated enol tautomerizes to a keto group.** In this reaction, catalyzed by pyruvate kinase, PEP is converted to the three-carbon α-keto acid pyruvate. Again, in this process, a molecule of ATP is generated via substrate-level phosphorylation.

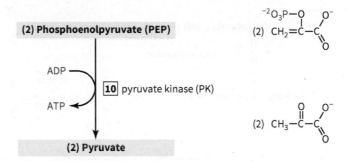

TABLE 6.2 Standard Free Energies of Hydrolysis of Some Molecules with a High Phosphate Transfer Potential (Strongly Negative $\Delta G^{\circ\prime}$ of Hydrolysis)

Intermediate	kJ/mol
Phosphoenolpyruvate	−61.9
1, 3–bisphosphoglycerate to 3-phosphoglycerate + Pi	−49.3
Phosphocreatine	−43.0
ATP to AMP and PPi	−45.6
ADP to AMP and Pi	−32.8
Acetyl-CoA	−31.4
ATP to ADP and Pi	−30.5
Glucose-3-phosphate	−20.9
PPi to 2 Pi	−19.2
Fructose-6-phosphate	−15.9
AMP to adenosine and Pi	−14.2
Glucose-6-phosphate	−13.8
Glycerol-3-phosphate	−9.2

Step 10 marks the end of the energy-yielding phase of glycolysis. Because two 3-carbon intermediates were generated in the energy-investment phase, and pass through steps 6–10, the payoff phase generates two molecules of NADH/H$^+$ in step 6 and four molecules of ATP, two in step 7 and two in step 10. Overall, given that two molecules of ATP were used in the investment phase of glycolysis, and two molecules of NADH/H$^+$ and four molecules of ATP were generated in the payoff phase, each round of this metabolic pathways provides a net gain of two NADH/H$^+$ and two ATP.

Energetics of glycolysis

The relative free energies of reaction for each of the steps of the glycolytic pathway are shown in **Figure 6.18**. Upon analyzing the relative values and signs of these reactions, some patterns begin to emerge. Steps where energy is invested, such as where molecules of ATP are expended, have strong negative $\Delta G^{\circ\prime}$ values, indicating spontaneous reactions under standard state conditions. These reactions are driven by the reactions of intermediates that can readily transfer a phosphate group, such as ATP, 1,3-bPG, or PEP. Isomerization reactions typically have $\Delta G^{\circ\prime}$ values close to zero, indicating that these reactions are near equilibrium in the standard state.

In the cell, these reactions (especially the concentrations of the reactants) are *not* found in standard state conditions. Also, changing the concentrations of reactants changes the ΔG of a reaction, making a reaction favorable that would normally be unfavorable. This is vividly illustrated in the fourth step of glycolysis, in which F-1,6-bP is cleaved to DHAP and GAP. Under standard state conditions, $\Delta G^{\circ\prime}$ has a strong positive value (23 kJ/mol), indicating that this reaction proceeds strongly in the reverse direction, favoring F-1,6-bP. However, in a cell undergoing glycolysis, the concentrations of F-1,6-bP, DHAP, and GAP are such that the reaction proceeds in the forward direction ($\Delta G = -0.23$ kJ/mol).

Free energies (ΔG values) can be used to predict *favorability* or *direction* of a reaction, but they have no bearing on the *rate* of a reaction (how quickly it proceeds). Several reactions of glycolysis, most notably the isomerization of DHAP to GAP, are near equilibrium in the cell, but they proceed at rates near the upper limit of enzymatic reactions.

Regulation of glycolysis: an overview

Metabolic pathways can be thought of as being analogous to a factory. If we wanted to control how fast the assembly line is going, we could do this in several ways. We could control how much material comes into the assembly line, or we could control how much leaves at the other end, causing backups. Alternatively, we could try to control the process through the slower or irreversible steps of the assembly. Thus, if one step takes longer than the other steps, it may be an ideal place to control the entire process.

The situation in the factory is similar to that of a metabolic pathway. Three different steps in glycolysis (steps 1, 3, and 10) are committed steps or irreversible steps, both good places to control flux through a pathway (**Figure 6.19**). Glycolysis is regulated to some extent at each

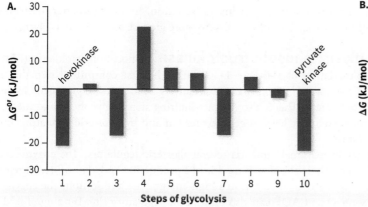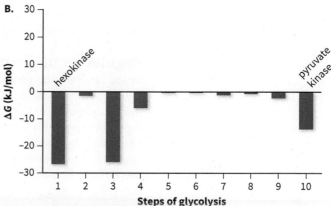

FIGURE 6.18 Energetics of glycolysis. **A.** In the standard state six of the steps of glycolysis are energetically unfavorable and would proceed in the reverse direction. **B.** Due to the concentrations of metabolites found in the cell, these reactions proceed favorably with much lower ΔG values.

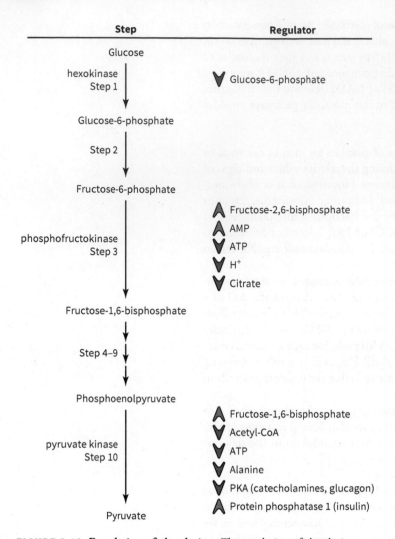

Step	Regulator

Glucose

hexokinase
Step 1   ∨ Glucose-6-phosphate

Glucose-6-phosphate

Step 2

Fructose-6-phosphate

                          ∧ Fructose-2,6-bisphosphate
                          ∧ AMP
phosphofructokinase   ∨ ATP
Step 3   ∨ H⁺
                          ∨ Citrate

Fructose-1,6-bisphosphate

Step 4–9

Phosphoenolpyruvate

                          ∧ Fructose-1,6-bisphosphate
                          ∨ Acetyl-CoA
pyruvate kinase   ∨ ATP
Step 10   ∨ Alanine
                          ∨ PKA (catecholamines, glucagon)
                          ∧ Protein phosphatase 1 (insulin)

Pyruvate

FIGURE 6.19 Regulation of glycolysis. The regulation of glycolysis occurs at several points in the pathway and through a variety of mechanisms, including both post-translational modification and allosteric regulation.

of these points. Multiple control points allow different metabolites to enter the pathway at different places, for example, glucose coming from outside the cell or glucose-1-phosphate derived from glycogen. Hexokinases regulate the phosphorylation of glucose and entry into glycolysis. The major enzymes that are regulated in the glycolytic pathway are phosphofructokinase and pyruvate kinase. They each provide some measure of regulation, depending on the tissue and metabolic state in question, and each can be regulated in several different ways. Enzyme activity can be modulated by post-translational modifications, such as phosphorylations. Shorter-term regulation can be accomplished through allosteric regulators, whereas changes in the expression of a gene coding for an enzyme can account for longer-term changes in the flux or flow of metabolites through a pathway.

Regulation of glycolysis: hexokinases

Trapping of glucose in the cell and the entry into glycolysis is accomplished by one of several enzymes, generically categorized as hexokinases. In mammals, there are four hexokinase isoforms designated as I, II, III, and IV. The latter is also known as glucokinase. All tissues express at least one of these enzymes. The II isoform is increased in cancer cells and may be partially responsible for the Warburg effect, which is an increased level of glycolysis seen in some tumors.

Hexokinases I, II, and III are almost ubiquitously expressed, and all have a high affinity for glucose at sub-physiological concentrations, that is, below 1 mM. The normal range of plasma glucose is 5.0–7.8 mM. These hexokinases are strongly inhibited by the product G-6-P and obey Michaelis-Menten kinetics.

Hexokinase IV (glucokinase) differs from the other isoforms in several ways. First, it is expressed in only a few tissues that regulate carbohydrate metabolism, including the liver, insulin-secreting β cells of the pancreas, and the hypothalamus. Second, it has a far lower affinity for glucose than the other members of the family. Despite being called "glucokinase," hexokinase IV can also phosphorylate other hexoses. Third, it is an allosteric enzyme and therefore does not obey Michaelis-Menten kinetics; however, the active sites in glucokinase are half saturated at approximately 8 mM glucose. Finally, hexokinase IV is not inhibited by the product of the reaction; this enables the pancreatic β cell to use G-6-P as a sensor of plasma glucose, which it uses to trigger insulin secretion. In addition, after meals, the liver uses different levels of hexokinase I and IV to absorb more glucose and store it as glycogen.

Regulation of glycolysis: phosphofructokinase

Phosphofructokinase (PFK) which catalyzes the phosphorylation of F-6-P to F-1,6-bP, is the first committed step that is unique to glycolysis. The G-6-P formed in step 1 can alternatively be directed to glycogen biosynthesis, a stored form of carbohydrate. This is also the rate-limiting step of glycolysis, and the key regulatory step. PFK is regulated at the level of gene expression and by allosteric regulators, but not directly by phosphorylation.

PFK is a homotetramer in mammals and has several allosteric regulators. The enzyme has two binding sites for ATP: a catalytic site that is used in the production of F-1,6-bP and a regulatory site. Binding of ATP signals that the cell is in a high-energy state and inhibits the enzyme. In contrast AMP and to some extent ADP can also bind in the regulatory site. Binding of AMP signals that the cell is in a low-energy state and stimulates the enzyme by stabilizing the active (R) conformation.

PFK is also inhibited by low pH. Anaerobic metabolism produces lactic acid, inhibiting PFK; this inhibition stops glycolysis and prevents further accumulation of damaging acidic by-products.

Glycolysis produces precursors for other metabolic pathways, such as the citric acid cycle. High levels of citrate indicate that these precursors are in abundance. Citrate is another inhibitor of PFK.

The strongest activator of PFK is fructose-2,6-bisphosphate (F-2,6-bP), which is not part of glycolysis but is made by a second enzyme, PFK-2 (**Figure 6.20**). When glucose is abundant, F-6-P levels are high. PFK-2 generates F-2,6-bP from F-6-P; in turn, the F-2,6-bP activates PFK. This is an example of **feed forward activation**, a type of regulation where the products of a reaction act to further activate the pathway. PFK-2 is a bifunctional enzyme composed of both the kinase that generates F-2,6-bP and a second domain containing the phosphatase that breaks it down. The switch between these two activities is largely regulated by phosphorylation, mediated by PKA. However, the regulation of this enzyme differs in different tissues. Hormones such as glucagon or epinephrine act to increase levels of cAMP in the liver cell, activating PKA, which in turn phosphorylates PFK-2. Phosphorylation of PFK-2 favors the phosphatase activity of this enzyme, breaking down F-2,6-bP, inhibiting PFK and slowing glycolysis. AMP-dependent protein kinase (AMPK) can phosphorylate a different site in the cardiac and inducible isoforms of PFK-2, increasing kinase activity and leading to increased levels of F-2,6-bP and therefore increased glycolysis. All other isoforms of PFK-2, such as the one found in skeletal muscle, lack the AMPK phosphorylation site and are not thought to be regulated by AMPK.

The Tp53 induced glycolysis and apoptosis regulator (TIGAR) is a multifunctional protein that inhibits glycolysis by acting as a fructose bisphosphatase, reducing cellular levels of both F-2,6-bP and F-1,6-bP. Originally identified as a protein that prevented the development of cancer, TIGAR blunts glycolysis and shuttles monosaccharides toward the pentose phosphate pathway. This lowers reactive oxygen species (ROS) in the cell (oxygen-containing molecules such as hydrogen peroxide that have been implicated in DNA damage). Expression of TIGAR also appears to regulate DNA damage-induced *apoptosis* (programmed cell death), although the mechanism by which TIGAR does this is still unknown. TIGAR is downregulated in some cancers, which may help explain some of the etiology of the Warburg effect.

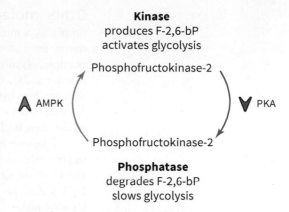

Kinase
produces F-2,6-bP
activates glycolysis

Phosphofructokinase-2

AMPK   PKA

Phosphofructokinase-2

Phosphatase
degrades F-2,6-bP
slows glycolysis

FIGURE 6.20 Regulation of phosphofructokinase-2 (PFK-2). PFK-2 has both kinase and phosphatase activities. The activity of PFK-2 is regulated by PKA and AMPK.

Regulation of glycolysis: pyruvate kinase

Pyruvate kinase (PK) catalyzes the transformation of PEP into pyruvate in the final step of glycolysis. Isoforms of the enzyme are found in different tissues, but this section focuses on the differences between the liver-specific isoform and those found in other tissues: predominantly skeletal muscle, the brain, and erythrocytes. All of the isoforms are regulated at the level of gene expression, and allosterically by small molecules. The liver isoform is also regulated by phosphorylation.

The liver form of PK is activated by insulin and inhibited by glucagon and epinephrine. All of these hormones act at least partially via a protein kinase A (PKA)-mediated mechanism. For example, glucagon and epinephrine inhibit PK via protein kinase A-mediated phosphorylation. Insulin, in contrast, acts by stimulating phosphodiesterase to decrease levels of cAMP and blunt the action of protein kinase A. Insulin also stimulates phosphoprotein phosphatase I, promoting dephosphorylation of PK, which is also allosterically stimulated by F-1,6-bP. Binding of F-1,6-bP increases the affinity of the PK for its substrate, PEP. PK is also inhibited by ATP, acetyl-CoA, and alanine. When glucose is abundant, levels of F-1,6-bP are high and PK is stimulated; when glucose is scarce, the reaction is inhibited. These molecules increase the rate of conversion of PEP to pyruvate when glycogenic metabolites are high. When there is abundant energy or when there are high levels of alanine, the rate is decreased.

The regulation of the other isoforms of PK is less complicated; they are simply stimulated by F-1,6-bP. This gives the organism a means of differentiating liver metabolism from the rest of the organ systems. When glucose is low and glucagon levels are high, PK and therefore glycolysis will be inhibited in the liver, but not in the other tissues of the body including the brain, erythrocytes, and muscle tissue.

Other molecules also use the glycolytic pathway

Glycolysis is often seen simply as a means of breaking down glucose and generating ATP or pyruvate for subsequent energy production. Although these are important functions of the pathway, there are others. For example, glycolysis provides a pathway for other carbohydrates and glycerol to enter metabolism (**Figure 6.21**). Glucose, like all monosaccharides, is relatively reactive, forming acetal linkages with random hydroxyl groups. In addition, changes in the concentration of glucose can influence the osmolarity of plasma or cytosol. Glycolysis provides a way to regulate and dispose of glucose, minimizing both of these potential problems.

Galactose is phosphorylated by galactokinase to form galactose-1-phosphate, which then reacts with uridinediphosphate glucose (UDP-glucose) to yield glucose-1-phosphate (G-1-P) and UDP-galactose. The G-1-P released here is used in the final step of the pathway. The enzyme UDP-galactose epimerase catalyzes the isomerization (epimerization) of UDP-galactose to UDP-glucose. This inversion of the stereochemistry at C-4 occurs by first oxidizing the carbon

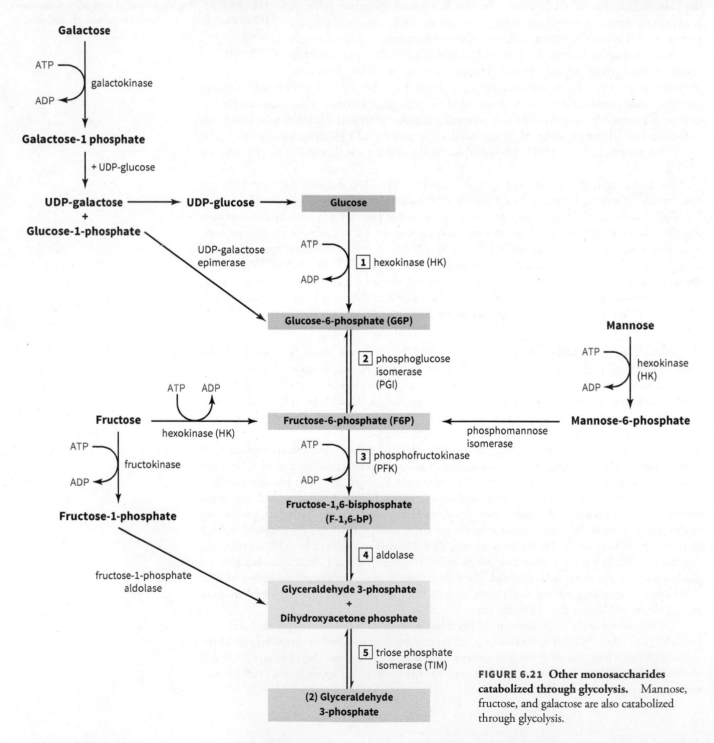

FIGURE 6.21 Other monosaccharides catabolized through glycolysis. Mannose, fructose, and galactose are also catabolized through glycolysis.

to a planar carbonyl group, followed by reduction of the carbonyl to the new stereochemistry. Finally, phosphoglucomutase isomerizes the G-1-P to G-6-P, where it can enter glycolysis.

Mannose is phosphorylated by hexokinase to form mannose-6-phosphate (M-6-P), which is then acted on by phosphomannose isomerase to form F-6-P. From there, it proceeds through glycolysis.

Fructose is metabolized via one of two different pathways, depending on the tissue in which it occurs. In muscle, fructose is phosphorylated by hexokinase to F-6-P, which can then directly enter glycolysis. In liver, the scenario is somewhat different. Liver has higher levels of glucokinase than hexokinase, and the affinity of glucokinase for fructose is low. Here, fructose is phosphorylated by fructokinase to form F-1-P. However, F-1-P cannot be phosphorylated further and enter glycolysis; rather, it is acted on by an aldolase to yield DHAP and glyceraldehyde. As discussed above, the DHAP is isomerized to GAP by triose phosphate isomerase and enters glycolysis, but the glyceraldehyde generated has several paths it can follow. Glyceraldehyde kinase can phosphorylate the glyceraldehyde to GAP, which can enter glycolysis. Glyceraldehyde can also reenter glycolysis as DHAP. Glyceraldehyde is first reduced to glycerol by alcohol dehydrogenase. The glycerol can be phosphorylated on C-3 by glycerol kinase to form glycerol-3-phosphate, which can be oxidized by glycerol phosphate dehydrogenase to form DHAP. DHAP in turn is isomerized by triose phosphate isomerase to GAP and enters glycolysis. The catabolism of fructose to GAP and entry into glycolysis has recently been referred to as fructolysis by some researchers.

Worked Problem 6.4 — Glycolysis and galactose?

Imagine that you are studying a novel strain of bacteria that can grow on galactose. You assume that the metabolism of galactose in this strain occurs through glycolysis. Using radiolabeled carbohydrates, design an experiment to show how galactose is being metabolized.

Strategy We need a means of following galactose through metabolism. Using a radiolabeled galactose in which one carbon has been labeled with C^{14} should allow us to trace this molecule as it is metabolized.

Solution Galactose can enter glycolysis through several different pathways. In mammals, galactose is phosphorylated to galactose-1-phosphate. A uridine monophosphate group is transferred from UDP-glucose to galactose, resulting in UDP-galactose and G-1-P. UDP-galactose is then epimerized to UDP-glucose, which liberates G-1-P as this cyclic pathway continues. If galactose were using a similar pathway to enter glycolysis, we would expect to see the radiolabel accumulate in those pools.

Follow-up question If galactose (rather than glucose) is entering glycolysis, what would be the product?

Summary

- Metabolic pathways consist of enzymatically catalyzed steps through which molecules are transformed.
- Glycolysis is the pathway through which glucose is converted into two molecules of the three-carbon α-keto acid pyruvate.
- Steps 1 and 3 of glycolysis are energetically possible due to the ability of ATP to donate phosphate groups to glycolytic intermediates.
- F-1,6-bP is cleaved into two 3-carbon intermediates (DHAP and GAP).
- DHAP is isomerized to GAP.
- GAP is changed into pyruvate in the final four steps of the pathway, producing a molecule of ATP in the process.
- Regulation of glycolysis occurs through both gene expression and allosteric regulation.

Concept Check

1. Describe in general what it means for something to be called a metabolic pathway.
2. Describe the reactions of glycolysis at the chemical level. Show how the structures and names of the glycolytic intermediates relate to one another.
3. Explain which steps of glycolysis require energy and which yield energy. Explain the energetics of glycolysis.

6.4 Gluconeogenesis

The cells and organ systems of the body differ in terms of metabolism. For example, some run almost exclusively on glucose as fuel. These include erythrocytes, the brain, and the central nervous system. During times of fasting or starvation, plasma glucose and glycogen stores may not be able to meet the demands of these organs. To deal with such situations, organisms have evolved the ability to produce glucose from other metabolites—a process known as **gluconeogenesis**.

Not all molecules can be made into glucose. Pyruvate, lactate, glycerol, and some amino acids all contribute to gluconeogenesis (**Figure 6.22**). However, acetyl-CoA cannot undergo this process. Once pyruvate has been decarboxylated to acetyl-CoA, its fate is sealed. It can enter the citric acid cycle, undergo fatty acid biosynthesis, or contribute to ketone body formation, but it cannot form carbohydrates.

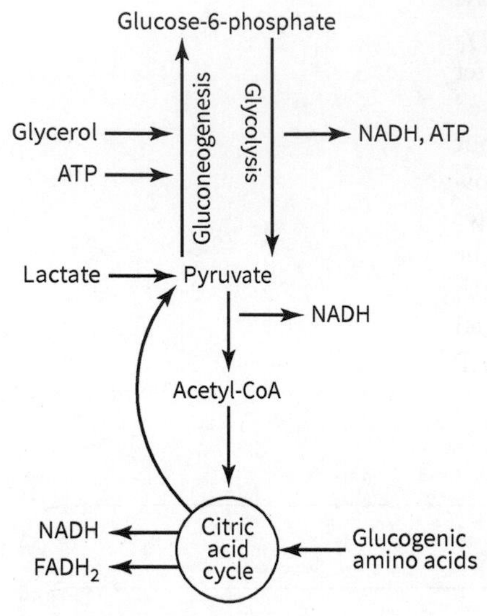

6.4.1 Gluconeogenesis differs from glycolysis at four reactions

We might expect gluconeogenesis to simply be the glycolytic pathway running in reverse. This is partially the case, but several reactions of glycolysis are energetically unfavorable when running in the reverse direction (**Figure 6.23**). These steps are the formation of glucose-6-phosphate (G-6-P), the formation of fructose-1,6-bisphosphate (F-1,6-bP), and the conversion of phosphoenolpyruvate (PEP) to pyruvate. These are also the most highly regulated steps of glycolysis. Different enzymes catalyze these reactions and couple them to other reactions, overcoming the energetic barriers. Likewise, different pathways for glycolysis and gluconeogenesis are controlled independently of one another: when one is turned on, the other can be turned off. The other reactions of glycolysis have ΔG values close to zero, and thus can be more readily tipped in the forward or reverse direction by modulating the concentrations of substrates or products.

Gluconeogenesis is a biosynthetic pathway, but it is often thought of as occurring during times of fasting, starvation, or in different disease states. As a synthetic pathway, gluconeogenesis requires energy and is in fact rather costly. To synthesize a single molecule of glucose from pyruvate requires six energetic equivalents of ATP (four ATP and two GTP).

Gluconeogenesis occurs predominantly in the liver but also in the cortex of the kidney and small intestine. The glucogenic fuel sources for these tissues vary. Lactate and alanine are used by the liver while lactate and glutamine are used by the kidneys, but the function of providing glucose for tissues in times of need is the same.

Conversion of pyruvate to phosphoenolpyruvate   The first step of gluconeogenesis involves the synthesis of PEP from pyruvate. This two-step process involves two different enzymes. First, pyruvate is converted to oxaloacetate by pyruvate carboxylase; this process occurs in the mitochondria and requires one molecule of ATP. The next step, generation of PEP from oxaloacetate, requires the enzyme phosphoenolpyruvate carboxykinase (PEPCK). A molecule of GTP is expended in this process, and the CO_2 fixed in the first step is lost. The cellular localization of PEPCK varies depending on the species. In humans, the enzyme is found in both mitochondria and cytosol. Specific transporters shuttle PEP out of the mitochondria, but no such transporters exist for oxaloacetate. In species that lack mitochondrial forms of PEPCK, oxaloacetate must first be transaminated into aspartate or reduced into malate, both of which can be transported to cytosol for PEP biosynthesis.

Following the synthesis of PEP, steps 4–9 of glycolysis are reversible and will proceed in the reverse direction if the concentration of glycolytic products exceeds substrates. All these steps (as well as the two remaining steps of gluconeogenesis) occur in the cytosol.

Liberation of glucose   The conversion of F-1,6-bP to F-6-P is carried out by fructose-1,6-bisphosphatase. Like glucose-1-phosphate, glucose-6-phosphate is trapped and unable to leave the cell. The conversion of G-6-P to glucose is catalyzed by glucose-6-phosphatase, an enzyme that is only found in tissues that export glucose for use elsewhere. Liver expresses this enzyme; as a result, it releases the glucose synthesized in the liver for use by other tissues (particularly the

FIGURE 6.22 Pathways that flow through gluconeogenesis. Glucogenic precursors include lactate, pyruvate, glycerol, and glucogenic amino acids.

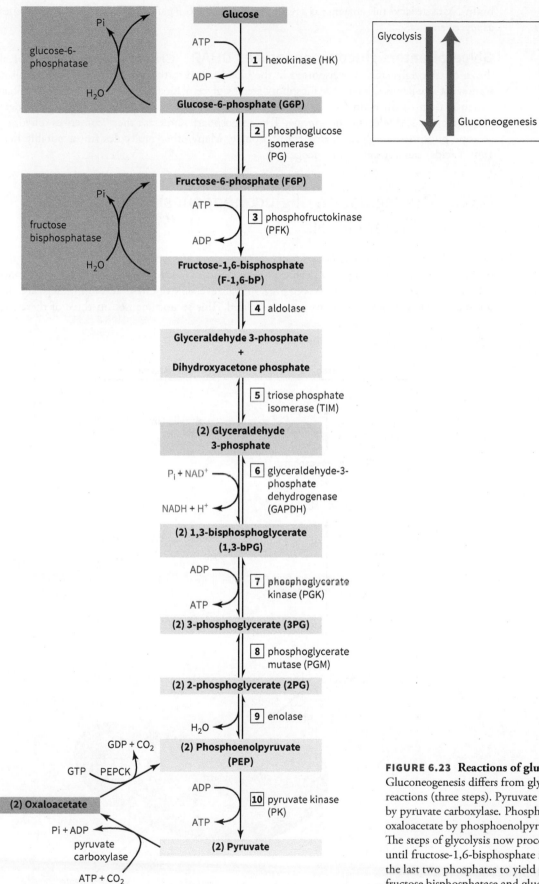

FIGURE 6.23 Reactions of gluconeogenesis.
Gluconeogenesis differs from glycolysis by four different reactions (three steps). Pyruvate is converted to oxaloacetate by pyruvate carboxylase. Phosphoenolpyruvate is formed from oxaloacetate by phosphoenolpyruvate carboxykinase (PEPCK). The steps of glycolysis now proceed in the reverse direction until fructose-1,6-bisphosphate is synthesized. Removal of the last two phosphates to yield glucose is accomplished by fructose bisphosphatase and glucose-6-phosphatase.

brain). Muscle lacks this enzyme; as a result, it fixes any G-6-P generated through gluconeogenesis as glycogen.

Glycerol enters gluconeogenesis as DHAP   Glycerol liberated from the breakdown of triacylglycerols is transported in the bloodstream to the liver, where it is metabolized, somewhat like hexoses. First, it is phosphorylated by glycerol kinase to yield glycerol-3-phosphate, which is then oxidized on C-2 to dihydroxyacetone phosphate by glycerol-3-phosphate dehydrogenase, an FAD-dependent enzyme. Triose phosphate isomerase then isomerizes dihydroxyacetone phosphate to glyceraldehyde-3-phosphate. Many other molecules (most notably many amino acids) can also enter gluconeogenesis.

6.4.2 The regulation of gluconeogenesis takes place at several different levels

The regulation of gluconeogenesis occurs through compartmentalization, allosteric regulation, phosphorylation, and changes in gene expression (**Figure 6.24**). The steps specific to gluconeogenesis are all points of control. As seen previously, irreversible or rate-determining steps are often the places where control over a pathway is exerted. This section focuses on three of these steps

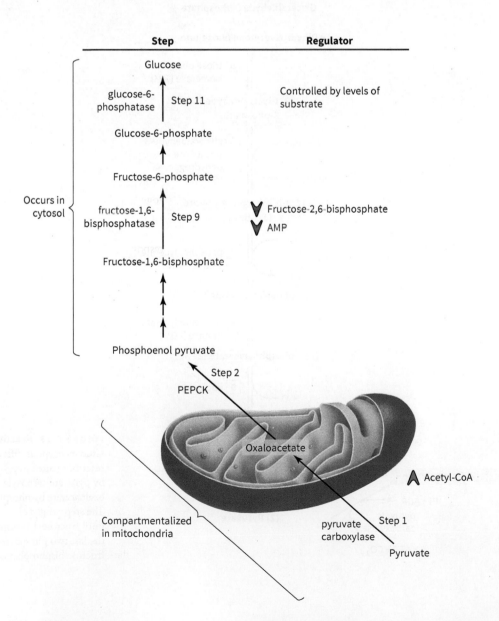

FIGURE 6.24 Regulation of gluconeogenesis. The regulation of gluconeogenesis is in many ways the reciprocal of glycolysis. In addition, the first two reactions of gluconeogenesis are compartmentalized to the mitochondria.

(pyruvate carboxylase, phosphoenolpyruvate carboxykinase [PEPCK], and fructose-1,6-bisphosphatase). However, control of gluconeogenesis and glycolysis is reciprocal. Thus, when the energetic level of the cell is low, glycolysis proceeds and gluconeogenesis is inhibited, but when plasma glucose is low, the reverse occurs—that is, gluconeogenesis proceeds and glycolysis is inhibited.

The first level of control is compartmentalization. Pyruvate carboxylase and PEPCK are found in the mitochondria. These reactions are kept separate from the reactions occurring in cytosol by the mitochondrial membranes.

Allosteric control is employed in the regulation of both pyruvate carboxykinase, which is stimulated by high levels of acetyl-CoA and fructose-1,6-bisphosphatase, which is inhibited by elevated levels of AMP and F-2,6-bP. High AMP levels indicate to the cell that energy levels are low and that energy should be replenished if possible. F-2,6-bP is a positive regulator of glycolysis, and it reciprocally controls gluconeogenesis.

6.4.3 Coordinated regulation of gluconeogenesis and glycolysis through hormones

Gluconeogenesis and glycolysis are tightly regulated and the coordination between two pathways is achieved by employing most of the enzymes that play a major role in glycolysis. The enzymes involved in seven reversible reactions also take part in the gluconeogenesis pathway. This pathway mainly occurs in the liver of mammals as an alternative source of glucose during the depletion of glycogen stores. Changes in the availability of substrates are responsible for most changes in metabolism either directly or indirectly acting *via* changes in hormone secretion. Three mechanisms are responsible for regulating the activity of enzymes in carbohydrate metabolism: (1) changes in the rate of enzyme synthesis, (2) covalent modification by reversible phosphorylation, and (3) allosteric effects.

Fructose 2,6-bisphosphate: A potential key player   Fructose 2,6-bisphosphate plays a unique role in the regulation of gluconeogenesis and glycolysis in the liver. Gluconeogenesis is stimulated by a decrease in the concentration of fructose 2,6-bisphosphate, which serves as a potent positive allosteric effector and, deactivates phosphofructokinase-1 and inhibits fructose-1,6-bisphosphatase. Inhibition of fructose-1,6-bisphosphatase is achieved by increasing K_M for fructose-1,6-bisphosphate. Its concentration is modulated through covalent modification (hormonal) and allosteric (substrate). The allosteric control of fructose 6-phosphate regulates the bifunctional enzyme by stimulating the kinase activity and inhibiting phosphatase. The bifunctional enzyme, phosphofructokinase-2 phosphorylates fructose 6-phosphate to form fructose 2,6-bisphosphate and the breakdown is possible due to the enzyme's fructose-2,6-bisphosphatase activity. Therefore, when the glucose level is high, the increased concentration of fructose-2,6-bisphosphate stimulates glycolysis by inhibiting fructose-1,6-bisphosphatase and activating phosphofructokinase-1 activity. Contrary to this, when the glucose level is low, the hormone, glucagon activates the cAMP-dependent protein kinase enzyme by stimulating the production of cAMP, which in turn activates fructose-2,6-bisphosphatase by phosphorylation and inactivates phosphofructokinase-2. Thus, the regulatory mechanism ensures the glucose release is achieved through glycogenolysis in the liver by glucagon rather than glycolysis.

Hormonal control of glucose homeostasis   The maintenance of glucose levels in the blood is one of the most finely regulated homeostatic mechanisms that involves hormones, the liver, and extrahepatic tissues. Blood glucose concentration is regulated by negative feedback pathways that are modulated by two separate hormones: insulin and glucagon. Both these hormones are produced in special cells called islet cells, or islets of Langerhans found in clusters throughout the pancreas. Islet cells make up a very small percentage of the pancreas (≤ 1–2%); the remainder of the organ is an exocrine gland producing digestive enzymes and bicarbonate ions. This tiny number of endocrine cells is exceedingly important. Each islet contains two kinds of cells: alpha cells, which produce glucagon, and beta cells, which produce insulin.

Insulin and glucagon function in an antagonistic manner. The result is a precise control of blood glucose levels. The insulin pathway is activated when blood glucose levels are too high. High blood glucose levels, as a result of a carbohydrate-rich diet or through gluconeogenesis, stimulate beta cells in the pancreas to release insulin. Insulin causes an increased uptake of glucose from the blood, promotes the conversion of glucose into triglycerides in the liver, fat, and muscle cells. It also increases the cellular rate of glycolysis by metabolizing glucose into smaller components that can be used for the synthesis of other compounds.

Contrary to the insulin pathway, the glucagon pathway is activated when blood glucose levels are too low. Low blood glucose levels, due to strenuous exercise or starvation, stimulate the alpha cells in the pancreas to produce glucagon. Glucagon stimulates the liver to convert stored glycogen into glucose followed by its release into the blood, a process called glycogenolysis. These two hormones, counter-regulate their levels to maintain glucose homeostasis. A decrease in insulin or glucose level stimulates the secretion of glucagon, while an increase in insulin or glucose level suppresses glucagon secretion. This results in a continuous cycle, with insulin and glucagon constantly monitoring blood glucose levels and regulating their secretion to maintain the level as nearly constant as possible. The main function of insulin is the removal of excess blood glucose through glycolysis. Since the function of glucagon is opposite to that of insulin, it stimulates the addition of glucose to the bloodstream. Thus, it targets cells with high concentrations of energy stored as glycogen, including the liver and skeletal muscles. It also stimulates gluconeogenesis from fats, thereby adipose tissue cells being another target of glucagon. The enzymes involved in the utilization of glucose (i.e., those of glycolysis and lipogenesis) become more active when there is a superfluity of glucose, and under these conditions, the enzymes responsible for gluconeogenesis have low activity. The secretion of insulin, in response to increased blood glucose, enhances the synthesis of the key enzymes in glycolysis. Likewise, it antagonizes the effect of the glucocorticoids and glucagon-stimulated cAMP, which induce synthesis of the key enzymes responsible for gluconeogenesis. The effects of insulin and glucagon on glucose homeostasis are summarized in **Tables 6.3** and **6.4**.

TABLE 6.3 Effects of Insulin on Glucose Homeostasis

	Metabolic Effect	Target Tissue		Target Enzyme
Increased	glucose uptake	Muscle	Increased	glucose transport
	glucose uptake	Liver		glucokinase activity
	glycogen synthesis	Muscle and liver		glycogen synthase activity
	glycolysis, acetyl -CoA production	Muscle and liver		phosphofructokinase-1 and pyruvate dehydrogenase complex
	fatty acid synthesis	Liver		acetyl-CoA carboxylase activity
	triacylglycerol synthesis	Adipose		lipoprotein lipase activity
Decreased	glycogen breakdown	Muscle and liver	Decreased	glycogen phosphorylase activity

TABLE 6.4 Effects of Glucagon on Glucose Homeostasis

Metabolic Effect	Target Tissue	Target Enzyme	Impact on Metabolism
Increased glycogen breakdown	Liver	Increased glycogen phosphorylase activity	Conversion of glycogen to glucose
Increased gluconeogenesis	Liver	Decreased fructose 1,6-bisphosphatase	Conversion of non-carbohydrate molecules (amino acids, glycerol, and oxaloacetate) to glucose
Increased mobilization of fatty acids	Adipose	Increased triacylglycerol lipase	Decreased glucose usage by liver and muscle
Decreased glycogen synthesis	Liver	Decreased glycogen synthase	Decreased storage of glucose
Decreased glucose breakdown	Liver	Decreased phosphofructokinase-1	Decreased usage of glucose

Hormones, glucagon, and to a lesser extent epinephrine are responsive to cause a decrease in blood glucose by inhibiting glycolysis and stimulating gluconeogenesis in the liver. This effect is achieved through an increase in the concentration of cAMP. This, in turn, activates cAMP-dependent protein kinase, leading to the phosphorylation and inactivation of pyruvate kinase. These hormones also affect the concentration of fructose 2,6-bisphosphate and therefore glycolysis and gluconeogenesis are maintained at equilibrium.

Insulin plays a central role

In addition to the direct effects of hyperglycemia in enhancing the uptake of glucose into the liver, the hormone insulin plays a central role in regulating blood glucose. It produced by the beta cells of the pancreas in response to hyperglycemia. The beta cells are freely permeable to glucose *via* glucose transporter 2 (GLUT2) for undergoing glycolysis through phosphorylation by glucokinase. Therefore, increasing blood glucose increases metabolic flux through glycolysis, the citric acid cycle, and the generation of ATP. The increase in [ATP] inhibits ATP-sensitive K^+ channels, causing depolarization of the beta cell membrane, which increases Ca^{2+} influx *via* voltage-sensitive Ca^{2+} channels, stimulating exocytosis of insulin. Thus, the concentration of insulin in the blood parallels that of blood glucose. Other substances causing the release of insulin from the pancreas include amino acids, free fatty acids, ketone bodies, glucagon, and secretin. Epinephrine and norepinephrine block the release of insulin. Insulin lowers blood glucose immediately by enhancing glucose transport into adipose tissue and muscle by recruitment of glucose transporter 4 (GLUT 4) from the interior of the cell to the plasma membrane. Although it does not affect glucose uptake into the liver directly, insulin does enhance long-term uptake as a result of its actions on enzymes controlling glycolysis, glycogenesis, and gluconeogenesis.

Glucagon counters the actions of insulin

Glucagon also enhances gluconeogenesis from amino acids and lactate. In all these actions, glucagon acts *via* the generation of cAMP. Both hepatic glycogenolysis and gluconeogenesis contribute to the hyperglycemic effect of glucagon, whose actions oppose those of insulin. Most of the endogenous glucagon and insulin are cleared from the circulation by the liver.

Role of other hormones in glucose homeostasis

The anterior pituitary gland secretes hormones that tend to elevate the blood glucose and therefore antagonize the action of insulin. These are growth hormone, ACTH (corticotrophin), and possibly other "diabetogenic" hormones. Growth hormone secretion stimulated by hypoglycemia causes decreased glucose uptake in muscles. Some of this effect may not be direct, since it stimulates mobilization of free fatty acids from adipose tissue which inhibits glucose utilization. The glucocorticoids (11-oxysteroids) secreted by the adrenal cortex increase gluconeogenesis. This is a result of enhanced hepatic uptake of amino acids and increased activity of aminotransferases and key enzymes of gluconeogenesis. In addition, glucocorticoids inhibit the utilization of glucose in extrahepatic tissues. In all these actions, glucocorticoids act in a manner antagonistic to insulin. Epinephrine is secreted by the adrenal medulla as a result of stressful stimuli such as fear, excitement, hemorrhage, hypoxia, hypoglycemia, etc. Epinephrine promotes glycogenolysis in the liver and muscle owing to the stimulation of phosphorylase *via* the generation of cAMP. In muscle, glycogenolysis results in increased glycolysis, whereas in the liver, glucose is the main product leading to an increase in blood glucose.

Worked Problem 6.5   Potential side effects

Metformin is a drug that is used to treat type 2 diabetes, and it works in part by inhibiting gluconeogenesis. A potential danger of taking metformin is lactic acidosis, the accumulation of excess lactic acid in the body. Why could taking metformin lead to lactic acidosis?

Strategy   Review the steps of gluconeogenesis. How is lactate linked to gluconeogenesis?

Solution   Lactate can act as a substrate for gluconeogenesis. If gluconeogenesis is inhibited or blocked, lactate levels in blood can increase, leading to lactic acidosis.

Follow-up question   If you were going to try to design a drug to inhibit gluconeogenesis, which step would you target and why?

Summary

- Gluconeogenesis is the synthesis of new glucose molecules. Only some metabolic precursors can be made into glucose.
- Gluconeogenesis differs from glycolysis at three steps. This means that both pathways can be independently controlled and are thermodynamically favorable.
- The regulation of gluconeogenesis occurs through compartmentalization, allosteric regulation, phosphorylation, and changes in gene expression.
- Gluconeogenesis and glycolysis are tightly regulated by enzymes of glycolysis and specific hormones.
- The concentration of fructose 2,6-bisphosphate varies with the level of glucose.
- Insulin and glucagon are the two major hormones that maintain glucose homeostasis and are counteractive.

Concept Check

1. Describe how glucose can be synthesized from different metabolic precursors.
2. Give an account of enzymes that are involved in the conversion of pyruvate to glucose.
3. Explain how the reactions of gluconeogenesis allow the cell to synthesize glucose under energetically unfavorable conditions and how compartmentalization facilitates this process.
4. Explain how gluconeogenesis is regulated, outlining which steps are regulated, and the molecules that regulate those steps.
5. Describe the actions of insulin and glucagon in regulating gluconeogenesis and glycolysis.

6.5 The Fates of Pyruvate

The pyruvate created at the end of glycolysis can have several different metabolic fates, depending on the tissue of origin and metabolic state of the organism. This section details five different fates of pyruvate, four of which occur in animals (**Figure 6.25**, **Table 6.5**).

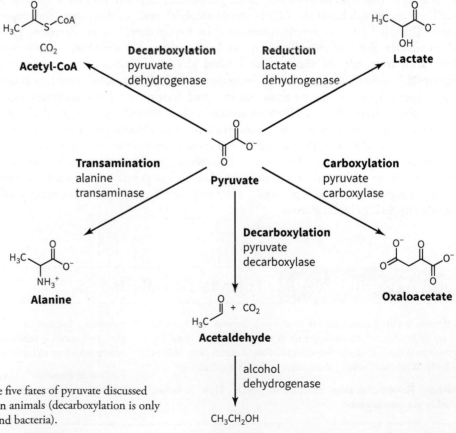

FIGURE 6.25 Fates of pyruvate. There are five fates of pyruvate discussed in this section, four of which are carried out in animals (decarboxylation is only carried out by microbes, such as some yeast and bacteria).

TABLE 6.5 Enzymes That Use Pyruvate as a Substrate

Enzyme	Pathway	Product
Alanine transaminase	Transamination	Alanine
Pyruvate dehydrogenase	Entry to citric acid cycle	Acetyl-CoA + CO_2
Pyruvate decarboxylase	Anaerobic glycolysis (microbes)	Acetaldehyde
Pyruvate carboxylase	Gluconeogenesis	Oxaloacetate
Lactate dehydrogenase	Anaerobic glycolysis (animals)	Lactate

6.5.1 Pyruvate can be decarboxylated to acetyl-CoA by pyruvate dehydrogenase

Under aerobic conditions, most pyruvate is converted to acetyl-CoA. This reaction yields one additional molecule of $NADH/H^+$ and one molecule of acetyl-CoA, which can be used for either fatty acid biosynthesis or the citric acid cycle. Pyruvate dehydrogenase, the enzyme catalyzing this reaction, is the committed step between carbohydrate metabolism and lipid metabolism. Although this is a decarboxylation reaction, and CO_2 is released from pyruvate, it is catalyzed by pyruvate dehydrogenase rather than by pyruvate decarboxylase (which produces acetaldehyde in facultative anaerobes). Once decarboxylated to acetyl-CoA, the carbons cannot be reconstituted to carbohydrates.

Pyruvate dehydrogenase is a large multienzyme complex that is found in the mitochondrial matrix (**Figure 6.26**). The functional unit is comprised of three enzymes: a pyruvate dehydrogenase (E1), a dihydrolipoyl transacetylase (E2), and a dihydrolipoyl dehydrogenase (E3).

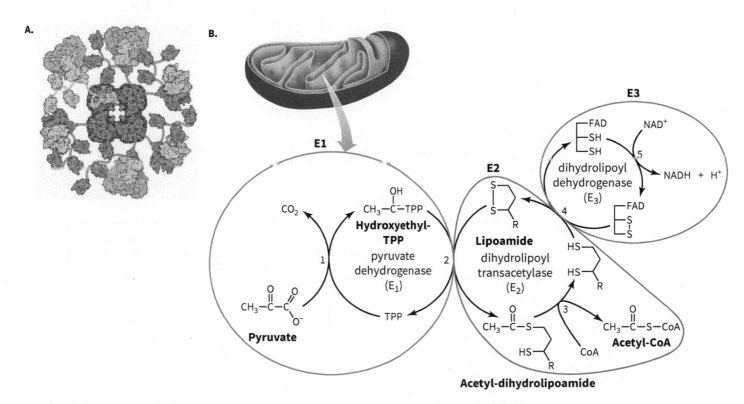

FIGURE 6.26 Topology of pyruvate dehydrogenase.   A. Pyruvate dehydrogenase is a large multienzyme complex with up to 60 subunits in some species. **B.** The basic reactions of pyruvate dehydrogenase are catalyzed by three enzymes: pyruvate dehydrogenase (E1), dihydrolipoyl transacetylase (E2), and dihydrolipoyl dehydrogenase (E3). The reactions that happen in each enzyme of the complex and their relationship to each other are shown.

(Source: Image from the RCSB PDB September 2012 Molecule of the Month feature by David Goodsell (doi: 10.2210/rcsb_pdb/mom_2012_9))

Sixty copies of E2 are found in the core of the complex, surrounded by a coating of 30 copies of E1, and 12 copies each of E3 and an E3 binding protein that links E2 and E3. In addition, E1 is a tetramer with $\alpha_2\beta_2$ stoichiometry, and E3 is a homodimer. Clearly, the topology of the collective assembly is complex.

The mechanism by which acetyl-CoA is generated from pyruvate is equally elegant and complex, employing both thiamine pyrophosphate (TPP) and lipoamide prosthetic groups (**Figure 6.27**). These cofactors and mechanisms are conserved throughout biochemistry and will be seen in other reactions. First, E1 removes a proton from TPP, generating a carbanion. This structure containing both positive and negative charges in the same molecule is known as a ylid. The carbanion then attacks the carbonyl carbon of pyruvate, resulting in protonation of the carbonyl oxygen. Electrons then flow out of the carboxyl moiety of pyruvate into the positively charged thiazole ring of TPP, resulting in decarboxylation of pyruvate. Further assisting this step is the resonance stabilization of the intermediate that exists following departure of the CO_2. The remaining molecule is hydroxyethyl-TPP.

The hydroxyethyl carbanion on TPP next reacts with the disulfide bond found in the lipoamide group of E2, resulting in a cleavage of the disulfide. One sulfur links to the carbanion; the other abstracts a proton to form a thiol. The former carbanion carbon is then oxidized back to

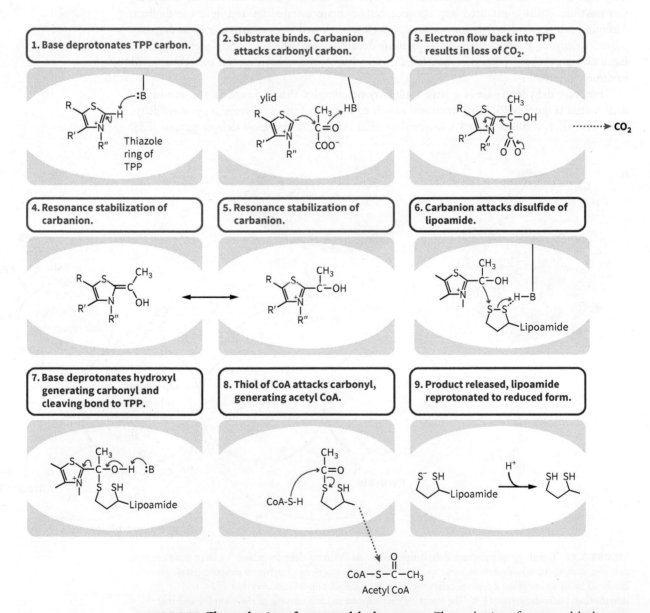

FIGURE 6.27 The mechanism of pyruvate dehydrogenase. The mechanism of pyruvate dehydrogenase proceeds through three different phases, each catalyzed by a different subunit. The two phases which generate acetyl-CoA are shown.

a carbonyl, forming a thioester and cleaving the bond to TPP in the process. The new thioester is transferred from lipoamide to coenzyme A, forming acetyl-CoA and leaving lipoamide as a dithiol. Lipoamide is on a long flexible arm, which permits it to move between distant sites in an enzyme complex. This theme occurs in other areas of biochemistry, most notably with the biotin coenzyme in fatty acid biosynthesis.

The lipoamide arm moves to the E3 subunit where the dithiol form of lipoamide is reoxidized to the disulfide, regenerating the active site of the enzyme. The two electrons and two protons are used to reduce a disulfide bond in E3. Finally, the disulfide in E3 is oxidized as the electrons and protons are transferred to FAD to form $FADH_2$. $FADH_2$ is oxidized by transfer of the electrons to NAD^+, yielding $NADH/H^+$.

Pyruvate dehydrogenase is directly regulated through product inhibition and phosphorylation (**Figure 6.28**). In terms of product inhibition, both NADH and acetyl-CoA inhibit the enzyme. In terms of phosphorylation, it is inactivated by pyruvate dehydrogenase kinase, a process that is reversed by a specific phosphatase (pyruvate dehydrogenase phosphatase). The phosphatase itself is highly regulated, being stimulated by insulin, phosphoenolpyruvate, and AMP, but inhibited by ATP, NADH, and acetyl-CoA. Muscle-specific isoforms of the enzyme are also stimulated by Ca^{2+}. These regulators make biochemical sense, in that molecules that indicate a low energetic state push the system toward increased formation of acetyl-CoA, whereas those that are plentiful in energetic abundance do the opposite. In active muscle tissue, Ca^{2+} ions are released and thus increase the rate of conversion of pyruvate into acetyl-CoA. But why does insulin, a hormone associated with the fed state, stimulate the conversion of pyruvate to acetyl-CoA? In this instance, the acetyl-CoA is destined not for the citric acid cycle but for fatty acid biosynthesis.

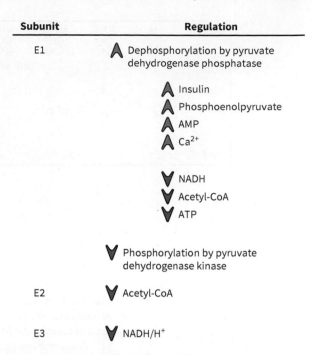

FIGURE 6.28 Regulation of pyruvate dehydrogenase. E1, E2, and E3 indicate the subunits of the pyruvate dehydrogenase complex. The numbers on arrows represent the individual reactions. Pyruvate dehydrogenase is regulated through substrate availability and product inhibition and through phosphorylation of the E1 subunit. Pyruvate dehydrogenase kinase and phosphatase regulate the activity of the E1 subunit and are regulated by many small molecules including ATP, pyruvate, and acetyl-CoA.

6.5.2 Pyruvate can be converted to lactate by lactate dehydrogenase

In vigorously active muscle, there is often a greater need for energy (ATP) than can be generated by oxidative metabolism (ATP synthesis in the presence of oxygen). In this state of anaerobic metabolism, muscle cannot make enough ATP, so muscle pyruvate is generated as glycolysis proceeds to generate ATP. Unfortunately, this is a limited process. The muscle cell (myocyte) will quickly deplete its supply of NAD^+; once this has happened, glycolysis will grind to a halt. To keep glycolysis moving forward and generating ATP, a means is needed to reset these electron carriers. One way in which animal metabolism resets these carriers is through the conversion of pyruvate into lactate. This process temporarily allows glycolysis to proceed, but only in the short term. Imagine someone who has sprinted for as long as she or he can and reached the point of collapse, or someone who has been rock climbing and has completely spent his or her arm strength hanging onto a cliff face; in both cases, we can see both the need for limited anaerobic metabolism and the aftereffects of prolonged muscle activity in the absence of oxygen.

Lactate is produced in tissues by the action of lactate dehydrogenase (LDH), which exists in five different isoforms (**Figure 6.29**). Interestingly, there are only two different genes for the enzyme: LDHA, which encodes the muscle-specific isoform, and LDHB, which encodes the heart-specific isoform. The heart isoform is referred to as "H," and the muscle and liver isoform as "M." The holoenzyme consists of a combination of four subunits which can be all M, all H, or a mix of the two isoforms. The products of LDHA and LDHB are found in different ratios (M4, M3H, M2H2, MH3, and H4) in different tissues, with M4 being found in muscle and the liver, and H4 in the heart. The M4 isoform has a lower affinity for pyruvate and is not inhibited by pyruvate, whereas the H4 isoform has a higher affinity for pyruvate and is allosterically inhibited by pyruvate. Therefore, the H isoform is better adapted to function in the conversion of lactate to pyruvate, and the M isoform is better suited to the reverse reaction.

A.

Isoform	H4	MH3	M2H2	M3H	M4
Heart	X	x			
RBC	x	X			
Brain	x	X	x		
Muscle			x	x	X
Liver				x	X

B.

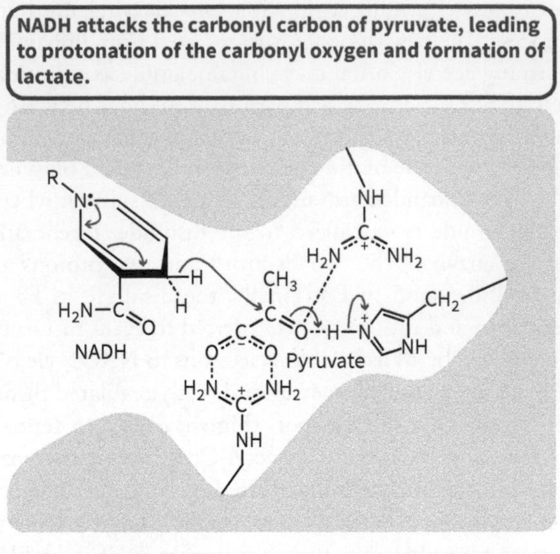

> NADH attacks the carbonyl carbon of pyruvate, leading to protonation of the carbonyl oxygen and formation of lactate.

FIGURE 6.29 Lactate dehydrogenase. A. Monomers of lactate dehydrogenase assemble and function as tetramers. The two genes for this enzyme are differentially expressed, leading to isoforms with different combinations of the M and H subunits, which are found in different tissues. A large X indicates the most prevalent combination in those tissues, and a small x indicates less prevalent combinations. **B.** Pyruvate is reduced to lactate in lactate dehydrogenase by transfer of electrons from NADH. Two arginines in the active site play a key role in the orientation of the pyruvate while an active site histidine acts as a proton donor to form the hydroxyl group on lactate.

The mechanism of lactate dehydrogenase employs two arginine residues in the active site. These residues play key roles in the orientation of the pyruvate substrate. Electrons flow from NADH to the carbonyl carbon of pyruvate and into the carbonyl oxygen. An active site histidine acts as a proton donor to complete formation of the hydroxyl group, forming lactate.

Lactate produced in active muscle can diffuse from the myocyte into the bloodstream. In the liver, lactate can be oxidized back to pyruvate by LDH (at the cost of one molecule of NADH/H$^+$). The liver can use this pyruvate to produce glucose via gluconeogenesis; it can then release the glucose for use by other tissues (presumably back to skeletal muscle). This is an example of how different organs have evolved and adapted to metabolically complement each other.

6.5.3 Pyruvate can be transaminated to alanine

Throughout this chapter, we have seen examples of how different organs and tissues evolved to have specialized functions. This situation is readily apparent on a macroscopic level. The functions of intestine and bone are clearly different. Similar dramatic differences are seen in metabolism at the biochemical level. One such difference is the means by which pyruvate and alanine are metabolized in liver and muscle.

Pyruvate can be transaminated into alanine. Transamination reactions exchange an amine and hydrogen for an oxygen, generating a carbonyl group. Pyruvate undergoes a transamination with the amino acid glutamate to generate the amino acid alanine. Several amino acids (serine, cysteine, threonine, and glycine) also proceed through an alanine intermediate during catabolism.

Alanine is released from muscle and passes through the circulation to the liver. Hence, alanine is serving two important metabolic functions in this situation. First, it is shuttling carbons to the liver for pyruvate production. Second, it is shuttling potentially toxic amino groups to the liver to be detoxified into urea. In the liver, a second transamination reaction generates pyruvate from the carbon skeleton of alanine. The liver typically uses this pyruvate to generate glucose via gluconeogenesis. The resulting ammonia is eventually detoxified via the urea cycle.

Again, alanine metabolism illustrates the level of specialization of metabolism in different tissues that perform specific tasks. By shifting to the liver the metabolic burden of detoxifying ammonia (via transamination, oxidative deamination, and urea synthesis) and generating glucose from pyruvate (via gluconeogenesis), muscle is freed to perform other tasks such as motility and mobility.

6.5.4 Pyruvate can be carboxylated to oxaloacetate by pyruvate carboxylase

Pyruvate can enter gluconeogenesis. The first step in this process is the formation of oxaloacetate from pyruvate and CO_2, a process known as carboxylation. The energy for this reaction is provided by the hydrolysis of ATP. This happens in the mitochondria, and the oxaloacetate produced is transaminated into aspartate, pumped out of the mitochondria, then transaminated back into oxaloacetate in the cytosol, where it serves as a substrate for PEPCK, and proceeds through gluconeogenesis. However, there is another role for oxaloacetate: as part of the citric acid cycle, it serves as a metabolic crossroads for several molecules, including the catabolism of several amino acids. The formation of oxaloacetate in this situation can act as an **anaplerotic** reaction, which are reactions that build up levels of citric acid cycle intermediates and restore levels of oxaloacetate.

6.5.5 Microbes can decarboxylate pyruvate into acetaldehyde

Under short-term, limited anaerobic conditions, animals can survive by temporarily converting pyruvate to lactate, allowing glycolysis to proceed and produce ATP. This is, at best, a stopgap measure. In contrast, yeast and some other microbes can survive in the absence of oxygen. When oxygen is available, it is used as a terminal electron acceptor. Water is formed, and the reducing equivalents NADH and $FADH_2$ are reset to NAD^+ and FAD. In the absence of oxygen, organisms need some way to produce ATP. Although ATP can be made through glycolysis, this requires a mechanism for oxidizing NADH and $FADH_2$ back to NAD^+ and FAD. In yeast, this is accomplished by the oxidative decarboxylation of pyruvate catalyzed by pyruvate decarboxylase (**Figure 6.30**). Pyruvate decarboxylase uses pyruvate produced at the end of glycolysis to produce

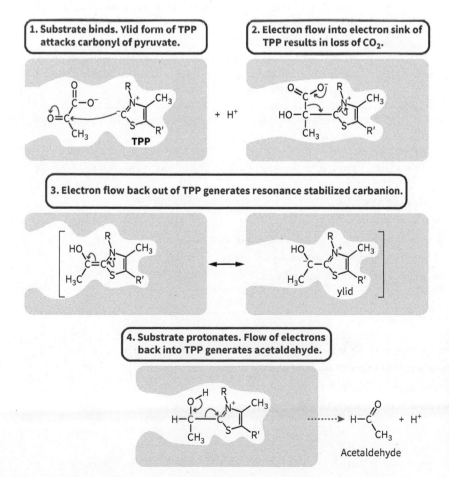

FIGURE 6.30 The mechanism of pyruvate decarboxylase. The mechanism of pyruvate decarboxylase is also a TPP-dependent mechanism. As we have previously seen, pyruvate forms an adduct through the ylid form of TPP (1). The nitrogen in the TPP acts as an electron sink, facilitating the decarboxylation of pyruvate (2). The resulting carbanion is stabilized through resonance. The anion protonates, permitting electron flow back from the hydroxyl group to reform the ylid (3). In the process, acetaldehyde is released (4).

acetaldehyde and CO_2. This is a TPP-dependent enzyme with a Mg^{2+} cofactor. The mechanism begins with the formation of a covalent adduct between pyruvate and the ylid form of TPP. The nitrogen in the TPP acts as an electron sink drawing electrons into the thiazole ring and away from pyruvate. This facilitates the decarboxylation of pyruvate and cleavage of the carbon–carbon bond, releasing CO_2. The resulting carbanion is stabilized through resonance. Electron flow back into the ylid electron sink causes formation of the carbonyl group and cleavage of the carbon–carbon bond between the substrate and TPP. Acetaldehyde is released, and the ylid form of the enzyme is regenerated.

Alcohol dehydrogenase then generates ethanol from the acetaldehyde. The cofactor in this redox reaction is $NADH/H^+$, which is reset to NAD^+ in the process. Effectively, acetaldehyde takes the place of oxygen as the terminal electron acceptor in this series of reactions.

Many strains of yeast found in nature can tolerate concentrations of ethanol as high as 14% v/v. These strains are responsible for the relatively high concentrations of ethanol found in some wines. Generation of spirits with higher concentrations of ethanol requires chemical concentration (distillation). Over the past 40 years, yeast strains that can tolerate higher concentrations of ethanol have been selected for or generated in the laboratory. Although these strains have found some use in the food and beverage industry, they are of potentially much greater value in the production of biofuels (notably bioethanol).

Worked Problem 6.6   Fates of pyruvate in a strain of bacteria

A strain of bacteria form pyruvate after metabolizing galactose through glycolysis. These bacteria grow under aerobic conditions without any apparent inhibition, using galactose as a carbon source. They can also grow under anaerobic conditions, albeit slowly and with the gradual acidification of the media. Suggest one mechanism through which these bacteria are conducting anaerobic metabolism.

Strategy   The question tells us that the bacteria can degrade galactose through glycolysis and live under both aerobic and anaerobic conditions. If the media these cells live in acidifies under anaerobic conditions, what is one likely fate of the pyruvate produced in glycolysis?

Solution   The bacteria may be producing lactate (lactic acid) with the pyruvate that results from glycolysis. Remember that the electron carrier $NADH/H^+$ needs to be reset for glycolysis to proceed. Under aerobic conditions, this is usually accomplished through the use of oxygen and the formation of water. Under anaerobic conditions, the bacteria need another mechanism to oxidize NADH back to NAD^+. Microbes such as yeast often reduce pyruvate to acetaldehyde and CO_2, and then to ethanol; however, many other microbes will produce other wastes, such as lactate.

Follow-up question   How could you test your hypothesis about the production of lactate using both a chemical approach and a genetic approach?

Summary

- Pyruvate has distinct fates depending on the type of cell (or organ) and the metabolic state of the organism.
- Pyruvate can be decarboxylated to acetyl-CoA by pyruvate dehydrogenase.
- Pyruvate can be converted to lactate by LDH.
- Pyruvate can be transaminated to alanine via alanine aminotransferase.
- Pyruvate can be carboxylated to oxaloacetate by pyruvate carboxylase.
- Microbes can decarboxylate pyruvate into ethanol.

Concept Check

1. Describe the fates of pyruvate in both higher organisms and in some microbes, and explain:
   ○ what pyruvate is transformed into in each of these cases.
   ○ the function of pyruvate in each of these cases.
   ○ pyruvate's role in the organism in a biochemical sense.

Bioinformatics Exercises

Exercise 1 Drawing Monosaccharides

Exercise 2 Introduction to Glycomics

Exercise 3 Oligosaccharides, Polysaccharides, and Glycoproteins

Exercise 4 Glycolysis and the KEGG Database

Exercise 5 Metabolism and the BRENDA and KEGG Database

Problems

6.1 Properties, Nomenclature, and Biological Functions of Monosaccharides

1. Describe these carbohydrates as

   a. aldose or ketose.

   b. triose, tetrose, pentose, hexose, or heptose

   c. furanose or pyranose

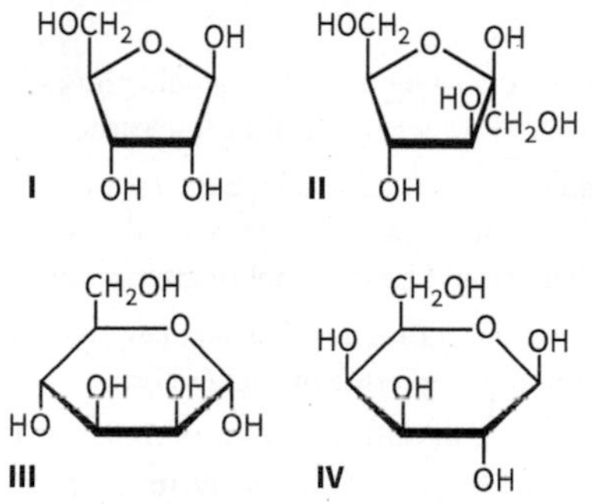

2. Given the following Fischer projections, draw the Haworth projection for either a furanose or pyranose.

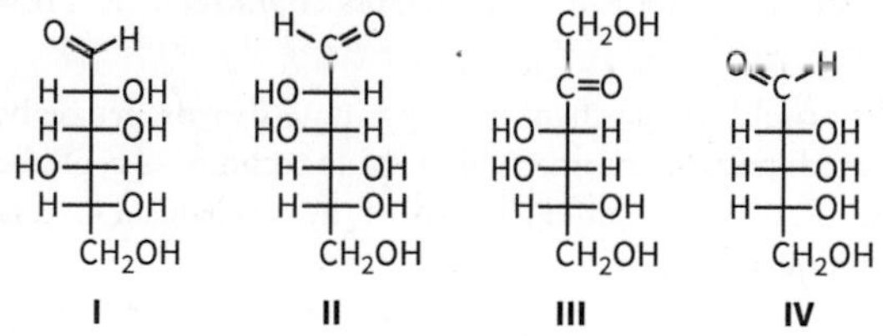

3. Given the following Haworth projections, draw the Fischer projections.

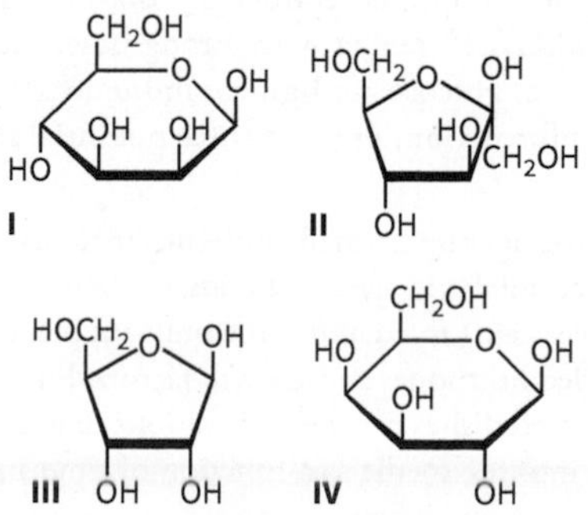

4. Draw the structures below as pyranoses.

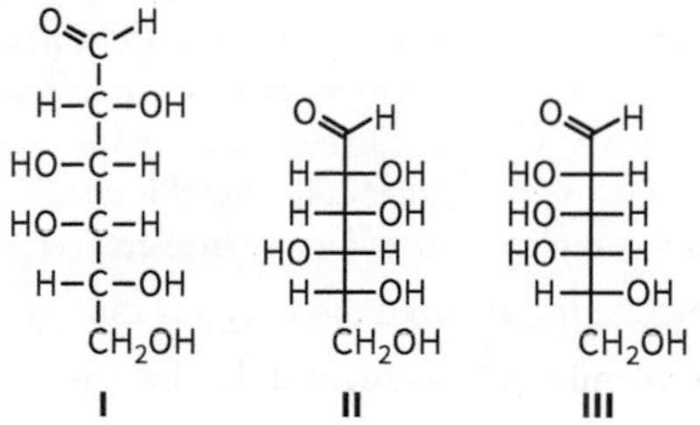

5. Draw the following structures as a furanose.

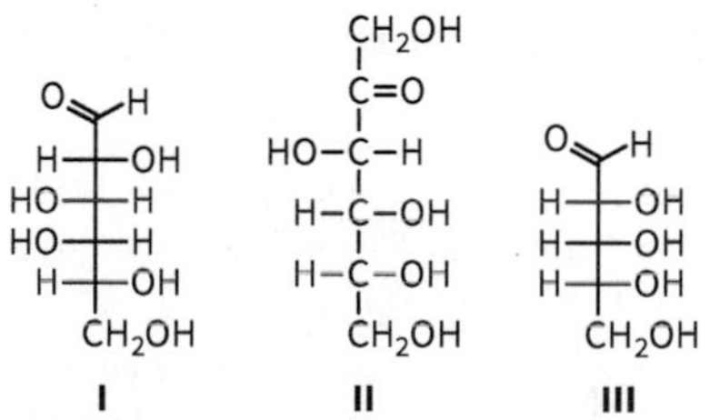

6. Draw the following structures in a chair or envelope projection, noting which anomer you believe would be most stable and explaining why.

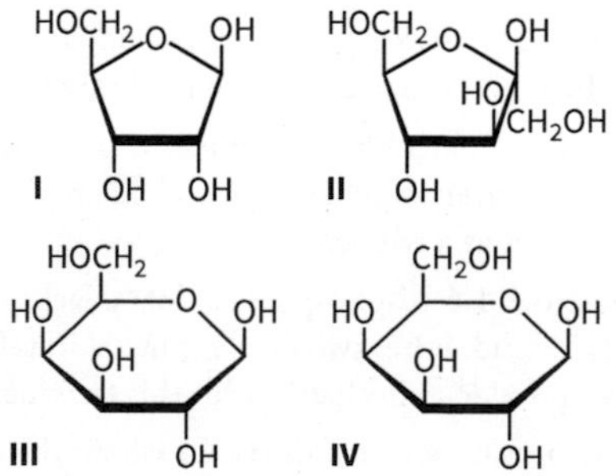

7. Draw the structures of the following modified monosaccharides.

   i. *N*-acetylgalactosamine

   ii. mannitol

   iii. glucuronic acid

   iv. alpha-D-glucopyranoside

6.2 Properties, Nomenclature, and Biological Functions of Complex Carbohydrates

8. Describe the constituents and linkages in the following diasacchrides.

   a. sucrose

   b. maltose

   c. lactose

   d. cellobiose

9. Elucidate the structure of chitin. Can humans digest chitin? Explain why or why not, using biochemical evidence.

10. Currency is printed on high-quality cellulose-based paper. Most paper consists of lower levels of cellulose and higher levels of starches and other carbohydrates. Pens used to detect counterfeit notes contain an iodine-based dye. What should the dye look like on an authentic note?

11. In making simple cheese, milk is treated with vinegar (acetic acid) to denature proteins; this creates a curd (the solid fraction containing fats and precipitated proteins) and whey (the liquid fraction). Based on this simple procedure, if the curd were separated from the whey, could someone who is lactose-intolerant consume this cheese (the curd) without suffering the complications of lactose intolerance?

12. Why is fructose metabolized more rapidly than glucose?

13. In organic chemistry, it is observed that increasing the temperature of a liquid–solid extraction increases the yield. In many cases, solvents are refluxed with a solid to extract the desired materials. Why is this not done in most biochemical extractions?

6.3 Glycolysis and an Introduction to Metabolic Pathways

14. In the reaction catalyzed by aldolase, the bond broken is between C-3 and C-4 of the substrate. Which functional groups are present on these two carbons (C-3 and C-4) in the products?

15. The energetics of glycolysis can be described using a roller coaster analogy. Initial energy is invested in glycolysis through phosphorylation (and hydrolysis of ATP). This is analogous to pulling the roller coaster up the hill. If glycolysis were a roller coaster, how would you envision the rest of the ride?

16. How important are enzymes of glycolysis for the survival of strains bacterial or yeast?

17. Drug companies are seeking compounds that act on F-2,6-bP. Why is this molecule being considered as a potential target?

18. Overexpression of glucokinase enzyme in the liver of a transgenic animal might affect monosaccharide metabolism in the body. How could it bring changes as a whole?

19. Bacterial fructose-1,6-bisphosphate aldolase lacks the critical lysine in the active site; instead, it has two divalent metal ions bound. Based on this information, provide a mechanism for this enzyme.

20. If there were a mutation to aldolase in which the active site aspartate was mutated to tyrosine, how might the mutation affect the kinetics of the aldolase? Would the enzyme be catalytically competent? Which kinetic parameters would you expect to be affected and how?

6.4 Gluconeogenesis

21. How could breakdown of triacylglycerols (a glycerol backbone with three fatty acids esterified to it) contribute to gluconeogenesis?

22. The amino acid glutamate is a glucogenic amino acid, meaning it can contribute carbons to gluconeogenesis. If the amino group was removed from glutamate, how could the remaining carbon skeleton contribute to gluconeogenesis?

23. If a molecule of pyruvate were labeled using ^{14}C on all three carbons and went through gluconeogenesis, where would those labels be found in the final molecule of glucose?

24. Coffee and many diet soft drinks and energy drinks contain caffeine. Caffeine is a phosphodiesterase inhibitor and blocks the breakdown of cAMP. How does caffeine affect glycolysis and gluconeogenesis?

25. Most tissues, including muscle, the kidneys, and the liver, are capable of some of the reactions of gluconeogenesis, but only liver tissue exports glucose. What does this say about the role of gluconeogenesis in the liver?

26. If the serine phosphorylated by PKA on PFK-2 were mutated to an aspartate, how might this affect gluconeogenesis?

27. How does phosphorylation increase the reactivity of glucose?

6.5 The Fates of Pyruvate

28. Pyruvate stands at a metabolic crossroads. Draw the structure of pyruvate and describe how pyruvate is made into the five different products.

29. If a person were on a high-fat diet, would you expect pyruvate levels to be high in liver? If so, where would those carbons originate?

30. If a molecule of pyruvate was labeled on the carboxyl carbon with ^{14}C and was used to make each of the five products discussed in the fates of pyruvate, where would be the label in each product?

31. Describe how the molecules that regulate pyruvate dehydrogenase make sense in the bigger picture of metabolism.

32. Describe how expression of the two different isoforms of LDH allows the different organs of the body to cooperate under hypoxic states, that is, under low oxygen concentration.

33. In describing the mechanism of pyruvate dehydrogenase, the designation "B" is used several times to designate a base. What types of functional groups, amino acids, or cofactors could serve as a base in this reaction?

34. How could the mechanism of pyruvate dehydrogenase be probed and assayed spectrophotometrically? What techniques could be used to test this mechanism? Could all parts of the mechanism be tested using this technique?

Data Interpretation

35. Imagine that you have isolated a slimy matrix secreted by a species of algae. The sample is retained by dialysis membrane with a pore size of 10 kDa. If treated with strong base, the sample yields a mixture of acetate, glucose, and glucosamine in a 1:1:1 molar ratio. Based on this information, propose three potential structures for this molecule.

36. You are investigating a small molecule that may potentially have use as a novel antidiabetic agent. The assay system consists of primary hepatocytes (liver cells) in a culture medium supplemented with serum to provide needed hormones and growth factors, but it is devoid of glucose. You incubated dishes of cells with various molecules and measured glucose release into the media as a function of time and treatment. The data obtained are shown below.

A. vehicle control

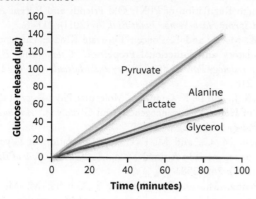

B. 10 mM drug

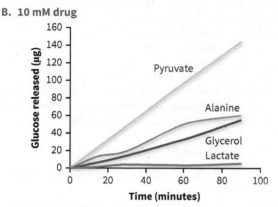

a. What do these data show?

b. What is the effect of this drug?

c. Why were hepatocytes chosen as the experimental model system? Would you expect a different result with other cell types?

d. Why was the assay run in media and serum rather than in saline or a buffered salt system? How could the use of media or serum complicate the assay?

e. If this drug had a molecular weight of 451 g/mol and a 10 mM concentration was needed to achieve the desired effect, would it ever make a good drug?

Experimental Design

37. Design an experiment to check the effect of a drug molecule on glucose metabolism in a diabetic mice model. What enzyme or hormone assays would you perform to evaluate the effect of the drug and how will you validate your results?

38. You have isolated a white crystalline solid from tree sap that gives a positive for Molisch and Seliwanoff's test. Research the Molisch test and design a confirmatory test to ascertain the properties of this molecule.

39. You are investigating the kinetic properties of glyceraldehyde-3-phosphate dehydrogenase. Provide two ways by which you could ensure that this enzyme is not hindered by exhausting the supply of NAD^+.

40. Design an experiment to test whether gluconeogenesis is increased in mice treated with an experimental drug. How would this experiment differ if you wanted to test the effect of the drug in cultured cells or in humans?

41. There are multiple fates of pyruvate. If you were studying pyruvate entering the citric acid cycle, what considerations would you need to take into account to ensure that pyruvate was not lost to other pathways?

Ethics and Social Responsibility

42. Insulin preparations mostly are now recombinant human, practically eliminating the common allergic reactions to the drug when it was extracted from animal sources. The insulin extracted from animals also caused religious concern in some patients. Many analogs are available that were created by modifying the human insulin molecule which alters absorption rates, duration, and time of action. Is it right to switch to modified agents? Do these cause any influence on the immune system and other metabolic pathways?

43. Significant public and private funds are being directed toward the development of cellulosic ethanol (fuel ethanol derived from nonfood sources). Should more public or private funds be spent on this endeavor? Investigate how much oil companies and governments worldwide are investing in this research. Do you feel this is a good distribution of funds? Find a student who disagrees with you and debate the points about this spending.

44. Suppose a country uses limited petrochemicals and instead fuels its cars with ethanol derived from sugarcane. What are the ethical considerations of fueling a car with ethanol derived from sugarcane versus petrochemicals?

45. Agriculture has progressed to the point where a 5 kilogram bag of sugar can be purchased for only a few dollars. Humans have a clear preference for sweet foods; this is built into our biology. At the same time, modern lifestyles have made it possible for people to exercise less and less. One could argue that we have progressed to the point where humans are the first species that needs to consciously think about food consumption and exercise. Other species hoard food for consumption over the winter or hibernate, and burn off fat they have stored over the summer months. Have humans lost the ability to seasonally regulate food consumption and metabolic output?

Suggested Readings

6.1 Properties, Nomenclature, and Biological Functions of Monosaccharides

Gabius, H. J., ed. *The Sugar Code.* Weinheim, Germany: GmbH & Co. KGaA, 2009.

Gard, H.G., M. K. Cowman, and C. A. Hales, eds. *Carbohydrate Chemistry, Biology, and Medical Applications.* Amsterdam, The Netherlands: Elsevier Sciences, 2008.

McNaught, A. D. "Nomenclature of Carbohydrates." *Advances in Carbohydrate Chemistry and Biochemistry.* 52 (1997): 43–177.

Pigman, W., and D. Horton. *The Carbohydrates: Chemistry and Biochemistry,* 2nd ed. New York, New York: Academic Press, 1972.

Stick, R.V., and Spencer, W. *Carbohydrates, Essential Molecules of Life.* 2nd ed. Amsertdam, The Netherlands: Elsevier Science, 2008.

Taylor, M. E., and K. Drickamer. *Introduction to Glycobiology.* New York, New York: Oxford University Press, 2011.

Thisbe, K. Lindhorst. *Essentials of Carbohydrate Chemistry and Biochemistry.* Weinham, Germany: GmbH & Co. KGaA, 2007.

Varki, A. *Essentials of Glycobiology.* 2nd ed. New York, New York: Cold Spring Harbor Press, 2008.

6.2 Properties, Nomenclature, and Biological Functions of Complex Carbohydrates

Cornish-Bowden, A. "Provisional International Union of Pure and Applied Chemistry and International Union of Biochemistry Joint Commission on Biochemical Nomenclature Symbol for Specifying the Conformation of Polysaccharide Chains." *Provisional Pure and Applied Chemistry.* 55, no. 8 (1983): 1269–72.

Genauer, C. H., and H. F. Hammer. "Maldigestion and Malabsorption." Chap. 101 in *Sleisenger & Fordtran's Gastrointestinal and Liver Disease,* edited by M. Feldman, L. S. Friedman, and M. H. Sleisenger. Philadelphia, Pa.: Saunders Elsevier, 2010.

Gupta, B., A. Arorab, S. Saxena, and Alam M. Sarwar. "Preparation of Chitosan–Polyethylene Glycol Coated Cotton Membranes for Wound Dressings: Preparation and Characterization." *Polymers for Advanced Technologies* 20, no.1 (2008): 58–65.

"Lactose intolerance." *The National Digestive Diseases Information Clearinghouse (NDDIC).* NIH Publication No. 09–2751. 2009. https://www.metsol.com/assets/sites/2/Foods_containing_lactose.pdf

Shahidi, F., J. Kamil, V. Arachchi, and Y. J. Jeon. "Food Applications of Chitin and Chitosans." *Trends in Food Science & Technology* 10 (1999): 37–51.

6.3 Glycolysis and an Introduction to Metabolic Pathways

Baggetto, L. G. "Deviant Energetic Metabolism of Glycolytic Cancer Cells." *Biochemie* 74, no. 11 (1992): 959–74.

Bayley, J. P., and P. Devilee. "The Warburg Effect in 2012." *Curr Opin Oncol.* 24, no.1 (2012): 62–7.

Cárdenas, M. L., A. Cornish-Bowden, and T. Ureta. "Evolution and Regulatory Role of the Hexokinases." *Biochim. Biophys. Acta.* 1401 (1998): 242–264.

de Souza A. C., G. Z. Justo, D. R. de Araújo, and A. D. Cavagis. "Defining the Molecular Basis of Tumor Metabolism: A Continuing Challenge Since Warburg's Discovery." *Cell Physiol Biochem.* 28, no. 5 (2011): 771–92.

Fothergill Gilmore, L. A., and P. A. M. Michels. "Evolution of Glycolysis." *Progress in Biophysics and Molecular Biology* 59, no. 2 (1993): 105–235.

Gatenby, R.A., and R. J. Gillies. "Why Do Cancers Have High Aerobic Glycolyis?" *Nature Reviews Cancer* 4, no. 11 (2004): 891–899.

Heinisch, J. "Isolation and Characterization of the Two Structural Genes Coding for Phosphofructokinase in Yeast." *Mol. Gen Genet.* 202 (1986): 75–82.

Hers, H. G., and L. Hue. "Gluconeogenesis and Related Aspects of Glycolysis." *Annual Review of Biochemistry* 52 (1983): 617–653.

Hue L., and M. H. Rider. "Role of Fructose 2,6-Bisphosphate in the Control of Glycolysis in Mammalian Tissues." *Biochemical Journal* 245, no. 2 (1987): 313–324.

Kim, J., and S. D. Copley. "Why Metabolic Enzymes Are Essential or Nonessential for Growth of Escherichia coli K12 on Glucose." *Biochemistry* 46, no. 44 (2007): 12501–11.

Kolwicz, S. C. Jr., and R. Tian. "Glucose Metabolism and Cardiac Hypertrophy." *Cardiovasc Res.* 90, no. 2 (2011):194–201.

Koppenol, W. H., P. L. Bounds, and C. V. Dang. "Otto Warburg's Contributions to Current Concepts of Cancer Metabolism." *Nat Rev Cancer.* 11, no. 5 (2011): 325–37.

Mor, I., E. C. Cheung, and K. H. Vousden. "Control of Glycolysis through Regulation of PFK1: Old Friends and Recent Additions." *Cold Spring Harb Symp Quant Biol.* 76 (2011): 211–6.

Munoz, M. E., and E. Ponce. "Pyruvate Kinase: Current Status of Regulatory and Functional Properties." *Comparative Biochemistry and Physiology Part B: Biochemistry and Molecular Biology* 135 (2003): 197–218.

Pilkis, S. J., and D. K. Granner. "Molecular Physiology of the Regulation of Hepatic Gluconeogenesis and Glycolysis." *Annual Review of Physiology* 54 (1992): 885–909.

Scrutton, M. C., and M. F. Utter. "Regulation of Glycolysis and Gluconeogenesis in Animal Tissues." *Annual Review of Biochemistry* 37 (1968): 249–258.

Sola-Penna, M., D. Da Silva, W. S. Coelho, M. M. Marinho-Carvalho, and P. Zancan. "Regulation of Mammalian Muscle Type 6-Phosphofructo-1-kinase and Its Implication for the Control of the Metabolism." *IUBMB Life* 62, no. 11 (2010): 791–6.

Sugden, M. C., and M. J. Holness. "The Pyruvate Carboxylase-Pyruvate Dehydrogenase Axis in Islet Pyruvate Metabolism: Going Round in Circles?" *Islets* 3, no. 6 (2011): 302–19.

Wilson, J. E. "Isozymes of Mammalian Hexokinase: Structure, Subcellular Localization and Metabolic Function." *Journal of Experimental Biology* 206, no. 12 (2003): 2049–57.

6.4 Gluconeogenesis

Exton, J. H. "Gluconeogenesis" *Metabolism-Clinical and Experimental* 21, no. 10 (1972): 945–990.

Friedmann, N. "Hormonal Regulation of Hepatic Gluconeogenesis." *Physiological Reviews* 64, no. 1 (1984): 170–259.

Pilkis, S. J., and D. K. Granner. "Molecular Physiology of the Regulation of Hepatic Gluconeogenesis and Glycolysis." *Annual Review of Physiology* 54 (1992): 885–909.

Pilkis, S. J., M. R. El Maghrabi, and T. H. Claus. "Hormonal Regulation of Hepatic Gluconeogenesis and Glycolysis." *Annual Review of Biochemistry* 57 (1988): 755–783.

6.5 The Fates of Pyruvate

Dienel, G. A., and L. Hertz. "Glucose and Lactate Metabolism During Brain Activation." *Journal of Neuroscience Research* 66, no. 5 (2001): 824–838.

Gladden, L. B. "Lactate Metabolism: A New Paradigm for the Third Millennium." *Journal of Physiology-London* 558, no. 1 (2004): 5–30.

Pronk, J. T., H. Y. Steensma, and J. P. vanDijken. "Pyruvate Metabolism in *Saccharomyces cerevisiae*." *Yeast* 12, no.16 (1996): 1607–1633.

Stacpoole, P. W. "The Pyruvate Dehydrogenase Complex as a Therapeutic Target for Age-Related Diseases." *Aging Cell* 11, no. 3 (2012): 371–377.

Sugden, M. C., and M. J. Holness. "The Pyruvate Carboxylase-Pyruvate Dehydrogenase Axis in Islet Pyruvate Metabolism: Going Round in Circles?" *Islets* 3, no. 6 (2011): 302–319.

Yadav, A. K., A. B. Chaudhari, and R. M. Kothari. "Bioconversion of Renewable Resources into Lactic Acid: An Industrial View." *Critical Reviews in Biotechnology* 31, no.1 (2011): 1–19.

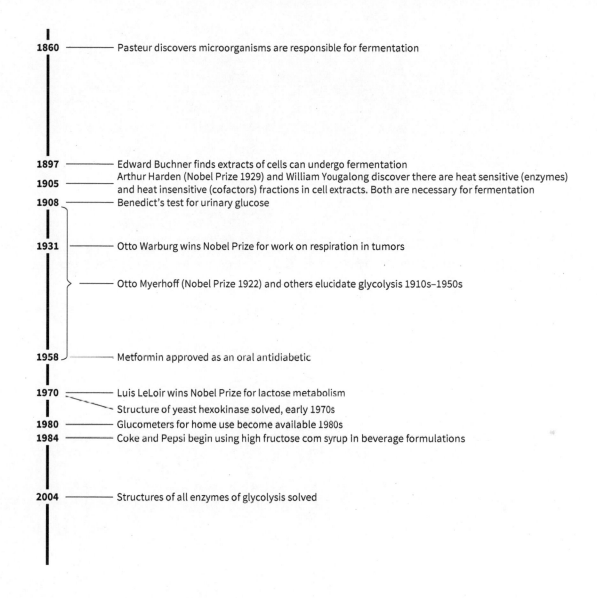

1860 ———— Pasteur discovers microorganisms are responsible for fermentation

1897 ———— Edward Buchner finds extracts of cells can undergo fermentation
1905 ———— Arthur Harden (Nobel Prize 1929) and William Yougalong discover there are heat sensitive (enzymes) and heat insensitive (cofactors) fractions in cell extracts. Both are necessary for fermentation
1908 ———— Benedict's test for urinary glucose

1931 ———— Otto Warburg wins Nobel Prize for work on respiration in tumors

———— Otto Myerhoff (Nobel Prize 1922) and others elucidate glycolysis 1910s–1950s

1958 ———— Metformin approved as an oral antidiabetic

1970 ———— Luis LeLoir wins Nobel Prize for lactose metabolism
———— Structure of yeast hexokinase solved, early 1970s
1980 ———— Glucometers for home use become available 1980s
1984 ———— Coke and Pepsi begin using high fructose com syrup In beverage formulations

2004 ———— Structures of all enzymes of glycolysis solved

CHAPTER 7

The Common Catabolic Pathway

Electron Transport in Context

Money, it has been said, makes the world go round. Under a barter system, trade is difficult to conduct over distances with any predictability. Paper money is not always the best medium for trade either. For example, it would be inconvenient or even dangerous to be paid weekly in cash or to make a large purchase, such as a car or house, using cash. Therefore, alternatives to money, such as credit cards or checks, have been developed. Money in all of its forms provides the potential to build things and do work.

This is analogous to the situation in the cell. Using the energy from one reaction to perform another is difficult if there is no common intermediate. Hence, the work of the cell—in the form of glycolysis, and now the citric acid cycle—gets paid in checks (NADH and $FADH_2$), which are cashed into ATP during oxidative phosphorylation.

Reducing equivalents (essentially electron carriers such as NAD^+) capture electrons in the citric acid cycle. These electrons move down a series of redox active centers, also electron carriers, ultimately forming water when they react with molecular oxygen. As the electrons move through these centers, the energy of their transfer is used to pump protons out of the mitochondrial matrix. This proton gradient (lower pH outside, higher pH inside) powers a tiny molecular motor—ATP synthase, the enzyme responsible for most of a cell's ATP production.

Chapter Outline

Common Themes

Evolution's outcomes are conserved.	• The citric acid cycle, electron transport, and ATP synthase are conserved throughout evolution.
	• Cytochrome *c* is used as a marker gene to track evolution and demonstrate degrees of relatedness between species.
	• ATP synthase has evolved for specialized functions such as ion transport or flagellar motility.
Structure determines function.	• The structure of the intermediates in the citric acid cycle dictates in part how these molecules react.
	• The structure of the mitochondria and the electron transport apparatus allows for the development of a proton gradient across the mitochondrial membrane.
	• ATP synthase acts as a molecular motor, using the energy of the proton gradient to drive ATP production.
	• The entire electron transport chain and ATP synthase reside in one supercomplex (the respirasome), which allows for substrate channeling.
Biochemical information is transferred, exchanged, and stored.	• Substrate availability, feedback inhibition and allosteric regulators control specific steps of the citric acid cycle.
Biomolecules are altered through pathways involving transformations of energy and matter.	• Energetically unfavorable reactions become favorable when coupled with reactions that have a highly negative ΔG value. Although we typically think of this in terms of ATP hydrolysis, other highly favorable reactions (such as those involving thioesters) are common in biochemistry. Likewise, reactions that have $\Delta G°$ values close to zero can be tipped to move in either direction by changing the concentrations of substrates or products.
	• Electrons in bonds of intermediates of the citric acid cycle are transferred to oxygen through a series of electron carriers, ultimately forming water. These electrons transfer from one electron carrier to another in the different electron transport complexes.
	• Electrons from the reduced form of nicotine adenine dinucleotide (NADH) or flavin adenine dinucleotide ($FADH_2$) are transferred through a series of redox active electron carriers, ultimately being consumed in the production of water.
	• The energy of electrons flowing through the electron transport complexes is used to pump protons out of the mitochondrial matrix.
	• ATP synthase uses the energy of the proton gradient to drive ATP production.

7.1 The Citric Acid Cycle

Acetyl-CoA is an important metabolic intermediate in several pathways. Chapter 6 describes how glucose and other monosaccharides are catabolized into pyruvate. Under aerobic conditions, the fate of most pyruvate is decarboxylation into acetyl-CoA. Likewise, Chapter 9 describes how fatty acids are catabolized into acetyl-CoA. Under normal conditions most of this acetyl-CoA enters the **citric acid cycle** (also known as the tricarboxylic acid cycle or the Krebs cycle) (**Figure 7.1**). The chemical bonds of acetyl-CoA are still energetically rich. The acetyl group of acetyl-CoA is ultimately oxidized in the citric acid cycle into CO_2, and the electrons obtained in these reactions are used to reduce the electron carriers NAD^+ and FAD to NADH and $FADH_2$. The electrons are used in subsequent reactions in the **electron transport chain** to generate a proton gradient, which subsequently drives ATP production in a process known as **oxidative phosphorylation**.

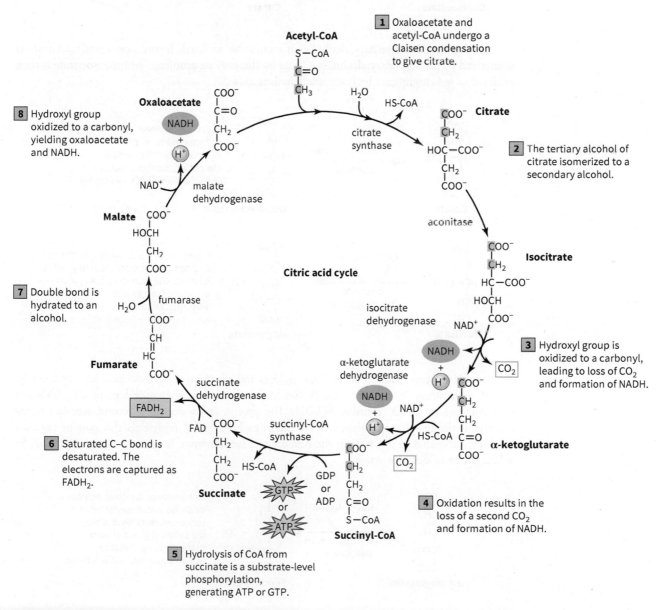

FIGURE 7.1 Citric acid cycle. The citric acid cycle (also known as the tricarboxylic acid cycle or the Krebs cycle) is the central hub of metabolism. While it oxidizes the carbons of acetyl-CoA to CO_2, it captures the electrons as NADH and $FADH_2$ for use in electron transport. Various pathways feed into and out of the cycle, linking central metabolism with other pathways.

(Source: Karp, *Cell and Molecular Biology: Concepts and Experiments*, 7e, copyright 2013, John Wiley & Sons. This material is reproduced with permission of John Wiley & Sons, Inc.)

A second and equally important function of the citric acid cycle is to serve as a central metabolic hub. Many molecules, such as the carbon skeletons of amino acids, are also metabolized through the citric acid cycle.

7.1.1 There are eight reactions in the citric acid cycle

The citric acid cycle begins with a condensation of acetyl-CoA with oxaloacetate, generating the six-carbon tricarboxylic acid citrate.

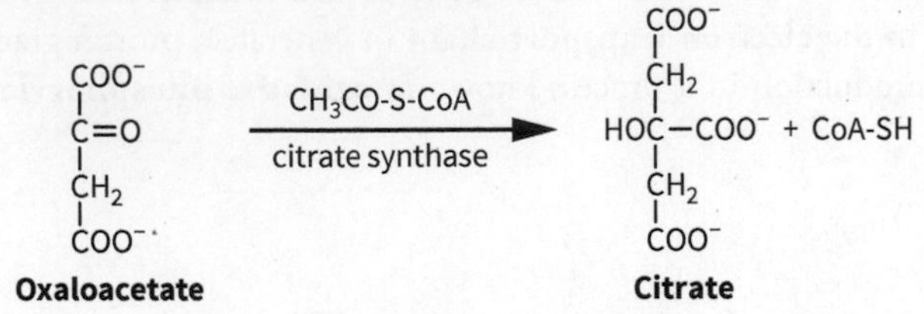

Citrate is formed from acetyl-CoA and oxaloacetate in an aldol condensation. Acetyl-CoA forms the enolate.

Citrate contains a tertiary alcohol that cannot be oxidized. In the next reaction, citrate is isomerized to the secondary alcohol isocitrate by the enzyme aconitase, and the isocitrate is then oxidized to α-ketoglutarate by isocitrate dehydrogenase.

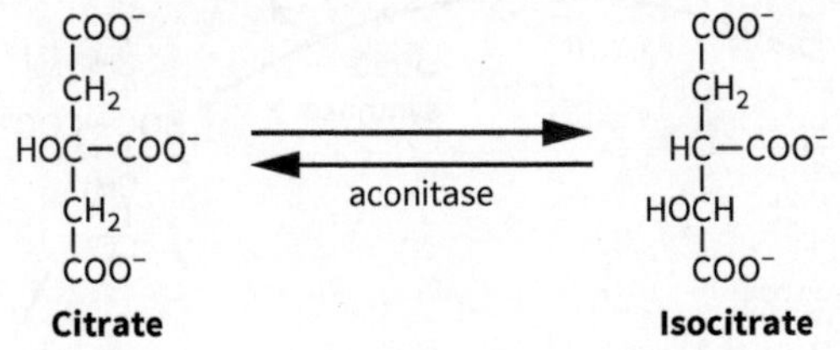

This is a combined dehydration-hydration reaction, which isomerizes the tertiary alcohol citrate to the secondary alcohol isocitrate.

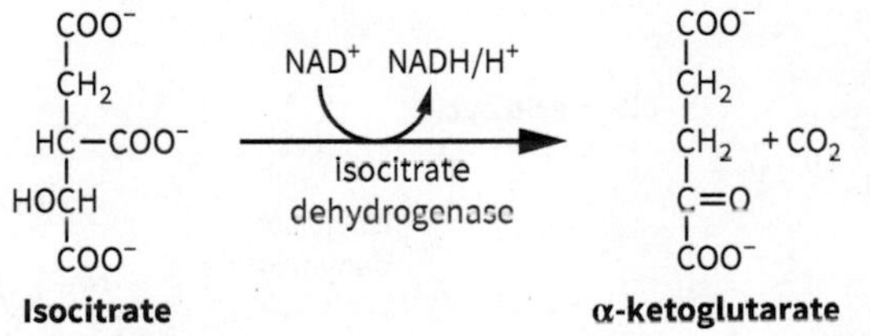

The secondary alcohol in isocitrate is oxidized (dehydrogenated) to give the intermediate oxalosuccinate, which decarboxylates to give α-ketoglutarate.

In this process, a molecule of CO_2 is lost, and electrons from isocitrate are captured by NAD^+, generating $NADH/H^+$. Recall that in these reactions the equivalent of a hydride ion is transferred to NAD^+ forming NADH. The proton is released as a second reaction product and maintains charge balance. Thus the term $NADH/H^+$ represents this pair of reaction products generated in this way. In the next step, α-ketoglutarate is oxidized to succinyl-CoA by α-ketoglutarate dehydrogenase.

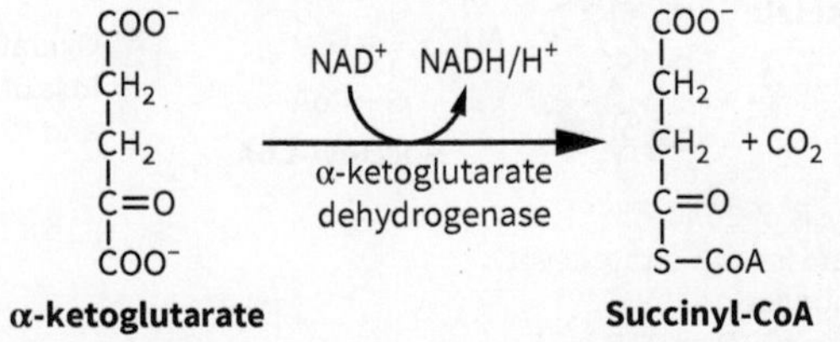

Alpha-ketoglutarate undergoes oxidative decarboxylation, a concerted reaction in which the carboxyl group closest to the carbonyl is lost as CO_2 and succinyl-CoA is formed.

Again, a molecule of $NADH/H^+$ is generated; however, in addition to the loss of CO_2, some energy is retained by the creation of a thioester bond between succinate and CoA. Some of this energy is used in the next reaction: the generation of succinate from succinyl-CoA.

The multienzyme complex α-ketoglutarate dehydrogenase is homologous to pyruvate dehydrogenase, seen in section 6.5. There are three different subunits to the enzyme: E1, a thiamine pyrophosphate (TPP) dependent decarboxylase; E2, a dihydrolipoyl transferase; and E3, a dihydrolipoyl dehydrogenase which employs NAD^+ and FAD. The reactions and chemistry that occur between these two oxidative decarboxylations are also conserved.

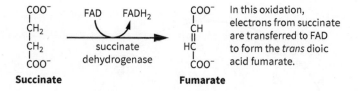

<table>
<tr><td>

COO⁻
|
CH₂
|
CH₂
|
C=O
|
S—CoA
Succinyl-CoA

</td></tr>
</table>

Succinyl-CoA → Succinate; GDP or ADP + Pᵢ + H₂O → GTP or ATP; succinyl-CoA synthetase

This oxidation is a substrate-level phosphorylation, a reaction that directly generates ATP or GTP.

Two isoforms of succinyl-CoA synthetase exist. One will form a molecule of guanosine-5′-triphosphate (GTP) from guanosine diphosphate (GDP) and inorganic phosphate (P_i) while the other will form ATP from ADP and P_i. Because GTP is a nucleoside triphosphate, it is energetically equivalent to ATP. Also, because GTP or ATP was generated directly by phosphorylating a nucleoside, this is a **substrate-level phosphorylation**, as opposed to an oxidative phosphorylation (one involving the oxidation of metabolites). Succinate and the remaining molecules in the cycle all have four carbons. The remaining reactions recycle these four carbons back to oxaloacetate, with two of the steps forming reducing equivalents. Succinate is a four-carbon dioic acid; there are no other functional groups, and the two central carbons are simply methylene units.

The remaining three reactions work together to incorporate a keto group onto one of the two central carbons. This keto group in oxaloacetate is necessary to provide the carbonyl group for the condensation reaction between acetyl-CoA and oxaloacetate in the first step of the cycle. The simplest way to think of this is to picture it as a series of simple organic reactions. An alkane can be oxidized into an alkene, which can be hydrated to an alcohol; the resultant alcohol can then be oxidized to a carbonyl group. The citric acid cycle works in the same way.

Succinate has two saturated central carbon atoms. These are desaturated to yield the *trans*-unsaturated dioic acid, fumarate.

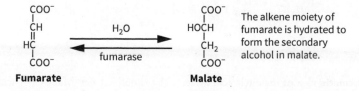

Succinate → Fumarate; FAD → FADH₂; succinate dehydrogenase

In this oxidation, electrons from succinate are transferred to FAD to form the *trans* dioic acid fumarate.

Only the *trans* isomer is formed by succinate dehydrogenase. The natural selection of fumarate may have evolved due to steric hindrance, but it is more likely that the *trans* fumarate was selected over the *cis* isomer for other reasons. The *cis* isomer is more reactive and might participate in unwanted Diels-Alder (2 + 4 cycloaddition) reactions. The reducing equivalent involved in this two-electron reduction is FAD bound to succinate dehydrogenase; however, this FAD is tightly bound to the enzyme and quickly passes these electrons to the mobile electron carrier ubiquinone (Q) producing reduced ubiquinone (QH_2) for the electron transport chain. Fumarate is hydrated to malate, generating a hydroxyl group on one of the two carbons formerly found in the double bond, with water being the source of the hydroxyl group in this reaction.

Fumarate → Malate; H₂O; fumarase

The alkene moiety of fumarate is hydrated to form the secondary alcohol in malate.

As in most hydration reactions, energy is neither consumed nor produced to an appreciable extent (ΔG is close to zero). Finally, malate is oxidized to oxaloacetate: a four-carbon dioic acid with a keto group adjacent (α) to one of the carboxyls.

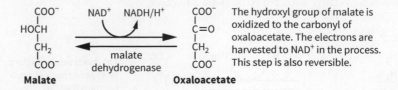

Again NAD$^+$ acquires electrons and is reduced to NADH. The cycle starts again using oxaloacetate.

To briefly review, the citric acid cycle takes a two-carbon intermediate (acetyl-CoA) and condenses it with a four-carbon intermediate (oxaloacetate), forming a six-carbon tricarboxylic acid (citrate). Citrate cannot be metabolized further and so is isomerized to isocitrate (moving from a tertiary to a secondary alcohol). Isocitrate is oxidized at this alcohol, first to α-ketoglutarate (five carbons) and then to succinyl-CoA (four carbons). At each of these oxidations, energy is harvested in the form of NADH/H$^+$. Succinyl-CoA is cleaved to liberate succinate, generating a molecule of GTP or ATP in the process. The final three reactions—dehydrogenation, hydration, and oxidation—regenerate oxaloacetate.

Desaturations often employ the cofactor FAD (as succinate dehydrogenase does in this reaction), whereas the second oxidation (malate dehydrogenase) uses NAD$^+$. Overall, the cycle generates two molecules of CO$_2$, one molecule of GTP or ATP, three molecules of NADH/H$^+$, and one molecule of FADH$_2$ (**Figure 7.2**).

Many of the cofactors employed in the citric acid cycle or other biochemical pathways are the vitamins and minerals we consume in our diet. A brief history of these molecules is found in **Societal and Ethical Biochemistry: Vitamins in human health and disease.**

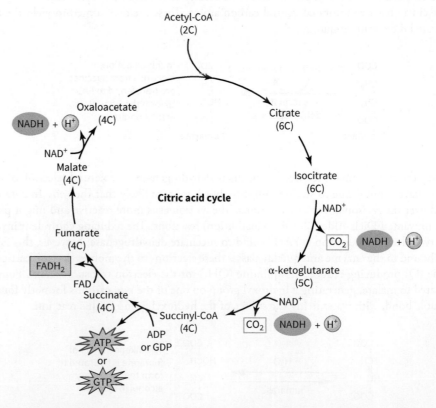

FIGURE 7.2 Schematic view of the citric acid cycle. This simplified view of the citric acid cycle shows the number of carbons in each intermediate as well as where energy is siphoned off as NADH, FADH$_2$, or GTP or ATP.

Societal and Ethical Biochemistry

Vitamins in human health and disease

There are several molecules such as NAD$^+$, FAD, biotin, and cobalamin that do not readily fit into the simple categories of amino acid, carbohydrate, fatty acid, and nucleotide. These molecules are derived from essential nutrients known as vitamins.

The term "vitamin" was coined by the Polish–American biochemist Casimir Funk, who merged the terms "vital" and "amine" (at the time, the compounds under investigation all contained amine moieties). Vitamins are small organic molecules produced by organisms, often by bacteria or plants.

Thiamine (vitamin B1)

Riboflavin (vitamin B2)

Nicotinamide (vitamin B3)

Hydroxycobalamin (X=OH) (vitamin B12)

Pantothenic acid (vitamin B5)

Pyridoxol (vitamin B6)

Biotin (vitamin B7)

Ascorbic acid (vitamin C)

Folic acid (vitamin B9)

Tocopherol (vitamin E)

Currently, 13 vitamins are recognized as necessary for human health: A, C, E, D, K, and the eight B vitamins. Humans cannot synthesize most of these compounds, except for small amounts of vitamins D and K. Vitamin B12 is generally obtained from animal sources in the diet; thus, vegetarians often have to take a vitamin B12 supplement. Vitamins themselves are not categorized as exclusive chemical species but rather as families of related molecules that can be biochemically converted from one form to another. Hence, "vitamin A" contains the active molecule retinal, but also contains retinol, lutein, and several other provitamins. Vitamins can be crudely categorized as being either fat soluble (A, D, E, and K) or water soluble (the B vitamins and vitamin C). Some of the chemical names of the vitamins in this list may be familiar as molecules encountered previously or as precursors for molecules studied; examples include vitamin B7 (biotin) seen in CO_2 binding, and vitamins B3 and B2 (niacin and riboflavin), which are precursors to NAD$^+$ and FAD.

When these compounds are lacking in the diet, it is hard if not impossible to perform the reactions they help to catalyze. This explains why a shortage or lack of vitamins in the diet leads to vitamin deficiencies and disease.

Vitamin deficiencies hold an odd niche both in medicine and in military history. Since the beginning of time, people have suffered from vitamin deficiencies based on poor diet or lack of particular vitamins at certain times of the year. Under the obligation and confinement of military service, these effects become more pronounced and obvious. British sailors in the 1700s often developed scurvy, which is caused by vitamin C deficiency. At that time, the diet of sailors was comprised of grains, fish, and dried meat, all of which lack vitamin C. Vitamin C is needed for the development of collagen. The vitamin is essential in the formation of hydroxyproline, one of the modified amino acids of collagen. Lacking vitamin C, sailors suffered from anemia, general lethargy, failure to heal wounds, loss of teeth and hair, bleeding from mucus membranes, and jaundice. Scurvy was often fatal, and it was of great concern to the Royal Navy. The Scottish physician James Lind published a paper in 1753 in which he proposed treating scurvy with citrus fruits. The Royal Navy adopted the idea and began using limes as part of a sailor's daily rations. Limes were useful because they were resistant to rot, but by themselves they were difficult to stomach, so sailors took a mixture of lime juice with cane sugar and rum, which is now the basis of the daiquiri cocktail. British sailors became known as "limeys," which entered the vernacular. Although the consumption of citrus helped to prevent scurvy in sailors, it was another 180 years before vitamin C was finally identified as the essential nutrient needed to prevent this disease.

Vitamin B12 deficiency leads to beriberi, a painful neurological and muscle disorder that can be fatal. Takaki Kanehiro, a physician in the Japanese Imperial Navy, noted that enlisted sailors who ate a diet of white rice suffered from this disease, whereas officers aboard the same ships who ate a Western diet did not. This led to an experiment in which sailors on one ship were given a diet of white rice only, and those on a second ship were given a diverse diet. Among the sailors on the traditional diet of white rice, there were 161 cases of beriberi and 25 deaths, whereas, among the sailors on the diverse diet, there were 14 cases of beriberi and no deaths. Although the results showed that the disease was probably linked to the diet, the results were unfortunately misinterpreted as being due to a protein deficiency and not to low vitamin B12 levels.

The United States also has a military link to vitamin deficiencies. On entering World War I and instituting the draft, many recruits were unfit for basic training because of malnutrition leading to rickets, pellagra, and goiters. This led to two major civilian initiatives. First, many foods in America (especially foods made with wheat flour) were "enriched"; that is, they were supplemented with B vitamins. The idea was that because Americans consume bread with most meals, adding vitamins to the bread would reduce malnutrition. The other initiative was the recommended daily allowance (RDA) of vitamins and minerals. Interestingly, the National Institutes of Health (NIH) does not oversee the RDA. Rather, the United States Department of Agriculture (USDA) does, an organization whose role is to promote the production and consumption of U.S. agricultural products both at home and abroad. The RDA has changed over the years. Initially, its purpose was research into the daily consumption of vitamins needed to prevent disease in young, adult Caucasian males. Stemming from that research, scientists began to question whether the absence of disease equates to health. More recently, the RDA has been amended to reflect the findings from more detailed research, for example, into how quickly vitamins and minerals pass through the body. Also, the values now take into account the different nutritional needs of various other groups of people: women, teenagers, children, pregnant women, and the elderly.

Vitamins are often conflated with minerals. However, minerals (calcium, chlorine, cobalt, copper, iodine, iron, magnesium, molybdenum, phosphorus, potassium, selenium, sodium, sulfur, and zinc) differ from vitamins in that they are all the ions of inorganic elements and are typically found in the diet as simple salts or as cofactors in the proteins of food. As discussed, these molecules are important in several enzymes and proteins. Other minerals that may have roles in biochemistry include arsenic, boron, chromium, and silicon, although the dietary need for any of these elements is likely to be exceedingly low as elevated levels of many of these compounds is toxic.

Today, many people consume vitamins as a dietary supplement. Although the efficacy of dietary supplements has not been wholeheartedly embraced by the medical field, most (but not all) physicians believe that limited supplements are not harmful to health. When taken in large quantity, however, supplements can lead to health problems. Several case studies have been published of people believing that it was healthy to consume massive amounts of carrots, carrot juice, or vitamin A (in some reports, patients consumed upwards of several pounds of carrots a day). Often, these patients have an orange skin tone, due to either excess carotenoid consumption or jaundice, and the effects can sometimes be fatal. Autopsies of these patients often reveal fatty liver and cirrhosis as a result of the vitamin A burden placed on their livers.

Exactly how much Americans spend on vitamins and minerals is unknown, but estimates range from $22 to $40 billion a year, more than the total amount the U.S. government spends on biomedical research. These numbers do not include nutritional supplements and alternative health practices, each of which is estimated at twice what is spent on vitamins and minerals.

Clearly, vitamins are big business, and many stores specialize in these compounds; however, what is less clear is whether vitamin supplements are necessary and in what doses. Questions also remain about the bioavailability of these molecules. Do vitamins in tablet form have the same bioavailability as those found naturally in food?

In many ways, the central importance of the citric acid cycle and its resulting popularity has been to its detriment. Many students first encounter the citric acid cycle in middle or high school, where it has been stripped of all chemical information and is merely a series of boxes. This is often repeated in biology classes leading all the way into biochemistry. The citric acid cycle is a series of chemical reactions; if you wish to memorize the pathway you can, but eventually your memory will fail, and the names and shapes, without any underlying chemistry or structure, are meaningless. However, if you take the time to study the reactions involved and look at the chemistry occurring, understand why a reaction has to proceed in the way that it does, and identify in detail what the structures are, it should make the cycle a lot easier to learn and much easier to understand. This approach applies to all biochemical pathways and all metabolism, not just the citric acid cycle (**Table 7.1**).

The citric acid cycle is an example of a **metabolon**: a group of enzymes performing reactions with a common purpose. These enzymes may be localized in an organelle or part of an organelle. This differs from a pathway in that the individual enzymes of the metabolon may function in multienzyme complexes, and there may be interactions between individual proteins or complexes of proteins. In a metabolon, **substrate channeling** may occur, that is, the diversion of the product of one enzymatic reaction directly into a subsequent reaction, to increase reaction rate and efficiency. As purification techniques and biophysical and analytical techniques improve, enzymes and reactions that were formerly thought to function in isolation are increasingly being identified as part of a larger system. Metabolons have been associated with glycolysis, glycogenolysis, fatty acid biosynthesis, and the electron transport chain. In this regard, these complexes are similar to the multienzyme systems involved in cellular processes such as DNA replication, transcription, translation, signal transduction, and vesicular trafficking.

TABLE 7.1 Reactions of the Citric Acid Cycle

Reaction	Structure	Enzyme	$\Delta G^{\circ\prime}$ (kJ/mol)	ΔG (kJ/mol)
Acetyl-CoA + oxaloacetate + $H_2O \rightarrow$ CoASH + citrate		Citrate synthase	−31.4	−53.9
Citrate $\rightarrow$ isocitrate		Aconitase	+6.7	+0.8
Isocitrate + $NAD^+ \rightarrow \alpha$-ketoglutarate + NADH + CO_2		Isocitrate dehydrogenase	−8.4	−17.5
α-ketoglutarate + CoASH + $NAD^+ \rightarrow$ succinyl-CoA + NADH + CO_2		α–ketoglutarate dehydrogenase	−30	−43.9
Succinyl-CoA + GDP or ADP + $P_i \rightarrow$ succinate + GTP or ATP + CoASH		Succinyl-CoA synthetase	−3.3	~0
Succinate + FAD $\rightarrow$ fumarate + $FADH_2$		Succinyl-Co A dehydrogenase	+0.4	~0
Fumarate + $H_2O \rightarrow$ malate		Fumarase	−3.8	~0
Malate + NAD $\rightarrow$ oxaloacetate + NADH + H^+		Malate dehydrogenase	+29.7	~0

Worked Problem 7.1   The citric acid cycles

Some view the citric acid cycle as two separate linear pathways that have joined through evolution. Can you identify these two pathways? What would be their function if they were independent of one another?

Strategy Examine Figure 7.1. Where are reasonable places to divide the citric acid cycle in half? Do any of the reactions seem to chemically fit together better than others?

Solution We can view the citric acid cycle as having two parts. One part, an initial oxidative part, is where citrate is gradually oxidized to α-ketoglutarate and then to succinyl-CoA. A secondary oxidative part is where succinate is gradually oxidized back into oxaloacetate to continue the cycle.

Evolutionary biochemists who study primitive bacteria have hypothesized instances where the second half of the citric acid cycle runs in reverse, that is, from oxaloacetate to malate to fumarate to succinate. This path would enable anaerobic bacteria to regenerate NAD^+ to keep glycolysis moving forward and generating ATP. The first half of the citric acid cycle could keep running to generate intermediates for amino acid production or other metabolic pathways.

Follow-up question If the two proposed pathways did evolve earlier in the history of life on Earth, what does this say about the other reactions of the citric acid cycle, the proteins that catalyze these reactions, and the genes that code for these proteins?

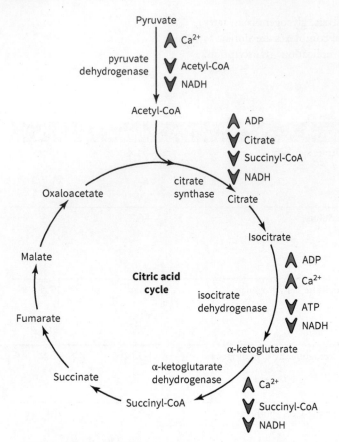

FIGURE 7.3 Regulation of the citric acid cycle. The citric acid cycle is regulated primarily by levels of substrates and products (such as citrate and succinate). The main allosteric regulator is ADP, but Ca^{2+}, NADH, and ATP regulate steps as well.

7.1.2 The citric acid cycle is regulated at multiple places and by several different mechanisms

In any metabolic pathway, we anticipate that the committed steps—those in which metabolites enter or leave the pathway or the rate-determining steps—are regulation points through the pathway. The citric acid cycle is no different, but because of its cyclic nature and the multiple substrates that feed into it, there are several enzymes and mechanisms of control. See **Figure 7.3** and Table 7.1.

The energetics of the citric acid cycle Examining the reactions of the citric acid cycle shows that, as expected, several of the reactions have large negative standard free energies ($\Delta G^{\circ\prime}$). The cleavage of the thioester bond in acetyl-CoA by citrate synthase helps to drive this reaction and give it a highly negative ΔG value. Likewise, the release of CO_2 in the reactions catalyzed by isocitrate dehydrogenase and α-ketoglutarate dehydrogenase increases the entropy of the system and helps to drive ΔG down, thus making the reaction more favorable.

Note that two values are provided in Table 7.1, ΔG and $\Delta G^{\circ\prime}$. Recall from Chapter 1 that $\Delta G^{\circ\prime}$ is established under standard concentrations and conditions which vary widely from the actual concentrations and conditions in the cell (section 1.1). In the cell, or in any reaction that is running under nonstandard conditions, the free energy change of a reaction, denoted here as ΔG, is equal to the standard free energy change, plus a term describing the reaction concentrations and temperature: $\Delta G^{\circ\prime} + RT \ln Q$, where Q is the reaction quotient (the concentration of the products over the reactants raised to the power of the coefficient preceding each substance in the balanced chemical equation).

The other reactions of the citric acid cycle have standard free energy values relatively close to zero. Therefore, in the cell under physiological conditions, these reactions are driven in the forward direction by manipulating concentrations of the substrate and product.

Substrate availability Flux through the citric acid cycle is largely regulated by levels of substrates. Increased levels of substrates generally increase flux through the pathway and most of the reactions are inhibited by their products. Therefore, a stoppage at any point in the pathway can lead to inhibition of most of the prior steps. NADH is a product of three of the four dehydrogenases of the citric acid cycle, and it inhibits all three of these enzymes: isocitrate dehydrogenase, α-ketoglutarate dehydrogenase, and malate dehydrogenase.

Several other products inhibit the enzymes that generate them. For example, succinyl-CoA inhibits α-ketoglutarate dehydrogenase, and citrate inhibits citrate synthase.

Allosteric regulation Because the citric acid cycle is central in the production of energy in the cell, it is also expected that molecular indicators of a low energy state will stimulate the pathway. High levels of adenosine diphosphate (ADP) indicate that energy (in the form of ATP) is in demand in the cell. Although the total concentration of all adenine nucleotides ([ATP] + [ADP] + [AMP]) in the cell is fairly constant, the balance between them shifts, based on the energetic state of the cell. Two reactions of the citric acid cycle are activated by increased concentrations of ADP: citrate synthase and isocitrate dehydrogenase.

Whether ATP inhibits the enzymes of the citric acid cycle is somewhat unclear. High concentrations of ATP will inhibit isolated enzymes from specific tissues (for example, isocitrate dehydrogenase from the heart) but not others (α-ketoglutarate dehydrogenase). Further complicating matters is the general observation that, in the transition from fully fed and rested to maximally exerted, total ATP concentrations change little (by less than 10%). Because of this relatively small change in concentration, it is unlikely that ATP can act as an allosteric regulator. ATP inhibition of the citric acid cycle has not been observed in isolated mitochondria. This highlights another complicating factor: the setting in which the enzymes function. In the mitochondrial matrix, these proteins are present at high concentration. In the purified state, devoid of the other molecules involved in the citric acid cycle, these enzymes may not behave as they do *in vivo*. Finally, structural data have not found an ATP binding site in either individual subunits or complexes of the enzymes of the citric acid cycle. The ATP/ADP ratio or energetic state of the cell is clearly involved in the regulation of the citric acid cycle, but which enzymes are interacting and involved is a matter of debate.

Ca^{2+} is an allosteric regulator that activates isocitrate dehydrogenase and α-ketoglutarate dehydrogenase. Calcium ions are important signaling molecules in the cell: they moderate vesicle fusion and muscle cell contraction (both of which require energy). Therefore, it makes chemical sense that these same signals also directly tell the citric acid cycle to feed more molecules toward ATP biosynthesis.

A final allosteric regulator that is worthy of mention is succinyl-CoA. Although this molecule inhibits the enzyme that makes it (α-ketoglutarate dehydrogenase), it also serves as an inhibitor of citrate synthase.

There are several other ways in which products of the citric acid cycle can increase or decrease flux though the pathway. Citrate inhibits the glycolytic enzyme phosphofructokinase, slowing glycolysis, which thus slows the production of pyruvate (and therefore of acetyl-CoA for the citric acid cycle). Pyruvate dehydrogenase (the committed step in the conversion of pyruvate to acetyl-CoA) is inhibited by its products NADH and acetyl-CoA and is activated allosterically by Ca^{2+}. The regulation of this step directly prevents the formation of acetyl-CoA (see section 6.5).

Hypoxia describes low oxygen levels in a cell, organ, or tissue. A family of transcription factors, known as hypoxia inducible factors (HIFs), is involved in the regulation of cellular processes such as angiogenesis (growth of new blood vessels), glycolysis, and the electron transport chain. HIFs respond to low oxygen levels. At high levels of oxygen, a hydroxyl group is added to a proline residue on HIF. The hydroxylation initiates assembly of a complex of proteins on HIF, including a ubiquitin ligase. Ubiquitin is a small protein that marks proteins for destruction through proteosomal degradation. At low levels of oxygen (hypoxia), there is less hydroxylation of HIF, which allows the transcription factor to move to the nucleus where it binds to DNA, triggering a cascade of transcriptional events that increase the amount of oxygen available to tissues.

Hydroxylation of prolines promotes interaction of HIF with the E3 ubiquitin ligase complex. The citric acid cycle intermediates succinate and fumarate inhibit prolyl 4-hydroxylases; therefore, high levels of these intermediates block the hydroxylase and thus block the interaction with E3 ubiquitin ligase, and the ubiquitination and degradation of HIF.

7.1.3 Anaplerotic reactions of the citric acid cycle replenish intermediates

If the only function of the citric acid cycle were to oxidize acetyl-CoA to CO_2, we could ignore the levels of the other intermediates in the cycle. However, numerous pathways converge on the citric acid cycle, either contributing to or depleting levels of the intermediates. Such pathways include biosynthesis and degradation of amino acids, gluconeogenesis, and biosynthesis of porphyrin and nucleotides. **Anaplerosis** is the scientific term for "filling up." Anaplerotic reactions replenish levels of citric acid cycle intermediates through a number of mechanisms, four of which are highlighted here: one that replenishes α-ketoglutarate, one that replenishes succinyl-CoA, and two that replenish oxaloacetate (**Figure 7.4**).

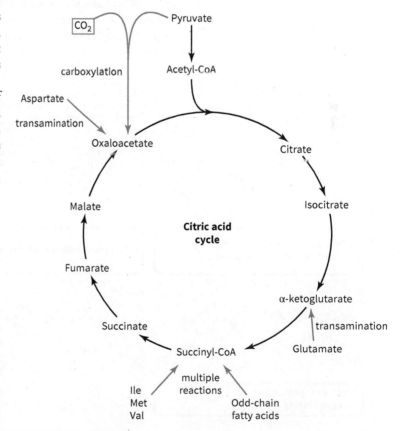

FIGURE 7.4 Central role of the citric acid cycle in metabolism. Numerous pathways converge on the citric acid cycle. Anaplerotic reactions such as the carboxylation of proprionyl-CoA to succinyl-CoA build up intermediates of the cycle, while cataplerotic reactions (transaminations with α-ketoglutarate, for example) deplete the cycle of intermediates.

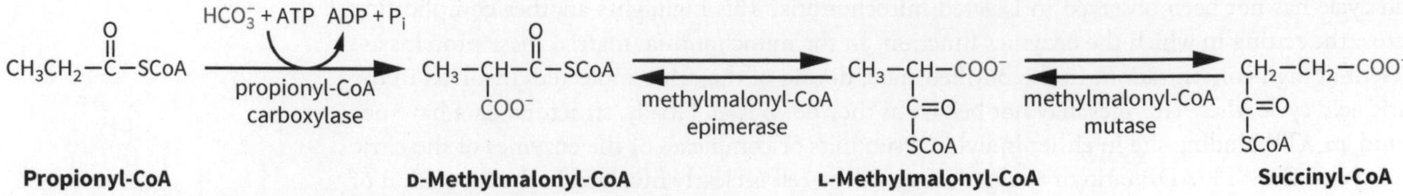

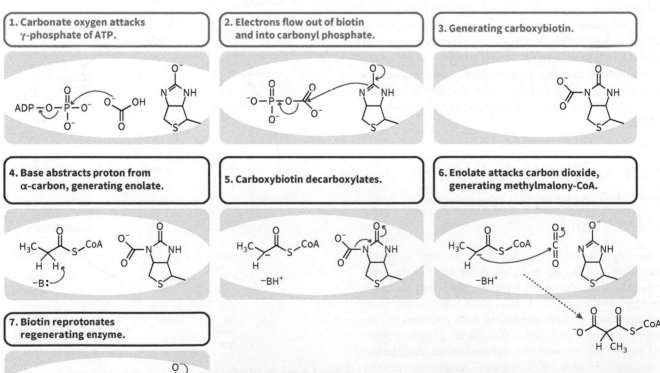

FIGURE 7.5 Conversion of propionyl-CoA to succinyl-CoA. Propionyl-CoA is derived from the catabolism of some amino acids and from fatty acids with odd numbers of carbons. Propionyl-CoA is carboxylated to methylmalonyl-CoA in a biotin-dependent process. Methylmalonyl-CoA is isomerized to succinyl-CoA by a vitamin B12 (cobalamin)-dependent enzyme.

Alpha-ketoglutarate

Levels of α-ketoglutarate can be replenished by removal of the amine moiety from the amino acid glutamate. This reaction is catalyzed by glutamate dehydrogenase and is a key reaction of anaplerosis, the detoxification of amine groups, and the catabolism of amino acids. Amine groups are often shuttled to α-ketoglutarate (a process called transamination) to generate glutamate and a new α-keto acid from the old amino acid. The new molecule of glutamate can be transported in the blood to the liver for deamination to α-ketoglutarate. The resulting amino group is detoxified via the urea cycle, which is discussed in section 11.3.

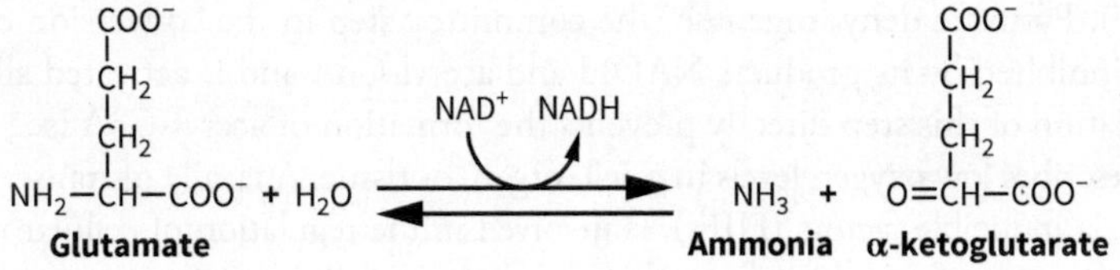

Succinyl-CoA

Levels of succinyl-CoA can be replenished through several intermediates that converge on the three-carbon fatty acid propionyl-CoA. Propionyl-CoA can be generated via catabolism of odd-chain fatty acids, branched-chain amino acids (leucine, isoleucine, and valine), or the amino acids threonine and methionine (via α-ketobutyrate) (**Figure 7.5**).

In propionyl-CoA carboxylase, a molecule of CO_2 is added to propionyl-CoA in a reaction employing a biotin functional group (**Figure 7.6**). Enzymes that add CO_2 to a molecule often

1. Carbonate oxygen attacks γ-phosphate of ATP.

2. Electrons flow out of biotin and into carbonyl phosphate.

3. Generating carboxybiotin.

4. Base abstracts proton from α-carbon, generating enolate.

5. Carboxybiotin decarboxylates.

6. Enolate attacks carbon dioxide, generating methylmalony-CoA.

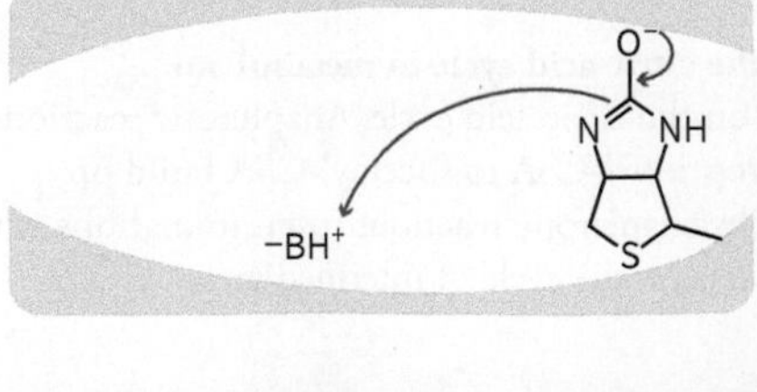

7. Biotin reprotonates regenerating enzyme.

FIGURE 7.6 Mechanism of propionyl-CoA carboxylase. Most carboxylases have a similar mechanism. Carbonate attacks the γ phosphate of ATP generating a carbonyl phosphate. The carbonyl phosphate is then attacked by one of the ring nitrogens of biotin. Phosphate departs and carboxybiotin is left. Carboxybiotin donates a CO_2 which is then attacked by the carbanionic form of propionyl-CoA. The product reprotonates to form malonyl-CoA.

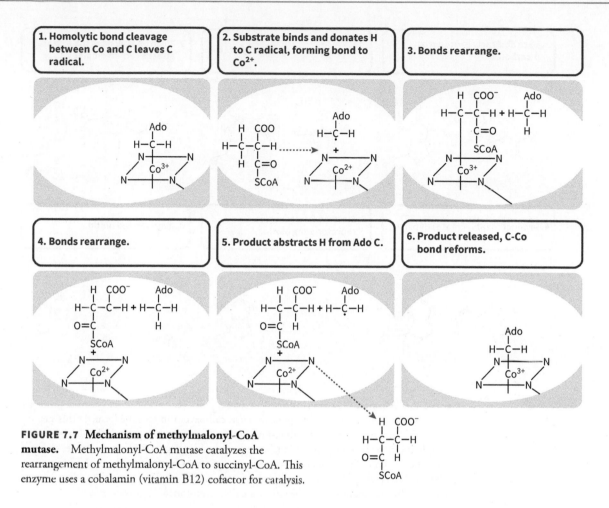

FIGURE 7.7 Mechanism of methylmalonyl-CoA mutase. Methylmalonyl-CoA mutase catalyzes the rearrangement of methylmalonyl-CoA to succinyl-CoA. This enzyme uses a cobalamin (vitamin B12) cofactor for catalysis.

employ a biotin functional group. In this particular reaction, CO_2 first reacts with the γ phosphate of ATP to form a carboxyphosphate. The lone pair of electrons on one of the ring nitrogens of biotin attacks the carbon of the carboxyphosphate, forming a carboxybiotin adduct as phosphate leaves. Next, the propionyl-CoA attacks the carboxyl carbon, facilitated by resonance stabilization and the anionic character of the α carbon of the propionate group. Methylmalonyl-CoA is released as the product. The S isoform is racemized into the R isomer by methylmalonyl CoA racemase (also known as methylmalonyl-CoA epimerase). Methylmalonyl-CoA mutase converts methylmalonyl-CoA to succinyl-CoA through a process that depends on cobalamin (vitamin B12). As with the biotin-dependent carboxylation described earlier in this chapter, this isomerization reaction also occurs in fatty acid biosynthesis (**Figure 7.7**).

Oxaloacetate
Oxaloacetate can be synthesized from either pyruvate or the amino acid aspartate (**Figure 7.8**). Pyruvate can be acted on by pyruvate carboxylase to generate the oxaloacetate used in gluconeogenesis. Like propionyl-CoA carboxylase, pyruvate carboxylase is biotin dependent. Oxaloacetate generated by this process is retained in the mitochondrial matrix for use in the citric acid cycle, whereas oxaloacetate used for gluconeogenesis is first reduced to malate, then exported from the mitochondrial matrix and reoxidized back to oxaloacetate.

In the same way that α-ketoglutarate can be generated from glutamate, oxaloacetate can be generated by the transamination of aspartate (**Figure 7.9**). Aspartate aminotransferase catalyzes the exchange of amino groups between aspartate and α-ketoglutarate, generating oxaloacetate and glutamate. Both of these enzymes employ a pyridoxal phosphate cofactor to carry out catalysis.

The reactions that replenish citric acid cycle intermediates also highlight the role of the citric acid cycle as a central metabolic hub. Many molecules are either broken down into citric acid cycle intermediates or synthesized from them. Taken a step further, not all metabolites need to completely participate in the citric acid cycle. For example, the carbon skeletons of amino acids that enter the citric acid cycle as α-ketoglutarate can exit as succinyl-CoA and contribute to heme biosynthesis.

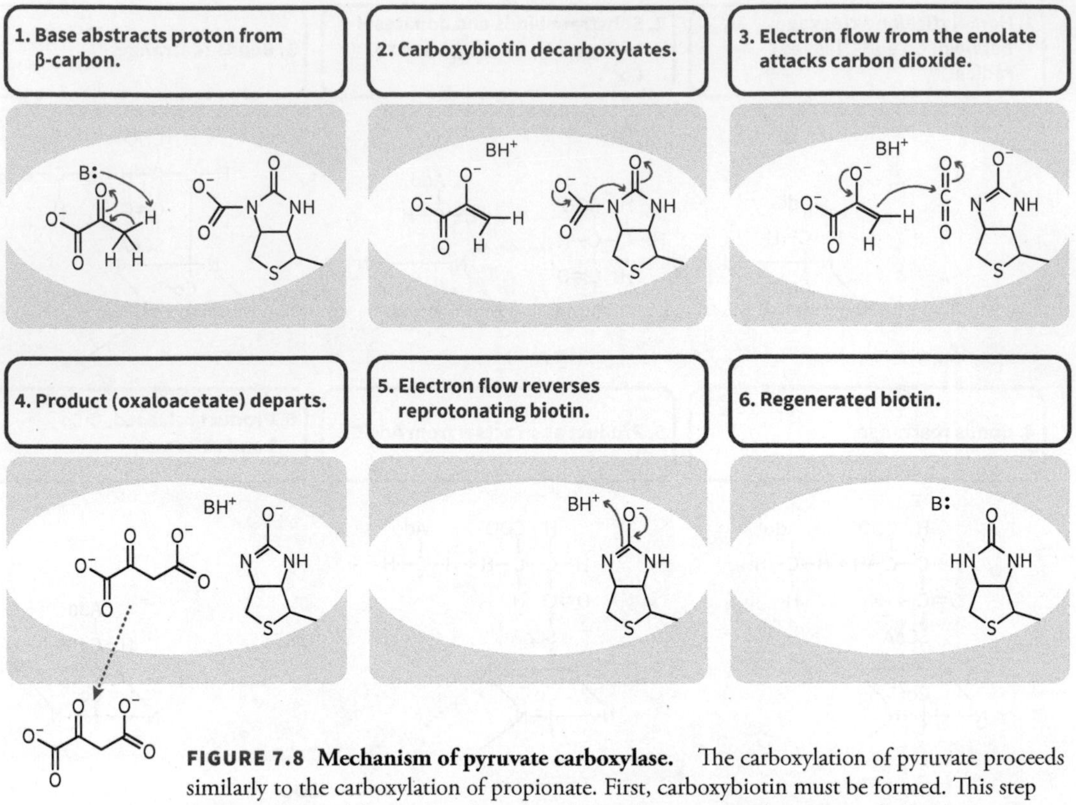

1. Base abstracts proton from
   β-carbon.

2. Carboxybiotin decarboxylates.

3. Electron flow from the enolate
   attacks carbon dioxide.

4. Product (oxaloacetate) departs.

5. Electron flow reverses
   reprotonating biotin.

6. Regenerated biotin.

FIGURE 7.8 Mechanism of pyruvate carboxylase. The carboxylation of pyruvate proceeds similarly to the carboxylation of propionate. First, carboxybiotin must be formed. This step requires ATP and formation of a carboxyl phosphate to provide energy and a leaving group to couple the carboxyl group to biotin (shown previously in Figure 7.5). In the model shown, the terminal carbon of pyruvate is deprotonated to generate an enolate. The enolate, in turn, attacks the CO_2 carbon, forming a new carboxyl group off C-3. The ring ketone of biotin is regenerated by protonation from the same base that deprotonated pyruvate earlier.

FIGURE 7.9 Transamination of aspartate. Aspartate can undergo a transamination reaction, effectively swapping an amino group with α-ketoglutarate to yield glutamate and oxaloacetate.

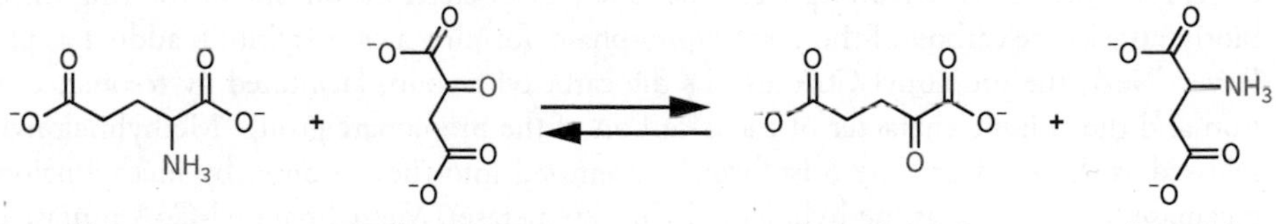

Glutamate + **Oxaloacetate** ⇌ **α-ketoglutarate** + **Aspartate**

Worked Problem 7.2   Anaplerotic reactions

Acetyl-CoA contributes directly to the formation of citrate, but acetate is not considered an anaplerotic intermediate. Explain why this is the case.

Strategy Review the reactions of the citric acid cycle. What does "anaplerosis" mean?

Solution Anaplerotic reactions "build up" levels of citric acid cycle intermediates, replenishing them. The intermediates most often discussed include α-ketoglutarate, succinyl-CoA, and oxaloacetate. The reactions of the citric acid cycle themselves are not thought of as anaplerotic because they do not lead to an increase in the concentration of that intermediate; the product of the reaction is immediately used by the next step. Because the formation of citrate from acetyl-CoA and oxaloacetate is part of the citric acid cycle, this step does not really elevate to the concentrations of intermediates and it is not anaplerotic.

Follow-up question In several neurodegenerative processes, fluctuations of glutamate and glutamine levels lead to low energy levels in the cell by slowing down the citric acid cycle. Suggest how depletion of cellular glutamate could reduce flux through the citric acid cycle.

Summary

- The citric acid cycle is an oxidative pathway that enables the organism to harvest electrons from carbon-containing intermediates (with the electrons captured in the reducing equivalents NAD^+ and FAD, and energy captured in the form of a molecule of GTP or ATP).
- The cycle is:
  - regulated by levels of substrates and allosteric regulators such as ADP and citrate.
  - the entry point of many molecules other than acetyl-CoA into metabolism, including many amino acid skeletons.
  - the source of the building blocks of many molecules, including many amino acids.
- Anaplerotic reactions act to build up levels of citric acid cycle intermediates that may have been drained by other metabolic needs.

Concept Check

1. Trace the reactions of the citric acid cycle in terms of the chemistry occurring in each step, the structures of the intermediates, and in terms of where NADH and $FADH_2$ are harvested.
2. Show which steps of the citric acid cycle are regulated and how they are regulated. (This should make biochemical sense: when energy (in the form of ATP) is abundant, the cycle is slowed, but when there is a lack of energy, flux through the pathway is increased.)
3. If $[4\text{-}^{14}C]$ oxaloacetate were used as the substrate for citric acid cycle, which carbon(s) of succinate would be labeled?

7.2 The Electron Transport Chain

In the metabolism covered so far in this chapter, a lot of chemistry has occurred, many carbons have been oxidized and numerous electrons have been harvested, but relatively little ATP has been generated. In the catabolism of glucose, glycolysis generated two molecules of ATP and captured four electrons in two molecules of NADH (**Figure 7.10**). Decarboxylation of pyruvate to form acetyl-CoA generated another two molecules of NADH, and oxidation of two molecules of acetyl-CoA into CO_2 produced another six molecules of NADH, two molecules of $FADH_2$, and two molecules of GTP or ATP. Thus, starting from a single molecule of glucose, at this point, metabolism has generated four ATP or their equivalent, but ten NADH, and two $FADH_2$.

Although relatively little ATP has been generated, the chemical energy of reactions has been stored in electron carriers, primarily NADH. This section describes the electron transport chain and how it uses that energy and those electrons to pump protons out of the mitochondrial matrix. Section 7.3 shows how the energy of the proton gradient created in this step drives ATP synthase.

The electron transport chain occurs in the matrix and inner mitochondrial membrane of eukaryotic cells. Protons are pumped out of the mitochondrial matrix and NADH and $FADH_2$ are oxidized back to NAD^+ and FAD. Four complexes—named I, II, III, and IV—are involved in this process. One of these complexes we have already seen. Complex II is succinate dehydrogenase from the citric acid cycle and was discussed in section 7.1.1. This section outlines the means by which electrons are transported through the complexes and discusses the structure and functions of each of the four complexes (**Figure 7.11**).

7.2.1 Electron transport occurs through a series of redox active centers from higher to lower potential energy

As glucose is metabolized through glycolysis, it is converted from one intermediate to the next. Almost all of these intermediates are stable

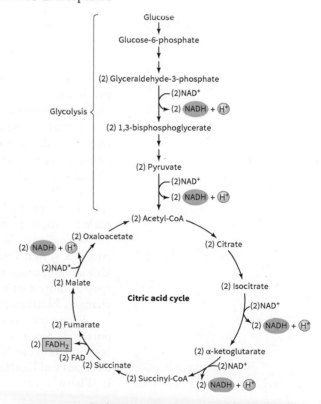

FIGURE 7.10 Reducing equivalents come from multiple metabolic pathways. The reducing equivalents derived from glycolysis, the citric acid cycle, and other metabolic pathways such as β-oxidation (section 9.2) are oxidized in the electron transport chain.

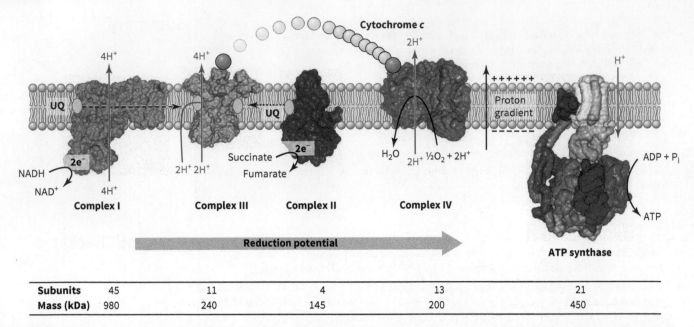

| Subunits | 45 | 11 | 4 | 13 | 21 |
| Mass (kDa) | 980 | 240 | 145 | 200 | 450 |

FIGURE 7.11 Overview of the electron transport chain.   The electron transport chain consists of four complexes that oxidize NADH and FADH$_2$ back to NAD$^+$ and FAD. The electrons from these complexes are passed through a series of redox centers ultimately being consumed in the formation of water in complex IV. The passage of electrons through these complexes provides energy to pump protons from the mitochondrial matrix to the intermembrane space. This proton gradient is used to synthesize ATP by the ATP synthase. Molecular weights and number of subunits provided are for the mammalian electron transport chain.

(Source: (Complex I) Data from PDB ID 5LDW Zhu, J., Vinothkumar, K.R., Hirst, J. (2016) Structure of mammalian respiratory complex I. *Nature* 536: 354-358. (Complex III) Data from PDB ID 3CX5 Solmaz, S.R., Hunte, C. (2008) Structure of complex III with bound cytochrome *c* in reduced state and definition of a minimal core interface for electron transfer. *J. Biol. Chem.* 283: 17542-17549. (Complex II) Data from PDB ID 1NEN Yankovskaya, V., Horsefield, R., Tornroth, S., Luna-Chavez, C., Miyoshi, H., Leger, C., Byrne, B., Cecchini, G., Iwata, S. (2003) Architecture of succinate dehydrogenase and reactive oxygen species generation *Science* 299: 700-704. (Complex IV) Data from PDB ID 5B1A Yano, N., Muramoto, K., Shimada, A., Takemura, S., Baba, J., Fujisawa, H., Mochizuki, M., Shinzawa-Itoh, K., Yamashita, E., Tsukihara, T., Yoshikawa, S. (2016) The Mg^{2+}-containing Water Cluster of Mammalian Cytochrome *c* Oxidase Collects Four Pumping Proton Equivalents in Each Catalytic Cycle. *J. Biol. Chem.* 291: 23882-23894. (ATP synthase) Data from PDB ID 5ARA Zhou, A., Rohou, A., Schep, D.G., Bason, J.V., Montgomery, M.G., Walker, J.E., Grigorieff, N., Rubinstein, J.L. (2015) Structure and conformational states of the bovine mitochondrial ATP synthase by cryo-EM. *Elife* 4: e10180-e10180)

molecules that can be isolated. Hence, it is possible to imagine glycolysis as an assembly line in which one molecule is transformed to the next, or as a cafeteria where one chooses a lunch, a drink, and finally a dessert from a range of alternatives. No matter how we envision these reactions occurring, the analogy falls woefully short when considering electron transport, which can seem more akin to a gymnastic routine. A gymnast cannot simply stop during a routine and hold a pose for a prolonged period; rather, the routine is dynamic and controlled with great precision. Similarly, it is the precise control of electron flow that permits electron transport to occur.

As with a gymnast performing a routine, electrons flow smoothly down a gradient of reduction potentials, moving from one electron carrier to another, and changing the oxidation state of that carrier as they do so (Table 7.2). In some instances, this change can be blocked and the reduced intermediate isolated, but often it cannot. Disruption of the complex would lead to oxidation or reduction of other carriers, and a loss of the signal. Unlike molecules, electrons cannot be isotopically labeled; instead, they are often tracked spectrophotometrically as they perform chemistry on a chromophore, or as their spin state changes. Electron spin can be determined using electron paramagnetic resonance (EPR).

Each electron carrier involved in the electron transport chain has a different reduction potential, dependent on both the carrier and its environment. Therefore, changes in protein conformation can help to fine-tune the affinity of the carrier for electrons and keep the flow of electrons unidirectional. The electron carriers found in the electron transport chain are listed in (Figure 7.12).

- Flavins, such as FAD (flavin adenine dinucleotide) and FMN (flavin mononucleotide (FMN), employ the tricyclic isoalloxazine ring to bind electrons on the N-5 position. Flavins can transport different numbers of electrons: none (fully oxidized), one (partially reduced), or two (fully reduced).

TABLE 7.2 Reduction Potentials of Electron Carriers Found in Biochemistry

Component	Reduction Potential ($E^{\circ\prime}$), V	Component	Reduction Potential ($E^{\circ\prime}$), V
NADH	−0.315	Ubiquinone	0.045
Complex I (NADH:coenzymeq oxidoreductase, ~900 kDa, 45 subunits)		**Complex III (UQ-cytochrome c oxdoreductase, 450 kDa, 9–11 subunits)**	
FMN	−0.380	Heme b_H	0.030
[2Fe-2S]N1a	−0.370	Heme b_L	−0.030
[2Fe-2S]N1b	−0.250	[2Fe-2S]	0.280
[4Fe-4S]N3, 4, 5, 6a, 6b 7	−0.250	Heme c_1	0.215
[2Fe-2S]N2	−0.150	Cytochrome c	0.235
Succinate	0.031	**Complex IV (cytochrome c oxidase, ~410 kDa 8–13 subunits)**	
Complex II (succinate-CoQ oxidoreductase, ~140 kDa, 4 subunits)		Heme a	0.210
FAD	−0.040	Cu_A	0.245
[2Fe-2S]	−0.030	Cu_B	0.340
[4Fe-4S]	-0.245	Heme a_3	0.385
[3Fe-4S]	−0.060	O_2	0.815
Heme b_{560}	−0.080		

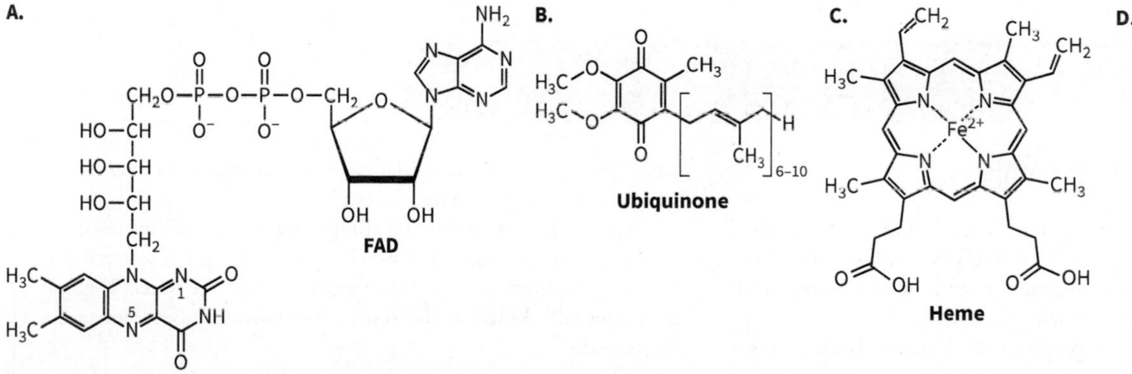

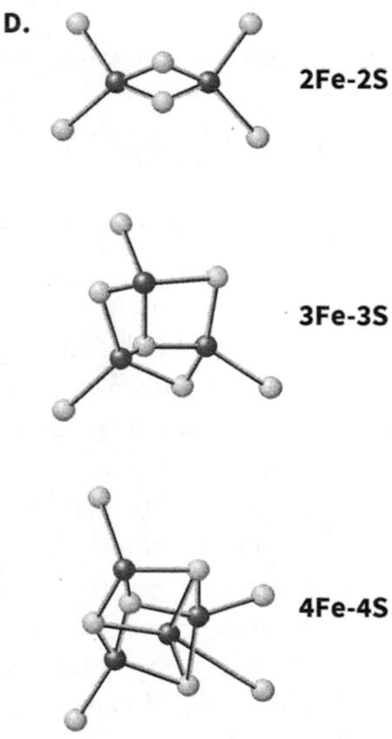

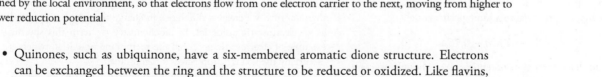

FIGURE 7.12 Electron carriers in the electron transport chain. Shown are some of the electron carriers employed by the electron transport chain. **A.** In FAD, the reduction occurs on the N-1 and N-5 positions. **B.** In ubiquinone (also known as coenzyme Q 10), the carbonyl groups are reduced to hydroxyl groups. **C.** Shown is an example of a heme group. The iron ion in the core of the organic complex (a protoporphyrin ring) carries the electrons. **D.** In iron-sulfur centers, the electrons are delocalized throughout the complex. The reduction potentials of each of these centers can be tuned by the local environment, so that electrons flow from one electron carrier to the next, moving from higher to lower reduction potential.

- Quinones, such as ubiquinone, have a six-membered aromatic dione structure. Electrons can be exchanged between the ring and the structure to be reduced or oxidized. Like flavins, quinones can be fully reduced (dihydroquinone), partially reduced (a semiquinone), or fully oxidized (quinone).

- Heme groups employ an active metal center, usually iron in electron transport, bound to a porphyrin ring in a planar, heterocyclic cofactor. The iron ion is bound by the four nitrogens of the porphyrin ring. Heme proteins bind to the porphyrin ring through either cysteine or histidine groups. Amino acid side chains of the protein (typically histidine or cysteine residues) also chelate (bind) the metal ion center above or below the plane of the porphyrin ring. Although this section focuses on heme proteins in the context of electron transport, heme groups also figure prominently in oxygen transport and in binding and catalysis. Heme groups bind a single electron, with the central iron ion fluctuating between Fe^{2+} and Fe^{3+} states.

- **Iron-sulfur centers** are found in iron-sulfur proteins. There are three different types: 2Fe-2S, 3Fe-3S, and 4Fe-4S. In each of these centers, there is a specific geometric arrangement of sulfur and iron atoms. Each iron-sulfur center can bind a single electron, delocalized across the entire center. Cysteine residues bind the iron-sulfur center in place in the protein. Often, multiple iron-sulfur centers are found in a single protein or protein complex.

- Copper centers found in cytochrome oxidase also act in electron transport. There are two such centers: Cu_A, which has two Cu^{2+} ions, and Cu_B, which has a single Cu^{2+} ion. The single electron transferred between these centers is delocalized over both copper ions in Cu_A, but it is bound exclusively by the single copper center in Cu_B.

The electrons flowing through the electron transport chain can be pictured as following little arrows as they move, but in reality, the electrons use the quantum mechanical property of **tunneling** to "hop" from one center to another. Tunneling is the ability of a small particle with sufficient energy, such as an electron, to pass through a barrier that is otherwise impenetrable. This may seem irrational on a macroscopic scale, but the probability of it happening can be predicted using quantum mechanics, and observed results validate the predictions.

While this section focuses on the electron carriers commonly found in electron transport, there are other electron carriers, many of which employ similar chemistry, although they may use different elements to carry these electrons. For example, plastocyanin proteins are small electron carriers used in photosynthesis; they are analogous to cytochromes in several ways but use copper ions rather than a heme group to store electrons. Ferredoxin is also an important electron carrier in several reactions; it employs an iron-sulfur center to store electrons. Other examples include the magnesium-containing center in chlorophyll, and the manganese and calcium atoms in the water-splitting enzyme of photosystem II. Clearly, biology has evolved numerous ways to transport and harness the power of electrons. A review of reduction potential and some of the roles it plays in biochemistry can be found in **Biochemistry: Electrochemistry in biochemistry**.

Biochemistry

Electrochemistry in biochemistry

Electrochemistry is important in biochemistry but is often poorly understood. In this chapter, electrochemistry figures prominently in two different ways: reduction potential and membrane potential. This section examines each in detail.

Reduction potential is a property of all materials but is often encountered in general chemistry as a series of reduction reactions of metals and metal ions. It is the property of all materials to gain or lose an electron or electrons; however, by convention, it is expressed as a reduction (gain of electrons). Reduction potential is measured in volts and is what we typically think of in terms of a battery or wet electrochemical cell. Materials with different reduction potentials will pass electrons from one material to another in a lower energy state (from a more negative reduction potential to a more positive one).

$$Zn^{2+} + 2e^- \rightarrow Zn(s) - 0.762$$

All molecules have reduction potentials, even though we commonly think of metals in relation to this type of electrochemistry. In biochemistry, we encounter reduction potentials in electron transport. As electrons pass from one electron carrier to another in the electron transport chain, they move down a series of reduction potentials, culminating in the reduction of oxygen (the formation of water).

$$O_2(g) + 4H^+ + 2e^- \rightarrow 2H_2O(1) + 1.229 \text{ V}$$

The electrochemical potential of protons, also known as the proton motive force or pmf, is the ability of a proton gradient to do work. Electrochemical potential has two components: the *chemical potential* and the *electrostatic potential*.

Chemical potential is the ability of a system of chemicals to do work. Imagine a system in which there is a beaker of water and a dye diffusing from the top of the beaker through the water. The dye molecules will diffuse as the system spontaneously becomes more disordered.

Electrostatic potential is the potential energy of a charged particle in an electric field, divided by the charge on the particle. The units of electrostatic potential are therefore joules/coulomb or volts. In biochemistry, we often see this as a disequilibrium of ions, an uneven concentration of ions across an impermeable or semipermeable membrane. In effect, this is the electronic component known as a capacitor. By segregating more ions on one side of the membrane than another, we create a difference in charge that can be expressed as an electrostatic potential. In biochemistry we term this specific example of electrostatic potential the membrane potential, which is essentially the voltage drop across a membrane.

The combination of membrane potential (electrostatic potential) and chemical potential results in the electrochemical potential, that is, the overall potential of an ion gradient to do work. This chapter discusses the electrochemical potential of the inner mitochondrial membrane, and in biochemistry there are frequent references to the electrochemical potential of cells, such as neurons or cardiomyocytes; however, most cells have an electrochemical potential across their membranes. Most cell types do not have a *polarizable* membrane with pumps and channels to create and use a gradient of ions such as those found in neurons, but most cell membranes are polarized.

Cytosolic NADH is shuttled into the mitochondrial matrix
Glycolysis occurs in the cytosol and produces NADH, but complex I (NADH dehydrogenase) oxidizes NADH in the mitochondrial matrix. For NADH to be oxidized, it must be imported to the mitochondrial matrix. The inner mitochondrial membrane is impermeable to NADH, but the cell has evolved several means of transporting other reduced molecules into the mitochondria where they are oxidized (**Figure 7.13**).

The **glycerophosphate shuttle** converts the glycolytic intermediate dihydroxyacetone phosphate (DHAP) to glycerol-3-phosphate. On the inner mitochondrial membrane, a flavoprotein (flavoprotein dehydrogenase) transfers electrons from glycerol-3-phosphate to the FAD center, generating $FADH_2$. The electrons are then passed on to ubiquinone, where they can continue to participate in electron transport. In the process of this reaction, the glycerol-3-phosphate is converted back to DHAP, which can continue to participate in glycolysis.

The **malate-aspartate shuttle** is another means of transporting electrons into the mitochondria. In this shuttle, two exchange proteins (the α-ketoglutarate-malate carrier and the

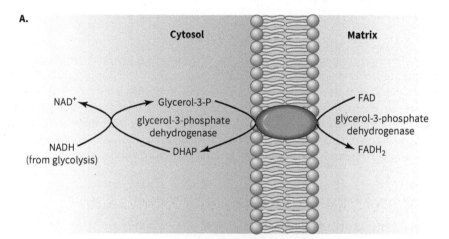

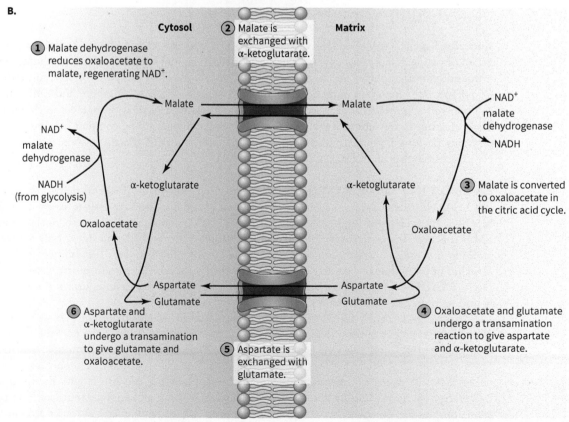

FIGURE 7.13 Mitochondrial shuttles. A. The glycerol phosphate shuttle uses DHAP to oxidize NADH, generating glycerol-3-phosphate in the process. These electrons are then transferred to FAD when glycerol-3-phosphate is oxidized back to dihydroxyacetone phosphate. **B.** In the malate-aspartate shuttle, glutamate and oxaloacetate undergo a transamination reaction to generate aspartate and α-ketoglutarate. In the cytosol another transamination regenerates glutamate and oxaloacetate. The oxaloacetate is reduced to malate, which is shuttled back into the mitochondrial matrix where it is oxidized to oxaloacetate, delivering electrons to the matrix in the process.

Worked Problem 7.3 | Electron flow

Working in the laboratory, you isolated an electron-transfer complex from a microbe that contains the following cofactors: a quinone, a heme complex, Cu^{2+}, and Zn^{2+}. Which of these cofactors are likely to be involved in electron transport? How might electrons flow through this complex?

Strategy Examine Table 7.2. Although all of these cofactors are found in biochemistry, some are more common in electron transport. Redox potentials depend partially on the carrier itself and partially on the microenvironment created by the protein around the center. What are the commonly accepted values for the centers listed in this problem?

Solution It is likely that the three-electron carriers are the quinone, the heme, and the copper ion. Zinc ions figure prominently

in biochemistry, and they are often used as an example when discussing electrochemistry in a general chemistry class, but zinc ions do not usually participate directly in redox chemistry in biochemistry.

In terms of the order, examination of the reduction potentials of the centers shows that the quinone has the lowest value, followed by the heme, and then the copper ion. This is one potential pathway; however, as we have seen, the environment of these redox centers can be altered by their environment.

Follow-up question How could you use spectroscopy to test your hypothesis about the flow of electrons?

---

aspartate-glutamate carrier) transport molecules across the inner mitochondrial membrane. The α-ketoglutarate-malate carrier exports α-ketoglutarate from the matrix and imports malate. Once in the mitochondrial matrix, malate can be converted to oxaloacetate through the citric acid cycle, generating NADH. Oxaloacetate can be transaminated to aspartate (exchanging amino groups with glutamate to generate α-ketoglutarate). The aspartate is exchanged from the matrix for glutamate. In the cytosol, aspartate is transaminated with α-ketoglutarate to yield oxaloacetate and glutamate. Unlike the glycerophosphate shuttle, the malate–aspartate shuttle is reversible.

7.2.2 Complex I (NADH dehydrogenase) transfers electrons from NADH to ubiquinone via a series of iron-sulfur centers

Complex I is the site of NADH oxidation to NAD^+. This 980 kDa, L-shaped complex is comprised of two large domains: a helical domain buried in the inner mitochondrial membrane, and a second globular domain on the matrix side. In eukaryotes, this complex contains more than 43 polypeptide chains, a tightly bound flavin mononucleotide (FMN) group, and nine different iron-sulfur centers that are used to transfer electrons from NADH, ultimately to the membrane electron carrier **ubiquinone**. The transmembrane domain of the complex is important in pumping protons out of the mitochondrial matrix, and the second domain is where reduction of NADH and transport of electrons occurs (**Figure 7.14**).

Electron transfer in complex I The precise mechanism of NADH dehydrogenase is still being elucidated, but several important aspects of the enzyme have yielded clues about its function. First, electrons are transferred from NADH in the mitochondrial matrix to the bound FMN cofactor. The isoalloxazine ring of FMN has three different oxidation states: fully reduced, semiquinone (one extra electron), and fully oxidized (two extra electrons). The ability of FMN to function in more than one oxidation state enables the transfer of electrons from the two-electron carrier (NADH). From FMN, the electrons flow through several (up to nine) iron-sulfur centers, including both Fe2-S2 and Fe4-S4.

Ultimately, electrons are transferred to the two-electron carrier ubiquinone. As with FMN, ubiquinone can exist in three different oxidation states: fully reduced (Q), the partially oxidized semiquinone (Q^-), and fully oxidized (QH_2) (**Figure 7.15**). In addition to its multiple oxidation states, ubiquinone has another feature essential to electron transport: its hydrophobic isoprenoid tail anchors the carrier in the inner mitochondrial membrane and prevents it (and the electrons it is carrying) from diffusing throughout the matrix. Instead, ubiquinone transfers electrons from complex I to complex III. As explained in section 7.2.3, ubiquinone is also used to transfer

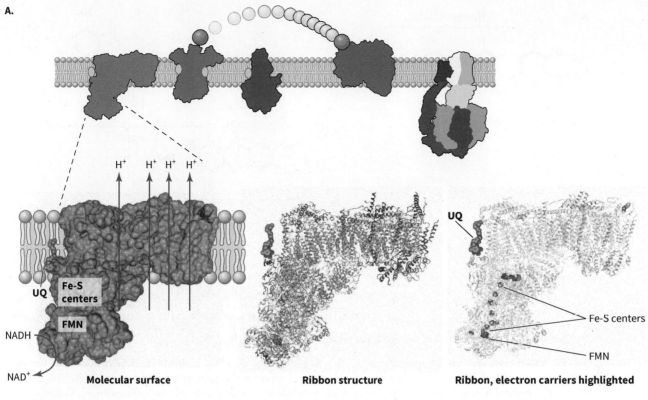

A.

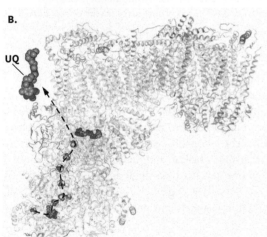

Molecular surface   Ribbon structure   Ribbon, electron carriers highlighted

B.

FIGURE 7.14 Complex I. **A.** Complex I (NADH dehydrogenase) consists of a large membrane-bound domain where proton pumping occurs and a cytosolic arm responsible for electron transport. Electrons are trafficked from NADH to a flavin center (FMN), through a series of iron-sulfur centers ultimately being used to reduce ubiquinone. A ribbon diagram of the NADH dehydrogenase shows the overall structure and the numerous transmembrane α-helical domains. Movement of electrons through the complex causes a conformational change that results in the pumping of four protons from the matrix to the cytosolic side of the inner mitochondrial membrane. **B.** A transparent view of complex I showing the electron carriers FMN and Fe-S centers and the site of ubiquinone binding.

(Source: Data from PDB ID 5LDW Zhu, J., Vinothkumar, K.R., Hirst, J. (2016) Structure of mammalian respiratory complex I. *Nature* 536: 354-358)

electrons from complex II to complex III. A different soluble carrier (cytochrome *c*) is used to transfer electrons from complex III to complex IV (see section 7.2.4).

Complex I pumps protons out of the mitochondrial matrix

As mentioned in section 7.1, electron transport serves two important functions: it oxidizes electron carriers back to their initial states, and it uses the energy of transferring these electrons to pump protons out of the mitochondrial matrix against their concentration gradient, that is, moving protons from lower H^+ concentration inside the matrix to higher H^+ concentration outside the matrix. Instead of wasting this surplus energy as heat, the electron transfer reactions are coupled to proton transfers across the inner mitochondrial membrane. The protons are transferred against their concentration gradient, and therefore the proton transfers on their own would be energetically unfavorable. However, the overall process, which combines the favorable electron transport processes with the unfavorable proton pumping processes, adds up to a net favorable combined process. In other words, the surplus energy from electron transfer drives the formation of the proton gradient.

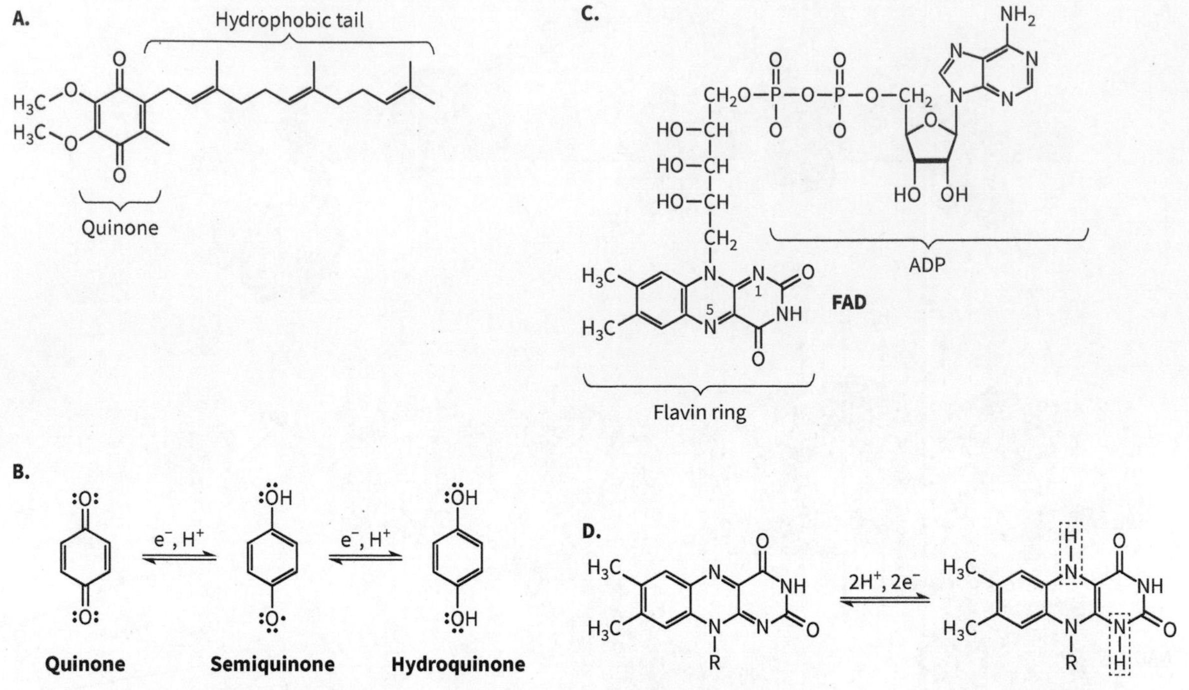

FIGURE 7.15 Redox states of ubiquinone and FAD.   A. Ubiquinone has a long, hydrophobic isoprenoid tail that anchors it in the inner mitochondrial membrane, but the redox chemistry happens at the quinone end of the molecule. **B.** Ubiquinone can be fully oxidized (the quinone), partially reduced (the semiquinone), or fully reduced (the hydroquinone). **C.** FAD consists of a tricyclic flavin ring connected to the polyol ribitol and then ADP. **D.** The redox chemistry occurs on the nitrogens at positions 1 and 5 on the flavin ring.

The cell uses the proton gradient to synthesize ATP. Complex I contributes to the proton gradient by pumping four protons out of the mitochondrial matrix. Analysis of a bacterial isoform of this protein (a simpler version than the mitochondrial isoform) reveals several features that help to elucidate the process. The bacterial isoform has three domains, each of which has architecture similar to Na^+/H^+ antiporters. These subunits are strongly helical, and they are buried in the inner mitochondrial membrane; they are connected to one another by a single helical segment that is orthogonal (rotated 90°) to the transmembrane helices. This single helix is thought to move back and forth like a piston, opening and shutting the ion channels and pumping protons out of the matrix as electrons flow through the complex.

7.2.3 Complex II is the citric acid cycle enzyme succinate dehydrogenase

Complex II is succinate dehydrogenase from the citric acid cycle (this enzymatic complex is discussed earlier in this chapter). This enzyme catalyzes the oxidation of succinate (a four-carbon dioic acid) to fumarate (a *trans* unsaturated dioic acid). In the process, two electrons are removed and transferred to FAD, generating $FADH_2$. Ultimately, the electrons from this reaction are transferred to ubiquinone and shuttled to complex III. No protons are pumped out of the matrix by complex II (**Figure 7.16**).

Complex II is comprised of four protein subunits: two transmembrane and two hydrophilic. The two hydrophilic proteins contain the FAD group (covalently attached to the protein via a histidine side chain), and an iron-sulfur protein containing three different iron-sulfur centers (2Fe-2S, 4Fe-4S, 3Fe-4S). Transfer of electrons from the iron-sulfur centers to ubiquinone is mediated by a heme center.

Electron transfer in complex II   As with complex I, the mechanism of electron carrier oxidation is not completely clear; however, X-ray crystallographic studies have provided some

A.

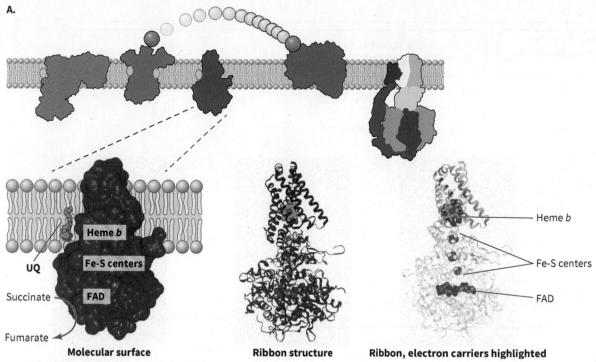

Molecular surface   Ribbon structure   Ribbon, electron carriers highlighted

B.

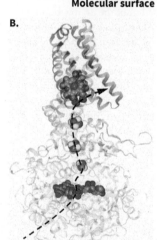

FIGURE 7.16 Complex II. Complex II is succinate dehydrogenase. **A.** Complex II does not transfer any protons from the matrix to the intermembrane space. Instead, it transfers electrons from the bound FAD cofactor through a series of iron-sulfur centers to a heme *b* group where they are transferred to ubiquinone. The panel on the right shows the arrangement of the electron carrying complexes in complex II. **B.** A transparent view of complex II showing the FAD, Fe-S, and heme centers.

(Source: Data from PDB ID 1NEN Yankovskaya, V., Horsefield, R., Tornroth, S., Luna-Chavez, C., Miyoshi, H., Leger, C., Byrne, B., Cecchini, G., Iwata, S. (2003) Architecture of succinate dehydrogenase and reactive oxygen species generation *Science* 299: 700-704)

potential leads as to how these reactions occur (**Figure 7.17**). There are two potential mechanisms for the formation of fumarate and the transfer of electrons to FAD: a concerted, E2 type of elimination reaction or a stepwise reaction (E1) with an enolate intermediate. If the reaction were to proceed via an E2 mechanism, a base (presumably a basic residue, such as histidine or deprotonated arginine) would deprotonate one of the α carbons of succinate. Electrons would then flow through the carbons, generating the double bond through the transfer of a hydride (H⁻) to the N-5 atom of the FAD group (generating FADH$_2$). The coplanar geometry needed for an E2 reaction helps to explain why the product (fumarate) is in a *trans* configuration. In the proposed E1 mechanism, a base abstracts an α proton (as it did in the E2); however, the electrons flow into one of the carbonyl oxygens, generating an enolate. Electron flow is then reversed, and FAD acquires a hydride ion to give the products. There is some experimental data to support both of these mechanisms.

Electrons from FADH$_2$ are passed through the iron-sulfur centers to the heme *b* group. It is unclear at this point whether the heme group is actually oxidized in this reaction (existing as a short-lived intermediate and essentially holding the electrons as they are passed), or helps to stabilize the semiquinone form of the ubiquinone as electrons are moving down the chain.

E1 mechanism

1. Base abstracts proton. Electrons flow into carbonyl forming enolate transition state.	2. Electron flow then reverses and a hydride is acquired by FAD.	3. Fumarate is formed and released.

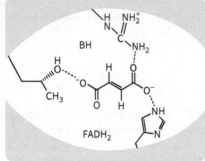

E2 mechanism

1. Concerted loss of proton and formation of double bond through a periplanar transition state.	2. Fumarate is formed and released.

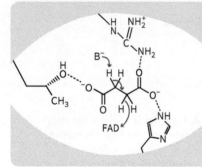

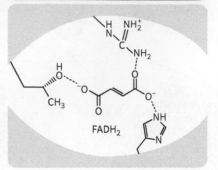

FIGURE 7.17 Two plausible mechanisms for succinate dehydrogenase. Two potential mechanisms for the formation of fumarate by succinate dehydrogenase exist. In the first, proton abstraction by base, formation of the double bond, and loss of a hydride ion to FAD forming $FADH_2$ occur in a concerted fashion through a classic E2 mechanism. The second proposed mechanism is basically an E1 reaction in which a basic residue abstracts a proton, and electrons flow into the carbonyl oxygen, forming an enolate transition state. Electron flow then reverses, and a hydride (H^-) is acquired by FAD. Experimental data exists that partially validates each mechanism.

Although this complex is called complex II, the reactions it carries out do not follow the reactions of complex I in any organized way. Both complex I and complex II independently generate ubiquinone, which transfers electrons to complex III.

7.2.4 Complex III is ubiquinone/cytochrome *c* reductase

Complex III takes electrons from the membrane-bound electron transporter ubiquinone and transfers them to the soluble electron carrier **cytochrome *c*** (**Figure 7.18**). This presents the cell with a challenge, in that ubiquinone can carry two electrons, whereas cytochrome *c* and the carriers in the remainder of the electron transport chain can carry only one. Electrons are passed through several heme centers and an iron-sulfur center (termed a Rieske center after its discoverer John Rieske) as they are transferred to cytochrome *c*. The net result of the oxidation of one molecule of ubiquinone is the formation of two molecules of reduced cytochrome *c* and the pumping of four protons out of the mitochondrial matrix.

Complex III contains 11 subunits, three of which are cytochrome proteins containing heme groups, and one of which contains an iron-sulfur center (the Rieske center). The entire complex is dimeric; it has an α-helical region that spans the mitochondrial membrane and a globular complex on the matrix side of the inner mitochondrial membrane.

Electron transport in complex III and the Q cycle As mentioned in the previous section, ubiquinone can carry two electrons, while the remainder of the electron transport

A.

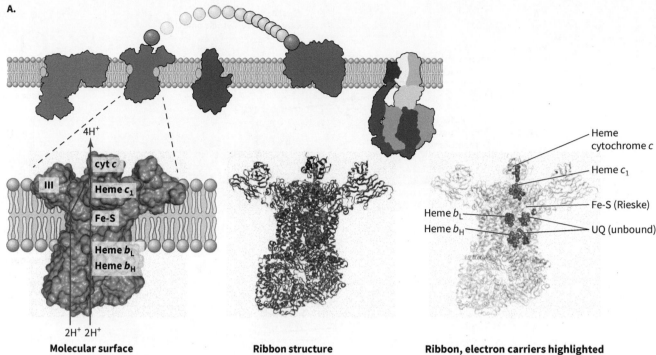

Molecular surface   **Ribbon structure**   **Ribbon, electron carriers highlighted**

B.

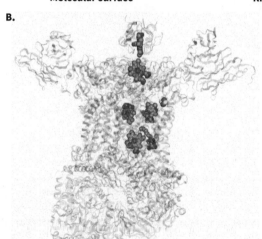

FIGURE 7.18 Complex III.   Complex III is the cytochrome *c*/ubiquinone oxidoreductase. **A.** Electrons in this complex are transferred from the two-electron carrier ubiquinone to the one-electron carrier cytochrome *c*. In the process, protons are pumped from the matrix, contributing to the proton gradient. **B.** The transparent view of complex III shows the several different heme groups that figure prominently in this process.

(Source: Data from PDB ID 3CX5 Solmaz, S.R., Hunte, C. (2008) Structure of complex III with bound cytochrome *c* in reduced state and definition of a minimal core interface for electron transfer. *J.Biol.Chem.* 283: 17542-17549)

chain can carry only one. Therefore, there needs to be some way to move from transporting electrons in pairs to transporting them singly; this mechanism is the **Q cycle** (**Figure 7.19**).

There are two ubiquinone binding sites in complex III: the Q_p site for the more positively charged intermembrane space, and the Q_n site for the more negative or matrix side. Fully reduced ubiquinone binds to complex III in the Q_p site. An electron is transferred first to the Rieske center and then to a heme (heme c_1), before passing to cytochrome *c*. In this process, the ubiquinone becomes partially reduced to the semiquinone form, and two protons are released to the inter-membrane space side of the inner mitochondrial membrane.

The second electron is then transferred from the semiquinone to another heme (heme b_L). This electron tunnels to the final heme in this complex (heme b_H). H and L refer to high ($E^{\circ\prime} = 0.050$) and low ($E^{\circ\prime} = -0.100$) electron-transfer potential, with H being found closer to the matrix side of the membrane. The electron is passed from heme b_H to a different molecule of fully oxidized ubiquinone bound at the Q_n site, reducing it to the semiquinone form.

The second half of the Q cycle begins as the first did. A new molecule of fully reduced ubiquinone binds in the Q_p site and undergoes a one-electron reduction, passing that electron off to a

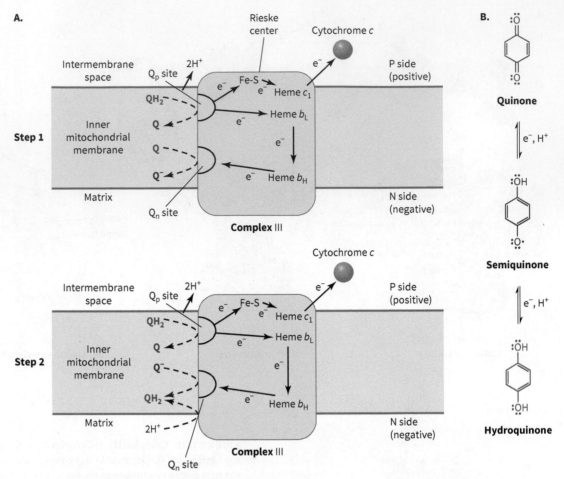

FIGURE 7.19 Q cycle and electron transport. The Q cycle switches electron transport from transporting pairs of electrons to transporting single electrons. **A.** In step 1 of the Q cycle, fully reduced ubiquinone (QH2) binds to complex III in the Q_p site. Two protons from coenzyme Q are released to the intermembrane space. One electron is transferred to cytochrome c while the other is shuttled via heme b_L and heme b_H to a fully oxidized molecule of ubiquinone (Q) bound in the Q_n site, generating a semiquinone (Q^-). In step 2 of the Q cycle, a new molecule of fully reduced ubiquinone binds in the Q_p site. Electrons are transferred and protons are pumped as in the first half; however, this results in the formation of a new molecule of fully reduced ubiquinone in the Q_n site. This molecule is released and can be subsequently oxidized in another cycle. **B.** Shown are the oxidation states of ubiquinone.

(Source: Illustration by Nicholas and Ferguson from *Plant Physiology and Development*, 6e, edited by Lincoln Taiz, Eduardo Zeiger, Ian Max Moller, and Angus Murphy, Copyright © 2014 by Sinauer Associates. Reprinted by permission of Oxford University Press, rights conveyed through Copyright Clearance Center, Inc.)

Rieske center, then to heme c_1 and finally to cytochrome c, releasing two protons on the intermembrane space side. The second electron from the semiquinone bound at the Q_p site is transferred to heme b_L and heme b_H. Heme b_H then reduces the semiquinone bound at the Q_n site to form fully reduced ubiquinone, which is released into the inner mitochondrial membrane. This may seem like a redundant pathway because it regenerates one reduced ubiquinone for each two that enter the cycle, but in fact, it provides a relatively simple way to switch from carrying one electron to carrying two. The Q pool is the term for the combination of oxidized and reduced forms of ubiquinone found in the mitochondrial membrane.

7.2.5 Cytochrome c is a soluble electron carrier

Electrons from complex III are transferred to cytochrome c. This small heme protein (12 kDa) carries a single electron in the heme group (**Figure 7.20**). The heme group itself is bound to the protein via cysteine side chains, thus differing slightly from the heme groups discussed previously. The critical heme cofactor is found buried in the core of the protein, and it is inaccessible to solvent

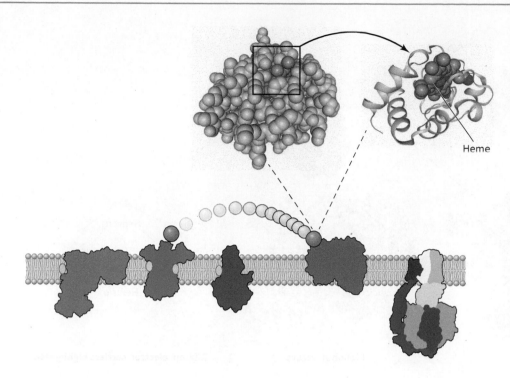

FIGURE 7.20 Cytochrome *c*.
Cytochrome *c* is a small soluble heme protein that acts to shuttle electrons from complex III to IV. The heme electron carrier is almost completely buried in cytochrome *c*.

(Source: Data from PDB 3CYT Takano, T., Dickerson, R.E. (1980) Redox conformation changes in refined tuna cytochrome *c. Proc Natl Acad.Sci.* USA 77:6371-6375)

or substrates. The surrounding protein provides an environment that regulates the affinity of the heme group for its passenger electron, and permits loading and unloading of the electron by binding to complexes III and IV at an appropriate location and distance from other electron carriers. Although the heme group is responsible for carrying the electron, the surrounding protein dictates the path it will follow in moving from complex III to complex IV.

How can the electron reduce the heme group, when the heme group is buried in the core of the protein? The answer probably lies in quantum mechanics. The electron being transported is thought to tunnel from one electron carrier to another, allowing it to cross barriers that normally seem impassable, and appear on the other side.

Cytochromes are an ancient family of proteins and are highly conserved throughout evolution. Their small size (approximately 100 amino acids) and ubiquitous nature (they are found in nearly all species) have facilitated their use in studies of evolution and genetics, especially cladistics: the placement of organisms on the tree of life.

7.2.6 In complex IV, oxygen is the terminal electron carrier

In mammals, **complex IV** (also known as cytochrome oxidase) is a 410 kDa homodimer, each half of which contributes 13 subunits to the structure (**Figure 7.21**). A single molecule of the phospholipid cardiolipin is found at the interface of the dimer. Complex IV takes four electrons from four copies of cytochrome *c* and uses them to reduce a molecule of molecular oxygen (O_2) to water (H_2O). To do this, it uses four different electron carriers: a bimolecular copper center (Cu_A), two heme groups (bound in cytochrome *a* and cytochrome a_3), and a second copper center (Cu_B).

Complex IV is comprised of 28 membrane-spanning α helices. The entire complex is predominantly helical, with two pieces of β sheet structure per monomer. Cytochrome *c* is thought to bind negatively charged residues near the β barrel structure on the intermembrane face of the complex through lysine side chains on cytochrome *c*.

Electron transport through complex IV and water formation   Once cytochrome *c* binds to complex IV, electrons are transferred from the heme of cytochrome *c* to the Cu_A of the complex, through the heme of cytochrome *a*, to cytochrome a_3 and then to Cu_B. In this final step, the electrons do not appear to completely move onto the second copper center (Cu_B); rather, they are shared between the copper, heme, and oxygen. This is where the formation of water occurs.

A.

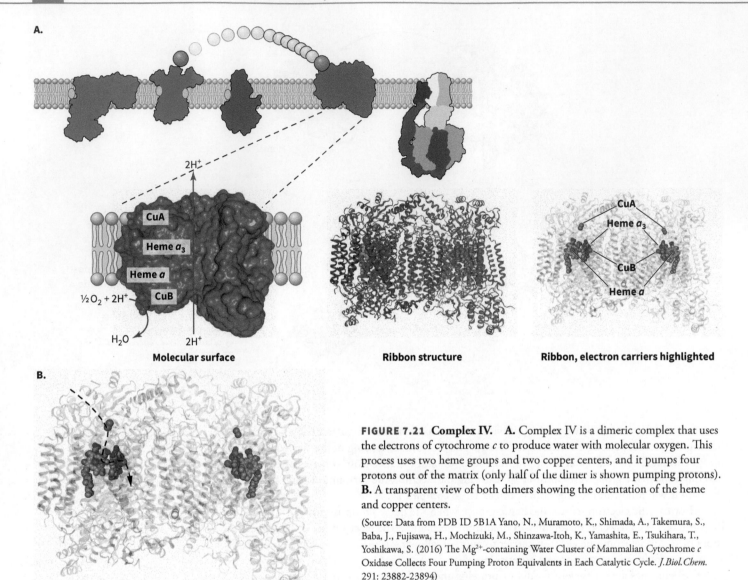

Molecular surface   **Ribbon structure**   **Ribbon, electron carriers highlighted**

B.

FIGURE 7.21 Complex IV. A. Complex IV is a dimeric complex that uses the electrons of cytochrome *c* to produce water with molecular oxygen. This process uses two heme groups and two copper centers, and it pumps four protons out of the matrix (only half of the dimer is shown pumping protons). **B.** A transparent view of both dimers showing the orientation of the heme and copper centers.

(Source: Data from PDB ID 5B1A Yano, N., Muramoto, K., Shimada, A., Takemura, S., Baba, J., Fujisawa, H., Mochizuki, M., Shinzawa-Itoh, K., Yamashita, E., Tsukihara, T., Yoshikawa, S. (2016) The Mg^{2+}-containing Water Cluster of Mammalian Cytochrome *c* Oxidase Collects Four Pumping Proton Equivalents in Each Catalytic Cycle. *J. Biol. Chem.* 291: 23882-23894)

The binuclear Cu_A center includes two cysteines, two histidines, a methionine and a glutamate residue. The cysteine residues in this complex give an overall geometry that is similar to a 2Fe-2S complex. Electrons tunnel from the heme of cytochrome *c* and are delocalized across the copper atoms in the Cu_A complex. Both heme groups have distal histidines in the fifth coordinate position to the iron atom of the heme group. The heme of cytochrome a_3 is coordinated to molecular oxygen in the sixth position. The other end of the oxygen molecule is coordinated to the Cu_B copper. In turn, the Cu_B copper is held in place by three histidines, one of which has a highly unusual covalent bond through the side chain to an adjacent tyrosine residue. This tyrosine is also thought to play a key role in catalysis. The bond formed by one of the ring carbons of tyrosine to histidine moves the tyrosine side chain, positioning it to become a donor of both H^+ and an electron during the reduction and cleavage of O_2. Formation of the bond to the nitrogen of the histidine lowers the pK_a of the phenolic proton, enabling proton donation; it also lowers the reduction potential, facilitating the formation of a tyrosyl radical.

The mechanism of water formation begins with the enzyme in the fully oxidized state (O), with the iron bound to a hydroxyl ion, the copper bound to water, and the tyrosine side chain in the deprotonated (phenolate) state (**Figure 7.22**). Both a proton and electron are transferred to the center, generating Cu^+ and a protonated phenol group, and water is released. A second proton and electron are transferred to the center. Water is formed again, this time on the iron

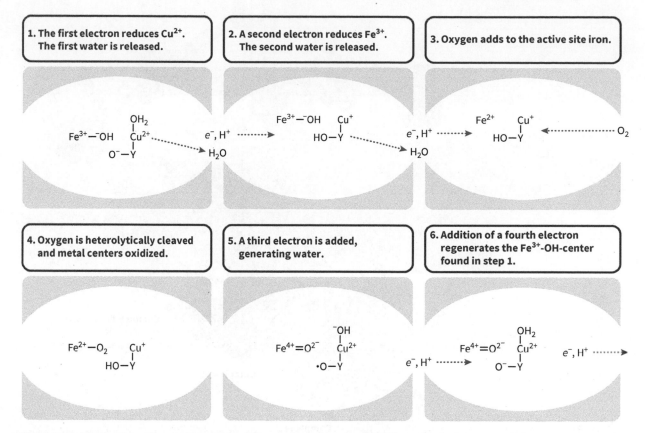

FIGURE 7.22 Mechanism of water formation. Water is formed in complex IV between the heme a$_3$ and Cu$_B$ centers. For each molecule of oxygen bound, four electrons and four protons are used to produce two molecules of water, released in the first two steps of the mechanism.

center, and is then released, leaving behind the Fe^{2+} state. The enzyme is now in the fully reduced state (R). Next, oxygen binds to the iron in the heme group and electrons rearrange themselves such that the iron forms a Fe^{4+} center coordinated to an O^{2-} atom. The second oxygen is transferred to the copper center, forming a Cu^{2+} oxidation state. This oxygen atom also acquires a proton from the tyrosine, generating a tyrosyl radical and a hydroxyl group coordinated to the Cu^{2+} center. Another proton and a third electron are transferred to the center, with the electron joining the radical to form a phenolate ion and the proton forming water on the Cu^{2+} center. Finally, a fourth and final electron is added, reducing the iron center from Fe^{4+} to Fe^{3+}, which is complexed to a hydroxyl ion, returning the enzyme to the fully oxidized state.

Proton pumping in complex IV   As electrons are used to reduce oxygen to water, four protons are pumped out of complex IV into the mitochondrial matrix. Two channels in complex IV are involved: the K and D channels (each named for a key lysine or a key aspartate residue). Both are thought to use a proton wire, a series of bound water molecules, to rapidly transfer protons to where they are needed. Although the exact mechanism of proton pumping remains unclear, it is likely that slight changes in structure as electrons move through the redox centers alter the pK_a values of the side chains lining the channel, causing protons to bind to or be released from water molecules lining the channel; the net result is a loss of protons on the intermembrane space side.

7.2.7 The entire complex working as one: the respirasome

To this point, this chapter has described the parts of electron transport, pieced together each of the electron carriers and complexes, and shown how electrons travel through these complexes. However, in reality, the story may be more complicated. Although each complex is enzymatically

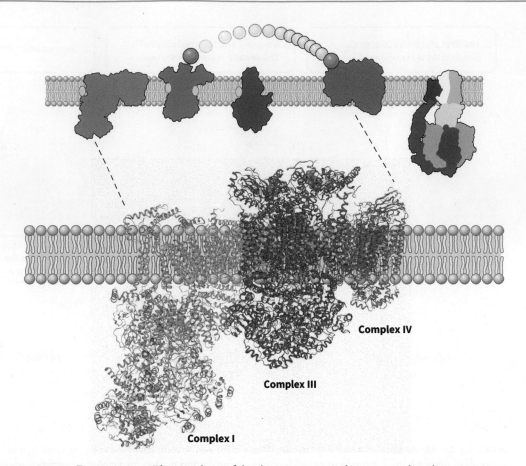

FIGURE 7.23 Respirasome.   The complexes of the electron transport chain are not thought to operate as single units but rather as a composite working in unison to facilitate reactions. This is termed the respirasome. Shown is a single unit comprised of complexes I, III, and IV. In this assembly, complex II is not seen, but other combinations of complexes have been observed.

(Source: Data from PDB ID 5J4Z Letts, J.A., Fiedorczuk, K., Sazanov, L.A. (2016) The architecture of respiratory supercomplexes. *Nature* 537: 644-648)

active on its own, data now suggest that the individual complexes aggregate together into one supercomplex, termed the **respirasome** (**Figure 7.23**). Also, although this chapter depicts the stoichiometry of the complexes as 1:1:1:1, there may be other ratios (a ratio of 1:1:3:7 has been proposed). Aggregating complexes would assist in substrate channeling (where substrate concentrations are higher at some locations, or where substrate can be directed from one active site to another). Likewise, keeping intermediates in close proximity may help in moving electrons via quantum mechanical tunneling.

The respirasome may be one of the largest biochemical structures ever proposed, but it has parallels in biochemistry. Other structures that have a higher order of organization, and effectively concentrate chemical reactions and chemical reactants in a small area, include the ribosome, replisome, transcriptional complex, and signal transduction complexes, among others.

Many poisons inhibit electron transport or uncouple the proton gradient from electron transport   Not surprisingly, compounds that inhibit electron transport are potent poisons. There are multiple places in the electron transport chain where poisons can bind. Many poisons have been characterized in terms of where they bind and block electron transport. Although highly toxic, these molecules are powerful research tools in the laboratory. Some molecules (for example, demerol and amytal) inhibit electrons passing through the iron-sulfur centers of complex I but do not provide their clinical effect via this particular mechanism. Nevertheless, this inhibition may provide clues to some of the side effects or more toxic effects seen in these compounds, especially when taken in excess (**Figure 7.24, Table 7.3**).

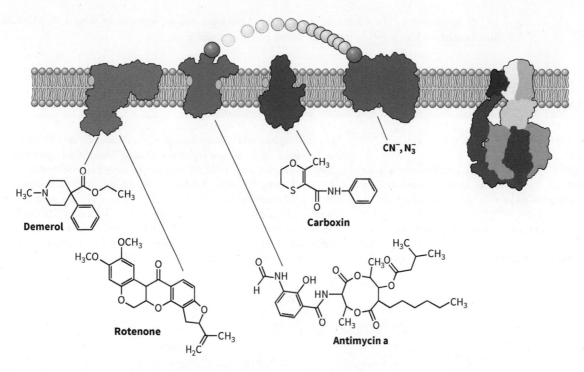

FIGURE 7.24 Poisons of electron transport. There are numerous compounds that can inhibit electron transport. Some act as poisons and block electron transport irreversibly, whereas others bind less tightly and have found use as pharmaceuticals.

TABLE 7.3 Drugs and Poisons and Where They Act in the Electron Transport Chain

Drug or Poison	Complex	Carrier Bound
Rotenone (insecticide)	I	Fe-S
Amytal (amobarbital, barbituates)		
Demerol (meperidine)		
Carboxin (fungicide)	II	Ubiqinone binding site
2-thionyltrifluoroacetone		
Antimycin A1	III	Cyt b_H in the Q_n site
Cyanide (CN^-)	IV	Heme cyt a_3
Azide (N_3^-)		
Carbon monoxide (CO)		
2,4-dinitrophenol (DNP)	ATP synthase	Uncoupling agent
Dicumarol		
FCCP		

Worked Problem 7.4   Mechanism of an antifungal agent

The antifungal agent myxothiazol binds to complex III and inhibits electron transport. Spectroscopic studies show that the absorbance spectra of cytochrome c_1, heme b_L, and cytochrome c all remain in the oxidized state (not reduced) when the drug is added. How might myxothiazol act?

Strategy Examine Figure 7.18. How do electrons travel through complex III? What is the purpose of the Q cycle?

Solution Crystallography shows that myxothiazol binds in the Q_p site, blocking the binding of ubiquinone and blocking transfer of electrons to the downstream electron carriers.

Follow-up question Cyanide (CN^-) binds tightly to the iron centers in some heme groups, but not all. Why?

A discussion of the effects of some proton gradient uncouplers can be found in **Medical Biochemistry: Uncoupling of the electron transport chain.**

Medical Biochemistry

Uncoupling of the electron transport chain

Imagine we had a magic pill to burn fat. We could consume whatever calories we wished and they would all burn away in our sleep. Although this sounds too good to be true, people are still searching for this magic pill.

Historically, there are multiple reports of munitions workers developing illnesses after being exposed to toxins in the course of their work. From the late 1880s until the late 1930s, munitions were often made with picric acid (2,4,6 trinitrophenol). Picrates are unstable molecules that form dangerous explosives. During World War II, picrate-based explosives were replaced with the more stable trinitrotoluene (TNT). During this period (and especially during wartime), working conditions were less than optimal, and munitions workers were frequently exposed to chemicals, including picrates. The exposure led to a set of symptoms including yellowed skin (picrates have a characteristic yellow color), elevated body temperature and respiration rate, profuse sweating, and weight loss in some cases resulting in death.

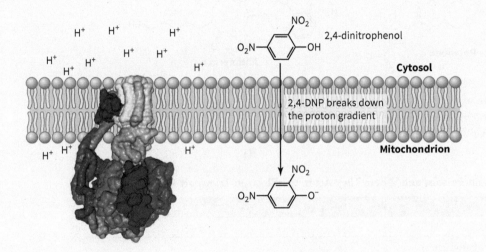

Breakdown products of picrates, particularly 2,4 dinitrophenol (DNP), are proton ionophores, molecules that can move specific ions across an otherwise impermeable barrier such as a plasma membrane. In the case of DNP, a combination of aromaticity and electron-withdrawing nitro groups lowers the pK_a of the phenolic hydroxyl proton to 4.1. In the intermembrane space, the deprotonated phenolate acquires a proton. Once protonated it becomes less soluble in aqueous solution and more soluble in lipid; it then diffuses across the inner mitochondrial membrane, where it deprotonates. Thus, DNP transports protons back into the mitochondria.

In light of this data, the symptoms and fate of those poisoned with DNP becomes clear. DNP uncouples electron transport from ATP production. The cell goes into overdrive, burning carbohydrates and lipids, sending acetyl-CoA through the citric acid cycle, passing electrons through NADH and $FADH_2$ to the electron transport chain and through the chain to form water. All this effort to pump protons out of the matrix is for nothing, because DNP simply shuttles them back in. A lot of heat is generated and oxygen consumed, but no ATP is made. Those affected either overheat or expire due to a lack of ATP for basic cellular needs.

In spite of these side effects, researchers have attempted to employ a controlled use of DNP as a means of losing weight. Following publication of a series of papers describing the potential weight-loss benefits of DNP, drugs appeared in the 1930s with varying levels of the compound. Unfortunately, people vary in their response to DNP and it has a relatively narrow therapeutic index (the dose of drug that is beneficial is very close to the dose that is harmful). DNP was officially withdrawn from diet pills in 1938, and it is not currently approved to treat any disease. However, it is available on the Internet and used illicitly by bodybuilders, boxers, and wrestlers to lose weight quickly. Web sites claim that this compound can be used safely, but the evidence, including deaths from DNP use, suggests otherwise.

Interestingly, nature has provided us with a potential alternative to DNP. Babies and hibernating animals express a protein in brown adipose tissue known as uncoupling protein-1 (UCP-1): a proton channel that is found in the inner mitochondrial membrane and serves to let protons back into the mitochondria. As seen in patients with DNP poisoning, such movement of protons can lead to increased burning of fat for energy and increased body temperature (which in the case of weight loss could be the desired effect). UCP-1 is responsible for non-shivering thermogenesis, which helps to keep babies and hibernating animals warm. The protein is only expressed in brown adipose tissue, but research is ongoing as to how this protein could be used to increase fat burning in obese or overweight humans.

(Source: Data from PDB ID 5ARA Zhou, A., Rohou, A., Schep, D.G., Bason, J.V., Montgomery, M.G., Walker, J.E., Grigorieff, N., Rubinstein, J.L. (2015) Structure and conformational states of the bovine mitochondrial ATP synthase by cryo-EM. *Elife* 4: e10180-e10180)

Summary

- Electrons harvested from catabolic reactions are used in the electron transport chain to pump protons out of the mitochondrial matrix, generating a proton gradient.

- Electrons are passed from one redox active center to another center with higher reduction potential.

- The redox active centers used in electron transport include iron-sulfur centers, heme groups, and organic electron carriers such as quinones.

- The energetic equivalent of NADH is transported into the mitochondrial matrix via either the glycerophosphate or malate–aspartate shuttle.

- Four macromolecular protein complexes are involved in electron transport:

  ○ Complex I is NADH dehydrogenase. It oxidizes NADH to NAD^+, passing electrons to ubiquinone and pumping four protons out of the matrix for each molecule of NADH.

  ○ Complex II is succinate dehydrogenase. This enzyme of the citric acid cycle oxidizes succinate to fumarate, capturing the two electrons in a molecule of $FADH_2$. Electrons from $FADH_2$ are transferred through several redox centers to ubiquinone.

  ○ Complex III is ubiquinone oxidase. Complex III uses ubiquinone to take electrons from complexes I and II and other catabolic processes, and it transfers them to the soluble electron carrier cytochrome c. Four additional protons are transferred to the intermembrane side of the inner mitochondrial membrane for each molecule of ubiquinone.

  ○ Complex IV is cytochrome c oxidase. Complex IV oxidizes cytochrome c, transferring electrons to oxygen to generate water. Two final protons are pumped out of the mitochondrial matrix, one for each molecule of cytochrome c.

- Many toxic compounds act by blocking specific steps of electron transport.

Concept Check

1. Analyze how the structure of each complex gives rise to its function.

2. Which prosthetic groups (cofactors) listed can accept or donate either one or two electrons because of the stability of the semiquinone state?

3. Explain why electrochemistry is used to describe the reactions in this chapter more than reactions in glycolysis, for example, and the net effect of all of these reactions.

4. Explain how different poisons can inhibit or block electron or proton transport.

7.3 ATP Biosynthesis

Until this point, a great deal of chemistry has been used in breaking glucose down into CO_2, but little ATP has been produced. The enzyme that is largely responsible for ATP production is known by many different names, the most common of which are **ATP synthase** or the **F_0/F_1 ATPase**. This multimeric enzyme uses the electrochemical energy of the proton gradient established by the electron transport chain to synthesize ATP from ADP and inorganic phosphate, P_i.

7.3.1 The structure of ATP synthase underlies its function

The general topology and structure of ATP synthases are conserved throughout biology. For simplicity, this section focuses on the bacterial form of the enzyme. ATP synthase can be divided into two complexes: F_0 and F_1. The membrane-bound F_0 complex is comprised of three different subunits (1 a, 2 b, and 10–15 c). The c subunits are arrayed in a disc that floats in bacterial plasma membrane or the inner mitochondrial membrane in eukaryotes. The a subunit acts as an adapter between the c subunits and the homodimer of b subunits. The b subunit spans the membrane and has an extended arm that cradles the F_1 complex (**Figure 7.25**).

The F_1 complex has nine subunits: three α, three β, one γ, one Δ, and one ϵ. A hexameric globular domain composed of all the α and β subunits in an alternating pattern forms the catalytic portion of the complex. The γ subunit forms an axle that passes through the center of the α-β hexameric complex and connects it to the F_0 complex via the c subunits. The Δ subunit caps the α-β hexamer and links it to the b homodimer of the F_0 complex. Finally, the ϵ subunit associates with the γ and c subunits, linking the complexes together.

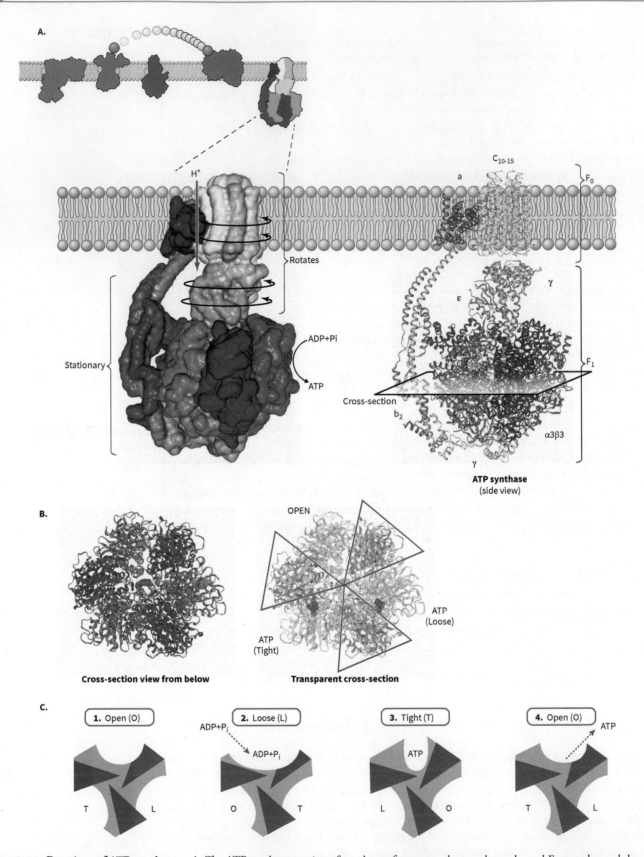

FIGURE 7.25 Reactions of ATP synthase.   A. The ATP synthase consists of two larger fragments, the membrane-bound F_0 complex and the catalytic F_1 complex. **B.** As the F_1 complex rotates past the b subunits, the conformation of a and b change, resulting in binding of ADP and P_i and formation of ATP. **C.** The complex consists of three active sites that progress through three different states: open, where nothing is bound; loose, where ADP and P_i bind; and tight, where the change in conformation generates ATP.

(Source: (A) Data from PDB ID 5ARA Zhou, A., Rohou, A., Schep, D.G., Bason, J.V., Montgomery, M.G., Walker, J.E., Grigorieff, N., Rubinstein, J.L. (2015) Structure and conformational states of the bovine mitochondrial ATP synthase by cryo-EM. *Elife* 4: e10180-e10180 (B) Data from PDB ID 1BMF Abrahams, J.P., Leslie, A.G., Lutter, R., Walker, J.E. (1994) Structure at 2.8 A resolution of F1-ATPase from bovine heart mitochondria. *Nature* 370: 621-628)

Mitochondrial forms of this enzymatic complex have retained the same architecture and function but are more complex, comprising at least 17 different subunits, each with a specialized function.

7.3.2 ATP synthase acts as a molecular machine driving the assembly of ATP molecules

We can visualize ATP synthase working as a tiny machine, in which protons slip down their electrochemical gradient back into the mitochondrial matrix. This movement of protons provides the energy for the synthase to generate ATP. As the protons move between the a and b subunits of the F_0 complex, they cause the ring of c subunits to rotate, which in turn causes the γ and ϵ subunits to also move. As the γ subunit passes through the hexamer of α and β subunits, a conformational change occurs, releasing newly generated ATP and binding to ADP and P_i. With each turn of the complex, a new molecule of ATP is released.

The mechanism of ATP synthase has not been observed directly, but there is much evidence to support the individual steps described above. Perhaps the most striking evidence comes from a paper that appeared in the journal *Nature* in 2004, in which investigators used chimeric proteins to fuse a fluorescent actin arm onto the F_0 subunit. A second modification of the F_1 subunit made it possible to mount the complex on a glass microscope slide. When ATP was added to this system, the position of the actin arm changed, providing evidence for the proposed mechanism (**Figure 7.26**).

The energetics of the proton motive force   The energy for ATP synthesis comes from the electrochemical potential created by the uneven distribution of protons across the inner mitochondrial membrane. This proton gradient is generated by protons being pumped out of the

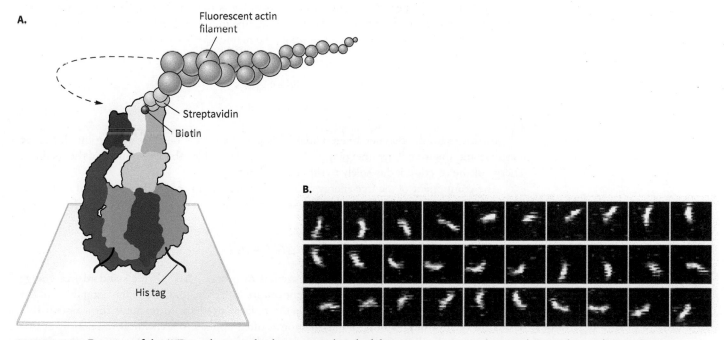

FIGURE 7.26 Rotation of the ATP synthase can be demonstrated in the laboratory.   A. In 2004, researchers used recombinant DNA techniques to produce an ATP synthase with two modifications, a histidine tag (His tag) on the a subunit and a biotinylation of the c subunit. Avidin binds very tightly to biotin. Therefore, a streptavidin tag fused to a fluorescent actin arm was used as a reporter. **B.** The modifications to the ATPase were minor enough to permit function of the complex but enable the use of the fluorescent actin arm as a reporter to follow rotation of the c subunit. The arm made a one-third rotation for every ATP molecule hydrolyzed. The modifications to the ATPase were minor enough to permit function of the complex but enable the use of the fluorescent actin arm as a reporter to follow rotation of the c subunit. The arm made a 1/3 rotation for every ATP molecule hydrolyzed. 4YXW 1

(Source: (A) Pratt and Cornely (2018). *Essential Biochemistry*, 4e (Wiley). Reproduced with permission of John Wiley & Sons, Inc. (B) Yoh Wada, Yoshihiro Sambongi, and Masamitsu Futai (2000). Biological nano motor, ATP synthase F_0F_1: from catalysis to $\gamma\epsilon c_{10-12}$ subunit assembly rotation. *Biochimica et Biophysica Acta (BBA) - Bioenergetics* Volume 1459, Issues 2–3: 499–505. https://doi.org/10.1016/S0005-2728(00)00189-4. Reproduced with permission of Elsevier.)

mitochondrial matrix during electron transport. The electrochemical potential can be separated into a chemical potential (a higher concentration of protons on the outside of the membrane than on the inside) and an electric potential (the added positive charge that accumulates on the outside of the membrane). These terms can be discussed quantitatively by describing each of the components (the chemical and the electrical) and by mathematically calculating their combined potential.

The electrical potential difference across a membrane is measured in volts or millivolts, and it is analogous to a weight lifted to some height or a car on the top of a hill. When describing electrical potentials, an analogy to water is often used, with the potential difference seen as analogous to water flowing over a waterfall or traveling through a pipe. The height of the waterfall or the diameter of the pipe relate to the potential voltage difference across a membrane or in an electrical circuit. The amount of water flowing equates to the number of electrons or ions moving and is the amperage of the system (a different quantity than voltage). There may be tall waterfalls with little water flowing over them (equating to high voltage and low current) or short waterfalls with a flood of water moving across them (equating to low voltage and high current). In biological systems, voltages are typically measured in tens of millivolts, and amperages in the order of picoamps. These values may seem small, but they can be readily measured across the plasma membrane of a cell. Also, given the size of the cell, the current density per unit area can be quite high.

The electrical component of the electrochemical potential can be envisioned as the voltage drop across the membrane. This is the potential difference or membrane potential:

$$\Delta\Psi_m = \Psi_{in} - \Psi_{out}$$

where $\Delta\Psi_m$ is the membrane potential, and Ψ_{in} and Ψ_{out} are the voltages on the inside and outside of the membrane, respectively. Electrical potential energy is related to the Gibbs free energy by the following constants:

$$\Delta G = n\mathscr{F}\Delta\Psi_m$$

where ΔG is the change in free energy of a process, n is the number of electrons per mole of product, $\mathscr{F}$ is the Faraday constant (96,485 coulombs mol^{-1} or 96.48 kJ/V·mol), and $\Delta\Psi_m$ is the potential difference. If the difference across the membrane is –0.040 volts, the free energy of the process is:

$$= (1)(96.48 \text{ kJ/V} \bullet \text{mol})(-0.040 \text{ V})$$

$$= -3.90 \text{ kJ/mol}$$

In this example, the membrane potential is positive on the outside, making the difference negative and the overall free energy negative, and thus favorable. This component of the total free energy of the reaction is due solely to the electrical potential.

The component of the free energy due to the difference in concentration created by pumping protons across the membrane also can be described using a version of the Gibbs free energy expression:

$$\Delta G = \Delta G^{\circ\prime} + RT \ln Q$$

in which, ΔG is the free energy for any reaction or process, $\Delta G^{\circ\prime}$ is the standard free energy exchange, R is the gas law constant, T is temperature in Kelvin, and $\ln Q$ is the natural log of the reaction quotient. The reaction quotient is the ratio of products to reactants for a system that is not at equilibrium. Although this equation is often applied to a chemical reaction, it can be applied just as easily to the imbalance of chemicals across a membrane. Here, it can be used to describe the concentration of protons on either side of the membrane:

$$\Delta G = RT \ln ([H_{in}^+]/[H_{out}^+])$$

Since the proton concentration is often discussed in terms of pH, it is easy to transform the previous equation, changing from a natural log to a base-10 log by multiplying by 2.303.

$$\Delta G = 2.303 \, RT \log([H_{in}^+]/[H_{out}^+])$$

Substituting pH for the negative log of $[H^+]$ gives

$$\Delta G = -2.303 \, RT \, (\text{pH}_{in} - \text{pH}_{out})$$

or

$$\Delta G = -2.303 \, RT \, \Delta\text{pH}$$

where ΔpH is the difference in pH from the inside to the outside of the membrane $(\text{pH}_{in} - \text{pH}_{out})$.

If the change in pH from the inside to the outside of a membrane was 1 pH unit, the free energy available to a system from this chemical gradient would be:

$$\Delta G = -2.303 \, (8.315 \times 10^{-3} \text{ kJ/mol} \cdot \text{K}) \, (310 \text{ K}) \, (1.0) = -5.93 \text{ kJ/mol}.$$

As discussed earlier in this section, this is only one component of the total free energy. The combined total of the electrical potential and the chemical potential exclusively for protons (disregarding other ions) is referred to as the **proton motive force** (pmf). The free energy expression for pmf is:

$$\Delta G = -2.303 \, RT \, \Delta\text{pH} + n\mathscr{F}\Delta\Psi_m$$

At body temperature (37°C, 310 K), the equation simplifies to:

$$\Delta G = -5.9 \, \Delta\text{pH} + n\mathscr{F}\Delta\Psi_m$$

This equation gives the free energy available for work generated by the proton gradient across the membrane. The amount of energy required to generate this gradient is therefore the same magnitude with the opposite sign (ΔG of the reaction is positive).

Worked Problem 7.5   Calculating the energy obtained from the pmf

In specific bacteria, there is a 120 mV potential drop across the membrane. The pH is 6.5 on the inside of the membrane and 7.2 on the outside. How much free energy is available for ATP biosynthesis?

Strategy Bacteria lack mitochondria but synthesize ATP using electron transport and an ATP synthase. What equation and information do we need to be able to calculate the free energy?

Solution   If we use the free energy expression for pmf,

$$\Delta G = -5.9 \, \Delta\text{pH} + n\mathscr{F}\Delta\Psi_m$$

we need to know the pH gradient and voltage drop across the membrane. The difference across the membrane $(\text{pH}_{in} - \text{pH}_{out})$ is:

$$6.5 - 7.2 = -0.7 \text{ pH units.}$$

The voltage across the membrane is 120 mV or 0.120 V. The constants in the expression are $n = 1$ (the charge on a proton) and $F = 96.485 \text{ kJ} \cdot \text{V}^{-1} \cdot \text{mol}^{-1}$. The free energy is therefore:

$$\Delta G = -5.9 \, (-0.7) + (1) \, (96.485 \text{ kJ} \cdot \text{V}^{-1} \cdot \text{mol}^{-1}) \, (0.120 \text{ V})$$

$$\Delta G = 4.13 + 11.58$$

$$\Delta G = 15.71 \text{ kJ} \cdot \text{mol}^{-1}$$

The equation tells us that the free energy required to create the proton gradient is 16.45 kJ $\cdot$ mol^{-1}. It takes energy to generate the gradient; therefore, the sign on the free energy is positive. The energy produced by the gradient is of the same magnitude but has the opposite sign (-16.45 kJ $\cdot$ mol^{-1}).

Follow-up question   If the free energy (ΔG) of the following reaction was 55 kJ/mol, how many protons would it take to synthesize one molecule of ATP?

$$\text{ADP} + \text{P}_i \rightarrow \text{ATP} + \text{H}_2\text{O}$$

ATP is transported out of the mitochondrial matrix by a translocase

ATP fuels many diverse reactions in the cell, and it is probably used in every organelle. Therefore, ATP must be released from the mitochondria to travel to its site of use. In return, ADP must travel to the mitochondrial matrix to be phosphorylated to ATP. The proteins responsible for this exchange are ATP:ADP translocases, 30 kDa monomeric transmembrane proteins found in the inner mitochondrial membrane.

The transmembrane domain of the translocase has six membrane-spanning α helices that form a cone-shaped depression. Binding of substrate in the cone causes the helices to move so that the opening shuts on one side of the membrane and opens on the other. After shuttling ATP from the matrix to the intermembrane space, the binding site will collect a molecule of ADP and reverse its conformation, completing the process. The binding site at the bottom of the funnel contains a conserved arginine-rich nucleotide-binding sequence. The ATP-ADP translocase is highly abundant in the inner mitochondrial membrane, comprising more than 12% of the protein found there. Once through the inner mitochondrial membrane, ATP is thought to pass unrestricted through pores in the outer mitochondrial membrane.

7.3.3 Other ATPases serve as ion pumps

There are other ATP synthases, the most common of which are also often characterized as ATPases. These ATPases use ATP hydrolysis to perform their proscribed function in reverse. Based on their structure and function, ATPases can be broadly sorted into five categories: F-type, A-type, V-type, P-type, and E-type.

F-type ATPases   F-type ATPases are those that have already been described in this section (F_0/F_1 ATPase). They are called "F" because of the isolation of a *factor* involved in phosphorylation of ADP.

A-type ATPases   A-type ATPases are found in the single-celled microorganisms *A*rchaea (hence the name "A-type"). They serve a similar function and have a similar architecture to F-type ATPases.

V-type ATPases   V-type ATPases are found in *v*acuoles. These complexes work in reverse compared to ATP synthase, using the energy of ATP hydrolysis to pump protons, or in some instances Na^+ ions, into a vacuole. They have similar architecture to F-type ATPases. V-type ATPases have been implicated in several cellular processes including receptor-mediated endocytosis, protein trafficking, active transport of metabolites, and neurotransmitter release.

P-type ATPases   P-type ATPases are pumps that use ATP hydrolysis to move ions from one side of a phospholipid bilayer to another. They differ from the F-type, A-type, and V-type enzymes based on their architecture and proposed mechanism of action. Typically, P-type ATPases have a single transmembrane subunit that is responsible for both ATPase and ion-pumping activity. A *p*hospho-enzyme intermediate (an aspartyl phosphoanhydride) is conserved in all P-type ATPases.

E-type ATPases   E-type ATPases are *e*xtracellular ATPases, and they can be characterized further into ecto-ATPases, CD39-like ATPases, and ecto-ATP/ADPases. All these enzymes hydrolyze extracellular ATP but may also hydrolyze ADP or other nucleotide triphosphates. Typically, E-type ATPases are glycoproteins that depend on divalent metal ions for activity. They may play a role in platelet aggregation, transplant rejection, and parasite survival.

7.3.4 Inhibitors of the ATPases can be powerful drugs or poisons

Given their central importance in energy production, it might seem that inhibition of ATP synthases and ATPases would be harmful. Several poisons (such as cyanide) work by inhibiting electron transport, thereby blocking ATP production in the cell. Often, however, the difference between a substance acting as a poison or a drug is a question of dosage. ATPases are also the target of several drugs and are an important field of research.

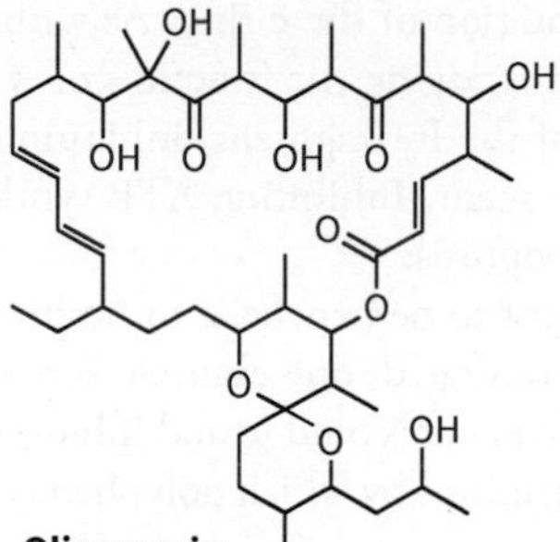

Oligomycin

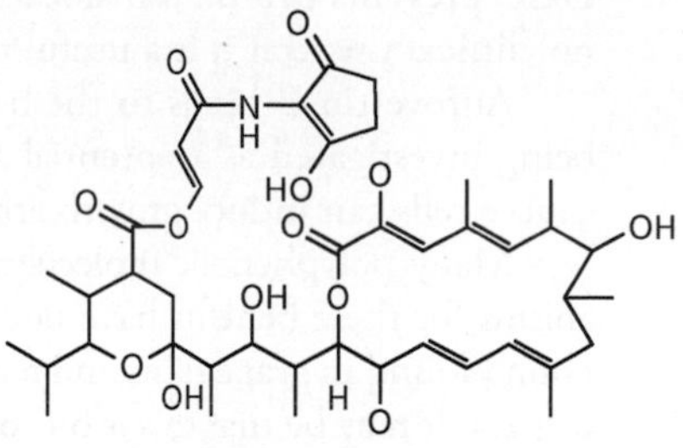

Balifomycin

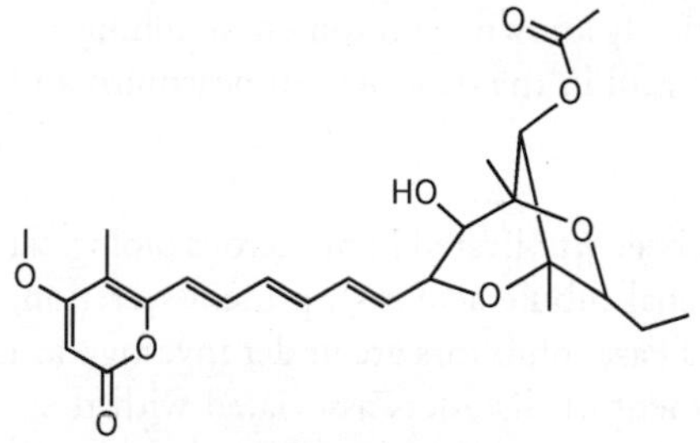

Aurovertin B 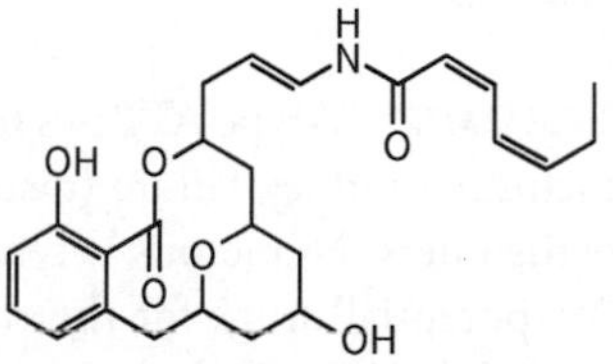   **Apicularen** 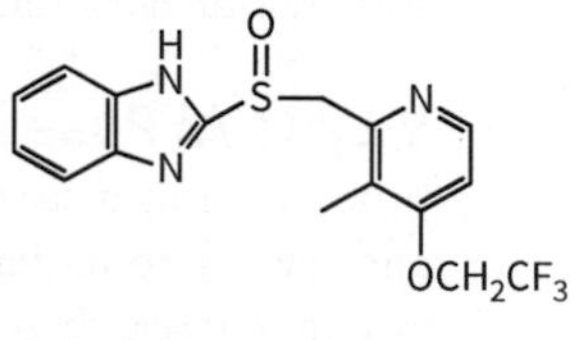   **Lansoprazole**

ATP synthase inhibitors

Oligomycin is a natural antibiotic isolated from the bacterium *Streptomyces*. Oligomycin is a member of a family of molecules termed macrolide antibiotics that contain a large, macrocyclic lactone that the bacterium probably secretes to poison any bacterial competitors. Oligomycin acts by binding at the interface of the a and c subunits of ATP synthase, blocking proton translocation and rotation of the ring and the associated F_1 complex, hence blocking ATP production. Although oligomycin is too toxic to be used in the clinic to treat patients with bacterial infections, similar molecules are used (such as the macrolide antibiotics azithromycin and arithromycin); however, these molecules act via a different mechanism (blocking protein biosynthesis). Oligomycin may not have a clinical use, but it is commonly used in the research laboratory for the study of electron transport and energy production.

The reagent dicyclohexylcarbodiimide (DCC) is commonly used in solid-phase peptide synthesis but also happens to be an inhibitor of ATP synthase. In theory, DCC could couple with any carboxylate group found in the protein; in practice, it preferentially couples to an aspartate residue in the c subunit. When administered to functional (coupled) mitochondria,

---

Worked Problem 7.6 | Steely hair disease

Menkes syndrome (also known as steely hair disease) is a genetic disorder that effects plasma concentrations of Cu^{2+}. Although the copper is absorbed from the small intestine, in this syndrome it does not leave the enterocytes and enter the plasma. The copper deficiency manifests itself in several ways (kinky hair is one obvious way) but is usually fatal in early childhood. The causal mutation in Menkes syndrome is in a gene coding for an ATPase. Propose a mechanism for how a mutation of the gene coding for this protein might impair copper absorption. What class of ATPase might be involved?

Strategy   Recall the structure of the intestine: enterocytes lining the intestinal lumen absorb nutrients and secrete them into the lymph or plasma, depending on the type of nutrient. If copper can enter these cells but not leave, it may be that the ions are somehow impaired from leaving the cell, or they become trapped in an organelle. What role might an ATPase play in this trafficking? Consider the different classes of ATPases.

Solution   The gene coding for the ATPase ATP7A protein is mutated in Menkes syndrome. This enzyme is a P-type ATPase that is involved in movement of divalent metal ions in the cell. In the normal cell, the enzyme resides in the Golgi; however, when copper levels rise, it translocates to the membrane surface and pumps copper ions out of the cell and into the plasma. Exactly how and where the mutation (or mutations) prevents this process from happening is still being investigated.

Follow-up question   The copper binding motif in ATP7A employs the sequence:

–DGMHCKSCV–

Think about which side chains interact with divalent metal ions. Which of these residues might be important for copper binding?

DCC prevents proton translocation and rotation of the c ring. As with oligomycin, DCC is of no clinical use, but it is a useful reagent for studying the function of ATP synthase.

Aurovertin B binds to the b subunit of the F_0 fragment, inhibiting ATP biosynthesis. It is being investigated as a potential anticancer agent. Inhibiting ATP synthesis in actively growing cancer cells can induce growth arrest and apoptosis.

Many polyphenolic molecules are thought to be beneficial to health, although distinct mechanisms for these benefits have not been demonstrated; one example is resveratrol, a phenolic compound found in grape skins and red wine. Polyphenols bind to and inhibit ATP synthase to differing degrees. It may be that this is one of the mechanisms by which polyphenols are beneficial.

P-type ATPase inhibitors   The hydrogen/potassium ATPase (H^+/K^+ ATPase) is responsible for the acidification of the stomach. Drugs such as omeprazole (Prilosec) and lansoprazole (Prevacid) inhibit the action of this ATPase, and are generically known as proton pump inhibitors. These drugs and others like them have been hugely successful in the treatment of heartburn and gastroesophageal reflux disease.

V-type ATPase inhibitors   V-type ATPases have been implicated in numerous biological processes and diseases including kidney failure (distal renal tubule acidosis), protein secretion, and several bone-density disorders. Numerous V-type ATPase inhibitors are under investigation as either research tools or potential drugs for the treatment of disorders associated with these proteins. Examples of such inhibitors include macrocyclic lactones (archazolid), plecomacrolides (bafilomycin and concanamycin), and benzolactones (apicularen and lobatamide). V-type ATPase inhibitors are produced by soil microbes and, as discussed for oligomycin, are used as a molecular defense against other microbes. Discovery of these molecules is another example of how studies in a field of biochemistry that may seem eccentric or obscure can yield important findings that have widespread implications for science and medicine (**Table 7.4**).

TABLE 7.4 Drugs That Affect ATPases

Drug	Class of ATPase	Effect
Apoptolidin	ATP synthase (F_0/F_1 ATPase)	Induces apoptosis; research tool
Bafilomycin B1	V-type inhibitor	Microbial antibiotic; research tool
Concanamycin C		Microbial antibiotic; research tool
Oligomycin B	F-type inhibitor	Microbial antibiotic; research tool
Ouabain	Na^+/K^+ ATPase inhibitor	Used in hunting (arrow poison); neuroscience research
Omeprazole/Lansoprazole/esomeprazole	H^+/K^+ ATPase inhibitor	Over-the-counter acid production blocker (proton pump inhibitor)

Summary

- The cell uses a tiny molecular motor to synthesize ATP.
- The energy of the proton gradient across the inner mitochondrial membrane can be described mathematically.
- Proteins related to ATP synthase work in reverse as ATPases, to generate ion gradients in vesicles or to pump ions across a cell membrane.
- Inhibitors of ATP synthase or ATPases can be potent poisons or beneficial drugs.

Concept Check

1. Describe the overall conformation of the ATP synthase.
2. If the pH of the matrix is 7.7, what is the pH of the intermembrane space if the ΔG for transport of H^+ is 20 kJ/mol at 37°C with $\Delta\gamma = 170$ mV?
3. Describe some of the other proteins that use ATP to generate an ion gradient (these are often thought of as ATP synthases running in reverse).
4. Explain how inhibition of ATP synthase or ATPases can be used in the laboratory or clinic.

Bioinformatics Exercises

Exercise 1 The Citric Acid Cycle and the KEGG Database

Exercise 2 Viewing and Analyzing Complexes I–IV

Exercise 3 Diversity of the Electron-Transport Chain and the KEGG Database

Problems

7.1 The Citric Acid Cycle

1. Examine these three citric acid cycle intermediates. Each has been labeled with C^{14}. Following one round of the citric acid cycle, where would the C^{14} be found?

2. Examine these three citric acid cycle intermediates. Each has been labeled with O^{18}. Following one round of the citric acid cycle, where would the O^{18} be found?

3. Without looking at a table of free energy values, what would you predict the free energies of reaction to be for the various reactions of the citric acid cycle? Which would be strongly positive, which negative, and which near zero?

4. Identify which reactions act to build up citric acid cycle intermediates, and explain the chemistry of these reactions.

5. Name the regulators of the citric acid cycle. At what point does each regulator affect the reactions? Which regulators are feedback inhibitors, and which are allosteric inhibitors?

6. Examine the structure of these enzymes using the Protein Data Bank (PDB). Where is the active site of each enzyme, and where does the regulator bind? Why are structures available for some but not all citric acid cycle enzymes in the PDB?

7. Could fluorescent molecules be used instead of radioisotopes in elucidation of citric acid cycle? How could you use a fluorescently tagged analog of isocitrate to help identify the next step in the citric acid cycle?

8. Based solely on the reactions of the citric acid cycle, could this pathway be used to fix CO_2 into acetyl-CoA or other molecules?

9. Examine the energetics of the citric acid cycle. Could this pathway be used to fix CO_2 into acetyl-CoA or other molecules?

10. If the citric acid cycle is so important to all of metabolism, why isn't there any redundancy to the pathway?

11. While some notable mutations exist, there are relatively few diseases that arise from complete deficiencies of any of the enzymes in the citric acid cycle. Why is this the case?

12. An individual with a shortage of B vitamins (which include thiamine and riboflavin) may feel fatigued because of decreased pyruvate dehydrogenase activity. How does reduced production of pyruvate impact the citric acid cycle?

7.2 The Electron Transport Chain

13. Structures for the complexes of the electron transport chain can be found in the PDB. Look up some of these structures in the PDB. Which species are they from, and why?

14. Many of the structures of the electron transport chain and ATP biosynthesis have been imaged with electron microscopy. Why has such a low-resolution technique been used, when it is usually associated with cell biology and used to examine larger structures?

15. What are the compounds that can block electron transport and how? Is is possible to block one single step of electron completely and not kill the organism? Explain.

16. Many biochemical reactions run at ΔG values close to zero but are reversible with slight changes in concentration. Is electron flow reversible through the electron transport chain? Is there a reason why electrons need to flow in the direction they do?

17. Describe the redox active centers found in each of the complexes I, II, III and IV.

18. Some organisms use molecules as electron carriers other than the ones discussed. If there were a protein with a Cu center, a heme group, and a manganese-containing heme analog, and based on the data in Table 7.2, predict the order in which they might function in electron transport. Which carriers would be used, and when?

19. Describe three techniques that can be used to show that the electron transport chain is found on the matrix side of the inner mitochondrial membrane.

20. Cytochrome c is often used as a marker for evolutionary change. Can it be used for all organisms? What are the strengths and limitations of using this single gene? What assumptions about evolutionary change must be made to use this gene successfully?

21. We may not think of gases as reagents in reactions, but several enzymes in biochemistry fix CO_2 while others use O_2 for a variety of processes. Describe the cofactors and chemistry involved in some of these processes.

22. In mitochondrial electron transport, there are four enzymes or enzyme complexes that funnel electrons to coenzyme Q (ubiquinone) from molecular sources other than reduced coenzyme Q. One of these is directly involved in lipid metabolism.

   a. What are the names of the three enzymes or enzyme complexes that are not directly involved in lipid metabolism?

   b. All four of these enzymes and enzyme complexes have a common location and one structural feature that is functionally significant (but is unrelated to the proteins' location). What is this common structural feature?

   c. Of these four enzymes and enzyme complexes, which one is distinct from the other three in its role in generating the intermediate energy form that is used to drive mitochondrial ATP synthesis?

23. What is wrong with this proposed mechanism?

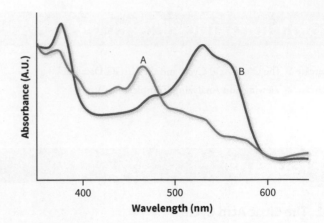

Oxaloacetate

24. What will be observed if a poison that prevents the transfer of electrons from the last [Fe–S] cluster of complex I to coenzyme Q is added to a suspension of actively respiring mitochondria?

7.3 ATP Biosynthesis

25. Both picric acid and DNP can act as uncoupling agents. Based on your knowledge of organic chemistry , which is likely to be the better uncoupling agent? How does treatment of liver cells with DNP cause increase in mitochondrial electron transport or decreases ATP activity?

26. What shuttle mechanism transfers cytosolic NADH into the mitochondria with a loss of reductive power? What are the number of ATP molecules synthesized?

27. Compare the structures of the different ATP synthases and ATPases discussed in section 7.3. How are they similar, and how are they different? Based on the structures alone, which would you expect to be part of the same evolutionary tree?

Questions 28–32 are concerned with ATPases that acidify vesicular structures in the cell.

28. How much ATP would you have to hydrolyze to cause a pH drop of 7.2 to 4.5 in a 0.1-micron diameter lysosome?

29. What assumptions would you need to make in performing this calculation?

30. If that change took two seconds, what would be the rate of acidification?

31. What is the turnover number of this enzyme if no pump was involved?

32. What is the turnover number of this enzyme if five pumps were involved?

Data Interpretation

33. A new graduate student in your laboratory has added radiolabeled citrate to cultured cells in an attempt to generate radiolabeled succinate. Instead, when checked, the label was found almost everywhere in the cell. The student believes he "got a bad batch" of radiolabeled citrate. What has probably happened instead? Why has the label shown up in other pools?

34. The same student is also attempting to use radiolabeled succinyl-CoA to produce radiolabeled GTP. Will this work?

35. The cobalt center in vitamin B12 fluctuates between the Co^{2+} and Co^{3+} states. These two states can be detected spectrophotometrically and have different visible range spectra as well as EPR signals. Co^{3+} is red colored and has no unpaired electrons, whereas Co^{2+} is yellow and has unpaired electrons. Which spectra (A or B) would match these oxidation states? Would you expect the molecule that generated each of the following spectra to give an EPR signal? Why or why not?

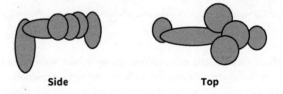

36. You isolated a purported electron transport complex from fungi. The individual complexes have the following molecular weights:

Complex	Mass	Complex	Mass
I	950 kDa	III	130 kDa
II	250 kDa	IV	190 kDa

Size exclusion chromatography of the sample reveals a single peak with an expected molecular weight of 1.5 to 1.8 MDa (1500 to 1800 kDa). Electron microscopy of the same sample gives the following images:

Side   **Top**

Propose a stoichiometry for the subunits in this complex. Could you algebraically predict the stoichiometry without the electron microscopy data?

Experimental Design

37. The proton motive force used to synthesize ATP has both a concentration and charge dependence. Because of the charge on the proton, these values cannot be completely divorced from one another. How could you tease out the chemical versus electrical component of pmf in mitochondria?

38. Using electron transport inhibitors, design an experiment to demonstrate that cytochrome *c* transports electrons from complex III to complex IV.

39. The FAD cofactor found in succinate dehydrogenase/complex II is tightly bound to the protein. Design an experiment to test this hypothesis.

40. You have identified a new protein in a species of algae. Based on sequence similarity to ATPases, you believe it may be involved in acidifying a subcellular compartment in the algae. Design an experiment to test this hypothesis.

Ethics and Social Responsibility

41. Colony collapse syndrome is a recently occurring phenomenon in which large numbers of bees have mysteriously vanished or died. Neonicotinoid pesticides used in many applications such as treating crops and flea and tick sprays for dogs and cats, are far less toxic than other insecticides (such as rotenone) but h been implicated as the causal agent in bee death. Research this issue and debate the use of neonicotinoid pesticides from both perspectives.

42. In the 1930s, many diet pills contained proton uncoupling agents such as DNP. Occasionally these drugs resurface, and there are stories of people using them with negative side effects. If these or similar molecules could be rendered safe, should people take them, or should weight be controlled by other means, such as diet and exercise?

43. Cyanide is a potent electron transport inhibitor, binding irreversibly to cytochromes. Cyanide fishing is a process in which fishermen dump cyanide into waters to stun tropical fish. The fish are scooped up in nets and sold to aquarium hobbyists. However, the health of the fish is often adversely affected by the cyanide poisoning, and the coral reef is killed by the residual cyanide. What are the socioeconomic circumstances that lead to this type of fishing? Is one group more to blame than another?

Suggested Readings

7.1 The Citric Acid Cycle

Brunengraber, H., and C. R. Roe. "Anaplerotic Molecules: Current and Future." *Journal of Inherited Metabolic Disease* 29, no. 2–3 (April-June 2006): 327–31.

Fandrey, J., T. A. Gorr, and M. Gassmann. "Regulating Cellular Oxygen Sensing by Hydroxylation." *Cardiovascular Research* 71, no. 4 (September 1, 2006): 642–51.

Kaelin, W. G., Jr., and P. J. Ratcliffe. "Oxygen Sensing by Metazoans: The Central Role of the HIF Hydroxylase Pathway." *Molecular Cell* 30, no. 4 (May 23, 2008): 393–402.

Lambeth, D. O., K. N. Tews, S. Adkins, D. Frohlich, and B. I. Milavetz. "Expression of Two Succinyl-CoA Synthetases with Different Nucleotide Specificities in Mammalian Tissues." *Journal of Biological Chemistry*, 279 (2004): 36621–4.

Lowenstein, J. M., ed. *Methods in Enzymology: Volume 13 Citric Acid Cycle*. Boston, Massachusetts: Academic Press, 1969.

Sazanov, L. A., and J. B. Jackson. "Proton-Translocating Transhydrogenase and NAD- and NADP-Linked Isocitrate Dehydrogenases Operate in a Substrate Cycle which Contributes to Fine Regulation of the Tricarboxylic Acid Cycle Activity in Mitochondria." *FEBS Letters* 344, no. 2–3 (May 16, 1994): 109–16.

Stobbe, M. D., S. M. Houten, A. H. van Kampen, R. J. Wanders, and P. D. Moerland. "Improving the Description of Metabolic Networks: The TCA Cycle as Example." *The FASEB Journal* 26, no. 9 (2012): 3625–36.

Sugden, M. C., and M. J. Holness. "The Pyruvate Carboxylase-Pyruvate Dehydrogenase Axis in Islet Pyruvate Metabolism: Going Round in Circles?" *Islets* 3, no. 6 (November–December 2011): 302–19.

Wan, B., K. F. LaNoue, J. Y. Cheung, and R. C. Scaduto, Jr. "Regulation of Citric Acid Cycle by Calcium." *Journal of Biological Chemistry* 264, no. 23 (1989): 13430–39.

7.2 The Electron Transport Chain

Bertini I., G. Cavallaro, and A. Rosato. "Evolution of Mitochondrial-Type Cytochrome c Domains and of the Protein Machinery for Their Assembly." *Journal of Inorganic Biochemistry* 101, no. 11–12, (2007): 1798–811.

Crofts, A. R., S. Hong, N. Ugulava, B. Barquera, R. Gennis, M. Guergova-Kuras, and E. Berry. "Pathways for Proton Release during Ubihydroquinone Oxidation by the bc1 Complex." *Proceedings of the National Academy of Sciences of the United States of America* 96, (1999): 10021–6.

Doherty, M. K., S. L. Pealing, C. S. Miles, R. Moysey, P. Taylor, M. D. Walkinshaw, G. A. Reid, and S. K. Chapman. "Identification of the Active Site Acid/Base Catalyst in a Bacterial Fumarate Reductase: A Kinetic and Crystallographic Study." *Biochemistry* 39, no. 35 (September 5, 2000): 10695–701.

Dudkinaa, N. V., M. Kudryashevb, H. Stahlbergb, and E. J. Boekemaa. "Interaction of Complexes I, III, and IV within the Bovine Respirasome by Single Particle Cryoelectron Tomography." *Proceedings of the National Academy of Sciences of the United States of America* 108, no. 37 (September 13, 2011): 15196–200.

Horsefield, R., V. Yankovskaya, G. Sexton, W. Whittingham, K. Shiomi, S. Ōmura, B. Byrne, G. Cecchini, and S. Iwata. "Structural and Computational Analysis of the Quinone-Binding Site of Complex II (Succinate-Ubiquinone Oxidoreductase): A Mechanism of Electron Transfer and Proton Conduction During Ubiquinone Reduction." *Journal of Biological Chemistry* 281, no. 11 (2006): 7309–16.

Johnson, D. C., D. R. Dean, A. D. Smith, and M. K. Johnson. "Structure, Function, and Formation of Biological Iron-Sulfur Clusters." *Annual Review of Biochemistry* 74 (2005): 247–81.

Lancaster, C. R., and A. Kröger. "Succinate: Quinone Oxidoreductases: New Insights from X-Ray Crystal Structures." *Biochimica et Biophysica Acta* 1459, no. 2–3 (August 15, 2000): 422–31.

Mowat, C. G., R. Moysey, C. S. Miles, D. Leys, M. K. Doherty, P. Taylor, M. D. Walkinshaw, G. A. Reid, and S. K. Chapman. "Kinetic and Crystallographic Analysis of the Key Active Site Acid/Base Arginine in a Soluble Fumarate Reductase." *Biochemistry* 40, no. 41 (2001): 12292–8.

Oyedotun, K. S., and B. D. Lemire. "The Quaternary Structure of the *Saccharomyces cerevisiae* Succinate Dehydrogenase: Homology Modeling, Cofactor Docking, and Molecular Dynamics Simulation Studies." *Journal of Biological Chemistry* 279, no. 10 (2004): 9424–31.

Palsdottir, H., Gomez-Lojero, B. L. Trumpower, and C. Hunte. "Structure of the Yeast Cytochrome bc1 Complex with a Hydroxyquinone Anion Q_o Site Inhibitor Bound." *Journal of Biological Chemistry* 278, no. 33 (2003): 31303–11.

Schägger, H., and K. Pfeiffer. "The Ratio of Oxidative Phosphorylation Complexes I–V in Bovine Heart Mitochondria and the Composition of Respiratory Chain Supercomplexes." *Journal of Biological Chemistry* 276, no. 41 (2001): 37861–7.

Solomon, Edward I., Uma M. Sundaram, and Timothy E. Machonkin. "Multicopper Oxidases and Oxygenases." *Chemical Reviews* 96, no. 7 (1996): 2563–606.

Sun, F., X. Huo, Y. Zhai, A. Wang, J. Xu, D. Su, M. Bartlam, and Z. Rao. "Crystal Structure of Mitochondrial Respiratory Membrane Protein Complex II." *Cell* 121 (2005): 1043–57.

Tran, Q. M., R. A. Rothery, E. Maklashina, G. Cecchini, and J. H. Weiner. "The Quinone Binding Site in *Escherichia coli* Succinate Dehydrogenase Is Required for Electron Transfer to the Heme b." *Journal of Biological Chemistry* 281, no. 43 (2006): 32310–7.

Yankovskaya, V., R. Horsefield, S. Törnroth, C. Luna-Chavez, H. Miyoshi, C. Léger, B. Byrne, G. Cecchini, and S. Iwata. "Architecture of Succinate Dehydrogenase and Reactive Oxygen Species Generation." *Science* 299, no. 5607 (2003): 700–4.

Zaunmüller, T., D. J. Kelly, F. O. Glöckner, and G. Unden. "Succinate Dehydrogenase Functioning by a Reverse Redox Loop Mechanism and Fumarate Reductase in Sulphate-Reducing Bacteria." *Microbiology* 152 (Pt 8) (August 2006): 2443–53.

Zhang, Z., L. Huang, V. M. Schulmeister, Y. I. Chi, K. K. Kim, L. W. Hung, A. R. Crofts, E. A. Berry, and S. H. Kim. "Electron Transfer by Domain Movement in Cytochrome bc1." *Nature* 392, (1998): 677–84.

7.3 ATP Biosynthesis

Huss, M., F. Sasse, B. Kunze, R. Jansen, H. Steinmetz, G. Ingenhorst, A. Zeeck, and H. Wieczorek. "Archazolid and Apicularen: Novel Specific V-ATPase Inhibitors." *BMC Biochemistry* 6 (August 4, 2005): 13.

Huss, M., and H. Wieczorek. "Inhibitors of V-ATPases: Old and New Players." *Journal of Experimental Biology* 212, (Pt 3) (February 2009): 341–6.

Ōmura, S. *Macrolide Antibiotics: Chemistry, Biology, and Practice.* 2nd ed. Boston, Massachusetts: Academic Press, 2002.

Plesner, L. "Ecto-ATPases: Identities and Functions." *International Review of Cytology* 158, (1995): 41–214.

Plesner, Liselotte, Terence L. Kirley, and Aileen F. Knowles, eds. *Ecto-ATPases: Recent Progress on Structure and Function.* New York, New York: Springer, 1997.

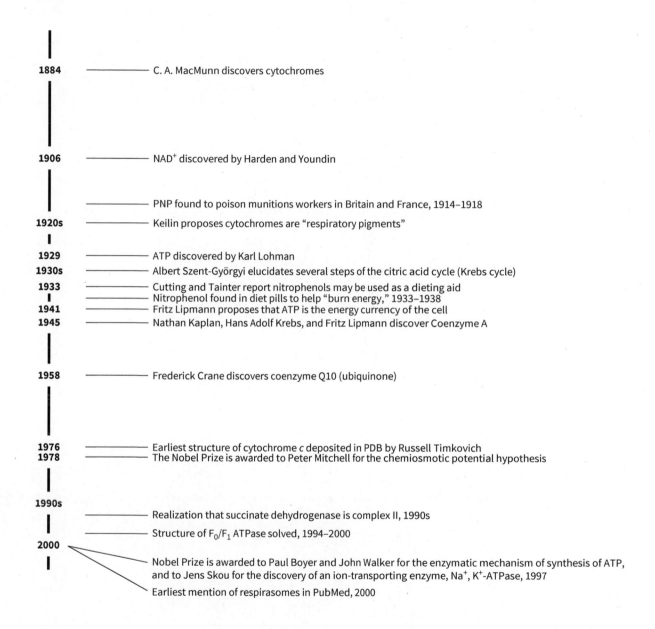

1884 ———————— C. A. MacMunn discovers cytochromes

1906 ———————— NAD$^+$ discovered by Harden and Youndin

———————— PNP found to poison munitions workers in Britain and France, 1914–1918
1920s ———————— Keilin proposes cytochromes are "respiratory pigments"

1929 ———————— ATP discovered by Karl Lohman
1930s ———————— Albert Szent-Györgyi elucidates several steps of the citric acid cycle (Krebs cycle)
1933 ———————— Cutting and Tainter report nitrophenols may be used as a dieting aid
———————— Nitrophenol found in diet pills to help "burn energy," 1933–1938
1941 ———————— Fritz Lipmann proposes that ATP is the energy currency of the cell
1945 ———————— Nathan Kaplan, Hans Adolf Krebs, and Fritz Lipmann discover Coenzyme A

1958 ———————— Frederick Crane discovers coenzyme Q10 (ubiquinone)

1976 ———————— Earliest structure of cytochrome *c* deposited in PDB by Russell Timkovich
1978 ———————— The Nobel Prize is awarded to Peter Mitchell for the chemiosmotic potential hypothesis

1990s

———————— Realization that succinate dehydrogenase is complex II, 1990s

———————— Structure of F$_0$/F$_1$ ATPase solved, 1994–2000
2000

———————— Nobel Prize is awarded to Paul Boyer and John Walker for the enzymatic mechanism of synthesis of ATP, and to Jens Skou for the discovery of an ion-transporting enzyme, Na$^+$, K$^+$-ATPase, 1997

———————— Earliest mention of respirasomes in PubMed, 2000

CHAPTER 8

Carbohydrates II

Polysaccharides in Context

Carbohydrates are nearly as diverse as proteins, and what they lack in functional group diversity and catalytic power, they make up for in overall diversity of form and structure. Starch and cellulose are polysaccharides, but so is glycogen, which acts as a store of energy in animals. Substances as diverse as mucus and connective tissue are largely comprised of polysaccharides, or proteins decorated with lengthy carbohydrate chains. These molecules are key to growth, development, and immunity; they also impart structure to the organism as a major component of cartilage and connective tissue. The many roles that carbohydrates play in biochemistry are just beginning to be understood.

This chapter starts with a discussion of the metabolism of glycogen. It highlights the regulation of glycogen metabolism, particularly how it differs from tissue to tissue, and its response to different chemical stimuli (hormones). Section 8.2 looks at the pentose phosphate pathway, which generates NADPH for lipid biosynthesis and redox needs of the cell, generates ribose-5-phosphate for nucleotide biosynthesis, and provides a way to catabolize other carbohydrates in the cell. Sections 8.3 and 8.4 discuss the properties and chemistry of the extracellular matrix: Section 8.3 focuses on polysaccharides in glycoproteins, glycolipids, proteoglycans, and peptidoglycans, and section 8.4 discusses matrix proteins and their assembly. The final section also looks at biofilms, analogous structures to the extracellular matrix made by some strains of microbes.

Chapter Outline

Common Themes

Evolution's outcomes are conserved.	• The pentose phosphate pathway evolved, in part, from the more primordial CO_2 fixation pathway used by photosynthetic bacteria and plants.
	• The eukaryotic extracellular matrix has its evolutionary origins in the secreted molecules made by microbes. Certain features of these structures, termed biofilms, are similar to those of the extracellular matrix, for example, group immunity due to secreted antibiotics, changes in phenotype and structure, and cell–cell communication.
	• Similar molecules (secreted glycoproteins and proteoglycans) are found in biofilms and the extracellular matrix.
	• Both biofilms and the extracellular matrix may have evolved from the bacterial cell wall, which has some similar features on a single-cell basis.
Structure determines function.	• The branched structure of glycogen provides numerous points at which glucose can be rapidly mobilized or incorporated.
	• The structure of extracellular matrix proteins and polysaccharides relates to their function. These molecules are often soft gels, but others provide some of the strongest structures in biochemistry. Many extracellular proteins are fibrous or filamentous, and they provide numerous locations for cross-linking.
	• Alterations of what we consider "normal" protein chemistry are often seen in the extracellular matrix. Such modifications include hydroxylated prolines and lysines, and cross-linked lysine and glutamine residues. These modifications lead to increased structural stability of the matrix.
Biochemical information is transferred, exchanged, and stored.	• Glycogen metabolism is tightly regulated by numerous kinases. These pathways are coordinated by hormones in the body.
	• Xylulose-5-phosphate, a product of the pentose phosphate pathway, is an important regulator of the transcription factor ChREBP (the carbohydrate response element-binding protein).
	• Hydroxy proline residues seen in collagen are also found in the transcription factor hypoxic inducible factor (HIF). When oxygen levels are low, HIF is not hydroxylated and is able to activate gene expression.
Biomolecules are altered through pathways involving transformations of energy and matter.	• As discussed in other chapters, energetically unfavorable reactions become favorable when coupled to reactions with a large, negative ΔG value (the hydrolysis or transfer of favorable leaving groups). We typically think of this in terms of adenosine-5′-triphosphate (ATP) hydrolysis or phosphate transfer; however, other reactions (such as hydrolysis of a thioester thioesters), are common in biochemistry. Similarly, reactions that have $\Delta G°$ values close to zero can be tipped to move in the forward or reverse direction by changing the concentrations of substrates or products.
	• Addition of carbohydrates to a molecule involves energetic activation through the use of uridine-5′-triphosphate (UTP), to produce uridine diphosphate (UDP)-carbohydrates.
	• Liberation of glucose monomers from glycogen is energetically favorable.
	• Isomerization of monosaccharides found in the pentose phosphate pathway occurs with ΔG values close to zero and can be tipped in the forward or reverse direction by the concentration of substrates.

8.1 Glycogen Metabolism

If we think of carbohydrates as a form of stored energy analogous to money, then glycogen is money in the bank. Glycogen is a stored form of glucose. Certain cells and tissues in the body (particularly the brain and erythrocytes, and to some extent the kidneys) require glucose for energy. For a variety of reasons, the energy source of these tissues is limited to glucose. Glycogen acts as a buffer, storing glucose when glucose is plentiful and releasing it at other times as the body needs it.

Glycogen is a highly branched polymer of α-1,4 and α-1,6 linked glucose monomers, with a single molecule of the protein **glycogenin** at the core (**Figure 8.1**). In the same way that the starch amylopectin stores energy in plants, glycogen stores energy in animals. Cells cannot store monomeric glucose. The anomeric carbon in monomeric glucose could react with free hydroxyl groups, randomly glycosylating molecules in the cell. In addition, a significant amount of water is needed to hydrate the glucose, which would generate a high osmotic pressure in the cell. Glycogen is not transported in the body; instead, it is synthesized and degraded in situ. Skeletal muscle stores most of the body's glycogen (75%); the remaining glycogen is stored in the liver. Muscle glycogen cannot release glucose for use by the rest of the body and so uses that glucose in muscle cells, but the liver can and does release glucose, so that it can fuel other tissues. The synthesis of glycogen is termed **glycogenesis** and the breakdown is **glycogenolysis**.

8.1.1 Glycogenesis is glycogen biosynthesis

Like the polymers discussed in other chapters (DNA, RNA, and proteins), glycogen is a polymer. However, glycogen differs from these other polymers in that glycogen is highly branched and has numerous termini—an important evolutionary adaptation. Biosynthesis and degradation can only occur at termini; hence, the large branched molecule of glycogen can be built up or broken down more rapidly than a linear molecule.

Glucose enters any cell via a glucose transporter, and it is rapidly phosphorylated to glucose-6-phosphate (G-6-P). Both of these steps were introduced in the discussion of glycolysis. When G-6-P levels are high, glycogenesis begins with G-6-P being isomerized into glucose-1-phosphate by the enzyme **phosphoglucomutase**. The addition of glucose monomers to glycogen is energetically unfavorable. Instead, carbohydrate metabolism is fueled by the formation of a molecule that has the ability to transfer glucose molecules to the growing chain.

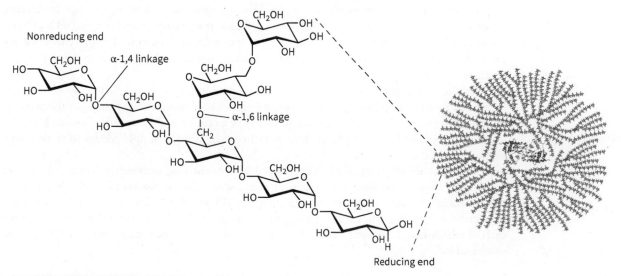

FIGURE 8.1 Glycogen is a storage form of glucose. Glycogen granules can contain tens of thousands of α-1,4 and α-1,6 linked glucose monomers and have a single molecule of the protein glycogenin in their core.

(Source: From Häggström, Mikael (2014). "Medical gallery of Mikael Häggström 2014". Wiki Journal of Medicine 1(2), DOI: 10.15347/wjm/2014.008. ISSN 2002–4436.)

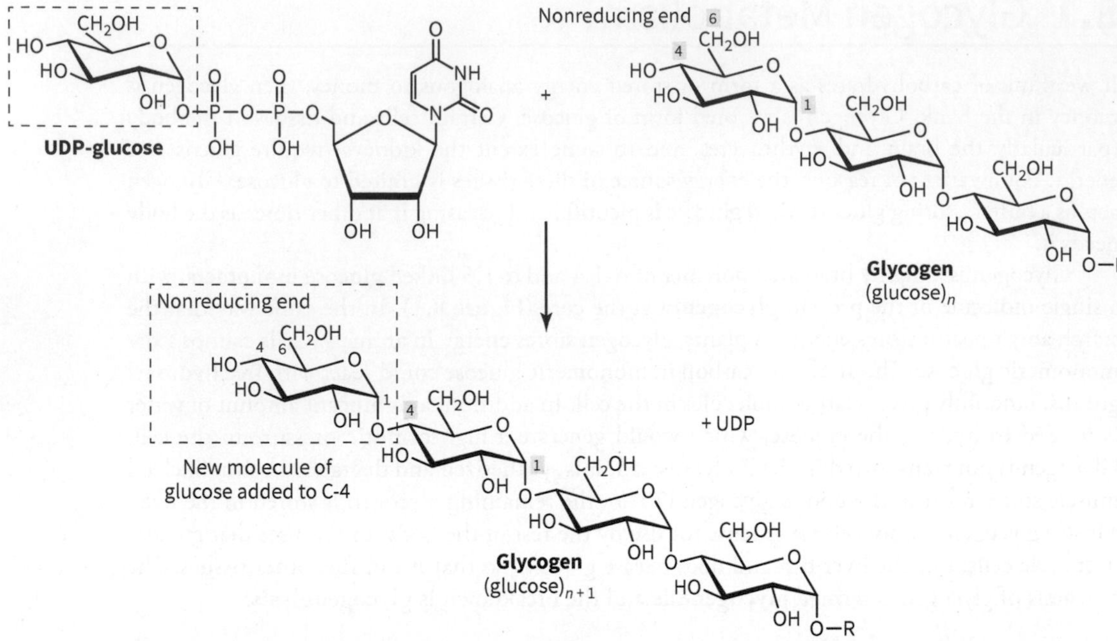

FIGURE 8.2 Synthesis of glycogen.   New glucose monomers are added to glycogen by glycogen synthase. The energy needed to drive the reaction comes from coupling glucose first to UDP. The bond to UDP joins the anomeric carbon of glucose (C-1) to C-4 of the growing glycogen chain.

That molecule is uridine diphosphate-glucose or UDP-glucose (**Figure 8.2**). In many reactions, we think of ATP as providing energy for a reaction, such as transfer of a phosphate group, to occur. Here, the nucleotide phosphate is UDP instead of ATP, and the group that is transferred is a molecule of glucose.

UDP-glucose is formed when the phosphate on glucose-1-phosphate attacks the α phosphate of UTP, liberating pyrophosphate and generating UDP-glucose. Although this reaction is not in and of itself energetically favorable, it is coupled to the hydrolysis of pyrophosphate into two molecules of inorganic phosphate, which is extremely favorable ($\Delta G = -33$ kJ/mol), making the overall net reaction favorable. Using UDP-glucose, **glycogen synthase** can add glucose monomers to the hydroxyl on C-4 of a growing glycogen molecule via an α-1,4 glycosidic bond (an acetal).

The initiation of a completely new molecule of glycogen proceeds through a slightly different pathway. The glycogenin protein at the core of glycogen is a glycosyl transferase that adds glucose molecules (as UDP-glucose) to an active site tyrosine molecule. The new glycogen chain continues to be extended until it is between 4 and 13 glucose monomers long, at which point glycogen synthase takes over.

Although glycogenin has only been found in mammalian systems, there may be paralogs (genes produced by duplication which evolve new but related functions) that regulate the priming of other polysaccharides in other species. This may be a rare instance of mammalian biochemistry informing microbial or plant biochemistry, rather than using a simple model to learn about human systems.

The branches in glycogen are incorporated by a **branching enzyme** (amylo α-1,4 → α-1, 6-transglycosylase) that transfers a chain of about seven glucose molecules to the hydroxyl of C-6 in either the same chain or an adjacent chain (**Figure 8.3**). The branch point is at least four carbons from the end of the chain. The resulting reaction leaves behind two new termini, each of which has an exposed hydroxyl group on C-4 to which new molecules of UDP-glucose can be added.

8.1.2 Glycogenolysis is glycogen breakdown

Because energy is required to form glycogen, the reverse reaction (liberation of glucose monomers) occurs spontaneously without input of additional energy.

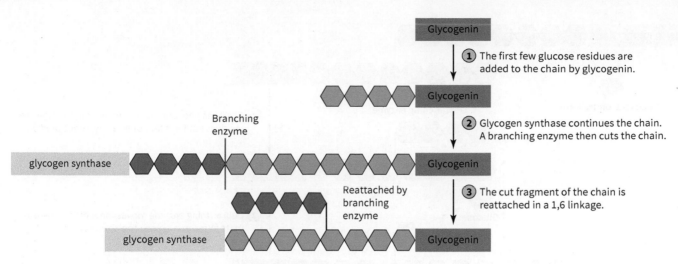

FIGURE 8.3 Branching enzyme. Branching enzyme transfers a 7-mer of glucose residues to the hydroxyl on C-6 of a glucose residue at least four glucose residues from the branch point.

Worked Problem 8.1 | Synthesizing polysaccharides

You identified a new strain of grass which, like some other grasses, stores polymers of fructose (inulin). When extracts of this grass are incubated with UDP-fructose or fructose alone, no new inulin is formed; however, incubation with sucrose and UDP-glucose or fructose and UDP-glucose produces new inulin. Based on this information and what you know of glycogen metabolism, how might the inulin polymers form?

Strategy What is the structure of glycogen? How is glycogen synthesized? What are the building blocks of glycogen, and how are they formed?

Solution Inulin is a β-1,2 linked polymer of fructose monomers. Formation of glycosidic bonds by the direct combination of two monosaccharides is energetically unfavorable. In glycogen, as in most carbohydrates, the reaction is made favorable by using UDP-glucose as a donor of one of the monosaccharides. This is still the case with inulin, but by an indirect route. Synthesis of inulin

involves fructose monomers, derived from sucrose, being added to the growing inulin chain. In the example of the grass, energy from UDP-glucose is used to drive sucrose production. Then fructose is transferred in a transglycosylation reaction to the growing inulin polymer.

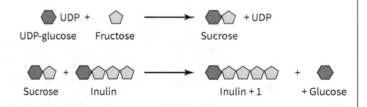

Follow-up question Secreted polysaccharides (known as exopolysaccharides) found in the extracellular matrix are often repeats of two monosaccharides. How might these form?

Glycogen is degraded one glucose unit at a time by **glycogen phosphorylase**. The enzyme removes glucose molecules at their α-1,4 linkages, liberating them into the cytosol as G-1-P, and leaving the glycogen shortened by one unit. Glycogen phosphorylase, which contains a pyridoxal phosphate cofactor, uses inorganic phosphate to complete this reaction. In this reaction, the inorganic phosphate serves as the nucleophile to attack C-1 of the terminal glucose monomer to generate glucose-1-phosphate. The phosphorylase cannot cleave branched termini. Instead, it stops four glucose residues before a branch point, and **glycogen debranching enzyme** removes three of those four residues, transferring them to the free C-4 terminus of another branch and leaving one glucose monomer at the branch point. The debranching enzyme then liberates this glucose using a hydrolysis reaction (**Figures 8.4, 8.5**).

G-1-P has two possible fates, depending on the cell type. In muscle, phosphoglucomutase can interconvert G-1-P and G-6-P. Therefore, G-1-P liberated from glycogen can be isomerized and enter into glycolysis, in the same way as any other molecule of G-6-P. In the liver, the story is somewhat different. G-1-P is still isomerized to G-6-P, but the liver also expresses glucose-6-phosphatase, which can cleave the phosphate from G-6-P, liberating glucose. Under these conditions, hexokinase and glucokinase cannot keep up with the glucose-6-phosphatase;

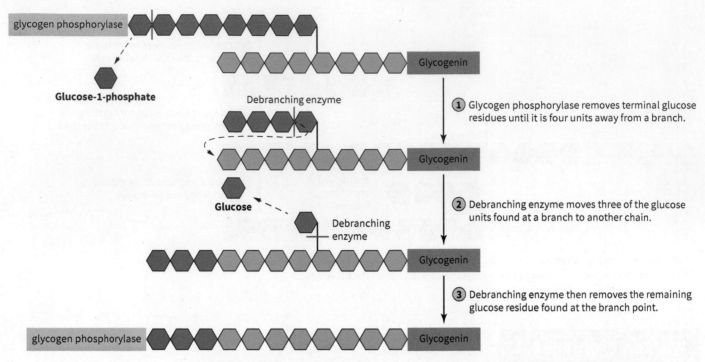

FIGURE 8.4 Glycogen breakdown. Glycogen is degraded by glycogen phosphorylase. Phosphorylase removes glucose residues and releases them to the cytosol as glucose-1-phosphate. Phosphorylase stops four glucose residues before a branch point. At that fork, glycogen debranching enzyme removes three of the four residues and transfers them to the end of another chain. The debranching enzyme then removes the branch glycogen and releases it as free glucose.

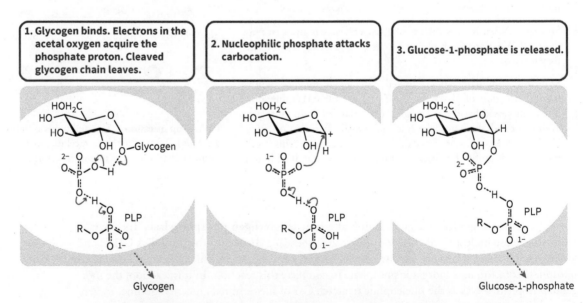

FIGURE 8.5 Mechanism of glycogen phosphorylase. Glycogen phosphorylase has an active-site pyridoxal phosphate used differently than we have seen previously. The phosphate moiety of PLP primes a molecule of inorganic phosphate for nucleophilic attack by the anomeric oxygen of the terminal glucose of glycogen. This results in cleavage of the anomeric linkage and formation of glucose-1-phosphate.

this results in elevated levels of cytosolic glucose, which diffuses out of the cell, providing glucose for other tissues that need it.

Mutations to the enzymes involved in the synthesis or breakdown of glycogen lead to glycogen storage disorders (**Table 8.1**).

TABLE 8.1 Representative Glycogen Storage Disorders

Disease	Mutation	Hypoglycemia	Hepatomegaly	Hyperlipidemia	Muscle symptoms
Type I (Von Gierke)	Glucose-6-phosphatase	Yes	Yes	Yes	None
Type II (Pompe)	Acid α–glucosidase	No	Yes	No	Weakness
Type III (Cori/Forbes)	Debranching enzyme	Yes	Yes	Yes	Myopathy
Type V (McArdle)	Muscle glycogen phosphorylase	No	No	No	Exercise induced cramping
Type IV (Tarui)	Muscle phosphofructokinase	No	No	No	Exercise induced cramping
Type 0	Glycogen synthase	Yes	No	No	Occasional cramping

Worked Problem 8.2   Glycogen storage disorder

Cori's disease is a rare autosomal recessive disorder in which the activity of debranching enzyme is defective. What effect would this have on glycogen metabolism? What might you expect for a phenotype?

Strategy What role does debranching enzyme play in glycogen metabolism? How would the lack of this enzyme affect whole-body metabolism? What would change, and why?

Solution Cori's disease is one of a family of more than 12 glycogen storage disorders. In these diseases, some aspect of glycogen metabolism is defective, and the body reacts to compensate. In the case of Cori's disease (also known as glycogen storage disease type III)

the absence of debranching enzyme means that glycogen cannot be debranched; thus, glycogen stores cannot be mobilized and the glycogen chains keep growing. This results in low fasting plasma glucose levels, elevated ketone bodies but high levels of glycogen storage in the liver and muscle (and, in some cases, the heart). These elevated levels result from the inability to catabolize glycogen properly. The clinical result of this disease is enlarged liver (hepatomegaly) and skeletal muscle weakness.

Follow-up question Why are patients with this disorder often treated with high-protein diets?

8.1.3 The regulation of glycogenesis and glycogenolysis

The regulation of glycogenesis and glycogenolysis is complex but can be distilled down to the regulation of two key enzymes: glycogen phosphorylase and glycogen synthase (**Figure 8.6**). Both of these enzymes are acutely regulated via the action of kinases and phosphatases and several allosteric regulators. As might be expected, the two proteins are reciprocally regulated. Protein kinase A is one of the main regulators of both glycogen synthase and glycogen phosphorylase, but other kinases and phosphatases also play a role in regulating glycogenesis and glycogenolysis.

Most kinases are regulated through a signaling cascade. This section discusses two of these cascades and then puts them into the context of glycogen metabolism. Among the various hormones involved in the cascades are insulin, which signals the fed state, and glucagon, which signals the fasted state. In these states, glycogen will either be synthesized or degraded, as necessary. Glycogen storage is also regulated by the catecholamine hormones epinephrine and norepinephrine. These hormones signal the "fight-or-flight" response, that is, the hormonal response that animals use to remove themselves from danger. Under these conditions, energy is required by the organism, so glycogen is broken down into glucose-1-phosphate.

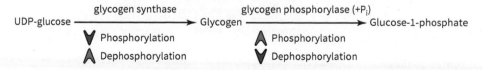

FIGURE 8.6 Overview of the regulation of glycogen metabolism. Glycogen levels are regulated by the balance of two enzymes, glycogen synthase and glycogen phosphorylase. Phosphorylation of glycogen synthase blocks the action of this enzyme and conversely activates glycogen phosphorylase. Dephosphorylation of glycogen synthase activates that enzyme and deactivates glycogen phosphorylase.

G protein coupled receptors

The peptide hormone glucagon and the catecholamines epinephrine and norepinephrine act via seven-transmembrane G protein-coupled receptors (GPCRs) as shown in **Figure 8.7**. Catecholamines or glucagon can bind to these receptors, which in turn activate a G protein. Tissue-specific expression of these receptors helps to fine-tune and integrate whole-body glycogen metabolism.

G proteins then activate adenylate cyclase. This membrane-bound enzyme catalyzes the conversion of ATP into the second messenger cyclic adenosine monophosphate (cAMP), which has several targets in the cell. Glycogen metabolism is dependent on the cAMP-dependent protein kinase (PKA). PKA is a tetrameric enzyme that is made up of a dimer of dimers. It has two regulatory subunits that bind the cAMP, and two catalytic subunits that use ATP to catalyze the phosphorylation of numerous target proteins.

Insulin receptors

Insulin signaling occurs via a different mechanism than GPCRs (**Figure 8.8**). The hormone insulin binds to an insulin receptor, causing dimerization of that receptor and autophosphorylation on specific tyrosine residues (this is a receptor tyrosine kinase, or RTK). The binding results in nucleation of scaffolding proteins and kinases, which assemble into a macromolecular signaling complex. Among the kinases are PI_3 kinase, which initiates a phosphoinositide signaling cascade. PI_3 kinase phosphorylates phosphatidylinositol 4,5-bisphosphate into phosphatidylinositol 3,4,5-trisphosphate. These second messengers activate the kinase Akt. Akt in turn phosphorylates multiple proteins, including glycogen synthase kinase-3 (GSK-3), inactivating it.

Insulin also acts to stimulate glycogenesis and block glycogenolysis by stimulating protein phosphatase 1 (PP1).

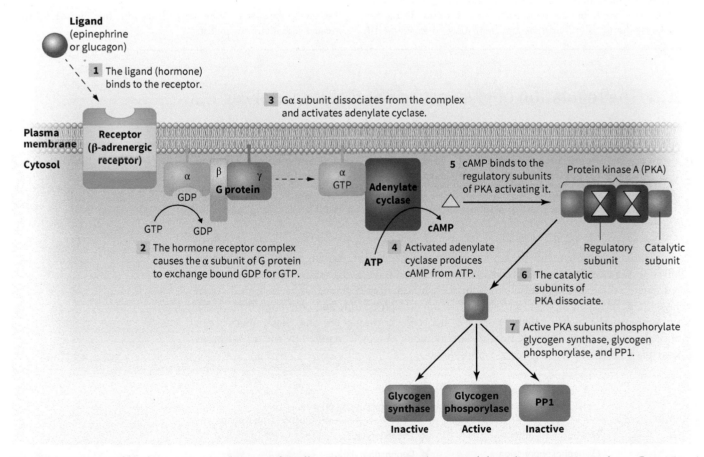

FIGURE 8.7 Action of glucagon and epinephrine on the cell.   Glucagon and epinephrine signal through a seven-transmembrane G protein-coupled receptor to activate a heterotrimeric G protein. Once bound to GTP, the Gα subunit will activate adenylate cyclase, which will produce cAMP from ATP. The second messenger cAMP will in turn activate protein kinase A, which can phosphorylate many targets in the cell, including glycogen synthase, glycogen phosphorylase, and protein phosphatase 1.

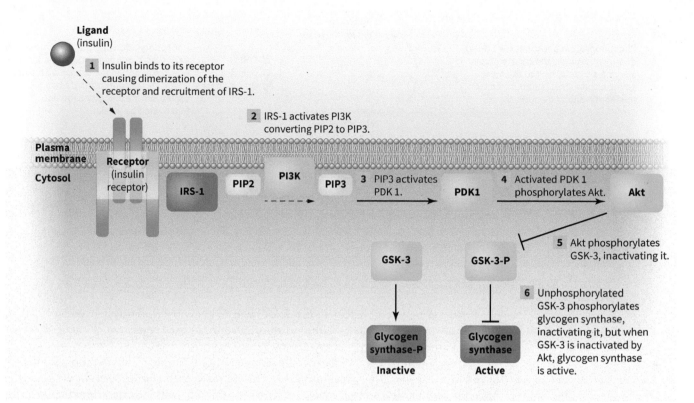

FIGURE 8.8 Insulin signaling. Insulin regulates both metabolism and growth. In metabolism, insulin signals through an insulin receptor to activate the phosphoinositide cascade via PI_3 kinase. This ultimately results in the activation of Akt, which phosphorylates and inhibits glycogen synthase kinase (GSK-3). Insulin also affects glycogen synthesis and degradation by activating phosphoprotein phosphatase 1 (not shown).

Bringing signals together: the balance of covalent modification and allosteric regulators

Both glycogen synthase and glycogen phosphorylase are regulated by covalent modification and by allosteric regulators. In the case of these enzymes, the covalent modifications consist of the phosphorylation or dephosphorylation of key serine residues, and the allosteric regulators are ATP and glucose-6-phosphate.

The a/b naming convention is used for both glycogen synthase and glycogen phosphorylase. In this convention the enzyme can be modeled as existing in one of two states, an active state "a" and a less active state "b." Phosphorylation or dephosphorylation dictates whether an enzyme is in the a or b state, but the regulation of glycogen synthase and glycogen phosphorylase are the reciprocal of each other. This means that glycogen synthase is most active (the a state) while dephosphorylated, but that glycogen phosphorylase is most active (the a state) when phosphorylated.

The activity of glycogen synthase and glycogen phosphorylase is also modified by allosteric regulators. The effectiveness of these regulators varies with regard to both the molecule itself (ATP vs. glucose-6-phosphate) as well as the phosphorylation state of the enzyme.

Bringing signals together: the effect on glycogen synthase

Glycogen synthase is activated by dephosphorylation and inactivated by phosphorylation (**Figure 8.9**). Therefore, any pathways that phosphorylate glycogen synthase will inactivate it, and any that dephosphorylate the enzyme will activate it. Glycogen synthase is phosphorylated by glycogen synthase kinase-3 or GSK-3, which is regulated by Akt. When Akt phosphorylates GSK-3, it inactivates GSK-3, thus favoring the dephosphorylated or active state of glycogen synthase. Glycogen synthase can be also directly phosphorylated by PKA and several other kinases.

Glycogen synthase is also regulated by phosphoprotein phosphatase 1 (PP1). The phosphatase removes phosphates from glycogen synthase, resulting in its activation. Interestingly, PP1 is

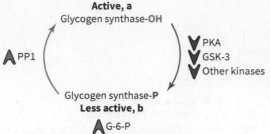

Phosphoprotein phosphatase 1 (PP1) dephosphorylates the b state of the enzyme and returns it to the more active a state. This is stimulated by the **insulin** signaling pathway.

Phosphorylation of glycogen synthase by PKA, GSK-3, or other kinases results in the enzyme adopting the less active b state. These kinases are stimulated by **epinepherine** and **glucagon** and inactivated by insulin.

Glucose-6-phosphate is an allosteric activator of glycogen synthase.

FIGURE 8.9 Regulation of glycogen synthesis.   Phosphorylation of glycogen synthase through any one of a number of possible kinases converts it to the inactive form. Dephosphorylation by phosphoprotein phosphatase 1 activates the enzyme. Phosphoprotein phosphatase itself is activated by phosphorylation by PKA.

also regulated by phosphorylation: it is inactivated by phosphorylation (catalyzed by PKA) and allosterically activated by G-6-P. This provides an added level of regulation of glycogen concentrations in the cell.

Glycogen synthase is also allosterically activated by G-6-P, which binds to the phosphorylated form of the enzyme. This makes metabolic sense in that, when glucose levels in the plasma are high, intracellular levels of glucose (and therefore of G-6-P) will increase, triggering the cell to store some of this G-6-P as glycogen.

Bringing signals together: the effect on glycogen phosphorylase

Glycogen phosphorylase is activated by phosphorylation and deactivated by dephosphorylation (**Figure 8.10**). The regulation of glycogen phosphorylase is similar to that of glycogen synthase (although slightly less complex). Phosphorylation of glycogen phosphorylase is catalyzed by phosphorylase kinase. Phosphorylase kinase is activated by phosphorylation (mainly by PKA) and also by calcium ions.

Dephosphorylation of glycogen phosphorylase is catalyzed by PP1, as previously discussed.

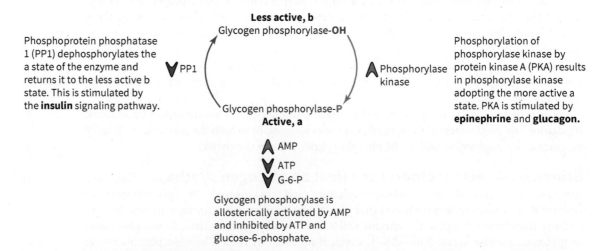

Phosphoprotein phosphatase 1 (PP1) dephosphorylates the a state of the enzyme and returns it to the less active b state. This is stimulated by the **insulin** signaling pathway.

Phosphorylation of phosphorylase kinase by protein kinase A (PKA) results in phosphorylase kinase adopting the more active a state. PKA is stimulated by **epinephrine** and **glucagon**.

Glycogen phosphorylase is allosterically activated by AMP and inhibited by ATP and glucose-6-phosphate.

FIGURE 8.10 Regulation of glycogen phosphorylase.   The activity of glycogen phosphorylase is regulated by both allosteric regulators (AMP, ATP, and glucose-6-phosphate) and phosphorylation state. A significant difference is that glycogen phosphorylase is active in the phosphorylated state. The phosphorylation of this enzyme is catalyzed by phosphorylase kinase, which in turn is regulated by PKA and Ca^{2+}. The dephosphorylation of glycogen phosphorylase is catalyzed by phosphoprotein phosphatase 1 (PP1).

PP1 is anchored in a complex with glycogen synthase and glycogen phosphorylase

PP1 is anchored within the cytosol in a complex with other proteins that regulate glycogen metabolism. These proteins include glycogen phosphorylase, phosphorylase kinase, and glycogen synthase. The protein that anchors the complex together with the phosphatase is G_M, which targets molecules to glycogen. G_M is itself phosphorylated at two different sites: one linked to insulin signaling and the other to catecholamine or glucagon signaling. Insulin-dependent phosphorylation of G_M results in activation of PP1 and dephosphorylation of the other members of the complex (phosphorylase kinase, glycogen phosphorylase, and glycogen synthase), leading to an increase in glycogen synthase activity, a decrease in glycogen phosphorylase activity, and an increase in overall glycogen storage.

Proteins that act as an organizing structure for other proteins are generically referred to as scaffolding proteins. **Scaffolding proteins** such as G_M are found in cell signaling, often with proteins that anchor kinases close to substrates, creating a locally elevated level of substrates and enzymes, and facilitating the signaling process.

Tissues respond to stimuli differently

Different tissues express different receptors, and thus may respond differently to a particular hormone (**Table 8.2**). An example of this is glycogen metabolism in the liver and muscle. Glucagon and epinephrine both elicit glycogenolysis in these tissues, but examination of the function of these tissues reveals clues as to what is happening in the underlying biochemistry. In a fight-or-flight response, muscle needs glucose to help remove the organism from danger, so muscle glycogen is depleted and used in glycolysis and the citric acid cycle to make ATP. In the liver, glycogen is mobilized, and glucose leaves the liver and travels to muscle, where again it is used to make ATP. In contrast, muscle uses its glucose in situ and does not release it. Also, muscle lacks glucagon receptors, whereas these receptors are present in the liver. When the body releases glucagon, which signals that blood sugar is low, muscle cannot respond, whereas the liver responds through glycogenolysis, releasing glucose.

TABLE 8.2 The Effect of Hormones on Glucose Metabolism

| | Insulin | | Glucagon | | Epinepherine | |
	muscle	liver	muscle	liver	muscle	liver
Glycogenesis	A	A	—	∀	∀	∀
Glycogenolysis	∀	∀	—	A	A	A

Worked Problem 8.3   Glycogen signals

A dog is brought into the veterinary clinic with weight loss. Analysis of blood samples indicates increased plasma glucose and decreased levels of plasma amino acids. Subsequent analysis reveals a pancreatic tumor and significantly elevated levels of glucagon in the plasma. How would elevated levels of glucagon affect glycogen metabolism?

Strategy In this scenario, we have an organism with elevated glucagon levels. How and where does glucagon signal in the body, and what is its ultimate effect on glycogen metabolism?

Solution Glucagon is a polypeptide (29 amino acids long) that is secreted from the α cells of the pancreas. It acts as an antagonist to insulin, raising blood glucose levels and promoting energy mobilization. Glucagon acts by binding to a G protein-coupled receptor that activates adenylate cyclase and elevates cAMP. This results in the activation of PKA, which will phosphorylate multiple targets in the cell, including phosphorylase kinase. Phosphorylase kinase is activated by phosphorylation and, in turn, activates phosphorylase,

causing glycogen breakdown and releasing more glucose into the bloodstream. PKA also targets glycogen synthase kinase and glycogen synthase. In the case of these last two enzymes, phosphorylation will block their activity, preventing the synthesis of more glycogen. This explains the elevated plasma glucose levels but not decreased concentrations of amino acids. Decreasing plasma amino acid concentrations are caused by an increasing contribution of these amino acids to gluconeogenesis to maintain plasma glucose levels in the absence of glycogen.

Glucagon receptors are predominantly expressed in the liver, but not in muscle.

This signaling cascade occurs in the context of many other processes. It is highly likely that there is cross talk between the glucagon signaling pathway and other pathways.

Follow-up question An energy drink contains both sucrose and caffeine (discussed in section 5.2). How might this combination affect glycogen metabolism?

Summary

- Glycogen is a highly branched storage form of glucose found in animals.
- Glycogen is built up from activated molecules of glucose (UDP-glucose) by the enzyme glycogen synthase.
- Glycogen phosphorylase catalyzes the release of glucose-1-phosphate (G-1-P) from glycogen.
- A complex series of kinases regulate glycogen storage. Insulin promotes storage of glycogen through a kinase cascade that results in phosphorylation of phosphoprotein phosphatase 1 (PP1) and dephosphorylation of glycogen synthase and glycogen phosphorylase. Glucagon and catecholamine hormones promote glycogen breakdown and mobilization, predominantly through PKA-mediated phosphorylation of glycogen synthase (inactivating it) and of glycogen phosphorylase (activating it).
- Glycogen synthase and glycogen phosphorylase are also regulated by allosteric mechanisms. Glycogen synthase is allosterically activated by glucose-6-phosphate. Glycogen phosphorylase is allosterically inhibited by glucose-6-phosphate and ATP, but activated by AMP.
- Changes to glycogen metabolism are tissue specific and are coordinated both by hormones and by kinase cascades.

Concept Check

1. What intermediates are required during the synthesis of glycogen?
2. Explain how glycogen formation is initiated and the reactions involved in the synthesis and breakdown of glycogen.
3. What proteins are activated by glucagon? What is the function of each of these proteins?
4. Discuss in molecular detail the kinases described in this section and the roles they play in glycogen metabolism.

8.2 The Pentose Phosphate Pathway

The pentose phosphate pathway integrates the synthesis of three different types of molecules. Section 6.1 noted that there are no essential carbohydrates. As long as we have glucose we can produce all the other carbohydrates we need. The pentose phosphate pathway provides a mechanism by which we can produce those different monosaccharides including ribose—the sugar used to produce nucleotides. There are pathways in metabolism that require reducing equivalents to synthesize molecules. The pentose phosphate pathway also provides those in terms of NADPH. Finally, the pentose phosphate pathway provides a means through which we can have a replenishable supply of antioxidants. Some of the NADPH produced is used to reduce the antioxidant peptide **glutathione**, helping keep the cell in the reduced state. Both the rearrangement of carbohydrates and the generation of NADPH use the **pentose phosphate pathway**, which is also known as the pentose phosphate shunt, the hexose monophosphate shunt, or the phosphogluconate pathway. The pathway can be divided into an oxidative phase that produces NADPH, and a second, nonoxidative phase that alters the structure of carbohydrates. The entire pentose phosphate pathway takes place in the cytosol.

8.2.1 The oxidative phase of the pentose phosphate pathway produces NADPH and ribulose-5-phosphate

The starting material for the pentose phosphate pathway is glucose-6-phosphate (G-6-P). The enzyme glucose-6-phosphate dehydrogenase catalyzes the first step in the pathway (**Figure 8.11**). This committed step is also the rate-determining step. The enzyme catalyzes a dehydrogenation, which is technically an oxidation of G-6-P to 6-phosphogluconate. The electrons from C-1 of G-6-P are passed to NADP⁺ to form NADPH. The hemiacetal carbon of G-6-P is oxidized from a hydroxyl group to a carbonyl group forming a lactone. Next, gluconolactonase catalyzes hydrolysis of the lactone (the cyclic ester) to give the carboxylic acid found on the C-1 and the

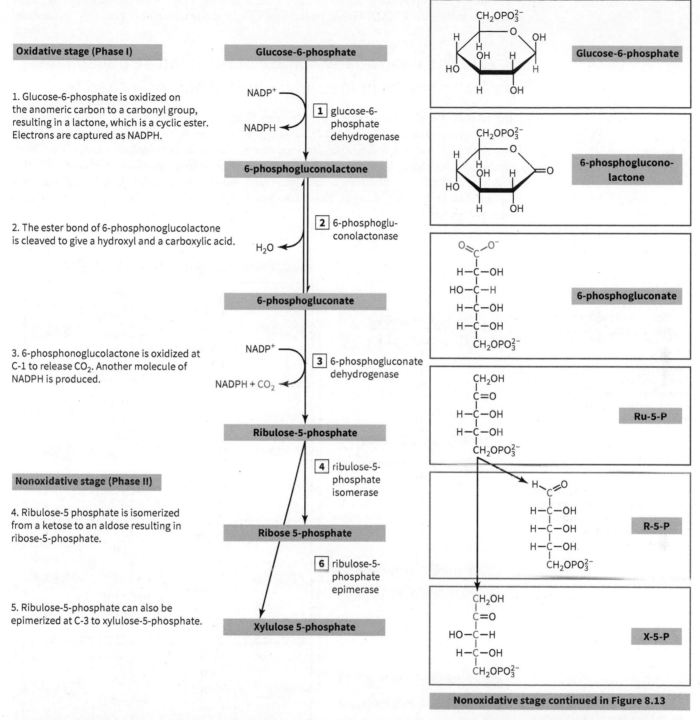

Oxidative stage (Phase I)

1. Glucose-6-phosphate is oxidized on the anomeric carbon to a carbonyl group, resulting in a lactone, which is a cyclic ester. Electrons are captured as NADPH.

2. The ester bond of 6-phosphonoglucolactone is cleaved to give a hydroxyl and a carboxylic acid.

3. 6-phosphonoglucolactone is oxidized at C-1 to release CO_2. Another molecule of NADPH is produced.

Nonoxidative stage (Phase II)

4. Ribulose-5 phosphate is isomerized from a ketose to an aldose resulting in ribose-5-phosphate.

5. Ribulose-5-phosphate can also be epimerized at C-3 to xylulose-5-phosphate.

FIGURE 8.11 Oxidative steps of the pentose phosphate pathway. The first three steps of the pentose phosphate pathway are referred to as phase I or the oxidative phase. In these steps, glucose-6-phosphate is oxidized first to a lactone, which is then opened to a carboxylic acid. Decarboxylation (loss of CO_2) generates a pentose (ribulose-5-phosphate). This process generates two molecules of NADPH. In the final step of this phase, ribulose-5-phosphate undergoes one of two different isomerizations to produce ribose-5-phosphate or xylulose-5-phosphate, as noted by the split arrows.

hydroxyl on C-5. The open-chain form of each carbohydrate is used in the rest of the pathway. The enzyme 6-phosphogluconate dehydrogenase oxidizes the carbohydrate further, oxidizing the C-3 to a keto group. The C-1 departs as CO_2, in a β keto-acid decarboxylation, producing the ketopentose ribulose-5-phosphate. The electrons removed in this reaction are shuttled to $NADP^+$, generating a second molecule of NADPH.

This is the end of the oxidative phase. So far, beginning with G-6-P, the pathway has generated two molecules of NADPH, one molecule of CO_2, and one molecule of ribulose-5-phosphate.

8.2.2 The nonoxidative phase of the pentose phosphate pathway results in rearrangement of monosaccharides

The second phase of the pentose phosphate pathway interconverts ribulose-5-phosphate with other monosaccharides (**Figure 8.12**). This provides a means to generate ribose from other carbohydrates and a way for other monosaccharides to enter carbohydrate metabolism. First, ribulose-5-phosphate isomerizes via an enediol intermediate to ribose-5-phosphate. This reaction is similar to phosphoglucose isomerase, which interconverts G-6-P and fructose-6-phosphate (F-6-P) in glycolysis. The ribose-5-phosphate produced in this step is used in nucleotide biosynthesis.

Nonoxidative stage

6. Transketolase transfers a two carbon group from xylulose-5-phosphate to ribose-5-phosphate, resulting in a molecule of glyceraldehyde-3-phosphate and sedoheptulose-7-phosphate.

7. Transaldolase transfers a three carbon group from sedoheptulose-7-phosphate to glyceraldehyde-3-phosphate resulting in fructose-6-phosphate and erythrose-4-phosphate.

8. A second reaction of transketolase combines xylulose-5-phosphate and erythrose-4-phosphate to yield fructose-6-phosphate and glyceraldehyde-3-phosphate.

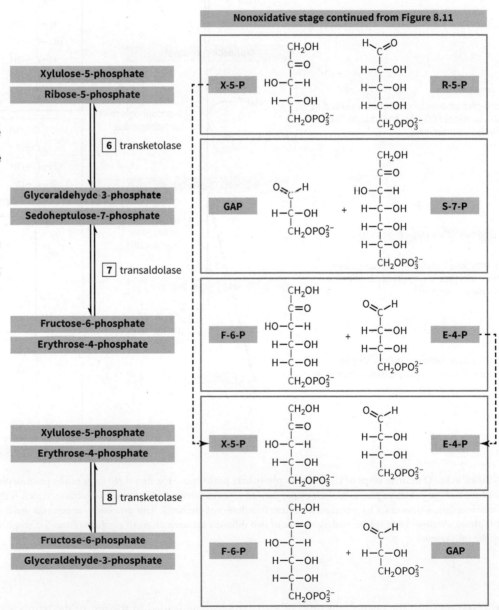

FIGURE 8.12 Nonoxidative reactions of the pentose phosphate pathway. The second phase of the pentose phosphate pathway is the nonoxidative phase. In this phase, aldoses and ketoses react to generate a new aldose and a new ketose. Dashed arrows indicate the branch points in this pathway (aldoses and ketoses of different steps combine to form new products).

Ribulose-5-phosphate can also be acted on by phosphopentose epimerase to yield xylulose-5-phosphate. Epimers are a type of isomer that differ only in the stereochemistry about a single chiral center (glucose and galactose are 4-epimers), whereas isomers differ in other ways, such as the placement of a carbonyl group (glucose is an aldose, and fructose is a ketose). Inversion of stereochemistry of C-3 on ribulose-5-phosphate occurs through an enediolate intermediate.

The final three reactions of the pentose phosphate pathway interconvert ribose-5-phosphate and xylulose-5-phosphate with other monosaccharides. These two molecules can undergo a transketolase reaction, to give sedoheptulose-7-phosphate and glyceraldehyde-3-phosphate (G-3-P). Subsequently, the two molecules can undergo a transaldolase reaction, to give erythrose-4-phosphate and F-6-P. Finally, another molecule of xylulose-5-phosphate can react with erythrose-4-phosphate to give F-6-P and G-3-P.

There are several patterns that help to make sense of this part of the pathway. First, the total number of carbons that go into any reaction must equal the total number that comes out. For example, in the first reaction of the nonoxidative stage of the pentose phosphate pathway in (reaction 6) Figure 8.12, two 5-carbon substrates go in (ribose-5-phosphate and xylulose-5-phosphate) and one 7-carbon and one 3-carbon intermediate come out (sedoheptulose-7-phosphate and glyceraldehyde-3-phosphate). Altogether, the following reactions happen:

$$5+5 \rightarrow 7+3 \rightarrow 4+6 \text{ (reactions 6 and 7)}.$$

In the final reaction, we have

$$4+5 \rightarrow 6+3 \text{ (reaction 8)}.$$

The other pattern that may be useful is that each reaction contains both an aldose and a ketose and each reaction results in the formation of a new aldose and a ketose.

The mechanisms of transketolase and transaldolase

Two enzymes figure prominently in the nonoxidative phase of the pentose phosphate pathway: transketolase and transaldolase. When looking at the products and reactants, these reactions look similar and thus might be expected to employ a similar mechanism. In fact, they use different mechanisms: one proceeding through a **thiamine pyrophosphate (TPP)** mediated mechanism and the other through a **Schiff base**.

In transketolase, TPP can deprotonate to form a TPP carbanion, which attacks the carbonyl carbon of the ketose substrate, forming an adduct with the TPP (**Figure 8.13**). This addition compound collapses into aldose as electrons flow into the thiazole ring nitrogen of TPP. The aldose product is released, and the new aldose substrate binds. Electron flow now reverses from the ring nitrogen, back out through glycolaldehyde, attacking the carbonyl carbon of the aldose substrate and forming a new carbon–carbon bond. The product deprotonates, generating a new carbonyl group on the ketose product and regenerating TPP.

In transaldolase, an active site lysine forms a Schiff base with the ketose substrate (**Figure 8.14**). This base protonates, resulting in electron flow into the base nitrogen, formation of an enzyme–substrate adduct, and the release of the aldose product. Next, an aldose substrate binds. Electron flow reverses out of the protonated Schiff base, attacking the carbonyl carbon of the aldose substrate. The base deprotonates and is hydrolyzed to yield the ketose product and regenerate the active site lysine.

8.2.3 Regulation of the pentose phosphate pathway

The pentose phosphate pathway occurs in the cytosol, together with glycolysis, most of gluconeogenesis, and several other pathways. Glucose-6-phosphate dehydrogenase catalyzes the committed step and entry into the pathway. If we consider only the oxidative phase of the pathway, this is also the rate-determining step. Because NADPH is used for biosynthetic pathways and NAD$^+$ is used in degradative ones, NADPH is found mainly in the reduced form (99 NADPH: 1NADP$^+$), whereas NAD$^+$ is found mainly in the oxidized form (99 NAD$^+$: 1NADH). The activity of glucose-6-phosphate dehydrogenase is regulated by levels of the substrate NADP$^+$.

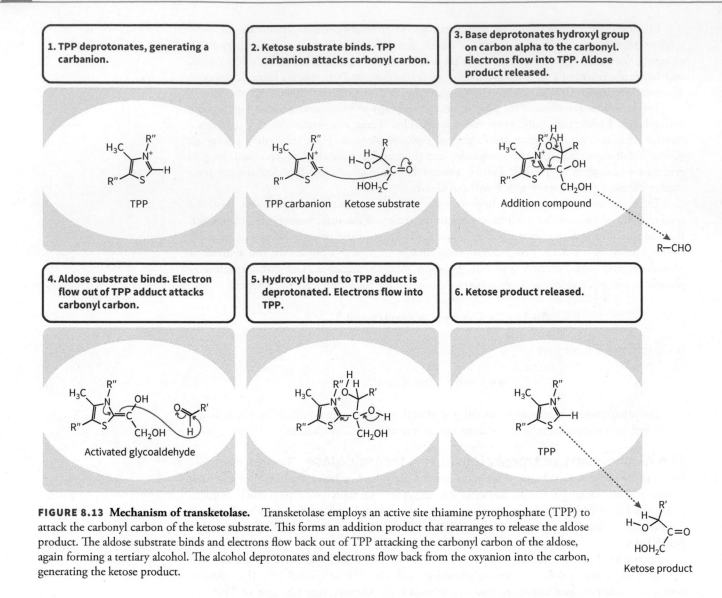

FIGURE 8.13 Mechanism of transketolase. Transketolase employs an active site thiamine pyrophosphate (TPP) to attack the carbonyl carbon of the ketose substrate. This forms an addition product that rearranges to release the aldose product. The aldose substrate binds and electrons flow back out of TPP attacking the carbonyl carbon of the aldose, again forming a tertiary alcohol. The alcohol deprotonates and electrons flow back from the oxyanion into the carbon, generating the ketose product.

The pentose phosphate pathway also has the ability to convert molecules such as ribose into hexoses for entry into glycolysis and to oxidize carbohydrates into CO_2, although these are thought to play relatively minor roles compared to NADPH and ribose-5-phosphate production.

8.2.4 The pentose phosphate pathway in health and disease

Different cells have a greater need for the products of the pentose phosphate pathway than others, such as:

- Cells that are rapidly dividing, which require both NADPH (for biosynthetic reducing equivalents) and ribose (for production of nucleotides and RNA and DNA).
- Cells of the liver, lactating mammary gland, and gonadal tissue involved in fatty acid or sterol biosynthesis that require NADPH.
- Macrophages and neutrophils, the phagocytic cells of the immune system, which require NADPH to produce the superoxide (O_2^{2-}) used in killing pathogens.

As a result, any of the cells noted in these examples have high levels of expression of the enzymes of the pentose phosphate pathway. There is one other cell type that also requires high levels of pentose phosphate pathway enzymes. Red blood cells (erythrocytes) require high levels

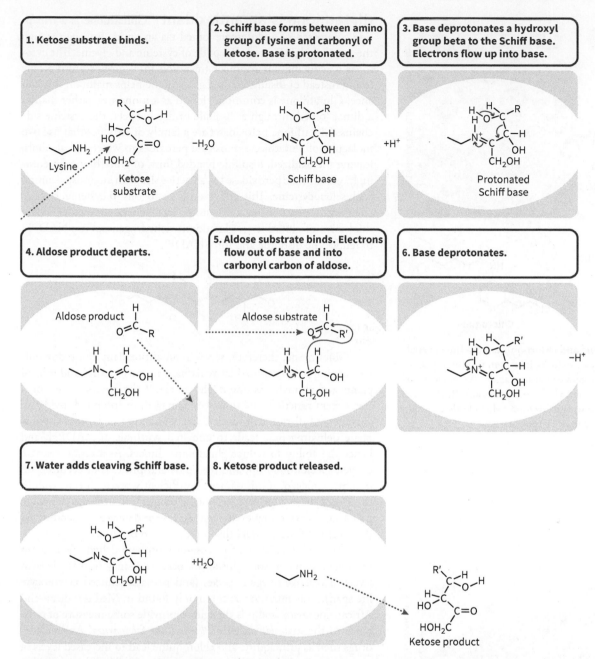

1. Ketose substrate binds.

2. Schiff base forms between amino group of lysine and carbonyl of ketose. Base is protonated.

3. Base deprotonates a hydroxyl group beta to the Schiff base. Electrons flow up into base.

4. Aldose product departs.

5. Aldose substrate binds. Electrons flow out of base and into carbonyl carbon of aldose.

6. Base deprotonates.

7. Water adds cleaving Schiff base.

8. Ketose product released.

FIGURE 8.14 Mechanism of transaldolase. Transaldolase catalyzes a similar reaction to transketolase but uses a very different mechanism. Rather than a thiamine pyrophosphate-dependent mechanism, transaldolase employs an active site lysine, which forms a Schiff base. In the first step of the mechanism, binding of the ketose substrate leads to a reaction of the active site lysine with the carbonyl carbon of the ketose, generating the Schiff base. The base is protonated and electrons flow down through an adjacent hydroxyl group, cleaving a carbon–carbon bond and leading to the release of the aldose product. Next, the aldose substrate binds and electron flow reverses from the Schiff base nitrogen up through the double bond to attack the carbonyl carbon of the aldose. Deprotonation of the Schiff base and addition of water regenerates the lysine and forms the final ketose product.

of NADPH produced by the pentose phosphate pathway to help neutralize reactive oxygen species. In erythrocytes, as in macrophages and neutrophils, there are high levels of reactive oxygen species such as hydrogen peroxide, due to high concentrations of oxygen within these cells and the potential for side reactions as molecular oxygen is carried by the heme group. In macrophages and neutrophils, hydrogen peroxide is synthesized and used to destroy pathogens and invaders. In all three cell types, the redox active tripeptide glutathione is used as a means of neutralizing reactive oxygen species that are free in the cell before they can cause oxidative damage to proteins and lipids. NADPH is used to reduce glutathione following oxidation.

A.

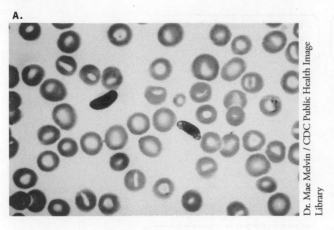

Dr. Mae Melvin / CDC Public Health Image Library

B.

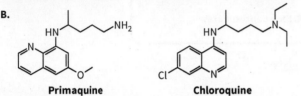

Primaquine   Chloroquine

FIGURE 8.15 Primaquine and chloroquine. A. The malarial parasite *Plasmodium falciparum* (dark images in this micrograph) degrades hemoglobin as a source of amino acids. The heme waste products crystalize into a nontoxic form (hemozoin). **B.** Primaquin and chloroquine block the recrystallization, which leads to accumulation of heme by-products toxic to the parasite.

A.

JIANG HONGYAN/Shutterstock

B.

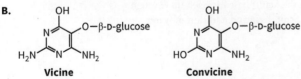

Vicine   Convicine

FIGURE 8.16 Fava beans. A. Fava beans (*Vicia faba*) are also known as broad beans. Despite their association with favism, they are a common component of the Mediterranean diet. **B.** Fava beans contain the glycosides vicine and covicine. When metabolized, these molecules lose their carbohydrate moieties. These compounds are toxic and induce hemolysis either by direct action on the erythrocyte membrane or by producing hydrogen peroxide, which in turn causes cell damage and death.

Glutathione is an antioxidant Glutathione is a tripeptide comprised of a glutamate joined via an amide linkage through the γ-carboxyl group to a dipeptide of cysteine and glycine. The cytosol is a reducing environment (meaning thiol groups are typically found instead of disulfides), and glutathione helps maintain this balance. Glutathione is commonly found as a monomer, rather than as a dimer joined through a disulfide bridge between the cysteine side chains. Glutathione peroxidases are a family of enzymes that use two molecules of glutathione to oxidize peroxides to water and alcohols, leaving the oxidized, disulfide-bonded form of glutathione. Interestingly, glutathione peroxidase has an active site with an unusual amino acid: selenocysteine. This side chain is analogous to cysteine, but has selenium in place of the sulfur atom. To regenerate the reduced free thiol form of glutathione, glutathione dehydrogenases use NADPH to reduce the disulfide and form NADP⁺.

Glutathione, therefore, serves as an antioxidant and redox buffer, helping to prevent oxidative damage in most tissues and cells of plants and animals. As one might expect, if the processes by which this system functions are mutated, damaged, or inhibited, oxidative damage and disease will result. Loss of function of glucose-6-phosphate dehydrogenase leads to impaired synthesis of NADPH, and hence the ability to reduce glutathione dimers. As a result, reactive oxygen species build up, and the lipids and proteins of erythrocytes become damaged. If this damage is sufficiently severe, it can result in the loss of the erythrocyte, eventually resulting in hemolytic anemia, a shortage of erythrocytes due to their loss or damage. Depending on the severity of the anemia, this condition can be fatal.

A partial deficiency of glucose-6-phosphate dehydrogenase can be advantageous in some instances, because it may lead to elevated levels of reactive oxygen species (and perhaps reduced erythrocyte life span). This mutation is commonly found in Mediterranean and African countries, and it is thought to provide some measure of resistance to the malarial parasite *Plasmodium falciparum*. Antimalarial drugs such as primaquine and chloroquine lead to increased levels of peroxides, which damage the erythrocyte in a similar fashion, making it less favorable to the parasite (**Figure 8.15**). Deficiencies in glucose-6-phosphate dehydrogenase can potentiate the effects of these drugs. Such deficiencies are thought to be the most common genetic defect in humans, affecting as many as 400 million people worldwide, although many people with the deficiency are asymptomatic.

Individuals with glucose-6-phosphate dehydrogenase deficiency can also have a reaction to other environmental factors. One such example is favism, a hemolytic response to consumption of fava beans (also known as broad beans, *Vicia faba*) (**Figure 8.16**). Fava beans are delicious and are commonly found in the diets of Mediterranean people but contain several molecules of the alkaloid glycosides class, such as vicine and covicine, which can damage erythrocytes by a mechanism similar to some antimalarial drugs.

Chronic granulomatous disease Chronic granulomatous disease is another disorder affected by glucose-6-phosphate dehydrogenase activity. One of the functions of NADPH is to act as an electron donor in the generation of large amounts of certain biologically useful reactive oxygen species, particularly in the phagocytic cells of the immune system. When the enzyme activity is low or absent, the ability to generate NADPH is limited. Without the ability to produce an oxidative burst of these chemicals to damage or kill intruders, these cells have a limited capacity to defend against pathogens, thus weakening the immune system and resulting in chronic infections.

The synthesis of reactive oxygen species and the response of phagocytic cells encompass a complex series of reactions involving multiple proteins. Production of NADPH is only one piece of that pathway. If any of the steps in peroxide production are in any way inhibited or defective, then chronic granulomatous disease can result. Hence, there are several classifications of the disease based on which enzymes are affected. The classifications may have different severities and pathophysiologies, but all have a similar set of symptoms.

Worked Problem 8.4   Rapid cell growth

How would the pentose phosphate pathway respond if cells were rapidly dividing?

Strategy What are the metabolic needs of a growing cell? What does the pentose phosphate pathway provide cells?

Solution The pentose phosphate pathway serves multiple functions. One function is to generate reducing equivalents in the form of NADPH. The growing cell could use NADPH in the synthesis of new fatty acids for phospholipid biosynthesis or new amino acids. The pentose phosphate pathway also provides ribose-5-phosphate for nucleotides (for DNA and RNA biosynthesis), and for NAD⁺, FAD, and other ribose-containing cofactors. In any growing cell,

there is a need for a balance between these metabolites; however, in a cell that is rapidly dividing, the need for ribose-5-phosphate for nucleotides will probably outweigh the need for reducing equivalents. In this case, the oxidative phase of the pathway can be bypassed by siphoning G-3-P and F-6-P out of glycolysis. Even though the final steps of the pathway are reversible, if these carbons are used for ribose-5-phosphate biosynthesis, they cannot be used in glycolysis or the citric acid cycle for energy.

Follow-up question Based on the pathways we have been discussing, propose how an organism could live off ribose as its sole carbon source.

8.2.5 Xylulose-5-phosphate is a master regulator of carbohydrate and lipid metabolism

Previously when we have discussed the regulation of metabolism we have focused on short-term regulation of enzyme activity modulated via phosphorylation or allosteric regulators. A third way that metabolism is regulated is by increasing the levels of enzymes through increased transcription of the genes that code for them. The proteins that regulate gene transcription are termed transcription factors. One such transcription factor is the **carbohydrate response element-binding protein (ChREBP)**, a transcription factor that responds to high levels of carbohydrates and regulates transcription of genes involved in carbohydrate and lipid metabolism. The most notable genes regulated by ChREBP are those for pyruvate kinase, as well as acetyl-CoA carboxylase and fatty acid synthase, both discussed in lipid metabolism.

In the absence of high levels of carbohydrate, ChREBP is found in the cytosol in the inactive form. When levels of G-6-P in the cell are high, the pentose phosphate pathway is active, generating NADPH for fatty acid biosynthesis. As flux through this pathway increases, it produces higher levels of xylulose-5-phosphate, which activates a phosphatase (protein phosphatase 2A). This dephosphorylates specific phosphoserine and phosphothreonine residues in ChREBP, activating the protein. The now-active transcription factor translocates from the cytosol to the nucleus, where it increases the expression of genes needed to store energy.

In summary, when G-6-P is plentiful in the cell, the pentose phosphate pathway product xylulose-5-phosphate signals to ChREBP via a phosphatase to increase the expression of the genes of glycolysis and fatty acid biosynthesis in order to store energy. If G-6-P levels are low, due to either a lack of glucose or high rates of flux through glycolysis and the citric acid cycle, the transcription factor remains phosphorylated in the cytosol, and gene transcription is not initiated (**Figure 8.17**).

FIGURE 8.17 Carbohydrate response element binding protein (ChREBP). The carbohydrate response element binding protein is a cytosolic transcription factor. In the inactive state, it is found phosphorylated on key residues (catalyzed by both PKA and AMPK). When levels of xylulose-5-phosphate rise, they activate phosphoprotein phosphatase 2A, which dephosphorylates the transcription factor. This leads to its translocation to the nucleus and activation of genes involved in energy storage.

(Source: Wessner, *Microbiology*, 2e, copyright 2017, John Wiley & Sons. This material is reproduced with permission of John Wiley & Sons, Inc.)

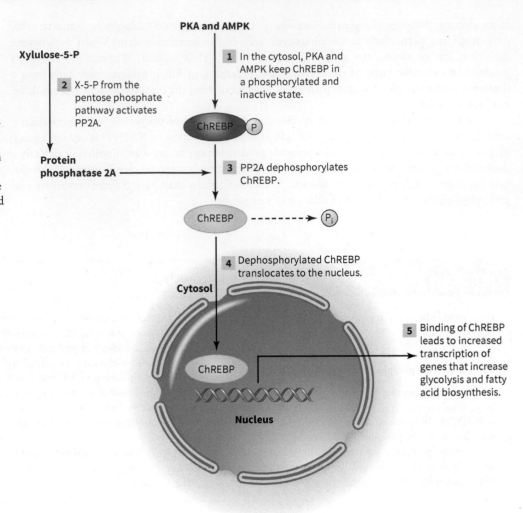

Xylulose-5-P

PKA and AMPK

1 In the cytosol, PKA and AMPK keep ChREBP in a phosphorylated and inactive state.

2 X-5-P from the pentose phosphate pathway activates PP2A.

ChREBP P

Protein phosphatase 2A

3 PP2A dephosphorylates ChREBP.

ChREBP ------> Pi

4 Dephosphorylated ChREBP translocates to the nucleus.

Cytosol

ChREBP

5 Binding of ChREBP leads to increased transcription of genes that increase glycolysis and fatty acid biosynthesis.

Nucleus

Worked Problem 8.5 | Fructose redux

High-fructose corn syrup has been targeted as a cause of obesity. Section 6.3 discussed how fructose is catabolized via glycolysis (fructolysis). How could fructose be catabolized via the pentose phosphate pathway, and what might be the ramifications of elevated fructose levels in the cell?

Strategy To answer this question, we need to know how fructose is metabolized in glycolysis and fructolysis, and how the pentose phosphate pathway works. We should also think about other products of the pentose phosphate pathway, and how they may contribute to the overall metabolism of the cell or organism.

Solution F-6-P is one product of the pentose phosphate pathway. Section 6.2 explained that monosaccharides diffuse into the cell down a concentration gradient. Once inside the cell, the monosaccharide is phosphorylated, either through fructokinase (to yield F-1-P) or hexokinase (to yield F-6-P). F-6-P is usually catabolized via glycolysis or fructolysis; however, the transketolation and transaldolation reactions of the pentose phosphate pathway are reversible. Therefore, if F-6-P is in abundance, it could react with G-3-P

to give xylulose-5-phosphate and erythrose-4-phosphate. Another molecule of F-6-P could then react with the erythrose-4-phosphate to yield sedoheptulose-7-phosphate and glyceraldehyde-3-phosphate. As described in section 8.2, in the forward direction of the pathway these two molecules can react together to give ribose-5-phosphate and another molecule of xylulose-5-phosphate. Thus, from one molecule each of fructose and of glyceraldehyde-3-phosphate, it is possible to generate two molecules of xylulose-5-phosphate and hence alter gene expression via ChREBP.

Although it is unlikely that most fructose is metabolized through this pathway, a demonstration that elevated fructose leads to altered gene expression could validate this hypothesis. There are many ways to catabolize fructose in the cell, all of which occur simultaneously; therefore, discerning increased flux through a particular pathway is difficult.

Follow-up question How could you test whether exposure to fructose alters gene expression? What would you use for controls, both positive and negative?

Summary

- The pentose phosphate pathway coordinates three divergent needs of the cell: ribose synthesis, NADPH synthesis, and carbohydrate metabolism.

- In the first (oxidative) phase of the pentose phosphate pathway, G-6-P is oxidized into 6-phosphogluconate, and then into ribulose-5-phosphate and CO_2. One molecule of NADPH is produced at each step.

- In the nonoxidative second phase of the pentose phosphate pathway, a molecule of ribulose-5-phosphate is isomerized to xylulose-5-phosphate. The subsequent reactions of the second phase are transaldolase and transketolase reactions that allow the rearrangement of these two pentoses into other aldoses and ketoses.

- The pentose phosphate pathway provides a way for carbohydrates to enter metabolism other than through glycolysis or gluconeogenesis.

- The NADPH produced in the pentose phosphate pathway is used in lipid biosynthetic pathways. The NADPH is also important in the oxidative response of some cells of the immune system and in maintaining a reducing environment in the cell by reducing oxidized glutathione dimers. Disruptions of these pathways can result in disease.

- Xylulose-5-phosphate is an important metabolic regulator; it activates protein phosphatase 2A, which in turn dephosphorylates and activates the transcription factor ChREBP, activating several genes involved in energy storage.

Concept Check

1. Discuss the two stages of pentose phosphate pathway. Outline the reactions taking place in each and products formed.

2. Explain how the flux through the pentose phosphate pathway coordinates diverse aspects of metabolism and provides for the different needs of the cell.

3. Why does the pentose phosphate pathway take place in the cytosol?

8.3 Carbohydrates in Glycoconjugates

Carbohydrates are important as structural molecules, as ligands for protein receptors, and in signaling. They can be linked to proteins or lipids on the surface of the cell, and they are also found in macromolecular assemblies outside the cell. This section discusses four main classes of carbohydrate modifications, known as glycoconjugates. These include:

- **Glycoproteins**—membrane-bound or extracellular proteins with some amount of carbohydrate modification.

- **Glycolipids**—membrane phospholipids with an attached carbohydrate moiety.

- **Proteoglycans**—extensive mesh nets of polysaccharides joined to fibrous proteins.

- **Peptidoglycans**—lengthy chains of polysaccharides cross-linked by peptides and found in bacterial cell walls.

Section 8.4 discusses the assembly of these molecules into the extracellular matrix.

8.3.1 Glycoproteins

Glycoproteins are proteins with attached carbohydrate modifications. In eukaryotic cells, these modifications are made initially in the endoplasmic reticulum and are then continued in the Golgi complex. Most glycosylated proteins are secreted by the cell to the outside, or retained to lysosomes, or found as integral plasma membrane proteins; few glycosylated proteins are found in the cytosol or in other organelles.

The carbohydrates found in glycoproteins can be either *N*-linked or *O*-linked. In *N*-linked glycoproteins, the carbohydrates are linked via an amide to the side chain of the amino acid asparagine. The **consensus recognition sequence**, the protein sequence that is recognized and

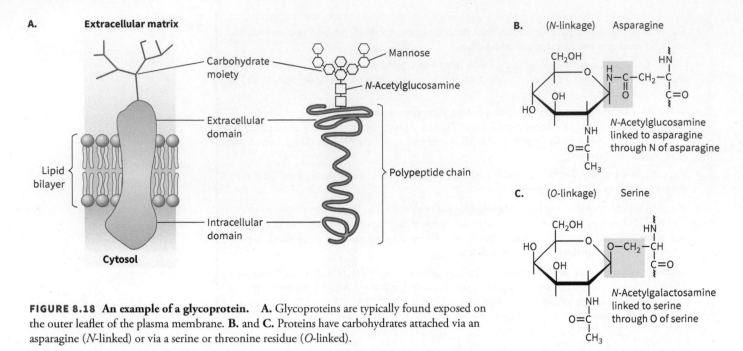

FIGURE 8.18 An example of a glycoprotein. **A.** Glycoproteins are typically found exposed on the outer leaflet of the plasma membrane. **B.** and **C.** Proteins have carbohydrates attached via an asparagine (*N*-linked) or via a serine or threonine residue (*O*-linked).

glycosylated, is (Asn-X-Ser/Thr) where X is any amino acid. *O*-linked glycoproteins are linked via an acetal linkage between the sugar and the hydroxyl group of a serine, or threonine residue, or hydroxylated residues (hydroxylysine and hydroxyproline) (**Figure 8.18**).

Numerous types of monosaccharides can be attached to proteins (**Figure 8.19**). In addition to some of the more familiar monosaccharides discussed in section 6.1 (glucose, mannose, galactose), there are aldoses (fucose and xylulose) and acetylated amino sugars. This last group includes

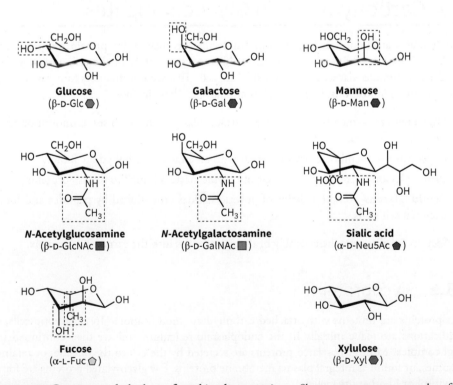

FIGURE 8.19 Common carbohydrates found in glycoproteins. Shown are common examples of monosaccharides found in glycoproteins and their abbreviations. Note the amino sugars *N*-Acetylglucosamine, *N*-Acetylgalactosamine, and neuraminic acid (*N*-Acetyl sialic acid). Dashed boxes highlight differences among these structures.

N-Acetylglucosamine (GlcNAc), *N*-Acetylgalactosamine (GalNAc), and *N*-Acetylneuraminic acid (Neu5Ac or NANA). Neuraminic acid and its derivatives are types of sialic acid. It has been estimated that half of all proteins are glycosylated and these modifications range from less than 1% to 60% carbohydrate by mass.

The roles of glycosylations in proteins are still being elucidated. Some regulate the activity of enzymes or stability of proteins, and may be involved in protein folding. Some modified carbohydrates, such as mannose-6-phosphate, are involved in protein trafficking, that is, the organized movement and marshaling of proteins in the cell. Prokaryotes have limited ability to glycosylate proteins, and among eukaryotic species, there is variation in ability to glycosylate proteins. This difference is important if using bacteria or an insect cell line to produce protein for study, where improper or missing glycosylation can (and does) lead to misfolded or inactive proteins.

Examples of glycoproteins   There are thousands of glycoproteins, and their functions in biochemistry are still being elucidated. This section discusses one example in which carbohydrate modification is important, the ABO blood groups.

ABO blood group antigens   The ABO blood group antigens are a series of carbohydrate modifications predominantly on a transmembrane protein found on the extracellular surface of erythrocyte plasma membranes (**Figure 8.20**). In the ABO system there are four blood groups: A, B, AB, and O. People with the O blood type have glycoproteins bound to a core pentasaccharide. Those with type A blood have an additional molecule of *N*-Acetylglucosamine, whereas those with type B have an additional molecule of galactose. Those with type AB blood have a combination of proteins with the A glycosylation and proteins with the B glycosylation. People with blood type O are known as *universal donors* because their blood can be given to anyone; their proteins lack these modifications and just have the core modification. People with blood type AB type are known as *universal recipients* because they can receive blood from anyone; their immune system recognizes both modifications as those of the recipients.

Receiving the wrong blood type elicits an immune response against the blood cells that are recognized as foreign; this can be fatal.

Blood types are usually listed as an ABO letter or letters followed by a (+) or (−) symbol known as the rhesus factor, Rh. The Rh is a different protein that is coded for by a different gene and is positive or negative, indicating the presence or absence of this protein.

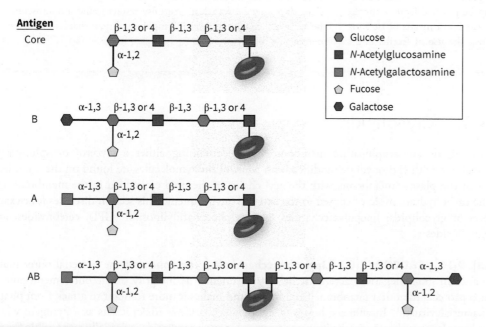

FIGURE 8.20 ABO blood group antigens.   The ABO blood group antigens result from carbohydrate modifications on the surface of erythrocytes. People with the O antigen have the core carbohydrate shown. Those with the A blood type have an extra molecule of *N*-Acetylgalactosamine, while those with the B blood type have an extra molecule of galactose. People with the AB blood group have a mixture of glycoproteins with both modifications found.

One group of proteins that bind to and recognize specific glycosylations are lectins, plant glycoproteins that are discussed in **Biochemistry: Lectins.**

Biochemistry

Lectins

Glycoproteins are proteins with attached carbohydrates; in contrast, lectins are proteins that recognize and bind tightly to specific carbohydrate sequences.

Although lectins were first identified in plants more than 100 years ago, little is known about their function. Lectins may be important in cell adhesion to an extracellular matrix or in the defense of the host against a pathogen. In animals, lectins bind to mannose-6-phosphate residues on glycosylated proteins in the Golgi complex, and they assist in the sorting of these proteins. This process, known as protein trafficking, is a field of intense investigation.

Lectins can be monomeric or multimeric proteins, and many are polyvalent (can bind to multiple carbohydrate chains at once). They are globular and have an average size of 40 to 75 kDa. Most lectins have at least one metal ion (either Mg^{2+} or Ca^{2+}). The lectin concanavalin A from jack beans was found to have an almost identical amino acid sequence to another protein found in the same plants, differing only at the amino and carboxyl termini. Concanavalin A exhibits a property called circular permutation. In this process, after translation and folding, the amino and carboxyl termini of the protein end up in close proximity to one another and then join via a process that is unclear, creating a circular loop through the amino acid backbone. Some other part of the backbone is cleaved, generating new amino and carboxyl termini.

In the laboratory, lectins have been used in affinity purification techniques (discussed in the techniques section). Many of the best characterized lectins come from legumes, an abundant starting material for purification. The lectin can be purified from a sample, affixed to a column or resin, and then used as a means of isolating or purifying other glycosylated proteins. The use of lectins in the laboratory decreased when advances in antibody production and molecular biology made it possible to add affinity tags to proteins of interest. However, more recently, fluorescently labeled lectins and advances in fluorescent technology have helped lectins regain their place in the laboratory.

Lectins have other applications. Many are categorized as agglutinins—compounds that make blood cells clump together. This happens as the lectins bind to the carbohydrate moiety of cell-surface glycoproteins. Because this binding is carbohydrate specific, lectins can be used to help identify the antigens, in this case, carbohydrate moieties, on the cell surface and perform blood typing.

Although lectins are a useful research tool, not all lectins are benign. One of the first lectins to be purified in 1888 was ricin from castor beans. Ricin binds to glycoproteins on the cell and thus gains entry to the cell through endocytosis. Once inside the cell, ricin is a potent inhibitor of ribosome function. One of the ricin subunits is a glycosidase that cleaves an *N*-glycosidic bond within the 28S subunit of the ribosome, removing an adenine. This irreversibly damages the ribosome, blocking protein synthesis. The turnover number for this enzyme is 1,777 per minute; that is, the enzyme can inactivate as many as 1,777 ribosomes per minute. Inactivation of the ribosomes results in cell death, and exposure to ricin is often fatal. The lethal dose in humans can be as little as 1.75 µg, which can be obtained from as few as five castor beans. Ricin is suspected of having been used as a poison in espionage. In one of the more infamous assassinations of the Cold War, Bulgarian dissident Georgi Markov was poisoned with a microscopic ricin-tipped dart fired into his leg from an umbrella as he waited for a bus in London. Four days later he was dead. Autopsy results found a microscopic pellet in his leg containing a 0.2 mg dose of ricin that had failed to completely dissolve.

8.3.2 Glycolipids

Glycolipids are amphipathic membrane lipids containing either a glycerol or sphingosine backbone with carbohydrate modifications. Most of these molecules are found on the outer leaflet of the plasma membrane, with the acyl chains of the lipid embedded in the membrane and the carbohydrate moiety exposed to the external environment. This section discusses four examples of glycolipids: lipopolysaccharides, glycosylphosphatidylinositol (GPI), cerebrosides, and gangliosides.

Lipopolysaccharide   **Lipopolysaccharide** is a component of the bacterial outer membrane. In gram-negative bacteria, it helps to maintain the integrity of the outer membrane; it also protects the outer membrane from lipases and makes it more resistant to attack from phages (bacterial viruses). In animals, however, lipopolysaccharide is toxic; it acts as a pyrogen, which induces fever and activates the immune system. Lipopolysaccharide is partially responsible for the septic shock that is sometimes seen in advanced infections.

Lipopolysaccharide has three parts: an outer polysaccharide termed the O antigen, a core oligosaccharide, and lipid A, which is a bacterial lipid comprised of a dimer of glucosamine molecules to which multiple (usually six) fatty acids are attached via ester or amide linkages.

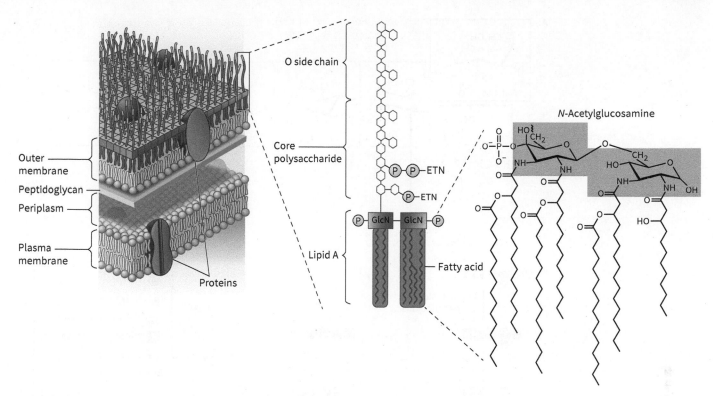

FIGURE 8.21 Structure of lipopolysaccharide. Lipopolysaccharide is found in the outer membrane of bacteria. It consists of lipid A bound to a core series of carbohydrates, two ethanolamine groups (ETN), and a lengthy carbohydrate tail.

(Source: Wessner, Microbiology, 2e, copyright 2017, John Wiley & Sons. This material is reproduced with permission of John Wiley & Sons, Inc.)

The core oligosaccharide acts as an adapter between lipid A and the O antigen. The core can vary considerably, but it often contains one, two, or three monosaccharides attached to lipid A through a glycosidic linkage to the glucosamine group. The outer O antigen also varies from species to species but typically consists of linear combinations of galactose, mannose, rhamnose, and other hexoses. The O antigen is also linked via a glycosidic bond to the core carbohydrate (**Figure 8.21**).

Glycosylphosphatidylinositol
Glycosylphosphatidylinositol, or **GPI** anchors are used by eukaryotic cells to anchor proteins to the outer leaflet of the plasma membrane (**Figure 8.22**). GPI contains a molecule of phosphatidyl inositol, a linker polysaccharide composed of glucosamine and several mannose residues, and a molecule of phosphoethanolamine attached to the protein. As in other carbohydrate-containing molecules, the polysaccharides may be modified with other groups (in this case, other fatty acids or carbohydrates).

Proteins to be affixed to a GPI anchor are translated into the ER, where a hydrophobic sequence at the protein's carboxy terminus attaches the protein to the ER membrane. This sequence is cleaved, and the resulting carboxy terminus is linked via an amide to the amine moiety of phosphoethanolamine, forming the GPI anchor.

There are many different types of GPI-anchored proteins, for example, enzymes, adhesion molecules, and proteins found in cell–cell recognition. Several important signal transduction pathways employ GPI-anchored proteins to tether them to the plasma membrane.

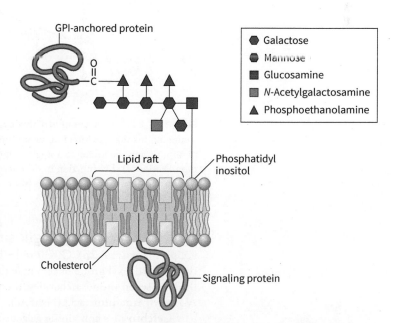

FIGURE 8.22 Structure of a glycosyl phosphoinositide (GPI) anchor. GPI anchors are used in eukaryotic cells to anchor proteins in the outer leaflet of the plasma membrane. The anchor consists of a molecule of phosphatidyl inositol to which a mannose-rich carbohydrate linker is attached. The mannose adapter is also linked to a phosphoethanolamine group, which is bound to a protein via an amide link to the carboxy terminus of the protein.

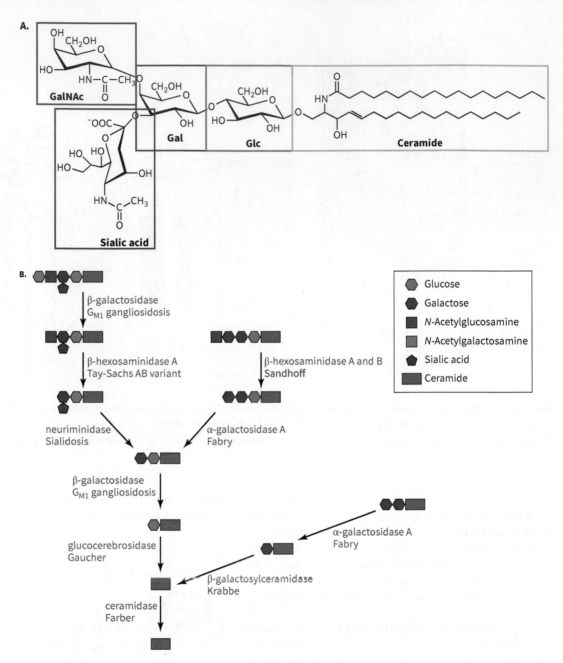

FIGURE 8.23 Structures of cerebrosides and gangliosides. Cerebrosides and gangliosides are glyco-sphingolipids that are found in higher concentrations in nervous tissue. **A.** Cerebrosides have the phospholipid ceramide joined to a single simple carbohydrate, while gangliosides have an oligosaccharide modification. **B.** Dysfunction in the enzymes (shown in blue) that produce or break down these molecules, lead to a variety of human diseases (shown in red).

Cerebrosides and gangliosides

Cerebrosides and gangliosides are glycosylated sphingolipids (**Figure 8.23**). Cerebrosides have a single sugar, usually glucose or galactose, attached to the free hydroxyl group of ceramide through a glycosidic linkage. Gangliosides have an additional three to seven monosaccharides; these are often not linear and include at least one molecule of *N*-Acetylneuraminic acid (Neu5Ac).

Cerebrosides containing galactose (galactocerebrosides) and gangliosides are highly enriched in the central nervous system but are found to some extent in most tissues. Cerebrosides and gangliosides are internalized and degraded in lysosomes. There are several inborn errors of metabolism (Gaucher's, Tay-Sachs, Nieman-Pick type C, and Fabry's diseases) that involve a deficiency in cerebroside or ganglioside degradation. In these disorders, lysosomes fill with glycolipids; the result is developmental disorders and, in many cases, death (**Figure 8.24**).

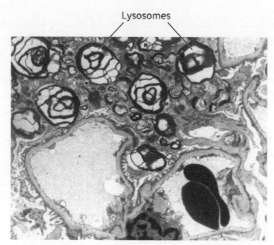

Lysosomes

H. Liapis, K. Foster, E. Theodoropoulou, et al. *Phenotype/Genotype Correlations in the Ultrastructure of Monogenetic Glomerular Diseases*, Journal Ultrastructural Pathology Volume 28, 2004 - issue 4, Taylor and Francis

FIGURE 8.24 Lysosomal storage disorder. The lysosomes in the micrograph are filled with whorls of membrane, indicative of a lysosomal storage disorder.

The Neu5Ac and other sialic acid derivatives found in gangliosides also have a role in infectious disease. For example, influenza virus has a protein on the surface of the virus called hemagglutinin that binds to sialic acid residues on the surface of cells. The virus uses this protein to gain entry and infect the cell. Recently, drugs have been developed that attempt to treat the spread of the flu virus through attacking these pathways. These are discussed in **Medical Biochemistry: Neuraminidase inhibitors**.

Medical Biochemistry

Neuraminidase inhibitors

There is a disease that accounts for more than half a million deaths each year. It mutates out of control at such a rate that vaccination programs and health organizations constantly work to stay ahead of it. In any given year, this disease may infect as many as 20% of the population. It is easily transmitted through the air or contaminated surfaces, can jump from species to species, and may be transmitted and spread through agriculture and migrating birds. In 1918, an outbreak of this disease killed some 50 to 100 million people, more than half of whom were in their 20s and 30s. That outbreak killed more than 1% of the world's population and caused more deaths in its first 25 weeks than the AIDS pandemic did in its first 25 years. The disease is influenza, commonly known as the "flu."

Influenza is caused by the influenza A virus (*Orthomyxoviridae*), a single-stranded RNA virus that attacks the epithelial cells of the mucus membranes and respiratory tract. Release of proinflammatory cytokines such as tumor necrosis factor α (TNF-α) and γ-interferon cause some of the other symptoms of influenza, including fever, muscle aches, and overall tiredness. The influenza virus has an envelope—a lipid bilayer studded with glycoproteins—that surrounds an RNA core. The RNA is in about eight pieces, each of which codes for one or two genes. Two important glycoproteins, hemagglutinin and neuraminidase, are on the surface of the virus particle, or the virion. These two proteins are reflected in the name of the virus, such as H5N1 or H2N2. There are 17 different subtypes of hemagglutinin and 9 different subtypes of neuraminidase.

Hemagglutinin is a carbohydrate-binding glycoprotein; essentially, it is a lectin. Hemagglutinin has two main roles in viral pathogenesis. It binds to sialic acid residues on the cell surface (particularly red blood cells and the epithelial cells of the upper respiratory tract); this binding fastens the virion to the host cell. The host cell uses endocytosis to internalize pieces of membrane, and viruses bound to sialic acid are internalized in this process. Inside the cell, these endosomes are processed as they normally would be. That is, proton pumps begin to acidify the endosome in preparation for fusion with other vesicles and transformation into a lysosome. However, in this instance, acidification of the contents is a trap. As the endosome is acidified, hemagglutinin changes in structure, forming a single α helix more than 60 amino acids long. The formation of this helix exposes a short hydrophobic peptide sequence at the end, which is thought to act as a molecular grappling hook, fastening into the endosomal membrane and facilitating fusion with the virion, releasing the viral contents into the cytoplasm of the host cell.

Inside the host cell, ribosomes bind and translate the viral RNA into viral proteins, including an RNA-dependent RNA polymerase (viral polymerase). The viral polymerase and messages translocate to the nucleus, where the new viral messages are transcribed. These new messages are transported back to the cytosol, where they assemble with riboproteins and other viral proteins at the cell membrane. Viruses bud from the cell but are still bound to the membrane by hemagglutinin.

Neuraminidase, the other glycoprotein of interest in this story, is a hydrolytic enzyme that cleaves sialic acid residues from the surface of the viral envelope and the cell. The removal of sialic acid enables the virus to leave the cell and infect other cells nearby; it also prevents viruses from binding to one another.

One line of drug development for influenza is based on identifying and exploiting differences between the biology of the pathogen and the infected person. In the case of the influenza virus, the presence of neuraminidase is one of the differences. Drugs that inhibit neuraminidase act by blocking the function of the enzyme; that is, the drugs block cleavage of sialic acid groups. The four drugs for influenza currently on the market are competitive inhibitors of neuraminidase. The infected cells can still make mature viral particles, but the drugs render those particles unable to release and propagate the infection. Although these drugs are not cures for influenza, they tend to lessen the severity of the infection and reduce the time with symptoms by one day.

8.3.3 Proteoglycans and non-proteoglycan polysaccharides

Proteoglycans are a major component of the extracellular matrix. They have two major parts: a glycosaminoglycan, or GAG, which is a large polysaccharide consisting in part of amino sugars, and a protein core. One significant difference between glycoproteins and proteoglycans is the overall amount of carbohydrate. Although glycoproteins can have extensive carbohydrate modifications, they are still predominantly proteins. Proteoglycans, on the other hand, have far more carbohydrate than protein. Cross-linking in proteoglycans between proteins and carbohydrate chains creates an extensive macromolecular network. One non-proteoglycan carbohydrate worthy of mention is hyaluronic acid. This lengthy polysaccharide is comprised of amino sugars but lacks the protein component found in proteoglycans.

Despite the amino groups on the carbohydrate moiety of proteoglycans, the molecule often carries an overall negative charge. This is due to oxidized carbohydrates via carboxylic acid groups, or sulfates being coupled to the carbohydrate. These negatively charged groups have sodium counter ions, which draw significant amounts of water into the structure. The ability to absorb water serves various functions, depending on the proteoglycan. Most proteoglycans are viscous, and many act as gels. Hence, some of these molecules can act as lubricants, keeping joints moving freely. Other proteoglycans form a much firmer gel and resist compaction, acting as shock absorbers. The layer of carbohydrate presents a partial barrier to both pathogens and cancerous cells, preventing or slowing their progression through a tissue (**Figure 8.25**).

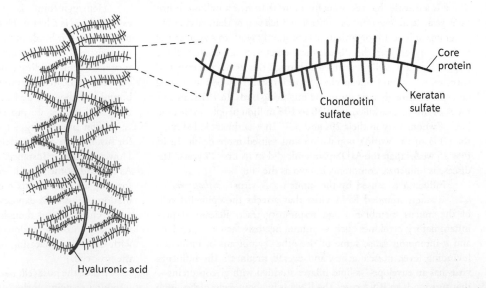

FIGURE 8.25 Structure of a proteoglycan.   Proteoglycans consist of lengthy carbohydrate polymers linked to a core protein. Often these are all organized on a central molecule of the polysaccharide hyaluronic acid.

(Source: Karp, *Cell and Molecular Biology: Concepts and Experiments*, 7e, copyright 2013, John Wiley & Sons. This material is reproduced with permission of John Wiley & Sons, Inc.)

Examples of glycosaminoglycans

Examples of common glycosaminoglycans found in humans include:

- **Heparan sulfate**—this linear polysaccharide is comprised of repeating disaccharides, usually glucuronic acid and *N*-Acetylglucosamine. The polysaccharide is sulfated at several locations in each disaccharide unit. Heparin is a related glycosaminoglycan that is enriched in iduronic acid and *N*-Acetylglucosamine. Heparan sulfate is found in the extracellular matrices of most animals.

Heparan sulfate

Glucosamine sulfate

Glucuronic acid sulfate

- **Chondroitin sulfate**—this linear polysaccharide is comprised of repeating glucuronic acid and *N*-Acetylgalactosamine disaccharides. As with heparan, it is multiply sulfated on each residue (typically on the 4 and 6 position of each *N*-Acetylglucosamine). Isomerization of glucuronic acid molecules into iduronic acid produces the molecule dermatan sulfate (a common component of epithelial tissues, such as skin and blood vessels). Chondroitin sulfate is found in materials of high tensile strength, such as tendons, cartilage, and the aorta. Many people currently take supplements of glucosamine and chondroitin for joint health, although their efficacy is debated.

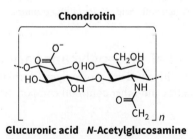

Chondroitin

Glucuronic acid   *N*-Acetylglucosamine

- **Keratan sulfate**—this linear polysaccharide is a polymer of galactose and *N*-Acetylglucosamine. It is typically less sulfated than chondroitin or heparan sulfate, having a single sulfate on C-6 of either monosaccharide. Keratan sulfate is found in bone, horn, and cornea.

Keratan sulfate

Galactose   *N*-Acetylglucosamine sulfate

These three polysaccharides have a far greater degree of variability than is seen in other classes of biological molecules. The sequence of a nucleotide or protein, although not set in stone, is highly invariant. Considerable biochemical effort is placed on making sure that DNA is copied with high fidelity—few to no mistakes—or that proteins are synthesized without error. The same does not seem to hold for proteoglycans. Depending on the tissue, organism, or even sample, there can be variability in characteristics such as degrees of sulfation, incorporation of other monosaccharides, isomerization or epimerization of monomers in the chain, and length of the chain. Whether this variability has a distinct purpose in biochemistry and what it is selected for are unclear, although it is known that tampering with these ratios can be harmful. In 2008, a batch of heparin was contaminated with chondroitin that had been chemically oversulfated. Patients given this heparin died rapidly from anaphylaxis, the severe allergic response commonly associated with bee stings or peanuts in those who are severely allergic. Further characterization revealed that this modified polysaccharide can elicit a release of the vasodilator bradykinin. Although the mechanism of the release of bradykinin is still unknown, this example shows that the interactions of cells with glycosaminoglycans can be harmful, and strongly suggests that the degree of sulfation of these polysaccharides is not random.

Hyaluronic acid is a non-proteoglycan matrix polysaccharide   Hyaluronic acid differs from the other polysaccharides discussed here, in that it is not a part of a proteoglycan. Rather, it is secreted as a lengthy linear polysaccharide of glucuronic acid and *N*-Acetylglucosamine. Molecules of hyaluronic acid can be huge, containing up to 25,000 monomers and ranging in molecular weight from several kilodaltons up to 20 million daltons. The average size ranges from 3 to 5 million daltons, depending on the tissue. Hyaluronic acid is synthesized by a complex in the plasma membrane and excreted from the cell as it is synthesized (whereas other glycosaminoglycans are added in the ER and Golgi complex).

Glucuronic acid   *N*-Acetylglucosamine

The protein components of proteoglycans vary widely   Numerous matrix proteins have been found to be involved in proteoglycan formation. We will discuss several fibrous proteins found in the matrix and other common components of proteoglycans in the next section.

8.3.4 Peptidoglycans

Peptidoglycans are the building blocks of the cell wall in bacteria. They are composed of a dimeric repeat of *N*-Acetylglucosamine (GlcNAc) and *N*-Acetylmuramic acid (MurNAc) in a β-1,4 linkage (**Figure 8.26**). Extending from each molecule of *N*-Acetylmuramic acid is a short (4 to 5 residue) peptide, linked through the oxygen on C-3. The amino acids found in the peptide vary by species, but include the L isomers of alanine and lysine, and the D isomers of glutamine, glutamic acid, and alanine. Please recall that the L isomers of amino acids are the common form found in proteins; D isomers are unusual and are found only in a few locations, such as the bacterial cell wall.

Cross-linking of these peptides with a pentaglycine linker by a transpeptidase provides a rigid net that gives the bacteria structure. Several antibiotics act by taking advantage of the chemistry of the bacterial cell wall. For example, penicillin and its derivatives irreversibly inhibit the transpeptidase, blocking formation of the peptide linkage between peptidoglycan chains in gram-positive bacteria (**Figure 8.27**). The chains of peptidoglycan in gram-negative bacteria use a different mechanism to cross-link chains; as a result, they are not sensitive to penicillin. Other derivatives of penicillin ("illin" class antibiotics) have improved properties; for example, they have greater bioavailability and stability and are more widely tolerated. These derivatives include ampicillin, amoxicillin, and methicillin, as well as second-generation drugs such as carbenicillin and piperacillin. Newer types of penicillin-class drugs include carbapenems and cephalosporins. Although these drugs look quite different from the original compound, they all retain the β-lactam ring, which is key in the inhibitory mechanism.

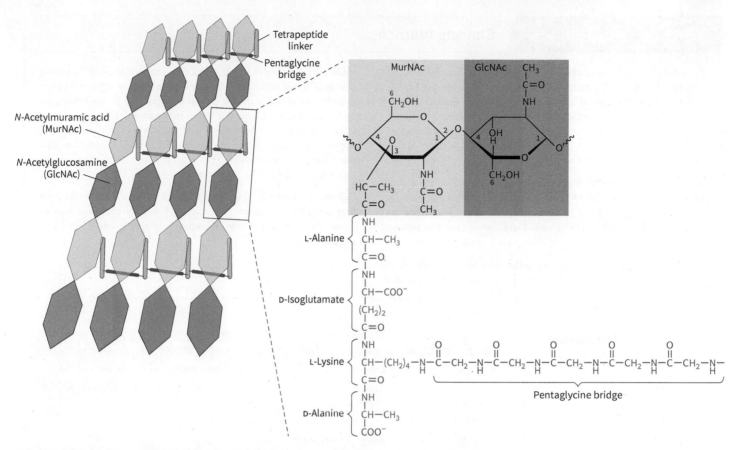

FIGURE 8.26 Structure of the peptidoglycan cell wall in gram-positive bacteria. Peptidoglycans consist of lengthy linear carbohydrate chains cross-linked by short peptides. While the carbohydrate and amino acid composition varies slightly across different phyla of bacteria, the overall structure is conserved.

(Source: Wessner, *Microbiology*, 2e, copyright 2017, John Wiley & Sons. This material is reproduced with permission of John Wiley & Sons, Inc.)

Nature also exploits the bacterial cell wall. Many organisms express lysozyme, a naturally occurring β-glycosidase that cleaves the linkage between *N*-Acetylglucosamine and *N*-Acetylmuramic acid. This effectively dissolves the bacterial wall, leading to bacterial lysis. Lysozyme is present in human tears (helping to prevent bacterial infections of the eye), and it is abundant in the egg white of chicken eggs, where it helps to protect the yolk and embryo from bacterial contamination.

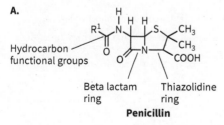

A.

Hydrocarbon functional groups

Beta lactam ring   Thiazolidine ring

Penicillin

FIGURE 8.27 Structure of penicillin G and other lactam antibiotics. **A.** Lactam antibiotics like penicillin act by inhibiting transpeptidase, the enzyme that catalyzes the cross-linking of polypeptides in the peptidoglycan cell wall. The lactam ring is essential in this mechanism. **B.** The structures of other first-generation penicillin analogs differ in the organic functional groups that are fixed to the β-lactam and fit differently in the active site of transpeptidase. Cephalosporin is a second-generation penicillin analog. While this structure differs in some ways, it retains the β-lactam ring essential for this inhibition.

B.

Benzylpenicillin   **Amoxicillin**   **Ampicillin**   **Cephalosporin**

Worked Problem 8.6   Sharing matrices

Experimental treatments are underway for people with severe muscle damage. In one such treatment, cells are scraped from pig bladder, and the tissue is sterilized and then transplanted into the damaged muscle. In time, new muscle grows into the transplanted matrix.

Explain why the transplanted pig bladder tissue did not turn into a new bladder. Also, why does the body fail to mount an immune response against this foreign object?

Strategy   Consider the matrix the cells are living in. How could these cells be prompted to grow and to grow properly? Why does one tissue (bladder) grow in one part of the body and a different tissue (muscle) in another? On a basic level, what causes an immune response?

Solution   When scraped free of cells, sterilized, and dried, the pig bladder resembles a piece of parchment paper or dried skin.

Essentially, it is just that, the remaining connective tissue or matrix devoid of living cells. The cells are what make this organ a bladder; most tissues of the body are rich in collagen and in the other proteoglycans discussed here and in the next section. There is too little molecular difference between these matrices to elicit an immune response, but there is an as-of-yet undetermined signal that tells muscle cells, or some type of pro-muscle stem cell, to migrate into this matrix and begin to regenerate the tissue.

Follow-up question   Pigs lacking α-1,3 galactosyltransferase have been suggested as a source of tissues for transplantation into human recipients. How might this mutation help to prevent rejection of these tissue grafts?

Summary

- Glycoproteins, glycolipids, proteoglycans, and peptidoglycans are all examples of glycosylated molecules.
- Glycoproteins are mainly extracellular proteins to which carbohydrate chains have been added.
- In glycoproteins, carbohydrates can be either O-linked (via a serine or threonine residue) or N-linked (through an asparagine residue).
- The ABO blood group antigens are examples of glycoproteins.
- Glycolipids are membrane lipids with attached carbohydrate groups.
- Lipopolysaccharide, glycosyl phosphoinositide (GPI) anchors, cerebrosides, and gangliosides are all examples of glycolipids.
- Proteoglycans are comprised of a protein linked to a glycosaminoglycan (a large carbohydrate polymer).
- Common examples of glycosaminoglycans, which are often sulfated, include:
  - heparan sulfate (glucuronic acid and N-Acetylglucosamine),
  - chondroitin sulfate (galactose and N-Acetylglucosamine),
  - keratan sulfate (iduronic acid and N-Acetylglucosamine).
- Hyaluronic acid is a non-proteoglycan glycosaminoglycan found in the extracellular matrix. Its structure consists of a repeating dimer of glucuronic acid and N-Acetylglucosamine.
- Gram-negative bacteria have a peptidoglycan cell wall made of chains of polysaccharides cross-linked by short peptides.
- The antibiotic penicillin blocks the cross-linking of the peptides. As a result, the bacterial cell wall loses integrity and the bacteria lyse.

Concept Check

1. Describe the four types of glycosylated molecules mentioned in this section with examples.
2. What are the components of the four matrix polysaccharides discussed? What are the sources of those carbohydrates?

8.4 Extracellular Matrices and Biofilms

Cells in eukaryotic systems are held together in a fibrous mesh of secreted polymers. These polymers (proteins and polysaccharides) form a gel that helps to give structure to and protect the underlying cells. Many prokaryotic systems also employ a secreted matrix of proteins and polysaccharides for similar protection and structure. This section discusses the secreted matrices of both eukaryotic and prokaryotic cells, and some recent discoveries and applications in this field.

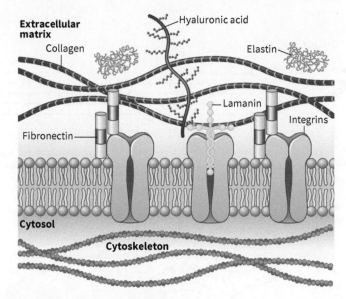

FIGURE 8.28 Schematic view of the extracellular matrix.   Matrix glycoproteins and proteoglycans form a gel-like mesh for the cell.

8.4.1 Eukaryotic extracellular matrix proteins

The extracellular matrix of a tissue is made of proteins, glycoproteins, and proteoglycans in a fibrous, gel-like mesh (**Figure 8.28**). The composition and function of the matrix has not been fully elucidated; however, more than 100 proteins have been found in the matrix, and these have diverse functions, including structure, ligand binding, immune responses, and regulation of growth and development. In addition, the matrix varies in composition from tissue to tissue and as the organism develops; it can also change in response to disease.

In addition to the glycosaminoglycans discussed in section 8.3 (heparan sulfate, chondroitin sulfate, keratan sulfate, and hyaluronic acid), several fibrous proteins are important in the matrix. Examples of some of the more common fibrous proteins, discussed below, are collagen, elastin, fibronectin, and laminin.

Collagen   Collagen is the protein found in the highest abundance in the extracellular matrix. It is present in almost every tissue of the body, and it has been estimated that as much as 25% of the protein in the body (by mass) is collagen. So far, 28 different types of collagen protein have been identified in the body, coded for by at least 42 different genes.

The structure of all collagens is similar. Collagens are helical, but unlike the α helix commonly seen in biochemistry, collagen adopts a structure known as the **collagen triple helix**. A detailed examination of the structure of collagen I (the most prevalent collagen) shows that it is comprised of three polypeptide chains: two copies of α1 and one copy of α2. These three chains wrap around each other, forming a left-handed helix (α helices are right handed). Another significant difference between collagen and other proteins is the amino acid composition. Collagen is enriched in three amino acids: glycine, proline, and lysine. In addition, the proline and lysine residues are often chemically modified into hydroxyproline and hydroxylysine.

Collagen synthesis begins with translation of collagen mRNA. A signal sequence directs this protein to the ER for processing. In the ER, the signal sequence is removed, prolines and lysines are hydroxylated, and hydroxylysines are often glycosylated; the collagen monomers form into tropocollagen, adopting the collagen triple helix structure. Glycosylation continues in the Golgi complex as these proteins are processed. Once secreted, the termini of the helices are truncated by collagen peptidases. The enzyme lysyl oxidase catalyzes the final assembly of tropocollagen molecules. Cross-linking into the mature form by lysyl oxidase is discussed in the next section; hydroxylation of proline residues is a hallmark of collagen and is discussed here (**Figure 8.29**).

The hydroxylation of proline is one of the key reactions in the formation of collagen. It requires L-ascorbate (vitamin C); thus, a deficiency of vitamin C causes disease, in the form of scurvy, as discussed in section 7.1. In scurvy, an inability to form extracellular matrix

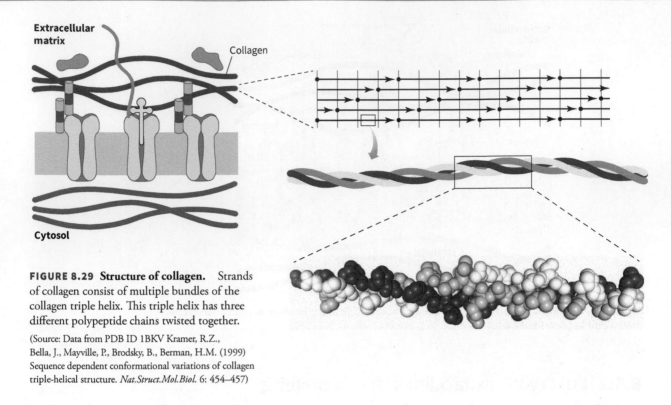

FIGURE 8.29 Structure of collagen. Strands of collagen consist of multiple bundles of the collagen triple helix. This triple helix has three different polypeptide chains twisted together.

(Source: Data from PDB ID 1BKV Kramer, R.Z., Bella, J., Mayville, P., Brodsky, B., Berman, H.M. (1999) Sequence dependent conformational variations of collagen triple-helical structure. *Nat.Struct.Mol.Biol.* 6: 454–457)

(connective tissue) leads to hair and tooth loss, bleeding from mucus membranes, and failure to heal wounds. Prolonged vitamin C deficiency is often fatal.

Prolyl hydroxylase is the enzyme responsible for the hydroxylation of proline. This enzyme is a tetramer with the stoichiometry α2β2. The α subunits are a mixed-function oxidase that hydroxylates proline using molecular oxygen, iron, and ascorbic acid; the β subunits are identical to the protein disulfide isomerase (another enzyme involved in protein modification and remodeling). As proline is being hydroxylated, a second oxidation reaction is also occurring, in which α-ketoglutarate is oxidized to succinate and CO_2. In the citric acid cycle, the oxidation of α-ketoglutarate to succinate and CO_2 proceeds through succinyl-CoA, generating a molecule of GTP in the process. The energy released in the α-ketoglutarate oxidation in prolyl hydroxylase is instead used to drive the hydroxylation of proline.

$$\text{Proline} + \text{α-ketoglutarate} \xrightarrow[\substack{\text{prolyl} \\ \text{hydroxylase}}]{\substack{\text{Ascorbate,} \\ O_2,\ Fe^{2+},\ CO_2}} \text{4-hydroxyproline} + \text{Succinate}$$

Hydroxyproline residues are best known for their role in collagen, but they are also important in other proteins such as elastin (discussed below), prions, the ion-channel poison conotoxin, and the RNA-silencing complex protein argonaut. Finally, hydroxylation of a key proline in hypoxic inducible factor (HIF), discussed in section 7.1, directs the protein to ubiquitination and proteosomal degradation. Although not all of these reactions are catalyzed by the same enzyme, they all result in hydroxyproline residues.

Elastin Elastin and fibrillin are the two proteins responsible for the elastic fibers found mainly in the skin and arteries. Elastin forms the core of these fibers, while fibrillin forms a mantle of microfibers coating the central core. Elastin is comprised largely of hydrophobic residues. It is enriched in proline and glycine, but these residues are not hydroxylated as they are in collagen. There are also repeated stretches that are rich in lysine and alanine. The lysine residues in this region are cross-linked to one another or to fibrillin via two different routes (**Figure 8.30**).

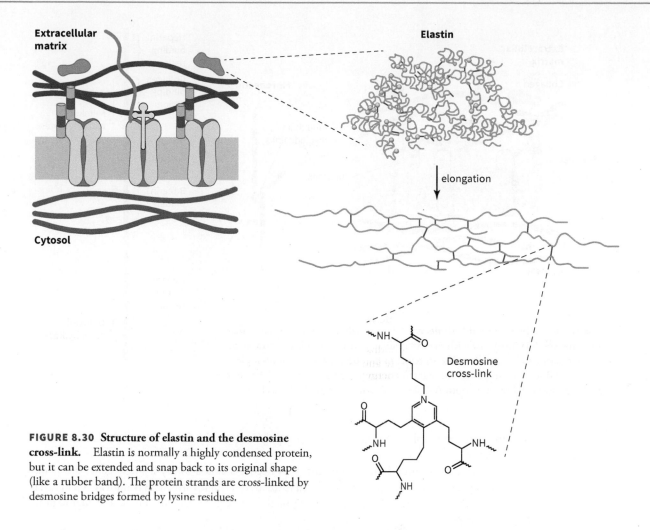

FIGURE 8.30 Structure of elastin and the desmosine cross-link. Elastin is normally a highly condensed protein, but it can be extended and snap back to its original shape (like a rubber band). The protein strands are cross-linked by desmosine bridges formed by lysine residues.

Transglutaminase catalyzes cross-linking between glutamine residues of fibrillin and lysine residues in elastin (**Figure 8.31**). Commercially, transglutaminase is used in the food industry as "meat glue" to join pieces of meat together. Ammonia (NH_3) is split out in the process, and an unusual peptide bond is formed between the γ-carboxamide of glutamine and the ε-nitrogen of lysine. **Lysyl oxidase** polymerizes monomers of a soluble pro-protein (tropoelastin) secreted from cells near the matrix into mature elastin. In the lysyl oxidase reaction, the side chain of lysine is oxidatively deaminated, and subsequent reactions (potentially uncatalyzed spontaneous reactions) form cross-links to other amino acids. Although tropoelastin is about 75 kDa, mature elastin fibrils can be millions of daltons in size.

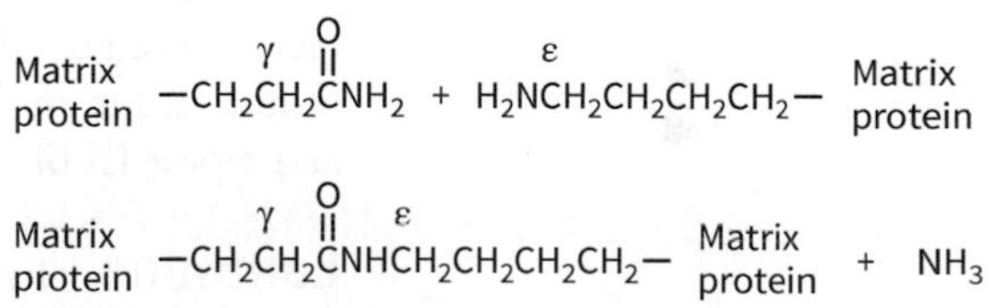

$$\text{Matrix protein} \quad -CH_2CH_2\overset{\gamma}{\overset{O}{\underset{\|}{C}}}NH_2 \; + \; \overset{\varepsilon}{H_2N}CH_2CH_2CH_2CH_2- \quad \text{Matrix protein}$$

$$\text{Matrix protein} \quad -CH_2CH_2\overset{\gamma}{\overset{O}{\underset{\|}{C}}}\overset{\varepsilon}{N}HCH_2CH_2CH_2CH_2- \quad \text{Matrix protein} \; + \; NH_3$$

FIGURE 8.31 Transglutaminase. In the extracellular matrix, transglutaminase generates an unusual amide bond between the side chains of lysine and glutamine.

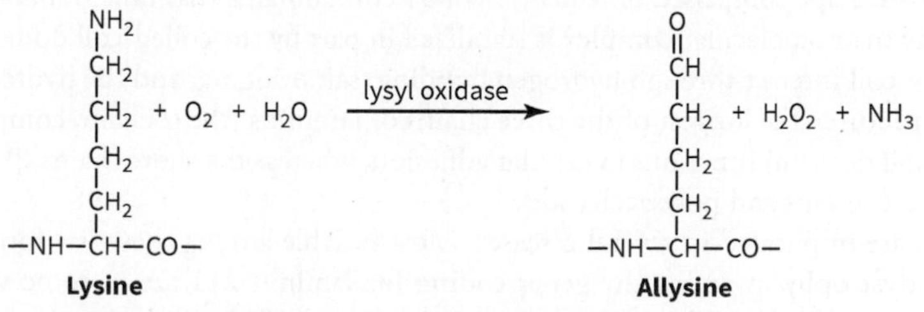

$$\begin{array}{c} NH_2 \\ | \\ CH_2 \\ | \\ CH_2 \\ | \\ CH_2 \\ | \\ CH_2 \\ | \\ -NH-CH-CO- \end{array} + O_2 + H_2O \quad \xrightarrow{\text{lysyl oxidase}} \quad \begin{array}{c} O \\ \| \\ CH \\ | \\ CH_2 \\ | \\ CH_2 \\ | \\ CH_2 \\ | \\ -NH-CH-CO- \end{array} + H_2O_2 + NH_3$$

Lysine   **Allysine**

Fibronectin

Fibronectin is a dimeric, 440-kDa glycoprotein found in the extracellular matrix. However, unlike the other proteins discussed in this section, it is also found in a soluble form in the circulation where it functions in the clotting cascade (**Figure 8.32**). Fibronectin acts as an adapter between the cell and the matrix. Integrins, integral membrane glycoproteins, bind

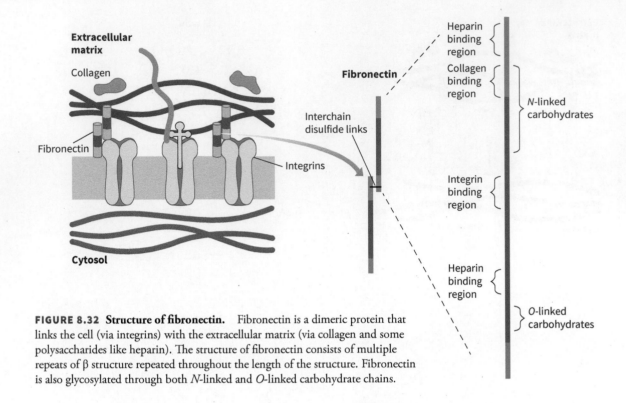

FIGURE 8.32 Structure of fibronectin. Fibronectin is a dimeric protein that links the cell (via integrins) with the extracellular matrix (via collagen and some polysaccharides like heparin). The structure of fibronectin consists of multiple repeats of β structure repeated throughout the length of the structure. Fibronectin is also glycosylated through both *N*-linked and *O*-linked carbohydrate chains.

to fibronectin and link it to other proteins and polysaccharides of the extracellular matrix to the cytoskeleton.

As with several other fibrillar proteins, fibronectin is coded for by a single gene, but it has multiple repeats in the protein sequence, each formed by a single exon. This is indicative of partial gene-duplication events. Differential splicing of this gene gives rise to soluble and filamentous forms of the protein.

The fibronectin molecule is a homodimer that contains multiple copies of three distinct repeats, each of which has two antiparallel β sheets. Up to 15 of these different features are found in fibronectin in different combinations. Repeat I binds to matrix proteins (collagen and fibrin), and repeat III binds to the cell at least partially via interactions with the integrins.

Laminins

Laminins are a family of 15 different proteins that function in the basal lamina, which is a layer of the basement membrane, a fibrous layer of connective tissue found under the epithelial layers (**Figure 8.33**). Laminins are heterotrimeric, consisting of one subunit each of α, β, and γ. There are five different α isoforms, four β isoforms, and three γ isoforms that exist and assemble into the mature holoprotein. Laminin 211, for example, has one copy each of the α2, β1, and γ1 subunits.

Laminins are large proteins ranging from 900 kDa to over 1.5 MDa. They form a four-armed cruciform shape comprised of α-helical coiled-coil domains that link globular bulges. The assembly of the macromolecular complex is stabilized in part by the coiled-coil domain. Repeated domains in the coil interact through hydrogen bonding, salt bridging, and the hydrophobic effect to form the structure. The longest of the three chains of laminins (the α chain, comprised mainly of the coiled-coil domain) functions in cellular adhesion, whereas the short chains (β and γ) attach to other matrix proteins and polysaccharides.

Laminins are implicated in several diseases. Most notable among these is a form of congenital muscular dystrophy in which the genes coding for laminin 211 are in some way mutated. More than 28 mutations have been characterized, ranging from simple point mutations to large deletions; all of these mutations result in defective proteins. The rest of the basal lamina seems unaffected, but the end result for the patient is muscle cell death, inflammation, and the invasion of other cell types (such as adipocytes) into the muscle tissue (**Figure 8.34**).

Because many molecules in the matrix are large and polyvalent, there are opportunities for multiple interactions between molecules. Matrix DB is software that tracks the interactions of the

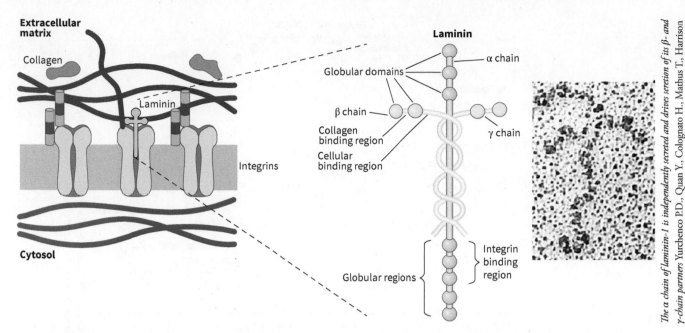

The α chain of laminin-1 is independently secreted and drives secretion of its β- and γ-chain partners Yurchenco P.D., Quan Y., Colognato H., Mathus T., Harrison D., Yamada Y., O'Rear J.J. 1997 Proceedings of the National Academy of Sciences of the United States of America, Sept 16;94(19): pp. 10189-10194. Copyright 1997 National Academy of Sciences, U.S.A.

FIGURE 8.33 Structure of the laminins. Like fibronectin, the laminins help link integrins to the extracellular matrix. The laminins consist of three polypeptide chains twisted around each other. Salt bridges and hydrophobic interactions hold these chains together. Laminins interact between integrins and fibronectin and collagen.

macromolecules of the extracellular matrix, based on what has been observed in the literature (**Figure 8.35**). More than 2,200 protein–protein and protein–glycosaminoglycan interactions have been mapped. The software also takes into consideration the multimeric nature of some members; for example, collagens and laminins function as multimers, enabling multiple interactions with other macromolecules. A group of molecules that interact with each other has been termed an interactome.

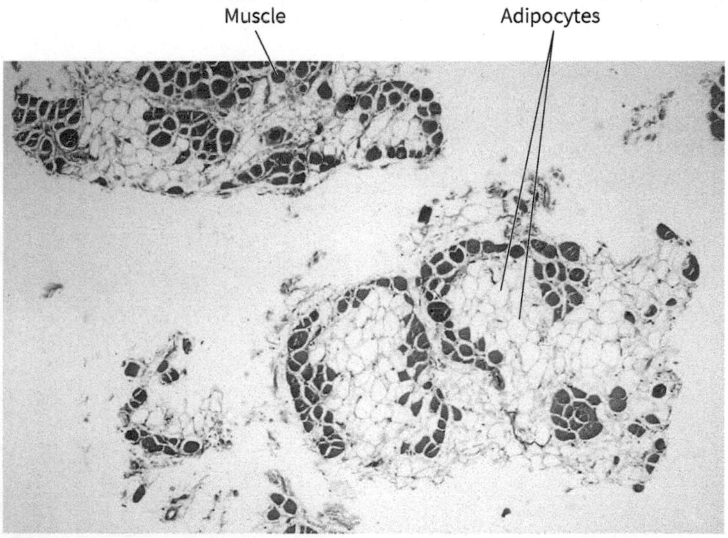

Cell culture and tissue engineering Cells

differ in the affinity with which they adhere to different surfaces. Over time, researchers discovered that surfaces usually need some sort of preparation to permit cell adherence. In plastic tissue-culture dishes, cultured cells adhere relatively poorly to untreated polystyrene. Thus, manufacturers treat the surface with an ionized gas or plasma creating oxidative modifications, for example, removing phenolic groups and incorporating hydroxyl, carbonyl, and carboxylic acid moieties. This simple treatment greatly improves the adherence of cells.

As different cell types have been adapted for culture, different surfaces have been developed, for example, a surface of biological molecules such as collagen, fibrin, fibronectin, or polymers of lysine. Such surfaces provide more of a native biological substrate for the cells to grow on. For some cell lines, this is still not enough; for example, embryonic stem cells require a layer of feeder cells to grow on. The feeder layer is a layer of cells grown to confluence: that is, the cells are grown to a contiguous monolayer of cells, then treated with pulses of electricity, radiation, or the DNA crosslinker and chemotherapeutic agent mitomycin C to stop the cells from dividing further (**Figure 8.36**). The feeder cells are metabolically active and thus still able to secrete the cytokines and hormones that the embryonic stem cells need to survive.

FIGURE 8.34 Dysfunction of the laminins results in muscular dystrophy. Shown is a micrograph of a muscle tissue biopsy from a patient with muscular dystrophy. Note the large number of adipocytes (large open cells) in contrast to the muscle cells. Several different diseases (known as laminopathies) can arise from mutations of the laminins.

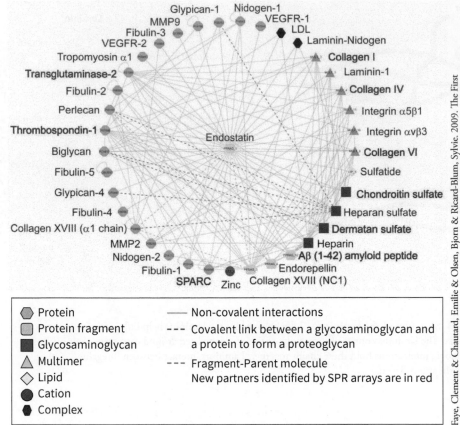

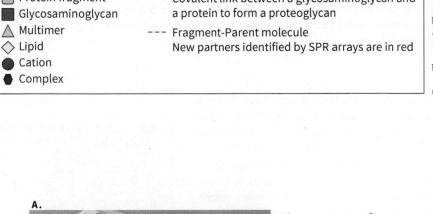

Faye, Clement & Chautard, Emilie & Olsen, Bjorn & Ricard-Blum, Sylvie. 2009. The First Draft of the Endostatin Interaction Network. *The Journal of Biological Chemistry.* 284. 22041-7. 10.1074/jbc.M109.002964.

FIGURE 8.35 Interactions of matrix proteins can be studied *in silico*. Shown is an image from Matrix DB, a web-based tool, that illustrates interactions of different molecules found in the extracellular matrix. This is a graphical interpretation of the interactome of these molecules.

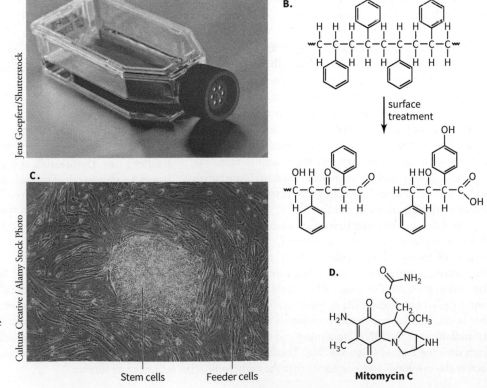

FIGURE 8.36 Cultured cells. A. Cells in culture are often grown in flasks. **B.** The surface of the flask is treated to generate a more hydrophilic surface to which cells can adhere. **C.** Embryonic stem cells (the colony in the center of the micrograph) live on a layer of feeder cells. **D.** These "zombie cells" are metabolically active but are not capable of division due to treatment with mitomycin c.

Despite these techniques for improved adhesion, there are still several draw-backs to cell culture. Eukaryotic cells prefer living systems to flat plastic surfaces. Primary cultures of cells, that is, cells that are directly derived from living tissue as opposed to transformed cultured cells, lose their phenotype after time in culture. For example, over time a hepatocyte (liver cell) will stop synthesis of the proteins that are typically found in the liver, making this model less valid for the study of liver processes.

The study of recreating tissues or organs lost through injury is *tissue engineering,* the growth of cultured cells in a three-dimensional matrix. Tissue engineering involves a combination of materials, matrices, and cells working to give a final product that has the desired mechanical, immunological, and biocompatible properties. In tissue engineering, the scaffold, which is the matrix to which the cells attach, needs to provide structure for the cells to grow. At the same time, the scaffold must allow the cells or host to reabsorb the matrix after the cells take hold. Hence, the scaffold is often made of proteins, such as collagen, or polyesters of lactic acid, glycolic acid, or caprolactone. As these molecules hydrolyze, they release monomers that are absorbed and used by the cell.

To date, there have been limited but promising successes with engineered tissues being transplanted into humans. For the most part, the tissues have been epithelial in nature; they include skin grafts, vascular tissue grafts, and artificial bladders (**Figure 8.37**).

BRIAN WALKER/APTN/ASSOCIATED PRESS

FIGURE 8.37 Engineered tissues. Recent advances in tissue engineering include human bladders grown in culture.

Worked Problem 8.7 Characterizing matrix proteins

Sodium dodecyl sulfate polyacrylamide gel electrophoresis (SDS-PAGE) is a common technique used in the analysis of proteins. Why is it often difficult to characterize matrix proteins using SDS-PAGE?

Strategy How is SDS-PAGE performed, and what information does it provide?

Solution SDS-PAGE is discussed in detail in the techniques appendix. In SDS-PAGE, a protein sample is denatured by heating in the presence of the detergent SDS. Often, a sulfhydryl reducing agent such as β-mercaptoethanol (β-ME) or dithiothreitol (DTT) is added to cleave disulfide bonds. The SDS imparts a negative charge, and the proteins can be separated in an electric field through a gel matrix such as polyacrylamide (hence the "PAGE" in SDS-PAGE). The movement through this field varies inversely with molecular weight; that is, larger proteins move more slowly than smaller ones. Typical ranges over which proteins can be separated run from several thousand up to several hundred thousand daltons, with 10 to 200 kDa being most common.

Analysis of matrix proteins by SDS-PAGE poses a range of problems. Often, matrix proteins are cross-linked to one another through covalent bonds that are not disulfides and are thus not susceptible to cleavage by β-ME or DTT. Also, matrix proteins form a polymer that is much too large to be analyzed using this technique. In addition, matrix proteins are often glycosylated, and glycosylations alter the migration of proteins in SDS-PAGE. Therefore, SDS-PAGE is a powerful technique but not necessarily the right one for analysis of matrix proteins.

Follow-up question How could some matrix proteins be processed so that they are suitable for analysis by SDS-PAGE?

8.4.2 Biofilms are composed of microbes living in a secreted matrix

Biofilms are associations of microbes (of a single species or multiple species) living in a secreted matrix (**Figure 8.38**). Biofilms are often found at an air–liquid or solid–liquid interface. The slime at the bottom of a pond can be an example of a biofilm, as can dental plaque, an infection in a tissue, or aggregations of microbes on a biomedical implant. They are also the primary methods used in sewage and wastewater treatment to remove contaminants and by the aquarium industry to remove ammonia from the water in fish tanks. Biofilms are responsible for fouling pumps and corrosion of equipment exposed to marine environments. Biofilms have been associated with almost 80% of chronic human infections, and they are thought to be responsible for tens of thousands of deaths a year. Hence, biofilms are of significant commercial, medical, and scientific interest.

Biofilms often form as thin films one microbe thick, but in some cases can grow to be several centimeters thick and as large as several meters across. Organisms in a biofilm have a different

Bacteria   Secreted matrix

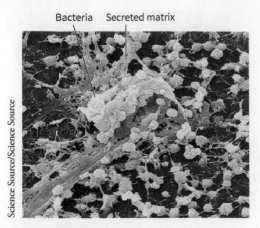

Science Source/Science Source

FIGURE 8.38 Structure of a biofilm. In a biofilm, bacteria live in a secreted matrix of proteins and polysaccharides. The film can consist of only one species or can have multiple species living in symbiosis. This is an example of a biofilm obtained from the inner surface of a needle-less connector.

phenotype and thus express different genes (or express the same genes but to differing degrees) than they would when floating freely in suspension or living singly on a surface.

Biofilm development
Biofilms undergo a distinct series of steps as they develop. In the earliest stage, single microbes begin to adhere to a surface or interface. As more members join the surface, they begin to alter gene expression and adhere irreversibly. At this point the microbes begin to secrete extracellular polysaccharides that form a matrix around the organisms. As the film begins to mature, more microbes are recruited; also, microbes already present in the film start to divide, increasing the size of the film. Matrix secretion continues and the film thickens. In the final stage of development, parts of the film may rupture, releasing its inhabitants to disperse and generate new films elsewhere (**Figure 8.39**).

Bacteria can communicate with one another using chemical messengers through mechanisms similar to those found in eukaryotic cells. This approach, termed quorum sensing, enables bacteria to determine the density of like organisms in the immediate area and alter their gene expression in response. Quorum-sensing signaling molecules have been shown to have a role in biofilm initiation in numerous species.

The ability of microbes to form biofilms is almost universal. When in a biofilm, the growth of the bacteria slows because the diffusion of small molecules may be limited. Although it may seem as though the bacteria are putting themselves at a disadvantage in this situation, the biofilm provides a stable environment. For example, bacteria are often more antibiotic resistant in a biofilm. Also, some members of the colony may themselves secrete antibiotics that stop competitors from growing and competing for resources. The development of a biofilm is an evolutionary adaptation; the bacteria work together and expend energy for both their own good and that of the film. These observations have been extended beyond bacteria. Fungal biofilms have now been described in the scientific literature.

Given the diversity of microbes, the extracellular matrix forming the biofilm might be expected to be equally diverse, and it is. Common polymers found in biofilms include alginates (polymers of mannuronic and guluronic acids), chitin (polymers of *N*-Acetylglucosamine), and polymers rich in glucose or mannose, or several other common monosaccharides and disaccharides. In addition to exopolysaccharides, the matrix contains proteins, including lectin-like proteins that bind to polysaccharides, and amyloid-like fibrils, unusual polymers of amino acids such as polyglutamate, and extracellular DNA.

Biofilm decontamination and prevention
Removal of a biofilm is not as simple as washing off surface microbes, because the microbes forming the film are cloistered in the matrix of the film and the film itself may be attached to the surface. There is significant ongoing research

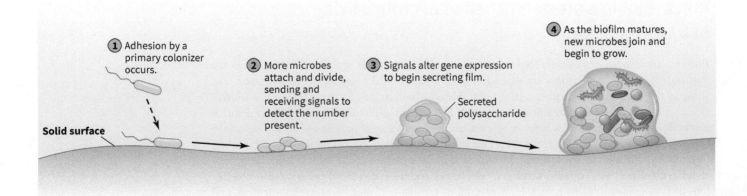

FIGURE 8.39 Life cycle of a biofilm.   Biofilms house microbes in a secreted matrix.

(Source: Wessner, *Microbiology*, 2e, copyright 2017, John Wiley & Sons. This material is reproduced with permission of John Wiley & Sons, Inc.)

into what combinations of detergents, antibiotics, and hydrolytic enzymes can disrupt biofilms, yet the films are notoriously hard to remove. One method being investigated is blasting films away using a jet of oxygen–helium plasma. This technique generates short-lived oxidative intermediates that are thought to eradicate the film without penetrating or damaging the tissue underneath. The treatment of surfaces with ozone is also being investigated. This technique may be useful in some cases but is not viable in a living organism. The ozone oxidizes the matrix polysaccharides, generating smaller fragments and creating hydrophilic modifications; it also kills the microbes in the film. This technique is often used in aquariums, pools, and spas.

An alternative to removing biofilms is to find ways to prevent their growth in the first place. Smooth surfaces with hydrophobic coatings discourage bacterial settling and biofilm initiation. Also, some polymers have been impregnated with antibiotics, hydrophobic oils, nanoparticles of silver, or positively charged groups such as *N*-alkylpyridinium salts, and these techniques seem to be effective at reducing biofilm formation. In biomedical applications, low-energy sound waves have been found to prevent microbes from binding to the surface of devices such as catheters.

Biofilms in human health
Given that most microbes can form biofilms, it stands to reason that these structures will have a role in human health. This section discusses two biofilms implicated in health and disease: *Pseudomonas aeruginosa* and dental plaque.

Pseudomonas aeruginosa
Pseudomonas aeruginosa is a common environmental pathogen. This pathogen is usually not a danger to healthy individuals, but it participates in opportunistic infections, for example, in the lungs of individuals with cystic fibrosis (**Figure 8.40**). In cystic fibrosis, a defective chloride transporter (cystic fibrosis transmembrane conductance regulator, CFTR) causes numerous pathophysiologies leading to the formation of viscous, sticky mucus in the lungs. *Pseudomonas* infections can be treated with antibiotics; however, by their mid-20s, most people with cystic fibrosis harbor antibiotic-resistant strains. Often, people with cystic fibrosis are on prophylactic antibiotics and require percussive therapy, which involves thumping on the chest several times a day to clear mucus. Newer therapies for lung disease in cystic fibrosis include gene therapy to replace the defective CFTR, attempts to trigger other chloride channels to open (e.g., the experimental drug Desnufosol), or lung transplant.

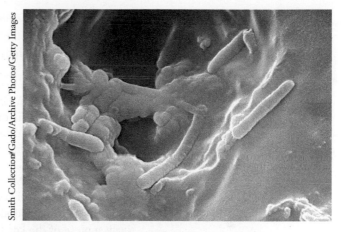

FIGURE 8.40 Micrograph of *Pseudomonas aeruginosa*. *Pseudomonas* in the lung can form biofilms secreting viscous mucus that help bacteria survive. *Pseudomonas* is responsible for one of the opportunistic infections that takes hold in cystic fibrosis.

Smith Collection/Gado/Archive Photos/Getty Images

Dental plaque
Dental plaque is a biofilm composed of microbes that live on the surface of teeth and gingival, or gum, tissues. The primary microbes are strains of *Streptococcus*; however, more than a thousand different strains of bacteria and yeast are thought to exist in dental plaque. The acidic waste products of these bacteria can damage the surface of the teeth, causing tooth decay and gum disease.

Although it is clear that bacteria secrete some of the matrix, it appears that another component of the matrix is salivary protein secreted by the human host perhaps to contain or otherwise control the growth or damage from the bacteria in the plaque.

The balance between health and disease in plaque is not dictated by the presence or absence of the biofilm; everyone has dental plaque. The difference between someone with advanced periodontal disease and someone with healthy teeth is more an issue of the types of microbes in the film and the extent of the plaque. Brushing and flossing are mechanical ways to control biofilms; various components of toothpaste and mouthwash also help to control the growth of mouth flora.

Symbiotic microbes: kefir and kombucha
Kefir and kombucha are ancient drinks that come from different parts of the world but have a similar biochemical makeup. Kefir is a fermented milk beverage from eastern Europe, whereas kombucha is a fermented tea drink thought to have originated in Mongolia and western China. While we may think of fermented beverages such as beer, mead, cider, and wine as being intoxicating, kefir lacks alcohol (lactate is the final fermentation product), and kombucha has only a low concentration of alcohol (0.5%).

Stephanie Frey/iStock/Getty Images

Brian Hagiwara/Photolibrary/Getty Images

FIGURE 8.41 Kefir grains and kombucha mushrooms.   **A.** Kefir grains are colonies of different strains of bacteria and yeast that live in a secreted extracellular matrix. **B.** Kombucha "mushrooms" grow in tea. While they have different microorganisms and matrix components, the theme is similar to to kefir grains.

Similar to beer and wine, kefir and kombucha probably arose in antiquity as a way of preserving leftover milk or tea. Today, these drinks are consumed worldwide, both for their taste and their health benefits.

Whereas beer and wine are primarily the products of yeast-based fermentations, kefir and kombucha result from a combination of yeast and bacteria living symbiotically in an exopolysaccharide matrix. In kefir, these microbes form structures called kefir grains: lumpy, white-to-yellow aggregates ranging in diameter from less than a millimeter to over a centimeter. In kombucha, the bacteria and yeast form a mat that floats at the air–liquid interface, resembling a jellyfish or mushroom, and is thus referred to as a kombucha mushroom. In well-developed kombucha cultures, the leathery mat can be several centimeters thick (**Figure 8.41**).

In both kefir and kombucha, the cultures are maintained by removing some of the culture medium or liquid, and adding fresh milk or tea, depending on the beverage being made. Alternatively, a piece of the kombucha mat (termed a "baby") or some kefir grains can be added to fresh liquid.

While this section has focused on the extracellular matrix of animal and microbial cells, plants too have an extracellular matrix, typically one rich in cellulose. The stored carbohydrates found in plant material or other sources, such as algae, can be used by humans as a source of fuel for their machines. These sources of energy, termed biofuels, are discussed in **Societal and Ethical Biochemistry: Biofuels.**

Societal and Ethical Biochemistry

Biofuels

All animals are chemotrophs; that is, they require energy from food to sustain their metabolic processes. However, only humans use external energy to alter their environment. Even ancient people extended the geographic range in which they could live by using fire to keep warm. Today, we take advantage of automation and technology such as transportation, refrigeration, and modern agricultural methods for many of the services that are essential to our daily lives.

As global economies develop and the global population increases, there is a greater demand for devices that use fuel and

therefore a greater demand for fuel. Building devices that are more fuel efficient makes sense, but this resolves only part of the problem. There has also been a shift in the past two centuries from fuels such as wood, whale oil, and dung to coal, oil, and gas. Some energy sources are considered to be carbon neutral; for example, the CO_2 released by burning wood is used by growing trees to make new glucose in photosynthesis and then transformed into cellulose, so there is no net increase in atmospheric carbon (although there is a carbon cost in harvesting and transporting the wood). Energy can come from other sources: in the past, energy came from wind and water, using windmills and watermills; today energy comes from sources such as wind turbines, hydropower, nuclear, and solar. Some of these options are growing in popularity as alternatives to fossil fuels. However, the greatest need today for fuels worldwide is for cars and trucks, and the leading candidates to replace the petroleum products used in vehicles are biofuels, that is, products derived from biological sources. Biofuels currently fall into two categories, biodiesel and ethanol.

Biodiesel is formed by esterifying methanol or other short-chain alcohols to fatty acids. By forming an ester (which is far less polar than the carboxyl group) the resulting molecule has a lower boiling point, higher volatility, and lower viscosity, and is more hydrophobic—all desirable properties for this type of fuel. These fuels are often formed from vegetable oils, such as corn or soybean; they can also be made from recycled vegetable oils. Producing biodiesel is relatively easy; the reaction is a simple Fisher esterification using an acid or base catalyst. The difficulty in using biofuels is quality control. Although it is relatively simple for an individual to produce ten gallons of biodiesel in the garage, homemade biodiesel is highly variable in quality and may harm the performance of the vehicle, in contrast to the quality fuel bought at the pump. This challenge does not exist in commercial biofuels where large-scale engineering and quality control address these concerns.

Ethanol can be made easily by fermenting starches and sugars. In the United States, corn is the main source of ethanol. In some parts of the country, gasoline has up to 10% ethanol as an additive; in other parts, fuel (referred to as E85) contains 85% ethanol and 15% petroleum products.

To be a successful alternative, it would be beneficial if there were other ways to generate ethanol. Plant biomass that is left behind, such as corn husks and stalks, have a lot of unused cellulose that is broken down by microbes in the soil. Switchgrass (*Panicum virgotum*) is another plant that has recently been targeted as a source of cellulose for biofuel production. The fuel produced from plant biomass and switchgrass is cellulosic ethanol; essentially, this is the same as ethanol obtained by the fermentation of starches or sucrose.

Ethanol was being produced from wood chips more than 100 years ago. In the late 1800s, chemists were generating ethanol by treating the wood chips with a dilute solution of sulfuric acid and heating it for several hours. More recently, to assist in the cleavage of cellulose into sugar, enzymes obtained from molds or fungi have been included in the process. Although yeast is still the main microbe used to convert sugar to ethanol in the fermentation, the yeast and other microbes involved in ethanol production have now been genetically engineered to tolerate higher ethanol concentrations, higher temperatures, and lower pH values.

The advantages of biofuels are many, but they are not a panacea. Even though these biofuels use carbon that was recently in the atmosphere, and ideally they could be produced locally cutting down the cost of transport, they are not free. There are energetic and monetary costs to growing the plants and processing them into ethanol or biodiesel. In addition, the fuels are not necessarily good for the environment just because they are derived from plants. For example, the growing and harvesting of sugar cane for ethanol production in Brazil is far from environmentally friendly. A large acreage must be dedicated to growing the sugar cane, the crop is frequently treated with large amounts of fertilizer and pesticides, and slash-and-burn techniques are used to harvest the cane. On the other hand, most vehicles in Brazil run on 97% ethanol, and the country imports almost no oil. While it may seem strange to run a car using processed vegetable oil or corn in the gas tank, fossil fuels are actually biofuels, just from a different era.

Summary

- Collagen, elastin, fibronectin, and laminin are all examples of large, secreted matrix glycoproteins.

- Hydroxyproline is found in collagen and elastin. Its synthesis is catalyzed by prolyl hydroxylase, a mixed-function oxidase that uses molecular oxygen, α-ketoglutarate, and a catalytic iron center.

- Cross-linking of lysines and glutamine residues in matrix proteins is common. These reactions are catalyzed by lysyl oxidase and transglutaminase.

- Matrix proteins are involved in development, immunity, and wound healing; they also give structure to the organism.

- The interactome describes how large matrix proteins interact with other macromolecules.

- The field of tissue engineering attempts to grow new tissues from inert matrices and cells.

- Biofilms are groups of microbes growing in a secreted matrix; they provide similar functions to the extracellular matrix in eukaryotic organisms, including scaffolding and anchoring the cells, directing their growth, and in some cases development.

- Bacteria in biofilms exhibit a different phenotype and group immunity through secreted antibiotics.

- Considerable effort is spent on controlling or eliminating biofilms from surfaces, or preventing their formation.

- Kefir and kombucha are fermented drinks created by colonizing microbes. These microbes live in symbiosis with each other in a secreted matrix.

Concept Check

1. Name and describe the molecules involved in the extracellular matrix.
2. How are biofilms commonly removed?
3. Explain how the structure of collagen I is different from other proteins. What changes in the structure lead to scurvy?
4. Give examples of how changes in the extracellular matrix cause or contribute to disease.

Bioinformatics Exercises

Exercise 1 Glycogen Phosphorylase Structure and Activity

Exercise 2 Glucagon Control of Glycogen and Glucose Metabolism

Problems

8.1 Glycogen Metabolism

1. Glycogen is stored glucose. If a plant stored fructose as a polymer for energy, would you anticipate that it would cost energy to synthesize the stored form of these molecules?

2. Some glycogen-storage diseases are treated by giving patients cornstarch. Why?

3. Discuss which hormones are important in the regulation of glycogen stores and how they function.

4. Glycogen granules are large cytoplasmic bodies that consist of large number of glycogen molecules and enzymes involved in its metabolism. In which tissues do you anticipate they will be present?

5. Muscle does not provide glucose for the brain during times of starvation. Why?

6. The brain is fueled almost exclusively by glucose. Using this fact, would it be a good idea to express glucose-1-phosphatase in muscle? What type of phenotype would be exhibited by a mouse expressing glucose-1-phosphatase in muscle?

7. Muscle does not express the glucagon receptor. If muscle did express the glucagon receptor, would muscle become responsive to glucagon? What would the response be? What would you hypothesize as a phenotype for a transgenic mouse expressing the glucagon receptor in muscle?

8.2 The Pentose Phosphate Pathway

8. Glucose-6-phosphate dehydrogenase catalyzes the first step of the pentose phosphate pathway. This enzyme is highly specific for $NADP^+$; the K_M for NAD^+ is about a thousand times that of $NADP^+$. What does this mean in terms of affinity of the enzyme for NAD^+ and affect rate?

9. Describe the roles that xylulose-5-phosphate and ChREBP play in carbohydrate metabolism.

10. Examining the mechanism of transaldolase, is there any other amino acid that could be substituted for lysine in the active site and retain activity?

11. If high levels of fructose in the diet were influencing the pentose phosphate pathway, how might this alter flux through this pathway? What kinetic information would you need to know to determine flux through this path?

12. In what metabolic situation would you need more of one metabolite of the pentose phosphate pathway than another?

13. Wernicke-Korsakoff syndrome is a degenerative brain disorder caused due to prolonged deficiency of thiamine in body. Would PPP be affected in this disorder?

14. Favism is caused by a mutation in the gene coding for glucose-6-phosphate dehydrogenase. This gene is on the X chromosome. Does this mean that more men or more women will be affected by favism? Is this a sex-linked trait?

15. Glutathione peroxidase has an active site selenocysteine rather than cysteine. How would the change from sulfur to selenium produce similar chemistry, and in what ways would the chemistry differ?

8.3 Carbohydrates in Glycoconjugates

16. Where in the cell would you expect to find GPI-anchored proteins?

17. How could you free a protein from a GPI anchor?

18. Tamiflu, was one of the anti-virals that were being excessively prescribed during COVID-19 pandemic. In context of glycolipids and what you have read about influenza, anticipate its mechanism of action.

19. Analyze the monosaccharides and disaccharides that make up the matrix polysaccharides discussed in this chapter. Based on your knowledge of glycogen metabolism, propose reactions by which these polysaccharides could be formed.

20. What are the components of the four matrix polysaccharides discussed in this chapter? What are the sources of those carbohydrates?

21. In the ABO blood type system, how many monosaccharides are found in each of the oligosaccharides for the various blood types?

22. Several inborn errors of metabolism arise from dysfunctional glycolipid metabolism. Describe three of these disorders on the chemical level and how these diseases manifest themselves in patients.

23. Look up the mechanism of action of penicillin. How do penicillin and other drugs inhibit the cross-linking of the peptidoglycan coating? Why are these drugs more effective on some bacteria than others?

8.4 Extracellular Matrices and Biofilms

24. Prolyl hydroxylase has an iron redox active center. Could copper substitute for the iron? Why or why not?

25. How could you test experimentally the proposed mechanism for prolyl hydroxylase? What technique would be most useful?

26. Describe the special properties of the molecules found in the extracellular matrix (for example, hydroxyproline), how these modifications are made, and how they give rise to function.

27. Could a kombucha mushroom produce kefir, or vice versa? Would you expect those cultures to be the same or to have the same matrix?

28. Explain the series of steps involved in the development of biofilms starting with adherence of single microbes to a surface or interface.

29. Electron microscopy is a technique commonly used to analyze matrix proteins. Why is this technique chosen instead of others?

30. Lysyl oxidase catalyzes cross-links between lysine residues in the matrix. Examine the structure of desmosine and isodesmosine. Can you determine the source of the ring carbons? Could any amino acid other than lysine participate in this reaction? What forces are helping to stabilize this cross-link?

31. Cross-linking is commonly seen in disulfide bonds but rarely in lysines and glutamines. Suggest why it is that we only see these cross-linking reactions in the matrix.

Data Interpretation

32. Question 32 pertains to a series of three electron micrographs obtained from a paper that examines the effect of the drug cantospermine on glycogen metabolism. (Molyneux et al., PNAS 82 pp 93–97 1985.)

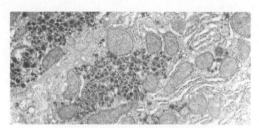

R Saul, J J Ghidoni, R J Molyneux, and A D Elbein/
Castanospermine inhibits alpha-glucosidase activities and alters glycogen distribution in animals/Published January 1, 1985, National Academy of Sciences

The image is a control image of an electron micrograph of a rat liver stained with alkaline lead citrate and uranyl acetate.

   a. Lead acetate and uranyl acetate are not used in light microscopy. Why are they used in electron microscopy?

   b. Which organelles can you identify in this image? How do they compare to illustrations you have seen in print?

   c. Why do some structures in this image appear darker than others?

   d. What are the darkest images in this micrograph?

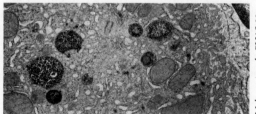

This second image has been prepared in the same way, but it is from a rat given the drug castanospermine.

   e. How has the image changed?

   f. Which organelles can you recognize now?

   g. How has the staining changed?

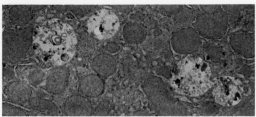

In this third image, castanospermine-treated liver samples were digested with amylase after the cells had been fixed and sliced.

   h. What role does amylase serve in the body?

   i. What reaction does amylase catalyze?

   j. How is amylase being used in this instance?

   k. How does this image differ from the other two?

   l. What does this data show?

   m. What is the effect of treating animals with castanospermine?

Experimental Design

33. Kefiran, the exopolysaccharide produced by bacteria in kefir grains, has a ratio of glucose to galactose of about 1:1. Based on what you know of other polysaccharides, what would you propose is the structure of kefiran? How might this molecule be synthesized by these bacteria? How could you test your hypotheses for the structure, and how could it be synthesized?

34. Metabolic requirements of a cancer cell are different from that of a normal cell, explain how will you investigate the involvement of PPP in cultured cancer cell. Based on your theoretical knowledge of cancer metabolism how would you expect PPP to be different?

35. Propose a way to measure glycogen levels in a dish of cultured cells.

36. Design an experiment to determine the degree of glycosylation, the amount of carbohydrate, in a membrane glycoprotein you have been studying.

37. How could you show that two matrix proteins were interacting in the extracellular matrix?

Ethics and Social Responsibility

38. This chapter discussed tissue engineering. Currently, tens of thousands of people are on waiting lists for organ transplants. Are you an organ donor? If not, what keeps you from becoming one?

39. Given the involvement of biofilms in human disease, do you think that more funding should be dedicated toward basic research into biofilms? Propose a source of this funding.

40. Many athletes who run marathons take advantage of carbo-loading. That is, using a dietary plan, they deplete muscle glycogen leading up to a big race and then supposedly build large stores to use on race day. Using drugs or artificially increasing your red blood cell count is considered illegal in sporting events. Why is dietary manipulation of glycogen stores not illegal? At what point does training become doping?

41. Usually *in vitro* studies (the ones that involve use of cultured cells) do not require any ethical clearance. What ethical guidelines would you draw for *in vitro* work?

42. Many of the inherited diseases mentioned in this chapter are incurable, despite advances in science. Often, we know what is wrong in a particular disease but cannot cure it. Does this situation give hope to people with such disorders, or does it simply make the problem more frustrating?

Suggested Readings

8.1 Glycogen Metabolism

Chaikuad, A., D. S. Froese, G. Berridge, F. von Delft, U. Oppermann, and W. W. Yue. "Conformational Plasticity of Glycogenin and Its Maltosaccharide Substrate During Glycogen Biogenesis." *Proceedings of the National Academy of Sciences of the United States of America* 108 (2011): 21028–33.

Gibbons, B. J., P. J. Roach, and T. D. Hurley. "Crystal Structure of the Autocatalytic Initiator of Glycogen Biosynthesis, Glycogenin." *Journal of Molecular Biology* 319, no. 2 (2000): 463–77.

Juhlin-Dannfelt, A. C., S. E. Terblanche, R. D. Fell, J. C. Young, and J. O. Holloszy. "Effects of Beta-Adrenergic Receptor Blockade on Glycogenolysis During Exercise." *Journal of Applied Physiology* 53, no. 3 (September 1982): 549–54.

Whelan, W. "Pride and Prejudice: The Discovery of the Primer for Glycogen Synthesis." *Journal of Protein Science* 7, no. 9 (1998): 2038–41.

8.2 The Pentose Phosphate Pathway

Cappellini, M. D., and G. Fiorelli. "Glucose-6-Phosphate Dehydrogenase Deficiency." *Lancet 371*, no. 9606 (2008): 64–74.

Dentin R., J. Girard, and C. Postic. "Carbohydrate Responsive Element Binding Protein (ChREBP) and Sterol Regulatory Element Binding Protein-1c (SREBP-1c): Two Key Regulators of Glucose Metabolism and Lipid Synthesis in Liver." *Biochimie* 87, no. 1 (2005): 81–6.

Glogauer, M. "Disorders of Phagocyte Function." Chap. 175 in *Cecil Medicine*, 23rd ed. Edited by L. Goldman and D. Ausiello. Philadelphia, PA: Saunders Elsevier, 2007.

Iizuka, K., and Y. Horikawa. "ChREBP: A Glucose-Activated Transcription Factor Involved in the Development of Metabolic Syndrome." *Endocrine Journal* 55, no. 4 (2008): 617–24.

Uyeda, K., and J. J. Repa. "Carbohydrate Response Element Binding Protein, ChREBP, a Transcription Factor Coupling Hepatic Glucose Utilization and Lipid Synthesis." *Cell Metabolism* 4, no. 2 (2006): 107–10.

Uyeda K., H. Yamashita, and T. Kawaguchi. "Carbohydrate Responsive Element-Binding Protein (ChREBP): A Key Regulator of Glucose Metabolism and Fat Storage." *Biochemical Pharmacology* 63, no. 12 (2002): 2075–80.

8.3 Carbohydrates in Glycoconjugates

Adam, A., N. Montpas, D. Keire, A. Désormeaux, N. J. Brown, F. Marceau, and B. Westenberger. "Bradykinin Forming Capacity of Oversulfated Chondroitin Sulfate Contaminated Heparin in Vitro." *Biomaterials* 31, no. 22 (2010): 5741–8.

Guerrini, M., D. Beccati, Z. Shriver, A. Naggi, K. Viswanathan, A. Bisio, I. Ishan Capila, et al. "Oversulfated Chondroitin Sulfate Is a Contaminant in Heparin Associated with Adverse Clinical Events." *Nature Biotechnology* 26 (2009): 669–75.

Han, L., M. Monné, H. Okumura, T. Schwend, A. L. Cherry, D. Flot, T. Matsuda, and L. Jovine. "Insights into Egg Coat Assembly and Egg-Sperm Interaction from the X-Ray Structure of Full-Length ZP3." *Cell* 143, no. 3 (2010): 404–15.

Hang, H. C., and C. R. Bertozzi. "The Chemistry and Biology of Mucin-Type O-Linked Glycosylation." *Bioorganic & Medicinal Chemistry* 13, no. 17 (September 1, 2005): 5021–34.

Knudson, C. B., and W. Knudson. "Cartilage Proteoglycans." *Seminars in Cell and Developmental Biology* 12, no. 2 (2001): 69–78.

Monné, M., L. Han, T. Schwend, S. Burendahl, and L. Jovine. "Crystal Structure of the ZP-N Domain of ZP3 Reveals the Core Fold of Animal Egg Coats." *Nature* 456, no. 7222 (2008): 653–7.

Peter-Katalinić, J. "Methods in Enzymology: O-Glycosylation of Proteins." *Methods in Enzymology* 405 (2005): 139–71.

Rankin, T., and J. Dean. "The Zona Pellucida: Using Molecular Genetics to Study the Mammalian Egg Coat." *Reviews of Reproduction* 5, no. 2 (2000): 114–21.

Schrager, J. "The Chemical Composition and Function of Gastrointestinal Mucus." *Gut* 11, no. 5 (1970): 450–6.

Tian, E., and K. G. Ten Hagen. "Recent Insights into the Biological Roles of Mucin-Type O-Glycosylation." *Glycoconjugate Journal* 26, no. 3 (April 2009): 325–34.

Wassarman, P. M. "Zona Pellucida Glycoproteins." *Journal of Biological Chemistry* 283, no. 36 (2008): 24285–9.

8.4 Extracellular Matrices and Biofilms

Alkawareek, M. Y., Q. T. Algwari, G. Laverty, S. P. Gorman, W. G. Graham, D. O'Connell, and B. F. Gilmore. "Eradication of *Pseudomonas aeruginosa* Biofilms by Atmospheric Pressure Non-Thermal Plasma." *PLoS One* 7, no. 8 (2012): 44289.

Blanc, P. J. "Characterization of the Tea Fungus Metabolites." *Biotechnology Letters* 18, no. 2 (1996): 139–42.

Cao, Y., J. P. Vacanti, K. T. Paige, J. Upton, and C. A. Vacanti. "Transplantation of Chondrocytes Utilizing a Polymer-Cell Construct to Produce Tissue-Engineered Cartilage in the Shape of a Human Ear." *Plastic and Reconstructive Surgery* 100, no. 2 (1997): 297–302.

CNN. "Ricin Dart Fired by Umbrella Killed Bulgarian Envoy." Retrieved from http://edition.cnn.com/2008/US/02/29/ricin.cases/index.html

Colognato, H., and P. D. Yurchenco. "Form and Function: The Laminin Family of Heterotrimers." *Developmental Dynamics* 218, no. 2 (2000): 213–34.

Costerton, J. W., P. S. Stewart, and E. P. Greenberg. "Bacterial Biofilms: A Common Cause of Persistent Infections." *Science* 284 (1999): 1318–22.

Csiszar, K. "Lysyl Oxidases: A Novel Multifunctional Amine Oxidase Family." *Progress in Nucleic Acid Research and Molecular Biology* 70 (2001): 1–32.

Endo, Y., and K. Tsurugi. "The RNA N-Glycosidase Activity of Ricin A-Chain. The Characteristics of the Enzymatic Activity of Ricin A-Chain with Ribosomes and with rRNA." *Journal of Biological Chemistry* 263, no. 18 (1988): 8735–9.

Epstein, A. K., T. S. Wong, R. A. Belisle, E. M. Boggs, and J. Aizenberg. "Liquid-Infused Structured Surfaces with Exceptional Anti-Biofouling Performance." *Proceedings of the National Academy of Sciences of the United States of America* 109, no. 33 (2012): 13182–7.

Garrote, G. L., A. G. Abraham, and G. L. De Antoni. "Chemical and Microbiological Characterization of Kefir Grains." *Journal of Dairy Research* 68, no. 4 (2001): 639–52.

Gorres, K. L., and R. T. Raines. "Prolyl 4-Hydroxylase." *Critical Reviews in Biochemistry and Molecular Biology* 45, no. 2 (2010): 106–24.

Hall-Stoodley, L., J. W. Costerton, and P. Stoodley. "Bacterial Biofilms: From the Natural Environment to Infectious Diseases." *Nature Reviews Microbiology* 2 (2004): 95–108.

Larivier, J. W., P. Kooiman, and K. Schmidt. "Kefiran a Novel Polysaccharide Produced in Kefir Grain by *Lactobacillus brevis*." *Archives of Microbiology* 59, no. 1–3 (1967): 269–70.

Lopez, D., H. Vlamakis, and R. Kolter. "Biofilms." In *Cold Spring Harbor Perspectives in Biology* (June 2010). doi: 10.1101/cshperspect.a000398

Marsh, P. D. "Contemporary Perspective on Plaque Control." *British Dental Journal* 212, no. 12 (2012): 601–6.

Mayser, P., S. Frommer, C. Leitzmann, and K. Gründer. "The Yeast Spectrum of the Tea Fungus Kombucha." *Mycoses* 38, no. 7–8 (1995): 289–95.

Miller, M. B., and B. L. Bassler. "Quorum Sensing in Bacteria." *Annual Review of Microbiology* 55 (2001): 165–199.

National Science Foundation (U.S.A.) (2004). The Emergence of Tissue Engineering as a Research Field. Retrieved from https://www.nsf.gov/pubs/2004/nsf0450/start.htm

Nerem, R. M. *Principles of Tissue Engineering*, 2nd ed. Edited by Joseph Vacanti, R. P. Lanza, and Robert S. Langer. Boston, MA: Academic Press, 2000.

Saul, R., J. J. Ghidoni, R. J. Molyneux, and A. D. Elbein. "Castanospermine Inhibits Alpha-Glucosidase Activities and Alters Glycogen Distribution in Animals." *Proceedings of the National Academy of Sciences of the United States of America* 82 no. 1 (January 1985): 93–7.

Seneviratne, C. J., C. F. Zhang, L. P. Samaranayake, and J. Chin. "Dental Plaque Biofilm in Oral Health and Disease." *Journal of Dental Research* 14, no. 2 (2011): 87–94.

Simova, E., D. Beshkova, A. Angelov, T. S. Hristozova, G. Frengova, and Z. Spasov. "Lactic Acid Bacteria and Yeasts in Kefir Grains and Kefir Made from Them." *Journal of Industrial Microbiology and Biotechnology* 28, no. 1 (2002): 1–6.

Singh, P. K., A. L. Schaefer, M. R. Parsek, T. O. Moninger, M. J. Welsh, and E. P. Greenberg. "Quorum-Sensing Signals Indicate That Cystic Fibrosis Lungs Are Infected with Bacterial Biofilms." *Nature* 407, no. 6805 (2002): 762–4.

Stoodley, P., K. Sauer, D. G. Davies, and J. W. Costerton. "Biofilms as Complex Differentiated Communities." *Annual Review of Microbiology* 56, (2002): 187–209.

Tada, S., Y. Katakura, K. Ninomiya, and S. Shioya. "Fed-Batch Coculture of *Lactobacillus kefiranofaciens* with *Saccharomyces cerevisiae* for Effective Production of Kefiran." *Journal of Bioscience and Bioengineering* 103, no. 6 (2007): 557–62.

Ullrich, M. *Bacterial Polysaccharides: Current Innovations and Future Trends*. Cambridge, MA: Caister Academic Press, 2009.

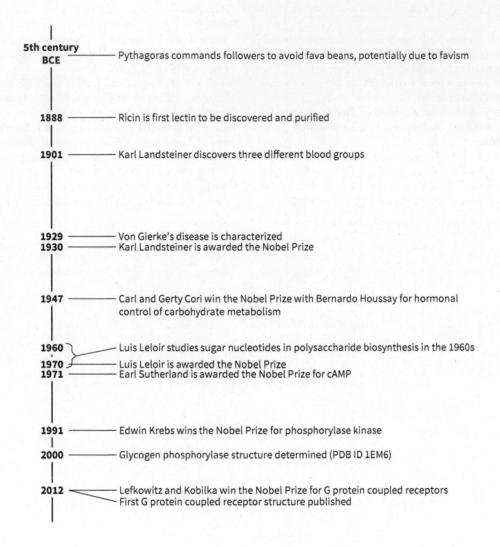

5th century BCE — Pythagoras commands followers to avoid fava beans, potentially due to favism

1888 — Ricin is first lectin to be discovered and purified

1901 — Karl Landsteiner discovers three different blood groups

1929 — Von Gierke's disease is characterized
1930 — Karl Landsteiner is awarded the Nobel Prize

1947 — Carl and Gerty Cori win the Nobel Prize with Bernardo Houssay for hormonal control of carbohydrate metabolism

1960 — Luis Leloir studies sugar nucleotides in polysaccharide biosynthesis in the 1960s
1970 — Luis Leloir is awarded the Nobel Prize
1971 — Earl Sutherland is awarded the Nobel Prize for cAMP

1991 — Edwin Krebs wins the Nobel Prize for phosphorylase kinase

2000 — Glycogen phosphorylase structure determined (PDB ID 1EM6)

2012 — Lefkowitz and Kobilka win the Nobel Prize for G protein coupled receptors
First G protein coupled receptor structure published

CHAPTER 9

Lipids I

Fatty Acids, Steroids, and Eicosanoids; Beta-Oxidation and Fatty Acid Biosynthesis

Lipids in Context

Lipids are perhaps the most maligned class of biological molecules. Many people consider fats the bane of those trying to lose weight. While many people may not know that cholesterol and steroids are lipids, they do associate them with negative thoughts. Cholesterol is viewed as a cause of blocked arteries, heart attacks, and stroke, whereas steroids are synonymous with disgraced athletes and are blamed for acne, mood swings, and unwarranted aggression. Yet, for all their bad press, we cannot live without these molecules: lipids are essential not only to human life but to *all* life.

Lipids are a diverse set of biomolecules unified by a single property: hydrophobicity. However, there is no single unifying structure among lipids. Lipids include cholesterol, corticosteroids, eicosanoids, fatty acids, and phospholipids and have a range of functions:

- Phospholipids in lipid bilayers separate the inside of the cell from the outside environment and form the defining perimeter of most organelles. They also can play important roles in signaling.

- Cholesterol is another membrane component in some cells where it helps maintain the fluidity of the lipid bilayer.

- Steroid hormones made from cholesterol are critical for sexual development, reproduction, and regulation of mineral balance in higher organisms.

- Fatty acids stored as triacylglycerols in adipose (fat) tissue serve to insulate and cushion vital organs.

- Eicosanoids and certain other lipids act as signaling molecules, regulating functions such as blood pressure, pain, inflammation, and labor and delivery.

This chapter begins with descriptions of the properties and functions of several of the major classes of lipids. We then discuss the degradation and biosynthesis of fatty acids, and the biosynthesis of ketone bodies and cholesterol. In each of these pathways, the discussion highlights the steps in the pathway, the mechanisms of reactions where applicable, and the regulation of the pathway. Finally, the chapter closes with the biosynthesis of eicosanoids and endocannabinoids, two important classes of signaling molecules.

Chapter Outline

9.1 Properties, Nomenclature, and Biological Functions of Lipid Molecules

9.2 Fatty Acid Catabolism

9.3 Fatty Acid Biosynthesis

9.4 Ketone Body Metabolism

9.5 Steroid Metabolism

9.6 Eicosanoid and Endocannabinoid Metabolism

Common Themes

Evolution's outcomes are conserved.	• Comparison of the major enzymes of fatty acid biosynthesis (acetyl-CoA carboxylase ACC and fatty acid synthase FAS) gives insights as to how this pathway has evolved. • Analysis of different species reveals that genes coding for the proteins and enzymes responsible for the synthesis of certain molecules, such as some steroids and eicosanoids, have evolved more recently; hence, these molecules may address needs that are lacking in simpler species.
Structure determines function.	• The structure of different lipids partially dictates their function; for example, amphipathic phospholipids form bilayers, whereas nonpolar triacylglycerols provide an anhydrous energy source. • The hydrophobic effect causes amphipathic lipids to form structures such as bilayers, micelles, and vesicles. • The structure of many enzymes in these pathways is conserved both within a single organism and between different species. • A common set of chemical reactions are used in the metabolism of lipids and in biosynthesis of different lipids.
Biochemical information is transferred, exchanged, and stored.	• Lipids can serve as important signaling molecules and second messengers.
Biomolecules are altered through pathways involving transformations of energy and matter.	• Reactions that are energetically unfavorable become favorable when coupled to reactions with large, negative ΔG values; we often think of this in terms of ATP hydrolysis, but other bonds with a large negative ΔG of hydrolysis, such as thioesters, are common in lipid biochemistry. Likewise, as we have seen elsewhere, reactions that have $\Delta G°$ values close to zero can be tipped to move in the forward or reverse direction by changing the concentrations of substrates or products. • Fatty acids and ketone bodies are synthesized and degraded using similar reactions. Different biochemical pathways and compartmentalization result in biosynthesis of fatty acids or β-oxidation. This is similar to the situation seen with glycolysis and gluconeogenesis. • Similar chemical patterns are found in β-oxidation, fatty acid synthesis, and ketone body synthesis and breakdown. Some of these reactions were discussed in the citric acid cycle.

9.1 Properties, Nomenclature, and Biological Functions of Lipid Molecules

This section provides an overview of the main categories of lipids: fatty acids, neutral lipids, phospholipids, steroids, bile salts, and eicosanoids.

9.1.1 Fatty acids are a common building block of many lipids

Fatty acids are components of neutral lipids, phospholipids, and eicosanoids. Typically, fatty acids are unbranched long-chain carboxylic acids, containing an even number of carbon atoms, generally between 12 and 26 with 16 and 18 carbons being the most common (**Figure 9.1**). Fatty acids are either **saturated** (no double bonds), **monounsaturated** (one double bond), or **polyunsaturated** (multiple double bonds). Unless otherwise noted, all double bonds in fatty acids can be assumed to be in the *cis* conformation. Figure 9.1 illustrates the notation system used to describe fatty acids. The number of carbon atoms and the number of double bonds are separated by a colon. The position of a double bond is designated by a superscript Greek delta, Δ, followed by the position of the first carbon of each double bond; for example, a double bond between C-9 and C-10 would be denoted Δ^9. Common polyunsaturated fatty acids do not have conjugated double bonds and instead have a single methylene unit between double bonded carbons.

Because of the way fatty acids are synthesized, elongated (extended beyond 16 carbons in length), and desaturated (how double bonds are incorporated), many fatty acids are also characterized by the position of the double bond farthest from the carboxyl group. The carbon farthest from the carboxyl group within that double bond is given the designation ω (omega, the last letter in the Greek alphabet); hence, we find ω-**3**, ω-**6**, ω-7, or ω-9 fatty acids, counting from the terminal carbon. The melting points of fatty acids vary widely, but some common trends emerge. As the fatty acid gets longer, the melting point increases. However, incorporation of double bonds greatly decreases the melting point (**Table 9.1**).

Fatty acids are typically known by their common name (stearic acid), or an International Union of Pure and Applied Chemistry [IUPAC] name such as octadecanoic acid, 18:0.

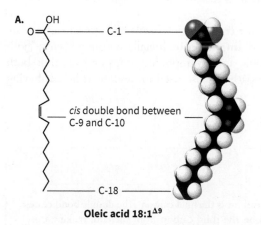

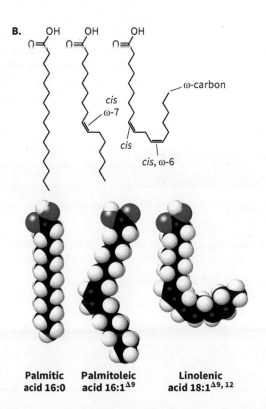

FIGURE 9.1 Structure of oleic acid and major categories of fatty acids. **A.** Fatty acids are long-chain, even-numbered carboxylic acids. They may be unsaturated (like oleic acid) in which instance they almost always have *cis* double bonds. **B.** Palmitic, palmitoleic, and linolenic acids are all fatty acids. Palmitic (16:0) is a saturated fatty acid, palmitoleic (16:1$^{\Delta 9}$) is a monounsaturated fatty acid, and linolenic (18:2$^{\Delta 9,12}$) is a polyunsaturated fatty acid. Note how the inclusion of *cis* double bonds introduces kinks into the carbon chain. Highly polyunsaturated fatty acids are kinked to the point of being almost circular in shape.

Oleic acid 18:1$^{\Delta 9}$

Palmitic acid 16:0   Palmitoleic acid 16:1$^{\Delta 9}$ Linolenic acid 18:1$^{\Delta 9,12}$

TABLE 9.1 Some Common Fatty Acids

Description	Common Name	IUPAC Name	Source	Melting Point	ω-Nomenclature
12:0	Lauric acid	Dodecanoic acid	Bay leaves	44.2	–
14:0	Myristic acid	Tetradecanoic acid	Coconut oil	53.9	–
16:0	Palmitic acid	Hexadecanoic acid	Palm oil	63.1	–
18:0	Stearic acid	Octadecanoic acid	Beef tallow	69.6	–
20:0	Arachidic acid	Eicosanoic acid	Peanut oil	76.5	–
$16:1^{\Delta 9}$	Palmitoleic acid	cis-9-hexadecenoic acid	Animal fats	1	ω-7
$18:1^{\Delta 9}$	Oleic acid	cis-9-octadecaenoic acid	Olive oil	13.4	ω-9
$18:2^{\Delta 9, 12}$	Linoleic acid	cis-, cis-9, 12-octadecadienoic acid	Safflower oil	−5	ω-6
$18:3^{\Delta 9, 12, 15}$	α-Linolenic	cis-, cis-, cis-9, 12, 15-octadecadienoic acid	Safflower oil	−11	ω-3
$20:3^{\Delta 8, 11, 14}$	DGLA, dihomogamma linolenic acid	Eicosatrienoic acid	Fish oil (rare)	~ −5	ω-6
$20:4^{\Delta 5,8,11,14}$	Arachidonic acid	Eicosatetraenoic acid		−49.5	ω-6
$20:5^{\Delta 5,8,11,14,17}$	EPA	Eicosapentaenoic acid		~ −50	ω-3
$22:6^{\Delta 4,7,10,13,16,19}$	DHA	Docosahexaenoic acid		−44	ω-3

Fatty acids: acid-base properties   Because fatty acids contain a carboxyl group, they act as weak acids. The pK_a of the carboxyl group is about 4.5, similar to other carboxylic acids such as amino acids or the intermediates of the citric acid cycle. Fatty acids are generally ionized (approximately 99.9%) at pH 7.4. In the laboratory, to increase the percentage of ionized species and thus make the molecule more soluble in aqueous solutions, the acid is often reacted with a strong base to generate a carboxylic acid salt or a carboxylate. Carboxylic acids, the protonated species, are distinct from carboxylates, the deprotonated species, and it is not surprising that the two forms have different properties and reactivity.

Fatty acids are typically found as esters or amides. Amides are referred to as peptides in protein chemistry but as amides in the lipid literature. Both ester and amide linkages are frequently observed in lipid chemistry. The term **free fatty acid** is reserved for fatty acids that are not part of an ester or an amide.

With a long, nonpolar hydrophobic tail at one end of the molecule and a polar carboxylate group at the other end, fatty acids are considered **amphipathic**, literally meaning having "both feelings," because they contain both hydrophobic and hydrophilic groups and are thus both water-loving and water-hating. The term amphipathic is also used to describe α helices having both a polar and a nonpolar face.

Worked Problem 9.1	Characterizing a fatty acid

Linolenic acid is $18:3^{\Delta 9, 12, 15}$. What is the omega designation of linolenic acid?

Strategy   Examine Figure 9.1 Think about what the omega designation tells us. It may be useful to draw the structure of linolenic acid.

Solution   Linolenic acid has the following structure:

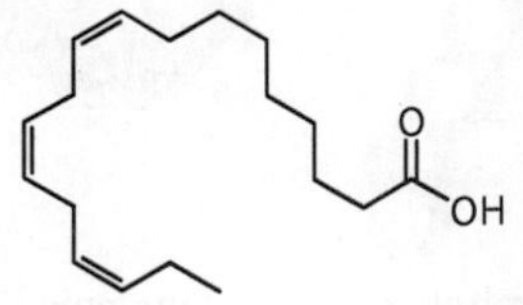

The omega carbon is the final carbon. The double bond closest to that carbon is on the third carbon from the end, making this a ω-3 fatty acid.

Follow-up question   Using Table 9.1, identify this fatty acid:

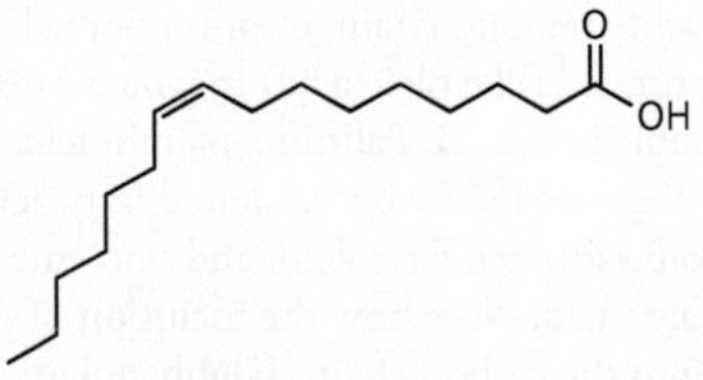

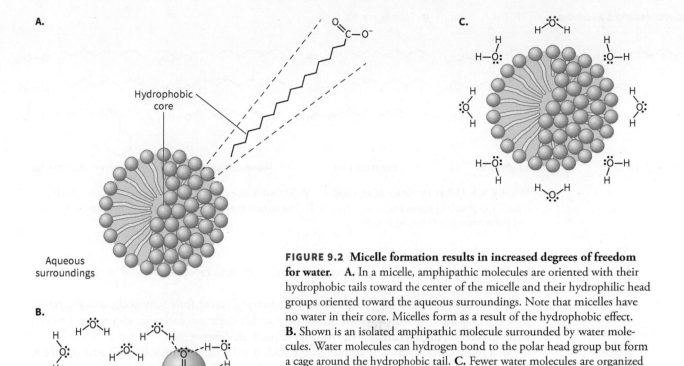

FIGURE 9.2 Micelle formation results in increased degrees of freedom for water. **A.** In a micelle, amphipathic molecules are oriented with their hydrophobic tails toward the center of the micelle and their hydrophilic head groups oriented toward the aqueous surroundings. Note that micelles have no water in their core. Micelles form as a result of the hydrophobic effect. **B.** Shown is an isolated amphipathic molecule surrounded by water molecules. Water molecules can hydrogen bond to the polar head group but form a cage around the hydrophobic tail. **C.** Fewer water molecules are organized around a micelle than around a similar number of monomeric amphiphiles. Despite the apparent gain of organization and loss of entropy in the organization of a micelle, the gain in entropy as a result of liberating water makes the overall ΔS positive. The free energy (ΔG) of the assembly of micelles is therefore negative, and the process is spontaneous.

Fatty acids: assembly into higher order structures Like other amphipathic molecules, fatty acids exhibit interesting properties when dissolved in water. At a very low concentration, fatty acids are found as single molecules in solution, but at higher concentrations, they aggregate into a roughly spherical structure known as a **micelle** (**Figure 9.2**). Within the micelle, the hydrophobic tails are oriented toward the waterless core of the sphere and the hydrophilic head groups are oriented toward the solvent. This structure maximizes the interaction of the polar head groups with water while shielding the hydrophobic tails. Similar to protein folding, when hydrophobic molecules are clustered together, water molecules are effectively freed from having to cage these molecules. Formation of a micelle appears to result in an apparent gain of organization and loss of entropy; however, the reverse is true when looking at the larger picture. The gain in entropy as a result of liberating water makes the overall ΔS *positive*. The free energy (ΔG) of the assembly of micelles is therefore negative, and the process is spontaneous. This is another illustration of the **hydrophobic effect**.

The concentration at which micelles form is called the **critical micellar concentration** (or CMC). The CMC is a property of all amphipathic molecules, including all detergents and many lipids and differs for each molecule. For fatty acids, the CMC is typically in the range of 1 to 6 µM; below this concentration, fatty acids are found as monomers in solution, and above it, they spontaneously assemble into micelles. The micelles exist in a dynamic equilibrium with monomeric (single) fatty acids in solution. Increasing the concentration of fatty acids above the CMC increases the *number* of micelles found in a sample but not the *size* of the micelles. When working with an amphipathic molecule in the laboratory, it is useful to know the CMC. Below the CMC, amphipathic molecules will partition into membranes; at concentrations above the CMC, membranes will begin to dissolve.

9.1.2 Neutral lipids are storage forms of fatty acids or cholesterol

Because fatty acids are carboxylic acids and are amphipathic, they can act as detergents and affect pH—properties that could be harmful to the organism. Therefore, fatty acids are stored as **neutral**

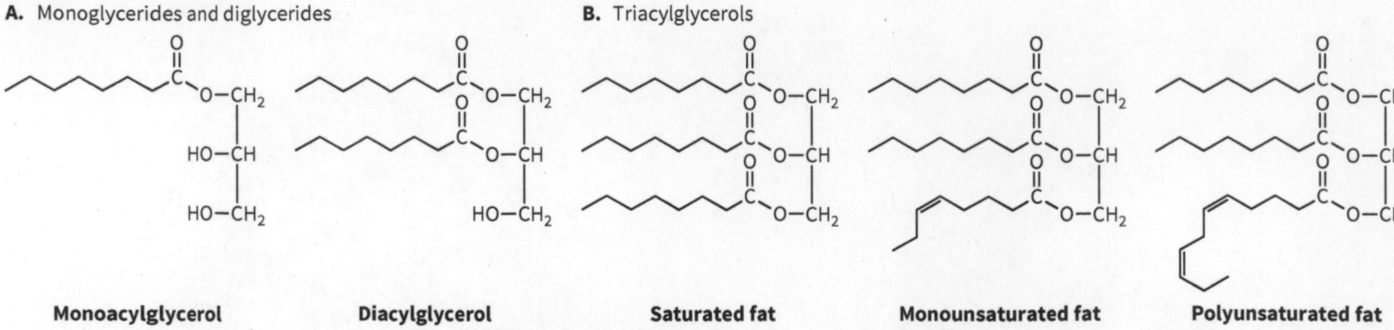

A. Monoglycerides and diglycerides

Monoacylglycerol   Diacylglycerol

B. Triacylglycerols

Saturated fat   Monounsaturated fat   Polyunsaturated fat

FIGURE 9.3 Other examples of glycerides. A. Shown is an example of a monoglyceride and a diglyceride, or monoacylglycerol and diacylglycerol. **B.** The second example shows saturated, monounsaturated, and polyunsaturated triacylglycerols.

lipids, which lack charged groups because the carboxylic acid group is esterified to either glycerol or cholesterol.

Fatty acids esterified to glycerol form a **monoacylglycerol** (one fatty acid), **diacylglycerol** (two fatty acids), or **triacyclglycerol** (three fatty acids); these molecules are also referred to as a **monoglyceride**, **diglyceride**, or **triglyceride**, respectively (**Figure 9.3**).

If a glyceride contains only saturated fatty acids (no double bonds), it is called a saturated fat; if it contains at least one fatty acid that has a single double bond, it is termed a monounsaturated fat; and if it contains at least one fatty acid with multiple double bonds, it is known as a polyunsaturated fat. Glycerides can be crudely categorized as fats or oils based on their melting points; fats are solid at room temperature, whereas oils are liquid at room temperature. A saturated fat is solid at room temperature because the straight chains of the fatty acids allow the molecules to pack tightly together. Saturated fats commonly come from animal origins, such as tallow from cows or lard from pigs. Unsaturated fats are liquid at room temperature because the kinked *cis* double bonds prevent the molecules from packing tightly together. Common sources of unsaturated fats are vegetables, such as corn, soy, or canola, and cold-water fish, such as salmon.

Although plant oils are readily available, they are less desirable for some food and cooking applications than solid fats because of their physical properties. For example, solid fats are preferred for baked and fried products. Hence, vegetable oils used in processed foods are often hydrogenated, that is, chemically treated with hydrogen gas and a catalyst to convert some of the unsaturated fatty acids into saturated ones. The complete conversion of all the *cis* double bonds in a vegetable oil results in a product that has too high a melting point and a waxy taste and feel. Therefore, oils are often partially hydrogenated. This partial reaction can result in the naturally occurring *cis* fatty acids being isomerized to *trans* fatty acids, forming ***trans* fats**. The properties of *trans* fats are similar to those of saturated fats; they are metabolized differently than either saturated or *cis* fatty acids. Although low quantities of *trans* fatty acids do occur naturally in some bacteria and ruminants, organisms have not evolved to metabolize large quantities of *trans* fats. It is possible to remove most of the *trans* fatty acids from processed foods. This is done by replacing the partially **hydrogenated fats** with blends of saturated and unsaturated fats that have similar physical properties to the hydrogenated fats.

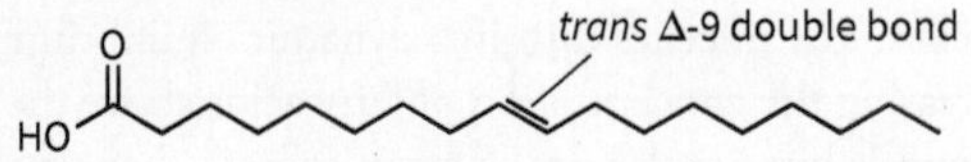

trans Δ-9 double bond

Fatty acids esterified to cholesterol yield **cholesteryl esters**, the main storage form of cholesterol. Cholesteryl esters form a large part of the plaque in arteries that leads to atherosclerosis and heart disease. Certain cells in the body also store cholesteryl esters as a cholesterol reserve, used to produce steroid hormones.

Fatty acids and cholesterol are not the only molecules stored as esters. Retinyl palmitate, a storage form of vitamin A, is an ester of retinol and palmitic acid. The esterification protects the hydroxyl group of retinol from oxidation to an aldehyde or carboxylic acid until the vitamin is needed in the body. Vitamin E is often administered in an esterified form, as tocopheryl acetate or succinate; the esterification prevents oxidation of the alcohol and thus helps to stabilize the

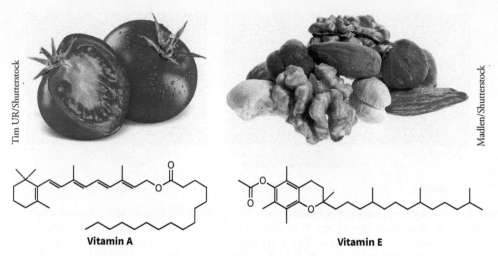

Vitamin A　　　　　　　　　　　**Vitamin E**

FIGURE 9.4 Structures of vitamins A and E.　Both vitamins A and E are fat-soluble vitamins and contain an ester moiety. For vitamin A (retinyl palmitate), this serves to protect the hydroxyl group from oxidation to an aldehyde or carboxylic acid until needed in the body. Vitamin E (tocopheryl acetate) is also frequently found as an ester—in this case an acetate. Again, this form of the vitamin is protected from premature oxidation by the ester. Vitamin A is found in foods such as tomatoes. Nuts are a rich source of vitamin E.

compound (**Figure 9.4**). The ester bond is cleaved in the stomach, and the free alcohol (tocopherol) is absorbed and used by the body.

Another class of neutral lipids worth mentioning is the **waxes**. These molecules are long-chain fatty acids esterified to a long-chain alcohol termed a fatty alcohol; waxes are lengthy and hydrophobic (**Figure 9.5**). Organisms produce and secrete these molecules as protective coatings that provide some degree of natural waterproofing and prevent desiccation. The best known include bee's wax; carnauba wax; jojoba oil; lanolin; and spermaceti, the wax found in the head of sperm whales. Lanolin, also known as wool grease, is secreted by sheep to assist in waterproofing their wool and skin. Found in many skin creams and lotions, lanolin differs from the other waxes mentioned above because the alcohol forming the ester is cholesterol rather than a fatty alcohol.

9.1.3 Phospholipids are important in membrane formation

Phospholipids are lipid molecules that contain a phosphate moiety. The hydrophilic nature of the phosphate group and the hydrophobic nature of the aliphatic lipid make these molecules amphipathic. Phospholipids are considered **polar lipids** and are the main structural component of membranes. Many phospholipids are also involved in signal transduction and act as lipid signaling mediators.

Phospholipids can broadly be divided into two categories, depending on the type of backbone they contain. **Glycerophospholipids** have a glycerol backbone, whereas **sphingolipids** have a sphingosine backbone. Because of their structure and geometry, some of these molecules can form micelles, but many will form other more complex structures, such as lipid bilayers, monolayers, or vesicles, a bilayer surrounding an aqueous compartment (**Figure 9.6**).

Triaconatyl palmitate
(bee's wax)

FIGURE 9.5 Structure of waxes.　Waxes, such as triaconatyl palmitate, consist of long-chain fatty acids esterified to long-chain alcohols. The highly hydrophobic nature of these molecules makes them useful as protectants against desiccation.

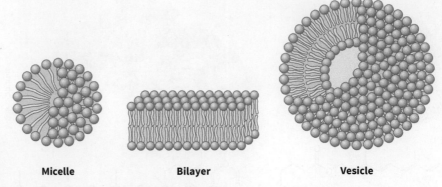

FIGURE 9.6 Examples of phospholipid structures. Phospholipids can form a variety of structures, depending on the head group, type of acyl chain, and number of acyl chains (one versus two). These structures can include some of the ones we have seen previously, for example, micelles and bilayers. Phospholipids can also form structures such as vesicles, a bilayer structure containing an aqueous core.

Glycerophospholipids

Glycerophospholipids feature two fatty acyl chains and a phosphoalcohol esterified to a glycerol backbone. Fatty acids are found esterified in the first two positions on the glycerol backbone. A phosphate group is in the final position, attached through a phosphoester linkage. Also connected to the phosphate is an alcohol, again through a phosphoester linkage. The alcohol gives rise to the common names of these lipids (**Figure 9.7**).

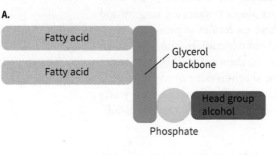

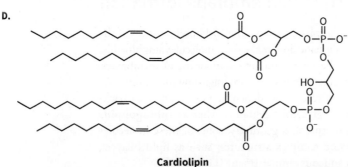

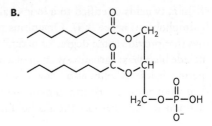

FIGURE 9.7 Structure of glycerophospholipids.
A. and **B.** Glycerophospholipids feature two fatty acyl chains and a phosphoalcohol esterified to a glycerol backbone. The length and degree of unsaturation of the fatty acids may vary. The most significant difference between these molecules occurs in the head group alcohol. **C.** The five common most head groups of phospholipids. At physiological pH both the amine on the choline head group as well as the phosphate moiety are ionized making the molecule zwitterionic. **D.** Cardiolipin is a glycerophospholipid comprised of two molecules of phosphatidate connected by a central glycerol molecule. It is prevalent in mitochondrial membranes.

FIGURE 9.8 Composition of lipid bilayers. The lipid composition of a membrane varies from cell type to cell type and from organelle to organelle. Shown is the composition for three different membranes. The lipids shown are cholesterol, phosphatidylethanolamine (PE), phosphatidylcholine (PC), sphingomyelin (SM), phosphatidylserine (PS), cardiolipin (CL), and all other lipids.

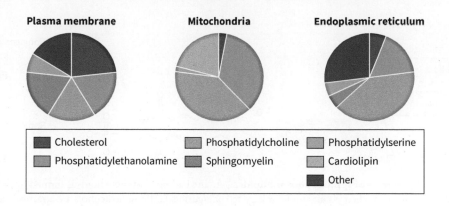

Phosphatidylcholine and phosphatidylethanolamine both contain positively charged amine groups in addition to the negatively charged phosphate, and like amino acids they are zwitterionic at neutral pH. Phosphatidylinositol, phosphatidylserine, and phosphatidate are all important in signaling cascades. These molecules are termed anionic phospholipids or acidic phospholipids due to their net negative charge.

Phosphatidylcholine and phosphatidylethanolamine also form the bulk of the lipid in cell membranes (typically over 50%), although membrane composition varies within the cell (such as the plasma membrane compared to mitochondrial or ER membrane), within the organism (a hepatocyte compared to a rod or cone cell in the eye), and between different species (erythrocytes from a human and a cow) (**Figure 9.8**). Phospholipid concentrations are also different in the two bilayers of the plasma membrane. Typically, the cytosolic leaflet of the plasma membrane is enriched in acidic phospholipids: phosphatidate, phosphatidylserine, phosphatidylglycerol, and phosphatidylinositol.

A phospholipid containing a single acyl chain and a free hydroxyl group is termed a **lysophospholipid**.

Sphingolipids

Sphingolipids contain many of the same fatty acids and alcohols as glycerophospholipids but differ in the backbone alcohol. The backbone alcohol in sphingolipids is sphingosine, which has two free hydroxyls, an amine, and a long (16 carbons) monounsaturated hydrocarbon tail. The phosphate is esterified through the primary hydroxyl most distal from the fatty-acid tail. Fatty acids are attached via an amide linkage to the amine, and the secondary hydroxyl between the amide and the fatty tail is not usually modified in any way.

One of the first sphingolipids identified was sphingomyelin, a major component of the myelin sheath coating some neurons. More recently, sphingolipids have gained attention for their possible involvement in the regulation of apoptosis and anoikis (programmed cell death). In particular, ceramide, sphingosine, and sphingosine-1-phosphate generated from sphingomyelin have been shown to promote apoptosis (**Figure 9.9**).

Worked Problem 9.2   Candy chemistry

Some candy bar labels state that the product contains soy lecithin, an emulsifier. Lecithin is another term for phosphatidylcholine. Describe how soy lecithin works as an emulsifier.

Strategy   To answer this question we need to know what lecithin is and how an emulsifier works. We discussed phosphatidylcholine in this section, and we previously discussed emulsifiers. Drawing the structure of phosphatidylcholine will help us to explain which properties of this molecule make it an emulsifier.

Solution   Emulsifiers are amphipathic molecules that act to suspend one type of molecule in another, such as a hydrophobic

molecule in an aqueous solvent. Soy lecithin is phosphatidylcholine obtained from soybeans. This phospholipid forms a micelle-like structure, coating and solubilizing fatty molecules in chocolate, for example. Within the micelles, the hydrophobic acyl chains are oriented toward the core of the particle and the polar head groups to the outside.

Follow-up question   We have previously seen how other amphipathic molecules can act as detergents. Could soy lecithin be used as a detergent?

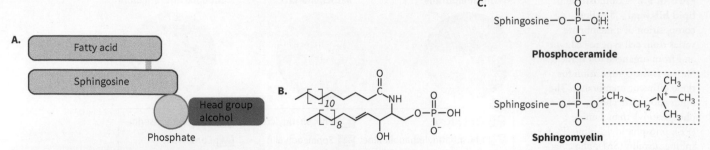

FIGURE 9.9 Structural examples of sphingophospholipids. **A.** The backbone of these lipids is formed by the long-chain fatty alcohol sphingosine. **B.** A single fatty acid is attached via an amide linkage to sphingosine. **C.** Phosphoalcohols are attached via a phosphoester bond to the more terminal hydroxyl of sphingosine.

9.1.4 All steroids and bile salts are derived from cholesterol

Steroids are a group of lipids with diverse functions but a common skeleton consisting of four fused rings. Variety among the steroids comes from incorporation of hydroxyl groups, carbonyl groups, and hydrocarbons around the ring system. **Cholesterol** is the most common steroid, and the one from which most other steroids are derived (**Figure 9.10**). It is an alcohol containing a single –OH group but is otherwise lacking in functional groups. The other steroids introduced in this section are steroid hormones (both sex hormones and corticosteroids) and bile salts.

In the cell, cholesterol is found within membranes and is oriented so that the hydroxyl moiety is away from the membrane core. Cholesterol is the master steroid, and it is found almost exclusively in animal tissue. Other organisms, such as plant sterols, have related compounds or employ other biochemistry, including modulating the fatty acyl composition of membranes to achieve the functions of cholesterol in animals. The fluidity of membranes is essential for a variety of functions including transport across the membrane, diffusion of proteins within the membrane, and membrane integrity; therefore, the concentration of cholesterol needs to be tightly controlled. At low concentrations, cholesterol maintains membrane fluidity by decreasing the interactions between lipids and proteins in the membrane; however, at higher concentrations, it decreases this fluidity.

Many steroids are also important signaling molecules. The two main classes of steroid hormones are the sex hormones and the corticosteroids. Cholesterol is converted into pregnenolone, and from that into a class of progestagens. The glucocorticoids, mineralocorticoids, and androgens are all derived from progestagens. Estrogens are derived from androgens, notably testosterone. All of these molecules share a common steroid skeleton, but they differ in the functional groups connected to that skeleton. A discussion of the role of steroid hormones in fertility and how hormones can be used to regulate fertility can be found in **Medical Biochemistry: Oral contraception.**

Bile salts, such as cholic acid or taurocholic acid, are detergents synthesized from cholesterol by the liver, and stored in the gall bladder. The body uses bile (a combination of cholesterol, bile salts, and amphipathic proteins) to solubilize dietary lipids in the gut; the bile increases the surface area and facilitates the enzymatic degradation of the lipids.

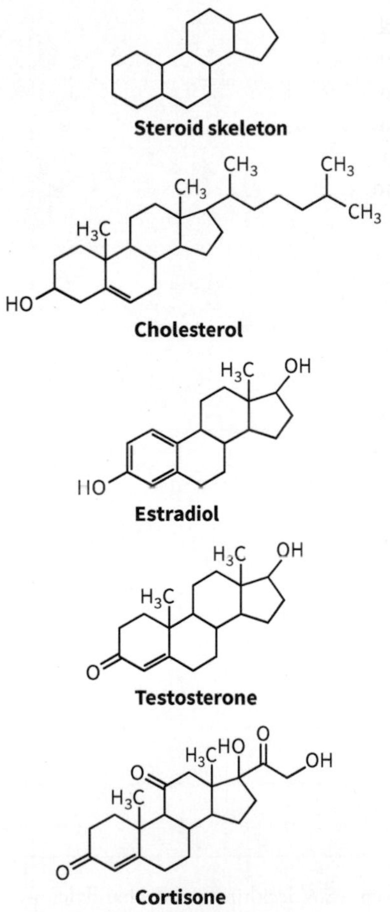

FIGURE 9.10 Steroids. All steroids have a common system of fused rings, and many steroids can be recognized by this characteristic shape. Shown are the steroid hormones estradiol, testosterone, and cortisone. Note the subtle differences in these structures that produce significant effects in growth and development or regulate fluid balance.

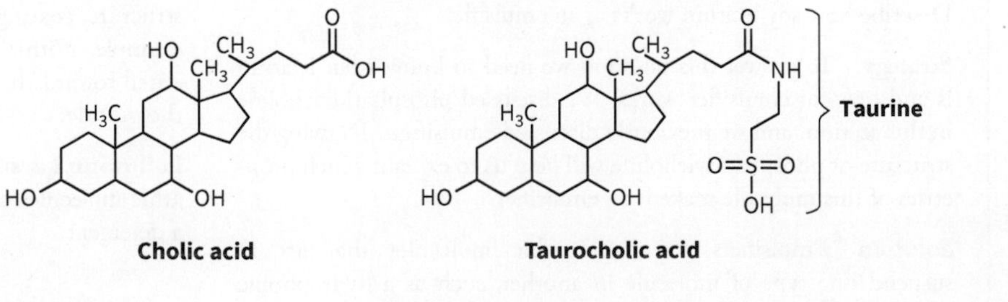

Medical Biochemistry

Oral contraception

Few drugs are as effective, successful, and controversial as oral contraceptives. Since their introduction in 1960, these drugs have been used by hundreds of millions of women worldwide. When used correctly, they are over 99.7% effective in preventing pregnancy, second only to surgical sterilization. The sheer number of people taking these drugs, combined with the issues surrounding their use—sexuality, morality, changing societal views, human rights, insurance coverage, and control of reproduction—continue to make these drugs among the most widely discussed and debated. It is little wonder that oral contraceptives are the one drug known as "The Pill."

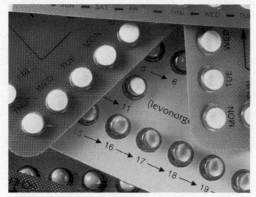

Adam Hart-Davis/Science Photo Library/Getty Images

From a biochemical perspective, human fertility is regulated by a cascade of both peptide and steroid hormones. Elevated progesterone levels from the ovary signal to the uterine lining (the endometrium) to prepare for an implanted embryo, should the egg become fertilized. If a fertilized egg does not implant in the endometrium, the cells in the ovary secreting hormones (the corpus luteum) die off, resulting in a drop in estrogen and progesterone levels. This also has the effect of shedding the endometrial lining, resulting in menstruation. However, if the egg is fertilized and implants in the endometrium, it begins to secrete human chorionic gonadotropin (a peptide hormone), which keeps the ovary secreting progesterone until later in the pregnancy when the placenta begins to play a larger role.

Oral contraceptives work by fooling the body into thinking that it is pregnant and thereby blocking or delaying ovulation. This is the only scientifically established mechanism of action of oral contraceptives; such methods do not work if an egg has been fertilized or if the woman is already pregnant. The drugs act by delivering a synthetic progesterone analog that blocks ovulation. The analog is needed because progesterone itself is poorly absorbed orally. Many oral contraceptives also contain estrogens, added to reduce adverse effects of the progesterone, such as breakthrough bleeding.

Progesterone

Norethindrone

Levonorgestrel

Mifepristone

The development of oral contraceptives is a fascinating example of how biochemistry, and science in general, can influence history. It had been proposed in the early part of the twentieth century that steroid hormones might be able to regulate ovulation, and hence fertility, but these molecules were difficult and expensive to obtain from animal sources. The combined skills of chemists, biologists, and botanists led to the discovery of a species of Mexican yams that produced a precursor steroid that can be chemically converted in the lab into progesterone or progesterone analogs. This breakthrough, more than any other, permitted the development of oral contraception.

There are several alternative delivery systems for synthetic progesterone. In addition to the oral route, such contraception can be delivered via transdermal patches, subcutaneous implants, and intrauterine devices.

Recently, new drugs have arrived on the market, advertised as emergency birth control or the "morning after pill." It has been known for many years that taking larger than normal doses of regular birth control pills can block ovulation. The difference in these new formulations is that they are now in a single pill, and they can be purchased from a pharmacy without a clinical consultation. The medical value of this approach and the societal issues accompanying it continue to be debated.

9.1.5 Eicosanoids are potent signaling molecules derived from 20-carbon polyunsaturated fatty acids

The term **eicosanoids** broadly describes molecules that are derived from 20-carbon polyunsaturated fatty acids. These fatty acids are arachidonic acid ($20:4^{\Delta 5,8,11,14}$), eicosapentaenoic acid ($20:5^{\Delta 5,8,11,14,17}$) and dihomogamalinolenic acid ($20:3^{\Delta 11,14,17}$). Eicosanoids such as the prostaglandins and leukotrienes are potent signaling molecules that mediate distinct and diverse functions in the body. Although best known as mediators of pain and inflammation, they also help to regulate blood pressure, sleep–wake cycles, labor, and delivery. Nonsteroidal anti-inflammatory drugs (NSAIDs) block the production of these signaling molecules.

Arachidonic acid   **Prostaglandin E2**   **Leukotriene A4**

Summary

- Fatty acids are common components of many lipids. Their properties vary with their length and number of double bonds.
- Some lipids can be stored as esters. Triacylglycerols are stored fatty acids esterified to glycerol; cholesteryl esters are stored cholesterol esterified to a fatty acid.
- There are two major categories of phospholipids: glycerophospholipids (with a glycerol backbone) and sphingolipids (with a sphingosine backbone). Both are found in membranes and other similarly complex structures.
- Steroids are lipids with a common skeleton of four fused rings. Cholesterol, sex hormones, and bile salts are steroids. The properties and functions of these molecules vary depending on the functional groups attached to the skeleton. Steroids have diverse functions, including acting as a structural component of membranes, acting as signaling molecules, or acting as detergents.
- Eicosanoids are signaling molecules derived from 20-carbon polyunsaturated fatty acids.

Concept Check

1. List the different types of lipids discussed. Provide examples of each type and discuss their properties.
2. Explain how phosphatidylcholine and phosphatidylserine are synthesized.
3. Think about the lipid molecule's structure in a schematic way; for example, a triacylglycerol has three fatty acids esterified to a glycerol backbone.

9.2 Fatty Acid Catabolism

The body's main store of energy is comprised of fatty acids esterified to glycerol as triacylglycerols. The highly reduced state and hydrophobic nature of triacylglycerols make them an ideal stored energy source. In contrast to carbohydrates, fats do not require water for hydration because they are stored in an anhydrous state. Compared to proteins and carbohydrates, the fats have the highest amount of energy per gram—9 Calories compared to the 4.5 Calories for proteins and carbohydrates.

The breakdown of fatty acids is known as **beta (β)-oxidation** because it occurs at the second carbon from the carboxyl functional group, that is, at the *beta* carbon or C-3. Beta-oxidation occurs in nearly every cell of the body; in some instances, for example, in the heart, it provides most of the energy requirements of the tissue. The oxidation occurs in the mitochondria. The process of moving fatty acids into the mitochondria requires them to be energetically activated. Activation for any molecule is achieved by coupling it to another molecule, which results in a product with different properties than the original. For example, in carbohydrate metabolism, glucose molecules need to be phosphorylated in the activation stage of glycolysis, before cleavage and energy return occur in subsequent stages. In the case of fatty acids, activation occurs through coupling to coenzyme A in the cytosol to form acyl-CoA. Once the acyl-CoA is formed, the reactions of β-oxidation are thermodynamically spontaneous, and no further energetic investment is required.

The energy to drive the formation of acyl-CoA comes from the hydrolysis of ATP into AMP and pyrophosphate (PPi), and the subsequent hydrolysis of PPi into two molecules of inorganic phosphate, P_i. The combined large negative $\Delta G°$ value of ATP and PPi hydrolysis is necessary to form the thioester linkage and makes this reaction essentially irreversible.

9.2.1 Fatty acids must be transported into the mitochondrial matrix before catabolism can proceed

Long-chain acyl-CoA (more than 16 carbons) are unable to penetrate the inner mitochondrial membrane. Instead, fatty acids are shuttled across the inner mitochondrial membrane as acylcarnitines. The fatty acid is transferred from CoA to carnitine (4-trimethylamino-3-hydroxybutarate). Carnitine palmitoyl transferase I (CPT I) catalyzes the transesterification from the thioester of the CoA to the ester of carnitine (**Figure 9.11**). Although CPT I catalyzes the transesterification of all long-chain acyl-CoA, the enzyme was first identified as catalyzing the transesterification of palmitoyl-CoA, and the name now represents the transesterification of all long-chain acyl-CoA. The formation of acylcarnitine by CPT I is thought to be the rate-determining step in β-oxidation. The transport of acylcarnitine into the mitochondria is mediated by carnitine carrier protein, also known as carnitine:acylcarnitine translocase, a membrane protein that exchanges free carnitine in the mitochondrial matrix for acylcarnitine in the intermembrane space. Once inside the mitochondrial matrix, a second transferase (CPT II) completes the reverse reaction to produce free carnitine and regenerate the acyl-CoA.

9.2.2 Fatty acids are oxidized to acetyl-CoA by β-oxidation

Each cycle of β-oxidation shortens a fatty acid by two carbon atoms. In each instance, a molecule of acetyl-CoA is removed from the fatty acid. Again, the group being oxidized is the β carbon. The process occurs in four steps:

1. **Formation of a double bond between the α and β carbons.** In the first step, a *trans*-Δ-2 double bond is incorporated between the α and β carbons in the fatty acyl-CoA. This is an oxidation reaction; the enzyme involved is acyl-CoA dehydrogenase, which uses FAD as a cofactor. FAD is often employed as a cofactor when a reaction incorporates a double bond, as happens in the citric acid cycle when succinate is oxidized to fumarate.

2. **Hydration of the double bond to an alcohol.** The second step of the pathway is a hydration reaction, in which enoyl-CoA hydratase adds a molecule of water in a hydration reaction to form an alcohol on the β carbon, yielding 3-L-hydroxyacyl-CoA. The addition is stereospecific, and only the l isomer is formed.

3. **Oxidation of the alcohol to a ketone.** The third step is another oxidation reaction, as 3-L-hydroxyacyl-CoA is converted to β-ketoacyl-CoA by 3-L-hydroxyacyl-CoA dehydrogenase, an NAD^+-dependent enzyme. A molecule of $NADH/H^+$ is generated at this step.

4. **Cleavage of the carbon–carbon bond and formation of a new acyl-CoA linkage.** In the final step, a molecule of acetyl-CoA is cleaved from the β-ketoacyl-CoA by the enzyme β-ketothiolase. Although similar to other thiolases encountered in this chapter, this enzyme is specific for degradative pathways and preferentially uses the β-keto substrate. If the resulting acyl-CoA is four carbons or more in length, the molecule will undergo repeated cycles of β-oxidation until it has been oxidized into acetyl-CoA units. The final cycle of β-oxidation yields two molecules of acetyl-CoA.

In summary, the oxidation of a fatty acid proceeds in a stepwise fashion of oxidation, hydration, oxidation, and thiolysis that is similar to the citric acid cycle. If we examine the final three steps of the citric acid cycle in detail (see section 7.1), we see that the chemistry is nearly the same as it is in β-oxidation. A saturated carbon–carbon bond is desaturated to an alkene, the alkene is hydrated to an alcohol, and the alcohol is oxidized to a ketone. Understanding how this stepwise oxidation occurs on the chemical level will help to clarify the overall pathway and the intermediates, enzymes, mechanisms, and energetics of the reactions. By repeating the four steps of β-oxidation multiple times, lengthy fatty acids can be broken into two-carbon acetyl-CoA units (**Figure 9.12**).

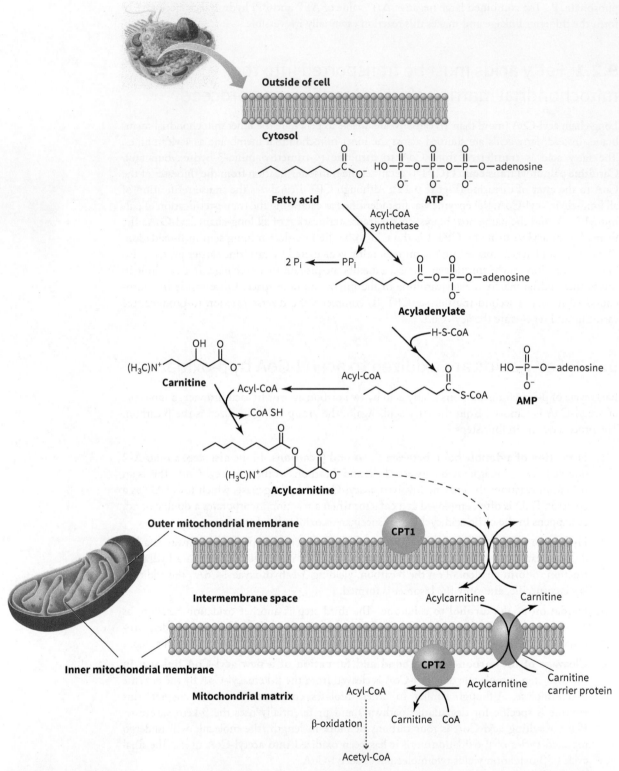

FIGURE 9.11 Activation and delivery of fatty acids to the mitochondria. Transfer of fatty acids into the mitochondrial matrix is a multistep process. First, fatty acids must be energetically activated by ATP and coupled to coenzyme A. Acyl-CoA is formed in a two-step process. A mixed anhydride is formed between the fatty acid and ATP. The energetic equivalent of two molecules of ATP (two phosphoanhydride bonds) are consumed in this process. Second, coenzyme A attacks the fatty acyl side of the anhydride to yield acyl-CoA and AMP. Energetic activation of molecules by ATP is also seen in glycolysis, glycogenesis, and protein synthesis. Following activation, the fatty acyl chains are transesterified to carnitine by carnitine palmitoyl transferase I (CPTI). The carnitine carrier protein can then shuttle these fatty acids into the mitochondrial matrix where carnitine palmitoyl transferase II (CPTII) can catalyze the reverse reaction (formation of a new acyl-CoA inside the mitochondrial matrix). Free carnitine is shuttled back through the membrane by carnitine carrier protein.

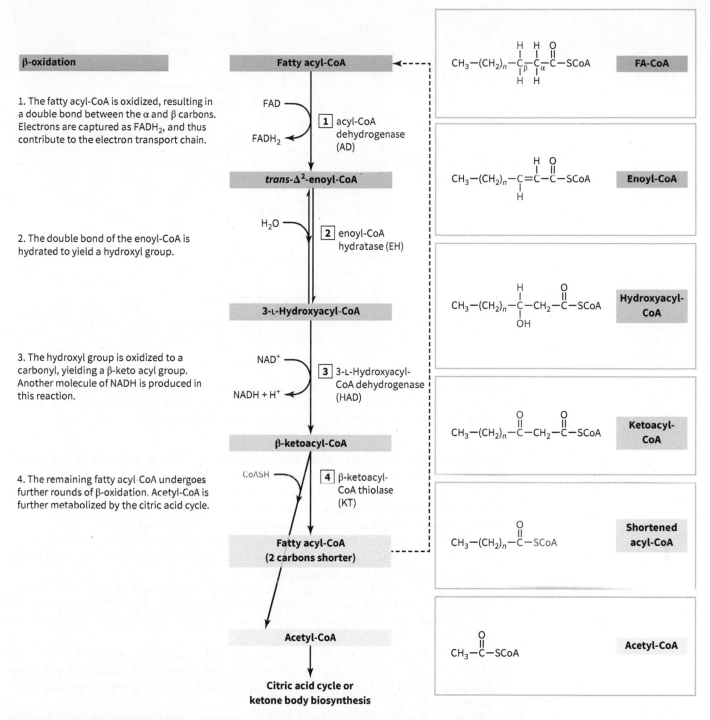

β-oxidation

1. The fatty acyl-CoA is oxidized, resulting in a double bond between the α and β carbons. Electrons are captured as $FADH_2$, and thus contribute to the electron transport chain.

2. The double bond of the enoyl-CoA is hydrated to yield a hydroxyl group.

3. The hydroxyl group is oxidized to a carbonyl, yielding a β-keto acyl group. Another molecule of NADH is produced in this reaction.

4. The remaining fatty acyl-CoA undergoes further rounds of β-oxidation. Acetyl-CoA is further metabolized by the citric acid cycle.

Fatty acyl-CoA

FAD
FADH$_2$
1 acyl-CoA dehydrogenase (AD)

trans-Δ²-enoyl-CoA

H$_2$O
2 enoyl-CoA hydratase (EH)

3-L-Hydroxyacyl-CoA

NAD$^+$
NADH + H$^+$
3 3-L-Hydroxyacyl-CoA dehydrogenase (HAD)

β-ketoacyl-CoA

CoASH
4 β-ketoacyl-CoA thiolase (KT)

Fatty acyl-CoA (2 carbons shorter)

Acetyl-CoA

Citric acid cycle or ketone body biosynthesis

FIGURE 9.12 Beta-oxidation of fatty acids. In β-oxidation, the β carbon is oxidized in a stepwise fashion to form acetyl-CoA. The remaining acyl-CoA is shortened by two carbons on each cycle of β-oxidation. On each pass, one molecule each of $FADH_2$ and NADH/H$^+$ is generated. These electron carriers pass electrons to the electron transport chain and are reduced to be used again.

The metabolic purpose of β-oxidation goes beyond the generation of acetyl-CoA; in two of the steps, the oxidation also leads to the controlled harvesting of electrons (as $FADH_2$ in the generation of the double bond, and as NADH/H$^+$ in the oxidation of the hydroxyl to a ketone). These electrons are used in oxidative phosphorylation to generate a proton gradient across the inner mitochondrial membrane, ultimately resulting in the generation of ATP, as seen in section 7.2.

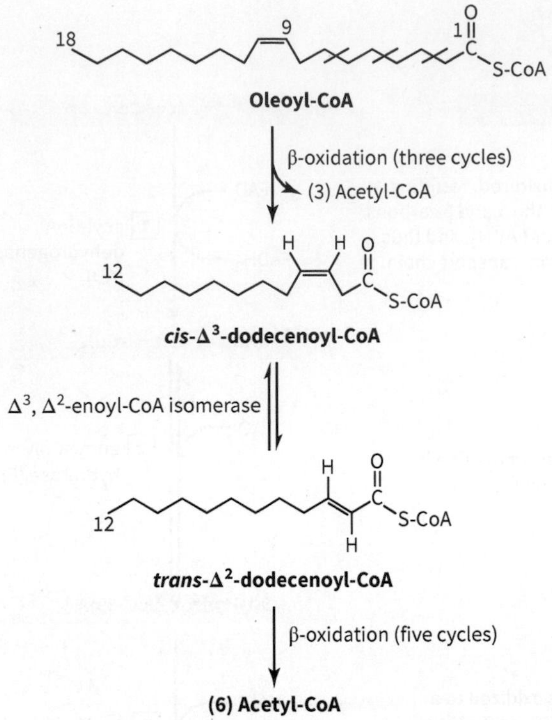

FIGURE 9.13 Oxidation of unsaturated fatty acids from the Δ-3 position. The initial steps of β-oxidation proceed without difficulty until the unsaturation is at the Δ-3 position. The bond can be isomerized to a *trans*-Δ-2 double bond, and β-oxidation proceeds as usual.

Double bonds in unsaturated fatty acids need to be moved or removed before oxidation

Several problems arise when trying to metabolize unsaturated fatty acids (**Figure 9.13**). If the fatty acids are naturally occurring, then the bonds are almost always *cis* double bonds. Often, these originate on an odd-numbered carbon, most commonly in the C-9 position, although sometimes in the C-5 position as well. In this instance, β-oxidation proceeds through several rounds until a β–γ double bond exists (a *cis*-Δ-3 double bond, between the C-3 and C-4 positions). This *cis*-Δ-3 double bond is isomerized to a *trans*-Δ-2 double bond by 3,2 enoyl-CoA isomerase, and β-oxidation proceeds.

If a double bond is found between even-numbered carbons, the first step of β-oxidation, formation of a *trans*-Δ-2 double bond, results in a conjugated system (**Figure 9.14**). In mammals, the Δ-4 double bond is reduced, resulting in a *trans*-Δ-3 double bond, which is isomerized back to *trans*-Δ-2 and, again, β-oxidation proceeds.

Occasionally, the 3,2 enoyl-CoA isomerase will sporadically generate a Δ-3, Δ-5 conjugated pair of double bonds. This scenario is rectified in the same way that a Δ-4 double bond is; that is, the pair of double bonds is reduced to a single Δ-3 double bond, followed by isomerization to Δ-2.

To review, if there is a double bond beginning at an odd-numbered carbon (Δ-3) it can be migrated to the Δ-2 position by an isomerase. However, if there is a Δ-4 double bond, β-oxidation begins and the two double bonds are rearranged to a single *trans*-Δ-3 double bond, which is then isomerized to Δ-2.

Odd-numbered fatty acids result in propionate

A small percentage of the fatty acids found in nature have an odd number of carbons. Such fatty acids are most commonly found in bacteria and ruminant tissues. They are degraded through β-oxidation until a three-carbon fatty acyl group (propionyl-CoA) is formed. Propionyl-CoA is then carboxylated to form D-methylmalonyl-CoA (the enantiomer of the L form found in fatty acid biosynthesis). The D isomer is converted to the L form by methylmalonyl-CoA epimerase. In a rearrangement reaction, methylmalonyl-CoA mutase converts the L isomer to succinyl-CoA, which enters the citric acid cycle.

9.2.3 Beta-oxidation is regulated at two different levels

Beta-oxidation is primarily regulated through the availability of substrate. Elevated levels of free fatty acids in the cellular environment (such as the extracellular fluid) result in increased uptake of free fatty acids into cells. Likewise, hydrolysis of a cell's triacylglycerol stores increases levels of free fatty acids in the cytosol. The increased fatty acid concentrations from both extracellular and intracellular sources lead, in turn, to elevated levels of cytosolic acyl-CoA. Thus, the pathways that dictate the breakdown of triacylglycerols, freeing up fatty acids, and the transport and trafficking of free fatty acids inside or outside the cell, have a major impact on the levels of fatty acid substrates available for β-oxidation. In general, pathways that regulate triacylglycerol breakdown include the hormones insulin or glucagon and two kinases, protein kinase A (PKA) and AMP dependent protein kinase (AMPK) and are discussed in section 10.6.

The other way in which the cell regulates β-oxidation is through the compartmentalization of substrate. Importing fatty acylcarnitines from the cytosol into the mitochondria is a point of control; CPT I is the protein responsible for transport across the outer mitochondrial membrane. CPT I is inhibited by malonyl-CoA, a metabolite produced in the cytosol that is used in the *biosynthesis* of fatty acids. High levels of malonyl-CoA indicate that the cell is also producing metabolites for fatty acid biosynthesis. Therefore, the cell is in an energetically rich state and does not need to burn fatty acids in β-oxidation. Inhibition of CPT I by malonyl-CoA blocks fatty acids from entering the mitochondria and therefore blocks β-oxidation. This is another example of control being exerted on a rate-determining and committed step.

The energetics of fatty acid oxidation   Fatty acid oxidation releases significant stores of energy. Let us examine the energetics of the oxidation of a single molecule of the 18-carbon fatty acid stearate (18:0). The activation of stearate costs a cell the energetic equivalent of two molecules of ATP since two phosphoanhydride bonds are cleaved. Stearate then goes through eight rounds of β-oxidation, with the final round yielding two molecules of acetyl-CoA. Each round of β-oxidation yields one molecule of NADH and one of $FADH_2$. Collectively, therefore, the oxidation of stearate has produced:

- eight molecules of NADH,
- eight molecules of $FADH_2$, and
- nine molecules of acetyl-CoA.

NADH and $FADH_2$ are oxidized to NAD^+ and FAD in the electron transport chain. As we have seen in section 7.3, the energetic yield for the oxidation of NADH and $FADH_2$ is approximately 2.5 ATP/NADH and 1.5 ATP/$FADH_2$. The acetyl-CoA will be further oxidized in the citric acid cycle, with each molecule of acetyl-CoA yielding the energetic equivalent of the hydrolysis of ten molecules of ATP. Collectively, this means that the catabolism of a single molecule of stearate will provide the energetic equivalent of the hydrolysis of 122 molecules of ATP (**Table 9.2**).

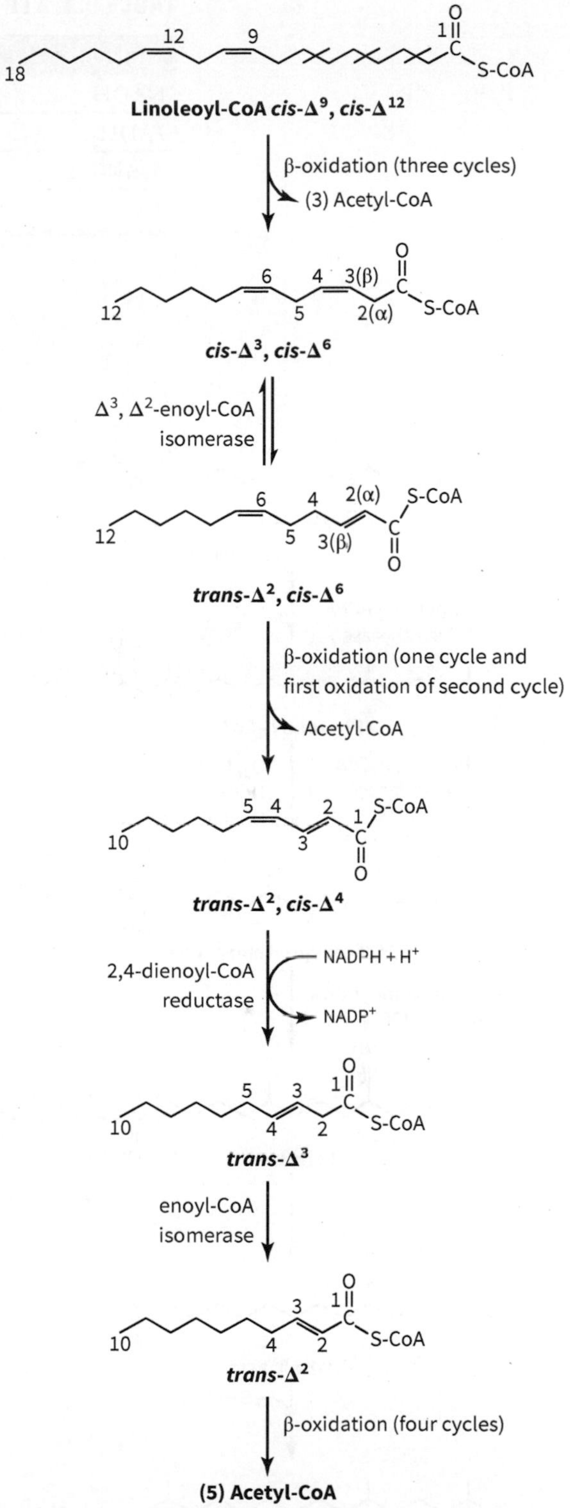

9.2.4 Other modes of fatty acid oxidation

beta (β)-oxidation is a major pathway for the oxidation of fatty acids, alpha (α)-oxidation and omega (ω)-oxidation are two other types that occur to metabolize fatty acids.

FIGURE 9.14 Oxidation of polyunsaturated fatty acids from the Δ-4 position.   If a Δ-4 double bond is encountered, β-oxidation begins, resulting in a conjugated Δ-2, Δ-4 system. This is reduced to a single Δ-3 double bond, which can then be isomerized to Δ-2 as previously seen, and oxidation proceeds.

TABLE 9.2 ATP Produced by the Oxidation of One Molecule of Stearate

Molecule	Number per Stearate	ATP per Molecule	Total ATP
NADH	8	2.5	20
FADH$_2$	8	1.5	12
Acetyl-CoA	9	10	90
Total			122

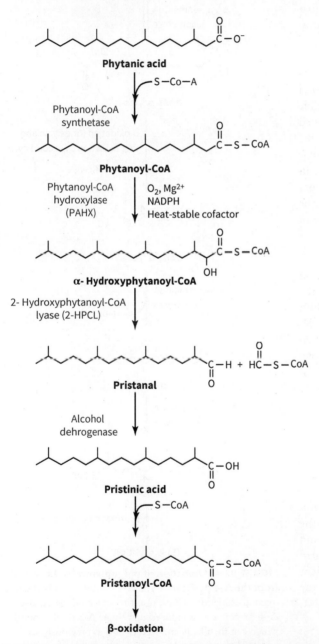

FIGURE 9.15 Alpha (α)-oxidation of fatty acid Alpha-oxidation occurs in those fatty acids that have a methyl group (−CH$_3$) at the β-carbon, which blocks β-oxidation. The carbon unit is removed in the form of CO$_2$ without any production of ATP.

Alpha (α)-oxidation of fatty acids

The α-oxidation sequence of 3-methyl-branched fatty acids starts with activation to the corresponding CoA-ester. Subsequently, this acyl-CoA-ester undergoes a 2-hydroxylation by the peroxisomal phytanoyl-CoA hydroxylase (PAHX). The initial hydroxylation reaction is catalyzed by a mitochondrial enzyme monooxygenase that requires O$_2$, Mg^{2+}, NADPH, and a heat-stable cofactor. Conversion of hydroxy fatty acid to CO$_2$ and the next lower unsubstituted acid appears to occur in the endoplasmic reticulum and requires O$_2$, Fe^{2+}, and ascorbate. In the third step, the peroxisomal 2-hydroxyphytanoyl-CoA lyase (2-HPCL) splits the carbon-carbon bond of the 2-hydroxy-intermediate into a 2-methyl ($n-1$) aldehyde and formyl-CoA, which is subsequently converted to formate and CO$_2$. Finally, the aldehyde is dehydrogenated by aldehyde dehydrogenase to the corresponding acid, which, after its conversion to the acyl-CoA ester, could be available as a substrate for β-oxidation (**Figure 9.15**). The important significance of α-oxidation is the catabolism of branched-chain fatty acids and oxidation of methylated fatty acids. The reaction is also a route for the synthesis of hydroxy fatty acids. The α-hydroxy fatty acid can be further oxidized and decarboxylated to a fatty acid one carbon shorter than the original.

Omega (ω)-oxidation of fatty acids

The ω-oxidation occurs in both animal and plant bacterial systems. It is mainly responsible for the oxidation of alkanes and is an alternative pathway to β-oxidation, that instead of involving β-carbon, involves the oxidation of the ω-carbon (the carbon most distant from the carboxyl group of fatty acid). This type of oxidation is important when the carboxyl end is unavailable or for the formation of ω-hydroxy fatty acids. Enzymes for ω-oxidation are located in the smooth endoplasmic reticulum of liver and kidney cells rather than in the mitochondria as with β-oxidation.

The mechanism entails an initial hydroxylation of the terminal methyl group to a primary alcohol. In animal's cytochrome P$_{450}$ system, hydroxylase is responsible for this alkane hydroxylation. While in bacteria, rubridoxin is the intermediate electron carrier that feeds electrons to ω hydroxylase system. The immediate product, RCH$_2$OH is oxidized to an aldehyde by alcohol dehydrogenase, which

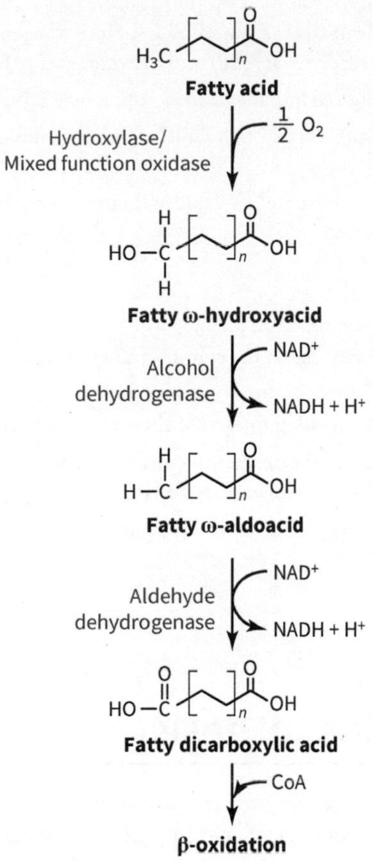

FIGURE 9.16 Omega (ω)-oxidation of fatty acid   Omega-oxidation serves as a subsidiary pathway for β-oxidation of fatty acids when its blocked. It is observed that ω- and (ω −1)-oxidation of fatty acids are related to energy metabolism.

in turn is oxidized to a carboxylic acid by aldehyde dehydrogenase enzyme producing a fatty acid with a carboxyl group at each end of the system. At this point, either end can be attached to coenzyme A, or the molecule can enter the mitochondrion and undergo β oxidation by the normal route. The methyl group is converted to a hydroxymethyl group which subsequently is oxidized to the carboxyl group, thus forming dicarboxylic acid. Once formed, the dicarboxylic acid may be shortened from either end of the molecule, by the β-oxidation sequence, to form acetyl-CoA (**Figure 9.16**).

Worked Problem 9.3   Following carbons in β-oxidation

If a molecule of palmitate is radiolabeled with C^{14} on the C-9 and C-10 positions, will the C^{14} be released in one or two molecules of acetyl-CoA?

Strategy   Think about which carbons are oxidized in β-oxidation. Beta-oxidation takes place on the β carbon and effectively removes carbons from the fatty acid as two carbon fragments.

Solution   Examine how palmitate proceeds through β-oxidation; the fatty acid is shortened by two carbons on each round of oxidation.

On the fifth round, the acetyl-CoA fragment containing C-9 and C-10 will be liberated. Thus, the C^{14} will come off in a single molecule of acetyl-CoA.

Follow-up question   If this same molecule of palmitate was completely oxidized through β-oxidation, how many ATP could be produced from the acetyl-CoA, NADH, and $FADH_2$ generated in these steps?

Summary

- Before catabolism, fatty acids must be activated by coupling to coenzyme A. The activation both requires energy (supplied by hydrolysis of ATP) and provides energy for subsequent steps.
- Activated fatty acids must be shuttled into the mitochondrial matrix (as acylcarnitines) for catabolism.
- Beta-oxidation is the stepwise process through which fatty acids are oxidized on the β carbon into two-carbon acetyl-CoA units.
- Beta-oxidation is regulated by substrate availability, compartmentalization within the cell, and inhibition of key steps in the pathway.

Concept Check

1. Illustrate how fatty acids are activated in the cytosol, transported to the mitochondrial matrix, and catabolized through β-oxidation.

2. Describe chemically what is happening in each of these steps and why these steps are necessary.

3. Analyze the thermodynamics of the reactions (which reactions help drive the entire pathway forward or have ΔG values close to zero and are reversible).

4. Describe how β-oxidation is regulated at the molecular level. What are some other modes of oxidation?

9.3 Fatty Acid Biosynthesis

We may associate fats and therefore fatty acids with adipose tissue. While the majority of the body's fats are stored in adipose tissue, fatty acid biosynthesis occurs most frequently in the liver or other tissues.

In this section, we first outline the process of fatty acid biosynthesis, using the example of palmitate (16:0), and then we show how longer or unsaturated fatty acids form from palmitate. The enzymes involved in the synthesis of palmitate are typically found in the cytosol; however, the enzymes involved in the elongation and desaturation of fatty acids are commonly found in the endoplasmic reticulum. Carbohydrate metabolism employs separate compartments (the cytosol and the mitochondria) for some of the steps of glycolysis and gluconeogenesis. This is done to ensure that the steps of these pathways are energetically favorable and to provide an additional means of regulating these pathways. Fatty acid metabolism employs a similar mechanism. While β-oxidation occurs in the mitochondria, fatty acid biosynthesis occurs in the cytosol.

Fatty acid biosynthesis requires an input of energy, which is provided by the reducing equivalent NADPH and the transfer potential of the thioester bonds found in the starting material, malonyl-ACP. ACP is acyl carrier protein, a small protein with a phosphopantetheine group. ACP in fatty acid biosynthesis acts in an analogous fashion to coenzyme A in fatty acid degradation and the citric acid cycle. Although fatty acids are built up from two-carbon subunits, and the reactions are similar, the process is not a simple reversal of β-oxidation.

Much of the early research on fatty acid biosynthesis was conducted in the bacterium *E. coli*, which differs from the mammalian enzymes and pathways described in this section. The differences are discussed as they arise in the text. Although *E. coli* is used as a model organism, different bacterial species may have other variations in fatty acid metabolism. Both the differences and similarities help illustrate how fatty acid metabolism continues to evolve.

9.3.1 Two major enzyme complexes are involved in fatty acid biosynthesis

Two, major, multifunctional enzyme complexes are responsible for the reactions of palmitate biosynthesis: **acetyl-CoA carboxylase (ACC)** and **fatty acid synthase (FAS)**. ACC synthesizes malonyl-CoA, a three-carbon building block that is the immediate precursor for fatty acid biosynthesis and a substrate for FAS. Regulation of fatty acid biosynthesis occurs through the regulation of both enzymes, but the rate-determining step is catalyzed by ACC.

Acetyl-CoA carboxylase synthesizes malonyl-CoA

ACC is a large, multifunctional enzyme complex that has three main domains: biotin carboxylase (BC), carboxyl transferase (CT), and the biotin binding (BB) domain (**Figure 9.17**). In bacteria such as *Escherichia coli*, the domains and any substrate binding sites are coded for by different genes, which give rise to different proteins that function in a multi-protein complex. In mammals, two different genes give rise to two different isoforms of ACC, each of which has all the binding sites and catalytic properties of the enzyme. The two isoforms, ACC1 and ACC2, catalyze similar reactions but have very different biological functions. ACC1 is found in all cells but particularly in lipogenic tissues, such as liver and adipose tissue; it is believed to be the isoform responsible for generating malonyl-CoA for fatty acid biosynthesis. This enzyme is associated with the cytoplasmic surface of the endoplasmic reticulum and releases its product (malonyl-CoA) into the cytosol. In contrast,

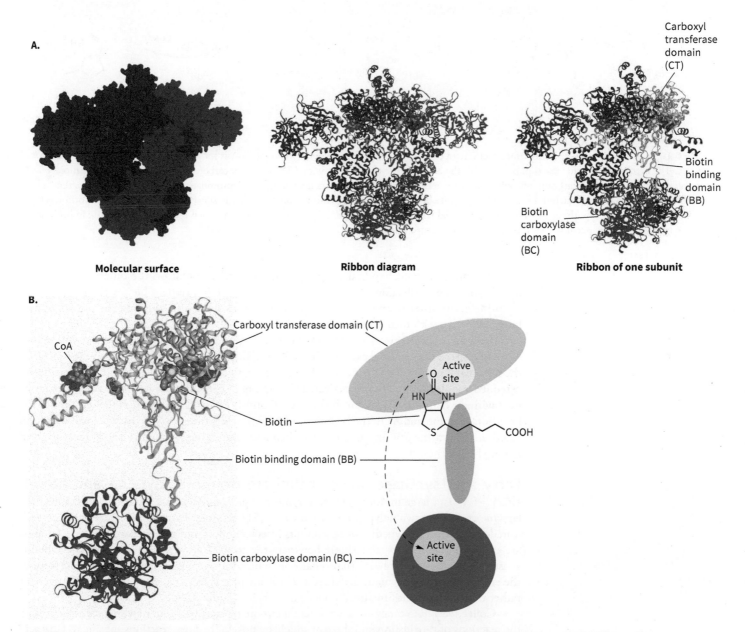

FIGURE 9.17 Structure of the biotin carboxylase domain of human acetyl-CoA carboxylase I from yeast.   **A.** ACC1 is a dimer (shown as a molecular surface and as a ribbon diagram in violet and purple). In the third structure, only one subunit is shown and the three key domains are colored blue (biotin caroxylase domain, BC), green (biotin binding domain, BB), and gold (carboxyl transferase domain, CT). **B.** The same three domains are shown in greater detail. The flexibility of the biotin-lysine arm (shown bound in the CT active site) allows this group to translocate between the CT and BC active sites (shown in pale blue and pale gold in the schematic view). ACC2 has a different structure and is not shown.

(Source: (A, B) Data from PDB ID 5CSL Wei, J., Tong, L. (2015) Crystal structure of the 500-kDa yeast acetyl-CoA carboxylase holoenzyme dimer. *Nature* 526: 723–727)

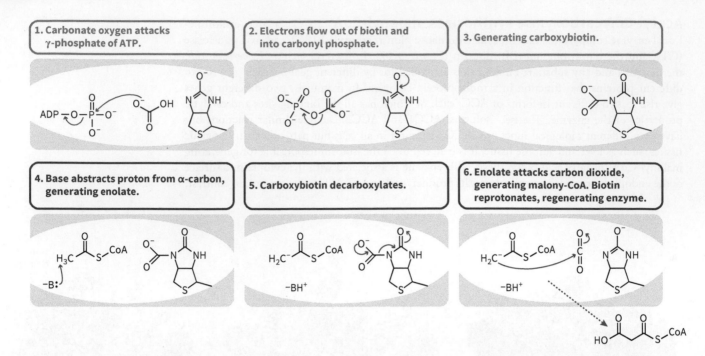

FIGURE 9.18 The overall reaction and mechanism of malonyl-CoA formation. Malonyl-CoA is formed by the carboxylation of acetyl-CoA. The mechanism of this reaction begins with the activation of bicarbonate with ATP to form a carbonyl phosphate. The amine moiety of biotin then attacks the carbonyl phosphate forming a carbamate. An enolate generated by the deprotonation of acetyl-CoA attacks the carbamate carbon to form malonyl-CoA and regenerate biotin. The lengthy combination of the lysine side chain and the arm on biotin has led to the swinging arm hypothesis in which the combined length of these two units provides a means by which biotin can participate in both of these reactions as they occur at two distantly separated active sites.

ACC2 contains a mitochondrial targeting sequence and is found on the outer mitochondrial membrane. The production of mitochondrial malonyl-CoA is responsible for the inhibition of carnitine acyltransferase and is therefore more likely involved with blunting the β-oxidation of fatty acids than with their synthesis.

Both ACC isoforms synthesize malonyl-CoA through a multistep process, with biotin (a B vitamin) as the activated carrier (**Figure 9.18**). The biotin is attached via an amide linkage to the terminal ε-nitrogen of a lysine side chain. The length of the lysine–biotin complex provides an extended and flexible arm for the activated CO_2 group to be transferred from one active site to the next without being released from the protein. Biotin is carboxylated through the reaction of the bicarbonate anion and ATP. The hydrolysis of ATP provides the energy to drive the carboxylation reaction. The carboxybiotinylated enzyme then adds the carboxyl group to acetyl-CoA to form malonyl-CoA.

Fatty acid synthase makes palmitate from malonyl-CoA and acetyl-CoA
FAS is a large multifunctional enzyme complex. The homodimeric human form of the enzyme consists of two polypeptides, each with 2,511 residues. Extensive research has been dedicated to determining whether these subunits interact and, if so, how. In mammals, the subunits are arranged in a head-to-tail fashion; however, it is not clear how strongly, if at all, the subunits assist one another in catalysis. The scenario may be completely different for the *E. coli* form of the enzyme because enzymatic activities in *E. coli* are provided by six, discrete individual proteins rather than a long multifunctional polypeptide chain (**Figure 9.19**).

Collectively, FAS catalyzes seven different enzymatic reactions, four of which are analogous to the reactions of β-oxidation and three of which are novel. The three reactions specific to fatty acid biosynthesis, described below, occur in the first steps of the pathway. Like glycolysis and gluconeogenesis, biology exploits reactions with $\Delta G°$ values that are close to zero, such as the hydration of a double bond to form a hydroxyl group. In glycolysis those reactions can be viewed as nearly reversible, and the value of ΔG can be influenced by the concentration of products and reactants in the reaction. In fatty acid synthase, substrate channeling (passing of substrate from one subunit

A.

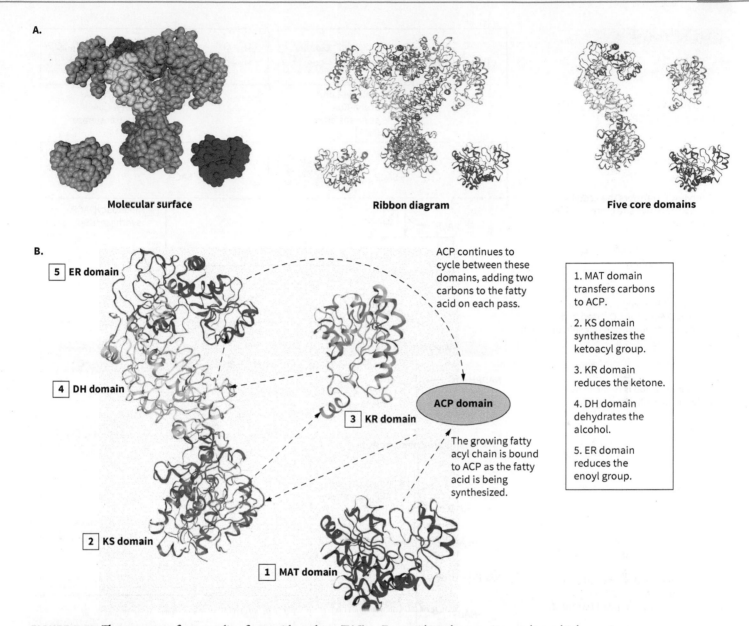

Molecular surface   Ribbon diagram   Five core domains

B.

5 ER domain

ACP continues to
cycle between these
domains, adding two
carbons to the fatty
acid on each pass.

4 DH domain

3 KR domain

ACP domain

The growing fatty
acyl chain is bound
to ACP as the fatty
acid is being
synthesized.

2 KS domain

1 MAT domain

1. MAT domain
transfers carbons
to ACP.

2. KS domain
synthesizes the
ketoacyl group.

3. KR domain
reduces the ketone.

4. DH domain
dehydrates the
alcohol.

5. ER domain
reduces the
enoyl group.

FIGURE 9.19 The structure of mammalian fatty acid synthase (FAS).   Fatty acid synthase carries out the multiple reactions necessary to synthesize fatty acids. In prokaryotes, these are performed by a multi-subunit protein. **A.** In mammals, these are found in a single polypeptide. The holoenzyme is a homodimer with each half containing all of the necessary subunits for fatty acid synthesis. Shown in **B.** are five of the essential domains (MAT, KS, KR, DH, and ER). The growing fatty acid, attached to the ACP domain, flips between these sites as the nascent fatty acid grows. When it achieves 16 carbons in length (palmitate), it is cleaved by ketothiolase (KT, not shown).

(Source: (A, B) Data from PDB ID 2CF2 Maier, T., Jenni, S., Ban, N. (2006) Architecture of Mammalian Fatty Acid Synthase at 4.5 A Resolution. *Science* 311: 1258)

to the next) artificially raises the effective concentration of the substrate and increases the rate and efficiency of the entire process (**Figure 9.20**).

1. **Transfer of activated carbon units to acyl carrier protein.** To start synthesizing fatty acids, malonyl-CoA and acetyl-CoA must be transferred to fatty acid synthase; one to the condensing domain via a thioether to a cysteine residue, and the other to the ACP domain through the phosphopantetheinyl group. In the first reaction of fatty acid synthesis, acetyl-CoA is transferred from the thioester of CoA to the ACP domain of FAS. The name ACP is something of a misnomer. In bacteria such as *E. coli*, ACP is a separate protein, but the ACP functionality is only one domain of the FAS complex in birds and mammals. The transfer of this group to the thioester of ACP is catalyzed by malonyl-CoA/acetyl-CoA acyltransferase (MAT).

Fatty acid biosynthesis

1. Acetyl-CoA is transferred to acyl carrier protein (ACP).

2. Acetyl-ACP transfers the acetyl group to the active site cysteine of the KS domain.

3. Malonyl-CoA is transferred to acyl carrier protein (ACP).

4. Malonyl-ACP and acetyl-S-KS are condensed into a single molecule of acetoacetyl-ACP. The CO_2 used in the formation of malonyl-CoA is lost in this step.

5. The β-keto group is reduced by NADPH/H⁺ to a β-hydroxyl group.

6. The alcohol is dehydrated leaving an α-β double bond.

7. The double bond is reduced by a second molecule of NADPH/H⁺ leaving the newly extended fatty acid.

8. Steps 3 through 7 are repeated six more times to generate a palmitoyl-ACP.

9. Palmitoyl-ACP is cleaved by a thioesterase to release palmitate and ACP.

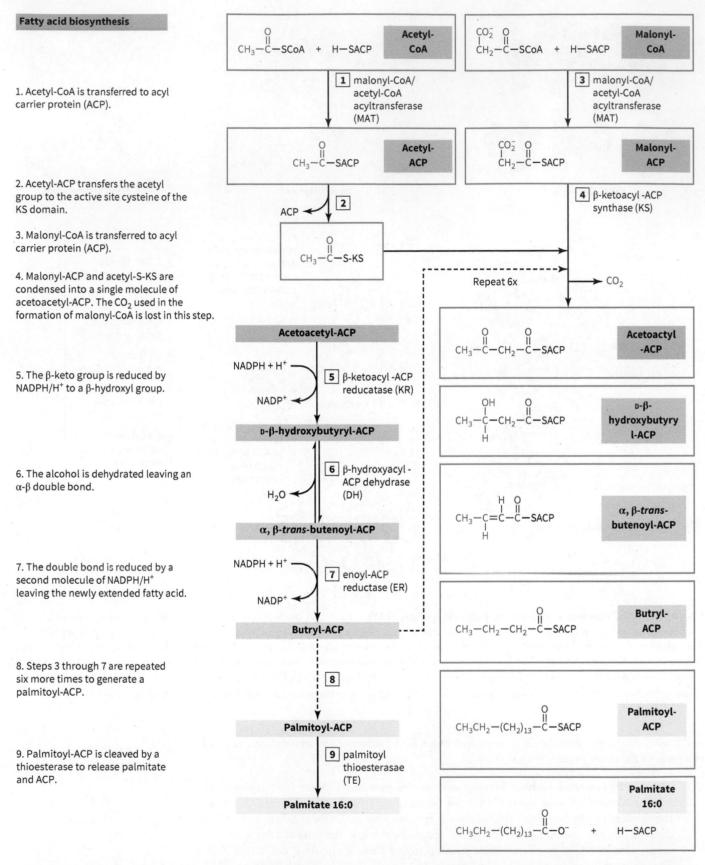

FIGURE 9.20 Reactions of fatty acid synthase. The reactions of fatty acid synthase are largely the reverse of β-oxidation with several notable exceptions. Acyl carrier protein is the handle used to manipulate the intermediates and NADPH is the reducing equivalent. Two carbon units are added through malonyl-ACP (carbon dioxide is lost to form the enolate required in this step).

2. **Transfer of acetyl group to the KS domain of FAS.** The acetyl group does not stay bound to ACP for long. Instead, it is transferred to the catalytic domain of the β-ketoacyl-ACP synthase (KS) domain. In the remaining steps of this round of fatty acid biosynthesis, the growing fatty acyl chain stays bound to ACP, but the growing chain appears to be transferred to the KS domain.

3. **Transfer of malonyl group from malonyl-CoA.** The active donor of carbon units for fatty acid biosynthesis is malonyl-CoA made by ACC1. The first step in using this is the transfer of the thioester bond from CoA to the cysteine of ACP.

4. **Condensation of malonyl- and acetyl-ACP.** These two activated carbon carriers are then condensed together in a Claisen condensation catalyzed by the β-ketoacyl-ACP synthase (KS) domain to form acetoacetyl-ACP. The CO_2 used to synthesize malonyl-CoA is lost at this step. The resulting acetoacetyl group remains on ACP, which in turn delivers this substrate to the other reaction centers. Therefore, the final two carbons of any fatty acid come from acetyl-CoA while the others are derived from malonyl-CoA.

5. **Reduction of the β-keto group by NADPH/H⁺ to a hydroxyl group.** The growing fatty acid has been elongated by two carbons, but it exists as a diketone instead of a hydrocarbon. It is reduced by a series of stepwise reactions. The first of these is the reduction of the ketone on the β carbon to a hydroxyl group catalyzed by the β-keto-ACP reductase (KR) domain. This enzyme requires NADPH/H⁺ to reduce the carbonyl of the β-keto group to a hydroxyl group.

6. **Dehydration of the alcohol to an α-β double bond.** The alcohol generated in the previous step is now dehydrated to a double bond by 3-hydroxyl acyl-ACP dehydratase (DH). This step requires no added input of energy and has a ΔG value close to zero.

7. **Reduction of the double bond by NADPH/H⁺.** The final step in the addition of the acetyl group is the reduction of the double bond to a saturated hydrocarbon chain by NADPH/H⁺. This is carried out by the enoyl-ACP reductase (ER) domain. This step finalizes the elongation of the growing acyl chain by two carbons.

8. **Steps 3 through 7 are repeated six more times to generate palmitoyl-ACP.** Six more two-carbon units are added to the growing fatty acyl chain as malonyl-ACP groups. Each undergoes all of the reactions of fatty acid synthesis to generate the 16-carbon fatty acid palmitate.

9. **Palmitoyl-ACP is cleaved by a thioesterase to release palmitate and ACP.** In the final reaction, palmitoyl thioesterase cleaves palmitate from ACP releasing a free fatty acid (palmitate) and acyl carrier protein. Palmitate may be acted on by elongases and desaturases to yield longer fatty acids or ones that contain double bonds.

While the steps of fatty acid biosynthesis employ reactions that are almost the reverse order of β-oxidation reactions, there are several significant differences (**Figure 9.21**).

First, the reducing equivalent used in steps 4 and 6 is NADPH/H⁺ rather than NADH and $FADH_2$. NADPH differs from NADH in that an additional phosphate has been esterified to the 2′ position of the ribose ring attached to the adenine group. Whereas NAD⁺/NADH is typically used in catabolic reactions, NADP/NADPH is often used in anabolic reactions, including fatty acid biosynthesis, cholesterol biosynthesis, and the reactions of the Calvin cycle in photosynthesis. NADPH is also the reducing equivalent used by the cytochromes, enzymes used by the liver to detoxify drugs and xenobiotics, and some of the reactions of bile salts and eicosanoid synthesis.

Second, the carbon units used in fatty acid biosynthesis are bound to ACP rather than CoA. In both ACP and CoA, acetate groups are bound via a thioester bond to the cysteamine end of a phosphopantetheine group (**Figure 9.22**). However, in contrast to CoA (in which the phosphopantetheine group is attached to AMP), the phosphopantetheine in the much larger ACP is linked to a serine residue.

A third difference between β-oxidation and fatty acid biosynthesis is the stereochemistry of the hydroxyl-bearing carbon. The D isomer is seen in biosynthesis (D-β-hydroxyacyl-ACP), whereas the L isomer (L-β-hydroxyacyl-CoA) is seen in β-oxidation.

The sequence of reactions in FAS is repeated six more times. In each round an additional two carbons are added onto the growing fatty acyl chain until the 16-carbon fatty acid palmitate is synthesized and palmitate is released.

Redox active carbon

NADPH

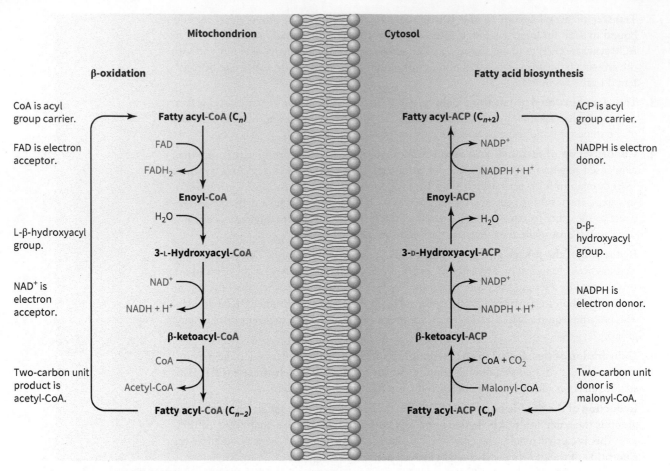

FIGURE 9.21 Comparison of β-oxidation and fatty acid biosynthesis. The five main differences between β-oxidation and fatty acid biosynthesis are shown.

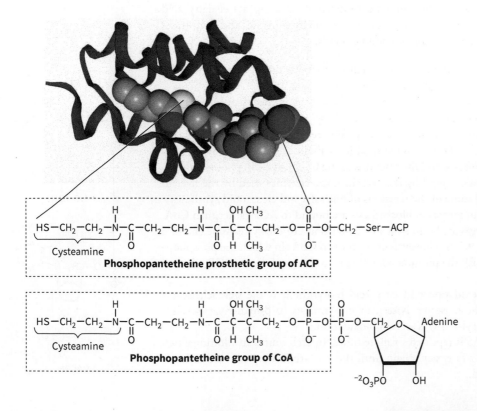

FIGURE 9.22 Structure of the acyl carrier protein (ACP). The phosphopantetheine group (space-filling) shown in the center is covalently bound to a serine residue in the protein. In the image, the phosphopantetheine group is bound to a hexanoyl group in the cavity of ACP. Note that ACP is a small protein. It is much larger than the phosphopantetheine group (and much larger than coenzyme A).

(Source: Data from PDB ID 2FAC Roujeinikova, A., Simon, W.J., Gilroy, J., Rice, D.W., Rafferty, J.B., Slabas, A.R. (2007) Structural Studies of Fatty Acyl-(Acyl Carrier Protein) Thioesters Reveal a Hydrophobic Binding Cavity that Can Expand to Fit Longer Substrates. *J. Mol. Biol.* 365: 135–145)

The reactions of β-oxidation and fatty acid biosynthesis may seem new to you, but in many ways these reactions are ones that you have covered in organic chemistry. The critical step of these reactions is discussed in **Biochemistry: The Claisen condensation**.

Biochemistry

The Claisen condensation

There are remarkably few ways in nature to generate carbon–carbon bonds, but the Claisen condensation is one exception. It is a classic reaction of organic chemistry and at the core of lipid biochemistry. The starting material for the reaction are two esters, and the final product is a β-diketone or, in the instance of lipid metabolism, a β-keto ester.

An enolate is necessary for the Claisen reaction to proceed. To form an enolate, there has to be a labile proton on the α carbon.

In the laboratory, a Claisen condensation is achieved using a strong base to remove the α proton and form the enolate. Enolate formation is the rate-limiting step in the reaction, and it is resonance-stabilized through the adjacent carbonyl group. In enzymatic reactions, an equivalent base forms in the active site in the enzyme to abstract the proton. One other limitation is that the base, in this case a proton acceptor in the active site of the enzyme, cannot be a strong nucleophile because it would add to the ester or a carbonyl carbon. This limitation is not typically a problem in biological systems.

A slightly different pathway is taken to obtain the enolate in fatty acid biosynthesis. Rather than generate the enolate by deprotonation, malonyl-CoA is decarboxylated to generate an enolate that can attack the ester carbon bound to fatty acid synthase, forming a covalently bound intermediate, before eliminating the synthase and the β-keto ester.

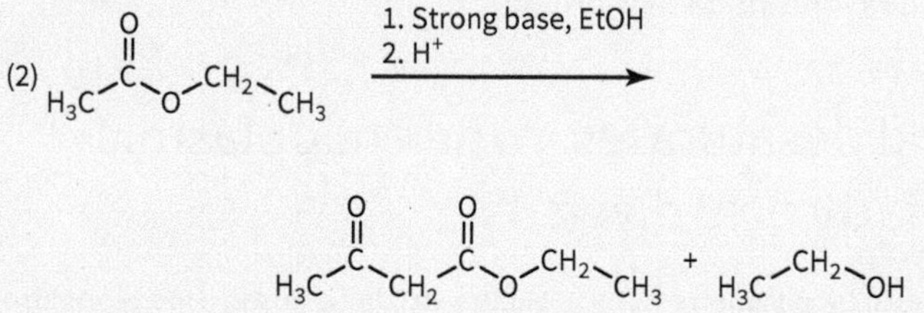

Another condition necessary for the reaction is that there needs to be a suitable leaving group adjacent to the enolate. Again, in the laboratory this is typically an alkoxide generated by one of the esters. In the instance of lipid biosynthesis, the leaving group is either the pantothenate group of coenzyme A or ACP.

Claisen condensations are found in fatty acid biosynthesis, ketone body biosynthesis, and the metabolism of some amino acids. The reverse reaction, a retro Claisen, or Claisen ester cleavage, is employed in β-oxidation. Aldol condensations and decarboxylation reactions of β-keto acids follow a similar scheme—formation of a resonance-stabilized carbanion intermediate, followed by nucleophilic attack.

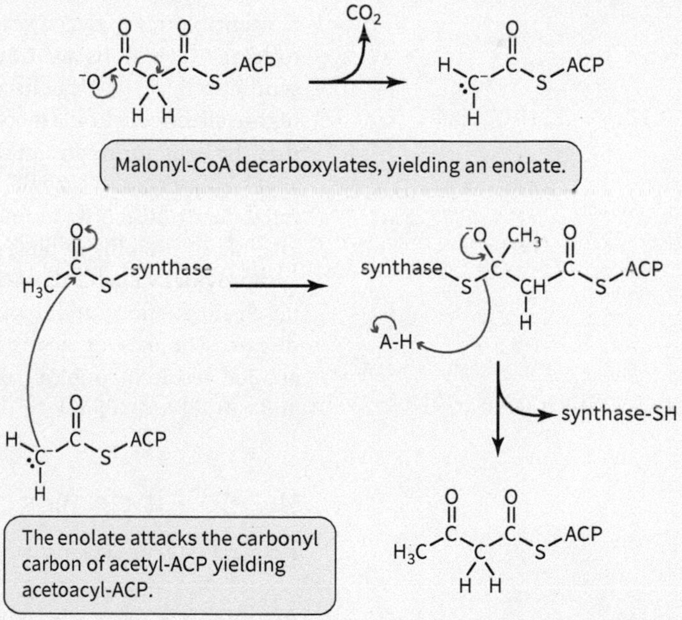

Malonyl-CoA decarboxylates, yielding an enolate.

The enolate attacks the carbonyl carbon of acetyl-ACP yielding acetoacyl-ACP.

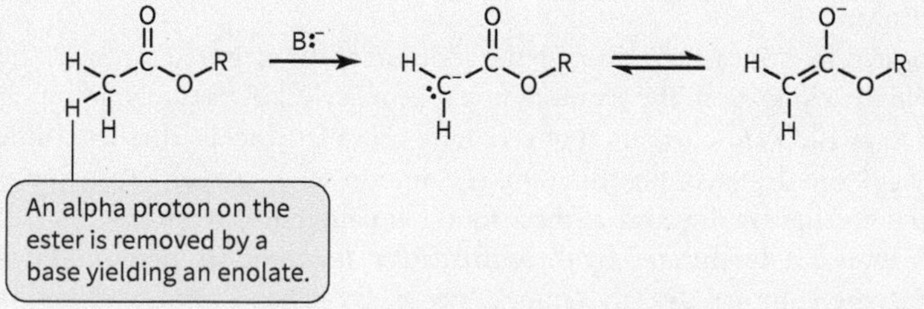

An alpha proton on the ester is removed by a base yielding an enolate.

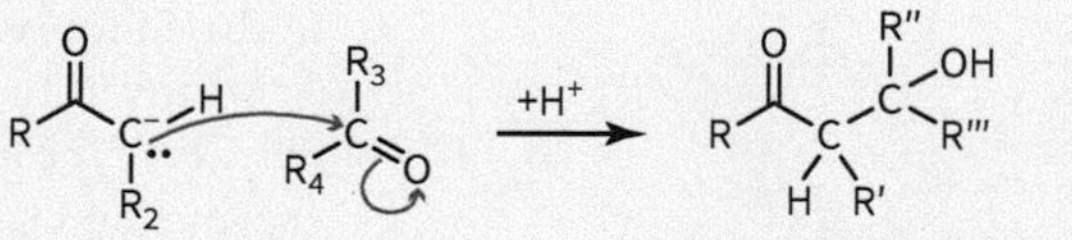

Aldol condensation

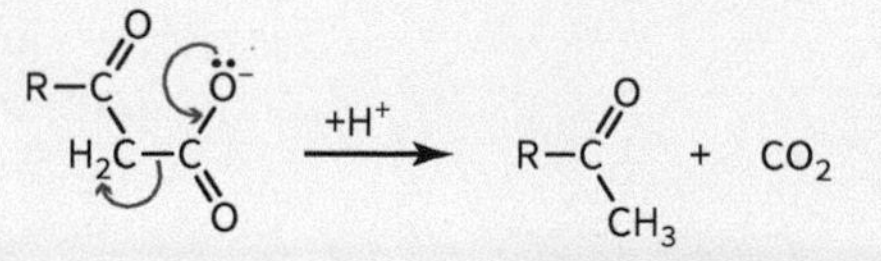

Decarboxylation reaction

A.

B.

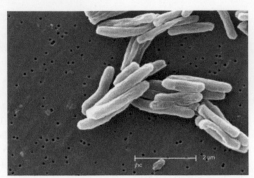

Isoniazid

FIGURE 9.23 Isoniazid and tuberculosis. **A.** Tuberculosis is a bacterial infection that often affects the lungs. It can be diagnosed on an X-ray. **B.** Isoniazid is a fatty acid synthesis inhibitor used to treat persistent bacterial infections (such as *Mycobacterim tuberculosis*, shown in the micrograph). It binds to and blocks enoyl-ACP reductase, blocking subsequent steps in fatty acid biosynthesis.

Sutthaburawonk/Getty Images

CDC/ Ray Butler, MS

Fatty acid biosynthesis differs in bacteria and mammals, presenting an opportunity to develop new antibiotics to treat diseases. An example of this is tuberculosis. Isoniazid is an antibiotic used to treat tuberculosis and other persistent bacterial infections. The skeleton of isoniazid was first synthesized in 1912, but its antibacterial properties were not discovered until 1951. The drug was highly effective, but the mechanism of action was unknown. In the early 1970s, it was discovered that the drug forms an adduct with NADP that inhibits the enoyl-ACP reductase, blocking fatty acid biosynthesis in bacteria, but not in mammals, resulting in slowing bacterial growth or in bacterial death. Recently, antibiotic-resistant tuberculosis has emerged (**Figure 9.23**). The structural analysis of proteins in these strains shows how individual mutations can lead to drug resistance and provides valuable data for designing new drugs. People often question why scientists spend so much time, effort, and resources on basic research that is not directly related to treating human disease. The development of isoniazid is a simple illustration that demonstrates how basic and applied research coupled together can unexpectedly lead to new treatments for human diseases and can serve as a guide in the development of new drugs.

9.3.2 Elongases and desaturases in the endoplasmic reticulum increase fatty acid diversity

As discussed above, the synthesis of palmitate (16:0) occurs in the cytosol, but elongation of palmitate to longer fatty acids and desaturation to unsaturated fatty acids occurs predominantly in the endoplasmic reticulum.

Fatty acids are elongated by two carbons through the addition of the activated carbons of malonyl-CoA. The first step in the elongation, the synthesis of a β-keto acyl-CoA, is catalyzed by various enzymes and is referred to as ELOVL (elongation of very long-chain fatty acids) that have different preferences for chain length and degree of unsaturation. A common set of enzymes then completes the process through reactions that are the same as those found in palmitate biosynthesis.

In mammals, fatty acids are desaturated by direct oxidative desaturation, predominantly by Δ-9 desaturase. This enzyme complex directly removes two electrons and associated hydrogens from stearoyl-CoA (18:0) to create oleyl-CoA ($18:1^{\Delta 9}$). The complex consists of three proteins: an NADH-cytochrome b5 reductase, cytochrome b5, and the desaturase. Electrons from both the fatty acid and NADH are transferred to the redox active iron centers in the cytochrome, and then to the desaturase. The final electron acceptor is molecular oxygen, generating two molecules of water. Inclusion of NADH in the reaction provides the two additional electrons needed to generate the second molecule of water, using up both atoms of oxygen in the process.

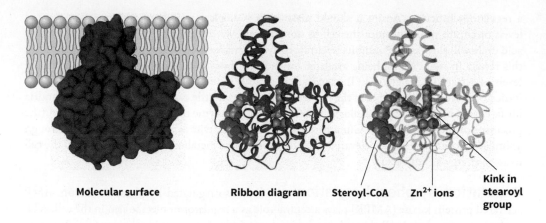

Molecular surface   **Ribbon diagram**   **Steroyl-CoA**   **Zn²⁺ ions**   **Kink in stearoyl group**

FIGURE 9.24 Structure of Δ-9 desaturase. Shown is the structure of Δ-9 desaturase from mouse (*Mus musculus*) in a molecular surface and ribbon depiction. The Δ-9 desaturase employs a pair of iron ions to generate the double bond, in this structure they have been replaced with zinc (shown in purple). Note the kink that the substrate binding site places on stearate. This ensures that the reaction happens with the proper stereochemistry.

(Source: Data from PDB ID 4YMK Bai, Y., McCoy, J.G., Levin, E.J., Sobrado, P., Rajashankar, K.R., Fox, B.G., Zhou, M. (2015) X-ray structure of a mammalian stearoyl-CoA desaturase. *Nature* 524: 252–256)

Because mammals lack desaturase activity at positions beyond the C-9 (for example, at Δ-12 or Δ-15), they are unable to add double bonds any farther down the fatty acid chain. Hence, fatty acids with double bonds in the ω-3 and ω-6 positions are considered essential fatty acids and must be supplied in the diet, typically in the form of $18:2^{\Delta 9,12}$ linoleic acid and $18:3^{\Delta 9,12,15}$α-linolenic acid. Both of these polyunsaturated fatty acids are commonly found in plants, where Δ12 or Δ15 desaturases are responsible for their synthesis. Humans can elongate and desaturate these fatty acids into the more highly unsaturated ω-3 and ω-6 fatty acids needed for eicosanoid biosynthesis, using a combination of Δ-4, Δ-5, Δ-6, and Δ-9 desaturases and elongases (**Figure 9.24**).

9.3.3 The formation of malonyl-CoA by acetyl-CoA carboxylase is the regulated and rate-determining step of fatty acid biosynthesis

As seen in other pathways, the rate-determining or committed step in a pathway is often a place where control mechanisms are found. In fatty acid biosynthesis, acetyl-CoA carboxylase (ACC) catalyzes the rate-determining step and is the main point of regulation (**Figure 9.25**).

Allosteric regulators Acetyl-CoA carboxylase is allosterically regulated by several molecules. For example, it is inactivated by acyl-CoA, a level of control that makes sense on both a chemical and biological level as an example of feedback inhibition. ACC synthesizes malonyl-CoA, which is used to synthesize fatty acyl-CoA. The allosteric regulation of ACC by fatty acyl-CoA provides a link to the energy level of the cell. If concentrations of acyl-CoA increase, their production is decreased, given that increased levels of acyl-CoA indicate either that triacylglycerols are being catabolized or that there is a block to their synthesis.

A second allosteric regulator of ACC is citrate, which binds to ACC1 and causes the enzyme to polymerize into lengthy strands (although the mitochondrial form of the enzyme does not polymerize). The individual complexes of ACC1 are less active than the polymer. Hence, although citrate does not participate in the conversion of acetyl-CoA to malonyl-CoA, it is still performing

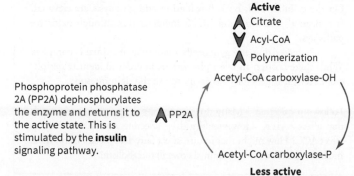

Active
∧ Citrate
∨ Acyl-CoA
∧ Polymerization
Acetyl-CoA carboxylase-OH

Phosphoprotein phosphatase 2A (PP2A) dephosphorylates the enzyme and returns it to the active state. This is stimulated by the **insulin** signaling pathway.

∧ PP2A

∨ AMPK

Phosphorylation of ACC I by AMP kinase (AMPK) or protein kinase A (PKA) results in ACC adopting the less active state. PKA is stimulated by **epinephrine** and **glucagon.**

Acetyl-CoA carboxylase-P
Less active

FIGURE 9.25 Regulation of ACC. ACC is regulated by phosphorylation and dephosphorylation, as well as polymerization, and two allosteric regulators (citrate and acyl-CoA).

a regulatory function. Again, it should make both biological and chemical sense that elevated levels of citrate would prompt the cell to store energy. When energetic levels are high, the citric acid cycle will slow as ATP inhibits isocitrate dehydrogenase, and levels of citrate will rise. In turn, this results in more citrate being available for the tricarboxylic acid transport system, increasing levels of citrate in the cytosol. Thus, citrate is serving multiple functions; it acts as a chemical indicator of the energetic state of the mitochondria, helps shuttle carbons out of the mitochondria for fatty acid biosynthesis, and allosterically activates the enzyme that produces malonyl-CoA (the precursor of those fatty acids). Glutamate and other dicarboxylic acids also activate ACC through a similar mechanism. Hence, elevated levels of several small metabolites converge to signal the cell to store energy as fatty acids.

Regulation by phosphorylation

ACC is also regulated by phosphorylation. AMP activated protein kinase (AMPK) plays a central role as a regulator of metabolism in the cell. ATP is often converted to AMP through biochemical reactions. As we saw in Chapter 6, ATP can be regenerated from AMP through substrate level phosphorylation or the F_0/F_1 ATPase found in mitochondria. Concentrations of ATP and AMP therefore reflect the energetic state of the cell with high concentrations of ATP (and low concentrations of AMP), signifying an energy-rich state while the converse (high AMP/low ATP) indicates an energy deficit.

High concentrations of AMP activate AMPK, which will phosphorylate ACC1 and thus inactivate it. As with the other regulatory mechanisms discussed, this should make metabolic sense. When concentrations of AMP are high, such as during exercise, the cell needs energy and should not be spending it converting acetyl-CoA to malonyl-CoA for fatty acid biosynthesis. Likewise, high levels of malonyl-CoA will limit β-oxidation by inhibiting carnitine palmitoyl transferase I and decreasing fatty acid import. Hence, in low-energy states, it makes sense to constrain the enzyme involved in malonyl-CoA synthesis. In these instances, biosynthetic pathways such as fatty acid or cholesterol biosynthesis are blunted, whereas those that yield energy (β-oxidation) are increased. PKA can also phosphorylate ACC, although the biological significance of this event is unclear.

Regulation of ACC gene expression

A final level of regulation that is worthy of mention is that of ACC gene expression. Two transcription factors play notable roles. First, ACC is partially regulated by the sterol response element binding protein (SREBP). This transcription factor binds to a sterol response element in the promoter region of the ACC gene, activating transcription. This transcription factor is partially responsible for activating many genes involved in lipid metabolism. The other transcription factor involved in the regulation of ACC is the carbohydrate response element binding protein (ChREBP), which was discussed in section 8.2. In response to a high-carbohydrate diet, ChREBP activates genes containing a carbohydrate response element (ChoRE); this includes both the ACC and FAS genes.

Further regulation of lipid metabolism occurs in triacylglycerol storage and mobilization. This will be discussed further in Chapter 10.

Worked Problem 9.4 — Predicting a knockout phenotype

There are two isoforms of ACC: ACC1 and ACC2. If we were to knock out either of these genes in mice, what would you predict for a phenotype? How would lacking these genes affect the animal and its metabolism?

Strategy What are the isoforms of ACC, and what functions do they serve? What would happen if one or both were missing?

Solution The ACC1 knockout mouse is unable to produce fatty acids, a mutation that is lethal to an embryo early in its development. The phenotype of the ACC2 knockout is a bit more complicated. Because ACC2 regulates CPT I, knocking out ACC2 leads to an increase in fatty acid transport into the mitochondria and increased β-oxidation. ACC2 knockout mice develop normally, have a normal appearance, and reproduce with vigor (a basic sign of overall animal health); however, compared to normal mice, they can consume more food and burn more fat through β-oxidation, and they are resistant to weight gain, diet-induced obesity, and type 2 diabetes mellitus. On the cellular level, both β-oxidation and glycolysis are elevated. It appears that eliminating ACC2 increases flux through oxidative pathways.

Characterizing any genetically manipulated animal is complex. While there is often an overt phenotype, in order to identify peripheral effects, it is generally necessary to ask the right questions.

Follow-up question Many drug companies are pursuing compounds that affect AMPK. How would a drug affecting AMPK be likely to affect ACC? How might that in turn affect fatty acid metabolism? What might be the effects on the animal's overall metabolism?

Summary

- Fatty acid biosynthesis occurs primarily in the cytosol and involves two important multifunctional enzyme complexes (ACC and FAS).
- The first step in fatty acid biosynthesis is the production of the three-carbon intermediate malonyl-CoA. Acetyl-CoA carboxylase synthesizes malonyl-CoA from acetate in the cytosol.
- FAS synthesizes palmitate (16:0) from malonyl-CoA and acetyl-CoA, using energy provided by NADPH.
- Longer and unsaturated fatty acids are produced from palmitate in the endoplasmic reticulum.
- Fatty acid biosynthesis is regulated by substrate availability, compartmentalization within the cell, and inhibition of key steps in the pathway itself. This is accomplished through both allosteric regulators and phosphorylation of key enzymes.

Concept Check

1. Describe the steps involved in fatty acid biosynthesis.
2. Explain the significant differences in the steps involved in β-oxidation and fatty acid biosynthesis.
3. Indicate which electron carriers are involved in biochemistry and how they affect the chemistry of the reactions.
4. What are the two sources of acetyl-CoA for fatty acid biosynthesis?
5. Describe how fatty acids longer than 16 carbons or containing double bonds are formed.

9.4 Ketone Body Metabolism

Fatty acids are an excellent source of stored energy, and evolution has produced elegant means for synthesizing, storing, and degrading these molecules. However, the organism must be protected from the detergent-like properties of fatty acids, properties that result from the amphipathic nature of fatty acids and would denature proteins and disrupt membranes. Fatty acids are released from adipose tissue and are mainly transported in the blood bound to albumin but are sometimes associated with plasma lipoproteins. In binding and sequestering fatty acids, albumin prevents them from acting as detergents. Although this system works well in delivering fatty acids to tissues such as the heart, the liver, and muscle, it is unable to deliver fatty acids to the brain.

The brain is protected and shielded by the blood–brain barrier, a semipermeable structure consisting of endothelial cells tightly joined together with an extracellular matrix that work together to exclude many metabolites. Large molecules and proteins including fatty acids bound to albumin are unable to cross the blood–brain barrier. Therefore, the brain cannot absorb or use fatty acids from the circulation.

Typically, the brain obtains most of its energy from glucose, which is transported across the blood–brain barrier. However, under fasting conditions, when there is low plasma glucose, or in certain disease states such as uncontrolled diabetes mellitus when the brain may be unable to use glucose as a fuel source, an alternative is needed. Under these conditions, the liver synthesizes **ketone bodies (acetone, acetoacetate** and β-**hydroxybutyrate**) to provide energy to the brain (**Figure 9.26**). Two of these three molecules (acetoacetate and β-hydroxybutyrate) are small, water-soluble equivalents of fatty acids, which can also permeate the blood–brain barrier.

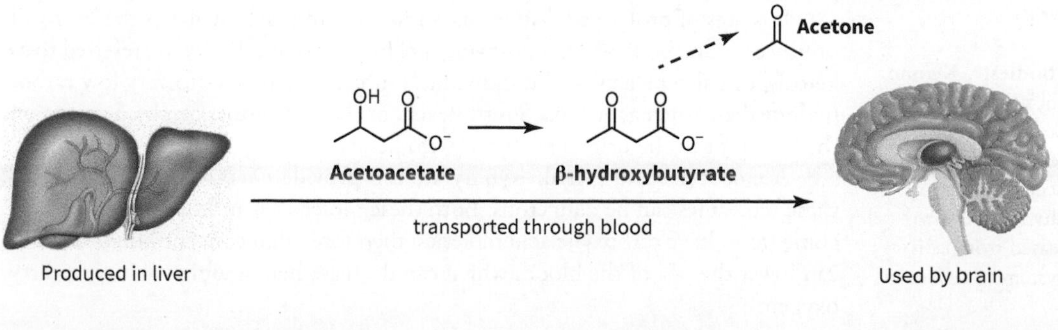

FIGURE 9.26 Structure of ketone bodies. Shown are acetone, acetoacetate, and β-hydroxybutyrate. These are produced by the liver. The body, especially the brain, can use acetoacetate and β-hydroxybutyrate for fuel. Acetone is a breakdown product formed by the spontaneous decarboxylation of acetoacetate.

(Source: (Liver) Tortora; Derrickson, *Principles of Anatomy & Physiology*, 15e, copyright 2016, John Wiley & Sons. This material is reproduced with permission of John Wiley & Sons, Inc. (Brain) Ireland, *Visualizing Human Biology*, 4e, copyright 2013, John Wiley & Sons. This material is reproduced with permission of John Wiley & Sons, Inc.)

9.4.1 Ketone bodies are made from acetyl-CoA

Ketone bodies are synthesized when glucose levels are low but acetyl-CoA levels are high. This occurs when fatty acids in the liver have been broken down to acetyl-CoA, but there are insufficient intermediates of the citric acid cycle to catabolize the acetate.

The process begins by the condensation of two molecules of acetyl-CoA into a molecule of acetoacetyl-CoA (**Figure 9.27**). The reaction is catalyzed by the enzyme thiolase (which is also involved in β-oxidation). The energy for the reaction is provided by the cleavage of the thioester bond between coenzyme A and acetate.

A third molecule of acetyl-CoA condenses with acetoacetyl-CoA. This reaction is catalyzed by β-hydroxy-β-methylglutaryl-CoA synthase (HMG-CoA synthase) to produce β-hydroxy-β-methylglutaryl-CoA (HMG-CoA). The first two reactions of ketone body synthesis are also the first two reactions of cholesterol biosynthesis.

Next, HMG-CoA is acted on by HMG-CoA lyase, which cleaves acetyl-CoA and liberates acetoacetate, a molecule with two metabolic fates. It can be secreted directly by the liver or can be reduced by β-hydroxybutyrate dehydrogenase in the liver to yield the D isomer of β-hydroxybutyrate. This too is released by the liver and can be used as a fuel by the brain. Beta-hydroxybutyrate has a chemical advantage over acetoacetate. While acetoacetate can undergo a spontaneous decarboxylation reaction to form acetone and CO_2, the more highly reduced state of β-hydroxybutyrate prevents this reaction from occurring. Although acetone and CO_2 are not toxic in the concentrations normally found in people undergoing ketone body production, these molecules are no longer usable, and the energy they contain is lost to metabolism.

9.4.2 Ketone bodies can be thought of as a water-soluble fuel source used in the absence of carbohydrates

Once released from the liver, ketone bodies travel through the circulatory system to tissues where they are metabolized. Ketone body metabolism occurs primarily in the brain, although heart and skeletal muscle also have the ability to use ketone bodies.

The first step in the use of ketone bodies is the conversion of β-hydroxybutyrate to acetoacetate by β-hydroxybutyrate dehydrogenase. Acetoacetate undergoes a substitution reaction with succinyl-CoA, forming acetoacetyl-CoA and succinate. The liver lacks 3-ketoacyl-CoA transferase, the enzyme needed for this reaction and thus cannot use ketone bodies for fuel. Coupling to coenzyme A energetically activates acetoacetate for subsequent cleavage. In the next reaction, acetoacetate reacts with CoA and thiolase to produce two molecules of acetyl-CoA, which can be further metabolized through the citric acid cycle to generate NADH and $FADH_2$ for the electron transport chain to use in ATP production (**Figure 9.28**).

In the discussion of the citric acid cycle, it was noted that the conversion of succinyl-CoA to succinate also yields a molecule of GTP or ATP. Therefore, the use of succinyl-CoA effectively costs a molecule of GTP or ATP.

In states of prolonged fasting, starvation, or untreated diabetes the liver will produce ketone bodies to help provide fuel for the brain. This state, referred to as ketosis, can also be achieved if individuals subject themselves to very low carbohydrate diets, such as Atkins, South Beach, or Paleo. Ketones can be detected on the breath or in the urine of subjects with ketosis.

Acetoacetate and β-hydroxybutyrate can provide energy, but high levels of these molecules can be dangerous. Both these molecules, and the fatty acids they come from, have carboxylic acid moieties; therefore, high concentrations of them can lower the pH of the blood, which can decrease hemoglobin's ability to carry oxygen.

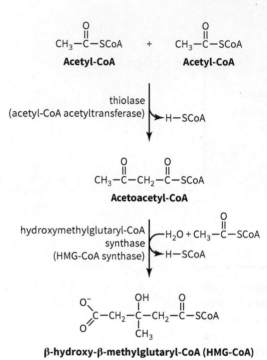

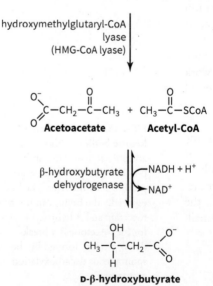

FIGURE 9.27 Synthesis of ketone bodies. Ketone body synthesis proceeds in the cytosol when levels of acetyl-CoA are elevated. Acetyl-CoA molecules condense in a series of Claisen condensations to form β-hydroxy-β-methylglutaryl-CoA. Acetoacetate can be reduced by β-hydroxybutyrate dehydrogenase to give β-hydroxybutyrate, which is released from cells and travels through the circulatory system to tissues in energetic need.

Although high levels of ketone bodies are associated with disease states or states of starvation, low levels of ketones are found in the bloodstream at concentrations that vary over the course of the day. It is probable that a low level of ketone body production is a part of normal physiology.

Regulation of ketone body production

As with other pathways, ketone body biosynthesis is regulated through hormonal control, allosteric regulators, and substrate availability. These three levels of control work together to regulate ketone body synthesis by regulating fatty acid transport into the mitochondrial matrix. Increases in the catecholamine hormones epinephrine and norepinephrine and decreases in insulin elevate triacylglycerol breakdown, leading to elevated concentrations of fatty acids available for β-oxidation. Glucagon decreases levels of hepatic malonyl-CoA by inhibiting acetyl-CoA carboxylase (ACC), resulting in carnitine palmitoyl transferase I (CPT I) activation, which is the rate-determining step in β-oxidation and ketone body biosynthesis. Finally, oxaloacetate helps regulate ketone body biosynthesis. If oxaloacetate levels are low due to increased gluconeogenesis, there will be less oxaloacetate available to enter the citric acid cycle. In this instance, acetyl-CoA will instead be used for acetoacetate biosynthesis. This pathway also liberates CoA molecules from acetyl-CoA, which allows β-oxidation to continue.

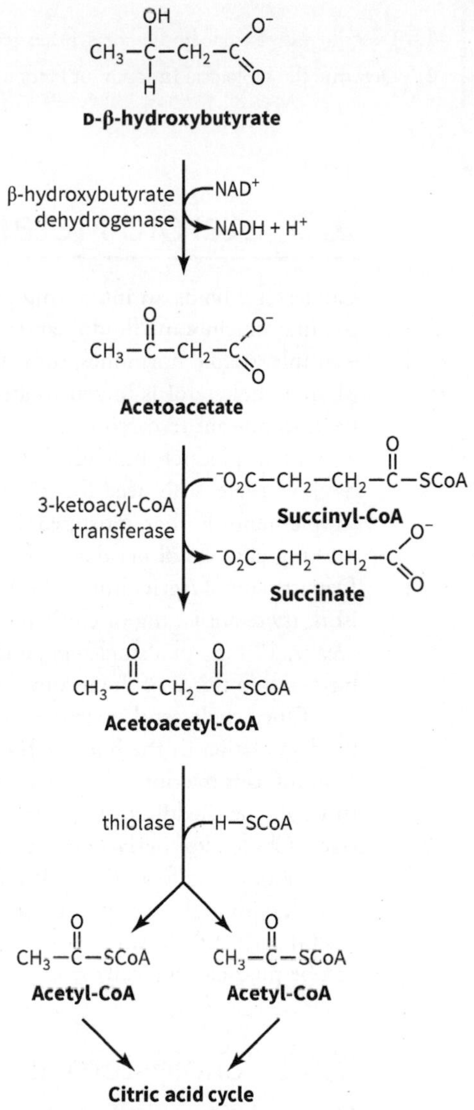

FIGURE 9.28 Breakdown of ketone bodies provides energy for tissues in energetic need. Tissues, such as those found in the brain, can use β-hydroxybutyrate and acetoacetate for fuel. Beta-hydroxybutyrate is oxidized back to acetoacetate by β-hydroxybutyrate dehydrogenase (the same enzyme that catalyzed the reduction in ketone body production). Acetoacetate can combine with succinyl-CoA to liberate succinate and form acetoacetyl-CoA. Acetoacetyl-CoA is cleaved by thiolase to give two molecules of acetyl-CoA, again in a reverse of the synthetic pathway.

Worked Problem 9.5   Ketones as a weight-loss agent

Ketone bodies are often elevated during periods of weight loss. Imagine that a drug is developed that increases the rate of acetoacetate decarboxylation into acetone. Would this drug be a good option for people who wish to lose weight?

Strategy Think about the chemical pathways involved in ketone body production and breakdown and the function of ketone bodies. Also think about why ketone bodies are produced during weight loss.

Solution Ketone bodies are produced by the liver during fasting; their function is to provide energy to brain tissue when glucose is not available. In some dietary states, carbohydrate intake is curtailed and, as a result, the liver increases ketone body production. Some fraction of acetoacetate spontaneously decarboxylates to form CO_2 and acetone. These carbons are either exhaled or are lost in the urine; they cannot be used for energy. Therefore, we might conclude that a molecule that increases this pathway would lead to carbons being wasted, and hence weight loss.

Follow-up question In the preceding question, we concluded that increased acetoacetate breakdown would be beneficial. Can you suggest why this might *not* be the case?

Summary

- Ketone bodies (acetone, acetoacetate, and β-hydroxybutarate) are small, soluble molecules made from acetyl-CoA.
- Ketone bodies provide energy to tissues, most often the brain, that need it in times of fasting.

Concept Check

HMG-CoA

HMG-CoA reductase ⟨ 2 NADPH / 2 NADP⁺ / CoA ⟩

Mevalonate

mevalonate-5-phosphotransferase ⟨ ATP / ADP ⟩

Phosphomevalonate

phosphomevalonate kinase ⟨ ATP / ADP ⟩

5-pyrophosphomevalonate

pyrophospho-mevalonate decarboxylase ⟨ ATP / ADP + Pᵢ + CO₂ ⟩

Isopentenyl pyrophosphate

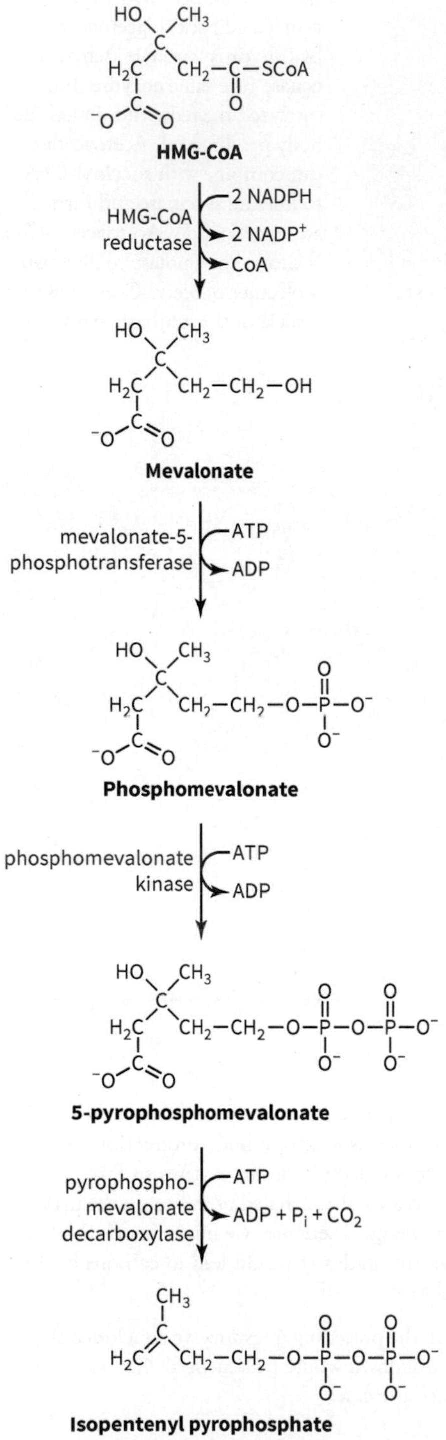

FIGURE 9.29 Mevalonate pathway. HMG-CoA is reduced to mevalonate. In this reaction, the thioester to CoA is broken and the carboxyl is reduced to a hydroxyl. Mevalonate is sequentially phosphorylated to form 5-pyrophosphomevalonate, which is decarboxylated to form the five-carbon isoprenoid isopentenyl pyrophosphate.

9.5 Steroid Metabolism

Cholesterol holds an interesting place in biochemistry. Animals require cholesterol to maintain membrane fluidity and to act as a precursor of several essential molecules, for example, steroid hormones, some fat-soluble vitamins, and bile salts. However, elevated plasma cholesterol is linked to atherosclerosis and heart disease. As a result, there has been significant research into the regulation of cholesterol levels. Cholesterol is largely an animal product, but humans can synthesize all that they need (nearly 1 gram per day), so there is no need for those who do not eat animal products to take cholesterol supplements. Rather, most people consume too much cholesterol in the diet. Less than 300 mg cholesterol per day is the intake currently recommended by the United States Department of Agriculture (USDA) and the American Heart Association. For comparison, there are 12 mg of cholesterol in a glass of low fat milk, 90 mg in a scoop of ice cream, 190 mg in a single egg, and over 205 mg in an average fast-food bacon cheeseburger, two-thirds of the recommended daily intake.

Once cholesterol is synthesized or taken in through the diet, it does not undergo biodegradation in the body; rather, cholesterol and its derivatives are lost or excreted. A significant fraction of the daily requirement for cholesterol is to replace what is lost through the shedding of epithelial cells in skin and the lining of the gastrointestinal tract. Cholesterol derivatives also form bile salts that act as detergents in the gastrointestinal tract and form the basis for steroid hormones. Cholesterol levels are tightly regulated, both in the cell and in the body as a whole, but both loss of cholesterol through the GI tract and blocking biosynthesis of new cholesterol are ways that cholesterol levels can be pharmacologically manipulated.

9.5.1 Cholesterol is synthesized in the liver through the addition of energetically activated isoprene units

Similar to fatty acids and ketone bodies, cholesterol is synthesized from acetyl-CoA; however, the synthesis of cholesterol from its immediate precursors is somewhat more modular than the synthesis of these other molecules. Energetically activated five-carbon alkenes are made from acetyl-CoA and then undergo condensation reactions to eventually form cholesterol. The first series of reactions used to form the two 5-carbon intermediates are known as the mevalonate pathway.

The first two reactions in the synthesis of cholesterol are the condensation of three molecules of acetyl-CoA. The condensation of the first two molecules, which is catalyzed by thiolase, forms acetoacetyl-CoA, the same reaction employed in the first step of ketone body biosynthesis, as shown in Figure 9.27. Although this is the same intermediate used in ketone body biosynthesis, the enzymes responsible for the synthesis of cholesterol precursors are found in the cytosol, while those responsible for the synthesis of ketone bodies are found in the mitochondria. It is the compartmentalization of these enzymes that regulates the metabolic fate of the products, determining whether they will form cholesterol or ketone bodies. The second reaction is addition of a third molecule of acetyl-CoA (catalyzed by HMG-CoA synthase) to form β-**hydroxy-β-methylglutaryl-CoA**, which is abbreviated as **HMG-CoA**. The name of this compound should help in deducing the structure. Both the hydroxyl group and the methyl group reside on the β carbon of glutarate. Glutarate is the deaminated form of the amino acid glutamate and is a five-carbon dioic acid. Both reactions involved in the synthesis of HMG-CoA are reversible. The next step is the reduction of HMG-CoA to **mevalonate** (**Figure 9.29**). This step requires NADPH to produce the primary alcohol found in mevalonate. Unlike the first two steps, the reduction of HMG-CoA is irreversible and results in the cleavage of the thioester bond to CoA. The enzyme responsible for this committed step in cholesterol biosynthesis is **HMG-CoA reductase**. Like many of the enzymes found in the biosynthesis of cholesterol, HMG-CoA reductase is a transmembrane protein found in the membrane of the endoplasmic reticulum and is the target of many of the most widely used and effective cholesterol-lowering drugs.

In cholesterol biosynthesis, mevalonate units are fused into activated isoprene units. These steps require energy from ATP. Mevalonate is first phosphorylated to 5-phosphomevalonate, which is further phosphorylated to give 5-pyrophosphomevalonate. This molecule gives rise to two important intermediates: $\Delta 3$ isopentenyl pyrophosphate and dimethyl allyl pyrophosphate. These molecules are both isoprenes, five-carbon alkenes that are coupled to readily hydrolyzable pyrophosphate groups. They are also isomers, differing only in the placement of the double bond.

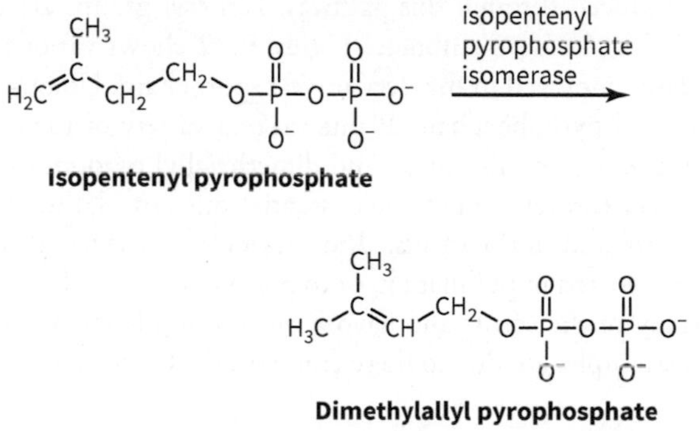

Isopentenyl pyrophosphate

isopentenyl pyrophosphate isomerase

Dimethylallyl pyrophosphate

The next steps in the synthesis of cholesterol involve the condensation of these isoprene units (**Figure 9.30**). The two 5-carbon isomers ($\Delta 3$ isopentenyl pyrophosphate and dimethyl allyl pyrophosphate) are condensed by prenyl transferase to form a one 10-carbon molecule, geranyl pyrophosphate. Loss of pyrophosphate provides the energy for this reaction. In this pathway, as with glycolysis and β-oxidation, an early investment of energy provides the energy for subsequent reactions. Another five-carbon molecule ($\Delta 3$ isopentenyl pyrophosphate) is added to geranyl pyrophosphate to yield the 15-carbon farnesyl pyrophosphate, again, with energy provided by the loss of a pyrophosphate group. Condensation of two farnesyl pyrophosphate groups produces squalene, a 30-carbon intermediate. The final two pyrophosphates are lost in this reaction, which also requires reducing equivalents, provided by NADPH/H$^+$.

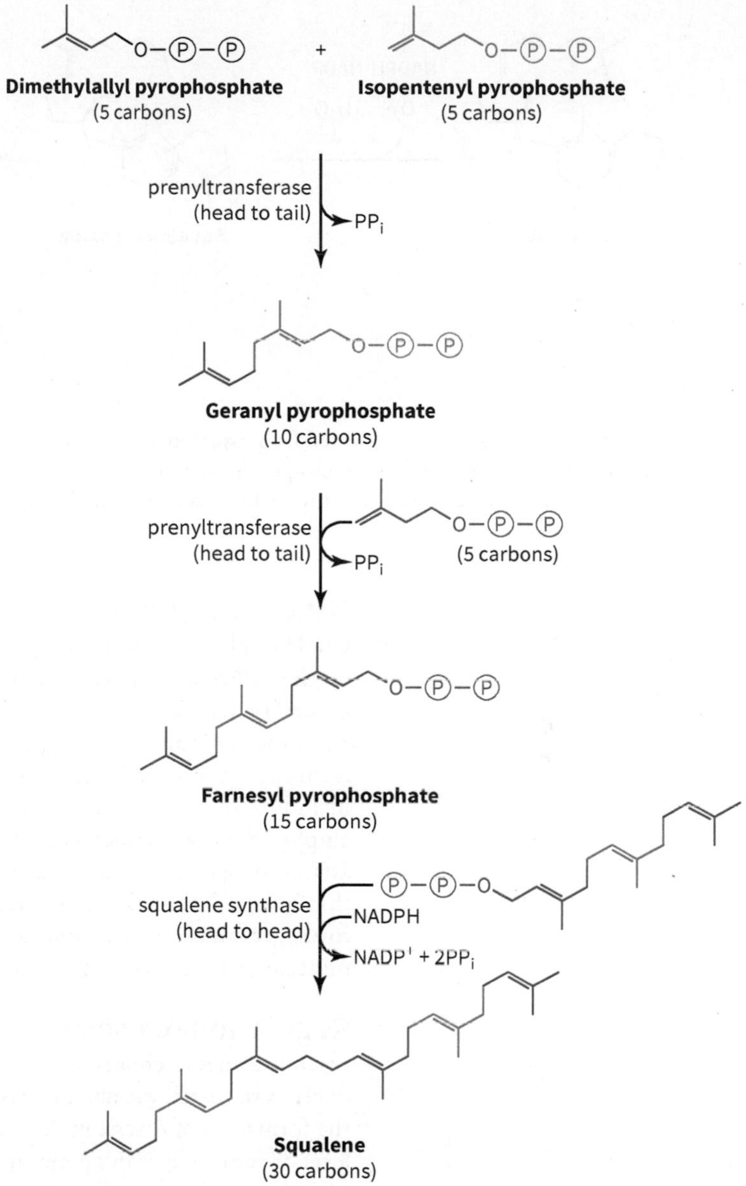

Squalene is highly branched, but it lacks the characteristic rings of a steroid. Two further steps are required to form that distinctive structure (**Figure 9.31**). First, squalene monooxygenase produces an epoxide between C-2 and C-3 of squalene. This process requires reducing equivalents that come from NADPH/H$^+$. Next, in one of the more amazing reactions in biochemistry, an enzyme known simply as cyclase catalyzes a concerted reaction in which all four rings of the steroid nucleus are formed, and the epoxide is cleaved to yield the molecule lanosterol. Nineteen additional steps are required to form the final cholesterol molecule.

Cholesterol is found at significant concentrations only in animal tissues. Plants and fungi employ different steroids as signaling molecules and vary the fatty acid composition of membranes to regulate fluidity.

Dimethylallyl pyrophosphate
(5 carbons)

+

Isopentenyl pyrophosphate
(5 carbons)

prenyltransferase
(head to tail)   PP$_i$

Geranyl pyrophosphate
(10 carbons)

prenyltransferase
(head to tail)   PP$_i$ (5 carbons)

Farnesyl pyrophosphate
(15 carbons)

squalene synthase
(head to head)   NADPH   NADP$^+$ + 2PP$_i$

Squalene
(30 carbons)

FIGURE 9.30 Synthesis of squalene from isoprenes. Squalene is formed by sequential condensation reactions fueled by the hydrolysis of phosphate groups. A molecule of each five-carbon isoprene condenses to form the ten-carbon geranyl pyrophosphate. An additional molecule of isopentenyl pyrophosphate adds to form the 15-carbon farnesyl pyrophosphate (pyrophosphate is lost in each of the first two reactions). Finally, two molecules of farnesyl pyrophosphate condense in a head-to-head fashion to form squalene.

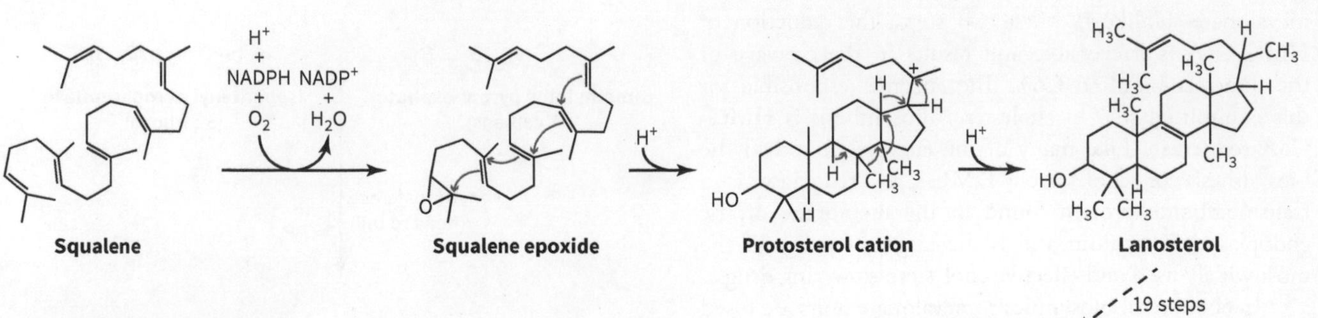

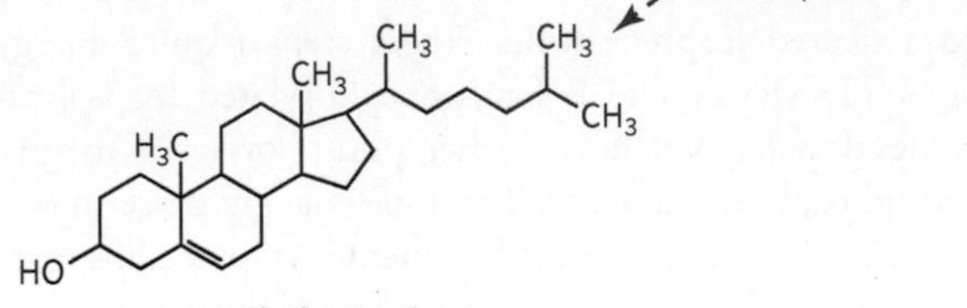

Squalene   **Squalene epoxide**   **Protosterol cation**   **Lanosterol**

19 steps

Cholesterol

FIGURE 9.31 Cholesterol is formed from squalene. Cyclase catalzyes the single step cyclization of squalene into lanosterol. An additional 19 steps are needed to form the final cholesterol molecule.

Other unsaturated molecules are also produced from mevalonate

Cholesterol is not the only molecule produced through this pathway. Farnesyl groups are used by the cell to anchor some proteins in the plasma membrane. **Figure 9.32** shows various isoprenes, such as ubiquinone, an antioxidant employed in the electron transport chain. Ubiquinone and other quinones are made from farnesyl pyrophosphate. Plants make a variety of molecules from the five-carbon building blocks isoprenyl pyrophosphate and dimethylallyl pyrophosphate, including the carotenoids retinol and beta carotene, numerous essential oils, and latex. Plants employ these molecules as both pigments and antioxidants. The molecules provide scents to attract some animals, for instance, those involved in pollination, or to poison others, such as those that would devour the entire plant. Many of the scents and flavors we commonly encounter in roses, geraniums, lemon, and clove, for example, are due to these compounds. Collectively, these molecules are termed terpenoids.

Regulation of cholesterol biosynthesis
The cell is faced with a complex problem when it comes to cholesterol biosynthesis. It needs to tightly regulate the levels of cholesterol production to match membrane biosynthesis, to generate isoprenoids, and to provide precursors for the formation of diverse molecules with a range of functions. Thus, the cell needs to exert a high level of control over this pathway.

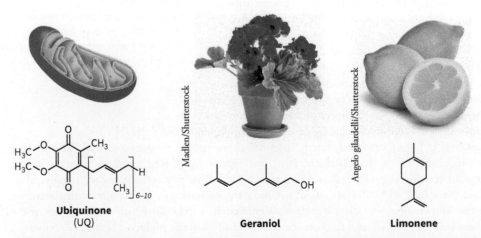

Ubiquinone (UQ)   **Geraniol**   **Limonene**

FIGURE 9.32 Structures of some other isoprenes. Isoprenes form many diverse structures, especially in plant biochemistry. Shown are ubiquinone (an electron carrier found in mitochondria), geraniol (a scented compound found in geraniums and citronella), and limonene (responsible for lemon/lime flavor). Latex rubber (not pictured) is another example of isoprene chemistry.

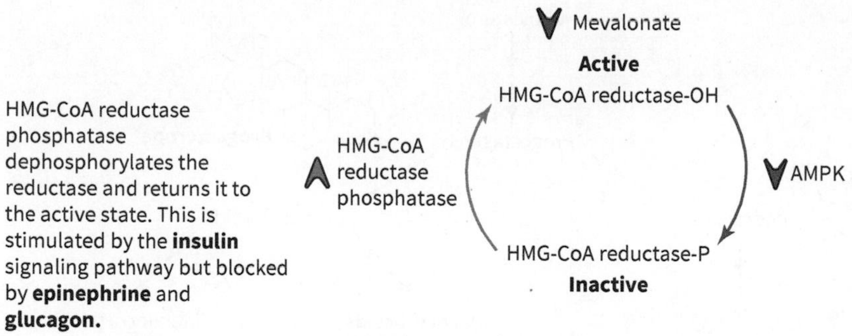

HMG-CoA reductase phosphatase dephosphorylates the reductase and returns it to the active state. This is stimulated by the **insulin** signaling pathway but blocked by **epinephrine** and **glucagon.**

FIGURE 9.33 Regulation of cholesterol biosynthesis. Cholesterol biosynthesis is largely regulated at the level of HMG-CoA reductase. Phosphorylation of this enzyme results in deactivation, while dephosphorylation produces the opposite effect. Also shown are several of the kinases and phosphatases known to regulate HMG-CoA reductase.

Regulation of cholesterol biosynthesis happens at several levels (**Figure 9.33**). Nearly all of these control steps affect the level of HMG-CoA reductase or the committed step in cholesterol biosynthesis, and they act over different timeframes.

Allosteric control

The most rapid control of cholesterol biosynthesis happens through a negative feedback control loop. Mevalonate, the product of HMG-CoA reductase, can directly inhibit cholesterol biosynthesis. Therefore, if the cell has high levels of mevalonate, the reaction is blocked, providing the most rapidly acting level of control.

Post-translational modification

The next level of regulation is through post-translational modification of HMG-CoA reductase. The enzyme is inactivated by phosphorylation, a reaction mediated by AMP kinase. Although the key phosphorylation and dephosphorylation events are directly regulated by AMPK and a specific HMG-CoA reductase phosphatase, these proteins and their activities are regulated on a broader level by several kinases and phosphatases, including PKA and PP2A. Catecholamines and glucagon inhibit cholesterol biosynthesis by inhibiting phosphatases. In contrast, insulin activates protein phosphatases and therefore activates HMG-CoA reductase.

HMG-CoA reductase can be ligated to ubiquitin and targeted for degradation by a proteasome. As described above, high levels of cholesterol can bind to HMG-CoA reductase through the sterol-sensing domain. When cholesterol is bound, the structure of HMG-CoA reductase is altered, promoting ubiquitination. Therefore, another level of regulation of HMG-CoA reductase is protein degradation. If cholesterol levels are moderately high, cholesterol will bind the protein and inhibit activity; however, at higher concentrations, the equilibrium will shift and the cholesterol-HMG-CoA reductase complex will predominate, increasing the likelihood that the complex will be ligated to ubiquitin and targeted for degradation.

Transcriptional regulation

The final means of regulation available to the cell is to regulate the transcription of the HMG-CoA reductase gene. This occurs through the sterol response element binding protein (SREBP) transcription factor.

The cell can obtain cholesterol from external sources via low-density lipoproteins (LDL). Additional regulation of the cell's free cholesterol levels happens through the LDL receptor, HMG-CoA synthase, and acyl-CoA cholesterol acyltransferase (ACAT, the enzyme responsible for the synthesis of cholesteryl esters).

The enzymes involved in the production of isoprenoids have a slightly higher affinity for their substrates; that is, they have lower K_m values than the enzymes leading to sterol production, meaning that isoprenoids can still be synthesized when mevalonate concentrations are low.

9.5.2 Steroid hormones are derived from cholesterol

Steroid hormones are all derivatives of cholesterol (**Figure 9.34**). The first steps in steroid biosynthesis are the oxidative cleavage of the cholesterol side chain and the oxidation of the hydroxyl group to yield progesterone, the master sex hormone. While the cleavage of the side chain occurs in the mitochondria, most other reactions of steroid hormone biosynthesis happen in the smooth endoplasmic reticulum. Many of the enzymes involved are part of the cytochrome P450 family of enzymes, a group of oxidases that employ heme cofactors to carry out reactions.

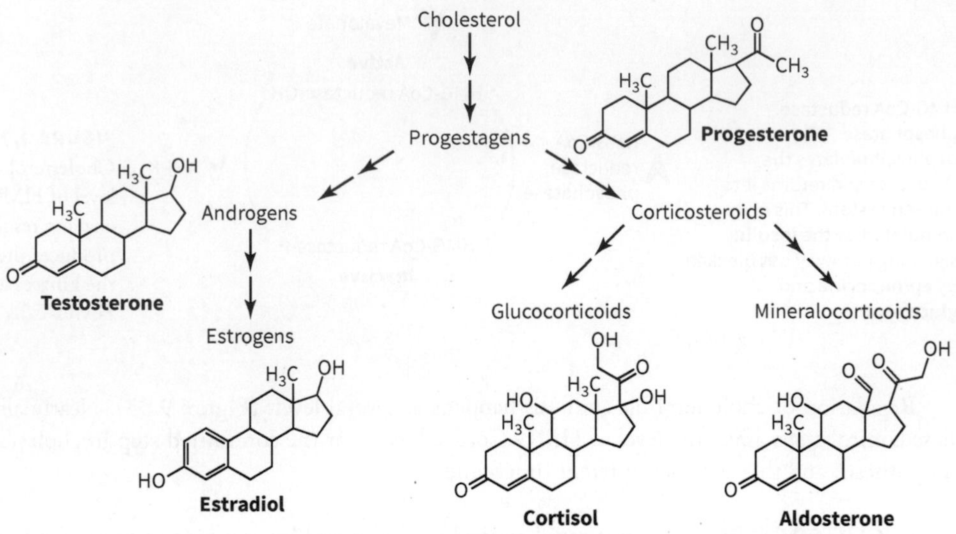

FIGURE 9.34 Steroid hormone biosynthesis. Steroid synthesis is a highly branched and multistep series of pathways. Following synthesis of progesterone, there is a split between the sex hormones and the corticosteroids. This intermediate leads to the production of all other steroid hormones through the action of numerous enzymes, many of which employ cytochrome P450 to accomplish the chemistry in question.

Worked Problem 9.6   Why this step?

Given all of the ways to control cholesterol levels, why would drug companies target HMG-CoA reductase as a way to regulate cholesterol levels?

Strategy Examine how cholesterol is synthesized. Is there anything different about HMG-CoA reductase?

Solution HMG-CoA reductase is an attractive molecule to target, for several reasons. First and foremost, it is the committed, irreversible, and rate-determining step in cholesterol biosynthesis. Such steps are often checkpoints for controlling flux through pathways. Also, HMG-CoA reductase is acted on by many different regulators. Any of these could, in theory, also be used to regulate the enzyme

(and thus the pathway), but modulating these molecules can have undesirable effects. Some of the signaling pathways that affect cholesterol biosynthesis (such as AMPK) also affect other physiological processes. Likewise, modulating other candidate molecules (such as mevalonate) directly can be problematic and may also have undesirable side effects.

Follow-up question Many proteins are joined to a farnesyl group (farnesylated) or otherwise prenylated. HMG-CoA reductase inhibitors may inhibit prenylation of some proteins. What does this say about how strongly these drugs inhibit the pathway? What does this also say about the cell's need for cholesterol versus the cells' need to prenylate proteins?

Progesterone has multiple fates. It can be further modified by hydroxylation and reduction to give the mineralocorticoids and glucocorticoids, such as aldosterone and cortisol. Progesterone also can be hydroxylated, leading to the synthesis of androgens. Androgens are the group of hormones associated with male sexual characteristics but are found in both males and females. Subsequent oxidations and reductions produce the rest of the androgens series, including testosterone. Estrogens are all derived from the androgens. The key step in this conversion is the aromatization of one of the rings of the steroid skeleton. Aromatization of the ring leads to estradiol if the precursor was testosterone.

As with most pathways, regulation of steroid hormone biosynthesis is tightly controlled at several levels. Pituitary signals trigger tissues to initiate production of steroid hormones. Chronic control of these steps occurs at the level of expression of the genes coding for proteins in the specific pathway of interest. At the acute level, a key regulatory step is the availability of cholesterol for conversion to progesterone. This step occurs in the inner mitochondrial membrane. The translocation of cholesterol from the outer to the inner mitochondrial membrane is regulated by the steroidogenic acute regulatory protein (StAR).

StAR is a 30 kDa phosphoprotein that has a hydrophobic channel with a curved β sheet structure (**Figure 9.35**). The molecule consists of a sheet topped by two helices and several turns. The hydrophobic channel is large enough to hold a single cholesterol molecule, but the exact mechanism of transfer—shuttling single molecules of cholesterol, acting as a channel for cholesterol, or facilitating temporary fusion of the inner and outer mitochondrial membranes—has yet to be elucidated.

Levels of precursor hormones can also affect levels of downstream hormones in steroid synthesis. For example, elevated levels of progesterone in the menstrual cycle or during pregnancy perturb levels of mineralocorticoids, resulting in changes in Na^+/K^+ mineral balance (and hence lead to water retention). Abuse of steroids by athletes can also lead to bloating. An extreme and dramatic illustration of the effects of precursor hormone levels is seen in diseases of the adrenal gland that affect mineralocorticoid production, which can lead to severe swelling and edema.

Steroid hormones are metabolized by cytochrome enzymes in the liver. Specifically, cytochrome P450 oxidases oxidize the steroid skeleton, adding hydroxyl groups to make the molecule more water soluble. The oxidized steroids are eliminated from the liver with bile.

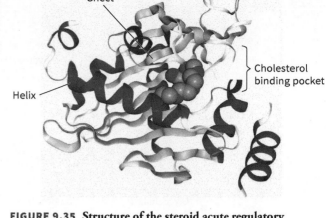

FIGURE 9.35 Structure of the steroid acute regulatory protein (StAR) domain. StAR binds cholesterol and regulates its transport from the outer to inner mitochondrial membrane where it is acted on by cytochrome enzymes involved in the synthesis of steroid hormones. Cholesterol molecules bind inside the pocket formed by the helix to the left and the β sheet structure.

(Source: Data from PDB ID 3FO5 Thorsell, A.G., Lee, W.H., Persson, C., Siponen, M.I., Nilsson, M., Busam, R.D., Kotenyova, T., Schuler, H., Lehtio, L. (2011) Comparative structural analysis of lipid binding START domains. *Plos One* 6: e19521-e19521)

9.5.3 Bile salts are steroid detergents used in the digestion of fats

Bile salts are amphipathic molecules that act as detergents, helping to solubilize dietary fats in the small intestine, and facilitating their digestion and breakdown into free fatty acids and glycerol before absorption by the intestine. The amphipathic nature of bile salts can be difficult to visualize when examining the structure. Cholesterol is a planar molecule; once derivatized into a bile salt, the molecule has two faces: one polar, the other nonpolar. As a result, these detergents help to solubilize lipids but are gentle enough not to denature enzymes or disrupt plasma membranes of the cells lining the gastrointestinal tract. As with fatty acids, the deprotonated form of the molecule is referred to as a bile salt, while the protonated form is termed a bile acid.

Bile salts can be categorized as conjugated or unconjugated. Conjugation in this sense does not refer to the number and position of double bonds but rather to the covalent addition of an amino acid, typically either glycine or taurine, to the carboxyl group. Addition of such groups typically makes the bile salt much more soluble in water and thus a better emulsifier.

Bile salts are synthesized in the liver (**Figure 9.36**). The rate-determining step in their synthesis is the action of cholesterol 7-hydroxylase, a cytochrome P450 enzyme that uses heme and molecular oxygen to incorporate a hydroxyl group on the C-7 of the cholesterol ring to form 7-α-hydroxycholest-4-en-3-one. The ring is often further hydroxylated at the C-12 position. Finally, the branched tail of cholesterol is oxidized to a carboxylic acid. In many instances, a highly hydrophilic group (glycine or taurine) is joined via an amide linkage to this carboxyl group.

Bile salts are secreted from the liver to the gall bladder, which stores bile (a mixture of bile salts, free cholesterol, and proteins) until it is needed for digestion. The body secretes upwards of 30 g of bile salts a day. Most of the bile salts in this mixture are reabsorbed via an intestinal, or ileal, bile acid transporter, and recycled to the liver. Bile acid sequestrants, that is, compounds that bind to bile salts in the small intestine and sequester them, can be effective at lowering plasma cholesterol, which is the precursor of bile salts. Two examples of these medications are cholestyramine and colsevelam.

Cholestyramine

Colesevelam

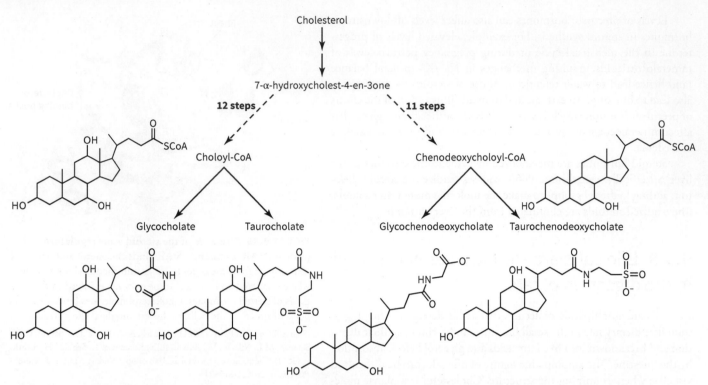

FIGURE 9.36 Bile salt biosynthesis.   Bile salts are made through multistep pathways. Two initial reactions change cholesterol into 7-α-hydroxycholest-4-en-3one. At this point the pathways diverge. The path shown to the left leads to choloyl-CoA (cholate) (12 steps). The pathway to the right leads to chenodeoxycholoyl-CoA (chenodeoxycholate) and requires one fewer step. Both of these bile salts can be conjugated (coupled to an amino acid) to give the secondary bile salts glycocholate, taurocholate, glycochenodeoxycholate, and taurochenodeoxycholate.

Summary

- Steroids are a class of lipids containing a planar systems of four fused rings.
- Cholesterol, the principal steroid from which all others are derived, is synthesized through the mevalonate pathway.
- Steroid hormones are derived from cholesterol and include the sex hormones, glucocorticioids, and mineralocorticoids.
- Bile is a biological detergent composed of cholesterol, proteins, and bile salts.

Concept Check

1. Identify what makes a molecule a steroid.
2. Explain in general terms how steroids and bile salts are made in the body.
3. Describe functions of steroids and isoprenes in biochemistry and biology.
4. Describe the mevalonate pathway, cholesterol biosynthesis, and the regulation of cholesterol biosynthesis.
5. Explain in general terms how steroid hormones are thought to act in the cell.

9.6 Eicosanoid and Endocannabinoid Metabolism

The eicosanoids and **endocannabinoids** are a large family of potent signaling molecules derived from 20-carbon polyunsaturated fatty acids. The best characterized of the precursors is arachidonic acid (AA, $20:4^{\Delta5,8,11,14}$); others include eicosapentaenoic acid (EPA $20:5^{\Delta5,8,11,14,17}$) and dihomogamalinolenicacid (DGLA, $20:3^{\Delta8,\,11,14}$).

In the cell, these fatty acids are typically esterified to the central hydroxyl group of phosphatidylcholine or phosphatidylethanolamine. Phospholipids with these fatty acids are found almost exclusively on the cytosolic side of the plasma membrane where fatty acids are released by the action of specific phospholipases.

Following its release, the fatty acid can be derivatized in several different ways as shown in for arachidonate.

9.6.1 Eicosanoids are classified by the enzymes involved in their synthesis

In the linear pathway, arachidonate is acted on by a **lipoxygenase** (LOX) to form **leukotrienes**, as shown in **Figure 9.37**. These molecules are linear, and they differ from the other eicosanoids in that they lack rings. Leukotrienes are so named because they were first identified in white blood cells and have at least three conjugated double bonds.

In the cyclic pathway, arachidonate is acted on by one of the **cyclooxygenase** enzymes (COX I or COX II) to form prostaglandin H_2 or PGH_2. The name reveals some of the features of these molecules. PG refers to **prostaglandin**, which has a single five-membered ring. PGI refers to prostacyclin, which has a second, oxygen-containing heterocyclic ring. TX refers to a thromboxane, which has a six-membered ring and also contains an oxygen atom. The next letter refers

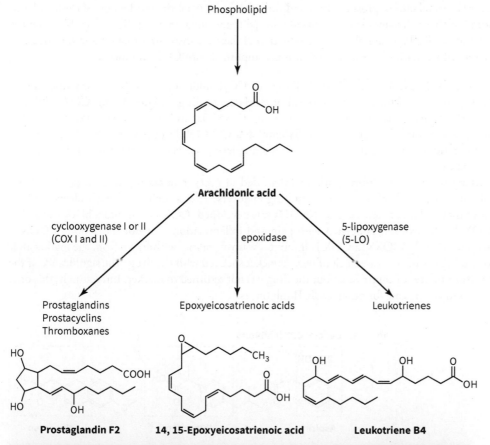

Drug	Inhibits	Used for
NSAIDS (aspirin, ibuprofen, ketoprofen, naproxyn)	COX I and II	Relief of pain, fever, inflammation
COX II inhibitors (rofecoxib, valdecoxib, celecoxib)	COX II	Chronic inflammation
Prostaglandin creams and suppositories	—	Induction of labor, vasodilation, ulcers
Leukotriene receptor agonist (inhibitors) (montelukast, zarfirlukast)	Leukotriene receptor	Asthma
5-LO inhibitors (zileuton)	5-LO	Asthma

FIGURE 9.37 Major pathways of eicosanoid biosynthesis. Arachidonate is released from phospholipids by a phospholipase. Once free, arachidonate can be acted on by cyclooxygenase (to yield the prostaglandins, prostacyclins, and thromboxanes), epoxidase (to give the epoxyeicosatrienoic acids) or 5-lipoxygenase (to yield the leukotrienes). All are important mediators of vasodilation, pain, and inflammation.

to the order in which it was discovered, and the number refers to the number of double bonds in the molecule outside the ring. Therefore PGH$_2$ is a prostaglandin (PG) that was discovered eighth (H) and has two double bonds.

Following production of PGH$_2$, pathways diverge. All of the other prostaglandins, prostacyclins, and thromboxanes are derived from PGH$_2$. It is generally thought that a single type of eicosanoid is produced in a given cell or tissue. The different eicosanoids are usually lipophilic and are thought to act as local signaling molecules, working over short distances in the body and binding to specific receptors. In some instances, eicosanoids can act as ligands for transcription factors such as the peroxisome proliferator-activated receptor (PPAR), although they more often act through a G protein coupled receptor (GPCR). Many details of eicosanoid signaling have yet to be elucidated (Figure 9.37).

There are two major isoforms of COX enzymes: COX I and COX II. COX I is constitutively expressed; that is, it is found at constant low levels in many cells and tissues. The products of COX I are used in the organism's day-to-day signaling pathways. Examples of such pathways include regulation of blood pressure, blood clotting, immune responses, labor and delivery, and sleep–wake cycles. In contrast, expression of COX II can be induced. When injury or inflammation damages a tissue, the gene coding for the COX II enzyme is activated and the enzyme is produced. Hence, this protein is involved with processes we typically think of as abnormal, such as pain and inflammation.

To summarize, the COX enzymes catalyze the production of PGH$_2$, which is the precursor of all other prostaglandins, prostacyclins, and thromboxanes. Also, the production of elevated levels of some of these eicosanoids is associated with pain and inflammation. Therefore, blocking the production of PGH$_2$ could block the production of these downstream products and thus decrease pain and inflammation—something that is accomplished with COX inhibitors.

NSAIDS are COX inhibitors

Forms of COX inhibitors have been used since antiquity; they are one of the most widespread of all classes of drugs (Figure 9.38). COX inhibitors include the nonsteroidal anti-inflammatory drugs (NSAIDs) aspirin, ibuprofen, ketoprofen, and naproxyn. The role of acetaminophen (Tylenol) as a COX inhibitor is still debated; it is used as an antipyretic (fever reducer) and analgesic (pain reliever) but is not commonly used as an anti-inflammatory drug.

Many people suffer from conditions that used to necessitate taking large doses of NSAIDs for significant periods of time, often resulting in adverse effects such as stomach ulcers. To help counter these effects, a new class of NSAIDs was developed. COX II inhibitors block the action of COX II, the isoform associated with pain and inflammation, but not the action of COX I. However, in 2004, VIOXX, a COX II inhibitor, was voluntarily withdrawn from the market after reports of a possible increased risk of heart attack associated with the drug. The significance of the risk is still a matter of controversy, but the drug has not returned to market. Interestingly, this does not seem to be the case for other COX II inhibitors.

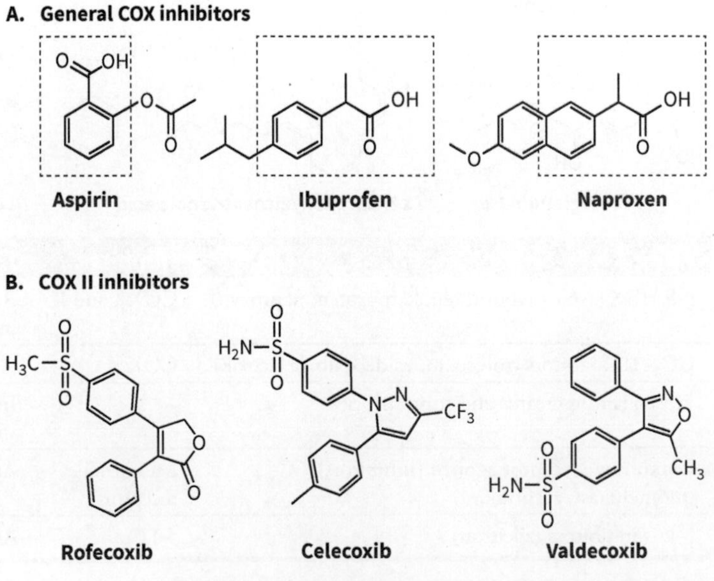

A. General COX inhibitors

Aspirin   Ibuprofen   Naproxen

B. COX II inhibitors

Rofecoxib   Celecoxib   Valdecoxib

FIGURE 9.38 Structures of COX inhibitors. Shown are **A.** the structures of the general COX inhibitors aspirin, ibuprofen, naproxen, and **B.** the COX II inhibitors rofecoxib, celecoxib, and valdecoxib. Note the structural similarities of these compounds.

9.6.2 Endocannabinoids such as anandamide are also arachidonate derivatives

A second class of signaling molecules derived from arachidonate are the endocannabinoids. The best characterized of these is anandamide (*N*-Arachidonyl ethanolamine), "ananda," named from the Sanskrit word for bliss. This molecule and other related structures are endogenous ligands for both the cannabinoid and vanilloid receptors. They all contain derivatives of arachidonic acid and are highly insoluble in water. Endocannabinoids are small lipophilic signaling molecules that are thought to act over short ranges.

The cannabinoid receptors (CB1 and CB2) are transmembrane G protein coupled receptors. They were first identified as the receptors that bind to Δ9 tetrahydrocannabinol (THC), the active compound in marijuana. Activation of these receptors results in hypotension, decreased gastrointestinal tract activity, increased appetite (also known as orexia), and decreased pain perception or nociception. The effects on the central nervous system are complex and probably involve more than simply a single G protein signaling pathway.

The vanilloid receptor, transient receptor potential cation channel subfamily V member 1 (TRPV1), is a nonspecific cation channel found in the peripheral nervous system and involved in nociception. The channel is stimulated by heat, low pH, and anandamide, but perhaps the best known ligand is capsaicin, the compound responsible for the burning taste of hot peppers.

Worked Problem 9.7   NSAIDs in pregnancy

Pregnant women are advised to avoid NSAIDs in the third trimester (the last third) of pregnancy. Why?

Strategy To answer this question, we need to examine the nature and functions of NSAIDs and the changes that occur in pregnancy that might be relevant for the use of NSAIDs.

Solution Although they are available over the counter and are typically thought of as innocuous, NSAIDs, as with any drug, can be dangerous in some circumstances. NSAIDs are general COX inhibitors, and they block the production of the prostaglandins and

prostacyclins associated with pain and inflammation. Prostaglandins of the PGE and PGF family are important in labor and delivery, specifically in uterine contractions and the softening of the uterine opening, referred to as cervical ripening. Inability to produce prostaglandins during this critical phase of pregnancy can lead to stalled labor and complications.

Follow-up question Prostaglandin creams are often used to induce labor. Why are the prostaglandins not administered orally or intravenously?

Both the endocannabinoids and their receptors are the focus of intense scrutiny from researchers and pharmaceutical companies. Due to their role in the regulation of pain, blood pressure, and appetite, these molecules are potential targets for drug development (**Figure 9.39**).

Because the endocannabinoids are all highly lipophilic, they are thought to be generated in the plasma membrane and to function there as well. Unlike other receptors we have seen, the ligand-binding pocket for CB1 and CB2 is in the region of the protein that is buried within the plasma membrane.

Drugs that modulate the cannabinoid receptors are of potential interest to medicine. One such compound, rimonabant, shown in Figure 9.39, is a reverse agonist to the CB1 receptor; that is, it binds to the receptor but dulls the response to that receptor rather than activating or blocking it. Rimonabant was developed as an anorectic or appetite suppressant to help combat obesity. It drives down appetite but also blunts other urges; for example, it blocks cravings for nicotine and has therefore been studied as a smoking cessation aid. However, the drug was withdrawn soon after its introduction to the market due to concerns about adverse effects including increased depression and risk of suicide. Scientists are now focused on developing drugs that have similar positive effects in modulating cannabinoid receptors but with fewer adverse effects.

We've now seen a variety of ways in which lipids can act as signaling molecules. A discussion of how some of these molecules are used as drugs is found in **Societal and Ethical Biochemistry: Performance enhancing drugs**.

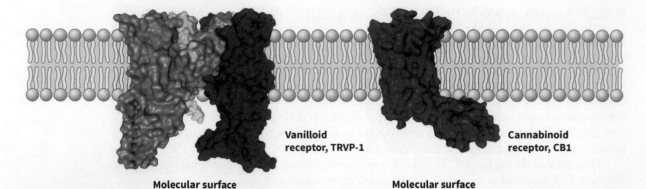

Compound	Receptor	Source	Effect	Structure
Anandamide or arachidonylethanolamine	Both	Endogenous ligand	Many, feelings of wellness	
Capsaicin	TRPV-1	Hot peppers	Burning sensations, pain relief	
Δ-9 tetrahydrocannabinol (THC)	CB1 and 2	Marijuana	Appetite stimulant, pain relief, antiemetic, treatment of glaucoma	
Rimonabant	CB1 (reverse agonist)	Synthetic	Appetite suppressant	
Vanillin	TRPV-1, likely others	Naturally occurring sources or synthetic	Agent responsible for vanilla flavor, biological effects unclear	

FIGURE 9.39 Structure of some endocannabinoids and vanilloids   Shown are the structures of the endogenous ligand for the cannabinoid receptors (anandamide or arachidonyl ethanolamine), and two compounds found in plants (the vanilloid receptor ligand capsaicin and Δ-9 tetrahydrocannabinol, the active component in marijuana). Rimonabant is an inverse agonist of the CB1 receptor. The CB1 receptor is shown in blue, and the vanilloid receptor (TRPV-1) is in red and orange.

(Source: (Vanilloid) Data from PDB ID 5IRX C Gao, Y., Cao, E., Julius, D., Cheng, Y. (2016) TRPV1 structures in nanodiscs reveal mechanisms of ligand and lipid action. *Nature* 534: 347–351. (Cannabinoid) Data from PDB ID 5TGZ Hua, T., Vemuri, K., Pu, M., Qu, L., Han, G.W., Wu, Y., Zhao, S., Shui, W., Li, S., Korde, A., Laprairie, R.B., Stahl, E.L., Ho, J.H., Zvonok, N., Zhou, H., Kufareva, I., Wu, B., Zhao, Q., Hanson, M.A., Bohn, L.M., Makriyannis, A., Stevens, R.C., Liu, Z.J. (2016) Crystal Structure of the Human Cannabinoid Receptor CB1. *Cell* 167: 750–762.e14)

Societal and Ethical Biochemistry

Performance enhancing drugs

Throughout history, people have been aware that physical and mental performance can be enhanced biochemically. For example, in ancient times, athletes and warriors would ingest herbal concoctions, animal organs, wine, or opium in the hope of besting their opponents. During the twentieth century, sports medicine was more evidence-based, and it became clear that some drug combinations could measurably improve performance, albeit often with adverse effects for the athlete. To protect athletes, rules were enacted to limit, eliminate, or curtail the use of these compounds. At the same time, there was an increase in the development and use of drugs to give greater improvements in performance and avoid detection from regulatory agencies.

Doping may have first had a significant impact in the 1960s in what was then East Germany. Records at international competitions were shattered by athletes who owed more to the laboratory than the practice field. Steroids, one of the main drugs of choice of the times, became synonymous with cheating, grotesquely overgrown musculature, hair growth, and terrible acne. Other, less obvious adverse effects of steroids include impotence, gynecomastia (breast enlargement in males), gonadal atrophy, and uncontrollable bouts of aggression. This was seen in both male and female athletes abusing these drugs.

There are multiple classes of performance enhancing drugs that are used by athletes to help them excel. We tend not to think of ibuprofen or caffeine as performance enhancing, but without these drugs many people would have a harder time going on their daily run or training for a marathon. Performance enhancing drugs fall into the main categories of painkillers and stimulants, sedatives (used, for example, in shooting sports to steady a hand), diuretics (used by bodybuilders to increase muscle definition, and by boxers and wrestlers to lose water and thus qualify for a lower weight class), and those that promote muscle growth. Anabolic steroids such as methandrostenolone, which falls into this latter class, are synonymous with drugs in sport.

Methandrostenolone
(anabolic steroid)

BMS-564-929
(selective androgen receptor modulator)

When people think of anabolic steroids, testosterone and its adverse effects is one of the first steroids that comes to mind. In addition to muscle growth, testosterone promotes hair growth, changes in voice, and acne. Women taking testosterone will experience virilization, or the development of male secondary sexual characteristics; men will experience testicular atrophy. Although testosterone will promote muscle growth, athletes are often looking for a specific type of muscle growth, for example, bulk versus definition or speed versus endurance. They also want a drug that is harder to detect than testosterone. In response to these challenges, biochemists developed several different approaches.

The pharmaceutical industry, as common practice, makes numerous derivatives of a parent compound and tests each to see which gives the desired effects. Some of these drugs have legitimate uses in veterinary medicine or as male hormone replacement therapy, but when abused these drugs can provide gains in performance.

A second approach used in drug design employs molecules that are not active as taken but are metabolized or activated by the body. For example, androst-4-ene-3α,17α-diol, or "andro" for short, is an endogenous testosterone precursor used by several professional baseball players in the 1990s. This drug was subsequently banned because of its testosterone elevating properties. Using this strategy, levels of particular precursors or important regulatory steps in the biosynthesis can be bypassed, leading to elevated levels of the hormone of interest. Additionally, some of these precursors may also have androgenic properties.

A new class of drugs that is emerging is the selective androgen receptor modulators (SARMs). These compounds are not steroids in the truest sense because they lack the steroid skeleton; rather, they act as specific agonists or antagonists of the androgen receptor. These drugs can be taken orally, have tissue-specific effects, and cause fewer adverse effects; thus, they may be suitable for treatment of diseases such as wasting, osteoporosis, and prostate cancers. However, they are also being abused by athletes.

All of these drugs work via a similar mechanism. They first bind to steroid receptors in the cytosol; the hormone-receptor complex then translocates to the nucleus, where it binds to specific DNA promoter sequences and affects gene transcription.

To counter some of the adverse effects of anabolic steroids and decrease the chances of detection, masking agents are often used. Such agents include inactive steroid precursors, taken to give the appearance of normal steroid precursor ratios in tests, and diuretics that assist in the elimination of compounds from the body.

Steroids are detected through blood or urine analysis. Typically, the analysis combines some sort of extraction or concentration step (such as a solid-phase extraction) with a chromatographic one (gas chromatography). As new types of drugs or combinations or dosages and patterns of use (termed cycles) are developed, new testing methods are also developed. Recent detection methods employ mass spectrometry and look for unnatural isotopic ratios in the body.

Whether or not steroid abuse is detected, biology will eventually catch up with those taking steroids; that is, the user will eventually experience the adverse effects of taking steroids. Steroid hormones and other performance enhancing drugs, such as erythropoietin, often act through changing levels of gene expression. Thus, there is a danger of eliciting an abnormal growth response, namely, cancer. Lyle Alzado, a former professional football player, attributed the brain tumor that eventually killed him to his rampant steroid use.

© Peter Read Miller/ASSOCIATED PRESS

Time & Life Pictures/Getty Images

Summary

- Eicosanoids are powerful signaling molecules derived from 20-carbon long polyunsaturated fatty acids. Their production can be blocked through the use of NSAIDs.
- Endocannabinoids are a second group of signaling molecules produced from arachidonic acid.

Concept Check

1. Explain how are eicosanoids classified by the enzymes involved in their synthesis.
2. List the processes that eicosanoids mediate.
3. Describe how NSAIDs block production of some eicosanoids.

Bioinformatics Exercises

Exercise 1 Drawing and Naming Fatty Acids

Exercise 2 Introduction to Lipidomics

Exercise 3 Viewing and Analyzing Fatty Acid Synthase

Exercise 4 Lipid Metabolism and the KEGG Database

Problems

9.1 Properties, Nomenclature, and Biological Functions of Lipid Molecules

1. Examine the molecules in the figure.

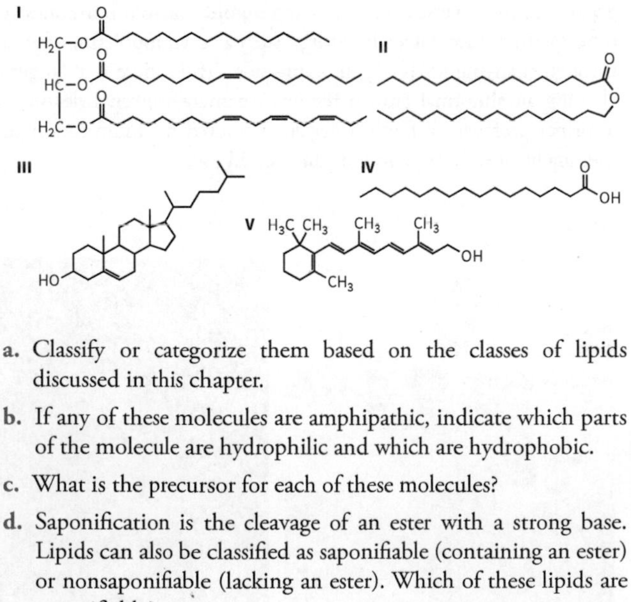

a. Classify or categorize them based on the classes of lipids discussed in this chapter.

b. If any of these molecules are amphipathic, indicate which parts of the molecule are hydrophilic and which are hydrophobic.

c. What is the precursor for each of these molecules?

d. Saponification is the cleavage of an ester with a strong base. Lipids can also be classified as saponifiable (containing an ester) or nonsaponifiable (lacking an ester). Which of these lipids are saponifiable?

2. Put each of the following series in order of increasing melting point.

a. stearate, palmitate, arachidate

b. oleate, stearate, linoleate

c. tripalmitin, tristearin, triolean

3. Examine the fatty acids in the figure used in problem 1 and:

a. Describe each using the delta nomenclature, such as $18:1\Delta^9$.

b. Describe each using the omega nomenclature, such as ω-9.

c. Give each an IUPAC name, such as octadecenoic acid, and a common name, for example, oleic acid.

4. Based on the structure and chemistry of phospholipids, how would they align and embed themselves in the lipid bilayer?

5. Lyso-phospholipids are glycerophospholipids lacking one of the two acyl chains. Based on the shape of these molecules, would it be likely that a solution of lyso-PC would form a bilayer in solution or would it form some other structure?

6. What are the advantages of storing energy in a reduced state?

7. Membrane phospholipids are often shown in cartoon form as a circle with two sticks coming out of it. Based on that model, propose cartoon stick figures for the following molecules:

I. lyso-phosphatidylethanolamine

II. cholesterol

III. cholesteryl ester

IV. 1-palmitoyl-2-linolyl-*sn*-glycero-3-phosphocholine

9.2 Fatty Acid Catabolism

8. When a molecule of steric acid (18:0) is catabolized through β-oxidation, how many molecules of acetyl-CoA and ATP' are generated? How many $FADH_2$? How many molecules of NADH/H$^+$?

9. How have organisms evolved to ensure that β-oxidation and fatty acid biosynthesis do not occur at the same time? How is this similar to carbohydrate metabolism?

10. Carnitine was popular for a short time as a dietary supplement to help people "burn fat." Do you think this approach would work?

11. How many ATP can be produced by the complete oxidation of palmitoleic acid, a 16-carbon monounsaturated fatty acid, considering that the fatty acid must first be activated?

12. Both oral contraceptives and anabolic steroids are sex hormones. Would you anticipate that the effects of one could replace the other? In other words, could anabolic steroids block ovulation or could oral contraceptives cause increased muscle growth?

9.3 Fatty Acid Biosynthesis

13. Describe in both general and specific terms how fatty acid biosynthesis is regulated.

14. In the formation of acetoacetyl-ACP, the CO_2 used to synthesize malonyl-CoA is lost, as shown in Figure 9.19. Would this help make the reaction more or less energetically favorable? Why? What thermodynamic parameters would this influence?

15. AICAR (5-amino-4-carboxamide ribonucleotide) is an experimental compound used in the laboratory but banned in competitive athletic events. It has been called "exercise in a pill." AICAR has been shown in the laboratory to selectively activate AMP kinase (AMPK). Describe what effect this might have on lipid metabolism.

16. C75 is a synthetic molecule that acts in the brain to block the feeding response. It acts in part by inhibiting FAS and blocking the malonyl-CoA mediated inhibition of CPT I. C75 also decreases the phosphorylation, and hence the activity, of AMPK. Taking these three effects into account, how is lipid metabolism affected at the cellular level?

17. Significant effort over the past six decades has been spent on determining the structure of both ACC and FAS. Why have scientists and funding agencies spent so much time and resources on these enzymes?

9.4 Ketone Body Metabolism

18. Bovine ketosis (elevated plasma ketone bodies) is a common disease among cattle that have recently given birth. Why would these animals become ketotic?

19. Ketones, actually acetoacetate, are often tested for in urine rather than in blood. How does assaying urine rather than blood change the test? Can the same assay be used for blood, plasma, and urine? What compounds and interfering substances need to be considered? How would taking readings from urine versus blood change the information the assay provides?

20. Acetoacetate can spontaneously decarboxylate in blood.

   a. What are the products of this reaction?

   b. Would the reaction proceed more rapidly under acidic or basic conditions, and would this matter?

21. Once released from the liver, ketone bodies travel through the circulatory system to tissues that can metabolize them—most notably the brain, although heart and skeletal muscle also have the ability to use ketone bodies. In these tissues, D-β-hydroxybutyrate is converted to acetoacetate by β-hydroxybutyrate dehydrogenase. What does this conversion tell us about the equilibrium and energetics of this reaction? Can heart and skeletal muscle synthesize ketone bodies? Why or why not?

22. What kind of diet is likely to promote the formation of ketone bodies?

23. Why is it unlikely that prokaryotes or plants would make ketone bodies?

9.5 Steroid Metabolism

24. Ketone bodies in urine are frequently measured with a dipstick assay. A plastic stick is dipped in urine, and a purple color develops in proportion to the amount of ketone present.

   a. Typically, these assays detect acetone and acetoacetate but not β-hydroxybutyrate. Would this complicate the interpretation of the result? Would the assay still be a valid measure of ketone production?

   b. The reagent in many commercially available ketone test strips is sodium nitroprusside. Does the ability of the test to detect acetate or acetoacetate and not β-hydroxybutyrate provide information on how these compounds would react?

   c. What does ketone dipstick assay tell you about normal levels of these compounds in urine? What other types of compounds might interfere?

   d. Why does the complex turn purple in the presence of ketones?

   e. If this test was used to assay ketones in the urine of a diabetic animal, would you expect the glucose in the urine to interfere with the result?

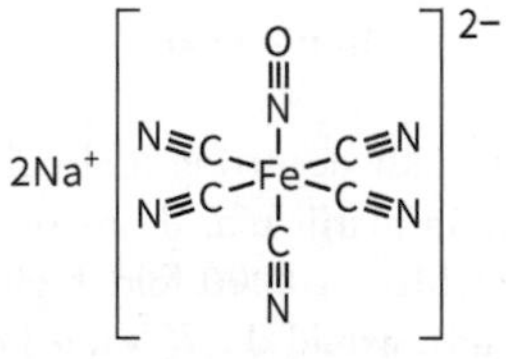

Prussian blue

25. Explain the difference between a sex hormone and a bile salt based on their structure.

26. Gallstones are small pebble-like aggregates of cholesterol and other bile components found in the gall bladder, the bile storage organ found below the liver. Although some small stones are comprised of bilirubin (a colored pigment), typical stones contain at least 80% cholesterol. The gall bladder can fill with these stones, creating a painful and dangerous situation that requires surgery. Two new therapies are being tested to remove gallstones. The first involves orally consuming bile salts, such as ursodeoxycholic acid and chenodeoxycholic acid; the second involves injecting methyl-tertbutyl ether into the gall bladder. Explain how each of these treatments would work.

27. One of the functions of cholesterol is regulating the fluidity of the plasma membrane of cells. Many species of cold water fish have low levels of cholesterol. How might such species maintain the fluidity of their membranes?

28. Pyrophosphomevalonate is synthesized by the sequential addition of phosphate groups to mevalonate. Draw the structures of mevalonate, 5-phosphomevalonate, and pyrophosphomevalonate. Where are the phosphate groups in pyrophosphomevalonate? Why are both groups not added simultaneously?

29. ADVAIR is an inhaled medication used to treat asthma. It is comprised of two compounds: fluticasone propionate and salmeterol. The structure of fluticasone is shown below. What type of molecule is this? How might it function?

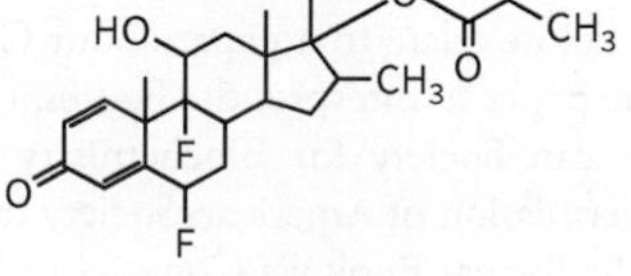

Fluticasone
(ADVAIR)

9.6 Eicosanoid and Endocannabinoid Metabolism

30. Why is it a poor idea for a pregnant woman to take drugs such as ibuprofen in the third trimester of pregnancy?

31. Rimonabant is a reverse agonist for the cannabinoid receptor CB-1. Several studies suggest that the CB-1 receptor ligand binding site is buried in the plasma membrane. If this is the case, would this change the way we think of typical receptor–ligand or enzyme–substrate interactions? What types of weak forces would be involved in the binding of this drug to the receptor?

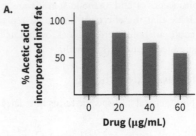

Rimonabant

32. There are many naturally occurring molecules related to THC, the active compound found in marijuana. If one of these molecules bound to the CB-1 receptor with a 10,000-fold higher affinity than anandamide (the native ligand), would the K_d value be higher or lower?

33. Capsaicin is the molecule that provides the heat in hot peppers.

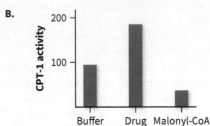

Capsaicin

It binds to vanilloid receptors and mediates signals of heat and pain to the brain. Drinking water fails to quench the heat of spicy foods, but eating dairy foods, especially vanilla ice cream), works relatively well. Explain why.

34. Aspirin is known to prevent heart attack. Explain the possible biochemical reason for this. What could be the biochemical reason behind this?

35. C17 fatty acids are often included as internal standards in high performance liquid chromatography (HPLC) or gas chromatography mass spectrometry (GC-MS) experiments.

    a. What is the purpose of an internal standard?

    b. Why is C17 often included as the internal standard?

36. Thin layer chromatography (TLC) is widely used technique for separating different lipid classes. TLC plates are typically coated with silica gel, a polar chromatography medium. If you wanted to resolve cholesteryl stearate and lysophosphatidate, would you obtain better results using a nonpolar mobile phase, such as hexanes, or a polar one, perhaps methanol?

Data Interpretation

37. The questions below relate to a paper about CPT activity. You do not need to read the paper to interpret the figures. (*Journal of Biological Chemistry* by American Society for Biochemistry & Molecular Biol. Reproduced with permission of American Society for Biochemistry and Molecular Biol in the format Book via Copyright Clearance Center).

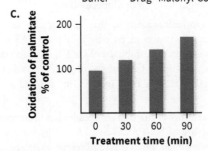

    a. What is shown in panel A? Why were so many concentrations of drug used? Would using one concentration of inhibitor and looking at different time points provide the same information?

    b. Based on panel B, how is the drug affecting CPT I activity?

    c. Describe what is happening in panel C. Are these data consistent with the observations in panels A and B?

38. Use the following figure to answer the questions.

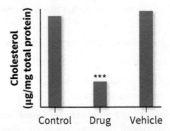

    a. What is the effect of the drug on cellular cholesterol levels?

    b. Why are the values on the *y*-axis not simply "cholesterol"?

    c. Based on this experiment, can you tell how the drug is working?

    d. Based on this experiment, can you hypothesize how the drug is working?

    e. What is the vehicle, and why did it elevate cholesterol levels?

The next series of experiments were run with radiolabeled cholesterol precursors.

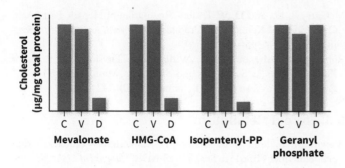

f. Based on these data, how would you propose the drug is functioning?

g. The graph shows no error bars or statistical analysis. Can you tell anything about the significance of the data without those analyses?

39. A collaborator has developed an assay in which she stimulates eicosanoid production in cultured cells by treating them with Indian cobra (*Naja naja atra*) venom. You attempt the assay but find that the cells have clearly begun to lyse. What is one potential conclusion that relates these two observations?

Experimental Design

40. Palmitic acid has a pK_a of 4.75. What percentage of this fatty acid would be ionized in the stomach at pH 2.0? In the plasma at pH 7.2? In a lysosome at pH 4.5?

41. Using radiolabeled carnitine, design an experiment to show that fatty acids must be first esterified to carnitine before they are transferred into the mitochondrial matrix.

42. Malonyl-CoA regulates β-oxidation by regulating fatty acid transport into mitochondria. How could you determine which CPT enzyme is being blocked by malonyl-CoA?

43. Before degradation, fatty acids must be activated by coupling them to coenzyme A. How could you show that coupling to CoA is an ATP-dependent process?

44. Design an experiment to show that humans can synthesize oleate (18:1$^{\Delta9}$) but not linolenate (18:2$^{\Delta9,12}$).

45. You are studying steroid metabolism in a species of plant. You hypothesize that some of the enzymes involved in squalene biosynthesis are in a single multi-enzyme complex, akin to fatty acid synthase. How would you test this hypothesis?

Ethics and Social Responsibility

46. Palm oil, also known as palm kernel oil, is a common ingredient in processed foods. Palm oil comes from tropical plantations and is typically raised in a manner that is not currently sustainable. Is it ethical to eat such foods?

47. Several COX II inhibitors come with a black box warning. These drugs have been linked to an increased risk of heart attack. All drugs have some level of risk and some side effects, some of which are positive, some adverse. What is an acceptable level of risk for a COX II inhibitor compared to a drug used to treat (a) metastatic bone cancer and (b) erectile dysfunction? Is the acceptance of risk the same in all cases? Should it be the same in all cases?

48. Babies born prematurely often have significant problems in breathing because they have not produced sufficient surfactant (a detergent-like combination of phospholipids and amphipathic proteins) to break the surface tension of water on the surface of the alveoli in their lung. Should government agencies devote more funding toward developing new treatments for these babies?

49. Sex hormones are commonly used as pharmaceuticals, often as birth control pills or in hormone replacement therapy. These hormones are excreted from the body and end up in waste water. Based on what you know of the function of these molecules, would you anticipate that their presence in waste water presents an environmental concern? What are the scientific questions pertinent to this discussion?

Suggested Readings

9.1 Properties, Nomenclature, and Biological Functions of Lipid Molecules

Ehnholm, Christian, ed. *Cellular Lipid Metabolism.* New York, New York: Springer, 2009.

Fahy, E., S. Subramaniam, R. C. Murphy, M. Nishijima, C. R. Raetz, T. Shimizu, F. Spener, et al. "Update of the LIPID MAPS Comprehensive Classification System for Lipids." *Journal of Lipid Research* 50 (April 2009): S9–14.

Gurr, I. M., ed. *Lipid Biochemistry: An Introduction.* 3rd ed. New York, New York: Springer, 1980.

The AOCS Lipid Library. Retrieved from http://lipidlibrary.aocs.org/index.html

Vance, D. E., and J. E. Vance, eds. *Biochemistry of Lipids, Lipoproteins, and Membranes.* 5th ed., New York, New York: Elsevier, 2008.

9.2 Fatty Acid Catabolism

Frayn, K. N. "Fat as a Fuel: Emerging Understanding of the Adipose Tissue-Skeletal Muscle Axis." *Acta Physiological (Oxford)* 199, no. 4 (August 2010): 509–18.

Goepfert, S., and Y. Poirier. "Beta-Oxidation in Fatty Acid Degradation and Beyond." *Current Opinion in Plant Biology* 10, no. 3 (June 2007): 245–51.

Houten, S. M., and R. J. Wanders. "A General Introduction to the Biochemistry of Mitochondrial Fatty Acid β-Oxidation." *Journal of Inherited Metabolic Disease* 33, no. 5 (October 2010): 469–77.

Wanders, R. J., and H. R. Waterham. "Biochemistry of Mammalian Peroxisomes Revisited." *Annual Review of Biochemistry* 75 (2006): 295–332.

9.3 Fatty Acid Biosynthesis

Baron, A., T. Migita, D. Tang, and M. Loda. "Fatty Acid Synthase: A Metabolic Oncogene in Prostate Cancer?" *Journal of Cellular Biochemistry* 91, no. 1 (January 1, 2004): 47–53.

Burdge, G. C., and P. C. Calder. "Conversion of α-Linolenic Acid to Longer-Chain Polyunsaturated Fatty Acids in Human Adults." *Reproduction Nutrition Development* 45, no. 5 (September–October 2005): 581–97.

Girard, J., D. Perdereau, F. Foufelle, C. Prip-Buus, and P. Ferré. "Regulation of Lipogenic Enzyme Gene Expression by Nutrients and

Hormones." *Federation of American Societies for Experimental Biology* 8, no. 1 (January 1994): 36–42.

Jakobsson, A., R. Westerberg, and A. Jacobsson. "Fatty Acid Elongases in Mammals: Their Regulation and Roles in Metabolism." *Progress in Lipid Research* 45, no. 3 (May 2006): 237–49.

Jitrapakdee, S., and J. C. Wallace. "The Biotin Enzyme Family: Conserved Structural Motifs and Domain Rearrangements." *Current Protein & Peptide Science* 4, no. 3 (June 2003): 217–29.

Kim, K. H. "Regulation of Mammalian Acetyl-Coenzyme A Carboxylase." *Annual Review of Nutrition* 17, (1997): 77–99.

Long, Y. C., and J. R. Zierath. "AMP-Activated Protein Kinase Signaling in Metabolic Regulation." *Journal of Clinical Investigation* 116, no. 7 (July 2006): 1776–83.

Ratledge, C. "Fatty Acid Biosynthesis in Microorganisms Being Used for Single Cell Oil Production." *Biochimie* 86, no. 11 (November 2004): 807–15.

Ruderman, N., and M. Prentki. "AMP Kinase and Malonyl-CoA: Targets for Therapy of the Metabolic Syndrome." *Nature Reviews Drug Discovery* 3, no. 4 (April 2004): 340–51.

Wakil, S. J., and L. A. Abu-Elheiga. "Fatty Acid Metabolism: Target for Metabolic Syndrome." *Journal of Lipid Research* 50 (April 2009): S138–43.

9.4 Ketone Body Metabolism

Baird, G. D. "Primary Ketosis in the High-Producing Dairy Cow: Clinical and Subclinical Disorders, Treatment, Prevention, and Outlook." *Journal of Dairy Science* 65, no. 1 (January 1982): 1–10.

Bergman, E. N. "Energy Contributions of Volatile Fatty Acids from the Gastrointestinal Tract in Various Species." *Physiological Review* 70, no. 2 (April 1990): 567–90.

Krebs, H. A. "The Regulation of the Release of Ketone Bodies by the Liver." *Advances on Enzyme Regulation* 4 (1966): 339–54.

McGarry, J. D., and D. W. Foster. "Regulation of Hepatic Fatty Acid Oxidation and Ketone Body Production." *Annual Review of Biochemistry* 49 (1980): 395–420.

Robinson, A. M., and D. H. Williamson. "Physiological Roles of Ketone Bodies as Substrates and Signals in Mammalian Tissues." *Physiological Reviews* 60, no. 1 (January 1980): 143–87.

9.5 Steroid Metabolism

Brown, M. S., and J. L. Goldstein. "Cholesterol Feedback: From Schoenheimer's Bottle to Scap's MELADL." *Journal of Lipid Research* 50 (April 2009): S15–27.

Goldstein, J. L., and M. S. Brown. "Regulation of the Mevalonate Pathway." *Nature* 343, no. 6257 (February 1, 1990): 425–30.

Russell, D. W., and K. D. Setchell. "Bile Acid Biosynthesis." *Biochemistry* 31, no. 20 (May 26, 1992): 4737–44.

9.6 Eicosanoid and Endocannabinoid Metabolism

Dennis, E. A., S. G. Rhee, M. M. Billah, and Y. A. Hannun. "Role of Phospholipase in Generating Lipid Second Messengers in Signal Transduction." *Federation of American Societies for Experimental Biology* 5, no. 7 (April 1991): 2068–77.

Funk, C. D. "Prostaglandins and Leukotrienes: Advances in Eicosanoid Biology." *Science* 294, no. 5548 (November 30, 2001): 1871–5.

Pertwee, R. G., A. C. Howlett, M. E. Abood, S. P. Alexander, V. Di Marzo, M. R. Elphick, P. J. Greasley, et al. "International Union of Basic and Clinical Pharmacology. LXXIX. Cannabinoid Receptors and Their ligands: Beyond CB_1 and CB_2." *Pharmacological Reviews* 62, no. 4 (December 2010): 588–631.

Shimizu, T., and L. S. Wolfe. "Arachidonic Acid Cascade and Signal Transduction." *Journal of Neurochemistry* 55, no. 1 (July 1990): 1–15.

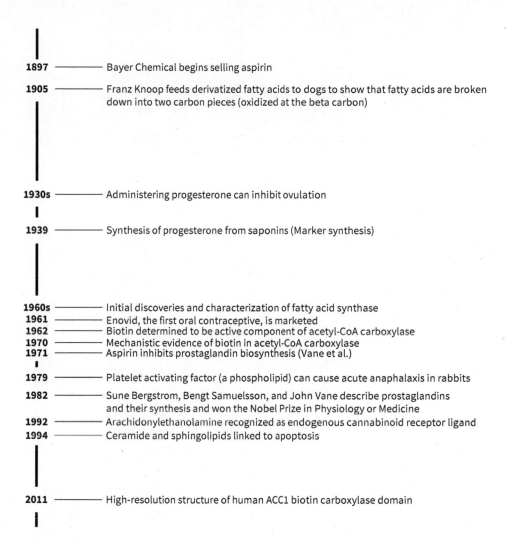

1897 ———————— Bayer Chemical begins selling aspirin

1905 ———————— Franz Knoop feeds derivatized fatty acids to dogs to show that fatty acids are broken down into two carbon pieces (oxidized at the beta carbon)

1930s ——————— Administering progesterone can inhibit ovulation

1939 ———————— Synthesis of progesterone from saponins (Marker synthesis)

1960s ——————— Initial discoveries and characterization of fatty acid synthase
1961 ———————— Enovid, the first oral contraceptive, is marketed
1962 ———————— Biotin determined to be active component of acetyl-CoA carboxylase
1970 ———————— Mechanistic evidence of biotin in acetyl-CoA carboxylase
1971 ———————— Aspirin inhibits prostaglandin biosynthesis (Vane et al.)

1979 ———————— Platelet activating factor (a phospholipid) can cause acute anaphalaxis in rabbits

1982 ———————— Sune Bergstrom, Bengt Samuelsson, and John Vane describe prostaglandins and their synthesis and won the Nobel Prize in Physiology or Medicine

1992 ———————— Arachidonylethanolamine recognized as endogenous cannabinoid receptor ligand

1994 ———————— Ceramide and sphingolipids linked to apoptosis

2011 ———————— High-resolution structure of human ACC1 biotin carboxylase domain

CHAPTER 10

Lipids II

Metabolism and Transport of Complex Lipids

Complex Lipids in Context

The United States imports large quantities of oil for domestic use. If this supply were stopped, it could disrupt life across the country. To act as a buffer in the event of a disruption to supplies, the U.S. maintains a large reserve of crude oil, known as the Strategic Petroleum Reserve (SPR). The SPR currently holds oil, which is enough to supply the total demands of the country for almost two months; however, not all of the supply is immediately accessible. Even at the maximum rate of pumping, it would take five months for the total amount of oil to be used. This is because the petroleum is stored as crude oil, rather than gasoline, heating oil, diesel, or jet fuel. The crude oil would need to be refined into those products before it could be used; it would also need to be transported from the SPR to a refinery and then to its final destination.

The SPR is in many ways a good analogy for the neutral lipid reserves of an organism. Animals store neutral lipids in specialized organelles in dedicated tissues that regulate the use of those lipids. Stored lipids provide energy for the organism in times of need and the starting materials for the synthesis of membranes. There are specific mechanisms for lipid mobilization—for example, for delivery to or release from tissues.

This chapter focuses on the synthesis and degradation of phospholipids and neutral lipids, and on the functions, regulation, transport, and trafficking of these molecules at the levels of both the cell and the organism.

Chapter Outline

Common Themes	
Evolution's outcomes are conserved.	• As seen in other chapters, pathways for synthesis of molecules common to all of life, such as phospholipids, are conserved throughout evolution.
	• As animals evolved, they adapted to different environments and challenges. Analysis of genes from different animals reveals that genes coding for proteins involved in neutral lipid transport and storage evolved more recently, and thus may address needs that are lacking in more primitive species.
Structure determines function.	• Lipoproteins and lipid storage droplets have a similar structure: an amphipathic coating of phospholipids and proteins surrounding a nonpolar core.
	• The proteins that coat lipoproteins or lipid storage droplets adhere to the surface of these structures via amphipathic helices or hydrophobic patches on the surface of the protein.
	• Many of the enzymes involved in lipid metabolism require a protein cofactor for optimal activity.
	• In higher organisms, nonpolar molecules require some type of soluble carrier to move them through the circulation.
Biochemical information is transferred, exchanged, and stored.	• The storage and mobilization of neutral lipid stores is tightly regulated by the actions of several hormones.
	• The regulation of the biosynthesis of neutral lipids and glycerophospholipids is coordinated through phosphatidate (PA) and lysophosphatidate (lyso PA) via lipins and acyltransferases.
Biomolecules are altered through pathways involving transformations of energy and matter.	• The metabolism of both glycerophospholipids and neutral lipids converge at PA and lyso PA.
	• Lipoproteins provide an elegant and specific system for delivering both dietary and *de novo* lipids to different tissues.
	• Lipid storage droplets are analogous to lipoproteins both in form and function. Lipolysis (the regulated breakdown of triacylglycerols in the lipid droplet) results in the mobilization of energy through the release of free fatty acids and glycerol into the plasma.

10.1 Phospholipid Metabolism

Phospholipids are amphipathic lipid molecules containing a phosphate group and can be broadly categorized by their backbone alcohol. Glycerophospholipids use glycerol as their backbone, whereas sphingolipids employ sphingosine. Both have a single phosphate esterified to the backbone alcohol. A second, head group alcohol, also esterified to the phosphate, imparts many of the physical and chemical properties to the phospholipid. In addition to being characterized by their backbone, phospholipids are also categorized and named by their head group (**Figure 10.1**).

The most common glycerophospholipids, discussed here, are phosphatidylcholine (PC), phosphatidylethanolamine (PE), phosphatidylserine (PS), phosphatidylinositol (PI), phosphatidate (PA, also referred to as phosphatidic acid), phosphatidylglycerol (PG), and cardiolipin (CL). The sphingolipids discussed in this chapter are ceramide, sphingomyelin, sphingosine, and glycosphingolipids.

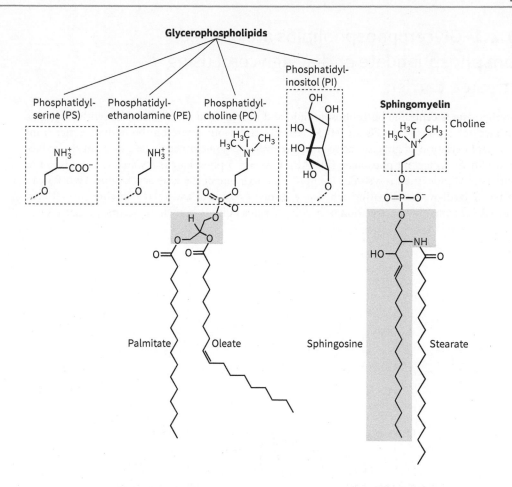

FIGURE 10.1 The structure of some common phospholipids. Shown are the glycerophospholipids phosphatidylserine, phosphatidylethanolamine, phosphatidylcholine, and phosphatidylinositol and the sphingolipid sphingomyelin.

Biosynthesis of phospholipids in eukaryotes occurs mainly in the endoplasmic reticulum (ER). However, phospholipids can be modified or remodeled in the plasma membrane; the type of modification varies depending on the location of the enzymes that catalyze the modifications (**Figure 10.2**). In prokaryotes phospholipid biosynthesis occurs primarily in the inner leaflet of the plasma membrane.

The synthesis of phospholipids is an anabolic process which requires energy that can come in several forms but is typically in the form of an activated carrier (as seen with glycogen synthesis and fatty acid metabolism, for example). In the case of phospholipid synthesis, the carrier is the nucleotide **cytidine diphosphate (CDP)**.

This section and section 10.2 highlight how phospholipid metabolism and neutral lipid metabolism are inextricably linked at several points.

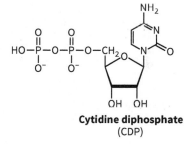

Cytidine diphosphate
(CDP)

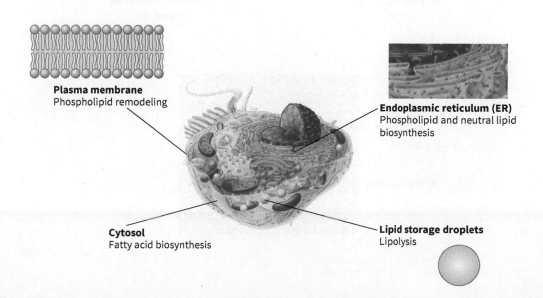

Plasma membrane
Phospholipid remodeling

Endoplasmic reticulum (ER)
Phospholipid and neutral lipid biosynthesis

Cytosol
Fatty acid biosynthesis

Lipid storage droplets
Lipolysis

FIGURE 10.2 Locations of lipid biosynthesis and degradation in the cell. Lipids are made and degraded in different organelles in the cell.

10.1.1 Glycerophospholipids are derived from phosphatidate or diacylglycerol using activated carriers

Phosphatidate is central to both phospholipid and neutral lipid metabolism. Its synthesis begins with glycerol-3-phosphate (**Figure 10.3**). The glycerol-3-phosphate can be derived either from the phosphorylation of glycerol, as seen in the liver and some other tissues, or from the reduction of dihydroxyacetone phosphate (DHAP). Glycerol-3-phosphate is acylated to form lyso PA. Lysophospholipids, such as lyso PA, are lipids that lack one of the fatty acyl chains. Lyso PA can also be derived by first acylating dihydroxyacetone phosphate and then reducing the acylated form. PA is formed by the acylation of lyso PA. At this point, a branching occurs, in that PA can

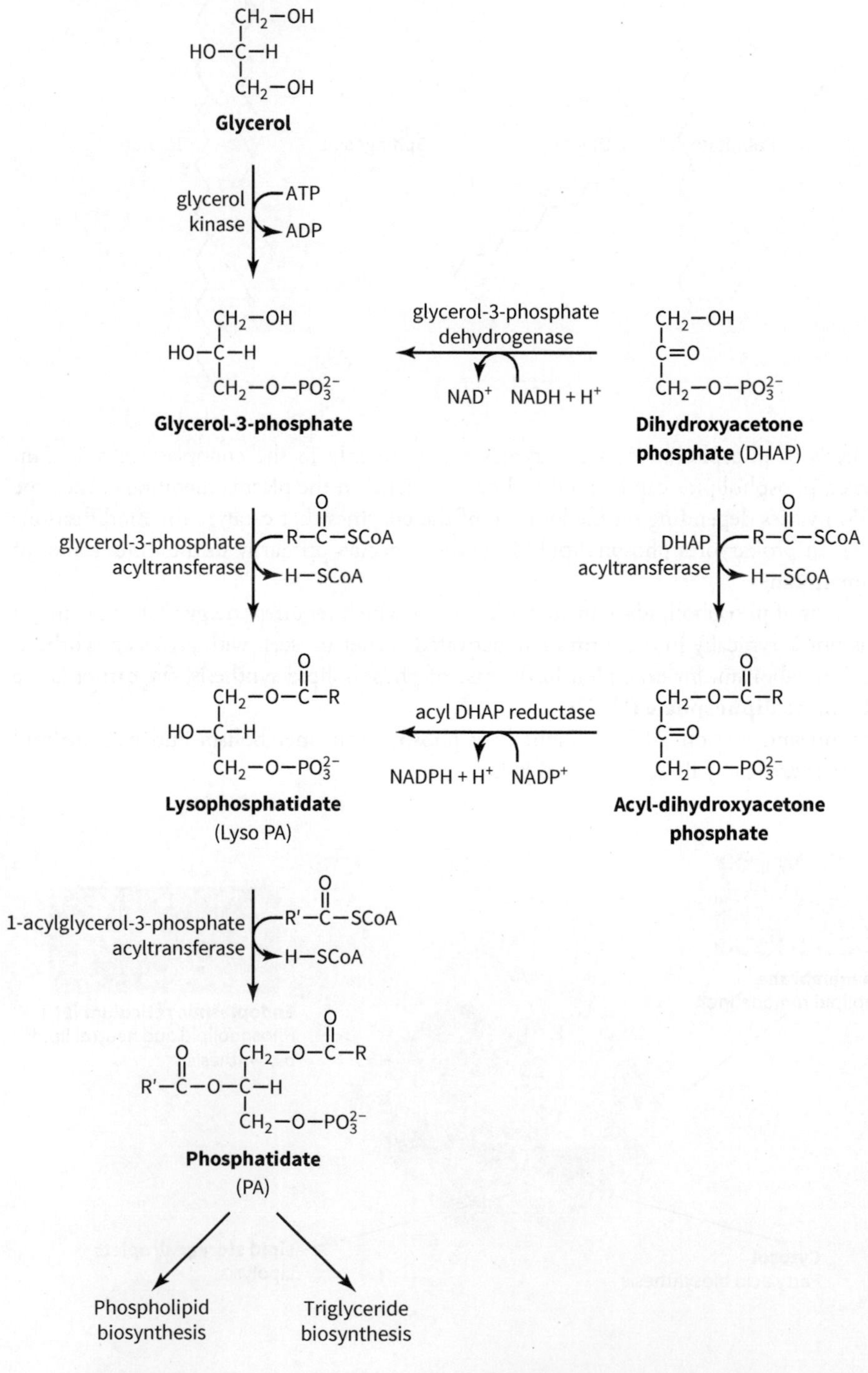

FIGURE 10.3 Phosphatidate (phosphatidic acid) biosynthesis. Phosphatidate is derived from either dihydroxyacetone phosphate or glycerol-3-phosphate.

be used as a building block for triacylglycerol biosynthesis or can be shuttled into phospholipid biosynthesis.

Phosphatidate can undergo an exchange reaction with CTP, catalyzed by CDP-diacylglycerol (DG) synthase, to yield CDP-DG and pyrophosphate, PP$_i$, which is further hydrolyzed into two molecules of inorganic phosphate, P$_i$ (**Figure 10.4**). Like other molecules joined to nucleotides,

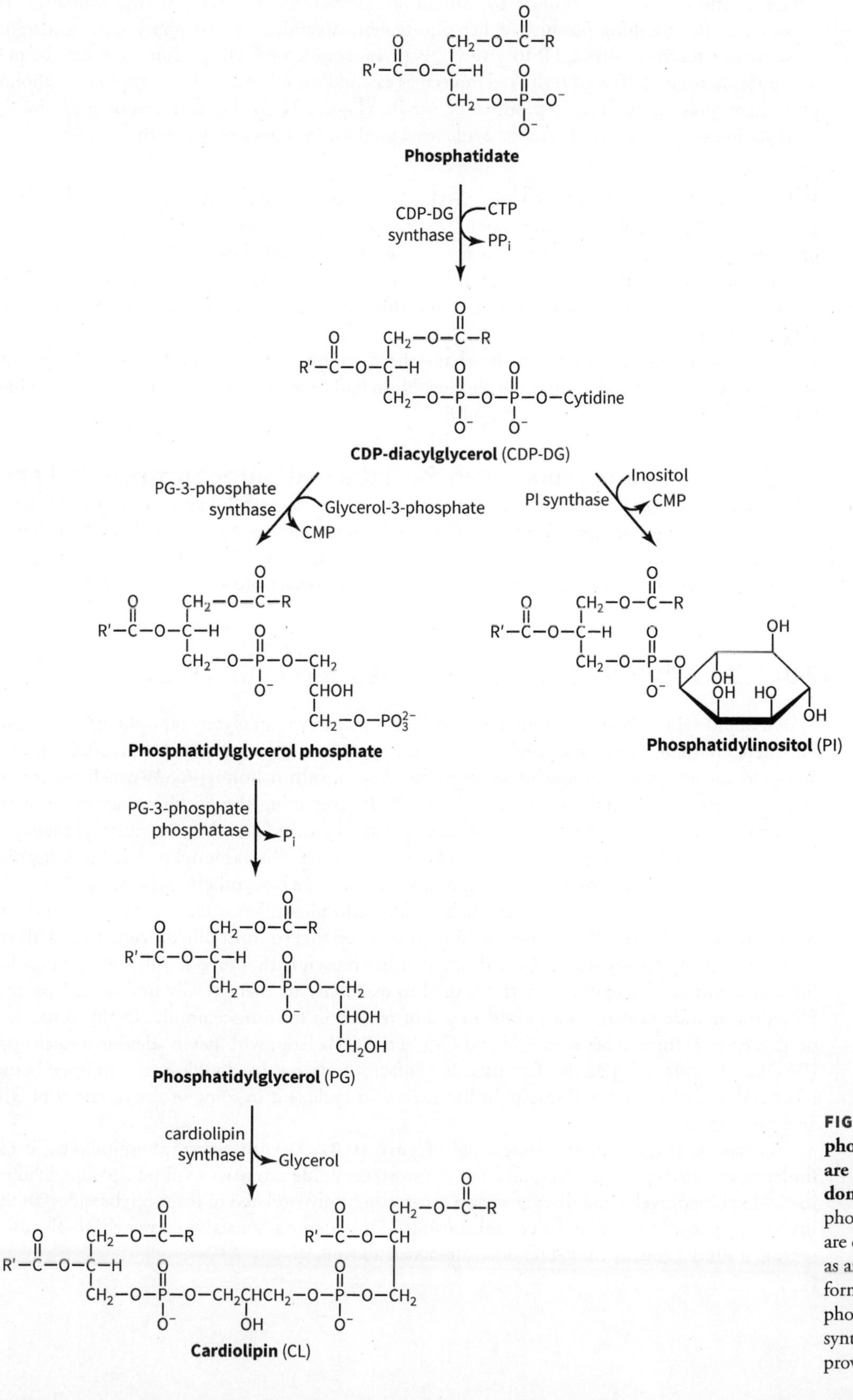

FIGURE 10.4 Phosphatidylinositol, phosphatidylglycerol and cardiolipin are built from activated diacylglycerol donors. The alcohol groups in phosphatidylinositol and diacylglycerol are coupled to CDP-DG, which acts as an activated carrier of diglyceride to form the phosphodiester bond in those phospholipids. In the mammalian synthesis of cardiolipin the alcohol is provided by phosphatidylglycerol.

CDP-DG is an active carrier of DG, and it undergoes exchange reactions with other molecules, such as inositol, to give phosphatidylinositol and CDP, glycerol-3-phosphate to give phosphatidylglycerol and CDP, and phosphatidylglycerol to give cardiolipin and CDP. In each of these cases, CDP acts as a carrier of DG.

Phosphatidylcholine and phosphatidylethanolamine also employ CDP in their biosynthesis, but in each case, CDP activates a different group. Ethanolamine is phosphorylated by ethanolamine kinase, and choline by choline kinase with ATP serving as the source of the phosphate. The resulting product, either phosphoethanolamine or phosphocholine, undergoes an exchange reaction with CTP to give CDP-ethanolamine or CDP-choline, respectively, plus pyrophosphate (PP_i). The CDP-linked molecules can add the ethanolamine phosphate or choline phosphate moieties to DG, to produce PC or PE (**Figure 10.5**). PC can also be made by the methylation of PE. In these reactions S-adenosyl methionine serves as the methyl donor.

Phosphatidylethanolamine can also be used as a phospholipid precursor

Phosphatidylethanolamine is one of the most common phospholipids in all organisms; it also serves as a precursor for several other phospholipids. We have just seen how PE can contribute to PC biosynthesis, but PE can also undergo head group exchange with serine, to form phosphatidylserine and ethanolamine. This reaction is catalyzed by PE serine transferase (**Figure 10.6**).

Some other phospholipids can also change head groups. For example, PC can exchange with serine to yield phosphatidylserine and choline. Phosphatidylserine can be decarboxylated by phosphatidylserine decarboxylase to produce PE.

Long-chain hydrocarbons can be attached through other linkages

While most often phospholipids have fatty acids attached via ester linkages to the glycerol backbone, other linkages are also observed. Ether lipids attach a long-chain fatty alcohol (analogous in structure to a fatty acid) to the glycerol backbone. In addition, ether lipids containing alcohols with double bonds between C-1 and C-2, termed plasmalogens, are also found in nature.

10.1.2 Sphingolipids are synthesized from ceramide

Sphingolipid synthesis follows a somewhat different pathway than glycerophospholipid synthesis, but there are many similarities such as the building up of molecules in a modular fashion from activated carriers. The synthesis of sphingolipids begins with palmitoyl-CoA, which reacts with serine to form 3-ketosphinganine (**Figure 10.7**). In examining the backbone structure of the sphingolipids, it may be helpful to recall the structure of serine and that the carboxyl moiety of serine is lost as CO_2 in the formation of 3-ketosphinganine. The carbonyl of 3-ketosphinganine is reduced to a hydroxyl group forming sphinganine. This is a biosynthetic pathway similar to the reactions of other anabolic pathways, such as fatty acid biosynthesis; hence, the electron donor for the reaction is NADPH. Dihydrosphingosine is acylated to form dihydroceramide, with the acyl donor being an acyl-CoA. As with many other biosynthetic processes, including phospholipid biosynthesis, an activated carrier is used to overcome an energetically unfavorable process. Dihydroceramide undergoes a second round of reduction to form ceramide. In this reaction, a double bond is formed between C-4 and C-5 in the acyl chain, with flavin adenine dinucleotide (FAD) as the cofactor. FAD is often used as a cofactor to introduce double bonds in several other biochemical pathways, for example, in the citric acid cycle and in some of the reactions of fatty acid biosynthesis.

Ceramide is the central sphingolipid (**Figure 10.8**). As with glycerophospholipids, it can undergo an exchange of head groups. For example, ceramide can react with phosphatidylcholine to yield sphingomyelin and diacylglycerol; it can also be glycosylated to form cerebrosides. In this instance, glucose is added to the ceramide using CDP-glucose as an active carrier. CDP-glucose is serving a similar role as CDP-DG does (discussed above).

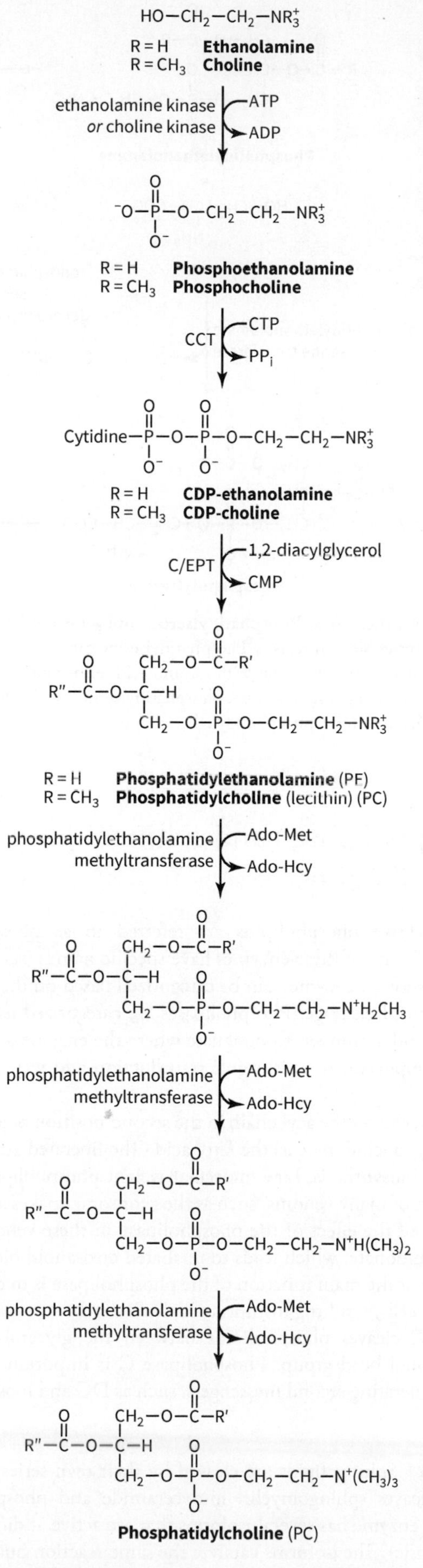

CCT=
CTP:phosphoethanolamine cytidyltransferase
or
CTP:phosphocholine cytdyltransferase

C/EPT=
CDP-ethanolamine:1,2-diacylglycerol phosphoethanolamine tranferase
or
CDP-choline:1,2-diacylglycerol phosphocholine transferase

Ado-Met=S-adenosylmethionine
Ado-Hcy=S-adenosylhomocysteine

FIGURE 10.5 Phosphatidylethanolamine and phosphatidylcholine biosynthesis employs an activated alcohol. Phosphatidylethanolamine is synthesized from diacylglycerol and an activated alcohol; in this case, the alcohol is activated by CTP resulting in CDP-ethanolamine or CDP-choline. Phosphatidylcholine can also be synthesized by the successive trimethylation of phosphatidylethanolamine. In this instance, *S*-adenosyl methionine serves as the methyl donor.

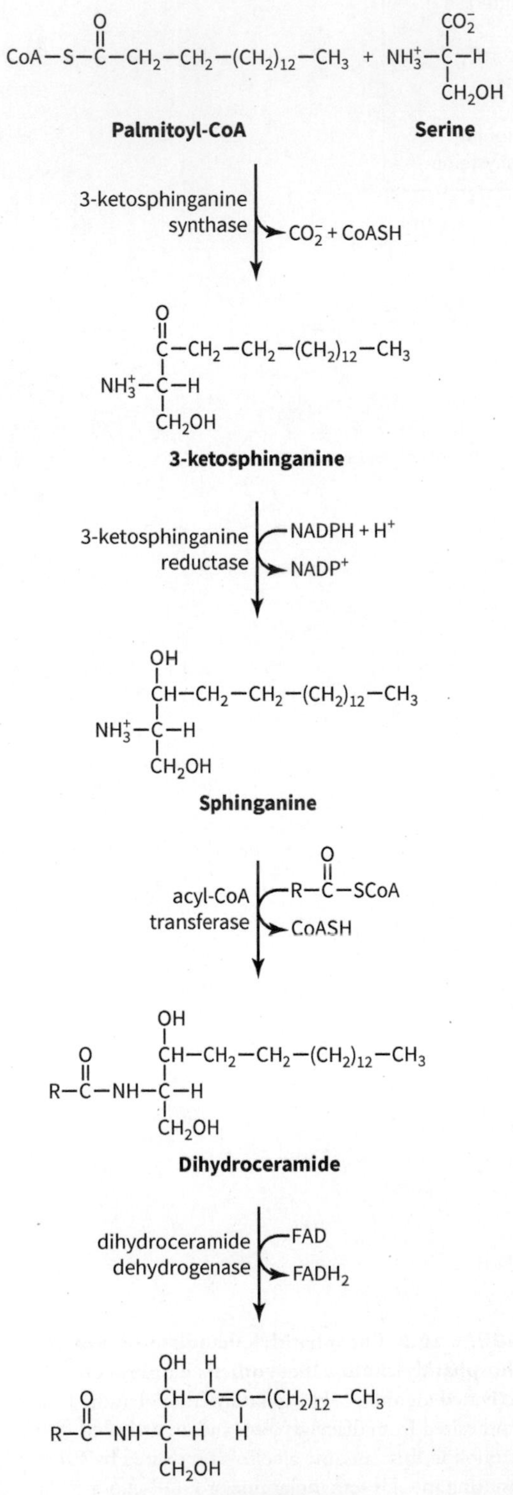

$$CoA-S-\overset{\overset{O}{\|}}{C}-CH_2-CH_2-(CH_2)_{12}-CH_3 \quad + \quad \overset{\overset{CO_2^-}{|}}{\underset{\overset{|}{CH_2OH}}{\overset{|}{NH_3^+-C-H}}}$$

Palmitoyl-CoA   **Serine**

3-ketosphinganine synthase $\searrow CO_2^- + CoASH$

$$\overset{\overset{O}{\|}}{C}-CH_2-CH_2-(CH_2)_{12}-CH_3$$
$$NH_3^+-\overset{|}{\underset{|}{C}}-H$$
$$CH_2OH$$

3-ketosphinganine

3-ketosphinganine reductase $\left\langle \begin{array}{l} NADPH + H^+ \\ NADP^+ \end{array} \right.$

$$\overset{OH}{\underset{}{CH}}-CH_2-CH_2-(CH_2)_{12}-CH_3$$
$$NH_3^+-\overset{|}{\underset{|}{C}}-H$$
$$CH_2OH$$

Sphinganine

acyl-CoA transferase $\left\langle \begin{array}{l} R-\overset{O}{\overset{\|}{C}}-SCoA \\ CoASH \end{array} \right.$

$$R-\overset{O}{\overset{\|}{C}}-NH-\overset{\overset{OH}{|}}{\underset{\overset{|}{CH_2OH}}{\overset{|}{C}-H}}\begin{array}{l}CH-CH_2-CH_2-(CH_2)_{12}-CH_3\end{array}$$

Dihydroceramide

dihydroceramide dehydrogenase $\left\langle \begin{array}{l} FAD \\ FADH_2 \end{array} \right.$

$$R-\overset{O}{\overset{\|}{C}}-NH-\overset{\overset{OH\ H}{|\ \ |}}{\underset{\overset{|\ \ |}{CH_2OH}}{\overset{|\ \ |}{C}-H\ H}}\begin{array}{l}CH-C=C-(CH_2)_{12}-CH_3\end{array}$$

Ceramide

FIGURE 10.7 Sphingolipid biosynthesis. Several common sphingolipids are intermediates in the biosynthesis of ceramide.

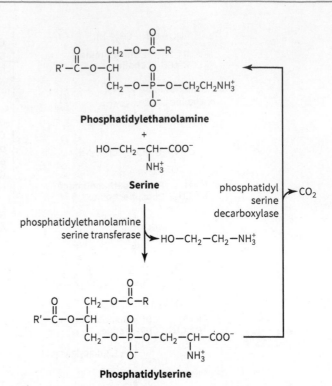

Phosphatidylethanolamine

+

$$HO-CH_2-\overset{\overset{|}{CH}-COO^-}{\underset{NH_3^+}{}}$$

Serine

phosphatidylethanolamine serine transferase $\searrow HO-CH_2-CH_2-NH_3^+$

phosphatidyl serine decarboxylase $\succ CO_2$

Phosphatidylserine

FIGURE 10.6 Phosphatidylserine and phosphatidylethanolamine biosynthesis. Phosphatidylserine can be derived through an exchange reaction of the amino acid serine and phosphatidylethanolamine. Likewise, phosphatidylserine can be decarboxylated to form phosphatidylethanolamine.

10.1.3 Phospholipases and sphingolipases cleave at specific sites

The enzymes that cleave phospholipids are referred to as phospholipases (**Figure 10.9**). While some of these enzymes have specific names (such as lipin) related to their discovery, all enzymes can be categorized based on the substrates they cut and the point of cleavage. Phospholipases are categorized using a capital letter, or a letter and a number, to designate where the enzyme acts. Two of these enzymes play important roles in signal transduction and were previously mentioned.

Phospholipase A_2 cleaves the acyl chain in the second position of a phospholipid, typically freeing arachidonate as the fatty acid. The liberated arachidonate is used in eicosanoid biosynthesis. Low molecular weight phospholipases A_2 are common components of many venoms, such as those of some snakes and of bees (**Figure 10.10**). Part of the effect of the phospholipase in these venoms is due to the release of arachidonate, which leads to unabated eicosanoid biosynthesis; however, it is likely that the main function of the phospholipase is to destroy the plasma membrane of cells, resulting in rupture and cell death.

Phospholipase C cleaves phospholipids between the glycerol backbone and the phosphoalcohol head group. Phospholipase C is important in several signaling cascades, generating second messengers, such as DG and inositol-1,4,5-trisphosphate.

Sphingolipases Sphingolipids are cleaved by their own series of lipases. Sphingomyelinase cleaves sphingomyelin into ceramide and phosphocholine (**Figure 10.11**). This enzyme has several isoforms that are active at different pHs (neutral, acidic, or basic). The isoforms catalyze the same reaction but are found

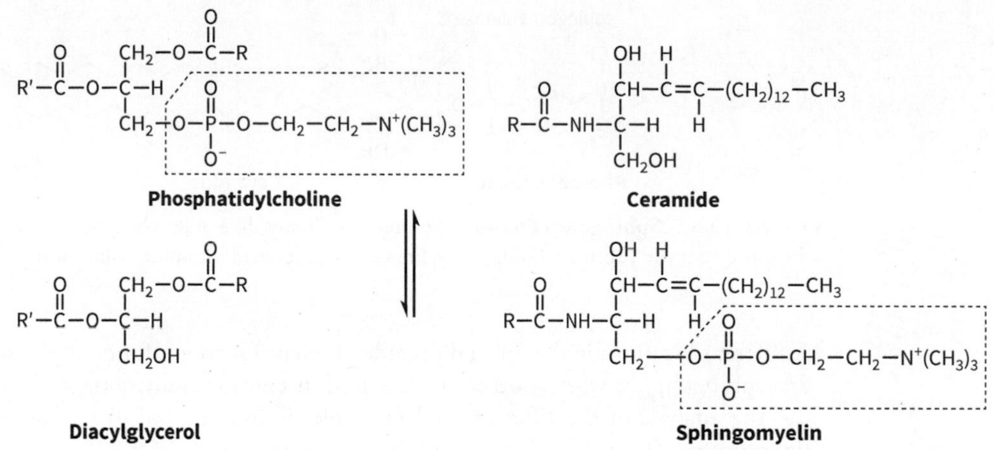

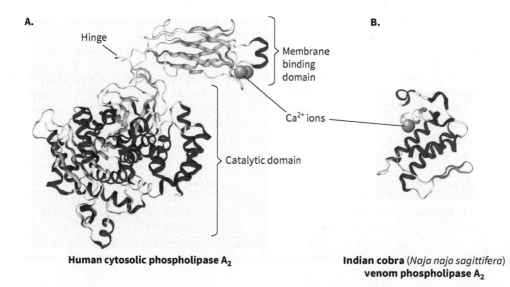

Phosphatidylcholine

Diacylglycerol

Ceramide

Sphingomyelin

FIGURE 10.8 Sphingomyelin can undergo head group exchange to produce other sphingolipids. Ceramide can be generated by the action of sphingomyelinases on sphingomyelin. It can undergo exchange reactions with other glycerophospholipids to exchange phospho head groups (highlighted with a dashed line) and generate diacylglycerol. In this instance, ceramide exchanges with phosphatidylcholine to yield sphingomyelin and diacylglycerol.

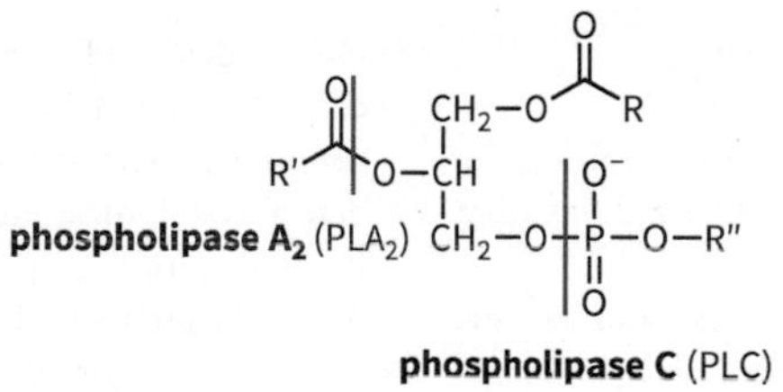

phospholipase A₂ (PLA₂)

phospholipase C (PLC)

FIGURE 10.9 Phospholipase cut sites. Phospholipases are named and categorized depending on where they cleave the phospholipid molecule. Phospholipases A$_2$ (PLA$_2$) cleaves the ester linkage in the 2 position of the glycerol backbone. Phospholipase C (PLC) cleaves on the backbone side of the phosphodiester linkage.

A.

Hinge

Membrane binding domain

Ca^{2+} ions

Catalytic domain

Human cytosolic phospholipase A₂

B.

Indian cobra (*Naja naja sagittifera*) **venom phospholipase A₂**

FIGURE 10.10 Structure of two different phospholipases A$_2$. **A.** Shown is the structure of a human cytosolic phospholipase A$_2$. The protein has two domains; the smaller of the two is a C2 domain and is important for binding Ca^{2+} and the lipid bilayer. This domain is primarily β sheet. The larger of the two domains is the catalytic domain. The C2 domain also acts as a lid to close down on the active site preventing substrate from binding in the active site before activation has occurred. Binding of Ca^{2+} by the C2 domain results in translocation of the enzyme from the cytosol to the membrane where liberation of arachidonate initiates prostaglandin production. **B.** The structure of a venom phospholipase from *Naja naja sagittifera* (Indian cobra) is shown. In contrast to the human cytosolic PLA$_2$, this is a much smaller molecule and is not as strictly regulated. Venom phospholipases lack the lid structure seen in other lipases.

(Source: (A) Data from PDB ID 1CJY Dessen, A., Tang, J., Schmidt, H., Stahl, M., Clark, J.D., Seehra, J., Somers, W.S. (1999) Crystal structure of human cytosolic phospholipase A2 reveals a novel topology and catalytic mechanism. *Cell* 97: 349–360) (B) Data from PDB ID 1YXH Jabeen, T., Singh, N., Singh, R.K., Ethayathulla, A.S., Sharma, S., Srinivasan, A., Singh, T.P. (2005) Crystal structure of a novel phospholipase A(2) from *Naja naja sagittifera* with a strong anticoagulant activity *Toxicon* 46: 865–875)

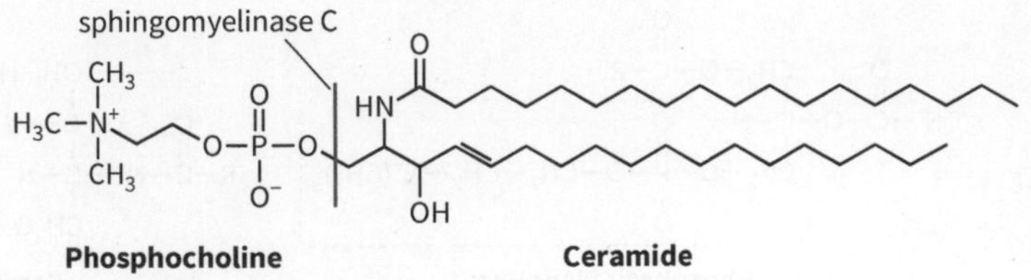

Phosphocholine　　　　　　　　**Ceramide**

FIGURE 10.11 Sphingomyelinases. Sphingomyelinases, like other phospholipases, are categorized by where they cut the substrate. Sphingomyelinase C cuts to yield phosphocholine and ceramide.

in different locations in the cell and provide different functions; for example, some modify structural sphingolipids, whereas others are involved in signal transduction.

The analysis of the different lipid molecules found in a cell, tissue, or organism is termed "lipodomics." This field is discussed in **Biochemistry: Lipidomics**.

Biochemistry

Lipidomics

Just as genomics is the study of the genome, and proteomics is the study of the proteome, lipidomics is the study of the lipidome, that is, the lipids of an organism, tissue, or cell.

As techniques and computational power increase and become more accessible, "–omics-based" approaches are another investigative tool for the researcher. Rather than asking a single specific question, such as "Do the levels of dioleoyl phosphatidylcholine in this cell increase when it is treated with drug X?," it is possible to ask a much broader question, such as "When cells are treated with drug X, what happens to the levels of each and every lipid in the cell?" The strength of this approach should be apparent. As with other bioinformatic studies, there is a plethora of potentially useful data, but sifting through that data continues to be challenging.

Lipid samples are, in many ways, simpler to manipulate than other types of biomolecules. Being small, largely hydrophobic, organic molecules, lipids can be isolated from tissues via simple liquid extraction with organic solvents. Care must be taken to avoid oxidation of sensitive species such as aldehydes. Likewise, the concentrations of some bioactive lipids may be at low levels. Analytical techniques, such as solid phase extraction, can be employed to concentrate these trace lipids.

One of the main workhorses of lipidomics is mass spectrometry. Although traditional mass spectrometry (GC-MS or FAB) has been used for decades to assist in the characterization of lipids, newer advances in ionization techniques have fewer drawbacks and can be used to detect whole molecules and polar lipids; these techniques are commonly used in lipidomics. The two most common ionization techniques, electrospray ionization (ESI) and matrix assisted laser desorption ionization (MALDI), permit the ionization of samples isolated from high performance liquid chromatography (HPLC) or other chromatographic systems. Recent advances in MALDI techniques have permitted the direct desorption of lipids from a slice of tissue, so called *in situ* mass spectrometry. These ionization methods can be coupled with other ion separation and detection techniques for analysis.

It used to be difficult to identify a sample based on spectral data. The spectra of known molecules were contained in incredibly large tomes of data, and an investigator had to sift through this material to characterize a sample. In the case of mass spectrometry, analysis of fragmentation patterns gave indications as to the functional groups present. Advances in computational power and software have made it easier to translate the raw mass spectrometry data into candidate lipids from a database. Likewise, computers have facilitated the development of metabolic maps of lipidomic data. Thus, advances in techniques make these experiments possible, and computers and software have made it much easier to interpret data. An example of this is the lipidomics profile of tears. Thanks to lipidomic studies, it is now known that tears contain hundreds of different lipids and that the composition of tears changes in different disease states. The individual roles and functions of many of these lipids remain unknown.

Clearly the future of these techniques holds considerable promise for lipid profiling.

Worked Problem 10.1　　Along came a spider . . .

Many venoms contain phospholipases that are important in the mechanism of action of venom. Your research laboratory has been characterizing venom from a newly discovered species of spider of the family *Sicariidae*, a group that includes the brown recluse spider. One of the researchers treated an extract of cellular lipids with the spider venom, performed thin layer chromatography (TLC) on the sample, and visualized them. The results of this experiment are shown below. Based on this information, propose one component of the venom.

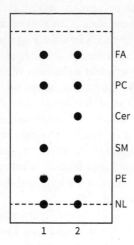

Lane 1 = control lipid
Lane 2 = venom treated lipid

Abbreviations: Cer, ceramide; FFA, free fatty acids;
NL, neutral lipid; PC, phosphatidylcholine;
PE, phosphatidylethanolamine; SM, sphingomyelin.

Strategy   Compare the lipid composition between the experimental venom-treated sample and the control. How have the lipids been affected by treatment with the venom? Based on the discussions in section 10.1, propose one or more enzymes that might be in the sample.

Solution   Many venoms contain phospholipases. TLC is a simple and inexpensive way to separate and analyze phospholipids. In these samples, treatment of the cellular lipid extract with the venom led to the loss of sphingomyelin and the accumulation of ceramide. These results indicate that the enzyme is probably a sphingomyelinase acting as a phospholipase C (see Figure 10.11). To verify this finding, the venom would have to be separated into its component proteins to check that it does not contain combinations of other enzymes that yield similar results to those seen with a sphingomyelinase.

Follow-up question   Describe the biological ramifications of this reaction in an organism bitten by this newly discovered species of spider.

Summary

- Glycerophospholipids have a glycerol backbone; sphingolipids have a sphingosine backbone.
- Phospholipases cleave glycerophospholipids at specific sites; sphingolipases do the same for sphingolipids.
- Glycerophospholipids and sphingolipids are built up and broken down in a modular fashion, using activated carriers.
- CDP is often used as the active carrier in biosynthesis of phospholipids. Glycerophospholipids are mainly derived from either DG or PA; sphingolipids are mainly derived from ceramide.

Concept Check

1. Describe how the phospholipids are synthesized and interconverted.
2. What is meant by *de novo* phospholipid biosynthesis?

10.2 Digestion of Triacylglycerols

Triacylglycerols represent a particularly rich source of energy, providing 9 kcal/g of energy, which is twice that of carbohydrates and proteins. Triacylglycerols are therefore an excellent way for an organism to store energy. They are also hydrophobic and unlike proteins or carbohydrates are stored in the absence of water. This too is an advantage for an organism that is mobile because storing energy as triglycerides means the organism can store energy without having water to keep these molecules in solution. However, due to their hydrophobic nature, digestion and transport of triacylglycerols is not as simple as that of carbohydrates and proteins.

10.2.1 Triacylglycerol digestion begins in the gastrointestinal tract

There are two locations in the body where triacylglycerol breakdown occurs. Each has a different class of lipase associated with it.

In animals, minor amounts of lipid degradation occur in the mouth and stomach. These reactions are catalyzed by lingual lipase, produced on the surface of the tongue, and gastric lipase, produced in the stomach. These lipases degrade triacylglycerols into diacylglycerol and free fatty acid. They are most active at low pH values and are characterized as acidic lipases. While these lipases are clearly important in digestion, they are not as efficient or as important as pancreatic lipase. Neither lingual lipase nor gastric lipase requires other protein cofactors for maximal activity.

Lipid digestion begins in earnest in the upper small intestine, referred to as the duodenum (**Figure 10.12**). Bile, comprised of bile salts, cholesterol and apolipoproteins, is secreted from the gallbladder, through the common bile duct, and into the duodenum. Bile breaks up and solubilizes dietary triacylglycerols in the intestinal lumen. In effect, the components of bile act as detergents, breaking large globules of dietary lipid into smaller micelles of lipid coated with bile. This emulsification greatly increases the surface area of the lipid and thus provides pancreatic lipase greater access to the triacylglycerol in the micelles.

The main step in the degradation of dietary lipids is the hydrolysis of the ester linkage between the fatty acids of a triacylglycerol and the glycerol backbone. This reaction is catalyzed by pancreatic lipase, which is secreted from the acinar cells of the pancreas and delivered through the pancreatic duct to the common bile duct and the duodenum. While the pancreas may be best known for the pancreatic islets that produce the endocrine hormone insulin, the exocrine tissues of the pancreas, including the acinar cells, form 80% of the mass of the pancreas and are responsible for the production of hydrolytic enzymes.

There are three major differences in the two lipases of the upper gastrointestinal tract, referred to as acidic lipases, and the single lipase of the lower gastrointestinal tract, pancreatic lipase.

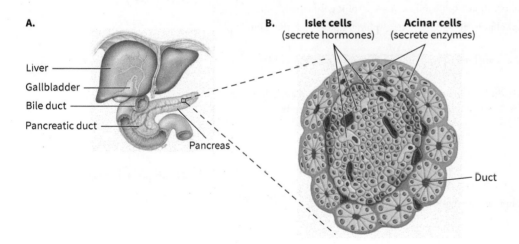

A.

Liver
Gallbladder
Bile duct
Pancreatic duct
Pancreas

B. **Islet cells** (secrete hormones) **Acinar cells** (secrete enzymes)

Duct

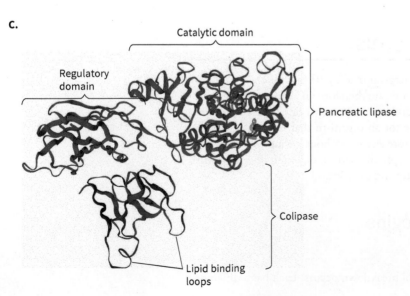

C.

Catalytic domain

Regulatory domain

Pancreatic lipase

Colipase

Lipid binding loops

FIGURE 10.12 Colipase and pancreatic lipase. A. Pancreatic lipase is made in the acinar cells of the pancreas. These cells form 90% of the mass of the pancreas and also secrete proteolytic enzymes. **B.** Acinar cells are different from the islet cells that produce the hormones insulin and glucagon. **C.** Pancreatic lipase consists of two domains, a larger, amino-terminal catalytic domain (right) and a smaller carboxy-terminal regulatory domain (left). The regulatory domain interacts with colipase, which is responsible for binding to lipid micelles. Colipase has three hydrophobic loops, termed "fingers" (two of which are visible), through which lipid binding occurs.

(Source: (A, B) Tortora; Derrickson, *Principles of Anatomy & Physiology*, 13e, copyright 2012, John Wiley & Sons. This material is reproduced with permission of John Wiley & Sons, Inc.
(C) Data from PDB ID 1N8S van Tilbeurgh, H., Sarda, L., Verger, R., Cambillau, C. (1992) Structure of the pancreatic lipase-procolipase complex. *Nature* 359: 159–162)

- Lingual and gastric lipases are active at acidic pH values (pH 3–6), whereas pancreatic lipase is active at alkaline pH. As partially digested food, referred to as chyme, leaves the stomach and moves into the duodenum and further down the digestive tract, the pH rises. As this happens, lingual and gastric lipases lose activity, and pancreatic lipase becomes more active.

- Lingual and gastric lipases are regulated differently. Acidic lipases do not require a protein cofactor for optimal activity, whereas pancreatic lipase requires a protein cofactor, known as colipase to achieve maximal catalysis.

- Lingual and gastric lipases have different substrates. In contrast to the acidic lipases, pancreatic lipase can act on either triacylglycerol (hydrolyzing it to diacylglycerol and a free fatty acid) or diacylglycerol itself (producing monoacylglycerol and a second molecule of free fatty acid). Therefore, pancreatic lipase can degrade one molecule of triacylglycerol into two molecules of free fatty acids and one molecule of 2-monoacylglycerol, all of which can be absorbed by the small intestine and metabolized.

Some lipases require protein cofactors for optimal activity The cofactor colipase is a small protein that binds to both pancreatic lipase and the micelle. In effect, colipase helps to anchor pancreatic lipase to the micelle and increase the availability of substrate. It also binds pancreatic lipase via the enzyme's regulatory domain in the non-catalytic carboxy-terminus in an active conformation, helping to stabilize the active state of the enzyme. Other enzymes that have hydrophobic substrates, such as lipoxygenases, have domains that include structures similar to the carboxy-terminus of pancreatic lipase. These domains help the enzyme to bind to the micelle, lipoprotein, or membrane. It is unclear whether these enzymes also require protein cofactors for optimal activity. As discussed in sections 10.3 and 10.6, several lipases including lipoprotein lipase, hormone sensitive lipase, and adipocyte triacylglycerol lipase require protein cofactors for maximal activity.

10.2.2 Dietary lipids are absorbed in the small intestine and pass into lymph before entering the circulation

Lipid micelles composed of monoacylglycerols, fatty acids, and cholesterol in the small intestine are absorbed by enterocytes. Within the enterocytes, the absorbance of these molecules is thought to occur predominantly by passive diffusion, although specific transporters may be involved in the absorbance of cholesterol and some fatty acids. Bile appears to be necessary for this process to occur; in the absence of bile, the diffusion of lipid to the enterocytes is so slow that it leads to lipid malabsorption or the inability to absorb dietary fat. The luminal side of these cells is the brush border, a surface covered in countless microvilli—tiny fingers of membrane that massively increase the surface area and facilitate absorption. Several properties help to drive diffusion of molecules into the enterocyte. For example, the fatty acids and 2-monoacylglycerols produced by the action of pancreatic lipase are rapidly absorbed by the enterocyte. Although it is unclear how these molecules cross the plasma membrane, it is generally accepted that, once they do, the two lipids are quickly reunited in the endoplasmic reticulum (ER). Likewise, the rapid formation of triacylglycerols and secretion into lymph all act to establish a concentration gradient of lipids across the enterocyte and thus facilitate diffusion.

In the enterocyte, the newly reassembled triacylglycerols from dietary lipids are packaged into a large particle called a **chylomicron** (plural **chylomicra**), which is used to transport dietary lipids in the circulation via the lymph. Chylomicra are a type of lipoprotein composed of a spherical assembly of amphipathic proteins and phospholipids surrounding a neutral lipid core. By entering the lymph first, lipids bypass the liver on their entry into the circulation (**Figure 10.13**).

Short-chain fatty acids (two to four carbons in length) are not re-esterified and are either directly metabolized by the enterocyte or enter the circulation through capillaries in the small intestine. Medium-chain fatty acids are re-esterified into triacylglycerols but are not packaged into chylomicra, and like short-chain fatty acids, enter the circulation directly through capillaries.

In enterocytes, triacylglycerols are formed by the 2-monoacylglycerol pathway. Although this pathway predominates in enterocytes, it is not how triacylglycerols are usually synthesized in other tissues, such as adipose or liver. In the 2-monoacylglycerol pathway, free fatty acids from the diet

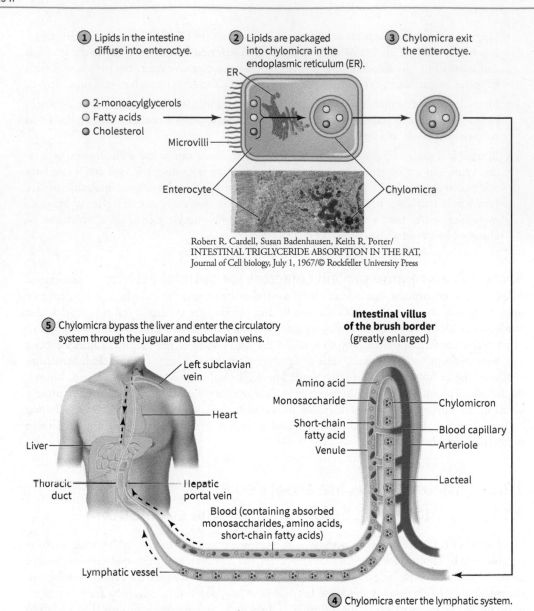

① Lipids in the intestine diffuse into enteroctye.

② Lipids are packaged into chylomicra in the endoplasmic reticulum (ER).

③ Chylomicra exit the enteroctye.

ER

● 2-monoacylglycerols
○ Fatty acids
● Cholesterol

Microvilli

Enterocyte

Chylomicra

Robert R. Cardell, Susan Badenhausen, Keith R. Porter/
INTESTINAL TRIGLYCERIDE ABSORPTION IN THE RAT,
Journal of Cell biology, July 1, 1967/© Rockfeller University Press

⑤ Chylomicra bypass the liver and enter the circulatory system through the jugular and subclavian veins.

Intestinal villus of the brush border (greatly enlarged)

Left subclavian vein

Heart

Liver

Thoracic duct

Hepatic portal vein

Amino acid
Monosaccharide
Short-chain fatty acid
Venule

Chylomicron
Blood capillary
Arteriole

Lacteal

Blood (containing absorbed monosaccharides, amino acids, short-chain fatty acids)

Lymphatic vessel

④ Chylomicra enter the lymphatic system.

FIGURE 10.13 Assembly and secretion of chylomicra into lymph. The microanatomy of the brush border helps explain the function of these cells. There is abundant surface area through which nutrients can be absorbed and secreted into the bloodstream (in the case of some nutrients) or the lymph, in the case of fats. Dietary lipids are reassembled into triacylglycerols in the endoplasmic reticulum (ER) of the brush border cells and packaged into chylomicra. These are secreted into the lymph, which then enter the circulatory system via the subclavian vein, bypassing the liver.

(Source: (ER) Alters, Biology: *Understanding Life*, 1e, copyright 2006, John Wiley & Sons. This material is reproduced with permission of John Wiley & Sons, Inc. (Steps 4,5) Tortora; Derrickson, *Principles of Anatomy & Physiology*, 13e, copyright 2012, John Wiley & Sons. This material is reproduced with permission of John Wiley & Sons, Inc.)

are first activated by coupling to coenzyme-A (forming an acyl-CoA). The energy required to form the acyl-CoA thioester linkage is provided by the hydrolysis of ATP, in a reaction catalyzed by acyl-CoA synthetase. The activated acyl chain is transferred to a molecule of 2-monoacylglycerol, in a reaction catalyzed by acyl-CoA:Monoacylglycerol acyltransferase (MGAT), forming a 1, 2-diacylglycerol. A third acyl chain is added by the action of acyl-CoA:DG acyltransferase (DGAT) (**Figure 10.14**). The acyl-CoA synthetase, MGAT, and DGAT are integral membrane proteins that are imbedded in the endoplasmic reticulum of the enterocyte where synthesis of triacylglycerol generally occurs.

As discussed later in this chapter, this is only one of the two major ways through which the organism can synthesize triacylglycerols, and 2-monoacylglycerol is only one of several substrates that can be used for triacylglycerol synthesis.

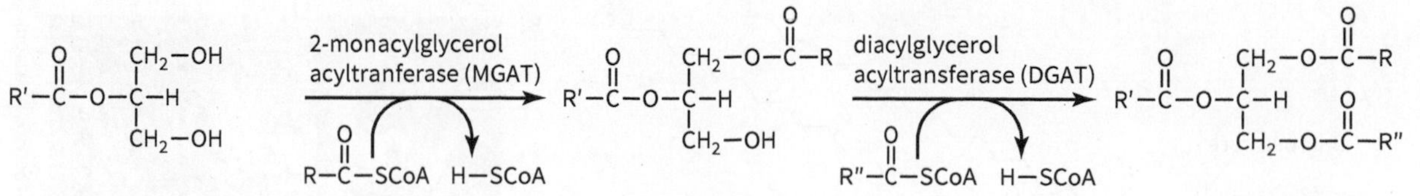

FIGURE 10.14 Triacylglycerol synthesis in the small intestine occurs through the 2-monoacylglycerol pathway. In the cells of the intestine, triacylglycerols are produced from 2-monoacylglycerol. This is different than in the other cells of the body.

10.2.3 Several molecules affect neutral lipid digestion

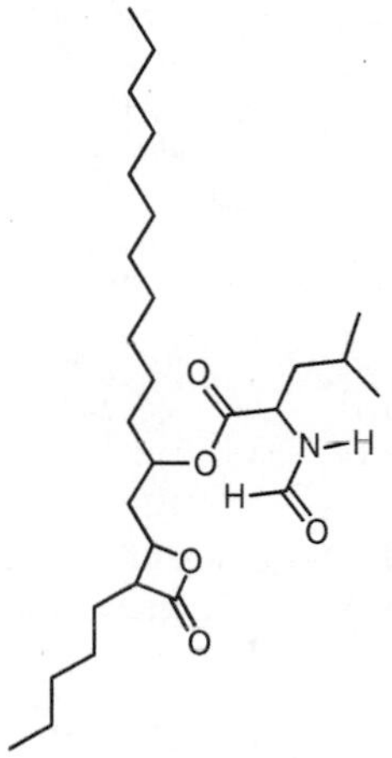

If triacylglycerols need to be broken down into fatty acids and monoglycerides for efficient absorption, then blocking their breakdown would also block their absorption. Lipase inhibitors are a class of dietary drugs that act as irreversible inhibitors of pancreatic (and to some extent gastric) lipases. The common trade name for this drug is orlistat. Orlistat is a synthetic analog of lipstatin, a naturally occurring molecule made by *Streptomyces toxytricini*. Examination of the structure reveals a long hydrophobic acyl chain, which may be important in substrate recognition, and a β-lactone ring. Beta-lactones are often employed as irreversible enzyme inhibitors. Once these molecules undergo nucleophilic attack, the strained β-lactone ring springs open and the nucleophile (often a serine in the active site of the enzyme) is irreversibly derivatized.

Orlistat is not a panacea for weight loss because it has an unpleasant side effect. When pancreatic lipase is inhibited and dietary fats remain undigested in the gastrointestinal tract, lipids leave the body as oily diarrhea (steatorreah).

Orlistat

Worked Problem 10.2   Making use of dietary triacylglycerols

Describe in your own words how the carbons found in the fatty acyl chains of a triacylglycerol molecule move from your digestive system to your circulatory system.

Strategy You can answer this question on several different levels. At a basic level, describe the way that triacylglycerols are degraded, absorbed, synthesized into new triacylglycerols, and secreted into the lymph. At the chemical level, describe each of those steps in molecular detail, discussing the enzymes and cofactors involved, and the microenvironments of the individual triacylglycerol molecule.

Solution A molecule of dietary triacylglycerol could be solubilized in a micelle containing bile salts in the small intestine, then degraded into free fatty acids and a molecule of monoglyceride by pancreatic lipase and its cofactor, colipase. The lipids are then absorbed by the enterocyte, where they are resynthesized into triacylglycerol

in the ER. To accomplish this, fatty acids must be activated (by acyl-CoA transferase) into fatty acyl-CoAs, then incorporated into a molecule of monoacylglycerol (by MGAT). The final fatty acid is added by DGAT. The "rebuilt" molecule of triacylglycerol is packaged into a chylomicron and secreted into the lymph. The lymph empties directly into the circulation. By entering the circulation via this pathway, the lipids bypass the liver on the first pass through the circulation, unlike dietary proteins and carbohydrates, which enter through the hepatic vein.

Follow-up question Many nutritional supplements contain short-chain or medium-chain triglycerides for "quick energy." How would the digestion of those molecules, such as trihexanoate, differ from a "normal" dietary fat, such as glycerol trihexadecanoate, also known as tripalmitate?

Olestra: an artificial fat
More than 12 artificial sweeteners are approved for sale, but there is only one artificial fat on the market. Olestra is a synthetic product designed to mimic the effects of fats and oils in cooking and to retain the texture of fats when eaten (a property termed mouth feel), but without the caloric value associated with triacylglycerols.

Olestra is made from sucrose and fatty acids. Recall that sucrose (table sugar) is a disaccharide of glucose and fructose and it has eight free hydroxyl groups available for hydrogen bonding;

A.

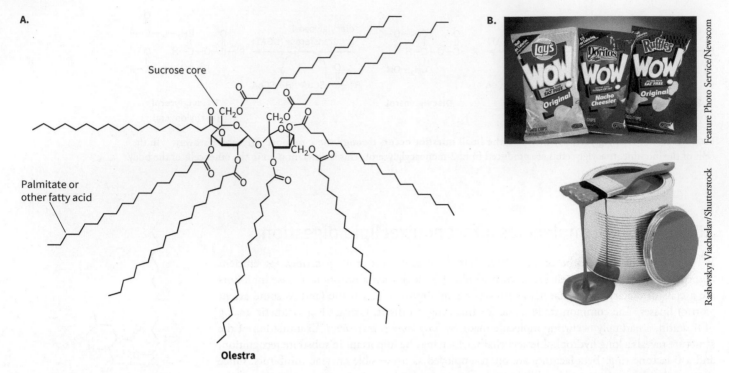

Sucrose core

CH₂O

CH₂O

CH₂O

Palmitate or
other fatty acid

Olestra

B.

Feature Photo Service/Newscom

Rashevskyi Viacheslav/Shutterstock

FIGURE 10.15 Structure of olestra. A. Olestra is synthesized from sucrose and oleic acid (or other fatty acids). Due to its complex structure, it is unable to be bound by esterases and degraded in the gut. **B.** While originally developed as a food ingredient, olestra is now being marketed as a paint solvent.

hydrogen bonding in sucrose is responsible for its crystalline solid structure as well as its high boiling point and high solubility in water. Recall also that fatty acids are amphipathic molecules comprised of a carboxyl group attached to a long aliphatic hydrocarbon chain. They can aggregate into water-soluble micelles, but once esterified to glycerol, as is the case in triacylglycerols, fatty acids become highly insoluble, behaving more as fats or oils. In olestra, fatty acids are esterified to the hydroxyl groups of sucrose, creating a molecule that has the physical properties of a fat in terms of boiling point and melting point, hydrophobicity, and insolubility in water. However, the molecule does not have the structure of a triacylglycerol molecule; hence, lipases are unable to bind to this molecule, and it passes through the digestive tract unscathed (**Figure 10.15**).

Unfortunately, olestra's resistance to degradation has some unintended side effects. To taste and feel like fat in the mouth, olestra has to melt at near or just below body temperature. This means that when it leaves the digestive tract at the other end, it does so in the form of an oily liquid; like orlistat, olestra produces the unpleasant side effect of steatorreah.

Interestingly, olestra has found new applications outside food and nutritional science. It is now being developed and marketed as a lubricant for small machines and as a hydrophobic solvent for some paints and stains. In both of these instances, it replaces petrochemicals and creates a potentially green alternative to traditional materials. Finally, olestra is being used as an experimental treatment to remove hydrophobic toxins, such as dioxin and other polychlorinated biphenyls (PCBs), from people exposed to these poisons.

Summary

- The degradation of dietary lipids begins in the mouth and stomach, but the process starts in earnest in the small intestine.
- Lingual, gastric, and pancreatic lipases all contribute to lipid digestion.
- Dietary lipids are re-esterified to neutral lipids, packaged into chylomicra and secreted into the lymph.
- Enterocytes employ a different pathway for triacylglycerol biosynthesis than elsewhere in the body.
- Several drugs block the intestinal breakdown of triacylglycerols.
- Artificial fats work because they cannot be broken down into fatty acids and absorbed.

1. Name the proteins involved in the steps for digestion of triglycerols and where they function.
2. Explain which of the steps involved in degradation of dietary lipids require an input of energy (ATP hydrolysis) and, for each step, explain why an energy input is or is not required.
3. Why does carbohydrate digestion begin in the mouth, protein digestion in the stomach, but lipid digestion in the intestine?

10.3 Transport of Lipids in the Circulation

Lipids are hydrophobic molecules that are generally insoluble in water. This presents a challenge to a multicellular organism; how can lipids and other important hydrophobic molecules be transported in an aqueous environment? This section discusses two major means of transport: the transport of aggregates of lipid and protein known as **lipoproteins** and the transport of hydrophobic molecules bound to a hydrophilic protein.

10.3.1 Lipoproteins have a defined structure and composition, and transport lipids in the circulation

Most lipoproteins have the same generic structure. With few exceptions, they are a spherical assembly consisting of a monolayer of phospholipid and cholesterol, surrounding a hydrophobic core of triacylglycerols and cholesteryl esters. Within the phospholipid–cholesterol monolayer, embedded proteins termed **apolipoproteins** act as ligands for receptors and as cofactors for lipases and other enzymes. Apolipoproteins help solubilize the lipoprotein's hydrophobic contents, and they are also important in the formation of lipoproteins (**Table 10.1**).

While the structure has not been solved for all apolipoproteins, many use amphipathic α helices to interact with the lipid. These helices have a hydrophobic face embedded in the lipid bilayer and a hydrophilic face exposed to the aqueous surroundings. These structures can be predicted with the help of a helical wheel diagram. A helical wheel diagram depicts the amino acids of the helix while looking down the helical shaft. The side chains are positioned every 100 degrees around the outside of the wheel. Using this depiction, we can see if a helix has a face that is polar, nonpolar, or some other interesting property. In the case of the helices found in apolipoproteins, there is a nonpolar face that is embedded in the lipoprotein, while the remainder of the amino acids are either polar or charged. Even though the diagrams alone are insufficient to predict the secondary structure of a protein, helical wheel diagrams do show the topology and faces of a known or presumed helical part of the protein (**Figure 10.16**).

There are several major classes of lipoprotein including chylomicra, **high density lipoproteins** (HDL), **low density lipoproteins** (LDL), and **very low density lipoproteins** (VLDL); classes are categorized based on their ratio of protein to lipid.

TABLE 10.1 Lipoproteins, Density, Composition, Origin, Fate

Lipoprotein	Density	Composition	Origin	Fate
Chylomicra	<1.006	B48, E, CII	Intestine	Loses TAG to muscle and adipose to become remnant
Chylo remnants	<1.006	B48, E	Blood	Binds to liver remnant receptor
VLDL	<1.006	B100, E, CII	Liver	Gets converted to IDL/LDL
IDL	1.006–1.019	B100, E	Blood	Gets converted to LDL
LDL	1.019–1.063	B100	Blood	Binds to tissues with LDLr, or oxidizes and forms plaques
HDL	1.063–1.31	AI, E, AIV, CII, CIII	Liver	Returns cholesterol to liver and steroidogenic tissues

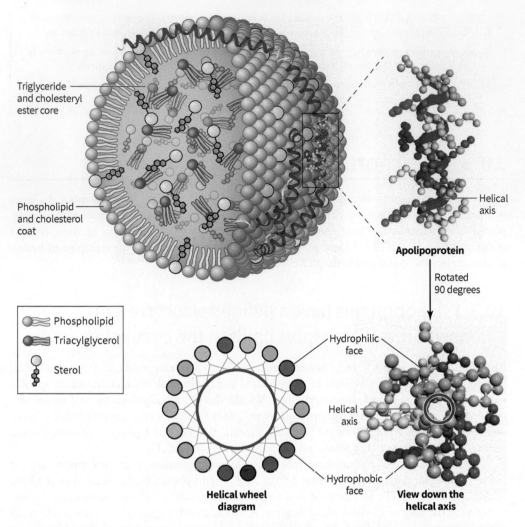

FIGURE 10.16 Structure of lipoproteins and helical wheel diagrams. All lipoproteins have a similar structure. A protein and phospholipid coat surrounds a hydrophobic core of triacylglycerol and cholesteryl esters. The coat proteins, such as apo AI, impart many of the properties of the particle by acting as ligands for receptors and as cofactors for enzymes. Along with apo AI, several other apolipoproteins use amphipathic α helixes to bind to the lipid bilayer. A segment of amino acid side chains are shown in a ball and stick depiction. These helices have a hydrophobic face and a hydrophilic face. This is best visualized by looking down the helical axis. A helical wheel diagram is an effective way to analyze helical structures and look for faces with properties such as hydrophobicity. In a helical wheel diagram the amino acids are plotted every 100 degrees of rotation around a central "wheel." In this view, hydrophobic residues are shown on one side of the helix, and hydrophilic ones are found on the other.

(Source: Data from PDB ID 3R2P Mei, X., Atkinson, D. (2011) Crystal Structure of C-terminal Truncated Apolipoprotein A-I Reveals the Assembly of High Density Lipoprotein (HDL) by Dimerization. *J. Biol. Chem.* 286: 38570–38582)

10.3.2 The trafficking of lipoproteins in the blood can be separated conceptually into three different paths

This section discusses individual lipoprotein species in the context of their metabolism. Although aspects of this metabolism happen concurrently, here the metabolism is split into three pathways to simplify and clarify the discussion.

Chylomicra deliver dietary lipids from the intestine to muscle and adipose   Section 10.2.2 discussed how lipids consumed in a fatty or mixed meal, comprised of fats, carbohydrates, and proteins, are broken down to the point where the lipids have been catabolized into free fatty acids and monoacylglycerols in the small intestine, absorbed by the intestinal enterocytes, and resynthesized into triacylglycerols in the endoplasmic reticulum (ER)

of those cells. The triacylglycerols are then packaged into the largest of the lipoproteins, the chylomicra, processed in the Golgi apparatus, and secreted. Chylomicra are large (upwards of half a micron in diameter) and coated with apolipoproteins AI, CII, and B48. Their size is attributable to a massive core of neutral lipid formed from dietary fats. This makes the particles buoyant and the least dense of all of the lipoproteins. If blood is drawn following a fatty meal, the particles will float to the surface, forming a creamy layer of fat on top of the blood cells (**Figure 10.17**).

The blood supply from the intestine first passes through the liver via the hepatic portal vein. Chylomicra are spared this fate by draining into the lymph. Since lymph drains into the venous side of the circulatory system through the thoracic duct, chylomicra avoid the liver in their first pass through the circulation. This avoids giving the liver an initial lipid burden and instead delivers lipids straight to muscle and adipose tissues.

The chylomicra enter muscle and adipose tissues by passing through capillary beds in those tissues. The endothelial cells in these beds are studded with lipoprotein lipase (LPL), which degrades the triacylglycerol found in the chylomicra. The fatty acids and glycerides produced are absorbed by myocytes (cells in muscle) and adipocytes (cells in adipose tissue), and are then either burned for energy or stored for future use. Lipoprotein lipase is activated by apo CII on the surface of the chylomicra; this apolipoprotein acts as a cofactor for the enzyme.

As the lipid contents of the chylomicron are degraded and lost, the chylomicron shrinks in size and increases in density (because it loses more lipid than protein). The resulting particle is a **chylomicron remnant**, that is, a particle that has lost much of its dietary triacylglycerols but is still enriched in dietary cholesterol. As the particle is metabolized, it loses apo AI and CII from the surface, and gains apo E (apo B48 is unchanged and is thus found in both chylomicra and remnants). The apo E on the chylomicron remnant binds to a remnant receptor on the surface of the hepatocyte, and the remnant is removed from the circulation through receptor-mediated endocytosis. Once internalized in a lysosome, chylomicrons are degraded and the components (lipids and amino acids) are used for other cellular processes, such as synthesis of membranes or new proteins (**Figure 10.18**).

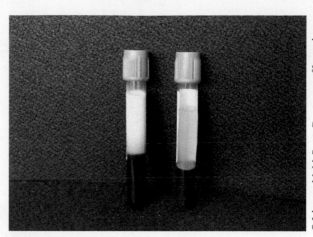

FIGURE 10.17 Chylomicra in blood samples. Shown are two blood samples, one from a normal patient (right) and one from a hyperlipidemic patient (left). Note the milky appearance of the plasma in the patient with hyperlipidemia.

Postprandial *de novo* lipids are secreted from the liver to deliver lipids to adipose and muscle

If we presume that a meal includes some carbohydrates and proteins in addition to lipids, then after eating, or postprandially, other metabolic processes will happen at the same time as the chylomicra are delivering triacylglycerols to muscle and adipose. Some carbohydrates are absorbed by the organism and burned for energy, whereas others are used to build up stores of glycogen. Proteins are catabolized into amino acids, and the amino acids not needed for synthetic purposes are deaminated, and their carbon skeletons broken down into acetyl-CoA or substrates for gluconeogenesis. Because many organisms (humans included) can rapidly build up carbohydrate stores and are not typically deficient in amino acids, most of the diet is quickly converted into triacylglycerols. People may consume a low-fat diet, but 20 minutes after drinking a can of cola, the drink consumed has more in common chemically with a stick of butter than with a sugary soft drink.

Excess carbohydrates and amino acids are converted into triacylglycerols in the liver. These triacylglycerols are said to be synthesized *de novo* (from new) to differentiate them from dietary triacylglycerols and are packaged with dietary cholesterol into large lipoproteins. Due to their high lipid content, these particles are particularly buoyant and are therefore called VLDL; they are coated by apo B100, apo E, and apo CII. The function of VLDL is to deliver lipids synthesized in the liver to muscle for energy and to adipose tissue for storage. As the VLDLs travel through the circulation, they interact with lipoprotein lipase with apo C II acting as a cofactor. The triacylglycerols are hydrolyzed and absorbed by tissues similar to the way in which chylomicra deliver dietary fat.

Because triacylglycerols in VLDL are hydrolyzed and lost, the particle shrinks in size. Again, because the particle is losing lipid rather than protein, it increases in density, converting first

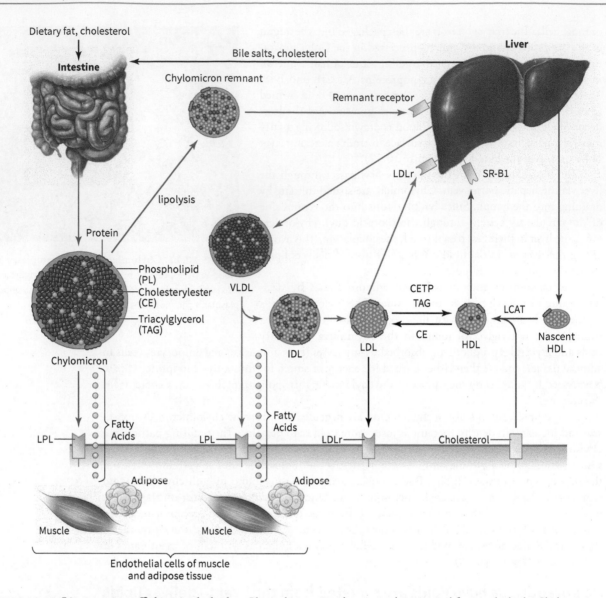

FIGURE 10.18 Lipoprotein trafficking in the body. Plasma lipoproteins have several origins and fates in the body. Chylomicra from the intestine transport dietary lipids to the endothelial lining of muscle and adipose tissue. Here LPLs is activated by its cofactor apo CII to hydrolyze triacylglycerols into free fatty acids and glycerol, which are absorbed by the tissues. VLDLs are released from the liver and transport fats made in the liver to muscle and adipose also using LPL to hydrolyze and deliver these molecules. As VLDLs lose triacylglycerols, they become smaller and denser, ultimately losing apo CII and forming LDL. HDLs emerge from the liver as discs of phospholipid surrounded by apo AI. Lecithin-cholesterol acyltransferase (LCAT) forms cholesteryl esters, producing mature HDLs, which are removed from circulation.

(Source: (Intestine, liver) Tortora; Derrickson, *Principles of Anatomy & Physiology*, 13e, copyright 2012, John Wiley & Sons. This material is reproduced with permission of John Wiley & Sons, Inc.)

to **intermediate density lipoprotein** (IDL) and finally to LDL, which has several different fates. First, LDLs are important in delivering cholesterol to tissues that need it, because the particles are enriched in cholesteryl esters (CEs), and can be bound by an LDL receptor (LDLr), which is expressed on the surface of tissues that are in need of cholesterol. The LDL particles are internalized by receptor-mediated endocytosis. Unfortunately, many people consume more cholesterol than their body needs. Over time, excess cholesterol accumulates in LDL particles, and the plasma concentration of these particles increases. The components of these particles are subject to oxidation. Both LDL and oxidized LDL are considered to be pro-atherogenic; that is, they contribute to the development of atherosclerosis, the hardening and narrowing of the arteries due to inflammation and accumulated cholesterol.

Because the vasculature of some arteries, especially the coronary arteries and carotids, is highly permeable to adjust for fluctuations in blood pressure, LDL particles pass from the lumen of the blood vessel into the subintimal space between the endothelial lining of the vessel and the

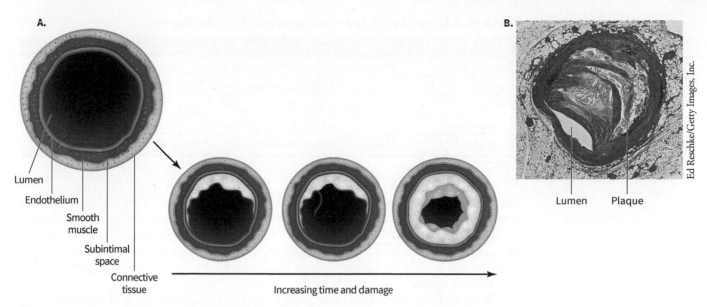

FIGURE 10.19 Development of atherosclerosis.   A. As we age, LDLs begin to accumulate in the subintimal space between the lining of the artery and the smooth muscle layer. Development of too much plaque can lead to tears and ruptures of the lining, resulting in a clotting response and heart attack or stroke. **B.** A micrograph of an artery showing plaque accumulation.

structural layers of muscle and connective tissue (**Figure 10.19**). LDL can become trapped in this space and accumulation of these particles will elicit an inflammatory response, drawing macrophages to the site and forming a plaque. Gradually, over the course of years, this damage increases, with more LDL adding to the lesion and more cells attempting to contain the damage. As this occurs, the vessel becomes more occluded and less flexible. Eventually, the endothelial layer tears, exposing the plaque to the circulation system and initiating a clotting response. The clots formed can lead to reduced blood flow (ischemia), which in turn can lead to tissue damage, myocardial infarction (heart attack), or cerebral ischemia (stroke).

High density lipoproteins are secreted by the liver and return cholesterol to the liver from the circulation and peripheral tissues
There is a process to counter the buildup of cholesterol in the circulation or in oxidized LDL. **Reverse cholesterol transport** is the removal of cholesterol from the circulation and peripheral tissues to the liver. To accomplish this, the liver secretes a small, protein-rich lipoprotein: HDL. These lipoproteins are initially found as a phospholipid disc surrounded by an annulus or ring of apolipoproteins (typically apo AI and apo E). Due to the high ratio of protein to lipid in HDL, the density is higher than that of the other lipoproteins discussed.

As the HDLs move through the circulation, they constantly collide with erythrocytes, other lipoproteins, and endothelial cells absorbing free cholesterol by diffusion. This is an effective means of absorbing free cholesterol, but the process would be limited if it were the only means of trapping cholesterol. The enzyme lecithin-cholesterol acyltransferase (LCAT) is found in plasma. It transesterifies an acyl chain from a molecule of PC (lecithin) in the HDL coat to free cholesterol, forming a molecule of cholesteryl ester (and lyso PC) (**Table 10.2**). Because the newly formed molecule of cholesteryl ester is more hydrophobic and less soluble in the lipid bilayer, large amounts of cholesteryl ester build up and form a core in the center of the HDL particles. As more cholesteryl ester is formed, the core of the particle grows, and the HDL changes from a disc to a sphere. Cholesteryl ester is trapped in the core of the HDL particle and is not free to exchange back with the surroundings.

When the HDL particle has increased in size and decreased in density, it has several fates. As shown in Figure 10.19, cholesteryl esters can be transferred to LDL by cholesteryl ester transfer protein (CETP). These LDLs can bind to receptors and be cleared from the circulation. Mature HDLs

TABLE 10.2 Lipases, Acyltransferases, and Cofactors

Enzyme	Protein Cofactor
Pancreatic lipase	Colipase
Lipoprotein lipase	Apo CII
Hepatic lipase	Apo AI
LCAT	Apo AI
Hormone sensitive lipase	Perilipin 1A
ATGL	CGI-58

TABLE 10.3 Drugs That Affect Plasma Lipids and How They Function

Drug	Function
Statins	HMG-CoA reductase inhibitors, blocks endogenous cholesterol biosynthesis
Niacin	Increases HDL, mechanism unknown
Fibrates	Transcription factor (PPAR-α) agonists decrease VLDL and increase HDL levels
Colesevelam	Bile acid sequestrant (blocks bile acid reabsorbtion in intestine)
Ezetimibe	Cholesterol-binding resin (binds cholesterol in intestine)

filled with cholesteryl ester binds to scavenger receptor B1 (SR-B1) on the surface of the liver, macrophages, and steroidogenic cells. The binding results in the delivery of cholesterol (but not HDL proteins) to these tissues. Likewise, HDL can bind to hepatic lipase. This lipase is highly homologous with lipoprotein lipase but, like SR-B1, is found in liver and steroidogenic tissues. Also, the uptake of HDL cholesteryl esters, like that of SR-B1, is independent of protein uptake and does not require the lipase to be catalytically active.

Because HDLs are involved in removing cholesterol from the circulation before it can accumulate in arteries, increased levels of plasma HDL cholesterol are thought to be anti-atherogenic; that is, the increased levels protect against the development of atherosclerosis. Numerous pharmaceutical approaches can be taken to try to modify or reduce plasma lipids. **Table 10.3** lists examples of some of the more popular drugs used to treat high cholesterol.

Elevated plasma lipids (especially triacylglycerols and LDL cholesterol) are associated with heart disease, stroke, kidney failure, and atherosclerosis. Oftentimes, elevated plasma lipids are treated pharmacologically. A discussion of this is found in **Medical Biochemistry: The pharmacological treatment of atherosclerosis.**

Medical Biochemistry

The pharmacological treatment of atherosclerosis

Heart disease, stroke, and atherosclerosis combined are the leading cause of death in the Western world. Clearly, these are complex diseases in which multiple aspects of genetics, environment, diet, and physiology contribute to the occurrence and eventual outcome. Studies suggest that aberrant lipid transport and storage, especially cholesterol and cholesteryl esters, are the culprits in the etiology of these disorders.

Much effort has been devoted to the development of pharmacological interventions by modifying lipid metabolism in some way in order to stave off the development of these diseases. Approaches include blocking *de novo* cholesterol biosynthesis, blocking absorption of cholesterol in the gut, and modulating gene expression, for example, by elevating levels of anti-atherogenic proteins—proteins involved in the reversal of the disease process.

The statins have been one of the most successful classes of drugs over the past 25 years. Statins are the generic trade name for a group of molecules that block the production of new cholesterol in the body by competitively inhibiting the rate-determining and committed step in cholesterol biosynthesis: hydroxymethylglutaryl-CoA reductase (HMG-CoA reductase). These drugs are therefore known as HMG-CoA reductase inhibitors. Accompanying decreased cholesterol production is a simultaneous increase in liver LDL receptor levels and reverse cholesterol transport as the liver adjusts to compensate, helping to create a more favorable lipoprotein profile. In this case, statins cause changes seemingly independent of cholesterol levels. These pleotropic effects, that is, effects that stem from a single change affecting multiple targets, include decreased oxidative stress and inflammation, and improved endothelial tissue function. Clinically, statins act to lower total plasma cholesterol by lowering the production of new cholesterol; although not a panacea, statins are widely used and highly effective. Annual sales of these drugs are in the tens of billions of dollars.

A second class of compounds developed to regulate cholesterol is bile acid sequestrants or bile acid binding resins. Much of the body's cholesterol stores are used each day in the production of bile, and this bile is reabsorbed and recycled back through the circulation via bile acid transporters. Bile acid sequestrants help to lower cholesterol by binding bile in the small intestine and thus preventing its absorption.

Niacin (also known as vitamin B3 or nicotinic acid) is also used to treat high cholesterol levels. Ingestion of large doses of niacin elevates HDL levels. The mechanism by which this works is still under investigation. Consumption of large doses of niacin can lead to skin flushing. Research is underway into niacin analogs that could achieve the positive HDL-elevating effects without this side effect.

The final class of compounds is the fibrates, lipid-lowering drugs that work predominantly to lower triglyceride levels and elevate HDL concentrations. PPAR-α is a transcription factor that effects the expression of many genes, including genes coding for the HDL proteins apo AI and apo AII, and the enzyme lipoprotein lipase.

Fibrates are PPAR-α agonists, and they increase fatty acid uptake; they also increase the activation of those fatty acids to fatty acyl-CoA and their β-oxidation. Overall, the picture is one of elevated levels of HDL (via increased levels of HDL proteins), increased catabolism of triacylglycerols, and lower levels of VLDL.

Often, statins, fibrates, niacin, and resins are used in combination to achieve the desired pharmacological effect. However, many of these drugs are metabolized using the same enzymes and pathways in the liver; therefore, clinicians and pharmacists need to be aware of drug interactions.

Other drug treatments for high plasma cholesterol are under development. Compounds which block the degradation of the LDL receptor hold significant promise as future treatments for hypercholesterolemia.

Worked Problem 10.3   It runs in the family

Familial hypercholesterolemia (FH) is a genetic disorder in which affected individuals have extremely high LDL levels. Normal levels in a healthy person are 50 to 80 mg of LDL cholesterol/deciliter, but patients with FH have levels of more than 1,000 mg/dl. Based on your knowledge of lipoprotein metabolism, propose a defect that could result in these high levels.

Strategy   At the core of FH are highly elevated plasma cholesterol levels. These could be caused by increased levels of cholesterol entering the circulation (either via increased synthesis or by increased secretion) or less cholesterol leaving the circulation. What types of mutations could lead to these effects? Which of these is more likely to happen?

Solution   Several mutations can cause FH, but all result from a loss of function, in this case, the ability to remove LDL from circulation. This can result from mutation of either the genes coding for the LDL receptor or the ligand of that receptor (apo B100).

Follow-up question   Would you anticipate that HDL levels would be changed in a patient affected by FH? Make a defendable hypothesis regarding how the conditions found in FH might affect HDL levels.

10.3.3 Brain lipids are transported on apo E-coated discs

The brain is separated from the rest of the circulation by the blood–brain barrier, a semipermeable membrane that selectively prevents entry of many molecules or assemblies, including lipoproteins, into the brain. As a result, brain tissue must synthesize its own cholesterol. Although neurons can synthesize cholesterol, they need to be able to export excess cholesterol; to do so they secrete small disk-shaped HDL-like particles coated with apo E. There are three different genotypes of the apo E allele, apo E2, 3, and 4, coding for the three different variants of apo E. These isoforms differ at positions 112 and 158 in the human protein. The E2 isoform has cysteine in both positions, the E4 isoform has arginine in both positions, and the E3 isoform has a cysteine in position 112 and an arginine in position 158. Although these variants have not been correlated with any type of cardiovascular disease or changes in lipoprotein metabolism, apo E4 has been correlated with an earlier age of onset of Alzheimer's disease. Having one copy of the apo E4 allele lowers the average age from 84 to 77 years for onset of Alzheimer's disease; having two copies lowers it to 69 years. Exactly how apo E4 is implicated in the etiology of this disease is unclear; however, it may be that apo E interacts with and assists in the clearance of A-β, one of the misfolded proteins found in β-amyloid plaques in the brains of patients with Alzheimer's disease. It could be that the apo E4 isoform of the protein is unable to clear these misfolded proteins and as such contributes to disease (**Figure 10.20**).

10.3.4 Fatty acids and hydrophobic hormones are transported by binding to carrier proteins

Free fatty acids released from adipose tissue are transported through the blood to the heart, liver, and muscle to be catabolized for energy. Because of their detergent-like nature, unesterified fatty acids could be harmful to the organism, denaturing proteins and solubilizing membranes. To render them harmless, fatty acids are transported bound to serum albumin.

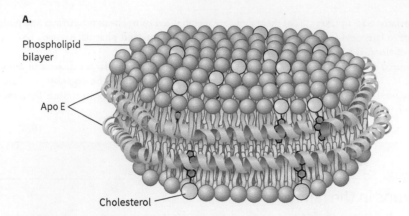

A.

Phospholipid bilayer

Apo E

Cholesterol

B.

Isoform	Isoform-specific amino acid difference	
	112	158
Apo-E2	Cys	Cys
Apo-E3	Cys	Arg
Apo-E4	Arg	Arg

FIGURE 10.20 Brain lipids are transported in phospholipid discs.   A. Lipids in the central nervous system are shuttled on apo E-coated discs. These lipoproteins are similar to nascent HDL secreted by the liver but lack a neutral lipid core. **B.** Polymorphisms at positions 112 and 158 in apo E are linked to an increased risk of Alzheimer's disease.

A globular protein, serum albumin is made in and secreted by the liver. It binds non-specifically to fatty acids and other molecules, including some drugs through its many hydrophobic pockets. Crystallographic data suggest that as many as 12 binding sites may exist, although it is likely that only one or two sites are occupied at any time. Albumin is conserved in higher organisms and is mostly α helical with a molecular weight of 60 to 68 kDa, depending on the species. Plasma concentrations of serum albumin may be as high as 5%, making it one of the more concentrated proteins in the body. Its high concentration is thought to help albumin maintain the osmotic pressure of the blood. Albumin is acidic, with negatively charged amino acids outnumbering positively charged ones. The isoelectric point (pI) of bovine serum albumin (BSA) is 4.7.

Drug molecules undergo complex interactions in the body. Many drugs bind to albumin in the blood and are thus rendered ineffective because they are sequestered away from their site of action. Although tight binding to albumin effectively lowers the concentration of a free drug, it may also increase the half-life of the drug in the body. Also, competition among different drug molecules (or between drug molecules and other hydrophobic molecules) to bind to albumin may lead to drug interactions by affecting a drug's free or functional concentration.

In the laboratory, albumin is often included in buffers, where it acts as a fatty acid acceptor or as an extra source of protein that can protect purified proteins from degradation due to contaminating proteases. Albumin provides a source of protein for enzymes to degrade, outcompeting any contaminating proteases that may remain. Albumin is also often used to coat plastics or hybridization membranes to decrease nonspecific binding of other proteins or nucleic acids (**Figure 10.21**).

Albumin is only one of several proteins that bind, sequester, and transport hydrophobic molecules; others include sex hormone–binding globulin, transcortin, thyroxine-binding globulin, and transthyretin. Each of these proteins is globular and contains a pocket that is responsible for binding the hydrophobic molecule it carries. The interactions of these proteins with the molecules they carry are analogous to the interactions between enzymes and substrates. In some cases, the interaction is highly specific, such as sex hormone–binding globulin interacting tightly and specifically with testosterone and estrogens but weakly, if at all, with corticosteroids. In other cases, the interaction is less specific; for example, transcortin binds to cortisone but also to aldosterone and progesterone. **Table 10.4** summarizes the molecules these proteins have been found to transport.

Carrier proteins have important functions. As alluded to with albumin, they bind and facilitate the transport of their hydrophobic cargo, and they bind to and thus block the effect of signaling molecules, as illustrated below for thyroid hormones and oral contraceptives.

Thyroid hormones, including thyroxine (T3) and triiodothironine (T4), are key chronic regulators of metabolism and energetic balance, that is, the balance between energy storage (as fat) and energy burning. When diagnosing a metabolic disease in a patient, a clinician will often ask the laboratory to assay for free T3 or free T4 (i.e., the fraction of each hormone that is free or unbound, which represents the amount of hormone that is functional in the bloodstream). Various conditions can affect the levels of thyroxine-binding globulin and hence indirectly influence the levels of T3 and T4 in the blood.

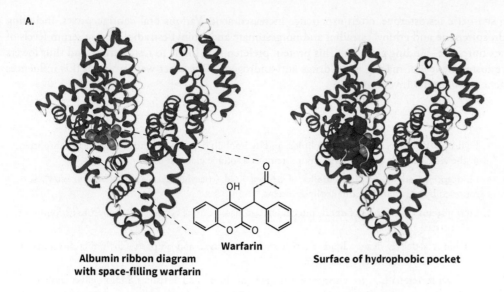

A.

Warfarin

**Albumin ribbon diagram
with space-filling warfarin**

Surface of hydrophobic pocket

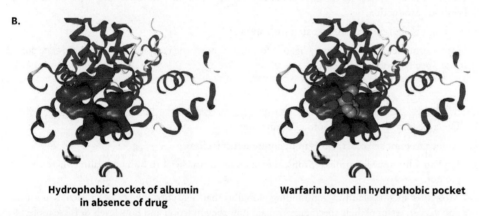

B.

**Hydrophobic pocket of albumin
in absence of drug**

Warfarin bound in hydrophobic pocket

FIGURE 10.21 Albumin bound to the anticoagulant drug warfarin. **A.** Serum albumin is a largely globular protein and is lacking in β sheet structures. It has several hydrophobic pockets, one of which is shown in a molecular surface. **B.** A close-up view of one of these hydrophobic pockets is shown in the absence and presence of the hydrophobic drug warfarin. These pockets can also bind to unesterified (free) fatty acids. Binding of drug molecules can decrease their efficacy, increase their half-life, and generally change their pharmacological properties (their pharmacokinetics).

(Source: (A, B) Data from PDB ID 2BXD Ghuman, J., Zunszain, P.A., Petitpas, I., Bhattacharya, A.A., Otagiri, M., Curry, S. (2005) Structural Basis of the Drug-Binding Specificity of Human Serum Albumin. *J. Mol. Biol.* 353: 38)

TABLE 10.4 Shuttle Proteins and Their Cargoes

Protein	Cargo
Sex hormone–binding globulin (SHBG)	Estrogen and testosterone
Transcortin (cortisol-binding globulin)	Glucocorticoids, progesterone, corticosteroids
Thyroxine-binding globulin	Triiodothironine (T3) and thyroxine (T4)
Transthyretin	Thyroxine (T4) and retinol
Serpin (thyroid-binding globulin)	Thyroxine and cortisol

Like many drugs, oral contraceptives are also transported in part by carrier proteins. A desirable side effect of some oral contraceptives is decreased incidence of acne. This condition is caused by a complex interplay of many factors, but one molecule implicated in the development of acne is testosterone. People with elevated testosterone levels, due either to natural biology or abuse of a synthetic testosterone, often experience increased acne. Various oral contraceptives, including drospirenone and ethinyl estradiol and norgestimate and ethinyl estradiol elevate serum levels of sex hormone–binding globulin. This protein preferentially binds to testosterone and thus lowers testosterone levels, making these drugs anti-androgenic. The exact way in which this influences acne is still under investigation.

Summary

- Hydrophobic molecules, including lipids, and hydrophobic hormones and vitamins are transported in the circulation bound to either lipoproteins or a soluble globular protein.
- Lipoproteins are spherical assemblies of neutral lipid (triacylglycerol and cholesteryl ester), surrounded by phospholipid and apolipoproteins.
- Lipoproteins have distinct fates in the circulation, based on the types of apolipoproteins found on their surface.
- Chylomicra deliver dietary lipids from intestine to muscle and adipose. Chylomicron remnants deliver dietary cholesterol to the liver.
- The liver secretes VLDL to transport fats made in the liver to other tissues. LDLs deliver cholesterol to tissues that need it, but it can accumulate and become oxidized, increasing risks of heart attack and stroke.
- HDLs bind cholesterol in the circulation and returns it to the liver or tissues involved in steroid hormone production.
- Brain lipids are bound to discs coated with apo E.
- Many hormones, including many ligands for nuclear hormone receptors (steroids), are transported inside specific globular proteins.

Concept Check

1. Describe how and where lipoproteins function.
2. Name enzymes and coactivators in lipoprotein metabolism.
3. Explain why some lipoprotein profiles are considered beneficial to human health, whereas others are not.
4. Molecules such as albumin and binding globulins that transport only a few molecules at a time, and vary in terms of their specificity. Explain how they function and how levels of these molecules influence the effectiveness of the molecules they bind.

10.4 Entry of Lipids into the Cell

Lipids gain entry into the cell in several ways. The process by which free fatty acids are absorbed is still debated, but other processes, such as receptor-mediated endocytosis of LDL, are well established.

10.4.1 Fatty acids can enter the cell via diffusion or by protein-mediated transport

As mentioned in section 10.3, lipoproteins are degraded by the lipases, lipoprotein lipase and hepatic lipase. The breakdown products of these lipases—free fatty acids and monoacylglycerols and diacylglycerols—are rapidly absorbed by cells. This poses an interesting question: How can the charged carboxyl head group of the fatty acid diffuse across the nonpolar lipid bilayer? One hypothesis proposes that microenvironments may exist in which the pH is low enough to protonate the fatty acid head group (fatty acids have a pK_a of 4 to 5, so they can be protonated at near physiological pH), thus reducing its polarity and allowing the molecule to diffuse across the membrane. In the cytosolic leaflet of the membrane, fatty acids are rapidly esterified to other lipids; again, this prevents them from acting as detergents and disrupting cellular functions (**Figure 10.22**).

An alternative hypothesis is that single molecules of long-chain fatty acids enter the cell via fatty acid transport proteins (FATPs), a family of six transmembrane proteins. Although the structure of these proteins and the mechanism they use to facilitate fatty acid movement is unknown, it is clear that FATPs increase fatty acid transport in a concentration-dependent manner; that is, more protein leads to increased transport. Additionally, the transport can be saturated and is specific for both long-chain and very-long-chain fatty acids (as opposed to shorter fatty acids or other hydrophobic molecules). The expression of these proteins is also tissue specific and responsive to cellular signals, such as insulin.

It is highly likely that FATPs work in conjunction with other proteins. They co-localize on the plasma membrane with CD-36, a protein hypothesized to bind fatty acids from plasma, and pass them to FATP. In some FATP family members, fatty acid transport and activation to a fatty acyl-CoA are coupled on the same protein, although mutation of the acyl-CoA ligase activity does not seem to impede lipid transport. As more is learned about this mechanism, it is likely that other proteins implicated in lipid transport, such as fatty acid binding protein (FABP) or the caveolins, may also play a role. Hence, it appears that more than one mechanism for fatty acid internalization may be at work.

Not all lipids are assimilated one molecule at a time. For example, LDLs bind to a receptor on a hepatocyte or other cell in need of cholesterol. The cell internalizes these particles and uses the cholesterol contents of the LDL. The process is termed receptor-mediated endocytosis and is the focus of section 10.4.2.

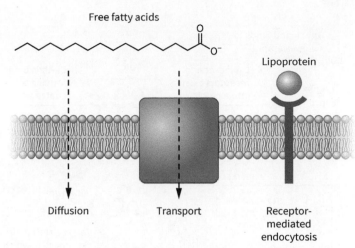

FIGURE 10.22 Lipids gain access to the cell via at least three different mechanisms. Single fatty acid molecules can gain entry to the cell either by diffusing through the plasma membrane or via a transport protein. Bulk lipids (lipoproteins) enter the cell through receptor-mediated endocytosis.

10.4.2 Lipoprotein particles and many other complexes enter the cell via receptor-mediated endocytosis

In receptor-mediated endocytosis, a ligand such as the apolipoprotein apo B100 binds to its receptor, in this case the LDL receptor. Binding of multiple ligands to receptors initiates a response on the cytosolic side of the membrane (**Figure 10.23**). The cytoplasmic tail of the LDL receptor interacts with adaptor proteins, such as AP2. As the name suggests, these proteins are adaptors between the receptor and its ligand, **clathrin**, a structural protein and key component of receptor-mediated endocytosis. The receptors begin to cluster in groups called **patches**, which then invaginate, forming a structure known as a **coated pit**. The cytosolic coating on this pit is largely comprised of clathrin.

As the clathrin network forms, additional proteins become involved, including motor proteins such as dynamin. The patch gradually invaginates into a coated pit; the pits are visible using electron microscopy, and the coating comes from the clathrin skeleton. Finally, the coated pit pinches off from the membrane and becomes a **coated vesicle**. The vesicle is trafficked inside the cell, where its clathrin and adaptor proteins are removed and recycled to form new coated pits. The uncoated vesicle fuses with other vesicles, including those containing proton pumps. These pumps are ATPases that use ATP hydrolysis to pump protons from the cytosol to the lumen of the vesicle. This acidifies the vesicle, resulting in protonation of key amino acids, conformational changes to proteins and ultimately the release of LDL from the LDL receptor.

LDL receptors are sorted away from the rest of the vesicular proteins and pinched off to be recycled back to the plasma membrane in an endosome, a vesicular trafficking organelle. Acidification of the vesicle also activates acidic lipases and proteases that break LDL into the constituent components.

AP2, the adaptor protein complex found in receptor-mediated endocytosis, is not related to the adipocyte fatty acid binding protein (AFABP), formerly known as AP2, or to the transcription factor ap2. These similar names highlight a problem that is often encountered when studying proteins with different or redundant names. Often, proteins, genes, or diseases are renamed when more information becomes available.

A.

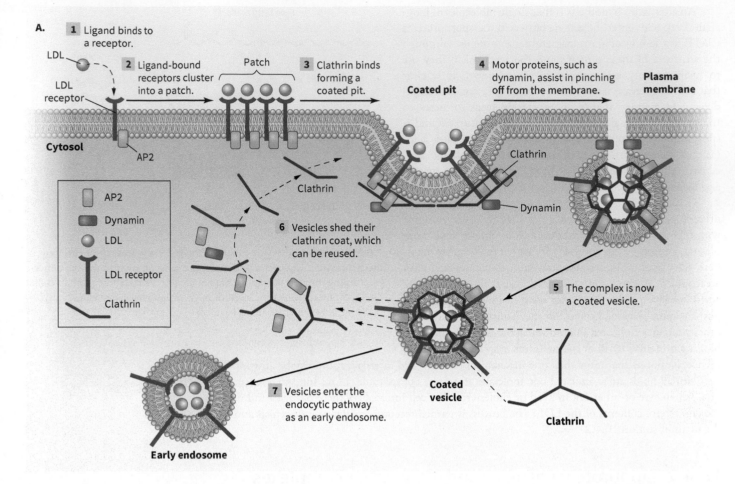

FIGURE 10.23 Receptor-mediated endocytosis is one of several trafficking pathways in the cell. **A.** In receptor-mediated endocytosis, vesicles coated by the protein clathrin are pinched off the plasma membrane and processed into endosomes. **B.** A space-filling structure of a clathrin triskelion shows it is comprised of three heavy chains and three light chains. These chains are pitched at an angle such that the entire structure is slightly puckered forming a tripod. **C.** A clathrin-coated vesicle is made up of 108 heavy and 108 light chains. A single triskelion of three heavy and three light chains is highlighted.

(Source: (B) Data from PDB ID 3LVH Wilbur, J.D., Hwang, P.K., Ybe, J.A., Lane, M., Sellers, B.D., Jacobson, M.P., Fletterick, R.J., Brodsky, F.M. (2010) Conformation switching of clathrin light chain regulates clathrin lattice assembly. *Dev. Cell* 18: 841–848 (C) Data from PDB ID 1XI4 Fotin, A., Cheng, Y., Sliz, P., Grigorieff, N., Harrison, S.C., Kirchhausen, T., Walz, T. (2004) Molecular model for a complete clathrin lattice from electron cryomicroscopy *Nature* 432: 573–579)

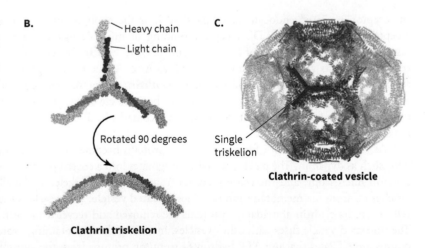

In addition to the adaptor proteins, other associated proteins are also found in coated pits. Many of these other proteins have domains or catalytic activities that modify the plasma membrane, facilitating the curvature of the plasma membrane, and formation of the coated pit.

Clathrin is found in the cell as a trimer of dimers. Each dimer is comprised of a rod-like light chain and a thicker, l-shaped heavy chain. The heavy and light subunits associate together, then three of these dimers bind through their light chains to form a shape termed a triskelion, as shown in Figure 10.23. The triskelion is not planar; rather, the arms extending from the central vertex are slightly bent toward each other, creating a gently sloping pyramidal shape overall. This shape helps the polymerized clathrin to form a basket-like structure. The clathrin triskelions join together to form a cage of repeating hexagons and pentagons that is similar to the pattern on a soccer ball.

The arms of the triskelion are flexible and have polyvalent interactions with all other proteins; thus, each protein contributes to the structure giving the coated pit the ability to assemble or disassemble in a rapid and cooperative fashion.

The architecture of these proteins and their chemistry contribute to the overall strength and dynamics of the interaction. The proteins are polyvalent, containing multiple receptors and clathrin-binding sites. Like antibodies, these proteins increase the likelihood of a second binding event and increase the overall strength of the interaction. This can be compared to the scene in many action films where someone is clinging to the edge of a cliff or ledge for dear life. If holding on with only one arm, people fall if they lose their grip; however, if holding on with two arms, they will still be able to hang on with one arm, even if the other slips. For both movie stars and proteins, having two arms of similar length makes it easier to create two points of contact.

After assembly of the coated pit, the vesicle needs to pinch off and split away from the membrane. The associated protein amphiphysin dimerizes around the neck of the vesicle and recruits dynamin to the complex. Dynamin is a GTPase that coils around the neck of the budding vesicle and extends as GTP is hydrolyzed, as shown in Figure 10.23.

Following the separation of the vesicle from the membrane, clathrin and the adaptor proteins are rapidly removed. Enzymes that modify phospholipids weaken the interactions between the coat components and the membrane coating via modification of head groups.

This section discussed receptor-mediated endocytosis in the context of the LDL receptor; however, there are more than 900 different receptors coupled to 7-transmembrane G proteins, and many of these other receptors are also internalized via clathrin-coated pits. Examples include the aquaporins, receptors involved in heavy metal transport, and opioid receptors in the brain.

Worked Problem 10.4   Need for GTP

How could you demonstrate that GTP hydrolysis is required for receptor-mediated endocytosis to proceed?

Strategy To demonstrate that receptor-mediated endocytosis depends on GTP hydrolysis, we would need a model system for studying the process and an experimental way to test the hypothesis.

Solution We could use a cultured cell system for the study. Many such cell lines are commonly used in thousands of laboratories worldwide. Some means of measuring receptor-mediated endocytosis would also be needed. Since it has already been demonstrated that LDL particles are internalized by receptor-mediated endocytosis, LDL particles that have been labeled with a fluorescent or radioactive lipid could be used, and their accumulation in an endosomal compartment could be followed as a way to quantitate the process. Several compounds act as GTPase inhibitors and are commercially available. Thus, control cells could be compared with cells treated with a GTPase inhibitor to look at whether inhibiting a GTPase activity blocks or dampens the endocytic response.

Follow-up question What type of phenotype would you predict for a clathrin knockout mouse?

Summary

- Lipids can enter the cell through either diffusion or endocytosis.
- Cells use receptor-mediated endocytosis to engulf LDL particles.
- Specific adaptor protein complexes bind to receptors and to the structural protein clathrin.
- The GTPase dynamin is important in the pinching off of the clathrin-coated vesicle.
- Following endocytosis, the vesicle is uncoated and most proteins are recycled.
- The new endosome is acidified and trafficked to a lysosome, where the cholesterol contents are liberated for other cellular processes.

Concept Check

1. Name the types of lipids that are absorbed by diffusion, and explain what helps to make that process favorable.
2. Describe the arguments supporting the hypothesis that specific proteins are involved in lipid absorption.
3. Outline the steps of receptor-mediated endocytosis and the proteins involved, and explain how the structure of these proteins contributes to their function.

10.5 Neutral Lipid Biosynthesis

High levels of free fatty acids are harmful to the cell. Although highly reduced and energetically rich, free fatty acids are amphipathic and are therefore not the best way to store energy. Instead, fatty acids are esterified to a glycerol backbone for storage; however, this pathway (*de novo* triglyceride biosynthesis) differs from that of dietary lipids (the 2-monacylglycerol pathway).

10.5.1 Triacylglycerols are synthesized by different pathways depending on the tissue

Triacylglycerol biosynthesis occurs predominantly in two tissues: liver and adipose. Liver tissue makes new fatty acids via *de novo* synthesis from acetyl-CoA, which is derived from glucose through glycolysis and decarboxylation of pyruvate. These new fatty acids are esterified into triacylglycerols and exported as VLDL. Adipose tissue absorbs free fatty acids generated by the action of lipoprotein lipase on the triacylglycerol-rich lipoprotein particles, chylomicra and VLDL; adipose tissue then re-esterifies these fatty acids into new triacylglycerols for storage. In both liver and adipose tissue, the enzymes required for triacylglycerol biosynthesis are found in the ER.

In the enterocyte, 2-monoacylglcerol is used as the source of the glycerol backbone in triacylglycerol biosynthesis; however, this is not the pathway used by other tissues. In liver and adipose tissue, the starting material is glycerol-3-phosphate.

There are several different pathways by which liver and adipose tissue can obtain glycerol-3-phosphate, the most common being glycolysis. We previously saw in Figure 10.3 how dihydroxyacetone phosphate (DHAP) can be reduced by glycerol-3-phosphate dehydrogenase to form glycerol-3 phosphate. Liver also contains a glycerol kinase that can directly phosphorylate glycerol into glycerol-3-phosphate. Because adipose tissue lacks significant levels of glycerol kinase, glycerol cannot be recycled in the adipocyte; this ensures that glycerol is exported when the cells are undergoing lipolysis (triacylglycerol breakdown).

The enzyme glycerol-3-phosphate acyltransferase acts on glycerol-3-phosphate to yield lysophosphatidate (lyso PA). The fatty acid is esterified to C-1 of the glycerol backbone. Fatty acids are not esterified directly; instead, they are first activated by coupling them to coenzyme A (producing fatty acyl-CoA), a process that requires the hydrolysis of ATP and provides the energy for the subsequent acyl transfers (**Figure 10.24**).

A second acyl chain is esterified to C-2 of glycerol by the action of lysophosphatidic acid acyltransferase (LPAT). This generates phosphatidate (PA). PA is then dephosphorylated to diacylglycerol (DG) by phosphatidic acid phosphatase (a phospholipase C). Finally, the remaining free hydroxyl group in DG is esterified to the third acyl chain via the action of diacylglycerol acyltransferase (DGAT).

Excess free cholesterol in the cell is stored as cholesteryl ester. In the ER, the enzyme responsible for this action is acyl-CoA: cholesterol acyltransferase (ACAT). Although this reaction produces cholesteryl ester, it differs from the synthesis of cholesteryl esters in the plasma. In plasma, the reaction is catalyzed by lecithin: cholesterol acyltransferase (LCAT), using phosphatidylcholine (PC or lecithin) as the acyl donor.

10.5.2 Triacylglycerol metabolism and phosphatidate metabolism are enzymatically linked

Phosphatidate and diacylglycerol, two intermediates in triacylglycerol biosynthesis, are also important intermediates in phospholipid biosynthesis (**Table 10.5**). The branch point in these pathways is the dephosphorylation of phosphatidate to diacylglycerol. This reaction is catalyzed by a phosphatidate phosphatase known as lipin, which helps coordinate cross talk between these two pathways.

Lipins are a family of related proteins, coded for by different genes. Each protein functions as a phosphatidate phosphatase, generating diacylglycerol from phosphatidate. Therefore, the lipins help to coordinate neutral lipid biosynthesis with the biosynthesis of phosphatidylcholine and phosphatidylethanolamine. Lipins also possess an amino acid sequence, termed a nuclear

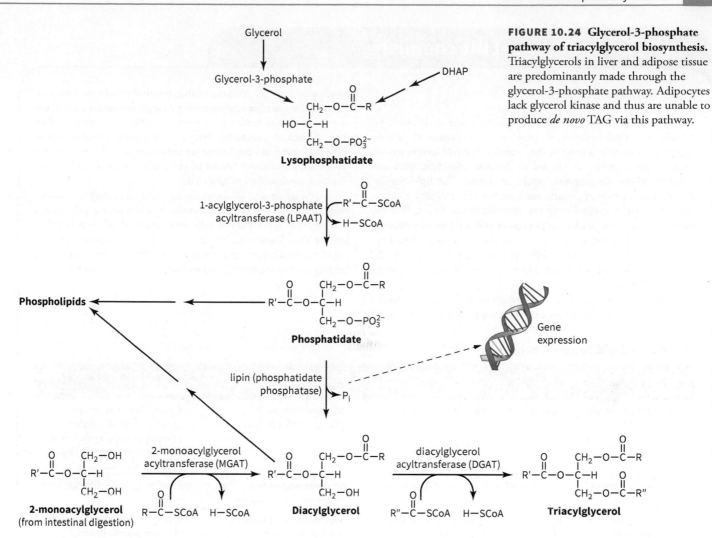

FIGURE 10.24 Glycerol-3-phosphate pathway of triacylglycerol biosynthesis. Triacylglycerols in liver and adipose tissue are predominantly made through the glycerol-3-phosphate pathway. Adipocytes lack glycerol kinase and thus are unable to produce *de novo* TAG via this pathway.

TABLE 10.5 Enzymes in This Chapter That Use Phosphatidate or Lysophosphatidate as a Substrate or Product: A Metabolic Crossroads

Protein	Enzymatic Activity	Other Ascribed Functions
Lipin	PA phosphatase/transcriptional co-activator	Regulates switch between glycerophospholipid and neutral lipid metabolism; also, functions in transcription in a complex with PPARs
DAG kinase	Phosphorylates DAG into PA	Functions prominently in signaling
MAG kinase	Phosphorylates MAG into lyso PA	Unknown
Endophilin-1	Lyso PA acyltransferase	Vesicular budding and fusion; plays role in membrane remodeling during receptor-mediated endocytosis

localization sequence, which allows them to traffic to the nucleus. Once inside, lipins can form a complex with several transcription factors. This complex helps to regulate the genes involved in fatty acid biosynthesis and β-oxidation. Thus, lipin is not only a key enzyme in lipid metabolism, it also acts to help regulate gene expression of proteins involved in lipid metabolic pathways. Mutation or deletion of the lipin genes results in several clinical syndromes, or lipodystrophies, in which irregular lipid metabolism leads to fatty liver, muscle damage and myoglobin accumulation in the blood, peripheral nerve damage or neuropathy, insulin resistance, inflammation, and anemia. A discussion of other diseases that result from problems with lipid metabolism can be found in **Societal and Ethical Biochemistry: Lipid storage disorders.**

Societal and Ethical Biochemistry

Lipid storage disorders

The inability to properly metabolize molecules frequently results in disease. When the molecule in question is common to all of biochemistry, such as a fatty acid, the mutation is rarely seen in nature because the organism would fail to develop. However, there are instances where the mutation results in disease. The lipid storage disorders are a group of diseases resulting from the inability to properly synthesize or catabolize several complex lipids. Often, in these disorders, lipids accumulate in lysosomes and are part of a broader category of diseases, referred to as lysosomal storage disorders. Depending on the mutation, different tissues are affected and different pathologies are observed. Many of these disorders are severe, and several are fatal.

Lipid storage disorders arise from the inability to properly metabolize a specific lipid molecule. This happens due to the loss of function of a specific enzyme in that metabolic pathway. These are recessive disorders because possessing a single functional copy of the gene is all that is needed to synthesize functional enzymes. Often these diseases are autosomal; that is, the genes implicated are found on an autosome, although some are sex-linked with the genes found on a sex chromosome. Several of the more common lipid storage disorders are described in Table 10.6.

Although some of these diseases are rare, they are clearly worth studying. Not only do advances in understanding the causes and effects of these diseases help those affected and their families, but the basic knowledge obtained contributes to our understanding of metabolic pathways, genetics, and biochemistry as a whole, helping to advance science and medicine in ways we cannot always predict.

TABLE 10.6 Lipid Storage Disorders

Disorder	Proteins Implicated	Location in the Cell	Pathophysiology	Manifestation
Niemann Pick disease (types A and B)	Acid sphingomyelinase	Lysosome	Accumulation of sphingomyelin in lysosomes	Accumulation of lipid in greatly enlarged lysosomes of many tissues. Type A fatal by 18 months; type B fatal by early adulthood.
Niemann Pick type C	NPC1 or NPC2	Lysosome	Accumulation of free cholesterol in lysosomes	Fatal by early adulthood
Cholesteryl ester storage disease (CESD) or Wolman's disease	Lysosomal acid lipase	Lysosome	Accumulation of cholesteryl esters in lysosomes	Enlarged liver (hepatomegaly), elevated plasma lipids (hyperlipidemia)
Gaucher's disease	Glucocerebrosidase	Lysosome	Accumulation of glucocerebrosides in lysosomes	Hepatomegaly, enlarged spleen, monocyte dysfunction
Tay-Sach's disease	Hexosaminidase A	Lysosome	Accumulation of gangliosides in neurons	Physical and mental deterioration, death by age 4
Fabry's disease	α galactosidase A	Lysosome	Accumulation of ceramide trihexoside	Pain, fatigue, kidney, heart, and skin complications
Chanarin-Dorfman syndrome	CGI-58	Lipid storage droplet	Increased storage of triacylglycerols in multiple tissues	Hepatomegaly, ichthyosis (scaly skin), cataracts, mild cognitive impairment

Worked Problem 10.5   Different paths

Triacylglycerol biosynthesis occurs through different pathways in the enterocyte (section 10.2) and the adipocyte (section 10.5). Why might the organism use two different means of synthesizing triacylglycerols?

Strategy   Compare the biosynthetic pathways and the functions of the two tissues. How might these give clues as to why the pathways are different?

Solution   While it is not possible to definitively state why there are two different means of triacylglycerol biosynthesis, we can identify two areas that provide clues. The first is the substrate available to each tissue. In the enterocyte, triacylglycerol is synthesized from 2-monoacylglycerol, whereas in the adipocyte or liver, the glycerol-3-phosphate is available through carbohydrate metabolism.

Follow-up question   In performing triacylglycerol breakdown (lipolysis) assays on isolated adipocytes in the laboratory, would it make a difference if glycerol or fatty acids were being measured?

Summary

- Triacylglycerol is synthesized by different pathways in different tissues.
- Triacylglycerol biosynthesis occurs in the endoplasmic reticulum.
- Phosphatidate (synthesized from glycerol-3-phosphate) is a key intermediate in the synthesis of triacylglycerol.
- Cross talk between phosphatidate and triacylglycerols occurs at several places in metabolism. The lipins may be involved in this cross talk.

Concept Check

1. Describe how triacylglycerol biosynthesis differs between an adipocyte and an enterocyte.
2. Describe the cross talk between phospholipid metabolism and triacylglycerol metabolism.

10.6 Lipid Storage Droplets, Fat Storage, and Mobilization

Neutral lipid molecules, such as triacylglycerols and cholesteryl esters, are too hydrophobic to be soluble in the cytosol. These molecules can be found in the lipid bilayer, but they will disrupt the bilayer at relatively low concentrations. Instead, neutral lipids are found in lipid storage droplets.

10.6.1 Bulk neutral lipids in the cell are stored in a specific organelle, the lipid storage droplet

Although triacylglycerol synthesis occurs in the endoplasmic reticulum (ER), storage of fats occurs in specialized organelles termed **lipid storage droplets**. These organelles are distinct from all others in the cell in that, instead of being bound by a phospholipid bilayer and an aqueous core, they have a phospholipid monolayer coating a core of hydrophobic triacylglycerols or cholesteryl esters.

Several mechanisms have been postulated for formation of lipid droplets. In this mechanism, triacylglycerols synthesized in the endoplasmic reticulum (ER) begin to form an inclusion, termed a "lens," between the leaflets of the ER membrane. These inclusions then either bud and are pinched off or are cleaved from both leaflets of the membrane to form new lipid storage droplets. Small lipid droplets could then fuse into larger ones through an unknown process. In the adipocyte, these droplets can comprise 95% of the volume of the cell and may exceed 100 microns in diameter.

All lipid storage droplets are coated with a variety of proteins involved in the trafficking and metabolism of the fatty contents. Chief among these are the perilipins, a family of proteins arising from five related genes that are important regulators of lipid metabolism. The best characterized of these proteins is perilipin 1A, which is found in adipose tissue (**Figure 10.25**).

10.6.2 Specific phosphorylation of lipases and lipid droplet proteins regulate lipolysis (triacylglycerol breakdown)

Lipolysis, the regulated mobilization and breakdown of triacylglycerols (TAG), is triggered by hormonal signals. In the basal or unstimulated state, perilipin is found on the lipid droplet surface, bound together with another regulator, the colipase CGI-58. Adipocyte triacylglycerol lipase (ATGL), the enzyme that catalyzes the first step of lipolysis, is distributed on the lipid storage droplet and in the cytosol. Hormone sensitive lipase (HSL) and monoacylglycerol lipase (MGL), the enzymes that catalyze the remaining two reactions, are found in the cytosol. Lipolysis is quiescent, occurring at low background levels.

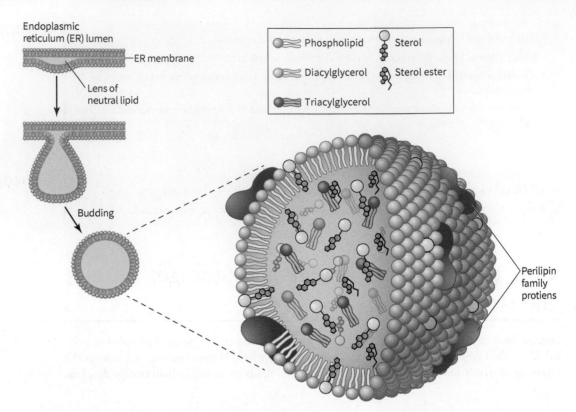

Endoplasmic reticulum (ER) lumen

ER membrane

Lens of neutral lipid

Budding

Phospholipid   Sterol

Diacylglycerol   Sterol ester

Triacylglycerol

Perilipin family protiens

FIGURE 10.25 Lipid droplet formation in the phospholipid bilayer.   There are several potential models for lipid droplet biogenesis. In this model, neutral lipids aggregate between the leaflets of the endoplasmic reticulum (ER) membrane until the droplet is large enough for budding to occur. The lipid droplet has a monolayer of phospholipid surrounding it.

Lipolytic stimuli to the adipocyte usually come in the form of a catecholamine signal. A hormone signal, often one of the catecholamines (epinephrine, norepinephrine, and dopamine) or glucagon, binds to a β-adrenergic receptor on the surface of the adipocyte. These receptors are coupled to a heterotrimeric G protein, and binding initiates the signaling cascade that activates the cAMP-dependent protein kinase, protein kinase A (PKA). We saw previously PKA function in a variety of metabolic pathways, including glycogen metabolism. PKA has two major substrates in the lipolytic cascade: HSL and perilipin A. Phosphorylation of these targets causes translocation of HSL to the surface of the lipid droplet and structural changes in perilipin proteins at the droplet surface, leading to a marked increase in lipolysis.

The distinct regulatory and catalytic cascade in lipolysis is still being elucidated; however, it is clear that three lipases are involved.

- Adipose triacylglycerol lipase (ATGL) is the enzyme which removes the first fatty acid from triacylglycerol resulting in diacylglycerol (DAG) and a free fatty acid. ATGL is stimulated through binding to its cofactor CGI-58. Release of CGI-58 from perilipin initiates this reaction and also starts triacylglycerol breakdown.

- The removal of the second fatty acid is catalyzed by HSL, resulting in monoacylglycerol (MAG) and a second free fatty acid. HSL is activated through interactions with perilipin, and both proteins must be phosphorylated by PKA for optimal activity.

- The third reaction is catalyzed by monoacylglycerol lipase (MGL), which cleaves the remaining ester resulting in free glycerol and a third molecule of fatty acid. This enzyme is cytosolic and is not thought to be regulated by other proteins.

Free fatty acids and glycerol generated in lipolysis are released into the circulation, where they are transported to the liver, muscle, and the heart for use. This is typically β-oxidation for fatty acids and gluconeogenesis for glycerol.

The proteins coating triacylglycerol-filled lipid storage droplets in adipose tissue are also found in cells active in steroid hormone production. In the Leydig cells of the testis, or in the cells of the adrenal cortex, perilipins coat cholesteryl ester-rich lipid droplets and, presumably, regulate interaction of their contents with lipases through a similar mechanism to triacylglycerol lipolysis (**Figure 10.26**).

Neutral lipids in plants have evolved some different, and in some cases highly specialized, functions. Plants with seeds that dry out as part of the maturation process (dessicatable seeds)

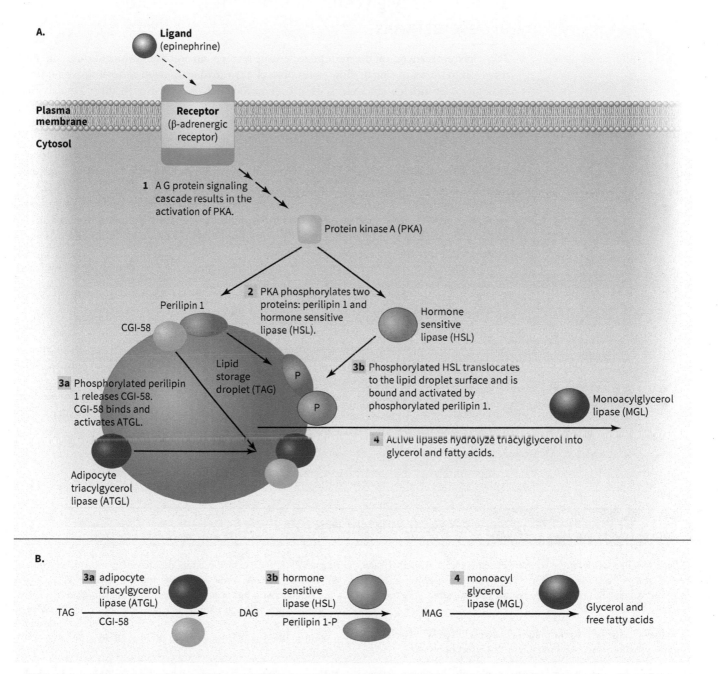

FIGURE 10.26 Signaling cascades implicated in lipolysis. Lipolysis is acutely regulated via the action of PKA on lipases and perilipin 1. **A.** Binding of a catecholamine or glucagon signal activates PKA, which has several targets in the adipocyte. Phosphorylation of perilipin 1 causes release of CGI-58, which can then bind and activate ATGL, catalyzing the release of the first fatty acid from triacylglycerol. Phosphorylation of HSL results in translocation to the lipid droplet surface. There, interaction of HSL with phosphorylated perilipin 1 results in cleavage of the second fatty acid from diacylglycerol. Hydrolysis of the third fatty acid is catalyzed by the cytosolic enzyme monoglyceride lipase. **B.** The summary of reactions and cofactors in triglyceride breakdown.

contain oil droplets coated with a layer of structural proteins called the oleosins. Like the perilipins, oleosins are thought to provide anchorage sites for enzymes and cofactors involved in metabolism of the oil body, a seed organelle analogous to the lipid storage droplet. Oleosins are thought to prevent oil droplets from coalescing in the dried seed. Because plant oils such as canola, corn, palm, and soy are potential sources of food, energy (biodiesel), and chemical feedstocks, there is considerable interest in the function of oleosins and their potential as a target for genetic modification.

10.6.3 Triacylglycerol metabolism is regulated at several levels

Triacylglycerols are one of the most prized energy stores of an organism, which devotes significant resources to the regulation of those stores. As is clear to anyone trying to either lose or gain weight, biology sometimes has other plans. This section briefly reviews how the organism regulates levels of triacylglycerols.

Triacylglycerol storage happens in two different time frames: acute storage and mobilization occurring over minutes to hours, and chronic storage occurring over weeks or longer.

Acute storage and mobilization of triacylglycerol occurs via the manipulation of levels of cAMP. When the organism needs to mobilize energy, it releases lipolytic hormones, such as the catecholamines epinephrine, norepinephrine, and dopamine, or the peptide hormone glucagon. These hormones act through receptors to activate adenylate cyclase, elevating levels of the second messenger cAMP. This second messenger activates protein kinase A, which then activates a lipolytic response through the phosphorylation of the perilipins and HSL. Lowering levels of cAMP by the action of phosphodiesterases inactivates PKA.

Insulin acts acutely on triacylglycerol metabolism by increasing protein phosphatase 1 activity, which in turn dephosphorylates HSL and perilipin 1A.

AMP kinase (AMPK) also plays a role in the regulation of lipolysis. HSL is also a substrate of AMPK; however, phosphorylation of HSL by AMPK inactivates the protein. Given that AMPK is active when cellular concentrations of AMP are high, and the energetic state of the cell is low, this may seem paradoxical; however, this may be a means by which the cell mediates lipolysis. Excess fatty acids require ATP for activation and reincorporation into complex lipids. AMPK is thought to help prevent lipolysis from proceeding unabated.

Chronic lipid storage levels are partially affected by both levels of gene expression and diet. Several transcription factors are important in the expression of genes coding for many of the proteins discussed here.

Worked Problem 10.6　　Coffee for weight loss?

Many weight-loss products contain caffeine (3,5,7-trimethyl xanthine), the world's most common phosphodiesterase inhibitor. Based on your knowledge of lipolysis, would caffeine affect this process, and if so, how?

Strategy　Caffeine is a phosphodiesterase inhibitor. What is the role of phosphodiesterase, and how might this affect lipolysis?

Solution　As the name suggests, phosphodiesterases cleave phosphodiesters. Here, the phosphodiester in question is cAMP. Phosphodiesterases act on cAMP to form AMP, which cannot activate PKA. Hence, by inhibiting the enzyme that breaks down the cAMP, levels of cAMP will be elevated and will activate PKA.

Anyone who has used strong coffee or energy drinks to help them stay awake all night knows that caffeine has many physiological functions. In theory it leads to increased lipolysis, although it is more likely that in weight loss products it serves as a general stimulant.

Follow-up question　Name and describe two techniques by which you could determine whether two of the proteins involved in lipolysis interact with one another. The techniques appendix may help you.

Summary

- Neutral lipids are stored in specific organelles termed lipid storage droplets.
- The surface of lipid storage droplets is coated with proteins, among them the perilipins.
- Phosphorylation of lipases and perilipins regulate lipolysis, the controlled breakdown of triacylglycerols.
- Triglyceride stores are regulated acutely by catecholamines and insulin through the modulation of cAMP levels, which affects PKA phosphorylation of lipases and perilipin. AMPK may also play a role.
- Long-term triglyceride stores are regulated by several transcription factors.

Concept Check

1. Describe the signaling pathways implicated in lipolysis.
2. Why are at least three different lipases involved in the mobilization of lipid stores?

10.7 Lipid Rafts as a Biochemical Entity

The term "raft" probably conjures up an image of a loose bundle of logs, crudely tied together to provide adventure when sailing down a river or as a method for castaways to escape from a desert island. Rafts may keep people afloat, but the strength of the structures and just how long they last are debatable. **Lipid rafts** are in many ways similar to these makeshift structures.

10.7.1 Lipid rafts are loosely associated groups of sphingolipids and cholesterol found in the plasma membrane

Lipid bilayers are comprised of different types of phospholipids, cholesterol, and multiple proteins, as discussed earlier, but this is only part of the picture. The lipids found in the lipid bilayer are not homogenously mixed. Lipid rafts are organized microdomains of the lipid bilayer. They are small (10 to 200 nm), loose aggregates of lipid, enriched in sphingomyelin and cholesterol, floating in the outer (exoplasmic) face of the lipid bilayer. Cholesterol appears to be a critical component of the raft, and depletion or manipulation of membrane cholesterol disrupts the raft. The acyl chain content of a raft is more saturated than the rest of the membrane; therefore, the raft is assumed to be less fluid than the surrounding phospholipid sea. Also, the increase in saturated lipids in the raft is thought to result in closer packing of the lipids, leading to stronger interactions of head groups through hydrogen bonding or coulombic interactions in the polar regions of the lipid head groups. The overall shape of the molecules involved in raft formation may also be important. Sphingomyelin and cholesterol form cone-shaped structures that complement each other to facilitate close molecular packing. The nature of the forces involved in raft formation and stabilization are still being elucidated.

Lipid rafts were originally identified as detergent-resistant extracts of membranes. When membranes were extracted at low temperatures (around 4°C) with non-ionic detergents such as Triton-X-100, these small structures remained. This led to some controversy in the field, with some researchers suggesting that these structures were artifacts of the experiment. More recently, numerous fluorescent techniques have been used to investigate rafts; for example, fluorescence resonance energy transfer microscopy (FRET) and fluorescence correlation/cross correlation spectroscopy (FCS/FCCS). These techniques, coupled with newly developed fluorescent probes and compounds that can alter the chemical composition of the lipid bilayer, are being used to study lipid rafts in terms of structure and lifetime, that is, how long the structure stays together (currently thought to be in the millisecond-to-second time scale) (**Figure 10.27**).

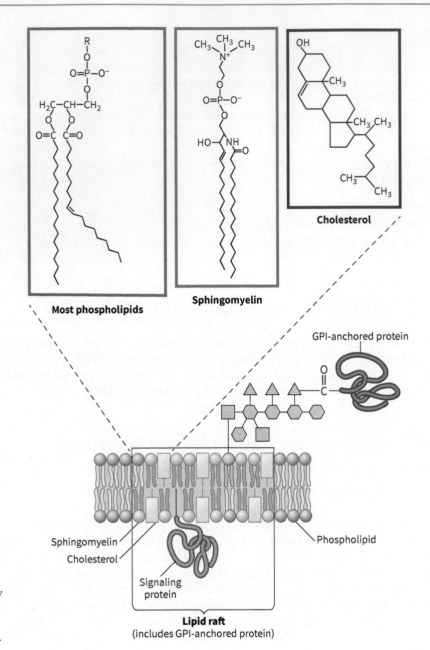

FIGURE 10.27 Lipid rafts and microdomains form in the plasma membrane. Lipid rafts are microdomains enriched in cholesterol and sphingolipids or glycosphingolipids. The geometry of cholesterol and sphingolipids allows them to pack more tightly together resulting in a patch of membrane that is less fluid than the surroundings.

10.7.2 Lipid rafts have been broadly grouped into two categories: caveolae and non-caveolar rafts

Caveolae are flask-shaped invaginations of the plasma membrane. They were originally identified in the 1950s from electron microscopy as cave-like invaginations of the plasma membrane. The caveolins, which are groups of three cholesterol-binding proteins, associate with these invaginations and form coated vesicles that are pinched off from the plasma membrane. Caveolae and caveolins are the best characterized raft-associated proteins, and they have been implicated in membrane trafficking. The rafts found in caveolae are enriched in cholesterol and sphingomyelin.

Non-caveolar rafts are flat structures found in the plane of the plasma membrane. In addition to sphingomyelin and cholesterol, these rafts are enriched in glycolipids, which may provide points at which cellular proteins or pathogens may attach.

Numerous proteins associate with lipid rafts. Palmitoylated proteins, such as those found in hedgehog, a key developmental regulator in *Drosophila* and humans, are common in rafts, as are

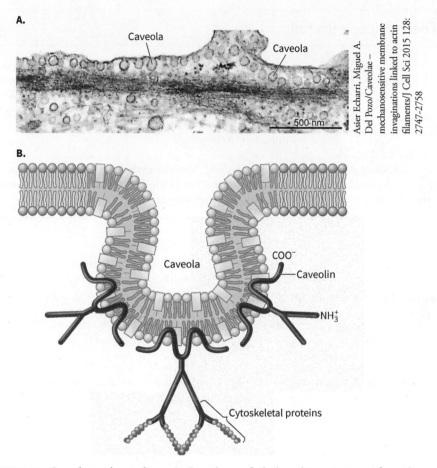

Asier Echarri, Miguel A. Del Pozo/Caveolae – mechanosensitive membrane invaginations linked to actin filaments/J Cell Sci 2015 128: 2747-2758

FIGURE 10.28　Caveolae and caveolins　**A.** Caveolae are flask-shaped invaginations of membrane coated with the caveolins. These structures differ from clathrin-coated pits, partially in their protein composition. **B.** Caveolins are integral membrane proteins associated with caveolae. These are dimeric proteins that bind to sphingomyelin and cholesterol-rich rafts through their carboxy-terminus and interact with the cytoskeleton through their amino-terminus.

GPI anchored proteins and proteins that bind to the membrane through glycolipids. This may be due to the saturated hydrocarbon tails of these lipids associating with the raft, although it is likely that protein factors are also important. Mutations of key residues at the lipid–protein interface will displace many proteins from rafts.

The importance of lipid rafts has now been demonstrated in many biological processes, and they are now part a rapidly expanding field of research. Two examples of these processes are cell signaling and the biology of pathogenic agents (**Figure 10.28**).

Cell signaling

Multiple signaling pathways are associated with lipid rafts. In each of these instances, the proteins of the pathway interact with the raft through glycolipid or GPI anchors, or through the raft itself via a membrane-spanning helix. Rafts may be involved in the clustering of these proteins and in assembly of macromolecular signaling complexes. Insulin signaling is one example of a pathway that employs lipid rafts.

Pathogenicity

Lipid rafts are important in the pathogenicity of various infectious agents, including some bacteria, viruses, prions, and protozoa. In these cases, the carbohydrate moieties of glycolipids are often the sites of attachment or budding for viruses, including human immunodeficiency virus-1 or HIV-1. They are also sites of action for bacterial toxins, such as anthrax; they may also be important for pathogen entry into the cell and survival such as *Plasmodium falciparum*, the causal agent in malaria.

A better understanding of rafts and their biological roles may make it possible to improve treatments for certain diseases by exploiting the pathogen's dependence on these structures.

Worked Problem 10.7   The trouble with rafts

You are investigating a protein complex that may be associated with lipid rafts. What complications might arise in investigating lipid rafts?

Strategy   What techniques have been used previously to examine lipid rafts? What are the strengths and limitations of these techniques?

Solution   Lipid rafts are relatively small and transient. Thus, the raft itself may contain relatively few molecules of cholesterol or sphingolipid (50 to several hundred) and only a few protein molecules. Rafts were originally identified using electron microscopy.

This technique can be used because of the size of rafts, but it fails to capture their transient and dynamic nature, since it requires the examination of dead, fixed material in a vacuum. Newer techniques employ fluorescent imaging of cells or cell membranes. These techniques (FRET and FCS/FCCS) can capture interactions in live cells occurring over microseconds.

Follow-up question   Methyl-β-cyclodextrins are a class of carbohydrate molecules that can be used to deplete the cell of cholesterol. How might these compounds be useful in a study of lipid rafts?

Summary

- Lipid rafts are microdomains of the plasma membrane enriched in cholesterol and sphingolipids.
- Lipid rafts have been implicated in membrane trafficking, protein interaction with the plasma membrane, and organization of signaling complexes.

Concept Check

1. Describe a lipid raft. How large are these rafts, and what are their proposed components?
2. Describe the forces involved in tethering raft lipids together.

Problems

10.1 Phospholipid Metabolism

1. Identify common themes in phospholipid metabolism.

2. Phospholipases cut phospholipids.

   a. Draw the structure of palmitoyl oleoyl phosphatidylcholine.

   b. What would the products be if that lipid were cut with phospholipase A$_2$?

   c. What would the products be if that lipid were cut with phospholipase C?

3. Choose a phospholipid and examine the thermodynamics of the steps in its biosynthesis. Based on your knowledge of the reactions involved, would the ΔG values for these reactions be positive, negative, or near zero? What is your reasoning for the values you generated?

4. Describes the sequence of events for the synthesis of phosphatidylcholine.

10.2 Digestion of Triacylglycerols

5. Shown is the structure of pancreatic lipase. Describe the topology of this molecule in your own words. Is it mostly α helix, β sheet, or some combination of the two? Can you see any potential domains to the protein? Are there any particularly striking features that you have not seen in other proteins? Can you propose a function for any of those

features? (Source: Data from PDB ID 1N8S van Tilbeurgh, H., Sarda, L., Verger, R., Cambillau, C. (1992) Structure of the pancreatic lipase-procolipase complex. *Nature* 359: 159–162)

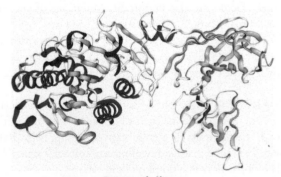

Pancreatic lipase

6. It is recommended that food made with olestra must be supplemented with vitamin K. What is the basis for this?

7. Trace the path a molecule of triacylglycerol would follow to enter the lymph.

8. Many lipases employ a catalytic triad (asp-his-ser), similar to the protease chymotrypsin, in their mechanism. Propose a mechanism for a lipase that cleaves an ester linkage.

9. Examine the structure of orlistat. Based on the structure, what weak forces may be involved in binding to pancreatic lipase? Which amino acids in pancreatic lipase would be involved in binding?

Orlistat

10. Orlistat and penicillin both contain β-lactam or lactone functional groups.

 a. What is a β-lactam or lactone?

 b. Can orlistat be used as an antibiotic? Can penicillin be used as a weight loss agent? Why or why not?

 c. Does the common β-lactam or lactone structure tell you anything about the mechanism of action of these drugs? What does it tell you about the enzymes they inhibit?

11. Methyl *tert*-butyl ether (MTBE) is being used experimentally as a treatment for gallstones. Examine the structure and properties of these molecules. Describe how this would work.

10.3 Transport of Lipids in the Circulation

12. Several enzymes, for example, hepatic lipase and lipoprotein lipase, can bind to lipoproteins. There are mutant variants of these enzymes that bind lipoproteins but are not catalytically active. Cells expressing these mutants can still absorb cholesterol. What does that tell us about the functions of these proteins in cholesterol absorption?

13. Heart disease is one of the main cause of death among other disease conditions. Why would such a flawed system of transport evolve? What are the environmental factors at work in the selection of genes coding for the proteins discussed in lipoprotein metabolism?

14. Why is fatty acid free bovine serum albumin (BSA) often included in lipase assays?

10.4 Entry of Lipids into the Cell

15. How could you show that the adaptor protein AP2 has a phosphatidylinositol 4,5 bisphosphate (PIP2) binding domain?

16. Antisense is a technique used to silence genes and decrease protein expression in a cell or organism. It is described in the techniques index. Using antisense, if you knocked down each of these proteins, what would you expect to happen? Where would molecules pile up or the pathway arrest?

 a. LDLr

 b. AP2

 c. clathrin

 d. dynamin

17. Clathrin is a structural protein and key component of receptor-mediated endocytosis. How could you calculate

 a. the ratio of clathrin subunits in the holoprotein?

 b. the number of clathrin molecules on the surface of an endosome?

10.5 Neutral Lipid Biosynthesis

18. Describe the synthesis of triacylglycerols in all three tissues, and explain how they differ.

19. A low-carbohydrate diet proclaims that, without glucose, adipose tissue is unable to store triacylglycerol. Is this true?

20. Is energy required for triacylglycerol biosynthesis? When, where, and how (in what form) is this energy provided?

21. What is the role of PA in neutral lipid metabolism?

22. Cholesteryl ester is the storage form of cholesterol. Based on your knowledge of lipid metabolism, does the synthesis of cholesteryl ester cost or provide energy? What might be the source of that energy in this reaction?

10.6 Lipid Storage Droplets, Fat Storage, and Mobilization

23. Which tissues are involved in the process of β-oxidation and for gluconeogenesis? Why catabolic (β-oxidation) and anabolic (gluconeogenesis) pathways occur at the same time?

24. Describe the interplay of lipases, perilipins, and other proteins in lipolysis.

25. Why are at least three different lipases involved in the mobilization of lipid stores?

26. The perilipins are neither acylated nor farnesylated. Propose a mechanism by which they tether to lipid storage droplets.

27. Lipid storage droplets are the only organelle known that is bound by a phospholipid monolayer. Why is there only a monolayer?

10.7 Lipid Rafts as a Biochemical Entity

28. Describe one process in which rafts have been implicated.

29. How many types of lipid rafts are there? What makes them more stable?

30. One of the complicating factors encountered when studying lipid rafts is their transient nature. If a structure lasts for only a short time (seconds or less), what techniques could be used to study its formation?

31. Examine the structures of lipids that are suspected to be found in a lipid raft. What types of interactions (weak forces) would you expect to help stabilize this structure? Characterize the strengths of those interactions.

32. Examine the overall shape of cholesterol and sphingomyelin shown in Figure 10.27. The shapes of these structures can be described as a cone or inverse cone. In your own words, describe how the packing of these molecules could occur in raft formation. If a caveolae or coated pit were forming, how would the composition of that membrane have to change? What changes would need to be made to the lipid composition to achieve that conformation?

Data Interpretation

33. A paper you are reading contains this figure of a micrograph and western blots. Please use the figure and caption to answer the questions below.

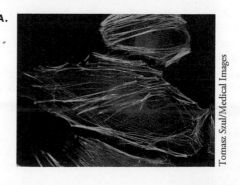

A.

A. Immunofluorescence image of POI (protein of interest) immunostained with Oregon Green 488 (green) and nuclei stained with DAPI (blue). **B.–F.** Western blots of sucrose ultracentrifugation gradients stained for the following proteins: **B.**, POI; **C.**, oxidative phosphorylation complex IV (mitochondria); **D.**, GAPDH (cytosol); **E.**, Golgin-97 (golgi); **F.**, disulfide isomerase (endoplasmic reticulum).

    a. Where is the POI found in the cell? Which organelle?

    b. What is the point of having two methods of determining the location of the protein?

    c. Why were these proteins chosen?

34. A paper hypothesizes that a specific protein cofactor is needed for optimal lipase activity. Both proteins are expressed in adipose tissue. Several knockdown constructs (RNAi) were prepared to silence expression of the cofactor and the following data were obtained, as shown in the table below. Check the appendix for descriptions of experimental techniques you may not have encountered before.

Values are standard error of the mean (S.E.M.) for *n* = 6 samples.

    a. Why were five constructs used?

    b. What is the purpose of the scrambled construct?

    c. Why were western blots run for both cofactor and lipase if the RNAi construct was only designed to silence the cofactor?

    d. Why were western blots performed rather than reverse transcriptase polymerase chain reaction (RT-PCR), northern blots, or some other technique?

    e. Which RNAi construct was most effective at silencing the expression of the cofactor?

    f. Why are glycerol and fatty acid release different?

    g. What does the statistical analysis of this data tell you?

    h. What are your conclusions from this figure?

35. You are investigating some of the enzymatic properties of a novel cobra venom. Samples of cellular lipids were treated with venom or a mixture of venom and EGTA. The resulting lipids were subjected to analysis by thin-layer chromatography.

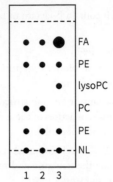

Lane 1 = control lipids (no treatment)
Lane 2 = lipids + venom + EGTA
Lane 3 = lipids + venom

    a. What enzymatic activity does the enzyme apparently have?

    b. What are the products of the reaction?

    c. What does the EGTA data tell you about the reaction?

Experimental Design

36. PE can be synthesized by several different pathways. How could you demonstrate that a cellular pool of PE was derived from phosphatidylserine rather than from some other source?

Construct	WB: cofactor	WB: lipase	Glycerol release (mmol/min/10⁻⁶ cells)	Fatty acid release (mmol/min/10⁻⁶ cells)
Scrambled	⬛	⬛	4.8 ± 0.2	0.7 ± 0.3
1	⬛	⬛	5.0 ± 0.3	1.2 ± 0.3
2	⬛	⬛	4.6 ± 0.1	0.9 ± 0.1
3	⬭	⬛	4.9 ± 0.4	0.8 ± 0.2
4	⬛	⬛	5.1 ± 0.2	1.2 ± 0.3

WB, western blot

37. The laboratory you are working in is investigating a family with very low levels of HDL and increased plasma triacylglycerols, but unexpectedly low risk of heart disease. What questions could you ask to start to investigate this paradox?

38. There are several points at which phospholipid metabolism and neutral lipid metabolism cross paths. Several proteins have recently been identified that have catalytic activity. Design an experiment to test whether a novel protein you have discovered has phosphatidate phosphatase activity.

39. The laboratory you are working in is studying how a newly discovered peptide hormone is absorbed by the cell. How could you test to see whether the receptor hormone complex is removed from the membrane via receptor-mediated endocytosis?

40. Design an experiment to test the hypothesis that the phosphorylation of perilipin A is required for maximal lipolysis.

Ethics and Social Responsibility

41. Many aspects of science that we now accept as dogma were once controversial or even heretical. The existence of lipid rafts was long considered a controversial topic. Can you think of other scientific topics in the current media that are considered controversial? What is your opinion on these topics? How have you formed these opinions? In other words, are your opinions based on science or punditry?

42. Olestra is a synthetic product designed to mimic the effects of fats and oils in cooking but without the caloric value associated with triacylglycerols. It was developed in 1968 but did not gain approval from the Food and Drug Administration (FDA) until 1996, some 28 years later, as concerns over its use were reviewed and tested. Based on what you know about olestra and lipid metabolism, write arguments from the perspective of the manufacturing company in support of approving olestra for public consumption, and from the perspective of a consumer watchdog group citing concerns about the approval.

43. One fear the FDA had about the widespread approval of olestra was that people would consume more olestra-rich foods in the belief that they were "healthy" food items. Do you think development of molecules such as olestra make people more lax with regard to their dietary habits?

44. Is it ethical to work on a project to develop a molecule like orlistat when millions of people around the world do not have enough to eat each day?

Suggested Readings

10.1 Phospholipid Metabolism

Ehnholm, Christian, ed. *Cellular Lipid Metabolism.* New York, New York: Springer, 2009.

Fahy, E., S. Subramaniam, R. C. Murphy, M. Nishijima, C. R. Raetz, T. Shimizu, F. Spener, et al. "Update of the LIPID MAPS Comprehensive Classification System for Lipids." *Journal of Lipid Research* 50 (April 2009): S9–14.

Gurr, I. M., ed. *Lipid Biochemistry: An Introduction.* 3rd ed. New York, New York: Springer, 1980.

Gurr, I. M., J. L. Harwood, and K. N. Frayne. *Lipid Biochemistry: An Introduction.* 5th ed. New York, New York: Wiley-Blackwell, 2002.

Kent, C. "Eukaryotic Phospholipid Biosynthesis." *Annual Review of Biochemistry* 64 (1995): 315–43.

Simmons, K., ed. *The Biology of Lipids: Trafficking, Regulation and Function.* Woodbury, New York: Cold Spring Harbor Laboratory Press, 2011.

The AOCS Lipid Library. Retrieved from http://lipidlibrary.aocs.org/index.html

Vance, D. E., and J. E. Vance, eds. *Biochemistry of Lipids, Lipoproteins, and Membranes.* 5th ed. New York, New York: Elsevier, 2008.

Wenk, Markus. "The Emerging Field of Lipidomics." *Nature Reviews Drug Discovery* 4 (2005).

10.2 Digestion of Triacylglycerols

Chapus, C., M. Rovery, L. Sarda, and R. Verger. "Minireview on Pancreatic Lipase and Colipase." *Biochimie* 70 (1988): 1223–34.

Cheng, D., T. C. Nelson, J. Chen, S. G. Walker, J. Wardwell-Swanson, R. Meegalla, R. Taub, et al. "Identification of Acyl Coenzyme A: Monoacylglycerol Acyltransferase 3, an Intestinal Specific Enzyme Implicated in Dietary Fat Absorption." *Journal of Biological Chemistry* 278, no. 16 (2003): 13611–4.

Hibuse, T., N. Maeda, T. Funahashi, K. Yamamoto, A. Nagasawa, W. Mizunoya, K. Kishida, et al. "Aquaporin 7 Deficiency Is Associated with Development of Obesity Through Activation of Adipose Glycerol Kinase." *Proceedings of the National Academy of Sciences of the United States of America* 102, no. 31 (August 2, 2005): 10993–8.

Li, C., T. Tan, H. Zhang, and W. Feng. "Analysis of the Conformational Stability and Activity of *Candida antarctica* Lipase B in Organic Solvents: Insight from Molecular Dynamics and Quantum Mechanics/Simulations." *Journal of Biological Chemistry* 285, no. 37 (2010): 28434–41.

Lowe, M. E. "Structure and Function of Pancreatic Lipase and Colipase." *Annual Review of Nutrition* 17 (1997): 141–58.

Lowe, M. E., J. L. Rosenblum, and A. W. Strauss. "Cloning and Characterization of Human Pancreatic Lipase cDNA." *Journal of Biological Chemistry* 264 (1989): 20042–8.

Persson, B., G. Bengtsson-Olivecrona, S. Enerback, T. Olivecrona, and H. Jornvall. "Structural Features of Lipoprotein Lipase. Lipase Family Relationships, Binding Interactions, Non-Equivalence of Lipase Cofactors, Vitellogenin Similarities and Functional Subdivision of Lipoprotein Lipase." *European Journal of Biochemistry* 179 (1989): 39–45.

van Tilbeurgh, H., S. Bezzine, C. Cambillau, R. Verger, and F. Carriere. "Colipase: Structure and Interaction with Pancreatic Lipase." *Biochimica et Biophysica Acta* 1441, no. 2–3 (1999): 173–84.

Weibel, E. K., P. Hadvary, E. Hochuli, E. Kupfer, and H. Lengsfeld. "Lipstatin, an Inhibitor of Pancreatic Lipase, Produced by *Streptomyces toxytricini*. 1. Producing Organism, Fermentation, Isolation and Biological Activity." *Journal of Antibiotics (Tokyo)* 40, no. 8 (August 1987): 1081–5.

Winkler, F. K., A. D'Arcy, and W. Hunziker. "Structure of Human Pancreatic Lipase." *Nature* 343 (1990): 771–4.

10.3 Transport of Lipids in the Circulation

Hamilton, R. L., M. C. Williams, C. J. Fielding, and R. J. Havel. "Discoidal Bilayer Structure of Nascent High Density Lipoproteins from Perfused Rat Liver." *Journal of Clinical Investigation* 58, no. 3 (September 1976): 667–680.

Mahley, R. W., T. L. Innerarity, S. C. Rall, Jr., and K. H. Weisgraber. "Plasma Lipoproteins: Apolipoprotein Structure and Function." *Journal of Lipid Research*, 25, (1984): 1277–1294.

Panáková, D., H. Sprong, E. Marois, C. Thiele, and S. Eaton. "Lipoprotein Particles Are Required for Hedgehog and Wingless Signaling." *Nature* 435, (2005): 58–65.

Vance, J. E., and H. Hayashi. "Formation and Function of Apolipoprotein E-Containing Lipoproteins in the Nervous System." *Biochimica et Biophysica Acta* 1801, no. 8 (August 2010): 806–18.

10.4 Entry of Lipids into the Cell

Fernandes, John. *Inborn Metabolic Diseases: Diagnosis and Treatment.* 4th ed. Berlin, Germany: Springer, 2006.

Goldstein, J. L., and M. S. Brown. "The LDL Receptor." *Arteriosclerosis, Thrombosis, and Vascular Biology* 29, no. 4 (April 29, 2009): 431–8.

Kikuchi, A., H. Yamamoto, and A. Sato. "Selective Activation Mechanisms of Wnt Signaling Pathways." *Trends in Cell Biology* 19, no. 3 (March 19, 2009): 119–29.

McMahon, H. T., and E. Boucrot. "Molecular Mechanism and Physiological Functions of Clathrin-Mediated Endocytosis." *Nature Reviews Molecular Cell Biology* 12, no. 8 (July 22, 2011): 517–33.

Reider, A., and B. Wendland. "Endocytic Adaptors—Social Networking at the Plasma Membrane." *Journal of Cell Science* 124, Pt. 10 (May 15, 2011): 1613–22.

Sorkin, A. "Cargo Recognition During Clathrin-Mediated Endocytosis: A Team Effort." *Current Opinion in Cell Biology* 16, no. 4 (2004): 392–9.

Tien, A. C., A. Rajan, and H. J. Bellen. "A Notch Updated." *Journal of Cell Biology* 184, no. 5 (March 9, 2009): 621–9.

Zwang, Y., and Y. Yarden. "Systems Biology of Growth Factor-Induced Receptor Endocytosis." *Traffic* 10, no. 4 (April 10, 2009): 349–63.

10.5 Neutral Lipid Biosynthesis

Athenstaedt, K., and G. Daum. "The Life Cycle of Neutral Lipids: Synthesis, Storage and Degradation." *Cellular and Molecular Life Sciences* 63, no. 12 (2006): 1355–1369.

Buhman, K. K., H. C. Chen, and R. V. Farese, Jr. "The Enzymes of Neutral Lipid Synthesis." *Journal of Biological Chemistry* 276, no. 44 (November 2, 2001): 40369–72.

Csaki, L. S., and K. Reue. "Lipins: Multifunctional Lipid Metabolism Proteins." *Annual Review of Nutrition* 30 (August 21, 2010): 257–72.

Reue, K., and D. N. Brindley. "Thematic Review Series: Glycerolipids. Multiple Roles for Lipins/Phosphatidate Phosphatase Enzymes in Lipid Metabolism." *Journal of Lipid Research* 49, no. 12 (December 2008): 2493–503.

Yen, C. L., S. J. Stone, S. Koliwad, C. Harris, and R. V. Farese, Jr. "Thematic Review Series: Glycerolipids. DGAT Enzymes and Triacylglycerol Biosynthesis." *Journal of Lipid Research* 49, no. 11 (November 2008): 2283–2301.

10.6 Lipid Storage Droplets, Fat Storage, and Mobilization

Bickel, P. E., J. T. Tansey, and M. A. Welte. "PAT Proteins, an Ancient Family of Lipid Droplet Proteins That Regulate Cellular Lipid Stores." *Biochimica et Biophysica Acta* 1791, no. 6 (June 2009): 419–40.

Brasaemle, D. L. "Thematic Review Series: Adipocyte Biology. The Perilipin Family of Structural Lipid Droplet Proteins: Stabilization of Lipid Droplets and Control of Lipolysis." *Journal of Lipid Research* 48, no. 12 (December 2007): 2547–59.

Olofsson, S. O, P. Boström, L. Andersson, M. Rutberg, M. Levin, J. Perman, and J. Borén. "Triglyceride Containing Lipid Droplets and Lipid Droplet-Associated Proteins." *Current Opinion in Lipidology* 19, no. 5 (October 2008): 441–7.

Walther, T. C., and R. V. Farese, Jr. "The Life of Lipid Droplets." *Biochimica et Biophysica Acta* 1791, no. 6 (June 2009): 459–66.

10.7 Lipid Rafts as a Biochemical Entity

Anderson, Richard G. W. "The Caeolae Membrane System." *Annual Review of Biochemistry* 67 (1998): 199–225.

Calder, P. C., and P. J. Yaqoob. "Lipid Rafts—Composition, Characterization, and Controversies." *Nutrition* 137, no. 3 (March 2007): 545–7.

Fielding, C. J., ed. *Lipid Rafts and Caveolae: From Membrane Biophysics to Cell Biology.* New York, New York: Wiley-VCH, 2006.

Glebov, Oleg O., Nicholas A. Bright, and Benjamin J. Nichols. Flotillin-1 defines a clathrin-independent endocytic pathway in mammalian cells *Nature Cell Biology* 8 (2006): 46–54.

Morrow, I. C., and R. G. Parton. "Flotillins and the PHB Domain Protein Family: Rafts, Worms and Anaesthetics." *Traffic* 6, no. 9 (September 2005): 725–40.

1950s	Caveolae is first described in electron micrographs
1953	Fibrates are first synthesized
1967	Fibrates first used to treat high levels of plasma lipids
1968	Olestra is developed
1971	First HMG-CoA reductase inhibitor (statin) is isolated from *Penicillium* fungus
1984	Brown and Goldstein win the Nobel Prize for regulation of cholesterol metabolism
1987	Mercor, the first statin approved for clinical use
1991	Perilipin is identified
1994	Sirtuins are first described in the literature
1996	Olestra is approved for sale to the public
1997	Lipid rafts are first characterized
1999	Pancreatic lipase inhibitor (orlistat) is approved for clinical use
2003	Earliest mention of the term "lipidomics" in the biomedical literature
2005	Statin sales exceed $19.7 billion in the United States alone
2012	Statin sales exceed $29 billion worldwide
2014	Total number of papers in medical publishing mentioning sirtuins breaks 3,500

CHAPTER 11

Amino Acid and Amine Metabolism

Amine Metabolism in Context

Most human endeavors involve waste. In almost all we do, we create some by-product that is unusable or needs to be directed toward some other function. A simple example is building a house. Whether we build it out of wood or stone, inevitably, there is scrap left behind. This scrap material may be wasted (put into a landfill), or it may be directed toward a different function, such as building a tree house or being processed into other products.

Biological systems are in many ways analogous to these examples. Chemotrophs are organisms that take in all the nutrients and energy they need in chemical form from their environment. The metabolism of some of these molecules produces toxic by-products that must be neutralized, recycled, or eliminated from the system.

Amino acids and xenobiotics are two types of molecules that pose challenges when metabolized. We do not typically store amino acids in the same way that we store carbohydrates or lipids, but we can catabolize the carbon skeleton of amino acids into other metabolites. This requires detoxifying the amino acid by removing the amino group, but direct removal of an amino group would generate ammonia, a highly toxic by-product. Animals have evolved different means of coping with this load; for example, many higher animals synthesize urea from the ammonia through a process called the urea cycle. Xenobiotics include molecules that are strange or foreign to the organism, including the compounds that give food and drink their characteristic flavors. Xenobiotics also include most drug molecules. Metabolism of xenobiotics is critical to modern medicine and pharmaceutical science.

Common Themes

Evolution's outcomes are conserved.	• Both the pathways and structures found in amino acid catabolism are conserved throughout evolution.
	• Specific proteins (cytochromes) encountered in this chapter are coded for by genes that can be used to track evolutionary change.
	• Organisms evolved specialized pathways to detoxify ammonia and spare water when excreting nitrogenous wastes.
	• Carnivorous organisms upregulated certain metabolic pathways to compensate for a diet rich in protein.
	• Pathways and whole-organ systems evolved to help organisms detoxify potentially harmful molecules.
Structure determines function.	• Many of the enzymes involved in amino acid metabolism are allosterically regulated.
	• Many xenobiotics are detoxified by making them more water soluble, which decreases their ability to enter the cell and increases the likelihood that they will be eliminated in the urine.
Biochemical information is transferred, exchanged, and stored.	• Allosteric regulators generated in other parts of metabolism regulate many aspects of amino acid catabolism.
	• Covalent modification (phosphorylation or adenylation) is used to regulate some pathways.
Biomolecules are altered through pathways involving transformations of energy and matter.	• The pathways of amino acid catabolism are coordinated with one another and with whole-body metabolism.
	• Reactions where one functional group is exchanged for another, such as a transamination, have little to no associated energetic cost.
	• Catabolic pathways are generally oxidative; that is, the products are more oxidized than the substrates, and yield energy, often as NADH, $FADH_2$, or NADPH.
	• The synthesis of urea is a biosynthetic pathway and a reduction of the carbon atom of urea. As a result, urea biosynthesis requires the hydrolysis of several molecules of ATP to make the reaction proceed in the forward direction.
	• Most amino acids use similar chemistry (transamination) to eliminate the amino group before catabolism of the carbon skeleton.
	• Similar types of reactions are seen in the citric acid cycle, β-oxidation, and catabolism of several amino acids.

11.1 Digestion of Proteins

All animals must consume proteins in their diet to provide a source of amino acids. Under most circumstances, the organism does not absorb whole proteins but breaks these larger molecules or protein polymers down into smaller subunits—di- and tripeptides and amino acid monomers. As with lipid and carbohydrate digestion, enzymes begin to break down protein molecules in the stomach, but most of the digestion and all of the absorption occurs in the small intestine.

11.1.1 Protein digestion begins in the stomach

The stomach is highly acidic with a pH of approximately 1. The acid-producing parietal cells lining the stomach employ ATPases in the plasma membrane to pump H^+ into the stomach lumen (**Figure 11.1**). Chief cells secrete proteases and peptidases, which begin to degrade proteins. These enzymes are synthesized in an inactive state, as **zymogens**, also known as proenzymes; in this way, enzymes are prevented from degrading the cell's proteins and hence from damaging the cell itself. Zymogens require some type of modification for activation. In this instance, proteolytic enzymes are activated by the acidic environment of the stomach. Consequently, the zymogen pepsinogen is converted to pepsin while still in the stomach.

Zymogens are not the only way in which the organism protects itself from its own digestive enzymes. Mucus, which is a viscous combination of glycoproteins, provides a barrier to help keep proteolytic enzymes away from the cells lining the stomach. Cells that secrete mucus also secrete bicarbonate ions, which prevent acidic damage. Depletion of this barrier, due to infection or injury, can lead to damage of the stomach lining causing an ulcer.

Stomach acid is controlled pharmacologically in different ways   Whether due to one's diet or stress, excess stomach acid is a common complaint. In response, science and medicine developed some simple treatments, and others more strategic, for excess stomach acid.

The oldest and still the most common treatment for excess stomach acid is to neutralize it with a weak base using antacid tablets. These tablets often contain calcium carbonate ($CaCO_3$) or magnesium carbonate ($MgCO_3$). Because they are weak bases, they work instantly by neutralizing stomach acid. In addition, these preparations provide Ca^{2+} or Mg^{2+}, necessary nutrients in the diet. Although antacids are generally effective, they have a chalky taste that some people find displeasing. In addition, stoichiometric amounts of base need to be consumed to neutralize the excess stomach acid.

Two newer approaches work at the level of acid production. Omeprazole (Prilosec) and lansoprazole (Prevacid) are examples of proton-pump inhibitors (PPIs). These drugs block the production of acid in the stomach by irreversibly inhibiting the action of the ATPase found on the surface of parietal cells in the gastric mucosa. This ATPase uses the energy of ATP hydrolysis to pump one proton into the stomach in exchange for a potassium ion. Prevacid and Nexium (the next-generation of Prilosec) are some of the best-selling drugs and are widely used; however, no matter how benign they may appear, these inhibitors are pharmaceuticals and taking them is not without risk. Although these drugs are well-tolerated and side effects are minor, they may increase susceptibility to disease, such as infection with *Clostridium difficile* (*C. diff*), block absorbance of some micronutrients, and interact with other medications. The United States Food and Drug Administration (FDA) recommends that patients consume a two-week course of proton-pump inhibitors no more than three times per year.

Omeprazole   **Lansoprazole**

A third means by which gastric acid can be controlled is through blocking the biochemical signals that cause the parietal cell to secrete acid. Histamine is the chemical signal used by the body to trigger these cells. Histamine is also the molecule that causes swelling and stuffy noses during allergy season, but its effect here is different. There are at least three classes of histamine receptors: H_1, which is ubiquitously expressed; H_2, which is most strongly expressed in the gastric mucosa; and H_3, which is found at the highest concentration in nervous tissue. All are examples of G protein coupled receptors. Inhibitors of the H_2 receptor, such as cimetidine (Tagamet) and ranitidine (Zantac) are H_2 receptor antagonists; they act by blocking the signals that stimulate the parietal cell to secrete acid.

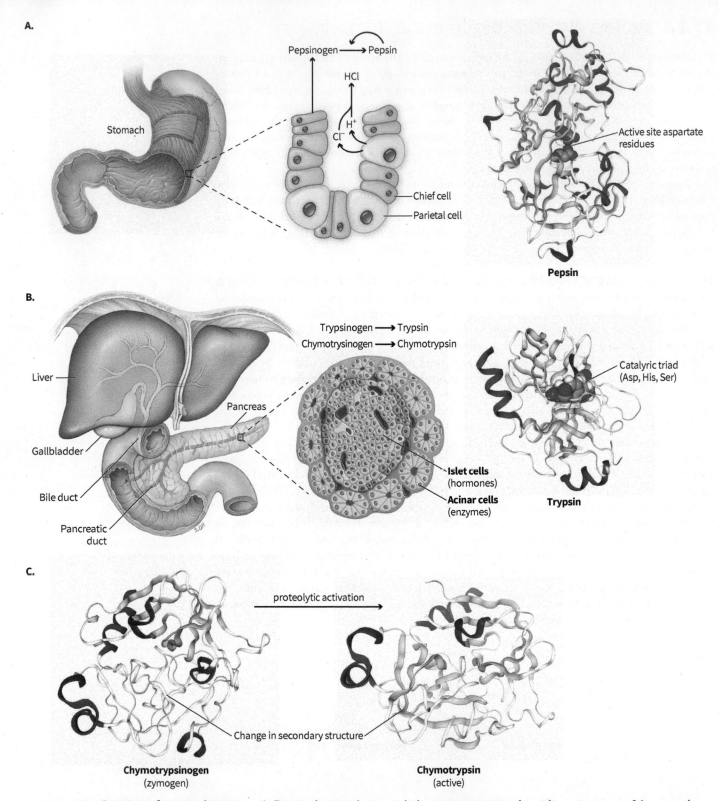

FIGURE 11.1 Overview of protein digestion. **A.** Protein digestion begins with the enzyme pepsin in the acidic environment of the stomach. **B.** The pancreatic enzymes trypsin and chymotrypsin continue protein degradation in the small intestine. **C.** Proteolytic activation of enzymes in the stomach leads to changes in protein structure.

(Source: (A, Stomach) Tortora; Derrickson, *Principles of Anatomy & Physiology*, 13e, copyright 2012, John Wiley & Sons. This material is reproduced with permission of John Wiley & Sons, Inc. (A) Data from PDB ID 1PSN Fujinaga, M., Chernaia, M.M., Tarasova, N.I., Mosimann, S.C., James, M.N. (1995) Crystal structure of human pepsin and its complex with pepstatin. *Protein Sci.* 4: 960–972. (B, Pancreas) Tortora; Derrickson, *Principles of Anatomy & Physiology*, 13e, copyright 2012, John Wiley & Sons. This material is reproduced with permission of John Wiley & Sons, Inc. (B) Data from PDB ID 1S81 Transue, T.R., Krahn, J.M., Gabel, S.A., DeRose, E.F., London, R.E. (2004) X-ray and NMR characterization of covalent complexes of trypsin, borate, and alcohols. *Biochemistry* 43: 2829–2839. (C, Left) Data from PDB ID 2CGA Wang, D., Bode, W., Huber, R. (1985) Bovine chymotrypsinogen A X-ray crystal structure analysis and refinement of a new crystal form at 1.8 A resolution. *J. Mol. Biol.* 185: 595–624. (C, Right) Data from PDB ID 4Q2K Chua, K.C.H., Pietsch, M., Zhang, X., Hautmann, S., Chan, H.Y., Bruning, J.B., Gutschow, M., Abell, A.D. (2014) Macrocyclic Protease Inhibitors with Reduced Peptide Character. *Angew. Chem. Int. Ed. Engl.*)

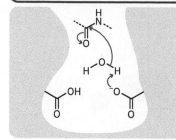

1. Substrate binds. Aspartyl residues in active site polarize water molecule that attacks peptide carbonyl carbon.

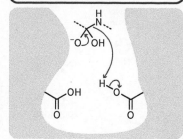

2. A tetrahedral intermediate forms. This intermediate collapses, resulting in cleavage of the peptide bond.

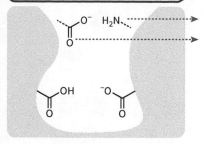

3. Products are released.

FIGURE 11.2 Pepsin is an aspartyl protease. Pepsin is an aspartyl protease using a pair of aspartate residues in the active site to polarize a water molecule, increasing its nucleophilic character and enabling attack on the carbonyl carbon of the peptide (amide) bond. The resulting tetrahedral intermediate collapses, resulting in bond cleavage.

11.1.2 Protein digestion continues in the small intestine, aided by proteases

Additional degradation of proteins takes place in the small intestine. Pancreatic enzymes such as trypsin and chymotrypsin further break down proteins into short peptides or free amino acids. The mechanism of action for these enzymes was discussed in section 4.3. The pancreatic enzymes are proteases; that is, they cleave peptide (amide) linkages between amino acids. Trypsin and chymotrypsin are serine proteases; they employ a classic catalytic triad (Asp-His-Ser) with the deprotonated serine acting as the nucleophile. Both trypsin and chymotrypsin are secreted as the zymogens trypsinogen and chymotrypsinogen, which are both activated by proteolytic cleavage. This cleavage can come from active trypsin itself or from enteropeptidase, another serine protease secreted from the intestinal mucosa. Pepsin is an aspartyl protease; it uses a pair of aspartate residues to coordinate a water molecule to attack the carbonyl carbon of the peptide bond (**Figure 11.2**).

Following cleavage by proteases, peptidases cleave the smaller units into individual amino acids, dipeptides, or tripeptides. The peptidases include aminopeptidase and carboxypeptidase, which cleave amino acids from the amino or carboxyl side of short oligopeptides (**Table 11.1**).

11.1.3 Amino acids are absorbed in the small intestine

The small intestine is also where absorption of amino acids occurs. The brush border of the small intestine is lined with epithelial enterocytes and embedded in their plasma membranes are transporters and peptidases (**Figure 11.3**). Transmembrane peptidases are integral membrane proteins that break dipeptides and tripeptides into free amino acids; most peptides longer than three amino acids are not absorbed. The epithelial enterocytes also have transporters that move the amino acids from the intestinal lumen to the bloodstream. These transporters are symporters that rely on the cotransport of sodium ions to drive amino acid absorption. Dipeptides and tripeptides are transported using the peptide transporter—a symporter that uses a proton

TABLE 11.1 Digestive Enzymes of the Gastrointestinal Tract

Zymogen	Enzyme	Location	Mechanism	Substrate
Pepsinogen	Pepsin	Stomach	Aspartyl protease	C side of phe, trp, tyr
Trypsinogen	Trypsin	Small intestine	Serine protease	C side of lys, arg
Chymotrypsinogen	Chymotrypsin	Pancreas/small intestine	Serine protease	C side of phe, tyr, trp

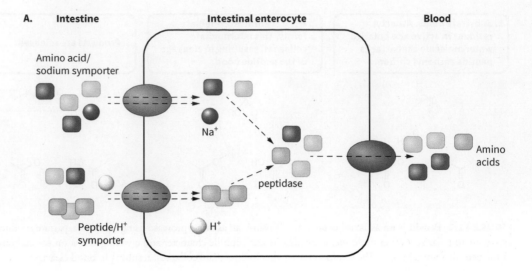

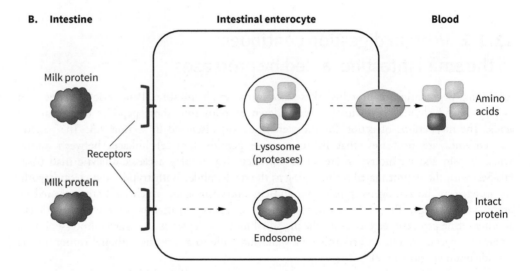

FIGURE 11.3 Amino acid absorption. **A.** Amino acids are transported into enterocytes via transporters. Each of these transporters uses the potential energy of chemical gradients, such as sodium, to drive amino acid transport. The amino acids are transported out of the enterocyte and into the bloodstream down their concentration gradient. **B.** Whole proteins are endocytosed into the enterocyte and then either degraded into amino acids in a lysosome or passed through the enterocyte and secreted into the circulation.

gradient rather than a sodium ion gradient to drive peptide transport into the cell. Once inside the cell, peptides are cleaved into amino acids by cytosolic peptidases.

There is one exception to the rule that proteins are degraded to amino acids before entering the circulation. In neonatal mammals, whole milk proteins are either endocytosed by the enterocyte and degraded in lysosomes, or the proteins pass unscathed through the enterocyte and enter the circulation in their native conformation. This mechanism is thought to grant neonates the ability to absorb antibodies from their mother's milk (an example of passive immunity) and protect them from infection in their first weeks of life. It is unclear at what point enterocytes stop importing whole proteins, known as the process of closure. In some animals, such as rats, closure can be as late as weaning, although it is likely that the switch is a combination of genetic and environmental factors.

Amino acids are secreted from enterocytes and transported in blood. Cells express amino-acid transporters on their surface and thus take up amino acids from the circulation. The transporters involved in this process are still being investigated, but several tissues deserve special mention. The kidney filters all amino acids out of blood and then reabsorbs them from the filtrate. As with the intestine, sodium ions drive this process, and many of the amino acid transporters expressed in enterocytes are also expressed in the kidney. The brain is protected from the circulation by the blood-brain barrier, a semipermeable membrane that permits the exchange of only some metabolites. This barrier is not freely permeable to amino acids. The adverse effects of a breakdown in parts of the blood-brain barrier can be seen in diseases such as brain cancer or stroke, or in physical trauma such as head injury. Failure to properly absorb amino acids is seen in several inborn errors of metabolism (**Table 11.2**).

TABLE 11.2 Examples of the Inborn Errors of Amino Acid Metabolism

Disorder	Pathway	Effect or Symptoms
Urea cycle disorders	Urea cycle	Six disorders related to deficiencies of the enzymes of the urea cycle. Patients experience hyperammonemia and nerve damage. Often severe to fatal.
Maple-syrup urine disease	Branched-chain amino acid metabolism (α-keto acid dehydrogenase)	Deficiency of branched-chain α-keto acid dehydrogenase leads to elevated levels of branched-chain amino acids and branched-chain α-keto acids, causing neurological damage and death. The disease is named for the sweet maple odor produced by the catabolite sotolon found in the urine of affected subjects.
Phenylketonuria (PKU)	Tyrosine metabolism (phenylalanine hydroxylase)	A deficiency in phenylalanine hydroxylase leads to elevated levels of phenylpyruvate, which is found in the urine of affected subjects. If left untreated, the disease can lead to mental retardation and neurological damage, but the disease is currently diagnosed in infants and successfully treated with a low phenylalanine diet.
Albinism	Tryosine metabolism (tyrosinase)	Tyrosinase deficiency results in the inability to convert tyrosine into the pigment melanin. Subjects lack pigmentation in skin, hair, and eyes.
Hyperprolinemia	Proline catabolism	A deficiency in one of the enzymes that catabolize protein leads to proline levels 3 to 10 times normal, resulting in neurological problems.

11.1.4 Amino acids serve many biological roles in the organism

Although we generally think of amino acids as the building blocks of proteins, they also serve many other important roles. Half of the amino acids can act as neurotransmitters, such as glutamate, or the precursors of neurotransmitters; for example, histidine is the precursor of histamine. Amino acids serve as metabolic intermediates that feed into other pathways, act as allosteric regulators, and are involved in the metabolism of other amino acids. Like most biological molecules, the function of amino acids is often not as simple as first appears.

Worked Problem 11.1   Acid reflux

If someone has heartburn and takes two over-the-counter calcium-based antacid tablets, how much acid will be neutralized? What will be the final pH of the stomach?

Strategy This question has several parts and requires several assumptions to be made. What is the initial pH of the stomach, the volume of the stomach (this can range from less than 100 milliliters to over 4 liters), and the dose and type of antacid? Also, are the stomach contents buffering any reaction that may take place?

Solution We can simplify the calculations by assuming that, for this example, the volume of the stomach is about 1 liter, pH is 1.0, and the stomach is currently not buffered by any contents (few biological buffers are active at such a low pH value).

Over-the-counter antacids typically contain 500 mg of calcium carbonate ($CaCO_3$) per tablet. Thus, two tablets of antacid contain 1 g of $CaCO_3$. Using the molecular mass of the compound (100 g/mol), we can calculate the number of moles in one gram:

$$1.00 \text{ g } CaCO_3/(100 \text{ g/mol}) = (0.01 \text{ mol})$$

The $CaCO_3$ will dissociate into Ca^{2+} and CO_3^{-2} ions in solution. The carbonate ion is the conjugate base of carbonic acid (H_2CO_3), which is a diprotic acid; hence, 1 mole of carbonate can react with 2 moles of protons. Therefore, two tablets of antacid can neutralize 0.02 mol of acid.

To calculate the pH change to the stomach, we will use an approach similar to that used for solving a titration problem. To find the molarity of acid in the stomach, we will take the negative antilog (10^{-x}) of the pH value:

$$10^{-1} = 0.1$$

At pH = 1, the concentration of acid in the stomach is 0.1 M. If the volume of the stomach is 1 liter, the stomach will contain 0.1 mol of acid:

$$0.1 \text{ M} = 0.1 \text{ mol/L}$$

$$0.1 \text{ mol/L (1 L)} = 0.1 \text{ mol}$$

We neutralized 0.02 mol of this acid and therefore have 0.08 mol of acid remaining:

$$0.1 \text{ mol acid} - 0.02 \text{ mol} = 0.08 \text{ mol}$$

To determine the new pH of the stomach, we must first convert this number of moles to a concentration. The negative log of this value will be the new pH:

$$(0.08 \text{ mol acid/1 L}) = 0.08 \text{ M acid}$$

$$-\log(0.08) = 1.1$$

Thus, consumption of the two tablets raises the stomach pH by 0.1 pH units (although they are still useful clinically).

Carbonic acid and bicarbonate (HCO_3^-) are weak bases, with pK_a values of 3.6 and 10.3, respectively. Why are these compounds not acting as buffers? The low pH of the stomach (more than 1 pH unit away from a pK_a value) prevents the buffers from working. Clearly, in other organs, including blood, bicarbonate is a critical buffer.

Follow-up question   The parietal cells of the stomach produce acid to maintain stomach pH at or near 1.0. Proton-pump inhibitors act to block these proteins and limit acid production. If we model the activity of proton pumps in the stomach using classical Michaelis-Menten kinetics, how might proton-pump inhibitors affect this process? What would be the final effect on pH?

Summary

- Proteins are degraded into short peptides and free amino acids by proteases and peptidases in the stomach and small intestine.
- Parietal cells in the stomach secrete acid to provide a low pH for optimal proteolytic activity in the stomach. Several classes of acid-lowering drugs target these cells.
- Enterocytes in the lining of the small intestine absorb amino acids, dipeptides, and tripeptides. Specific transporters facilitate this process.
- Amino acids are secreted into the bloodstream and absorbed by cells.

Concept Check

1. Explain how proteins are degraded and how amino acids are assimilated by the organism.
2. How do proteins, fatty acids and carbohydrates in a balanced meal get digested and absorbed?
3. Parietal cells in the stomach secrete acid. What is the purpose of acid secretion? List some of the ways by which acid production is affected.

11.2 Amino Acid Deamination and Transamination

Following a meal, excess carbohydrates can either be used to build up glycogen stores or be catabolized through glycolysis, converted into fatty acids, and stored as triacylglycerols. The energy found in excess dietary amino acids is also stored as fat. In times of extreme need, the organism can catabolize proteins found in muscle. In both storage and breakdown, excess amino acids offer a distinct metabolic challenge to the organism.

The amine moiety central to all amino acids is not found in carbohydrates and lipids and must be removed from the amino acids in order to convert them to fatty acids for storage or catabolized through the citric acid cycle.

To complicate the matter further, ammonia (NH_4^+) formed by the removal of a primary amine is somewhat toxic to the organism and must either be metabolized further or quickly eliminated. Rather than deaminate amino acids in peripheral tissues, amine groups are transported on amino acids, and the amines are detoxified in the kidney or the liver.

11.2.1 Ammonia can be removed from an amino acid in two different ways

Alpha-keto acids such as pyruvate or α-ketoglutarate are commonly found in biochemistry and play key roles in amine metabolism. Ammonia can be removed from an amino acid through **transamination**, in which the amine is transferred to an α-keto acid, or **oxidative deami-**

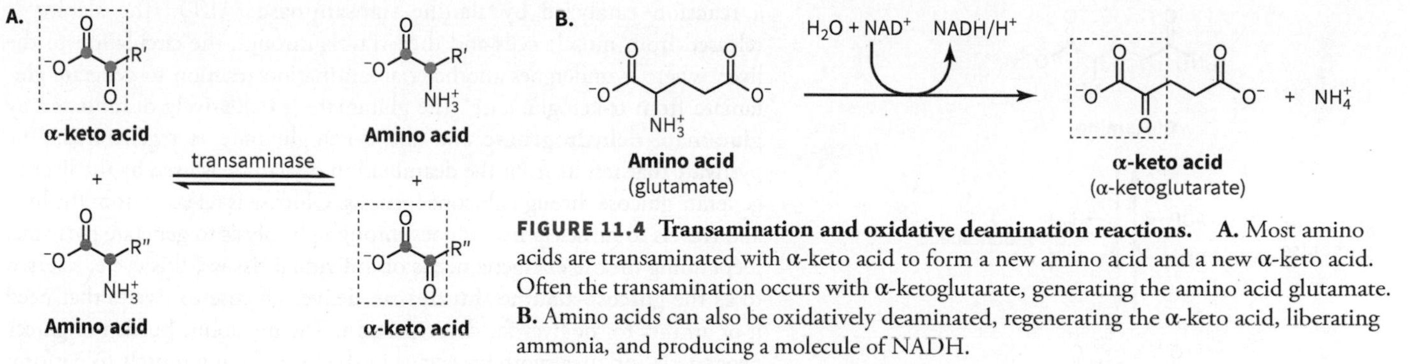

A.

α-keto acid   +   Amino acid

⇌ transaminase ⇌

Amino acid   +   α-keto acid

B.

Amino acid (glutamate)

$H_2O + NAD^+$ $NADH/H^+$

→

α-keto acid (α-ketoglutarate) $+ NH_4^+$

FIGURE 11.4 Transamination and oxidative deamination reactions. **A.** Most amino acids are transaminated with α-keto acid to form a new amino acid and a new α-keto acid. Often the transamination occurs with α-ketoglutarate, generating the amino acid glutamate. **B.** Amino acids can also be oxidatively deaminated, regenerating the α-keto acid, liberating ammonia, and producing a molecule of NADH.

nation, in which the amine is removed as ammonia. Both reactions involve an α-keto acid, which is characterized by having a carbonyl adjacent to a carboxylic acid moiety (**Figure 11.4**). Alpha-keto acids may contain other functional groups, such as other carboxylic acids or hydroxyl groups; they also vary in the number of carbons they contain.

Transamination generates a new amino acid and a new α-keto acid. Most amino acids have corresponding transaminases, but none has been identified for lysine and threonine. The α-keto acid employed by many of these enzymes is either α-ketoglutarate or pyruvate, and they generate the amino acids glutamate and alanine, respectively, when transaminated. Therefore, the cell also uses this mechanism as a simple means of synthesizing amino acids from their corresponding α-keto acids.

In oxidative deamination, the source of the oxygen is water and a molecule of $NADH/H^+$ is generated. The most common amino acid used is glutamate, which is interconverted to α-ketoglutarate. The resulting free ammonia is converted to other less-toxic nitrogenous compounds such as uric acid (in birds and reptiles) or **urea** (in mammals).

11.2.2 The glucose–alanine shuttle moves nitrogen to the liver and delivers glucose to tissues that need it

Transamination reactions provide the organism with an elegant means of shuttling carbon and nitrogen metabolites to and from tissues that need them (**Figure 11.5**). There are several metabolic states in which muscle will catabolize protein in order to provide glycolytic intermediates and glucose. These include starvation and uncontrolled diabetes mellitus, both instances of where amino acids are burned for energy. In muscle, alanine is generated by the transamination of pyruvate,

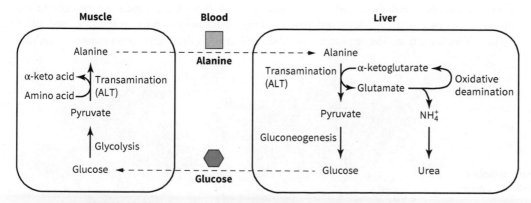

FIGURE 11.5 Glucose–alanine shuttle. The glucose–alanine shuttle allows amines and pyruvate from muscle tissue to be shuttled to the liver to provide substrates for gluconeogenesis and the urea cycle. In turn, muscle tissue is provided with glucose, and muscle glycolysis proceeds.

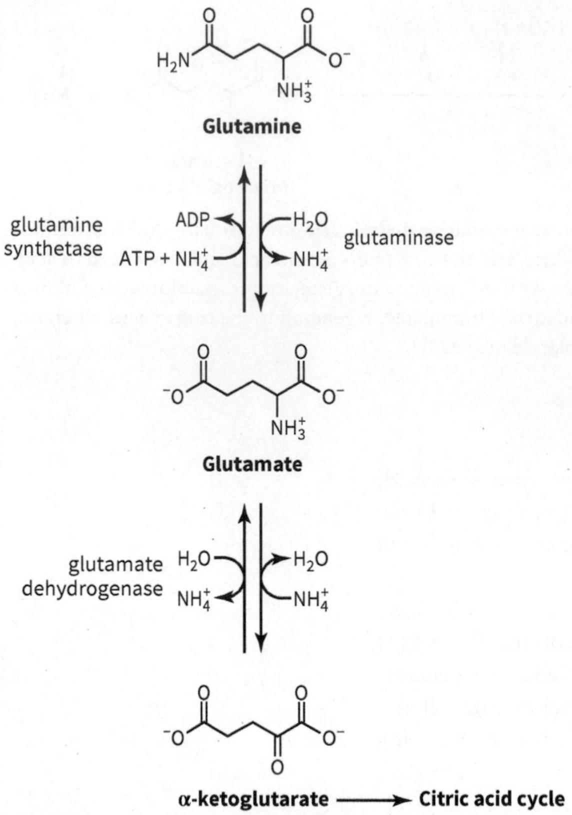

Glutamine

glutamine synthetase — ADP ← | → H_2O — glutaminase
ATP + NH_4^+ ← | → NH_4^+

Glutamate

glutamate dehydrogenase — H_2O ← | → H_2O
NH_4^+ ← | → NH_4^+

α-ketoglutarate ⟶ **Citric acid cycle**

FIGURE 11.6 Overview of glutamine metabolism.
Glutamine and glutamate are important molecules in amine metabolism. Several enzymes are involved in the synthesis and breakdown of these metabolites.

a reaction catalyzed by alanine transaminase (ALT). The alanine is released from muscle cells and then travels through the circulation to the liver, where it undergoes another transamination reaction to generate glutamate from α-ketoglutarate. The glutamate is oxidatively deaminated by **glutamate dehydrogenase** and the α-ketoglutarate is regenerated. The pyruvate that results from the deamination of alanine is used by the liver to generate glucose through gluconeogenesis. Glucose is released from the liver and travels to muscle where it goes through glycolysis to generate pyruvate. Depending on the energetic needs of individual tissues, this cycle, referred to as the **glucose-alanine shuttle**, can deliver glucose to tissues that need it or amines to the liver for detoxification. The metabolic burden of gluconeogenesis and urea synthesis occurs in the liver, freeing muscle to perform physical work.

11.2.3 Glutamine is also important in nitrogen transport

Although alanine is used to shuttle amines to the liver, it is not the only mechanism in the body for moving amino groups (**Figure 11.6**). Glutamine is the amino acid found in the highest concentration in the blood; it also serves to detoxify and transport ammonia. The enzyme responsible for the synthesis of glutamine from glutamate and ammonium ion in the liver is glutamine synthetase.

Glutamine synthetase regulates glutamine levels

Glutamine synthetase is a dodecamer (12 subunits) of identical subunits arranged in two hexagonal rings stacked on top of one another. Each subunit has an active site that is accessible from both the top and the bottom. ATP enters from the top, and glutamate and ammonia enter from the interface between the two hexameric rings. Each active site contains two divalent metal ions (either Mg^{2+} or Mn^{2+}), located in the center of the active site; these ions are important in glutamate binding and catalysis (**Figure 11.7**).

Glutamine synthetase is acutely regulated by covalent modification and numerous allosteric regulators. We often think of phosphorylation as a covalent modification that can alter the activity or localization of an enzyme; glutamine synthetase is not phosphorylated and is instead adenylated on a specific tyrosine residue. In adenylation, an adenosine monophosphate (AMP) group is added through a phosphodiester linkage to a tyrosine residue. Addition of the bulky adenylate group blocks the activity of that subunit of the enzyme. However, because there are 12 subunits, regulation can be fine-tuned by adjusting the level of adenylation.

Glutamine synthetase is also controlled by allosteric regulators. These small molecules coordinate glutamine metabolism with other pathways, and they offer a regulatory checkpoint for the synthesis of glutamine. Nine different regulators bind to glutamine synthetase, and all of them inhibit its action. The regulators are AMP, amino acids that figure prominently in cellular

FIGURE 11.7 Structure of glutamine synthetase.
Glutamine synthetase is a homododecamer, 12 subunits composed of two 6-membered rings.

(Source: Data from PDB ID 1LGR Liaw, S.H., Jun, G., Eisenberg, D. (1994) Interactions of nucleotides with fully unadenylylated glutamine synthetase from Salmonella typhimurium. *Biochemistry* 33: 11184–11188)

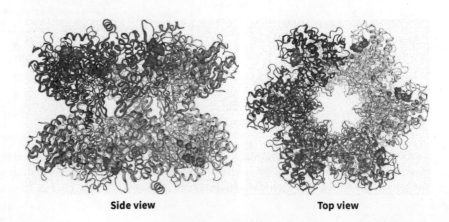

Side view   **Top view**

amino acid metabolism (alanine, glycine, and serine), carbamoyl phosphate, and downstream products of glutamine (cytidine triphosphate [CTP], glucosamine-6-phosphate, histidine, and tryptophan). While this might seem like a curious list, examination of the molecules in question reveals that each of these molecules either provides a readout on the metabolic state of the cell or is a product made from glutamine.

Glutamine has several advantages as a nitrogen carrier. First, it carries two equivalents of nitrogen: one as an amine in the common portion of the amino acid and one as a carboxamide group in the side chain. Because the amide side chain of glutamine is not charged, glutamine can transport at least one of these moles of nitrogen in an uncharged and detoxified state.

Glutaminase generates glutamate and ammonia

Glutamine serves as a starting material for several important intermediates; thus, it can be metabolized to eliminate ammonia in several different ways. In the kidney and liver, glutamine can be deamidated to yield glutamate and free ammonia; **glutaminase** is the main enzyme responsible for this reaction. In the liver, ammonia generated in this way is consumed in urea synthesis, whereas in the kidney it is directly eliminated in the urine. The liver also contains glutamine synthetase; however, glutamine synthetase and glutaminase are found in different cells within the liver. The presence of glutamine synthetase allows the liver to sequester excess ammonia as glutamine if urea biosynthesis becomes overwhelmed. In addition, having these two distinct pathways provides the liver with a way to detoxify amines in situ via hepatic formation of urea or to shift the metabolic burden of disposal to the kidney, where glutamine is deamidated and disposed of as ammonia.

Under conditions of starvation or a high-protein diet, glutaminase gene expression is upregulated, as might be expected. Other factors that regulate the kidney and liver isoforms of this protein are not clearly understood, although kinases involved in cell growth and differentiation (MAP kinases) have been implicated in the regulation of both isoforms.

As discussed in the next section, glutamine serves several important roles in the central nervous system. Glutamate is an important neurotransmitter but can be deactivated by adding an amino group to form glutamine. Neurons are highly sensitive to ammonia. Glutamate can act as a buffer for ammonia or can be used to alter levels of glutamine in these cells.

Glutamate dehydrogenase liberates a second mole of ammonia

Glutamine carries two amine moieties, one on the side chain and the other in the α-amino group common to all amino acids. After deamination to glutamate, the second amine can still be liberated. Glutamate dehydrogenase catalyzes the oxidation of the amine of glutamate to an imine, which quickly hydrolyzes in water to liberate a second ammonia group and produce α-ketoglutarate, shown in Figure 11.6. Glutamate dehydrogenase is found in the mitochondrial matrix of the kidney, the liver, and nervous tissue.

Glutamate dehydrogenase is regulated by numerous allosteric regulators, which can bind at several sites on each monomer. For example, ADP and NAD⁺ bind in the second coenzyme-binding site, adjacent to the hinge. Binding stabilizes the enzyme in the active conformation, in which the active site is accessible to substrate. In contrast, ATP, GTP, leucine, and NADH stabilize the closed conformation by binding either in the substrate site or in the second coenzyme-binding site. Some data indicate that there may be a third binding site, which could explain some of the more complex effects observed. These regulators should make metabolic sense. Those that signal the cell is in a low energy state (NAD⁺ and ADP) activate the enzyme to produce more α-ketoglutarate, a key citric acid cycle intermediate. Conversely, regulators that indicate the cell has sufficient energy (ATP, GTP, and NADH) inhibit the enzyme. As with glutamine synthetase, the regulators of this enzyme balance the need for biochemical building blocks with the energetic demands of the cell.

Ammonia toxicity arises as a result of glutamate dehydrogenase. This enzyme generates free ammonia and α-ketoglutarate from glutamate, but it can also perform the reverse reaction. If ammonia levels are high, this can lead to depletion of mitochondrial α-ketoglutarate and disturbances of the citric acid cycle and ATP production. In addition, this reaction requires NADPH, which will potentially deplete if the reaction proceeds unabated. This reaction happens at ammonia concentrations as low as 100 μM, too low to substantially influence plasma pH but high enough to cause cell or organismal death.

The causal parasite in malaria (*Plasmodium falciparum*) infects erythrocytes and employs glutamate dehydrogenase to synthesize NADPH, which is used as an electron source for antioxidative enzymes (glutathione reductase and thioredoxin reductase). Because the erythrocyte lacks glutamate dehydrogenase, small molecules that target this enzyme are potential candidate molecules for antimalarial drug development.

Why are there two different mechanisms (transamination and oxidative deamination) for the metabolism of amines? First, having two mechanisms instead of just one provides a safe way of shuttling potentially toxic metabolites around the organism. Second, transamination provides a pathway for synthesis of amino acids and α-keto acids by exchanging amino groups with α-keto acids. Third, transamination and the glucose–alanine shuttle help to concentrate metabolic tasks to specific organs, such as urea production in the liver, allowing for more specific organ function. Muscle does not make urea and does not need to because muscle and the liver have evolved specialized functions. Nitrogen from amino acid catabolism in muscle is sent to the liver for detoxification. In addition, muscle gains glucose that is made by the liver.

The question of which of these two pathways is used on a particular amino acid at a particular time is more complex. The interplay of the energetic state of the organism, the need for specific amino or α-keto acids, and the availability of those and other metabolites in a given tissue shift the balance between transamination and oxidative deamination.

We may also ask why alanine and glutamine are the most common nitrogen carriers in the blood. Again, the answers are imbedded in the multiple functions these compounds play in metabolism. Alanine generated in muscle, either by protein breakdown or by transamination of pyruvate obtained through glycolysis, moves toxic nitrogenous wastes to the liver for detoxification. In addition, the pyruvate that results from transamination provides a source of carbons for gluconeogenesis. Glutamine can transport two amines, one in the side chain as an amide and the other as the amino group common to all amino acids. The amide is not protonated and bears no charge; hence, this is another way of transporting nitrogenous wastes safely, with no risk of affecting plasma pH. The deamidation of glutamine in either the liver or kidney yields glutamate, which can be used as an amino acid, or can be further transaminated or oxidatively deaminated to give α-ketoglutarate. This intermediate provides another link between the citric acid cycle and the rest of metabolism. In addition, this process can keep glycolysis moving forward under anaerobic conditions for a limited time.

Worked Problem 11.2   Oxidative deamination

Draw the α-keto acid that would result from the oxidative deamination of each of these amino acids.

Serine   **Glutamate**   **Isoleucine**

Strategy   Examine the structures of each amino acid. Which amine will be oxidized? What will be found in its place in an α-keto acid?

Solution

Follow-up question   What are the products of these transamination reactions? Would energy be consumed in these reactions?

Summary

- Amino acids all contain a potentially toxic amino group that must be removed and detoxified.
- Amino groups can be shuttled between amino acids and α-keto acids, resulting in a new α-keto acid and a new amino acid.
- The glucose–alanine shuttle is a means of delivering ammonia to the liver for detoxification while delivering glucose back to tissues that need them.
- Amines can also be removed as ammonia, in a process known as oxidative deamination; one molecule of NADH/H⁺ is produced in the process.
- Glutamine is one of the main carriers of ammonia in the blood.

1. Describe how and where the first steps in amino acid catabolism occur and give the products of these reactions.

2. Explain which of the reactions in the first steps of amino acid catabolism are energetically "free," that is, neither costing nor producing energy, and which produce energy.

3. Which enzyme is involved in the oxidative deamination of glutamate and what are the products formed?

11.3 The Urea Cycle

Ammonia is a toxic substance. Normal blood levels range from 10 to 40 micromoles/L. High levels of ammonia elevate blood pH, making it more difficult for hemoglobin to deliver oxygen to metabolically active tissues, and more difficult to transport carbon dioxide (CO_2) in the blood. Many organisms are slightly tolerant of a small drop in plasma pH, for example, from pH 7.4 to 7.2, but less tolerant of a similar rise in plasma pH, for instance, from pH 7.4 to 7.6. Furthermore, ammonia is a neurotoxin and can permeate the blood-brain barrier. Once it enters the brain, ammonia can undergo an amination reaction with α-ketoglutarate to form glutamate. This depletes the neurons of citric acid cycle intermediates, effectively blocking a major pathway of energy production in the cells and thus leading to cell death. Elevated levels of glutamate, itself a neurotransmitter, lead to increased glutamine production. This depletes the cells of glutamate by shunting the molecule toward glutamine production and away from the production of γ amino butyric acid (GABA), a protective neurotransmitter. Also, high levels of ammonia may help to break down the H⁺ gradient in mitochondria, another potentially toxic effect. An organism that lacks a means of coping with ammonia will not survive for long.

11.3.1 Ammonia detoxification begins with the synthesis of carbamoyl phosphate

The first step in the detoxification of ammonia is the condensation of ammonia with CO_2 to from carbamoyl phosphate. This is a reaction catalyzed by carbamoyl phosphate synthetase and requiring ATP (**Figure 11.8**). The reaction occurs in three distinct stages:

- CO_2 reacts with ATP to form a carboxyl phosphate.
- The phosphate is exchanged for a molecule of ammonia, creating a carbamate.
- The carbamate is phosphorylated by a second molecule of ATP to produce carbamoyl phosphate.

There are different isoforms of carbamoyl phosphate synthetase. This section focuses on the mitochondrial form of the enzyme: carbamoyl phosphate synthetase I. This isoform is involved in the production of carbamoyl phosphate for the urea cycle. Carbamoyl phosphate synthetase II, the cytosolic form of the enzyme, is responsible for the production of building blocks for pyrimidine biosynthesis.

11.3.2 The urea cycle synthesizes urea and other metabolic intermediates

Carbamoyl phosphate is not the terminal destination for nitrogenous waste. Mammals convert carbamoyl phosphate into the far less toxic urea using a pathway termed the **urea cycle**. This cycle employs several amino acids, such as ornithine and citrulline, which have not been mentioned previously. They are not included in the 20 amino acids commonly found in proteins; nevertheless, they are amino acids and are important in amine metabolism.

The reactions of the urea cycle   The urea cycle begins with carbamoyl phosphate condensing with ornithine to form citrulline. The phosphate on carbamoyl phosphate is the

FIGURE 11.8 Carbamoyl phosphate synthesis.
Bicarbonate anion attacks the phosphorous center of the γ-phosphate of ATP, generating an ATP-carboxyl phosphate. ADP departs, and ammonia binds. The lone pair of electrons on ammonia attacks the carbonyl carbon, generating a carbamate. ATP binds and, as before, the negatively charged oxygen attacks the γ-phosphate of ATP. ADP leaves, and carbamoyl phosphate is generated.

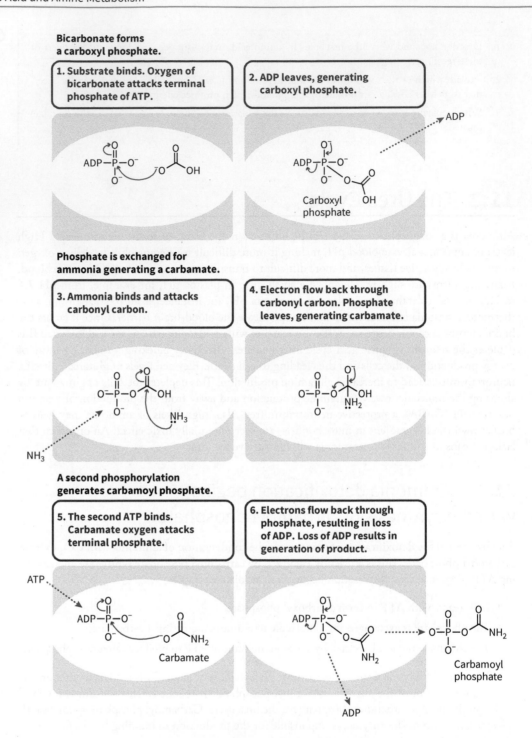

Bicarbonate forms a carboxyl phosphate.

1. Substrate binds. Oxygen of bicarbonate attacks terminal phosphate of ATP.

2. ADP leaves, generating carboxyl phosphate.

Phosphate is exchanged for ammonia generating a carbamate.

3. Ammonia binds and attacks carbonyl carbon.

4. Electron flow back through carbonyl carbon. Phosphate leaves, generating carbamate.

A second phosphorylation generates carbamoyl phosphate.

5. The second ATP binds. Carbamate oxygen attacks terminal phosphate.

6. Electrons flow back through phosphate, resulting in loss of ADP. Loss of ADP results in generation of product.

leaving group in this reaction, and it provides the driving energy for the reaction to proceed. Examining the structures of ornithine and carbamoyl phosphate can help in determining the rest of the steps in urea production. As we have seen in other pathways, comparing structures and functional groups of molecules in a pathway helps connect a chemical logic to the steps in the urea cycle. The structure of ornithine might look familiar: ornithine is effectively lysine minus a methylene group. Coupling carbamoyl phosphate to ornithine could be achieved in several ways, but the one employed by nature is to couple the carbamoyl carbon to the primary amine on the side chain of ornithine. This reaction is analogous to the second step of carbamoyl phosphate synthesis in which the attack of ammonia on carboxyl phosphate forms the initial carbamate. In both reactions, the phosphate is the leaving group (**Figure 11.9**).

In the next step of the pathway, citrulline links with aspartate to form argininosuccinate, a reaction that is catalyzed by argininosuccinate synthase. Again, the carbon involved in the chemistry of the reaction is the carbamoyl carbon, and the nitrogen comes from the amine of aspartate.

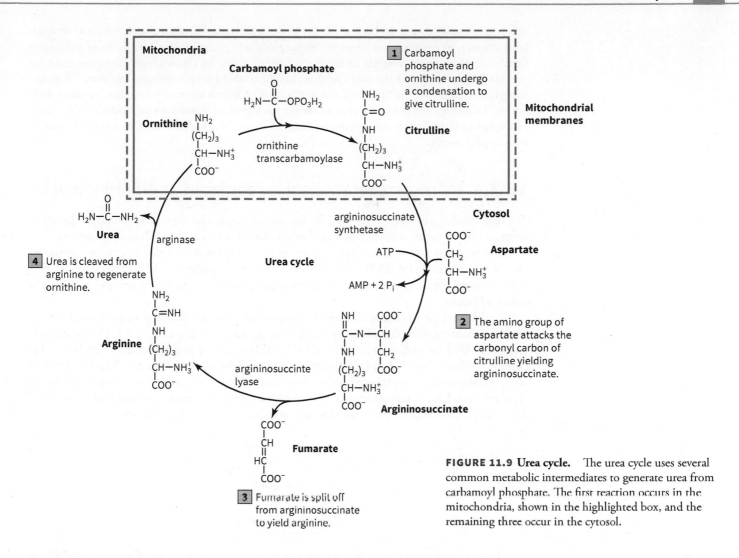

FIGURE 11.9 Urea cycle. The urea cycle uses several common metabolic intermediates to generate urea from carbamoyl phosphate. The first reaction occurs in the mitochondria, shown in the highlighted box, and the remaining three occur in the cytosol.

A molecule of ATP is hydrolyzed into AMP and PPi in the reaction (providing the energetic equivalent of two molecules of ATP being hydrolyzed).

Argininosuccinate is cleaved by argininosuccinase (also known as argininosuccinate lyase) to give two molecules, both of which should be familiar at this point: the amino acid arginine and the citric acid cycle intermediate fumarate (a *trans* unsaturated dioic acid, four carbons in length). Tracking the structures of molecules throughout the pathway or considering the structure of arginine can help in recalling the intermediates in these pathways. The final step of the urea cycle is the formation of urea from arginine, in a reaction catalyzed by arginase. Water serves as the source of oxygen for the new carbamate, and ornithine is regenerated.

The urea cycle occurs in two subcellular compartments. The first two reactions (formation of carbamoyl phosphate and condensation of carbamoyl phosphate with ornithine to form citrulline) occur in the mitochondrial matrix. This location takes advantage of free ammonia generated by the oxidative deamination of amino acids in the mitochondria without releasing ammonia into the cytosol, where it may be able to enter the bloodstream. Citrulline is exported from the matrix to the cytosol via the ornithine transporter, which has two isoforms (ORC 1 and ORC 2), and also transports arginine, citrulline, and lysine. The remaining steps of the urea cycle occur in the cytosol. Ornithine regenerated following the cleavage of urea is imported back into the mitochondrial matrix to propagate the cycle.

The urea cycle can be studied as a separate and discrete entity, but in reality it always occurs in a sea of metabolites. The aspartate employed in the urea cycle is the same as the aspartate discussed in other chapters. However, partitioning of metabolites can direct them to different metabolic fates and proscribe different metabolic functions. For example, fumarate generated in the urea cycle (in the cytosol) is chemically the same as fumarate generated in the citric acid cycle (in the mitochondrial matrix), but partitioning by the mitochondrial membranes leads to different metabolic fates. In both pools, fumarate is converted to malate and then, via fumarase (a hydratase), to oxaloacetate. However, in the citric acid cycle, oxaloacetate serves as a substrate for citrate production, whereas in the cytosol it can either be transaminated into aspartate (to provide intermediates for the urea cycle) or decarboxylated into phosphoenolpyruvate for gluconeogenesis. Likewise, the intermediates in the urea cycle can serve as precursors for other molecules. Thus, arginine can serve as a source for the signaling hormone nitric oxide (NO), being converted to citrulline in the process, and ornithine can be decarboxylated to form polyamines, such as putrescene.

11.3.3 Nitrogen metabolism is regulated at different levels

The urea cycle is primarily responsible for the detoxification and disposal of ammonia. Levels of ammonia rise in response to the oxidative deamination of amino acids, particularly glutamate.

The regulation of the urea cycle occurs at two different levels, through either carbamoyl phosphate synthetase I (CPS I) or the other enzymes of the pathway. CPS I is regulated allosterically, whereas the other enzymes of the pathway are primarily regulated by levels of substrate, as discussed below.

Carbamoyl phosphate synthetase I is responsible for the first step in urea synthesis, in which carbamoyl phosphate is synthesized from ammonia, CO_2, and ATP. CPS I is allosterically activated by *N*-Acetylglutamate (NAG). This regulator is synthesized by *N*-Acetylglutamate synthase, which employs glutamate and acetyl-CoA as substrates. The synthesis of *N*-Acetylglutamate is positively regulated by levels of substrate, glutamate, and the allosteric regulator arginine. Concentrations of both glutamate and arginine are elevated when there are excess amino acids, thus increasing urea production via *N*-Acetylglutamate.

***N*-Acetylglutamate** (NAG)

The other enzymes of the urea cycle are acutely regulated by the levels of substrate available to them. Other means of control, such as phosphorylation or allosteric mechanisms, have not been observed for these enzymes; however, they are also regulated at transcriptional level. For example, consumption of a diet rich in protein leads to upregulation of the genes coding for these enzymes.

11.3.4 Mechanisms for elimination of nitrogenous wastes differ between mammals and non-mammals

So far, this section has focused on human nitrogen metabolism. Several factors must be taken into consideration when thinking about nitrogen metabolism in other organisms. Urea biosynthesis is energetically expensive; neutralizing a single molecule of ammonia requires the energetic equivalent of the hydrolysis of three ATP molecules and sufficient water for excretion via the kidneys. Hence, not all organisms use this method to dispose of nitrogenous wastes.

Ammonotelic organisms such as insects, crustaceans, aquatic invertebrates, and some fish and amphibians typically eliminate wastes as ammonia secreted through the skin. These species adapted to the ample water in their environment and have no need to expend energy on urea biosynthesis. This indicates that evolution of the genes encoding proteins found in the urea cycle have evolved more recently than the appearance of these organisms.

Birds and reptiles employ a different means of eliminating nitrogenous wastes. Termed **uricotelic** organisms, they synthesize uric acid and eliminate it in their waste, which is technically not urine (**Figure 11.10**). Birds and reptiles have evolved adaptations to different environmental challenges. For example, to facilitate their ability to fly, birds have evolved ways to conserve weight. Using urea to eliminate nitrogenous wastes would necessitate a urinary bladder to store

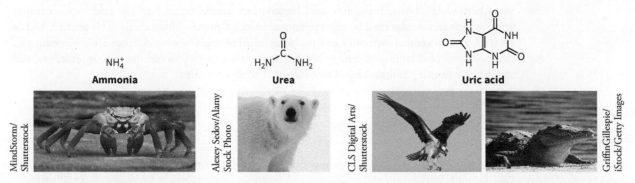

FIGURE 11.10 Nitrogenous wastes. Different organisms produce different types of nitrogenous wastes. Aquatic organisms such as sponges, corals, jellyfish, and crustaceans secrete ammonia. Mammals produce urea, which is secreted in the urine, whereas birds and some reptiles secrete uric acid in their waste.

the urine; there would be also a need to carry the water weight of the urine itself. By producing uric acid, birds avoid this necessity. Reptiles face a different challenge. Many reptiles, in particular the diapsids, such as crocodiles, lizards, and snakes, come from arid or saltwater environments where fresh water is in short supply. These animals expel nitrogenous waste as uric acid to spare water rather than weight.

11.3.5 Some mammals have adapted to high- or low-protein diets

Up until now, our discussion of urea biosynthesis presumed that the organism in question was human and consumed a relatively protein-rich diet, consuming about 15% of calories from animal or plant protein. However, not all mammals fall into this category; some consume a carnivorous diet with exceedingly high amounts of protein, whereas others (most herbivores) consume far less. Both groups of animals have evolved adaptations to assist in their management of nitrogen.

Humans and all higher animals are unable to "fix" nitrogen, that is, to use N_2 gas in the air to produce reduced forms of nitrogen, such as the nitrogen of amino acids. In order to synthesize the nonessential amino acids, we must have a source of reduced nitrogen, such as an amine, to begin the process.

Carnivores Obligate carnivores consume all their calories from animal sources. In the wild, this means a diet with a high concentration of protein and little carbohydrate and fat. Zebras and antelopes, the prey of large carnivores, are generally lean animals that do not exhibit the marbling seen in a good beef steak, and are far lower in fat than the meat that humans would typically eat. The domestic cat is an example of an obligate carnivore, as are all felids (the cat family), pinnipeds (walruses and seals), and weasels. The diet of the cat has led to some important and interesting traits in nitrogen metabolism. For example, the urea cycle in cats is always functioning at a high level, and the control mechanisms discussed previously do not apply in this case. Thus, cats are always at the ready to catabolize a high-protein diet, but they cannot shut down the urea cycle and cannot spare their own protein when food is scarce.

This difference in diet and metabolism leads to some interesting complexities in carnivore metabolism. Cats are unable to synthesize the amino acid arginine. If cats consume a diet that contains enough arginine to fuel the urea cycle, this does not present a problem; however, a single meal devoid of arginine can make a cat hyperamonemic (high plasma ammonia levels, which affect respiration and the nervous system). This condition can be fatal. To combat this, many commercial cat foods are supplemented with arginine.

Herbivores Herbivores face a different problem; in the wild these organisms consume a low-protein diet that consists mainly of carbohydrates. Dietary proteins are at a premium for these organisms, and they have evolved mechanisms to help conserve the nitrogen found in such proteins. In herbivorous mammals, urea is shunted back to the intestine in a similar fashion as uric acid in some herbivorous birds. There, intestinal bacteria can reduce the urea or uric acid back to amino acids, which can be absorbed by the animal and used for protein biosynthesis. This ancient, highly evolved, and efficient system can cause problems when animals become domesticated. For example, cattle or horses fed large quantities of high-protein grain that they would not encounter in the wild can develop health problems related to an elevated nitrogen burden.

Worked Problem 11.3   Follow that nitrogen

Parts of the urea cycle occur in the mitochondria and parts in the cytosol. How could you demonstrate that this is the case? Design an experiment to show which aspects of the urea cycle happen in the mitochondria and which happen in the cytosol.

Strategy   We will need a source of tissue or cells capable of undergoing the urea cycle and some means of tracking the intermediates. We will also need some means of separating the mitochondrial and cytosolic fractions of the cell. Radiolabeled tracers have been helpful in deducing metabolic pathways in the past, but another approach would be to use stable isotopes (such as N^{15} labeled ammonia) and follow those using mass spectrometry.

Solution   First, we would fractionate the cells into a mitochondrial and a cytosolic fraction. To each fraction we would add labeled ammonia ($N^{15}H_4^+$). If necessary, we could supplement these fractions

with exogenous levels of the required cofactors or precursors (ATP, citrulline, or aspartate). Samples would be incubated for different times, and the products examined by mass spectrometry to see what new molecules the labeled ammonia has been converted into. If the urea cycle in the sample is functioning as described in this chapter (highly likely), then we should see no product from the cytosolic fraction; that is, the ammonia would not be metabolized. However, in the mitochondrial fraction, some of the N^{15} ammonia would be converted to N^{15} carbamoyl phosphate. Both fractions would have to be combined to obtain N^{15} urea as a final product.

Follow-up question   Zinc is an important cofactor in some enzymes. How could you demonstrate which steps in nitrogen metabolism are zinc dependent?

Summary

- Mammals detoxify ammonia to urea through a process termed the urea cycle.
- Ammonia is bound to CO_2 to form carbamoyl phosphate; the equivalent of two molecules of ATP is consumed in this reaction.
- The intermediates of the urea cycle include ornithine and citrulline, two amino acids not found in proteins. The urea cycle also contains amino acids commonly found in proteins, such as arginine and aspartate, and citric acid cycle intermediates such as fumarate.
- The urea cycle synthesizes urea, helps to metabolize amino acids, and provides citric acid cycle intermediates.
- Animals other than humans have evolved different mechanisms to help them adapt to their diet. These adaptations include elevated levels of urea cycle enzymes in obligate carnivores or the means to recycle nitrogen back through the gut in some herbivores.

Concept Check

1. Describe how nitrogenous wastes are detoxified in the organism.
2. What are the factors that regulate the urea cycle? What are the intermediates and final products of the urea cycle?
3. Explain how animals other than humans metabolize nitrogen.

11.4 Metabolism of Amino Acid Carbon Skeleton

Following removal of the amino group from an amino acid, several different fates can befall the remaining carbon skeleton. This section describes how different amino acid skeletons are broken down for energy or long-term storage as fat. Many amino acid derivatives are important signaling molecules, cofactors, or metabolic intermediates. Those pathways are not discussed here, except in terms of where these metabolites lead.

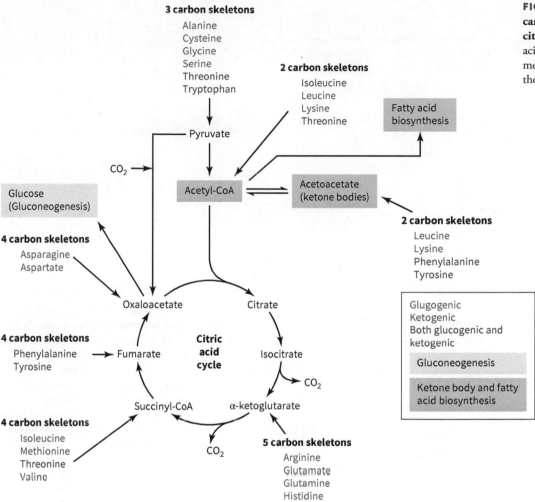

FIGURE 11.11 Amino acid carbon skeletons enter the citric acid cycle. All amino acid breakdown products yield metabolites that are consumed in the citric acid cycle.

Amino acids can be categorized as **glucogenic**, producing breakdown products which undergo gluconeogenesis, or **ketogenic**, producing breakdown products which can only produce ketone bodies (**Figure 11.11**). There are only two solely ketogenic amino acids: leucine and lysine. Some amino acid skeletons, such as threonine, have their carbons catabolized into both glucogenic and ketogenic substrates but are considered glucogenic amino acids because they contribute to glucose production. In addition, any amino acid can be broken down and stored as fat; when consumed in excess, this is the fate of many amino acids.

Although numerous reactions are involved in amino acid catabolism, most of them can be ordered into a few categories. The first reaction is generally the removal of the amino group, often through a transamination to an α-keto acid as discussed in section 11.2. The reactions of amino acid catabolism are often similar to those seen in β-oxidation, in that carbon skeletons are desaturated (an FAD-dependent process), hydrated to an alcohol, and then oxidized to a carbonyl (an NAD$^+$-dependent process). Hydroxylation of rings or ring opening often involves molecular oxygen and the dioxygenase class of enzymes. Finally, carbon groups can be added or lost as CO_2, or removed by tetrahydrofolate-dependent reactions (**Figure 11.12**).

Like fatty acids and carbohydrates, the carbon skeleton of an amino acid is catabolized via the citric acid cycle; however, in contrast to fatty acids and carbohydrates, there are multiple places where the carbon skeletons of amino acids can contribute. For example, amino acids can contribute to seven different metabolites that either lead into the citric acid cycle or are part of the cycle itself (α-ketoglutarate, succinate, fumarate, oxaloacetate, pyruvate, acetyl-CoA, and acetoacetate). The last two metabolites on this list, acetyl-CoA and acetoacetate, cannot be converted into glucose. Skeletons metabolized through these two can only lead to ketone body production, not glucose. The remaining metabolites, pyruvate, α-ketoglutarate, succinyl-CoA,

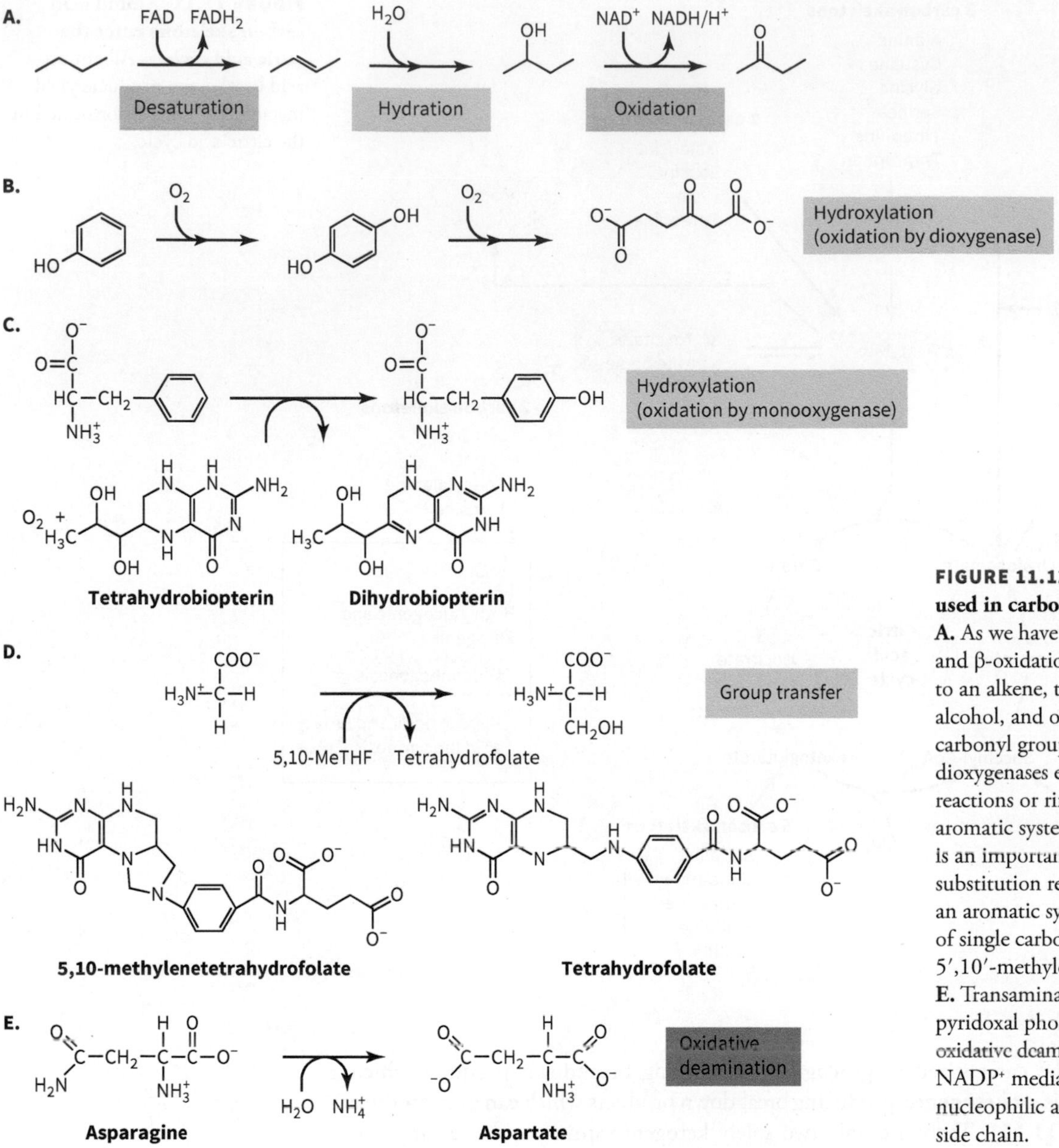

A.

Desaturation   Hydration   Oxidation

B.

Hydroxylation
(oxidation by dioxygenase)

C.

Hydroxylation
(oxidation by monooxygenase)

Tetrahydrobiopterin   **Dihydrobiopterin**

D.

Group transfer

5,10-MeTHF   Tetrahydrofolate

5,10-methylenetetrahydrofolate   **Tetrahydrofolate**

E.

Oxidative
deamination

Asparagine   **Aspartate**

H$_2$O   NH$_4^+$

FIGURE 11.12 Types of reactions used in carbon skeleton metabolism.
A. As we have seen in the citric acid cycle and β-oxidation, alkanes can be oxidized to an alkene, then hydrated to form an alcohol, and oxidized again to form a carbonyl group. **B.** Mono- and dioxygenases either perform substitution reactions or ring opening reactions on aromatic systems. **C.** Tetrahydrobiopterin is an important cofactor in substitution reactions occurring in an aromatic system. **D.** The donor of single carbons in this reaction is 5′,10′-methylenetetrahydrofolate. **E.** Transamination reactions employ pyridoxal phosphate as the cofactor but oxidative deaminations use NAD$^+$ or NADP$^+$ mediated mechanisms or nucleophilic attack by a threonine side chain.

fumarate, and oxaloacetate can all contribute to gluconeogenesis. A summary is given in **Table 11.3**. This section discusses the catabolism of amino acids in terms of their end product, starting with three-carbon amino acids.

11.4.1 Three-carbon skeletons produce pyruvate

The metabolism of three-carbon animo acid skeletons is shown in **Figure 11.13**.

- *Threonine* is cleaved into acetaldehyde and then converted into acetyl-CoA, a ketogenic product. Threonine can also undergo cleavage to form glycine, which leads to pyruvate.

- *Glycine* undergoes addition of a hydroxymethyl group to form serine. N^5, N^{10}–Methylenetetrahydrofolate (N^5, N^{10}–Methylene-THF) serves as the donor of the hydroxymethyl group.

- *Alanine* can be transaminated with α-ketoglutarate to give pyruvate and glutamate. This is a pyridoxal phosphate (PLP)-dependent reaction.

- *Serine* is dehydrated by serine dehydratase, also a PLP-dependent enzyme. The resulting product spontaneously hydrolyzes into ammonia and pyruvate.

- *Cysteine* can be acted on by various pathways to eliminate the sulfur as H$_2$S, SO$_3^{2-}$, or SCN$^-$. Oxidative deamination of the product yields pyruvate.

TABLE 11.3 Amino Acid Catabolism

Amino Acid	Catabolized Through	Glucogenic or Ketogenic	Amino Acid	Catabolized Through	Glucogenic or Ketogenic
Alanine	Pyruvate (3 carbon)	Glucogenic	Phenylalanine	Fumarate	Glucogenic
Serine			Tyrosine		
Cysteine			Aspartate		
Glycine					
Threonine					
Tryptophan					
Aspartate	Oxaloacetate (4 carbon)	Glucogenic	Leucine	Acetyl-CoA	Ketogenic
Asparagine			Isoleucine		Both
			Threonine		Both
Glutamate	α-keto glutarate (5 carbon)	Glucogenic	Leucine	Acetoacetate	Ketogenic
Glutamine			Lysine		Ketogenic
Proline			Phenylalanine		Both
Arginine			Tyrosine		Both
Histidine			Tryptophan		Both
Valine	Succinyl-CoA	Glucogenic			
Isoleucine					
Methionine					

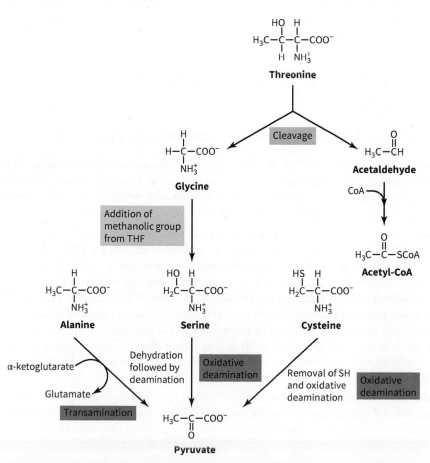

FIGURE 11.13 Metabolism of three-carbon skeletons. The three-carbon amino acids cysteine, alanine, and serine all feed into pyruvate through oxidative loss of ammonia. Threonine is cleaved into acetaldehyde or acetyl-CoA and glycine. Glycine is carboxylated into serine.

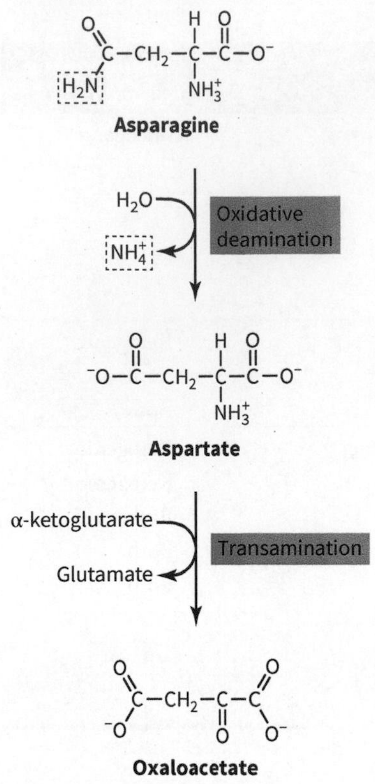

FIGURE 11.14 Metabolism of four-carbon skeletons. Asparagine is deaminated into aspartate. The amino group that is removed is highlighted with a dashed box. Aspartate is transaminated into oxaloacetate.

11.4.2 Four-carbon skeletons produce oxaloacetate

The metabolism of four-carbon animo acid skeletons is shown in **Figure 11.14**.

- *Aspartate* undergoes a transamination reaction with α-ketoglutarate to give oxaloacetate and glutamate.

- *Asparagine* can be hydrolyzed by asparaginase to ammonia and aspartate. As discussed in section 11.3, asparagine also participates in the urea cycle, in which case the carbons end up as fumarate, another citric acid cycle intermediate.

11.4.3 Five-carbon skeletons produce α-ketoglutarate

The metabolism of five-carbon animo acid skeletons is shown in **Figure 11.15**.

- *Histidine* is deaminated to urocanate. Urocanate undergoes two successive reactions involving water. First, a hydratase generates 4-imidazolone-5-propionate, and then a second molecule of water is added to form *N*-Formaminoglutamate. The terminal formamino group is transferred to **tetrahydrofolate** to yield glutamate.

- *Arginine* is hydrolyzed to urea and ornithine. Ornithine undergoes a transamination with α-ketoglutarate to yield glutamate and glutamate-γ-semialdehyde, which is oxidized to glutamate.

- *Proline* is oxidized to form pyrroline-5-carboxylate. This spontaneously decomposes in a ring-opening reaction to form glutamate-5-semialdehyde; from there it is metabolized to glutamate.

- *Glutamine* undergoes oxidative deamination to give ammonia and glutamate.

- *Glutamate* can undergo a number of different transamination reactions to form α-ketoglutarate. The other amino acids in this category are also catabolized into glutamate and thus enter the citric acid cycle.

11.4.4 Methionine, valine, and isoleucine produce succinyl-CoA

The metabolism of methionine, valine, and isoleucine is shown in **Figure 11.16**:

- *Methionine* is demethylated via **S-Adenosylmethionine** to form homocysteine. Homocysteine undergoes a condensation reaction with serine to form cystathione. Cystathione is cleaved into cysteine and deaminated to α-ketobutyrate, which is then decarboxylated to form propionyl-CoA, which is metabolized as above.

- *Valine* loses its amine moiety via transamination and is decarboxylated to generate a short branched-chain acyl-CoA. It is desaturated in an FAD-dependent reaction to generate an α–β double bond, and then hydrated to yield β-hydroxybutyryl-CoA. Coenzyme-A is lost, and the hydroxyl group is oxidized to a ketone (methylmalonate semialdehyde). These reactions are similar to the steps of β-oxidation. Loss of CO_2 and oxidation yields propionyl-CoA, which is converted to succinyl-CoA by successive carboxylation and isomerization. These final few steps of the pathway are the same ones used in the catabolism of odd-chain fatty acids in β-oxidation. Beta oxidation cleaves the odd-chain fatty acid to a three-carbon propionyl-CoA, which is carboxylated to methylmalonyl-CoA and then isomerized to succinyl-CoA.

- *Isoleucine* is metabolized in an analogous fashion to valine. It is deaminated, desaturated, hydrated, and oxidized, but in this case the product of the reactions is α-methylacetoacetyl-CoA. Loss of acetyl-CoA yields propionyl-CoA, which is converted to succinyl-CoA.

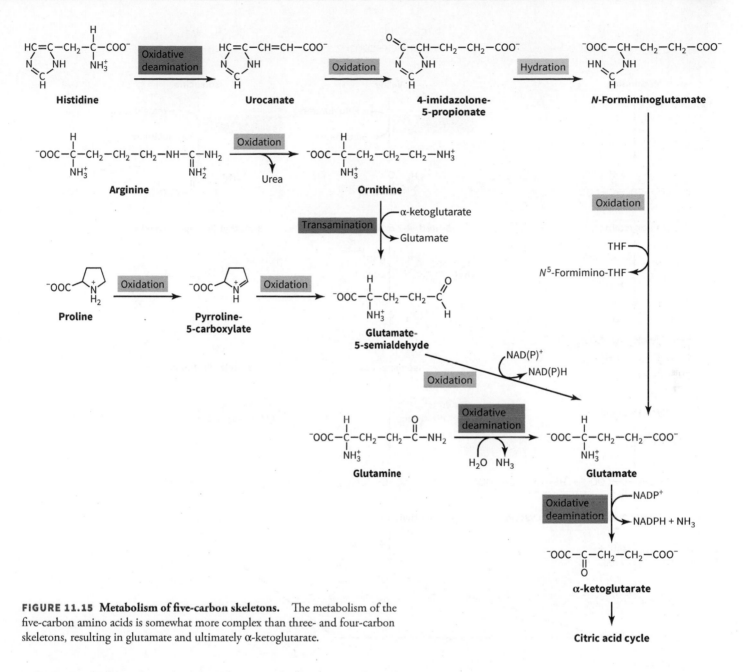

FIGURE 11.15 Metabolism of five-carbon skeletons. The metabolism of the five-carbon amino acids is somewhat more complex than three- and four-carbon skeletons, resulting in glutamate and ultimately α-ketoglutarate.

11.4.5 Other amino acids produce acetyl-CoA, acetoacetate, or fumarate

The metabolism of amino acids that yield acetyl-CoA, acetoacetate, or fumarate is shown in **Figure 11.17**.

- *Leucine* is metabolized like the other branched-chain amino acids, isoleucine and valine. It is deaminated and desaturated but then carboxylated to form β-methylglutaconyl-CoA. This molecule is then hydrated to form β-hydroxy-β-methylglutaryl-CoA or HMG-CoA, discussed in section 9.5. HMG-CoA is acted on by HMG-CoA lyase to generate acetyl-CoA and acetoacetate.

- *Phenylalanine* is hydroxylated to tyrosine in a reaction that is **tetrahydrobiopterin** dependent.

- *Tyrosine* is transaminated to yield *p*-hydroxyphenylpyruvate. Parahydroxyphenylpyruvate dioxygenase catalyzes a reaction resulting in the loss of CO_2, hydroxylation of the phenyl ring, and rearrangement of the product to give homogentisate. A second dioxygenase causes ring opening and formation of 4-maleylacetoacetate. An isomerization yields 4-fumarylacetoacetate, which is hydrolyzed to fumarate and acetoacetate.

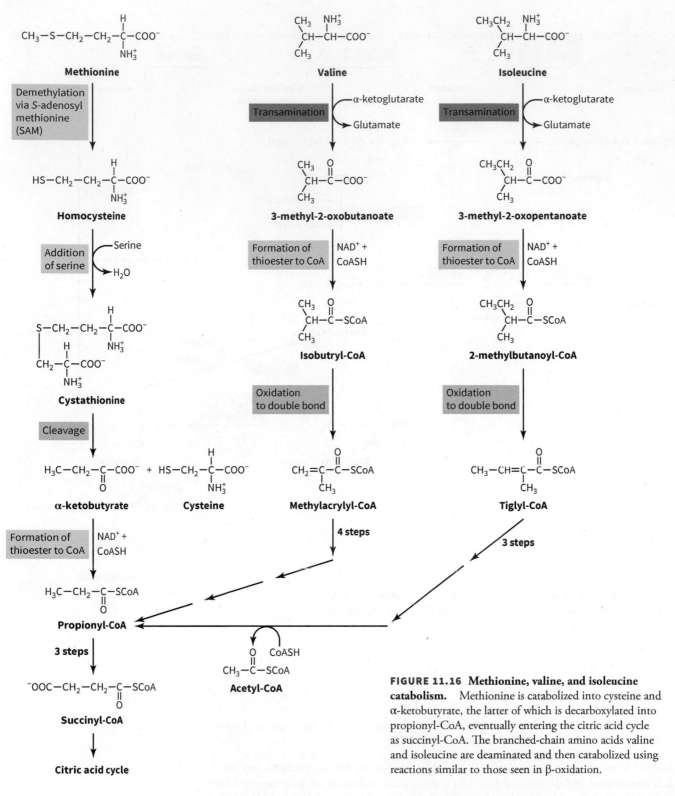

FIGURE 11.16 Methionine, valine, and isoleucine catabolism. Methionine is catabolized into cysteine and α-ketobutyrate, the latter of which is decarboxylated into propionyl-CoA, eventually entering the citric acid cycle as succinyl-CoA. The branched-chain amino acids valine and isoleucine are deaminated and then catabolized using reactions similar to those seen in β-oxidation.

- *Lysine* undergoes a condensation reaction with α-ketoglutarate to form saccharopine. Glutamate is cleaved from saccharopine in an oxidation reaction to form α-aminoadipate-6-semialdehyde. A second oxidation gives α-aminoadipate, and a transamination results in α-ketoadipate. Carbon dioxide (CO_2) is lost to yield glutaryl-CoA. At this point in the pathway, similarities are seen with both branched-chain amino acid catabolism and β-oxidation. A double bond is incorporated to yield glutaconyl-CoA, another molecule of CO_2 is lost to yield crotonyl-CoA, and the double bond is hydrated to yield β-hydroxybutyryl-CoA. One final oxidation gives acetoacetyl-CoA, which can be converted to acetoacetate as discussed previously.

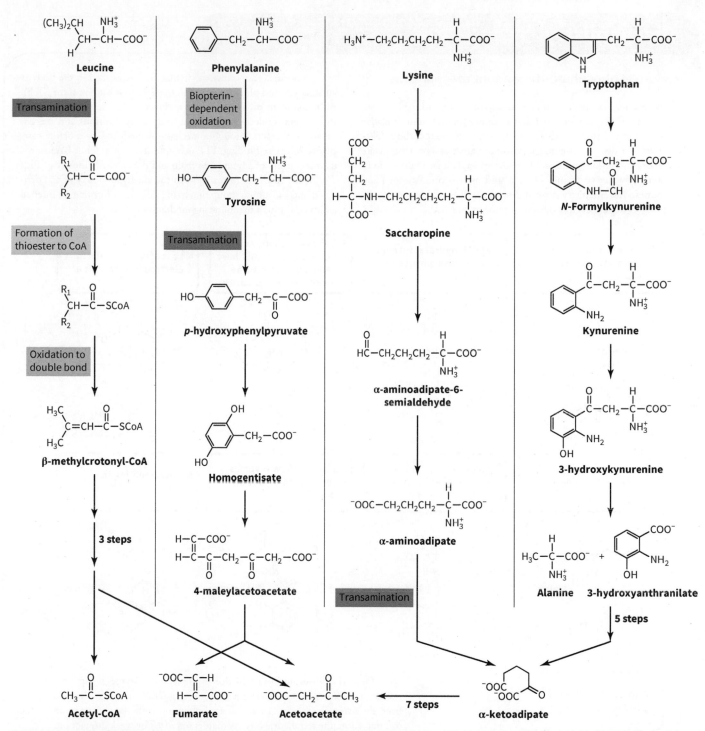

FIGURE 11.17 Catabolism of leucine, tyrosine, phenylalanine, lysine, and tryptophan.   These amino acids are catabolized through multistep pathways resulting in the production of acetyl-CoA, fumarate, and acetoacetate.

- *Tryptophan* requires the most steps to catabolize. The first step is the opening of the five-membered nitrogen-containing indole ring, catalyzed by a dioxygenase. This is decarboxylated to yield kynurenine, which is hydroxylated on the C-3 of the aromatic ring that is ortho to the amino group. The next reaction cleaves this molecule in two, leaving alanine and 3-hydroxyanthranilate. From here, the molecule undergoes a series of oxidations and rearrangements culminating in the loss of ammonia and generation of α-ketoadipate. From this point, the reactions proceed as they would for lysine.

A common motif in amino acid metabolism is the use of **pyridoxal phosphate**-dependent mechanisms. The chemistry of these mechanisms is discussed in **Biochemistry: Pyridoxal phosphate-dependent mechanisms.**

Biochemistry

Pyridoxal phosphate-dependent mechanisms

Many reactions in metabolism occur adjacent to a carboxyl carbon. For example, this chapter highlights transamination and oxidative deamination reactions, both of which occur on the α carbon. Likewise, we can deprotonate and reprotonate an α carbon to generate a stereoisomer of an amino acid, or decarboxylate an amino acid to generate a bioactive amine, often found as neurotransmitters and hormones. The chemistry surrounding these diverse reactions is governed by a single cofactor: pyridoxal phosphate (PLP).

Pyridoxal, also known as vitamin B_6, is absorbed in the body as an aldehyde and phosphorylated in the liver to the active form (PLP). Although many different enzymes and mechanisms are associated with PLP, several underlying principles unify these reactions and mechanisms. PLP reacts through a mechanism mediated by a Schiff base. Before substrate binding, PLP is bound through a Schiff base to the ε-amino group of a lysine side chain and the aldehyde moiety of PLP. The Schiff base here forms between the aldehyde moiety of the pyridoxal and the amino group of a lysine side chain forming an internal aldimine through the following mechanism:

---

1. Amino group of lysine in active site attacks carbonyl of PLP.

2. PLP dehydrates forming Schiff base, an internal aldamine.

3. Substrate binds. Amino group attacks Schiff base generating an external aldamine with substrate.

4. Loss of proton leads to electron flow to amine.

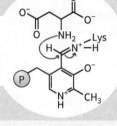

H_2O

5. Eliminating lysine.

6. Electrons flow into ring system of PLP, resulting in deprotonation.

7. Electron flow reverses through quinoid form of PLP.

8. Reprotonating the Schiff base carbon.

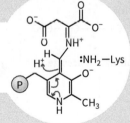 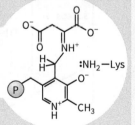

9. The α-keto product is produced.

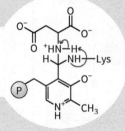

When the substrate binds (in this example, the enzyme is aspartate transaminase and the substrate is aspartate), the Schiff base is exchanged from the internal aldimine to an external aldimine with the substrate. In this transition, the tetrahedral intermediate is stabilized by weak interactions with the oxyanion on the PLP ring or amino acid side chains in the enzyme (interactions with arginine side chains are shown).

Electrons flow from the substrate down through the Schiff base and into the ring nitrogen, which acts as an electron sink, drawing the electrons through the complex, forming a quinonoid intermediate.

Electrons flow back up through the ring and abstract a proton from an adjacent amino acid, forming a ketamine intermediate. Finally, water enters, displacing the Schiff base and forming the α-keto acid and pyridoxamine phosphate. Next, a new α-keto acid (typically α-keto glutarate) binds, and the reverse reaction proceeds.

Not all PLP-mediated mechanisms employ the Schiff base aspect of the molecule. Glycogen phosphorylase uses the phosphate group of PLP in catalyzing the breakdown of glycogen.

11.4.6 Biosynthesis of Amino Acids

Humans can produce only 11 of the 20 amino acids—alanine, arginine, asparagine, aspartic acid, cysteine, glutamic acid, glutamine, glycine, proline, serine, and tyrosine—those which are synthesized by simple pathways. These are collectively called the **nonessential amino acids** and are not required to be part of the diet. The remaining nine amino acids—histidine, isoleucine, leucine, lysine, methionine, phenylalanine, threonine, tryptophan, and valine—that must be obtained from the diet are called the **essential amino acids**.

Amino acids are synthesized from their respective metabolic precursors that are intermediates of the metabolic pathways such as glycolysis, the citric acid cycle (Figure 11.11), or the pentose phosphate pathway.

- α-Ketoglutarate produces glutamate, glutamine, proline, and arginine.
- 3-Phosphoglycerate produces serine, glycine, and cysteine.
- Oxaloacetate produces aspartate, asparagine, methionine, threonine, and lysine.
- Pyruvate produces alanine, valine, leucine, and isoleucine.
- Phosphoenolpyruvate and erythrose 4-phosphate produce tryptophan, phenylalanine and tyrosine.
- Ribose-5-phispshate produces histidine.

Nitrogen gets into these pathways in the form of glutamate and glutamine. Apart from the six precursors mentioned above, there are some other significant intermediates obtained from the pathways of amino acid and nucleotide synthesis.

Worked Problem 11.4 · Complete proteins

Shown below is the typical amino acid breakdown of a soy protein sample. Based on the amino acid metabolism discussed so far and presuming one could eat enough, could a person survive on soy protein alone?

Amino Acid	% Mass of Dry Weight Protein	Amino Acid	% Mass of Dry Weight Protein
Arginine	1.12	Lysine	1.04
Alanine	0.71	Methionine	0.23
Aspartate	1.91	Phenylalanine	0.81
Cysteine	0.14	Proline	0.81
Histidine	0.45	Serine	0.84
Glutamate	2.96	Threonine	0.64
Glycine	0.70	Tyrosine	0.41
Isoleucine	0.72	Valine	0.63
Leucine	1.28		

Strategy Plant proteins are often considered an incomplete source of protein because they lack specific amino acids sufficient in quantity to prevent malnutrition. Therefore, combinations of proteins are often consumed to preserve health. To simplify the problem, we can ignore for a moment the fact that humans are unable to synthesize ten of the amino acids, and we will focus solely on the carbon skeletons.

Solution Could a diet of soy protein provide enough glucose through gluconeogenesis to keep someone alive? Examining the ratios of glucogenic to ketogenic amino acids, it is clear that all of the glucogenic amino acids are present. Thus, if soy were the sole source of food, and carbohydrate and lipid metabolism were the only concern, a person could apparently survive for some time before other dietary deficiencies would appear. Soy protein differs somewhat from other plant proteins in being a complete protein, containing all of the essential amino acids in sufficient quantity.

Follow-up question Would supplementing the above diet with lysine from corn be helpful to the situation?

Summary

- Following deamination, the side chains of amino acids are catabolized into either glucogenic or ketogenic substrates.
- Glucogenic amino acids can be converted into glucose via gluconeogenesis.
- Ketogenic amino acids can only be converted into lipids (fatty acids or ketone bodies).
- Amino acids with three, four, or five carbons enter into metabolism as pyruvate, oxaloacetate, and α-ketoglutarate, respectively.
- Other amino acids yield fumarate, succinyl-CoA, acetoacetate, or acetyl-CoA.
- The biosynthesis of nonessential amino acids involves relatively simple pathways, whereas those forming the essential amino acids are generally more complex.

11.5 The Detoxification of Other Amines and Xenobiotics

Not all molecules we consume fall into the simple categories of carbohydrate, fat, protein, or nucleic acids. All foods contain compounds that provide color, flavor, and aroma. These molecules are broadly categorized as **xenobiotics**. These compounds may have evolved to attract insects or to repel predators, but they are not molecules found in human metabolism. Hence, they need to be eliminated from the body.

In addition, other organisms produce toxins that we may consume, either intentionally or unintentionally. For example, while molds and fungi produce molecules that are toxic in high concentrations, many people enjoy the strong flavor of blue cheese or the savory flavor of some mushrooms, flavors caused by these same molds and fungi. Small amounts of these molecules in the diet may not be harmful, but these also need to be eliminated from the body.

In the course of our metabolism, some alkenes or alcohols become oxidatively degraded to form aldehydes or carboxylic acids. These molecules may be toxic if allowed to accumulate, so they too require some form of detoxification and removal.

Finally, modern medicine has provided us with a wide array of compounds that can relieve problems as minor as headaches or as significant as cancer. The metabolism and elimination of these drugs is critical to their function.

The metabolism of foreign molecules consumed in the diet, toxins, and drugs is collectively known as **xenobiotic metabolism**, the process by which the body detoxifies and eliminates these foreign molecules. The overall scheme of xenobiotic metabolism is to create more water-soluble molecules. This decreases the likelihood of xenobiotics entering a cell through the plasma membrane and increases the likelihood of their being eliminated in the urine. Xenobiotics may also be cleaved into smaller fragments and then modified with small molecules that are recognized by receptors, which remove these molecules from the cell. Many cancers that are multidrug resistant (MDR) owe that resistance to the ability of the patient to detoxify or export anticancer drugs from their site of action.

Xenobiotic metabolism can be divided into two phases: phase I, oxidation, and phase II, conjugation to larger molecules such as glutathione or glucuronic acid (**Figure 11.18**).

11.5.1 Phase I metabolism makes molecules more hydrophilic through oxidative modification

In phase I, xenobiotics undergo reactions that make them more hydrophilic, either through oxidation or cleavage reactions, such as an ester cleavage that unmasks more polar moieties (**Figure 11.19**). These reactions are carried out by several families of enzymes, including dehydrogenases, hydrolases, flavin monooxidases (FMO), cytochrome P450 enzymes, and **monoamine oxidases (MAOs)**. The latter two are discussed below.

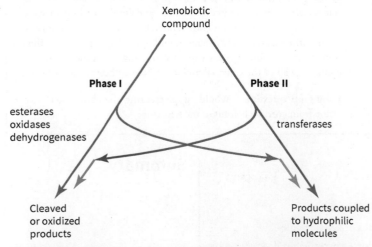

FIGURE 11.18 Phase I and phase II metabolism. Xenobiotic metabolism can be classified as either phase I or phase II. In phase I, oxidative reactions make molecules smaller and more water soluble. In phase II, molecules are coupled to soluble groups to increase their likelihood of being excreted in the urine. Molecules also may be metabolized through both processes.

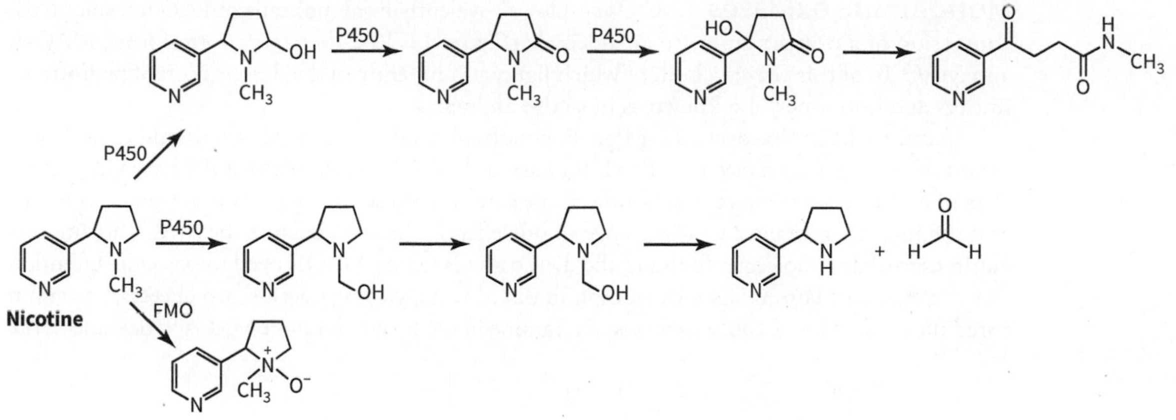

FIGURE 11.19 Examples of phase I metabolism.   Nicotine is acted on by several enzymes to give a range of metabolites. Note that the final products may appear quite different from the starting metabolite.

Cytochrome P450 enzymes

Cytochrome P450 mixed function oxidases are a broad family of enzymes that employ NADPH and molecular oxygen (O_2) to oxidatively modify substrate molecules. These oxidases are membrane-associated enzymes commonly found in either the inner mitochondrial membrane or the endoplasmic reticulum (ER), and they are most strongly expressed in the liver. There are many cytochrome P450 enzymes: humans have nearly 60 genes, and over 11,000 genes are known in other species. Each enzyme has a broad range of substrate specificities; however, individual drug molecules may be acted on by specific cytochromes. The enzymes implicated in the metabolism of a drug provide important information used in developing the dosages of that compound.

The functions of these enzymes can be broadly divided into those involved in detoxifying xenobiotics and those involved in lipid metabolism.

There is a standard nomenclature system for cytochrome P450 genes. All are abbreviated as CYP, followed by a numeral indicating the gene family, a capital letter indicating the subfamily, and a second number indicating the gene, for example, CYP3A4. Some genes have retained other names, such as thromboxane A2 synthase or aromatase, in addition to the standard nomenclature.

Previously we discussed cytochrome proteins in terms of the electron transport chain. The cytochromes of the electron transport chain are small, low molecular weight cytochromes. Both low molecular weight cytochromes and cytochrome P450 enzymes are heme proteins. While both catalyze redox chemistry, cytochrome *c* and cytochrome P450 enzymes serve very different functions in biology.

Cytochrome P450 enzymes are globular, predominantly α helical proteins found in the ER or mitochondria. They all have a heme prosthetic group bound by coordination of the central heme iron to a cysteine on the distal side of the ring. The active site of the enzyme (CYP3A4 in this example) is remarkably small, and the structure undergoes relatively little conformational change upon substrate binding. It is likely that this contributes to the broad substrate specificity of this enzyme. Key hydrophobic residues line the active site and play a role in controlling the steric interactions of the enzyme with its many substrates. CYP3A4 is one of the most highly expressed cytochromes and is involved in the detoxification of more than 200 common pharmaceuticals.

NADPH supplies the electrons used to reduce CYP, although numerous mechanisms and proteins are involved in the transfer of these electrons. Typically, these involve flavoproteins or small cytochromes such as cyt b5. The mechanisms involved for other cytochromes vary depending on the substrates and electron donor pairs, although all employ a heme group to bind the reactive oxygen species.

Phase I metabolism often results in products that are still metabolically active. Several drugs are supplied in a biologically inactive form that only become active after undergoing phase I metabolism. For example, the HMG-CoA reductase inhibitor Mevacor (lovastatin), used to treat overactive high plasma cholesterol, must be cleaved by esterases to generate the active form of the molecule.

Monoamine oxidases   MAOs employ a covalently fixed molecule of FAD to catalyze the conversion of a primary amine to an aldehyde (**Figure 11.20**). There are two isoforms, MAO-A and MAO-B, of this enzyme, both of which figure prominently in the deactivation of neurotransmitters and hormones, also known as bioactive amines.

Because MAOs deactivate these signaling molecules, inhibition of MAOs should result in elevation of the signaling molecules. This is the case, and MAO inhibitors (MAOIs) are an important class of antidepressant molecules. However, the extensive substrate range for these enzymes means that the inhibition results in a wide range of side effects. Likewise, because the MAOs are important in detoxifying molecules found in the diet, patients taking MAOIs need to pay close attention to what they eat. For example, consumption of red wine, yeast extracts, sharp cheeses, or certain cured meats can elevate concentrations of tyramine in the blood. Without MAOs to degrade it, the

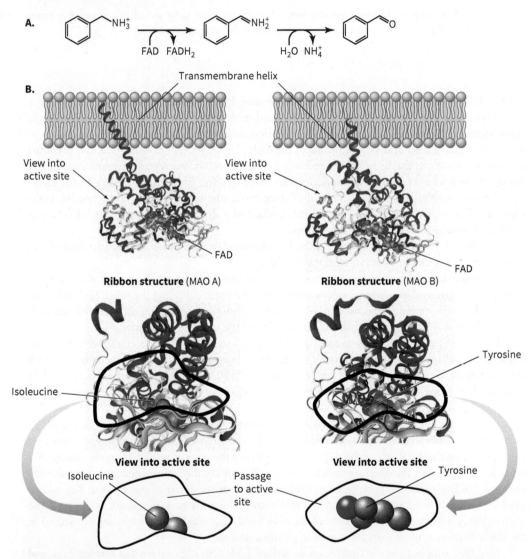

FIGURE 11.20 Monoamine oxidases are FAD-dependent monoxidases.   **A.** Monoamine oxidases are responsible for inactivating many neurotransmitters and other bioactive amines. **B.** MAO-A and B employ a similar mechanism but differ in the access to the active site. MAO-A contains an isoleucine guarding the passage while MAO-B has the active site blocked by a much bulkier tyrosine residue. This regulates substrate specificity.

(Source: (Left) Data from PDB ID 2Z5X Son, S.Y., Ma, J., Kondou, Y., Yoshimura, M., Yamashita, E., Tsukihara, T. (2008) Structure of human monoamine oxidase A at 2.2-A resolution: The control of opening the entry for substrates/inhibitors. *Proc. Natl. Acad. Sci. Usa* 105: 5739–5744. (Right) Data from PDB ID 4CRT Esteban, G., Allan, J., Samadi, A., Mattevi, A., Unzeta, M., Marco-Contelles, J., Binda, C., Ramsay, R.R. (2014) Kinetic and Structural Analysis of the Irreversible Inhibition of Human Monoamine Oxidases by Ass234, a Multi-Target Compound Designed for Use in Alzheimer's Disease. *Biochim. Biophys. Acta* 1844:1104)

usually harmless tyramine can build up, resulting in greatly increased blood pressure and creating a potentially fatal condition known as hypertensive crisis.

Tyramine

MAOIs also illustrate a mechanism of drug–drug interactions. Over-the-counter cold medications often include molecules that elicit similar effects to MAOIs. One example is the combination of the MAOI phenelzine (Nardil) and the expectorant dextromethorphan. Both of these medications elevate levels of serotonin, so if taken together they can result in dangerously high levels of serotonin, referred to as serotonin syndrome. Although cases of this interaction are rare, serotonin syndrome can also be fatal. Patients exhibit restlessness, hallucination, increased body temperature, tachycardia (increased heart rate), nausea, and rapid changes in blood pressure. Similar effects can be seen in subjects who abuse cough syrups that contain dextromethorphan. As a result of these and other similar interactions, patients taking MAOIs are told not to take any over-the-counter cold medicines.

The isoforms MAO-A and MAO-B differ in their overall topology. Both have a covalently linked FAD cofactor and are anchored in the outer mitochondrial membrane by a C-terminal hydrophobic α helix; however, the shape of the substrate-binding pocket differs. MAO-A has a single large hydrophobic cavity for substrate binding, whereas MAO-B has two smaller funnel-shaped cavities separated by an isoleucine residue, a molecular gatekeeper. Rotation of this isoleucine side chain moves it out of the way, enabling the two chambers to accommodate substrates of varying size. Another difference is that MAO-B contains a tyrosine residue at position 326, whereas MAO-A contains an isoleucine at an analogous position in the active site (335). This simple difference changes the inhibitor specificity for the isoforms.

Most of the chemistry that occurs in MAOs happens in the flavin ring, but three amino acid side chains also figure prominently in the reaction. A conserved lysine residue binds to both a water molecule and the N-5 position of the flavin ring, and two conserved tyrosine residues sandwich substrates, potentially helping with substrate binding and increasing the nucleophilicity of the substrate amine. Mutation of any of these residues alters enzymatic activity.

11.5.2 Phase II metabolism couples molecules to bulky hydrophilic groups

In phase II metabolism, molecules are conjugated to other chemical groups, the most common of which are glucuronic acid, glutathione, glycine, methyl groups, and sulfate. Glycine, methyl groups, and sulfate should need little explanation, but glucuronic acid and glutathione may be less familiar.

Glucuronic acid   **Glutathione**

Glucuronic acid is an oxidized form of glucose in which the C-6 has been enzymatically oxidized from an alcohol to a carboxylic acid. Conjugation to an oxidized xenobiotic occurs between the C-1 of the glucuronic acid and a hydroxyl or amino group in the xenobiotic, with UDP-glucuronic acid acting as the active donor.

UDP-glucuronosyl transferases (UGTs)

Biphenylamine   **Glucuronic acid derivative**

Glutathione (GSH) is a tripeptide of glycine, cysteine, and glutamate; however, the glutamic acid is attached via an unusual peptide bond formed between the side chain of the glutamate (the γ-carboxyl group) and the free amine of the cysteine. Glutathione helps to provide a reducing environment in the cell and is found in its oxidized state as a glutathione dimer. In some cell types, such as liver cells, the concentration of glutathione can be quite high, upwards of 5 mM in a 90:10 ratio of reduced monomer:oxidized dimer. When used in phase II metabolism, the thiol moiety is the active group, undergoing conjugation to electrophilic centers in the target xenobiotic. Attaching this bulky, highly soluble, anionic group to xenobiotics helps to make the xenobiotics less able to permeate into cells and more water soluble; hence, they are more easily passed through the kidney and into the urine.

A group of transferases are responsible for covalently modifying molecules in phase II. The best characterized of these are the **glutathione-*S*-transferases (GSTs)**.

Glutathione-*S*-transferases

GSTs are the enzymes that conjugate glutathione to an electrophilic center of a substrate molecule. They are expressed differentially in all tissues of the body, with expression levels of the different isoforms varying in response to factors such as tissue, environmental challenge, disease states (for example, some cancers), sex, and age.

This section focuses on the role that GSTs play in phase II metabolism, but other enzymes serve similar purposes. For example, *N*-Acetyltransferases, methyltransferases, UDP-glucuronosyl transferases and sulfotransferases are all used by the liver to derivatize xenobiotics. The result of these modifications is to render the xenobiotic inactive through the addition of more polar or charged groups. This prevents the molecule from crossing membranes and makes it more likely to be eliminated in the urine. Although this part of xenobiotic metabolism is termed phase II, some molecules undergo these reactions without first being modified in phase I.

Some examples of how phase I and II metabolism are used in the analysis of many drugs are discussed in **Societal and Ethical Biochemistry: Drug testing.**

Societal and Ethical Biochemistry

Drug testing

Throughout history, people have used or abused molecules for a variety of reasons. Examples include molecules that allow one to wake up or to be more alert (caffeine or amphetamines), to relax (alcohol or tetrahydrocannabinol, THC), to relieve pain (ibuprofen or morphine), to have spiritual experiences (nicotine or hallucinogens), to escape reality, or to fit in socially (all of the above). Some of these molecules (caffeine, nicotine, ibuprofen) are deemed relatively safe, and their use is legal. Others (alcohol, THC, narcotics, opioids, hallucinogens) are either highly regulated or illegal.

Governments have a strong mandate to create and enforce laws prohibiting the use of many recreational drugs, and to keep people who use such drugs from working in positions where they may cause harm if they are in an impaired or drug-dependent state. Hence, scientists have developed tests for illegal compounds.

Drug testing offers an interesting window into both drug metabolism and pharmacology. Often the substance assayed is urine, although in some instances blood or even breath (in the case of ethanol use) can be assayed. Each of these tests has different strengths and weaknesses. Breath can be assayed easily and gives an indication of the amount of alcohol in blood. Urine testing can provide an average of the metabolites excreted over the past several hours, but the substances assayed may be metabolic breakdown products and also may be dilute. Blood gives perhaps the most "accurate" picture, but the amounts that can be assayed are limited.

As small organic molecules, drugs can generally be identified using the methods found in analytical laboratories. Often, the substances being tested for in urine or blood need to be concentrated using solid-phase extraction. Molecules are extracted from the sample by passing them over a hydrophobic interaction column—a column coated with hydrophobic molecules like phenyl groups or chains of carbons of varying length. The drugs and their derivatives are more nonpolar than water and adhere to the column matrix. The column is eluted with a nonpolar solvent, such as hexane, and analyzed using either high performance liquid chromatography (HPLC) or mass spectrometry. These techniques can be highly sensitive, detecting nanogram-to-picogram quantities of material. The molecules found can be either positively identified by mass or by comparison to known standards.

It is helpful to look at the metabolism of a single drug as an example of what can happen in metabolism and excretion. Caffeine (1,3,7-trimethylxanthine) is a common molecule that many people use to help start their day. It is found naturally in many plants, including coffee, tea, and cocoa. A typical dose of caffeine in an 8-ounce cup of drip coffee is about 100 to 175 mg, but some energy drinks contain more than twice that concentration. In the body, the liver works to detoxify caffeine and make it more water soluble so that it can be eliminated in the urine by the kidney.

The first step in caffeine metabolism is the removal of one of the methyl groups, resulting in paraxanthine (1,7-dimethylxanthine, the major metabolite), theobromine (3,7-dimethylxanthine), or theophylline (1,3-dimethylxanthine), all of which are examples of phase I metabolism.

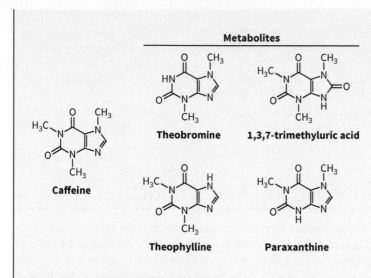

Metabolites

Caffeine

Theobromine

1,3,7-trimethyluric acid

Theophylline

Paraxanthine

Each of these molecules can be detected in urine, but each can also be further metabolized. Theobromine, for example, can be further demethylated into methyl xanthines and methyl uric acid. Hence, there can be multiple breakdown products for a single drug, any one of which can be telling when it comes to drug testing. Although testing for caffeine and its metabolites is relatively innocuous, testing for metabolites or derivatives of other substances (for example, morphine) is important in medicine, society, and law.

Worked Problem 11.5 — Metabolism of an anticancer drug

The anticancer drug tamoxifen is often used to treat some types of breast cancer. Its structure is shown along with the structure of one of its metabolites. Based on these structures, propose a pathway through which tamoxifen is metabolized by the liver.

Strategy Examine the structure of tamoxifen and the structure of the metabolite. What chemical changes have been made to these structures? Of the reactions discussed in this section, which reactions could lead to these changes and which could not?

Tamoxifen

4-hydroxytamoxifen

Solution Tamoxifen is hydroxylated by a cytochrome P450 enzyme. We can eliminate phase II metabolism as a potential pathway because the molecule is not coupled to a larger hydrophilic molecule, such as glucuronic acid or glutathione. Likewise, we can rule out some of the reactions of phase I metabolism because there are no esters to be cleaved or monoamines to be oxidized. This leaves the cytochromes as likely candidates for metabolizing these molecules. In the case of tamoxifen, the specific isoform implicated is CYP2D6, although others are involved in producing other metabolites. In the course of designing new drugs, structure–activity relationships provide some guidance as to how a drug may be metabolized, but the specific route used to detoxify a drug must be experimentally observed.

Follow-up question One of the metabolites is *N*-Desmethyl tamoxifen. What reactions and enzymes might be responsible for this metabolite of tamoxifen?

Summary

- There are many foreign molecules, or xenobiotics, in the body that need to be eliminated.
- Xenobiotics are typically cleaved into smaller pieces, made more hydrophilic, or coupled to larger hydrophilic molecules to decrease their ability to cross the plasma membrane and increase their ability to be excreted in urine.
- Phase I metabolism involves cleaving the xenobiotic or oxidatively damaging it to make it more water soluble.
- The cytochrome p 450 enzymes employ a heme functional group to generate reactive oxygen species that react with substrate molecules.
- MAOs are flavoproteins that use a covalently bound flavin group to oxidize amines. These enzymes act in neurotransmitter and hormone metabolism and in detoxifying reactions.
- Phase II metabolism involves coupling molecules to larger, water-soluble molecules such as glucuronic acid and the tripeptide glutathione.
- GSTs are enzymes that couple glutathione to substrate molecules to be excreted.

Concept Check

1. What are xenobiotics? Discuss their origin and modes of entry into the human body.
2. Describe how xenobiotics are metabolized and eliminated, including the two main pathways that are used.
3. Describe the chemistry of the three major enzymatic systems involved in the detoxification of xenobiotics discussed in this section (CYP, MAO, and GST).
4. Explain why understanding how a drug is metabolized is crucial to understanding how it functions.

11.6 The Biochemistry of Renal Function

Nitrogenous wastes, drugs, and xenobiotic metabolites need to be secreted from organisms. In fish and higher organisms, kidneys filter this waste and remove it from the circulation. In addition to filtering out wastes, the kidneys also need to be able to retain metabolites such as amino acids and glucose that the organism needs. This section discusses renal or kidney function, the filtration of wastes and retention of metabolites, as well as three other related renal functions: acid–base balance, CO_2 synthesis, and regulation of plasma osmolarity. These functions collectively allow the kidney to eliminate wastes as well as regulate fluid retention, pH, and concentrations of electrolytes in the body. The role of hormones on the kidney and the associated adrenal gland is also discussed.

There are many different xenobiotics, and an organism may be exposed to different molecules depending on its diet. The exposure can change through seasonal fluctuations, location, or progression through the life cycle. Likewise, as xenobiotics are detoxified, multiple products can be formed, meaning that many different molecules need to be expelled from the organism. Using specific pumps or channels for each molecule would be inefficient and complicated, yet these molecules must be removed to avoid their rapid buildup in the circulation. Organisms have evolved such that *all* small molecules are filtered out of the blood; an elegant mechanism of molecular pumps and diffusion is then used to recover and retain needed metabolites and water.

Kidney function is critical to health. One means of ascertaining kidney function is an analysis of the chemicals found in blood. This is discussed in **Medical Biochemistry: Clinical chemistry**.

Medical Biochemistry

Clinical chemistry

Biochemistry provides a wealth of tools used by clinicians and investigators to examine the metabolism of a patient or test subject. One of the simplest and most widely used assays is a CHEM 7 or CHEM 20, also known as blood work. In this assay, a sample of blood is drawn, usually in the fasted state, and tested for about 7 to 20 different parameters. A common panel of assays is shown in **Table 11.4**. Collectively, the results give the physician or investigator clues (some of them strong clues) as to what is happening at the chemical level and possibly causing disease or a disorder. Several of the compounds were mentioned in this chapter and merit further discussion.

LDH is lactate dehydrogenase, encountered in section 6.5 (fates of pyruvate). In general, LDH is localized to specific tissues: heart, kidney, liver, muscle. If LDH is released from those tissues, it will not be excreted by the kidney and will accumulate. LDH is therefore a marker for general tissue damage.

This chapter discusses alanine transaminase (ALT) and aspartate aminotransferase (AST) and their role in amine and amino acid metabolism. Both of these enzymes are localized to the liver. Therefore, like LDH, elevated levels of these enzymes in blood are a marker of liver damage. Similarly, gamma-glutamyl transpeptidase (gamma-GT) is another liver enzyme that is a marker for liver damage. Among its multiple functions, gamma-GT catalyzes the transfer of the gamma-linked glutamate from glutathione to an acceptor molecule.

BUN is a measure of the urea level in blood. Elevated BUN levels typically indicate problems with renal function and decreased urea clearance rates in the kidney. Creatinine, an important breakdown product of creatine in muscle, is discussed in chapter 12. Like urea, creatinine is typically cleared from circulation by the kidney, and elevated levels indicate renal damage.

One final molecule that is often included on chemistry panels is creatine phosphokinase (CPK). This enzyme is specifically found in liver and muscle tissue, where it functions in creatine metabolism. Elevated levels of CPK indicate muscle damage or, more frequently, heart attack.

TABLE 11.4 Clinical Chemistry Panel

Metabolite	Normal Concentration	Metabolite	Normal Concentration
Normal results		**Normal results**	
Albumin	3.9–5.0 g/dL	Gamma-GT	0–51 IU/L
Alkaline phosphatase	44–147 IU/L	Glucose test	64–128 mg/dL
ALT	8–37 IU/L	LDH	105–333 IU/L
AST	10–34 IU/L	Phosphorus—serum	2.4–4.1 mg/dL
BUN	7–20 mg/dL	Potassium	3.7–5.2 mEq/L
Calcium—serum	8.5–10.9 mg/dL	Serum sodium	136–144 mEq/L
Serum chloride	101–111 mmol/L	Total bilirubin	0.2–1.9 mg/dL
CO_2	20–29 mmol/L	Total cholesterol	100–240 mg/dL
Creatinine	0.8–1.4 mg/dL	Total protein	6.3–7.9 g/dL
Direct bilirubin	0.0–0.3 mg/dL	Uric acid	4.1–8.8 mg/dl

11.6.1 Molecules smaller than proteins are filtered out of the blood by the glomerulus

The basic filtration unit of the kidney is the **nephron**. Each human kidney contains 1 to 1.5 million nephrons. In other animals, kidneys operate similarly, but particular functions depend on the species and the environment for which they have adapted. The kidney can broadly be divided into two parts: the cortex or outer coating and the medulla or core. The nephron bridges these two parts. Each nephron can be subdivided into two structures: one that filters the blood and another to reabsorb water and other essential molecules. The remaining liquid, the urine, flows through the ureter and is collected in the urinary bladder.

The glomerulus and Bowman's capsule The filtration aspect of kidney function is carried out in the **glomerulus**, a network of highly porous capillaries. Large objects, such as cells, are retained in the capillaries, while small molecules are filtered out. The filtration unit consists of fenestrated endothelial cells; these are cells with small pores that form the capillary and associated basement membrane. The spaces are small enough to allow molecules with a Stokes radius of less than 2 nm (an approximate molecular weight of 7 kDa) to be filtered out. Anything with a Stokes radius greater than 3.5 nm is almost completely retained (serum albumin, with a molecular weight of 64 kDa and a Stokes radius of 3.5 nm, is retained). To keep the blood liquid, about 20% of plasma is also retained, but everything else is lost to the filtrate.

In certain cases of kidney damage or persistent high blood pressure, proteins can leak from the glomerulus into the filtrate and eventually into the urine, a condition known as **proteinuria**.

The filtrate lost from the capillary network in the glomerulus enters the **Bowman's capsule**—a glove that envelops the glomerulus and collects the filtrate. At this point, the filtrate contains water, electrolytes, amino acids, and glucose in about the same concentration as found in blood, but it lacks cells and proteins. If the filtrate itself were to be excreted in humans, the organism would lose these essential molecules and would excrete nearly 200 liters of fluid a day. Water and metabolites are reabsorbed in the second phase of renal function.

11.6.2 Water, glucose, and electrolytes are reabsorbed in the proximal convoluted tubule, loop of Henle, and distal convoluted tubule

The initial stages of reabsorption occur in the **proximal convoluted tubule**. The proximal convoluted tubule reabsorbs glucose and amino acids using a sodium ion coupled symporter (**Figure 11.21**).

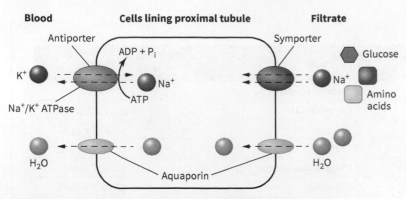

FIGURE 11.21 Ion pumps and aquaporins in the proximal tubule. Small organic molecules such as glucose, amino acids, and some metabolites are transported back into the cells lining the proximal tubule using specific Na^+ symporters. The sodium gradient used to drive this transport is created by Na^+/K^+ ATPases. Water diffuses back into the cell from the filtrate and out of the cell into the blood through aquaporins.

The gradient of sodium ions, which helps to drive the absorption of glucose and amino acids, is created by a sodium/potassium antiporter on the plasma side of the cell. The energy to drive this process is substantial; 10%–15% of the organism's resting energy demands go to maintaining these gradients and to reabsorbing nutrients and electrolytes from the filtrate. The energy comes from the hydrolysis of ATP and is thus considered active transport.

The gradient created by reabsorbing sodium ions, glucose, and amino acids also draws water back from the filtrate into the blood. Water moves through an aquaporin, its own protein channel. This system normally absorbs most of the glucose and amino acids as well as 70% of the water lost in the initial filtration step. If plasma glucose is high, as is the case in diabetes, the glucose transporters are unable to reabsorb all the glucose, and some of it appears in the urine. The remaining glucose in the filtrate also shifts the osmotic gradient across the cells of the proximal convoluted tubule, so that water is also retained in the filtrate. This results in increased urinary volume and the increased frequency of urination seen in people with diabetes.

The proximal convoluted tubule absorbs glucose and amino acids; the filtrate that remains is still relatively high in electrolytes and contains a considerable volume of water. If this were urine, the total volume in humans would be over 20 liters a day and be nearly as concentrated in electrolytes as seen in plasma. This difficulty is overcome in two ways. The first lies in the anatomy of the nephron. As the filtrate is concentrated during its passage through the proximal convoluted tubule, it descends into a structure called the **loop of Henle** (Figure 11.22). The descending arm of this loop is permeable to water, whereas the ascending arm is impermeable to water. The loop also contains sodium pumps capable of setting up a 200 mOsM gradient between the filtrate and the extracellular fluid surrounding the loop. However, because the descending arm is water permeable, water diffuses out to maintain a small gradient between the fluid in the lumen of the loop and the interstitial fluid. The net result of this is a large gradient between the interstitial fluid at the entry point of the loop and the bottom of the loop (from 300 to 1200 mOsM).

Intertwined with the loop of Henle is a network of capillaries known as the **vasa recta**. In the capillaries surrounding the descending limb of the loop of Henle, Na^+ and Cl^- ions are absorbed, and water leaves due to the high concentration at the bend of the loop. However, in the vessels surrounding the ascending loop, the reverse occurs because the plasma is now hypertonic compared to the extracellular fluid; thus, electrolytes are lost from the plasma, and water is reabsorbed until an isotonic concentration is reached.

The story is not complete, however. The urine that leaves the proximal convoluted tubule is still quite dilute. In the final stage of reabsorption, water is removed from the collecting duct to yield urine. There are no pumps for water molecules; rather, water is removed as it is in the other steps of reabsorption, by passive diffusion. This is regulated in several different ways. First, as the dilute urine passes down the collecting duct, it passes through the intracellular region by the loop of Henle. Pores in the cells of the collecting duct allow water to diffuse out of the duct in response to the high salt concentration around the loop of Henle. This porosity is acutely regulated by antidiuretic hormone. High levels of antidiuretic hormone make this duct more porous. Significant quantities of water escape the duct, resulting in the production of concentrated urine. In the absence of the hormone, the pores are closed and large quantities of dilute urine are produced. On a molecular level, antidiuretic hormone is a peptide hormone that acts via a 7-transmembrane helix spanning receptor to elevate cytosolic levels of cAMP. Elevated cAMP concentrations activate protein kinase A (PKA), which phosphorylates proteins on the surface of vesicles containing aquaporins, causing them to migrate to the plasma membrane of endothelial cells lining this duct and facilitate water transport (Figure 11.23).

The other major hormone system that regulates water reabsorption is the renin–angiotensin system. Aldosterone is a steroid hormone made in the adrenal gland. It acts in the distal convoluted

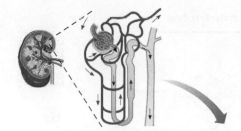

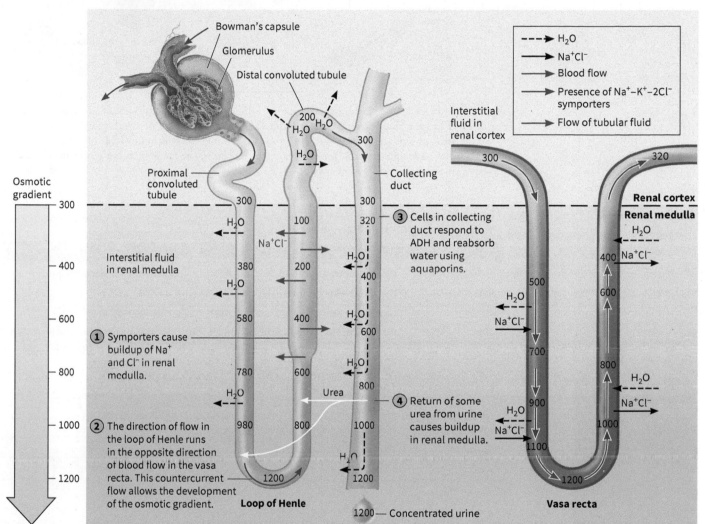

FIGURE 11.22 Ion balance and water reabsorption. The proximal tubule, loop of Henle, and distal tubule are wrapped by the vasa recta. The proximal tubule and descending loop of Henle are permeable to water, increasing the concentration of the filtrate. In the distal tubule, sodium and chloride ions are pumped out generating dilute filtrate. Values are measured in milliOsmoles (mOsm), the millimoles of solute per liter that contribute to the osmotic pressure of a solution. In the collecting duct, more ions and metabolites (Na^+, Cl^-, bicarbonate, and urea) are reabsorbed or diffused into the vasa recta, bringing water with them. This results in concentrated urine.

(Source: Tortora; Derrickson, *Principles of Anatomy & Physiology*, 13e, copyright 2012, John Wiley & Sons. This material is reproduced with permission of John Wiley & Sons, Inc.)

tubule to increase expression of sodium pumps, thereby increasing retention of sodium and water. The renin–angiotensin system signals to the adrenal gland to release aldosterone.

Not all molecules leave the kidney by diffusion. Some are pumped out via generic anion or cation pumps (**Table 11.5**). These molecules recognize small organic ions and pump them into the cells lining the proximal convoluted tubule. This process is driven by gradients of Na^+ diffusing into and dicarboxylic acids diffusing out of the cell.

Bicarbonate is reabsorbed in the proximal convoluted tubule through a multistep mechanism. Carbonic anhydrase is found in two locations: the cytosol and the membrane lining the

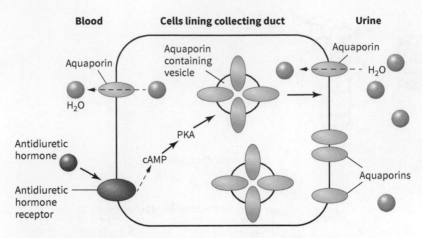

FIGURE 11.23 Antidiuretic hormone. The peptide antidiuretic hormone signals through a heterotrimeric G protein coupled receptor to activate protein kinase A. PKA phosphorylation of vesicle proteins causes fusion with the luminal side of the plasma membrane. This presents more aquaporins to the membrane and results in the reabsorption of more water.

TABLE 11.5 Transport of Metabolites in the Kidney

Metabolite	Where Reabsorbed	How Reabsorbed
H_2O	Proximal convoluted tubule	Osmotic gradient, via aquaporins
Na^+	Distal convoluted tubule/collecting duct	Na^+/K^+ ATPase
K^+	Distal convoluted tubule/collecting duct	Na^+/K^+ ATPase
Cl^-	Distal convoluted tubule/collecting duct	Cl^- channels (passive) and $Na^+/K^+/Cl^-$ cotransporters
CO_2	Collecting duct	Passive diffusion
HPO_4^{-2}	Proximal convoluted tubule	Na^+ symporter
Glucose	Proximal convoluted tubule	Na^+ symporter
Amino acids	Proximal convoluted tubule	Na^+ or H^+ symporter
Organics	Proximal convoluted tubule	Na^+ or H^+ symporter
Urea	Collecting duct	Passive diffusion

distal convoluted tubule. In the filtrate, carbonic anhydrase acts to form CO_2 and water. The CO_2 diffuses into the cells lining the tubule, where a second molecule of carbonic anhydrase reforms carbonic acid (**Figure 11.24**). The acid dissociates and the resulting HCO_3^- is pumped from the endothelial cell into the bloodstream. To maintain cellular pH and keep the process going, protons are pumped back into the filtrate using a sodium-dependent antiporter.

Surprisingly, perhaps not all the urea is eliminated in the filtrate. In addition to allowing water to pass, the aquaporins found in the collecting duct are permeable to urea. Increased urea absorption results in increased water retention and decreased urinary volume.

Kidney function is ascertained clinically using several different tests. The total amount of urea found in blood (blood urea nitrogen, BUN) and creatinine (a muscle catabolite) are typically used as indirect measures of kidney function. If these molecules are not cleared from the circulation by the kidney, they will accumulate, leading to elevated levels. Accumulation of proteins in the urine, particularly albumin, is also indicative of kidney damage. An injection of the polysaccharide inulin is sometimes administered as a final test of renal function. This molecule is neither retained in the circulation nor reabsorbed; as such, it can be used as a measure of how well filtration is occurring, such as the glomerular filtration rate.

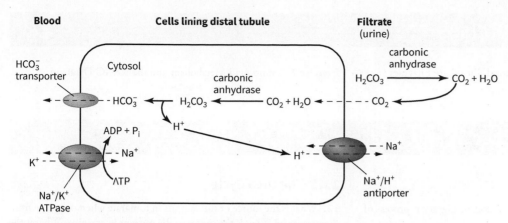

FIGURE 11.24 Acid–base balance and CO₂ balance in the kidney. Protons are pumped from cells lining the distal tubule into the filtrate. This pushes the equilibrium to the right, favoring CO_2 and H_2O. CO_2 diffuses into the cells where the reverse reaction (also catalyzed by carbonic anhydrase) occurs. This generates carbonic acid (H_2CO_3), which ionizes into bicarbonate and a proton. Bicarbonate (HCO_3^-) diffuses out of the cell into the blood side of the cell. Finally, Na^+ ions are pumped from the cell by a Na^+/K^+ ATPase. This mechanism helps the body conserve sodium ions as well as provide the gradient that helps keep bicarbonate retention possible.

Worked Problem 11.6   Active or passive

The kidney offers an interesting example of the interdependence of many biological systems. Which molecules are dependent on sodium gradients for transport?

Strategy Think about the processes that go on in the kidney. Which of these require some investment of energy or a gradient for transport to occur?

Solution Glucose, amino acids, urea, and protons are either directly or indirectly dependent on sodium ions for transport.

Follow-up question Glucose is reabsorbed in the kidney by a glucose transporter. How could you show that this process can be saturated?

Summary

- Kidneys serve several functions, including filtration of wastes from plasma, hormone secretion, acid–base balance, secretion of ammonia, and maintenance of osmolarity.
- Filtration of the blood occurs in the glomerulus and Bowman's capsule.
- All cells and all molecules or complexes larger than a medium-sized protein are retained in the glomerulus.
- Key molecules, water, electrolytes, and amino acids are reabsorbed in the proximal convoluted tubule, loop of Henle, and distal convoluted tubule.
- The kidney uses a combination of active and passive transport to reabsorb these molecules.
- By balancing water and sodium reabsorption, the kidney is able to regulate the osmolarity of the plasma.
- By balancing ammonia biosynthesis and carbonate reabsorption, the kidney regulates acid–base balance.

Concept Check

1. Explain the process by which the kidney filters out wastes from blood.
2. Describe how different molecules are reabsorbed into the bloodstream and where in the kidney these processes occur.
3. Describe the function of the hormones involved in the regulation of reabsorption of electrolytes and water from the filtrate.
4. Describe other functions performed by the kidney such as acid–base balance.
5. What are the biochemical measures for ascertaining kidney function clinically?

Bioinformatics Exercises

Exercise 1 Viewing and Analyzing Pyridoxal Phosphate Enzymes

Exercise 2 Amino Acid Metabolism and the KEGG Database

Problems

11.1 Digestion of Proteins

1. What are zymogens? Discuss their function in digestion process of proteins in stomach.

2. Pepsin is an aspartyl protease. How might mutation of the active site aspartate residue to a glutamate residue affect the enzyme? What kinetic parameters (e.g., V_{max}, k_{cat}, K_M) would be affected, and how? What would the mechanism of the new mutant look like? If one of the active site aspartates were mutated to asparagine, how would that affect the enzyme?

3. Trypsin is a serine protease with a classic catalytic triad. How would mutation of the active site serine to each of the following amino acids affect catalysis and kinetic parameters of the enzyme?

    a. serine to alanine

    b. serine to aspartate

    c. serine to threonine

4. In the course of a hypothetical meal, the stomach expands from 0.25 liters to 1.25 liters. If we assume that no buffering occurs and no acid is produced from the stomach contents, how much acid must be secreted to maintain pH = 1.0? How many grams of ATP would be consumed in the production of that acid?

5. The digestive enzyme renin is found in juvenile goats, cows, camels, and sheep. Renin, which is used in cheese-making, is a protease. Following heating and renin treatment, the curds coagulate and separate away from the whey. Why would treating a protein with a protease make it coagulate, when it might be expected that smaller pieces would be more soluble?

6. The pK_a of the amino and carboxyl groups on isoleucine are 9.6 and 2.36, respectively. Under physiological conditions (pH 7.4), what percentage of isoleucine is ionized?

11.2 Amino Acid Deamination and Transamination

7. Alanine has a hydrophobic side chain. Would you anticipate solubility of this amino acid in plasma to be a problem?

8. Draw the structure resulting from the transamination of these amino acids.

    I. serine

    II. arginine

    III. glutamate

    IV. phenylalanine

9. What is the reason for funneling all amino residues to glutamate?

10. We have not discussed the energetics of the transaminase reaction. What would you predict for a ΔG value for a transaminase? Strongly positive, strongly negative, or close to zero? Explain your reasoning.

11. A mutation in GDH of patients with HI/HA (hypoglycemia/hyperammonemia) results in a decrease in allosteric control of GDH via GTP. What will be observed in the patients?

11.3 The Urea Cycle

12. What is the source of the carbons in fumarate when they arise from the urea cycle? What is the source of the carbons in arginine? What is the source of the nitrogen atoms in arginine?

13. Ornithine can be decarboxylated to form polyamines such as putrescene (1,4-diaminobutane). What would the structure look like if ornithine was decarboxylated? Give a general description of the chemistry of the decarboxylation. What cofactors, electron donors, electron acceptors, or nucleophiles may be involved in this chemistry? What would you predict for a value of the free energy of reaction (positive, negative, near zero)? Explain your reasoning.

14. Citrulline, ornithine, and other metabolites are now commonly found in energy or sports recovery drinks. What is the reason for adding these metabolites to energy drinks? What might they be doing, and does it work? What drawbacks would these molecules have to overcome to do what they are described as being able to do?

15. What is the correct sequence of reactants for carbamoyl phosphate synthetase? In the mechanism, which group undergoes nucleophilic attack by ammonia?

16. Cats require arginine for the urea cycle but cannot synthesize it, so many commercial cat foods are supplemented with arginine. What is the likely effect if a cat has a single meal devoid of arginine?

11.4 Metabolism of Amino Acid Carbon Skeleton

17. We tend to focus on the fates of the carbon atoms in amino acid carbon skeleton scavenging. Where in these pathways is energy required or released? What other thermodynamic principles allow these reactions to proceed in the forward direction?

18. The catabolism of carbon skeletons of amino acids is organized by how they enter the common catabolic pathways (acetyl-CoA and the citric acid cycle), but they could also be organized by type of reactions. If you were to reorganize these pathways, what other means could you use to organize them?

19. In obligate carnivores, are carbon skeletons truly scavenged? What is the fate of these carbons?

20. Many humans are vegetarians or vegans whereas others eat large quantities of meat or fish. Can those people adapt their nitrogen metabolism to their specific diet, as other species discussed in this chapter have?

11.5 The Detoxification of Other Amines and Xenobiotics

21. Agonists are drugs that bind to receptors and activate them. The β-adrenergic receptor agonist propranolol is metabolized in the liver to 4-hydroxypropranolol. Would you expect this to be phase I or phase II metabolism? Based on the structure of propranolol, propose a structure of 4-hydroxypropranolol. Of the other metabolites shown in the figure, which pathways would you categorize as phase I and which as phase II?

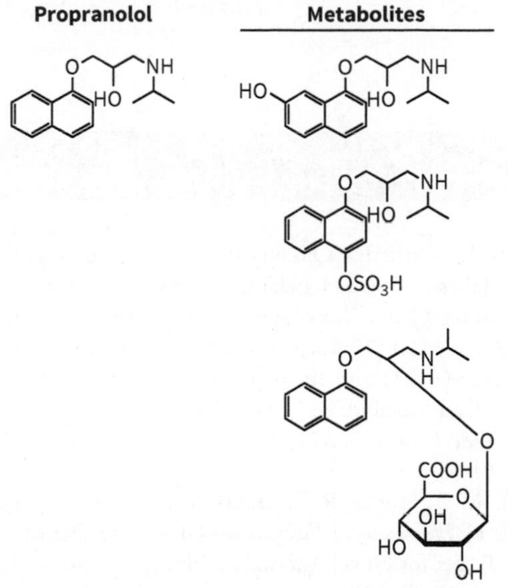

Propranolol   **Metabolites**

22. Which reactions of phase I and phase II metabolism require energy, and where does this energy come from (in the molecular form)?

23. MAO inhibitors (MAOIs) are an important class of antidepressant molecules. Patients taking MAOIs need to pay close attention to their diet and when taking any over the counter drugs. Explain.

11.6 The Biochemistry of Renal Function

24. Cells with large numbers of Na⁺ pumps often have more mitochondria. Why might this be the case?

25. About 68% of a small molecule is cleared in one pass of the kidney. What does this tell us about how this molecule is being treated in the kidney?

26. How many passes would it take to clear 80% of inulin from the circulation?

27. If 1.0 kg of NaCl is filtered each day by the kidney, how many sodium ions are there? What is the energetic potential of separating that number of sodium ions across a plasma membrane?

28. On average, 180 liters of plasma are filtered each day. If humans had to expend one molecule of ATP for every molecule of water retained, approximately how many molecules of ATP would be required? If 36 molecules of ATP are produced by the aerobic catabolism of one molecule of glucose, how many grams of glucose would be required?

29. Crush syndrome (traumatic rhabdomyolysis) is a phenomenon frequently observed after earthquakes or building collapses. In this syndrome, people who have had a body part, such as a leg trapped or crushed, enter acute kidney failure following release of the trapped body part. Based on your knowledge of kidney function and general biochemistry, what might be some of the agents released from the damaged tissues that could contribute to this organ failure?

Data Interpretation

30. A protease you are studying cleaves between basic residues. It is insensitive to metal chelating agents such as EDTA and EGTA and to

thiol-derivitizing agents, such as *N*-Ethylmaleamide, but is inhibited by organic phosphonates, such as phenylmethyl sulfonyl fluoride. Two residues appear to be critical for catalysis: a serine and a glutamate. Based on this collective data, which of the proteases noted in this text would be similar, and how might this protein function?

31. Given the following data, what is the sequence of this peptide?

Treatment with	Yields (No Stoichiometry or Sequence Implied)
Carboxypeptidase A	No products
Trypsin	No products
Chymotrypsin	(Met, Ile, Ser, Tyr) Trp (Ala, Lys)
Thrombin	No products
Dabsyl Cl/HCl	Lys, Ser
O-iodosobenzoate	Trp, (Ala, Lys) (Trp, Tyr, Ile, Met, Ser)
CNBr	(Met, Ser) (Trp, Tyr, Ile, Lys, Ala)

If you need help with how these reagents work, consult the techniques appendix.

32. Melamine is a component of some plastics. Because of its high nitrogen content and low price, it has been illegally used in the past to make samples of protein look as though they have higher levels of protein. Why would melamine give a positive test for protein when tested with a Kjeldahl assay but not with a BCA or Bradford assay?

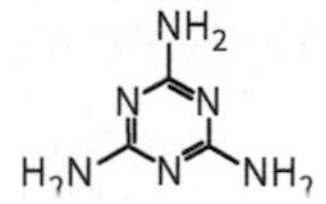

Melamine

33. A newborn presents in the clinic with vomiting, poor feeding, lethargy, and seizures. The child smells faintly of dirty socks. Urinalysis reveals elevated levels of isovaleryl glycine. Plasma chemistry indicates elevated levels of isovaleryl carnitine. What is the disorder, and where is the odor, isovaleryl glycine, and isovaleryl carnitine originating?

Experimental Design

34. You discover a novel protease in a species of fungi. How would you go about determining where the protease cuts substrate proteins?

35. Some reactions of the urea cycle occur in the mitochondria and others in the cytosol. How would you design an assay to test where in the cell a process is occurring, especially if the process occurs partly in one organelle and partly in another?

36. In an organism that does not have a liver, how could you test to see where in that organism urea biosynthesis is occurring? What controls would you need to use? What type of assay would you need to perform, and what would need to be included in that assay?

37. You discover a novel inhibitor of glucose transport in the kidney. For getting approval for you new drug, you need to show how it is metabolized and cleared from the body. How would you test to see how your drug is metabolized and excreted?

38. If the drug referred to in the previous question inhibits glucose transport, how might it affect other aspects of renal function? How would you test your hypothesis?

Ethics and Social Responsibility

39. Affirm, deny, or qualify this statement: "Like many over-the-counter drugs, proton pump inhibitors are overused to the point of abuse."

40. A friend decides to consume a vegetarian diet but only wants to eat French fries. How would you use your biochemical knowledge to

explain to her that this is probably not the healthiest course of action? A few weeks later your friend has decided she will supplement her diet with lentils, which she explains are high in protein. What might you recommend to her at this point?

Suggested Readings

11.1 Digestion of Proteins

Barrett, G. C., and D. T. Elmore. *Amino Acids and Peptides*. Cambridge, United Kingdom: Cambridge University Press, 1998.

Benos, Dale J. "Developmental Biology of Membrane Transport Systems." Retrieved from http://books.google.com/books/about/Developmental_biology_of_membrane_transp.html?id=GVjnebaPMMgCAcademic Press

Brix, Klaudia, and Walter Stöcker, eds. *Proteases: Structure and Function*. New York, New York: Springer, 2013.

Bröer, S. "Amino Acid Transport Across Mammalian Intestinal and Renal Epithelia." *Physiological Reviews* 88, no. (January 2008): 49–86.

Chiba, Isamu, and Takao Kami, eds. *Serine Proteases: Mechanism, Structure and Evolution*. Hauppauge, New York: Nova Science Publishers, 2012.

Lendeckel, Uwe, and Nigel M. Hooper, eds. *Proteases in Gastrointestinal Tissues (Proteases in Biology and Disease)*. New York, New York: Springer, 2010.

11.2 Amino Acid Deamination and Transamination

Brosnan, M. E., and J. T. Brosnan. "Hepatic Glutamate Metabolism: A Tale of 2 Hepatocytes." *The American Journal of Clinical Nutrition* 90, no. 3 (September 2009): 857S–861S.

Curthoys, Norman P., and Malcolm Watford. "Regulation of Glutaminase Activity and Glutamine Metabolism." *Annual Review of Nutrition* 15 (1995): 133–159.

Eisenberg, David S., Harindarpal S. Gill, Gaston M. U. Pfluegl, and Sergio H. Rothstein. "Structure-Function Relationships of Glutamine Synthetases." *Biochimica et Biophysica Acta* 1477 (2000): 122–145.

Gill, H. S., and D. Eisenberg. "The Crystal Structure of Phosphinothricin in the Active Site of Glutamine Synthetase Illuminates the Mechanism of Enzymatic Inhibition." *Biochemistry* 40 (2001): 1903–1912.

Haussinger, D., and W. Gerok. "Hepatocyte Heterogeneity in Glutamate Uptake by Isolated Perfused Rat Liver." *European Journal of Biochemistry* 136 (1983): 421–5.

Liaw, Shwu-Huey, I. Kuo, and D. Eisenberg. "Discovery of the Ammonium Substrate Site on Glutamine Synthetase, a Third Cation Binding Site." *Protein Science* 4 (1995): 2358–65.

Peterson, P. E., and T. J. Smith. "The Structure of Bovine Glutamate Dehydrogenase Provides Insights into the Mechanism of Allostery." (1999) *Structure* 7 (1999): 769–82.

Smith, T. J., P. E. Peterson, T. Schmidt, J. Fang, C. A. Stanley. "Structures of Bovine Glutamate Dehydrogenase Complexes Elucidate the Mechanism of Purine Regulation." *Journal of Molecular Biology* 307 (2001): 707.

Thangavelu, K., Catherine Qiurong Pan, Tobias Karlberg, Ganapathy Balaji, Mahesh Uttamchandani, Valiyaveettil Suresh, Herwig Schüler, Boon Chuan Low, and J. Sivaramana. "Structural Basis for the Allosteric Inhibitory Mechanism of Human Kidney-Type Glutaminase (KGA) and Its Regulation by Raf-Mek-Erk Signaling in Cancer Cell Metabolism." *Proceedings of the National Academy of Sciences of the United States of America* 109, no. 20 (May 15, 2012): 7705–7710.

Werner, C., M. T. Stubbs, R. L. Krauth-Siege, G. Klebe. "The Crystal Structure of *Plasmodium Falciparum* Glutamate Dehydrogenase, a Putative Target for Novel Antimalarial Drugs." *Journal of Molecular Biology* 349 (2005): 597.

11.3 The Urea Cycle

Hoar, William Stewart, Anthony P. Farrell, David J. Randall, Patricia A. Wright, and Paul M. Anderson. *Fish Physiology: Nitrogen Excretion*. Vol. 20. Cambridge, MA: Academic Press, 2001.

Kaneko, Jiro J., John W. Harvey, and Michael Bruss. *Clinical Biochemistry of Domestic Animals*. Retrieved from http://books.google.com/books/about/Clinical_Biochemistry_of_Domestic_Animal.html?id=spsD4WQbL0QC

Prosser, C. Ladd. *Comparative Animal Physiology, Part 2*. Hoboken, New Jersey: John Wiley and Sons, 1991.

Singer, M. A. "Do Mammals, Birds, Reptiles and Fish Have Similar Nitrogen Conserving Systems?" *Comparative Biochemistry and Physiology—Part B: Biochemistry & Molecular Biology* 134, no. 4 (April 2003): 543–58.

Walsh, Patrick J., and Patricia A. Wright. "Nitrogen Metabolism and Excretion." Boca Raton, Florida: CRC Press, 1995.

11.4 Metabolism of Amino Acid Carbon Skeleton

Abelson, John N., Melvin I. Simon, John R. Sokatch, and Robert Adron Harris. *Branched-Chain Amino Acids, Part B, Volume 324 (Methods in Enzymology)*. Amsterdam, Netherlands: 2000.

Bender, David A. *Amino Acid Metabolism*. 3rd ed. Hoboken, New Jersey: John Wiley & Sons, 2012.

Jorgensen, H., W. C. Sauer, and P. A. Thacker. "Amino Acid Availabilities in Soybean Meal, Sunflower Meal, Fish Meal and Meat and Bone Meal Fed to Growing Pigs." *Journal of Animal Science* 58, no. 4 (1984).

Kaplan, Nathan P., Nathan P. Colowick, Herbert Tabor, and Celia White Tabor, eds. *Metabolism of Amino Acids and Amines, Part B, Volume 17B*. Amsterdam, Netherlands: 1971.

Sharma, Virender K., and Steven E. Rokit. *Reactive Intermediates in the Oxidation of Amino Acids, Peptides, and Proteins*. Hoboken, New Jersey: Wiley, 2013.

11.5 The Detoxification of Other Amines and Xenobiotics

Angelucci, Francesco, Paola Baiocco, Maurizio Brunori, Louise Gourlay, Veronica Morea, and Andrea Bellelli. "Insights into the Catalytic Mechanism of Glutathione S-Transferase: The Lesson from *Schistosoma haematobium*." *Structure* 13, no. 9 (September 1, 2005): 1241–6.

Cornish, H., and A. A. Christman. "A Study of the Metabolism of Theobromine, Theophylline, and Caffeine in Man." *Journal of Biological Chemistry* 228 (1957): 315–24.

De Montellano, P. R. O. *Cytochrome P450: Structure, Mechanism, and Biochemistry.* New York, New York: Springer, 2005.

Dick, C. W., and R. J. Baker. "Phylogenetic Relationships Among Megabats, Microbats, and Primates." *Proceedings of the National Academy of Sciences of the United States of America* 88 (1991): 10322–26.

Finberg, J. P., and M. B. Youdim. "Selective MAO A and B Inhibitors: Their Mechanism of Action and Pharmacology." *Neuropharmacology* 22, spec. no. 3 (Mar 1983): 441–6.

Guengerich, F. P. "Mechanisms of Cytochrome P450 Substrate Oxidation: Minireview." *Journal of Biochemical and Molecular Toxicology.* 21 (2007):163–168.

Hayes, J. D., and D. J. Pulford. "The Glutathione S-transferase Supergene Family: Regulation of GST and the Contribution of the Isoenzymes to Cancer Chemoprotection and Drug Resistance." *Critical Reviews in Biochemistry and Molecular Biology* 30, no. 6 (1995): 445–600.

Kot, M., and W. A. Daniel. "The Relative Contribution of Human Cytochrome P450 Isoforms to the Four Caffeine Oxidation Pathways: An in Vitro Comparative Study with cDNA-expressed P450s Including CYP2C Isoforms." *Biochemical Pharmacology* 76, no. 4 (August 15, 2008): 543–51.

Szklarz, G., and J. R. Halpert. "Molecular Basis of p450 Inhibition and Activation: Implications for Drug Development and Drug Therapy." *Drug Metabolism and Disposition* 26, no. 12 (1998).

11.6 The Biochemistry of Renal Function

Boron, W., and E. Boulpaep. *Medical Physiology.* 3rd ed. Philadelphia, Pennsylvania: Elsevier, 2016.

Hamburger, Jean. *Structure and Function of the Kidney.* Philadelphia, PA: W. B. Saunders Co., 1971.

Lote, C. J. *Principles of Renal Physiology.* 4th ed. New York, New York: Springer, 2000.

More, Ed. *Biochemistry Kidney Functions* (INSERM Symposium). Amsterdam, Netherlands: Elsevier, 1982.

Valtin, Heinz, and James A. Schafer. *Renal Function.* 3rd. ed. Philadelphia, Pennsylvania: Lippincott Williams & Wilkins, 1995.

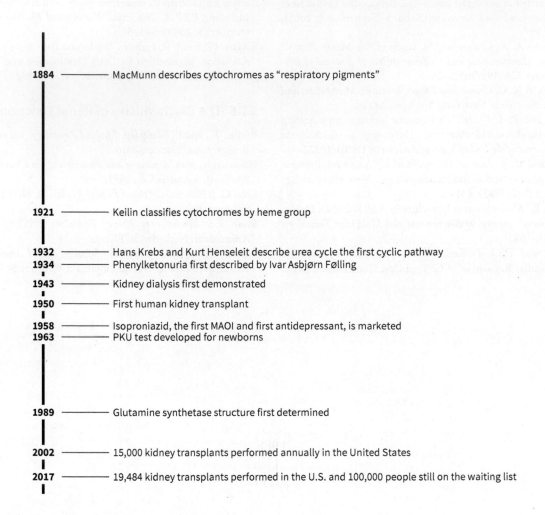

1884 ——————— MacMunn describes cytochromes as "respiratory pigments"

1921 ——————— Keilin classifies cytochromes by heme group

1932 ——————— Hans Krebs and Kurt Henseleit describe urea cycle the first cyclic pathway
1934 ——————— Phenylketonuria first described by Ivar Asbjørn Følling
1943 ——————— Kidney dialysis first demonstrated
1950 ——————— First human kidney transplant
1958 ——————— Isoproniazid, the first MAOI and first antidepressant, is marketed
1963 ——————— PKU test developed for newborns

1989 ——————— Glutamine synthetase structure first determined

2002 ——————— 15,000 kidney transplants performed annually in the United States

2017 ——————— 19,484 kidney transplants performed in the U.S. and 100,000 people still on the waiting list

CHAPTER 12

Regulation and Integration of Metabolism

Metabolism in Context

Running any business, even a small one, is a very complicated proposition. Will the business be able to keep enough inventory to meet customers' demands? How will it store that inventory, and how can the business coordinate between manufacturers, warehouses, and retail outlets? The small business in this analogy can represent any biochemical pathway. Studied on its own in terms of its steps, how it is regulated, and where it occurs, any biochemical pathway can be simply viewed as stand-alone, unconnected, and isolated. This chapter unites and explores the complex interplay of these different pathways. To return to the small business analogy, this chapter is less about a single business and more about economics; how all the small businesses, or pathways, interact with one another to form the economy or a biochemical network. A perturbation in one area will affect another area and when one sector, such as energy or housing, is faring poorly, the entire economy suffers.

First, we look at the different players in the metabolic economy, that is, the molecules commonly found at the intersection of different pathways and the organs and tissues that metabolize these molecules. Next, we discuss how the organs communicate with each other, and the vital role of the brain in overall regulation of metabolism. Finally, the chapter examines the role of metabolism in disease, explaining how aberrant metabolism can cause disease and conversely how disease can turn metabolism against the organism.

Chapter Outline

12.1 A Review of the Pathways and Crossroads of Metabolism

12.2 Organ Specialization and Metabolic States

12.3 Communication between Organs

12.4 Metabolic Disease

Common Themes

Evolution's outcomes are conserved.	• Many of the pathways we will see in this chapter are conserved throughout evolution. • Organs have evolved specific functions to help cope with metabolic challenges. A comparison of humans with other organisms illustrates some of these differences. • Specific mechanisms have evolved for organisms to cope with the ebb and flow of energy throughout the day, the year, the seasons, and the life span. • The explosion of diseases related to metabolic syndromes reflects how humans as a species have been unable to adapt to recent changes in diet and lifestyle, compared to the conditions adapted to over the past 40,000 years.
Structure determines function.	• The structure of simple organic molecules, such as catecholamines, steroids, and peptides, are critical in their interaction with receptors. • The chemical bonds in triacylglycerols are highly reduced, increasing the amount of energy that can be obtained through oxidation. • Triacylglycerols are hydrophobic molecules and are stored without water. This is not the case with carbohydrates or proteins. • The structure and chemistry of some metabolites—ketone bodies, acetate, glucose—can directly or indirectly cause damage to molecules, cells, and tissues when concentrations are elevated, as happens in some disease states.
Biochemical information is transferred, exchanged, and stored.	• Metabolism is regulated hormonally through endocrine hormones and neuronally through direct innervation of some tissues from the brain. • The hypothalamus is important to the integration of metabolism resulting in cravings, such as appetite, and complex behaviors, like eating. • Hormonal signals that regulate metabolism are constantly in flux.
Biomolecules are altered through pathways involving transformations of energy and matter.	• Under different metabolic states, a metabolite may be transformed for specific purposes. For example, alanine formed from pyruvate in muscle can be shuttled to the liver for detoxification of the amine moiety and generation of glucose from the carbon backbone. Similarly, the liver uses acetyl-CoA to form ketone bodies that can provide fuel for the brain. • The fundamental laws of thermodynamics, such as the conservation of energy, underlie and apply to all of metabolism.

12.1 A Review of the Pathways and Crossroads of Metabolism

An important distinction between metabolism as a whole and the organic reactions of metabolism is the context under which these reactions occur. For instance, if the metabolites and enzymes that participate in metabolic pathways were isolated and recombined in a beaker, the end products might be the same, but many of the important levels of control would be lost since metabolic pathways are interdependent. This section briefly reviews the major catabolic and anabolic pathways covered in previous chapters, identifies key metabolites that help to shuttle metabolites between pathways, and provides a more integrated discussion of the roles of these metabolites.

12.1.1 The pathways of metabolism are interconnected

As we have discussed, different metabolic pathways build up or break down different metabolites reviewed in **Figure 12.1**. These pathways are highly conserved throughout biology.

All of the major metabolic reactions for lipids, carbohydrates, and proteins take place within the cell and are almost evenly divided between the cytosol and the mitochondria (**Figure 12.2**). Two pathways, the urea cycle and gluconeogenesis, take place in both locations. With the exception of the electron transport chain, none of these pathways require molecular oxygen, and each of these pathways can operate either in the aerobic or anaerobic state.

Beta oxidation catabolizes fatty acids   Fatty acids are broken down to acetyl-CoA through β-oxidation. This pathway occurs in the mitochondria in most cells of the body. The brain is an exception in that it can undergo low levels of β-oxidation, but it cannot obtain fatty acids from the blood for this process due to the blood-brain barrier. Most of the acetyl-CoA produced by β-oxidation is further oxidized in the citric acid cycle; however, if concentrations elevate, some is diverted to ketone body production in the liver.

Fatty acid biosynthesis generates fatty acid from acetyl-CoA   Fatty acid biosynthesis occurs in the cytosol, producing fatty acids from reduced building blocks

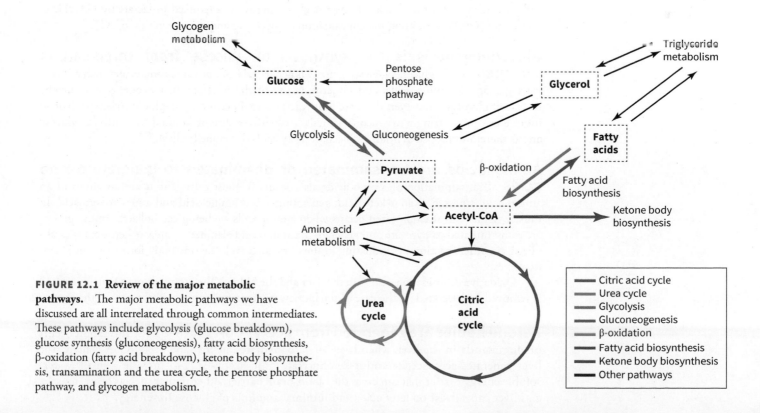

FIGURE 12.1 Review of the major metabolic pathways. The major metabolic pathways we have discussed are all interrelated through common intermediates. These pathways include glycolysis (glucose breakdown), glucose synthesis (gluconeogenesis), fatty acid biosynthesis, β-oxidation (fatty acid breakdown), ketone body biosynthesis, transamination and the urea cycle, the pentose phosphate pathway, and glycogen metabolism.

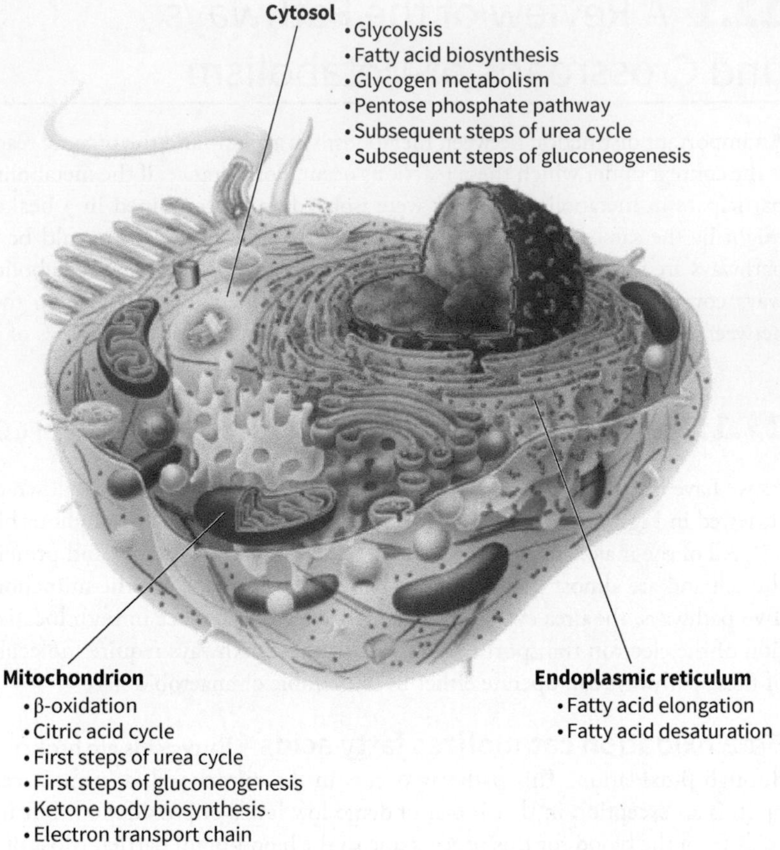

FIGURE 12.2 Overview of where metabolism occurs in the cell. The major metabolic reactions we have discussed all take place within the cell. These reactions are almost evenly divided between the cytosol and the mitochondria. Two pathways (the urea cycle and gluconeogenesis) take place in both locations. Fatty acid biosynthesis occurs in the cytosol, while elongation and desaturation happen in the endoplasmic reticulum.

Cytosol
- Glycolysis
- Fatty acid biosynthesis
- Glycogen metabolism
- Pentose phosphate pathway
- Subsequent steps of urea cycle
- Subsequent steps of gluconeogenesis

Mitochondrion
- β-oxidation
- Citric acid cycle
- First steps of urea cycle
- First steps of gluconeogenesis
- Ketone body biosynthesis
- Electron transport chain

Endoplasmic reticulum
- Fatty acid elongation
- Fatty acid desaturation

(acetyl-CoA). Fatty acyl-CoAs can be esterified to glycerol-3-phosphate to form lysophosphatidate, the branching point between phospholipid and triacylglycerol metabolism.

Glycolysis catabolizes glucose to pyruvate   Glucose is degraded through glycolysis into pyruvate. This process occurs in the cytosol of all known cell types. In the absence of sufficient oxygen for the electron transport chain, pyruvate is reduced to lactate by NADH, to regenerate NAD^+, permitting the continuation of glycolysis and the generation of ATP.

Gluconeogenesis is the synthesis of glucose from three-carbon precursors   In gluconeogenesis, glucose is generated from precursor molecules such as pyruvate, glucogenic amino acids, and glycerol. Many of the reactions in gluconeogenesis are the reverse of glycolysis; however, there are four reactions that occur only in gluconeogenesis. Not all molecules can contribute to gluconeogenesis. Acetyl-CoA cannot be used to synthesize glucose and is therefore trapped as lipidic molecules (fatty acids or ketone bodies).

Amino acids are transaminated or deaminated to generate α-keto acids   Transamination of amino acids also occurs in most cells. The α amino group of an amino acid is passed to an α-keto acid, generating a new amino acid and a new α-keto acid. In muscle tissue, this reaction often occurs when amino acids are being catabolized. Amino groups are transaminated to pyruvate and generate the amino acid alanine; the new α-keto acid is catabolized via other reactions, culminating in the citric acid cycle, ketone body formation, or gluconeogenesis.

Oxidative deamination occurs in the liver and the kidney. The amino group of an amino acid is removed and secreted, in the case of the kidney, or detoxified to urea, in the case of the liver.

Ketone bodies are generated from acetyl-CoA   Ketone body formation occurs predominantly in the liver, when levels of carbohydrates are low and acetyl-CoA is high. The liver generates acetoacetate and α-ketobutyrate from acetyl-CoA to provide a small, water-soluble energy carrier that can cross the blood-brain barrier; other tissues such as heart, muscle, and liver can subsist on fatty acids and minimal amounts of glucose for energy.

12.1.2 Several metabolites are at the intersection of multiple pathways

By this stage, many metabolites will be familiar. At the mention of fatty acids or glucose, you should have a mental picture of the structure of the molecules, their biological functions, and how they are built up and broken down. Hopefully, you can also place these molecules in some sort of metabolic context, although this is harder to do because the role of metabolites is often unclear. Scientific opinion about the function of these molecules evolves as research generates new findings. This section reviews and examines the role of many of the major metabolites in the context of metabolism in the body (**Table 12.1**).

- Glucose is blood sugar, the main carbohydrate source of fuel in the body. It is stored in and mobilized from glycogen. Glucose is the main fuel source for some tissues such as the brain and the kidney and the sole fuel source for some others, including erythrocytes and sperm. It is catabolized in all tissues via glycolysis and can be produced from noncarbohydrate sources through gluconeogenesis in the liver.

- Fatty acids in the blood come from **adipose tissue** (fat tissue) and are transported bound to albumin. They are stored in limited amounts in the liver and muscle, and can fuel these tissues and the heart but not the brain. Due to their reduced state, fatty acids are a committed fuel—once reduced to a fatty acid, the carbons can be burned through β-oxidation, the citric acid cycle and oxidative phosphorylation, or converted to ketone bodies.

TABLE 12.1 Metabolites at a Crossroads

Metabolite	Structure	Made In	Stored In	Metabolized By
Glucose		Liver (gluconeogenesis)	Liver and muscle (glycogen)	All tissues (glycolysis)
Fatty acids		Liver (fatty acid biosynthesis)	Adipose tissue (triacyglycerol)	Liver and muscle (β-oxidation)
Pyruvate		All tissues, product of glycolysis	Transient	All tissues (citric acid cycle)
Lactate		Muscle, anaerobic product of glycolysis	Transient	Muscle (pyruvate) Liver (gluconeogenesis)
Alanine		Muscle, transamination	Transient	Liver (transamination and gluconeogenesis)
Glutamine		Muscle, transamination	Transient	Liver and kidney (transamination, oxidative deamination, urea cycle)
Ketone bodies		Liver, ketone body biosynthesis	Transient	Brain (used as fuel for citric acid cycle)
Glycerol		Liver, by-product of glycolysis	Adipose tissue (triacylglycerol)	Liver (gluconeogenesis)

- Pyruvate is found at the intersection of four pathways in mammals. It can be used in muscle tissue to generate lactate under anaerobic conditions, to keep glycolysis proceeding. In transamination reactions, pyruvate can accept an amino group from other amino acids, producing alanine. Pyruvate can be oxidized into acetyl-CoA for the citric acid cycle and oxidized to generate reducing equivalents for the electron transport chain. Finally, pyruvate can be carboxylated to provide oxaloacetate. Although this reaction can be anaplerotic (it replenishes the oxaloacetate of the citric acid cycle), it is often used instead in gluconeogenesis to provide glucose for other tissues.

- Lactate is generated in anaerobically exercising muscle as a way to oxidize NADH to NAD⁺. It can accumulate in muscle, causing other physiological effects, such as a drop in pH and an associated increase in oxygen delivery by hemoglobin; alternatively, it can be transported through the blood to the liver, where it is converted to pyruvate and used in gluconeogenesis.

- Alanine is made in muscle from pyruvate. It serves a dual role in metabolism, shuttling what would otherwise be a toxic ammonia molecule to the liver for detoxification by the urea cycle and, at the same time, transferring carbon atoms to the liver for use in gluconeogenesis.

- Glutamine is another amino acid from muscle that transports amines from muscle to the liver for urea biosynthesis. If the liver cannot keep up, as is seen in some disease states such as hepatitis and cirrhosis, the kidney can deaminate glutamine into first glutamic acid, then α-ketoglutarate. The ammonia is released into the urine, and the α-ketoglutarate can participate in renal gluconeogenesis.

- Ketone bodies are formed from acetyl-CoA in times of low glucose availability. These molecules are made by the liver, and they provide an alternative energy source for the brain.

- Glycerol forms the backbone of triacylglycerol molecules. It is released from adipose tissue and transported to the liver, where it participates in synthesis of liver triacylglycerols or gluconeogenesis.

As we can see, several metabolites converge on multiple pathways, and measuring the rate of flux through these pathways may not be as simple as once envisioned. The study of flux through a pathway is considered in **Biochemistry: Flux through a pathway**.

Biochemistry

Flux through a pathway

Flux is the rate of flow or passage through a pathway. When we think of pathways as being analogous to an assembly line, we can model flux simply as the rate of production over time. This is typically given the term J:

$$J = \frac{dq}{dt}$$

where q is the quantity of product made.

If a metabolic pathway such as glycolysis were simply analogous to an assembly line, this approximation might fit, but as we have seen, metabolism is a complex interplay of many factors transpiring in the compartments of the cell. There are numerous places that metabolic intermediates can enter or exit from a metabolic pathway, the pathways themselves are complex, and there are sometimes competing reactions or other enzymes that can catalyze the same reaction (e.g., glucokinase and hexokinase in glycolysis).

Modern approaches to metabolic flux use computational models that take some of these factors into account. There are several mathematical models including metabolic control analysis, flux balance analysis, choke point analysis, and dynamic metabolic simulation that are used in these studies. Each of these models examines the concentration of metabolites and rates of enzymatic reactions under different conditions and assumptions, such as steady state or a single enzyme as the rate determining step, to calculate flux through the pathway. Disruptions are introduced into the system to see how flux changes. Why might it be important to measure, calculate, or understand flux through a pathway? When we consider that pathways are not isolated but are rather interconnected, we begin to see how flux through a pathway can influence all of metabolism. This may have important implications in allosteric regulation of individual enzymes in a pathway, athletic performance, or progression of disease. Likewise, if we are engineering a biological system to produce a specific product—an intermediate in a pathway, an engineered protein, or the organism itself—we might be concerned as to what conditions are optimal for growth. All of these parameters can be considered using flux analysis.

Worked Problem 12.1   Follow that nitrogen...again!

A mouse previously fed a diet containing amino acids labeled with the stable isotope ^{15}N was moved to the fasted state. Based on your current knowledge of metabolism, which labeled compounds would you expect to find in the blood and urine of the mouse in the fasted state?

Strategy   Levels of amino acids increase following a meal but drop relatively soon thereafter. If an animal has been fasted, it has received no food for enough time to ensure that there are no dietary amino acids in the bloodstream; in addition, levels of stored carbohydrates will have dropped. If we presume the fast has gone on long enough to catabolize some protein (most likely from muscle, the largest store of amino acids in the body), what will be the result of muscle protein catabolism?

Solution   We may expect to see three pools of ^{15}N in the blood, corresponding to alanine, glutamine, and urea. The alanine and glutamine act as transporters of nitrogen from muscle to the liver, where the amine moiety of the amino acids is detoxified through urea synthesis. The liver secretes urea for elimination by the kidney. In the urine, we may expect to see one or two pools of ^{15}N. We would definitely expect to see urea, but we might also expect to see some levels of ammonia synthesized by oxidative deamination of amino acids (glutamine) by the kidney.

Follow-up question   Name and describe two methods for detecting ^{15}N.

Summary

- The major anabolic and catabolic pathways are all intertwined.
- Many metabolites serve to link different metabolic pathways or organ systems. This has arisen as specialized tissues have evolved that can shift the metabolic burden from one tissue to another.
- Pyruvate, alanine, glutamine, glycerol, and lactate all serve to shift the metabolic burden from one tissue to another.

Concept Check

1. Describe what happens in each of the metabolic pathways, and the major products in each case.
2. Describe other ways molecules can enter or leave these pathways.
3. Several metabolites converge on multiple pathways and facilitate more complex level of metabolic organization. Identify key metabolites that help to shuttle metabolites between pathways and explain how these molecules function.

12.2 Organ Specialization and Metabolic States

Organisms struggle throughout their lifespan to maintain **homeostasis** or the balance between energy-poor and energy-rich states. Considering only the pathways studied in previous chapters, as many as 100 different enzymes catalyze the reactions involved. A similar number of regulatory proteins, hormones, receptors, kinases, phosphatases, and other proteins are also involved in the regulation of metabolism. Thus, biochemistry is not just a study of catalysis but also a study of regulation.

In a complex organism, the situation is even more complicated because distinct cell types have evolved with differing metabolic functions. Although the core processes of metabolism (glycolysis, β-oxidation, the citric acid cycle, and the electron transport chain) are clearly conserved, cells have evolved elegant mechanisms of control and levels of specialization. For example, muscle cells, adipocytes, and neurons are all capable of conducting glycolysis, but the conditions under which the glycolysis will occur are regulated differently in the three cell types, with each having evolved a distinct function to fill a distinct need. The specialization of different cell types has further evolved into discrete organ systems with specific metabolic functions. This section first examines the function of some of the major organs of the body in the context of metabolism, then focuses on the distinct metabolic states that the organism may pass through in its lifecycle and the demands these states put on each system.

12.2.1 Different organs play distinct roles in metabolism

Organs play distinct roles in metabolism. Some, such as the liver and adipose tissue, generate energy for other tissues, whereas others, such as the brain, cannot respond to an energy demand from any other tissues. This section discusses and reviews the functions of these tissues in metabolism as a whole and how they work in conjunction with one another.

The liver is central to the metabolism of proteins, carbohydrates, and lipids

Whereas the brain is the master regulator of metabolism, the liver is the caretaker (**Figure 12.3**). Although the liver is indeed involved in the detoxification of various metabolites, it is even more important in the balance of carbohydrate, lipid, and amino acid metabolism.

The liver can conduct all of the major metabolic pathways. The rest of this section discusses the metabolism of three major metabolites: amino acids, lipids, and carbohydrates.

Most urea production occurs in the liver. Amino acids from the diet or from the breakdown of muscle tissue are shuttled to the liver for oxidative deamination, or for transamination with α-keto acids into other amino acids. The liver also receives alanine from muscle, created by the transamination of other amino acids to pyruvate. In the liver, these amino acids are detoxified; the amino group is oxidatively removed and used in urea biosynthesis. From here, urea is released into the bloodstream and excreted by the kidney. The carbon skeletons of amino acids are used to make new glucose via gluconeogenesis, fatty acids, or ketone bodies.

The liver also receives fats from chylomicron remnants. These lipids are repackaged with other lipids made from glucose and fatty acids, and they are secreted into the bloodstream as very-low-density lipoprotein (VLDL), which delivers triacylglycerols to muscle and adipose tissue. The liver can also accept free or (unesterified) fatty acids and glycerol from adipose tissue. These fatty

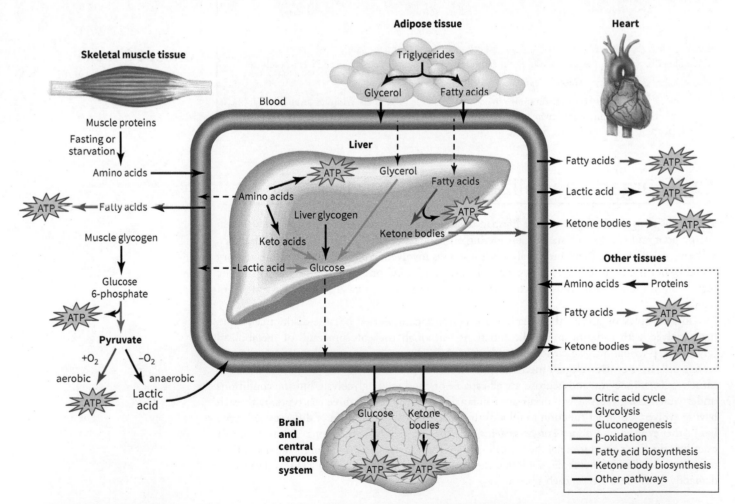

FIGURE 12.3 Overview of where metabolism occurs in the body.   Specific organs evolved to fulfill different metabolic jobs. The heart and the brain do not store any energy but rather depend on other organs for fuel. The liver and muscle store carbohydrate as glycogen. Adipose is the predominant triacylglycerol store, although muscle and the liver can store triacylglycerol as well. The liver is responsible for the urea cycle, gluconeogenesis, and ketone body biosynthesis.

(Source: Tortora; Derrickson, *Principles of Anatomy & Physiology*, 13e, copyright 2012, John Wiley & Sons. This material is reproduced with permission of John Wiley & Sons, Inc.)

acids are used to make ketone bodies to fuel the brain, while the glycerol fraction is used for gluconeogenesis. As with ketone bodies, the glucose produced in the liver is not retained but rather
is secreted to the blood for other tissues to use.

The liver and muscle are the two major stores of glycogen in the body. About 25% of the
body's glycogen stores are found in the liver, and the remaining 75% are found in muscle tissue.
Because the liver expresses the enzyme glucose-1-phosphatase, it can produce glucose from the
glucose-1-phosphate liberated in glycogenolysis and hence can release this glucose to the bloodstream for use by other tissues. Any glucose made in **hepatocytes (liver cells)** via gluconeogenesis
will also be released.

Muscle catabolizes molecules for energy

Muscles allow organisms to move to
either obtain energy or get out of harm's way. Unlike the liver, muscle does not normally contribute to the metabolic needs of other organs; instead, it makes demands on other organs such as the
liver and adipose tissue. In rare worst-case scenarios, muscle can catabolize amino acids for fuel.

While all tissues can use amino acids to synthesize proteins, excess amino acids are catabolized. Amino acids that are catabolized in muscle are often transaminated to pyruvate, generating alanine. Alanine travels through the blood to the liver, where it is transaminated to another
α-keto acid or oxidatively deaminated to regenerate pyruvate. In the liver, pyruvate is used for
gluconeogenesis, producing new glucose that is returned to muscle and burned as fuel. This is the
glucose–alanine cycle.

Muscle can use fatty acids from chylomicra, VLDLs, or free fatty acids released from adipose
tissue and transported in the blood by albumin. Muscle has the ability to store small quantities of
these molecules in lipid storage droplets, although most of the lipid used by muscle is burned for
energy. Muscle does not generate appreciable quantities of ketone bodies.

The other primary muscle fuel is carbohydrate. Muscle readily undergoes glycolysis,
and it contains three-fourths of the body's glycogen reserves. Unlike the liver, muscle lacks
glucose-6-phosphatase. As a result, any glucose released from glycogen is retained by muscle and
burned in that cell.

Muscle can broadly be categorized as red or white. Red muscle, characterized metabolically
as oxidative muscle, is slow-twitch muscle. Its cells are enriched in myoglobin and mitochondria
and are therefore redder in color. White muscle, characterized metabolically as glycolytic muscle,
is fast-twitch muscle; its primary fuel is glucose. White muscle cells contain less myoglobin and
fewer mitochondria, and it has a higher amount of collagen. Section 12.4 discusses the roles of
these different types of muscle and the use of an alternative fuel.

We might think that a well-trained athlete has lean muscle devoid of fat, but depending on
the type of athlete, the opposite may be found. This is discussed in **Medical Biochemistry: The
athlete's paradox**.

Medical Biochemistry

The athlete's paradox

Well-conditioned endurance athletes have low percentages of body
fat. The muscles of these athletes have an increased capacity to burn
fatty acids; that is, these muscles are oxidative and have a greater
sensitivity to insulin. We might anticipate these states in a trained
athlete, but what is surprising is that the muscles of these athletes
have elevated levels of stored triacylglycerols. This is known as the
athlete's paradox.

Normally, levels of muscle triacylglycerols correlate directly
with insulin resistance. Overweight subjects have more triacylglycerol deposition and more insulin resistance than lean individuals; obese individuals have even more stored triacylglycerols and

increased insulin resistance. In conditioned athletes, there is elevated
triacylglycerol storage, but this fails to correlate with insulin resistance. Obese and insulin-resistant patients who undertake a chronic
exercise program also see selective partitioning of triacylglycerol into
muscle, coupled with improved insulin sensitivity. Because of this
disconnect, and the implications for type 2 diabetes, the athlete's
paradox is a fascinating question that has ramifications for health.

There are several theories about what causes the athlete's paradox. One of the more promising explanations is that in athletes
the muscle triacylglycerol turns over more rapidly than normal. This
theory is supported by findings that levels of diacylglycerol and ceramide are also elevated in athletes, as are levels of several enzymes and
regulatory proteins.

Adipose tissue stores fat for future use

Like the liver, adipose tissue takes care of the rest of the body. Adipose tissue was once thought of as an inert warehouse where energy was stored as triacylglycerol. However, more recent studies have shown that adipose tissue is both a thriving, metabolically active tissue and an endocrine tissue, secreting several molecules that regulate appetite and metabolism. Thus, there are two tissues that help to take care of the energetic needs of the other organs of the body: the liver, which provides in the short term, and adipose tissue, which functions in both long-term storage and the near future.

Adipose tissue does little in terms of amino acid metabolism. While it uses some glucose to produce glycerol for triacylglycerol metabolism and uses glycolysis and oxidative phosphorylation to make ATP, it cannot store carbohydrates as glycogen. Instead, adipose tissue tends to store and regulate the release of free fatty acids and glycerol from triacylglycerols.

Adipocytes, the fat-storing cells of adipose tissue, acquire fatty acids from either chylomicrons or VLDL, storing these lipids in triacylglycerols in a single large lipid storage droplet. When stimulated, the adipocyte will release free fatty acids and glycerol. Due to the absence of glycerol kinase, the adipocyte cannot recycle glycerol once all three fatty acyl chains have been hydrolyzed. Instead, glycerol is released from the adipocyte and travels to the liver where it is used in gluconeogenesis to make glucose. Fatty acids are transported in the blood by serum albumin to muscle and liver tissue. Muscle will burn these fatty acids for fuel, whereas the liver will convert them to ketone bodies.

Adipose tissue is organized into discreet deposits, effectively organs or parts of organs, termed fat pads. There are eight major fat deposits in the body. As anyone who has tried to gain or lose weight knows, not all fat pads are metabolically equivalent; they burn or store fat at different rates. It may be easier, for example, to lose weight in your arms and legs compared to the abdomen. Studies of how individual fat pads respond to metabolic stimuli are the focus of research.

The blood transports metabolites

The blood serves as a transport system for many of the metabolites discussed, including glucose, free fatty acids, pyruvate, lactate, alanine, glycerol, and ketone bodies. Although many molecules are transported in the blood, others cannot be transported in this way. For example, ATP and glycogen are isolated in the cell; also, once formed inside an adipocyte or **myocyte**, triacylglycerol is not secreted back into blood; the liver secretes triacylglycerols in lipoprotein complexes.

Blood cells use glucose as their main fuel. The phagocytic cells of the immune system, such as macrophages, require additional NADPH formed in the pentose phosphate pathway to generate reactive oxygen species.

The kidney secretes toxic metabolites but can also carry out gluconeogenesis

The kidney is important for the excretion of toxic metabolites from the body. In the context of metabolism, the main focus is on the metabolite urea. Although the kidney can make some urea and eliminate some nitrogen directly, most urea is made in the liver and transported to the kidney via the blood. Due to the large chemical gradients used in filtering the plasma, the kidney has a high need for energy; in the resting state, it consumes 10% of the body's energetic need.

During liver failure or in extreme states of starvation, the liver has a limited ability to produce glucose via gluconeogenesis. In such situations, the kidney may be able to provide as much as half of the gluconeogenesis needed. Glutamine from muscle tissue is transported to the kidney, where it can undergo oxidative deamination, first to glutamate and then to α-ketoglutarate. The ammonia groups released in these reactions are not recaptured as urea but rather are directly secreted into the urine as ammonia. The α-ketoglutarate formed is anaplerotic and enters the citric acid cycle; from there it can participate in gluconeogenesis via oxaloacetate.

The brain is unable to store energy

The brain plays two roles in metabolism: Like all tissues, it has a metabolic need; it also regulates both appetite and metabolism (discussed in section 12.3).

The brain is the least self-sufficient of any of the organs; it requires a constant supply of energy and cannot survive for more than a few minutes without it. Two other factors work against the brain. First, almost the entire brain is important and cannot be scavenged for energy. Whereas we can lose 50% or more of our muscle mass and still move around unhindered, the same is not true of neurological function; any loss of brain or nervous tissue has significant consequences.

Second, the brain is protected and separated from the rest of the body by the blood-brain barrier, a semipermeable membrane that protects the central nervous system. This barrier is permeable to glucose and many amino acids but not to free fatty acids. One further metabolic complication is that the brain uses several amino acids as signaling molecules. As discussed in section 11.3, accumulation of ammonia in the brain can lead to depletion of α-ketoglutarate, blocking energy production in the neuron and leading to the death of the cell. For all of these reasons, glucose is the brain's fuel of choice. In the resting state, the brain needs about 120 g of glucose a day, roughly 20% of the body's resting energetic needs.

The brain's dependency on glucose could pose a significant problem for the organism. If the brain were dependent on glucose alone for fuel, the consequences for the organism would be severe if glycogen supplies were depleted, unless there was a proteinaceous tissue that could be catabolized to generate new glucose via gluconeogenesis. Nature has evolved a relatively simple and ingenious means for solving this problem. Instead of importing fatty acids, the liver converts fatty acids to ketone bodies. Once ketone bodies enter the cell, the reactions that were used to synthesize β-hydroxybutyrate are reversed, producing two molecules of acetyl-CoA that can be metabolized for energy by the brain.

12.2.2 The organism shifts between different metabolic states depending on access to food

Having discussed the various roles that organs play in metabolism, the next point to consider is the different metabolic states the organism is likely to encounter over its life span, such as when the organism is fed, when it is fasted, or when it may be starved (**Figure 12.4**). Some of these states are encountered several times a day, whereas others may never be experienced. The states discussed in this section are time dependent; that is, it takes time to eat food, digest it, and then become hungry again. They are discussed also in terms of both human and mouse metabolism because the mouse is the most commonly used model organism in metabolic studies. Other organisms may eat far less frequently or have other adaptations that help to compensate for dietary excess or need. For example, reptiles may eat as infrequently as once a month, whereas hibernating animals store large amounts of energy between spring and fall, then burn those calories off during their winter hibernation.

In each of the states considered, we will assume that the organism has consumed in the diet a combination of proteins, carbohydrates, and lipids. In metabolic studies, this is known as a **mixed meal**, which typically consists of some defined mixture; for example, 20% of calories from fats, 20% from proteins, and 60% from carbohydrates (**Figure 12.5**). In what is known as

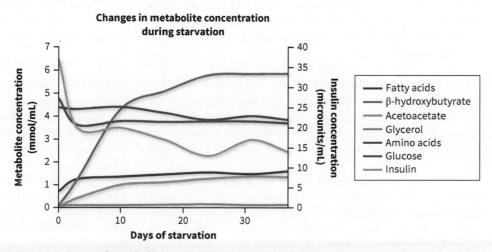

FIGURE 12.4 States of metabolism. As starvation progresses, plasma glucose levels drop and plateau as protein is spared. When this happens, levels of ketone bodies rise and replace some of the need generated by the absence of glucose. All concentrations are measured using the left-hand axis, with the exception of insulin, which uses the right-hand axis.

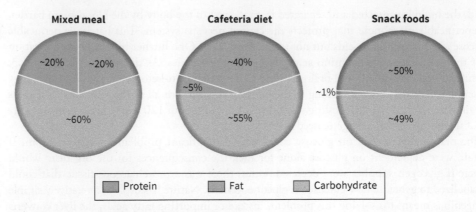

FIGURE 12.5 Compositions of different diets.   An average diet in a mixed meal may consist of 60% calories from carbohydrate, 20% of calories from protein, and 20% of calories from fat. A cafeteria diet may consist of 55% calories from carbohydrates, 5% of calories from protein, and 40% calories from fat. Some snack foods have less than 1% calories from protein, 50% of calories from fat, and the remainder from carbohydrates. In many instances these carbohydrates come from simple sugars.

the **cafeteria diet**, there is a high concentration of saturated fats and simple sugars, as one might expect to find in a cafeteria; in this case, the figures are 40% of calories from fat, 5% from protein, and 55% from simple sugars. Variants of the cafeteria diet are used to study obesity and heart disease. In a diet filled with snack foods, people consume large quantities of lipids and carbohydrates with minimal protein.

People often undertake diets to lose weight or for other health reasons. These can range from moderate caloric restriction plans to severe limitations on some nutrients, such as diets low in sodium or cholesterol, to the absolutely bizarre diet regimens. People may lose a few pounds consuming nothing but cabbage soup, grapefruit, or boiled kelp for a few weeks, but such diets are certainly not healthy. Moreover, restriction plans rarely produce meaningful weight loss one year after they cease. The only weight loss systems that have proven efficacious are those that teach people to consume healthy food in moderation and to combine that plan with a reasonable daily exercise program.

In recent years, high protein–low carbohydrate diets have become popular. These diets allow consumption of high amounts of fat and protein but without carbohydrate. Part of the notion behind these diets is that consuming protein and fat provides a feeling of fullness, so the person then eats less. Biochemically, a diet low in carbohydrates forces the body to burn glycogen quickly, resulting in rapid weight loss; this is followed by a period of **ketosis**, presumably to mobilize the body's fat stores. Undoubtedly, some of these diets help people to lose weight, but it is unclear how the weight loss is achieved. Most studies find that people tire of eating steak with butter at every meal, and simply eat less. The long-term health ramifications of eating a high-fat, high-protein diet with the elevated dietary cholesterol and sodium levels such a diet typically entails is still being debated.

Fed and postprandial states occur immediately after eating   As its name suggests, the **fed state** is one in which the organism has just finished feeding and has food in its stomach. An overlapping state is the **postprandial state**, that is, the periods during which metabolites are absorbed from the small intestine. These terms are often used interchangeably; clearly, they have a high degree of similarity and can be difficult to delineate outside a laboratory setting.

In the fed and postprandial states, amylases have begun to break dietary polysaccharides (starches, amylose, amylopectin) into monosaccharides. Proteases have begun to cleave proteins into amino acids or dipeptides and tripeptides. As the partially digested material (known as chyme) leaves the stomach, it encounters other proteases and lipases. These molecules are absorbed by the body and enter the circulation. Glucose and amino acids enter the blood, while dietary lipids are processed into chylomicrons and secreted first into the lymph and then from there into the bloodstream. This allows lipids to bypass exposure to the liver on their first pass through the circulation.

TABLE 12.2 Metabolic States

State	Fuel
Fed	Mixed from diet
Postprandial	Fatty acids made from dietary sources in the liver; glucose
Postabsorptive	Fatty acids from adipose tissue; glucose from glycogen stores
Re-fed	Mixed from diet; use of liver gluconeogenesis to replenish glycogen
Fasted	Carbohydrates depleted, fatty acids, ketone bodies, gluconeogenesis
Starved	Ketone bodies and fatty acids, limited gluconeogenesis

In the fed and postprandial states, elevated levels of glucose stimulate the pancreas to secrete the peptide hormone **insulin**. In the healthy organism, increased insulin elicits a signaling response in cells, causing more glucose transporters to move to the cell surface. This allows cells to absorb more glucose and either burn it or store it for future use (**Table 12.2**).

In the fed state, lipids are transported through the circulation on chylomicrons. Lipids are exposed to lipoprotein lipase in the capillary beds in adipose tissue and muscle tissue. Apolipoprotein CII is found on the surface of the chylomicrons and acts as a cofactor, stimulating the lipase to degrade triacylglycerols in the chylomicrons and absorb the released fatty acids and monoglycerides. The remaining particle, a chylomicron remnant, is enriched in dietary cholesterol; it binds to a chylomicron-remnant receptor on the surface of the hepatocyte and is internalized by receptor-mediated endocytosis.

As cells absorb and become satiated with glucose, there are two ways to store the carbon atoms from the glucose: as glycogen or as fatty acids (after oxidation to acetyl-CoA). Dietary amino acids also cannot be stored in large quantities. Most dietary protein, over 90% in some diets, is catabolized into carbon skeletons and stored as fat, with the liver being the location for both of these pathways. Yet the liver is not adapted to storing large quantities of fat and will malfunction if it is forced to do so. Instead, the liver synthesizes triacylglycerols de novo from carbohydrates and amino acid skeletons, packages these triacylglycerols plus cholesterol esters into VLDL, and secretes these into the circulation. In the circulation, VLDL delivers triacylglycerols to adipose tissue and muscle tissue for storage or energy. The VLDL loses some of its triacylglycerol core, shrinks in size and increases in relative density, becoming an LDL particle. LDL can deliver cholesterol to tissues as needed by binding an LDL receptor and being internalized through receptor-mediated endocytosis.

Collectively, the fed and postprandial states constitute the **absorptive state**, that is, the state in which nutrients are absorbed by the body. The act of eating can take anything from a few minutes to several hours. In humans, the postprandial state is often defined as from 0 to 5 hours after eating a mixed meal, although the actual time depends on factors such as the composition of the diet. In other animals, these numbers may vary considerably, depending on the organism's biology. For example, an owl or a snake that swallows prey whole may have that meal in the stomach for much longer than is the case with someone eating a pizza. It is more difficult to determine when rodents, such as rats and mice, have recently eaten and are therefore in the postprandial state. In some studies, experiments with small rodents are carried out early in the morning, based on the assumption that these nocturnal animals have recently eaten. It is common practice in the laboratory to develop a feeding protocol in an attempt to dictate the feeding time for animals on a study.

The postabsorptive state follows digestion but precedes hunger
The postabsorptive state follows the postprandial one. In the former, there is a fall in glucose from dietary sources and postprandial lipids because these molecules become sequestered in tissues that will burn or store them. As plasma glucose is depleted, the pancreatic hormone **glucagon** signals for glycogen to be mobilized from the liver to help maintain euglycemia, that is, the normal concentration of plasma glucose of 70–120 mg/dl. In addition, epinephrine and other **catecholamines** signal adipose tissue to mobilize fatty acids from triacylglycerols for fuel. As the postabsorptive state continues, the liver begins to use glycerol from adipose tissue, and alanine and lactate from muscle, to produce new glucose by gluconeogenesis. Glucose produced via

gluconeogenesis and glucose liberated from glucose-1-phosphate via glycogenolysis are released from the liver and used by other tissues, predominantly the brain. In this state, insulin levels are low, and glucagon and epinephrine levels are high. This promotes gluconeogenesis in the liver, and the breakdown of glycogen and use of fatty acids as fuel by the liver and muscle.

As the postabsorptive state continues, the liver begins to use acetyl-CoA from fatty acids to produce ketone bodies. Elevated ketone levels are associated with disease states or weight loss, but ketone bodies are a normal part of metabolism and are found at low levels in plasma at all times. Ketone bodies increase each night as we sleep and fall when we eat breakfast.

Many humans in the developed world eat several meals each day and have snacks in between. Some researchers argue that this situation means that many people never fully enter a postabsorptive or fasted state.

The re-fed state describes a re-fueling state

When someone in the postabsorptive state has consumed a meal, the metabolism does not immediately return to the fed or postprandial state. As levels of glucose, amino acids, and dietary fats in chylomicrons increase, the liver remains refractory or resistant to the effects of glucose and insulin, allowing other tissues to "feed" first, replenish glycogen levels, and return to a glycolytic state. Indeed, the liver continues to undergo gluconeogenesis but uses this glucose for liver glycogenesis. As glucose levels increase, the liver again begins to respond and returns to a postprandial state, absorbing glucose and synthesizing fatty acids and triglycerides for VLDL.

In the fasted state, energy conservation begins

Somewhere between 12 and 48 hours after consuming a meal the body is in the **fasted state**, which is marked by a lack of dietary lipids in the circulation. Chylomicrons and their remnants are absent, VLDL concentration is low, and LDL and HDL predominate. Levels of free fatty acids from adipose tissue increase in the blood, while levels of fatty acids and triacylglycerols increase in the liver. Liver glycogen stores have been largely depleted, and the liver undergoes high levels of gluconeogenesis and ketone body production. In this state, the body begins to shift into a metabolic state where it can preserve itself for the longer term in the absence of other food sources. Liver, muscle, and heart tissues can all use fatty acids to meet their energetic needs; however, the brain also needs glucose due to the blood-brain barrier. The liver can use glycerol from adipose tissue or most amino acid skeletons as a source of carbon for gluconeogenesis, but using these fuels is not without consequence. Amino acids cannot be stored; therefore, any use of amino acids for fuel is observed as a loss of protein mass, and wasting begins.

The starved state attempts to preserve life

Somewhere between 24 and 48 hours after a meal, the body moves from the fasted to the **starved state**, altering metabolism to spare protein in muscle and use ketone bodies as the predominant fuel for the brain. This happens in part through upregulation of mitochondrial hydroxymethylglutaryl-CoA synthase (mitochondrial HMG-CoA synthase), the rate-determining step in ketone body biosynthesis. As the starved state progresses, the brain also switches from using glucose to using ketone bodies as its main source of fuel. This too helps spare muscle tissue because the metabolic demand of the brain drops from its normal requirement of 120 g of glucose a day to only 40 g in the starved state. The details of this metabolic switch are still under investigation (Table 12.3).

Theoretically, the starved state can last for some time. For example, a 100 kilogram individual with 28% body fat has about 28 kilos of triacylglycerol available to burn. In the absence of other food, this fuel could last for 155 days, or slightly over 5 months, presuming the burning of 180 g a day. When triacylglycerol stores are fully depleted, the individual begins to burn whatever fuels remain, demyelinating nerves and damaging organs as protein is catabolized from the heart, liver, and kidneys. By this point, even if the individual returns to a normal diet, damage can persist, sometimes permanently. However, this situation rarely occurs because someone undergoing prolonged starvation usually succumbs to an opportunistic infection due to a weakened immune system, or to heart failure precipitated by a lack of potassium in the diet (**Figure 12.6**).

TABLE 12.3 Fuel Metabolism in Starvation

Fuel	Fuel Consumption, g/24 h	
	Day 3	Day 40
Fuel use by brain		
Glucose	120	40
Ketone bodies	50	100
Glucose use by other tissues	50	40
Fuel mobilization		
Adipose tissue lipolysis	180	180
Muscle protein degradation	75	20
Fuel output by the liver		
Glucose	150	80
Ketone bodies	150	150

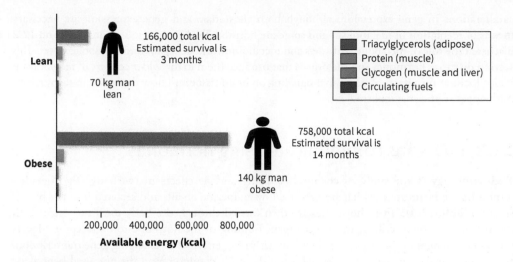

FIGURE 12.6 Energy reserves in normal and obese states. Shown are the energetic reserves of a 70 kg and a 140 kg man. In the obese state, more than four times as many calories are stored, and the majority of them are stored as triacylglycerols.

Worked Problem 12.2   I could eat …

Think about the last thing you had to eat and when you ate it. What metabolic state are you currently in? Are your levels of amino acids, glucose, dietary fats, fatty acids, and ketone bodies likely to be elevated or depressed?

Strategy Examine what you have eaten over the past 12 hours and at what times. How does your food intake compare with the metabolic states discussed?

Solution Derek, a student taking biochemistry, skipped his breakfast this morning due to an early class. For lunch, about two hours ago he ate an apple and two granola bars. This meal was relatively low in protein but mixed in terms of carbohydrate and fat content. Given that he was coming off what amounts to a short (approximately 14 hour) fast, the carbohydrates that entered his circulation probably went first to replenishing muscle glycogen. However, since it has been two hours since he last ate, any remaining carbohydrates have probably been taken up by his liver and either stored as glycogen or transformed into triacylglycerols. So at this point, he probably has low levels of ketone bodies and fatty acids in his blood and moderate levels of glucose, amino acids, and VLDL.

Follow-up question Presuming you eat nothing more for the next 24 hours, how will the levels of metabolites change?

Summary

- Different organs have evolved specialized functions and roles in metabolism.
- Some organs store metabolites: Liver and muscle tissues store glycogen, and adipose tissue stores triglycerides; muscle can serve as a source of amino acids in times of energetic need.
- Some organs have specialized metabolic needs and cannot store energy. The brain requires a constant supply of energy in the form of glucose or ketone bodies.
- The organism passes through multiple metabolic states as it goes through a day, year, or lifetime.
- Specific biochemical adaptations have evolved to help the organism survive in the absence of a constant flow of nutrients.

Concept Check

1. Describe the roles each organ plays: explain which organs act as energy storehouses and for which molecules, and which molecules are released and where they go.
2. Explain what metabolic process occurs in each state from a biochemical perspective, when these switches in metabolism occur, and how the alterations in metabolism help the organism survive in the absence of food.
3. What biochemical adaptations take place during starvation or in absence of constant flow of nutrients?

12.3 Communication between Organs

As we have seen recurrently throughout metabolism, signaling molecules or hormones and signaling cascades, especially ones involving kinases, are prevalent in biochemistry. These types of regulatory mechanisms are well-adapted to balance metabolism in the short term. Signaling molecules do not offer the acute regulation of allosterism, nor do they take as long to effect change as alterations in gene expression, although both allosterism and gene expression are necessary means of regulation of metabolism and some are hormonally dependent. Sections 12.1 and 12.2 discussed the role of metabolic pathways and metabolites, and of organs and metabolic states. This section discusses hormones and the organs that produce them in the wider context of metabolism; it also looks at the effects of hormonal signaling on brain tissue and how the brain, in turn, affects both appetite and metabolism.

12.3.1 Organs communicate using hormones

Endocrinology is the study of chemical signaling and its effects in the body. The signaling molecules are **hormones**, which are produced by endocrine organs and secreted into the bloodstream (**Table 12.4**). These hormones are then carried through the body until they meet with the receptor on or in cells of the target organ. Binding of the hormone to its receptor triggers metabolic changes in the target organ such as activating enzymes or changing gene transcription. Hormones can be categorized in several ways; the most common, and the one used here, is by structure. These include peptide hormones, hormones derived from tyrosine, and steroid hormones. (Source: Data from PDB ID 1GCN Sasaki, K., Dockerill, S., Adamiak, D.A., Tickle, I.J., Blundell, T. (1975) X-ray analysis of glucagon and its relationship to receptor binding. *Nature* 257: 751–757)

Glucagon
(peptide hormone)

Thyroxine
(derived from tyrosine)

Epinephrine
(derived from tyrosine)

Estrogen
(steroid hormone)

Peptide hormones   **Peptide hormones** can be as few as eight amino acid residues in length, but others are on the order of small proteins. For example, human insulin is 51 amino acids long. Peptide hormones are often synthesized from an mRNA message as part of a larger protein, known as a prohormone. The prohormone is cleaved in the endoplasmic reticulum to generate the active peptide, which is then secreted. Peptide hormones are important in the regulation of most physiology, and they are among the most powerful signaling molecules. Insulin and glucagon are examples of peptide hormones discussed in previous chapters, but other peptide hormones important in metabolism include **adiponectin**, cholecystokinin (also known as CCK-PZ), gastrin, **ghrelin**, **leptin**, and **PYY(3-36)**. Neuropeptides (short peptides that act as neurotransmitters, signaling between neurons) are synthesized and function in a similar manner. The list of peptide hormones continues to grow.

TABLE 12.4 Hormones: Where They Are Made and How They Elicit Responses

Hormone	Source	Released by	Target Tissue	Receptor Type	Effect
T3 and T4	Thyroid	Thyroid stimulating hormone (TSH)	All	Nuclear receptor (transcription factor)	Gene expression (increase metabolic rate)
Insulin	Pancreas	High glucose	All	Receptor tyrosine kinase	Glucose uptake, gene expression
Glucagon		Low glucose	Liver	7 TM GPCR	Energy mobilization from stores
Catecholamines (dopamine, epinephrine, norepinephrine)	Adrenals	Nervous inputs to adrenal, stress	All	7 TM GPCR	Energy mobilization from stores
Corticosteroids (minerals and glucocorticoids)		Adrenocorticotropic hormone, stress	All	Nuclear receptor (transcription factor)	Energy mobilization (gluconeogenesis) Water reabsorption from the kidney
Leptin	Adipose tissue			Receptor tyrosine kinase	Satiety
Adiponectin		Under investigation	Hypothalamus/ arcuate		
Ghrelin					Appetite
GLP-1					Satiety
PYY(3-36)	Gut				Appetite
Cholecystokinin			Pancreas, gut, brain	7 TM GPCR	Stomach emptying, appetite
Vasopressin	Posterior pituitary	Plasma osmolarity	Kidney		Water reabsorbtion in the kidney
Oxytocin		Innervation of pituitary	Brain		Human bonding

Catecholamines and thyroid hormones are derived from tyrosine

The second class of hormones can be broadly categorized as tyrosine derivatives; they include the catecholamines and thyroid hormones.

The catecholamines **epinephrine**, dopamine, and **norepinephrine** are small organic molecules that are named because of their structural similarity to catechol (1,2 dihydroxybenzene). The catecholamines are synthesized in a multistep pathway (**Figure 12.7**).

First, tyrosine is hydroxylated to L-Dihydroxyphenylalanine (L-DOPA) in a reaction catalyzed by tyrosine hydroxylase. This is the rate-limiting step in catecholamine biosynthesis and is dependent on a tetrahydrobiopterin cofactor for activity. Molecular oxygen is the source of the oxygen in the hydroxyl group. The L-DOPA is decarboxylated by L-DOPA decarboxylase, leaving dopamine. As with many enzymes that catalyze reactions centered on the α carbon of amino acids, this enzyme is dependent on pyridoxal phosphate (PLP).

In the next step, dopamine is hydroxylated to yield norepinephrine. While the hydroxylations discussed in previous chapters were the product of the hydration of a double bond, here, the benzylic position from the aromatic ring forms a more stable carbocation and is directly hydroxylated by dopamine β-hydroxylase. Again, molecular oxygen is required. In addition, this enzyme uses ascorbic acid and employs a similar mechanism to proline hydroxylase, although dopamine hydroxylase employs an active-site copper ion instead of iron and succinate.

Norepinephrine undergoes *N*-Methylation to form epinephrine, in a reaction catalyzed by phenylethanolamine *N*-Methyltransferase, with *S*-Adenosylmethionine as the source of the methyl group.

The **thyroid hormones**, including triiodothyronine (T3) and thyroxine (T4), are also derived from tyrosine, although their pathways differ markedly from that of catecholamine formation. Thyroid hormone biosynthesis begins with the synthesis of the protein thyroglobulin, a large (669 kDa) protein, in the endoplasmic reticulum. Thyroglobulin is processed in the thyroid and undergoes substitution reactions on the aromatic tyrosine rings of thyroglobulin, resulting in monoiodotyrosine or diiodotyrosine.

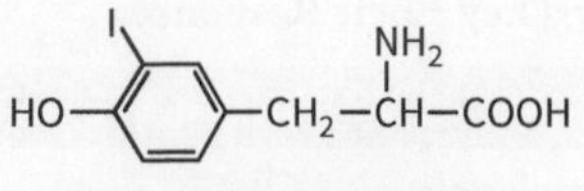

Monoiodotyrosine (MIT)

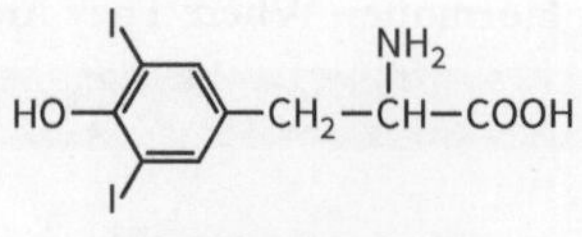

Diiodotyrosine (DIT)

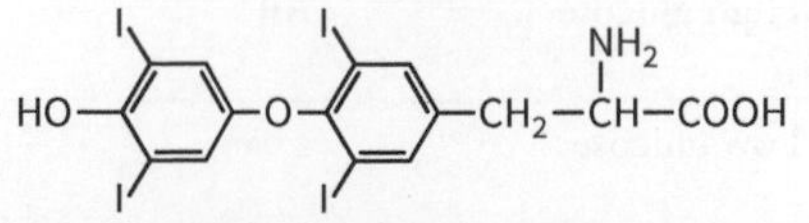

3,5,3′,5′-tetraiodothyronine (L-Thyroxine; T4)

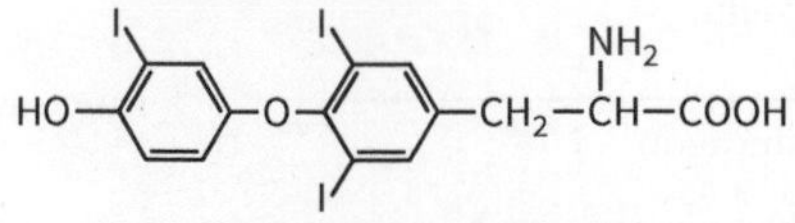

3,5,3′-triiodothyronine (T3)

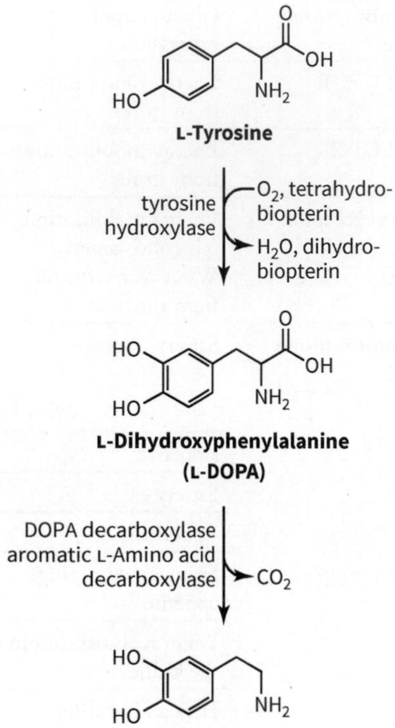

FIGURE 12.7 Biosynthesis of catecholamines. Catecholamines are synthesized in a cascade. First, tyrosine is hydroxylated into L-DOPA. DOPA is decarboxylated into dopamine. Hydroxylation of dopamine yields norepinephrine. *N*-Methylation of norepinephrine gives the final product of the pathway, epinephrine.

Once derivatized, the iodinated tyrosyl side chains undergo a conjugation reaction, forming dimers that are still attached to thyroglobulin. The modified thyroglobulin is endocytosed by follicular cells in the thyroid. In the lysosome, it undergoes proteolytic cleavage to release T4 and some T3. After release from the follicular cell, these hormones travel to other tissues, where they are further modified, either by decarboxylation or by deiodination catalyzed by deiodinases, a family of selenium-dependent enzymes. These other forms of thyroid hormone may be more active or have more distinct effects in different cells and tissues of the body.

Thyroid hormones act through nuclear hormone receptors, a family of proteins that includes steroid receptors. These proteins are found inside the cell where they bind to the hormone and translocate to the nucleus to change gene expression.

Steroid hormones are derived from cholesterol
Steroid hormones, such as **cortisol** and the sex hormones, are hydrophobic hormones that are derived from cholesterol and made in gonadal tissue or the adrenal glands. Like thyroid hormones, steroids act through nuclear hormone receptors, proteins that bind to the hormone and act as transcription factors, to alter gene expression. Because this process is lengthier, involving RNA and protein synthesis, steroid hormones typically function over longer periods of time—hours to days rather than seconds to minutes (**Figure 12.8**).

Hormones have distinct mechanisms of action
Both the catecholamines and the thyroid hormones affect almost every cell of the body. The catecholamines act primarily through transmembrane receptors. Despite being mainly hydrophobic, thyroid hormones are not membrane permeable because the iodo groups on the aromatic rings are sufficiently electron-withdrawing to lower the pK_a of the phenolic proton, leaving the hormone with two negative charges at physiological pH. Instead, there are hormone transporters that usher T3 and T4 into the cytosol. Thyroid hormones bind to specific receptors in the cytosol that translocate to the nucleus and act as transcription factors, influencing gene expression. Steroid hormones and other hydrophobic signaling molecules also act via nuclear hormone receptors.

Organs produce and respond to specific hormonal signals
Endocrinology focuses on the endocrine organs, including the pancreas, thyroid, adrenals, gonads, and pituitary; however, it is now generally accepted that most, if not all, organs have endocrine function and can secrete hormonal signals. The brain, pancreas, adrenal glands, thyroid, gut, and adipose tissue are all important in the regulation of metabolism. The function of these organs and the hormones that they secrete are found in **Table 12.5**.

12.3.2 Hormonal signals to the brain regulate appetite and metabolism

Advances in neuroscience and biochemistry have shown that a few select regions of the hypothalamus regulate the intersection of metabolism, appetite, and complex behaviors such as feeding.

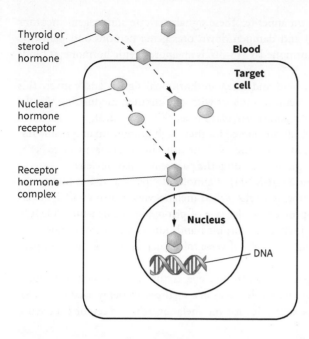

FIGURE 12.8 Action of thyroid and steroid hormones. Thyroid and steroid hormones work by binding to a nuclear hormone receptor in the cytosol. Binding of the hormone to the receptor activates the receptor, facilitating translocation to the nucleus, binding of DNA, and activation of specific genes.

TABLE 12.5 Hormonal Disorders

Disease	Hormone Affected	Symptoms
Diabetes insipidus	Vasopressin	Mutation of vasopressin receptor leads to less sodium reabsorbtion in the kidney and excessive urination.
Diabetes mellitus	Insulin	Signaling defect results in inability to properly take glucose into cells.
Addison's disease	Corticosteroids	Low levels of corticosteroids leads to mineral imbalance and hypoglycemia.
Cushing's disease	Adrenocorticotropic hormone (ACTH)	Increased levels of ACTH leads to elevated corticosteroids and weight gain.
Hashimoto's disease	Thyroid hormones	Autoimmune attack on thyroid protein(s) results in depressed thyroid hormone levels and weight gain.
Graves' disease	Thyroid hormones	Elevated thyroid hormones result in weight loss and increase in metabolic rate.

The hypothalamus coordinates hormonal signals and neuronal responses

The **hypothalamus** is a roughly spherical organ, about 1 cm in diameter, with a notch on the bottom side that divides the organ into two symmetric halves. The notch is the third ventricle, a fluid-filled cavity that runs through the midline of the brain. The regions of the hypothalamus lining the third ventricle are the sites of action of many of the hormones discussed here, and the blood-brain barrier is more porous in this region, permitting passage of these hormones. This section focuses on two regions of the hypothalamus: the arcuate nucleus and the paraventricular nucleus.

The **arcuate nucleus** is a cluster of cells less than a millimeter across. In the 1950s it was discovered that lesions of the brain in the region of the arcuate nucleus led to voracious appetite; hence, it was hypothesized that this tiny region of the brain integrated or controlled appetite. It contains two types of neurons pertinent to this discussion. These neurons are broadly classified by the types of signals they release: one class of neuron releases **neuropeptide Y (NPY)**, agouti-related peptide (AgRP), and γ-amino butyric acid (GABA); the other neuron releases the neuro-peptides pro-opiomelanocortin (POMC) and cocaine-amphetamine related transcript (CART). The NPY/AgRP/GABA neurons are **orexigenic**; that is, they promote appetite. The POMC/CART neurons are **anorexigenic**; they blunt appetite. These different classes of neuron have different receptors to receive communications from the rest of the body. Both classes receive signals from leptin and insulin, which blunt appetite. These signals activate anorexigenic neurons and pathways (POMC/CART neurons) and dampen down orexigenic neurons and pathways (NPY/AgRP/GABA neurons). The gut hormone PYY(3-36) is an anorexigenic hormone; it binds and inhibits NPY/AgRP/GABA neurons.

The neurons of the arcuate nucleus bind and integrate hormonal signals and convert this information into neuronal signals. These neurons interact with one another, modulating signals; they also interact with other neurons in the paraventricular nucleus (**Figure 12.9**).

The neurotransmitters of the NPY family are named for their high proportion of tyrosine residues, as many as five in a 36-amino-acid-long peptide. NPY neurons in the arcuate secrete NPY, which signals to neurons in a variety of regions, including the paraventricular nucleus.

Alpha-melanocyte stimulating hormone (α-MSH) is a molecule that was discovered for its role in the stimulation of melanin in pigmented cells, called melanocytes. It is one of eight different signaling molecules derived from proteolytic cleavage of pro-opiomelanocortin (POMC). Other molecules derived from POMC include endorphins and adrenocorticotropic hormone (ACTH). The products of pro-opiomelanocortin play diverse roles, such as the regulation of pain and blood pressure.

The receptor for α-MSH in the paraventricular nucleus is the melanocortin-4 receptor (MC4R). This receptor is interesting due to its involvement in appetite and energy balance, as well as in reproductive physiology. Male knockout mice for the melanocortin-4 receptor experience

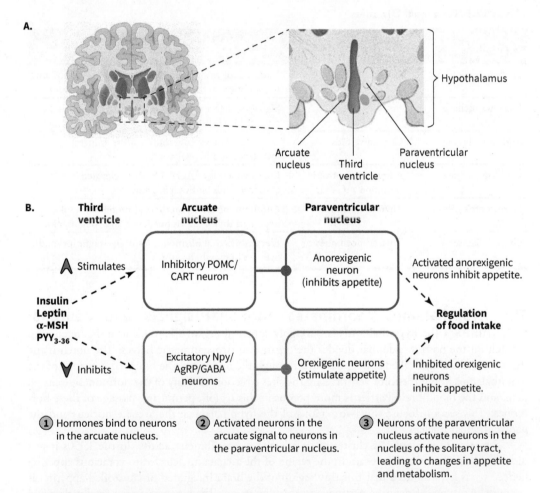

FIGURE 12.9 Innervation of the arcuate nucleus. **A.** A detailed representation of the communications between the third ventricle, arcuate, and the periventricular nucleus is shown. **B.** The neurons of the arcuate respond to leptin and insulin and produce either the neurotransmitters NPY/AgRP/GABA or POMC and CART. NPY/AgRP/GABA neurons are orexigenic and signal to eat more. POMC/CART neurons are anorexigenic and blunt appetite.

sexual dysfunction (decreased motivation and decreased copulatory performance) compared to wild-type siblings. Melanocortin-4 receptor agonists, either drugs or hormones, cause the opposite effect and have been tested with favorable results in both men and women with sexual dysfunction. Therefore, any drug developed to regulate appetite via melanocortin-4 receptor may also affect sexual behavior, and vice versa.

Neurons in other areas of the hypothalamus also are innervated by the arcuate nucleus. Signals from these are thought to regulate appetite and energy balance.

Although this section focuses on the hormonal basis of feeding and energy balance, it is only one part of a complex story. Humans eat for a variety of reasons. It is probably fair to say that many people often eat even though they are not hungry. Eating is usually pleasurable and, as such, has a large hedonic component. For example, we eat hot dogs at sporting events and popcorn at the movies, and we feast on holidays, often simply because of culture and habit. The mere mention of a plate of warm chocolate chip cookies or pizza can stimulate the appetite of someone who would not otherwise be hungry. Clearly, there are other signals that collectively weigh in to alter the decision to eat or not, for example, visual signals, olfactory signals, memories of pleasurable meals past and emotional states. Interestingly, the opposite is also true. Someone who becomes violently ill or suffers with a bout of food poisoning carries an aversion to those foods in her or his memory. The sight, smell, or even the mere mention of them can cause a physiological response, making someone feel ill at the thought of consuming them again.

12.3.3 Transcription factors and histone acetylases and deacetylases regulate metabolism in the longer term

We have discussed acute regulation of metabolism via allosteric control and modifications of enzymes, but alterations in gene expression are also important in the regulation of metabolism. The sirtuins, a group of seven proteins that can act as deacetylases, regulate genes by modifying histones and may directly affect metabolic pathways by deacetylating enzymes.

Hormones regulate expression of genes controlling metabolism

Hormones regulate gene expression through several different mechanisms. This section discusses two of these mechanisms, nuclear hormone receptors and resident nuclear transcription factors, and how they respond to hormonal signals.

Nuclear hormone receptors are small proteins found in the cytosol or nucleus of the cell. Upon binding to a ligand, they become active and translocate to specific DNA sequences, where they bind and activate gene transcription. These ligands are largely hydrophobic: steroid and thyroid hormones. Often, the hormone–receptor complexes dimerize with other nuclear hormone receptors to fine-tune gene activation. Therefore, binding of the hormone to the receptor causes activation, resulting in a receptor that can bind to specific DNA sequences and activate specific genes.

The cyclic AMP response element binding protein (CREB) is an example of a resident nuclear transcription factor (**Figure 12.10**). This transcription factor resides permanently in the nucleus but is activated by phosphorylation. The phosphorylation of CREB may be catalyzed by several different enzymes, but protein kinase A (PKA) was one of the first enzymes identified. Catecholamines or glucagon act through a 7-transmembrane helix receptor coupled to a heterotrimeric G protein to activate PKA resulting in phosphorylation of CREB. Once phosphorylated, CREB dimerizes with either another copy of itself or with a related transcription factor; it then binds to the cyclic AMP response element, a short stretch of DNA that is found in the promoter region of cAMP responsive genes. Binding of the transcription factor complex initiates binding of other proteins, including coactivators and histone-modifying enzymes, all of which act to alter the levels of gene expression.

Sirtuins broadly regulate gene expression through histone modification

Sirtuins are found in organisms ranging from bacteria to humans and may regulate the expression of genes involved in metabolism and aging. Originally identified in yeast, these proteins have an NAD^+-dependent histone deacetylase activity, removing acetyl groups from lysine and increasing the interaction of histone proteins with DNA (**Figure 12.11**). This activity results in the breakdown of NAD^+ into nicotinamide and O-Acetyl-ADP ribose. Nicotinamide is an inhibitor of sirtuin activity.

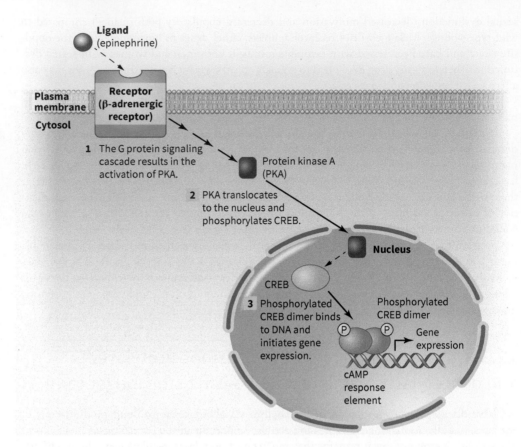

FIGURE 12.10 Action of CREB. CREB is the cyclic AMP response element binding protein. This resident nuclear transcription factor responds to PKA phosphorylation to bind a cyclic AMP response element in promoter regions of some genes, activating transcription.

The sirtuin story is not a typical one from a scientific perspective. It has been known for decades that, in many organisms, severe caloric restriction results in dramatically increased life span, but for many years the causal factor was unknown. At the same time scientists were fascinated with the French paradox; that is, the observation that people in some parts of France live longer in spite of having greater risk factors for premature mortality. It was unclear what aspect of the French diet or lifestyle was responsible for this increased longevity. The answer in both cases has been linked to the sirtuins, but the results of many of these studies are controversial. Overexpression of sirtuins results in increased life span in several model organisms. In addition, resveratrol, a component of red wine, seems to activate sirtuins. The apparent benefits of sirtuins in relation to longevity made them desirable targets for the treatment of several diseases found in aging, such as type 2 diabetes, cancer, and neurodegenerative disorders. Again controversies arose. The effects of the sirtuins are not the same in all organisms and, more important, when transgenic organisms with increased life span were bred into a common background, in some instances the observed benefit was lost; that is, life spans returned to normal.

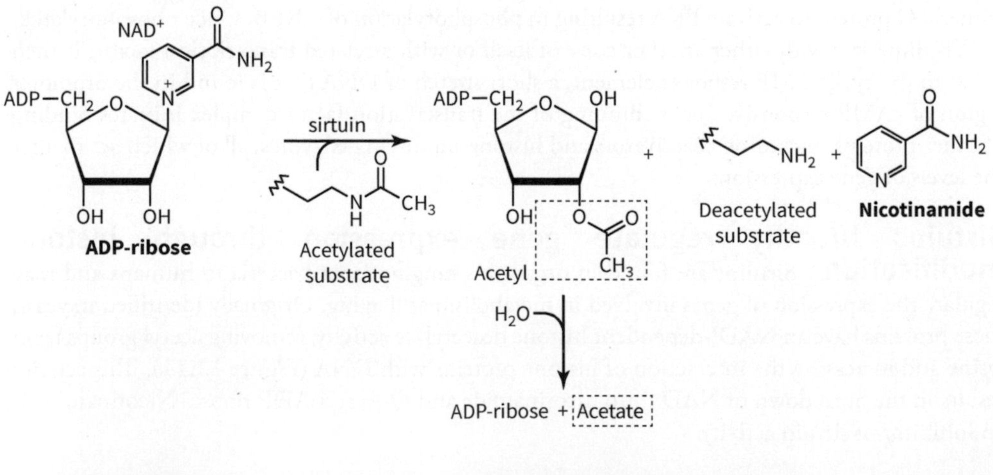

FIGURE 12.11 Sirtuins reactions. The sirtuins are a family of NAD^+-dependent deacetylases. These proteins have been implicated in many cellular processes including the regulation of metabolism and aging. Sirtuins deacetylate lysines generating nicotinamide, and 2'-O-Acetyl-ADP-ribose (OAADPr). This reaction begins with binding of acetyllysine, then NAD^+. Nicotinamide is released followed by deacetylated lysine and OAADPr.

The sirtuin story is still unfolding. Clearly sirtuins have a role in the cell. One of the more interesting observations is that the sirtuins deacetylate many proteins, potentially preserving function in the event that a protein has been acetylated accidentally. This includes several enzymes in metabolic pathways we have studied including acetyl-CoA synthetase, long-chain acyl-CoA dehydrogenase, and 3-hydroxy-3-methylglutaryl-CoA synthase 2. Given that several of the sirtuins are found in the mitochondrial matrix, where there are high levels of acetyl-CoA (an acetyl group donor), it may be that at least one of these proteins serves to maintain and protect mitochondrial proteins. Further research will be needed to clarify the role of the sirtuins (if any) in aging and longevity.

Worked Problem 12.3   Obese mice

There are six different naturally occurring obese mouse models. Two of these are the obese (ob/ob) and diabetic (db/db) phenotypes. One of these mice has a defect in leptin production; the other has a defect in the leptin receptor. Propose a simple experiment to determine which mouse is which.

Strategy Leptin is a soluble proteinaceous signaling molecule synthesized in adipose tissue. It acts in the brain to regulate satiety and metabolism. If you had access to a supply of leptin or you were able to synthesize this hormone, how might this factor into your experiment? What would be valid controls for the experiment?

Solution One simple means of testing this hypothesis would be to inject both animals with leptin. The mice that were defective in leptin production should show a decrease in appetite and in weight because the leptin would replace the hormone that was naturally absent. On the other hand, mice deficient in the leptin receptor would be unaffected by injections of leptin.

There are several problems that can be encountered in an experiment such as this. Extensive preliminary experiments would be needed to determine the time it takes for the hormone to act, what level of hormone to inject, and the time taken for the body to eliminate the hormone. When it was first proposed that there was a hormone that regulated satiety in these mice, the factor itself was unknown. A clever, but somewhat disturbing, experiment used to check this was a parabiosis study: two mice, one lacking the hormone and one lacking the receptor, were surgically conjoined so that they had a common circulation. The mouse lacking the ability to make leptin lost weight while the other remained unchanged.

Follow-up question It has been proposed that in some people leptin injections could help with weight loss. Based on what you know of other metabolic diseases and how other hormones signal, discuss this proposal. Does it have merit? What are the strengths and weaknesses of this therapy? Are there ethical ramifications of such a treatment?

Summary

- The hormones discussed in this chapter can broadly be divided structurally into peptides, tyrosine derivatives, and steroids.
- Specific organs synthesize, secrete, and respond to these hormonal signals.
- Some organs such as the pancreas, thyroid, and adrenals are classically thought of as endocrine organs, but more recently it has been found that most tissues secrete some type of signaling molecule; these molecules include adiponectin and leptin from adipose tissue, and others from the gut.
- Many hormones signal to the brain to regulate appetite and metabolism. These hormones act through the hypothalamus.
- Hormones from the circulation act in the arcuate nucleus, a region of the brain that receives hormonal signals, that is, ghrelin, insulin, leptin, and PYY(3-36), and transmits signals to other parts of the hypothalamus, specifically, the paraventricular nucleus. From the paraventricular nucleus, the arcuate nucleus signals are processed in other parts of the brain to regulate complex behaviors such as appetite and satiety.
- Metabolism is regulated in the longer term by transcription factors and histone-modifying proteins such as the sirtuins.

Concept Check

1. Describe differences among the chemical signals discussed and the ways in which they are synthesized.
2. Describe where signals such as catecholamines and peptide hormones are made and where they act in the body.
3. Where are the hormones leptin and PYY3-36 released from and where are their receptors located? What kind of signals are sent by them?

12.4 Metabolic Disease

Homeostasis represents a metabolic ideal. By its very nature, metabolism is a study in energy balance. Concentrations of nutrients and signaling molecules are not fixed; they vary from hour to hour, day to day, season to season, and year to year. To prosper, the organism has to be able to cope and change with these fluctuations. Unfortunately, organisms cannot always cope with changing conditions. Hunger and famine have always plagued humankind and were one of the leading killers in the twentieth century. Curiously, in Western society, famine has now been replaced by diseases stemming from nutritional excess. Many modern practices result in people living in a state outside homeostasis.

This section starts with a discussion of overnutrition (consumption of too many calories), obesity, diabetes, and metabolic syndrome. It also explores diseases that stem from a lack of nutrition or an inability to properly regulate metabolic pathways, conditions seen in acquired immunodeficiency syndrome (AIDS) and some cancers. The section then discusses metabolic changes resulting from alcohol consumption and concludes by considering several metabolic states that are important in biochemistry but are not truly disease states, such as the metabolic alterations that occur during exercise and hibernation.

12.4.1 Diseases of excess: obesity, diabetes, and metabolic syndrome

If food cost is represented as a percentage of personal income, the United States has the cheapest food supply of any country on Earth, and the easiest access (**Figure 12.12**). Consequently, a high percentage of people eat too much, although even in the United States, 5.4% of households, including 9.8% of households with children, suffer from hunger and do not get enough to eat each day. Unfortunately, access to an almost unlimited food supply, coupled with minimal physical effort in everyday life, differs greatly from the conditions that prevailed as we evolved, resulting in considerable health risks for people today.

Obesity is the most common disease in America   Body mass index (**BMI**) is an indicator of how overweight an individual is. It is calculated by dividing mass in kilograms by height in meters squared. An optimal BMI is in the range of 18.5 to 24.9. Currently, 32.3% of Americans are overweight, having a BMI of over 25 kg/m². A further 36.5% of Americans are considered obese (BMI > 30 kg/m²). Collectively, this means that 68.8% of Americans are in an unhealthy weight range.

BMI is not a perfect measure of how overweight someone is. It fails to take into consideration people of different builds or ethnicities and fails to consider the distribution of fat in the body

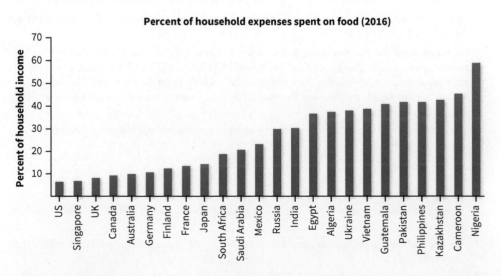

FIGURE 12.12 Percent of household income spent on food.   In 2016, Americans spent less on food as a percentage of household income than any other country in the world. (Source: From U.S. Department of Agriculture.)

(abdominal fat versus whole body fat). Nevertheless, BMI positively correlates with weight and is one measure that is used to determine obesity. Over the past years the number of people who are overweight or obese has exceeded.

Fat storage is an important evolutionary adaptation that allows us to survive in times when food is not available; however, an excess of fat storage can be problematic. Although obesity is not in itself a terminal disease, it is a major contributor to type 2 diabetes, heart disease, and some cancers.

Obesity can have hormonal causes, but it is generally the result of overnutrition and lack of exercise. The effect of obesity is significant. In the obese state, adipose tissue becomes leaky, releasing free fatty acids into the plasma in the absence of lipolytic stimuli. These elevated plasma fatty acids have been implicated in the development of type 2 diabetes. The influx of fatty acids to muscle cells shifts metabolism in that tissue toward β-oxidation and away from glycolysis. At the same time, the liver is also undergoing greater levels of β-oxidation and of gluconeogenesis due to elevated fatty acids. Collectively, this contributes to **hyperglycemia**, that is, elevated plasma glucose.

Obesity also leads to hormonal changes. In the obese state there are elevated levels of stress hormones, such as cortisol and epinephrine, as well as several pro-inflammatory cytokines, which are signaling molecules secreted by the immune system. At the same time, insulin levels are often elevated, but tissues are less responsive to these elevated levels, a prediabetic condition known as insulin resistance. Leptin levels are elevated, and adiponectin levels are decreased. For reasons that are not fully clear, hormones regulated by the hypothalamus are also aberrant. For example, thyroid releasing hormone (TRH) is often elevated. Paradoxically, levels of thyroid hormones may or may not be elevated, although hypothyroidism (low levels of thyroid hormones) is associated with obesity. Growth hormone levels are low because its release is blunted, whereas levels of both androgens and estrogens are elevated. Overall, this presents a rich and complex regulatory issue that underlies many of the problems and complications of obesity.

Obesity can be treated pharmacologically in several ways (**Figure 12.13**). These include the fat absorption inhibitor Orlistat and several antidepressants that act to blunt appetite. The drug metformin, an insulin sensitizer, increases the sensitivity and responsiveness of cells to insulin; it has been used in some instances to treat obesity, although it is more commonly used to treat diabetes. Newer experimental treatments include PYY(3-36), growth hormone, leptin, and various drugs that target the hypothalamus. These drugs may have the advantage of truly targeting metabolism rather than simply appetite, although there is always the potential for adverse effects.

Diabetes is a disease of insulin signaling Diabetes has been recognized as a disease since 1500 BCE. There are several different types of diabetes. These diseases as well as other hormonal disorders are discussed in Table 12.5. Diabetes insipidus results from an imbalance of vasopressin and thus the inability to reabsorb sodium from urine, which in turn causes a loss of sodium and water. The outcome causes copious urination and loss of electrolytes. Diabetes insipidus is unrelated to the other three forms of diabetes, which are types of **diabetes mellitus** that arise from complications in insulin signaling. It is diabetes mellitus that people typically think of

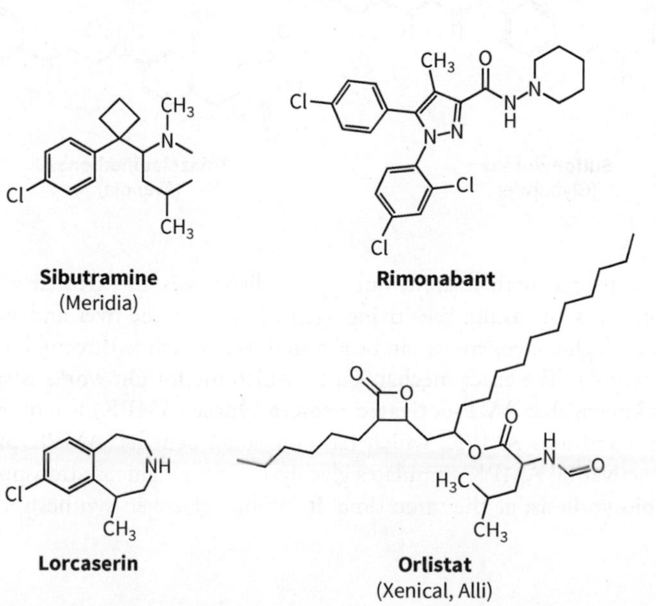

Sibutramine
(Meridia)

Rimonabant

Lorcaserin

Orlistat
(Xenical, Alli)

FIGURE 12.13 Drugs that target obesity. Drugs that target obesity can be categorized into those that block lipid absorption and those that suppress appetite. Several of these drugs have been withdrawn due to undesirable side effects. Sibutramine (Meridia) is an SSRI and appetite suppressant and was withdrawn in 2010. Rimonabant is a reverse agonist of the cannabinoid receptor CB1, an appetite suppressant, and was withdrawn in 2009. Lorcaserin is a 5-HT 2c receptor agonist and an appetite suppressant. Orlistat (Xenical, Alli) is a pancreatic lipase inhibitor and blocks fat absorption.

when they hear the term diabetes. Diabetes mellitus can be broadly divided into two categories: **type 1 diabetes mellitus**, previously known as juvenile diabetes, and **type 2 diabetes mellitus**, previously known as adult onset diabetes. A third form of the disease, gestational diabetes, occurs in pregnant mothers.

Type 1 diabetes mellitus is an autoimmune disease that typically strikes children and young adults between the ages of 12 and 21. The immune system of the patient recognizes the pancreatic β cells as foreign and destroys them, resulting in the loss of insulin. The causes of type 1 diabetes are not totally clear, with both genetic and environmental factors apparently affecting the development of the disease. Currently, 5% of people with diabetes in the United States have type 1 disease.

In type 2 diabetes, the problem arises not from a lack of insulin but an inability of cells to properly respond to this signal. The defect usually lies not at the level of the insulin receptor but is a post-receptor defect in signaling. The disease begins to manifest itself as a prediabetic condition termed **insulin resistance**. In the case of insulin resistance, plasma insulin concentrations become higher and higher, until the pancreas can no longer keep up and the subject goes into β cell failure, resulting in type 2 diabetes. The causes of type 2 diabetes are also unknown, but several risk factors have been identified. Although there is a genetic component to the disease, environmental factors seem to be more important, including dietary excess, obesity, and sedentary lifestyles.

Gestational diabetes is similar to type 2 diabetes and is thought to arise from elevated levels of pregnancy hormones influencing insulin signaling. The disease often reverts after delivery, but women who have experienced gestational diabetes are at increased risk for developing type 2 diabetes later in life.

At the root of all types of diabetes is the body's inability to use glucose. Without insulin or insulin signaling, glucose transporters are not mobilized to the cell surface, so glucose cannot enter the cell. Sensing that glucose is low or unavailable, the body shifts its metabolism to the starved state. Fatty acids are mobilized from adipose tissue, the liver produces both ketone bodies and glucose through gluconeogenesis, and muscle begins to waste as amino acids are catabolized to provide glucogenic precursors.

Multiple drugs are available for treatment of diabetes
Several pharmacological interventions are available for treatment of diabetes. Arguably one of the greatest discoveries in the history of endocrinology, and a turning point in the history of medicine, was the discovery and isolation of insulin. Today, numerous formulations of insulin are available. They include slow- or fast-acting insulin, in which single amino acid mutations have been made to recombinant human insulin, to promote or discourage crystallization. As the crystals dissolve in the bloodstream, the concentration of active insulin increases.

Biguanides
(Metformin)

Sulfonylureas
(Glyburide)

Thiazoladinediones
(Avandia)

All types of diabetes can be treated with insulin, but type 2 diabetes is also treated with oral antidiabetic agents. Metformin is an insulin sensitizing agent; it acts on the liver and suppresses gluconeogenesis. In diabetes, gluconeogenesis can be elevated by as much as threefold, but metformin can reduce this to twofold. The exact mechanism by which metformin works is still being elucidated; however, it is known that **AMP-activated protein kinase (AMPK)** is required for activity, and that metformin activates AMPK, which has additional benefits. AMPK promotes catabolic pathways, and activating AMPK stimulates glycolysis through glucose transport, β-oxidation and ketone body biosynthesis; at the same time, it inhibits glycogen synthesis and triglyceride biosynthesis.

Sulfonylureas, such as glyburide, are drugs that act as insulin secretagogues; that is, they promote the secretion of insulin. Sulfonylureas work by inhibiting the ATP-dependent K⁺ pump required for insulin secretion. This causes a depolarization of the β cell membrane resulting in the opening of Ca^{2+} channels. The increase in cellular Ca^{2+} causes fusion of vesicles of insulin with the plasma membrane.

A final class of oral antidiabetic drugs is the **thiazoladinediones (TZDs)**, such as rosiglitizone. TZDs are **peroxisome proliferator-activated receptor (PPAR)** agonists. These drugs bind to the PPAR class of transcription factor, activating them. The transcription factor PPAR-γ regulates many genes involved in lipid metabolism and fat storage. Therefore, by activating these genes, cells are more able to store lipid. This may seem like a paradox: why give a drug that makes people store more fat to treat a disease with obesity as a risk factor? However, TZDs apparently help to store fatty acids and make adipose tissue less "leaky," lowering levels of plasma fatty acids.

Although there are many treatments for diabetes, it is still a highly prevalent disease. In 2016 it was estimated that 9.3% of the population of the United States is directly affected by diabetes: over 29 million Americans are diagnosed with the disease, and an additional 8.1 million are undiagnosed. The American Diabetes Association reports that diabetes is the leading cause of blindness and kidney failure and accounts for over 65,000 amputations each year. Diabetes also contributes to cardiovascular disease and stroke.

Metabolic syndrome is an amalgam of diverse symptoms underlying many diseases

It has long been known that there are significant links between diabetes, obesity, heart disease, stroke, high blood pressure (hypertension), elevated plasma lipids (hyperlipidemia), and elevated plasma glucose (hyperglycemia). Collectively, these conditions have been named "syndrome X" or **metabolic syndrome**. Although these conditions had a relatively minor role in human health before the industrial revolution, in modern times, metabolic syndrome is a major part of the etiology (the study of causation of disease) of five of the top ten causes of death in the United States: heart disease, cancer, stroke, diabetes, and kidney disease (**Figure 12.14**). In most people, the root cause of metabolic syndrome is simple: overnutrition compounded by a sedentary lifestyle. If those two circumstances were reversed, most cases of metabolic disease could be ameliorated or remedied. Despite research and advances in understanding, the reality is that changes to diet and lifestyle would help more than most pharmaceutical intervention ever aspires to achieve.

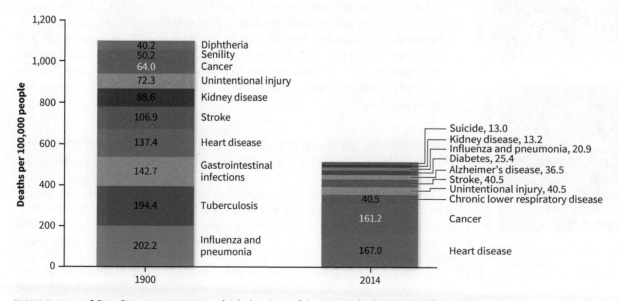

FIGURE 12.14 Mortality rates in 1900 and 2014. Five of the top ten leading causes of mortality in the United States are affected by the metabolic syndrome. These causes include heart disease, diabetes, cerebrovascular disease, kidney disease, and some cancers. (Source: Data from CDC.)

12.4.2 Diseases of absence: starvation, cachexia, and cancer

Although obesity and diabetes are clearly epidemic health problems, at the opposite extreme, metabolic wasting diseases and starvation are also a cause for concern. Despite increased caloric consumption, there are many disease states in which patients waste away, a condition known as **cachexia**; these diseases include AIDS, cancer, and tuberculosis. Although a secondary effect of these diseases and not in and of itself terminal, cachexia is a risk factor and does contribute to the mortality in these diseases.

Two possible causes have been suggested for cachexia, and both may contribute to the disease state. The first possible cause relates to the fact that anorexia is observed in some but not all instances of cachexia, which suggests that there may be an alteration of metabolic rate independent of food intake. The hypothesis is that there may be a disconnect in a metabolic feedback loop, possibly involving adiponectin, ghrelin, leptin, or PYY(3-36), and leading to a decrease in appetite but an increase in metabolic rate.

The second plausible cause states that elevated levels of cytokines (the signaling molecules of the immune system), either coming from a tumor or secreted in response to a tumor, alter gene expression, resulting in wasting. The cytokines implicated in this mechanism include tumor necrosis factor-α (TNF-α), interleukin-6 (IL-6), interferon-γ (INF-γ), and the hormones angiotensin II and the glucocorticoids. Secretion of these factors contributes to the loss of skeletal muscle mass due to both decreased muscle protein synthesis and increased degradation. Additionally, these muscles burn more energy through increased thermogenesis due in part to increased expression of uncoupling proteins.

Cancer cells exhibit rampant glycolysis   The Warburg effect is the observation that cancer cells in many tumors exist in a highly glycolytic state, using glucose as their sole energy source. In addition, these cells do not produce acetyl-CoA from pyruvate; instead, they ferment pyruvate to lactate in the cytosol, even in the presence of oxygen. In terms of *why* this adaptation occurs, one hypothesis is that it may allow cells to grow at low oxygen concentrations inside the tumor; another hypothesis is that cancer cell proliferation occurs because the cells need to use the products of the glycolysis to synthesize the molecules necessary for cell growth. In terms of *how* it occurs, the tumor cell may produce proteins or silence genes to somehow disable mitochondria and thus prevent them from initiating **apoptosis**.

The Warburg effect may also have a direct impact on gene expression. For example, in colon cancer cells, research has shown that by switching from other fuels, such as fatty acids to glucose, butyrate will accumulate, inhibiting histone deacetylases and affecting gene expression through epigenetic changes. Sirtuins also may be important in the regulation of the Warburg effect. Sirtuins have been shown to regulate the stability of transcription factor HIF-1, which regulates the expression of several key proteins in glycolysis.

Kwashiorkor and marasmus result from starvation and malnutrition   Two other states related to starvation merit discussion: kwashiorkor and marasmus.

Kwashiorkor is a type of starvation marked by severe protein deficiency. It is most commonly seen in small children and infants who are abruptly weaned from milk or formula onto a high-carbohydrate diet, as is often seen during a famine. Because of the lack of protein in the diet, there is fatty liver, edema, and distention of the abdomen (**Figure 12.15**).

Marasmus is the general starvation that comes from lack of sufficient calories. Victims of marasmus often obtain calories but at levels below those needed for basic survival. Likewise, meals may come at irregular intervals, complicating the metabolic picture. The result is general wasting, dehydration, and malnutrition.

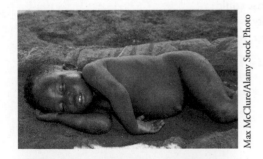
Max McClure/Alamy Stock Photo

Mike Goldwater/Alamy Stock Photo

FIGURE 12.15 Kwashiorkor and marasmus.
A. Kwashiorkor results from severe protein deficiency. The abdomens of affected children are distended from fluid retention and edema.
B. Marasmus results from overall caloric insufficiency leaving the victim emaciated.

12.4.3 Diseases of indulgence: alcohol overconsumption

Ethanol is a naturally occurring molecule and a small part of the diet in all cultures of the world. Although some cultures abstain from consuming alcoholic beverages, many foods contain small amounts of naturally occurring ethanol; for example, ordinary orange juice is typically 0.5% ethanol by volume. Other cultures have used ethanol historically to preserve foods and beverages; for example, in grape-growing countries, wine is a means of preserving grape juice. Likewise, beer and other fermented drinks were thought to rise in popularity, partially from the need for a beverage that was not contaminated with pathogenic microbes. The hops and other herbs in fermented drinks contain microbial growth inhibitors. Of course, ethanol is consumed today for numerous reasons. The National Institutes of Health (NIH) estimates that in the United States one out of every three people abstain from all alcoholic beverages, or drink in very limited amounts (a few drinks per year), and that one person in six has a drinking problem. Even if people abstain from alcoholic beverages, they still need to metabolize the ethanol found in some foods.

Ethanol is largely metabolized by the liver. Alcohol dehydrogenase catalyzes the oxidation of ethanol to acetaldehyde, a process that is NAD$^+$ dependent (**Figure 12.16**). Acetaldehyde is oxidized to acetate in the mitochondria by aldehyde dehydrogenase, generating a second molecule of NADH. Acetate generated in the mitochondria is typically coupled to coenzyme A and oxidized through the citric acid cycle, although this is not a committed fate. Free acetate can be detected in the cytosol as well as the plasma. Acetate contributes to both fatty acid and cholesterol biosynthesis and may serve as a fuel for the brain when consumed chronically in excess.

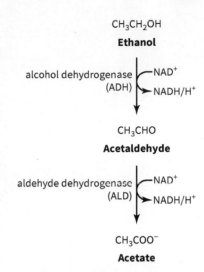

FIGURE 12.16 Ethanol metabolism. Most ethanol is oxidized to acetaldehyde by alcohol dehydrogenase. Acetaldehyde has a short half-life and is quickly converted into the less reactive acetate by aldehyde dehydrogenase.

The biochemistry of hangovers   Many people overconsume ethanol on occasion; the ethanol permeates most tissues of the body and rapidly increases in concentration as consumption continues. Ethanol is generally categorized as a depressant because it decreases respiration and slows reflexes, which alters perceptions and affects judgment. The legal blood alcohol concentration in most states is 0.08% v/v. A concentration above 0.25% typically results in unconsciousness, whereas one above 0.4% can result in coma or death. Even if these excessive levels are not reached, there can still be aftereffects. In the drunken state (blood alcohol of 0.08% to 0.2%) ethanol concentrations are in the millimolar level, but the K_m for most of the isozymes of alcohol dehydrogenase are also in the low millimolar to micromolar level, so ethanol is eliminated relatively rapidly.

Acetaldehyde, the first product of ethanol consumption is toxic, but levels rarely elevate to the point of toxicity. The toxicity of acetaldehyde is perhaps best illustrated by populations sensitive to alcohol consumption. In many Asian populations, a point mutation in aldehyde dehydrogenase renders the enzyme inactive. In people with this mutation, consumption of even small amounts of ethanol rapidly leads to flushing and hangover-like symptoms, including headaches and nausea. Pharmacological inhibition of aldehyde dehydrogenase can produce similar results. Disulfiram, the active ingredient in the drug Antabuse, used to treat chronic alcoholism, or any one of several related toxins derived from fungi, result in inhibition of aldehyde dehydrogenase. Consumption of small amounts of ethanol while taking Antabuse results in acute hangover-like symptoms.

Disulfiram
(Antabuse)

People suffering from a hangover often feel weak. As ethanol is metabolized, the ratio of NADH/NAD$^+$ is increased; that is, more NADH is produced in the oxidation of ethanol to acetate. Changes in this ratio result in decreased gluconeogenesis in the liver. As a result, plasma glucose levels drop (hypoglycemia) and lactate levels from muscle increase. Elevated lactate in conjunction with increased acetate can result in **acidosis**, or decreased plasma pH.

Other hangover symptoms result from dehydration. Ethanol blocks the release of vasopressin from the **posterior pituitary**, resulting in decreased sodium reabsorption in the kidney and increased urination. This effect, compounded by nausea and vomiting from stomach irritation, can account for some of the dizziness and thirst felt in a hangover.

Chronic alcohol consumption causes other health effects

Chronic alcohol consumption carries with it other significant health effects. Alcoholism often results in malnutrition due to the general failure to consume a nutritious diet, the loss of water-soluble electrolytes and vitamins through increased urination, and the inability to absorb certain B vitamins in the intestine.

The chronic consumption of ethanol provides a higher than normal ratio of $NADH/NAD^+$. As mentioned, this decreases gluconeogenesis and shifts metabolism to ethanol and its final product acetate; it also contributes to increased fatty acid biosynthesis and hypoglycemia. Because the brain and body tissues are starved of glucose, lipolytic signals trigger mobilization of fatty acids out of adipose tissue for ketone body production. This results in increased fat accumulation in the liver, referred to as fatty liver, and in the blood, referred to as hyperlipidemia. Fatty liver is a reversible process but is the beginning of the scarring of the liver and the development of cirrhosis and hepatocarcinoma, or liver cancer. The result of this is a decrease of plasma glucose and a significant elevation of ketone bodies, lactate, amino acids, and lipids, both free fatty acids and VLDL/LDL. Perhaps the only metabolic benefit in those addicted to alcohol is an elevated HDL level.

Another complication of ethanol metabolism is the formation of adducts between ethanol, acetaldehyde, or acetate and different molecules of the body. Acetaldehyde in particular is quite reactive; it can randomly add to molecules, altering their structure, function, or activity. The molecules of the cell also can be altered through reactive oxygen species, that is, reactive by-products of oxidative metabolism that randomly modify and damage proteins, membranes, and DNA. These species are clearly causal factors in hepatic cell death and liver scarring. Over time, oxidative damage to DNA is responsible for hepatocarcinoma. Increased risk of developing several cancers, including liver and breast cancer, has been linked to alcohol consumption.

A final complication of chronic ethanol consumption is hypoxia or low oxygen concentration. Ethanol puts a metabolic burden on the liver, increasing flux through several pathways that require molecular oxygen. This creates a need for more oxygen in the liver and can create such a demand that regions of the liver distant from arteries face hypoxic conditions. In this state, the cells generate signaling molecules that leave them further depleted of oxygen. In addition, decreased plasma pH decreases the ability of the hemoglobin to carry oxygen; this in combination with a malnourished state can lead to anemia in the person with alcoholism. Collectively, these effects contribute to the hypoxic state, which is another source of damage to the tissues. Ethanol can also result in changes in gene expression, although it is unclear whether this is directly due to the ethanol itself or to downstream effects of cell signaling.

As damage to the liver continues, levels of proinflammatory cytokines, such as TNF-α, increase. In cirrhosis, cells apoptose, or in some instances become necrotic, and are replaced by connective tissue. Over time, this combination can be so severe as to drive people with alcoholism into liver failure.

12.4.4 Other metabolic states

Several other metabolic states warrant commentary. First are those found in trained athletes. Metabolic states in exercise physiology can be categorized by the type of exercise performed: these are sprinting exercise, when running at full speed for less than 400 meters, and distance exercise, any exercise greater than sprinting (**Table 12.6**). Sports that require short bursts of energy, from table tennis to powerlifting, fall into the sprinting category, while anything longer is a distance activity.

The delineation between these two types of exercise can be clearly seen if one examines elite athletes' biochemistry. Runners covering short distances (up to 200 meters) have a similar speed. The runner needs to cover a short distance to gain the momentum to achieve that speed; however, once the

TABLE 12.6 Speed of Track and Field Events

Distance, meters	Time, sec	Velocity, m/s
100	9.58	10.438
150	14.35	10.453
200	19.19	10.422
300	30.85	9.724
400	43.03	9.296
600	72.81	8.241
800	100.91	7.928

speed is attained, runners can continue at that speed for about 200 meters or a little less than 20 seconds. Thus, the world record in the 100 meters is a bit more than double the speed of a top football recruit running the 40 meters, and about half of the world record in the 200 meters. At greater distances, such as 400 and 800 meters, the speed does not simply double again but drops significantly. Biochemistry explains the basis of this difference.

The biochemistry of the sprint

A sprint happens almost instantaneously. There is only one fuel that can be used so quickly: ATP. The initial stages of a sprint use the muscle cell's ATP store to rapidly generate explosive muscle movement, but the ATP store lasts only a few seconds. Glycolysis will rapidly take over, producing more ATP and generating lactate from the pyruvate product; therefore, sprinting is largely considered an anaerobic activity. A mechanism is needed to restore these ATP levels. There is a buffer for ATP, that is, a means of rapidly regenerating ATP in the absence of other energy production. Creatine kinase can transfer a phosphate group from creatine phosphate to ADP, or vice versa; under states where ATP is depleted, creatine phosphate can help rapidly replenish levels, but levels of creatine phosphate are also limited: there is only enough to use for a few seconds. Athletes attempt to elevate levels of creatine phosphate by taking creatine supplements, but the value of this practice is questionable.

Creatine that escapes from muscle is catabolized in the liver to creatinine and secreted by the kidneys.

Distance events require a different metabolic approach

In marathon running, we see the opposite extreme of exercise physiology. Marathon runners who are trained to run at this distance (42.2 km) also largely burn carbohydrates through glycolysis and oxidative phosphorylation, mostly from stored glycogen. Most distance athletes consume some carbohydrate drink or gel while running. In some cycling events, for example, the Tour de France, half of the total calories consumed in the course of the day are taken in on the bike. If athletes do not replenish these carbohydrates in some way during a race, they risk total depletion of carbohydrate reserves and resulting muscle exhaustion. When this occurs, muscles shut down and the athlete cannot use them. This is referred to as "hitting the wall," and stops runners in their tracks until they replenish some carbohydrates, either by consuming carbohydrates or through gluconeogenesis from protein or glycerol sources.

Many animals undergo periods of metabolic dormancy

Another interesting metabolic state is **hibernation**, the winter period when some mammals sleep and enter torpor, in which metabolic rate, respiration rate, and body temperature all drop. Most animals have evolved mechanisms to downregulate their metabolism in times of energy scarcity or depleted resources. Torpor may refer to the metabolic state seen in hibernation, but in some organisms it can occur daily or due to environmental conditions, such as low temperature or lack of food. Torpor is thought to preserve the organism and let it conserve energy. Another, similar process to hibernation is aestivation. **Aestivation** is a process undergone by organisms, such as mollusks and arthropods, in which they become dormant and slow their metabolism in warm or dry months. Physiological changes also help these organisms to conserve water. A third state is **brumation**, in which animals are still awake but have slowed their metabolic rate. Ectotherms, such as fish and reptiles, undergo brumation; they are unable to regulate their body temperature and therefore do not alter it as is done in hibernation. Thus, brumation differs from hibernation and aestivation. Other than these gross observations, metabolic links between brumation and hibernation have not been clearly established.

Studies of animal dormancy have direct implications for human health. In preparation for hibernation, many animals model obesity by eating excessively and storing calories as fat. In addition, some animals become insulin resistant before hibernation, similar to type 2 diabetes.

With changes in feeding in the periods proceeding dormancy, many birds will overeat prior to migration. Humans have noticed this since antiquity and have also noted how this changes the different aspects of birds as a human food source, most notably the amount of fat that accumulates in the liver. One particular example of this is discussed in **Societal and Ethical Biochemistry: Foie gras.**

Societal and Ethical Biochemistry

Foie gras

Few foods are as curious or controversial as foie gras, a delicacy famed in French cuisine for its unusual rich buttery flavor and texture. Foie gras is often lightly seared or grilled, or is blended into a spread or pate. However, the production of foie gras has raised ethical concerns.

Foie gras is the liver of a duck or goose that has been specifically fattened. In fact, ducks and geese are gavaged or force fed as much grain as they can hold, through a special funnel inserted into the esophagus of the bird. This process is repeated several times a day for two to three weeks. As this happens, the bird's liver engorges and becomes fatty, enlarging up to ten times its normal size, and producing the buttery texture sought after in foie gras.

In the biochemical sense, the liver of these animals is storing fat produced predominantly from carbohydrates found in grain. Although this is a metabolic extreme, it is not totally unnatural. In the wild, migratory birds will engorge themselves before starting their migration, and their livers will swell with fat.

Because of the process by which foie gras is produced, many localities now consider serving the product unethical. For example, the state of California, the city of Chicago, and several countries in Europe, have sought to ban foie gras. Nevertheless, gavage of animals has gone on since the time of 2500 BCE, and France recognizes foie gras as part of its cultural heritage.

Alternative production methods are under investigation. Appetite is controlled in part through the hypothalamus, so one approach has been to surgically damage the regions of the hypothalamus involved in the regulation of appetite. Geese that have undergone this process consume three or four times more food, and they grow fatty livers that are similar to those of gavaged birds, but this technique has not grown in popularity, perhaps due to the resources needed to perform it. A second technique capitalizes on the natural behavior of geese and ducks. While preparing to migrate, the geese and ducks are given free access to a diet rich in acorns and olives. This results in a smaller liver, but one that still has the flavors, textures, and levels of fat that chefs and diners prefer. It is not yet clear whether this product will be as popular with consumers as foie gras.

Worked Problem 12.4 — Characterizing a knockout mouse

You are asked to help characterize a knockout mouse made in your laboratory. A gene coding for a fat storage protein was deleted in this mouse, which is lean and has normal food consumption and activity, but has elevated plasma insulin and an impaired **glucose tolerance test (GTT)**.

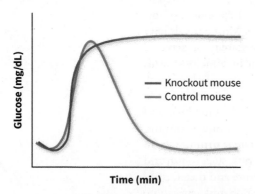

How would you characterize these results?

Strategy Examine the data at hand. What does a GTT determine, and how does it work? What is the significance of the elevated insulin level?

Solution This mouse is a lean model for type 2 diabetes mellitus. It needs elevated insulin levels to maintain normal metabolism because its cells respond poorly to the insulin signal. The GTT is a way to ascertain how well the organism can respond to a glucose challenge. Glucose levels rise in both animals, but glucose clearance is delayed in the knockout mouse; the explanation for diabetes is not as apparent, but it may be due to an impairment in triacylglycerol storage, which can result in elevated plasma fatty acids and contribute to an insulin resistant phenotype.

Follow-up question Given that the mouse has reduced adipose tissue mass, elevated insulin levels, and normal activity and food consumption, predict the levels of other hormones in this organism. In other words, how has this mutation affected hormone levels, based on the phenotype observed?

Summary

- Obesity resulting from overnutrition and lack of exercise is at epidemic proportions in Western society. It is a significant risk factor for the development of type 2 diabetes mellitus.

- Type 2 diabetes mellitus arises from cells of the body failing to appropriately respond to insulin secreted from the pancreas. The cause is still unknown, but it is likely that the disease is polygenic, resulting from the interplay of multiple proteins.

- Metabolic syndrome or syndrome X is a group of related pathophysiologies that contribute to many of the most common and debilitating diseases, including heart disease, stroke, diabetes, and some cancers.

- Cancer, AIDS, and other diseases can lead to cachexia, a wasting of the body. This may be due to damaged feedback loops in hormonal signaling, elevated cytokines, or other unknown causes.
- The Warburg effect is the observation that tumors are highly glycolytic and undergo anaerobic metabolism in the presence of oxygen. Sirtuins, HIF-1, and epigenetic modifications may play a role in this syndrome.
- Starvation can be exhibited as kwashiorkor or marasmus.
- Excessive ethanol consumption results in decreased hepatic gluconeogenesis and elevated fatty acid metabolism. Chronic ethanol consumption results in a hypoxic state in the liver, causing liver damage or cirrhosis, and eventual liver failure.
- Other metabolic states resulting from diet, ethanol consumption, and exercise or hibernation alteration in metabolism have been extensively studied.

Concept Check

1. Explain the term metabolic syndrome. What is the root cause and possible ways for reversing it?

2. Describe what is happening in obesity, starvation, cancer, and alcoholism at the molecular level. Explain the symptoms of these conditions and complications, if any.

3. Explain what treatments are available for these diseases and how they act on the molecular level.

Bioinformatics Exercises

Exercise 1 Insulin Secretion and the KEGG Database

Exercise 2 Insulin Control of Glucose Metabolism and the KEGG Database

Problems

12.1 A Review of the Pathways and Crossroads of Metabolism

1. Glycerol kinase and glucose-6-phosphatase are two enzymes that tailor metabolic responses to specific tissues. How do these two enzymes provide specialized metabolic responses? What do these two enzymes do to specialize the metabolic response? What other enzymes act this way?

2. The urea cycle is an example of a pathway that occurs in a specific tissue that requires a specialized metabolic response. What does the urea cycle do, and where does it do it? What are other examples of specific pathways, and where are they found?

12.2 Organ Specialization and Metabolic States

3. Why is re-fed state also a re-fueling stage? Explain why the liver undergo gluconeogenesis when in the re-fed state.

4. There are two hydroxymethylglutaryl-CoA (HMG-CoA) synthases in the cell. Where are they found? In which pathways do they function? One of these synthases is under control of SREBP and the other of a PPRE. Hypothesize as to which molecule controls which synthase.

12.3 Communication between Organs

5. Propose a mechanism for the decarboxylation of L-DOPA to dopamine in a pyridoxal phosphate-dependent mechanism.

6. Explain the synthesis of epinephrine from L-DOPA. Why not insert a double bond in dopamine and then hydrate it to form teh alchol?

7. Examine the name phenylethanolamine *N*-Methyltransferase. What is the substrate for this enzyme? What does this enzyme do? What is the source of the methyl group, and where does it end up in the final molecule? Draw the structure of the final compound.

8. Thyroid follicular cells use a sodium transporter to bring iodide ions into the cell. What type of transport mechanism is this?

9. Explain how the hypothalamus functions in the regulation of metabolism, how transcription factors assist in the regulation of metabolism, and the role of sirtuins.

10. Propose a mechanism for the proteolytic cleavage of T3 and T4 from thyroglobulin.

11. Glucagon is a short helical peptide.

   a. How could you determine the tertiary structure of glucagon?

   b. In what ways would it be easier to determine the structure of glucagon than of a larger protein?

   c. In what ways would it be harder to determine the structure?

12. Some hormonal disorders include Addison's disease, Cushing's disease and Hashimoto's disease. Identify the hormones affected and key symptoms of these diseases.

13. How has completion of the human genome assisted in the discovery of new hormones?

14. Many hormones are peptides and are degraded in the stomach; hence, they are not orally available. Why is this an evolutionary advantage?

15. Based on the discussion in the chapter, what are the symptoms of diabetes insipidus and diabetes mellitus.

16. Hyperthyroidism and Graves' disease are treated with radioactive iodine, most notably ^{131}I.

    a. How does this treatment work?

    b. What type of radiation does ^{131}I emit?

    c. What are the possible side effects of ^{131}I treatment?

17. What classes of hormones are corticotropin releasing hormone and adrenocorticotropic hormone? What are the functions of these hormones?

18. Adiponectin levels are increased by treatment with thiazoladinediones. Correlate these two observations, and explain them.

19. Goiter results from insufficient iodine in the diet. What are the metabolic symptoms of this disorder?

12.4 Metabolic Disease

20. Diabetes insipidus is a disease caused by a mutation to a gene coding for the vasopressin receptor. Describe how this mutation would affect vasopressin signaling and the vasopressin signaling loop, and lead to the symptoms of the disease.

21. Often, menstruation ceases in women who are professional athletes or bodybuilders or who have anorexia. Propose a role for hormonal signaling to explain this observation.

22. What metabolite is used in unusually high amounts by cancerous tumors?

23. Would the phenotype of the MC4R knockout mouse be lean or obese? Would you anticipate any problems maintaining a colony of these mice?

24. Bremalanotide is a hexapeptide analog of α-MSH. It was developed as a tanning agent but is now marketed for the treatment of sexual dysfunction. What is the reason for this?

25. List and discuss the health effects of chronic alcohol consumption.

26. Explain why insulin cannot be taken orally; and steroid hormones can be applied topically or delivered by a patch.

27. Lysine can spontaneously form adducts with acetaldehyde. Draw this structure. What type of functional group is it?

28. What factors would influence adduct formation? Would you anticipate that some proteins would have more acetaldeyde adducts than others?

29. Both serum and urinary creatinine are typically measured as part of a panel of clinical tests. What organ function is being measured in these tests?

30. Why is creatine kinase used as a marker for heart attacks? How does this test work?

31. Would it benefit a golfer, baseball player, football player or weightlifter to supplement with creatine?

Data Interpretation

32. Only 5% of people with diabetes currently have type 1 diabetes, whereas 15 years ago, 10% had it. Why is the number dropping?

33. In the literature, there are several cases of a disorder known as hamburger thyrotoxicosis. In this disorder, subjects suffer from sleeplessness, nervousness, weight loss, headache, and fatigue. These cases have been linked to eating hamburger contaminated with thyroid tissue. Examine the graphs shown below.

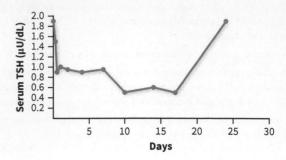

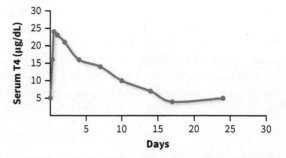

a. Do the data in these graphs substantiate or invalidate the hypothesis that an acute event caused the change in hormone levels?

b. What is an alternative hypothesis, and how might you test it (in either a human or animal model)?

c. Why does thyroid stimulating hormone drop?

d. Given what is known above, what would be an appropriate control for this study?

e. Is there an inherent weakness in the study?

Experimental Design

34. Engineered forms of insulin have point mutations that allow the molecule to form microcrystals that slowly dissolve, providing a "slow release" form of the hormone. How would you test to see whether an engineered protein was forming microscopic crystals? Design an assay to see how quickly these crystals would dissolve.

35. You identify a short (40 amino acid long) peptide that you hypothesize regulates appetite. How could you test to see which residues of the peptide are critical for function?

36. Following one week on a diet enriched in certain fats, mice have elevated plasma ketone bodies compared to control animals. How could you determine whether the effect on ketone bodies is due to increased gene expression, increased enzyme activity, or some other regulatory mechanism?

Ethics and Social Responsibility

37. In some countries, prisoners are punished by being fed a "starving ration" or by receiving no food for one in every three days. Is this ethical? Is it healthy? Is it punishment?

38. Political prisoners or those wishing to make a political statement sometimes undertake hunger strikes, in some instances literally starving themselves to death. Given that these people cannot resist medical intervention at some point, is it ethical to let them die? Why is starvation—a slow, degenerative and painful death—considered acceptable but other forms of suicide are not?

39. Numerous college students (both male and female) suffer from eating disorders. Seek out your campus student center and see what resources they have on eating disorders. Share this information with your friends and classmates.

40. Diabetes and obesity are diseases that carry with them a huge cost in terms of diminished quality of life; leading a healthier lifestyle can help to prevent or delay the onset of many of the problems associated with these diseases. These diseases are also expensive to manage and treat. Given this information, how would you develop a company policy to limit the liability associated with this risk?

Suggested Readings

Ehnholm, Christian, ed. *Cellular Lipid Metabolism.* New York, NY: Springer, 2009.

Fahy E., S. Subramaniam, R. C. Murphy, M. Nishijima, C. R. Raetz, T. Shimizu, F. Spener, G. van Meer, M. J. Wakelam, and E. A. Dennis. "Update of the LIPID MAPS Comprehensive Classification System for Lipids." *Journal of Lipid Research* 50 (April 2009): S9–14.

Hers, H. G., and L. Hue. "Gluconeogenesis and Related Aspects of Glycolysis." *Annual Review of Biochemistry* 52 (1983): 617–53.

Houten, S. M., and R. J. Wanders. "A General Introduction to the Biochemistry of Mitochondrial Fatty Acid β-oxidation." *Journal of Inherited Metabolic Diseases* 33, no. 5 (October 2010): 469–77.

Pilkis, S. J., and D. K. Granner. "Molecular Physiology of the Regulation of Hepatic Gluconeogenesis and Glycolysis." *Annual Review of Physiology* 54 (1992): 885–909.

Scrutton, M. C., and M. F. Utter. "Regulation of Glycolysis and Gluconeogenesis in Animal Tissues." *Annual Review of Biochemistry* 37 (1968): 249–58.

Wakil, S. J., and L. A. Abu-Elheiga. "Fatty Acid Metabolism: Target for Metabolic Syndrome." *Journal of Lipid Research* 50 (April 2009): S138–43.

Walsh, Patrick J., and Patricia A. Wright. "Nitrogen Metabolism and Excretion." Boca Raton, Florida: CRC Press, 1995.

Frayn, K. N. "Fat as a Fuel: Emerging Understanding of the Adipose Tissue-Skeletal Muscle Axis." *Acta Physiologica* 199, no. 4 (August 2010): 509–18.

Grundy, S. M., H. B. Brewer, Jr., J. I. Cleeman, S. C. Smith, Jr., and C. Lenfant. "Definition of Metabolic Syndrome: Report of the National Heart, Lung, and Blood Institute/American Heart Association Conference on Scientific Issues Related to Definition." *Circulation* 109, no. 3 (January 27, 2004): 433–8.

Gurr, M. I., J. L. Harwood, and K. N. Frayn. *Lipid Biochemistry: An Introduction.* Hoboken, New Jersey: Blackwell Science Ltd., 2002.

McIntosh, Brian, Phillip Gardiner, and Alan McComas. *Skeletal Muscle: Form and Function.* 2nd ed. Champagne, Illinois: Human Kinetics, 2005.

Symonds, Michael E. E., ed. *Adipose Tissue Biology.* Philadelphia, PA: Springer Science and Business Media, LLC, 2012.

Batterham, R. L., H. Heffron, S. Kapoor, J. Chivers, K. Chandarana, H. Herzog, C. W. Le Roux, E. L. Thomas, J. D. Bell, D. J. Withers. "Critical Role for Peptide YY in Protein-Mediated Satiation and Body-Weight Regulation." *Cell Metabolism* 4, no. 3 (2006): 223–33.

Batterham, R. L., M. A. Cowley, C. J. Small, H. Herzog, M. A. Cohen, C. L. Dakin, A. M. Wren, et al. "Gut Hormone PYY(3-36) Physiologically Inhibits Food Intake." *Nature* 418, no. 6898 (August 8, 2002): 650–4.

Boron, W. F. *Medical Physiology: A Cellular and Molecular Approach.* Philadelphia, PA: Elsevier-Saunders, 2003.

DeGroot, Leslie Jacob. *Endocrinology.* Edited by J. E. McGuigan. Philadelphia, PA: Saunders, 1989.

Gromada, J., I. Franklin, and C. B. Wollheim. "Alpha-Cells of the Endocrine Pancreas: 35 Years of Research but the Enigma Remains." *Endocrine Reviews* 28, no. 1 (February 2007): 84–116.

Guarente, L. "Sirtuins, Aging, and Medicine." *New England Journal of Medicine* 364, (2011): 2235–44.

Havel, P. J., J. O. Akpan, D. L. Curry, J. S. Stern, R. L. Gingerich, and B. Ahren. "Autonomic Control of Pancreatic Polypeptide and Glucagon Secretion During Neuroglucopenia and Hypoglycemia in Mice." *American Journal of Physiology* 265, no. 1 Pt 2 (July 1993): R246–54.

Holst, J. J. "Glucagon and Glucagon-Like Peptides 1 and 2." *Results and Problems in Cell Differentiation* 50, (2010): 121–35.

Lihn, A. S., S. B. Pedersen, and B. Richelsen. "Adiponectin: Action, Regulation and Association to Insulin Sensitivity." *Obesity Reviews* 6, no. 1 (February 2005): 13–21.

Lombard, David B., and Richard A. Miller. "Ageing: Sorting Out the Sirtuins." *Nature* 483 (2011): 166–7.

Minor, R. K., J. W. Chang, and R. de Cabo. "Hungry for Life: How the Arcuate Nucleus and Neuropeptide Y May Play a Critical Role in Mediating the Benefits of Calorie Restriction." *Molecular and Cellular Endocrinology* 299, no.1 (2009): 79–88.

Murphy, K. G., and S. R. Bloom. "Gut Hormones and the Regulation of Energy Homeostasis." *Nature* 444, no. 7121 (2006): 854–9.

Myers, M. G., M. A. Cowley, and H. Münzberg. "Mechanisms of Leptin Action and Leptin Resistance." *Annual Review of Physiology* 70 (2008): 537–56.

Wu, Y., and R. J. Koenig. "Gene Regulation by Thyroid Hormone." *Trends in Endocrinology and Metabolism* 11, no. 6 (August 2000): 207–11.

Yamauchi, T., J. Kamon, Y. Ito, A. Tsuchida, T. Yokomizo, S. Kita, T. Sugiyama, et al. "Cloning of Adiponectin Receptors That Mediate Antidiabetic Metabolic Effects." *Nature* 423, no. 6941 (2003): 762–9.

Zhang, J., and M. A. Lazar. "The Mechanism of Action of Thyroid Hormones." *Annual Review of Physiology* 62 (2000): 439–66.

Bray, G. A. "Pathophysiology of Obesity." *The American Journal of Clinical Nutrition* 55 (1992): 488S–945.

Gatenby, R. A., and R. J. Gillies. "Why Do Cancers Have High Aerobic Glycolysis?" *Nature Reviews Cancer* 4, no. 11 (2004): 891–9.

Hansen, B. C., and G. A. Bray. "The Metabolic Syndrome: Epidemiology, Clinical Treatment, and Underlying Mechanisms." *Humana Press* 2010.

Lean, M. E. "Pathophysiology of Obesity." *The Proceedings of the Nutrition Society* 59, no. 3 (2000): 331–6.

Orth, Jeffrey D., Ines Thiele, and Bernhard Ø. Palsson. "What Is Flux Balance Analysis?" *Nature Biotechnology* 28, no. 3 (2010): 245–8.

Padwal, R. S., and S. R. Majumdar. "Drug Treatments for Obesity: Orlistat, Sibutramine, and Rimonabant." *The Lancet* 369, no. 9555 (2007): 6–12, 71–77.

Pi-Sunyer, F. Xavier. "The Obesity Epidemic: Pathophysiology and Consequences of Obesity." *Obesity Research* 10 (2002): 97S–104S.

Scacchi, M., A. I. Pincelli, and F. Cavagnini. "Growth Hormone in Obesity." *International Journal of Obesity and Related Metabolic Disorders* 23, no. 3 (1999): 260–71.

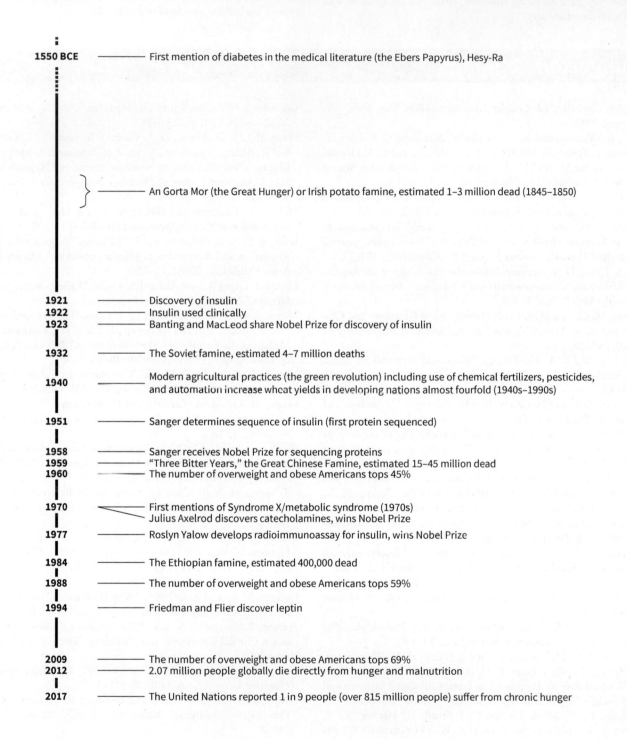

1550 BCE ——————— First mention of diabetes in the medical literature (the Ebers Papyrus), Hesy-Ra

}——————— An Gorta Mor (the Great Hunger) or Irish potato famine, estimated 1–3 million dead (1845–1850)

1921 ——————— Discovery of insulin
1922 ——————— Insulin used clinically
1923 ——————— Banting and MacLeod share Nobel Prize for discovery of insulin

1932 ——————— The Soviet famine, estimated 4–7 million deaths

1940 ——————— Modern agricultural practices (the green revolution) including use of chemical fertilizers, pesticides, and automation increase wheat yields in developing nations almost fourfold (1940s–1990s)

1951 ——————— Sanger determines sequence of insulin (first protein sequenced)

1958 ——————— Sanger receives Nobel Prize for sequencing proteins
1959 ——————— "Three Bitter Years," the Great Chinese Famine, estimated 15–45 million dead
1960 ——————— The number of overweight and obese Americans tops 45%

1970 ——————— First mentions of Syndrome X/metabolic syndrome (1970s)
——————— Julius Axelrod discovers catecholamines, wins Nobel Prize

1977 ——————— Roslyn Yalow develops radioimmunoassay for insulin, wins Nobel Prize

1984 ——————— The Ethiopian famine, estimated 400,000 dead

1988 ——————— The number of overweight and obese Americans tops 59%

1994 ——————— Friedman and Flier discover leptin

2009 ——————— The number of overweight and obese Americans tops 69%
2012 ——————— 2.07 million people globally die directly from hunger and malnutrition

2017 ——————— The United Nations reported 1 in 9 people (over 815 million people) suffer from chronic hunger

CHAPTER 13

Nucleotide and Deoxynucleotide Metabolism

Nucleotides and Deoxynucleotides in Context

Most of us recognize such iconic structures as the pyramids of Egypt, the Great Wall of China, or the Taj Mahal and marvel at their size and beauty, but we rarely appreciate what materials these masterpieces are made from or how they were made. Specialists like engineers or architects know the sources of various building materials and their properties. They understand the processes by which these raw materials were transformed and the labor it took to erect these structures. They know the engineering principles that have kept them upright for hundreds or thousands of years. Understanding materials and building processes can help experts to protect and repair these ancient structures as well as to innovate in new forms of architecture, such as modern skyscrapers made of glass and steel.

Just as these great monuments were formed from individual blocks of stone, life itself is made up of building blocks of nucleotides. We recognize life in its many forms all around us, but biochemistry helps us appreciate the structural components DNA and RNA, how they are synthesized, and their role in sustaining health. Nucleotides and deoxynucleotides not only form the building blocks of DNA and RNA, but they are also key components of the electron carriers NAD$^+$ and FAD and serve as reaction intermediates in many biochemical reactions.

This chapter discusses the synthesis of purines and pyrimidines and how these molecules are recycled via salvage pathways; we also examine the synthesis of deoxynucleotides and discuss how these molecules are catabolized. Because these pathways have ramifications for human health, several disorders of nucleotide metabolism are also discussed as well as how analogs of these molecules can be used as therapeutic agents.

Chapter Outline

Common Themes

Evolution's outcomes are conserved.	• Because nucleotides are fundamental to biochemistry and found in all organisms, the pathways governing their synthesis and breakdown are conserved throughout evolution.
	• Because life as we know it is thought to have evolved from an RNA-based world in which there was little or no DNA or protein, purines and pyrimidines must also have been found in the primordial world.
Structure determines function.	• Substrate channeling in carbamoyl phosphate synthetase increases the efficiency of this complex.
	• What prokaryotes accomplish in a metabolic pathway, eukaryotes accomplish via a single multi-enzyme complex.
Biochemical information is transferred, exchanged, and stored.	• The synthesis of purines and pyrimidines is tightly regulated by other nucleotides via product inhibition and allosteric regulatory mechanisms.
	• Purines and pyrimidines form the basis of nucleotides and deoxynucleotides, the building blocks of RNA and DNA.
Biomolecules are altered through pathways involving transformations of energy and matter.	• The synthesis of purines and pyrimidines is costly in terms of both starting materials and energetic requirements. Therefore, salvage pathways have evolved to help recycle these molecules.
	• As with other types of biochemicals, small changes made to a common intermediate, such as inosine monophosphate, lead to structural and functional diversity in the final products (adenosine and guanosine monophosphate).

13.1 Purine Biosynthesis

Our discussion of nucleotide metabolism begins with biosynthesis of the **purines** adenosine and guanosine (**Table 13.1**). Purines are produced as nucleosides, synthesized off a molecule of ribose, and subsequently phosphorylated into adenosine triphosphate (ATP) and guanosine triphosphate (GTP). ATP is commonly known as the cell's energy currency, and adenosine is also a component of several important molecules in biochemistry, including NAD^+, $NADP^+$, and FAD. GTP has a free energy of hydrolysis similar to that of ATP and can just as readily act as a phosphate donor. GTP is also important in cell signaling, protein biosynthesis, and cytoskeletal stability. ATP and GTP are made from adenosine monophosphate (AMP) and guanosine monophosphate (GMP), both of which are derived from inosine monophosphate (IMP), which is synthesized in the *de novo* **biosynthetic pathway**. Later we will discuss an alternative pathway, the **salvage biosynthetic pathway**, which can build purines through recycling of hypoxanthine. There are equivalent *de novo* and salvage pathways that build pyrimidines, discussed in section 13.3.

13.1.1 The *de novo* biosynthetic pathway builds the purine backbone from phosphoribose

The common purine backbone is the base hypoxanthine, which is converted to IMP while coupled to ribose-5-phosphate. The IMP is then further derivatized to produce AMP and GMP.

TABLE 13.1 Building Blocks of Nucleic Acids

Base	Structure	Ribonucleoside	Ribonucleotide (5′ monophosphate)
Adenine (A)		Adenosine	Adenosine 5′ monophosphate (AMP), adenylate
Guanine (G)		Guanosine	Guanosine 5′ monophosphate (GMP), guanylate
Cytosine (C)		Cytidine	Cytidine 5′ monophosphate (CMP), cytidylate
Uracil (U)		Uridine	Uridine 5′ monophosphate (UMP), uridylate

Base	Structure	Deoxyribonucleoside	Deoxyribonucleotide (5′ monophosphate)
Adenine (A)		Deoxyadenosine	Deoxyadenosine 5′ monophosphate (dAMP), deoxyadenylate
Guanine (G)		Deoxyguanosine	Deoxyguanosine 5′ monophosphate (dGMP), deoxyguanylate
Cytosine (C)		Deoxycytidine	Deoxycytidine 5′ monophosphate (dCMP), deoxycytidylate
Thymine (T)		Deoxythymidine or thymidine	Deoxythymidine or thymidine 5′ monophosphate (AMP), deoxythymidylate or thymidylate

Inosine consists of the base hypoxanthine joined to ribose. Note that inosine is the only ribonucleotide that deviates from the common nomenclature.

Several of the reactions of purine metabolism employ the cofactor tetrahydrofolate (THF), previously encountered in amino acid metabolism. The chemistry of folate, a water-soluble B vitamin that is important in human health, is discussed in **Biochemistry: Folate**.

The synthesis of IMP is an 11-step process (**Figure 13.1**). The pathway begins with the addition of pyrophosphate (PP$_i$) to ribose-5-phosphate forming **phosphoribosyl**

Biochemistry

Folate

Several methods from organic chemistry can be used to synthesize carbon–carbon bonds. These include Suzuki coupling, Michael additions, Grignard reactions, or Wittig reactions. Chapters 6 and 9 discussed examples of Diels–Alder, Claisen, and aldol reactions in biochemistry. Biochemistry also provides a way to add single carbons to a molecule. In this chapter, the reactions use different forms of folate, which serve as donors of formyl, methyl, methylene, or hydride groups.

Folate is reduced in the liver to the active forms dihydrofolate (DHF) and tetrahydrofolate (THF) by two successive reduction reactions, both of which are NADPH dependent. These reactions transfer hydride groups to the bicyclic pterin ring structure of folate. THF can be converted to 5,10-methylene-THF through the transfer of a methylene group from either a glycine or serine donor. Here, the methylene group bridges between N-5 on the pterin ring and N-10 proximal to the phenyl group. Although this chapter focuses on the use of folate metabolites in nucleotide synthesis, these reactions also show the roles that folate plays in amino acid metabolism. 5,10-methylene-THF can be used as an active donor of this group in dTMP biosynthesis, or the methylene group can be further reduced to 5-methyl-THF in an NADH-dependent reaction. This folate metabolite is used in the synthesis of methionine from homocysteine (a cysteine analog containing an extra methylene unit). Finally, 5,10-methylene-THF can be converted to 10-formyl-THF through oxidation of the methylene unit by methylene-THF cyclohydrolase.

Medicinally, folate occupies an interesting niche. Humans cannot synthesize folate, yet it is required for nucleotide biosynthesis and therefore for life itself. Molecules that inhibit folate production in microbes (such as prontosil) can be used to effectively target these organisms. Likewise, drugs can target folate metabolism in mammalian cells that are dividing rapidly. These molecules (methotrexate or aminopterin) can be effective anticancer or immunosuppressive agents.

Humans are unable to synthesize folate and must consume it in their diet. The liver changes folate into active forms (DHF and THF). It is a water-soluble vitamin (vitamin B$_9$, although it is typically known simply as "folate" or "folic acid"). Folate content is highest in leafy greens, whole grains, beans and legumes, and animal tissue. Because it is water soluble, folate must be consumed daily, although it takes nearly a month without consumption for a person to become folate deficient. Those who are folate deficient are generally anemic and have lower white blood cell counts. Folate has been shown to greatly reduce neural tube defects such as spina bifida. Because of this, and of how well folate is tolerated in the diet, pregnant women or women seeking to become pregnant are strongly advised to take folate supplements.

Folate is not the only single-carbon donor available to biochemistry. S-adenosyl-L-methionine (AdoMet) is also used to donate methyl groups to a molecule. In the process this molecule becomes S-adenosyl-L-homocysteine (AdoHcy). AdoHcy levels have become a focus of research in recent years, as they have become associated with the levels of oxidative stress in an organism.

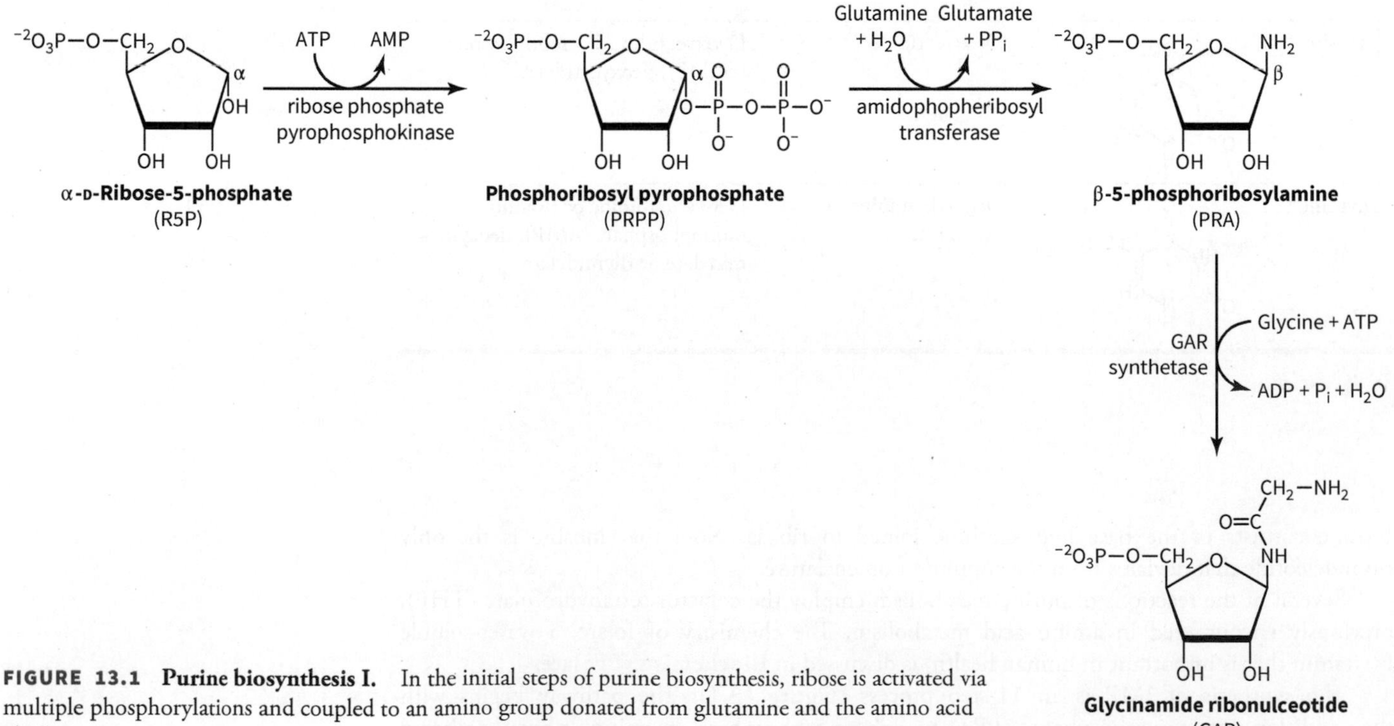

FIGURE 13.1 Purine biosynthesis I. In the initial steps of purine biosynthesis, ribose is activated via multiple phosphorylations and coupled to an amino group donated from glutamine and the amino acid glycine.

pyrophosphate (PRPP). This intermediate will not only be used in the synthesis of IMP but also in several other subsequent pathways to generate nucleotides.

Next, PRPP is amidated to form β-5-phosphoribosylamine. The amino group comes from the carboxamide group of glutamine and yields glutamate as a side product. The donated amino group becomes the N-9 position of the purine ring.

Glycine forms an amide with the amine of the phosphoribosylamine to produce glycinamide ribonucleotide (GAR), with glycine contributing the C-4, C-5, and N-7 positions of the growing purine. This reaction is driven by the hydrolysis of ATP. GAR is formylated on the free amine to yield formylglycinamide ribonucleotide (FGAR), as shown in **Figure 13.2**. The formyl group donor in this reaction is 10-formyl-THF, and the formyl group carbon becomes the C-8 of the purine.

The carbonyl oxygen of the amide linkage in FGAR is replaced by a second amino group, producing formylglycinamidine ribonucleotide (FGAM). The source of the amino group in this reaction is also glutamine, and the reaction is also driven by ATP hydrolysis.

The five-membered imidazole ring of the purine undergoes an intramolecular condensation to form 5-aminoimidazole ribonucleotide (AIR). This reaction consumes another molecule of ATP.

AIR is carboxylated on the N-3 position; this rearranges to add the carboxyl to the C-4 position, forming carboxyaminoimidazole ribonucleotide (CAIR). Another molecule of ATP is consumed.

Aspartate forms an amide linkage via the new carboxyl group to produce 5-aminoimidazole-4-(*N*-succinylocarboxamide) ribonucleotide (SAICAR), and another molecule of ATP is consumed **Figure 13.3**).

Fumarate is cleaved from SAICAR to yield 5-aminoimidazole-4-carboxamide ribonucleotide (AICAR).

AICAR is formylated on the N-3 amino group to produce 5-formylaminoimidazole-4-carboxamide ribonucleotide (FAICAR). As seen earlier, the formyl group is donated by 10-formyl-THF.

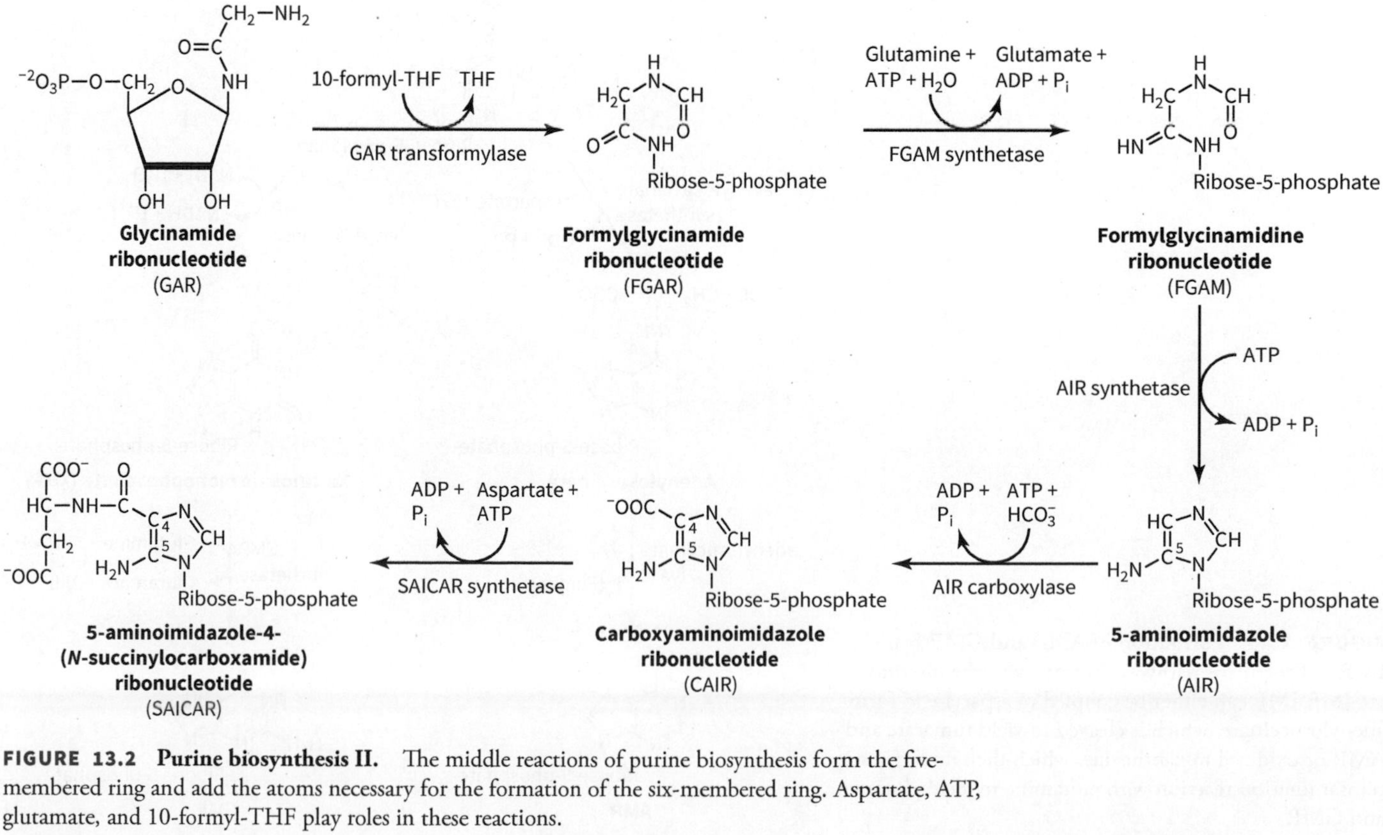

FIGURE 13.2 Purine biosynthesis II. The middle reactions of purine biosynthesis form the five-membered ring and add the atoms necessary for the formation of the six-membered ring. Aspartate, ATP, glutamate, and 10-formyl-THF play roles in these reactions.

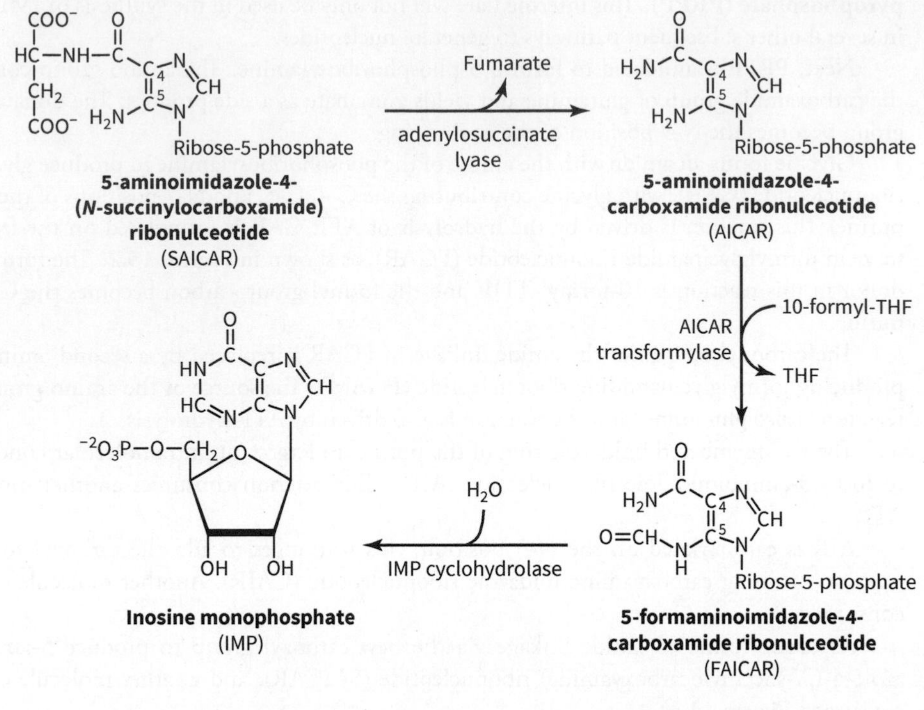

FIGURE 13.3 Purine biosynthesis III.
In the final steps of purine biosynthesis, inosine monophosphate is formed.

The six-membered ring is formed through a condensation reaction of the N-1 amino group and the C-2 formyl group to produce inosine monophosphate (IMP). Water is released in this final step.

After formation of IMP, the purine pathway forks into two branches: one leading to AMP and one to GMP (**Figure 13.4**). Both branches employ amino acids as cofactors in the reactions.

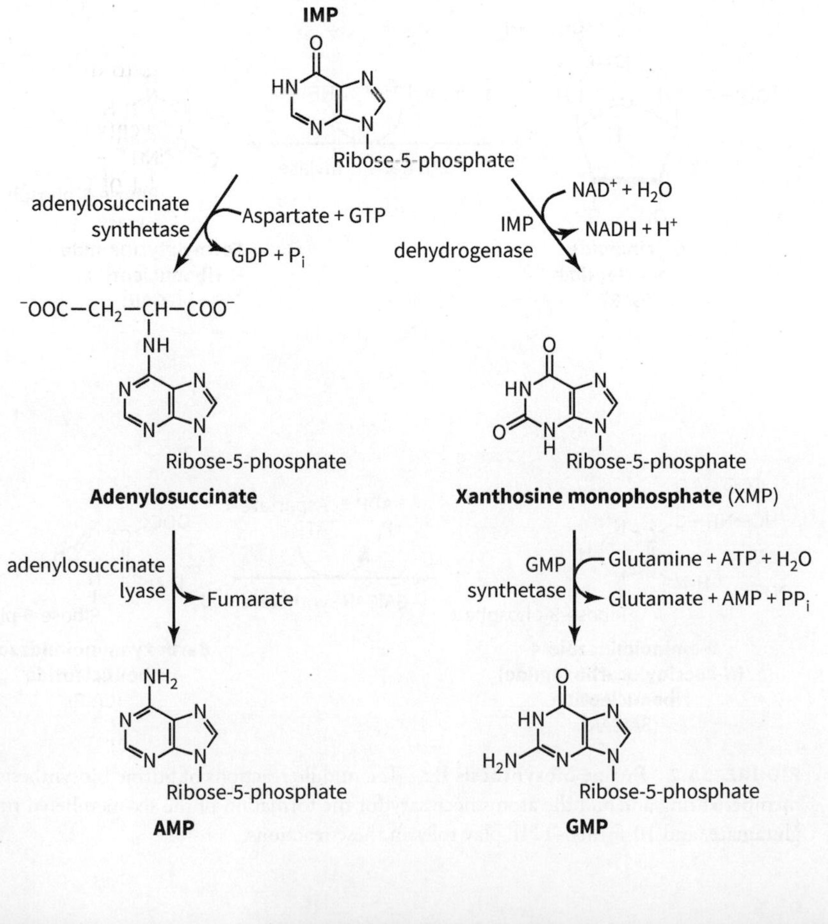

FIGURE 13.4 Formation of AMP and GMP from IMP. The purine pathway diverges with the intermediate IMP. IMP can either be coupled to aspartate to form adenylosuccinate, which is cleaved to yield fumarate and AMP, or oxidized to xanthosine, which then undergoes a transamination reaction with glutamine to yield glutamate and GMP.

The branch leading to AMP begins with the addition of aspartate to the C-6 position of the purine skeleton to form adenylosuccinate. Hydrolysis of a molecule of GTP provides the energy for this reaction. In the second step, adenylosuccinate is cleaved to yield AMP and fumarate.

The branch of the pathway leading to GMP begins with the oxidation of the C-2 to form xanthosine monophosphate. Next, the C-2 is transaminated using the terminal side chain amino group from glutamine to form the final products GMP and glutamate. ATP is a cofactor in this reaction, and it is hydrolyzed into AMP and PP$_i$.

13.1.2 The salvage biosynthetic pathway recycles hypoxanthine into adenosine and guanosine

The cost of synthesizing purines is relatively high, both in terms of the number of steps in the pathway and in the required number of molecules of ATP (six) and tetrahydrofuran (two). Rather than waste energy producing molecules that are hydrolyzed from the ribose backbone and lost, salvage pathways employ two enzymes—one for adenine (**adenine phosphoribosyltransferase**, or **APRT**) and the other for guanine and hypoxanthine (**hypoxanthine guanine phosphoribosyltransferase**, or **HGPRT**) (**Figure 13.5**). In salvage pathways, the base is transferred to PRPP to generate AMP, GMP or IMP, and PP$_i$. As seen in previous reactions, liberation and subsequent hydrolysis of PP$_i$ provides the energy to drive this reaction.

The purine salvage pathways are significant for more than just sparing metabolites, which is demonstrated by the fact that deficiencies in these pathways lead to disease. **Lesch–Nyhan syndrome** is caused by absent or dysfunctional HGPRT, which prevents the salvage of guanine and hypoxanthine. As a result, **uric acid** (the final breakdown product of purines) accumulates in all tissues of the body, causing severe developmental problems, moderate intellectual disability, poor muscle control, and kidney problems. Perhaps the most unusual effect of this syndrome is that children with it exhibit self-mutilation in the form of destructive biting of the lips and fingers. Because the gene for HGPRT is found on the X chromosome, it affects only males. The drug **allopurinol** decreases uric acid levels and relieves some of the symptoms of the syndrome, but it does not reduce the neurological signs, suggesting that more is at play in this disorder than simple uric acid accumulation.

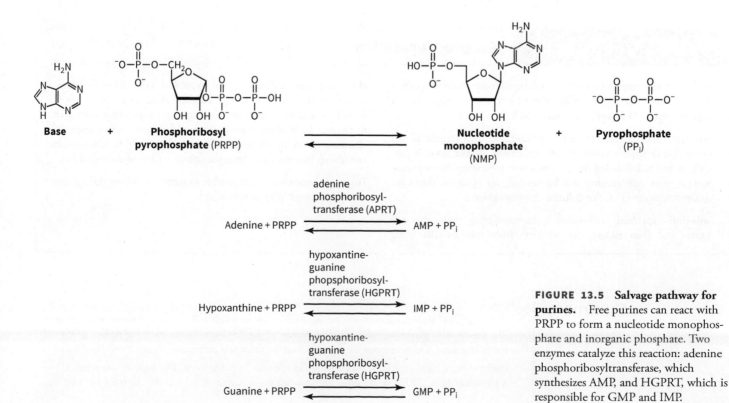

FIGURE 13.5 Salvage pathway for purines. Free purines can react with PRPP to form a nucleotide monophosphate and inorganic phosphate. Two enzymes catalyze this reaction: adenine phosphoribosyltransferase, which synthesizes AMP, and HGPRT, which is responsible for GMP and IMP.

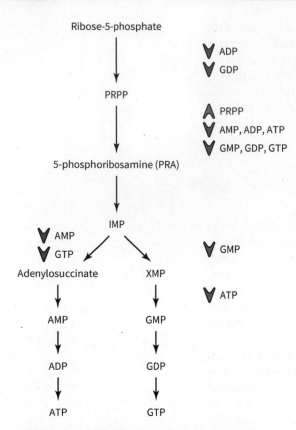

FIGURE 13.6 Regulation of purine biosynthesis.
Purine biosynthesis is generally regulated via the levels of
PRPP. High levels of PRPP activate the pathway, whereas
high levels of nucleotide phosphates serve to inhibit it.
This figure also shows the reciprocal regulation of ATP by
GTP and GTP by ATP.

13.1.3 Purine biosynthesis is tightly regulated

As seen in other pathways, the steps of purine biosynthesis that are tightly regulated occur either at the beginning or end of the pathway or at branch points.
Purine synthesis is regulated at the first step (formation of PRPP), the second
step (formation of 5-phosphoribosamine [PRA]), and at the steps leading to
AMP and GMP biosynthesis (**Figure 13.6**).

The first step of the pathway, synthesis of PRPP by ribose phosphate
pyrophosphokinase, is under a feedback inhibition mechanism, the inhibition being mediated by adenosine diphosphate (ADP) and guanosine diphosphate (GDP). Recall that **feedback inhibition** is when a product of a reaction
inhibits the reaction itself.

The second step, synthesis of PRA by amidophosphoribosyltransferase,
is considered the rate-determining step of the pathway. As with the first step,
it is subjected to feedback inhibition, but the mechanism of regulation is
somewhat different. This enzyme is inhibited by any of the purine nucleotides
(ATP, ADP, AMP, GTP, GDP, or GMP), but the adenyl and guanyl nucleotides bind to different sites on the enzyme. Thus, each of these nucleotides
can influence the enzyme independently or in concert with one another, acting synergistically. The enzyme is also stimulated by PRPP in a feed-forward
regulatory mechanism. Feed-forward regulation occurs when a downstream
product of a reaction activates a reaction.

The branch point where IMP leads to either AMP or GMP is also regulated. GTP is a substrate for AMP biosynthesis (at adenylosuccinate synthetase), and ATP is a substrate for GMP synthesis (at GMP synthetase); it
is these steps that are regulated. Higher concentrations of either nucleotide
triphosphate, ATP or GTP, will increase the rate of production of the other.
These enzymes are also feedback inhibited by their products, in this case
acting as competitive inhibitors and thus balancing production of AMP
and GMP.

Worked Problem 13.1   Recovery operation

Design an experiment that uses radiolabeled tracer molecules to test
the hypothesis that hypoxanthine can be recovered through the salvage pathway and incorporated into AMP.

Strategy   Adding radiolabeled hypoxanthine to a sample of cell
lysate that is actively making new nucleotides should make it possible to find radiolabeled AMP. A means of separating the two compounds from one another will be needed, for example, thin-layer
chromatography (TLC) or column chromatography.

Solution   Incubate radiolabeled hypoxanthine with the cell
lysate, and then isolate the products from one another using
chromatography. Unincorporated hypoxanthine could easily be isolated from its derivatives using TLC, and AMP could be separated
from IMP and adenylosuccinate using a cation-exchange resin such
as Dowex. Each of these products would require further chemical
characterization for ribose and phosphate but could be identified as
containing hypoxanthine from the incorporation of the radiolabel.

Follow-up question   Is it possible to recover xanthine through the
salvage pathway? Why or why not?

Summary

- In the *de novo* biosynthetic pathway, purine nucleotides are synthesized from a phosphoribose group.
  The amino acids glutamine, glycine, and aspartate all participate in the reaction and donate atoms to
  the purine base. Other carbons are donated from formate and carbonate.

- In the salvage biosynthetic pathway, the bases adenine, hypoxanthine, and guanine are rejoined to
  a phosphoribose. PRPP is the other substrate in these reactions, which are catalyzed by adenine

phosphoribosyltransferase or hypoxanthine guanine phosphoribosyltransferase. Deficiencies in the salvage pathway can lead to diseases such as Lesch–Nyhan syndrome.

- Purine biosynthetic pathways branch at IMP and are regulated by levels of the starting material PNPP at the second step of the pathway (formation of PRA) and through the enzymes leading to formation of AMP and GMP.

Concept Check

1. Illustrate how purines are synthesized through the *de novo* pathway.
2. What steps in purine biosynthesis are regulated by feedback inhibition by ADP and GDP?
3. Distinguish the origin of each of the atoms in the purine bases.
4. Discuss the salvage pathway for purines.

13.2 Pyrimidine Biosynthesis

The other class of bases besides purines is the monocyclic **pyrimidines**: cytosine, thymidine, and uracil.

Cytosine   **Thymidine**   **Uracil**

Like purines, pyrimidines are made from amino acids and carbonate, but unlike purines, pyrimidines are synthesized as free bases rather than ribosides. Just as the purines AMP and GMP are made from a common intermediate (IMP), cytosine, thymidine, and uracil are made from uridine monophosphate (UMP).

13.2.1 The *de novo* pathway of pyrimidine biosynthesis generates UMP

The *de novo* pyrimidine biosynthesis pathway builds the base orotate from carbamoyl phosphate and aspartate. Orotate is coupled to PRPP to generate orotidine monophosphate, which is further derivatized into UMP, cytidine-5-monophosphate (CMP), and thymidine monophosphate (TMP).

The synthesis of UMP requires six steps (**Figure 13.7**). The process begins with carbamoyl phosphate, which is synthesized from bicarbonate, glutamine, and ATP. As discussed in Chapter 11, carbamoyl phosphate is involved in the first step in urea biosynthesis, detoxifying ammonia liberated from amino acids. In nucleotide biosynthesis, carbamoyl phosphate is produced by **carbamoyl phosphate synthetase II (CPSII)**, which uses the amino group from the side chain of glutamine in the reaction. CPSII is found in the cytosol and is one of the rate-determining enzymes of pyrimidine biosynthesis.

The carbamoyl phosphate produced in the first step is transferred to the amino group of aspartate, forming carbamoyl aspartate. The loss of the phosphate-leaving group from carbamoyl phosphate drives this reaction.

Next, carbamoyl aspartate is dehydrated to dihydroorotate in a cyclization reaction. Water is lost in the formation of the amide bond, resulting in the six-membered ring dihydroorotate.

Dihydroorotate is dehydrogenated to orotate, generating a double bond between C-4 and C-5. The reducing agent used in this reaction is quinone, resulting in dihydroquinone.

Next, orotate is coupled to PRPP, forming orotidine-5′-monophosphate (OMP). The loss of PP_i makes this reaction energetically favorable. PRPP is the same intermediate used in the initial steps of purine biosynthesis.

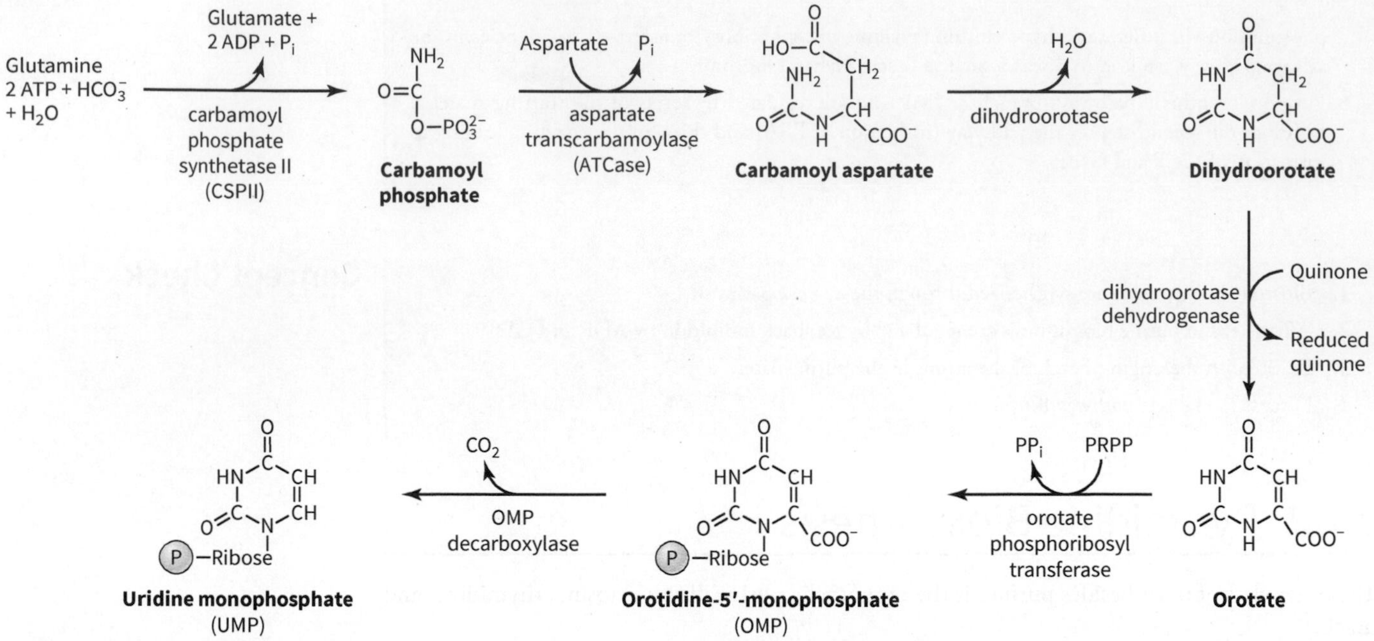

FIGURE 13.7 Pyrimidine biosynthesis. Pyrimidine biosynthesis forms a free base (orotate), which is then joined to PRPP.

Finally, orotidine monophosphate is decarboxylated on C-5 to form UMP. UMP then undergoes two successive phosphorylations to give uridine-5-triphosphate (UTP). This nucleoside triphosphate (NTP) is the basis of the remaining two purine nucleoside phosphates.

Cytidine-5-triphosphate (CTP) is produced from UTP. CMP is synthesized from UTP and glutamine by the action of CTP synthetase. Again, the terminal side chain amino group from glutamine is used as the substrate in this transamination reaction (**Figure 13.8**).

13.2.2 The pyrimidine salvage pathway converts CMP back into UMP

We have considered the salvage pathway for purines and seen how these bases can be recycled back into nucleotides. By contrast, because pyrimidines are made as free bases that are then transferred to PRPP, some of the enzymatic machinery necessary for recycling or salvage of these molecules is already found in the *de novo* pathway (**Figure 13.9**).

Cytidine and deoxycytidine can be deaminated to uridine and deoxyuridine (in humans) by cytidine deaminase, but using this pathway exclusively instead of recycling these bases would be energetically wasteful to the organism and could be detrimental to rapidly dividing cells because of a buildup of toxic ammonia. Additionally, pyrimidine metabolism needs to be roughly balanced with purine metabolism to provide the building blocks of DNA and RNA.

There is another important difference between purine and pyrimidine metabolism to consider. As we will see in the next section, purines are catabolized into uric acid, which is relatively insoluble. The products of pyrimidine breakdown are more soluble and can be reassimilated into lipid metabolism. This also provides a clue to why the salvage pathway for pyrimidines is necessary: it avoids wasting chemically and energetically expensive bases, prevents the accumulation of toxic by-products, and provides NTPs for rapidly dividing cells.

13.2.3 Pyrimidine biosynthesis is regulated allosterically

Aspartate transcarbamoylase (ATCase) is a well-studied enzyme and a model for allosteric regulation. Allosteric enzymes do not follow Michaelis–Menten kinetics; instead, they have sigmoidal, or

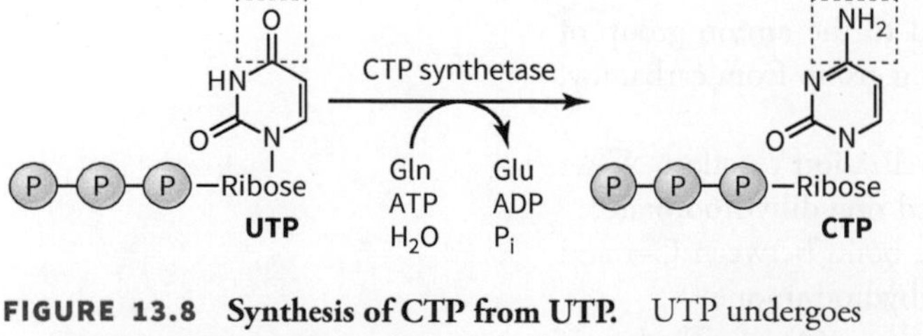

FIGURE 13.8 Synthesis of CTP from UTP. UTP undergoes a transamination reaction with the side chain amino group of glutamine to form CTP. The energy required for this reaction comes from the hydrolysis of a molecule of ATP.

S-shaped, activity curves that are shifted to the left upon stimulation or to the right upon inhibition.

Pyrimidine biosynthesis regulation in *E. coli* The *Escherichia coli* (*E. coli*) isoform of ATCase is the one that is most often studied, and it is the regulated and rate-determining step in *E. coli* pyrimidine biosynthesis; however, the same is not true in all species.

ATCase catalyzes the formation of carbamoyl aspartate from aspartate and carbamoyl phosphate. This enzyme is a dodecamer comprised of two trimers of catalytic subunits arranged in a central core, surrounded by three dimers of regulatory subunits around the equator of the complex (**Figure 13.10**). ATCase has two notable allosteric regulators: ATP and CTP. ATP binds to the regulatory subunits, which stabilizes the R state of the enzyme and increases enzymatic activity, thereby shifting the activity curve to the left. CTP is a downstream product of ATCase and an allosteric inhibitor. It binds to a different site on the regulatory subunit, which stabilizes the T state and slows enzymatic activity, thereby shifting the curve to the right. In other species of bacteria, UTP is the allosteric regulator of ATCase that inhibits activity.

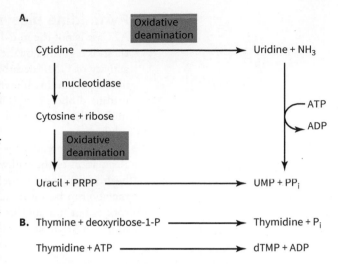

FIGURE 13.9 Salvage pathway for pyrimidines. A. Uracil can be salvaged by coupling to PRPP. Cytosine or cytidine can be deaminated to uracil or uridine and salvaged. **B.** Thymine can be salvaged by coupling to 1-deoxyribose.

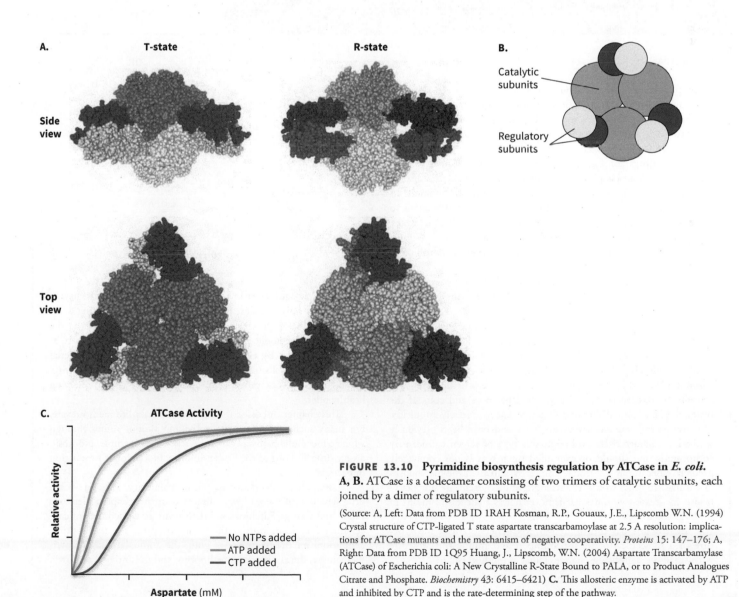

FIGURE 13.10 Pyrimidine biosynthesis regulation by ATCase in *E. coli*. A, B. ATCase is a dodecamer consisting of two trimers of catalytic subunits, each joined by a dimer of regulatory subunits.

(Source: A, Left: Data from PDB ID 1RAH Kosman, R.P., Gouaux, J.E., Lipscomb W.N. (1994) Crystal structure of CTP-ligated T state aspartate transcarbamoylase at 2.5 A resolution: implications for ATCase mutants and the mechanism of negative cooperativity. *Proteins* 15: 147–176; A, Right: Data from PDB ID 1Q95 Huang, J., Lipscomb, W.N. (2004) Aspartate Transcarbamylase (ATCase) of Escherichia coli: A New Crystalline R-State Bound to PALA, or to Product Analogues Citrate and Phosphate. *Biochemistry* 43: 6415–6421) **C.** This allosteric enzyme is activated by ATP and inhibited by CTP and is the rate-determining step of the pathway.

Pyrimidine biosynthesis regulation in mammals
By contrast, in mammals ATCase is not the rate-determining step and is not highly regulated. Instead, pyrimidine biosynthesis is regulated by the level of carbamoyl phosphate in the cytosol (via CPSII) and through the activity of OMP decarboxylase (part of the UMP synthase complex). Carbamoyl phosphate for use in pyrimidine biosynthesis is produced in the cytosol by CPSII. This enzyme is inhibited by uridine diphosphate (UDP) and UTP and activated by ATP and PRPP. OMP decarboxylase is inhibited by its immediate product, UMP, and by the downstream product, CMP, via product inhibition.

The differences observed in humans and other species are more than a biochemical curiosity. Regulation of the pathways of purine and pyrimidine biosynthesis is critical for the growth and health of any organism. Investigators are currently attempting to exploit differences in nucleotide metabolism between humans and other organisms to develop drugs that target pathways in one species but not another to treat infection. A discussion of how drugs like these are developed is found in **Medical Biochemistry: Drug design.**

Medical Biochemistry

Drug design

Many people hold the erroneous belief that physicians develop new drugs while working in isolation. Instead, companies (either large pharmaceutical companies or small startups) develop and bring drugs to the market. There are several ways that drugs are found and developed.

Many drugs come to us from plants or fungi that have been used by indigenous people since antiquity. Ethanol, tobacco, caffeine, codeine, and morphine are all naturally occurring compounds. In addition, it had been known since the 1800s that modifications to these naturally occurring molecules could produce different effects, some desirable (such as the acetylation of salicylic acid to form aspirin), and others less so (such as the acetylation of morphine to make heroin).

There are currently two major ways that drug companies develop new compounds: library screens and structure-based drug design.

All drug companies have libraries of compounds, usually hundreds of thousands of small molecules cataloged and chemically characterized but with unknown biological effect. In a library screen, an assay is developed (blocking cell division, for example), and the compounds in the library are screened to see whether they will achieve the desired effects. This method has pros and cons. If the system is automated, dishes of cells can be screened relatively quickly. The compounds are already in existence and have been previously studied or characterized. Although this type of screen can identify compounds that should be examined further, it is not completely predictive because it is a much simpler system than the metabolism of animals. In other words, the strength of the system (its simplicity) is also its greatest weakness. It is nevertheless a successful means of finding new drugs.

In structure-based drug design, researchers examine the structure of a molecule (folate, for example) and mimic this structure with small but significant changes (such as methotrexate).

Folate

Methotrexate

These changes result in a molecule that can bind to an enzyme as the native molecule does but acts differently, either acting as an inhibitor of the enzyme or acting as a substrate that results in the blockade of a metabolic process or pathway, generally referred to as an antimetabolite. By exploiting the differences between bacterial and mammalian cells (or healthy versus diseased cells), these drugs can be used to block processes in undesired cells while preserving healthy ones.

This chapter discussed the folate antimetabolite methotrexate, but many such molecules exist, providing various routes through which a metabolic pathway can be attacked. For example, prontosil, a sulfa drug and one of the first antibiotics, blocks folate production in bacteria.

Using their knowledge of purine chemistry, structure-based drug design, and the pathways discussed in this chapter, Gertrude Ellion and George Hitchings and their team developed the anticancer drug 6-mercaptopurine, the immunosuppressive azathioprine, allopurinol for the treatment of gout, the antimalarial agent pyrimethamine, the antibiotic trimethoprim, and the antiviral acyclovir.

Drug	Target Organism	Targeted Step in Metabolism	Medical Use
6-mercaptopurine	Humans	HGPRT substrate, metabolites slow purine production leading to decreased DNA synthesis	Leukemia, lymphoma, arthritis, inflammatory bowel disease
Allopurinol	Humans	Xanthine oxidase inhibitor	Gout
Azathioprine	Humans	Phosphoribosyl transferase inhibitor (ultimately DNA synthesis)	Immunosuppression following organ transplantation
Pyrimethamine	Protozoans (*Plasmodium falciparum* or *Toxoplasmosis gondii*)	DHF reductase inhibitor	Antimalarial treatment of toxoplasmosis in immunocompromised patients
Trimethoprim	Bacteria	DHF reductase inhibitor	Urinary tract infections
Acyclovir	Retroviruses	Reverse transcriptase (viral replication)	Herpes, HIV, shingles

Worked Problem 13.2 Predicted kinetics

You are characterizing a novel inhibitor of bacterial ATCase. Based on structural similarities, you hypothesize that the novel factor will act as an allosteric inhibitor of the enzyme. What would a graph of reaction rate versus substrate concentration look like?

Strategy ATCase is an allosteric enzyme. What do the kinetics of an allosteric reaction look like?

Solution The graph of reaction rate versus substrate concentration would produce an S-shaped, or sigmoidal, curve. In these enzymes, the cooperativity between the subunits leads to different activities at different substrate concentrations. In the presence of an allosteric inhibitor, the curve would retain the same shape but be shifted to the right, meaning that it would take more substrate to achieve the same reaction rate.

Follow-up question Could this novel inhibitor of bacterial ATCase be developed as a potential antibiotic? What considerations would you have to take into account to develop it as an antibiotic?

Summary

- Pyrimidines are synthesized in the cytosol from carbamoyl phosphate and aspartate. *De novo* pyrimidine biosynthesis results in UTP, which can be aminated to form CTP.
- In the salvage pathway of pyrimidine biosynthesis, CMP is converted back into UMP via the action of cytosine deaminase.
- The synthesis of pyrimidines in *E. coli* is regulated through ATCase. CTP acts as an allosteric inhibitor, whereas ATP acts as an allosteric activator. In mammals, pyrimidine synthesis is regulated at two points: the synthesis of carbamoyl phosphate by CPSII and the synthesis of UMP by OMP decarboxylase. CPSII is activated by ATP and PRPP and inhibited by UDP and UTP. OMP decarboxylase is inhibited by UMP and CMP.

Concept Check

1. Describe the steps of the *de novo* pathway for pyrimidine biosynthesis, including the chemical reactions and origins of each of the atoms in the pyrimidine bases.
2. Discuss the salvage pathway for pyrimidine biosynthesis.
3. Describe how and where pyrimidine biosynthesis is regulated in *E. coli* and in mammals.

13.3 Deoxyribonucleotide Biosynthesis

So far, this chapter has discussed the synthesis of ribonucleotides, which are used in the synthesis of all forms of RNA. The cell also uses the potential energy in the phosphoanhydride bonds of ribonucleotides to transfer phosphate groups, drive the biosynthesis of lipids and carbohydrates, and act as the energy currency of the cell. Ribonucleotides are also the components of several important cofactors, including NAD$^+$, FAD, and NADP$^+$. These cofactors are not, however, the

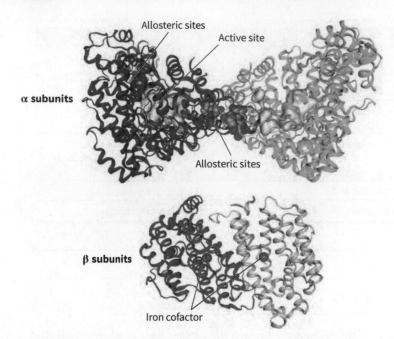

α subunits

Allosteric sites

Active site

Allosteric sites

β subunits

Iron cofactor

FIGURE 13.11 Ribonucleotide reductase. Ribonucleotide reductase can adopt several different quaternary structures in the cell. Shown is the $\alpha_2\beta_2$ arrangement with the two catalytic subunits (α) at the top and the two regulatory subunits (β) at the bottom.

(Source: Top: Data from PDB ID 3HNE Fairman, J.W., Wijerathna, S.R., Ahmad, M.F., Xu, H., Nakano, R., Jha, S., Prendergast, J., Welin, R.M., Flodin, S., Roos, A., Nordlund, P., Li, Z., Walz, T., Dealwis, C.G. (2011) Structural basis for allosteric regulation of human ribonucleotide reductase by nucleotide-induced oligomerization. *Nat. Struct. Mol. Biol.* **18**: 316–322; Bottom: Data from PDB ID 2VUX Welin, M., Moche, M., Andersson, J., Arrowsmith, C.H., Berglund, H., Busam, R.D., Collins, R., Dahlgren, L.G., Edwards, A.M., Flodin, S., Flores, A., Graslund, S., Hammarstrom, M., Herman, M.D., Johansson, A., Johansson, I., Kallas, A., Karlberg, T., Kotenyova, T., Lehtio, L., Nilsson, M. E., Nyman, T., Persson, C., Sagemark, J., Schueler, H., Svensson, L., Thorsell, A.G., Tresaugues, L., van Den Berg, S., Weigelt, J., Wikstrom, M., Nordlund, P., Structural Genomics Consortium (SGC) (2008). Human ribonucleotide reductase, subunit M2 B. To be published.)

building blocks of DNA. An important difference between RNA and DNA is the lack of the hydroxyl group in the 2′ position of the deoxyribonucleotides used to synthesize DNA. Deoxyribonucleotides are synthesized from ribonucleotides using the enzyme **ribonucleotide reductase**.

There are three classes of ribonucleotide reductase, all of which employ a free radical for catalysis; however, they differ in how they generate this radical. This section focuses on the most common form (class I) from humans, although the enzyme is highly conserved throughout evolution.

13.3.1 The simplest active structure of ribonucleotide reductase is a dimer of dimers

Ribonucleotide reductases have two types of subunits, alpha (α) and beta (β) (**Figure 13.11**). The α subunit contains the active site and two allosteric sites, whereas the β subunit contains a tyrosine residue that is critical for the reaction.

In their simplest form, ribonucleotide reductases are tetrameric proteins comprising a dimer of dimers, giving an overall topology of $\alpha_2\beta_2$. As discussed in the following paragraphs, the complex can form other ratios, depending on the concentrations of substrates and products. The α subunits twist around each other, forming an α_2 dimer to which the β subunits bind.

The β subunits also contain a pair of metal ions (typically either Fe^{3+} or Mn^{3+}), bridged by an oxygen atom. These cofactors are necessary for generation of the tyrosyl radical.

13.3.2 Ribonucleotide reductase reduces the 2′ carbon of the NDPs

The substrates of the ribonucleotide reductase are the nucleotide diphosphates ADP, GDP, cytidine diphosphate (CDP), and UDP. The action of the enzyme produces the deoxynucleotides dADP, dCDP, and dGDP; dTDP is produced through the methylation of dUDP, discussed later.

Ribonucleotide reductase has a pair of thiols in its active site (**Figure 13.12**). The reaction proceeds through the reduction of the 2′ carbon; the hydroxyl group formerly on the 2′ carbon is released as water, and the enzyme is oxidized, with both former thiols now forming a disulfide linkage.

To regenerate the reduced form of ribonucleotide reductase, electrons in the disulfide must be donated to an acceptor; there are two ways by which this occurs. In the first, an electron is passed to reduced **glutaredoxin**, a protein that also has a redox-active thiol pair. Loss of the electrons to ribonucleotide reductase generates the oxidized form of glutaredoxin. Glutaredoxin is reset to the reduced form using two equivalents of glutathione, a redox-active tripeptide that is important in many redox reactions and helps to maintain a reducing environment within the cell. A third enzyme, **glutathione reductase**, regenerates glutathione in the cell using electrons from NADPH.

The second way in which some organisms reduce ribonucleotide reductase is through the redox-active protein **thioredoxin**. As with glutaredoxin, there is a redox-active thiol pair in thioredoxin that becomes oxidized when electrons are lost to ribonucleotide reductase. The oxidized thioredoxin is reduced by **thioredoxin reductase**, using $FADH_2$ as a cofactor. As with glutathione reductase, thioredoxin reductase is reduced by NADPH to regenerate the enzyme.

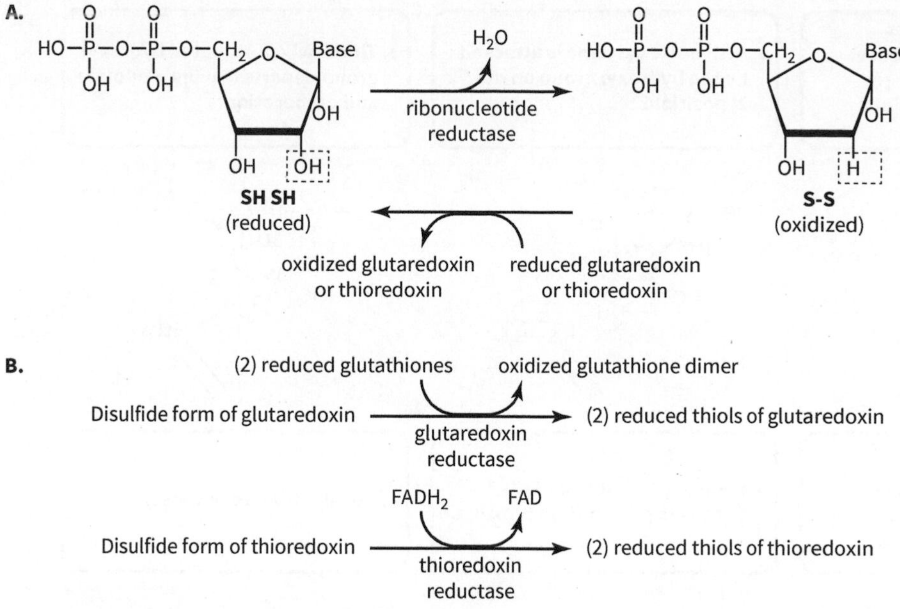

FIGURE 13.12 Ribonucleotide reductase reaction. A. Ribonucleotide reductase uses a pair of active cysteines to remove the hydroxyl group from the 2′ position of the ribose ring. The thiols are oxidized in the course of the reaction. **B.** To reduce the thiols and regenerate the enzyme, electrons are passed to the disulfide, either from glutaredoxin or thioredoxin, each of which also has an active cysteine pair. Glutaredoxin or thioredoxin must also be reduced, either by glutathione reductase or thioredoxin reductase. Both of these enzymes ultimately use NADPH as the source of reducing equivalents.

13.3.3 Ribonucleotide reductase employs a free radical and a series of redox-active thiols to synthesize deoxyribonucleotides

The mechanism of ribonucleotide reductase commences in the β subunit (**Figure 13.13**). The local environment (including the two iron atoms and oxygen) polarizes a tyrosine side chain, resulting in the formation of a tyrosyl radical. In some species, this radical is up to 35 Å from the active site on the α subunit. The radical is then transferred through a proton-coupled electron transfer pathway. The electron travels through this pathway to the α subunit, where it reduces a thiol side chain of a cysteine residue to generate a thiyl radical (–RS·). The radical abstracts a proton from the 3′ position of the ribose ring, regenerating the thiol of the cysteine side chain and generating a new radical on the ribose ring.

Next, the positioning of the ribose ring facilitates attack of the lone pair of electrons from the 2′ hydroxyl oxygen on a different thiol in the active site. This results in protonation of the hydroxyl group and generation of a thiolate anion from the cysteine side chain. The protonated hydroxyl group is lost, leaving a carbocation in the 2′ position of the ring, stabilized through resonance with the radical on the 3′ position.

A third cysteine residue then comes into play. The thiolate generated in the previous step attacks an adjacent thiol on the third cysteine. This forms a disulfide bond between the two and results in an attack of the thiol on the carbocation, regenerating a carbon–hydrogen bond (effectively a hydride-transfer reaction).

In the final step of the mechanism, the hydrogen atom (hydrogen radical) is transferred from the thiol of the first cysteine back to the ribose ring, forming a new carbon–hydrogen bond and regenerating the thiyl radical.

13.3.4 Regulation of ribonucleotide reductase balances deoxynucleotide concentrations and impacts the rate of DNA synthesis

Ribonucleotide reductase is regulated at the level of transcription and by allosteric regulators. The yeast isoform of the enzyme also has a protein inhibitor, but this has not been observed in humans. Both α and β subunits of ribonucleotide reductase are involved in the regulation of function. Expression of the β subunit of the enzyme varies throughout the cell cycle, peaking with

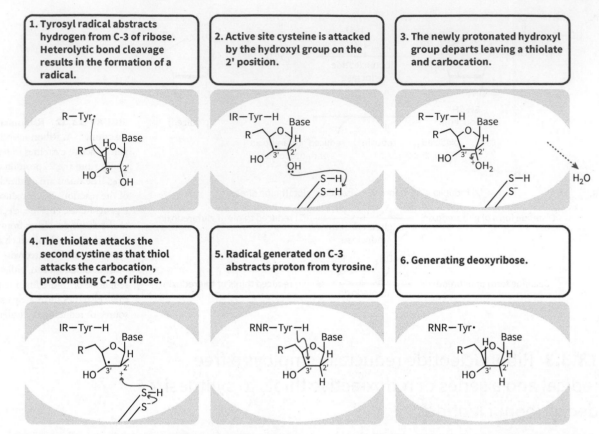

1. Tyrosyl radical abstracts hydrogen from C-3 of ribose. Heterolytic bond cleavage results in the formation of a radical.

2. Active site cysteine is attacked by the hydroxyl group on the 2' position.

3. The newly protonated hydroxyl group departs leaving a thiolate and carbocation.

4. The thiolate attacks the second cystine as that thiol attacks the carbocation, protonating C-2 of ribose.

5. Radical generated on C-3 abstracts proton from tyrosine.

6. Generating deoxyribose.

FIGURE 13.13 **The mechanism of ribonucleotide reductase.** Ribonucleotide reductase uses a tyrosyl radical and a pair of cysteine residues to protonate the 2' hydroxyl group of the ribose ring. The protonated hydroxyl group departs, resulting in 2' deoxyribose.

DNA synthesis at S phase. Because the enzyme requires the β subunit for activity, regulation of expression of this subunit can regulate the overall level of active enzyme.

Why does the cell require tight regulation of dNTP production? Ribonucleotide reductase catalyzes the production of the building blocks of DNA, low levels of which will prevent the cell from replicating. Furthermore, if any single deoxynucleotide is absent, the cell will be unable to divide or repair its DNA. Conversely, high levels of dNPs would be wasteful. Complicating matters further, ratios of dATP to dGTP and dCTP to thymidine triphosphate (dTTP) need to be balanced because an imbalance increases the likelihood that an incorrect nucleotide will be inserted during replication—in other words, an imbalance is mutagenic. Therefore, levels of each product need to alter the substrate specificity of the enzyme to stimulate the production of the partner deoxyribonucleotide. This finer level of regulation calls for an allosteric regulation of activity.

There are two regulatory sites on the α subunit termed the S (for specificity) and A (for activity) sites (**Figure 13.14**). The S site binds to several different NTPs or dNTPs and, as mentioned earlier, alters the substrate specificity of the enzyme. Binding of dATP or ATP to this site increases the specificity of the enzyme for CDP and UDP, whereas binding of dGTP increases the specificity for ADP, and binding of dTTP increases the specificity for GDP.

The A site binds to either ATP or dATP and regulates activity. When dATP is bound, the enzyme becomes less active. Interestingly, the enzyme binds 100 times more tightly to dATP than to ATP. This helps to explain why dATP is found in lower abundance than ATP, but the mechanism of this specificity is not known.

Binding of dATP has another effect on ribonucleotide reductase. Previously, we discussed how the α and β subunits of ribonucleotide reductase form an active tetrameric complex with an $\alpha_2\beta_2$ stoicheometry. Structural studies indicate that binding of dATP can cause other ratios to form, including $\alpha_4\beta_4$ or $\alpha_6\beta_2$. Also, binding of ATP can induce higher order forms (such as $\alpha_6\beta_6$). The implications of these complex higher order forms is not yet clear, but it may be that binding

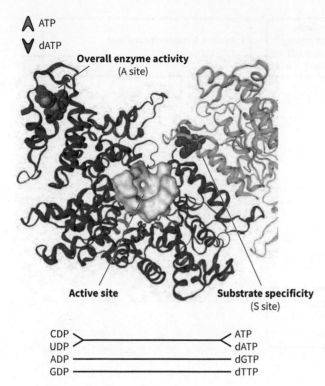

FIGURE 13.14 **Regulation of ribonucleotide reductase.** Ribonucleotide reductase is stimulated by ATP and inhibited by dATP at the allosteric site. When active, if the regulatory subunit binds to either ATP or dATP, the catalytic site will recognize UDP or CDP. When dGTP is bound by the regulatory subunit, the catalytic site will recognize ATP. Finally, when the regulatory site is bound to dTTP, the catalytic site will recognize GTP. This complex regulatory mechanism ensures balance between the concentrations of deoxynucleotides.

(Source: Data from PDB ID 3HNE Fairman, J. W., Wijerathna, S. R., Ahmad, M. F., Xu, H., Nakano, R., Jha, S., Prendergast, J., Welin, R. M., Flodin, S., Roos, A., Nordlund, P., Li, Z., Walz, T., Dealwis, C. G. (2011) Structural basis for allosteric regulation of human ribonucleotide reductase by nucleotide-induced oligomerization. *Nat. Struct. Mol. Biol.* 18: 316–322)

of either ATP or dATP results in different multimeric forms: an active form bound to ATP and an inactive form bound to dATP.

Morpheeins are proteins that exist in one of several oligomeric forms. The switch between these forms is dependent on binding a small molecule such as dATP. Ribonucleotide reductase is one enzyme that may fit the morpheein model of allosteric regulation.

Just as NTPs can be generated by transfer of a phosphoryl group from any other NTP to a NDP, the same reaction happens with dNTPs. In fact, any of the NTPs or dNTPs can transfer a phosphoryl group to a ribonucleoside-5-diphosphate (NDP) or dNDP, in a reaction catalyzed by **nucleoside diphosphate kinase**.

13.3.5 Thymidine is synthesized from dUMP

Thymidine is synthesized *de novo* through the methylation of deoxyuridate (dUMP) by thymidylate synthase. The methyl group donor in this reaction is 5,10-methylene-THF (**Figure 13.15**).

The mechanism of thymidylate synthase

The proposed mechanism of thymidylate synthase is shown in **Figure 13.16**. First, an active site thiolate anion attacks the C-6 of the uridylate ring, forming a covalent intermediate. Electrons flow up and into the carboxyl oxygen from the C-4, forming an enolate. Electrons from the double bond between C-4 and C-5 attack the methylene carbon of 5,10-methylene-THF. This carbon exists in equilibrium with an iminium cation form of THF. Both substrates are now in a covalent complex with the enzyme. Electron flow back down the oxygen regenerates the carboxyl group and forms a link between C-5 and the methylene group. Next, a base abstracts a proton from C-5. This results in electron flow up into the C-5–methylene bond and from the methylene group back up into the THF ring. Electron flow from the ring nitrogen of THF results in the transfer of a hydride ion (H–)

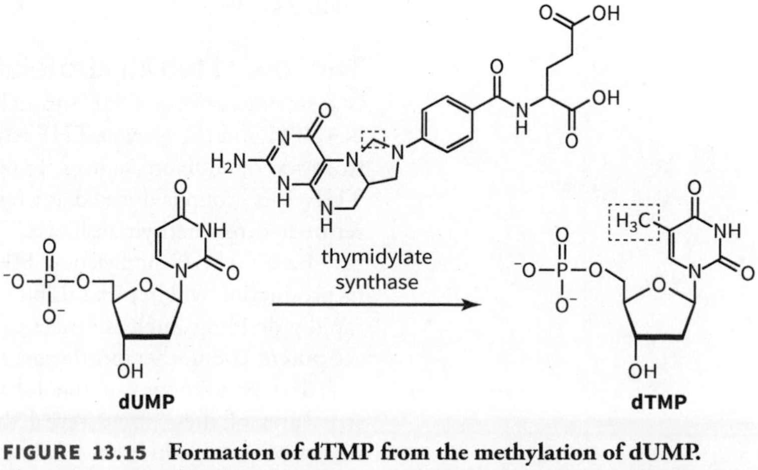

FIGURE 13.15 **Formation of dTMP from the methylation of dUMP.** Thymidine monophosphate (dTMP) is synthesized through the methylation of deoxyuridate (dUMP) by thymidylate synthase. 5,10-methylene-THF is the source of the donated carbon (shown in the dashed box).

1. **Active site Cys forms adduct with uracil ring.**

2. **Protonation of methylene-THF results in cleavage of the THF ring containing active methylene group. Tautomeric form of uracil attacks methylene group.**

3. **Adduct between uracil and THF collapses.**

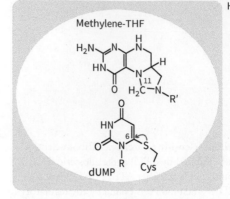

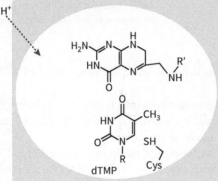

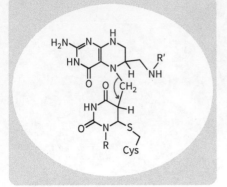

4. **Protonation of methylene group and electron flow back through uracil ring cleaves bond to cystine.**

5. **Release of dTMP product and DHF regenerates enzyme. Thiolate protonates.**

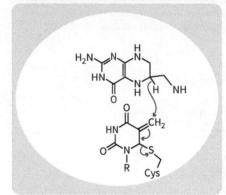

FIGURE 13.16 The mechanism of thymidylate synthase. The methylation of dUMP by thymidylate synthase proceeds through a covalent adduct of dUMP to cystine. Methylene-THF provides a methylene group, which is protonated in the final steps of the mechanism.

to the methylene group of the pyrimidine, generating a methyl group and DHF. The resulting electron flow cleaves the carbon–sulfur bond, regenerating the thiolate and releasing the dTMP product.

In humans, free thymine does not appear to be metabolized through a salvage pathway. However, thymidine can be phosphorylated to form thymidine monophosphate (dMTP) by thymidine kinase.

The role of tetrahydrofolate in nucleotide metabolism
5,10-methylene-THF is regenerated from DHF through two successive reactions (**Figure 13.17**). The first employs NADPH and the enzyme THF reductase to generate THF from DHF. THF can acquire carbons from several different sources. In mammals, a methylene group is added to form 5,10-methylene-THF. This group is donated as a hydroxymethyl group from serine, and the enzyme responsible is serine hydroxymethyltransferase.

Because 5,10-methylene-THF is so central to nucleotide production, compounds that inhibit its production will in turn inhibit nucleotide production. This is most pronounced in cells that are rapidly dividing, such as cancer cells, and is the reason why drugs that inhibit THF metabolism are potent chemotherapeutic agents; these drugs are classified as folate antimetabolites.

The best known of the folate antimetabolites are **methotrexate** and **aminopterin**. The structures of these drugs reveal their mechanism of action: they are competitive inhibitors of **dihydrofolate reductase**, binding tightly and blocking the regeneration of THF. These drugs indiscriminately kill any cells that are rapidly dividing, including endothelial cells, bone marrow cells, and cells of the immune system. This helps to explain some of the side effects associated with

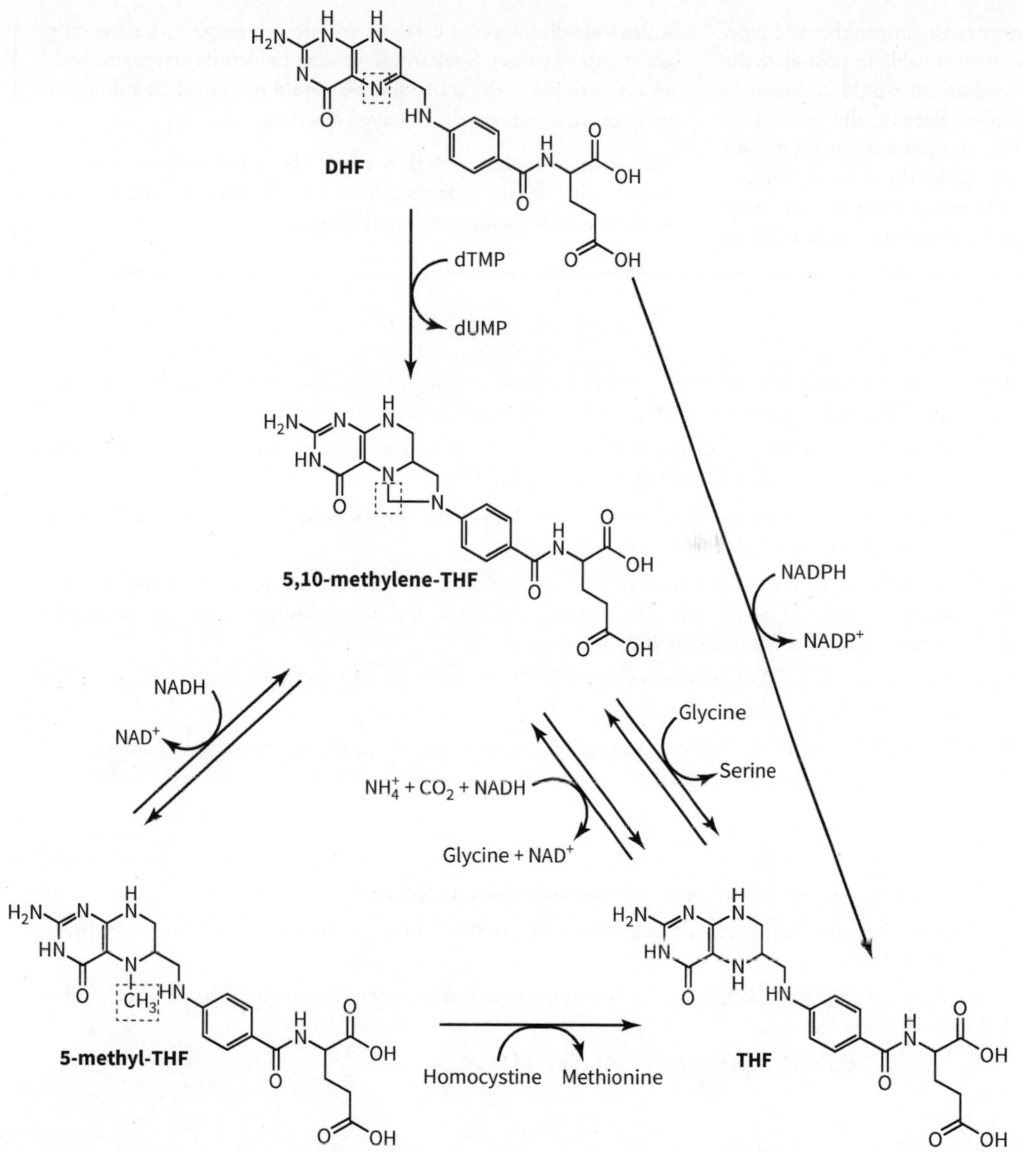

FIGURE 13.17 Regeneration of 5,10-methylenetetrahydrofolate. THF is regenerated from DHF by THF reductase, an NADPH-dependent enzyme. The methylene group is obtained from serine, generating glycine as a product.

cancer treatment, including hair loss, skin changes, nausea, mouth sores, diarrhea, anemia, and compromised immune system. Despite these side effects, folate antimetabolites are highly effective. In the 1950s, before the discovery of drugs such as methotrexate, childhood leukemia was a death sentence; today the survival rate of some of these cancers is as high as 95%.

Worked Problem 13.3   Amino acid analogs

We often think of thiols as being analogous to hydroxyl groups with regard to their chemistry. If a double mutant of ribonucleotide reductase had a pair of serine residues instead of a pair of cysteines in the active site, would you anticipate that it would be catalytically active? Explain your reasoning.

Strategy Because sulfur is below oxygen in the periodic table, and because cysteine is a structural analog of serine, we might anticipate that the chemistry these two side chains can perform is equivalent and redundant. However, the stronger basicity of the alkoxide versus the thiolate may alter the chemistry enough to stop the reaction

from proceeding. Recall from organic chemistry that the thiol is more acidic than the hydroxyl and is as many as 9 pK_a units lower, which is a billion times more acidic. Also recall that sulfur is less electronegative than oxygen and that therefore the thiolate is a better nucleophile than the alkoxide. These properties will also influence the reactivity of the mutant.

Solution The double mutant is probably not an active catalyst. Although the oxygen–hydrogen bond of the serine side chain is more polar than the sulfur–oxygen bond of cysteine, the pK_a of the hydroxyl proton is higher by several orders of magnitude. Likewise, if the mechanism proceeded in a manner analogous to the wild type, and if one of the serines in the active site were able to protonate the 2′ hydroxyl of the ribose ring, the mechanism would continue to produce some form of peroxide complex. Because the peroxide is more highly reactive than the disulfide, the potential is greater for an unfavorable side reaction that would render the enzyme useless.

There are enzymatic reactions employing peroxide, but these reactions usually generate or degrade hydrogen peroxide (such as superoxide dismutase or catalase) and do not employ a catalytically active pair of serines. Similarly, if we were to identify an enzyme with an active serine in the active site, we would need to elicit a different mechanism to explain the observed reaction.

Follow-up question   Methionine is the other sulfur-containing amino acid. Would substitution of a methionine for the cysteine result in a catalytically competent enzyme?

Summary

- Deoxyribonucleotides are synthesized from ribonucleotides via the action of ribonucleotide reductase. Ribonucleotide reductase is comprised of multimers of α and β subunits.
- The oxidized enzyme is reduced back to the active state by either glutaredoxin or thioredoxin. In either case, the electrons ultimately come from NADPH.
- The mechanism of ribonucleotide reductase employs a tyrosyl free radical and reduces the 2′ carbon of ribose using a pair of redox-active disulfides.
- Allosteric regulation of ribonucleotide reductase controls the activity, substrate specificity, and oligomerization of the enzyme. The allosteric regulation of ribonucleotide reductase is essential for management of concentrations and ratios of dNTPs.
- The transfer of a phosphate from other NTPs or dNTPs, catalyzed by phosphoribonucleotide kinase, produces dNTPs.
- Thymidine is synthesized *de novo* through the methylation of dUMP by thymidylate synthase.

Concept Check

1. Illustrate the mechanism by which ribonucleotides are reduced.
2. Describe how ribonucleotide reductase is regulated and how the multimeric structure contributes to activity.
3. What are the two mechanisms by which the organisms may regenerate the reduced form of ribonucleotide reductase?
4. How is NADPH necessary for production of DNA?

13.4 Catabolism of Nucleotides

As with several other classes of complex biological molecules, nucleotides are both built up and broken down in a modular fashion. In general, nucleotides are first dephosphorylated to nucleosides and then deglycosylated to yield the purine or pyrimidine base. This deglycosylation step is catalyzed by a phosphorylase akin to the one used in the breakdown of glycogen. It employs inorganic phosphate ($H_2PO_4^-$) in the reaction to generate ribose-1-phosphate. Finally, the purine or pyrimidine base is catabolized and eliminated. As with amino acids, these bases contain amines that need to be detoxified and eliminated by the body. For pyrimidines, this means transamination to β-keto acids or oxidative deamination. For purines, amino groups that are attached to the bicyclic ring system are oxidatively deaminated, but the bicyclic ring system remains intact. The **xanthine** base undergoes one final oxidation to become uric acid, which is also eliminated in the urine (or in solid waste in birds and reptiles). This section examines each of these steps in detail.

13.4.1 Purines are catabolized to nucleosides and then uric acid

Purine bases are freed from nucleotides through two successive reactions (**Figure 13.18**). They are first dephosphorylated to nucleosides by **nucleotidase**, then to purine bases and ribose-1-phosphate by purine nucleotide phosphorylase. The catabolism of AMP or adenosine differs slightly,

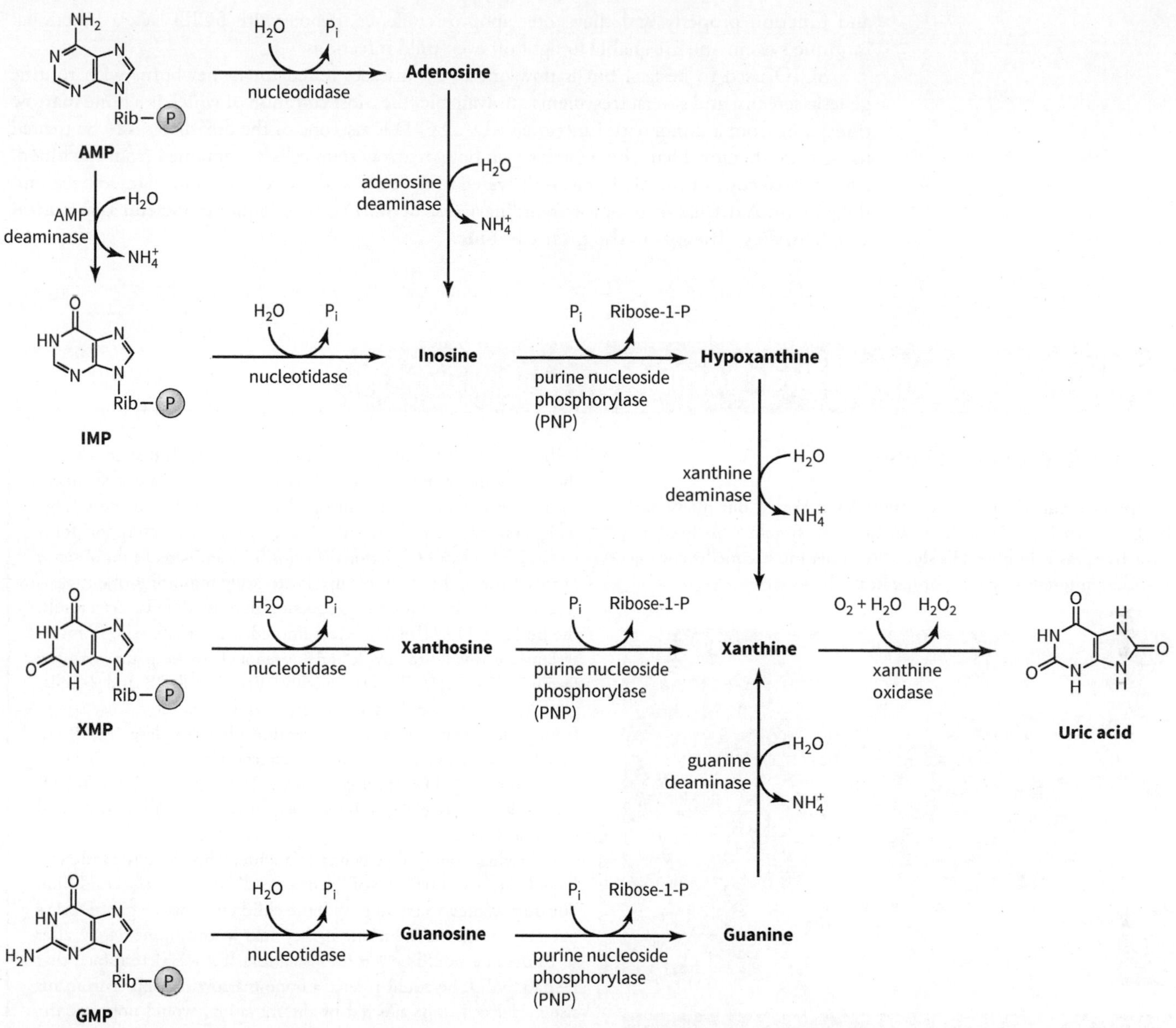

FIGURE 13.18 Catabolism of purines. Purines are dephosphorylated and deglycosylated to free bases. The bases are deaminated or oxidized to xanthine, which is oxidized to uric acid by xanthine oxidase. Despite the complexity of this diagram, there are only four types of reactions (deamination, dephosphorylation, phosphorolysis, and oxidation), which are carried out by relatively few enzymes.

in that either of these bases is first deaminated (either by AMP deaminase or by adenosine deaminase) to IMP or inosine before removal of the ribose. Recall that hypoxanthine is the base that forms inosine. This leaves three bases: hypoxanthine, xanthine, and guanine. Hypoxanthine is oxidized to xanthine via the action of **xanthine oxidase**, while guanine is deaminated, also resulting in xanthine. Xanthine is oxidized again by xanthine oxidase to give the final breakdown product, uric acid.

Diseases caused by defects in nucleotide catabolism
Severe combined immunodeficiency syndrome (SCID) is a hereditary disease resulting from a defect in the adenosine catabolic pathway, typically **adenosine deaminase (ADA)**. Infants with this condition cannot catabolize adenine. In the absence of ADA, levels of dATP rise, allosterically inhibiting ribonucleotide reductase and decreasing formation of the other deoxynucleotides. As a result, DNA synthesis is impaired in rapidly growing cells and tissues. The disease manifests itself in the cells of the immune system, which grow rapidly compared to many other body systems. Without enough nucleotides for DNA biosynthesis, the cells of the immune system are not able to grow

and function properly, and they enter apoptosis. Hence, people with SCID lack a functional immune system and are unable to fight off even mild infections.

SCID used to be fatal but is now one of 29 diseases tested for in newborns with routine genetic screens, and several treatments are available, the most common of which is a bone marrow transplant from a donor with functional ADA. SCID is also one of the first diseases to be treated using gene therapy. Here, the patient's own bone marrow stem cells are obtained from the blood, a functional copy of the ADA gene is delivered to the cells, and the cells are transplanted back into the patient. A discussion of one extraordinary case of SCID can be found in **Societal and Ethical Biochemistry: The boy in the plastic bubble.**

Societal and Ethical Biochemistry

The boy in the plastic bubble

Few people will recognize the name David Vetter, but many recall the boy born in 1971 without an immune system who lived in a germ-free plastic bubble. His short life under intense media coverage provides an interesting case for medical ethics.

Bettmann/Getty Images

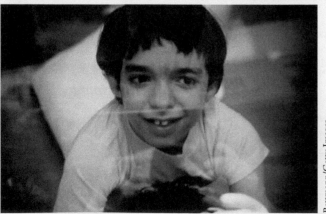
Bettmann/Getty Images

David's parents lost their first child shortly after birth from severe combined immunodeficiency, or SCID—a hereditary disease caused by the lack of a functional immune system. As a result, simple microbes that would normally not pose a threat easily overwhelmed the infant.

Like many genetic diseases, several different mutations can lead to SCID. David's disease (the most common form) was due to an inability to produce interleukins, the protein signals that coordinate the cells of the immune system. This form of the disease is X-linked. Two other common forms of the disease affect the purine salvage pathways and are not sex-linked. The second-most-common form comes from a lack of functional adenosine deaminase. In the absence of this protein, dATP builds up allosterically inhibiting ribonucleotide reductase, blocking the production of other dNTPs. As a result, the building blocks of DNA are limited, and cells of the immune system are unable to divide quickly enough to keep up with the immune challenge. The most highly affected cells (B and T cells) enter apoptosis. The third-most-common form of the disease comes from a deficiency in purine nucleotide phosphorylase (PNP). In the absence of this protein, dGTP concentrations increase relative to the other dNTPs, leading to lower levels of dCTP and dTTP. This induces apoptosis in T-cells, also resulting in a compromised immune system.

David's parents already had a daughter who, because of the sex-linked nature of this form of the disease, did not have the condition, but they wanted to try to have more children. Doctors gave David's parents a potential solution. If they had a son, immediately after birth the boy would be put into isolation. If it was determined that he had SCID, he could receive a bone marrow transplant from his sister. Her cells, not affected by the mutation, would populate the baby's bone marrow and provide the newborn with an immune system.

According to plan, David was put into a sterile incubator immediately after birth, and tests revealed that he did have SCID. In preparation for the planned bone marrow transplant, doctors found that David's sister was not a compatible donor, and at the time there were no bone marrow donor registries. Now, what was intended to be a short-term solution (a stay in a sterile incubator) became David's prognosis for life.

As David grew, he was profiled in the news media as a medical miracle. A baby that would have otherwise had a life span of weeks to months grew up in a safe, sterile environment. David was profiled on news programs and in print. Everyone knew David. Yet few people knew David's pain. David could look outside and see children playing but was unable to join them. Being confined to a small space, he could not do many of the things ordinary kids did and struggled with ordinary concepts like wind or the change of the seasons. NASA went as far as to build David a special suit so that he could explore the world a bit more, but the suit was difficult to use and required a bulky cart to filter his air. David's wish on his 11th birthday was to be able to go outside and see the stars, something he had never done before. David understood that he could not do some of these things and that he was different, but these were clearly extremely difficult conditions for a growing child.

David was also under the care of his physicians, who appreciated that David was not only a patient but also a unique case of what happens when a person is born without a functional immune system. He was subjected to multiple studies, all in the name of helping his condition. The value of these individual studies can be debated, but his samples provided valuable information about how the immune system functions and the etiology of SCID.

As David grew, he faced increasing bouts of depression and anxiety. Here again he was a subject of study. Psychologists seeking to help David and his unique condition interviewed and worked with him, but again, some of this work was used in research studies. David had become not only a patient, but a research subject.

When David was 12, doctors proposed a new solution. Researchers had discovered a new way to transplant bone marrow that would allow David's sister to become a donor. The operation was attempted, but the outcome was tragic. Unknowingly, David's sister had a latent virus (Epstein-Barr virus) in her bone marrow. Although this virus caused her no difficulties, it generated tumors in David's nascent, transplanted immune system, and he developed lymphoma. When it became clear that he was very ill, David was removed from his bubble for the first time in his life. He died shortly thereafter.

Although SCID is still a grave concern, it is now largely a treatable disease. It occurs in 1 in 50,000 to 100,000 live births in the general population but is higher in some groups such as the Navajo and Apache tribes of Native Americans. It is one of 29 genetic diseases that many states commonly screen for at birth. Some 90% of babies born with SCID are cured using either a bone marrow transplant or transfusions of enzymes or immune system proteins. Newer, experimental treatments include gene therapy or modifying stem cells to replace the defective gene and then reintroducing these cells back into the patient.

David's case is interesting and still discussed in ethics classes for several reasons. His treatment occurred in the 1970s, before informed consent—the concept that patients are active decision makers in their care and understand the ramifications—was as commonly practiced as it is today. Likewise, there were no standing hospital ethics committees as there are now. It is not clear that David's parents understood the ramifications of what would happen if David were born with SCID, or what would happen if the original bone marrow transplant could not be performed. Likewise, there was some hubris on the part of David's doctors to think that everything would go as planned. If it did not, were they really planning on keeping a human being in a plastic bubble for his entire life?

Often, people make claims about the indirect benefits of biomedical research. For example, they may claim that research made possible through David's illness has provided the groundwork that has saved potentially tens of thousands of lives. This claim should be examined carefully. The research team that initiated David's care only published three papers on his condition and moved on to other opportunities within a few years of his birth. What has proven valuable are David's direct contributions. Samples from David and several other patients have been used to identify the mutation that causes his disease, providing new knowledge about how interleukins are formed and work in the immune system. Likewise, David's development of Epstein-Barr virus and the link to the lymphoma that eventually caused his death also helped to form a foundational link between viruses and cancer. This information affects not only people with SCID but also those with cancer, AIDS, and other blood disorders and has informed some of our knowledge of bone marrow transplants.

Clearly, David Vetter's case is a complex and interesting topic for reflection and discussion. Above all, this case causes us to reflect on quality-of-life issues. What is the difference between being alive and living?

Uric acid   Uric acid serves several purposes in biochemistry and is involved in several diseases. We have already discussed the role of elevated uric acid levels in Lesch–Nyhan syndrome in section 13.1.

An examination of the structure of uric acid may raise the question of why this is an acid. Most acids in biochemistry have carboxyl or phosphate groups, but uric acid differs in that it contains two carbonyl groups flanked on either side by amino groups. It bears a structural resemblance to urea but has an important difference. Because these groups are found attached to a larger cyclic structure, charge can be delocalized, and the carbonyl group can tautomerize to a hydroxyl group and form an enolate-like structure. Both of these groups can ionize in uric acid, having pK_a values of 5.4 and 10.3, respectively.

Uric acid can hydrogen bond through multiple functional groups, but the molecule is not particularly water soluble at only 66 mg/L compared to a similarly sized molecule like tryptophan, with a solubility of 13,600 mg/L. This attribute is important in the use of uric acid by the body and in several pathophysiologies, and normal concentrations of uric acid hover near the solubility limit.

Uric acid is eliminated by the kidneys. If levels of uric acid are elevated, there is the possibility that crystals of uric acid or its salt, monosodium urate, will form and accumulate, resulting in urate-based kidney stones (**Figure 13.19**). These stones can also form nucleation sites for other types of crystals (oxalate salts) to form, which again leads to kidney stones.

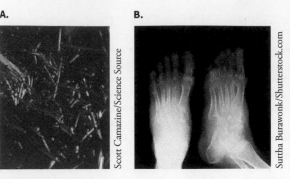

FIGURE 13.19 Disorders related to uric acid accumulation. When uric acid crystals accumulate in the joints, they lead to gout, whereas in the kidney they can lead to kidney stones. **A.** Uric acid crystals can be seen in synovial fluid. **B.** In this X-ray of someone suffering from gout, note the disturbances around the joints of the big toe. **C.** This urate-containing kidney stone has a core of uric acid coated with calcium oxalate salts.

Uric acid crystals can form in other parts of the body as well. Elevated uric acid concentrations in blood contribute to the growth of crystals of uric acid. When crystals grow in the synovial fluid of joints, they can cause pain and inflammation, known as gout, which affects as much as 4% of the population in developed countries.

Gout has some basis in genetics and some in environment. It can be partially controlled by reducing or eliminating foods in the diet that contribute to the production of uric acid, namely, meats and legumes. Also, pharmaceutical interventions can be used, including allopurinol and **probenecid**. Allopurinol is a competitive inhibitor of **xanthine oxidase**. It competes with xanthine for the active site of xanthine oxidase and slows the rate of uric acid production.

Allopurinol   **Probenecid**

Probenecid acts to increase uric acid secretion by the kidney. In the kidney, all small molecules are filtered out of the blood in the glomerulus and then reabsorbed via transport or diffusion. Probenecid competes with uric acid for the organic ion transporter, blocking reabsorption of uric acid from the glomerular filtrate. It was originally developed for a different purpose but one with a similar mechanism. Probenecid binds to and inhibits a similar transporter in the proximal tubule that pumps other organics (drugs) into the filtrate, decreasing their elimination and increasing the plasma concentration. This drug was originally used in World War II to treat patients who were also taking penicillin. In order to conserve the supply of penicillin, probenecid was used to alter the pharmacokinetics of the drug to decrease elimination and increase plasma concentrations, meaning less penicillin could be used to obtain the same effects. In 2005, concerns about influenza epidemics and drug shortages also prompted the use of probenecid to extend drug supplies.

Uric acid metabolism differs throughout the animal kingdom (**Figure 13.20**). Humans and great apes lack uricase, an enzyme that degrades uric acid to 5-hydroxyisourate and then to **allantoin**. Other animals, including diapsids (saltwater crocodiles and birds), eliminate nitrogenous wastes as uric acid in their solid waste. Here, the crystallization of uric acid into a solid is a benefit to the organism. Water is spared in these organisms and is not needed for elimination of nitrogenous wastes. In the case of crocodiles, this saves water, whereas in birds it saves weight.

It is still unclear what other roles uric acid may play in biochemistry. This redox-active molecule readily acts as a reducing agent or antioxidant. It has been suggested that uric acid is second only to ascorbate (vitamin C) in terms of its effectiveness as an antioxidant.

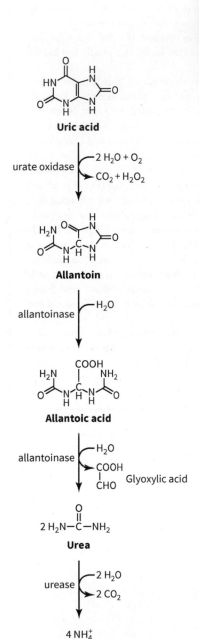

FIGURE 13.20 Uric acid metabolism in other species. Not all species catabolize nucleotides similarly. Primates are rare among mammals in that they lack urate oxidase and therefore are unable to break down uric acid. Some birds and reptiles use uric acid as a means of disposing of nitrogen catabolized from amino acids as well.

13.4.2 Pyrimidines are catabolized to malonyl-CoA and methylmalonyl-CoA

The pyrimidine nucleotides are catabolized by the liver. In short, pyrimidine catabolism is almost the reverse of pyrimidine biosynthesis. It results in β-alanine or β-aminoisobutyrate, amino acids that can be transaminated and coupled to CoA to yield malonyl-CoA and methylmalonyl-CoA, respectively.

The pathway begins in a manner analogous to the purine catabolic pathway (**Figure 13.21**). CMP, dTMP, and UMP are dephosphorylated by nucleotidase. Cytidine is then deaminated by

FIGURE 13.21 Pyrimidine catabolism.
Uracil or thymine undergoes a ring opening, loss of ammonia, and finally catabolism via lipid catabolic pathways (via malonyl-CoA or methyl-malonyl-CoA).

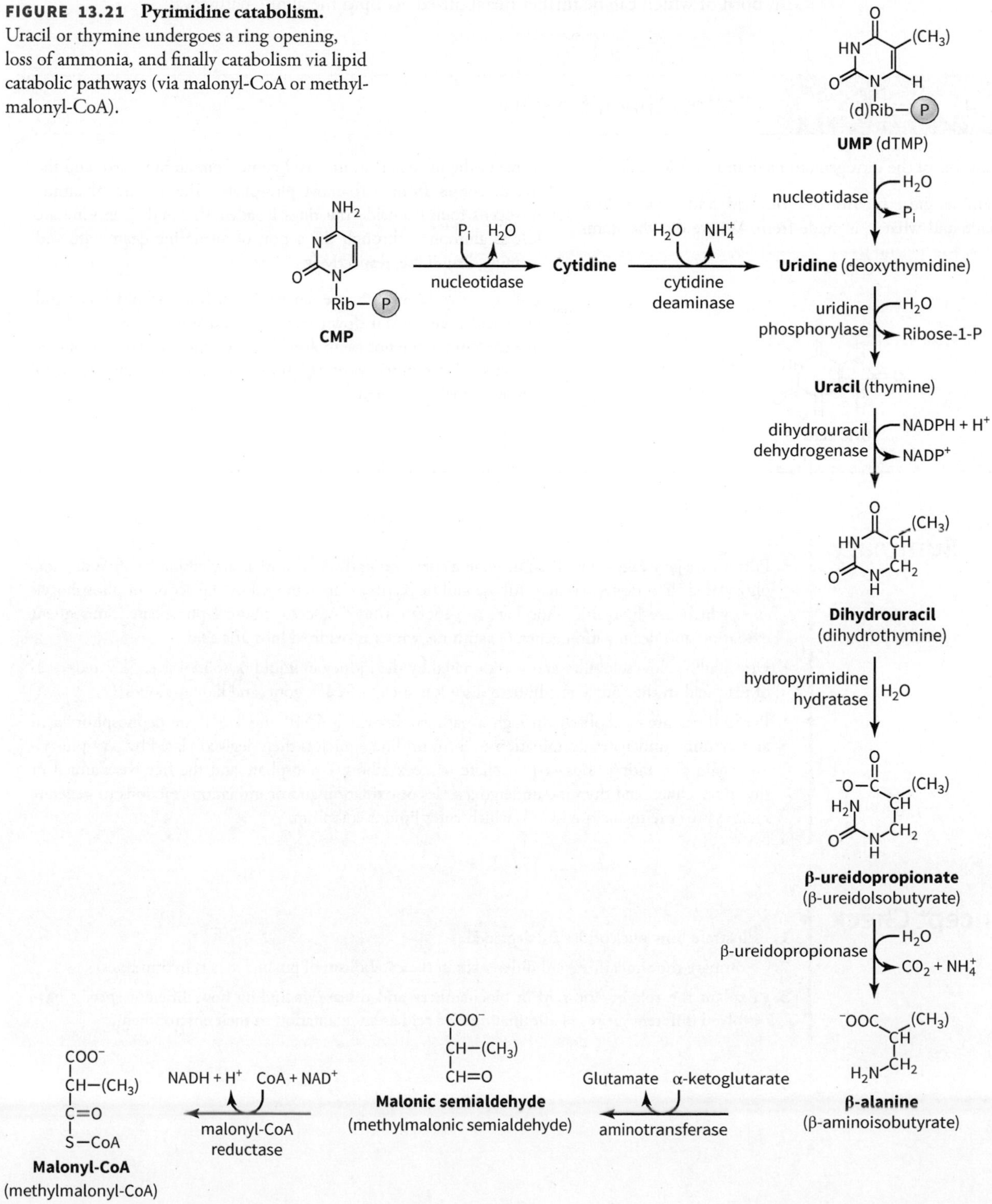

cytidine deaminase to uridine, where the pathways converge. Next, the ribose (or deoxyribose for deoxythymidine) is removed by uridine phosphorylase. This reaction is again analogous to glycogen phosphorylase, in that it uses inorganic phosphate to generate a ribose-1-phosphate product that can be recycled in the biosynthesis of pyrimidine nucleotides or PRPP. The double bond of bases uracil and thymine is hydrogenated by dihydrouracil dehydrogenase. Next, the ring is opened through the addition of water to the linkage formed between the C-4 and the N-3. β-uriedopropionase cleaves both ammonia and CO_2 from this molecule, leaving β-alanine (from uracil) or β-aminoisobutyrate (from thymine). The next steps of pyrimidine catabolism are analogous to reactions seen in amino acid metabolism. β-alanine is transaminated with α-ketoglutarate to yield glutamate, while the reaction of β-aminoisobutyrate with α-ketoglutarate gives malonic semialdehyde. Conjugation of these products with CoA gives malonyl-CoA and methylmalonyl-CoA, both of which can be further metabolized via lipid metabolic pathways.

Worked Problem 13.4 — Follow that nitrogen

What are the sources of the nitrogen atoms found in uric acid?

Strategy Examine the structure of uric acid and look at how uric acid is made and what it is made from. Where did the atoms originate?

Solution

Uric acid

Three of the nitrogens in uric acid come from amino acids, and the fourth comes from carbamoyl phosphate. The additional amine nitrogens found outside the rings in adenosine and guanosine are lost as ammonia, through the action of adenosine deaminase and guanine deaminase, respectively.

Follow-up question Some molecules, such as carbohydrates and fatty acids, are broken down into CO_2 and water, whereas others, like cholesterol, are not catabolized but are simply lost by the organism in small amounts every day. Why, therefore, does the organism bother to make uric acid?

Summary

- Purines are generally catabolized through a common pathway in which nucleotides are first dephosphorylated then deglycosylated. Ribose and deoxyribose are removed by the action of phosphorylases, which use inorganic phosphate to generate ribose or deoxyribose-1-phosphate. Subsequent oxidation and deamination generate xanthine, which is oxidized into uric acid.

- Uric acid has low solubility and is eliminated by the kidney in liquid or solid waste. Elevated levels of uric acid in the body contribute to disorders such as SCID, gout, and kidney stones.

- Pyrimidines are catabolized through a pathway in which CMP and UMP are dephosphorylated and cytidine undergoes deamination to form uridine, which is then deglycosylated by phosphorylase, again generating ribose-1-phosphate or deoxyribose-1-phosphate and the free bases uracil or thymine. Uracil and thymine undergo a series of oxidation and deamination reactions to generate malonyl- or methylmalonyl-CoA, which enter lipid metabolism.

Concept Check

1. Illustrate how nucleotides are degraded.

2. Compare the similarities and differences in the catabolism of purines versus pyrimidines.

3. Explain the role of uric acid in biochemistry and disease, including how different species have evolved different means of eliminating uric acid as an adaptation to their environment.

Bioinformatics Exercises

Exercise 1 Viewing and Analyzing Thymidylate Synthase

Exercise 2 Nucleotide Metabolism and the KEGG Database

Problems

13.1 Purine Biosynthesis

1. What are the direct sources of purine ring atoms in the *de novo* synthesis of IMP? After formation of IMP, the purine pathway forks into two branches. What are the end products of both branches and what are the cofactors used?

2. What does the role of glycine and aspartate as the building blocks of purines say about the composition of the primordial soup and the RNA world?

3. Would IMP, AMP, adenylosuccinate, or hypoxanthine be the most highly retained by a cation exchange resin like Dowex? Why?

4. Based on your knowledge of amino acid chemistry, how might a bird or reptile synthesize uric acid? Where would the nitrogen atoms come from? Would this deplete any pools of amino acids? (Hint: The atoms of uric acid come from hypoxanthine via guanine and adenine nucleotides.)

5. Describe how and where purine biosynthesis is regulated.

6. Uric acid doesn't look like most acids in biochemistry. Which protons are acidic?

7. The symptoms of Lesch–Nyhan syndrome are caused by a deficiency of which enzyme? In this syndrome which purine base is not salvaged and what are its consequences?

8. Lesch–Nyhan syndrome is a sex-linked disorder only observed in boys, but some forms of SCID are found in both males and females. What does this tell you about the location of the causal genes in these disorders? Would you expect these disorders to be dominant or recessive?

9. Von Gierke disease is a glycogen-storage disorder. In this disease, glucose-6-phosphatase activity is deficient, resulting in increased glucose-6-phosphate. Based on your knowledge of purine metabolism, how might this mutation result in elevated purine biosynthesis?

13.2 Pyrimidine Biosynthesis

10. If a drug is acting as an ATCase inhibitor, what would be the net overall effect on the production of all the different nucleotides?

11. Describe how and where pyrimidine biosynthesis is regulated in mammals.

12. How does the regulation of pyrimidine *de novo* synthesis get controlled in mammals allosterically?

13.3 Deoxyribonucleotide Biosynthesis

13. If any of the NTPs or dNTPs can be used as substrate for nucleoside diphosphate kinase, what does that say about the substrate-binding pocket of this enzyme?

14. Describe how structural analogs of folate (folate antimetabolites) can be used as anticancer drugs.

15. If any NTP can be used to phosphorylate an NDP, what does this say about the thermodynamics of the reaction?

16. How could a mutation to a gene involved in glutathione metabolism affect the cell's ability to replicate?

17. When in the cell cycle would you expect dNTP levels to be highest? Why must the synthesis of the dNTPs controlled and how?

18. What is the strategy adopted by ribonucleotide reductase to afford a proper balance of dNTPs for the cell?

19. Ribonucleotide reductase can be separated into two dimers, an α_2 dimer and a β_2 dimer. Would you expect either of these two dimers independently to have catalytic activity? Explain.

20. What are the cofactors required in the conversion of dUMP to dTMP by thymidylate synthase? Why can the thymidylate synthase reaction be considered unique?

21. Low-dose methotrexate is used to treat autoimmune diseases such as rheumatoid arthritis and Crohn's disease. Describe how this drug would provide relief for people with these conditions.

22. Some chemotherapeutic treatments involve giving the patient a lethal dose of an antimetabolite drug and then rescuing the patient with calcium folinate. Explain how this therapy would work.

13.4 Catabolism of Nucleotides

23. SCID can be treated with bone marrow transplants, stem cells, or gene therapy. How does that work? Why can't Lesch–Nyhan syndrome be treated the same way?

24. Draw connections between nucleotide metabolism and other metabolic pathways.

25. Why are the carbons of purine degradation eliminated as uric acid and not oxidized and expelled as CO_2?

26. Why are the carbons of pyrimidine degradation converted into malonyl- or methylmalonyl-CoA?

27. Elevated uric acid levels are seen in Lesch–Nyhan syndrome. Why?

28. What is the final common product obtained from both purine degradation and pyrimidine degradation?

29. Compare the energetic cost of eliminating a molecule of ammonia as urea versus as uric acid. Is one more energetically economical than another? What other factors are involved in the elimination of nitrogenous wastes?

30. Recombinant forms of the enzyme urate oxidase are available for therapeutic purpose. Identify the condition/ disease in which this might be a useful therapy.

31. In this chapter we have discussed several types of molecules that influence nucleotide metabolism. Some of these are antitumor agents, some immunosuppressive, and some antimalarial, whereas others block

the breakdown of nucleotides. What are the similarities and differences among these molecules? How is it that such a diverse set of diseases and conditions can be treated using molecules that all affect nucleotide metabolism?

Data Interpretation

32. The anticancer drug methotrexate acutely increases risk of sunburn. Provide a molecular explanation for this observation.

33. You have isolated a strain of bacteria that has a mutation in one of the enzymes of purine metabolism. Given this table of data, what enzyme might you anticipate is defective? Explain in your own words.

Metabolite	Concentration in wild type (μM)	Concentrations in mutant (μM)
5-phosphorobosylamine	120	265
AIR	130	135
AICAR	120	130
IMP	155	165
Adenylosuccinate	70	270

34. You are studying a novel isoform of ATCase from a species of *Lactobacillus*. When inhibited with orotidine monophosphate (OMP), you obtain the following curves:

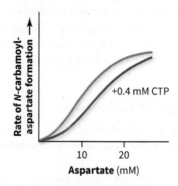

How is OMP acting in this reaction? Explain your findings in your own words.

35. You have been tasked with developing a defined food for laboratory mice. The food you have developed so far contains all the required amino acids, as well as fats, carbohydrates, and minerals, but you have been asked to define which vitamins are added. This batch contains vitamins A, C, D, E, K, and folate. Despite this, these mice exhibit signs of folate deficiency. Explain this observation.

Experimental Design

36. Presume a problem exists where an enzyme was overexpressed. Hypothesize a phenotype for cells or organisms with this mutation for each of the following enzymes.

a. ribonucleotide reductase

b. xanthine oxidase

c. OMP decarboxylase

d. ribose phosphate pyrophosphokinase

37. Presume a problem exists where an enzyme was mutated. Hypothesize a phenotype for cells or organisms with each of the following mutations.

a. A mutation to ribonucleotide reductase that leaves it unable to bind to ATP in the regulatory subunit.

b. A mutation to ribonucleotide reductase in which the radical generating tyrosine has been replaced by serine.

c. A mutation resulting in an increase of the K_M value for glutamine by CPSII.

d. A mutation to dihydrouracil dehydrogenase that leaves it nonfunctional.

38. What would be the best technique to separate each of the following molecules?

a. adenine and AMP

b. AMP and ATP

c. adenosine from AMP

d. dATP from a trimer of dATP molecules

39. What sort of technique would you use to test a newborn baby for an impairment in purine or pyrimidine metabolism? How would this differ from an adult or from a nonhuman source?

Ethics and Social Responsibility

40. Gout is a disease that is traditionally associated with excess. Although there is a dietary component to gout, and some of gout's symptoms can be treated with diet, a larger component of the disease is genetic in nature. Can you think of other diseases where people's lifestyle is blamed for their poor health? For the disease you are thinking of, does this association bear up under scientific scrutiny? In other words, is there truly an association (such as the link between smoking and cancer), or is it imagined? Why is it that some of these associations persist?

41. Diseases such as SCID and Lesch–Nyhan syndrome are dramatic but very rare. Studying people with these disorders provides an interesting insight into human metabolism and development, but is it ethical? If you are a parent of a child with a rare disease like one of these, could you ethically opt out of a study knowing that it could help your child and others in the future?

42. After reading about the David Vetter case (the boy with SCID who had to live in a sterile environment), do you believe any of the following people acted unethically? If so, state how.

- David's doctors
- David's parents
- David's psychologist
- The news media

43. Clearly, David Vetter was affected psychologically by having to live in his bubble, and yet it kept him alive for 12 years. Did he live less of a life than an ordinary kid his age? Why or why not? Debate your answer with another student holding the opposing view.

Suggested Readings

13.1 Purine Biosynthesis

Martin, W., and M. J. Russell. "On the Origin of Biochemistry at an Alkaline Hydrothermal Vent." *Philosophical Transactions of the Royal Society B-Biological Sciences* 362, no. 1486 (October 29, 2007): 1887–1925.

Zalkin, H., and J. E. Dixon. "De-Novo Purine Nucleotide Biosynthesis." *Progress in Nucleic Acid Research and Molecular Biology* 42 (1992): 259–287.

Zhang, Y., M. Morar, and S. E. Ealick. "Structural biology of the purine biosynthetic pathway." *Cellular and Molecular Life Sciences* 65, no. 23 (November 2008): 3699–3724.

13.2 Pyrimidine Biosynthesis

Callahan, B. P., and B. G. Miller. "OMP Decarboxylase—An Enigma Persists." *Bioorganic Chemistry* 35, no. 6 (2007): 465–469.

Christopherson, R. I., and S. D. Lyons. "Potent Inhibitors of De Novo Pyrimidine and Purine Biosynthesis as Chemotherapeutic-Agents." *Medicinal Research Reviews* 10, no. 4 (1990): 505–548.

Radzicka, A., and R. Wolfenden, "A Proficient Enzyme." *Science* 267, no. 5194 (1995): 90–93.

Shambaugh, G. E., III. "Pyrimidine Biosynthesis." *American Journal of Clinical Nutrition* 32, no. 6 (1979): 1290–1297.

Traut, T. *Allosteric Regulatory Enzymes.* New York: Springer, 2007.

Traut, T. W., and B. R. S. Temple. "The Chemistry of the Reaction Determines the Invariant Amino Acids during the Evolution and Divergence of Orotidine 5'-Monophosphate Decarboxylase." *The Journal of Biological Chemistry* 275 (2000): 28675–28681.

Vincenzetti, S., S. Pucciarelli, F. M. Carpi, D. Micozzi, V. Polzonetti, P. Natalini, I. Santarelli, P. Polidori, and A. Vita. "Site Directed Mutagenesis as a Tool to Understand the Catalytic Mechanism of Human Cytidine Deaminase." *Protein & Peptide Letters* 20, no. 5 (May 2013): 538–549.

13.3 Deoxyribonucleotide Biosynthesis

Christopherson, R. I., S. D. Lyons, and P. K. Wilson. "Inhibitors of *de Novo* Nucleotide Biosynthesis as Drugs." *Accounts of Chemical Research* 35, no. 11 (November 2002): 961–971.

Fairman, J. W., et al. "Structural Basis for Allosteric Regulation of Human Ribonucleotide Reductase by Nucleotide-Induced Oligomerization." *Nature Structural and Molecular Biology* 18, no. 3 (March 2011): 316–322.

Fox J. T., and P. J. Stover. "Folate-Mediated One-Carbon Metabolism." In *Folic Acid and Folates*, edited by G. Litwack, pp. 1–44. San Diego, CA: Academic Press, 2008.

Hofer, A., M. Crona, D. T. Logan, and B. M. Sjöberg. "DNA Building Blocks: Keeping Control of Manufacture." *Critical Reviews in Biochemistry and Molecular Biology* 47, no. 1 (January–February 2012): 50–63.

Jordan, A., and P. Reichard. "Ribonucleotide Reductases." *Annual Review of Biochemistry* 67 (1998): 71–98.

Maden, B. E. H. "Tetrahydrofolate and Tetrahydromethanopterin Compared: Functionally Distinct Carriers in C-1 Metabolism." *Biochemical Journal* 350 (September 15, 2000): 609–629.

13.4 Catabolism of Nucleotides

Becker, B. F. "Towards the Physiological-Function of Uric-Acid." *Free Radical Biology and Medicine* 14, no. 6 (1993): 615–631.

Berg, L. J. "The 'Bubble Boy' Paradox: An Answer That Led to a Question." *Journal of Immunology* 181 (2008): 5815–5816.

Borges, F., E. Fernandes, and F. Roleira. "Progress towards the Discovery of Xanthine Oxidase Inhibitors." *Current Medicinal Chemistry* 9, no. 2 (2002): 195–217.

Celus, A. C. *On Medicine.* Accessed at http://penelope.uchicago.edu/Thayer/E/Roman/Texts/Celsus/4*.html.

Choi, H. K., D. B. Mount, and A. M. Reginato. "Pathogenesis of Gout." *Annals of Internal Medicine* 143, no. 7 (2005): 499–516.

el Kouni, M. H. "Potential Chemotherapeutic Targets in the Purine Metabolism of Parasites." *Pharmacology & Therapeutics* 99, no. 3 (2003): 283–309.

Emmerson, B. T. "Drug Therapy—The Management of Gout." *New England Journal of Medicine* 334, no. 7 (1996): 445–451.

Fox, I. H. "Purine Ribonucleotide Catabolism—Clinical and Biochemical Significance." *Nutrition and Metabolism* 16, no. 2 (1974): 65–78.

Laverty, G., and E. Skadhauge. "Adaptive Strategies for Post-Renal Handling of Urine in Birds." *Comparative Biochemistry and Physiology A Molecular & Integrative Physiology* 149, no. 3 (2008): 246–254.

McVicker, S. "Bursting the Bubble." *Houston Press*, Houston, TX, April 10, 1997.

Nyhan, W. L. "Disorders of Purine and Pyrimidine Metabolism." *Molecular Genetics and Metabolism* 86, no. 1–2 (2005): 25–33.

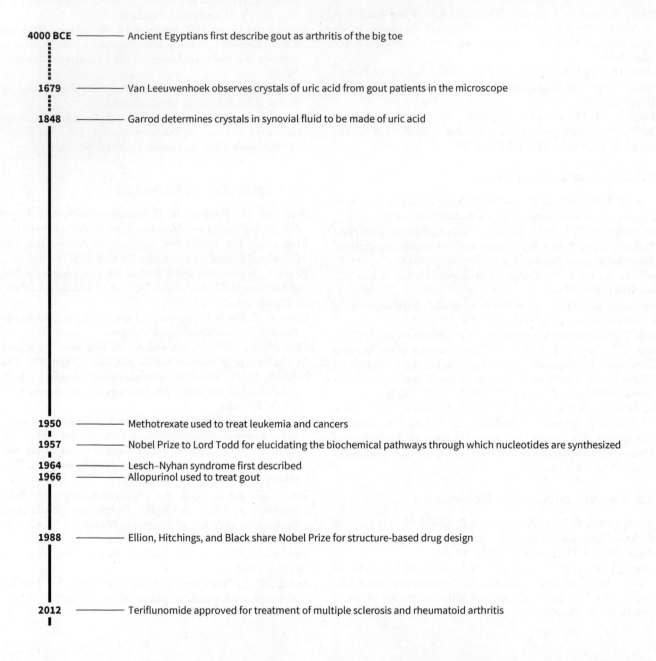

4000 BCE ——————— Ancient Egyptians first describe gout as arthritis of the big toe

1679 ——————— Van Leeuwenhoek observes crystals of uric acid from gout patients in the microscope

1848 ——————— Garrod determines crystals in synovial fluid to be made of uric acid

1950 ——————— Methotrexate used to treat leukemia and cancers

1957 ——————— Nobel Prize to Lord Todd for elucidating the biochemical pathways through which nucleotides are synthesized

1964 ——————— Lesch–Nyhan syndrome first described

1966 ——————— Allopurinol used to treat gout

1988 ——————— Ellion, Hitchings, and Black share Nobel Prize for structure-based drug design

2012 ——————— Teriflunomide approved for treatment of multiple sclerosis and rheumatoid arthritis

CHAPTER 14

DNA Replication, Damage, and Repair

DNA in Context

Human culture depends on sharing information, from weather forecasts to cupcake recipes to the position of military troops. To ensure that information is accurate, it must be transmitted with fidelity—with few errors. Ancient cultures transferred information orally from one generation to the next through storytelling. As stories were told and retold, however, people remembered things incorrectly or altered a story to fit their needs, which meant this mode of transmission had a high error rate. In the Middle Ages, written language was a more effective means of distributing information, but it still encountered problems when it came to high-fidelity transmission. As monks spent an unimaginable amount of time copying texts by hand, they introduced errors that altered the original information. In the 1400s, Gutenberg's printing press finally meant that works could be printed and distributed with fewer errors and in a quicker time frame. This technology has advanced to today's printers, copiers, and scanners.

This chapter explains how prokaryotic and eukaryotic cells copy their DNA messages with high fidelity. The copying must be done quickly enough to allow the cell or organism to propagate and survive. Also, the process must allow for recognizing and removing errors where possible and coordinate the packing and unpacking of DNA into more complex structures (chromosomes). We also discuss the types of errors and damage that can occur to DNA and how the cell tries to repair them. We conclude with a discussion of homologous recombination, a method by which an organism can increase genetic diversity either in its offspring or in the cells of its own immune system.

Learning about DNA replication may seem as daunting a task as copying a book by hand, but when we examine it on the molecular level, the complexity gives way to elegance and beauty.

Chapter Outline

Common Themes

Evolution's outcomes are conserved.	• DNA replication, mutation and damage to DNA, and homologous recombination all account for changes to the genome. • These changes can, in turn, result in alterations of gene expression or changes to the amino acid sequence coded for by a gene.
Structure determines function.	• The structure of DNA, both at the chromosomal level and at the chemical level, provides important clues to how DNA is replicated. • Understanding the chemistry of DNA reveals why some types of DNA damage occur and some of the ramifications of such damage.
Biochemical information is transferred, exchanged, and stored.	• DNA is perhaps the ultimate molecule for storing information. The code that is used to transmit the information remains largely unchanged throughout evolution, but the information that is encoded changes gradually as species evolve from generation to generation. • Some organisms (typically bacteria) can exchange DNA with one another, acquiring new traits. Viruses can infect cells of other organisms, in some cases incorporating viral DNA into the genome of these cells. • Damage to DNA can block the transfer of information or alter the code of the original molecule. • Homologous recombination is a means of generating genetic diversity, either in offspring or in the molecules of the immune system.
Biomolecules are altered through pathways involving transformations of energy and matter.	• DNA is a highly organized structure, and it needs to stay that way to retain its function. • Because DNA is such an organized structure, synthesizing and maintaining it requires a considerable amount of energy to overcome the decrease in entropy. • The energy that is required comes from hydrolysis of nucleotide triphosphates and deoxynucleotide triphosphates.

14.1 Challenges of DNA Replication

To review from previous chapters, DNA is a large polymer comprised of deoxyribonucleotides (**Figure 14.1**). These monomers contain a specific base—either a purine (adenine or guanine) or a pyrimidine (cytosine or thymine)—joined by a glycosidic linkage to the sugar deoxyribose at the $1'$ position. The hydroxyl group on the $5'$ position is linked to a phosphate group. These phosphates form phosphodiesters between the $5'$ carbon of one deoxynucleotide and the $3'$ carbon of the next. These polymeric strands can be over 100 million bases in length. Two of these strands run in opposite directions to one another and twist together to form the familiar double helix of DNA.

Even this brief review of the structure of DNA immediately highlights some of the difficulties the cell faces in trying to replicate this molecule. This section discusses some of those challenges and how the cell responds to them.

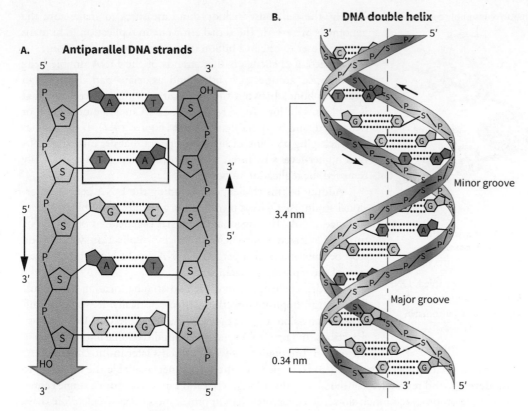

FIGURE 14.1 Structure of DNA. DNA consists of two strands. **A.** The backbone of the strands is comprised of sugar phosphates. **B.** The strands run in opposite directions (an antiparallel configuration) and are twisted together into a double helix with a major and minor groove. The bases form hydrogen bonds with one another to help hold the two strands together.

(Source: Snustad; Simmons, *Principles of Genetics*, 6e, copyright 2012, John Wiley & Sons. This material is reproduced with permission of John Wiley & Sons, Inc.)

14.1.1 The structure of DNA creates several challenges for replication

The structure of a DNA molecule poses major challenges to replication: its size and complexity, its stable double-helical structure, and the antiparallel orientation of the DNA strands. Eukaryotes face the added challenge of replicating much more DNA in a short period of time.

First, the size and complexity of DNA makes it challenging to replicate. Temporarily saving the information the DNA encodes while synthesizing a new strand is not an option. For comparison, imagine trying to read this page, remember the words, and then retype it from memory. This means the transfer of information from a parent DNA molecule to its replicated copy has to occur at the same time as synthesis of the daughter strand.

Replicating such a complex molecule also increases the likelihood of introducing errors, and DNA replication must have high fidelity for an organism to thrive. In the human genome with 3.3 billion base pairs, an error rate of one in a million would mean 3,300 mistakes per cell division. Given that some cells in the human body may divide as rapidly as once every 20 hours, and the average life span of a human is about 78 years, some cells could divide as many as 34,000 times over the course of an individual's life. At a rate of 3,300 errors per division, more than a third of an individual's genome (more than a billion errors) would be randomized, which clearly cannot be the case. Consider, for example, how this error rate would affect the accuracy of a biochemistry textbook.

Error rates in printed material:
1 in 100: ~1 error per sentence
1 in 1×10^3: ~1 error per page
1 in 1×10^6: ~1 error per book
1 in 1×10^9: ~1 error per thousand books

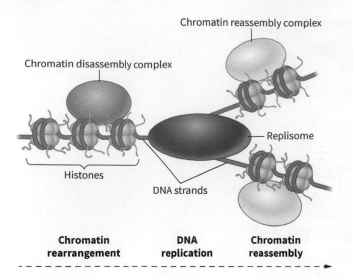

FIGURE 14.2 Chromatin remodeling. Prior to being copied, DNA needs to be separated from histones (chromatin rearrangement). After replication, it is then quickly repackaged with histones (chromatin reassembly).

Clearly, the cell must include some measures to make sure the genome is preserved. The actual error rate in replication in humans may be closer to one in a billion (three errors per cell division).

A second challenge is that strands of the DNA double helix are wrapped around each other and associate partially through hydrogen bonds between the base pairs, forming a structure that is stable in solution. To be replicated, DNA must be unwound, or denatured, and temporarily stabilized (**Figure 14.2**). If the cell has any packaging proteins, or histones, present that organize the DNA into higher-level structures, these proteins need to be temporarily removed for replication to occur.

Adding to this challenge, unwinding the DNA generates torsional strain. As DNA is unwound, it generates a bubble of single-stranded DNA. The torsional strain that is generated on the DNA must be released somewhere; otherwise, replication would soon stop because knots of supercoiled DNA would block the progression of the replication machinery.

A third challenge is that the two strands forming the double helix run in opposite directions. Mechanistically, it would be difficult to envision an enzyme that could synthesize DNA in both directions. In fact, DNA synthesis is unidirectional and proceeds in the 5′ to 3′ orientation, as we will discuss later in the chapter.

Finally, the entire genome needs to be copied in a timely manner to enable the organism to grow, develop, and reproduce as needed. In prokaryotes, replication proceeds from a single center. The replication process is much slower in eukaryotes than in prokaryotes because eukaryotes have about 1,000 times more DNA to replicate. Proceeding from a single origin, eukaryotic DNA replication would be much too slow for the organism to thrive. To solve this problem, replication in eukaryotes begins in tens of thousands of centers. The replication of the genome at these many sites must be coordinated to ensure that it happens at the appropriate time in the cell's life cycle. In the next section, we look in greater detail at how cells overcome these challenges.

14.1.2 DNA replication addresses the challenges posed by its structure

The process of DNA replication must address each of the challenges posed by the complex structure of DNA.

Because a DNA molecule is so large and complex, the transfer of information from a parent DNA molecule to its replicated copy has to occur at the same time as synthesis of the daughter strand to maintain high fidelity. A classic 1958 experiment by Meselson and Stahl shows how this happens. In this experiment, cells undergoing replication were incubated with heavy isotopes of N^{15}-labeled nucleotides, and the resulting DNA was analyzed using ultracentrifugation. Results indicated that all the new copies of DNA produced through replication had equivalent densities, with one heavy and one light strand. That is, each of the double-stranded daughter DNA helices contained one parent strand (with the heavy label) and one newly synthesized strand (without the heavy label), with the parent strand serving as a **template** for the nascent strand. The Meselson–Stahl experiment demonstrated that DNA replication is **semiconservative**, in that one strand of the parent DNA is conserved in each of the daughter DNA helices.

To replicate DNA, the strands of the double helix need to be unwound and temporarily prevented from reannealing so that the information coded for by the bases are available to the cellular machinery carrying out replication. To accomplish this, packaging proteins—histones in eukaryotes and HU proteins in prokaryotes—must be removed or shuffled. As the DNA is unwound, a structure termed a replication bubble forms in a chromosome (**Figure 14.3**). DNA synthesis occurs at either end of the replication bubble, at locations termed **replication forks**. As the replication fork slides and copies DNA, the DNA builds up torsional strain. This strain is released by a class of enzymes termed topoisomerases, which we will discuss further in the next section.

The cellular machinery that is responsible for synthesizing new DNA, the **replisome**, is comprised of several enzymes and proteins that unwind the double helix, synthesize DNA, and

A. Prokaryotic cell

Origin

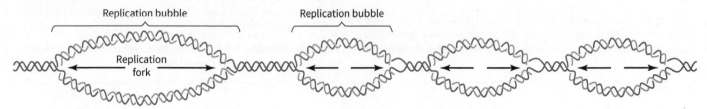

Replication fork

B. Eukaryotic cell

Replication bubble   Replication bubble

Replication
fork

FIGURE 14.3 Replication bubbles.   A. In prokaryotic cells, there are relatively few origins of replication, or distinct sequences where replication bubbles initiate. **B.** In eukaryotic cells, numerous replication bubbles form in each chromosome in response to cell cycle–dependent kinases.

(Source: (A) Snustad; Simmons, *Principles of Genetics*, 6e, copyright 2012, John Wiley & Sons. This material is reproduced with permission of John Wiley & Sons, Inc. (B) Alters, *Biology*, 1e, copyright 2006, John Wiley & Sons. This material is reproduced with permission of John Wiley & Sons, Inc.)

proofread the newly synthesized strand. **Proofreading** in this instance is a method of checking the newly synthesized strand for errors and fixing those errors. Other nuclear components are capable of detecting DNA damage and inducing repair as well.

Although the two strands of DNA run in opposite directions, DNA synthesis proceeds only in the 5′ to 3′ direction. As DNA is unwound into a replication fork, the **leading strand** can be copied continuously as the double helix unwinds. Meanwhile, the **lagging strand** is synthesized in short segments called **Okazaki fragments** 5′ to 3′ in the reverse direction and then joined together. There are two very good chemical reasons for this directionality: energy and proofreading.

First, the assembly of the DNA polymer and formation of the phosphodiester bonds require energy. In the instance of DNA or RNA, this energy is provided by the hydrolysis of pyrophosphate (PPi) and the subsequent hydrolysis of pyrophosphate to inorganic phosphate from the incoming deoxynucleotide triphosphate (dNTP). This happens as the incoming dNTP is attacked by the 3′ hydroxyl group of the lengthening chain. If the triphosphate were found on the 3′ carbon, it could, in theory, be attacked by an incoming 5′ hydroxyl, resulting in elongation. However, although this might have worked for DNA, the adjacent 2′ hydroxyl in RNA could also provide a place for a nucleophilic attack, resulting in chain termination. Therefore, this system is unfavorable and may have been selected against early in prebiotic conditions.

The second reason that nucleotides are added to the 3′ end has to do with proofreading. Let us again assume that there were conditions in which the growing DNA molecule contained the pyrophosphate-leaving group. If an incorrect nucleotide were incorporated, there would be no means by which the cell could proofread and edit out the incorrect base. If the terminal nucleotide were incorrectly base-paired and a proofreading mechanism removed it, there would be only a 3′ phosphate remaining, not the triphosphate needed to drive the reaction to the next nucleotide. Therefore, this mechanism too has been selected against at the chemical level.

The genomes of most eukaryotes are larger than those of prokaryotes, in some cases as much as 100,000 times larger, and contain multiple chromosomes. Therefore, replication must proceed from multiple sites simultaneously for it to occur in a timely manner. Bacteria usually have one well-defined sequence where replication of their single chromosome begins. In eukaryotes, instead of a single location, there are tens to hundreds of thousands of sequences where replication begins. Unlike in bacteria, there is no single consensus sequence shared among the sites. Consensus sequences are conserved sequences of nucleotides or amino acids that are recognized by a specific binding protein.

Finally, the coordination of DNA replication and cellular division (in effect, the entire life cycle of the cell) is coordinated by a group of proteins termed the cyclins. These signaling proteins work through a kinase cascade (the cyclin-dependent kinases) to initiate events such as DNA replication and chromatin remodeling. The cyclins are discussed in detail in Chapter 21.

Worked Problem 14.1 | Interpreting a classic experiment

Results of the Meselson–Stahl experiment indicated that both helices of DNA created in the first round of replication had equivalent densities with one heavy and one light strand. In subsequent rounds of DNA replication, the product of intermediate density remained, but a new product comprised of two strands of the less dense DNA were observed. What would the ratio of the densities be after four generations?

Strategy The molecules resulting from replication would have different densities, depending on which method was involved in the replication. How might these results look when analyzed for density?

Solution After one round of replication, both daughter helices would have their density increased due to the heavy isotope. However, in the second round, only half the DNA (2 out of 4 products) would have the heavier density fraction (from the original parent strands). In the third round, the number of copies would double to 8, but only 2 would retain the originally labeled strands. In the fourth round, there would be 16 copies of the DNA, but only 2 would retain the heavy label. Therefore the answer would be 2:16 or 1:8.

Follow-up question Write an equation to describe how much N^{15} label would remain after 11 generations.

Summary

- The structure of DNA reveals several challenges to replication, including maintaining a low error rate in the transfer of information, unwinding the helix and removing organizing proteins, relieving torsional strain as the DNA is copied, and overcoming the antiparallel orientation of the DNA strands. Eukaryotes face the additional challenge of replicating much more DNA in a timely manner.

- Replication forms a bubble in the DNA strand; at either end of the replication bubble is a replication fork. The replisome is the name given to the replication complex, which contains proteins that unwind the helix, copy and proofread the new DNA sequence, and organize the DNA as replication is occurring.

- DNA synthesis proceeds only in the 5′ to 3′ direction because of the structure of the nucleotides being added and the necessity of being able to proofread the copy.

- DNA replication begins at discrete sites within the chromosome. Bacteria have a single site, whereas in eukaryotes there are tens of thousands of sites at which replication initiates.

- The cyclins are a group of signaling proteins that regulate DNA replication and cell division in eukaryotes.

Concept Check

1. The structure of DNA complicates its replications. Do you agree? Give reason in support of your answer.
2. Explain how the cell compensates for challenges presented by its structure.
3. Explain the basic process of DNA replication.

14.2 DNA Replication in Prokaryotes

This section looks at the replication machinery used by prokaryotes, focusing on *Escherichia coli* (*E. coli*) as an example. At least nine different proteins in the *E. coli* replisome work in concert to replicate the cell's DNA. In the next section, we will focus on eukaryotic DNA replication, which has many similarities to the process in prokaryotes but differs in several important ways. Although there are some variations in how DNA is replicated among prokaryotes, *E. coli* illustrates the general principles involved. Replication begins at a replication fork and has two main parts: initiation and elongation.

14.2.1 Initiation of replication begins at an *ori* sequence

In *E. coli*, DNA replication begins at a specific DNA sequence termed the ***oriC*** (**origin of replication**) sequence. This sequence contains between seven and nine repeats of a consensus sequence rich in thymine and adenine. The initiation complex protein DnaA binds to adenosine triphosphate (ATP) and then binds to the *oriC* sequence. When all sites are occupied by DnaA, the pre-replication complex is complete. Binding of DnaA in this region bends and destabilizes the double helix, resulting in exposure of single-stranded DNA. The active DnaA complex also recruits DnaC (the helicase loader) and multimers of DnaB, which form the helicase. DnaC uses the energy of ATP hydrolysis to catalyze the formation of the DnaB hexamer on the lagging strand. Finally, the primase DnaG binds and forms **primers** on both strands. After this step, the origin is ready for replication.

Replication of DNA in prokaryotes is regulated through several mechanisms, including the levels of both ATP and the pre-replication complex protein DnaA. DnaA binds to both adenosine diphosphate (ADP) and ATP with equal affinity. Therefore, the ratio of ATP to ADP needs to be sufficient to generate enough competent DnaA–ATP complexes to bind to *oriC*. Also, levels of the DnaA–ATP complex need to be sufficient to bind all of the binding sites in the *oriC* sequence. Finally, replication can also be regulated by blocking access of the *oriC* sequence to DnaA.

When DnaA is sequestered on the plasma membrane of the cell, it is unable to form a pre-replication complex. However, when liberated, this protein again becomes competent to form a replication complex. SeqA is a protein that will bind to methylated DNA sequences. Four of these repeats are found in the *oriC* sequence. Thus, replication is also regulated by levels of methylation and of SeqA protein.

14.2.2 Elongation replicates the *E. coli* genome

Following priming, the replisome forms at the replication fork (**Figure 14.4**). The replisome is comprised of numerous proteins and multi-subunit enzymes that work together to replicate DNA. Elongation is the term used to describe the synthesis of DNA following the priming step.

Proteins and enzymes of the replisome   **Topoisomerase** relieves the torsional strain that is induced when the double helix is unwound. There are multiple types of topoisomerases; the predominant isoform involved in DNA replication in *E. coli* is topoisomerase II, also known as gyrase. This enzyme cleaves both of the phosphodiester backbones; uses the energy of ATP hydrolysis to pass a second strand between the cleaved ends; and then rejoins the bonds that were originally broken, relaxing the supercoils.

Helicase is formed from the hexameric subunits of DnaB. It unwinds the DNA in the replication fork to provide single-stranded templates for polymerase in a process that requires ATP hydrolysis.

Single-stranded binding proteins (SSBs) are small (12 kDa) homotetrameric proteins that have single-stranded DNA wrapped around them. The binding of SSBs helps to prevent formation of secondary structures in the displaced strand and to block nucleophilic attack on the phosphodiester backbone.

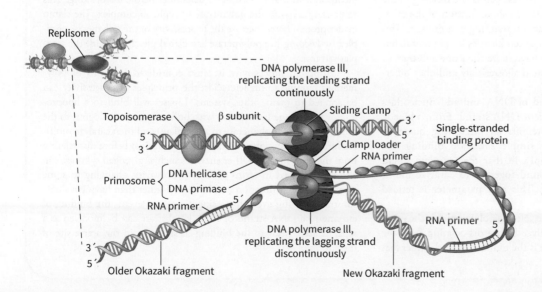

FIGURE 14.4 Proteins of the *E. coli* replisome. The replisome is a protein complex that duplicates DNA at a replication fork. It consists of gyrase (topoisomerase II), helicase, polymerases, primase, sliding clamp proteins, and a clamp loader.

(Source: Snustad; Simmons, *Principles of Genetics*, 6e, copyright 2012, John Wiley & Sons. This material is reproduced with permission of John Wiley & Sons, Inc.)

Primase is responsible for the synthesis of the RNA primer to which DNA polymerases bind. It is found on the leading strand when transcription is initiated and on the lagging strand every several hundred bases before generation of an Okazaki fragment.

Sliding clamps (also known as β clamps) are ring-shaped protein complexes that bind to DNA and act as dynamic scaffolding to tether the catalytic pieces of the replisome to the DNA being replicated. The clamps are dimers of the protein β and slide along each of the two daughter strands of DNA, bound to DNA polymerase III (pol III). The sliding clamps help to keep DNA pol III bound to DNA and increase the **processivity** of the enzyme, namely, the number of nucleotides it can incorporate per binding event.

The **clamp loader** consists of the γ complex—a complex of five different subunits that take an open clamp protein and load it onto the lagging strand at a primed piece of DNA. The clamp loader then flips the new clamp and primed lagging strand up into one of the units of DNA pol III.

DNA polymerase III (pol III) is the enzyme responsible for most of the DNA synthesis in *E. coli*. Each of its three subunits has its own function: DNA pol α is the polymerase, DNA pol ε contains the 3′ to 5′ **exonuclease** activity, and DNA pol θ increases the enzyme's proofreading function. Two copies of pol III are found in the replisome.

DNA polymerase I (pol I) is the polymerase that removes the RNA primer (using a 5′ to 3′ exonuclease activity) and fills in the gap that is left following primer removal with complementary DNA. Following DNA pol I action, both strands have been synthesized, although there are nicks, single-stranded cuts in the sugar–phosphate backbone, in the phosphodiesterase backbone of the lagging strand. As we will see later, DNA pol I is also involved in proofreading and DNA repair. If the holoenzyme is cleaved by being treated with the protease subtilisin, the proofreading domain is removed. The remaining polymerase is termed the **Klenow fragment** and is often used in molecular biology experiments.

DNA ligase catalyzes the formation of phosphodiester linkages between the pieces on the lagging strand. The *E. coli* isoform of this enzyme uses the unusual hydrolysis of NAD^+ instead of ATP to provide the energy for this reaction. Other isoforms of ligase serve similar roles in DNA repair.

Also worthy of mention is τ, a protein that forms a dimeric structure that links the two units of DNA pol III to the clamp loader complex. This protein acts as scaffolding to hold the other units of the replisome together.

Polymerases catalyze multiple reactions once they bind to DNA. This is termed processivity and is discussed in **Biochemistry: Processivity of an enzyme.**

Biochemistry

Processivity of an enzyme

The first step of an enzyme-catalyzed reaction is the binding of the enzyme to the substrate or substrates. In some fraction of the cases, the enzyme is able to catalyze a reaction resulting in a product. This simple model of basic Michaelis–Menten kinetics is very useful, but it is not universal. Some enzymes have more than one substrate or product. Other enzymes are regulated allosterically and obey different kinetic rules.

Polymerases bind to a strand of DNA and add nucleotides one at a time to the growing nascent DNA strand. From a kinetic perspective, DNA replication presents an interesting question: Clearly the standard model of simple substrate binding and catalysis does not completely apply, in that the template strand and polymerase seem to stay bound together to catalyze multiple reactions before dissociating. This new parameter is termed processivity.

The processivity of an enzyme (often a polymerase) is the number of catalytic events it can catalyze before dissociating from the template. For DNA polymerase, it is the number of nucleotides that can be incorporated before the complex dissociates and must reassociate to continue. A highly processive polymerase may be able to incorporate as many as 10,000 nucleotides before dissociating. This value can vary with the polymerase or replicon complex. The clamp loader protein helps increase the processivity of the polymerase complex by keeping the polymerase associated physically with the template strand.

Although processivity is often considered in terms of polymerases binding to nucleic acids, the principles of processivity can be applied to many other systems. Lipases will bind to a lipoprotein or membrane (association with bulk substrate, analogous to the binding of DNA polymerase to a template) and then catalyze multiple individual reactions (liberation of fatty acids) before dissociating from the lipid surface. Other enzymes such as glycogen synthase can also be described in terms of their processivity, changing to some extent both their function and kinetics. Rather than having a single substrate-binding event for a single catalytic event, the binding of enzyme to a DNA strand or to bulk substrate can be modeled as a separate step, prior to the binding of substrate in the active site of the enzyme.

Subunits of the replisome working together   As the replisome slides into the replication fork, the unit of DNA pol III on the leading strand continually copies DNA. On the lagging strand, the piece of template that is exposed must be translated in the opposite direction due to the orientation of the template. The model that is currently favored for this is the **trombone model**. As helicase exposes single-stranded template, primase lays down a short (7 to 12 nucleotide) RNA primer. SSBs bind exposed single-stranded DNA to preserve the template. As the RNA–DNA hybrid passes the clamp loader, a new clamp is loaded onto the strand and is then flipped into the active site of DNA pol III on the lagging strand. DNA pol III extends from this primer, synthesizing an Okazaki fragment. Such fragments are typically 200 to 2,000 nucleotides long. The clamp then dissociates from the strand, which is released from that unit of DNA pol III. Next, DNA pol I binds, filling in any gaps between the end of one Okazaki fragment and the primer of the previous fragment. DNA pol I also removes the RNA primer from the previous fragment. Finally, DNA ligase seals the nick that is left by DNA pol I. Elongation continues in both directions around the entire *E. coli* chromosome.

Termination   Replication terminates through a combination of specific terminator sequences (Ter sequences) and a Ter-binding protein named Tus (terminus utilization substance). Collectively this is termed the **Ter–Tus system**. This protein blocks the advancing helicase–polymerase complex and terminates replication.

When replicated, both circular daughter chromosomes are joined in links termed concatamers, which are connected like the links of a chain. Topoisomerase IV decatenates the links to produce two separate daughter chromosomes.

14.2.3 Topoisomerases relieve strain in the DNA helix

Among the challenges of DNA replication discussed earlier is the torsional tension that builds up as the double helix is unwound. This results in supercoils formed in the DNA. Topoisomerase I and II both relieve strain in the replicating DNA helix and return it to the relaxed state, but they employ different mechanisms.

Type I topoisomerase   Type IA and IB topoisomerases form a covalent DNA–enzyme adduct through an active-site tyrosine (**Figure 14.5**). Type IA topoisomerases break both strands

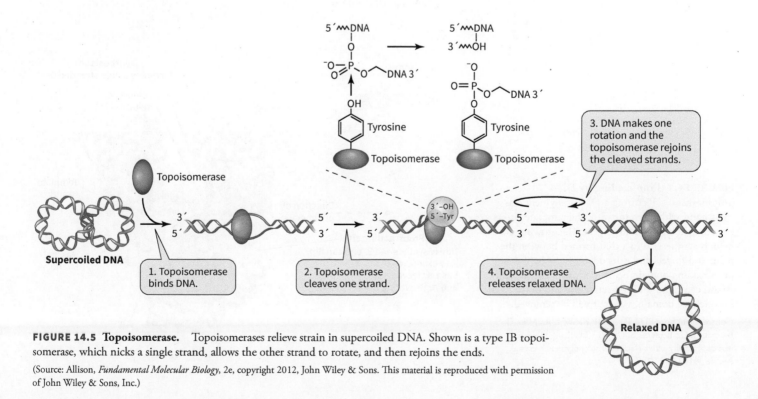

FIGURE 14.5 Topoisomerase.   Topoisomerases relieve strain in supercoiled DNA. Shown is a type IB topoisomerase, which nicks a single strand, allows the other strand to rotate, and then rejoins the ends.

(Source: Allison, *Fundamental Molecular Biology*, 2e, copyright 2012, John Wiley & Sons. This material is reproduced with permission of John Wiley & Sons, Inc.)

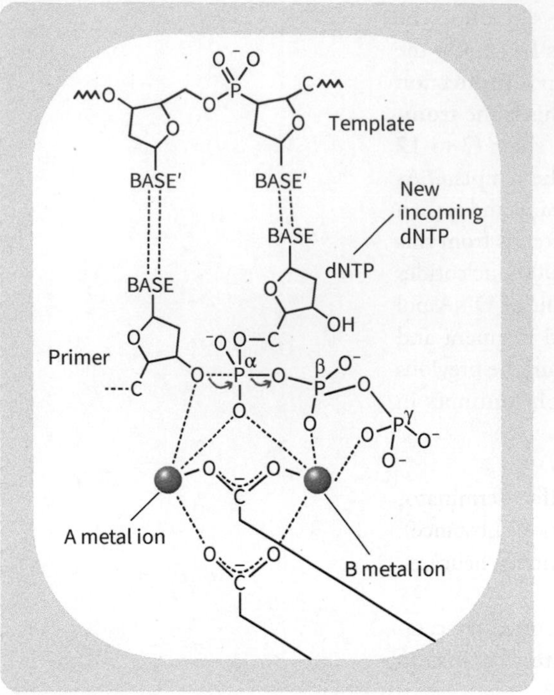

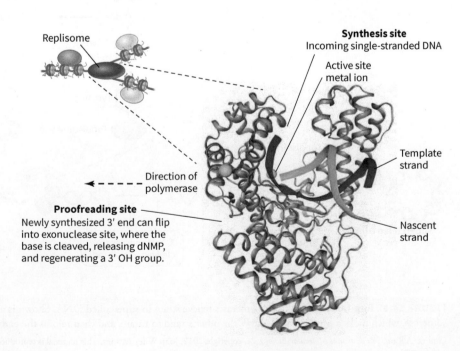

FIGURE 14.6 The mechanism of DNA polymerases. DNA polymerases employ a dyad of metal ions coordinated by anionic amino acids, in this case the aspartate residues. The metal ions interact with the phosphates of the nucleotide being added, priming the α phosphate for nucleophilic attack and stabilizing the charge on the departing pyrophosphate leaving group.

of the helix and allow the passage of an adjacent piece of the DNA helix before re-ligating the original break. By contrast, type IB enzymes generate a nick in one strand, which allows rotation around the bonds of the other strand, thus relieving strain. Type IB enzymes are capable of relieving multiple turns in one catalytic cycle.

Type II topoisomerase Type II topoisomerases bind and cleave both strands of DNA. Like the type IA enzymes, they then bind a distal loop of DNA, which is passed through the two cleaved fragments. This process requires the hydrolysis of ATP. The fragments are then re-ligated. Looping the distal piece through the temporary gap relieves the strain.

14.2.4 DNA polymerases synthesize and proofread DNA

In DNA polymerases, the overall structure and mechanism are conserved among species. The structure of DNA polymerases is like a right hand, with fingers, palm, and thumb cradling the DNA being replicated. The active site also contains two divalent metal ions (typically Mg^{2+} ions) that are critical for catalysis (**Figure 14.6**). The metal ions are coordinated in part by two conserved aspartate residues and in part by phosphate groups on the incoming dNTP. The "A" metal ion activates the 3′ hydroxyl for nucleophilic attack on the α phosphate, whereas the "B" metal ion stabilizes the charge on the departing oxygen and chelates the pyrophosphate group generated from the β and γ phosphates.

DNA pol III is not only responsible for most DNA synthesis in prokaryotes, but it is also responsible for proofreading the DNA sequence (**Figure 14.7**). The polymerase holoenzyme is comprised of two large domains: the polymerase domain and a proofreading domain. The polymerase domain rides along the single-stranded DNA template, incorporating nucleotides as they hydrogen bond in the proper place. The proofreading domain possesses a 3′ to 5′ exonuclease activity and removes the terminal nucleotide from the growing DNA chain.

FIGURE 14.7 Proofreading by DNA polymerases. Picturing a DNA polymerase as a right hand with palm forward, the template enters between the thumb and fingers. The growing double helix is synthesized in a cleft formed between the palm and fingers and exits down the wrist. In the proofreading domain below the palm, the nascent strand can flip out of the active site and down to the exonuclease domain, where it will be removed.

(Source: Data from PDB ID 3BDP Kiefer, J. R., Mao, C., Braman, J. C., Beese, L. S. (1998) Visualizing DNA replication in a catalytically active *Bacillus* DNA polymerase crystal. *Nature* 391: 304–307)

If an incorrect nucleotide has been incorporated into the growing chain, it will have improper geometry due to incorrect base pairing. Because of this mismatch, the polymerase will stall, and because the terminal nucleotide is not tightly bound, the mismatched nucleotide can flip out of the active site of the polymerase and into the active site of the exonuclease. The terminal nucleotide will be removed, and chain synthesis will resume, presumably inserting the correct nucleotide the second time around. Using this mechanism, the polymerase sometimes mistakenly removes the correct nucleotide. This may occur as much as 5% of the time. Even so, this proofreading mechanism enables the replication process to have a 1,000-fold lower error rate than polymerases lacking the exonuclease domain.

Worked Problem 14.2   Clogged gears

Novobiocin is an aminocoumarin antibiotic. It is a natural product of *Streptomyces*, which is a potent inhibitor of bacterial gyrase. How does this antibiotic function?

Strategy   Review the function of bacterial gyrase. How would inhibition of this enzyme affect the organism?

Solution   When bacterial gyrase is inhibited, bacteria are unable to relieve strain in the DNA ahead of the replication fork. Thus, torsional strain builds up, and the replisome stalls. Bacteria are unable to replicate their DNA and, as a result, are unable to reproduce. This does not necessarily kill the bacteria outright, but it does prevent the spread of infection. Ideally, the host's immune system would be able to contain the remaining pathogens.

Follow-up question   Would there be any benefit to inhibiting human topoisomerases pharmacologically? Explain your reasoning.

Summary

- In *E. coli*, DNA replication is initiated at the *oriC* sequence, a short, 150-nucleotide sequence that is rich in adenine and thymine base pairs.
- Prokaryotic DNA replication employs multiple proteins and enzymes functioning in a complex termed the replisome. These include gyrase, helicase, polymerase, single-stranded binding protein, primase, ligase, and structural proteins that coordinate the catalytic ones or increase the processivity of polymerases.
- The individual proteins and enzymes involved in replication act in concert in the replisome, the multiprotein complex that replicates DNA of subunits.
- Topoisomerases use a range of mechanisms to relieve strain in the replicating helix.
- DNA polymerases employ conserved aspartates and divalent metal ions to catalyze the addition of deoxynucleotides to the growing chain. DNA poll III has subunits that act as 3′ to 5′ exonucleases and are involved in proofreading of the nascent DNA sequence.

Concept Check

1. In the cell, ATP is also required for DNA replication. What steps of replication require ATP?
2. Describe the individual proteins involved in prokaryotic replication and their function.
3. Describe how the proteins and enzymes involved in replication act in the replisome.
4. Discuss enzyme processivity and its characteristics.

14.3 DNA Replication in Eukaryotes

DNA replication in eukaryotes is different from the process in prokaryotes in several important ways: how replication initiates, which proteins are involved in replication, and how histones are moved to enable replication. The final part of this section discusses how the ends of chromosomes, termed telomeres, function and the role of telomerase in maintaining DNA. As we will see, not all DNA is maintained through replication.

14.3.1 In eukaryotes replication originates in multiple sites

Because of the size and complexity of the eukaryotic genome, multiple origins of replication are needed; however, replication must occur only once before cell division. The replication also needs to be coordinated across tens of thousands of sites within the genome simultaneously.

Replication licensing To find and coordinate these multiple origins of replication requires a process termed **replication licensing**, which primes the origins of replication to bind to the replisome and begin replication (**Figure 14.8**). Licensing occurs between late M and G2 phases of the cell cycle and is then canceled as DNA is replicated in S phase. This ensures that DNA replication can only occur once per cell cycle.

Licensing has two phases. First is the assembly of the **pre-replication complex (pre-RC)** on the origin sequence or autonomously replicating sequence (ARS). This begins with the binding of the **origin of replication complex (ORC)**, Cdc6 and Ctd1, which in turn promote the binding of hexamers of the **minichromosome maintenance (MCM) complex** proteins (specifically, MCM2 through MCM7) on the DNA strand. This protein complex acts as the core of the ATP-dependent helicase that unwinds the DNA double helix and provides single-stranded DNA for the binding of the rest of the replisome. The second phase of licensing is the activation of the MCM2–7 complex to "fire" and initiate DNA replication. This requires the association of MCM2–7 with two other proteins, Cdc45 and GINS, which are suggested to close the six members of the MCM2–7 complex around the DNA helix and initiate unwinding.

Licensing and firing are regulated by signaling proteins, termed the cyclins, and the kinases that require these factors, called cyclin-dependent kinases. Cyclin-dependent signaling cascades are discussed in detail in Chapter 21.

In contrast to prokaryotes, specific DNA sequences are not required in eukaryotes, but the members of the pre-RC are needed. This complex has been shown to initiate from several different sequences. Bioinformatics studies are currently investigating whether there is a common denominator to these sites.

Replication sites It appears that replication does not occur randomly throughout the eukaryotic genome but instead is centered in replication factories. These are places where multiple initiation complexes are found at once and where there are higher concentrations of the proteins involved in replication.

Fewer than 20% of these licensed sites are used as origins of replication. This seems to be stochastic (random) in nature but may provide the cell with a flexible mechanism to overcome several challenges at once. For example, by having multiple sites available for replication, DNA can be replicated no matter which genes are being transcribed at any given point in time. Having multiple sites available affords the cell flexibility in terms of how quickly DNA can be replicated. In embryogenesis, replication rates are higher, probably because of increased numbers of sites. Finally, having this level of flexibility provides the cell with a means to tailor replication rates under poor growth

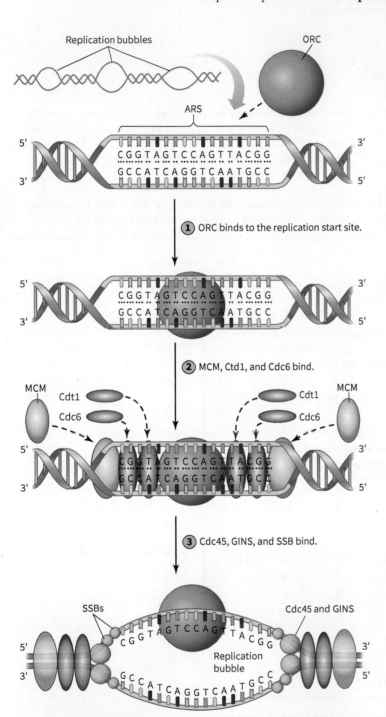

FIGURE 14.8 Replication licensing and initiation of replication in eukaryotes. Replication licensing describes the preparation of a site in the chromosome for replication to begin. In eukaryotes, this involves the binding of proteins to the ARS site.

(Source: Wessner, *Microbiology*, 1e, copyright 2013, John Wiley & Sons. This material is reproduced with permission of John Wiley & Sons, Inc.)

conditions or replicative stress, both of which typically lead to an increase in the number of origins of replication.

There is a positive correlation between transcriptionally active DNA and replication origins, possibly due to the availability of DNA to the replication machinery. It seems that making the DNA more accessible to the ORC or replisome would have the same effect as it would in transcription. Nucleosomes that have been modified to decrease the interactions between histones and DNA are more likely to form origins of replication. Several enzymes are involved in chromatin remodeling, including histone acetylases and methyltransferases such as the proteins HBO-1 and PR-Set7.

14.3.2 DNA replication systems in eukaryotes are more complex than in prokaryotes

The overall scheme of DNA replication is conserved between prokaryotes and eukaryotes; however, eukaryotic systems are somewhat more complicated. These proteins also have different names in eukaryotes, as noted in **Figure 14.9** and summarized in **Table 14.1**.

The eukaryotic helicase complex (the Cdc45/MCM/GINS helicase complex) is formed by the hexameric MCM2–7 complex that is assembled during initiation, Cdc45, and the tetrameric complex GINS. Each of these latter two protein complexes acts as a go-between, enabling interaction between the MCM2–7 complex and polymerases.

Eukaryotic polymerases replicate DNA and are more varied and specific than their prokaryotic equivalents. The three main eukaryotic polymerases are identified with Greek letters rather than Roman numerals: DNA pol α contains the primase activity necessary for initiation of transcription on either strand, DNA pol δ is the main polymerase found on the lagging strand, and

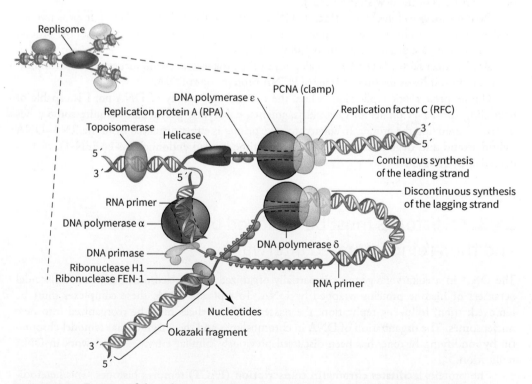

FIGURE 14.9 Proteins of the eukaryotic replisome.   The human replisome is analogous to the prokaryotic replisome shown in Figure 14.4 but with several differences. It consists of topoisomerase; the multi-subunit helicase; several polymerases; primase; single-stranded binding proteins; replication factor C (RFC); sliding clamp proteins (PCNA); and a clamp loader.

(Source: Snustad; Simmons, *Principles of Genetics*, 6e, copyright 2012, John Wiley & Sons. This material is reproduced with permission of John Wiley & Sons, Inc.)

TABLE 14.1　A Comparison of Replisome Proteins in *E. coli* and Humans

Function	*E. coli*	Human
Relief of torsional strain	Gyrase	Topoisomerase II
Helicase-unwinding of double helix	DnaB	MCM2–7
Single-strand maintenance	Single-strand binding protein (SSB)	Replication protein A (RPA)
Priming	Primase (DnaG)	Primase/pol α
Sliding clamp	β protein	Proliferating cell nuclear antigen (PCNA)
Clamp loading	γ complex	Replication factor C (RFC)
Strand elongation	DNA pol III	DNA pol δ and DNA pol ε
RNA primer removal	DNA pol I	Flap endonuclease-1 (FEN-1) or RNase H
Ligation of Okazaki fragments	Ligase	Ligase

DNA pol ε is the main polymerase found on the leading strand. DNA pol δ and ε have 3′ to 5′ exonuclease activities involved in proofreading. Following synthesis of the RNA primer and an extension of about 20 nucleotides from that primer by DNA pol α, the clamp loader switches polymerases (a process known as polymerase switching) to DNA pol δ, which extends the rest of the lagging strand up to the next Okazaki fragment. Eleven other DNA polymerases have been discovered that participate in the replication and repair of both genomic and mitochondrial DNA.

Replication protein A (RPA) is a heterotrimeric protein that is analogous in function to single-stranded binding protein. These proteins bind to the exposed DNA on the lagging strand before synthesis of the new strand occurs.

Proliferating cell nuclear antigen (PCNA) is analogous to clamp protein β in prokaryotes. PCNA functions to help tether the polymerase to the DNA strand. PCNA increases the processivity of polymerase ε and δ more than 1,000-fold.

Replication factor C (RFC) is analogous to the prokaryotic clamp loader protein complex. In eukaryotes, it loads the homotrimer of PCNA proteins onto DNA.

Unlike prokaryotic replication, where the exonuclease activity of DNA pol I is capable of degrading the RNA primer of Okazaki fragments, eukaryotic polymerases require some other means of removing the primer. In eukaryotes, the primer is either displaced from the RNA–DNA hybrid strand and degraded one nucleotide at a time by flap endonuclease-1 (FEN-1), or it is degraded by RNase H.

14.3.3　Histones must be removed before replication and then reincorporated afterward

The DNA in a eukaryotic genome is partially organized into nucleosomes, which are discoidal octamers of histone proteins wrapped by DNA. To replicate DNA, these complexes must be removed; then, following replication, the nascent DNA helices must be reorganized into new nucleosomes. The organization of DNA in chromosomes and the enzymes that remodel chromatin by modifying histones has been discussed previously. Similar enzymes are at work in DNA replication.

The protein **facilitates chromatin transcription (FACT)** removes histones. This heterodimeric protein binds to the H2A–H2B dimer, stabilizing the free-histone complex. It does this in part by interacting with H2B and blocking the interaction of that histone with DNA, which leads to the unwinding of that portion of the nucleosome. This, in turn, enhances the gradual winding and unwinding of DNA around the histone core by breaking the interaction of the first 30 base pairs with the histone.

Immediately following replication, proteins such as chromatin assembly factor-1 (CAF-1) and Rtt106 orchestrate the formation of new nucleosomes.

CAF-1 is a heterotrimeric protein that catalyzes the formation of new nucleosomes following the replication fork. It joins an (H3–H4)$_2$ heterotetramer to the nascent DNA double helix. It then adds two H3–H4 dimers to complete the nucleosome.

Rtt106 is a histone chaperone that acts as a histone acetylase to promote formation of (H3–H4)$_2$ heterotetramers. These structures are thought to form the basis of new nucleosomes and promote association between the nascent DNA helix and the newly assembling nucleosome core.

Replication of the mitochondrial chromosome   The mitochondrial chromosome is small (16 kb) and circular, encoding the 13 genes essential for mitochondrial function. Mitochondrial DNA replication is unclear and currently debated, but it is known that there are specific polymerases (pol γ) and specific proteins (the helicase TWINKLE and mitochondria-specific single-stranded binding proteins) that may be analogous to those proteins in the replication of genomic DNA. Likewise, little is known about the initiation of mitochondrial DNA replication. Although clearly regulated, mitogenesis, the replication of mitochondria, occurs independent of cell division and therefore probably functions under different control systems.

14.3.4 Telomeres present a problem for DNA replication

Telomeres are the ends of the linear chromosomes found in eukaryotes, literally the "end pieces" (**Figure 14.10**). Early in the development of a model of DNA replication, it became clear that the ends of chromosomes present a conundrum. If the replication complex runs off the end of a linear chromosome, replication could occur on the leading strand, but the discontinuous nature of replication on the lagging strand would leave exposed Okazaki fragments. There would be no way for DNA pol III to remove these fragments; hence, every time DNA was replicated, a chromosome would shrink by several hundred bases. This would not present a major problem if a chromosome only needed to undergo a few divisions, but given the need to faithfully replicate the chromosome through generations, this presents a significant problem. Clearly, something else must be responsible for the maintenance of telomeres.

The structure of the telomere   Telomeres do not end with an exposed 3′ terminus. Instead, the DNA component of the telomere loops back on itself, forming a large loop several thousand nucleotides in length referred to as a T-loop, T for "telomere" (**Figure 14.11**). The loop is closed by the base pairing of the terminus of the DNA farther up the strand. This displaces one strand of the double helix, generating a second loop structure termed the D-loop, D for "displacement."

The DNA sequence of the telomere consists of numerous repeats of a single sequence (TTAGGG) and contains anywhere from several hundred bases to several kilobases, depending on the species. In mammals, the telomeric DNA is bound by a complex of six proteins (TRF1, TRF2, RAP1, TIN2, POT1, and TPP1), termed the **shelterin complex**. An analogous system is found in other eukaryotes, such as yeast. TIN2, TPP1, and RAP1 bind to telomeric DNA, and anchor TRF1 and 2 and POT1. The function of the shelterins is not completely known, but they have been shown to provide structural integrity to the telomere and recruit telomerase, the enzyme responsible for telomere maintenance and replication.

Functions of telomeres   Telomeres have two important functions in the cell. First, as mentioned earlier, they allow DNA replication to occur without loss of DNA at the telomere. Without these ends and the enzyme telomerase (discussed later), cells would lose some DNA at each replication. Second, telomeres protect the

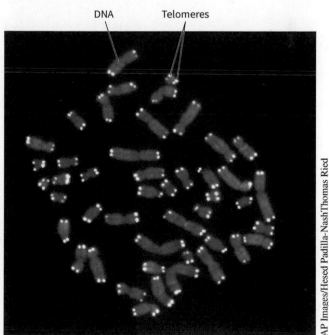

DNA   Telomeres

FIGURE 14.10 Telomeres.   In this metaphase spread of chromosomes, the telomeres are stained to distinguish them from the rest of the DNA. Note that telomeres are just the tip of each chromatid.

AP Images/Hesed Padilla-Nash Thomas Ried

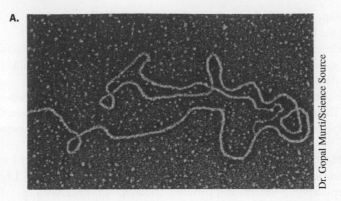

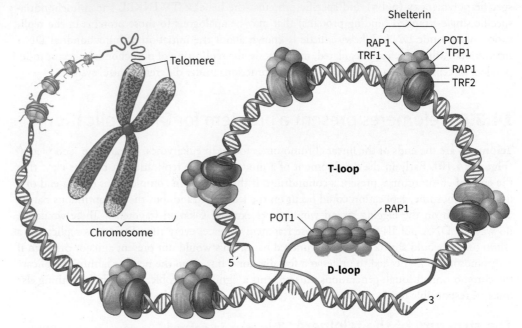

FIGURE 14.11 Structure of the telomere. **A.** In a T-loop, the terminal telomeric sequence has invaded the parent strand upstream, forming a large loop several thousand bases in length. **B.** This diagram of the T-loop structure shows the D-loop where the single-stranded 3′ end invades the parent. The shelterin complex consists of the DNA-binding proteins TRF1 and 2, Rap1, and TIN2 and may also contain TPP1 and POT1. The single strand displaced in the D-loop is bound by POT1.

(Source: From Jason C. Kovacic, Pedro Moreno, Vladimir Hachinski, et al, Cellular Senescence, Vascular Disease, and Aging, Volume 123; Issue 15, Copyright 2011, Wolters Kluwer Health, Inc. Reproduced with the permission of Wolters Kluwer Health, Inc.)

chromosome from the cell's own DNA repair machinery. In cells that have been manipulated to lack either telomeres or some members of the shelterin complex, DNA repair mechanisms recognize the telomere as a piece of damaged DNA and attempt to fuse the end of the chromosome lacking a telomere with another chromosome. Chromosome fusion can result in oncogenesis (events that initiate cancer) or fatal defects for the cell.

Telomere maintenance Telomeres are reproduced and maintained by telomerases, enzymes that regenerate the telomere (**Figure 14.12**). How is the telomeric repeat regenerated, given the problem of replicating the lagging strand? Telomerase is a combination of both a reverse transcriptase (writing DNA from an RNA code) and a terminal transferase (adding nucleotides to the end of a nucleic acid sequence). The other important feature of telomerase is that the enzyme contains within the active site a strand of RNA that helps direct and code for the growing telomere. After the telomerase has added bases to lengthen the telomere, the enzyme slides down the

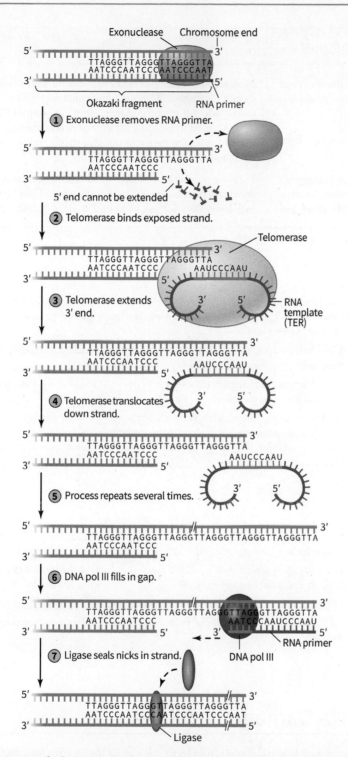

FIGURE 14.12 Action of telomerase.   Telomerase contains a short RNA in the active site that it uses as a template to lay down the hexameric repeat of the telomere.

(Source: Wessner, *Microbiology*, 1e, copyright 2013, John Wiley & Sons. This material is reproduced with permission of John Wiley & Sons, Inc.)

nascent strand and continues to extend the strand. Telomerase is processive, meaning that it is capable of repeating this reaction through multiple cycles.

Following the action of telomerase, DNA pol III can extend both strands of the new telomere, filling in any remaining gaps. The fragments are joined by ligase as they would be normally during replication.

Worked Problem 14.3 | Missing pieces

If an organism had a mutation that resulted in a dysfunctional PCNA protein, how might this affect replication? How would this affect the health of the organism?

Strategy What is the role of PCNA in DNA replication? How might lacking this protein change how DNA is replicated?

Solution PCNA is analogous to the clamp protein in prokaryotes in that it increases the processivity of DNA polymerases. In other words, it keeps the polymerase tethered to the DNA strand and

increases the number of bases that can be added before the polymerase dissociates. Without a functional copy of PCNA, the polymerase will fall off prematurely, greatly slowing DNA replication. This might be seen during organismal development or, if the organism makes it to maturity, in cells that divide rapidly such as those of the blood or immune system.

Follow-up question Based on its role in replication, would you anticipate that malignant cancer cells would have higher than normal, lower than normal, or normal levels of telomerase?

Summary

- Replication in eukaryotes originates at multiple sites found throughout the genome. These sites are coordinated through replication licensing, which primes the origins of replication to bind to the replisome and begin replication.
- The proteins involved in eukaryotic replication are analogous to prokaryotic proteins but have more specialized functions.
- Chromatin remodeling is the process of removing histones prior to DNA replication and then reorganizing histones following replication.
- The presence of telomeres at the ends of linear chromosomes explains why chromosomes do not shrink when replicated. Telomeres also prevent the cellular DNA repair machinery from fusing chromosomes nonspecifically. Telomerase is the enzyme that maintains the end of chromosomes by adding nucleotides to the telomere.

Concept Check

1. Compare prokaryotic and eukaryotic DNA replication, including the different proteins involved.
2. How is the eukaryotic genome well equipped with the demand for increased replication rates during specific situations like embryogenesis?
3. Describe what happens to histones during DNA replication.
4. Explain why telomeres of eukaryotic chromosomes require special treatment during replication and how they are replicated.

14.4 DNA Damage and Repair

DNA is unique among biological molecules in that it carries the instructions essential for the life of the organism. Although many organisms are diploid, carrying two copies of each autosome, many others are haploid and have only a single copy. In the latter case, any damage to the DNA code may be lethal for the organism. Functional loss of a gene essential for the life of a cell would cause the loss of that cell. Likewise, loss of a gene that is essential for reproduction of the entire organism could prevent the organism from reproducing and passing on its other genes. Finally, damage to DNA in a eukaryote can result in alterations to genes coding for the regulation of growth and development, possibly leading to cancer.

This section first describes the different types of DNA damage (also known as DNA lesions) and the agents that can cause them. Then we compare how prokaryotes and eukaryotes detect and repair DNA damage.

TABLE 14.2 DNA-Damaging Agents, Damage Caused, and Repair Mechanisms

Damaging Agent	Lesion	Repair Mechanism
Replication errors	Base mismatch Insertions Deletions	Mismatch repair (MMR)
Oxygen radicals Hydrolysis Alkylating agents	Abasic sites Single-strand breaks 8-oxoguanine lesions	Base excision repair (BER)
UV light Chemicals	Bulky adducts Pyrimidine dimers	Nucleotide excision repair (NER), photolyase
Ionizing radiation X-rays Antitumor drugs	Double-strand breaks Single-strand breaks Intrastrand cross-links Interstrand cross-links	Non-homologous end joining (NHEJ) Homologous recombination (HR)

14.4.1 DNA damage occurs in various ways through different agents

An examination of the structure of DNA and a consideration of how DNA is replicated provide valuable clues on how and where DNA damage can occur and how it is repaired (**Table 14.2**). DNA can be damaged during replication or by chemicals, radiation, or viruses. Collectively, these are termed **mutagens**—compounds that mutate DNA. The phosphodiester backbone of DNA provides structure to each strand, but the information coded for by DNA is held in the base pairs in the core. Damage to the DNA code can include insertions and deletions, alkylation of a base, oxidative damage, creation of thymine dimers, and viral damage. Damage can lead to **mutation**, which is a permanent change in the DNA sequence.

Insertions and deletions   One type of DNA damage involves **insertions** (as extra bases are inserted, sometimes multiple times) or **deletions** (as bases are skipped) during replication. If this occurs in a coding sequence, it will result in a frameshift mutation, in which the reading frame of the message is shifted, resulting in the loss of the code for the protein from that point in the gene forward. Frameshift mutations often result in shorter, prematurely terminated products due to generation of a stop codon in the frameshift.

Insertions are generated when molecules slip between bases. DNA is not typically thought of as a hydrophobic molecule, but it does have a hydrophobic core where the base pairs meet and the bases stack. Although there is little room in this core, some molecules can become inserted between the bases. This is referred to as **base pair intercalation**, and the molecules that do this are termed **base pair intercalators** (**Figure 14.13**). Intercalators are similar to the double helix in width, about three or four six-membered rings wide. Because they are planar or nearly planar, they are able to fit into the space between the bases. Most are hydrophobic aromatic molecules,

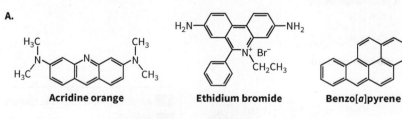

A.

Acridine orange   **Ethidium bromide**   **Benzo[a]pyrene**

B.

Molecules of ethidium bromide are intercalated between base pairs.

FIGURE 14.13 Base pair intercalation.
A. Shown are the structures of the base pair intercalators acridine orange, ethidium bromide, and benzo[a]pyrene. **B.** Intercalators interact with DNA using a variety of weak interactions, including base pair stacking, π-π stacking, and the hydrophobic effect.

and examples include the fluorescent dye ethidium bromide and the polycyclic aromatic hydrocarbon benzo[*a*]pyrene, a product of hydrocarbon combustion. These molecules interact with DNA through a variety of weak interactions we have seen previously: base stacking, π-π stacking, van der Waals forces, and the hydrophobic effect.

Base pair intercalators do not usually damage DNA on their own, but they cause polymerases and the replisome to slip, introducing insertions or deletions during replication.

Alkylations   Many compounds generated in the cell or in the environment can react with the bases of DNA, causing chemical modification of those bases. Here, the types of damage that occur are typically alkylations, in which alkyl groups are directly added to the N and O atoms of bases, forming adducts (**Figure 14.14**). More than 20 common DNA adducts have been identified. A common type of agent that has been shown to form DNA adducts is the nitrosamines, a group of molecules commonly found in tobacco smoke. Nitrosamines also form whenever nitrates are found in conjunction with amines. Sodium nitrate is a common preservative in many cured meats, including bacon. The high temperature at which bacon is cooked also favors nitrosamine formation. The antioxidants ascorbate (vitamin C) or erythorbate (an isomer of vitamin C) added to cured meats reduce but do not completely prevent the formation of nitrosamines.

Certain drugs can also alkylate DNA. Mustard gas is a chemical warfare agent that was used during World War I. This compound causes terrible blisters and chemical burns on the skin and mucous membranes of the victim. A secondary effect is that people subjected to gas attacks became immunocompromised and have depressed levels of white blood cells, a condition called leukopenia. The effects of mustard agents are due to alkylation of DNA, which in turn reduces the white blood cell count.

Alkylation can also be applied to therapeutic treatments. In the 1940s, it was found that chemical analogs of mustard gas (nitrogen mustards or sulfur mustards) could shrink the size of tumors in mice. Since the 1950s, alkylating agents such as cyclophosphamide and streptozotocin have been used as chemotherapeutic agents to treat cancers and some autoimmune diseases; they are also used to prepare recipients for bone marrow transplants.

Oxidative damage   Certain types of DNA damage can result in the complete loss of a base, called an abasic site. Such damage occurs due to chemistry at the glycosidic linkage between the deoxyribose ring and the purine or pyrimidine base. Although this could be caused directly by an oxidant, more often such damage is caused by ionizing radiation. When a particle of high-energy radiation (an α particle, a high-energy β particle, or a γ photon) passes through the cell, it can interact with any of the molecules found there, including DNA, and cause damage. However, the amount of water in a cell means that either water or molecular oxygen usually absorbs the energy and undergoes a reaction. Although this protects the cell from damaging radiation, damage to water or molecular oxygen molecules can result in formation of highly reactive H^+, OH^-,

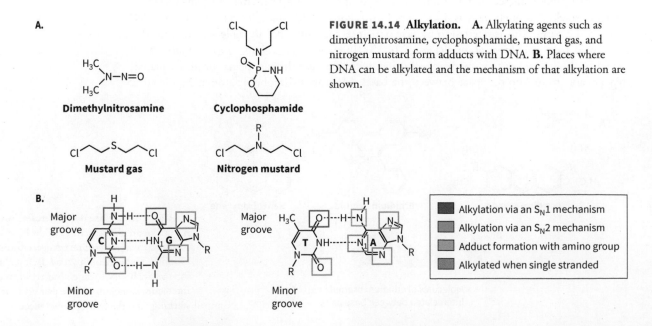

FIGURE 14.14 Alkylation.   **A.** Alkylating agents such as dimethylnitrosamine, cyclophosphamide, mustard gas, and nitrogen mustard form adducts with DNA. **B.** Places where DNA can be alkylated and the mechanism of that alkylation are shown.

H_2O_2, or O_2 radicals, all of which can react with DNA to oxidize bases or generate abasic sites. It is in part to combat this problem that the cell contains enzymes such as superoxide dismutase to catalyze the breakdown of these molecules or antioxidant molecules such as glutathione that will react with these compounds, or antioxidant vitamins such as retinal (vitamin A), ascorbate (vitamin C), and tocopherols (vitamin E).

Thymine dimers Ultraviolet (UV) light also damages DNA (**Figure 14.15**). Molecules with double bonds have electrons in π bonds. If a molecule has conjugated double bonds, it has a system of π electrons that allows them to interact with longer wavelengths of radiation. The greater the extension of the π network, the longer the wavelength that will be absorbed. Most UV radiation from the sun is absorbed by ozone (O_3) in the ozone layer of Earth's atmosphere, which absorbs this radiation and reemits the energy as heat. Without ozone, more UV would reach the ground and cause damage.

The π bond network in thymine is of a size that allows it to interact with UV light in the 280 to 315 nm range (UVB radiation), with the UV photons exciting electrons in the π bond. In the case of thymine, if there is an adjacent molecule of thymine in the same strand, the potential exists to form new covalent bonds, bridging the two bases with a cyclobutane ring and linking them together in a dimer. This causes a bulge in the DNA helix, which makes it difficult for enzymes and proteins to bind the DNA, resulting in errors in replication, transcription, and translation.

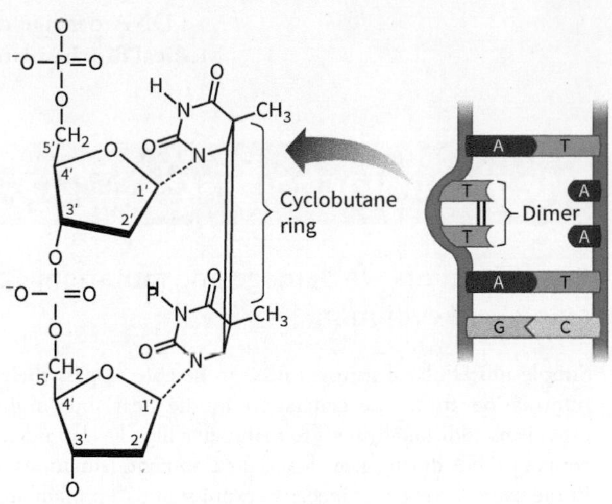

FIGURE 14.15 Effects of ultraviolet light. Ultraviolet light can induce the formation of thymine dimers, a cyclobutane linkage between the two pyrimidines.

(Source: Wessner, *Microbiology*, 1e, copyright 2013, John Wiley & Sons. This material is reproduced with permission of John Wiley & Sons, Inc.)

Double-strand breaks The final type of damage that can occur is double-strand breaks, in which the phosphodiester backbone breaks on both strands.

If the breaks are across from one another or there are not enough base pairs between the breaks to keep the strands joined, what results is a double-strand break. Ionizing radiation and topoisomerase inhibitors (including the antibiotic and chemotherapeutic agent bleomycin) all induce double-strand breaks, although the mechanism by which these breaks occur is not always clear.

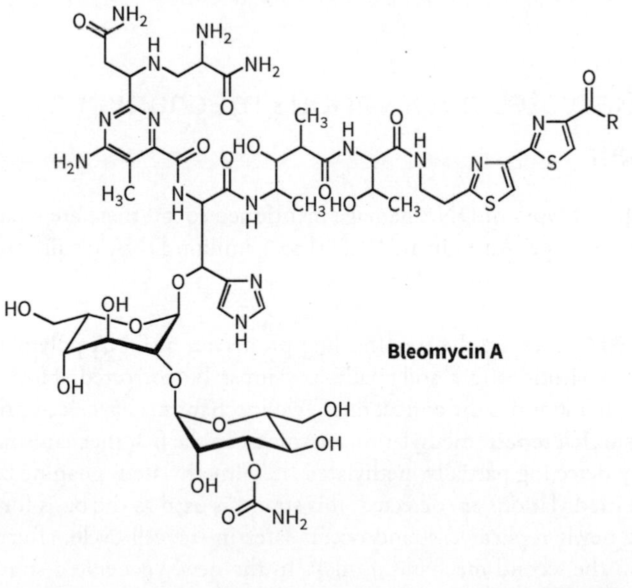

Bleomycin A

Viral damage Damage can also occur when retroviruses write their genetic code back into a copy of the cell's DNA. Thousands of viruses have incorporated viral DNA into the genome over time, and the cell has no means for detecting or removing this information from the genome once encoded. Benign viral information is simply replicated from generation to generation; however, some viruses have the ability to transform cells, activating oncogenes and causing cancer. The best-known example is the human papilloma virus (HPV), which causes over 12,000 cases of cervical cancer in women each year in the United States, resulting in 3,400 deaths. A preventative vaccine called Gardasil can reduce the rates of HPV infection, and HPV-related diseases (cervical cancer, genital warts, and genital cancers) have dropped by 70%–100% in vaccinated populations.

DNA damage can have serious consequences. The results of DNA damage are discussed in **Medical Biochemistry: The impact of DNA damage and mutation: Cancer, aging, and evolution.**

Medical Biochemistry

The impact of DNA damage and mutation: Cancer, aging, and evolution

Simple unicellular organisms need to be able to pass their genetic information from one generation to the next, but multicellular organisms additionally need to maintain a high level of fidelity when copying DNA during growth and development. Mutations resulting in the overexpression or incorrect expression of a protein are usually harmless to a multicellular organism, but in some cases, they can be disastrous. Failure to copy the genetic code faithfully has been implicated in cancer and aging but is also one of the driving forces in evolution.

Let us examine two genes in a liver cell, one coding for hemoglobin and one for a signaling protein involved in cell proliferation. All cells in the human body have the gene for hemoglobin, but in liver cells it is not expressed. This means that if this gene were mutated or damaged, there would be no noticeable impact to the organism. However, if the other gene were damaged and this somehow triggered the cell to begin dividing, there would be the potential for the development of cancer. The current model is not quite as simple as this and assumes that there must be multiple mutations to genes regulating cell growth or mutations to both copies of a gene (known as the two-hit hypothesis).

As we learned in this chapter, chemicals, radiation, or viruses that damage DNA can lead to oncogenesis, or the development of cancer. The cells that are most directly exposed to mutagens (for example, lung cells in smokers) are more likely to suffer damage and therefore more likely to become cancerous. Likewise, some cancers run in families. In such cases, there may already be a damaged allele coding for a growth regulator that is inherited and passed down through generations. This also explains why cancer affects a greater proportion of older people. As people age, they accumulate a greater amount of DNA damage and mutations. When these mutations affect the genes involved in cell growth, it creates the potential for cancer.

DNA damage has also been implicated in the aging process. Telomerases maintain the ends of chromosomes and enable cells in culture and cancer cells to keep dividing. We have also seen that telomere shortening is linked to the ability of cells to divide (known as the Hayflick limit). However, the aging process is complex, and many factors contribute to it.

Finally, although we typically think of DNA mutation in somatic cells, there are also mutations that occur to the cells of the germ line. Here also, recombination comes into play, increasing the diversity of successive generations. The rate of mutations in these cells determines the rate of evolution, or how rapidly a species changes through generations or over time.

14.4.2 Prokaryotes have various mechanisms for DNA repair

For each of the different types of DNA damage mentioned so far, there are repair mechanisms in place to remedy the damage. An estimated 1,000 to 1 million DNA repairs are made in the life of an average cell.

Mismatch repair Despite the proofreading properties of DNA polymerases, mismatches (incorrectly paired nucleotides) are still made and must be corrected. How does the cellular machinery know which strand is the correct one? Two mechanisms have been proposed for identifying errors for **mismatch repair**: methylation and nicks. In the first mechanism, the parent DNA can be identified by detecting partially methylated (hemimethylated) guanine bases of the parent strand. Once these methylations are detected, this strand is used as the basis for comparison. The methylation of the newly replicated strand occurs later in the cell cycle, after mismatch repairs have been made. In the second mechanism, nicks in the newly generated strand (perhaps nicks not yet sealed by ligase) are the marker for the nascent strand.

The best studied of the mismatch repair systems is the Mut system of *E. coli* (**Figure 14.16**). The proteins that are critical for the detection and repair of mismatches are the Mut proteins (MutH, MutL, and MutS).

Several models describe mismatch repair. In one such model, homodimers of MutS encircle and ride along the DNA double helix, searching for mismatches. The mismatch would be identified due to changes in the conformation of the helix such as a bulge generated by two purines at the mismatch site. Binding of ATP increases the processivity of MutS. Upon encountering a

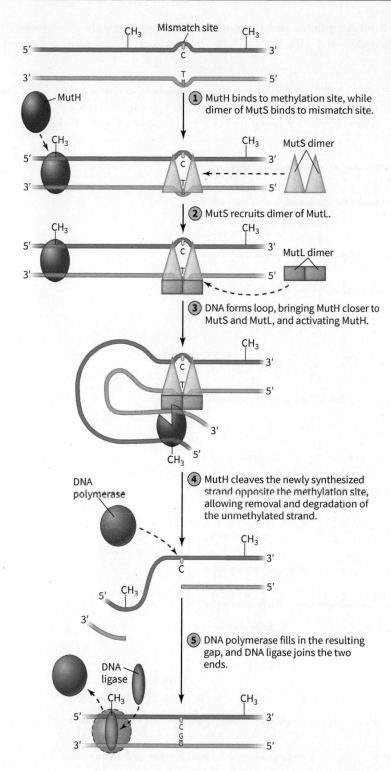

① MutH binds to methylation site, while dimer of MutS binds to mismatch site.

② MutS recruits dimer of MutL.

③ DNA forms loop, bringing MutH closer to MutS and MutL, and activating MutH.

④ MutH cleaves the newly synthesized strand opposite the methylation site, allowing removal and degradation of the unmethylated strand.

⑤ DNA polymerase fills in the resulting gap, and DNA ligase joins the two ends.

FIGURE 14.16 Mismatch repair pathway. In the Mut mismatch repair system, MutS recognizes and binds to mismatches. This recruits and anchors MutL and MutH. MutH is an endonuclease that nicks and excises the daughter strand. The damaged strand is removed, and a new one is inserted by DNA pol III and ligase.

(Source: Wessner, *Microbiology*, 1e, copyright 2013, John Wiley & Sons. This material is reproduced with permission of John Wiley & Sons, Inc.)

mismatch, MutS pauses and recruits a homodimer of the adapter protein MutL. The latter also requires ATP; it causes MutS to bind more tightly and recruits and activates MutH, a restriction **endonuclease** that recognizes the methylated strand of DNA and incises the daughter strand at a specific sequence (dGATC). This sequence may be as far as 1 kb from the mismatch. Beginning at the nick generated by MutH, the DNA is unwound by the helicase UvrD, exposing the single-stranded parent chain. The daughter strand is degraded by exonucleases (ExoI, ExoVII, ExoX, and RecJ). The original parent strand is bound by SSBs, and a replacement is generated by DNA pol III and ligase.

Mismatch repair in other organisms may function in a similar way but without heterodimers of MutS and MutL homologs and without MutH. All these systems are the focus of ongoing study.

Base excision repair

A different means of repair, called **base excision repair**, is needed in instances where a base has been chemically modified (e.g., 8-oxoguanine, 7-methyladenine, or hypoxanthine), uracil has been mistakenly substituted, or a base is missing (**Figure 14.17**). In these

FIGURE 14.17 Base excision repair. In the base excision repair system, the damaged base is first removed by a glycosylase, resulting in an abasic site. Next, a nuclease nicks the damaged strand at the abasic site. The repair can continue by either the short-patch or long-patch pathway. In short-patch, the single abasic site is removed, and a new nucleotide is incorporated by DNA pol β. In long-patch, DNA pol δ or ε binds and synthesizes a sequence about 10 bases long, displacing the old strand as it is made. The flap of the displaced strand is removed by FEN-1, and the nick is sealed by ligase.

(Source: Snustad; Simmons, *Principles of Genetics*, 6e, copyright 2012, John Wiley & Sons. This material is reproduced with permission of John Wiley & Sons, Inc.)

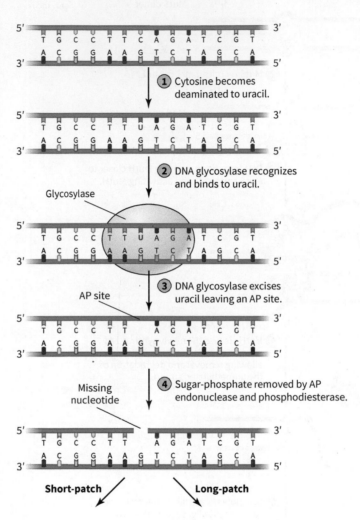

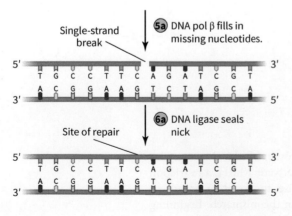

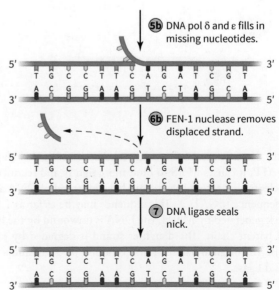

instances, the base is first excised from the DNA helix by a **DNA glycosylase**, and the chemically damaged base will presumably be lost; catabolized; or, at the least, not reincorporated into nucleotides and then DNA. The removal of the base leaves an abasic site, that is, a hole in the helix where a base is missing. Recalling nucleotide metabolism or DNA synthesis, it is unlikely that the correct base can be reinserted and the glycosidic linkage to the deoxyribose backbone re-formed. Instead, abasic sites (also referred to as apurinic or apyrimidinic or **AP sites**) are recognized by an **AP endonuclease**. This enzyme nicks the DNA at the AP site, leaving a free 3′ hydroxyl group.

At this point, there are two different types of repair that can occur: short-patch or long-patch. In short-patch repair, the deoxyribose phosphate lacking the base is removed by 5′ deoxyribose phosphate lyase. This enzyme removes the phosphoribose, leaving a gap, which is filled in by DNA pol β (in humans); the resulting nick is then sealed by ligase. DNA pol β inserts only a single base before dissociating from the DNA helix.

In long-patch repair, DNA pol δ or ε binds in conjunction with PCNA and synthesizes a new piece of DNA, extending from the 3′ end of the nick. This long-patch is up to ten bases in length and displaces the old strand as it is made. Flap endonuclease-1 (FEN-1) degrades this flap, and the remaining nick is sealed with ligase.

Nucleotide excision repair

The third means the cell has of detecting and repairing DNA damage is **nucleotide excision repair (NER)** (**Figure 14.18**). NER plays an important role in detecting and eliminating bulky DNA adducts such as thymine dimers or bases to which a large group has been added. In NER, a lesion is detected, and nicks are made on either side of the lesion. The undamaged strand is used as a template in the synthesis of the patch.

In prokaryotes, NER is carried out by the UvrABC system, which is comprised of four proteins: UvrA, UvrB, UvrC, and UvrD (also known as helicase II). A dimer of UvrA and B slides along the DNA helix and detects the lesion. When this happens, UvrA leaves, and a new dimer of UvrB and UvrC forms. UvrB cuts the phosphodiester backbone four nucleotides downstream of the damage, and UvrC cuts eight nucleotides upstream, generating a 12-nucleotide fragment that is removed by UvrD. The repair is completed by DNA pol I and ligase.

In eukaryotes, there are two systems for recognizing damage, involving 11 different proteins (**Figure 14.19**): global genomic nucleotide excision repair (GG-NER) and transcription-coupled nucleotide excision repair (TC-NER). The chief difference between the two is that TC-NER occurs when an RNA polymerase stalls during transcription, whereas GG-NER is ongoing. Each of these two detection systems is associated with disease (xeroderma pigmentosum with defective GG-NER and Cockayne syndrome with defective TC-NER). In both diseases, patients suffer damage to skin and photosensitivity, although differences in the two disorders reveal that these pathways play other roles in health and development.

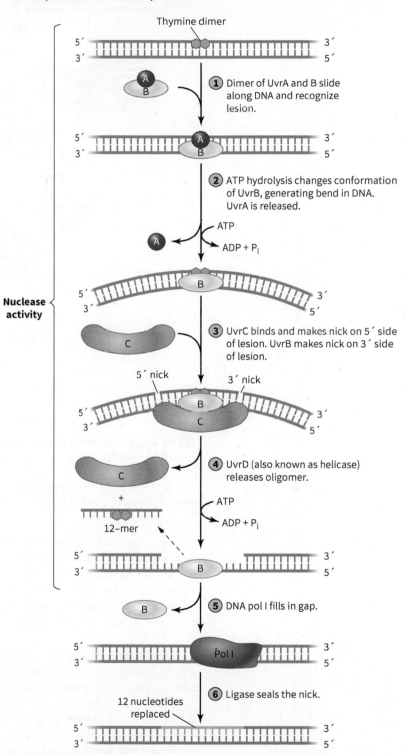

FIGURE 14.18 Nucleotide excision repair. Nucleotide excision repair is one way the cell has of dealing with thymidine dimers or bulky DNA adducts. In the UvrABC system, a bulky lesion is first detected by UvrA and B, which recruit UvrC. The complex cuts on either side of the bulky lesion. The repair is completed by DNA pol I and ligase.

(Source: Snustad; Simmons, *Principles of Genetics*, 6e, copyright 2012, John Wiley & Sons. This material is reproduced with permission of John Wiley & Sons, Inc.)

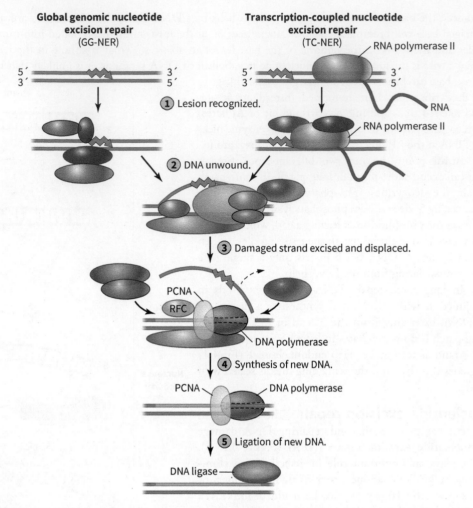

FIGURE 14.19 Nucleotide excision repair in eukaryotes. Nucleotide excision repair in eukaryotes is detected by one of two different means. In GG-NER, a complex of proteins recognizes the lesion, whereas in TC-NER, the stalling of RNA polymerase II triggers recognition. In both cases, after recognition the DNA is unwound by transcription factor II H, and approximately 25 nucleotides are excised around the lesion. The repair is patched by one of several DNA polymerases and sealed by ligase.

Once detected, the lesion is removed through the action of transcription factor II H. This multiprotein complex is comprised of ten different subunits (including an endonuclease, a helicase, and an ATPase), which cut the damaged strand on the 3′ side of the lesion. A second endonuclease cuts approximately 25 to 30 nucleotides away, releasing the damaged fragment.

Following excision, RFC loads PCNA, which recruits DNA polymerases (δ, ε, or κ) that fill in the gap. FEN-1 and ligase complete the repair.

Thymine dimer repair **Photolyase** is an enzyme that uses the energy of a photon of light to hydrolyze thymine dimers generated through exposure to UV light (**Figure 14.20**). This type of repair is also known as direct repair because the nucleotides involved remain in place.

E. coli photolyase contains two cofactors: a molecule of the fully reduced hydroquinone form of FADH⁻ that is directly involved in the reduction and cleavage of the cyclobutane ring that forms between the adjacent thymines, and a molecule of methylenetetrahydrofolate. The methylenetetrahydrofolate acts as a kind of photoantenna, absorbing photons of far blue light (300 to 450 nm) and liberating an electron. This electron tunnels through three tryptophan residues to the FADH⁻ and catalyzes the cleavage of the cyclobutane ring in less than a nanosecond.

Although photolyase enzymes are common in most organisms, they are absent in mammals, which employ other repair mechanisms.

Double-strand break repair There are two different mechanisms by which a **double-strand break repair (DSBR)** can be made: homologous recombination (discussed in the

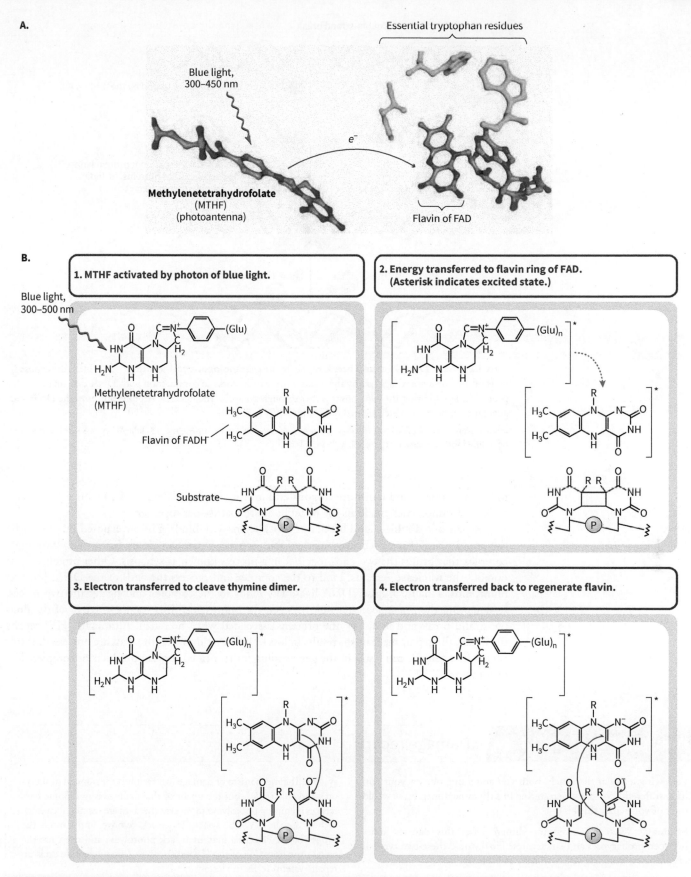

FIGURE 14.20 The mechanism of photolyase. **A.** Illustrated are the key elements of the photolyase active site. **B.** Photolyase employs methylenetetrahydrofolate (MTHF) and a flavin ring to cleave the cyclobutane ring found in thymine dimers. This proceeds through a radical mechanism (a single electron transfer).

(Source: (A) Data from PDB ID 1DNP Park, H. W., Kim, S. T., Sancar, A., Deisenhofer, J. (1995) Crystal structure of DNA photolyase from *Escherichia coli. Science* 268: 1866–1872. Copyright 1995. Reproduced with permissions of American Association for the Advancement of Science – AAAS.)

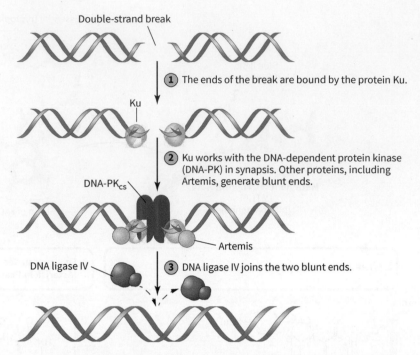

Double-strand break

1 The ends of the break are bound by the protein Ku.

Ku

2 Ku works with the DNA-dependent protein kinase (DNA-PK) in synapsis. Other proteins, including Artemis, generate blunt ends.

DNA-PK$_{cs}$

Artemis

DNA ligase IV

3 DNA ligase IV joins the two blunt ends.

FIGURE 14.21 Double-strand break repair by non-homologous end joining. In NHEJ, the exposed ends of the break are recognized by the protein Ku, which works in conjunction with DNA-dependent protein kinase to bring the two ends together through synapsis. The nuclease Artemis processes the DNA and generates blunt ends, which are then joined by DNA ligase.

(Source: Karp, *Cell and Molecular Biology: Concepts and Experiments*, 7e, copyright 2013, John Wiley & Sons. This material is reproduced with permission of John Wiley & Sons, Inc.)

following section) and **non-homologous end joining (NHEJ)** (**Figure 14.21**). NHEJ takes two ends of chromosomes and joins the phosphodiester backbones together.

Repair of a double-strand break by NHEJ begins with binding of the exposed DNA ends by the protein Ku. This protein recruits the DNA-dependent protein kinase (DNA-PK), forming a complex that brings the two ends together in a process known as synapsis. Other proteins, most notably the **nuclease** Artemis, bind to the complex and process the ends of the DNA, forming blunt ends. Finally, a ligase (DNA ligase IV) binds to the complex and ligates the two pieces, regenerating the phosphodiester link between the blunt ends. This results in repair of the double-strand break and release of the proteins associated with the repair. Although NHEJ repairs the break, the repair mechanism results in loss of information when the ends are processed. If this occurs in a gene, it can result in the gene coding for truncated protein products or complete loss of function.

Worked Problem 14.4   Damage report

Following a day at the beach, both you and a microbe on your skin have suffered UV exposure resulting in a thymine dimer. How would each of you repair this lesion?

Strategy What is a thymine dimer? Does this damage differ between prokaryotes and eukaryotes? How would these two types of organisms detect and repair this type of damage?

Solution Thymine dimers result from UV exposure. A cyclobutane ring is formed between adjacent thymine bases on the same DNA strand. This changes the alignment of the bases, causing a kink in the DNA helix that is detected by the DNA base excision repair

system. The mechanism of damage formation is the same in prokaryotes and eukaryotes, but the repair of this damage is not. Some bacteria can employ photolyase to reverse the damage (as noted earlier). Others use the UVRabc system to detect, remove, and reinsert the proper bases. Placental mammals lack photolyase and must use the eukaryotic homolog of the UVRabc system (the nucleotide excision repair system) to fix the lesion.

Follow-up question Scientists hypothesize that life on Earth was not possible before the development of the ozone layer. Why might this have been the case?

Summary

- DNA can be damaged through chemicals, radiation, or viruses. Specific types of DNA damage can include insertions or deletions, alkylation of DNA, oxidative damage, thymine dimers, or double-strand breaks.
- Chemical damage can prevent DNA replication or generate problems in coding for information, which can lead to cancer, other illnesses, or death of an organism.
- DNA repair mechanisms identify damage (although not necessarily mutations), remove the damaged area, and regenerate the strand. Specific mechanisms of repair include mismatch repair, base excision repair, nucleotide excision repair, thymine dimer repair, and double-strand break repair.

Concept Check

1. Describe with examples, the different types of DNA damage. Discuss and compare the significance of each type.
2. Enumerate the types of DNA repair mechanisms, and describe the basics of how each functions.
3. Describe the differences between prokaryotic and eukaryotic DNA repair mechanisms.

14.5 Homologous Recombination

Nature is amazingly diverse, even within a single species. A superficial examination of a single species such as humans reveals that there is enormous variability in outward appearance, even among members of the same family. Much of this variability is due to homologous recombination, which is the focus of this section.

14.5.1 Homologous recombination generates genetic diversity

Homologous recombination is effectively the trading or swapping of different pieces of chromosomes. It has been known since the early 1900s that chromosomes in germ cells (the cells that become gametes) from eukaryotic organisms will cross over, exchanging different segments of chromosomes after replication but before mitosis (**Figure 14.22**). This effectively generates two daughter chromosomes that have components of both parent chromosomes but are not complete copies of either. In germ cells this generates genetic diversity in offspring, but in other cells it also provides a mechanism for DNA repair and is responsible for generating the diversity seen in the proteins of the immune system. Recombination of some kind is seen in all organisms and is employed by some viruses as a way to integrate into the host chromosome.

This diversity not only helps make us individuals, but it can also impact how society views and develops treatments. This is described in **Societal and Ethical Biochemistry: Personalized medicine: Knowing your code.**

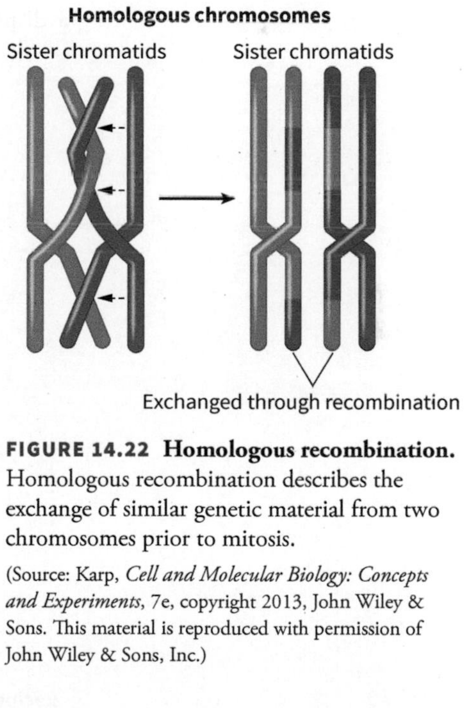

Homologous chromosomes

Sister chromatids   Sister chromatids

Exchanged through recombination

FIGURE 14.22 Homologous recombination. Homologous recombination describes the exchange of similar genetic material from two chromosomes prior to mitosis.

(Source: Karp, *Cell and Molecular Biology: Concepts and Experiments*, 7e, copyright 2013, John Wiley & Sons. This material is reproduced with permission of John Wiley & Sons, Inc.)

14.5.2 Recombination in prokaryotes is mediated by Rec proteins

Recombination in prokaryotes proceeds through a double-strand break repair process (**Figure 14.23**). The process has two parts: **branch migration** (formation of a link between the damaged chromosome and a homologous chromosome) and resolution (separation of the two intertwined chromosomes).

Following a double-strand break, the end of the DNA is recognized by the RecBCD complex, which has several enzymatic activities. Initially, the complex employs an ATP-dependent helicase activity to unwind the DNA and degrade the sequence with an endonuclease. This goes on from the end of the break to an eight-base sequence (5'-GCTGGTGG-3') known as Chi. At

Societal and Ethical Biochemistry

Personalized medicine: Knowing your code

Although modern medicine offers a wide variety of treatments, a single drug may affect different people in different ways due to polymorphisms in genes and differing degrees of gene expression.

A simple analogy can be drawn with artificial sweeteners. Overall, most people would agree that sucrose (table sugar), saccharin (Sweet'n Low), aspartame (NutraSweet), and sucralose (Equal) all taste sweet, but the degree to which people find each of these substances sweet might vary widely. Likewise, about 25% of the population finds that saccharin has a strong metallic aftertaste. To some extent, the same variability is true for every molecule from antihistamines to narcotics; any of these molecules may have different effects in individual members of the population.

One advantage to knowing aspects of your own DNA sequence such as which alleles of specific genes you have is that medical treatments can be tailored specifically for your genotype. Such personalized medicine at the molecular level is already in use in many cancer centers for a limited but growing number of diseases.

When a patient has been diagnosed with breast cancer, for example, she immediately undergoes genotyping for about 30 genes that code for proteins that might make the tumor resistant to chemotherapy or result in a more aggressive cancer. Her doctors consult these results before beginning any type of therapy. Should the results show that the tumor may be resistant to some types of chemotherapy, oncologists can immediately pursue second-line courses of action. By tailoring the treatment to the individual, doctors can bypass ineffective treatments and improve outcomes.

Chi, the complex binds to DNA, forming a single-stranded loop that binds to the protein RecA. The latter increases the interactions of the single strand with the complementary strand on a homologous chromosome. The single-stranded DNA–RecA protein complex then displaces one of the strands on the other copy of the DNA, in a process termed strand invasion, and then forms a displacement loop. Following displacement, branch migration occurs, where the new strand

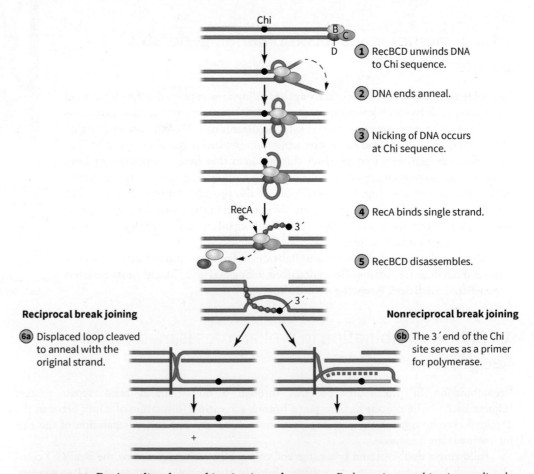

FIGURE 14.23 RecA-mediated recombination in prokaryotes. Prokaryotic recombination mediated by the RecBCD complex of proteins can proceed through either reciprocal or nonreciprocal break joining. In both mechanisms RecBCD catalyzes the unwinding and nicking of DNA.

displaces one of the old strands, re-forming the double helix in the process. RecA helps to stabilize this complex, resulting in regions thousands of base pairs in length that have crossed over.

Following strand invasion and branch migration, the complex may complete recombination in one of two ways. The displaced loop of DNA may be cleaved and anneal to the original strand in a process known as reciprocal break joining. Alternatively, the 3′ terminus of the Chi sequence can act as a primer for DNA polymerase, which can fill in the gap in a process known as nonreciprocal break joining.

Reciprocal break joining results in an X-shaped structure (termed a **Holliday junction**, after its discoverer) that needs to be broken before the DNA molecules can separate. This is resolved by the cleavage of the junction by the RuvABC complex. This complex is an endonuclease that recognizes the sequence 5′-(A/T)TT(G/C)-3′. The resulting nicks are sealed by ligase.

Nonreciprocal break joining results in two chromosomes, one of which is recombinant and the other of which is a copy of one of the parent chromosomes.

14.5.3 Recombination in eukaryotes can proceed through several different models

Several different mechanisms for homologous recombination in eukaryotes have been proposed. The two best characterized are the double-strand break repair (DSBR) mechanism and the synthesis-dependent strand-annealing (SDSA) path (**Figure 14.24**). Both pathways begin with a double-strand break in one of the sister chromatids. These ends are bound by the MRN complex. The 5′ blunt ends are resected (cut back) by a nuclease (Sae2) that leaves a 3′ overhang. Sgs1 helicase denatures the DNA, leaving single strands, which are bound by the protein Rad51. The single-stranded DNA–Rad51 complex mediates strand invasion in the sister chromatid, displacing one of the two strands of the other chromosome and generating a D-loop. The invading strand is used as a primer for DNA polymerases to extend and synthesize a new DNA strand, using the invaded strand as the template. At this point, the two pathways diverge.

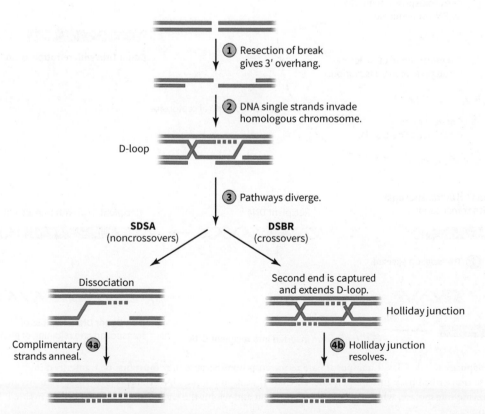

FIGURE 14.24 Homologous recombination in eukaryotes. Homologous recombination in eukaryotes can proceed through either the double-strand break repair (DSBR) mechanism or the synthesis-dependent strand-annealing (SDSA) path.

In SDSA, the synthesis of DNA results in the sliding of the Holliday junction down the strand in branch migration (as with the recombination and repair mechanisms discussed previously). This displaces the newly synthesized strand from the template, which can anneal to the other single strand of the original broken chromosome. Here again, the DNA repair mechanisms (DNA polymerase and ligase) will complete the repair. SDSA results in one sister chromatid that has recombined and one that is unchanged.

In DSBR, both strands of the cut chromosome are involved in the formation of Holliday junctions. Both strands undergo DNA synthesis and branch migration before resolution and separation of the sister chromatids. Exactly how this process occurs is still under investigation, but it is known that nicking endonucleases that only cleave one strand are involved in the separation of the two chromatids. In DSBR, both strands are recombinant, exchanging sequence with each other.

14.5.4 Transposons are hopping genes

Some pieces of DNA are highly mobile and can move throughout the genome, effectively hopping from one chromosome to another (**Figure 14.25**). These pieces of hopping DNA are termed transposable elements, or **transposons**, and were originally identified in corn (*Zea mays*), where they constitute 90% of the genome. These elements were later found in all species, including humans, where they comprise 50% of the genome.

There are two main classes of transposons: class I and class II.

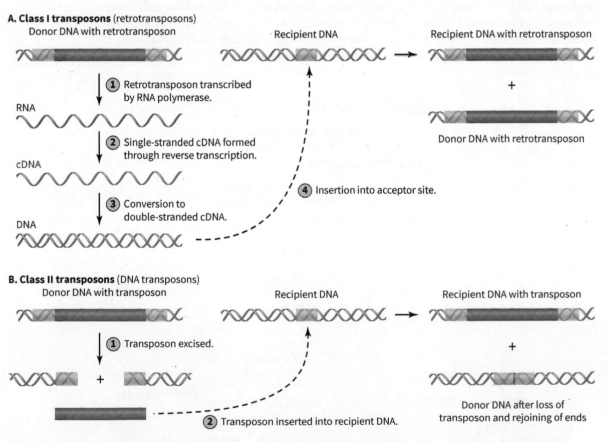

FIGURE 14.25 Transposons. A. Class I transposons are retrotransposons because they must first be transcribed into RNA and then reverse transcribed back into the genome. **B.** Class II transposons are DNA transposons. These elements are recognized by the enzyme transposase, which cuts out the transposon and rejoins the ends. The transposon is then inserted elsewhere in the genome.

(Source: Karp, *Cell and Molecular Biology: Concepts and Experiments*, 7e, copyright 2013, John Wiley & Sons. This material is reproduced with permission of John Wiley & Sons, Inc.)

Class I transposons

Class I transposon, also known as retrotransposons, must first pass through an RNA element that is reverse-transcribed into a new location in the genome. The process has many similarities to retroviral infection of a cell. The three main classes of retrotransposons are those with long terminal repeats (LTRs), long interspersed nucleotide elements (LINEs), and short interspersed nucleotide elements (SINEs). Of these, LTRs and LINEs have sequences that code for reverse transcriptase, whereas SINEs do not.

LINEs are approximately 6 kb in length and typically code for two proteins: a reverse transcriptase/integrase (noted previously) and an RNA-binding protein. These are transcribed by RNA polymerase II (discussed in Chapter 15). There are about 500,000 LINEs in the human genome, constituting about 17% of the total sequence.

SINEs are shorter than LINEs, averaging 100 to 600 bp in length. These are transcribed by RNA polymerase III. There are about 1.5 million copies of SINEs in the human genome (~11% of the total sequence). SINEs do not code for a reverse transcriptase; instead, they rely on the LTR or LINE reverse transcriptase for **transposition**. SINEs also do not code for protein but for RNA molecules that are similar in sequence to tRNA or rRNA. The function of these molecules and the function of these transposable elements are still under investigation.

The most common SINEs in humans and all primates are Alu sequences. Alu sequences are about 300 bp long and are named for the restriction enzyme (Alu I) that cleaves in this site. The function of Alu sequences in human biology is still unclear.

Class II transposons

Class II transposons are directly "cut" from one part of the genome and "pasted" into another. They differ from retrotransposons in that they do not have to pass through an RNA intermediate before reintegration into the genome. Instead, DNA transposon hopping is specifically catalyzed by transposases, enzymes that recognize a sequence framing the transposable element, cut it out, and reinsert it in a new location. Approximately 2%–3% of the human genome is comprised of class II transposons.

Transposases (such as the proteins Tcn1/mariner, Tn5, or sleeping beauty) have a conserved mechanism that is similar to the mechanism of DNA polymerase, RNase H, and retroviral integrase. The active site contains two metal ions chelated by three conserved negatively charged amino acids (two aspartate and one glutamate residue) termed a DDE motif (**Figure 14.26**). These residues and metal ions are responsible for cleavage of the phosphodiester backbone of the transposon at a specific sequence (5'-TACA-3') and integration of the transposon into the new site, coded for by the sequence (5'-TA-3'). The missing nucleotides on the recipient strand are filled in by the cell's DNA repair machinery. On the donor strand, the strands are filled in to

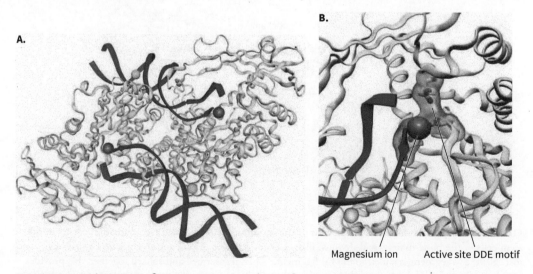

Magnesium ion   Active site DDE motif

FIGURE 14.26 Active site of transposase.   **A.** A dimer of transposases is bound to a piece of DNA. **B.** The active site of transposase contains three conserved anionic amino acids termed a DDE motif, which chelate a magnesium ion (shown in space-filling depiction) that is involved in the cleavage of the phosphodiester backbone.

(Source: (A, B) Data from PDB ID 1MUH Davies, D. R., Goryshin, I. Y., Reznikoff, W. S., Rayment, I. (2000) Three-dimensional structure of the Tn5 synaptic complex transposition intermediate. Science 289: 77–85)

generate blunt ends and joined through DSBR mechanisms. This provides a short, repeated sequence (in effect, a scar of the original transposon) termed an excision footprint.

Although the function of transposons is unknown, they may play significant roles in evolution. Movement of fragments of DNA around the genome can lead to gene duplications, disruptions of some genes, or chromosomal rearrangements.

14.5.5 VDJ recombination generates genetic diversity in the immune system

The cells of the immune system that are responsible for production of antibodies and T-cell receptors have a unique job because each must be able to produce a unique protein, either an antibody or a receptor. The proteins must have the variety and diversity needed to detect foreign proteins in the body but retain common structural elements needed for immune system function. The cell needs to synthesize the protein reproducibly and pass this information from one generation of cell to the next in the event that the cell is stimulated to divide and proliferate. To generate these proteins, these cells use a special process called somatic recombination, which differs from the recombination events seen in gametes (**Figure 14.27**). This process is also termed variable,

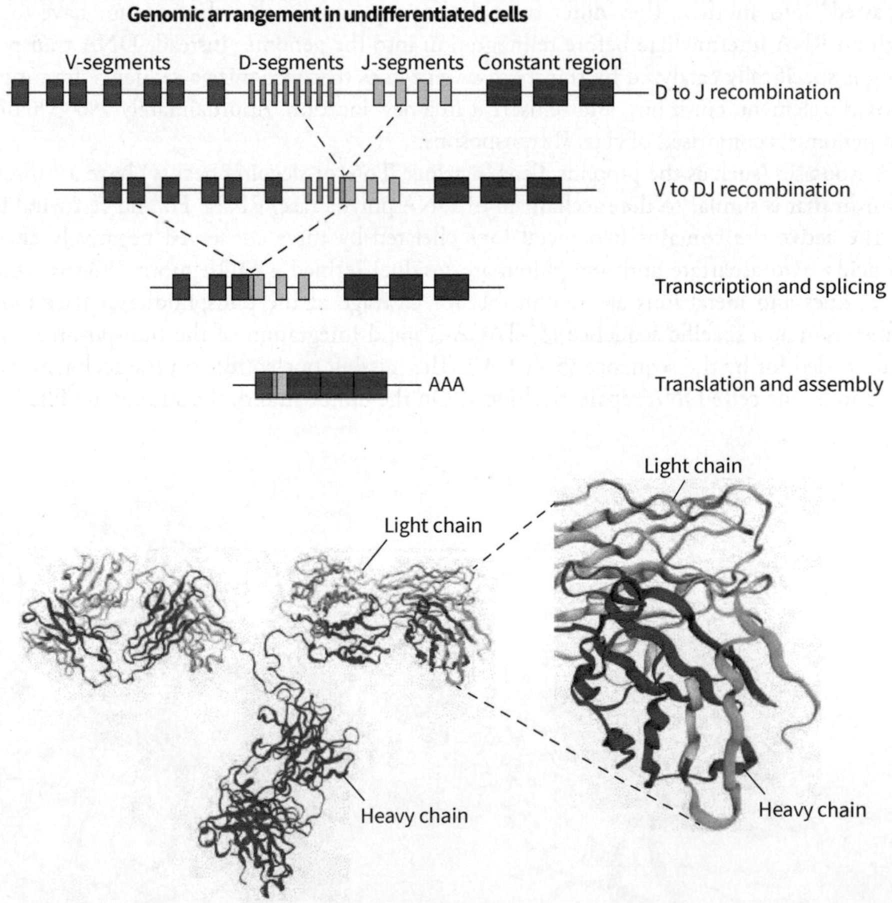

FIGURE 14.27 Recombination in the cells of the immune system.   VDJ recombination occurs in the stem cells of the bone marrow and thymus. First, a D and J segment join, and the intervening DNA is lost. Next, a V segment is joined to the DJ fragment. When transcribed and spliced together, this provides a wide array of possible coding combinations for the variable termini of immunoglobulins and T-cell receptors. The recombination of the light chain is not shown in detail here.

(Source: Data from PDB ID 1IGT Harris, L. J., Larson, S. B., Hasel, K. W., McPherson, A. (1997) Refined structure of an intact IgG2a monoclonal antibody. Biochemistry 36: 1581–1597)

diverse, and joining (VDJ) replication, based on the three original names for the pieces of DNA involved in the recombination event.

Depending on the type of protein produced by the cell (immunoglobulin G or a T-cell receptor), there are different regions that code for V, D, and J. For example, in a cell producing an immunoglobulin G heavy chain, there are 44 copies of V, 27 copies of D, and 6 copies of J that can be combined to give the final coding sequence. The first recombination event joins a D and a J unit together, and the intervening DNA is lost from the cell. In the second recombination event, a V unit is added to the DJ fragment, and intervening DNA is again lost. By recombining fragments in this way, the cell is able to generate great diversity; it is estimated that the number of different types of immunoglobulins that can be generated is on the order of 10^{12}.

Worked Problem 14.5   Synthesis or not

How could you demonstrate that DNA synthesis is required for homologous recombination to occur in a mammalian system?

Strategy   If DNA synthesis were required for homologous recombination, we should be able to use what we know about DNA synthesis to test this hypothesis. Other than the template and primer, what else would be needed?

Solution   We could test the hypothesis using dNTPs radiolabeled on the α phosphate adjacent to the 2-deoxyribose ring. If these were

incorporated, it would result in a radiolabeled product that could be easily separated from the monomeric dNTPs by either precipitation or size-exclusion chromatography. If the labeled dNTPs were not incorporated, it could be assumed that the radionucleotides had not been incorporated.

Follow-up question   Is there a way to conduct this study without using radiolabeled nucleotides? What are the strengths and weaknesses of that experimental technique?

Summary

- Homologous recombination is the process by which similar DNA sequences can exchange with one another to repair DNA and increase diversity in gametes and cells of the immune system.

- The Rec proteins mediate recombination by unwinding DNA, binding to single-stranded DNA, and promoting strand invasion. Holliday junctions are the cross-shaped structures DNA forms during recombination. They are "resolved" (cleaved) in prokaryotes by the RuvABC complex.

- Recombination in eukaryotes can proceed by either the double-strand break repair pathway or the synthesis-dependent strand-annealing pathway.

- Transposons are short regions of chromosomes that can move throughout the genome. Class I transposons, including LTRs, LINES, and SINES, are retrotransposons that must pass from DNA through an RNA intermediate before being reverse-translated back into the genome. Class II transposons are DNA transposons bounded by sequences that are cut by transposase and then reintegrated elsewhere in the genome.

- VDJ recombination is the process through which the proteins of the immune system (antibodies and T-cell receptors) generate the diversity required in that system. Three different components (V, D, and J) are recombined to form the mature gene.

Concept Check

1. Explain the evolutionary benefit of homologous recombination.
2. Describe different mechanisms of recombination, including the RecABC system, the DSBR path, and the SDSA path.
3. Explain the consequences of the two ways in which holiday junctions resolve.
4. Describe the VDJ system and how it provides a mechanism for immune system diversity in terms of protein structure and function.

Bioinformatics Exercises

Exercise 1 Melting Temperature and the GC Content of Duplex DNA

Exercise 2 Viewing and Analyzing DNA-Binding Proteins

Exercise 3 DNA Replication and the KEGG Database

Exercise 4 Viewing and Analyzing Uracil–DNA Glycosylase in DNA Repair

Problems

14.1 Challenges of DNA Replication

1. List the similarities and differences between prokaryotic and eukaryotic cells that impact cell replication.

2. Is DNA replication unidirectional? If yes, specify the direction in terms of DNA strands. Explain how does the second strand then replicates.

3. Hydrolysis of a triphosphate nucleotide or diphosphate nucleotide molecule is required for the formation of an unfavorable phosphodiester bond between two nucleotide molecules. Explain the chemistry involved in this step. Also, explain why the incoming group's triphosphate is located on 5′ carbon of the pentose sugar.

4. If the *E. coli* genome is 4.6×10^6 base pairs and is replicated in 30 minutes, what is the reaction rate of helicase in terms of nucleotides per minute?

5. Different topoisomerases either prevent supercoiling of DNA during replication or separate the newly copied DNA molecules. Based on section 14.1, are these reactions carried out by the same enzyme or different enzymes?

6. Explain how and why ATP/ADP ratio will affect the assembly of the replication complex.

7. Why is it unlikely from a thermodynamic perspective that DNA synthesis would run 3′ to 5′? Could a polymerase run in the 3′ to 5′ direction? Why or why not?

14.2 DNA Replication in Prokaryotes

8. Describe initiation of DNA replication in prokaryotes.

9. Biochemically, would it make better sense for histones to be methylated or demethylated prior to replication? Why?

10. The *oriC* sequence is rich in adenine and thymine. Why does this make sense chemically? Could a sequence rich in guanine and cytosine work as well?

11. A prokaryotic replisome typically contains two molecules of DNA pol III, but only one molecule of DNA pol I. Why?

12. We have discussed the structure if DNA at several different levels, including the nucleotide sequence, double helix, nucleosome, and chromosome. What techniques were used to determine these structures?

13. If you mutated a polymerase so that the active site bound nucleoside diphosphates instead of nucleoside triphosphates, would you anticipate that this enzyme would be active? How might you have affected the kinetics of the enzyme?

14.3 DNA Replication in Eukaryotes

14. Explain why telomeres and telomerase are necessary for the current model of DNA replication. Most cancer cells express elevated levels of telomerase. Why?

15. How does the cell ensure that DNA replication must occur once per cell cycle?

16. Would you anticipate that telomerase levels are higher in short-lived organisms like the shrew or long-lived organisms like the giant sequoia, or are they the same? Explain your answer.

14.4 DNA Damage and Repair

17. In double-strand breaks, the phosphodiester backbone breaks on both strands. What are the agents that induce double-strand breaks?

18. Nitrosamine (NR_2NO) and formaldehyde (CH_2O) can both form adducts with DNA. How might each of the types of damage caused by these be repaired by the cell? How would this differ between eukaryotic and prokaryotic cells?

19. Why does exposure to UV light in a few instances cause a bulge in the DNA double helix?

20. Cisplatin is a chemotherapeutic drug used to treat several different cancers, including testicular cancer. It forms adducts to guanine residues that often result in cross-linking of two guanines, similar to a thymine dimer. How might this damage be repaired?

21. Which type of DNA repair is used to remove and replace a single modified base that cannot be converted back to the normal base by a direct repair process?

22. Compare the processivities of DNA polymerase β and δ. Could they substitute for one another? How has natural selection influenced the evolution of the genes coding for each of these proteins?

23. Cockayne's syndrome is a disease that stems from damage to the nucleotide excision repair system. Research this disorder. How does the disease differ from xeroderma pigmentosum? Which genes and proteins are affected? What do differences and similarities between the two disorders tell you about the broader roles of these proteins in DNA repair?

14.5 Homologous Recombination

24. *E. coli* strains that are deficient in RecA are commonly used in molecular cloning experiments. Why might it be to a scientist's advantage to use these strains?

25. What are the odds of randomly encountering AT or TACA in the genome? If these sequences are not random, what does that say about them?

26. The quaternary structure of RecA is a linear polymer of RecA subunits that binds to single-stranded DNA. How could this help facilitate recombination?

27. Conserved enzymes such as cytochromes are often used to track evolution. Could Alu sequences be used instead? How would the information obtained differ?

28. Describe what transposons are and how they function in the genome. Discuss two ways that transposons could be used in mutagenesis.

29. The active site of recombinase is similar to some other enzymes (such as some polymerases) that we have seen in that it uses two metal ions for binding and catalysis. What are the common divalent metal ions employed by enzymes that have a DNA substrate? Which other metals could work, and which could not? Propose a reason why these ions work and others do not.

30. Explains why researchers often use *E. coli* strains that are RecA minus (recA-) when transforming a cloned gene into a plasmid. What does this explain about the behavior of a RecA strain?

31. How would an alteration to the DDE sequence or to the types of metal ions influence the kinetics of the recombinase? What kinetic parameters (V_{max}, K_M, etc.) would be affected?

32. Relate evolution, aging, and cancer to DNA replication, repair, recombination, and telomere maintenance.

Data Interpretation

Questions 33 to 37 refer to the following data from Beck and Brubaker, *J. Bacteriol.* 1973. In that paper, strains of *E. coli* were grown in media containing ³H thymine for 24 hours then transferred to unlabeled media. The percentage of radiolabel that was found as material precipitated by trichloroacetic acid is shown.

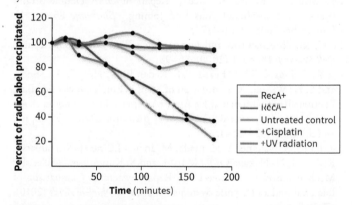

33. Why would radiolabel precipitate in the presence of trichloroacetic acid? What has happened to the thymine?

34. Why is the radiolabel no longer in the insoluble fraction following UV exposure?

35. Why is the radiolabel no longer in the soluble fraction following exposure to cisplatin?

36. What does the insoluble radiolabel say about the type of lesion cisplatin generates?

37. Why are cells expressing the RecA protein retaining more radiolabel? The RecA protein is part of the RecBCD repair system.

Experimental Design

38. Design an experiment to measure how often the correct nucleotide is excised from a piece of replicating DNA.

39. If you wished to repeat the Meselson–Stahl experiment (section 14.1) but did not have access to an ultracentrifuge, is there a different way you could run the experiment and detect the results?

40. Design an experiment to determine if the exonuclease function of DNA pol III is necessary for replication.

41. Design two experiments, the first to determine how damaging a chemical is to DNA *in vitro* and the second to see how mutagenic the same chemical is *in vivo*.

Ethics and Social Responsibility

42. DNA evidence is often used to help convict or exonerate the accused. Knowing this, is it a good idea to sequence everyone's DNA to have in a database? How is DNA any different in this regard than fingerprints?

43. What do you understand about personalizing medicine? State your views on how it might or might not revolutionize modern medicine.

44. When someone gets DNA sequenced to determine if they are at risk for a disease or carriers of a potential birth defect, it may also provide some information about other family members. Although counseling is often offered to the person tested, nothing is offered for family members. Is this ethical? What provisions should be made in these instances?

45. Transgenic, or knockout, organisms are a highly valuable resource for researchers. Do you feel it is ethical to make genetically modified animal or plant model systems for use in the laboratory? Explain your position.

Suggested Readings

14.1 Challenges of DNA Replication

Bell, S. D., M. Méchali, and M. L. DePamphilis, eds. *DNA Replication*. Cold Spring Harbor, NY: Cold Spring Harbor Press, 2013.

Cook, P. R. "Molecular Biology—The Organization of Replication and Transcription." *Science* 284, no. 5421 (June 11, 1999): 1790–1795.

Lohman, T. M., and K. P. Bjornson. "Mechanisms of Helicase-Catalyzed DNA Unwinding." *Annual Review of Biochemistry* 65 (1996): 169–214.

Waga, S., and B. Stillman. "The DNA Replication Fork in Eukaryotic Cells." *Annual Review of Biochemistry* 67 (1998): 721–751.

Wang, J. C. *Untangling the Double Helix: DNA Entanglement and the Action of the DNA Topoisomerases.* Cold Spring Harbor, NY: Cold Spring Harbor Laboratory Press, 2009.

14.2 DNA Replication in Prokaryotes

Crooke, E., D. S. Hwang, K. Skarstad, B. Thöny, and A. Kornberg. "*E. coli* Minichromosome Replication: Regulation of Initiation at OriC." *Research in Microbiology* 142, no. 2–3 (1991): 127–130.

Kamada, K., T. Horiuchi, K. Ohsumi, N. Shimamoto, and K. Morikawa. "Structure of a Replication-Terminator Protein Complexed with DNA." *Nature* 383 (1996): 598–603.

Steitz, T. A. "DNA Polymerases: Structural Diversity and Common Mechanisms." *The Journal of Biological Chemistry* 274 (1999): 17395–17398.

Wang, J. C. "Cellular Roles of DNA Topoisomerases: A Molecular Perspective." *Nature Reviews Molecular Cell Biology* 3 (June 2002): 430–440.

14.3 DNA Replication in Eukaryotes

Brown, T. A., C. Cecconi, A. N. Tkachuk, C. Bustamante, and D. A. Clayton. "Replication of Mitochondrial DNA Occurs by Strand Displacement with Alternative Light-Strand Origins, Not via a Strand-Coupled Mechanism." *Genes & Development* 19 (2005): 2466–2476.

DeLange, T. "T-Loops and the Origin of Telomeres." *Nature Reviews Molecular Cell Biology* 5, no. 4 (2004): 323–329.

Greider, C. W. "Telomerase Is Processive." *Molecular and Cellular Biology* 11, no. 9 (September 1991): 4572–4580.

Greider, C. W., and E. H. Blackburn. "Identification of a Specific Telomere Terminal Transferase Activity in Tetrahymena Extracts." *Cell* 43 (1985): 405–413.

Greider, C. W., and E. H. Blackburn. "A Telomeric Sequence in the RNA of *Tetrahymena* Telomerase Required for Telomere Repeat Synthesis." *Nature* (London) 337 (1989): 331–337.

Holt, I. J., and A. Reyes. "Human Mitochondrial DNA Replication." *Cold Spring Harbor Perspectives in Biology* 1, no. 12 (2012): a012971.

Hondele, M., T. Stuwe, M. Hassler, F. Halbach, A. Bowman, E. T. Zhang, B. Nijmeijer, C. Kotthoff, V. Rybin, S. Amlacher, E. Hurt, and A. G. Ladurner. "Structural Basis of Histone H2A–H2B Recognition by the Essential Chaperone FACT." *Nature* 499, no. 7456 (July 4, 2013): 111–114.

Hseih, F. K., O. I. Kulaeva, S. S. Patel, P. N. Dyer, K. Luger, D. Reinberg, and V. M. Studitsky. "Histone Chaperone FACT Action during Transcription through Chromatin by RNA Polymerase II." *Proceedings of the National Academy of Science USA* 110, no. 19 (May 2013): 7654–7659.

Krude, T. "Chromatin Assembly during DNA Replication in Somatic Cells." *European Journal of Biochemistry* 263, no. 1 (1999): 1–5.

Leonard, A. C., and M. Méchali. "DNA Replication Origins." *Cold Spring Harbor Perspectives in Biology* 5, no. 10 (2013): a010116.

Lu, G.-L., J. D. Bradley, L. D. Attardi, and E. H. Blackburn. "*In vivo* Alteration of Telomere Sequences and Senescence Caused by Mutated *Tetrahymena* Telomerase RNAs." *Nature* (London) 344 (1990): 126–132.

Mcintosh, D., and J. J. Blow. "Dormant Origins, the Licensing Checkpoint, and the Response to Replicative Stresses." *Cold Spring Harbor Perspectives in Biology* 4, no. 10 (2012).

McKinney, E. A., and M. T. Oliveira. "Replicating Animal Mitochondrial DNA." *Genetics and Molecular Biology* 36, no. 3 (2013): 308–315.

Nandakumar, J., and T. R. Cech. "Finding the End: Recruitment of Telomerase to the Telomere." *Nature Reviews Molecular Cell Biology* 14, no. 2 (February 2013): 69–82.

Nishitani, H., and Z. Lygerou. "Control of DNA Replication Licensing in a Cell Cycle." *Genes to Cells* 7, no. 6 (June 2002): 523–534.

Orphanides, G., G. LeRoy, C.-H. Chang, D. S. Luse, and D. Reinberg. "FACT, a Factor That Facilitates Transcript Elongation through Nucleosomes." *Cell* 92, no. 1 (1998): 105–116.

Reinberg, D., and R. J. Sims, III. "de FACTo Nucleosome Dynamics." *The Journal of Biological Chemistry* 281, no. 33 (2006): 23297–23301.

Stewart, L., M. R. Redinbo, X. Qiu, W. G. J. Hol, and J. J. Champoux. "A Model for the Mechanism of Human Topoisomerase I." *Science* 279, no. 5356 (March 1998): 1534–1541.

Stoeber, K., T. D. Tlsty, L. Happerfield, G. A. Thomas, S. Romanov, L. Bobrow, E. D. Williams, and G. H. Williams. "DNA Replication Licensing and Human Cell Proliferation." *Journal of Cell Science* 114 (2001): 2027–2041.

Truong, L. N., and X. Wu. "Prevention of DNA Re-replication in Eukaryotic Cells." *Journal of Molecular Cell Biology* 3, no.1 (2011): 13–22.

Vijayraghavan, S., and A. Schwacha. "The Eukaryotic Mcm2–7 Replicative Helicase." *Subcellular Biochemistry* 62 (2012): 113–134.

14.4 DNA Damage and Repair

Balakrishnan L., and R. A. Bambara. "Flap Endonuclease 1." *Annual Review of Biochemistry* 82 (2013): 119–138.

Brettel, K., and M. Byrdin. "Reaction Mechanisms of DNA Photolyase." *Current Opinion in Structural Biology* 20, no. 6 (December 2010): 693–701.

Cerritelli, S. M., and R. J. Crouch. "Ribonuclease H: The Enzymes in Eukaryotes." *FEBS Journal* 276, no. 6 (March 2009): 1494–505.

Drabløs F., E. Feyzi, P. A. Aas, C. B. Vaagbø, B. Kavli, M. S. Bratlie, J. Peña-Diaz, M. Otterlei, G. Slupphaug, and H. E. Krokan. "Alkylation Damage in DNA and RNA—Repair Mechanisms and Medical Significance." *DNA Repair* (Amst) 3, no. 11 (November 2004): 1389–1407.

Essen, L. O., and T. Klar. "Light-Driven DNA Repair by Photolyases." *Cellular and Molecular Life Sciences* 63, no. 11 (June 2006): 1266–1277.

Fortini P., and E. Dogliotti. "Base Damage and Single-Strand Break Repair: Mechanisms and Functional Significance of Short- and Long-Patch Repair Subpathways." *DNA Repair* 6, no. 4 (2007): 398–409.

Jiricny, J. "Postreplicative Mismatch Repair." *Cold Spring Harbor Perspectives in Biology* 5, no. 4 (2013): a012633.

Lees-Miller, S. P., and K. Meek. "Repair of DNA Double Strand Breaks by Non-Homologous End Joining." *Biochimie* 85, no. 11 (November 2003): 1161–1173.

Li, G. M. "Mechanisms and Functions of DNA Mismatch Repair." *Cell Research* 18, no. 1 (2008): 85–98.

Liu Y., R. Prasad, W. A. Beard, P. S. Kedar, E. W. Hou, D. D. Shock, and S. H. Wilson. "Coordination of Steps in Single-Nucleotide Base Excision Repair Mediated by Apurinic/Apyrimidinic Endonuclease 1 and DNA Polymerase β." *Journal of Biological Chemistry* 282, no. 18 (2007): 13532–13541.

Morita R., S. Nakane, A. Shimada, M. Inoue, J. Iino, T. Wakamatsu, K. Fukui, N. Nakagawa, R. Matsui, and S. Kuramitsu. "Molecular Mechanisms of the Whole DNA Repair System: A Comparison of Bacterial and Eukaryotic Systems." *Journal of Nucleic Acids* (2010): 179594.

Sandigursky, M., and W. A. Franklin. "The Post-Incision Steps of the DNA Base Excision Repair Pathway in *Escherichia coli*: Studies with a Closed Circular DNA Substrate Containing a Single U:G Base Pair." *Nucleic Acids Research* 26, no. 5 (1998): 1282–1287.

Sarasin, A., and R. Monier. "DNA Repair Pathways and Associated Human Diseases." *Biochimie* 85, no. 11 (2003): 1041.

Wang, H., and J. B. Hays. "Signaling from DNA Mispairs to Mismatch-Repair Excision Sites Despite Intervening Blockades." *The EMBO Journal* 23, no. 10 (2004): 2126–2133.

Wold, M. S. "Replication Protein A: Heterotrimeric, Single-Stranded DNA-Binding Protein Required for Eukaryotic DNA Metabolism." *Annual Review of Biochemistry* 66, no. 1 (1997): 61–92.

Yang, W. "Surviving the Sun: Repair and Bypass of DNA UV Lesions." *Protein Science* 20, no. 11 (2011): 1781–1789.

14.5 Homologous Recombination

Collier, L. S., C. Carlson, S. Ravimohan, A. J. Dupuy, and D. A. Largaespada. "Cancer Gene Discovery in Solid Tumours Using Transposon-Based Somatic Mutagenesis in the Mouse." *Nature* 436 (July 14, 2005): 272–276.

Ding, S., X. Wu, G. Li, M. Han, Y. Zhuang, and T. Xu. "Efficient Transposition of the PiggyBac (PB) Transposon in Mammalian Cells and Mice." *Cell* 122 (August 12, 2005): 473–483.

Dupuy, A. J., K. A. Kagi, D. A. Largaespada, N. G. Copeland, and N. A. Jenkins. "Mammalian Mutagenesis Using a Highly Mobile Somatic Sleeping Beauty Transposon System." *Nature* 436 (July 14, 2005): 221–226.

Goodsell, D. S. "The Molecular Perspective: Double-Stranded DNA Breaks." *The Oncologist* 10, no. 5 (May 2005): 361–362.

Grindley, N. D. F., K. L. Whiteson, and P. A. Rice. "Mechanisms of Site-Specific Recombination." *Annual Review of Biochemistry* 75 (2006): 567–605.

Horne, J., H. T. Lawless, W. Speirs, and D. Sposato. "Bitter Taste of Saccharin and Acesulfame-K." *Chemical Senses* 27, no. 1 (2002): 31–38.

Ivics, Z., P. B. Hackett, R. H. Plasterk, and Z. Izsvák. "Molecular Reconstruction of Sleeping Beauty, a Tc1-Like Transposon from Fish, and Its Transposition in Human Cells." *Cell* 91, no. 4 (November 1997): 501–510.

Ivics, Z., M. A. Li, L. Mátás, J. D. Boeke, A. Bradley, and Z. Izsvák. "Transposon-Mediated Genome Manipulation in Vertebrates." *Nature Methods* 6, no. 6 (2006): 415–422.

Kramerov, D. A., and N. S. Vassetzky. "Origin and Evolution of SINEs in Eukaryotic Genomes." *Heredity* (Edinb) 107, no. 6 (December 2011): 487–495.

Reed, D. R., S. Li, X. Li, L. Huang, M. G. Tordoff, R. Starling-Roney, K. Taniguchi, D. B. West, J. D. Ohmen, G. K. Beauchamp, and A. A. Bachmanov. "Polymorphisms in the Taste Receptor Gene (Tas1r3) Region Are Associated with Saccharin Preference in 30 Mouse Strains." *The Journal of Neuroscience* 24, no. 4 (2004): 938–946.

Reznikoff, W. S. "The Tn5 Transposon." *Annual Review of Microbiology* 47 (1993): 945–963.

TIMELINE CHAPTER **14**

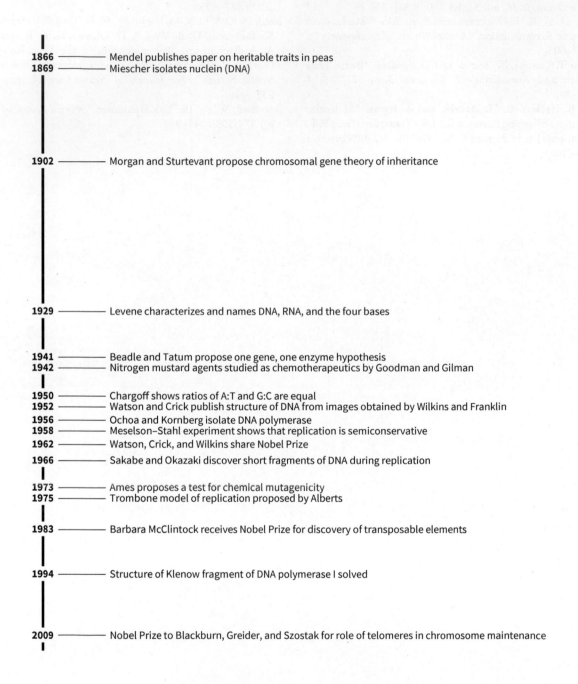

1866 —————— Mendel publishes paper on heritable traits in peas
1869 —————— Miescher isolates nuclein (DNA)

1902 —————— Morgan and Sturtevant propose chromosomal gene theory of inheritance

1929 —————— Levene characterizes and names DNA, RNA, and the four bases

1941 —————— Beadle and Tatum propose one gene, one enzyme hypothesis
1942 —————— Nitrogen mustard agents studied as chemotherapeutics by Goodman and Gilman

1950 —————— Chargoff shows ratios of A:T and G:C are equal
1952 —————— Watson and Crick publish structure of DNA from images obtained by Wilkins and Franklin
1956 —————— Ochoa and Kornberg isolate DNA polymerase
1958 —————— Meselson–Stahl experiment shows that replication is semiconservative
1962 —————— Watson, Crick, and Wilkins share Nobel Prize
1966 —————— Sakabe and Okazaki discover short fragments of DNA during replication

1973 —————— Ames proposes a test for chemical mutagenicity
1975 —————— Trombone model of replication proposed by Alberts

1983 —————— Barbara McClintock receives Nobel Prize for discovery of transposable elements

1994 —————— Structure of Klenow fragment of DNA polymerase I solved

2009 —————— Nobel Prize to Blackburn, Greider, and Szostak for role of telomeres in chromosome maintenance

CHAPTER 15

RNA Synthesis and Processing

RNA Synthesis in Context

I have a friend who has an amazing collection of music. She began collecting works from different artists back in the 1980s in the form of records, 12-inch discs of stamped vinyl with music physically encoded on them. As technology advanced, cassette tapes and players were invented, making music more portable than it was on records, which can only be played on a turntable. Also, it was possible to create a mix tape by recording different pieces of music onto a single tape, for example, to suit a particular mood, for a party, or to let someone know you were thinking of them. Today, people make playlists to load on their phones so they can listen at the gym or in the car.

The process of creating a mix tape or playlist has some similarities to transcription of information in the cell from the DNA in the genome (like a record collection on vinyl) to RNA (like a playlist). Synthesis of the RNA message effectively condenses the massive amount of information contained in the genome into smaller parts, which can then be spliced together to provide the code necessary for what is going on in the cell at that point in time.

This chapter describes how both prokaryotes and eukaryotes synthesize RNA from DNA in a process known as transcription. The RNA includes not only messenger RNA (mRNA), but also transfer, ribosomal, and several other types of small RNA molecules. The chapter also discusses how these nascent molecules are processed into mature transcripts and then exported from the nucleus.

Today's technologies have largely rendered mix tapes obsolete, but the concept of the mix tape is hard to escape—as if it were in your DNA.

Chapter Outline

Common Themes

Evolution's outcomes are conserved.	• Separation of transcription from translation, compartmentalization in the nucleus, the introduction of introns into the genome, and mRNA processing are major events in evolution; these significant steps help delineate prokaryotes from eukaryotes at the biochemical level.
	• The lack of conservation between RNA and DNA polymerases indicates that proteins coding for these genes arose and evolved independent of one another.
	• RNA polymerases are conserved at the gene and amino acid level, indicating a diversification of and descent from a single progenitor gene. These proteins are also structurally conserved and employ common mechanisms.
	• The protein structures found in the nuclear pore complex (NPC) are conserved and are found throughout eukaryotic biochemistry. The NPC must have coevolved with the nuclear envelope to retain its function.
Structure determines function.	• Although RNA and DNA polymerases are functionally related, their structures differ.
	• Many of the functions of the complex structures formed by RNA molecules are parallel to those of proteins (e.g., catalysis).
	• The protein structures found in the NPC serve similar roles in other aspects of biochemistry, for example, in receptor-mediated endocytosis.
	• The intrinsically disordered phenylalanine–glycine (FG) proteins found in the center of the NPC hinder the passage of most molecules but facilitate the transit of molecules and complexes that recognize the FG motif.
	• Transcription and processing of RNA in eukaryotes appears to be coupled with export from the nucleus. These processes are catalyzed by complexes consisting of multiple molecules of protein and RNA.
Biochemical information is transferred, exchanged, and stored.	• RNA molecules can act as messages (coding for information), adapters (assisting in the translation of a message), and catalysts (catalyzing the transcription and translation of a message).
	• RNA molecules are thought to have been the key progenitors of life as it evolved from a prebiotic state.
	• The central dogma of biochemistry is that DNA molecules code for information that is passed through an RNA intermediate and translated into protein.
	• RNA-based genomes found in retroviruses violate that central dogma by having an initial step in which RNA is reverse-transcribed into DNA.
	• Splicing provides a means to increase the diversity of proteins without increasing the total number of genes.
Biomolecules are altered through pathways involving transformations of energy and matter.	• The synthesis, processing, and export of RNA molecules require a large energy investment by the organism.
	• Reactions such as those in splicing preserve energy by performing energetically neutral transesterification reactions.

15.1 Transcription: RNA Synthesis

The central dogma of biochemistry is that DNA molecules code for information, which is then transcribed into an RNA intermediate before being translated into protein. This section deals with the transcription part of this process. Synthesis of an RNA message from a DNA template through **transcription** is an important regulatory step in gene expression and requires a lot of energy and resources. An RNA intermediate also allows some species to diversify proteins through differential splicing of mRNA messages, discussed later in section 15.2.

Eukaryotic genes that are actively undergoing transcription are organized into transcription factories, that is, regions of the nucleus that contain multiple copies of RNA polymerase (pol) II and loops of DNA coding for genes actively undergoing transcription. These regions are often clustered around the nuclear envelope and are directly tied to the enzymes involved in splicing and processing of nascent RNAs.

The process of transcription differs slightly between prokaryotes and eukaryotes, but the overall steps are similar. This section will describe RNA polymerases and the steps involved in transcription for both prokaryotes and eukaryotes in parallel, comparing and contrasting as we go. The remaining two sections of the chapter focus on processes exclusive to eukaryotes.

15.1.1 Different types of RNA polymerase catalyze synthesis of different types of RNA

Recall from Chapter 2 that the cell uses multiple types of RNA, including **ribosomal RNA (rRNA)**, **transfer RNA (tRNA)**, and **messenger RNA (mRNA)**. Found in both prokaryotes and eukaryotes, these three types of RNA are the best characterized. In this chapter we will encounter two additional types of RNA: **Small nuclear RNAs (snRNAs)** are responsible for splicing eukaryotic mRNAs, and **microRNAs (miRNAs)** affect gene expression.

Bacteria have a single **RNA polymerase** that is responsible for transcription of all RNAs, both coding and **noncoding sequences**. In eukaryotes, specific RNA polymerases have evolved to synthesize the different classes of RNA. RNA pol I synthesizes the rRNAs in the cell, except for 5S rRNA. RNA pol II makes up 80% of the RNA polymerases in the cell and is the best studied of these enzymes; this enzyme synthesizes mRNA, miRNA, and most snRNA. RNA pol III is responsible for the synthesis of 5S rRNA, tRNA, and several other small RNAs, such as the U6 RNA component of the spliceosome.

Plants have two other RNA polymerases (IV and V) that participate in transcription of short interfering RNAs (siRNAs), which function in post-transcriptional regulation of gene expression. There is a high degree of conservation between these polymerases and RNA pol II.

In eukaryotes, RNA polymerases are large (~500 kDa) structures that contain 12 subunits (**Figure 15.1**). The replisome employs multiple proteins to initiate replication, unwind the DNA helix, relieve strain, and prime and polymerize the new strand. In contrast, the enzymes that complete these tasks in RNA polymerases are part of a single complex. The 12 subunits are collectively responsible for unwinding a small region of the DNA helix, initiating the synthesis of the new RNA molecule, ensuring that the nucleotides added to the growing chain are correctly matched, and generating the new bond to the growing chain.

Other protein factors are responsible for directing the polymerase to bind at a particular point and begin transcription. The carboxy terminal domain (CTD) of RNA pol II contains 52 repeats of the sequence Tyr-Ser-Pro-Thr-Ser-Pro-Ser, with the serines in the 2 and 5 positions of this repeat phosphorylated. This repeated domain is an important adapter between the polymerase and several other proteins, including core transcription factors and an adapter protein complex termed Mediator.

Polyphosphorylation of the CTD by one of the core transcription factors signals the initiation of transcription. The enzymes involved in modifications of mRNAs such as capping and splicing also bind through the CTD.

RNA polymerases and DNA polymerases both employ an active site Mg^{2+} ion for catalysis, and they use similar mechanisms to polymerize nucleotides. Nevertheless, there is no structural homology between these classes of enzymes, and they appear to have evolved independently from one another.

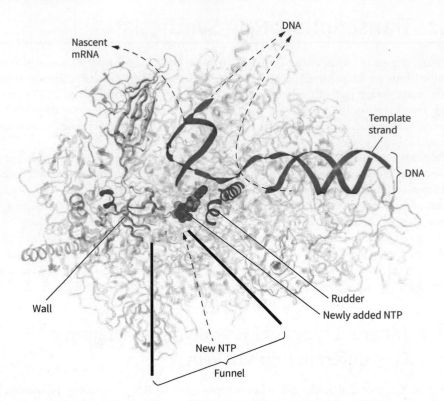

FIGURE 15.1 Structure of RNA polymerase II. RNA polymerase II is a hetero-dodecameric enzyme consisting of 12 different subunits. The template DNA is bound in the polymerase by the combination of a clamp and lid structure. The helix then encounters a beta-sheet structure termed the wall, which introduces an 80° bend to the DNA and facilitates the local denaturation of the double helix. The template strand of the DNA is diverted to the active site by the rudder. RNA is synthesized in the active site, which employs an Mg²⁺ ion for catalysis. New nucleotide triphosphates (NTPs) enter through a funnel on the lower side of the structure. The template and new RNA strand exist as a hybrid for approximately nine bases, at which point they diverge, and RNA exits.

(Source: Data from PDB ID 2E2H Wang, D., Bushnell, D. A., Westover, K. D., Kaplan, C. D., Kornberg, R. D. (2006) Structural basis of transcription: role of the trigger loop in substrate specificity and catalysis. *Cell* 127: 941–954)

Scientists consider RNA to be older than life itself and believe it is one of the molecules that could be synthesized on the young Earth. This is discussed in **Biochemistry: The RNA world.**

Biochemistry

The RNA world

One of the most intriguing questions in science is "How did life begin?" Approaching this question from a biochemical rather than a religious or spiritual perspective poses an interesting conundrum. Evolution clearly explains how eukaryotes arose from prokaryotes and how species have diversified and prospered in nearly all environments on Earth, but it does not explain how life began when Earth was between 100 million and 1 billion years old.

Currently, life on Earth consists almost exclusively of lipids surrounding aqueous compartments in which DNA molecules code for information, RNA acts as a carrier of that information and as a catalyst, and proteins carry out the bulk of the functions of the cell. The current view of early life on Earth is that RNA may have

been the progenitor molecule for all of life because of its ability to both code for information and act as a catalyst. RNA serves as the active component of the ribosome, and many other RNA-dependent processes are slow to evolve, suggesting their involvement in early life. Additionally, many of the common cofactors found in biochemistry (ATP, NADH, FADH₂, and acetyl-CoA) are derived from nucleotides, suggesting that nucleotides were early players in the game of life.

For life to begin, we must assume that the biochemical building blocks of life are present and can assemble into polymers and that these polymers can be organized into cells. In each step, catalysis would facilitate chemical reactions, yielding more rapid reactions and enabling the accumulation of the necessary molecules in the necessary time frame for life to begin.

We might think that the accumulation of the building blocks of life would be one of the highest hurdles to overcome, but this is actually not the case. In 1953, Stanley Miller and Howard Urey created an experiment to simulate the early Earth and determine whether those conditions could create biochemicals. They filled a flask with the water and gases that were likely present in the early atmosphere (CH_4, H_2O, NH_3, and H_2), heated the flask, and passed an electric current through it to simulate lightning. To their amazement, within 24 hours they had generated substantial quantities of several amino acids.

More recent analysis of the Miller–Urey experiment shows that, in fact, far more compounds are generated, including many of the common amino acids and several intermediary metabolites. Other researchers have since shown how nucleotides can also easily form from simple molecules that would have been found on the early Earth and how inorganic minerals and clays can serve as catalysts in the formation of some of these compounds.

Significant advances have also been made in answering the question of how molecules reproduce. For example, certain small RNA molecules have been found to catalyze the formation of copies of themselves, and phospholipid molecules can spontaneously form sheets and micelles, structures that could easily entrap replicating molecules and their precursors.

Other models of abiogenesis (the origin of life) favor ideas such as life arising from a complex mixture of molecules or from catalytic inorganic minerals. It may be that, as with many things in life, it was not one thing but many things that contributed to the origin of life.

Stanley Miller

Bettmann/Getty Images

Harold C. Urey

U.S. Department of Energy

15.1.2 Transcription involves five stages

Transcription in eukaryotes generally has five stages: preinitiation, initiation, promoter clearance, elongation, and termination. These stages depend on signals that direct RNA polymerase to bind at the appropriate location on the DNA helix and begin transcribing RNA. These signals are derived in part from the DNA sequence and in part from proteins called **transcription factors** that bind to these sequences. Transcription factors fall into two categories: core **promoters** (general transcription factors such as the bacterial sigma factors) and specific transcription factors, discussed in Chapter 17.

Stage 1: Preinitiation   There are two major parts to the bacterial **preinitiation complex**: the RNA polymerase and a core promoter. The core promoter is also known as a **sigma (σ) factor**, a protein that works in conjunction with RNA polymerase to form the **holoenzyme** complex. As will be described in Chapter 17, various sigma factors can upregulate genes in times of stress or in response to environmental conditions.

Eukaryotes have a core promoter sequence that starts about 1 to 60 bases upstream of the transcriptional start site (noted as –1 to –60). This site binds to RNA polymerase and the core transcription factors that recruit RNA polymerase (**Figure 15.2**). One of the best-known core promoter sequences is the **TATA box**, that is, a short sequence (TATAAA) found 25 bases before the start site (at –25) that is bound by **TATA binding protein (TBP)**. TBP is part of a larger core transcription factor termed TFIID (TF for transcription factor, II for RNA pol II, and D for the order in which it was discovered), which is comprised of TBP and 14 other proteins termed **TBP-associated factors (TAFs)**. Chapter 17 discusses the various protein motifs that are found in transcription factors. Many of these structures bind by fitting an alpha helix into the major groove of DNA. In contrast, TBP employs a cupped antiparallel beta sheet to encircle the DNA double helix and then binds in the minor groove, after which six other complexes bind to form the preinitiation complex.

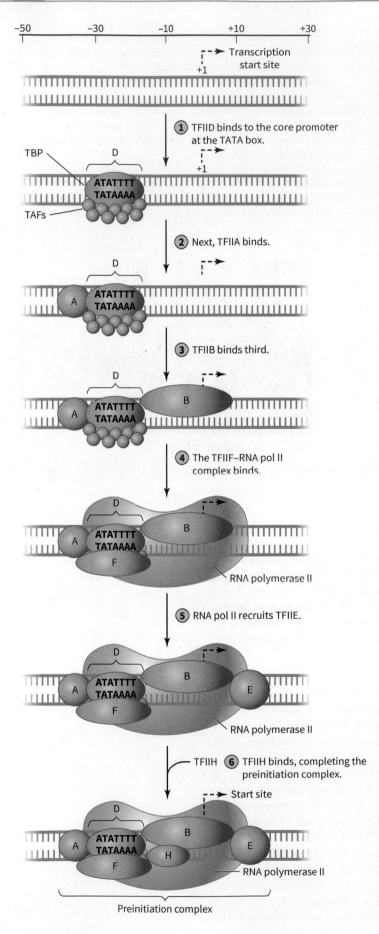

-50 -30 -10 +10 +30

Transcription start site
+1

① TFIID binds to the core promoter at the TATA box.
+1

TBP

D

ATATTTT
TATAAAA

TAFs

② Next, TFIIA binds.

D

A ATATTTT
TATAAAA

③ TFIIB binds third.

D

A ATATTTT
TATAAAA B

④ The TFIIF–RNA pol II complex binds.

D

A ATATTTT
TATAAAA B

F

RNA polymerase II

⑤ RNA pol II recruits TFIIE.

D

A ATATTTT
TATAAAA B E

F

RNA polymerase II

TFIIH ⑥ TFIIH binds, completing the preinitiation complex.

Start site

D

A ATATTTT
TATAAAA B E

H

F

RNA polymerase II

Preinitiation complex

FIGURE 15.2 Steps in the initiation of transcription in eukaryotes. Initiation consists of the sequential binding of a series of core transcription factors and the unwinding of DNA. It ends when the complex clears the start site and elongation begins.

(Source: (Top) Snustad; Simmons, *Principles of Genetics*, 6e, copyright 2012, John Wiley & Sons. This material is reproduced with permission of John Wiley & Sons, Inc. (Bottom) Karp, *Cell and Molecular Biology: Concepts and Experiments*, 7e, copyright 2013, John Wiley & Sons. This material is reproduced with permission of John Wiley & Sons, Inc.)

Next, TFIIB binds to TBP and to a GC-rich sequence found downstream of the TATA box. This serves to orient TFIIB, which in turn orients the polymerase complex. Thus, TFIIB dictates the direction the polymerase should travel, which strand should be copied, and where the polymerase should bind with respect to the transcription start site (the +1 site). TFIIA also binds to TBP, stabilizing the first three factors in the complex (TFIIA, TFIIB, and TFIID).

The next molecule to bind is RNA pol II, in a complex with TFIIF. Binding this molecule recruits TFIIE, which in turn recruits TFIIH. The latter has multiple functions, among them an ATP-dependent helicase function that is partially responsible for forming the transcription bubble and a kinase that polyphosphorylates the carboxy terminal domain of RNA pol II. The phosphorylation of RNA pol II and the helicase activity of TFIIH lead to unwinding of the DNA double helix and binding of the template strand by RNA pol II, assisted by TFIIE. At this point, low levels of transcription can occur; however, the expression of most genes is either increased or decreased by other transcription factors, corepressors, and coactivators that bind upstream of the core transcription complex. These proteins interact with RNA pol II through the aptly named protein, Mediator.

Mediator is a large (1.2 MDa), boomerang-shaped complex of 30 proteins that links distal transcription factors and the carboxy terminal domain of RNA pol II to modulate the activity of the polymerase. The other transcription factors, coactivators, and corepressors typically bind anywhere from −100 to −1,000 base pairs from the transcriptional start site in the promoter region of the gene. Multiple transcription factors can each bind to its own particular sequence and act through Mediator and the carboxy terminal domain of RNA pol II to influence transcription. Now the preinitiation complex is comprised of the promoter region of the gene and multiple protein factors, including RNA polymerase, the core promoter complex, Mediator, transcription factors, coactivators, and corepressors. These molecules work together to regulate expression.

Stage 2: Initiation In bacteria, the binding of RNA polymerase to a sigma factor completes the initiation complex (the RNA polymerase holoenzyme), and transcription commences. In eukaryotes, initiation begins with the phosphorylation of the carboxy terminal domain of RNA pol II by TFIIH. At this point, the polymerase complex is competent and begins synthesizing RNA. In both instances, initiation involves the unwinding of the double helix to expose a short section of single-stranded DNA in the active site of the polymerase.

Stage 3: Promoter clearance

As RNA synthesis commences, the polymerase leaves the transcription initiation complex and proceeds down the DNA, transcribing the message as it goes. This process, termed promoter clearance, occurs in both prokaryotes and eukaryotes. As synthesis of RNA begins, multiple short (ten-nucleotide) transcripts are often produced in a process termed abortive initiation that occurs prior to promotor clearance. The role that abortive initiation plays in promoter clearance is still unclear. In eukaryotes, promoter clearance coincides with recruitment of the capping enzyme, discussed in section 15.2.

Stage 4: Elongation

In the elongation phase, the polymerase adds to the growing RNA. Nucleotides are added by base pairing with the DNA template, with uracil (U) incorporated instead of thymine (T). The energy to form each new phosphodiester linkage in the growing chain comes from the loss and subsequent degradation of pyrophosphate.

As with DNA polymerases, RNA polymerase can proofread messages to ensure that they are accurate copies of the template. Because creating an exact copy is less important for mRNA than for DNA, however, the error rate of RNA polymerases (about 1 in 100,000) is much higher than that of DNA polymerases.

In addition, the proofreading mechanism of RNA polymerase differs from that of DNA polymerase. The RNA polymerase sits in one of three states: (1) In the pre-translocation state, the nucleotide binding site is bound by a nucleotide; (2) in the post-translocation state, the nucleotide binding site is empty; and (3) in the backtracked state, the enzyme has moved in the reverse direction by one nucleotide. RNA pol II is in equilibrium between moving in the forward and reverse directions along the DNA template. Nucleotide binding in the active site of the polymerase favors movement in the forward direction, whereas encountering DNA damage or an incorrectly incorporated nucleotide favors movement in the reverse direction, which would allow TFIIS to cleave the last nucleotide incorporated.

In both prokaryotes and eukaryotes, multiple copies of RNA polymerase can bind to the same promoter and transcribe the DNA message into RNA simultaneously (**Figure 15.3**).

Stage 5: Termination

Bacteria employ two types of termination of transcription: one that depends on the protein factor **Rho** and one that is Rho independent. In Rho-independent termination, the end of the RNA transcript has a stretch of nucleotides that forms a GC hairpin loop, followed by a sequence of uridines. The formation of the GC hairpin is thought to dislocate the message from the template and pull it free from the polymerase, terminating transcription and dislodging the enzyme from the template (**Figure 15.4**).

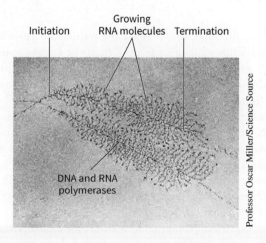

Initiation   Growing RNA molecules   Termination

DNA and RNA polymerases

Professor Oscar Miller/Science Source

FIGURE 15.3 Numerous transcripts synthesized from a gene at one time. This electron micrograph shows RNA polymerases transcribing DNA into RNA. Multiple copies of the polymerase are seen on each gene. In each instance, one copy of the polymerase travels down the DNA molecule with an elongating RNA molecule extending from it as it goes.

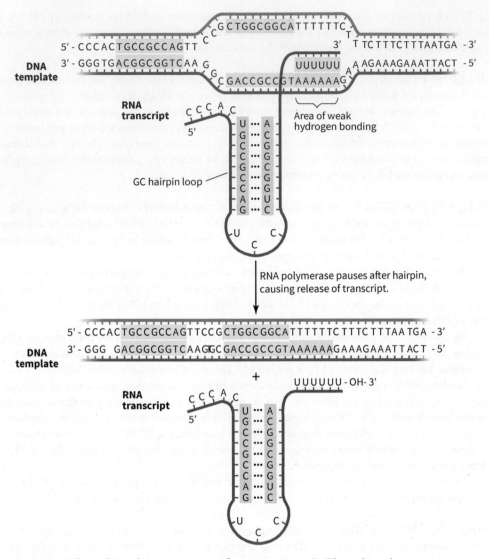

FIGURE 15.4 Rho-independent termination of transcription. In Rho-independent termination, the polymerase encounters a GC-rich sequence followed by a stretch of adenines. The base pairing in the GC-rich sequence forms a hairpin loop that pulls the mRNA out of the active site of the polymerase, terminating transcription.

(Source: Snustad; Simmons, *Principles of Genetics*, 6e, copyright 2012, John Wiley & Sons. This material is reproduced with permission of John Wiley & Sons, Inc.)

In Rho-dependent termination, Rho binds to the terminator pause sequence of the message, a GC-rich sequence that follows the open reading frame or coding sequence of the RNA. Rho wraps around the RNA and acts as an ATP-dependent helicase, riding along the nascent message. When Rho finally reaches the RNA polymerase, its helicase function unwinds the newly synthesized end of the RNA from the DNA template, causing dissociation.

Termination of transcription is not understood as well in eukaryotes, and it differs depending on the polymerase. RNA pol I ceases transcription when a specific termination factor similar to Rho binds to the message and dissociates it from the template. Genes transcribed by RNA pol III have a specific termination sequence that forms a hairpin structure, in a process analogous to Rho-independent termination in bacteria. Termination of genes transcribed by RNA pol II (typically mRNA) is less clear but is thought to be coupled with the addition of a poly(A) tail following the open reading frame, described further in section 15.2.

Different cell types express different genes, which leads to the phenotype of these cells and tissues. Most cells have specific profiles of gene expression, but some can be prompted to turn into other cells—a process termed differentiation. One group of cells that is capable of turning into many different types of cells are stem cells, which are profiled in **Societal and Ethical Biochemistry: Stem cells.**

Societal and Ethical Biochemistry

Stem cells

In the past 15 years, few topics in biomedical science have been as controversial or misunderstood as the study and use of stem cells.

Numerous types of cells exist in the body, varying from neurons to blood cells to osteocytes. Each cell arises from a progenitor cell, and its lineage can be traced back to one of the three different types of cells (ectoderm, endoderm, and mesoderm) and then to the original zygote that these three cell types were derived from. Stem cells are cells that have not yet been terminally differentiated; they can be prompted (either in the body or in the laboratory) to differentiate into other cells or tissues. Cells that can differentiate into any of the cells of the body are referred to as totipotent, whereas those that can differentiate into the three germ layers are referred to as pluripotent. Cells that can only differentiate into related forms are termed multipotent. Totipotent stem cells are obtained from embryonic tissue early in development (within four to five days of fertilization) and are referred to as embryonic stem cells.

The original lines of human embryonic stem (hES) cells were obtained from human embryos created using *in vitro* fertilization techniques. This source has led to much controversy. Those who believe that life begins at conception consider the use of these cells to be unethical because to them the destruction of these embryos represents a loss of life. Some groups concerned with the ethics of using hES cells favor the use of existing lines over the generation of new ones, whereas other groups wish to defund stem cell research altogether.

In fact, much stem cell research today does not rely on cells obtained from human embryos. We now know that most tissues contain stem cells with varying degrees of pluripotency and that such cells can be easily obtained from some tissues. Also, it has become possible to reprogram otherwise terminally differentiated cells back into stem cells; such cells are referred to as induced pluripotent stem (iPS) cells. Such cells can be derived from most tissues or body fluids (including endothelial cells found in urine) without the ethical questions that surround hES cells.

Unfortunately, opponents of stem cell research and even some lawmakers may be unaware of the difference between stem cells originating from embryos and those from other sources like donated blood. Therefore, some people still think that iPS cells should be banned simply because they are stem cells. This is an example of public knowledge lagging behind scientific advances. Scientists and physicians can help to educate the general public to facilitate an informed discussion about the use of stem cells and other technologies.

Given the many model systems available, why is there such interest in studying stem cells? First, understanding the differentiation of human cells yields important data about how humans grow and develop on the molecular level, which can lead to better understanding of human biology and disease. Second, seeing how human cells grow and differentiate can improve our understanding of situations in which cells can no longer grow and divide (as in aging) or grow out of control (as in cancer). Stem cells provide valuable models for both of these processes.

Finally, and perhaps most important, stem cells hold promise in the field of regenerative medicine. Because such cells are in many ways a blank canvas that can be differentiated into many cell types, they may one day provide a means to treat diseases that involve organ or tissue degeneration. For example, it may become possible to grow an organ outside the body that can then be used for transplantation; to inject cells that seek out a target tissue or wound and help repair damage; or to take stem cells that carry a genetic defect, genetically engineer them to remove that defect, and then transplant the healthy cells back into the body.

Many potential applications of stem cells still need to be studied and developed, but at some level stem cell technologies have already arrived. In the United States alone, over 18,000 lifesaving stem cell transplants are performed each year in the form of bone marrow transplants.

15.1.3 RNA polymerase inhibitors can be powerful drugs or poisons

Numerous molecules can serve as inhibitors of RNA polymerases. Many of these are useful research tools or beneficial drugs, but some are highly toxic. This section discusses three RNA polymerase inhibitors with different mechanisms of action and uses (**Table 15.1**).

Of the thousands of mushrooms and fungi found in the world, fewer than 75 are known to be toxic. Of these, one of the deadliest is *Amanita phalloides*, the death cap, shown in **Figure 15.5**. This mushroom produces several toxins, among them α-amanitin, a bicyclic octapeptide that is a potent inhibitor of RNA polymerases. In particular, α-amanitin inhibits RNA pol II with a K_I in the picomolar range. Ingestion of the toxin results in liver and kidney failure, leading to death within five to ten days. X-ray structures of RNA pol II bound to α-amanitin show that the binding of the peptide constrains the bridge helix of the enzyme, resulting in a dramatic slowing of translocation.

Rifampicin is another naturally produced cyclic compound. The organism that produces it is a strain of the bacterium *Streptomyces*, and the cyclic backbone is predominantly formed from

TABLE 15.1 RNA Polymerases and Their Inhibitors

Polymerase	RNA	Inhibitors
RNA pol I	5.8, 18, and 28 S rRNA	None
RNA pol II	mRNA and snRNA	α-amanitin, 8-hydroxyquinoline, and lomofungin
RNA pol III	5 S rRNA, tRNA, and U6	α-amanitin at high concentrations
Prokaryotic RNA polymerase	All prokaryotic RNAs	Rifampicin

hydrocarbon molecules. Rifampicin is also an inhibitor of RNA polymerases, but it is specific for prokaryotic polymerases and therefore can be used to treat bacterial infections. Although the drug is potentially effective against a wide spectrum of bacterial pathogens, it has potentially significant side effects, including liver damage. Therefore, it is reserved for use against some of the more highly resistant strains of disease-causing bacteria: tuberculosis, leprosy, and methicillin-resistant *Staphylococcus aureus* (MRSA).

The final molecule discussed in this section is 8-hydroxyquinoline, a small molecule that is a divalent metal chelator. The mechanism of action of 8-hydroxyquinoline is thought to involve chelation of the twin Mg^{2+} ions from RNA pol, but it is unclear whether this is how it functions *in vivo* or why it would not therefore chelate all Mg^{2+} ions in the organism. Nevertheless, 8-hydroxyquinoline is used in topical creams and ointments for treatment of fungal infections such as those caused by *Trichoderma* and *Myrothecium*.

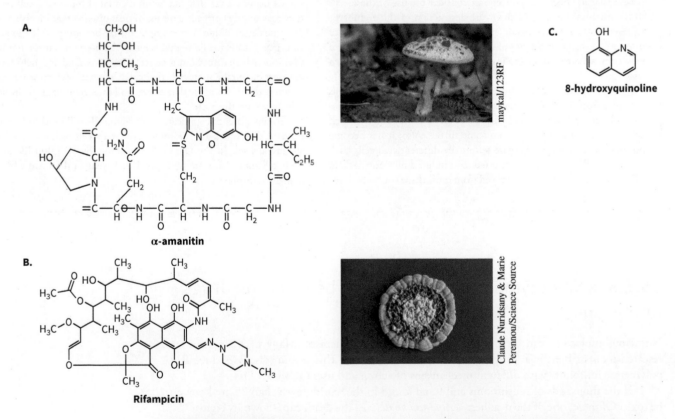

A. α-amanitin

B. Rifampicin

C. 8-hydroxyquinoline

maykal/123RF

Claude Nuridsany & Marie Perennou/Science Source

FIGURE 15.5 Inhibition of RNA polymerases.   A. The structure of α-amanitin. The death cap mushroom, *Amanita phalloides*, is the source of α-amanitin. **B.** The structure of the antibiotic rifampicin, a natural product made by *Streptomyces*. **C.** The structure of 8-hydroxyquinoline, which inhibits polymerases by chelating the divalent metal ions in the active site of the polymerase.

(Source: (A, Left) From David A. Bushnell, Patrick Cramer, Structural basis of transcription: α-Amanitin–RNA polymerase II cocrystal at 2.8 Å resolution Feb 5; 99(3): 1218–1222, Copyright 2002. Reproduced with permission of National Academy of Sciences.)

15.1.4 Retroviruses employ reverse transcription

Retroviruses are viruses that have an RNA genome that must be written into DNA in the host cell for the virus to reproduce. The enzyme responsible for this is **reverse transcriptase (RT)**, which reverse-transcribes an RNA template (the viral genome) into a complementary DNA (cDNA) copy. Other viral enzymes called integrases then insert these cDNA sequences into the host genome, where they reside (**Figure 15.6**).

As mentioned in Chapter 2, due to the 2′ hydroxyl group on the ribose ring, RNA molecules are inherently more prone to degradation than DNA. Thus, it is often useful to produce a DNA copy of an RNA molecule in the laboratory. RT is now commercially available and enables biochemists and molecular biologists to produce cDNA copies of an RNA template. Today, this approach is widely used in biochemistry and molecular biology.

Because humans lack RT, the retroviral enzyme has become a target for pharmaceutical companies wishing to produce drugs to combat infection by viruses such as human immunodeficiency virus (HIV). Two main categories of drugs used to combat HIV infection are RT inhibitors: nucleoside analog RT inhibitors (NARTIs) and non-nucleoside RT inhibitors (NNRTIs). Both classes of drugs inhibit RT, but they do it through different mechanisms. NARTIs are analogs of the nucleotide substrates that the enzyme uses, and they act as competitive inhibitors of the enzyme. As their name suggests, NNRTIs are not nucleoside analogs; instead, they inhibit the enzyme through noncompetitive mechanisms. Some of these inhibitors (such as rilpivirine) can bind to different sites within RT, making them less vulnerable to mutations to the enzyme that could lead to drug resistance.

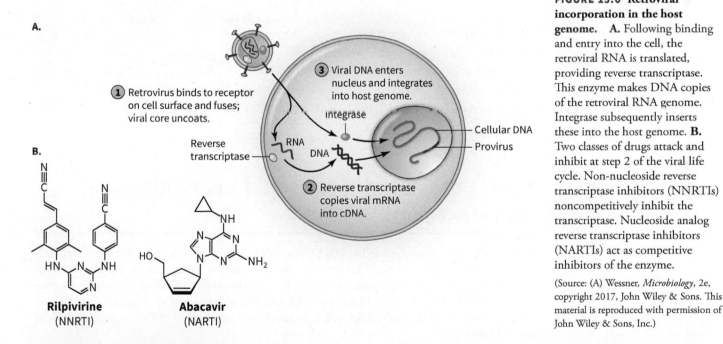

A.

1. Retrovirus binds to receptor on cell surface and fuses; viral core uncoats.

Reverse transcriptase

RNA

DNA

3. Viral DNA enters nucleus and integrates into host genome.

integrase

2. Reverse transcriptase copies viral mRNA into cDNA.

Cellular DNA
Provirus

B.

Rilpivirine
(NNRTI)

Abacavir
(NARTI)

FIGURE 15.6 Retroviral incorporation in the host genome. A. Following binding and entry into the cell, the retroviral RNA is translated, providing reverse transcriptase. This enzyme makes DNA copies of the retroviral RNA genome. Integrase subsequently inserts these into the host genome. **B.** Two classes of drugs attack and inhibit at step 2 of the viral life cycle. Non-nucleoside reverse transcriptase inhibitors (NNRTIs) noncompetitively inhibit the transcriptase. Nucleoside analog reverse transcriptase inhibitors (NARTIs) act as competitive inhibitors of the enzyme.

(Source: (A) Wessner, *Microbiology*, 2e, copyright 2017, John Wiley & Sons. This material is reproduced with permission of John Wiley & Sons, Inc.)

Worked Problem 15.1   Transcription question

ATP is required for synthesis of RNA, both as a substrate for RNA pol II and as a cofactor for TFIIH. Design an experiment to test the hypothesis that the function of TFIIH is ATP dependent.

Strategy Think about the sort of model you could use to conduct this experiment and what variables and controls you would use to test your hypothesis.

Solution This experiment requires a model that includes all of the core transcriptional components (basal transcription factors, Mediator, and RNA pol II) and a template. Usually, the template would contain all four bases (A, C, G, and T) and would synthesize an mRNA with A, C, G, and U. To determine if the complex requires ATP to function, you would need to omit ATP from the system and see whether mRNA synthesis still occurred. This would require a DNA template that lacked thymine bases because ATP would be incorporated in the growing RNA if there were thiamine in the template sequence.

Follow-up question Does this experiment reveal the function of TFIIH? How might you test the hypothesis that TFIIH acts as a helicase?

Summary

- Transcription is the production of RNA molecules from a DNA template. It is thought to occur at transcription factories, localized regions of the nucleus where actively transcribed genes, polymerases, and other proteins involved in transcription are concentrated.

- Bacteria employ a single RNA polymerase for transcription. Eukaryotes use three different polymerases, depending on the type of RNA being produced: RNA pol I synthesizes most rRNAs; RNA pol II synthesizes mRNA, miRNA, and most snRNA; and RNA pol III synthesizes 5S rRNA, tRNA, and U6 snRNA.

- Transcription has five stages: preinitiation, initiation, promoter clearance, elongation, and termination. In bacteria, transcription begins when the RNA polymerase binds a sigma factor to the DNA and terminates at the dissociation of the template and transcript. Termination can occur either by formation of a hairpin loop (Rho-independent termination) or binding of the helicase Rho (Rho-dependent termination).

- In eukaryotes, formation of the initiation complex involves at least seven major complexes in addition to the polymerase. Termination in eukaryotes is not well understood.

- Several molecules in nature are inhibitors of RNA polymerases. Some of these molecules can be used as antifungal or antibacterial agents; others are potent poisons.

- Reverse transcriptase (RT) is a retroviral enzyme that back-translates an RNA message into DNA. Several anti-HIV therapies target this enzyme. Researchers use RT in the laboratory to make DNA copies of RNA molecules.

Concept Check

1. Describe where in the cell RNA synthesis happens and how this process differs between bacteria and humans.
2. What is the structure of the nucleic acid complex formed during RNA polymerization?
3. Outline the steps involved in eukaryotic RNA synthesis.
4. Describe how transcription is terminated in prokaryotes.
5. Give examples of how inhibitors of RNA pol or RT can be useful in medicine and research.

15.2 Processing of Nascent Eukaryotic RNA Messages

Prokaryotic organisms transcribe DNA into RNA in the cytosol and then immediately begin to transcribe the RNA messages into protein; there is little or no processing of RNA molecules. Conversely, eukaryotic organisms modify their RNA molecules, often while they are being transcribed. Modifications to mRNA include capping, polyadenylation, and splicing; modifications to tRNA and rRNA generally involve other types of chemical modification of bases (**Figure 15.7**).

Given that prokaryotes do not modify RNA in any significant way, why have these processes evolved in eukaryotes? The function of some modifications is unclear, but capping is known to assist in the export of RNA from the nucleus and polyadenylation to increase the stability of RNA. This section discusses the different modifications made to RNA and their effects.

15.2.1 Messenger RNA molecules receive a 5'-methylguanine cap

Messenger RNA molecules are modified before transcription has been completed. Following synthesis of the first 25 to 30 nucleotides of the nascent mRNA, a **5'-methylguanine cap** is added in three separate steps (**Figure 15.8**). First, RNA triphosphatase removes the γ-phosphate from the 5' nucleotide. Next, guanylyltransferase transesterifies a molecule of GMP to the terminal phosphate, leaving a structure with the sequence GpppN, in which N is the first nucleotide of

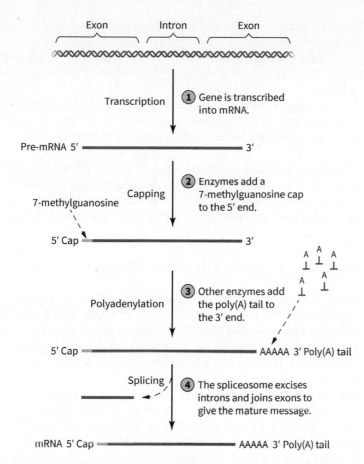

FIGURE 15.7 Messenger RNA processing in eukaryotes. Messenger RNAs undergo the addition of a 7-methylguanine cap to the 5′ end of the message, addition of a poly(A) tail to the 3′ end, and splicing to remove introns prior to export from the nucleus.

(Source: Wessner, *Microbiology*, 2e, copyright 2017, John Wiley & Sons. This material is reproduced with permission of John Wiley & Sons, Inc.)

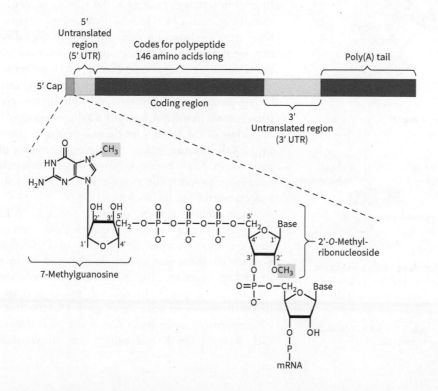

FIGURE 15.8 Capping of mRNA molecules. Messenger RNAs have a cap structure consisting of a guanine nucleotide that is methylated at the 7 position of the purine ring. This group is attached through a triphosphate linkage to the 5′ nucleotide of the message. In addition to the cap, the 2′ hydroxyl of the nucleotides in either the first or second position can also be methylated.

(Source: Karp, *Cell and Molecular Biology: Concepts and Experiments*, 7e, copyright 2013, John Wiley & Sons. This material is reproduced with permission of John Wiley & Sons, Inc.)

the message. The donor molecule in this step is GTP, and pyrophosphate is lost and degraded in the course of the reaction. In mammals, these first two steps are catalyzed by a multifunctional enzyme complex. Finally, the cap is methylated on the 7 position of the ring by guanine-7-methyltransferase. Here, the methyl group donor is S-adenosyl methionine, and S-adenosyl homocysteine is formed.

The capping of messages is coupled to their transcription. The multifunctional capping enzyme in mammals is bound to the phosphorylated form of RNA pol II. Loss of the phosphate group on serine 5 of RNA pol II early in elongation signals the dissociation of the capping enzyme complex.

The 5′ methylguanine cap serves several purposes. Capped mRNAs bind to the cap-binding complex (CBC), which is involved in mRNA export from the nucleus; these mRNAs are resistant to degradation by exonucleases and are recognized in the cytosol by eIF4F, a protein complex involved in ribosome binding and the initiation of translation (discussed in Chapter 16).

15.2.2 Messenger RNA molecules receive a poly(A) tail

Another step in mRNA processing is polyadenylation, the addition of a long tail of adenosine (A) residues at the 3′ end of the message (**Figure 15.9**). This **poly(A) tail** is not coded for by a template but is instead added enzymatically. The modification occurs past the end of the open reading frame for the message but typically within several hundred nucleotides of the stop codon. Addition of the poly(A) tail is signaled by the sequence AAUAAA, and the tail begins 10 to 30 nucleotides later.

In mammals, polyadenylation requires five different proteins, four of which are involved in the recognition and cleavage of the site; the remaining protein catalyzes the addition of the poly(A) tail. Polyadenylation is thought to have effects similar to addition of the cap sequence, rendering mRNA more resistant to attack by nucleases and increasing its half-life in the cell. The modification may also affect translation. Differences in the extent of polyadenylation may give mRNA molecules longer or shorter half-lives in the cell.

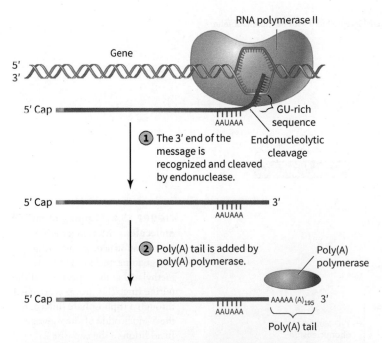

FIGURE 15.9 Polyadenylation of mRNA molecules. Polyadenylation is a multistep process involving at least five different proteins. Following synthesis, the message undergoes cleavage by an endonuclease and then enzymatic addition of the poly(A) tail.

(Source: Snustad; Simmons, *Principles of Genetics*, 6e, copyright 2012, John Wiley & Sons. This material is reproduced with permission of John Wiley & Sons, Inc.)

15.2.3 The nascent mRNA is spliced into different messages

In common usage, splicing is the joining of two ends of something, such as a rope or a piece of film or tape. When films were produced on tapes rather than digitally, scenes that were not needed in the final film were cut out, and the remaining pieces of tape were spliced together to create the final product. In biology, **splicing** refers to cutting out certain pieces of RNA (introns) and rejoining the remaining parts (exons) in mRNA to create a spliced message.

Most of the splicing in the cell occurs in the nucleus in the **spliceosome**, which is a dynamic complex of RNA and protein. The spliceosome works with the transcriptional machinery to splice the message as it is being synthesized and exported from the nucleus.

The coding regions of most genes in eukaryotes are interrupted by introns. These regions were initially thought of as noncoding segments of DNA, but it is now known that some introns do code and are important in certain proteins. The boundary between exons and introns is not specifically defined, but certain sequences in the intron favor splicing (**Figure 15.10**). The 5′ end of the intron usually has a donor site with the sequence GU, whereas the 3′ end usually has the sequence AG.

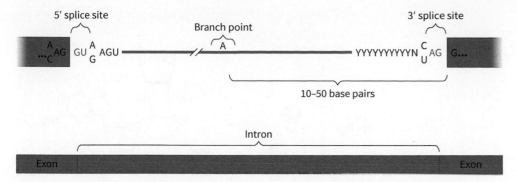

FIGURE 15.10 Sequence information coding for splicing within introns. Introns are demarcated with conserved sequences at their termini: GU at the 5′ end and AG at the 3′ end. Introns also have a branch point, followed by a stretch of pyrimidines, shown as YYYYY, called the polypyrimidine tract.

In between are two other elements: a splice branch point consisting of a single A residue and a stretch of pyrimidines referred to as the polypyrimidine tract, comprising a region rich in C and U bases.

Splicing is a two-step process in which complexes formed between the spliceosome and mRNAs lead to the activation of the 2′ hydroxyl of the branch point nucleotide and attack on the phosphate between the first base of the intron (usually that of the GU sequence) and the last base of the first exon (**Figure 15.11**). This forms a looped structure termed a lariat and frees the 3′ terminus of the first exon. Next, the 3′ hydroxyl at the terminus of the first exon attacks the phosphate between the last base of the intron and the first base of the second exon to splice the two exons. The resulting intron retains the looped lariat structure until it is degraded and the nucleotides are recycled.

The spliceosome

Although the large and dynamic spliceosome is not yet completely understood, two experimental approaches have been used to partially elucidate the structure of the complex and its proposed mechanisms: examining molecules with similar function or examining fragments of the spliceosome.

This first approach to understanding the mechanism of the spliceosome is by comparing it to similar molecules that are better understood. The reactions performed by spliceosomes are similar in many ways to those of the self-splicing class II introns, a group of introns in which the RNA folds into a complex structure and becomes a ribozyme capable of undergoing self-splicing. In the structure of class II introns, four metal ions are critical for catalysis: a pair of divalent metal ions (potentially Mg^{2+} ions) and a pair of K^+ ions (**Figure 15.12**). The two divalent metal ions activate a water molecule that is used in the attack on the phosphate backbone of the mRNA being spliced. This mechanism is similar to that of restriction endonucleases, although in this instance it is the folds of the RNA molecule that orient the substrate and catalytic centers rather than amino acids. A similar mechanism may be at work in the spliceosome.

A second approach has been to examine pieces of the spliceosome. For example, the U5 complex includes the protein Prp8 in complex with Aar2. An examination of the structure of this catalytic fragment of the spliceosome also reveals an active site

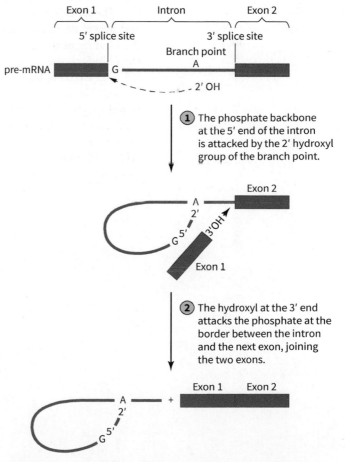

FIGURE 15.11 Splicing of mRNA molecules. The splicing mechanism for mRNAs consists of two separate transesterification reactions. In the first reaction, the phosphate backbone at the 5′ end of the intron is attacked by the 2′ hydroxyl group of the branch point. This leaves a 3′ hydroxyl group at the end of the first exon. This hydroxyl then attacks the phosphate at the border between the 3′ end of the intron and the next exon, joining the two exons.

(Source: Karp, *Cell and Molecular Biology: Concepts and Experiments*, 7e, copyright 2013, John Wiley & Sons. This material is reproduced with permission of John Wiley & Sons, Inc.)

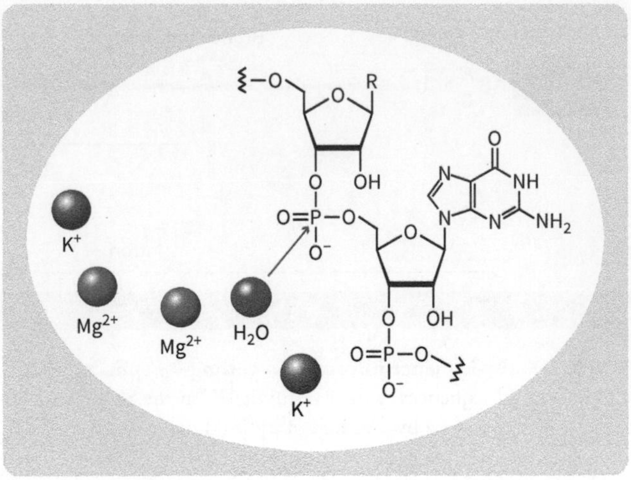

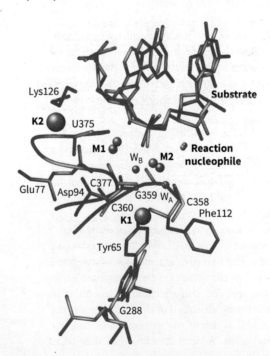

FIGURE 15.12 Mechanism of the spliceosome. The spliceosome has four metal ions bound in the active site, two potassium ions, and two divalent metal ions that are critical for catalysis. The divalent metal ions polarize a nucleophile, most likely a water molecule, which then attacks the phosphate backbone.

with chelated divalent metal ions and a retained fold, a structure that is conserved between the spliceosome and restriction endonucleases (**Figure 15.13**). Both experimental approaches have reached similar conclusions about the topology and chemistry of the spliceosome.

The spliceosome is a complex formed from five different snRNAs (U1, U2, U4, U5, and U6) and dozens of different proteins in a four-step process (**Figure 15.14**). Its formation begins with the binding of the U1 ribonucleoprotein complex to the 5′ end of the intron. The U2 complex then binds to the 3′ end of the intron, forming the prespliceosome (complex A). Next, the U4, U5, and U6 complex binds, generating the precatalytic spliceosome (complex B). Following this, U1 and U4 are lost, activating the precatalytic complex. Complex C forms in the first catalytic step through the formation of the lariat structure branch point and transesterification. The second catalytic step yields the postspliceosomal complex. The excised intron, now in a lariat structure, is released, as are the U2, U5, and U6 complex and the spliced mRNA. In each of these steps, several ATP- or GTP-dependent proteins either bind or catalyze the exchange of one group for another using the energy of NTP hydrolysis.

FIGURE 15.13 Structure of the spliceosome conserved with restriction endonucleases. Shown is a stereo close-up of the active site of both the spliceosome and the restriction endonuclease BamHI. The active site has a conserved topology containing two potassium ions (K), two divalent metal ions (M), and a nucleophile (an activated water molecule).

(Source: From Marcia M, Pyle AM, "Visualizing group II intron catalysis through the stages of splicing", Cell Press, Copyright 2012. Reproduced with permission of Elsevier Inc.)

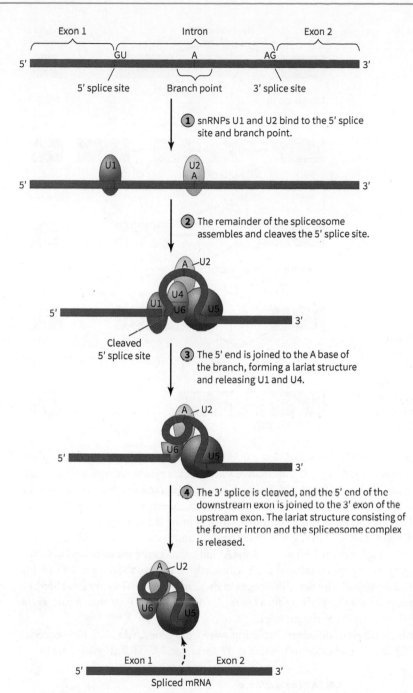

FIGURE 15.14 The spliceosome as a dynamic ribonu-cleoprotein splicing assembly. The spliceosome assembles at an intron to catalyze the splicing of two exons. Many of these steps require hydrolysis of NTP cofactors.

(Source: Snustad; Simmons, *Principles of Genetics*, 6e, copyright 2012, John Wiley & Sons. This material is reproduced with permission of John Wiley & Sons, Inc.)

Although most RNA is spliced by spliceosomes, there are other types of RNA and other mechanisms through which splicing can occur. These include self-splicing RNAs and enzyme-dependent splicing pathways in tRNAs.

Differential splicing   The human genome has fewer than 20,000 genes but more than 100,000 different proteins. Most of this diversity comes from the **differential splicing** of genes, which creates different combinations of introns and exons to produce related protein structures. Over 95% of genes undergo differential splicing.

There are five different mechanisms through which genes may be differentially spliced (**Figure 15.15**): exon skipping, mutually exclusive exons, alternative donor sites, alternative acceptor sites, and intron retention. There are also variations within these mechanisms. In the case of intron retention, the intron may code for a string of amino acids followed by a stop codon, or it may lack a stop codon and continue to code for amino acids. Should the intron lack a stop codon, it needs to retain the reading frame of the overall protein to retain the code for the exons that are coded for downstream of this region.

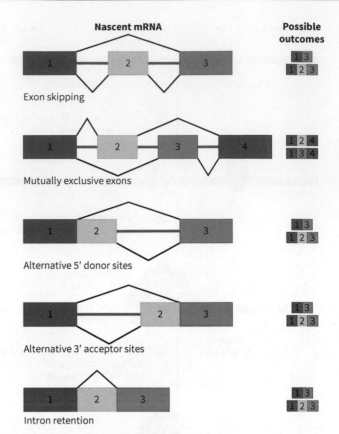

FIGURE 15.15 Examples of differential splicing. Five different arrangements of introns and exons have been proposed to result in differential splicing. In these examples, the exons are shown in alternating colored boxes. The column on the right shows the different combinations that are possible. The black lines above and below the sequences indicate two different ways the sequence can be spliced.

Two other mechanisms of RNA modification can result in different protein products: alternative start sites or multiple poly(A) signals. Alternative start sites are promoter regions that result in differential binding of RNA pol II and a different 5′ end of the message. Such sites may be extensive and incorporate multiple new exons and introns in the message. Multiple poly(A) signals are found past the stop codon in the 3′ untranslated region of the message. Why do these sites exist, and what impact do they have, given that they are found outside the coding region? There are several plausible explanations. First, the poly(A) tail protects the message and acts as a molecular timer. The longer the poly(A) tail and the 3′ untranslated region, the longer the half-life of the message in the cell. It may also be that differences in these adenylation sites alter the lifetime of the message. Second, there may be binding sites for miRNAs in the 3′ region that would again influence the half-life or translation of the message.

One of the first observations of differential splicing was in the gene CALCA, which encodes mRNAs for both the 32 amino acid calcitonin and the 37 amino acid CGRP (**Figure 15.16**).

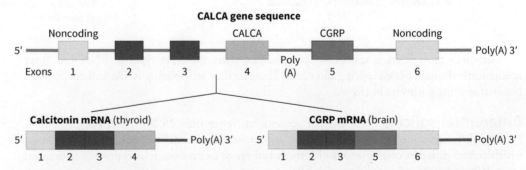

FIGURE 15.16 Differential splicing in calcitonin. The CALCA gene is differentially spliced to produce calcitonin in the thyroid and CGRP in the brain. This gene has six exons: (1) a 5′ noncoding exon, (2 and 3) two exons common to both transcripts, (4 and 5) two mutually exclusive exons, and (6) a 3′ noncoding exon followed by a poly(A) tail.

(Source: From James R. Roesser, Kurt Liittschwager, and Stuart E. Leff. Regulation of Tissue-specific Splicing of the Calcitonin/Calcitonin Gene-related Peptide Gene by RNA-binding Proteins. Vol. 263, No. 11, Issue of April 15, pp. S36 3375,1993. http://www.jbc.org/content/268/11/8366.full.pdf. Copyright 1993. Reproduced with permission of The American Society for Biochemistry and Molecular Biology, Inc.)

TABLE 15.2 Splicing Disorders in Human Health

Disorder	Cause	Effects
Familial isolated growth hormone deficiency	Lack of growth hormone	Diminutive stature and impaired immune system
Spinal muscular atrophy (SMA)	Lack of a fully functional protein, leading to gradual death of motor neurons	Muscular atrophy, eventually resulting in paralysis and death
Frontotemporal dementia and parkinsonism linked to chromosome 17 (FTDP-17)	Lack of a functional protein, leading to loss of neurons in the frontal and temporal lobes	Neurodegeneration, dementia, and parkinsonism
Fraser syndrome	Mutation in one of three different proteins	Developmental problems such as eyes that are completely covered by skin and usually malformed (cryptophthalmos), fusion of the skin between the fingers and toes (cutaneous syndactyly), and abnormalities of the genitalia and the urinary tract
Atypical cystic fibrosis	Mutations to a chloride channel	A milder form of cystic fibrosis that may only affect a single organ system
Retinitis pigmentosa	Mutations to any one of 50 genes in the photoreceptors of the retina, leading to cell death	Blindness
Myotonic dystrophy	Mutation to one of two proteins	Inability to relax muscles, cataracts, intellectual disabilities, and heart problems

Following splicing, several proteins remain bound to the mature mRNA, including the transcription-export (TREX) complex (discussed in section 15.3) and the exon junction complex (EJC) proteins. These proteins are involved in export of the mRNA from the nucleus and in translation. EJC proteins are also involved in the quality control of splicing. There are multiple places and multiple mechanisms for overseeing message formation to ensure that messages have been properly spliced before they are exported from the nucleus. If the EJC complex is bound to a message downstream of a stop codon, the cell degrades that message and prevents its export and translation. The importance of accurate splicing is demonstrated in the diseases caused by an inability to properly splice a message (Table 15.2). One such example of this is the disease spinal muscular atrophy (SMA), discussed in **Medical Biochemistry: Spinal muscular atrophy and mRNA splicing**.

Medical Biochemistry

Spinal muscular atrophy and mRNA splicing

Spinal muscular atrophy (SMA) is a group of neuromuscular degenerative disorders caused by an autosomal recessive mutation. One in 50 adults carries a mutation capable of causing SMA, and it is the second-most-common genetically inherited cause of infant death in populations of European descent (cystic fibrosis being the first). Worldwide, between 1 in 6,000 and 10,000 live births have some form of SMA.

SMA is caused by the absence of the survival motor neuron (SMN) protein. In the absence of this protein, motor neurons die because they are unable to grow properly and therefore fail to develop the correct axonal connections. People with SMA gradually lose control of motor functions, and in severe cases, the disease often results in digestive and respiratory difficulties in infants. People with the most severe type of SMA (SMA I) have a life expectancy of less than four years. People with less severe forms of the disease can have a much better prognosis; they do not experience symptoms until later in life, and many have a normal life expectancy.

The genetics and biochemistry of SMA are somewhat more complicated than in simple autosomal recessive disorders. Humans have two highly related genes, termed SMN1 and SMN2, both of which are found on chromosome 5. SMN1 is located near the telomere of this chromosome and, in healthy individuals, codes for the normal SMN protein. This message is eight exons long, and any deleterious mutations to this gene cause SMA. However, the severity of the disease is determined in part by the second gene, SMN2, which is found near the centromere and is nearly identical to SMN1, with one critical difference: The SMN2 gene has a C to T conversion that results in differential splicing of the message. Instead of a full eight exons, exon 7 is spliced out in about 80%–90% of transcripts, resulting in a dysfunctional protein, SMNΔ7. However, the remaining 10%–20% of transcripts produce functional copies of SMN. Complicating matters further, SMN2 varies in copy number, with some people having as many as four copies of the gene.

Higher concentrations of the SMN protein are correlated with cell survival and a less severe disease phenotype. Therefore, research into treating people with SMA has focused on increasing transcription or altering the splicing of SMN2. Sodium vanadate, trichostatin A, and aclarubicin have been shown to enhance SMN2 expression by inducing the signaling kinase STAT5 in cellular and mouse models

of the disease. These approaches elevate transcription and protein levels of SMN but do not affect splicing.

Other approaches address the splicing question. One research group has created a bifunctional U7-snRNA construct that binds to the 3′ end of exon 7 and recruits necessary splicing factors. This approach has been successful in both cultured cell and mouse models, although it does not result in complete reversal of the disease.

The most recent approaches to treating SMA are some of the most promising and exciting. Genetically engineered viruses are being used to deliver functional copies of the gene to patients. This has the potential to be an actual cure for the disease, but it is one of the most expensive treatments ever developed at over $2 million for a one-time treatment.

It is unclear in evolutionary history when introns and spliceosomes formed. Likewise, the functions of introns are still debated. Introns and splicing clearly provide a mechanism for introducing variation and diversity to a population of proteins without having to replicate genes. Also, the extra DNA they provide to the genome affords a measure of protection against viruses, base insertions, or base deletions that might otherwise interrupt and disrupt a gene with an exon.

15.2.4 Transfer RNA molecules have several uncommon modifications

Transfer RNAs are small (73 to 94 nucleotide) adapter RNA molecules that ribosomes use to add amino acids to a growing protein; they are synthesized by RNA pol III. Nearly 500 genes code for tRNAs, and over 300 code for tRNA-derived pseudogenes, that is, regions of DNA clearly conserved with a functional tRNA gene but unable to produce a tRNA due to a mutation. The structure of tRNA molecules is discussed in greater detail in Chapter 16 in relation to their function in protein synthesis, but several modifications that tRNAs undergo are discussed here (**Figure 15.17**).

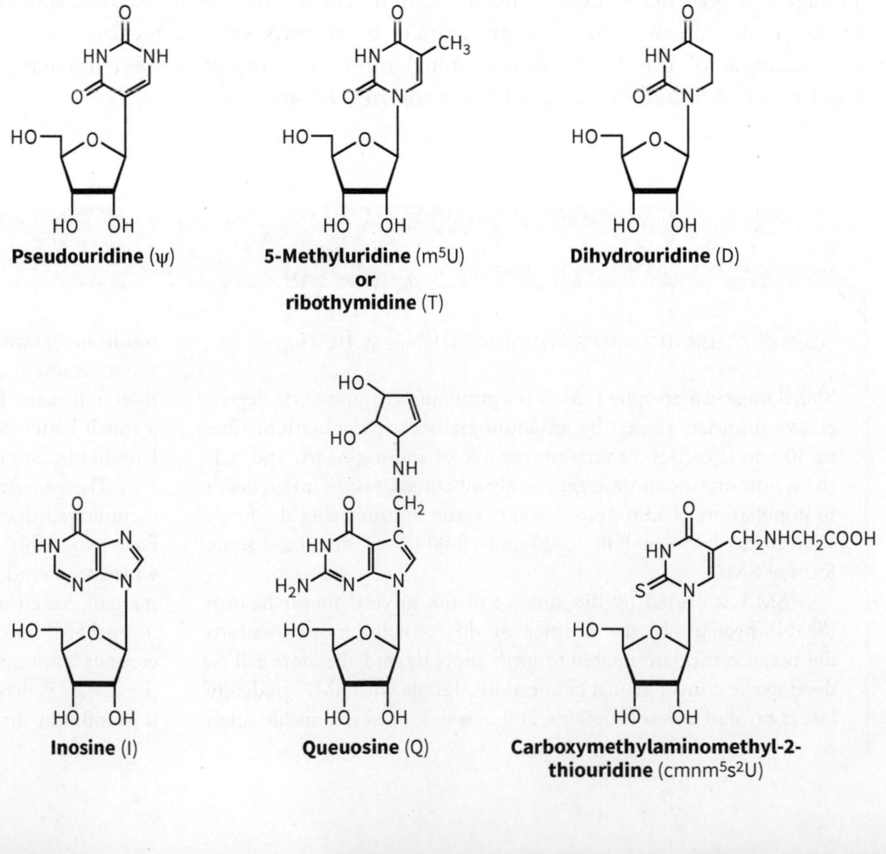

Pseudouridine (ψ)

5-Methyluridine (m^5U) **or ribothymidine** (T)

Dihydrouridine (D)

Inosine (I)

Queuosine (Q)

Carboxymethylaminomethyl-2-thiouridine ($cmnm^5s^2U$)

FIGURE 15.17 Examples of the modifications made to tRNA molecules. All tRNA molecules are known have some form of modification. Examples of some of the more than 20 different modifications found to date are shown.

Following synthesis of a nascent tRNA molecule, the 5′ leader and 3′ trailer sequences are removed by endonucleases: RNase P in the case of the 5′ sequence and tRNase Z for the 3′ sequence. Some tRNA molecules contain an intron and thus need to be spliced; this splicing is catalyzed not by the spliceosome but by a splicing endonuclease. The 3′ end of tRNA molecules contains a common sequence (CCA-3′) that is added by nucleotidyl transferase. Like the 3′ tail of mRNA, this is not coded for by a template but is added enzymatically, one nucleotide at a time. No fewer than 25 different chemically modified residues have been observed in different tRNAs. These modifications range from methylations to thiolations and additions of isopentenyl or carboxymethylamino groups. The most common modification is the inclusion of pseudouridine, an isomer of uracil. The implications of these modifications are unclear, but most are thought to impart structural stability to the tRNA.

| **Worked Problem 15.2** | **Effect of a mutation on a noncoding region** |

There are numerous places that a mutation might occur in a chromosome. How might a mutation to a noncoding region affect gene expression?

Strategy Mutations are often thought of as affecting the coding sequence of a gene. What other parts of a gene are there, and what do they do? Would altering those sequences matter?

Solution Mutations to noncoding regions can have serious implications for gene expression. For example, a mutation in an intron that led to the loss of a splicing site could in turn lead to a mutant or truncated protein and thus perhaps a complete or partial loss of function.

Follow-up question Would a mutation to a gene coding for a tRNA be lethal? Explain your reasoning.

Summary

- In eukaryotes, mRNA molecules undergo several different types of modifications: capping, polyadenylation, and splicing.

- During capping, mRNA molecules undergo a three-step process in which they receive a 7-methylguanine cap that is attached through a series of three phosphates (5′-7mGpppN), where N is the first base of the mRNA.

- Polyadenylation is when a poly(A) tail is added to mRNAs on the 3′ end of the message, with the addition of the adenosines being enzymatic rather than being found in the template; however, a polyadenylation signal (5′-AAUAAA-3′) is found in the message.

- Splicing involves cutting out introns and rejoining exons in mRNA to create a spliced message. Splicing is catalyzed by the spliceosome, a large, dynamic complex that is comprised of five different RNA complexes and dozens of proteins and proceeds through two different transesterifications. First, the 5′ end of the intron is transesterified to the 2′ hydroxyl group of a specific adenosine residue in the intron, and second, the free 3′ end of the first exon is joined to the second to splice out the intron. Differential splicing allows for the great diversity of proteins produced by a much smaller number of genes.

- A 3′ tail sequence (CCA-3′) is added enzymatically to all tRNA molecules, which also have numerous chemically modified bases, including pseudouridine, isopentenyl groups, carboxymethylamino groups, and thiolations.

Concept Check

1. Name and briefly describe the major classes of RNA modifications in eukaryotes, and explain their purpose.

2. Explain how and where RNA modifications are made.

3. Describe the steps involved in the assembly of the spliceosome.

4. What is the purpose of the poly(A) tail on eukaryotic mRNA?

15.3 RNA Export from the Nucleus

In eukaryotic cells, the genome resides in the nucleus. This is also where transcription occurs, but translation—the synthesis of proteins from the mRNA message—occurs in the cytosol or endoplasmic reticulum. For translation to happen, the mRNA, tRNA, and rRNA must all make their way out of the nucleus.

The nucleus is protected and separated from the rest of the cell by a double membrane perforated with several thousand pores. As discussed previously in relation to studies of the plasma membrane or mitochondria, the main function of the largely impermeable lipid bilayer of a membrane-bound structure is to protect the contents of the organelle or cell. In contrast, the protein pores imbedded in the membrane permit or facilitate the passage of particular molecules into or out of the structure. In this case, the nucleus must import nucleotides and deoxynucleotides, inorganic ions, water, and many other small molecules, as well as proteins. At the same time, the nucleus must keep out viruses or pathogens that could jeopardize the stability of the genome and its function. The only way into or out of the nucleus is through the nuclear pore complex.

15.3.1 Transit into and out of the nucleus is regulated by the nuclear pore complex

The **nuclear pore complex (NPC)** is a cylindrical macromolecular assembly that spans the nuclear envelope and regulates passage of molecules into and out of the nucleus (**Figure 15.18**). It has an estimated mass of 120 MDa (120,000 kDa) and is comprised of repeats of over 30 different proteins termed nucleoporins; it has a total of over 400 subunits. The complex has eight-fold symmetry and several distinguishing structural features. On the nuclear side of the NPC are eight intranuclear filaments, connected through the distal ring to the nuclear basket. The top of the basket, associated with the inner nuclear membrane, is the nuclear ring, which is bound to the core scaffold or luminal ring that bridges the inner and outer nuclear membranes. Bound to the outer membrane is the cytoplasmic ring, which has eight cytoplasmic filaments attached to it. Overall, the pore resembles a basketball hoop but with a thickened rim and fingers extending from both ends. The filaments can extend as far as 100 nm into the cytosol on one side and the nucleus on the other. The channel itself is not an open structure but contains numerous proteins that regulate the passage of molecules through the pore.

The NPC has to be large enough to allow the passage of rRNA molecules. The diameter of the entire NPC is approximately 120 nm (~30 times the diameter of an average globular protein), but the diameter of the pore is only a fraction of that. Complexes as large as 39 nm have been transported through the NPC, indicating that some viruses could pass through the nuclear envelope intact. Molecules smaller than 60 kDa generally diffuse through the pore, although, as discussed later, some smaller molecules are directed either into or out of the nucleus by signal sequences or chaperone proteins.

The core scaffold of the NPC is formed by nucleoproteins that fold almost exclusively into either β-propeller or α-solenoid structures. This structure is conserved between membrane-associated pore-forming structures, including the clathrin and adaptin proteins (discussed in Chapter 10) and the protein trafficking coat protein complex (COP) proteins, discussed in Chapter 22. In addition, NPCs are structurally and functionally conserved throughout the eukaryotes, but they are not as well conserved at the nucleotide or amino acid level.

The pore of the NPC is lined with proteins that contain phenylalanine-glycine-rich repeats and are thus termed FG nucleoproteins. Curiously, these proteins are thought to be unfolded in the native state, forming a tangle of threads that clogs the pore and generates unfavorable conditions for molecules that do not interact with the FG repeat.

Finally, the NPC is a dynamic structure. The number of pores varies with cell type, stage of development of the organism, and stage of the cellular life cycle. In all cases, it is elevated when cells are undergoing significant growth or protein synthesis. Conversely, NPCs and the entire nuclear envelope are disassembled and reassembled during cell division.

A.

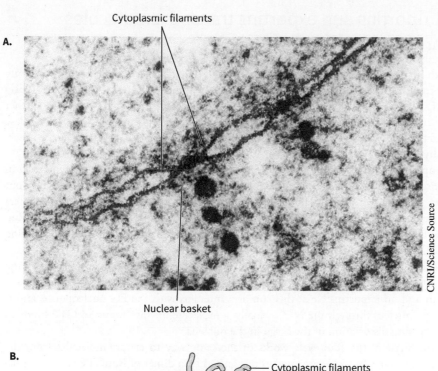

Cytoplasmic filaments

Nuclear basket

CNRI/Science Source

B.

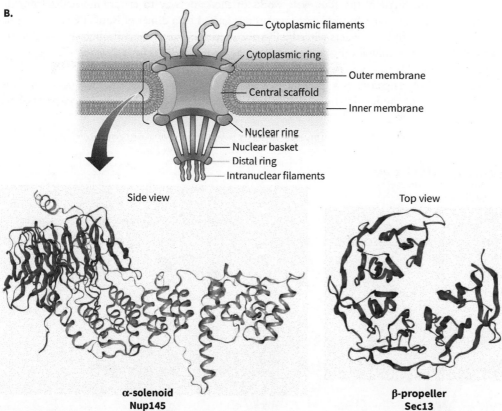

Cytoplasmic filaments

Cytoplasmic ring

Outer membrane

Central scaffold

Inner membrane

Nuclear ring

Nuclear basket

Distal ring

Intranuclear filaments

Side view

Top view

α-solenoid
Nup145

β-propeller
Sec13

FIGURE 15.18 Selective passage of proteins and RNA molecules through the nuclear pore complex.
A. An electron micrograph of the nuclear membrane and a nuclear pore complex indicates the cytoplasmic filaments and nuclear basket. **B.** The nuclear pore complex (NPC) consists of a basket-shaped structure on the nuclear side of the pores, a central scaffold lined with FG proteins, and a cytosolic ring and associated filaments. Small ions and metabolites can pass freely through the NPC, but passage of proteins and nucleic acids is selective. The NPC is lined with a cylinder formed by eight heptameric repeats (56 subunits each) of the proteins Sec13 and Nup145, which form β-propeller and α-solenoid structures.

(Source: (B, Top) From Peters R., "Functionalization of a nanopore: the nuclear pore complex paradigm". 2009 Oct; 1793(10):1533–9. doi: 10.1016/j.bbamcr.2009.06.003. Epub 2009 Jul 9. Copyright 2009. Reproduced with permission of Elsevier Inc. (B, Bottom) Data from PDB ID 3BG0 Hsia, K. C., Stavropoulos, P., Blobel, G., Hoelz, A. (2007) Architecture of a coat for the nuclear pore membrane. Cell 131: 1313–1326)

15.3.2 Importins and exportins transport molecules in the Ran cycle

Transport both into and out of the nucleus occurs through the NPC, largely mediated by a family of proteins termed the karyopherins and the GTPase Ran (**Figure 15.19**). We have previously encountered GTPases in cell signaling in Chapter 5 and will see them again when we further explore signal transduction in Chapter 21. Although Ran is not a signaling pathway, the GTPase activity of Ran provides a function analogous to GTPases found in those pathways. As we will see, Ran uses the hydrolysis of GTP as a switch to alter the function of the protein.

The karyopherins can be broadly divided into two classes: **importins**, which are involved in moving proteins into the cell nucleus, and **exportins**, which direct movement of molecules out of the nucleus. Both classes of karyopherins interact with Ran. The karyopherins act as adapters that bind to their cargo through either a nuclear localization sequence (NLS, a cationic amino acid sequence, KKKRK) or a nuclear export sequence (NES, of which there are several, including L-X(2,3)-LIVFM-X(2,3)-L-X-LI).

Molecules can be imported into and exported from the nucleus through the **Ran cycle**. In the import phase of the Ran cycle, a complex forms on the cytoplasmic side of the envelope between a cargo protein and an importin. Next, this complex interacts with the FG nucleoporins and is shuttled into the nucleus through the NPC. Finally, a molecule of Ran bound to GTP binds to the importin, causing dissociation of the cargo in the nucleus.

The reverse phase of the Ran cycle works in the same way to export molecules from the nucleus. First, a molecule targeted for export is recognized by a dimer of RanGTP and an exportin. Next, this complex interacts with the FG nucleoporins and transits through the NPC to the cytosol. Finally, the complex dissociates in the cytosol when GTP is hydrolyzed to GDP and P_i, a process enhanced by the presence of the Ran GTPase-activating protein (GAP).

This hydrolysis generates a gradient in which most of the Ran in the nucleus is bound to GTP, but most of the Ran in the cytosol is bound to GDP. As is the case in some aspects of cell

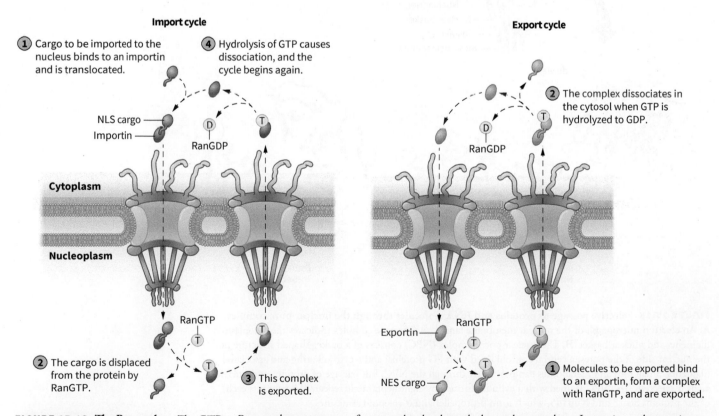

FIGURE 15.19 The Ran cycle. The GTPase Ran regulates transport of most molecules through the nuclear envelope. Importins and exportins chaperone the proteins being transported. Hydrolysis of GTP causes complex dissociation.

signaling (see Chapters 5 and 21), GTPase activity acts as a molecular timer, being active in the GTP-bound state but inactive upon hydrolysis. Also similar to cell signaling is the presence of GAP.

15.3.3 Different classes of RNA molecules employ different protein adapters for transport

There are different mechanisms for exporting different types of RNA molecules through the nuclear envelope (**Figure 15.20**). Export of tRNAs, miRNAs, snRNAs, and rRNAs involves exportins and the Ran cycle, and in each case the export requires mature RNA that has undergone modification and splicing. Export of mRNAs is not dependent on exportins and the Ran cycle, but it does require splicing and processing of the mRNA.

Export of tRNA and miRNA   Mature tRNAs are bound by a complex of exportin and RanGTP and exported through the NPC. On the cytoplasmic side of the nuclear envelope, GTP hydrolysis results in dissociation of the tRNA from the complex. The nuclear exclusion sequence on exportin-t is a tertiary structure that is not found as contiguous amino acids in the protein.

Like tRNA, miRNA is bound in the nucleus by a complex of exportin and RanGTP. The exportin in this case is exportin 5. Following release into the cytosol, the miRNA is cleaved by the nuclease Dicer and processed by the RNA-induced silencing complex (RISC) into mature miRNA. The processing of miRNA and siRNA is discussed in the techniques appendix.

Export of snRNA   The mechanism for exporting snRNA is more complex than for tRNA or miRNA because it involves an adapter protein that binds to both the snRNA and the export protein. Export of mRNA and rRNA also uses an adapter complex between the RNA and exportin. In the case of snRNA, the adapter proteins that bind to both the RNA and exportin are CBC and PHAX. Phosphorylation of PHAX is required for binding of exportin 1 (CRM1) and RanGTP to form the mature transport complex. In the cytosol, hydrolysis of GTP and dephosphorylation of PHAX cause dissociation of the complex and release of the snRNA.

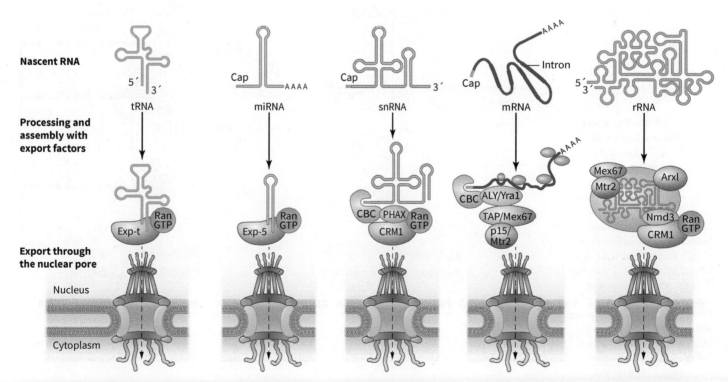

FIGURE 15.20 Different exportins involved in nuclear export.   Transfer RNAs and miRNAs are exported simply through binding to exportins (Exp-t or Exp-5) and the RAN cycle. Small nuclear RNAs and rRNAs employ adapter proteins to bind to exportins and RAN. Messenger RNAs are transported by several different mechanisms.

(Source: From Köhler A, Hurt E., "Exporting RNA from the nucleus to the cytoplasm". Nat Rev Mol Cell Biol. 2007 Oct;8 (10):761-73. Copyright 2007. Reproduced with permission of Springer Nature Limited.)

Why is snRNA exported from the nucleus when the splicing of pre-mRNA occurs in the nucleus? In fact, snRNA must be imported back into the nucleus later to form the mature spliceosome. It is unclear why this occurs, but it may be related to processing or proofreading of the snRNA.

Export of rRNA

The ribosome is comprised of four different rRNA molecules and at least 70 different proteins. The structure of the ribosome is discussed in greater detail in Chapter 16, but the basic structure has one large and one small subunit. The sizes of ribosomes and their RNA components were originally determined using analytical ultracentrifugation. In eukaryotes, the large subunit is termed 60S and the small subunit 40S. S is the Svedberg unit, which combines the dimensions and density of the particle as determined through ultracentrifugation.

Before being exported, the rRNA and proteins are assembled into 60S and 40S subunits in the nucleoli. The mechanism for exporting the 60S subunit is similar to that used for snRNAs in that adapter proteins are involved in the interaction of the RNA with the exportin–RanGTP complex. In the case of the 60S subunit, this is the adapter protein Nmd3, which interacts with CRM1 and facilitates binding of RanGTP.

In yeast, the proteins Arx1 and the complex Mex67-Mtr2 are known to interact with the 60S subunit. These proteins are thought to assist with transport of the 60S subunit through the NPC.

The pathways dictating how the 40S subunit is exported are still being examined, but several of the same proteins (CRM1 and the Mex67-Mtr2 complex) are thought to be involved.

Export of mRNA

Export of mRNA molecules presents particular challenges because these molecules differ in their size, sequence, and tertiary structure. This may be part of the reason why export of RNA molecules is not dependent on the Ran cycle. Several models have been proposed for export of mRNAs. One of these is the TREX-dependent pathway, in which the TREX complex of proteins (comprising the multiprotein complex THO, as well as ALY and UAP56) forms at the 5′ end of the mRNA at the CBC, as the mRNA is being capped and spliced (**Figure 15.21**).

FIGURE 15.21 Transport across the nuclear envelope by mRNA. There are several models for mRNA transport. Shown is the TREX pathway in both yeast (**A**) and metazoans (multicellular animals, **B**). Nascent mRNAs are capped and bound by cap-binding protein (CBP) and spliced. CBP is bound by the TREX complex. This causes recruitment of Tap/p15 in mammals or Mex 67-Mtr2 in yeast. These proteins result in transport across the nuclear membrane, where they are released.

(Source: From Köhler A, Hurt E., "Exporting RNA from the nucleus to the cytoplasm". Nat Rev Mol Cell Biol. 2007 Oct;8 (10):761–73. Copyright 2007. Reproduced with permission of Springer Nature Limited.)

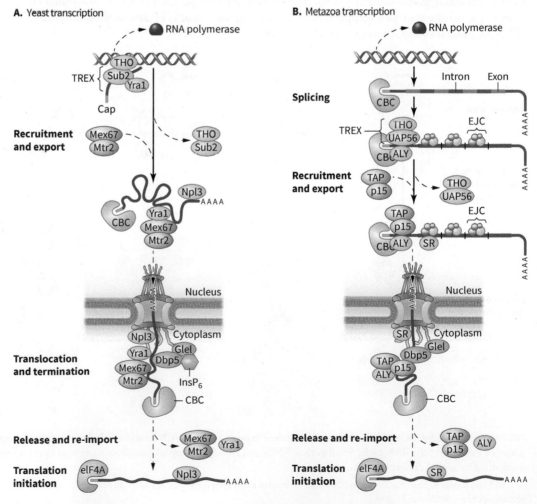

Formation of this complex recruits the export protein complex TAP/p15, which is analogous to the yeast Mex 67-Mtr2 discussed in rRNA transport. Although TAP/p15 does not interact with RanGTP and is unrelated to the exportins, it serves a similar function, interacting with the FG-rich nucleoporins and facilitating export from the nucleus.

The pathways described here are conserved throughout metazoans (complex organisms). Pathways that are homologous but slightly different are found in yeast.

Worked Problem 15.3   Pore things

A 2002 paper by Panté and Kann (*Mol. Biol. Cell.* 2002 13(2) 425–434) described an experiment in which gold nanoparticles were coated with importins, and their transit across the nuclear envelope was measured by electron microscopy. The findings are shown here.

A.

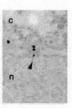
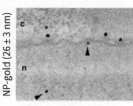

NP-gold (22 ± 2 nm)

B.

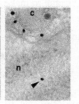

NP-gold (26 ± 3 nm)

C.

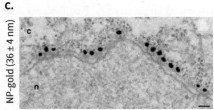

NP-gold (36 ± 4 nm)

The American Society for Cell Biology

In each photo, "c" denotes cytoplasm and "n" nucleoplasm. Each particle has a protein coating approximately 7 nm thick. Based on these data, how large a pore would you predict in the NPC?

Strategy   Gold particles are visible on electron microscopy because metals appear opaque in an electron micrograph. What is the diameter of each particle?

Solution   Gold particles can be seen on either side of the nuclear envelope in the 22 and 26 nm particles but not in the 36 nm particles; however, the actual diameter of these particles needs to take into account the protein coating. If the protein coat is 7 nm thick, the actual diameter of the particles becomes 36, 40, and 50 nm (i.e., the diameter of the nanoparticle + 2 × the diameter of the protein coat). Because we can see particles on either side of the envelope in the two smaller-sized proteins but not in the largest, we know that the size of the pore must be somewhere between 40 and 50 nm.

Follow-up question   Design an experiment to show that exportin must bind to the RanGTP complex before binding to tRNA.

Summary

- The NPC is a massive multiprotein pore that penetrates the nuclear envelope and selectively permits passage of key molecules into and out of the nucleus.

- In the import phase of the Ran cycle, importins bind to cargo proteins and bring them through the NPC to the nucleus, where binding of RanGTP to the importin causes dissociation of the cargo. In the export phase of the Ran cycle, an exportin–RanGTP complex binds to a cargo molecule and exports it through the NPC, where hydrolysis of GTP causes release of the cargo.

- Transfer RNA and miRNA are exported using the Ran cycle. Ribosomal RNA and snRNA require an additional adapter protein to interface between the RNA and the exportin before being exported. Several different mechanisms have been proposed for the export of mRNA, which lacks the common structural elements found in other RNA molecules.

Concept Check

1. Describe the NPC and its main structural features.

2. Explain why the nuclear localization sequence cannot be simply an electrostatic interaction between the protein and DNA.

3. Explain how the five different types of RNA discussed pass through the pore, and compare the different mechanisms by which they do so.

Bioinformatics Exercises

Exercise 1 Viewing and Analyzing RNA Polymerase

Exercise 2 RNA Polymerase, Transcription, and the KEGG Database

Problems

15.1 Transcription: RNA Synthesis

1. There are six core transcription factors: TFIIA, TFIIB, TFIID, TFIIE, TFIIF, and TFIIH. List the functions of each and where they function in the cell.

2. How much energy in terms of ATP equivalents does it take to make a 1.8-kb mRNA?

3. Identify the amino acids in α-amanitin and the unusual linkages present.

4. Would you anticipate α-amanitin would be water or fat soluble? Why?

5. Imagine that a new drug is proposed which is hypothesized to selectively bind the transcription factor TBP. Describe how RNA polymerases will be affected.

6. If translocation of RNA pol II is inhibited upon binding of α-amanitin but NTP binding is not, how would this impact kinetics of the reaction?

7. Compare prokaryotic and eukaryotic RNA and DNA polymerases in terms of their structures and functions. How are they similar and different? How do their structures influence their activities?

8. Describe how bacteria could become resistant to rifampicin. What general or specific types of mutations would have to occur to impart resistance to this drug?

9. Rifampicin has an intense red color. Explain why this might be the case.

10. Rifampicin imparts a red color to the urine and tears of users, which dissipates within a few hours. What does this suggest about its metabolism?

11. Trace the pi-bond network in rifampicin.

12. List the different isoforms of RNA polymerase and how they differ in terms of function.

13. Is 8-hydroxyquinoline used internally or externally? What does the mechanism of action suggest about fungal biology or the permeability of the drug to fungal cells?

15.2 Processing of Nascent Eukaryotic RNA Messages

14. What are the modifications that occur in all eukaryotic mRNA?

15. Why is GTP the donor in the capping reaction? Could other nucleotides, nucleosides, or bases work? Why or why not?

16. What are the different mechanisms involved in differential splicing?

17. What types of mutations could result from errors of splicing?

18. Explain the sorts of modifications that are made to tRNA molecules, and contrast these with the modifications made to mRNA molecules.

19. Does the branch point in splicing of the intron have to be an A? If so, why? Does the chemistry of the base matter here? How?

20. Are the reactions of mRNA splicing truly transesterification reactions? Why is that important energetically? Is another mechanism possible?

21. The 2′ OH on ribose is what gives RNA its inherent instability compared to DNA. What value does that 2′ OH have in biochemistry?

22. Why does reverse transcriptase that facilitates reverse transcription has no exonuclease activity?

23. How would an siRNA directed against exon 1, 2, 5, or 6 of calcitonin affect protein levels? Review how siRNA is used experimentally in the techniques appendix if necessary.

24. How many amino acids does each exon of calcitonin code for?

25. What protein/RNA complex is responsible for the removal of introns? Would you expect retroviruses to contain introns? Explain.

26. A new antibiotic was discovered which strongly inhibited mRNA precursor transcripts and snRNA transcripts. Which RNA polymerase is expected to be inhibited by this antibiotic?

27. Why are there so many tRNAs? Explain your answer.

15.3 RNA Export from the Nucleus

28. Illustrate the Ran cycle and the ways in which it is similar to GTPases, discussed previously.

29. Why is it interesting that FG proteins are unfolded in the native state?

30. What does the folding state of the FG proteins suggest?

31. HIV Rev binds to exportin 1. Propose a mechanism by which this mRNA exits the nucleus.

32. What is the function of nuclear pore complex (NPC)? Explain why NPCs are structurally and functionally conserved, but not at the nucleotide level.

Data Interpretation

Questions 33 to 38 pertain to calcitonin (*Homo sapiens* gene *CALCA* gene ID 796).

33. Using GenBank, look up the sequence for calcitonin, find where it splices, and show intron–exon boundaries. Can you find conserved nucleotides in the introns (GU, A, polypyrimidine tract, etc.)?

34. Of the different types of splicing discussed in the chapter, what is the pattern for calcitonin?

35. If an mRNA were annealed to a genomic DNA and electron micrographs were taken, what would it look like?

36. What sizes of message would you calculate for this gene?

37. What would the banding pattern of a northern blot look like for calcitonin? If you are unfamiliar with northern blots, see the techniques appendix.

38. What would the banding pattern look like for a Southern blot if the DNA was digested with EcoRI, PstI, or Bgl II and probed with the cDNA of the α splice variant? If you are unfamiliar with Southern blots, see the techniques appendix.

Experimental Design

39. What techniques are best to detect splice variants at the mRNA level?

40. A student has given a drug to cells in culture expecting to block mRNA synthesis, but it has not worked. What are some possible problems with this design?

41. How could you determine if a gene of interest is improperly splicing in a mutant organism?

42. What is the best technique currently available to determine if a chemical modification has been made to a select tRNA?

43. Design an experiment to test the hypothesis that a certain modification made to a tRNA molecule occurs in the cytosol instead of the nucleus.

Ethics and Social Responsibility

44. What newborn genetic disease tests does your area hospital conduct routinely? Are any of these splicing disorders?

45. In a very few countries, most antibiotics require a prescription. In most of the other countries, many of these drugs are available over the counter (OTC). List the merits and disadvantages of making drugs available only by prescription versus OTC.

46. Consumption of some antibiotics such as rifampicin renders other drugs, such as oral contraceptives, ineffective. How might this work?

47. Many diseases (such as SMA) are now known to be inherited. How would you advise a family about having children if they were carriers of SMA and had a one in four chance that their child would be affected? Would your advice change if the disease were something like ALS, a neuromotor disease that is equally fatal but does not affect people until later in life?

Suggested Readings

15.1 Transcription: RNA Synthesis

Gilbert, W. "The RNA World." *Nature* 319, no. 6055 (February 1986): 618.

Maynard Smith, J., and E. Szathmáry. *The Major Transitions in Evolution.* Oxford, UK: Oxford University Press, 1995.

Mitchell, J. A., and P. Fraser. "Transcription Factories Are Nuclear Subcompartments That Remain in the Absence of Transcription." *Genes & Development* 22 (1996): 20–25.

Papantonis, A., and P. R. Cook. "Transcription Factories: Genome Organization and Gene Regulation." *Chemical Reviews* 113, no. 11 (2013): 8683–8705.

Rieder, D., Z. Trajanoski, and J. G. McNally. "Transcription Factories." *Frontiers in Genetics* 3 (2012): 221–230.

15.2 Processing of Nascent Eukaryotic RNA Messages

Black, D. L. "Mechanisms of Alternative Pre-Messenger RNA Splicing." *Annual Review of Biochemistry* 72 (2003): 291–336.

Darnell, J. E., Jr. "Reflections on the History of Pre-mRNA Processing and Highlights of Current Knowledge: A Unified Picture." *RNA* 19, no. 4 (2013): 443–460.

Faustino, N. A., and T. A. Cooper. "Pre-mRNA Splicing and Human Disease." *Genes & Development* 17 (2003): 419–437.

Galej, W. P., C. Oubridge, A. J. Newman, and K. Nagai. "Crystal Structure of Prp8 Reveals Active Site Cavity of the Spliceosome." *Nature* 493 (January 31, 2013): 638–643.

Hocine, S., R. H. Singer, and R. Luhrmann. "RNA Processing and Export." *Cold Spring Harbor Perspectives in Biology* 2, no. 12 (2010): a000752.

Hoskins, A. A., and M. J. Moore. "The Spliceosome: A Flexible, Reversible Macromolecular Machine." *Trends in Biochemical Sciences* 37, no. 5 (2012): 179–188.

Konarska, M. M., R. A. Padgett, and P. A. Sharp. "Recognition of Cap Structure in Splicing In Vitro of mRNA Precursors." *Cell* 38, no. 3 (1984): 731–736.

Kornblihtt A. R., I. E. Schor, M. Alló, G. Dujardin, E. Petrillo, and M. J. Muñoz. "Alternative Splicing: A Pivotal Step between Eukaryotic Transcription and Translation." *Nature Reviews Molecular Cell Biology* 14, no. 3 (2013): 153–165.

Marcia, M., and A. M. Pyle. "Visualizing Group II Intron Catalysis through the Stages of Splicing." *Cell* 151, no. 3 (2012): 497–507.

Matlin, A. J., F. Clark, and C. W. Smith. "Understanding Alternative Splicing: Towards a Cellular Code." *Nature Reviews Molecular Cell Biology* 6, no. 5 (2005): 386–398.

Padgett, R. A. "New Connections between Splicing and Human Disease." *Trends in Genetics* 28, no. 4 (2012): 147–154.

Pan, Q., O. Shai, L. J. Lee, B. J. Frey, and B. J. Blencowe. "Deep Surveying of Alternative Splicing Complexity in the Human Transcriptome by High-Throughput Sequencing." *Nature Genetics* 40, no. 12 (2008): 1413–1415.

Roesser, J. R., K. Liittschwager, and S. E. Leff. "Regulation of Tissue-Specific Splicing of the Calcitonin/Calcitonin Gene-Related Peptide Gene by RNA-Binding Proteins." *Journal of Biological Chemistry* 268, no. 11 (April 15, 1993): 8366–8375.

Sleeman, J. "Small Nuclear RNAs and mRNAs: Linking RNA Processing and Transport to Spinal Muscular Atrophy." *Biochemical Society Transactions* 41, no. 4 (2013): 871–875.

Toh, Y., H. Hori, K. Tomita, T. Ueda, and K. Watanabe. "Transfer RNA Synthesis and Regulation." In eLS. Chichester, UK: John Wiley & Sons, 2009. http://www.els.net.

Toor, N., K. S. Keating, S. D. Taylor, and A. M. Pyle. "Crystal Structure of a Self-Spliced Group II Intron." *Science* 320, no. 5872 (April 4, 2008): 77–82.

van der Feltz, C., K. Anthony, A. Brilot, and D. A. Pomeranz Krummel. "Architecture of the Spliceosome." *Biochemistry* 51, no. 16 (2012): 3321–3333.

Will, C. L., and R. Luhrmann. "Spliceosome Structure and Function." *Cold Spring Harbor Perspectives in Biology* 3, no. 7 (2011): a003707.

15.3 RNA Export from the Nucleus

Björk, P., and L. Wieslander. "Mechanisms of mRNA Export." *Seminars in Cell and Developmental Biology* 32 (August 2014): 47–54.

Carmody, S. R., and S. R. Wente. "mRNA Nuclear Export at a Glance." *Journal of Cell Science* 122 (2009): 1933–1937.

Chook, Y. M., and G. Blobel. "Karyopherins and Nuclear Import." *Current Opinion in Structural Biology* 11, no. 6 (December 2001): 703–715.

Cullen, B. R. "Nuclear RNA Export Pathways." *Molecular and Cellular Biology* 20, no. 12 (2000): 4181.

Grünwald, D., R. H. Singer, and M. Rout "Nuclear Export Dynamics of RNA-Protein Complexes." *Nature* 475, no. 7356 (2011): 333–341.

Katahira, J., and Y. Yoneda. "Nucleocytoplasmic Transport of MicroRNAs and Related Small RNAs." *Traffic* 12, no. 11 (2011): 1468–1474.

Köhler A., and E. Hurt. "Exporting RNA from the Nucleus to the Cytoplasm." *Nature Reviews Molecular Cell Biology* 8, no. 10 (October 2007): 761–773.

Lei, E. P., and P. A. Silver. "Protein and RNA Export from the Nucleus." *Developmental Cell* 2 (March 2002): 261–272.

Panté, N., and M. Kann. "Nuclear Pore Complex Is Able to Transport Macromolecules with Diameters of ~39 nm." *Molecular Biology of the Cell* 13, no. 2 (2002): 425–434.

Peters, R. "Functionalization of a Nanopore: The Nuclear Pore Complex Paradigm." *Biochimica et Biophysica Acta* 1793, no. 10 (2009): 1533–1539.

Schwartz, T. "Functional Insights from Studies on the Structure of the Nuclear Pore and Coat Protein Complexes." *Cold Spring Harbor Perspectives in Biology* 5, no. 7 (July 2013): pii: a013375.

Stewart, M. "Nuclear Export of mRNA." *Trends in Biochemical Sciences* 35, no. 11 (2010): 609–617.

Strambio-De-Castillia, C., M. Niepel, and M. P. Rout. "The Nuclear Pore Complex: Bridging Nuclear Transport and Gene Regulation." *Nature Reviews Molecular Cell Biology* 11, no. 7 (2010): 490–501.

Tu, L. C., and S. M. Musser. "Single Molecule Studies of Nucleocytoplasmic Transport." *Biochimica et Biophysica Acta* 1813, no. 9 (2011): 1607–1618.

Wälde, S., and R. H. Kehlenbach. "The Part and the Whole: Functions of Nucleoporins in Nucleocytoplasmic Transport." *Trends in Cell Biology* 20, no. 8 (2010): 461–469.

Wente, S. R., and M. P. Rout. "The Nuclear Pore Complex and Nuclear Transport." *Cold Spring Harbor Perspectives in Biology* 2, no. 10 (2010): a000562.

1868 ——————— Miescher discovers nucleic acids DNA and RNA

1959 ——————— Nobel Prize to Ochoa for *in vitro* synthesis of RNA using polynucleotide phosphorylase
1960 ——————— Loe, Stevens, and Hurwitz independently discover RNA polymerase
1961–62 ——————— Nirenberg, Khorana, and Holley crack mRNA genetic code for protein, sequence tRNA, Nobel Prize 1968
1970 ——————— Baltimore and Temin discover reverse transcriptase, Nobel Prize 1975

1977 ——————— Sharp and Roberts discover split genes, Nobel Prize 1993
1978 ——————— Gilbert coins terms *intron* and *exon*
1980s ——————— Cech and Altman discover catalytic RNA molecules involved in splicing, Nobel Prize 1989
1986 ——————— Ecker demonstrates antisense RNA in petunia plants

1998 ——————— Fire and Mello publish work on RNAi in *C. elegans* (flatworm), Nobel Prize 2006

CHAPTER 16

Protein Biosynthesis

Protein Biosynthesis in Context

It seems to be part of human nature to keep secrets and to try to uncover the secrets of others. In some ways, science can be seen as an attempt to find out and make sense of nature's secrets.

The meaning of the hieroglyphics found on ancient Egyptian tombs and artifacts remained a mystery until the discovery of the Rosetta Stone in 1799. This stone fragment contained a decree issued by King Ptolemy, with parallel text in Ancient Greek and Ancient Egyptian, allowing archeologists to translate hieroglyphics for the first time. The discovery of the stone was a significant leap in knowledge because it enabled us to decode what was once enigmatic.

Some 50 years ago, the messages encoded in nucleic acids were equally enigmatic, and no Rosetta Stone was available to crack this code. Instead, the genetic code was deciphered through years of experimentation. As with the Rosetta Stone, elucidation of the genetic code represented a significant leap in knowledge.

This chapter begins with a discussion of the genetic code and how the information contained in it codes for proteins; it then discusses the molecules and individual steps in the synthesis of proteins. These include detailed descriptions of tRNA and the ribosome, which is largely composed of rRNA. The chapter concludes by introducing some molecules that inhibit protein synthesis; some of these molecules are highly toxic, whereas others are used as chemotherapeutic agents or antibiotics.

Chapter Outline

16.1 The Genetic Code

16.2 Machinery Involved in Protein Biosynthesis

16.3 Mechanism of Protein Biosynthesis

Evolution's outcomes are conserved.	• The broad themes of this chapter—the functions of mRNA, tRNA, and rRNA in protein biosynthesis—are conserved throughout biology, suggesting an ancient origin.
	• The genetic code is almost universal, suggesting that it too was selected early in the history of life.
	• Deviations from the genetic code are minor and are conserved during evolution.
Structure determines function.	• The conserved structures found in tRNA and rRNA are necessary for protein biosynthesis.
	• Often, changes to a single base in the 2 or 3 positions of a codon give a conservative change or no change at all to the amino acid being encoded.
Biochemical information is transferred, exchanged, and stored.	• The information that has passed from DNA to RNA is used to synthesize proteins in translation.
	• Proteins regulate gene expression and epigenetic modifications to DNA as well as regulating genes via signaling cascades.
Biomolecules are altered through pathways involving transformations of energy and matter.	• To be coupled to a tRNA, amino acids must first be chemically activated via coupling to ATP.
	• Free amino acids are not added to a protein; rather, the amino group of the aminoacyl-tRNA undergoes nucleophilic attack on the ester linkage of the growing peptide bound to the ribosome.

16.1 The Genetic Code

The information that is coded for by DNA is transcribed into RNA. In addition, DNA can serve as a template in its own replication and repair and as a template for RNA biosynthesis. Similarly, messenger RNA (mRNA) contains the code for proteins to be synthesized. In the case of protein synthesis, there are two complications: extracting the information stored within the code and synthesizing a protein (made of amino acids) from a message (made of nucleic acids). This section describes the genetic code and the decoding of the information found within mRNA. Sections 16.2 and 16.3 discuss the RNA molecules involved in protein biosynthesis and the steps involved in protein production.

16.1.1 The genetic code translates nucleic acids into amino acids

The **genetic code** is the name given to the sequences of RNA that code for individual amino acids (**Figure 16.1**). If we think of the deoxyribonucleotides of DNA or the nucleotides of RNA as a long string of letters on a page, then the code assembles these letters into words and sentences that communicate meaning.

The genetic code **translates** nucleic acids into amino acids. Scientists cracked the code using synthetic mRNAs and radiolabeled amino acids. The first mRNA to be tried was polyuridine (5′–UUUUUUUUU–3′), which generated the polypeptide polyphenylalanine, indicating

1st letter

	U	C	A	G	
					2nd letter
U	Phenylalanine	Serine	Tyrosine	Cysteine	U
	Phenylalanine	Serine	Tyrosine	Cysteine	C
	Leucine	Serine	**stop**	**stop**	A
	Leucine	Serine	**stop**	Tryptophan	G
C	Leucine	Proline	Histidine	Arginine	U
	Leucine	Proline	Histidine	Arginine	C
	Leucine	Proline	Glutamine	Arginine	A
	Leucine	Proline	Glutamine	Arginine	G
A	Isoleucine	Threonine	Asparagine	Serine	U
	Isoleucine	Threonine	Asparagine	Serine	C
	Isoleucine	Threonine	Lysine	Arginine	A
	(start) Methionine	Threonine	Lysine	Arginine	G
G	Valine	Alanine	Aspartic acid	Glycine	U
	Valine	Alanine	Aspartic acid	Glycine	C
	Valine	Alanine	Glutamic acid	Glycine	A
	Valine	Alanine	Glutamic acid	Glycine	G

3rd letter

FIGURE 16.1 Genetic code. The genetic code is made up of three-letter codons, most of which code for a specific amino acid. One codon (AUG) is the start codon and codes for methionine. Three codons (UAA, UAG, and UGA) are stop codons, signaling the termination of translation.

(Source: Karp, *Cell and Molecular Biology: Concepts and Experiments*, 7e, copyright 2013, John Wiley & Sons. This material is reproduced with permission of John Wiley & Sons, Inc.)

that UUU is the codon for the amino acid phenylalanine. Poly(A) and poly(C) were the next to be tried, and these generated polylysine and polyproline (AAA codes for lysine and CCC for proline).

Subsequent experiments revealed that the genetic code is comprised of three-nucleotide "words" that do not overlap or have gaps between them. With only four nucleotides in mRNA and 20 amino acids to be produced, a single nucleotide cannot code for a single amino acid. Using two nucleotides to code for a single amino acid would generate only 16 different combinations ($4^2 = 16$), which is still not enough to have a single unit of code for each amino acid. Using three nucleotides as the unit of the code generates 64 different combinations ($4^3 = 64$), more than enough to cover the 20 amino acids. The three-nucleotide units that make up the genetic code are termed **codons**. To continue our language analogy, each word in the genetic code is three letters long.

The genetic code has just two forms of punctuation: where to start a sentence and where to stop. The message to code for a protein always begins with a **start codon** (AUG) and ends with one of the three **stop codons** (UAA, UAG, or UGA). Of the 64 possible codons, 61 code for amino acids; three are stop codons, signaling the termination of protein synthesis; and one is the start codon, coding for methionine, which starts every protein. As the only codon for methionine, AUG also serves to code for that amino acid wherever it appears in a protein.

The string of code between a start codon and a stop codon (the sentence in our analogy) is called an **open reading frame (ORF)**. Because each codon is comprised of three bases, there are three possible reading frames in the forward direction for any message, but only one of these will generate the correct protein. Incorrect reading frames generate the wrong protein and frequently terminate prematurely. As discussed later in section 16.3, one of the functions of the initiation of translation is to ensure that the reading frame is correct.

16.1.2 Codon degeneracy and wobble benefit protein synthesis

Many amino acids are coded for by multiple codons, termed **degenerate codons**. The term degeneracy indicates that these codons provide built-in redundancy in the code, not that they show moral decline. Often, degenerate codons differ only in the third position or sometimes in both the second and the third positions. The lack of consistency in the third position of such codons is termed **wobble**. Furthermore, some organisms favor certain degenerate codons over others, called **codon bias**.

Degeneracy and wobble have several benefits. Because there are multiple codons for a single amino acid and the third base is less important in coding, mutations to that position have less effect in that they are often neutral or silent. In eight of the amino acids, altering the base in the third position of the codon gives the same amino acid. Most of the other 12 amino acids are coded for by at least two codons.

Furthermore, there is structural organization among codons so that mutations are often conservative. For example, codons with A in the second position code for hydrophilic amino acids, whereas codons with U in the second position code for hydrophobic amino acids. Finally, wobble speeds up protein synthesis because the weaker hydrogen bonding that occurs in the third position means that the corresponding tRNA is loosely bound and can thus dissociate rapidly.

16.1.3 The genetic code is almost universal

The genetic code used to translate mRNA into protein is almost the same among all organisms, but there are slight differences, even within a single organism. For example, the genetic code used by mitochondria to translate the small mitochondrial genome differs between species by as many as four codons. In addition, codons other than AUG can serve as the start codon. Other differences from the universal genetic code have been noted in yeast, flatworms, and some microbes. Also, some species (typically among the Archaea) have codons that are used by the amino acids selenocysteine (UGA) and pyrrolysine (UAG) (**Figure 16.2**).

Nonetheless, these differences are minor in comparison to the number of codons that are unchanged and the number of species that use the universal genetic code. Changes to the common genetic code are rare in part because of the success of the genetic code itself, which was selected for early in evolutionary history. Because these changes are so rare, they can be traced back through evolution to determine relationships among species.

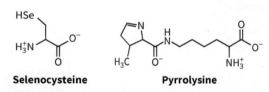

FIGURE 16.2 Structure of the unusual amino acids selenocysteine and pyrrolysine. Selenocysteine and pyrrolysine are rarely encountered in proteins but do occur in some species, including mammals.

Worked Problem 16.1   Secret code

What peptide would be generated from this DNA sequence?

    5′–TATGCCAGCGAGGGTCTGACC–3′ coding strand
    3′–ATACGGTCGCTCCCAGACTGG–5′ template strand

Strategy   In this DNA sequence, one strand codes and the other acts as a template. The strands given here are labeled, but in the laboratory, it is necessary to either search for an ORF or otherwise experimentally verify which strand is being transcribed and translated. The mRNA is made using the coding strand as a template. How would this strand be translated? Where would the translation begin?

Solution   The DNA translates into the sequence:

    5′–UAUGCCAGCGAGGGUCUGACC–3′

To translate this sequence, we need to find an ORF. Searching for the start codon, we find AUG beginning on the second nucleotide. This provides the reading frame of the protein and enables us to use the genetic code to read through the rest of the sequence.

    5′–U **AUG** CCA GCG AGG GUC **UGA** CC–3′
       H₃N⁺-Met-Pro-Ala-Arg-Val-COO⁻

Follow-up question   What changes to the sequence would be needed to convert any polar amino acids in the original sequence to alanine?

Summary

- The genetic code is comprised of three-base sequences termed codons. Each codon either codes for an amino acid or provides the signal to stop translation. There is one start codon and three stop codons. The code was solved by translating synthetic RNA molecules, and it was found that there are no gaps between codons and no overlap of codons.
- Most amino acids are coded for by more than one codon. The redundancy these degenerate codons provide allows a greater tolerance for mutations. The variability in these codons, typically in the third position, is called a wobble. The code deviates from the norm in only a few instances, for example, in mitochondria, yeast, flatworms, and some microbes.

Concept Check

1. What are the main features of genetic code?
2. Explain the advantages of degeneracy and wobble.
3. Explain where and potentially why the genetic code deviates from what is universally accepted.

16.2 Machinery Involved in Protein Biosynthesis

Previous chapters have alluded to the final step of the central dogma, in which the message that is now encoded in mRNA is translated into proteins. This section reviews some of the important structural elements found in tRNA, rRNA, and mRNA and explains how these elements are used in translation.

16.2.1 Messenger RNA contains structural information

We often think of mRNA as the sequence that codes for protein. Although this is partially correct, there are important parts of the mRNA that do not code for amino acids. In addition to the coding sequence (the reading frame), which is bordered by the start codon (AUG) at one end and a stop codon (UGA, UAG, UAA) at the other, mRNAs contain 5′ and 3′ untranslated regions.

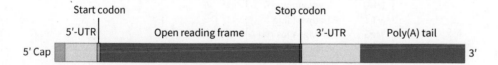

A 5′ **untranslated region (UTR)** extends from the 5′-methylguanine cap of mRNA to the start codon (AUG). In prokaryotes, this region is short (<10 nucleotides) and contains a ribosome-binding site known as the **Shine–Dalgarno sequence** (AGGAGGU). In eukaryotes, this region has an average size of 250 bases, but it can range from several hundred to several thousand bases. It, too, contains a ribosome binding site, in this case termed a **Kozak sequence**, which overlaps with the start codon (ACC**AUG**G). Eukaryotic mRNAs also frequently form secondary structures in the 5′-UTR that may influence gene expression (for example, in the regulation of ferritin translation by iron-response elements, discussed in Chapter 17).

Following the stop codon is the 3′-UTR, which is involved in the stability and translation of the message, partly through binding of microRNAs (miRNAs) or regulatory proteins that bind to AU-rich regions. The 3′-UTR is also the location of the poly(A) tail of the mRNA and is usually about 700 bases in length.

16.2.2 Transfer RNAs act as adapters for the genetic code

Transfer RNA molecules are adapters that interface between the mRNA code and the ribosome to provide an amino acid for the growing polypeptide chain. Often, the focus of the molecular structure of the tRNA is on the **anticodon loop**, which is the region of the tRNA that interacts with the mRNA codon, binding to it and ensuring that the correct amino acid is incorporated.

The tRNA has to do more than just interact with the codon, however; it must also interact with the ribosome to add the amino acid to the protein. The coupling of the tRNA to the correct amino acid is catalyzed by an **aminoacyl-tRNA synthetase**, an enzyme that needs to recognize both the specific tRNA and the specific amino acid and then catalyze the formation of the new ester bond, despite there being only small structural differences in the substrates. The entire structure of the tRNA, not just the anticodon loop, is important in its function.

All tRNAs have a common shape (**Figure 16.3**). Between 75 and 95 nucleotides in length, tRNAs have a secondary structure that is reminiscent of a three-leafed clover due to hydrogen bonding between base pairs. In each of the three hairpin loops, there is a double-stranded region that twists into a helical shape and a loop.

FIGURE 16.3 Structure of tRNA. A. Transfer RNA molecules base pair with each other to form a clover-leaf pattern that folds into an L or 7 shape in three dimensions. Several of the bases in tRNA are modified. **B.** When folded, the anticodon loop is found at the bottom of the structure, and the amino acid is coupled to the adapter stem at the other end.

(Source: (A) Snustad; Simmons, *Principles of Genetics*, 6e, copyright 2012, John Wiley & Sons. This material is reproduced with permission of John Wiley & Sons, Inc. (B, Left) Snustad; Simmons, *Principles of Genetics*, 6e, copyright 2012, John Wiley & Sons. This material is reproduced with permission of John Wiley & Sons, Inc. (B, Middle/Right) Data from PDB ID 2TRA Westhof, E., Dumas, P., Moras, D. (1988) Restrained refinement of two crystalline forms of yeast aspartic acid and phenylalanine transfer RNA crystals. *Acta Crystallogr., Sect.A* 44: 112-123)

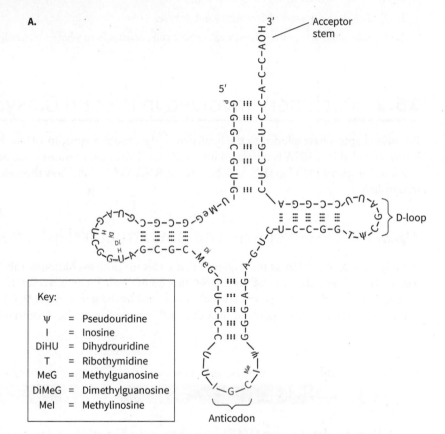

Key:

Ψ	=	Pseudouridine
I	=	Inosine
DiHU	=	Dihydrouridine
T	=	Ribothymidine
MeG	=	Methylguanosine
DiMeG	=	Dimethylguanosine
MeI	=	Methylinosine

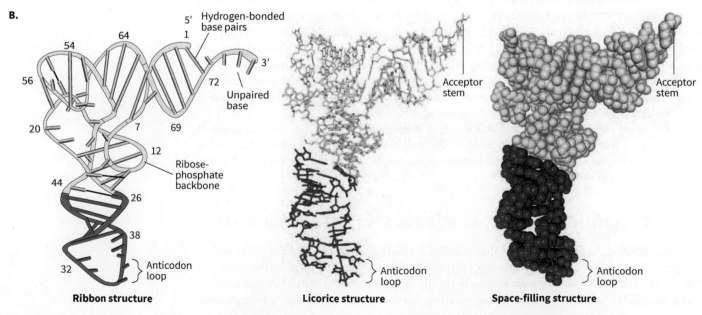

Ribbon structure   Licorice structure   Space-filling structure

Transfer RNA molecules have several characteristics that distinguish them from other nucleic acids. Both the sugar–phosphate backbone and the bases that comprise tRNA molecules are highly modified. For example, the 5′ end of tRNA is phosphorylated as part of the maturation of the molecule. Other modifications include bases not normally found in mRNA (e.g., inosine, pseudouridine, and dihydrouridine) and methylations of other bases. Some of these modifications occur in the nucleus, whereas others occur after export from the nucleus.

Inosine **Pseudouridine** **Dihydrouridine** **5-Methylcytosine**

The first loop that is formed is termed the D-loop, a four base–pair hairpin that frequently contains the base dihydrouridine, as shown in Figure 16.3. At the bottom of the second hairpin is the anticodon loop, the region of the tRNA that is recognized by the mRNA codon. For binding to occur, the anticodon loop and the codon of the mRNA must align in an antiparallel orientation to each other, just as the strands of DNA are in an antiparallel alignment. The next loop is the short variable loop. This structure may contain methylated bases such as methyl-guanine or methylcytosine. Moving counterclockwise, the third loop or leaf of the clover is the T arm, which contains the TΨC loop. The TΨC loop includes the conserved sequence TΨC, where Ψ is the modified base pseudouridine. The final structural item of note is the **acceptor stem**, which is the region of tRNA that is esterified to an amino acid. All tRNA molecules end with the sequence CCA-OH, and it is this hydroxyl group that is used in the esterification reaction. The CCA sequence is not coded for by the tRNA gene but is added enzymatically as part of post-transcriptional processing.

In the tertiary structure of the tRNA, the leaves of the clover fold onto each other, generating an L or 7 shape. Again, tRNAs are typically depicted with the anticodon loop at the bottom of the image and the 3′-hydroxyl acceptor stem at the top.

Types of tRNA molecules   The genetic code includes 64 codons, and 20 different amino acids are commonly found in proteins. Some tRNAs can bind to more than one codon, and there are 49 different tRNAs that differ in the anticodon loop. Multiple tRNAs that can couple to one amino acid (e.g., two different tRNA molecules bind to different codons for serine) are termed **isoacceptors**. Surprisingly, in humans there are 503 genes that code for tRNAs scattered throughout the genome, including 274 different tRNA species. These molecules mainly differ not in the anticodon loop but elsewhere in the tRNA molecule. As an example, the amino acid alanine has four different codons that are recognized by one isoacceptor, but there are about 40 alanine tRNA molecules that differ elsewhere in the sequence. These are coded for by 46 distinct alanine tRNA genes, scattered across 11 different chromosomes. New computational methods continue to identify additional tRNA genes.

16.2.3 Ribosomal RNA molecules provide both structure and catalysis

As its name suggests, ribosomal RNA (rRNA) is found in the **ribosome**; hence, this section describes the features of the entire ribosome. At low resolution, ribosomes are somewhat egg shaped and are comprised of two subunits, termed the large and small subunits. Prokaryotes and eukaryotes have similar ribosomes but with several important differences.

Prokaryotic ribosomes are about 20 nm (200 Å) in diameter and consist of a **30S subunit** and a **50S subunit** (**Figure 16.4**). These subunits come together to form a **70S ribosome**. Analytical ultracentrifugation experiments on ribosomes and their subunits yield sedimentation coefficients, a rough measure of particle size. This technique is the basis for Svedberg units (S), and these values are not additive. By mass, the prokaryotic ribosome is about 65% RNA and

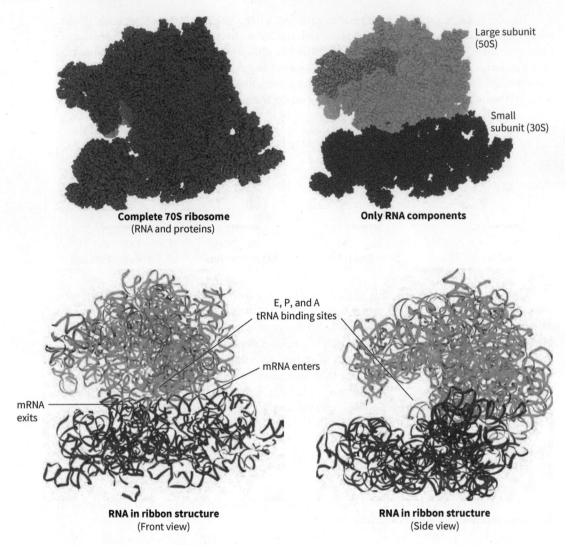

Complete 70S ribosome
(RNA and proteins)

Only RNA components

Large subunit
(50S)

Small
subunit (30S)

E, P, and A
tRNA binding sites

mRNA enters

mRNA
exits

RNA in ribbon structure
(Front view)

RNA in ribbon structure
(Side view)

FIGURE 16.4 Prokaryotic ribosome. Shown is the structure of the 70S ribosome. In this view, the 50S large subunit is on the top, and the 30S small subunit is on the bottom. Messenger RNA enters from the right and exits from the left side of the structure. The three binding sites for tRNA (the E, P, and A sites) are found in the front of the structure.

(Source: Data from PDB ID 4V4A Vila-Sanjurjo, A., Ridgeway, W. K., Seymaner, V., Zhang, W., Santoso, S., Yu, K., Cate, J. H. D. (2003) X-ray crystal structures of the WT and a hyper-accurate ribosome from Escherichia coli. *Proc.Natl.Acad.Sci.USA* 100: 8682-8687)

35% protein. The 30S subunit has a single 16S RNA (1,540 bases) that is complexed with 21 different proteins. The 50S subunit contains 31 proteins and two RNAs: a 5S of 120 bases and a 23S of 2,900 bases. The proteins are labeled by subunit and number (e.g., S4 or L23).

Eukaryotic ribosomes are slightly larger, measuring 25 nm (250 Å) in diameter, and they are composed of a **40S subunit** and a **60S subunit**, which together form an **80S ribosome**. The small subunit contains 33 proteins and a 1,900-base 18S RNA. In contrast, the large subunit contains 46 proteins and three RNAs: a 5S of 120 bases, a 5.8S of 160 bases, and a 23S of 4,700 bases. Eukaryotic ribosomes are about 50% protein and 50% RNA by mass.

In both prokaryotes and eukaryotes, RNA molecules themselves are responsible for catalysis. The proteins found on ribosomes are thought to help provide structure and regulation, but they do not contribute catalytically. It comes as a surprise to some that RNA can be catalytic, but there are numerous instances of RNA performing duties commonly associated with proteins.

The fully assembled active ribosome consists of the two subunits clamped around the mRNA template. The template is bound by the small subunit and is topped by the large subunit, but a groove separates the two subunits. The large subunit has three characteristic binding sites for tRNAs (see again Figure 16.4): the **aminoacyl (A) site**, where the new, loaded aminoacyl-tRNA binds and is ready to be added to the protein; the **peptide (P) site**, where the growing polypeptide is bound and the peptide bonds are formed; and the **exit (E) site**, where the now unoccupied tRNA is found before it is ejected from the ribosome. The newly translated peptide progresses up through a tunnel in the large subunit and is eventually released.

Ribosomes may be found free in the cytosol or bound to a membrane—the endoplasmic reticulum (ER) in eukaryotes or the plasma membrane in prokaryotes. The structure that forms when multiple ribosomes in the cytosol begin to translate an mRNA is termed a polysome. Ribosomes bound to membranes are oriented such that the protein being translated is either synthesized through a pore into the lumen of the ER, or it is released into the ER membrane to become a membrane protein. The Sec proteins are partially responsible for secretion of proteins or generation of integral membrane proteins in bacteria.

The function of the ribosome in translation is discussed in section 16.3.

Worked Problem 16.2 — The shape of tRNA

Examine this sequence from a tRNA molecule:

5′–GΨUGG(3mC)UUGAHACCAACAU–3′

Can you identify any secondary structures that may form? How might you determine whether these structures exist *in vitro* or *in vivo*?

Strategy Examine the sequence shown. Ψ is pseudouridine, 3mC is 3-methylcytosine, and H is a modified adenosine base. Are there any regions that might form secondary structures?

Solution The sequence shown is a serine tRNA from chicken (*Gallus gallus*) and has the structure shown in the diagram.

Follow-up question Would mutations to the tRNA occurring outside the anticodon loop be potentially harmful to the organism? Explain your reasoning.

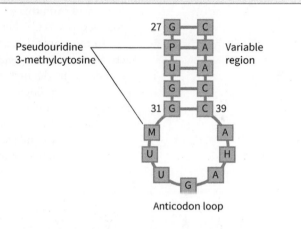

5′–GΨUGG(3mC)UUGAHACCAACAU–3′

Summary

- Messenger RNAs contain 5′ and 3′ untranslated regions. The 5′-UTR contains a Shine–Dalgarno sequence in prokaryotes or a Kozak sequence in eukaryotes. These sequences aid in the recognition of the start codon.

- All 247 different types of tRNA have a characteristic cloverleaf structure, but they are enzymatically modified in several locations, including methylations or alterations of bases. Isoreceptors are tRNAs that bind to the same amino acid but can recognize more than one codon.

- Several rRNA molecules contribute to the ribosome, which is comprised of two subunits: a small subunit consisting of a single RNA and about 30 proteins, and a larger subunit consisting of three different rRNAs and about 40 additional proteins. The ribosome has three different sites bound by tRNA molecules: the A (aminoacyl), P (peptide), and E (exit) sites. The catalytic properties of the ribosome are generated through RNA, not amino acids. Ribosomes may synthesize proteins free in the cytosol, into the lumen of the ER (eukaryotes), or into the plasma membrane (prokaryotes).

Concept Check

1. How do the three types of RNA work together in protein biosynthesis?
2. Describe the modifications made to mRNA, tRNA, and rRNA molecules and what role those modifications may play in their function.
3. Identify and name the common pieces of mRNA and tRNA molecules.

16.3 Mechanism of Protein Biosynthesis

Protein biosynthesis can be broken into four major steps: First is **activation**, where amino acids are coupled to tRNAs. Second is **initiation**, where the ribosome and associated proteins, termed initiation factors, bind to the start of the message. Third is **elongation**, where additional amino acids are added to the growing polypeptide chain. Fourth and last is **termination**, where the new protein is released, and the individual pieces of the translation complex are released for reuse. This section reviews each of these steps and provides a short introduction to protein folding and insertion of proteins into a membrane.

16.3.1 Activation is the coupling of amino acids to tRNA

To synthesize proteins, the cell needs a source of building blocks. These are not simply amino acids but are **aminoacyl-tRNAs**. The coupling of the amino acid to the tRNA provides an adapter that the ribosome uses in decoding the mRNA message; it also generates an ester linkage between the amino acid and the tRNA. This ester is essential for protein synthesis because the transfer of the amino acid from this state to form the new peptide bond is energetically favorable; that is, it has a more negative ΔG value compared to the formation of the peptide between two free amino acids.

In most organisms, there are 20 aminoacyl-tRNA synthetases, one for each amino acid. The activation of amino acids is a two-step process, with both steps catalyzed by aminoacyl-tRNA synthetase. In the first step of the reaction, amino acids react with ATP to form an aminoacyl-adenylate or aminoacyl-AMP, shown in **Figure 16.5**. Pyrophosphate (PP$_i$) is released in this reaction and subsequently hydrolyzed. Thus, the energetic equivalent of the hydrolysis of two molecules of ATP is consumed in the production of the aminoacyl-adenylate (a mixed anhydride). In the second step, the aminoacyl-adenylate reacts with the 2′ or 3′ hydroxyl group of the terminal ribose ring of the specific tRNA to form an aminoacyl-tRNA. Here again, the donation of the aminoacyl group from the mixed anhydride makes this reaction energetically favorable.

Although in most organisms there is one aminoacyl-tRNA synthetase for every amino acid, this is not the case in all organisms. Some bacteria, for example, use a single synthetase to charge tRNAs coding for both glutamate and glutamine with the amino acid glutamate. A second enzyme performs a transamination reaction on the glutamine-specific tRNAs to generate the correct amino acid–tRNA pair.

All aminoacyl-tRNA synthetases catalyze similar reactions and use similar substrates, and thus it might be assumed that they arose from a single family of related enzymes, coded for by a conserved family of genes. This is not the case. Although aminoacyl-tRNA synthetases are highly conserved throughout evolution, the study of their sequences and structures presents an interesting picture of early evolutionary history. There are two major structural classes of aminoacyl-tRNA synthetases that seem to have arisen independently in evolution and do not share any common sequences or structures (**Figure 16.6**). Lateral gene transfer must have happened at several places early in evolution to account for the sharing of these gene families and the differences between bacterial, archaeal, and eukaryotic tRNA synthetases that are observed in current organisms.

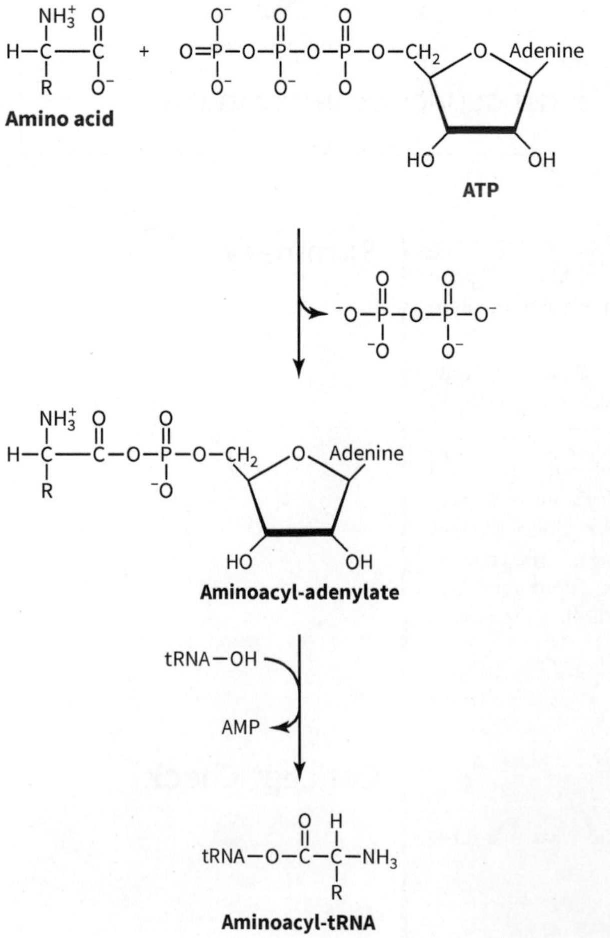

FIGURE 16.5 Coupling reaction of amino acids to tRNA. Amino acids are activated by coupling to their cognate tRNA. This reaction takes place in two steps. First, the amino acid reacts with ATP to generate an aminoacyl-adenylate, and pyrophosphate is lost and degraded. Next, the mixed anhydride bond between the amino acid and AMP is attacked by the 2′ or 3′ OH group from the ribose ring of the tRNA. AMP departs, and the amino acid is esterified to the tRNA.

Class I enzymes
(monomeric)

Class II enzymes
(multimeric)

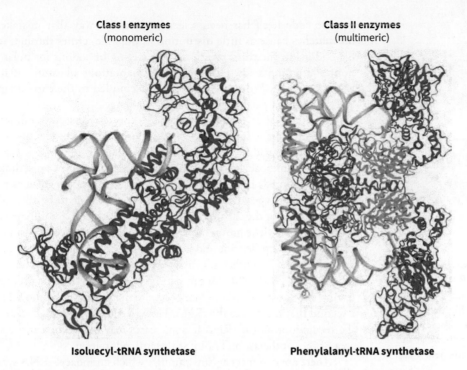

Isoluecyl-tRNA synthetase

Phenylalanyl-tRNA synthetase

FIGURE 16.6 Classes of aminoacyl-tRNA synthetases.   Aminoacyl-tRNA synthetases are organized into two classes by their structure. Shown are isoleucyl-tRNA synthetase, a class I monomeric enzyme, and phenylalanyl-tRNA synthetase, a class II multimeric enzyme.

(Source: (Left) Data from PDB ID 1FFY Silvian, L. F., Wang, J., Steitz, T. A. (1999) Insights into editing from an ile-tRNA synthetase structure with tRNAile and mupirocin. *Science* 285: 1074-1077; (Right) Data from PDB ID 1EIY Goldgur, Y., Mosyak, L., Reshetnikova, L., Ankilova, V., Lavrik, O., Khodyreva, S., Safro, M. (1997) The crystal structure of phenylalanyl-tRNA synthetase from thermus thermophilus complexed with cognate tRNAPhe. *Structure* 5: 59-68)

Aminoacyl-tRNA synthetases can be categorized as class I or class II. Class I enzymes are mono- or dimeric and contain a Rossmann fold, which is a combination of six planks of beta sheet and two alpha helices commonly found in nucleotide binding proteins (**Figure 16.7**). These enzymes make contact with the tRNA over a wide area and join the amino acid to the 2′ OH of the ribose ring. These amino acids spontaneously migrate to the 3′ position in a transesterification reaction to give the final product. The active site in class I enzymes is more exposed and, perhaps as a result, these enzymes generally couple the bulkier amino acids: leucine, isoleucine, valine, cysteine, methionine, tyrosine, tryptophan, arginine, glutamine, and glutamate.

Class II enzymes are multimeric proteins with a common antiparallel beta sheet fold (**Figure 16.8**). These proteins bind the tRNA from the opposite side to class I enzymes, generally making less contact. The active site of class II enzymes is generally more deeply buried in the structure than it is in class I enzymes and, as a result, these enzymes couple the less bulky amino acids: histidine, proline, serine, threonine, aspartate, asparagine, lysine, glycine, alanine, and phenylalanine. However, the aminoacyl-tRNA synthetase for lysine is a class I enzyme in some organisms.

The role of aminoacyl-tRNA synthetases in achieving a low error rate
The three previous chapters have shown that the cell devotes a significant portion of its resources to ensuring that the genetic code is faithfully copied, any errors or damage are repaired, and the messages made from that code are transcribed with as high fidelity as possible. Translation poses several challenges. As discussed previously, the

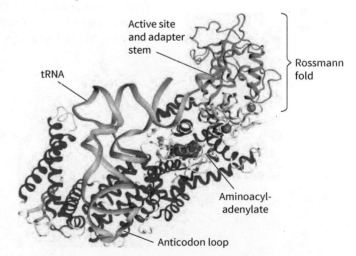

FIGURE 16.7 Class I aminoacyl-tRNA synthetases.   Shown is the valyl-tRNA synthase from *Thermus thermophilus* complexed to its cognate tRNA. In this view, the anticodon loop is found at the bottom left and is partially occluded by turns. The active site and the adapter stem are found at the other end of the tRNA (lower left). Here the active site is formed by a Rossmann fold comprised of two alpha helices and four strands of beta sheet.

(Source: Data from PDB ID 1GAX Fukai, S., Nureki, O., Sekine, S., Shimada, A., Tao, J., Vassylyev, D. G., Yokoyama, S. (2000) Structural basis for double-sieve discrimination of L-valine from L-isoleucine and L-threonine by the complex of tRNA(Val) and valyl-tRNA synthetase. *Cell* 103: 793-803)

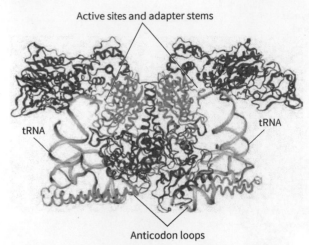

FIGURE 16.8 Class II aminoacyl-tRNA synthetases.
Shown is the structure of the phenylalanyl-tRNA synthetase complexed with its cognate tRNA. The enzyme is a heterotetramer with an a2b2 stoichiometry. In this view, the symmetric halves of the enzyme are each complexed to phenylalanyl-tRNAs with the anticodon loop found at the bottom and the adapter arm at the top near the beta sheet containing the active site.

(Source: Data from PDB ID 1EIY Goldgur, Y., Mosyak, L., Reshetnikova, L., Ankilova, V., Lavrik, O., Khodyreva, S., Safro, M. (1997) The crystal structure of phenylalanyl-tRNA synthetase from thermus thermophilus complexed with cognate tRNAPhe. *Structure* 5: 59-68)

genetic code itself has been selected in such a way that mistakes or mismatches cause as little disruption as possible, either through wobble codons in the third position or clustering of codons for polar and nonpolar amino acids. However, these adaptations all assume that the tRNA being used in protein synthesis is coupled to the correct amino acid, which is a particularly complex challenge.

One of the essential keys to the success of aminoacyl-tRNA synthetase is substrate specificity, both in terms of identifying the specific amino acid to couple to the tRNA and in binding the correct tRNA. The aminoacyl-tRNA synthetase needs to be able to discriminate between hundreds of different tRNA molecules and recognize as many as six different types of anticodon loops. It also needs to be able to differentiate among the 20 amino acids, some of which have differences as slight as a methylene group (aspartate and glutamate) or the positioning of a methyl group (leucine and isoleucine). Despite this challenging task, aminoacyl-tRNA synthetases catalyze these reactions with an error rate of less than 1 in 20,000. We can simply think of the tRNA synthetase as having two substrate binding sites: one site that binds the amino acid and a second that recognizes the tRNA. However, we need to appreciate that the recognition of the tRNA in some cases may be spread across a wide area of the synthetase and not just the anticodon loop.

There are two mechanisms through which aminoacyl-tRNA synthetases ensure high fidelity in charging tRNAs. The first mechanism that helps to ensure these enzymes are catalyzing the correct reaction is highly specific binding in the active site of the enzyme. An example of this is the selectivity of tyrosyl-tRNA synthetase for tyrosine over phenylalanine (**Figure 16.9**). These two amino acids differ only by the phenolic hydroxyl group found in tyrosine. Tyrosyl-tRNA synthetase has an aspartate side chain in the active site that can hydrogen-bond with the tyrosine, strengthening the interactions of the enzyme and substrate. Phenylalanine cannot undergo this interaction and thus binds far less tightly. Hydrogen bonds are thought of as a weak interaction, and most of the hydrogen bonds that exist between a substrate and enzyme only contribute 0.5 to 1.5 kcal/mol of binding energy. However, in the instance where a hydrogen bond is forming with a charged group, the strength of the bond is much higher, on the order of 3.5 to 4.5 kcal/mol. This change in binding energy is reflected in a change of the free energy of the transition state ($\Delta\Delta G^{\ddagger}$), resulting in a thousand-fold change in enzymatic rate. As discussed previously, small changes in structure can result in large changes in catalytic activity.

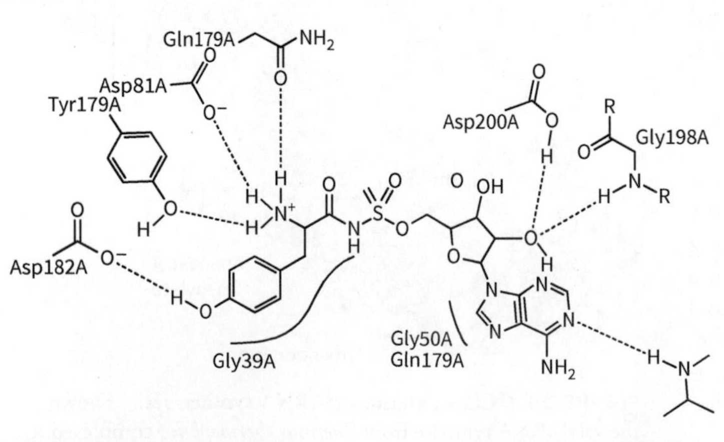

The second mechanism that helps to ensure that aminoacyl-tRNA synthetases are catalyzing the correct reaction is that several of these enzymes also have a proofreading function via a second site adjacent to the active site (**Figure 16.10**). The flexible adapter stem of the tRNA allows the coupled amino acid to flip out of the active site and into this proofreading site. Similar topology is seen in enzymes that employ a lengthy arm (e.g., the lysine–biotin combination used in carboxylation reactions) to separate active sites. If the substrate fits in the proofreading site, then the amino acid is hydrolyzed, and a new aminoacyl-AMP must be added. For example, if the tRNA for threonine is incorrectly loaded with serine, the serine can fit into the proofreading site of the synthetase and will be hydrolyzed. Threonine is too large to fit into the site, and therefore it will not be cleaved. This example shows that the proofreading site for the threonyl-tRNA synthetase must also be specific; it cannot accept threonine or the enzyme would undergo a futile cycle, coupling and uncoupling threonine. Likewise, the proofreading site cannot be so general that it accepts

FIGURE 16.9 Active site of tyrosyl-tRNA synthetase. The active site of tyrosyl-tRNA synthetase is able to discriminate between tyrosine and phenylalanine. Shown is a modified substrate (a non-hydrolyzable thiolate analog of the tyrosyl-AMP) and the interactions the enzyme makes with this substrate.

(Source: Data from PDB ID 1VBN Kobayashi, T., Sakamoto, K., Takimura, T., Sekine, R., Vincent, K., Kamata, K., Nishimura, S., Yokoyama, S. (2005) Structural basis of nonnatural amino acid recognition by an engineered aminoacyl-tRNA synthetase for genetic code expansion. *Proc.Natl.Acad.Sci. USA* 102: 1366-1371)

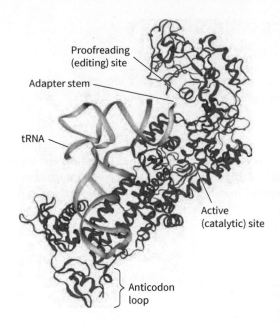

Proofreading (editing) site

Adapter stem

tRNA

Active (catalytic) site

Anticodon loop

FIGURE 16.10 Proofreading function of aminoacyl-tRNA synthetases. Many aminoacyl-tRNA synthetases have a proofreading (or editing) site adjacent to the active (catalytic) site. The adapter stem of the tRNA can flip back and forth between the active site and the proofreading site. If an amino acid that has been incorrectly coupled can also bind in the proofreading site, it is cleaved, and the enzyme must bind a new amino acid and ATP.

(Source: Data from PDB ID 1FFY Silvian, L. F., Wang, J., Steitz, T. A. (1999) Insights into editing from an ile-tRNA synthetase structure with tRNAile and mupirocin. *Science* 285: 1074-1077)

every amino acid. Therefore, the substrate-binding site that initially forms the aminoacyl-AMP and the substrate-specificity site have evolved to work together to keep the error rate low.

Activation is an ongoing process. We will assume for the remainder of this chapter that all of the necessary tRNAs are available for protein biosynthesis at all times, although Chapter 17 includes an example of how having low concentrations of an aminoacyl-tRNA can be used as part of a transcriptional control mechanism.

Amino acids are joined to tRNAs through ester linkages. The bonds that join many biological molecules are discussed in **Biochemistry: Esters, amides, and thioesters.**

Biochemistry

Esters, amides, and thioesters

Carboxylic acids are a common functional group in biochemistry. Often, they are part of amino acids or fatty acids, but they can also be found in other molecules. Carboxylic acids are more reactive than most other functional groups. As seen in this and other chapters, carboxylic acids can react with alcohols to form esters, with amines to form amides (termed peptides when encountered in proteins), and with thiols to form thioesters; however, the energetics of these reactions are worth reviewing.

In organic chemistry, esters can be formed by many different pathways, although most use strong conditions or reactive reagents such as acid chlorides, strong dehydrating agents, or oxidation of ketones with peroxides such as the Baeyer–Villiger reaction. In biochemistry, there are still energetic constraints to overcome, but instead of using harsh conditions, catalysts (enzymes) usually assist in overcoming energetic barriers. Although enzymes lower activation energies, they do not permit energetically unfavorable reactions to proceed. Instead, esters are often generated in biochemistry through a transesterification reaction with other esters, for example, thioesters from coenzyme A (CoA) or from an anhydride such as the esterification reaction of an aminoacyl-adenylate to the hydroxyl of tRNA, as seen in this chapter. In these instances, the generation of the ester from substrates in a relatively higher energy state makes the reaction favorable overall.

As with esters, amides (and peptides) are formed in the laboratory under harsh conditions. The Schotten–Baumann reaction is analogous to the formation of an ester from an acid chloride and an alcohol. In biochemistry, most amides—including peptides and amide linkages found in lipids—are formed through aminolysis of an ester or thioester (in the case of long-chain acyl-CoAs). The ester or thioester is attacked by an amine, generating the amide and eliminating the alcohol or thiol (CoASH).

Reactions analogous to those seen with esters and amides are found with thioesters. In the laboratory, a thioester can be generated in the Mitsunobu reaction, using an alcohol, thiocarboxylic acid, triphenyl phosphine, and an azodicarboxylate (effective, though hardly biological). In the cell, thioesters are formed from free carboxylic acids and a thiol (such as a fatty acid and CoA), through the energy released in ATP hydrolysis. Here, the energetic equivalent of two phosphoanhydride linkages is used to drive the reaction. Thioesters are also formed in the synthesis of acetyl-CoA and succinyl-CoA, through reactions involving decarboxylation steps. In both of these reactions, the loss of CO_2 drives the reaction, resulting in the formation of the thioester.

16.3.2 Initiation is the binding of the ribosome and the beginning of protein biosynthesis

All of the pieces needed for protein synthesis are now in place, but the components must be assembled in the correct order to begin protein synthesis. This step, termed initiation, differs slightly between prokaryotes and eukaryotes, so we will treat them separately.

Initiation in prokaryotes Initiation in prokaryotes begins with the 30S ribosomal subunit (**Figure 16.11**), which is activated by three different proteins termed **initiation factors** (IF-1, IF-2, and IF-3) and GTP. The subunit binds to the mRNA near the start codon, recognizing the Shine–Dalgarno sequence. At this point, a special initiator codon (f-Met-tRNA) binds to the

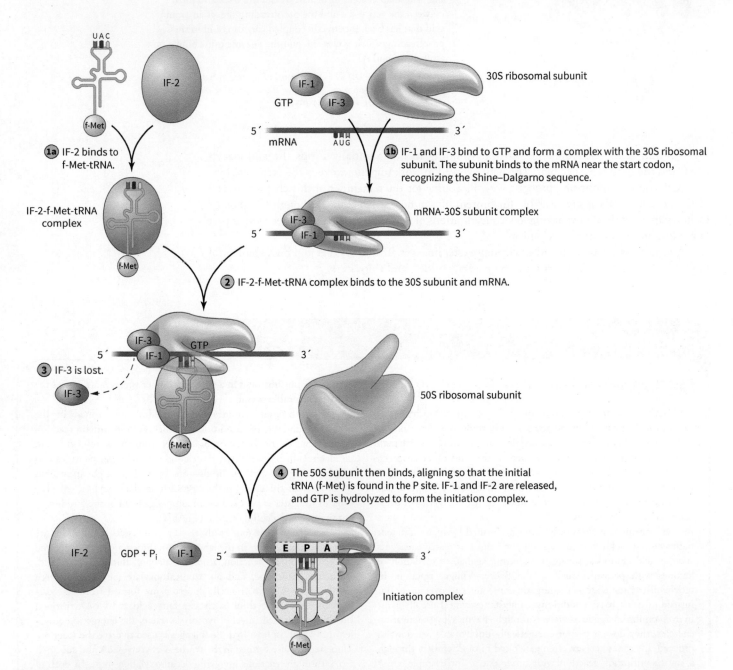

FIGURE 16.11 Initiation of translation in prokaryotes. Translation in prokaryotes begins with the formation of several complexes. First, f-Met tRNA binds to IF-2, IF-1 binds to GTP, and IF-3 forms a complex with the 30S subunit and mRNA. These form the 30S initiation complex and cause release of IF-3. Finally, IF-2 hydrolyzes GTP and the 50S subunit binds, causing release of both IF-1 and IF-2 and forming the 70S initiation complex.

(Source: Snustad; Simmons, *Principles of Genetics*, 6e, copyright 2012, John Wiley & Sons. This material is reproduced with permission of John Wiley & Sons, Inc.)

start codon, and IF-3 is lost. Formylated methionine, **f-Met**, is a methionine molecule in which a formyl group has been added to the amino terminus. It is formed by methioninyl-tRNA formyl-transferase, which adds the formyl group after the methionine has been coupled to the tRNA. The 30S initiation complex is the complex formed by the 30S subunit, f-Met-tRNA, IF-1 and IF-2, and the mRNA. The 50S subunit then binds, aligning so that the initial tRNA (f-Met) is found in the P site. This binding causes IF-1 and IF-2 to be released and GTP to be hydrolyzed to GDP and P_i.

Formylmethionine (f-Met)

Initiation in eukaryotes

Initiation of translation in eukaryotes is similar to the process in prokaryotes, with several key differences (**Figure 16.12**). The small ribosomal subunit (the 40S subunit) is bound by the eukaryotic initiation factors eIF-2, eIF-3, and eIF-4 along with Met-tRNA. The initial amino acid in eukaryotes is simply methionine, and it is not formylated; eIF-2 is complexed to GTP, and its role is analogous to that of IF-2 in prokaryotes (binding and escorting the Met-tRNA to the small subunit). This entire preinitiation complex assembles at the 5′ cap and begins scanning; that is, it moves down the 5′-UTR until it encounters the start codon. When this occurs, eIF-2 hydrolyzes the GTP that it was bound to and dissociates. In turn, this dissociation signals the departure of the other initiation factors and the binding of the large subunit (the 60S in eukaryotes). As in prokaryotes, the initial tRNA is bound in the P site, whereas all subsequent tRNAs bind in the A site.

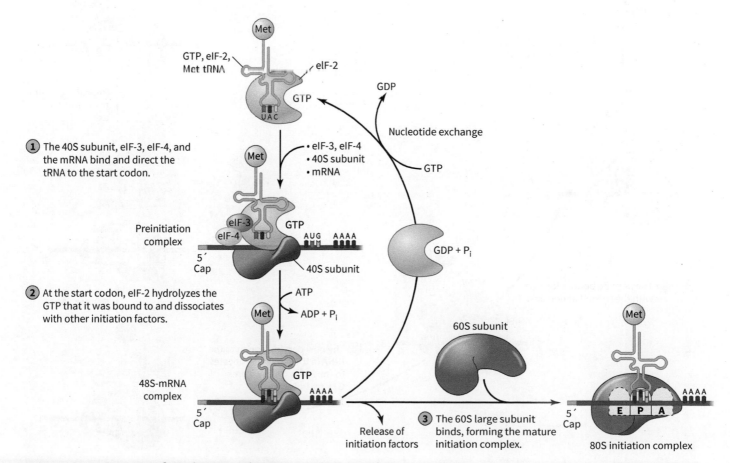

FIGURE 16.12 Initiation of translation in eukaryotes. Initiation in eukaryotes is analogous to initiation in prokaryotes. The 40S subunit forms a complex with eIF-1 and eIF-3. Bound to GTP, eIF-2 escorts the initiator tRNA (Met) to the 40S subunit. The mRNA binds along with the subunits of eIF-2, eIF-3, and eIF-4. At this point, the mRNA is scanned until the AUG is found and recognized by the 43S complex. Finally, the 60S subunit binds, displacing the other eIFs and forming the mature initiation complex.

(Source: Allison, *Fundamental Molecular Biology*, 2e, copyright 2012, John Wiley & Sons. This material is reproduced with permission of John Wiley & Sons, Inc.)

16.3.3 Elongation is the addition of more amino acids to the growing polypeptide chain

Elongation begins with the formation of the first peptide bond (**Figure 16.13**). The aminoacyl-tRNA for the second amino acid in the growing protein binds in the A site.

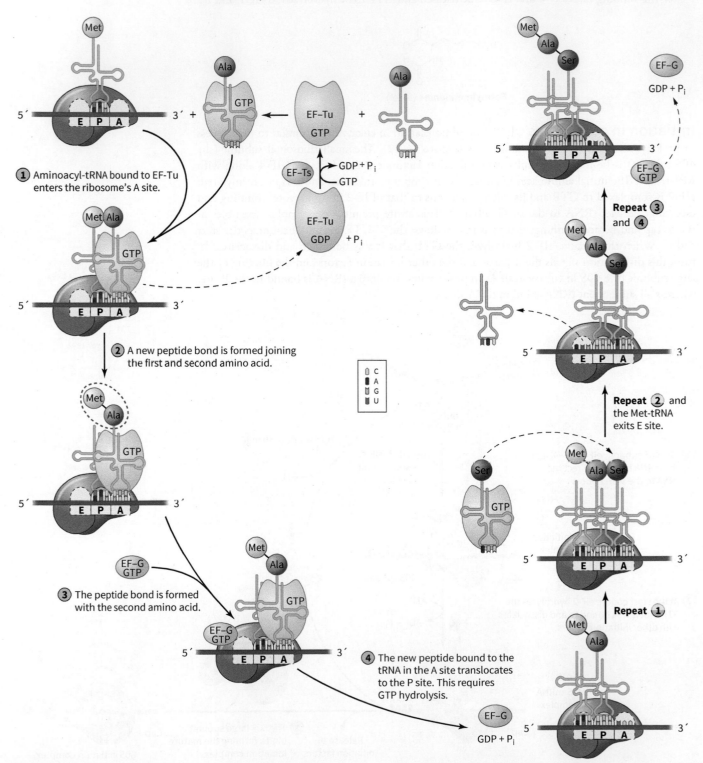

FIGURE 16.13 Elongation in eukaryotes. Aminoacyl-tRNAs are escorted to the A site of the ribosome by EF-Tu bound to GTP. GTP is hydrolyzed, and EF-Tu departs. The new peptide bond is formed between the new amino acid and the peptide, resulting in the peptide being transferred to the tRNA bound in the A site. EF-G in the GTP-bound state now associates with the ribosome. GTP hydrolysis results in translocation of the ribosome down the mRNA. Release of EF-G, GDP, and P_i leaves the A site empty and ready to accept a new aminoacyl-tRNA.

(Source: Snustad; Simmons, *Principles of Genetics*, 6e, copyright 2012, John Wiley & Sons. This material is reproduced with permission of John Wiley & Sons, Inc.)

The carboxyl carbon of the methionine is attacked by the amino nitrogen of the second amino acid, releasing the methionine from its tRNA and forming a dipeptide between methionine and the second amino acid. This new peptide is still joined to the second tRNA through the ester linkage of the second amino acid's carboxyl group to the tRNA.

Following the formation of the first peptide bond, the ribosome moves over by one codon. This puts the unoccupied f-Met-tRNA in the E site and the tRNA that carried the second amino acid (and now bears the dipeptide) in the P site. This movement leaves the A site unoccupied, and it requires the hydrolysis of GTP: by EF-G in prokaryotes or by the analogous eEF-2 in eukaryotes.

Next, a new aminoacyl-tRNA binds in the A site. This binding is assisted by an **elongation factor**, a protein that assists in the elongation process. In this case, the elongation factor is EF-Tu, a GTPase that binds to aminoacyl-tRNAs in the cytosol and directs their binding to the A site of the ribosome. If the binding of the codon and anticodon is correct, the ribosome will change conformation slightly. This change activates the GTPase activity of EF-Tu, which in turn causes the release of the EF-Tu-GDP complex and P_i. EF-Tu can exchange GDP for GTP in the cytosol before binding a new aminoacyl-tRNA. As discussed in other chapters (for example, in cell signaling and nuclear export), this exchange is catalyzed by a guanine nucleotide exchange factor; in this instance, the exchange factor is EF-Ts.

The selection of the new tRNA is directed by the base pairing of the codon of the mRNA and the anticodon of the tRNA. As with the formation of the first peptide bond, the 23S rRNA positions the growing peptide adjacent to the new aminoacyl-tRNA. The new peptide bond is then formed by the attack of the free amino group of the new amino acid on the ester linkage between the tRNA in the P site and the growing peptide, increasing the size of the peptide by one amino acid and leaving a free tRNA. Elongation factor G (EF-G) catalyzes the ratcheting of the ribosome down by one codon, to place the unoccupied tRNA in the E site and the growing peptide in the P site.

Section 16.3.1 noted that the synthesis of proteins proceeds with high fidelity due to the high degree of specificity employed by the aminoacyl-tRNA synthetase and the proofreading capacity of that enzyme. There are two other places in protein synthesis where a system of checks enables the proofreading of the final protein product. One place is the binding and incorporation of the aminoacyl-tRNA, and the other is a final quality control step that the ribosome employs following peptide bond synthesis.

In addition to facilitating the binding of the aminoacyl-tRNA to the A site, EF-Tu is important in matchmaking between the codon of the mRNA and the anticodon of the tRNA. This is thought to happen on three different levels. First, through an unknown mechanism, the binding of EF-Tu to incorrectly coupled aminoacyl-tRNA pairs occurs at a lower than predicted rate. Second, EF-Tu binding gives protein synthesis a brief pause while anticodon recognition goes on. If the binding is incorrect, this pause gives the aminoacyl-tRNA time to diffuse away from the A site; conversely, if the binding is correct, the GTPase activity of EF-Tu causes hydrolysis of GTP. There is a second pause while the EF-Tu-GDP complex and P_i diffuse away. This second pause also provides time for an incorrectly paired aminoacyl-tRNA to diffuse out of the A site.

The third place where proofreading occurs is after the formation of the peptide bond. Codon–anticodon mismatch in the P site can lead to termination by one of two methods. First, binding of release factors RF-2 and RF-3, typically encountered when the ribosome reaches a stop codon, causes dissociation of the ribosomal subunits from the mRNA or the detection of a mismatch in the P site. Second, the detection of a mismatch in the P site causes an alteration in the ribosomal structure such that there is a loss of specificity and increased potential for incorrect recognition of the new aminoacyl-tRNA binding in the A site. This in turn amplifies errors and leads to a greater chance of premature termination by binding RF-2 and RF-3.

In summary, errors can be detected at three places in translation: at the level of the aminoacyl-tRNA synthetase, at the point when the aminoacyl-tRNA-EF-Tu-GTP complex binds to the ribosome, and after peptide bond formation.

The process of elongation continues, adding and proofreading one amino acid at a time until a stop codon is reached. The addition of new amino acids is a much slower process than the synthesis of either DNA or RNA. This is in part due to the increased complexity of having to select the correct amino acid out of 20 different options instead of four different nucleotides. Protein synthesis in *Escherichia coli* (*E. coli*) typically proceeds at the rate of about 20 amino acids per second, compared to 750 nucleotides per second during replication. The process is even slower in metazoans, which incorporate only four to five amino acids per second.

16.3.4 The final step in protein synthesis is termination of translation

Binding of the stop codon in the A site signals the termination of translation. Rather than a tRNA binding, a protein termed a **release factor** binds instead (**Figure 16.14**). In prokaryotes there are two primary release factors: RF-1 (which recognizes UAG) and RF-2 (which recognizes

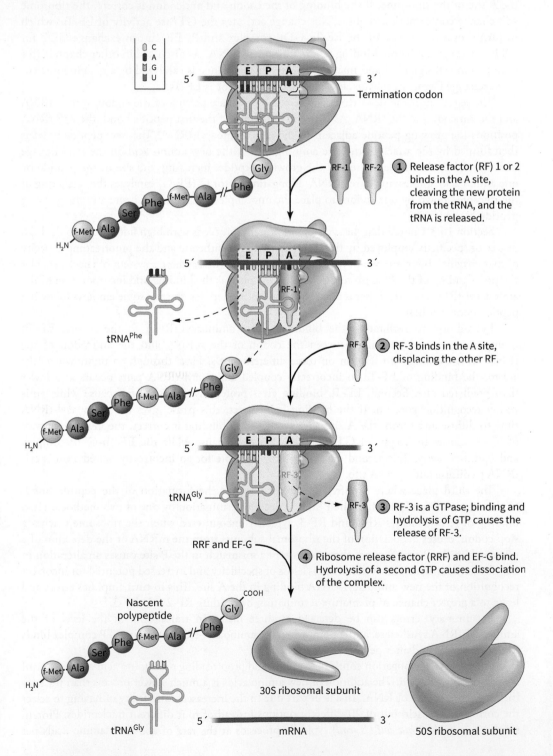

FIGURE 16.14 Termination of translation in prokaryotes. When a stop codon is encountered, one of the release factors, RF-1 or RF-2, binds. These proteins recruit release factor 3 (RF-3) bound to GDP, which in turn causes release of the new protein. Exchange of the GDP for GTP causes release of RF-1 and RF-2. Hydrolysis of GTP releases RF-3 from the ribosome. Next, RRF and EF-G (bound to GTP) bind. Hydrolysis of GTP causes dissociation of the ribosomal subunits. Finally, IF-3 binds, causing dissociation of the remaining tRNA.

(Source: Snustad; Simmons, *Principles of Genetics*, 6e, copyright 2012, John Wiley & Sons. This material is reproduced with permission of John Wiley & Sons, Inc.)

1 Release factor (RF) 1 or 2 binds in the A site, cleaving the new protein from the tRNA, and the tRNA is released.

2 RF-3 binds in the A site, displacing the other RF.

3 RF-3 is a GTPase; binding and hydrolysis of GTP causes the release of RF-3.

4 Ribosome release factor (RRF) and EF-G bind. Hydrolysis of a second GTP causes dissociation of the complex.

UGA). Both proteins recognize UAA. Binding of either of these factors causes cleavage of the ester bond between the new protein and the tRNA in the P site and release of the new protein. A third release factor, RF-3, binds to the complex, displacing the other release factor. RF-3 is a GTPase; binding and hydrolysis of GTP causes the release of RF-3 but leaves the ribosome bound to the mRNA. Next, another GTPase termed ribosome release factor (RRF) binds in conjunction with EF-G. Hydrolysis of GTP causes the dissociation of the complex.

Termination in eukaryotes is thought to proceed through a similar pathway, but eukaryotes have only a single release factor (eRF-1) that recognizes all three stop codons.

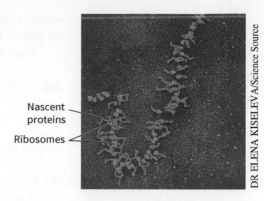

FIGURE 16.15 Polysomes. Polysomes consist of multiple ribosomes translating the same message at once.

16.3.5 Proteins may be made on free polysomes or translated into a membrane

Protein synthesis can be broadly classified into one of two different categories. Some 10% of proteins are synthesized in the cytosol and largely reside in the cytosol or have an amino acid sequence that targets them to some other organelle (e.g., a nuclear localization sequence). These proteins are synthesized by multiple ribosomes on a single free mRNA in the cytosol, an assembly referred to as a **polysome** (Figure 16.15).

Proteins found imbedded in membranes, inside the lumen of most organelles, or secreted from the organism are produced in the rough endoplasmic reticulum (ER). The ER carries out many jobs in the cell and can be divided into rough (bound to ribosomes) or smooth (without ribosomes). The synthesis of the protein proceeds largely as has already been described in this section but with one significant difference. Proteins that are targeted to the rough ER have, at or near their amino terminus, a specific amino acid sequence termed the **signal peptide**, a largely hydrophobic, alpha-helical region, 5 to 30 amino acids long (Figure 16.16). This sequence is recognized and bound as the nascent protein is being synthesized by a small protein termed the **signal recognition particle (SRP)**. The SRP–peptide complex binds to a second protein, the SRP receptor, which is a transmembrane protein pore found in the rough ER. The SRP threads the signal peptide

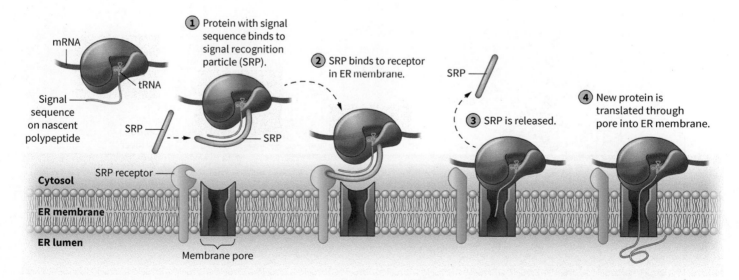

FIGURE 16.16 Peptide synthesis into ER.   Proteins that are to be translated into the ER contain an amino terminal signal peptide between 5 and 30 amino acids in length. Following synthesis of the signal peptide, the ribosome pauses, allowing for the signal recognition particle (SRP) to bind to the signal peptide. The SRP–signal peptide complex is recognized and bound by the SRP receptor in the ER membrane. Protein complexes, such as Sec 61, aid in the translation of the protein into the membrane. The signal peptide is typically removed by a protease, such as signal peptidase.

(Source: Karp, *Cell and Molecular Biology: Concepts and Experiments*, 7e, copyright 2013, John Wiley & Sons. This material is reproduced with permission of John Wiley & Sons, Inc.)

sequence through the SRP receptor, and the nascent protein is translated through the pore. In the ER lumen, the signal peptide is recognized and cleaved by an enzyme termed signal peptidase.

The trafficking of proteins to different organelles is covered in greater detail in Chapter 22.

16.3.6 Nascent proteins fold and undergo processing

As proteins emerge from the ribosome, they fold almost exclusively into a single stable conformation. About 90% of all proteins do this without any assistance; the remainder requires some type of protein, termed a **chaperone**, to bind and stabilize some domains as the protein folds.

Chaperones are a structurally diverse group of functionally related proteins that recognize and bind to denatured proteins, preventing their aggregation and facilitating folding into their correct conformation. One example of a chaperone is the *E. coli* protein GroEL (**Figure 16.17**); the eukaryotic analog of this protein is heat shock protein 60. GroEL, a tetradecamer, works in conjunction with another protein, GroES, the eukaryotic analog of which is Hsp10. Together, GroEL and GroES form a cylindrical chamber consisting of two rings of seven subunits each of GroEL, capped by an assembly of six to eight molecules of GroES.

Denatured proteins are bound by hydrophobic patches on the inside of the cylindrical GroEL complex. Each of the subunits of GroEL has an ATP-binding site. Binding of ATP induces a structural change that has several ramifications. First, the change in structure reorients the GroEL monomers to hide the hydrophobic patch and expose a now totally hydrophilic chamber. Second, this new structure recruits the GroES lid, which traps the denatured protein inside. Third, the hydrophilic nature of the chamber promotes burying of hydrophobic residues in the core of the folding protein, essentially isolating these structures and promoting the hydrophobic effect.

The process of protein folding for the 90% of proteins that do not require chaperones is spontaneous and does not require any additional expenditure of energy. The weak forces discussed in Chapter 3 all contribute to the formation of the final structure.

As discussed in Chapters 3, 5, and 8, many proteins undergo **post-translational modification**, for example, through the removal of signal peptides (as mentioned previously), phosphorylation, glycosylation, acetylation, prenylation, or farnesylation. Protein folding will be discussed in greater detail in Chapters 18 and 21.

In this section we have discussed protein synthesis. Shorter fragments (peptides) may be produced by cleaving them off a larger protein. Other peptides may be made enzymatically. One such peptide is discussed in **Societal and Ethical Biochemistry: Harmful algal blooms**.

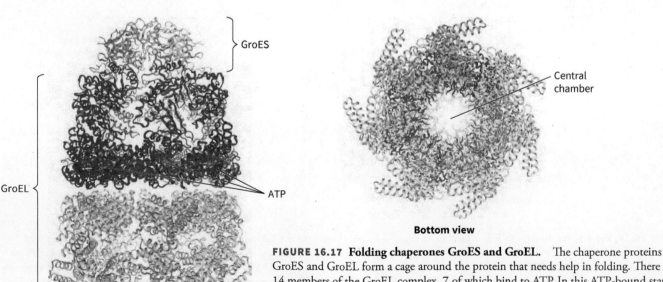

Side view

Bottom view

FIGURE 16.17 Folding chaperones GroES and GroEL. The chaperone proteins GroES and GroEL form a cage around the protein that needs help in folding. There are 14 members of the GroEL complex, 7 of which bind to ATP. In this ATP-bound state, the chamber they form is hydrophobic; however, upon ATP hydrolysis, the structure changes, becoming hydrophilic.

(Source: Data from PDB ID 3ZQ1 Chen, D.-H., Madan, D., Weaver, J., Lin, Z., Schroder, G. F., Chiu, W., Rye, H. S. (2013) Visualizing GroEL/ES in the Act of Encapsulating a Folding Protein. *Cell* 153: 1354)

Societal and Ethical Biochemistry

Harmful algal blooms

Imagine having to live for three days without running water. In August 2014, the city of Toledo, Ohio (population 500,000), took the unprecedented precaution of telling residents that they could not drink or otherwise use the municipal water supply. Toledo's water comes from Lake Erie, and in the days leading up to the ban,

Microcystis aeruginosa growing in lake water.

Satellite image of algae growing along the edges of the lake.

there had been a bloom of algae (a harmful algal bloom [HAB]), leaving the lake the color and clarity of pea soup. Residents were cautioned that drinking the water could be fatal and even boiling the water would not make it safe.

When conditions are favorable, algae can grow almost out of control, producing an algal bloom. As with any cell, algae need nutrients. In the case of the Toledo bloom, the nutrients were provided by agricultural runoff (nitrogenous wastes from farming) and phosphorous runoff (from farms, lawns, and golf courses) entering the water. As temperatures climb in the summer months, algae grow and can persist until the weather cools or other conditions change. Climate change accelerates these blooms. HABs are commonly found in lakes and harbors but can be found in almost any body of water.

The algae themselves are relatively easy to remove from the water supply by precipitation and filtration and are not by themselves toxic. However, they secrete small molecules that inhibit the pathways of other organisms in their niche (their competitors or predators), in effect poisoning these neighbors. In the Toledo algae bloom, the culprit was *Microcystis aeruginosa*, a species of cyanobacteria.

The toxin secreted by *Microcystis* is microcystin, a cyclic peptide. Several of the amino acids in microcystin and the peptide linkages that hold the molecule together will be familiar, but some of the amino acids are irregular, for example, dehydroalanine or the aromatic 3-amino-9-methoxy-2,6,8-trimethyl-10-phenyldeca-4,6-dienoic acid (ADDA). Microcystins are persistent toxins, and a quick examination of the structure (see the

Microcystin

figure) indicates why this is the case. In water contaminated with this molecule, the peptide backbone will not be cleaved by simply boiling the water; instead, the peptide is typically filtered out using activated charcoal. Biologically, microcystins are transported to the liver, where they accumulate and act as protein phosphatase 1 and 2 inhibitors. This leads to hepatotoxicity (liver toxicity) and can lead to liver failure and death. The current accepted guideline for levels of this molecule in water is 1 µg/L (1 part per billion [ppb]).

The contamination that led to the 2014 Toledo water ban is an example of a modern scientific question that can only be resolved through input from numerous experts in diverse fields (environmental science; meteorology; hydrology; agricultural science; toxicology; microbiology; medicine; and, of course, biochemistry).

16.3.7 Inhibitors of protein biosynthesis may be lifesaving drugs or deadly toxins

Chapter 7 discussed the beta lactam class of antibiotics; these drugs, which include penicillin, work by inhibiting key steps in bacterial cell wall biosynthesis. Chapter 11 discussed sulfa drugs and methotrexate, which are folate antimetabolites—molecules that inhibit folate metabolic pathways and also act as antibiotics or chemotherapeutic agents. Antibiotics in a third class exploit the differences between the prokaryotic and the eukaryotic ribosome and inhibit protein biosynthesis in bacteria. Members of this class are generically referred to with the suffix "-mycin," and they include several structurally different groups of molecules. These molecules can also be broadly divided according to whether they bind to the 30S subunit, the 50S subunit, or both (**Figure 16.18**).

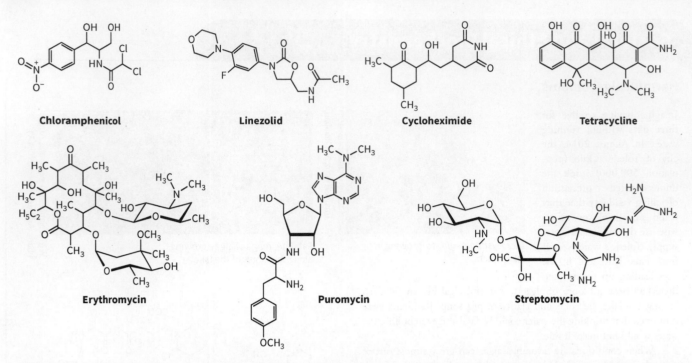

FIGURE 16.18 Examples of protein synthesis inhibitors. Shown are representative drugs that inhibit protein synthesis. Chloramphenicol, linezolid, and erythromycin all bind to the 50S subunit. Tetracycline, puromycin, and streptomycin all bind to the 30S subunit. Cycloheximide binds to the ribosomal complex of either eukaryotes or prokaryotes.

Drugs that bind the 30S subunit

Aminoglycosides (including streptomycin, kanamycin, gentamycin, and neomycin) consist of two carbohydrate rings joined to a central cyclohexane ring. Each ring has numerous amino modifications, which often include the substitution of an amino group for a hydroxyl group. These modifications result in different interactions (e.g., hydrogen bonding) between the drug and ribosome and can account for the specificity or effectiveness of the drug.

Aminoglycosides inhibit protein synthesis by binding to the 16S rRNA found in the small subunit of the ribosome, causing misreading of the code and inhibition of initiation and translocation. In addition, these molecules appear to inhibit translational proofreading, also leading to premature termination.

Like many other members of the -mycin class, aminoglycosides are produced naturally by bacteria from the genus *Streptomycetes* or the genus *Micromonospora*. *Streptomyces* is a common soil bacterium that should not be confused with *Streptococcus*, the human pathogen that causes strep throat, among other diseases. Over 70% of all antibiotics have been isolated from this genus.

Tetracycline and doxycycline are members of the **tetracycline** class of antibiotics, and as that name suggests, they contain four rings, each of which has six members. Like the aminoglycosides, tetracyclines bind to the 16S RNA in the 30S subunit and compete in the A site for the binding of aminoacyl-tRNAs, effectively blocking binding of tRNA.

Drugs that bind the 50S subunit

Originally isolated from a strain of *Streptomyces*, erythromicin and its derivatives (e.g., azithromycin and clarithromycin) are **macrolide** antibiotics. Molecules in this family are comprised of a macrocyclic lactone (a large ring linked by an ether), typically decorated with several carbohydrate groups. Macrolides share the -mycin name and similarly inhibit protein biosynthesis, but these are a separate class of drugs, used to treat different infections and are metabolized differently. Other members of the macrolide family vary structurally in terms of substituents attached to the ring; this variation affects the drugs' effectiveness against pathogens and how they are metabolized.

Erythromycin and its derivatives act by binding to the 50S subunit of the bacterial ribosome and blocking translocation sterically. The tRNA that is unable to translocate to the P site is bound to a peptide, resides in the A site, and is not released for subsequent binding of a new tRNA. With the protein unable to elongate, protein synthesis terminates, and the organism is severely compromised, although it is debated whether these antibiotics are bactericidal (able to kill bacteria) or only bacteriostatic (inhibiting bacterial growth). By preventing the pathogenic bacteria from making proteins, these antibiotics enable the host's immune system to fight off the infection and destroy the foreign cells.

Chloramphenicol is a small organic molecule that is also derived from *Streptomyces*. It too interferes with substrate binding, in this case by binding to the A2451 and A2452 residues in the 23S rRNA of the 50S ribosomal subunit and thereby preventing peptide bond formation.

Oxazolidones are the most recent addition to the antibiotic arsenal and are the only molecules discussed in this section that are not naturally occurring. All members of this class of drugs have a common structural feature (the 2-oxoazolidone ring) and share the common suffix "-zolid" (e.g., linezolid). Oxazolidones work by binding the 23S tRNA of the 50S subunit and sterically blocking formation of the initiation complex with the 30S subunit. Several members of this class have been approved, and several more are in the pipeline. Oxazolidones may cause significant side effects, including bone marrow suppression, nerve damage, and lactic acidosis. Nonetheless, they are one of the few compounds that can be used to treat infections such as methicillin-resistant *Staphylococcus aureus* (MRSA).

Several other classes of drugs act by inhibiting protein synthesis through other mechanisms. For example, puromycin is an aminonucleoside, in effect a tyrosine tRNA analog consisting of a modified adenosine nucleoside linked to a methoxy tyrosine via a nonconventional amide at the 2′ position of the ribose ring. This molecule also binds in the A site. Attack of the drug's peptide group causes coupling of puromycin to the nascent peptide and premature chain termination. The exact mechanism of coupling is still unclear, but the presence of the amide linkage at the 3′ position (instead of the ester linkage found in aminoacyl-tRNAs) may offer extra stability and resistance to hydrolysis that prevent removal of the modification.

A final drug worthy of mention is **cycloheximide**. This drug blocks translocation and therefore elongation, but it works only in eukaryotic cells, not prokaryotic cells, so it is ineffective against bacterial infections. Although cycloheximide was investigated as an antifungal and chemotherapeutic agent, the drug is also a mutagen and teratogen and has been abandoned as a treatment for disease. It is, however, a useful reagent in the laboratory when there is a need to block protein synthesis.

Ribosome-inactivating proteins

Ricin (discussed in Chapter 3), abrin, and viscumin are lectins; that is, they are carbohydrate-binding proteins found in certain plants. The sources of these three lectins are castor beans, rosary pea, and mistletoe, respectively (**Figure 16.19**). Lectins gain access to cells via their carbohydrate-binding property, but they also possess enzymatic activity in that they are ribosome-inactivating proteins (RIPs) and endoglycosylases, which depurinate a key adenine base (A4324) in the 28S RNA of the 60S subunit of the ribosome.

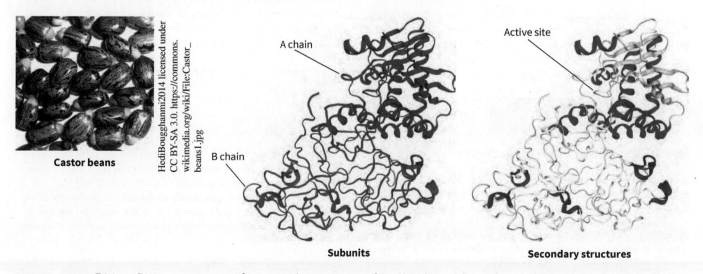

HediBougghanmi2014 licensed under CC BY-SA 3.0. https://commons.wikimedia.org/wiki/File:Castor_beans1.jpg

Castor beans

A chain

B chain

Subunits

Active site

Secondary structures

FIGURE 16.19 Ricin. Ricin, a protein toxin from castor beans, consists of A and B chains. The catalytic A chain removes the adenine base, thereby inactivating the ribosome. The A chain has both alpha helices and beta sheets, but the B chain is almost devoid of these structures.

(Source: Data from PDB ID 2AAI Rutenber, E., Katzin, B. J., Ernst, S., Collins, E. J., Mlsna, D., Ready, M. P., Robertus, J. D. (1991) Crystallographic refinement of ricin to 2.5 A. *Proteins* 10: 240–250)

Removal of this adenine permanently inactivates the ribosome, leaving the cell (and eventually the organism) unable to make proteins. As a result, these lectins are potent poisons. The lethal dose for abrin is about 0.1 μg/kg, which is about 70 μg for the average adult, an amount smaller than a few grains of sand.

Drugs have traditionally been small organic compounds, but newer generations of drugs are proteins, hormones, or even genes. These drugs are called biologics and are discussed in **Medical Biochemistry: Biologics: peptide- and protein-based drugs**.

Medical Biochemistry

Biologics: peptide- and protein-based drugs

A century ago, most drugs were natural products isolated from plant extracts, for example, opiates or quinine. As chemistry advanced, it became possible to further characterize and modify these compounds. Scientists generated morphine from the opium poppy and aspirin from willow bark. Advances in microbiology in the early part of the twentieth century led to the isolation of molecules from microbes, including penicillin and erythromycin. Again, these molecules were typically obtained from a natural source, but they could be produced in mass quantities thanks to advances in science. Advances in molecular biology and biochemistry since the 1980s have led to the production of biologics—biological molecules (typically proteins) that are used as pharmacological agents. Examples of biologics are insulin, hormones, recombinant vaccines, and monoclonal antibody treatments.

In 1921, insulin was isolated by Frederick Banting in the laboratory of Henry Macleod; this was the first peptide hormone to be identified and the first protein that could be used to treat a disease—diabetes. For decades, insulin was isolated from the pancreas of pigs, but this was not a perfect solution. For example, some people objected to the use of porcine insulin on religious or personal grounds; also, some people had allergic reactions to this type of insulin and instead had to use insulin obtained from other sources, such as dogs or horses. In the 1980s, insulin was cloned and expressed in mammalian cells. Currently, most insulin that is used medicinally is recombinant human insulin or mutated forms of human insulin.

Other biologics are peptide-based hormones. Recombinant erythropoietin (EPO alfa) is used to treat anemias associated with some chemotherapeutic treatments, and human growth hormone (marketed as Nutropin or several other names) is used to treat muscle wasting and some congenital deficiencies.

Several vaccines are now made by expressing components of a virus, such as the coat proteins. An example of this is the vaccine against hepatitis C. Such vaccines are advantageous because there are no potentially infectious virions in the vaccine.

Monoclonal antibodies are antibodies generated in the laboratory from the immortalization of a single B cell. Because they are derived from a single cell, the antibodies produced are all the same (clonally derived) and bind a single location in the target protein. These antibodies can be generated to target molecules in humans that block what might otherwise be normal function in healthy tissue. Examples of these drugs include the breast cancer treatment trastuzumab (Herceptin), the rheumatoid arthritis treatment etanercept (Enbrel), and the psoriasis treatment ustekinumab (Stelara). ZMapp is an experimental monoclonal antibody treatment for Ebola virus disease.

Biologics have made incredible advances in medicine, but they also present some challenges compared to routine pharmaceuticals. Protein-based drugs are not as shelf stable as small organics and also often have a shorter biological half-life. Most cannot be taken orally or transdermally and so must be injected. They can also be expensive to make; for example, monoclonal antibodies need to be synthesized in cultured mammalian cells. Because these molecules are often recombinant proteins or monoclonal antibodies, differences can arise in the processing and glycosylation of these molecules, which can affect activity. Also, due to their complex structure and nature of synthesis, the exact structure and purity of the molecule may not be completely reproducible, leading to variability between generic and brand-name compounds. Despite these drawbacks, these drugs and others like them can be lifesavers.

Worked Problem 16.3   What does it cost?

How much energy does it take to make a protein?

Strategy   Review the process of translation. Which steps require energy in terms of nucleotide triphosphates? How many phosphate equivalents are needed per amino acid of protein synthesized?

Solution   Four phosphate equivalents are needed for each amino acid, plus one to start the process. One molecule of GTP is hydrolyzed in the formation of the initiation complex. The energetic equivalent of two molecules of ATP is required to activate an amino acid (in the activation reaction, ATP is used to make an aminoacyl-AMP and PP_i). When an aminoacyl-tRNA is marshalled to the A site in the ribosome by EF-Tu, a molecule of GTP is hydrolyzed to GDP and P_i; in the translocation step, EF-2 (also known as EF-G or eEF-2) hydrolyzes a single molecule of GTP to GDP and P_i.

This process is thus quite costly to the cell. The synthesis of an average-sized protein (450 amino acids) is the energetic equivalent of 1801 ATP; nevertheless, the cell must synthesize these complex molecules.

Follow-up question   Propose a mechanism of action for puromycin. How might the structure of puromycin contribute to its antibiotic properties?

Summary

- In activation, the synthesis of proteins begins with the coupling of amino acids to the correct tRNA by aminoacyl-tRNA synthases. Amino acids are first energetically activated by coupling with ATP to form an aminoacyl-AMP. The aminoacyl-AMP is then joined to the correct tRNA through an ester linkage with the carboxyl group of the amino acid. Aminoacyl-tRNA synthetases have tight substrate specificity and a proofreading function to ensure that the correct amino acid and tRNA have been joined.

- In initiation, the halves of the ribosome join around the mRNA, assisted by initiation factors. The first codon (AUG) is bound by the methionine tRNA in the P site of the ribosome.

- In elongation, the rest of the amino acids are added one at a time, assisted by elongation factors. Aminoacyl-tRNAs bind in the A site, the growing peptide is transferred to the new amino acid, and the ribosome ratchets down the mRNA. As amino acids are added, the ribosome has a proofreading function that gives one final opportunity for the protein to be checked before it is released.

- In termination, the translation of the message ends at the stop codon. Release factors bind to the ribosome and separate the ribosomal subunits.

- Proteins may be synthesized in the cytosol on a structure termed a polysome, or they may be associated with a membrane or the ER. A signal peptide in the nascent protein binds to the signal recognition particle and targets proteins to the ER.

- Some 90% of proteins fold on their own; the remainder require a chaperone, such as the GroEL/GroES complex, for correct folding. Following folding, proteins may be post-translationally modified.

- Many molecules inhibit the synthesis of proteins; some of these are natural in origin and act as a defense mechanism for the organism that produces them. Scientists have isolated many of these inhibitors for use as antibiotics or as tools in basic research.

Concept Check

1. Give an overview of the four major steps of protein synthesis. Explain what happens in each step.

2. Discuss the mechanisms that contribute to high fidelity in protein synthesis.

3. Describe how proteins can be inserted into membranes or the ER.

4. Describe how different molecules can inhibit the individual steps of protein biosynthesis and why they make effective drugs.

Bioinformatics Exercises

Exercise 1 Viewing and Analyzing Transfer RNA

Exercise 2 Ribosomes, Protein Processing, and the KEGG Database

Problems

16.1 The Genetic Code

1. If there were five different bases in mRNA instead of four, how many codons would be possible, assuming that there were still three nucleotides in a codon?

2. If the assignments of the 20 amino acids to the 64 codons of the genetic code were random instead of there being one genetic code that has been conserved throughout evolution, how many different combinations would there be?

3. The human genome has 3.3 billion bases, of which 3% codes for protein. If a codon were four letters instead of three, what percentage of the genome would have to code to generate the amount of protein?

4. The genetic code is not completely conserved but varies slightly in some organisms. What do these differences say about the evolution of those species (e.g., the Archaea versus metazoans)?

5. One type of sickle cell anemia is caused by a single point mutation in the hemoglobin gene, in which a glutamic acid is replaced by valine in the final protein. Identify the codons might result in this disease.

16.2 Machinery Involved in Protein Biosynthesis

6. How do the modifications made to eukaryotic mRNA during its synthesis feature in the function of the molecule in translation?

7. How do the Shine–Dalgarno and Kozak sequences differ in terms of sequence versus functionality?

8. Describe the general three-dimensional structure of a tRNA.

9. What sorts of modifications are made to the bases of tRNA? How do these alterations change the secondary and tertiary structure of the molecule?

10. Examine the structure of a ribosome using a structure viewer. Which rRNA molecules are found in each subunit? What are the rough

shapes of each? How are they oriented with respect to one another in the mature ribosome?

11. Why does the aminoacyl-tRNA synthetase for isoleucine rarely incorporate valine despite the similar structure of these two amino acids?

16.3 Mechanism of Protein Biosynthesis

12. List out the events of protein synthesis in correct order.

13. The aminoacyl-AMP intermediate is a stable intermediate in the coupling reaction of amino acids to tRNA. Based on this data alone, draw a Cleland diagram to model this reaction.

14. In prokaryotes, translation begins as soon as transcription is initiated and the 5′ end of the message is exposed. Why can't this occur in eukaryotes?

15. Describe the preparation of fMet-tRNA for translation in prokaryotes. Translate the following mRNA into a peptide.

5′-CAUGCCAAGGGCGGCGCGCCGAUGA-3′

16. Discuss current theories of how proteins fold.

17. Draw a model of translation using images rather than words. How is this model similar to or different from the illustrations seen in this chapter or online? How would you know if your model is drawn to scale?

18. Examine the structure of streptomycin and the stereochemistry of the rings. Can you identify these molecules? What is the likely source of the amino groups?

19. Classify these antibiotic drugs based on their structure. Identify salient functional groups in each.

I.

II.

III.

20. Examine the structures of the antibiotics shown in Problem 19. Based on what you have seen previously in this text, which biochemical building blocks were likely used in the synthesis of these molecules?

21. Many antibiotics are modified versions of a common skeleton. Examine the structure of the macrolide family of antibiotics. How do you think these modifications were made? What were the sources of these atoms (i.e., what donor groups did they come from)?

22. Peptides are synthesized from the amino or N-terminus to the carboxy or C-terminus. Explain this in molecular detail.

23. Chloramphenicol interferes with substrate binding, thereby preventing peptide bond formation. Chloramphenicol is available as a palmitate ester for oral use and as succinyl ester for intravenous use. Propose reasons why might the compound be supplied in these two forms.

24. Structure-based drug design is a method of designing molecules based on making minor structural modifications to a common skeleton compound. Nature has used a similar approach in the alterations made to some families of molecules. Examine the members of the macrolide class of antibiotics. What is the core structure?

25. Linezolid can cause lactic acidosis as a result of suppressed mitochondrial function. Explain how this might happen.

26. Puromycin is a reversible inhibitor of both serine peptidases and metalloproteases. Examine the structure of puromycin, and propose a mechanism through which puromycin could inhibit these enzymes.

Puromycin

27. Examine the structure of puromycin given in Problem 26.

   a. Identify parts of the structure that are similar to molecules we have already studied?

   b. How is this structure similar to tRNA?

Cycloheximide

   c. What might be the source of these molecules?

28. Examine the structure of cycloheximide. Cycloheximide breaks down in alkaline solutions. Where is it cleaved? Propose a mechanism for the hydrolysis.

29. Examine the structure of cycloheximide. Why is this drug named cycloheximide and not cyclohexamide?

30. A 1.4-mg dose of ricin is considered lethal for most humans. Ricin has a molecular weight of 65 kDa. How many molecules of ricin are in a lethal dose?

Data Interpretation

Questions 31 to 33 pertain to the data in **Table 16.1** and Figure 16.9. These data refer to the change in the binding energy of substrates between wild type and mutant tyrosyl-tRNA synthetase. All of the data relate the binding of the substrate (ATP and tyrosine) to the surrounding topology of the active site. $\Delta\Delta G_T$ refers to the change in the binding energy of the substrates (a measure of how tightly the substrates are bound in the active site) due to the difference seen.

TABLE 16.1 Effect of mutation on substrate binding

Wild type	Mutant	Substrate	$\Delta\Delta G_T$ kcal/mol
Phe34	Tyr34	Tyr	0.52
Gly35	Cys35	ATP	1.14
Ala51	Cys51	ATP	0.47
Gly48	Asn48	ATP	0.77
Gly48	His48	ATP	0.96
Phe169	Tyr169	Tyr	3.72
Glu195	Gln195	Tyr	4.49
Gly35	Ser35	ATP	−0.04
Ala51	Thr51	ATP	−0.44

31. Using the structure from Figure 16.9 as a guide and the data from Table 16.1, reconstruct the active site for the proteins in Table 16.1.

32. Which data in Table 16.1 stand out as anomalous to you? Why?

33. Which mutations are these? Would other mutations cause as much of a change in $\Delta G^\ddagger$?

Experimental Design

34. What are some of the challenges that you might face when searching for tRNA genes in a genome the size of the human genome?

35. What techniques could you use to determine the structure of a tRNA?

36. Would it be possible to engineer an aminoacyl-tRNA synthetase to couple a different amino acid to a tRNA? What considerations would you need to make in engineering that enzyme? What would you start with, and how might you change it?

37. Design an experiment to test the hypothesis that eIF-2 associates with aminoacyl-tRNAs only in the GTP-bound state.

Ethics and Social Responsibility

38. Should antibiotics receive patent protection? How are considerations different for patent protections on a lifestyle drug (e.g., a hair loss treatment)?

39. Pharmaceutical companies are frequently accused of being more interested in developing drugs for chronic conditions (e.g., diabetes) instead of drugs like antibiotics. Is this a fair accusation? Examine the issues surrounding the development of new drugs. Should these companies be obligated to develop new antibiotics?

40. Several movies and television programs have featured ricin poisoning, promoting awareness of this method of poisoning to the general public. Is this ethical? Would it be ethical to instruct someone in how to purify ricin? Explain your reasoning.

Suggested Readings

16.1 The Genetic Code

Agris, P. F. "Decoding the Genome: A Modified View." *Nucleic Acids Research* 32, no. 1 (2004): 223–238. doi: 10.1093/nar/gkh185.

Crick, F. H. C., L. Barnett, S. Brenner, and R. J. Watts-Tobin. "General Nature of the Genetic Code for Proteins." *Nature* 192 (1961): 1227–1232. doi:10.1038/1921227a0.

Gaston, M. A., R. Jiang, and J. A. Krzycki. "Functional Context, Biosynthesis, and Genetic Encoding of Pyrrolysine." *Current Opinion in Microbiology* 14, no. 3 (June 2011): 342–349. doi: 10.1016/j.mib.2011.04.001.

Khorana, H. G. "Polynucleotide Synthesis and the Genetic Code." *Federation Proceedings* 24, no. 6 (November–December 1965): 1473–1478.

Maraia, R. J., and J. R. Iben. "Different Types of Secondary Information in the Genetic Code." *RNA* 20, no. 7 (2014): 977–984. doi: 10.1261/rna.044115.113.

Nirenberg, M. "Protein Synthesis and the RNA Code." *Harvey Lectures* 59 (1965): 155–158.

Ochoa, S. "Translation of the Genetic Message." *Bulletin de la Société de chimie biologique* (Paris) 49, no. 7 (July 1967): 721–737.

Tsugita, A., and H. Fraenkel-Conrat. "The Aminoacid Composition and C-Terminal Sequence of a Chemically Evoked Mutant of TMV." *Proceedings of the National Academy of Sciences* 46 (1070): 636–642.

16.2 Machinery Involved in Protein Biosynthesis

Babendure, J. R., J. L. Babendure, J. H. Ding, and R. Y. Tsien. "Control of Mammalian Translation by mRNA Structure Near Caps." *RNA* 12, no. 5 (2006): 851–861.

Ben-Shem A., N. Garreau de Loubresse, S. Melnikov, L. Jenner, G. Yusupova, and M. Yusupov. "The Structure of the Eukaryotic Ribosome at 3.0 Å Resolution." *Science* 334, no. 6062 (February 2011): 1524–1529.

Chan, P. P., and T. M. Lowe. "GtRNAdb: A Database of Transfer RNA Genes Detected in Genomic Sequence." *Nucleic Acids Research* 37 (2009): D93–D97.

Ibba, M., and D. Söll. "Aminoacyl-tRNAs: Setting the Limits of the Genetic Code." *Genes & Development* 18 (2004): 731–738.

Klinge, S., F. Voigts-Hoffmann, M. Leibundgut, S. Arpagaus, and N. Ban "Crystal Structure of the Eukaryotic 60S Ribosomal Subunit in Complex with Initiation Factor 6." *Science* 334, no. 6058 (November 2011): 941–948.

Kutter, C., G. D. Brown, Â. Gonçalves, M. D. Wilson, S. Watt, A. Brazma, R. J. White, and D. T. Odom. "Pol III Binding in Six Mammalian Genomes Shows High Conservation among Aminoacid Isotypes, Despite Divergence in tRNA Gene Usage." *Nature Genetics* 43, no. 10 (August 28, 1011): 948–955.

Lander, E., et al. (2001). "Initial Sequencing and Analysis of the Human Genome." *Nature* 409, no. 6822 (2001): 860–921. doi:10.1038/35057062. PMID 11237011.

Lowe, T. M., and S. R. Eddy "tRNAScan-SE: A Program for Improved Detection of Transfer RNA Genes in Genomic Sequence." *Nucleic Acids Research* 25 (1997): 955–964.

Phizicky, E. M., and A. K. Hopper. "tRNA Biology Charges to the Front." *Genes & Development* 24, no. 17 (September 1, 2010): 1832–1860. doi: 10.1101/gad.1956510.

Raina, M., and M. Ibba. "tRNAs as Regulators of Biological Processes." *Frontiers in Genetics* 5 (June 11, 2014): 171. doi: 10.3389/fgene.2014.00171.

16.3 Mechanism of Protein Biosynthesis

Beissinger, M., and J. Buchner. "How Chaperones Fold Proteins." *Biological Chemistry* 379, no. 3 (March 1998): 245–259.

Fekkes, P., and A. J. M. Driessen. "Protein Targeting to the Bacterial Cytoplasmic Membrane." *Microbiology and Molecular Biology Reviews* 63, no. 1 (March 1999): 161–173.

Fersht, A. R., J.-P. Shi, J. Knill-Jones, D. M. Lowe, A. J. Wilkinson, D. M. Blow, P. Brick, P. Carter, M. M. Y. Waye, and G. Winter. "Hydrogen Bonding and Biological Specificity Analysed by Protein Engineering." *Nature* 314 (1985): 235–238.

Gingold, H., and Y. Pilpel. "Determinants of Translation Efficiency and Accuracy." *Molecular Systems Biology* 7 (April 12, 2011): 481. doi: 10.1038/msb.2011.14.

Jakubowski, H., and E. Goldman. "Editing of Errors in Selection of Aminoacids for Protein Synthesis." *Microbiology Reviews* 56, no. 3 (September 1992): 412–429.

Kirino, Y., and T. Suzuki. "Human Mitochondrial Diseases Associated with tRNA Wobble Modification Deficiency." *RNA Biology* 2, no. 2 (April 2005): 41–44.

Korostelev, A. A. "Structural Aspects of Translation Termination on the Ribosome." *RNA* 17, no. 8 (August 2011): 1409–1421. doi: 10.1261/rna.2733411.

Lundblad, R., and F. M. MacDonald, eds. *Handbook of Biochemistry and Molecular Biology*, 4th ed. Boca Raton, FL: CRC Press, 2010.

Moody, P., and A. J. Wilkinson. *Protein Engineering*. Oxford, UK: IRL Press, 1990.

Park, S. G., P. Schimmel, and S. Kim. "Aminoacyl tRNA Synthetases and Their Connections to Disease." Proceedings of the National Academy of Sciences USA. 105, no. 32 (August 12, 2008): 11043–11049. doi: 10.1073/pnas.0802862105.

Ramakrishnan, V. "Ribosome Structure and the Mechanism of Translation." *Cell* 108, no. 4 (February 22, 2002): 557–572.

Rodnina, M. V., M. Beringer, and W. Wintermeyer. "How Ribosomes Make Peptide Bonds." *Trends in Biochemical Sciences* 32, no. 1 (2007): 20–26. doi:10.1016/j.tibs.2006.11.007. PMID 17157507.

Saibil, H. R. "Chaperone Machines in Action: Current Opinion in Structural Biology 18, no. 1 (February 2008): 35–42.

Saier, M. H., Jr., P. K. Werner, and M. Müller. "Insertion of Proteins into Bacterial Membranes: Mechanism, Characteristics, and Comparisons with the Eucaryotic Process." *Microbiology Reviews* 53, no. 3 (September 1989): 333–366.

Winter, G., A. R. Fersht, A. J. Wilkinson, M. Zoller, and M. Smith. "Redesigning Enzyme Structure by Site-Directed Mutagenesis: Tyrosyl tRNA Synthetase and AT Binding." *Nature* 299 (1982): 756–758.

Woese, C. R., G. J. Olsen, M. Ibba, and D. Söll. "Aminoacyl-tRNA Synthetases, the Genetic Code, and the Evolutionary Process." *Microbiology and Molecular Biology Reviews* 64, no. 1 (March 2000): 202–236.

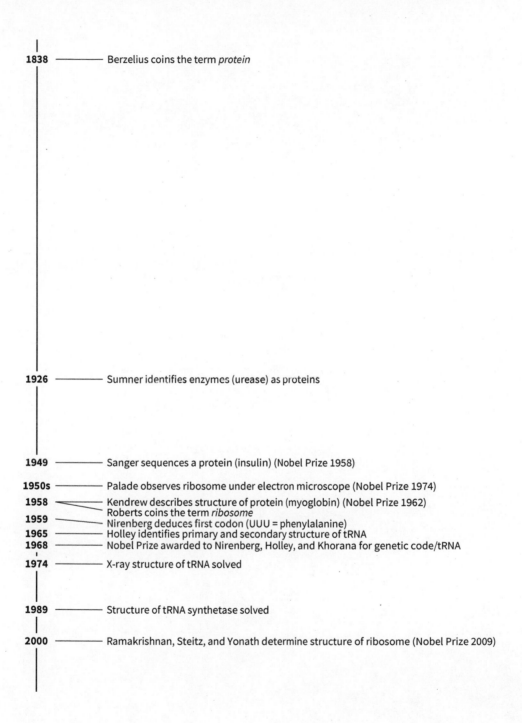

1838 ——————— Berzelius coins the term *protein*

1926 ——————— Sumner identifies enzymes (urease) as proteins

1949 ——————— Sanger sequences a protein (insulin) (Nobel Prize 1958)

1950s ——————— Palade observes ribosome under electron microscope (Nobel Prize 1974)

1958 ——————— Kendrew describes structure of protein (myoglobin) (Nobel Prize 1962)
——————— Roberts coins the term *ribosome*

1959 ——————— Nirenberg deduces first codon (UUU = phenylalanine)

1965 ——————— Holley identifies primary and secondary structure of tRNA

1968 ——————— Nobel Prize awarded to Nirenberg, Holley, and Khorana for genetic code/tRNA

1974 ——————— X-ray structure of tRNA solved

1989 ——————— Structure of tRNA synthetase solved

2000 ——————— Ramakrishnan, Steitz, and Yonath determine structure of ribosome (Nobel Prize 2009)

CHAPTER 17

Control and Regulation of Gene Expression

Gene Regulation in Context

Many people know that DNA is an important molecule, and some may be able to describe DNA as the blueprint of life or the genetic code. Those who have studied science may even be aware that DNA molecules code for all the proteins of the cell. Relatively few people, however, appreciate all the functions of DNA.

We can use an architectural analogy to provide a general view of these other functions. Imagine the organism as a skyscraper, in which each cell of the organism is a room. DNA serves as a highly detailed blueprint for that building, a copy of which is found in every room. The plans for this building need to describe the exact materials used to construct every aspect of the structure. However, DNA provides far more information than what is contained in a blueprint. Although a detailed architectural drawing would provide information about what types of doorknobs to use and where to put electrical outlets, it would not tell the builders when to order and install those doorknobs and outlets. In contrast, DNA provides the when in addition to the where. That is, the DNA sequence contains information that the cell uses to transcribe genes in response to environmental changes and challenges as the organism proceeds through its life cycle.

This chapter examines the similarities in different DNA sequences that regulate genes and how DNA-binding motifs found in proteins interact with DNA in both specific and nonspecific ways. Next, it considers how prokaryotes control gene expression both globally and on a gene-by-gene basis, using sigma factors, operons, and riboswitches. Finally, it explores how eukaryotes organize and consolidate DNA into chromosomes and modify gene expression to adapt to their environment. We also discuss biochemical aspects of several human diseases that arise as a result of aberrations in gene expression and the ways they are currently treated or may be treated in the future.

Chapter Outline

17.1 DNA–Protein Interactions

17.2 Regulation of Gene Expression in Prokaryotes

17.3 Regulation of Gene Expression in Eukaryotes

Evolution's outcomes are conserved.	• Organisms face environmental or developmental challenges and must adapt and respond in order to prosper and pass on their genes. • Like protein and nucleic acid structures, the control mechanisms themselves are often conserved. All the common mechanisms of gene regulation (operons, ligand-responsive transcription factors, riboswitches, attenuators, antiterminators, and miRNAs) are found in multiple species and often throughout biology.
Structure determines function.	• Several different protein structural domains are involved in DNA–protein interactions and, consequently, in DNA-binding proteins used for gene regulation. These domains are structurally conserved. • Like proteins, nucleic acids (specifically RNAs) can form complex but organized tertiary structures. • There are many weak interactions (hydrogen bonding, London dispersion forces, dipole–dipole interactions, and ionic interactions) between molecules. The sum total of the interactions between macromolecules (nucleic acids and proteins) and solvents forms the basis of how these molecules bind to each other. Slight perturbations to this chemistry lead to changes in how tightly this binding occurs and cause molecules to either associate with or dissociate from one another, ultimately resulting in changes to gene expression.
Biochemical information is transferred, exchanged, and stored.	• The nucleic acid sequences (and therefore nucleic acid structures) involved in the regulation of genes are conserved throughout evolution. In some instances, the control sequences are more tightly regulated than the coding sequences of the genes themselves.
Biomolecules are altered through pathways involving transformations of energy and matter.	• Many of the macromolecular interactions involved in the regulation of transcription can be attributed to chemical modifications of proteins, nucleic acids, or both.

17.1 DNA–Protein Interactions

It has long been known that cells regulate genes using a combination of specific sequences of DNA and protein molecules that bind DNA. Interactions between DNA and proteins are key to the regulation of **gene expression**. How do these structures regulate transcription on the molecular and chemical levels? In other words, what is the function of these molecules and interactions? Answering these questions provides a starting point for interpreting these chemical changes and their biological ramifications in development and disease.

17.1.1 DNA structure determines the interactions that regulate genes

The three-dimensional structure of the DNA helix limits the ways in which DNA can interact with proteins to control gene expression. The iconic structure of DNA is the double helix, with hydrogen-bonded bases in the core of the helix and antiparallel anionic sugar–phosphate

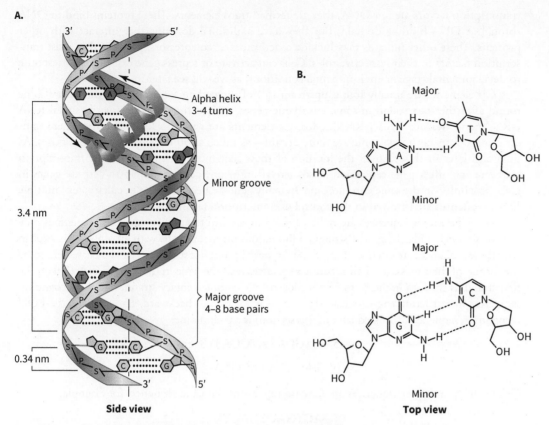

FIGURE 17.1 Interactions of DNA with the alpha helix via the major groove. **A.** Proteins can interact with the double helix of DNA via an alpha helix. Because both the DNA double helix and alpha helix are rodlike and somewhat rigid, there is a limited region through which they can interact, typically consisting of four to eight base pairs of DNA and the amino acids on three to four turns of an alpha helix. **B.** Both the major and minor grooves of DNA have functional groups that can interact with DNA-binding proteins. This top view shows the functional groups on the major and minor groves through which proteins can interact.

(Source A: Snustad: *Principles of Genetics*, 6e; Copyright 2012, John Wiley & Sons. This material is reproduced with permission of John Wiley & Sons, Inc.)

backbones on the outside. Collectively, the structure resembles a ladder that has been twisted like a spring.

Proteins can bind to DNA through interactions with the sugar–phosphate backbone, but the specificity of the interaction occurs through interactions with bases via an alpha helix. A closer examination of the DNA helix structure shows two grooves spiraling around the outside of the helix. The narrower of the two is the **minor groove**, and the wider is the **major groove**. The major groove is about as wide as the diameter of an alpha helix, and the two structures fit neatly together. Both grooves can participate in DNA–protein interactions, but the diameter of the major groove allows alpha helices of proteins to specifically interact with DNA. A collection of weak forces accounts for this specificity, including hydrogen bonding between the bases and amino acid side chains, steric effects, and the hydrophobic effect. There is often a focus on the interactions of the base pairs within DNA, but other functional groups on the bases (not involved in base pairing) are exposed to either the major or the minor groove (**Figure 17.1**).

17.1.2 DNA promoter sequences regulate gene expression

Specific noncoding DNA sequences are called **promoter sequences**. Because these are parts of DNA that regulate genes, they are termed **cis elements**. Promoter sequences are recognized and bound by proteins called **transcription factors** that regulate gene transcription either positively or negatively. **Positive regulation** involves upregulating or activating gene expression, and a gene that is able to be turned on is **inducible**. **Negative regulation** involves downregulating or repressing gene expression, and a gene that is able to be turned off is **repressible**. Because transcription factors are not DNA, they are termed trans elements. These proteins bind to DNA through a DNA-binding domain, but they have additional domains that interact with other proteins. These other proteins may include coactivators or corepressors, proteins that assist transcription factors in their respective roles. Cells can activate or repress gene expression in response to developmental stage or environmental conditions, as we will see later in the chapter.

Cis elements are usually found upstream (5′) of the coding sequence, but they can also be found after the start codon in some eukaryotic genes, such as those coding for ribosomal RNA (rRNA) and transfer RNA (tRNA). Most cis elements are within 5 kb of the site where transcription occurs, although some are more distant—as much as 50 kb distant in globin genes. As discussed later in this chapter, the location of these sequences with regard to the transcription start site can affect gene expression. Also, activation of one gene may help to activate genes in close proximity on the same chromosome. In many genes, especially those in eukaryotes, multiple DNA sequences are involved in the control of transcription factors.

DNA promoter sequences are rather short compared to the rest of the gene, sometimes only four to six nucleotides long. The structural limitations on interactions with an alpha helix explain why these sequences are often so short. Several types of promoter sequences can be characterized by the format of their messages. **Palindromic sequences** read the same in the 5′ to 3′ direction on the first strand as they do in the 3′ to 5′ direction on the complementary strand, like the recognition sequences in restriction enzymes read the same way forward or backward. For example, the POU recognition sequence (discussed later in this section) is a palindrome:

$$5'-ATGCATATGCAT-3'$$

$$3'-TACGTATACGTA-5'$$

Other DNA promoter sequences are **tandem repeats** of the same sequence, for example:

$$5'-ACTGACTGACTG-3'$$

Note that transcription factors typically bind sequences that are longer than this example. Sequence repeats provide multiple binding sites for a DNA-binding protein, which greatly increases the binding affinity. This is a theme that we have frequently seen in biochemistry, for example, with antibodies binding an antigen. Repeated DNA sequences are often recognized by multimeric proteins that bind the same sequence with repeated subunits.

Promoter sequences can be highly specific. For example, the difference between the Oct-1 and Pit-1 promoter sequences is only two nucleotides out of eight:

Oct-1   ATGCA<u>A</u>AT

Pit-1   ATG<u>AA</u>T<u>A</u>T

However, just replacing these two nucleotides changes expression radically. Genes controlled by Pit-1 are expressed exclusively in the pituitary, whereas those controlled by Oct-1 are expressed in B cells. This specificity means that a single substitution mutation in this sequence could be harmful to the health of an organism.

Transcription factors recognize specific nucleotide sequences. We have seen analogous patterns in proteins that recognize other proteins and in proteins that recognize nucleic acids. Many proteins have short stretches of amino acids that may be recognized by a protease or kinase. Also, different proteins recognized by a single enzyme may all include the same short sequence. For example, proteins recognized by protein kinase A (PKA) contain the short sequence

Arg-Arg or Lys-Any AA-Ser or Thr

(R-R/K-X-S/T)

in which the final amino acid (Ser or Thr) is phosphorylated. There is some variability to the sequence. In some instances, the kinase recognizes either an arginine or a lysine residue, and it will phosphorylate either serine or threonine. However, researchers have been able to deduce the common sequence, termed a **consensus sequence**, that this kinase recognizes. Promoter regions of DNA also contain consensus sequences, but these are sequences of nucleotides, not amino acids. As seen in the example of Oct-1 and Pit-1, some consensus sequences are tightly conserved, and slight deviations will result in radical changes in activity. However, in other cases, such as the recognition sequence of PKA, there is more tolerance for variation.

17.1.3 Transcription factors use specific protein motifs to interact with DNA

Many different proteins bind DNA, but the transcription factors known to date use only a few distinct motifs to interact with the DNA helix. As mentioned previously, most binding occurs specifically in the major groove of DNA. Although the beta sheet is important in several DNA-binding motifs (for example, zinc fingers), it is usually the direct interactions of alpha-helical regions of the protein that dictate binding. This finding may change as the structures of more transcription factors are elucidated.

This section focuses on seven particular DNA-binding domains or motifs: helix-turn-helix, helix-loop-helix, leucine zipper, zinc finger, homeodomain, winged helix, and POU (**Table 17.1**). These motifs contain many conserved features and are pertinent in eukaryotic and prokaryotic gene regulation and human disease. Additional DNA-binding motifs are known, however, and more are being elucidated each year.

The helix-turn-helix motif   The **helix-turn-helix** motif was one of the earliest motifs to be discovered and is one of the most common. It functions in DNA binding in both eukaryotes and prokaryotes. The domain consists of two alpha helices, each about 20 amino acids in length, connected by a short turn four to six amino acids long. The helices fit in the major groove, interacting through hydrogen bonding between the amino acid side chains that protrude from the helices and the exposed amino and carbonyl groups on the major groove side of the helix. This motif is found in many DNA-binding proteins and transcription factors, including several discussed later in this chapter, such as the *trp* repressor and cAMP-receptor protein (CRP).

The helix-loop-helix motif   Similar to the helix-turn-helix motif is the **helix-loop-helix**. In this arrangement, the two alpha helices that make contact and bind to DNA are separated by a large loop (12 to 28 amino acids) rather than a short turn (2 to 4 amino acids). Both helices, as well as the turn region, make contact with the DNA double helix. Helix-loop-helix motifs are found in the transcription factors ADD1 and MyoD.

The leucine zipper motif   The **leucine zipper** motif is also known as the basic leucine zipper (bZIP) motif or transcription factor. The topology of this motif includes two helices joined to each other by a series of leucines that repeat every seven amino acids along the alpha helix. This creates a gently pitched face of the branched, hydrophobic leucine side chains on each alpha helix. These side chains intercalate with one another like the teeth of a zipper (hence the name), knitting the two helices together. At the end of the zipper, the helices diverge, and each fits into the major groove of DNA. Leucine zipper motifs are found in the transcription factors cAMP response element–binding (CREB), c-Jun, and Fos.

The zinc finger motifs   The **zinc finger** family of DNA-binding motifs actually describes six different motifs. The feature that unites this family is the use of a divalent zinc ion (Zn^{2+}) that is complexed to a combination of histidine or cysteine residues to stabilize the structure. These motifs are relatively small (23 to 28 amino acids) and may contain both beta sheet and alpha helix. Zinc finger motifs are often repeated in a protein, and a sequence of two or three repeated motifs is sometimes found.

The most common zinc finger structure encountered is the Cys_2His_2 motif. In this configuration, a single zinc ion is chelated by two cysteines and two histidines (as the name would suggest). The consensus amino acid sequence for this motif is

$$X_2\text{-Cys-}X_{2,4}\text{-Cys-}X_{12}\text{-His-}X_{3,4,5}\text{-His}$$

where X is any amino acid, and the subscript number refers to the number of amino acids present.

In this arrangement, the histidine residues are found in the alpha-helical region and the cysteine residues in the turn between the two strands of antiparallel beta sheet. As seen in other transcription factors, the helical region of the Cys_2His_2 zinc finger is the part that is responsible for DNA binding. The many Cys_2His_2 zinc finger proteins that have been identified can and do bind to different DNA sequences using other parts of the sequence.

TABLE 17.1 Key DNA Binding Motifs

Motif/Domain	Description/Examples	Protein Data Bank ID
Helix-turn-helix motif	A pair of alpha helices connected by a short loop. Variants of this motif are repeated in several other classes of transcription factors (helix-loop-helix, homeodomain, winged helix, and POU). **Example:** Oct-1	Source: Data from PDB ID 2CGP Passner, J.M., Steitz, T.A. (1997) The structure of a CAP-DNA complex having two cAMP molecules bound to each monomer. *Proc.Natl. Acad.Sci.USA* **94**: 2843–2847
Helix-loop-helix motif	Two helices connected by a loop of greater than five amino acids. Both helices make contact with the DNA major groove. **Examples:** MyoD, ADD1	Source: Data from PDB ID 1AM9 Parraga, A., Bellsolell, L., Ferre-D'Amare, A.R., Burley, S.K. (1998) Co-crystal structure of sterol regulatory element binding protein 1a at 2.3 Å resolution. *Structure* **6**: 661–672
Leucine zipper motif	Two alpha helices held together by intercalated leucine side chains. The distal ends of these helices intersect with the major groove of DNA. **Examples:** CREB, C/EBP	Source: Data from PDB ID 1NWQ Miller, M., Shuman, J.D., Sebastian, T., Dauter, Z., Johnson, P.F. (2003) Structural Basis For DNA Recognition By The Basic Region Leucine Zipper Transcription Factor CCAAT/Enhancer Binding Protein Alpha. *J.Biol.Chem.* **278**: 15178–15184
Zinc finger motif	A helix and sheet motif held together by at least one zinc ion. Multiple subclasses that differ in how the zinc ligand is bound (Cys_2His_2, Zn_2/Cys_6, and Zn_2/Cys_8). **Examples:** PPARs, steroid receptors	Source: Data from PDB ID 3DZY Chandra, V., Huang, P., Hamuro, Y., Raghuram, S., Wang, Y., Burris, T.P., Rastinejad, F. (2008) Structure of the intact PPAR-gamma-RXR-nuclear receptor complex on DNA. *Nature* **456**: 350–356
Homeodomain	Three helices, consisting of 60 amino acids; a variant of the helix-turn-helix. **Example:** HOX	Source: Data from PDB ID 1PUF Laronde-Leblanc, N.A., Wolberger, C. (2003) Structure of HOXA9 and PBX1 bound to DNA: HOX hexapeptide and DNA recognition anterior to posterior. *Genes Dev.* **17**: 2060–2072
Winged helix domain	Four helices topped by a double-stranded beta sheet, consisting of about 100 amino acids. **Example:** FOX	Source: Data from PDB ID 6EL8 Newman, J.A., Aitkenhead, H.A., Pinkas, D.M., von Delft, F., Arrowsmith, C.H., Edwards, A., Bountra, C., Gileadi, O. Crystal structure of the Forkhead domain of human FOXN1 in complex with DNA. *To be published*.
POU domain	A homeodomain and a specific POU domain, consisting of 150 amino acids. **Examples:** POU (pituitary specific, octomer, Unc-86)	Source: Data from PDB ID 1PUF Laronde-Leblanc, N.A., Wolberger, C. (2003) Structure of HOXA9 and PBX1 bound to DNA: HOX hexapeptide and DNA recognition anterior to posterior. *Genes Dev.* **17**: 2060–2072

Zinc finger motifs are found in steroid and nuclear hormone receptors, including the peroxisome proliferator-activated receptors (PPARs), RXR, and the estrogen receptor.

Homeodomain proteins Homeobox (HOX) genes code for proteins containing a sequence of about 60 amino acids called the **homeodomain** that binds DNA promoter sequences. The domain is folded into three alpha-helical segments, two of which form a helix-turn-helix motif. The third alpha helix is involved in stabilizing the interactions of the other two helices. Like most transcription factors, HOX proteins bind DNA through multiple weak interactions; some of these are nonspecific (for example, salt bridging of basic amino acids to phosphates in the DNA backbone), whereas others are specific, such as HOX proteins recognizing the DNA consensus sequence 5′-ATTA-3′. We will say more about HOX genes and their important role in development when we discussion control of gene expression in eukaryotes in section 17.3.

The winged helix domain The **winged helix** family of transcription factors binds DNA with a unique binding domain consisting of 100 amino acids. The motif responsible for protein–DNA interactions is sometimes termed the fork head domain because it is found in fork head box (FOX) proteins. This motif interacts via an alpha-helical segment but also contains three other helices and a two-stranded antiparallel beta sheet. This gives the whole molecule the appearance of having a wing, hence the name winged helix. FOX proteins bind DNA at a single location and act in a monomeric fashion. In addition, they are regulated by phosphorylation through either PKB/Akt or other downstream kinases from the PI3K signaling pathway. The FOX family is comprised of over 40 proteins organized into 17 different subgroups. Because FOX genes are responsible for proliferation and differentiation, mutations in these in these genes have been implicated in many cancers and developmental disorders.

The POU domain POU proteins are a family of transcription factors, named for the three original family members: pituitary specific, octamer, and unc-86. As with winged helix and homeodomains, they are found exclusively in eukaryotes. POU family members have a DNA-binding domain of about 150 amino acids that comprises both a homeodomain and a specific POU motif linked by a flexible loop. This loop allows both subdomains to interact with DNA at the same time and to bind a longer stretch of DNA. The consensus sequence for the pituitary gland–specific transcription factor Pit-1 is ATGAATAA/T.

Many transcription factors will change their subcellular localization or function based on post-translational modification, ligand binding, or binding to partner proteins. Specifics of these interactions are described later in this chapter.

Worked Problem 17.1 — DNA structure and protein interactions

You have identified a DNA-binding protein that specifically interacts with the sequence 5′-AT-3′. Where will the protein interact on DNA, and how will it bind specifically with this sequence? What will the topology of the DNA be where the protein binding occurs, and what functional groups will be involved?

Strategy If the protein is binding to the sequence 5′-AT-3′, we should examine the topology of the major groove. What functional groups are there in the major groove that could interact with proteins? Using Figure 17.1 as a guide and knowledge of the conformation of the double helix, we can piece together the functional groups in this binding cleft.

Solution If we examine the functional groups in the major groove of the sequence 5′-AT-3′, we arrive at the following:

A. An A::T base pair

B. Top view

C. Side view

In this view, the major groove is shown across the top of the AT base pair (panel A). Three groups can participate in hydrogen bonding (shown in circles or squares). To simplify this exercise, we will not examine the effect of the methyl group on the thymine ring. Because the sequence we are examining is 5′-AT-3′, the groups available for interactions are flipped in the second base pair (panel B). Panel C shows a simplified view of what would be available for interaction in the major groove: two hydrogen bond donors and four acceptors.

If an alpha helix was interacting with this stretch of DNA, we could hypothesize that several polar amino acids could interact with these groups. Similarly, we could assume that many other amino acids (the nonpolar ones) would not interact. These hypotheses would need to be tested experimentally.

This example illustrates that it is possible to work out which functional groups are available for DNA–protein interactions, that there are many possibilities, and that it is difficult (and, in fact, currently not possible) to predict exactly which amino acids will interact with these groups.

Follow-up question   What will the topology be for the sequence 5′-TAAT-3′?

Summary

- Proteins and enzymes interact with the DNA helix through functional groups in the major and minor groove and through the sugar–phosphate backbone.
- Specific DNA sequences are involved in the regulation of gene expression. These can include palindromic sequences and tandem repeats. Small changes to some sequences have a major impact on gene expression; others have more tolerance for variation.
- Specific protein motifs (DNA-binding domains) are involved in the recognition of the DNA helix, including the helix-turn-helix, helix-loop-helix, leucine zipper, and zinc finger motifs, as well as the homeodomain and the winged helix and POU domains. Many of these domains employ alpha helices to interact with the major groove of DNA.

Concept Check

1. Describe the physical features of the DNA helix that are involved in DNA–protein interactions.
2. Identify the types of specific DNA promoter sequences involved in regulation of gene expression.
3. Compare specific motifs or domains that proteins employ in DNA–protein interactions. What do you mean by of the helix-turn-helix motif?

17.2 Regulation of Gene Expression in Prokaryotes

Both prokaryotes and eukaryotes must regulate gene expression to adapt to changing environmental conditions. Gene regulation presents fewer challenges for prokaryotes than for eukaryotes, however, because prokaryotic organisms are unicellular and relatively simple. Nevertheless, prokaryotes use several mechanisms to regulate expression both globally and specifically, including sigma factors, operons, and riboswitches.

17.2.1 Sigma factors regulate large groups of genes

RNA polymerase is a multi-subunit complex that includes both catalytic and regulatory subunits. One of these subunits is a **sigma factor (σ)** responsible for recognition of the promoter site where transcription begins. Following initiation, once the holoenzyme has bound and transcription has begun, the sigma factor dissociates from the holoenzyme and stays bound to the promoter, and RNA synthesis proceeds. Globally, the cell can change gene expression by changing the types of sigma factors it produces and thus can respond to and survive under different environmental challenges.

The factor sigma 70 (σ70), named for its molecular weight in *Escherichia coli* (*E. coli*), is thought to be the default promoter found under normal, or basal (unstimulated), conditions. It has been associated with genes that are commonly expressed and are not highly regulated or inducible under special circumstances, sometimes called **housekeeping genes**. This term is also used in eukaryotic cells to describe genes that are constitutively expressed—that is, continually expressed at an unchanging rate.

When bacteria face various environmental challenges (for example, changes in nutrient levels, osmolarity, or temperature), they can respond by synthesizing different sigma factors, and as many as 14 such factors have been identified. For example, when presented with increased temperature in its environment, the cell produces s32, a sigma factor with a molecular weight of 32 kDa. This factor binds to the heat shock promoter sequences, activating transcription in these genes, which in turn produce proteins that the cell uses to respond to this challenge. Sigma factors are involved in the pathogenicity, or virulence, of bacteria and can globally influence the genes being expressed. Particular sigma factors can upregulate the genes involved in resistance to stressful environments, such as heat, low pH, or the high concentrations of hydrogen peroxide sometimes seen in abscessed tissue. For example, a common problem associated with cystic fibrosis is bacterial infections that produce significant amounts of mucus. A specific sigma factor has been found that upregulates genes involved in producing secreted polysaccharides (alginate) associated with this mucus.

17.2.2 Operons regulate small groups of genes that code for proteins with a common purpose

Often, the cell needs to respond to a challenge that requires a more focused approach than simply changing the types of sigma factors synthesized. One such instance would be a shift in the availability of different food sources. An organism would be at a disadvantage if it were forced to produce all the enzymes needed to metabolize a diverse array of carbohydrates at all times or if it were unable to switch from one carbon source to another (for example, from glucose to lactose). Operons provide one way in which prokaryotes can respond to such a challenge.

An **operon** is an organization of genes that code for proteins involved in a common purpose, such as the metabolism of a specific carbohydrate or amino acid (**Figure 17.2**). Within an operon, the promoter region serves as a control site to regulate the synthesis of a single mRNA that codes for all the genes for proteins involved in the specific metabolic pathway. The promoter region binds to RNA polymerase. The **operator** region binds to repressor proteins encoded by genes upstream of the promoter to regulate transcription of the structural genes that follow in the message. The operon allows the regulation (either positive or negative) of related genes to be accomplished by a single control region, rather than multiple control sequences regulating multiple independent genes. The end result is a single **polycistronic message**, that is, a message that contains multiple start and stop sites and codes for more than one protein. Operons also have regions upstream of the control sites that code for a regulatory protein (the repressor). Often, these proteins will bind a metabolite involved in the pathway specific to the operon and regulate the expression of the operon structural genes either positively or negatively. By binding different regulatory elements, genes can be either up- or downregulated, as appropriate. The remainder of this section examines two examples of operons: the *lac* operon and the *trp* operon.

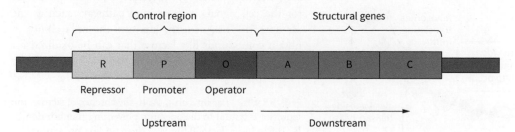

FIGURE 17.2 Structure of the operon.   Operons have a series of structural genes (A, B, and C) coding for enzymes or transporters translated into a single polycistronic message. Upstream of the structural genes is the control region, which consists of the repressor, promoter, and operator regions. The repressor gene codes for the repressor protein, which binds to the operator region, blocking transcription or binding of RNA polymerase to the promoter sequence.

The *lac* operon Lactose is a disaccharide, a β–1,4 linked dimer of glucose and galactose that is commonly found in milk. To catabolize lactose, *E. coli* must absorb it into the cell and cleave the glycosidic linkage between the glucose and galactose. Glucose can then be catabolized directly via glycolysis, whereas galactose requires further epimerization before entering the same pathway.

The *lac* **operon** codes for three structural genes: *lacZ*, *lacY*, and *lacA*. The gene *lacZ* codes for a β-galactosidase that is responsible for breaking the lactose into its component monosaccharides (**Figure 17.3**); *lacY* codes for a lactose permease, a membrane protein that facilitates the transport of lactose into *E. coli*; and *lacA* codes for a transacetylase, an enzyme that esterifies an acyl group to the free hydroxyl on the 6 position of galactose. Although the structure of the transacetylase and the kinetics of this reaction have been studied, its biological ramifications remain unclear.

Upstream of the structural genes is one other coding sequence, *lacI*. This gene codes for the *lac* **repressor (LacI)**, a helix-turn-helix DNA-binding protein, which is constitutively expressed, that is, continually expressed under most conditions. Hence, the *lac* repressor protein is present most of the time. This protein can bind either of two key molecules in this regulatory system: lactose or a specific stretch of DNA found in the operon.

Between the two coding sequences are the control regions of the operon—the promoter and operator regions—bound by the RNA polymerase and *lac* repressor, respectively.

In the absence of lactose in the environment, the *lac* repressor is synthesized, and its conformation is such that the protein can bind to the operator region of the operon. In this state, the promoter is unable to bind RNA polymerase and transcribe the structural genes *lacZ*, *lacY*, and *lacA*. However, if lactose is abundant in the environment, the *lac* repressor can bind to lactose; this frees the operator region, which permits RNA polymerase to bind at the promoter, allowing the transcription of the genes required for further lactose metabolism. This is an example of negative regulation.

There is a second means by which the cell regulates this operon. Many organisms employ cyclic adenosine monophosphate (cAMP) as an indicator of the energetic state of the cell, and this signal has been conserved throughout evolution. High levels of cAMP activate pathways that mobilize energy (such as glucose mobilization from glycogen or neutral lipid breakdown) through the signaling pathway. Bacteria also use cAMP to regulate fuel sources. In *E. coli*, if fuel such as glucose is absent, cAMP levels are high. Once bound to cAMP, the regulatory protein known as the **cAMP receptor protein (CRP)** employs a helix-turn-helix motif to interact with DNA (**Figure 17.4**). CRP is also known as the catabolite activator protein (CAP).

CRP functions as a homodimer, and each monomer has a cAMP-binding site. In the cAMP bound state, the CRP–cAMP complex binds to specific sequences upstream of the operator region. Binding introduces a nearly 90° bend in the DNA helix, which in turn facilitates the binding of RNA polymerase and initiation of transcription in the presence or absence of the *lac* repressor (LacI) (**Figure 17.5**). Therefore, a low energetic state for the cell signals it to activate pathways such as the *lac* operon. This, in turn, increases the cell's ability to obtain compounds like lactose that can be oxidized for energy. The activation of genes involved in this metabolic

A. Absence of inducer: Repression

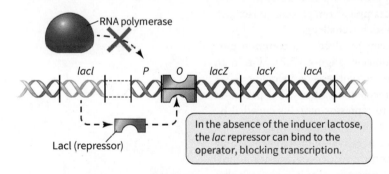

In the absence of the inducer lactose, the *lac* repressor can bind to the operator, blocking transcription.

B. Presence of inducer: Induction

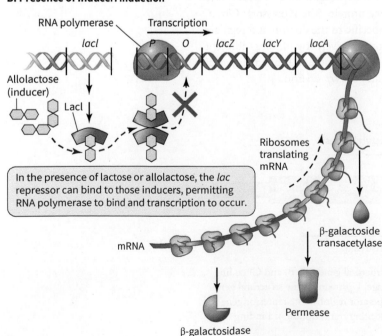

In the presence of lactose or allolactose, the *lac* repressor can bind to those inducers, permitting RNA polymerase to bind and transcription to occur.

FIGURE 17.3 *Lac* **operon.** **A.** In the absence of lactose, the repressor is bound to the operator, preventing transcription. **B.** The inducer (lactose) binds to the repressor, preventing repressor binding to the operator sequence. This allows binding of RNA polymerase and transcription of the structural genes (Z, Y, and A).

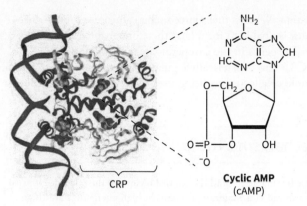

FIGURE 17.4 Cyclic AMP and CRP. The structure of cyclic AMP (cAMP) is a key indicator of the energetic state of the cell. The structure of the cAMP receptor protein (CRP) is shown bound to cAMP and DNA. The molecule binds DNA as a homodimer using a helix-turn-helix motif.

(Source: Data from PDB ID 2CGP Passner, J. M., Steitz, T.A. (1997) The structure of a CAP-DNA complex having two cAMP molecules bound to each monomer. *Proc. Natl. Acad. Sci. USA* 94: 2843–2847)

Cyclic AMP (cAMP)

CRP

Glucose concentration	Lactose concentration	cAMP concentration	Operon status	Level of *lacZ*, *lacY*, and *lacA* transcription	Lactose metabolized?
Low	High	High	*P O lacZ*	High	Yes
Low	Low	High	*P O lacZ*	Low	No
High	Low	Low	*P O lacZ*	Low	No
High	High	Low	*P O lacZ*	Low	No

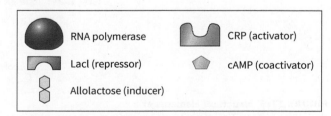

RNA polymerase CRP (activator)

LacI (repressor) cAMP (coactivator)

Allolactose (inducer)

FIGURE 17.5 Role of CRP in the *lac* operon. When the cell is low in energy, cAMP levels rise. Then cAMP binds to CRP and acts as an allosteric regulator. The CRP–cAMP complex binds to DNA and facilitates binding of RNA polymerase, increasing transcription of the *lac* operon. In the absence of cAMP, the equilibrium is shifted, CRP no longer binds to the DNA, and transcription is not activated.

pathway is an example of positive regulation. When a fuel source such as glucose is present, cAMP levels drop, and the CRP–cAMP complex is unable to form (and therefore unable to bind DNA). This is known as **catabolite repression**. The genes in the *lac* operon are therefore inducible.

Two molecules of cAMP bind to CRP and allosterically regulate the interaction of this protein with DNA. This is a further example of allosterism being used as a means of fine-tuning a protein's activity.

Worked Problem 17.2 — Prokaryotic control mechanisms: the *lac* operon

If both glucose and lactose were available to them, which fuel source would bacteria use? How would the *lac* operon function in this instance?

Strategy Assuming the bacteria can use either the disaccharide lactose or the monosaccharide glucose as a carbon source for energy and that both carbohydrates are abundant, we might assume that the bacteria would preferentially use glucose because glucose can be directly catabolized by glycolysis, whereas lactose cannot.

Solution Based on the assumption that the cell will preferentially use glucose, the genes of the *lac* operon will be downregulated. Lactose will still be present and will bind the *lac* repressor, which will

therefore be unable to bind the control regions of the operon. However, if the cell can maintain a relatively high energetic state (high levels of ATP), it will have relatively low levels of cAMP. This means that cAMP will not bind to CRP, so the CRP–cAMP complex will be unable to form and bind DNA (and upregulate transcription). Transcription of the operon will thus remain relatively low, despite the absence of the *lac* repressor.

Follow-up question If a mutant strain of bacteria had a gene coding for CRP that was bound to DNA in the presence or absence of cAMP (that is, constitutively bound), how would this affect the expression of genes in the *lac* operon?

The *trp* operon Operons are typically thought of as controlling carbohydrate metabolism, but there are operons with other functions. One example is the **_trp_ operon**, which regulates production of the enzymes involved in the synthesis of the amino acid tryptophan (**Figure 17.6**). In contrast to the *lac* operon (an inducible and positively regulated system), the *trp* operon is negatively regulated and repressible. In addition, the *trp* operon employs a process known as **attenuation** to control transcription.

Tryptophan is the largest and most hydrophobic of the amino acids; its synthesis from chorismate takes five steps and requires five different genes. In addition to its use as a building block for proteins, tryptophan is used by bacteria as a precursor for the production of NAD^+.

Clearly, it would be energetically favorable for an organism to use other sources of tryptophan if available. However, it would also benefit the organism to be able to produce its own tryptophan in times of scarcity. Therefore, tight control of the ability to make tryptophan is an evolutionary advantage for an organism.

The *trp* operon in *E. coli* contains five genes under the control of a common promoter: *trpE*, *trpD*, *trpC*, *trpB*, and *trpA*. The genes code for two multi-subunit enzymes (anthranilate synthase and tryptophan synthase) and one monomeric enzyme (*N*-(5′-phosphoribosyl) anthranilate isomerase), also known as indole-3-glycerol phosphate synthase. Two of these enzymes catalyze

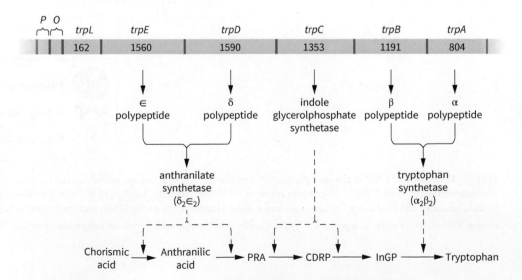

FIGURE 17.6 Structural elements of the *trp* operon in *E. coli*. This map shows the major structural elements of the tryptophan operon and the proteins they encode.

(Source: Snustad; Simmons, *Principles of Genetics*, 6e, copyright 2012, John Wiley & Sons. This material is reproduced with permission of John Wiley & Sons, Inc.)

multiple reactions. Upstream of these genes is the control sequence. As with the *lac* operon, a structural gene codes for a control protein (in this case, the *trp* repressor), and there are operator and promoter sequences between the gene coding for the repressor and the structural genes of the operon. One important difference in the *trp* operon is the leader sequence (*trpL*) that immediately precedes the structural genes.

The *trp* operon is a repressible promoter. In an environment rich in tryptophan, the organism has no need to produce its own tryptophan or the enzymes required for the biosynthesis of tryptophan. The *trp* repressor protein is always being made. This dimeric protein has a binding site for tryptophan in each of its subunits. When tryptophan is present at a sufficiently high concentration, it binds to the *trp* repressor. The tryptophan–*trp* repressor complex changes conformation and binds, using a helix-turn-helix motif, to the operator sequence of the *trp* operon. Binding of the repressor–tryptophan complex blocks binding of RNA polymerase, preventing transcription and ultimately resulting in the production of fewer copies of the enzymes involved in tryptophan biosynthesis.

When the organism encounters lower concentrations of tryptophan, the equilibrium among the operator, the repressor protein, and tryptophan shifts. Without tryptophan bound to the *trp* repressor, the protein dissociates from the operator site, permitting binding of RNA polymerase and transcription of the structural genes of the operon.

There is a second means of control in the *trp* operon. The leader sequence of the operon (the *trpL* gene) employs attenuation to exert further regulatory control over the gene (**Figure 17.7**).

A. The tryptophan (*trp*) operon and regulatory region of mRNA

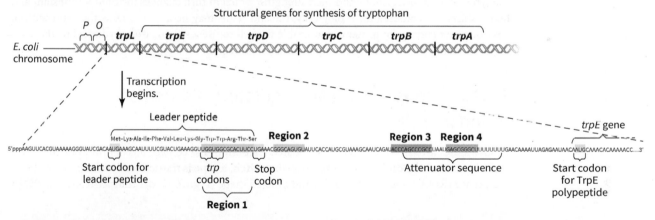

B. Attenuation of transcription by high levels of tryptophan

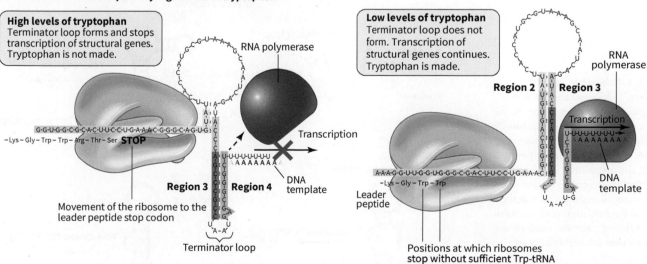

FIGURE 17.7 Attenuation in the *trp* operon. **A.** The *trp* operon control regions are in the *trpL* gene. **B.** When tryptophan is abundant, levels of Trp-tRNA are high. This prevents formation of the 2,3 hairpin and instead favors formation of the 3,4 hairpin, which results in premature termination. When tryptophan levels are low, stalling of the ribosome permits formation of the 2,3 hairpin, which prevents formation of the 3,4 terminator structure.

(Source: Wessner, *Microbiology*, 2e, copyright 2017, John Wiley & Sons. This material is reproduced with permission of John Wiley & Sons, Inc.)

The *trpL* gene contains several adjacent tryptophan codons. Although tryptophan (like most amino acids) is found in most proteins, it is also the amino acid of lowest abundance. Encountering several tryptophan codons in rapid succession gives the cell another means of regulating this operon. The *trpL* sequence is capable of forming complex secondary structures. Four different regions can contribute to the formation of stem loop structures. As the ribosome begins translating the mRNA, it encounters the first of several tryptophan codons. If tryptophan is abundant, charged tryptophan-tRNAs will be available for incorporation into the growing polypeptide, and the ribosome will continue translating through the leader sequence. As this happens, a stem loop structure is formed between regions 3 and 4; this causes termination of transcription (attenuation) and blocks the translation of the downstream genes. However, if there is a shortage of tryptophan-tRNAs, when the ribosome encounters the multiple tryptophan codons in the mRNA, it will stall as it waits for a tryptophan-tRNA. During this pause, a loop forms between regions 2 and 3. This loop is not a terminator sequence in that it allows the gene to be fully transcribed. Therefore, if there is plenty of tryptophan available for protein synthesis, synthesis of the genes involved in tryptophan biosynthesis are blocked. Conversely, if there is insufficient tryptophan, stalling of the ribosome facilitates the formation of different mRNA secondary structures and results in translation of the genes involved in tryptophan biosynthesis.

These complex interactions between large macromolecules can be thought of in basic chemical terms. The various transcription factors (sigma factors, *lac* repressor, CRP, and *trp* repressor) bind to a specific DNA sequence with a certain affinity, so there is an equilibrium constant between free DNA and transcription factor and the complex. Binding of a small ligand either facilitates a change in the conformation of the transcription factor or stabilizes a different conformation. This changes the types of weak interactions and their strengths, which in turn changes the binding constant and hence shifts the equilibrium. Biologically, the end result may be as complex as the expression of a set of genes coding for proteins that enable the cell to use a specific carbohydrate, but the steps involved can be broken down to simple chemical principles.

17.2.3 Riboswitches are regulatory elements found in mRNA

Riboswitches are the most recently elucidated transcriptional control mechanism, but they were probably one of the earliest to evolve. In a **riboswitch**, a gene is transcribed into an mRNA; however, that mRNA folds into a complex and ordered shape (**Figure 17.8**). Transfer RNA and rRNA

FIGURE 17.8 Schematic view of riboswitches. Riboswitches have two distinct regions: an aptamer region, which folds into a complex structure that binds to a metabolite, and an expression platform, which codes for a key protein in the synthesis of the metabolite. In the presence of the metabolite, the conformation of the aptamer is changed, resulting in a change in expression of the structural gene. This can occur either through altering transcription (**A**), in which case antiterminators are formed, or translation (**B**), in which case the Shine–Dalgarno sequence is masked through secondary structures.

(Source: Wessner, *Microbiology*, 2e, copyright 2017, John Wiley & Sons. This material is reproduced with permission of John Wiley & Sons, Inc.)

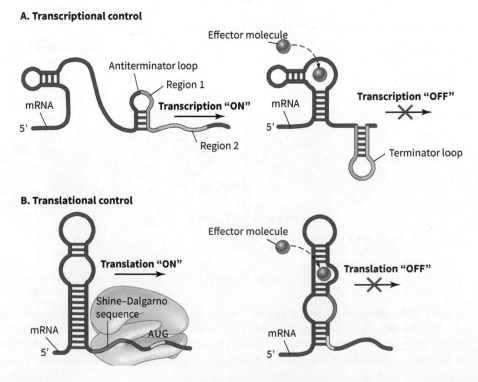

A. Transcriptional control

Effector molecule
Antiterminator loop
Region 1
mRNA
Transcription "ON"
5′
Region 2

mRNA
Transcription "OFF"
5′
Terminator loop

B. Translational control

Effector molecule
Translation "ON"
Shine–Dalgarno sequence
mRNA
AUG
5′

Translation "OFF"
mRNA
5′

molecules can fold into highly ordered structures, just as proteins do. The structures adopted by riboswitches bind specifically and selectively to a small molecule, akin to an enzyme binding a substrate or a protein binding a small ligand. The small molecule is bound by a region of the RNA termed an **aptamer**. The other domain of the riboswitch is termed the **expression platform**—the domain that codes for protein and any up- or downstream sequences responsible for regulating ribosome binding or message stability.

Binding of the small molecule causes changes in the structure of the riboswitch or stabilizes the small ligand–mRNA complex through multiple weak interactions; again, this is analogous to an enzyme–substrate or protein–ligand interaction. This binding changes the way the mRNA is translated. This is an example of translational control of gene expression. Usually, the change is to block the translation of the message into the enzyme involved in the synthesis or transport of the small ligand. Therefore, the riboswitch acts as a direct sensor of small molecules. In the control mechanisms examined previously, the small molecule or ligand is bound to a protein adapter that interacts between the small molecule and the nucleic acid (in those cases, DNA). However, in the case of the riboswitch, the direct binding to small molecules obviates the need for an adapter, as well as for the various genes and steps involved in the production of that adapter.

Several different mechanisms through which riboswitches might act have been postulated or demonstrated. This section discusses two common mechanisms (blocking of transcriptional termination through the use of antiterminators and blocking of translational initiation by sequestering the ribosome binding site) and one less common mechanism (self-cleaving ribozymes).

Antiterminators **Antiterminators** are stretches of RNA sequences that form into GC-rich hairpin, or stem-loop, structures. Once folded, these structures prevent intrinsic termination, also known as premature or rho-independent termination. When ribosomes transcribing a message encounter an antiterminator, they typically pause, in a similar way to that seen in attenuation; however, in contrast to an encounter with a **terminator** loop, antiterminators do not cause the ribosome to abort transcription.

The flavin mononucleotide (FMN) riboswitch is one example of a riboswitch acting as an antiterminator (**Figure 17.9**). FMN is found in the 5′ region of the genes involved

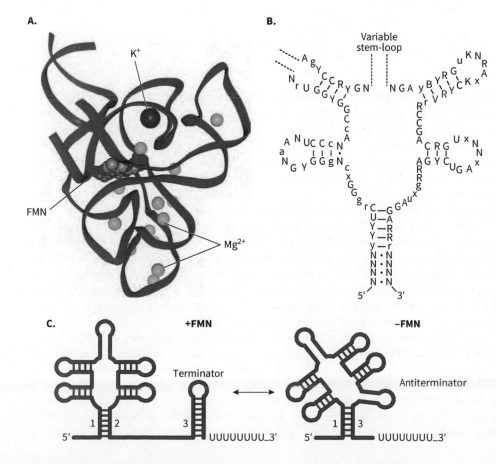

FIGURE 17.9 FMN riboswitch as an antiterminator. **A.** The bound FMN is nearly completely enshrouded by the RNA in the folded state. Note the riboflavin, numerous Mg^{2+} cofactors, and the K^+ ion. **B.** This schematic shows the base pairing in the aptamer region of the FMN. **C.** In the presence of bound FMN, a terminator forms, blocking translation. In the absence of FMN, an antiterminator loop forms, and transcription proceeds.

(Source: (A) Data from PDB ID 3F4E Serganov, A., Huang, L., Patel, D. J. (2009) Coenzyme recognition and gene regulation by a flavin mononucleotide riboswitch. *Nature* **458**: 233–237)

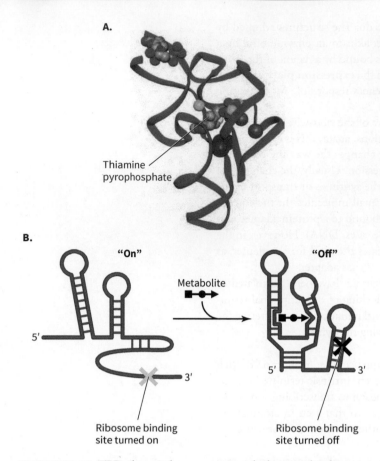

A.

Thiamine
pyrophosphate

B.

"On"

Metabolite

"Off"

5'

3'

5'   3'

Ribosome binding
site turned on

Ribosome binding
site turned off

FIGURE 17.10 TPP riboswitch as a sequestered ribosome-binding site. **A.** The structure of the TPP riboswitch is bound in the center of the RNA. **B.** Binding of a metabolite, in this case TPP, to the aptamer region forms a secondary structure that masks the ribosome-binding site (indicated with a star), blocking transcription.

(Source: (A) Data from PDB ID 2GDI Serganov, A., Polonskaia, A., Phan, A.T., Breaker, R. R., Patel, D.J. (2006) Structural basis for gene regulation by a thiamine pyrophosphate-sensing riboswitch. *Nature* **441**: 1167–1171)

in flavin mononucleotide biosynthesis. FMN is used as a cofactor in many redox reactions in metabolism and is typically found tightly bound to the enzymes catalyzing those reactions. When excess free FMN is present, the aptamer domain of the FMN riboswitch will bind to FMN, resulting in the formation of a terminator loop. Ribosomes that initiate translation on these mRNAs are terminated early and are unable to synthesize functional proteins. In the absence of FMN, an antiterminator structure is formed, and translation proceeds.

Sequestered ribosome-binding sites
Riboswitches can also regulate gene expression by masking the ribosome-binding site, a mechanism called a sequestered ribosome-binding site. If initiation factors are unable to bind to the Shine–Dalgarno sequence, the initiation complex cannot form, and the message cannot be translated. Upon binding to the ligand, riboswitches of this class stabilize a structure that prevents initiation of translation and thus prevents protein synthesis.

An example of this class of riboswitches is the TPP riboswitch (also known as the THI element) (**Figure 17.10**). TPP is thiamine pyrophosphate, the metabolically active form of thiamine (vitamin B1). This multifunctional cofactor is used in a series of important metabolic reactions, including isomerizations, decarboxylations, and deaminations. When the cell has low levels of TPP, the aptamer domain is free, and it forms a secondary structure with an exposed Shine–Dalgarno sequence; however, when TPP is abundant, a different secondary structure is stabilized, the Shine–Dalgarno sequence is imbedded in a secondary structure, and initiation fails to occur.

TPP is the only riboswitch that has been found in eukaryotes to date.

Self-cleaving ribozyme
Some RNA molecules have catalytic properties, catalyzing, for example, splicing reactions and the synthesis of new proteins and ribozymes. To carry out such reactions, the RNA molecules must fold into the correct conformation. In the self-cleaving ribozyme mechanism, the functional ribozyme will only form if the aptamer is bound to its ligand. Once it has attained the correct conformation, the substrate for the ribozyme is found farther downstream in the same molecule. Cleaving of the mRNA results in a nonfunctional message and prevents translation.

An example of the self-cleaving riboswitch mechanism is the glucosamine–6-phosphate (GlcN6P) riboswitch (**Figure 17.11**). This motif is found in the *glmS* gene, which codes for the enzyme glutamine-fructose–6-phosphate amidotransferase. This enzyme uses glutamine and fructose–6-phosphate to generate glutamine–6-phosphate, a crucial component of the bacterial cell wall. On binding of GlcN6P, the message undergoes a self-cleaving event that results in a nonfunctional mRNA. The binding of GlcN6P is highly specific, in that the addition of either glucose–6-phosphate or glucosamine to the message fails to elicit cleavage. A similar mechanism is used in other riboswitches to cause changes in splicing of mRNAs.

Because RNA molecules or DNA molecules coding for RNA molecules are relatively easy to synthesize and introduce into cells, researchers are interested in engineering riboswitches to manipulate gene expression. These methods may be used in the future for gene therapy and as novel antibiotics.

The direct binding and regulation of an RNA function by a small ligand that is not a protein presents an interesting paradigm. Other roles for RNA molecules in the cell include catalysis, information storage, protein synthesis, and ligand binding, and these structures and functions have been conserved. Thus, RNA molecules are leading candidates as one of the original molecules of the primordial soup from which life arose. Present-day RNA molecules that bind small organic metabolites and can regulate mRNA function provide further evidence that life on Earth may have started with RNA-based life forms.

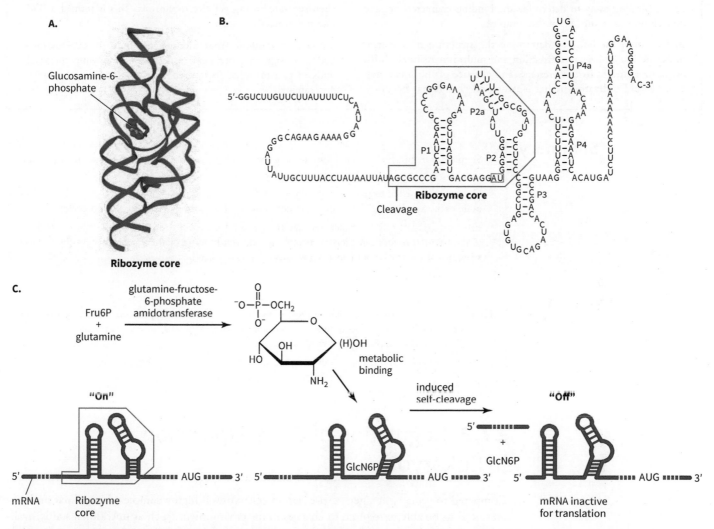

FIGURE 17.11 Glucosamine-6-phosphate riboswitch as a self-cleaving ribozyme. **A.** The structure of the ribozyme core is shown in the presence of glucosamine-6-phoshpate. **B.** The schematic shows the base pairing in the riboswitch. **C.** In the absence of glucosamine-6-phosphate, the riboswitch fails to form and the gene encoding glucosamine-6-phosphate synthase is translated. As levels of glucosamine-6-phosphate increase, there is increased binding of glucosamine-6-phosphate to the RNA, stabilizing the riboswitch. This results in a catalytically active ribozyme, which undergoes self-cleavage, preventing subsequent translation.

(Source: (A) Data from PDB ID 3B4A Klein, D.J., Been, M.D., Ferre-D'Amare, A.R. (2007) Essential Role of an Active-Site Guanine in glmS Ribozyme Catalysis. *J.Am.Chem. Soc.* 129: 14858–14859)

Worked Problem 17.3   Testing a potential riboswitch

You have identified a gene that you hypothesize may be acting as a riboswitch. How could you demonstrate that the mRNA is

a. binding to your small organic metabolite of interest?

b. undergoing a structural change upon binding of the metabolite?

Strategy Assume that we have access to a source of our mRNA molecule and the small molecule that may bind to it. We need to

identify one technique that can determine whether two molecules bind together and another that can detect a structural change.

Solution **a.** There are several different ways to determine whether a small molecule is binding or interacting with a larger one (for example, a receptor or binding protein). One approach would be to use a means of detecting the small molecule such as a radiolabel or a fluorescent label and a means of separating the free molecule from its bound form like size-exclusion chromatography, dialysis, or gel electrophoresis. We could mix the labeled small molecule with the mRNA; if binding occurs, the small molecule would behave as though it were the size of the larger molecule (i.e., it would exit a size-exclusion column earlier, be retained in dialysis, or run differently in an electrophoresis experiment). We could then use a ligand-binding assay to determine the binding characteristics and dissociation constant (K_d) of the complex.

b. Common methods to determine the structure of a macromolecule include X-ray diffraction and multidimensional NMR. Structural studies in the presence and absence of the small molecule would indicate whether a structural change has occurred.

More simply, we could digest samples of the mRNA with ribonuclease in the presence and absence of the small molecule. If binding of the small molecule elicits a conformational change, we would expect different regions of the mRNA to become accessible to the ribonuclease and thus give a different pattern of cleavage when separated using gel electrophoresis. This is termed a ribonuclease protection assay.

Follow-up question   Your riboswitch binds to galactosamine (GalN). Design an experiment to determine how your riboswitch binds to galactosamine (compared to its binding to galactose or glucosamine).

Summary

- Sigma factors bind to specific promoter sequences and regulate large groups of genes involved in a response to a stimulus such as heat shock.
- Operons are used to regulate groups of genes dedicated to a common purpose (for example, a metabolic pathway); they can be regulated either positively (inducibly) or negatively (repressibly).
- Riboswitches are mRNA molecules that contain an aptamer, a region that folds into a complex tertiary structure and can bind to a small ligand. The binding of the ligand changes the tertiary structure of the mRNA and affects how well the message can be translated.

Concept Check

1. Explain which methods provide a more global means of control (expression of many genes) and which provide a level of fine-tuning (expression of a single gene) and how they achieve this, providing several examples of each method.
2. What are operons? Describe their structure and function using *lac* operon as example.

17.3 Regulation of Gene Expression in Eukaryotes

Compared to prokaryotes, gene expression in eukaryotes is highly complex. Unicellular prokaryotes need to be able to respond to changes in the environment (such as nutritional status, availability of a specific metabolite, temperature, and osmolarity of their surroundings) and to cycle through a comparatively simple life cycle (regulation and coordination of replication). In contrast, most eukaryotes need to respond to all of the aforementioned challenges and also temporally and spatially regulate the genes involved in their growth and development. Gene expression changes often during development as the organism moves from a fertilized zygote through maturity and into senescence. It can also change in response to long-term environmental conditions and sometimes in response to disease states.

17.3.1 Eukaryotic DNA is organized into chromosomes

The regulation of eukaryotic gene expression is more complicated in part because the basic structure of a eukaryotic genome is more complex. Placed end to end, all the DNA molecules from a single human cell would stretch two meters. Each cell of the organism must consolidate that DNA into tightly packed and organized bundles that nevertheless allow access to the information coded in those DNA strands. The cell uses histones to help bundle the DNA molecules, discussed in greater detail in the following section. The nucleosome is an octamer of histones that forms a core wrapped with DNA (**Figure 17.12**).

As discussed previously, the DNA molecule is helical in nature. When wrapped around nucleosomes, DNA is twisted further, resulting in a supercoiled structure. As the DNA supercoils, nucleosomes pack together and form a superhelical structure known as a 30-nanometer

A.

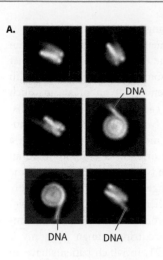

DNA

DNA DNA

B.

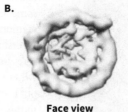

Face view **Side view**

FIGURE 17.12 Cryo-EM structure of a nucleosome. DNA is spooled onto histones to form nucleosomes. **A.** Shown are electron micrographs of histones in different orientations. **B.** The composite structure obtained from averaging thousands of those images.

(Source: Reprinted with Permissions from Springer Nature: Silvija Bilokapic et al. Nature Structural & Molecular Biology 25, pages 101–108, 2018)

fiber (based on the dimensions of the structure as determined by electron microscopy). The 30-nanometer fibers are further organized into loops by scaffolding proteins. This is probably the structure that chromosomes adopt in interphase (the growth period between division cycles). The iconic images of highly condensed chromosomes are only observed at metaphase.

The number of chromosomes an organism has can vary greatly (**Table 17.2**). The fruit fly, *Drosophila melanogaster*, has only four pairs, whereas corn has 10 pairs, humans have 23 pairs, and the king crab has about 104 pairs. The number of chromosomes does not correlate with the size of the genome (in terms of base pairs) or the complexity of the genome (in terms of its organization).

Telomeres (discussed in Chapter 14) are the terminal regions of the chromosome. These zones are not coded for by DNA; rather, they are added during DNA replication by telomerase. Telomeres have garnered attention due to their potential roles in the **pluripotency** of a cell (that is, the ability of the cell to keep dividing or to reach a predetermined number of replications). Cell division is important in both cancer and aging; hence, research on telomeres is of interest to the scientific and medical community and to the general public.

Centromeres are comprised of highly repeated sequences of DNA in the central regions of a chromosome, as is shown in **Figure 17.13**. These sequences hold the two copies (**sister chromatids**) of each chromosome together after DNA replication but before cell division (mitosis). Centromeres interact with **spindle fibers**—the actin and microtubule cages that assist in separation and segregation of chromosomes during the anaphase segment of mitosis—and are essential for proper cell division.

TABLE 17.2 A Comparison of Genomes

Species	Number of chromosomes	Base pairs of DNA	Number of genes
Human (*Homo sapiens*)	46	2.9×10^9	~21,000
Saccharomyces cerevisiae	30	12×10^6	5,700
Rice (*Oryza sativa*)	24	3.9×10^8	~37,500
Corn (*Zea mays*)	20	2.0×10^9	~32,000
Fruit fly (*Drosophila melanogaster*)	8	1.23×10^8	~17,000
Flatworm (*Caenorhabditis elegans*)	7	1.00×10^8	~22,000
Colonial amoeba (*Dictyostelium discoideum*)	6	3.4×10^7	~9,000
Mustard plant (*Arabidopsis thaliana*)	5	1.15×10^8	~27,000
E. coli K12	1	4.6×10^6	4,376
HIV-1 (RNA base pairs)	1	9.7×10^3	10
Epstein-Barr virus (causes mononucleosis)	1	1.7×10^3	80
Human mitochondrion	1	1.6×10^3	37

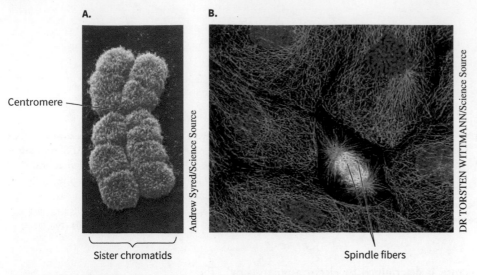

A.

Centromere

Andrew Syred/Science Source

Sister chromatids

B.

DR TORSTEN WITTMANN/Science Source

Spindle fibers

FIGURE 17.13 Electron micrographs of the metaphase chromosome and spindle fibers.
A. An electron micrograph of a fully condensed metaphase chromosome. The centromere is found in the middle of the image. **B.** A cell undergoing mitosis. The spindle fibers are shown guiding the separation of the metaphase chromosomes.

Changes to the number or arrangement of chromosomes can alter gene expression on that or other chromosomes and have serious health consequences for an organism. Such changes can take the form of a translocation (where one part of a chromosome is moved to or swapped with a segment of another chromosome), an inversion (where part of a chromosome is rearranged internally), or a polysomy (where there are more than two copies of a chromosome). The most commonly occurring human polysomy is trisomy 21, in which patients have an extra copy of chromosome 21. This manifests itself as Down syndrome, which causes varying levels of difficulty in physical and mental development. Recent research into this polysomy has focused on several microRNAs coded for by chromosome 21. Table 17.3 summarizes some common polysomies and chromosomal rearrangements and how these changes correlate to disease. Fluorescence *in situ* hybridization (FISH) is a technique that can be used to determine where a gene is found in the genome or whether a major rearrangement of DNA has occurred. Fluorescence techniques are discussed further in **Biochemistry: The use of fluorescence in biochemistry.**

TABLE 17.3 Chromosomal Disorders

Disorder	Description	Frequency	Symptoms	Prognosis
Down syndrome (Trisomy 21)	A third copy of chromosome 21	1 in 733 live births; frequency increases with maternal age.	Characteristic appearance, delayed social and mental development	Increased risk of heart defects and leukemias; shorter-than-normal life expectancy (55 years)
Edwards syndrome (Trisomy 18)	A third copy of chromosome 18	1 in 3,000 live births; frequency increases with maternal age.	Small at birth, heart defects, severe intellectual disability	Most children die within the first month of life; 10% live a year.
Patau syndrome (Trisomy 13)	A third copy of chromosome 13	1 in 10,000 to 21,000 live births; many affected fetuses miscarry or are stillborn.	Multiple and severe organ defects and deformities	Eighty percent of children die within the first month of life.
Klinefelter syndrome (XXY syndrome)	An extra X chromosome	1 in 650 males	Phenotypically male with decreased testosterone and elevated FSH and LH levels	Normal life expectancy; may have decreased fertility
Triple X syndrome (XXX syndrome)	An extra X chromosome	1 in 1,000 females	Phenotypically female; mild or no effects	Normal life expectancy and fertility
t(8;14)(q24;q32)	A translocation of the *c-myc* gene from chromosome 8q24 to the immunoglobulin heavy chain locus chromosome 14q32	300 cases annually in the United States; far more common in Africa and in HIV-positive patients	Burkitt's lymphoma	A five-year survival rate of 50%
t(2;3)(q13;p25)	A translocation of the PAX8 gene (paired box gene 8 on chromosome 2q13 to PPARγ 1 on chromosome 3p25)	5,500 cases annually	Follicular thyroid cancer	95% survival rate

Biochemistry

The use of fluorescence in biochemistry

Of the fundamental changes that have facilitated advances in biochemistry over the past generation, few can compare with the growing use of fluorescent molecules called fluorophores. Often when people hear the term *fluorescent*, they think of cheap glow-in-the-dark items or compact fluorescent light bulbs. Although some of these are truly fluorescent, others glow as a result of different physical phenomena.

Fluorescence is an intrinsic property typically associated with highly conjugated organic molecules (molecules with alternating sets of single and double bonds). As organic molecules contain more conjugated double bonds, they interact with longer wavelengths of light. With just a few conjugated double bonds, the molecule will absorb light in the ultraviolet (UV) range (think of benzene rings or the conjugated ring systems of the amino acids tryptophan and tyrosine). As more double bonds are found in a molecule, it will begin to interact with longer and longer wavelengths of light. The electrons of these molecules (β-carotene, for example) will absorb photons of light with frequencies (energies) relative to the degree of conjugation.

Once energy is absorbed, electrons in these molecules need to dissipate that energy in some form, most often as heat, such as when a dark object warms as it absorbs sunlight. Sometimes that energy triggers a chemical reaction, such as when inks and dyes fade in sunlight or when thymine dimers form when DNA is exposed to UV light. In other cases, the energy can be emitted as a new photon of light, leading to fluorescence or phosphorescence. Because the electron exists in an excited state for some brief period of time, it can evolve some energy as heat or undergo other relaxation events to reach lower energy states. The light that is emitted has less energy than the light that was absorbed and hence has a lower frequency (and longer wavelength). Therefore, the light is shifted to the longer wavelengths of the spectrum and is said to be red shifted.

Because the excited molecule can lose energy by collisions with other molecules (a process known as quenching), the phenomenon is environmentally dependent. For example, the fluorescent properties of a molecule will change as the molecule goes from an aqueous to hydrophobic environment.

Advances in optics and detection systems have permitted very sensitive levels of detection of fluorophores. Typical fluorescent detection systems are a thousand times more sensitive than colorimetric systems but a thousand times less sensitive than those that employ radioactivity.

In addition to advances in the optics used to detect fluorophores, there are now thousands of different fluorophores available, each with different functional groups or conjugated to different biological molecules for use in studies. The properties of these fluorophores have been fine-tuned by adding or removing double bonds to attain the desired absorbance and emission properties. There are virtually no wavelengths of the ultraviolet, infrared, or visible spectrum that are not covered.

Alongside small organics, biochemists have naturally occurring fluorescent proteins such as green fluorescent protein (GFP) at their disposal. In nature, organisms such as the jellyfish *Aequorea victoria* produce these molecules (ostensibly to signal to each other). In the laboratory, the cDNA coding for GFP can be used as a reporter gene or fused to the cDNA coding for a protein of interest. Once expressed, these molecules fluoresce when excited with the proper wavelength of light, revealing their location in the cell or organism.

Fluorescent molecules play a significant role in biochemistry today. Fluorescent dyes are the workhorse of high-throughput DNA sequencing and contributed greatly to sequencing the human genome. Immunofluorescence microscopy provides rapid subcellular localization of proteins, complexes, and organelles, all while leaving the greater context of the cell intact. Fluorescent dyes can measure molecules as small as a calcium ion or complexes as large as mitochondria and can do it at sub-nanomolar concentrations. Fluorescence *in situ* hybridization (FISH) is used to detect chromosomal translocations and genetic aberrations. Coupling of fluorophores in fluorescence resonance energy transfer (also known as Förster resonance energy transfer) is used to detect direct protein–protein interactions.

Translocation in chromosome 1 of a cancer cell

Chromosome

Centromeres

Look at Sciences/Medical Images

Instead of small organic particles, the latest generation of fluorophores employs nanoparticles (also known as quantum dots) made of cadmium or zinc selenide. The photoproperties of these small complexes arise from their small structure. Because conjugated double bonds are not needed, the particles never bleach or fade and are orders of magnitude brighter than the dyes used only a few years ago. These dyes are gaining favor in techniques such as live cell imaging.

Clearly, the basic physical phenomenon of fluorescence has spawned numerous advances, and more keep coming.

17.3.2 Histones and other proteins organize and give structure to the chromosome

Histones are the major structural proteins bound to DNA and are key members of the nucleosome. These proteins are found exclusively in the nuclei of eukaryotic cells. They constitute a superfamily of at least 77 proteins and can be separated into core histones (families H2A, H2B,

H3, and H4) and linker histones (the H1 family). As the name suggests, core histones form the core of the nucleosome, whereas the location and function of linker histones outside the core are uncertain.

The core histones form an octameric structure. Proteins H3 and H4 form a tetrameric structure (a dimer of dimers), as do H2A and H2B. Each of these complexes forms a disclike structure. The two discs then associate with each other to form a short cylinder. A groove runs around the outside of this cylinder, where the DNA wraps around the cylinder 1.7 times (about 147 base pairs of DNA) to form the nucleosome.

Each histone protein is small (~14 to 33 kDa) and highly basic. For example, in H4 in humans, 27 of the 103 residues are basic, and the protein has a pI of 10.2. Histones have a long helix capped by a helix-turn-helix domain at either end. At the end of each protein is a highly modified tail that is key to the regulation of histone function.

Histone proteins are highly conserved, with relatively few mutations or polymorphisms. This indicates that histones solve a particular challenge that provides the organism with an evolutionary advantage.

Several other proteins are also important in both organization of the chromosome and separation of sister chromatids during mitosis. The **structural maintenance of chromosome (SMC) proteins** are coded for by a family of six related genes. These proteins are responsible for higher-order chromosomal structure and function, that is, for condensation of 30-nanometer fibers into larger diameter structures and cohesion of sister chromatids. In contrast to histones, SMC protein homologs are found in bacteria. SMC proteins have a simple yet distinctive structure and are composed of two ATPase domains connected by lengthy (40 to 50 nm, ~300 amino acids long) coiled-coil domains (an alpha-helical domain in which multiple helices are coiled around one another like a braid or rope). These, in turn, are connected by a hinge region.

SMC proteins 1 and 3 are complexed with two other proteins (Scc1 and 3) to form the ring-shaped **cohesin complex** (**Figure 17.14**). Cohesins act to bind the sister chromatids together before separation in anaphase. Hydrolysis of ATP by the ATPase domains of the SMC proteins can mediate opening and closing of the ring. It is unclear how this complex interacts with DNA, but it seems likely that the ring encloses both copies of the sister chromatids before separation.

Condensins are proteins that have a similar structure to cohesins in that they contain a pair of SMC proteins (2 and 4) and form a ring in concert with other proteins. Condensin rings appear to contain five members, although the identity of the other proteins is unclear. There are two different classes of condensin complexes, at least one of which contains a cyclin-dependent kinase (CDK), and another interacts with topoisomerase II. The condensins are all involved in sister-chromatid separation during anaphase.

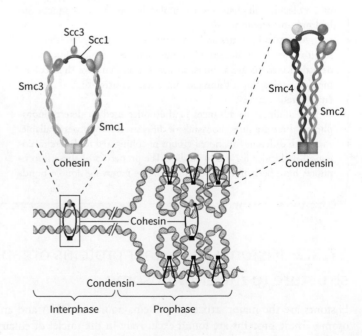

FIGURE 17.14 Cohesins and condensins in the cell cycle. Condensins assemble on the chromosomes in prophase following replication. They stay associated with the daughter chromatids until after separation in anaphase and assist in organizing the DNA into higher-order structures (loops). Following cell division, the condensins depart from the chromosomes, and the chromosomes return to their uncondensed state.

(Source: Karp, *Cell and Molecular Biology: Concepts and Experiments*, 7e, copyright 2013, John Wiley & Sons. This material is reproduced with permission of John Wiley & Sons, Inc.)

17.3.3 Epigenetic gene regulation can affect expression by modifying histones or DNA

Genetics is the study of heritable traits and is often taken to mean the study of genes. However, environmental factors are now known to also affect the expression of those genes. The study of modifications that affect gene expression but not DNA sequence is termed **epigenetics**.

Such modifications may occur over the course of an organism's development. One example is the differentiation of cells as an organism develops from pluripotent stem cells (able to differentiate into any of the cells of the organism) to **totipotent cells** (those able to differentiate into a limited number of cell types) to **terminally differentiated cells** (those unable to differentiate from one cell type to another). Modifications in gene expression may also occur in response to environmental stress. For example, survivors of the Dutch famine of 1944 were found to have increased risk of obesity, type 2 diabetes, and cardiovascular disease. Later research showed that the children and even grandchildren of women who survived the famine also suffered from these disorders at a higher rate. This finding suggests that the susceptibility to these conditions that resulted from experiencing famine was somehow passed down through these women's genes. A final example is when identical twins or congenic animals look or behave differently owing to environmental effects, despite having identical DNA. Although their genes are identical, expression of those genes has been affected by the individuals' environment and life experiences.

Two ways that cells can regulate gene expression are modifications to histones or to DNA.

Post-translational histone modifications Cells can control gene expression by regulating histone–DNA interactions to make the DNA more or less accessible to transcription factors and RNA polymerase. Histones are enriched in basic residues: arginine; histidine; and, in particular, lysine. The positively charged histones interact with the negatively charged DNA through coulombic interactions (analogous to a salt bridge in a protein) and other weak interactions; thus, hydrogen bonding occurs between the histones and DNA, dipoles from the alpha helices of the histones interact with the DNA helix, and there are hydrophobic-effect interactions between the members of the complex. Any change in histone–DNA interactions has the potential to affect gene transcription. Even relatively minor modifications (e.g., a phosphorylation) can radically change the function of a protein, and numerous modifications of histones can occur (**Table 17.4** and **Figure 17.15**).

TABLE 17.4 Histone Modifications

Modification	Explanation	Result
Mono-, di-, trimethylation	Addition (or loss) of one to three methyl groups on lysine or arginine residues	Typically decreases histone–DNA interactions and activates chromatin for transcription
Acetylation	Addition (or loss) of acetate groups on lysine residues	Typically decreases histone–DNA interactions and activates chromatin
Phosphorylation	Phosphorylation of serine, threonine, and tyrosine residues	Key step in some regulatory events, which may initiate methylation or DNA repair mechanisms.
SUMOylation	Ligation to SUMO, a small protein (12 kDa) similar to ubiquitin, but which doesn't target protein for degradation	Varies, often inhibits transcription
Ubiquitination	Ligation to ubiquitin, an 8.5 kDa protein (usually to lysine)	May mark histone for degradation; also signals DNA repair and transcriptional repression
ADP-ribosylation	Addition of ADP-ribose to lysine residues	Significant modification both in terms of charge and size; results in disruption of the nucleosome
Deimination/ citrullination	Loss of methylamine from methyl arginine/formation of citrulline	Varies, alters expression levels
Biotinylation	Addition of a biotin group to a lysine residue	Found in repressed regions of heterochromatin; probably silences genes

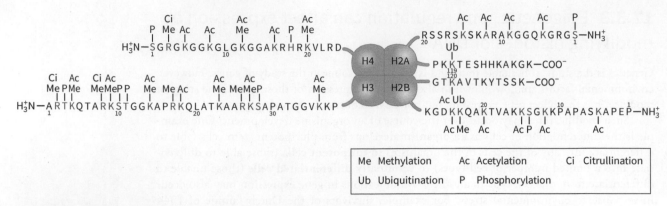

FIGURE 17.15 Histone modifications.   Histones, especially the tails of histones, are among the most highly modified proteins. This schematic shows some of the known modifications of these proteins.

Several enzyme classes modify histones to either increase or decrease histone–DNA interactions. **Histone acetyltransferases (HATs)** and **histone deacetylases** add acetyl groups to or remove them from specific lysine residues in histone tails. Acetylases use acetyl-CoA to form an amide linkage between the free amine of the lysine and the acetate group. **Histone methyltransferases (HMTs)** transfer one to three methyl groups to lysine and arginine residues, also in the tails of histone proteins. *S*-adenosylmethionine is the activated methyl donor for these reactions (**Figure 17.16**). Collectively, the proteins and enzymes involved in the changes to histones and DNA are known as the **chromatin-remodeling complex**.

One particular chromatin-remodeling complex is worthy of mention. The **switch/sucrose non-fermentable (SWI/SNF) complex** (pronounced *switch-sniff*) is an ATP-dependent histone-remodeling complex originally found in yeast; homologs are found in other eukaryotes, including humans (**Figure 17.17**). The SWI/SNF complex is large, consisting of 11 proteins with a combined molecular weight of about 1.5 megadaltons. The complex breaks down ATP, uses that

FIGURE 17.16 Histone methyltransferase (HMT) proteins.

A. This is the structure of the human lysine *N*-methyltransferase domain of the SETMAR protein. **B.** This HMT specifically methylates Lys3 and Lys36 of histone H3 in humans using *S*-adenosylmethionine. The structure has an architecture called a beta clip that is common in methylases and AdoMet binding proteins.

(Source: (A) Data from PDB ID 3B05 Nagai, T., Unno, H., Janczak, M. W., Yoshimura, T., Poulter, C.D., Hemmi, H. (2011) Covalent modification of reduced flavin mononucleotide in type-2 isopentenyl diphosphate isomerase by active-site-directed inhibitors. *Proc.Natl.Acad.Sci.USA* **108**: 20461–20466)

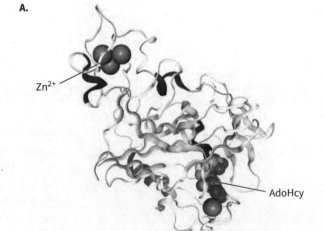

A.

B.

S-Adenosylmethionine
(AdoMet)

S-Adenosylhomocysteine
(AdoHcy)

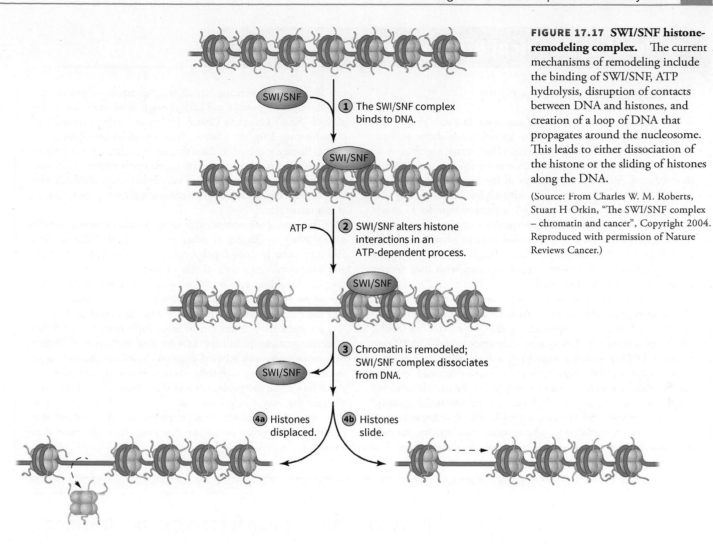

FIGURE 17.17 SWI/SNF histone-remodeling complex. The current mechanisms of remodeling include the binding of SWI/SNF, ATP hydrolysis, disruption of contacts between DNA and histones, and creation of a loop of DNA that propagates around the nucleosome. This leads to either dissociation of the histone or the sliding of histones along the DNA.

(Source: From Charles W. M. Roberts, Stuart H Orkin, "The SWI/SNF complex – chromatin and cancer", Copyright 2004. Reproduced with permission of Nature Reviews Cancer.)

energy to pinch a loop of DNA away from a nucleosome, and then slides that nucleosome farther down the DNA molecule (like sliding a bead down a string). Several proteins in the SWI/SNF complex contain **bromodomains**, four-helix bundle domains that interact with acetyllysines in histone tails.

By modifying histones after translation, cells change the way in which histones bind to DNA and thus change the level of sequestration of the DNA. By loosening or tightening the hold histones have on DNA, the cell makes DNA more or less accessible to the transcription machinery of the cell.

DNA methylation

In addition to proteins, DNA can be modified to regulate gene expression. Some bacteria methylate DNA—either on the N-6 position of adenine residues or on the C-5 position of cytosine residues—to block the action of specific restriction enzymes. Eukaryotes have several systems for modifying DNA to alter gene expression, for example, methylations of DNA on C-5 of cytosine bases by DNA methyltransferases. The cytosines are found in sequences called **CpG islands** because they are rich in CG repeats, with a phosphoester linkage (designated by "p") between the two. Methylation does not affect base pairing because the cytosines still form hydrogen bonds with guanines. However, the methyl group is found in the major groove and can affect how the methylated DNA interacts with proteins such as histones or transcription factors. CpG islands are common in the promoter regions of genes and near transcriptional start sites.

Both epigenetics and genetics contribute to the phenotype of an organism. We are just beginning to scratch the surface of how epigenetics functions in gene expression, but we know far more about genetics. In recent years, several companies have offered genetic testing to provide information about your ancestry or risk of developing diseases, discussed further in **Societal and Ethical Biochemistry: The business of gene sequencing.**

Societal and Ethical Biochemistry

The business of gene sequencing

It is human nature to be curious and want to know things. This curiosity is compounded when the knowledge is about us, especially when there are gaps in our personal information such as where our families came from or what diseases we may be susceptible to in the future. Since the completion of the Human Genome Project and with the advent of new technologies, it has become simple and affordable to have our DNA sequenced in order to answer some of these questions. Several companies now promise to tell you which small town your ancestors hailed from or what diseases your great-great-grandparents likely suffered. But how truthful are these claims? How do these services actually operate, what data do they obtain, and how is that data used?

To establish a genetic history, researchers typically track polymorphisms throughout the genome such as variability in the copy number of a tandem repeat or single-nucleotide polymorphisms (SNPs). Direct-to-consumer services such as Ancestry.com and 23andMe use Illumina DNA-sequencing technology to sequence approximately 80,000 polymorphisms in each sample of saliva submitted by a consumer. Results are then compared to an SNP database to identify other people with these same polymorphisms to determine which groups of SNPs link a person with certain geographic regions or diseases.

At-home gene testing may be an interesting activity, but there are some problems with the claims these companies make. It is true that DNA sequencing can allow us to trace our genetic history through our parents and back through other ancestors. It has allowed some adoptees to identify biological family members and helped reunite long-lost relatives. However, given the dynamics of human history, it is hard to claim that an individual's genotype geographically locates their distant ancestors. Furthermore, these analyses are based on a comparison of data only from individuals who have already been tested by a given company, meaning the results are at best incomplete.

There are also problems with using these analyses to identify susceptibility to diseases or other personal traits. Some services claim to have identified polymorphisms associated with complex behavioral traits such as fear of public speaking. Although the polymorphisms may be linked with these traits, clearly other factors, including personal experiences, play a role. More concerning, the data showing you may be at risk for cancer or a carrier of a potential birth defect can be more frightening than helpful, especially outside the context of a medical consultation. People may make important medical decisions based on incomplete or inaccurate information. Finally, there is reason to be concerned about how these companies use and share your personal data and whether that may pose privacy risks.

More than anything, these considerations demonstrate that genomics is entering a new phase that poses more questions than answers.

17.3.4 Eukaryotic transcription factors can be classified based on their mechanism of action

In eukaryotes, transcription factors that selectively bind to the promoter region of genes to activate them are a major means for control of gene expression. Transcription factors can be regulated by phosphorylation or by binding to small ligands, other transcription factors, or parts of the transcriptional machinery. It is estimated that nearly 10% of the human genome codes for transcription factors.

In terms of structure, a transcription factor has three distinct regions: a DNA-binding domain (discussed earlier in the chapter), a trans-acting domain, and a signal-sensing domain. The trans-acting domain is responsible for interactions with other proteins such as coregulators, which may act either as coactivators or corepressors. The signal-sensing domain is the part of the protein that regulates function; this is the part that binds to ligands or may be phosphorylated.

Because these transcription factors act on DNA, they need to function in the nucleus. However, many transcription factors are found in the cytosol and have to be activated and translocated to the nucleus before they can affect gene expression.

Transcription factors can be classified based on their function into six different categories: general, developmentally regulated, nuclear hormone, ligand dependent, resident nuclear, and cytosolic translocating to the nucleus. Representative transcription factors from these categories are discussed here to illustrate their characteristics, properties, and functions. Often, the example given is just one member of a family of transcription factors. Each factor may have other (as yet undiscovered) modes of action, and there may be many other proteins that act in a similar fashion. Table 17.5 summarizes and compares several of these transcription factors.

General: C/EBP   CCAAT/enhancer-binding protein (C/EBP) is a general transcription factor that acts as a master regulator of gene expression in many tissues (**Figure 17.18**). It contains a leucine zipper domain that recognizes the CCAAT box, a common promoter region in many

TABLE 17.5 Representative Eukaryotic Transcription Factors

Factor	Binding Motif	Regulated By	Class and Function
C/EBP	Leucine zipper	Phosphorylation (among others)	General transcription factor. Plays role in differentiation, cancer, and inflammation; heterodimerizes with other leucine zipper transcription factors
HOX	Homeodomain	Chemical gradients	Developmentally regulated. Key players in anterior–posterior body development
PPAR	Zinc finger	Ligand binding and phosphorylation	Nuclear hormone receptor. Important in tissue-specific (e.g., adipose, liver, or muscle) gene expression and metabolism; heterodimerizes with RXR
SREBP	Helix-loop-helix and leucine zipper	Ligand binding, proteolytic processing	Ligand dependent. Binds sterol in ER, processed in Golgi apparatus and translocates to nucleus; active SREBP to sterol response elements
CREB	Leucine zipper	Phosphorylation	Resident nuclear. Responds to cAMP levels (via PKA); important in nervous system development
NF-κB	Rel homology	Cytosolic inhibitor protein	Cytosolic translocating to the nucleus. Plays key role in inflammation and immune responses; involved in cell proliferation; anti-apoptotic

eukaryotic genes. C/EBP binds to DNA as a dimer, interacting with other proteins through the leucine zipper motif. However, C/EBP does not need to dimerize with other copies of C/EBP or even with other C/EBP isoforms. There are six members in the C/EBP family, designated α to ζ. C/EBP can form homodimers but has been shown to dimerize with other leucine zipper proteins, such as the Jun/Fos family and CREB.

C/EBPβ is regulated by phosphorylation via mitogen-activated protein (MAP) kinase, at least in part, which results in changes in translocation of the transcription factor to the nucleus or its ability to interact with DNA. Various other members of the C/EBP family are also regulated by phosphorylation but through a different mechanism. Phosphorylation of key residues of C/EBPα (by an unknown kinase) primes the protein for coupling to ubiquitin. As with many proteins, ubiquitination targets C/EBP for proteasomal degradation. Thus, phosphorylation of C/EBPα and ubiquitination decreases the activity of the protein, resulting in decreased gene expression. In the case of C/EBPα, one effect of this is a reduction the ability of stem cells to differentiate into adipocytes.

As might be expected for a transcription factor that acts as a general regulator, the C/EBP family regulates a diverse set of genes found in many tissues. C/EBP is important in cell differentiation, as progenitor cells change into different cell types. As such, it is a key player in the development of adipocytes, myeloid cells, and osteoclasts, and dysfunction of C/EBP has been implicated in diseases such as obesity, type 2 diabetes, leukemia, and osteoporosis. C/EBP is an **anti-proliferative** transcription factor (that is, it promotes differentiation), but it slows cellular replication (in the same way that nuclear anti-proliferation treaties among nations are designed to stop the spread and reduce the number of nuclear weapons). Defects in C/EBP function have been implicated in several cancers. Finally, C/EBP functions as a master regulator of genes involved in metabolism. Numerous hepatic and adipocyte genes coding for key proteins in metabolic pathways contain CCAAT sequences in their promoter regions. This includes other transcription factors (for example, PPARγ), hormones (PEPCK and leptin), and enzymes (glycogen synthase and stearoyl-CoA desaturase).

C/EBP is not dependent on binding any small ligands for activity.

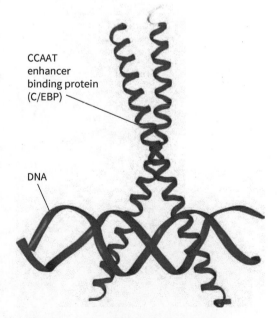

CCAAT enhancer binding protein (C/EBP)

DNA

FIGURE 17.18 CCAAT/enhancer-binding protein (C/EBP). C/EBP dimerizes and binds DNA using a leucine zipper motif. This transcription factor has been implicated in a diverse array of processes.

(Source: Data from PDB ID 1NWQ Miller, M., Shuman, J.D., Sebastian, T., Dauter, Z., Johnson, P.F. (2003) Structural Basis for DNA Recognition by the Basic Region Leucine Zipper Transcription Factor CCAAT/enhancer Binding Protein Alpha. *J. Biol. Chem.* 278: 15178–15184)

Developmentally regulated: HOX genes

Homeobox (HOX) genes are a group of genes coding for transcription factors, all of which contain a homeodomain. The HOX genes are key regulators of pattern development and anterior–posterior body development in animals (**Figure 17.19**). Originally identified and characterized in the fruit fly (*Drosophila*), the genes are conserved among all metazoans, including humans.

The homeodomain comprises a helix-turn-helix motif stabilized by a third helix. The consensus sequence for the homeodomain is

**RRRKRTA-YTRYQLLE-LEKEFLF-NRYLTRRRRIELAHSL-
NLTERHIKIWFQN-RRMK-WKKEN**

This sequence is highly conserved across the animal kingdom; again, this suggests that mutations to the protein are likely to be harmful to the organism and that there is considerable selective pressure to maintain this sequence. Structural studies indicate that these proteins can bind DNA as a monomer, but they may also interact with other HOX transcription factors, and they can form a dimeric or multimeric complex with DNA to regulate gene transcription.

HOX proteins bind to this short DNA sequence:

$$5'–TAAT–3'$$

A comparison of the DNA-binding domain of two HOX proteins (for example, antennapedia, which controls leg development, and bithorax, which controls anterior–posterior development) illustrates how small changes either in the promoter region bound by the HOX protein or to the protein itself can manifest as significant changes in gene expression.

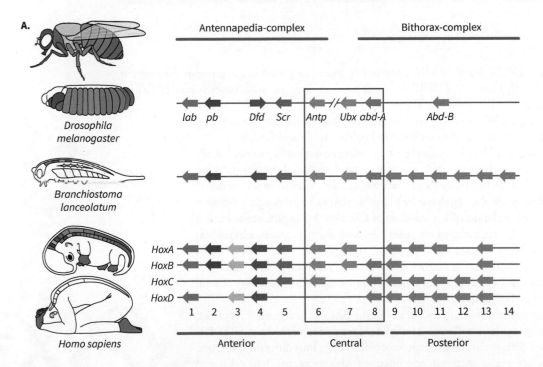

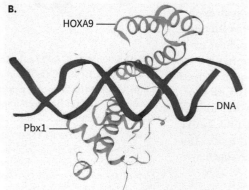

FIGURE 17.19 HOX genes in development.   A. HOX genes are conserved throughout metazoan evolution and have common functions in species as diverse as *Drosophila* and humans. Each arrow represents a single HOX gene and where it is found relative to other family members, numbered 1 to 14. Color coding of the anatomical drawings indicates where each gene affects development. **B.** The structure of the HOXA9 and Pbx1 heterodimer bound to DNA.

(Source: (A) Stefanie D. Hueber; BioEssays. Copyright 1984. Reproduced with permissions of John Wiley & Sons Inc. (B) Data from PDB ID 1PUF Laronde-Leblanc, N.A., Wolberger, C. (2003) Structure of HOXA9 and PBX1 bound to DNA: HOX hexapeptide and DNA recognition anterior to posterior. *Genes Dev.* **17**: 2060–2072)

HOX genes are notable for their role in development and for the dramatic phenotypes that result from faulty expression of these genes. For example, in the fly, mutations of the *antennapedia (Antp)* gene result in the development of a set of legs where the antenna should be, and mutations of the *ultrabithorax (Ubx)* gene can result in a second set of wings forming (essentially a second thorax). Clearly, these are complex, multi-step processes that are more complicated than turning a single gene on or off. Some HOX proteins respond to gradients of small soluble molecules, and the position of HOX genes in the genome regulates how these genes are expressed. Many (if not all) of these genes and mutations have been discovered through genetic rather than biochemical techniques. Hence, although some outcomes (for example, a headless fly embryo) are clear, the mechanisms behind these mutations have yet to be elucidated.

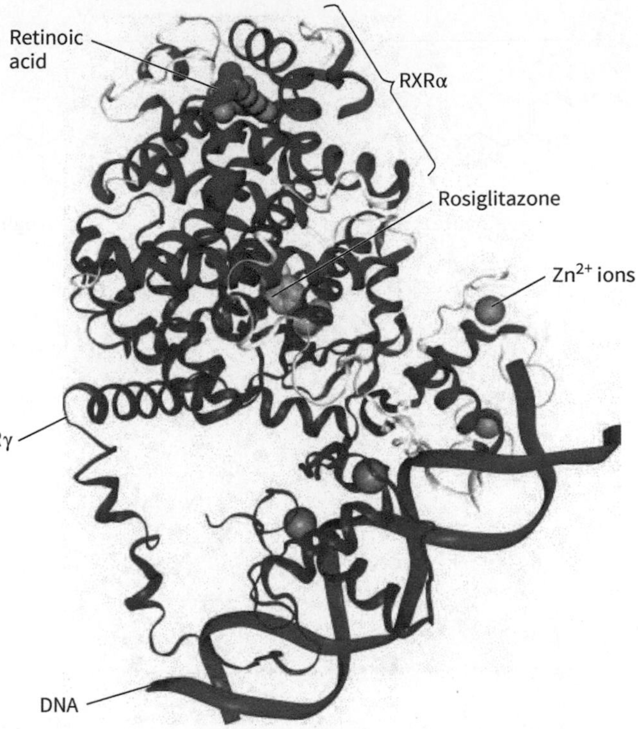

Nuclear hormone receptors: PPARs

Peroxisome proliferator activated receptors (PPARs) are a family of three related nuclear hormone receptors (α, β/δ, and γ) that are important in the differentiation and development of specific tissues (e.g., adipose, liver, or muscle) and the metabolic response of those tissues (**Figure 17.20**). The PPAR family was originally identified as being responsible for increasing lipid oxidation in the presence of long-chain unsaturated fatty acids, especially ω–3 and ω–6 polyunsaturated fatty acids. Each PPAR recognizes a different sequence based on the dimeric binding partner for the PPAR. For example, PPARα will form a heterodimer with the retinoid X receptor α (RXRα), and it recognizes a direct repeat of the sequence 5′-AGGTC-3′, separated by a single nucleotide (X), as shown:

5′–AGGTCXAGGTC–3′

FIGURE 17.20 PPAR transcription factors. The structure of a PPARγ–RXRα complex bound to DNA, 9-*cis*-retinoic acid, and rosiglitazone. Note that each transcription factor has its own ligand and that both transcription factors bind to DNA.

(Source: Data from PDB ID 3DZY Chandra, V., Huang, P., Hamuro, Y., Raghuram, S., Wang, Y., Burris, T.P., Rastinejad, F. (2008) Structure of the intact PPAR-gamma-RXR-nuclear receptor complex on DNA. *Nature* 456, 350–356)

This sequence is recognized by several other transcription factors that dimerize with RXRα, including the T3 and vitamin D3 receptors. However, the distance between the binding sites varies, with four nucleotides in the case of the T3 receptor and five in the case of the vitamin D3 receptor. These elements are collectively known as direct repeat 1, 4, or 5 elements (DR1, DR4, or DR5), with the number signifying the number of nucleotides in between the repeats. Again, several important aspects of transcription factors are seen. Although fundamental to the DNA–transcription factor interaction, the primary sequence of the DNA molecule is not as important as the three-dimensional shape of the molecule. Also, this illustrates how transcription factors (in this case, RXRα) can bind with different partners, ultimately resulting in a finer degree of control of gene expression.

The PPARs interact with DNA via two zinc finger domains. Members of this family also possess a large ligand-binding domain, linked by a hinge region. PPARs function as heterodimers with RXR, linked predominantly through the ligand-binding domains.

PPARs are regulated by both ligand binding and phosphorylation. The native ligand for PPARs is still a matter of debate, although polyunsaturated fatty acids and eicosanoids are probable candidates (prostaglandin J2 has been shown to stimulate PPARγ and leukotriene B4 to inhibit the function of PPARα). In relation to phosphorylation, PKA and protein kinase C (PKC) and the MAP kinases ERK, p38, and JNK have all been shown to phosphorylate PPARs. The precise ramifications of these phosphorylation events are unclear, but there are several interesting hypotheses. One is that phosphorylations may change the affinity of the transcription factor for its ligand or the interaction of the transcription factor complex with DNA. Alternatively, it may target the transcription factor for coupling to ubiquitin and subsequent degradation or translocation of the PPAR to the nucleus.

In the inactive state, PPARs are found in the cytosol or nucleus. Upon ligand binding, the PPAR dimerizes with RXR. This complex then binds DNA and can affect gene transcription.

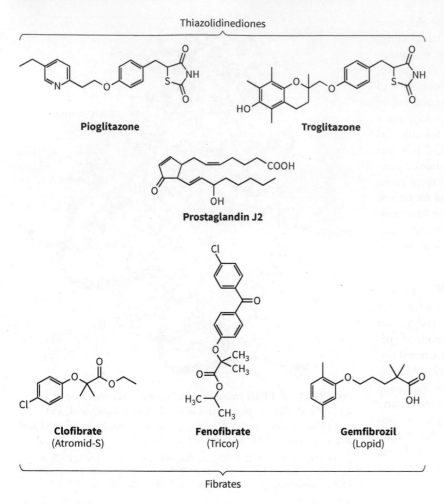

Thiazolidinediones

Pioglitazone

Troglitazone

Prostaglandin J2

Clofibrate
(Atromid-S)

Fenofibrate
(Tricor)

Gemfibrozil
(Lopid)

Fibrates

FIGURE 17.21 Ligands for the PPARs.   Prostaglandin J2 (PGJ2) is thought to be the native ligand for the PPARs. A number of very successful drugs have been developed that bind to PPARs and modulate their activity. These include the thiazolidinediones (pioglitazone and troglitazone), which activate PPARγ and are antidiabetic agents, and the fibrates (clofibrate, fenofibrate, and gemfibrozil), which activate PPARα and are used to treat hyperlipidemia (high-plasma triacylglycerols and cholesterol).

However, the story is complicated by the fact that RXR can bind to retinoids, 9-*cis*-retinoic acid being the endogenous ligand (although other retinoids can bind as well). In addition, there are several coactivators (PGC–1, the PPARγ coactivator; SRC–1, the steroid receptor coactivator) and corepressors (NCoR, the nuclear receptor corepressor; SMRT, the silencing mediator of retinoid and thyroid receptors) of the PPARs that can affect gene transcription. Last, PPARs are competing with other transcription factors (for example, the T3 and vitamin D3 receptors mentioned earlier) for RXR binding. The interactions of PPARs and their role in gene transcription are the subject of numerous research studies.

The PPARs are the targets of several drugs or pharmaceuticals (**Figure 17.21**), such as the fibrate class (for example, clofibrate, gemfibrozil, and fenofibrate) of cholesterol-lowering drugs that are used to treat hypercholesterolemia (high-plasma cholesterol). These drugs act by binding and inhibiting PPARα; hence, they are PPARα agonists and are thought to downregulate genes involved in the biosynthesis of lipids.

The thiazolidinediones (TZDs or glitazones) are a class of oral antidiabetic agents used in the management of type 2 diabetes. These drugs (rosiglitazone, pioglitazone, and troglitazone) bind and activate PPARγ; that is, they act as PPAR agonists. In effect, these drugs make tissues more responsive to glucose by turning on fat genes and, paradoxically, making the tissue able to store more fat.

Collectively, these classes of drugs are known as PPAR modulators. They are hugely successful compounds from both a medical and a financial standpoint.

Ligand dependent: SREBP   As its name suggests, the **sterol regulatory element–binding protein (SREBP)** is a transcription factor that binds to a small ligand (in this instance, a sterol) and to a sterol-response element (SRE)—the promoter sequence of DNA upstream of genes that are responsive to sterols (**Figure 17.22**). There is some flexibility in the sequence that SREBP recognizes. The sterol response element sequence is

5'–TCACNCCAC–3'

where N is any nucleotide. But SREBP will also recognize

5'–CANNTG–3'

This variability is based on several factors. First, SREBP is a member of the helix-loop-helix/leucine zipper family of transcription factors. It can form homodimers or heterodimers with other family members that have the same DNA-binding domain. However, in place of an arginine (found in almost all other family members), SREBP has a tyrosine. This single change enables SREBP to recognize the SRE.

SREBP activation is different from the other transcription factors we have considered in that it is activated by proteolytic cleavage. In the inactive form, SREBP is found inserted into the endoplasmic reticulum (ER) membrane in a hairpinlike structure. The cytosolic DNA-binding domain is found in the amino terminus of the protein and is connected by a transmembrane helix

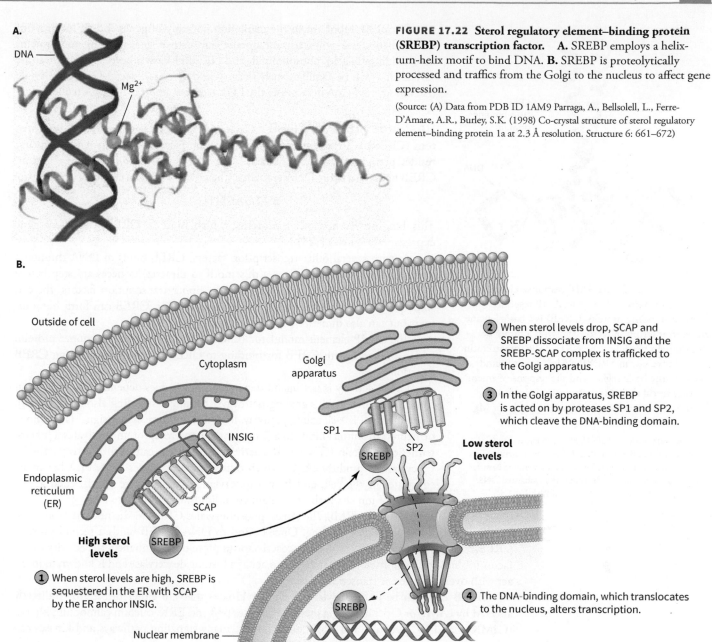

FIGURE 17.22 Sterol regulatory element–binding protein (SREBP) transcription factor. **A.** SREBP employs a helix-turn-helix motif to bind DNA. **B.** SREBP is proteolytically processed and traffics from the Golgi to the nucleus to affect gene expression.

(Source: (A) Data from PDB ID 1AM9 Parraga, A., Bellsolell, L., Ferre-D'Amare, A.R., Burley, S.K. (1998) Co-crystal structure of sterol regulatory element–binding protein 1a at 2.3 Å resolution. Structure 6: 661–672)

to a loop on the luminal side of the ER. A second transmembrane helix connects this loop to a regulatory domain in the cytosol.

In the ER, SREBP interacts with the SREBP cleavage–activating protein (SCAP), which has several important functions. SCAP has a sterol-binding domain that acts as a sterol sensor, and when sterols are plentiful, SCAP also binds to a third protein, called INSIG. The INSIG protein is found only in the ER, and it anchors SCAP (and indirectly SREBP) in the ER when sterols are available. Once sterol levels drop, SCAP dissociates from INSIG, and the SCAP–SREBP complex is trafficked to the Golgi complex apparatus via a COPII-coated vesicle.

In the Golgi apparatus, SREBP is acted upon by two aptly named proteases: site 1 protease and site 2 protease. The first cleavage (site 1) occurs in the lumenal loop (i.e., the loop in the lumen of the ER) between the two transmembrane helices; it generates two fragments, each of which contains a cytosolic domain and a transmembrane helix. The second cleavage (carried out by the site 2 protease) occurs within the transmembrane helix of the original amino terminal fragment of SREBP. This second cleavage frees the active form of SREBP, which is then translocated to the nucleus, where it can activate gene transcription (including transcription of genes involved in sterol synthesis).

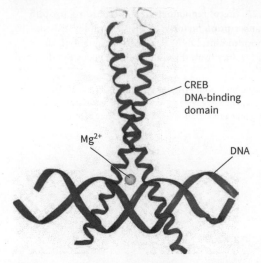

CREB
DNA-binding
domain

Mg²⁺

DNA

FIGURE 17.23 Cyclic AMP response element–binding protein. The cyclic AMP response element–binding protein (CREB) is a basic leucine zipper transcription factor. Shown is a dimer of this domain bound to the CRE of the somatostatin gene. As we saw in SREBP, a Mg²⁺ ion is bound between the basic region and the response element. Studies reveal that this magnesium increases binding by over 25-fold, suggesting a role for Mg²⁺ in these transcription factors.

(Source: Data from PDB ID 1DH3 Schumacher, M.A., Goodman, R.H., Brennan, R.G. (2000) The structure of a CREB bZIP.somatostatin CRE complex reveals the basis for selective dimerization and divalent cation-enhanced DNA binding. *J. Biol. Chem.* **275**: 35242–35247)

SREBP is important in the regulation not only of genes coding for proteins involved in sterol synthesis and uptake but also of genes coding for proteins involved in fatty acid, phospholipid, and NADPH biosynthesis. Among these are several that will be familiar, such as fatty acid synthase, acetyl-CoA carboxylase (ACC), stearoyl-CoA desaturase, the LDL receptor, and HMG-CoA reductase.

Resident nuclear: CREB   Cyclic AMP response element–binding protein (CREB) is a resident nuclear transcription factor that (as its name suggests) resides permanently in the nucleus. Upon phosphorylation and dimerization, CREB binds to a short DNA sequence (the cAMP-response element, or CRE):

$$5'–TGACGTCA–3'$$

This binding recruits other proteins, which bind to CREB and affect gene expression (**Figure 17.23**).

As with several other transcription factors, CREB binds to DNA through a leucine zipper motif. It also uses this motif to dimerize, a necessary step before DNA binding. And, as with other leucine zipper transcription factors, there is some promiscuity with regard to the binding partner. CREB can form homodimers but can also dimerize with other leucine zipper proteins.

The cAMP-element modulator (CREM) is a related leucine zipper protein, which competes with CREB for binding to CRE, effectively inactivating CREB function.

CREB activity is not ligand dependent, but it does depend on phosphorylation; hence, many cell signaling pathways that affect signaling also affect CREB. Consider the PKA-signaling pathway, in which an extracellular signal (a hormone or neurotransmitter) binds to a 7-transmembrane receptor and activates a heterotrimeric G protein (GPCR). The activated G protein can in turn activate adenylyl cyclase to produce cAMP, which binds to and activates PKA. PKA has many substrates in the cell, and its phosphorylation causes acute effects, such as the opening of ion channels or breakdown of lipids, but it can also affect gene expression through CREB. The kinase phosphorylates CREB in the nucleus, leading to CREB dimerization and DNA binding. Once bound to DNA, CREB binds a second protein, CREB-binding protein (CBP), which is a coactivator (a protein that enhances the activity of the factor it binds to). CBP itself has been found to act as a histone deacetylase and is known to interact with over 60 different transcription factors.

CREB can also be phosphorylated by other kinases that are either directly or indirectly affected by cAMP or Ca²⁺, including the MAP kinases p38 and ERK, calcium calmodulin kinases (CaMK) I and IV, p90S6K, and RSK1 and 2. Thus, many signaling pathways and kinases can affect gene expression by acting through CREB.

Just as adding a phosphate to CREB activates it, the reverse (dephosphorylation) inactivates CREB. The proteins implicated in the dephosphorylation of CREB are protein phosphatases 1 and 2A (PP1 and PP2A).

Both CREB and CREM are necessary for proper development and function of the nervous system. Several genes are regulated at least in part by CREB, including genes coding for growth factors (brain-derived neurotrophic factor), neurotransmitters and hormones (somatostatin), ion channels (Na⁺/K⁺ ATPase α), and signal transduction (neuronal nitric oxide synthetase). These are only a few of the hundreds of genes known to be regulated at least in part by CRE.

Despite similarities in their names and interaction between the three, CEB/P, SREBP, and CREB are different transcription factors with different functions.

Cytosolic translocating to the nucleus: NF-κB   Nuclear factor kappa B (NF-κB) is a cytosolic protein that, upon activation, translocates to the nucleus to affect gene transcription (**Figure 17.24**). It was originally identified as a key factor in the transcription of kappa light chains of immunoglobulins, hence the name. NF-κB binds to the κB promoter sequence:

$$5'–GGGRNYYYCC–3'$$

where R is a purine, Y is a pyrimidine, and N is any unspecified base.

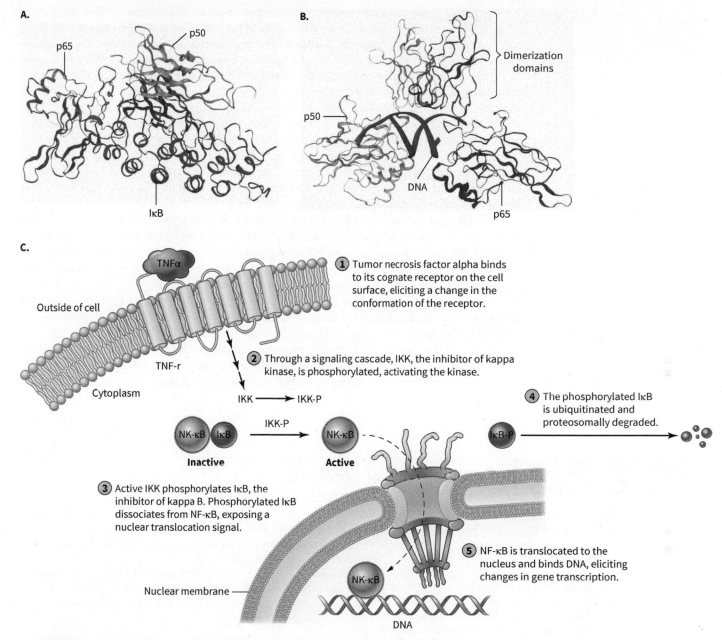

FIGURE 17.24 Nuclear factor kappa B (NF-κB) **A.** The structure of NF-κB is bound in a complex to its cytosolic inhibitor IκB. IκB is a β-hairpin/α-hairpin repeat (an ankyrin repeat). The NF-κB dimer consists of two beta-rich structures: p50 and p65. **B.** The p50/p65 heterodimer is shown bound to DNA. The dimerization domains participate in binding using Rel homology domains, β-rich immunoglobulinlike folds. **C.** Details of the NF-κB signaling pathway are shown.

(Source: (A) Data from PDB ID 1IKN Huxford, T., Huang, D.B., Malek, S., Ghosh, G. (1998) The crystal structure of the IkappaBalpha/NF-kappaB complex reveals mechanisms of NF-kappaB inactivation. *Cell* **95**: 759–770. (B) Data from PDB ID 1VKX Chen, F.E., Huang, D.B., Chen, Y.Q., Ghosh, G. (1998) Crystal structure of p50/p65 heterodimer of transcription factor NF-kappaB bound to DNA. *Nature* **391**: 410–413)

NF-κB binds to DNA using a Rel homology (RH) domain—an immunoglobulinlike β-barrel fold that fits into the major groove of the target DNA sequence. Because NF-κB is active as a dimer, two such domains pinch the DNA helix. NF-κB can form homodimers or heterodimers with other NF-κB or Rel proteins. NF-κB is in the same family as Rel proteins and has many functional similarities.

To generalize, NF-κB is a transcription factor that responds to inflammatory signals and helps to regulate cell proliferation and survival. There are many ways to activate NF-κB. It responds to signals as diverse as stress, proinflammatory cytokines, or UV light. One representative pathway is the response to the proinflammatory cytokine tumor necrosis factor alpha (TNFα). In this scenario, a trimer of TNFα molecules on the outside of the cell binds to a TNFα receptor, eliciting a conformational change inside the cell. As with many other signaling pathways, this conformational change leads to the development of a nucleation site for scaffolding proteins and kinases. In this instance, a complex of proteins activates the inhibitor of kappa B kinase (IKK) through phosphorylation. Active IKK phosphorylates a cytosolic complex of NF-κB and its inhibitor, the aptly named inhibitor of κB (IκB). Phosphorylation occurs on the IκB protein. Once phosphorylated, IκB dissociates from NF-κB and is then ubiquitinated and marked for degradation by a proteasome. Nuclear localization signal sequences in NF-κB are now exposed, and NF-κB is translocated to the nucleus to bind to DNA and affect gene transcription. Hence, NF-κB is not directly affected by phosphorylation or ligands, but its binding partner, IκB, is affected by phosphorylation. The implications of this are that multiple extracellular ligands and receptors can have an effect on NF-κB signaling by acting upstream of IKK.

NF-κB was originally identified in the cells of the immune system, but it is also involved in many cancers, in neuronal plasticity (the ability of neurons to strengthen or change their connections to other neurons), and memory. NF-κB regulates the body's immune response, and dysfunction of this transcription factor has been found in some autoimmune diseases. In general, NF-κB regulates cell survival and proliferation and is antiapoptotic.

NF-κB is considered to be a fast-acting transcription factor—similar to nuclear hormone receptors, c-Jun and the STAT family—because no new proteins need to be synthesized to respond to the signal.

In this section, we have described transcription factors that regulate genes in a variety of circumstances and with a variety of mechanisms. However, most genes in eukaryotes have multiple promoter and enhancer sites in the 5′ control regions, meaning that most genes have multiple factors that affect their expression.

Worked Problem 17.4   Identification of a transcription factor binding site

You hypothesize that a gene you have been characterizing is partially regulated by a cytosolic transcription factor similar to NF-κB. How can you determine whether this protein is involved in binding promoter sequences for your gene?

Strategy   There are several ways to analyze promoter sequences. Classically, this would have been done using a reporter gene assay with a variety of truncated promoters or a footprinting assay. A more modern approach would be to use a chromatin immunoprecipitation (ChIP) assay. This assay is described in the Techniques Appendix.

Solution   To use a ChIP assay, we would need an antibody to the transcription factor in question. Cells are treated with formaldehyde to covalently cross-link DNA and any proteins that are bound to the DNA. The DNA is then fragmented using sonication, and the complex is immunoprecipitated with the antibody to the transcription factor and antibody-binding beads. Unbound proteins are washed off the precipitate, and the cross-linking is reversed by heating the sample. Finally, the DNA sequence is identified by using PCR with specific primers or is sequenced to identify the actual nucleotide sequence.

Follow-up question   If this transcription factor were a nuclear hormone receptor (as is the case with steroid receptors or the PPARs), how could you use a ChIP assay to see whether it binds DNA in the presence or absence of a drug you hypothesize to be acting as a ligand?

17.3.5 MicroRNAs regulate gene expression at the message level

This chapter has explained how, in prokaryotes, regulation of transcription involves a complex interplay of DNA, RNA, and proteins, and how each of these molecules can affect the regulation of gene expression. In eukaryotes, this chapter has focused so far on DNA and proteins; however, RNA molecules are also involved.

The RNA molecules that function in the regulation of gene expression are **microRNAs (miRNAs)**. These are short stretches of RNA that, once processed, act to silence gene expression at the message level by hybridizing with complementary mRNA and eliciting sequestration or degradation of the message. The human genome is now thought to contain over a thousand such miRNAs, which may control 60% of human genes.

MiRNAs arise either from a primary-miRNA (pri-miRNA) synthesized by RNA pol II (or, in some instances, pol III) or from an intron spliced out of an existing message (**Figure 17.25**). Multiple miRNAs can be found in a single pri-miRNA. The nascent miRNA forms a hairpin structure that is recognized by a nuclear protein complex termed **Pasha** (known in humans as DGCR8); this complex, in turn, is bound and cleaved by the RNAse **Drosha** to liberate a short pre-miRNA, the immediate precursor to the miRNA.

The pre-miRNAs are exported from the nucleus by the protein exportin-5, in a process that depends on hydrolysis of GTP bound to the RAN protein. Once in the cytosol, the pre-miRNA is acted on by a second RNAse, called **Dicer**, which cleaves the loop region of the pre-miRNA to leave a short (22 nucleotides), double-stranded miRNA. This miRNA then associates with a large complex of proteins—including Dicer and the Argonaute proteins—to form the **RNA-induced silencing complex (RISC)**. **Argonaute proteins** bind and orient mature, single-stranded miRNA molecules with the appropriate target mRNA. The Argonaute proteins themselves possess RNAse activity and may act to cleave the target mRNA, effectively silencing the gene, or they may simply remain bound to the mRNA–RISC complex, preventing translation.

For reasons that are unclear, there is a high degree of conservation among the miRNA sequences of different species. In many cases, the miRNAs are more tightly conserved than the

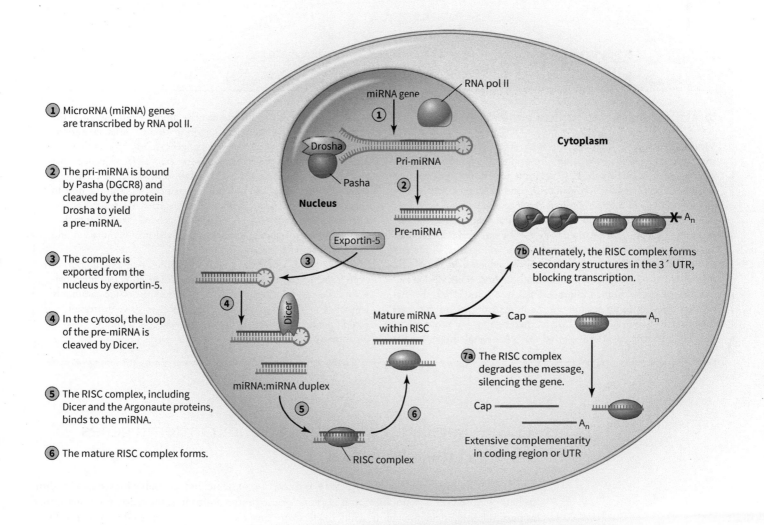

FIGURE 17.25 MicroRNA processing.   MicroRNAs enable the cell to regulate gene expression at the post-transcriptional level.

(Source: Allison, *Fundamental Molecular Biology*, 2e, copyright 2012, John Wiley & Sons. This material is reproduced with permission of John Wiley & Sons, Inc.)

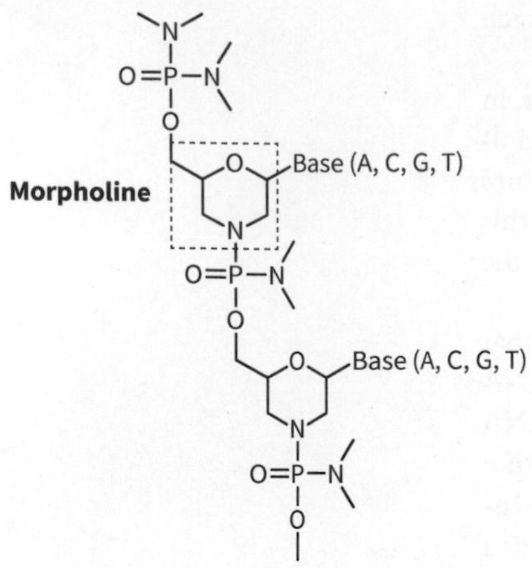

Morpholine

Base (A, C, G, T)

Base (A, C, G, T)

FIGURE 17.26 Morpholino structure. Morpholinos are one means of silencing genes. The morpholino consists of a synthetic RNA analog that replaces the ribose ring with morpholine. Unlike shRNA techniques, morpholinos block translation by sterically interfering with ribosome binding.

coding regions of the genes they regulate. This indicates that the miRNAs were selected for early in evolution and are under considerable evolutionary pressure to remain as they are.

Even though miRNAs are involved in disease, they also hold great potential as treatments. Given that oligonucleotides and their chemical analogs (for example, morpholinos) can be synthesized and manipulated easily, medicine may soon be able to regulate defective genes precisely at the mRNA level (**Figure 17.26**).

In the laboratory, **RNA interference (RNAi) techniques** such as short interfering RNA (siRNA), short hairpin RNA (shRNA), and morpholinos silence genes by taking advantage of the cell's native machinery (Dicer, Argonaute, and the RISC complex). These techniques are simple, inexpensive, and readily adaptable to different experimental conditions and to many genes. Solid-phase synthesis and robotic nucleotide synthesizers have made it inexpensive and easy to procure specific DNA, RNA, or modified oligonucleotide analogs. Using these techniques, it has become routine to silence a single gene in a cell or organism or to silence a whole spectrum of genes to see how this affects a pathway of interest. Such techniques are rapidly changing the face of biochemical research.

Viruses use multiple approaches to regulate gene expression, discussed further in **Medical Biochemistry: Viral mechanisms to control gene expression.**

Medical Biochemistry

Viral mechanisms to control gene expression

Viruses are parasites that use the host organism's own expression machinery to replicate. In addition, viruses affect the host's gene expression, plausibly to prevent the host organism from detecting or having a negative impact on virus propagation. To accomplish these feats, viruses employ a variety of mechanisms to regulate gene expression.

Viruses infect cells and integrate the viral genome into the host genome. Viral proteins are often coded for by one polycistronic message with a single control element similar to an operon. In the case of most viruses, the virus has a promoter sequence in the 5' end of the viral genome that recognizes multiple host transcription factors. In the case of human immunodeficiency virus (HIV), the viral promoter region binds strongly to the human transcription factors NF-κB, AP–1, NFAT, and C/EBP, assisted by the viral proteins Vpr and Tat. Following synthesis of viral mRNA, several viruses employ

antiterminator loops to further regulate protein synthesis. Both the respiratory syncytial virus (RSV) and the vaccinia virus use antiterminator loops to regulate expression of genes found early in their life cycles. Likewise, viral genomes code for short RNA sequences that also serve in a regulatory capacity (akin to miRNAs in eukaryotic cells). Finally, viruses can mediate epigenetic changes in the host to generate conditions favorable for viral propagation or latency. Kaposi's sarcoma–associated virus and Epstein-Barr virus both manipulate host genomes to repress host genes involved in inflammation and immunity to remain latent.

Because of both their ubiquitous expression and high levels of expression, several viral promoters—including HIV, cytomegalovirus (CMV), and simian virus 40 (SV40) promoters—are commonly employed in molecular biology experiments to drive expression of ectopic genes (genes that are not normally expressed in those cells) in organisms or cultured cells.

17.3.6 Regulation of the ferritin and transferrin genes occurs at the mRNA level

Iron has an important role in biochemistry as the center of heme, indispensable for oxygen binding in hemoglobin and myoglobin. Iron is also one of the central atoms in some redox reactions, either as part of heme, in an iron sulfur center in electron transport, or in cytochrome P450 proteins. Deficiencies of iron result in lowered red blood cell count, a condition known as anemia.

Despite its necessity in human health, iron is also toxic. Due in part to its ability to facilitate and carry out redox reactions, many transition metal ions (including iron) can wreak chemical havoc in the cell, generating reactive oxygen species, oxidizing proteins and lipids, and damaging DNA.

Most organisms that employ iron use an array of proteins to segregate and control iron stores, for example, using **ferritin** (the iron-free precursor of which is apoferritin) to sequester iron in the circulation or in the cells or a **transferrin receptor** to bind iron and transport it into cells (**Figure 17.27**).

The regulation of these genes happens at the mRNA level. Genes coding for both ferritin and the transferrin receptor contain an **iron-response element (IRE)**. This sequence forms a hairpin loop in the mRNA. In ferritin it is found in the 5′ untranslated region; in the transferrin receptor, several such repeats are found in the 3′ untranslated region. IREs are bound by a specific protein known as an **IRE-binding protein (IRP)** (**Figure 17.28**). The IRP can bind to either the IRE or iron (in an Fe_4S_4 cluster) but not to both. When iron concentrations are low in the organism, the IRP is bound not to iron, but instead to the mRNA through the IRE. This produces two different effects in the cell. In ferritin (where the IRE is at the 5′ end of the message), binding of the message to the IRP blocks translation of the ferritin message. This results in lower concentrations of ferritin protein and makes more iron available to the organism. The transferrin receptor uses a different mechanism to regulate translation. Again, when iron is low, the IRP binds to the IREs in the transferrin receptor mRNA. However, because these regions are downstream of the initiation of translation and the coding region, protein synthesis still proceeds. Hence, in the absence of iron, the cell attempts to obtain iron by synthesizing the receptor to transport it into the cell. In contrast, when iron concentrations are high, the IRP binds iron rather than IRE. This shift in equilibrium results in free mRNA molecules. In the case of the ferritin receptor, the lack of IRP binding destabilizes the mRNA, rapidly degrading it. When iron in the cell is plentiful, there is no need to synthesize more receptors to increase the concentration further.

Although these mechanisms are found in eukaryotic cells and employ adapter proteins, they are similar to some of the operon and riboswitch mechanisms discussed previously with regard to prokaryotic gene regulation.

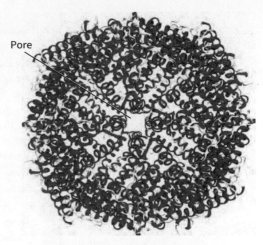

FIGURE 17.27 Structure of ferritin. The ferritin holoprotein is a 24mer consisting of two types of subunits (heavy and light) organized in a sphere. Iron is stored in the Fe^{3+} state as ferrihydrite $(FeO(OH))_8[FeO(H_2PO_4)]$ in the core. As iron is oxidized (from the Fe^{3+} to Fe^{2+}), it forms a hydrate $(Fe^{2+} \cdot 6(H_2O))$ and is released through one of the pores in the protein (shown in the center of the image).

(Source: Data from PDB ID 2FHA Hempstead, P.D., Yewdall, S.J., Fernie, A.R., Lawson, D.M., Artymiuk, P.J., Rice, D.W., Ford, G.C., Harrison, P.M. (1997) Comparison of the three-dimensional structures of recombinant human H and horse L ferritins at high resolution. *J. Mol. Biol.* **268**: 424–448)

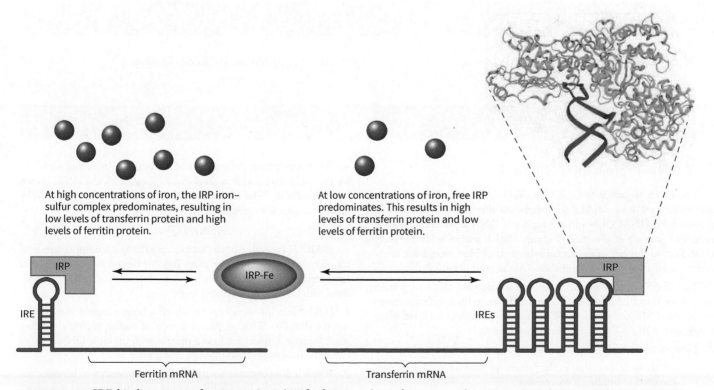

At high concentrations of iron, the IRP iron–sulfur complex predominates, resulting in low levels of transferrin protein and high levels of ferritin protein.

At low concentrations of iron, free IRP predominates. This results in high levels of transferrin protein and low levels of ferritin protein.

FIGURE 17.28 IRE-binding protein function. In coding for ferritin and transferrin genes, the system responds in an attempt to maintain constant concentrations of iron for the cell.

(Source: Data from PDB ID 3SNP Walden, W.E., Selezneva, A.I., Dupuy, J., Volbeda, A., Fontecilla-Camps, J.C., Theil, E.C., Volz, K. (2006) Structure of dual function iron regulatory protein 1 complexed with ferritin IRE-RNA. *Science* **314**: 1903–1908)

Summary

- Eukaryotic genomic DNA is organized into chromosomes. Changes to the number or arrangement of chromosomes can have serious consequences for the health of an organism.

- Histones help the cell bundle DNA and proteins into chromosomes. Histones can be classified as core histones (which form the core of the nucleosome) or linker histones (which are outside the core). SMC proteins are responsible for higher-order chromosomal structure and function.

- Epigenetics are changes that affect gene expression but not the DNA sequence, for example, by modifying proteins or DNA. Histones have multiple post-translational modifications including methylations, acetylations, and phosphorylations regulated by enzymes. Methylation can affect how DNA interacts with histones and transcription factors.

- Transcription factors are proteins that bind to DNA promoter sequences and affect gene expression. Transcription factors can be classified as general, developmentally regulated, nuclear hormone, ligand dependent, resident nuclear, and cytosolic translocating to the nucleus.

- MicroRNAs are short RNA molecules that regulate gene expression at the message level by binding to mRNA and targeting it for degradation.

- Some genes (e.g., those for ferritin and transferrin) are regulated through the interaction of mRNA with specific protein products.

Concept Check

1. Describe the specific challenges that eukaryotes face and how they have evolved regulatory mechanisms to cope with these challenges.

2. Explain how DNA is organized in eukaryotes and how chemical modifications to DNA and proteins (specifically histones) regulate gene expression.

3. Describe the six general classes of transcription factors discussed in this section, and discuss the similarities and differences in their function.

Bioinformatics Exercises

Exercise 1 Viewing and Analyzing the *lac* Repressor

Exercise 2 Apoptosis and the KEGG Database

Problems

17.1 DNA–Protein Interactions

1. The DNA sequence 5′-GGCGCGATG–3′ is methylated. On the molecular level, how would this methylation affect the interactions of a protein with DNA? Draw out the region of DNA affected in the presence and absence of the methyl group. Which amino acids are likely candidates to interact in the unmethylated state? How would this interaction be affected by the incorporation of the methyl group?

2. The BLAST search algorithm is used to compare DNA or protein sequences. Can BLAST be used to search for transcription factor recognition sites? Why or why not? What are the difficulties incurred when trying to use BLAST to search through sequences?

3. How would the following mutations affect gene expression?

   a. an Ala to Ile mutation to the protein–protein interacting domain involved in binding a corepressor

   b. a Ser to Ala mutation of a phosphorylation site for PKA

   c. a Lys to Pro in one of the helical regions of the DNA-binding domain

4. When we examine the sequence of a helical DNA-binding motif, we see that only certain amino acids make contact with the DNA, shown highlighted here. Why are these residues distributed as they are? Why are the residues not contiguous?

RNRAYQVAS

5. SREBP is a member of the helix-loop-helix/leucine zipper family of transcription factors. In place of an arginine (found in almost all other family members), SREBP has a tyrosine. How might this affect interactions with DNA at the chemical level?

6. POU transcription factors have both a homeodomain and a POU specific domain. What is the advantage of having multiple binding sites? Propose a model for how a protein with multiple DNA-binding domains could have arisen through evolution.

7. Why is it that nearly any stretch of DNA can interact with histones, but a transcription factor like CREB can bind to only a very short sequence?

8. DNA is a large polyanion. Explain how the positively charged proteins be drawn across the cell to DNA. Identify the sites/regions where proteins interact with DNA.

17.2 Regulation of Gene Expression in Prokaryotes

9. In prokaryotes, what can affect the controlling gene expression relative to the concentration of a metabolite or catabolite?

10. The metabolism of galactose is controlled by an operon. It consists of three structural genes (coding for epimerase, galactose transferase, and galactokinase), two operator sites, a promoter region, and a repressor gene. The operon is also controlled by the CRP protein as it is for the *lac* operon. Based solely on this information, generate a model for the *gal* operon similar to Figure 17.5.

11. Flavin mononucleotide (FMN) is known to participate in a riboswitch. How does it contribute to the activity?

12. How is the tight control of tryptophan synthesis advantageous evolutionarily in prokaryotes?

17.3 Regulation of Gene Expression in Eukaryotes

13. Histone proteins are highly basic with high pI values. Transcription factors also bind DNA but have pI values that are typically much lower. Provide an explanation as to why this is the case.

14. Histones are highly modified proteins, and we have discussed some of their modifications. What is the likely source of the donor for each of the following modifications? If the donor group itself is not part of the bond, what is the source of energy for the group transfer?

   a. methylation

   b. ubiquitination

15. We have seen in previous chapters that steroid hormones can have an impact on gene expression. How can hormones such as insulin that never enter the cell also affect gene expression?

16. In this chapter we discussed nuclear hormone receptors, which include steroid hormone receptors. How does these interact with the DNA?

17. Plasmids engineered in the laboratory have been hugely helpful in biochemical and molecular biological studies. An expression plasmid is used to induce mammalian cells to ectopically express a cDNA. In this technique a cDNA is cloned in frame downstream of a promoter, and the resulting construct is transfected into mammalian cells for functional studies or to overexpress the cDNA of interest. In many instances, a viral promoter such as CMV or HIV is employed. Why are these promoters chosen?

18. NIH 3T3-L1 cells are a fibroblastic cell line that can be induced to differentiate into an adipocytelike cell line (3T3-L1 adipocytes). A cocktail of dexamethasone, insulin, and isobutylmethyl xanthine (DMX) is often employed to induce differentiation and turn on adipocyte-specific genes. What are these three molecules, and how do they work? Describe in molecular detail how these compounds could affect gene expression and, in this case, differentiation.

19. Many of the participants in the NF-κB cascade have had their structures solved earlier than the components of the SREBP cascade. Why might this be the case?

20. How does the transcriptional machinery gain access to the nucleosome-sequestered DNA?

21. Sometimes genes are used as markers of evolution to show how closely related two species are. Would histone genes (or the proteins they code for) be a good choice to use for this technique? Explain.

22. Tumor promoters are compounds that do not cause cancer but do promote the growth of tumors. Vitamin A in high dosed was once hypothesized to act as a tumor promoter. Explain.

23. Which of these factors correlates most strongly with organismal complexity: number of chromosomes, size of the genome in terms of base pairs, number of genes, or number of proteins?

24. It is estimated that humans have over 100,000 proteins. How can there be over 100,000 proteins in an organism with only 21,000 genes?

25. Consider structures of methylarginine, methylamine and citrulline. Propose a mechanism for deiminase, the enzyme that removes methylamine from methylarginine, resulting in citrulline

26. Why is citrullination an irreversible modification? Why is it you can modify arginine but you cannot make the reverse modification to citrulline?

27. What are the different ways in which the family of HOX genes is conserved? Explain in terms different aspects of HOX genes discussed in this chapter.

28. Describe in general terms how changes in transcription can affect development and metabolism. Substantiate your statements with examples from the chapter.

29. Describe the ways, other than promoters, that nucleic acids use to regulate gene expression.

30. A certain transcription factor (TF) has three zinc finger domains. Write a mathematical expression to describe the interaction of the three domains with DNA. Several assumptions will simplify the calculation. HINT: Protein–ligand interactions will be discussed in depth in Chapter 4.

Data Interpretation

31. A certain transcription factor (TF) has been determined to bind to the sequence 5′-CCAAGG–3′. An electrophoretic mobility shift assay (EMSA) was performed to determine which base pairs are essential for interactions. Data are shown here:

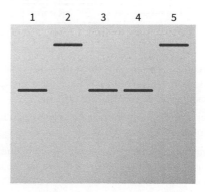

Lane 1, control sequence 5′-CCAAGG–3′ without TF

Lane 2, control sequence 5′-CCAAGG–3′ with TF

Lane 3, sequence A, 5′-CCATGG–3′ with TF

Lane 4, sequence B, 5′-CCAATG–3′ with TF

Lane 5, sequence C, 5′-CCAAGT–3′ with TF

  a. What is an EMSA assay?

  b. What is the purpose of running the control sequence in the presence and absence of TF?

  c. What conclusions can be drawn from these data?

32. A hypothetical operon involved in the synthesis of an amino acid "X" is 'ON' (transcribing) in the presence of low levels of "X" and 'OFF' (not transcribing) in presence of high level of "X". The symbols a, b and c (in the table below) represents a structural gene for the synthesis of X (X-synthase), the operator region, and the gene encoding the repressor—but not necessarily in that order. From the following data, in which superscripts denote wild type or defective genotype, identify which of these genes for X-synthase, operator region, and the repressor.

Strain	Genotype	X-synthase activity in the presence of	
		Low level of 'X'	High level of 'X'
1.	$a^-b^+c^+$	Detected	Detected
2.	$a^+b^+c^-$	Detected	Detected
3.	$a^+b^-c^-$	Not detected	Not detected
4.	$a^+b^+c^+$ / $a^-b^-c^-$	Detected	Not detected
5.	$a^+b^+c^-$ / $a^-b^-c^+$	Detected	Not detected
6.	$a^-b^+c^+$ / $a^+b^-c^-$	Detected	Detected

33. You are studying a gene expressed in muscle tissue that may play a role in the development of type 2 diabetes. You have found a paper discussing the regulation of expression of this gene. A reporter gene assay was conducted to determine which parts of the promoter are important. The data are shown here:

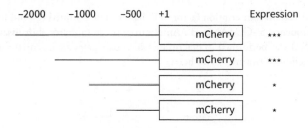

a. What was the reporter gene chosen for this experiment? Why was this reporter gene chosen? What other choices are there? How is expression detected in this system?

b. What parts of the promoter appear to be important for regulation? Explain.

c. Would there be another way to ask the same questions? What assays would you need to perform? What are the strengths and weaknesses of this assay?

Experimental Design

34. Transgenes often employ different promoters to attain tissue-specific expression. A common problem in neuroscience is ensuring that the cells being studied are indeed the ones the investigator hopes they are. Classically, this was accomplished by performing functional studies (typically electrophysiology or patch-clamp studies) on the cells of interest and then fixing and staining the cells after the study was complete to determine if the cells were expressing the protein of interest. This was time consuming and frustrating to the investigator. Often, only one in three cells studied were expressing the proteins of interest; the remainder of the data was on other cells and needed to be disregarded.

Kv3.2 is a voltage-gated potassium channel found in mouse hippocampus. You are interested in conducting studies only on live cells expressing Kv3.2. Design an experiment to help you identify which cells in this region of brain are expressing Kv3.2. HINT: Use a fluorescent transgene as a reporter. What sort of promoter would you want to use?

35. Your principal investigator has a hypothesis that a key lysine residue in histone H3 is acetylated at one stage in development. Design two experiments to test this hypothesis.

36. Your laboratory has been studying a novel leucine zipper transcription factor that is expressed in adipose tissue. You have an antibody raised against this transcription factor. A potential collaborator is interested in the role this TF may play in the regulation of a gene she is studying. She has the entire gene sequenced, including 5 kb upstream of the transcriptional start site. Propose a ChIP assay to see if the TF is binding to promoter sequences in this gene. Presuming that it does bind, provide a means of determining the specific sequence to which it is binding.

37. You have two samples of DNA, one bound to nucleosomes and one stripped free of any associated protein. Would you anticipate one or the other to have a higher melting point? Which one? Propose a hypothesis based on your knowledge of DNA and histone proteins. Design an experiment to test your hypothesis, including appropriate controls.

38. Your friend is studying the regulation of a gene in corn (*Zea mays*). His data indicate that the regulatory step may be at the mRNA level. He hypothesizes that an unknown protein is binding to the mRNA to block translation. Design a footprinting assay to determine if the protein is binding and, if so, where in the mRNA.

39. You have identified a protein from cell lysates that copurifies with DNA. You have hypothesized that this protein is a DNA-binding protein that recognizes a short stretch of DNA in the promoter region of a gene you are studying. How can you test to see if this protein is binding specifically to your promoter region?

Ethics and Social Responsibility

40. In Victorian times, selective breeding ran amok. Many breeds of dogs, varieties of roses and tulips, and strains of mice were bred into existence, often as hobbies of the breeders. Although this selective breeding resulted in aesthetically pleasing plants and animals, inbreeding often led to congenital defects and increased risk of disease. Was this practice ethical? Is selective breeding ethical today? Argue both for and against selective breeding.

41. Is selectively breeding strains of mice into existence by crossing animals with desired traits any different ethically than genetically manipulating them in the laboratory? Explain your answer.

42. Cara's class is doing a laboratory project in which members of the class are screened for a mutation in CFTR, a chloride transporter that is the causal gene in cystic fibrosis. The mutation rate in the general population for this particular mutation is around 1 in 19, but based on a family history of the disease, Cara believes her odds may be higher.

a. If Cara's sister is a known carrier, what are her odds of being a carrier?

b. If there are 41 people in Cara's class, what are the chances that no one will be a carrier?

c. Is Cara obligated to participate in the study? Explain your position to someone who disagrees with you.

43. Tens of billions of dollars are spent on PPAR agonists for the treatment of elevated plasma triglycerides (fibrates) and type 2 diabetes (thiazolidinediones or TZDs). Could this money be better spent in other areas, such as for other diseases, prevention, education, or scientific research? Discuss both sides of the issue.

Suggested Readings

General

Latchman, D. *Gene Control.* New York: Garland Science, 2010.

Skloot, R. *The Immortal Life of Henrietta Lacks.* New York: Broadway Books, 2011.

17.1 DNA–Protein Interactions

Branden, C., and J. Tooze. *Introduction to Protein Structure,* 2nd ed. New York: Garland Science, 1999.

Higgins, P. J., ed. *Transcription Factors: Methods and Protocols.* New York: Humana Press, 2010.

Ladunga, I., ed. *Computational Biology of Transcription Factor Binding.* New York: Humana Press, 2010.

Lesk, A. M. *Introduction to Protein Architecture.* London: Oxford University Press, 2001.

Matthews, J. M., and M. Sunde. "Zinc Fingers—Folds for Many Occasions." *IUBMB Life* 54, no. 6 (December 2002): 351–355.

Miller, M. "The Importance of Being Flexible: The Case of Basic Region Leucine Zipper Transcriptional Regulators." *Current Protein and Peptide Science* 10, no. 3 (June 2009): 244–269.

Schreiter, E. R., and C. L. Drennan. "Ribbon-Helix-Helix Transcription Factors: Variations on a Theme." *Nature Reviews Microbiology* 5, no. 9 (September 2007): 710–720.

Shapiro, J. A. "Genome Organization and Reorganization in Evolution: Formatting for Computation and Function." *Annals of the New York Academy of Sciences* 981 (December 2002): 111–134.

17.2 Regulation of Gene Expression in Prokaryotes

Beckwith, J. R. "Regulation of the *lac* Operon. Recent Studies on the Regulation of Lactose Metabolism in *Escherichia coli* Support the Operon Model." *Science* 156, no. 3775 (May 5, 1967): 597–604.

de Crombrugghe B., S. Busby, and H. Buc. "Cyclic AMP Receptor Protein: Role in Transcription Activation." *Science* 224, no. 4651 (May 25, 1984): 831–838.

Fondi, M., G. Emiliani, and R. Fani. "Origin and Evolution of Operons and Metabolic Pathways." *Research in Microbiology* 160, no. 7 (September 2009): 502–512.

Kercher, M. A., P. Lu, and M. Lewis. "*Lac* Repressor–Operator Complex." *Current Opinion in Structural Biology* 7, no. 1 (February 1997): 76–85.

Merino, E., R. A. Jensen, and C. Yanofsky. "Evolution of Bacterial *trp* Operons and Their Regulation." *Current Opinion in Microbiology* 11, no. 2 (April 2008): 78–86.

Nudler, E., and A. S. Mironov. "The Riboswitch Control of Bacterial Metabolism." *Trends in Biochemical Sciences* 29, no. 1 (January 2004): 11–17.

Osbourn, A. E., and B. Field. "Operons." *Cell and Molecular Life Sciences* 66, no. 23 (December 2009): 3755–3775.

Roderick, S. L. "The *lac* Operon Galactoside Acetyltransferase." *Comptes Rendus Biologies* 328, no. 6 (June 2005): 568–575.

Serganov, A. "Determination of Riboswitch Structures: Light at the End of the Tunnel?" *RNA Biology* 7, no. 1 (January–February 2010): 98–103.

Vilar, J. M., C. C. Guet, and S. Leibler. "Modeling Network Dynamics: The *lac* Operon, a Case Study." *Journal of Cell Biology* 161, no. 3 (May 2003): 471–476.

Wagner, R. *Transcription Regulation in Prokaryotes.* London: Oxford University Press, 2000.

Wilson, C. J., H. Zhan, L. Swint-Kruse, and K. S. Matthews. "The Lactose Repressor System: Paradigms for Regulation, Allosteric Behavior and Protein Folding." *Cell and Molecular Life Sciences* 64, no. 1 (January 2007): 3–16.

Zhang, J., M. W. Lau, and A. R. Ferré-D'Amaré. "Ribozymes and Riboswitches: Modulation of RNA Function by Small Molecules." *Biochemistry* 49, no. 43 (November 2010): 9123–9131.

17.3 Regulation of Gene Expression in Eukaryotes

Albini, S., and P. L. Puri. "SWI/SNF Complexes, Chromatin Remodeling and Skeletal Myogenesis: It's Time to Exchange!" *Experimental Cell Research* 316, no. 18 (November 2010): 3073–3080.

Brivanlou, A. H., and J. E. Darnell. "Signal Transduction and the Control of Gene Expression." *Science* 295, no. 5556 (2002): 813–818.

Brown, M. S., and J. L. Goldstein. "The SREBP Pathway: Regulation of Cholesterol Metabolism by Proteolysis of a Membrane-Bound Transcription Factor." *Cell* 89, no. 3 (May 1997): 331–340.

Carey, M. F., C. L. Peterson, and S. T. Smale. *Transcriptional Regulation in Eukaryotes: Concepts, Strategies, and Techniques,* 2nd ed. Cold Spring Harbor, NY: Cold Spring Harbor Laboratory Press, 2009.

Courey, A. J. *Mechanisms in Transcriptional Regulation.* Malden, MA: Blackwell Publishers, 2008.

Deaton, A. M., and A. Bird. "CpG Islands and the Regulation of Transcription." *Genes & Development* 25, no. 10 (May 2011): 1010–1022.

Desvergne, B., and W. Wahli. "Peroxisome Proliferator-Activated Receptors: Nuclear Control of Metabolism." *Endocrine Reviews* 20, no. 5 (October 1999): 649–688.

Hinojos, C. A., Z. D. Sharp, and M. A. Mancini. "Molecular Dynamics and Nuclear Receptor Function." *Trends in Endocrinology and Metabolism* 16, no. 1 (January–February 2005): 12–18.

Krumlauf, R. "Hox Genes in Vertebrate Development." *Cell* 78, no. 2 (July 1994): 191–201.

Latchman, D. S. *Eukaryotic Transcription Factors.* Amsterdam: Elsevier/Academic Press, 2008.

Li, G., and D. Reinberg. "Chromatin Higher-Order Structures and Gene Regulation." *Current Opinion in Genetics & Development* 21, no. 2 (April 2011): 175–186.

Maiese, K., ed. *Forkhead Transcription Factors: Vital Elements in Biology and Medicine.* New York: Landes Bioscience and Springer Science+Business Media, 2009.

Martin-Subero, J. I., and M. Esteller. "Profiling Epigenetic Alterations in Disease. *Advances in Experimental Medicine and Biology* 711 (2011): 162–177.

Pérez E., W. Bourguet, H. Gronemeyer, and A. R. de Lera. "Modulation of RXR Function through Ligand Design." *Biochimica et Biophysica Acta* 1821, no. 1 (January 2012): 57–69.

Suh, N., and R. Blelloch. "Small RNAs in Early Mammalian Development: From Gametes to Gastrulation." *Development* 138, no. 9 (May 2011): 1653–1661.

Tsukada, J., Y. Yoshida, Y. Kominato, and P. E. Auron. "The CCAAT/Enhancer (C/EBP) Family of Basic-Leucine Zipper (bZIP) Transcription Factors Is a Multifaceted Highly-Regulated System for Gene Regulation." *Cytokine* 54, no. 1 (April 2011): 6–19.

Valencia-Sanchez, M. A., J. Liu, G. J. Hannon, and R. Parker. "Control of Translation and mRNA Degradation by miRNAs and siRNAs." *Genes & Development* 20, no. 5 (May 2006): 515–524.

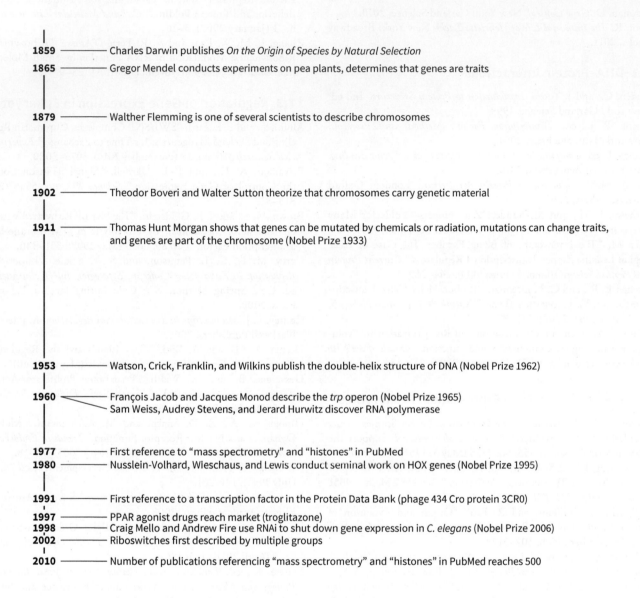

1859 — Charles Darwin publishes *On the Origin of Species by Natural Selection*

1865 — Gregor Mendel conducts experiments on pea plants, determines that genes are traits

1879 — Walther Flemming is one of several scientists to describe chromosomes

1902 — Theodor Boveri and Walter Sutton theorize that chromosomes carry genetic material

1911 — Thomas Hunt Morgan shows that genes can be mutated by chemicals or radiation, mutations can change traits, and genes are part of the chromosome (Nobel Prize 1933)

1953 — Watson, Crick, Franklin, and Wilkins publish the double-helix structure of DNA (Nobel Prize 1962)

1960 — François Jacob and Jacques Monod describe the *trp* operon (Nobel Prize 1965)
Sam Weiss, Audrey Stevens, and Jerard Hurwitz discover RNA polymerase

1977 — First reference to "mass spectrometry" and "histones" in PubMed
1980 — Nusslein-Volhard, Wieschaus, and Lewis conduct seminal work on HOX genes (Nobel Prize 1995)

1991 — First reference to a transcription factor in the Protein Data Bank (phage 434 Cro protein 3CR0)
1997 — PPAR agonist drugs reach market (troglitazone)
1998 — Craig Mello and Andrew Fire use RNAi to shut down gene expression in *C. elegans* (Nobel Prize 2006)
2002 — Riboswitches first described by multiple groups
2010 — Number of publications referencing "mass spectrometry" and "histones" in PubMed reaches 500

CHAPTER 18

Determination of Macromolecular Structure

Macromolecular Structure in Context

A hammer may well be one of the first tools ever created. In its most basic design, a hammer is just a blunt object at the end of a stick or arm. Hammers can vary in size and weight from those small and light enough to use to eat a crab to those large and heavy enough to shatter boulders. However, all hammers have a similar function, that is, to hit something with force. A trip to the tool aisle of a home improvement center would yield a large variety of hammers, and we could propose a job for each based on its size and shape. Likewise, if an archeological dig unearthed a stick attached to a stone, we could propose that it was used as a hammer based on its overall shape.

Thus, knowing the structure of an object provides important information about its function. If we were to find an object that was similar to a hammer but had a sharp head instead of a blunt one, we might propose a new function (chopping) and give that object a new name (an axe). The hammer and the axe share a common region or domain (the handle), which has a similar function in both cases.

Similarly, elucidating the structure of a protein or other biological macromolecule provides us a wealth of information and a basis for comparison. Determining the structure of an enzyme, for example, can provide the topology of the substrate-binding site, indicate sites where other molecules may interact, and suggest which amino acids are in the active site and thus involved in catalysis. The structure of a receptor or enzyme can provide a map for structure-based drug design or help to identify the native ligand or substrate for the molecule.

This chapter begins with the basics of determining molecular structure and then discusses three different techniques—electron microscopy, diffraction, and nuclear magnetic resonance—that are currently used to determine the structure of macromolecules and macromolecular complexes.

Chapter Outline

Common Themes

Evolution's outcomes are conserved.	• Determination of structure has shown that many proteins are conserved through evolution, not only at the level of nucleic or amino acid sequence but also at the level of structure. • Conservation at the structural level can occur even when nucleic and amino acid sequences or structures diverge. • Protein structures can show convergent or divergent evolution.
Structure determines function.	• Different laboratory techniques enable scientists to examine macro-molecules or complexes of macromolecules at different resolutions and different magnifications. • Elucidation of a structure provides important information about the function of the molecule and the mechanism through which it works. • Comparison of structural data can often provide new insights into the function of a protein.
Biochemical information is transferred, exchanged, and stored.	• Elucidation of the structure of DNA has provided key information that relates the structure of the molecule to its function (information storage) and mechanism of replication.
Biomolecules are altered through pathways involving transformations of energy and matter.	• The basic rules of thermodynamics govern protein structure.

18.1 An Introduction to Structure Determination

Some structures can be seen quite easily; for example, in general or cell biology, light microscopy is often sufficient to show the finer structures of an insect wing or even a cell or tissue. Other structures can be deduced; for example, in general chemistry, drawing a Lewis structure or performing a simple analysis can indicate whether a functional group is present. However, as we progress to finer levels of detail, determination of structure becomes more complex (**Figure 18.1**).

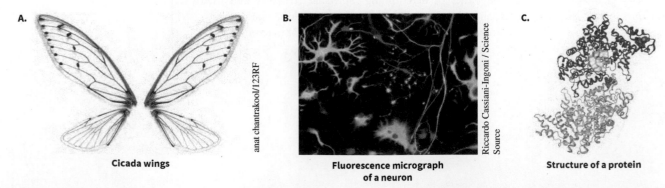

A. Cicada wings

anat chantrakool/123RF

B. Fluorescence micrograph of a neuron

Riccardo Cassiani-Ingoni / Science Source

C. Structure of a protein

FIGURE 18.1 Structure determination.   Discussion of biological structures has evolved over time from **A.** gross anatomy and form to **B.** the fine structures seen in cells, all the way down to **C.** the structures of molecules such as this protein. In all instances, the structure is conserved through evolution.

(Source: Data from PDB ID 3HNE Fairman, J.W., Wijerathna, S.R., Ahmad, M.F., Xu, H., Nakano, R., Jha, S., Prendergast, J., Welin, R.M., Flodin, S., Roos, A., Nordlund, P., Li, Z., Walz, T., Dealwis, C.G. (2011) Structural basis for allosteric regulation of human ribonucleotide reductase by nucleotide-induced oligomerization. *Nat. Struct. Mol. Biol.* **18:** 316–322)

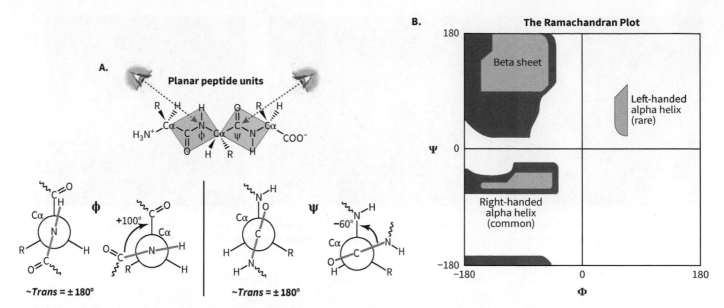

FIGURE 18.2 Phi and psi bond angles on a Ramachandran plot. **A.** Phi and psi are the torsional angles found on either side of the peptide bond. Establishing these angles gives the general shape of the protein. **B.** Graphing these angles for each amino acid with phi on the *x*-axis and psi on the *y*-axis results in a Ramachandran plot. Regions of this plot correspond to the bond angles seen in the alpha helix and beta sheet. Therefore, these plots can be used to determine where amino acid residues can be found in a protein.

(Source (A): Adapted from drawings created by Jason D. Kahn, Univ. Maryland.)

In organic chemistry, **nuclear magnetic resonance (NMR)** spectroscopy is used to find the connectivity of atoms in a larger molecule, such as one with perhaps a dozen carbons or heteroatoms. NMR can also probe a molecule's three-dimensional (3D) shape, called its conformation. In biochemistry, the situation is far more complex. A biological molecule may well have a molecular weight of 50,000 daltons or more and be comprised of thousands of carbons and heteroatoms. Such molecules—typically proteins and nucleic acids but also some carbohydrates—are complex, but they have a defined chemical composition and a defined or definable shape or conformation.

What does it mean to determine the 3D structure of a macromolecule, and how can it be done? In a protein, the peptide bond is planar, and the overall shape of the peptide backbone can be determined if the torsional bond angles around the peptide bond—the phi (ϕ) and psi (Ψ) angles—are known; in effect, this produces a low-resolution image of the protein (**Figure 18.2**). This analysis will not give the positions of side chains, although sometimes they can be inferred by identifying weak interactions, such as salt bridges, hydrogen bonds, or steric hindrances, that force a residue into or out of a pocket. A plot of the psi and phi bond angles for each amino acid residue is known as a Ramachandran plot, and specific regions of that plot correspond to the alpha-helical and beta-sheet elements of secondary structure.

This section defines some general terms and concepts related to determining the structure of a macromolecule. The remaining sections explore specific techniques used to determine macromolecular structure.

18.1.1 Resolution is a key feature of structure determination

In determining the structure of a molecule, **resolution** refers to the ability to discern from a distance whether we are looking at a single point or two points. For example, if you see a brick building in the distance, you cannot tell where one brick ends and another begins. As you move closer the building, however, you will eventually be able to distinguish individual bricks. At this point, you are said to be able to "resolve" those bricks (**Figure 18.3**).

FIGURE 18.3 Bricks as an example of resolution. **A.** We may surmise, based on prior knowledge and our observations, that the building is made of brick, but we cannot resolve individual bricks at this distance. **B.** Examination of a brick wall at a closer distance allows us to resolve the individual bricks; that is, we are able to see where one brick ends and another begins.

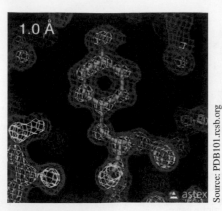  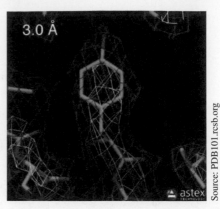

FIGURE 18.4 Amino acids as an example of resolution. The three images of a hexagonal tyrosine side chain are shown at resolutions ranging from 1 to 3 Å. At 3 Å resolution, the entire side chain is an irregular blob, but at higher resolution more details (including hydrogen atoms) can be seen.

In light microscopy, the resolution is given by the formula

$$r = \frac{0.61\lambda}{NA}$$

where NA is the numerical aperture of the microscope (a term that describes the range of angles over which light can enter a lens), and λ is the wavelength of light.

Therefore, the resolution limit between two objects is a function of two things: the quality of the lens and the wavelength of light. Even with the best lenses, the resolution of a light microscope depends on the wavelength of light that is used. Visible light ranges from 400 nm to 700 nm in wavelength; hence, resolution in even the best light microscope is limited to 0.22 μm—far too large for the resolution of molecular structures. To investigate such structures, biochemists use forms of radiation that have a higher energy and therefore a shorter wavelength, such as high-energy electrons in electron microscopy and X-rays in X-ray crystallography.

In practice, resolution is related to how much detail we can observe in a structure (**Figure 18.4**). At low resolution (~10 Å), a molecule of hemoglobin may look like a simple sphere. At higher resolution (~5 Å), alpha helices can be seen as rods, but it is still not possible to determine the position of the amino acid side chains. This level of resolution was used in determining the earliest protein structures and producing the first images of DNA. The possible positions of some side chains or nucleotides of DNA can be deduced based on the sequence and chemical composition of the molecule, but their specific positions are a point of conjecture. At higher resolutions, it becomes possible to see the carbons and heteroatoms of nucleotides and amino acids (at ~2 Å) and finally hydrogen atoms (at ~1.5 Å). Increasing the resolution even further allows the determination of other aspects of protein structure, for example, the oxidation state of a metal ion or the identity of a previously unknown metal cofactor. To date, the highest resolution achieved has been less than 1 Å.

18.1.2 Contrast is important in some techniques for structure determination

A second parameter that is often discussed in determining a structure is **contrast**, that is, the difference in brightness or hue between two objects that makes them stand out or become visible (**Figure 18.5**). Thus, a black object on a white background has high contrast, whereas a pale gray object on a white background has lower contrast. Contrast is important in visualizing objects, for example, in electron microscopy.

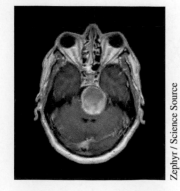

FIGURE 18.5 An example of contrast. Administration of a gadolinium-based contrast agent to a patient enhances the MRI signal and clearly reveals the damage caused by an aneurysm in the patient's brain.

18.1.3 The means of depicting structures have evolved along with advances in structure determination

In the past, as structures were determined, physical models were constructed using clay, metal, or plastic (**Figure 18.6**). Often, electron densities were drawn on Plexiglas and the sheets layered one on

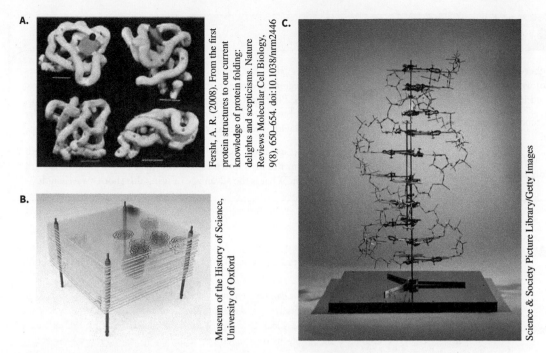

Fersht, A. R. (2008). From the first protein structures to our current knowledge of protein folding: delights and scepticisms. Nature Reviews Molecular Cell Biology, 9(8), 650–654. doi:10.1038/nrm2446

Museum of the History of Science, University of Oxford

Science & Society Picture Library/Getty Images

FIGURE 18.6 Examples of early models. **A.** The original model of myoglobin was constructed with metal rods and modeling clay. **B.** The electron density of penicillin was drawn out on glass sheets that were stacked together to generate a 3D image. **C.** The original model of DNA was made with metal rods.

top of the other to generate a 3D structure. Since the early 1990s, computers have been used to depict molecular structures, and today this can be done using tablets and handheld devices. However, these programs simply create an image of a known protein or macromolecule; a great deal more computing power is required for the initial structure determination. Nevertheless, computers can quickly produce images in which specific residues are highlighted or present different views (such as space-filling versus ball-and-stick models) to highlight certain aspects of a protein.

The structures of all solved macromolecules are stored in the Protein Data Bank (PDB), a publically accessible database. The PDB also contains the coordinates for the solved structure, published references, and information about how the structure is related to other molecules.

Structural biology continues to change as new technologies emerge. Currently, physical models are routinely made using 3D printers, and virtual reality headsets now allow interaction with molecules (**Figure 18.7**).

Although it may be tempting to accept such a vivid model as fact without scrutiny or analysis, such structures deserve to be as closely scrutinized as any other piece of data. Furthermore, determining a structure is not the end of the story. Structures may be dynamic, that is, able to interact with numerous partners.

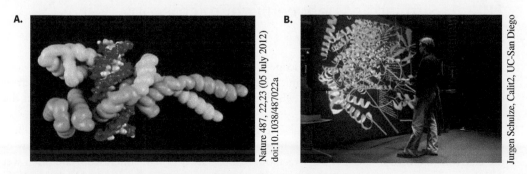

Nature 487, 22.23 (05 July 2012) doi:10.1038/487022a

Jurgen Schulze, Calit2, UC-San Diego

FIGURE 18.7 Evolution of molecular modeling. Modern molecular modeling uses computer-generated images. **A.** Physical models can be generated using 3D printers. **B.** Molecules can also be viewed using virtual reality walls or headsets.

Worked Problem 18.1   Light microscopy

The images produced by higher-quality light microscopes often appear blue because filters remove the red end of the visible light spectrum. Why is blue light used in light microscopy?

Strategy   What are the differences between red and blue light? Why would blue be preferable?

Solution   The resolution of any optical technique is limited by the wavelength of the radiation used. The longer the wavelength, the lower the resolution. Hence, shorter-wavelength (higher-energy) regions of the electromagnetic spectrum are used to resolve molecules. Blue light has a shorter wavelength than red light; thus, the use of blue light in microscopy makes it possible to view smaller structures with greater clarity.

Follow-up question   CD players use red lasers with a wavelength of 650 nm. Blu-ray players use a blue–violet laser with a wavelength of 405 nm. What is the advantage of using this shorter wavelength?

Summary

- The structure of a macromolecule often provides crucial information about that molecule's function or mechanism of action.
- Resolution refers to the level of detail that can be seen in a structure. A technique that provides high resolution makes it possible to see fine details such as the side chains of amino acids or the hydrogens on those side chains. Techniques that provide only low resolution can also be valuable, for example, to illustrate the gross interactions of subunits on a macromolecular complex. Contrast—the ability of a structure to stand out from its background—is a term usually reserved for microscopy or electron microscopy.
- In the past, once a structure was determined, a model was constructed using, for example, layers of Plexiglas, metal rods, or clay. Today, models are generated and viewed on a computer or sometimes printed on a 3D printer. The PDB is the international repository of macromolecular structures (mostly proteins and nucleic acids).

Concept Check

1. Describe how resolution and contrast impact macromolecular structure determination.
2. Discuss how models were constructed in the past and how they are generated today.

18.2 Electron Microscopy

It would be convenient if there were a device similar to a microscope that could be used to determine the shapes of macromolecules, in the way that light microscopy can be used to visualize cells and some subcellular structures. However, as outlined in the previous section, this is not possible because the smallest wavelength of light is too long compared to the size of molecules. Currently, the best option for determining the crude shape of a macromolecule is **electron microscopy (EM)**.

18.2.1 Similarities between light microscopy and electron microscopy

In many ways, using an electron microscope is similar to using a light microscope. The obvious difference between the two is that light microscopy uses photons of visible light, whereas electron microscopy uses high-energy electrons, which creates several important differences between the two techniques (**Figure 18.8**).

Light passes through air easily. High-energy electrons, on the other hand, interact strongly with air, generating heat. To prevent this heat generation, the path that electrons must follow in the electron microscope is under high vacuum. Samples used in light microscopy are often wet and, in some cases, still alive; this is not possible in a vacuum, where water would boil or quickly evaporate at ambient temperatures. Therefore, samples used in the electron microscope must be fixed and dry.

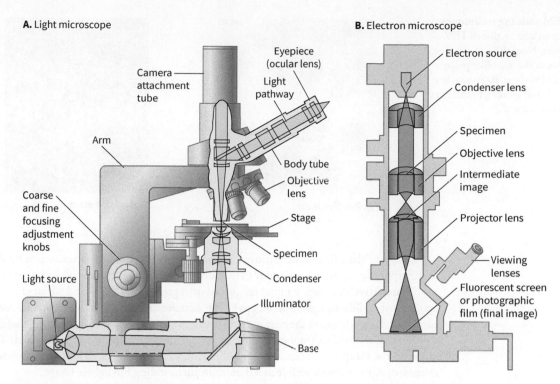

A. Light microscope

Eyepiece (ocular lens)

Light pathway

Camera attachment tube

Arm

Body tube

Objective lens

Coarse and fine focusing adjustment knobs

Stage

Specimen

Condenser

Light source

Illuminator

Base

B. Electron microscope

Electron source

Condenser lens

Specimen

Objective lens

Intermediate image

Projector lens

Viewing lenses

Fluorescent screen or photographic film (final image)

FIGURE 18.8 Light microscopy versus electron microscopy.   The principles of electron microscopy are analogous to those of light microscopy, but instead of a beam of light, high-energy electrons are focused using magnets and pass through a sample. The electrons that pass through the sample are focused either onto a phosphorescent screen or a detector.

(Source: Black, *Microbiology: Principles and Explorations*, 8e, copyright 2012, John Wiley & Sons. This material is reproduced with permission of John Wiley & Sons, Inc.)

Samples in light microscopy are often stained with colored dyes to enhance contrast or to identify specific molecules or cell types. Electron microscopy also employs contrast agents, but due to the nature of electrons interacting with the sample, these agents are often heavy metals such as gold, lead, uranium, osmium, platinum, or tungsten.

Light microscopy uses lenses for magnification and to focus the beam of light. In contrast, electron microscopy employs positively charged plates (electromagnetic lenses) to focus and direct the path of the negative electrons. A basic light microscope may only employ a few lenses to focus the beam of light, whereas electron microscopes often use more than 20 electromagnetic lenses to focus the electron beam.

Finally, in light microscopy, the viewer's eye or a camera captures the photons of light passed through the sample. In electron microscopy, images are either generated by projection of the electrons onto a phosphorescent screen or captured by a camera. In both types of microscopy, various techniques are used to process the acquired images.

18.2.2 Sample preparation is essential to visualization in electron microscopy

In light microscopy, visible light passes through a sample. Regions of the sample that are optically dense due to the composition or amount of material absorb more light and appear dark, whereas regions that are less dense permit the passage of that light to the viewer's eyes or a camera. Transmission electron microscopy (TEM) is similar, but the beam of electrons—analogous to the light source in light microscopy—will readily pass through most biological samples. The interaction between electrons and atoms is proportional to the square of the atomic number (Z) of the atoms. Therefore, an electron will interact over 6,600 times more strongly with an atom of lead (Pb, atomic number 82) than with an atom of carbon (C, atomic number 6). This is why most samples to be studied with electron microscopy are stained with some type of heavy metal.

Of the different electron microscopy techniques used in structure determination, the most common are positive staining, negative staining, and shadowing (**Figure 18.9**).

FIGURE 18.9 EM staining techniques.
A. Positively stained and sectioned HIV particles are shown. Note the cone-shaped core of the virus and the knoblike projections from the viral surface. **B.** Shown is a negative stain of *E. coli* glutamyl synthetase stained with uranyl acetate.

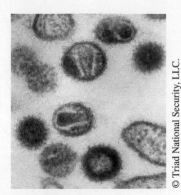

A. Positive stain

© Triad National Security, LLC.

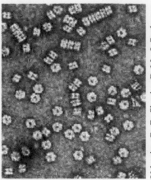

B. Negative stain

Valentine, R. C., Shapiro, B. M., & Stadtman, E. R. (1968). Regulation of glutamine synthetase. XII. Electron microscopy of the enzyme from Escherichia coli. Biochemistry, 7(6), 2143–2152. doi:10.1021/bi00846a017

Many of the different metals used as stains are **positive stains**. These stains bind generically to different classes of molecules; for example, osmium tetroxide binds phospholipids.

In **negative staining**, samples are placed on a grid (a tiny metal and Mylar circle that is analogous to the slide in light microscopy). The samples (for example, lipoproteins) are allowed to adhere to the grid, where they are quickly dried. A stain such as phosphotungstic acid (tungsten), ammonium molybdate (molybdenum), or uranyl acetate (uranium) is then added. The dye does not stain the sample itself but rather puddles around the structure. When viewed with an electron microscope, the metal is visible as a dark stain surrounding the lighter image.

In **shadowing**, which is similar to negative staining, a metal source is evaporated at a known angle to the sample. As the metal deposits on the sample, it accumulates more heavily on the side facing the source, in the way that blown snow or sand builds up around an object such as a car. When viewed in the electron microscope, the height of the object can be calculated based on the angle of the source in relation to the sample and the length of the shadow. Some newer techniques rotate the sample while coating (referred to as sputter coating), generating a metal-coated sample that can also be imaged.

In **cryogenic electron microscopy (cryo-EM)**, samples of purified macromolecules are prepared by flash freezing them in vitrified water. **Vitrification** is the process of creating a glassy solid, effectively an amorphous solid of ice, by instantly freezing the sample in water at very low temperatures. This prevents the formation of ice crystals and allows for the study of the macromolecules in the sample.

University of Victoria

FIGURE 18.10 The Hitachi HF-3300 eV STEHM. This modern, high-powered electron microscope is capable of imaging some molecules down to the picometer range.

18.2.3 Advances in electron microscopy have dramatically improved resolution

In light microscopy, changing the magnification of the scope involves changing lenses. The same is true for electron microscopy in that the magnification can be changed by focusing the electron beam more tightly. However, there is an important distinction between the two types of microscopy in this respect. All light microscopes use visible light for magnification, and, as mentioned previously, blue light has a shorter wavelength and therefore higher resolving power than red light. In electron microscopy, the "color" of the electron beam (effectively the wavelength of the electron) is a function of how strongly the electrons are accelerated. The first electron microscopes used electrons that were 60, 120, or 180 kiloelectron volts (kEv, a measure of energy; 1 kEv = 1.6×10^{-16} J). Newer microscopes routinely use electrons that are 300 to 400 kEv. These higher energies produce electrons with shorter wavelengths, which permit higher resolution. The use of higher-energy electrons combined with advances in microscope design have resulted in the development of extraordinary instruments. One of the world's most powerful microscope is the Hitachi HF-3300 eV scanning transmission electron holography microscope (STEHM) at the University of Victoria in British Columbia, Canada (**Figure 18.10**). This microscope is over 4.7 m tall, weighs 7 tons, and has 65 electromagnetic lenses. It is housed in a room specially designed to dampen vibrations. This microscope is capable of a resolution of 35 picometers (pm), a magnification of over 30 million-fold, which allows for the imaging of individual atoms in some materials.

Cryo-EM relies heavily on sampling large numbers of images, averaging these images together, and processing these data to generate a 3D image of a macromolecule. Advances in the microscopes themselves, direct detection systems, image analysis software, and computational power have made this possible. Cryo-EM allows researchers to image macromolecules in molecular detail (~2 Å resolution) in only a few hours, where only 4 to 5 Å resolution was available as recently as five years ago.

Worked Problem 18.2   A case for electron microscopy

SecA is an ATP-dependent molecular motor that is involved in translocating proteins out of *E. coli*. It has been shown to interact with both the ribosome and a membrane complex containing two other proteins. In addition, high resolution X ray shows both monomeric and dimeric forms of SecA. How could electron microscopy help to elucidate the role of SecA?

Strategy   What might images of SecA interacting with a ribosome tell you?

Solution   In 2014, Sing et al. used high-resolution electron microscopy to show SecA bound to the large subunit (50S) of the bacterial ribosome. Specifically, SecA binds to either side of the exit tunnel through which the nascent protein emerges. Both one and two copies of SecA were observed, suggesting a role for dimeric SecA on the ribosome.

In this instance, electron microscopy made it possible to identify how many copies of SecA were binding to the ribosome and where they were binding—something that would be difficult to achieve with other imaging techniques.

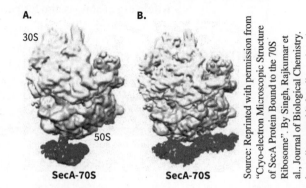

Source: Reprinted with permission from "Cryo-electron Microscopic Structure of SecA Protein Bound to the 70S Ribosome". By Singh, Rajkumar et al., Journal of Biological Chemistry. https://www.ncbi.nlm.nih.gov/pmc/articles/PMC3945378/

Follow-up question   Metallic stains and contrast agents were not used in the acquisition of these images. Instead, contrast was enhanced through imaging techniques. Why might staining this structure with metals be inadvisable? How would you find a copy of this paper to see how the authors imaged the structure?

18.2.4 Electron microscopy for next generation macromolecular structure determination

Electron microscopy (EM) has allowed researchers to gain an understanding of macromolecular complexes such as such as the ribosome, viruses, molecular chaperones, etc., and medium to low-resolution structures of biological molecules and processes. Over the past decade, the advancement in the field has given rise to the need for the development of robust instrumentation, in-situ platforms, and data-driven experimentation. Researchers and technologies across the globe are focusing on the development of next-generation electron microscopy to adapt to upcoming transformative research tasks and optimize analytical workflows. Transmission electron microscope (TEM) and the scanning electron microscope (SEM) are the two main techniques used to obtain the structure of macromolecular complexes ranging from ~200 kDa to hundreds of MDa.

Transmission electron microscope (TEM)   TEM is typically used to examine tissue sections, internal cell structure, and molecules. It produces its two-dimensional image by transmitting electrons through the specimen. Electrons cannot penetrate through the thick surface, so specimens are prepared as extremely thin slices (70–90 nm thick), by embedding them in a block of plastic and then cutting with a glass or diamond knife. These are then stained or coated with chemicals such as lead acetate, phosphotungstate, uranium, palladium, and gold. This is essential to enhance contrast in the image in TEM. The section is placed on a copper grid in the chamber. A beam of electrons passes through it via the electromagnetic condenser. The electron beam that comes out of it spreads and projects over a fluorescent screen by the electromagnetic lens. It is used to study the interior of cells, the structure of protein molecules and their arrangement in cell membranes, and the organization of macromolecules in viruses, etc.

Scanning electron microscope (SEM) SEM is used to create an image of the surface of the specimen, produced from secondary electrons emitted back from the object's surface. SEM is a more recent invention than TEM. In SEM, the specimen is first supercooled in liquid propane at –180°C and then dehydrated in alcohol at –70°C. It is then coated with a thin layer of heavy metal such as gold and palladium for creating a reflecting surface. Then, it is subjected to a narrow electron beam, which rapidly scans the surface of the specimen and causes the release of a shower of secondary electrons and other types of radiation from the specimen surface. These emitted radiations are collected by the detector which generates an electronic signal, and these signals are scanned in the same manner as in a television system to produce an image on a cathode ray tube. It is used to obtain detailed images of the surface of the cells and whole organisms, as also for particle counting and size determination.

Summary

- Electron microscopy is analogous to light microscopy, but in electron microscopy, high-energy electrons rather than photons of visible light are used to visualize a sample. Electron microscopy employs magnetic plates or rings as lenses and a phosphorescent screen or camera for visualization.

- Samples in electron microscopy need to be dried and fixed. They are typically stained with heavy metals such as tungsten, gold, osmium, or uranium.

- Higher-energy electrons produce higher resolution in electron microscopy. Cryo-EM differs from other EM techniques in that the samples are flash frozen in vitreous water, and multiple images are analyzed to generate the final structure.

Concept Check

1. Describe the technique of electron microscopy and explain how it compares with light microscopy.
2. Name the different stains and types of staining used in electron microscopy.
3. Explain why an electron microscope is capable of higher resolution than a light microscope.

18.3 X-Ray Diffraction and Neutron Scattering

Advances in electron microscopy are enabling the examination of ever-smaller structures. However, atomic-level resolution of molecules is largely accomplished through two other techniques—X-ray diffraction and NMR—both of which enable investigators to discern the structure of molecules with angstrom or sub-angstrom resolution.

X-ray diffraction is also known as X-ray crystallography or X-ray scattering. In this technique, X-rays are focused on a crystal or fiber of a biological material (typically a protein or nucleic acid). The molecules in the crystal scatter the X-rays, and the resulting pattern allows for reconstruction of the coordinates of the atoms of the crystal. This section discusses the steps involved in this process.

18.3.1 A two-dimensional example illustrates the principles of diffraction

Diffraction is a phenomenon that occurs with all waves. In the simplest case, single-slit diffraction, a wave passes through a single aperture (slit or hole) that is small compared to the wavelength of the wave. As it emerges, the wave is bent around the slit and emerges on the other side as if were emanating from a point source. The equation that describes this is

$$d \sin \theta_{min} = \lambda$$

where d is the size of the slit, λ is the wavelength, and θ_{min} is the angle of the first minimum seen (**Figure 18.11**). A closer examination of the diffraction pattern shows multiple higher-order maxima and minima.

If there are two slits rather than a single slit, the waves emanating from them interact and interfere with one another, called double-slit diffraction. Constructive interference is when waves are in phase with one another, meaning that either or both of their peaks or troughs align. Destructive interference is when waves are out of phase, meaning neither their peaks nor troughs align (**Figure 18.12**). The two waves are shifted in relation to each other, resulting in the complex pattern observed. Both slits contribute to all the peaks and troughs observed. In addition, if the two slits are closer together, the peaks that are observed will be farther apart, and vice versa.

The wave intensity patterns observed in Figure 18.11 are basically combinations of different sine wave functions with varying phase intensity, shifted with regard to each other. These curves (and any wave) can be represented as a product of different sine waves. The wave intensity patterns observed in the two-slit experiment are also a product of two waves. These two functions can be broken into their original wave patterns, in a process called deconvolution. The mathematical manipulation that is used to do this is a Fourier transform, discussed in a later section in relation to how to extract the coordinates of atoms in a protein from a complex diffraction pattern.

18.3.2 Diffraction from a crystal occurs in three dimensions

How does diffraction from a crystal of atoms compare to diffraction from slits or holes? Babinet's principle states that diffraction from a small, opaque object (such as a molecule) is identical to diffraction from a hole of the same size and shape, but it differs in intensity. Therefore, objects such as molecules can diffract electromagnetic radiation if the molecules are on the order of the size of the wavelength of the radiation (e.g., X-rays). The value of using a crystal over a single molecule is that the repeating nature of the crystal produces many scattering centers, all in the same orientation. This greatly increases the signal generated compared to what would be obtained from a single scattering center.

Diffraction from a crystal is similar to diffraction from a slit (**Figure 18.13**). The Bragg equation relates the scattering of X-rays by atoms to the position of those atoms in the crystal. A crystal can be thought of as having scattering centers in planes

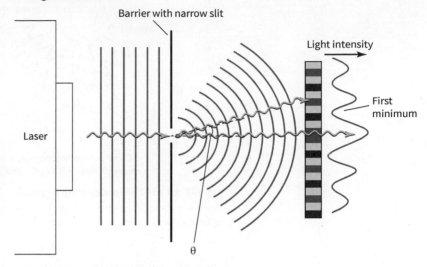

A. Single-slit diffraction

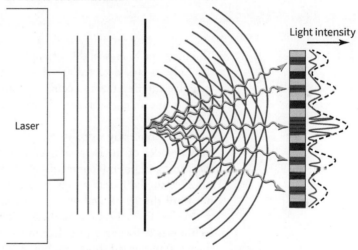

B. Double-slit diffraction

FIGURE 18.11 Diffraction experiments.   A. Waves that pass through a slit that is small compared to the wavelength of the wave will be diffracted, or bent, as they pass through that opening. This is observed with any waves, including electromagnetic radiation. **B.** When two slits are present, both can diffract the wave in question. This leads to a complex pattern as the waves interfere with one another.

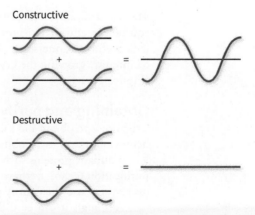

FIGURE 18.12 Constructive and destructive interference.   When the peaks of a wave are found at the same point, they are said to be in phase with one another. These waves interfere constructively and increase intensity. Conversely, waves that are out of phase with one another interfere destructively and decrease intensity.

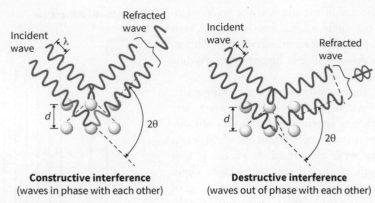

FIGURE 18.13 Diffraction from a crystal. The atoms in a crystal can diffract X-rays in the same way that a slit does. Bragg's equation relates the distance between scattering centers (spheres) with the angle of diffraction and wavelength of radiation used.

separated by distances (d). The extra length the X-ray must travel is an integer (n) times the wavelength of the X-ray (λ):

$$2d \sin \theta = n\lambda$$

The incident X-rays in this example hit all atoms and are scattered in all directions; however, the only X-rays detected are those that constructively interfere with one another. For constructive interference to occur, the path length of the X-ray must vary by the parameters set forth in the Bragg equation. That is, $2d \sin \theta$ must equal an integer times the wavelength of the X-ray.

The pioneering work of the father–son team of Lawrence and Charles Wilson Bragg in the early decades of the twentieth century enabled the determination of thousands of structures of inorganic compounds but relatively few biological ones. The added complexity of biological structures required different means of data interpretation and more computational time; hence, it was not until the 1940s and 1950s that researchers began to determine the structures of biological molecules. Even then, it took the development of computers and advances in protein purification techniques before structure determination became common. In 1981, the PDB included fewer than 100 published structures, but by 2013 it had surpassed 100,000 structures, and it has grown exponentially each year since.

18.3.3 The determination of structures by X-ray diffraction can be broken into four steps

The determination of a macromolecular structure by X-ray diffraction requires dedicated equipment and investigators, and it can easily take several years. The process is complex, and it is useful to understand the four main steps involved: obtaining and purifying the protein of interest, growing the crystal, bombarding it with X-rays, and analyzing the collected data (**Figure 18.14**).

Obtaining and purifying the protein   The first step in determining macromolecular structure is to obtain and isolate the protein of interest. The first protein structures determined—hemoglobin and myoglobin—were abundant, being purified from sperm whale and horse. For less abundant proteins such as transcription factors, advances in molecular biology and protein purification techniques have assisted and improved structure determination by making it easier to overexpress and then purify proteins in bacteria, yeast, or mammalian cells.

Although it can be difficult to isolate an entire protein, large domains or fragments of a protein may lend themselves to purification and crystallization, and these can provide important data. Here again, advances in molecular techniques have made the generation of multiple fragments of a protein and expression of those fragments a relatively straightforward process.

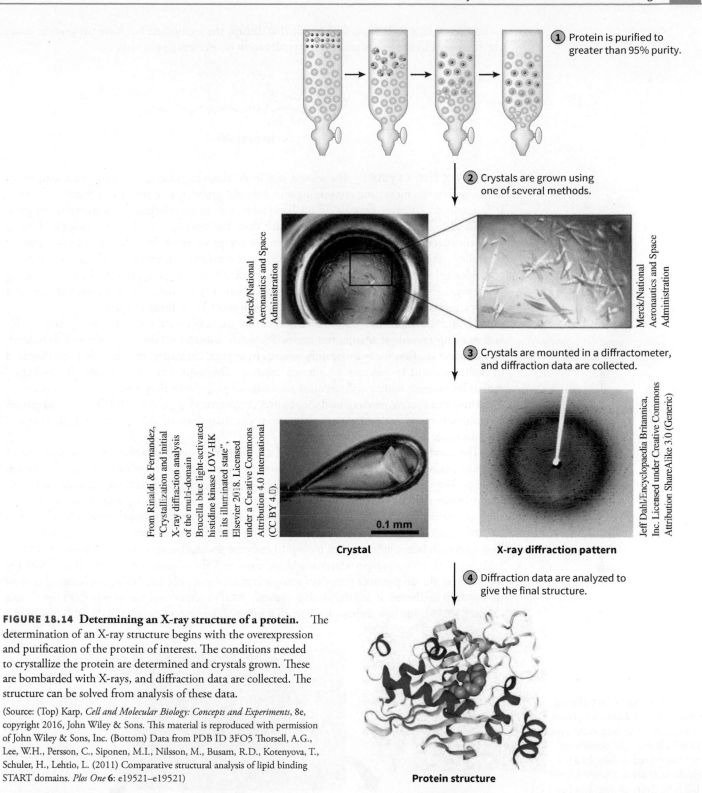

① Protein is purified to greater than 95% purity.

② Crystals are grown using one of several methods.

Merck/National Aeronautics and Space Administration

Merck/National Aeronautics and Space Administration

③ Crystals are mounted in a diffractometer, and diffraction data are collected.

From Rinaldi & Fernandez, "Crystallization and initial X-ray diffraction analysis of the multi-domain Brucella blue light-activated histidine kinase LOV-HK in its illuminated state", Elsevier 2018. Licensed under a Creative Commons Attribution 4.0 International (CC BY 4.0).

0.1 mm

Crystal

Jeff Dahl/Encyclopaedia Britannica, Inc. Licensed under Creative Commons Attribution ShareAlike 3.0 (Generic)

X-ray diffraction pattern

④ Diffraction data are analyzed to give the final structure.

FIGURE 18.14 Determining an X-ray structure of a protein. The determination of an X-ray structure begins with the overexpression and purification of the protein of interest. The conditions needed to crystallize the protein are determined and crystals grown. These are bombarded with X-rays, and diffraction data are collected. The structure can be solved from analysis of these data.

(Source: (Top) Karp, *Cell and Molecular Biology: Concepts and Experiments*, 8e, copyright 2016, John Wiley & Sons. This material is reproduced with permission of John Wiley & Sons, Inc. (Bottom) Data from PDB ID 3FO5 Thorsell, A.G., Lee, W.H., Persson, C., Siponen, M.I., Nilsson, M., Busam, R.D., Kotenyova, T., Schuler, H., Lehtio, L. (2011) Comparative structural analysis of lipid binding START domains. *Plos One* **6**: e19521–e19521)

Protein structure

Biological molecules to be used in crystallization need to be as uniform and pure as possible—typically over 99.95% pure. Furthermore, any post-translational modifications to a protein (particularly glycosylations) that result in a diverse population of structures need to be eliminated lest they change the structure of the protein. Hence, proteins are often treated with phosphatases and glycosylases to remove any phosphate or carbohydrate modifications before crystallization.

Because crystallization usually occurs in aqueous conditions, membrane proteins, especially transmembrane proteins, have proven more difficult to isolate and crystallize than others.

Gentle detergents such as octyl glucoside that disrupt the membrane but leave the protein intact are often used in the purification and crystallization of membrane proteins.

Octyl glucoside

Growing the crystal

The second step in determining the structure of a macromolecule is also one of the most time consuming and difficult: growing a crystal that can diffract X-rays. It is common to grow crystals of small organic molecules in organic chemistry or to grow crystals of sugar to make rock candy. In each case, the process is similar. A molecule of interest is dissolved in a warm solvent, and then the temperature of the solution is cooled past its solubility point so that crystals form. In organic chemistry or in making rock candy, scratched glassware or an irregular surface can facilitate initiation of crystal growth. In recrystallizing small organic molecules, the choice of solvent is often important. Crystallization is used as a purification step because only like molecules pack together to form a crystal.

There are similarities and differences between growing crystals of proteins or nucleic acids and growing crystals of smaller molecules. Biological molecules exist in a watery environment; hence, water is often the main or only solvent. In organic chemistry, molecules may be dissolved in boiling solvent to generate a saturated solution. This approach is not suitable for biological molecules because boiling will denature them, destroying the tertiary and quaternary structures. Therefore, the process needs to be done at lower temperatures (e.g., at 4°C). Finally, the structures of biological macromolecules are far more complex than those of simple organic molecules and are thus more difficult to crystallize.

The conditions for growing crystals of any biological macromolecule must be determined experimentally. Because the molecules are complex, it can take weeks or even months for crystals to form. Therefore, hundreds of crystallization reactions are sometimes set up at once, with variations in salt type and concentration, pH, and the addition of small organics. Other molecules, such as polyethylene glycol, are often added to help concentrate the macromolecule of interest. When organic molecules are crystallized, the sample is often concentrated through evaporation. This approach is not applicable to biological macromolecules because the concentration of other solutes in the crystallization reaction will increase, and those solutes may crystallize before the macromolecule. To promote growth of a single crystal, vapor diffusion (using the hanging drop or sitting drop methods) or microdialysis is typically used to remove solvent slowly from the sample (**Figure 18.15**). In these techniques, a small droplet of the protein to be crystallized is placed in a

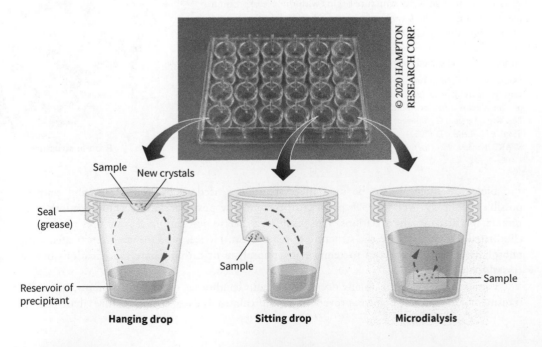

FIGURE 18.15 Crystallizing proteins. Proteins are grown using three common techniques in which multiple conditions are attempted at once in a multi-well plate. Shown are the hanging drop, sitting drop, and microdialysis methods. In each, the solvent is slowly removed from a concentrated sample of protein. The three differ in the geometry of how the experiment is set up and in how solvent is removed (evaporation or diffusion through a dialysis membrane). Different pH and salt concentrations are used to favor growth of a single high-quality crystal.

© 2020 HAMPTON RESEARCH CORP.

Hanging drop   **Sitting drop**   **Microdialysis**

sealed receptacle that contains a saturated solution of salt or polyethylene glycol. Water diffuses through the vapor phase into the more concentrated solution, gradually and slowly increasing the concentration of the protein and (if all goes as planned) facilitating crystal growth.

The crystals used in modern crystallography can be as small as 0.05 to 0.1 mm wide, but they must nevertheless be of sufficient quality to diffract. Thus, they should have few imperfections and cracks and should not be twinned (a condition where two crystals grow together and share a common face). Needle-shaped crystals diffract poorly. Some proteins, for reasons that are still unclear, fail to crystallize.

X-ray diffraction
The device used to conduct the X-ray experiment is an X-ray diffractometer (**Figure 18.16**). This instrument consists of an X-ray source, a mount for the sample, and a detector.

There are several ways to obtain X-rays for a diffraction experiment. The classical method is to bombard a metal target (usually copper) with high-energy electrons. Depending on the energy of the electrons and the metal used, X-rays of different wavelengths can be obtained. The metal surface is rotated during the experiment to limit heating. This type of X-ray generator is termed a rotating anode X-ray generator, and it generates monochromatic X-rays, that is, those with a single wavelength.

The second means of generating X-rays employs a cyclotron or synchrotron particle accelerator (**Figure 18.17**). Electrons are accelerated to high voltages and move in circular paths until X-rays are emitted. The X-rays are ejected from the accelerator through a beamline and can then be used in a diffraction experiment. There are several advantages to synchrotron X-rays. The rays are polychromatic (that is, having a range of wavelengths) and can thus provide more highly resolved structures than monochromatic X-rays. Also, the X-ray source emanating from a synchrotron is brighter (has a greater intensity), so the crystal requires shorter exposure times. Obtaining the necessary data can require several days of exposure using a rotating anode X-ray generator but as few as 10 minutes with a synchrotron X-ray generator. The drawback of particle accelerators is that they can cost up to billions of dollars to construct and are therefore scarce, with only six such facilities in the United States in 2020.

Crystals to be diffracted are held in the X-ray beam by a device that permits exact movement of the crystal in three dimensions. The entire sample chamber is cooled to low temperatures (–80° C) to prevent thermal damage when the crystal is hit by X-rays. Although the geometry varies among instruments, all diffractometers must detect the diffracted spots in a 360° sphere around the crystal. Therefore, the crystal, the detector, or both must move.

Finally, there is a detector. In the past this was a piece of film that was exposed by the diffracted X-rays. Current instruments use electronic X-ray detectors (charge-coupled device sensors) that identify the angle of scatter from the source (position) and the intensity of the scattered beam at that point. Modern instruments also often include shielding to ensure that those operating the instrument are not inadvertently irradiated.

X-rays are directed at the crystal, and the detector reads the position and intensity of the spots generated by constructive interference of the diffracted X-rays (also known as **reflections**). Often, more than one full read of the diffraction pattern is necessary to obtain a full dataset. The earliest experiments in the 1950s involved the analysis

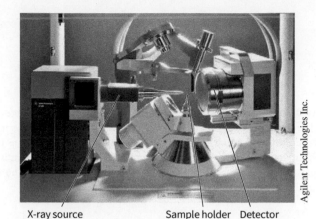

X-ray source   Sample holder   Detector

FIGURE 18.16 X-ray diffractometer. An Agilent GV1000 X-ray diffractometer is shown with the Atlas S2 CCD detector. The X-ray source is shown on the left, the sample holder in the middle, and the detector on the right.

Agilent Technologies Inc.

A.

Berkeley UNIVERSITY OF CALIFORNIA

B.

Particle beam

1 Electron gun
2 Linear accelerator
3 Booster synchrotron
4 Storage ring
5 Beamlines
6 Experiment stations

FIGURE 18.17 Synchrotron. A synchrotron is a type of particle accelerator commonly used in particle physics and X-ray crystallography. **A.** Shown is the Advanced Light Source at Lawrence Berkeley National Laboratory. **B.** In a synchrotron, electrons are accelerated by magnets to high speeds to facilitate the emission of other wavelengths of electromagnetic radiation, including X-rays. These X-rays are focused down one of several beamlines tangential to the synchrotron where they can be used in experiments.

(Source: From National Synchrotron Light Source II. Public domain.)

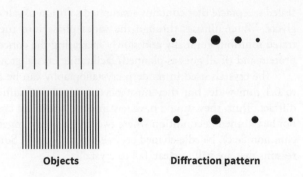

FIGURE 18.18 Diffraction patterns of repeating arrays.   Like the lines on the left, repeating arrays of molecules seen in a crystal diffract X-rays in a pattern that is the reciprocal of what is observed in real space. On the right, as in a two-slit diffraction pattern, objects closer together produce a pattern that is scattered farther apart, and objects oriented vertically produce a pattern arrayed horizontally.

Objects   **Diffraction pattern**

of over 13,000 reflections, and modern, high-resolution experiments may analyze 100,000 to several million reflections.

Analysis of diffraction data: the Fourier transform   The overall goal of using X-ray diffraction is to determine the structure (electron density) of the molecule in question from the scattered electrons. To do this, two problems need to be overcome. First, the data obtained is not simply a picture of the molecule itself. When X-rays are diffracted, as mentioned previously, slits that are closer together yield diffraction spots that are farther apart (and vice versa), and a vertical array of spots give rise to a horizontal diffraction pattern. In both instances, the pattern observed is in reciprocal, or inverse, space (**Figure 18.18**).

To extract the data from reciprocal space into real space, a **Fourier transform** is required to allow one function (frequency) to be represented as another (time). A Fourier transform is a mathematical equation that breaks the electron density into two parts: the information obtained from the magnitude and the information obtained from the phase of the scattered X-rays. The Fourier transform is presented in greater detail in **Biochemistry: Fourier series and Fourier transforms**. The advanced mathematics employed in this step are relatively straightforward using modern computational tools. The intensity data are simply obtained from the magnitude of the spot. However, identifying the phase of the scattered X-rays is more problematic, and this constitutes the second problem that needs to be solved, which is addressed in the next section.

Biochemistry

Fourier series and Fourier transforms

Cookbooks tell a chef how to put a recipe together. The cookbook contains all the information (ingredients, times, directions) to make a certain recipe. Sometimes a commercial restaurant has a dish or sauce or drink that is so good that people want to make it at home. In order to re-create the recipe for a final product such as a flavored coffee, we would start by breaking it down into components: coffee, cream, vanilla, nutmeg, and sugar. This is a good analogy for the mathematical processes known as Fourier series and Fourier transforms.

Any periodic function can be represented as the summary of sine and cosine functions. This is the essence of the Fourier summation. If we have enough of the right simple equations (ingredients) added together, we can make something more complex (a coffee drink).

A Fourier transform is somewhat more complicated. The Fourier transform seeks to determine which equations would contribute to the complex function—in effect to figure out the ingredients that make up the coffee drink. The Fourier transform, F, is accomplished by representing the function as a series of complex exponential expressions of the form $e^{2\pi ix\theta}$ and taking the integral of the resulting function:

$$F(\theta) = \int_{-\infty}^{\infty} f(x)e^{-i2\pi x\theta}dx$$

This allows us to obtain both the amplitude (x) and phase of the wave (the starting value of θ).

The net result of this is that a complex function can be expressed in terms of another variable. An expression that was a function of time, $f(t)$, for example, can now be expressed as a function of frequency, $f(\nu)$. In effect, we have gone from the time domain, t, to the frequency domain, $1/t$.

Fourier transforms come up twice in this chapter. X-ray diffractometry uses Fourier transforms to convert the diffraction pattern from reciprocal space to real space. In X-ray diffractometry, the solution of the Fourier transform is often represented as a series of sums:

$$\rho(xyz) = 1/V \sum_h \sum_k \sum_l F(hkl)\, e^{-2\pi i(hx + ky + lz) + i\phi(hkl)}$$

where ρ(*xyz*) is the electron density in real space, $1/V$ is a scaling factor that relates the volume of the unit cell to the electron density, and the remaining terms relate the intensity of the diffracted X-rays in the diffraction pattern (*F*(*hkl*)). The phase information, which must be obtained through other methods, is the final term in the exponential expression *i*φ(*hkl*).

The other place that Fourier transforms surface in this chapter is in the acquisition of modern NMR signals. Classically, we could sweep a field of radio waves across the NMR sample and see where they would absorb and generate a signal. This is analogous to finding a radio station by changing the channel and listening to see if there is a station or song you want to hear at each frequency. This would mean basically running the NMR experiment at every frequency and seeing where absorbances occur. This would be laborious and time consuming to do manually, and it would leave us with a single dataset for each frequency. Instead, modern NMR (and most spectral techniques) employs a Fourier transform. A short pulse of radio

waves excites all nuclei in the sample, which then process at the same time as they return to the unexcited state. This signal, which is a convolution of all the processing frequencies expressed as a function of time, is referred to as free induction decay (FID). Because it is simply a summation of these periodic functions, it can also be deconvoluted and expressed as intensities at different frequencies through the easily recognizable NMR spectrum.

Fourier transforms have numerous other uses in science and technology. Many computer files are compressed to reduce bandwidth or storage space. In this process, a signal for a sound or image file can be deconvoluted into its component waves through a Fourier transform. Then, only the waves that significantly contribute to the signal are retained. This maintains the majority of the information in the file but decreases the file size significantly. When music or video files (MP3 or MP4) or image files (JPEGs) are compressed, the compression is due in part to the use of a Fourier transform.

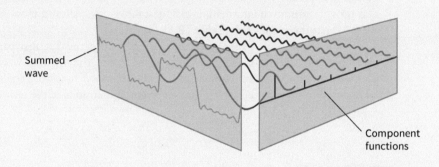

Summed wave

Component functions

Analysis of diffraction data: solving the phase problem

The X-rays diffracted by a crystal provide two pieces of data: how "bright" the spot is (the intensity of the scattered X-rays) and the position of the reflection (the angle of the scattered beam). Also known is the wavelength of the X-rays, which is unchanged. However, one important piece of data is missing: the phase of the scattered X-rays. When scattered, these beams are phase shifted with regard to one another, and that phase information needs to be determined in order to find the coordinates of the scattering centers.

There are several ways to solve the **phase problem**, depending on the complexity of the crystal structure. In the examination of macromolecular structures such as proteins and nucleic acids, the most common approach is **multiple isomorphous replacement (MIR)**. In MIR, data are first collected from the native crystal, and then a crystal is soaked in a solution of heavy atoms (e.g., mercury, gold, platinum, or iodine). The heavy atoms diffuse into the crystal and bind, ideally at only a few (three to eight) regular locations. For the technique to work, the structure of the crystal should not change (it should remain isomorphous). Scatter data are collected from the crystal bound to the heavy atoms. Because only a few atoms have been added to each unit cell, the overall scatter pattern (the location of the spots) should not change, but the intensity of the spots should increase in some cases and decrease in others. The difference between the intensities of the spots can be used to find the positions of the heavy atoms. The Patterson function (the square of the electron density function) is used to locate the heavy atoms in the unit cell and thus help to solve the phase problem. Based on the position of the heavy atoms, it is possible to estimate the phase of the X-rays and then calculate the position of other atoms in the crystal. Modern software can perform multiple rapid calculations to find the best fit.

As data are interpreted and the phase of the scattered X-rays determined, the initial structure determination can take place. The diffraction data are now transformed into electron density data. The amino acid or nucleotide sequence of the molecule is traced through the electron density, giving the initial model. In the 1960s to 1980s, these models were constructed of wire, metal rods, clay, or Plexiglas. Today, they are constructed on a computer using software designed for the task.

The initial model can be used to help refine the structure and redefine the phase of the X-rays, leading to clearer and higher-resolution structures. During this refinement, bond lengths and bond angles are checked to ensure that they are within expected limits. If only a few bond lengths or angles are outside the norm, it may simply be the nature of the structure, but if many are, it suggests a problem with data analysis and a need for further refinement.

18.3.4 A table of data describes the quality of an X-ray crystallography structure

Once a structure has been determined, we check whether the data from the experiment are of high enough quality to ensure that the structure is an accurate model of nature. A structure or model may look appealing, but data are needed to substantiate it. Therefore, most scientific publications that provide structural information include a table of statistics to show that the model fits within accepted natural parameters (Table 18.1).

These statistics tables are often divided into two parts. The first part gives some of the raw data for the native crystal and the different MIR experiments that were run. For example, this part will contain the total number of reflections (spots) that were acquired, the types of heavy atoms used in phasing the crystal, the shape and dimensions of the unit cell, the completeness of

TABLE 18.1 Crystallographic Data

Data collection	
Space group	P2 2_1 2_1
Cell dimensions	
a, b, c (Å)	60.2, 70.3, 80.1
α, β, γ (°)	90, 90, 90
Resolution (Å)	48.7–2.20 (2.26–2.20)
R_{sym}	0.124 (0.550)
$I/\sigma I$	14 (3.9)
Completeness (%)	99.0 (97.8)
Redundancy	6.9 (7.1)
Refinement	
Resolution (Å)	37.5–2.2
Number of reflections	14,775
R_{work}/R_{free}	20.1/24.6
(Number of atoms)	
Protein	2436
Ligand/Mg^{2+} ion	58/4
Water	140
(B factors)	
Protein	24.8
Ligand/ion	27.2
Water	28.8
RMSD	
Bond lengths (Å)	0.008
Bond angles (°)	1.05
Ramachandran statistics	
Most favorable region (%)	96.0
Allowed region (%)	4.0
Disallowed region (%)	0

the dataset, and how highly resolved the structure was (this value often has a range, with different structures being obtained at different resolutions). The second half of the table includes the statistical analysis that describes the quality of the structure and the computational methods used in solving the phase problem. This part of the table usually contains the total number of atoms in the structure, the total number of solvent atoms, and any other groups that were part of the crystal but are not part of the molecule of interest.

The table contains data on the **root-mean-square deviation (RMSD)** for both bond angles and bond lengths. Each of these values represents how much the average bond length or bond angle deviates from the norm across the crystal. Low values (<0.01 Å and 1.0°) are desirable because atoms in any molecule have accepted bond lengths (~1.54 Å for a C–C single bond) and bond angles (109.5° for a tetrahedral bond angle) and should not exhibit significant variability. High values indicate a likelihood that there was a problem in solving the structure.

Reliability (R) factors are another means of determining the quality of a structure. There are several means of calculating an R factor—sometimes called R_{work}— but the calculation can almost be thought of as a percent error. The R factor itself is not a true measure of error but rather of how closely the structure predicts the observed diffraction data. In an R factor, the sum of the absolute values of the observed X-ray intensities is subtracted from the absolute value of the predicted X-ray intensities and divided by the sum of the absolute value of the predicted values:

$$R = \frac{\sum \| F_{obs} | - | F_{calc} \|}{\sum | F_{obs} |}$$

For large molecules such as proteins and nucleic acids, the value of the R factor varies between 0.6 and 0.2, with lower values indicating greater reliability.

The final value that most tables include is a **B factor** (also known as a temperature factor or Debye–Waller factor). The B factor describes how much of the scattering intensity is lost due to thermal motion of the scattering center. In a macromolecule such as a protein, the B factor indicates that there are regions of the crystal where the protein exhibits more motion, and therefore the structures are harder (or sometimes impossible) to determine. Each atom of the structure has a B factor assigned to it, and these values are averaged to obtain the mean B factor for the entire structure. B factors are measured in Å², and they range from 30 Å² to 60 Å². As with the R factor, lower numbers indicate greater confidence in the data.

To validate a structure, it is also common practice to map its phi and psi bond angles to a Ramachandran plot to verify that the elements of secondary structure fall within predicted angles for alpha helix and beta sheet.

18.3.5 Neutron scattering is similar to X-ray diffraction

Neutron scattering is similar to X-ray diffraction and generates similar structural data, but the two techniques have several important differences. First, neutron scattering uses neutrons rather than X-rays, generally with wavelengths of 0.04 to 3.00 nm. Second, X-rays interact with the clouds of electrons around nuclei, whereas neutrons interact with and are scattered by the nuclei themselves. Neutrons interact differently with each isotope, but they generally interact less as the atomic number of the nuclei increases. This differs from electrons and X-rays, which interact more as the square of the atomic number increases. Thus, compared with X-rays, neutrons interact much more strongly with small nuclei such as carbon and hydrogen.

Neutron scattering has several advantages over X-ray diffraction. Because the scattering centers are small compared to the size of the atom, it is not necessary to take into account any effects of the shape of the electron cloud when interpreting data. Scattering by nuclei also means that there is no loss of signal as the scattering angle increases. As already discussed, in an X-ray experiment, the highest angle-scattering data yield the highest resolution structure, but they are also the poorest quality data. In the neutron-scattering experiment, there is no loss of quality at high angles, making it possible to determine higher-resolution structures, although this can also lead to complications in the calculation of the Fourier transform.

Worked Problem 18.3   Quality of a structure

Interpret and discuss this table of data from an X-ray structure determination:

X-Ray Refinement Data

Total atoms: protein/solvent/ligand	4472/472/21
R factor	0.23
RMSD bond angles (°)	1.073
RMSD bond lengths (Å)	0.006
B factor (Å^2)	32

Strategy   What do the different refinements listed in the table mean? What are the commonly accepted values for high-quality structures?

Solution   The data suggest that this is a structure of reasonably high quality. Based on the statistical parameters discussed in the text (RMSD for bond lengths and angles, R factor, and B factor), the values are all within ranges that suggest the structure was accurately determined. Higher values (>0.6 for R factors or >40 for B factors) would indicate that there was some variability or difficulty in fitting the electron density to the structure. Overall, however, based on these analyses, the data appear sound. A further test would be to map the coordinates of the amino acids to a Ramachandran plot to see where the phi and psi bond angles fall on the diagram.

Follow-up question   What would be the advantages and limitations of using neutron scattering rather than X-ray diffraction to determine a structure?

Summary

- When any wave passes through an opening or around an object that is small compared to the wavelength of the wave, that wave is diffracted. When multiple scattering centers are present, the diffracted waves interfere with one another. Most interference is destructive, resulting in decreased amplitude or annihilation of the signal; however, some interference is constructive, resulting in increased amplitude of the signal.

- Because the wavelength of X-rays is similar to the size of atomic structures, X-rays will be scattered by a crystal of a macromolecule, which can be used to calculate the coordinates of the atoms in the macromolecule. Bragg's equation describes the relationship between the wavelength of the X-ray, the angle of scatter, and the distance between planes in the crystal.

- In X-ray diffraction, X-rays are generated and focused on a crystal of a macromolecule. A detector records the position and intensity of the refracted X-rays. Two problems arise in determining the structure: translating the diffracted X-rays observed in reciprocal space back into real space (employing a Fourier transform) and determining how the X-rays have been phase shifted through the scattering process (solving the phase problem).

- The quality of a published X-ray structure can be determined by several statistical calculations usually found in a table of data that includes the RMSD, R factors, and B factors.

Concept Check

1. Describe the theory behind X-ray diffraction.
2. Describe the steps in an X-ray diffraction experiment.
3. Some X-ray diffractometers use X-rays of various wavelengths, while other use monochromatic X-rays. Discuss the techniques used for generating these.
4. Explain how it is possible to determine the quality of an X-ray structure based on a table of scatter data.

18.4 Nuclear Magnetic Resonance

Nuclear magnetic resonance (NMR), also known as NMR spectroscopy, is widely used in organic chemistry to solve the structure of small molecules. In introductory organic chemistry, the aim is often simply to determine the connectivity of atoms, but the actual conformation of a molecule can also be obtained using a more complex NMR experiment.

NMR is also used in biochemistry, but due to the size and complexity of biological macromolecules, other steps must be taken to solve the structure. These include using higher magnetic fields to increase the resolving power of the experiment and using various spectral techniques to obtain the necessary conformational information.

18.4.1 NMR relies on the quantum mechanical property of nucleus spin

The protons and neutrons in the nucleus of an atom have the quantum mechanical property of spin. If there are an uneven number of protons and neutrons, the nucleus itself exhibits this property. In the simplest example of the hydrogen nucleus, which consists of a single proton, two possible and energetically equal (degenerate) states exist in similar numbers: $+\frac{1}{2}$ and $-\frac{1}{2}$ (**Figure 18.19**). If an external magnetic field is applied to these nuclei, they will align either with or against the magnetic field. Aligning with the magnetic field is a lower-energy state than aligning against it. This creates two different energies, one higher ($-\frac{1}{2}$) and one lower ($+\frac{1}{2}$). The difference between the two states is due in part to the spin of the nucleus and in part to the strength of the magnetic field. When subjected to the magnetic field, slightly more nuclei (~1%) will be in the lower energy state.

The nuclei in the magnetic field of the NMR experiment are not static. If the nuclei absorb electromagnetic radiation of a particular wavelength in the radio range, they can be knocked out of alignment with the magnetic field and resonate or flip as they realign with the field. Because the nuclei are charged and processing, they induce a current in a loop of wire that can be detected as a free induction decay (FID) (**Figure 18.20**).

The difference between the two populations of nuclei and the ability to detect changes in them are not in themselves useful in an NMR experiment. However, the electronic environment immediately surrounding the nucleus affects the spin slightly, shifting the frequency of radio waves required to flip the nucleus. This difference in absorbance can be detected as a chemical shift (δ) that represents the splitting of the resonant peak due to electrons surrounding the nucleus. The NMR output, therefore, gives a spectrum of peaks shifted and split by the groups attached to the atom of interest.

NMR spectroscopy may seem reserved for the laboratory, but simple variations of NMR are used in the clinic, as described in **Medical Biochemistry: Medical imaging**.

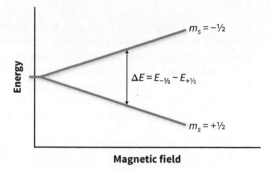

FIGURE 18.19 Spin and magnetic fields. Nuclei with the property of spin are split by an external magnetic field, aligning either with or against the field. This gap is increased with the strength of the field.

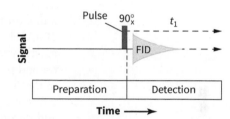

FIGURE 18.20 NMR experiment. In a basic NMR experiment, a sample is excited with a short, narrow-band pulse of radio waves oriented 90° to the sample. As the sample returns to the ground state, the nuclei process and emit energy, which is detected as a free induction decay (FID). The Fourier transform of this FID gives the familiar NMR spectrum.

(Source: From Horst Joachin Schirra.)

Medical Biochemistry

Medical imaging

Since the 1890s, if you broke your leg, an X-ray could show the physician the position and severity of the break. The same could be said of any bone break, but a tear to a ligament or soft tissue injury cannot be clearly seen on an X-ray. Physicians require other means to image these types of injuries.

One of the greatest advances in visualizing injuries has been magnetic resonance imaging, or MRI, shown in the photo.

An MRI is simply an NMR that measures different relaxation times for the protons in water in different tissues of the body. The spin-lattice relaxation time (t_1) is a comparison of the time it takes for nuclei to relax relative to the surroundings (the lattice). This technique is very valuable in imaging soft tissues, such as delineating the difference between gray and white matter in the brain. The spin-spin relaxation time (t_2) generally measures how long it takes for nuclei to return to their original state. This technique is useful in imaging tissues with different water content (e.g., bone, ligament, adipose, and muscle) or pockets of fluid. Both types of MRI imaging are used to generate a slice through the body. These can be used to reconstruct a three-dimensional image of the tissue of interest, such as the brain, shown here.

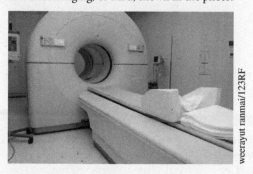

weerayut ranmai/123RF

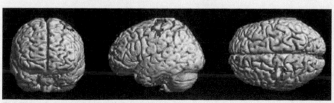

Pr Michel ZANCA/ISM/Medical Images.com

There are other means of doing medical imaging. CT scans use a series of X-rays taken through the body to reconstruct transverse sections. There have been vast improvements in the resolution of ultrasound technology, which uses the reflection of sound waves to image tissues. Positron image tomography (PET) scans are becoming more common. PET scans measure the emission of positrons (a type of radioactive decay) through a tissue. This requires the consumption of a short-lived positron emitter such as [18]fluorodeoxyglucose that accumulates in tissues that are actively using glucose (such as tumors).

Advances in imaging technology are helping diagnose not only broken bones but also cancer, atherosclerosis, and disorders of the central nervous system. Medical imaging has already transformed medicine and continues to improve, offering new hope for the future.

18.4.2 An NMR spectrometer generates a magnetic field

In an NMR spectrometer, a magnet holds the sample and provides the magnetic field that the nuclei align with or against (**Figure 18.21**). In modern NMR instruments, this is usually a superconducting cryomagnet that is held at temperatures near 0 K using a combination of liquid helium and liquid nitrogen. Such magnets have field strengths between 7 and 21 tesla (nearly a million times Earth's magnetic field) and can help to provide resolutions of between 300 and 900 MHz. The NMR sample fits into a small tube typically 2 to 3 mm in diameter and 20 cm in length, which is inserted into the magnet. The core of the magnet also contains the pulse generator and probe. These parts of the instrument generate a radio frequency signal and detect changes in the electromagnetic field.

Finally, the NMR spectrometer has a console that is used to control the entire experiment and deconvolute the signal into readable spectra. In modern instruments, this is accomplished using a Fourier transform.

18.4.3 NMR spectra are interpreted by analyzing peaks

In a simple NMR experiment, a sample (e.g., ethanol) is placed into an NMR tube and subjected to a simple analysis (**Figure 18.22**). The resulting spectrum has a standard peak (often tetramethylsilane, or TMS, fixed at 0) and three different peaks for ethanol, corresponding to the three different types of protons found: those of the methyl group, those of the methylene group, and the single proton of the hydroxyl group. The area under each peak (shown as the green integral above the spectrum) gives the relative number of protons in each peak. The peaks themselves are shifted based on the electronic environment of the molecule. The methyl and methylene peaks are found in the $\delta = 1$ to 4 range. The peaks for the methyl group are where one might expect to find them based on standard observations ($\delta \approx 0.9$), but the peaks for the methylene group are shifted farther to the left (downfield) than might

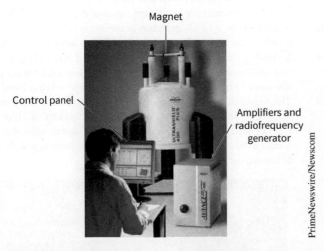

FIGURE 18.21 NMR spectrophotometer. An NMR spectrometer consists of a control panel, a cabinet of amplifiers and radio frequency generators, and the magnet. The magnet in modern NMR spectrometers is a superconducting magnet that must be maintained at very low temperatures. Higher-field NMR spectrometers employ larger magnets.

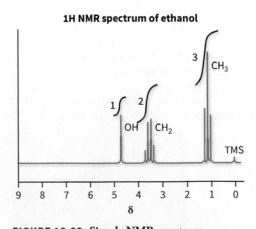

FIGURE 18.22 Simple NMR spectrum. Shown are the three peaks generated by a sample of ethanol and a TMS standard. Note that each of the different types of protons (OH, CH_2, and CH_3) have different splitting patterns and are shifted from 0 in the spectrum.

be expected given that methylene groups are typically found around $\delta = 1.2$. The observed shift is due to the neighboring oxygen found in the hydroxyl group. This final proton is even farther downfield due to the electronic environment around the electronegative oxygen.

The splitting of the peaks provides additional information about the electronic environment. The general type of splitting seen in this experiment, called J-coupling, is due to bond effects. The simplest rule for interpretation of these spectra is the $n + 1$ rule, where n is equal to the number of nearest neighbors to the peak in question. The general limit of J-coupling is three bonds. Returning to the spectrum in Figure 18.22, because the methylene protons are three bonds away, they split the methyl proton signal into $2 + 1 = 3$ peaks, forming a triplet. Similarly, the methylene protons are split by their nearest neighbors: the methyl protons. This produces $3 + 1 = 4$ peaks, forming a quartet. Finally, the spectrum gives a single peak for the hydroxyl proton. Why is this signal not split by the neighboring methylene, or why does this proton fail to split the methylene group into five peaks rather than four? The answer is probably due to the exchange of the hydroxyl proton with solvent. Modified experimental conditions may show stronger interactions.

The process described here is the basis for most simple NMR experiments. Spectra are analyzed based on the number of peaks, the position of the peaks, the number of equivalent nuclei in each peak, and the splitting of the peaks.

18.4.4 The Overhauser effect describes the through-space effects of one nuclei on others

Other phenomena can influence the electronic environment of the nuclei and can be useful in NMR. One such phenomenon is the Overhauser effect, or **nuclear Overhauser effect (NOE)**, whereby the spin of one nucleus can influence the spin of another. Whereas splitting in a traditional NMR experiment is a through-bond effect, the NOE is a through-space effect. In other words, one active nucleus can influence the spin of another merely by being close to it, even without a bond. The NOE is also distance dependent and varies with the inverse of the distance to the 6th power ($1/r^6$). Thus, it is a weak interaction but one that can be readily used to determine whether one nucleus is close (within 5 Å) to another.

Qualitatively, an NOE experiment can be thought of as stemming from a traditional NMR experiment. In the example of analyzing a simple NMR spectrum of 3-ethylidenequinuclidine to determine whether two peaks are within 5 Å of each other, it is possible to perform an NOE, saturating the peak of interest. This is accomplished by adding a radio frequency signal that matches the absorbance of one of the peaks, thereby eliminating or saturating it. This should increase the signal at peaks that are within the range of the effect. In the example shown in **Figure 18.23**, the stereochemistry of the compound in question can be determined by irradiating the peak corresponding to the methyl group and seeing what effect this has on the proton attached to the tertiary carbon. If a strong increase is seen in the signal of the proton attached to the tertiary carbon, the effect could be attributed to the methyl protons being found *cis* to that proton.

An NOE is not symmetric or reciprocal. Irradiating the methyl protons does not yield the same spectrum as irradiating the proton on the tertiary carbon because of different distances and interactions through space.

18.4.5 Multidimensional NMR techniques are used to probe three-dimensional structure

In a traditional, one-dimensional (1D) NMR experiment, samples are aligned in a magnetic field, a pulse of radio waves is administered at 90° to the axis of the field, and the signal is collected as the sample returns to the original state, a process termed relaxation. Modern instruments can pulse samples at different angles. Because nuclei relax at different rates based on their electronic environment, different relaxation times are used in obtaining these spectra, in what are termed **multidimensional NMR** experiments. The rest of this section discusses the three types of these experiments that are most commonly used in determination of biological structure: correlation spectroscopy, total correlation spectroscopy, and nuclear Overhauser effect spectroscopy.

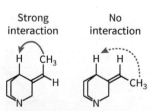

3-Ethylidenequinuclidine

FIGURE 18.23 The nuclear Overhauser effect. The nuclear Overhauser effect (NOE) describes the effect of one spin-active nucleus on another through space. Here, the effect of the protons in the methyl group can produce an Overhauser effect on the designated proton if found in the correct stereochemistry, as shown in the structure on the left. In the structure on the right, the methyl group is too distant to cause the effect.

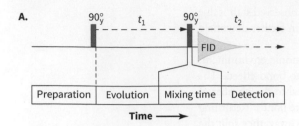

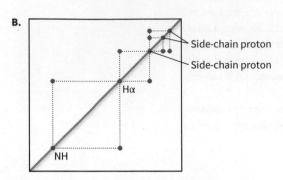

FIGURE 18.24 Correlation spectroscopy.
A. The COSY spectrum is a 2D NMR technique. Instead of a simple pulse 90° to the sample, there is a waiting period (t_1) followed by a second pulse (t_2) and data acquisition. **B.** All nuclei within three bonds of each other give a cross peak.

Correlation spectroscopy

Correlation spectroscopy (COSY) was the first type of multidimensional NMR spectroscopy to be used (**Figure 18.24**). In a traditional COSY experiment, the sample is usually subjected to a 90° pulse, given time (termed t_1) to relax, and then subjected to another 90° pulse (t_2), after which the signal is acquired.

The second pulse given in the COSY experiment enables direct identification of nuclei that are within three bonds of each other, through a process termed cross polarization. A COSY spectrum looks quite different from a traditional 1D NMR experiment. The 1D spectrum is found on the diagonal between the *x*- and *y*-axes. Peaks that are off the diagonal correlate the two peaks found on the diagonal to one another. In other words, off-diagonal peaks show which nuclei of the 1D spectrum are within two to three bonds of each other; the stronger the peak, the stronger the interaction. The phi bond angle in proteins can be determined by the strength of the coupling between the α carbon proton and the peptide proton in a protein, using the Karplus equation, discussed at the end of this section.

Total correlation spectroscopy

Total correlation spectroscopy (TOCSY) is useful for determining all the nuclei in a spin system, that is, a group of nuclei that are all coupled to one another through bonds (**Figure 18.25**). In biochemistry, this could be in a protein (where the nuclei of amino acids are isolated from one another by peptide bonds), in a polysaccharide (where the individual nuclei of the monosaccharide are isolated by the glycosidic linkages), or in a nucleotide (where individual sugars and bases are isolated from each other by the phosphate groups of the nucleotide backbone). As with COSY, off-diagonal peaks show interactions between nuclei found on the 1D spectrum (the diagonal), except that in this case, the off-diagonal peaks show not only the peaks observed through three bonds (the COSY spectrum) but also through all nuclei in the spin system.

Nuclear Overhauser effect spectroscopy

Nuclear Overhauser effect spectroscopy (NOESY) is another multidimensional technique (**Figure 18.26**). It uses the Overhauser effect to show all through-space interactions of a nucleus with others in the molecule. As with any Overhauser effect, nuclei (protons) need to be within 5 Å for signals to be seen. Because these protons can be distant from one another in the primary sequence but must be close in actual distance, they can provide critical information necessary for structure identification. For example, two amino

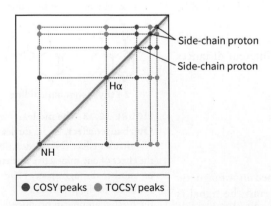

FIGURE 18.25 Total correlation spectroscopy. The pulse pattern in the NMR experiment in TOCSY enables the detection of all peaks coupled with one another in a spectrum. In other words, a TOCSY spectrum enables the identification of all the peaks in an amino acid of a peptide or protein. This spectrum contains all the peaks shown in the previous COSY spectrum and added TOCSY peaks.

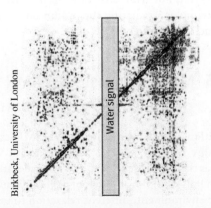

FIGURE 18.26 Nuclear Overhauser effect spectroscopy. NOESY spectra identify Overhauser effects as cross peaks off the 1D spectrum generated by two peaks interacting through space. These spectra can be complicated, to say the least.

acids in a generic protein (serine 21 and phenylalanine 341) are distant from each other in the amino acid sequence, but a strong Overhauser effect of the methyl group from phenylalanine 341 and the proton on the carbon of serine 21 place them within 5 Å of each other in the tertiary structure.

The Karplus equation   The Karplus equation describes the strength of the coupling constant of the torsional angle, phi (ϕ), between the alpha proton and amide proton in a peptide backbone by the relationship:

$$J(\phi) = A \cos^2 \phi + B \cos \phi + C$$

where J is the coupling constant measured in Hz or ppm, and the values of A, B, and C are adapted for the individual molecule, substituents, and atoms being used. In one example derived using the protein ubiquitin, these values were 12 (A), -1.4 (B), and 1.7 (C). The value of the coupling constant is strongest when there is the most overlap between the orbital ($\phi = 0$ or 180) and weakest when there is little overlap ($\phi = 90$).

18.4.6 NMR is developing as a technology

Recent advances in NMR have made it possible to determine the structure of components of the bacterial cell wall in intact cells. Because NMR is conducted in solution, scientists have used it to probe protein–ligand interactions and to study protein dynamics and nucleotide folding. NMR is now commonly used in some hospitals to identify HDL and LDL cholesterol levels quickly. NMR is a technology that is of growing importance in biochemistry research. Whether research teams use NMR or other tools, properly crediting those involved in the project can be complicated, as discussed in **Societal and Ethical Biochemistry: Giving credit where credit is due.**

Societal and Ethical Biochemistry

Giving credit where credit is due

The general public often thinks of scientists laboring for long hours alone in the laboratory until a breakthrough is made. Although science requires a high level of commitment, it is rare for a scientist to work on a research question alone. The more typical situation is to have multiple laboratories, each with multiple investigators, all working toward a single goal. In this regard, modern scientific work is less a solo performance and more like a band or orchestra.

Similarly, the public may think of science in terms of discoveries and breakthroughs, but in reality, most discoveries arise from a wide body of published work. By publishing information, scientists make the results available for the greater scientific community to corroborate (or refute) and to use in their own research. Publications and how they are viewed and used (cited) by other scientists are therefore a measure of accomplishment in the laboratory.

Because of the weight publications carry in the career of a scientist and the number of people involved in a scientific endeavor, there are often questions about who should be included as an author on a scientific paper. In other fields (for example, religion or literature), it is difficult to imagine a group of people with diverse talents contributing to a single published article, but in science the opposite is true. Often, an array of scientists at varying points in their career will all contribute to a scientific paper, and they all deserve authorship in some capacity.

The rules for authorship of a scientific paper vary among disciplines and laboratories. In general, the investigator who has contributed most to answering the scientific research question is the primary or lead author. Some papers include a footnote to indicate that particular authors contributed equally, but there is currently no accepted or effective method for giving scientists equal credit in joint authorship. In biochemistry, it is common for the director of the laboratory (the professor or principal investigator), who may be the most established author on the paper, to be listed last in the order of authors. Other authors are usually listed in the order of how much they contributed to the project, and some articles list numerous authors whose contributions may vary considerably.

It is generally accepted that scientists involved in a project should be authors on a paper, but the question is, what level of contribution is needed for authorship? Should those who shared ideas at a laboratory meeting be included? What about the technicians or undergraduate students who conduct experiments but did not design those experiments or the laboratory manager who orders supplies and manages the animals? Should the principal investigator, as the main driving force of scientific inquiry in the laboratory, be listed as the first author rather than the last? The answers to such questions are rarely straightforward, but the importance of publications in a career path means that authorship needs to be discussed and agreed on as early as possible in a collaboration or research project to avoid disagreements or conflicts.

Worked Problem 18.4   Interpretation of a 2D NMR spectrum

Begin to interpret this simple COSY spectrum of a dipeptide. Which peaks in the 1D spectrum are associated with one another?

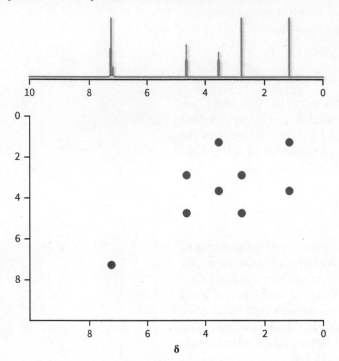

Strategy The peaks in the 1D spectrum are overlapping peaks at $\delta = 7.5$ (areas of 1, 2, and 2), a triplet at $\delta = 4.7$ (area of 1), a doublet of doublets at $\delta = 3.6$ (area of 1), a doublet at $\delta = 3.4$ (area of 2) and a doublet at $\delta = 1.7$ (area of 3).

Solution With the peaks numbered 1 through 5 from right to left, peaks 2 and 4 have an off-diagonal peak (indicating that these protons couple with one another) and another off-diagonal peak between peaks 3 and 5. Either side of the diagonal could have been used to find this peak because the spectrum is symmetrical. Peak 1 appears not to be coupled with any of the others.

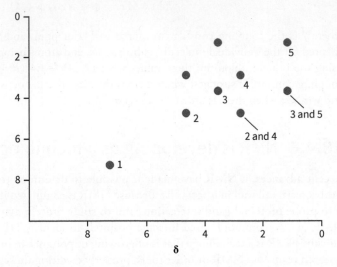

Follow-up question Based on the chemical shifts, the areas, and the splitting patterns, which amino acids are in the dipeptide? Which can be ruled out?

Summary

- Nuclei that exhibit the quantum-mechanical property of spin will align with or against an applied magnetic field and can be prompted to flip from one state to another if they absorb radiation of the appropriate frequency (radio waves). Because molecules relax differently based on their electronic environment, it is possible to deduce structural information from the differences in these signals through NMR.

- Simple 1D NMR uses J-coupling to determine splitting patterns and deduce the structure of the molecule. Spectra are analyzed based on the number of peaks, the position of the peaks, the number of equivalent nuclei in each peak, and the splitting of the peaks.

- The nuclear Overhauser effect describes the interactions of nuclei over a distance instead of through bonds.

- Multidimensional NMR, including COSY, TOCSY, and NOESY, uses different pulse patterns to probe both through-bond and through-space effects on molecules with complex structures. Off-diagonal peaks are used to correlate signals that occur between peaks on the diagonal and indicate interactions between these nuclei.

Concept Check

1. Describe the basic theory of NMR.

2. Explain the schematic diagrams describing the 2D NMR experiments shown in Figure 18.24A. What is happening at each step?

3. Describe the nuclear Overhauser effect and how it is used in 2D NMR spectroscopy such as NOESY.

4. Describe the other 2D techniques used in this chapter: TOCSY and COSY.

Bioinformatics Exercises

Exercise 1 Structural Homology of Globins

Problems

18.1 An Introduction to Structure Determination

1. Define resolution. Explain its significance in structure determination.

2. The resolution of an imaging technique must be matched to the size of the objects that are to be observed. If you were to map your classroom with 1-meter resolution, which features could you see? Which could you not see at that resolution? How would this differ if you had 1-centimeter or 1-millimeter resolution?

3. Define contrast. Why is contrast important in determining some structures?

4. How can structural data be validated? For example, how can one determine if an enzyme is a functional multimer in the cell?

18.2 Electron Microscopy

5. Electron microscopes are usually installed in a vibration-proof room. Even small vibrations such as a truck passing outside can make acquiring data difficult. Why is this the case?

6. Why are magnets used to focus the electron beam in electron microscopy?

7. What is characteristic of sample preparation in case of cryo-EM? Technological advancements have now made it possible to improve the resolution of cryo-EM, such that large biomolecular assemblies can be structurally analyzed. What information can be obtained from these studies?

8. How does sample preparation potentially influence an electron microscopy experiment?

9. Why is a Ramachandran plot not typically used to interpret electron microscopy data?

10. Why are stains such as phosphotungstic acid and uranyl acetate used in electron microscopy instead of organic dyes like the ones commonly used in light microscopy?

18.3 X-Ray Diffraction and Neutron Scattering

11. Why would modern computers facilitate X-ray crystallography compared to what was done in the 1950s? Which steps have been improved and which have not?

12. How have other areas of biochemistry and molecular biology facilitated X-ray crystallography?

13. List the steps that are needed to determine the 3D structure of a protein by X-ray crystallography.

14. How does sample preparation potentially influence the structure observed in X-ray crystallography?

15. Describe the difficulties involved in determining structures bound to membranes or proteins that are highly glycosylated. What approaches have been used to help solve these structures?

16. Using simple geometry, derive the Bragg equation to show that the distance between refracting planes is proportional to the wavelength of radiation used.

17. What are the acceptable values of RMSD, R factor and B factor for high quality X-ray structures of proteins?

18. The formula that is used to calculate the B factor is $B = 8\pi^2 Ui^2$, where Ui^2 is the mean squared displacement of the atom. What is the mean squared displacement of a carbon atom at B factor values of 10, 30, and 60 Å^2? For comparison, the diameter of a carbon atom is 3.4 Å.

19. Multiple isomorphous replacement (MIR) is used to help solve the phase problem in X-ray structures. How does MIR work? Could other atoms such as selenium be used to solve the phase problem? What about sulfur? Explain.

20. Interpret this table of X-ray data. What do each of these values tell you about the quality of the data?

TABLE 1 Crystallographic Data, Phasing, and Refinement Statistics

	Native	KI	HgCl$_2$	K$_2$PtCl$_4$
Data collection and phasing				
Space group	P6$_3$22	P6$_3$22	P6$_3$22	P6$_3$22
Unit cell	114.3, 97.5, 131.0	114.2, 97.3, 131.1	114.4, 97.6, 131.2	114.3, 97.7, 131.1
Resolution	2.7	2.9	3.1	3.1
R_{merge}[a]	0.071	0.077	0.081	0.069
Completeness	99.9	98.7	95.1	97.1
Unique reflections	12,421	9171	10471	7141
Phasing power[b]		0.059	0.062	0.061
Number of sites		4	5	4
Figure of merit[c]	0.41			
Refinement				
R_{work}[d]	0.224			
R_{free}[e]	0.267			
Cross-validated Luzzati error	0.47			
RMSD, bonds	0.009			
RMSD, angles	1.2			
Mean B	56.4			
Wilson B	55.2			
Protein atoms	1497			
Solvent atoms	0			
BME atoms	6			

[a] $R_{merge} = \sum |I(k) - \langle I(k) \rangle| / \sum I(k)$.
[b] Phasing power is $\sum |F_H| / \sum |E_{ms}|$ for acentric reflections, where E is the lack of closure.
[c] MIRAS figure of merit prior to density modification.
[d] $R_{work} = \sum |F_{obs} - kF_{calc}| / \sum |F_{obs}|$ summed over all reflections used in refinement.
[e] R_{free} is the R value calculated for a test set of reflections, comprising a randomly selected 5% of the data, not used during refinement.

21. Interpret this table of X-ray data. What does each of these values tell you about the quality of the data?

Data collection			(Number of atoms)	
Space group	P 2 2$_1$ 2$_1$		Protein	2436
Cell dimensions			Ligand/Mg^{2+} ion	58/4
a, b, c (Å)	60.2, 70.3, 80.1		Water	140
α, β, γ (°)	90, 90, 90		B-factors	
Resolution (Å)	48.7–2.20 (2.26–2.20)		Protein	24.8
R_{sym}	0.124 (0.550)		Ligand/ion	27.2
$I/\sigma I$	14 (3.9)		Water	28.8
Completeness (%)	99.0 (97.8)		RMSD	
Redundancy	6.9 (7.1)		Bond lengths (Å)	1.05
			Bond angles (°)	0.008
Refinement			Ramachandran statistics	
Resolution (Å)	37.5–2.2		Most favorable region (%)	96.0
(Number of reflections)	14,775		Allowed region (%)	4.0
R_{work}/R_{free}	20.1/24.6		Disallowed region (%)	0

22. Calculate the Bragg angles for the scattering of an X-ray with a wavelength of 1.45 Å by planes that are 3.5, 13.5, and 30 Å apart.

18.4 Nuclear Magnetic Resonance

23. Explain why high-field NMR instruments are used in determining the structures of large molecules. What are the characteristic components used in these instruments?

24. NMR structures in the PDB do not list the reliability factor, R, and the temperature factor, B, as is customary for X-ray structures. Explain.

25. The following nuclei are all spin active and are used in NMR experiments. Which would be more useful in biochemical experiments? Why?

Nuclei	Abundance
H^1	99.98%
Li^7	7.42%
C^{13}	1.108%
N$_{15}$	0.37%
O^{17}	0.037%
F^{19}	100%
P^{31}	100%

26. NMR can be used for determining the 3D structure of proteins in solution. What are the insights obtained into the protein structure from the NMR data?

27. Compare the structures generated using EM, X-ray diffraction, and NMR. Which of these is the highest resolution? Which is the best reflection of what is going on in the cell?

28. Is there a chance that new nuclei will be identified that can be used in NMR experiments? Explain.

29. A 60-MHz NMR is very useful for determining the structure of simple organic compounds but not for proteins. Why is a higher-field instrument necessary for protein structure determination?

Data Interpretation

30. Before the advent of computers and calculators, scientists routinely relied on a table of logarithms and a slide rule. Look up these instruments. Explain how they could be used for calculations. Keep in mind that more than 20 structures were determined this way, by hand and without help from computers.

Experimental Design

31. In Chapter 3 we discussed several different examples of proteins. For each of these proteins, discuss the merits and drawbacks of using each of the three techniques (EM, diffraction, or NMR) to determine their structure.

a. chymotrypsin

b. hemoglobin

c. immunoglobulin γ

d. insulin

e. myosin

Ethics and Social Responsibility

32. Some labs include undergraduate researchers as authors on papers, whereas others do not. What are the guidelines for establishing authorship? Should a principal investigator (the professor that directs the lab), who thinks up the big ideas but does not conduct any experiments, be given authorship? What about a technician who runs experiments but does not interpret any data?

33. A current common convention in biochemistry is for the lead author on a paper to be the one who has done the most work on the study and for the principal investigator (the lab director or professor) to be the last author. Other authors are listed in order of contribution. Do you agree with this ordering? This is not a universal convention. Should all fields of science agree on a single protocol for establishing authorship?

34. Authorship rights vary throughout disciplines and change over time. Some projects like the Human Genome Project or the Large Hadron Collider have hundreds of authors. If you were the director of a large project overseeing tens or hundreds of scientists, how would you establish authorship? What guidelines would you use?

35. Biochemical research is usually done by large research groups. The members have a variety of responsibilities and contributions to the project. Read the section *Giving credit where is due* in this chapter. Describe a fair way of listing authors on the research paper published by a particular research group at a university.

36. Authorship fights can be bitter, damage careers, or end collaborations and friendships. How can these conflicts be eliminated?

Suggested Readings

18.1 An Introduction to Structure Determination

Cantor, C. R., and P. R. Schimmel. *Biophysical Chemistry Part II: Techniques for the Study of Biological Structure and Function.* New York: W. H. Freeman and Company, 1980.

Van Holde, K. D. *Physical Biochemistry.* Englewood Cliffs, NJ: Prentice Hall, 1985.

18.2 Electron Microscopy

Bozzola, J. J., and L. D. Russell. *Electron Microscopy*, 2nd ed. Sudbury, MA: Jones and Bartlett Publishers, 1998.

Hendricks G. M. "Metal Shadowing for Electron Microscopy." *Methods in Molecular Biology* 1117 (2014): 73–93.

Kuo, J., ed. *Electron Microscopy: Methods and Protocols* (Methods in Molecular Biology series), 3rd ed. Totowa, NJ: Humana Press, 2013.

Newbery, D. E., D. C. Joy, P. Echlin, C. E. Fiori, and J. I. Goldstein. (1986). *Advanced Scanning Electron Microscopy and X-Ray Microanalysis.* New York: Plenum Press, 1986.

18.3 X-Ray Diffraction and Neutron Scattering

Bergfors, T. M. *Protein Crystallization*, 2nd ed. La Jolla, CA: International University Line Publishers, 2009.

Blow, D. *Outline of Crystallography for Biologists.* Oxford, UK: Oxford University Press, 2002.

Cowtan, K. *Kevin Cowtan's Book of Fourier Transforms.* Heslington, York, UK: University of York, Department of Chemistry, 2014 (http://www.ysbl.york.ac.uk/~cowtan/fourier/fourier.html).

Drenth, J. *Principles of Protein X-Ray Crystallography.* New York: Springer, 2006.

Lattman, E. E., and P. J. Loll. *Protein Crystallography: A Concise Guide.* Baltimore, MD: Johns Hopkins University Press, 2008.

Pynn, R. "Neutron Scattering: A Primer." *Los Alamos Science* (Summer 1990): 1–32.

Rhodes, G. *Crystallography Made Crystal Clear: A Guide for Users of Macromolecular Models*, 3rd ed. San Diego, CA: Academic Press, 2002.

Rupp, B. *Biomolecular Crystallography: Principles, Practice and Applications to Structural Biology.* New York: Garland Science, 2009.

Sherwood, D., and J. Cooper. *Crystals, X-rays and Proteins: Comprehensive Protein Crystallography.* Oxford, UK: Oxford University Press, 2010.

18.4 Nuclear Magnetic Resonance

Cavanagh, J., W. J. Fairbrother, A. G. Palmer, III, M. Rance, and N. J. Skelton. *Protein NMR Spectroscopy, Principles and Practice*, 2nd ed. San Diego, CA: Academic Press, 2005.

Evans, J. N. S. *Biological NMR Spectroscopy.* Oxford, UK: Oxford University Press, 1995.

Keeler, J. *Understanding NMR Spectroscopy*, 2nd ed. New York: Wiley, 2010.

Macomber, R. S. *A Complete Introduction to Modern NMR Spectroscopy.* New York: Wiley Interscience, 1997.

Marion D. "An Introduction to Biological NMR Spectroscopy." *Molecular & Cellular Proteomics* 12, no. 11 (November 2013): 3006–3025.

Rule, G. S, and T. K. Hitchens. *Fundamentals of Protein NMR Spectroscopy.* Dordrecht: Springer, 2006.

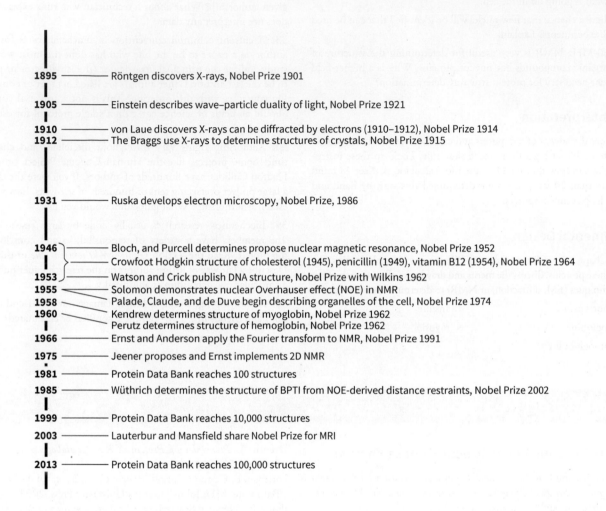

1895 —————— Röntgen discovers X-rays, Nobel Prize 1901

1905 —————— Einstein describes wave–particle duality of light, Nobel Prize 1921

1910 —————— von Laue discovers X-rays can be diffracted by electrons (1910–1912), Nobel Prize 1914
1912 —————— The Braggs use X-rays to determine structures of crystals, Nobel Prize 1915

1931 —————— Ruska develops electron microscopy, Nobel Prize, 1986

1946 —————— Bloch, and Purcell determines propose nuclear magnetic resonance, Nobel Prize 1952
        —————— Crowfoot Hodgkin structure of cholesterol (1945), penicillin (1949), vitamin B12 (1954), Nobel Prize 1964
1953 —————— Watson and Crick publish DNA structure, Nobel Prize with Wilkins 1962
1955 —————— Solomon demonstrates nuclear Overhauser effect (NOE) in NMR
1958 —————— Palade, Claude, and de Duve begin describing organelles of the cell, Nobel Prize 1974
1960 —————— Kendrew determines structure of myoglobin, Nobel Prize 1962
        —————— Perutz determines structure of hemoglobin, Nobel Prize 1962
1966 —————— Ernst and Anderson apply the Fourier transform to NMR, Nobel Prize 1991
1975 —————— Jeener proposes and Ernst implements 2D NMR
1981 —————— Protein Data Bank reaches 100 structures
1985 —————— Wüthrich determines the structure of BPTI from NOE-derived distance restraints, Nobel Prize 2002
1999 —————— Protein Data Bank reaches 10,000 structures
2003 —————— Lauterbur and Mansfield share Nobel Prize for MRI
2013 —————— Protein Data Bank reaches 100,000 structures

CHAPTER 19

Protein Production, Folding, and Engineering

Protein Folding and Engineering in Context

Thanks to modern biochemistry and molecular biology techniques, it is now possible to engineer virtually endless changes into a protein and produce these altered proteins in significant quantities. However, the mechanisms by which proteins fold into a stable conformation remain enigmatic.

When humans discover something new, our first instinct is often to take it apart and see how it works. People may take apart a computer or car, hack an operating system, or figure out the ingredients in a recipe. These modifications can have benefits; for example, upgrading a driver in a computer may improve the machine's performance, or adding just a little cilantro to a guacamole recipe may improve the flavor.

In many ways, biochemistry is no different. When scientists discover a new protein or enzyme, they are keen to determine its function. Hence, they modify parts of the protein—the domains or individual amino acids—in an effort to alter the protein's behavior or performance. This may provide important data about enzyme function or create a modified protein that is an effective pharmaceutical. More recently, scientists have been able to insert artificial amino acids into a protein, again with the hope of determining or altering protein function.

This chapter begins by discussing various methods for producing peptides and proteins, including chemical methods of peptide synthesis and biological overexpression of proteins. It then discusses some of the common tools used to modify proteins in the laboratory, including epitope tags and chimeric proteins. The chapter concludes with a discussion of protein folding, explaining what is currently known about how an amino acid sequence becomes the stable tertiary or quaternary configuration of a native protein.

Chapter Outline

Common Themes

Evolution's outcomes are conserved.	• Among over 100,000 known protein structures in the Protein Data Bank (PDB), there are only 1,300 conserved folds, indicating that these structures are selected for by evolution. • Directed evolution (randomizing a nucleotide sequence and placing artificial selective pressures on enzyme function) can be used to help generate altered aminoacyl-tRNA synthetases.
Structure determines function.	• Altering the structure of a protein by swapping out a domain or altering a single amino acid can result in substantial changes to protein function. • Unnatural amino acids can be used as probes of both protein structure and function, depending on the modifications made.
Biochemical information is transferred, exchanged, and stored.	• The flow of information from DNA to RNA to protein can be used to generate altered DNA sequences, which in turn code for proteins that include specific changes. • Other pieces of DNA (promoter elements) can be used to drive overexpression of a mutant DNA sequence.
Biomolecules are altered through pathways involving transformations of energy and matter.	• Protein folding is driven by both entropic and enthalpic processes such as the formation of weak interactions through the peptide backbone and amino acid side chains and the entropic gains made through the hydrophobic effect. • An energy funnel is a means of depicting how proteins fold to the most stable state.

19.1 Protein Production

There are many different ways to determine the function or characterize the properties of a protein. One way is to purify the protein, isolating it from all other regulators and interactors. This approach makes it possible to answer questions that would be technically difficult or impossible to resolve in a complex system. However, in its purified state, a protein might lack important regulators or subtle nuances of function.

A second way to study a protein, which overcomes some of the problems of working on purified material, is to either leave the protein *in situ*—that is, within its complex biological system—or introduce it into a cell, tissue, or organism. This is usually done by genetic engineering in which nucleic acids that code for a protein of interest are introduced into a cell or organism, although purified protein is sometimes used. Chapter 2 discussed the expression of proteins in transgenic organisms and blocking expression of a protein using knockout and silencing techniques. It may be helpful to review the concepts from that chapter for a reminder of how these techniques can be used to probe a biochemical system.

Both types of protein analysis—in a purified state or a complex system—provide important data and contribute to answering the larger question of the molecule's role in biology. This section discusses two ways to produce proteins for study: synthesis in the laboratory using the tools of chemistry and production in a cell using the tools of biology. The subsequent purification of such proteins and peptides is covered in Chapter 3.

19.1.1 Solid-phase peptide synthesis is a means of chemically synthesizing proteins

To a chemist, a protein is a condensation polymer of bifunctional monomers. Each monomer has two functional groups (an amine and a carboxyl group) that can participate in the polymerization reaction, forming an amide (peptide) bond in the process. If all amino acids had the same side chain, a polymer could be created by simply mixing the amino acids in a flask under the appropriate conditions. However, the reality is more complex. There are 20 different amino acids that need to be incorporated in the correct order to produce a functional protein. Even a single amino acid substitution can lead to a complete loss of protein activity.

Complicating matters further, the side chains of some amino acids have functional groups (hydroxyl, carboxyl, carboxamide, and amino groups) that could, in theory, participate in a polymerization reaction. Clearly, the chemical synthesis of a peptide or protein is not as straightforward as simply mixing and polymerizing amino acids.

The solution to this problem is a technique termed **solid-phase peptide synthesis** (**SPPS**) (**Figure 19.1**). In SPPS, the growing peptide is coupled to an inert resin matrix: a sandlike

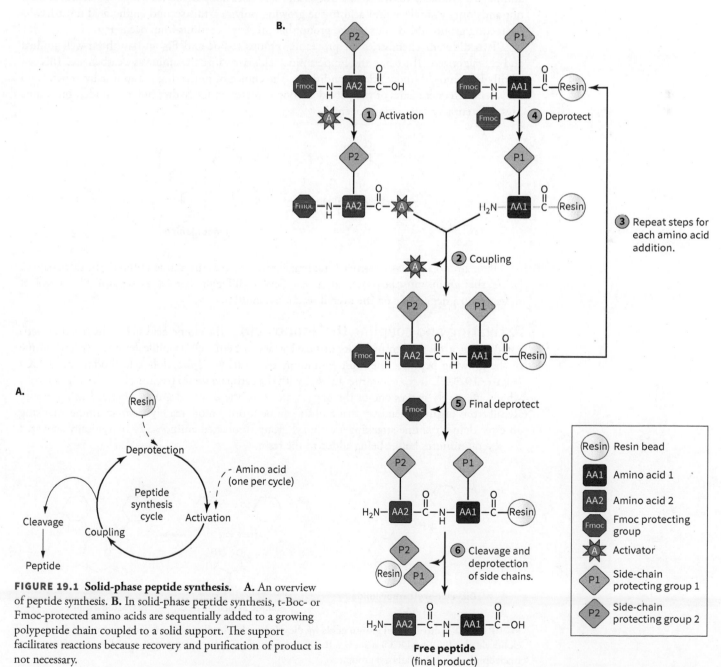

FIGURE 19.1 Solid-phase peptide synthesis. A. An overview of peptide synthesis. **B.** In solid-phase peptide synthesis, t-Boc- or Fmoc-protected amino acids are sequentially added to a growing polypeptide chain coupled to a solid support. The support facilitates reactions because recovery and purification of product is not necessary.

compound comprising millions of microscopic polystyrene beads. Attaching the growing chain to a solid surface means that the desired amino acids remain while unwanted materials are washed away. Amino acids are then added sequentially to the growing chain until the full-length peptide is achieved.

As amino acids are added in SPPS, they must be chemically protected to prevent unwanted reactions among them. The basic steps in SPPS are activating the protected amino acid and coupling it to the previous amino acid in the peptide chain. Before the process can be repeated, the next terminal amino acid of the peptide must be deprotected. The growing peptide is washed between certain steps to remove unwanted material. Once the peptide is complete, it is cleaved from the resin solid. This process, described in detail in the following subsections, is repeated for each amino acid added to the growing peptide.

Protecting the amino acid to be added

Unlike protein synthesis in a cell, which proceeds from amino to carboxy terminus, SPPS begins with the carboxy terminal amino acid and works backward. The first amino acid is coupled to the resin bead through an ester that joins the amino acid's carboxyl group to a hydroxyl group on the polystyrene resin. Next, the second amino acid is added. To ensure that this amino acid does not polymerize with other copies of itself but adds only a single amino acid to the growing polymer, this second amino acid must have a protecting group added to its amino group that prevents reactions from occurring.

Two different chemistries for protecting groups (t-Boc and Fmoc) have been well studied and characterized. The t-Boc and Fmoc amino acids derivatize the amine as a carbamate. This can readily be removed under acidic conditions. The choice of protecting group usually depends on the overall characteristics of the peptide; t-Boc is preferred for hydrophobic peptides, but Fmoc uses a potentially safer set of reagents.

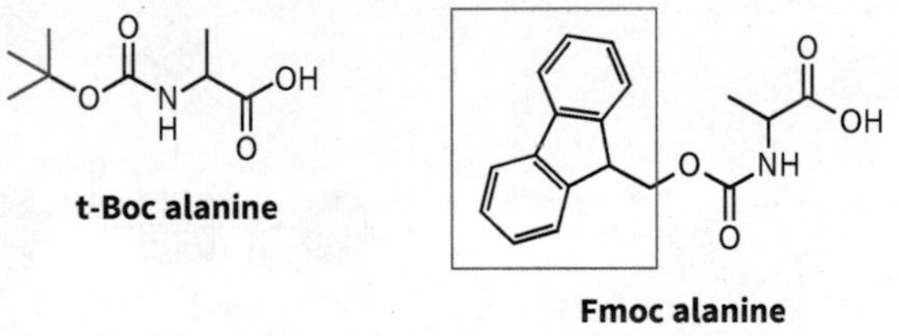

t-Boc alanine

Fmoc alanine

If an amino acid has a reactive functional group, such as the amino group in the side chain of lysine, that group must be protected as well. Several different chemistries are available to protect these groups, depending on the overall reaction conditions.

Activating and coupling the amino acid

The amino acid to be added is then activated using a catalyst to make its carboxyl group a better electrophile for the exposed amine of the growing peptide. The most commonly used catalyst is dicyclohexylcarbodiimide (DCC) (**Figure 19.2**). This reagent forms an ester with the carboxyl group (technically an *O*-acylisourea), which effectively makes one of the oxygen atoms of the carboxyl group a better leaving group. Recall from organic chemistry that amides can be formed more readily from an amine attacking an ester than an amine attacking a carboxyl group. Protected amino acids are typically activated for several minutes before being added to the resin.

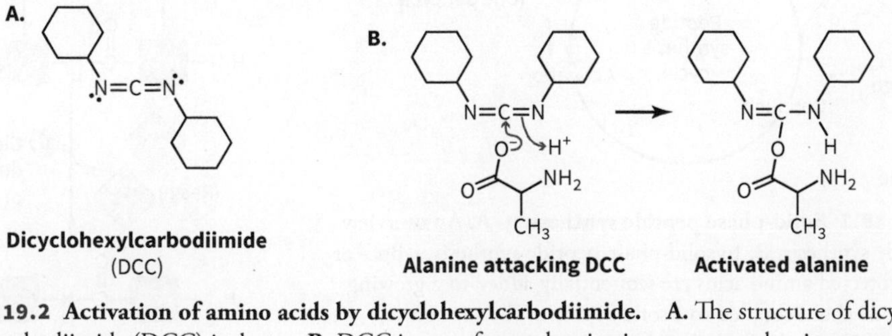

A.

Dicyclohexylcarbodiimide (DCC)

B.

Alanine attacking DCC   **Activated alanine**

FIGURE 19.2 Activation of amino acids by dicyclohexylcarbodiimide.   **A.** The structure of dicyclohexylcarbodiimide (DCC) is shown. **B.** DCC is one of several activating agents used to increase the nucleophilicity of the carboxyl group.

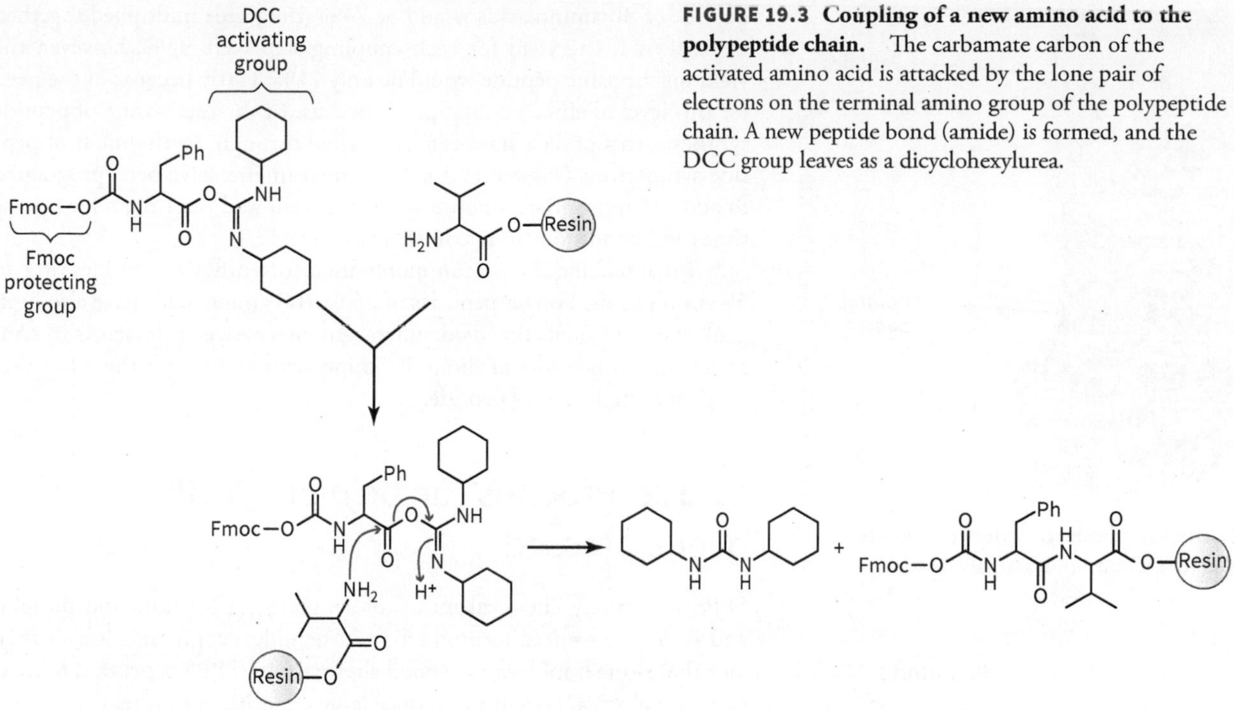

FIGURE 19.3 Coupling of a new amino acid to the polypeptide chain. The carbamate carbon of the activated amino acid is attacked by the lone pair of electrons on the terminal amino group of the polypeptide chain. A new peptide bond (amide) is formed, and the DCC group leaves as a dicyclohexylurea.

Next, the activated amino acid is coupled to the growing peptide chain on the resin bead (**Figure 19.3**). The coupling reaction takes from five minutes to an hour, depending on reaction conditions. Following the coupling reaction, any unreacted amino acids and remaining catalyst are washed away from the resin solid using the solvent dimethylformamide (DMF). After washing, the peptide is left with its newly added amino acid still bound to the resin bead.

Deprotection of the terminal amino acid
To initiate the next round of synthesis, the amino terminus of the peptide must have its protecting group removed (**Figure 19.4**). How this is done depends on the type of protecting group. Anhydrous hydrofluoric acid (HF) is used to remove t-Boc protecting groups. HF is highly corrosive; it requires special equipment for handling and to ensure the process remains anhydrous. Piperidine is used to remove Fmoc protecting groups; again, this reagent must be handled with care, but it is considerably less dangerous than HF. The use of the solid phase facilitates these reactions. Following deprotection, the removed protecting groups and the reagents used to remove them are rinsed away from the solid resin, leaving the peptide still bound. The whole process of protecting, activating, coupling, and deprotecting is then repeated for the next amino acid in the sequence. When all of the amino acids have been added, the completed peptide is cleaved from the resin surface.

Challenges in synthesizing peptides
Clearly, many different reactions need to happen to synthesize a peptide. A poor yield at any step will greatly limit the final yield of the product. If the yield for each coupling step of a peptide synthesis were 99%, the final yield for a

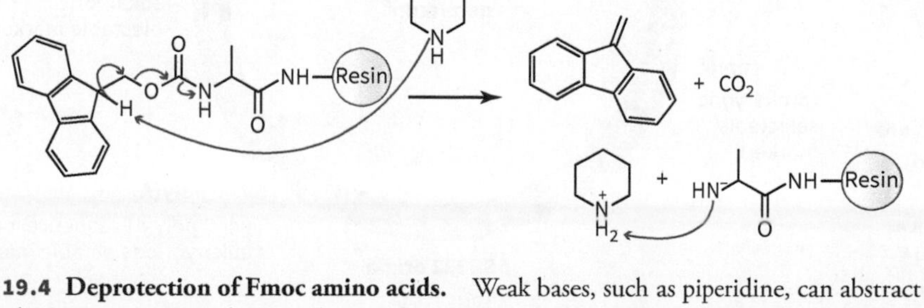

FIGURE 19.4 Deprotection of Fmoc amino acids. Weak bases, such as piperidine, can abstract a proton from the Fmoc group, resulting in electron flow up into the peptide group. This results in the collapse of the carbamate into the deprotected amine, a fluorenyl piperidine adduct, and carbon dioxide.

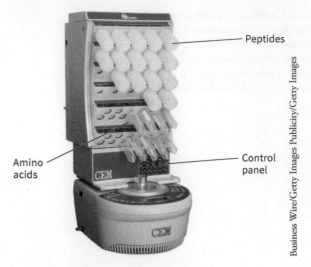

Business Wire/Getty Images Publicity/Getty Images

FIGURE 19.5 Peptide synthesizer.   Modern peptide synthesizers complete reactions under highly controlled conditions.

peptide of 30 amino acids would be 74% (the yields multiplied together, or 0.99^{30}). If the yield for each coupling step were 95%, however, the yield for the same peptide would be only 21%. Partly because of the need for this level of efficiency and partly because of the rote nature of peptide synthesis, this process has been automated through development of peptide synthesizers (**Figure 19.5**). These instruments have been programed to add the appropriate amount of each reagent and react them for precise times and conditions to maximize peptide yield.

These techniques are commonly used to synthesize peptides of 2 to 50 amino acids. Longer peptides of up to 100 amino acids have also been generated, although the yield suffers. An alternative approach is to construct shorter peptides of about 50 amino acids and couple them together to form a single, larger peptide.

19.1.2 Proteins can be biologically overexpressed

SPPS is a purely chemical means of synthesizing peptides and proteins and is not well suited to production of peptides or proteins longer than 30 amino acids. To produce these longer molecules, a good alternative to SPPS is **protein overexpression**, which is the use of a biological system to produce large quantities of protein.

Protein expression requires generating a piece of DNA that encodes the directions for producing the protein along with regulatory elements that ensure the protein is expressed. Also required is an organism or cell in which to express the protein and some means of getting this piece of DNA into the cell. This section examines two elements of this process: expression vectors and organisms used for protein expression.

Expression vectors   **Expression vectors** are pieces of DNA that encode both the message for the protein that is to be expressed and the regulatory information needed to express that protein at a high level. These vectors are synthetic structures, fragments of DNA spliced together in the laboratory to synthesize high levels of protein.

Expression vectors have several important features (**Figure 19.6**). There is an *ori* (or origin of replication) sequence that enables the plasmid to replicate within the bacteria. There is also a promoter region, which depends on the type of cell that expresses the protein. For example,

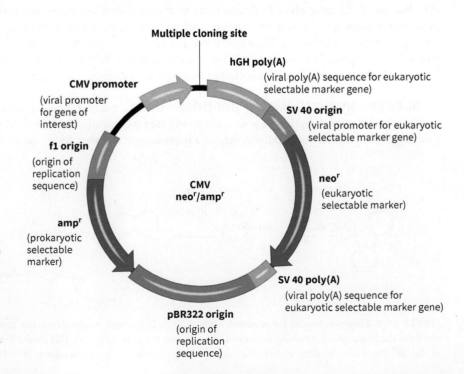

FIGURE 19.6 Expression plasmids.   Plasmids are one type of expression vector. A plasmid needs to have a selectable marker for the cell type in which it is going to be used, an *ori* sequence to drive replication in bacteria, a cloning site to add in the cDNA of interest, and a promoter specific to the cell type being used to express the protein. A polyadenylation sequence is often included as well.

in bacterial cells, the RecA promoter is common. RecA is a bacterial gene that codes for a DNA repair protein, but that function is not important in this situation, where the RecA promoter is used because it ensures a high level of expression in bacteria.

A common promoter in yeast is TEF1. As with RecA, TEF1 is organism specific; that is, genes driven by the TEF1 promoter are expressed strongly in yeast but not in other species. As with RecA, TEF1 is a constitutive promoter, meaning that the promoter is always on and is not inducible. Rec A and TEF1 are both strong promoters, leading to high levels of expression. Other promoters have medium or low levels of expression, which is useful in certain situations, for example, where expression of high levels of a particular protein would be lethal in the expression system.

In mammalian cells, a strong viral promoter is sometimes used, for example, the promoter regions of cytomegalovirus (CMV), simian virus 40 (SV40), or human immunodeficiency virus (HIV). These systems do not produce actual virus or even viral proteins; they are simply using the viral directions (promoter) to drive expression of the protein of interest.

The promoter sequence is followed by the cDNA coding for the protein of interest. Generally, this cDNA is inserted as a single coding region without any introns. In part this is because bacterial systems cannot splice messages and in part because it tends to make the production of the expression construct easier; fewer pieces need to be joined together, so the total size is smaller.

Depending on the cell type being used to express the protein, there may also be a polyadenylation sequence such as the one that follows the cDNA in the human growth hormone. Where a polyadenylation sequence is included, it is thought to serve a similar function to its role in nature; that is, it increases levels of expression, mRNA stability, and half-life in the cell.

Plasmids are often used as the means of constructing and delivering DNA to the cell in which proteins will be expressed, but there are alternative methods. The processes used to deliver plasmids are typically transfection (delivering DNA chemically) or electroporation (using pulses of electricity to drive plasmids into cells). For cells that are hard to transfect (including some mammalian cells), retroviral expression systems are used to deliver RNA to cells. The viral machinery then reverse-transcribes this RNA into DNA in the cell's genome. These techniques are discussed in the Techniques Appendix as well as in Chapter 2.

19.1.3 Different organisms can be used for protein expression

Several organisms are available for protein expression (**Figure 19.7**). Because all organisms use a single genetic code, the directions for producing a human protein can be used to make that

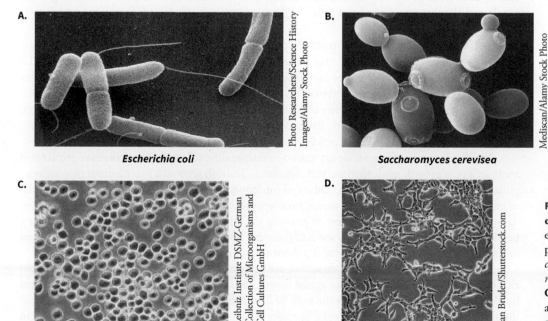

FIGURE 19.7 Organisms for expressing proteins. Shown are examples of cells used to express protein in the lab. **A.** *Escherichia coli* is a gut bacterium. **B.** *Saccharomyces cerevisiae* is brewer's yeast. **C.** *Spodoptera frugiperda* (Sf9) cells are ovarian fibroblasts from the armyworm. **D.** HEK cells come from a human embryonic kidney.

protein in bacterial cells; however, the production of *mature* proteins can be more complex. Different cell types may modify proteins differently, adding (or failing to add) glycosylations or other post-translational modifications. These altered modifications may, in turn, affect protein folding or function. Whether this matters will depend on the purpose of the expression. If a protein is to be used for a structural study (for example, to grow crystals), the altered modifications may be less important than if the molecule is being developed as a drug. Therefore, the choice of cell type can be important in the expression of a protein product.

Although all organisms use the same genetic code, many amino acids are coded for by more than one codon, and some organisms preferentially use certain codons over others through codon bias. **Codon optimization** is the matching of codons in a cDNA with the preferred codons of the host organism to achieve maximum protein expression. Such optimization is particularly useful when industrial-scale expression is required.

Bacterial, yeast, and insect cells

Bacterial cells are the simplest expression system to work with. The most common and well-studied bacterial system is *Escherichia coli*, which grows rapidly in an inexpensive broth, produces a high protein yield, and can easily be transformed through either chemical means or electroporation. However, *E. coli* does have some disadvantages. For example, multi-subunit and multi-domain proteins often fail to fold properly in *E. coli* and instead aggregate in inclusion bodies, creating granules of misfolded protein. It is then necessary to recover the proteins and often to refold them to their native conformations, a tedious and arduous task. Also, because proteins are not typically secreted from *E. coli*, the cells must often be lysed to harvest the protein.

Yeast provides an alternative to bacterial expression. Like bacteria, yeast grow quickly and only require a simple broth; however, unlike bacteria, yeast are eukaryotic. This eliminates some of the protein-folding and modification issues that can arise with *E. coli*. Also, yeast secrete proteins more readily than bacterial strains, a feature that can facilitate certain overexpression and purification schemes. However, if an overexpressed protein is not secreted, then the yeast must be lysed before purification, adding an extra step to the process. Common strains of yeast used for protein expression include *Saccharomyces cerevisiae* and *Pichia pastoris*.

A third alternative is the baculovirus system, a strain of virus that is pathogenic to insects. First, the baculovirus is engineered to code for the protein of interest. Then the virus is used to infect Sf21 or Sf9 cells, lines of ovarian cells derived from *Spodoptera frugiperda*, the fall armyworm. As with yeast and bacterial cells, Sf21 cells grow in suspension in a simple broth. Expression in a baculovirus system is strong, and this system avoids the modification and secretion issues found in bacterial systems because Sf21 cells are eukaryotic. At the end of the viral life cycle, the cells lyse, releasing the protein product into the medium. Hence, there is no need to rupture the cells to obtain the protein of interest; however, the protein must be purified from both the growth media and the remains of the lysed cells.

Mammalian cells

In some instances, the folding or post-translational modifications made to a protein necessitate the use of mammalian cells for expression. Human lines include HeLa (named for their source, Henrietta Lacks) and HEK, human embryonic kidney. Animal lines include CHO, Chinese hamster ovary cells, and CV-1, African green monkey kidney fibroblasts. These cell lines have gained popularity over others for their ease of transfection and growth. Also, because such systems involve mammalian proteins being expressed in a host cell that is also mammalian, the cells are more likely to process the protein in the same way as the native cell.

The disadvantage of large-scale expression in mammalian cells is that the cells have greater and more stringent growth needs than simpler organisms. The cells grow in a rich medium that includes glucose, pyruvate, vitamins, amino acids, and nucleotides. They may also require hormones such as insulin or supplementation with serum from another mammalian species such as bovine serum. This can add substantially to the cost of growing the cells and complicate the purification of any proteins after expression. However, some of these cell lines (particularly CHO) have been adapted to grow in serum-free conditions with supplementation of the required growth factors.

Bacterial, yeast, insect, and mammalian cell systems are by far the most popular for protein expression. However, there are other available systems; for example, proteins have been expressed in plants (banana or tobacco) or in the milk of livestock species. Again, there are advantages and disadvantages to each of these systems, but they illustrate the diversity and flexibility of protein expression systems.

Worked Problem 19.1 | Producing a protein for study

You are working on a clinical study in which cancer patients are to be administered a fragment of an interleukin (a small signaling protein produced by white blood cells) that is 30 amino acids long. You have been tasked with producing 2.5 g of the pure fragment for the study. Describe in detail what options you have for producing this quantity of peptide.

Strategy What techniques are available to produce a peptide 30 amino acids long? What are the strengths and weaknesses of each approach, keeping in mind the quantity of peptide needed?

Solution To synthesize a protein using SPPS, it would be necessary to couple sequentially a protected amino acid to a solid resin, deprotect that amino acid, and then repeat with each amino acid in sequence. The chemistry used to protect the amino acids (t-Boc or Fmoc) would dictate the reagents used to deprotect and ultimately cleave the peptide from the resin. The synthesis could be done

manually, but an automated peptide synthesizer would be easier and probably provide a higher yield. It would be fairly straightforward to synthesize 2.5 g of peptide using this approach, but it would require careful quality control (usually mass spectrometry) to ensure that the peptide was pure and had the correct sequence.

An alternative would be to use a bacterial expression system, for example, taking a cDNA coding for the peptide of interest, cloning it directionally into an expression vector, and using that vector to transform *E. coli*. The bacteria could then be cultured to produce the protein of interest. In this scenario, the protein would have to be isolated from either the bacteria or the bacterial broth.

Follow-up question Would your answer be different if you were asked to produce 2.5 g of a full-length protein of 165 amino acids? Why or why not?

Summary

- Proteins can be synthesized chemically or using biological expression systems.
- Solid-phase peptide synthesis (SPPS) is a form of chemical protein synthesis carried out on an immobilized resin. In SPPS, an amino acid is activated and coupled to the growing peptide, excess reagent is washed away, the terminal amino acid is deprotected and rinsed, and then the process is repeated to form the full peptide. One of two types of protecting groups (t-Boc or Fmoc) is used in the synthesis.
- To express a protein in a living system, a segment of DNA called an expression vector is generated that contains a region coding for the protein of interest and other regions coding for promoters. Promoter regions used in expression vectors differ depending on the cell type in which the protein will be expressed.
- Proteins can be expressed in various different systems, including bacteria, yeast, insect cells, and mammalian cells. Each expression system has advantages and disadvantages related to ease of use, cost, and problems with protein folding or modifications.

Concept Check

1. Compare and contrast the different protein-production systems discussed in the section and explain when each should be used.
2. Describe the steps involved in chemically synthesizing proteins using solid-phase peptide synthesis (SPPS). Explain the significance of use of solid matrix and protecting group.
3. Describe the types of expression vector available, including an explanation of the purpose of the promoter region and why it differs among expression vectors.
4. Explain why different organisms are chosen to express proteins and how a protein expressed in a bacterial system might differ from one expressed in a mammalian system.

19.2 Protein Folding

By this point in your study of biochemistry you should be familiar with what a protein looks like. The linear polymer of amino acids that makes up a protein is rarely in an extended and disordered conformation; instead, proteins are highly organized. Generally, a protein will assemble into a single shape or conformation, be relatively condensed, and be either globular or fibrillar in nature (**Figure 19.8**).

The process by which a protein goes from a linear polymer to a tightly folded structure is termed **protein folding**. Protein folding is the topic of this section, and it remains one of the great unanswered questions of biochemistry.

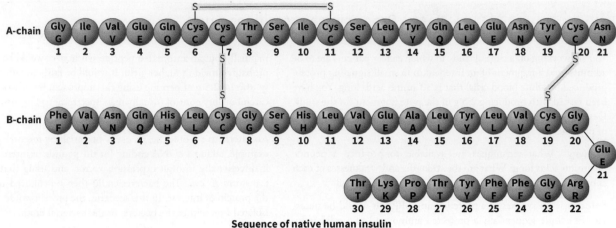

Sequence of native human insulin

fold

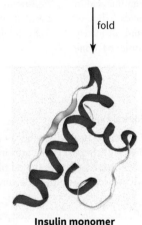

Insulin monomer

FIGURE 19.8 Protein folding. Protein folding describes how proteins adopt a single native conformation. The information needed is found in the amino acid sequence of the protein, and 90% of proteins do this without assistance in less than a millisecond.

(Source: Data from PDB ID 1ZEH Whittingham, J.L., Edwards, D.J., Antson, A.A., Clarkson, J.M., Dodson, G.G. (1998) Interactions of phenol and m-cresol in the insulin hexamer, and their effect on the association properties of B28 pro --> Asp insulin analogues. *Biochemistry* 37: 11516–11523)

19.2.1 There is debate about the mechanisms by which proteins fold

It is well known that proteins are made from amino acids and that they have a complex tertiary conformation held together by weak interactions. How proteins form this tertiary structure is not as well understood. Why, for example, do two specific amino acids in a protein hydrogen-bond with one another instead of with the many other amino acids they could bind to? Two observations can help in thinking about this problem: Levinthal's paradox and work conducted on ribonuclease A (RNAse A).

Levinthal's paradox What if protein folding were random, and the protein simply went through a series of different conformations (termed a random walk) until it found the most stable one? Cyrus Levinthal asked this question in 1969. He reasoned that if a protein had 100 amino acids, it would have 99 peptide bonds and therefore 99 phi and 99 psi angles. If each of these angles had only three possible conformations, then the number of total conformations for the protein would be 3^{198} (or 2.95×10^{94}) different combinations, that is, three different combinations for each of the 198 different bond angles. Given that a transition state is about 10^{-12} seconds, the time needed to proceed through all of these different states would be 2.95×10^{82} seconds, or 9.3×10^{74} years, that is, 6×10^{64} times longer than the age of the universe.

In fact, even this large number is an undercount of the true number of possible combinations. First, a typical protein is likely to be more than 100 amino acids in length (~450 or so). Second, each psi and phi bond angle can actually assume hundreds if not thousands of different conformations rather than just three. Third, the side chains of the amino acids must be considered. When all of these parameters are taken into account, it would require an astronomical amount of time for a protein to find the most stable (folded) state by trying different combinations at random. This is termed **Levinthal's paradox**. In reality, 90% of proteins fold spontaneously, without assistance, and within microseconds to milliseconds.

Clearly, there is something going on other than random folding. A clue to what this might be is provided by the evolutionary biologist Richard Dawkins. As an analogy to protein folding, Dawkins chose a sentence from Hamlet: "Methinks it is like a weasel." Randomly punching keys on a keyboard, it would take 10^{40} keystrokes to correctly type that sentence. However, if the keys were hit at random, but any correct letters were retained at each position each time, the total number of keystrokes would drop to only a few thousand, a far more reasonable number.

Correct:	`Methinks it is like a weasel`
Random:	`Hgqwnmpl `**`iv`**` er asv`**`e`**` w ghnhgr` (2 correct)
Retaining correct letters:	`Rghfds`**`k`**`a `**`iy`**` wq pym`**`e`**` `**`l`**` rfved`**`l`** (4 correct)

This analogy suggests that retaining "correct" or "allowed" conformations—that is, those elements of secondary structure that were stable—would make it possible to achieve the overall correct conformation much more rapidly than would randomly trying combinations (a random walk).

Ribonuclease A and protein folding

The second series of clues comes from work conducted on RNAse A. This small, stable protein was chosen for study in part because it was readily obtained and assayed. However, the results obtained with RNAse A have been repeated with and extended to other proteins.

When RNAse A is treated with heat, detergents, disulfide-reducing agents such as β-mercaptoethanol, or chaotropic agents such as urea, the weak forces stabilizing this protein are disrupted. As a result, it denatures, loses its tertiary structure, and becomes disordered (as discussed in Chapter 3). The curve that results from graphing the temperature or concentration of denaturant on the x-axis against the percentage of protein denatured on the y-axis is sigmoidal (**Figure 19.9**). The denaturation of DNA or the cooperativity of different subunits of an allosterically regulated enzyme both generate curves with a similar shape. The sigmoidal curve indicates that the protein denaturation occurs over an abrupt transition and is cooperative in nature. Proteins are never half-folded; rather, they exist in either a folded or unfolded state.

These experiments can be reversed; for example, solutions of denatured proteins could be cooled or the denaturants dialyzed away to refold the protein. In such situations, the same curve is found. Therefore, both protein folding and unfolding are cooperative processes. In addition, these data indicate another point that is often taken for granted: the information required to fold a protein properly is found within the protein sequence itself.

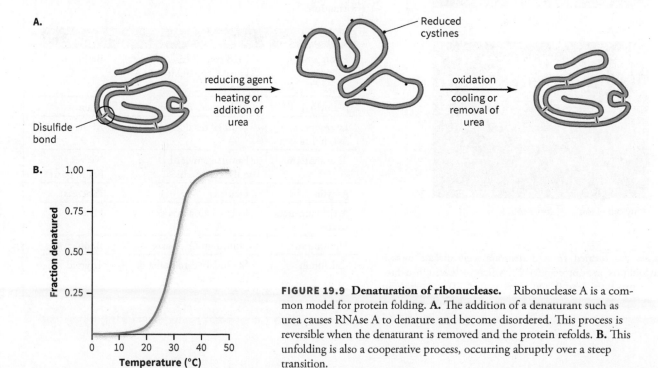

FIGURE 19.9 Denaturation of ribonuclease. Ribonuclease A is a common model for protein folding. **A.** The addition of a denaturant such as urea causes RNAse A to denature and become disordered. This process is reversible when the denaturant is removed and the protein refolds. **B.** This unfolding is also a cooperative process, occurring abruptly over a steep transition.

The folding of proteins is not simply a question of interest to science. It is now known that many diseases develop as a result of protein misfolding. These are discussed in **Medical Biochemistry: Protein folding and disease.**

Medical Biochemistry

Protein folding and disease

Chapter 3 discussed protein structures and introduced protein folding as well as diseases related to protein folding, such as kuru, mad cow disease, and Creutzfeldt–Jakob disease. Over the past 20 years, as understanding of these diseases and of protein folding has grown, so has the appreciation that several diseases, including several prominent neurodegenerative disorders, are the result of protein misfolding.

Alzheimer's disease is a progressive neurodegenerative disorder marked by cognitive difficulty and loss of short-term memory and motor skills, eventually leading to death. The causes of the disease are not completely understood and are probably polygenic (multiple genes contribute to the observed phenotype), but several key observations have been made. One of these observations is the accumulation of plaques of protein found in the brains of Alzheimer's patients.

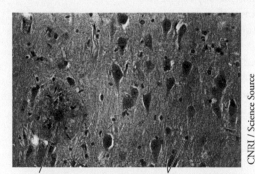

Plaque   Vacuoles

CNRI / Science Source

Amyloid plaque   Nerve cell

selvanegra/Getty Images

These plaques are formed from a neuronal cell surface protein termed amyloid precursor protein (APP), which is cleaved into three fragments by proteases (the β and γ secretases). The β fragment

is a short peptide 36 to 43 amino acids long. Characterization of β-amyloid structures has been difficult. The molecule fails to crystallize and appears amorphous. For reasons that are still unclear, this peptide can form oligomers or polymers that aggregate together to produce the amyloid plaques that are one of the hallmarks of this disease.

Alzheimer's is a type of disease known as amyloidosis; other diseases in this class are Parkinson's disease and Huntington's disease. Although the causes of many of these diseases are unclear, certain breeds of dog, including Shar-Peis, beagles, and English foxhounds, have a predisposition to amyloidosis, which suggests a genetic component. Currently, over 20 diseases and over 40 proteins are known to contribute to amyloid formation; the diseases include some forms of type 2 diabetes, cancer, and heart disease.

In all forms of disease related to amyloid formation, a functional protein is partially degraded in what appears to be a normal cellular process, but instead of complete degradation, a fragment of the protein adopts a different structure. These structures are rich in beta sheets and have hydrophobic faces that can associate with one another. As a result, the fragments begin to associate and form oligomers and polymers, at some point becoming insoluble plaques. It is currently unclear whether the plaques themselves or some step along the way is toxic to the cell or tissue, but the end result (accumulation of plaque and associated organ and tissue damage) is similar.

There are currently few treatments for people suffering with these diseases. Research is focused on identifying what changes to the protein result in misfolding, what aspects of this process lead to tissue damage, and what can be done to prevent plaque formation.

Protein	Disease	Tissue
Prion protein	Mad cow disease, scrapie, kuru, Creutzfeldt–Jakob disease	Brain
β-amyloid	Alzheimer's disease	Brain
Lysozyme, apo A-I, fibrinogen	Familial renal amyloidosis	Kidney
Transthyretin	Familial amyloid polyneuropathies	Nerves, brain, heart
Amylin	Diabetes	Pancreas
Atrial natriuretic factor	Isolated atrial amyloidosis	Heart
Huntingtin	Huntington's disease	Brain
Calcitonin	Medullary carcinoma of the thyroid	Thyroid

19.2.2 Protein folding can be analyzed using thermodynamics

Proteins are stabilized by weak forces, mainly hydrogen bonding, the hydrophobic effect, salt bridges, cation–π interactions, and disulfide bonds. The most stable structure would presumably be the one that has as many of these weak forces as possible. This situation is best depicted by an energy funnel, a three-dimensional graph of the types of interactions found against the free energy of the structure (**Figure 19.10**). The structure with the lowest free energy is the most stable and is therefore most likely to be the native structure. Unfolded structures are located higher in the funnel, outside this energy well.

Protein folding can be analyzed by looking at the thermodynamics of protein folding, breaking the problem into several parts. We can use the Gibbs free energy expression ($\Delta G = \Delta H - T\Delta S$) to dissect this problem. For the protein on its own, entropy (ΔS) decreases as the protein moves toward a single conformation because there are fewer microstates for the protein to adopt as it folds. On its own, this would be unfavorable because of that decrease in entropy. In contrast, examining the effects of solvent shows that burying hydrophobic residues in the core of the protein and the associated freeing of water molecules (the hydrophobic effect) has a substantial positive entropy, which makes a corresponding contribution to the free energy of folding. The hydrophobic effect is probably the main contributor to protein folding, but enthalpic contributions (ΔH) are also important; these are made through the formation of weak interactions among backbone groups and side chains, hydrogen bonding being the predominant force. Recall that breaking bonds requires energy ($+\Delta H$), but forming them releases energy ($-\Delta H$). Most reactions require both breaking and forming bonds, and it is the net sum of these energies that gives the energy of the reaction. Overall, ΔS and ΔH combine to give an overall ΔG of folding for the protein that is about 25 to 50 kcal/mol.

The folding topology can also be represented by other models. For example, the spherical model has two concentric spheres: a larger, external sphere modeling the completely unfolded state and a smaller, internal sphere signifying the folded state. As the protein folds and achieves a lower energy state, it moves from the external sphere to the internal one. This model can be used to illustrate both the folding funnel and the golf-course model (discussed in Worked Problem 19.2) and can take into account both the entropy of folding (in this case, the number of different microstates the protein can adopt) and the energetics of folding with regard to the decrease in free energy as bonds are formed.

Not all proteins can fold on their own. Previous chapters discussed chaperones and heat shock proteins, a family of proteins that form a temporary hydrophobic cage around nascent proteins, enabling them to rearrange and form a stable conformation. About 10% of proteins need some assistance in forming a stable conformation.

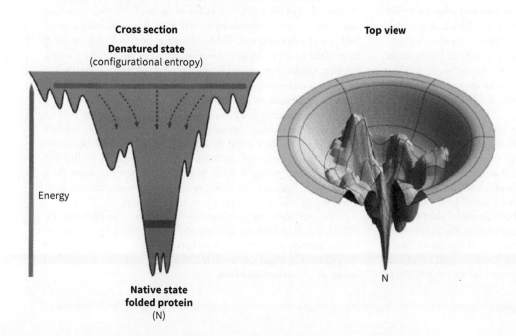

Cross section

Denatured state
(configurational entropy)

Top view

Energy

Native state
folded protein
(N)

N

FIGURE 19.10 Protein folding as depicted by an energy funnel. One way that protein folding can be depicted is using an energy funnel, a three-dimensional surface in which contacts are graphed on one axis, entropy on a second, and free energy on a third. The protein folds as the free energy decreases and comes to the lowest and most stable state. Shown are a cross section and a top-down view of one of these folding funnels.

(Source: Reprinted with permission from Tracking Chaperone-Mediated Folding Using Force Spectroscopy, João Manuel das Eiras Nunes, 2012.)

Although proteins in their native state are intrinsically stable, it is remarkable how they are at once both tolerant and intolerant of change. In the laboratory, it is often possible to produce mutant or truncated proteins that fold properly. Likewise, the amino acid sequence of a protein may seem sacrosanct and alterations to it harmful, but nature may suggest otherwise. Aligning a protein sequence from different species reveals how minor or even large-scale changes can occur that do not readily affect protein function. Finally, PCR-based techniques and phage display have been used with numerous proteins to remove various core or surface amino acids or to determine which residues are key for activity or for protein–protein interactions. Often, numerous conservative changes can be made without loss of function or binding, indicating that proper folding has taken place.

19.2.3 Computational modelling to explore the protein folding and structure

Knowing the structure of a protein is valuable and informative, but discerning this is a substantial undertaking. The stakes are so high in this fast-moving field that some researchers may be tempted to falsify results, as discussed in **Societal and Ethical Biochemistry: Scientific misconduct or fraud**.

Societal and Ethical Biochemistry

Scientific misconduct or fraud

In 2009, the University of Alabama at Birmingham concluded that a researcher there had faked the structures of almost a dozen proteins, falsifying results that appeared in ten scientific papers that were cited more than 450 times. All of these structures had been deposited in the Protein Data Bank, and one of them may have contributed to wasting time and resources in the search for a drug to treat dengue fever.

Many ethical pitfalls can trip up an investigator. Other chapters have discussed conflicts of interest, informed consent, use of animals in research, authorship, and plagiarism, but perhaps the most important ethical issue is scientific misconduct or fraud.

Scientific misconduct is a broad term that can cover many different violations of ethics, including intentional falsification of data (for example, to mislead a reviewer, colleague, or funding agency), stealing ideas from others when reviewing grants or manuscripts, giving a biased review of a grant or manuscript, or negligence that has resulted in someone being misled. It can be as small as the intentional mislabeling of a control or as large as fabricating data for a grant application.

Although misconduct is relatively rare, its ramifications are important and wide reaching. Several investigators have been found guilty of misconduct that continued over many years. In one case, an investigator was a recipient of over $2.9 million received through fraudulent grant applications. In other cases, multiple papers had to be retracted and protein structures removed from databases. In the meantime, these fraudulent reports had been used and cited by other members of the scientific community, potentially misleading those researchers or complicating their interpretation of legitimate data.

Generally, it is a series of papers that creates a trail of misconduct, but even a single paper or discovery can have vast ripple effects.

Perhaps the best recent example is the now-discredited work of Andrew Wakefield, who falsely associated the measles, mumps, and rubella (MMR) vaccine with autism in 1990. Even though that work was discredited and retracted, it led to a significant drop in MMR vaccination and is still being cited by the anti-vaccine movement as a reason not to immunize children, putting their health and lives at risk.

It may seem curious that people think they can get away with scientific fraud and that irregularities in their data or difficulties in repeating the work will go unnoticed. It is possible that people get caught up in a spiral of manufacturing data, lying to cover their tracks, and then having to repeat the process to keep making "progress." Nevertheless, many fraudulent investigators are eventually discovered, and the ramifications for them can be severe. Common consequences are retraction of an article by the publishing journal and loss of job and career. Where there has been a deliberate effort to mislead grant reviewers, some investigators have had to return funding and are then ineligible to apply for funding for several years. In at least one instance, the violations were severe enough to warrant criminal investigations and prison time.

Modern science is a competitive field. It is difficult to publish high-quality work, and funding rates for grant proposals are low (~21% in 2018, depending on the funding agency). Furthermore, there are strong motivations in biomedical research to be the first to discover something. A study may lead to a new treatment for a disease, a change in public policy, or discovery of a blockbuster drug. These things can affect people's lives in drastic ways.

Despite these temptations, there is *never* a justification for falsifying data or ignoring a questionable experimental procedure or result. Science only works if scientists can trust each other and if the public trusts that the work is being done ethically. Misconduct brings all of this into question.

Computational modeling can help predict protein folding and determine protein structures from primary amino acid sequences. Several different computational processes are commonly used.

Predictions of the structure of a protein solely from the amino acid sequence are termed ***de novo* structure predictions**. These are similar to the *ab initio* calculations used in organic or inorganic chemistry to predict the structure and properties of a small molecule (<1000 daltons). *Ab initio* calculations use quantum mechanics to predict the properties of a small molecule, and they are currently not applicable to larger, more complex systems such as proteins because the necessary computational power is not yet available.

The most common *de novo* structure prediction that is performed on a biological system is a **molecular dynamics (MD) calculation**. Multiple algorithms are available for use in a molecular dynamics simulation, but all have roughly the same outline:

1. Atoms are assigned initial positions.

2. Forces are applied to those atoms based on the functional groups they have formed (for example, hydroxyl groups will form hydrogen bonds).

3. Those forces are used to calculate accelerations toward or away from each other over short periods of time.

4. The calculation is repeated multiple times until the most stable state is found.

Some MD calculations take into account solvent, temperature, or other parameters. One shortcoming of MD calculations is that they typically focus more on enthalpic contributions (bonding) than on entropic ones, such as the organization of the protein molecule and the freeing of organized or bound water molecules. The energies calculated are typically enthalpies rather than free energies. There are hundreds of different variants of MD, each of which optimizes one parameter over another.

There are several options for calculating the structure of a protein when the structure of a related protein, such as one with a similar primary amino acid sequence, is already known. These techniques are called template-based methods, and they rely on related structures that have already been solved by other methods, such as X-ray diffraction and nuclear magnetic resonance (NMR). Template-based methods include homology modeling, threading, and fold prediction.

Homology modeling

Homology modeling involves making use of a protein that is highly related at the amino acid level but has a solved structure that can serve as a framework for the unknown protein. First, a sequence alignment search identifies proteins that are related to the unknown protein at the amino acid level. A second search identifies which (if any) of the related proteins have solved structures. Once a solved structure is found, the computer fits the amino acid sequence of the unknown protein to the existing structure. A refining calculation then takes into account differences between the two sequences and modifies the modeled structure accordingly, for example, moving structures slightly to account for steric clashes introduced by bulky amino acid side chains.

Threading

As with homology modeling, **threading** involves searching a database of existing structures, but threading is used to find structures for which there are no homologous protein structures. Instead, the search is for shorter sequences that form discrete folds (conserved elements of tertiary structure). Of the more than 100,000 proteins found in the PDB, there are relatively few types of folds (~1,300). Threading programs look for sequence similarity between these folds to identify regions of homology and then build up a structure based on these elements.

Fold prediction

Fold prediction is similar to threading but takes into consideration shorter pieces of a protein rather than the whole. It involves finding short stretches of an amino acid sequence and comparing them to a database of existing folds such as a Greek key motif or a zinc finger. The two most commonly used databases are SCOP (Structural Classification of Proteins) and CATH (Class, Architecture, Topology, Homology). If there is significant homology, the program assigns the fold to that region and assumes the function that piece of the protein may serve. For example, a zinc finger domain would indicate that there might be a DNA-binding domain.

Computational modeling of proteins has grown in recent times as the processing speed of computers has increased and the internet has made those computers more accessible. However,

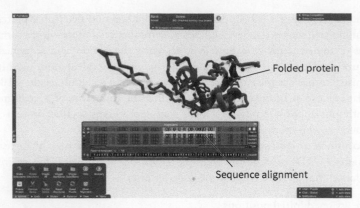

Folded protein

Sequence alignment

FIGURE 19.11 Foldit.   Foldit is a web-based protein folding game that is being used to crowdsource protein folding problems. Here players are attempting to determine the structure of a monkey virus protein. (Source: University of Washington)

such modeling is still computationally heavy, requiring large amounts of processing time. One interesting avenue that is being explored combines game theory, crowdsourcing, and molecular dynamics to solve structures. Foldit is a web-based "game" in which players attempt to fold proteins into native configurations. It is an excellent way to review basic protein chemistry and protein folding, but its main value lies in its predictive ability. Once players have learned the basic rules of the game (the rules of protein folding), they are given real-life structures that have not been solved. In 2011, players were challenged to solve the structure of a retroviral protease. Crystals of this protein had been obtained, but due to difficulties in data interpretation, the structure had not been solved. Using the data obtained from gameplay of several thousand players, investigators were able to solve the structure of the protease (**Figure 19.11**). More recently, players were able to modify a computationally derived enzyme to increase its activity by 18-fold.

Worked Problem 19.2   Levinthal's golf course

Protein folding energy funnels are used to depict the energy landscape as a protein folds. What would an energy funnel look like if Levinthal's paradox were true?

Strategy   Examine the energy funnels shown in Figure 19.10. How would the number of contacts or bonds increase and entropy decrease as a protein folded under the guidelines set forth in Levinthal's paradox?

Solution   The energy funnel that is used to describe Levinthal's paradox is often called a golf-course model. In Levinthal's golf course,

the "ball" (the unfolded protein) moves at random around the green (a flat surface) without forming any favorable bonds or significantly gaining entropy until it chances upon a hole. When the protein falls into the hole, all of the correct bonds are formed, and the overall entropy of the system increases as the protein folds into its native conformation. (Note that this is simply a mental exercise and is not considered a viable model for folding.)

Follow-up question   The champagne glass (a broad depression leading to an energy well) is another energy funnel. What would this representation depict?

Summary

- Levinthal's paradox illustrates that proteins must retain "correct" folds as they assemble rather than folding at random. Studies on the small molecule RNAse A reveal that proteins denature and refold abruptly. The sigmoidal shape of the graph indicates that protein folding is cooperative and that information necessary for proper folding is contained within the protein sequence itself.

- Many common neurodegenerative diseases, including Alzheimer's, Parkinson's, and Huntington's diseases, are caused by protein misfolding that leads to amyloidosis.

- Proteins are stabilized by weak forces, mainly hydrogen bonding, the hydrophobic effect, salt bridges, cation–π interactions, and disulfide bonds. Protein folding can be depicted as an energy funnel, in which unfolded states are found higher up in the funnel, and the native state is at the bottom in the most energetically stable state.

- Molecular dynamics calculations are used in *de novo* protein prediction. Template-based modeling approaches include homology modeling, threading, and fold prediction.

Concept Check

1. In context of experiments conducted on RNAse A, explain the mechanism involved in protein folding.
2. Discuss how protein misfolding can lead to some common neurodegenerative diseases.
3. Explain how protein folding can be analyzed in terms of thermodynamics.
4. Discuss the computational techniques used to predict protein folding.

19.3 Protein Engineering

Biochemistry has grown tremendously as a field of study over the past 30 years, in part because of advances in the ability to manipulate and alter DNA coding sequences. This situation has far-reaching ramifications. At a basic level, proteins that were once scarce and difficult to obtain can now be overexpressed in culture, generating grams of starting material. Perhaps more important, it is now possible to alter an overexpressed protein, for example, adding or deleting domains that may regulate the protein or facilitate its purification. It is also possible to probe the function of a protein one amino acid at a time, surgically mutating select amino acids to determine which ones are critical for activity or regulation of an enzyme or receptor. In the most recent adaptation of these alterations, amino acids that are not usually found in nature, such as halogenated amino acids, can be incorporated at key locations in a protein to enable physical or chemical measurements that were previously unattainable.

This section discusses several common modifications that are made to proteins and describes how these modifications can be made in the laboratory.

19.3.1 Site-directed mutagenesis is a way to alter precisely the amino acid sequence of a protein

Site-directed mutagenesis is a widely used technique for making small alterations to a protein's coding sequence. Typically, these mutations affect only a single amino acid, and they may be as small as a single nucleotide change, although it is often advantageous to modify three to six bases.

Design of a mutagenesis experiment A mutagenesis experiment may seem straightforward, but good design can facilitate subsequent experiments. Two factors need to be taken into consideration: the desired alteration in the amino acid sequence and the benefits to be gained by any additional changes, for example, incorporation of novel restriction sites in the DNA sequence.

Imagine that a protein is thought to be phosphorylated by protein kinase A on a particular serine residue. Mutation of that serine to an alanine would be a relatively minor change in terms of the size of the amino acid side chain (CH_2OH to CH_3). However, the mutation would prevent protein kinase A from phosphorylating the protein because of the absence of the hydroxyl group in alanine. Imagine also that examination of the mRNA for this protein produces the following sequence:

mRNA ... AGA-AAA-CGU-AAA-UCU-UUU-GGU-GCU
Protein ... Arg- Lys- Arg- Lys- Ser- Phe- Gly- Ala

Both serine and alanine codons are degenerate, which means these amino acids are coded for by multiple codons. Of the six different codons for serine, the one being used here is UCU (refer Figure 16.1). Hence, of the four different codons for alanine, it would be best to use GCU because it requires the smallest change from the serine codon UCU; namely, the two codons differ by only one nucleotide.

The second thing to consider in the design of the mutant is whether the alterations made to the DNA sequence can help in other areas of the experiment, for example, by incorporating restriction sites in the cDNA or deleting such sites. Restriction sites can be used as a preliminary confirmation that the mutation has occurred; they can also be useful in subsequent cloning steps. The first step in this method is to analyze the mutant nucleotide sequence and its flanking regions to determine which restriction enzymes, if any, would recognize the sequence. This is most conveniently done with sequence analysis software. Using such tools, investigators can submit a nucleotide sequence and quickly gauge whether any novel sites have been introduced.

If no novel site has been introduced, but one is desired, it may be possible to exploit the property of degeneracy to introduce a site. Introducing silent mutations that alter the nucleotide sequence but not the amino acid sequence can provide or delete a restriction site.

There is currently no automated way to screen for such mutations. Instead, the investigator must determine whether fragments of restriction sites can be introduced through silent mutations. For example, the recognition sequence of the restriction enzyme EcoRI is G^AATTC, in which the enzyme cuts after the G on both strands, leaving a four-base overhang. Thus, if the sequence GATTTC is present in a protein, and it is possible to make the T to A mutation without altering the amino acid sequence, that sequence might be useful in subsequent screening or cloning steps. Clearly, this process can be laborious, and it does not always yield results, but it can be of great help when successful.

Current techniques used in site-directed mutagenesis

Site-directed mutagenesis has benefitted greatly from advances in nucleic acid chemistry, most notably the ease with which short sequences of nucleotides (primers) can now be synthesized and the use of the polymerase chain reaction (PCR). This section discusses three types of mutagenesis experiments: cassette mutagenesis, PCR-based methods, and whole-plasmid mutagenesis.

In **cassette mutagenesis**, two complementary DNA strands are chemically synthesized, each containing the mutation of interest (**Figure 19.12**). The strands are mixed together, heated to melt any preexisting secondary structure, and allowed to anneal into a short, double-stranded sequence of DNA. This fragment is then cloned into a plasmid containing the cDNA of interest. The cloned fragment must be in the same reading frame as the fragment that was removed; that is, the fragment must not disrupt the order of codons or how they are read. Cassette mutagenesis is a relatively straightforward technique, and most of the products created have the desired mutation, but the process is complicated by several factors. First, the sequence to be mutated needs to have two restriction sites positioned close enough together for a cassette of reasonable size (generally 10 to 50 nucleotides) to be cut out. Second, oligonucleotide synthesizers often produce a range of side products in addition to the nucleotide sequence that was requested. These side products can also complicate the experiment, and it may be helpful to purify the desired oligonucleotide to facilitate subsequent mutagenesis steps.

Some cDNAs lack the appropriate restriction sites for cassette mutagenesis. In such cases, other techniques would be more appropriate, the most widely used being **PCR-based mutagenesis** methods. The simplest of such methods is to introduce a mutation using two primers: one normal primer and one mutagenic primer that contains mismatches where the primer will not anneal to the template. The design of the primers used must take this mismatch into consideration, and the mutagenic primer must have a homologous region downstream (in the 3′ direction) of the mismatch that is long enough for the polymerase to extend from. Typically, this region must be at least eight nucleotides long. The primers must also cover an area that encompasses restriction sites flanking the mutation to allow the cloning of the mutated fragment into the cDNA.

If the region to be mutated is too distant from a restriction site to include it in the same primer, a different technique must be used (**Figure 19.13**). In this case, two separate PCR reactions are run, each with a mutagenic primer and a primer that is some distance from the mutation but includes a restriction site. The PCR reaction is run with each of these "half reactions," and each half reaction generates a mutagenic product. These products are mixed together in a denatured state, allowed to anneal to one another, and then extended, with each PCR product serving as both a template for one reaction and a primer for the other. The resulting product is a larger fragment containing the desired mutation incorporated into a larger piece of the cDNA. This cDNA contains two restriction sites that can be used to clone the fragment into the native cDNA. Sequencing the final product verifies that the required mutation (and no other mutations) has been incorporated.

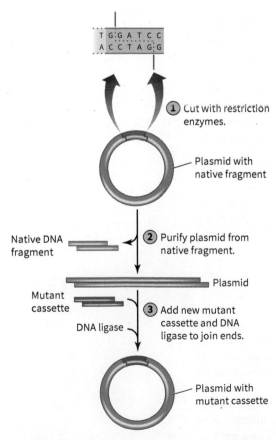

FIGURE 19.12 Cassette mutagenesis. In cassette mutagenesis, a mutant fragment of DNA is synthesized and then cloned into the cDNA of interest.

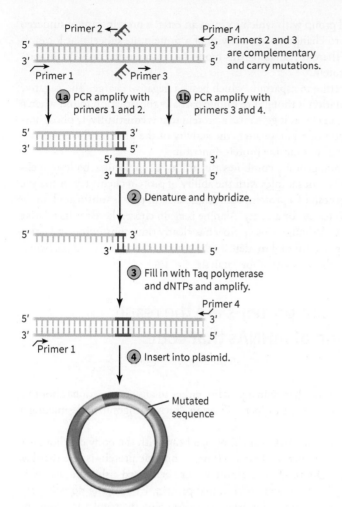

FIGURE 19.13 PCR-based mutagenesis. PCR-based mutagenesis methods employ mutagenic primers, which contain enough homology to anneal to the template but also mismatches that generate the mutation. Pairs of mutated primers can be used in sequential PCR reactions to generate a PCR product with a mutation closer to the center of the product.

(Source: Allison, *Fundamental Molecular Biology*, 2e, copyright 2012, John Wiley & Sons. This material is reproduced with permission of John Wiley & Sons, Inc.)

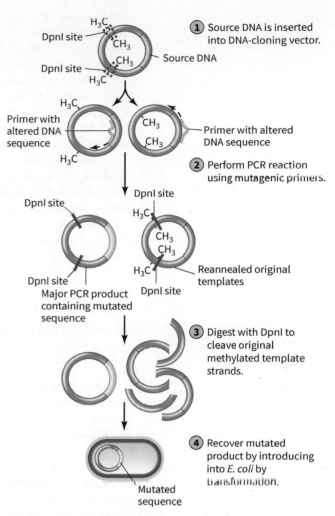

FIGURE 19.14 Whole-plasmid mutagenesis. In whole-plasmid mutagenesis, two primers overlap the region to be mutated but face opposite directions. First, the entire plasmid is amplified. Next, the template is degraded using the restriction enzyme DpnI that recognizes methylated DNA. The template, being bacterial in origin, is methylated, whereas the products are not.

(Source: Wessner, *Microbiology*, 2e, copyright 2017, John Wiley & Sons. This material is reproduced with permission of John Wiley & Sons, Inc.)

Another option is **whole-plasmid mutagenesis** (Figure 19.14). This is a variation of a PCR-based technique that uses two complementary mutagenic primers to polymerize an entire copy of the plasmid template. The template is then degraded using a restriction enzyme (DpnI) that only recognizes methylated DNA. Hence, it recognizes the original template DNA (which is of bacterial origin and therefore methylated) but not the unmethylated strands that were synthesized in the reaction (which are therefore preserved). This step leaves behind a copy of the original plasmid with the mutation incorporated, which can be used to transform bacteria and generate more of the plasmid for subsequent experimentation.

Common examples of site-directed mutagenesis

In theory, any amino acid could be mutated into any other; however, in the laboratory, some amino acid substitutions are more common than others. Perhaps the most commonly incorporated mutation is the substitution of an alanine for a serine residue. This can be done for several reasons. First, as has been noted several times in earlier chapters, serine residues are often important nucleophiles in enzymatic mechanisms. The loss of the hydroxyl group when serine changes to alanine means there is no longer a nucleophile in the mechanism, and catalysis usually ceases. Second, serine residues are important as acceptors of phosphate groups, acyl groups, carbohydrate chains, or ubiquitin modifications. Without a hydroxyl group with which to form an ester, a protein cannot undergo these modifications. This can lead to changes in the activity of a protein, its subcellular localization, half-life, or overall function. Therefore, making this modification can often provide important data about the function of a protein.

A related mutation is that of serine to aspartate, which bears a negative charge. The negative charge introduced through this mutation is thought to mimic partially the situation found when a phosphate is esterified to the serine; in effect, it generates a protein that is constitutively phosphorylated. In situations where the action of a kinase alters the activity of the protein, this mutation could be used as an experimental tool to examine protein function.

Another technique, alanine mutagenesis, combines the ability of molecular biology techniques to generate easily large numbers of samples with the ability of protein chemistry to analyze protein function. In alanine mutagenesis of a protein, each amino acid in turn is substituted for an alanine, and the resulting protein is tested for activity. Alanine is again chosen here for its relative chemical inertness and its small size. Although this approach is clearly time consuming and complex, alanine mutagenesis can be a powerful tool to identify the amino acids involved in particular aspects of protein function, such as those essential for catalysis.

19.3.2 Chimeric or fusion proteins are the result of in-frame combination of mRNAs that code for different proteins

It is often desirable to make a large-scale change in a protein, either to answer questions about its location in the cell or its function or to assist in its purification. Such changes include making a chimera or a fusion protein.

Chimeric proteins are named for the mythical chimera, a beast with the body of a lion, the head of a goat, and the tail of a serpent (**Figure 19.15**). Thus, a chimeric protein is one that has different domains grafted together. Often, chimeric proteins are generated using the catalytic domain of one protein and the regulatory domain of a related protein. Exchanging domains can allow investigators to determine their functions; for example, exchanging the regulatory domains of hepatic and lipoprotein lipase gives the chimeric proteins the substrate specificity of the donor protein.

Fusion proteins are similar to chimeric proteins but are generated by the in-frame splicing of cDNAs for two unrelated proteins. Many different fusion proteins are commonly used, and they are generally grouped by function: fluorescent proteins, ligand-binding proteins, and enzymes.

Fluorescent proteins   Proteins with fluorescent properties include green fluorescent protein (GFP), enhanced GFP (EGFP), enhanced cyan fluorescent protein (ECFP), yellow

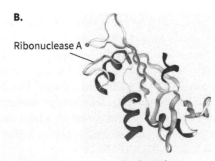

Ribonuclease A

Angiogenin

FIGURE 19.15 Chimera and chimeric proteins.   A. The chimera is a mythical beast with the body of a lion, the head of a goat, and the tail of a serpent. **B.** Chimeric proteins, such as this engineered human angiogenin, contain domains from different proteins. In this case, the catalytic loop of angiogenin has been replaced by the analogous domain from ribonuclease A.

(Source: (B) Data from PDB ID 1UN5 Holloway, D.E., Chavali, G.B., Hares, M.C., Baker, M.D., Subbarao, G.V., Shapiro, R., Acharya, K.R. (2004) Crystallographic Studies on Structural Features That Determine the Enzymatic Specificity and Potency of Human Angiogenin: Thr44, Thr80 and Residues 38-41. *Biochemistry* **43**: 1230)

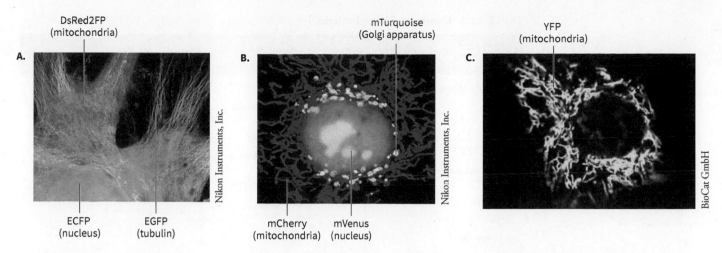

FIGURE 19.16 Fluorescent proteins. Fluorescent proteins can be used to tag another protein that will then migrate to the organelle where it is normally functional in the cell. **A.** The fluorescent protein tags EGFP, ECFP, and DsRed2FP function in tubulin, the nucleus, and mitochondria, respectively. **B.** Fluorescent proteins mCherry, mTurquoise, and mVenus have been fused to proteins found in mitochondria, Golgi, and the nucleus, respectively, to illuminate these different regions of the cell. **C.** These mitochondria are labeled with the yellow fluorescent protein tag (YFP).

fluorescent protein (YFP), the red fluorescent proteins mCherry and DsRedFP, and the blue and yellow proteins mTurquoise and mVenus (**Figure 19.16**). When fused with another protein, fluorescent proteins can be used as **tags** in living cells to indicate where the protein localizes in the cell or how it behaves when cells are given a chemical signal. For example, some proteins reside in the cytosol until the cell receives a hormone signal, at which point they translocate to the cell membrane; a GFP or mCherry fusion protein can be used to indicate this change in living cells.

Ligand-binding proteins

Many proteins bind to small ligands. A fusion protein generated using such a protein will retain its binding properties and can therefore be used in purification or pulldown assays (assays used to determine whether two or more proteins interact) or in microscopy. Antibodies generated against all of these proteins are commercially available, providing more tools that can be used to investigate protein function. Among the proteins that are commonly used to generate a fusion protein are the bacterial protein streptavidin, which binds to biotin (**Figure 19.17**); maltose binding protein (MBP), which binds to the carbohydrates maltose or amylose; the viral coat protein hemagglutinin (HA), which binds to the carbohydrate sialic acid; protein A and protein G, two bacterial proteins that are recognized and bound by all IgG antibodies; and the cellulose-binding domain (CBD) of any one of numerous proteins.

Enzymes

Several enzymes are also used in the generation of fusion proteins, mainly as reporter genes but also in purification or pulldown assays. Such proteins include chloramphenicol acetyltransferase (CAT), luciferase, glutathione *S*-transferase (GST), alkaline phosphatase (ALP), horseradish peroxidase (HRP), and dihydrofolate reductase (DHFR). Some of these enzymes will be familiar from pathways discussed previously, for example, GST in Chapter 11 and DHFR in Chapter 13. Again, the choice of enzyme varies with the design of the experiment. Fusion proteins such as ALP, CAT, luciferase, and HRP are used because they generate a product that can be detected in an experiment, whereas GST and DHFR are used because they bind tightly to a substrate.

Generally, experiments can be designed to take advantage of other aspects of a fusion protein. For example, an antibody-coated

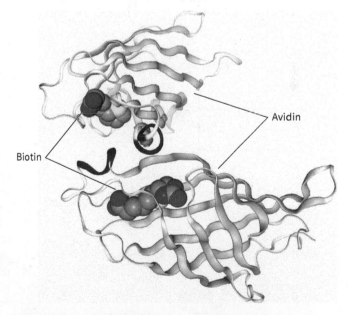

FIGURE 19.17 Avidin and biotin. Biotin is a small molecule that is used as a cofactor in several reactions we have studied. It is bound tightly by avidin, a small protein. Shown is a dimer of avidin with biotin bound in the center cavity of a β-barrel motif. The dissociation constant between avidin and biotin is 1×10^{-15} M, among the tightest bindings seen in nature.

(Source: Data from PDB ID 2AV1 Gagnon, S.J., Borbulevych, O.Y., Davis-Harrison, R.L., Baxter, T.K., Clemens, J.R., Armstrong, K.M., Turner, R.V., Damirjian, M., Biddison, W.E., Baker, B.M. (2005) Unraveling a Hotspot for TCR Recognition on HLA-A2: Evidence Against the Existence of Peptide-independent TCR Binding Determinants. *J.Mol.Biol.* **353**: 556–573)

TABLE 19.1 Common Fusion Proteins

Fluorescent/ Chemiluminescent	Ligand Binding	Enzymes
Green fluorescent protein (GFP)	Streptavidin	Chloramphenicol acetyltransferase (CAT)
Enhanced GFP (EGFP)	Maltose-binding protein (MBP)	Luciferase
Enhanced cyan fluorescent protein (ECFP)	Hemagglutinin (HA)	Glutathione *S*-transferase (GST)
Yellow fluorescent protein (YFP)	Cellulose-binding domain (CBD)	Alkaline phosphatase (ALP)
mCherry	Protein A and protein G	Horseradish peroxidase (HRP)
DsRedFP	FLAG® and 3xFLAG	Dihydrofolate reductase (DHFR)
mTurquoise	c-Myc	Hemagglutinin A (HA)
mVenus	Histidine (6-His)	
Luciferase		

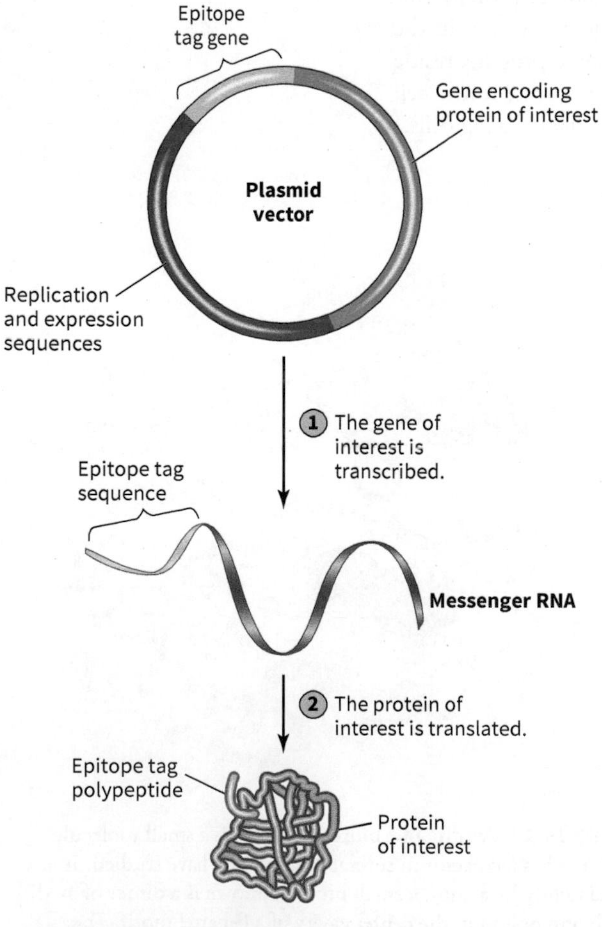

FIGURE 19.18 Epitope tags. Epitope tags are short sequences of amino acids that are attached to either the amino or carboxy terminus of proteins. When the protein is produced in the cell, these additional amino acids form a handle that can be recognized by antibodies or other binding proteins. These tags facilitate immunoblots and immunocytochemistry, and they can be used in immunoprecipitation or pulldown assays.

bead can be used to bind to a GFP fusion protein. Likewise, GST will bind to a bead coated with glutathione. **Table 19.1** indicates some of the different fusion proteins that are commonly available to investigators.

19.3.3 Epitope tags are modifications made to proteins to assist in detection or purification

As discussed in previous chapters, antibodies are molecules generated by an organism's immune system in response to antigens (molecules recognized as foreign). Antibodies bind antigens with high affinity. They can be generated in an animal such as a rabbit by injecting a small amount of antigen (purified peptide or protein) into that animal; the antibodies can then be isolated from the animal's blood. In some cases, however, it is not feasible to obtain an antibody against a specific protein.

Antibodies actually recognize and bind to small parts of an antigen, termed **epitopes**. An **epitope tag** is a short amino acid sequence that is incorporated into a protein of interest, which is used to detect or purify that protein via a commercially available antibody generated against the epitope tag (**Figure 19.18**). Because the tags are generally small (<20 amino acids), they are thought (or hoped) not to interfere with the function of the native protein. Epitope tags are usually placed at the amino or carboxy terminus of a protein, but in theory they can be placed anywhere in the sequence, provided they do not interfere with the native structure of the protein and can be bound by an antibody.

As with fusion proteins, the molecular instructions for including an epitope tag are encoded in the cDNA, and care must be taken to ensure that the tag is in the proper reading frame with the rest of the protein. There are several common epitope tags, each with their own advantages. Two peptides—FLAG (DYKDDDDK) and the related 3xFLAG (DYKDHDGDYKDHDIDYKDDDK)—are highly immunogenic and have produced antibodies with high binding affinities. Because these sequences are not found in nature, they can be used in a wide variety of model systems. In addition, the highly charged nature of the FLAG epitope means that it is more likely to be found on the surface of a protein and therefore to be able to interact with an anti-FLAG antibody.

A.

| EcoR I | | BamH I | Kpn I | RBS | | Sgf I | Asc I |

CTATAGGGCGGCCGGGAATTCGTCGACTGGATCCGGTACCGAGGAGATCTGCCGCCGCGATCGCCGGCGCGCCAGATCT

| Hind III | | Nhe I | Rsr II | Mlu I | | Not I | | Xho I | | Myc tag |

CAAGCTTAACTAGCTAGCGGACCG ACG CGT ACG CGG CCG CTC GAG CAG AAA CTC ATC TCT GAA GAG
                                       T   R   T   R   P   L   E   Q   K   L   I   S   E   E

| | | | EcoR V | | | 6-His tag | | | | Pme I | Fse I |

GAT CTG GCA AAT GAT ATC CTG CAT CAC CAT CAC CAT CAC GTT TAA ACGGCCGGCCGCGG
 D   L   A   N   D   I   L   H   H   H   H   H   H   V   Stop

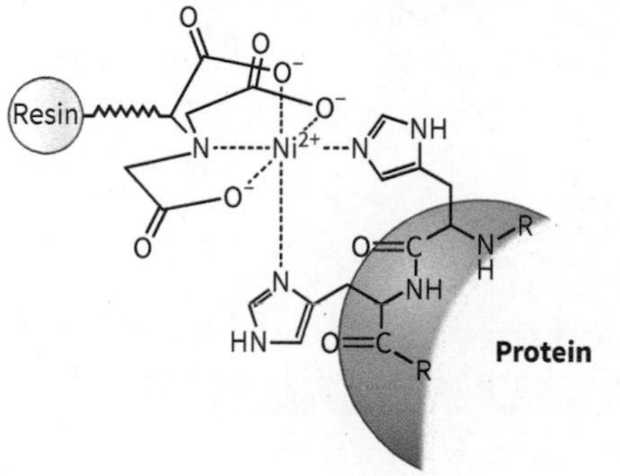

FIGURE 19.19 His-tagged proteins. **A.** The 6-His tag is a series of six histidine residues attached to either the amino or carboxy terminus of a protein. **B.** The imidazole rings of histidine can chelate to divalent metal ions, specifically Ni^{2+} and Zn^{2+}. Shown here is a protein binding to Ni^{2+}, which can be useful in purification schemes.

(Source: (A) Reprinted with permission from pCMV6-AC-Myc-His, mammalian vector with C-terminal Myc-His tag, 10ug, OriGene Technologies. (B) Reprinted with permission from GoldiBlot™ His-Tag Western Blot Kit, Nanoprobes, Inc.)

The c-Myc (pronounced see-mick, with the sequence EQKLISEEDL) and hemagglutinin A (HA, with the sequence YPYDVPDYA) epitope tags are small fragments of naturally occurring proteins; again, antibodies are used to recognize the tagged protein. The full-length c-Myc and HA proteins can also be used in the generation of fusion proteins. Investigators need to be cautious and understand the differences between the two. Although the epitope tag is simply a short peptide used to identify a protein of interest, c-Myc is a transcription factor, and hemagglutinin is a viral protein that binds and recognizes sialic acid groups on the cell surface. These molecules may retain their functionality in a fusion protein, altering the outcome of the experiment.

Another epitope tag is the His (or 6-His) tag, which, as its name suggests, is comprised of six histidine residues in a row (**Figure 19.19**). Although antibodies can be generated against the His epitope, this tag is more commonly used for other reasons. The imidazole rings of histidine will chelate metal ions, binding them with relatively high affinity. A protein labeled with the His tag will bind to a bead coated in divalent metal ions (typically Ni^{2+} or Zn^{2+}). This technique, termed **immobilized metal affinity chromatography** (**IMAC**), is the basis of many purification schemes.

In some instances, an investigator may wish to remove an epitope tag or fusion protein. This can be achieved by engineering a specific protease cleavage site between the epitope tag and the native protein.

Finally, there are other molecules that can be coupled to a protein (or nucleic acid) and then detected with antibodies or other proteins. For example, organic molecules such as fluorescent dyes, biotin, or the plant steroid digoxigenin can be coupled to a molecule and then detected using an affinity technique (essentially a western blot or immunocytochemistry); these techniques are discussed in Chapter 3 and in the Techniques Appendix.

19.3.4 Amber codon suppression is a way to incorporate unnatural amino acids into proteins

Biochemistry has been highly conserved throughout evolution, and although cells may chemically modify some amino acids, there are generally only 20 to choose from. At times, this situation can be limiting to an investigator. For example, it might be useful to incorporate an amino acid that could be cross-linked to a neighboring protein upon exposure to light, a halogenated amino acid that could be used to probe an enzymatic mechanism, or an amino acid that would give a therapeutic protein increased solubility or half-life.

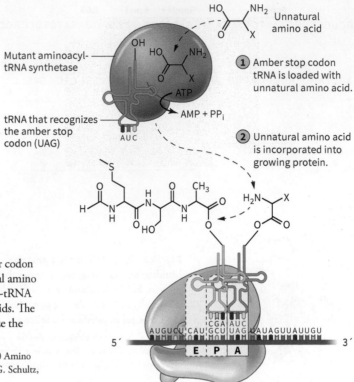

FIGURE 19.20 Amber codon suppression. In amber codon suppression, a stop codon (UAG) is used to add unnatural amino acids to a protein. This is done using a mutant aminoacyl-tRNA synthetase and chemically generated, unnatural amino acids. The key to this system is the synthetase, which has to recognize the unnatural amino acid specifically over others.

(Source: Reprinted with permission from Beyond the Canonical 20 Amino Acids: Expanding the Genetic Lexicon, Travis S. Young and Peter G. Schultz, 2010, *Journal of Biological Chemistry*.)

Amber codon suppression is an emerging technique in which unnatural amino acids can be strategically incorporated into a protein in a living cell (**Figure 19.20**). Initially, investigators used cell-free systems in which an unnatural amino acid was substituted for the amber stop codon, UAG, via a chemically coupled aminoacyl-tRNA. More recent developments of this technique have made it possible to incorporate unnatural amino acids in living cells. Generally, this requires two steps. First, the amber codons need to be edited out of the genome, except where the unnatural amino acid is to be used. In a typical *E. coli* genome, the UAG stop codon is used infrequently: ~30 to 300 times in the entire genome, depending on the strain. Strains have been engineered in which release factor-1 (RF-1), the termination factor that recognizes the UAG codon and causes termination of translation, has been knocked out. Additionally, these strains have been engineered to introduce genes with altered stop codons in situations where the loss of the UAG codon would be lethal. As a result, these strains grow normally but use only two of the three possible stop codons.

To couple the unnatural amino acid to the amber tRNA, a mutant aminoacyl-tRNA synthetase is used. The tRNA, unnatural amino acid, synthetase, and codon are referred to as an **orthogonal set** because they must be unable to communicate in any way with the native machinery of the cell. Otherwise, the system might incorporate the unnatural amino acid elsewhere, or other amino acids might be incorporated into the protein of interest, either of which would lead to experiment failure.

The key component in this translation system is the tRNA synthetase that will accept the unnatural amino acid. The synthetase is generated through a process termed **directed evolution**, in which the region around the amino acid binding site for the tRNA is randomly mutated. This is often done using a PCR-based technique with random or degenerate primers. This step generates a large number (>10^8) of mutagenized copies of the synthetase gene, which can then be used to transform *E. coli*. The *E. coli* are also transformed with a plasmid encoding an antibiotic resistance gene (for example, chloramphenicol acyltransferase, or CAT), which has been mutated to require incorporation of the unnatural amino acid early in its coding sequence. Bacteria are grown in the presence of the unnatural amino acid and the antibiotic (in this case, the antibiotic chloramphenicol). If the mutant aminoacyl-tRNA synthetase is able to couple the unnatural amino acid to the proper tRNA and that amino acid is properly inserted into the antibiotic resistance gene, the cells will be able to grow. However, if any of the steps in this process fail, the cell will be unable to produce the protein, become susceptible to the antibiotic, and die.

In addition to this positive selection, a second, negative selection is also made. In this second screen, the RNAse barnase is inserted into the cell to kill the cells if it is transcribed and translated into protein. Barnase also contains a mutation that incorporates the unnatural amino acid, but the cells are not provided with the unnatural amino acid. If the synthetase conjugates the wrong amino acid to the tRNA or if the ribosome permits another tRNA to bind to the amber codon, the toxic protein will be synthesized, and the cells will die. This provides an additional means of ensuring that the mutant tRNA synthetase functions with high specificity and fidelity.

Over the past few years, this and related methods have been used to generate over 100 different aminoacyl-tRNA synthetases that will recognize different unnatural amino acids (**Figure 19.21**). We may ask why we might want to add groups like alkynes or azo groups to amino acids. One reason would be to use these in subsequent experiments, like those described in **Biochemistry: Click chemistry**.

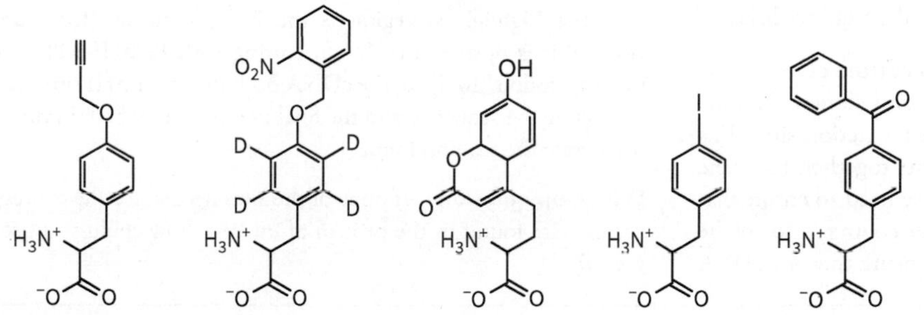

FIGURE 19.21 Unnatural amino acids. Some of over 100 different unnatural amino acids that have been incorporated into proteins.

Biochemistry

Click chemistry

One of the more interesting technologies to emerge in recent times is click chemistry. These reactions were so named by Nobel Laureate Barry Sharpless because the reactants "click" together with high yield under conditions that are less than ideal for most reactions.

The unifying principles of click chemistry state that the reactions are favorable under biological conditions (large negative ΔG values in aqueous solvents at room temperature), have high yield and atom economy (meaning most of the atoms in the reactant end up in the product), are stereochemically pure, and are both simple yet flexible enough to be employed in multiple scenarios.

If we consider most of the reactions studied in organic chemistry or carried out in a laboratory course, we should instantly see the advantages of click chemistry to a biochemist. Many of the reactions done in organic chemistry require harsh reagents (acid chlorides, anhydrides, strong acids and bases, and reducing and oxidizing reagents) in nonpolar solvents at temperatures beyond the stability of many biological molecules and structures. Due to the chiral nature of most biological molecules, it is important to pay attention to the stereochemistry of any reaction to be performed on a living system. Finally, given the diversity of molecules in a biological system, it would be a great advantage to have a single modular system that could be employed to answer multiple questions about the molecules involved. This is the strength of click chemistry.

The reaction that most people associate with click chemistry is currently the Huisgen cycloaddition of an azide to an alkyne, but many other reactions also fit the definition, for example, the Diels–Alder 4 + 2 cycloaddition, which you may recall from organic chemistry. Other reactions where there is nucleophilic addition to a strained ring system (such as epoxides) can also be thought of as click reactions. Nature employs a similar reaction when the lactam ring of penicillin opens following nucleophilic attack by the bacterial transpeptidase.

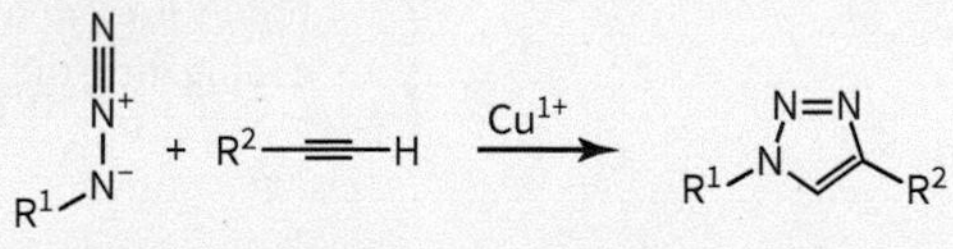

Huisgen cycloaddition of an azide to an alkyne

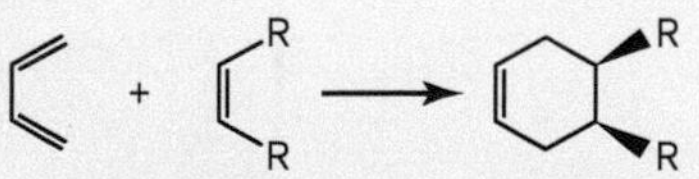

Diels–Alder 4+2 cycloaddition

In biochemistry, reagents are currently commercially available to label proteins, nucleic acids, lipids, and carbohydrates with click functional groups (examples include azo or alkyne groups). These kits can be used to generate reagents that could be used to identify or label other molecules. Other kits have been developed to detect apoptosis and identify lipid peroxidation, cell proliferation, and protein synthesis. The possibilities for click chemistry are just beginning to become apparent.

Worked Problem 19.3   Generating a fusion protein

You have been asked to express a particular protein with a C-terminal 6-His epitope tag. The mRNA coding for this protein has been cloned into a vector with a Sal I site at the 5′ end of the cDNA and a BamHI site at the 3′ end. It ends in the sequence 5′…TGG CCA TGG CTG GGA TCC ATC-3′.

You also have an expression vector for making a C-terminal 6-His epitope tag, which has the following multiple cloning site: AgeI, SalI, NcoI, and BamHI. The reading frame for the His sequence begins immediately after the BamHI site.

Propose a strategy to generate the fusion protein.

Strategy   Examine both sequences. The location of unique restriction sites is shown. Would using one of these to clone the sequence help?

Solution   The simplest strategy would be to cut the cDNA insert out of the first vector using SalI and BamHI and clone it into the vector containing the His tag using the same restriction sites. This strategy should work for cloning the fragments together, but there are two concerns about this approach. First, we need to ensure that the stop codon for the cDNA is not between the coding region of the cDNA and the His tag. Second, we need to ensure that the cDNA

and the tag are in the same reading frame, that is, that cloning the two together has not induced a shift of the reading frame by introducing one or two nucleotides. (Additions or deletions of three are acceptable, provided that added or deleted amino acids are not important.) To check both the stop codon and the reading frame, we should verify the nucleotide sequence:

> **DNA** 5′… TGG CCA TGG CTG GGA TCC ATC-3′
>
> **mRNA** … ACC GGU ACC GAC CCU AGG UAG
>
> **Protein**   Thr- Gly- Thr- Asp- Pro- Arg- STOP

The highlighted region is the recognition site for BamHI. Because it is upstream of UAG, cutting with BamHI will remove the stop codon. To ligate the cDNA back into a BamHI site, we will regenerate the sequence and the final two codons, which in turn will regenerate the reading frame.

Follow-up question   If none of the sites needed for the expression vector were found in the protein of interest, how could you use the vector?

Summary

- Site-directed mutagenesis is a means of directly and specifically altering an amino acid sequence, typically involving a polymerase extending from a mutagenic primer. Incorporation of silent mutations into a mutagenic site can be used to introduce novel restriction sites, which facilitates screening of mutant constructs.

- Chimeric or fusion proteins are created by joining two messages together. This can generate a protein with the regulatory region of one protein and the functional domain of another. Fluorescent proteins, ligand-binding proteins, or enzymes are often used in the generation of fusion proteins.

- Epitope tags are smaller modifications made to proteins; the addition of an epitope produces a handle to which antibodies can bind. 3xFLAG, HA, and c-Myc are examples of epitope tags.

- Amber codon suppression is a means of incorporating unnatural amino acids into proteins; in this technique, the amber (UAG) stop codon is used to insert an unnatural amino acid. Directed evolution is used to mutate the UAG aminoacyl-tRNA synthetase into an enzyme capable of specifically conjugating the unnatural amino acid to the UAG tRNA.

Concept Check

1. Describe the major changes that can be made to proteins.

2. Explain how amber codon suppression can be used to insert an unnatural amino acid into a protein.

3. Explain why an investigator would want to make a site-directed mutant of a protein or a fusion protein. Describe the process for generating these.

Bioinformatics Exercises

Exercise 1 Capstone: Structure and Function of Thyroid Hormone Receptors

Problems

19.1 Protein Production

1. If you wished to synthesize a peptide eight amino acids long, and each step of the solid-phase synthesis had a 99% yield, what would be your overall yield?

2. You are synthesizing a 12mer peptide using SPPS. If your deprotecting reactions ran at 99% yield, what would your coupling efficiency need to be for your overall yield to be 60%?

3. If you wanted an overall yield of 55% and if each reaction ran at 90% efficiency, how long a peptide could you make?

4. Propose mechanisms for the steps involved in SPPS.

5. What sort of protecting groups would be used for side chains of amino acids? How would these protecting groups be different than the protecting groups on terminal amine?

6. How might you ascertain peptide purity at the end of SPPS? If you synthesized a peptide 15 amino acids long using SPPS, how could you purify that peptide?

7. In the step of deprotection of terminal amino acid in SPPS, what reagents would you use to remove the protecting groups?

8. What is the function of each of the following parts of an expression plasmid?

   a. multiple cloning site

   b. polyadenylation sequence

   c. promoter

   d. *ori* sequence

   e. selectable marker

9. Any cDNA that is to be expressed must be cloned into an expression vector in the correct orientation (with the 5′ end of the coding region nearest the promoter). Given the following plasmid and cDNA, which combinations of restrictions enzymes could be used to directionally clone the cDNA into the plasmid? Which combinations could not? Explain your answers.

Vector cloning site: 5′ EcoRI, HindIII, PstI, BglII, NotI, ScaI 3′

cDNA: 5′ HindIII, BglII-coding sequence of the cDNA-HindIII, EcoRI, ScaI 3′

10. An insect expression system uses a baculovirus that is engineered to encode the protein of interest. The virus is used to infect ovarian cells from *Spodoptera frugiperda*, the fall armyworm. What are the advantages of using this system to produce engineered proteins?

11. What are the advantages of expressing a protein in *E. coli* compared to human cells?

12. Could you use an *E. coli* expression vector to express protein in human cells or a human expression vector to express protein in *E. coli*? Why or why not?

19.2 Protein Folding

13. A common calculation performed to illustrate Levinthal's paradox is to consider a protein of 101 amino acids and 100 "bonds" each with three conformations.

   a. How many conformations would there be with these parameters?

   b. If a transition state is 1×10^{-12} seconds, how long would it take to proceed through all of the states?

14. Repeat the calculation that Richard Dawkins made regarding correctly conserved sequences. How many attempts would it take to type "Methinks it is like a weasel" by hitting keys at random? What variables did Dawkins use to get 10^{40}?

15. Explain how each of the following reagents would denature a protein and how each could be reversed:

   a. heat

   b. β-mercaptoethanol

   c. urea

   d. guanidinium hydrochloride

   e. pH

16. Proteins can be denatured using β-mercaptoethanol, but they are less likely to fold properly when it is removed. Why might this be the case?

Questions 17 to 20 pertain to following peptide:

NH_3^+-Met-Ala-Ser-Ile-Phe-Asp-Cys-Pro-Gln-Val-Gly-Cys-His-Phe-COO⁻

   1   2   3   4   5   6   7   8   9   10   11   12   13   14

17. Identify which side chains of the peptide interact through the following forces:

   a. salt bridges

   b. hydrogen bonds

   c. hydrophobic effect

   d. disulfide bonds

   e. cation–π interactions

18. As this peptide folds, where might these side chains be found?

19. Generate a model of this peptide either on paper or using a physical model such as a molecular modeling kit or pipe cleaners.

20. Identify any elements of secondary structure in your model.

21. The thermodynamic stabilization of the protein in folded state is partly due to the hydrophobic effect that accompanies protein folding. Where will be the hydrophobic effect manifested in proteins?

22. When we cook eggs or meat, the proteins present are denatured. Why do they not refold when they are cooled back down?

23. We often think of a salt bridge forming between two residues, one positively and one negatively charged, but in reality all the ions of a protein interact with each other, albeit weakly. If we were to model this, how many ion pairs would exist for a protein that had 23 cationic and 28 anionic amino acids, presuming all of them were ionized?

24. We often think of protein folding in terms of entropy, or bonds formed. How could you calculate the entropy of protein folding?

19.3 Protein Engineering

25. Why might you want to add an epitope tag to a protein?

26. Why does families of proteins with homologous tertiary structures that have diverged from a common ancestor show difference in amino acid sequence?

27. Altered forms of insulin are termed insulin analogs. These hormones have been engineered with specific changes to alter their properties. Aspart is an insulin analog in which aspartate is substituted for histidine. This does not affect receptor binding, but it prevents the analog from forming inactive dimers and hexamers and as a result is fast acting. Describe how this change might have been generated and how the mutant protein was produced.

28. Subtilisin is a bacterial protease that is commonly used as an additive in laundry detergent. In its native state, it has a methionine that becomes oxidized, leading to enzyme inactivation. How might you engineer the enzyme to prevent inactivation?

Experimental Design

Questions 29 to 32 pertain to the following sequences:

DNA sequence:

5′-TTCGCGGCTCTGAGGCCAGAGATCGGGCCC-3′

3′-AAGCGCCGAGACTCCGGTCTCTAGCCCGGG-5′

RNA sequence:

5′-AAGCGCCGAGACUCCGGUCUCUAGCCCGGG-3′

Protein: -Lys- Arg- Arg- Asp- Ser- Gly- Leu- STOP

29. This protein has a putative protein kinase A site in which the serine is phosphorylated. Propose a conservative mutation that would render this site inactive.

30. Design a primer to mutate the serine found at this site to an alanine using the least number of changes possible.

31. Design a primer to mutate the serine found at this site to an aspartate residue using the least number of changes possible.

32. Examine the cloning site for the plasmid found in Figure 19.19. Design a plan to generate a C-terminal fusion protein of this protein with the 6-His tag.

Ethics and Social Responsibility

33. What are the mechanisms for reporting academic misconduct at your institution? Where can you find about reporting this at your institute?

34. Would you expect the rules around academic misconduct to be similar or different at other institutions? Explain.

35. Depositing falsified protein structure data into the Protein Data Bank, PDB is a huge disservice to the scientific community. Express your thoughts on this.

36. Whistleblower is the term used for someone who reports an instance of a violation to superiors or oversight committees. In many instances whistleblowers face some form of repercussion for reporting the violation, even when they are in the right. What protections do whistleblowers have at your institution or elsewhere?

Suggested Readings

19.1 Protein Production

Benoiton, N. L. *Chemistry of Peptide Synthesis.* Boca Raton, FL: CRC Press, 2005.

Bösze, Z., and L. Hiripi. "Recombinant Protein Expression in Milk of Livestock Species." *Methods in Molecular Biology* 824 (2012): 629–41. doi:10.1007/978-1-61779-433-9_34.

Chan, W. C., and P. D. White, eds. *Fmoc Solid Phase Peptide Synthesis: A Practical Approach.* Oxford, UK: Oxford University Press, 2000.

Gellissen, G., ed. *Production of Recombinant Proteins: Novel Microbial and Eukaryotic Expression Systems.* Weinheim, Germany: Wiley-VCH, 2005.

Mergulhão, J. F., D. K. Summers, and G. A. Monteiro. "Recombinant Protein Secretion in *Escherichia coli.*" *Biotechnology Advances* 23, no. 3 (2005): 177–202.

Pednnock, G. D., C. Shoemaker, and L. K. Miller. "Strong and Regulated Expression of *Escherichia coli* Beta-Galactosidase in Insect Cells with a Baculovirus Vector." *Molecular and Cellular Biology* 4, no. 3 (March 1984): 399–406.

Smith, G. E., M. D. Summers, and M. J. Fraser. "Production of Human Beta Interferon in Insect Cells Infected with a Baculovirus Expression Vector." *Molecular and Cellular Biology* 3, no. 12 (December 1983): 2156–65.

19.2 Protein Folding

Alm, E., and D. Baker. "Prediction of Protein-Folding Mechanisms from Free-Energy Landscapes Derived from Native Structures." *Proceedings of the National Academy of Sciences USA* 96, no. 20 (September 1999): 11305–10.

Anfinsen, C. B. "Principles That Govern the Folding of Protein Chains." *Science* 181, no. 4096 (1973): 223–30.

Best, M. D. "Click Chemistry and Bioorthogonal Reactions: Unprecedented Selectivity in the Labeling of Biological Molecules." *Biochemistry* 48, no. 28 (July 2009): 6571–84.

Bicout, D. J., and A. Szabo. "Entropic Barriers, Transition States, Funnels, and Exponential Protein Folding Kinetics: A Simple Model." *Protein Science* 9, no. 3 (March 2000): 452–65.

Bogatyreva, N. S., and A. V. Finkelstein. "Cunning Simplicity of Protein Folding Landscapes." *Protein Engineering, Design and Selection* 14, no. 8 (August 2001): 521–3.

Dawkins, R. *The Blind Watchmaker.* New York: Norton, 1987.

Dill, K. A., and H. S. Chan. "From Levinthal to Pathways to Funnels." *Nature Structural & Molecular Biology* 4, no. 1 (January 1997): 10–19.

Fändrich, M. "On the Structural Definition of Amyloid Fibrils and Other Polypeptide Aggregates." *Cellular and Molecular Life Sciences* 64, no. 16 (August 2007): 2066–78.

Fersht, A. *Structure and Mechanism in Protein Science: A Guide to Enzyme Catalysis and Protein Folding.* New York: W. H. Freeman, 1998.

Fiser, A. "Template-Based Protein Structure Modeling." *Methods in Molecular Biology* 673 (2010): 73–94. doi:10.1007/978-1-60761-842-3_6.

Hinchliffe, A. *Molecular Modeling for Beginners.* New York: Wiley, 2008.

Källberg, M., H. Wang, S. Wang, J. Peng, Z. Wang, J. Lu, and J. Xu. "Template-Based Protein Structure Modeling Using the RaptorX Web Server." *Nature Protocols* 7 (2012): 1511–22. doi:10.1038/nprot.2012.085.

Knowles, R. P., M. Vendruscolo, and C. M. Dobson. "The Amyloid State and Its Association with Protein Misfolding Diseases." *Nature Reviews Molecular Cell Biology* 15, no. 6 (June 2014): 384–96. doi:10.1038/nrm3810.

Kukol, A. *Molecular Modeling of Proteins* (Methods in Molecular Biology). Totowa, NJ: Humana Press, 2014.

Levinthal, C. "Are There Pathways for Protein Folding?" *Journal de Chimie Physique* 65 (1968): 44–45.

Levinthal, C. "How to Fold Graciously." In *Mossbauer Spectroscopy in Biological Systems, Proceedings of a Meeting Held at Allerton House,*

Monticello, IL, edited by P. Debrunner, J. C. M. Tsibris, and E. Münck (pp. 22–24). Urbana: University of Illinois Press, 1969.

Ramírez-Alvarado, M., J. S. Merkel, and L. Regan. "A Systematic Exploration of the Influence of the Protein Stability on Amyloid Fibril Formation *in vitro*." *Proceedings of the National Academy of Sciences* 97, no. 16 (2000): 8979–84.

Reynaud, E. "Protein Misfolding and Degenerative Diseases." *Nature Education* 3, no. 9 (2010): 28.

Sears, P., and C.-H. Wong. "Engineering Enzymes for Bioorganic Synthesis: Peptide Bond Formation." *Biotechnology Progress* 12, no. 4 (1996): 423–33.

19.3 Protein Engineering

Hopp, T. P., K. S. Prickett, V. L. Price, R. T. Libby, C. J. Match, D. P. Cerretti, D. L. Urdal, and P. J. Conlon. "A Short Polypeptide Marker Sequence Useful for Recombinant Protein Identification and Purification." *Bio/Technology* 6, no. 10 (1988): 1204–10. doi:10.1038/nbt1088–1204.

Li, X., and C. C. Liu. "Biological Applications of Expanded Genetic Codes." *Chembiochem* 15, no. 16 (November 2014): 2335–41. doi:10.1002/cbic.201402159.

Lutz, S., and U. T. Bornscheuer, eds. *Protein Engineering Handbook*, Vol. 3. New York: Wiley-VCH, 2012.

Mahalik, S., A. K. Sharma, and K. J. Mukherjee. "Genome Engineering for Improved Recombinant Protein Expression in *Escherichia coli*." *Microbial Cell Factories* 13, no. 1 (December 2014): 177.

Mukai, T., A. Hayashi, F. Iraha, A. Sato, K. Ohtake, S. Yokoyama, and K. Sakamoto. "Codon Reassignment in the *Escherichia coli*

Genetic Code." *Nucleic Acids Research* 38, no. 22 (2010): 8188–95. doi:10.1093/nar/gkq707.

Ohtake, K., A. Sato, T. Mukai, N. Hino, S. Yokoyama, and K. Sakamoto. "Efficient Decoding of the UAG Triplet as a Full-Fledged Sense Codon Enhances the Growth of a prfA-Deficient Strain of *Escherichia coli*." *Journal of Bacteriology* 194, no. 10 (May 2012): 2606–13.

Park, S. J., and J. R. Cochran, eds. *Protein Engineering and Design*. Boca Raton, FL: CRC Press, 2009.

Tomme, P., R. A. J. Warren, R. C. Miller, Jr., D. G. Kilburn, and N. R. Gilkes. "Cellulose-Binding Domains: Classification and Properties." Chapter 10 in *Enzymatic Degradation of Insoluble Carbohydrates* (ACS Symposium Series, Vol. 618), edited by J. N. Saddler and M. H. Penner. Oxford, UK: Oxford University Press, 1996.

Wals, K., and H. Ovaa. "Unnatural Amino Acid Incorporation in *E. coli*: Current and Future Applications in the Design of Therapeutic Proteins." *Frontiers in Chemistry* 2 (April 2014): 15.

Wells, J. A., M. Vasser, and D. B. Powers. "Cassette Mutagenesis: An Efficient Method for Generation of Multiple Mutations at Defined Sites." *Gene* 34, no. 2–3 (1985): 315–23.

Xiao, H., A. Chatterjee, S.-H. Choi, K. M. Bajjuri, S. C. Sinha, and P. G. Schultz. "Genetic Incorporation of Multiple Unnatural Amino Acids into Proteins in Mammalian Cells." *Angewandte Chemie International Edition* 52, no. 52 (2013): 14080–83.

Young, T. S., and P. G. Schultz. "Beyond the Canonical 20 Amino Acids: Expanding the Genetic Lexicon." *Journal of Biological Chemistry* 285, no. 15 (April 2010): 11039–44. doi:10.1074/jbc.R109.091306.

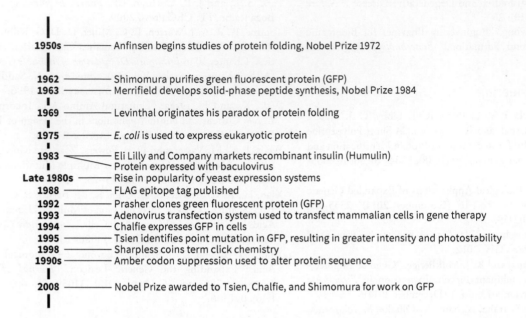

1950s	Anfinsen begins studies of protein folding, Nobel Prize 1972
1962	Shimomura purifies green fluorescent protein (GFP)
1963	Merrifield develops solid-phase peptide synthesis, Nobel Prize 1984
1969	Levinthal originates his paradox of protein folding
1975	*E. coli* is used to express eukaryotic protein
1983	Eli Lilly and Company markets recombinant insulin (Humulin)
	Protein expressed with baculovirus
Late 1980s	Rise in popularity of yeast expression systems
1988	FLAG epitope tag published
1992	Prasher clones green fluorescent protein (GFP)
1993	Adenovirus transfection system used to transfect mammalian cells in gene therapy
1994	Chalfie expresses GFP in cells
1995	Tsien identifies point mutation in GFP, resulting in greater intensity and photostability
1998	Sharpless coins term click chemistry
1990s	Amber codon suppression used to alter protein sequence
2008	Nobel Prize awarded to Tsien, Chalfie, and Shimomura for work on GFP

CHAPTER 20

Bioinformatics and Omics

Bioinformatics in Context

If you want to know whether people like chocolate cookies, you could simply ask. Provided you pose this question to a large enough group, you will be able to determine what proportion of the population likes chocolate cookies.

To gather more information, you could generate a survey aimed at teasing out the details of dessert preferences. For example, do people prefer cupcakes or raspberry bars over cookies? Chocolate icing or vanilla? By posting the survey online, you could obtain a wealth of data, which you could then analyze to identify patterns or similarities.

Today's computing power, combined with the vast reach of the internet, makes it possible to quickly and easily ask many questions and generate massive datasets; however, it is important to be smart about what we ask in a survey, how we gather and process the data, and how we interrogate or probe the resulting dataset.

The dessert preference survey is an analogy for the revolution that has occurred in bioinformatics and omics. Where in the past we would ask a single question of a biological system, we can now ask many questions at once.

This chapter begins with a brief description of bioinformatics and omics, focusing on the differences between these and classical biochemical approaches. It then discusses the analytical tools that generate large datasets. It briefly reviews DNA sequencing and high-throughput DNA sequencing from Chapter 2. Next it considers the techniques used to generate data about the transcriptome. This chapter additionally focuses on mass spectrometry, a tool that is more commonly used in the characterization of the other types of biological molecules and their modifications. The final section discusses the tools used to analyze large datasets.

Chapter Outline

Evolution's outcomes are conserved.	• The chemical underpinnings of evolution can be analyzed using bioinformatic techniques.
	• Comparisons of nucleic acid and protein sequences across species illustrate the beauty of evolution and demonstrate the conservation of changes over time.
Structure determines function.	• Bioinformatic techniques can be used to ascribe functions to a protein based solely on the similarity of its amino acid sequence or three-dimensional structure to those of known proteins.
	• Post-translational modifications of proteins or modifications of other molecules such as lipids can lead to functional changes. Bioinformatics can be used to observe such modifications in numerous types of molecules simultaneously.
Biochemical information is transferred, exchanged, and stored.	• Bioinformatic analysis of nucleic acid sequences in different species illustrates how genes are transferred or exchanged. It also indicates how these genes evolve across generations.
Biomolecules are altered through pathways involving transformations of energy and matter.	• Bioinformatic techniques such as mass spectrometry can be used to identify post-translational modifications. Analysis of these modifications across a proteome can provide important biochemical information about the state of a cell, tissue, organ, or organism.

20.1 Introduction to Bioinformatics

Bioinformatics is a relatively new and rapidly expanding field of science that involves generating and using large datasets to answer multiple questions simultaneously. This section explains what a bioinformatics question is and what a bioinformatics approach involves.

20.1.1 A classical approach to a problem can be reformulated using a bioinformatics approach

Consider a simple experiment in which a drug binds to a transcription factor and alters gene expression, for example, one of the thiazolidinedione (TZD) class of drugs binding to the transcription factor PPARγ. TZDs are a popular class of drugs commonly used to treat type 2 diabetes. A classical approach might ask whether the PPARγ gene is upregulated when cells are incubated with a TZD and use a technique such as a northern blot to detect any change in gene expression (**Figure 20.1**). In a northern blot, the RNA (either total RNA or mRNA) of a sample is isolated and separated on an agarose gel. It is then transferred to a nylon membrane, and the RNA of interest is identified using a radiolabeled probe. If upregulation has occurred, more mRNA will be present in the band corresponding to the PPARγ gene.

It is relatively easy to analyze several samples simultaneously in a northern blot, together with the appropriate controls. Thus, rather than just asking, "Does TZD upregulate the gene of interest?" we could ask several related questions such as, "What are the effects of different concentrations of drug or of using different cell types?" It would also be possible to administer TZD to an animal and isolate RNA from different tissues to look for gene expression changes in those tissues.

A. Northern blot

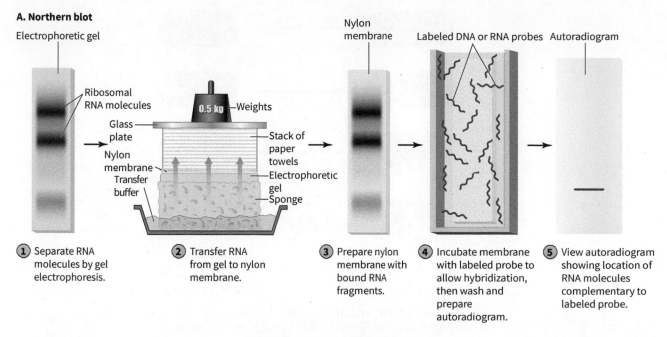

Electrophoretic gel

Ribosomal
RNA molecules

Glass
plate

Nylon
membrane

Transfer
buffer

0.5 kg — Weights

Stack of
paper
towels

Electrophoretic
gel

Sponge

Nylon
membrane

Labeled DNA or RNA probes   Autoradiogram

1. Separate RNA molecules by gel electrophoresis.

2. Transfer RNA from gel to nylon membrane.

3. Prepare nylon membrane with bound RNA fragments.

4. Incubate membrane with labeled probe to allow hybridization, then wash and prepare autoradiogram.

5. View autoradiogram showing location of RNA molecules complementary to labeled probe.

B. Dot blot

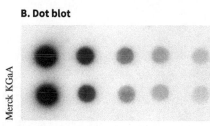

Merck KGaA

FIGURE 20.1 Review of northern blotting.   **A.** In a northern blot, mRNAs are resolved based on size in an agarose gel and transferred to a nylon membrane. The membrane is then treated with a radiolabeled oligonucleotide probe, which is the complement to the RNA. The membrane is exposed to film, and the radiolabel indicates where the probe has bound. Northern blotting can be used to detect mRNA, but it is commonly used to quantify levels of mRNA expression for an individual gene. **B.** A dot blot skips the first two steps of the northern blot and simply spots RNA samples on a membrane prior to development.

(Source: (A) Wessner, *Microbiology*, 2e, copyright 2017, John Wiley & Sons. This material is reproduced with permission of John Wiley & Sons, Inc.)

If it became apparent that the mRNA message was the same size each time the experiment was run and that the probe bound only the mRNA of interest, the gel component of the northern blot could be omitted. Instead, a dot or slot blot could be used, in which the isolated RNA is spotted directly onto a nylon membrane and probed. Dot blots are quicker and easier than northern blots. Also, about 20 samples can be run on a 14 × 20 cm northern blot, whereas hundreds or even thousands of samples can be tested in a single dot blot experiment. The dot blot begins to allow us to answer multiple questions simultaneously.

These scenarios involve a sample that contains all of the RNAs and a single probe. However, if specific, isolated samples were available, they could be spotted onto the membrane, one gene per spot. Next, the mixture of RNAs could be labeled and incubated with the membrane; the appearance of a spot would indicate gene expression. This experiment is similar to a microarray, discussed later in section 20.2. This basic, single experiment allows us to ask not only "Does this drug alter the expression of the gene of interest?" but also "What does this drug do to the expression levels of *all genes at once*?" When we ask about all genes rather than one gene, we move from a classical to a bioinformatic experiment.

20.1.2 Bioinformatics or omics experiments can be used to test all aspects of biochemistry

The earliest bioinformatics techniques focused on DNA; however, all classes of biological molecules can now be studied using a bioinformatics approach (**Figure 20.2**).

Genomics is the study of the **genome**, that is, of all of the genes of an organism and their interactions. The Human Genome Project, which sequenced the 3.3 billion base pairs of the human genome for the first time, was one of the most ambitious scientific collaborations in history. Read more about how such research is funded in **Societal and Ethical Biochemistry: Sources of funding for scientific research.**

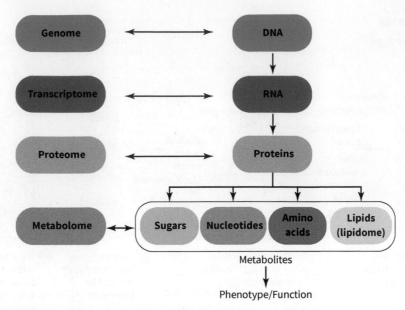

FIGURE 20.2 Omics studies. This flowchart shows some of the major types of omics studies now conducted and how they relate to one another.

Societal and Ethical Biochemistry

Sources of funding for scientific research

Completed in 2003, the government-funded Human Genome Project took 13 years and almost $3 billion. Because the research was publicly funded, the data generated was made freely available to the scientific community and launched countless other projects. Clearly, the ability to conduct scientific research that is not tied to expectations of profit is invaluable to scientific progress. There are currently several means by which biochemistry and biomedical research are funded. The two largest funding agencies in the United States are publicly funded government agencies: the National Institutes of Health (NIH) and the National Science Foundation (NSF).

The NIH is a collection of 27 institutes that fund biomedical research, which includes research aimed at understanding and alleviating disease and generally improving human health. In 2020 the NIH had a budget of $41.5 billion. To put this into perspective, in 2019, U.S. citizens spent $30.2 billion on alternative health care (which is not evidence based) and more than $60 billion on weight loss.

About 10% of the NIH budget is used internally at the NIH main campus in Bethesda, Maryland, or at several other institutes. A further 80% or more is given as grants to colleges, universities, hospitals, nonprofit organizations, and some businesses to fund biomedical research.

The NSF budget is primarily used to fund basic science questions that are not related to human health but do have broader impact. Thus, the NSF funds biochemical research related to biotechnology and bioengineering, development of new tools for studying biochemical systems, alternative fuels, protein structure and function, and basic questions of science. The 2020 NSF budget (which goes to all branches of science, not just biochemistry) was $7.1 billion.

Funding opportunities also exist in the private sector. Many charities and professional organizations fund researchers, including the American Heart Association, the American Diabetes Association, and the American Cancer Society. These grants are often smaller than government grants but are nevertheless important in helping new investigators or funding disease-specific research questions that have previously gone unfunded.

Several funding agencies merit special mention. The Howard Hughes Medical Foundation (HHMF) is a multibillion-dollar trust set up by the estate of Howard Hughes that funds investigators through different funding mechanisms around the world. In 2006, it established the Janelia Farm research campus, where scientists with diverse professional backgrounds work on common questions pertaining to aging and neuroscience. Other charitable organizations that fund biomedical research include Britain's Wellcome Trust (established by the estate of Sir Henry Wellcome), Scripps Research, and the Bill and Melinda Gates Foundation.

How do these numbers compare globally? In 2012, the U.S. government spent a total of about $125.6 billion on all types of research, whereas private industry spent $279.6 billion, nonprofit organizations spent $14.5 billion, and academia spent $12 billion. This amounts to about 2.7% of the country's gross domestic product (GDP). Overall, the United States leads the world in total research expenditure, but in terms of percentage of GDP, it lags behind South Korea, Israel, Japan, and Sweden. China is second in terms of overall expenditures, with $337.5 billion spent on research and development compared to $405.3 billion for the United States.

The suffix "-omics" comes from the Greek for "all," so the term **genomics** literally indicates "all the genes." The genome was characterized first, and fields of study devoted to other biological information followed this nomenclature. Hence, they all end in -ome, and their study is -omics. Taken as a group, the **omics** represent a scientific approach that considers all members of a certain class of molecules or interactions collectively. Because all the cells of an organism contain the same copies of DNA, the genome is consistent across an entire organism. The same cannot be said for the other classes of information, so for these other groups of data, we may consider all the molecules in a certain cell or tissue rather than in the whole organism.

The **transcriptome** is the collection of all of the RNA molecules in an organism, cell, or tissue. Unlike the genome, the transcriptome varies across an organism and over its life span as genes are expressed in various cells and tissues and at different stages of development or disease.

The **proteome** is the collection of all of the proteins in an organism, cell, or tissue. Like the transcriptome, it can change with the age of an organism, stage of development, tissue or cell type, or disease state. In research, the transcriptome is often used as a surrogate for the proteome, but their data are not interchangeable. The transcriptome is responsive to the up- or downregulation of genes, but the transcription levels of genes may or may not correlate positively with protein levels. Also, the half-lives of proteins vary, which may confound the interpretation of experimental results. It has now become clear that many genes are regulated after transcription or even after translation. Thus, the results of a study of the transcriptome cannot necessarily be extrapolated to the proteome.

The **kinome** draws its name from the term kinase. It consists of all the phosphorylation modifications made in the cell, including those made to proteins, lipids, carbohydrates, and metabolites. The kinome is highly transient, and it can change abruptly upon stimulation.

The **lipidome** is the collection of all the hydrophobic molecules found in a sample. Lipidomic studies often identify molecules that were previously unknown to science or are unusual metabolites. For example, studies of the lipidome of tears have identified over 100 lipid components, many of which have unknown origins and functions.

The **glycome** is the collection of carbohydrates, encompassing both free carbohydrates in solution and carbohydrate modifications to molecules in an organism, tissue, or cell. Because of the number of carbohydrate modifications and molecules that can, in theory, be modified (most proteins and lipids), the glycome is highly complex.

The **metabolome** is the collection of all the small metabolites of an organism, tissue, or cell. Of the different classes of molecules mentioned, the metabolome and the kinome are the ones that fluctuate most in the short term, for example, throughout the day in humans.

A final -ome that is worthy of mention does not quite fit with those already listed. The **interactome** includes all interactions of the molecules of the cell. For instance, when an enzyme with a protein cofactor and a lipid substrate docks to another molecule, all three of these molecules interact with the enzyme. The interactome is useful for study in that it depicts how components of the cell actually work in conjunction with one another. Its limitation is that, unlike the other bioinformatics approaches, there is no automated way to test for these interactions. Instead, the data substantiating them must be generated step by step, experiment by experiment, and paper by paper. Thus, although interactomes can be hugely useful, they can be slow to develop.

20.1.3 Bioinformatics experiments differ in the volume of data generated and the methods used to process the data

Bioinformatics experiments generate large amounts of data. For example, we could compare a gene sequence from an unknown species of soil bacterium to a database of DNA sequences. The bacterial DNA sequence might be hundreds to thousands of bases long, and we would be comparing it to a database of 180 million sequences stretching over 180 billion bases of DNA. Clearly, effective and efficient analysis of data of that size requires powerful computers and software tools. **Figure 20.3** depicts how different branches of science interact to give rise to bioinformatics.

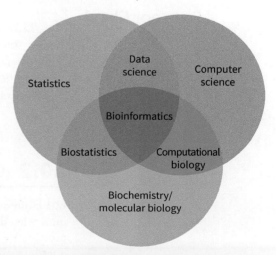

FIGURE 20.3 Bioinformatics studies. A Venn diagram illustrates how computer science, biochemistry and molecular biology, and statistics combine in bioinformatics studies.

The field of bioinformatics has developed in part due to advances in technology over the past 20 years, particularly in software, computing, and statistical analysis packages that enable investigators to analyze data or develop software for such analysis. Furthermore, the internet has enabled investigators to access large databases and freely share data on collective servers, building off the work of prior investigations from around the world.

Other advances in technology that have facilitated bioinformatics studies include mass spectrometry techniques that make it possible to analyze large molecules such as proteins, high-throughput nucleic acid sequencing techniques, development and commercialization of novel fluorophores, and the availability of reproducible two-dimensional acrylamide gels.

Worked Problem 20.1   **Designing a bioinformatics experiment**

Statins are HMG-CoA reductase inhibitors, widely used drugs that inhibit the committed step in cholesterol biosynthesis (discussed in Chapter 10). Design two experiments—one with a classical approach and one based on bioinformatics—to determine the effect of novel statin HMG-CoA reductase inhibitor on the expression of liver genes.

Strategy   What techniques are used to study gene expression? How would these differ if we were examining a single gene or all the genes expressed in the liver?

Solution   One classical approach would be to use a northern blot to see whether any genes were up- or downregulated. One bioinformatics approach would be to use a microarray to compare the transcriptome of the livers of control animals to those exposed to the drug.

Follow-up question   Would you anticipate that the transcriptome of the pancreas would show similar changes to the liver when exposed to the drug? Explain your reasoning.

Summary

- Bioinformatics uses modern techniques and computing power to answer multiple questions simultaneously.

- Omics study the totality of a certain biological molecule, unit, or modification, including the genome (genes), transcriptome (RNA), proteome (proteins), glycome (carbohydrates), lipidome (lipids), kinome (phosphorylations), and metabolome (small metabolites). Study of the interactome, which includes all direct interactions among components of a biological system, has not yet achieved the level of automation found in the other omics techniques.

- Bioinformatics studies differ from classical biochemical studies in both the volume of data generated and the statistical analysis needed to extract information from that data.

Concept Check

1. Explain how bioinformatics studies differ from classical studies. What factors lead to the development of bioinformatics as a distinct field of study?

2. Describe the types of molecules or modifications that can be studied using bioinformatics approaches.

3. Describe both classical (revisit Chapter 2) and modern methods used to sequence DNA.

20.2 Generating Bioinformatic Data

In a bioinformatics approach, the type of molecule being studied and the questions asked determine what data are gathered and the method of gathering it. For example, questions about which proteins are present in a sample or which mRNAs are found in a tissue are quite different from questions about the phosphorylation state of those same proteins or the relative abundance of those mRNAs. Bioinformatics studies are also driven by the techniques available. This section introduces some of these techniques, grouping them according to the type of molecule studied and the techniques employed.

20.2.1 High-throughput sequencing is used to study the genome

Chapter 2 discussed DNA sequencing in detail. To review briefly, the earliest approach was to read a DNA sequence one base at a time using chain termination. In this technique, four separate sequencing reactions are run, each containing a polymer, a template, a polymerase, deoxynucleotide triphosphates, and *one* dideoxy nucleotide triphosphate (ddATP, ddCTP, ddGTP, or ddTTP). Incorporation of a dideoxynucleotide means that the chain cannot be elongated further, but it also indicates that the chain must terminate with that nucleotide (A, C, G, or T). The four reactions are resolved on an acrylamide gel, and the bases are read one at a time. Up to 200 bases can be read in one reaction, but the process is time consuming.

More recent advances in sequencing include fluorescent sequencing and high-throughput parallel sequencing.

Fluorescent sequencing

Fluorescent sequencing uses the same dideoxy chemistry as earlier methods but employs fluorescent dideoxy terminators (**Figure 20.4**). Because four

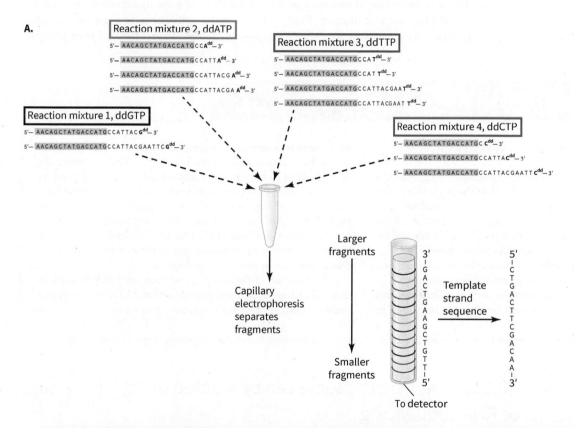

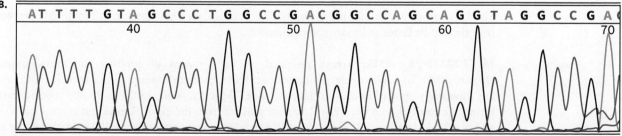

FIGURE 20.4 Fluorescent DNA sequencing.   **A.** In fluorescent DNA sequencing, DNA, a polymerase, and a specific primer are mixed together in the presence of dexoynucleotides and four different fluorescent nucleotides. Once incorporated, the fluorescent nucleotides are unable to participate in chain growth and are referred to as chain terminators. The fluorescent DNA is read by a DNA sequencer, which separates the samples using capillary electrophoresis and reads the fluorescent signal. **B.** An example of fluorescent sequencing data is shown.

different fluorophores are used, all four reactions can be conducted in the same tube, and because the detection system can read the four different fluorophores, it can be automated. In the early 2000s, fluorescent sequencing was often used to read lengths of about 700 bases per reaction, and today it can be used for even longer sequences. It was this technology that made possible the completion of many genome projects, including the sequencing of the mouse and human genomes.

High-throughput parallel sequencing   More recently, **high-throughput parallel sequencing** methods can completely sequence and assemble entire genomes (billions of nucleotides) within days. They employ hundreds of thousands or even billions of reactions, read optically from a microscope slide. Short lengths of DNA (<250 bp) are sequenced, and these short sequences are then assembled into the whole genome using existing genomes as scaffolding and advanced computational techniques. As we will see in the next section, high-throughput parallel sequencing can also be a secondary step in some studies of the transcriptome.

Knowing the sequence of an entire genome provides enormous insight into the nucleotides and proteins that can be found in an organism, how those molecules are related to others evolutionarily, and possibly even when and where those molecules are expressed and how they may function. However, the sequence of a genome is not the entire story. Genomics cannot answer many of the more fundamental questions of protein function or gene dysregulation that underlie biochemical processes and diseases. Read more about the application of these technological advances to medicine in **Medical Biochemistry: Personalized medicine: Beyond your genome**.

Medical Biochemistry

Personalized medicine: Beyond your genome

Chapter 14 discussed how medicine is increasingly using a patient's DNA sequence to profile cancer risk or select the best cancer therapy. Although this is a significant step in terms of personalized medicine, we also know that we are not merely the collection of our genes. Environment and epigenetic changes also affect who we are biochemically and how our cells and bodies behave. How could epigenomic or kinomic data be used to predict our health and healthcare outcomes? In other words, what are we beyond our genome?

Genomic and nucleic acid methods such as microarrays and RNA-Seq (discussed in the next section) are currently being used in the clinic to treat patients. Proteomic and other omic approaches do not yet have clinical applications, but they are likely to eventually. The U.S. Food and Drug Administration is working to "respond to, anticipate, and help drive scientific developments in personalized therapeutics and diagnostics." Advances in pharmacogenomics allow prediction of the behavior or effect of a drug based on someone's genetic makeup, discussed in Chapter 14. The development of new biomarkers, such as a protein that may be elevated or misfolded in a disease state, can be used to aid in diagnosis.

Examination of the kinomes and proteomes of individuals as they age or become sick will provide a wealth of information that could aid diagnosis and treatment in the near future.

20.2.2 The transcriptome can be studied using microarrays or RNA sequencing

Similar to genomics techniques, methods of generating data about the transcriptome have evolved over time to be faster and more comprehensive.

Microarrays   **Microarrays** provided an early means of probing the transcriptome (**Figure 20.5**). In a microarray, thousands of cDNAs, each corresponding to a specific transcript, are robotically spotted onto a microscope slide. Typically, some 3,000 to 30,000 specific transcripts are spotted onto the array. RNAs are purified from the experimental and control samples, and a cDNA copy is made from those RNAs using reverse transcriptase. Next, each sample of RNA is labeled using a fluorescent dye. Often, the control sample is labeled with a green dye and the experimental sample with a red dye. The two samples are combined in overall equal proportions and hybridized to the array. If equivalent amounts of the control and experimental RNA are present, both should bind equally well to their specific spot on the array; hence, the signal generated will have equivalent intensity of both green and red dyes. If the experimental sample contains more of one particular transcript than the control, the red signal will be stronger than the

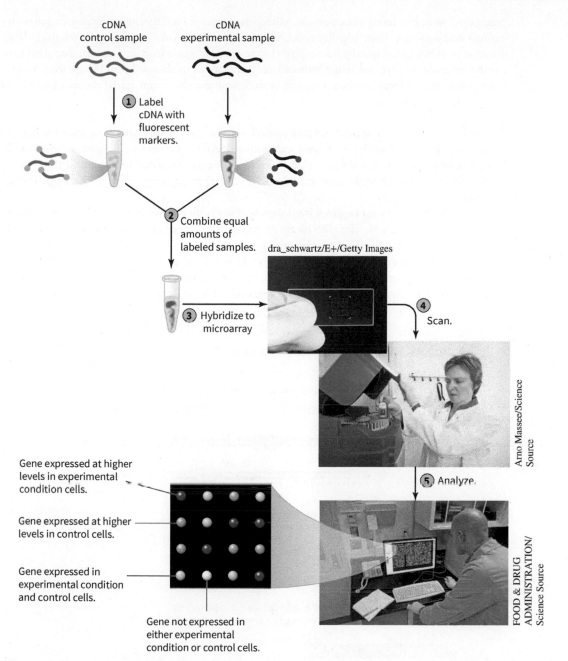

FIGURE 20.5 Expression microarray. A microarray consists of thousands of sequences, each representing a single gene spotted on a glass microscope slide. In the microarray experiment, RNA samples from control and experimental samples are isolated and labeled with fluorescent dyes (red for one sample and green for another). The labeled RNAs are then mixed and hybridized to the microarray. Spots where there is more binding of the red-labeled probe than green will show up red (indicating higher levels of that mRNA), whereas the opposite (more green than red) would indicate less of that mRNA in the sample. Spots in which both probes are bound with similar intensity would be colored yellow by the computer.

(Source: Wessner, *Microbiology*, 2e, copyright 2017, John Wiley & Sons. This material is reproduced with permission of John Wiley & Sons, Inc.)

green signal in that spot. Conversely, if a gene is downregulated in the experimental sample, the green signal will be stronger than the red. Microarrays are read in a scanner that can identify the intensity, color, and location of each spot on the array. Because the location of each individual spot is known, investigators can quickly determine which genes are up- or downregulated.

Microarrays are powerful experiments that have become cheaper and more accessible as libraries have been sequenced and as technologies for labeling cDNAs and producing arrays have improved. However, microarrays still have limitations. For example, because they rely on hybridization, they may produce false positives for highly related genes. Microarrays have a relatively high background (many genes are expressed at some level), which may make it difficult to identify transcripts with low levels of expression. Microarrays cannot identify single-nucleotide polymorphisms and can only rarely identify splice variants. Finally, microarrays require a relatively large amount of RNA and a specific microarray chip to start. If one is studying human liver, this may not be an issue, as there are many human liver microarrays to choose from, but generation of a microarray from a rarely studied organism would require a significant initial outlay of resources to develop the array itself.

Serial analysis of gene expression

Some of the limitations of microarrays are addressed with **serial analysis of gene expression (SAGE)**. Compared to microarrays, SAGE has the advantage that you do not need to know the sequence ahead of time, and the quantification of gene expression is more exact. SAGE sequences fragments of mRNA using classical technologies.

The SAGE experiment begins with isolation of the cell's mRNA and its reverse transcription into cDNA (**Figure 20.6**). The cDNAs are enzymatically cleaved and then amplified using PCR to generate short tags 11 to 15 nucleotides long with sticky ends. Next, these fragments are joined into a single **concatemer** containing hundreds of tags. These pieces can be sequenced to give a snapshot of which genes are being expressed in the cell at a particular point in time. Tags that are more prevalent are from the genes that are expressed at higher levels in the cell.

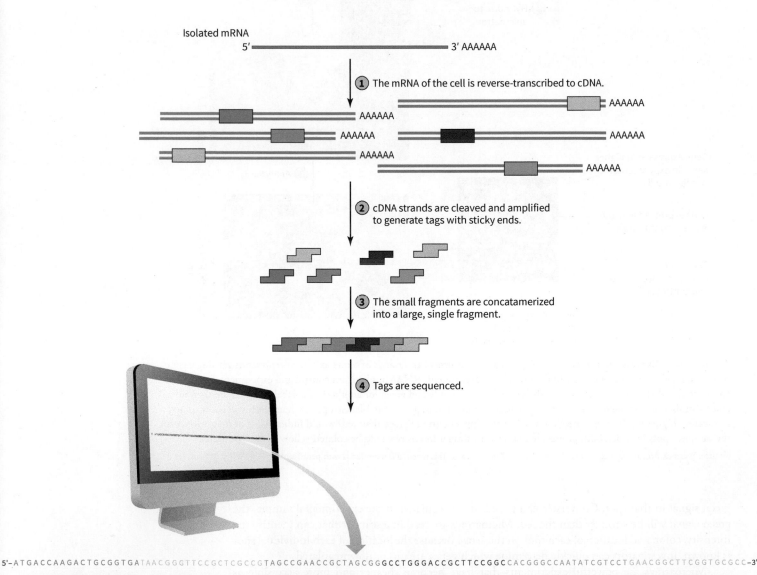

FIGURE 20.6 Serial analysis of gene expression (SAGE). In SAGE, short, unique fragments of mRNA are isolated from a sample and concatamerized into several large messages. This is usually accomplished using PCR and specific SAGE primers. These larger compilations of DNA are sequenced. The frequency of a SAGE tag in the sequence correlates to the expression level of that gene.

(Source: Wessner, *Microbiology*, 2e, copyright 2017, John Wiley & Sons. This material is reproduced with permission of John Wiley & Sons, Inc.)

SAGE is an older technique that is more expensive and time consuming than modern methods. A SAGE experiment also produces less data. Using music as an analogy for gene sequences, imagine breaking your entire music collection into clips about 12 seconds long. If we could use SAGE to sequence the music, these fragments would be assembled in random order into several long pieces of music, each containing portions of hundreds of songs. Analysis of these long pieces of music would show, on average, how much of each different style of music or particular artist was in your collection. It would be possible to identify a song but not whether verses were missing or had been rearranged. Thus, the dataset would be useful but incomplete. In comparison, modern methods of RNA sequencing would allow you to regenerate the entire music collection by reassembling all the fragments of each song.

RNA sequencing

RNA sequencing (RNA-Seq)—also known as whole transcriptome shotgun sequencing (WTSS)—is the current method of choice for analysis of the transcriptome.

An RNA-Seq experiment begins with isolation of the cell's mRNA, separating it from other types of RNA (**Figure 20.7**). This is often accomplished using a poly(A) sequence, although other techniques have been developed for the isolation of bacterial mRNAs. Because eukaryotic mRNAs have a poly(A) tail, they can be isolated using a poly(T)-coated resin. The mRNA sample is then fragmented into shorter lengths (~250 bp) and reverse-transcribed into cDNA. These fragments are ligated to adapter arms and sequenced using any of the high-throughput methods mentioned earlier. The data generated are sometimes assembled and analyzed using the organism's genome as a template. Alternatively, because multiple sequences are typically obtained from a single transcript, data can sometimes be assembled *de novo* based on overlapping sequences.

RNA-Seq provides information about which genes are being actively transcribed at a particular moment and about the relative abundance of the transcript, which indicates how strongly a gene is being transcribed. Thus, for a transcription factor expressed at low levels, the cell will contain only a few copies of the relevant mRNA, whereas for an abundant structural protein such as actin, the cell will contain many copies of the relevant mRNA. Sequence information also reveals mutations (including single-nucleotide substitutions, insertion–deletions, and gene fusions) and the prevalence of different splice variants.

Investigators can choose among these techniques and many others to find the one that best meets their needs. **Table 20.1** gives an overview of the strengths and weaknesses of each technique discussed here.

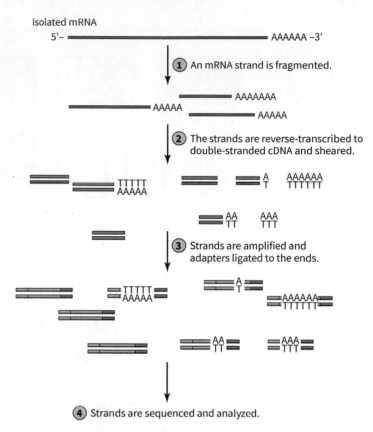

FIGURE 20.7 Initial steps of RNA-Seq. RNA-Seq can be conducted using several different methods. In this method, mRNA is reverse-transcribed into cDNA and sheared into smaller fragments. These fragments are mixed with adapter arms and ligated. The samples can then be sequenced using high-throughput parallel technologies.

(Source: From RNA-Seq, in SOLAS by Hugues Richard and Marcel Schulz. Max Planck Institute for Molecular Genetic.)

TABLE 20.1 A Comparison of Bioinformatic Techniques

Technology	Microarray	SAGE	RNA-Seq
Resolution	Several to 100 bp	Single base	Single base
Throughput	High	Low	High
Detection of mutations and splice variants	No	Limited	Yes
Sensitivity (expression detection range)	Difficult to identify genes with low levels of expression	Varies with expression level	High
RNA needed for analysis	Relatively large sample plus a specific microarray chip	Small sample	Small sample
Cost	High	Medium	Low

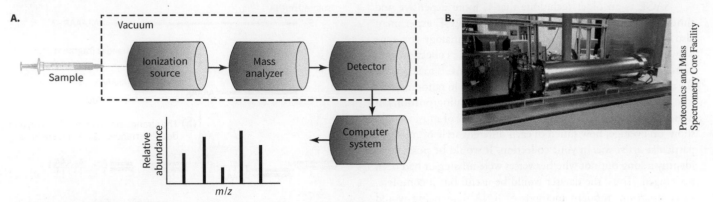

FIGURE 20.8 MALDI-TOF   A. A schematic diagram illustrates the different steps in mass spectrometry. **B.** The Bruker ultrafleXtreme MALDI-TOF is shown with cover removed.

20.2.3 Mass spectrometry is used to quantify proteins, lipids, carbohydrates, and metabolites

Similar to the impact of high-throughput sequences on the field of genomics, advances in mass spectrometry have revolutionized proteomics, lipidomics, glycomics, kinomics, and metabolomics. Mass spectrometry can be used to determine the mass of a molecule in three main steps: ionization, resolution of different mass-to-charge ratios, and detection (**Figure 20.8**).

Ionization   In the mass spectrometer, a volatile sample is ionized, often through collision with an ionized gas or plasma. The charged molecules are then separated from one another based on their mass-to-charge ratios. The molecules are resolved in an electromagnetic field and travel until they reach a sensor that detects the charge, either through the collision of the charged molecule or through induction of a current as the charged species passes.

This basic process presents some challenges to the biochemist in terms of determining molecular mass. Often, biological molecules are large and in an aqueous environment; few are volatile or lend themselves to injection into the gaseous phase. However, if the mass of a molecule is known to a high degree of precision, mass spectrometry can indicate whether that molecule has been modified (for example, phosphorylated), what that molecule is, or how prevalent it is in the sample.

The greatest recent advances in mass spectrometry have been in relation to how biological molecules are ionized. The two main developments are the soft ionization techniques—**electrospray ionization (ESI)** and **matrix-assisted laser desorption ionization (MALDI)**—which are described in detail in **Biochemistry: Mass spectrometer ionization techniques**. These techniques make it possible to ionize and analyze even large protein or carbohydrate molecules.

Biochemistry

Mass spectrometer ionization techniques

ESI and MALDI are now commonly used to characterize biological molecules. Both ESI and MALDI are considered soft ionization techniques, in contrast to fast atom bombardment or other older, classical techniques used to characterize small organics. The advantage of ESI or MALDI is that proteins remain largely intact.

In ESI, proteins or other molecules are separated using chromatography, for example, a reverse phase column. As proteins elute from this column, they pass through an atomizer that generates microscopic droplets of protein in solvent (water). These droplets pass through a charged ring that gives them a static-electric charge. As the droplets enter the mass spectrometer, they encounter a vacuum. This causes the loss of solvent molecules and a concentration of charge on the surface of the protein. The charged molecule can now be separated and resolved based on its mass-to-charge ratio, from which the mass of the protein can be determined.

In MALDI, a solution of protein is mixed with a matrix of cinnamic acid or one of its derivatives in organic solvent. The mixture is applied to a metal plate and irradiated with a laser, often in the UV range (266 to 355 nm). The matrix absorbs the laser light, resulting in ionization and ablation of the protein from the surface of the plate and into the mass spectrometer. MALDI is often coupled with a technique for separating the ions termed time of flight (TOF), in which the transition of ions through a magnetic field in the mass spectrometer is measured. When combined, these techniques are often referred to as MALDI-TOF.

Resolution   Following ionization, molecules are resolved by their mass-to-charge ratio; this is usually achieved by passage through an electromagnetic field. With larger molecules, such as those found in biochemical systems, two different technologies are often used. The first is a **quadrupole mass analyzer** (**Figure 20.9**), in which four parallel metal rods surround the ion beam.

A. Quadrupole mass analyzer

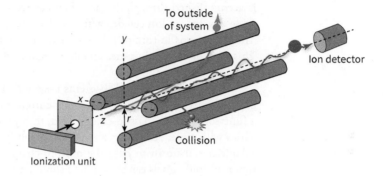

B. Time-of-flight mass analyzer

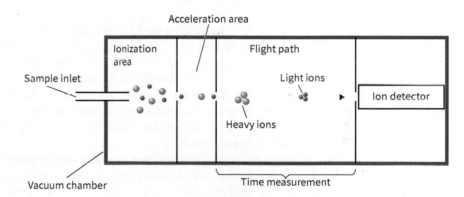

C. Fourier transform mass spectrometer

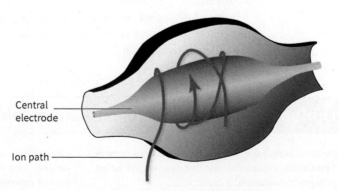

FIGURE 20.9 Types of mass analyzers.   A. The quadrupole mass analyzer consists of four cylindrical electrodes arrayed in a square. The electron beam travels through the center of this bundle along the axis of the four electrodes. Ions can be separated based on their mass-to-charge ratio by modulating the magnetic field generated by the quadrupole. **B.** In a time-of-flight (TOF) mass analyzer, ions are generated in an antechamber and introduced into a mass analyzer. Heavier ions bearing the same charge travel more slowly through the instrument and are detected later. **C.** In a Fourier transform mass spectrometer or Orbitrap, ions enter from the side and move about a central electrode according to their mass-to-charge ratio. The ions induce a current in a coil of wire, generating a signal that can be interpreted using a Fourier transform.

(Source: (A): From Introduction to LC-MS Part6, Shimadzu Corporation. Reproduced with permission of Shimadzu Corporation; (B): From Mass Spectrometry, Sam Adam-Day; (C): Gbdivers. Licensed under CC BY SA 3.0.)

These rods are charged with an electromagnetic field that can be quickly oscillated on or off, which tunes the instrument to allow the passage of particular mass-to-charge ratios. Often, three quadrupoles are arrayed in an instrument in an arrangement termed a **triple quad**. A triple quad offers higher sensitivity and greater signal-to-noise ratios than a single-quad system.

The second type of ion separation system that is widely used is termed time-of-flight (TOF) mass spectrometry. In this technique, an electromagnetic field is used to accelerate all ions to the same potential (voltage). If the ions all have the same charge, the separation (velocity) will be based solely on their mass, with lighter ions hitting the detector first. The time it takes for the ions to migrate to the detector is measured, and this in turn can be used as a measure of mass. Newer forms of ion resolution are the Orbitrap or **Fourier transform mass spectrometer (FTMS)**. In these instruments, ions injected into a magnetic field orbit a central electrode and oscillate with a frequency that is proportional to their mass-to-charge ratio.

Detection   The final stage of mass spectrometry is detection. In most instruments, detection is accomplished through the use of an **electron multiplier** or **microchannel plate detector**. In these devices, an ion collides with a surface subjected to a high voltage, ejecting several electrons. These electrons in turn collide with a second surface, resulting in the ejection of more electrons. This cascade results in a measurable current at the far end of the detector that can be converted into a signal or peak.

The Orbitrap or FTMS systems use another means of detection. In these instruments, ions oscillate past a coil of wire, inducing a current in that coil. Because the movement of the ions is a function of their mass-to-charge ratio, they periodically induce the current; that is, the current varies with time. Amplifiers are used to magnify these signals, which can be deconvoluted using a Fourier transform to yield the mass-to-charge ratio. One advantage of this type of detector is that multiple signals can be acquired simultaneously, giving higher precision and lower noise than other instruments.

Other mass spectral techniques   Samples can be prepared for mass spectrometry in various ways. For example, they can simply be separated using some form of liquid chromatography (LC), in which case the technique is termed LC-MS. Proteins can also be isolated on either an SDS-PAGE or two-dimensional PAGE gel in which proteins are separated in the first dimension by isoelectric point and then subjected to SDS-PAGE. Proteins may be eluted from the chromatography column and run in their native state, or they can be digested using a protease into smaller fragments before being characterized, an approach known as peptide mass fingerprinting.

Alternatively, **tandem mass spectrometry (MS-MS)** can be used to characterize a protein. In this technique, a protein is first isolated based on its mass-to-charge ratio in one mass spectrometer, and then a sample of the ions is diverted into a second mass spectrometer. This second instrument is equipped with some means of fragmenting the protein, for example, **collision-induced dissociation (CID)**. In CID, the proteins collide with molecules of an inert gas such as argon, which causes them to fragment. The fragments are analyzed and the masses reconstructed to identify the protein.

Worked Problem 20.2   Analysis of gene expression

Design an experiment to examine how gene expression is altered in a mouse model of pancreatic cancer. Assume that you already have a mouse model of pancreatic cancer and a control.

Strategy   In order to study gene expression, we need to examine the transcriptome via one of the options available.

Solution   Using RNA-Seq, we could generate short cDNAs from control and diseased tissue and sequence these fragments.

Comparison back to the mouse genome would indicate which genes are expressed, and the prevalence of the message would indicate the level of expression.

Follow-up question   What are the strengths and weaknesses of this approach compared with SAGE or a microarray?

Summary

- Fluorescent sequencing and high-throughput parallel sequencing techniques can be used to study the genome.
- Several techniques quantify the transcriptome. In a microarray, labeled RNA samples compete for binding to specific cDNAs spotted onto a microscope slide to indicate transcription levels. In SAGE, short mRNA tags are concatamerized and sequenced to provide the identity of the gene and level of transcription. RNA-Seq provides a more complete dataset than SAGE, enabling investigators to identify polymorphisms and mutations in a sequence.
- Mass spectrometry is used in the generation of data in proteomics, lipidomics, glycomics, kinomics, and metabolomics; it involves ionization, resolution, and detection. The two main means of ionizing large nonvolatile molecules are ESI and MALDI. Ions are resolved using either a quadrupole or time-of-flight instrument and detected using an electron multiplier or by inducing a current in a coil of wire.

Concept Check

1. Explain the main techniques by which genomic data are generated. Name and explain the main techniques used to study the genome.
2. Discuss the strengths and weaknesses of using microarrays, SAGE, or RNA-Seq to probe a transcriptome.
3. Describe the process of mass spectrometry, including why mass spectrometry is often the method of choice for bioinformatics.

20.3 Analyzing Bioinformatic Data

Once a large dataset has been generated in a bioinformatics experiment, a computer-based means of analyzing that data is needed. Statistical analysis of the dataset can establish some idea of its validity, and data mining involves searching through a database to find particular information.

20.3.1 Several computational tools are used in bioinformatic data analysis

The collective data acquired in a bioinformatics study are typically stored in a database. That database may be on a local server or desktop computer, accessible only by the people working on the project, but it is more likely to be part of a publicly accessible compilation of the data of tens of thousands of investigators. Bioinformatics Exercises are included at the end of each chapter to give students practice in accessing these databases and using the tools found there.

Hundreds of databases are available, containing information ranging from nucleotide and protein sequences to the taxonomy of bacteria and mitochondrial proteomics. This section discusses the largest and most frequently used databases.

GenBank is the international repository of all information on nucleotides and all translated protein sequences (**Figure 20.10**). Hosted in the United States, GenBank is part of the National Library of Medicine, a branch of the National Institute of Biotechnology, which in turn is part of the National Institutes of Health. Two other organizations—the European Molecular Biology Laboratory (EMBL) and the DNA Data Bank of Japan (DDBJ)—collaborate with GenBank, and together the three organizations maintain copies of the collective database, which is updated daily.

Protein structures are collected and maintained by the Protein Data Bank (PDB). This is also an international collaboration between Europe, Japan, and the United States that is updated daily.

Investigators can download and create local copies of these databases but usually simply access them online. The groups that maintain the databases also host commonly used tools for data mining and bioinformatics, and many of these tools also change daily as new advances are made and new programs are added. Because the URLs for these databases and tools may change, we recommend accessing them through a simple browser search.

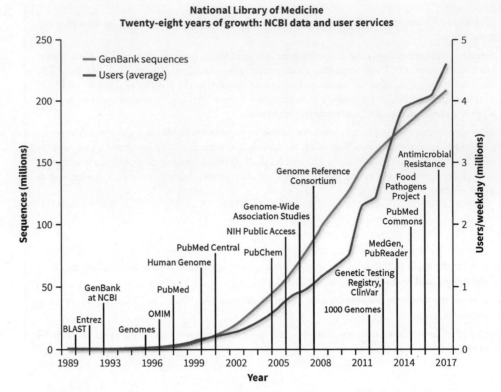

FIGURE 20.10 GenBank. GenBank is the world repository of biological sequence information, curated by the U.S. National Center for Biotechnology Information (NCBI), the DNA Data Bank of Japan (DDBJ), and the European Molecular Biology Laboratory (EMBL). The graphs depicts how the number of sequences and users has grown over time.

(Source: From Congressional Justification FY 2019, U.S. National Library of Medicine. Public Domain.)

Sometimes the tools available are insufficient to answer an investigator's question. In such cases, the investigator may need to develop software tools to extract and analyze data. This is becoming easier as more professionals, no matter their background or field, learn to write code and develop proficiency in programming.

Five programming languages are common in the development of bioinformatics tools: Java, JavaScript, Perl, R, and Python. Each of these languages has different strengths and weaknesses as well as different syntax and philosophy. For example, the central tenet of Perl is, "There's more than one way to do it," whereas Python's is, "There should be one, and preferably only one, obvious way to do it." These languages are freely available, and there are many materials and communities that assist with programming. Simplifying matters even further, members of the programming community have shared pieces of code (termed applets, modules, or packages) that solve many programming problems. These units can be joined together to generate a functional tool that can be used to answer bioinformatics questions.

20.3.2 Search algorithms are important in bioinformatics

To obtain meaningful information from a large dataset, it is necessary to interrogate the data. Often, the questions relate to **sequence alignment**, which involves lining up sequences of biological information. This helps to determine which sequences are useful for identifying a related sequence, establishing phylogeny among sequences, recognizing tertiary motifs or patterns, or predicting structure.

Sequence alignments can be broadly categorized as global or local. **Global alignments** attempt to align every member of a sequence and are more useful when there is a high degree of homology across the entire sequence. **Local alignments** take into account that there may be regions of higher or lower homology and preferentially align regions with higher homology. These different approaches can result in vastly different sequence alignments, and software is available for screening and comparing sequences with different levels of complexity or length.

Overview of search algorithms   The software for screening and comparing uses a **search algorithm** (a series of steps designed to locate specific data from a larger pool) to decide how a sequence best aligns or to locate a sequence in a database. These algorithms have certain common characteristics. First, they employ **dynamic programming**, a technique wherein a complex problem is broken into simpler pieces. Second, they need some basis for making comparisons among different amino acids. This is accomplished using a **substitution matrix**, a table that has one set of data (the different amino acids) along the side of the table and the same data along the base of the table. At each block in the table is a number that provides a score correlating the extent of the match or mismatch found at each point.

   There are numerous ways to generate a substitution matrix. The most popular matrix (and the one used in BLAST searches, discussed later) is the BLOSUM62 (BLOcks SUbstitution Matrix), which is constructed from proteins with less than 62% homology (**Figure 20.11**). In a BLOSUM62 matrix, positive scores indicate a match, and negative scores indicate a mismatch. The score is also positive if an amino acid is conserved—that is, substituted for by another amino acid with the same properties, such as aspartate for glutamate—although not as high as it would be if there were a match. Furthermore, not all amino acids are found at the same prevalence in a protein. If an amino acid is rare, an aligning sequence is given a higher score if it is conserved. For example, tryptophan, a less common amino acid, is given a score of 11 in a BLOSUM62 matrix. The values in a BLOSUM62 matrix were derived by examining conserved regions of related proteins and then calculating the likelihood that one amino acid would be substituted for another. These values were used to calculate a log of the odds score for each pair of amino acids.

Needleman–Wunsch algorithm   The first algorithm developed to compare sequences was the Needleman–Wunsch algorithm, which uses dynamic programming to make global alignments. In this algorithm, two sequences are aligned by generating a **similarity matrix**, that is, a table in which one sequence is listed along the top and one down the side. Scores are entered in each cell of the table. The algorithm then steps through the table, attempting to find the path at each step that provides the lowest score. An insertion or deletion (a gap) introduces a penalty score from 1 to 10.

Smith–Waterman algorithm   Smith and Waterman modified the Needleman–Wunsch algorithm to generate a new algorithm that provides local alignments. Whereas the Needleman–Wunsch algorithm compares the entire sequence, the Smith–Waterman algorithm compares fragments of each sequence and optimizes the scoring to enable the alignment of shorter sequences of higher homology with larger gaps found in between. This is accomplished by taking negative values found in the Needleman–Wunsch similarity matrix and changing them to 0, which in turn

	Ala	Arg	Asn	Asp	Cys	Gln	Glu	Gly	His	Ile	Leu	Lys	Met	Phe	Pro	Ser	Thr	Trp	Tyr	Val
Ala	4																			
Arg	-1	5																		
Asn	-2	0	6																	
Asp	-2	-2	1	6																
Cys	0	-3	-3	-3	9															
Gln	-1	1	0	0	-3	5														
Glu	-1	0	0	2	-4	2	5													
Gly	0	-2	0	-1	-3	-2	-2	6												
His	-2	0	1	-1	-3	0	0	-2	8											
Ile	-1	-3	-3	-3	-1	-3	-3	-4	-3	4										
Leu	-1	-2	-3	-4	-1	-2	-3	-4	-3	2	4									
Lys	-1	2	0	-1	-3	1	1	-2	-1	-3	-2	5								
Met	-1	-1	-2	-3	-1	0	-2	-3	-2	1	2	-1	5							
Phe	-2	-3	-3	-3	-2	-3	-3	-3	-1	0	0	-3	0	6						
Pro	-1	-2	-2	-1	-3	-1	-1	-2	-2	-3	-3	-1	-2	-4	7					
Ser	1	-1	1	0	-1	0	0	0	-1	-2	-2	0	-1	-2	-1	4				
Thr	0	-1	0	-1	-1	-1	-1	-2	-2	-1	-1	-1	-1	-2	-1	1	5			
Trp	-3	-3	-4	-4	-2	-2	-3	-2	-2	-3	-2	-3	-1	1	-4	-3	-2	11		
Tyr	-2	-2	-2	-3	-2	-1	-2	-3	2	-1	-1	-2	-1	3	-3	-2	-2	2	7	
Val	0	-3	-3	-3	-1	-2	-2	-3	-3	3	1	-2	1	-1	-2	-2	0	-3	-1	4
	Ala	Arg	Asn	Asp	Cys	Gln	Glu	Gly	His	Ile	Leu	Lys	Met	Phe	Pro	Ser	Thr	Trp	Tyr	Val

FIGURE 20.11 BLOSUM62 matrix.   The BLOSUM62 matrix is a means of weighing the strength of a retained or changed amino acid in a protein–protein sequence comparison.

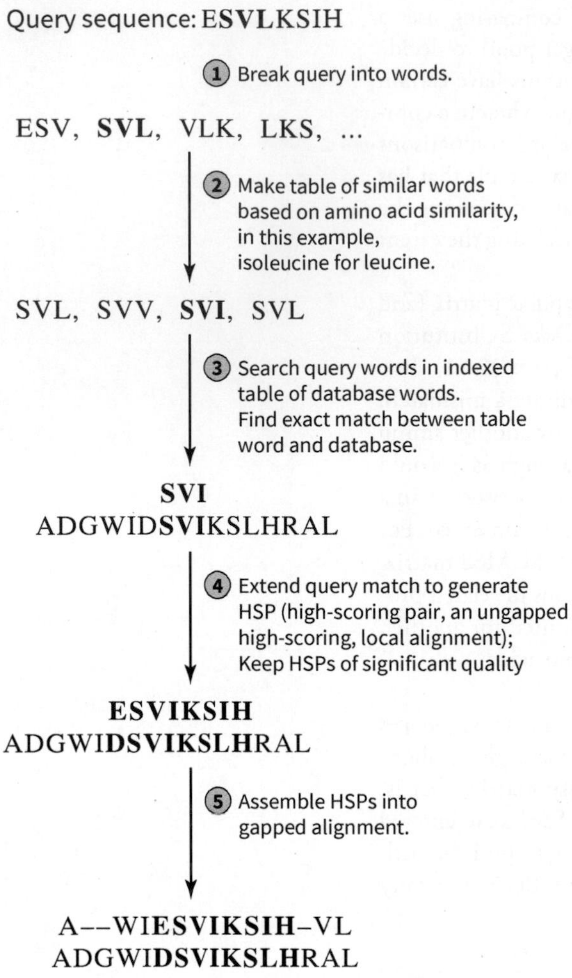

Query sequence: E**SV**LKSIH

① Break query into words.

E**SV**, **SVL**, VLK, LKS, ...

② Make table of similar words based on amino acid similarity, in this example, isoleucine for leucine.

SVL, **SVV**, **SVI**, **SVL**

③ Search query words in indexed table of database words. Find exact match between table word and database.

SVI
ADGWID**SVI**KSLHRAL

④ Extend query match to generate HSP (high-scoring pair, an ungapped high-scoring, local alignment); Keep HSPs of significant quality

ESVIKSIH
ADGWID**SVIKSLH**RAL

⑤ Assemble HSPs into gapped alignment.

A--WI**ESVIKSIH**–VL
ADGWID**SVIKSLH**RAL

FIGURE 20.12 BLAST search using the Altschul algorithm. To ensure maximum "hits" in minimal time, the algorithm takes the query sequence and chops it into short "words" of approximately three amino acids (or 11 nucleotides) in length. These words can be used to generate a second list of related words. The lists are scanned to remove words of low complexity. Then the library of words is used to probe the database and generate exact hits. These hits are then used to extend the matches and generate high-scoring segment pairs and ultimately a gapped alignment of the sequences.

gives stronger scores for short regions of highly similar sequences (local alignments flanked by gaps).

BLAST algorithm

The Needleman–Wunsch and Smith–Waterman algorithms were large advances in bioinformatics, but both are computationally time consuming when used to compare large sequences. A breakthrough came with the advent of the **basic local alignment search tool (BLAST)**, which employs the Altschul algorithm. This is now the most common tool used in bioinformatics.

The BLAST algorithm works by first removing low-complexity sequences with simple repeats, or stretches of the same sequence, from the query sequence (**Figure 20.12**). It then divides the query sequence into short "words," typically three amino acids for proteins and 11 nucleotides for DNA. These words are given a score based on their complexity, and sequences of low complexity are removed from the comparison. The remaining set of high-complexity words is used to scan the database for exact matches in what is termed a **seeded search**, with the short words being the seeds. From these exact matches, the other words are used to extend the sequence on either side. This generates a list of **high-scoring segment pairs (HSPs)**, which are subjected to further analysis. Scores above a preset threshold are subjected to statistical analysis and then merged to give longer sequences.

Using BLAST to locate sequence alignments

It is relatively easy to use BLAST to search for sequences with homology. Starting with the BLAST page at the National Center for Biotechnology Information (NCBI), the user has several options (**Figure 20.13**), for example, choosing what organism to run a BLAST search against or conducting a basic BLAST search against all sequences. There are five basic types of BLAST (**Table 20.2**). The first, BLASTn, uses a nucleotide sequence to query a nucleotide database. It has several variants, including megaBLAST, which allows for alignment of lengthy sequences. The second, BLASTp, uses a protein sequence to search a protein database. Submitting a nucleotide sequence to BLASTp or a protein sequence to BLASTn would not

FIGURE 20.13 BLAST homepage. All of the different BLAST options can be accessed from the BLAST homepage.

(Source: From U.S. National Library of Medicine. Public Domain.)

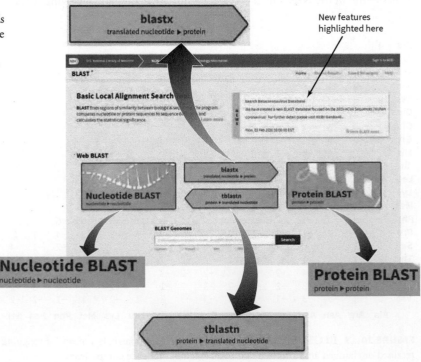

TABLE 20.2 Types of BLAST

Type	Used for
BLASTn	Nucleotide sequences; megaBLAST variant used for long sequences
BLASTp	Protein sequences
BLASTx	Probing a protein database with a nucleotide sequence
tBLASTn	Probing a translated nucleotide database using a protein query
tBLASTx	Probing a translated nucleotide database using a translated nucleotide query

return a result. The other three BLAST options—BLASTx, tBLASTn, and tBLASTx—use translations of nucleotides. These are more complex and require more computing time than a standard BLASTn, but they can often provide results within minutes, even for complex searches. In a BLASTx search, a protein database is probed using a translated nucleotide sequence. All six reading frames of the nucleotide sequence are translated and compared to the protein database. In a tBLASTn search, a translated nucleotide database is probed using a protein query, and in tBLASTx a translated nucleotide database is probed using a translated nucleotide query.

As BLAST has grown in popularity, more specialized BLAST searches have been developed. They include a means to search for specific primer sequences in a DNA strand (primer-BLAST), to align two specific sequences (bl2seq), and to screen sequences for vector contamination (vecscreen). There are many other tools for searching or examining bioinformatics data hosted by NCBI or other institutes such as ExPASy, the Swiss bioinformatics resource portal.

Submitting data for a BLAST search   Once a method has been chosen, data can be submitted by several different means. For example, sequences can be submitted as text using standard characters for nucleotides or amino acids; these are termed **bare sequences**. Also, sequences can be submitted using the FASTA format, in which a descriptor line denoted by a ">" symbol is found preceding the sequence, as shown in this example from NCBI:

>gi|129295|sp|P01013|OVAX_CHICK GENE X PROTEIN (OVALBUMIN-RELATED)
QIKDLLVSSSTDLDTTLVLVNAIYFKGMWKTAFNAEDTREMPFHVTKQESK
PVQMM CMNNSFNVATLPAE

The FASTA format does not support blanks, blank lines, or spaces in the sequence, whereas bare code does allow for the inclusion of spaces and numbers.

Another option is to load sequences by accession number or GI number. These are specific serial numbers assigned to each sequence when it is deposited in GenBank. GI numbers are typically denoted by the letters gi and the pipe character "|" followed by a series of numbers, for example, gi|129295. Accession numbers may have some combination of letters preceding the number such as p01013. In both cases, the syntax is specific, and only a single "word" (termed a token) is permitted. If any spaces are included, the file name is invalid. Finally, if there are multiple sequences to examine or large stretches of sequence, sequence files can be directly uploaded to the BLAST page.

Accessing results of a BLAST search   Once the BLAST is submitted, results are usually returned within seconds to minutes. The first piece of data returned is a graphical representation of the results (**Figure 20.14**). The graph consists of a series of horizontal lines that correspond to regions of nucleotide or protein sequences that have homology to the submitted sequence. The color of the sequence is related to the score (from 0 to over 200), which indicates the strength of the match. The higher the score, the stronger the match; however, the score is affected both by the degree of homology and the length of the sequence. Short sequences will always provide lower scores, even if they are complete matches.

As the cursor is moved over the graphic, names of the individual sequences appear. These individual lines of data are described in a table following the first graphic. This table has the hyperlinked match entry, followed by a row of statistics generated by BLAST (**Figure 20.15**). The statistics include the maximum and total score, the query cover (how much of the query sequence was able to be aligned with the database sequence), the expectation (E) value, the percentage

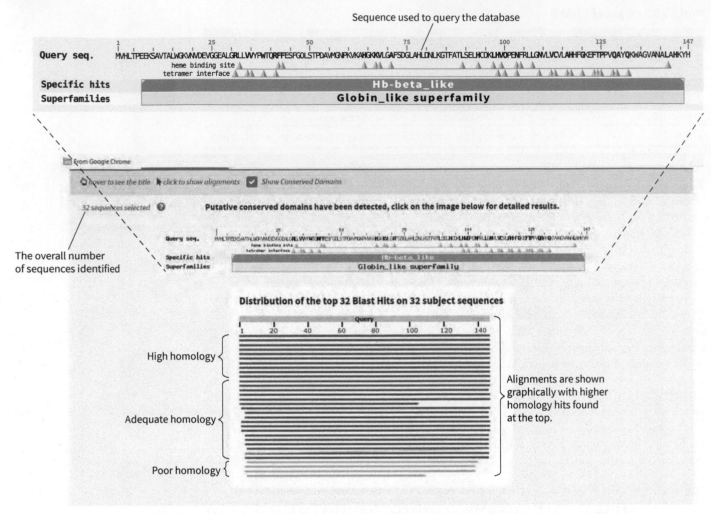

FIGURE 20.14 Graphical depiction of BLAST results. In these results of a BLAST search, each horizontal line represents a single hit (sequence) generated by the search. The color indicates the strength of each hit, and the continuity of the line indicates where the alignment has occurred.

(Source: From U.S. National Library of Medicine. Public Domain.)

identity (the percentage of complete matches), and the accession number of the sequence that was identified.

Most of these scores are self-explanatory, but some merit discussion. The maximum or **total score** indicates the quality of the alignment. The total score can be calculated in several different ways, and some of the BLAST parameters can be altered using advanced options. In general, the longer the match, the greater the identity or similarity; also, the fewer the mismatches or gaps, the higher the score will be. The **E value** is a statistical assessment of the likelihood that the sequences in the database would be identified by chance. Therefore, a lower value indicates a greater likelihood that the sequences are biologically relevant and not random. Strong E values can range from 10^{-20} to 10^{-130}, extremely low numbers.

The final piece of the BLAST result is a pairwise alignment of each result with the query sequence. This can be obtained by scrolling down the screen or by clicking on the link in either the graphical or tabular representation of the results.

Clustal alignment tool

A common tool that allows the comparison and alignment of two or more sequences is the Clustal program. Variants of this tool can be used to draw phylogenetic trees. Clustal first compares the sequences, finds the two that are most highly related, and aligns them in what is termed a **pairwise alignment**. Next, it constructs a guide tree, which directs the sequential addition of each new sequence to the alignment based on its similarity to the initial pair. This approach is termed a hierarchical or progressive alignment. Clustal can be used to align either nucleotide or protein sequences for both global and local alignments **(Figure 20.16)**.

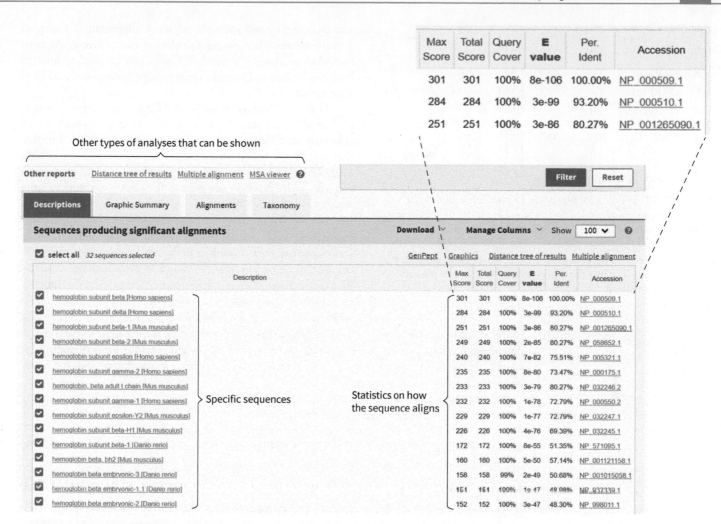

FIGURE 20.15 Table of BLAST results. This table corresponds to the data found in Figure 20.14. In this table, we see the species and sequence that generated the result, the max and total scores, the amount of coverage the hit has in common with the query, the E value (a statistical analysis of the quality of the hit), the percent identity between the two, and the accession number of the hit.

(Source: From U.S. National Library of Medicine. Public Domain.)

Other sequence alignment tools

T-Coffee (tree-based consistency objective function for alignment evaluation) is another tool that is used for aligning multiple sequences (**Figure 20.17**). It differs from Clustal in that it first pairs each set of sequences and then uses a library of these pair scores to generate the overall alignment. In effect, T-Coffee uses the broader local environment to ensure that parts of the sequence that may have less similarity do

	Insulin B-chain	Insulin A-chain
Human	FVNQHLCGSHLVEALYLVCGERGFFYTPKT	GIVEQCCTSICSLYQLENYCN
Great apes	FVNQHLCGSHLVEALYLVCGERGFFYTPKT	GIVEQCCTSICSLYQLENYCN
Macaque	FVNQHLCGSHLVEALYLVCGERGFFYTPKT	GIVEQCCTSICSLYQLENYCN
Rabbit	FVNQHLCGSHLVEALYLVCGERGFFYTPKS	GIVEQCCTSICSLYQLENYCN
Dog	FVNQHLCGSHLVEALYLVCGERGFFYTPKA	GIVEQCCTSICSLYQLENYCN
Horse	FVNQHLCGSHLVEALYLVCGERGFFYTPKA	GIVEQCCTGICSLYQLENYCN
Pig	FVNQHLCGSHLVEALYLVCGERGFFYTPKA	GIVEQCCTSICSLYQLENYCN
Cat	FVNQHLCGSHLVEALYLVCGERGFFYTPKA	GIVEQCCASVCSLYQLEHYCN
Rat	FVKQHLCGPHLVEALYLVCGERGFFYTPKS	GIVDQCCTSICSLYQLENYCN
Mouse	FVKQHLCGPHLVEALYLVCGERGFFYTPKS	GIVDQCCTSICSLYQLENYCN
Cattle	FVNQHLCGSHLVEALYLVCGERGFFYTPKA	GIVEQCCASVCSLYQLENYCN

Insulin B-chain positions: 1 5 10 15 20 25 30
Insulin A-chain positions: 1 5 10 15 20

FIGURE 20.16 Example of a Clustal alignment. In this example, sequences of the same protein (insulin) from multiple species were aligned with one another. Divergent residues are highlighted. These proteins are highly conserved with one another.

```
SeqA   GARFIELD   THE   LAST   FAT   CAT
       ||||||||   |||   ||||   |||              Weight = 88
SeqB   GARFIELD   THE   FAST   CAT


SeqA   GARFIELD   THE   LAST   FAT   CAT
       ||||||||   |||        \  |||   \\\
SeqC   GARFIELD   THE   VERY   FAST  CAT       Weight = 77
       ||||||||   |||         ||||   ||
SeqB   GARFIELD   THE               FAST  CAT


Seq1   GARFIELD   THE   LAST   FAT   CAT
                  |||          |||   |||
SeqD              THE          FAT   CAT       Weight = 100
                  |||          || \  \\\
SeqB   GARFIELD   THE          FAST  CAT
```

FIGURE 20.17 T-Coffee alignment.   In a T-Coffee alignment, individual sequences are paired first to give a library of paired sequences through which the final group alignment is made.

(Source: From Cédric Notredame et. al. T-coffee: a novel method for fast and accurate multiple sequence alignment, Volume 302, Issue 1, 8 September 2000, Pages 205-217. Edited by J. Thornton. doi:10.1006/jmbi.2000.4042. © 2000 Elsevier. Reproduced with permission from Elsevier.)

not skew the overall multiple sequence alignment. T-Coffee is a more accurate but slower calculation than Clustal. As with the other methods discussed, T-Coffee can be used to analyze both nucleotide and protein sequences for both global and local alignments.

Hidden Markov models (HMMs) are statistical models that allow an investigator to make a conceptual toolkit that encompasses different sources of divergent information. Imagine that you have a friend on a distant campus and you text or talk to that friend every day. Each day, your friend does one of three activities—she plays soccer, goes to the library, or works in the laboratory—but her choice of what she does depends entirely on whether she has an exam the next day. Using a hidden Markov model, you could predict whether or not she has an exam. First, you would assign probabilities to having an exam or not (a 10% overall chance on any given day) and to each of the three activities depending on two different states: exam or no exam. For example, let us say that if there is an exam the next day, there is an 85% chance that the friend went to the library, a 10% chance that she was in the laboratory, and a 5% chance that she played soccer. If she had no exam the next day, these probabilities would be different (for example, 50/25/25). From these probabilities, it would be possible to construct a statistical model that would provide the likelihood that the friend had an exam the next day or not.

Eddy (*Nature Biotech*, 2004) uses the example of a program that could identify the splice site of a gene. Let us presume that in exons there is an equal probability (25%) for each nucleotide (A, C, G, and T) that introns are AT rich (40% for each AT and 10% for GC) and that the 5′ splice site is almost always a G (95% G, 5% A). These probabilities can be combined in a model to predict where the 5′ splice site is likely to be found.

20.3.3 Interaction maps indicate how proteins interact

A final example of a bioinformatics tool is an interaction map. Such maps show which proteins interact with one another or the conditions under which they interact. As mentioned in section 20.1, interactomes are typically not well suited for automated generation because currently the interactions need to be annotated by hand and entered into the database.

There are several ways to depict interaction maps. In some maps, individual proteins are located on the edge of a circle, and lines crossing the circle indicate an interaction. In such maps, different colored lines indicate interactions that occur through phosphorylation or other modifications. Interaction maps can also show interactions between molecules other than proteins. An interaction map of glycosaminoglycans is shown in Figure 8.35. In other types of interaction maps, proteins are shown as circles connected by lines (**Figure 20.18**). Cytoscape and Medusa are software packages that can be used to display interaction data visually.

Worked Problem 20.3   **Characterizing a new gene**

You have identified a protein that may be an important regulator of a pathway you are studying. You have a cDNA that codes for this protein, and you know the sequence of that cDNA. Give one method for identifying the gene.

Strategy   We could use the cDNA sequence to search a database such as GenBank to find known matches. Based on that data, we could determine whether the gene, a protein, or a homolog has been previously identified.

Solution   We could simply use BLAST to compare the sequence against a database. If we know the organism, we could use the genome or some other library such as expressed sequence tags (ESTs) or mRNAs from that organism to identify the sequence.

Follow-up question   If we had only the amino acid sequence, could we still use BLAST to identify the gene?

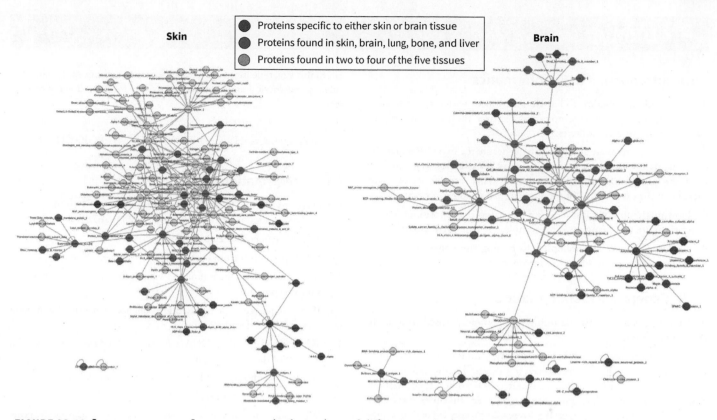

Skin

Proteins specific to either skin or brain tissue
Proteins found in skin, brain, lung, bone, and liver
Proteins found in two to four of the five tissues

Brain

FIGURE 20.18 Interactome map. Interactomes can be depicted several different ways. In this view, proteins found in skin and brain are shown. If a protein is specific to either tissue, it is shown in red. Blue nodes represent proteins found in skin, brain, lung, bone, and liver, whereas gray indicates proteins found in two to four of the five tissues listed.

(Source: From U.S. National Library of Medicine. Public Domain.)

Summary

- Data from most bioinformatics projects are stored in large, publicly accessible databases such as GenBank. Many programming tools for bioinformatics can be easily written in the programming languages Perl, Java, JavaScript, R, or Python.

- Sequence alignments can be global (lower homology over larger areas) or local (higher homology over shorter stretches). In the BLAST algorithm, sequences are chopped into short words, and a subset of these is used to probe the database. From early hits on the database, regions of larger, more complex similarity can be identified. The E value of a BLAST search gives the likelihood of a true match. The Clustal and T-Coffee programs are used to make pairwise comparisons of two or more sequences.

- Interaction maps provide visual depictions of the interactions of molecules in a cell, tissue, or organism.

Concept Check

1. Explain what GenBank is and how is it useful.
2. Compare the different programs used for multiple sequence alignments.
3. Explain what interaction maps are and how they are read.

Bioinformatics Exercises

Exercise 1 Using BLAST to Identify an Unknown Protein

Exercise 2 Gene Alignment and the Endosymbiotic Theory

Exercise 3 Conformations of DNA

Exercise 4 Sequence Alignments of Subunit 1 of Cytochrome c Oxidase

Exercise 5 Globin Sequence Alignment and Evolution

Exercise 6 Evolution and Structures of Serine Proteases

Exercise 7 Ion Channels

Exercise 8 Capstone: Translating, Identifying, and Analyzing an Unknown Bacterial Gene

Exercise 9 Capstone: Sequence Alignment and Structure of Bacterial Sigma Factors

Problems

20.1 Introduction to Bioinformatics

1. Of the bioinformatics studies we mentioned, which are most time sensitive in terms of how samples change over the time necessary to harvest the molecules of interest?

2. Biological systems are always in flux. What are the two classes of molecules fluctuate the most on a short time scale? What concerns might there be in the examination of these two classes? What other concerns might there be in the examination of a kinome or metabolome, for example?

3. How might an investigator ensure that sample compositions or concentrations are not changing as the samples are being harvested?

4. To date, no one has attempted to catalog the epigenetic changes found in specific genes. Propose a reason why.

20.2 Generating Bioinformatic Data

5. Compare and contrast the techniques commonly used to quantify the transcriptome.

6. Mass spectrometers used in proteomics studies are often high-end instruments requiring specialized training and dedicated operators and lab space for their use. Because of this need for specialization, many investigators collaborate with a laboratory that uses mass spectrometry. If you were to undertake a research project involving a component that required mass spectrometry, what aspects of the project could you perform prior to sending samples to a collaborator for mass spectrometry?

20.3 Analyzing Bioinformatic Data

7. If you are not familiar with programming, there are several places where you can learn. Scratch is an easy introduction to programming. The system seems simplistic and is easy enough to use for children and high school students but is also used in introductory programming classes. Go to the Scratch website (https://scratch.mit.edu), and complete one of their simple projects. Could you use this software to generate a simple animation of a biochemical process?

8. Use the NCBI site (https://www.ncbi.nlm.nih.gov/) and search for NP_001278826.1, without choosing a particular database. What are the access links obtained from your search? What further information do you obtain by clicking on the protein link?

9. ExPASy is a bioinformatics website that contains numerous tools. Among them is the ProtParam tool, https://web.expasy.org/protparam/.

Go to this site, type in the amino acid sequence without spaces, and calculate some physical and chemical parameters for the following peptide sequence:

10   20   30

FVNQHLCGSH   LVEALYLVCG   ERGFFYTPKT

Data Interpretation

For questions 10 to 20, use one of the following GI numbers or a GI number for a sequence of your choice: gi|254826781.

10. Has this sequence been identified?

11. What species is it from or most likely from?

12. If the sequence is for a protein, what is the cDNA sequence that codes for this protein?

13. If the sequence is a nucleotide sequence, what is the protein sequence once it is decoded?

14. What is the genomic DNA sequence 1 kb upstream and 1 kb downstream of the beginning of this gene?

15. In what chromosome and band is this gene located?

16. Are any genomic markers linked to the gene?

17. What genes are immediately upstream and downstream of the gene, and how far apart are they?

18. Are any single nucleotide polymorphisms (SNPs) found in this gene?

19. Identify two proteins found in the same species that are conserved at the protein level.

20. Identify two sequences that are conserved at the DNA level and found in different species.

Ethics and Social Responsibility

21. Societal and Ethical Biochemistry: *Sources of funding for scientific research*, presents the comparative spend of U.S. government on scientific research with other countries, private industry, academia and nonprofit organizations. Do you agree with how much the U.S. spends on scientific research? Explain.

22. Go to the NIH Web site, and examine how the NIH divides its budget to investigate different diseases. Would you propose any changes to how the NIH divides its resources? Again, substantiate your opinion with facts and data.

Suggested Readings

20.1 Introduction to Bioinformatics

Bakalar, N. "The Alternative Medical Bill: $30.2 Billion." *New York Times* (June 27, 2016). https://www.nytimes.com/2016/06/28/health/alternative-complementary-medicine-costs.html.

Battelle. "Battelle R&D Magazine Annual Global Funding Forecast Predicts R&D Spending Growth Will Continue While Globalization Accelerates." Press Release (December 16, 2011). http://www.battelle.org/media/press-releases/battelle-r-d-magazine-annual-global-funding-forecast-predicts-r-d-spending-growth-will-continue-while-globalization-accelerates.

Bessant, C., D. Oakley, and I. Shadforth. *Building Bioinformatics Solutions*, 2nd ed. Oxford, UK: Oxford University Press, 2014.

Choudhuri, S. *Bioinformatics for Beginners: Genes, Genomes, Molecular Evolution, Databases and Analytical Tools*. San Diego, CA: Academic Press, 2014.

Edwards, D., J. Stajich, and D. Hansen, eds. *Bioinformatics: Tools and Applications*. New York: Springer, 2009.

Hodgman, T. C., A. French, and D. R. Westhead. *Bioinformatics*. New York: Taylor & Francis, 2010.

Korpelainen, E., J. Tuimala, P. Somervuo, M. Huss, and G. Wong. *RNA-Seq Data Analysis: A Practical Approach* (Mathematical and Computational Biology). Boca Raton, FL: Chapman & Hall/CRC Press, 2014.

Lesk, A. *Introduction to Bioinformatics*. Oxford, UK: Oxford University Press, 2014.

National Institutes of Health. "Budget." http://www.nih.gov/about/budget.htm.

Wilson, P. "Americans Spend $33.9 Billion a Year on Alternative Medicine." *Consumer Reports* (August 3, 2009). http://www.consumerreports.org/cro/news/2009/08/americans-spend-33-9-billion-a-year-on-alternative-medicine/index.htm.

20.2 Generating Bioinformatic Data

Amaratunga, D., and J. Cabrera. *Exploration and Analysis of DNA Microarray and Other High-Dimensional Data*, 2nd ed. Hoboken, NJ: Wiley, 2014.

Brown, S. M. *Next-Generation DNA Sequencing Informatics*. Woodbury, NY: Cold Spring Harbor Press, 2013.

Caister, J. X. *Next-Generation Sequencing: Current Technologies and Applications*. San Diego, CA: Academic Press, 2014.

Eidhammer, I., H. Barsnes, G. E. Eide, and L. Martens. *Computational and Statistical Methods for Protein Quantification by Mass Spectrometry*. Chichester, UK: Wiley, 2013.

Gross, J. H. *Mass Spectrometry: A Textbook*, 2nd ed. New York: Springer, 2011.

Lovrić, J. *Introducing Proteomics: From Concepts to Sample Separation, Mass Spectrometry and Data Analysis*. Chichester, UK: Wiley, 2011.

Paša-Tolic, L., and M. S. Lipton. *Mass Spectrometry of Proteins and Peptides: Methods and Protocols*, 2nd ed. (Methods in Molecular Biology). New York: Humana Press, 2014.

Russell, S., and L. A. Meadows. *Microarray Technology in Practice*. San Diego, CA: Academic Press, 2008.

Uttamchandani, M., and S. Q. Yao. *Small Molecule Microarrays: Methods and Protocols* (Methods in Molecular Biology). New York: Humana Press, 2010.

Velculescu, V. E., L. Zhang, B. Vogelstein, and K. W. Kinzler. "Serial Analysis of Gene Expression." *Science* 270 (1995): 484–87.

Velculescu, V. E., L. Zhang, W. Zhou, J. Vogelstein, M. A. Basrai, D. E. Bassett, P. Heiter, B. Vogelstein, and K. W. Kinzler. "Characterization of the Yeast Transcriptome." *Cell* 88, no. 2 (1997): 243–51.

Wang, Z., M. Gerstein, and M. Snyder. "RNS-Seq: A Revolutionary Tool for Transcriptomics." *Nature Reviews Genetics* 10, no. 1 (2009): 57–63. doi:10.1038/nrg2484.

20.3 Analyzing Bioinformatic Data

Altschul, S. F., W. Gish, W. Miller, E. W. Myers, and D. J. Lipman. "Basic Local Alignment Search Tool." *Journal of Molecular Biology* 215, no. 3 (1990): 403–10.

Chautard, E., M. Fatoux-Ardore, L. Ballut, N. Thierry-Mieg, and S. Ricard-Blum. "MatrixDB, the Extracellular Matrix Interaction Database." *Nucleic Acids Research* 39, Database issue (2011): D235–40.

Christiansen, R., B. D. Foy, and L. Wall. *Programming Perl: Unmatched Power for Text Processing and Scripting*. Sebastopol, CA: O'Riley Media, 2012.

Eddy, S. R. "What Is a Hidden Markov Model?" *Nature Biotechnology* 22 (2004): 1315–6.

Haverbeke, M. *Eloquent JavaScript: A Modern Introduction to Programming*. San Francisco, CA: No Starch Press, 2014.

Higgins, D. G., and P. M. Sharp. "CLUSTAL: A Package for Performing Multiple Sequence Alignment on a Microcomputer." *Gene* 72, no. 1 (1988): 237–44.

Hoffman, P. *Perl for Dummies*, 4th ed. Indianapolis, IN: Wiley, 2003.

Jones, N. C., and P. A. Pevzner. *An Introduction to Bioinformatics Algorithms* (Computational Molecular Biology). Cambridge, MA: MIT Press, 2004.

Lutz, M. *Learning Python*, 5th ed. Sebastopol, CA: O'Riley Media, 2013.

Model, M. L. *Bioinformatics Programming Using Python: Practical Programming for Biological Data*. Sebastopol, CA: O'Riley Media, 2009.

Mount, D. M. *Bioinformatics: Sequence and Genome Analysis*, 2nd ed. Woodbury, NY: Cold Spring Harbor Laboratory Press, 2004.

Notredame, C., D. G. Higgins, and J. Heringa. "T-Coffee: A Novel Method for Fast and Accurate Multiple Sequence Alignment." *Journal of Molecular Biology* 302, no. 1 (2000): 205–17.

Schildt, H. *Java: A Beginner's Guide*. New York: McGraw-Hill, 2014.

Simoncelli, T., et al. *Paving the Way for Personalized Medicine: FDA's Role in a New Era of Medical Product Development*. Washington, DC: U.S. Food and Drug Administration, 2013. http://www.fda.gov/downloads/ScienceResearch/SpecialTopics/PersonalizedMedicine/UCM372421.pdf.

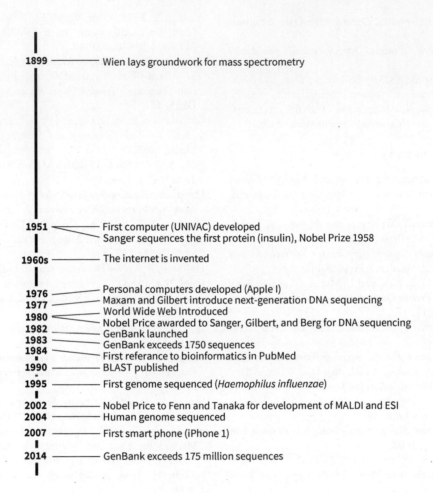

1899 —— Wien lays groundwork for mass spectrometry

1951 —— First computer (UNIVAC) developed
Sanger sequences the first protein (insulin), Nobel Prize 1958

1960s —— The internet is invented

1976 —— Personal computers developed (Apple I)
1977 —— Maxam and Gilbert introduce next-generation DNA sequencing
World Wide Web Introduced
1980 —— Nobel Price awarded to Sanger, Gilbert, and Berg for DNA sequencing
1982 —— GenBank launched
1983 —— GenBank exceeds 1750 sequences
1984 —— First referance to bioinformatics in PubMed
1990 —— BLAST published

1995 —— First genome sequenced (*Haemophilus influenzae*)

2002 —— Nobel Price to Fenn and Tanaka for development of MALDI and ESI
2004 —— Human genome sequenced

2007 —— First smart phone (iPhone 1)

2014 —— GenBank exceeds 175 million sequences

CHAPTER 21

Signal Transduction

Cell Signaling in Context

Chapter 5 described communication among people as an analogy for how cells communicate with each other through metabolic signals. Although this analogy works for simple signal transduction, the process of cell signaling, like that of communicating with friends, can be far more complicated.

In today's technology-saturated world, communication among friends as well as strangers has been both facilitated and complicated by social media. Rather than communicating directly with one person via phone messages or letters, we can now communicate instantaneously with a much broader group of connections using Instagram, Twitter, Facebook, or other social media. If you posted a message to one of these platforms, you could get different responses from different followers. For example, your friends might have a different reaction to a photo from last weekend than your parents. In addition, these accounts are often linked so that a post made to one account automatically appears in another (a Tweet appearing on Instagram, for example). This allows the message to reach different groups of followers and get even more diverse responses.

The kind of complex communication that occurs through social media is a better analogy than a phone message or letter for the signaling pathways we discuss in this chapter. Similar to how posts on one platform may appear on another, activation of one pathway can sometimes lead to the activation of a divergent but related process through cross-talk.

Previous chapters discussed signaling pathways that are fundamental to almost all of biochemistry and respond largely to metabolic signals. In contrast, the pathways discussed in this chapter are more specialized, with most occurring only in eukaryotes and some only in fungi or animals. They regulate gene expression, cell growth, and differentiation of cells.

Chapter Outline

21.1 A Review of Signal Transduction

21.2 An Overview of Regulation of Signal Transduction

21.3 Six New Signal Transduction Pathways

Common Themes

Evolution's outcomes are conserved.	• Signaling pathways are conserved throughout evolutionary history.
	• Specific conserved domains found in signaling molecules, such as a kinase domain or a growth factor receptor, can be found in different signaling pathways.
Structure determines function.	• Many signaling molecules have conserved domains such as Src homology 2 (SH2) and kinase domains.
	• Alterations in the structure of molecules due to covalent modifications such as addition of phosphates or the binding of coactivators can lead to translocation of a signaling protein from one part of the cell to another or to activation or inactivation of a signaling molecule.
Biochemical information is transferred, exchanged, and stored.	• Signal transduction allows for the transmission of information among cells and different parts of an organism or among organisms.
	• Signal transduction pathways signal through transcription factors to regulate gene expression.
Biomolecules are altered through pathways involving transformations of energy and matter.	• Transfer of phosphate groups to or from signaling molecules such as kinases leads to conformational changes that activate or inactivate these proteins.

21.1 A Review of Signal Transduction

Signal transduction is the means by which cells communicate with one another. Why might cells need to communicate with one another? All cells need to be able to detect, respond to, and react to their environment, whether they are single-celled organisms or part of a complex metazoan.

Chapter 5 and subsequent chapters focused on three signaling pathways and their roles in metabolism: protein kinase A, insulin, and AMP-dependent protein kinase, each of which has diverse effects that include activation of gene transcription. These chapters focused mainly on the metabolic ramifications of signaling, for example, altering the activity of enzymes such as glycogen synthase or hormone-sensitive lipase. Those examples illustrate how a relatively small number of signaling molecules can generate diverse effects in different tissues through differential expression of receptors. For example, the signaling molecule glucagon has no effect on muscle glycogen because that tissue lacks the glucagon receptor, but it does stimulate glycogenolysis in the liver.

This chapter looks beyond regulation of enzymatic activity to examine some other facets of cell signaling, particularly in growth and differentiation of cells and tissues. One of the advantages of specializing tissue function into storage (adipose), movement (muscle), and collected metabolic pathways (liver) is that an organism can have dedicated systems for these different metabolic roles. A complex organism needs to first generate these tissues and systems, then orchestrate their growth and differentiation, and finally coordinate their response to different metabolic states. Collectively, all this information is significant not only in terms of basic biochemistry but also in immunology, developmental biology, and the processes involved in diseases such as cancer.

21.1.1 Signal transduction follows certain basic principles

Signal transduction is generally accomplished through chemical signals, although some specialized cells can detect and respond to light. For communication to occur, three things need to be in place: a signal (a chemical compound released by other cells somewhere in the organism or by other organisms in the environment), a receptor for the signal, and a means by which the cell can respond (**Figure 21.1**). Signals are often hormones, and receptors are often transmembrane proteins.

The binding of the signal by the receptor on the extracellular side is specific, and it elicits a conformational change in the receptor that causes a change on the intracellular side of the protein. The change in the conformation typically activates enzymes such as kinases, and it can stimulate protein signaling pathways directly or be amplified by other enzymes and second messenger signaling molecules such as Ca^{2+} or cyclic AMP (cAMP). Recall that a **second messenger** is a chemical message made in response to a signal that amplifies and propagates it inside the cell. The end result of signaling is often a change or modulation in enzymatic function (discussed in Chapter 5) or a change in gene transcription (discussed here).

One pathway can often communicate with another through **cross-talk**. This usually occurs through a common signaling molecule that is active in both pathways.

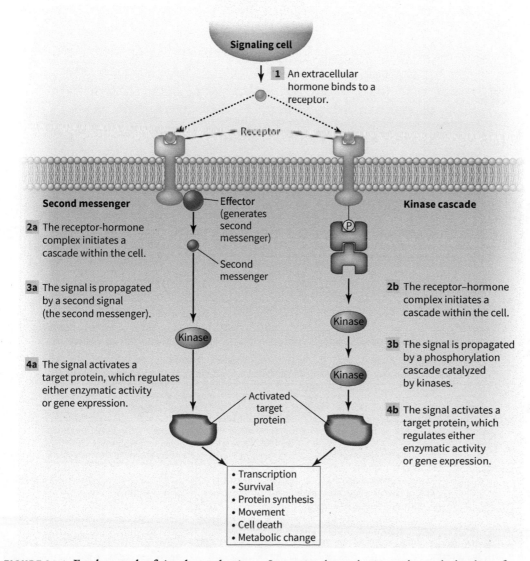

FIGURE 21.1 Fundamentals of signal transduction. In any signal transduction pathway, the binding of a signaling cell to a receptor initiates a cascade within the cell that ultimately results in alterations to enzyme activity or gene expression. Sometimes the signal is first propagated and amplified through a second messenger.

21.1.2 The PKA signaling pathway involves a second messenger

The **protein kinase A (PKA)** signaling pathway is an example of a G protein-coupled pathway involving a second messenger (**Figure 21.2**). PKA can be activated by a signal from a catecholamine such as dopamine, epinephrine, or norepinephrine. The catecholamine binds to a β-adrenergic receptor, activating a heterotrimeric G protein in which the Gα subunit exchanges GDP for GTP. The binding of GTP activates this G protein, which in turn activates adenylate cyclase, a membrane-bound enzyme that generates the second messenger cAMP from ATP.

PKA is a tetrameric enzyme consisting of two regulatory and two catalytic subunits. Each of the regulatory subunits can allosterically bind two molecules of cAMP and release the two active catalytic subunits. The enzyme has hundreds, if not thousands, of substrates in the cell. PKA is important in the phosphorylation (and activation) of glycogen phosphorylase kinase, which in turn activates glycogen phosphorylase, resulting in glycogen breakdown. PKA also mobilizes energy from fat stores by stimulating lipolysis in adipocytes through the phosphorylation of perilipin 1 and hormone-sensitive lipase.

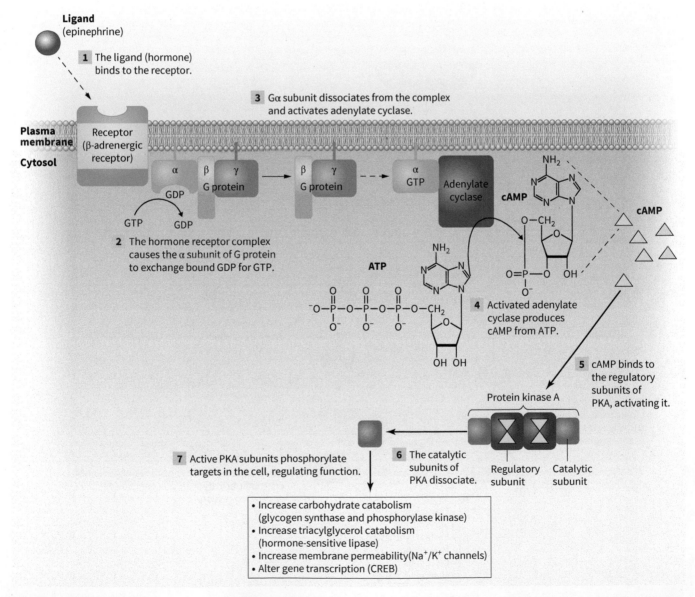

FIGURE 21.2 PKA signaling pathway. In the PKA signaling pathway, an extracellular signal binds to a 7-transmembrane G protein-coupled receptor. This signals through a heterotrimeric G protein to activate adenylate cyclase, which produces the second messenger cAMP, which finally activates PKA.

21.1.3 The insulin signaling pathway has multiple functions

The peptide hormone **insulin** is an iconic signaling molecule. Insulin has multiple functions in the body. The intake of food causes insulin to be released from the pancreas. The hormone then binds to the insulin receptor, resulting in a signaling cascade that causes vesicles containing glucose transporters to fuse with the plasma membrane, which in turn leads to greater uptake of plasma glucose into the cells. Insulin signaling also activates protein phosphatase, which blunts the breakdown of carbohydrates and lipids. Finally, insulin is a growth factor and can activate transcription factors leading to increased growth, protein synthesis, and activation of gene transcription.

Insulin is an example of a growth factor that functions through a receptor tyrosine kinase (**Figure 21.3**). Insulin signaling begins with the binding of insulin to its receptor. This binding causes the receptor to dimerize and become a functional kinase; that is, the insulin receptor is itself a tyrosine kinase. The receptor then autophosphorylates, with each half of the receptor thought to

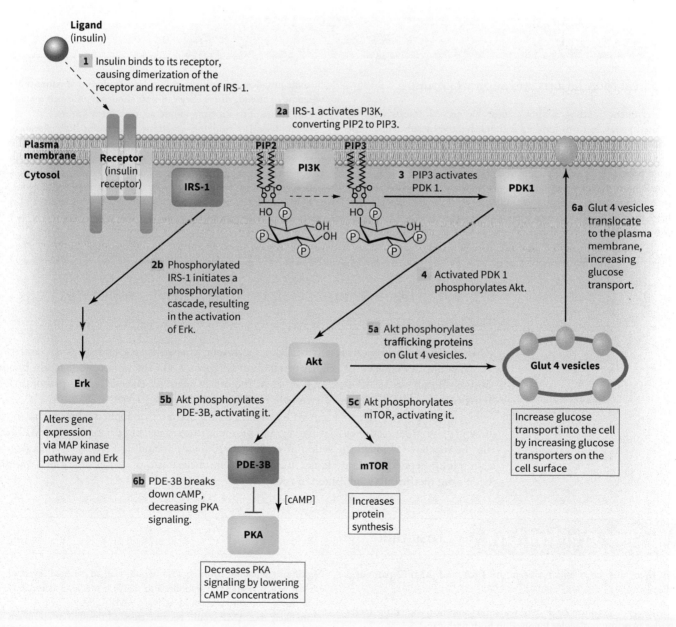

FIGURE 21.3 Insulin signaling pathway. Insulin binds to its receptor, causing dimerization and autophosphorylation of the receptor. Next, the insulin receptor substrate IRS-1 binds. IRS is a scaffolding protein that can bind to several other proteins, which leads to the MAP kinase pathway (Erk) and alterations in gene expression. IRS also activates PI3 kinase, which leads to activation of the Akt pathway and translocation of vesicles of glucose transporters to the cell surface.

phosphorylate key tyrosine residues on the other half. Once phosphorylated, the receptor recruits other factors, beginning with the insulin receptor substrate (IRS), which acts as a scaffolding protein from which other factors can nucleate.

At this point, insulin signaling bifurcates. It can activate the MAP kinase pathway (Erk, discussed later), ultimately resulting in cell division and growth. Alternatively, it can activate phosphatidyl inositol 4,5-bisphosphate kinase (PI3 kinase) and the phosphoinositide cascade, thus affecting metabolism.

When activated by the insulin receptor, PI3 kinase phosphorylates phosphatidyl inositol 4,5-bisphosphate (PIP2) to generate phosphatidyl inositol 3,4,5-trisphosphate (PIP3). PIP3 activates the protein-dependent kinase PDK1, which can interact with several different kinases, including protein kinase C (PKC) and Akt, both of which are discussed in the next section.

Many people are familiar with insulin because of the role it plays in diabetes. Diabetes was one of the first diseases for which a support network was established not only for patients but also for friends and family of those with the disease. These support networks are discussed in **Medical Biochemistry: Medical foundations and support groups.**

Medical Biochemistry

Medical foundations and support groups

Government funding supports most biomedical research in the United States, but this funding is supplemented by substantial support from nongovernmental medical foundations. Organizations like the American Heart Association or American Diabetes Association have hundreds of thousands of members, including clinicians and researchers. In recent years, these organizations have opened up their membership to an even larger audience, for example, people who have these disorders or their friends and caregivers. They even publish journals and electronic publications written for a lay audience.

In some instances, scientific meetings have sessions that patients are welcome to attend, where they can hear about progress being made toward a cure or treatment and interact with investigators.

For scientists studying questions with biomedical relevance, interactions with those affected by a disease are a reminder of why their work is important. In addition, patients get a ringside seat to watch progress being made on a condition that affects them at a personal level. This helps to give patients and friends a realistic view of how scientific research is conducted and how slow progress can be, while still providing hope for a cure.

21.1.4 The AMP kinase signaling pathway regulates cell metabolism

AMP-activated protein kinase (AMPK) is a cytosolic kinase that responds to AMP levels and acts as an acute regulator of cellular metabolism (**Figure 21.4**). The total concentration of adenosine phosphates (AMP + ADP + ATP) in the cell is relatively constant. ATP predominates, but breakdown of ATP due to energetic need leads to increased levels of AMP. The elevated concentrations of AMP activate AMPK, which in turn activates catabolic pathways that break down triglycerides, activates β-oxidation, and promotes the storage of carbohydrates. AMPK can also be regulated through the action of several other kinases, including PKA, Akt, and calcium/calmodulin-dependent kinase kinase (CaMKK). One mechanism of action of the oral antidiabetic drug metformin is inhibition of AMPK.

Worked Problem 21.1   Cross-talk

Is there any cross-talk between the PKA and AMPK pathways? Explain.

Strategy Examine both pathways for common signaling molecules. Are any molecules involved in both pathways?

Solution AMPK cannot directly affect the PKA pathway. If AMP becomes elevated, AMPK will phosphorylate targets that activate pathways to increase the cell's energetic state (increase ATP levels).

Significant signaling through PKA would lead to elevated levels of cAMP, which would be catabolized to AMP; however, it is unlikely that this small amount could activate AMPK. Nevertheless, it is more likely that AMP would become elevated through other means such as depletion of cellular ATP through metabolism.

Follow-up question Do these two pathways affect enzyme activity, gene transcription, or both?

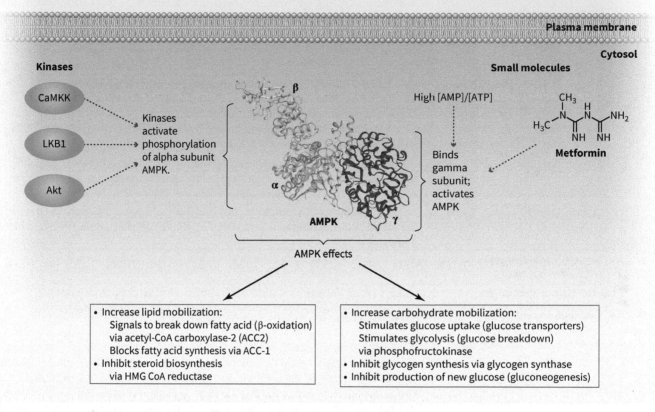

FIGURE 21.4 AMPK signaling pathway. AMP-activated protein kinase (AMPK) is activated by high levels of AMP but also by calcium/calmodulin-dependent protein kinase (CaMKK). AMPK regulates several metabolic pathways, including glycogen biosynthesis, gluconeogenesis, β-oxidation, and fatty acid biosynthesis.

(Source: Data from PDB ID: 4CFF Xiao, B., Sanders, M.J., Carmena, D., Bright, N.J., Haire, L.F., Underwood, E., Patel, B.R., Heath, R.B., Walker, P.A., Hallen, S., Giordanetto, F., Martin, S.R., Carling, D., Gamblin, S.J. (2013) Structural Basis of Ampk Regulation by Small Molecule Activators. *Nat. Commun.* **4**: 3017)

Summary

- Signal transduction is the process by which cells communicate with one another. It can lead to changes in metabolism or gene expression or affect the development of the organism. For signal transduction to occur, a chemical signal needs to be received by a receptor, and the cell needs some means of responding to that signal inside the cell.

- Signaling through PKA is activated by the second messenger cAMP. This molecule is produced by the enzyme adenylate cyclase, which is activated by a heterotrimeric G protein.

- Insulin signals through the insulin receptor. When bound to insulin, the two halves of the receptor dimerize and autophosphorylate. This causes binding of the insulin receptor substrate (IRS) and activation of two different pathways. The insulin receptor can influence gene expression via MAP kinase or alter glucose levels in the cell via the kinase Akt.

- When the energy state of the cell is low, AMP kinase is activated by binding to AMP, which activates pathways that mobilize energy stores.

Concept Check

1. Describe how cells communicate, including the role of a second messenger.
2. Describe the steps of the three signal transduction pathways discussed here: protein kinase A, insulin, and AMP-activated protein kinase.
3. Discuss the signals used by protein kinase A and AMP-activated protein kinase signaling pathways and the effects mediated by them.

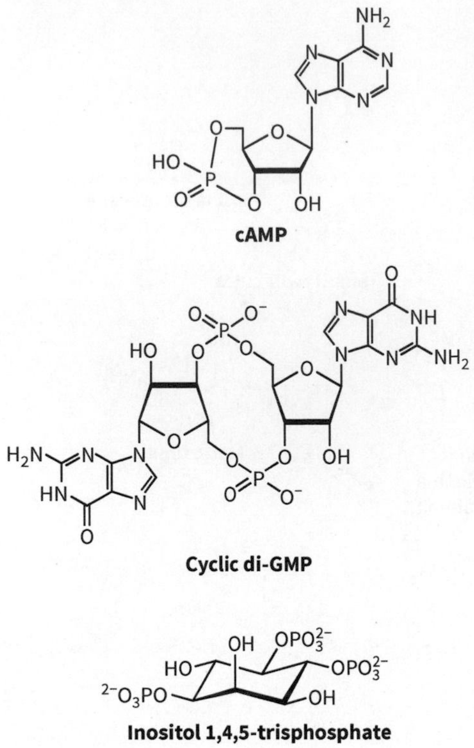

cAMP

Cyclic di-GMP

Inositol 1,4,5-trisphosphate

FIGURE 21.5 Common second messengers. Common second messengers in the cell include cAMP, cyclic di-GMP, and inositol 1,4,5-tri phosphate.

21.2 An Overview of Regulation of Signal Transduction

The three signaling pathways reviewed from Chapter 5—protein kinase A, insulin, and AMP-dependent protein kinase—give an indication of how a hormone signal can affect an enzyme in a metabolic pathway. There are many other signals, and cells need to be able to generate a range of responses based on these stimuli, including changes in gene expression. The result may simply be up- or downregulation of certain genes, but it could also be cell proliferation and mitogenesis (increased cell division), or differentiation, such as a stem cell becoming a liver cell. Signaling could also trigger a cell to enter apoptosis (programmed cell death).

Clearly, the effects of signaling are complex and require both fine-tuning and tight regulation. Failure to regulate a step could be disastrous, for example, leading to unabated growth, aberrant gene expression, or premature apoptosis. Signaling problems that affect development can lead to significant defects and are often fatal to the organism.

Certain patterns of organization, function, and structure in signal transduction are useful in elucidating new pathways and provide greater understanding of existing pathways.

21.2.1 Second messengers regulate certain pathways

As discussed in section 21.1, PKA is an example of a molecule that is regulated by a second messenger (cAMP). Second messengers are generally small, soluble signals that are generated or released within the cell in response to an extracellular signal. Other second messengers include cyclic guanosine monophosphate (cGMP), cyclic di-GMP, calcium ions (Ca^{2+}), and various differentially phosphorylated forms of the cyclic alcohol inositol (**Figure 21.5**).

21.2.2 Kinase cascades are an alternative to second messengers

Other signaling pathways are regulated via **kinase cascades**. In a kinase cascade, a kinase becomes active and phosphorylates a second kinase downstream of the first. As many as four or five layers of kinases regulating kinases are seen in biochemistry. It may seem redundant to have so many levels of regulation; however, each level provides opportunities for cross-talk between pathways, adds levels of control, and potentially amplifies the signal.

We may ask how it is known that proteins in these pathways bind and interact with one another. There are several methods through which these interactions can be detected, either directly or indirectly. These are discussed in **Biochemistry: Detection of protein–protein interactions.**

Biochemistry

Detection of protein–protein interactions

This chapter discusses numerous pathways in which one protein interacts with another. Some of these proteins are enzymes with distinct catalytic capabilities such as a kinase or phosphatase. In some cases, two proteins interact with each other, for example, when one phosphorylates the other, but in other cases, a direct protein–protein interaction is less clear.

There are several ways an investigator can tell whether two proteins interact directly. A technique such as immunofluorescence (IF) microscopy, where proteins are stained with specific fluorescent antibody tags, could seem like a candidate for detecting protein–protein interactions. In fact, the resolution of IF is generally too low at hundreds of nanometers to fractions of a micron. The best that IF can do is to co-localize two proteins to the same organelle or structure.

Co-immunoprecipitation (Co-IP) is a technique that provides a much closer approximation of how proteins interact. A protein of interest is precipitated using an antibody and a microscopic bead of agarose or other material that binds specifically to the antibody. The complex of the bead plus antibody, the protein of interest, and any associated material is separated on SDS-PAGE and analyzed using immunoblotting. The identification of a second protein *could* indicate a protein–protein interaction, but it could also indicate that the second protein binds to the first through some sort of intermediate or is part of a larger complex.

A more direct way to probe protein–protein interactions is to use fluorescence resonance energy transfer (FRET, also known as Förster resonance energy transfer). In this technique, two proteins hypothesized to interact directly are fluorescently tagged with different fluorophores. The choice of fluorophore is important because in the FRET experiment one fluorophore is excited, but the technique looks for emission from the second. If the two tagged proteins are in close proximity to one another (within 5 Å), and the fluorophores are properly aligned with each other, then the excited state energy of the first will induce excitation of the second through resonance, resulting in emission from the second fluorophore. This experiment is more complex than IF or Co-IP, but it makes a much stronger case that two proteins are physically interacting with each other.

Another technique that is commonly employed to ascertain protein–protein interactions is the two-hybrid system. This experiment can be conducted in bacteria, yeast, or mammalian cells, although the most common approach uses yeast. One protein (the bait) is coded for by a plasmid and expressed as a fusion protein with one half of a dimeric transcription factor. The cDNAs coding for the other proteins (the targets) are found in a library of constructs. These proteins are also expressed as fusion proteins with the complementary half of the transcription factor. If the two halves of the transcription factor are brought together by a protein–protein interaction, then a reporter gene will be expressed through the activity of the transcription factor.

If two proteins interact directly, it may be possible to cross-link them chemically. Several protein chemistry reagents enable the cross-linking of proteins by chemically derivatizing specific groups such as primary amines. If these groups are close together, it may be possible to link them. Subsequent analysis using western blotting or mass spectrometry would be necessary to identify which proteins are cross-linked. Cross-linking chemistry has improved greatly, but the technique still has drawbacks, most notably the need to hit the "sweet spot" in the experiment between cross-linking the proteins of interest and cross-linking everything else in a sample.

Finally, if two proteins interact directly and can be crystallized together, the gold standard for detecting protein–protein interaction is to determine the structure of the complex using either X-ray diffraction or NMR spectroscopy. Although potentially more laborious than other techniques, this is the only approach that provides a picture of exactly how two proteins interact.

Regulation of kinases by inhibition Although there are numerous kinases, there are relatively few mechanisms by which such enzymes are regulated. One mechanism involves the use of an inhibitory protein that binds to the kinase and either keeps it in an inactive state or blocks access of the active site to substrates. This mechanism is seen in PKA, where the regulatory domains of PKA bind and inactivate the catalytic domains of the enzyme (**Figure 21.6**).

Other inhibitors block the interaction of one component of the signaling pathway with a downstream effector; one example is β-arrestin blocking access of G proteins to the β-adrenergic receptor. A different type of regulation is found in enzymes such as protein kinase C (PKC). PKC does not bind to a regulatory protein; instead, it contains a **pseudosubstrate domain** that is a structural homolog of its substrate but lacks the essential serine, threonine, or tyrosine residue necessary for phosphorylation. Thus, although the domain binds in the active site of the kinase, it cannot be phosphorylated. While bound, it blocks binding of substrate, acting as a built-in inhibitor that stays bound until an activation event. Finally, phosphorylation or binding of a second messenger elicits a conformational change that results in release of the pseudosubstrate domain and allows binding and phosphorylation of the true substrate.

Deactivation of kinase cascades by phosphatases Like all signal transduction pathways, kinase cascades must have a way to be turned off. Phosphatases cleave phosphates from proteins and thus play a key role in reversing the action of kinases and shutting down kinase signaling cascades. For example, glycogen phosphorylase is activated by phosphorylation by PKA, increasing the breakdown of glycogen.

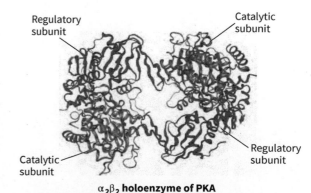

Regulatory subunit

Catalytic subunit

Catalytic subunit

Regulatory subunit

$\alpha_2\beta_2$ **holoenzyme of PKA**

FIGURE 21.6 Inhibition of PKA. PKA is an example of a protein kinase whose activity is kept in check by binding to a regulatory domain of a second protein. Shown is the $\alpha_2\beta_2$ holoenzyme of PKA, which consists of two regulatory and two catalytic subunits.

(Source: Data from PDB ID 3TNP Zhang, P., Smith-Nguyen, E.V., Keshwani, M.M., Deal, M.S., Kornev, A.P., Taylor, S.S. (2012) Structure and allostery of the PKA RIIbeta tetrameric holoenzyme *Science* **335**: 712–716)

Under high-glucose conditions, insulin signaling activates protein phosphatase 1 (PP1). PP1 dephosphorylates glycogen phosphorylase, inactivating it and stopping the release of glucose from glycogen breakdown.

21.2.3 SH2 domains are common structural motifs found in signaling proteins

Many signaling proteins contain **Src homology 2 (SH2)** domains (**Figure 21.7**). These domains are named for the protein Src (pronounced "sark"), one of the signaling proteins in which this domain is found. Src is a signaling kinase that was originally found in cells infected with the Rous sarcoma virus. The Src protein is partially responsible for the transformation of healthy cells infected with the virus into the cancerous state.

SH2 domains bind specifically but reversibly to phosphorylated tyrosine residues. Many signaling proteins are protein tyrosine kinases or receptor tyrosine kinases. If a tyrosine becomes phosphorylated by one of these kinases, a downstream protein can bind to it via an SH2 domain.

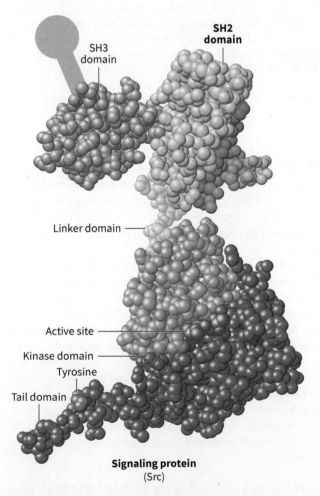

FIGURE 21.7 The SH2 domain. Src homology 2 (SH2) domains are named for the signaling protein Src. SH2 domains bind tightly to phosphorylated tyrosine domains. This is a common motif found in signaling proteins.

(Source: Image from the RCSB PDB July 2003 Molecule of the Month feature by David Goodsell)

Worked Problem 21.2 | Kinase sites

A certain kinase recognizes a conserved or canonical substrate sequence consisting of two positively charged residues, any residue, followed by a serine or threonine. A related kinase has a domain with the following amino acid sequence:

Ser-Ile-Tyr-Arg-Arg-Gly-Ala-Arg-Arg-Trp-Arg-Lys-Leu

Propose a function for this domain.

Strategy Compare the amino acid sequence to the canonical sequence. How does the kinase normally act on the substrate? Is there anything in the new sequence that appears similar?

Solution The domain may be a pseudosubstrate domain. Comparison of the two sequences indicates the substitution of an alanine for a serine or threonine normally found in the substrate. This leaves the enzyme unable to phosphorylate the domain, but the remaining sequence is largely unchanged, indicating that it could still fit into the active site of the kinase.

Follow-up question You have identified an SH2 domain in a protein with no known catalytic activity or function. Based solely on the presence of this domain, what might we predict about this protein?

Summary

- Signal transduction pathways can be broadly divided into those that mediate signals via second messengers and those that mediate signals through kinase cascades.

- One class of kinases uses a pseudosubstrate domain of the same polypeptide to mask the active site when the enzyme is in the inactive form; the other class binds to a separate polypeptide chain (a different subunit), which sterically blocks the active site. Kinase cascades are deactivated in part by phosphatases.

- SH2 domains are commonly conserved structural elements that bind to the phosphorylated tyrosine residues.

Concept Check

1. Describe the difference between signal transduction pathways that are mediated by second messengers and those that employ kinase cascades.
2. Discuss the structural motifs described in this section and what they can do.

21.3 Six New Signal Transduction Pathways

Along with the three pathways just discussed, numerous other signaling pathways can compete or cooperate to modulate the cell's metabolic state, managing which genes the cell is expressing and whether the cell is senescent or dividing. This section discusses six of these pathways, summarized in **Table 21.1**. Some, like JAK-STAT and MAP kinase, are expressed in all cells; others, such as the toll-like receptors (TLRs), are more specialized and restricted to specific cell types. As with other pathways, there are numerous points in these six pathways where proteins can modulate the action of one another, amplify signals, or make an effect more general or more specific, depending on the pathway and the situation.

TABLE 21.1 Key Signal Transduction Pathways

Pathway	Signal	Response
JAK-STAT	Interferons (IFNs) Cytokines	Initiation of gene expression
MAP kinase	Mitogens	Regulation of transcription factors
Toll-like receptor	Lipopolysaccharides	Inflammatory response Developmental changes
PKC	Tumor promotors	Arrest of cell growth Apoptosis
CDK	Cyclins	Cell cycle regulation
PKG	Nitric oxide (NO)	Regulation of kinases

21.3.1 The JAK-STAT pathway is involved in inflammation

The enzyme Janus kinase (JAK) and the signal transducer and activator of transcription (STAT) were identified independently in the early 1990s in studies examining the cellular response to interferons (IFNs), protein signals produced by the immune system in response to infections. It was not until 1994 that investigators determined that JAK and STAT work in a concerted fashion in a single signaling pathway, which they termed the **JAK-STAT pathway**. Although JAK can phosphorylate other proteins and STAT can be phosphorylated by other kinases, the name JAK-STAT has persisted.

Janus kinase (JAK) is named after the Roman god Janus, who had two faces—one facing forward and another on the back of his head. This allowed him to see in all directions while guarding Rome. Janus kinase is equally two faced, possessing two similar kinase domains. One domain is responsible for the transfer of phosphate groups from ATP to the tyrosine of the target protein. The other domain is a pseudokinase that is highly related structurally to the first but lacks catalytic activity. This second domain is thought to regulate the overall activity of the kinase. JAK also possesses several SH2 domains, which are probably responsible for interactions with phosphorylated receptors, STAT, or other proteins.

In its inactive state, STAT is a monomeric protein found in the cytosol. Once phosphorylated, STAT will form homodimers through the interaction of its SH2 domains, translocate to the nucleus, and bind to DNA, thereby altering gene transcription.

The JAK-STAT pathway is found in animals and the colonial amoeba *Dictyostelium discoideum*, and it is expressed in every cell of the organism. It is not found in bacteria, fungi, or plants. JAK-STAT is important in the immune response to pathogens, but it generally functions in any condition in which proinflammatory cytokines are elevated, including cancer and obesity.

The JAK-STAT pathway is activated by several different signals and receptors, the most common of which are the cytokines, small signaling peptides such as the IFNs and interleukins. Leptin and several other peptide hormones can also signal via the JAK-STAT pathway through their cognate receptors.

Steps involved in JAK-STAT signaling   The first step in JAK-STAT signaling is that an extracellular ligand such as a cytokine binds to its receptor, which has similarities with growth factor receptors such as the insulin receptor (**Figure 21.8**). Upon binding of ligand, the receptor dimerizes. In growth factor receptors, the receptor itself has a kinase domain. In contrast, in JAK-STAT signaling, binding of ligand to the receptors and the ensuing dimerization elicits binding of JAK, which in turn phosphorylates the key tyrosine residues on both the receptor and the transcription factor STAT.

One of the genes transcriptionally activated by STAT is suppressor of cytokine signaling proteins (SOCS). SOCS inhibits JAK-STAT by binding to JAK and preventing its interaction with STAT. This is an example of a **feedback inhibition loop** similar to the long-term regulation of signaling pathways discussed in other chapters.

Phosphorylation of receptor tyrosine residues by JAK can induce binding of other proteins to the receptor. Such proteins include adapter proteins such as Shp, which can cross-talk to the PI3K pathway and the MAP kinase pathway.

The STAT proteins can also interact with certain growth factor receptors. These receptors have an intrinsic kinase domain that can phosphorylate the key tyrosine in STAT, leading to dimerization and activation.

JAK-STAT signaling in science and medicine   The JAK-STAT pathway has been implicated in inflammation, cancer, obesity, and the immune response. One example of the JAK-STAT pathway is cytokines signaling to adipose tissue in obesity. In the obese state, adipocytes (fat cells), preadipocytes, and resident macrophages in adipose tissue secrete proinflammatory cytokines such as tumor necrosis factor alpha (TNF-α), interleukin-6 (IL-6), and interferon gamma (IFN-γ). These signals act on preadipocytes through the JAK-STAT pathway to promote expression of the transcription factor PPARγ and differentiation into adipocytes. Stimulation of this pathway also promotes lipid storage, angiogenesis (growth of new blood vessels), extracellular matrix remodeling, and clearance of the debris left over from cells that have undergone apoptosis. Because of these relationships between proinflammatory signals and obesity, it is thought that drugs that attempt to silence these pathways might be effective in the treatment of obesity or its complications.

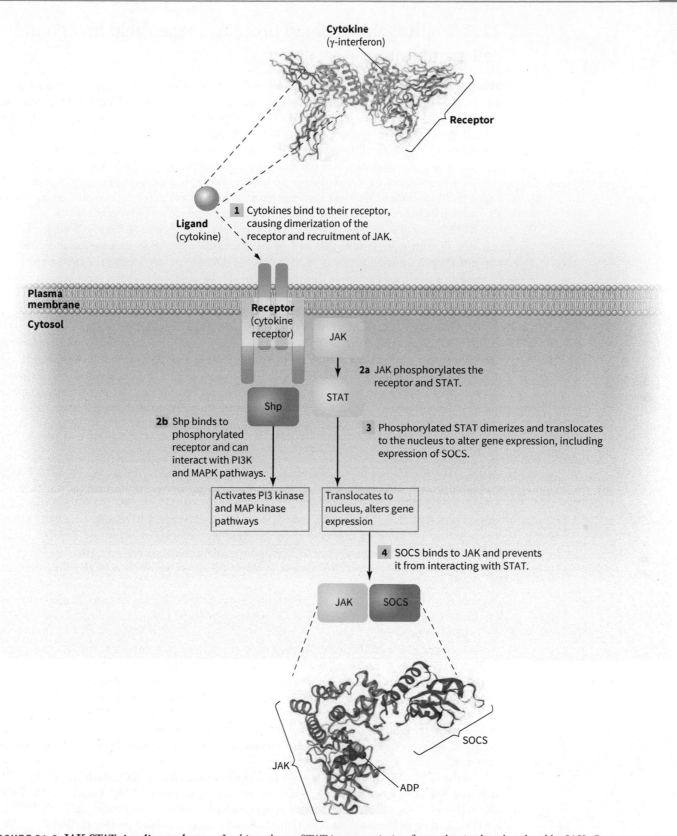

FIGURE 21.8 JAK-STAT signaling pathway. In this pathway, STAT is a transcription factor that is phosphorylated by JAK. Once phosphorylated, STAT will dimerize and move to the nucleus, where it binds to DNA and alters gene expression.

(Source: (Top) Data from PDB ID 1FG9 Thiel, D.J., le Du, M.H., Walter, R.L., D'Arcy, A., Chene, C., Fountoulakis, M., Garotta, G., Winkler, F.K., Ealick, S.E. (2000) Observation of an unexpected third receptor molecule in the crystal structure of human interferon-gamma receptor complex. *Structure Fold. Des.* **8**: 927–936; (Bottom) Data from PDB ID 6C7Y Liau, N.P.D., Laktyushin, A., Lucet, I.S., Murphy, J.M., Yao, S., Whitlock, E., Callaghan, K., Nicola, N.A., Kershaw, N.J., Babon, J.J. (2018) The molecular basis of JAK/STAT inhibition by SOCS1. *Nat Commun* **9**: 1558–1558)

21.3.2 Mitogen-activated protein kinases help to regulate cell proliferation

Mitogen-activated protein (MAP) kinases are a family of kinases that respond to **mitogens**, that is, growth factors and proinflammatory signals that lead to cell proliferation (mitogenesis). MAP kinases are important in the regulation of cell proliferation and therefore in processes such as healing, development of the organism, and the immune response. They also contribute to the unmitigated growth of cells in cancer. Hence, MAP kinases are an active field of research.

There are three major branches to the MAP kinase pathway: the ERK, JNK, and p38 pathways. These branches were each discovered independently and subsequently identified as MAP kinase family members. The three branches have similar patterns of regulation and activation, and all function in the cellular response to stress as well as in growth and development.

MAP kinases themselves do not bind to ligands; instead, they are activated through a kinase cascade. The upstream ligands that lead to the activation of MAP kinases are highly varied. As with the insulin signaling pathway, the MAP kinase cascade can be activated through the binding of insulin to the insulin receptor. The cascade can also be activated through other growth factors and their receptors. Additional factors that can stimulate this cascade include low pH, osmotic or heat shock, oxidative stress, radiation, and other cellular stressors.

MAP kinases activate gene transcription by activating transcription factors. Each member of the MAP kinase family can phosphorylate a wide number of transcription factors. In general, the ERK kinases activate transcription factors that lead to increased growth, development, and proliferation, including c-Fos, STAT, and c-Jun. JNK and p38 kinases stimulate transcription factors that in part regulate inflammation, apoptosis, differentiation, and proliferation, such as CREB, ATF, and cMyc. The ERKs can also stimulate translation through the activation of kinases such as p90 S6 kinase or RSK that phosphorylate or activate initiation factors like eIF4E, either directly or indirectly.

Steps involved in MAP kinase signaling

The MAP kinase pathway has two distinguishing features: the cascade aspect of the pathway and regulation of the kinases through phosphorylation. MAP kinases are regulated through phosphorylation by MAP kinase kinases, also known as MAPKKs or MEKs, which are in turn regulated by phosphorylation by a MAP kinase kinase kinase (MAP3K or MEKK). There are multiple isoforms of MEK and MEKK; hence, different MAP kinase pathways can communicate with or modulate one another.

Many enzymes are regulated through a single phosphorylation, but MAP kinases require multiple residues to be phosphorylated for activation. Also, whereas many kinases can be broadly defined as serine threonine kinases or tyrosine kinases, MAP kinases have a broader substrate specificity and generally phosphorylate the hydroxyl group of any of these three amino acids in their substrate protein.

An example of a MAP kinase pathway is signaling through a growth factor receptor that is a receptor tyrosine kinase (**Figure 21.9**). On binding to its ligand, the receptor dimerizes and autophosphorylates, recruiting two proteins: Grb2 and SOS. These two proteins act as adapters for the low-molecular-weight G protein Ras. Like the heterotrimeric G proteins seen in PKA signaling, the low-molecular-weight G proteins exchange GDP in the inactive state for GTP in the active state. However, in contrast to heterotrimeric G proteins, these proteins are smaller, cytosolic, and monomeric. They are also unrelated to other G proteins in terms of structure or sequence.

In the GTP-bound state, Ras is active and can bind and activate Raf, which is a MEKK. The activated Raf phosphorylates MEK, which in turn phosphorylates a MAP kinase such as ERK. The activated ERK in turn phosphorylates and activates transcription factors such as CREB or c-MYC, leading to increased transcription of S6 and thus to increased protein synthesis.

The MAP kinase pathway is almost ubiquitous, being found in most eukaryotes, including plants, fungi, and animals. When depicted in illustrations, the cross-talk between the various branches of the MAP kinase pathway and other pathways often appears more like a web than a straightforward pathway. Specific aspects of MAP kinase pathways are modulated either directly or indirectly by growth factor receptors, TLRs, cAMP, Ca^{2+}, and PKC. Overall, a large number of signals converge on the MAP kinases.

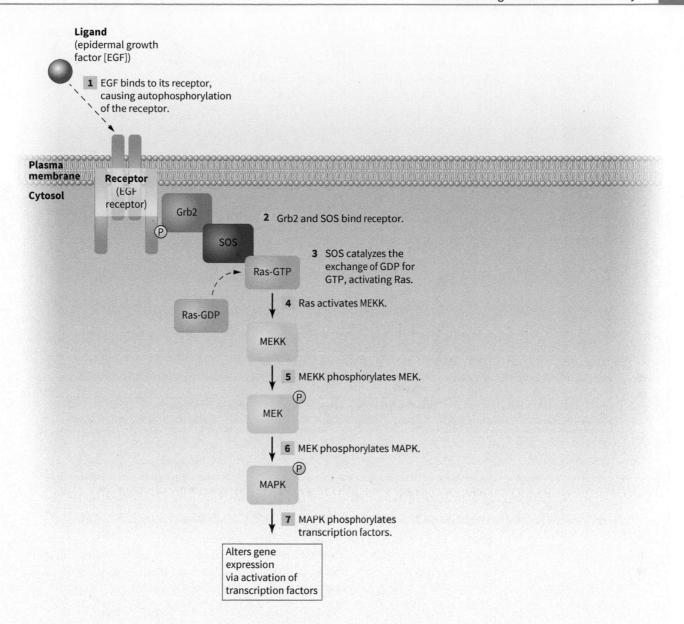

FIGURE 21.9 MAP kinase signaling pathway. The MAP kinase pathway can be activated through several different methods. In this example, binding of the epidermal growth factor (EGF) to the epidermal growth factor receptor causes autophosphorylation and binding of Grb2 and SOS. These, in turn, activate the small monomeric G protein Ras. Ras exchanges GDP for GTP and activates a MEKK. MEKK phosphorylates a MEK, which in turn phosphorylates a MAP kinase. MAP kinases activate downstream targets such as transcription factors, activating gene expression.

MAP kinase signaling in science and medicine

Because MAP kinases are central to cell growth and the inflammatory response, blocking these pathways pharmacologically has been used in treatments for cancer or inflammation. Although there are no drugs currently available that target MAP kinases directly, numerous drugs are available that block tyrosine kinases involved in the MAP kinase pathway. The best known of these drugs is imatinib mesylate (Gleevec), which inhibits the Bcl-Abl fusion protein found in acute myelogenous leukemia (**Figure 21.10**). This drug has been shown to inhibit several other kinases, including Src and the receptors for platelet-derived growth factor (PDGF). It is currently used to treat certain blood disorders and gastrointestinal cancers. Other drugs have been developed to target Ras because this protein is mutated and constitutively active in 25% of human cancers. Drugs such as tipifarnib (Zarnestra) inhibit Ras activation by blocking the farnesylation of

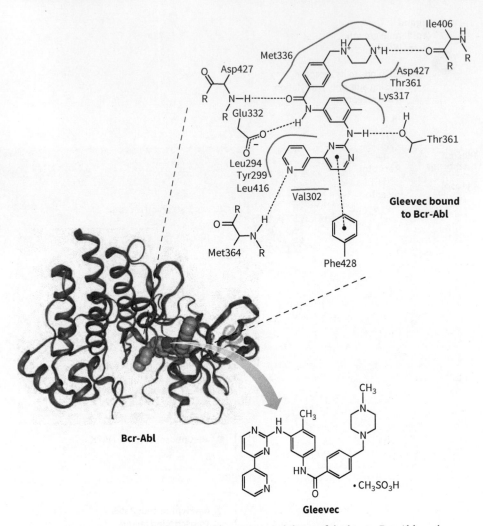

FIGURE 21.10 Structure of Gleevec.   Gleevec is an inhibitor of the kinase Bcr-Abl used to treat chronic myelogenous leukemia. It has doubled the five-year survival rate for this disease since its discovery.

(Source: Data from PDB ID 3GVU Salah, E., Ugochukwu, E., Barr, A., Mahajan, P., Shrestha, B., Savitsky, P., Knapp, S. The crystal structure of human ABL2 in complex with GLEEVEC. *To be published.*)

the protein rather than its catalytic activity. Vemurafenib (Zelboraf), dabrafenib (Tafinlar), and encorafenib (Braftovi) are drugs that act directly on mutant forms of Ras (most notably the V600E mutation) to inhibit the kinase. These drugs are used for treatment of malignant melanoma.

Despite advances in computational modeling and cultured cell models, animals still play a key role in the development of these and all drugs. Their use is discussed in **Societal and Ethical Biochemistry: The use of animals in biomedical research.**

Societal and Ethical Biochemistry

The use of animals in biomedical research

Biochemists use the tools of chemistry to ask questions of biological systems. Often, this necessitates the use of animals in research, but for various reasons some people are opposed to using animals in this way. Biomedical research carried out on animals can improve or even save human lives.

It is important to ask whether animals are necessary for a specific biochemical study, but there are still experiments for which animals are critical. For example, in toxicity studies during drug development, there is no alternative to animal data. However, in most instances, animals are not necessary because research can be conducted using bacteria, yeast, plants, cultured cells, or samples prepared from any of these systems. In many ways, such systems are simpler to study and maintain in the laboratory than animals. Science has made significant progress using these tools, but for now, many complex diseases that afflict human beings are too complex to study with these simple samples. For example, studying aspects of kidney function generally requires use of a functioning kidney.

If an animal system is needed, why not simply use humans? In fact, humans or human samples are often used in the laboratory with strict precautions to ensure the samples do not harbor pathogens or contaminate the researchers. Nevertheless, use of humans is not always feasible. For example, although humans have kidneys, few would voluntarily give one up simply for the good of science, especially when a mouse's kidney could be used instead. Also, the history of medicine is filled with stories of investigators failing to inform study participants of all the risks. Finally, some experiments are clearly unethical to perform on people. An institutional review board (IRB) considers the ethics of human experimentation and grants permission for experiments involving humans.

The choice to use an animal model for study is not a simple one. Even small animals are expensive to use in a laboratory setting.

Investigators must demonstrate that animals are necessary for the proposed study, ensure that they suffer as little as possible, and use as few animals as possible.

Animals used in biomedical research come under the oversight of an animal care and use committee comprised of veterinarians, nonscientists, and members of the community, as well as researchers. Investigators must apply for permission to use animals, describing housing and living conditions as well as how they will be used in the experiments. Guidelines are in place to minimize the suffering and discomfort of animals. Veterinarians are employed to carry out routine checks on the animals, and the committee regularly inspects both the laboratories and the animal facility.

Those opposed to animal research sometimes portray it as cruel and senseless. However, the mechanisms in place ensure that animals are used as little as possible and that when animals must be used, strict policies prevent harm to animal subjects and ensure that any suffering is minimized.

21.3.3 Toll-like receptor signaling is critical to the innate immune response

Innate immunity is the aspect of an organism's immune system that is present from birth. In contrast, adaptive immunity is developed through exposure to pathogens. Because many organisms can respond to a bacterial infection, and many bacteria elicit similar responses in different organisms, it was hypothesized in the late 1800s that bacteria must secrete toxins that elicit the innate immune response. The receptors that bind these toxins and several endogenous signals are called toll receptors.

Toll receptors were first identified through their function in a different process. *Drosophila* embryos harboring a mutation in the toll genes have a developmental defect that prevents establishment of a proper dorsal–ventral axis (**Figure 21.11**). Basically, these embryos have difficulty distinguishing their front side from their back. The name for the toll gene comes from the German investigators who first observed the mutant phenotype in the 1980s. On seeing the embryos, they exclaimed, "*Das ist ja toll!*" which roughly translates to "That's great!"

Some ten years after their discovery, it was found that the toll receptors are also responsible for *Drosophila*'s resistance to fungal infection. Thirteen different but related toll receptors are found in humans; however, in humans (and in most mammals), these receptors are thought to be more important in immunity than in development. Because the family now includes multiple, related members, the name **toll-like receptor (TLR)** has been adopted.

TLRs have been shown to bind to lipopolysaccharides, which belong to a class of molecule known as endotoxins that includes bacterial lipids, carbohydrates, and proteins. These molecules are part of the bacterial cell wall (where they have other functions), but they also cause fever and inflammation in response to infection. TLRs bind not only to lipopolysaccharides but also to the bacterial proteins fibrillin and profilin and to several heat shock proteins.

Signaling through TLRs activates gene transcription, ultimately resulting in increased concentrations of IFNs and other proinflammatory cytokines. These compounds signal to other cells of the organism (in particular those of the immune system) to mount an attack on the pathogens generating the signal.

Fungal infection

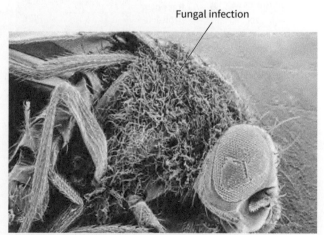

Lemaitre, B., Nicolas, E., Michaut, L., Reichhart, J.-M., & Hoffmann, J. A. (1996). The Dorsoventral Regulatory Gene Cassette spätzle/Toll/cactus Controls the Potent Antifungal Response in Drosophila Adults. Cell, 86(6), 973–983

FIGURE 21.11 Toll mutation in *Drosophila*. Toll mutations in *Drosophila* have several phenotypes. These genes were originally identified because they affected the dorsal–ventral axis of the fly embryo. Shown is a different manifestation of a toll mutation. Flies with these mutations have defective immune systems and are susceptible to fungal infections, which is the cause of the hairy coating on this dead fly.

The steps involved in signaling through toll-like receptors
Binding of ligand (lipopolysaccharide) to a TLR causes the receptor to dimerize (**Figure 21.12**). This dimerization is somewhat analogous to an insulin or growth factor receptor binding to substrate or cytokines acting through the JAK-STAT pathway. However, in contrast to some of these other

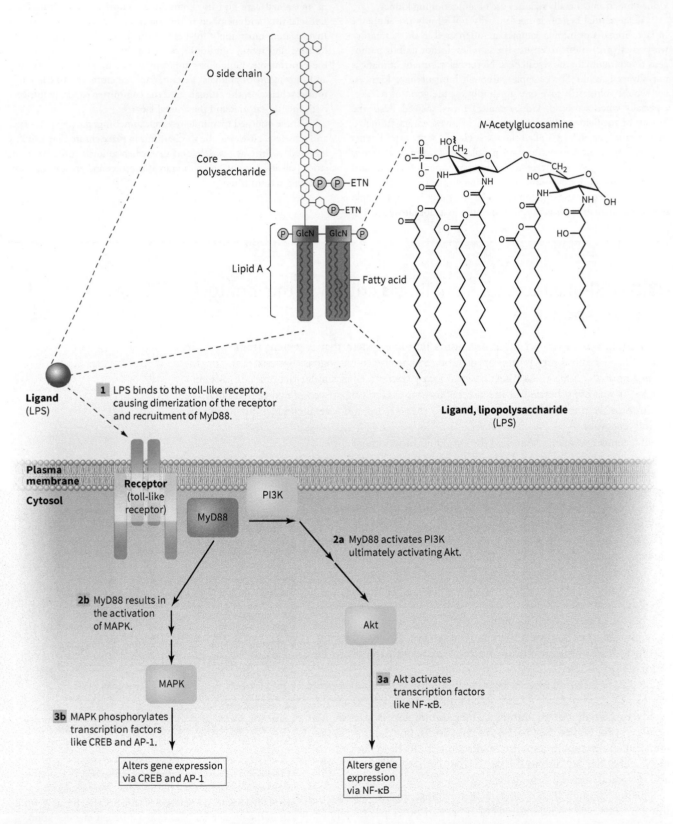

FIGURE 21.12 Toll-like receptor signaling. Toll-like receptors (TLRs) can signal through multiple pathways. Lipopolysaccharide (LPS) is a component of some bacterial cell walls and a ligand for TLRs. Binding of LPS causes receptor dimerization, similar to the insulin or cytokine pathways. The adaptor protein MyD88 signals through several different intermediates to activate either MAP kinases or NF-kB.

receptors, TLRs are homodimers. Dimerization of the TLR leads to interaction with one of two different proteins, which in turn leads to activation of two different pathways. These proteins can be thought of as analogous in some ways to the insulin receptor substrate (IRS) proteins found in insulin signaling.

TLRs exhibit considerable cross-talk with other pathways. Most interact with MyD88, which initiates a kinase cascade that activates PI3K, ultimately activating Akt. MyD88 also activates the MAP kinase cascade. Both of these pathways in turn activate specific transcription factors. Akt activates the transcription factor NF-κB by phosphorylating the regulatory protein Iκ-B, leading to its degradation. The MAP kinases phosphorylate and activate CREB and AP-1, which also translocate to the nucleus and activate transcription of genes coding for proinflammatory cytokines.

The other protein that binds to the active receptor is TRIF, which recruits other kinases. These kinases in turn activate the IFN regulatory factors (IRFs). Once phosphorylated, IRFs translocate to the nucleus and specifically regulate the production of the IFNs.

Toll-like receptors in science and medicine   Because of the roles they play in immunity and infection, TLRs are the targets of several drugs. Blocking TLRs could, in theory, blunt the immune system's response. This may be valuable in treating sepsis (a life-threatening immune response to infection) or autoimmune diseases such as rheumatoid arthritis or inflammatory bowel disease. Likewise, there may be clinical benefit to activating TLRs when a patient has been diagnosed with cancer or received a vaccine. Resiquimod is an agonist of TLRs 7 and 8 and is used as an antiviral and antitumor compound in dermatology (**Figure 21.13**).

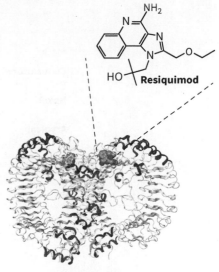

Toll-like receptor 8
(TLR8)

FIGURE 21.13 Resiquimod. Resiquimod is a toll-like receptor agonist that has been used in dermatology.

(Source: Data from PDB ID 3W3L Tanji, H., Ohto, U., Shibata, T., Miyake, K., Shimizu, T. (2013) Structural reorganization of the Toll-like receptor 8 dimer induced by agonistic ligands. *Science* **339**: 1426–1429)

21.3.4 PKC and the phosphoinositide cascade affect gene expression and enzyme activity

PKC was first identified as a calcium-dependent kinase that was not dependent on cyclic nucleotides such as cAMP for activation. It was also found to be responsive to some lipids or lipid analogs. Investigators later recognized its importance in a wide variety of effects on the cell and organism, mitigated through activation of transcription or enzymes. PKC can be activated via several types of receptors, but the common second messengers include Ca^{2+} and diacylglycerol.

Signaling through PKC can elicit both an alteration in gene expression and changes in enzyme activity. The responses of cells vary, depending on the type of cell and its location in the organism. Examples of cellular responses to PKC stimulation include changes in permeability of the cell membrane, secretion, migration, muscle contraction, proliferation, apoptosis, and hypertrophy.

Fifteen different PKC isoforms are found in humans. These isoforms can be broadly classified as conventional, novel (or unconventional), and atypical (**Figure 21.14**). They vary with regard to how they are regulated, which in turn depends on their structure. The conventional PKCs are activated by Ca^{2+} and diacylglycerol; the novel PKCs require diacylglycerol but not Ca^{2+}; and the atypical PKCs require neither diacylglycerol nor Ca^{2+} but instead the phospholipid phosphatidyl serine. PKC isoforms contain two distinct domains that impart these differences in regulation. The C1 domain is present, in different forms, in all three classes of PKC; this domain is responsible for diacylglycerol binding in the conventional and novel forms. The C2 domain is also present in all three classes of proteins. It is responsible for Ca^{2+} binding in the conventional isoforms but not in the other two classes.

The steps involved in protein kinase C signaling   An example of PKC signaling is activation of PKC through 7-transmembrane G protein-coupled receptor (7TM-GPCR) signaling (**Figure 21.15**). A ligand such as epinephrine binds to a 7TM-GPCR such as the α-adrenergic receptor. As in PKA signaling, this class of receptor is associated with a heterotrimeric G protein. However, in contrast to the PKA pathway, this combination of G proteins activates phospholipase C (PLC) rather than adenylate cyclase. The activated PLC cleaves between the glycerol backbone and the phosphoalcohol head group of some phospholipids. In this example, the PLC isoform specifically cleaves phosphatidyl inositol 4,5-bisphosphate (PIP2), resulting in the production of diacylglycerol and inositol 1,4,5-trisphosphate (IP3). The latter is soluble in the cytosol and will diffuse away from the plasma membrane. This second messenger can activate

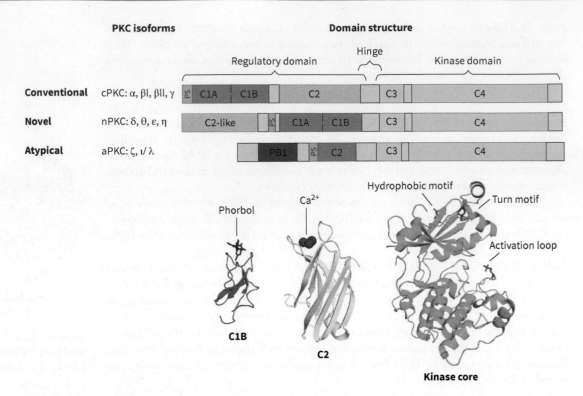

FIGURE 21.14 Subclasses of PKC.   Protein kinase C can be divided into three categories (conventional, novel, and atypical) based on its domain composition. Each of these classes has different conserved domains including kinase, C1B, and C2 domains. Each class member is conserved but differs in sequence.

(Source: Adapted from Newton AC, Antal CE, Steinberg SF. "Protein kinase C mechanisms that contribute to cardiac remodelling." (2016) Clinical Science 130(17):1499–1510.)

specific calcium channels in the endoplasmic reticulum, elevating the concentration of Ca^{2+} in the cytosol. This activates cytosolic PKC, causing it to translocate to the plasma membrane. There, PKC encounters the diacylglycerol product generated in the previous step, which activates the enzyme, leading to phosphorylation of multiple substrates.

The PKC pathway can activate several branches of the MAP kinase pathway. Some isoforms of PKC can activate adenylate cyclases, leading to elevated levels of cAMP and activation of PKA-dependent processes as well as activation of phosphodiesterases. The product of these phosphodiesterases (AMP) can also activate AMP kinase (AMPK). Finally, some growth factor receptors are coupled to PLC, which, when activated, can generate reactive signaling molecules (phosphoinositides and diacylglycerol) that in turn can activate PKC.

PKC itself can activate transcription through the action of several transcription factors, most notably NF-κB.

PKC signaling in science and medicine   **Tumor promoters** are not by themselves damaging to DNA or carcinogenic; rather, they facilitate the growth of tumors once a carcinogenic event has occurred. This is the equivalent of releasing the brakes in a car parked at the top of a hill; once something causes the car to start rolling, there are no brakes to hold it back. The opposite of tumor promotion is tumor suppression. When **tumor suppressors** are turned on, these genes generate proteins that either suppress growth of cells or direct cells toward apoptosis. Examples of tumor suppressor genes include *TP53*, which is mutated in over half of all cancers, and *BRCA1* and *BRCA2*, the genes implicated in several forms of breast cancer. In contrast, oncogenes (genes implicated in cancer) such as *MEK*s and *RAF* regulate expression of transcription factors that trigger growth and cell survival.

PKC is the target of several compounds that act as tumor promoters, the best known of which is tetradecanoyl phorbol acetate (TPA), also referred to as phorbol myristyl acetate (PMA) (**Figure 21.16**). These phorbol esters are naturally occurring compounds that bind to and strongly activate PKC.

Many herbal remedies that are viewed as benign can actually be tumor promoters. One such remedy, *Croton tiglium*, is known in Chinese herbal medicine as *ba dou* or "purging croton."

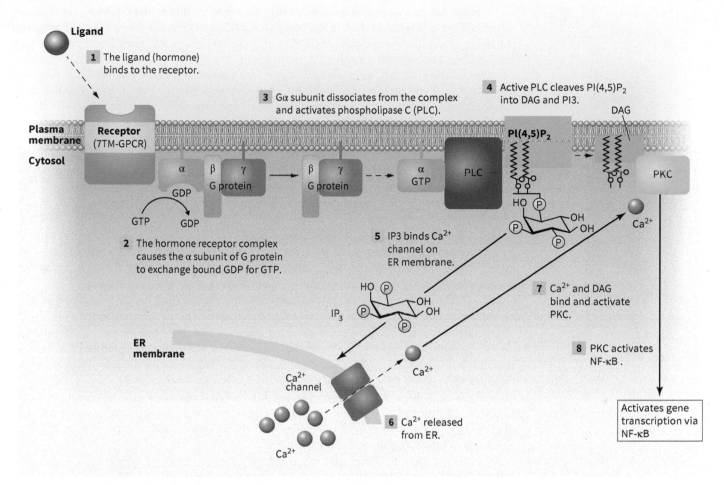

FIGURE 21.15 Example of protein kinase C signaling. In this example, a hormone binds to a 7TM-GPCR, which signals through a heterotrimeric G protein to activate phospholipase C. PLC produces two signaling molecules, diacylglycerol and IP3. IP3 is soluble and activates a calcium channel in the endoplasmic reticulum to elevate cytosolic Ca^{2+} levels. This isoform of PKC is responsive to Ca^{2+} and translocates to the plasma membrane where it is activated by diacylglycerol.

The seeds and oil they contain can be used as a purgative to relieve a variety of gastrointestinal maladies or skin disorders. In a diluted form, the oil is used to produce a warming effect in the treatment of sore muscles and joints. Current uses include "rejuvenating" face peels. However, croton oil contains as much as 12% phorbol ester by mass. This herbal treatment clearly has the biological effect of relaxing muscles or causing skin peeling, but at the same time it is highly toxic and tumor promoting, acting through stimulation of PKC. Although "natural," this treatment is certainly not benign.

How does TPA/PMA activate PKC? An examination of the structure of TPA/PMA provides clues. The acetate and myristoyl groups of TPA/PMA are structurally similar to diacylglycerol. Structural analysis of PKC isoforms bound to TPA/PMA confirms that this is where and how the tumor promoter binds and activates the enzyme. TPA/PMA activates PKC so effectively that it is often used in research settings to activate PKC signaling specifically.

21.3.5 The cyclin-dependent kinases regulate cell growth and replication

Studies in the 1980s on cell division identified a group of proteins whose concentrations varied throughout the cell cycle. These proteins, which came to be known as the cyclins, are critical regulators of cell growth and replication and act via **cyclin-dependent kinases (CDKs)**.

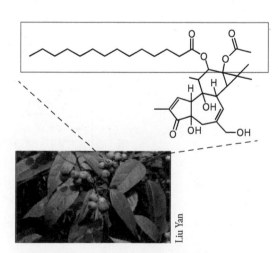

FIGURE 21.16 Phorbol esters. Phorbol esters are naturally occurring toxins found in the seeds of several species of the croton plant. These molecules fit in the diacylglycerol binding site of PKC and strongly activate it. The highlighted portion of the molecule is thought to act like diacylglycerol in the activation of PKC.

Cyclins can be grouped based whether they are expressed during the G_1, M, G_2, or S phases in the cell cycle (**Figure 21.17**). In contrast, the CDKs are expressed constitutively throughout the cell cycle. Mutations to cyclins or CDKs can lead either to a lack of growth and division or to rampant and uncontrolled growth.

The CDKs are small, serine-threonine protein kinases that regulate progression throughout the cell cycle. Two events need to occur for a CDK to be activated. First, the CDK binds to its cognate cyclin, and then a specific threonine adjacent to the active site is phosphorylated by a CDK-activating kinase (CAK). The phosphorylation leads to a conformational change that activates the CDK.

CDKs can be inhibited or negatively regulated by phosphorylation of specific tyrosine residues on a glycine-rich loop (the G loop). This reduces the affinity of CDK for its substrates. CDKs can also be negatively regulated by INK4 and Kip proteins. INK4 binds to specific CDKs, preventing formation of a cyclin–CDK complex and blocking phosphorylation by CAK. In contrast, Kip proteins bind to an already active cyclin–CDK complex, blocking activity. Cdc-25 is a phosphatase that removes the phosphate that activates CDKs.

There are many CDKs, with 20 occurring in mammals alone. They mostly fall into two groups based on their effects: those that affect transcription and those that affect cell cycle progression.

The CDKs that regulate the cell cycle are found in the nucleus, where they phosphorylate and activate transcription factors or regulatory elements such as the retinoblastoma protein, Rb. Other targets for the CDKs in the nucleus include the condensins and the lamins. Condensins are the proteins that regulate the formation of the DNA replication complex and are partially responsible for assembling metaphase chromosomes. The lamins are the nuclear equivalent of the cytoskeleton.

CDKs that affect transcription typically bind to and phosphorylate splicing factors, the C-terminal domain (CTD) of RNA polymerase II, or the mediator complex. These molecules are discussed in Chapters 14 and 15. As with most of the kinases discussed in this chapter, there are hundreds of substrate proteins for the CDKs.

The structure and regulation of CDKs
Like all protein kinases, CDKs have a bilobed structure consisting of an N-terminal lobe rich in beta sheets and a C-terminal lobe rich in alpha helices (**Figure 21.18**). The two lobes are joined by a hinge, and the active site is found at the interface. In the case of the CDKs, cyclin recognition and binding occurs on a face formed between the lobes. Two other structural features of note are the G loop and the C helix. The G loop is a glycine-rich loop that includes two threonines that are important in inhibition. The C helix contains the sequence PSTAIRE, which folds down over the active site in the inactivated state. Binding of cyclin moves the C helix up and out, exposing the active site and a critical threonine. Phosphorylation of this threonine stabilizes the open, active conformation.

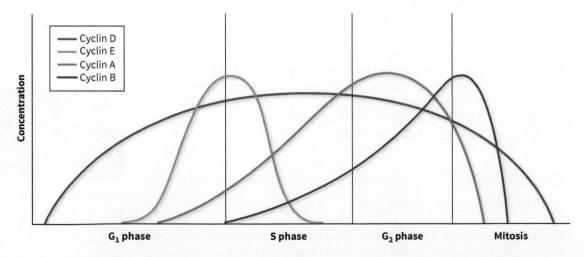

FIGURE 21.17 Cyclins. The cyclins are small protein moderators that bind and activate cyclin-dependent kinases. The levels of these proteins vary throughout the cell cycle.

The CDKs are highly conserved throughout the eukaryotes. They are part of a superfamily of kinases that includes the MAP kinases and glycogen synthase kinase 3β (GSK-3β). However, although they are conserved and structurally related, the CDKs are not thought to be regulated by pathways other than the ones previously mentioned. In contrast to kinases such as the MAP kinases, which have significant cross-talk with each other and can be activated by other pathways, the CDKs appear to function somewhat in isolation from other signaling cascades.

CDK signaling in science and medicine   Because CDKs act as master regulators of the cell cycle, they might be useful targets for anticancer drugs. Many drugs that target CDKs are in development, including some to treat aggressive lung and breast cancers, and several have passed into clinical trials.

21.3.6 Nitric oxide signals via protein kinase G

Drugs such as nitroglycerine and amyl nitrite have been known for over a hundred years to produce physiological effects including vasodilation and smooth muscle relaxation. Research in the

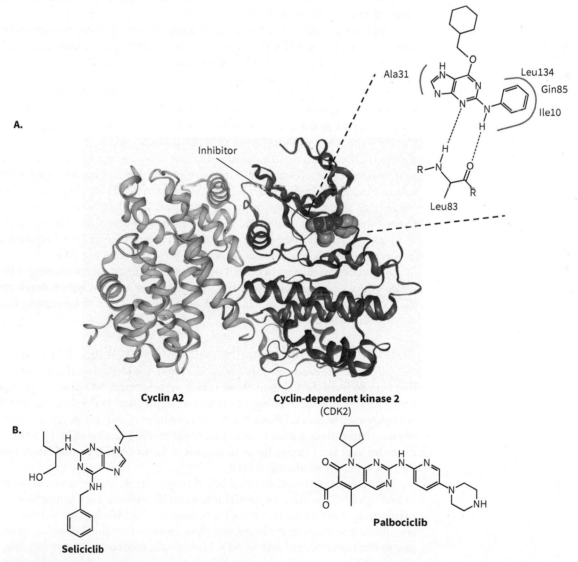

FIGURE 21.18 Structure of CDK2 bound to cyclin A2 and an inhibitor.   **A.** CDKs are activated by binding to their respective cyclin. An inhibitor is shown bound to the kinase. **B.** Several inhibitors of CDKs are used as drugs. Shown are Seliciclib (roscovitine) and Palbociclib (Ibrance), CDK inhibitors used in experimental treatments or treatment of some cancers.

(Source: Data from PDB ID 1H1Q Davies, T.G., Bentley, J., Arris, C.E., Boyle, F.T., Curtin, N.J., Endicott, J.A., Gibson, A.E., Golding, B.T., Griffin, R.J., Hardcastle, I.R., Jewsbury, P., Johnson, L.N., Mesguiche, V., Newell, D.R., Noble, M.E.M., Tucker, J.A., Wang, L., Whitfield, H.J. (2002) Structure-Based Design of a Potent Purine-Based Cyclin-Dependent Kinase Inhibitor. *Nat.Struct.Mol.Biol.* **9**: 745)

late 1960s on cAMP led to studies of the effects of other cyclic nucleotides, including cGMP, and thus to the identification and characterization of the enzymes involved in cGMP synthesis. This led to the key and fortuitous discovery that sodium azide, a poison and common laboratory biocide, stimulates production of cGMP. Following this discovery, scientists re-examined several drugs containing nitro groups and several other compounds (nitroglycerine and sodium nitroprusside) that elicited production of cGMP. Each of these molecules contained labile nitrogen oxide (NO). The final piece of the puzzle was that NO, a component of cigarette smoke, elicited the same response.

NO is a gaseous signaling molecule. It should not be confused with nitrous oxide (N_2O, used as laughing gas) or nitrogen dioxide (NO_2, a poisonous by-product of combustion and a component of pollution). NO is produced through the reduction of arginine, catalyzed by NO synthase (NOS).

There are three known isoforms of NOS in mammals. They vary in whether they are found in the endothelium or in neurons and whether they are constitutively or inducibly expressed. This section focuses on the endothelial form of the enzyme, which regulates vasodilation.

Endothelial NOS (eNOS) is a homodimer (**Figure 21.19**). Each subunit has two major domains. One domain catalyzes the oxidation of the guanidino group of arginine to form NO, whereas the other provides the electrons for the reaction. Five different electron carriers are involved in the synthesis of NO.

The reaction that eNOS catalyzes is unusual in that it involves two successive oxidation reactions, each requiring NADPH and oxygen. Overall, this reaction consumes two moles of O_2 and one and a half moles of NADPH and produces NO and citrulline as products. The actual steps involved are somewhat more complicated.

There are five electron carriers in eNOS: NADPH, FAD, FMN, heme, and tetrahydrobiopterin. These carriers pass electrons from NADPH to heme, where two successive but different oxygen-dependent oxidation reactions occur. Unlike the heme-containing enzymes discussed previously, there is no proximal histidine in eNOS; rather, a proximal cysteine coordinates the fifth position of the iron in the heme group. Addition of an electron and an oxygen to this group results in the formation of an electrophilic iron-oxo species that is often seen in cytochrome P450 mechanisms. This group hydroxylates the arginine to form NOH arginine. Next, another electron and oxygen add to the Fe^{3+}. One of the electrons from NOH arginine is hypothesized to contribute to the formation of this complex. It is thought that, rather than cleaving to give water and the complex found in the previous step, the peroxo form attacks the NOH arginine directly, resulting in a tetrahedral intermediate that collapses to form citrulline and NO.

NO elicits production of cGMP, which is an important second messenger in vasodilation and is also critical in vision, olfaction, and some aspects of nervous system development. The role of cGMP is to activate PKG, which, as with other kinases discussed here, phosphorylates numerous downstream targets.

An example of nitric oxide and PKG signaling

NO is soluble and can move freely through the plasma membrane of cells. From the blood, it diffuses into smooth muscle cells that help to form the lining of blood vessels. In these cells, NO binds to guanylate cyclase, activating it. This enzyme is analogous in function to the adenylate cyclase seen in PKA signaling, but guanylate cyclase uses GTP as substrate and produces cGMP as a product. There are two isoforms of guanylate cyclase: a soluble form that binds to and is activated by NO and a transmembrane form that acts as a receptor for atrial natriuretic factor (ANF), a vasodilatory hormone. Both isoforms respond by producing cGMP.

Soluble guanylate cyclases bind NO through a heme moiety, much as oxygen binds to heme in hemoglobin, but there are several important differences. Like hemoglobin, the heme in guanylate cyclase has a proximal histidine (**Figure 21.20**). Binding of NO forms a temporary six-coordinate iron atom in the heme, with four coordination sites belonging to the porphyrin ring, one to the histidine, and one to NO. However, in contrast to oxygen binding, NO generates a stronger NO–Fe bond, breaking the bond between iron and histidine and leaving a five-coordinate iron atom. When histidine is not bound to the iron in the heme group, it moves away, resulting in enzyme activation.

Next, cGMP activates PKG (**Figure 21.21**). Upon binding of cGMP, the conformation of the heterodimer changes, resulting in activation of the catalytic subunit. However, the catalytic

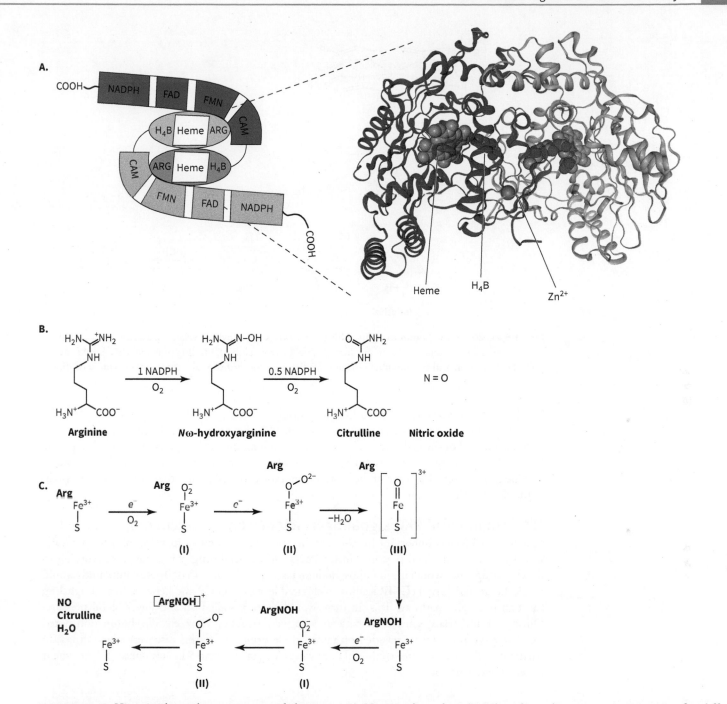

FIGURE 21.19 Nitric oxide synthase structure and chemistry.   **A.** Nitric oxide synthase (NOS) is a homodimeric enzyme containing five differ-ent electron carriers. The final oxidation of arginine occurs at the active site heme group. **B.** NOS catalyzes two successive oxidations of arginine to form nitric oxide and citrulline. **C.** The mechanism of eNOS is thought to be similar to those seen in cytochrome P450 mechanisms. The heme–oxygen complex hydroxylates the arginine to form NOH arginine. Next, another electron and oxygen add to the Fe^{3+}. Finally, the peroxo form of the complex attacks the NOH arginine directly, resulting in a tetrahedral intermediate that collapses to form citrulline and nitric oxide.

(Source: Data from PDB ID 1NSE Raman, C.S., Li, H., Martasek, P., Kral, V., Masters, B.S., Poulos, T.L. (1998) Crystal structure of constitutive endothelial nitric oxide syn-thase: a paradigm for pterin function involving a novel metal center. *Cell* **95**: 939–950)

subunit remains bound rather than dissociating as it does with kinases such as PKA. In smooth muscle cells, PKG phosphorylates numerous targets. It probably elicits its relaxing effect by phos-phorylating and acting on proteins to lower muscle cell Ca^{2+} levels. This can have two direct effects: blocking muscle cell contraction at the level of myosin and stimulating calcium-sensitive potassium channels that regulate muscle cell contraction by hyperpolarizing the cell.

Cross-talk between the NO/PKG pathway is not as apparent as cross-talk in other pathways. The only known "receptor" for NO is guanylate cyclase. Although there are membrane-bound

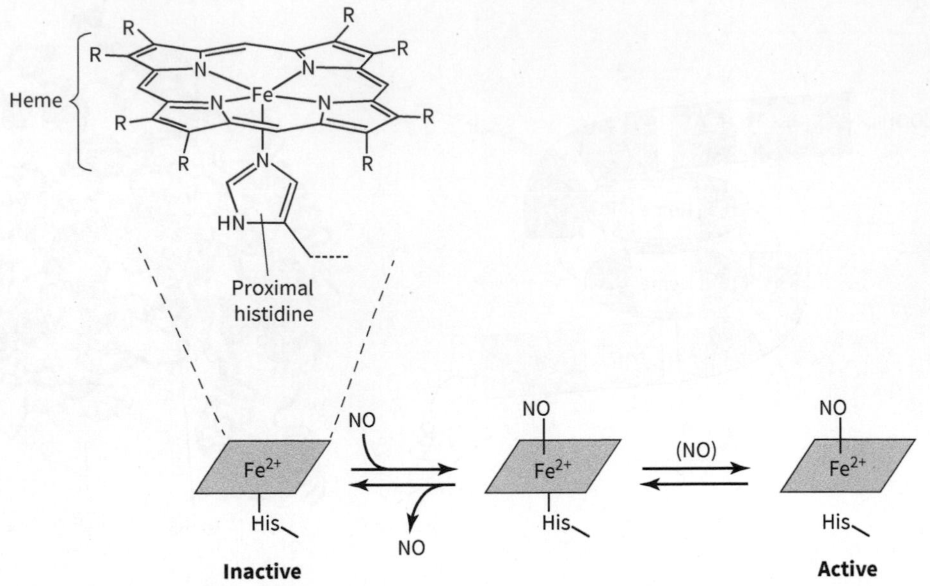

FIGURE 21.20 Heme bound to NO. Nitric oxide binds to guanylate cyclase via a heme group. Unlike some other heme proteins (for example, hemoglobin and myoglobin), binding of nitric oxide breaks the bond between iron and the proximal histidine, leading to a conformational change and enzyme activation.

isoforms of this enzyme, these too have a relatively narrow specificity for ligands such as ANF. Both isoforms of guanylate cyclase produce cGMP, the only molecule that activates PKG. External control of the pathway can be exerted through modulation of phosphodiesterases specific for cGMP. This may include insulin signaling to modulate the expression of phosphodiesterases. It is likely, however, that some of the downstream targets of PKG and other kinases can cross-talk at that level.

Nitric oxide and PKG signaling in science and medicine
As discussed, NO signaling to PKG is important in vasodilation, the relaxation and expansion of arteries associated with lowering blood pressure. In the late 1990s, several drug companies were working to develop drugs that would inhibit phosphodiesterases specific for cGMP. These compounds would block the breakdown of cGMP, leading to elevated levels of cGMP and hence increased signaling through the PKG pathway. This, in turn, would cause vasodilation and lower blood pressure. During clinical trials, some male patients who were suffering from erectile dysfunction noted that their symptoms were reversed on taking these drugs. Since this discovery, these drugs (in particular, sildenafil citrate, commonly known as Viagra) became a hugely successful treatment for erectile dysfunction.

Worked Problem 21.3   Taking signaling to heart

Many people with the heart condition known as unstable angina take either nitroglycerine or nitrates for heart pain caused by constricted blood vessels. Propose a mechanism for how these drugs function.

Strategy   We have studied several signaling pathways; one of these has mentioned vasodilation. What can you determine from the structures of the drugs discussed?

Solution   Both nitroglycerine and other drugs known as nitrosovasodilators act by generating NO, which in turn activates guanylate cyclase in the vasculature and thus activates PKG. The activated PKG acts through a calcium channel to dilate the muscle layer and lower blood pressure.

Follow-up question   One side effect of phosphodiesterase inhibitors such as Viagra is a change in vision, especially with regard to color perception. In some people, images appear to have a blue tint. What does this tell us about how color is perceived and how vision works?

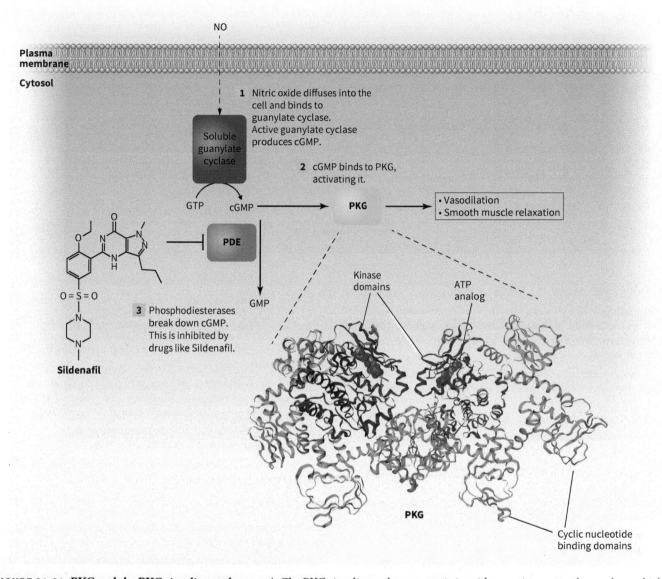

FIGURE 21.21 PKG and the PKG signaling pathway.   A. The PKG signaling pathway uses nitric oxide to activate guanylate cyclase, which makes cGMP, which in turn activates PKG. Phosphodiesterases break down cGMP. Inhibition of this enzyme by drugs like Sildenafil leads to elevated levels of cGMP and activation of PKG. PKG has both a kinase domain and nucleotide binding domains.

(Source: Data from PDB ID 5DZC El Bakkouri, M., Kouidmi, I., Wernimont, A.K., Amani, M., Hutchinson, A., Loppnau, P., Kim, J.J., Flueck, C., Walker, J.R., Seitova, A., Senisterra, G., Kakihara, Y., Kim, C., Blackman, M.J., Calmettes, C., Baker, D.A., Hui, R. (2019) Structures of the cGMP-dependent protein kinase in malaria parasites reveal a unique structural relay mechanism for activation. *Proc. Natl. Acad. Sci. USA* **116**: 14164–14173)

Summary

- In the JAK-STAT pathway, a receptor activates JAK, which phosphorylates the transcription factor STAT, leading to dimerization, translocation to the nucleus, and initiation of gene expression.

- The MAP kinase pathway is actually three multilayered kinase cascades that regulate transcription by phosphorylating transcription factors. Numerous pathways converge on MAP kinase.

- The TLRs were first identified in development but are now accepted to play roles in immune system regulation.

- PKC is a calcium-dependent kinase that responds to tumor promotors such as phorbol esters. It regulates various processes, including the arrest of cell growth and apoptosis.

- CDKs are kinases that become active when they bind to a cyclin, a small protein that triggers the progression of the cell cycle. CDKs phosphorylate several targets, which activate the next step of the cell cycle.

- NO is a gaseous signaling molecule that is synthesized from arginine. NO binds to guanylate cyclase, activating it and producing cGMP, which in turn activates PKG. PKG then phosphorylates multiple targets, including calcium channels, which results in vasodilation.

Concept Check

1. Compare and contrast the six different signaling pathways described in this section.
2. Describe the basic roles of each pathway and the implications it has for biochemistry.
3. Sketch out each of these pathways. Indicate how these pathways could cross-talk with one another.

Bioinformatics Exercises

Exercise 1 G Protein-Coupled Receptors and Receptor Tyrosine Kinases

Exercise 2 Biosignaling and the KEGG Database

Problems

21.1 A Review of Signal Transduction

1. *Dictyostelium discoideum* (dicty) is a colonial amoeba used as a model for development in the laboratory. What are the signaling needs of dicty? What is the dicty life cycle like, and why is it a good model for studying cell signaling and development?

2. What is the order of events to afford cellular change with beta-adrenergic response? What are the likely cellular outcomes of catecholamine binding to this receptor?

3. What are the transmembrane domains of 7-transmembrane G protein-coupled receptors like? Describe them in chemical detail.

21.2 An Overview of Regulation of Signal Transduction

4. Are SH2 domains phosphorylated? Explain.

5. What are the advantages of protein phosphorylation in regard to biochemical signaling?

21.3 Six New Signal Transduction Pathways

6. There are actually seven STAT proteins in mammals, each coded for by a different but highly related gene. *Dictyostelium discoideum* also has seven different STATs coded for by different genes. What does this tell us about the evolution of this pathway?

7. STAT knockout mice are immune compromised. Propose a reason why this might be.

8. Two mammalian cell lines were found to express either epidermal growth factor receptor (EGFR) alone (cell line A) or both EGFR and Ras (cell line B). These cell lines were treated with epidermal growth factor (EGF) and protein phosphorylation was examined in the membrane and cytosolic fractions using anti-phosphotyrosine and anti-phosphoserine antibodies. Where does the EGF-dependent tyrosine phosphorylation detect?

9. How could you tell if the receptors found in the JAK-STAT pathway were similar either structurally or functionally to growth factor receptors?

10. What is common to both the insulin receptor and JAK/STAT receptor?

11. STATs bind DNA as a dimer. Why is this advantageous? What might this suggest about the STAT binding site in DNA?

12. Compare and contrast the PKG and PKA signaling pathways.

13. How is SOCS similar to other regulatory mechanisms we have seen, for example, the PKA pathway? How is it different from acute regulation of some pathways?

14. Illustrate how signaling through a cytokine receptor could activate MAP kinase signaling.

15. MAP kinases are generally categorized as protein kinases in that they are capable of phosphorylating serine, threonine, or tyrosine residues. How might the mechanisms for these three reactions differ?

16. What is farnesylation? Why is it important, and how might this impact Ras signaling?

17. Based on your knowledge of toll-like receptors, why might infection during pregnancy lead to birth defects?

18. Toll-like receptors have been implicated in allergies to nickel metal. Based on your knowledge of amino acid chemistry and how nickel might interact with amino acids, propose a molecular basis for this reaction based on function of the TLR pathway.

19. Phosphorylation of tyrosine residues in the G loop of CDKs reduces the affinity of the CDK for its substrates. How would this be reflected in the kinetics of the enzyme?

20. How does the phorbol esters stimulate the phosphoinositol pathway that leads to tumor promotion?

21. PDK1 and Akt are kinases that we did not discuss in depth in this chapter. Research how these kinases interact with the kinases in the pathways we did discuss. Illustrate a pathway that shows the interaction of PDK1 and Akt with the other pathways.

22. Unlike MAP kinases, which have significant cross-talk with each other and can be activated by other pathways, the CDKs appear to function somewhat in isolation from other signaling cascades. Explain why this may be the case.

23. Describe the activation of a G protein? And also, conditions in which the state of the G protein is unactivated?

24. Why might sulfur be important in the mechanism of NO synthase compared to nitrogen or any another atom? Why is No freely soluble through plasma membrane?

25. Classify each kinase pathway discussed in this section based on whether it uses

   a. Second messengers, a kinase cascade, or both.

   b. A separate inhibitory subunit or a pseudosubstrate domain.

Experimental Design

26. Design an experiment to show that SOCS binds to JAK and not STAT.

27. How could you demonstrate that the interaction of JAK and SOCS is phosphorylation dependent?

Ethics and Social Responsibility

28. Gleevec is a successful anti-leukemic that works by inhibiting the MAP kinase pathway. It can improve life expectancy from nine months to upward of five years. However, it is very expensive, costing approximately $10,000 per month, which is one-fourth of median annual income in the country. Is it ethical to sell a drug for tis much money?

29. Do you think it is ethical to use animals in research or education? Why or why not?

Suggested Readings

21.1 A Review of Signal Transduction

Dennis, E. A., S. G. Rhee, M. M. Billah, and Y. A. Hannun. "Role of Phospholipase in Generating Lipid Second Messengers in Signal Transduction." *FASEB Journal* 5, no. 7 (1991): 2068–77.

Gomperts, B. D., I. M. Kramer, and P. E. R. Tatham. *Signal Transduction*. San Diego, CA: Academic Press, 2003.

Milligan, G., ed. *Signal Transduction: A Practical Approach*. New York: Oxford University Press, 1999.

Schönbrunner, N., J. Cooper, and G. Krauss. *Biochemistry of Signal Transduction and Regulation*, translated by G. Weinheim. New York: Wiley-VCH, 2000.

Sitaramayya, A., ed. *Introduction to Cellular Signal Transduction*. Boston, MA: Birkhauser, 1999.

21.2 An Overview of Regulation of Signal Transduction

Cantley, L. C., T. Hunter, R. Sever, and J. Thorner, eds. *Signal Transduction: Principles, Pathways, and Processes*. Cold Spring Harbor, NY: Cold Spring Harbor Laboratory Press, 2014.

Gomperts, B. D., I. M. Kramer, and P. E. R. Tatham. *Signal Transduction*. San Diego, CA: Academic Press, 2003.

Milligan, G., ed. *Signal Transduction: A Practical Approach*. New York: Oxford University Press, 1999.

Pawson, T. "Protein Modules and Signaling Networks." *Nature* 373, no. 6515 (1995): 573–80.

21.3 Six New Signal Transduction Pathways

Akira, S., K. Takeda, and T. Kaisho. "Toll-Like Receptors: Critical Proteins Linking Innate and Acquired Immunity." *Nature Immunology* 2 (2001): 675–80.

Alexander, W. S., and D. J. Hilton. "The Role of Suppressors of Cytokine Signaling (SOCS) Proteins in Regulation of the Immune Response." *Annual Review of Immunology* 22 (2004): 503–29. doi: 10.1146/annurev.immunol.22.091003.090312.

Bonifacino, J. S., E. C. Dell'Angelica, and T. A. Springer. "Immunoprecipitation." *Current Protocols in Molecular Biology* (2001): 10.16.1–29.

Burkhard, K., and P. Shapiro. "Use of Inhibitors in the Study of MAP Kinases." *Methods in Molecular Biology* 661 (2010): 107–22.

Darnell, J. E., I. M. Kerr, and G. R. Stark. "JAK-STAT Pathways and Transcriptional Activation in Response to IFNs and Other Extracellular Signaling Proteins." *Science* 264, no. 5164 (June 3, 1994): 1415–21.

Fruhbeck, G. "Intracellular Signaling Pathways Activated by Leptin." *Biochemical Journal* 393, Part 1 (2006): 7–20. doi: 10.1042/BJ20051578.

Hecker, E. "Cocarcinogenic Principles from the Seed Oil of *Croton tiglium* and from Other Euphorbiaceae." *Cancer Research* 28 (1968): 2338–49.

Hennessy, E. J., A. E. Parker, and L. A. O'Neill. "Targeting Toll-Like Receptors: Emerging Therapeutics?" *Nature Reviews Drug Discovery* 9, no. 4 (2010): 293–307.

John, P. C., M. Mews, and R. Moore. "Cyclin/Cdk Complexes: Their Involvement in Cell Cycle Progression and Mitotic Division." *Protoplasma* 216, no. 3–4 (2001): 119–42.

Johnson, G. L., and R. Lapadat. "Mitogen-Activated Protein Kinase Pathways Mediated by ERK, JNK, and p38 Protein Kinases." *Science* 298, no. 5600 (2002): 1911–12.

Krishnan, J., G. Lee, and S. Choi. "Drugs Targeting Toll-Like Receptors." *Archives of Pharmacal Research* 32, no. 11 (2009): 1485–502.

Lee, M., and P. Nurse. "Complementation Used to Clone a Human Homologue of the Fission Yeast Cell Cycle Control Gene *cdc2*." *Nature* 327 (1987): 31–35.

Malumbres, M. "Cyclin-Dependent Kinases." *Genome Biology* 15 (2014): 122.

Manning, G., D. B. Whyte, R. Martinez, T. Hunter, and S. Sudarsanam. "The Protein Kinase Complement of the Human Genome." *Science* 298, no. 5600 (2002): 1912–34.

Marletta, M. A. "Nitric Oxide Synthase Structure and Mechanism." *Journal of Biological Chemistry* 268, no. 17 (1993): 12231–4.

Mellor, H., and P. J. Parker. "The Extended Protein Kinase C Superfamily." *Biochemical Journal* 332, Pt. 2 (1998): 281–92.

Mesa, R. A. "Tipifarnib: Farnesyl Transferase Inhibition at a Crossroads." *Expert Review of Anticancer Therapy* 6 (2006): 313–9.

Morgan, D. O. "Cyclin-Dependent Kinases: Engines, Clocks, and Microprocessors." *Annual Review of Cell and Developmental Biology* 13 (1997): 261–91.

Morgan, D. O. "Principles of CDK Regulation." *Nature* 374 (1995): 131–33.

Morgan, D. O. *The Cell Cycle: Principles of Control*. London: New Science Press, 2007.

Nathan, C., and Q.-W. Xie. "Nitric Oxide Synthases: Roles, Tolls, and Controls." *Cell* 78, no. 6 (1994): 915–8.

Nishizuka, Y. "The Role of Protein Kinase C in Cell Surface Signal Transduction and Tumor Promotion." *Nature* 308 (April 19, 1983): 693–8. doi:10.1038/308693a0.

Richard, A. J., and J. M. Stephens. "The Role of JAK-STAT Signaling in Adipose Tissue Function." *Biochimica et Biophysica Acta* 1842, no. 3 (March 2014): 431–9.

Schindler, C., D. E. Levy, and T. Decker. "JAK-STAT Signaling: From Interferons to Cytokines." *Journal of Biological Chemistry* 282, no. 28 (2007): 20059–63. doi: 10.1074/jbc.R700016200.

Schmidt, M., B. Raghavan, V. Müller, T. Vogl, G. Fejer, S. Tchaptchet, S. Keck, C. Kalis, P. J. Nielsen, C. Galanos, J. Roth, A. Skerra, S. F. Martin, M. A. Freidenberg, and M. Goebeler. "Crucial Role for Human Toll-Like Receptor 4 in the Development of Contact Allergy to Nickel." *Nature Immunology* 11, no. 9 (2010): 814–9. doi: 10.1038/ni.1919.

Shuai, K., and B. Liu. "Regulation of JAK-STAT Signaling in the Immune System." *Nature Reviews Immunology* 3, no. 11 (2003): 900–11. doi: 10.1038/nri1226.

Shuai, K., G. R. Stark, I. M. Kerr, and J. E. Darnell. "A Single Phosphotyrosine Residue of STAT91 Required for Gene Activation by Interferon-Gamma." *Science* 261, no. 5129 (September 24, 1993): 1744–6.

Shuai, K., A. Ziemiecki, A. F. Wilks, A. G. Harpur, H. B. Sadowski, M. Z. Gilman, and J. E. Darnell. "Polypeptide Signaling to the Nucleus through Tyrosine Phosphorylation of JAK and STAT Proteins." *Nature* 366, no. 6455 (December 9, 1993): 580–3.

Stuehr, D. J. "Mammalian Nitric Oxide Synthases." *Biochimica et Biophysica Acta* 1411, no. 2–3 (May 1999): 217–30.

Takeda, K., and S. Akira. "Toll-Like Receptors in Innate Immunity." *International Immunology* 17, no. 1 (2005): 1–14. doi:10.1093/intimm/dxh186.

Takeda, K., T. Kaisho, and S. Akira. "Toll-Like Receptors." *Annual Review of Immunology* 21 (2003): 335–76.

Wilks, A. F., A. G. Harpur, R. R. Kurban, S. J. Ralph, G. Zürcher, and A. Ziemiecki. "Two Novel Protein-Tyrosine Kinases, Each with a Second Phosphotransferase-Related Catalytic Domain, Define a New Class of Protein Kinase." *Molecular and Cellular Biology* 11, no. 4 (April 1991): 2057–65.

Wong, S., and O. N. Witte. "The BCR-ABL Story: Bench to Bedside and Back." *Annual Review of Immunology* 22 (2004): 247–306.

Yu, H., D. Pardoll, and R. Jove. "STATs in Cancer Inflammation and Immunity: A Leading Role for STAT3." *Nature Reviews Cancer* 9, no. 11 (2009): 978–809.

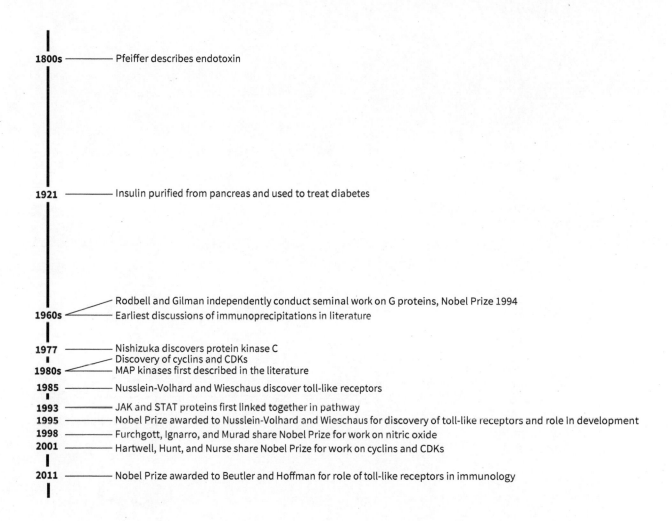

1800s ———— Pfeiffer describes endotoxin

1921 ———— Insulin purified from pancreas and used to treat diabetes

1960s ———— Rodbell and Gilman independently conduct seminal work on G proteins, Nobel Prize 1994
———— Earliest discussions of immunoprecipitations in literature

1977 ———— Nishizuka discovers protein kinase C
———— Discovery of cyclins and CDKs
1980s ———— MAP kinases first described in the literature
1985 ———— Nusslein-Volhard and Wieschaus discover toll-like receptors
1993 ———— JAK and STAT proteins first linked together in pathway
1995 ———— Nobel Prize awarded to Nusslein-Volhard and Wieschaus for discovery of toll-like receptors and role In development
1998 ———— Furchgott, Ignarro, and Murad share Nobel Prize for work on nitric oxide
2001 ———— Hartwell, Hunt, and Nurse share Nobel Prize for work on cyclins and CDKs

2011 ———— Nobel Prize awarded to Beutler and Hoffman for role of toll-like receptors in immunology

CHAPTER 22

Protein Trafficking

Protein Trafficking in Context

Many purchases today are made online. Instead of visiting a store and picking up what we need, we simply click a button, and it is delivered to our door. Think for a minute about how this process occurs and the steps involved. An item made or stored in one place needs to be identified and separated from everything else in the warehouse, packaged up, labeled or sorted, and delivered to whomever has ordered it. This has to happen with a high degree of fidelity, or our packages will end up on someone else's doorstep.

This process is somewhat analogous to the cellular processes ensuring that molecules, especially proteins, are delivered to the organelles where they function. Proteins have signaling peptides or carbohydrate modifications that are recognized by other proteins. Proteins are sorted based on these modifications and packaged into vesicles, which are moved about the cell by molecular motors moving on cytoskeletal proteins. All of these steps ensure that proteins get where they need to go, and if any of them go wrong, disease can result.

This chapter examines three related macromolecular processes: sorting of proteins in the cell (protein trafficking), movement of molecules in the cell (molecular motors), and fusion with the target membrane. Each of these complex systems involves numerous proteins.

Evolution's outcomes are conserved.	• The processes involved in protein trafficking, movement, and secretion provide examples of both convergent and divergent evolution.
	• These processes use proteins that contain several conserved structural motifs or domains encountered previously in other proteins.
Structure determines function.	• In motor proteins, the length of the arm determines the length of the step the protein can take. This correlates with the distance between binding sites for the motor on the microtubule or microfilament.
	• The distinct domains found in many of the proteins discussed here are found in other proteins with similar functions, such as ATPases.
	• Membrane budding and fusion are mediated in part by proteins that bind membranes using transmembrane helices or acyl chains. These groups pull on membranes and cause the bending that is necessary to disrupt the lipid bilayer and initiate fusion.
Biochemical information is transferred, exchanged, and stored.	• The information needed to sort proteins for trafficking is found in the amino acid sequence of the protein.
	• Chemical modifications made to proteins alter how they are trafficked in the cell.
Biomolecules are altered through pathways involving transformations of energy and matter.	• G proteins become active when they have exchanged GDP for GTP.
	• Other proteins use the energy obtained in ATP hydrolysis to move unidirectionally in the cell or to separate the SNARE proteins following vesicular fusion.

22.1 Molecular Aspects of Protein Trafficking

All cells are complex systems, but eukaryotic cells, with their myriad organelles, present a particular challenge. Protein synthesis proceeds either in the cytosol or at the surface of the endoplasmic reticulum (ER), and the newly synthesized proteins then need to be directed to the region of the cell where they will function. The process by which proteins are packaged, are sorted, and find their place in the cell is known as **protein trafficking**. Prior chapters have already touched on interactions related to protein trafficking. Chapter 3 covered the anchoring of proteins in a membrane, an organelle, or a region of the cell through post-translational modifications such as farnesylation or palmitoylation. Chapter 9 explored interactions with phospholipid head groups or lipid rafts.

Although this section focuses on trafficking in eukaryotic cells, bacteria also have systems for sorting proteins into different compartments such as the cytosol, membrane, and cell wall.

22.1.1 Signal sequences direct proteins to specific organelles

The assembly of the ribosome around the mRNA and the first stages of protein synthesis occur in the cytosol. The next stages continue either in the cytosol, in the membrane of the ER, or sometimes in the membrane of other organelles, such as mitochondria.

Proteins that move to a location outside the cytosol have a short peptide sequence called a **signal sequence** that directs the nascent protein to an organelle. **Table 22.1** provides some of these sequences. Specific signal sequences have been discovered for the matrix, the inner and outer

TABLE 22.1 Signal Peptides

Organelle	Signal Peptide
ER retention	KEDL
ER lumen (secretion)	MMSFVSLLLVGILFWATEAEQLTKCEVFQ
Mitochondrial matrix	MLSLRQSIRFFKPATRTLCSSRYLL
Nuclear localization sequence	PKKKRKV
Nuclear exclusion sequence	IDMLIDLGLDLSD

mitochondrial membranes, and the intermembrane space of the mitochondria; the inner and outer membranes of the chloroplast; the membrane and space of the thylakoid and stroma; and the peroxisome. Various proteins facilitate this protein movement; for example, the TOM proteins translocate nascent proteins across the outer mitochondrial membrane. In some cases, more than one signal sequence is necessary to position a protein properly.

In the case of proteins synthesized into the membrane of the ER, the signal peptide of the nascent protein interacts with the signal recognition complex in the ER membrane. Following synthesis, these nascent proteins are embedded in the ER membrane by a single transmembrane alpha helix. **Signal peptidase** is a protease that resides in the ER lumen; it cleaves the signal peptide of some proteins, resulting in loss of the transmembrane domain and formation of a soluble protein within the ER lumen.

Within the ER, proteins fold into their correct conformation. This folding may involve chaperone proteins or protein disulfide isomerase, an enzyme involved in the formation of the correct disulfide bonds in a folding protein. Proteins to be retained in the ER contain a specific four-amino-acid sequence: lysine-aspartate-glutamate-leucine, known by its single-letter abbreviation, KDEL. The KDEL sequence is bound by a receptor and retained within the ER.

Most proteins are then moved on to the Golgi apparatus through vesicular transport.

22.1.2 Many proteins are trafficked using vesicles

Most proteins are packaged into membrane-bound **transport vesicles** for trafficking around the cell. The trafficking of proteins out of the cell is known as **anterograde transport**. In contrast, **retrograde transport** is movement from the Golgi apparatus back to the ER. Several proteins, termed coat proteins, are involved in the formation of these vesicles (**Figure 22.1**). One such protein is clathrin, discussed

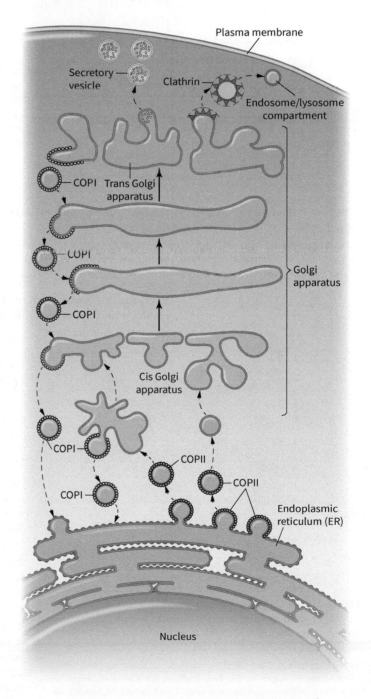

FIGURE 22.1 Major coat proteins. The three major coat proteins found in the cell are clathrin, COPI, and COPII. Clathrin-coated vesicles move among the trans Golgi apparatus, the plasma membrane, and endosomal compartments. COPI-coated vesicles move cargo from the Golgi apparatus to the ER, whereas COPII-coated vesicles move in the reverse direction.

(Source: Karp, *Cell and Molecular Biology: Concepts and Experiments*, 8e, copyright 2016, John Wiley & Sons. This material is reproduced with permission of John Wiley & Sons, Inc.)

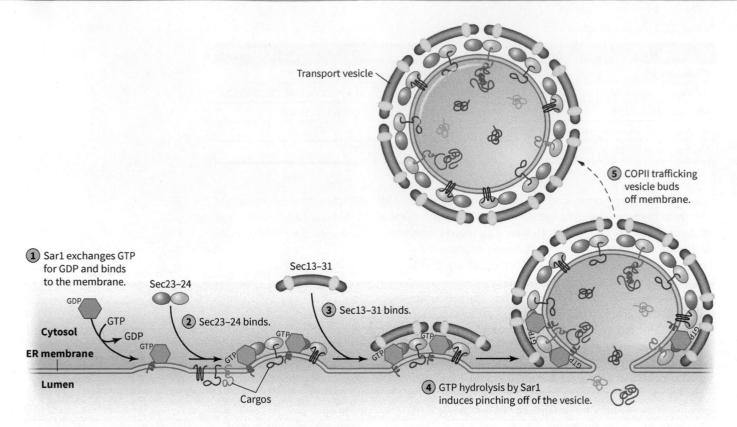

FIGURE 22.2 Assembly of COPII.   The assembly of a complex of COPII and the budding of a vesicle off the ER membrane requires several of the Sec proteins.

(Source: Adapted from D'Arcangelo, J. G., Stahmer, K. R., & Miller, E. A. (2013). Vesicle-mediated export from the ER: COPII coat function and regulation. *Biochimica et Biophysica Acta (BBA) - Molecular Cell Research*, 1833(11), 2464–2472. 2013 Elsevier.)

previously in the context of receptor-mediated endocytosis of lipoproteins. Clathrin is one of the main coat proteins found trafficking among the trans Golgi apparatus, the plasma membrane, and endosomal compartments.

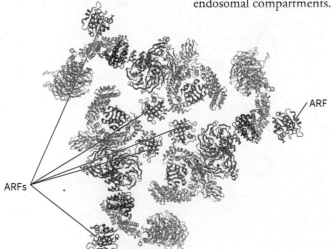

FIGURE 22.3 COPI structure and organization.   COPI is a complex of seven proteins activated by the G protein ARF. In this structure, there is a trimer of ARF molecules surrounded by the six different COPI proteins.

(Source: Data from PDB ID 5A1U Dodonova, S.O., Diestelkoetter-Bachert, P., Von Appen, A., Hagen, W.J.H., Beck, R., Beck, M., Wieland, F., Briggs, J.A.G. (2015) Vesicular Transport. A Structure of the Copi Coat and the Role of Coat Proteins in Membrane Vesicle Assembly. *Science* **349**: 195)

The coat protein complex   The **coat protein complex (COP)** contains two different protein complexes that are important in the movement of cargo vesicles in the cell: COPI and COPII. COPII is involved in anterograde transport and is partially responsible for formation of vesicles bound for the Golgi apparatus from the ER. Some details of the mechanism by which these vesicles move to the Golgi apparatus and fuse have been discovered (**Figure 22.2**). The COPII complex is comprised of two heterodimers: Sec23–Sec24 and Sec13–Sec31. These proteins work in conjunction with Sar1, a small GTPase. Vesicle formation begins with the association of Sar1 with the ER through a hydrophobic tail. Next, COPII proteins begin to bind, forming a larger group and budding a vesicle off the ER membrane. More proteins join to form a larger structure that buds off the ER. This process appears to resemble the pinching of clathrin-coated vesicles from the plasma membrane.

In contrast to COPII, COPI-coated vesicles undergo retrograde transport. The COPI complex is heptameric, but it is divided into a trimeric and a tetrameric substructure and contains conserved structural motifs that are also found in the proteins in the COPII complex. Whereas COPII complexes are recruited to the membrane surface by Sar1, an ADP-ribosylation factor (ARF) recruits and assembles COPI vesicles (**Figure 22.3**). As with the GTPase, the activity of ARF is dependent upon binding to GTP. Specific proteins are important in regulating the function of these GTPases and ARFs: the guanine nucleotide exchange factors (GEFs) and the GTPase activating proteins (GAPs).

As is often the case in biochemistry, we recognize the true importance of these proteins when they are not working properly. When protein trafficking is disrupted, for example because of genetic mutation, it can lead to catastrophic disease, as discussed in **Medical Biochemistry: Protein trafficking disorders**.

Medical Biochemistry

Protein trafficking disorders

It is not surprising that mutations to genes coding for proteins involved in protein trafficking, movement, and vesicular fusion are harmful and potentially disease causing. There are over 30 recognized inherited disorders of trafficking. The underlying etiologies (causes) of these diseases are all now known to result from mutations to specific genes.

Muscular dystrophies are a family of diseases in which there is a loss of muscle mass and weakness due to an absence of muscle proteins. These impairments may be systemic or occur only in specific muscle groups. Other diseases that could also lead to weakness and loss of muscle tissue include neuromuscular disorders, motor neuron disorders, mitochondrial disorders, myopathies (in which the muscle proteins fail to function normally), and ion channel disorders. All told, there have been nearly 80 such diseases identified.

One of these diseases is limb-girdle muscular dystrophy (LGMD), which affects the muscles of the hips, shoulders, and upper arms and legs. Mutations to any of over 33 different genes can cause LGMD. These can be divided into two groups: autosomal dominant and autosomal recessive. Any one of these proteins could have mutations that lead to a loss of function. Dysferlin, a protein involved in membrane repair in muscle cells (sarcolemma repair) has over 300 different mutations that have been identified, each causing disease. Several genes identified in LGMD (including *TRAPPC11* and *PODC1*) encode trafficking proteins.

Although their symptoms can be treated, there are currently no cures for many of these diseases. In the future, gene or stem cell therapy may offer some hope.

Protein Trafficking Disorders

Disorder	Gene	Symptoms
Charcot–Marie–Tooth disease (CMT)	*RAB7*	Neuronal disorder that leads to neuropathy and loss of muscle control, resulting in muscle wasting
Griscelli syndrome (GS) type I	*MYO5A*	Albinism and immunodeficiency leading to death in early childhood
Sensorineural deafness	*MYO7*	Damaged or deficient hair cells in cochlea, leading to deafness
CEDNIK syndrome (cerebral dysgenesis, neuropathy, ichthyosis, and keratoderma)	*SNAP29*	Neuronal and skin disorders, cerebral dysgenesis, and facial dysmorphism
Limb-girdle muscular dystrophy (LGMD)	*DYSF*	Weakness and muscle wasting in the muscles of the arms and legs, especially the shoulders, hips, upper arms, and thighs

Other trafficking proteins: Rabs, GEFs, GAPs, and GDI In addition to COP, other proteins are involved in either formation of vesicles or association of vesicles with motor proteins. Rabs are a family of small monomeric GTPases related to the signaling protein Ras; they bind and act as adapters between motor proteins and the vesicle. Various Rabs are involved in most aspects of both anterograde and retrograde vesicular transport.

As with other G proteins, Rabs must bind GTP to be active. The exchange of GDP for GTP is enhanced by GEFs (guanine nucleotide exchange factors). Conversely, GAPs (GTPase activating proteins) increase the GTPase activity of Rabs and blunt their trafficking functions. Finally, once the Rab has served its purpose and assisted in the delivery of cargo to the plasma membrane, GDP dissociation inhibitor (GDI) binds to the Rab, blocks exchange of GDP for GTP, and assists in the return of the Rab back to the Golgi apparatus. A summary of these trafficking proteins can be found in **Table 22.2**.

TABLE 22.2 Some Trafficking Proteins

Name	Function
ADP-ribosylation factor (ARF)	Guanine nucleotide-binding proteins acting as molecular switches in protein trafficking
Coat protein complex I (COPI)	The group of proteins that participate in retrograde transport from the Golgi apparatus toward the nucleus
Coat protein complex II (COPII)	The group of proteins that participate in anterograde transport from the nucleus to the membrane of the cell; includes the pairs Sec23–Sec24 and Sec13–Sec31
GTPase-activating protein (GAP)	Proteins that increase the GTPase activity of a Rab, slowing trafficking
GDP-dissociation factor (GDI)	Proteins that bind to Rabs and block the exchange of GTP for GDP, inactivating the Rab and causing its return to the Golgi apparatus
Guanine nucleotide-exchange factor (GEF)	Proteins that enhance the exchange of GDP for GTP in a Rab, regulating its activity
Rabs	Small monomeric GTPases that act as adapters between motor proteins and cargo
Sar1	A GTPase that recruits Sec proteins (COPII) to the membrane surface

22.1.3 Proteins are sorted and modified in the Golgi apparatus

Passage through the Golgi apparatus sorts proteins into those that will be secreted, embedded in the plasma membrane, or trafficked to a lysosome. Vesicles of proteins (referred to as **cargo**) arrive from the ER at the cis face of the Golgi apparatus, pass through the cisternae, and move out of the trans Golgi network. As proteins move through the Golgi apparatus, several post-translational modifications can occur, particularly glycosylations.

Different modifications occur in the different compartments of the Golgi apparatus, as discussed in Chapter 8. Generally speaking, mannose residues added in the ER are removed in the cis Golgi apparatus. Further removal of mannose and addition of *N*-acetylglucosamine residues happen in the medial Golgi apparatus, whereas galactose and sialic acid are added in the trans Golgi apparatus. Phosphorylations and sulfations also occur in the Golgi apparatus. These modifications affect where the protein is trafficked.

Addition of mannose-6-phosphate groups to specific proteins in the Golgi apparatus targets proteins to the lysosome. Such proteins are directed into specific regions of the Golgi apparatus through interaction with a mannose-6-phosphate receptor.

Worked Problem 22.1   Bad traffic

Explain how a mutation to an asparagine residue in a protein could result in accumulation of this protein in the Golgi apparatus.

Strategy What is the structure of asparagine? What roles can asparagine serve in protein function?

Solution One role for asparagine is to act as the site of attachment for *N*-linked carbohydrates. If asparagine were mutated to some other amino acid, *N*-linked glycosylation could not take place. Recall that asparagine is the only amino acid that can participate in *N*-linked glycosylation; glutamine cannot. The inability to glycosylate a protein on this residue may prevent it from being properly recognized and sorted, for example, by a mannose-6-phosphate receptor.

Follow-up question You have identified a Rab with a point mutation that is unable to hydrolyze GTP. Hypothesize how this might affect protein trafficking in the cell.

Summary

- Signal peptides within proteins direct those proteins to specific organelles or the ER. Signal peptidase is an enzyme in the ER that cleaves the signal peptide, releasing the protein to the lumen of the ER.
- Proteins transit through the ER to the Golgi on COPII-coated vesicles. The GTPase Sar1 is necessary for vesicle formation. Trafficking back to the ER is accomplished using COPI-coated vesicles. Here, an ARF is necessary for vesicle formation.
- Modifications to proteins, such as the addition of manose-6-phosphate groups, help direct them to different destinations in or out of the cell.

Concept Check

1. Explain how proteins use signal sequences to traffic proteins to different regions of the cell.
2. Describe in general terms how proteins move through the ER and Golgi apparatus and into vesicles.

22.2 Molecular Motor Proteins

Molecules in the cell generally move randomly by diffusion. However, in the same way that people follow the aroma of a freshly baked pie into the kitchen or leave a smoke-filled room where it is hard to breathe, there are times when organisms benefit from deliberate movement toward or away from something. Similarly, individual cells may translocate, or molecules and assemblies of molecules in a cell may move in ways that go beyond diffusion. These motions require energy (e.g., through ATP hydrolysis) and are mediated in part via **molecular motors proteins**.

Several classes of molecules generate motion in the cell. This section focuses on three groups of proteins that transport complexes in the cell along the "tracks" of cytoskeletal proteins, each of which has numerous members with specialized functions: the myosins, kinesins, and dynesins. These three groups share several properties. First, all three travel along a cytoskeletal element—myosin along filaments made of the protein actin and the kinesins and dynesins along microtubules made of the protein tubulin. Second, they all use the energy of ATP hydrolysis to generate motion. Third, they tend to move along the cytoskeleton in a stepwise fashion similar to walking.

Each of these three motor proteins binds to a cytoskeletal element and then catalyzes numerous steps before falling off. This is another example of **processivity**, or the property of some enzymes to bind to a bulk substrate or strata and catalyze numerous reactions before completely dissociating. Other highly processive enzymes include nucleotide polymerases and lipases.

22.2.1 Myosins move along actin microfilaments

Myosins were the first molecular motor to be characterized. These proteins are best known for acting in conjunction with actin in muscle cell contraction, but various members of this family have other functions. This section discusses two particular myosins: myosin II, which is responsible for muscle cell contraction, and myosin V, which is responsible for vesicular transport.

Myosin II (biochemistry of muscle cell contraction) Each molecule of myosin II has a globular head (the ATPase motor domain), a flexible neck, and a long, alpha-helical tail comprising a coiled-coil domain (**Figure 22.4**). Two of these heavy chains dimerize through the coiled-coil domain. Each heavy chain is also associated with two light chains at the neck domain. The light chains are thought to have a regulatory function, although the exact mechanism for this is unclear. In muscle fibers, myosins are grouped together and associate through their coiled-coil tails to form large bundles. These groups of myosins cause muscle cell contraction by pulling on actin fibers.

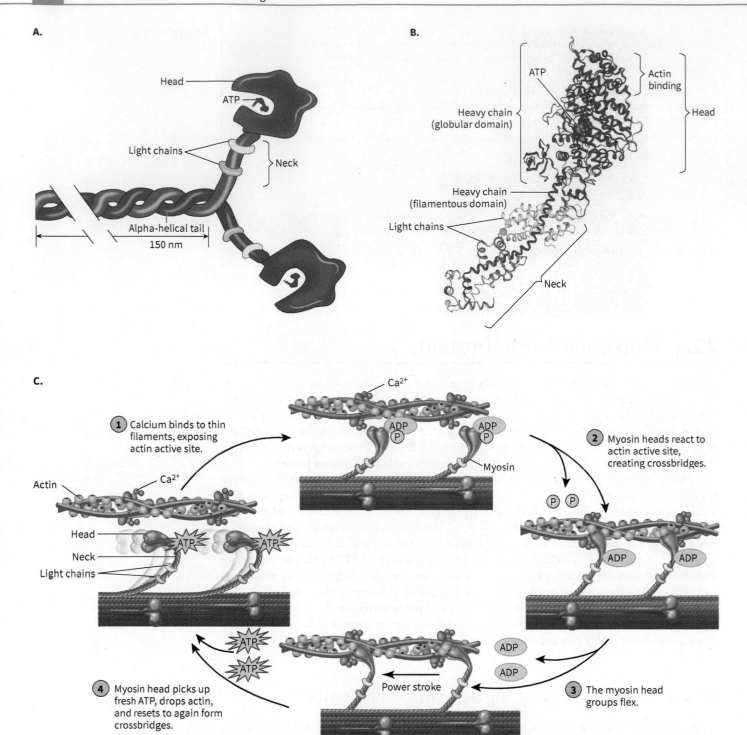

FIGURE 22.4 Myosin II   A. and B. The heavy chain consists of a globular domain containing the actin-binding region, the ATP-binding site, and an extended filamentous domain. Myosin heavy chains bundle together joined by two other subunits (the light chains). **C.** Myosin functions in movement.

(Source: (A) Karp, *Cell and Molecular Biology: Concepts and Experiments,* 7e, copyright 2013, John Wiley & Sons. This material is reproduced with permission of John Wiley & Sons, Inc. (B) Data from PDB ID 1S5G Risal, D., Gourinath, S., Himmel, D.M., Szent-Gyorgyi, A.G., Cohen, C. (2004) Myosin subfragment 1 structures reveal a partially bound nucleotide and a complex salt bridge that helps couple nucleotide and actin binding. *Proc.Natl.Acad.Sci.Usa* **101**: 8930–8935; (C) Ireland, *Visualizing Human Biology,* 4e, copyright 2013, John Wiley & Sons. This material is reproduced with permission of John Wiley & Sons, Inc.)

The power stroke of myosins proceeds as follows. Myosin bound to ADP is associated with actin. Myosin exchanges ADP for ATP and dissociates from the actin filament. Next, myosin hydrolyzes the γ phosphate from ATP. This causes a structural change in the myosin that results in tight binding to actin. The phosphate produced in ATP hydrolysis is then released, causing a second conformational change. This second change rocks the myosin head backward, causing

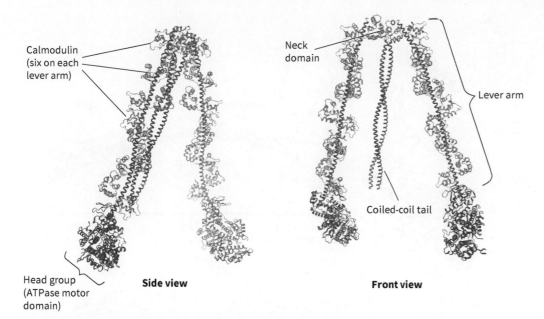

Calmodulin
(six on each
lever arm)

Neck
domain

Lever arm

Coiled-coil tail

Head group
(ATPase motor
domain)

Side view   **Front view**

FIGURE 22.5 Structure of myosin V.   Myosin V is a dimeric molecule consisting of a globular head group, a rigid lever arm, a coiled-coil tail, and a linker group (not shown) that binds to cargo. Because of its long arms, myosin V takes large steps when walking. Myosin V is shown here in the inactive state in which the tail domain that binds to cargo is bent down toward the motor domains instead of extended above.

(Source: Data from PDB ID 2DFS Liu, J., Taylor, D.W., Krementsova, E.B., Trybus, K.M., Taylor, K.A. (2006) Three-dimensional structure of the myosin V inhibited state by cryoelectron tomography. *Nature* **442**: 208–211)

leverage across the tail and contraction. Binding to actin is necessary for myosin to release phosphate and ADP; in the absence of actin, these molecules stay bound in the active site. Likewise, in the absence of ATP, the complex stays bound to actin and fails to enter the dissociated state.

Contractility in muscle fibers is regulated by myocyte calcium levels. Here, the calcium does not act on myosin but rather on troponin C, another actin-binding protein. Troponin C acts in conjunction with tropomyosin to block myosin-binding sites sterically. Binding of calcium to troponin C and allosteric modulation of tropomyosin cause dissociation of the troponin C–tropomyosin complex. This exposes myosin-binding sites on actin and initiates the contraction.

Myosin V (biochemistry of vesicular transport)   Like myosin II, myosin V has a globular head group, flexible neck, and elongated coiled-coil tail (**Figure 22.5**). However, myosin V acts as a dimer rather than a bundle. Instead of pulling a bundle of myosin fibers, the ends of the myosin V tails are bound to a cargo. That cargo may be a group of proteins, a vesicle, or even an organelle such as a lysosome or mitochondria. In addition, there is greater separation of the ATPase motor domains in myosin V than in myosin II. This separation occurs through a rigid lever arm found extending from the neck domain to the coiled coil. Each lever arm is bound by six copies of the protein calmodulin. These proteins stabilize the structure and coordinate the activity of the motor domains. Because of the lengthy arms on myosin V, the protein moves in large steps of 36 nm. By comparison, other myosins move in much smaller steps of 5.5 nm.

22.2.2 Kinesins move along tubulin microtubules

Kinesins move along tubulin microtubules in the same direction as myosins, that is, toward the positive or growing end of the microtubule (**Figure 22.6**). In general, kinesin-mediated transport is anterograde, moving material away from the microtubule organizing center or center of a eukaryotic cell. Kinesins are important in the movement of many different types of vesicles in the cell but also in the separation of chromosomes during mitosis and the destabilization of microtubules following cell division.

Kinesins are similar to myosins in structure. A single tetrameric complex contains two kinesin heavy chains and two light chains. As with myosin, the heavy chains have a globular head

FIGURE 22.6 Structure and walking of kinesin. **A.** Like myosin, kinesin has a globular head group, a neck, a coiled-coil domain, and a tail that binds to cargo. **B.** The steps involved in kinesin walking are shown. Differences in conformation due to ATP binding and hydrolysis lead to differences in affinity for the motor subunits and the β-tubulin binding site on the microtubule.

(Source: (A, Left) Data from PDB ID 2Y5W Kaan, H.Y.K., Hackney, D.D., Kozielski, F. (2011) The Structure of the Kinesin-1 Motor-Tail Complex Reveals the Mechanism of Autoinhibition. *Science* **333**: 883; (A, Right) Karp, *Cell and Molecular Biology: Concepts and Experiments*, 8e, copyright 2016, John Wiley & Sons. This material is reproduced with permission of John Wiley & Sons, Inc. (B) Wessner, *Microbiology*, 2e, copyright 2017, John Wiley & Sons. This material is reproduced with permission of John Wiley & Sons, Inc.)

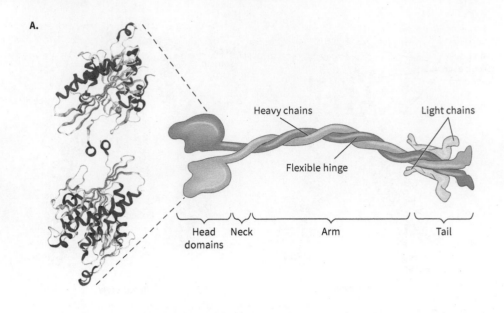

A.

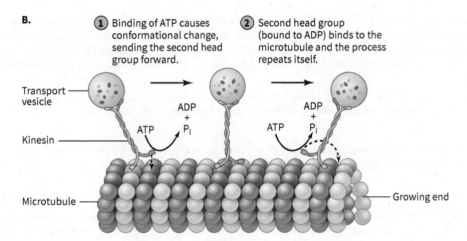

B.

domain, a neck linker sequence, a long arm interrupted by several flexible hinge regions, and a short tail sequence. The head domain is a mixed α/β structure similar to the GTPase Ras. This domain contains the ATPase motor, whereas the arm is an alpha-helical coiled-coil domain. The two heavy chains dimerize through the interactions of the arm. The light chains of kinesin associate with the tails of the heavy chains, but they do not associate with one another. Instead, the light chains act as adapters, interacting between the heavy chains and the cargo. They associate with some Rabs, scaffolding proteins, and transmembrane receptors.

The kinesin light chain plays an important trafficking role in neuronal axons. For example, deletion of this gene in *Drosophila* results in neuronal degradation and death of the embryo. Kinesin light chains have also been shown to interact with proteins implicated in neurodegenerative diseases such as Alzheimer's and Huntington's diseases. The precise role of these interactions in the development and progression of these diseases is not yet clear, but this example illustrates how an understanding of motor proteins can contribute to our knowledge of biological processes in health and disease.

Before binding to a cargo protein or vesicle, kinesin is found in the cytosol in an inactive state bound to ADP. In this state, the arm domain is folded at the hinge domains so that the tail domains of the heavy chains interact with the motor domains. This compact state prevents the protein from binding to cargo, and the ATPase activity of the motor domains is decreased. Extending of the protein is induced by phosphorylation of kinesin light chains by AMP kinase. A similar regulatory mechanism is found in myosin V.

Although their structures are similar, the action of kinesin is different from that of myosin. In the inactive state, the motor domains of kinesin are bound to ADP. When kinesin encounters a microtubule, several things happen. First, the motor domain binds to the microtubule, releasing

ADP and binding to ATP. The binding of ATP changes the conformation of the motor domain, causing the neck to zipper forward. This results in the second motor domain being thrown forward, putting that group in a favorable position to bind to the microtubule, which is about 8 nm away. ATP is hydrolyzed in the first motor domain, releasing phosphate and resulting in a conformational change that releases the neck domain. At this point, the second motor domain exchanges ADP for ATP, and the cycle continues.

Unlike myosin "walking," the steps of kinesins occur around a central pivot point closer to the two motor domains. The movement is therefore more like a penguin waddling than a pair of legs striding.

22.2.3 Dyneins move cargo back to the ER

In contrast to kinesins, **dyneins** move microtubule traffic in the negative direction, back toward the microtubule organizing center (retrograde transport).

Dynein is a large (~1.5 megadalton) multisubunit complex (**Figure 22.7**). The largest subunits are the heavy chains. A dynein complex has two heavy chains, each over 500 kDa. Each of these chains has a small globular domain that interacts with the microtubule, a short arm comprising a coiled-coil domain, and a large heptameric head comprising seven separate domains: six ATPase domains from the AAA gene family and one domain of unknown structure. Joined to the ATPase domain of the heavy chain are another ten intermediate and light subunits, the functions of which are uncertain, although some of these subunits direct associations of dynein with its cargo. Also associated with these subunits is the protein complex **dynactin**, which again acts as an adapter between dynein and its cargo.

The movement of dynein has not been as clearly elucidated as that of myosin and kinesin. The stem region of the dynein complex joins the heavy chain at one of the six AAA domain repeats. Four of these domains bind ATP, but it is thought that the other two domains and the C terminus of the heavy chain do not bind ATP. The arm that binds to microtubules is found between repeats 4 and 5 of the AAA domains. The power stroke in dynein is thought to begin with a conformational change of the microtubule-binding tip and association of this domain with a microtubule. This changes the conformation of the coiled-coil arm, releases ADP from the first AAA domain, and causes the head to rotate and the arm to move by about 15 nm.

Dynein is responsible for the movement of large vesicles and organelles around the cell. For example, it is involved in the movement of chromosomes, Golgi apparatus, ER, lysosomes, and endosomes.

22.2.4 Other proteins also generate motion

In addition to the molecular motors discussed here, other proteins generate motion, some of which have been discussed previously (**Figure 22.8**). One example is the F_0/F_1 ATPase that functions in mitochondria to generate ATP. This protein uses the energy of the proton gradient to rotate as ATP is generated. It can also run in reverse, hydrolyzing ATP to generate rotational motion. Bacteria and some cells with flagella or cilia use analogs of the F_0/F_1 ATPase to generate such motion.

Several proteins that act on DNA result in movement. For example, topoisomerases and helicases unwind the double helix through rotational movement. The hinge-shaped SMC proteins that are partially responsible for the separation of sister chromatids during mitosis generate motion in the opening and closing of the molecule through the hydrolysis of ATP.

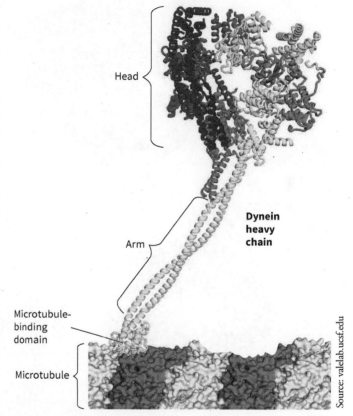

Source: valelab.ucsf.edu

FIGURE 22.7 Structure of dynein. Shown in detail is the head domain, coiled-coil arm, and microtubule-binding domain of dynein. The light chain and stem domain, which attach to cargo, are not shown. The head domain of dynein consists of six repeats of the AAA ATPase and the linker. As ATP is hydrolyzed, the circular unit ratchets forward, moving the arm relative to the stem. Animated models of the walking motion of dynein indicate that it may be a more wobbly motion than myosin or kinesin.

(Source: From DYNEIN-KINESIN COMPARISON. Reproduced with permission from Ronald D. Vale.)

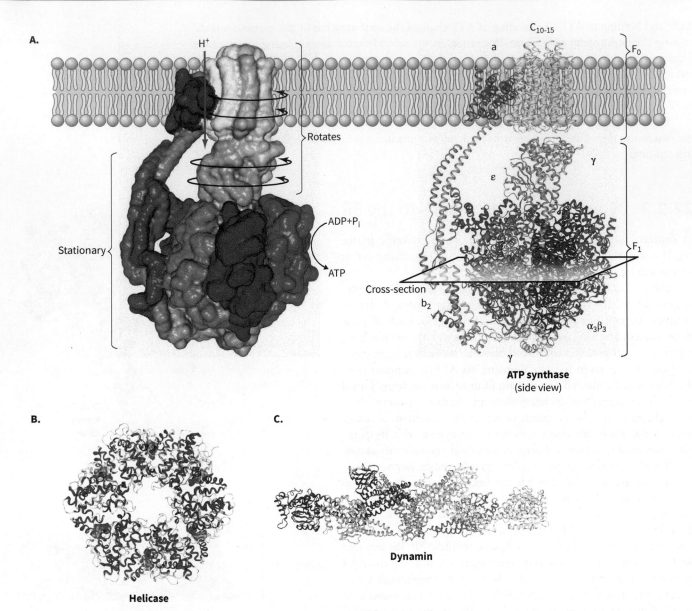

FIGURE 22.8 Other examples of molecular motors. The F_0/F_1 ATPase (**A**), helicase (**B**), and dynamin (**C**) are all examples of molecular motors we have seen previously.

(Source: (A) Data from PDB ID 5ARA Zhou, A., Rohou, A., Schep, D.G., Bason, J.V., Montgomery, M.G., Walker, J.E., Grigorieff, N., Rubinstein, J.L. (2015) Structure and conformational states of the bovine mitochondrial ATP synthase by cryo-EM. *Elife* **4**: e10180-e10180; (B) Data from PDB ID 6QI8 Munoz-Hernandez, H., Pal, M., Rodriguez, C.F., Fernandez-Leiro, R., Prodromou, C., Pearl, L.H., Llorca, O. (2019) Structural mechanism for regulation of the AAA-ATPases RUVBL1-RUVBL2 in the R2TP co-chaperone revealed by cryo-EM. *Sci Adv* **5**: eaaw1616-eaaw1616; (C) Data from PDB ID 5A3F Reubold, T.F., Faelber, K., Plattner, N., Posor, Y., Ketel, K., Curth, U., Schlegel, J., Anand, R., Manstein, D.J., Noe, F., Haucke, V., Daumke, O., Eschenburg, S. (2015) Crystal Structure of the Dynamin Tetramer. *Nature* **525**: 404)

A different kind of molecular motion is generated by the protein **dynamin**, previously discussed in the context of endocytosis. In receptor-mediated endocytosis, a basket of clathrin forms around a patch of clustered receptors. Dynamin is a spiral-shaped protein that assembles around the neck of this structure; it also possesses GTPase activity. As dynamin hydrolyzes GTP, the protein twists, extends, and constricts, a bit like a boa constrictor, pinching off the coated vesicle from the rest of the membrane.

The proteins of the cytoskeleton, notably actin and tubulin, are polar polymers, which means they have two different ends that grow through the addition of subunits to only one end of the growing cytoskeletal element. The growth of these cytoskeletal elements can also generate directional motion, but this is different from the other processes discussed in this section.

Finally, there is the possibility of synthesizing molecules to operate as molecular motors, which could have amazing applications in medicine and engineering (see **Biochemistry: Synthetic molecular motors**).

Biochemistry

Synthetic molecular motors

In his famous address to the American Physical Society in 1959, titled "There's Plenty of Room at the Bottom," Nobel laureate Richard Feynman postulated that significant gains could be made in both basic science and engineering by miniaturizing things. Since that time, massive advances in the miniaturization of computer and electronic components have, in many ways, more than accomplished the challenge Feynman set forth in that address. And we continue to make progress on the miniaturization of devices that use energy (thermal, light, or chemical) to generate motion.

This chapter discusses biological motors such as proteins or assemblies of proteins that use the chemical energy stored in ATP to generate linear motion along a cytoskeletal track. They also include the rotational engines of the F_0/F_1 ATPase in mitochondria that generate ATP using the energy of the proton gradient.

Recent advances in material science and nanotechnology make it possible to synthesize chemically molecules capable of motion. Among these are molecules that undergo *cis-trans* isomerism upon exposure to light, randomly rotate along an axis due to Brownian motion, or are capable of moving along a track. There is even a tiny molecular car driven by a paddle that flaps in response to the light-driven isomerization of an alkene group.

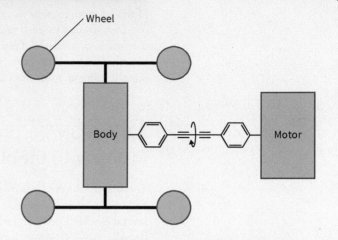

Although these devices may seem whimsical, they represent significant achievements in synthesis and advance our understanding of how molecules operate on the nanoscale. In the future, molecular machines may be used to efficiently generate macroscopic motion or purposefully deliver nanoscale payloads to specific cellular addresses. The possibilities are endless.

Worked Problem 22.2 — Myosin structure function

Examine the structure of myosin V. How do the different features of that molecule contribute to its function? How might changing those features change its function?

Strategy Myosin is a dimer of two subunits, each consisting of a head group responsible for binding actin, a flexible neck domain, and an elongated coiled-coil tail through which the protein dimerizes and cargo is bound (see again Figure 22.5). How might altering the physical properties of this molecule affect its function?

Solution The neck domains of myosin V serve as levers. If they were shortened or if they were floppy instead of rigid, it would have a negative impact on this protein's function.

Follow-up question Based on the structures of the motor proteins found in this section, would you anticipate their genes to be conserved? Why or why not?

Summary

- Motor proteins use the power of ATP hydrolysis to move cargo around the cell on tracks made of actin or tubulin.

- Myosin is a muscle fiber protein but also acts as a molecular motor, using ATP hydrolysis to pull actin filaments along in the anterograde direction.

- Kinesin is similar in structure and function to myosin but uses microtubules made of tubulin.

- Dynein is structurally different from myosin or kinesin and moves vesicles along microtubules in the retrograde direction.

Concept Check

1. Explain the importance of molecular motor proteins to a cell. What are the common properties in the three groups of motor proteins discussed, responsible for this function?

2. Compare the structure and function of myosin II and myosin V.

3. Describe the three classes of molecular motors discussed and how they function.

4. Describe some of the proteins, other than molecular motor proteins, that can generate motion and their mode of action.

22.3 Vesicular Fusion

This section discusses the final step in the movement of a cargo out of a cell: the **vesicular fusion** of the transport vesicle to its target membrane. This could occur via the fusion of a vesicle of neurotransmitter or integral membrane proteins with the plasma membrane. It might also occur via the fusion of a vesicle of enzymes to a lysosome. Vesicular fusion triggers exocytosis of the vesicle contents, either into the target membrane or outside the cell in the form of secretion. To return to our shipping analogy from the chapter opener, this step is the delivery of the package.

22.3.1 SNARE proteins are essential mediators of vesicular fusion with the plasma membrane

The fusion of vesicles is mediated by the **SNARE proteins**. SNARE stands for SNAP receptor, where SNAP is the soluble *N*-ethylmaleimide-sensitive (NSF) attachment protein. Over time, several SNARE proteins have undergone name changes as their functions have been elucidated. Broadly, SNARE proteins can be grouped into two classes: the v-SNAREs, which are bound to vesicles, and the t-SNAREs (target SNAREs), which are bound to the target membrane with which the vesicle will merge. The interaction of the two classes of SNAREs is responsible for vesicular fusion.

Because of the importance of vesicular fusion to the plasma membrane in neurotransmission, the SNARE proteins are some of the most highly characterized proteins, and their roles in neuronal signaling and release of neurotransmitter have been studied extensively. The devastating symptoms of botulism and tetanus result when bacterial neurotoxins affect these SNAREs (see **Societal and Ethical Biochemistry: Botulinum toxin: medicine or poison?**).

Societal and Ethical Biochemistry

Botulinum toxin: medicine or poison?

Among the different types of food poisoning, few are as terrifying as botulism, caused by botulinum toxin produced by the bacteria *Clostridium botulinum*. This anaerobic bacillus or its spores are commonly found in the soil, and they can survive in poorly preserved or prepared food. Once ingested, the bacterium can thrive in the intestine, where it secrets a lethal toxin that causes paralysis. Doses of less than 200 ng of botulinum toxin can kill an adult. Thankfully, there are fewer than 200 cases of botulism per year in the United States, often traced to canned or fermented foods that were improperly prepared at home.

Botulinum toxin causes paralysis by preventing release of acetylcholine at the synapse between a motor neuron and a muscle. The toxin itself consists of a heavy and a light chain, which must be internalized by the neuron. Once inside, the small subunit of the toxin acts as a metalloprotease to cleave specific proteins, notably SNAP-25, blocking vesicular fusion with the plasma membrane and acetylcholine release.

Botulinum toxin poisoning is clearly a serious danger, but the purified toxin has several clinical uses. Anyone who has had a cramp knows how painful a muscle spasm can be. If such spasms are out of control or occur due to some pathology, they can be debilitating. Injections of small doses of botulinum toxin to the muscle inhibit release of acetylcholine and prevent the spasm. Botulinum toxin was originally approved by the FDA in 1989 for the treatment of blepharospasm (nervous ticks of the eye muscles) and strabismus (being cross-eyed). Since then, uses have expanded to include hyperhidrosis (excessive sweating), migraines, and cerebral palsy.

The most common use of botulinum toxin is actually cosmetic. The commercial compound marketed as Botox is most commonly injected into the corners of the eye or mouth to counteract crow's feet and frown lines, wrinkles associated with aging.

22.3.2 The structure of the SNARE proteins serves their function

The SNARE protein family has over 60 members. Although these show some diversity in terms of structure, several conserved domains or features have been identified (**Figure 22.9**). Given the function of the SNARE proteins in mediating vesicle fusion, we could easily hypothesize several of these features. For example, the SNARE proteins need some means to associate with a membrane,

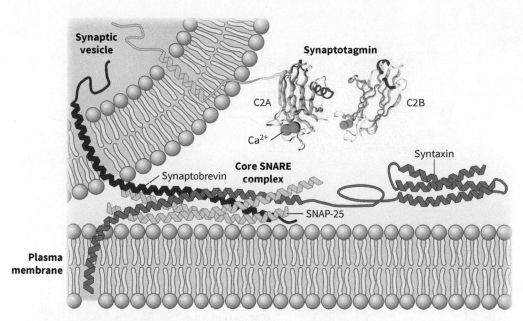

FIGURE 22.9 Structure of the SNARE protein complex.   Shown are the three members of the coiled-coil domain (SNAP-25, synaptobrevin, and syntaxin) and the calcium sensor (synaptotagmin). Note the C2 domains of synaptotagmin, which are responsible for calcium binding.

(Source: Data from PDB ID 6ANK Voleti, R., Tomchick, D.R., Sudhof, T.C., Rizo, J. (2017) Exceptionally tight membrane-binding may explain the key role of the synaptotagmin-7 C2A domain in asynchronous neurotransmitter release. *Proc. Natl. Acad. Sci. U.S.A.* **114**: E8518–E8527)

associate with each other (a v-SNARE with a t-SNARE), and merge the two membranes. Finally, some means of regulating this process is required so that it happens only when needed.

The SNARE proteins are integral membrane proteins, anchored in their C termini. Most members of the SNARE family are anchored via a single transmembrane helix, but almost a third are tethered through palmitoylation of cysteine residues. Palmitoyl groups are 16-carbon-long fatty acids that provide a hydrophobic anchor to the plasma membrane.

This tethering is thought to provide two important features. First, SNAREs that tether through palmitoylation can have this group added or removed as necessary, altering the association of the SNARE with the membrane and thus altering the function. This feature may have a role in synaptic plasticity, or the changes made to the synapse due to stimuli. Second, having a palmitoyl group instead of a transmembrane helix anchors the protein in only one leaflet of the lipid bilayer. This feature may have a role in the fusion of one membrane (the vesicle) with another (the plasma membrane) by only involving one leaflet at a time.

All SNARE proteins also contain a SNARE domain: a 60-amino-acid-long alpha-helical region that participates in the formation of a coiled-coil structure. In turn, that structure is involved in the association of the different SNARE proteins with one another.

22.3.3 Vesicular fusion involves the formation and disassociation of the SNARE complex

This section explores the mechanism of vesicular fusion through the example of synaptic vesicles fusing with the plasma membrane in a neuron. The signal for fusion to occur in this case is an increase of cytosolic Ca^{2+}, triggered by neuronal depolarization (**Figure 22.10**). Four proteins are involved in the formation of the fusion complex: three in the core SNARE complex and one that acts as a calcium sensor. The core complex is formed by the v-SNARE synaptobrevin (also known as VAMP) and the t-SNARE proteins syntaxin and SNAP-25. Synaptotagmin is the Ca^{2+} sensor.

Synaptotagmin contains a single transmembrane domain linked to two domains discussed previously. These are the C2A (calcium-binding) and C2B (PIP3-binding) domains of protein kinase C (PKC). In synaptotagmin, these domains play a similar role. Binding of calcium increases the affinity of synaptotagmin for syntaxin and helps to form the SNARE complex.

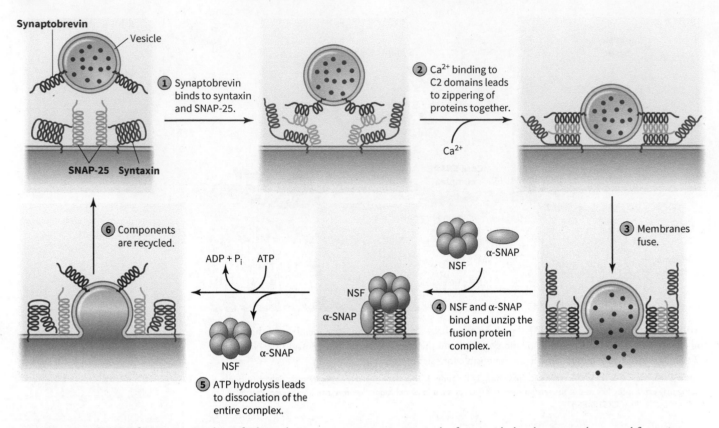

FIGURE 22.10 SNARE function. Binding of calcium by synaptotagmin initiates vesicular fusion with the plasma membrane and formation of the coiled-coil structure, leading to membrane fusion. NSF and its cofactor, α-SNAP, bind to disentangle the SNARE complex, using ATP hydrolysis to fuel the unwinding.

(Source: Josep Rizo & Thomas C. Südhof *Nature Reviews Neuroscience* 3, 641–653 (August 2002). © 2002 Springer Nature. Reproduced with permission from Springer Nature.)

The three SNAREs form a coiled-coil domain parallel to the plane of the plasma membrane. Four helices are involved in this structure: two from SNAP-25 and one each from synaptobrevin and syntaxin. The core of this complex is similar in structure to a leucine zipper, flanked by the interactions of glutamine and arginine residues.

One currently accepted model for the mechanism of fusion is the so-called **zippering** hypothesis. It is based on the idea that the formation of the coiled-coil domain gradually pulls the proteins (and therefore the vesicle and membrane) closer together. As the coiled coil twists, this may apply torque to the transmembrane domains, generating a bridge or pore through which membranes fuse. The physical chemistry of the lipid rearrangement that allows for this fusion is still under debate, but several interesting phenomena have been noted. If SNARE recombinant proteins are inserted into synthetic vesicles, their mere presence is enough to cause vesicular fusion; no added energy (in terms of ATP hydrolysis, for example) is necessary for fusion. From an energetic standpoint, this means that the formation of the SNARE complex from individual proteins is favorable ($\Delta G < 0$).

Once fusion has occurred, the SNARE complex needs to be disassembled. This is accomplished by the AAA ATPase NSF (a hexameric protein first identified as being sensitive to the derivatizing agent *N*-ethylmaleimide) and its cofactor α-SNAP (**Figure 22.11**). Previously in this chapter we discussed the use of an AAA ATPase by the protein dynein to accomplish translocation. NSF uses the energy released in ATP hydrolysis to unzip the SNARE complex and recycle the components. These observations support a hypothesis that the overall driving force for vesicular fusion is the energy released in ATP hydrolysis, which is used to separate the members of the SNARE complex. That is, formation of the energetically less favorable state (individual SNARE proteins) through ATP hydrolysis provides the energy for a subsequent favorable reaction (formation of the SNARE complex and vesicle fusion).

The mechanisms of vesicular fusion are at once fascinating and complex. It is a rapidly evolving field of science and one that is likely to yield important findings in the near future.

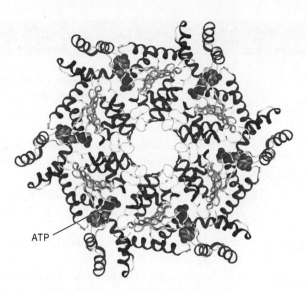

ATP

FIGURE 22.11 Structure of NSF. Like the structure of the motor domain of dynein, NSF is an AAA ATPase. It contains a hexameric repeat of these ATPase structures and uses the energy of ATP hydrolysis to separate the SNARE complex.

(Source: Data from PDB ID 1NSF Yu, R.C., Hanson, P.I., Jahn, R., Brunger, A.T. (1998) Structure of the ATP-dependent oligomerization domain of N-ethylmaleimide sensitive factor complexed with ATP. *Nat. Struct. Mol. Biol.* **5:** 803–811)

Worked Problem 22.3   Rusty nails and chipped beakers

Tetanus toxin works through a mechanism similar to that of botulinum toxin, but it attacks the interneurons that block activation of motor neurons. Tetanus poisoning often occurs when someone has a deep puncture wound as you might get from a nail or chipped beaker. How does botulinum toxin work on the molecular level? Based on this information, describe the symptoms of tetanus toxin poisoning.

Strategy Review how botulinum toxin functions. Tetanus toxin is similar but acts on the neurons that silence motor neurons. How might this manifest itself in a person or animal?

Solution *Clostridium tetani* is the soil bacterium that secretes tetanus toxin. Like botulinum toxin, tetanus toxin has a light chain that acts as a metalloprotease, cleaving specific proteins of the SNARE complex, in this instance, synaptobrevin/VAMP. In the absence of this protein, the interneuron is unable to secrete neurotransmitters and signal to the motor neuron to silence it. Hence, the motor neuron has a higher-than-normal level of activity. The end result of this is increased firing of the motor neuron and uncontrolled muscle contraction, leading to spasm and lockjaw.

Follow-up question Tetanus vaccine is one of the most important vaccines for laboratory workers to have. Why is there an added risk for exposure to tetanus when working in the laboratory?

Summary

- Vesicles fuse with the plasma membrane through the action of the SNARE proteins. Vesicle- or v-SNAREs reside on vesicles, whereas t-SNAREs reside on the target to which the vesicle is fusing, such as the plasma membrane.
- Binding of synaptotagmin by Ca^{2+} initiates vesicular fusion. Syntaxin, synaptobrevin, and SNAP-25 all interact to form a coiled-coil domain comprised of four alpha helices. NSF uses the energy of ATP hydrolysis to unfold the mature SNARE complex.

Concept Check

1. Differentiate the two main classes of SNARE proteins.
2. Identify similarities and patterns in the structure of the SNARE proteins which serves their function.
3. Discuss how the SNARE proteins work together to cause fusion between vesicles and the plasma membrane.

Problems

22.1 Molecular Aspects of Protein Trafficking

1. Protein disulfide isomerase and the KDEL sequence are used as markers in immunofluorescence and cell fractionation studies. Why are these used as markers for ER?

2. How do proteins leave the ER in COPII-coated vesicles?

3. Small monomeric G proteins have been discussed previously. Where and how do they function?

22.2 Molecular Motor Proteins

4. What would the curve of troponin C binding to myosin look like in the presence and absence of calcium?

5. How do kinesin light chains contribute to the kinesin-mediated transport?

6. Compare the protein sequences and structures of myosin and kinesin. What do these data tell us about the evolution of these proteins?

7. What happens in the co-translational translocation of a protein that spans the membrane multiple times?

8. What properties of molecular motor proteins make them processive?

9. What properties of lipases or nucleotide polymerases increase their processivity?

22.3 Vesicular Fusion

10. Several SNAREs are palmitoylated on cysteine groups. How is the palmitate attached in this case, and how are palmitoyl groups usually attached to a protein?

11. Describe the structure of coiled-coil domains and how they function in the motor proteins and SNAREs.

12. Coiled-coil domains contain a heptad repeat, which is a repeat of polar and hydrophobic residues that facilitate the formation of a coiled-coil domain. Based on your knowledge of protein structure, how would this pattern of amino acids help to stabilize the coiled-coil domain?

13. How does a coiled-coil differ from a four-helix bundle which is a common element of tertiary protein structure?

14. What are the likely sources of a reservoir of Ca^{2+} for synaptotagmin to use in vesicle fusion?

15. Draw a simple model of the formation of the coiled-coil domain in the SNARE complex. What are the energetic difficulties that must be overcome for this complex to form?

16. Dissect the thermodynamics of membrane fusion. At what points in the process are there gains or losses in terms of enthalpy, entropy, and free energy?

17. Refute or defend this statement: The SNARE proteins are simply enzymes.

18. Does the ability of the components of the SNARE complex to react together spontaneously affect rate? Explain your answer.

19. Propose a mechanism for the metalloprotease activity of botulinum toxin.

Data Interpretation

20. The structure of the coiled coil found in the SNARE complex is PDB ID# 1SFC. Look up this structure. What weak forces stabilize the coiled coil?

Experimental Design

21. A commonly accepted model of molecular motor proteins is one in which the cytoskeletal binding element moves hand over hand. Earlier models proposed that the complex shuffled along the microtubule or filament. Propose an experiment to test those models.

Ethics and Social Responsibility

22. Most procedures performed with Botox are cosmetic. Are cosmetic procedures ever medically warranted, or do they pose an unnecessary risk? Should health insurance cover cosmetic procedures?

Suggested Readings

22.1 Molecular Aspects of Protein Trafficking

Bean, A. J., ed. *Protein Trafficking in Neurons*. Amsterdam: Elsevier/Academic Press, 2007.

Brandizzi, F., and C. Barlowe. "Organization of the ER–Golgi Interface for Membrane Traffic Control." *Nature Reviews Molecular Cell Biology* 14 (June 2013): 382–92.

De Matteis, A., A. Budnik, and D. J. Stephens, eds. "ER Exit Sites—Localization and Control of COPII Vesicle Formation." *FEBS Letters* 583, no. 23 (December 3, 2009): 3796–803.

Dupré, D. J., T. E. Hebert, and R. Jockers. *GPCR Signalling Complexes: Synthesis, Assembly, Trafficking and Specificity*. Dordrecht: Springer, 2012.

Fielding, C. J., ed. *Lipid Rafts and Caveolae: From Membrane Biophysics to Cell Biology*. Weinheim: Wiley-VCH, 2006.

Henderson, B., and A. G. Pockley, eds. *Cellular Trafficking of Cell Stress Proteins in Health and Disease*. Dordrecht: Springer, 2012.

Jena, B. P. *NanoCellBiology of Secretion: Imaging Its Cellular and Molecular Underpinnings*. New York: Springer, 2012.

Rothblatt, J., P. Novick, and T. H. Stevens, eds. *Guidebook to the Secretory Pathway*. Oxford, UK: Oxford University Press, 1995.

Segev, N., ed., with A. Alonson, J. G. Donaldson, and G. S. Payne. *Trafficking inside Cells: Pathways, Mechanisms, and Regulation*. Austin, TX: Landes Bioscience, 2009.

Stenmark, H., and V. M. Olkkonen. "The Rab GTPase Family." *Genome Biology* 2, no. 5 (2001): reviews3007.1–7. doi:10.1186/gb-2001-2-5-reviews3007.

St. George-Hyslop, P. H., W. C. Mobley, and Y. Christen, eds. *Intracellular Traffic and Neurodegenerative Disorders*. Berlin: Springer, 2009.

Szul, T., and E. Sztul. "COPII and COPI Traffic at the ER-Golgi Interface." *Physiology* 26, no. 5 (2011): 348–64.

Tang, B. L. *Membrane Trafficking* (Methods in Molecular Biology), 2nd ed. New York: Humana Press, 2015.

Tzfira, T., and V. Citivsky, eds. *Nuclear Import and Export in Plants and Animals*. Georgetown, TX: Landes Bioscience, 2005.

Yarden, Y., and G. Tarcic. *Vesicle Trafficking in Cancer*. New York: Springer, 2013.

22.2 Molecular Motor Proteins

Browne, W. R., and B. L. Feringa. "Making Molecular Machines Work." *Nature Nanotechnology* 1 (2006): 25–35.

Burgess, S. A., M. L. Walker, H. Sakakibara, P. J. Knight, and K. Oiwa. "Dynein Structure and Power Stroke." *Nature* 421 (2003): 715–18.

Credi, A., S. Silvi, and M. Venturi, eds. *Molecular Machines and Motors: Recent Advances and Perspectives*. Cham, Switzerland: Springer, 2014.

Gissen, P., and E. R. Maher. "Cargos and Genes: Insights into Vesicular Transport from Inherited Human Disease." *Journal of Medical Genetics* 44 (2007): 545–55. doi:10.1136/jmg.2007.050294.

Goodsell, D. S. *Our Molecular Nature: The Body's Motors, Machines, and Messages*. New York: Copernicus, 1996.

Joachim, C., and G. Rapenne. *Single Molecular Machines and Motors: Proceedings of the 1st International Symposium on Single Molecular Machines*. Cham, Switzerland: Springer Verlag, 2015.

Kolomeisky, A. B. *Motor Proteins and Molecular Motors*. Boca Raton, FL: CRC Press, 2015.

Mehta, A. D., R. S. Rock, M. Rief, J. A. Spudich, M. S. Mooseker, and R. E. Cheney. "Myosin-V Is a Processive Actin-Based Motor." *Nature* 400, no. 6744 (1999): 590–3.

Morin, J.-F., Y. Shirai, and J. M. Tour. "En Route to a Motorized Nanocar." *Organic Letters* 8 (2006): 1713–16.

Roberts, A. J., T. Kon, P. J. Knight, K. Sutoh, and S. A. Burgess. "Functions and Mechanics of Dynein Motor Proteins." *Nature Reviews Molecular Cell Biology* 14 (2013): 713–26.

Schliwa, M., ed. *Molecular Motors*. Weinheim: Wiley-VCH, 2003.

Schliwa, M., and G. Woehlke. "Molecular Motors." *Nature* 422 (2003): 759–65.

Schmidt, H., E. S. Gleave, and A. P. Carter. "Insights into Dynein Motor Domain Function from a 3.3-Å Crystal Structure." *Nature Structural & Molecular Biology* 19 (2012): 492–7.

Setou, M., D.-H. Seog, Y. Tanaka, Y. Kanai, Y. Takei, M. Kawagishi, and N. Hirokawa. "Glutamate-Receptor-Interacting Protein GRIP1 Directly Steers Kinesin to Dendrites." *Nature* 417 (May 2, 2002): 83–87.

Shiroguchi, K., and K. Kinosita. "Myosin V Walks by Lever Action and Brownian Motion." *Science* 316, no. 5838 (May 25, 2007): 1208–12.

Sperry, A. O. *Molecular Motors: Methods and Protocols* (Methods in Molecular Biology). New York: Humana Press, 2007.

Trybus, K. "Myosin V from Head to Tail." *Cellular and Molecular Life Sciences* 65, no. 9 (2008): 1378–89.

Vale, R. D., and R. A. Milligan. "The Way Things Move: Looking under the Hood of Molecular Motor Proteins." *Science* 288 (2000): 88–95.

Verhey, K. J., D. Meyer, R. Deehan, J. Blenis, B. J. Schnapp, T. A. Rapoport, and B. Margolis. "Cargo of Kinesin Identified as Jip Scaffolding Proteins and Associated Signaling Molecules." *Journal of Cell Biology* 152, no. 5 (2001): 959–70.

Youell, J., and K. Firman. *Molecular Motors in Bionanotechnology*. Singapore: Pan Stanford Publishing, 2013.

22.3 Vesicular Fusion

Chapman, E. R. "Synaptotagmin: A Ca²⁺ Sensor That Triggers Exocytosis?" *Nature Reviews Molecular Cell Biology* 3, no. 7 (2002): 498–508.

Fernández-Chacón, R., A. Königstorfer, S. H. Gerber, J. García, M. F. Matos, C. F. Stevens, et al. "Synaptotagmin I Functions as a Calcium Regulator of Release Probability." *Nature* 410, no. 6824 (2001): 41–49.

Linial, M., A. Grasso, and P. Lazarovici, eds. *Secretory Systems and Toxins*. Amsterdam: Harwood Academic Publishers, 1998.

Meriney, S. D., J. A. Umbach, and C. B. Gunderson. "Fast, Ca²⁺-Dependent Exocytosis at Nerve Terminals: Shortcomings of SNARE-Based Models." *Progress in Neurobiology* 121 (October 2014): 55–90.

Riso, J., and T. C. Südhof. "The Membrane Fusion Enigma: SNAREs, Sec1/Munc18 Proteins, and Their Accomplices—Guilty as Charged?" *Annual Review of Cell and Developmental Biology* 28 (2012): 279–308.

Söllner, T., M. K. Bennett, S. W. Whiteheart, R. H. Scheller, and J. E. Rothman. "A Protein Assembly-Disassembly Pathway in Vitro That May Correspond to Sequential Steps of Synaptic Vesicle Docking, Activation, and Fusion." *Cell* 75, no. 3 (19936): 409–18.

Sutton, R. B., D. Fasshauer, R. Jahn, and A. T. Brunger. "Crystal Structure of a SNARE Complex Involved in Synaptic Exocytosis at 2.4 Å Resolution." *Nature* 395, no. 6700 (September 24, 1998): 347–53.

Zimmerberg, J., S. A. Akimov, and V. Frolov. "Synaptotagmin: Fusogenic Role for Calcium Sensor?" *Nature Structural & Molecular Biology* 13, no. 4 (2006): 301–3.

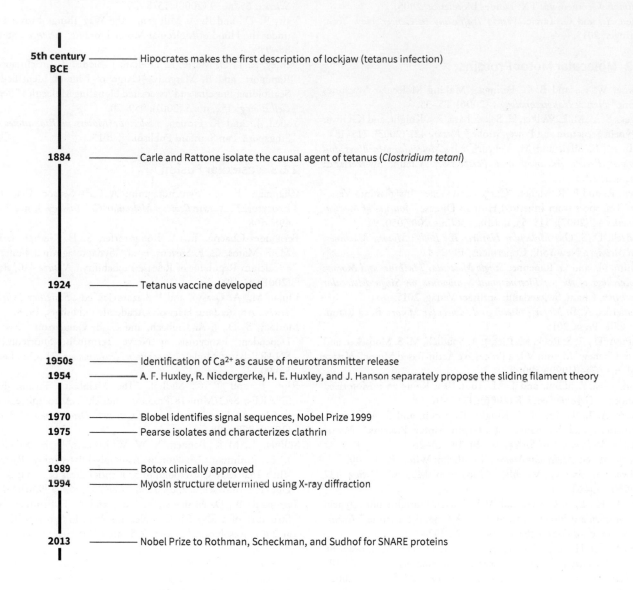

5th century BCE —————— Hipocrates makes the first description of lockjaw (tetanus infection)

1884 —————— Carle and Rattone isolate the causal agent of tetanus (*Clostridium tetani*)

1924 —————— Tetanus vaccine developed

1950s —————— Identification of Ca^{2+} as cause of neurotransmitter release
1954 —————— A. F. Huxley, R. Niedergerke, H. E. Huxley, and J. Hanson separately propose the sliding filament theory

1970 —————— Blobel identifies signal sequences, Nobel Prize 1999
1975 —————— Pearse isolates and characterizes clathrin

1989 —————— Botox clinically approved
1994 —————— Myosin structure determined using X-ray diffraction

2013 —————— Nobel Prize to Rothman, Scheckman, and Sudhof for SNARE proteins

CHAPTER 23

Photosynthesis and Nitrogen Fixation

Photosynthesis and Nitrogen Fixation in Context

Scientifically, we know that mass is conserved in chemical reactions, but we may forget that if we burn tons of coal, the mass of the coal still exists in gaseous form. This chapter is about how the invisible becomes visible when gases are chemically reduced into new products through photosynthesis and nitrogen fixation.

In photosynthesis, plants and cyanobacteria use the energy of the sun to drive carbon fixation, producing carbohydrates from carbon dioxide. In the process, they also make ATP and NADPH and produce gaseous O_2. A waste product of photosynthesis, this oxygen has made our lives, the lives of all other anaerobes, and indeed most life on Earth possible through production of ozone to protect us from the ultraviolet rays of the sun. Likewise, plants are the only way we have of producing carbohydrates because gluconeogenesis requires a partially oxidized precursor and cannot use highly reduced sources of carbon.

In nitrogen fixation, bacteria remove dinitrogen gas (N_2) from the air and reduce it into ammonia (NH_3) for use in other biochemical processes. Again, higher organisms reap the benefits of the work of these microbes. Humans are unable to reduce nitrogen for use in the synthesis of amino acids, nucleotides, and cofactors ourselves, and so we rely on bacteria to do these jobs for us.

Structure, energy transfer, evolution, conservation of themes, motifs, and domains of proteins, and transformation of one molecule into another have served as organizing principles for this book. This chapter provides a big-picture review of many of the topics studied in earlier chapters. Photosynthesis and nitrogen fixation are in a way biochemistry in a microcosm.

Chapter Outline

23.1 Photosynthesis

23.2 Nitrogen Fixation

Common Themes

Evolution's outcomes are conserved.	• The evolution of photosynthesis affected all organisms on the planet.
	• Nitrogen fixation is used by relatively few organisms, but the products of this reaction are essential for all organisms.
	• The pathways for both photosynthesis and nitrogen fixation are ancient but not as old as life itself.
	• Like most other complex and essential pathways, photosynthesis has been refined over time but is highly conserved.
Structure determines function.	• As with the electron transport centers in mitochondria, the proteins involved in photosynthesis use many redox-active metal ions and metal clusters, some of which are unique to this pathway.
	• Many of the functional domains seen in the proteins described here have similar functions in other proteins, for example, tethering to a membrane or protein–protein interactions.
	• The chemical structures of pigment molecules (for example, metal ions or a significant number of conjugated double bonds) allow them to interact with photons of light with energies and wavelengths in the visible range.
Biochemical information is transferred, exchanged, and stored.	• The chloroplast, like the mitochondrion, has a small, compact genome that is used to produce a limited number of proteins. Both mitochondria and chloroplasts demonstrate evidence of endosymbiosis.
Biomolecules are altered through pathways involving transformations of energy and matter.	• Photosynthesis uses solar energy to cleave water molecules into protons, electrons, and oxygen. Electrons excited to higher energy states are then used to generate NADPH and a proton gradient to form ATP.
	• The light-independent reactions of photosynthesis (the Calvin cycle) use some energy from the pathway's light-dependent reactions to fix carbon dioxide into carbohydrates.

23.1 Photosynthesis

Photosynthesis is often one of our first introductions to biochemistry. Most children learn in elementary school that plants use the energy of the sun to make carbohydrates from carbon dioxide (CO_2). They may even learn the overall reaction of photosynthesis:

$$6CO_2 + 6H_2O + light \Rightarrow C_6H_{12}O_6 + 6O_2$$

As students progress in their education, they may learn other aspects of photosynthesis, including the parts of the cell where this occurs and the dependence of this process on different wavelengths of light. They may even know that there are different phases to photosynthesis. This chapter connects some of these molecular dots and draws important parallels to systems discussed previously to create a richer picture of photosynthesis.

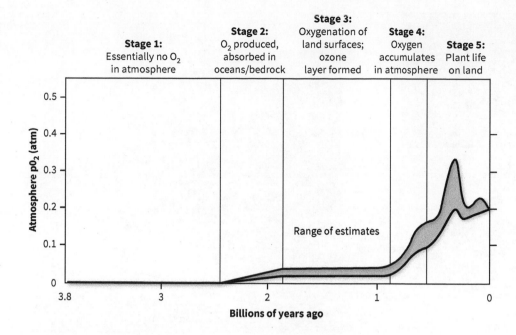

Stage 1:
Essentially no O_2
in atmosphere

Stage 2:
O_2 produced,
absorbed in
oceans/bedrock

Stage 3:
Oxygenation of
land surfaces;
ozone
layer formed

Stage 4:
Oxygen
accumulates
in atmosphere

Stage 5:
Plant life
on land

Range of estimates

FIGURE 23.1 Timeline of Earth's geologic history. Earth is approximately 4.5 billion years old. The first atmosphere formed about 4.3 billion years ago, and the first evidence of life dates from approximately 3.8 billion years ago. Atmospheric oxygen levels did not increase until approximately 2.5 billion years ago.

23.1.1 An overview of photosynthesis reveals several important reactions

Photosynthesis is the process through which plants and other photosynthetic organisms such as cyanobacteria use solar energy to produce ATP and NADPH, reduce CO_2 to glucose, and produce oxygen as a by-product. Although photosynthesis is usually associated with plants, most of today's plants only evolved into their current forms over the past 75 million years (**Figure 23.1**). The dinosaurs came and went before flowering plants evolved, with cyanobacteria and single-celled algae the predominant photosynthetic organisms from 3.5 billion to 2 billion years ago.

Photosynthetic organisms have shaped Earth more than any volcano or asteroid strike. As these ancient photosynthetic organisms thrived, they released oxygen into Earth's atmosphere. The results of this are apparent in the fossil record. Oxygen is highly reactive. When it is not being constantly generated, it quickly (in geological terms) reacts with other elements. During times when oxygen was plentiful, iron ores rapidly oxidized, forming rust-colored bands of minerals. Conversely, when oxygen was less abundant, the iron-rich ores were far more reduced when they were deposited and thus formed greenish-colored bands.

Most of the oxygen on Earth has been generated through photosynthesis. Without plants, algae, and cyanobacteria, Earth would run out of oxygen gas in less than 20,000 years. Although oxygen production benefits aerobes, this is not the main function of photosynthesis. Rather, the process can be broken into two phases: the light-dependent reactions, which use the energy of photons from the sun to split water into molecular oxygen (O_2) and protons to produce ATP and NADPH; and the light-independent reactions, which use some of that ATP and NADPH to fix CO_2, reducing it into glucose. Both sets of reactions occur in a specialized organelle: the chloroplast.

23.1.2 Photosynthesis takes place in the chloroplast

A typical algal cell has a single **chloroplast**, whereas some plants have close to 100. Like the nucleus and mitochondria, chloroplasts are surrounded by a double membrane (**Figure 23.2**). In between the outer and inner membrane of the chloroplast is the intermembrane space. Inside the inner membrane is the **stroma** (analogous to the matrix in mitochondria) and a membranous structure termed the **thylakoid**. The thylakoid membrane delineates the thylakoid space, called the lumen, from the stromal space.

The light-dependent reactions of photosynthesis occur in the thylakoid membrane and lumen, whereas the light-independent reactions occur in the stroma. The chloroplast is in many

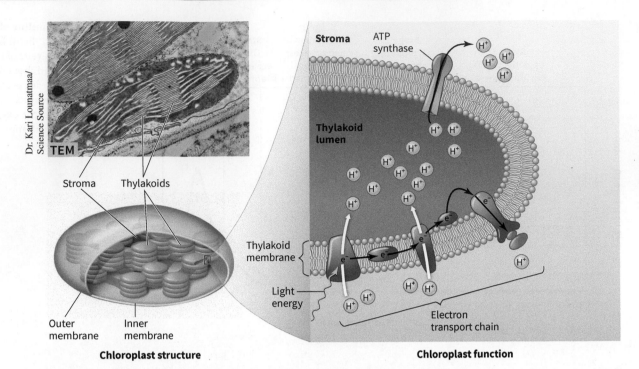

FIGURE 23.2 Structure and function of the chloroplast. The chloroplast is bound by two membranes, and there is a membranous structure called the thylakoid found in its center. The thylakoid membrane is where the light-dependent reactions of photosynthesis occur. The light-independent reactions (Calvin cycle) occur in the stromal space.

(Source: Wessner, *Microbiology*, 2e, copyright 2017, John Wiley & Sons. This material is reproduced with permission of John Wiley & Sons, Inc.)

ways parallel to mitochondria. In both systems, the cell tightly regulates the passage of molecules through these membranes. Both mitochondria and chloroplasts have a small piece of DNA that codes for several proteins. Like mitochondria, thylakoids are used as an example of endosymbiotic evolution, the theory that eukaryotes arose from engulfed prokaryotes.

23.1.3 The light-dependent reactions of photosynthesis generate ATP, NADPH, and oxygen

The light-dependent reactions of photosynthesis use photons of light to mobilize electrons. In the process, O_2 is generated, and the energy captured is used to drive the production of ATP and NADPH (**Figure 23.3**).

Four proteins or complexes are involved in photon capture and electron transport in photosynthesis. Many of these processes are similar in structure and function to those seen in redox enzymes or the electron transport chain. The process begins in **photosystem II** (so named only because it was discovered after photosystem I). This multiprotein complex serves several roles. First, it captures photons of light and uses them to eject an electron to the electron carrier plastoquinone. In the process of transferring electrons, photosystem II splits a molecule of water. Splitting two molecules of water generates one molecule of O_2 and four free protons in the thylakoid lumen. Plastoquinone is a quinone similar in structure and function to coenzyme Q10 (ubiquinone), which is found in the electron transport chain. Plastoquinone shuttles the electrons to the cytochrome b_6f complex, moving two protons from the stromal space to the thylakoid lumen in the process. Electrons are passed through various redox centers in the cytochrome b_6f complex (similar to the electron transport chain in mitochondria). Eventually, the electrons reach the soluble electron carrier plastocyanin, which is analogous to cytochrome C in the electron transport chain. Plastocyanin transfers electrons from the cytochrome b_6f complex to photosystem I; a second photon is captured, and the energy of that capture is used to further excite the electron. Electrons are passed through this complex to ferredoxin and ultimately to ferredoxin NADP+ reductase, which is found on the stromal face of the thylakoid membrane. The energy of the electrons is then used to reduce NADP+ to NADPH. The overall pathway of electron flow

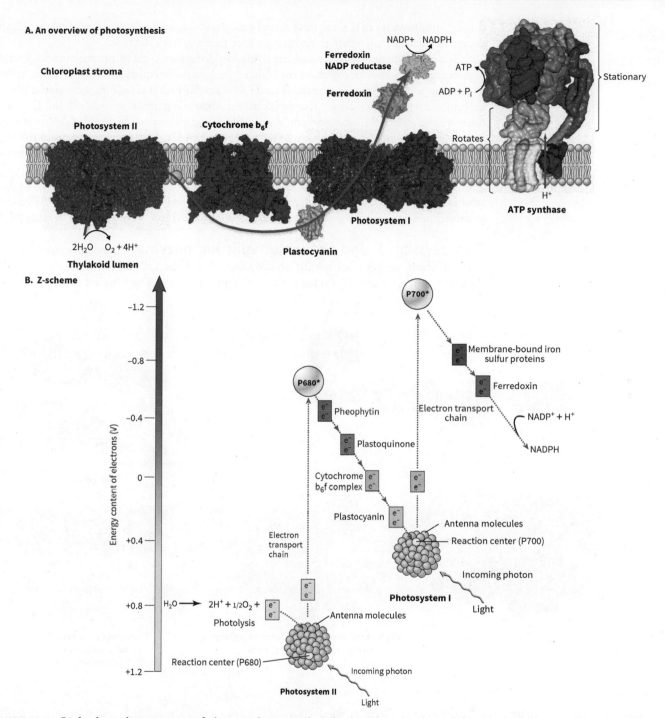

FIGURE 23.3 **Light-dependent reactions of photosynthesis. A.** The light-dependent reactions of photosynthesis begin at photosystem II. Photons excite an electron to a higher energy state. To fill the gap left by the excited electron, water is split into oxygen and protons. The electron is passed through plastoquinone to the cytochrome b_6f complex and then to photosystem II, where it is again excited to a higher energy state. The electron is passed through ferredoxin to ferredoxin reductase, where it is used to produce NADPH. The generated proton gradient is used to power an F_0/F_1 ATPase to make ATP. **B.** This is shown schematically in the Z-scheme.

(Source: A. (Photosystem II) Data from PDB ID 4YUU Ago, H., Adachi, H., Umena, Y., Tashiro, T., Kawakami, K., Kamiya, N., Tian, L., Han, G., Kuang, T., Liu, Z., Wang, F., Zou, H., Enami, I., Miyano, M., Shen, J.-R. (2016) Novel Features of Eukaryotic Photosystem II Revealed by Its Crystal Structure Analysis from a Red Alga *J.Biol.Chem.* **291**: 5676–5687; (Cytochrome b_6f complex) Data from PDB ID 1VF5 Kurisu, G., Zhang, H., Smith, J.L., Cramer, W.A. (2003) Structure of the Cytochrome B6F Complex of Oxygenic Photosynthesis: Tuning the Cavity. *Science* **302**: 1009–1014; (Plastocyanin) Data from PDB ID 1AG6 Xue, Y., Okvist, M., Hansson, O., Young, S. (1998) Crystal structure of spinach plastocyanin at 1.7 A resolution. *Protein Sci.* **7**: 2099–2105; (Photosystem I) Data from PDB ID 1JB0 Jordan, P., Fromme, P., Witt, H.T., Klukas, O., Saenger, W., Krauss, N. (2001) Three-dimensional Structure of Cyanobacterial Photosystem I at 2.5 A Resolution. *NATURE* **411**: 909–917; (Ferredoxin) Data from PDB ID 1A70 Binda, C., Coda, A., Aliverti, A., Zanetti, G., Mattevi, A. (1998) Structure of the mutant E92K of [2Fe-2S] ferredoxin I from Spinacia oleracea at 1.7 A resolution. *Acta Crystallogr.,Sect.D* **54**: 1353–1358; (Ferredoxin NADP+ reductase) Data from PDB ID 1GJR Hermoso, J., Mayoral, T., Faro, M., Gomez-Moreno, C., Sanz-Aparicio, J., Medina, M. (2002) Mechanism of Coenzyme Recognition and Binding Revealed by Crystal Structure Analysis of Ferredoxin-Nadp(+) Reductase Complexed with Nadp(+). *J.Mol.Biol.* **319**: 1133; (ATP synthase) Data from PDB ID 4YXW Bason, J.V., Montgomery, M.G., Leslie, A.G., Walker, J.E. (2015) How release of phosphate from mammalian F1-ATPase generates a rotary substep. *Proc.Natl.Acad.Sci.USA* **112**: 6009–6014; B. Karp, *Cell and Molecular Biology: Concepts and Experiments*, 8e, copyright 2016, John Wiley & Sons. This material is reproduced with permission of John Wiley & Sons, Inc.)

through photosynthesis is sometimes referred to as a Z-scheme because a plot of relative electron energy as a function of where electrons are in the pathway has a Z shape.

Both the splitting of water and the oxidation–reduction cycle of plastoquinone generate a proton gradient across the thylakoid membrane. Because this membrane is generally impermeable to protons, a chemiosmotic potential is established similar to that seen in mitochondria; however, an important difference exists. The thylakoid membrane is permeable to Mg^{2+} and Cl^-, which dissipate the electrochemical gradient. As a result, the gradient used by the thylakoid membrane is simply the chemical concentration of the protons (the pH). This is advantageous to the organism because, without this, the gradient would need to be nearly 1,000 times more concentrated to generate enough potential to drive ATP synthesis. As might be expected, the final step of the light-dependent cycle of photosynthesis is the synthesis of ATP from ADP and P_i via a proton-gradient-driven ATP synthase that is similar in structure and function to the F_0/F_1 ATPase used in mitochondria. Both the proton gradient and the F_0/F_1 ATPase were discussed in Chapter 7.

Photosystem II and the water-splitting enzyme   The light-dependent reactions of photosynthesis begin with photosystem II and the splitting of water (**Figure 23.4**). Photosystem II is a large (~700 kDa) dimeric complex, each half of which is comprised of more

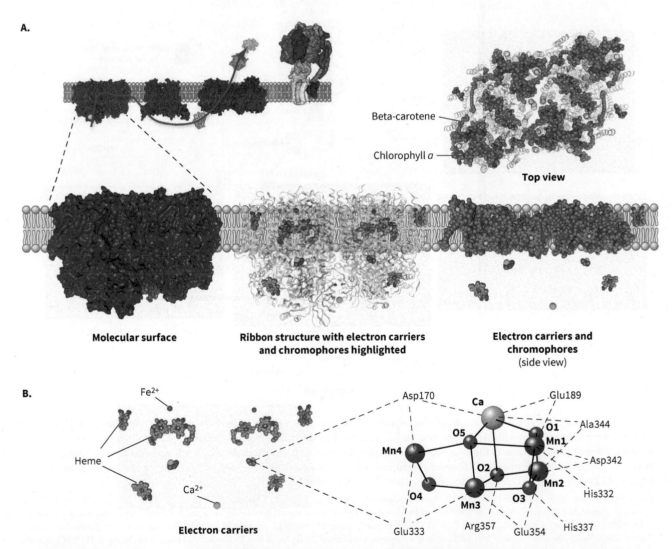

FIGURE 23.4 Photosystem II and water splitting.   A. Photosystem II consists of a large transmembrane region (the antenna complex) made up of several proteins and numerous pigments and electron transporters. Adjacent to this is the water-splitting enzyme domain, which consists of a largely helical domain in the center of the protein and a β barrel structure in the thylakoid lumen. **B.** This region of the protein contains a manganese center, which performs the catalytic splitting of water.

(Source: Data from PDB ID 4YUU Ago, H., Adachi, H., Umena, Y., Tashiro, T., Kawakami, K., Kamiya, N., Tian, L., Han, G., Kuang, T., Liu, Z., Wang, F., Zou, H., Enami, I., Miyano, M., Shen, J.-R. (2016) Novel Features of Eukaryotic Photosystem II Revealed by Its Crystal Structure Analysis from a Red Alga *J.Biol. Chem.* **291**: 5676–5687)

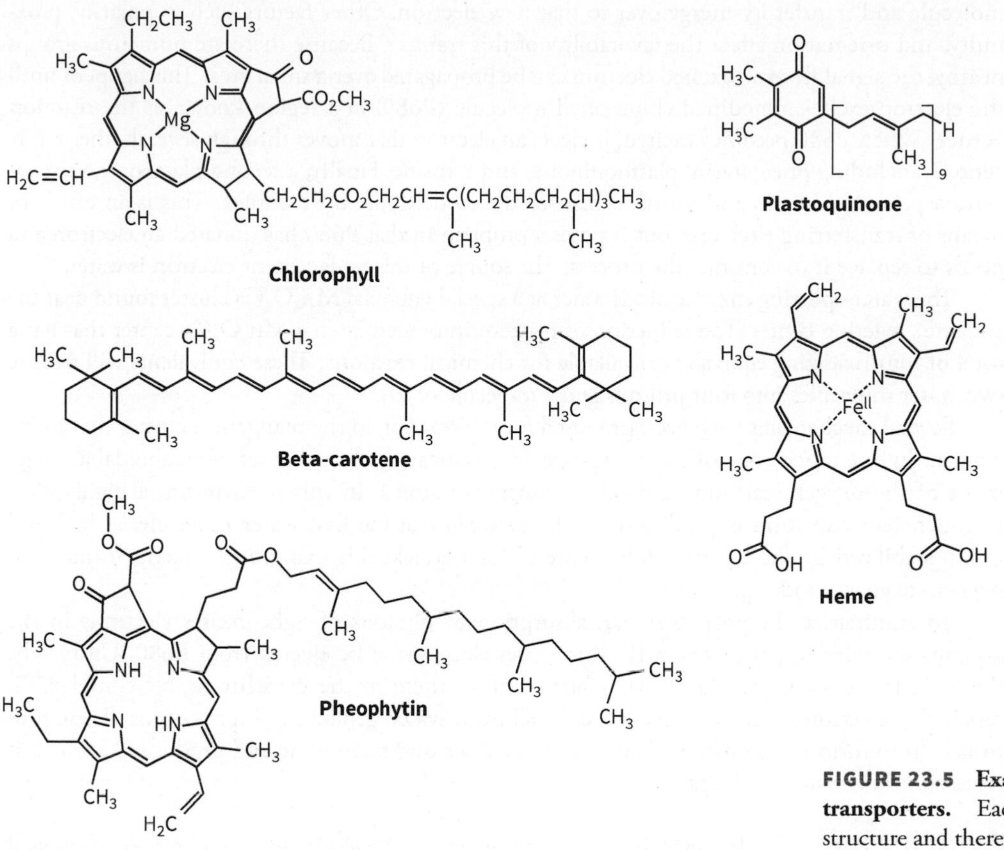

FIGURE 23.5 Examples of plant pigments and electron transporters.   Each of these pigments has a different structure and therefore absorbs at a different wavelength.

than 20 proteins and 75 cofactors. Among these cofactors are numerous structural molecules, pigments, and electron carriers (**Figure 23.5**). The cofactors include membrane phospholipids, chlorophyll, β-carotene, pheophytin, plastoquinone, heme, Fe^{2+} and Ca^{2+} ions, and the Mn_4CaO_4 cluster where water is cleaved into protons, electrons, and oxygen. When other species consume plants, they use some of these molecules as vitamins or cofactors for their own enzymes. Chemicals from plants can also be used in medicine (see **Medical Biochemistry: Plants as sources of pharmaceuticals**).

When a photon of light encounters photosystem II, it is likely to be absorbed by one of the numerous pigment molecules clustered in a structure termed the **antenna complex**. These pigments span the visible spectrum. This excites an electron in the pigment, moving it from the ground state to an excited state. If the pigment were isolated (e.g., in a test tube), it would probably release this energy as heat and fall back to the ground state, but that is not the case in this complex. Because there are numerous pigment molecules in close proximity to one another, those molecules can undergo resonance energy transfer (similar to the technique FRET). Hence, an electron that exists in an excited state in one molecule can excite a neighbor in a nearby

Medical Biochemistry

Plants as sources of pharmaceuticals

Until a little over 100 years ago, most pharmaceuticals came from plants. The best known of these are perhaps tobacco (nicotine), tea and coffee (caffeine), willow bark (salicylic acid), opium (morphine and codeine), and cinchona bark (quinine). Herbal specialists know of hundreds of plants that can produce responses in the human body. Some of these have been well studied and are used to produce actual drugs. For example, the ornamental plant foxglove (*Digitalis*)

produces digoxin, an aminoglycoside that acts as an ATPase inhibitor. This compound is used clinically to increase cardiac output. Other plant products are well known but have questionable efficacy. For example, many people use coneflower (*Echinacea*) to treat colds, but the effectiveness of this treatment and the nature of the active molecule is still being investigated. The National Institutes of Health Office of Complementary and Integrative Health is one group that funds studies examining such compounds.

molecule and transfer its energy over to that new electron. Other factors such as polarity, proximity, and orientation affect the favorability of this transfer. Because there are numerous groups nearby, the signal from an excited electron can be propagated over a small area. This happens until the electron excites a modified chlorophyll molecule (P680) in a region known as the **reaction center**. When P680 becomes excited, it ejects an electron that moves through several other redox centers, including pheophytin, plastoquinone, and tyrosine. Finally, a second plastoquinone captures a pair of electrons and shuttles them to the cytochrome b_6f complex. This is an excellent means of transferring electrons, but it poses a problem in that P680 has donated an electron and needs to replace it to continue the process. The source of this replacement electron is water.

The water-splitting enzyme binds water at a special cuboidal Mn_4O_4Ca cluster found near the enzyme's reaction center. The reduction of plastoquinone results in a Mn_4O_4Ca center that has a total of four oxidizing equivalents available for chemical reactions. These equivalents will oxidize two water molecules into four protons and a molecule of O_2.

Several mechanisms have been proposed for the reaction in the manganese cluster that forms oxygen. In one model, one of the manganese ions is located off the corner of a cuboidal arrangement of the oxygen, calcium, and other manganese atoms. In this mechanism, a highly electron-deficient oxo form is produced via the oxidation of the first water molecule. A hydroxyl group stabilized by the coordination sphere of Ca^{2+} attacks this oxo group directly, forming an oxygen–oxygen bond.

To summarize the process so far, absorption of photons of light excites electrons in the antenna complex of photosystem II. This causes electrons to be ejected from P680. Ultimately, these electrons reach plastoquinone, which shuttles them to the cytochrome b_6f complex. To repair this electron hole, electrons are removed from water, generating four protons. These protons help to form the proton gradient across the thylakoid membrane and produce O_2, which is released into the surroundings.

Cytochrome b_6f

The cytochrome b_6f complex is a 225-kDa dimeric complex of integral membrane proteins. Each half of the complex consists of four heavy subunits, four light subunits, and seven redox-active cofactors. These cofactors include four heme groups (one of which is unique to this complex), a Rieske iron-sulfur center (2Fe:2S), a chlorophyll molecule, and a β-carotene molecule (**Figure 23.6**).

The electron transport chain of mitochondria poses a dilemma in that electrons are transferred in pairs in the first part of the chain but later transferred singly. Photosynthesis poses a similar problem. Plastoquinone carries two electrons, but plastocyanin only carries one. The solution to this conundrum in the electron transport chain is to employ a half-oxidation step known as the Q cycle. Similarly, cytochrome b_6f employs a Q cycle to switch from carrying electrons in pairs to carrying them singly (**Figure 23.7**).

In the oxidation of plastoquinone, protons are removed from that carrier and released into the thylakoid lumen. This is the other reaction that generates the proton gradient that drives ATP biosynthesis through the F_0/F_1 ATPase.

Plastocyanin

Plastocyanin is a small (10.5 kDa) soluble protein found on the lumenal side of the thylakoid membrane (**Figure 23.8**). The protein consists of a single, eight-sheet β barrel structure with one associated alpha helix. Plastocyanin transfers a single electron from the cytochrome b_6f complex to photosystem I. To do this, it employs a single Cu^{2+} ion chelated by two histidine residues, a cysteine and a methionine, found at one end of the barrel. Unlike some of the metal clusters studied previously in which the charge of the electron is delocalized, the Cu^{2+} bears the entire charge in plastocyanin, switching from a Cu^{2+} to a Cu^{1+} state upon reduction.

Photosystem I

Electrons from plastocyanin are shuttled into **photosystem I**. This system uses energy captured from sunlight to raise the electrons to a higher energy state and transfer them to ferredoxin.

As with photosystem II, photosystem I is a complex of integral membrane proteins (**Figure 23.9**). Photosystem I has three identical sections, each of which contains 12 proteins, 10 of which have membrane-spanning alpha helices. In addition, each section of photosystem I has 96 molecules of chlorophyll, 22 molecules of β-carotene, 2 molecules of phylloquinone, 3 molecules of phosphatidyl glycerol, 1 molecule of diglyceride, 3 4FE:4S clusters, and a calcium ion. As with photosystem II, these pigments form an antenna complex to capture photons. Once the

A.

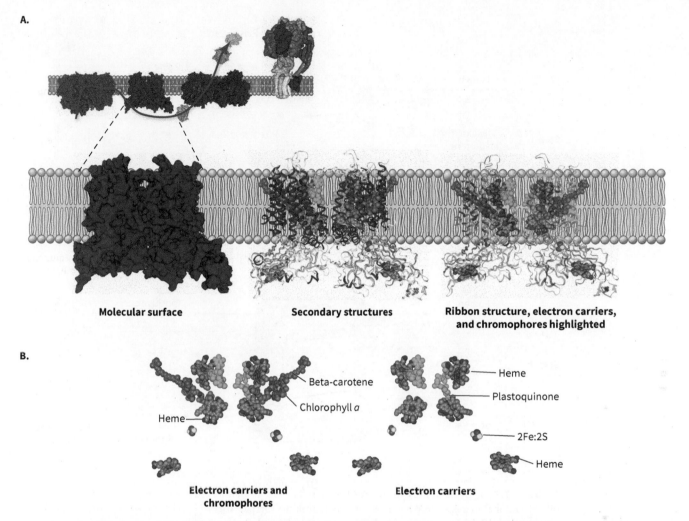

Molecular surface   **Secondary structures**   **Ribbon structure, electron carriers, and chromophores highlighted**

B.

Heme

Beta-carotene

Chlorophyll *a*

Heme

Plastoquinone

2Fe:2S

Heme

Electron carriers and chromophores   **Electron carriers**

FIGURE 23.6 Cytochrome b$_6$f complex.   A. As with the other main centers in photosynthesis, the cytochrome b$_6$f complex is comprised of numerous proteins and is an integral membrane protein with numerous alpha-helical regions. **B.** The complex transfers electrons from plastoquinone to plastocyanin. It is symmetric with two globular beta-sheet domains that interact with plastocyanin.

(Source: Data from PDB ID 1VF5 Kurisu, G., Zhang, H., Smith, J.L., Cramer, W.A. (2003) Structure of the Cytochrome B6F Complex of Oxygenic Photosynthesis: Tuning the Cavity. *Science* **302**: 1009–1014)

electrons are captured, their resonant energy is transferred to the reaction center. In photosystem II, this was a P680 reaction center, whereas in photosystem I, it is a P700 reaction center. The P700 center consists of a dimer of chlorophyll molecules that have a maximal absorbance (λ_{max}) of 700 nm. The complex uses the energy of captured photons to elevate the electrons to a higher energy state as they move through. The potential difference across the P700 complex is −1.2 volts, one of the highest values seen in biochemistry.

The reactions in photosystem I seem simple. By using the energy of light to excite electrons to higher energies, the system takes a process that would normally move in the reverse direction and makes it energetically favorable and spontaneous in the forward direction. In the absence of photosystem I, electrons could not be spontaneously transferred from plastocyanin to ferredoxin but would instead spontaneously flow from ferredoxin to plastocyanin.

Electrons are transferred out of the P700 complex to reduce phylloquinone, which in turn passes the electrons on through a series of three 4Fe:4S clusters and then to ferredoxin. Here, the electrons have two possible fates. In **linear electron transport**, electrons are transferred to ferredoxin NADP$^+$ reductase, and the process continues as described in the next section to produce NADPH. In **cyclic electron transport**, however, electrons can be passed from ferredoxin *back* through plastoquinone and b$_6$f to plastocyanin and again to the P700 complex. We may ask why this process exists. In this state, the cell can still generate a proton gradient and produce ATP, but it does not produce NADPH. Therefore, the cell can use this pathway when NADPH levels are high, but the cell still has an energetic need for ATP.

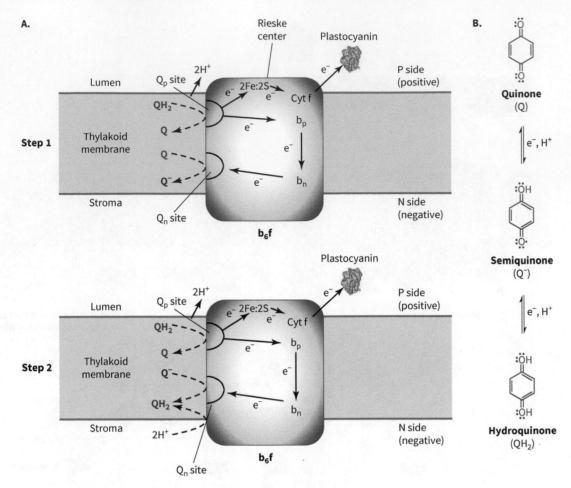

FIGURE 23.7   Q cycle in the cytochrome b$_6$f complex.   A. To transfer electrons from a two-electron carrier (plastoquinone) to a one-electron carrier (plastocyanin), the complex uses a Q cycle analogous to the one used in mitochondria. **B.** The redox states of plastoquinone are identical to ubiquinone.

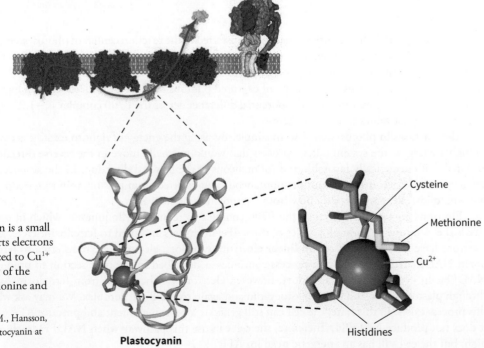

FIGURE 23.8   Plastocyanin.   Plastocyanin is a small soluble copper-binding protein that transports electrons in photosynthesis. A single Cu^{2+} ion is reduced to Cu^{1+} as it transports the electron. Shown is a view of the copper ion bound by two histidines: a methionine and a cysteine.

(Source: Data from PDB ID 1AG6 Xue, Y.,Okvist, M., Hansson, O., Young, S. 1998) Crystal structure of spinach plastocyanin at 1.7 A resolution. *Protein Sci.* 7: 2099–2105)

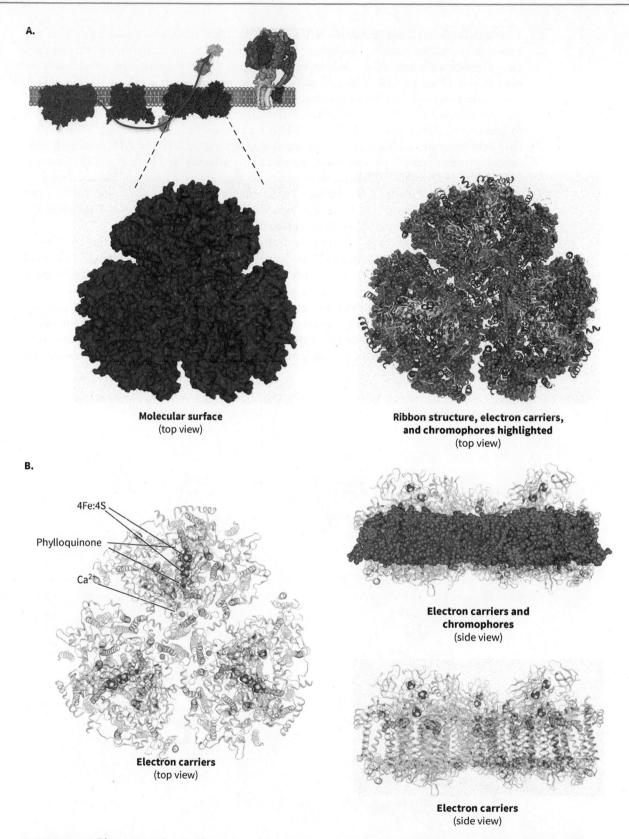

A.

Molecular surface
(top view)

**Ribbon structure, electron carriers,
and chromophores highlighted**
(top view)

B.

4Fe:4S

Phylloquinone

Ca²⁺

Electron carriers
(top view)

**Electron carriers and
chromophores**
(side view)

Electron carriers
(side view)

FIGURE 23.9 Photosystem I. **A.** Photosystem I is a trimeric complex largely embedded in the thylakoid membrane through alpha-helical structures. In addition to the numerous proteins, each subunit has over 100 cofactors. **B.** Shown in the middle and lower right is a cross section through the membrane of the photosystem I complex with space-filling models of the cofactors. The lower panels show the same figure with identical orientation, but the chromophores are left out, and the alpha-helical proteins and electron carriers are shown. These include three iron-sulfur centers, two molecules of phylloquinone, and a calcium ion.

(Source: Data from PDB ID 1JB0 Jordan, P., Fromme, P., Witt, H.T.,Klukas, O., Saenger, W., Krauss, N. (2001) Three-dimensional Structure of Cyanobacterial Photosystem I at 2.5 A Resolution. *NATURE* **411**: 909–917)

Ferredoxin and ferredoxin NADP$^+$ reductase

Ferredoxins are a family of proteins that employ redox-active iron-sulfur centers to transport electrons. In this way, they resemble two of the other small soluble electron carrier proteins discussed previously: cytochromes and plastocyanin (**Figure 23.10**). Although certain members of this family employ a 3Fe:4S or 4Fe:4S cluster, the chloroplast variant of ferredoxin uses a 2Fe:2S center.

Ferredoxin NADP$^+$ reductase is the enzyme that uses the electrons carried by ferredoxin to generate NADPH from NADP$^+$. NADPH is used in biosynthetic reactions, such as those of fatty acid biosynthesis. The next section explains how the NADPH and ATP generated in the light-dependent phase of photosynthesis are used in the reduction and fixation of CO_2. To reduce NADP$^+$, ferredoxin NADP$^+$ reductase requires two electrons and a proton. Because ferredoxin carries only a single electron, two molecules of ferredoxin must be used. To accomplish this two-step reduction, ferredoxin NADP$^+$ reductase uses an FAD cofactor. (Recall that FAD-containing enzymes catalyze either one or two electron reductions.)

Ferredoxin NADP$^+$ reductase consists of two domains: a β barrel domain that contains the FAD cofactor and a mixed alpha-helical and beta-sheet (α/β) domain that is responsible for binding to NADP$^+$ (**Figure 23.11**). Ferredoxin binds to the α/β domain near to the flavin group, and the active site is found at the interface between the two domains. Ferredoxin is active as a monomer in solution, but it can dimerize and associate with membrane-bound proteins via a polyproline II helix. This binding appears to be pH dependent, and it may influence the activity of the reductase. In the dark, when the light-dependent reactions of photosynthesis are not operating and protons are not being pumped, this mechanism provides a means for the reductase to be retained in an inactive state.

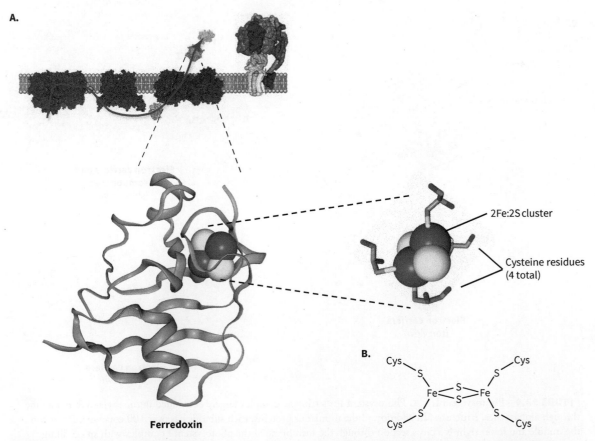

FIGURE 23.10   Structure of ferredoxin. A. Like cytochrome C and plastocyanin, ferredoxin is a small soluble electron carrier. It binds to an electron through its iron-sulfur center and transfers the electron from photosystem I to ferredoxin reductase to make NADPH. **B.** The 4Fe:4S center in ferredoxin is bound in place by four cysteine residues.

(Source: Data from PDB ID 1A70 Binda, C., Coda, A., Aliverti, A., Zanetti, G., Mattevi, A. (1998) Structure of the mutant E92K of [2Fe-2S] ferredoxin I from Spinacia oleracea at 1.7 A resolution. *Acta Crystallogr.,Sect.D* **54**: 1353–1358)

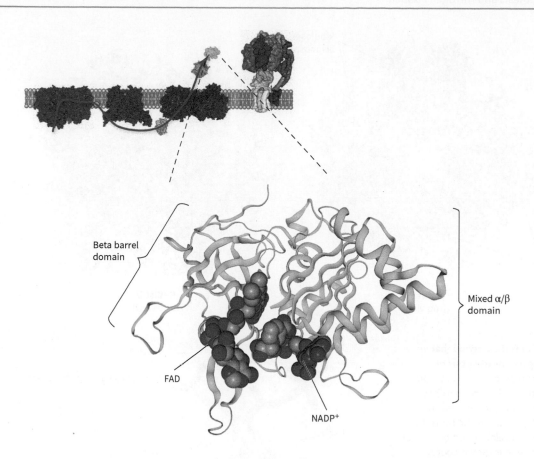

FIGURE 23.11 Structure of ferredoxin NADP$^+$ reductase. Ferredoxin reductase has two main domains: a β barrel domain that binds to FAD and a mixed α/β domain. The binding of ferredoxin occurs at the interface of the two domains.

(Source: Data from PDB ID 1GJR Hermoso, J., Mayoral, T., Faro, M., Gomez-Moreno, C., Sanz-Aparicio, J., Medina, M. (2002) Mechanism of Coenzyme Recognition and Binding Revealed by Crystal Structure Analysis of Ferredoxin-Nadp(+) Reductase Complexed with Nadp(+). *J. Mol. Biol.* **319**: 1133)

23.1.4 The light-independent reactions of photosynthesis fix CO$_2$ and generate carbohydrates

The light-independent reactions of photosynthesis, known as the **Calvin cycle** after one of their discoverers, use the ATP and NADPH generated in the light-dependent part of the process to reduce CO$_2$ into carbohydrates that the plant can use for building blocks or store as energy for future use. These reactions were formerly known as the "dark reactions" of photosynthesis, but they do not actually require dark conditions; rather, they do not need photons to accomplish the chemistry as the light-dependent reactions do.

The overall reaction of the Calvin cycle is

$$3CO_2 + 6NADPH + 5H_2O + 9ATP \rightarrow \text{glyceraldehyde-3-phosphate (G3P)}$$
$$+ 2H^+ + 6NADP^+ + 9ADP + 8P_i$$

Glucose is often thought of as being the product of photosynthesis, but technically this is not the case. Rather, the Calvin cycle produces 3-phosphoglycerate (3PG), which the cell can easily use in gluconeogenesis to generate glucose. These reactions occur in the stroma of the chloroplast.

There are two different phases to the Calvin cycle: first, the fixation of CO$_2$ and formation of 3PG and second, the regeneration of ribulose-1,5-bisphosphate, the substrate used to fix CO$_2$ and generate 3PG.

Fixing CO$_2$ and reducing the carbons

The enzyme that catalyzes the fixation of CO$_2$ is ribulose-1,5-bisphosphate carboxylase/oxygenase (RuBisCO) (**Figure 23.12**). Variants of two enzymes previously seen in gluconeogenesis catalyze the other two steps. RuBisCO takes a molecule of CO$_2$ and uses it to carboxylate a molecule of ribulose-1,5-bisphosphate. The resulting compound is unstable and rapidly breaks down, creating two molecules of 3-phosphoglycerate (3PG). These molecules are acted on by 3-phosphoglycerate kinase, which uses ATP to generate 1,3-bisphosphoglycerate. In turn, the latter is reduced to glyceraldehyde-3-phosphate (GAP) by GAP dehydrogenase in an NADPH-dependent mechanism. These steps are analogous to those in gluconeogenesis.

A.

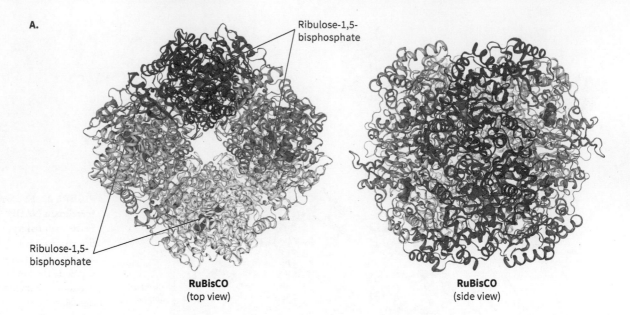

RuBisCO
(top view)

RuBisCO
(side view)

B.

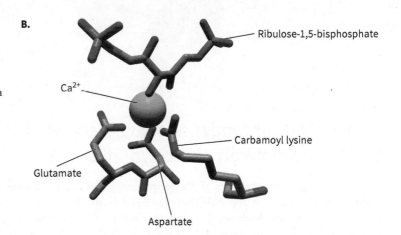

FIGURE 23.12   RuBisCO. A. RuBisCO is the enzyme that catalyzes the first steps of the light-independent reactions of photosynthesis. In plants, RuBisCO forms an octameric complex. Each subunit consists of a heavy and light chain that coordinates ribulose-1,5-bisphosphate via a calcium or magnesium ion. **B.** Substrate binding by RuBisCO and stabilization of the transition state enolate is accomplished by the use of a calcium ion (shown in the center). The carbamoylated lysine residue is shown coordinating the calcium from the lower right. Aspartate and glutamate residues also coordinate to the calcium ion.

(Source: Data from PDB ID 1RCX Taylor, T.C., Andersson, I. (1997) The structure of the complex between rubisco and its natural substrate ribulose 1,5-bisphosphate. *J.Mol.Biol.* **265**: 432-444; Data from PDB ID 1RXO Taylor, T.C., Andersson, I. (1997) The structure of the complex between rubisco and its natural substrate ribulose 1,5-bisphosphate. *J.Mol.Biol.* **265**: 432-444)

RuBisCO has been well characterized, in part because of its important role in the Calvin cycle. In bacteria, it is a dimeric enzyme consisting of a single heavy and a single light chain. In contrast, in plants and algae, RuBisCO is a complex of eight heavy and eight light chains. Despite the multiple interactions of the subunits in this protein, RuBisCO does not appear to be allosterically regulated.

The active site of RuBisCO has several interesting features. First, it has an active-site magnesium ion. This ion is important in stabilizing the transition state; however, it needs a modified amino acid to hold it in the correct position. The modified amino acid is a lysine that has been carboxylated to form a carbamate. This is *not* the molecule of CO_2 that is used as a substrate in the reaction. Rather, this molecule acts as a regulator of activity. RuBisCO activase is an enzyme that forms the carbamate on the lysine to regulate activity. RuBisCO is also regulated by pH. When the light-dependent reactions of photosynthesis are running rapidly, the pH of the stroma rises due to proton pumping across the thylakoid membrane and into the thylakoid lumen. In addition, Mg^{2+} comes out of the thylakoid. Both of these conditions (lower pH favoring protonation of the amino group of lysine and the increased concentration of Mg^{2+} in the stroma) favor the formation of the carbamate, increasing activity of RuBisCO.

Previous chapters have highlighted several reactions through which a carboxyl group is added to a molecule. These include the biotin-dependent reactions of acetyl-CoA carboxylase, propionyl-CoA carboxylase, and pyruvate carboxylase. Each of these reactions is basically an aldol condensation in which the carbonyl carbon of CO_2 acts as the electrophile. Examination of the mechanism of RuBisCO does not give any surprises (**Figure 23.13**). Following binding of ribulose-5-phosphate, a base abstracts a proton from the carbon alpha to the keto group. The keto

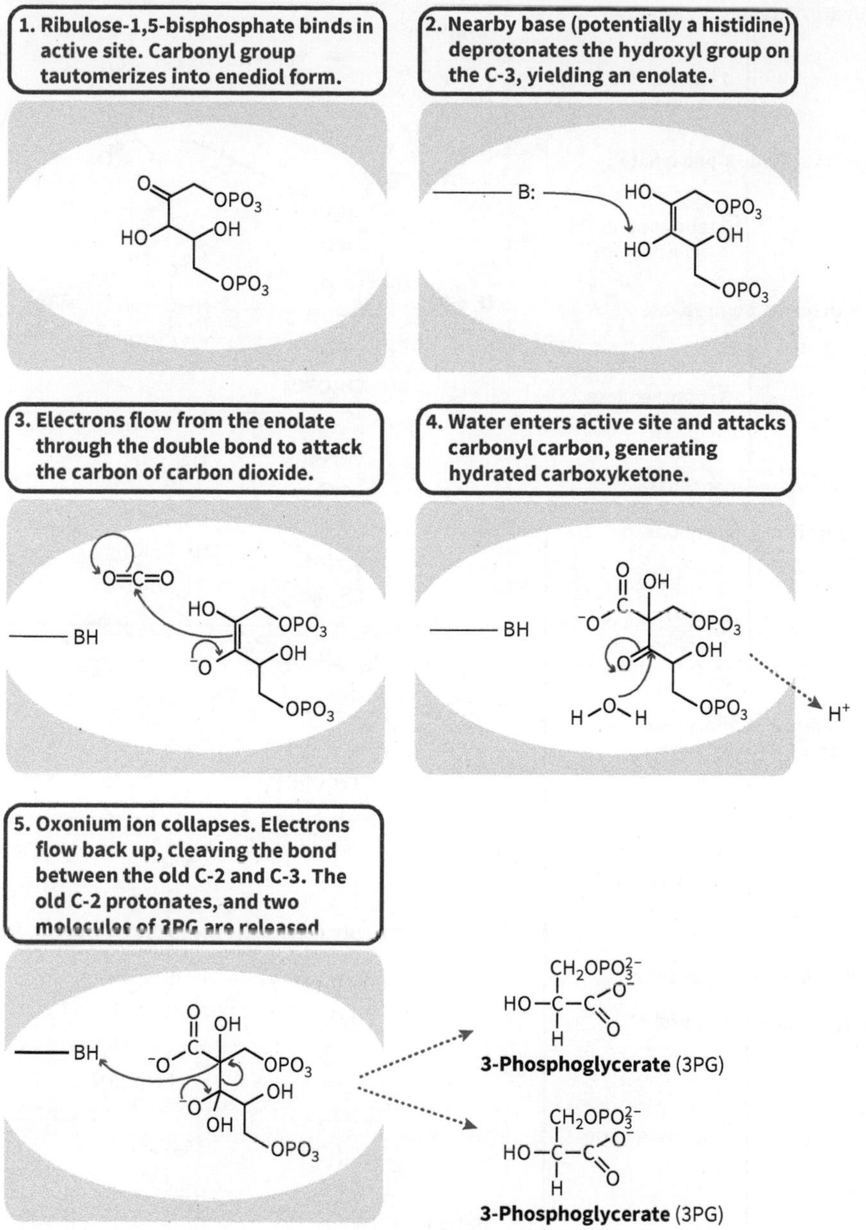

1. **Ribulose-1,5-bisphosphate binds in active site. Carbonyl group tautomerizes into enediol form.**

2. **Nearby base (potentially a histidine) deprotonates the hydroxyl group on the C-3, yielding an enolate.**

3. **Electrons flow from the enolate through the double bond to attack the carbon of carbon dioxide.**

4. **Water enters active site and attacks carbonyl carbon, generating hydrated carboxyketone.**

5. **Oxonium ion collapses. Electrons flow back up, cleaving the bond between the old C-2 and C-3. The old C-2 protonates, and two molecules of 3PG are released.**

3-Phosphoglycerate (3PG)

3-Phosphoglycerate (3PG)

FIGURE 23.13   The mechanism of RuBisCO.   Following substrate binding and deprotonation of the two-carbon hydroxyl, the enolate form of ribulose-1,5-bisphosphate attacks the carbon of carbon dioxide. This forms a new bond to C-2 of the substrate. Next, water attacks the carbonyl group of the substrate, resulting in cleavage of the bond between the old C-2 and C-3 and release of two molecules of 3PG.

group is oriented close to the magnesium ion in the active site, which in turn stabilizes the newly generated enolate. Electrons in the enolate flow from the double bond to the carbonyl carbon of CO_2. This results in a new carbon–carbon bond between C-2 of ribulose-5-phosphate and CO_2. The final step of the RuBisCO reaction is a retro-Claisen condensation in which the new enolate is the leaving group.

RuBisCO is the rate-determining step of the Calvin cycle. It turns fewer than 10 times per second, but it makes up for this leisurely pace with total numbers. RuBisCO is both the predominant enzyme in photosynthetic plants (comprising as much as half of the total protein mass in chloroplasts) and the most prevalent enzyme on Earth.

Regenerating ribulose-1,5-bisphosphate   At this point, the 3PG formed in the Calvin cycle could be shuttled into the cytosol of the plant and used to make glucose and other carbohydrates via gluconeogenesis. However, a mechanism for regeneration of ribulose-1,5-bisphosphate is needed (**Figure 23.14**). The last eight reactions of the Calvin cycle may look familiar. Essentially, they are the same as the reactions found in glycolysis, gluconeogenesis, and the pentose phosphate pathway. These pathways differ only in that the final product of the Calvin cycle is the regeneration of ribulose-1,5-bisphosphate, and the enzymes involved are chloroplast specific, encoded for by chloroplast-specific genes. In both pathways, the laws of thermodynamics apply.

1. Dihydroxyacetone phosphate (DHAP) and glyceraldehyde 3-phosphate (GAP) can be isomerized to one another by triose phosphate isomerase.

2. Aldolase catalyzes the condensation of GAP and DHAP into fructose-1,6-bisphosphate. Fructose-1,6-bisphosphatase removes the phosphate at the 1 position, leaving fructose-1,6-phosphate (F-6-P).

3. Transketolase transfers a carbon from F-6-P to GAP, generating xylulose-5-phosphate (X-5-P) and erythrose-4-phosphate (E-4-P).

4. X-5-P is acted on by phosphopentose epimerase to generate ribulose-5-phosphate.

5. E-4-P and DHAP are acted on by aldolase to yield sedoheptulose-1,7-bisphosphate. A phosphatase cleaves the phosphate on the 1 position, leaving sedoheptulose-7-phosphate (S-7-P).

6. Transketolase transfers a two-carbon unit from S-7-P to GAP, generating X-5-P and ribose-5-phosphate (R-5-P).

7. X-5-P is acted on by phosphopentose epimerase to generate ribulose-5-phosphate (Ru-5-P). Ribose-5-phosphate is isomerized to ribulose-5-phosphate.

8. Ribulose-5-phosphate is phosphorylated by phosphoribulokinase to yield ribulose-1,5-bisphosphate (Ru-1,5-bP).

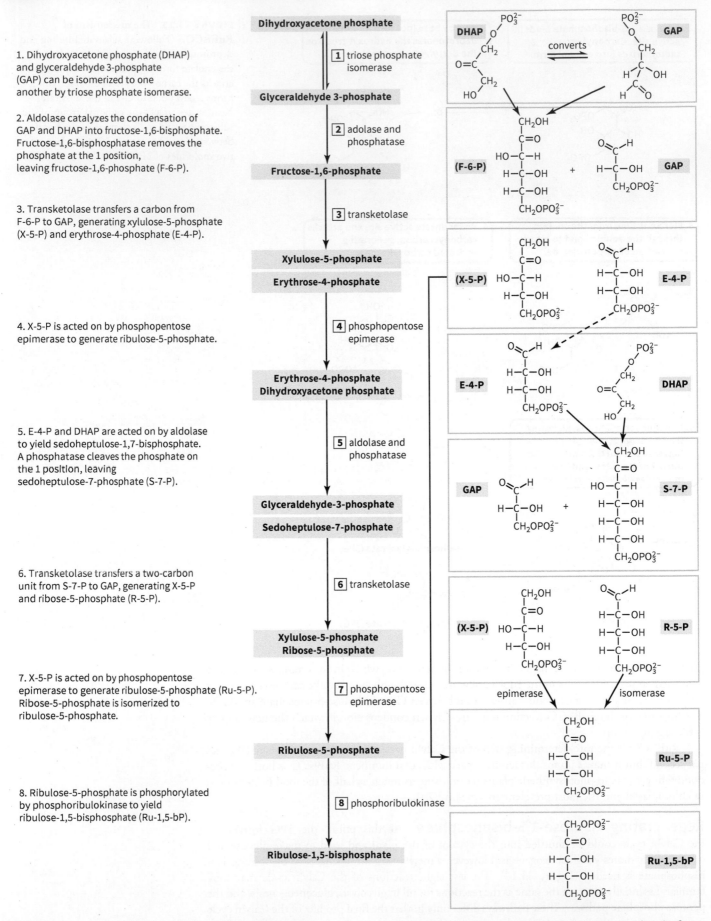

Dihydroxyacetone phosphate

1 triose phosphate isomerase

Glyceraldehyde 3-phosphate

2 adolase and phosphatase

Fructose-1,6-phosphate

3 transketolase

Xylulose-5-phosphate
Erythrose-4-phosphate

4 phosphopentose epimerase

Erythrose-4-phosphate
Dihydroxyacetone phosphate

5 aldolase and phosphatase

Glyceraldehyde-3-phosphate
Sedoheptulose-7-phosphate

6 transketolase

Xylulose-5-phosphate
Ribose-5-phosphate

7 phosphopentose epimerase

Ribulose-5-phosphate

8 phosphoribulokinase

Ribulose-1,5-bisphosphate

FIGURE 23.14 Steps of the second phase of the Calvin cycle. In the second phase of the Calvin cycle, the 3PG that was generated in the first phase undergoes transfer reactions to regenerate ribulose-1,5-bisphosphate. These reactions are analogous to the ones that happen in the pentose phosphate pathway, employing transaldolases and transketolases, but these are chloroplast-specific isomers of these enzymes.

As seen throughout our studies of metabolism, reactions such as isomerizations are generally energetically neutral, with ΔG values close to 0. Other reactions, such as addition of a phosphate to a hydroxyl group, usually require a molecule with significant transfer potential to drive the reaction forward. In general, such a molecule has a large negative ΔG of hydrolysis, an example being ATP.

The regeneration of ribulose-1,5-bisphosphate begins with GAP and a series of reactions that are similar to those found in gluconeogenesis. Triose phosphate isomerase isomerizes the GAP to dihydroxyacetone phosphate (DHAP). Aldolase then catalyzes an aldol condensation between a molecule of GAP and dihydroxyacetone phosphate to generate fructose-1,6-bisphosphate. The enzyme fructose-1,6-bisphosphatase removes the phosphate from the 1 position, leaving fructose-6-phosphate. The next series of reactions are similar to those found in the pentose phosphate pathway. Transketolase removes a two-carbon intermediate from fructose-6-phosphate, leaving the four-carbon erythrose-4-phosphate. The remaining two-carbon intermediate is attached to GAP to yield xyulose-5-phosphate, which can be acted on by phosphopentose epimerase to yield ribulose-5-phosphate. Aldolase combines erythrose-4-phosphate and dihydroxyacetone phosphate to yield sedoheptulose-1,7-bisphosphate. A phosphatase (sedoheptulose-1,7-bisphosphatase) cleaves the phosphate from the 1 position, leaving sedoheptulose-7-phosphate. Transketolase removes a two-carbon group from sedoheptulose-7-phosphate and joins it to another molecule of GAP to yield a further molecule of xyulose-5-phosphate. The remainder of the sedoheptulose-7-phosphate is released as ribose-5-phosphate. As mentioned previously, xyulose-5-phosphate can be epimerized into ribulose-5-phosphate, which in turn can be acted on by phosphopentose isomerase, again yielding ribulose-5-phosphate. Finally, ribulose-5-phosphate is phosphorylated on C-1 by phosphoribulokinase to yield the starting material of the Calvin cycle, ribulose-1,5-bisphosphate.

The Calvin cycle is regulated at two points. As mentioned in the previous section, ferredoxin reductase is regulated in part by light levels. Regulation of the reductase affects levels of NADPH, which are essential for the reactions to proceed. The other point of regulation occurs with RuBisCO. If this enzyme is inactive, there will be no 3-phosphoglycerate, and again, the reactions cannot proceed.

The pathway just described is referred to as the **C3 pathway** because three-carbon intermediates are formed first. Other plants, including the economically important crops corn, sugarcane, and millet, use the C4 pathway.

The **C4 pathway** is used by some species of plants to fix CO_2 (**Figure 23.15**). This pathway is named for the four-carbon intermediates used in this pathway. Because it is not highly selective for CO_2, RuBisCO can use O_2 as a substrate, which it does as much as 25% of the time. When it does, this results in **photorespiration**, using O_2 to generate 3PG and 2-phosphoglycolate instead of a second molecule of 3PG. 2-Phosphoglycolate is an inhibitor of several Calvin cycle enzymes that must be further metabolized at an energetic cost to recover those carbons. To avoid this problem, C4 plants have evolved cellular structures with higher carbon dioxide levels and lower oxygen levels than C3 plants. This involves two cell types: mesophyll cells in which the light-dependent reactions of photosynthesis occur and bundle sheath cells in which RuBisCO is found. CO_2 levels are increased in bundle sheath cells by importing four-carbon molecules such as malate and aspartate and decarboxylating them to release CO_2 for RuBisCO. Depending on the species, decarboxylation and subsequent reactions result in the formation of either pyruvate or alanine, which are cycled back to the mesophyll cells. In the mesophyll cells, alanine or pyruvate can be

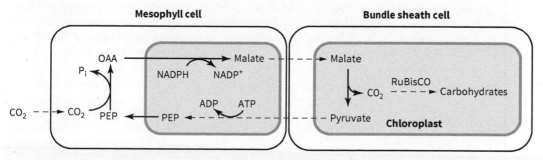

FIGURE 23.15 C4 carbohydrate synthesis. Because RuBisCO can add oxygen instead of CO_2 to ribulose-1,5-bisphosphate, some plants have evolved mechanisms to increase the concentration of CO_2 near RuBisCO. These plants first fix CO_2 as a four-carbon intermediate (malate or aspartate) and then import this to the cell where RuBisCO is found. In that second cell (bundle sheath cells), the CO_2 carrier is decarboxylated, and the CO_2 is used by RuBisCO.

carboxylated to regenerate malate or aspartate. If we follow the metabolism of malate, we find that the imported pyruvate is first phosphorylated to phosphoenolpyruvate, which is then carboxylated to oxaloacetate using the enzyme phosphoenolpyruvate carboxylase. Oxaloacetate, in turn, is oxidized in an NADPH-dependent mechanism to generate malate, and the cycle continues.

We might ask why RuBisCO has this odd quirk and why plants have evolved this somewhat complex work-around. RuBisCO is an ancient enzyme that evolved before oxygen levels in the atmosphere were as high as they are now. Therefore, the selective pressures on RuBisCO have changed over time, but the RuBisCO gene has not evolved to keep pace. Instead, in C4 plants we see common metabolites and elements of other pathways (transporters, oxidoreductases, and transaminases) employed to solve the problem. Plants are highly diverse, and it appears that different mechanisms of C4 metabolism have independently evolved multiple times in an example of convergent evolution. Plants that have evolved this mechanism need to expend more ATP and NADPH to generate glucose, but they do so at a much higher rate due to retention of the carbons not lost to photorespiration.

Worked Problem 23.1   Copper deficiency

Imagine that you have identified a strain of cyanobacteria that can conduct photosynthesis and grow normally in the absence of copper. Identify why this might be an interesting phenomenon and how the plants might be compensating for the lack of copper.

Strategy   How and where is copper used in photosynthesis? What barrier is formed by the absence of copper? What alternatives would biology use to compensate for this situation?

Solution   Plastocyanin is a small, blue, copper-binding protein that transports electrons from the cytochrome b$_6$f complex to photosystem I. In the absence of copper, plastocyanin is unable to function and transport electrons. One alternative situation could be that the cyanobacteria evolved another mechanism through which electrons could be transported. To be able to do this, the carrier would need to have a reduction potential somewhere between the potential of the b$_6$f complex and photosystem I.

Follow-up question   Would a plant with a mutation that prevented it from producing plastocyanin be viable? Explain your answer.

Summary

- Photosynthesis is the means by which plants, algae, and some bacteria and archaea use solar energy to produce ATP and NADPH and reduce CO_2 to carbohydrates. Most O_2 found on Earth is a product of photosynthesis.

- The reactions of photosynthesis take place in the chloroplast, which has a double membrane surrounding the stroma, inside of which is a membranous structure called the thylakoid.

- The light-dependent reactions of photosynthesis generate ATP, NADPH, and oxygen. Photosystem II is a multiprotein complex that splits water and passes the electrons to plastoquinone, generating four proteins in the thylakoid lumen and a molecule of O_2. An antenna complex of chlorophyll, β-carotene, and other molecules captures photons and excites electrons. The splitting of water occurs at a unique cuboidal Mn_4O_4Ca cluster. Plastoquinone shuttles electrons to the cytochrome b$_6$f complex, where they are passed through a series of heme and iron-sulfur centers to the soluble, copper-containing protein plastocyanin, which carries electrons. The cytochrome b$_6$f complex uses a Q cycle to transfer from a two-electron carrier (plastoquinone) to a one-electron carrier (plastocyanin). Plastocyanin transfers electrons to photosystem I, which excites them to higher energy levels and passes them to ferredoxin. The antenna complex of photosystem I consists of over 100 molecules. The electrons fed to ferredoxin are used by ferredoxin reductase to generate NADPH. Ferredoxin reductase uses a flavin cofactor to perform the two-electron reduction of NADP$^+$ to NADPH. The splitting of water and the oxidation of plastoquinone generate a proton gradient across the thylakoid membrane, which is used to drive an F_0/F_1 ATPase.

- The light-independent reactions of photosynthesis, called the Calvin cycle, produce carbohydrates from CO_2 and water using the energy harvested in the light-dependent cycle, making 3PG, and resetting the ribulose-5-phosphate needed to keep the reaction running. RuBisCO adds a molecule of CO_2 to the alpha carbon of the keto group of ribulose-1,5-bisphosphate in an aldol condensation and then completes a retro-Claisen condensation, with the enolate acting as the leaving group, producing two molecules of 3PG. In the last eight reactions of the Calvin cycle, transaldolases and transketolases swap different two-carbon and three-carbon groups to regenerate ribulose-5-phosphate, which is then acted on by phosphoribulokinase to regenerate ribulose-1,5-bisphosphate. C4 plants use a different pathway to fix CO_2, carboxylating phosphoenolpyruvate into oxaloacetate, which then joins the last eight steps of the Calvin cycle.

1. Explain how photosynthetic organisms trap electrons and use this energy to drive reactions.
2. Describe photosynthesis in terms of its thermodynamics or energetics.
3. Discuss where and how oxygen is produced by photosynthesis.
4. Molecules involved in photosynthesis include chlorophyll, β-carotene, phylloquinone, phosphatidyl glycerol, diglyceride, iron-sulfur clusters, and calcium. Describe the general functions of each of these.

23.2 Nitrogen Fixation

The previous section discussed how some organisms (notably plants and cyanobacteria) use sunlight to fix CO_2 via photosynthesis, generating ATP, glucose, and O_2 in the process. This section describes how some species of bacteria and archaea can fix another gas—molecular nitrogen (N_2)—and convert it to ammonia. To put this process in context, it is useful to first examine the geological history of Earth's atmosphere.

23.2.1 Nitrogen levels in Earth's atmosphere changed over time

Initially, Earth had no atmosphere because light gases such as hydrogen and helium were lost to the early solar system. As the planet solidified and cooled, heavier gases escaped Earth's core and collected to form the earliest atmosphere. This early atmosphere consisted of water, ammonia, methane, and nitrogen. This was a reducing atmosphere, with most of the atoms in a reduced state and little or no O_2 present. These conditions prevailed from about 4.5 billion to 4 billion years ago (the Hadean Eon). As Earth continued to cool, water condensed and formed seas. Chapter 1 discussed the Miller–Urey experiment, in which a flask filled with the gases of the primordial atmosphere was warmed and sparked with an electric arc to simulate the effect of lightning, forming many amino acids. About 3.5 billion years ago, the earliest organisms—microscopic archea and protobacteria—evolved from this primordial soup. These organisms were anaerobic and probably lived in the seas. At the time, Earth had no ozone layer, meaning that ultraviolet radiation could damage the DNA of surface organisms. These earliest organisms obtained carbon either from methane or CO_2, oxygen from water, and nitrogen from ammonia. By this point in evolutionary history, it is likely that these organisms were already undergoing photosynthesis to fix CO_2 and harvest the energy of the sun. Although we often connect photosynthesis with plants, most of the plants known today did not evolve for another 3.3 billion years. Early photosynthesis was carried out by cyanobacteria.

Photosynthesis altered life on Earth forever because of its oxygen by-product. For hundreds of millions of years, photosynthetic cyanobacteria captured protons, split water, fixed carbon, and released oxygen. The oxidation levels of iron-containing minerals in ancient rocks demonstrate that oxygen levels varied at different points in Earth's history. As levels of oxygen rose, organisms evolved means of living and even thriving in an oxygen-rich atmosphere, with some organisms using O_2 as an electron acceptor in the electron transport chain.

The preponderance of oxygen in the atmosphere had another important ramification for early life on Earth. In the presence of ammonia or other reduced amines, oxygen will quickly oxidize the amines to various nitrogen oxides. Over time, the relatively inert N_2 eventually became predominant thanks to its stability. Over a geological timespan, the ammonia that had been present in the atmosphere and dissolved in the oceans was replaced by nitrogen gas (**Figure 23.16**).

A lack of reduced ammonia presents a problem for biochemistry. All of the oxygen we breathe and the carbohydrates we eat originate from plants; thus, at some level, we obtain all our energy from the sun. However, to produce amino acids, nitrogen gas must be transformed into something that can be used in a biosynthetic pathway. The agents of this transformation are nitrogen-fixing bacteria and cyanobacteria.

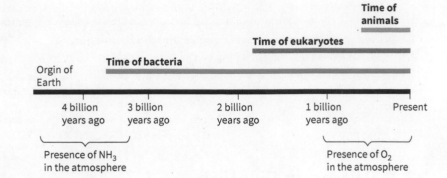

FIGURE 23.16   Ammonia in early Earth's atmosphere.   Ammonia levels had dropped to insignificant levels by 3.2 billion years ago. Oxygen did not begin to accumulate until 2.5 billion years ago.

23.2.2 The nitrogen cycle describes nitrogen chemistry in the environment

The first section of this chapter focused on the carbon cycle, and this section looks at the nitrogen cycle (**Figure 23.17**). In the nitrogen cycle, atmospheric nitrogen is fixed by specific organisms in the environment, termed **diazotrophs**. These are often symbiotic bacteria of the genuses *Rhizobia* and *Frankia* but also independently living cyanobacteria.

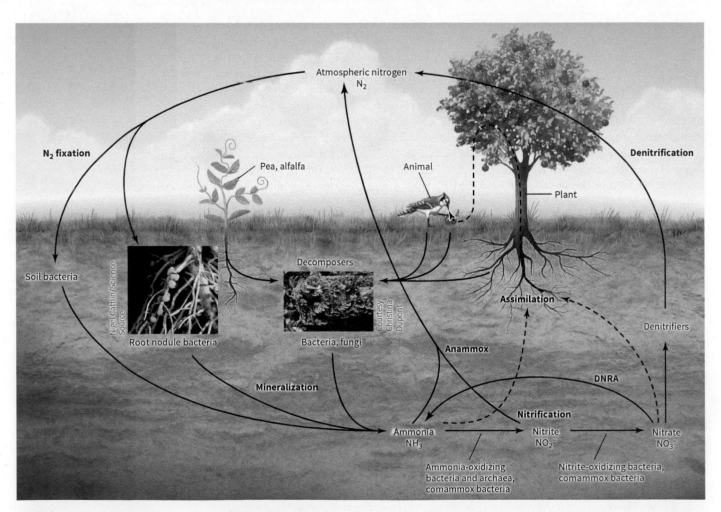

FIGURE 23.17   Global nitrogen cycle.   Nitrogen is fixed by anaerobic bacteria that live in the soil or in the root nodules of some plants. These plants provide the nitrogen for all other life on the planet. Typically, in the short term, the nitrogen cycles between ammonia and nitrogen oxides, but if left with molecular oxygen, nitrogen oxides will eventually reform N_2.

(Source: Wessner, *Microbiology*, 2e, copyright 2017, John Wiley & Sons. This material is reproduced with permission of John Wiley & Sons, Inc.)

Many of these symbionts live in the root nodules of legumes. The plant provides nutrients for the bacteria, protects them from other microbes, and limits their exposure to oxygen. In turn, the bacteria supply the plant with ammonia for the production of amino acids, nucleotides, and other biomolecules. Understanding the importance of nitrogen to plant growth was key to developing nitrogen-based crop fertilizers that revolutionized global agriculture (see **Societal and Ethical Biochemistry: Fritz Haber—good and evil**).

When plants die, they are degraded by various organisms (fungi, bacteria, and animals) that can use the amine-containing molecules in the synthesis of other molecules. The waste products of the catabolism of these amino groups are often ammonia, uric acid, or urea. These molecules can

Societal and Ethical Biochemistry

Fritz Haber—good and evil

We take for granted in the developed world that food is plentiful and crops can be grown at high yield and with ease. Part of the reason for this is the use of fertilizer. It has been appreciated since antiquity that plants require some level of nutrients for growth. For example, certain Native American tribes buried fish with particular seeds to encourage their growth. The mechanism by which feeding plants affected their growth was not known, but it was evident that plants that could not easily fix their own nitrogen benefitted from having nitrogen supplementation in the soil. Until the early 1900s, the main means of fertilizing soils was through the use of bat or bird guano. A few parts of the world (notably Chile) had natural deposits of nitrate-containing minerals that could be mined and used as a source of nitrate for fertilizer. This approach was expensive and time consuming, and it relied on the cooperation of governments around the world.

This situation changed in the early 1900s due to the work of German chemist Fritz Haber. An interesting character, Haber struggled with his early education and went on to have difficulties working at his father's textile company. In contrast, he was highly successful in academic positions, where he studied topics ranging from dyes to electrochemical reactions to corrosion. Later, Haber helped to develop the thermodynamic understanding of reactions known as the Born–Haber cycle, familiar to students of general chemistry. Haber is perhaps best known for the Haber–Bosch process, which combined small amounts of hydrogen gas, an iron catalyst, and the nitrogen found in air at high pressure and temperature to produce ammonia. For the first time, fertilizer could be produced in nearly limitless amounts using air as a starting material. For this discovery, Haber was awarded the Nobel Prize in Chemistry in 1918. Today, the process is used to produce over 100 million tons of ammonia-based fertilizer globally each year. It is estimated that Haber's discovery feeds over half of the world's population and has made massive advances in agriculture possible. In this regard, Haber was a hero who saved millions from starvation.

Ammonia and nitrates are also used in the production of explosives, but Haber seemed to have no moral qualms about how his advances in the laboratory were used. He was fervently nationalistic and a proud German. As World War I started, Haber was quoted as saying, "During peace time a scientist belongs to the world, but during war he belongs to his country."

The technological history of World War I is at once fascinating and terrifying. Although some units were still using horses and swords, World War I saw the first use of tanks, automatic weapons, and airplanes. Haber was asked to head a special chemical unit.

On April 22, 1915, as evening fell across the town of Ypres in western Belgium, he ordered his soldiers to carry 5,700 cylinders of chlorine gas to the front lines and open them. The prevailing breeze carried a toxic cloud of chlorine across the battlefield to the unsuspecting allied troops. Thousands died within minutes as the chlorine reacted with their eyes and mucous membranes to form hydrochloric acid, which caused horrific burns.

The use of chemical weapons was already outlawed; nevertheless, British and French forces felt the need to retaliate in kind. The French employed their best chemist (Victor Grignard, the same chemist studied in organic chemistry classes today) to help develop chemical weapons. Over the next three years there was a chemical arms race. Because of this, World War I is sometimes known as the "chemist's war." Chlorine gas, tear gas, and burning sulfur led to the use of mustard gas, a vesicant, or blister agent, that causes chemical burns and blisters to the skin and lungs. Another chemical used was phosgene, which causes lung damage. By the end of the war, over 21 different chemical weapons had been used, leading to 1.2 million military and 250,000 civilian casualties and over 100,000 military deaths.

Haber was noted as saying that death was death, and it made no difference whether you were poisoned or blown up. Although Haber had no personal issue with his work, those around him did. His first wife Clara (also a chemist) committed suicide after confronting him about what had happened at Ypres. Many other scientists found his work unethical and shunned him.

Haber was decorated for his work during the war, but his success in this arena was relatively short lived. Following the war, National Socialism swept Germany and with it a rise in anti-Semitism. Although by this time Haber was a practicing Lutheran, he had Jewish parents. He left Germany in 1933, traveled throughout Europe, and died in Switzerland in 1934.

The academic groups and companies Haber helped to set up continued their work after he had left. One of these (Degesch) developed the pesticide and fumigant Zyklon B, the poison gas that was used in the Holocaust to kill over 1 million people at Auschwitz.

Often, we cannot foresee how and to what end an invention or discovery will be used, but scientists always need to be mindful of their work. Biochemistry provides amazing opportunities to improve human health and relieve human suffering but also the possibility of causing immeasurable suffering and death. Whatever Haber's motivations or his beliefs about how he was benefitting his country, we cannot ignore the atrocities that were committed with his inventions. Perhaps Haber's greatest legacy is not his gift of fertilizer but the complexities of the rest of his life that should make us think carefully.

be used directly by some organisms or converted into nitrites (NO_2^-) and nitrates (NO_3^-). These can be used by plants but are often toxic to animals and some microbes, which is why nitrates are commonly used as food preservatives. Also, other bacteria can use nitrates and nitrites, reducing these compounds back to nitrogen gas. Alternatively, the compounds can react with oxygen to form volatile nitrogen oxides, leading to the formation of nitrogen dioxide (NO_2), which undergoes photochemical degradation into nitrogen gas.

23.2.3 Nitrogenase employs unique structural features to catalyze the reduction of nitrogen gas

Nitrogenase is the enzyme responsible for producing reduced nitrogen (NH_3) from N_2 gas. The nitrogenase complex is made up of two proteins: the Fe protein (which provides electrons for the reactions to occur) and the MoFe protein (which catalyzes the reduction of molecular nitrogen) (**Figure 23.18**). Two MoFe proteins come together to form a heterotetramer with two α and two

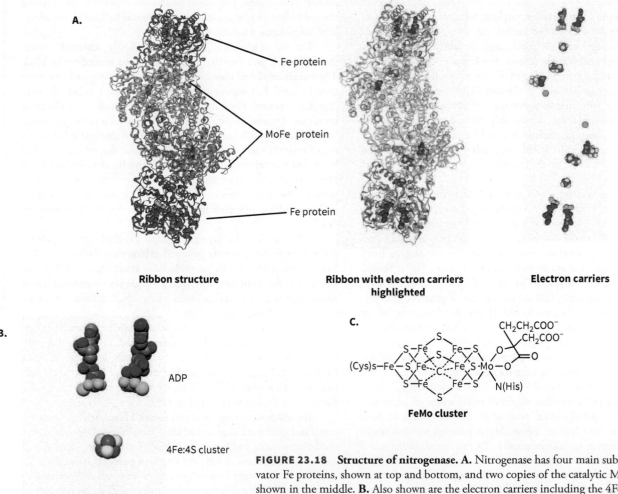

FIGURE 23.18   Structure of nitrogenase. A. Nitrogenase has four main subunits: two activator Fe proteins, shown at top and bottom, and two copies of the catalytic MoFe protein, shown in the middle. **B.** Also shown are the electron carriers including the 4Fe:4S, 8Fe:7S, and FeMo clusters. Note the unusual geometry of these cofactors. **C.** The FeMo cofactor consists of two separate metal clusters enclosing what is believed to be a single carbon atom. Reduction of N_2 is thought to occur at the molybdenum atom. As the Fe protein binds, it transfers an electron that flows down the reduction potential of electron carriers as shown, eventually reaching the nitrogen and reducing it. The hydrolysis of two molecules of ATP is necessary to cause the proper conformational change and prompt the electrons to flow in the direction opposite to how they would flow in the absence of this protein.

(Source: Data from PDB ID 1N2C Schindelin, H., Kisker, C., Schlessman, J.L., Howard, J.B., Rees, D.C. (1997) Structure of ADP x AlF4(-)-stabilized nitrogenase complex and its implications for signal transduction. *Nature* **387**: 370–376)

β subunits. The Fe proteins bind at the poles of this complex and feed in the electrons necessary for the reaction.

Nitrogen gas is one of the most stable molecules encountered in biology, with a bond enthalpy of $\Delta H = -946$ kJ/mol. Because of the stability of N_2 gas and the triple bond that holds it together, six electrons are needed to cleave the triple bond and generate two molecules of ammonia. Two more electrons are needed to produce the hydrogen gas (H_2) that is a by-product of the reaction. To drive this reaction forward, two molecules of ATP are hydrolyzed for each electron employed in the reduction:

$$N_2 + 8H^+ + 8e^- + 16ATP \longrightarrow 2NH_3 + H_2 + 16ADP + 16P_i$$

Cofactor structure and discovery To complete the reduction of nitrogen, the Fe and MoFe proteins use several metallic cofactors, two of which are unique. The first electron carrier employed by the Fe protein has a traditional 4Fe:4S center. Such centers have been discussed in the context of electron transport and in several redox-active enzymes. Perhaps the best known of these is the enzyme aconitase, which catalyzes the isomerization of citrate to isocitrate in the citric acid cycle. As with other iron-sulfur centers, electrons are thought to be delocalized across the atoms in the cluster; individual atoms are not oxidized or reduced by the loss or gain of electrons. Understanding the principles of inorganic chemistry can help explain these clusters (see **Biochemistry: Inorganic chemistry**).

Biochemistry

Inorganic chemistry

Most biochemistry courses require students to have studied organic chemistry. Biochemists use analytical techniques to separate and analyze compounds, physical techniques to determine structures, and a knowledge of organic chemistry to explain many reactions and mechanisms. However, inorganic chemistry (the study of elements other than carbon) is also vital to biochemistry. Many mechanisms covered in this text, particularly in this chapter, have metal cofactors or reaction centers where many of the reactions involve metal ions or clusters of metals and nonmetals.

Particular aspects of inorganic chemistry that are pertinent to biochemistry are worth reviewing. Of all the elements encountered in biochemistry, the ones that are often unfamiliar to the student are sulfur; phosphorus; and transition metals such as iron, calcium, zinc, copper, manganese, magnesium, and, occasionally, some of the other 3d group elements (chromium, cobalt, or vanadium). In terms of electronic structure, each of these elements includes both 3p and 3d orbitals, meaning that they often contain expanded octets and can make more than four covalent bonds. Many of these elements can also form metal clusters, combining with oxygen or sulfur to form a collective group that can act as a single entity. In these instances, oxidations or reductions of the cluster will result in electron density being delocalized throughout the cluster rather than focused on a single atom. In other cases, metal ions can act alone, being bound in place by various amino acid side chains such as histidine or cysteine. Finally, metals can be found in a larger functional group such as the porphyrin ring of a heme group. In all these instances, the interactions of the metal with other atoms can change the reactivity of the metal center. It can also modulate the strength of bonds to the metal, resulting in changes in conformation of a protein with a change in the chemistry of the metal center. Metals are involved in redox chemistry, in catalysis as Lewis acids, and in stabilizing anionic transition states.

Inorganic chemistry may not be the first thing that comes to mind as a foundation for biochemistry, but its value in understanding many biochemical processes is clear.

From the 4Fe:4S center, electrons flow into a group known as the P cluster. This unique 8Fe:7S cluster is bound to nitrogenase through three cysteine sulfhydryl groups from the α subunit (Cysα63, Cysα88, and Cysα154) and three from the β subunit (Cysβ70, Cysβ95, and Cysβ153). Four of these groups (Cysα63, Cysα154, Cysβ70, and Cysβ153) ligate one iron atom each; the other two (Cysα88 and Cysβ95) ligate two iron atoms each. The P cluster differs from some iron-sulfur clusters in that it undergoes a significant structural shift when oxidized, with half the cluster adopting a more open conformation. This changes the ligand binding and introduces two new ligands: the backbone amide group of Cysα88 and the serine oxygen of Serβ188. Both of these interactions are unique to this pathway. The ramifications of the structural change are not yet fully understood, but it is likely that this conformational change is involved in electron flow from the P cluster to the FeMo cluster.

The FeMo cluster (also known as the M cluster) has the formula $MoFe_7S_9C$-homocitrate. It is completely buried in the α subunit, 14 Å away from the P cluster, and it is unusual in several ways. First, it can be viewed as two distinct clusters—a 4Fe:3S cluster and a Mo:3Fe:3S cluster—bridged by three additional sulfur atoms. Second, at the molybdenum end of the group is a homocitrate group. Finally, in the middle, between the two clusters, there is a small, light atom that has been determined to be a carbon atom. The complex is bound to the protein through coordination of a histidine (His442) to the molybdenum atom at one end of the molecule and a cysteine (Cys225) to an iron atom at the opposite end.

The structure of nitrogenase was first determined in 1992. At that time, it was understood that there was a metal cluster at the core of the protein, but the resolution was not high enough to give a detailed structure of the FeMo cluster. Over the next 30 years, the structure and mechanism of nitrogenase was further explored. One fundamental question that arose was the placement of the atom in the center of the core. Early experiments found no electron density in the core of the complex because of the resolution of the experiment, a loss of structural detail in the Fourier analysis of the diffracted X-rays, and the size and lightness of the atom involved. As techniques improved, it became clear that there was an atom in the center of the cluster but that the hole provided by the other groups meant that it had to be a light atom (probably carbon, oxygen, or nitrogen). High-resolution X-ray diffraction studies and density field theory calculations indicated that the atom might be a nitrogen. However, subsequent studies with even higher resolution X-ray diffraction, X-ray emission spectroscopy, and electron spin echo envelope modulation (ESEEM) showed it to be a carbon atom donated by *S*-adenosylmethionine.

The structure of nitrogenase was elucidated through studies in many laboratories, involving possibly hundreds of investigators, over more than 20 years. It is a fascinating illustration of how science can be so reluctant to yield its secrets and yet so beautiful and elegant when it does.

The assembly of the P and FeMo cofactors is beyond the scope of this text, but it complements the structures in terms of complexity and elegance. Several excellent reviews of this topic are listed in the Suggested Readings at the end of this chapter.

The nitrogenase reaction in detail

The exact mechanism of the nitrogenase reaction is still under debate, but some of its more general aspects are well known. The electrons used in the reaction originate from central metabolism and are transferred using either ferredoxin or flavoredoxin, depending on the organism in question. These electrons are transferred to the 4Fe:4S cluster of the Fe protein, as shown in Figure 23.18. The Fe protein binds to two ATP molecules, causing a conformational change that results in the transfer of electrons through the P cluster and to the FeMo cofactor, where catalysis occurs. Hydrolysis of the two ATPs induces a further conformational change that releases the bound Fe protein. This cycle is repeated seven more times to transfer the necessary electrons.

The actual mechanism of nitrogen reduction is still not clear, but it likely involves a stepwise addition of electrons and protons to the nitrogen while it is bound to a metal atom (possibly the molybdenum) and proceeding through a nitride (N^{3-}) intermediate.

Nitrogenase cleaves the nitrogen triple bond and generates ammonia; however, N_2 is not its only substrate. Nitrogenase can also reduce several other small alkenes and alkynes, including azide (N_3^-), cyanide (CN^-), thiocyanate (SCN^-), and acetylene (HCCH).

Given the importance of compounds containing reduced nitrogen in biochemistry, it may seem strange that so few organisms synthesize their own ammonia. There are several reasons for this. First, nitrogenase is inhibited by oxygen. Most dizaotrophs (organisms that fix nitrogen) are anaerobes, and the few that are aerobes employ special mechanisms to sequester any oxygen away from nitrogenase. Second, ammonia is toxic to most animals on several levels: it is reactive; raises pH; and acts as a neurotoxin in that it depletes glutamate, other neurotransmitters, and citric acid cycle intermediates by reacting to form glutamine. This is another instance in which organisms that have emerged more recently simply stood on the biochemical shoulders of their ancestors.

Why is the chemistry of nitrogenase such an important area of research? Beyond simple intellectual curiosity about this process, there is important science underlying these mechanisms. Nitrogenase catalyzes a reaction that is energetically difficult. The ammonia it produces can be used to grow crops for food or produce explosives for weapons. Although we have been able to synthesize ammonia chemically for over 100 years, today's chemical processes for making ammonia are energetically expensive. An understanding of how proteins such as nitrogenase work or how diazotrophs live can help to inform other research questions and advance science overall.

Worked Problem 23.2   Something in the air

One theory about past mass extinctions is that they were caused by a gamma ray burst from a nearby dying star. If a gamma ray burst sterilized the planet, eradicating all life, how would the composition of the atmosphere change over time?

Strategy   This chapter discussed how three gases in the atmosphere (N_2, O_2, and CO_2) are either produced or used by biological systems. What would happen in the absence of the organisms that produce these gases?

Solution   We know that most of the oxygen on Earth is produced by photosynthesis. In the absence of photosynthetic organisms, the bulk of this oxygen would react with other molecules in relatively short order (tens of thousands of years), and oxygen levels would

plummet. By contrast, nitrogen is stable and would probably remain the predominant gas in the atmosphere. Even if ozone were depleted, N_2 does not appreciably absorb UV light and undergo photochemistry. Nitrogen fixation by microbes would stop, but in the short term this would not appreciably affect nitrogen or ammonia levels in the atmosphere. Over longer periods of time, if biological nitrogen-containing molecules were exposed to oxygen, they would break down into nitrogen oxides and eventually into molecular nitrogen. Likewise, biological carbon exposed to oxygen would eventually oxidize to CO_2 and raise the atmospheric concentration of this gas.

Follow-up question   Is it surprising that so few organisms undertake nitrogen fixation? What does it mean from an evolutionary perspective that so few do?

Summary

- The early atmosphere of Earth was rich in reduced nitrogen in the form of ammonia. As oxygen levels increased, levels of ammonia dropped.
- Nitrogen is fixed (reduced to ammonia) by diazotrophs in part of the nitrogen cycle.
- Nitrogen reduction occurs via nitrogenase, which is comprised of two separate proteins: the Fe protein and the MoFe protein. The Fe protein shuttles electrons through a 4Fe:4S center to the MoFe protein. Two molecules of ATP are hydrolyzed in this process for each electron transferred. The overall reduction of one molecule of N_2 requires 8 protons, 8 electrons, and 16 molecules of ATP. The MoFe protein uses two unique iron-sulfur centers to transfer the electrons and catalyze the reduction: the P center and the FeMo cluster. The mechanism of nitrogenase is not clear, but it is accepted that electrons are passed out of the Fe protein through the 4Fe:4S center, to the P center, and then to the FeMo cluster. It is also likely that the nitrogen binds to the molybdenum group when reduction occurs.

Concept Check

1. Describe the nitrogen cycle. Discuss the role of diazotrophs in nitrogen fixation and that of nirogenase in nitrogen reduction.
2. Describe the overall reaction of nitrogenase and why so many ATP molecules are hydrolyzed.
3. Where does the ferredoxin used in nitrogen fixation come from? Explain your answer.
4. Explain why there is such strong research interest in how bacteria make ammonia.

Bioinformatics Exercises

Exercise 1 Viewing and Analyzing Photosystems I and II

Exercise 2 Photosynthesis and the KEGG Database

Problems

23.1 Photosynthesis

1. Compare the oxidative phosphorylation to photophosphorylation.

2. Compare photosynthesis and electron transport in mitochondria. How are these pathways similar, and how are they different?

3. Explain why plants appear green but photosystems I and II have λ_{max} of 700 and 680 nm. Also, justify why do photosystems I and II respond to different wavelengths of light.

4. Why doesn't plastoquinone serve as a proton-gradient uncoupler?

5. Plastocyanin is a small protein of 10.5 kDa. How large would an average or large protein be?

6. A houseplant has been mutated so that its chlorophyll is more efficient at fluorescence than normal. What could be the likely result?

7. Some plants have leaves that are red or purple in color. What would this mean in terms of the structure of the chromophores found in these plants? How might this affect photosynthesis?

8. Explain the movement of electrons in the Z-scheme of photosynthesis.

9. Why does cyanide inhibit electron transport, but not photosynthesis?

23.2 Nitrogen Fixation

10. Could a device such as a catalytic converter be used to produce NH_3 from N_2 and H_2 gas under conditions normally found in biological systems? Explain your answer.

11. Using a table of standard enthalpies of formation that you find online, explain why N_2 is more stable than nitrogen oxides such as NO_2.

12. In nitrogen cycle, explain the

   a. significance of leghaemoglobin in root nodules.

   b. function of nitrobacter in biological fixation.

13. Using standard bond enthalpies, illustrate the energetics of ammonia synthesis. Explain why is it necessary to hydrolyze 16 ATP to fix nitrogen as ammonia.

14. How might elevated levels of hydrogen affect the kinetics in the reaction of nitrogenase?

15. CO is a mixed inhibitor of nitrogenase. How would a Lineweaver–Burk plot appear in the presence and absence of CO?

16. Binding and hydrolysis of which molecule during a nitrogenase reaction leads conformational changes that cause the transfer of electrons from P cluster to the Fe–Mo cofactor? Describe the process.

17. Why does nitrogenase apparently add electrons one at a time instead of all at once?

18. Explain the salient features of nitrogenase, the cofactors involved, and the features that make those cofactors interesting.

19. Nitrogenase is found in anaerobic bacteria. What information does this provide about its evolutionary history? How does chemistry inform evolution in this instance?

20. What are the unusual electron carriers contained by nitrogenase? Although it is not completely understood, how do you think the movement of electron takes place?

Ethics and Social Responsibility

21. Based on what you have read, what is your view of Haber? How do you think you might have acted in this situation?

22. Two wrongs do not make a right, but are we ever absolved for one transgression because of a good action done elsewhere?

23. Examine the professional codes of conduct for the American Chemical Society, American Society for Biochemistry and Molecular Biology, and International Union of Biochemistry and Molecular Biology. What types of activities are prohibited by these societies?

24. Are modern weapons of mass destruction, such as nuclear, biological, and chemical weapons, more problematic ethically than traditional weapons? Why or why not?

Suggested Readings

23.1 Photosynthesis

Aliverti, A., C. M. Bruns, V. E. Pandini, P. A. Karplus, M. A. Vanoni, B. Curti, and G. Zanetti. "Involvement of Serine 96 in the Catalytic Mechanism of Ferredoxin-NADP+ Reductase: Structure—Function Relationship as Studied by Site-Directed Mutagenesis and X-Ray Crystallography." *Biochemistry* 34 (1995): 8371–9.

Aliverti, A., V. Pandini, A. Pennati, M. de Rosa, and G. Zanetti. "Structural and Functional Diversity of Ferredoxin-NADP+ Reductases." *Archives of Biochemistry and Biophysics* 474, no. 2 (2008): 283–91.

Alte, F., A. Stengel, J. P. Benz, E. Petersen, J. Soll, M. Groll, and B. Bölter. "Ferredoxin:NADPH Oxidoreductase Is Recruited to Thylakoids by Binding to a Polyproline Type II Helix in a pH-Dependent Manner." *Proceedings of the National Academy of Sciences U S A* 107, no. 45 (2010): 19260–5.

Arakaki, A K., E. A. Ceccarelli, and N. Carrillo. "Plant-Type Ferredoxin-NADP+ Reductases: A Basal Structural Framework and a Multiplicity of Functions." *FASEB Journal* 11, no. 2 (1997): 133–40.

Barber, J. "Photosystem II: The Engine of Life." *Quarterly Reviews of Biophysics* 36 (2003): 71–89.

Barber, J. "Photosystem II: The Water-Splitting Enzyme of Photosynthesis." *Cold Spring Harbor Symposia on Quantitative Biology 77* (2012): 295–307.

Bassham, J. A. "Mapping the Carbon Reduction Cycle: A Personal Retrospective." *Photosynthesis Research* 76, no. 1–3 (2003): 35–52.

Chitnis, P. R. "Photosystem I: Function and Physiology." *Annual Review of Plant Physiology and Plant Molecular Biology* 52 (2001): 593–626.

Ferreira, K. N., T. M. Iverson, K. Maghlaoui, J. Barber, and S. Iwata. (2004) "Architecture of the Photosynthetic Oxygen-Evolving Center." *Science* 303 (2004): 1831–8.

Fromme, P., and P. Mathis. "Unraveling the Photosystem I Reaction Center: A History, or the Sum of Many Efforts." *Photosynthesis Research* 80, no. 1–3 (2004): 109–24.

Golbeck, J. H. "Structure, Function and Organization of the Photosystem I Reaction Center Complex." *Biochimica et Biophysica Acta* 895, no. 3 (1987): 167–204.

Guskov A., J. Kern, A. Gabdulkhakov, M. Broser, A. Zouni, and W. Saenger. "Cyanobacterial Photosystem II at 2.9 Å Resolution and the Role of Quinones, Lipids, Channels and Chloride." *Nature Structural and Molecular Biology* 16, no. 3 (2009): 334–42.

Hasan, S. S., E. Yamashita, D. Baniulis, and W. A. Cramer. "Quinone-Dependent Proton Transfer Pathways in the Photosynthetic Cytochrome b6f Complex." *PNAS* 110, no. 11 (2013): 4297–302.

Jordan, P., P. Fromme, H. Tobias Witt, O. Kuklas, W. Saenger, and N. Krauss. "Three-Dimensional Structure of Cyanobacterial Photosystem I at 2.5 A Resolution." *Nature* 411 (2001): 909–17.

Kameda, H., K. Hirabayashi, K. Wada, and K. Fukuyama. "Mapping of Protein-Protein Interaction Sites in the Plant-Type [2Fe-2S] Ferredoxin." *PLoS One* 6 (2011): e21947–e21947.

Karapetyan, N. V., A. R. Holzward, and M. Rogner. "The Photosystem I Trimer of Cyanobacteria: Molecular Organization, Excitation Dynamics and Physiological Significance." *FEBS Letters* 460 (1999): 395–400.

Kurisu, G, M. Kusunoki, E. Katoh, T. Yamazaki, K. Teshima, Y. Onda, Y. Kimata-Ariga, and T. Hase. "Structure of the Electron Transfer Complex between Ferredoxin and Ferredoxin-NADP+ Reductase." *Nature Structural & Molecular Biology* 8, no. 2 (2001): 117–21.

Loll, B., J. Kern, W. Saenger, A. Zouni, and J. Biesiadka. "Towards Complete Cofactor Arrangement in the 3.0 A Resolution Structure of Photosystem II." *Nature* 438 (2005): 1040–4.

Paladini, D. H., M. A. Musumeci, N. Carrillo, and E. A. Ceccarelli. "Induced Fit and Equilibrium Dynamics for High Catalytic Efficiency in Ferredoxin-NADP(H) Reductases." *Biochemistry* 48, no. 24 (2009): 5760–8.

Talts, E., V. Oja, H. Rämma, B. Rasulov, A. Anijalg, and A. Laisk. "Dark Inactivation of Ferredoxin-NADP Reductase and Cyclic Electron Flow under Far-Red Light in Sunflower Leaves." *Photosynthesis Research* 94, no. 1 (2007): 109–20.

23.2 Nitrogen Fixation

Beinert, H., R. H. Holm, and E. Münck. "Iron-Sulfur Clusters: Nature's Modular, Multipurpose Structures." *Science* 277, no. 5326 (1997): 653–9.

Byer, A. S., E. M. Shepard, J. W. Peters, and J. B. Broderick. "Radical S-Adenosyl-L-Methionine Chemistry in the Synthesis of Hydrogenase and Nitrogenase Metal Cofactors." *Journal of Biological Chemistry* 290, no. 7 (2015): 3987–94. doi: 10.1074/jbc.R114.578161.

Einsle, O. "Nitrogenase FeMo Cofactor: An Atomic Structure in Three Simple Steps." *Journal of Biological Inorganic Chemistry* 19, no. 6 (2014): 737–45. doi: 10.1007/s00775-014-1116-7.

Fontecave, M. "Iron-Sulfur Clusters: Ever-Expanding Roles." *Nature Chemical Biology* 2 (2006): 171–4.

Hu, Y., and M. W. Ribbe. "Nitrogenase Assembly." *Biochimica et Biophysica Acta* 1827, no. 8–9 (2013): 1112–22. doi: 10.1016/j.bbabio.2012.12.001.

Johnson, D. C., D. R. Dean, A. D. Smith, and M. K. Johnson "Structure, Function, and Formation of Biological Iron-Sulfur Clusters." *Annual Review of Biochemistry* 74 (2005): 247–81.

MacLeod, K. C., and P. L. Holland. "Recent Developments in the Homogeneous Reduction of Dinitrogen by Molybdenum and Iron." *Nature Chemistry* 5 (2013): 559–65.

Seefeldt, L. C., B. M. Horrman, and D. R. Dean. "Mechanism of Mo-Dependent Nitrogenase." *Annual Review of Biochemistry* 78 (2009): 701–22. doi: 10.1146/annurev.biochem.78.070907.103812.

TIMELINE CHAPTER 23

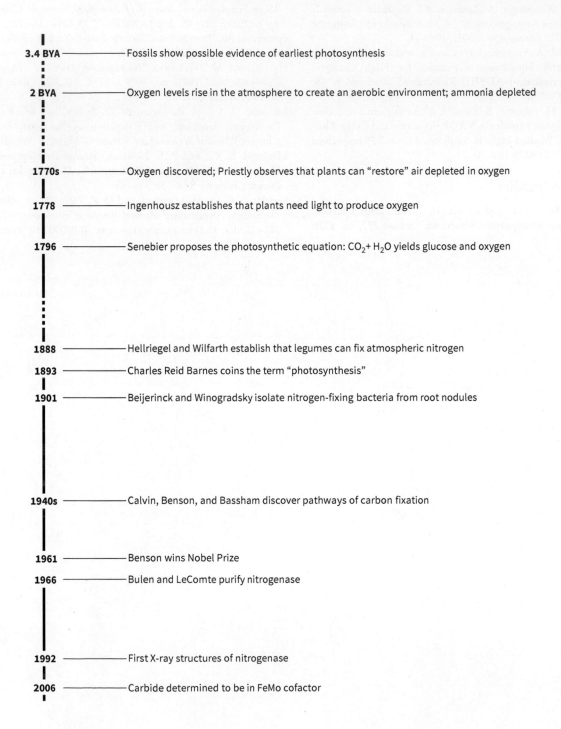

3.4 BYA —————— Fossils show possible evidence of earliest photosynthesis

2 BYA —————— Oxygen levels rise in the atmosphere to create an aerobic environment; ammonia depleted

1770s —————— Oxygen discovered; Priestly observes that plants can "restore" air depleted in oxygen

1778 —————— Ingenhousz establishes that plants need light to produce oxygen

1796 —————— Senebier proposes the photosynthetic equation: $CO_2 + H_2O$ yields glucose and oxygen

1888 —————— Hellriegel and Wilfarth establish that legumes can fix atmospheric nitrogen

1893 —————— Charles Reid Barnes coins the term "photosynthesis"

1901 —————— Beijerinck and Winogradsky isolate nitrogen-fixing bacteria from root nodules

1940s —————— Calvin, Benson, and Bassham discover pathways of carbon fixation

1961 —————— Benson wins Nobel Prize

1966 —————— Bulen and LeComte purify nitrogenase

1992 —————— First X-ray structures of nitrogenase

2006 —————— Carbide determined to be in FeMo cofactor

Affinity chromatography

When and why it is used: Affinity chromatography is one of the most common chromatography techniques, and it is used in nearly any biochemical isolation or purification.

How it works: In affinity chromatography, as in all chromatography, there is a mobile and a stationary phase. Molecules partition differentially between the two, leading to separation and enrichment of one molecule over another. In affinity chromatography, the solid phase can be more general such as a C8 or C18 matrix, meaning a matrix coated with hydrocarbon chains either 8 or 18 carbons long, or it can be more specific such as an antibody binding to a specific antigen or protein of interest. The mobile phase washes away molecules that have not bound as tightly, and the molecules of interest are eluted, or released from the solid phase, through a change in solvent polarity or pH, or outcompeted by addition of more ligand.

Strengths and weaknesses: There are many different types of affinity chromatography, and the technique can be easily scaled up to preparative levels or scaled down to run small analytical assays. Although the matrix costs can vary across the different types of techniques, affinity chromatography is a common and flexible technique used in labs worldwide.

Analytical ultracentrifugation

When and why it is used: Analytical ultracentrifugation is a technique that measures the ability of a pure protein or macromolecular complex to sediment when subjected to a high centrifugal force. It is used to deduce the hydrodynamic properties of a particle or molecule, such as the number of subunits that are associated with one another in solution.

How it works: A sample is applied to the top of an analytical centrifuge tube and spun in the centrifuge. The optics of the centrifuge detect the progression of the band of protein as it migrates to the bottom of the tube as the tube is spun. Based on how quickly the sample sediments, the composition of the liquid in the tube, and several other factors, the sedimentation coefficient, S, can be determined. S for any protein or complex is dependent on several factors including the size and dimensions of the complex.

Strengths and weaknesses: Analytical ultracentrifugation requires a dedicated instrument; an ordinary ultracentrifuge will not work. The information it provides, for example, the stoichiometry of subunits in solution, is of interest but can also be derived from other techniques such as size-exclusion chromatography or X-ray crystallography.

Chemical tests for carbohydrates

When and why it is used: Tests for carbohydrates are classic techniques in biochemistry. Although they have largely been replaced by enzymatic methods, the chemistry that underlies them is still the basis of several detection methods. These tests are often an excellent first step in the characterization of an unknown.

How it works: There are several assays for carbohydrates. Most take advantage of the observation that carbohydrates contain aldehyde and ketone groups and are subject to oxidation. Some common tests are:

Benedict's test consists of an alkaline solution of Cu^{2+} ions and citrate. The copper ions will react with free aldehydes, generating brick-red copper I oxide (cuprous oxide, Cu_2O) from the blue Cu^{2+} ions and oxidizing the aldehyde to a carboxylic acid. Ketoses will isomerize to aldoses under the assay's alkaline conditions. This test is therefore a general test for almost any monosaccharide or disaccharide.

Barfoed's test uses similar chemistry to Benedict's test with one important difference. The assay is run under mildly acidic conditions. Under these conditions, ketoses will not isomerize and disaccharides also fail to react. This test is more specific for aldoses.

Seliwanoff's test and *Bial's test* exploit different aspects of carbohydrate chemistry. In both of these tests, carbohydrates are dehydrated with acid to form a conjugated alkene, either furfural or hydroxymethyl furfural. Seliwanoff's reagent contains resorcinol, which undergoes a condensation reaction with hydroxymethylfurfural to give a red product. This test is used for ketohexoses or disaccharides containing a ketohexose. Bial's test contains orcinol instead of resorcinol and reacts with furfural to give a green product, indicative of pentoses.

Molisch's test uses somewhat similar chemistry to Seliwanoff's and Bial's tests. A sample of carbohydrates is mixed with α-napthol. The mixture is overlaid on a sample of concentrated H_2SO_4; a purple color is indicative of monosaccharides that have formed a colored product (furfural or hydroxymethylfurfural) with the napthol.

The *iodine test for starches* takes advantage of the observation that iodine will bind to starches and turn blue. A solution of molecular iodine in water (I_2), often mixed with an iodine salt such as KI to help stabilize the iodine, is added to a sample. If the sample turns blue (the color can range from purple to green), it is a positive indication for starch. The blue color is thought to arise from I_3^- ions binding in the core of the starch helix.

Strengths and weaknesses: These tests are rapid, inexpensive, and easy to perform. Benedict's test was the basis of home diabetes testing for over 80 years. The tests are usually qualitative, but some can be made quantitative with the use of standards. The tests are usually not specific; for example, they may reveal that a sample contains a reducing sugar, but they not explicitly identify which sugar. They also require more material than modern analytical techniques.

Cell disruption

When and why it is used: Most biological samples are found inside the cell or tissue. It is often necessary to disrupt those cells to gain access to those molecules for purification or characterization. Following disruption, proteins can be studied or isolated, or the resulting cellular lysate can be fractionated using differential ultracentrifugation, a process termed cell fractionation. Cell fractionation is used to identify where in the cell a specific protein or process occurs, by colocalizing that protein or activity with known markers.

How it works: Numerous disruption techniques are available, but they require care because overambitious disruption can damage organelles, complexes, or proteins. Some cells can be lysed simply by incubating them in a hypotonic solution. Depending on the sample, detergents or enzymes may be added to assist in the lysis. Shearing a sample of cells by passing them through a narrow orifice, such as a narrow-gauge needle, is another way to gently disrupt the cell. These gentle techniques often leave many cellular complexes intact and are used to study larger organelles and complexes.

A more vigorous means of disrupting cells involves physically grinding them in a blender, homogenizer, or bead beater. Blenders and homogenizers shear or physically cut the cells open using mechanical force. Bead beaters use glass or ceramic beads sealed in a container with cells and buffer. The container is shaken vigorously, and the cells rupture. A French press homogenizer passes cells through a narrow orifice at high pressure, greater than 5,000 psi. These techniques are used for cells that are more resistant to lysis such as those with a cell wall or protective coating, for example, spores.

In nitrogen cavitation, a sample is incubated with nitrogen or another inert gas at high pressure, greater than 20,000 psi, until the gas is dissolved in the cell. The pressure is rapidly released, and the gas boils out of the cell, rupturing cells and organelles.

Chemically, cells can be disrupted either using detergents to dissolve the plasma membrane or enzymes to degrade the cell wall or membrane. Here, care must be taken to choose detergents that will permeabilize the cell without denaturing the proteins of interest. Bacteria are often lysed using lysozyme, an enzyme commonly isolated from hen eggs. This enzyme degrades the polysaccharides found in the bacterial cell wall. Without this wall, the bacterium loses structural integrity and is more susceptible to hypotonic lysis.

Finally, ultrasonic disruption uses a titanium probe that vibrates at 20,000 Hz to disrupt cells. This is the most effective means of disrupting cells, but it is so strong that damage to macromolecules is a potential problem. Also, precautions must also be taken to avoid overheating the sample.

Strengths and weaknesses: There are many techniques available to disrupt cells, each with its own strengths and weaknesses. Depending on the desired outcome, cells can be gently lysed or blown apart.

ChIP (chromatin immunoprecipitation) assay

When and why is it used: Chromatin immunoprecipitation (ChIP) assays are used to identify proteins, such as histones or transcription factors, that are bound to specific genes or DNA sequences.

How it works: Proteins and DNA are reversibly cross-linked using formaldehyde. The sample is sonicated to shear the DNA into smaller fragments. An immunoprecipitation is performed using an antibody directed against a protein of interest, for example, a specific transcription factor. This precipitation will pull down the protein–antibody complex, along with the DNA cross-linked to the protein. The cross-linking is reversed by heating, and the DNA fragments are amplified using PCR and sequenced. There are multiple variants of this assay, which vary with regard to cross-linking conditions and detection methods.

Strengths and weaknesses: The ChIP assay is relatively simple to perform and can be adapted for high-throughput analysis. However, the ChIP assay often suffers from low sensitivity and often cannot give a target sequence with high resolution. Finally, ChIP assays cannot give

functional data; it is not possible to tell if the proteins bound to the DNA sequence are biologically significant.

Circular dichroism

When and why it is used: Circular dichroism (CD) is a technique used to analyze the ability of a sample to differentially absorb either left- or right-hand circular polarized light. The absorbance of circular polarized light is due to interactions with chiral structures. As such, it is used as a measure of alpha-helicity in proteins or the helical structures of nucleic acids.

How it works: The CD spectrophotometer measures the absorbance of polarized light. Basic units operate much like any other sort of absorbance spectrophotometer. There is a light source, typically in the UV range, the sample, and a detector. CD differs in that its filters circularly polarize the light before it reaches the sample. More advanced instruments enable investigators to alter experimental conditions, such as temperature, and measure changes in the spectra.

Strengths and weaknesses: CD spectroscopy is quick and simple to use and provides straightforward answers, but these answers are often limited. A CD experiment can tell you, for example, that a protein sample is 45% alpha helix, but it cannot tell you where that helix is found in the protein or what the rest of the structure looks like.

Click chemistry

When and why it is used: Click chemistry is a generic term used to describe small modular chemical reactions that proceed with high yield in a complex environment. Although these reactions do not occur in biochemical systems, they can be effectively used to label or otherwise react biological molecules under biological conditions.

How it works: There are several examples of click chemistry. One of the best studied examples of a click reaction is the copper catalyzed alkyne-azide cycloaddition (the Huisgen cycloaddition).

In this reaction, an alkyne and an azide undergo a cycloaddition, forming a five-membered ring, and Cu^+ serves as a catalyst. Because the reaction is specific and can occur at high yield in conditions typically found in a biological system, and neither the azo nor the alkyne functional moieties are typically found in nature, this reaction fits the description of a click reaction.

Either azo or alkyne functional groups are synthetically incorporated into a lipid, carbohydrate, nucleotide, or amino acid in the laboratory. These modified molecules (for example, an alkyne fatty acid) are then biosynthetically incorporated into a larger biological molecule, such as a phospholipid. This molecule can be labeled in a biological system using an azo click chemistry partner that is conjugated to a fluorescent dye or other molecule.

Strengths and weaknesses: The main advantage of click chemistry is the high yield of these reactions under what might normally be very difficult reaction conditions—generally biological conditions, at relatively low temperature, with numerous contaminants, and using an aqueous solvent. Click labels are very small compared to other labeling technologies. Numerous reagents are now commercially available for click labeling, and several assays, such as assays for cell proliferation and lipid peroxidation, are now commercially available. The weaknesses of this system are that the copper used can be mildly cytotoxic, and the click-labeled reagents need to be generated in such a way that they are bioavailable to the cell; that is, the cells need to be able to incorporate these molecules as they normally would for other biological processes.

CRISPR-Cas9

When and why it is used: CRISPR is an acronym for clustered regularly interspersed short palindromic repeats. CRISPR describes short sequences identified in prokaryotes that act in conjunction with the Cas9 nuclease to serve as a primitive immune system in these organisms. CRISPR-Cas9 is a flexible and easy means of editing a genome.

How it works: There are two components to the CRISPR-Cas9 system. In prokaryotes, the Cas9 nuclease recognizes specific DNA sequences and cleaves them. In these organisms, the DNA sequences that are recognized come from viruses that have previously infected the bacterium. Cas9 is directed to these sites through a guide RNA. It is relatively easy to transfect other cells using a cDNA that encodes the Cas9 nuclease and contains a guide RNA directed against the region of the genome that an investigator wishes to edit. This means that based on the design of the guide RNA, any gene can be altered using this technique, and it can be done in a live complex organism.

Strengths and weaknesses: CRISPR is a relatively new technology but one with incredible promise. CRISPR is a very simple and inexpensive technology to use, involving only transfection or infection of cells with a plasmid encoding Cas9 and the guide RNA. It effectively and efficiently edits the genome and can be altered to make several different types of changes, including deletions and insertions. People have raised ethical concerns that CRISPR could be used to edit human genomes, but these questions are often raised when novel technologies are developed.

Cryogenic electron microscopy (cryo-EM)

When and why it is used: Cryo-EM is an imaging technique used to determine the structure of macromolecules, including proteins and nucleic acids. Often these are larger complexes of molecules.

How it works: Cryo-EM is a transmission electron microscopy technique. In this technique, there are several important differences between cryo-EM and traditional EM techniques. First, the experiment is conducted at very low temperature and using a weaker electron beam to avoid damaging the molecules. Second, the images of thousands of proteins are analyzed to generate a high-resolution image of the molecule in question.

Strengths and weaknesses: Structure determination of macromolecules is still difficult to obtain. NMR is largely useful only for small molecules, and X-ray diffraction is often limited by the ability to grow crystals of a protein. Cryo-EM does not require crystallization and therefore presents a much easier way to determine the structure of a protein or complex. In addition, complexes that will not crystallize due to complexity or other factors can still be imaged using cryo-EM.

Density-gradient ultracentrifugation

When and why it is used: Ultracentrifugation is a classic technique used to separate the components found in cellular lysates or plasma lipoproteins. The technique is commonly employed in cell fractionation studies, where cells are lysed and the components separated based on density.

How it works: There are variations on this technique, but all use salts, sucrose, or colloidal solutions, such as Percoll (colloidal silica coated with polyvinyl pyrolidone) to adjust the density of the solution. Gradients are centrifuged until the molecular complexes pellet or band at their isopycnic density, the point where the density of the complex and the surrounding medium are equivalent. Typical centrifugation experiments span several hours to a day in length and employ forces of 20,000 to 2,000,000 × g to achieve separations. In differential centrifugation, the density of the centrifugation media is adjusted and the molecule, complex, or organelle of interest is floated or pelleted away from the other components of the lysate. In isopycnic centrifugation, samples are centrifuged until they float at the same density of the media in which they are suspended. Cellular lysates in buffer are overlaid, underlaid, or placed within a gradient of media of differing densities. Samples are centrifuged until the complex of interest forms a band at the density of the complex. Complexes migrate through a density gradient based on the mass, shape, and partial specific volume of the complex. These techniques easily separate organelles, plasma lipoproteins, and macromolecular complexes, but they are not commonly used to purify or isolate soluble proteins.

Preparative ultracentrifugation differs from analytical ultracentrifugation experiments, which measures the sedimentation rate of a pure protein to give hydrodynamic information about the protein. Analytical ultracentrifugation is performed under very different conditions and with a different instrument.

Strengths and weaknesses: Ultracentrifugation is a simple and common technique for the isolation of organelles and macromolecular complexes. The technique used to disrupt the cell prior to centrifugation must be chosen carefully to avoid destroying the finer structure of organelles and complexes.

Dialysis and ultrafiltration

When and why it is used: Dialysis and ultrafiltration employ microporous membranes to achieve separation of molecules based on size or on concentration of larger molecules.

How it works: Dialysis uses a dialysis membrane—a bag with pores large enough to permit passage of small molecules but not larger ones. A range of different cutoff sizes is available, ranging from hundreds of daltons to hundreds of thousands of daltons. A sample is put into the dialysis bag and the bag is bathed in a beaker containing buffer or other solutions. Over time, generally several hours, molecules below the molecular weight cutoff, which correspond to the size of the pore, will diffuse between the bag and the buffer in the beaker, while the larger molecules will be retained. This technique is typically used to remove a small molecule, such as a salt, or in some cases larger molecules up to the size of large proteins.

Ultrafiltration is often used to concentrate a sample of macromolecules. This technique also employs a dialysis membrane but in a different configuration. Here, the membrane is found at the bottom of a chamber that is filled with the sample to be filtered. Pressure is applied using either a pressurized gas or a centrifuge to drive the solution through the dialysis membrane. In effect, this works like catching pasta in a colander, in that larger molecules are retained while the solvent and smaller molecules are lost. Newer versions of this apparatus are designed to prevent the loss of all solvent and thus prevent the filter from running dry.

Strengths and weaknesses: Dialysis has been employed for decades and is affordable and easy to do. Newer geometries of dialysis membranes minimize sample loss, but there is always a fear of losing sample due to a leaky membrane.

Differential scanning calorimetry (DSC)

When and why it is used: Differential scanning calorimetry (DSC) is a calorimetric technique that measures the absorbance of energy required to change the temperature of a sample. In the simplest case,

DSC can detect phase transitions. Although it is frequently used to precisely detect how a protein and nucleic acid denature, DSC is used as a tool to measure protein stability and protein folding.

How it works: Highly pure samples are placed into the calorimeter. The device precisely measures the amount of energy input to the calorimetry chamber as well as the temperature. Based on these values, phase transitions or denaturation can be determined.

Strengths and weaknesses: DSC is a relatively common technique that is simple to perform. The experiment requires small and relatively pure samples for analysis but provides interesting and important thermodynamic data.

DNA (or RNA) footprinting assay

When and why is it used: Footprinting assays are used to determine which specific segments of a nucleic acid are protected from enzymatic degradation by nucleases. Classically, the footprint is thought of as being bound by proteins, such as histones or transcription factors, but it can also be due to secondary structure, as in a riboswitch.

How it works: Typically, fragments of nucleic acids, either DNA or RNA, are terminally labeled with radiolabeled nucleotides. These fragments are then partially digested by a nuclease such as DNAse I in the presence or absence of molecules thought to bind or otherwise protect the nucleic acid from degradation. The resulting fragments are separated by gel electrophoresis, usually acrylamide. In this instance, the fragments can be detected using autoradiography, but other techniques such as fluorescence, Southern blotting, or northern blotting can also be used.

Strengths and weaknesses: Footprinting assays can be highly informative, revealing which DNA sequences are protected with base-pair resolution, but finding the correct conditions for protein binding and digestion can be challenging.

DNA sequencing

When and why it is used: Classically, DNA sequencing is used to determine the nucleotide sequence in DNA. Newer techniques use parallel technologies to sequence entire genomes.

How it works: In first-generation DNA sequencing, there is a sequencing mixture that contains a DNA polymerase, a template to be sequenced, a primer, an oligonucleotide from which the polymerase can extend that will specifically bind the sequence to be sequenced, and radiolabeled deoxynucleotide triphosphates. This mixture is divided into four reactions; each reaction utilizes a small amount of a different dideoxynucleotide triphosphate. Because these dideoxynucleotides lack a hydroxyl group on C-3 of the deoxyribose ring, the polymerase is unable to extend from one of these nucleotides, blocking further elongation from that chain. Because a different dideoxynucleotide is in each reaction, the base found at the end of that chain can be identified. The products of the four reactions are resolved on a polyacrylamide electrophoresis gel that can resolve pieces of DNA that differ in length by a single base. A film exposed to this gel shows bands that give the order of the individual bases.

The next evolution of this technology employs terminator nucleotides tagged with fluorescent dye. Terminator or dideoxynucleotides lack a 3′ hydroxyl group and prevent further elongation of the DNA chain. Because the different nucleotides can be labeled with different fluorophores, only one reaction is needed. In addition, the strands are separated with a capillary column rather than a gel. This enables the entire system to be automated to run as many as 384 samples at once.

Using the latest techniques, it is possible to sequence an entire genome at once. These techniques use a combination of modern polymerase chain reaction (PCR), advances in microscopic techniques, and advanced data processing. Between hundreds of thousands and billions of individual sequencing reactions occur at once and are read on a microscope slide. Short sequences of 25 to 250 bases are gradually built up, analyzed for quality, and then assembled into a full genome.

Strengths and weaknesses: Knowing the nucleotide sequence of a piece of DNA is amazingly powerful. It can show which sequences code and which do not, what is coded for by the sequencing codes, and what control elements there are. Start-up costs can be high, but the cost of sequencing has dropped by several orders of magnitude as the technology has progressed. Another challenge is how to mine useful information from the huge volume of data that modern sequencing produces.

Ectopic protein expression

When and why it is used: In many instances, it is desirable to study proteins in different cell types or to produce significant quantities of protein. In such cases, ectopic protein expression, or overexpression, is a useful technique.

How it works: The cDNA of the gene of interest is cloned into a vector (such as a plasmid) and transduced or infected into the cells of interest. The cells are often put under some type of selection, usually treatment with an antibiotic, to select for those that have acquired the vector. If the vector has been engineered with the proper promoter sequence for the cell type that has acquired the DNA, the gene will be expressed. The protein may accumulate within the cell or be secreted, depending on the nature of the protein and how it is processed in the cell. This will also vary depending on the protein and the cell type.

Strengths and weaknesses: The ability of a cell to internalize a piece of DNA and produce protein is one of the most powerful tools of modern biochemistry. Overall, it is relatively easy to use DNA to get cells to produce a protein, but the process can be difficult when a specific protein needs to be expressed in a specific cell type or using a particular vector. Investigators must understand that not all cells process proteins in the same way or have the capacity to secrete these proteins. Likewise, when investigating proteins in cells that do not normally produce the protein in question, there can be complications, for example, other compensatory mechanisms or missing interaction partners.

Edman degradation

When and why it is used: The Edman degradation is a means of sequencing a protein by rounds of successive derivatization and cleavage of the amino terminus. Using this technique, it is possible to determine the sequence of an entire protein or identify a protein based on the sequence of the first few amino acids. The process was first described by Pehr Edman in 1950.

How it works: In an Edman degradation, the amino terminus of a protein is derivatized, usually with phenylisothiocyanate. This compound reacts with the free amino terminus to form a phenylthiocarbamoyl derivative, which is readily cleaved with acid to yield a thiazolinone. Subsequent treatment of the cleaved amino acid derivative with more acid gives a phenylthiohydantoin, which can be identified by high-performance liquid chromatography (HPLC) or mass spectrometry. The process is repeated until the protein is characterized.

Strengths and weaknesses: The Edman degradation was one of the first successful ways to determine the sequence of a protein. The process is simple and straightforward and has been used to develop automated sequencing devices. Unfortunately, in practice the characterization of a protein is complicated by incomplete reactions and usually only about 30 amino acids can be identified from a single amino terminus. Today, the amino acid sequence of a protein is more often determined from the mRNA sequence, and unknown proteins are identified by mass spectrometry.

Electron paramagnetic resonance (EPR)

When and why it is used: Electron paramagnetic resonance (EPR), also known as electron spin resonance (ESR), is a technique that is used to probe the spin states of electrons and thus detect unpaired electrons in a sample or reaction or to probe the redox states of a complex.

How it works: EPR is analogous to nuclear magnetic resonance (NMR). Samples are placed in a strong magnetic field. All electrons have a spin state associated with them known as the m quantum number. The electron either has a spin state of $+\frac{1}{2}$ or $-\frac{1}{2}$. Most electrons found in filled orbitals or in chemical bonds adhere to Hund's rule and must be spin-paired. However, unpaired electrons such as radicals and some redox states are not spin-paired. As a result, when probed with the correct wavelength of energy, typically microwaves in the EPR experiment, the system will absorb energy, giving a signal analogous to the NMR signal. The EPR signal is far simpler to interpret than an NMR signal because of the paucity of unpaired electrons in the sample.

Strengths and weaknesses: EPR can only be used when unpaired electrons are present; however, if they are, it is a relatively simple matter to obtain a spectrum. The normal complications in an NMR experiment such as complexity of signal, solvent effects, and overlap of peaks are not concerns in EPR.

Enzymatic assays

When and why it is used: Enzymatic assays are used to assay a sample for a specific molecule or metabolite. They do not typically require the isolation of the molecule to be tested; that is, they can be used on a crude mixture or biological sample, and they can be highly specific and sensitive.

How it works: An enzyme that catalyzes a reaction specific for the molecule of interest yields a product. For example, glucose oxidase reacts specifically with glucose to give glucuronic acid and hydrogen peroxide. One of these products is used in a secondary reaction to give a fluorescent, colored, or chemiluminescent product, which is detected spectrophotometrically. Hydrogen peroxide, for example, will react with dyes, fluorophores, or luminescent molecules to provide a signal. The assay is typically used with a standard curve to provide quantitative results.

Strengths and weaknesses: Enzymatic assays can be both highly specific and sensitive. For example, assays can detect glucose but not galactose or glucosamine, and they can detect this carbohydrate in the nanomolar range. With advances in molecular biology and protein expression, almost any enzyme can now be overexpressed; as a result, prices of enzymes have reduced dramatically, making these assays affordable, although not as inexpensive as chemical tests for functional groups, such as the tests used for carbohydrates. Enzymatic assays are usually easy to perform, especially the commercial kits or clinical tests, and little to no specific equipment is needed other than a detection system. However, results must be validated in some way to ensure that the biological samples are not being complicated by other molecules in the mixture.

Enzyme-coupled assay

When and why it is used: Often in the laboratory or clinic, the concentration of a specific metabolite must be obtained. However, the concentration of the metabolite may be low submicromolar to nanomolar, and the metabolite may be in a complex mixture with other biomolecules. Also, the molecule in question may be similar to other molecules in the mix.

One method commonly used to assay molecules in such situations is to employ an enzyme-based assay. Enzymes often function perfectly well in complex mixtures and can be highly specific for a given substrate.

How it works: Enzyme-based assays consist of a sample, a reaction mixture that includes an enzyme specific for the molecule in question, and some means of detecting that the reaction has happened. Often these enzymes are oxidoreductases, which cause a change in the redox state of NAD^+, $NADP^+$, or FAD, or generate H_2O_2. Assays that cause a change in the redox state of an electron carrier can be detected spectrophotometrically. Assays that generate H_2O_2 include fluorophores or dyes that react with the H_2O_2 and are read spectrophotometrically. The values obtained are compared to a set of standards to determine the concentration in the sample. If a specific oxidoreductase is not available for the molecule in question, other enzymes may be included to transform the unknown molecule into one that can be used to produce a signal.

Strengths and weaknesses: Enzyme-coupled assays can be highly specific, rapid, and sensitive. With advances that have been made in recombinant enzyme expression and purification, the cost of these assays has dropped considerably. Likewise, advances in fluorescent dye chemistry have resulted in assays that can easily measure unknowns in microliter-sized samples. One weakness of these assays is the nature of the detection system. If the detection system is based on the amount of $FADH_2$ or H_2O_2 produced for a signal, any component in the sample producing those compounds could affect the result.

Enzyme-linked immunosorbent assay (ELISA) and radioimmunoassay (RIA)

When and why they are used: ELISA and RIA are techniques used to quantify the amount of a single biomolecule—often a protein—in a complex mixture.

How they work: ELISA and RIA take advantage of an antibody specific to the molecule of interest for detection. There are several different variations on the assay.

In a direct ELISA assay, a mixture of proteins is bound to an assay plate or dish. An antibody directed against the molecule of interest is added and binds specifically to the protein of interest; all other antibodies are washed away. The antibody can then be detected using a second antibody that will recognize the first. This second antibody has an enzyme conjugated to it that catalyzes a chemical reaction, resulting in the development of a colored product. In a sandwich ELISA assay, the dish is coated with an antibody that recognizes the protein of interest. All other proteins are washed away, and the protein is detected with a second enzyme-coupled antibody against the protein.

In an RIA, a small amount of pure target molecule is radiolabeled with iodine (^{125}I). The antibody is bound to the assay plate with the radiolabeled target molecule. An unknown sample is added and competes with the antibody for the radiolabeled target. If more sample is added, more radiolabeled protein will be released. This can be quantified in a gamma counter and the amount of protein in the unknown sample calculated.

Strengths and weaknesses: Both ELISA and RIA are able to quantify the amount of a molecule in a mixture; there is no need to purify the molecule of interest, and the assays are sensitive into the nanomolar range. Kits are currently available for many molecules that make the assay simple and quick to conduct, but kits can be expensive, ranging from several hundred to a thousand dollars for 100 assays. RIA assays use radioactivity, typically isotopes of iodine, and thus require special handling.

Fluorescence anisotropy

When and why it is used: Fluorescence anisotropy is used to determine how quickly a fluorescent molecule (fluorophore) rotates in a system. In effect, it is a measure of rotational freedom and can be used in conjunction with other parameters to calculate the general axial ratios of a molecule, referred to as the Stokes radius.

How it works: Polarized light is used to excite a fluorophore, which captures the photon. The fluorophore keeps this photon as the molecule freely rotates. At some later time, microseconds to milliseconds, the photon is emitted. Because the fluorophore can move, the degree of polarization of the emitted photon is lost, and the loss is dependent on the movement—with molecules exhibiting greater movement also exhibiting greater loss of polarization. Because the polarization can be easily measured, this is an excellent means of calculating the degree of rotation of the fluorophore.

Strengths and weaknesses: Fluorescence anisotropy is typically conducted using specialized fluorimeters, purified samples, and fluorescent tags or molecules. Once the initial setup is completed, anisotropy experiments often yield interesting biophysical data about a system.

Fluorescence *in situ* hybridization (FISH)

When and why it is used: FISH is used to quickly identify where in the genome a gene or fragment of a chromosome is located.

How it works: FISH is a fluorescence microscopy technique. A spread of metaphase chromosomes is obtained from a biological sample, such as a cheek swab or blood sample. The chromosomes are fixed and stained with fluorescent dyes to indicate where both the DNA and gene or sequence of interest is found. The samples are analyzed under a fluorescence microscope. Because of the availability of different colored dyes and advanced microscope optics, it is very common to compare multiple colors simultaneously, meaning that multiple sequences can be identified. FISH commonly uses three to five different dyes; however, as many as 23 different colors (a different color for each chromosome) can be compared, leading to the other name for this technique: chromosome painting.

Strengths and weaknesses: FISH is a relatively easy, rapid, straightforward technique. Better dyes and staining kits have improved the technique. The cost of the microscope is often the limiting factor, although simple FISH experiments can be done with any fluorescence microscope. FISH can provide quick answers to simple questions, for instance, if there are multiple copies of a chromosome or whether or not a chromosomal rearrangement occurred. However, FISH is an optical technique and cannot provide information at the molecular level. For instance, it could be used to determine if a fragment of one chromosome has been fused to another, but it could not tell you exactly where in the sequence the fusion occurred.

Fluorescence microscopy or immunofluorescence

When and why it is used: Fluorescence microscopy, or immunofluorescence, is used to identify and localize proteins in the cell. It is also the foundation of several related techniques, including FRET (Förster or fluorescence resonance energy transfer), FRAP (fluorescence recovery after photobleaching), FISH (fluorescence *in situ* hybridization), and a host of other techniques.

How it works: Cells are labeled with a fluorophore, or a molecule that will fluoresce under certain wavelengths of light. The fluorophore can be a nonspecific dye, such as nucleic acid dyes or dyes for lipids and hydrophobic compartments; a dye that targets specific organelles, such as mitochondria or lysosomes; or a chimeric protein made by expressing the cDNA of a protein of interest fused to the cDNA from a naturally fluorescent protein such as green fluorescent protein (GFP), yellow fluorescent protein, or mCherry. Proteins may also be labeled with a fluorescent antibody. Often, two antibodies are used: the first one recognizes the protein of interest, whereas the second is fluorescently labeled and recognizes the first antibody. Using a combination of these techniques, multiple molecules can be detected at once.

Strengths and weaknesses: Fluorescence is a workhorse of modern experimental biology and has become one of the most common techniques of cell biology. It also provides invaluable information to the biochemist and offers additional means of localizing molecules in the cell, thus providing clues to the function of a molecule or complex. The technique requires a fluorescence microscope, the cost of which can be prohibitive, depending on the technique being employed; higher end confocal microscopes can cost upward of half a million dollars. Specific antibodies or cDNA constructs are also required to visualize each molecule of interest.

Fluorescence recovery after photobleaching (FRAP)

When and why it is used: FRAP is used to measure the mobility of a molecule in a plasma membrane; that is, it measures whether a molecule is floating freely in the membrane or is tethered to other molecules or in other ways restrained.

How it works: FRAP is performed using a fluorescence microscope fitted with special optics. First, a cell is fluorescently labeled using a fluorescent tag or dye. A laser is used to photobleach a spot on the cell. If the labeled molecules are small and untethered, they will quickly diffuse back into the spot, and the fluorescent signal seen at the spot will recover. If, however, the labeled molecule is anchored to some immobile element, the fluorescent signal will not return after photobleaching. Based on the rate of return, the diffusional coefficient of the molecule in question can be determined. The diffusional coefficient can provide other information; for example, if the diffusional coefficient correlates with a much slower time, it indicates that the labeled molecule may be associated with a much larger complex.

Strengths and weaknesses: FRAP is a relatively simple and rapid way to determine the lateral diffusional coefficient of a molecule in a membrane, but it can be expensive. Usually, FRAP is part of a confocal microscope setup that costs several hundred thousand dollars.

Fluorescence spectroscopy

When and why it is used: Fluorescence spectroscopy is a common optical technique that is used to directly detect some biological molecules. It is now currently used in conjunction with synthetic dyes and techniques (such as anisotropy, FRET, or immunofluorescence microscopy) as a detection method.

How it works: Fluorescence is a physical property of some compounds; generally, organic compounds with highly conjugated π-bond systems or some inorganic compounds. When these molecules interact with light of a specific wavelength, they absorb a photon, exciting an electron from the ground state to an excited energy state. That electron spends nanoseconds to milliseconds in the excited state before losing some energy as heat and dropping to a lower energy level. When the electron falls back to the ground state, it emits a photon of light; however, because this photon is at a lower energy level, it is at a longer wavelength and therefore a different color than the exciting photon.

Strengths and weaknesses: Fluorophores are typically on the order of a thousand times more sensitive than colorimetric molecules. Fluorescence is an environmentally dependent property and will vary based on the surroundings of the fluorophore. This property of fluorescence can also be used in experimental design.

Förster or fluorescence resonance energy transfer (FRET)

When and why it is used: FRET is a technique used to identify two different fluorophores that are within close proximity (5 Å) to one another. It is typically used to identify protein–protein interactions.

How it works: Molecules are tagged with two different fluorophores, often with two different fluorescent proteins such as GFP and mCherry. The FRET experiment is typically conducted using a fluorescence microscope with software that controls and analyzes the experiment. A fluorophore is excited with the correct wavelength of light; if a second fluorophore is within a specified distance and in the correct orientation, the energy of the excited state is transferred from the first fluorophore to the second, which will then emit a photon.

Strengths and weaknesses: Although FRET experiments require specialized equipment, there are few experiments that can detect protein–protein interactions. Nonetheless, there are multiple reasons why a FRET pair of fluorophores may not interact, giving a false negative result.

Glucose tolerance test (GTT)

When and why it is used: The glucose tolerance test (GTT) is used to test for diabetes or insulin resistance in an organism.

How it works: In a GTT, a fasted test subject is given a dose of glucose. This may be taken orally, in which case it is an oral GTT, or OGTT, or injected intraperitoneally, in which case it is an intraperitoneal GTT, or IGTT. As the glucose is absorbed, the plasma glucose level increases. This is measured over a specific time interval, for example every 15 minutes for 3 hours, and the glucose concentration will increase rapidly over the first 15 to 30 minutes. In the healthy state, glucose concentrations will decrease as insulin is released and cells absorb the glucose. This should result in a decrease in plasma glucose. However, if the subject has diabetes or insulin resistance, the time for the glucose levels to return to normal is increased, if it occurs at all over the time course of the experiment. The time required for glucose levels to return to normal can be compared to normal values to determine the degree of insulin resistance.

Strengths and weaknesses: With the advent of microscale glucose monitoring, glucose tolerance tests have become more common, less expensive, and easier to perform in the laboratory setting. The GTT is a powerful tool, but it must be coupled with some other assay, such as plasma insulin concentrations or insulin clamp, to determine where a defect has occurred in metabolism.

High-pressure liquid chromatography (HPLC)

When and why it is used: HPLC is used to resolve closely related molecules, for example, different fatty acids or isomers of small organic molecules. Although an analytical technique, HPLC can be adapted for use on a preparative scale.

How it works: HPLC is a liquid chromatography technique. The stationary phase of the chromatography system consists of a column filled with fine beads coated with the molecule that gives the stationary phase its characteristic properties. Coatings can be polar, aromatic, ionic, or nonpolar. Nonpolar columns are often referred to as reverse-phase columns and are often used to separate nonpolar molecules. The mobile phase, which consists of a mixture of solvents, varies depending on the conditions of the experiment. Gradient HPLC employs more than one mobile phase. As the experiment is run, solvents are gradually mixed together, for example, to shift from one pH to another, or to make a mobile phase more or less polar. Analytes can be detected by various methods including ultraviolet or visible range spectrometry, MS, or refractive index.

Strengths and weaknesses: HPLC is a common and reliable technique, and autosamplers permit the analysis of tens to hundreds of samples with minimal labor. However, the technique can be time consuming; depending on conditions, a single analysis can take anywhere from several minutes to more than an hour. Identification of specific analytes often requires other techniques to positively confirm their identity.

Homologous recombination for gene deletion (knockout organisms)

When and why it is used: Knockout organisms are used to ascertain the function of a protein throughout the organism's development and life by deleting the gene coding for that protein. Knockout mice are one of the most often studied knockout organisms.

How it works: The creation of a knockout organism through homologous recombination uses numerous experimental techniques. In short, the gene to be deleted must be identified, mapped using restriction enzymes, and subcloned. Next, a DNA construct is created with two regions of homology flanking a selectable marker. Longer regions increase the likelihood of homologous recombination. The construct is electroporated into mouse stem cells. These cells are selected using the selectable marker and are assayed to see whether the construct has in fact recombined correctly in the genome and deleted the gene of interest. The cells with the proper deletion are then injected into a developing embryo and implanted into a female mouse. The mice that are born are heterozygotes; they are crossed with one another to give homozygous knockout mice.

Strengths and weaknesses: The surgical deletion of a single gene can provide a wealth of information about that protein's functions in an organism. It may provide unexpected results in fields of science where one may not normally be focused. For example, deletion of the hormone-sensitive lipase gene also renders male mice infertile. However, recombination has several drawbacks. It can take more than a year to generate a knockout mouse. The process is highly involved and requires input from veterinarians, animal husbandry experts, cell biologists, and molecular biologists. These experiments are expensive and, depending on the laboratory, costs can easily range into hundreds of thousands of dollars. Finally, even if the deletion is achieved, it is possible that it will cause an embryonic lethality, in which case there is no mouse; a sterility, in which case there will be no more mice; or another protein or family of proteins that compensates for the gene that was deleted, in which case the mouse appears normal until it is tested or stressed in some way. Nevertheless, the opportunities to conduct science on what is basically a novel organism are exciting and almost limitless.

Ion-selective electrodes (ISEs)

When and why it is used: Ion-selective electrodes (ISEs) are electrochemical probes designed to detect specific solutes. Probes exist for many different molecules, but the most common example of an ISE is a pH probe.

How it works: ISEs consist of a glass tube with an ion-permeable window or membrane. This membrane is typically permeable only to the ion of interest. A wire (the electrode wire) is inserted into the glass tube. This wire is connected to one terminal of a galvanometer, while a reference wire is attached to the other terminal. The transmission of ions across the membrane is reflected and measured as current running across the membrane. This value can be converted to a chemical concentration.

Strengths and weaknesses: Newer ISEs are highly specific for solute and are robust compared to earlier generations, but they must still be handled with care. The main advantage of ISEs is that they are the only technique that can provide real-time monitoring of concentrations without further analysis. Electrodes exposed to conditions where living cells are present (for example, in bodies of water, in tissue culture, or within organisms) can be subject to biofouling, or the growth of cells on the electrode. Although the price of ISEs continues to drop, they can be prohibitively expensive for certain applications.

Immunoblotting (western blotting)

When and why it is used: Immunoblotting, or western blotting, is used to identify or quantitate amounts of a specific protein in a sample.

How it works: As the name implies, immunoblotting uses antibodies in the detection scheme. In this technique, a sample of proteins is separated using SDS-PAGE. The separated proteins are then transferred to a nylon or nitrocellulose membrane, typically performed using an electrophoretic technique. The blot is then coated with proteins that "block" the binding of any other proteins to the membrane, a step called blocking. Blocking ensures that antibodies used in the subsequent steps of the detection do not give high nonspecific backgrounds on the blot. Next, the blot is incubated in a sample of antibodies that bind to the protein of interest. The blot is washed, and a secondary antibody that recognizes the first antibody is added. This secondary antibody is coupled to an enzyme, protein, or fluorophore employed in the detection step. Finally, the blot is detected. Bands on the blot will vary in intensity depending on the amount of protein found.

Strengths and weaknesses: Immunoblotting is a standard technique used in many labs around the world. It has been optimized so that it can be done relatively quickly and inexpensively, and it is the basis for several other techniques such as IP, pulldown assays, and ChIP. Drawbacks include the need for specific antibodies to detect each protein.

Immunoprecipitation/Co-immunoprecipitation/Pulldown assays (IP/CoIP/Pulldown)

When and why it is used: Immunoprecipitations or pulldown assays are used to help identify proteins that are either directly or indirectly interacting with one another.

How it works: Cells containing a protein of interest are lysed and centrifuged to remove insoluble material. An antibody directed against the protein of interest is added. This antibody binds to the protein of interest. The antibody complex is then precipitated using a bead coated with either a second antibody that recognizes the first or a bacterial protein that binds nonspecifically to antibodies; protein A or protein G is commonly used. If the protein of interest is bound to other proteins, these should also precipitate with the antibody-bead complex. The precipitated proteins are eluted from the bead and analyzed using western blotting. Variants of this experiment can be conducted using fusion proteins (proteins that are joined together and encoded by a single cDNA) that bind to other ligands, for example, an avidin fusion protein binding to a biotinylated bead.

Strengths and weaknesses: The IP experiment is useful for demonstrating that proteins are found in a complex with one another. In general, the experiment is easy and inexpensive and lends itself well to large-scale experimentation. Investigators need to be careful not to overinterpret results. The IP shows that proteins are found in the same complex, but it does not necessarily show direct protein–protein interactions. One drawback of the experiment is that the investigator must hypothesize which proteins interact with the protein of interest in order to identify them. As improvements are made in IP and mass spectrometry technology, it may become easier to identify unknown proteins using this technique.

In situ hybridization

When and why it is used: An *in situ* hybridization, or simply *in situ*, is a technique that is used to analyze which cells are expressing a specific gene. It differs from some related techniques in that it examines expression at the mRNA level and does it on a cell-by-cell basis.

How it works: In an *in situ*, a complementary DNA or RNA probe is generated and binds to the RNA of interest. This probe may be fluorescent, radiolabeled, or detected with an antibody that generates a colored product. A section of tissue is fixed and permeabilized, and the probe is hybridized to the RNA of interest. If the probe is radiolabeled or generates a colored product, it must be developed by incubating the section with the appropriate chemicals. The sections are viewed under a microscope to determine which cells are expressing the gene of interest.

Strengths and weaknesses: *In situ* is one way of determining which cells of a sample express a gene of interest. It plays an important role in neuroscience, developmental biology, and some fields of microanatomy. The technique is relatively straightforward to complete, requires little specialized equipment, and yields specific results. Immunofluorescence microscopy asks similar questions at the protein level and can complement *in situ* studies.

Langmuir–Blodgett troughs (monolayer systems)

When and why it is used: A Langmuir–Blodgett trough is a device used to construct and study the physical properties of monolayers of amphiphiles including phospholipids and proteins that associate with them.

How it works: A Langmuir–Blodgett trough consists of a Teflon trough, a moveable barrier, and a Wilhelmy plate or Langmuir balance to measure surface pressure. A buffer solution is placed in the trough and an amphiphile is floated on the surface of the buffer. The barrier is used to adjust the area available to the amphiphile and therefore the packing of those molecules on the surface of the monolayer or film. The Wilhelmy plate indicates the surface pressure, which will change as a function of the packing of the surface or any change made to the surface, such as the binding or loss of molecules, or any changes in molecular conformation or packing.

Strengths and weaknesses: Langmuir–Blodgett troughs are useful in determining the binding characteristics of proteins to a phospholipid monolayer as well as the packing of membrane lipids, and they can provide important and interesting biophysical data. Systems are typically a more specialized piece of equipment and are highly sensitive, often requiring some type of isolation from background vibration.

Lectin-based techniques

When and why it is used: Lectins are carbohydrate-binding proteins. These proteins can be labeled using a fluorescent tag and then used to detect carbohydrate sequences in microscopy or western blotting. Alternatively, lectins can be employed in a purification scheme.

How it works: Lectins bind to short specific carbohydrate sequences using the same types of interactions that are found between a receptor and a ligand or an enzyme and a substrate. When originally identified, they were used in a way that was similar to the way antibodies are commonly used today, for example, in fluorescence microscopy, western blotting, and affinity chromatography.

Strengths and weaknesses: Lectins have seen a resurgence in their popularity with the advent of new fluorescent tags and further characterization of the specific carbohydrate each lectin recognizes. In some instances, lectins offer an alternative to antibodies. They are also still used in some clinical applications, such as blood typing.

Ligand-binding assays

When and why it is used: A ligand-binding assay is used to determine the affinity or binding constant (or dissociation constant) of two molecules, usually a ligand and a receptor. The ligand may be a small organic or inorganic compound or a larger molecule such as a protein.

How it works: Many different assays can be used to determine ligand binding. In most of these assays, the ligand is labeled either with a radiolabel or a fluorescent dye. In all of the assays, analytical techniques need to be used to identify which fraction of the ligand is bound to the receptor and which is unbound. Typically, this is performed by using a fixed concentration of the receptor, a fixed concentration of the labeled ligand, and a varying concentration of unlabeled ligand. In this format, the unlabeled ligand will compete with the labeled ligand for binding to the receptor. The higher the concentration of the more unlabeled ligand, the less the labeled ligand will bind to the receptor. The percentage of labeled ligand bound is graphed as a function of the log of the competing ligand, and a sigmoidal curve is observed.

Strengths and weaknesses: Ligand-binding assays require a labeled ligand, which either involves the use of radioactivity or a fluorescent tag. In the case of the latter, investigators must ensure that the modification to the ligand does not affect binding. Once developed, a ligand-binding assay is a workhorse in many laboratories and is a common technique in receptor biology and pharmaceutical development. Variants of ligand-binding assays include the enzyme-linked immunoassay (ELISA) and radioimmunoassay (RIA), both of which are used to measure the concentration of molecules.

Lipidomics

When and why it is used: Lipidomics is a relatively new field. Similar to other -omics based fields, such as genomics or proteomics, lipidomics seeks to identify and characterize all of the lipids in a sample.

How it works: Experimental lipid samples are often extracted away from cells, tissues, or media by an organic liquid–liquid extraction. The resulting lipids may be separated by thin layer chromatography or HPLC before characterization. The characterization is usually conducted using mass spectrometry with the samples being introduced via either ESI or MALDI; however, some samples are also identified using NMR. Depending on the separation and characterization scheme, the exact identity of lipids, including the percentages of individual phospholipid subclasses and their specific acyl chain composition, can be determined. Interestingly, several new lipids, in particular, signaling lipids found in low abundance, have been identified using this technique. The functions of these molecules remain unknown.

Strengths and weaknesses: A lipidomics experiment can generate massive amounts of data. Any given biological sample can contain many thousands of different lipid species, and the lipidomics experiment can enumerate each one. Even relatively pure samples, such as the lipoprotein subclasses, have significant diversity with regard to their lipid components. Lipids found in small quantities, although they are some of the more interesting molecules, are also often the hardest to isolate and characterize. Although the experiment can yield vast quantities of data, careful analysis and interpretation is needed. Another potential drawback is the cost incurred when trying to examine large numbers of samples by mass spectrometry.

Making solutions and buffers

When and why it is used: Buffers and solutions are central to most experimental biochemistry. All students of biochemistry must know how to calculate the correct concentrations in solutions and buffers and how to make these solutions properly.

How it works: Making solutions and buffers usually involves the dissolution of solids, liquids, or even gases in a solvent, typically water. Solutes are dissolved in a smaller amount of solvent and then diluted to the final volume. Similarly, when buffers are prepared, the salts and acids used in the buffer are dissolved in a small volume of liquid, the pH is adjusted to the proper value, and the solution is diluted to the final volume. This does not substantially affect the pH of the final product as long as the dilution factor is small compared to the buffering capacity of the solution.

Several measuring devices are commonly employed in the biochemistry laboratory. These include graduated cylinders and volumetric flasks for larger volumes, pipettes and burets for volumes of 1 to 50 mL, and micropipettes and syringes for smaller volumes (down to the nanoliter scale). Students should be reminded to use proper laboratory practice and accurately measure all liquids in the laboratory. The gradations on a beaker, capped test tube, or Erlenmeyer flask carry with them a generally large (±10%) error.

Strengths and weaknesses: There are really no strengths or weaknesses to making solutions and buffers because they are a basic part of laboratory life. However, some laboratories will purchase ready-made solutions; this may seem extravagant, but it can save time, a valuable commodity in a research endeavor. In addition, purchasing the solutions generally ensures that they are uniform, which in some cases, such as in diagnostic or clinical chemistry, may be warranted.

Mass spectrometry (MS)

When and why it is used: Mass spectrometry (MS) is used to determine the mass of biological molecules. Historically, it has been used to identify and characterize small organic molecules, but advances in technology now permit analysis of macromolecules and complexes of macromolecules. Based on the mass, the user can positively identify a molecule or determine whether a modification has been made. Because MS can identify both the molecule in question and any modifications directly, it is rapidly gaining in popularity.

How it works: The mass spectrometer can be thought of as a modular instrument. Before mass determination, samples may be isolated by chromatography, depending on the experiment and the instrumental configuration. The instrument itself consists of a means of introducing and ionizing the sample (the ion source), a resolution chamber (the mass analyzer), and a detector. Common means of introducing and ionizing small organic molecules include fast atom bombardment (FAB) or chemical ionization (CI). Larger molecules, such as proteins, are ionized using matrix-assisted laser desorption ionization (MALDI) or electrospray ionization (ESI). Resolution is accomplished by separating ions not solely by their mass but also by their mass-to-charge ratio (m/z). Magnetic sector, time-of-flight, quadrupole, and ion traps are the main categories of instruments used to resolve ions. In an electron multiplier, ions are detected by the impact of the ion on the surface of a detector. In an Orbitrap (a type of ion trap) ions are detected by generation of a current in a nearby coil of wire.

Strengths and weaknesses: Mass spectroscopy identifies unknown molecules far more quickly and easily than any other technique, often with unambiguous results. Likewise, it provides information about modifications of proteins that is difficult to obtain using other techniques. The primary drawback of MS is the cost. High-end instruments typically cost more than $300,000 and often require specialists to keep them running efficiently.

Microarrays

When and why it is used: Microarrays are used to obtain expression data for a large number of genes in a cell, tissue, organ, or organism at once.

How it works: To perform a microarray experiment, a microarray specific for the organism or tissue of interest and two RNA samples are needed for comparison. For example, these RNA samples could be one from a healthy tissue and one from a diseased tissue. A microarray is most often a glass microscope slide with anywhere from 3,000 to 20,000 microscopic spots of specific cDNAs arrayed on the slide; the location of each cDNA is known. These two RNA samples are labeled using two different-colored fluorescent tags: red for the experimental sample and green for the control sample. Care must be taken to ensure that the degree of labeling is similar in both samples. Next, the labeled RNA samples are mixed in equal amounts and hybridized to the slide. The slides are washed and then detected. If the original RNA samples have more of one gene being expressed in the experimental sample than in the control, the correlating spot on the microarray will light up red. If a gene is more strongly expressed in the control sample, it will light up green. By comparing the intensity differences of the two colors and knowing which genes are found at each spot, a complete map of how expression has been altered in the experimental sample can be determined.

Strengths and weaknesses: Microarrays provide data simultaneously on all of the transcripts of the cell. Expression of a transcript is measured relative to the other genes being compared. As a result, changes in gene expression levels with very low or high levels of expression may be harder to determine. The strength of microarrays lies in the total amount of data obtained, but this is in many ways also a liability. Microarrays provide clues regarding genes that warrant further investigation; however, microarrays do not usually provide critical information regarding a single gene. Likewise, the complexity, and in the past the cost, of the experiment usually kept experimental sample values low; $n = 3$ was not uncommon. The cost of microarrays has fallen dramatically since the technology was first developed.

Northern blotting

When and why it is used: Northern blotting is a technique that is used to identify and obtain the size of mRNAs. It is often used as a *de facto* measure of expression, although investigators should be cautioned that message levels do not always correlate with protein levels.

How it works: Total RNA or mRNA is isolated from a sample of interest. The RNA can be from an entire organism, organs, tissues, or cells. The RNA molecules are denatured and separated on an agarose gel similar to other nucleic acid separation techniques. The separated molecules are transferred to a membrane, typically nylon or PVDF, through osmosis; they are permanently affixed to the membrane by drying under vacuum or cross-linking with UV light. Next, unoccupied sites on the membrane are "blocked" by treating with a blocking agent, typically a solution of nucleic acids, proteins, or polymers that discourage the binding of other nucleic acids. The membrane is then probed with a radioactive cDNA generated against the gene or message of interest. Ideally, the radioactive probe will bind only to its complementary mRNA on the blot. The unbound probe is washed away, and the blot is exposed to film or a detection system for imaging. Bands corresponding to the message or messages are seen and are proportional to the amount of RNA found in the sample.

Strengths and weaknesses: Northern blotting is a widely used and reliable technique. An experienced investigator can easily survey multiple genes or multiple samples for expression in a relatively short amount of time, usually over several days. This technique is a workhorse for many labs, but it is gradually being replaced by modern technologies such as RT-PCR, RNA-Seq, and microarrays. These newer technologies have advantages in throughput (number of samples analyzed over time) and flexibility. Another drawback of northern blotting is its reliance on radioisotopes for detection. Newer northern blot techniques employ fluorescent detection, decreasing the amount of regulation needed and general work load.

Nuclear magnetic resonance (NMR)

When and why it is used: Nuclear magnetic resonance (NMR) is a spectral technique used to determine the structure of molecules.

How it works: A sample is placed into an NMR spectrometer. This device consists of a strong magnet, a radio frequency generator, a detector, and a control panel used to run the experiment. In the magnetic field of the instrument, the sample absorbs energy in the radio wave frequency and reemits this energy as it returns to the ground state. Molecules differ in how they absorb and reemit this energy based on the nature and electronic environment of the nuclei. Based on these differences, the connectivity or structure of the molecule can be determined.

Strengths and weaknesses: NMR is a very popular technique in organic chemistry but less so in biochemistry, due partially to the size and complexity of the molecules involved. A common organic molecule may contain 20 protons that need to be assigned in an NMR spectrum, but a simple protein could easily have a thousand protons. To assist in resolving these peaks, NMR instruments needed for structure determination in biochemistry often employ high magnetic fields and are expensive instruments costing millions of dollars.

Polymerase chain reaction (PCR)

When and why it is used: PCR is used to amplify, that is, make multiple copies of, DNA sequences. PCR and variations on this technique are an integral part of many modern approaches to molecular biology, such as sequencing, site-directed mutagenesis, and genotyping.

How it works: The PCR reaction consists of four components: a DNA template to be replicated, short oligonucleotides that flank the region to be amplified, deoxynucleotide triphosphates, and a polymerase that does not denature at high temperatures. The components are mixed in a buffer solution and placed into a thermocycler—a programmable heating and cooling system that precisely cycles the temperature through multiple cycles. The thermocycler heats the sample to melt the template DNA, then cools it to a temperature where primers will anneal to the template strand, and finally warms it to enable the polymerase to copy the template strand. After 25 to 30 cycles, the target sequence may be amplified by as much as a million-fold.

Strengths and weaknesses: PCR is easy to conduct and is relatively inexpensive. Commercial availability of polymerases from thermophilic organisms has made PCR widely accessible to many; the limiting factor in many cases is access to a thermocycler, which can be purchased for less than $4,000. The technique is rapid (less than three hours), easy to perform, and sensitive enough to detect a single molecule of DNA. However, the sensitivity means that even slight levels of contaminants will be amplified in the reaction, leading to false positives.

Protein assays

When and why they are used: Protein assays are used to determine the concentration of total protein in a sample. Many different methods can be used, depending on the purpose of the assay. In the clinic, protein assays are used to determine the presence or absence of protein in a sample such as urine. In the research laboratory, protein assays are a mainstay of many analytical and purification schemes. In those applications, knowing the concentration of proteins in a sample enables the investigator to make consistent comparisons between samples or track how well a purification scheme is going.

How they work: There are several different techniques available for protein determination:

UV spectrophotometry uses the absorbance of tyrosine and tryptophan residues at 280 nm to correlate to total protein in a sample.

The Bradford protein assay uses a blue dye (Coomassie Brilliant Blue G250) that non-specifically binds protein, causing a shift in the spectral characteristics of the dye. The assay is read using a visible-range spectrophotometer.

The *Lowry assay* and *BCA assay* use similar chemistry. Both of these assays take advantage of the observation that peptide bonds and potentially some side chains will stoichiometrically reduce Cu^{2+} to Cu^+. The reduced Cu^+ will react with a mixture of phosphotungstic and phosphomolybdic acid in a Lowry assay, or with the dye bicinchoninic acid in a BCA. These complexes have characteristic absorbances and can be determined using a visible range spectrophotometer as with the Bradford assay.

Finally, protein can be assayed on a much larger scale using the *Kjeldahl assay*, which destroys proteins by digesting them with sulfuric acid to produce ammonium sulfate. The ammonia is titrated to give the total amount of nitrogen in the sample. Thus, the assay is actually measuring nitrogen rather than protein, but on the scale of grams to kilograms it is relatively accurate, provided several corrections are taken into consideration.

Strengths and weaknesses: UV spectrophotometry is nondestructive but relatively insensitive. It also assumes a constant ratio of tyrosine and tryptophan to other amino acids in all proteins—a dangerous assumption when working with a purification scheme. In addition, other biological molecules (notably nucleosides and nucleic acids) also absorb in the UV range.

The Bradford is a classic assay and is more sensitive than UV but is incompatible with several common laboratory reagents.

The Lowry is more time consuming than the Bradford but is used as a standard assay in many laboratories worldwide. The BCA—the newest of these assays—has the fewest interfering substances and takes less time than the Lowry. It is also the most sensitive of the methods discussed.

Interpretation of the Kjeldahl assay can be complicated by other nitrogen sources in the sample. It can be time consuming, but it is the standard method for measuring protein levels in many industrial applications such as food and dairy science.

Pulsed-field gel electrophoresis (PFGE)

When and why it is used: PFGE is a variant on normal agarose gel electrophoresis. By varying the electric field, the resolution of very large (megabase) pieces of DNA is possible.

How it works: In a routine agarose gel electrophoresis, the electric field is applied in a single direction. This allows for the resolution of small pieces of DNA; when the pieces exceed approximately 20 kb, the ability to resolve them ceases, and larger pieces run as a single band. In PFGE, three different fields are used. The first one is oriented through the gel as in routine gel electrophoresis. Two other fields are oriented approximately 60 degrees to either side of the first field and are pulsed periodically, ranging from a few seconds every few minutes to an hour. When this happens, the very large molecules of DNA will react to the field at different rates, based on their size. This enables the resolution of these large molecules.

Strengths and weaknesses: PFGE is a common and relatively simple technique. A programmable power supply is needed. It is a central technique in genomic analysis, genotyping, or genetic fingerprinting of organisms. Due to the size of the fragments and the need to pulse the gel, PFGE experiments often have to run overnight (16 to 20 hours).

Quasi-elastic light scatter (QELS), Raleigh scattering, photon correlation spectroscopy, or dynamic light scattering (DLS)

When and why it is used: QELS is a technique that uses the ability of a particle, a protein, macromolecular complex, or organelle to scatter light as a function of its shape and size. QELS can be used to determine the shape and size of a pure sample of particles or the distribution of a diverse population of particles.

How it works: The QELS experiment shines laser light on a sample and detects the scattered light 90° to the incident beam. When measured over short microsecond to millisecond time scales, the signal will vary over time due to the Brownian motion of particles in solution. The variation in the signal can be correlated with the size and crude shape of the particle and the distribution of particle sizes.

Strengths and weaknesses: QELS is a simple but elegant technique that provides data regarding size of particles. QELS is nondestructive and easy to perform quickly. It does require a dedicated instrument with sensitive optics.

Real-time PCR (RT-PCR) or q-PCR

When and why it is used: Real-time, or quantitative, PCR is a technique that is used to measure the expression of a specific mRNA transcript.

How it works: RT-PCR is similar in many ways to PCR; RT-PCR employs a special thermocycler and relies on fluorescent detection methods. Like for PCR, there needs to be a template, generally a cDNA copy of the cell's mRNA, nucleotides, and a thermostable polymerase. RT-PCR differs in that it uses special primers. The primers contain a hairpin loop secondary structure in addition to the specific sequence that anneals to the template. At one end of the hairpin is a fluorophore, and adjacent to it is a fluorescence quencher. Because the quencher is near the fluorophore, fluorescence is quenched and there is no signal; however, as the primer is incorporated into PCR product, the quencher and fluorophore become separated and the signal is no longer quenched. Therefore, accumulation of PCR product can be measured as an accumulation of fluorescent signal; the RT-PCR thermocycler has the capability to measure this fluorescence each cycle.

Newer generation RT-PCR employs normal PCR conditions and primers but includes dyes that intercalate into the PCR product as it accumulates. These are also quantitated by the thermocycler.

Strengths and weaknesses: The price of RT-PCR and specific primers has fallen dramatically as technology has improved. RT-PCR enables investigators to measure expression levels within a few hours as compared to several days when using northern blotting. In addition, the technology is PCR-based; it is simple to run hundreds of samples in a single day or automate the system. One drawback of RT-PCR is that, as a PCR-based technique, care must be taken to ensure that samples are not contaminated, leading to artifactual results.

Reporter gene assay

When and why is it used: Reporter gene assays are used to detect the segments of a promoter region involved in the regulation of a specific gene.

How it works: A region of DNA upstream of a gene of interest is selected. Typically, this region is anywhere from 2 to 5 kilobases upstream of the start site. This piece of DNA is cloned in a vector in front of a reporter gene, a gene encoding a protein that can be easily detected. Classical reporter genes include chloramphenicol acyltransferase (CAT) or luciferase, but often today green fluorescent protein (GFP) is used. Different constructs using truncated segments of the promoter sequence are generated and tested to determine which regions are necessary for reporter gene expression.

Strengths and weaknesses: Reporter gene assays are relatively easy to conduct and can be informative, but they require the use of cultured cell systems, which may not accurately recreate the environment found in native tissue.

RNA silencing (shRNA, siRNA, RNAi)

When and why it is used: RNA silencing is a family of techniques, including shRNA, siRNA, and morpholino, that silence genes at the post-transcriptional level using the RNAi, or RNA interference, pathway.

How it works: RNA silencing techniques introduce RNA or DNA constructs to the cell. These constructs code for specific sequences that form double-stranded RNA structures for processing by the cell's native RNAi enzymes. Ideally these constructs will silence a specific mRNA.

Strengths and weaknesses: RNA silencing techniques are rapidly evolving as more is known about the science underlying the technology. These are effective and easy experiments to conduct, and due to their nature—transfecting cells with nucleic acids—they are relatively straightforward and easy to scale up. Many constructs developed to silence RNAs only do so partially, resulting in an intermediate phenotype. In and of itself this can be experimentally valuable, but it raises questions about whether the RNA of interest has been completely silenced and whether protein levels have been affected as predicted.

Salting in or salting out

When and why it is used: Salting in or salting out is simply precipitating (out) or increasing (in) the solubility of a macromolecule, usually a protein, through the addition of a salt.

How it works: Proteins are kept in solution through the interaction of water molecules with polar and hydrophilic parts of the protein surface. As increasing amounts of salt are added to a solution, the ions from the dissolved salt are hydrated and reorganize water molecules away from the protein. This causes the protein molecules to clump together and precipitate. Similarly, proteins are most soluble in weak salt solutions. Salting in occurs with weak salt solutions and happens partially due to ions neutralizing the charge on the surface of a protein, which in turn increases the activity of the solvent.

Strengths and weaknesses: Salting in or out is one of the simplest purification techniques and can quickly, easily, and inexpensively remove many contaminating molecules. The salt is typically used at a high concentration and must be removed in subsequent steps, which can be a consideration in some purification procedures. Conditions for salting in or out must be empirically determined.

Site-directed mutagenesis

When and why it is used: Site-directed mutagenesis is a means of surgically altering a DNA sequence that is often used to change a coding sequence.

How it works: There are several techniques available for generating site-directed mutants. The easiest and most straightforward technique involves the polymerase chain reaction (PCR) and mutagenic primers. In this reaction, one of the primers is a normal complement to the template strand, and the other primer contains the desired mutation. A stretch of nucleotides, which will base pair with the template, follows and provides a site for the polymerase to extend. Because the primer contains the mutation, the products generated from that template or amplified from that product will also contain the mutation. Modifications of this technique coupled with subcloning steps can be used to quickly generate a battery of specific mutants for any protein.

Strengths and weaknesses: Thanks to advances in primer synthesis and PCR, site-directed mutagenesis is now a very straightforward technique that can be accomplished in a very short time and on a very limited budget. Products generated using this technique should be sequenced to confirm that they have only the desired mutations.

Size-exclusion or germeation chromatography

When and why it is used: Size-exclusion chromatography is a technique used to separate molecules based on size. It is a common purification technique with many applications.

How it works: Size-exclusion chromatography employs a size-exclusion resin or medium. This resin consists of microscopic beads of a polymer; agarose and acrylamide are commonly used. The degree of polymerization is controlled to provide pores or channels in each bead that are on the order of the size of the molecules being separated. When a mixture of molecules is applied to the media, smaller molecules will fit into the pores while larger molecules will be excluded. This effectively gives the smaller molecules a larger volume to pass through and results in their retention, while larger molecules, being excluded, will elute from the media first. Variation in pore size means that depending on the type of media size, exclusion can be used for removing salt from larger molecules or single nucleotides from oligonucleotides. Larger molecules, such as proteins that differ in size or in some cases shape, can also be resolved from one another.

Strengths and weaknesses: Size-exclusion chromatography is a well-studied and widely used technique. It is relatively inexpensive and easy to learn. It works best when separating molecules that vary greatly in size, for example, a salt and a protein. Coupling this media with a high-pressure chromatography system increases the resolving power.

Sodium dodecyl sulfate polyacrylamide gel electrophoresis (SDS-PAGE)

When and why it is used: Sodium dodecyl sulfate polyacrylamide gel electrophoresis (SDS-PAGE) is a nearly 50-year-old technique that is used to separate proteins from a mixture in a polymer gel matrix. This technique and several popular variants can be used on their own or as the basis for other techniques, including immunoblotting or western blotting.

How it works: A sample of proteins is denatured in the presence of SDS, a detergent. Often, disulfide reducing agents such as β-mercaptoethanol or dithiothreitol are included. The detergent imparts a negative charge to the denatured proteins. An electric field is applied, and proteins are separated in a polyacrylamide gel matrix. The gel matrix has pores that are small enough to impede the movement of larger proteins; thus, proteins are separated largely by molecular weight. Modifications to this basic system, such as the use of a second "stacker" gel on top of the first with differing pH buffers in each, increases the resolution of the technique. There are numerous ways to detect proteins in the gel, including Coomassie Blue, silver staining, or fluorescent dyes. Additionally, the proteins can be transferred to membranes for immunological detection, known as immunoblotting; individual bands can be isolated from a gel and characterized by mass spectrometry.

Strengths and weaknesses: SDS-PAGE is a workhorse of modern biochemistry; it is easy to learn and relatively inexpensive. Drawbacks of SDS-PAGE include difficulties resolving some proteins from each other, and post-translational modifications such as acylation, phosphorylation, or glycation can alter the behavior of proteins. Finally, although polymerized acrylamide is nontoxic and is used in numerous consumer products, the monomeric form of acrylamide used to make the gel is a neurotoxin. Care should be used in the handling of this compound.

Solid-phase peptide synthesis (SPPS)

When and why it is used: Solid-phase peptide synthesis (SPPS) is used to chemically synthesize peptides. The technique is often used to produce antigens for antibody production.

How it works: SPPS takes advantage of solid-phase technology. The first amino acid in the synthesis, the carboxy terminal amino acid, is supplied bound to a plastic resin bead. Beads are incubated with a second amino acid that has a protective group on its amino terminus, and this amino acid couples to the bead. Unbound or unreacted amino acids are then washed off the bead. The newly formed dipeptide still bound to the bead is deprotected, and a third amino acid is added. The series is repeated until the desired peptide is complete. At that point, the peptide is released from the solid support using acid.

Strengths and weaknesses: SPPS is now commonly used in automated peptide synthesizers, and it is relatively inexpensive to order microgram quantities of peptides of up to 20 amino acids in length. For peptides of greater lengths, is it is generally more effective to use recombinant DNA techniques to overexpress a protein in bacteria or cultured eukaryotic cells.

Southern blotting

When and why it is used: Southern blotting is a means of identifying specific fragments of DNA. It has applications in genotyping some organisms and in cloning.

How it works: DNA, typically genomic DNA, is cleaved into smaller fragments using a restriction enzyme. These fragments are separated on an agarose gel. The DNA fragments are denatured and transferred to a nitrocellulose membrane. Next, a radiolabeled fragment of DNA that is complementary to the fragment of interest, termed a probe, is incubated with the membrane. The probe should anneal or base-pair specifically to the fragment of interest. The unbound probe is washed away, and the membrane is exposed to film or a detection system. Final analysis should show specific bands corresponding to the size of the fragment of interest.

Strengths and weaknesses: Southern blotting is a very common and simple technique used to identify specific DNA fragments. It often employs radiolabeled DNA, although some fluorescent techniques have been developed. Southern blots are somewhat time-consuming, taking a full day to perform and up to two weeks before results can be analyzed. As a result, PCR-based techniques have often replaced them.

Thin-layer chromatography (TLC)

When and why it is used: TLC is used to separate classes of small organic molecules. Generally these are lipids but occasionally other types of molecules based on their chemical properties. It is often coupled with subsequent detection or characterization steps to identify specific molecules of interest. Depending on the scale involved, TLC can be either an analytical or a preparative technique.

How it works: As the name implies, TLC is a chromatographic technique. The stationary phase is often a thin coating of silica gel bound to a glass plate, and the mobile phase is a solution of organic solvents, weak acids or bases, and water. A sample of the molecules is applied in a small spot to a TLC plate. Plates are developed in a TLC tank, a receptacle containing a small volume of the mobile phase at equilibrium with its vapor. As the mobile phase wicks up the TLC plate, it carries the analytes (the molecules in the sample) with it. Analytes that interact more strongly with the stationary than the mobile phase will not move up the plate as far as those that interact strongly with the mobile phase. By choosing appropriate mobile phases, a wide variety of separations can be conducted. Detection of samples can be accomplished by many different techniques including exposure to iodine vapor, charring, use of dyes or reagents specific to a functional group, radiometry, fluorescent detection, gas chromatography, or mass spectrometry.

Strengths and weaknesses: TLC has been in use for many years and is still employed in many laboratories. It is inexpensive, simple to perform, and versatile. Recent refinements of solid-phase compositions have resulted in "high-resolution TLC," which has higher resolving power than older techniques and allows the investigator to use smaller samples.

TLC lacks the resolving power of other analytical techniques, such as GC and HPLC; it is not commonly used to specifically identify molecules. For example, TLC cannot resolve two different fatty acids. It is also not typically thought of as a high-throughput technique (testing many samples at once) and does not lend itself well to automation. Finally, although TLC seems benign, continued exposure to silica gel on TLC plates can lead to silicosis, or white lung, a potentially lethal pulmonary condition. Facemasks or other types of personal protective equipment are advised when working continually in an environment of silica dust.

Tissue-specific knockout organisms: the Cre-Lox system

When and why it is used: Many proteins are expressed in multiple tissues. Tissue-specific knockout of a gene enables the investigator to determine the function of a gene in a specific tissue by deleting the gene only in the cells of that organ or tissue. These are similar to knockout mice discussed in the section "Homologous recombination for gene deletion (knockout organisms)."

How it works: Tissue-specific knockouts using the Cre-Lox system involve the production of both a knockout mouse and a transgenic mouse. The DNA construct for the knockout mouse consists of the gene of interest flanked by LoxP sites. These short sequences were originally found in bacteria. These sites are recognized by the enzyme Cre. Cre is a recombinase that recombines and deletes DNA found between the LoxP sites. Here, the deletion would remove all or a significant piece of the gene of interest, rendering it unable to transcribe. In wild type cells there is no Cre recombinase and the cells—and mouse—will be normal. However, if the mouse in question were to be crossed with a mouse expressing the Cre recombinase, the gene of interest would be deleted, creating a knockout. In order to produce a tissue-specific knockout, the expression of Cre is driven by a tissue-specific promoter. Therefore, from one "floxed" (flanked Lox) knockout mouse, a series of tissue-specific knockout mice can be created depending on the tissue-specific Cre transgenic mouse used for breeding.

Strengths and weaknesses: The Cre-Lox system provides a very useful model system for asking big-picture questions; for example, what happens when liver or muscle can no longer respond to insulin? However, a tissue-specific knockout is a very complex experiment requiring numerous scientists and significant resources. Once a single floxed mouse is made, it can be crossed to transgenic mice expressing Cre in a variety of tissues, providing multiple model organisms.

Transgenic organisms

When and why it is used: Transgenic organisms ectopically express a gene that has been added to an organism's genome. This allows, the function of the protein encoded by that gene to be studied in a somewhat native environment.

How it works: A piece of DNA consisting of the cDNA coding for the protein of interest, a promotor, and a polyadenylation sequence is injected into a fertilized oocyte. The piece of DNA integrates somewhere in the genome of the organism. If done early enough in development, the DNA replicates with each cellular division and will be passed on in the progeny.

Strengths and weaknesses: Ectopic expression provides a way to study the function of a protein in an organism. It can be done relatively inexpensively, often for several thousand dollars. Care must be taken, however, to ensure that the effects observed in the organism are due to the gene of interest being expressed either at very high levels or in tissues where it is not normally found.

Transmission electron microscopy (TEM)

When and why it is used: Traditionally associated with cell biology, transmission electron microscopy (TEM) is also a technique involved with imaging macromolecular complexes. TEM cannot provide the resolution that diffraction techniques such as neutron or X-ray diffraction can, nor can it provide measurement of some of the smaller molecular distances afforded by spectral techniques such as NMR or FRET. Nonetheless, it is very useful for examining larger complexes of proteins and seeing how complexes may interface with one another.

How it works: TEM works much like a light microscope but takes advantage of the shorter wavelength of electrons. When using light to examine a structure, the resolution of the instrument is limited by the wavelength of light. Shorter wavelengths, and therefore higher-energy particles, permit higher resolution and magnification. Whereas light microscopy is typically limited to approximately 400× magnification, the power and wavelength of electrons permit magnifications of over $1 \times 10^6 \times$ with near-Angstrom resolution. At these magnifications, it becomes easy to visualize macromolecular complexes of proteins, lipoproteins, or ribosomes.

A source of electrons analogous to the light source in conventional microscopy is focused on the sample. Whereas lenses are used to focus light, magnets and magnetic fields are used to focus the electron beam. In TEM, heavy metals are used as contrast agents, rather than dyes, which are often used to stain molecules or increase contrast in light microscopy. Because of the high energy of the electrons, typically between 100 and 300 kEv, the entire system is held under high vacuum. Images are projected onto a fluorescent screen or captured using a digital camera.

Strengths and weaknesses: TEM samples are imaged under vacuum and must therefore be dry. A contrast agent is used to make the sample stand out from the background. The sample preparation process can introduce artifacts that need to be controlled, such as the aggregation of samples. As with any experimental technique, it behooves the investigator to validate the findings using a different technique, in this case, size-exclusion chromatography, laser light scatter, or non-denaturing gel electrophoresis when using TEM. TEM are expensive instruments (a standard TEM costs over $150,000) and require highly trained technical help to keep the instrument functional; however, they are relatively simple to use, and data can be obtained quickly and easily.

Transposon mutagenesis

When and why it is used: Transposons are "jumping genes," segments of DNA that move from one chromosome to another. Transposon mutagenesis is a technique that is used either to disrupt genes, effectively knocking them out, or to insert a gene of interest into a genome. The technique is growing in popularity as commercially available kits enter the market and familiarity with the technique grows.

How it works: Transposon mutagenesis can be carried out in one of two different ways. In insertional activation, a transposon is randomly inserted across the genome. The insertion contains stop codons, often in all six reading frames. Should the transposon insert into a gene, that gene will be unable to properly code for mRNA and, ultimately, protein. These are considered loss-of-function experiments. Typically, a library of mutants is generated and used to screen for a particular phenotype. The sequence of the transposon can then be used as a primer to determine where in the genome the sequence has integrated. Purified transposonase makes it possible to use transposon mutagenesis *in vitro*. Alternately, the technique can be performed *in vivo* using a plasmid coding for transposonase.

The second type of mutagenesis involves insertion of a gene of interest into host cells. In this instance, the gene of interest (for example, a healthy copy of a defective gene in a patient) is cloned between the long terminal repeat sequences of the transposon. This piece of DNA can be delivered to cells through any one of several different transfection techniques, dependent on whether the DNA is delivered to an organism or cultured cells. Once inside the cell, the transposon is processed and integrated into the chromosome, and presumably the genes found in the transposon are expressed.

Strengths and weaknesses: Transposon technology is inexpensive and growing in popularity. It overcomes problems associated with viral infection of cells including immune response, random insertion of DNA, cost of producing high-quality virus, and naked DNA transfection or electroporation methods such as failure to integrate or low expression levels. The techniques are flexible and versatile and promise to be increasingly used in research and in the clinic. Transposons have been used in gene therapy trials in several mouse models of human disease, including hemophilia, sickle cell anemia, and lymphoma. Clinical trials with humans have already begun.

Visible-range spectroscopy

When and why it is used: Visible-range spectroscopy is one of the oldest and most common forms of spectroscopy. It is used to detect changes in the concentration of a light-absorbing compound known as a chromophore. You may be familiar with visible-range spectroscopy as a technique used to quantitatively measure dyes, but it can also be used to monitor chemical changes in naturally occurring chromophores in a biochemical reaction.

How it works: In visible-range spectroscopy, a sample is placed in a clear chamber called a cuvette, and visible light is shined on the sample. Based on the molecular structure, the photons of light may interact with the compounds in the sample, absorbing the light. The remaining light passes to a detector, either a photomultiplier tube or an array of light-sensitive diodes, where it is translated into an electronic signal. Biochemical samples that contain brightly colored or pigmented groups such as flavins and heme groups provide natural chromophores to be used in spectroscopy. As reactions occur and changes are made to the redox state of these molecules, there is an accompanying change in absorbance. Therefore, a change in the state of these molecules as a reaction progresses can be tracked spectrophotometrically.

Strengths and weaknesses: Visible-range spectroscopy is a classic technique and is the basis of many assays. It is hard to think of a laboratory that does not use it in some way. It is the perfect tool for many situations, but it is not as sensitive as fluorescence spectroscopy or the use of radiolabeled molecules.

X-ray diffraction/X-ray crystallography

When and why it is used: X-ray diffraction is a technique used to determine the structure of molecules.

How it works: Crystals of the molecule of interest are placed into a diffractometer, an instrument that generates X-rays, shines these X-rays on the crystal, and detects the angle and intensity of the diffracted spots. Mathematical analysis of the diffracted spots enables the identification of electron density in the original sample. Based on this, the structure of the molecule can be determined.

Strengths and weaknesses: X-ray structure analysis has arguably been one of the most transformative techniques of the last century. It provides detailed, high-resolution information on the structure of molecules that can yield important data related to molecular interactions and mechanism of action. The drawbacks of X-ray crystallography are the time and investment needed to yield these results. Proteins must be highly purified, and crystallization conditions are experimentally determined. Advances in diffractometer design and X-ray sources have improved the acquisition of data, and advances in computing have no doubt made the analysis of the data easier. Determination of a structure using X-ray diffraction has become much more accessible, but it is still one of the most difficult of biochemical experiments.

Yeast 2-hybrid system

When and why it is used: The yeast 2-hybrid system is one of only a few techniques that can be used to directly measure protein–protein interactions.

How it works: The yeast 2-hybrid system is a complementation assay that takes advantage of a reporter gene as an output. Transcription factors usually have two domains: a DNA binding domain and an activation domain that is responsible for transcriptional activation. If these two domains are expressed separately, expression will not occur; however, if they are in proximity to one another due to the interactions of other proteins, there is expression. The yeast 2-hybrid system can be used as a screening technique for protein–protein interactions. Two different types of plasmids are generated in the system. One plasmid has the DNA-binding domain of a transcription factor fused to a protein of interest. The other plasmid has the activation domain fused to a cDNA library coding for different proteins. Some of these may interact with the first fusion protein, termed the bait. Cells are transformed with both plasmids. Colonies are selected that will only grow if the gene regulated by the transcription factor is expressed. Therefore, colonies that grow indicate an interaction between the bait protein and some other protein. These colonies are isolated, and the genes coding for the second protein are sequenced and identified.

There are several variants of the original yeast 2-hybrid system that can be used to detect protein–DNA interactions. These employ different host systems such as *E. coli* or eukaryotic cells or employ different detection systems, including fluorescent detection.

Strengths and weaknesses: The yeast 2-hybrid system can identify interaction partners for the bait protein, but it is easy to over- or under-interpret the results. For example, the lack of an interaction in the 2-hybrid cannot necessarily be interpreted as the absence of an interaction *in vivo*.

ANSWERS TO WORKED PROBLEM FOLLOW-UP QUESTIONS

Chapter 1

Worked Problem 1.1 Factors that influence ΔG include the concentration of reactants and products, temperature, pH, and partial pressure of gases. The most common way ΔG can be driven to the negative in the cell is to increase the concentration of reactants and decrease the concentration of products. Altering ΔG will not affect reaction rate but altering concentrations of reactants will.

Worked Problem 1.2 The products would be an amine and a carboxylic acid.

Worked Problem 1.3 Basic conditions would favor deprotonation of the amino group, which would generate the nucleophile that would attack the carbonyl carbon. A protonated amine won't do that.

Worked Problem 1.4 Using the Henderson–Hasselbalch equation, the volumes needed would be 863 ml of sodium acetate and 137 ml of acetic acid.

Chapter 2

Worked Problem 2.1 Two techniques used to determine the structure of a macromolecule would be NMR and X-ray crystallography. Both techniques could determine the structure of the tRNA and could therefore be used to help determine function.

Worked Problem 2.2 Oxidatively damaged or alkylated bases can lead to improper base paring in the double helix. These modified bases are incorrectly copied during replication. These mutations could alter a coding sequence, a splice site, or a regulatory sequence, any of which would lead to changes in transcription and translation.

Worked Problem 2.3 This process would in many ways be the same. Two differences would be the type of plasmid or DNA construct used. If a plasmid were used instead of a virus or other means, it would need to have mammalian promoters to be expressed in a mammalian cell. The other significant way these would differ would be how the DNA was introduced to the cell. Transfection of mammalian cells differs from transduction of bacteria.

Chapter 3

Worked Problem 3.1 The pK_a of the carboxyl group of alanine has a pK_a around 2.34. Because the pK_a is more than one pH unit away from the pH of the buffer, we would expect it to be outside the buffer range. At a lower pH it could serve as a buffer.

Worked Problem 3.2 Myoglobin has a molecular weight of 16.7 kDa. An average protein has a MW of about 50 kDa with a range of 30 to 80. Myoglobin is a relatively small protein.

Worked Problem 3.3 Isoleucine is a medium-sized hydrophobic amino acid. If it were substituted with another similar amino acid, such as leucine or valine, this would probably not affect protein function. If other hydrophobic amino acids were substituted, such as a tryptophan or glycine, the change in size might create a steric clash. If a charged or polar amino acid were substituted, the polar or charged group might orient toward solvent instead of the nonpolar core and alter protein structure or stability.

Worked Problem 3.4 This protein is globular and mostly alpha-helical. It has a heme functional group, which is important in oxygen binding. If we were to compare it to other proteins we have seen in this chapter, we might most closely relate it to the oxygen transport protein hemoglobin. Unlike hemoglobin, this protein has only a single subunit. Therefore, although it could transport or bind oxygen, it could not do it in an allosteric fashion. This protein is the muscle-oxygen-binding protein myoglobin.

Worked Problem 3.5 The crude mixture could be treated with incrementally increasing salt concentrations (for example, 15%, 20%, 25%, and 30%) to see where the optimal separation of lysozyme from other proteins occurs.

Worked Problem 3.6 Potentially. If there were enough of a size difference between a compact folded form and an extended, unfolded protein, then they could be resolved by size exclusion.

Worked Problem 3.7 The peptide in the first part of the question differs from this one in that this peptide has several anionic residues (aspartate, D, and glutamate, E) in place of the positively charged ones. Therefore, a good choice for chromatography would be an anion-exchange resin such as DEAE.

Chapter 4

Worked Problem 4.1 Three examples here would include: α-ketoglutarate dehydrogenase, an oxidoreductase; lipase, a hydrolase; and T4 DNA ligase, an example of a ligase.

Worked Problem 4.2 To calculate k_{cat}, you would need $[E]_T$ in addition to V_{max}.

Worked Problem 4.3 Both glutamate and aspartate have a carboxyl group. They differ by the additional methylene group found in glutamate. It is likely that the aspartyl group would compensate for the loss of the glutamate group and, therefore, not effect the proposed mechanism.

Worked Problem 4.4 Allosteric enzymes or proteins with homotropic allosteric regulators are multimeric. When a substrate or molecule that is being transported binds to one site, it influences the affinity of the other sites.

Worked Problem 4.5 You would need to calculate a Hill plot to determine the cooperativity of the subunits. To do that, you would need the concentration of the ligand (the allosteric regulator) and the fraction of sites bound in the enzyme at any given ligand concentration. The Hill plot can be used to determine the Hill coefficient. If that number (the slope of the central portion of the line) is greater than 1, then the system is cooperative.

Worked Problem 4.6 If there were a mutation to bisphosphoglycerate mutase such that it couldn't make 2,3-bPG, then hemoglobin would lack 2,3-bPG. This would stabilize the R state of hemoglobin and increase hemoglobin's affinity for oxygen.

Worked Problem 4.7   The binding of sucralose to one binding site influences the binding of sucralose to other binding sites in the complex. This is analogous to one molecule of oxygen binding to hemoglobin and influencing oxygen binding at the other sites in the molecule. This is an example of homotropic binding.

Chapter 5

Worked Problem 5.1   The change in diffusional constant reflects a change in the structure of the protein. If the ligand is small, we would not expect the addition of ligand to change the diffusional coefficient directly, but a change to the conformation of the protein or a change in interaction partners for that protein might lead to a change in the diffusional coefficient.

Worked Problem 5.2   Caffeine causes increased alertness, flushing, increased heart rate, increased respiration, and increased sweating. Caffeine has multiple effects in the body, but it generally activates the sympathetic nervous system and acts as a stimulant. Many of these effects are due to caffeine-blocking phosphodiesterases, leading to increased cAMP levels and increased PKA signaling.

Chapter 6

Worked Problem 6.1

Worked Problem 6.2   Two glucose monomers joined by an α-1,4 linkage is maltose. It is able to be digested by humans, whereas cellobiose cannot be digested.

Worked Problem 6.3   Acetylated pectins have alcohols (typically methanol) found in ester linkages to the carboxyl groups on the pectin. The acetylation derivatizes the carboxyl and prevents formation of a cross-link through a metal ion.

Worked Problem 6.4   The products of glycolysis wouldn't be any different than those derived from glucose. The product would be two molecules of pyruvate.

Worked Problem 6.5   PEPCK would be a potential target to inhibit. It is highly regulated; it is isoform specific, so we could potentially target only one isoform; and it is found at the beginning of gluconeogenesis.

Worked Problem 6.6   A chemical approach would involve measurement of lactate. This could be done by enzymatic assay, HPLC, or mass spectrometry. A genetic approach might include deleting genes, such as lactate dehydrogenase, using CRISPR or other techniques.

Chapter 7

Worked Problem 7.1   This would imply that the reactions of the citric acid cycle, and perhaps the proteins and genes, existed prior to the evolution of the cycle. In other words, it suggests that this complex pathway did not evolve as a complete pathway but rather as a series of individual steps that evolved over time into the pathway we now know.

Worked Problem 7.2   In some cells glutamate is used as a neurotransmitter. It is produced by the transamination of α-ketoglutarate. Therefore, reactions that deplete glutamate could in turn lead to a depletion of α-ketoglutarate, which would slow the citric acid cycle.

Worked Problem 7.3   Some of the redox centers mentioned in the problem, most notably heme and Cu^{2+}, will change their absorbance or color if they are reduced or oxidized. This could be determined spectrophotometrically and used to detect which redox centers are being affected. Spectroscopy could provide a means of determining which centers are reduced and potentially the order in which they are reduced. This could be done by removing some of the centers or reconstituting the structure in the presence or absence of some of the other centers.

Worked Problem 7.4   To react and bind, cyanide has to be able to access the central iron atom. In some heme groups this iron is shielded by proteins and is inaccessible.

Worked Problem 7.5   In the solution to worked problem 7.5, we saw that it took 16.45 kJ/mol to generate the proton gradient. Therefore, it would take approximately three protons (45 kJ/mol; 16.45 kJ/mol = 2.7 protons) to synthesize 1 molecule of ATP.

Worked Problem 7.6   Copper II is a divalent metal ion with a +2 charge. Amino acids that frequently interact with divalent metals include histidine and cysteine.

Chapter 8

Worked Problem 8.1   Either the unit could be added as a disaccharide to the growing chain, or the enzymes incorporating new monosaccharides could have substrate specificity, which would necessitate that the alternating monosaccharides be added. This could either be due to the geometry of the enzyme's active sites or the geometry of the carbohydrates that are being added.

Worked Problem 8.2   If glycogen is unavailable to provide glucose for the individual, the glucose could be replaced in part through gluconeogenesis. In this instance, the glucose would come from the catabolism of amino acid skeletons, which would then undergo gluconeogenesis.

Worked Problem 8.3   Sucrose is a disaccharide, which will enter the cell either as fructose or glucose. Both of these will presumably be metabolized into glucose-6-phosphate, then glucose-1-phosphate, then UDP-glucose before incorporation into glycogen. Caffeine is a phosphodiesterase inhibitor. It blocks the breakdown of cAMP and in turn increases the activity of PKA. PKA will phosphorylate several proteins, most notably GSK-3, glycogen phosphorylase, and glycogen synthase. In these instances, it will inhibit glycogen synthesis and stimulate glycogen breakdown. In short, the two compounds will work against each other.

Worked Problem 8.4   Ribose could proceed through the pentose phosphate pathway and enter glycolysis either as fructose-6-phosphate or glyceraldehyde-3-phosphate.

Worked Problem 8.5   On one level this is a very simple question. Cells could be grown in the presence or absence of fructose, and the expression of several genes assayed using northern blotting or RT-PCR to determine if the carbohydrate affected the expression of these genes. However, this presumes the effect was large enough to see and the genes assayed were sensitive to fructose. This also assumes that fructose itself was not inadvertently providing some other effect (for example, increased energy). Another way to do the study would be to use two groups of organisms, such as mice or humans, and then ascertain gene expression on the whole genome, either by using a technique such as a microarray or RNA-Seq. These experiments would also have challenges, not the least of which would be obtaining RNA to assay and determining what constitutes a change in gene expression.

Worked Problem 8.6   Humans lack α-1,3-galactose linkages, whereas pigs do not. Because pigs have this modification, their tissues may cause an immune response in some transplant recipients. However, pigs lacking this modification would be less immunogenic and thus better sources for tissue donation.

Worked Problem 8.7 Matrix proteins can be solubilized using strong detergents, cross-links (chemically cleaved if at all possible), and carbohydrate modifications removed using glycosylases to facilitate their study.

Chapter 9

Worked Problem 9.1 This is $16:1^{\Delta 9}$ palmitoleic or 9-dodecahexaenoic acid.

Worked Problem 9.2 Yes, due to its amphipathic nature, lecithin could be used as a detergent but would probably work as well as a synthetic detergent.

Worked Problem 9.3 The yield would be 108 molecules of ATP from one molecule of palmitate. Examine Table 9.2 for a breakdown of where these ATP molecules originate.

Worked Problem 9.4 A drug that activated AMPK would lead to phosphorylation and inactivation of ACC1. This in turn would block malonyl-CoA production and fatty acid biosynthesis.

Worked Problem 9.5 Acetoacetate breakdown may not be beneficial because the energy found in acetoacetate would not reach its destination (tissues that need it). CO_2 may contribute to the acidification of the blood, and it is not clear whether acetone in high concentrations would damage proteins or organs, especially the kidney.

Worked Problem 9.6 The prenylation of proteins is not affected by this class of drugs. Therefore, we can assume that the enzymes involved are not completely inhibited as some molecules are made. This also indicates that protein prenylation occurs preferentially to cholesterol biosynthesis.

Worked Problem 9.7 Prostaglandin creams are used because they can deliver the drug directly to the tissues that need it. Oral or intravenous delivery of drugs would mean that the drug would need to be absorbed and not entirely metabolized to get to the tissues that need it. Likewise, these drugs could potentially have broader impacts on health if distributed throughout the body.

Chapter 10

Worked Problem 10.1 The phospholipases found in spider venom could lead to damage to the plasma membrane of cells, in turn leading to tissue damage. These enzymes would also cause the release of fatty acids, such as arachidonyl acid, which could contribute to uncontrolled eicosanoid production.

Worked Problem 10.2 Short- or medium-chain triglycerides are not hydrolyzed and metabolized in the intestine like other fats. Rather, they are left intact and enter the bloodstream where they travel directly to the liver for metabolism.

Worked Problem 10.3 One hypothesis would be that because the organism cannot take cholesterol up from the circulation via LDL, it would sense that it had low levels of cholesterol and therefore downregulate the production of HDL. This hypothesis would need to be tested.

Worked Problem 10.4 Because clathrin plays a central role in receptor-mediated endocytosis, we would expect that the organism could not properly undergo this process. However, because receptor-mediated endocytosis is probably involved in many cellular processes, we might not expect a complex organism like a mouse to be viable.

Worked Problem 10.5 It potentially would. Adipocytes lack glycerol kinase, so any glycerol that was liberated would be released as glycerol. Fatty acids, on the other hand, could be re-esterified to other lipids in the cell, giving an artificially low value of the lipid breakdown occurring.

Worked Problem 10.6 Two techniques that could be used to determine if two proteins are interacting would be FRET and co-immunoprecipitation. FRET is fluorescence or Forster's energy resonance transfer. In this technique both proteins are tagged with fluorophores with overlapping spectra. One fluorophore is excited, and if the proteins are in close proximity, the second fluorophore will emit a photon. In co-immunoprecipitation, an antibody is used to selectively precipitate a protein. Any proteins associated with this protein will co-precipitate and can be detected using western blotting or mass spectrometry.

Worked Problem 10.7 Methyl-β-cyclodextrin could be used to deplete a membrane of cholesterol. The cell could then be examined to see what effect if any this would have on raft formation.

Chapter 11

Worked Problem 11.1 Because these drugs are irreversible inhibitors, they would act like a suicide substrate, effectively lowering the concentration of enzyme as the inhibitor reacts.

Worked Problem 11.2 Transamination of these α-keto acids would regenerate the amino acids they were originally. Transamination reactions typically have ΔG values near zero.

Worked Problem 11.3 The enzymes could be isolated and the reactions performed in the presence and absence of EDTA to bind divalent metal ions. If the enzyme is able to catalyze the reaction in the absence of zinc, this indicates that zinc is not important.

Worked Problem 11.4 Lysine is an essential amino acid and one that contributes to both ketone and glucose production.

Worked Problem 11.5 Desmethyl tamoxifen is missing a methyl group on the amine (N). This would involve class I metabolism in which the substrate is oxidized. This could be due to a demethylase or oxidase such as a cytochrome enzyme or monoamine oxidase.

Worked Problem 11.6 To do this you would need some means of assaying glucose and a source of the transporter, presumably in a membrane or cell. Glucose transport would need to be measured as a function of glucose concentration. This would be graphed and if it saturates, the curve would look like the rectangular parabola found in Michaelis–Menten kinetics.

Chapter 12

Worked Problem 12.1 Two means of detecting ^{15}N are mass spectrometry and NMR. Both of these techniques are described in the appendix.

Worked Problem 12.2 If you consume no food for the next 24 hours, levels of plasma triacylglycerols should drop, but levels of free fatty acids and glycerol from adipose tissue should increase. Glucose may drop slightly, but this should be offset by glycogenolysis and gluconeogenesis from the liver. Ketone bodies should increase.

Worked Problem 12.3 Insulin is taken several times daily by millions of diabetics worldwide. Other hormones are delivered this way when their levels are low or elevated levels of a hormone are needed to combat a disease. This does not always work due to a post receptor defect in some diseases, but if we put that observation aside, it might work.

Worked Problem 12.4 With reduced adipose tissue mass, hormones such as leptin would presumably be decreased; however, based on the food consumption and body mass data, we might anticipate that leptin levels would be normal or some other mechanism might compensate. Elevated insulin levels would indicate that glucagon would probably be low as well.

Chapter 13

Worked Problem 13.1   No, once xanthine is formed from either the oxidation of hypoxanthine or the oxidative deamination of guanine, it is unable to be reduced back to either form. The fate of xanthine at that point is to be converted into uric acid and eliminated.

Worked Problem 13.2   Yes, in theory and potentially in practice. Bacterial ATCase differs in how it is regulated from the human form of the enzyme. It could be used as a drug if this difference could be exploited and a drug could be found that would inhibit the bacterial form of the enzyme but not the human one. Considerations would include how the drug would be absorbed, distributed in the body, metabolized, and excreted; these are called the ADME properties of the drug.

Worked Problem 13.3   No, methionine is a thioether, whereas cysteine exists as either a thiol or disulfide. Cystine is closer in reactivity to a hydroxyl-containing amino acid, such as serine.

Worked Problem 13.4   There are probably several reasons why uric acid is used to eliminate purines. First, purine bases themselves are potent signaling molecules or the basis of signaling molecules. Therefore, it is unlikely that the organism would use these as a means of disposal. Second, uric acid is more highly oxidized than xanthine and therefore production is presumably energetically favorable, requiring little input of energy from the organism. Third, some species use uric acid as a means of disposing of nitrogenous wastes. In those instances, the organism's energy expenditure in the synthesis of uric acid is worthwhile in terms of the advantage it gains in biological function (i.e., water or weight savings).

Chapter 14

Worked Problem 14.1   The amount of label left would be 2 parts in 2^{nth} generations of DNA. In this case it would be 2 in 2^{11}.

Worked Problem 14.2   Inhibiting human topoisomerase would lead to DNA damage and theoretically block DNA replication and cell division. This drug could be used to block rapidly dividing cells, such as those found in cancer.

Worked Problem 14.3   Higher than normal. Nearly all malignant cells express abnormally high levels of telomerase, potentially due to the high levels of DNA replication going on in these cells.

Worked Problem 14.4   Prior to the formation of the ozone layer, Earth was bombarded with UV radiation. This radiation would easily cross-link thymine molecules in DNA and could potentially lead to damage of other nucleotides. It is thought that the levels of UV radiation would have been high enough to prevent the establishment of life on Earth.

Worked Problem 14.5   If we were to obtain dNTPs in which the α phosphate groups were isotopically labeled, using nonradioactive isotopes of phosphorous or oxygen, for example, we would be able to detect these using mass spectroscopy in the product. A strength of this technique would be that we would not have to use radiolabel; a weakness would be that we would need access to a mass spectrometer capable of analyzing the results.

Chapter 15

Worked Problem 15.1   Depending on how the experiment was designed, it could identify one or even several functions of TFIIH. One way to test whether TFIIH acts as a helicase would be to incubate TFIIH with a segment of DNA in the presence and absence of ATP and examine the structure using cryo-EM. If the TFIIH complex were able to unwind the DNA, it might be visible using this technique.

Worked Problem 15.2   Most likely not, because even if the mutation left that tRNA nonfunctional, there are multiple tRNA genes that have redundant functionality and act as molecular "backups."

Worked Problem 15.3   One way to do this experiment would be to incubate radiolabeled tRNA with and without purified exportin and the RanGTP complex. Complexes could then be separated using size-exclusion chromatography or native gel electrophoresis and the size of the complex detected by finding the free and bound tRNA.

Chapter 16

Worked Problem 16.1   To change the codon for arginine (in this specific case AGG) to a codon encoding valine, we would need to change the initial A to a G and the second G to a C, giving the codon GCG. Proline (CCA) is also a polar amino acid, and its sequence would have to have the initial C modified to a G to encode for alanine (GCA).

Worked Problem 16.2   Yes, mutations occurring outside the anticodon loop would be potentially deleterious to the organism. This is because the tRNA molecule also requires the CAA sequence at the 3' hydroxyl end of the RNA to add amino acids. Transfer RNA must also be recognized by an enzyme; this recognition site may happen over a large area and not a small patch of the tRNA. Having said that, we could also hypothesize that there are many different tRNA molecules, and therefore potentially mutating one would still leave multiple tRNAs that could fulfill the role, leading to a silent mutation.

Worked Problem 16.3   Puromycin's structure looks similar to an amino acid coupled to a nucleotide. Specifically, it looks similar to the amino acid tyrosine joined to an adenine residue. Based on this and the knowledge of where the drug binds, we could presume that it is analogous enough to part of an aminoacyl-tRNA to bind and inhibit the synthesis of proteins.

Chapter 17

Worked Problem 17.1   Review the functional groups exposed for the sequence 5'-AT-3'. The functional groups found in 5'-TAAT-3' will be the same but arrayed as two repeats of the AT structure facing one another.

Worked Problem 17.2   The CRP binds to cAMP and signals that the cell requires energy and therefore should transcribe the *lac* operon and produce these proteins and enzymes. CRP binds upstream of the *lac* repressor and activates its expression. In the absence of lactose, there will still be minimal transcription of the operon due to the *lac* repressor binding to the operator region. In the mutation, this will become the new default state and the operon will be transcribed in the presence or absence of cAMP (or glucose, for that matter).

Worked Problem 17.3   The cDNA encoding for the aptamer region of the riboswitch could be fused to a reporter gene such as GFP. This mRNA could be incubated with the different carbohydrates to see how each would affect translation. Translation could be detected by measuring the accumulation of GFP.

Worked Problem 17.4   One would need to set up two conditions: one in the presence and the other in the absence of the drug. The cellular lysate would be incubated with or without the drug and the ChIP assay performed. The proteins would be cross-linked to the DNA, the DNA sheared into small fragments, cross-linking reversed, and DNA detected using PCR. If the drug affected binding, it should show up as a difference in the ChIP assay.

Chapter 18

Worked Problem 18.1   The blue light has a shorter wavelength, which means that smaller pieces of data can be encoded on the disc. This translates to a higher density of data and more data per disc.

Worked Problem 18.2   The authors may have been concerned that staining the image with metals would change the structure or lead to artifacts (data that appear real but are actually due to experimental conditions). To find a copy, you could search PubMed for the author, key terms, and year of publication.

Worked Problem 18.3   Neutron scattering is dependent on low-molecular-weight nuclei, such as those found in proteins. Also, there is no loss of signal with higher-order diffractions as there is with an X-ray experiment. A limitation of using neutron diffraction is the need for a high-energy source of neutrons.

Worked Problem 18.4   Peak 1 is shifted down to 7 to 8 ppm and has areas of 1, 2, and 2, indicating that it is likely an aromatic residue, probably phenylalanine. If we examine the other peaks, we have a doublet at 1.7 ppm with an area of 3. This is indicative of a methyl group. Those groups are found in alanine, threonine, valine, leucine, isoleucine, and methionine; however, there are not enough peaks to support many of these structures, leading us to conclude that the peak is probably alanine.

Chapter 19

Worked Problem 19.1   If you needed to synthesize a protein of 165 amino acids, you would probably need to use a cell-based expression system rather than SPPS.

Worked Problem 19.2   If the vector lacked the sites that were necessary, there are multiple options. The piece of DNA encoding the protein of interest could be mutated to include one of the sites. This could be done through site-directed mutagenesis. Another option would be vector hopping. This would involve cloning the fragment of interest into a third vector to acquire the desired restriction sites, then cutting out the fragment from that vector with the new restriction sites incorporated.

Worked Problem 19.3   The champagne glass (also known as a margarita glass) is broad at the top and gradually comes to a point in the bottom center. In this model of folding, the protein would spend time in numerous equivalent semi-stable states until it found the most stable configuration.

Chapter 20

Worked Problem 20.1   We would not anticipate that this drug would influence these two tissues the same way. The liver is the site of both cholesterol biosynthesis and drug detoxification. Therefore, we might anticipate that the genes encoding enzymes involved in lipid or cholesterol metabolism might be affected, but we might also anticipate that the genes involved in drug metabolism (cytochrome enzymes) would be affected. We would expect low levels of expression of these genes in the pancreas, and we might not anticipate that their levels of expression would be affected.

Worked Problem 20.2   RNA-Seq is a newer technique than SAGE or microarrays. In general, it is an easier experiment to perform; is lower in cost; and, most important, gives a richer dataset than either of the other two. SAGE and microarrays can tell you which genes are up- or down-regulated, but RNA-Seq can give information about splice variants that will most likely be missing from the other techniques. One weakness would be the quantity of data that one would then have to analyze and manipulate using bioinformatics.

Worked Problem 20.3   Yes, there are several ways to use an amino acid sequence in a BLAST search. This could include BLASTp or tBLASTn.

Chapter 21

Worked Problem 21.1   Protein kinase A is known to affect enzyme activity and gene transcription. AMP kinase is known to affect enzyme activity, but it is not yet known whether it affects gene expression.

Worked Problem 21.2   SH2 domains are Src homology 2 domains. These domains are commonly found in many signaling proteins and bind to phosphorylated tyrosine residues. Based on that knowledge, we can surmise that the new protein would also bind to phosphorylated tyrosine residues.

Worked Problem 21.3   Phosphodiesterase inhibitors block the breakdown of phosphodiesters, such as cAMP and cGMP, so some of these enzymes must be involved in the vision signaling pathway. In fact, if you consult a resource on the vision pathway, you will find that a cGMP phosphodiesterase plays a key role its function. Even slightly inhibiting this enzyme changes how we perceive color.

Chapter 22

Worked Problem 22.1   Rabs are GTPases that act as adapters between cargo and motor proteins. Rabs only interact with these proteins in their GTP-bound state and dissociate once GTP is hydrolyzed. A mutant Rab that can bind to GTP but not hydrolyze it will be stuck in the "on" position. This could have a negative impact on trafficking due to the loss of the ability to regulate this protein and therefore the entire process. Cargo could not be released when the complex reaches its destination.

Worked Problem 22.2   Yes, based on the conserved function and structure, we might anticipate that these protein amino acid sequences and genes are conserved; however, a further examination indicates that there is greater diversity among these structures and that they have evolved through different means (convergent evolution).

Worked Problem 22.3   Lab workers have an increased chance of puncture injuries from chipped glassware, animal bites, or other sharp objects. Lab workers and all healthcare workers receive the tetanus vaccine as part of a Tdap (tetanus, diphtheria, pertussis) vaccine. Lab and healthcare workers should also receive vaccines for infectious diseases such as hepatitis B, influenza, measles, mumps, rubella, meningitis, and chickenpox.

Chapter 23

Worked Problem 23.1   It is unlikely that this plant would be viable. Not all photosynthetic organisms have plastocyanin, but they all need some means of either transferring electrons from cytochrome b_6f to photosystem I or oxidizing plastoquinone to allow electron transport to occur in photosystem II. If neither of these conditions is met, the organism would probably not survive.

Worked Problem 23.2   Most organisms rely on some other organism for a source of nutrients or cofactors. As we have seen in this chapter, we all rely on other organisms for the air we breathe. The fact that most other organisms are unable to fix their own nitrogen from the air suggests that these organisms have adapted to obtaining fixed nitrogen from some other source—for example, by digesting another organism to obtain amino acids—and that this is not a major burden. It also suggests that the organisms that do fix nitrogen may have arrived at this adaptation early in evolution before oxygen accumulated, and they have not felt significant negative selective pressure on this trait over time.

Chapter 1

1. The three laws of thermodynamics are: (1) Energy is conserved (meaning energy can neither be created nor destroyed); (2) The entropy, or disorder, of the universe is always increasing; and (3) The entropy of a pure crystal at absolute zero is zero. The system is the sample we are studying, and the surroundings are everything else in the universe. Free energy is the amount of energy available in a reaction to do work.

3. Because the structures form spontaneously, ΔG must be favorable (negative). This is due to several factors, including the formation of multiple weak interactions between the different parts of a macromolecule with itself. This organization may seem to decrease the overall disorder of the system. However, one of the main factors that are important in the formation of complex structures is the advantage that is gained by liberating water molecules, termed the hydrophobic effect.

5. The polymerization of the monomer into the polymer would seemingly have a positive entropy (the system is becoming more ordered and has fewer microstates). However, in the synthesis of the polymer, if water molecules can be liberated from caging individual monomers, the overall entropy may be lower.

7. All amino acids have carboxyl and amino groups. Many other amino acids contain amide, hydroxyl, carboxamide, thiol, or hydrocarbon groups. All of these groups can participate in London dispersion interactions. Those with amino or hydroxyl groups can participate in hydrogen bonding, and charged groups can also participate in coulombic interactions.

9. most, middle, least for water, the order would be reversed for hexane

11.

13. deoxyribonucleotides

15. The bond from an OH–F would be shortest due to the electronegativity of the fluorine.

17. 1.0, –0.5, 1.9, 14.3, 13.3, 2.6, 2.1, 10.8, 10.49, 10.45

19. 4.28

21. 12.3 mL

23. 1 M MOPS; the higher concentration has the higher capacity to absorb or donate protons.

25. The pK_a is the pH at the half equivalence point (about 3.9). This solution would buffer from pH 2.9–4.9, one pH unit on either side of the pK_a value.

27. There are numerous ways to answer this question. Here is one answer. Make a saturated solution of tyrosine in 250 mM NaCl. Withdraw a known volume of the saturated solution. Dry off the water and weigh the remaining solid. The mass of the solid will include the mass of the sodium chloride as well as the mass of tyrosine in that volume of the saturated solution.

29. Dissolve 18.5 g KCl in approximately 900 mL water in a beaker. Bring volume to 1 L in a graduated cylinder or volumetric flask.

The answers for these questions will vary from individual to individual.

31. Potential answers might include genetically modified organisms, gene therapy in humans (CRISPR), and stem cell technology.

33. As a society, we have decided to regulate some compounds but not others. There are a variety of reasons why these regulations are in place, from protecting youth to preventing the abuse of some molecules.

35. Although the new drug is far easier to develop and test than the original, it is a different compound and may elicit different pharmacological properties than the original. This new pure compound has to be approved and regulated.

Chapter 2

1. Oswald Avery, Colin MacLeod, and Maclyn McCarty in a milestone experiment showed that the entity causing the transformation of R strains to S strain was DNA.

3. It is easy for the RNA to move from one place to other and it has faster rate of replication.

5. Purine and pyrimidine are generally insoluble in water, but the modifications made to them (carbonyl, hydroxyl, and amino groups outside the ring structures) make the compounds more soluble.

7. Check Figure 2.3 for the enol forms of each base. If the bases were enols instead of carbonyl groups, they could act as both hydrogen bond donors and acceptors instead of simply acceptors.

11. The exact weak forces would depend on the nucleotide in question but would include columbic interactions between charged species in the phosphate groups; hydrogen bonding in the base, sugar, and phosphate; and van der Waals/London dispersion interactions and dipole–dipole interactions throughout the molecule.

13.

15. DNA can interact with RNA, proteins, and small ions through all of the weak interactions we have discussed (coulombic interactions, hydrogen bonding, London dispersion forces, and the hydrophobic effect). Of these interactions, hydrogen bonding, and coulombic interactions are several orders of magnitude stronger than London dispersion forces. Dipole–dipole interactions fall somewhere in between.

17. Generally speaking, for DNA to be replicated there needs to be some mechanism for the DNA helix to be unwound and a single-stranded template exposed for synthesis to occur. As the replication forks slide away from each other and form a replication bubble, one strand (the leading strand) is continually synthesized while the other (the lagging strand) is synthesized in short stretches that are joined together.

19. Cross-linked thymine dimers are constrained and form a kink in the DNA backbone. This makes both replication and transcription more difficult for the cell.

21. Cells continually need to replace proteins that are damaged or to make new proteins as the organism grows and develops. Blocking mRNA or protein synthesis is almost always fatal to cells. Protein synthesis inhibitors can be powerful antibiotics or chemotherapeutic agents but can also be used to study how rapidly proteins are being synthesized or turning over in the cell.

23. Most organisms currently known to science employ a DNA genome and use RNA in the synthesis of proteins. Retroviruses have an RNA genome that needs to be reverse transcribed into the host genome as part of the viral life cycle.

25. Proteins, including restriction enzymes, often recognize specific DNA sequences through binding in the major groove of DNA with an alpha helix. Given that both the double helix and the alpha helix are relatively inflexible over the lengths being described, there is limited contact between the two structures. Sequences are often palindromic to assist in binding because multiple binding sites are better than a single site, or to assist in cutting, where the enzyme recognizes and cuts two sites, one on each strand. Sequences with a disrupted palindrome, and therefore greater distance between the binding sites, may be due to a dimeric protein.

27. A six-base sequence randomly made from four different nucleotides will occur 4^6 or once every 4,096 bases. On average it would cut 2.5 times (10,000/4,096).

29. Exposure to UV light can damage DNA, leading to the formation of thymine dimers. There are enzymes in the cell that can help reverse this process, but the same cannot be said of DNA in a gel.

31. The chimera was a creature with the body and head of a lion, the head of a goat, and the head of a serpent. The term "chimera" in biology refers to any organism in which there is a fusion of different organisms, akin to the chimera. In the production of a knockout mouse, the first pups are a chimera—mice consisting of cells derived from either the parents that generated the blastocyst or parents that generated the embryonic stem cells.

33. High throughput methods enable the sequencing of an entire genome or large quantity of DNA over a relatively short period of time. In addition, due to how the experiment is conducted, the price per nucleotide is much lower than in other techniques. However, if an investigator only needs to sequence a short stretch of DNA and not an entire genome, high throughput sequencing is not the correct approach and more traditional techniques provide reliable data quickly.

35. Potentially very different results. Introns and exons in eukaryotes would result in a much larger product than from one using a prokaryotic template.

37. The melting point would be the temperature at which one-half of the DNA is melted. In this instance it would be the midpoint of the vertical part of the curve.

39. EDTA is a compound that binds to divalent metal ions such as Ca^{2+} and Mg^{2+}. When EDTA is added, it binds to these metals, which usually interact with the phosphate backbone of DNA. If these metal ions are absent, then there is nothing to balance out the charges of the phosphate backbones, leading to repulsive forces between the two strands and a lower melting point.

41.

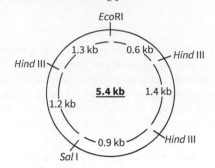

43. To do this one would need a piece of DNA (often a plasmid) that has the cDNA of the protein in question, flanked by sequences that the host cell would recognize to express the protein. Typically, these sequences involve a strong promoter and a polyadenylation sequence. These vectors also contain some sort of selectable marker, which allows for selection of cells that are expressing the plasmid over those that are not. DNA could be introduced into eukaryotic cells using lipofection, electroporation, or some other chemical means, like calcium phosphate precipitation. The cells would be selected using a selectable marker, for example, an antibiotic, to produce cells that express the protein.

45. You would need a means of measuring the melting or denaturation of DNA. This could be done with a technique such as differential scanning calorimetry or possibly by examining single- and double-stranded DNA on a gel. The melting of DNA would be studied in the presence and absence of the protein to see if the protein had an effect on the melting temperature.

47. Examine the coding sequence of the gene. Phosphorylations occur on serine, threonine, and tyrosine residues. The codon that codes for phosphorylated amino acid needs to be mutated to another amino acid. Often a serine is changed to an alanine. Synthesize a primer that is a match for the sequence in question but contains the proposed change. Use this primer in a PCR reaction to amplify the region of the gene. The amplified fragment will contain the mutation, which can be cloned into a full-length copy of the gene.

The answers for these questions will vary from individual to individual.

49. People tend to fear the unknown or the new. It may be that bacteria are perceived as less of a threat than macroscopic organisms because they are microscopic.

51. This is useful data to have, but caution needs to be exercised in how it is used or shared.

Chapter 3

1. Threonine and isoleucine.

3. None of the amino acids have enough conjugated double bonds to absorb in the visible spectra; however, several have enough conjugation to absorb in the ultraviolet range. These include tryptophan; tyrosine; and, to a lesser degree, phenylalanine.

5. Chiral structures rotate plane-polarized light. While we usually think of chiral molecules such as small organics, larger structures, such as α helices, are also chiral. β sheets are not chiral and therefore will not rotate plane-polarized light.

7. If we assume that the α amino and carboxyl groups are not interfering with the ionization, we can simply use the Henderson–Hasselbalch equation and pK_a to find the percent ionization. 98.3%, approximately 100%, 99%, 1.2%.

9. Huckle's rule states that a planar ring system with $4n+2$ π electrons is aromatic. Tyrosine, tryptophan, and phenylalanine are aromatic; each has $4n+2$ electrons (six for tyrosine and phenylalanine and 10 for tryptophan).

11. Nearly all amino acids can be found in both α helix and β sheet because those secondary structures are largely shaped by the structure of the backbone which is common to all amino acids (with the possible exception of proline).

13. 110 is a rounded number based on the average amino acid molecular weight. In an actual protein, the distribution of amino acids may vary and shift this number, although 110 is still a good first approximation.

15. ATP is used instead of free phosphate, for example, because of the free energy of the reaction. ATP is said to have high transfer potential for phosphate. In other words, the free energy of the reaction in which ATP transfers a phosphate to a hydroxyl group in a protein is negative compared to the phosphorylation of that same protein by a free phosphate group (which is positive). ATP is frequently used as a source of phosphate in phosphorylation reactions, but several other molecules can also be used. It is the gamma phosphate (the one most distal from the ribose ring) that is used in the reaction. In terms of equilibrium, there is a relatively high concentration of free phosphate in the cell, which could drive a reaction forward if it were the reactant in the reaction. Glycogen phosphorylase uses the energy of glucose hydrolysis from the glycogen main chain to generate glucose-1-phosphate. This happens without ATP.

17. Most of the structure of an enzyme is important for scaffolding the residues found in the active site. The rest of the enzyme also may have sites for localization in the cell or regulator binding sites.

19. The core of a protein is relatively oily and is enriched in hydrophobic amino acids. Charged amino acids in the core are often involved in a salt bridge. Polar and charged amino acids in the core of an enzyme are often involved with binding substrate or catalyzing reactions. The surface of proteins also contains many hydrophobic amino acids but also contains many polar and charged amino acids.

21. The region of immunoglobulins that binds to antigen is found at the tip of their variable regions generated through a special recombination event in the cells that produce these proteins. The rest of the protein is conserved among all cells of the organism.

23. The mutation changes a charged amino acid (glutamate) into a hydrophobic one (valine). The substitution of this valine generates a hydrophobic patch that facilitates assembly of the hemoglobin polymer. This only occurs in the deoxygenated state because of conformational changes of the hemoglobin molecule upon binding of oxygen.

25. The other amino acids in chymotrypsin are important in scaffolding the catalytic residues in place, forming the substrate binding site or other elements of the active site, and potentially regulating the enzyme.

27. **a.** Obtaining proteins from human tissue would require permission from either living or deceased donors. Many regulations govern the use of human tissues. Other sources such as animal or yeast would be easier to obtain, but proteins obtained from these sources might not be completely homologous to human proteins.

b. Cell culture would provide a large supply of protein from a controlled laboratory setting, but cells grown in culture display differences from those in tissue. These might lead to protein modifications or processing differences in culture that would need to be screened for.

c. The adrenal glands are very small and located at the top of the kidneys. They are difficult to remove, and each gland would only provide a small amount of protein. The liver is more accessible and would provide greater quantities of protein due to its larger size.

d. Mice can be easily cared for in a laboratory, have a relatively short gestation period for breeding, and require fewer resources to raise than donkeys. However, a large quantity of mice would be needed to obtain the same amount of tissue as a donkey could provide.

29. Taking into account that a dialysis is at one level simply a dilution, we can use $M_1V_1 = M_2V_2$ to calculate the volume: (1 M)(3 mL) = (0.145 M)(× mL). This would give us a volume of 20.7 mL.

31. The temperature, the volume of the sample, the concentration of the sample/contaminants, the viscosity, and the volume of dialysis buffer compared to the sample all affect the time it takes to reach equilibrium.

33. Proteins of higher molecular weight elute from size-exclusion columns before low-molecular-weight proteins.

35. No. The resolution range of the column cannot be increased simply by increasing the amount of resin or bead. These properties are determined by the beads themselves. Yes, potentially. If the column were to flow too slowly, peaks would diffuse together, leading to a loss of resolution. If the flow rate were too high, proteins would not have enough time to diffuse into the beads, again leading to a loss of resolution.

37. **a.** prolate 5:1:1

b. oblate 1:2:1

c. prolate 7:4:1

d. prolate 8:4:1

39. Proteins in HIC are often bound to the column under conditions of high salt concentration. Therefore, it makes sense to use HIC after a salting-out step when the salt concentration may already be high.

41. Amines are cations and have a positive charge. Sulfonates are anions and bear a negative charge under the conditions typically used for chromatography. Therefore, the amines would be anion exchangers and the sulfonates cation exchangers.

43. Presumably, putting a tag at the amino or carboxy terminus has a lower chance of disrupting a functional domain of the native protein. These tags are also less likely to be buried and are therefore accessible to use for methods such as protein purification.

45. You could make a recombinant protein by putting an engineered fragment of DNA encoding the His tag onto the end of a cDNA for the protein of interest and inserting into an expression vector of choice.

47. As an electric charge is applied to the protein, it migrates from where it started to a region of the gel that has a lower pH. It moves toward the cathode (the positive electrode) until the protein stops moving. The pH at which it stops moving will have negatively and positively charged groups equally protonated (there will be no net charge on the protein). This pH is called the isoelectric point (pI) of the protein. The protein is in the pH 4 to 5 range, indicating that the protein has more acidic groups (aspartate and glutamate) than basic amino acids.

49. The molecular weight of a protein can be estimated as we have done in this chapter or calculated based on the actual amino acid sequence. The mass as determined by SDS-PAGE is commonly used in many experimental techniques but may vary from the true value due to modifications of the protein (e.g., phosphorylations may make a protein run differently in SDS-PAGE). Size exclusion chromatography can measure the protein under native conditions instead of the denaturing ones found in SDS-PAGE. Therefore, the molecular weight found using column chromatography indicates that the protein is found as a dimer in solution (approximately twice the mass of what is observed on SDS-PAGE).

51. Metal affinity chromatography

53. The high-speed spin and ultrafiltration could be eliminated in terms of purification. It may be necessary to remove precipitates before the column is run.

55. If you had a technique to measure the folding of a protein (such as circular dichroism of an alpha-helical protein), you could measure the reversible folding and unfolding of a protein under different conditions.

57. First, disrupt the germinating seeds by crushing them, and remove the seed and cell components by centrifugation. Another crude protein purification step, such as salting out, might also be useful. The easiest way to purify this protein from the remainder of the unwanted proteins would be to use an affinity column with galactose-linked resin. The protein of interest will stick to the column, and others will not. The protein can then be eluted by flushing with free galactose in solution.

The answers for these questions will vary from individual to individual.

59. The definition of a drug is often more of a legal question than a biological one.

61. Find out how sulfur amino acids deteriorate health. What happens if you have too much sulfur in your body?

63. Different journals and granting agencies have different rules, but most will be disqualified from reviewing if they have a major professional role in the project, if they or a close family member have a financial interest in the project, or if they are from the same institution as the author.

Chapter 4

1. Active sites are often found in clefts such as the deep cleft between the two globular domains. There is a second notch at the upper right and a broad cuplike feature in the lower left that could also be the active site.

3. RNA molecules have hydroxyl, carbonyl, and amino groups that can all participate in catalysis. They lack thiols and carboxyl groups.

5. A mixture (or cocktail) of protease inhibitors is likely to contain molecules such as ethylene diamine tetraacetic acid (EDTA) which will bind to metal ions; inhibiting metalloproteases; competitive inhibitors of proteases (such as the peptide aprotinin); and suicide substrates such as phenyl methyl sulfonyl fluoride (PMSF), which inhibit serine proteases. These molecules will each inhibit different types of proteases and are broadly effective.

7. An examination of the structure of this drug reveals that it looks a lot like a ribonucleoside. Based on the structure, we would presume that it would be a reverse transcriptase inhibitor (RTI).

9. An examination of the structure reveals a modified peptide backbone with what appear to be the side chains of glycine, phenylalanine, valine, and several five-membered rings that might be analogous in some way to histidine or proline.

11. $K_M = 0.74$ mM and $V_{max} = 2.9$ mM/s

13. BS3 will react with the primary groups of proteins and amines through attacks of these amino groups on the carbonyl carbon at either end of the eight-carbon long arm. In both cases it will form a new amide linkage and the sulfosuccinimidyl group (the heterocyclic ring structure) will be lost.

15. $k_{cat} = 8 \times 10^7$ min^{-1}

17. Probably not. Zn^{2+} could be replaced with another divalent metal such as Ca^{2+} or Mg^{2+} and we might expect to retain activity, but the ions suggested in the problem will probably not have the correct size, electronegativity, or electronic structure to promote polarization of water as Zn^{2+} would.

19. Potentially less harmful. This mutation would affect substrate recognition and binding more than catalysis.

21. Yes, genes can still evolve. Other changes can be found elsewhere in the gene. These changes could be made to the coding sequence and affect other residues that scaffold the essential ones, or they could result in different regulatory mechanisms, for example. Other changes to the gene could affect how or when the gene is expressed or how the message is processed and translated.

23. This mutation would insert a small side chain (a methyl group) in place of the hydrogen atom found in glycine. Based on this small change we might predict that the enzyme would not be dramatically affected. If it were affected, it would be more likely that k_{cat} (catalysis) was affected compared to K_M (substrate binding).

25. The substrate binding pocket of chymotrypsin recognizes and binds to bulky hydrophobic residues. A mutation of a residue lining that pocket to a lysine might better accommodate a small negatively charged residue such as aspartate. This would most likely effect the binding of substrate (K_M) over k_{cat}.

27. Both allosterism and phosphorylation are reversible modifications whereas proteolysis is not. Allosterism takes place very rapidly (as fast as the enzyme can bind to the regulator), whereas phosphorylation and proteolysis can be slower (from milliseconds to minutes). The question of fine tuning a reaction rate is often a confusing one. In terms of most of these control mechanisms, an enzyme is either active or is not—there is no middle state. However, in terms of a population of enzyme molecules, each mechanism allows some level of fine tuning of activity as more or fewer enzymes are activated.

29. Using this equation, we calculate the relative strength of this interaction across the protein to be 34.6 kJ/mol. If this were happening in water, the strength would be 1.7 kJ/mol.

31. The macromolecular structure of the enzyme could be determined using electron microscopy, size exclusion chromatography, native gel electrophoresis, dynamic light scatter, or some other technique that determines if an enzyme is a monomer or multimer in the presence and absence of an allosteric regulator.

33. If the Hill coefficient were 1, there would be no cooperativity; like myoglobin, hemoglobin would be quickly saturated even at low levels of oxygen. If this were the case, there would be very little delivery of oxygen in the tissues because hemoglobin would remain mostly saturated at tissue concentrations of O_2. If the Hill coefficient were 1.5 or 2, hemoglobin would still display cooperativity but not to the same level, so oxygen delivery would occur but not to the same degree as normal hemoglobin.

35. The Hill equation assumes that the total number of binding sites will be filled and that these binding sites are all filled with ligand simultaneously.

37. An orthosteric inhibitor interferes directly with the protein-protein interface, disrupting binding, whilst an allosteric inhibitor induces a conformational change to the binding interface region of the protein that indirectly disables binding. Proteins/enzymes can be regulated allosterically by molecules that differ structurally from a substrate. They can be both activated and inhibited.

39. a. Blue whales breathe air and then dive deep into the ocean. Blue whale hemoglobin must efficiently deliver oxygen to the tissues in a low-oxygen environment; therefore, blue whale hemoglobin must be adapted to stabilize the T state through subunit electrostatic interactions or through increased levels of 2,3-bPG.

   b. Mountain goats live at high altitude, where pO_2 is lower. Mountain goat hemoglobin must be adapted to bind more oxygen at the lower pO_2 concentrations in the lungs while delivering adequate oxygen to the tissues. This can be done by stabilizing the T state through carbamylation with increased CO_2, alterations to the 2,3-bPG-binding site for increased affinity for 2,3-bPG, and mutations that stabilize the subunit interactions in the T state.

c. Cheetahs must deliver large amounts of oxygen to muscles when they are running. Running would also produce large amounts of CO_2, so hemoglobin delivery of oxygen would be enhanced by the Bohr effect. We would expect to find amino acids that can be protonated and carbamylated to stabilize the T state.

d. Migratory birds fly at high altitudes with lower pO_2 along with the necessary delivery of oxygen to muscles during flight. Hemoglobin in these birds would be adapted to stabilize the T state for greater oxygen delivery through subunit interactions and enhanced 2,3-bPG binding. In addition, bird hemoglobin must have increased affinity for O_2 in the lungs.

41.

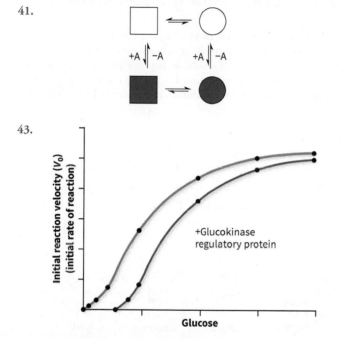

43.

45. It can serve as a chemotherapeutic agent because it inhibits synthesis of deoxyribonucleotides, which are essential for DNA replication. Without DNA replication, cancer cells cannot proliferate. In a similar manner, hydroxyurea would decrease immune response by inhibiting the proliferation of immune cells.

47. Yes, because the replacement of the bone marrow did not change the genetic makeup of the person. The recipient still has two copies of the sickle cell hemoglobin allele and will pass the allele on to any offspring.

49. a. These molecules could act like EDTA and chelate or form a complex with the iron.

b. Iron is highly reactive and can participate in many oxidation reactions if it is not bound in a protein or complex.

c. No, the problem is not the lack of iron; it is how the iron functions in hemoglobin.

d. In sickle cell anemia, the loss of blood cells is not accompanied by the loss of iron from hemoglobin due to the polymerization of the protein.

51. Based on the structure of the hemoglobin holoprotein complex, we might predict that the γ subunits would have similar structure to the β subunits and be largely alpha helical with a globin fold.

53. Antagonists bind to a receptor and block its activation. A weak dopamine receptor antagonist binds weakly to the dopamine receptor and blocks its activation.

55. This means that nicotine can bind to nicotinic acetylcholine receptors and activate them.

57. • A Lineweaver-Burke or double reciprocal plot reveals that V_{max} for this enzyme is 48.3 µmol/min/mg protein and a K_M of 1.19 mM.

• The inhibitor alters both V_{max} and K_m making this a mixed inhibitor.

• You would need the concentration of the inhibitor and enzyme.

59. This graph depicts allosteric regulation. ATP is an allosteric inhibitor, shifting the curve to the right. AMP seems to reverse some of the effect of ATP.

61. It is likely that this holoenzyme complex consists of three monomeric subunits of 66 kDa each that would give an approximate molecular weight of 198, relatively close to 200 kDa.

63. You will need some means of assaying the activity of the enzyme in both the phosphorylated and dephosphorylated state. You will also need some means of phosphorylating the enzyme (in other words, you need to know what kinase is involved in the phosphorylation of the enzyme and how to activate that kinase). The phosphorylation state of the enzyme could be determined using either mass spectrometry, incorporating a radiolabeled phosphate, studying the effect of kinase inhibitors, or using site directed mutagenesis to mutate the phosphorylated residue.

65. Perform kinetics experiments on the HIV protease in the presence and absence of the inhibitor and analyze the results using a Lineweaver-Burke plot.

67. Because sickle cell hemoglobin has an additional negative charge in the β subunit (Glu replaced uncharged Val), you should see a shift in migration in a native gel. The sickle cell hemoglobin will travel closer to the cathode (+ end) of the gel than native hemoglobin. This would only be seen with native proteins, not in SDS-PAGE.

69. If a thalassemia resulted in an unequal ratio of α to β subunits, SDS-PAGE could be used to detect this disorder.

The answers for these questions will vary from individual to individual.

71. Is the lack of antibiotics considered a threat to the public health of the nation? If so, how would you propose funding this research and who would conduct it?

73. Both catalase and the metal are catalysts and catalyze the same reaction. As a protein, catalase has to be purified from a biological source and has to be stabilized in some way. On the other hand, the metals used in catalysts are often heavy metals such as manganese. Both have their advantages and disadvantages.

75. One might anticipate antibiotics, basic pain relievers, and blood pressure medications, but the basic list is actually far more extensive.

77. Yes, bone marrow is replaceable, the procedure has become as painless as a blood draw, and it would provide another person with a chance at a better life.

79. You have a right to deny participation. It is your responsibility to understand the study to the best of your ability and make it clear to the researchers (and to any family members who might try to persuade you) that you do not wish to participate.

Chapter 5

1. If we assume the mass of a phospholipid to be approximately 1,000 g/mol and the mass of an average protein to be 50 kDa (50,000 g/mol), the approximate ratio is 50:1 lipid to protein.

3. Phospholipid head groups contain several different functional groups including phosphate, amino, carboxyl, and hydroxyl groups. In general, ionic or coulombic interactions and hydrogen bonding are two of the main ways that proteins interact with these groups. The amino acids involved in these interactions would include charged and polar amino acids.

5. The membrane consists of two sheets or leaflets of phospholipids. The two leaflets are arranged so that the acyl chains of one adjoin the acyl chains of the other, forming a sandwich of acyl chains with polar head groups on either side of the membrane. This structure is termed a phospholipid bilayer.

7. Leu, Gln, Asp

9. Integral membrane proteins are proteins that are somehow imbedded in the plasma membrane. In contrast, peripheral membrane proteins are associated with the surface of the membrane.

11. It may be possible that one organism engulfed another, but rather than one digesting the other the two organisms began a symbiotic relationship through the sharing of substrates and products. This is most likely an example of a punctuated equilibrium—a significant event that occurs comparatively rarely but puts one population at a significant advantage over another—rather than a series of gradual changes over time. It is unclear if this type of symbiosis occurred (or occurs) very rarely or relatively commonly.

13. Insulin signaling can directly influence gene expression through Akt via the activation of the MAP kinase cascade, ultimately leading to the phosphorylation of transcription factors. Insulin can also indirectly influence gene expression through the action of phosphodiesterase. Insulin signaling increases the activity of phosphodiesterase, breaking down cAMP. This in turn decreases the activity of PKA. Lower PKA activity means less phosphorylation of the transcription factor CREB and therefore decreased expression of the genes that transcription factor regulates.

15. There are many reasons why drugs are banned for athletic competition. While caffeine is widely regarded as a drug, most people also recognize the near universal availability of caffeine in the general population and the limited effect it has on performance. AICAR, on the other hand, is not recognized as a commonly found molecule and may have harmful side effects that are yet to be discovered.

17. Receptor antagonists bind to a receptor and block the function of the receptor. They bind with a similar K_M value, but binding fails to lead to the structural changes needed to activate the receptor. Caffeine is a xanthine derivative. As such, it shares a common skeleton (the purine skeleton) with adenine, which is also a xanthine derivative.

19. Akt is farther downstream in the signaling cascade and therefore is more specific than simply signaling through the insulin receptor. Signaling through the receptor may well cause a variety of changes including alterations to metabolism and gene expression.

21. Phosphodiesterase inhibitors block the breakdown of cAMP, allowing levels to increase and stimulate PKA. This means that there is a low level of cAMP present at all times in the cell due to basal level stimulation of the pathways leading to activation of adenylate cyclase. In most signaling pathways there is often a low level of background signal present.

23. The location of the protein could be identified by disrupting a sample of cells or tissue and separating out the components using density gradient ultracentrifugation. The location of the protein could be compared to other cellular proteins. A second means would be using immunofluorescence. Here an antibody generated against the protein would be used to visualize where in the cell the protein could be found.

25. Samples of membrane lipids can be isolated and characterized, and their physical properties (such as melting point) can be measured.

27. There are several ways to assay receptor–ligand interactions. One way would be to use a radiolabeled ligand (radiolabeled EGF). This ligand can be incubated with the receptor, then the receptor isolated away from the unbound ligand. This is often accomplished using size-exclusion chromatography, dialysis, or filtration. If the radiolabeled ligand binds to the receptor, the isolated receptor sample should now contain the radiolabeled ligand. In the second experiment, we could add EGF to a sample of cells expressing the insulin receptor and measure some output of the insulin signaling cascade (such as phosphorylation of the insulin receptor or activation of downstream kinases) using western blotting with phosphospecific antibodies. For more on this technique, please check the Techniques Appendix.

The answers for these questions will vary from individual to individual.

29. Students should examine the Controlled Substances Act and the updates and amendments that have been made to that act for information on this topic.

31. Clearly people should not judge something based purely on the name. As to whether the use of stem cells or cultured cells is somehow ethically tainted is clearly a matter of debate. Perhaps the greater issue here is to examine how practices were conducted in the past and try to project how they might be conducted in the future to avoid future ethical dilemmas.

Chapter 6

1. **Ia.** aldose   **IIa.** ketose   **IIIa.** aldose   **IVa.** aldose

    Ib. pentose   **IIb.** hexose   **IIIb.** hexose   **IVb.** hexose

    Ic. furanose   **IIc.** furanose   **IIIc.** pyranose   **IVc.** pyranose

3.

5.

7.

9. No, humans and most animals lack chitinase, the enzymes needed to degrade chitin; however, some fish and amphibians have chitinase and therefore can digest this material.

11. We would assume that lactose would remain soluble and partition with the whey. In this instance, we would assume that those who are lactose intolerant could consume these curds. In many cases, those who are lactose intolerant can consume some dairy products such as cheese.

13. Although refluxing a biological sample with solvent is done in some instances to increase extraction, increasing the temperature of an extraction of a biological sample will lead to denaturation of the higher-order structures of most biological molecules, including nucleic acids, proteins, and some carbohydrates, rendering the molecule useless or inactive.

15. The roller coaster would vary with the metabolic state of the cell (as the cell switched between glycolysis and gluconeogenesis), but for the most part the rest of the "ride" would be rather flat because many of these reactions have ΔG values in the cell close to 0 kJ/mol. The end of the ride would also finish lower than the starting point because ΔG for the entire pathway is negative.

17. F-2,6-bP is an important regulator of glycolysis, regulating phosphofructokinase, the key rate-determining enzyme in this pathway. Therefore, molecules that regulate the synthesis or degradation of F-2,6-bP will in turn regulate glycolysis.

19. Without a lysine residue, bacterial aldolase cannot generate a Schiff base. Instead, a plausible mechanism would use a metal ion to polarize a water molecule to attack the carbon–carbon bond of the aldol, resulting in bond cleavage.

21. Glycerol can contribute to gluconeogenesis through dihydroxyacetone phosphate. It is phosphorylated to glycerol-3-phosphate and then oxidized to DHAP. Fatty acids cannot contribute to gluconeogenesis.

23. They would be found on any of the carbons of glucose; the label would be scrambled by triose phosphate isomerase.

25. Although many of the reactions of gluconeogenesis are reversible and used in glycolysis, only the liver and kidneys are capable of producing glucose through gluconeogenesis. Only the liver releases glucose for other organs (typically the brain).

27. Phosphorylation is very important in glycolysis to trap glucose inside the cell, make ATP and facilitate enzyme binding.

29. If a person were on a high-fat diet, the diet would have low levels of carbohydrates. As a result, the liver would be producing more glucose through gluconeogenesis. This would potentially deplete levels of pyruvate from the liver. One source of these carbons would be the glycerol backbone of fats.

31. Pyruvate dehydrogenase is the committed step in the synthesis of acetyl-CoA from pyruvate. The molecules that regulate pyruvate dehydrogenase are all molecules that reflect the energetic state of the cell and therefore regulate flow of metabolites into the citric acid cycle.

33. This could be any molecule that can serve as a proton acceptor. Typically, these would be deprotonated carboxyl groups in aspartate and glutamate or deprotonated histidine or tyrosine residues.

35. These data would suggest that the molecule in question is a polysaccharide with a molecular weight of at least 10 kDa. The hydrolysis data suggest that the polysaccharide is a combination of glucose and glucosamine residues in a 1:1 ratio. It is likely that the acetate is found as a modification of the glucosamine residues (N-acetylglucosamine), but further experimentation would need to be conducted to test this hypothesis.

37. The shells could easily be composed of a carbohydratelike chitin. I would advise her to test the samples for carbohydrate or perform mass spectrometry on the sample to help determine the composition.

39. The Molisch test is a test for all carbohydrates in which a strong acid dehydrates an alcohol to an aldehyde, which reacts with napthol to give a purple-colored product. The positive Molisch test indicates that the sample is a carbohydrate. Other carbohydrate tests, such as the starch–iodine test, Benedict's test, or Bial's test, can be used to determine what types of carbohydrates are found in the sample.

41. This experiment would require two groups of mice—one treated with the drug and a control group. Both groups would need to be fasted to induce gluconeogenesis. A simple blood test to measure glucose would give an indication of glucose levels but not the origin. To measure gluconeogenesis, mice could be administered radiolabeled alanine or pyruvate, and as the label accumulated in glucose or in the steps of gluconeogenesis, it could be used to measure gluconeogenesis. A similar approach could be used in cultured cells. This would be made somewhat easier by the lack of different tissues and ease of administering different metabolites to the dish of cells. A different approach would be needed in humans. Gluconeogenesis would largely be determined by measuring plasma glucose without radiolabels.

The answers for these questions will vary from individual to individual.

43. How much public support should go to any endeavor? Is developing an alternative fuel in the public interest? How much public money is used now for either alternative or traditional fuels?

45. Arguably, most humans currently have access to foods that are not seasonally regulated. For examples, fresh fruit can be moved around the world in ships and planes and is routinely available; however, one can argue that humans have always stored some food for the winter or that for many generations trade has enabled us to move foods from far away. Likewise, humans, unlike many other species, do not have a limited geographic range on Earth.

Chapter 7

1. I. It was lost as CO_2 in the first round.

   II. Half of the label would be lost as CO_2 in the oxidation of isocitrate; the other half would be in one of the two carboxyl groups of α-ketoglutarate.

   III. It was lost as CO_2 in the first round (half in the oxidation of isocitrate, half in the oxidation of α-ketoglutarate).

3. We might expect that the reactions in which CO_2 is lost or in which there is a transfer of a group from CoA would have negative ΔG values. Other reactions, such as isomerizations or hydrations, would quite likely have ΔG values near zero.

5. Regulation of the citric acid cycle is depicted in Figure 7.4. The citric acid cycle is positively regulated by Ca^{2+} and negatively regulated by NADH, succinyl-CoA, ATP, and citrate. The key points of control are at citrate synthase (the entry point to the pathway), isocitrate dehydrogenase, and α-ketoglutarate dehydrogenase.

7. Radiolabeled tracers are effective because they do not have a significant impact on the reactions that the molecules are undergoing. In contrast, fluorescent tags are much larger than the intermediates in question in the citric acid cycle. As a result, the tag might interfere with how the molecule was able to react in the cycle. In theory this may have been able to work, but in practice it may have been more difficult.

9. If there were some way to overcome the energetic barriers found at the steps where carbon dioxide is released and thioesters are cleaved, most of the other steps of the cycle would have ΔG values near zero and could be easily reversible by altering concentrations of substrates in the cell.

11. Complete deficiencies of a key step in the citric acid cycle would result in an organism that is not viable and would probably never come into existence.

13. The species that are found vary from *E. coli* to humans and *Bos taurus* (cow). Some species like cows have been chosen because they are a rich source of protein for purification. Others, such as microbes, are easy to work with in the lab.

15. We might imagine that if we could block electron flow through complex II with some sort of poison, electrons could still flow from complex I to complex III and we might still be able to generate ATP. The problem with this line of reasoning is that we would rapidly deplete the supply of FAD and be unable to generate NADH through the citric acid cycle. Anaerobic organisms can operate without the citric acid cycle, but aerobes cannot.

17. The redox-active centers used in electron transport include iron-sulfur centers, heme groups, and organic electron carriers such as quinones. Four macromolecular protein complexes are involved in electron transport:

Complex I is NADH dehydrogenase. It oxidizes NADH to NAD^+, passing electrons to ubiquinone and pumping four protons out of the matrix for each molecule of NADH.

Complex II is succinate dehydrogenase. This enzyme of the citric acid cycle oxidizes succinate to fumarate, capturing the two electrons in a molecule of $FADH_2$. Electrons from $FADH_2$ are transferred through several redox centers to ubiquinone.

Complex III is ubiquinone oxidase. Complex III uses ubiquinone to take electrons from complexes I and II and other catabolic processes, and it transfers them to the soluble electron carrier cytochrome *c*. Four additional protons are transferred to the intermembrane side of the inner mitochondrial membrane for each molecule of ubiquinone.

Complex IV is cytochrome *c* oxidase. Complex IV oxidizes cytochrome *c*, transferring electrons to oxygen to generate water. Two final protons are pumped out of the mitochondrial matrix, one for each molecule of cytochrome *c*.

19. One means by which we could identify the location of the electron transport complex would be to fix and section cells and stain them with an antibody directed against one of the proteins of the electron transport chain. This could be detected using immunogold and electron microscopy to show the location. A second technique would involve isolating intact mitochondria and treating them with a protease. Proteins on the cytosolic side of the mitochondria would be exposed to the protease and degraded. Proteins in the mitochondrial matrix would be protected. This could be visualized using SDS-PAGE and western blotting using antibodies that would detect proteins of the electron transport chain. A third means would be to chemically derivatize the proteins that are on the cytosolic side of the mitochondria using a dye or fluorophore and show that the proteins of the electron transport chain are not stained.

21. Organisms that fix CO_2 often employ biotin to bind and polarize the molecule. Reactions with oxygen most often use heme groups again to bind and polarize the molecule for subsequent chemical reactions.

23. The mechanism has electrons flowing from a partially positively charged carboxyl carbon to the negatively charged enolate, whereas it should be the other way around.

25. Due to the electron-withdrawing nitro groups and resonance, the pK_a of picric acid is much lower (0.4) than that of 2,4-DNP (4.1). Therefore, it would be harder for picric acid to protonate and deprotonate and pass through the membrane, making DNP the better uncoupler.

27. Each of these ATPases and synthases can synthesize or hydrolyze ATP, and most are coupled to some sort of chemical gradient. Some, such as F, A, and V, have similar architecture, but others vary. Unlike the others, E-type lacks a transmembrane domain and P has a single subunit.

29. We would have to assume that a single molecule of ATP could be used to pump three protons; we would have to assume the lysosome was spherical, and that there was no leakage, buffering, or other likely possibilities.

31. The pump would pump 14 molecules per minute.

33. The citric acid cycle is the cell's central metabolic hub. Many other molecules are also metabolized through the citric acid cycle. What probably happened is the label was allowed to incorporate into these other molecules.

35. Spectra A absorbs maximally in the 450 to 500 nm range and is the yellow spectrum. Spectra B absorbs maximally in the 500 to 600 nm range and is the red spectrum. EPR detects unpaired electrons and would detect the Co^{2+} state.

37. One experiment would be to measure ATP production in the presence and absence of other ions to alter the charge distribution across the membrane.

39. Radiolabeled FAD could be incubated with the succinate dehydrogenase complex to see if it exchanges from bulk solution to the enzyme. If radiolabeled FAD was found to associate with the enzyme, this would be evidence that it exchanged with the protein.

The answers for these questions will vary from individual to individual.

41. What is the data that indicate neonicotinoids may be responsible or not, and is it coming from a reputable source?

43. How do these fish get to market? Who are the players in this process?

Chapter 8

1. Building up any molecule requires an input of energy. In the instance of polymers of monosaccharides, the energy is supplied in one way or another from a nucleotide triphosphate. When these bonds are hydrolyzed, more stable products are formed, and energy is released in the formation of those new molecules.

3. Glycogen storage is by insulin, glucagon, and catecholamine hormones, epinephrine and norepinephrine

5. Muscle lacks the phosphatase required to liberate free glucose from glucose-1-phosphate (the product of glycogen breakdown); as a result, glucose is unavailable to diffuse out of the myocyte as it does in the liver. Muscle keeps its carbohydrates and uses them for glycolysis instead of fueling the brain like the liver does.

7. If muscle expressed the glucagon receptor, it would become responsive to glucagon because the rest of the PKA signaling pathway is already in place. This would mean that a glucagon signal would elicit a response that is similar to those seen in other tissues. Glycogen would be catabolized, but muscle does not express glucose-1-phosphatase or glucose-6-phosphatase and would be unable to release the liberated glucose. Instead there would be an increased usage of glucose, futile cycling, or some pathophysiology resulting from increased glucose levels in the cell.

9. Xylulose-5-phosphate, a product of the pentose phosphate pathway, is an important regulator of the transcription factor ChREBP (the carbohydrate response element binding protein) that responds to high levels of carbohydrates and regulates transcription of genes involved in carbohydrate and lipid metabolism.

11. Fructose-6-phosphate can react with either glyceraldehyde-3-phosphate or erythulose-4-phosphate to form xylulose-5-phosphate and erythulose-4-phosphate or sedoheptulose-7-phosphate and glyceralde-hyde-3-phosphate, respectively. This could shift concentrations of those monosaccharides, altering flux through pathways like glycolysis or gene expression via xylulose-5-phosphate. To determine flux through the pathway, you would need to know the concentrations of the different intermediates and the rates of catalysis.

13. Yes, PPP will be affected because trasnketolase involved in the non-oxidative phase of the PPP pathway employs an active site with thiamine pyrophosphate to attack the carbonyl carbon of the ketose substrate, thus deficiency of thiamine will render the enzyme dysfunctional.

15. Selenium is just below sulfur in the periodic table and to some extent acts as sulfur or oxygen would. Sulfur is less electronegative and less of a nucleophile than oxygen, and selenium is even less so than sulfur.

17. Because a protein is covalently linked to the GPI anchor, it would need to be cleaved by a protease or the anchor degraded by a phospholi-pase (often phospholipase C in the cell) to separate the two.

19. These polysaccharides could be synthesized by the addition of acti-vated monosaccharides (analogous to CDP-glucose), or they could be added as disaccharide units using a transferase to join the disaccharide unit to the growing polysaccharide chain.

21. A: 6, B: 6, O: 5.

23. Penicillin blocks the cross-linking of the cell wall by irreversibly derivatizing the enzyme responsible for catalyzing the cross-linking. The sensitivity of bacteria to penicillin depends on the strain of bacteria and the presence or absence of a cross-linked cell wall (gram-positive versus gram-negative bacteria).

25. Because of the single electrons being transferred in prolyl hydroxy-lase, a technique such as EPR (electron paramagnetic resonance) would be useful to detect these states and changes.

27. No, the microbes in kombucha and kefir are different and secrete different matrix polysaccharides.

29. When using electron microscopy, we do not need to crystalize a protein. Elimination of that step can streamline the structure determi-nation process.

31. It is likely that in the cell lysine and glutamine residues are nec-essary for other aspects of protein chemistry, such as catalysis. Matrix proteins are either cross-linked in the Golgi apparatus prior to secretion or cross-linked in the matrix itself.

33. Based on what we know from other carbohydrates, the structure of kefiran would be an alternating linear polymer of glucose and galac-tose. It could be synthesized by the incorporation of lactose groups to the growing polysaccharide as UDP galactose. The structure could be determined through X-ray diffraction or through fragmentation of the polysaccharide into small, soluble di- and trisaccharides and analysis of the fragments using chemical means or mass spectrometry.

35. The cells could be fragmented and large molecules, such as glyco-gen, separated from small molecules, such as glucose. Once separated, glycogen could be hydrolyzed using an acid or base and the resulting glucose monomers analyzed.

37. Co-immunoprecipitation and fluorescence (or Forster's) resonance energy transfer (FRET), are two common means of determining protein–protein interactions.

The answers for these questions will vary from individual to individual.

39. Biofilms are one of hundreds of research areas that are currently ongoing around the world. Funding for this work comes from both the private (companies) and public (government) sectors.

41. Good record collection, keeping and management, transparency and appropriate representations of results.

Chapter 9

1. a. In order, these molecules are: triacylglycerol, wax, steroid, fatty acid, terpenoid (vitamin A).

   b. The steroid, fatty acid, and terpenoid are amphipathic. The polar functional groups on these molecules (the hydroxyl and carboxyl groups) are the water soluble parts.

   c. Each of these is built up from acetyl-CoA.

   d. the triacylglycerol and wax

3. a. The fatty acid in this problem is 16:0.

   b. This fatty acid has no double bonds, so it has no omega no-menclature.

   c. hexadecanoic acid

5. Because lyso-PC is more cone-shaped, it would more likely form some sort of micelle or sphere than a bilayer.

7.

A. Lyso-phosphatidylethanolamine   B. Cholesterol   C. Cholesteryl ester   D. Palmitoyl linolyl phosphatidylcholine

9. Beta-oxidation and fatty acid biosynthesis are compartmentalized in the cell. Beta-oxidation occurs in the mitochondria, whereas fatty acid biosynthesis occurs in the cytosol. This is similar to the compart-mentalization that is found in the reactions specific to glycolysis and gluconeogenesis. Other differences include the use of NAD^+ versus NADPH and the stereochemistry of different intermediates.

11. 104.5

13. In fatty acid biosynthesis, acetyl-CoA carboxylase (ACC) catalyzes the rate-determining step and is the main point of regulation.

15. AMPK acts on ACC1. Phosphorylation by AMPK inhibits ACC1 and would slow fatty acid synthesis by limiting the production of malonyl-CoA.

17. There are several reasons for this. First of all, the two enzyme com-plexes clearly play a fundamental role in lipid metabolism and therefore are germane to many diseases that have lipid metabolism as a central feature. Second, there are several important differences between eukar-yotic and prokaryotic fatty acid metabolism, and therefore molecules that would target these enzymes would be potentially good drugs for the treatment of infections.

19. Because urine and blood have different compositions, the assay for these compounds is different. Testing ketones (or any compound) in the urine is affected by when the sample is obtained (early in the morning after waking versus in the middle of the day) and can also be affected by how much urine the patient is producing. Depending on the fluid being tested, there may be interference due to pH, electrolytes, and proteins.

21. Because D-β-hydroxybutyrate is converted to acetoacetate in heart and muscle tissue, we know that the reaction has to proceed with a neg-ative ΔG value at these concentrations. Heart and muscle tissue cannot synthesize ketone bodies because of this energetic barrier and the lack of the enzymes required to synthesize them in these tissues.

23. The synthesis of ketone bodies is necessitated by the inability of the brain to catabolize fatty acids. If the brain could catabolize fatty acids, ketone body biosynthesis would be unnecessary. Because plants and prokaryotes lack barriers, such as the blood–brain barrier, it is unlikely that they use ketone body biosynthesis.

25. Steroids are lipids with a common skeleton of four fused rings. Cholesterol, sex hormones, and bile salts are steroids. The properties and functions of these molecules vary depending on the functional groups attached to the skeleton.

27. Rather than using cholesterol, cold water fish have high levels of polyunsaturated fatty acids. These fatty acids lower the melting point of the lipids and the entire membrane.

29. Based on the structure of fluticasone we can assume that it is acting as a steroid. Steroids act by binding to a cytosolic receptor, translocating to the nucleus where they act as transcription factors and alter gene expression.

31. Examining the structure of this molecule, we see several places where the drug could hydrogen bond. In a typical aqueous environment, a ligand binding site might have water molecules that would hydrogen bond to the ligand or ligand binding site. In a lipid bilayer this would not be the case. Both hydrogen bonding and hydrophobic interactions would play important roles.

33. Capsaicin is a hydrophobic molecule. When someone eats or drinks a food or beverage with excess lipids in it, the lipids will bind to the capsaicin and remove it from the mouth. Drinking water will not do this. Once the capsaicin leaves the mouth, the burning sensation ceases, but other effects mediated through vanilloid receptors in other parts of the body can still be felt.

35. a. An internal standard is a standard (a known molecule and concentration) that is added to an experiment. In this way, the standard can be used to track how efficient different aspects of the experiment are and to normalize data at the end.

   b. Most naturally occurring fatty acids contain even numbers of carbons. By using an odd number, it becomes easy to trace the standard.

37. a. The percent of acetic acid decreases in proportion to the amount of drug used. Different concentrations were used to show the effect of drug concentration on the outcome. Adding other time points may show similar data, but this adds the additional variable of time.

   b. The drug is activating CPT I.

   c. Treatment with the drug is increasing palmitate oxidation over time. This is consistent with the other data observed; CPT activity is one of the rate-determining steps of β-oxidation.

39. Cobra venom contains phospholipases. These phospholipases could release arachidonate, which would be used for eicosanoid production, but they could also damage the plasma membrane of cells to the point of causing cell lysis.

41. Incubate cells with radiolabeled carnitine. Fractionate cells at different time points and analyze for carnitine and acyl carnitine. Radiolabeled fatty acyl carnitines should appear in the cytosol before appearing in the mitochondrial matrix.

43. Incubate fatty acids, coenzyme A, and fatty acyl-CoA synthetase in the presence and absence of ATP and see if a product forms.

45. One way to test this would be to isolate the enzymatic activity responsible for each reaction. Use size-exclusion chromatography to see if they are running together as a single complex.

The answers for these questions will vary from individual to individual.

47. Empirically examine the case for and against medical marijuana. Outside historical and cultural biases, does this make medical and scientific sense?

49. How is this currently funded? How is this balanced versus funding other conditions?

Chapter 10

1. Phospholipids, are conserved throughout evolution. As animals evolved, they adapted to different environments and challenges. Analysis of genes from different animals reveals that genes coding for proteins involved in neutral lipid transport and storage evolved more recently, and thus may address needs that are lacking in more primitive species.

3. If we consider the synthesis of any molecule, there are steps, such as the formation of an ester, which seem energetically unfavorable if the substrates were simply an alcohol and fatty acid (ΔG would be positive). In the synthesis of biological molecules, nature generates activated donors, often through coupling one molecule to another. An example of this in phospholipid metabolism is the addition of fatty acids to a lyso lipid. Fatty acyl-CoAs are used as the donor to form this ester bond. This overall reaction has a negative ΔG value.

5. Pancreatic lipase has both elements of a helix and β sheet. There are two globular domains to this protein separated by what appears to be a flexible hinge, which may allow for movement of the two domains. The smaller of the two globular domains contains more β sheet and also binds to the colipase cofactor.

7. In the enterocyte, the newly reassembled triacylglycerols from dietary lipids are packaged into a large particle called a chylomicron, which is used to transport dietary lipids in the circulation via the lymph.

9. Orlistat has three hydrophobic moieties, an ester, and an amide. It is likely that binding of the hydrophobic groups occurs in some sort of hydrophobic pocket lined with hydrophobic residues. The ester and amide group hydrogen bond well and therefore could interact with polar residues capable of undergoing hydrogen bonding.

11. MTBE is a hydrophobic solvent. It is used to dissolve the gallstones and flush the solubilized bile from the gallbladder.

13. Numerous factors are involved in the evolution of such a complex system. Factors that should be considered include changes in diet and lifestyle in modern times compared to several hundred or several thousand years ago. Also most people who develop cardiovascular disease do so after they have passed on their genes to their offspring.

15. There are multiple ways to address this question. One way would be a phospholipid array. Different phospholipids, including PIP2, could be spotted to a nitrocellulose membrane. Next a mixture of proteins containing AP2 is added. AP2 is detected by treating the membrane as an immunoblot, that is, detecting the binding of AP2 with an antibody reporter complex.

17. We would need some means of separating the individual clathrin proteins from the holoprotein, such as a size exclusion column. The column could be run in native and denaturing conditions and the products compared. Under native conditions, we would isolate a single band for the clathrin holoprotein. Under denaturing conditions, we would expect to see two bands, one for the heavy subunit and one for the light subunit, but adding the molecular weight of these two would only reach a third of the mass of the holoprotein. If we know the length and angles of a clathrin trimer and the size of the endosome, we could reconstruct how many trimers are needed to cover the surface.

19. This is somewhat true. Adipose tissue synthesizes the triacylglycerol it stores from fatty acids obtained from chylomicra and VLDL and esterifies these to glycerol phosphate, which is typically obtained as a side reaction of glycolysis.

21. Phosphatidate is at a crossroads between neutral lipid metabolism and phospholipid metabolism. Removal of the phosphate group yields diacylglycerol, which is esterified to triacylglycerol.

23. Beta-oxidation occurs in nearly all tissues except the brain but predominantly in the heart and in muscle tissue. Gluconeogenesis occurs largely in the liver, which ensures that the liver can provide glucose for the brain. The other organs can use β-oxidation.

25. A different lipase hydrolyzes each of the three fatty acids from triacylglycerol.

27. Because the lipid storage droplet has a hydrophobic core and not an aqueous one, only a monolayer of phospholipids is necessary to coat the core.

29. Lipid rafts have been broadly grouped into two categories: caveolae and non-caveolar rafts. These small lipid rafts may be captured and stabilized by lipid-anchored, transmembrane proteins.

31. Lipid rafts are largely composed of sphingolipids and cholesterol. There are hydrophobic interactions between the ring systems of cholesterol and the fatty acyl chains of the sphingolipids. The head groups of the sphingolipids and the hydroxyl group of cholesterol can interact through hydrogen bonding.

33. a. Based on both the micrograph and the centrifugation gradient, the protein of interest appears to be in the endoplasmic reticulum.

   b. Multiple independent, redundant methods provide an added level of verification in any experiment.

   c. These proteins are commonly found in these organelles and are often used as indicators of protein localization.

35. a. The venom appears to break phosphatidylcholine into lyso-phosphatidylcholine.

   b. Lysophosphatidylcholine and free fatty acids

   c. EGTA seems to prevent the reaction from occurring. EGTA acts by binding divalent metal ions, such as Ca^{2+}, indicating that this ion may play a role in enzyme activity

37. An investigator could ask if the other levels of lipoproteins are normal. She or he could also ask if the enzymes, receptors, and apolipoproteins had normal activity or were in any way altered.

39. The hormone receptor could be labeled using an antibody labeled with gold beads. Electron microscopy images could be analyzed to note where in the cell the gold beads accumulated to determine if they used receptor-mediated endocytosis to enter the cell.

The answers for these questions will vary from individual to individual.

41. Current controversial topics in the media include genetic modification of crops or humans, effectiveness of alternative treatments, and what should or should not be in the diet.

43. Has the introduction of other foods, such as artificial sweeteners, increased consumption of foods containing those ingredients?

Chapter 11

1. Zymogens are inactive forms of enzymes. They are released in stomach as they do not damage to the tissues from which they are secreted. When it is required and the conditions are optimal, it is converted to the active form and acts on target tissues.

3. a. Loss of function. Loss of the hydroxyl group on serine prevents formation of a nucleophile to attack the peptide. The enzyme may still be able to bind substrate but will not be catalytically active.

   b. Probable loss of function. The aspartate is a much weaker nucleophile than the hydroxyl of serine and therefore will probably not be able to attack the peptide bond. Substrate binding may be unchanged, but catalysis (k_{cat}) will be much lower.

   c. Probable loss of function. Although the threonine contains a hydroxyl group, it is a secondary, not a primary, alcohol. This means that there are potentially steric concerns when attacking the peptide bond.

5. Cleaving a protein results in fragments that denature or denature more readily at a lower temperature. These denatured proteins form the curds of cheese. Any monomeric amino acids or short peptides that would be released are soluble.

7. No, the ionized amino and carboxyl groups on the free amino acid backbone ensure that this amino acid is quite soluble in the blood.

9. It is because glutamate plays a central role in nitrogen metabolism.

11. Increase in ammonia and NADH will be observed in patients.

13. Decarboxylation of ornithine would yield 1,4-diaminobutane, also known as putrescene. Decarboxylation of amino acids, or any reaction involving the α carbon, is often accomplished using a pyridoxal phosphate-dependent mechanism. Because CO_2 is being released in a decarboxylation, it probably contributes favorably to increasing entropy and therefore drives ΔG lower.

15. The correct sequence of reactants for carbamoyl phosphate synthetase is ATP and bicarbonate, then ammonia, then ATP. The carbonyl group of carboxyphosphate undergoes nucleophilic attack by ammonia.

17. Scavenging of amino acid skeletons is a catabolic process and reactions are typically near zero or negative in terms of ΔG values. Energy is captured in terms of NADH. Overall, ΔG for these reactions is negative for a cell undergoing amino acid catabolism. This also means that the flux through these pathways is dependent on the concentration of reactants and products.

19. Obligate carnivores (animals that only eat meat) eat a very high protein diet. Most of the amino acids they consume are either burned as energy via the citric acid cycle or used to make glucose via gluconeogenesis.

21.

Propranolol   **4-hydroxypropranolol**

4-hydroxypropranolol is an oxidative product and therefore part of phase I metabolism.

The sulfation and glucuronic acid derivatives are both examples of phase II metabolism.

Oxidized products come from phase I metabolism.

23. The remaining MAOIs in the system react or are excreted, and the body synthesizes new enzymes to replace the ones that have been derivatized.

25. The molecule is probably being excreted in the filtrate and not reabsorbed.

27. 1.0 kg of salt is 17.1 mol of salt, or moles of sodium ions. To separate 1 mol of ions requires 0.33 mol of ATP. If we consider that hydrolysis of a mole of ATP liberates 7.3 kcal, the overall energy required to pump this many sodium ions is approximately 41.2 kcal.

29. In crush syndrome, damaged tissues release proteins (often myoglobin) that denature in the circulation or kidney and precipitate, damaging kidney function.

31. Based on where these reagents cleave, the order would be N-terminus SerMetIleTyrTrpTrpAlaLys C-terminus.

33. The disease is caused by a deficiency in one of the enzymes of valine degradation. Valine is being deaminated and potentially coupled to

coenzyme A but is unable to be desaturated to methylacrylyl-CoA. Instead, it accumulates and the acylcarnitine is made as a side product. Short-chain fatty acids are often odiferous. The odor likely comes from isovalerate in urine. As isovaleryl carnitine accumulates in the circulation, it is detoxified in the liver to isovaleryl glycine, which is then eliminated in the urine where it breaks down.

35. Isolate mitochondria and cytosol. Attempt to regenerate some of the reactions of these pathways in the presence of only cytosol or mitochondria as a source of enzymes.

37. Give the drug to an animal model like a rat or pig and see how and when it is secreted in the animal's urine. Characterization of the derivatives found in the urine would provide keys as to how the drug was metabolized.

The answers for these questions will vary from individual to individual.

39. Are many over-the-counter drugs abused?

Chapter 12

1. Glycerol kinase (the enzyme that catalyzes the phosphorylation of glycerol in the liver) and glucose-6-phosphatase (the enzyme catalyzing the dephosphorylation of glucose-6-phosphate in the final step of gluconeogenesis in the liver) both tailor specific responses to a tissue.

5.

7. Phenylethanolamine *N*-Methyltransferase transfers a methyl group to the nitrogen (N) of phenylethanolamine. Based on some of the chemistry we have seen in this chapter and others that the methyl group donor is *S*-Adenosylmethionine. Folate can also act as a donor of methyl groups in some reactions.

9. Hypothalamus regulates the intersection of metabolism, appetite, and complex behaviors such as feeding. Hormones regulate gene expression through several different mechanisms such as nuclear hormone receptors and resident nuclear transcription factors. The cyclic AMP response element-binding protein (CREB) is an example of a resident nuclear transcription factor. CREB is the cyclic AMP response element-binding protein. This resident nuclear transcription factor responds to PKA phosphorylation to bind a cyclic AMP response element in promoter regions of some genes, activating transcription. The sirtuins, a group of seven proteins that can act as deacetylases, regulate genes by modifying histones and may directly affect metabolic pathways by deacetylating enzymes.

11. a. NMR or X-ray diffraction could provide a molecular structure.

b. Because glucagon is small, it would presumably be easier to crystalize it for X-ray diffraction or make it more soluble for an NMR experiment. It would also be less complex overall.

In this instance, both of these enzymes are expressed in the liver. This means that the specific pathways they are a part of can only happen in that tissue; for example, glycerol released from triacylglycerol hydrolysis in adipose tissue cannot be recycled in that tissue without first cycling through the liver to be phosphorylated and used in some other metabolic process (phospholipid or neutral lipid biosynthesis or gluconeogenesis). Likewise, glucose-6-phosphatase is only expressed in the liver. As such, the liver is the predominant tissue that releases glucose following gluconeogenesis (the kidney can also perform gluconeogenesis under some conditions). Muscle lacks this enzyme and is unable to export glucose-6-phosphate for use by other cells. Instead muscle tissue uses glucose from gluconeogenesis for energy via glycolysis or glycogen biosynthesis. There are numerous other enzymes that are either differentially expressed or differentially regulated to coordinate organismal metabolism.

3. When someone in the postabsorptive state has consumed a meal, the metabolism does not immediately return to the fed or postprandial state. As levels of glucose, amino acids, and dietary fats in chylomicrons increase, the liver remains refractory or resistant to the effects of glucose and insulin, allowing other tissues to "feed" first, replenish glycogen levels, and return to a glycolytic state. Indeed, the liver continues to undergo gluconeogenesis but uses this glucose for liver glycogenesis. As glucose levels increase, the liver again begins to respond and returns to a postprandial state, absorbing glucose and synthesizing fatty acids and triglycerides for VLDL.

c. Because the structure is small, it may have more motion than a larger protein and therefore be harder to determine using either of these techniques.

13. The human genome has allowed scientists to identify small peptide hormones or the genes for their precursors. It has also facilitated the discovery of their receptors.

15. Because thyroid hormone increases expression of genes involved in metabolism, low levels of thyroid hormone lead to lower expression of these proteins and metabolic rate. This manifests itself as weight gain, and people suffer from having low energy, among many other symptoms.

17. Corticotropin and adrenocorticotropic hormone are pituitary peptide hormones.

19. The symptoms of goiter would be the same as symptoms of low levels of thyroid hormone (hypothyroidism). Because thyroid hormone increases expression of genes involved in metabolism, low levels of thyroid hormone lead to lower expression of these proteins and metabolic rate. This manifests itself as weight gain, and people suffer from having low energy, among many other symptoms.

21. One underlying observation that relates these three groups would be very low body fat percentage. Body fat correlates with leptin levels. Lack of leptin signaling can lead to amenorrhea in females.

23. The melanocortin-4 receptor (MC4R) binds to melanocortins, especially α-MSH and POMC. Mutations to this receptor have been associated with obesity in humans. Based on that data, we would anticipate that a knockout mouse for MC4R would also be obese. Because this receptor also plays a role in sexual function, we might also anticipate that eliminating this receptor could lead to a lower reproductive rate in these mice.

25. Chronic alcohol consumption carries with it other significant health effects. Alcoholism often results in malnutrition due to the general failure to consume a nutritious diet, the loss of water-soluble electrolytes and vitamins through increased urination, and the inability to absorb certain B vitamins in the intestine.

27. Lys−N=CHCH$_3$

    Schiff base

29. Creatinine concentrations are used as a measure of kidney function. High plasma levels and low urinary levels indicate that the kidneys are not functioning normally.

31. Creatine is thought to act as a buffer to help replenish ATP stores in the short term. Therefore, it is thought to help athletes who undergo short bursts of activity that are almost exhausting. These would potentially include baseball, weightlifting, soccer, and football due to the sprinting activities that occur in each sport. In reality, the data regarding creatine use and performance are mixed, and there are no clear guidelines.

33. a. The data in this graph support the hypothesis that an acute event led to increased levels of thyroid hormone and a decreased level of thyroid stimulating hormone.

    b. An alternative hypothesis might be that some toxin that had been consumed is causing people to produce elevated levels of their own endogenous thyroid hormones. A test of thyroid function would indicate if the hormones in question were coming from their own thyroid or ones that were consumed.

    c. High levels of thyroid hormone act in a feedback loop to decrease levels of TSH.

    d. Someone in the same group or family who had not consumed hamburger, or the same people following a washout period where the thyroid hormone levels dropped, would be an appropriate control.

    e. This is more of a case study than an experimental study. It would be potentially difficult to ethically produce this effect in people.

35. You could make truncations or mutants of the protein in which regions were missing or deleted. These new proteins could be tested for function to find the specific region or regions that were important for binding.

The answers for these questions will vary from individual to individual.

37. This practice is considered unethical and unhealthy.

39. Campus health centers typically have information on eating disorders.

Chapter 13

1. The six-membered ring is formed through a condensation reaction of the N-1 amino group and the C-2 formyl group to produce inosine monophosphate. After formation of IMP, the purine pathway forks into two branches: one leading to AMP and one to GMP. Both branches employ amino acids as cofactors in the reactions.

3. Dowex is a cation-exchange resin, that is, a resin with cationic groups on its surface that will bind selectively to anions. The more strongly charged the anion, the more selectively it will adhere to the resin. Because IMP, AMP, and adenylosuccinate are all negatively charged, it is likely they will be retained by the resin while hypoxanthine would not be.

5. Purine synthesis is regulated at the first step (formation of PRPP), the second step (formation of 5-phosphoribosamine [PRA]), and at the steps leading to AMP and GMP biosynthesis.

7. HPRT1 gene mutations that cause Lesch Nyhan syndrome result in a severe shortage (deficiency) or complete absence of hypoxanthine phosphoribosyltransferase. When this enzyme is lacking, purines are broken down but not recycled, producing abnormally high levels of uric acid.

9. Elevated levels of glucose-6-phosphate are metabolized via the pentose phosphate pathway. This uses NADPH and perturbs concentrations of ribose-5-phosphate, resulting in changes to purine metabolism. In von Gierke disease, this leads to elevated levels of uric acid via elevated purine catabolism.

11. Regulation of pyrimidine *de novo* synthesis in mammals is through allosteric control by OMP decarboxylase.

13. It means that the binding site is not dependent on the nature of the base or the hydroxyl group in the 2′ position for substrate recognition. Instead, binding must occur through recognition of other groups: the remaining carbons and phosphates on the ribose or deoxyribose ring and the three phosphate groups.

15. It says that the free energy of the phosphate transfer should be equivalent from each NTP.

17. We might anticipate levels of dNTPs would be highest in G_1, when the cell is preparing for DNA replication and mitosis.

19. The mechanism of ribonucleotide reductase uses a radical generated in the β subunit to cause chemistry in the α subunit. Both are required in the correct configuration for chemistry to occur. Neither would be catalytically active on its own.

21. Low-dose methotrexate is used to treat autoimmune diseases because the drug targets cells that are causing inflammation. Although the exact mechanism of action is not clear, the drug acts as a folate antimetabolite and causes alterations to nucleotide metabolism in these cells. This blocks the production of immune cells, leading to a decreased immune response.

23. In bone marrow transplants, a patient's bone marrow is destroyed using chemotherapy or radiation, and then the bone marrow is replaced using a transfusion of bone marrow cells from a donor. Experimental treatments include using a patient's own stem cells or genetically manipulated cells (gene therapy) as the donor cells. Lesch-Nyhan syndrome cannot be treated this way because all the cells of the body, not just bone marrow or blood cells, are affected.

25. There are multiple reasons why purines are eliminated as uric acid. One reason is that if they were catabolized via other pathways, four molecules of ammonia would need to be detoxified. This would be energetically costly if it proceeded through the urea cycle.

27. Lesch-Nyhan syndrome is caused by a deficiency in hypoxanthine-guanine phosphoribosyl transferase; the salvage pathway is dysfunctional. This leads to elevated levels of purine catabolism, which in turn leads to elevated uric acid levels.

29. Synthesizing a molecule of urea costs the equivalent of four molecules of ATP. Uric acid is a catabolic product of purines and has no energetic cost associated with its synthesis; however, the synthesis of purines requires the equivalent of at least seven molecules of ATP, as well as tetrahydrofolate and several amino acids. Other factors that are important in thinking about disposal of nitrogenous wastes are how much waste will be generated and how it will be transported or eliminated. Urea requires water for disposal, which is an issue for animals that are conserving weight (birds) or that live in arid or saltwater environments (some reptiles). In these cases, uric acid is eliminated as a solid.

31. Each of these molecules has structural similarities to the substrates that these enzymes typically bind. Each of these molecules differs from the substrate structurally but each binds in the active site and competes with the native substrate for the enzyme (they are often competitive inhibitors). These drugs can be used to treat a variety of conditions because inhibiting cell division provides relief in each of these diseases.

33. Elevated levels of 5-phosphoribosamine and adenylosuccinate indicate that the defective enzyme might be adenylosuccinate lyase. This would lead to decreased levels of AMP. Because AMP serves as a negative regulator of the enzymes making these metabolites, low levels of AMP would increase the activity of those enzymes.

35. This diet is deficient in all B vitamins except folate. Because the cells lack other B vitamins, they couldn't generate FAD or NAD$^+$. These cofactors would be necessary for the reduction of folate. In other words, while the cells have folate, they wouldn't be able to use each molecule more than once.

37. a. Because binding of ATP in this site activates the enzyme, this mutation would result in an enzyme with decreased activity.

    b. In this instance the enzyme would be catalytically dead.

    c. If the K_M value increased, it would mean that the activity of this enzyme would likely be decreased or would require higher concentrations of glutamine to achieve the same level of activity.

    d. This would leave the organism unable to catabolize pyrimidines.

39. Babies could be tested for irregular levels of purine or pyrimidine metabolites compared to normal cells. This could easily be done using mass spectrometry. If a specific enzyme was suspected, a mutation in the gene could be identified using DNA sequencing. Babies differ from adults or nonhumans in that the amount of tissue available for testing is highly limited. Thankfully, both of these techniques use very limited sample sizes.

The answers for these questions will vary from individual to individual.

41. This may vary depending on the question being researched. Providing a blood or tissue sample may be minimally invasive and may help research greatly, but spending considerable time in a hospital may not.

43. I think David led a very different life than most people. I don't know that it was more or less of a life than that of someone who cannot leave a wheelchair or is unable to communicate, but his situation was different. I think because of the nature of David's disease it was simple to imagine him as healthy or normal, and therefore it is easy to compare him to a healthy child.

Chapter 14

1. Both prokaryotes and eukaryotes need to replicate DNA faithfully and with high fidelity (few to no errors) in a timely fashion with regard to their biology (it cannot take too long compared to the life of the cell). Differences between the two include the increased size and complexity of the eukaryotic genome (including having multiple chromosomes) and having multiple organelles and the membrane of the nucleus to reorganize upon division.

3. In order for the phosphodiester bond to be formed and the nucleotides to be joined, the tri-phosphate or di-phosphate forms of the nucleotide building blocks are broken apart to give off energy required to drive the enzyme-catalyzed reaction.

5. It is likely that that these are different enzymes based on the mechanism of either winding DNA more tightly or releasing tension in the DNA.

7. It is energetically unfavorable ($+\Delta G$) to form a sugar–phosphate linkage from monophosphates. The pyrophosphate leaving group is needed to help drive the reaction energetically.

9. Methylation of histones typically happens on amino groups. This makes the group bulkier and more hydrophobic. This decreases the interaction of the lysine of the histone with the anionic phosphate groups in the DNA backbone. Therefore, it makes sense that histones would become methylated to decrease the interactions of the histone with DNA prior to transcription.

11. DNA pol I replaces the RNA primers with DNA, which really only needs to be done repetitively on one strand, while both strands are worked on by the DNA pol III.

13. The enzyme would potentially still be able to bind substrate, but having only one phosphate to cleave would not provide the driving energy for the formation of the phosphodiester bonds of the DNA backbone. Therefore, it is likely that k_{cat} would be negatively impacted.

15. Replication licensing, which primes the origins of replication to bind to the replisome and begin replication, ensures that DNA is replicated only once during a single cell cycle.

17. Ionizing radiation and topoisomerase inhibitors induce double-strand breaks.

19. This is because of formation of covalent bonds between two adjacent thymine residues in the DNA double helix upon UV damage.

21. Base excision repair is used to remove and replace a single modified base.

23. Patients with Cockayne's syndrome suffer from microencephaly (small head size), impaired nervous system development, failure to thrive, and premature aging. They do not suffer from cancers or have an impaired immune system like patients with xeroderma pigmentosum. In Cockayne's syndrome, the genes affected are found in the transcription coupled nucleotide excision repair mechanism (TC-NER). In xeroderma pigmentosum, the genes impacted are found in global genomic nucleotide excision repair (GG-NER). These differences in both the genes affected and the disease phenotypes illustrate how these pathways differ in terms of their functions.

25. Based on the number of different bases (4) and the length of the sequences (2 or 4), we would expect the frequency of encountering an AT pair to be 1 in 16 (1 in 4 times 1 in 4 or 4^2) and the frequency of finding the TACA sequence to be 1 in 256 (4^4). If these sequences are found at other frequencies in the genome, it suggests that they are being selected for or against.

27. Alu sequences could be used as markers of evolution, but several important differences exist. First, Alu sequences are significantly shorter (~300 bp) than the gene for cytochromes (a few kb). Second, cytochromes have well-described functions and roles in metabolism, whereas the Alu sequence does not. Therefore, some of the selective forces on cytochromes are known, whereas for Alu they are not as clear.

29. Two common ions that are involved in protein–DNA interactions are Zn^{2+} and Mg^{2+}. In theory, other ions such as Cu^{2+} or Ni^{2+} could work but may have other issues (such as reacting differently in the active site). Ions are likely selected for based on their size and reactivity.

31. If the divalent metal ions were involved in both substrate binding and catalysis, then alterations to the DDE sequence would impact both V_{max}, k_{cat}, and K_M. If, however, only substrate recognition were affected, then we would only expect to see changes to K_M.

33. The thymine has been incorporated into DNA. Polymers like DNA and protein precipitate in the presence of trichloroacetic acid, whereas monomeric thymine would not.

35. Cisplatin has induced DNA damage and DNA repair mechanisms. This has also resulted in the liberation of thymine from the DNA polymer.

37. The RecA protein binds to single-stranded DNA and assists in recombination. It is likely that expression of this protein is resulting in increased recombination and therefore increased retention of undamaged parts of the sequence.

39. The Meselson–Stahl experiment could potentially be run in a modern setting using mass spectrometry to determine the difference in masses between the parent and daughter fragments.

41. The *in vitro* system could be constructed by incubating a piece of genomic or plasmid DNA with the chemical in a test tube and examining the DNA at different time points using a technique such as mass spectrometry. A similar technique could be used *in vivo*. The chemical would be incubated with cells or injected into an animal, and chemically derivatized DNA bases could be analyzed using mass spectrometry.

The answers for these questions will vary from individual to individual.

43. What is personalized medicine? How it will help doctors in overcoming difficulties faced today in treating patients with different set of genes?

45. Why a certain section of society oppose transgenic organisms? Is their concern genuine? How it can be a threat to the society in future?

Chapter 15

1. TFIIA binds to TBP; stabilizes the first four factors in the preinitiation complex (TFIIA, TFIIB, TFIIF, and TFIID) in the nucleus of the cell.

TFIIB binds to both TBP and to a GC-rich sequence in the DNA downstream of the TATA box; orients the polymerase with respect to the transcription start site in the preinitiation complex. The positioning dictates the DNA strand that serves as template.

TFIID consists of multiple proteins including the TATA binding protein (TBP) and TBP-associated factors (TAFs) that bind to the promoter region of DNA in the nucleus of the cell; essential to the formation of the preinitiation complex in the nucleus.

TFIIE recruits TFIIH to the preinitiation complex; facilitates the activity of TFIIH and the binding of RNA pol II to the template strand in the nucleus of the cell.

TFIIF forms a complex with RNA pol II and allows binding of pol II to the preinitiation complex; recruits TFIIE to the complex.

TFIIH has two enzymatic functions: an ATP-dependent helicase and a kinase that polyphosphorylates the CTD of RNA pol II; essential for unwinding of the DNA double helix and forming a transcription bubble in the nucleus of the cell.

3. The amino acids found in α-amanitin include tryptophan, cysteine, isoleucine, and proline. Several of these amino acids are hydroxylated, including proline and isoleucine. The structure is bicyclic, which occurs via two linkages: a joining of the carboxy-terminus to the amino group and a tryptothionine linkage between cysteine and tryptophan.

5. TBP (TATA box binding protein), a general transcription factor required for proper initiation of gene expression by RNA polymerase II. Drug binding to TBP would affect all RNA polymerase I, RNA polymerase II and RNA polymerase III.

7. All RNA and DNA polymerases require Mg^{2+} cofactors, use a DNA strand as template, and use nucleoside triphosphates to extend a growing nucleotide strand on the 3′ hydroxyl through nucleophilic attack on the α phosphate. Both DNA and RNA polymerases have proofreading functions, but they work through different mechanisms, and RNA polymerases have a higher error rate. Unlike DNA polymerases, RNA polymerases have subunits that have helicase and topoisomerase activity and do not require primers. DNA and RNA polymerases have no structural homology. Prokaryotes have a single RNA polymerase that transcribes both coding and noncoding sequences. Eukaryotes have distinct RNA polymerases that synthesize the different classes of RNA. The initiation complex involves multiple transcription factors that interact with each other and RNA polymerase in eukaryotes, but only the sigma factor is required in prokaryotes. Initiation also requires phosphorylation of the CTD in eukaryotic RNA pol II.

9. The structure of rifampicin has a highly conjugated network of double bonds. When a molecule has enough of these double bonds, it begins to absorb light in the visible range, giving it a red color.

11. Pi bonds will be found in all the double bonds of rifampicin and are delocalized in areas of conjugation.

13. The drug 8-hydroxyquinoline is used externally (topically). The drug must be permeable to fungal cells.

15. GTP in the cap must be methylated at the 7 position of the ring by guanine-7-methyltransferase. This methyl group is essential for binding of the cap-binding complex (CBC) and is also involved in ribosome binding. Because of the specificity of the interactions, it is unlikely that other nucleotides could be substituted.

17. The types of mutations that could result from splicing errors include frameshift mutations, missense mutations, and premature termination.

19. The branch point must be an A. The 2′ OH that carries out the nucleophilic attack to cleave the 5′ end of the intron is not unique to adenine and could be carried out by the 2′ OH on any ribonucleotide. However, because of the observed specificity, the stereochemistry of the adenine base must be important for interacting with the spliceosome complex.

21. The 2′ OH is very reactive and allows splicing and catalytic functions of RNA to occur.

23. An siRNA directed against exon 1 or 2 would interfere with proper translation of both the thyroid and neural proteins and substantially lower both protein levels. An siRNA against exon 5 or 6 would only affect the production of the neural protein.

25. Intronase is responsible for the removal of introns. Some retrovirus genomes contain introns, which allows for alternative splicing of the viral mRNA and the production of different proteins from a single genomic fragment.

27. One possible explanation is that there is some built-in redundancy in case of mutation or malfunction of any one tRNA gene. This ensures that there will always be a tRNA for each of the amino acids to enable proper protein synthesis.

29. The vast majority of proteins that have been studied are folded into a single stable conformation in the native state. FG proteins represent a new paradigm.

31. The mRNA must interact with RanGTP and exportin 1. The mRNA/RanGTP/exportin 1 complex can then interact with FG nucleoporins to move through the pore complex to the cytosol, where the complex dissociates.

33. There should be five introns and six exons with conserved nucleotides at the 5′ and 3′ splice points at the boundaries. There are A branch points and polypyrimidine tracts in each of the introns.

35. There would be loops in the genomic DNA that would not be hybridized to the mRNA. These loops would correspond to the introns and could be observed in the micrographs.

37. There would be two bands: the top band would correspond to the β splice variant (larger mRNA), and the bottom band would correspond to the α splice variant (smaller mRNA).

39. mRNA level: real-time PCR or northern blot protein level: SDS PAGE, western blot (immunoblot), or HPLC

41. You could use real-time PCR or northern blots to see the size of the mutant transcripts compared to the nonmutant.

43. One possibility would be to isolate the modified tRNA and determine whether it interacts with exportin through footprinting experiments. This would indicate that the modification occurred before export (in the nucleus) because the tRNA could still interact with exportin.

The answers for these questions will vary from individual to individual.

45. Because these drugs are only available by prescription in the United States, a person must visit a physician or other clinician who can prescribe medication. This could be particularly detrimental to people with limited access to healthcare. However, keeping these drugs available only by prescription means that they are only given to patients who have an actual bacterial infection that can be helped by antibiotics. It also gives healthcare providers the opportunity to instruct patients on proper use of the antibiotics. Making these available over the counter (OTC) could lead to more misuse of the drugs. For example, people might take them for illnesses that are not bacterial (such as viral infections like colds and flu) or perhaps not take a full course of antibiotics for a bacterial infection, leading to more antibiotic-resistant strains of pathogens.

47. One simple answer could be that the parents should be advised of the chances of a child having the disease and the variations in life expectancy that could occur with the severity of the disease so that they can make an informed decision on whether they would want to have children. This advice would apply even for a disease that might not present until later in life, assuming it is a life-threatening disorder without a cure.

Chapter 16

1. It would be 5^3 or 125 different codons.

3. Ninety-nine million bases code for protein in the human genome, with three bases comprising a codon. This totals 33 million codons within the human genome. If there were four bases in a codon instead of three, the 33 million codons would take 132 million nucleotides, increasing the percentage of the genome to 4% that encodes protein.

5. GAA to GUA

7. Both of these sequences are part of the 5′-UTR and aid in ribosome binding and recognition of the start codon. They differ in sequence. The Shine–Dalgarno sequence in prokaryotes is AGGAGGU and is upstream of the start codon. The Kozak sequence in eukaryotes is ACCAUGG and includes the start codon (AUG).

9. Modifications include bases not normally found in mRNA (inosine, pseudouridine, and dihydrouridine) and methylations of guanine and cytosine. Modified bases are found throughout the tRNA, but they are especially prevalent in the D-loop and TψC loop. The modified groups help stabilize the interactions between the two loops in the tertiary structure.

11. The enzyme contains a proofreading site that removes valine from the tRNA.

13. The Cleland diagram should show ATP and amino acid binding to enzyme E. This is followed by release of PP$_i$, leaving enzyme E bound to the aminoacyl-AMP intermediate. The diagram should then show tRNA binding, release of AMP, and finally release of aminoacyl-tRNA leaving the original enzyme E.

15. Formyl-met-tRNA is formed by the addition of methionine to tRNAfMet (formyl-methionine specific-tRNA) by the fMet-tRNA synthetase to produce methionine tRNA. Met-Pro-Arg-Ala-Ala-Arg-Asp-Stop

17. Answers will vary. Models should contain the combined information from Figures 16.13 through 16.15.

19. I. This antibiotic is a macrolide with methylated glycosyl groups.

   II. This antibiotic is an aminoglycoside with amino, hydroxyl, and methyl groups.

   III. This antibiotic is a tetracycline with hydroxyl and carbonyl groups.

21. Macrolide antibiotics are formed by chemical modifications of erythromycin. Modifications could be made by reaction of keto or hydroxyl groups of erythromycin such as the addition of methyl groups from methanol and the conversion of a ketone to an oximine through reaction with hydroxylamine.

23. The ester would form between the carboxylic acid of the palmitate and the hydroxyl group of the chloramphenicol. The drug might be supplied this way either to prevent reactions of the alcohol (for example, an oxidation to an aldehyde) or to decrease the solubility of the drug because of the hydrophobic palmitate group. The succinyl group would make the compound more soluble due the presence of additional carbonyl and carboxyl groups.

25. If mitochondrial function (respiration) were blocked, the cell might shift to anaerobic metabolism to produce ATP; this could result in increased levels of lactic acid.

27. Puromycin has a portion that looks like a nucleoside and a portion that looks like an amino acid. These are most likely from modifications of adenosine and tyrosine. The structure is similar to the aminoacyl end (3′ end) of an activated tRNA.

29. An imide bond has a nitrogen linking two carbonyl groups, as seen in one of the cycloheximide rings. An amide bond has a nitrogen bound to a single carbonyl group. There are no amide bonds present in the structure.

31. The amino acids Phe169 and Glu195 in the enzyme most likely correspond to Tyr175 and Asp182 in Figure 16.9, respectively. Mutations of these residues greatly change ΔG^+. The Phe34 mutation to Tyr has no real effect, so it is unclear where it would correspond to Figure 16.9, but it likely is not in the area of Tyr binding. Amino acids Ala51 and Gly48 most likely correspond to Asp200 and Gly98 in Figure 16.9; mutating these residues with H-bonding capabilities increases binding to ATP perhaps through better interactions with the ribose hydroxyls. Gly35 may be in the area of Gly50 and Gln179 in Figure 16.9, interacting either through backbone H-bonding to the adenine ring or perhaps in a different location acting to stabilize ATP phosphates.

33. Any mutation that enhanced binding through hydrogen-bond interactions would increase ΔG^+.

35. X-ray crystallography or cryo-EM could be used to determine the three-dimensional structure of tRNA.

37. One possibility would be to use a footprinting assay and compare the results of tRNA digestion in the presence eIF-2 with GTP bound and in the presence of just eIF-2. Another option would be to design a binding assay using the aminoacyl-tRNA as a ligand with eIF-2 in the presence of GDP and in the presence of GTP or in the presence of non-hydrolyzable analogs of GDP and GTP.

The answers for these questions will vary from individual to individual.

39. Drug companies do need to recoup their investment in developing new drugs and make a profit to stay in business, but pricing should not be so disparate in different areas of the world. The high prices of

name-brand drugs can prevent low-income patients from access to necessary treatments.

Chapter 17

1. a. The C and A residues could be methylated.

    b. The methylations on cytosines would be found in the major groove. The methylations on the adenines would also occur in the major groove but would not affect base pairing.

    c. Addition of a methyl group changes size (geometry) and hydrophobicity in the major groove of DNA and can change interactions with amino acid residues in the DNA binding sites of proteins, which can lead to altered affinity.

    d. Proteins with small hydrophobic residues such as alanine or valine in their DNA-binding sites would be more prone to interact with methylated bases. In the absence of methyl groups, especially in the case of methylated adenine, residues that are capable of hydrogen bonding, such as serine, threonine, and histidine, would interact with the major groove more readily.

3. a. Changing a small hydrophobic Ala to a larger hydrophobic Ile would change the geometry of the binding site and change the affinity of the protein to the corepressor, thus changing gene expression.

    b. Without Ser, phosphorylation could not occur, and this would alter gene expression because the transcription factor would always be in its unphosphorylated state.

    c. A proline in the helical region would disrupt the alpha helix so that a portion of the protein could no longer interact properly with DNA and prevent transcription factor binding.

    d. A mutation of Ser to Asp would prevent ubiquination of the protein. This would inhibit transcription factor activation by ubiquination.

5. It could be that the positively charged arginine in the basic helix-loop-helix domain is important for interactions of the transcription factor with the negatively charged DNA. If this residue were substituted for a tyrosine residue, it would decrease the strength of an interaction.

7. CREB interacts with DNA through specific interactions with the major groove, whereas histones interact with DNA nonspecifically through charges on the phosphate backbone.

9. CAP or CRP and riboswitches.

11. When FMN is present, it binds the aptamer domain of the FMN riboswitch. It results in the formation of a terminator loop.

13. Transcription factors use multiple types of weak interactions to bind to highly specific sequences in the major groove. Histones are positively charged proteins that interact nonspecifically with the phosphate backbone and therefore use coulombic interactions to mediate their interactions.

15. Hormones that bind outside the cell can cause changes inside the cell through the activation of kinases either via a kinase cascade or through a second messenger.

17. They are constitutive promoters that work in a variety of mammalian cell types and allow for a high level of gene expression.

19. Because SREBP is localized in the ER membrane and activation occurs in the ER in conjunction with other ER proteins, it would be more difficult to isolate than NF-κB, which is a cytosolic protein.

21. Histone genes would not be good markers of evolution. Histones from different species all interact with DNA in the same way, and histone structure is highly conserved. Mutations to histones would be detrimental to their function. Therefore, you would expect very little variation in genes of different species.

23. Number of proteins appears to correlate most strongly with organismal complexity. Crabs have more chromosomes than humans do, and grasses such as corn have more DNA. Several species have as many or nearly as many genes as humans do, but humans appear to splice those genes differentially to generate more proteins than other species.

25. One possible reaction mechanism would be a nucleophilic attack by an amino acid (perhaps Ser or Cys) on the guanidinium carbon of methyl arginine. Resolution of the tetrahedral intermediate could release methylamine. Hydrolysis of the alkyl-enzyme intermediate could then result in citrulline by forming a carbonyl.

27. All HOX genes contain the homeobox sequence that encodes for the homeobox DNA-binding domain, which is very highly conserved. HOX genes also consistently occur in clusters.

29. MicroRNAs regulate gene expression at the message level by binding to mRNA and targeting it for degradation. Some genes (e.g., those for ferritin and transferrin) are regulated through the interaction of mRNA with specific protein products.

31. a. EMSA is an electrophoresis technique that takes advantage of the fact that DNA bound to proteins will move more slowly through a gel than DNA alone. This enables researchers to determine whether proteins can bind to specific sequences (on small fragments) of DNA.

    b. This ensures that the transcription factor does in fact bind to the DNA sequence. If it does not bind, there will be no difference in migration between the two lanes. It also allows comparison of migration of the known sequence to the transcription to the other sequences.

    c. The AG in positions 4 and 5 of the sequence is essential for transcription-factor binding.

33. a. The reporter gene is mCherry. It encodes a red fluorescent protein that serves a visual product of gene expression. Expression is detected with fluorescence microscopy. Other choices for reporter genes are GFP and luciferase.

    b. The −1000 region of the promoter appears to be important for regulation. Only promoters with this region showed expression of the reporter gene.

    c. Another alternative might be to observe the mRNA transcripts produced with various promoters using western blots or real-time PCR. Limitations of the reporter gene assays would include low expression or constitutive expression.

35. Some possible ideas: Isolate histone H3 at various stages of development, and look for acetylation on the key lysine via HPLC or other techniques. Inhibit histone acetyltransferases, and see if the stage of development still occurs.

37. The DNA associated with nucleosomes would most likely have a higher melting point because the DNA would be stabilized by association with the nucleosome and less susceptible to breaking of the H bonds between strands. You could perform a melting point assay of each sample by measuring the absorbance at 260 nm at different temperatures. You would want to do a control of the histone proteins alone as well to be sure they are not affecting the absorbance.

39. You could do an EMSA assay using the protein and a synthetic strand of DNA that has the specific sequence of the promoter region and compare it to a nonspecific coding sequence.

The answers for these questions will vary from individual to individual.

41. Although selective breeding is a type of genetic manipulation, it is not as directed as laboratory manipulation. Genetic manipulation in the laboratory is more directed and specific—and is done for specific

purposes. Some would argue that laboratory manipulation is more ethical because the changes in genes are specific, the animals are not released to the wild, and they are bred solely for a research purpose. In contrast, selective breeding of mice by hobbyists or pet breeders would have no real purpose beyond aesthetics or novelty.

43. Because these conditions are so prevalent, many would argue that this is money well spent to help treat patients and give them better lives. However, others might argue that it would be better to spend the money on prevention methods and education so that fewer people suffer from these largely preventable ailments to begin with.

Chapter 18

1. Resolution is the smallest distance at which two objects can still be seen as two distinct objects rather than a single entity. The higher the resolution of the diffraction data, the more measurements are present to base the model on.

3. Contrast is the difference between the brightest whites in an image and the darkest blacks. It is important in determining some structures because a high-contrast image shows some of the relief between different elements of the structure.

5. Just as motion in a light photograph will lead to a blurry image, movement can generate blurry images in electron micrographs. This is further complicated by the fact that electron microscopes amplify small vibrations.

7. In cryogenic electron microscopy (cryo-EM), samples of purified macromolecules are prepared by flash freezing them in vitrified water. Advances in electron microscopy and computer technology make it possible to improve the resolution of cryo-EM, such that large biomolecular assemblies can be structurally analyzed, providing information about protein-ligand interactions and multimeric protein assemblies.

9. Typical electron microscopy data is not as high resolution as an X-ray structure, and therefore phi and psi bond angles are not determined.

11. In the 1950s, X-ray diffraction patterns were analyzed and structures were determined by making physical models by hand. Now computers are used to analyze diffraction patterns and determine structure. Computer software is also readily available to convert structure data into a computer-generated image.

13. Pure protein is crystallized through a variety of techniques until a single pure crystal is obtained. The crystal provides repeating patterns of atoms within the structure. X-rays are then focused on the crystal. When the X-rays hit the electrons in the molecule, they are diffracted (scattered). This produces a diffraction pattern of differing intensities that can then be analyzed to determine where atoms are located in space and where they are located with respect to each other. The structure of the protein can be determined by comparing this data with the known amino acids in the protein and the allowable angles of the peptide bonds.

15. Membrane proteins are often difficult to purify from membranes. In addition, the hydrophobic portions on the exterior of the protein that interact with membranes are not compatible with polar crystallization solvents. Mild detergents can be used to isolate these proteins, and solvent can be used to crystallize some of them. Glycosylation occurs post-translationally and can result in a diverse population of protein structures within a sample. This would not provide a unique repeating pattern necessary for proper crystal formation. Glycosylases are sometimes used to remove these residues so that a crystal can be produced.

17. Acceptable values for overall resolution of an X-ray structure are 0.5 Å to 3.0 Å, with lower values near 0.5 Å resolving individual atoms, and values closer to 3 Å showing overall protein structure.

Acceptable values for RMSD for angles are <0.01 Å for bond length and <1.0° for bond angles.

Acceptable values for R factor are 0.2 to 0.6, with lower values showing greater reliability.

Acceptable values for B factor are 30 Å^2 to 60 Å^2, with lower values being better.

19. MIR works by soaking the crystal in heavy atoms (mercury, gold, platinum, or iodine) and comparing scattering data with the native crystal. The difference between the intensities of the diffraction patterns can be used to locate the heavy atoms in the unit cell. Based on the position of the heavy atoms, it is possible to estimate the phase of the X-rays and then calculate the position of other atoms in the crystal. Other atoms would only work if they did not disrupt the crystal structure and only entered at a few locations. Because selenium is a heavy atom, it would theoretically work, but sulfur is not heavy and would probably have little difference in intensity from other common atoms in a protein.

21. The resolution of the refined structure of 2.2 Å is good, and overall protein folds and structure should be correct. After refinement, the B factors and RMSD for bond angles are in the acceptable ranges. However, the R factor and RMSD values for bond length are well out of range. This structure probably has errors even though the resolution is good.

23. Higher magnetic fields is used to increase the resolving power of the experiment. A high-field NMR instrument uses a superconducting cryomagnet cooled with liquid helium or nitrogen, radiofrequency pulse generator, probe, and a detector.

25. Nuclei that are useful in biochemical experiments include phosphorous 31, nitrogen 15, oxygen 17, carbon 13, and protons. Some of these are more useful than others depending on the experiment being performed and the abundance of the isotope.

27. The highest resolution is found in X-ray diffraction structures, followed by NMR, with cryo-EM giving the lowest resolution. Because NMR is carried out in solution, it is able to provide information about protein structure in its native environment. Because of the rapid freezing in cryo-EM, it would have a closer-to-native structure than X-ray structures, which depend on crystallization in solvents and only present a single static form of the protein.

29. A higher-field instrument is necessary to increase the resolution of the numerous nuclei found in biological molecules. Simple instruments also may lack the ability to conduct 2D experiments.

31. a. Aquaporins are large membrane proteins with large hydrophobic regions. Structural determination by NMR would be difficult because it is carried out in solution. X-ray crystallography could be done but would require that the protein be isolated from the membrane and crystallized with mild detergents. Cryo-EM would be effective for observing the large membrane protein.

b. This small globular protein is water soluble, and its structure could be determined by any of the three methods.

c. Because collagen is a fibrous protein that is insoluble and very large, NMR would not be a good method to determine its structure. X-ray diffraction and cryo-EM could be used.

d. Hemoglobin is a four-subunit globular protein, and its structure could be identified by any of the three methods.

e. This multimeric protein has some flexibility, and a structure of the entire protein that also depicts this flexibility could be determined by NMR, although the size of IgG is larger than the proteins usually elucidated with NMR. Cryo-EM and X-ray crystallography could also be used to identify the entire structure or pieces of the structure.

f. This small protein consists of two chains linked by a disulfide bond and is water soluble. Its structure could be determined by any of the three methods.

g. Myosin is a molecular motor consisting of four subunits. Because it is very large, its structure would best be determined by X-ray crystallography or cryo-EM.

The answers for these questions will vary from individual to individual.

33. This convention of ordering seems fair, with the person doing the most work receiving first author status. It would be beneficial if all fields of science agreed on the establishment of authorship because it would enable the reader of the paper to know both who did the majority of the work and who was the PI for the project.

35. The first author on the paper is the one who has done most of the work on the study and the principal investigator (lab director or professor) is the last author. Group members who contributed to some sections of the paper are listed. The order of listing the names of the authors is not important.

Chapter 19

1. If there were three steps for each amino acid, that would be 24 total reactions. 99% (0.99) to the 24th power is 78%.

3. If there were three steps for the addition of each amino acid, you could only add two amino acids ($0.90^6 = 0.53$ or 53%) to achieve this yield.

5. Aromatic or hydrocarbon protecting groups would keep the ionizable side chains from reacting. Examples are benzyl groups and tert-butyl groups for carboxyl side chains. These groups would need to be different from the group used for the terminal amino group so that the protecting group would not be removed in the deprotecting step before the coupling of the next amino acid.

7. A solid-phase protein synthesis reaction cycle includes deprotection of the peptide amino terminus using piperidine.

9. BglII and ScaI could be used to clone the cDNA in the proper direction. Although there is a HindIII site on either end of the cDNA, there is no way to ensure that the cDNA will insert in the proper direction using the HindIII site in the vector alone.

11. *E. coli* cells grow rapidly, can be easily transformed, produce large amounts of protein, and are inexpensive.

13. a. This would be 3^{100} or 5.1×10^{47}.

b. It would take 5.1×10^{35} seconds or 1.63×10^{28} years, more than a trillion times the age of the universe.

15. a. Heat increases molecular motion and disrupts hydrogen bonding and hydrophobic interactions. These may be reversed by lowering the temperature, but heat can also permanently denature proteins in many cases, for example, due to aggregation as when heating albumin in an egg white.

b. β-mercaptoethanol is a reducing agent that reduces disulfide bonds (linkages) in proteins, destabilizing protein structure. Removing β-mercaptoethanol will allow the disulfide bonds to reform.

c. Urea is a chaotropic agent that denatures proteins by destabilizing the noncovalent interactions among amino acid side chains and facilitating the entry of water into the protein, which further disrupts hydrophobic interactions. Removing urea will allow the protein to renature.

d. Guanidinium hydrochloride is another chaotropic agent that denatures proteins in a manner similar to urea. Removing guanidinium hydrochloride will allow the protein to renature.

e. Changing pH will cause changes in the protonation of ionizable amino acid side chains, disrupting salt bridges and hydrogen-bonding interactions and therefore structure. These changes are reversed when the pH is restored to normal, but extreme changes in pH can permanently denature a protein.

17. a. Asp, His

b. Ser, Asp, Cys, Gln, His

c. Met, Ala, Ile, Phe, Pro, Val, Gly

d. Cys

e. His, Phe

19. Answers will vary.

21. The hydrophobic effect in proteins is manifested in burial of the exposed hydrophobic side chains and on release of the solvating water molecules.

23. Each one would react with all of the others, both attracting those of opposite charge and repelling those of similar charge. This would give rise to 1,225 different interactions.

25. An epitope tag is a short amino acid sequence that is incorporated into a protein of interest, which is used to detect or purify that protein via a commercially available antibody generated against the epitope tag.

27. A cDNA for insulin could be obtained and used to make a site-directed mutant DNA in which the His codon is converted to a codon encoding Asp. This could be generated using PCR and a mutagenic primer or via cassette mutagenesis using two complementary DNA strands with the mutation.

29. Mutating the serine to an alanine would be a conservative mutation in terms of size, but it would remove the hydroxyl group necessary for phosphorylation. This would render the phosphorylation site inactive, and the serine would be unable to be phosphorylated.

31. You would need to change at least two nucleotides to change the Ser codon to Asp. One example would be (TCC to GAC), so a potential primer would be (nucleotide changes underlined):

$$5'-CTG\underline{C}TGCCA-3'$$
$$3'-GAC\underline{G}ACGGT-5'$$

The answers for these questions will vary from individual to individual.

33. Most institutions have a process in place that involves working with the provost or ombudsman. Search the institution website for terms like: academic misconduct, academic integrity, research integrity, code of conduct, conflict of interest.

35. It may lead to misleading protein-drug interactions when used in drug discovery, wasted financial resources, wasted time using erroneous structures, and retracted publications.

Chapter 20

1. The kinome and the metabolome would change most rapidly because kinases and metabolites are in greatest flux over time in response to the needs of the organism.

3. An investigator would need to take many different time samples and compare the composition/concentration results, ideally compared to a control that is from the same tissue but not subject to large changes over time. Keeping samples cold to prevent degradation and including enzyme inhibitors would also help.

5. See Table 22.1 for detailed comparison. RNA-Seq would be best for detecting polymorphisms because of its ability to distinguish different isoforms and alleles to single-base resolution. Microarrays would be best for detecting splice variants due to high throughput and sensitivity to changes in genome expression.

7. Students can access Scratch at: https://scratch.mit.edu.

9. Theoretical p*I* 6.90. Molecular weight 3430

11. Answer for gi|254826781: *Mus musculus* (house mouse)

13. Answer for gi|254826781:

MFPRETKWNISFAGCGFLGVYHIGVASCLREHAPFLV ANATHIYGASAGALTATALVTGACLGEAGANIIEVSKEARKRF LGPLHPSFNLVKTIRGCLLKTLPADCHERANGRLGISLT RVSDGENVIISHFSSKDELIQANVCSTFIPVYCGLIPPTLQ GVRYVDGGISDNLPLYELKNTITVSPFSGESDICPQDSSTN IHELRVTNTSIQFNLRNLYRLSKALFPPEPMVLREMCKQG YRDGLRFLRRNALLEACVEPKDLMTTLSNMLPVRLATAMM VPYTLPLESAVSFTIRLLEWLPDVPEDIRWMKEQTGSICQYLV MRAKRKLGDHLPSRLSEQVELRRAQSLPSVPLSCATYSE ALPNWVRNNLSLGDALAKWEECQRQLLLGLFCTNVAFPPD ALRMRAPASPTAADPATPQDPPGLPPC

15. Answer for gi|254826781: 7

17. Answer for gi|254826781: upstream: Rplp2 (~4 kb between); downstream: Cracr2b (~140 bp)

19. Answer for *Mus musculus*: perilipin-2 and adipose differentiation related protein, isoform CRA_

The answers for these questions will vary from individual to individual.

21. Yes, the U.S. federal government spends over $120 billion a year on basic research, applied research, and development.

Chapter 21

1. Dicty must respond to the changes in its environment through signaling pathways. It serves as a good model because its complete genome is known, and many genes are homologous to human genes. It has a short life cycle and can be cultivated in the lab. Over the course of its life cycle, dicty has both unicellular and multicellular stages. The multicellular stage has different cell types with different signaling pathways.

3. G protein-coupled receptors have seven membrane-spanning helical regions. These regions cluster together to provide the core of the receptor complex. Each of these helices has a hydrophobic face exposed to the lipid bilayer, but the remainder of each helix faces the other helices and interacts with those helices via weak interactions. These helices do not form a pore but rather transmit the signal that the receptor has bound to ligand on the cytosolic face to G proteins via a conformational change.

5. Phosphorylation is rapid and specific. It is enzymatically reversible. It can be amplified by cascade systems. It results in covalent modification of enzymes.

7. STAT knockout mice are immune compromised because the mice lack STAT, and this protein plays an important role in the immune response to cytokines such as interferon-γ. Without STAT, the cells cannot respond to these signals from the immune system and thus cannot fight off disease.

9. Structural similarity could be explored with BLAST searches for conserved regions/sequence homology, X-ray crystallography, or NMR. Functional similarity could be studied via ligand-binding studies using techniques like FRET and *in vitro* studies in model organisms.

11. Dimerization provides two identical binding sites and allows STAT to bind to DNA sequences with dyad symmetry (palindromes).

13. SOCS binds to JAK, forming a protein–protein interaction that prevents JAK from phosphorylating STAT. In a similar way, the protein regulatory subunits bind to PKA and prevent it from phosphorylating other proteins. This is different from the immediate signaling that would occur in the presence of increased concentrations of second messenger or another effector that does not rely on transcription of the inhibitory protein.

15. For all three reactions, the mechanism will involve a transfer of the phosphoryl group from ATP to the enzyme via a nucleophilic attack. The differences would be primarily in the transfer of the phosphoryl group from the enzyme to the Ser, Thr, and Tyr. All three can act as a nucleophile when stripped of their –OH proton by a base in the active site. The environment of the active site would need to differ to enable a microenvironment to facilitate a proton removal. The resulting nucleophiles could then be used to attack the phosphate.

17. Because there is cross-talk through toll signaling both in response to infections and in development, there is the possibility that infections encountered during pregnancy could affect the TLR pathway and therefore cause developmental problems (birth defects).

19. This would probably manifest itself as an increase in the K_m value of these substrates for the CDK. K_m is the inverse measure of affinity for a protein to a substrate. Therefore, as the K_m increases, the affinity decreases, as it does in this case.

21. PDK1 is phosphoinositide-dependent kinase 1. Akt is also known as protein kinase B. Activation of the insulin receptor leads to the formation of phosphatidylinositol 3,4,5-trisphosphate (PIP3). PIP3 interacts with both PDK1 and Akt. This allows PDK1 to phosphorylate Akt and activate it. Akt can then phosphorylate numerous targets, including MAPKKK and GSK, which prevents the degradation of cyclin downstream.

23. G proteins are molecular switches that are activated by receptor-catalyzed GTP for GDP exchange on the G protein alpha subunit. G proteins are inactivated by RGS proteins (for "Regulator of G protein signalling") that stimulate GTP hydrolysis (creating GDP, thus turning the G protein off).

25. **a.** JAK-STAT: kinase cascade

    MAP kinase: kinase cascade

    Toll-like receptor: kinase cascade

    PKC: both

    CDK: kinase cascade

    PKG: both

  b. JAK-STAT: separate subunits

    MAP kinase: pseudosubstrate

    Toll-like receptor: separate subunits

    PKC: pseudosubstrate

    CDK: separate subunits

    PKG: neither

27. This could be tested by generating proteins that were mutants unable to be phosphorylated. If the phosphorylation site tyrosine were mutated to a phenylalanine, the protein (JAK) would be unable to be phosphorylated and the necessity of this phosphorylation tested.

The answers for these questions will vary from individual to individual.

29. The ethics of using animals in research and education must balance the potential benefit to humanity against the suffering of animals. Therefore, it would seem ethical to use animals in research that could save human lives when there is no alternative. It may not be ethical to use them for education when human lives are not likely to be at stake.

Chapter 22

1. Protein disulfide isomerase is localized in the ER. Proteins with a KDEL sequence are also found in the ER and can also be used as a marker for ER.

3. Small, monomeric G proteins are found in signaling pathways such as the insulin signaling and MAP kinase pathways. They function as GTPase signaling molecules.

5. The light chains act as adapters, interacting between the heavy chains and the cargo. They associate with some Rabs, scaffolding proteins, and transmembrane receptors.

7. The co-translational pathway utilizes the signal recognition particle (SRP) to deliver secretory proteins to the ER membrane while they are still being synthesized by ribosomes.

9. Nucleotide polymerases have a high processivity because they can add large numbers of nucleotides per binding of template nucleic acid strand. Proteins associated with the strand increase processivity by enabling the polymerase to stay bound to the template while adding nucleotides. Lipases have domains that bind to the membrane or vesicle.

11. Coiled-coil domains of motor proteins allow attachment of the protein to other proteins (for example, myosin II forming muscle fibers or dynein binding to microtubules) or other cellular components (for example, kinesin attaching to transport vesicles). Coiled-coil domains of SNAREs permit interaction among the SNARE proteins, drawing the vesicle closer to the membrane.

13. A coiled-coil domain consists of two alpha-helical regions twisted around each other, whereas a four-helix bundle consists of four shorter alpha helices associated with one another.

15. Formation of the SNARE complex (Figure 24.10) is more ordered/entropically unfavorable, and that negative entropy must be overcome for this formation to be thermodynamically favorable. This is most likely accomplished through increased enthalpy due to the stability of the coiled-coil complex provided by the hydrophobic effect and other noncovalent interactions.

17. SNAREs do not have enzymatic activity of their own. They interact with each other via binding interactions. However, the uncoiling of SNAREs does require the action of an ATPase (NSF). This protein interacts with SNAREs but is not itself a SNARE.

19. The botulinum toxin has a metalloprotease that cleaves proteins of SNAP-25 protein. This prevents fusion of vesicles with the membrane so that neurotransmitters cannot be released. If we compare this metalloprotease to those found in Chapter 4, we see that the active site of the enzyme has a metal ion that polarizes a water molecule. This water molecule then acts as a nucleophile to attack the carbonyl carbon of the peptide bond in the substrate.

21. Shuffling implies that each hand moves independently. One method might be making a mutant protein in which one hand could either not bind or not hydrolyze ATP, while maintaining the function of the other. In a shuffling model, this mutant protein would still be able to show movement along the microtubule, but the hand-over-hand model could not.

Chapter 23

1. The key difference between oxidative phosphorylation and photophosphorylation is that ATP production is driven by electron transfer to oxygen in oxidative phosphorylation while sunlight drives ATP production in photophosphorylation.

3. Chlorophyll and β-carotene are pigments that act to absorb photons of light and help excite electrons. Phylloquinone and iron-sulfur clusters act to transport these electrons. Phosphatidyl glycerol and diglycerides are components of the membrane that help the photosystem pack together. Finally, calcium ions help regulate protein function in photosystem I and may play a role in catalysis. These plants appear green because they absorb light in the red range of the spectrum.

5. The average size of a protein is approximately 450 amino acids or about 50 kDa. The largest protein known is titin, which is approximately 34,000 amino acids, with a molecular weight of 3,900 kDa.

7. Plants that are red or purple in color lack the green pigment chlorophyll found in most plants. These plants would be unable to use light in the blue and red parts of the spectrum. These plants often grow in the shade of other plants and are exposed to different wavelengths of light. On the other hand, these plants could still absorb using their other pigments. Beta-carotene absorbs in the blue range, phycocyanin in the green, and phycoerythrin in the yellow.

9. In animals, cyanide binds to heme groups in complex IV, which blocks electron transport and therefore prevents ATP synthesis. This is not the only heme group in electron transport or photosynthesis. It is likely that the cyanide cannot access any of the heme groups in photosynthesis and therefore is unable to bind to them.

11. The energy required to cleave N_2 into two molecules of nitrogen is greater than the energy required to cleave a molecule of a nitrogen oxide (such as NO_2) into nitrogen and oxygen atoms.

13. Using standard bond enthalpies, we would take the sum of bonds broken minus the sum of bonds formed. The bonds broken would be three hydrogen–hydrogen bonds and one nitrogen–nitrogen triple bond. The bonds formed would be six nitrogen–hydrogen bonds. A general chemistry reference will tell us that the bond enthalpies are 436 kJ for a hydrogen–hydrogen bond, 941 kJ for a nitrogen–nitrogen bond, and 391 kJ for a nitrogen–hydrogen bond. Therefore, we find the overall enthalpy of the reaction is –97 kJ. If we are hydrolyzing 16 ATP molecules and if each one has a standard free energy of hydrolysis of approximately –32 kcal/mol, the energy required to fix nitrogen would be 16 times that value, or –512 kcal/mol. This value could be considerably off if we consider the equilibrium concentrations of intermediates in the cell instead of the standard free energy of hydrolysis.

15. A mixed inhibitor affects both substrate binding (K_m) and catalysis (k_{cat}). Therefore, the y-intercept ($1/V_{max}$) would be increased to give a smaller value of V_{max}, and the x-intercept ($-1/K_m$) would be decreased to give a larger value of K_m.

17. Electrons are added singly due to the nature of the electron carriers and reaction centers and the number needed to perform the chemistry. Some electron carriers, such as quinones, can exist in more than one oxidation state, but not all carriers can. Furthermore, just because a molecule *can* be found in a different oxidation state does not mean it *should* be used in that state. In some instances, the redox-active center may be too reactive and cause side reactions.

19. This indicates that the genes encoding nitrogenase evolved prior to the genes encoding the proteins involved in aerobic metabolism. This is an instance of evolution and chemistry informing one another in response to a complex situation. An understanding of both ancient atmospheric chemistry and evolution is necessary to paint a complete picture of what is going on.

The answers for these questions will vary from individual to individual.

21. Haber was a complex person, but he seemed to act without a moral compass.

23. Interestingly, only the International Union of Biochemistry and Molecular Biology prohibits the development of biological or chemical weapons.

2′-deoxyribose (deoxyribose) a ribose sugar that is lacking the hydroxyl group on the 2 position and found in DNA.

5′-methylguanine cap a modification made to eukaryotic mRNAs in which the 5′ end has a guanine residue added to it, which is subsequently methylated.

30S subunit the small subunit of the prokaryotic ribosome.

40S subunit the small subunit of the eukaryotic ribosome.

50S subunit the large subunit of the prokaryotic ribosome.

60S subunit the large subunit of the eukaryotic ribosome.

70S ribosome the complete prokaryotic ribosome, consisting of a small (30S) subunit and a large (50S) subunit.

80S ribosome the complete eukaryotic ribosome, consisting of a small (40S) subunit and a large (60S) subunit.

3_{10} helix a less common example of secondary protein structure, similar to an alpha helix but with tighter coils due to hydrogen bonding occurring in an $i+3$ pattern compared to the alpha-helical $i+4$.

Absorptive state the metabolic state in which metabolites are absorbed from the diet, typically up to six hours after eating.

Acceptor stem the part of the tRNA molecule where the amino acid binds.

Acetyl-CoA carboxylase (ACC) the enzyme that synthesizes malonyl-CoA from acetyl-CoA and CO_2.

Achiral characteristic of a molecule with no chiral center.

Acidosis a build-up of acid in the bloodstream, leading to decreased plasma pH.

Activation energy, E_a the energy required to overcome the energetic barrier between reactants (substrates) and products.

Activation the coupling of amino acids to tRNAs.

Active site the place on an enzyme where catalysis occurs.

Active transport transport across a membrane driven via the expenditure of some form of energy, including primary active transport, which is driven by the breakdown of ATP, and secondary active transport, when the transport of a molecule is coupled to gradients of other molecules.

Adenine a purine base found in both DNA and RNA.

Adenine phosphoribosyltransferase (APRT) the enzyme that catalyzes the transfer of free adenine to PRPP in the salvage pathway.

Adenosine deaminase (ADA) the enzyme that converts adenosine to inosine.

Adenylate cyclase the enzyme that produces cAMP from ATP.

Adipocyte a triacylglycerol-storing cell found in adipose tissue.

Adiponectin a peptide hormone produced in adipose tissue that acts in the brain to increase appetite.

Adipose tissue the tissue of the body that primarily store triacylglycerols.

Aestivation similar to hibernation, a dormant metabolic state organisms enter as a result of dry or hot conditions.

Affinity chromatography a purification method that separates proteins based on weak interactions between the molecule of interest and the binding resin.

Agonist a molecule that binds to a receptor and activates it.

Aldol an organic molecule containing a carbonyl group such as an aldehyde with a hydroxyl moiety β to the carbonyl.

Aldose a monosaccharide containing aldehyde groups.

Alginate an anionic polysaccharide comprised of polymers of glucuronic acid.

Allantoin a breakdown product of uric acid found in most mammals but not in apes or humans.

Allopurinol a xanthine oxidase inhibitor used to treat gout.

Allosteric regulation a regulatory mechanism in which a molecule binds somewhere on a protein or enzyme other than the active site and affects enzyme activity.

Allosteric regulator a molecule that regulates the function of a protein by binding at some site distant to the one where the protein's primary function is occurring.

Allosterism regulation of proteins (enzymes, receptors, and transport proteins) through binding at a site other than the active site.

Alpha helix an element of secondary protein structure, consisting of a right-handed twist of the peptide backbone with side chains radiating outward from the central region.

Amber codon suppression a technique through which unnatural amino acids are strategically incorporated into a protein in a living cell.

Amino acid one of a family of biological molecules that contain an amino group and a carboxy group and serve as the building blocks of proteins.

Aminoacyl (A) site the site in the ribosome where new amino acids bind.

Aminoacyl-tRNA a tRNA molecule with an amino acid bound to it.

Aminoacyl-tRNA synthetase the enzyme that recognizes and joins the tRNA to the appropriate amino acid.

Aminoglycoside any of a class of antibiotics that blocks protein synthesis in bacteria and is characterized by the presence of an amino sugar.

Aminopterin a folate antimetabolite, dihydrofolate reductase inhibitor, and chemotherapeutic agent.

Amino sugar a sugar in which an amino group has been substituted for a hydroxyl group.

Amino terminus the free amino end of a polypeptide where protein synthesis begins, continuing from there to the carboxy (C) terminus; also called the N terminus.

Ammonotelic organisms that secrete waste as ammonia.

AMP-activated protein kinase (AMPK) a kinase regulated by levels of AMP that coordinates metabolic processes in the cell.

Amphipathic characteristic of a molecule with both hydrophobic and hydrophilic groups.

Amylopectin a branched polymer of glucose that plants use to store energy.

Amylose a linear polymer of glucose that plants use to store energy; one of the polysaccharides in starch.

Anabolic reaction a reaction that builds up or synthesizes biological molecules, typically requiring an input of energy.

Anaplerosis a process that elevates levels of a citric acid cycle intermediate.

Anaplerotic reaction a reaction that replenishes citric acid cycle intermediates.

Anion-exchange resin a diethylaminoethyl resin containing a positively charged amino group.

Anomeric carbon the hemiacetal or hemiketal carbon in a carbohydrate; the new chiral center formed from the former carbonyl carbon.

Anomer the stereoisomer of the hemiacetal forms of a carbohydrate, with α and β anomers related to one another by the position of the hydroxyl group on the anomeric carbon (α on the bottom face, β on the top face).

Anorexigenic characteristic of hormones or neurotransmitters that blunt appetite.

Antagonist a molecule that binds to a receptor and blocks its action.

Antenna complex a structure containing a cluster of numerous pigment molecules involved in photosynthesis.

Anterograde transport the trafficking of proteins or cargos out of the cell; the opposite of retrograde transport.

Antibody a protein produced by the immune system that recognizes and binds molecules foreign to the host organism.

Anticodon loop the region of the tRNA that recognizes and binds the codon.

Antiparallel characteristic of two strands of a molecule (DNA or protein) that run in opposite directions (e.g., the double helix of DNA or strands of protein in a beta sheet).

Antiporter a transport protein that traffics two or more molecules across the lipid bilayer in opposite directions.

Anti-proliferative a compound or protein that acts to block cell proliferation (reproduction).

Antiterminator a region of GC-rich sequences that form hairpins in the mRNA.

AP endonuclease the enzyme that removes a damaged base, generating an AP site.

Apolipoprotein the protein component of a lipoprotein.

Apoprotein a protein lacking its cofactor or prosthetic group.

Apoptosis programmed cell death in which cells undergo a systematic disassembly of cellular components culminating in the fragmentation of DNA.

AP site (apurinic/apyrimidinic site) a site where a purine or pyrimidine has been removed from DNA.

Aptamer the region of a riboswitch that binds to a small metabolite, regulating expression of the gene.

Arcuate nucleus the region of the hypothalamus that integrates neuronal and hormonal signals that regulate metabolism and appetite.

Argonaute protein a protein involved in the RISC complex.

Aspartate transcarbamoylase (ATCase) the rate-determining enzyme in purine synthesis in *E. coli*, frequently used as a model of allosteric regulation.

ATP synthase (F_0/F_1ATPase) the mitochondrial complex that uses the energy of the proton gradient to synthesize ATP from ADP and P_i.

Attenuation a regulatory mechanism in which the secondary structure of the RNA molecule can lead to alterations in the expression of the gene, based on levels of an amino acid such as tryptophan.

β-hydroxy-β-methylglutaryl-CoA (HMG-CoA) a metabolic intermediate at the crossroads of ketone body and cholesterol metabolism.

Bare sequence a set of standard characters representing a sequence of nucleotides or amino acids.

Base excision repair a DNA repair mechanism in which a chemically damaged base is removed from the DNA double helix.

Base pair intercalation the process through which compounds slide in between bases in the double helix.

Base pair intercalator a chemical that fits in between bases in the DNA double helix.

Base stacking the collective term for the weak forces that stabilize the double helix of DNA, including the hydrophobic effect, London-dispersion, van der Waals, dipole–dipole, and π–π stacking.

Basic local alignment search tool (BLAST) software that employs the Altschul algorithm to compare large sequences of biological data.

Bed volume the volume of buffer as well as the beads in a chromatography column.

Beta sheet an element of secondary protein structure consisting of elongated planks of amino acids that hydrogen-bond to each other, generating a sheetlike structure; also called a beta-pleated sheet.

Beta oxidation the process through which fatty acids are catabolized into acetyl-CoA.

B factor a measure of movement, or thermal noise, in the analysis of a crystal structure, with higher B values indicating more movement in the crystal during data collection.

Bile salt a steroid-based detergent made by the liver and stored in the bile duct until needed.

Binding site the site on a protein where binding of a small molecule occurs, including the active site of an enzyme, the ligand binding site of a receptor, and the allosteric regulatory site on an enzyme.

Bioinformatics a field of science involving the generation and analysis of large biological

datasets to answer multiple questions simultaneously.

Body mass index (BMI) a measure of adiposity (obesity) calculated by dividing a person's body mass in kilograms by their height in meters squared.

Bowman's capsule the compartment that surrounds the glomerulus and collects the filtrate.

Branching enzyme the enzyme responsible for forming branches in the glycogen molecule.

Branch migration the movement of a Holliday structure along the chromosome before the strands separate.

Bromodomain a four-helix bundle domain found in SWI/SNF proteins.

Brønsted–Lowry acid any chemical species that can donate one or more protons in the form of hydrogen ions.

Brumation the hibernation-like state that reptiles enter in cold weather.

Buffer a combination of a weak acid or base with its conjugate salt that is resistant to changes in pH near the pK_a value.

Buffer capacity the ability of a buffer to resist changes in pH, which increases with the concentration of buffer.

C3 pathway part of the Calvin cycle that uses three-carbon intermediates to fix CO_2.

C4 pathway part of the Calvin cycle in some plants that uses four-carbon intermediates to fix CO_2.

Cachexia the wasting associated with cancer and some other diseases.

Cafeteria diet a diet enriched in fats and carbohydrates.

Calvin cycle the light-independent reactions of photosynthesis that reduce CO_2 into carbohydrates.

cAMP (cyclic AMP) one of several cyclic nucleotides that act as second messengers.

cAMP receptor protein (CRP) a protein that binds to cAMP and the *lac* operon and activates expression of the operon; also known as catabolite activator protein (CAP).

Carbamoyl phosphate synthetase II (CPSII) the enzyme responsible for synthesis of carbamoyl phosphate from glutamine, carbonate, and ATP in pyrimidine biosynthesis.

Carbohydrate a molecule with the basic formula $C_x(H_2O)_x$.

Carbohydrate response element binding protein (ChREBP) a transcription factor that binds to carbohydrates and regulates transcription of genes involved in carbohydrate and lipid metabolism.

Carboxy terminus the free carboxy end of a polypeptide where protein synthesis (begun at the amino terminus) ends; also called the C terminus.

Cargo vesicles of proteins that are transported throughout the cell.

Cassette mutagenesis a type of site-directed mutagenesis that is an efficient method for generating multiple mutations at defined sites using synthesized DNA and restriction-enzyme cloning.

Catabolic reaction a reaction that breaks down or degrades biological molecules typically resulting in the production of ATP.

Catabolite repression upregulation or activation of gene expression by a regulatory factor.

Catalyst a participant in a chemical reaction that increases rate but is not consumed in the reaction.

Catalytic subunit a subunit of an enzyme where catalysis occurs.

Catecholamine any of a group of hormones derived from tyrosine.

Cation-exchange resin a carboxymethyl resin containing a negatively charged carboxyl group.

Cation–π interaction any interaction between a positively charged amino acid and one containing an aromatic π bond network.

Caveola a flask-shaped invagination of the plasma membrane coated with the caveolins.

CCAAT enhancer-binding protein (C/EBP) a general transcription factor that acts as the master regulator of transcription in many tissues.

Cellulose a structural polymer of glucose made by plants and some microbes.

Centromere the central region of the eukaryotic chromosome, which holds the two sister chromatids together.

Channel any of a group of gated transmembrane proteins that permit regulated passage of molecules across the membrane.

Chaperone a protein that assists in the folding of a protein.

Chimera an organism consisting of cells derived from multiple organisms.

Chimeric protein a protein that has different domains grafted together.

Chiral characteristic of molecules with a chiral center, typically a carbon atom joined to four different groups; most biochemicals are chiral, including 19 of the 20 amino acids.

Chitin a structural polymer of glucosamine.

Chloramphenicol an antibiotic from *Streptomyces* that blocks protein synthesis in bacteria by binding to the 50S subunit.

Chloroplast the plant organelle where photosynthesis occurs.

Cholesterol a membrane lipid and the most common steroid, which regulates the fluidity of membranes.

Cholesteryl ester a storage form of cholesteryl consisting of a molecule of cholesterol joined to a fatty acid through an ester.

Chondroitin sulfate a linear polysaccharide composed of repeating glucuronic acid and *N*-acetylgalactosamine disaccharides.

Chromatin-remodeling complex a complex of combinations of proteins that are responsible for altering the interactions between histones and DNA.

Chylomicron a lipoprotein particle that forms in the intestine and delivers dietary triacylglycerols to muscle and adipose tissue; plural *chylomicra*.

Chylomicron remnant the lipoprotein that remains after chylomicra are stripped of dietary triacylglycerols.

Cis element a gene regulatory element found within the DNA.

Citric acid cycle part of the central catabolic pathway in which carbons derived from acetyl-CoA are gradually oxidized to carbon dioxide, in the process harvesting electrons on the carriers NADH and $FADH_2$; also known as the tricarboxylic acid cycle or the Krebs cycle.

Clamp loader the *E. coli* protein responsible for adding the sliding clamp protein to the replication fork.

Clathrin a cytosolic protein that forms a coating around membranous pits, aiding in endocytosis.

Cloning the production of an exact copy of a piece of DNA or an organism.

Coated pit an intermediate step in receptor-mediated endocytosis, formed when patched receptors are pinched off the plasma membrane.

Coated vesicle a vesicle of plasma membrane coated with the protein clathrin and adapter proteins.

Coat protein complex (COP) a protein complex important in the movement of cargo vesicles in the cell.

Codon bias the observation that some organisms prefer certain codons over others in proteins.

Codon optimization the matching of codons in a cDNA with the preferred codons of the host organism to achieve maximum protein expression.

Codons the three-nucleotide "words" that make up the genetic code.

Cofactor a part of an active holoprotein that is not comprised of amino acids, including metal ions, heme groups, and iron sulfur centers; also known as a prosthetic group.

Cohesin complex the ring-shaped protein assembly that assists in holding the chromatids together before separation in anaphase.

Coil a lengthy section of amino acid sequences that links elements of secondary structure; sometimes called a loop.

Collagen triple helix the tertiary structure of the matrix protein collagen, consisting of a left-handed helical twist of three collagen strands.

Collision-induced dissociation (CID) the fragmentation of a protein by collision with an inert gas such as argon.

Competitive agonist a molecule that competes with the native ligand in binding to the receptor.

Competitive inhibitor an inhibitor that competes with substrate for binding to the enzyme.

Complex I (NADH oxidase) the complex that transfers electrons from NADH to ubiquinone and pumps four protons out of the mitochondrial matrix.

Complex II (succinate dehydrogenase) the complex that takes electrons from $FADH_2$ from the citric acid cycle and transfers them to ubiquinone.

Complex III (ubiquinone/cytochrome *c* reductase) the complex that transfers electrons from reduced ubiquinones to cytochrome *c* resulting in electrons passing from two-electron carriers to a one-electron carrier.

Complex IV (cytochrome oxidase) the complex that transfers electrons from cytochrome *c* to oxygen to make water, in which each oxidized cytochrome *c* results in one proton being consumed in the production of water and an additional proton being pumped out of the mitochondrial matrix.

Concatemer short fragments of DNA linked together in a single, long strand containing hundreds of tags.

Concerted model a model of allosterism in which the entire complex exists in either the T (active) or R (relaxed) state; also known as the MWC model.

Consensus recognition sequence a binding site found in one protein and recognized by another.

Consensus sequence a core amino acid or nucleotide sequence that has a common function (for example, substrate binding).

Conservative substitution a change to a protein that results in one amino acid being substituted for a similar amino acid, for example the substitution of a leucine for isoleucine through evolution.

Contrast the difference in brightness or hue between two objects.

Copolymer a polymer in which the monomeric subunits alternate in the pattern [ABABAB], typically built using dimeric building blocks.

Correlation spectroscopy (COSY) a multidimensional NMR technique used to determine which nuclei are within three bonds of each other.

Cortisol a glucocorticoid steroid hormone involved in regulation of metabolism.

Covalent catalysis catalysis in which at least one step of the mechanism involves a covalent linkage of the enzyme to the substrate.

Covalent modification a modification made to a protein through a covalent bond, for example, phosphorylation.

CpG island a region of CG repeats in bacterial DNA in which cytosines are methylated.

CRISPR a genome editing technique in which a guide RNA directs an enzyme to make specific changes (insertions or deletions) to a DNA strand.

Critical micellar concentration (CMC) the concentration of an amphipathic molecule at which micelles form and below which amphiphiles are found as monomers.

Cross-talk one or more aspects of a signaling pathway that affects another.

Cryogenic electron microscopy (cryo-EM) a macromolecular structure-determination technique in which samples are frozen in vitreous ice and imaged using electron microscopy.

Cyclic AMP response element–binding protein (CREB) a transcription factor that responds to elevated levels of cAMP via phosphorylation by protein kinase A.

Cyclic electron transport the process of passing electrons back to earlier complexes during photosynthesis.

Cyclin-dependent kinase (CDK) a small serine-threonine protein kinase that regulates progression throughout the cell cycle.

Cycloheximide a drug that serves as a prokaryotic protein-synthesis inhibitor.

Cyclooxygenase the enzyme that converts arachidonic acid into prostaglandins.

Cytidine diphosphate (CDP) a nucleotide that acts as an active carrier in the synthesis of lipids.

Cytochrome *c* a small, heme-containing redox active protein that transports electrons from complex III to complex IV in the electron transport chain.

Cytochrome P450 mixed function oxidases a broad family of oxidases that use heme to catalyze oxidation reactions and function prominently in both xenobiotic and drug metabolism.

Cytosine a pyrimidine base found in both DNA and RNA.

Debranching enzyme the enzyme responsible for removing branches in glycogen breakdown.

Degenerate codon one of multiple codons that encode the same amino acid.

Deletion the removal of nucleotides from a DNA sequence.

***De novo* biosynthetic pathway** the pathway by which nucleotides are synthesized from simple building blocks.

***De novo* structure prediction** a method of protein structure prediction based solely on amino acid sequence.

Deoxynucleoside a nitrogenous base joined to a sugar (deoxyribose) through a glycosidic linkage.

Deoxynucleotide a nitrogenous base joined to a sugar (2-deoxyribose) phosphate via a glycosidic linkage.

Deoxy sugar a sugar lacking at least one hydroxyl group, for example, 2-deoxy ribose found in DNA.

Dextrose a term for glucose in commonly used clinical settings.

Diabetes mellitus an endocrine disease of glucose metabolism in which insulin signaling is deficient either due to an autoimmune attack on the pancreas or due to a post-insulin receptor signaling defect.

Diacylglycerol (diglyceride) a glycerol molecule with two fatty acids esterified to it.

Dialysis a classic method to desalt or change the buffer of a protein solution without precipitating the protein of interest.

Diazotroph an organism that fixes atmospheric nitrogen.

Dicer an RNAse that cleaves the loop region of hairpin RNA molecules.

Dielectric constant a means of quantifying the polarity of a solvent or medium.

Differential splicing the ability of eukaryotes to generate multiple proteins from a single gene through different arrangements of exons.

Diffraction the process by which waves on the same wavelength spread out when they come into contact with an opening or particle.

Diffusion controlled limit the limit at which some enzymes operate, determined by how fast substrate can diffuse into the active site.

Dihydrofolate reductase the enzyme responsible for the reduction of dihydrofolate to 5,10-methylenetetrahydrofolate.

Dipole–dipole interaction the weak force that describes the interaction of at least two dipoles in a molecule.

Directed evolution a protein engineering method that mimics natural selection to steer proteins or nucleic acids to a user-defined function.

Disaccharide a carbohydrate comprised of two monosaccharide units.

Disulfide bond the S–S linkage created by the oxidation of two cysteine side chains.

Ditopic a characteristic of protein or other molecule spanning both leaflets of the lipid bilayer, that is, transmembrane proteins.

DNA glycosylase the enzyme that cleaves bases from deoxyribose.

DNA ligase the enzyme that joins two ends of DNA together to seal a nick in the DNA strand, frequently used in the laboratory in DNA manipulations.

DNA polymerase I (pol I) the enzyme in *E. coli* that is responsible for DNA damage repair and replacement of the RNA primer on the lagging strand during DNA synthesis.

DNA polymerase III (pol III) the main DNA polymerase from *E. coli* that synthesizes DNA on the leading strand and lagging strand.

Domain a discretely folding protein structure that retains its shape independently from the rest of the protein.

Double-strand break repair (DSBR) a DNA repair mechanism used to repair a two-stranded break in the DNA.

Drosha an RNAse involved in microRNA processing.

Dynactin an adapter between motor protein dynein and its cargo.

Dynamic programming a computing technique that breaks a complex problem into smaller pieces.

Dynamin a molecular motor involved in receptor-mediated endocytosis.

Dynein a molecular motor that moves along microtubules, exhibiting retrograde transport.

EC number a serial number for an enzyme that describes its activity.

Eicosanoid any one of the signaling molecules that are derivatives of the 20-carbon unsaturated fatty acid arachidonate.

Electron microscopy (EM) a visualization technique analogous to light microscopy in which samples are visualized using high-energy electrons.

Electron multiplier the most common type of mass spectrometer detector, in which single-ion collision induces a cascade of electrons that gives a detectable signal or peak.

Electron transport chain the system by which electrons aare used to drive the production of a proton gradient, oxidizing electron carriers and making water in the process.

Electroporation a technique that uses pulses of electricity to introduce DNA into a cell.

Electrospray ionization (ESI) a soft ionization technique that involves spraying a sample at high voltage to create an ionized aerosol.

Elongation factor a protein that assists in the elongation step of protein synthesis.

Elongation the addition of amino acids to the growing polypeptide chain.

Elution volume the point at which a specific molecule elutes during size-exclusion chromatography; visualized by a peak on a chromatogram.

Endocannabinoid any one of the naturally occurring ligands of the cannabinoid receptor, for example, arachidonylethanolamine.

Endocrinology the study of hormones and their functions in the body.

Endonuclease an enzyme that cleaves between nucleotides in a DNA sequence.

Enthalpy the thermodynamic term associated with the energetics of a reaction or system, in biochemistry typically associated with the energy needed to break a bond.

Entropy the thermodynamic term associated with the amount of disorder or microstates of a system.

Enzyme one of tens of thousands of protein catalysts that catalyze most reactions in an organism.

Enzyme inhibitor a molecule (generally small) that lowers the rate of an enzymatically catalyzed reaction.

Enzyme kinetics the study of enzymatically catalyzed reactions and their rates.

Epigenetics the field that studies the effect of modifications to genes and how they influence gene expression.

Epimer a stereoisomer that differs at a single chiral center.

Epinephrine a type of catecholamine that stimulates the fight-or-flight response; also known as adrenalin.

Epitope a small part of an antigen that antibodies recognize and bind.

Epitope tag a short amino acid sequence incorporated into a protein of interest that is used to detect or purify that protein by high-affinity antibodies.

Equilibrium constant, K_{eq} the ratio of products to reactants at equilibrium.

Equilibrium the chemical state where the rates of the forward and reverse reactions are (not necessarily when the concentrations of reactants and products are equal).

E value the statistical assessment of the likelihood that a sequence in the BLAST database would be identified by chance.

Exit (E) site the site where empty tRNAs exit the ribosome.

Exon one of multiple regions of nascent eukaryotic mRNA that are spliced together into a mature mRNA.

Exonuclease an enzyme that cleaves a nucleotide from the end of the DNA sequence.

Exportin a protein that moves proteins out of the nucleus.

Expression platform the region of a riboswitch that contains the coding sequence of the mRNA.

Expression vector a piece of DNA that encodes the message for expression of protein and the regulatory information needed to express at high levels.

Facilitates chromatin transcription (FACT) the protein associated with the MCM helicase that assists in the remodeling of chromatin.

Fasted state the metabolic state in which an organism has not eaten for 12 to 24 hours.

Fatty acid synthase (FAS) the multifunctional enzyme that synthesizes the fatty acid palmitate from malonyl-CoA.

Fed state the metabolic state lasting from the time an organism has eaten until 12 hours after.

Feedback inhibition a regulatory mechanism in which the product of the reaction inhibits the steps of the pathway.

Feedback inhibition loop a regulatory mechanism where the product of the reaction inhibits the steps of the pathway.

Feed forward activation a regulatory process in which the product of a reaction regulates that reaction and makes the reaction go at a faster rate.

Ferritin an iron-binding protein.

Fischer projection a two-dimensional depiction of molecular structure in which crossed lines are used to represent chiral centers.

Fluorescent sequencing a gene sequencing technique that uses dideoxy chemistry with fluorescent dideoxy terminators.

f-Met (formylmethionine) the first amino acid in prokaryotic proteins.

Fold prediction a template-based method of protein structure prediction that involves finding short stretches of amino acid sequence

and comparing them to a database of existing folds.

Fold purification a measure of the amount of protein of interest in the crude fraction divided by the amount of protein of interest in the final sample.

Fourier transform a mathematical operation in which one function, such as frequency, is represented as another, such as time.

Fourier transform mass spectrometer (FTMS) a type of mass spectrometer in which ions injected into a magnetic field orbit around a central electrode and oscillate with a frequency proportional to their mass-to-charge ratio.

Free energy of the standard state (standard free energy state, $\Delta G^{\circ\prime}$) the free energy of a reaction in the standard state as opposed to the one in which the reaction occurs in the cell.

Free energy of the transition state (free energy of activation, $\Delta G^{\ddagger}$) the energy needed to form the transition state, which can be used to calculate the kinetics of a reaction.

Free fatty acid a fatty acid that is not esterified to any other group.

Functional group a group of atoms in molecules (for example, carbonyl, hydroxyl, or carboxyl groups) with common known functions, properties, and reactivity.

Furanose a monosaccharide in a five-membered ring (four carbons and one oxygen).

Fusion protein the protein that results when cDNA of two unrelated proteins is spliced together in-frame to produce a single polypeptide.

Gene expression the activation and regulation of transcription of a gene.

General acid-base catalysis an enzyme mechanism in which an amino acid side chain acts as a proton donor or acceptor.

Gene silencing any one of several techniques used to reduce or prevent gene expression.

Gene a sequence of DNA that codes for protein.

Genetic code a sequence of RNA that codes for individual amino acids.

Genome all the genes of an organism.

Genomics the study of the genome and gene interactions.

Ghrelin a peptide hormone from the gut that regulates appetite by sending orexigenic signals to the brain to stimulate appetite.

Gibbs free energy (ΔG) the amount of energy in a chemical reaction available to do work, calculated by considering enthalpy, entropy, and temperature; generally used to predict the favorability of a reaction or the direction in which it will proceed spontaneously.

Global alignment   a bioinformatic approach that attempts to align every member of a gene sequence, which is most useful if there is high homology across the entire sequence.

Globin fold   the tertiary structure used by hemoglobin and myoglobin.

Glomerulus   a network of porous capillaries that filter the blood, retaining things larger than proteins.

Glucagon   a pancreatic peptide hormone that is released in the fasted or starved state.

Glucogenic   characteristic of amino acids that can contribute to gluconeogenesis.

Gluconeogenesis   the metabolic pathway through which organisms generate glucose from simple building blocks.

Glucosamine   a common amino sugar consisting of glucose in which one hydroxyl group (at the number 2 position) has been replaced by an amino group.

Glucose–alanine cycle   the metabolic pathway in which alanine from muscle is used to shuttle amino groups to the liver where, the amino groups are detoxified via the urea cycle, and the carbon skeletons are used in gluconeogenesis.

Glucose–alanine shuttle   the metabolic process through which nitrogenous wastes generated in muscle are transferred to pyruvate, generating alanine, which in turn is transported to the liver where it is deaminated and the pyruvate presumably is used in gluconeogenesis.

Glucose tolerance test (GTT)   a test used to determine if an individual is diabetic or insulin resistant.

Glucuronic acid   an acidic carbohydrate that is covalently bound to molecules to increase their solubility and make them more prone to be excreted by the kidney.

Glutamate dehydrogenase   the enzyme that catalyzes the oxidation of glutamate, resulting in the formation of α-ketoglutarate.

Glutaminase   the enzyme responsible for removal of the side chain amino group from glutamine, generating glutamate.

Glutamine synthetase   the liver enzyme responsible for synthesizing glutamine from glutamate, free ammonia, and ATP.

Glutaredoxin   the enzyme that uses glutathione to reduce ribonucleotide reductase.

Glutathione   a small peptide comprised of glycine, cysteine, and aspartate that maintains a reducing atmosphere in the cell.

Glutathione reductase   the enzyme responsible for reducing oxidized glutathione dimers back into their sulfhydryl form.

Glutathione-*S*-transferase (GST)   any one of a family of enzymes that couple the tripeptide glutathione to molecules.

Glycerophosphate shuttle   the process through which the electrons from NADH are shuttled into the mitochondria via glycerophosphate.

Glycerophospholipid   any one of a class of phospholipids that use glycerol as the base alcohol.

Glycogen   a highly branched polymer of glucose found in animals and used as a source of stored energy.

Glycogenesis   the creation of new glycogen.

Glycogenin   the protein found at the core of glycogen that synthesizes glycogen in the initial steps of glycogen synthesis.

Glycogenolysis   the breakdown and mobilization of glycogen stores.

Glycogen phosphorylase   the enzyme that catabolizes glycogen, liberating glucose-1-phosphate.

Glycogen synthase   the enzyme responsible for adding new molecules of glucose to glycogen with UDP-glucose as the donor molecule.

Glycolipid   a phospholipid with a carbohydrate modification to it.

Glycolysis   the metabolic pathway through which glucose is broken down into pyruvate.

Glycome   the collection of all the carbohydrates in an organism, tissue, or cell, including free carbohydrates in solution and carbohydrate modifications to molecules.

Glycoprotein   a protein with a carbohydrate modification to it.

Glycosidic bond (linkage)   the acetal or ketal bond or linkage found between carbohydrates.

GPI anchor (glycosylphosphatidylinositol)   a glycolipid that anchors some proteins to the cell membrane, often to a lipid raft.

G protein   any one of several classes of signaling proteins that bind GTP, at which point, these proteins are in the active state.

Guanine   a purine base found in both DNA and RNA.

Haworth projection   a molecular representation of cyclic monosaccharides that depicts their three-dimensional structure.

HDL (high-density lipoprotein)   a lipoprotein that transports cholesterol back to the liver from the circulation.

Helicase   the protein responsible for unwinding the DNA double helix during replication.

Helix-loop-helix   a conserved DNA-binding motif consisting of a helix, a larger loop, and a second helix.

Helix-turn-helix   a conserved DNA-binding motif consisting of a helix, a turn, and a second helix.

Henderson–Hasselbach equation   the equation that relates pH, pK_a, and the ratio of conjugate acid to conjugate base.

Heparan sulfate   a linear polysaccharide comprised of repeating disaccharides, usually glucuronic acid and *N*-acetylglucosamine.

Hepatocyte   a liver cell.

Heterotropic regulation   allosteric regulation by a molecule that is not the substrate for the enzyme, for example, 2,3-bPG.

Hexose   a six-carbon monosaccharide.

Hibernation   the dormant metabolic state mammals enter in winter.

Hidden Markov model (HMM)   a statistical model used to generate a conceptual toolkit that encompasses different sources of divergent information.

High-scoring segment pair (HSP)   a subsegment of a pair of sequences that share a high level of similarity.

High-throughput parallel sequencing   a gene sequencing technique that employs hundreds of thousands to billions of short sequencing reactions simultaneously to sequence entire genomes within days.

Hill equation   the equation used to predict the cooperativity of an allosteric complex.

Histone acetyltransferase (HAT)   an enzyme that adds acetyl groups to histones.

Histone   a protein that helps organize DNA.

Histone deacetylase   an enzyme that removes acetyl groups from histones.

Histone methyltransferase (HMT)   an enzyme that transfers methyl groups to histones, decreasing interactions with DNA.

HMG-CoA reductase   the committed and rate-determining step in cholesterol biosynthesis.

Hofmeister series   a classification of anions and cations according to how they decrease or increase protein solubility.

Holliday junction   the X-shaped structure that forms when two homologous chromosomes exchange DNA during recombination.

Holoenzyme   an enzyme complex with all its parts (such as cofactors) complete.

Holoprotein   a complete protein (i.e., one with its complementary cofactors).

Homeodomain   a family of transcription factors originally identified in *Drosophila* but conserved throughout evolution; involved in differentiation of cells and tissues.

Homeostasis the concept that a living system has control systems in place to maintain conditions necessary for life, including systems to regulate pH, store and mobilize energy, and maintain temperature.

Homogenization a method for disrupting tissue to bring a sample into uniform distribution.

Homologous recombination the process by which two homologous genes or chromosomes can exchange strands.

Homology modeling a template-based method of protein structure prediction that solves the structure of a protein using a different protein with a solved structure and highly related amino acid sequence as a framework.

Homotropic regulation allosteric regulation by substrate molecules acting to regulate the activity at other sites in the complex, for example, oxygen binding to hemoglobin.

Hormone a chemical signal that is secreted in one part of the organism but works elsewhere.

Housekeeping gene a gene that is constitutively expressed, that is, always expressed in the cell.

Hydrodynamic property one of several properties of a hydrated particle in aqueous solution, including viscosity, sedimentation coefficient, diffusion coefficients, and hydrodynamic radius.

Hydrodynamic radius the radius of a molecule when modeled as a sphere; also known as the Stokes radius.

Hydrogenated fat any fat or oil in which double bonds have been removed by chemically treating them with hydrogen gas and a catalyst.

Hydrogen bonding the weak interaction that occurs between the lone pair of electrons on a very electronegative atom (oxygen, nitrogen, or fluorine) and a hydrogen bound to an electronegative atom.

Hydrophobic effect the phenomenon by which hydrophobic groups in a molecule aggregate together, liberating water molecules and increasing the overall entropy of the structure.

Hydrophobic interaction chromatography (HIC) an affinity chromatography method that uses resin coated with hydrophobic groups.

Hyperglycemia high blood sugar.

Hypothalamus the region of the brain that integrates neuronal and hormonal signals.

Hypoxanthine guanine phosphoribosyltransferase (HGPRT) the enzyme respon-

sible for coupling guanine or hypoxanthine to PRPP in the purine salvage pathway.

IDL (intermediate density lipoprotein) an intermediate that forms when VLDL are converted to LDL.

Immobilized metal affinity chromatography (IMAC) a variant of affinity chromatography in which proteins or peptides are separated based on their affinity to immobilized metal ions on the matrix.

Importin a protein that shuttles proteins into the nucleus.

Induced fit a description of enzymatic catalysis in which a substrate helps form the shape of the active site as it binds the enzyme.

Inducible able to be upregulated or activated through positive regulation.

Initial rate (v_0) the instantaneous rate of an enzymatically catalyzed reaction at $t = 0$; also known as initial velocity.

Initiation the assembly of the ribosome on the mRNA.

Initiation factor (IF) a protein that assists in the assembly of the ribosome on the start codon.

Insertion the addition of nucleotides to a DNA sequence.

Insulin a peptide hormone made by the pancreas that has effects on both metabolism and transcription using a receptor tyrosine kinase.

Insulin resistance a prediabetic state in which cells become refractory to the effect of insulin.

Integral membrane protein a membrane protein that cannot be removed from the lipid bilayer without the use of detergents.

Interactome the collection of all the interactions of the molecules of the cell.

Intron a region of a nascent eukaryotic mRNA that is spliced out to form a mature mRNA.

Inverse agonist a molecule that binds to the same site as an agonist but exerts the opposite effect.

Ion-exchange chromatography (IEX) an affinity chromatography method that uses coulombic effects to separate proteins.

IRE-binding protein (IRP) a trans-acting regulatory element that binds to either the IRE or iron in an iron–sulfur cluster.

Iron-response element (IRE) the cis regulatory element found upstream of the ferritin and transferrin receptor genes.

Iron-sulfur center a redox active cluster of iron and sulfur atoms found in electron transport, including Fe2-S2, Fe3-S3, and Fe4-S4.

Irreversible inhibitor *see* suicide substrate.

Isoacceptor one of multiple tRNAs that bind to the same amino acid.

Isoelectric point the pH at which a protein is electronically neutral.

Isoform one of multiple proteins with highly related structures that serve an identical function and are encoded by different genes.

JAK-STAT pathway a highly conserved signaling pathway generally functioning in any condition in which pro-inflammatory cytokines are elevated.

Keratan sulfate a linear polysaccharide that is a polymer of galactose and *N*-acetylglucosamine.

Ketoacidosis the acidification of the blood via the accumulation of ketone bodies.

Ketogenic characteristic of amino acids that are broken down into ketone bodies.

Ketone body one of a group of three small soluble molecules (acetone, acetoacetate, and β-hydroxybutyrate) that are made in the liver in times of energetic scarcity and can provide energy to tissues such as the brain when glucose is not available.

Ketose a monosaccharide containing a keto group.

Ketosis a metabolic state characterized by excessive ketone bodies in the blood.

Kinase an enzyme that adds a phosphate group to a molecule, typically donated from a molecule with a high transfer potential (such as ATP) and added to a hydroxyl group.

Kinase cascade a reaction in which a kinase becomes active and phosphorylates a second kinase downstream of the first.

Kinesin a molecular motor that moves along tubulin microtubules.

Kinome the collection of all the phosphorylation modifications made in the cell, including those made to proteins, lipids, carbohydrates, and metabolites.

Klenow fragment a proteolytic fragment of *E. coli* DNA pol I used in molecular biology.

K_M (1) The dissociation constant of the Michaelis complex. (2) The ratio of rate constants found, which form the Michaelis complex over those that break it down. (3) The concentration of substrate required for an enzyme to operate at $V_{max}/2$.

Knockout an organism in which a gene (or multiple genes) have been removed or mutated to prevent their expression.

Kozak sequence the region of the genetic message where the ribosome binds in eukaryotes.

***Lac* operon** an operon that regulates the expression of three genes involved in lactose metabolism; the first operon characterized.

Lac repressor (LacI) a protein that binds to either lactose or the cis elements of the *lac* operon, preventing gene expression.

Lactose milk sugar; a disaccharide composed of a molecule of glucose and galactose.

Lagging strand the DNA strand on which discontinuous DNA replication happens.

LDL (low-density lipoprotein) a lipoprotein that can deliver cholesterol from the liver to tissues that need it but can also become oxidized and accumulate in the linings of arteries causing atherosclerosis.

Leading strand the strand of DNA that is continuously copied in the 5′ to 3′ direction.

Le Châtelier's principle the observation that systems tend to move toward equilibrium.

Leptin a peptide hormone produced by fat tissue that regulates satiety (feelings of fullness).

Lesch–Nyhan syndrome a sex-linked genetic disorder in which HGPRT is dysfunctional, frequently resulting in self-mutilation among other abnormalities.

Leucine zipper a conserved DNA-binding motif consisting of two helices that associate by the zippering of leucine residues on the sides of the helices.

Leukotriene any molecule in a subclass of eicosanoid, a family of signaling molecules made from arachidonic acid by the enzyme lipoxygenase.

Levinthal's paradox the reasoning that it would require an astronomical amount of time for a protein to find the most stable (folded) state by trying different combinations at random.

Linear electron transport the process of transferring electrons to the next step in a linear fashion during photosynthesis.

Lineweaver–Burke plot a double-reciprocal plot, that is, one in which $1/[S]$ versus $1/v$ is plotted to determine the values of V_{max} and K_M.

Lipid bilayer a structure comprised of two monomolecular films of phospholipids arrayed to form a hydrophobic barrier, including most of the membranous structure in cells and organelles.

Lipidome the collection of all the hydrophobic molecules found in a sample.

Lipid raft a transient structure of weakly associated lipids, such as sphingolipids and cholesterol, found in the plasma membrane.

Lipid storage droplet the organelle that stores neutral lipids in the cell.

Lipopolysaccharide (LPS) a glycolipid found in the cell wall of some bacteria that is inflammatory in a cell or organism; also known as endotoxin.

Lipoprotein a spherical micelle of phospholipids and apolipoproteins surrounding a neutral lipid core that transports lipids in the circulation.

Lipoxygenase the enzyme that converts arachidonic acid into leukotrienes.

Local alignment a bioinformatic approach that preferentially aligns regions of a gene sequence with higher homology.

Lock and key one description of enzymatic catalysis in which a substrate fits an enzyme like a key in a lock.

London-dispersion force one in a class of weak intermolecular forces that all molecules exhibit.

Loop of Henle the continuation of the convoluted tubule in which partial permeability and sodium pumps set up a concentration gradient between the lumen of the tube and the extracellular fluid.

Lysophospholipid a glycerophospholipid that is missing a single acyl chain.

Lysyl oxidase the enzyme that cross-links lysine residues in the extracellular matrix.

Macrolide any of a class of antibiotics that blocks protein synthesis in bacteria and is characterized by a macrocyclic ring.

Major groove the larger of the two grooves on the DNA double helix where proteins bind specifically.

Malate–aspartate shuttle the process through which the electrons from NADH are shuttled into the mitochondria via malate and aspartate, with glutamate and α-ketoglutarate also playing important roles.

Maltose malt sugar; a disaccharide composed of two glucose molecules.

Matrix-assisted laser desorption ionization (MALDI) a soft ionization technique that involves ionizing a sample without fragmentation.

Messenger RNA (mRNA) a type of RNA that provides the code for protein production.

Metabolic syndrome a group of ten related pathophysiologies related to obesity and diabetes.

Metabolite an intermediate in a metabolic pathway.

Metabolome the collection of all the small metabolites of an organism, tissue, or cell.

Metabolon a metabolic process in which a group of enzymes function in concert to synthesize a product.

Metal affinity chromatography (MAC) an affinity chromatography method that separates proteins based on interactions between specific side chains and metal ions bound to a column.

Metal ion catalysis an enzyme mechanism in which a metal ion is participating in redox chemistry or acting as a Lewis acid or Lewis base.

Methotrexate a folate antimetabolite that acts as a dihydrofolate reductase inhibitor and was among the first successful chemotherapeutic drugs.

Mevalonate an intermediate in cholesterol metabolism.

Micelle the structure formed from the aggregation of amphipathic molecules such as fatty acids or detergents, which has a hydrophobic core and a hydrophilic coating.

Michaelis complex the complex formed between enzyme and substrate.

Michaelis–Menten equation $v_0 = V_{max}[S]/K_M + [S]$.

Michaelis–Menten kinetics the best-known model of enzyme kinetics, which relates rate (velocity) to substrate concentration.

Microarray a gene sequencing technique in which the cDNA of transcripts of interest are adhered to a microscope slide.

Microchannel plate detector a type of mass spectrometer detector that detects a signal through ion collision and a cascade of electrons.

MicroRNA (miRNA) a short stretch of RNA that, once processed, binds to mRNAs to silence them.

Minichromosome maintenance (MCM) complex the proteins that form a ring around the DNA helix at the origin of replication and act as a helicase to unwind the double helix.

Minor groove the smaller of the two grooves on the DNA double helix.

Mismatch repair a DNA repair mechanism in which a mismatch in the double helix is identified, the site is excised from the newly synthesized DNA strand, and the repair is made through the Mut proteins (MutS, MutL, and MutH).

Mitogen-activated protein (MAP) kinase one of a family of kinases that respond to mitogens.

Mitogen a growth factor and proinflammatory signal that leads to cell proliferation.

Mixed inhibitors an inhibitor that can bind to either free enzyme or *ES* complex.

Mixed meal a meal in which fats, proteins, and carbohydrates are consumed in proportion to each other.

Molecular dynamics (MD) calculation the most common *de novo* structure prediction that is performed on a biological system.

Molecular motor an enzyme that uses hydrolysis for directional motion.

Monoacylglycerol (monoglyceride) a glycerol molecule with a single fatty acid esterified to it.

Monoamine oxidase (MAO) any in a family of oxidases that oxidatively deaminate amines to carbonyl groups.

Monomer the building block of a polymer, for example, amino acids, nucleotides, and carbohydrates.

Monosaccharide the simplest carbohydrate.

Monotopic a protein or other molecule inserted into only one leaflet of the lipid bilayer.

Monounsaturated a characteristic of fatty acid with one double bond.

Motif a recurring combination of secondary structure in proteins, for example, a four-helix bundle or Greek key motif.

Multidimensional NMR a category of NMR techniques that employ multiple radio pulses applied at angles to one another to probe the sample.

Multimer a molecule made of several monomers.

Multiple isomorphous replacement (MIR) a technique in which multiple heavy atoms (metals) are infused into a crystal of a protein, which changes the intensity of the diffraction pattern but not the shape of the crystal or the location of the spots; used to solve the phase problem in X-ray diffraction experiments.

Mutagen a chemical or virus that causes a mutation to DNA.

Mutation a change in the sequence of a nucleic acid.

Mutorotation the process through which one anomer can interconvert to another by ring opening and closing.

Myocyte a muscle cell.

Myosin a molecular motor that moves along actin microfilaments.

Negative regulation a type of regulation in which proteins bind and repress the expression of an operon.

Negative stain an electron microscopy technique in which structures can be determined by stain puddling around them.

Nephron the basic filtration unit of the kidney.

Neuropeptide Y (NPY) a peptide neurotransmitter that regulates and stimulates appetite.

Neutral lipid a lipid without a charged group, typically triacylglycerol and cholesteryl ester.

N-linked carbohydrate a carbohydrate moiety connected to a protein through an asparagine residue.

Noncoding sequence a gene sequence that does not code for protein.

Noncompetitive agonist a molecule that does not compete with the native ligand for binding to the receptor.

Non-homologous end joining (NHEJ) a DNA repair mechanism in which two free ends of chromosomes are joined together without a template.

Norepinephrine a catecholamine hormone.

Nuclear factor kappa B (NF-κB) a transcription factor involved in the immune system and many cancers.

Nuclear magnetic resonance (NMR) a structure-determination technique in which nuclei are aligned in a magnetic field and irradiated with radio waves.

Nuclear Overhauser effect (NOE) the observation that the spin of one nucleus can influence the spin of another through space; used in multidimensional NMR experiments and determination of protein structure.

Nuclear Overhauser effect spectroscopy (NOESY) a type of multidimensional NMR spectroscopy that uses the Overhauser effect to identify the interactions of some nuclei.

Nuclear pore complex (NPC) the large, basketlike structure that forms pores in the nuclear membrane.

Nuclease an enzyme that degrades nucleic acids into nucleotide phosphates.

Nucleic acid a polymer of nucleotides (RNA) or deoxynucleotides (DNA).

Nucleoside a nitrogenous base joined to a sugar (ribose) through a glycosidic linkage with no phosphate.

Nucleoside diphosphate kinase the enzyme responsible for adding the third phosphate to a nucleoside diphosphate to generate a nucleoside triphosphate.

Nucleotidase the enzyme that cleaves phosphates from nucleotides to generate a nucleoside.

Nucleotide a nitrogenous base joined to a sugar (ribose) phosphate via a glycosidic linkage.

Nucleotide excision repair (NER) a DNA repair mechanism in which a thiamine dimer is removed and then the damaged strand is repaired by DNA polymerase.

Okazaki fragment one of the short, 200-base pair sequences of DNA formed on the lagging strand that are joined together in DNA synthesis.

Oligopeptide a molecule consisting of more than a few amino acids, typically between 8 and 20.

Oligosaccharide one in a class of short polymers of monosaccharides between four and twenty monosaccharides in length.

O-linked carbohydrate a carbohydrate moiety connected to a protein through a serine or threonine residue.

Omega-3 (ω-3) fatty acid a fatty acid with double bonds three carbons from the carbon at the end of the acyl chain.

Omega-6 (ω-6) fatty acid a fatty acid with double bonds six carbons from the carbon at the end of the acyl chain.

Omics the group of scientific approaches that considers all members of a certain class of molecules or interactions collectively.

Open reading frame (ORF) the region of genetic code between a start and stop codon.

Operator the region of DNA in an operon that binds to repressors.

Operon a regulatory arrangement in which a single regulatory domain controls multiple genes.

Orexigenic characteristic of neurons, hormones, or drugs that promote appetite.

oriC (origin of replication) the DNA sequence at which DNA replication initiates.

Origin of replication complex (ORC) the complex of proteins including the MCM proteins, Cdc6, and Ctd1 that bind to the pre-replication complex.

Orthogonal set in the context of protein chemistry, a set consisting of tRNA, unnatural amino acid, synthetase, and codon that are unable to communicate with the native machinery of the cell in any way.

Orthosteric regulation control of an enzyme through mechanisms that act at the native ligand-binding site.

Osmosis the diffusion of water from areas of lower concentrations of molecules to higher concentrations of molecules.

Oxazolidone any of a class of antibiotics that blocks protein synthesis in bacteria by blocking initiation.

Oxidative deamination the oxidative removal of an amino group as ammonia generating an α-keto acid and NADH.

Oxidative phosphorylation the process through which metabolites are oxidized and the energy of these processes is used to generate a proton gradient, which is then used to generate ATP.

Pairwise alignment an alignment of two of the most highly related sequences.

Palindromic sequence a DNA sequence that reads the same forward as backward.

Parallel characteristic of a structure in which the direction strands are all running in the same direction.

Partial agonist a molecule that binds to and activates the receptor but to a lesser extent than the native ligand.

Pasha a nuclear pore protein complex that recognizes and binds short hairpin RNAs.

Passive transport the transport of molecules across a membrane down a concentration gradient via diffusion.

Patch one of the clusters of receptors on the cell surface and one of the initial steps in receptor-mediated endocytosis.

PCR-based mutagenesis a method for generating site-directed mutations that utilizes mismatched primer pairs to introduce mutations to specific locations within DNA plasmids.

Pentose a five-carbon monosaccharide.

Pentose phosphate pathway the pathway through which the cell generates ribose, produces NADPH, and transforms monosaccharides.

Penultimate carbon the chiral center farthest from the anomeric carbon.

Peptide a chain of one or more amino acids.

Peptide bond the amide bond that links amino acids together in a protein.

Peptide hormone a hormone made from short chains of amino acids.

Peptide (P) site the site in the ribosome where the peptide emerges.

Peptidoglycan a molecule composed of peptides and carbohydrates found in the bacterial cell wall.

Peripheral membrane protein a membrane protein that can be dissociated from the lipid bilayer through the use of salts or changes in pH.

Peroxisome proliferator-activated receptor (PPAR) a transcription factor important in expression of many genes involved in lipid metabolism.

Phase problem determining how diffracted X-ray beams are phase-shifted with respect to each other.

Phosphatase an enzyme that removes a phosphate.

Phosphoglucomutase the enzyme that isomerizes glucose-6-phosphate and glucose-1-phosphate.

Phospholipid bilayer a structure of phospholipids found in the plasma membrane and many organelles in which the hydrophobic acyl chains of phospholipids are oriented toward the core of the bilayer and the hydrophilic head groups are oriented to either face of the bilayer; also known as a lipid bilayer.

Phospholipid one of a class of polar lipids that have glycerol at their core.

Phosphoribosyl pyrophosphate (PRPP) a ribose ring with phosphate at the 5′ carbon and pyrophosphate on the 1′ carbon.

Phosphorylation the addition of a phosphate to a hydroxyl group, typically of a protein.

Photolyase the enzyme that uses the energy of a photon of light to cleave a thymine dimer.

Photorespiration the process in which oxygen is used as a substrate for RuBisCO to generate 3PG and 2-phosphoglycolate.

Photosynthesis the process that uses solar energy to produce ATP and NADPH, reduce CO_2 to glucose, and produce oxygen.

Photosystem I the second photosystem in photosynthesis, which uses sunlight to produce electrons of higher energy states.

Photosystem II the first protein complex in photosynthesis, which is involved in photon capture and electron transport.

pH the scale used to measure the concentration of H^+ in solution.

Pi (π) helix a less common example of secondary structure, similar to an alpha helix but with looser coils due to hydrogen bonding occurring in an $i+5$ pattern compared to the alpha-helical $i+4$ pattern.

pK_a the negative log of the acid ionization constant, which can be used generically to discuss how acidic a proton is (how tightly a proton is bound to another atom).

Pluripotency the ability of a cell to divide into numerous different cell types.

Polar lipid a lipid molecule containing polar groups such as phosphates, typically found in lipid bilayers.

Poly(A) tail a tail of adenosine residues added enzymatically to the end of an mRNA.

Polycistronic message a message encoding multiple proteins.

Polymer a molecule made up of multiple subunits joined together, including proteins and nucleic acids.

Polymerase chain reaction (PCR) a laboratory technique in which short stretches of RNA can be amplified many times over, which is the basis of numerous other techniques.

Polypeptide a chain of amino acids between 20 and 100 amino acids in length.

Polysaccharide one of any polymers of monosaccharides.

Polysome the assembly of multiple ribosomes translating proteins on an mRNA in the cytosol.

Polyunsaturated having multiple double bonds.

Positive regulation a type of regulation in which proteins bind and activate gene expression.

Positive stain an electron microscopy technique in which stains bind directly to the molecules of interest.

Posterior pituitary the rear section of the pituitary gland that secretes two hormones, oxytocin and vasopressin.

Postprandial state the metabolic state found up to two hours after eating.

Post-translational modification changes such as glycosylation, acylation, or phosphorylation made to proteins following synthesis.

POU a family of transcription factors named for the original members of the family (pituitary specific, octamer, and unc-86).

Preinitiation complex the complete complex of transcription factors and RNA polymerase prior to transcription.

Pre-replication complex the complex of proteins including ORC, MCM, Cdc6, and Ctd1 that bind to the origin of replication sequence.

Primary protein structure the amino acid sequence of the protein.

Primase the enzyme that synthesizes an RNA primer from the template strand.

Primer a short segment of DNA or RNA from which a DNA polymerase can extend.

Probenecid an inhibitor of the transporter that reabsorbs uric acid from urine, leading to increased uric acid excretion but potentially also blocking the excretion of other drugs.

Processivity the property of an enzyme that catalyzes multiple reactions; a processive enzyme will bind to bulk substrate (such as a DNA sequence) and catalyze multiple reactions before dissociating.

Product the substance made in an enzymatic reaction.

Proliferating cell nuclear antigen (PCNA) the eukaryotic homolog of the β clamp. This protein complex increases processivity of the DNA polymerase.

Promoter sequence the segment of DNA upstream of the coding sequence of a gene that is responsible for regulation of expression.

Proofreading the process through which an enzyme, such as a DNA polymerase, ensures that the correct base has been incorporated into the growing chain.

Prostaglandin one of a subclass of eicosanoid, a family of signaling molecules made from arachidonate by the enzyme cyclooxygenase.

Protein folding the process by which a protein goes from a linear polymer to a tightly folded structure.

Protein kinase an enzyme that adds a phosphate to a protein.

Protein kinase A (PKA) the first protein kinase discovered, which catalyzes the phosphorylation of many proteins, altering their activity.

Protein a polymer of amino acids that serves functions ranging from structural to catalytic.

Protein overexpression the use of a biological system to produce large quantities of protein.

Protein trafficking the process by which proteins are packaged, are sorted, and find their place in the cell.

Proteinuria a condition in which protein is found in the urine, typically due to kidney damage.

Proteoglycan a part of the extracellular matrix composed of a protein core with significant carbohydrate modifications.

Proteolytic cleavage a break in the peptide backbone of a protein.

Proteome the collection of all the proteins in an organism, cell, or tissue.

Proton motive force (PMF) the combined total of the proton chemical gradient and the electrical gradient across the inner mitochondrial membrane.

Proximal convoluted tubule the structure in the kidney that reabsorbs fluid and metabolites after filtration.

Pseudosubstrate domain a structural homolog of its substrate but lacking essential residues necessary for phosphorylation.

Purine one of the class of bicyclic nitrogenous bases that are synthesized from a purine ring and include adenine and guanine; along with pyrimidines, one of the two categories of bases found in nucleic acids.

Pyranose a monosaccharide in a six-membered ring (five carbons and one oxygen).

Pyridoxal phosphate (PLP) a cofactor found in several reactions involving the α carbon of amino acids.

Pyrimidine one of the class of single-ring nitrogenous bases that are derivatives of pyrimidine and include uracil, cytosine, and thymine; along with purines, one of the two categories of bases found in nucleic acids.

Pyruvate a three-carbon α-keto acid found at the convergence of several metabolic pathways.

PYY(3-36) a peptide neurotransmitter that is involved in regulation of metabolism and appetite.

Q cycle the mechanism by which electrons are transferred in complex III from a two-electron carrier (ubiquinone) to a one-electron carrier (cytochrome *c*).

Quadrupole mass analyzer a device consisting of four parallel metal rods surrounding the ion beam that resolves mass-to-charge ratio in a mass spectrometer.

Quaternary protein structure how different subunits interact with one another to form a multisubunit complex.

Ran cycle the GTPase cycle that describes how molecules are transported into or out of the nucleus.

Rate constant, *k* the number that quantifies the rate of a chemical reaction used in rate laws, mathematical expressions which relate chemical rate or velocity to the concentration of reactants in the reaction.

Rate law a mathematical expression that describes how chemical reactions occur.

Reaction center the site of primary energy reactions during photosynthesis.

Receptor tyrosine kinase (RTK) one in a class of signaling receptors that upon binding of ligand (hormones such as insulin) will dimerize.

Reflection the total number of spots found in a diffraction experiment.

Regulatory subunit a subunit of an enzyme that regulates binding of substrate or catalysis.

Relaxed (R) state the more active state in allosteric control.

Release factor a protein that causes the release of the nascent protein and dissociation of the ribosome.

Reliability (R) factor a measure of how much a structure determined using X-ray diffraction differs from a predicted one, with lower R factors being generally more desirable.

Replication factor C (RFC) the eukaryotic homolog of the clamp loader complex, which assists in the trombone model and in loading polymerases to the DNA strand.

Replication fork the place where DNA is unraveled and replication occurs.

Replication licensing the process by which eukaryotic cells identify replication start sites in chromosomes.

Replication protein A (RPA) the eukaryotic homolog of single-stranded binding protein.

Replication the process through which DNA is replicated.

Replisome the collective complex responsible for replicating DNA.

Repressible able to be downregulated or repressed through negative regulation.

Resolution the shortest distance at which two objects can be distinguished as separate entities.

Respirasome the name of the combination of multiple electron transport complexes working in unison.

Restriction enzyme a bacterial enzyme that recognizes and cleaves specific DNA sequences, often employed in the laboratory in DNA manipulations.

Retrograde transport the trafficking of proteins or cargos inward to a central cellular location; the opposite of anterograde transport.

Retrovirus a virus with an RNA genome that uses RNA and viral enzymes to insert a genetic message into the host's genome.

Reverse cholesterol transport the process through which cholesterol is removed from the circulation and returned to the liver.

Reverse transcriptase (RT) the retroviral enzyme that synthesizes a viral RNA genome into the host's DNA.

Rho a protein cofactor involved in the termination of transcription that determines when the RNA polymerase should dissociate from the gene.

Ribonucleotide reductase the enzyme responsible for making 2′ deoxyribonucleotides out of ribonucleotides.

Ribose a five-carbon (a pentose) sugar that forms a five-membered ring (a furanose) and is used in the synthesis of ribonucleotides.

Ribosomal RNA (rRNA) the type of RNA that is responsible for protein synthesis and forms the majority of the structure of the ribosome.

Ribosome the protein–RNA complex responsible for the synthesis of proteins in the cell.

Riboswitch a regulatory element in which the mRNA has a regulatory region that can bind to a small metabolite (termed an aptamer) and the coding region of the mRNA (the expression platform).

RNA-induced silencing complex (RISC) the complex of Pasha, Drosha, Dicer, and Argonaute that processes miRNAs and silences genes.

RNA interference (RNAi) technique a laboratory technique used to silence genes by using the cell's native RNA processing machinery, including siRNA, shRNA, and morpholinos.

RNA polymerase one of several enzymes responsible for the synthesis of RNA from a DNA template.

RNA sequencing (RNA-Seq) a gene sequencing technique that uses high-throughput sequencing to determine the presence and relative abundance of a transcript at a particular moment in a sample; also known as whole transcriptome shotgun sequencing (WTSS).

Root-mean-square deviation (RMSD) a measure of how much experimentally determined atomic coordinates differ from predicted values.

S-Adenosylmethionine a donor of activated methyl groups.

Salt bridging the coulombic interactions between positively and negatively charged amino acids in a protein.

Salting in a method of protein purification that uses substantial changes to salt concentration to alter protein stability, causing some proteins to solubilize; the reverse of salting out.

Salting out a method of protein purification that uses substantial changes to salt concentration to alter protein stability, causing some proteins to precipitate out; the reverse of salting in.

Salvage biosynthetic pathway the pathway through which bases can be recovered and formed into nucleotides.

Saturated having no double bonds.

Scaffolding protein one of the group of proteins that act as to organize other proteins in the cell to coordinate a process such as signal transduction.

Schiff base the product resulting from the reaction of the nitrogen side chain of lysine with a carbonyl group.

Search algorithm in bioinformatics, a series of steps in a software program used to search a dataset for similarities or patterns.

Secondary active transport a cellular transport method that uses an electrochemical gradient to drive transport of a small molecule across a membrane.

Secondary protein structure basic common units of structure including alpha-helix, parallel and antiparallel beta-sheets, and turns or coils.

Second messenger a chemical message made or released inside the cell that amplifies and propagates the message relayed by a hormonal signal, for example Ca^{2+}, cAMP, and IP_3.

Seeded search a scan of databases for exact matches using high-complexity short sequences.

Semiconservative replication the copying of one DNA strand using the other strand as a template such that each of the final copies contains one copy of the original.

Sequence alignment lining up similar sequences of biological information.

Sequential model a model of allosterism in which the subunits of the complex transition between the T and R states one at a time; also known as the KNF model.

Serial analysis of gene expression (SAGE) a transcriptome analysis technique in which fragments of mRNA are sequenced using classical technologies.

Serine protease an enzyme that uses an active site serine in the hydrolysis of proteins.

Severe combined immunodeficiency (SCID) a family of diseases in which the patient is severely immunocompromised due to deficiencies in nucleotide synthesis.

Shadowing an electron microscopy staining technique in which a metal source is evaporated at a known angle to the sample.

Shelterin complex the protein complex responsible for protecting and maintaining telomeres and directing telomerase activity.

Shine–Dalgarno sequence the region of the genetic message where the ribosome binds in prokaryotes.

Sickle cell anemia an inherited disease in which a point mutation results in a single substitution of a valine for a glutamate in the β chain of hemoglobin, causing the protein to polymerize and cells to form a sickled shape under low-oxygen conditions.

Side chain the unique part of each amino acid along with a region that is common to all amino acids (the amino group, α carbon, and carboxyl).

Sigma (σ) factor a protein cofactor that recognizes and binds to the core promoter in prokaryotes.

Signaling cascade a pathway through which signals in the cell are relayed and amplified, often employing kinases.

Signal peptidase a protease in the ER lumen that cleaves the signal peptide from a protein.

Signal peptide a short, hydrophobic peptide that targets proteins to the endoplasmic reticulum.

Signal recognition particle (SRP) the protein that binds to the signal peptide and signals for translation into the ER membrane.

Signal sequence a short, N-terminal peptide sequence that directs nascent protein to the ER.

Signal transduction a means by which cells communicate with one another.

Similarity matrix a table used to identify similarities among DNA sequences in which one sequence of DNA is listed along the top and one down the side.

Single-stranded binding protein (SSB) a protein that binds to the exposed single strands of DNA at the replication fork.

Sister chromatid one of the pairs or arms of the chromosome.

Site-directed mutagenesis a technique used to make small alterations to a protein's coding sequence, ranging from a single nucleotide change to modification of three to six bases.

Size-exclusion chromatography a method of protein purification that separates proteins based on size; also known as gel filtration.

Sliding clamp the *E. coli* protein complex that encircles the DNA strand and acts as scaffolding for the replisome, anchoring the polymerase and increasing processivity; also known as a β clamp.

Small nuclear RNA (snRNA) a type of small RNA molecule found in the nucleus and responsible for splicing mRNA.

SNARE protein a complex of proteins that mediates the fusion of vesicles to the plasma membrane.

Solid-phase peptide synthesis (SPPS) a chemical condensation reaction that links amino acids via amide bonds to produce peptides.

Sonication a crude step of protein purification in which a sample is bombarded with ultrasonic waves to create a uniform distribution.

Specific activity a measure of purity that describes units catalyzed per unit time per milligram of enzyme.

Sphingolipid one of a class of phospholipids that use sphingosine as the base alcohol.

Spindle fiber a cytoskeletal element of the cell that centromeres interact with during cell division.

Spliceosome the protein–RNA complex responsible for recognizing and catalyzing the splicing of exons in mRNA.

Splicing the removal of introns and joining of different exons during mRNA processing.

Src homology 2 (SH2) domain a common structural motif found in signaling proteins that reversibly binds phosphorylated tyrosine residues.

Standard state a defined state for comparison with others, for example, the chemical standard state, $\Delta G°$, which defines temperature at 25°C (298 K), a pressure of 1 atmosphere and solutes with an activity of 1 (roughly corresponding to their molar concentrations).

Start codon the codon (AUG) that begins every message and encodes methionine in eukaryotes or formyl methionine in prokaryotes.

Starved state the metabolic state achieved when one has not eaten for more than 24 hours.

Steady state assumption the assumption that the enzyme is operating at a fixed rate, that is $d[ES]/dt = 0$.

Steroid hormone one of a category of lipid hormones that are derived from cholesterol.

Steroid one of a group of steroid hormones that work by altering gene expression.

Sterol regulatory element–binding protein (SREBP) a transcription factor found in the endoplasmic reticulum that is proteolytically processed to respond to elevated sterol levels.

Stop codon any of three codons (UAA, UAG, and UGA) that tell the ribosome to stop translating the mRNA message.

Stroma the fluid within the chloroplast; analogous to the matrix in mitochondria.

Structural maintenance of chromosomes (SMC) protein one of a group of six related proteins that organize and condense chromosomes.

Substitution matrix a table used to compare DNA sequences that contains one set of data along the side of the table and the same set of data along the base of a table.

Substrate channeling the direct diversion of the product of one reaction into a subsequent reaction.

Substrate-level phosphorylation a reaction that directly results in the transfer of a phosphate in an enzymatic reaction, along with oxidative phosphorylation, one of two means by which ATP can be produced.

Substrate the reactant in an enzymatic reaction.

Sucrose table sugar; a disaccharide comprised of glucose and fructose.

Sugar acid a monosaccharide in which one of the carbons has been oxidized to a carboxylic acid.

Sugar alcohol a monosaccharide in which the carbonyl carbon has been reduced to a hydroxyl group.

Suicide substrate an inhibitor of an enzymatic reaction that becomes irreversibly coupled to the enzyme, preventing further catalytic events from occurring; also known as an irreversible inhibitor.

Sulfonylurea one of a class of oral antidiabetic drugs that act by promoting insulin secretion.

Switch/sucrose non-fermentable (SWI/ SNF) complex a histone remodeling complex found in yeast.

Symporter a transport protein that transports a molecule in conjunction with another molecule.

Tag a peptide sequence added to a recombinant protein, for example, epitope tags and fluorescent tags such as GFP.

Tandem mass spectrometry (MS-MS) a technique for characterizing a protein that uses one mass spectrometer to isolate the protein followed by a second one to fragment and detect the protein for identification.

Tandem repeat a region of DNA in which a sequence is repeated in tandem; a binding site for proteins.

TATA binding protein (TBP) the protein that recognizes and binds to the TATA box.

TATA box a segment of DNA found immediately upstream of the transcription start site in genes to which promoter elements and transcription factors bind.

TBP-associated factor (TAF) a protein that recognizes and binds to the proteins binding the TATA box.

Telomere the end of eukaryotic chromosomes.

Template the strand of DNA that is copied.

Tense (T) state the less active state in allosteric control.

Terminally differentiated cell a cell that is unable to differentiate or change from one cell type to another.

Termination the completion of protein synthesis and the dissociation of the ribosome from the mRNA.

Terminator a hairpin structure found in mRNAs that causes termination of translation.

Tertiary protein structure how units of secondary structure assemble to form the shape of the protein.

Ter–Tus system the system in which the Tus protein (termination utilization substance) binds to Ter (termination) sequences to signal termination of DNA replication.

Tetracycline any of a class of antibiotics that blocks protein synthesis in bacteria and is characterized by four rings in their structure.

Tetrahydrobiopterin a cofactor used in aromatic amino acid metabolism.

Tetrahydrofolate a donor of activated carbon groups.

Thalassemia one of a family of inherited diseases in which one of the chains of hemoglobin is damaged or missing, resulting in its decreased ability to carry oxygen.

Thiamine pyrophosphate (TPP) an organic cofactor in some enzymes that figures prominently in transaldolase.

Thiazoladinedione (TZD) one of a class of oral antidiabetic drugs that generally act by activating genes involved in fat metabolism.

Thioredoxin a redox-active protein that accepts electrons from other enzymes, becoming reduced in the process.

Thioredoxin reductase the enzyme responsible for reducing thioredoxin using NADPH.

Threading a template-based method of protein structure prediction that uses conserved elements of tertiary structure to identify regions of homology to build a structure.

Thylakoid a membranous structure within the chloroplast.

Thymine a pyrimidine base found exclusively in DNA.

Thyroid hormone any of the iodinated hormones derived from tyrosine and made in the thyroid gland.

Toll-like receptor (TLR) a receptor that binds to lipopolysaccharides, activating gene transcription and resulting in increased concentrations of cytokines.

Topoisomerase a class of proteins that cleave the DNA strand and release torsional strain or supercoiling prior to re-ligating the DNA fragments.

Total correlation spectroscopy (TOCSY) a multidimensional NMR technique in which all nuclei coupled to each other through bonds can be identified.

Total score an indication of the quality of the sequence alignment in a BLAST search.

Totipotent cell a cell that can differentiate into a limited number of cell types.

Transamination the transfer of an amino group from an α-keto acid to generate a new amino acid and new α-keto acid.

Transcription the process by which RNAs are synthesized from a gene.

Transcription factor a protein that binds to the promoter region of a gene to regulate its transcription.

Transcriptome the collection of all the RNA molecules in an organism, cell, or tissue.

trans fat any of the fatty acids with a *trans* double bond resulting from the isomerization of *cis* double bonds during partial hydrogenation.

Transfection a technique that uses recombinant virus to introduce DNA into a eukaryotic cell.

Transferrin receptor a receptor that binds and imports iron from the circulation.

Transfer RNA (tRNA) a type of small, highly modified RNA molecule that serves as an adapter between the message and the growing protein during translation.

Transformation a technique that uses chemical methods to introduce DNA into a bacterium.

Transgenic organism an organism that has been manipulated to express genes found in other organisms.

Transglutaminase the enzyme that cross-links glutamine and lysine residues in the extracellular matrix.

Translation the synthesis of proteins from an mRNA template.

Transmembrane protein a protein that spans the plasma membrane.

Transporter protein proteins that are involved in the passage of small molecules across the plasma membrane; also known as a transporter.

Transport vesicle a protein-containing structure that is trafficked around the cell.

Transposition the act of segments of DNA moving from one location in the genome to another.

Transposon A segment of DNA that moves from one location in the genome to another.

Triacyclglycerol (triglyceride) a glycerol molecule with all three hydroxyl groups esterified to fatty acids.

Triple quad a device consisting of three sequential quadrupoles surrounding the ion beam that resolves the mass-to-charge ratio in a mass spectrometer.

Trisaccharides a short carbohydrate comprised of three monosaccharides.

Trombone model the current model of DNA synthesis in which the lagging strand of DNA moves like the slide of a trombone.

Trp operon an operon regulating the expression of multiple genes involved in the synthesis of tryptophan.

Tumor promoter a gene that facilitates the growth of tumors once a carcinogenic event has occurred.

Tumor suppressor a gene that generates proteins that either suppress growth of cells or direct cells toward apoptosis.

Tunneling a quantum mechanical explanation of how small particles such as electrons are able to move from one place to another.

Turnover number the number of reactions that an enzyme can catalyze per second, represented in Michaelis–Menten kinetics by k_{cat} or k_2.

Turns a short stretch of amino acids in a protein that links elements of secondary structure.

Type 1 diabetes mellitus an autoimmune disease in which the immune system attacks the insulin-producing β cells of the pancreas.

Type 2 diabetes mellitus the more common form of diabetes in which the organism gradually becomes refractory to the effects of insulin; often related to obesity.

Ubiquinone a redox active quinoid group attached to a lengthy isoprenoid tail that shuttles electrons between complexes I and III and complex II; also known as coenzyme Q, CoQ_{10}, or UQ.

Ultrafiltration a method of protein purification that uses a dialysis membrane and centrifugation to concentrate a protein sample.

Uncompetitive inhibitor an inhibitor that can only bind to the *ES* complex.

Uniporter a transport protein that moves a single molecule across a membrane in a flow that may be uni- or bidirectional.

Untranslated region (UTR) the region upstream of the start codon in the mRNA.

Uracil a pyrimidine base found exclusively in RNA.

Urea cycle the metabolic pathway that generates urea from carbamoyl phosphate.

Urea the metabolic waste product formed through the urea cycle consisting of two amino groups joined by a central carbonyl carbon and not ionized at neutral pH.

Uric acid the oxidation product of xanthine, used as a means of disposal of purine nucleotides and nitrogenous waste in some organisms; the causal molecule in gout.

Uricotelic characteristic of organisms that synthesize uric acid as a nitrogenous waste product.

Uridine diphosphate glucose (UDP-glucose) the source of glucose used in the synthesis of glycogen.

van der Waals force one in a class of weak intermolecular forces that all molecules exhibit.

Vasa recta a network of capillaries that reabsorb electrolytes and water.

Vesicular fusion the attachment of a vesicle to the membrane that triggers exocytosis of the vesicle contents; the final step of cargo transport.

Virus a simple intracellular pathogen that requires a host for replication.

Vitrification the process of instantly freezing water to form an amorphous solid; used in sample preparation for cryo-EM.

VLDL (very-low-density lipoprotein) a lipoprotein secreted by the liver to deliver triacylglycerols made in the liver to peripheral tissues.

V_{max} the theoretical maximal velocity at which an enzymatically catalyzed reaction can proceed.

Void volume the volume of the buffer around the beads in a chromatography column.

Wax a naturally hydrophobic coating, consisting of a fatty acid esterified to a fatty alcohol.

Whole-plasmid mutagenesis a variation of PCR-based mutagenesis that uses two complementary mutagenic primers to polymerize an entire copy of the plasmid template and incorporate the mutations.

Winged helix a family of transcription factors that are roughly wing-shaped; also known as fork head.

Wobble a lack of consistency in the nucleotide composition of degenerate codons, particularly in the third position.

Xanthine oxidase the enzyme that converts xanthine to uric acid.

Xanthine the purine that is the substrate from which adenine and guanine are made.

Xenobiotic metabolism how the body metabolizes foreign molecules prior to excretion.

Xenobiotic a molecule the body recognizes as foreign.

X-ray diffraction a structure-determination technique in which X-rays are scattered by a crystal or fiber of a macromolecule and then used to determine the positions of atoms in the crystal.

Zinc finger one of several conserved DNA-binding motifs consisting of a helix stabilized by the binding of zinc through either cystine or histidine residues.

Zippering a hypothesis for the mechanism of vesicular fusion in which coiled-coil domains pull proteins together, leading to fusion.

Zwitterion a class of molecules that bear both a positive and negative charge at the same time, including amino acids, some phospholipids, and many metabolites.

Zymogen an inactive enzyme that requires proteolytic cleavage for activation; also known as a proenzyme.

Numbers and Greek letters are alphabetized as if they were spelled out. Page numbers in **bold** refer to a major discussion of the entry. F after a page number refers to a figure. Positional and configuration designations in chemical names are ignored in alphabetizing.

H